The Elements

Element	Symbol	Atomic Number	Atomic Mass*
Actinium	Ac	89	(227)
Aluminum	Al	13	26.98
Americium	Am	95	(243)
Antimony	Sb	51	121.8
Argon	Ar	18	39.95
Arsenic	As	33	74.92
Astatine	At	85	(210)
Barium	Ba	56	137.3
Berkelium	Bk	97	(247)
Beryllium	Be	4	9.012
Bismuth	Bi	83	209.0
Bohrium	Bh	107	(267)
Boron	B	5	10.81
Bromine	Br	35	79.90
Cadmium	Cd	48	112.4
Calcium	Ca	20	40.08
Californium	Cf	98	(249)
Carbon	C	6	12.01
Cerium	Ce	58	140.1
Cesium	Cs	55	132.9
Chlorine	Cl	17	35.45
Chromium	Cr	24	52.00
Cobalt	Co	27	58.93
Copper	Cu	29	63.55
Curium	Cm	96	(247)
Darmstadtium	Ds	110	(281)
Dubnium	Db	105	(262)
Dysprosium	Dy	66	162.5
Einsteinium	Es	99	(254)
Erbium	Er	68	167.3
Europium	Eu	63	152.0
Fermium	Fm	100	(253)
Fluorine	F	9	19.00
Francium	Fr	87	(223)
Gadolinium	Gd	64	157.3
Gallium	Ga	31	69.72
Germanium	Ge	32	72.61
Gold	Au	79	197.0
Hafnium	Hf	72	178.5
Hassium	Hs	108	(277)
Helium	He	2	4.003
Holmium	Ho	67	164.9
Hydrogen	H	1	1.008
Indium	In	49	114.8
Iodine	I	53	126.9
Iridium	Ir	77	192.2
Iron	Fe	26	55.85
Krypton	Kr	36	83.80
Lanthanum	La	57	138.9
Lawrencium	Lr	103	(257)
Lead	Pb	82	207.2
Lithium	Li	3	6.941
Lutetium	Lu	71	175.0
Magnesium	Mg	12	24.31
Manganese	Mn	25	54.94
Meitnerium	Mt	109	(268)
Mendelevium	Md	101	(256)

Element	Symbol	Atomic Number	Atomic Mass*
Mercury	Hg	80	200.6
Molybdenum	Mo	42	95.94
Neodymium	Nd	60	144.2
Neon	Ne	10	20.18
Neptunium	Np	93	(244)
Nickel	Ni	28	58.70
Niobium	Nb	41	92.91
Nitrogen	N	7	14.01
Nobelium	No	102	(253)
Osmium	Os	76	190.2
Oxygen	O	8	16.00
Palladium	Pd	46	106.4
Phosphorus	P	15	30.97
Platinum	Pt	78	195.1
Plutonium	Pu	94	(242)
Polonium	Po	84	(209)
Potassium	K	19	39.10
Praseodymium	Pr	59	140.9
Promethium	Pm	61	(145)
Protactinium	Pa	91	(231)
Radium	Ra	88	(226)
Radon	Rn	86	(222)
Rhenium	Re	75	186.2
Rhodium	Rh	45	102.9
Roentgenium	Rg	111	(272)
Rubidium	Rb	37	85.47
Ruthenium	Ru	44	101.1
Rutherfordium	Rf	104	(263)
Samarium	Sm	62	150.4
Scandium	Sc	21	44.96
Seaborgium	Sg	106	(266)
Selenium	Se	34	78.96
Silicon	Si	14	28.09
Silver	Ag	47	107.9
Sodium	Na	11	22.99
Strontium	Sr	38	87.62
Sulfur	S	16	32.07
Tantalum	Ta	73	180.9
Technetium	Tc	43	(98)
Tellurium	Te	52	127.6
Terbium	Tb	65	158.9
Thallium	Tl	81	204.4
Thorium	Th	90	232.0
Thulium	Tm	69	168.9
Tin	Sn	50	118.7
Titanium	Ti	22	47.88
Tungsten	W	74	183.9
Uranium	U	92	238.0
Vanadium	V	23	50.94
Xenon	Xe	54	131.3
Ytterbium	Yb	70	173.0
Yttrium	Y	39	88.91
Zinc	Zn	30	65.41
Zirconium	Zr	40	91.22
		112**	(285)
		114	(289)
		116	(292)

*All atomic masses are given to four significant figures. Values in parentheses represent the mass number of the most stable isotope.

**The names and symbols for elements 112, 114, and 116 have not been chosen.

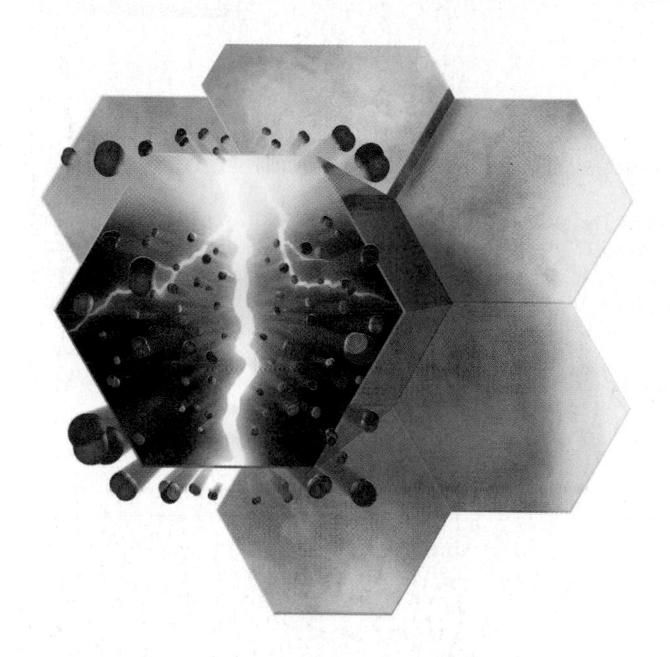

FOURTH EDITION

CHEMISTRY

The Molecular Nature of Matter and Change

Martin S. Silberberg

Consultants

Randy Duran
University of Florida—Gainesville

Charles G. Haas (Emeritus)
Pennsylvania State University

Arlan D. Norman
Western Washington University

Boston Burr Ridge, IL Dubuque, IA Madison, WI New York San Francisco St. Louis
Bangkok Bogotá Caracas Kuala Lumpur Lisbon London Madrid Mexico City
Milan Montreal New Delhi Santiago Seoul Singapore Sydney Taipei Toronto

The McGraw-Hill Companies

CHEMISTRY: THE MOLECULAR NATURE OF MATTER AND CHANGE, FOURTH EDITION

Published by McGraw-Hill, a business unit of The McGraw-Hill Companies, Inc., 1221 Avenue of the Americas, New York, NY 10020.

This book is printed on acid-free paper.

2 3 4 5 6 7 8 9 0 DOW/DOW 0 9 8 7 6 5
1 2 3 4 5 6 7 8 9 0 DOW/DOW 0 9 8 7 6 5 4

ISBN 0–07–255820–2
ISBN 0–07–296439–1 (Annotated Instructor's Edition)

Editorial Director: *Kent A. Peterson*
Sponsoring Editor: *Thomas D. Timp*
Senior Developmental Editor: *Donna Nemmers*
Developmental Editor: *Joan M. Weber*
Managing Developmental Editor: *Shirley R. Oberbroeckling*
Freelance Developmental Editor: *Karen Pluemer*
Senior Marketing Manager: *Tamara L. Good-Hodge*
Lead Project Manager: *Peggy J. Selle*
Lead Production Supervisor: *Sandy Ludovissy*
Lead Media Project Manager: *Judi David*
Senior Media Technology Producer: *Jeffry Schmitt*
Senior Designer: *David W. Hash*
Cover/Interior Designer: *Jamie E. O'Neal*
Cover Illustration: *Michael Goodman*
Page Layout/Special Features Designer: *Ruth Melnick*
Illustrations: *Federico/Goodman Studios; Furious Design*
Senior Photo Research Coordinator: *Lori Hancock*
Photo Research: *Chris Hammond/PhotoFind, LLC*
Supplement Producer: *Brenda A. Ernzen*
Compositor: *The GTS Companies/Los Angeles, CA Campus*
Typeface: *10.5/12 Times Roman*
Printer: *R. R. Donnelley, Willard, OH*

Library of Congress Cataloging-in-Publication Data

Silberberg, Martin S. (Martin Stuart), 1945–
Chemistry : the molecular nature of matter and change / Martin S. Silberberg. — 4th ed.
p. cm.
Includes bibliographical references and index.
ISBN 0–07–255820–2 (hard copy : alk. paper)
1. Chemistry. I. Title.

QD33.2.S55 2006
540—dc22

2004018874
CIP

www.mhhe.com

To Ruth and Daniel,
with all my love.

Brief Contents

Detailed Contents

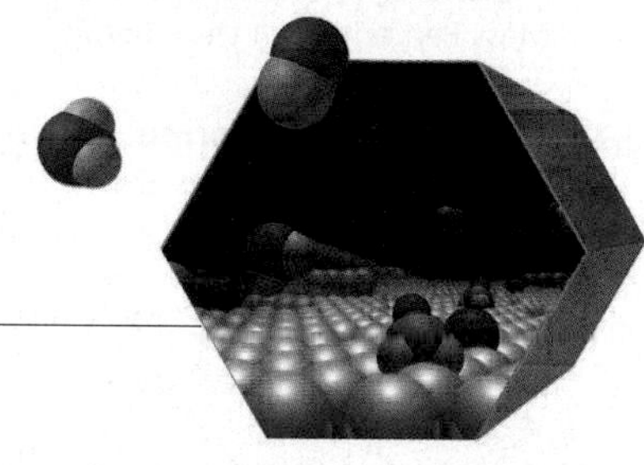

1 CHAPTER

Keys to the Study of Chemistry 1

2 CHAPTER

The Components of Matter 38

3 CHAPTER

Stoichiometry of Formulas and Equations 86

4 CHAPTER

The Major Classes of Chemical Reactions 134

5 CHAPTER

Gases and the Kinetic-Molecular Theory 176

6 CHAPTER

Thermochemistry: Energy Flow and Chemical Change 224

7 CHAPTER

Quantum Theory and Atomic Structure 256

8 CHAPTER

Electron Configuration and Chemical Periodicity 290

9 CHAPTER

Models of Chemical Bonding 328

10 CHAPTER

The Shapes of Molecules 365

11 CHAPTER

Theories of Covalent Bonding 398

12 CHAPTER

Intermolecular Forces: Liquids, Solids, and Phase Changes 424

13 CHAPTER

The Properties of Mixtures: Solutions and Colloids 488

INTERCHAPTER

A Perspective on the Properties of the Elements 541

14 CHAPTER

Periodic Patterns in the Main-Group Elements 552

15 CHAPTER

Organic Compounds and the Atomic Properties of Carbon 616

16 CHAPTER

Kinetics: Rates and Mechanisms of Chemical Reactions 672

17 CHAPTER

Equilibrium: The Extent of Chemical Reactions 722

18 CHAPTER

Acid-Base Equilibria 766

19 CHAPTER

Ionic Equilibria in Aqueous Systems 814

20 CHAPTER

Thermodynamics: Entropy, Free Energy, and the Direction of Chemical Reactions 863

21 CHAPTER

Electrochemistry: Chemical Change and Electrical Work 902

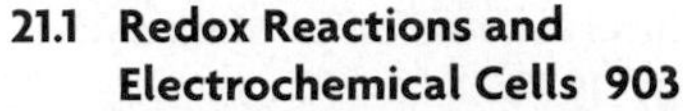

22 CHAPTER

The Elements in Nature and Industry 960

23 CHAPTER

The Transition Elements and Their Coordination Compounds 1002

24 CHAPTER

Nuclear Reactions and Their Applications 1044

LIST OF MARGIN NOTES

This icon in the text indicates a related application, study aid, or historical note appearing in the margin.

About the Author and Consultants

Martin S. Silberberg received a B.S. in Chemistry from the City University of New York and a Ph.D. in Chemistry from the University of Oklahoma. He then accepted a research position in analytical biochemistry at the Albert Einstein College of Medicine in New York City, where he studied neurotransmitter metabolism in Parkinson's disease. Following his years in research, Dr. Silberberg joined the faculty of Simon's Rock College of Bard, a liberal arts college known for its excellence in teaching small classes of highly motivated students. As head of the Natural Sciences Major and Director of Premedical Studies, he taught courses in general chemistry, organic chemistry, biochemistry, and liberal-arts chemistry. The close student contact afforded him insights into how students learn chemistry, where they have difficulties, and what strategies can help them succeed. In 1983, Dr. Silberberg applied these insights in a broader context by establishing a text writing, editing, and consulting company. Before writing his own text, he worked as a consulting and development editor on chemistry, biochemistry, and physics texts for several major college publishers. He resides with his wife and son in the Pioneer Valley near Amherst, Massachusetts, where he enjoys the rich cultural and academic life of the area and relaxes by cooking, gardening, and hiking.

Dr. Silberberg is very grateful to his consultants, whose advice and insight have helped optimize this text for a variety of students from diverse learning environments.

Randy Duran is Professor of Chemistry within the Butler Polymer Laboratory and Adjunct Professor of Materials Science at the University of Florida. In addition to teaching general chemistry and advanced courses in physical and polymer chemistry, Dr. Duran directs an active research program in the area of polymer surfaces and interfaces.

Charles G. Haas, emeritus Professor of Chemistry at The Pennsylvania State University, earned his Ph.D. at the University of Chicago under the supervision of Norman Nachtrieb. During his 38-year career at Penn State, he taught general chemistry, undergraduate and graduate inorganic chemistry, and courses in chemical education for public school teachers at all levels. His research focuses on the chemistry of transition metals and coordination compounds.

Arlan D. Norman served as Professor of Chemistry and Biochemistry at the University of Colorado at Boulder, where he taught general and inorganic chemistry for 33 years, concentrating on molecular graphics, modeling, and visualization techniques. He is now founding Dean of Western Washington University's newly formed College of Sciences and Technology.

Preface

IS CHEMISTRY CHANGING?

As in any dynamic, modern science, theories in chemistry are refined to reflect new data, established ideas are applied to new systems, and connections are forged with other sciences to uncover new information. But chemistry, as the science of matter and its changes, is central to so many other sciences—physical, biological, environmental, medical, and engineering—that it must evolve continuously to enable their progress. Designing "greener" ways to make plastics, fuels, and other commodities, monitoring our atmosphere and oceans to predict the effects of their changing compositions, comprehending our genetic makeup to develop novel medicines and disease treatments, and synthesizing nanomaterials with revolutionary properties are among the countless areas in which chemistry is evolving.

Nevertheless, the basic concepts of chemistry still form the essence of the course. The mass laws and the mole concept still apply to the amounts of substances involved in a chemical reaction. Atomic properties, and the periodic trends and types of bonding derived from them, still determine molecular structure—which in turn still governs the forces between molecules and the resulting physical behavior of substances. And the central concepts within the fields of kinetics, equilibrium, and thermodynamics still account for the dynamic aspects of chemical change.

The challenge for a modern chemistry text, then, is to do two jobs at once: present the fundamental principles clearly and show how they apply to the newly emerging areas of chemistry. Like chemistry itself, the Fourth Edition of ***Chemistry: The Molecular Nature of Matter and Change*** has evolved in important ways to meet this challenge. Read on to learn how we've enhanced this edition to continue to meet the changing needs of students and instructors.

WHAT SETS THIS BOOK APART?

Respected and emulated through three editions, ***Chemistry: The Molecular Nature of Matter and Change*** has set—and then raised—the standard for general chemistry texts. And while the content has been continually updated to reflect the changing impact of chemistry in the world, the mechanisms of the text—the teaching approaches that are so admired—have remained the same. Three hallmarks make this text the market leader:

1. Visualizing chemical models—macroscopic to molecular
2. Thinking logically to solve problems
3. Applying ideas to the real world.

Visualizing Chemical Models: Macroscopic to Molecular

Because chemistry deals with observable changes in the world around us that are caused by unobservable atomic-scale events, a size gap of mind-boggling proportions must be spanned. Throughout the text, concepts are explained first at the macroscopic level and then from a molecular point of view, with the text's groundbreaking illustrations placed next to the discussion to bring the point home for today's visually oriented students.

Thinking Logically to Solve Problems

The problem-solving approach, based on a widely accepted, four-step method, is introduced in Chapter 1 and employed *consistently* throughout the text. It encourages students to first plan a logical approach to a problem, and only then proceed to solve it mathematically. The check, a step universally recommended by instructors, fosters the habit of assessing the reasonableness and magnitude of the answer. For practice and reinforcement, each worked problem is followed immediately by a similar one, for which an abbreviated solution is given at the end of the chapter.

Applying Ideas to the Real World

An understanding of modern chemistry influences a person's attitudes about public policy issues, such as climate change, health care, and agricultural methods, and also explains everyday phenomena, such as the spring in a running shoe, the display of a laptop screen, and the fragrance of a rose. Today's students may enter one of the emerging chemistry-related, hybrid fields—biomaterials science or planetary geochemistry, for example—and their text should point out the relevance of chemical concepts to such career directions. Chemical Connections, Tools of the Laboratory, Galleries, and margin notes are up-to-date pedagogic features that complement the text content.

EMBRACING CHANGE: HOW WE EVALUATED YOUR NEEDS

Just as the applications of chemistry change, so do your needs in the classroom. Martin Silberberg and McGraw-Hill listened—and responded. They invited instructors like you from across the nation—with varying teaching styles, class sizes, and student background—to provide feedback through reviews and focus groups. This feedback was then used to carefully revise and mold this new edition of ***Chemistry: The Molecular Nature of Matter and Change***, resulting in new topic coverage, succinct and logical presentation, and expanded treatment in key areas.

WHAT'S NEW IN THE FOURTH EDITION?

Enhancements to this Edition

Annotated Instructor's Edition

Created to help you optimize use of the text, the annotated edition uses the margins to cite journal articles and lecture demonstrations, to display icons for transparencies, online resources, figures and tables, worked problems, and animations that are available to you, and to show content related to other fields (biological, engineering, environmental/green chemistry, and organic chemistry), as well as to indicate the general level of difficulty of each homework problem.

Integration of Organic and Biochemistry

In response to students' strong interest in biology-related material, applications of mainstream chemical concepts to bio-organic topics are incorporated into many chapters, including new discussions in Chapters 2, 3, 13, and 15 (see below). All of these discussions are clearly identified, so they can be skipped or included as you wish.

Integration of Green Chemistry

This strong new research direction to develop substances and processes with only benign effects on the environment receives coverage, including homework problems, in Chapters 3, 6, 13, 15, and 16 (see below). These discussions are identified and can be skipped or included as you wish.

Molecular-View Sample Problems

A new type of conceptual sample problem is introduced that finally teaches students how to look at simple molecular scenes and think through qualitative and quantitative problems based on them. These 10 new exercises—five sample problems and their five follow-up problems—appear in Chapters 2, 3, 5, 13, and 17.

Electron Density Art

Striking new art based on the latest quantum-mechanical calculations depicts the distribution of electron density in covalent bonds. These illustrations appear in Chapters 7, 9, 11, 14, and 16.

End-of-Chapter Problems

Over 275 new problems appear, many at the challenging level, with applications from biology, organic chemistry, engineering, and environmental science. Several new molecular-scene problems have also been included.

In This Chapter...

At the end of its introduction, each chapter includes this short preview that orients the student to the sequence of main topics and correlates with the chapter outline.

Changes to Individual Chapters

Chapter 2 briefly introduces the naming and structures of organic compounds and discusses the elements found in living systems in the context of the periodic table.

Chapter 3 introduces isomers in terms of molecular formulas and discusses percent yield in the multistep synthesis of medicines. The green-chemistry concept of atom economy is discussed, with a sample calculation.

Chapter 4 highlights the role of water in aqueous systems and emphasizes the activity series of metals in the context of redox reactions.

Chapter 6 brings up the green chemistry of fuels and new energy sources and updates the essay on climate change and alternative energy.

Chapter 9 has been greatly improved with art and discussions that clarify the importance of covalent bonding and trends in electronegativity. Coverage of foods and fuels has been revised and moved here from Chapter 6; it now complements the full treatment of bond energy. And the essay on IR spectroscopy now includes isomers, with new art.

Chapter 12 emphasizes the universal importance of dispersion forces and summarizes all intermolecular forces with new art.

Chapter 13 includes a major new section on how intermolecular forces stabilize the structures of proteins, nucleic acids, polysaccharides, and cell membranes, with striking new art. A new margin note discusses the role of greener solvents in syntheses.

Chapter 15 includes carbon-skeleton formulas in discussions (and problems) to prepare students for organic chemistry courses. In the biopolymer section, the structural hierarchy of proteins is covered, and a unique, new essay pays homage to the remarkable achievement of the Human Genome Project by describing the chemistry of DNA sequencing.

Chapter 16 describes the transition state with new art and updates coverage of catalytic hydrogenation and ozone depletion.

Chapter 20 has been completely rewritten to reflect a new approach to the coverage of entropy. The vague notion of "disorder" (with analogies to macroscopic systems) has been replaced with the idea that entropy is related to the dispersal of a system's energy and the freedom of motion of its particles. Also clarified are the characteristics of a reversible process and the connection between free energy and equilibrium.

Chapter 21 extends the consistent method for determining cell potential and updates coverage of batteries and fuel cells.

Appendices have been greatly expanded to include extensive tables of thermodynamic and equilibrium data.

Acknowledgments

My deepest appreciation goes to all the academic and research chemists who have generously contributed their time and expertise to improving the text. Once again, I was extremely fortunate to have the meticulous eyes of Dorothy B. Kurland reviewing every chapter and each new homework problem. The following experts helped me keep specific content areas up-to-date and accurate: Ronald J. Gillespie of McMaster University for advice on the new electron-density relief maps and Michel Rafat of the University of Manchester Institute of Science and Technology for providing data for these illustrations and many useful discussions regarding them; Deborah Exton of the University of Oregon for new coverage of green chemistry; Jonathan Kurland of Dow Chemical Company for essays on fuels (Chapter 6), ozone depletion (Chapter 16), acid rain (Chapter 19), and numerous other points of industrial and atmospheric chemistry; Chad Mirkin and Sungho Park of Northwestern University for advice on nanotechnology (Chapter 12); Steven A. Soper of Louisiana State University and Bruce A. Roe of the University of Oklahoma for the new essay on DNA sequencing (Chapter 15); Frank Lambert of Occidental College for insightful advice and comments on the coverage of entropy (Chapters 13 and 20); Stewart Strickler of the University of Colorado–Boulder for in-depth help in revising the coverage of entropy and free energy (Chapters 13, 20, 21); and John Newman of the University of California/Berkeley and Perla Balbuena of the University of South Carolina for the coverage of batteries and fuel cells (Chapter 21).

Special thanks go to Rich Bauer of Arizona State University and Sue Nurrenbern of Purdue University for their helpful reviews of the new molecular sample problems and to the professors who contributed many of the excellent new homework problems: Joseph Bularzik, Sarina Ergas of the University of Massachusetts–Amherst, Rich Langley of Stephen F. Austin State University, S. Walter Orchard of Tacoma Community College, Jeanette K. Rice of Georgia Southern University, and Marcy Whitney of the University of Alabama.

Modern texts are served by a battery of supplements, and this one is very lucky to have supplement authors so committed to accuracy and clarity for student and instructor. John Pollard of the University of Arizona researched and coordinated all of the annotations that appear in the Annotated Instructor's Edition of this text. Elizabeth Bent Weberg wrote a superb *Student Study Guide*. Richard H. Langley of Stephen F. Austin University diligently prepared the *Instructor's Solutions Manual* and *Student Solutions Manual*. S. Walter Orchard of Tacoma Community College updated the *Test Bank*. Christina Bailey of California Polytechnic University again provided the excellent PowerPoint Lecture Outlines that appear on the Digital Content Manager CD.

I am especially grateful for the support of the Board of Advisors, a select group of chemical educators dedicated to helping make this text the optimum teaching tool. The board contributed insightful comments during the revision that shaped this edition in many significant ways.

Board of Advisors

Ramesh D. Arasasingham
University of California, Irvine
Margaret Asirvatham
University of Colorado, Boulder
Christina A. Bailey
Cal Poly, San Luis Obispo
David S. Ballantine, Jr.
Northern Illinois University
Rich Bauer
Arizona State University
Tim Bays
United States Military Academy
Don Berkowitz
University of Maryland
Phil Brucat
University of Florida, Gainesville
Dominick Casadonte
Texas Tech University
Deborah Exton
University of Oregon
David Frank
Cal State University, Fresno
Russell Geanangel
University of Houston
Mark A. Griep
University of Nebraska, Lincoln
John M. Halpin
New York University
Richard H. Langley
Stephen F. Austin State University
Amy Lindsay
University of New Hampshire
Mike Lipschutz
Purdue University
One Sun (Trevor Roberti)
University of California, Santa Cruz
Lou Pignolet
University of Minnesota
John Pollard
University of Arizona
Jeanette Rice
Georgia Southern University
Bill Robinson
Purdue University
Reva Savkar
Northern Virginia Community College
Bob Scheidt
University of Notre Dame
Harvey Schugar
Rutgers University
Diane Smith
San Diego State University
John Vincent
University of Alabama
Sharon Vincent
Shelton State Community College

Included especially in this group of professors are all those who participated in focus groups, reviewed content, and class-tested the Third Edition:

Patricia G. Amateis
Virginia Institute of Technology
David Anderson
University of Colorado, Colorado Springs
Frank C. Andrews
University of California, Santa Cruz
Ramesh D. Arasasingham
University of California, Irvine
Todd L. Austell
University of North Carolina, Chapel Hill
Kishore K. Bagga
Central Community College, Hastings
Yiyan Bai
Houston Community College
David W. Ball
Cleveland State University
David S. Ballantine, Jr.
Northern Illinois University
Joseph Bariyanga
University of Wisconsin, Milwaukee
Rebecca E. Barlag
University of Cincinnati
Jeffrey M. Bartolin
University of Michigan
Tim Bays
United States Military Academy
Debbie J. Beard
Mississippi State University
Suely Meth Black
Norfolk State University
Bob Blake
Texas Tech University
Bob Bryant
University of Virginia
Brian Buffin
Western Michigan University
William Burns
Arkansas State University
Stuart Burris
Western Kentucky University
Kim C. Calvo
The University of Akron
C. Kevin Chambliss
Baylor University
Julia Chan
Louisiana State University
Tsun-Mei Chang
University of Wisconsin, Parkside
Allen Clabo
Francis Marion University
Ross D. Compton
Southwest Texas State University
Elzbieta Cook
Southern University and A&M College
Brandon Cruickshank
Northern Arizona University
Mark Cybulski
Miami University of Ohio
William M. Davis
University of Texas, Brownsville
Dru L. DeLaet
Southern Utah University
Judy Dirbas
Grossmont Community College
Daniel S. Domin
Tennessee State University
William Donovan
The University of Akron
Mark Draganjac
Arkansas State University
Stephen Drucker
University of Wisconsin, Eau Claire
Bill Durham
University of Arkansas
Karen E. Eichstadt
Ohio University
Robert J. Eierman
University of Wisconsin, Eau Claire
Tom Engel
University of Washington, Seattle
Jeffrey Evans
University of Southern Mississippi
Emmanuel Ewane
Houston Community College
David Farrelly
Utah State University
Debra A. Feakes
Texas State University, San Marcos
Gregory M. Ferrence
Illinois State University
Paul A. Flowers
University of North Carolina, Pembroke
Sonya J. Franklin
University of Iowa
Cheryl Baldwin Frech
University of Central Oklahoma
Daniel Lee Fuller
College of DuPage
Nancy J. Gardner
California State University, Long Beach
Russell Geanangel
University of Houston
Susan E. Geldert
University of Rhode Island
Amy Collins Gottfried
University of Michigan
Pierre Y. Goueth
Santa Monica College
Todor K. Gounev
University of Missouri, Kansas City
David Grainger
Colorado State University
Palmer Graves
Florida International University
Thomas J. Greenbowe
Iowa State University of Science and Technology
Mark A. Griep
University of Nebraska, Lincoln
John M. Halpin
New York University
Jessica A. Hessler
University of Michigan
Narayan S. Hosmane
Northern Illinois University
Michael O. Hurst, Sr.
Georgia Southern University
Chui Kwong Hwang
Evergreen Valley College
Michael A. Janusa
Stephen F. Austin State University
Andy Jorgensen
University of Toledo
Steven W. Keller
University of Missouri, Columbia
Colleen Kelley
Pima Community College, West
Richard Kiefer
College of William & Mary
Jim Klent
Ohlone College
Evguenii I. Kozliak
University of North Dakota
Mary Beth Kramer
University of Delaware
Robert M. Kren
University of Michigan, Flint
John Krenos
Rutgers University
Brian B. Laird
University of Kansas
Brian D. Lamp
Truman State University
Richard H. Langley
Stephen F. Austin State University
Robley Light
Florida State University
David Lippmann
Southwest Texas State University
James M. LoBue
Georgia Southern University
Jeffrey A. Mack
California State University, Sacramento
Donald E. Mencer, Jr.
Wilkes University
Gabriel Miller
New York University
Chad Mirkin
Northwestern University
John T. Moore
Stephen F. Austin State University
Mark E. Noble
University of Louisville
Susan Nurrenbern
Purdue University
John G. O'Brien
Truman State University
Michael Y. Ogawa
Bowling Green State University
One Sun (Trevor Roberti)
University of California, Santa Cruz
Jessica N. Orvis
Georgia Southern University

Robert A. Orwoll
College of William & Mary
Jason S. Overby
College of Charleston
Gilbert E. Pacey
Miami University
Gholam H. Pahlavan
Houston Community College (Southwest College)
Yasmin Patell
Kansas State University
M. Diane Payne
Villa Julie College
Karl P. Peterson
University of Wisconsin, River Falls
Lou Pignolet
University of Minnesota, Minneapolis
Al Pinhas
University of Cincinnati
John Pollard
University of Arizona
Cathrine E. Reck
Indiana University, Bloomington
Arnold L. Rheingold
University of California, San Diego
Jeanette K. Rice
Georgia Southern University
B. Ken Robertson
University of Missouri, Rolla
Bill Robinson
Purdue University
Jill Robinson
Indiana University
Jon Russ
Arkansas State University
E. Alan Sadurski
Ohio Northern University
Svein Saebo
Mississippi State University
Tris Samberg
Edmonds Community College
Jerry L. Sarquis
Miami University
Barbara Sawrey
University of California, San Diego
Harvey Schugar
Rutgers, The State University of New Jersey
William D. Scott, III
University of Mississippi
Thomas Selegue
Pima Community College
Susan M. Shih
College of DuPage
Alka Shukla
Southeast College (Houston Community College system)
Shyam S. Shukla
Lamar University
Steven Sincoff
Butte College
Tom Sorenson
Univeristy of Wisconsin, Milwaukee
Lothar Stahl
University of North Dakota
Paul B. Steinbach
Stephen F. Austin State University
Robert P. Stewart, Jr.
Miami University
Alan Stolzenberg
West Virginia University
Ryan Sweeder
University of Michigan
Greg Szulczewski
University of Alabama
Vicente Talanquer
University of Arizona
James G. Tarter
College of Southern Idaho
Jason R. Telford
University of Iowa
Larry C. Thompson
University of Minnesota, Duluth
Mark Thomson
Arkansas State University
Wayne Tikkanen
California State University, Los Angeles
Edmund L. Tisko
University of Nebraska, Omaha
John Todd
Bowling Green State University
William C. Trogler
University of California, San Diego
Martin Vala
University of Florida
Petr Vanýsek
Northern Illinois University
John Verkade
Iowa State University
Tito Viswanathan
University of Arkansas, Little Rock
Robert Walker
University of Maryland, College Park
Neil Weinstein
Sante Fe Community College
David D. Weis
Skidmore College
Gary D. White
Middle Tennessee State University
Charles A. Wilkie
Marquette University
Vickie M. Williamson
Texas A & M University
Donald R. Wirz
University of California, Riverside
Troy Wood
SUNY Buffalo
Kim Woodrum
University of Kentucky
Chris Yerkes
University of Illinois, Champaign
Linda S. Zarzana
American River College
Jin Z. Zhang
University of California, Santa Cruz

Class-Test Schools

Madeline Adamczeski
San Jose City College
Rebecca Barlag
University of Cincinnati
Jerry Sarquis
Miami University of Ohio
Lothar Stahl
University of North Dakota

So many exceptional publishing professionals played key roles in this new edition, and I owe them my warmest gratitude. Heading the McGraw-Hill Higher Education team with their friendship and support were Editorial Director Kent Peterson and Sponsoring Editor Thomas Timp. When my previous Development Editor Joan Weber moved to other projects, she deftly passed this one into the able hands of Development Editor Donna Nemmers; Donna handled innumerable text and supplement details with her unusual mix of warmth and competence. Project Manager Peggy Selle engineered countless production details with consummate experience and skill. Senior Designer David Hash supervised the striking new design by freelancers Jamie O'Neal and Michael Goodman. And Marketing Manager Tami Hodge developed exciting new ways to present the book and supplements to the academic community.

A wonderful group of expert freelancers contributed as well. Jane Hoover did a masterful copyediting job, and Katie Aiken and Janelle Pregler followed with superb proofreading. Michael Goodman again performed his magic on in-text molecular art and the cover. And the remarkable Karen Pluemer somehow managed to coordinate all the complex interactions with publisher and freelancers, while keeping me on schedule and sane.

Finally, in far more ways than I could ever begin to say here, I am indebted to my son Daniel and wife Ruth for their love and confidence. Daniel, now 16, has become an accomplished artist, and he drafted initial designs for several key pieces of art, including the cover, before the illustrators took over. As in past editions, Ruth played numerous essential roles—laying out the student-friendly pages of text, art, and tables, collaborating on style and design, editing new text and art, checking all page proofs, and, in general, helping author and publisher maintain the highest standards of quality and consistency throughout the project.

LEARNING CHEMISTRY JUST GOT EASIER

Success starts here.

Take this guided tour for *Chemistry: The Molecular Nature of Matter and Change,* Fourth Edition, and you'll see how to get the most out of your textbook!

Chapter Opener

The opener provides a thought-provoking figure and legend that relate to a main topic of the chapter.

Concepts and Skills to Review

This unique feature helps you prepare for the upcoming chapter by referring to key material from earlier chapters that you should understand *before* you start reading this one.

A Select Few Particles of solid lead(II) iodide swirling through a sodium nitrate solution are the products of this precipitation reaction, an example of one of a handful of major reaction classes. Distinguishing the key features of these classes and learning the role of water in many of them form the focus of the chapter.

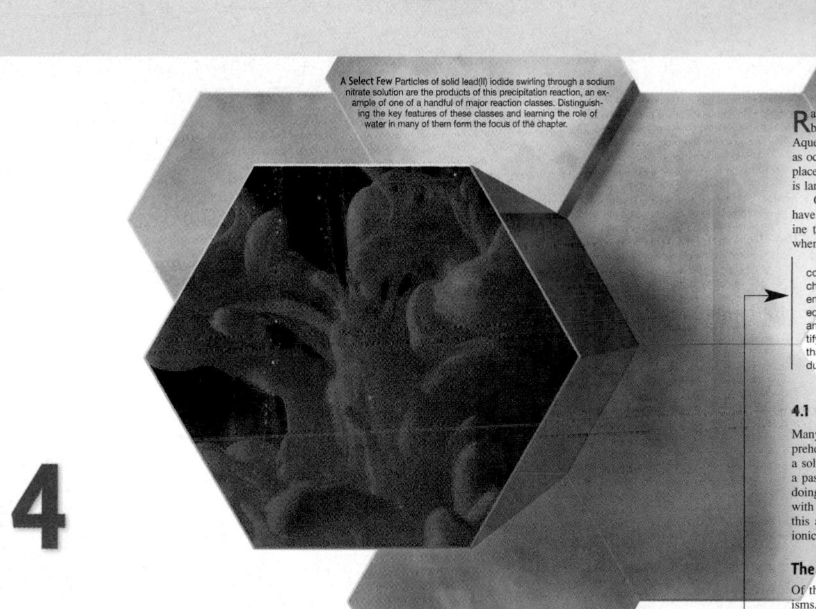

4

The Major Classes of Chemical Reactions

Rapid chemical changes occur among gas molecules as sunlight bathes the atmosphere or lightning rips through a stormy sky. Aqueous reactions go on unceasingly in the gigantic containers we know as oceans. And, in every cell of your body, thousands of reactions taking place right now allow you to function. Indeed, the amazing variety in nature is largely a consequence of the amazing variety of chemical reactions.

Of the millions of chemical reactions occurring in and around you, we have examined only a tiny fraction so far, and it would be impossible to examine them all. Fortunately, it isn't necessary to catalog every reaction, because when we survey even a small percentage of reactions, a few major patterns emerge.

IN THIS CHAPTER . . . We examine the underlying nature of the three most common reaction processes. One of our main themes is aqueous reaction chemistry, so we first investigate how the molecular structure of water influences its crucial role as a solvent in these reactions. We see how to use ionic equations to describe reactions. We focus in turn on precipitation, acid-base, and oxidation-reduction reactions, examining why they occur and how to quantify them. We classify several important types of oxidation-reduction reactions that include elements as reactants or products. The chapter ends with an introductory look at the reversible nature of all reactions.

Concepts & Skills to Review
before you study this chapter

- names and formulas of compounds (Section 2.8)
- nature of ionic and covalent bonding (Section 2.7)
- mole-mass-number conversions (Section 3.1)
- molarity and mole-volume conversions (Section 3.5)
- balancing chemical equations (Section 3.3)
- calculating amounts of reactants and products (Section 3.4)

4.1 THE ROLE OF WATER AS A SOLVENT

Many reactions take place in aqueous solution, and our first step toward comprehending these reactions is to understand how water acts as a solvent. The role a solvent plays in a reaction depends on its chemical nature. Some solvents play a passive role, dispersing the dissolved substances into individual molecules but doing nothing further. Water plays a much more active role, interacting strongly with the substances and, in some cases, even reacting with them. To understand this active role, we'll examine the structure of water and how it interacts with ionic and covalent solutes.

The Polar Nature of Water

Of the many thousands of reactions that occur in the environment and in organisms, nearly all take place in water. Water's remarkable power as a solvent results from two features of its molecules: *the distribution of the bonding electrons* and *the overall shape.*

Recall from Section 2.7 that the electrons in a covalent bond are shared between the bonded atoms. In a covalent bond between identical atoms (as in H_2, Cl_2, O_2, etc.), the sharing is equal, so no imbalance of charge appears (Figure 4.1A). On the other hand, in covalent bonds between nonidentical atoms, the sharing is unequal: one atom attracts the electron pair more strongly than the other. For reasons discussed in Chapter 9, an O atom attracts electrons more strongly than an H atom. Therefore, in each O—H bond in water, the shared electrons spend more time closer to the O atom (Figure 4.1B).

This unequal distribution of negative charge creates partially charged "poles" at the ends of each O—H bond. The O end acts as a slightly negative pole (represented by the red shading and the $\delta-$), and the H end acts as a slightly positive pole (represented by the blue shading and the $\delta+$). Figure 4.1C indicates the bond's polarity with a *polar arrow* (the arrowhead points to the negative pole and the tail is crossed to make a "plus").

The H—O—H arrangement forms an angle, so the water molecule is bent. The combined effects of its bent shape and its polar bonds make water a **polar molecule:** the O portion of the molecule is the partially negative pole, and the region midway between the H atoms is the partially positive pole (Figure 4.1D).

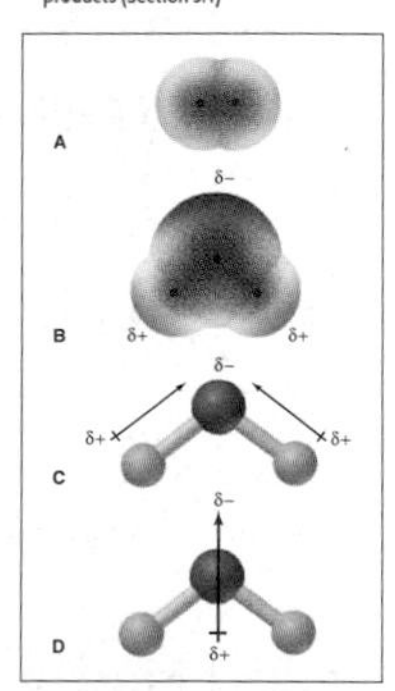

Figure 4.1 Electron distribution in molecules of H_2 and H_2O. A, In H_2, the identical nuclei attract the electrons equally. The central region of higher electron density (red) is balanced by the two outer regions of lower electron density (blue). **B,** In H_2O, the O nucleus attracts the shared electrons more strongly than the H nucleus. **C,** In this ball-and-stick model, a polar arrow points to the negative end of each O—H bond. **D,** The two polar O—H bonds and the bent shape give rise to the polar H_2O molecule.

Chapter Outline and *In This Chapter*

Each chapter begins with an outline that shows the sequence of topics and subtopics. And, at the end of the introduction, a paragraph called *In This Chapter* ties the main topics to the outline.

SOLVING PROBLEMS, STEP BY STEP

Using this clear and thorough problem-solving approach, you'll learn to think through chemistry problems logically and systematically.

Sample Problems

A worked-out problem appears whenever an important new concept or skill is introduced. The step-by-step approach is shown consistently for every sample problem in the text.

Steps

Plan

analyzes the problem so that you can use what is known to find what is unknown. This approach develops the habit of thinking through the solution *before* performing calculations.

Solution

shows the calculation steps *in the same order* as they are discussed in the plan.

Check

fosters the habit of going over your work quickly to make sure that the answer is reasonable, chemically and mathematically—a great way to avoid careless errors.

Comment

provides an additional insight, alternative approach, or common mistake to avoid.

Follow-up Problem

gives you immediate practice by presenting a similar problem.

NEW and Unique to This Text!

Conceptual (picture) problems apply this stepwise strategy to help you interpret molecular scenes and solve problems based on them.

Problem-Solving Roadmaps

are block diagrams—specific to the problem and shown alongside the Solution—that lead you visually through the needed calculation steps.

SAMPLE PROBLEM 5.2 Applying the Volume-Pressure Relationship

Problem Boyle's apprentice finds that the air trapped in a J tube occupies 24.8 cm^3 at 1.12 atm. By adding mercury to the tube, he increases the pressure on the trapped air to 2.64 atm. Assuming constant temperature, what is the new volume of air (in L)?

Plan We must find the final volume (V_2) in liters, given the initial volume (V_1), initial pressure (P_1), and final pressure (P_2). The temperature and amount of gas are fixed. We convert the units of V_1 from cm^3 to mL and then to L, rearrange the ideal gas law to the appropriate form, and solve for V_2. We can predict the direction of the change: since P increases, V will decrease; thus, $V_2 < V_1$. (Note that the roadmap has two parts.)

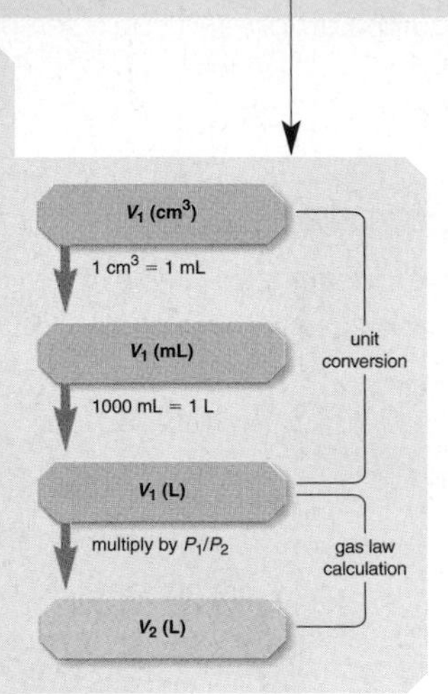

Solution Summarizing the gas variables:

$P_1 = 1.12$ atm $\quad P_2 = 2.64$ atm

$V_1 = 24.8\ cm^3$ (convert to L) $\quad V_2 =$ unknown $\quad T$ and n remain constant

Converting V_1 from cm^3 to L:

$$V_1 = 24.8\ cm^3 \times \frac{1\ mL}{1\ cm^3} \times \frac{1\ L}{1000\ mL} = 0.0248\ L$$

Arranging the ideal gas law and solving for V_2: At fixed n and T, we have

$$\frac{P_1V_1}{\cancel{n_1}\cancel{T_1}} = \frac{P_2V_2}{\cancel{n_2}\cancel{T_2}} \quad \text{or} \quad P_1V_1 = P_2V_2$$

$$V_2 = V_1 \times \frac{P_1}{P_2} = 0.0248\ L \times \frac{1.12\ atm}{2.64\ atm} = 0.0105\ L$$

Check As we predicted, $V_2 < V_1$. Let's think about the relative values of P and V as we check the math. P more than doubled, so V_2 should be less than $\frac{1}{2}V_1$ ($0.0105/0.0248 < \frac{1}{2}$).

Comment Predicting the direction of the change provides another check on the problem setup: To make $V_2 < V_1$, we must multiply V_1 by a number *less than* 1. This means the ratio of pressures must be *less than* 1, so the larger pressure (P_2) must be in the denominator, P_1/P_2.

FOLLOW-UP PROBLEM 5.2 A sample of argon gas occupies 105 mL at 0.871 atm. If the temperature remains constant, what is the volume (in L) at 26.3 kPa?

Brief Solutions to Follow-up Problems

5.1 P_{CO_2} (torr) $= (753.6\ mmHg - 174.0\ mmHg) \times \dfrac{1\ torr}{1\ mmHg}$

$= 579.6$ torr

P_{CO_2} (Pa) $= 579.6\ torr \times \dfrac{1\ atm}{760\ torr} \times \dfrac{1.01325\times10^5\ Pa}{1\ atm}$

$= 7.727\times10^4$ Pa

P_{CO_2} (lb/in^2) $= 579.6\ torr \times \dfrac{1\ atm}{760\ torr} \times \dfrac{14.7\ lb/in^2}{1\ atm}$

$= 11.2$ lb/in^2

5.2 P_2 (atm) $= 26.3\ kPa \times \dfrac{1\ atm}{101.325\ kPa} = 0.260$ atm

V_2 (L) $= 105\ mL \times \dfrac{1\ L}{1000\ mL} \times \dfrac{0.871\ atm}{0.260\ atm} = 0.352$ L

5.3 T_2 (K) $= 273\ K \times \dfrac{9.75\ cm^3}{6.83\ cm^3} = 390.$ K

5.4 P_2 (torr) $= 793\ torr \times \dfrac{35.0\ g - 5.0\ g}{35.0\ g} = 680.$ torr

(There is no need to convert mass to moles because the ratio of masses equals the ratio of moles.)

5.5 $n = \dfrac{PV}{RT} = \dfrac{1.37\ atm \times 438\ L}{0.0821\ \dfrac{atm{\cdot}L}{mol{\cdot}K} \times 294\ K} = 24.9\ mol\ O_2$

Mass (g) of $O_2 = 24.9\ mol\ O_2 \times \dfrac{32.00\ g\ O_2}{1\ mol\ O_2} = 7.97\times10^2\ g\ O_2$

Brief Solutions to Follow-up Problems

provide multistep solutions at the end of the chapter, not just a one-number answer at the back of the book. This fuller treatment is an excellent way for you to reinforce your problem-solving skills.

SAMPLE PROBLEM 17.14 Determining Equilibrium Parameters from Molecular Scenes

Problem For the reaction,

$$X(g) + Y_2(g) \rightleftharpoons XY(g) + Y(g) \quad \Delta H > 0$$

the following molecular scenes depict different reaction mixtures (X = green, Y = purple):

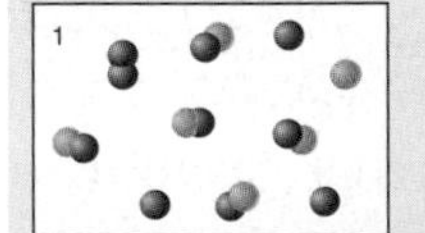

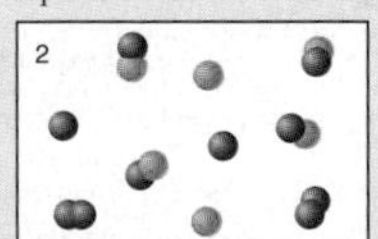

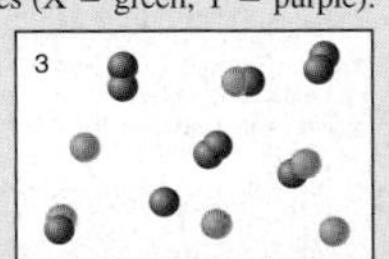

(a) If $K_c = 2$ at the temperature of the reaction, which scene represents the mixture at equilibrium?

(b) Will the reaction mixtures in the other two scenes proceed toward reactants or toward products to reach equilibrium?

(c) For the mixture at equilibrium, how will a rise in temperature affect $[Y_2]$?

Plan (a) We are given the balanced equation and the value of K_c and must choose the scene representing the mixture at equilibrium. We write the expression for Q_c, and for

DESIGNED TO HELP YOU "SEE" CHEMISTRY

Three-Level Illustrations

A Silberberg hallmark, these illustrations provide macroscopic and molecular views of a process that help you connect these two levels of reality with each other and with the chemical equation that describes the process in symbols.

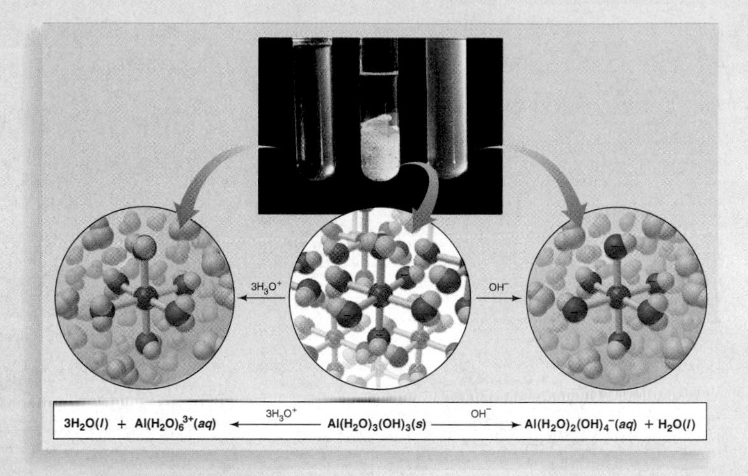

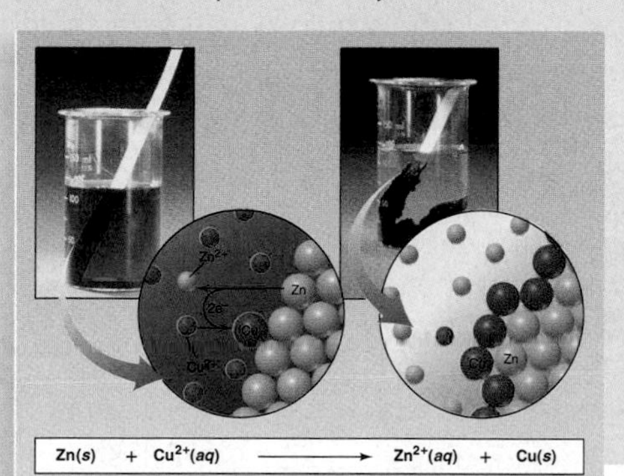

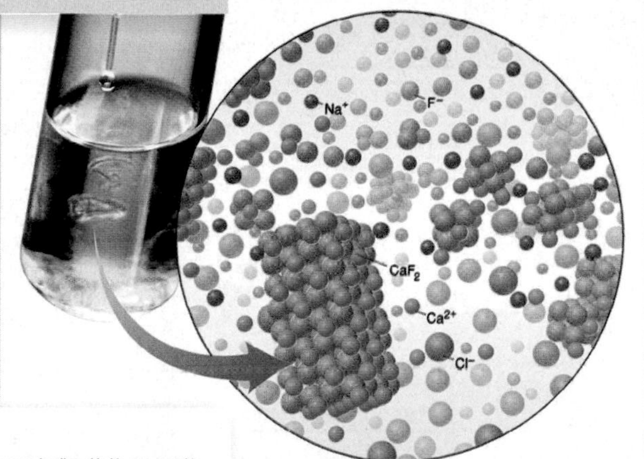

Cutting-Edge Molecular Models

Author and artist worked side by side and employed the most advanced computer-graphic software to provide accurate molecular-scale models and vivid scenes. Included in the Fourth Edition are many new pieces of molecular art as well as newly designed electron-density illustrations.

58 **Chapter 2** The Components of Matter

The formation of the binary ionic compound sodium chloride, common table salt, is depicted in Figure 2.12, from the elements through the atomic-scale electron transfer to the compound. In the electron transfer, a sodium atom, which is neutral because it has the same number of protons as electrons, *loses* 1 electron and forms a sodium cation, Na^+. (The charge on the ion is written as a *right superscript*.) A chlorine atom *gains* the electron and becomes a chloride anion, Cl^-. (The name change from the nonmetal atom to the ion is discussed in the next section.) Even the tiniest visible grain of table salt contains an enormous number of sodium and chloride ions. The oppositely charged ions (Na^+ and Cl^-) attract each other, and the similarly charged ions (Na^+ and Na^+, or Cl^- and Cl^-) repel each other. The resulting solid aggregation is a regular array of alternating Na^+ and Cl^- ions that extends in all three dimensions.

Animation: Formation of an Ionic Compound
Online Learning Center

The strength of the ionic bonding depends to a great extent on the net strength of these attractions and repulsions and is described by *Coulomb's law*, which can be expressed as follows: *the energy of attraction (or repulsion) between two particles is directly proportional to the product of the charges and inversely proportional to the distance between them.*

$$\text{Energy} \propto \frac{\text{charge 1} \times \text{charge 2}}{\text{distance}}$$

In other words, ions with higher charges attract (or repel) each other more strongly than ions with lower charges. Likewise, smaller ions attract (or repel) each other more strongly than larger ions, because their charges are closer together. These effects are summarized in Figure 2.13.

Ionic compounds are neutral; that is, they possess no net charge. For this to occur, they must contain equal numbers of positive and negative *charges* but not necessarily equal numbers of positive and negative *ions*. Because Na^+ and Cl^- each bear a unit charge (1+ or 1−), equal numbers of the ions are present in sodium chloride. However, in sodium oxide, for example, twice as many Na^+ ions balance the 2− charge of the oxide ions, O^{2-}.

Can we predict the number of electrons a given atom will lose or gain when it forms an ion? In the formation of sodium chloride, for example, why does each

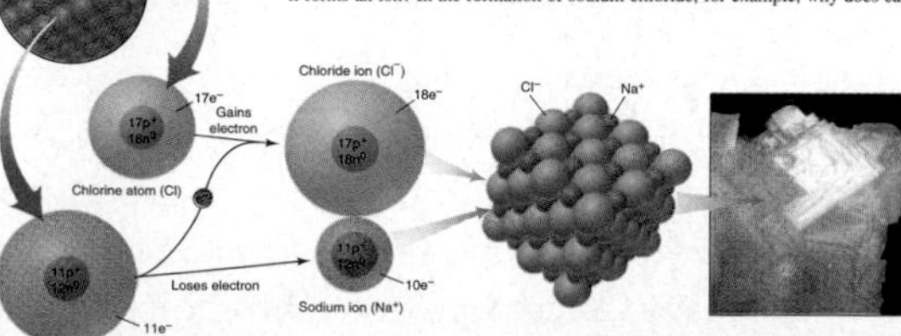

Figure 2.12 The formation of an ionic compound. A, The two elements as seen in the laboratory. **B,** The elements as they might appear on the atomic scale. **C,** The neutral sodium atom loses one electron to become a sodium cation (Na^+), and the chlorine atom gains one electron to become a chloride anion (Cl^-). (Note that when atoms lose electrons, they become ions that are smaller, and when they gain electrons, they become ions that are larger.) **D,** Na^+ and Cl^- ions attract each other and lie in a three-dimensional crystalline array. **E,** This cubic array is reflected in the structure of crystalline NaCl, which occurs naturally as the mineral halite, hence the name *halogens* for the Group 7A(17) elements.

Page Layout

Author and pager collaborated on page layout to ensure that all figures and tables are as close as possible to their related text. In some cases, art is wrapped by text so that you can read about concepts and see them depicted simultaneously.

MAKING CHEMISTRY "REAL" THROUGH MODERN APPLICATIONS

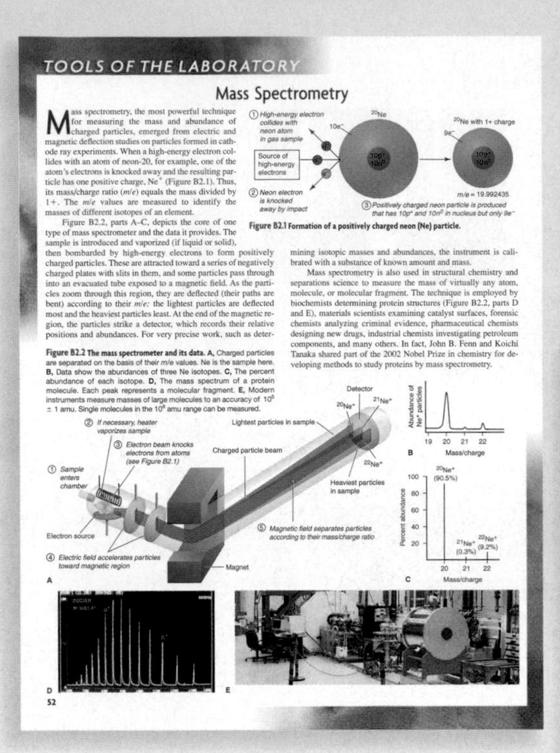

TOOLS OF THE LABORATORY

Mass Spectrometry

Mass spectrometry, the most powerful technique for measuring the mass and abundance of charged particles, emerged from electric and magnetic deflection studies on particles formed in cathode ray experiments. When a high-energy electron collides with an atom of neon-20, for example, one of the atom's electrons is knocked away and the resulting particle has one positive charge, Ne^+ (Figure B2.1). Thus, its mass/charge ratio (m/e) equals the mass divided by 1+. The m/e values are measured to identify the masses of different isotopes of an element.

Figure B2.2, parts A–C, depicts the core of one type of mass spectrometer and the data it provides. The sample is introduced and vaporized (if liquid or solid), then bombarded by high-energy electrons to form positively charged particles. These are attracted toward a series of negatively charged plates with slits in them, and some particles pass through into an evacuated tube exposed to a magnetic field. As the particles zoom through this region, they are deflected (their paths are bent) according to their m/e: the lightest particles are deflected most and the heaviest particles least. At the end of the magnetic region, the particles strike a detector, which records their relative positions and abundances. For very precise work, such as determining isotopic masses and abundances, the instrument is calibrated with a substance of known amount and mass.

Mass spectrometry is also used in structural chemistry and separations science to measure the mass of virtually any atom, molecule, or molecular fragment. The technique is employed by biochemists determining protein structures (Figure B2.2, parts D and E), materials scientists examining catalyst surfaces, forensic chemists analyzing criminal evidence, pharmaceutical chemists designing new drugs, industrial chemists investigating petroleum components, and many others. In fact, John B. Fenn and Koichi Tanaka shared part of the 2002 Nobel Prize in chemistry for developing methods to study proteins by mass spectrometry.

Figure B2.1 Formation of a positively charged neon (Ne) particle.

Figure B2.2 The mass spectrometer and its data. A, Charged particles are separated on the basis of their m/e values. Ne is the sample here. **B,** Data show the abundances of three Ne isotopes. **C,** The percent abundance of each isotope. **D,** The mass spectrum of a protein molecule. Each peak represents a molecular fragment. **E,** Modern instruments measure masses of large molecules to an accuracy of 10^5 ± 1 amu. Single molecules in the 10^8 amu range can be measured.

52

Tools of the Laboratory

These essays describe the key instruments and techniques that chemists use in modern practice to obtain the data that underlie their theories.

Chemical Connections

These extended essays show the interdisciplinary nature of chemistry by applying chemical principles discussed in the text to related scientific fields, including physiology, geology, biochemistry, engineering, and environmental science.

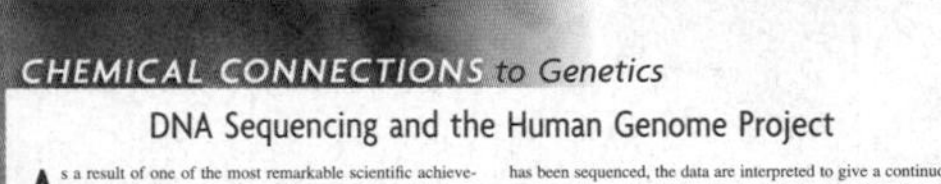

CHEMICAL CONNECTIONS to Genetics

DNA Sequencing and the Human Genome Project

As a result of one of the most remarkable scientific achievements in modern times, we now know our complete genetic makeup, the sequence of the 3 billion nucleotide base pairs in the DNA of the entire human genome! The potential benefits for medical science and for an understanding of our relationship with other creatures are profound.

As is true of so many large scientific endeavors, the Human Genome Project (www.genome.gov) began as a vision of forward-thinking scientists meeting at a conference—in 1986, at the Cold Spring Harbor Symposium in New York. In 1988, after much debate over feasibility—and even advisability—the U.S. Congress funded the $3-billion, 15-year program. Government scientists worked independently of, but in parallel with, private researchers at Celera Genomics, and in early 2003, the project was completed.

Central to this accomplishment is *DNA sequencing*, the process used to determine the identity and order of bases. Sequencing is an indispensable tool in molecular biology and biochemical genetics research. The process usually begins with a series of enzymatic and cloning steps to prepare the sample, but we will focus on the chemical steps that yield the final result.

An Overview of DNA Sequencing

The genome consists of DNA molecules; segments of their chains are the genes, the functional units of heredity. The DNA exists in each cell's nucleus as tightly coiled threads arranged, in the human, into 23 pairs of chromosomes (Figure B15.6). A given chromosome may have 100 million nucleotide bases, but the sequencing process can handle, at one time, DNA fragments only about 2000 bases long. Therefore, the chromosome is first broken into pieces by enzymes that cleave at specific sites. Then, through a variety of "amplification" methods, many copies of the individual DNA "target" fragments are made. After each target fragment has been sequenced, the data are interpreted to give a continuous sequence that represents the structure of the entire chromosome.

Steps in the Sequencing

Once a DNA target fragment is prepared, either of two methods are used to sequence it. We'll introduce the first and detail the second because it is much more common today.

1. *The Maxam-Gilbert chemical cleavage method* uses chemical reactions to break the DNA at specific locations. The total amount of the DNA target fragment is divided into four samples, and each is labeled with phosphorus-32, a radioactive "tag" that will make the final product easy to visualize. Each of the four samples is treated with reactants under conditions of temperature, concentration, and acidity that will cleave the DNA in characteristic ways. For instance, dimethylsulfate ($CH_3OSO_2OCH_3$) followed by heating cleaves the chain at a guanine (G). The cleavage products in each tube are then separated by size (length) through gel electrophoresis (discussed below). Photographic film placed on the gel is exposed by the radioactive tags in the cleavage products. The result is a series of dark bands, from which the sequence of the original DNA target fragment is determined.

2. *The Sanger chain-termination method* uses chemically altered bases to stop the growth of a complementary DNA chain at specific locations. As you've seen, the chain consists of linked 2′-deoxyribonucleoside monophosphate units (dNMP, where N represents A, T, G, or C). The link is a phosphodiester bond from the 3′-OH of one unit to the 5′-OH of the next. The free monomers used to construct the chain are 2′-deoxyribonucleoside triphosphates (dNTP). The Sanger method uses a modified monomer, called a *di*deoxyribonucleoside triphosphate (ddNTP), in which the 3′-OH group is also missing from the ribose unit (Figure B15.7). As soon as the ddNTP is incorporated into the growing chain,

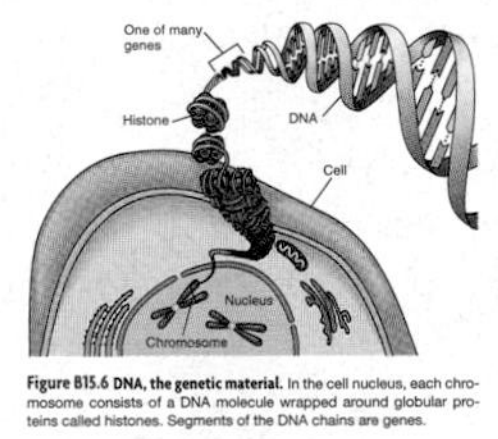

Figure B15.6 DNA, the genetic material. In the cell nucleus, each chromosome consists of a DNA molecule wrapped around globular proteins called histones. Segments of the DNA chains are genes.

662

NEW and Unique to this Text!

In Chapter 15, a Chemical Connections essay describes the chemistry behind the groundbreaking achievement of elucidating the sequence of the human genome.

GALLERY

Molecular Beauty: Odd Shapes with Useful Functions

Chemists see a strange beauty in the geometric intricacy of nature's most minute objects. The simplicity of spherical atoms disappears when they combine to form pentagons, helices, and countless other shapes. Moreover, many "beautiful" molecules have marvelous practical uses.

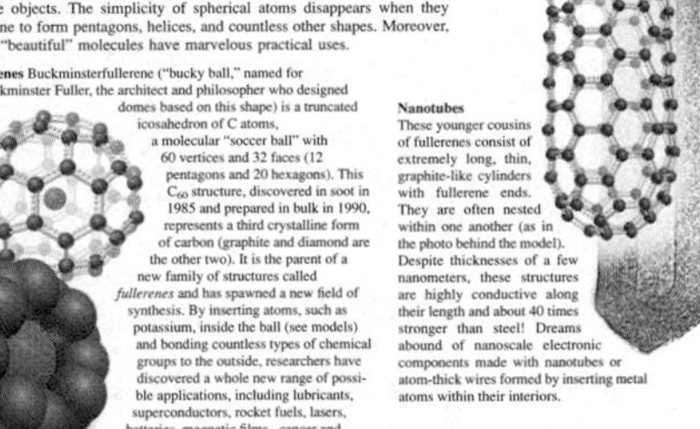

Fullerenes Buckminsterfullerene ("bucky ball," named for R. Buckminster Fuller, the architect and philosopher who designed domes based on this shape) is a truncated icosahedron of C atoms, a molecular "soccer ball" with 60 vertices and 32 faces (12 pentagons and 20 hexagons). This C_{60} structure, discovered in soot in 1985 and prepared in bulk in 1990, represents a third crystalline form of carbon (graphite and diamond are the other two). It is the parent of a new family of structures called *fullerenes* and has spawned a new field of synthesis. By inserting atoms, such as potassium, inside the ball (see models) and bonding countless types of chemical groups to the outside, researchers have discovered a whole new range of possible applications, including lubricants, superconductors, rocket fuels, lasers, batteries, magnetic films, cancer and AIDS drugs—the list grows daily!

Nanotubes These younger cousins of fullerenes consist of extremely long, thin, graphite-like cylinders with fullerene ends. They are often nested within one another (as in the photo behind the model). Despite thicknesses of a few nanometers, these structures are highly conductive along their length and about 40 times stronger than steel! Dreams abound of nanoscale electronic components made with nanotubes or atom-thick wires formed by inserting metal atoms within their interiors.

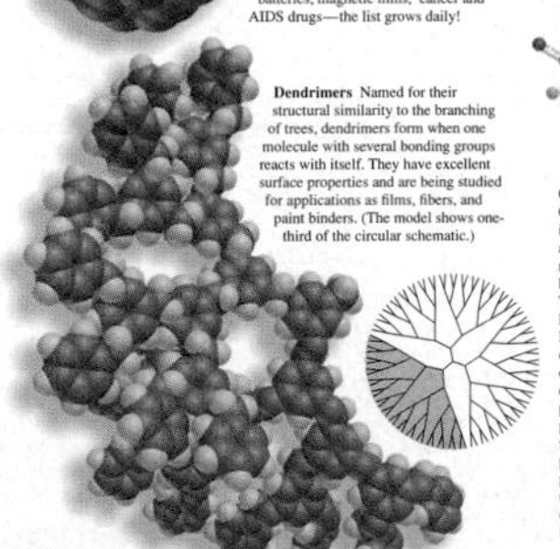

Dendrimers Named for their structural similarity to the branching of trees, dendrimers form when one molecule with several bonding groups reacts with itself. They have excellent surface properties and are being studied for applications as films, fibers, and paint binders. (The model shows one-third of the circular schematic.)

Cubanes Compressing carbon's ideal tetrahedral bond angle of 109.5° to 90° requires energy, which becomes stored in the bond. Thus, cyclobutane, a square of C atoms with two H atoms at each corner, is unstable above 500 K. Much more energy is stored in the bonds of cubane, a cube of C atoms with one H at each corner, synthesized in 1964. The bond-strain energy of cubane can yield enormous explosive power. Tetranitrocubane (shown here) is a very powerful explosive, and octanitrocubane, which was synthesized in 2000, is considered the most powerful non-nuclear explosive known. Other properties of cubanes, which are apparently unrelated to the bond energy, include attacking cancer cells and inactivating an enzyme involved in Parkinson's disease.

386

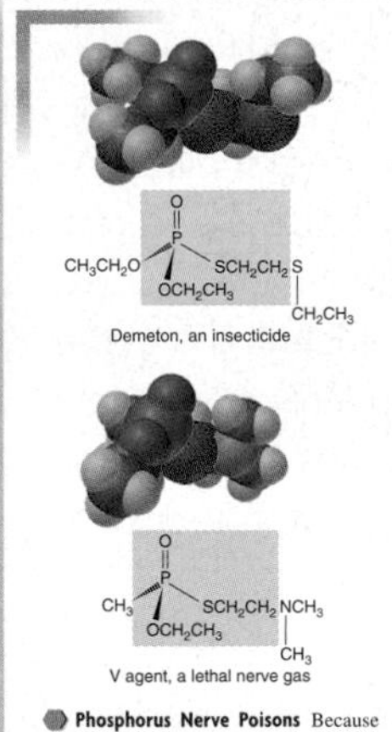

Phosphorus Nerve Poisons Because phosphorus is essential for all organisms, its compounds have many biological actions. In addition to their widespread use as fertilizers, phosphorus compounds are also used as pesticides *and* as military nerve gases. In fact, these two types of substances often have similar structures (see above) and modes of action. Both types affect the target organism by inactivating an enzyme involved in muscle contraction. They bind an amino acid critical to the enzyme's function, allowing the muscle to contract but not relax; the effect on the organism—insect or human—is often fatal. Chemists have synthesized a nerve-gas antidote whose shape mimics the binding site of the enzyme. The drug binds the poison, removing it from the enzyme, and is excreted.

Margin Notes

More than 135 short, lively explanations apply ideas presented in the text. You'll learn how water controls the temperatures of your body and our planet, how crime labs track illegal drugs, how gas behavior affects lung function, how fat-free chips and decaf coffee are made, and what the risks of nuclear radiation are, in addition to handy tips for memorizing relationships, and much more.

Galleries

These illustrated summaries of applications show how common and unusual substances and processes relate to chemical principles. You'll learn how a ballpoint pen works, why bubbles in a drink are round, why contact-lens rinse must have a certain concentration, why the strongest non-nuclear explosive is so powerful, and many other intriguing applications.

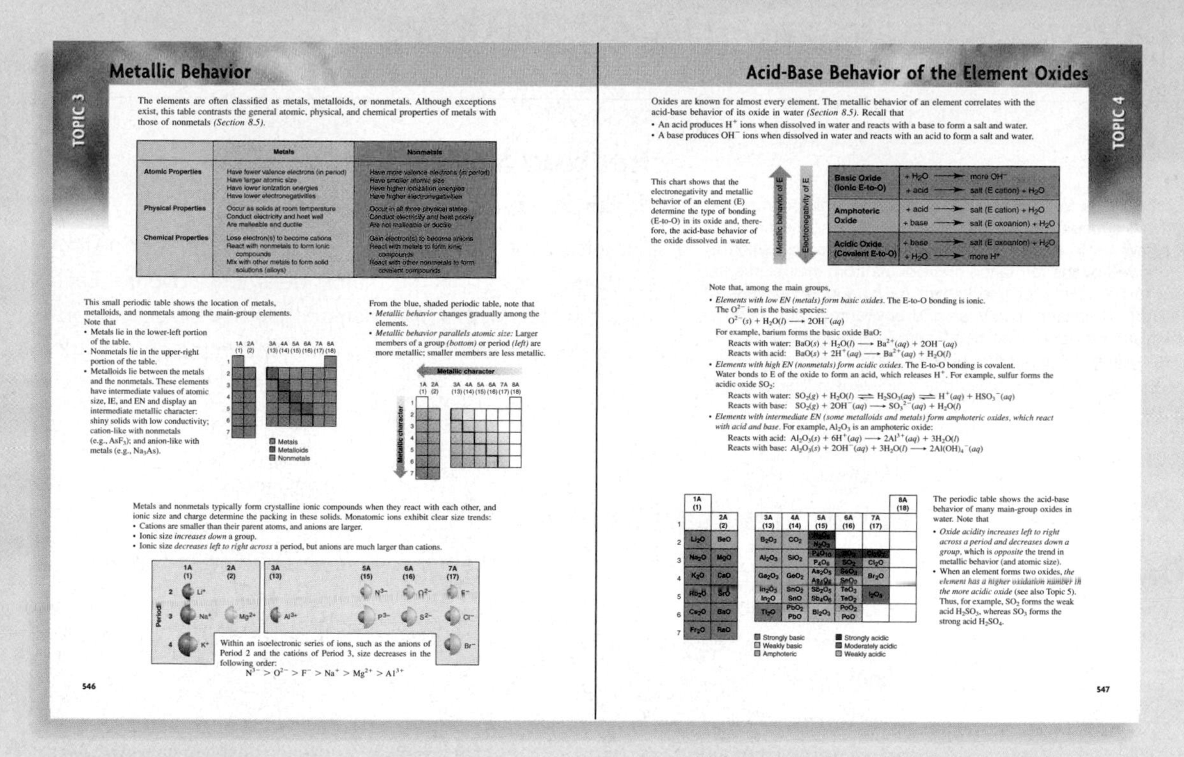

TOPIC 3
Metallic Behavior
The elements are often classified as metals, metalloids, or nonmetals. Although exceptions exist, this table contrasts the general atomic, physical, and chemical properties of metals with those of nonmetals (Section 8.5).
Metals
Nonmetals
Atomic Properties
Physical Properties
Chemical Properties
Metallic character
546
Acid-Base Behavior of the Element Oxides
TOPIC 4
Oxides are known for almost every element. The metallic behavior of an element correlates with the acid-base behavior of its oxide in water (Section 8.5). Recall that
• An acid produces H⁺ ions when dissolved in water and reacts with a base to form a salt and water.
• A base produces OH⁻ ions when dissolved in water and reacts with an acid to form a salt and water.
Basic Oxide (Ionic E-to-O)
Amphoteric Oxide
Acidic Oxide (Covalent E-to-O)
547

Interchapter

This multipage, illustrated Perspective on the Properties of the Elements reviews the major concepts from Chapters 7–13 that deal with atomic properties and their resulting effects on element behavior.

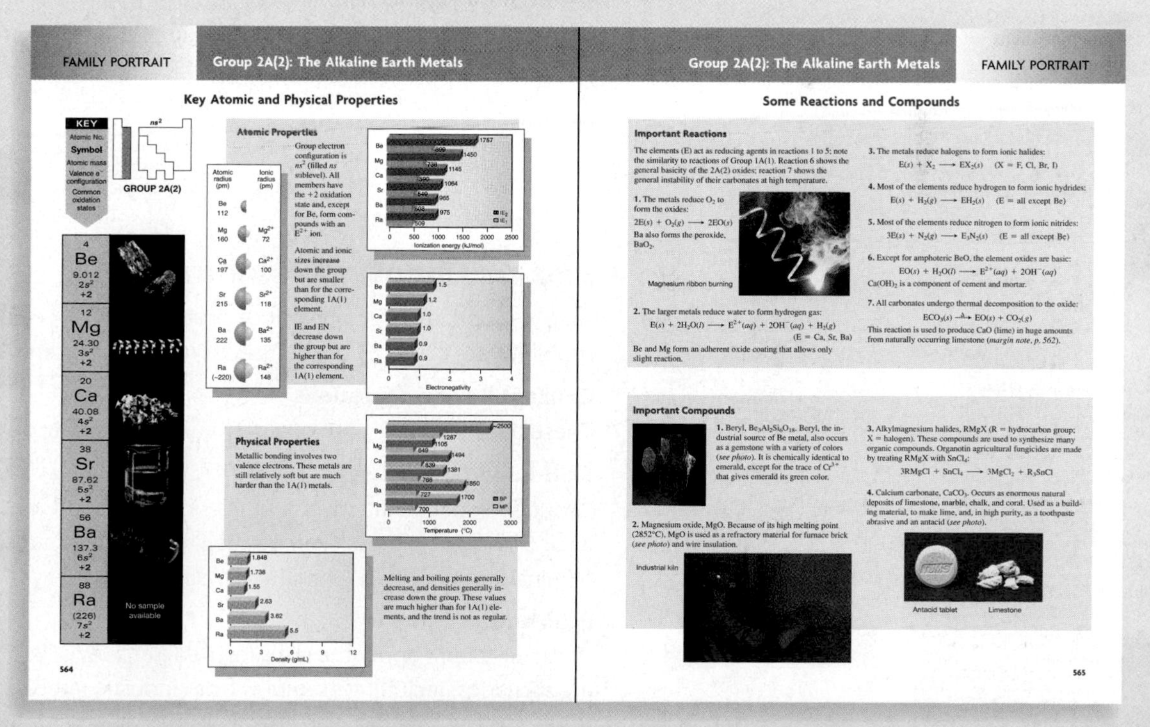

Family Portraits

These illustrated summaries in Chapter 14 detail the atomic, physical, and chemical properties of each main group of elements and describe some of their most important compounds.

LAYING A SOLID LEARNING FOUNDATION

SECTION SUMMARY

Standard states are chosen conditions for substances. When 1 mol of a compound forms from its elements with all substances in their standard states, the enthalpy change is ΔH_f^0. Hess's law allows us to picture a reaction as the decomposition of reactants to their elements, followed by the formation of products from their elements. We use tabulated ΔH_f^0 values to find ΔH_{rxn}^0 or use known ΔH_{rxn}^0 and ΔH_f^0 values to find an unknown ΔH_f^0. Because of dwindling resources and environmental concerns, such as global warming, chemists are developing new energy alternatives, including coal and biomass conversion, hydrogen fuel, and noncombustible energy sources.

Section Summaries and Chapter Perspective

Concise summary paragraphs conclude each section, immediately restating the major ideas just covered. Each chapter ends with a brief perspective that places its topics in the context of previous and upcoming ones.

For Review and Reference (Numbers in parentheses refer to pages, unless noted otherwise.)

Learning Objectives

Relevant section and/or sample problem (SP) numbers appear in parentheses.

Understand These Concepts

1. How Lewis structures depict the atoms and their bonding and lone electron pairs in a molecule or ion (10.1)
2. How resonance and electron delocalization explain bond properties in many compounds with double bonds adjacent to single bonds (10.1)
3. The meaning of formal charge and how it is used to select the most important resonance structure; the difference between formal charge and oxidation number (10.1)
4. The octet rule and its three major exceptions—molecules with a central atom that has an electron deficiency, an odd number of electrons, or an expanded valence shell (10.1)
5. How electron-group repulsions lead to molecular shapes (10.2)
6. The five electron-group arrangements and their associated molecular shapes (10.2)
7. Why double bonds and lone pairs cause deviations from ideal bond angles (10.2)
8. How bond polarities and molecular shape combine to give a molecule polarity (10.3)

Master These Skills

1. Using a stepwise method for writing a Lewis structure from a molecular formula (SPs 10.1 to 10.3 and 10.5)
2. Writing resonance structures for molecules and ions (SP 10.4)
3. Calculating the formal charge of any atom in a molecule or ion (10.5)
4. Predicting molecular shapes from Lewis structures (SPs 10.6 to 10.8)
5. Using molecular shape and electronegativity values to predict the direction of a molecular dipole (SP 10.9)

Key Terms

Section 10.1
Lewis structure (Lewis formula) (366)
resonance structure (resonance form) (369)
resonance hybrid (370)
electron-pair delocalization (370)
formal charge (371)
electron deficient (372)
free radical (373)
expanded valence shell (373)

Section 10.2
valence-shell electron-pair repulsion (VSEPR) theory (376)
molecular shape (376)
bond angle (376)
linear arrangement (377)
linear shape (377)
trigonal planar arrangement (377)
bent shape (V shape) (378)
tetrahedral arrangement (379)
trigonal pyramidal shape (379)
trigonal bipyramidal arrangement (380)
equatorial group (380)
axial group (380)
seesaw shape (380)
T shape (380)
octahedral arrangement (381)
square pyramidal shape (381)
square planar shape (381)

Section 10.3
molecular polarity (387)
dipole moment (μ) (387)

Key Equations and Relationships

10.1 Calculating the formal charge on an atom (371):
Formal charge of atom
$$= \text{no. of valence } e^- - (\text{no. of unshared valence } e^- + \tfrac{1}{2} \text{ no. of shared valence } e^-)$$

10.2 Ranking the effect of electron-pair repulsions on bond angle (380):
Lone pair–lone pair > lone pair–bonding pair > bonding pair–bonding pair

Highlighted Figures and Tables

These figures (F) and tables (T) provide a review of key ideas.

For Review and Reference

A rich catalog of study aids ends each chapter to help you review its content:

Learning Objectives

are listed, with section and/or sample problem numbers, to focus you on key concepts and skills.

Key Terms

are boldfaced within the chapter and listed here by section (with page numbers); they are defined again in the Glossary.

Key Equations and Relationships

are screened and numbered within the chapter and listed here with page numbers.

Highlighted Figures and Tables

are listed with their page numbers so that you can find them easily and review their essential content.

2.113 Each box contains a representation of a binary compound. Determine its name, formula, and molecular (formula) mass.

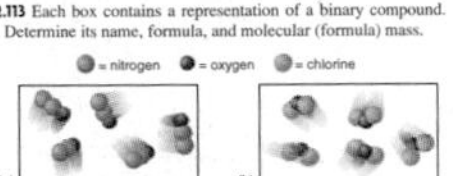

Mixtures: Classification and Separation

Concept Review Questions

2.114 In what main way is separating the components of a mixture different from separating the components of a compound?

2.115 What is the difference between a homogeneous and a heterogeneous mixture?

2.116 Is a solution a homogeneous or a heterogeneous mixture? Give an example of an aqueous solution.

Skill-Building Exercises *(grouped in similar pairs)*

2.117 Classify each of the following as a compound, a homogeneous mixture, or a heterogeneous mixture: (a) distilled water; (b) gasoline; (c) beach sand; (d) wine; (e) air.

2.118 Classify each of the following as a compound, a homogeneous mixture, or a heterogeneous mixture: (a) orange juice; (b) vegetable soup; (c) cement; (d) calcium sulfate; (e) tea.

2.119 Name the technique(s) and briefly describe the procedure you would use to separate each of the following mixtures into two components: (a) table salt and pepper; (b) table sugar and sand; (c) drinking water contaminated with fuel oil; (d) vegetable oil and vinegar.

2.120 Name the technique(s) and briefly describe the procedure you would use to separate each of the following mixtures into two components: (a) crushed ice and crushed glass; (b) table sugar dissolved in ethanol; (c) iron and sulfur; (d) two pigments (chlorophyll *a* and chlorophyll *b*) from spinach leaves.

Problems in Context

2.121 Which separation method is operating in each of the following procedures: (a) pouring a mixture of cooked pasta and boiling water into a colander; (b) removing colored impurities from raw sugar to make refined sugar; (c) preparing coffee by pouring hot water through ground coffee beans?

Comprehensive Problems

2.122 Helium is the lightest noble gas and the second most abundant element (after hydrogen) in the universe.
(a) The radius of a helium atom is 3.1×10^{-11} m; the radius of its nucleus is 2.5×10^{-15} m. What fraction of the spherical atomic volume is occupied by the nucleus (V of a sphere $= \frac{4}{3}\pi r^3$)?
(b) The mass of a helium-4 atom is 6.64648×10^{-24} g, and each of its two electrons has a mass of 9.10939×10^{-28} g. What fraction of this atom's mass is contributed by its nucleus?

2.123 From the following ions and their radii (in pm), choose a pair that gives the strongest ionic bonding and a pair that gives the weakest: Mg^{2+} 72; K^+ 138; Rb^+ 152; Ba^{2+} 135; Cl^- 181; O^{2-} 140; I^- 220.

2.124 Prior to 1961, the atomic mass standard was defined as $\frac{1}{16}$ of the mass of ^{16}O. Based on that standard: (a) What was the mass of carbon-12, given the modern atomic mass of oxygen is 15.9994 amu? (b) What was the mass of potassium-39, given its modern isotopic mass is 38.9637 amu?

2.125 Transition metals, located in the center of the periodic table, have many essential uses as elements and form many important compounds as well. Calculate the molecular mass of the following transition metal compounds:
(a) $[Co(NH_3)_6]Cl_3$ (b) $[Pt(NH_3)_4BrCl]Cl_2$
(c) $K_4[V(CN)_6]$ (d) $[Ce(NH_3)_6][FeCl_4]_3$

2.126 A rock is 5.0% by mass fayalite (Fe_2SiO_4), 7.0% by mass forsterite (Mg_2SiO_4), and the remainder silicon dioxide. What is the mass percent of each element in the rock?

2.127 Polyatomic ions are named by patterns that apply to elements in a given group. Using the periodic table and Table 2.5, give the name of each of the following: (a) SeO_4^{2-}; (b) AsO_4^{3-}; (c) BrO_2^-; (d) $HSeO_4^-$; (e) TeO_3^{2-}.

2.128 Scenes A–I depict various types of matter on the atomic scale. Choose the correct scene(s) for each of the following:
(a) A mixture that fills its container
(b) A substance that cannot be broken down into simpler ones
(c) An element with a very high resistance to flow
(d) A homogeneous mixture
(e) An element that conforms to the walls of its container and displays a surface
(f) A gas consisting of diatomic particles
(g) A gas that can be broken down into simpler substances
(h) A substance with a 2:1 number ratio of its component atoms
(i) Matter that can be separated into its component substances by physical means
(j) A heterogeneous mixture
(k) Matter that obeys the law of definite composition

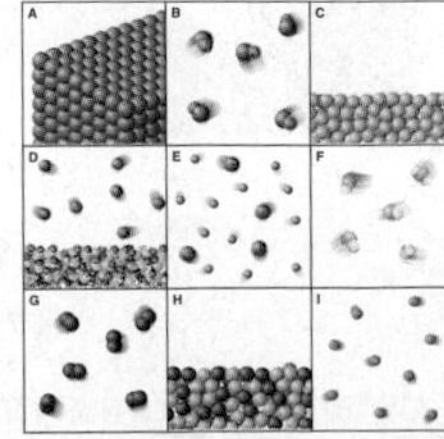

2.129 Nitrogen forms more oxides than any other element. The percents by mass of N in three different nitrogen oxides are (I) 46.69%; (II) 36.85%; (III) 25.94%. (a) Determine the empirical formula of each compound. (b) How many grams of oxygen per 1.00 g of nitrogen are in each compound?

End-of-Chapter Problems

An exceptionally large number of problems follow each chapter. Three types of problems are keyed by chapter section, and comprehensive problems follow:

Concept Review Questions

These questions test your qualitative understanding of key ideas.

Skill-Building Exercises

Grouped in pairs that cover a similar idea, with one of each pair answered in the back of the book, these exercises begin simply and increase in difficulty, gradually eliminating your need for multistep directions.

Problems in Context

These problems apply the skills learned in the Skill-Building Exercises to interesting scenarios, including examples from industry, medicine, and the environment.

Comprehensive Problems

Following the section-based problems is a large group of more challenging problems. These are based on concepts and skills from any section and/or earlier chapter and are filled with applications from related sciences.

SUPPLEMENTS FOR THE INSTRUCTOR

Multimedia Supplements

Digital Content Manager

Electronic art at your fingertips! This cross-platform CD-ROM provides you with artwork from the text in multiple formats. You can easily create customized classroom presentations, visually based tests and quizzes, dynamic content for a course website, or attractive printed support materials. Available on this CD-ROM are the following resources:

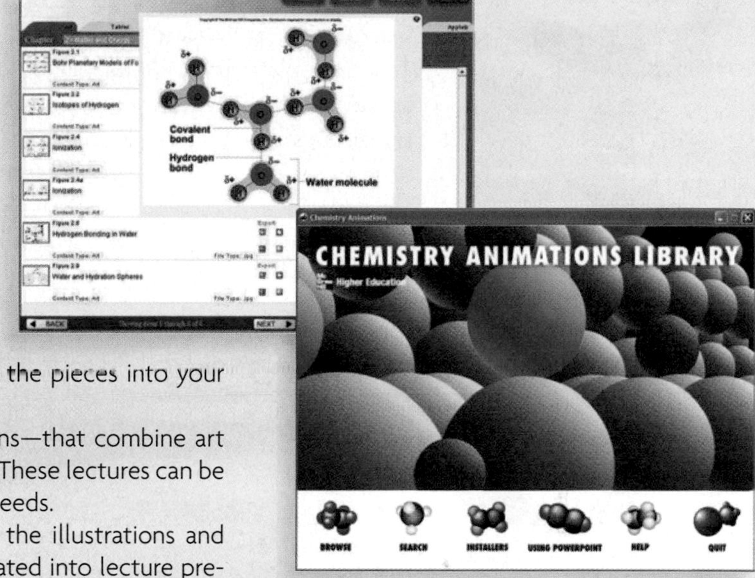

- **Active Art Library** These key art pieces—formatted as PowerPoint slides—illustrate difficult concepts in a step-by-step manner. The artwork is broken into small, incremental frames, allowing you to incorporate the pieces into your lecture in whatever sequence or format you desire.
- **PowerPoint Lecture Outlines** Ready-made presentations—that combine art and lecture notes—cover all of the chapters in the text. These lectures can be used as is or customized by you to meet your specific needs.
- **Art and Photo Library** Full-color digital files of all of the illustrations and many of the photos in the text can be readily incorporated into lecture presentations, exams, or custom-made classroom materials.
- **Worked Example Library** and **Table Library** Access the worked examples and visual tables from the text in electronic format for inclusion in your classroom resources.

Chemistry Animations DVD

This DVD contains more than 300 animations, several authored by Martin Silberberg. This easy-to-use DVD allows you to view the animations quickly and import them into PowerPoint to create multimedia presentations.

Chemistry Online Learning Center

www.mhhe.com/silberberg

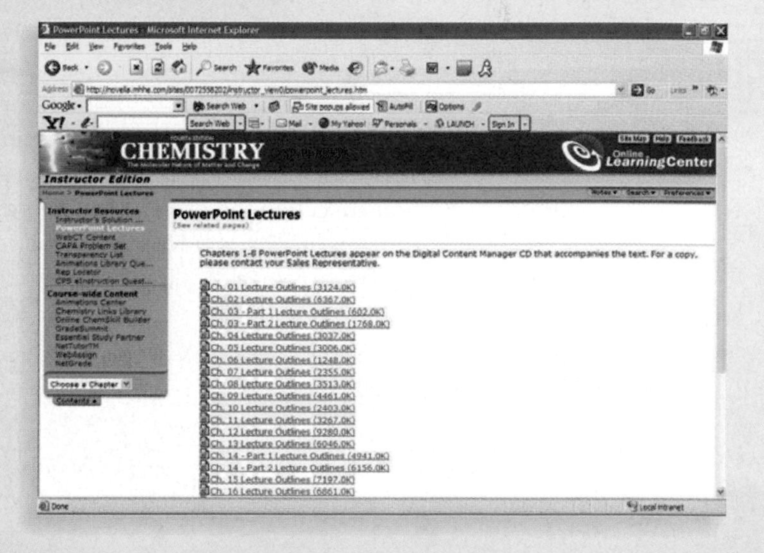

McGraw-Hill's *Chemistry* **Online Learning Center** contains many useful tools for empowering both students and instructors.

- **All assignments and questions are directly tied to text-specific materials** in *Chemistry*, Fourth Edition, but you can edit questions and algorithms, import your own content, and create announcements and due dates for assignments.
- A **secured Instructor Center** stores your course materials, saving you preparation time.
- ***Chemistry*** **Online Learning Center** provides you with access to these essential resources: PowerPoint lecture outlines, *Instructor's Solutions Manual,* animations, and more.
- **ChemSkill Builder** online, McGraw-Hill's powerful electronic homework system, gives students the tutorial practice they need to master concepts covered in your general chemistry course. *ChemSkill Builder* contains more than 1,500 algorithmically generated questions as well as interactive exercises, quizzes, animations, and study tools that correlate directly with each chapter of the text. A record of student work is maintained in an online gradebook so that homework can be easily assigned and factored into the course syllabus.

Course Management Software

With help from **WebCT**, **Blackboard,** or **WebAssign**, you can take complete control over your course content. These course cartridges also provide online testing and powerful student tracking features. The *Chemistry* Online Learning Center is available within all of these platforms. Contact your McGraw-Hill representative for more details.

Instructor's Testing and Resource CD-ROM

This cross-platform CD-ROM includes the *Instructor's Solutions Manual,* which is comprised of all answers from the textbook's end-of-chapter problems, as well as the *Test Bank,* which offers additional questions that can be used for homework assignments and/or the preparation of exams; both are available in Word and pdf formats. The user-friendly computerized test bank utilizes testing software to allow you to quickly create customized exams by sorting questions by format, editing existing questions or adding new ones, and scrambling questions for multiple versions of the same test.

Instructor's Solutions Manual

By Richard H. Langley of Stephen F. Austin State University

This supplement contains complete, worked-out solutions for all the end-of-chapter problems in the text. It can be found within the secure Instructor's Center, within the Online Learning Center.

Printed Supplements

Transparencies

This boxed set of 300 full-color transparency acetates features images from the text that are modified to ensure maximum readability in both small and large classroom settings.

Primis LabBase

By Joseph Lagowski of University of Texas at Austin

More than 40 general chemistry lab experiments are available in this database collection, some from the *Journal of Chemical Education* and others used by Professor Lagowski, enabling you to create your own custom laboratory manual.

General Chemistry Laboratory Manual

By Petra A. M. van Koppen of University of California, Santa Barbara

This definitive lab manual for the two-semester general chemistry course contains 21 experiments that cover the most commonly assigned experiments for the introductory level.

Cooperative Chemistry Laboratory Manual

By Melanie Cooper of Clemson University

This innovative guide features open-ended problems designed to simulate experience in a research lab. Working in groups, students investigate one problem over a period of several weeks, thus completing three or four projects during the semester, rather than one preprogrammed experiment per class. The emphasis here is on experimental design, analysis, problem solving, and communication.

LEARNING AIDS FOR STUDENTS

Multimedia Supplements

Online Learning Center ***www.mhhe.com/silberberg***

McGraw-Hill's ***Online Learning Center*** is your online home page for help.

Text-specific features to complement and solidify lecture concepts include:

- Online homework and quizzes (which are automatically graded and recorded for your instructor)
- Study tools that relate directly to each chapter of the text

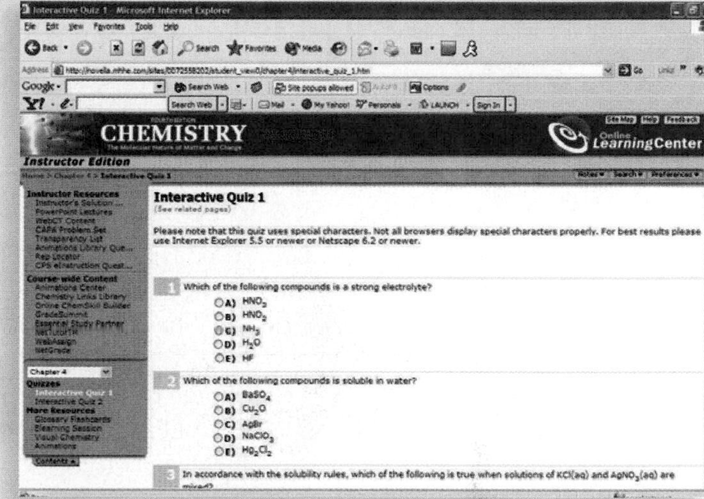

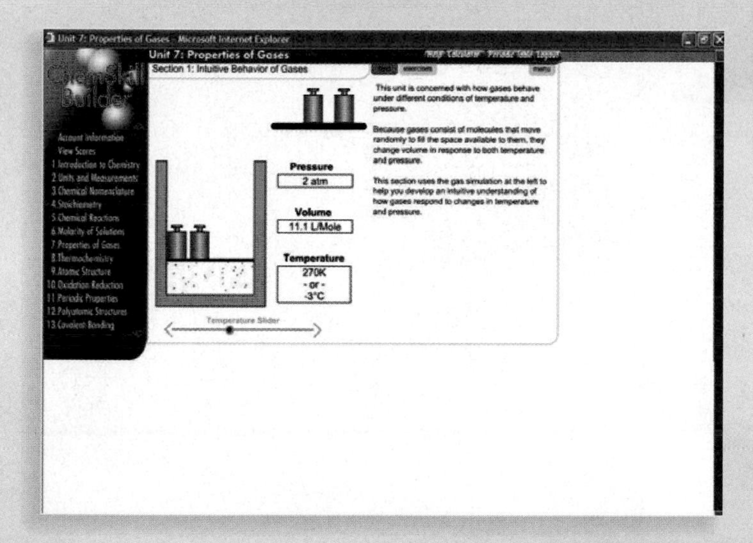

- **ChemSkill Builder** online, McGraw-Hill's powerful electronic homework system, gives you the tutorial practice you need to master concepts covered in your general chemistry course. *ChemSkill Builder* contains more than 1,500 algorithmically generated questions as well as interactive exercises, quizzes, animations, and study tools matched to each chapter of the text. A record of your work is maintained in an online gradebook so that your homework scores can be easily viewed.

Printed Supplements

Student Solutions Manual

By Richard H. Langley of Stephen F. Austin University

This supplement contains detailed solutions and explanations for all color-coded problems in the main text.

Student Study Guide

By Libby Weberg

This valuable ancillary is designed to help you recognize your learning style; understand how to read, classify, and create a problem-solving list; and practice problem-solving skills. For each section of a chapter, the author provides study objectives and a summary of the corresponding text. Following the summary are sample problems with detailed solutions. Each chapter has true-false questions and a self-test, with all answers provided at the end of the chapter.

Chemistry Resource Card

The resource card is a quick and easy source of information on general chemistry. Without having to consult the text, you have right at hand the periodic table and list of elements, tables for conversion factors, equilibrium and thermodynamic data, nomenclature, and key equations.

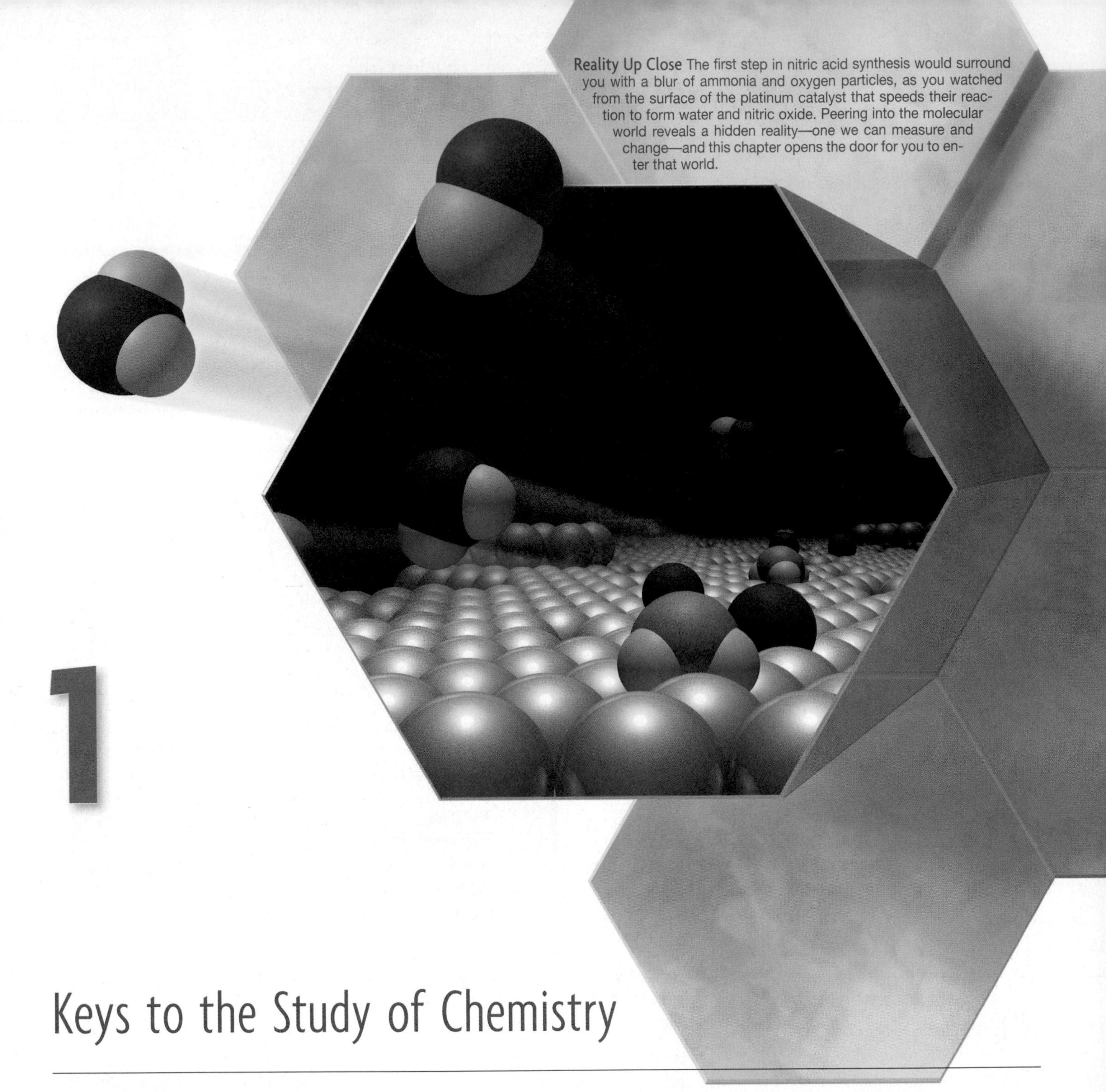

Reality Up Close The first step in nitric acid synthesis would surround you with a blur of ammonia and oxygen particles, as you watched from the surface of the platinum catalyst that speeds their reaction to form water and nitric oxide. Peering into the molecular world reveals a hidden reality—one we can measure and change—and this chapter opens the door for you to enter that world.

Keys to the Study of Chemistry

The science of chemistry stands at the forefront of change in the 21st century, creating "greener" energy sources to power society *and* sustain the environment, using breakthrough knowledge of the human genetic makeup to understand diseases and design medicines, and even researching the origin of life as we investigate our and nearby solar systems for signs of it. Addressing these and countless other challenges and opportunities depends on an understanding of the concepts you will learn in this course.

At this point, your interest in chemistry may be little more than satisfying a requirement for your major. You know that chemistry is a basic science with connections to many careers, but once this course is out of the way, who cares? You may wonder whether chemistry is important at all to you personally. Soon, you'll be wondering how any educated person can function today without knowing some chemistry!

In fact, the impact of chemistry on your daily life is mind-boggling. Consider the beginning of a typical day—perhaps this one—from a chemical point of view. Molecules align in the liquid crystal display of your clock, electrons flow through its circuitry to create a noise, and you throw off a thermal insulator of manufactured polymer. You jump in the shower to emulsify fatty substances on your skin and hair with treated water and formulated detergents. You adorn yourself in an array of processed chemicals—pleasant-smelling pigmented materials suspended in cosmetic gels, dyed polymeric fibers, synthetic footwear, and metal-alloyed jewelry. Today, breakfast is a bowl of nutrient-enriched, spoilage-retarded cereal and milk, a piece of fertilizer-grown, pesticide-treated fruit, and a cup of a hot aqueous solution of neurally stimulating alkaloid. After brushing your teeth with artificially flavored, dental-hardening agents dispersed in a colloidal abrasive, you're ready to leave, so you grab your laptop, an electronic device based on ultrathin, microetched semiconductor layers powered by a series of voltaic cells, collect some books—processed cellulose and plastic, electronically printed with light- and oxygen-resistant inks—hop in your hydrocarbon-fueled, metal-vinyl-ceramic vehicle, electrically ignite a synchronized series of controlled gaseous explosions, and you're off to class!

The influence of chemistry extends to the natural environment as well. The air, water, and land and the organisms that thrive there form a remarkably complex system of chemical interactions. While modern chemical products have enhanced the quality of our lives, their manufacture and use also pose increasing dangers, such as toxic wastes, acid rain, global warming, and ozone depletion. If our careless application of chemical principles has led to some of these problems, our careful application of the same principles is helping to solve them.

Perhaps the significance of chemistry is most profound when you contemplate the chemical nature of biology. Molecular events taking place within you right now allow your eyes to scan this page and your brain cells to translate fluxes of electric charge into thoughts. The most vital biological questions—How did life arise and evolve? How does an organism reproduce, grow, and age? What is the essence of health and disease?—ultimately have chemical answers.

This course comes with a bonus—the development of two mental skills you can apply to any science-related field. The first skill, common to all science courses, is the ability to solve problems systematically. The second is specific to chemistry, for as you comprehend its ideas, your mind's eye will learn to see a hidden level of the universe, one filled with incredibly minute particles hurtling at fantastic speeds, colliding billions of times a second, and interacting in ways that determine everything inside and outside of you. This first chapter holds the keys to help you enter this new world.

Concepts & Skills to Review

before you study this chapter

- exponential (scientific) notation (Appendix A)

IN THIS CHAPTER . . . We begin with fundamental definitions and concepts of matter and energy and their changes. Then, a brief discussion of chemistry's historical origins leads to an overview of how scientists build models to study nature. We consider chemical problem solving, including unit conversion, modern systems of measurement—focusing on mass, length, volume, density, and temperature—and numerical manipulations in calculations. A final essay examines how modern chemists work with other scientists for society's benefit.

1.1 SOME FUNDAMENTAL DEFINITIONS

The science of chemistry deals with the makeup of the entire physical universe. A good place to begin our discussion is with the definition of a few central ideas, some of which may already be familiar to you. **Chemistry** is *the study of matter and its properties, the changes that matter undergoes, and the energy associated with those changes.*

The Properties of Matter

Matter is the "stuff" of the universe: air, glass, planets, students—*anything that has mass and volume.* (In Section 1.5, we discuss the meanings of mass and volume in terms of how they are measured.) Chemists are particularly interested in the **composition** of matter, *the types and amounts of simpler substances that make it up.* A *substance* is a type of matter that has a defined, fixed composition.

We learn about matter by observing its **properties,** *the characteristics that give each substance its unique identity.* To identify a person, we observe such properties as height, weight, eye color, race, fingerprints, and, now, even DNA fingerprint, until we arrive at a unique identification. To identify a substance, chemists observe two types of properties, physical and chemical, which are closely related to two types of change that matter undergoes. **Physical properties** are those that a substance shows *by itself, without changing into or interacting with another substance.* Some physical properties are color, melting point, electrical conductivity, and density.

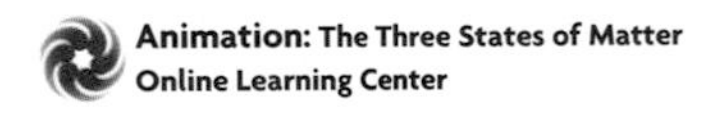
Animation: The Three States of Matter
Online Learning Center

A **physical change** occurs when a substance *alters its physical form,* ***not*** *its composition.* Thus, a physical change results in different physical properties. For example, when ice melts, several physical properties have changed, such as hardness, density, and ability to flow. But the sample has *not* changed its composition: it is still water. The photo in Figure 1.1A shows this change the way you

Figure 1.1 **The distinction between physical and chemical change.**

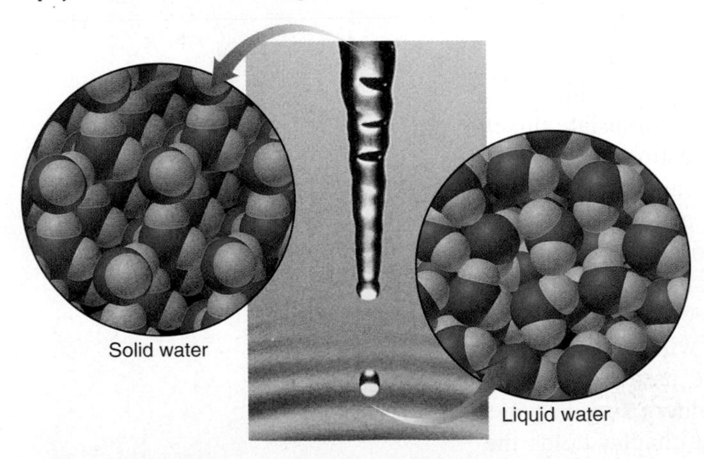

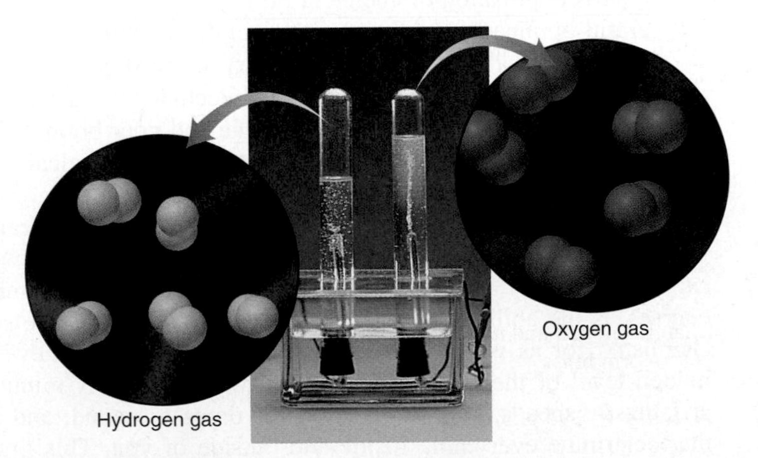

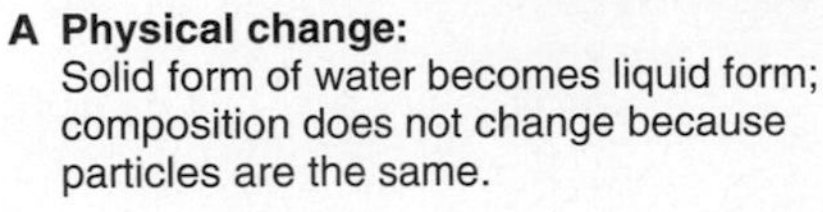
A Physical change:
Solid form of water becomes liquid form; composition does not change because particles are the same.

B Chemical change:
Electric current decomposes water into different substances (hydrogen and oxygen); composition does change because particles are different.

would see it in everyday life. In your imagination, try to see the magnified view that appears in the "blow-up" circles. Here we see the particles that make up the sample; note that the same particles appear in solid and liquid water.

Physical change (same substance before and after):

$$\text{Water (solid form)} \longrightarrow \text{water (liquid form)}$$

On the other hand, **chemical properties** are those that a substance shows *as it changes into or interacts with another substance (or substances).* Some examples of chemical properties are flammability, corrosiveness, and reactivity with acids. A **chemical change,** also called a **chemical reaction,** occurs when *a substance (or substances) is converted into a different substance (or substances).*

Figure 1.1B shows the chemical change (reaction) that occurs when you pass an electric current through water: the water decomposes (breaks down) into two other substances, hydrogen and oxygen, each with physical and chemical properties different from each other *and* from water. The sample *has* changed its composition: it is no longer water, as you can see from the different particles in the magnified view.

Chemical change (different substances before and after):

$$\text{Water} \xrightarrow{\text{electric current}} \text{hydrogen gas} + \text{oxygen gas}$$

A substance is identified by its own set of physical and chemical properties. Some properties of copper appear in Table 1.1.

Table 1.1 Some Characteristic Properties of Copper

Physical Properties	Chemical Properties
Reddish brown, metallic luster  Easily shaped into sheets (malleable) and wires (ductile)	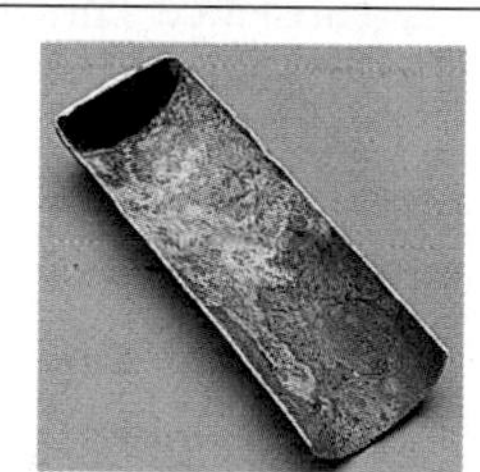 Slowly forms a basic, blue-green sulfate in moist air
Good conductor of heat and electricity  Can be melted and mixed with zinc to form brass	 Reacts with nitric acid (photo) and sulfuric acid
Density = 8.95 g/cm^3 Melting point = 1083°C Boiling point = 2570°C	 Slowly forms a deep-blue solution in aqueous ammonia

The Incredible Range of Physical Change Scientists often study physical change in remarkable settings, using instruments that allow observation far beyond the confines of the laboratory. Instruments aboard the *Voyager* and *Galileo* spacecrafts and the Hubble Space Telescope have measured temperatures on Jupiter's moon Io (shown here) that are hot enough to maintain lakes of molten sulfur and cold enough to create vast snowfields of sulfur dioxide and polar caps swathed in hydrogen sulfide frost. (On Earth, sulfur dioxide is one of the gases released from volcanoes and coal-fired power plants, and hydrogen sulfide occurs in swamp gas.)

The Three States of Matter

Matter occurs commonly in three physical forms called **states:** solid, liquid, and gas. As shown in Figure 1.2 for a general substance, each state is defined by the way it fills a container. A **solid** *has a fixed shape that does not conform to the container shape.* Solids are *not* defined by rigidity or hardness: solid iron is rigid, but solid lead is flexible and solid wax is soft. A **liquid** *conforms to the container shape but fills the container only to the extent of the liquid's volume;* thus, a liquid forms a *surface.* A **gas** *conforms to the container shape also, but it fills the entire container,* and thus, does *not* form a surface. Now, look at the views within the blow-up circles of the figure. The particles in the solid lie next to each other in a regular, three-dimensional array with a definite shape. Particles in the liquid also lie together but move randomly around one another. Particles in the gas usually have great distances between them, as they move randomly throughout the container.

Depending on the temperature and pressure of the surroundings, many substances can exist in each of the three physical states and undergo changes in state as well. As the temperature increases, solid water melts to liquid water, which boils to gaseous water (also called *water vapor*). Similarly, with decreasing temperature, water vapor condenses to liquid water, and with further cooling, the liquid freezes to ice. Many other substances behave in the same way: solid iron melts to liquid (molten) iron and then boils to iron gas at a higher temperature. Cooling the iron gas changes it to liquid and then to solid iron.

Thus, a physical change caused by heating can generally be reversed by cooling, and vice versa. This is *not* generally true for a chemical change. For example, heating iron in moist air causes a chemical reaction that yields the brown, crumbly substance known as rust. Cooling does not reverse this change; rather, another chemical change (or series of them) is required.

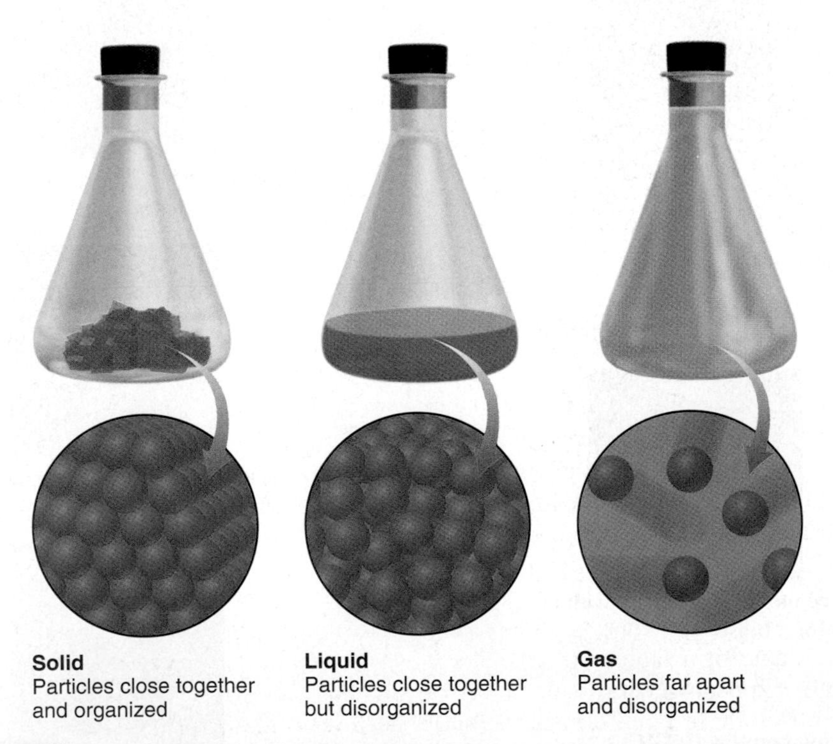

Figure 1.2 The physical states of matter. The magnified (blow-up) views show the atomic-scale arrangement of the particles in the three states of matter.

To summarize the key distinctions:

- A physical change leads to a different form of the same substance (same composition), whereas a chemical change leads to a different substance (different composition).
- A physical change caused by a temperature change can generally be reversed by the opposite temperature change, but this is not generally true of a chemical change.

The following sample problem provides more examples of these types of changes.

SAMPLE PROBLEM 1.1 Distinguishing Between Physical and Chemical Change

Problem Decide whether each of the following processes is primarily a physical or a chemical change, and explain briefly:
(a) Frost forms as the temperature drops on a humid winter night.
(b) A cornstalk grows from a seed that is watered and fertilized.
(c) Dynamite explodes to form a mixture of gases.
(d) Perspiration evaporates when you relax after jogging.
(e) A silver fork tarnishes slowly in air.

Plan The basic question we ask to decide whether a change is chemical or physical is, "Does the substance change composition or just change form?"

Solution (a) Frost forming is a physical change: the drop in temperature changes water vapor (gaseous water) in humid air to ice crystals (solid water).
(b) A seed growing involves chemical change: the seed uses substances from air, fertilizer, soil, and water and energy from sunlight to make complex changes in composition.
(c) Dynamite exploding is a chemical change: the dynamite is converted into other substances.
(d) Perspiration evaporating is a physical change: the water in sweat changes its form, from liquid to gas, but not its composition.
(e) Tarnishing is a chemical change: silver changes to silver sulfide by reacting with sulfur-containing substances in the air.

FOLLOW-UP PROBLEM 1.1 Decide whether each of the following processes is primarily a physical or a chemical change, and explain briefly:
(a) Purple iodine vapor appears when solid iodine is warmed.
(b) Gasoline fumes are ignited by a spark in an automobile engine cylinder.
(c) A scab forms over an open cut.

The Central Theme in Chemistry

Understanding the properties of a substance and the changes it undergoes leads to the central theme in chemistry: *macroscopic* properties and behavior, those we can see, are the results of *submicroscopic* properties and behavior that we cannot see. The distinction between chemical and physical change is defined by composition, which we study macroscopically. But it ultimately depends on the makeup of substances at the atomic scale, as the magnified views of Figure 1.1 show. Similarly, the defining properties of the three states of matter are macroscopic, but they arise from the submicroscopic behavior shown in the magnified views of Figure 1.2. Picturing a chemical scene on the molecular scale, even one as common as the flame of a laboratory burner (see margin), clarifies what is taking place. What is happening when water boils or copper melts? What events occur in the invisible world of minute particles that cause a seed to grow, a neon light to glow, or a nail to rust? Throughout the text, we return to this central idea: we study *observable* changes in matter to understand their *unobservable* causes.

Methane and oxygen form carbon dioxide and water in the flame of a lab burner.

The Importance of Energy in the Study of Matter

In general, physical and chemical changes are accompanied by energy changes. **Energy** is often defined as *the ability to do work.* Essentially, all work involves moving something. Work is done when your arm lifts a book, when an engine moves a car's wheels, or when a falling rock moves the ground as it lands. The object doing the work (arm, engine, rock) transfers some of the energy it possesses to the object on which the work is done (book, wheels, ground).

The total energy an object possesses is the sum of its potential energy and its kinetic energy. **Potential energy** is the *energy due to the* ***position*** *of the object.* **Kinetic energy** is the *energy due to the* ***motion*** *of the object.* Let's examine four systems that illustrate the relationship between these two forms of energy: (1) a weight raised above the ground, (2) two balls attached by a spring, (3) two electrically charged particles, and (4) a fuel and its waste products. A key concept illustrated by all four cases is that *energy is conserved: it may be converted from one form to the other, but it is not destroyed.*

Look at Figure 1.3A and consider a weight you lift above the ground. The energy you use to move the weight against the gravitational attraction of the Earth increases the weight's potential energy (energy due to its position). When the weight is dropped, this additional potential energy is converted to kinetic energy (energy due to motion). Some of this kinetic energy is transferred to the ground as the weight does work, such as driving a stake or simply moving dirt and pebbles. As you can see, the added potential energy does not disappear: it is converted to kinetic energy.

In nature, situations of lower energy are typically favored over those of higher energy: because the weight has less potential energy (and thus less total energy) at rest on the ground than held in the air, it will fall when released. We speak of the situation with the weight elevated and higher in potential energy as being *less stable,* and the situation after the weight has fallen and is lower in potential energy as being *more stable.*

To bring the concept somewhat closer to chemistry, consider the two balls attached by a relaxed spring in Figure 1.3B. When you pull the balls apart, the energy you exert to stretch the spring increases its potential energy. This change in potential energy is converted to kinetic energy when you release the balls and they move closer together. The system of balls and spring is less stable (has more potential energy) when the spring is stretched than when it is relaxed. When you wind a clock, the potential energy in the mainspring increases as the spring is compressed. As the spring relaxes, its potential energy is slowly converted into the kinetic energy of the moving gears and hands.

There are no springs in a chemical substance, of course, but the following situation is similar in terms of energy. Much of the matter in the universe is composed of positively and negatively charged particles. A well-known behavior of charged particles (similar to the behavior of the poles of magnets) results from interactions known as *electrostatic forces: opposite charges attract each other, and like charges repel each other.* When work is done to separate a positive particle from a negative one, the potential energy of the particles increases. As Figure 1.3C shows, that increase in potential energy is converted to kinetic energy when the particles move together again. Also, when two positive (or two negative) particles are pushed toward each other, their potential energy increases, and when they are allowed to move apart, that increase in potential energy is changed into kinetic energy. Like the weight above the ground and the balls connected by a spring, charged particles move naturally toward a position of lower energy, a situation that is more stable.

The chemical potential energy of a substance results from the relative positions and the attractions and repulsions among all its particles. Some substances

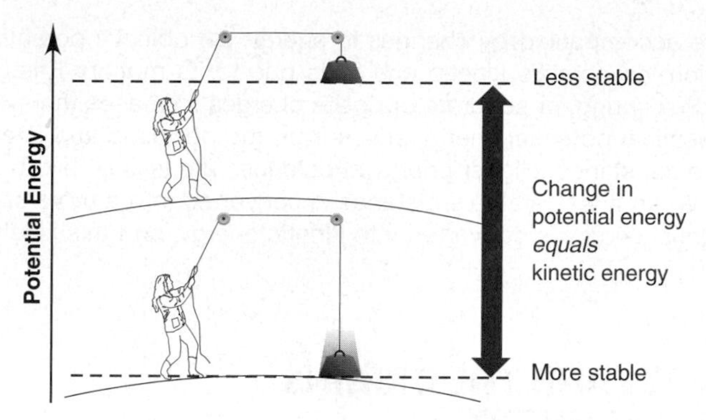

A **A gravitational system.** The potential energy gained when a weight is lifted is converted to kinetic energy as the weight falls.

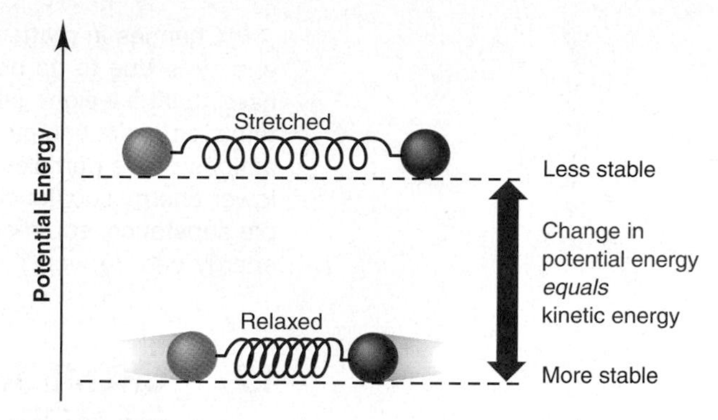

B **A system of two balls attached by a spring.** The potential energy gained when the spring is stretched is converted to the kinetic energy of the moving balls when it is released.

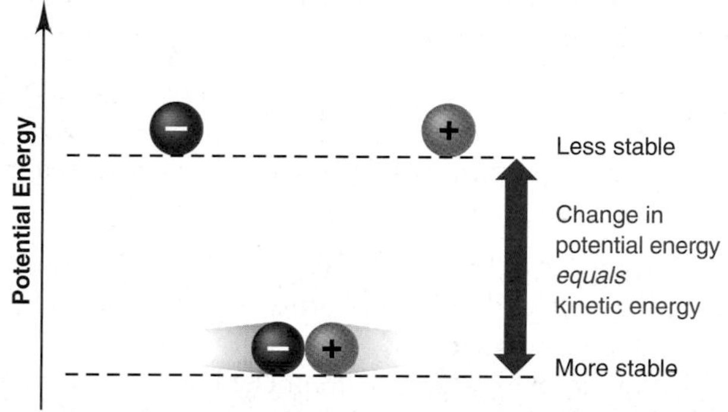

C **A system of oppositely charged particles.** The potential energy gained when the charges are separated is converted to kinetic energy as the attraction pulls them together.

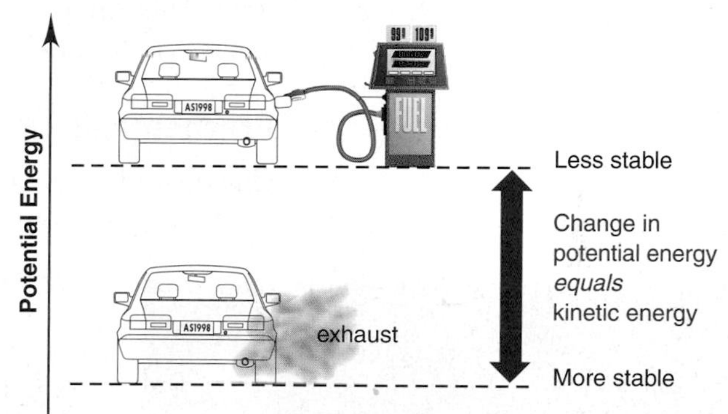

D **A system of fuel and exhaust.** A fuel is higher in chemical potential energy than the exhaust. As the fuel burns, some of its potential energy is converted to the kinetic energy of the moving car.

Figure 1.3 **Potential energy is converted to kinetic energy.** In all four parts of the figure, the dashed horizontal lines indicate the potential energy of the system in each situation.

are richer in this chemical potential energy than others. Fuels and foods, for example, contain more potential energy than the waste products they form. Figure 1.3D shows that when gasoline burns in a car engine, substances with higher chemical potential energy (gasoline and air) form substances with lower potential energy (exhaust gases). This difference in potential energy is converted into the kinetic energy that moves the car, heats the passenger compartment, makes the lights shine, and so forth. Similarly, the difference in potential energy between the food and air we take in and the waste products we excrete is used to move, grow, keep warm, study chemistry, and so on. Note again the essential point: *energy is neither created nor destroyed—it is always conserved as it is converted from one form to the other.*

SECTION SUMMARY

Chemists study the composition and properties of matter and how they change. Each substance has a unique set of physical properties (attributes of the substance itself) and chemical properties (attributes of the substance as it interacts with or changes to other substances). Changes in matter can be physical (different form of the same substance) or chemical (different substance). Matter exists in three physical states—solid, liquid, and gas. The observable features that distinguish these states reflect the arrangement of their particles. A change in physical state brought about by heating may be reversed by cooling. A chemical change can be reversed only by other chemical changes. Macroscopic changes result from submicroscopic changes.

Changes in matter are accompanied by changes in energy. An object's potential energy is due to its position; an object's kinetic energy is due to its motion. Energy used to lift a weight, stretch a spring, or separate opposite charges increases the system's potential energy. Chemical potential energy arises from the positions and interactions of the particles in a substance. Higher energy substances are less stable than lower energy substances. When a less stable substance is converted into a more stable substance, some potential energy is converted into kinetic energy, and this kinetic energy can do work.

1.2 CHEMICAL ARTS AND THE ORIGINS OF MODERN CHEMISTRY

Chemistry has a rich, colorful history. Even some concepts and discoveries that led temporarily along confusing paths have contributed to the heritage of chemistry. This brief overview of early breakthroughs and false directions provides some insight into how modern chemistry arose and how science progresses.

Prechemical Traditions

Chemistry has its origin in a prescientific past that incorporated three overlapping traditions: alchemy, medicine, and technology.

The Alchemical Tradition The occult study of nature practiced in the 1st century AD by Greeks living in northern Egypt later became known by the Arabic name **alchemy.** Its practice spread through the Near East and into Europe, where it dominated Western thinking about matter for more than 1500 years! Alchemists were influenced by the Greek idea that matter naturally strives toward perfection, and they searched for ways to change less valued substances into precious ones. What started as a search for spiritual properties in matter evolved over a thousand years into an obsession with potions to bestow eternal youth and elixirs to transmute "baser" metals, such as lead, into "purer" ones, such as gold. Greed prompted some later European alchemists to paint lead objects with a thin coating of gold to fool wealthy patrons.

Alchemy's legacy to chemistry is mixed at best. The confusion arising from alchemists' use of different names for the same substance and from their belief that matter could be altered magically was very difficult to eliminate. Nevertheless, through centuries of laboratory inquiry, alchemists invented the chemical methods of distillation, percolation, and extraction, and they devised apparatus that today's chemists still use routinely (Figure 1.4). Most important, alchemists encouraged the widespread acceptance of observation and experimentation, which replaced the Greek approach of studying nature solely through reason.

Figure 1.4 An alchemist at work. The apparatus shown here is engaged in distillation, a process still commonly used to separate substances.
This is a portion of a painting by the Englishman Joseph Wright entitled "The Alchymist, in search of the Philosopher's Stone, Discovers Phosphorus, and prays for the successful Conclusion of his Operation as was the custom of the Ancient Chymical Astrologers." Some suggest it portrays the German alchemist Hennig Brand in his laboratory, lit by the glow of phosphorus, which he discovered in 1669.

The Medical Tradition Alchemists greatly influenced medical practice in medieval Europe. Since the 13th century, distillates of roots, herbs, and other plant matter have been used as sources of medicines. Paracelsus (1493–1541), an active alchemist and important physician of the time, considered the body to be a chemical system whose balance of substances could be restored by medical treatment. His followers introduced mineral drugs into 17th-century pharmacy. Although many of these drugs were useless and some harmful, later practitioners employed other mineral prescriptions with increasing success. Thus began the indispensable alliance between medicine and chemistry that thrives today.

The Technological Tradition For thousands of years, people have developed technological skills to carry out changes in matter. Pottery making, dyeing, and especially metallurgy (begun about 7000 years ago) contributed greatly to experience with the properties of materials. During the Middle Ages and the Renaissance, such technology flourished. Books describing how to purify, assay, and coin silver and gold and how to use balances, furnaces, and crucibles were published and regularly updated. Other writings discussed making glass, gunpowder, and other materials. Some even introduced quantitative measurement, which had been lacking in alchemical writings.

Many creations of these early artisans are still unsurpassed today, and people marvel at them in the great centers of world art. Nevertheless, while the artisans' working knowledge of substances was expert, their approach to understanding matter shows little interest in exploring *why* a substance changes or *how to predict* its behavior.

The Phlogiston Fiasco and the Impact of Lavoisier

Chemical investigation in the modern sense—inquiry into the causes of changes in matter—began in the late 17th century but was hampered by an incorrect theory of **combustion,** the process of burning.

At the time, most scientists embraced the **phlogiston theory,** which held sway for nearly 100 years. The theory proposed that combustible materials contain varying amounts of an undetectable substance called *phlogiston,* which is released when the material burns. Highly combustible materials like charcoal contain a lot of phlogiston, and thus release a lot when they burn; similarly, slightly combustible materials like metals contain very little and thus release very little.

However, the theory could not answer some key questions from its critics: "Why is air needed for combustion, and why does charcoal stop burning in a closed vessel?" The theory's supporters responded that air "attracted" the phlogiston out of the charcoal, and that burning in a vessel stops when the air is "saturated" with phlogiston. When a metal burns, it forms its *calx,* which weighs more than the metal, so critics asked, "How can the *loss* of phlogiston cause a *gain* in mass?" Supporters proposed that phlogiston had negative mass! These responses seem ridiculous now, but they point out that the pursuit of science, like any other endeavor, is subject to human failings; even today, it is easier to dismiss conflicting evidence than to give up an established idea.

Into this chaos of "explanations" entered the young French chemist Antoine Lavoisier (1743–1794), who demonstrated the true nature of combustion. In a series of careful measurements, Lavoisier heated mercury calx, decomposing it into mercury and a gas, whose combined masses equaled the starting mass of calx. The reverse experiment—heating mercury with the gas—re-formed the mercury calx, and again, *the total mass remained constant.* Lavoisier proposed that when a metal forms its calx, it does not lose phlogiston but rather combines with this gas, which must be a component of air. To test this idea, Lavoisier heated mercury in a measured volume of air to form mercury calx and noted that only four-fifths of the air volume remained. He placed a burning candle in the remaining air, and it went out, showing that the gas that had combined with the mercury was necessary for combustion. Lavoisier named the gas *oxygen* and called metal calxes *metal oxides.*

Lavoisier's new theory of combustion made sense of the earlier confusion. A combustible substance such as charcoal stops burning in a closed vessel once it combines with all the available oxygen, and a metal oxide weighs more than the metal because it contains the added mass of oxygen. This theory triumphed because it relied on *quantitative, reproducible measurements,* not on the strange

Scientific Thinker Extraordinaire Lavoisier's fame would be widespread, even if he had never performed a chemical experiment. A short list of his other contributions: He improved the production of French gunpowder, which became a key factor in the success of the American Revolution. He established on his farm a scientific balance between cattle, pasture, and cultivated acreage to optimize crop yield. He developed public assistance programs for widows and orphans. He quantified the relation of fiscal policy to agricultural production. He proposed a system of free public education and of societies to foster science, politics, and the arts. He sat on the committee that unified weights and measures in the new metric system. His research into combustion clarified the essence of respiration and metabolism. To support these pursuits, he joined a firm that collected taxes for the king, and only this role was remembered during the French Revolution. Despite his devotion to French society, the father of modern chemistry was guillotined at the age of 50.

properties of undetectable substances. Because this approach is at the heart of science, many propose that the *science* of chemistry began with Lavoisier.

A Great Chemist Yet Strict Phlogistonist Despite the phlogiston theory, chemists made key discoveries during the years it held sway. Many were made by the English clergyman Joseph Priestley (1733–1804), who systematically studied the physical and chemical properties of many gases (inventing "soda water," carbon dioxide dissolved in water, along the way). The gas obtained by heating mercury calx was of special interest to him. In 1775, he wrote to his friend Benjamin Franklin: "Hitherto only two mice and myself have had the privilege of breathing it." Priestley also demonstrated that the gas supports combustion, but he drew the wrong conclusion about it. He called the gas "dephlogisticated air," air devoid of phlogiston, and thus ready to attract it from a burning substance. Priestley's contributions make him one of the great chemists of all time. He was a liberal thinker, favoring freedom of conscience and supporting both the French and American Revolutions, positions that caused severe personal problems throughout his later life. But, scientifically, he remained a conservative, believing strictly in phlogiston and refusing to accept the new theory of combustion.

SECTION SUMMARY

Alchemy, medicine, and technology established processes that have been important to chemists since the 17th century. These prescientific traditions placed little emphasis on objective experimentation, focusing instead on practical experience or mystical explanations. The phlogiston theory dominated thinking about combustion for almost 100 years, but in the mid-1770s, Lavoisier showed that oxygen, a component of air, is required for combustion and combines with a substance as it burns.

1.3 THE SCIENTIFIC APPROACH: DEVELOPING A MODEL

The principles of chemistry have been modified through time and are still evolving. Imagine how differently people learned about the material world tens of thousands of years ago. At the dawn of human experience, our ancestors survived through knowledge acquired by *trial and error:* which types of stone were hard enough to shape others, which types of wood were rigid and which were flexible, which hides could be treated to make clothing, which plants were edible and which were poisonous. Today, the science of chemistry, with its powerful *quantitative theories,* helps us understand the essential nature of materials to make better use of them and create new ones: specialized drugs, advanced composites, synthetic polymers, and countless other new materials (Figure 1.5).

Is there something special about the way scientists think? If we could break down a "typical" modern scientist's thought processes, we could organize them into an approach called the **scientific method.** This approach is not a stepwise checklist, but rather a flexible process of creative thinking and testing aimed at objective, verifiable discoveries about how nature works. It is very important to realize that there is no typical scientist and no single method, and that luck can

A

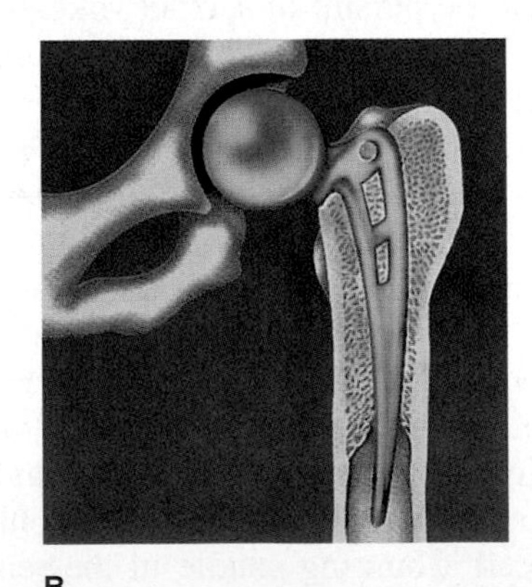
B

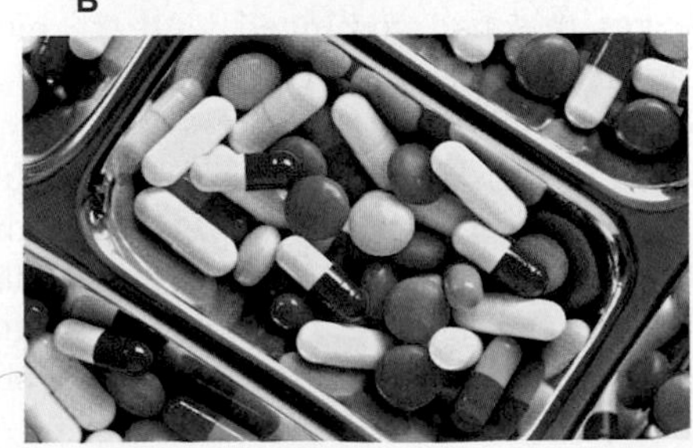
C

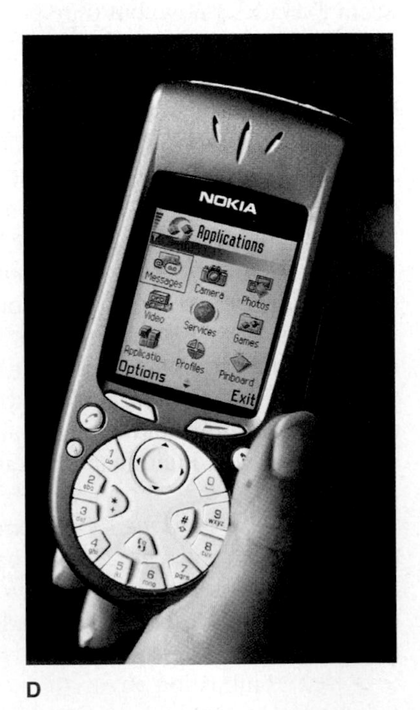

D

Figure 1.5 Modern materials in a variety of applications. A, Specialized steels in bicycles; synthetic polymers in clothing and helmets. **B,** High-tension polymers in synthetic hip joints. **C,** Medicinal agents in pills. **D,** Liquid crystal displays in electronic devices.

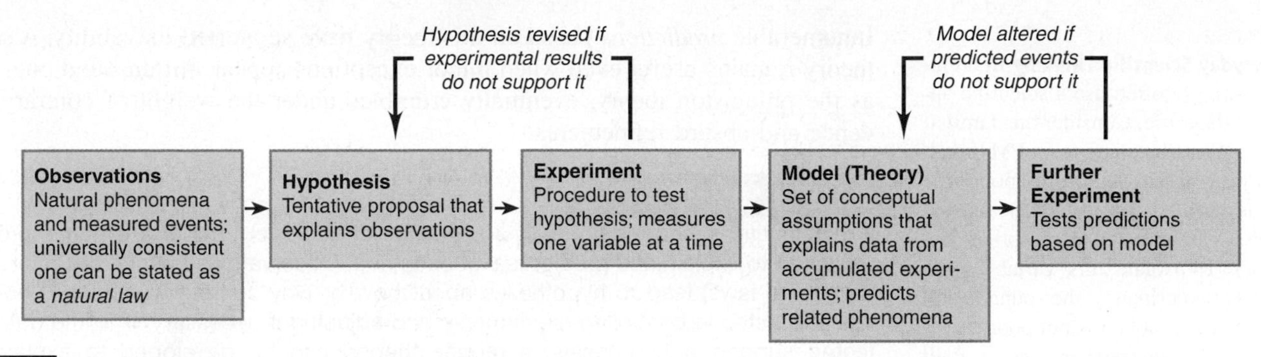

Figure 1.6 The scientific approach to understanding nature. Note that hypotheses and models are mental pictures that are changed to match observations and experimental results, *not* the other way around.

and often has played a key role in scientific discovery. In general terms, the scientific approach includes the following parts (Figure 1.6):

1. ***Observations.*** These are the facts that our ideas must explain. Observation is basic to scientific thinking. The most useful observations are quantitative because they can be compared and allow trends to be seen. Pieces of quantitative information are **data.** When the same observation is made by many investigators in many situations with no clear exceptions, it is summarized, often in mathematical terms, and called a **natural law.** The observation that mass remains constant during chemical change—made by Lavoisier and numerous experimenters since—is known as the law of mass conservation (discussed in Chapter 2).

2. ***Hypothesis.*** Whether derived from actual observation or from a "spark of intuition," a hypothesis is a proposal made to explain an observation. A valid hypothesis need not be correct, but it must be testable. Thus, a hypothesis is often the reason for performing an experiment. If the hypothesis is inconsistent with the experimental results, it must be revised or discarded.

3. ***Experiment.*** An experiment is a clear set of procedural steps that tests a hypothesis. Experimentation is the connection between our hypotheses about nature and nature itself. Often, hypothesis leads to experiment, which leads to revised hypothesis, and so forth. Hypotheses can be altered, but the results of an experiment cannot.

An experiment typically contains at least two **variables,** quantities that can have more than a single value. A well-designed experiment is **controlled** in that it measures the effect of one variable on another while keeping all others constant. For experimental results to be accepted, they must be *reproducible,* not only by the person who designed the experiment, but also by others. Both skill and creativity play a part in great experimental design.

4. ***Model.*** Formulating conceptual models, or **theories,** *based on experiments* is what distinguishes scientific thinking from speculation. As hypotheses are revised according to experimental results, a model gradually emerges that describes how the observed phenomenon occurs. A model is not an exact representation of nature, but rather a simplified version of nature that can be used to make *predictions* about related phenomena. Further investigation refines a model by testing its predictions and altering it to account for new facts.

Lavoisier's overthrow of the phlogiston theory demonstrates the scientific approach. *Observations* of burning and smelting led some to *hypothesize* that combustion involved the loss of phlogiston. *Experiments* by others showing that air is required for burning and that a metal gains mass during combustion led Lavoisier to propose a new *hypothesis,* which he tested repeatedly with quantitative *experiments.* Accumulating evidence supported his developing *model (theory)* that combustion involves combination with a component of air (oxygen).

Everyday Scientific Thinking In an informal way, we often use a scientific approach in daily life. Consider this familiar scenario. While listening to an FM broadcast on your stereo system, you notice the sound is garbled (observation) and assume it is caused by poor reception (hypothesis). To isolate this variable, you play a CD (experiment): the sound is still garbled. If the problem is not poor reception, perhaps the speakers are at fault (new hypothesis). To isolate this variable, you play the CD and listen with headphones (experiment): the sound is clear. You conclude that the speakers need to be repaired (model). The repair shop says the speakers check out fine (new observation), but the power amplifier may be at fault (new hypothesis). Replacing a transistor in the amplifier corrects the garbled sound (new experiment), so the power amplifier was the problem (revised model). Approaching a problem scientifically is a common practice, even if you're not aware of it.

Innumerable *predictions* based on this theory have supported its validity. A sound theory remains useful even when minor exceptions appear. An unsound one, such as the phlogiston theory, eventually crumbles under the weight of contrary evidence and absurd refinements.

SECTION SUMMARY

The scientific method is not a rigid sequence of steps, but rather a dynamic process designed to explain and predict real phenomena. Observations (sometimes expressed as natural laws) lead to hypotheses about how or why something occurs. Hypotheses are tested in controlled experiments and adjusted if necessary. If all the data collected support a hypothesis, a model (theory) can be developed to explain the observations. A good model is useful in predicting related phenomena but must be refined if conflicting data appear.

1.4 CHEMICAL PROBLEM SOLVING

In many ways, learning chemistry is learning how to solve chemistry problems, not only those in exams or homework, but also more complex ones in professional life and society. (The Chemical Connections essay at the end of this chapter provides an example.) This textbook was designed to help strengthen your problem-solving skills. Almost every chapter contains sample problems that apply newly introduced ideas and skills and are worked out in detail. In this section, we discuss the problem-solving approach. Because most problems include calculations, let's first go over some important ideas about measured quantities.

Units and Conversion Factors in Calculations

All measured quantities consist of a number *and* a unit; a person's height is "6 feet," not "6." Ratios of quantities have ratios of units, such as miles/hour. (We discuss the most important units in chemistry in the next section.) To minimize errors, try to make a habit of including units in all calculations.

The arithmetic operations used with measured quantities are the same as those used with pure numbers; in other words, units can be multiplied, divided, and canceled:

- A carpet measuring 3 feet (ft) by 4 ft has an area of

$$\text{Area} = 3\ \text{ft} \times 4\ \text{ft} = (3 \times 4)\ (\text{ft} \times \text{ft}) = 12\ \text{ft}^2$$

- A car traveling 350 miles (mi) in 7 hours (h) has a speed of

$$\text{Speed} = \frac{350\ \text{mi}}{7\ \text{h}} = \frac{50\ \text{mi}}{1\ \text{h}}\ (\text{often written } 50\ \text{mi}\cdot\text{h}^{-1})$$

- In 3 hours, the car travels a distance of

$$\text{Distance} = 3\ \cancel{\text{h}} \times \frac{50\ \text{mi}}{1\ \cancel{\text{h}}} = 150\ \text{mi}$$

Conversion factors are ratios used to express a measured quantity in different units. Suppose we want to know the distance of that 150-mile car trip in feet. To convert the distance between miles and feet, we use equivalent quantities to construct the desired conversion factor. The equivalent quantities in this case are 1 mile and the number of feet in 1 mile:

$$1\ \text{mi} = 5280\ \text{ft}$$

We can construct two conversion factors from this equivalency. Dividing both sides by 5280 ft gives one conversion factor (shown in blue):

$$\frac{1\ \text{mi}}{5280\ \text{ft}} = \frac{5280\ \cancel{\text{ft}}}{5280\ \cancel{\text{ft}}} = 1$$

And, dividing both sides by 1 mi gives the other conversion factor (the inverse):

$$\frac{1\ \cancel{\text{mi}}}{1\ \cancel{\text{mi}}} = \frac{5280\ \text{ft}}{1\ \text{mi}} = 1$$

It's very important to see that, since the numerator and denominator of a conversion factor are equal, multiplying by a conversion factor is the same as multiplying by 1. Therefore, *even though the number and unit of the quantity change, the size of the quantity remains the same.*

In our example, we want to convert the distance in miles to the equivalent distance in feet. Therefore, we choose the conversion factor with units of feet in the numerator, because it cancels units of miles and gives units of feet:

$$\text{Distance (ft)} = 150\ \cancel{\text{mi}} \times \frac{5280\ \text{ft}}{1\ \cancel{\text{mi}}} = 792{,}000\ \text{ft}$$

$$\text{mi} \Longrightarrow \text{ft}$$

Choosing the correct conversion factor is made much easier if you think through the calculation to decide whether the answer expressed in the new units should have a larger or smaller number. In the previous case, we know that a foot is *smaller* than a mile, so the distance in feet should have a *larger* number (792,000) than the distance in miles (150). The conversion factor has the larger number (5280) in the numerator, so it gave a larger number in the answer. The main goal is that *the chosen conversion factor cancels all units except those required for the answer.* Set up the calculation so that the unit you are converting *from* (beginning unit) is in the *opposite position in the conversion factor* (numerator or denominator). It will then cancel and leave the unit you are converting *to* (final unit):

$$\cancel{\text{beginning unit}} \times \frac{\text{final unit}}{\cancel{\text{beginning unit}}} = \text{final unit} \qquad \text{as in} \qquad \cancel{\text{mi}} \times \frac{\text{ft}}{\cancel{\text{mi}}} = \text{ft}$$

Or, in cases that involve units raised to a power,

$$(\cancel{\text{beginning unit}} \times \cancel{\text{beginning unit}}) \times \frac{\text{final unit}^2}{\cancel{\text{beginning unit}^2}} = \text{final unit}^2$$

$$\text{as in} \qquad (\cancel{\text{ft}} \times \cancel{\text{ft}}) \times \frac{\text{mi}^2}{\cancel{\text{ft}^2}} = \text{mi}^2$$

Or, in cases that involve a ratio of units,

$$\frac{\cancel{\text{beginning unit}}}{\text{final unit}_1} \times \frac{\text{final unit}_2}{\cancel{\text{beginning unit}}} = \frac{\text{final unit}_2}{\text{final unit}_1} \qquad \text{as in} \qquad \frac{\cancel{\text{mi}}}{\text{h}} \times \frac{\text{ft}}{\cancel{\text{mi}}} = \frac{\text{ft}}{\text{h}}$$

We use the same procedure to convert between systems of units, for example, between the English (or American) unit system and the International System (a revised metric system discussed fully in the next section). Suppose we know the height of Angel Falls in Venezuela (Figure 1.7) to be 3212 ft, and we find its height in miles as

$$\text{Height (mi)} = 3212\ \cancel{\text{ft}} \times \frac{1\ \text{mi}}{5280\ \cancel{\text{ft}}} = 0.6083\ \text{mi}$$

$$\text{ft} \Longrightarrow \text{mi}$$

Now, we want its height in kilometers (km). The equivalent quantities are

$$1.609\ \text{km} = 1\ \text{mi}$$

Since we are converting from miles to kilometers, we use the conversion factor with kilometers in the numerator in order to cancel miles:

$$\text{Height (km)} = 0.6083\ \cancel{\text{mi}} \times \frac{1.609\ \text{km}}{1\ \cancel{\text{mi}}} = 0.9788\ \text{km}$$

$$\text{mi} \Longrightarrow \text{km}$$

Notice that, since kilometers are *smaller* than miles, this conversion factor gave us a *larger* number (0.9788 is larger than 0.6083).

Figure 1.7 Angel Falls. The world's tallest waterfall is 3212 ft high.

If we want the height of Angel Falls in meters (m), we use the equivalent quantities 1 km = 1000 m to construct the conversion factor:

$$\text{Height (m)} = 0.9788 \cancel{\text{km}} \times \frac{1000\ \text{m}}{1\ \cancel{\text{km}}} = 978.8\ \text{m}$$

$$\text{km} \Longrightarrow \text{m}$$

In longer calculations, we often string together several conversion steps:

$$\text{Height (m)} = 3212\ \cancel{\text{ft}} \times \frac{1\ \cancel{\text{mi}}}{5280\ \cancel{\text{ft}}} \times \frac{1.609\ \cancel{\text{km}}}{1\ \cancel{\text{mi}}} \times \frac{1000\ \text{m}}{1\ \cancel{\text{km}}} = 978.8\ \text{m}$$

$$\text{ft} \Longrightarrow \text{mi} \Longrightarrow \text{km} \Longrightarrow \text{m}$$

The use of conversion factors in calculations is known by various names, such as the factor-label method or **dimensional analysis** (because units represent physical dimensions). We use this method in quantitative problems throughout the text.

A Systematic Approach to Solving Chemistry Problems

The approach we use in this text provides a systematic way to work through a problem. It emphasizes reasoning, not memorizing, and is based on a very simple idea: plan how to solve the problem *before* you go on to solve it, and then check your answer. Try to develop a similar approach on homework and exams. In general, the sample problems consist of several parts:

1. **Problem.** This part states all the information you need to solve the problem (usually framed in some interesting context).
2. **Plan.** The overall solution is broken up into two parts, *plan* and *solution,* to make a point: *think* about how to solve the problem *before* juggling numbers. There is often more than one way to solve a problem, so the plan shown in a given text problem is just one possibility. The plan will
   - Clarify the known and unknown. (What information do you have, and what are you trying to find?)
   - Suggest the steps from known to unknown. (What ideas, conversions, or equations are needed to solve the problem?)
   - Present a "roadmap" of the solution for many problems in early chapters (and in some later ones). The roadmap is a visual summary of the planned steps. Each step is shown by an arrow labeled with information about the conversion factor or operation needed.
3. **Solution.** In this part, the steps appear in the same order as they were planned.
4. **Check.** In most cases, a quick check is provided to see if the results make sense: Are the units correct? Does the answer seem to be the right size? Did the change occur in the expected direction? Is it reasonable chemically? We often do a rough calculation to see if the answer is "in the same ballpark" as the calculated result, just to make sure we didn't make a large error. *Always* check your answers, especially in a multipart problem, where an error in an early step can affect all later steps. Here's a typical "ballpark" calculation. You are at the music store and buy three CD's at $14.97 each. With a 5% sales tax, the bill comes to $47.16. In your mind, you quickly check that 3 times approximately $15 is $45, and, given the sales tax, the cost should be a bit more. So, the amount of the bill is in the right ballpark.
5. **Comment.** This part is included occasionally to provide additional information, such as an application, an alternative approach, a common mistake to avoid, or an overview.
6. **Follow-up Problem.** This part consists of a problem statement only and provides practice by applying the same ideas as the sample problem. Try to solve it *before* you look at the brief worked-out solution at the end of the chapter.

Of course, you can't learn to solve chemistry problems, any more than you can learn to swim, by reading about an approach. Practice is the key to mastery. Here are a few suggestions that can help:

- Follow along in the sample problem with pencil, paper, and calculator.
- Do the follow-up problem as soon as you finish studying the sample problem. Check your answer against the solution at the end of the chapter.
- Read the sample problem and text explanations again if you have trouble.
- Work on as many of the problems at the end of the chapter as you can. They review and extend the concepts and skills in the text. Answers are given in the back of the book for problems with a colored number, but try to solve them yourself first. Let's apply this approach in a unit-conversion problem.

SAMPLE PROBLEM 1.2 Converting Units of Length

Problem To wire your stereo equipment, you need 325 centimeters (cm) of speaker wire that sells for \$0.15/ft. What is the price of the wire?

Plan We know the length of wire in centimeters and the cost in dollars per foot (\$/ft). We can find the unknown price of the wire by converting the length from centimeters to inches (in) and from inches to feet. Then the cost (1 ft = \$0.15) gives us the equivalent quantities to construct the factor that converts feet of wire to price in dollars. The roadmap starts with the known and moves through the calculation steps to the unknown.

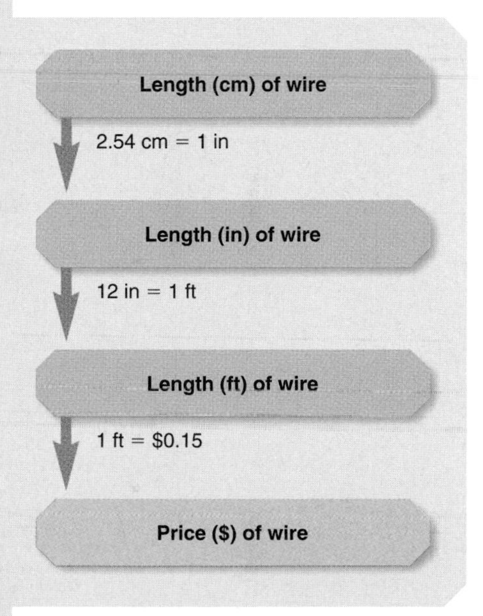

Solution Converting the known length from centimeters to inches: The equivalent quantities alongside the roadmap arrow are the ones needed to construct the conversion factor. We choose 1 in/2.54 cm, rather than the inverse, because it gives an answer in inches:

$$\text{Length (in)} = \text{length (cm)} \times \text{conversion factor} = 325\ \cancel{\text{cm}} \times \frac{1\ \text{in}}{2.54\ \cancel{\text{cm}}} = 128\ \text{in}$$

Converting the length from inches to feet:

$$\text{Length (ft)} = \text{length (in)} \times \text{conversion factor} = 128\ \cancel{\text{in}} \times \frac{1\ \text{ft}}{12\ \cancel{\text{in}}} = 10.7\ \text{ft}$$

Converting the length in feet to price in dollars:

$$\text{Price (\$)} = \text{length (ft)} \times \text{conversion factor} = 10.7\ \cancel{\text{ft}} \times \frac{\$0.15}{1\ \cancel{\text{ft}}} = \boxed{\$1.60}$$

Check The units are correct for each step. The conversion factors make sense in terms of the relative unit sizes: the number of inches is *smaller* than the number of centimeters (an inch is *larger* than a centimeter), and the number of feet is *smaller* than the number of inches. The total price seems reasonable: a little more than 10 ft of wire at \$0.15/ft should cost a little more than \$1.50.

Comment 1. We could also have strung the three steps together:

$$\text{Price (\$)} = 325\ \cancel{\text{cm}} \times \frac{1\ \cancel{\text{in}}}{2.54\ \cancel{\text{cm}}} \times \frac{1\ \text{ft}}{12\ \cancel{\text{in}}} \times \frac{\$0.15}{1\ \text{ft}} = \$1.60$$

2. There are usually alternative sequences in unit-conversion problems. Here, for example, we would get the same answer if we first converted the cost of wire from \$/ft to \$/cm and kept the wire length in cm. Try it yourself.

FOLLOW-UP PROBLEM 1.2 A furniture factory needs 31.5 ft^2 of fabric to upholster one chair. Its Dutch supplier sends the fabric in bolts of exactly 200 m^2. What is the maximum number of chairs that can be upholstered by 3 bolts of fabric (1 m = 3.281 ft)?

SECTION SUMMARY

A measured quantity consists of a number and a unit. Conversion factors are used to express a quantity in different units and are constructed as a ratio of equivalent quantities. The problem-solving approach used in this text usually has four parts: (1) devise a plan for the solution, (2) put the plan into effect in the calculations, (3) check to see if the answer makes sense, and (4) practice with similar problems.

How Many Barleycorns from His Majesty's Nose to His Thumb? Systems of measurement began thousands of years ago as trade, building, and land surveying spread throughout the civilized world. For most of that time, however, measurement was based on inexact physical standards. For example, an inch was the length of three barleycorns (seeds) placed end to end; a yard was the distance from the tip of King Edgar's nose to the tip of his thumb with his arm outstretched; and an acre was the area tilled by one man working with a pair of oxen in a day.

1.5 MEASUREMENT IN SCIENTIFIC STUDY

Almost everything we own—clothes, house, food, vehicle—is manufactured with measured parts, sold in measured amounts, and paid for with measured currency. Measurement is so commonplace that it's easy to take for granted, but it has a history characterized by the search for *exact, invariable standards.*

Our current system of measurement began in 1790, when the newly formed National Assembly of France, of which Lavoisier was a member, set up a committee to establish consistent unit standards. This effort led to the development of the *metric system.* In 1960, another international committee met in France to establish the International System of Units, a revised metric system now accepted by scientists throughout the world. The units of this system are called **SI units,** from the French **S**ystème **I**nternational d'Unités.

General Features of SI Units

As Table 1.2 shows, the SI system is based on a set of seven **fundamental units,** or **base units,** each of which is identified with a physical quantity. All other units, called **derived units,** are combinations of these seven base units. For example, the derived unit for speed, meters per second (m/s), is the base unit for length (m) divided by the base unit for time (s). (Derived units that occur as a ratio of two or more base units can be used as conversion factors.) For quantities that are much smaller or much larger than the base unit, we use decimal prefixes and exponential (scientific) notation. Table 1.3 shows the most important prefixes. (If

Table 1.2 SI Base Units

Physical Quantity (Dimension)	Unit Name	Unit Abbreviation
Mass	kilogram	kg
Length	meter	m
Time	second	s
Temperature	kelvin	K
Electric current	ampere	A
Amount of substance	mole	mol
Luminous intensity	candela	cd

Table 1.3 Common Decimal Prefixes Used with SI Units

Prefix*	Prefix Symbol	Word	Conventional Notation	Exponential Notation
tera	T	trillion	1,000,000,000,000	1×10^{12}
giga	G	billion	1,000,000,000	1×10^{9}
mega	M	million	1,000,000	1×10^{6}
kilo	k	thousand	1,000	1×10^{3}
hecto	h	hundred	100	1×10^{2}
deka	da	ten	10	1×10^{1}
—	—	one	1	1×10^{0}
deci	d	tenth	0.1	1×10^{-1}
centi	c	hundredth	0.01	1×10^{-2}
milli	m	thousandth	0.001	1×10^{-3}
micro	μ	millionth	0.000001	1×10^{-6}
nano	n	billionth	0.000000001	1×10^{-9}
pico	p	trillionth	0.000000000001	1×10^{-12}
femto	f	quadrillionth	0.000000000000001	1×10^{-15}

*The prefixes most frequently used by chemists appear in bold type.

you need a review of exponential notation, read Appendix A.) Because these prefixes are based on powers of 10, SI units are easier to use in calculations than are English units such as pounds and inches.

Some Important SI Units in Chemistry

Let's discuss some of the SI units for quantities that we use early in the text: length, volume, mass, density, temperature, and time. (Units for other quantities are presented in later chapters, as they are used.) Table 1.4 shows some useful SI quantities for length, volume, and mass, along with their equivalents in the English system.

Table 1.4 Common SI-English Equivalent Quantities

Quantity	SI	SI Equivalents	English Equivalents	English to SI Equivalent
Length	1 kilometer (km)	1000 (10^3) meters	0.6214 mile (mi)	1 mile = 1.609 km
	1 meter (m)	100 (10^2) centimeters	1.094 yards (yd)	1 yard = 0.9144 m
		1000 millimeters (mm)	39.37 inches (in)	1 foot (ft) = 0.3048 m
	1 centimeter (cm)	0.01 (10^{-2}) meter	0.3937 inch	1 inch = 2.54 cm (exactly)
Volume	1 cubic meter (m^3)	1,000,000 (10^6) cubic centimeters	35.31 cubic feet (ft^3)	1 cubic foot = 0.02832 m^3
	1 cubic decimeter (dm^3)	1000 cubic centimeters	0.2642 gallon (gal)	1 gallon = 3.785 dm^3
			1.057 quarts (qt)	1 quart = 0.9464 dm^3
				1 quart = 946.4 cm^3
	1 cubic centimeter (cm^3)	0.001 dm^3	0.03381 fluid ounce	1 fluid ounce = 29.57 cm^3
Mass	1 kilogram (kg)	1000 grams	2.205 pounds (lb)	1 pound = 0.4536 kg
	1 gram (g)	1000 milligrams (mg)	0.03527 ounce (oz)	1 ounce = 28.35 g

Length The SI base unit of length is the **meter (m).** The standard meter is now based on two quantities, the speed of light in a vacuum and the second. A meter is a little longer than a yard (1 m = 1.094 yd); a centimeter (10^{-2} m) is about two-fifths of an inch (1 cm = 0.3937 in; 1 in = 2.54 cm). Biological cells are often measured in micrometers (1 μm = 10^{-6} m). On the atomic-size scale, nanometers and picometers are used (1 nm = 10^{-9} m; 1 pm = 10^{-12} m). Many proteins have diameters of around 2 nm; atomic diameters are around 200 pm (0.2 nm). An older unit still in use is the angstrom (1 Å = 10^{-10} m = 0.1 nm = 100 pm).

How Long Is a Meter? The history of the meter exemplifies the ongoing drive to define units based on unchanging standards. The French scientists who set up the metric system defined the meter as 1/10,000,000 the distance from the equator (through Paris!) to the North Pole. The meter was later redefined as the distance between two fine lines engraved on a corrosion-resistant metal bar kept at the International Bureau of Weights and Measures in France. Fear that the bar would be damaged by war led to the adoption of an exact, unchanging, universally available atomic standard: 1,650,763.73 wavelengths of orange-red light from electrically excited krypton atoms. The current standard is even more reliable: 1 meter is the distance light travels in a vacuum in 1/299,792,458 second.

Volume Any sample of matter has a certain **volume (*V*)**, the amount of space that the sample occupies. The SI unit of volume is the **cubic meter (m^3).** In chemistry, the most important volume units are non-SI units, the **liter (L)** and the **milliliter (mL)** (note the uppercase L). Physicians and other medical practitioners measure body fluids in cubic decimeters (dm^3), which is equivalent to liters:

$$1\ L = 1\ dm^3 = 10^{-3}\ m^3$$

As the prefix *milli-* indicates, 1 mL is $\frac{1}{1000}$ of a liter, and it is equal to exactly 1 cubic centimeter (cm^3):

$$1\ mL = 1\ cm^3 = 10^{-3}\ dm^3 = 10^{-3}\ L = 10^{-6}\ m^3$$

A liter is slightly larger than a quart (qt) (1 L = 1.057 qt; 1 qt = 946.4 mL); 1 fluid ounce ($\frac{1}{32}$ of a quart) equals 29.57 mL (29.57 cm^3).

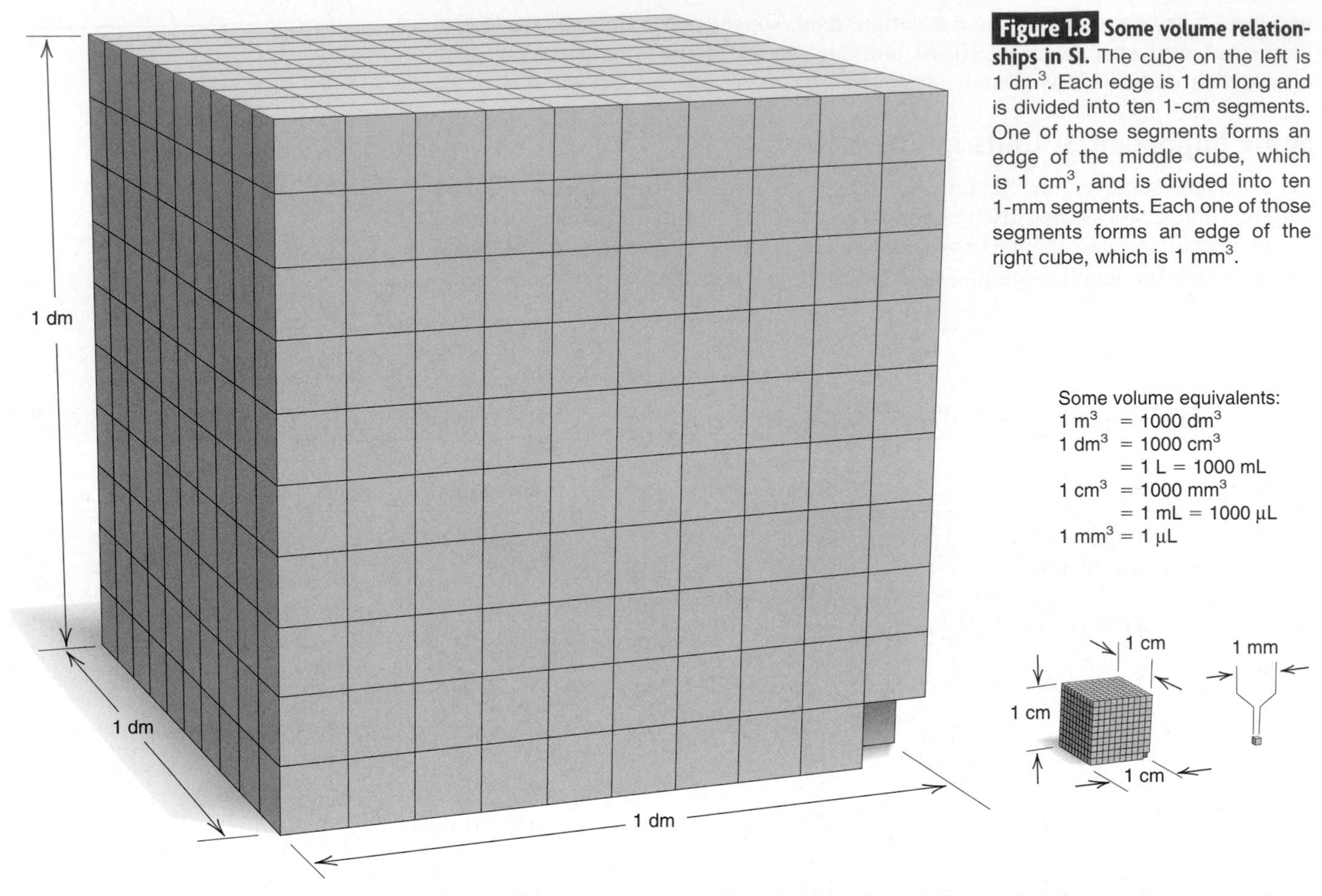

Figure 1.8 Some volume relationships in SI. The cube on the left is 1 dm^3. Each edge is 1 dm long and is divided into ten 1-cm segments. One of those segments forms an edge of the middle cube, which is 1 cm^3, and is divided into ten 1-mm segments. Each one of those segments forms an edge of the right cube, which is 1 mm^3.

Figure 1.8 is a life-size depiction of the 1000-fold decreases in volume from the cubic decimeter to the cubic millimeter. The edge of a cubic meter would be about 2.5 times the width of this textbook when open.

Figure 1.9 shows some of the types of laboratory glassware designed to contain liquids or measure their volumes. Many come in sizes from a few milliliters to a few liters. Erlenmeyer flasks and beakers are used to contain liquids. Graduated cylinders, pipets, and burets are used to measure and transfer liquids. Volumetric flasks and many pipets have a fixed volume indicated by a mark on the neck. In quantitative work, liquid solutions are prepared in volumetric flasks,

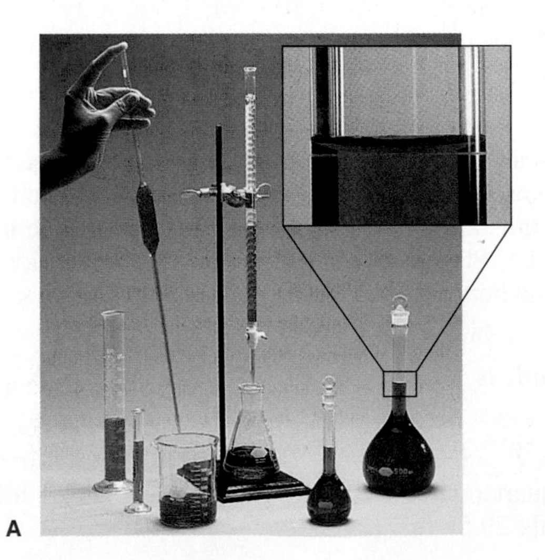

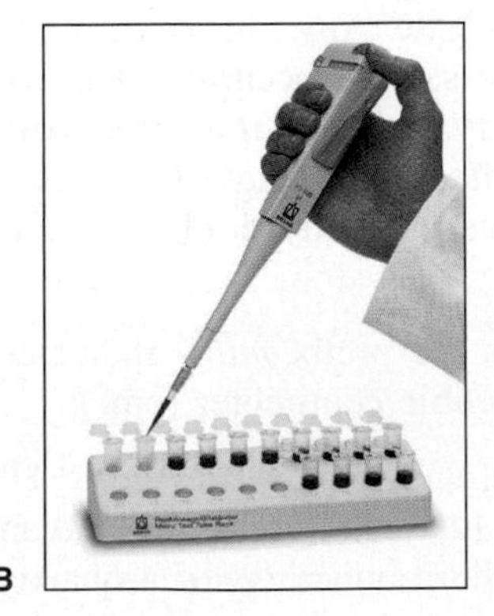

Figure 1.9 Common laboratory volumetric glassware. A, From left to right are two graduated cylinders, a pipet being emptied into a beaker, a buret delivering liquid to an Erlenmeyer flask, and two volumetric flasks. **Inset,** In contact with glass, this liquid forms a concave meniscus (curved surface). **B,** Automatic pipets deliver a given volume of liquid.

measured in cylinders, pipets, and burets, and then transferred to beakers or flasks for further chemical operations. Automatic pipets transfer a given volume of liquid accurately and quickly.

SAMPLE PROBLEM 1.3 Converting Units of Volume

Problem The volume of an irregularly shaped solid can be determined from the volume of water it displaces. A graduated cylinder contains 19.9 mL of water. When a small piece of galena, an ore of lead, is added, it sinks and the volume increases to 24.5 mL. What is the volume of the piece of galena in cm^3 and in L?

Plan We have to find the volume of the galena from the change in volume of the cylinder contents. The volume of galena in mL is the difference in the known volumes before and after adding it. The units mL and cm^3 represent identical volumes, so the volume of the galena in mL equals the volume in cm^3. We construct a conversion factor to convert the volume from mL to L. The calculation steps are shown in the roadmap.

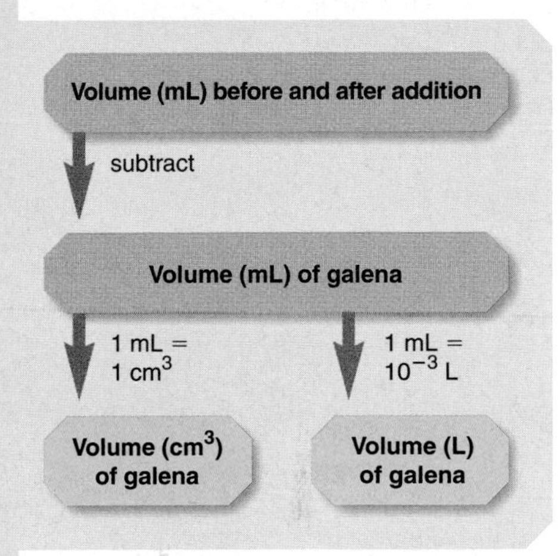

Solution Finding the volume of galena:

$$\text{Volume (mL)} = \text{volume after} - \text{volume before} = 24.5\text{ mL} - 19.9\text{ mL} = 4.6\text{ mL}$$

Converting the volume from mL to cm^3:

$$\text{Volume (cm}^3\text{)} = 4.6\ \cancel{\text{mL}} \times \frac{1\text{ cm}^3}{1\ \cancel{\text{mL}}} = \boxed{4.6\text{ cm}^3}$$

Converting the volume from mL to L:

$$\text{Volume (L)} = 4.6\ \cancel{\text{mL}} \times \frac{10^{-3}\text{ L}}{1\ \cancel{\text{mL}}} = \boxed{4.6\times10^{-3}\text{ L}}$$

Check The units and magnitudes of the answers seem correct. It makes sense that the volume expressed in mL would have a number 1000 times larger than the volume expressed in L, because a milliliter is $\frac{1}{1000}$ of a liter.

FOLLOW-UP PROBLEM 1.3 Within a cell, proteins are synthesized on particles called ribosomes. Assuming ribosomes are generally spherical, what is the volume (in dm^3 and μL) of a ribosome whose average diameter is 21.4 nm (V of a sphere $= \frac{4}{3}\pi r^3$)?

Mass The **mass** of an object refers to the quantity of matter it contains. The SI unit of mass is the **kilogram (kg),** the only base unit whose standard is a physical object—a platinum-iridium cylinder kept in France. It is also the only base unit whose name has a prefix. (In contrast to the practice with other base units, however, we attach prefixes to the word "gram," as in "microgram," rather than to the word "kilogram"; thus, we never say "microkilogram.")

The terms *mass* and *weight* have distinct meanings. Since a given object's quantity of matter cannot change, its *mass is constant.* Its **weight,** on the other hand, depends on its mass *and* the strength of the local gravitational field pulling on it. Because the strength of this field varies with height above the Earth's surface, the object's weight also varies. For instance, you actually weigh slightly less on a high mountaintop than at sea level.

Does this mean that if you weighed an object on a laboratory balance in Miami (sea level) and in Denver (about 1.7 km above sea level), you would obtain different results? Fortunately not. Such balances are designed to measure mass rather than weight, so this chaotic situation does not occur. (We are actually "massing" an object when we weigh it on a balance, but we rarely use that term.) Mechanical balances compare the object's unknown mass with known masses built into the balance, so the local gravitational field pulls equally on them. Electronic (analytical) balances determine mass by generating an electric field that counteracts the local gravitational field. The magnitude of the current needed to restore the pan to its zero position is then displayed as the object's mass. Therefore, an electronic balance must be readjusted with standard masses when it is moved to a different location.

SAMPLE PROBLEM 1.4 Converting Units of Mass

Problem International computer communications are often carried by optical fibers in cables laid along the ocean floor. If one strand of optical fiber weighs 1.19×10^{-3} lb/m, what is the mass (in kg) of a cable made of six strands of optical fiber, each long enough to link New York and Paris (8.84×10^{3} km)?

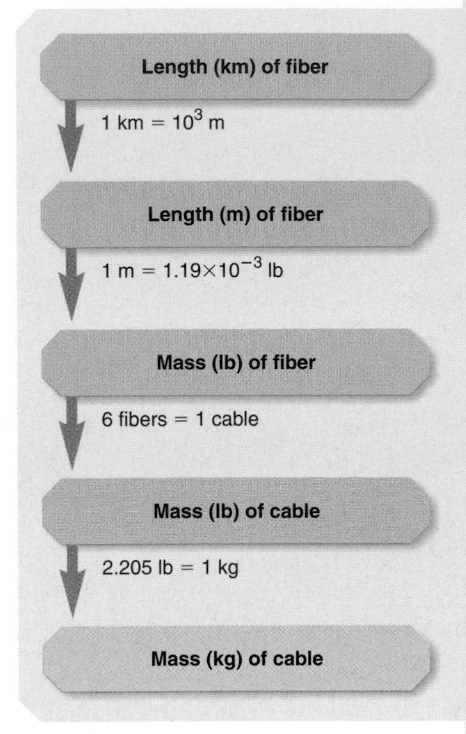

Plan We have to find the mass of cable (in kg) from the given mass/length of fiber, number of fibers/cable, and the length (distance from New York to Paris). One way to do this (as shown in the roadmap) is to first find the mass of one fiber and then find the mass of cable. We convert the length of one fiber from km to m and then find its mass (in lb) by using the lb/m factor. The cable mass is six times the fiber mass, and finally we convert lb to kg.

Solution Converting the fiber length from km to m:

$$\text{Length (m) of fiber} = 8.84\times10^{3}\ \cancel{\text{km}} \times \frac{10^{3}\ \text{m}}{1\ \cancel{\text{km}}} = 8.84\times10^{6}\ \text{m}$$

Converting the length of one fiber to mass (lb):

$$\text{Mass (lb) of fiber} = 8.84\times10^{6}\ \cancel{\text{m}} \times \frac{1.19\times10^{-3}\ \text{lb}}{1\ \cancel{\text{m}}} = 1.05\times10^{4}\ \text{lb}$$

Finding the mass of the cable (lb):

$$\text{Mass (lb) of cable} = \frac{1.05\times10^{4}\ \text{lb}}{1\ \cancel{\text{fiber}}} \times \frac{6\ \cancel{\text{fibers}}}{1\ \text{cable}} = 6.30\times10^{4}\ \text{lb/cable}$$

Converting the mass of cable from lb to kg:

$$\text{Mass (kg) of cable} = \frac{6.30\times10^{4}\ \cancel{\text{lb}}}{1\ \text{cable}} \times \frac{1\ \text{kg}}{2.205\ \cancel{\text{lb}}} = 2.86\times10^{4}\ \text{kg/cable}$$

Check The units are correct. Let's think through the relative sizes of the answers to see if they make sense: The number of m should be 10^3 larger than the number of km. If 1 m of fiber weighs about 10^{-3} lb, about 10^7 m should weigh about 10^4 lb. The cable mass should be six times as much, or about 6×10^4 lb. Since 1 lb is about $\frac{1}{2}$ kg, the number of kg should be about half the number of lb.

Comment Actually, the pound (lb) is the English unit of *weight,* not *mass.* The English unit of mass, called the *slug,* is rarely used.

FOLLOW-UP PROBLEM 1.4 An intravenous bag delivers a nutrient solution to a hospital patient at a rate of 1.5 drops per second. If a drop weighs 65 mg on average, how many kilograms of solution are delivered in 8.0 h?

Figure 1.10 shows the ranges of some common lengths, volumes, and masses.

Density The **density (*d*)** of an object is its mass divided by its volume:

$$\text{Density} = \frac{\text{mass}}{\text{volume}} \qquad \textbf{(1.1)}$$

Whenever needed, you can isolate mathematically each of the component variables by treating density as a conversion factor:

$$\text{Mass} = \text{volume} \times \text{density} = \cancel{\text{volume}} \times \frac{\text{mass}}{\cancel{\text{volume}}}$$

Or,

$$\text{Volume} = \text{mass} \times \frac{1}{\text{density}} = \cancel{\text{mass}} \times \frac{\text{volume}}{\cancel{\text{mass}}}$$

Under given conditions of temperature and pressure, *density is a characteristic physical property of a substance* and has a specific value, even though the separate values of mass and volume vary. Mass and volume are examples of **extensive properties,** those dependent on the amount of substance present. Density, on the other hand, is an **intensive property,** one that is independent of the amount of substance present. For example, the mass of a gallon of water is four times the mass of a quart of water, but its volume is also four times greater; therefore, the

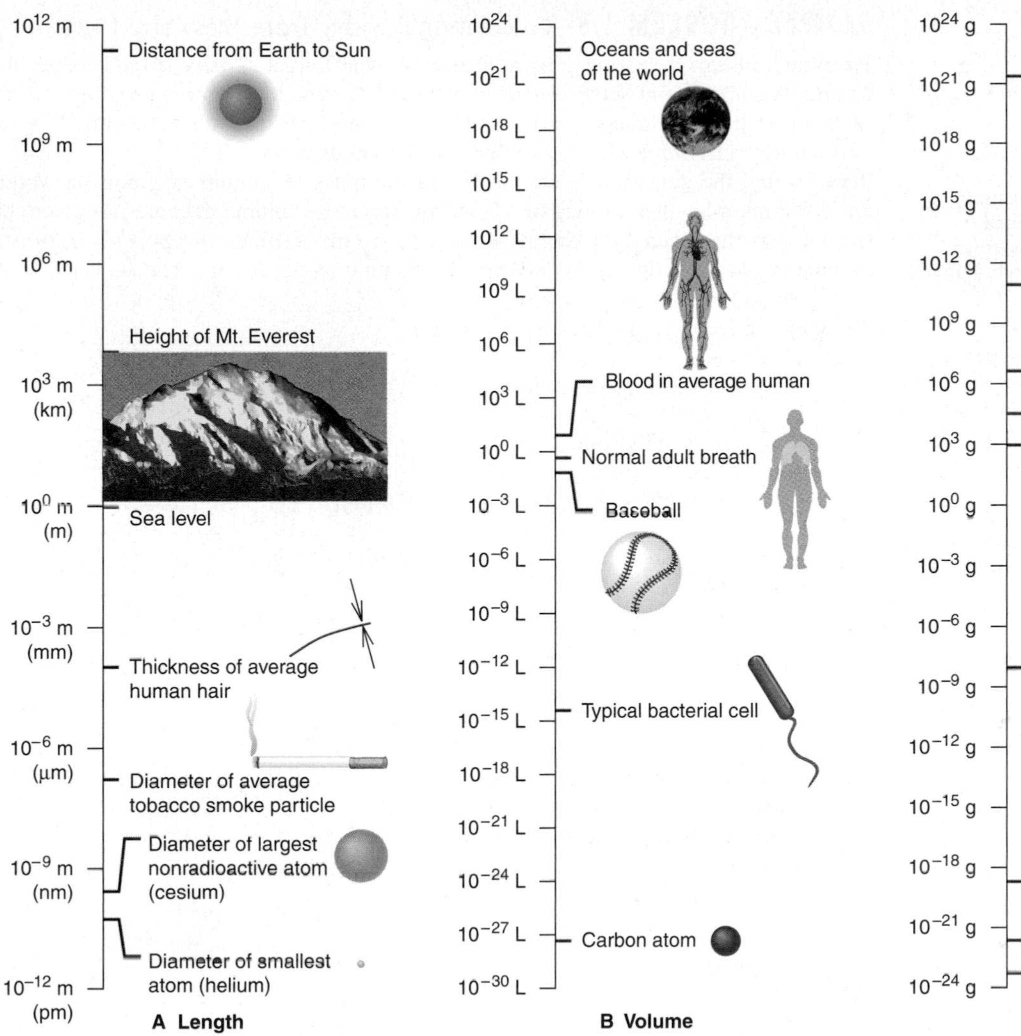

Figure 1.10 Some interesting quantities of length (A), volume (B), and mass (C). Note that the scales are exponential.

density of the water, the *ratio* of its mass to its volume, is constant at a particular temperature and pressure, regardless of the sample size.

The SI unit of density is the kilogram per cubic meter (kg/m^3), but in chemistry, density is typically given in units of g/L (g/dm^3) or g/mL (g/cm^3). For example, the density of liquid water at ordinary pressure and room temperature (20°C) is 1.0 g/mL. The densities of some common substances are given in Table 1.5. As you might expect from the magnified views of the physical states (see Figure 1.2), the densities of gases are much lower than those of liquids or solids.

Table 1.5 Densities of Some Common Substances*

Substance	Physical State	Density (g/cm^3)
Hydrogen	gas	0.0000899
Oxygen	gas	0.00133
Grain alcohol	liquid	0.789
Water	liquid	0.998
Table salt	solid	2.16
Aluminum	solid	2.70
Lead	solid	11.3
Gold	solid	19.3

*At room temperature (20°C) and normal atmospheric pressure (1 atm).

SAMPLE PROBLEM 1.5 Calculating Density from Mass and Length

Problem Lithium is a soft, gray solid that has the lowest density of any metal. It is an essential component of some advanced batteries, such as the one in your laptop. If a small rectangular slab of lithium weighs 1.49×10^3 mg and has sides that measure 20.9 mm by 11.1 mm by 11.9 mm, what is the density of lithium in g/cm^3?

Plan To find the density in g/cm^3, we need the mass of lithium in g and the volume in cm^3. The mass is given in mg, so we convert mg to g. Volume data are not given, but we can convert the given side lengths from mm to cm, and then multiply them to find the volume in cm^3. Finally, we divide mass by volume to get density. The steps are shown in the roadmap.

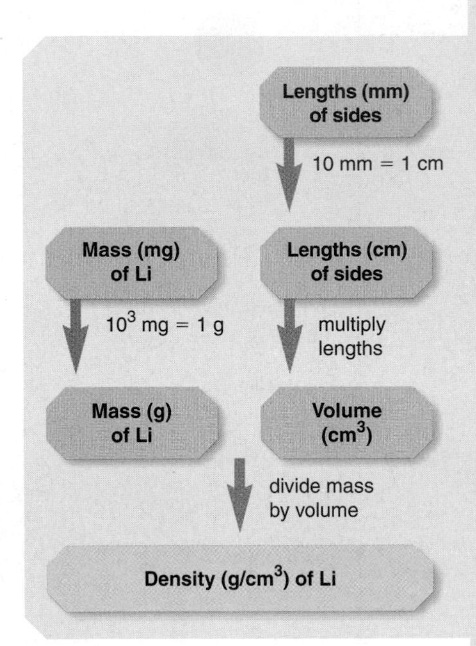

Solution Converting the mass from mg to g:

$$\text{Mass (g) of lithium} = 1.49 \times 10^3 \cancel{\text{mg}} \left(\frac{10^{-3}\ \text{g}}{1\ \cancel{\text{mg}}}\right) = 1.49\ \text{g}$$

Converting side lengths from mm to cm:

$$\text{Length (cm) of one side} = 20.9\ \cancel{\text{mm}} \times \frac{1\ \text{cm}}{10\ \cancel{\text{mm}}} = 2.09\ \text{cm}$$

Similarly, the other side lengths are 1.11 cm and 1.19 cm.
Finding the volume:

$$\text{Volume (cm}^3) = 2.09\ \text{cm} \times 1.11\ \text{cm} \times 1.19\ \text{cm} = 2.76\ \text{cm}^3$$

Calculating the density:

$$\text{Density of lithium} = \frac{\text{mass}}{\text{volume}} = \frac{1.49\ \text{g}}{2.76\ \text{cm}^3} = 0.540\ \text{g/cm}^3$$

Check Since 1 cm = 10 mm, the number of cm in each length should be $\frac{1}{10}$ the number of mm. The units for density are correct, and the size of the answer (~0.5 g/cm^3) seems correct since the number of g (1.49) is about half the number of cm^3 (2.76). Since the problem states that lithium has a very low density, this answer makes sense.

FOLLOW-UP PROBLEM 1.5 The piece of galena in Sample Problem 1.3 has a volume of 4.6 cm^3. If the density of galena is 7.5 g/cm^3, what is the mass (in kg) of that piece of galena?

Temperature There is a common misunderstanding about heat and temperature. **Temperature (*T*)** is a measure of how hot or cold a substance is *relative to another substance*. **Heat** is the energy that flows between objects that are at different temperatures. Temperature is related to the *direction* of that energy flow: when two objects at different temperatures touch, energy flows from the one with the higher temperature to the one with the lower temperature until their temperatures are equal. When you hold an ice cube, its "cold" seems to flow *into* your hand; actually, heat flows *from* your hand into the ice. (In Chapter 6, we will see how heat is measured and how it is related to chemical and physical change.) Energy is an *extensive* property (as is volume), but temperature is an *intensive* property (as is density): a vat of boiling water has more energy than a cup of boiling water, but the temperatures of the two water samples are the same.

In the laboratory, the most common means for measuring temperature is the **thermometer,** a device that contains a fluid that expands when it is heated. When the thermometer's fluid-filled bulb is immersed in a substance hotter than itself, heat flows from the substance through the glass and into the fluid, which expands and rises in the thermometer tube. If a substance is colder than the thermometer, heat flows outward from the fluid, which contracts and falls within the tube.

The three temperature scales most important for us to consider are the Celsius (°C, formerly called centigrade), the Kelvin (K), and the Fahrenheit (°F) scales. The SI base unit of temperature is the **kelvin (K);** note that the kelvin has no degree sign (°). Figure 1.11 shows some interesting temperatures in this scale. The Kelvin scale, also known as the *absolute scale,* is preferred in all scientific work, although the Celsius scale is used frequently. In the United States, the Fahrenheit scale is still used for weather reporting, body temperature, and other everyday purposes. *The three scales differ in the size of the unit and/or the temperature of the zero point.* Figure 1.12 shows the freezing and boiling points of water in the three scales.

The **Celsius scale,** devised in the 18th century by the Swedish astronomer Anders Celsius, is based on changes in the physical state of water: 0°C is set at water's freezing point, and 100°C is set at its boiling point (at normal atmospheric pressure). The **Kelvin (absolute) scale** was devised by the English physicist William Thomson, known as Lord Kelvin, in 1854 during his experiments on the expansion and contraction of gases. *The Kelvin scale uses the same size degree unit as the Celsius scale*—$\frac{1}{100}$ of the difference between the freezing and boiling points of water—*but it differs in zero point.* The zero point in the Kelvin scale, 0 K, is called *absolute zero* and equals −273.15°C. In the Kelvin scale, *all temperatures have positive values.* Water freezes at +273.15 K (0°C) and boils at +373.15 K (100°C).

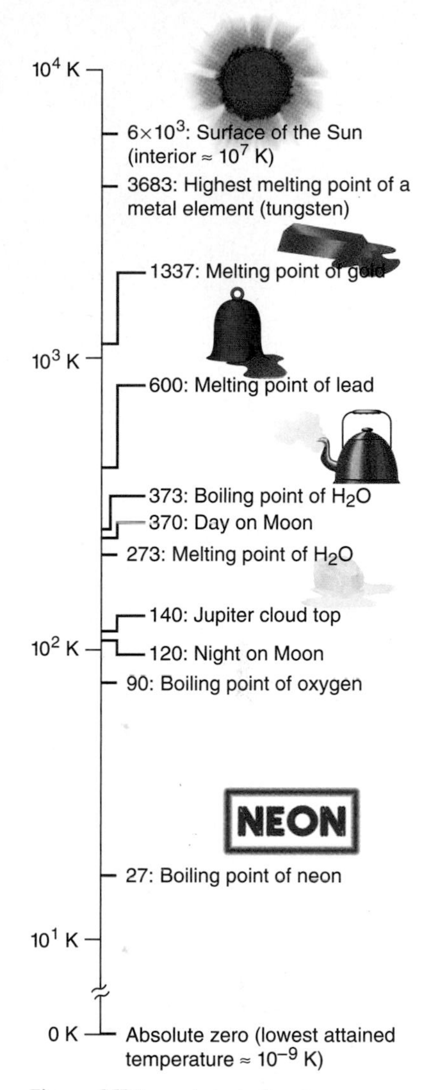

Figure 1.11 Some interesting temperatures.

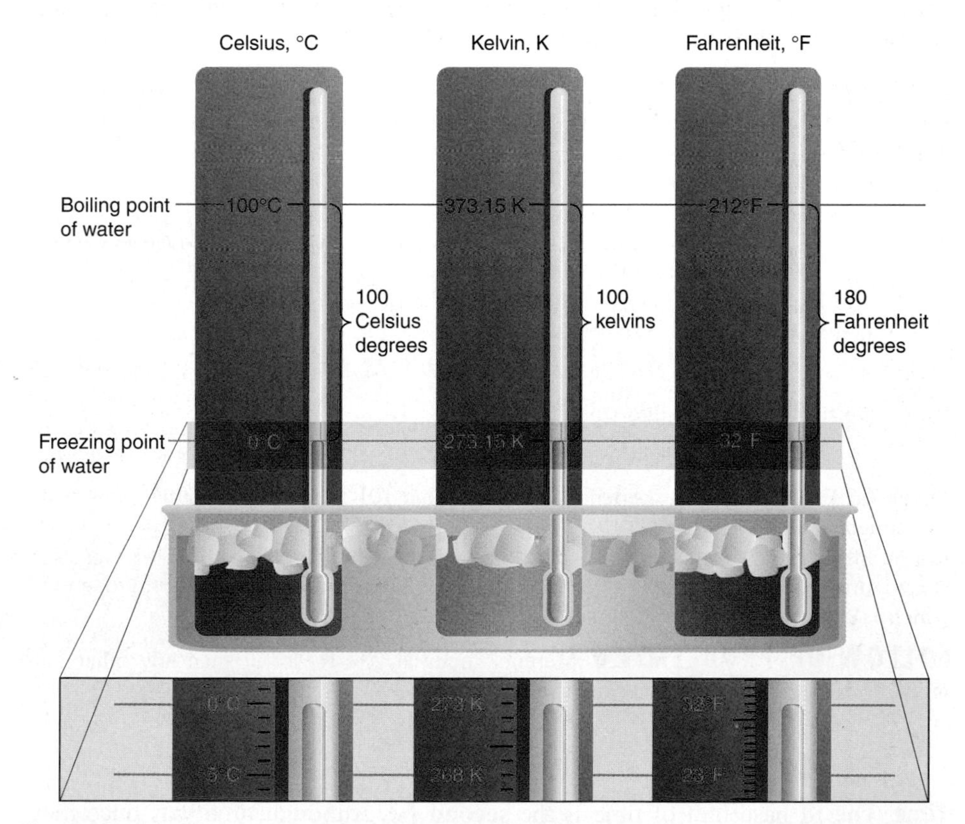

Figure 1.12 The freezing point and the boiling point of water in the Celsius, Kelvin (absolute), and Fahrenheit temperature scales. As you can see, this range consists of 100 degrees on the Celsius and Kelvin scales, but 180 degrees on the Fahrenheit scale. At the bottom of the figure, a portion of each of the three thermometer scales is expanded to show the sizes of the units. A Celsius degree (°C; *left*) and a kelvin (K; *center*) are the same size, and each is $\frac{9}{5}$ the size of a Fahrenheit degree (°F; *right*).

We can convert between the Celsius and Kelvin scales by remembering the difference in zero points: since 0°C = 273.15 K,

$$T \text{ (in K)} = T \text{ (in °C)} + 273.15 \quad \textbf{(1.2)}$$

Solving Equation 1.2 for T (in °C) gives

$$T \text{ (in °C)} = T \text{ (in K)} - 273.15 \quad \textbf{(1.3)}$$

The Fahrenheit scale differs from the other scales in its zero point *and* in the size of its unit. Water freezes at 32°F and boils at 212°F. Therefore, 180 Fahrenheit degrees (212°F − 32°F) represents the same temperature change as 100 Celsius degrees (or 100 kelvins). Because 100 Celsius degrees equal 180 Fahrenheit degrees,

$$1 \text{ Celsius degree} = \tfrac{180}{100} \text{ Fahrenheit degrees} = \tfrac{9}{5} \text{ Fahrenheit degrees}$$

To convert a temperature in °C to °F, first change the degree size and then adjust the zero point:

$$T \text{ (in °F)} = \tfrac{9}{5}T \text{ (in °C)} + 32 \quad \textbf{(1.4)}$$

To convert a temperature in °F to °C, do the two steps in the opposite order; that is, first adjust the zero point and then change the degree size. In other words, solve Equation 1.4 for T (in °C):

$$T \text{ (in °C)} = [T \text{ (in °F)} - 32]\tfrac{5}{9} \quad \textbf{(1.5)}$$

(The only temperature with the same numerical value in the Celsius and Fahrenheit scales is −40°; that is, −40°F = −40°C.)

SAMPLE PROBLEM 1.6 Converting Units of Temperature

Problem A child has a body temperature of 38.7°C.
(a) If normal body temperature is 98.6°F, does the child have a fever?
(b) What is the child's temperature in kelvins?
Plan (a) To find out if the child has a fever, we convert from °C to °F (Equation 1.4) and see whether 38.7°C is higher than 98.6°F.
(b) We use Equation 1.2 to convert the temperature in °C to K.
Solution (a) Converting the temperature from °C to °F:

$$T \text{ (in °F)} = \tfrac{9}{5}T \text{ (in °C)} + 32 = \tfrac{9}{5}(38.7\text{°C}) + 32 = 101.7\text{°F; yes, the child has a fever.}$$

(b) Converting the temperature from °C to K:

$$T \text{ (in K)} = T \text{ (in °C)} + 273.15 = 38.7\text{°C} + 273.15 = 311.8 \text{ K}$$

Check (a) From everyday experience, you know that 101.7°F is a reasonable temperature for someone with a fever.
(b) We know that a Celsius degree and a kelvin are the same size. Therefore, we can check the math by approximating the Celsius value as 40°C and adding 273: 40 + 273 = 313, which is close to our calculation, so there is no large error.

FOLLOW-UP PROBLEM 1.6 Mercury melts at 234 K, lower than any other pure metal. What is its melting point in °C and °F?

Time The SI base unit of time is the **second (s).** Although time was once measured by the day and year, it is now based on an atomic standard: microwave radiation absorbed by cesium atoms (Figure 1.13). In the laboratory, we study the speed of a reaction by measuring the time it takes a fixed amount of substance to undergo a chemical change. The range of reaction speed is enormous: a fast reaction may be over in less than a nanosecond (10^{-9} s), whereas slow ones, such as rusting or aging, take years. Chemists now use lasers to study changes that occur in a few picoseconds (10^{-12} s) or femtoseconds (10^{-15} s).

Figure 1.13 The cesium atomic clock. The accuracy of the best pendulum clock is to within 3 seconds per year and that of the best quartz clock is 1000 times greater. The most recent version of the atomic clock, NIST-F1, developed by the Physics Laboratory of the National Institute of Standards and Technology, is over 6000 times more accurate still, to within 1 second in 20 million years! Rather than using the oscillations of a pendulum, the atomic clock measures the oscillations of microwave radiation absorbed by gaseous cesium atoms: 1 second is defined as 9,192,631,770 of these oscillations. This new clock cools the cesium atoms with infrared lasers to around 10^{-6} K, which allows much longer observation times of the atoms, and thus much greater accuracy.

SECTION SUMMARY

SI units consist of seven base units and numerous derived units. Exponential notation and prefixes based on powers of 10 are used to express very small and very large numbers. The SI base unit of length is the meter (m). Length units on the atomic scale are the nanometer (nm) and picometer (pm). Volume units are derived from length units; the most important volume units in chemistry are the cubic meter (m^3) and the liter (L). The mass of an object, a measure of the quantity of matter present in it, is constant. The SI unit of mass is the kilogram (kg). The weight of an object varies with the gravitational field influencing it. Density (*d*) is the ratio of mass to volume of a substance and is one of its characteristic physical properties. Temperature (*T*) is a measure of the relative hotness of an object. Heat is energy that flows from an object at higher temperature to one at lower temperature. Temperature scales differ in the size of the degree unit and/or the zero point. In chemistry, temperature is measured in kelvins (K) or degrees Celsius (°C). Extensive properties, such as mass, volume, and energy, depend on the amount of a substance. Intensive properties, such as density and temperature, are independent of amount.

The Central Importance of Measurement in Science It's important to keep in mind *why* scientists measure things: "When you can measure what you are speaking about, and express it in numbers, you know something about it; but when you cannot measure it, . . . your knowledge is of a meager and unsatisfactory kind; it may be the beginning of knowledge, but you have scarcely, in your thoughts, advanced to the stage of science" (William Thomson, Lord Kelvin, 1824–1907).

1.6 UNCERTAINTY IN MEASUREMENT: SIGNIFICANT FIGURES

We can never measure a quantity exactly, because measuring devices are made to limited specifications and we use our imperfect senses and skills to read them. Therefore, every measurement we make includes some **uncertainty.**

The measuring device we choose in a given situation depends on how much uncertainty we are willing to accept. When you buy potatoes, a supermarket scale that measures in 0.1-kg increments is perfectly acceptable; it tells you that the mass is, for example, 2.0 ± 0.1 kg. The term "± 0.1 kg" expresses the uncertainty in the measurement: the potatoes weigh between 1.9 and 2.1 kg. For a large-scale reaction, a chemist uses a lab balance that measures in 0.001-kg increments in order to obtain 2.036 ± 0.001 kg of a chemical, that is, between 2.035 and 2.037 kg. The greater number of digits in the mass of the chemical indicates that we know its mass with *more certainty* than we know the mass of the potatoes.

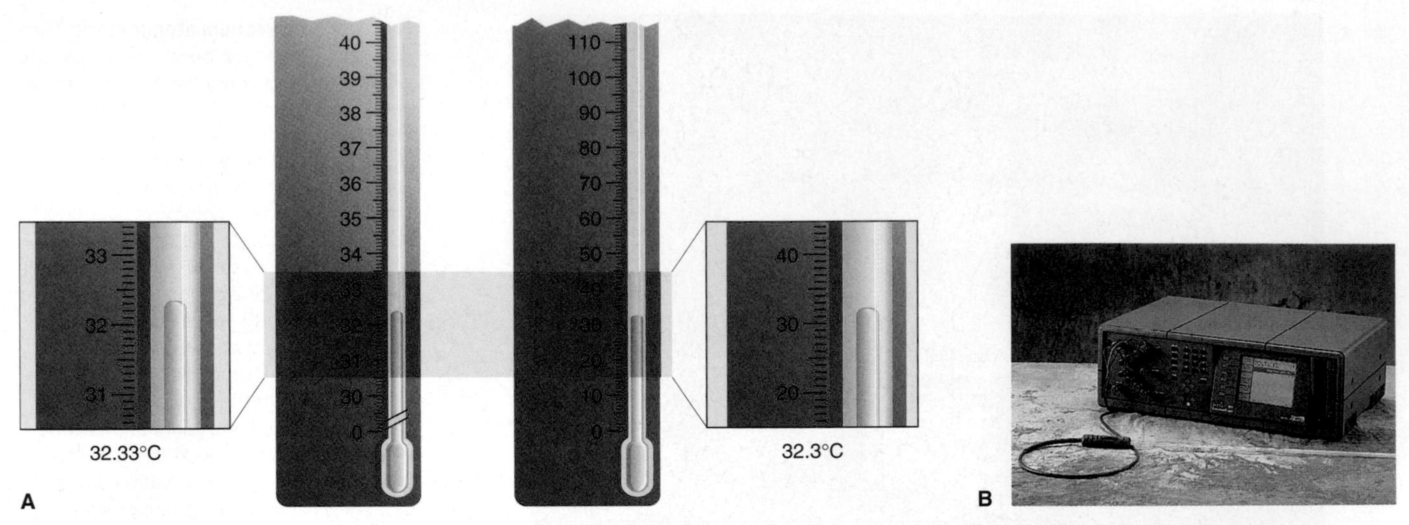

Figure 1.14 The number of significant figures in a measurement depends on the measuring device. A, Two thermometers measuring the same temperature are shown with expanded views. The thermometer on the left is graduated in 0.1°C and reads 32.33°C; the one on the right is graduated in 1°C and reads 32.3°C. Therefore, a reading with more significant figures (more certainty) can be made with the thermometer on the left. **B,** This modern electronic thermometer measures the resistance through a fine platinum wire in the probe to determine temperatures to the nearest microkelvin (10^{-6} K).

We *always estimate the rightmost digit* when reading a measuring device. The uncertainty can be expressed with the ± sign, but generally we drop the sign and *assume an uncertainty of one unit in the rightmost digit.* The digits we record in a measurement, both the certain and the uncertain ones, are called **significant figures.** There are four significant figures in 2.036 kg and two in 2.0 kg. *The greater the number of significant figures in a measurement, the greater is the certainty.* Figure 1.14 shows this point for two thermometers.

Determining Which Digits Are Significant

When you take measurements or use them in calculations, you must know the number of digits that are significant. In general, *all digits are significant, except zeros that are not measured but are used only to position the decimal point.* Here is a simple procedure that applies this general point:

1. Make sure that the measured quantity has a decimal point.
2. Start at the left of the number and move right until you reach the first nonzero digit.
3. Count that digit and every digit to its right as significant.

Sometimes, there may be a slight complication with zeros that *end* a number. Zeros that end a number and lie either after or before the decimal point *are* significant; thus, 1.030 mL has four significant figures, and 5300. L has four significant figures also. The problem arises if there is no decimal point, as in 5300 L. In such cases, we would assume that the zeros are *not* significant; exponential notation is needed to show which of the zeros, if any, were measured and therefore are significant. Thus, 5.300×10^3 L has four significant figures, 5.30×10^3 L has three, and 5.3×10^3 L has only two. In this and other modern texts (and research articles), a terminal decimal point is used when necessary to clarify the number of significant figures; thus, 500 mL has one significant figure, but 5.00×10^2 mL, 500. mL, and 0.500 L have three.

SAMPLE PROBLEM 1.7 Determining the Number of Significant Figures

Problem For each of the following quantities, underline the zeros that are significant figures (sf), and determine the number of significant figures in each quantity. For **(d)** to **(f)**, express each in exponential notation first.

(a) 0.0030 L **(b)** 0.1044 g **(c)** 53,069 mL
(d) 0.00004715 m **(e)** 57,600. s **(f)** 0.0000007160 cm^3

Plan We determine the number of significant figures by counting digits, as just presented, paying particular attention to the position of zeros in relation to the decimal point.

Solution (a) $0.003\underline{0}$ L has 2 sf

(b) $0.1\underline{0}44$ g has 4 sf

(c) $53{,}\underline{0}69$ mL has 5 sf

(d) 0.00004715 m, or 4.715×10^{-5} m, has 4 sf

(e) $57{,}6\underline{00}.$ s, or $5.76\underline{00}\times10^{4}$ s, has 5 sf

(f) $0.000000716\underline{0}$ cm^3, or $7.16\underline{0}\times10^{-7}$ cm^3, has 4 sf

Check Be sure that every zero counted as significant comes after nonzero digit(s) in the number.

FOLLOW-UP PROBLEM 1.7 For each of the following quantities, underline the zeros that are significant figures and determine the number of significant figures (sf) in each quantity. For **(d)** to **(f)**, express each in exponential notation first.

(a) 31.070 mg **(b)** 0.06060 g **(c)** 850.°C
(d) 200.0 mL **(e)** 0.0000039 m **(f)** 0.000401 L

Significant Figures in Calculations

Measurements often contain differing numbers of significant figures. In a calculation, we keep track of the number of significant figures in each quantity so that we don't claim more significant figures (more certainty) in the answer than in the original data. If we have too many significant figures, we **round off** the answer to obtain the proper number of them.

The general rule for rounding is that *the least certain measurement sets the limit on certainty for the entire calculation and determines the number of significant figures in the final answer.* Suppose you want to find the density of a new ceramic. You measure the mass of a piece on an analytical balance and obtain 3.8056 g; you measure its volume as 2.5 mL by displacement of water in a graduated cylinder. The mass has five significant figures, but the volume has only two. Should you report the density as 3.8056 g/2.5 mL = 1.5222 g/mL or as 1.5 g/mL? The answer with five significant figures implies more certainty than the answer with two. But you didn't measure the volume to five significant figures, so you can't possibly know the density with that much certainty. Therefore, you report the answer as 1.5 g/mL.

Significant Figures and Arithmetic Operations The following two rules tell how many significant figures to show based on the arithmetic operation:

1. *For multiplication and division.* The answer contains the same number of *significant figures* as in the measurement with the fewest significant figures. Suppose you want to find the volume of a sheet of a new graphite composite. The length (9.2 cm) and width (6.8 cm) are obtained with a meterstick and the thickness (0.3744 cm) with a set of fine calipers. The volume calculation is

$$\text{Volume (cm}^3) = 9.2 \text{ cm} \times 6.8 \text{ cm} \times 0.3744 \text{ cm} = 23 \text{ cm}^3$$

The calculator shows 23.4225 cm^3, but you should report the answer as 23 cm^3, with two significant figures, because the length and width measurements determine the overall certainty, and they contain only two significant figures.

2. *For addition and subtraction.* The answer has the same number of *decimal places* as there are in the measurement with the fewest decimal places. Suppose you measure 83.5 mL of water in a graduated cylinder and add 23.28 mL of protein solution from a buret. The total volume is

$$\text{Volume (mL)} = 83.5 \text{ mL} + 23.28 \text{ mL} = 106.8 \text{ mL}$$

Here the calculator shows 106.78 mL, but you report the volume as 106.8 mL, with one decimal place, because the measurement with fewer decimal places (83.5 mL) has one decimal place.

Rules for Rounding Off In most calculations, you need to round off the answer to obtain the proper number of significant figures or decimal places. Notice that in calculating the volume of the graphite composite above, we removed the extra digits, but in calculating the total protein solution volume, we removed the extra digit and increased the last digit by one. Here are rules for rounding off:

1. If the digit removed is *more than 5,* the preceding number is increased by 1: 5.379 rounds to 5.38 if three significant figures are retained and to 5.4 if two significant figures are retained.
2. If the digit removed is *less than 5,* the preceding number is unchanged: 0.2413 rounds to 0.241 if three significant figures are retained and to 0.24 if two significant figures are retained.
3. If the digit removed *is 5,* the preceding number is increased by 1 if it is odd and remains unchanged if it is even: 17.75 rounds to 17.8, but 17.65 rounds to 17.6. If the 5 is followed only by zeros, rule 3 is followed; if the 5 is followed by nonzeros, rule 1 is followed: 17.6500 rounds to 17.6, but 17.6513 rounds to 17.7.
4. *Always carry one or two additional significant figures through a multistep calculation and round off the final answer* ***only.*** Don't be concerned if you string together a calculation to check a sample or follow-up problem and find that your answer differs in the last decimal place from the one in the book. To show you the correct number of significant figures in text calculations, we round off intermediate steps, and this process may sometimes change the last digit.

Significant Figures and Electronic Calculators A calculator usually gives answers with too many significant figures. For example, if your calculator displays ten digits and you divide 15.6 by 9.1, it will show 1.714285714. Obviously, most of these digits are not significant; the answer should be rounded off to 1.7 so that it has two significant figures, the same as in 9.1. A good way to prove to yourself that the additional digits are not significant is to perform two calculations, including the uncertainty in the last digits, to obtain the highest and lowest possible answers. For (15.6 ± 0.1)/(9.1 ± 0.1),

$$\text{The highest answer is } \frac{15.7}{9.0} = 1.744444\ldots \quad \text{The lowest answer is } \frac{15.5}{9.2} = 1.684782\ldots$$

No matter how many digits the calculator displays, the values differ in the first decimal place, so the answer has two significant figures and should be reported as 1.7. Many calculators have a FIX button that allows you to set the number of digits displayed.

Significant Figures and Choice of Measuring Device The measuring device you choose determines the number of significant figures you can obtain. Suppose you are doing an experiment that requires mixing a liquid with a solid. You weigh the solid on the analytical balance and obtain a value with five significant figures. It would make sense to measure the liquid with a buret or pipet, which measures volumes to more significant figures than a graduated cylinder. If you chose the cylinder, you would have to round off more digits in the calculations, so the certainty in the mass value would be wasted (Figure 1.15). With experience, you'll choose a measuring device based on the number of significant figures you need in the final answer.

Figure 1.15 Significant figures and measuring devices. The mass (6.8605 g) measured with an analytical balance *(top)* has more significant figures than the volume (68.2 mL) measured with a graduated cylinder *(bottom).*

Exact Numbers Some numbers are called **exact numbers** because they have no uncertainty associated with them. Some exact numbers are part of a unit definition: there are 60 minutes in 1 hour, 1000 micrograms in 1 milligram, and 2.54 centimeters in 1 inch. Other exact numbers result from actually counting individual items: there are exactly 3 quarters in my hand, 26 letters in the English alphabet, and so forth. Since they have no uncertainty, *exact numbers do not limit the number of significant figures in the answer.* Put another way, exact numbers have as many significant figures as a calculation requires.

SAMPLE PROBLEM 1.8 Significant Figures and Rounding

Problem Perform the following calculations and round the answer to the correct number of significant figures:

(a) $\dfrac{16.3521\ \text{cm}^2 - 1.448\ \text{cm}^2}{7.085\ \text{cm}}$ **(b)** $\dfrac{(4.80\times10^4\ \text{mg})\left(\dfrac{1\ \text{g}}{1000\ \text{mg}}\right)}{11.55\ \text{cm}^3}$

Plan We use the rules just presented in the text. In **(a)**, we subtract before we divide. In **(b)**, we note that the unit conversion involves an exact number.

Solution (a) $\dfrac{16.3521\ \text{cm}^2 - 1.448\ \text{cm}^2}{7.085\ \text{cm}} = \dfrac{14.904\ \text{cm}^{\cancel{2}}}{7.085\ \cancel{\text{cm}}} = \boxed{2.104\ \text{cm}}$

(b) $\dfrac{(4.80\times10^4\ \cancel{\text{mg}})\left(\dfrac{1\ \text{g}}{1000\ \cancel{\text{mg}}}\right)}{11.55\ \text{cm}^3} = \dfrac{48.0\ \text{g}}{11.55\ \text{cm}^3} = \boxed{4.16\ \text{g/cm}^3}$

Check Note that in (a) we lose a decimal place in the numerator, and in (b) we retain 3 sf in the answer because there are 3 sf in 4.80. Rounding to the nearest whole number is always a good way to check: **(a)** $(16 - 1)/7 \approx 2$; **(b)** $(5\times10^4/1\times10^3)/12 \approx 4$.

FOLLOW-UP PROBLEM 1.8 Perform the following calculation and round the answer to the correct number of significant figures: $\dfrac{25.65\ \text{mL} + 37.4\ \text{mL}}{73.55\ \text{s}\left(\dfrac{1\ \text{min}}{60\ \text{s}}\right)}$

Precision, Accuracy, and Instrument Calibration

Precision and accuracy are two aspects of certainty. We often use these terms interchangeably in everyday speech, but in scientific measurements they have distinct meanings. **Precision,** or *reproducibility,* refers to how close the measurements in a series are to each other. **Accuracy** refers to how close a measurement is to the actual value.

Precision and accuracy are linked with two common types of error:

1. **Systematic error** produces values that are *either* all higher or all lower than the actual value. Such error is part of the experimental system, often caused by a faulty measuring device or by a consistent mistake in taking a reading.
2. **Random error,** in the absence of systematic error, produces values that are higher *and* lower than the actual value. Random error *always* occurs, but its size depends on the measurer's skill and the instrument's precision.

Precise measurements have low random error, that is, small deviations from the average. *Accurate measurements have low systematic error and, generally, low random error as well.* In some cases, when many measurements are taken that have a high random error, the *average* may still be accurate.

Suppose each of four students measures 25.0 mL of water in a pre-weighed graduated cylinder and then weighs the water *plus* cylinder on a balance. If the density of water is 1.00 g/mL at the temperature of the experiment, the actual mass of 25.0 mL of water is 25.0 g. Each student performs the operation four times, subtracts the mass of the empty cylinder, and obtains one of the four graphs

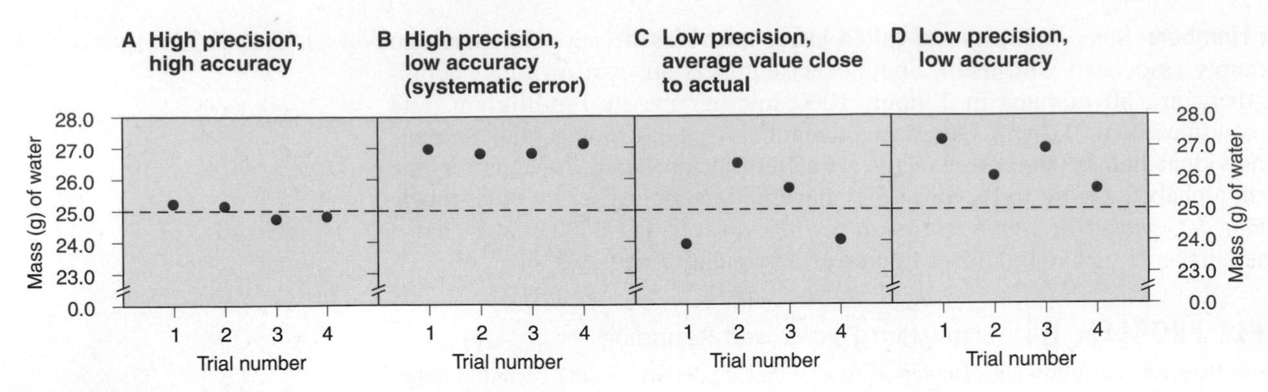

Figure 1.16 Precision and accuracy in a laboratory calibration. Each graph represents four measurements made with a graduated cylinder that is being calibrated (see text for details).

shown in Figure 1.16. In graphs A and B, the random error is small; that is, the precision is high (the weighings are reproducible). In A, however, the accuracy is high as well (all the values are close to 25.0 g), whereas in B the accuracy is low (there is a systematic error). In graphs C and D, there is a large random error; that is, the precision is low. Large random error is often called large *scatter.* Note, however, that in D there is also a systematic error (all the values are high), whereas in C the average of the values is close to the actual value.

Systematic error can be avoided, or at least taken into account, through **calibration** of the measuring device, that is, by comparing it with a known standard. The systematic error in graph B, for example, might be caused by a poorly manufactured cylinder that reads "25.0" when it actually contains about 27 mL. If you detect such an error by means of a calibration procedure, you could adjust all volumes measured with that cylinder. Instrument calibration is an essential part of careful measurement.

SECTION SUMMARY

Because the final digit of a measurement is estimated, all measurements have a limit to their certainty, which is expressed by the number of significant figures. The certainty of a calculated result depends on the certainty of the data, so the answer has as many significant figures as in the least certain measurement. Excess digits are rounded off in the final answer. The choice of laboratory device depends on the certainty needed. Exact numbers have as many significant figures as the calculation requires.

Precision (how close values are to each other) and accuracy (how close values are to the actual value) are two aspects of certainty. Systematic errors result in values that are either all higher or all lower than the actual value. Random errors result in some values that are higher and some values that are lower than the actual value. Precise measurements have low random error; accurate measurements have low systematic error and often low random error. The size of random errors depends on the skill of the measurer and the precision of the instrument. A systematic error, however, is often caused by faulty equipment and can be compensated for by calibration.

Chapter Perspective

This chapter has provided several keys for you to use repeatedly in your study of chemistry: descriptions of some essential concepts; insight into how scientists think; the units of modern measurement and the mathematical skills to apply them; and a systematic approach to solving problems. You can begin using these keys in the next chapter, where we discuss the components of matter and their classification and trace the winding path of scientific discovery that led to our current model of atomic structure.

Chemistry Problem Solving in the Real World

Learning chemistry is essential to many fields, including medicine, engineering, and environmental science. It is also essential to an understanding of complex science-related issues, such as the recycling of plastics, the reduction of urban smog, and the application of genetic cloning—to mention just three of many.

Any major scientific discipline such as chemistry consists of several subdisciplines that form connections with other sciences to spawn new fields. Traditionally, chemistry has five main branches—organic, inorganic, analytical, physical, and biological chemistry—but these long ago formed interconnections, such as physical organic and bioinorganic chemistry. Solving the problems of today requires further connections, such as ecological chemistry, materials science, atmospheric chemistry, and molecular genetics. The more complex the system under study is, the greater the need for interdisciplinary scientific thinking.

Environmental issues are especially complex, and one of the most intractable is the acid rain problem. Let's see how it is being approached by chemists interacting with scientists in related fields. Acid rain results in large part from burning high-sulfur coal, a major fuel used throughout much of North America and Europe. As the coal burns, the gaseous products, including an oxide of its sulfur impurities, are carried away by prevailing winds. In contact with oxygen and rain, this sulfur oxide undergoes chemical changes, yielding acid rain. (We discuss the chemical details in later chapters.) In the northeastern United States and adjacent parts of Canada, acid rain has killed fish, decimated forests, injured crops, and released harmful substances into the soil. Acid rain has severely damaged many forests and lakes in Germany, Sweden, Norway, and several countries in central and eastern Europe. And acidic precipitation has now been confirmed at both Poles!

Chemists and other scientists are currently working together to solve this problem (Figure B1.1). As geochemists search for low-sulfur coal deposits, their engineering colleagues design better ways of removing sulfur oxides from smokestack gases. Atmospheric chemists and meteorologists track changes through the affected regions, develop computer models to predict the changes, and coordinate their findings with environmental chemists at ground stations. Ecological chemists, microbiologists, and aquatic biologists monitor the effects of acid rain on microbes, insects, birds, and fish. Agricultural chemists and agronomists study ways to protect crop yields. Biochemists and genetic engineers develop new, more acid-resistant crop species. Soil chemists measure changes in mineral content, sharing their data with forestry scientists to save valuable timber and recreational woodlands. Organic chemists and chemical engineers convert coal to cleaner fuels. Working in tandem with this intense experimental activity are scientifically trained economic and policy experts who provide business and government leaders with the information to make decisions and foster "greener" approaches to energy use. With all this input, interdisciplinary understanding of the acid rain problem has increased enormously and certainly will continue to do so.

These professions are just a few of those involved in studying a single chemistry-related issue. Chemical principles apply to many other specialties, from medicine and pharmacology to art restoration and criminology, from genetics and space research to archaeology and oceanography. Chemistry problem solving has far-reaching relevance to many aspects of your daily life and your future career as well.

Figure B1.1 The central role of chemistry in solving real-world problems. Researchers in many chemical specialties join with those in other sciences to investigate complex modern issues such as acid rain. **A,** The sulfur oxide in power plant emissions will be reduced by devices that remove it from smokestack gases. **B,** Atmospheric chemists study the location and concentration of air pollutants with balloons that carry monitoring equipment aloft. **C,** Ecologists sample lake, pond, and river water and observe wildlife to learn the effects of acidic precipitation on aquatic environments.

For Review and Reference

(Numbers in parentheses refer to pages, unless noted otherwise.)

Learning Objectives

Relevant section and/or sample problem (SP) numbers appear in parentheses.

Understand These Concepts

1. The distinction between physical and chemical properties and changes (1.1; SP 1.1)
2. The defining features of the states of matter (1.1)
3. The nature of potential and kinetic energy and their interconversion (1.1)
4. The process of approaching a phenomenon scientifically, and the distinctions between observation, hypothesis, experiment, and model (1.3)
5. The common units of length, volume, mass, and temperature and their numerical prefixes (1.5)
6. The distinctions between mass and weight, heat and temperature, and intensive and extensive properties (1.5)
7. The meaning of uncertainty in measurements and the use of significant figures and rounding (1.6)
8. The distinctions between accuracy and precision and between systematic and random error (1.6)

Master These Skills

1. Using conversion factors in calculations (1.4; SPs 1.2–1.4)
2. Finding density from mass and volume (SP 1.5)
3. Converting among the Kelvin, Celsius, and Fahrenheit scales (SP 1.6)
4. Determining the number of significant figures (SP 1.7) and rounding to the correct number of digits (SP 1.8)

Key Terms

Section 1.1
chemistry (2)
matter (2)
composition (2)
property (2)
physical property (2)
physical change (2)
chemical property (3)
chemical change (chemical reaction) (3)
state of matter (4)
solid (4)
liquid (4)
gas (4)
energy (6)
potential energy (6)
kinetic energy (6)

Section 1.2
alchemy (8)
combustion (9)
phlogiston theory (9)

Section 1.3
scientific method (10)
observation (11)
data (11)
natural law (11)
hypothesis (11)
experiment (11)
variable (11)
controlled experiment (11)
model (theory) (11)

Section 1.4
conversion factor (12)
dimensional analysis (14)

Section 1.5
SI unit (16)
base (fundamental) unit (16)
derived unit (16)
meter (m) (17)
volume (*V*) (17)
cubic meter (m^3) (17)
liter (L) (17)
milliliter (mL) (17)
mass (19)
kilogram (kg) (19)
weight (19)
density (*d*) (20)
extensive property (20)
intensive property (20)
temperature (*T*) (22)
heat (22)
thermometer (22)
Kelvin (K) (23)
Celsius scale (23)
Kelvin (absolute) scale (23)
second (s) (24)

Section 1.6
uncertainty (25)
significant figures (26)
round off (27)
exact number (29)
precision (29)
accuracy (29)
systematic error (29)
random error (29)
calibration (30)

Key Equations and Relationships

1.1 Calculating density from mass and volume (20):

$$\text{Density} = \frac{\text{mass}}{\text{volume}}$$

1.2 Converting temperature from °C to K (24):

$$T \text{ (in K)} = T \text{ (in °C)} + 273.15$$

1.3 Converting temperature from K to °C (24):

$$T \text{ (in °C)} = T \text{ (in K)} - 273.15$$

1.4 Converting temperature from °C to °F (24):

$$T \text{ (in °F)} = \tfrac{9}{5}T \text{ (in °C)} + 32$$

1.5 Converting temperature from °F to °C (24):

$$T \text{ (in °C)} = [T \text{ (in °F)} - 32]\tfrac{5}{9}$$

Highlighted Figures and Tables

These figures (F) and tables (T) provide a review of key ideas. Entries in **color** contain frequently used data.

F1.1 The distinction between physical and chemical change (2)
F1.2 The physical states of matter (4)
F1.3 Potential energy and kinetic energy (7)
F1.6 The scientific approach (11)
T1.2 SI base units (16)
T1.3 Decimal prefixes used with SI units (16)
T1.4 SI-English equivalent quantities (17)
F1.8 Some volume relationships in SI (18)

Brief Solutions to Follow-up Problems

1.1 (a) Physical. Solid iodine changes to gaseous iodine.
(b) Chemical. Gasoline burns in air to form different substances.
(c) Chemical. In contact with air, torn skin and blood react to form different substances.

1.2 No. of chairs

$$= 3 \text{ bolts} \times \frac{200 \text{ m}^2}{1 \text{ bolt}} \times \frac{3.281 \text{ ft}}{1 \text{ m}} \times \frac{3.281 \text{ ft}}{1 \text{ m}} \times \frac{1 \text{ chair}}{31.5 \text{ ft}^2}$$
$$= 205 \text{ chairs}$$

1.3 Radius of ribosome (dm) $= \frac{21.4 \text{ nm}}{2} \times \frac{1 \text{ dm}}{10^8 \text{ nm}}$
$$= 1.07\times10^{-7} \text{ dm}$$

Volume of ribosome (dm^3) $= \frac{4}{3}\pi r^3 = \frac{4}{3}(3.14)(1.07\times10^{-7} \text{ dm})^3$
$$= 5.13\times10^{-21} \text{ dm}^3$$

Volume of ribosome (μL) $= (5.13\times10^{-21} \text{ dm}^3)\left(\frac{1 \text{ L}}{1 \text{ dm}^3}\right)\left(\frac{10^6 \ \mu\text{L}}{1 \text{ L}}\right)$
$$= 5.13\times10^{-15} \ \mu\text{L}$$

1.4 Mass (kg) of solution $= 8.0 \text{ h} \times \frac{60 \text{ min}}{1 \text{ h}} \times \frac{60 \text{ s}}{1 \text{ min}} \times \frac{1.5 \text{ drops}}{1 \text{ s}}$
$$\times \frac{65 \text{ mg}}{1 \text{ drop}} \times \frac{1 \text{ g}}{10^3 \text{ mg}} \times \frac{1 \text{ kg}}{10^3 \text{ g}}$$
$$= 2.8 \text{ kg}$$

1.5 Mass (kg) of sample $= 4.6 \text{ cm}^3 \times \frac{7.5 \text{ g}}{1 \text{ cm}^3} \times \frac{1 \text{ kg}}{10^3 \text{ g}}$
$$= 0.034 \text{ kg}$$

1.6 T (in °C) = 234 K − 273.15 = −39°C
T (in °F) $= \frac{9}{5}(-39°\text{C}) + 32 = -38°\text{F}$
Answer contains two significant figures (see Section 1.6).

1.7 (a) 31.0̲7̲0̲ mg, 5 sf (b) 0.060̲6̲0̲ g, 4 sf
(c) 85̲0̲.°C, 3 sf (d) 2.0̲0̲0̲$\times10^2$ mL, 4 sf
(e) 3.9$\times10^{-6}$ m, 2 sf (f) 4.0̲1̲$\times10^{-4}$ L, 3 sf

1.8 $$\frac{25.65 \text{ mL} + 37.4 \text{ mL}}{73.55 \text{ s}\left(\frac{1 \text{ min}}{60 \text{ s}}\right)} = 51.4 \text{ mL/min}$$

Problems

Problems with **colored** numbers are answered in Appendix E. Sections match the text and provide the numbers of relevant sample problems. Most offer Concept Review Questions, Skill-Building Exercises (grouped in pairs covering the same concept), and Problems in Context. Comprehensive Problems are based on material from any section or previous chapter.

Some Fundamental Definitions

(Sample Problem 1.1)

Concept Review Question

1.1 Scenes A and B depict changes in matter at the atomic scale:

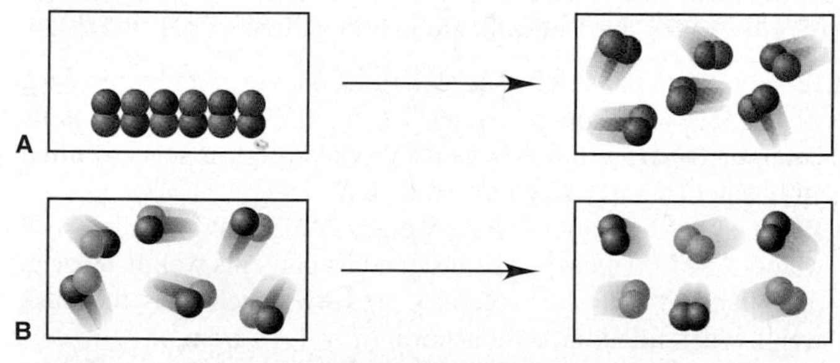

(a) Which show(s) a physical change?
(b) Which show(s) a chemical change?
(c) Which result(s) in different physical properties?
(d) Which result(s) in different chemical properties?
(e) Which result(s) in a change in state?

Skill-Building Exercises *(grouped in similar pairs)*

1.2 Describe solids, liquids, and gases in terms of how they fill a container. Use your descriptions to identify the physical state (at room temperature) of the following: (a) helium in a toy balloon; (b) mercury in a thermometer; (c) soup in a bowl.

1.3 Use your descriptions in the previous problem to identify the physical state (at room temperature) of the following: (a) air in your room; (b) vitamin tablets in a bottle; (c) sugar in a packet.

1.4 Define *physical property* and *chemical property*. Identify each type of property in the following statements:
(a) Yellow-green chlorine gas attacks silvery sodium metal to form white crystals of sodium chloride (table salt).
(b) A magnet separates a mixture of black iron shavings and white sand.

1.5 Define *physical change* and *chemical change*. State which type of change occurs in each of the following statements:
(a) Passing an electric current through molten magnesium chloride yields molten magnesium and gaseous chlorine.
(b) The iron in discarded automobiles slowly forms reddish brown, crumbly rust.

1.6 Which of the following is a chemical change? Explain your reasoning: (a) boiling canned soup; (b) toasting a slice of bread; (c) chopping a log; (d) burning a log.

1.7 Which of the following changes can be reversed by changing the temperature (that is, which are physical changes): (a) dew condensing on a leaf; (b) an egg turning hard when it is boiled; (c) ice cream melting; (d) a spoonful of batter cooking on a hot griddle?

1.8 For each pair, which has higher potential energy?
(a) The fuel in your car or the products in its exhaust
(b) Wood in a fireplace or the ashes in the fireplace after the wood burns

1.9 For each pair, which has higher kinetic energy?
(a) A sled resting at the top of a hill or a sled sliding down the hill
(b) Water above a dam or water falling over the dam

Chemical Arts and the Origins of Modern Chemistry

Concept Review Questions

1.10 The alchemical, medical, and technological traditions were precursors to chemistry. State a contribution that each made to the development of the science of chemistry.

1.11 How did the phlogiston theory explain combustion?

1.12 One important observation that supporters of the phlogiston theory had trouble explaining was that the calx of a metal weighs more than the metal itself. Why was that observation important? How did the phlogistonists respond?

1.13 Lavoisier developed a new theory of combustion that overturned the phlogiston theory. What measurements were central to his theory, and what key discovery did he make?

The Scientific Approach: Developing a Model

Concept Review Questions

1.14 How are the key elements of scientific thinking used in the following scenario? While making your breakfast toast, you notice it fails to pop out of the toaster. Thinking the spring mechanism is stuck, you notice that the bread is unchanged. Assuming you forgot to plug in the toaster, you check and find it is plugged in. When you take the toaster into the dining room and plug it into a different outlet, you find the toaster works. Returning to the kitchen, you turn on the switch for the overhead light and nothing happens.

1.15 Why is a quantitative observation more useful than a nonquantitative one? Which of the following are quantitative?
(a) The sun rises in the east.
(b) An astronaut weighs one-sixth as much on the Moon as on Earth.
(c) Ice floats on water.
(d) An old-fashioned hand pump cannot draw water from a well more than 34 ft deep.

1.16 Describe the essential features of a well-designed experiment.

1.17 Describe the essential features of a scientific model.

Chemical Problem Solving
(Sample Problem 1.2)

Concept Review Question

1.18 When you convert feet to inches, how do you decide which portion of the conversion factor should be in the numerator and which in the denominator?

Skill-Building Exercises *(grouped in similar pairs)*

1.19 Write the conversion factor(s) for (a) in^2 to cm^2; (b) km^2 to m^2; (c) mi/h to m/s; (d) lb/ft^3 to g/cm^3.

1.20 Write the conversion factor(s) for (a) cm/min to in/min; (b) m^3 to in^3; (c) m/s^2 to km/h^2; (d) gallons/h to L/s.

Measurement in Scientific Study
(Sample Problems 1.3 to 1.6)

Concept Review Questions

1.21 Describe the difference between intensive and extensive properties. Which of the following properties are intensive: (a) mass; (b) density; (c) volume; (d) melting point?

1.22 Explain the difference between mass and weight. Why is your weight on the Moon one-sixth that on Earth?

1.23 For each of the following cases, state whether the density of the object increases, decreases, or remains the same:
(a) A sample of chlorine gas is compressed.
(b) A lead weight is carried from sea level to the top of a high mountain.
(c) A sample of water is frozen.
(d) An iron bar is cooled.
(e) A diamond is submerged in water.

1.24 Explain the difference between heat and temperature. Does 1 L of water at 65°F have more, less, or the same quantity of energy as 1 L of water at 65°C?

1.25 A one-step conversion is sufficient to convert a temperature in the Celsius scale into the Kelvin scale, but not into the Fahrenheit scale. Explain.

Skill-Building Exercises *(grouped in similar pairs)*

1.26 The average radius of a molecule of lysozyme, an enzyme in tears, is 1430 pm. What is its radius in nanometers (nm)?

1.27 The radius of a barium atom is 2.22×10^{-10} m. What is its radius in angstroms (Å)?

1.28 What is the length in inches (in) of a 100.-m soccer field?

1.29 The center on your basketball team is 6 ft 10 in tall. How tall is the player in centimeters (cm)?

1.30 A small hole in the wing of a space shuttle requires a 17.7-cm^2 patch. (a) What is the patch's area in square kilometers (km^2)? (b) If the patching material costs NASA \$3.25/in^2, what is the cost of the patch?

1.31 The area of a telescope lens is 6322 mm^2. (a) What is the area of the lens in square feet (ft^2)? (b) If it takes a technician 45 s to polish 135 mm^2, how long does it take her to polish the entire lens?

1.32 Express your body weight in kilograms (kg).

1.33 There are 2.60×10^{15} short tons of oxygen in the atmosphere (1 short ton = 2000 lb). How many metric tons of oxygen are present (1 metric ton = 1000 kg)?

1.34 The average density of the Earth is 5.52 g/cm^3. What is its density in (a) kg/m^3; (b) lb/ft^3?

1.35 The speed of light in a vacuum is 2.998×10^8 m/s. What is its speed in (a) km/h; (b) mi/min?

1.36 The volume of a certain bacterial cell is 1.72 μm^3. (a) What is its volume in cubic millimeters (mm^3)? (b) What is the volume of 10^5 cells in liters (L)?

1.37 (a) How many cubic meters of milk are in 1 qt (946.4 mL)? (b) How many liters of milk are in 835 gallons (1 gal = 4 qt)?

1.38 An empty vial weighs 55.32 g. (a) If the vial weighs 185.56 g when filled with liquid mercury (d = 13.53 g/cm^3), what is its volume? (b) How much would the vial weigh if it were filled with water (d = 0.997 g/cm^3 at 25°C)?

1.39 An empty Erlenmeyer flask weighs 241.3 g. When filled with water (d = 1.00 g/cm^3), the flask and its contents weigh 489.1 g. (a) What is the flask's volume? (b) How much does the flask weigh when filled with chloroform (d = 1.48 g/cm^3)?

1.40 A small cube of aluminum measures 15.6 mm on a side and weighs 10.25 g. What is the density of aluminum in g/cm^3?

1.41 A steel ball-bearing with a circumference of 32.5 mm weighs 4.20 g. What is the density of the steel in g/cm^3 (V of a sphere = $\frac{4}{3}\pi r^3$; circumference of a circle = $2\pi r$)?

1.42 Perform the following conversions:
(a) 72°F (a pleasant spring day) to °C and K
(b) −164°C (the boiling point of methane, the main component of natural gas) to K and °F
(c) 0 K (absolute zero, theoretically the coldest possible temperature) to °C and °F

1.43 Perform the following conversions:
(a) 106°F (the body temperature of many birds) to K and °C

(b) 3410°C (the melting point of tungsten, the highest for any element) to K and °F
(c) 6.1×10^3 K (the surface temperature of the Sun) to °F and °C

Problems in Context

1.44 Anton van Leeuwenhoek, a 17th-century pioneer in the use of the microscope, described the microorganisms he saw as "animalcules" whose length was "25 thousandths of an inch." How long were the animalcules in meters?

1.45 The distance between two adjacent peaks on a wave is called the *wavelength*.
(a) The wavelength of a beam of ultraviolet light is 255 nanometers (nm). What is its wavelength in meters?
(b) The wavelength of a beam of red light is 683 nm. What is its wavelength in angstroms (Å)?

1.46 In the early 20th century, thin metal foils were used to study atomic structure. (a) How many in^2 of gold foil with a thickness of 1.6×10^{-5} in could have been made from 2.0 troy oz? (b) If gold cost \$20.00/troy oz at that time, how many cm^2 of gold foil could have been made from \$75.00 worth of gold (1 troy oz = 31.1 g; d of gold = 19.3 g/cm^3)?

1.47 A cylindrical tube 7.8 cm high and 0.85 cm in diameter is used to collect blood samples. How many cubic decimeters (dm^3) of blood can it hold (V of a cylinder = $\pi r^2 h$)?

1.48 Copper can be drawn into thin wires. How many meters of 34-gauge wire (diameter = 6.304×10^{-3} in) can be produced from the copper in 5.01 lb of covellite, an ore of copper that is 66% copper by mass? (*Hint:* Treat the wire as a cylinder: V of cylinder = $\pi r^2 h$; d of copper = 8.95 g/cm^3.)

Uncertainty in Measurement: Significant Figures

(Sample Problems 1.7 and 1.8)

Concept Review Questions

1.49 What is an exact number? How are exact numbers treated differently from other numbers in a calculation?

1.50 All nonzero digits are significant. State a rule that tells which zeros are significant.

1.51 A newspaper reported that the attendance at Slippery Rock's home football game was 5209. (a) How many significant figures does this number contain? (b) Was the actual number of people counted? (c) After Slippery Rock's next home game, the newspaper reported an attendance of 5000. If you assume that this number contains two significant figures, how many people could actually have been at the game?

Skill-Building Exercises *(grouped in similar pairs)*

1.52 Underline the significant zeros in the following numbers:
(a) 0.39 (b) 0.039 (c) 0.0390 (d) 3.0900×10^4

1.53 Underline the significant zeros in the following numbers:
(a) 5.08 (b) 508 (c) 5.080×10^3 (d) 0.05080

1.54 Round off each number to the indicated number of significant figures (sf): (a) 0.0003554 (to 2 sf); (b) 35.8348 (to 4 sf); (c) 22.4555 (to 3 sf).

1.55 Round off each number to the indicated number of significant figures (sf): (a) 231.554 (to 4 sf); (b) 0.00845 (to 2 sf); (c) 144,000 (to 1 sf).

1.56 Round off each number in the following calculation to one fewer significant figure, and find the answer:

$$\frac{19 \times 155 \times 8.3}{3.2 \times 2.9 \times 4.7}$$

1.57 Round off each number in the following calculation to one fewer significant figure, and find the answer:

$$\frac{9.8 \times 6.18 \times 2.381}{24.3 \times 1.8 \times 18.5}$$

1.58 Carry out the following calculations, making sure that your answer has the correct number of significant figures:
(a) $\dfrac{2.795\text{ m} \times 3.10\text{ m}}{6.48\text{ m}}$
(b) $V = \frac{4}{3}\pi r^3$, where r = 9.282 cm
(c) 1.110 cm + 17.3 cm + 108.2 cm + 316 cm

1.59 Carry out the following calculations, making sure that your answer has the correct number of significant figures:
(a) $\dfrac{2.420\text{ g} + 15.6\text{ g}}{4.8\text{ g}}$ (b) $\dfrac{7.87\text{ mL}}{16.1\text{ mL} - 8.44\text{ mL}}$
(c) $V = \pi r^2 h$, where r = 6.23 cm and h = 4.630 cm

1.60 Write the following numbers in scientific notation:
(a) 131,000.0 (b) 0.00047 (c) 210,006 (d) 2160.5

1.61 Write the following numbers in scientific notation:
(a) 281.0 (b) 0.00380 (c) 4270.8 (d) 58,200.9

1.62 Write the following numbers in standard notation. Use a terminal decimal point when needed:
(a) 5.55×10^3 (b) 1.0070×10^4
(c) 8.85×10^{-7} (d) 3.004×10^{-3}

1.63 Write the following numbers in standard notation. Use a terminal decimal point when needed:
(a) 6.500×10^3 (b) 3.46×10^{-5} (c) 7.5×10^2 (d) 1.8856×10^2

1.64 Convert the following into correct scientific notation:
(a) 802.5×10^2 (b) 1009.8×10^{-6} (c) 0.077×10^{-9}

1.65 Convert the following into correct scientific notation:
(a) 14.3×10^1 (b) 851×10^{-2} (c) 7500×10^{-3}

1.66 Carry out each of the following calculations, paying special attention to significant figures, rounding, and units (J = joule, the SI unit of energy; mol = mole, the SI unit for amount of substance):
(a) $\dfrac{(6.626\times10^{-34}\text{ J·s})(2.9979\times10^8\text{ m/s})}{489\times10^{-9}\text{ m}}$
(b) $\dfrac{(6.022\times10^{23}\text{ molecules/mol})(1.19\times10^2\text{ g})}{46.07\text{ g/mol}}$
(c) $(6.022\times10^{23}\text{ atoms/mol})(2.18\times10^{-18}\text{ J/atom})\left(\dfrac{1}{2^2} - \dfrac{1}{3^2}\right)$, where the numbers 2 and 3 in the last term are exact.

1.67 Carry out each of the following calculations, paying special attention to significant figures, rounding, and units:
(a) $\dfrac{8.32\times10^7\text{ g}}{\frac{4}{3}(3.1416)(1.95\times10^2\text{ cm})^3}$ (The term $\frac{4}{3}$ is exact.)
(b) $\dfrac{(1.84\times10^2\text{ g})(44.7\text{ m/s})^2}{2}$ (The term 2 is exact.)
(c) $\dfrac{(1.07\times10^{-4}\text{ mol/L})^2(2.6\times10^{-3}\text{ mol/L})}{(8.35\times10^{-5}\text{ mol/L})(1.48\times10^{-2}\text{ mol/L})^3}$

1.68 Which statements include exact numbers?
(a) Angel Falls is 3212 ft high.
(b) There are nine known planets in the Solar System.
(c) There are 453.59 g in 1 lb.
(d) There are 1000 mm in 1 m.

1.69 Which of the following include exact numbers?
(a) The speed of light in a vacuum is a physical constant; to six significant figures, it is 2.99792×10^8 m/s.
(b) The density of mercury at 25°C is 13.53 g/mL.
(c) There are 3600 s in 1 h.
(d) In 2003, the United States had 50 states.

Problems in Context

1.70 How long is the metal strip shown below? Be sure to answer with the correct number of significant figures.

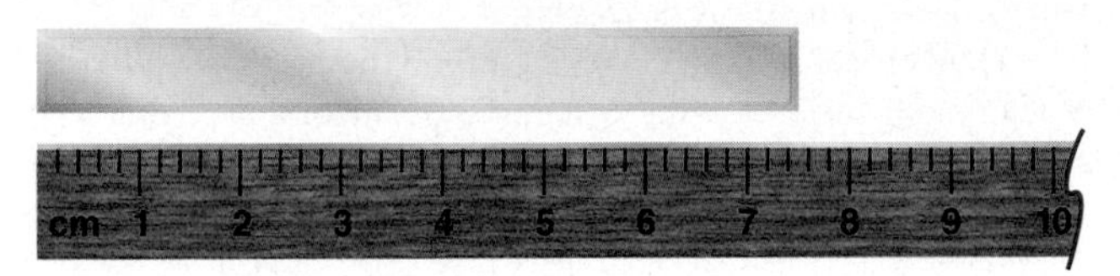

1.71 These organic solvents are used to clean compact discs:

Solvent	Density (g/mL) at 20°C
Chloroform	1.492
Diethyl ether	0.714
Ethanol	0.789
Isopropanol	0.785
Toluene	0.867

(a) If a 15.00-mL sample of CD cleaner weighs 11.775 g at 20°C, which solvent is most likely to be present?
(b) The chemist analyzing the cleaner calibrates her equipment and finds that the pipet is accurate to ±0.02 mL, and the balance is accurate to ±0.003 g. Is this equipment precise enough to distinguish between ethanol and isopropanol?

1.72 A gathering of 300 families attends a Founder's Day picnic in a small town, and the local Chamber of Commerce provides 3 bagels, 2 donuts, and 4 cans of iced tea to each family. After the picnic, 18 kg of garbage is collected. Which of the numbers—300, 3, 2, 4, 18—are exact numbers? Explain.

1.73 A laboratory instructor gives a sample of amino acid powder to each of four students, I, II, III, and IV, and they weigh the samples. The true value is 8.72 g. Their results for three trials are
I: 8.72 g, 8.74 g, 8.70 g II: 8.56 g, 8.77 g, 8.83 g
III: 8.50 g, 8.48 g, 8.51 g IV: 8.41 g, 8.72 g, 8.55 g
(a) Calculate the average mass from each set of data, and tell which set is the most accurate.
(b) Precision is a measure of the average of the deviations of each piece of data from the average value. Which set of data is the most precise? Is this set also the most accurate?
(c) Which set of data is both the most accurate and most precise?
(d) Which set of data is both the least accurate and least precise?

1.74 The following dartboards illustrate the types of errors often seen in measurements. The bull's-eye represents the actual value, and the darts represent the data.

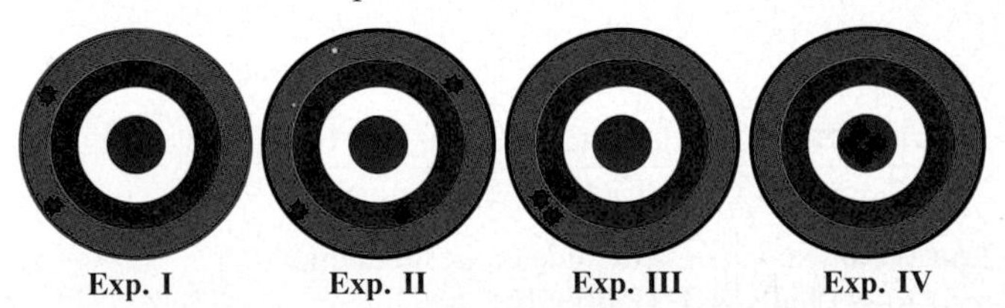

(a) Which experiments yield the same average result?
(b) Which experiment(s) display(s) high precision?
(c) Which experiment(s) display(s) high accuracy?
(d) Which experiment(s) show(s) a systematic error?

Comprehensive Problems

1.75 To make 2.000 gal of a powdered sports drink, a group of students measure out 2.000 gal of water with 500.-mL, 50.-mL, and 5-mL graduated cylinders. Show how they could get closest to 2.000 gal of water, using these cylinders the fewest times.

1.76 Two blank potential energy diagrams (see Figure 1.3, p. 7) appear below. Beneath each diagram are objects to place in the diagram. Draw the objects on the dashed lines to indicate higher or lower potential energy and label each case as more or less stable:

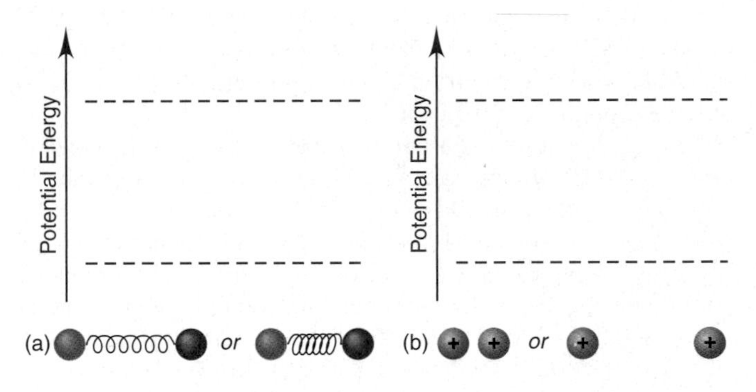

(a) Two balls attached to a relaxed *or* a compressed spring.
(b) Two positive charges near *or* apart from each other.

1.77 Stainless steels are a major component of fine surgical instruments. An engineer, testing a more energy-efficient process for preparing Stainless Steel 304 (d = 7.90 g/cm^3), makes a cube of the steel that is 15.0 mm on each edge and weighs it. If the new process works, what should the mass of the cube be?

1.78 Soft drinks are about as dense as water (1.0 g/cm^3); many common metals, including iron, copper, and silver, have densities around 9.5 g/cm^3. (a) What is the mass of the liquid in a standard 12-oz bottle of diet cola? (b) What is the mass of a dime? (*Hint:* A stack of five dimes has a volume of about 1 cm^3.)

1.79 Suppose your dorm room is 11 ft wide by 12 ft long by 8.5 ft high and has an air conditioner that exchanges air at a rate of 1200 L/min. How long would it take the air conditioner to exchange the air in your room once?

1.80 In 1933, the United States went off the international gold standard, and the price of gold increased from \$20.00 to \$35.00/troy oz. The twenty-dollar gold piece, known as the double eagle, weighed 33.436 g and was 90.0% gold by mass. (a) What was the value of the gold in the double eagle before and after the price change? (b) How many coins could be made from 50.0 troy oz of gold? (c) How many coins could be made from 2.00 in^3 of gold (1 troy oz = 31.1 g; d of gold = 19.3 g/cm^3)?

1.81 An Olympic-size pool is 50.0 m long and 25.0 m wide. (a) How many gallons of water (d = 1.0 g/mL) are needed to fill the pool to an average depth of 4.8 ft? (b) What is the mass (in kg) of water in the pool?

1.82 At room temperature (20°C) and pressure, the density of air is 1.189 g/L. An object will float in air if its density is less than that of air. In a buoyancy experiment with a new plastic, a chemist creates a rigid, thin-walled ball that weighs 0.12 g and has a volume of 560 cm^3.
(a) Will the ball float if it is evacuated?
(b) Will it float if filled with carbon dioxide (d = 1.830 g/L)?
(c) Will it float if filled with hydrogen (d = 0.0899 g/L)?
(d) Will it float if filled with oxygen (d = 1.330 g/L)?
(e) Will it float if filled with nitrogen (d = 1.165 g/L)?
(f) For any case that will float, how much weight must be added to make the ball sink?

1.83 Asbestos is a fibrous silicate mineral with remarkably high tensile strength. But it is no longer used because airborne asbestos particles can cause lung cancer. Grunerite, a type of asbestos, has a tensile strength of 3.5×10^4 kg/cm^2 (thus, a bar of grunerite with a 1-cm^2 cross-sectional area can hold up to 3.5×10^4 kg). The tensile strengths of aluminum and Steel No. 5137 are 2.5×10^4 lb/in^2 and 5.0×10^4 lb/in^2, respectively. Calculate the cross-sectional area (in cm^2) of bars of aluminum and Steel No. 5137 that have the same tensile strength as a bar of grunerite with a cross-sectional area of 25 mm^2.

1.84 According to the lore of ancient Greece, Archimedes discovered the displacement method of density determination while bathing and used it to find the composition of the king's crown. If a crown weighing 4 lb 13 oz displaces 186 mL of water, is the crown made of pure gold (d = 19.3 g/cm^3)?

1.85 Earth's oceans have an average depth of 3800 m, a total area of 3.63×10^8 km^2, and an average concentration of dissolved gold of 5.8×10^{-9} g/L. (a) How many grams of gold are in the oceans? (b) How many m^3 of gold are in the oceans? (c) If a recent price of gold was $370.00/troy oz, what is the value of gold in the oceans (1 troy oz = 31.1 g; d of gold = 19.3 g/cm^3)?

1.86 Brass is an alloy of copper and zinc. Varying the mass percentages of the two metals produces brasses with different properties. A brass called *yellow zinc* has high ductility and strength and is 34%–37% zinc by mass. (a) Find the mass range (in g) of copper in 174 g of yellow zinc. (b) What is the mass range (in g) of zinc in a sample of yellow zinc that contains 46.5 g of copper?

1.87 For the year 2002, worldwide production of aluminum was 24.4 million metric tons (t). (a) How many pounds of aluminum were produced? (b) What was its volume in cubic feet (1 t = 1000 kg; d of aluminum = 2.70 g/cm^3)?

1.88 Liquid nitrogen is obtained from liquefied air and is used industrially to prepare frozen foods. It boils at 77.36 K. (a) What is this temperature in °C? (b) What is this temperature in °F? (c) At the boiling point, the density of the liquid is 809 g/L and that of the gas is 4.566 g/L. How many liters of liquid nitrogen are produced when 895.0 L of nitrogen gas is liquefied at 77.36 K?

1.89 The speed of sound varies according to the material through which it travels. Sound travels at 5.4×10^3 cm/s through rubber and at 1.97×10^4 ft/s through granite. Calculate each of these speeds in m/s.

1.90 If a raindrop weighs 65 mg on average and 5.1×10^5 raindrops fall on a lawn every minute, what mass (in kg) of rain falls on the lawn in 1.5 h?

1.91 A jogger runs at an average speed of 5.9 mi/h. (a) How fast is she running in m/s? (b) How many kilometers does she run in 98 min? (c) If she starts a run at 11:15 am, what time is it after she covers 4.75×10^4 ft?

1.92 The Environmental Protection Agency (EPA) has proposed a new safety standard for microparticulates in air: for particles up to 2.5 μm in diameter, the maximum allowable amount is 50. μg/m^3. If your 10.0 ft × 8.25 ft × 12.5 ft dorm room just meets the new EPA standard, how many of these particles are in your room? How many are in each 0.500-L breath you take? (Assume the particles are spheres of diameter 2.5 μm and made primarily of soot, a form of carbon with a density of 2.5 g/cm^3.)

1.93 Earth's surface area is 5.10×10^8 km^2, and its crust has a mean thickness of 35 km and mean density of 2.8 g/cm^3. The two most abundant elements in the crust are oxygen (4.55×10^5 g/metric ton, t) and silicon (2.72×10^5 g/t), and the two rarest non-radioactive elements are ruthenium and rhodium, each with an abundance of 1×10^{-4} g/t. What is the total mass of each of these elements in Earth's crust (1 t = 1000 kg)?

1.94 Nutritional tables give the potassium content of a standard apple (about 3 apples/lb) as 159 mg. How many grams of potassium are in 4.25 kg of apples?

1.95 The three states of matter differ greatly in their viscosity, a measure of their resistance to flow. Rank the three states from highest to lowest viscosity. Explain in submicroscopic terms.

1.96 If a temperature scale were based on the freezing point (5.5°C) and boiling point (80.1°C) of benzene and the temperature difference between these points was divided into 50 units (called °X), what would be the freezing and boiling points of water in °X? (See Figure 1.12, p. 23.)

Parts of a Whole Like this part of a larger mosaic, any sample of matter we observe consists of smaller components that give rise to the sample's unique properties. In this chapter, we explore the nature of these components and of their components in turn.

2

The Components of Matter

Questioning what something is made of was a common practice even among the philosophers of ancient Greece, although we approach the question very differently today. They believed that everything was made of one or, at most, a few elemental substances (elements). Some believed this substance to be water, because rivers and oceans extend everywhere. Others thought it was air, which was "thinned" into fire and "thickened" into clouds, rain, and rock. Still others believed there were four elements—fire, air, water, and earth—whose properties accounted for hotness, wetness, sweetness, and all other characteristics of things.

Democritus (c. 460–370 BC), the father of atomism, took a different approach. He focused on the ultimate components of *all* substances, and his reasoning went something like this: if you cut a piece of, say, copper smaller and smaller, you must eventually reach a particle of copper so small that it can no longer be cut. Therefore, matter is ultimately composed of indivisible particles, with nothing between them but empty space. He called the particles *atoms* (Greek *atomos*, "uncuttable") and proclaimed: "According to convention, there is a sweet and a bitter, a hot and a cold, and according to convention, there is order. In truth, there are atoms and a void." However, Aristotle (384–322 BC), who elaborated the idea of four elements, held that it was impossible for "nothing" to exist, and his influence was so great that the concept of atoms was suppressed for 2000 years.

Finally, in the 17th century, the great English scientist Robert Boyle argued that an element is composed of "simple Bodies, not made of any other Bodies, of which all mixed Bodies are compounded, and into which they are ultimately resolved," a description that is remarkably similar to today's idea of an element, in which the "simple Bodies" are atoms. Boyle's hypothesis began the wonderful process of discovery, debate, and rediscovery that marks scientific inquiry, as exemplified by Lavoisier's work. Further studies in the 18th century gave rise to laws concerning the relative masses of substances that react with each other. Then, at the beginning of the 19th century, John Dalton proposed an atomic model that explained these mass laws and soon led to rapid progress in chemistry. By that century's close, however, further observation exposed the need to revise Dalton's model. A burst of creativity in the early 20th century gave rise to a picture of the atom with a complex internal structure, which led to our current model.

IN THIS CHAPTER . . . We compare the properties and composition of the three types of matter—elements, compounds, and mixtures—on the macroscopic and atomic scales. We examine the mass laws and Dalton's theory to explain them and then cover key experiments that led to our current model of the atom. Atomic structure is described, and then we see how elements are organized and classified in the periodic table. We discuss the two ways elements combine to form compounds, and learn how to derive compound names, formulas, and masses. Finally, we see how mixtures are classified and separated.

Concepts & Skills to Review

before you study this chapter

- physical and chemical change (Section 1.1)
- states of matter (Section 1.1)
- attraction and repulsion between charged particles (Section 1.1)
- meaning of a scientific model (Section 1.3)
- SI units and conversion factors (Section 1.5)
- significant figures in calculations (Section 1.6)

2.1 ELEMENTS, COMPOUNDS, AND MIXTURES: AN ATOMIC OVERVIEW

Matter can be broadly classified into three types—elements, compounds, and mixtures. An **element** is the simplest type of matter with unique physical and chemical properties. *An element consists of only one kind of atom.* Therefore, it cannot be broken down into a simpler type of matter by any physical or chemical methods. An element is one kind of **pure substance** (or just **substance**), matter whose composition is fixed. Each element has a name, such as silicon, oxygen, or copper. A sample of silicon contains only silicon atoms. A key point to remember is that the *macroscopic* properties of a piece of silicon, such as color, density, and combustibility, are different from those of a piece of copper because silicon atoms are different from copper atoms; in other words, *each element is unique because the properties of its atoms are unique.*

Most elements exist in nature as populations of atoms. Figure 2.1A shows atoms of a gaseous element such as neon. However, several elements occur naturally as molecules: a **molecule** is an independent structural unit consisting of two or more atoms chemically bound together (Figure 2.1B). Oxygen, for example, occurs in air as *diatomic* (two-atom) molecules.

A **compound** is a type of matter composed of *two or more different elements that are chemically bound together.* Be sure you understand that the elements in a compound are not just mixed together; rather, their atoms have joined chemically (Figure 2.1C). Ammonia, water, and carbon dioxide are some common compounds. One defining feature of a compound is that *the elements are present in fixed parts by mass* (fixed mass ratio). Because of this fixed composition, *a compound is also considered a substance.* Any molecule of the compound has the same fixed parts by mass because it consists of *fixed numbers* of atoms of the component elements. For example, any sample of ammonia is 14 parts nitrogen by mass plus 3 parts hydrogen by mass. Since 1 nitrogen atom has 14 times the mass of 1 hydrogen atom, 1 molecule of ammonia always consists of 1 nitrogen atom and 3 hydrogen atoms:

Ammonia is 14 parts N and 3 parts H by mass
1 N atom has 14 times the mass of 1 H atom
Therefore, ammonia has 1 N atom and 3 H atoms.

Another defining feature of a compound is that *its properties are different from those of its component elements.* Table 2.1 shows a striking example. Soft, silvery sodium metal and yellow-green, poisonous chlorine gas have very different properties from the compound they form—white, crystalline sodium chloride, or common table salt! Unlike an element, a compound *can* be broken down into simpler substances—its component elements. For example, an electric current breaks down molten sodium chloride into metallic sodium and chlorine gas. Note that this breakdown is a *chemical change,* not a physical one.

Figure 2.1D depicts a **mixture,** a group of two or more substances (elements and/or compounds) that are physically intermingled. In contrast to a compound, *the components of a mixture* ***can*** *vary in their parts by mass.* Since its composition is not fixed, a mixture is *not* a substance. A mixture of the two compounds sodium chloride and water, for example, can have many different parts by mass of salt to water. Since the components are physically mixed, not chemically combined, a mixture at the atomic scale is merely a group of the individual units that make up its component elements and/or compounds. Therefore, *a mixture retains many of the properties of its components.* Salt water, for instance, is colorless like water and tastes salty like sodium chloride. Unlike compounds, mixtures can be separated into their components by *physical changes;* chemical changes are not needed. For example, the water in salt water can be boiled off, a physical process that leaves behind the sodium chloride.

Figure 2.1 Elements, compounds, and mixtures on the atomic scale. A, Most elements consist of a large collection of identical atoms. **B,** Some elements occur as molecules. **C,** A molecule of a compound consists of characteristic numbers of atoms of two or more elements chemically bound together. **D,** A mixture contains the individual units of two or more elements and/or compounds that are physically intermingled. The samples shown here are gases, but elements, compounds, and mixtures occur as liquids and solids also.

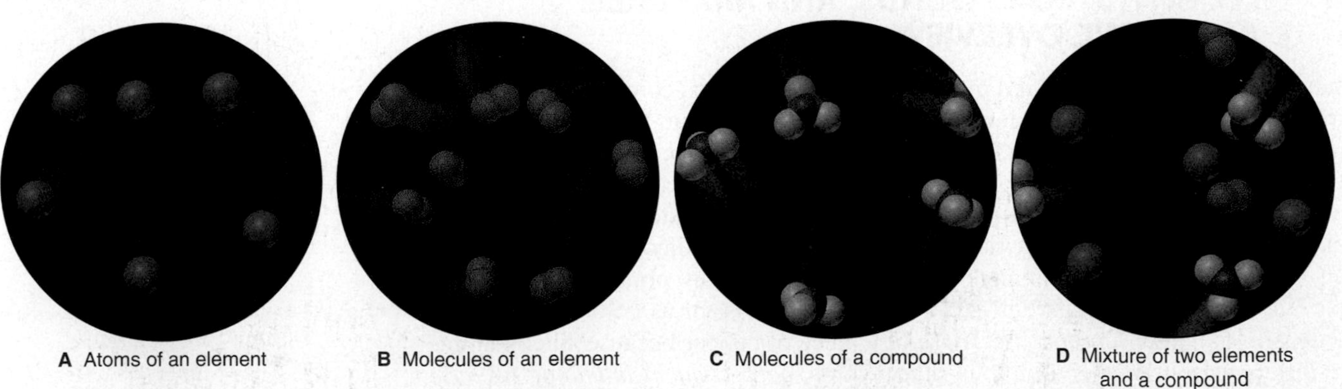

Table 2.1 Some Properties of Sodium, Chlorine, and Sodium Chloride

Property	Sodium	+	Chlorine	⟶	Sodium Chloride
Melting point	97.8°C		−101°C		801°C
Boiling point	881.4°C		−34°C		1413°C
Color	Silvery		Yellow-green		Colorless (white)
Density	0.97 g/cm^3		0.0032 g/cm^3		2.16 g/cm^3
Behavior in water	Reacts		Dissolves slightly		Dissolves freely

SECTION SUMMARY

All matter exists as either elements, compounds, or mixtures. An element consists of only one type of atom. A compound contains two or more elements in chemical combination; it exhibits different properties from its component elements. The elements of a compound occur in fixed parts by mass because each unit of the compound has fixed numbers of each type of atom. Elements and compounds are referred to as substances because their compositions are fixed. A mixture consists of two or more substances mixed together, not chemically combined. The components retain their individual properties and can be present in any proportions.

2.2 THE OBSERVATIONS THAT LED TO AN ATOMIC VIEW OF MATTER

Any model of the composition of matter had to explain two extremely important chemical observations that were well established by the end of the 18th century: the *law of mass conservation* and the *law of definite (or constant) composition.* As you'll see, John Dalton's atomic theory explained these laws and another observation now known as the *law of multiple proportions.*

Mass Conservation

The most fundamental chemical observation of the 18th century was the **law of mass conservation:** *the total mass of substances does not change during a chemical reaction.* The *number* of substances may change, and by definition their properties must, but the *total amount* of matter remains constant. Lavoisier had first stated this law on the basis of his combustion experiments, in which he found the mass of oxygen plus the mass of mercury equal to the mass of mercuric oxide they formed. Figure 2.2 illustrates mass conservation in a reaction that occurs in water.

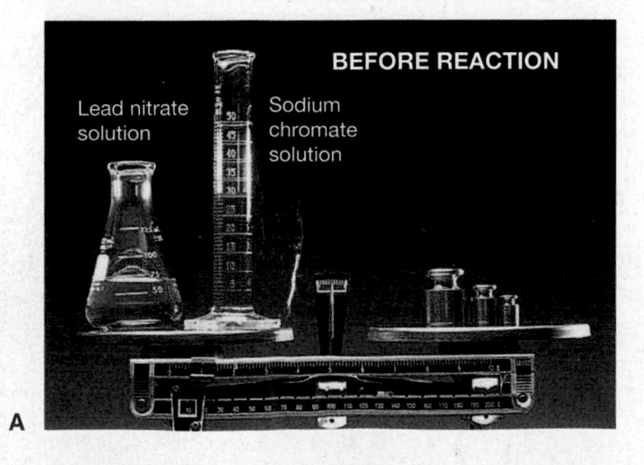

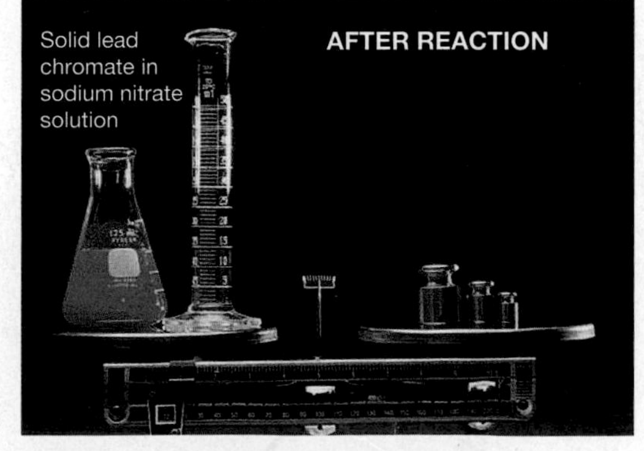

Figure 2.2 The law of mass conservation: mass remains constant during a chemical reaction. The total mass of lead nitrate solution and sodium chromate solution before they react **(A)** is the same as the total mass after they have reacted **(B)** to form lead chromate (yellow solid) and sodium nitrate solution.

Immeasurable Changes in Mass Based on the work of Albert Einstein (1879–1955), we now know that some mass *does* change into energy during a chemical reaction. But the amount is too small to measure, even by the best modern balance. For example, when 100 g of carbon burns in oxygen, carbon dioxide is formed, and only 0.000000036 g (3.6×10^{-8} g) of mass is converted to energy. The energy yields of chemical reactions are relatively so small that, for all practical purposes, mass *is* conserved. As you'll see later, however, energy changes in nuclear reactions are so large that mass changes are measured easily.

Even in a complex biochemical change within an organism, such as the metabolism of the sugar glucose, which involves many reactions, mass is conserved:

$$180\text{ g glucose} + 192\text{ g oxygen gas} \longrightarrow 264\text{ g carbon dioxide} + 108\text{ g water}$$
$$372\text{ g material before change} \longrightarrow 372\text{ g material after change}$$

Mass conservation means that, based on all chemical experience, *matter cannot be created or destroyed.*

Definite Composition

Another fundamental chemical observation is summarized as the **law of definite (or constant) composition:** *no matter what its source, a particular compound is composed of the same elements in the same parts (fractions) by mass.* The **fraction by mass (mass fraction)** is that part of the compound's mass contributed by the element. It is obtained by dividing the mass of each element by the total mass of compound. The **percent by mass (mass percent, mass %)** is the fraction by mass expressed as a percentage.

Let's examine the meanings of these ideas in terms of a box of marbles *(right)*. The box contains three types of marbles: yellow marbles weigh 1.0 g each, purple marbles 2.0 g each, and red marbles 3.0 g each. Each type makes up a fraction of the total mass of marbles, 16.0 g. The *mass fraction* of the yellow marbles is their number times their mass divided by the total mass: (3 × 1.0 g)/16.0 g = 0.19. The *mass percent* (parts per 100 parts) of the yellow marbles is 0.19 × 100 = 19% by mass. The purple marbles have a mass fraction of 0.25 and are 25% of the total by mass, and the red marbles have a mass fraction of 0.56 and are 56% by mass. Similarly, in a compound, each element has a *fixed* mass fraction (and mass percent).

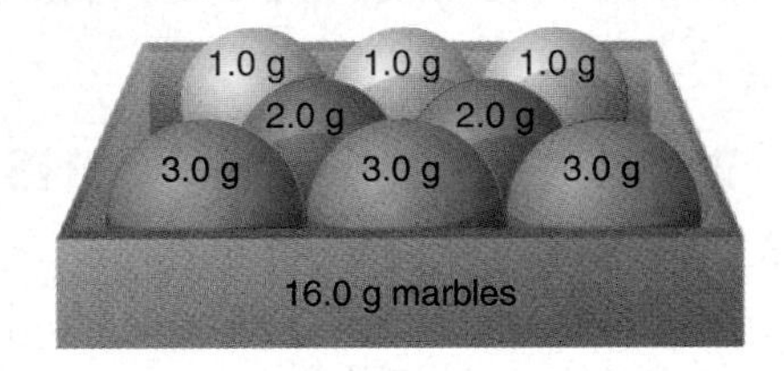

Consider calcium carbonate, the major compound in marble. The following results are obtained for the elemental mass composition of 20.0 g of calcium carbonate (for example, 8.0 g of calcium/20.0 g = 0.40 parts of calcium):

Analysis by Mass (grams/20.0 g)	Mass Fraction (parts/1.00 part)	Percent by Mass (parts/100 parts)
8.0 g calcium	0.40 calcium	40% calcium
2.4 g carbon	0.12 carbon	12% carbon
9.6 g oxygen	0.48 oxygen	48% oxygen
20.0 g	1.00 part by mass	100% by mass

As you can see, the sum of the mass fractions (or mass percents) equals 1.00 part (or 100%) by mass. The law of definite composition tells us that pure samples of calcium carbonate always contain the same elements in the same percents by mass (Figure 2.3).

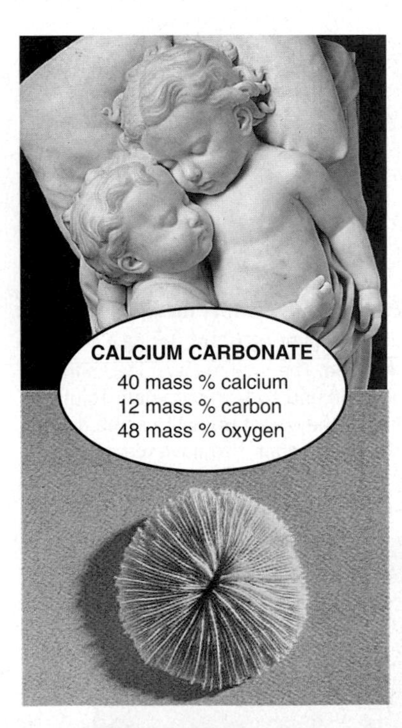

Figure 2.3 The law of definite composition. Calcium carbonate is found naturally in many forms, including marble *(top)*, coral *(bottom)*, chalk, and seashells. The mass percents of its component elements do not change regardless of the compound's source.

Since a given element always constitutes the same mass fraction of a given compound, we can use the mass fraction to find the actual mass of the element in any sample of the compound:

$$\text{Mass of element} = \text{mass of compound} \times \frac{\text{part by mass of element}}{\text{one part by mass of compound}}$$

Or, more simply, since mass analysis tells us the parts by mass, we can use that directly with *any* mass unit and skip the need to find the mass fraction first:

$$\text{Mass of element in sample} = \text{mass of compound in sample} \times \frac{\text{mass of element in compound}}{\text{mass of compound}} \quad \textbf{(2.1)}$$

SAMPLE PROBLEM 2.1 Calculating the Mass of an Element in a Compound

Problem Pitchblende is the most commercially important compound of uranium. Analysis shows that 84.2 g of pitchblende contains 71.4 g of uranium, with oxygen as the only other element. How many grams of uranium can be obtained from 102 kg of pitchblende?

Plan We have to find the mass of uranium in a known mass of pitchblende, given the mass of uranium in a different mass of pitchblende. The mass ratio of uranium/pitchblende is the same for any sample of pitchblende. Therefore, as shown by Equation 2.1, we multiply the mass (in kg) of pitchblende by the ratio of uranium to pitchblende that we construct from the mass analysis. This gives the mass (in kg) of uranium, and we just convert kilograms to grams.

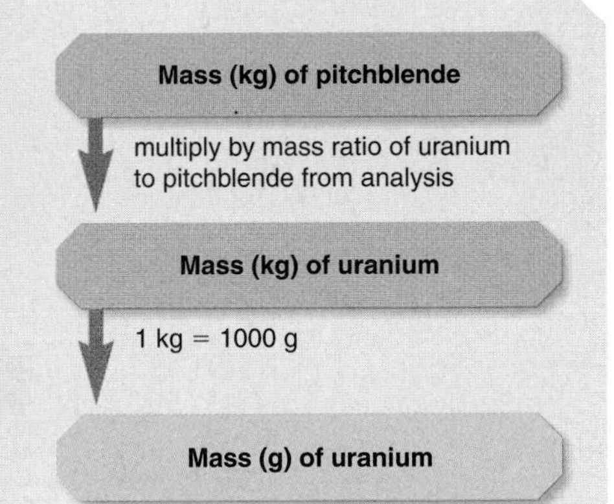

Solution Finding the mass (kg) of uranium in 102 kg of pitchblende:

$$\text{Mass (kg) of uranium} = \text{mass (kg) of pitchblende} \times \frac{\text{mass (kg) of uranium in pitchblende}}{\text{mass (kg) of pitchblende}}$$

$$\text{Mass (kg) of uranium} = 102\ \cancel{\text{kg pitchblende}} \times \frac{71.4\ \text{kg uranium}}{84.2\ \cancel{\text{kg pitchblende}}} = 86.5\ \text{kg uranium}$$

Converting the mass of uranium from kg to g:

$$\text{Mass (g) of uranium} = 86.5\ \cancel{\text{kg}}\ \text{uranium} \times \frac{1000\ \text{g}}{1\ \cancel{\text{kg}}}$$

$$= 8.65 \times 10^4\ \text{g uranium}$$

Check The analysis showed that most of the mass of pitchblende is due to uranium, so the large mass of uranium makes sense. Rounding off to check the math gives:

$$\sim 100\ \text{kg pitchblende} \times \frac{70}{85} = 82\ \text{kg uranium}$$

FOLLOW-UP PROBLEM 2.1 How many metric tons (t) of oxygen are combined in a sample of pitchblende that contains 2.3 t of uranium? (*Hint:* Remember that oxygen is the only other element present.)

Multiple Proportions

Dalton described a phenomenon that occurs when two elements form more than one compound. His observation is now called the **law of multiple proportions:** *if elements A and B react to form two compounds, the different masses of B that combine with a fixed mass of A can be expressed as a ratio of small whole numbers.* Consider two compounds that form from carbon and oxygen; for now, let's call them carbon oxides I and II. They have very different properties. For example, measured at the same temperature and pressure, the density of carbon oxide I is 1.25 g/L, whereas that of II is 1.98 g/L. Moreover, I is poisonous and flammable, but II is not. Analysis shows that their compositions by mass are

Carbon oxide I: 57.1 mass % oxygen and 42.9 mass % carbon
Carbon oxide II: 72.7 mass % oxygen and 27.3 mass % carbon

To see the phenomenon of multiple proportions, we use the mass percents of oxygen and of carbon in each compound to find the masses of these elements in a given mass, for example, 100 g, of each compound. Then we divide the mass of oxygen by the mass of carbon in each compound to obtain the mass of oxygen that combines with a fixed mass of carbon:

	Carbon Oxide I	Carbon Oxide II
g oxygen/100 g compound	57.1	72.7
g carbon/100 g compound	42.9	27.3
g oxygen/g carbon	$\frac{57.1}{42.9} = 1.33$	$\frac{72.7}{27.3} = 2.66$

If we then divide the grams of oxygen per gram of carbon in II by that in I, we obtain a ratio of small whole numbers:

$$\frac{2.66 \text{ g oxygen/g carbon in II}}{1.33 \text{ g oxygen/g carbon in I}} = \frac{2}{1}$$

The law of multiple proportions tells us that in two compounds of the same elements, the mass fraction of one element relative to the other element changes in *increments based on ratios of small whole numbers*. In this case, the ratio is 2:1—for a given mass of carbon, II contains *2 times* as much oxygen as I, not 1.583 times, 1.716 times, or any other intermediate amount. As you'll see next, Dalton's theory allows us to explain the composition of carbon oxides I and II on the atomic scale.

SECTION SUMMARY

Three fundamental observations known as the mass laws state that (1) the total mass remains constant during a chemical reaction; (2) any sample of a given compound has the same elements present in the same parts by mass; and (3) in different compounds of the same elements, the masses of one element that combine with a fixed mass of the other can be expressed as a ratio of small whole numbers.

2.3 DALTON'S ATOMIC THEORY

With almost 200 years of hindsight, it may be easy to see how the mass laws could be explained by an atomic model—matter existing in indestructible units, each with a particular mass—but it was a major breakthrough in 1808 when John Dalton (1766–1844) presented his atomic theory of matter in *A New System of Chemical Philosophy.*

Dalton's Revival of Atomism Although John Dalton, the son of a poor weaver, had no formal education, he established one of the most powerful concepts in science. Dalton began teaching science at 12 years of age, and later studied color blindness, a personal affliction still known as *daltonism.* In 1787, he began his life's work in meteorology, recording daily weather data until his death 57 years later. His studies on humidity and dew point led to a key discovery about the behavior of gases (Section 5.4) and eventually to his atomic theory. In 1803, he stated, "I am nearly persuaded that [the mixing of gases and their solubility in water] depends upon the mass and number of the ultimate particles. . . . An enquiry into the relative masses of [these] particles of bodies is a subject . . . I have lately been prosecuting . . . with remarkable success." The atomic theory was published 5 years later.

Postulates of the Atomic Theory

Dalton expressed his theory in a series of postulates. Like most great thinkers, Dalton incorporated the ideas of others into his own to create the new theory. As we go through the postulates, which are presented here in modern terms, let's see which were original and which came from others. (Later, we can examine the key differences between Dalton's postulates and our present understanding.)

1. All matter consists of **atoms,** tiny indivisible particles of an element that cannot be created or destroyed. (Derives from the "eternal, indestructible atoms" of Democritus more than 2000 years earlier and conforms to mass conservation as stated by Lavoisier.)
2. Atoms of one element *cannot* be converted into atoms of another element. In chemical reactions, the atoms of the original substances recombine to form different substances. (Rejects the alchemical belief in the magical transmutation of elements.)
3. Atoms of an element are identical in mass and other properties and are different from atoms of any other element. (Contains Dalton's major new ideas: unique mass and properties for all the atoms of a given element.)
4. Compounds result from the chemical combination of a specific ratio of atoms of different elements. (Follows directly from the fact of definite composition.)

How the Theory Explains the Mass Laws

Let's see how Dalton's postulates explain the mass laws:

- *Mass conservation.* Atoms cannot be created or destroyed (postulate 1) or converted into other types of atoms (postulate 2). Since each type of atom has a fixed mass (postulate 3), a chemical reaction, in which atoms are just combined differently with each other, cannot possibly result in a mass change.

- *Definite composition.* A compound is a combination of a *specific* ratio of different atoms (postulate 4), each of which has a particular mass (postulate 3). Thus, each element in a compound constitutes a fixed fraction of the total mass.
- *Multiple proportions.* Atoms of an element have the same mass (postulate 3) and are indivisible (postulate 1). Because different numbers of B atoms combine with each A atom in different compounds, the masses of element B that combine with a fixed mass of element A give a small, whole-number ratio.

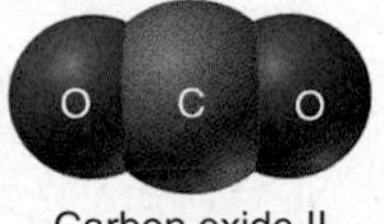

Figure 2.4 The atomic basis of the law of multiple proportions. Carbon and oxygen combine to form carbon oxide I (carbon monoxide) and carbon oxide II (carbon dioxide). The masses of oxygen in the two compounds relative to a fixed mass of carbon are in a ratio of small whole numbers.

The *simplest* arrangement consistent with the mass data for carbon oxides I and II in our earlier example is that one atom of oxygen combines with one atom of carbon in compound I (carbon monoxide) and that two atoms of oxygen combine with one atom of carbon in compound II (carbon dioxide) (Figure 2.4).

The Relative Masses of Atoms

After publication of the atomic theory, investigators tried to determine the masses of atoms from the mass fractions of elements in compounds. But an individual atom is so small that the mass of all the atoms of one element can be determined only relative to the mass of all the atoms of another element. But what should the mass standard be? Many conflicts arose, and they were resolved only decades later in an 1860 conference attended by over 140 of the world's leading chemists. There, a list of elements with accurate relative atomic masses was accepted—16 for oxygen, 12 for carbon, and so forth—and chemical formulas such as H_2O for water, NH_3 for ammonia, and CH_4 for methane were put into common use. Dalton's atomic model was crucial to this understanding because it originated the idea that masses of reacting elements could be explained in terms of atoms.

However, the model did not explain why atoms bond as they do: for example, why do two, and not three, H atoms bond with one O atom in water? Also, Dalton's "billiard ball" atom did not account for the charged particles that were being observed in experiments. Clearly, a more complex atomic model was needed.

Atoms? Humbug! Rarely does a major new concept receive unanimous acceptance. Despite the atomic theory's impact, several major scientists denied the existence of atoms for another century. In 1877, Adolf Kolbe, an eminent organic chemist, said, "[Dalton's atoms are] . . . no more than stupid hallucinations . . . mere table-tapping and supernatural explanations." The influential physicist Ernst Mach believed that scientists should look at facts, not hypothetical entities such as atoms. It was not until 1908 that the famous chemist and outspoken opponent of atomism Wilhelm Ostwald wrote, "I am now convinced [by recent] experimental evidence of the discrete or grained nature of matter, which the atomic hypothesis sought in vain for hundreds and thousands of years." He was referring to the discovery of the electron.

SECTION SUMMARY

Dalton's atomic theory explained the mass laws by proposing that all matter consists of indivisible, unchangeable atoms of fixed, unique mass. Mass is constant during a reaction because atoms form new combinations; each compound has a fixed mass fraction of each of its elements because it is composed of a fixed number of each type of atom; and different compounds of the same elements exhibit multiple proportions because they each consist of whole atoms. Studies stimulated by Dalton's model eventually led to consistent values for relative masses of atoms.

2.4 THE OBSERVATIONS THAT LED TO THE NUCLEAR ATOM MODEL

The path of discovery is often winding and unpredictable. Basic research into the nature of electricity eventually led to the discovery of *electrons,* negatively charged particles that are part of all atoms. Soon thereafter, other experiments revealed that the atom has a *nucleus*—a tiny, central core of mass and positive charge. In this section, we examine some key experiments that led to our current model of the atom.

Discovery of the Electron and Its Properties

Nineteenth-century investigators of electricity knew that matter and electric charge were somehow related. When amber is rubbed with fur, or glass with silk, positive and negative charges form—the same charges that make your hair crackle

The Familiar Glow of Colliding Particles The electric and magnetic properties of charged particles that collide with gas particles or hit a phosphor-coated screen have familiar applications. A "neon" sign glows because electrons collide with the gas particles in the tube, causing them to give off light. An aurora display occurs when Earth's magnetic field bends streams of charged particles coming from the Sun, which then collide with gases in the atmosphere. In a television tube or computer monitor, the cathode ray passes back and forth over the coated screen, creating a pattern that the eye sees as a picture.

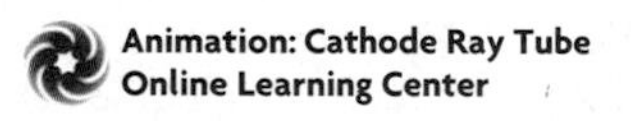
Animation: Cathode Ray Tube
Online Learning Center

and cling to your comb on a dry day. They also knew that an electric current could decompose certain compounds into their elements.

What they did not know, however, was how electricity could be studied in the *absence* of matter. Some investigators tried passing an electric current from a high-voltage source through nearly evacuated glass tubes fitted with metal electrodes that were sealed in place and connected to an external source of electricity. When the power was turned on, a "ray" could be seen striking the phosphor-coated end of the tube and emitting a glowing spot of light. The rays were called **cathode rays** because they originated at the negative electrode (cathode) and moved to the positive electrode (anode). Cathode rays typically travel in a straight line, but in a magnetic field the path is bent, indicating that the particles are charged, and in an electric field the path bends toward the positive plate. The ray is identical no matter what metal is used as the cathode (Figure 2.5). It was concluded that cathode rays consist of negatively charged particles found in all matter. The rays appear when these particles collide with the few remaining gas molecules in the evacuated tube. Cathode ray particles were later named *electrons*.

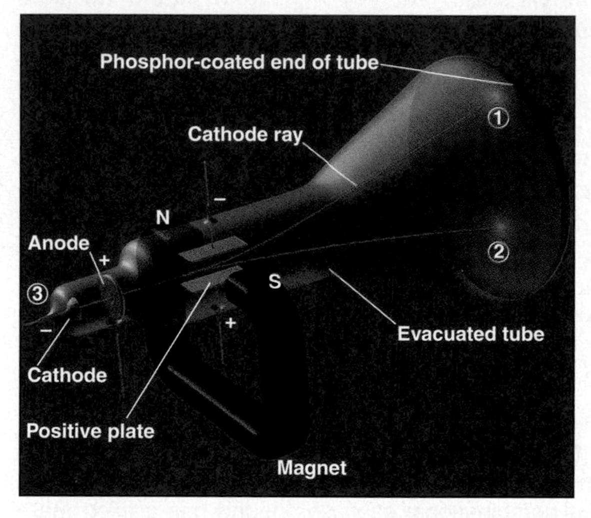

Figure 2.5 Experiments to determine the properties of cathode rays. A cathode ray forms when high voltage is applied to a partially evacuated tube. The ray passes through a hole in the anode and hits the coated end of the tube to produce a glow.

OBSERVATION	CONCLUSION
1. Ray bends in magnetic field	Consists of charged particles
2. Ray bends toward positive plate in electric field	Consists of negative particles
3. Ray is identical for any cathode	Particles found in all matter

Just over a century ago, in 1897, J. J. Thomson (1856–1940) used magnetic and electric fields to measure the ratio of the cathode ray particle's mass to its charge. By comparing this value with the mass/charge ratio for the lightest charged particle in solution, Thomson estimated that the cathode ray particle weighed less than $\frac{1}{1000}$ as much as hydrogen, the lightest atom! He was shocked because this implied that, contrary to Dalton's atomic theory, *atoms are divisible into even smaller particles*. Thomson concluded, "We have in the cathode rays matter in a new state, . . . in which the subdivision of matter is carried much further . . . ; this matter being the substance from which the chemical elements are built up." Fellow scientists reacted at first with disbelief, and some even thought Thomson was joking.

In 1909, the American physicist Robert Millikan (1868–1953) measured the *charge* of the electron. He did so by observing the movement of tiny droplets of the "highest grade clock oil" in an apparatus that contained electrically charged plates and an x-ray source (Figure 2.6). Here is a description of the basis of the experiment: X-rays knocked electrons from gas molecules in the air, and as an oil droplet fell through a hole in the positive (upper) plate, the electrons stuck to the drop, giving it a negative charge. With the electric field off, Millikan measured the mass of the droplet from its rate of fall. By turning on the field and

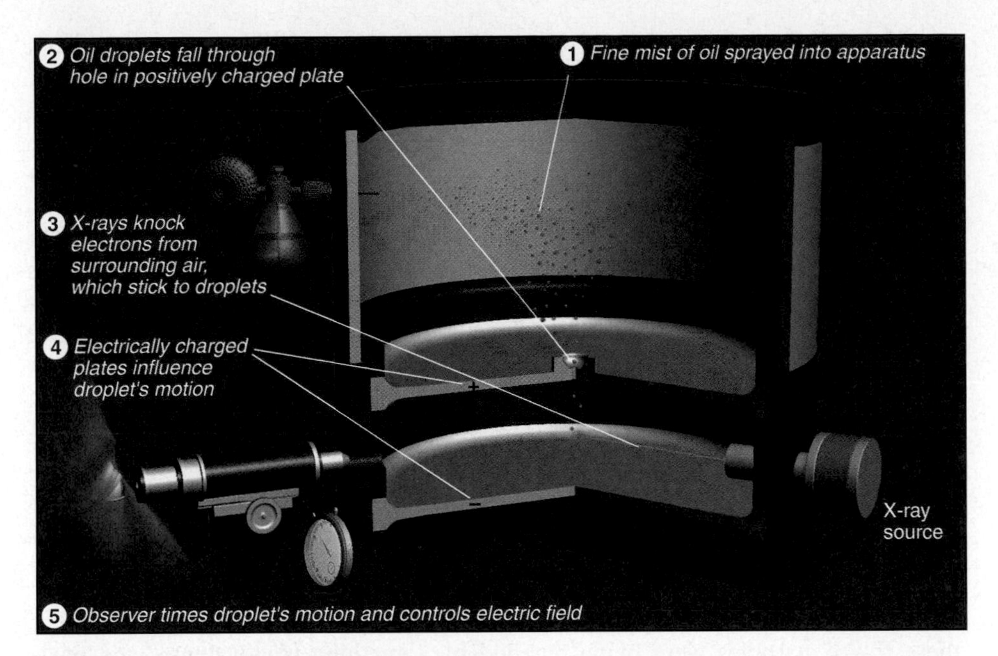

2 Oil droplets fall through hole in positively charged plate
1 Fine mist of oil sprayed into apparatus
3 X-rays knock electrons from surrounding air, which stick to droplets
4 Electrically charged plates influence droplet's motion
X-ray source
5 Observer times droplet's motion and controls electric field

Figure 2.6 Millikan's oil-drop experiment for measuring an electron's charge. The motion of a given oil droplet depends on the variation in electric field and the total charge on the droplet, which depends in turn on the number of attached electrons. Millikan reasoned that the total charge must be some whole-number multiple of the charge of the electron.

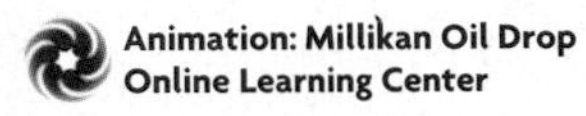

Animation: Millikan Oil Drop
Online Learning Center

varying its strength, he could make the drop fall more slowly, rise, or pause suspended. From these data, Millikan calculated the total charge of the droplet.

After studying many droplets, Millikan calculated that the various charges of the droplets were always some *whole-number multiple of a minimum charge.* He reasoned that different oil droplets picked up different numbers of electrons, so this minimum charge must be that of the electron itself. The value, which he calculated over 95 years ago, is within 1% of the modern value of the electron's charge, -1.602×10^{-19} C (C stands for *coulomb,* the SI unit of charge). Using the electron's mass/charge ratio from work by Thomson and others and this value for the electron's charge, let's calculate the electron's *extremely* small mass the way Millikan did:

$$\text{Mass of electron} = \frac{\text{mass}}{\text{charge}} \times \text{charge} = \left(-5.686\times10^{-12}\ \frac{\text{kg}}{\cancel{\text{C}}}\right)(-1.602\times10^{-19}\ \cancel{\text{C}})$$
$$= 9.109\times10^{-31}\ \text{kg} = 9.109\times10^{-28}\ \text{g}$$

Discovery of the Atomic Nucleus

Clearly, the properties of the electron posed problems about the inner structure of atoms. If everyday matter is electrically neutral, the atoms that make it up must be neutral also. But if atoms contain negatively charged electrons, what positive charges balance them? And if an electron has such an incredibly tiny mass, what accounts for an atom's much larger mass? To address these issues, Thomson proposed a model of a spherical atom composed of diffuse, positively charged matter, in which electrons were embedded like "raisins in a plum pudding."

Near the turn of the 20th century, French scientists discovered radioactivity, the emission of particles and/or radiation from atoms of certain elements. Just a few years later, in 1910, the New Zealand–born physicist Ernest Rutherford (1871–1937) used one type of radioactive particle in a series of experiments that solved this dilemma of atomic structure.

Figure 2.7 is a three-part representation of Rutherford's experiment. Tiny, dense, positively charged alpha (α) particles emitted from radium were aimed, like minute projectiles, at thin gold foil. The figure illustrates (A) the "plum-pudding" hypothesis, (B) the apparatus used to measure the deflection (scattering) of the α particles from the light flashes created when the particles struck a circular, coated screen, and (C) the actual result.

With Thomson's model in mind, Rutherford expected only minor, if any, deflections of the α particles because they should act as tiny, dense, positively charged "bullets" and go right through the gold atoms. According to the model, the embedded electrons could not deflect the α particles any more than a Ping-Pong ball could deflect a speeding baseball. Initial results confirmed this, but soon the unexpected happened. As Rutherford recalled: "Then I remember two or three days later Geiger [one of his coworkers] coming to me in great excitement and saying, 'We have been able to get some of the α particles coming backwards . . .' It was quite the most incredible event that has ever happened to me in my life. It was almost as incredible as if you fired a 15-inch shell at a piece of tissue paper and it came back and hit you."

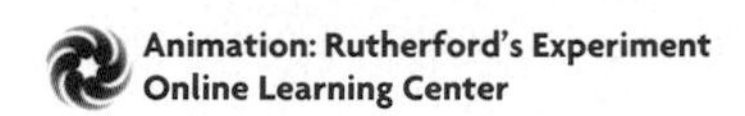
Animation: Rutherford's Experiment
Online Learning Center

The data showed that very few α particles were deflected at all, and that only 1 in 20,000 was deflected by more than 90° ("coming backwards"). It seemed that these few α particles were being repelled by something small, dense, and positive within the gold atoms. From the mass, charge, and velocity of the α particles, the frequency of these large-angle deflections, and the properties of electrons, Rutherford calculated that *an atom is mostly space occupied by electrons,* but in the center of that space is a tiny region, which he called the **nucleus,** that contains *all the positive charge and essentially all the mass of the atom.* He proposed that positive particles lay within the nucleus and called them *protons,* and then calculated the magnitude of the nuclear charge with remarkable accuracy. Rutherford's model explained the charged nature of matter, but it could not account for all the atom's mass. After more than 20 years, this issue was resolved when, in 1932, James Chadwick discovered the *neutron,* an uncharged dense particle that also resides in the nucleus.

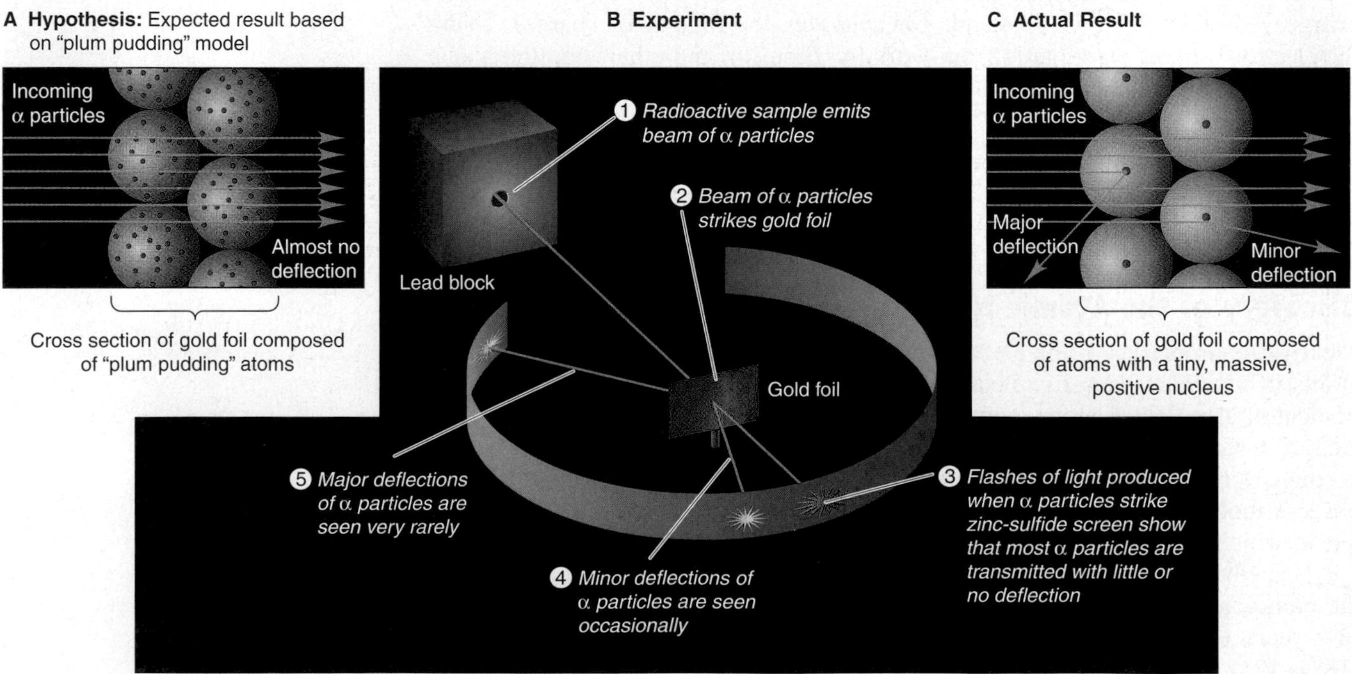
A Hypothesis: Expected result based on "plum pudding" model
Incoming α particles
Almost no deflection
Cross section of gold foil composed of "plum pudding" atoms
B Experiment
1 Radioactive sample emits beam of α particles
2 Beam of α particles strikes gold foil
Lead block
Gold foil
3 Flashes of light produced when α particles strike zinc-sulfide screen show that most α particles are transmitted with little or no deflection
4 Minor deflections of α particles are seen occasionally
5 Major deflections of α particles are seen very rarely
C Actual Result
Incoming α particles
Major deflection
Minor deflection
Cross section of gold foil composed of atoms with a tiny, massive, positive nucleus

Figure 2.7 Rutherford's α-scattering experiment and discovery of the atomic nucleus. A, HYPOTHESIS: Atoms consist of electrons embedded in diffuse, positively charged matter, so the speeding α particles should pass through the gold foil with, at most, minor deflections. **B,** EXPERIMENT: α Particles emit a flash of light when they pass through the gold atoms and hit a phosphor-coated screen. **C,** RESULTS: Occasional minor deflections and very infrequent major deflections are seen. This means very high mass and positive charge are concentrated in a small region within the atom, the nucleus.

SECTION SUMMARY

Several major discoveries at the turn of the 20th century led to our current model of atomic structure. Cathode rays were shown to consist of negative particles (electrons) that exist in all matter. J. J. Thomson measured their mass/charge ratio and concluded that they are much smaller and lighter than atoms. Robert Millikan determined the charge of the electron, which he combined with other data to calculate its mass. Ernest Rutherford proposed that atoms consist of a tiny, massive, positive nucleus surrounded by electrons.

2.5 THE ATOMIC THEORY TODAY

For nearly 200 years, scientists have known that all matter consists of atoms, and they have learned astonishing things about them. Dalton's hard, impenetrable spheres have given way to atoms with "fuzzy," indistinct boundaries and an elaborate internal architecture of subatomic particles. In this section, we examine our current model and begin to see how the properties of these particles affect the properties of atoms. Then we'll see how Dalton's theory stands up today.

Structure of the Atom

An *atom* is an electrically neutral, spherical entity composed of a positively charged central nucleus surrounded by one or more negatively charged electrons (Figure 2.8). The electrons move rapidly within the available atomic volume, held there by the attraction of the nucleus. The nucleus is incredibly dense: it contributes 99.97% of the atom's mass but occupies only about 1 ten-trillionth of its volume. (A nucleus the size of a period on this page would weigh about 100 tons, as much as 50 cars!) An atom's diameter ($\sim10^{-10}$ m) is about 10,000 times the diameter of its nucleus ($\sim10^{-14}$ m).

An atomic nucleus consists of protons and neutrons (the only exception is the simplest hydrogen nucleus, which is a single proton). The **proton (p^+)** has a positive charge, and the **neutron (n^0)** has no charge; thus, the positive charge of the nucleus results from its protons. The *magnitude* of charge possessed by a proton is equal to that of an **electron (e^-)**, but the *signs* of the charges are opposite.

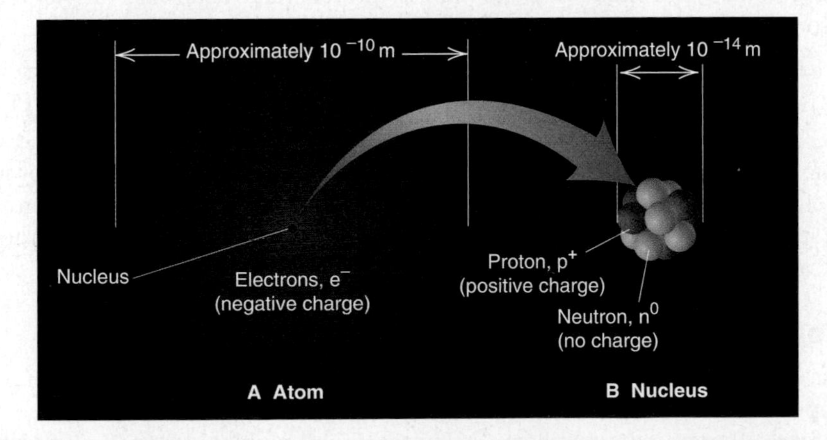

Figure 2.8 General features of the atom. A, A "cloud" of rapidly moving, negatively charged electrons occupies virtually all the atomic volume and surrounds the tiny, central nucleus. **B,** The nucleus contains virtually all the mass of the atom and consists of positively charged protons and uncharged neutrons. If the nucleus were actually the size in the figure (~1 cm across), the atom would be about 100 m across—slightly more than the length of a football field!

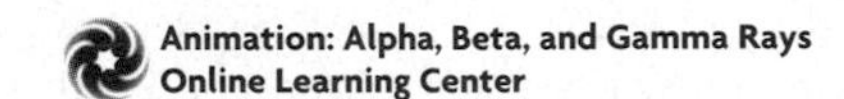
Animation: Alpha, Beta, and Gamma Rays
Online Learning Center

Table 2.2 Properties of the Three Key Subatomic Particles

	Charge		Mass		
Name (Symbol)	Relative	Absolute (C)*	Relative (amu)†	Absolute (g)	Location in Atom
Proton (p^+)	1+	$+1.60218\times10^{-19}$	1.00727	1.67262×10^{-24}	Nucleus
Neutron (n^0)	0	0	1.00866	1.67493×10^{-24}	Nucleus
Electron (e^-)	1−	-1.60218×10^{-19}	0.00054858	9.10939×10^{-28}	Outside nucleus

*The coulomb (C) is the SI unit of charge.
†The atomic mass unit (amu) equals 1.66054×10^{-24} g; discussed later in this section.

An atom is neutral because the number of protons in the nucleus equals the number of electrons surrounding the nucleus. Some properties of these three subatomic particles are listed in Table 2.2.

Atomic Number, Mass Number, and Atomic Symbol

The **atomic number (Z)** of an element equals the number of protons in the nucleus of each of its atoms. *All atoms of a particular element have the same atomic number, and each element has a different atomic number from that of any other element.* All carbon atoms ($Z = 6$) have 6 protons, all oxygen atoms ($Z = 8$) have 8 protons, and all uranium atoms ($Z = 92$) have 92 protons. There are currently 114 known elements, of which 90 occur in nature; the remaining 24 have been synthesized by nuclear processes.

The total number of protons and neutrons in the nucleus of an atom is its **mass number (A).** Each proton and each neutron contributes one unit to the mass number. Thus, a carbon atom with 6 protons and 6 neutrons in its nucleus has a mass number of 12, and a uranium atom with 92 protons and 146 neutrons in its nucleus has a mass number of 238.

The nuclear mass and charge are often included with the **atomic symbol** (or *element symbol*). Every element has a symbol based on its English, Latin, or Greek name, such as C for carbon, O for oxygen, S for sulfur, and Na for sodium (Latin *natrium*). The atomic number (Z) is written as a left *sub*script and the mass number (A) as a left *super*script to the symbol, so element X would be ${}^{A}_{Z}X$. Since the mass number is the sum of protons and neutrons, the number of neutrons (N) equals the mass number minus the atomic number:

$$\text{Number of neutrons} = \text{mass number} - \text{atomic number}, \quad \text{or} \quad N = A - Z \qquad (2.2)$$

Thus, a chlorine atom, which is symbolized as ${}^{35}_{17}Cl$, has $A = 35$, $Z = 17$, and $N = 35 - 17 = 18$. Each element has its own atomic number, so we know the atomic number from the symbol. For example, every carbon atom has 6 protons. Therefore, instead of writing ${}^{12}_{6}C$ for carbon with mass number 12, we can write ${}^{12}C$ (spoken "carbon twelve"), with $Z = 6$ understood. Another way to write this atom is carbon-12.

Naming an Element Element names have a variety of origins. Carbon comes from coal (Latin *carbo,* "ember"). Mercury (Hg) is named for the planet, but its symbol comes from its earlier name ***hy**dra**g**yrum* (Latin, "liquid silver"). Some names are based on a chemical action, such as hydrogen (Greek *hydros,* "water," and *genes,* "producer"), or on an obvious property, such as chlorine (Greek *chloros,* "yellow-green"). Sometimes, an element name refers to a country, as in germanium for Germany and americium for America. Ytterbium, yttrium, erbium, and terbium are all named for Ytterby, the Swedish town where these elements were discovered. Paying homage to great scientists led to the names of many synthetic elements, such as einsteinium for Albert Einstein and curium for Marie and Pierre Curie.

Isotopes and Atomic Masses of the Elements

All atoms of an element are identical in atomic number but not in mass number. **Isotopes** of an element are atoms that have *different numbers of neutrons* and therefore different mass numbers. For example, all carbon atoms ($Z = 6$) have 6 protons and 6 electrons, but only 98.89% of naturally occurring carbon atoms have 6 neutrons in the nucleus ($A = 12$). A small percentage (1.11%) have 7 neutrons in the nucleus ($A = 13$), and even fewer (less than 0.01%) have 8 ($A = 14$). These are carbon's three naturally occurring isotopes—${}^{12}C$, ${}^{13}C$, and ${}^{14}C$. Five other carbon isotopes—${}^{9}C$, ${}^{10}C$, ${}^{11}C$, ${}^{15}C$, and ${}^{16}C$—have been observed in the laboratory. Figure 2.9 depicts the atomic number, mass number, and symbol for four atoms, two of which are isotopes of the same element.

A key point is that the chemical properties of an element are primarily determined by the number of electrons, so *all isotopes of an element have nearly identical chemical behavior,* even though they have different masses.

SAMPLE PROBLEM 2.2 Determining the Number of Subatomic Particles in the Isotopes of an Element

Problem Silicon (Si) is essential to the computer industry as a major component of semiconductor chips. It has three naturally occurring isotopes: ^{28}Si, ^{29}Si, and ^{30}Si. Determine the numbers of protons, neutrons, and electrons in each silicon isotope.

Plan The mass number (*A*) of each of the three isotopes is given, so we know the sum of protons and neutrons. From the elements list on the text's inside front cover, we find the atomic number (*Z*, number of protons), which equals the number of electrons. We obtain the number of neutrons from Equation 2.2.

Solution From the elements list, the atomic number of silicon is 14. Therefore,

^{28}Si has $14p^+$, $14e^-$, and $14n^0$ (28 − 14)
^{29}Si has $14p^+$, $14e^-$, and $15n^0$ (29 − 14)
^{30}Si has $14p^+$, $14e^-$, and $16n^0$ (30 − 14)

FOLLOW-UP PROBLEM 2.2 How many protons, neutrons, and electrons are in **(a)** $^{11}_{5}Q$? **(b)** $^{41}_{20}X$? **(c)** $^{131}_{53}Y$? What element symbols do Q, X, and Y represent?

The mass of an atom is measured *relative* to the mass of an atomic standard. The modern atomic mass standard is the carbon-12 atom. Its mass is defined as *exactly* 12 atomic mass units. Thus, the **atomic mass unit (amu)** is $\frac{1}{12}$ the mass of a carbon-12 atom. Based on this standard, the ^1H atom has a mass of 1.008 amu; in other words, a ^{12}C atom has almost 12 times the mass of an ^1H atom. We will continue to use the term *atomic mass unit* in the text, although the name of the unit has been changed to the **dalton (Da);** thus, one ^{12}C atom has a mass of 12 daltons (12 Da, or 12 amu). The atomic mass unit, which is a unit of relative mass, has an absolute mass of 1.66054×10^{-24} g.

The isotopic makeup of an element is determined by **mass spectrometry,** a method for measuring the relative masses and abundances of atomic-scale particles very precisely (see the Tools of the Laboratory essay on the next page). For example, using a mass spectrometer, we measure the mass ratio of ^{28}Si to ^{12}C as

$$\frac{\text{Mass of } ^{28}\text{Si atom}}{\text{Mass of } ^{12}\text{C standard}} = 2.331411$$

From this mass ratio, we find the **isotopic mass** of the ^{28}Si atom, the mass of the isotope relative to the mass of the standard carbon-12 isotope:

$$\begin{aligned}\text{Isotopic mass of } ^{28}\text{Si} &= \text{measured mass ratio} \times \text{mass of } ^{12}\text{C}\\ &= 2.331411 \times 12 \text{ amu} = 27.97693 \text{ amu}\end{aligned}$$

Along with the isotopic mass, the mass spectrometer gives the relative abundance (fraction) of each isotope in a sample of the element. For example, the percent abundance of ^{28}Si is 92.23%. Such measurements provide data for obtaining the **atomic mass** (also called *atomic weight*) of an element, the *average* of the masses of its naturally occurring isotopes weighted according to their abundances.

Each naturally occurring isotope of an element contributes a certain portion to the atomic mass. For instance, as just noted, 92.23% of Si atoms are ^{28}Si. Using this percent abundance as a fraction and multiplying by the isotopic mass of ^{28}Si gives the portion of the atomic mass of Si contributed by ^{28}Si:

$$\text{Portion of Si atomic mass from } ^{28}\text{Si} = 27.97693 \text{ amu} \times 0.9223 = 25.8031 \text{ amu}$$
(retaining two additional significant figures)

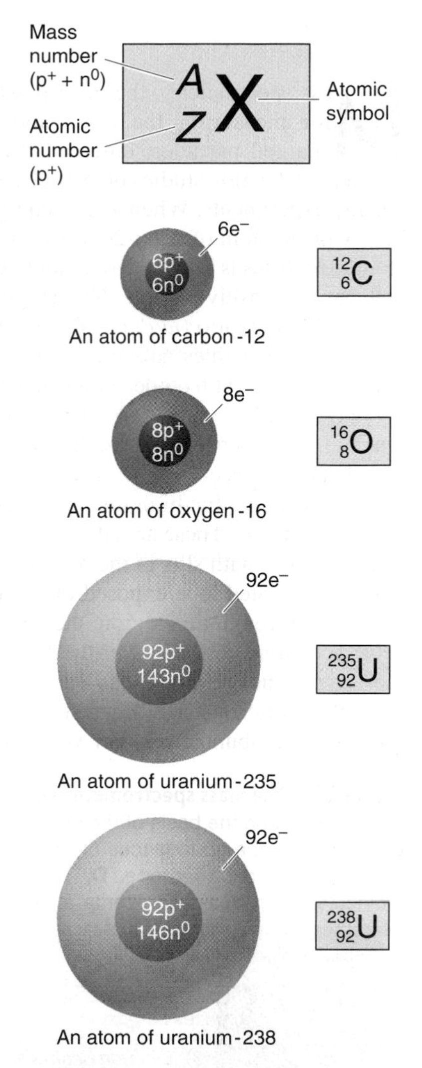

Figure 2.9 Depicting the atom. Atoms of carbon-12, oxygen-16, uranium-235, and uranium-238 are shown (nuclei not drawn to scale) with their symbolic representations. The sum of the number of protons (*Z*) and the number of neutrons (*N*) equals the mass number (*A*). An atom is neutral, so the number of protons in the nucleus equals the number of electrons around the nucleus. The two uranium atoms are isotopes of the element.

3. *All atoms of an element have the same number of protons and electrons, which determines the chemical behavior of the element.* We now know that isotopes of an element differ in the number of neutrons, and thus in mass number, but a sample of the element is treated as though its atoms have an *average* mass.
4. *Compounds are formed by the chemical combination of two or more elements in specific ratios.* We now know that a few compounds can have slight variations in their atom ratios, but this postulate remains essentially unchanged.

Even today, our picture of the atom is being revised. Although we are confident about the distribution of electrons within the atom (Chapters 7 and 8), the interactions among protons and neutrons within the nucleus are still on the frontier of discovery (Chapter 24).

SECTION SUMMARY

An atom has a central nucleus, which contains positively charged protons and uncharged neutrons and is surrounded by negatively charged electrons. An atom is neutral because the number of electrons equals the number of protons. An atom is represented by the notation $^{A}_{Z}X$, in which Z is the atomic number (number of protons), A the mass number (sum of protons and neutrons), and X the atomic symbol. An element occurs naturally as a mixture of isotopes, atoms with the same number of protons but different numbers of neutrons. Each isotope has a mass relative to the ^{12}C mass standard. The atomic mass of an element is the average of its isotopic masses weighted according to their natural abundances and is determined by modern instruments, especially the mass spectrometer.

2.6 ELEMENTS: A FIRST LOOK AT THE PERIODIC TABLE

At the end of the 18th century, Lavoisier compiled a list of the 23 elements known at that time; by 1870, 65 were known; by 1925, 88; today, there are 114 and still counting! These elements combine to form millions of compounds, so we clearly need some way to organize what we know about their behavior. By the mid-19th century, enormous amounts of information concerning reactions, properties, and atomic masses of the elements had been accumulated. Several researchers noted recurring, or *periodic,* patterns of behavior and proposed schemes to organize the elements according to some fundamental property.

In 1871, the Russian chemist Dmitri Mendeleev published the most successful of these organizing schemes, a table that listed the elements by increasing atomic mass, arranged so that elements with similar chemical properties fell in the same column. The modern **periodic table of the elements,** based on Mendeleev's earlier version, is one of the great classifying schemes in science and has become an indispensable tool to chemists. Throughout your study of chemistry, the periodic table will guide you through an otherwise dizzying amount of chemical and physical behavior.

Organization of the Periodic Table A modern version of the periodic table appears in Figure 2.10 and inside the front cover. It is formatted as follows:

1. Each element has a box that contains its atomic number, atomic symbol, and atomic mass. The boxes lie in order of *increasing atomic number* (number of protons) as you move from left to right.
2. The boxes are arranged into a grid of **periods** (horizontal rows) and **groups** (vertical columns). Each period has a number from 1 to 7. Each group has a number from 1 to 8 *and* either the letter A or B. A new system, with group

numbers from 1 to 18 but no letters, appears in parentheses under the number-letter designations. (Most chemists still use the number-letter system, so the text retains it, but shows the new numbering system in parentheses.)

3. The eight A groups (two on the left and six on the right) contain the *main-group,* or *representative, elements.* The ten B groups, located between Groups 2A(2) and 3A(13), contain the *transition elements.* Two horizontal series of *inner transition elements,* the lanthanides and the actinides, fit *between* the elements in Group 3B(3) and Group 4B(4) and are usually placed below the main body of the table.

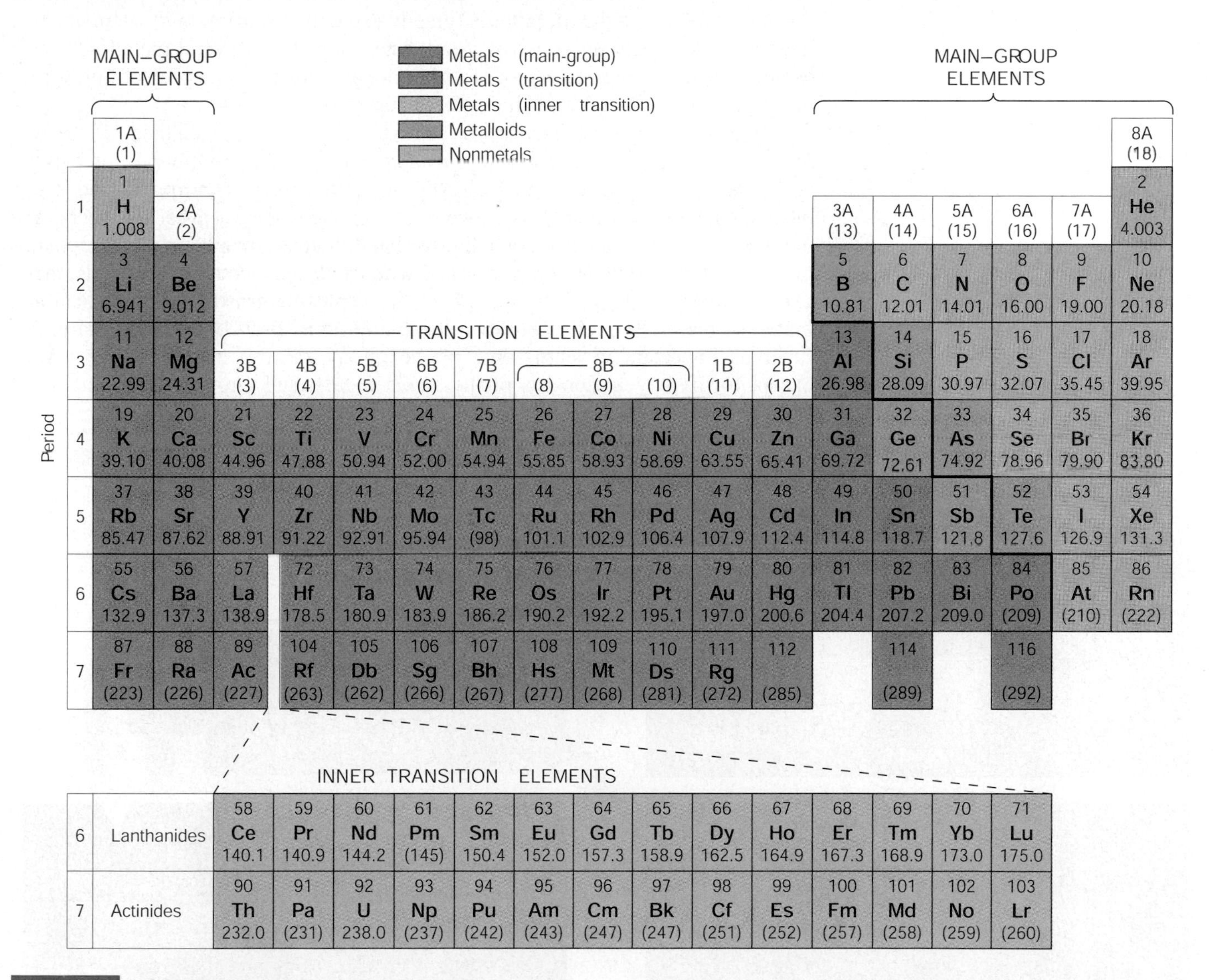

Figure 2.10 **The modern periodic table.** The table consists of element boxes arranged by *increasing* atomic number into groups (vertical columns) and periods (horizontal rows). Each box contains the atomic number, atomic symbol, and atomic mass. (A mass in parentheses is the mass number of the most stable isotope of that element.) The periods are numbered 1 to 7. The groups (sometimes called *families*) have a number-letter designation and a new group number in parentheses. The A groups are the main-group elements; the B groups are the transition elements. Two series of inner transition elements are placed below the main body of the table but actually fit between the elements indicated. Metals lie below and to the left of the thick "staircase" line [top of 3A(13) to bottom of 6A(16)] and include main-group metals *(purple-blue),* transition elements *(blue),* and inner transition elements *(gray-blue).* Nonmetals *(yellow)* lie to the right of the line. Metalloids *(green)* lie along the line. We discuss the placement of hydrogen in Chapter 14. As of late 2004, elements 112, 114, and 116 had not yet been named.

At this point in the text, the clearest distinction among the elements is their classification as metals, nonmetals, or metalloids. The "staircase" line that runs from the top of Group 3A(13) to the bottom of Group 6A(16) is a dividing line for this classification. The **metals** (three shades of blue) appear in the large lower-left portion of the table. About three-quarters of the elements are metals, including many main-group elements and all the transition and inner transition elements. They are generally shiny solids at room temperature (mercury is the only liquid) that conduct heat and electricity well and can be tooled into sheets (malleable) and wires (ductile). The **nonmetals** (yellow) appear in the small upper-right portion of the table. They are generally gases or dull, brittle solids at room temperature (bromine is the only liquid) and conduct heat and electricity poorly. Along the staircase line lie the **metalloids** (green; also called **semimetals**), elements that have properties between those of metals and nonmetals. Several metalloids, such as silicon (Si) and germanium (Ge), play major roles in modern electronics. Figure 2.11 shows examples of these three classes of elements.

Two of the major branches of chemistry can almost be defined by the elements that each studies. *Organic chemistry* studies the compounds of carbon, specifically those that contain hydrogen and often oxygen, nitrogen, and a few other elements. This branch is concerned with fuels, drugs, dyes, polymers, and the like. *Inorganic chemistry,* on the other hand, focuses mainly on the compounds of all the other elements. It is concerned with catalysts, electronic materials, metal alloys, mineral salts, and the like. With the explosive growth in biomedical and materials research, the line between these traditional branches is disappearing.

It is important to learn some of the group (family) names. Group 1A(1), except for hydrogen, consists of the *alkali metals,* and Group 2A(2) consists of

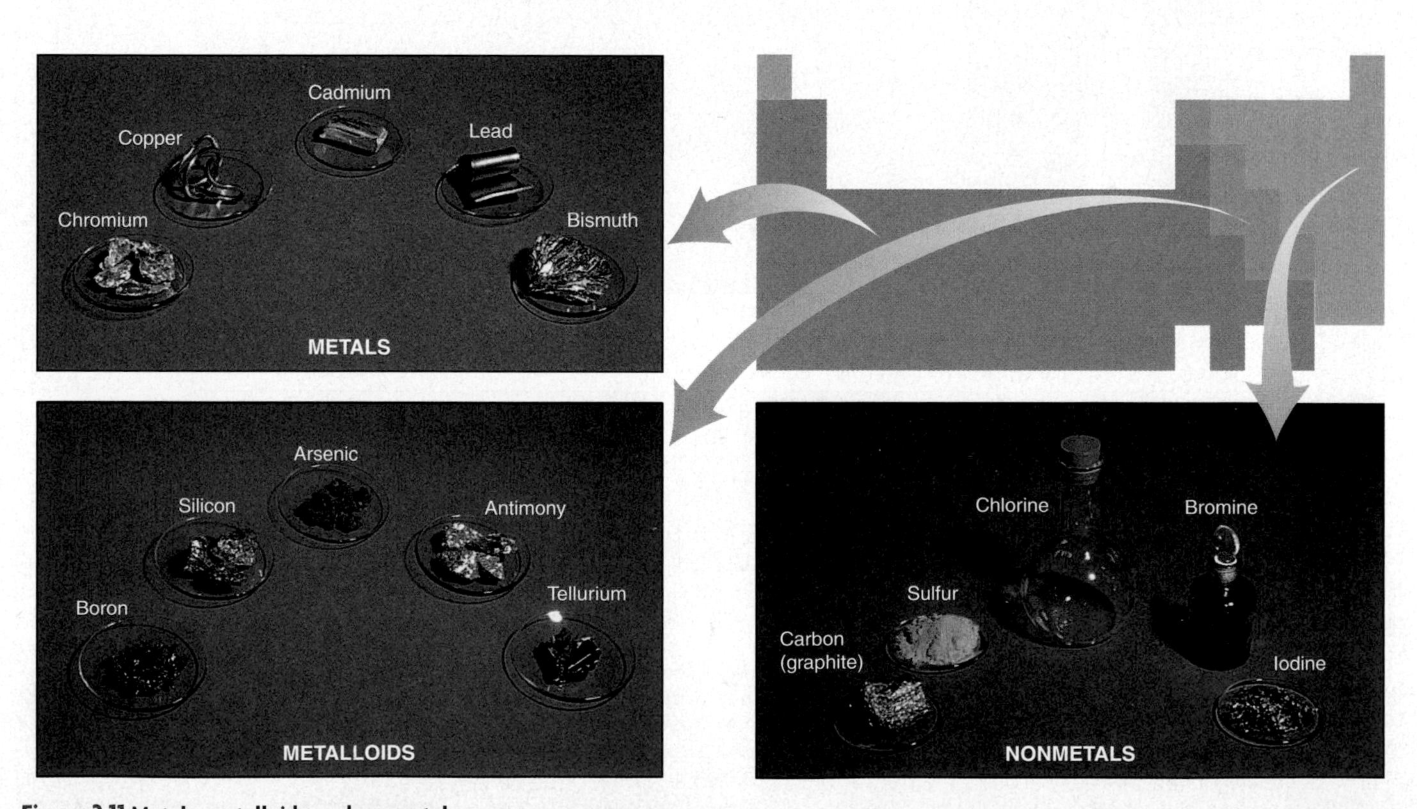

Figure 2.11 Metals, metalloids, and nonmetals.

the *alkaline earth metals*. Both groups of metals are highly reactive elements. The *halogens,* Group 7A(17), are highly reactive nonmetals, whereas the *noble gases,* Group 8A(18), are relatively unreactive nonmetals. Other main groups [3A(13) to 6A(16)] are often named for the first element in the group; for example, Group 6A is the *oxygen family.*

A key point that we return to many times is that, in general, *elements in a group have **similar** chemical properties and elements in a period have **different** chemical properties.* We begin applying the organizing power of the periodic table in the next section, where we discuss how elements combine to form compounds.

Animation: Formation of an Ionic Compound
Online Learning Center

SECTION SUMMARY

In the periodic table, the elements are arranged by atomic number into horizontal periods and vertical groups. Because of the periodic recurrence of certain key properties, elements within a group have similar behavior, whereas elements in a period have dissimilar behavior. Nonmetals appear in the upper-right portion of the table, metalloids lie along a staircase line, and metals fill the rest of the table.

2.7 COMPOUNDS: INTRODUCTION TO BONDING

The overwhelming majority of elements occur in chemical combination with other elements. In fact, only a few elements occur free in nature. The noble gases—helium (He), neon (Ne), argon (Ar), krypton (Kr), xenon (Xe), and radon (Rn)—occur in air as separate atoms. In addition to occurring in compounds, oxygen (O), nitrogen (N), and sulfur (S) occur in the most common elemental form as the molecules O_2, N_2, and S_8, and carbon (C) occurs in vast, nearly pure deposits of coal. Some of the metals, such as copper (Cu), silver (Ag), gold (Au), and platinum (Pt), may also occur uncombined with other elements. But these few exceptions reinforce the general rule that elements occur combined in compounds.

It is the electrons of the atoms of interacting elements that are involved in compound formation. Elements combine in two general ways:

1. *Transferring electrons* from the atoms of one element to those of another to form **ionic compounds**
2. *Sharing electrons* between atoms of different elements to form **covalent compounds**

These processes generate **chemical bonds,** the forces that hold the atoms of elements together in a compound. We'll introduce compound formation next and have much more to say about it in later chapters.

The Formation of Ionic Compounds

Ionic compounds are composed of **ions,** charged particles that form when an atom (or small group of atoms) gains or loses one or more electrons. The simplest type of ionic compound is a **binary ionic compound,** one composed of just two elements. It typically forms *when a metal reacts with a nonmetal.* Each metal atom loses a certain number of its electrons and becomes a **cation,** a positively charged ion. The nonmetal atoms gain the electrons lost by the metal atoms and become **anions,** negatively charged ions. In effect, the metal atoms *transfer electrons* to the nonmetal atoms. The resulting cations and anions attract each other through electrostatic forces and form the ionic compound. *All binary ionic compounds are solids.* A cation or anion derived from a single atom is called a **monatomic ion;** we'll discuss polyatomic ions, those derived from a small group of atoms, later.

The formation of the binary ionic compound sodium chloride, common table salt, is depicted in Figure 2.12, from the elements through the atomic-scale electron transfer to the compound. In the electron transfer, a sodium atom, which is neutral because it has the same number of protons as electrons, *loses* 1 electron and forms a sodium cation, Na^+. (The charge on the ion is written as a *right superscript*.) A chlorine atom *gains* the electron and becomes a chloride anion, Cl^-. (The name change from the nonmetal atom to the ion is discussed in the next section.) Even the tiniest visible grain of table salt contains an enormous number of sodium and chloride ions. The oppositely charged ions (Na^+ and Cl^-) attract each other, and the similarly charged ions (Na^+ and Na^+, or Cl^- and Cl^-) repel each other. The resulting solid aggregation is a regular array of alternating Na^+ and Cl^- ions that extends in all three dimensions.

Animation: Formation of an Ionic Compound
Online Learning Center

The strength of the ionic bonding depends to a great extent on the net strength of these attractions and repulsions and is described by *Coulomb's law,* which can be expressed as follows: *the energy of attraction (or repulsion) between two particles is directly proportional to the product of the charges and inversely proportional to the distance between them.*

$$\text{Energy} \propto \frac{\text{charge 1} \times \text{charge 2}}{\text{distance}}$$

In other words, ions with higher charges attract (or repel) each other more strongly than ions with lower charges. Likewise, smaller ions attract (or repel) each other more strongly than larger ions, because their charges are closer together. These effects are summarized in Figure 2.13.

Ionic compounds are neutral; that is, they possess no net charge. For this to occur, they must contain equal numbers of positive and negative *charges* but not necessarily equal numbers of positive and negative *ions.* Because Na^+ and Cl^- each bear a unit charge (1+ or 1−), equal numbers of the ions are present in sodium chloride. However, in sodium oxide, for example, twice as many Na^+ ions balance the 2− charge of the oxide ions, O^{2-}.

Can we predict the number of electrons a given atom will lose or gain when it forms an ion? In the formation of sodium chloride, for example, why does each

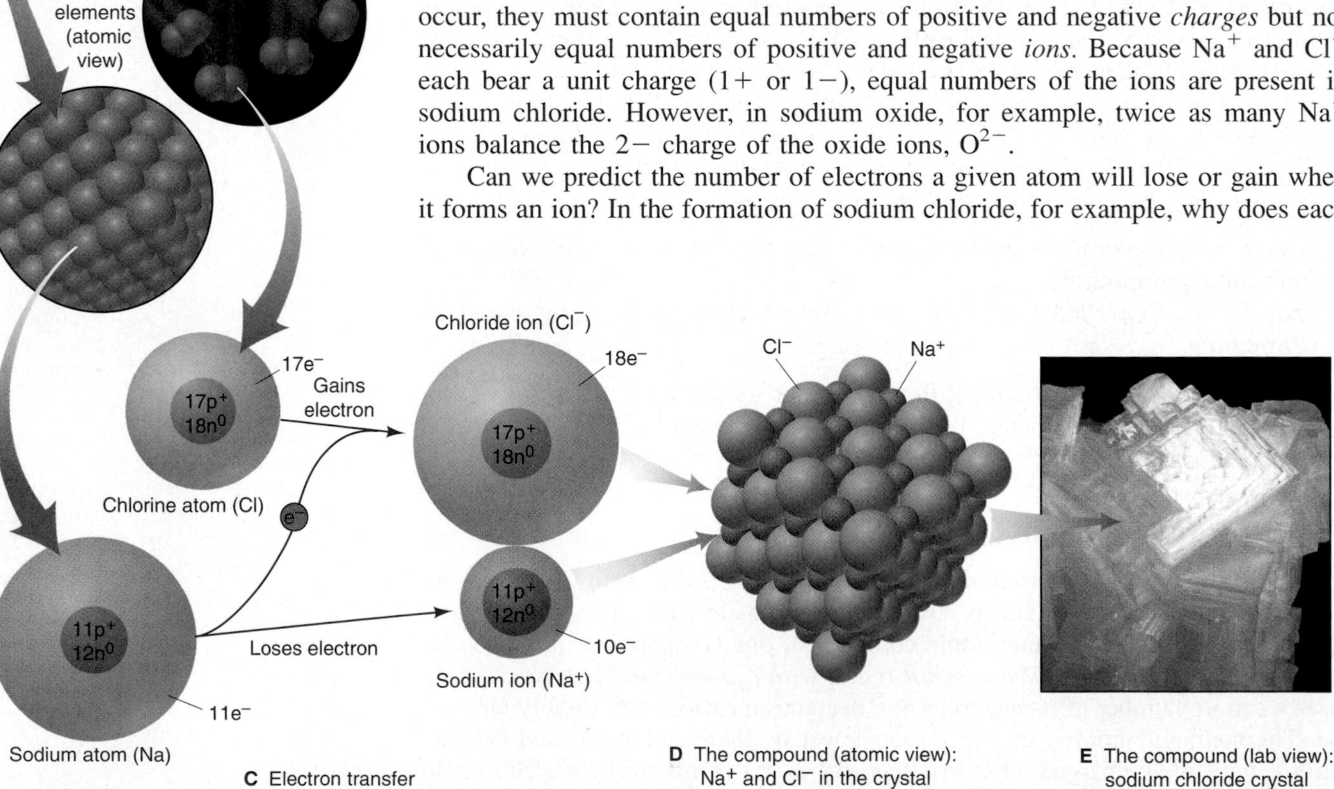

Figure 2.12 The formation of an ionic compound. A, The two elements as seen in the laboratory. **B,** The elements as they might appear on the atomic scale. **C,** The neutral sodium atom loses one electron to become a sodium cation (Na^+), and the chlorine atom gains one electron to become a chloride anion (Cl^-). (Note that when atoms lose electrons, they become ions that are smaller, and when they gain electrons, they become ions that are larger.) **D,** Na^+ and Cl^- ions attract each other and lie in a three-dimensional crystalline array. **E,** This cubic array is reflected in the structure of crystalline NaCl, which occurs naturally as the mineral halite, hence the name *halogens* for the Group 7A(17) elements.

sodium atom give up only 1 of its 11 electrons? Why doesn't each chlorine atom gain two electrons, instead of just one? For A-group elements, the periodic table provides an answer. We generally find that metals lose electrons and nonmetals gain electrons to *form ions with the same number of electrons as in the nearest noble gas* [Group 8A(18)]. Noble gases have a stability (low reactivity) that is related to their number (and arrangement) of electrons. A sodium atom ($11e^-$) can attain the stability of neon ($10e^-$), the nearest noble gas, by losing one electron. Similarly, by gaining one electron, a chlorine atom ($17e^-$) attains the stability of argon ($18e^-$), its nearest noble gas. Thus, when an element located near a noble gas forms a monatomic ion, *it gains or loses enough electrons to attain the same number as that noble gas.* Specifically, the elements in Group 1A(1) lose one electron, those in Group 2A(2) lose two, and aluminum in Group 3A(13) loses three; the elements in Group 7A(17) gain one electron, oxygen and sulfur in Group 6A(16) gain two, and nitrogen in Group 5A(15) gains three.

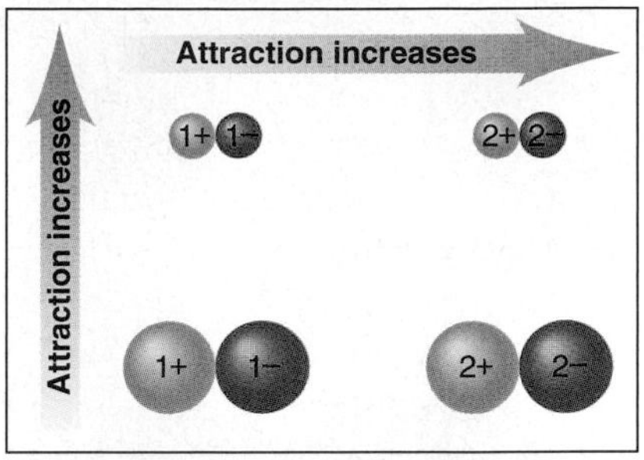

With the periodic table printed on a two-dimensional surface, as in Figure 2.10, it is easy to get the false impression that the elements in Group 7A(17) are "closer" to the noble gases than the elements in Group 1A(1). Actually, both groups are only one electron away from having the same number of electrons as the noble gases. To make this point, Figure 2.14 shows a modified periodic table that is cut and rejoined, with the noble gases in the center. Now you can see that fluorine (F; $Z = 9$) has one electron fewer and sodium (Na; $Z = 11$) has one electron more than the noble gas neon (Ne; $Z = 10$); thus, they form the F^- and Na^+ ions. Similarly, oxygen (O; $Z = 8$) gains two electrons and magnesium (Mg; $Z = 12$) loses two to form the O^{2-} and Mg^{2+} ions and attain the same number of electrons as neon.

Figure 2.13 Factors that influence the strength of ionic bonding. For ions of a given size, strength of attraction *(arrows)* increases with higher ionic charge *(left to right).* For ions of a given charge, strength of attraction increases with smaller ionic size *(bottom to top).*

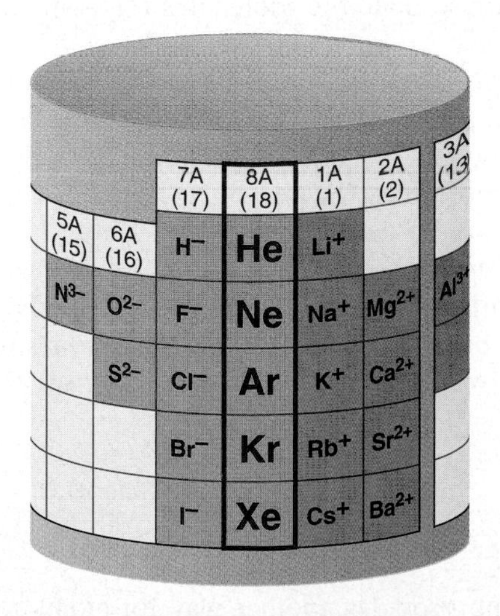

Figure 2.14 The relationship between ions formed and the nearest noble gas. This periodic table was redrawn to show the positions of other nonmetals *(yellow)* and metals *(blue)* relative to the noble gases and to show the ions these elements form. The ionic charge equals the number of electrons lost (+) or gained (−) to attain the same number of electrons as the nearest noble gas. Species in the same row have the same number of electrons. For example, H^-, He, and Li^+ all have two electrons. [Note that H is shown here in Group 7A(17).]

SAMPLE PROBLEM 2.4 Predicting the Ion an Element Forms

Problem What monatomic ions do the following elements form?

(a) Iodine ($Z = 53$) **(b)** Calcium ($Z = 20$) **(c)** Aluminum ($Z = 13$)

Plan We use the given Z value to find the element in the periodic table and see where its group lies relative to the noble gases. Elements in groups that lie *after* the noble gases *lose* electrons to attain the same number as the nearest noble gas and become positive ions; those in groups that lie *before* the noble gases *gain* electrons and become negative ions.

Solution (a) I^- Iodine ($_{53}I$) is a nonmetal in Group 7A(17), one of the halogens. Like any member of this group, it gains 1 electron to have the same number as the nearest Group 8A(18) member, in this case $_{54}Xe$.

(b) Ca^{2+} Calcium ($_{20}Ca$) is a member of Group 2A(2), the alkaline earth metals. Like any Group 2A member, it loses 2 electrons to attain the same number as the nearest noble gas, in this case, $_{18}Ar$.
(c) Al^{3+} Aluminum ($_{13}Al$) is a metal in the boron family [Group 3A(13)] and thus loses 3 electrons to attain the same number as its nearest noble gas, $_{10}Ne$.

FOLLOW-UP PROBLEM 2.4 What monatomic ion does each of the following elements form: **(a)** $_{16}S$; **(b)** $_{37}Rb$; **(c)** $_{56}Ba$?

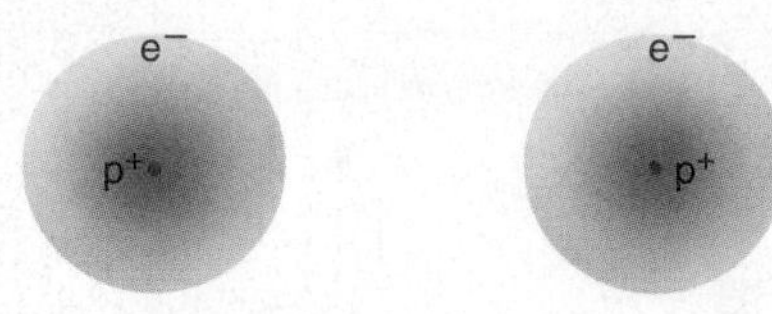

A No interaction

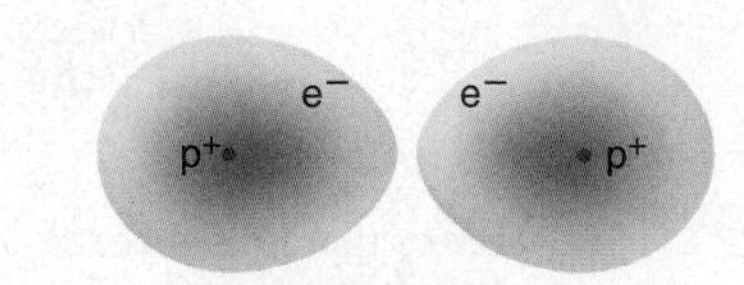

B Attraction begins

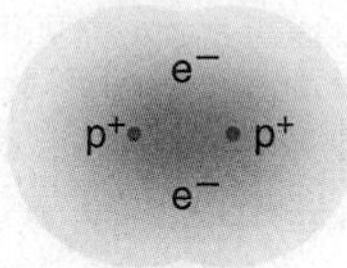

C Covalent bond

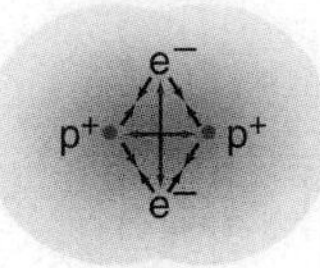

D Combination of forces

Figure 2.15 Formation of a covalent bond between two H atoms. A, The distance is too great for the atoms to affect each other. **B,** As the distance decreases, the nucleus of each atom begins to attract the electron of the other. **C,** The covalent bond forms when the two nuclei mutually attract the pair of electrons at some optimum distance. **D,** The H_2 molecule is more stable than the separate atoms because the attractive forces *(black arrows)* between each nucleus and the two electrons are greater than the repulsive forces *(red arrows)* between the electrons and between the nuclei.

The Formation of Covalent Compounds

Covalent compounds form when elements share electrons, which usually occurs between nonmetals. Even though relatively few nonmetals exist, they interact in many combinations to form a very large number of covalent compounds.

The simplest case of electron sharing occurs not in a compound but between two hydrogen atoms (H; $Z = 1$). Imagine two separated H atoms approaching each other, as in Figure 2.15. As they get closer, the nucleus of each atom attracts the electron of the other atom more and more strongly, and the separated atoms begin to interpenetrate each other. At some optimum distance between the nuclei, the two atoms form a **covalent bond,** a pair of electrons mutually attracted by the two nuclei. The result is a hydrogen molecule, in which each electron no longer "belongs" to a particular H atom: the two electrons are *shared* by the two nuclei. Repulsions between the nuclei and between the electrons also occur, but the net attraction is greater than the net repulsion. (We discuss the properties of covalent bonds in great detail in Chapter 9.)

A sample of hydrogen gas consists of these diatomic molecules (H_2)—pairs of atoms that are chemically bound and behave as an independent unit—*not* separate H atoms. Other nonmetals that exist as diatomic molecules at room temperature are nitrogen (N_2), oxygen (O_2), and the halogens [fluorine (F_2), chlorine (Cl_2), bromine (Br_2), and iodine (I_2)]. Phosphorus exists as tetratomic molecules (P_4), and sulfur and selenium as octatomic molecules (S_8 and Se_8) (Figure 2.16). At room temperature, covalent substances may be gases, liquids, or solids.

Atoms of different elements share electrons to form the molecules of a covalent compound. A sample of hydrogen fluoride, for example, consists of molecules in which one H atom forms a covalent bond with one F atom; water consists of molecules in which one O atom forms covalent bonds with two H atoms:

Hydrogen fluoride, HF

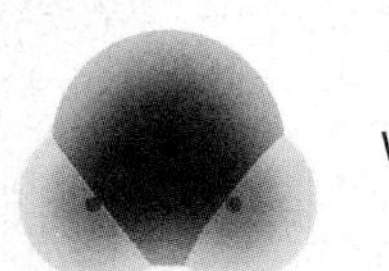

Water, H_2O

(As you'll see in Chapter 9, covalent bonding provides another way for atoms to attain the same number of electrons as the nearest noble gas.)

Figure 2.16 Elements that occur as molecules.

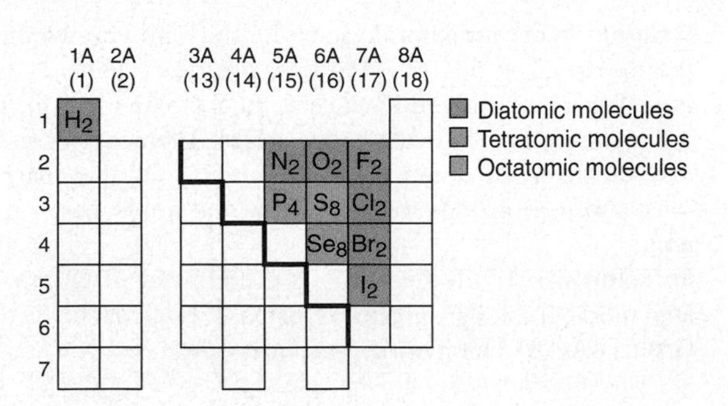

Distinguishing the Entities in Covalent and Ionic Substances There is a key distinction between the chemical entities in covalent and ionic substances. *Most covalent substances consist of molecules.* A cup of water, for example, consists of individual water molecules lying near each other. In contrast, under ordinary conditions, *no molecules exist in a sample of an ionic compound.* A piece of sodium chloride, for example, is a continuous array of oppositely charged sodium and chloride ions, *not* a collection of individual "sodium chloride molecules."

Another key distinction exists between the particles attracting each other. Covalent bonding involves the mutual attraction between two (positively charged) nuclei and the two (negatively charged) electrons that reside between them. Ionic bonding involves the mutual attraction between positive and negative ions.

Polyatomic Ions: Covalent Bonds Within Ions Many ionic compounds contain **polyatomic ions,** which consist of two or more atoms bonded *covalently* and have a net positive or negative charge. For example, the ionic compound calcium carbonate is an array of polyatomic carbonate anions and monatomic calcium cations attracted to each other. The carbonate ion consists of a carbon atom covalently bonded to three oxygen atoms, and two additional electrons give the ion its 2− charge (Figure 2.17). In many reactions, a polyatomic ion stays together as a unit.

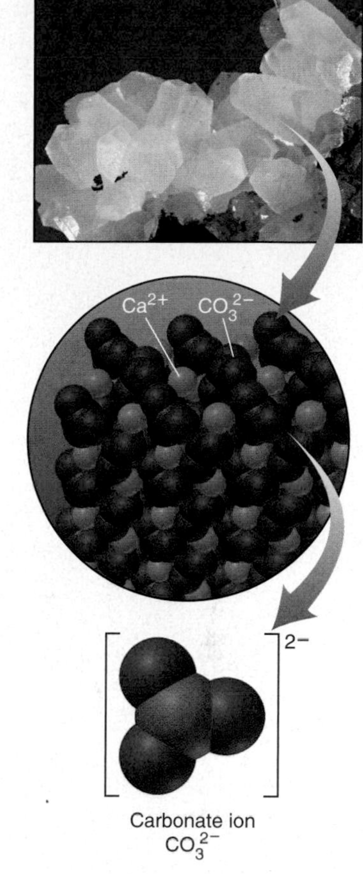

Figure 2.17 A polyatomic ion. Calcium carbonate is a three-dimensional array of monatomic calcium cations *(purple spheres)* and polyatomic carbonate anions. As the bottom structure shows, each carbonate ion consists of four covalently bonded atoms.

The Elements of Life

About one-quarter of all the elements have known roles in organisms. As you can see in Figure 2.18, metals, nonmetals, and metalloids are among these essential elements. But, except for some diatomic oxygen and nitrogen molecules inhaled into the lungs, none of the elements in organisms occurs in pure form; rather, they appear in compounds or as ions in solution.

The elements of life are often classified by the amount present in organisms. The four nonmetals carbon (C), oxygen (O), hydrogen (H), and nitrogen (N) are the *building-block elements* because they make up a major portion of biological molecules. Over 99% of the atoms in organisms are C, O, H, and N; in humans, they account for over 96% by mass of body weight. The nonmetals O and H make up the water in organisms, of course, and together with C occur in all four major classes of biological molecules—carbohydrates, fats, proteins, and nucleic acids. All proteins and nucleic acids also contain N.

The seven *major minerals* (or *macronutrients*) range from around 2% by mass for calcium (Ca) to around 0.14% by mass for chlorine (Cl). The alkali metals sodium and potassium and the halogen chlorine are dissolved in cell fluids as the

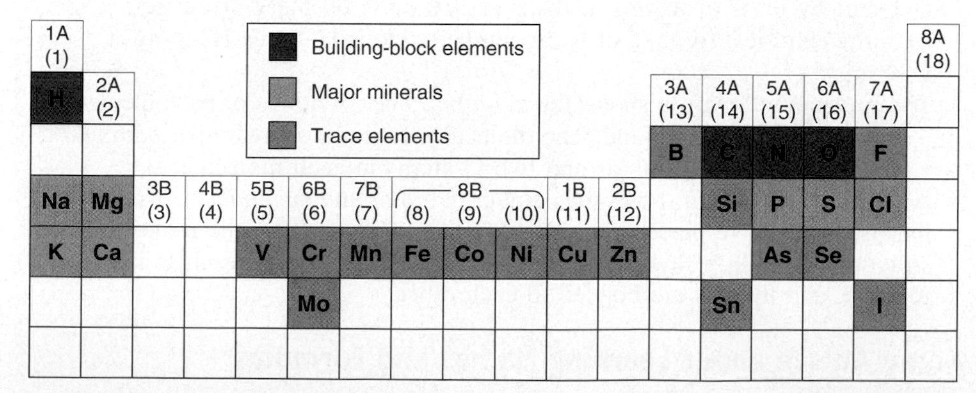

Figure 2.18 A biological periodic table. The building-block elements and major minerals are required by all organisms. Most organisms, including humans, require the trace elements as well. Many other elements (not shown) are found in organisms but have no known role.

ions Na^+, K^+, and Cl^-. The alkaline earth metals magnesium and calcium occur as Mg^{2+} and Ca^{2+}, most often bound to proteins or, in the case of calcium, in bones and teeth. Sulfur (S) occurs mostly in proteins, but phosphorus (P) also occurs in nucleic acids, many fats, and sugars, and as part of a polyatomic ion in bone and cell fluids.

The *trace elements* (or *micronutrients*) are present in much lower amounts, with iron (Fe) most abundant at only 0.005% by mass. Most are associated with protein functions. We look more closely at the trace elements in Chapter 23.

SECTION SUMMARY

Although a few elements occur uncombined in nature, the great majority exist in compounds. Ionic compounds form when a metal *transfers electrons* to a nonmetal, and the resulting positive and negative ions attract each other to form a three-dimensional array. In many cases, metal atoms lose and nonmetal atoms gain enough electrons to attain the same number of electrons as in atoms of the nearest noble gas. Covalent compounds form when elements, usually nonmetals, *share electrons.* Each covalent bond is an electron pair mutually attracted by two atomic nuclei. Monatomic ions are derived from single atoms. Polyatomic ions consist of two or more covalently bonded atoms that have a net positive or negative charge due to a deficit or excess of electrons. The elements in organisms are found as ions or, most often, bonded in large biomolecules. Four building-block elements (C, O, H, N) form these compounds, while seven other elements (major minerals, or macronutrients) are also common, and many others (trace elements, or micronutrients) occur in tiny amounts and play specific roles.

2.8 COMPOUNDS: FORMULAS, NAMES, AND MASSES

Names and formulas of compounds form the vocabulary of the chemical language; now it's time for you to begin speaking and writing this language. In this discussion, you'll learn the names and formulas of ionic and simple covalent compounds and how to calculate the mass of a unit of a compound from its formula.

Types of Chemical Formulas

In a **chemical formula,** element symbols and numerical subscripts show the type and number of each atom present in the smallest unit of the substance. There are several types of chemical formulas for a compound:

1. The **empirical formula** shows the *relative* number of atoms of each element in the compound. It is the simplest type of formula and is derived from the masses of the component elements. For example, in hydrogen peroxide, there is 1 part by mass of hydrogen for every 16 parts by mass of oxygen. Therefore, the empirical formula of hydrogen peroxide is HO: one H atom for every O atom.
2. The **molecular formula** shows the *actual* number of atoms of each element in a molecule of the compound. The molecular formula of hydrogen peroxide is H_2O_2; there are two H atoms and two O atoms in each molecule.
3. A **structural formula** shows the number of atoms and *the bonds between them;* that is, the relative placement and connections of atoms in the molecule. The structural formula of hydrogen peroxide is H—O—O—H; each H is bonded to an O, and the O's are bonded to each other.

Some Advice about Learning Names and Formulas

Perhaps in the future, systematic names for compounds will be used by everyone. However, many reference books, chemical supply catalogs, and practicing chemists still use many common (trivial) names, so you should learn them as well.

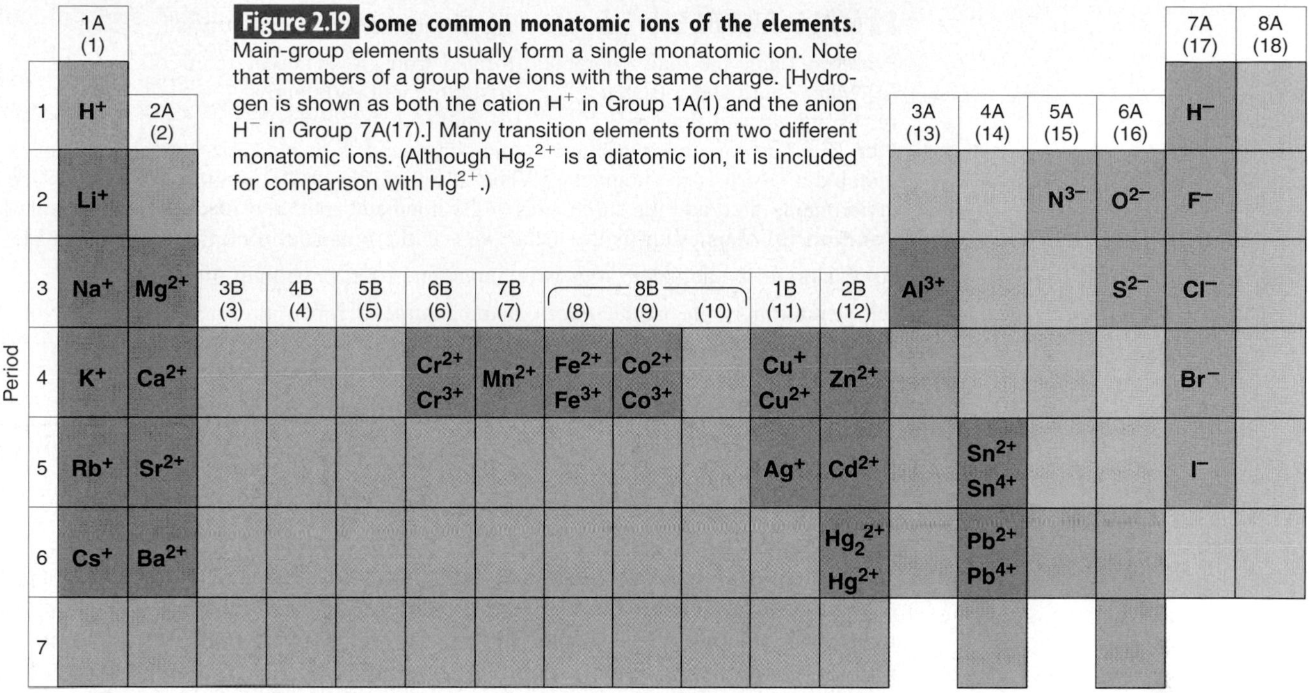

Figure 2.19 Some common monatomic ions of the elements. Main-group elements usually form a single monatomic ion. Note that members of a group have ions with the same charge. [Hydrogen is shown as both the cation H^+ in Group 1A(1) and the anion H^- in Group 7A(17).] Many transition elements form two different monatomic ions. (Although Hg_2^{2+} is a diatomic ion, it is included for comparison with Hg^{2+}.)

Period	1A (1)	2A (2)	3B (3)	4B (4)	5B (5)	6B (6)	7B (7)	8B (8)	8B (9)	8B (10)	1B (11)	2B (12)	3A (13)	4A (14)	5A (15)	6A (16)	7A (17)	8A (18)
1	H^+																H^-	
2	Li^+														N^{3-}	O^{2-}	F^-	
3	Na^+	Mg^{2+}											Al^{3+}			S^{2-}	Cl^-	
4	K^+	Ca^{2+}				Cr^{2+} Cr^{3+}	Mn^{2+}	Fe^{2+} Fe^{3+}	Co^{2+} Co^{3+}		Cu^+ Cu^{2+}	Zn^{2+}					Br^-	
5	Rb^+	Sr^{2+}									Ag^+	Cd^{2+}		Sn^{2+} Sn^{4+}			I^-	
6	Cs^+	Ba^{2+}										Hg_2^{2+} Hg^{2+}		Pb^{2+} Pb^{4+}				
7																		

Here are some points to note about ion formulas:

- Members of a periodic table group have the same ionic charge; for example, Li, Na, and K are all in Group 1A and all have a 1+ charge.
- For A-group cations, ion charge = group number: for example, Na^+ is in Group 1A, Ba^{2+} in Group 2A. (Exceptions in Figure 2.19 are Sn^{2+} and Pb^{2+}.)
- For anions, ion charge = group number minus 8: for example, S is in Group 6A ($6 - 8 = -2$), so the ion is S^{2-}.

Here are some suggestions about how to learn names and formulas:

1. Memorize the A-group monatomic ions of Table 2.3 (all except Ag^+, Zn^{2+}, and Cd^{2+}) according to their positions in Figure 2.19. These ions have the same number of electrons as an atom of the nearest noble gas.
2. Consult Table 2.4 (page 65) and Figure 2.19 for some metals that form two different monatomic ions.
3. Divide the tables of names and charges into smaller batches, and learn a batch each day. Try flash cards, with the name on one side and the ion formula on the other. The most common ions are shown in **boldface** in Tables 2.3, 2.4, and 2.5, so you can focus on learning them first.

Table 2.3 Common Monatomic Ions*

Charge	Formula	Name
Cations		
1+	H^+	hydrogen
	$\mathbf{Li^+}$	**lithium**
	$\mathbf{Na^+}$	**sodium**
	$\mathbf{K^+}$	**potassium**
	Cs^+	cesium
	$\mathbf{Ag^+}$	**silver**
2+	$\mathbf{Mg^{2+}}$	**magnesium**
	$\mathbf{Ca^{2+}}$	**calcium**
	Sr^{2+}	strontium
	$\mathbf{Ba^{2+}}$	**barium**
	$\mathbf{Zn^{2+}}$	**zinc**
	Cd^{2+}	cadmium
3+	$\mathbf{Al^{3+}}$	**aluminum**
Anions		
1−	H^-	hydride
	$\mathbf{F^-}$	**fluoride**
	$\mathbf{Cl^-}$	**chloride**
	$\mathbf{Br^-}$	**bromide**
	$\mathbf{I^-}$	**iodide**
2−	$\mathbf{O^{2-}}$	**oxide**
	$\mathbf{S^{2-}}$	**sulfide**
3−	N^{3-}	nitride

*Listed by charge; those in **boldface** are most common.

Names and Formulas of Ionic Compounds

All ionic compound names give the positive ion (cation) first and the negative ion (anion) second.

Compounds Formed from Monatomic Ions Let's first consider binary ionic compounds, those composed of ions of two elements.

- *The name of the cation is the same as the name of the metal.* Many metal names end in *-ium.*
- *The name of the anion takes the root of the nonmetal name and adds the suffix -ide.*

For example, the anion formed from brom*ine* is named brom*ide* (brom+ide). Therefore, the compound formed from the metal calcium and the nonmetal bromine is calcium bromide.

SAMPLE PROBLEM 2.5 Naming Binary Ionic Compounds

Problem Name the ionic compound formed from the following pairs of elements:
(a) Magnesium and nitrogen (b) Iodine and cadmium
(c) Strontium and fluorine (d) Sulfur and cesium

Plan The key to naming a binary ionic compound is to recognize which element is the metal and which is the nonmetal. When in doubt, check the periodic table. We place the cation name first, add the suffix *-ide* to the nonmetal root, and place the anion name last.

Solution (a) Magnesium is the metal; *nitr-* is the nonmetal root: magnesium nitride

(b) Cadmium is the metal; *iod-* is the nonmetal root: cadmium iodide

(c) Strontium is the metal; *fluor-* is the nonmetal root: strontium fluoride (Note the spelling is fl*uo*ride, not fl*ou*ride.)

(d) Cesium is the metal; *sulf-* is the nonmetal root: cesium sulfide

FOLLOW-UP PROBLEM 2.5 For the following ionic compounds, give the name and periodic table group number of each of the elements present: **(a)** zinc oxide; **(b)** silver bromide; **(c)** lithium chloride; **(d)** aluminum sulfide.

Ionic compounds are arrays of oppositely charged ions rather than separate molecular units. Therefore, we write a formula for the **formula unit,** which gives the *relative* numbers of cations and anions in the compound. Thus, ionic compounds generally have only empirical formulas.* The compound has zero net charge, so the positive charges of the cations must balance the negative charges of the anions. For example, calcium bromide is composed of Ca^{2+} ions and Br^- ions; therefore, two Br^- balance each Ca^{2+}. The formula is $CaBr_2$, not Ca_2Br. In this and all other formulas,

- The subscript refers to the element *preceding* it.
- The *subscript 1 is understood* from the presence of the element symbol alone (that is, we do not write Ca_1Br_2).
- The charge (without the sign) of one ion becomes the subscript of the other:

$$Ca^{2+} \quad Br^{1-} \qquad \text{gives} \qquad Ca_1Br_2 \qquad \text{or} \qquad CaBr_2$$

Reduce the subscripts to the smallest whole numbers that retain the ratio of ions. Thus, for example, from the ions Ca^{2+} and O^{2-} we have Ca_2O_2, which we reduce to the formula CaO (but see the footnote).

SAMPLE PROBLEM 2.6 Determining Formulas of Binary Ionic Compounds

Problem Write empirical formulas for the compounds named in Sample Problem 2.5.

Plan We write the empirical formula by finding the smallest number of each ion that gives the neutral compound. These numbers appear as *right subscripts* to the element symbol.

Solution
(a) Mg^{2+} and N^{3-}; three Mg^{2+} ions (6+) balance two N^{3-} ions (6−): Mg_3N_2

(b) Cd^{2+} and I^-; one Cd^{2+} ion (2+) balances two I^- ions (2−): CdI_2

(c) Sr^{2+} and F^-; one Sr^{2+} ion (2+) balances two F^- ions (2−): SrF_2

(d) Cs^+ and S^{2-}; two Cs^+ ions (2+) balance one S^{2-} ion (2−): Cs_2S

Comment Note that ion charges do *not* appear in the compound formula. That is, for cadmium iodide, we do *not* write $Cd^{2+}I_2^{\ -}$.

FOLLOW-UP PROBLEM 2.6 Write the formulas of the compounds named in Follow-up Problem 2.5.

*Compounds of the mercury(I) ion, such as Hg_2Cl_2, and peroxides of the alkali metals, such as Na_2O_2, are the only two common exceptions. Their empirical formulas are HgCl and NaO, respectively.

Table 2.4 Some Metals That Form More Than One Monatomic Ion*

Element	Ion Formula	Systematic Name	Common (Trivial) Name
Chromium	Cr^{2+}	chromium(II)	chromous
	$\mathbf{Cr^{3+}}$	**chromium(III)**	chromic
Cobalt	Co^{2+}	cobalt(II)	
	Co^{3+}	cobalt(III)	
Copper	$\mathbf{Cu^{+}}$	**copper(I)**	cuprous
	$\mathbf{Cu^{2+}}$	**copper(II)**	cupric
Iron	$\mathbf{Fe^{2+}}$	**iron(II)**	ferrous
	$\mathbf{Fe^{3+}}$	**iron(III)**	ferric
Lead	$\mathbf{Pb^{2+}}$	**lead(II)**	
	Pb^{4+}	lead(IV)	
Mercury	Hg_2^{2+}	mercury(I)	mercurous
	$\mathbf{Hg^{2+}}$	**mercury(II)**	mercuric
Tin	$\mathbf{Sn^{2+}}$	**tin(II)**	stannous
	Sn^{4+}	tin(IV)	stannic

*Listed alphabetically by metal name; those in **boldface** are most common.

Compounds with Metals That Can Form More Than One Ion Many metals, particularly the transition elements (B groups), can form more than one ion, each with a particular charge. Table 2.4 shows some examples. Names of compounds containing these elements include a *Roman numeral within parentheses* immediately after the metal ion's name to indicate its ionic charge. For example, iron can form Fe^{2+} and Fe^{3+} ions. The two compounds that iron forms with chlorine are $FeCl_2$, named iron(II) chloride (spoken "iron two chloride"), and $FeCl_3$, named iron(III) chloride.

In common names, the Latin root of the metal is followed by either of two suffixes:

- The suffix *-ous* for the ion with the lower charge
- The suffix *-ic* for the ion with the higher charge

Thus, iron(II) chloride is also called ferr*ous* chloride and iron(III) chloride is fer*ric* chloride. (You can easily remember this naming relationship because there is an ***o*** in *-ous* and *lower,* and an ***i*** in *-ic* and *higher.*)

SAMPLE PROBLEM 2.7 Determining Names and Formulas of Ionic Compounds of Elements That Form More Than One Ion

Problem Give the systematic names for the formulas or the formulas for the names of the following compounds: **(a)** tin(II) fluoride; **(b)** CrI_3; **(c)** ferric oxide; **(d)** CoS.

Solution (a) Tin(II) is Sn^{2+}; fluoride is F^-. Two F^- ions balance one Sn^{2+} ion: tin(II) fluoride is SnF_2. (The common name is stannous fluoride.)

(b) The anion is I^-, iodide, and the formula shows three I^-. Therefore, the cation must be Cr^{3+}, chromium(III): CrI_3 is chromium(III) iodide. (The common name is chromic iodide.)

(c) Ferric is the common name for iron(III), Fe^{3+}; oxide ion is O^{2-}. To balance the ionic charges, the formula of ferric oxide is Fe_2O_3. (The systematic name is iron(III) oxide.)

(d) The anion is sulfide, S^{2-}, which requires that the cation be Co^{2+}. The name is cobalt(II) sulfide.

FOLLOW-UP PROBLEM 2.7 Give the systematic names for the formulas or the formulas for the names of the following compounds: **(a)** lead(IV) oxide; **(b)** Cu_2S; **(c)** $FeBr_2$; **(d)** mercuric chloride.

Table 2.5 Common Polyatomic Ions*

Formula	Name
Cations	
NH_4^+	**ammonium**
H_3O^+	**hydronium**
Anions	
CH_3COO^- (or **$C_2H_3O_2^-$**)	**acetate**
CN^-	cyanide
OH^-	**hydroxide**
ClO^-	hypochlorite
ClO_2^-	chlorite
ClO_3^-	**chlorate**
ClO_4^-	**perchlorate**
NO_2^-	nitrite
NO_3^-	**nitrate**
MnO_4^-	**permanganate**
CO_3^{2-}	**carbonate**
HCO_3^-	**hydrogen carbonate** (or **bicarbonate**)
CrO_4^{2-}	chromate
$Cr_2O_7^{2-}$	**dichromate**
O_2^{2-}	peroxide
PO_4^{3-}	**phosphate**
HPO_4^{2-}	hydrogen phosphate
$H_2PO_4^-$	dihydrogen phosphate
SO_3^{2-}	sulfite
SO_4^{2-}	**sulfate**
HSO_4^-	hydrogen sulfate (or bisulfate)

*__Boldface__ ions are most common.

No. of O atoms (increasing upward)	Prefix	Root	Suffix
	per	*root*	ate
		root	ate
		root	ite
	hypo	*root*	ite

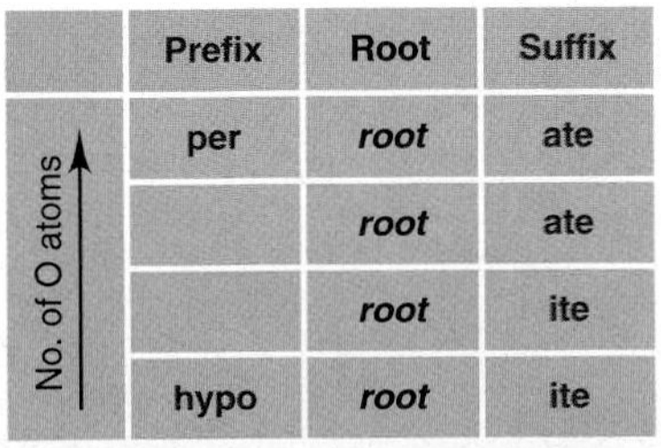

Figure 2.20 Naming oxoanions. Prefixes and suffixes indicate the number of O atoms in the anion.

Compounds Formed from Polyatomic Ions Ionic compounds in which one or both of the ions are polyatomic are very common. Table 2.5 gives the formulas and the names of some common polyatomic ions. Remember that *the polyatomic ion stays together as a charged unit.* The formula for potassium nitrate is KNO_3: each K^+ balances one NO_3^-. The formula for sodium carbonate is Na_2CO_3: two Na^+ balance one CO_3^{2-}. *When two or more of the same polyatomic ion are present in the formula unit, that ion appears in parentheses with the subscript written outside.* For example, calcium nitrate, which contains one Ca^{2+} and two NO_3^- ions, has the formula $Ca(NO_3)_2$. Parentheses and a subscript are *not* used unless *more than one* of the polyatomic ions is present; thus, sodium nitrate is $NaNO_3$, *not* $Na(NO_3)$.

Families of Oxoanions As Table 2.5 shows, most polyatomic ions are **oxoanions,** those in which an element, usually a nonmetal, is bonded to one or more oxygen atoms. There are several families of two or four oxoanions that differ only in the number of oxygen atoms. A simple naming convention is used with these ions.

With two oxoanions in the family:

- The ion with *more* O atoms takes the nonmetal root and the suffix *-ate.*
- The ion with *fewer* O atoms takes the nonmetal root and the suffix *-ite.*

For example, SO_4^{2-} is the sulf*ate* ion; SO_3^{2-} is the sulf*ite* ion; similarly, NO_3^- is nitr*ate*, and NO_2^- is nitr*ite*.

With four oxoanions in the family (usually a halogen bonded to O), as Figure 2.20 shows:

- The ion with *most* O atoms has the prefix *per-*, the nonmetal root, and the suffix *-ate*.
- The ion with *one fewer* O atom has just the root and the suffix *-ate*.
- The ion with *two fewer* O atoms has just the root and the suffix *-ite*.
- The ion with *least (three fewer)* O atoms has the prefix *hypo-*, the root, and the suffix *-ite*.

For example, for the four chlorine oxoanions,

ClO_4^- is *per*chlor*ate*, ClO_3^- is chlor*ate*, ClO_2^- is chlor*ite*, ClO^- is *hypo*chlor*ite*

Hydrated Ionic Compounds Ionic compounds called **hydrates** have a specific number of water molecules associated with each formula unit. In their formulas, this number is shown after a centered dot. It is indicated in the systematic name by a Greek numerical prefix before the word *hydrate*. Table 2.6 shows these prefixes. For example, Epsom salt has the formula $MgSO_4{\cdot}7H_2O$ and the name magnesium sulfate *hepta*hydrate. Similarly, the mineral gypsum has the formula $CaSO_4{\cdot}2H_2O$ and the name calcium sulfate *di*hydrate. The water molecules, referred to as "waters of hydration," are part of the hydrate's structure. Heating can remove some or all of them, leading to a different substance. For example, when heated strongly, blue copper(II) sulfate pentahydrate ($CuSO_4{\cdot}5H_2O$) is converted to white copper(II) sulfate ($CuSO_4$).

SAMPLE PROBLEM 2.8 Determining Names and Formulas of Ionic Compounds Containing Polyatomic Ions

Problem Give the systematic names for the formulas or the formulas for the names of the following compounds:
(a) $Fe(ClO_4)_2$ **(b)** Sodium sulfite **(c)** $Ba(OH)_2{\cdot}8H_2O$

Solution (a) ClO_4^- is perchlorate; since it has a 1− charge, the cation must be Fe^{2+}. The name is iron(II) perchlorate. (The common name is ferrous perchlorate.)

(b) Sodium is Na^+; sulfite is SO_3^{2-}. Therefore, two Na^+ ions balance one SO_3^{2-} ion. The formula is Na_2SO_3. **(c)** Ba^{2+} is barium; OH^- is hydroxide. There are eight (*octa-*) water molecules in each formula unit. The name is barium hydroxide octahydrate.

FOLLOW-UP PROBLEM 2.8 Give the systematic names for the formulas or the formulas for the names of the following compounds:
(a) Cupric nitrate trihydrate **(b)** Zinc hydroxide **(c)** LiCN

Table 2.6 Numerical Prefixes for Hydrates and Binary Covalent Compounds

Number	Prefix
1	mono-
2	di-
3	tri-
4	tetra-
5	penta-
6	hexa-
7	hepta-
8	octa-
9	nona-
10	deca-

SAMPLE PROBLEM 2.9 Recognizing Incorrect Names and Formulas of Ionic Compounds

Problem Something is wrong with the second part of each statement. Provide the correct name or formula.
(a) $Ba(C_2H_3O_2)_2$ is called barium diacetate.
(b) Sodium sulfide has the formula $(Na)_2SO_3$.
(c) Iron(II) sulfate has the formula $Fe_2(SO_4)_3$.
(d) Cesium carbonate has the formula $Cs_2(CO_3)$.

Solution (a) The charge of the Ba^{2+} ion *must* be balanced by *two* $C_2H_3O_2^-$ ions, so the prefix *di-* is unnecessary. For ionic compounds, we do not indicate the number of ions with numerical prefixes. The correct name is barium acetate.
(b) Two mistakes occur here. The sodium ion is monatomic, so it does *not* require parentheses. The sulfide ion is S^{2-}, *not* SO_3^{2-} (called "sulfite"). The correct formula is Na_2S.
(c) The Roman numeral refers to the charge of the ion, *not* the number of ions in the formula. Fe^{2+} is the cation, so it requires one SO_4^{2-} to balance its charge. The correct formula is $FeSO_4$.
(d) Parentheses are *not* required when only one polyatomic ion of a kind is present. The correct formula is Cs_2CO_3.

FOLLOW-UP PROBLEM 2.9 State why the second part of each statement is incorrect, and correct it:
(a) Ammonium phosphate is $(NH_3)_4PO_4$.
(b) Aluminum hydroxide is $AlOH_3$.
(c) $Mg(HCO_3)_2$ is manganese(II) carbonate.
(d) $Cr(NO_3)_3$ is chromic(III) nitride.
(e) $Ca(NO_2)_2$ is cadmium nitrate.

Acid Names from Anion Names Acids are an important group of hydrogen-containing compounds that have been used in chemical reactions since before alchemical times. In the laboratory, acids are typically used in water solution. When naming them and writing their formulas, we consider them as anions connected to the number of hydrogen ions (H^+) needed for charge neutrality. The two common types of acids are binary acids and oxoacids:

1. *Binary acid* solutions form when certain gaseous compounds dissolve in water. For example, when gaseous hydrogen chloride (HCl) dissolves in water, it forms a solution called hydrochloric acid. The name consists of the following parts:

   Prefix *hydro-* + nonmetal *root* + suffix *-ic* + separate word *acid*
   hydro + chlor + ic + acid

   or hydrochloric acid. This naming pattern holds for many compounds in which hydrogen combines with an anion that has an *-ide* suffix.
2. *Oxoacid* names are similar to those of the oxoanions, except for two suffix changes:
   - *-ate* in the anion becomes *-ic* in the acid
   - *-ite* in the anion becomes *-ous* in the acid

   The oxoanion prefixes *hypo-* and *per-* are kept. Thus,

   BrO_4^- is *per*brom*ate*, and $HBrO_4$ is *per*brom*ic* acid
   IO_2^- is iod*ite*, and HIO_2 is iod*ous* acid

SAMPLE PROBLEM 2.10 Determining Names and Formulas of Anions and Acids

Problem Name the following anions and give the names and formulas of the acids derived from them: **(a)** Br^-; **(b)** IO_3^-; **(c)** CN^-; **(d)** SO_4^{2-}; **(e)** NO_2^-.

Solution **(a)** The anion is bromide; the acid is hydrobromic acid, HBr.

(b) The anion is iodate; the acid is iodic acid, HIO_3.

(c) The anion is cyanide; the acid is hydrocyanic acid, HCN.

(d) The anion is sulfate; the acid is sulfuric acid, H_2SO_4. (In this case, the suffix is added to the element name *sulfur,* not to the root, *sulf-*.)

(e) The anion is nitrite; the acid is nitrous acid, HNO_2.

Comment We added *two* H^+ ions to the sulfate ion to obtain sulfuric acid because it has a 2− charge.

FOLLOW-UP PROBLEM 2.10 Write the formula for the name or name for the formula of each acid: **(a)** chloric acid; **(b)** HF; **(c)** acetic acid; **(d)** sulfurous acid; **(e)** HBrO.

Names and Formulas of Binary Covalent Compounds

Binary covalent compounds are formed by the combination of two elements, usually nonmetals. Several are so familiar, such as ammonia (NH_3), methane (CH_4), and water (H_2O), that we use their common names, but most are named in a systematic way:

1. The element with the lower group number in the periodic table is the first word in the name; the element with the higher group number is the second word. (*Exception:* When the compound contains oxygen and any of the halogens chlorine, bromine, and iodine, the halogen is named first.)
2. If both elements are in the same group, the one with the higher period number is named first.
3. The second element is named with its root and the suffix *-ide*.
4. Covalent compounds have Greek numerical prefixes (see Table 2.6) to indicate the number of atoms of each element in the compound. The first word has a prefix *only* when more than one atom of the element is present; the second word *usually* has a numerical prefix.

SAMPLE PROBLEM 2.11 Determining Names and Formulas of Binary Covalent Compounds

Problem **(a)** What is the formula of carbon disulfide? **(b)** What is the name of PCl_5? **(c)** Give the name and formula of the compound whose molecules each consist of two N atoms and four O atoms.

Solution **(a)** The prefix *di-* means "two." The formula is CS_2.

(b) P is the symbol for phosphorus; there are five chlorine atoms, which is indicated by the prefix *penta-*. The name is phosphorus pentachloride.

(c) Nitrogen (N) comes first in the name (lower group number). The compound is dinitrogen tetraoxide, N_2O_4.

FOLLOW-UP PROBLEM 2.11 Give the name or formula for **(a)** SO_3; **(b)** SiO_2; **(c)** dinitrogen monoxide; **(d)** selenium hexafluoride.

SAMPLE PROBLEM 2.12 Recognizing Incorrect Names and Formulas of Binary Covalent Compounds

Problem Explain what is wrong with the name or formula in the second part of each statement and correct it: **(a)** SF_4 is monosulfur pentafluoride. **(b)** Dichlorine heptaoxide is Cl_2O_6. **(c)** N_2O_3 is dinitrotrioxide.

Solution (a) There are two mistakes. *Mono-* is not needed if there is only one atom of the first element, and the prefix for four is *tetra-*, not *penta-*. The correct name is sulfur tetrafluoride.

(b) The prefix *hepta-* indicates seven, not six. The correct formula is Cl_2O_7.

(c) The full name of the first element is needed, and a space separates the two element names. The correct name is dinitrogen trioxide.

FOLLOW-UP PROBLEM 2.12 Explain what is wrong with the second part of each statement and correct it: **(a)** S_2Cl_2 is disulfurous dichloride. **(b)** Nitrogen monoxide is N_2O. **(c)** $BrCl_3$ is trichlorine bromide.

An Introduction to Naming Organic Compounds

Organic compounds typically have complex structural formulas that consist of chains, branches, and/or rings of carbon atoms bonded to hydrogen atoms and, often, to atoms of oxygen, nitrogen, and a few other elements. At this point, we'll look at one or two basic principles for naming them. Much more on the rules of organic nomenclature appears in Chapter 15.

Hydrocarbons, the simplest type of organic compound, contain *only* carbon and hydrogen. *Alkanes* are the simplest type of hydrocarbon; many function as important fuels, such as methane, propane, butane, and the mixture of alkanes in gasoline. The simplest alkanes to name are the *straight-chain alkanes* because the carbon chains have no branches. Alkanes are named with a *root,* based on the number of C atoms in the chain, followed by the suffix *-ane*. Table 2.7 gives the names, molecular formulas, and space-filling models (discussed shortly) of the first 10 straight-chain alkanes. Note that the roots of the four smallest ones are new, but those for the larger ones are the same as the Greek prefixes (Table 2.6).

Alkanes (and other organic compounds) *with* branches have a *prefix* in the name as well. *The prefix names the length of the branch and numbers the carbon atom in the main chain that the branch is attached to.* A prefix consists of a root plus the ending *-yl*. Thus, for example, the compound with a one-carbon ("meth") branch attached to the second carbon of the main chain of butane is *2-methylbutane,* where "2-methyl" is the prefix (see margin).

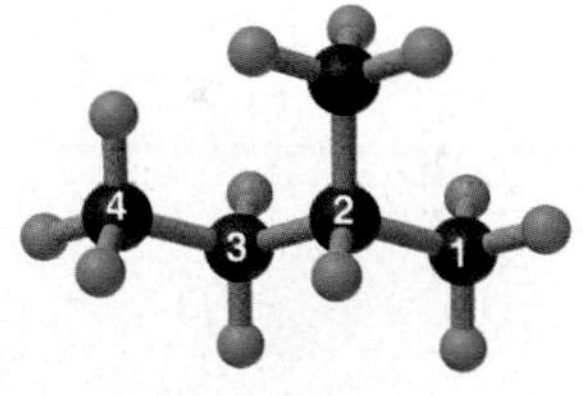

2-methylbutane

Organic compounds other than alkanes have names derived from a particular region of the molecule, called the *functional group,* which consists of one or a

Table 2.7 The First 10 Straight-Chain Alkanes

Name	Formula	Model
Methane	CH_4	
Ethane	C_2H_6	
Propane	C_3H_8	
Butane	C_4H_{10}	
Pentane	C_5H_{12}	
Hexane	C_6H_{14}	
Heptane	C_7H_{16}	
Octane	C_8H_{18}	
Nonane	C_9H_{20}	
Decane	$C_{10}H_{22}$	

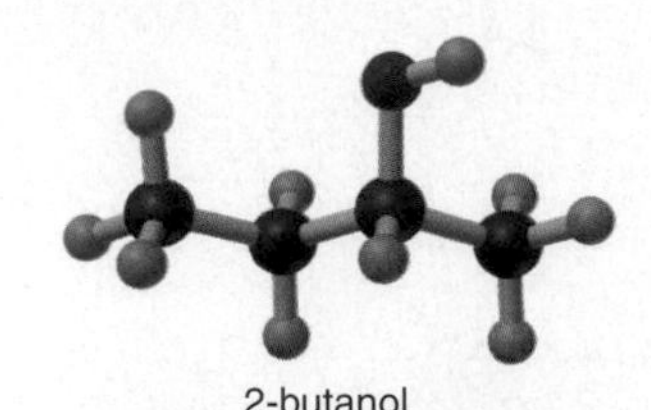
2-butanol

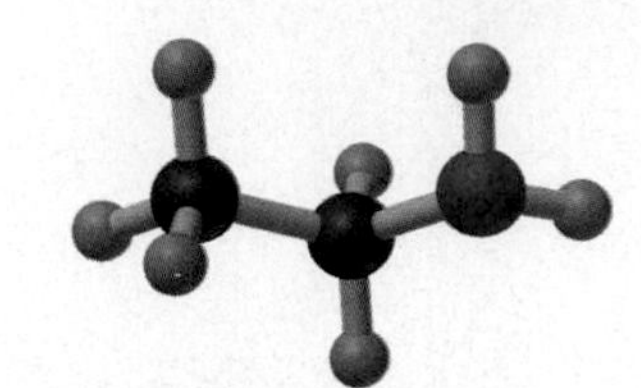
ethanamine

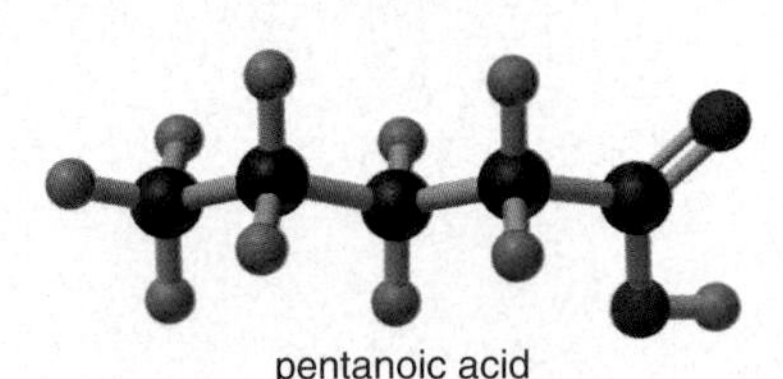
pentanoic acid

few atoms bonded in a specific way. *The functional group determines how the compound reacts.* The *alcohol* functional group is a hydroxyl group, O—H, in place of one of the H atoms in the hydrocarbon; an *amine* has an amino group, $—NH_2$; a *carboxylic acid* has a carboxyl group, —COOH; and so forth. Each functional group has its own suffix. Thus, the compound with a hydroxyl group attached to the second carbon in butane is called *2-butanol* (see margin); the compound with an amino group bonded to a two-carbon chain is *ethanamine;* the compound with a carboxyl group bonded to a four-carbon chain is *pentanoic acid* (the C of the carboxyl group counts as one of the carbons). We'll examine the types of reactions the different functional groups undergo in Chapter 15.

Molecular Masses from Chemical Formulas

In Section 2.5, we calculated the atomic mass of an element. Using the formula of a compound to see the number of atoms of each element and the periodic table, we calculate the **molecular mass** (also called *molecular weight*) of a formula unit of the compound as the sum of the atomic masses:

$$\text{Molecular mass} = \text{sum of atomic masses} \quad \textbf{(2.3)}$$

The molecular mass of a water molecule (using atomic masses to four significant figures from the periodic table) is

$$\begin{aligned}\text{Molecular mass of } H_2O &= (2 \times \text{atomic mass of H}) + (1 \times \text{atomic mass of O}) \\ &= (2 \times 1.008 \text{ amu}) + 16.00 \text{ amu} = 18.02 \text{ amu}\end{aligned}$$

Ionic compounds are treated the same, but because they do not consist of molecules, we use the term *formula mass* for an ionic compound. To calculate its formula mass, *the number of atoms of each element inside the parentheses is multiplied by the subscript outside the parentheses.* For barium nitrate, $Ba(NO_3)_2$,

$$\begin{aligned}&\text{Formula mass of } Ba(NO_3)_2 \\ &= (1 \times \text{atomic mass of Ba}) + (2 \times \text{atomic mass of N}) + (6 \times \text{atomic mass of O}) \\ &= 137.3 \text{ amu} + (2 \times 14.01 \text{ amu}) + (6 \times 16.00 \text{ amu}) = 261.3 \text{ amu}\end{aligned}$$

Note that atomic, not ionic, masses are used. Although masses of ions differ from those of their atoms by the masses of the electrons, electron loss equals electron gain in the compound, so electron mass is balanced.

SAMPLE PROBLEM 2.13 Calculating the Molecular Mass of a Compound

Problem Using data in the periodic table, calculate the molecular (or formula) mass of:
(a) Tetraphosphorus trisulfide **(b)** Ammonium nitrate

Plan We first write the formula, then multiply the number of atoms (or ions) of each element by its atomic mass, and find the sum.

Solution (a) The formula is P_4S_3.

$$\begin{aligned}\text{Molecular mass} &= (4 \times \text{atomic mass of P}) + (3 \times \text{atomic mass of S}) \\ &= (4 \times 30.97 \text{ amu}) + (3 \times 32.07 \text{ amu}) = \boxed{220.09 \text{ amu}}\end{aligned}$$

(b) The formula is NH_4NO_3. We count the total number of N atoms even though they belong to different ions:

$$\begin{aligned}&\text{Formula mass} \\ &= (2 \times \text{atomic mass of N}) + (4 \times \text{atomic mass of H}) + (3 \times \text{atomic mass of O}) \\ &= (2 \times 14.01 \text{ amu}) + (4 \times 1.008 \text{ amu}) + (3 \times 16.00 \text{ amu}) = \boxed{80.05 \text{ amu}}\end{aligned}$$

Check You can often find large errors by rounding atomic masses to the nearest 5 and adding: **(a)** $(4 \times 30) + (3 \times 30) = 210 \approx 220.09$. The sum has two decimal places because the atomic masses have two. **(b)** $(2 \times 15) + 4 + (3 \times 15) = 79 \approx 80.05$.

FOLLOW-UP PROBLEM 2.13 What is the formula and molecular (or formula) mass of each of the following compounds: **(a)** hydrogen peroxide; **(b)** cesium chloride; **(c)** sulfuric acid; **(d)** potassium sulfate?

In the next sample problem, we use molecular depictions to find the formula, name, and mass.

SAMPLE PROBLEM 2.14 Naming Compounds from Their Depictions

Problem Each box contains a representation of a binary compound. Determine its formula, name, and molecular (formula) mass.

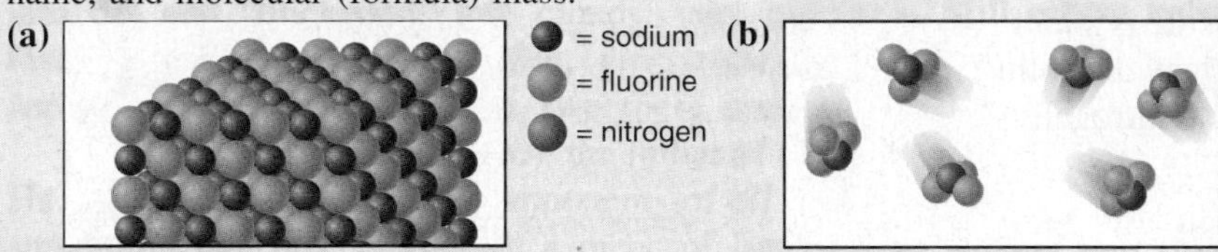

Plan Each of the compounds contains only two elements, so to find the formula, we find the simplest whole-number ratio of one atom to the other. Then we determine the name (see Sample Problems 2.7 and 2.11) and the mass (see Sample Problem 2.13).

Solution (a) There is one brown (sodium) for each green (fluorine), so the formula is NaF. A metal and nonmetal form an ionic compound, in which the metal is named first: sodium fluoride.

$$\text{Formula mass} = (1 \times \text{atomic mass of Na}) + (1 \times \text{atomic mass of F})$$
$$= 22.99 \text{ amu} + 19.00 \text{ amu} = 41.99 \text{ amu}$$

(b) There are three green (fluorine) for each blue (nitrogen), so the formula is NF_3. Two nonmetals form a covalent compound. Nitrogen has a lower group number, so it is named first: nitrogen trifluoride.

$$\text{Molecular mass} = (1 \times \text{atomic mass of N}) + (3 \times \text{atomic mass of F})$$
$$= 14.01 \text{ amu} + (3 \times 19.00 \text{ amu}) = 71.01 \text{ amu}$$

Check (a) For binary ionic compounds, we predict ionic charges from the periodic table (see Figure 2.14). Na forms a 1+ ion, and F forms a 1− ion, so the charges balance with one Na^+ per F^-. Also, ionic compounds are solids, consistent with the picture. (b) Covalent compounds often occur as individual molecules, as in the picture. Rounding in (a) gives 25 + 20 = 45; in (b), we get 15 + (3 × 20) = 75, so there are no large errors.

FOLLOW-UP PROBLEM 2.14 Each box contains a representation of a binary compound. Determine its name, formula, and molecular (formula) mass.

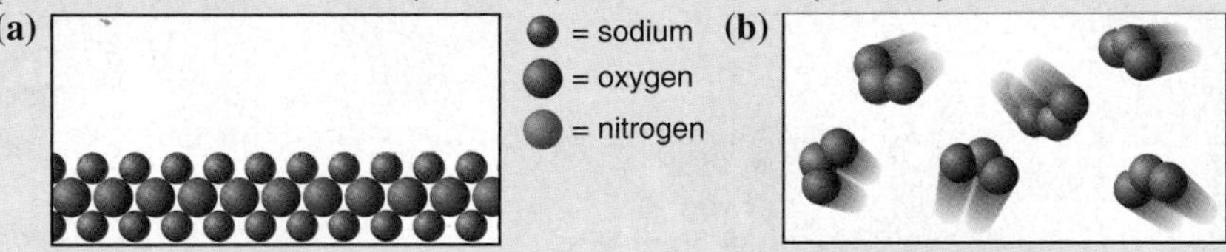

The Gallery on the next page shows some of the ways that chemists picture molecules and the enormous range of molecular sizes.

SECTION SUMMARY

Chemical formulas describe the simplest atom ratio (empirical formula), actual atom number (molecular formula), and atom arrangement (structural formula) of one unit of a compound. An ionic compound is named with cation first and anion second. For metals that can form more than one ion, the charge is shown with a Roman numeral. Oxoanions have suffixes, and sometimes prefixes, attached to the element root name to indicate the number of oxygen atoms. Names of hydrates give the number of associated water molecules with a numerical prefix. Acid names are based on anion names. Names of covalent compounds have the element that is leftmost or lower down in the periodic table first, and prefixes show the number of each atom. The molecular (or formula) mass of a compound is the sum of the atomic masses in the formula. Molecules are three-dimensional objects that range in size from H_2 to biological and synthetic macromolecules.

Picturing Molecules

The most exciting thing about learning chemistry is training your mind to imagine a molecular world, one filled with tiny objects of various shapes. Molecules are depicted in a variety of useful ways, as shown at right for the water molecule:

Chemical formulas show only the relative numbers of atoms.

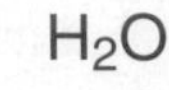

H:O:H

Electron-dot and *bond-line formulas* show a bond between atoms as either a pair of dots or a line.

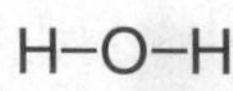

Ball-and-stick models show atoms as spheres and bonds as sticks, with accurate angles and relative sizes, but distances are exaggerated.

Space-filling models are accurately scaled-up versions of molecules, but they do not show bonds.

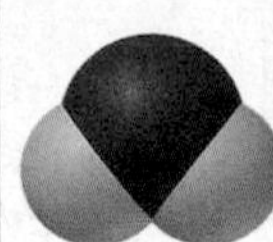

Electron-density models show the ball-and-stick model within the space-filling shape and color the regions of high *(red)* and low *(blue)* electron charge.

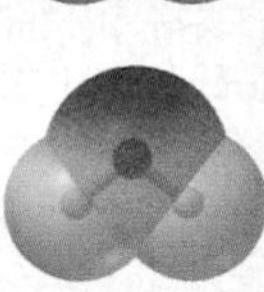

All molecules are minute, with their relative sizes depending on composition. A water molecule is small because it consists of only three atoms. Many air pollutants, such as ozone, carbon monoxide, and nitrogen dioxide, also consist of small molecules.

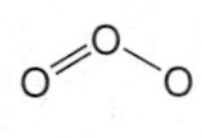

Ozone (O_3, 48.00 amu) contributes to smog; natural component of stratosphere that absorbs harmful solar radiation.

C≡O

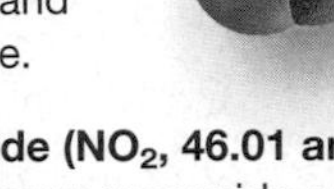

Carbon monoxide (CO, 28.01 amu), toxic component of car exhaust and cigarette smoke.

Nitrogen dioxide (NO_2, 46.01 amu) forms from nitrogen monoxide and contributes to smog and acid rain.

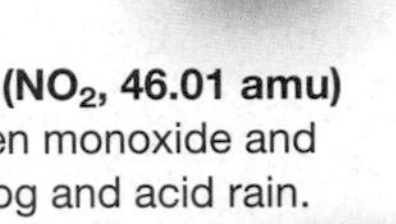

Many household chemicals, such as butane, acetic acid, and aspirin, consist of somewhat larger molecules. The biologically essential molecule heme is larger still.

Butane (C_4H_{10}, 58.12 amu), fuel for cigarette lighters and camping stoves.

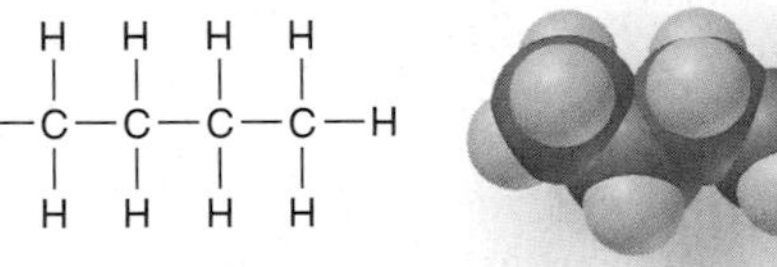

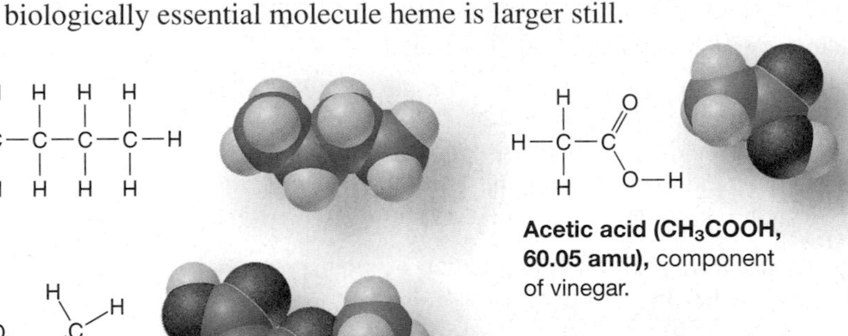

Acetic acid (CH_3COOH, 60.05 amu), component of vinegar.

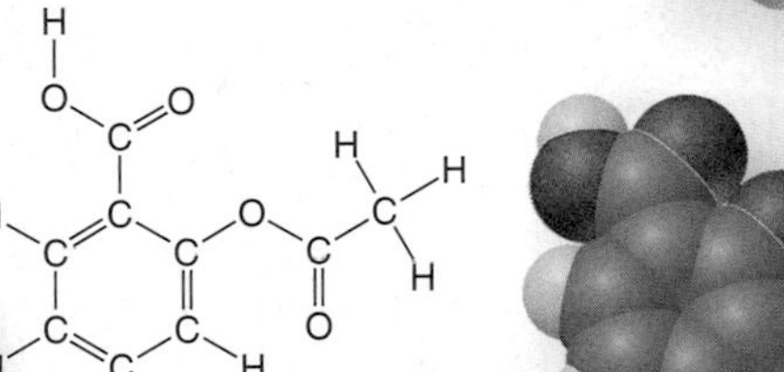

Aspirin ($C_9H_8O_4$, 180.15 amu), most common pain reliever in the world.

Heme ($C_{34}H_{32}FeN_4O_4$, 616.49 amu), part of the blood protein hemoglobin, which carries oxygen through the body.

Very large molecules, called macromolecules, can be synthetic, like nylon, or natural, like DNA, and typically consist of thousands of atoms.

Nylon-66 (~15,000 amu), relatively small, synthetic macromolecule used to make textiles and tires.

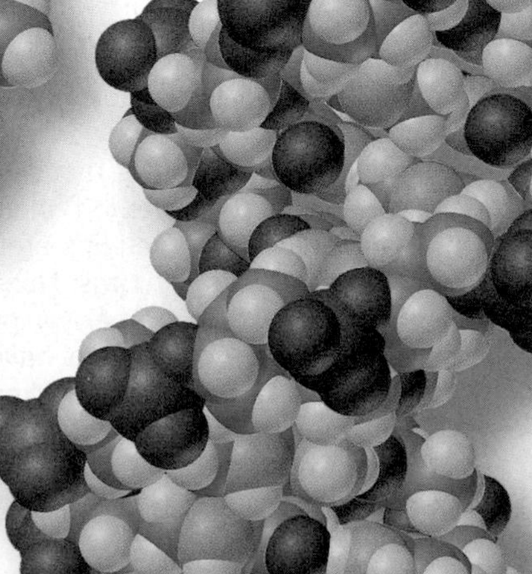

Deoxyribonucleic acid (DNA, ~10,000,000 amu), cellular macromolecule that contains genetic information.

2.9 MIXTURES: CLASSIFICATION AND SEPARATION

Although chemists pay a great deal of attention to pure substances, this form of matter almost never occurs around us. In the natural world, *matter usually occurs as mixtures.* A sample of clean air, for example, consists of many elements and compounds physically mixed together, including oxygen (O_2), nitrogen (N_2), carbon dioxide (CO_2), the six noble gases [Group 8A(18)], and water vapor (H_2O). The oceans are complex mixtures of dissolved ions and covalent substances, including Na^+, Mg^{2+}, Cl^-, SO_4^{2-}, O_2, CO_2, and of course H_2O. Rocks and soils are mixtures of numerous compounds—such as calcium carbonate ($CaCO_3$), silicon dioxide (SiO_2), aluminum oxide (Al_2O_3), and iron(III) oxide (Fe_2O_3)—perhaps a few elements (gold, silver, and carbon in the form of diamond), and petroleum and coal, which are complex mixtures themselves. Living things contain thousands of substances: carbohydrates, lipids, proteins, nucleic acids, and many simpler ionic and covalent compounds.

There are two broad classes of mixtures. A **heterogeneous mixture** has one or more visible boundaries between the components. Thus, its composition is *not* uniform. Many rocks are heterogeneous, showing individual grains and flecks of different minerals. In some cases, as in milk and blood, the boundaries can be seen only with a microscope. A **homogeneous mixture** has no visible boundaries because the components are mixed as individual atoms, ions, and molecules. Thus, its composition *is* uniform. A mixture of sugar dissolved in water is homogeneous, for example, because the sugar molecules and water molecules are uniformly intermingled on the molecular level. We have no way to tell visually whether an object is a substance (element or compound) or a homogeneous mixture.

A homogeneous mixture is also called a **solution.** Although we usually think of solutions as liquid, they can exist in all three physical states. For example, air is a gaseous solution of mostly oxygen and nitrogen molecules, and wax is a solid solution of several fatty substances. Solutions in water, called **aqueous solutions,** are especially important in chemistry and comprise a major portion of the environment and of all organisms.

Recall that mixtures differ fundamentally from compounds in three ways: (1) the proportions of the components can vary; (2) the individual properties of the components are observable; and (3) the components can be separated by physical means. In some cases, if we apply enough energy to the components of the mixture, they react with each other chemically and form a compound, after which their individual properties are no longer observable. Figure 2.21 shows such a case with a mixture of iron and sulfur.

In order to investigate the properties of substances, chemists have devised many procedures for separating a mixture into its component elements and compounds. Indeed, the laws and models of chemistry could never have been formulated without this ability. Many of Dalton's critics, who thought they had found compounds with varying composition, were unknowingly studying mixtures! The Tools of the Laboratory essay on the next two pages describes some of the more common laboratory separation methods.

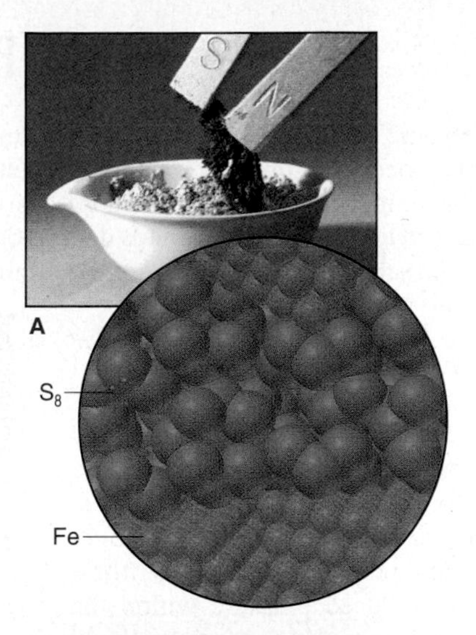

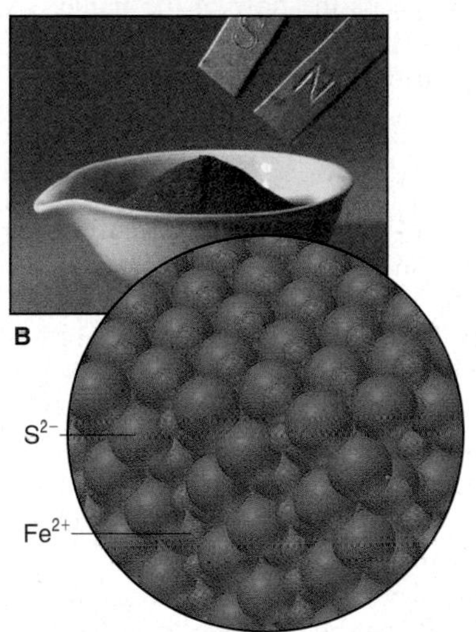

Figure 2.21 The distinction between mixtures and compounds. A, A *mixture* of iron and sulfur can be separated with a magnet because only the iron is magnetic. The blow-up shows separate regions of the two elements. **B,** After strong heating, the *compound* iron(II) sulfide forms, which is no longer magnetic. The blow-up shows the structure of the compound, in which there are no separate regions of the elements.

SECTION SUMMARY

Heterogeneous mixtures have visible boundaries between the components. Homogeneous mixtures have no visible boundaries because mixing occurs at the molecular level. A solution is a homogeneous mixture and can occur in any physical state. Mixtures (not compounds) can have variable proportions, can be separated physically, and retain their components' properties. Common physical separation processes include filtration, crystallization, extraction, chromatography, and distillation.

Basic Separation Techniques

Some of the most challenging and time-consuming laboratory procedures involve separating mixtures and purifying the components. Several common separation techniques are described here. All these methods depend on the *physical properties* of the substances in the mixture; no chemical changes occur.

Filtration separates the components of a mixture on the basis of *differences in particle size.* It is used most often to separate a liquid (smaller particles) from a solid (larger particles). Figure B2.3 shows simple filtration of a solid reaction product. In vacuum filtration, reduced pressure within the flask speeds the flow of the liquid through the filter. Filtration is a key step in the purification of the tap water you drink.

Figure B2.3 Filtration.

Crystallization is based on *differences in solubility.* The *solubility* of a substance is the amount that dissolves in a fixed volume of solvent at a given temperature. The procedure shown in Figure B2.4 applies the fact that many substances are more soluble in hot solvent than in cold. The purified compound crystallizes as the solution is cooled. Key substances in computer chips and other electronic devices are purified by a type of crystallization.

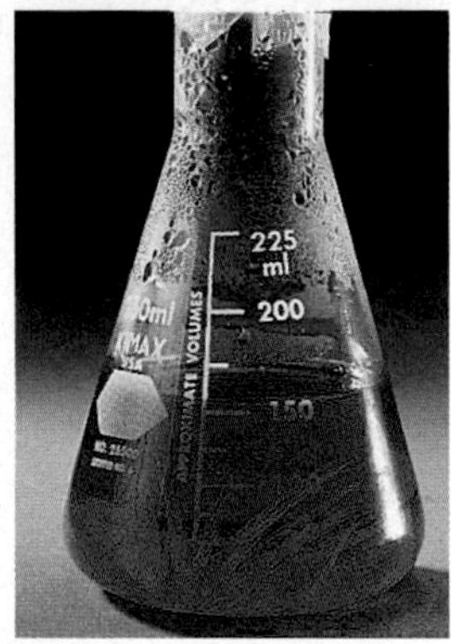
225
ml
200
150

Figure B2.4 Crystallization.

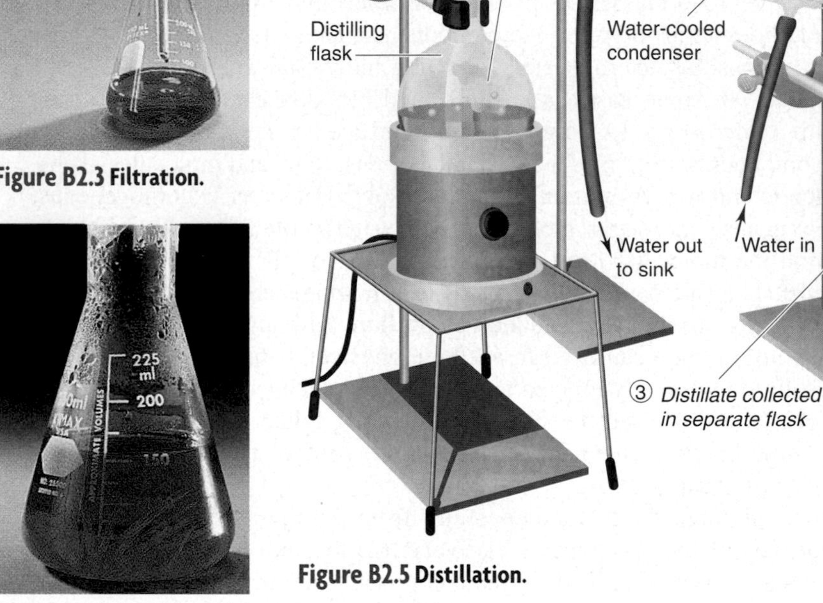
① Mixture is heated and volatile component vaporizes
Thermometer
② Vapors in contact with cool glass condense to form pure liquid distillate
Distilling flask
Water-cooled condenser
Water out to sink
Water in
③ Distillate collected in separate flask

Figure B2.5 Distillation.

Distillation separates components through *differences in **volatility,*** the tendency of a substance to become a gas. Ether, for example, is more volatile than water, which is much more volatile than sodium chloride. The simple distillation apparatus shown in Figure B2.5 is used to separate components with *large* differences in volatility, such as water from dissolved ionic compounds. As the mixture boils, the vapor is richer in the more volatile component, which is condensed and collected separately. Separating components with small volatility differences requires many vaporization-condensation steps (discussed in Chapter 13).

Extraction is also based on *differences in solubility.* In a typical procedure, a natural (often plant or animal) material is ground in a blender with a solvent that extracts (dissolves) soluble compound(s) embedded in insoluble material. This extract is separated further by the addition of a second solvent that does not dissolve in the first. After shaking in a separatory funnel, some components are extracted into the new solvent. Figure B2.6 shows the extraction of plant pigments from water into hexane, an organic solvent.

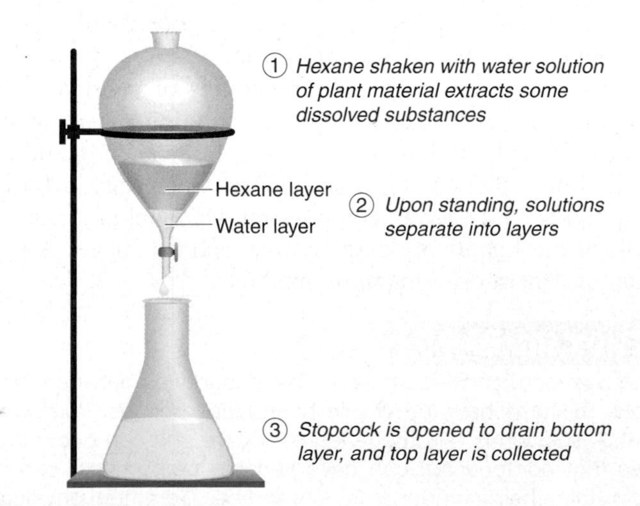
① Hexane shaken with water solution of plant material extracts some dissolved substances
Hexane layer
Water layer
② Upon standing, solutions separate into layers
③ Stopcock is opened to drain bottom layer, and top layer is collected

Figure B2.6 Extraction.

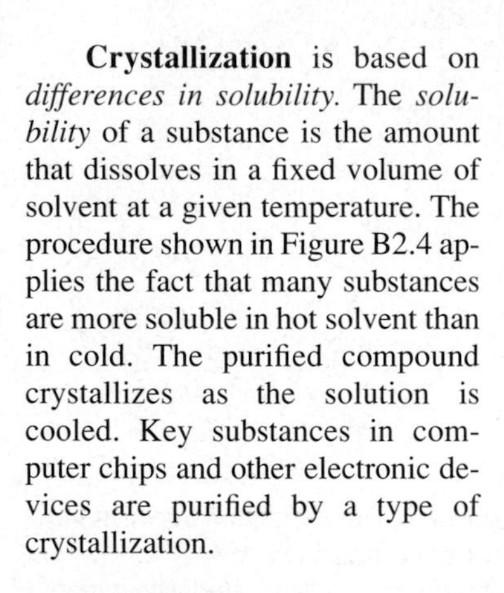

Chromatography is a third technique based on *differences in solubility*. The mixture is dissolved in a gas or liquid called the *mobile phase,* and the components are separated as this phase moves over a solid (or viscous liquid) surface called the *stationary phase*. A component with low solubility in the stationary phase spends less time there, thus moving faster than a component that is highly soluble in that phase. Figure B2.7 depicts the separation of a mixture of pigments in ink.

Many types of chromatography are used to separate a wide variety of substances, from simple gases to biological macromolecules. In *gas-liquid chromatography (GLC),* the mobile phase is an inert gas, such as helium, that carries the previously vaporized components into a long tube that contains the stationary phase (Figure B2.8, part A). The components emerge separately and reach a detector to create a chromatogram. A typical chromatogram has numerous peaks of specific position and height, each of which represents the amount of a given component (Figure B2.8, part B).

The principle of *high-performance (high-pressure) liquid chromatography (HPLC)* is very similar. However, in this technique the mixture is not vaporized, so a more diverse group of components, which may include nonvolatile compounds, can be separated (Figure B2.9).

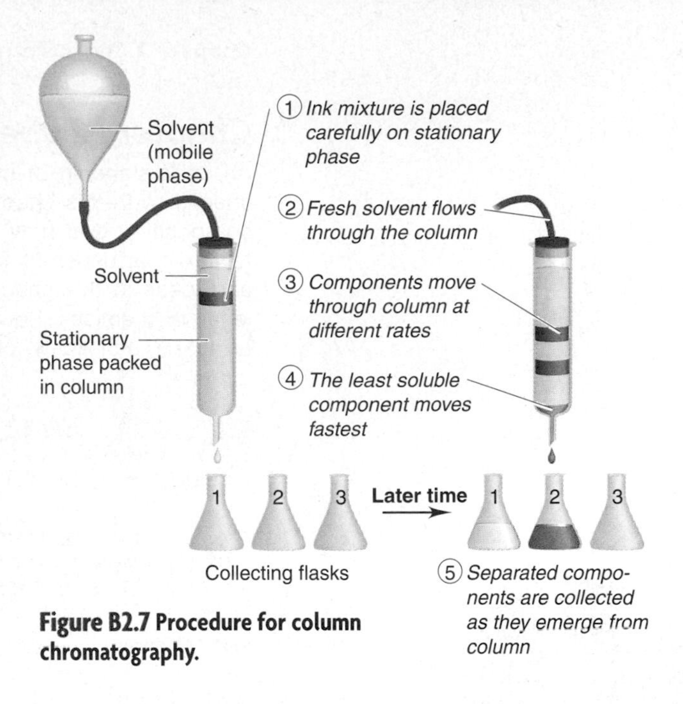

Figure B2.7 Procedure for column chromatography.

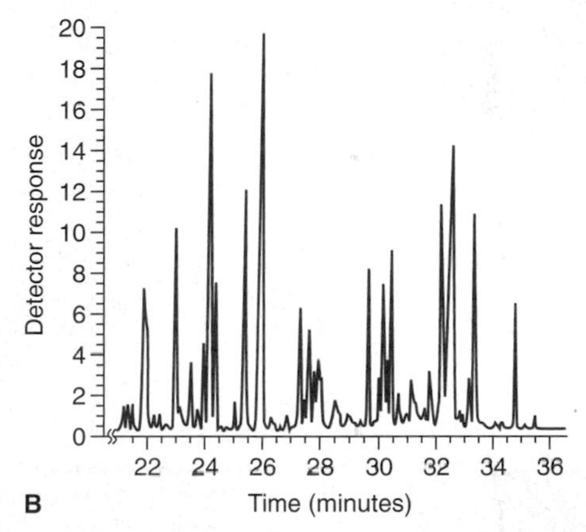

He gas
He gas
Before interacting with the stationary phase
He gas
He gas
A
After interacting with the stationary phase

Figure B2.8 Principle of gas-liquid chromatography (GLC). A, The mobile phase *(purple arrow)* carries the sample mixture into a tube packed with the stationary phase *(gray outline on yellow spheres),* and each component dissolves in the stationary phase to a different extent. A component *(red)* that dissolves less readily than another *(blue)* emerges from the tube sooner. **B,** A typical gas-liquid chromatogram of a complex mixture displays each component as a peak.

Figure B2.9 A high-performance liquid chromatograph.

Chapter Perspective

An understanding of matter at the observable and atomic levels is the essence of chemistry. In this chapter, you have learned how matter is classified in terms of its composition and how it is named in words and formulas, which are major steps toward that understanding. Figure 2.22 provides a visual review of many key terms and ideas in this chapter. In Chapter 3, we explore one of the central quantitative ideas in chemistry: how the observable amount of a substance relates to the number of atoms, molecules, or ions that make it up.

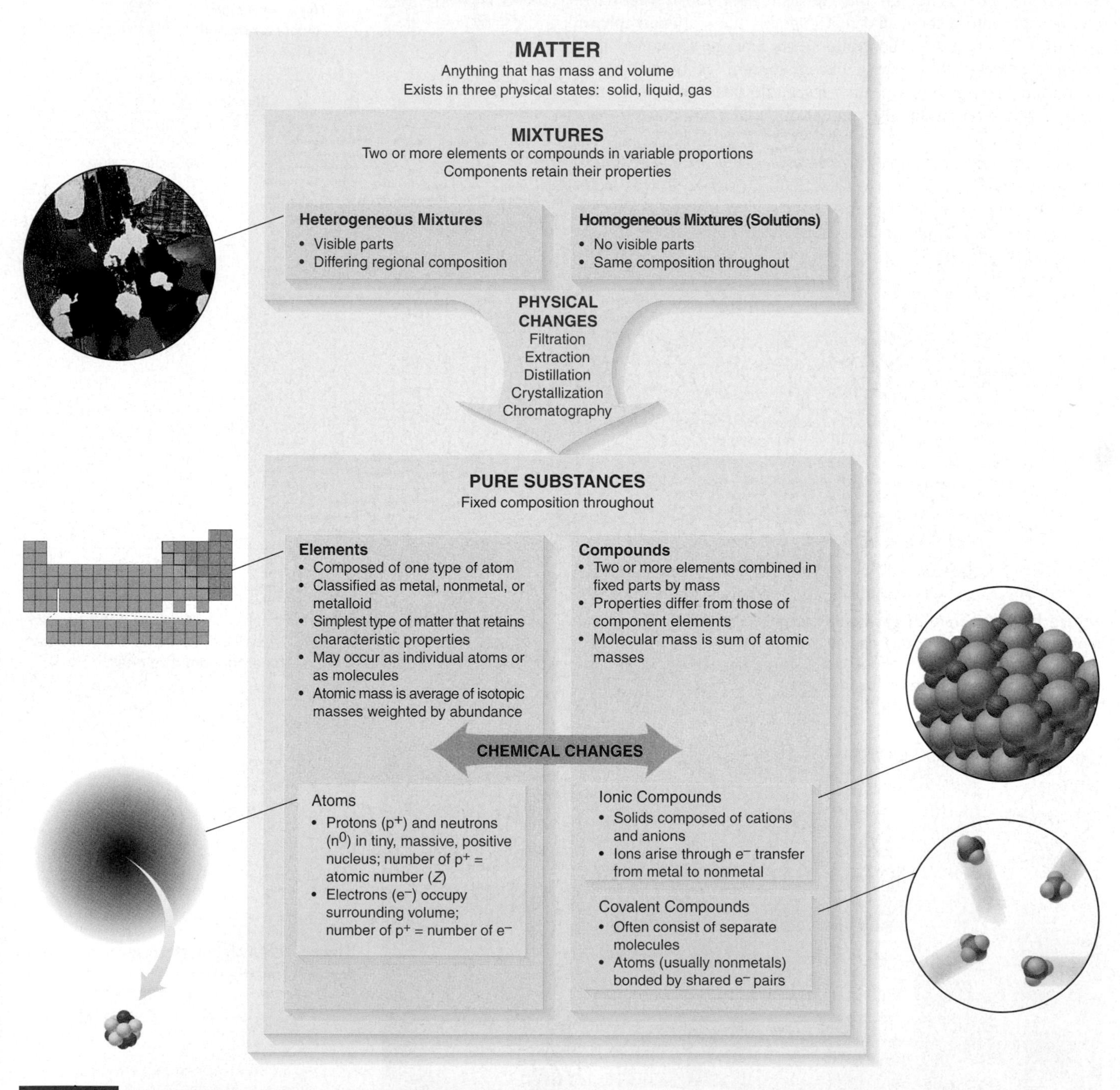

MATTER
Anything that has mass and volume
Exists in three physical states: solid, liquid, gas
MIXTURES
Two or more elements or compounds in variable proportions
Components retain their properties
Heterogeneous Mixtures
• Visible parts
• Differing regional composition
Homogeneous Mixtures (Solutions)
• No visible parts
• Same composition throughout
PHYSICAL CHANGES
Filtration
Extraction
Distillation
Crystallization
Chromatography
PURE SUBSTANCES
Fixed composition throughout
Elements
• Composed of one type of atom
• Classified as metal, nonmetal, or metalloid
• Simplest type of matter that retains characteristic properties
• May occur as individual atoms or as molecules
• Atomic mass is average of isotopic masses weighted by abundance
Compounds
• Two or more elements combined in fixed parts by mass
• Properties differ from those of component elements
• Molecular mass is sum of atomic masses
CHEMICAL CHANGES
Atoms
• Protons (p^+) and neutrons (n^0) in tiny, massive, positive nucleus; number of p^+ = atomic number (Z)
• Electrons (e^-) occupy surrounding volume; number of p^+ = number of e^-
Ionic Compounds
• Solids composed of cations and anions
• Ions arise through e^- transfer from metal to nonmetal
Covalent Compounds
• Often consist of separate molecules
• Atoms (usually nonmetals) bonded by shared e^- pairs

Figure 2.22 The classification of matter from a chemical point of view. Mixtures are separated by physical changes into elements and compounds. Chemical changes are required to convert elements into compounds, and vice versa.

For Review and Reference (Numbers in parentheses refer to pages, unless noted otherwise.)

Learning Objectives

Relevant section and/or sample problem (SP) numbers appear in parentheses.

Understand These Concepts

1. The defining characteristics of the three types of matter—element, compound, and mixture—on the macroscopic and atomic levels (2.1)
2. The significance of the three mass laws—mass conservation, definite composition, and multiple proportions (2.2)
3. The postulates of Dalton's atomic theory and how it explains the mass laws (2.3)
4. The major contribution of experiments by Thomson, Millikan, and Rutherford concerning atomic structure (2.4)
5. The structure of the atom, the main features of the subatomic particles, and the importance of isotopes (2.5)
6. The format of the periodic table and general location and characteristics of metals, metalloids, and nonmetals (2.6)
7. The essential features of ionic and covalent bonding and the distinction between them (2.7)
8. The types of mixtures and their properties (2.9)

Master These Skills

1. Using the mass ratio of element to compound to find the mass of an element in a compound (SP 2.1)
2. Using atomic notation to express the subatomic makeup of an isotope (SP 2.2)
3. Calculating an atomic mass from isotopic composition (SP 2.3)
4. Predicting the monatomic ion formed from a main-group element (SP 2.4)
5. Naming and writing the formula of an ionic compound formed from the ions in Tables 2.3 to 2.5 (SP 2.5 to 2.10 and 2.14)
6. Naming and writing the formula of an inorganic binary covalent compound (SP 2.11, 2.12, and 2.14)
7. Calculating the molecular mass of a compound (SP 2.13)

Key Terms

Section 2.1
element (39)
pure substance (39)
molecule (40)
compound (40)
mixture (40)

Section 2.2
law of mass conservation (41)
law of definite (or constant) composition (42)
fraction by mass (mass fraction) (42)
percent by mass (mass percent, mass %) (42)
law of multiple proportions (43)

Section 2.3
atom (44)

Section 2.4
cathode ray (46)
nucleus (48)

Section 2.5
proton (p^+) (49)
neutron (n^0) (49)
electron (e^-) (49)
atomic number (Z) (50)
mass number (A) (50)
atomic symbol (50)
isotope (50)
atomic mass unit (amu) (51)
dalton (Da) (51)
mass spectrometry (51)
isotopic mass (51)
atomic mass (51)

Section 2.6
periodic table of the elements (54)
period (54)
group (54)
metal (56)
nonmetal (56)
metalloid (semimetal) (56)

Section 2.7
ionic compound (57)
covalent compound (57)
chemical bond (57)
ion (57)
binary ionic compound (57)
cation (57)
anion (57)
monatomic ion (57)
covalent bond (60)
polyatomic ion (61)

Section 2.8
chemical formula (62)
empirical formula (62)
molecular formula (62)
structural formula (62)
formula unit (64)
oxoanion (66)
hydrate (66)
binary covalent compound (68)
molecular mass (70)

Section 2.9
heterogeneous mixture (73)
homogeneous mixture (73)
solution (73)
aqueous solution (73)
filtration (74)
crystallization (74)
distillation (74)
volatility (74)
extraction (74)
chromatography (75)

Key Equations and Relationships

2.1 Finding the mass of an element in a given mass of compound (42):

Mass of element in sample

$$= \text{mass of compound in sample} \times \frac{\text{mass of element}}{\text{mass of compound}}$$

2.2 Calculating the number of neutrons in an atom (50):

$$\text{Number of neutrons} = \text{mass number} - \text{atomic number}$$

or

$$N = A - Z$$

2.3 Determining the molecular mass of a formula unit of a compound (70):

$$\text{Molecular mass} = \text{sum of atomic masses}$$

Highlighted Figures and Tables

These figures (F) and tables (T) provide a review of key ideas. Entries in **color** contain frequently used data.

Brief Solutions to Follow-up Problems

2.1 Mass (t) of pitchblende

$$= 2.3\ \cancel{\text{t uranium}} \times \frac{84.2\ \text{t pitchblende}}{71.4\ \cancel{\text{t uranium}}} = 2.7\ \text{t pitchblende}$$

Mass (t) of oxygen

$$= 2.7\ \cancel{\text{t pitchblende}} \times \frac{(84.2 - 71.4\ \text{t oxygen})}{84.2\ \cancel{\text{t pitchblende}}} = 0.41\ \text{t oxygen}$$

2.2 (a) Q = B; $5p^+, 6n^0, 5e^-$
(b) X = Ca; $20p^+, 21n^0, 20e^-$
(c) Y = I; $53p^+, 78n^0, 53e^-$

2.3 $10.0129x + [11.0093(1 - x)] = 10.81$; $0.9964x = 0.1993$; $x = 0.2000$ and $1 - x = 0.8000$; % abundance of $^{10}B = 20.00\%$; % abundance of $^{11}B = 80.00\%$

2.4 (a) S^{2-}; (b) Rb^+; (c) Ba^{2+}

2.5 (a) Zinc [Group 2B(12)] and oxygen [Group 6A(16)]
(b) Silver [Group 1B(11)] and bromine [Group 7A(17)]
(c) Lithium [Group 1A(1)] and chlorine [Group 7A(17)]
(d) Aluminum [Group 3A(13)] and sulfur [Group 6A(16)]

2.6 (a) ZnO; (b) AgBr; (c) LiCl; (d) Al_2S_3

2.7 (a) PbO_2; (b) copper(I) sulfide (cuprous sulfide); (c) iron(II) bromide (ferrous bromide); (d) $HgCl_2$

2.8 (a) $Cu(NO_3)_2{\cdot}3H_2O$; (b) $Zn(OH)_2$; (c) lithium cyanide

2.9 (a) $(NH_4)_3PO_4$; ammonium is NH_4^+ and phosphate is PO_4^{3-}.
(b) $Al(OH)_3$; parentheses are needed around the polyatomic ion OH^-.
(c) Magnesium hydrogen carbonate; Mg^{2+} is magnesium and can have only a 2+ charge, so it does not need (II); HCO_3^- is hydrogen carbonate (or bicarbonate).
(d) Chromium(III) nitrate; the *-ic* ending is not used with Roman numerals; NO_3^- is nitr*ate*.
(e) Calcium nitrite; Ca^{2+} is calcium and NO_2^- is nitr*ite*.

2.10 (a) $HClO_3$; (b) hydrofluoric acid; (c) CH_3COOH (or $HC_2H_3O_2$); (d) H_2SO_3; (e) hypobromous acid

2.11 (a) Sulfur trioxide; (b) silicon dioxide; (c) N_2O; (d) SeF_6

2.12 (a) Disulfur dichloride; the *-ous* suffix is not used.
(b) NO; the name indicates one nitrogen.
(c) Bromine trichloride; Br is in a higher period in Group 7A(17), so it is named first.

2.13 (a) H_2O_2, 34.02 amu; (b) CsCl, 168.4 amu; (c) H_2SO_4, 98.09 amu; (d) K_2SO_4, 174.27 amu

2.14 (a) Na_2O. This is an ionic compound, so the name is sodium oxide.

Formula mass

$$= (2 \times \text{atomic mass of Na}) + (1 \times \text{atomic mass of O})$$
$$= (2 \times 22.99\ \text{amu}) + 16.00\ \text{amu} = 61.98\ \text{amu}$$

(b) NO_2. This is a covalent compound, and N has the lower group number, so the name is nitrogen dioxide.

Molecular mass

$$= (1 \times \text{atomic mass of N}) + (2 \times \text{atomic mass of O})$$
$$= 14.01\ \text{amu} + (2 \times 16.00\ \text{amu}) = 46.01\ \text{amu}$$

Problems

Problems with **colored** numbers are answered in Appendix E. Sections match the text and provide the numbers of relevant sample problems. Most offer Concept Review Questions, Skill-Building Exercises (grouped in pairs covering the same concept), and Problems in Context. Comprehensive Problems are based on material from any section or previous chapter.

Elements, Compounds, and Mixtures: An Atomic Overview

Concept Review Questions

2.1 What is the key difference between an element and a compound?

2.2 List two differences between a compound and a mixture.

2.3 Which of the following are pure substances? Explain.
(a) Calcium chloride, used to melt ice on roads, consists of two elements, calcium and chlorine, in a fixed mass ratio.
(b) Sulfur consists of sulfur atoms combined into octatomic molecules.
(c) Baking powder, a leavening agent, contains 26% to 30% sodium hydrogen carbonate and 30% to 35% calcium dihydrogen phosphate by mass.
(d) Cytosine, a component of DNA, consists of H, C, N, and O atoms bonded in a specific arrangement.

2.4 Classify each substance in Problem 2.3 as an element, compound, or mixture, and explain your answers.

2.5 Explain the following statement: The smallest particles unique to an element may be atoms or molecules.

2.6 Explain the following statement: The smallest particles unique to a compound cannot be atoms.

2.7 Can the relative amounts of the components of a mixture vary? Can the relative amounts of the components of a compound vary? Explain.

Problems in Context

2.8 The tap water found in many areas of the United States leaves white deposits when it evaporates. Is this tap water a mixture or a compound? Explain.

2.9 Samples of illicit "street" drugs often contain an inactive component, such as ascorbic acid (vitamin C). After obtaining a sample of cocaine, government chemists calculate the mass of vitamin C per gram of drug sample, and use it to track the drug's distribution. For example, if different samples of cocaine obtained on the streets of New York, Los Angeles, and Paris all contain 0.6384 g of vitamin C per gram of sample, they very likely come from a common source. Is this street sample a compound, element, or mixture? Explain.

The Observations That Led to an Atomic View of Matter

(Sample Problem 2.1)

Concept Review Questions

2.10 Why was it necessary for separation techniques and methods of chemical analysis to be developed before the laws of definite composition and multiple proportions could be formulated?

2.11 To which classes of matter—element, compound, and/or mixture—do the following apply: (a) law of mass conservation; (b) law of definite composition; (c) law of multiple proportions?

2.12 In our modern view of matter and energy, is the law of mass conservation still relevant to chemical reactions? Explain.

2.13 Identify the mass law that each of the following observations demonstrates, and explain your reasoning:
(a) A sample of potassium chloride from Chile contains the same percent by mass of potassium as one from Poland.
(b) A flashbulb contains magnesium and oxygen before use and magnesium oxide afterward, but its mass does not change.
(c) Arsenic and oxygen form one compound that is 65.2 mass % arsenic and another that is 75.8 mass % arsenic.

2.14 (a) Does the percent by mass of each element in a compound depend on the amount of compound? Explain. (b) Does the mass of each element in a compound depend on the amount of compound? Explain.

2.15 Does the percent by mass of the elements in a compound depend on the amounts of the elements used to make the compound? Explain.

Skill-Building Exercises *(grouped in similar pairs)*

2.16 State the mass law(s) demonstrated by the following experimental results, and explain your reasoning:
Experiment 1: A student heats 1.00 g of a blue compound and obtains 0.64 g of a white compound and 0.36 g of a colorless gas.
Experiment 2: A second student heats 3.25 g of the same blue compound and obtains 2.08 g of a white compound and 1.17 g of a colorless gas.

2.17 State the mass law(s) demonstrated by the following experimental results, and explain your reasoning:
Experiment 1: A student heats 1.27 g of copper and 3.50 g of iodine to produce 3.81 g of a white compound, and 0.96 g of iodine remains.
Experiment 2: A second student heats 2.55 g of copper and 3.50 g of iodine to form 5.25 g of a white compound, and 0.80 g of copper remains.

2.18 Fluorite, a mineral of calcium, is a compound of the metal with fluorine. Analysis shows that a 2.76-g sample of fluorite contains 1.42 g of calcium. Calculate the (a) mass of fluorine in the sample; (b) mass fractions of calcium and fluorine in fluorite; (c) mass percents of calcium and fluorine in fluorite.

2.19 Galena, a mineral of lead, is a compound of the metal with sulfur. Analysis shows that a 2.34-g sample of galena contains 2.03 g of lead. Calculate the (a) mass of sulfur in the sample; (b) mass fractions of lead and sulfur in galena; (c) mass percents of lead and sulfur in galena.

2.20 Magnesium oxide (MgO) forms when the metal burns in air. (a) If 1.25 g of MgO contains 0.754 g of Mg, what is the mass ratio of magnesium to oxide? (b) How many grams of Mg are in 435 g of MgO?

2.21 Zinc sulfide (ZnS) occurs in the zincblende and wurtzite crystal structures. (a) If 2.54 g of ZnS contains 1.70 g of Zn, what is the mass ratio of zinc to sulfide? (b) How many kilograms of Zn are in 2.45 kg of ZnS?

2.22 A compound of copper and sulfur contains 88.39 g of metal and 44.61 g of nonmetal. How many grams of copper are in 5264 kg of compound? How many grams of sulfur?

2.23 A compound of iodine and cesium contains 63.94 g of metal and 61.06 g of nonmetal. How many grams of cesium are in 38.77 g of compound? How many grams of iodine?

2.24 Show, with calculations, how the following data illustrate the law of multiple proportions:
Compound 1: 47.5 mass % sulfur and 52.5 mass % chlorine
Compound 2: 31.1 mass % sulfur and 68.9 mass % chlorine

2.25 Show, with calculations, how the following data illustrate the law of multiple proportions:
Compound 1: 77.6 mass % xenon and 22.4 mass % fluorine
Compound 2: 63.3 mass % xenon and 36.7 mass % fluorine

Problems in Context

2.26 Dolomite is a carbonate of magnesium and calcium. Analysis shows that 7.81 g of dolomite contains 1.70 g of Ca. Calculate the mass percent of Ca in dolomite. On the basis of the mass percent of Ca, and neglecting all other factors, which is the richer source of Ca, dolomite or fluorite (see Problem 2.18)?

2.27 The mass percent of sulfur in a sample of coal is a key factor in the environmental impact of the coal because the sulfur combines with oxygen when the coal is burned and the oxide can then be incorporated into acid rain. Which of the following coals would have the smallest environmental impact?

	Mass (g) of Sample	Mass (g) of Sulfur in Sample
Coal A	378	11.3
Coal B	495	19.0
Coal C	675	20.6

Dalton's Atomic Theory

Concept Review Questions

2.28 Which of Dalton's postulates about atoms are inconsistent with later observations? Do these inconsistencies mean that Dalton was wrong? Is Dalton's model still useful? Explain.

2.29 Use Dalton's theory to explain why potassium nitrate from India or Italy has the same mass percents of K, N, and O.

The Observations That Led to the Nuclear Atom Model

Concept Review Questions

2.30 Thomson was able to determine the mass/charge ratio of the electron but not its mass. How did Millikan's experiment allow determination of the electron's mass?

2.31 The following charges on individual oil droplets were obtained during an experiment similar to Millikan's. Determine a charge for the electron (in C, coulombs), and explain your answer: -3.204×10^{-19} C; -4.806×10^{-19} C; -8.010×10^{-19} C; -1.442×10^{-18} C.

2.32 Describe Thomson's model of the atom. How might it account for the production of cathode rays?

2.33 When Rutherford's coworkers bombarded gold foil with α particles, they obtained results that overturned the existing (Thomson) model of the atom. Explain.

The Atomic Theory Today

(Sample Problems 2.2 and 2.3)

Concept Review Questions

2.34 Define *atomic number* and *mass number.* Which can vary without changing the identity of the element?

2.35 Choose the correct answer. The difference between the mass number of an isotope and its atomic number is (a) directly related to the identity of the element; (b) the number of electrons; (c) the number of neutrons; (d) the number of isotopes.

2.36 Even though several elements have only one naturally occurring isotope and all atomic nuclei have whole numbers of protons and neutrons, no atomic mass is a whole number. Use the data from Table 2.2 to explain this fact.

Skill-Building Exercises *(grouped in similar pairs)*

2.37 Argon has three naturally occurring isotopes, ^{36}Ar, ^{38}Ar, and ^{40}Ar. What is the mass number of each? How many protons, neutrons, and electrons are present in each?

2.38 Chlorine has two naturally occurring isotopes, ^{35}Cl and ^{37}Cl. What is the mass number of each isotope? How many protons, neutrons, and electrons are present in each?

2.39 Do both members of the following pairs have the same number of protons? Neutrons? Electrons?
(a) $^{16}_{8}$O and $^{17}_{8}$O (b) $^{40}_{18}$Ar and $^{41}_{19}$K (c) $^{60}_{27}$Co and $^{60}_{28}$Ni
Which pair(s) consist(s) of atoms with the same *Z* value? *N* value? *A* value?

2.40 Do both members of the following pairs have the same number of protons? Neutrons? Electrons?
(a) $^{3}_{1}$H and $^{3}_{2}$He (b) $^{14}_{6}$C and $^{15}_{7}$N (c) $^{19}_{9}$F and $^{18}_{9}$F
Which pair(s) consist(s) of atoms with the same *Z* value? *N* value? *A* value?

2.41 Write the $^{A}_{Z}$X notation for each atomic depiction:

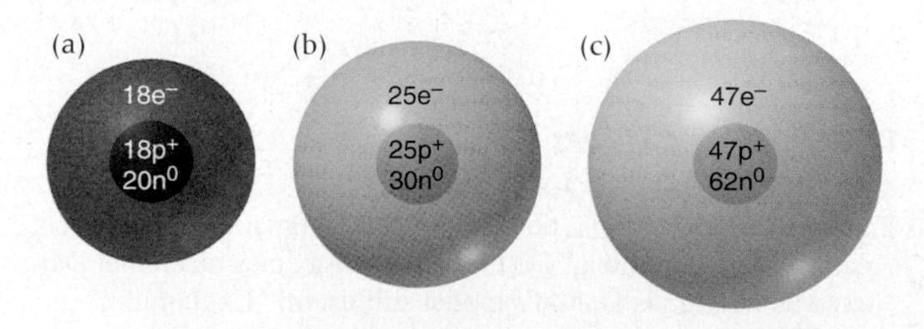

2.42 Write the $^{A}_{Z}$X notation for each atomic depiction:

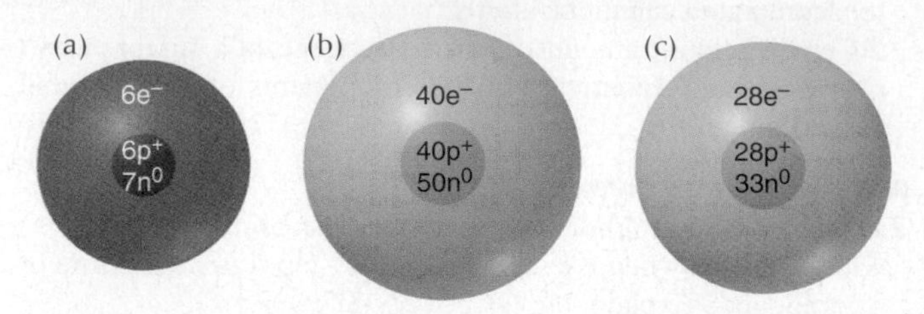

2.43 Draw atomic depictions similar to those in Problem 2.41 for (a) $^{48}_{22}$Ti; (b) $^{79}_{34}$Se; (c) $^{11}_{5}$B.

2.44 Draw atomic depictions similar to those in Problem 2.41 for (a) $^{207}_{82}$Pb; (b) $^{9}_{4}$Be; (c) $^{75}_{33}$As.

2.45 Gallium has two naturally occurring isotopes, ^{69}Ga (isotopic mass 68.9256 amu, abundance 60.11%) and ^{71}Ga (isotopic mass 70.9247 amu, abundance 39.89%). Calculate the atomic mass of gallium.

2.46 Magnesium has three naturally occurring isotopes, ^{24}Mg (isotopic mass 23.9850 amu, abundance 78.99%), ^{25}Mg (isotopic mass 24.9858 amu, abundance 10.00%), and ^{26}Mg (isotopic mass 25.9826 amu, abundance 11.01%). Calculate the atomic mass of magnesium.

2.47 Chlorine has two naturally occurring isotopes, ^{35}Cl (isotopic mass 34.9689 amu) and ^{37}Cl (isotopic mass 36.9659 amu). If chlorine has an atomic mass of 35.4527 amu, what is the percent abundance of each isotope?

2.48 Copper has two naturally occurring isotopes, ^{63}Cu (isotopic mass 62.9396 amu) and ^{65}Cu (isotopic mass 64.9278 amu). If copper has an atomic mass of 63.546 amu, what is the percent abundance of each isotope?

Elements: A First Look at the Periodic Table

Concept Review Questions

2.49 How can iodine ($Z = 53$) have a higher atomic number yet a lower atomic mass than tellurium ($Z = 52$)?

2.50 Correct each of the following statements:
(a) In the modern periodic table, the elements are arranged in order of increasing atomic mass.
(b) Elements in a period have similar chemical properties.
(c) Elements can be classified as either metalloids or nonmetals.

2.51 What class of elements lies along the "staircase" line in the periodic table? How do their properties compare with those of metals and nonmetals?

2.52 What are some characteristic properties of elements to the left of the elements along the "staircase"? To the right?

2.53 The elements in Groups 1A(1) and 7A(17) are all quite reactive. What is a major difference between them?

Skill-Building Exercises *(grouped in similar pairs)*

2.54 Give the name, atomic symbol, and group number of the element with the following *Z* value, and classify it as a metal, metalloid, or nonmetal:
(a) $Z = 32$ (b) $Z = 16$ (c) $Z = 2$ (d) $Z = 3$ (e) $Z = 42$

2.55 Give the name, atomic symbol, and group number of the element with the following *Z* value, and classify it as a metal, metalloid, or nonmetal:
(a) $Z = 33$ (b) $Z = 20$ (c) $Z = 35$ (d) $Z = 19$ (e) $Z = 12$

2.56 Fill in the blanks:
(a) The symbol and atomic number of the heaviest alkaline earth metal are ____________ and ____________.
(b) The symbol and atomic number of the lightest metalloid in Group 5A(15) are ____________ and ____________.
(c) Group 1B(11) consists of the *coinage metals*. The symbol and atomic mass of the coinage metal whose atoms have the fewest electrons are ____________ and ____________.
(d) The symbol and atomic mass of the halogen in Period 4 are ____________ and ____________.

2.57 Fill in the blanks:
(a) The symbol and atomic number of the heaviest noble gas are ____________ and ____________.
(b) The symbol and group number of the Period 5 transition element whose atoms have the fewest protons are ____________ and ____________.
(c) The elements in Group 6A(16) are sometimes called the *chalcogens*. The symbol and atomic number of the only metallic chalcogen are ____________ and ____________.
(d) The symbol and number of protons of the Period 4 alkali metal atom are ____________ and ____________.

Compounds: Introduction to Bonding

(Sample Problem 2.4)

Concept Review Questions

2.58 Describe the type and nature of the bonding that occurs between reactive metals and nonmetals.

2.59 Describe the type and nature of the bonding that often occurs between two nonmetals.

2.60 How can ionic compounds be neutral if they consist of positive and negative ions?

2.61 Given that the ions in LiF and in MgO are of similar size, which compound has stronger ionic bonding? Use Coulomb's law in your explanation.

2.62 Are molecules present in a sample of BaF_2? Explain.

2.63 Are ions present in a sample of SO_3? Explain.

2.64 The monatomic ions of Groups 1A(1) and 7A(17) are all singly charged. In what major way do they differ? Why?

2.65 Describe the formation of solid magnesium chloride ($MgCl_2$) from large numbers of magnesium and chlorine atoms.

2.66 Describe the formation of solid potassium sulfide (K_2S) from large numbers of potassium and sulfur atoms.

2.67 Does potassium nitrate (KNO_3) incorporate ionic bonding, covalent bonding, or both? Explain.

Skill-Building Exercises *(grouped in similar pairs)*

2.68 What monatomic ions do potassium ($Z = 19$) and iodine ($Z = 53$) form?

2.69 What monatomic ions do barium ($Z = 56$) and selenium ($Z = 34$) form?

2.70 For each ionic depiction, give the name of the parent atom, its mass number, and its group and period numbers:

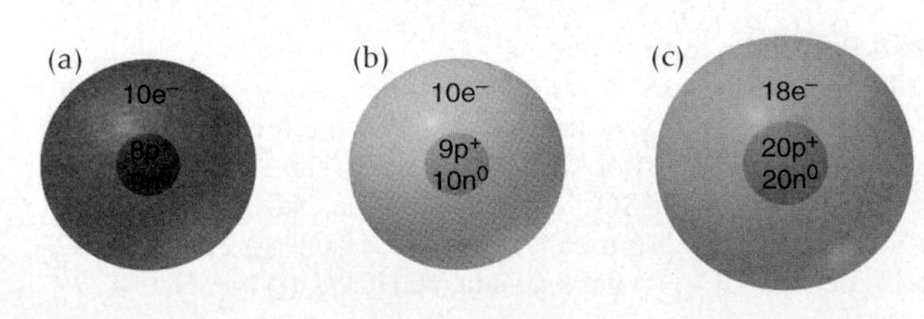

2.71 For each ionic depiction, give the name of the parent atom, its mass number, and its group and period numbers:

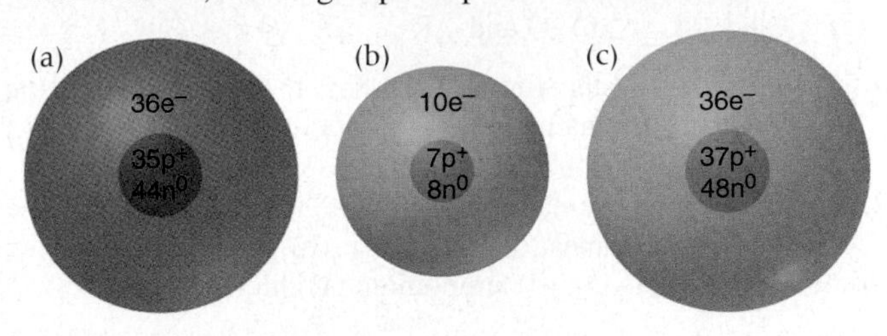

2.72 An ionic compound forms when lithium ($Z = 3$) reacts with oxygen ($Z = 8$). If a sample of the compound contains 5.3×10^{20} lithium ions, how many oxide ions does it contain?

2.73 An ionic compound forms when calcium ($Z = 20$) reacts with iodine ($Z = 53$). If a sample of the compound contains 7.4×10^{21} calcium ions, how many iodide ions does it contain?

2.74 The radii of the sodium and potassium ions are 102 pm and 138 pm, respectively. Which compound has stronger ionic attractions, sodium chloride or potassium chloride?

2.75 The radii of the lithium and magnesium ions are 76 pm and 72 pm, respectively. Which compound has stronger ionic attractions, lithium oxide or magnesium oxide?

Compounds: Formulas, Names, and Masses

(Sample Problems 2.5 to 2.14)

Concept Review Questions

2.76 What is the difference between an empirical formula and a molecular formula? Can they ever be the same?

2.77 How is a structural formula similar to a molecular formula? How is it different?

2.78 Consider a mixture of 10 billion O_2 molecules and 10 billion H_2 molecules. In what way is this mixture similar to a sample containing 10 billion hydrogen peroxide (H_2O_2) molecules? In what way is it different?

2.79 For what type(s) of compound do we use Roman numerals in the name?

2.80 For what type(s) of compound do we use Greek numerical prefixes in the name?

2.81 For what type of compound are we unable to write a molecular formula?

Skill-Building Exercises *(grouped in similar pairs)*

2.82 Write an empirical formula for each of the following:
(a) Hydrazine, a rocket fuel, molecular formula N_2H_4
(b) Glucose, a sugar, molecular formula $C_6H_{12}O_6$

2.83 Write an empirical formula for each of the following:
(a) Ethylene glycol, car antifreeze, molecular formula $C_2H_6O_2$
(b) Peroxodisulfuric acid, a compound used to make bleaching agents, molecular formula $H_2S_2O_8$

2.84 Give the name and formula of the compound formed from the following elements: (a) lithium and nitrogen; (b) oxygen and strontium; (c) aluminum and chlorine.

2.85 Give the name and formula of the compound formed from the following elements: (a) rubidium and bromine; (b) sulfur and barium; (c) calcium and fluorine.

2.86 Give the name and formula of the compound formed from the following elements:
(a) $_{12}L$ and $_{9}M$ (b) $_{11}L$ and $_{16}M$ (c) $_{17}L$ and $_{38}M$

2.87 Give the name and formula of the compound formed from the following elements:
(a) $_3Q$ and $_{35}R$ (b) $_8Q$ and $_{13}R$ (c) $_{19}Q$ and $_{53}R$

2.88 Give the systematic names for the formulas or the formulas for the names: (a) tin(IV) chloride; (b) $FeBr_3$; (c) cuprous bromide; (d) Mn_2O_3.

2.89 Give the systematic names for the formulas or the formulas for the names: (a) Na_2HPO_4; (b) potassium carbonate dihydrate; (c) $NaNO_2$; (d) ammonium perchlorate.

2.90 Give the systematic names for the formulas or the formulas for the names: (a) CoO; (b) mercury(I) chloride; (c) $Pb(C_2H_3O_2)_2{\cdot}3H_2O$; (d) chromic oxide.

2.91 Give the systematic names for the formulas or the formulas for the names: (a) $Sn(SO_3)_2$; (b) potassium dichromate; (c) $FeCO_3$; (d) copper(II) nitrate.

2.92 Correct each of the following formulas:
(a) Barium oxide is BaO_2.
(b) Iron(II) nitrate is $Fe(NO_3)_3$.
(c) Magnesium sulfide is $MnSO_3$.

2.93 Correct each of the following names:
(a) CuI is cobalt(II) iodide.
(b) $Fe(HSO_4)_3$ is iron(II) sulfate.
(c) $MgCr_2O_7$ is magnesium dichromium heptaoxide.

2.94 Give the name and formula for the acid derived from each of the following anions:
(a) hydrogen sulfate (b) IO_3^- (c) cyanide (d) HS^-

2.95 Give the name and formula for the acid derived from each of the following anions:
(a) perchlorate (b) NO_3^- (c) bromite (d) F^-

2.96 Many chemical names are similar at first glance. Give the formulas of the species in each set: (a) ammonium ion and ammonia; (b) magnesium sulfide, magnesium sulfite, and magnesium sulfate; (c) hydrochloric acid, chloric acid, and chlorous acid; (d) cuprous bromide and cupric bromide.

2.97 Give the formulas of the compounds in each set: (a) lead(II) oxide and lead(IV) oxide; (b) lithium nitride, lithium nitrite, and lithium nitrate; (c) strontium hydride and strontium hydroxide; (d) magnesium oxide and manganese(II) oxide.

2.98 Give the name and formula of the compound whose molecules consist of two sulfur atoms and four fluorine atoms.

2.99 Give the name and formula of the compound whose molecules consist of two iodine atoms and seven oxygen atoms.

2.100 Correct the name to match the formula of the following compounds: (a) calcium(II) chloride, $CaCl_2$; (b) copper(II) oxide, Cu_2O; (c) stannous fluoride, SnF_4; (d) hydrogen chloride acid, HCl.

2.101 Correct the formula to match the name of the following compounds: (a) iron(III) oxide, Fe_3O_4; (b) chloric acid, HCl; (c) mercuric oxide, Hg_2O; (d) dichlorine pentaoxide, Cl_2O_7.

2.102 Give the number of atoms of the specified element in a formula unit of each of the following compounds, and calculate the molecular (formula) mass:
(a) Oxygen in aluminum sulfate, $Al_2(SO_4)_3$
(b) Hydrogen in ammonium hydrogen phosphate, $(NH_4)_2HPO_4$
(c) Oxygen in the mineral azurite, $Cu_3(OH)_2(CO_3)_2$

2.103 Give the number of atoms of the specified element in a formula unit of each of the following compounds, and calculate the molecular (formula) mass:
(a) Hydrogen in ammonium benzoate, $C_6H_5COONH_4$
(b) Nitrogen in hydrazinium sulfate, $N_2H_6SO_4$
(c) Oxygen in the mineral leadhillite, $Pb_4SO_4(CO_3)_2(OH)_2$

2.104 Write the formula of each compound, and determine its molecular (formula) mass: (a) ammonium sulfate; (b) sodium dihydrogen phosphate; (c) potassium bicarbonate.

2.105 Write the formula of each compound, and determine its molecular (formula) mass: (a) sodium dichromate; (b) ammonium perchlorate; (c) magnesium nitrite trihydrate.

2.106 Calculate the molecular (formula) mass of each compound: (a) dinitrogen pentaoxide; (b) lead(II) nitrate; (c) calcium peroxide.

2.107 Calculate the molecular (formula) mass of each compound: (a) iron(II) acetate tetrahydrate; (b) sulfur tetrachloride; (c) potassium permanganate.

2.108 Give the formula, name, and molecular mass of the following molecules:

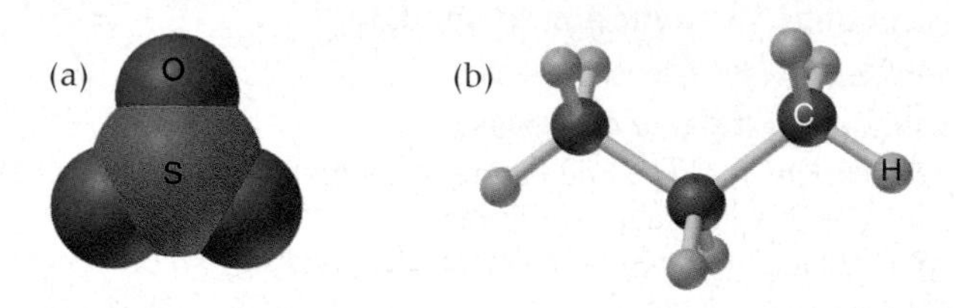

2.109 Give the formula, name, and molecular mass of the following molecules:

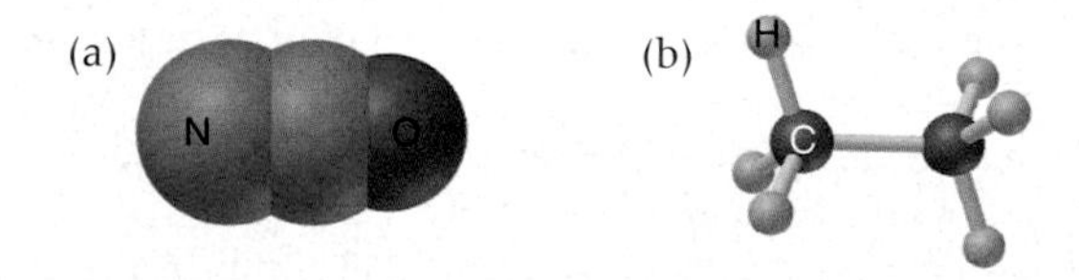

2.110 Give the name, empirical formula, and molecular mass of the molecule depicted in Figure P2.110.

2.111 Give the name, empirical formula, and molecular mass of the molecule depicted in Figure P2.111.

Figure P2.110 Figure P2.111

Problems in Context

2.112 Before the use of systematic names, many compounds had common names. Give the systematic name for each of the following: (a) blue vitriol, $CuSO_4{\cdot}5H_2O$; (b) slaked lime, $Ca(OH)_2$; (c) oil of vitriol, H_2SO_4; (d) washing soda, Na_2CO_3; (e) muriatic acid, HCl; (f) Epsom salts, $MgSO_4{\cdot}7H_2O$; (g) chalk, $CaCO_3$; (h) dry ice, CO_2; (i) baking soda, $NaHCO_3$; (j) lye, NaOH.

2.113 Each box contains a representation of a binary compound. Determine its name, formula, and molecular (formula) mass.

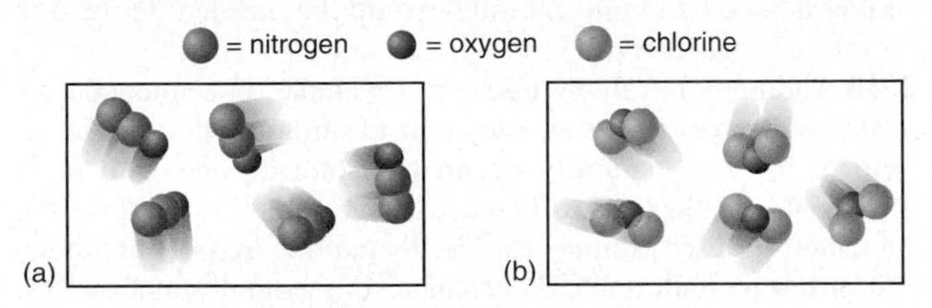

Mixtures: Classification and Separation

Concept Review Questions

2.114 In what main way is separating the components of a mixture different from separating the components of a compound?

2.115 What is the difference between a homogeneous and a heterogeneous mixture?

2.116 Is a solution a homogeneous or a heterogeneous mixture? Give an example of an aqueous solution.

Skill-Building Exercises *(grouped in similar pairs)*

2.117 Classify each of the following as a compound, a homogeneous mixture, or a heterogeneous mixture: (a) distilled water; (b) gasoline; (c) beach sand; (d) wine; (e) air.

2.118 Classify each of the following as a compound, a homogeneous mixture, or a heterogeneous mixture: (a) orange juice; (b) vegetable soup; (c) cement; (d) calcium sulfate; (e) tea.

2.119 Name the technique(s) and briefly describe the procedure you would use to separate each of the following mixtures into two components: (a) table salt and pepper; (b) table sugar and sand; (c) drinking water contaminated with fuel oil; (d) vegetable oil and vinegar.

2.120 Name the technique(s) and briefly describe the procedure you would use to separate each of the following mixtures into two components: (a) crushed ice and crushed glass; (b) table sugar dissolved in ethanol; (c) iron and sulfur; (d) two pigments (chlorophyll *a* and chlorophyll *b*) from spinach leaves.

Problems in Context

2.121 Which separation method is operating in each of the following procedures: (a) pouring a mixture of cooked pasta and boiling water into a colander; (b) removing colored impurities from raw sugar to make refined sugar; (c) preparing coffee by pouring hot water through ground coffee beans?

Comprehensive Problems

2.122 Helium is the lightest noble gas and the second most abundant element (after hydrogen) in the universe.
(a) The radius of a helium atom is 3.1×10^{-11} m; the radius of its nucleus is 2.5×10^{-15} m. What fraction of the spherical atomic volume is occupied by the nucleus (V of a sphere $= \frac{4}{3}\pi r^3$)?
(b) The mass of a helium-4 atom is 6.64648×10^{-24} g, and each of its two electrons has a mass of 9.10939×10^{-28} g. What fraction of this atom's mass is contributed by its nucleus?

2.123 From the following ions and their radii (in pm), choose a pair that gives the strongest ionic bonding and a pair that gives the weakest: Mg^{2+} 72; K^+ 138; Rb^+ 152; Ba^{2+} 135; Cl^- 181; O^{2-} 140; I^- 220.

2.124 Prior to 1961, the atomic mass standard was defined as $\frac{1}{16}$ of the mass of ^{16}O. Based on that standard: (a) What was the mass of carbon-12, given the modern atomic mass of oxygen is 15.9994 amu? (b) What was the mass of potassium-39, given its modern isotopic mass is 38.9637 amu?

2.125 Transition metals, located in the center of the periodic table, have many essential uses as elements and form many important compounds as well. Calculate the molecular mass of the following transition metal compounds:
(a) $[Co(NH_3)_6]Cl_3$ (b) $[Pt(NH_3)_4BrCl]Cl_2$
(c) $K_4[V(CN)_6]$ (d) $[Ce(NH_3)_6][FeCl_4]_3$

2.126 A rock is 5.0% by mass fayalite (Fe_2SiO_4), 7.0% by mass forsterite (Mg_2SiO_4), and the remainder silicon dioxide. What is the mass percent of each element in the rock?

2.127 Polyatomic ions are named by patterns that apply to elements in a given group. Using the periodic table and Table 2.5, give the name of each of the following: (a) SeO_4^{2-}; (b) AsO_4^{3-}; (c) BrO_2^-; (d) $HSeO_4^-$; (e) TeO_3^{2-}.

2.128 Scenes A–I depict various types of matter on the atomic scale. Choose the correct scene(s) for each of the following:
(a) A mixture that fills its container
(b) A substance that cannot be broken down into simpler ones
(c) An element with a very high resistance to flow
(d) A homogeneous mixture
(e) An element that conforms to the walls of its container and displays a surface
(f) A gas consisting of diatomic particles
(g) A gas that can be broken down into simpler substances
(h) A substance with a 2:1 number ratio of its component atoms
(i) Matter that can be separated into its component substances by physical means
(j) A heterogeneous mixture
(k) Matter that obeys the law of definite composition

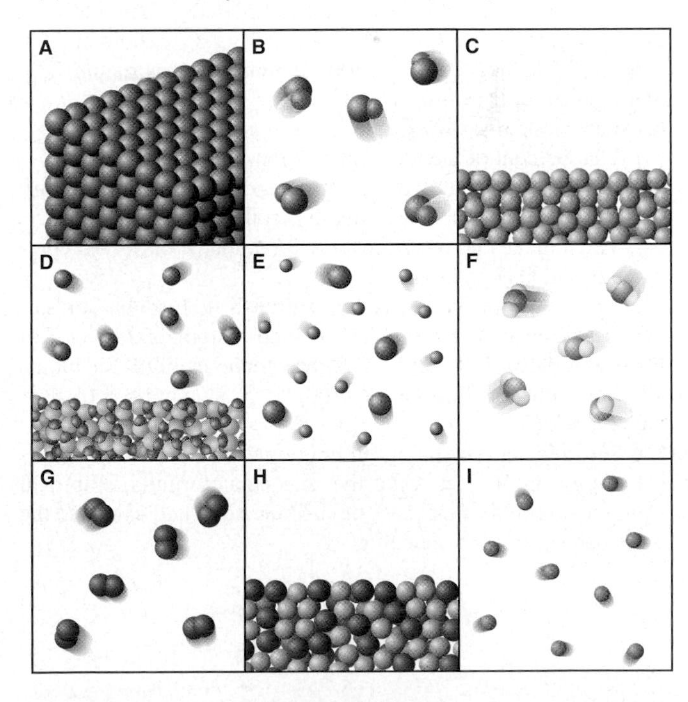

2.129 Nitrogen forms more oxides than any other element. The percents by mass of N in three different nitrogen oxides are (I) 46.69%; (II) 36.85%; (III) 25.94%. (a) Determine the empirical formula of each compound. (b) How many grams of oxygen per 1.00 g of nitrogen are in each compound?

2.130 Boron has two naturally occurring isotopes, ^{10}B (19.9%) and ^{11}B (80.1%). Although the B_2 molecule does not exist naturally on Earth, it has been produced in the laboratory and been observed in stars. (a) How many different B_2 molecules are possible? (b) What are the masses and percent abundances of each?

2.131 Which of the following pictures is (are) consistent with the fact that compounds of nitrogen *(blue)* and oxygen *(red)* exhibit the law of multiple proportions? Explain.

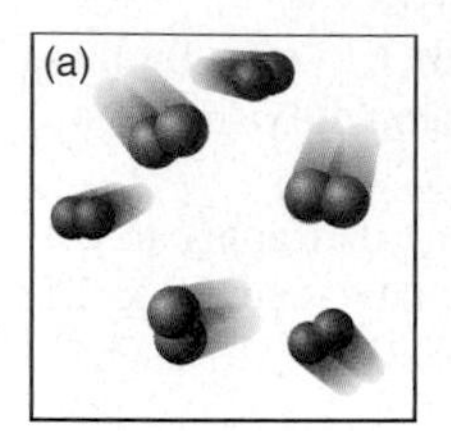

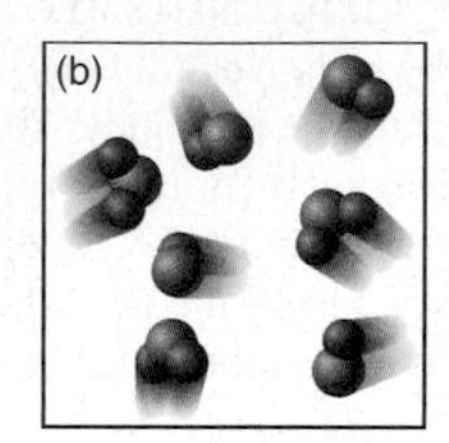

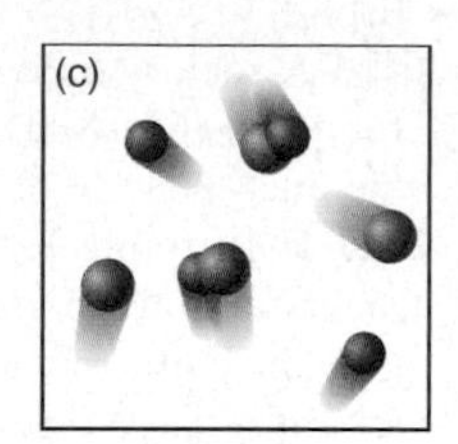

2.132 The number of atoms in 1 dm^3 of aluminum is nearly the same as the number of atoms in 1 dm^3 of lead, but the densities of these metals are very different (see Table 1.5). Explain.

2.133 You are working in the laboratory preparing sodium chloride. Consider the following results for three preparations of the compound:

Case 1: 39.34 g Na + 60.66 g Cl_2 ⟶ 100.00 g NaCl

Case 2: 39.34 g Na + 70.00 g Cl_2 ⟶ 100.00 g NaCl + 9.34 g Cl_2

Case 3: 50.00 g Na + 50.00 g Cl_2 ⟶ 82.43 g NaCl + 17.57 g Na

Explain these results in terms of the laws of conservation of mass and definite composition.

2.134 The seven most abundant ions in seawater make up more than 99% by mass of the dissolved compounds. They are listed in units of mg ion/kg seawater: chloride 18,980; sodium 10,560; sulfate 2650; magnesium 1270; calcium 400; potassium 380; hydrogen carbonate 140.
(a) What is the mass % of each ion in seawater?
(b) What percent of the total mass of ions is sodium ion?
(c) How does the total mass % of alkaline earth metal ions compare with the total mass % of alkali metal ions?
(d) Which makes up the larger mass fraction of dissolved components, anions or cations?

2.135 When barium (Ba) reacts with sulfur (S) to form barium sulfide (BaS), each Ba atom reacts with an S atom. If 2.00 cm^3 of Ba reacts with 1.50 cm^3 of S, are there enough Ba atoms to react with the S atoms (d of Ba = 3.51 g/cm^3; d of S = 2.07 g/cm^3)?

2.136 Succinic acid *(below)* is an important metabolite in biological energy production. Give the molecular formula, empirical formula, and molecular mass of succinic acid, and calculate the mass percent of each element.

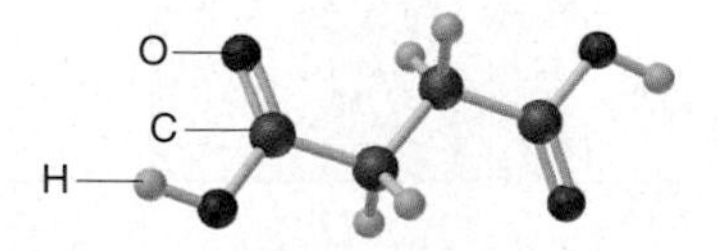

2.137 Fluoride ion is poisonous in relatively low amounts: 0.2 g of F^- per 70 kg of body weight can cause death. Nevertheless, in order to prevent tooth decay, F^- ions are added to drinking water at a concentration of 1 mg of F^- ion per L of water. How many liters of fluoridated drinking water would a 70-kg person have to consume in one day to reach this toxic level? How many kilograms of sodium fluoride would be needed to treat a 7.00×10^7-gal reservoir?

2.138 Antimony has many uses—for example, in semiconductor infrared devices and as an alloy in lead storage batteries. The element has two naturally occurring isotopes, one with mass 120.904 amu, the other with mass 122.904 amu. (a) Write the ^A_ZX notation for each isotope. (b) Use the atomic mass of antimony from the periodic table to calculate the natural abundance of each isotope.

2.139 Nitrous oxide (N_2O) is a potent greenhouse gas. It enters the atmosphere from fertilizer breakdown, car exhaust, and many other sources. Some studies have shown that the isotope ratios of ^{15}N to ^{14}N and of ^{18}O to ^{16}O in N_2O depend on the source. Thus, measuring the relative abundance of molecular masses in a sample of N_2O can help determine the source
(a) What different molecular masses are possible for N_2O?
(b) The percent abundance of ^{14}N is 99.6%, and that of ^{16}O is 99.8%. Which molecular mass of N_2O is least common, and which is most common?

2.140 The scene below represents a chemical reaction in a container. Which of the mass laws—mass conservation, definite composition, or multiple proportions—is (are) illustrated?

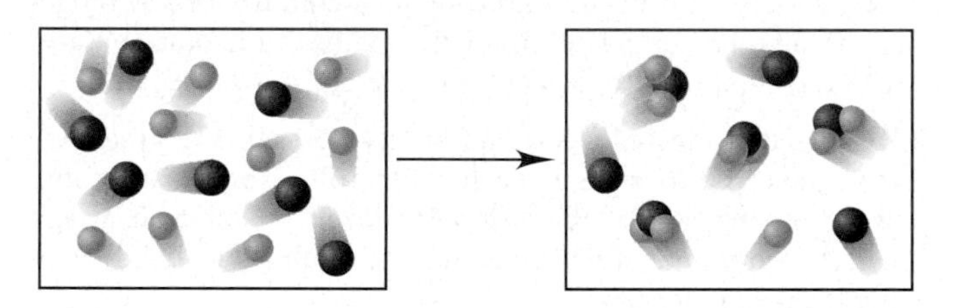

2.141 The two isotopes of potassium with significant abundance in nature are ^{39}K (isotopic mass 38.9637 amu, 93.258%) and ^{41}K (isotopic mass 40.9618 amu, 6.730%). Fluorine has only one naturally occurring isotope, ^{19}F (isotopic mass 18.9984 amu). Calculate the formula mass of potassium fluoride.

2.142 Boron trifluoride is used extensively in the synthesis of organic compounds. When it is analyzed by mass spectrometry (see Tools of the Laboratory, p. 52), several different 1+ ions form, including ions representing the whole molecule as well as molecular fragments formed by the loss of one, two, and three F atoms. Given that boron has two naturally occurring isotopes, ^{10}B and ^{11}B, and fluorine has one, ^{19}F, calculate the masses of all possible 1+ ions.

2.143 Nitrogen monoxide (NO) is a bioactive molecule in blood. Low NO concentrations cause respiratory distress and the formation of blood clots. Doctors prescribe nitroglycerin, $C_3H_5N_3O_9$, and isoamyl nitrate, $(CH_3)_2CHCH_2CH_2ONO_2$, to increase NO. If each compound releases one molecule of NO per atom of N, calculate the mass percent of NO in each medicine.

2.144 Nuclei differ in their stability, and some are so unstable that they undergo radioactive decay. The ratio of the number of neutrons to number of protons (N/Z) in a nucleus correlates with its stability. Calculate the N/Z ratio for (a) ^{144}Sm; (b) ^{56}Fe; (c) ^{20}Ne; (d) ^{107}Ag. (e) The radioactive isotope ^{238}U decays in a series of nuclear reactions that includes another uranium isotope, ^{234}U, and three lead isotopes, ^{214}Pb, ^{210}Pb, and ^{206}Pb. How many neutrons, protons, and electrons are in each of these five isotopes?

2.145 TNT (trinitrotoluene; *below*) is used as an explosive in construction. Calculate the mass of each element in 1.00 lb of TNT.

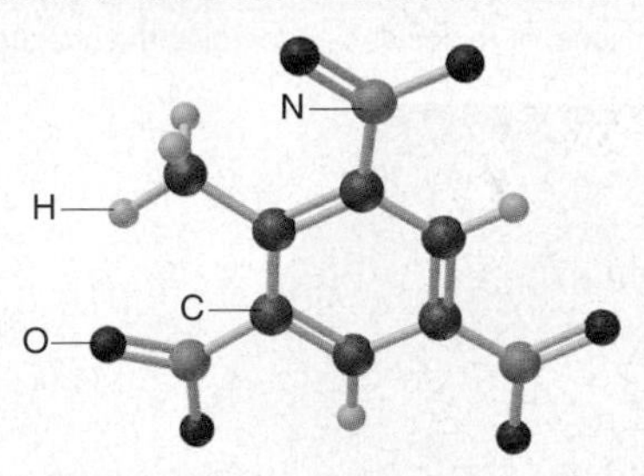

2.146 Carboxylic acids react with alcohols to form esters, which are found in all plants and animals. Some are responsible for the flavors and odors of fruits and flowers. What is the percent by mass of carbon in each of the following esters?

Name	Formula	Odor
Isoamyl isovalerate	$C_4H_9COOC_5H_{11}$	apple
Amyl butyrate	$C_3H_7COOC_5H_{11}$	apricot
Isoamyl acetate	$CH_3COOC_5H_{11}$	banana
Ethyl butyrate	$C_3H_7COOC_2H_5$	pineapple

2.147 The anticancer drug Platinol (Cisplatin), $Pt(NH_3)_2Cl_2$, reacts with the cancer cell's DNA and interferes with its growth. (a) What is the mass % of platinum (Pt) in Platinol? (b) If Pt costs \$19/g, how many grams of Platinol can be made for \$1.00 million (assume that the cost of Pt determines the cost of the drug)?

2.148 Grignard reagents, which have the general formula $CH_3—(CH_2)_x—MgBr$, are essential in the synthesis of organic compounds. They contain a C—Mg bond, with the Mg so positive that the reagent exists as an ionic compound. (a) Calculate the mass percent of Mg if $x = 0$. (b) Calculate the mass percent of Mg if $x = 5$. (c) Calculate the value of x if the mass percent of Mg is 16.5%.

2.149 In a sample of any metal, spherical atoms pack closely together, but the space between them means that the density of the sample is less than that of the atoms themselves. Iridium (Ir) is one of the densest elements: 22.56 g/cm^3. The atomic mass of Ir is 192.22 amu, and the mass of the nucleus is 192.18 amu. Determine the density (in g/cm^3) of (a) an Ir atom and (b) an Ir nucleus. (c) How many Ir atoms placed in a row would extend 1.00 cm (radius of Ir atom = 1.36 Å; radius of Ir nucleus = 1.5 femtometers (fm); V of a sphere $= \frac{4}{3}\pi r^3$)?

2.150 Dimercaprol ($HSCH_2CHSHCH_2OH$) is a complexing agent developed during World War I as an antidote to arsenic-based poison gas and used today to treat heavy-metal poisoning. Such an agent binds and removes the toxic element from the body.
(a) If each molecule binds one arsenic (As) atom, how many atoms of As could be removed by 250. mg of dimercaprol?
(b) If one molecule binds one metal atom, calculate the mass % of each of the following metals in a metal-dimercaprol complex: mercury, thallium, chromium.

2.151 Choose the box color(s) in the periodic table below that match(es) the following:

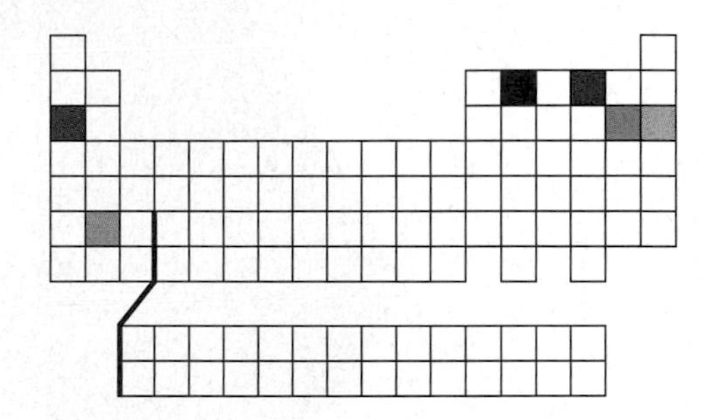

(a) Four elements that are nonmetals
(b) Two elements that are metals
(c) Three elements that are gases at room temperature
(d) Three elements that are solid at room temperature
(e) One pair of elements likely to form a covalent compound
(f) Another pair of elements likely to form a covalent compound
(g) One pair of elements likely to form an ionic compound with formula MX
(h) Another pair of elements likely to form an ionic compound with formula MX
(i) Two elements likely to form an ionic compound with formula M_2X
(j) Two elements likely to form an ionic compound with formula MX_2
(k) An element that forms no compounds
(l) A pair of elements whose compounds exhibit the law of multiple proportions
(m) Two elements that are building blocks in biomolecules
(n) Two elements that are macronutrients in organisms

2.152 Which of the following steps in an overall process involve(s) a physical change and which involve(s) a chemical change?

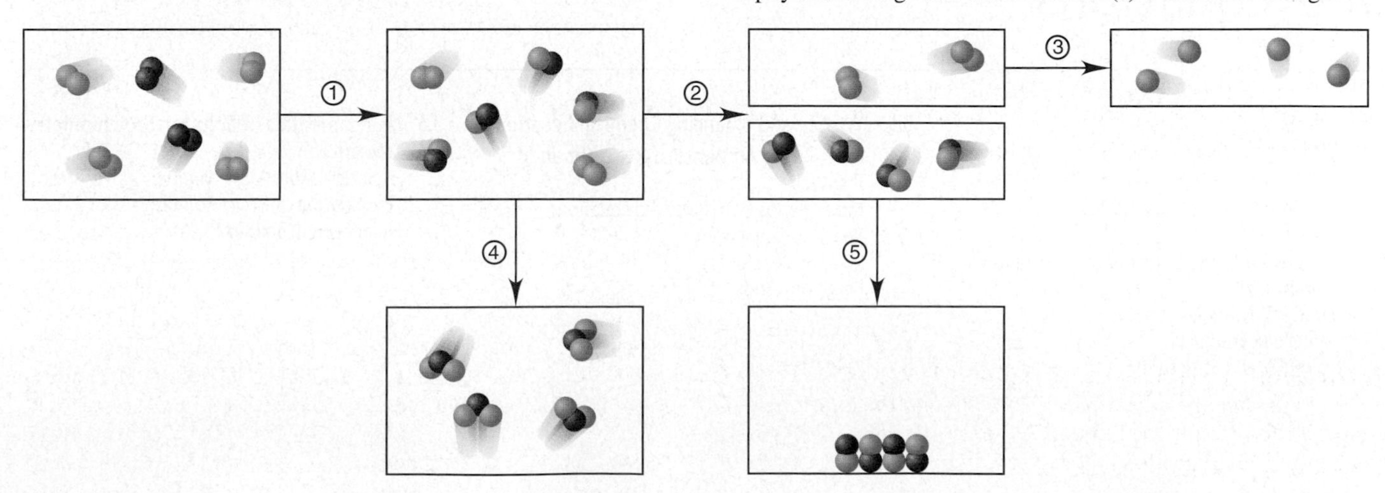

Mass and Number During any chemical reaction, such as this one between potassium and water to form hydrogen gas and a solution of potassium hydroxide, total mass is constant but individual masses change. As you'll see in this chapter, weighing provides a means for knowing not only the *mass* of each substance involved in a reaction, but also the *number* of atoms, ions, or molecules undergoing the change.

3

Stoichiometry of Formulas and Equations

Chemistry is a practical science. Just imagine how useful it could be to determine the formula of a compound from the masses of its elements or to predict the amounts of substances consumed and produced in a reaction. Suppose you are a polymer chemist preparing a new plastic: how much of this new material will a given polymerization reaction yield? Or suppose you're a chemical engineer studying rocket engine thrust: what amount of exhaust gases will a test of this fuel mixture produce? Perhaps you are on a team of environmental chemists examining coal samples: what quantity of air pollutants will this sample release when burned? Or, maybe you're a biomedical researcher who has extracted a new cancer-preventing substance from a tropical plant: what is its formula, and what quantity of metabolic products will establish a safe dosage level? You can answer such questions and countless others like them with a knowledge of **stoichiometry** (pronounced "stoy-key-AHM-uh-tree"; from the Greek *stoicheion,* "element or part," and *metron,* "measure"), the study of the quantitative aspects of chemical formulas and reactions.

IN THIS CHAPTER . . . We relate the mass of a substance to the number of chemical entities comprising it (atoms, ions, molecules, or formula units). We convert the results of mass analysis into a chemical formula and distinguish the types of chemical formulas and their relation to molecular structures. Reading, writing, and thinking in the language of chemical equations are applied to finding the amounts of reactants and products involved in a reaction. These methods are also applied to reactions that occur in solution.

Concepts & Skills to Review

before you study this chapter

- isotopes and atomic mass (Section 2.5)
- names and formulas of compounds (Section 2.8)
- molecular mass of a compound (Section 2.8)
- empirical and molecular formulas (Section 2.8)
- mass laws in chemical reactions (Section 2.2)

3.1 THE MOLE

All the concepts and skills discussed here depend on an understanding of the *mole* concept, so let's begin with this central idea. In daily life, we typically measure things out by counting or by weighing, with the choice based on convenience. It is more convenient to weigh beans or rice than to count individual pieces, and it is more convenient to count eggs or pencils than to weigh them. To measure out such things, we use mass units (a kilogram of rice) or counting units (a dozen pencils). Similarly, daily life in the laboratory involves measuring out substances to prepare a solution or "run" a reaction. However, an obvious problem arises when we try to do this. The atoms, ions, molecules, or formula units are the entities that react with one another, so we would like to know the numbers of them that we mix together. But, how can we possibly count entities that are so small? To do this, chemists have devised a unit called the mole to *count chemical entities by weighing them.*

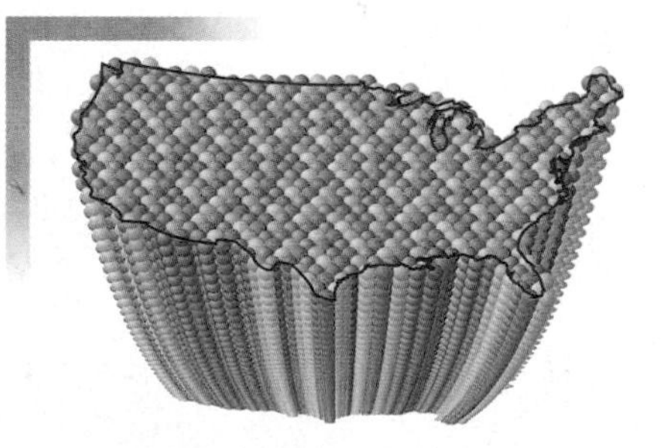

Imagine a Mole of . . . A mole of any ordinary object is a staggering amount: a mole of periods (.) lined up side by side would equal the radius of our galaxy; a mole of marbles stacked tightly together would cover the United States 70 miles deep. However, atoms and molecules are not ordinary objects: a mole of water molecules (about 18 mL) can be swallowed in one gulp!

Defining the Mole

The **mole** (abbreviated **mol**) is the SI unit for amount of substance. It is defined as *the amount of a substance that contains the same number of entities as there are atoms in exactly 12 g of carbon-12.* This number is called **Avogadro's number,** in honor of the 19th-century Italian physicist Amedeo Avogadro, and as you can tell from the definition, it is enormous:

$$\text{One mole (1 mol) contains } 6.022\times10^{23} \text{ entities (to four significant figures)} \quad (3.1)$$

Thus,

1 mol of carbon-12	contains	6.022×10^{23} carbon-12 atoms
1 mol of H_2O	contains	6.022×10^{23} H_2O molecules
1 mol of NaCl	contains	6.022×10^{23} NaCl formula units

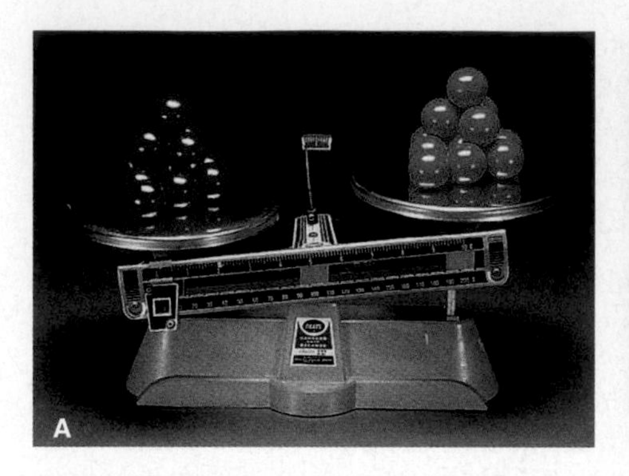

Figure 3.1 Counting objects of fixed relative mass. A, If marbles had a fixed mass, we could count them by weighing them. Each red marble weighs 7 g, and each yellow marble weighs 4 g, so 84 g of red marbles and 48 g of yellow marbles each consists of 12 marbles. Equal numbers of the two types of marbles always have a 7:4 mass ratio of red:yellow marbles. **B,** Because atoms of a substance have a fixed mass, we can weigh the substance to count the atoms; 55.85 g of Fe (left pan) and 32.07 g of S (right pan) each consists of 6.022×10^{23} atoms (1 mol of atoms). Any two samples of Fe and S that contain equal numbers of atoms have a 55.85:32.07 mass ratio of Fe:S.

However, the mole is not just a counting unit, like the dozen, which specifies only the *number* of objects. The definition of the mole specifies the number of objects in a fixed *mass* of substance. Therefore, *1 mole of a substance represents a fixed number of chemical entities* ***and*** *has a fixed mass.* To see why this is important, consider the marbles in Figure 3.1A, which we'll use as an analogy for atoms. Suppose you have large groups of red marbles and yellow marbles; each red marble weighs 7 g and each yellow marble weighs 4 g. Right away you know that there are 12 marbles in 84 g of red marbles or in 48 g of yellow marbles. Moreover, because one red marble weighs $\frac{7}{4}$ as much as one yellow marble, any given *number* of red and of yellow marbles always has this 7:4 *mass* ratio. By the same token, any given *mass* of red and of yellow marbles always has a 4:7 *number* ratio. For example, 280 g of red marbles contains 40 marbles, and 280 g of yellow marbles contains 70 marbles. As you can see, the fixed masses of the marbles allow you to count marbles by weighing them.

Atoms have fixed masses also, and the mole gives us a practical way to determine the number of atoms, molecules, or formula units in a sample by weighing it. Let's focus on elements first and recall a key point from Chapter 2: the atomic mass of an element (which appears on the periodic table) is the weighted average of the masses of its naturally occurring isotopes. For purposes of weighing, all atoms of an element are considered to have this atomic mass. That is, all iron (Fe) atoms have an atomic mass of 55.85 amu, all sulfur (S) atoms have an atomic mass of 32.07 amu, and so forth.

The central relationship between the mass of one atom and the mass of 1 mole of those atoms is that *the atomic mass of an element expressed in* ***amu*** *is numerically the same as the mass of 1 mole of atoms of the element expressed in* ***grams.*** You can see this from the definition of the mole, which referred to the number of atoms in "**12** g of carbon-**12.**" Thus,

1 Fe atom	has a mass of	55.85 amu	and	1 mol of Fe atoms	has a mass of	55.85 g
1 S atom	has a mass of	32.07 amu	and	1 mol of S atoms	has a mass of	32.07 g
1 O atom	has a mass of	16.00 amu	and	1 mol of O atoms	has a mass of	16.00 g
1 O_2 molecule	has a mass of	32.00 amu	and	1 mol of O_2 molecules	has a mass of	32.00 g

Moreover, because of their fixed atomic masses, we know that 55.85 g of Fe atoms and 32.07 g of S atoms each contains 6.022×10^{23} atoms. As with marbles of fixed mass, one Fe atom weighs $\frac{55.85}{32.07}$ as much as one S atom, and 1 mol of Fe atoms weighs $\frac{55.85}{32.07}$ as much as 1 mol of S atoms (Figure 3.1B).

A similar relationship holds for compounds: *the molecular mass (or formula mass) of a compound expressed in* ***amu*** *is numerically the same as the mass of 1 mole of the compound expressed in* ***grams.*** Thus, for example,

1 molecule of H_2O	has a mass of	18.02 amu	and	1 mol of H_2O (6.022×10^{23} molecules)	has a mass of	18.02 g
1 formula unit of NaCl	has a mass of	58.44 amu	and	1 mol of NaCl (6.022×10^{23} formula units)	has a mass of	58.44 g

To summarize the two key points about the usefulness of the mole concept:

- The mole maintains the *same mass relationship* between macroscopic samples as exists between individual chemical entities.
- The mole relates the *number* of chemical entities to the *mass* of a sample of those entities.

A grocer cannot obtain 1 dozen eggs by weighing them because eggs vary in mass. But a chemist *can* obtain 1 mol of copper atoms (6.022×10^{23} atoms) simply by weighing 63.55 g of copper. Figure 3.2 shows 1 mol of some familiar elements and compounds.

Figure 3.2 One mole of some familiar substances. One mole of a substance is the amount that contains 6.022×10^{23} atoms, molecules, or formula units. From left to right: 1 mol (172.19 g) of calcium sulfate dihydrate, 1 mol (32.00 g) of gaseous O_2, 1 mol (63.55 g) of copper, and 1 mol (18.02 g) of liquid H_2O.

Molar Mass

The **molar mass ($\mathcal{M}$)** of a substance is the mass per mole of its entities (atoms, molecules, or formula units). Thus, molar mass has units of grams per mole (g/mol). Table 3.1 summarizes the meanings of mass units used in this text.

Table 3.1 Summary of Mass Terminology*

Term	Definition	Unit
Isotopic mass	Mass of an isotope of an element	amu
Atomic mass (also called atomic weight)	Average of the masses of the naturally occurring isotopes of an element weighted according to their abundance	amu
Molecular (or formula) mass (also called molecular weight)	Sum of the atomic masses of the atoms (or ions) in a molecule (or formula unit)	amu
Molar mass ($\mathcal{M}$) (also called gram-molecular weight)	Mass of 1 mole of chemical entities (atoms, ions, molecules, formula units)	g/mol

*All terms based on the ^{12}C standard: 1 atomic mass unit = $\frac{1}{12}$ mass of one ^{12}C atom.

The periodic table is indispensable for calculating the molar mass of a substance. Here's how the calculations are done:

1. **Elements.** You find the molar mass of an element simply by looking up its atomic mass in the periodic table and then noting whether the element occurs naturally as individual atoms or as molecules.

- *Monatomic elements.* For elements that occur as collections of individual atoms, the molar mass is the numerical value from the periodic table expressed in units of grams per mole.* Thus, the molar mass of neon is 20.18 g/mol, the molar mass of iron is 55.85 g/mol, and the molar mass of gold is 197.0 g/mol.
- *Molecular elements.* For elements that occur naturally as molecules, you must know the molecular formula to determine the molar mass. For example, oxygen exists normally in air as diatomic molecules, so the molar mass of O_2 molecules is twice that of O atoms:

$$\text{Molar mass } (\mathcal{M}) \text{ of } O_2 = 2 \times \mathcal{M} \text{ of O} = 2 \times 16.00 \text{ g/mol} = 32.00 \text{ g/mol}$$

The most common form of sulfur consists of octatomic molecules, S_8:

$$\mathcal{M} \text{ of } S_8 = 8 \times \mathcal{M} \text{ of S} = 8 \times 32.07 \text{ g/mol} = 256.6 \text{ g/mol}$$

*The mass value in the periodic table is unitless because it is a *relative* atomic mass, given by the atomic mass (in amu) divided by 1 amu ($\frac{1}{12}$ mass of one ^{12}C atom in amu):

$$\text{Relative atomic mass} = \frac{\text{atomic mass } (\cancel{\text{amu}})}{\frac{1}{12} \text{ mass of } ^{12}\text{C } (\cancel{\text{amu}})}$$

Therefore, you use the same number for the atomic mass (weighted average mass of one atom in amu) and the molar mass (mass of 1 mole of atoms in grams).

Combining the steps for each of the other two elements in glucose:

$$\text{Mass \% of H} = \frac{\text{mol H} \times \mathcal{M} \text{ of H}}{\text{mass of 1 mol glucose}} \times 100 = \frac{12 \text{ mol H} \times \frac{1.008 \text{ g H}}{1 \text{ mol H}}}{180.16 \text{ g}} \times 100$$

$$= \boxed{6.714 \text{ mass \% H}}$$

$$\text{Mass \% of O} = \frac{\text{mol O} \times \mathcal{M} \text{ of O}}{\text{mass of 1 mol glucose}} \times 100 = \frac{6 \text{ mol O} \times \frac{16.00 \text{ g O}}{1 \text{ mol O}}}{180.16 \text{ g}} \times 100$$

$$= \boxed{53.29 \text{ mass \% O}}$$

Check The answers make sense: the mass % of O is greater than the mass % of C because, even though there are equal numbers of moles of each in the compound, the molar mass of O is greater than the molar mass of C. The mass % of H is small because the molar mass of H is small. The total of the mass percents is 100.00%.

(b) Determining the mass (g) of carbon

Plan To find the mass of C in the glucose sample, we multiply the mass of the sample by the mass fraction of C from part (a).

Solution Finding the mass of C in a given mass of glucose (with units for mass fraction):

$$\text{Mass (g) of C} = \text{mass of glucose} \times \text{mass fraction of C} = 16.55 \text{ g glucose} \times \frac{0.4000 \text{ g C}}{1 \text{ g glucose}}$$

$$= \boxed{6.620 \text{ g C}}$$

Check Rounding shows that the answer is "in the right ballpark": 16 g times less than 0.5 parts by mass should be less than 8 g.

Comment 1. A *more direct approach* to finding the mass of element in any mass of compound is similar to the approach we used in Sample Problem 2.1 (p. 43) and eliminates the need to calculate the mass fraction. Just multiply the given mass of compound by the ratio of the total mass of element to the mass of 1 mol of compound:

$$\text{Mass (g) of C} = 16.55 \text{ g glucose} \times \frac{72.06 \text{ g C}}{180.16 \text{ g glucose}} = 6.620 \text{ g C}$$

2. From here on, you should be able to determine the molar mass of a compound, so that calculation will no longer be shown.

FOLLOW-UP PROBLEM 3.3 Ammonium nitrate is a common fertilizer. Agronomists base the effectiveness of fertilizers on their nitrogen content.
(a) Calculate the mass percent of N in ammonium nitrate.
(b) How many grams of N are in 35.8 kg of ammonium nitrate?

SECTION SUMMARY

A mole of substance is the amount that contains Avogadro's number (6.022×10^{23}) of chemical entities (atoms, molecules, or formula units). The mass (in grams) of a mole has the same numerical value as the mass (in amu) of the entity. Thus, the mole allows us to count entities by weighing them. Using the molar mass ($\mathcal{M}$, g/mol) of an element (or compound) and Avogadro's number as conversion factors, we can convert among amount (mol), mass (g), and number of entities. The mass fraction of element X in a compound is used to find the mass of X in any amount of the compound.

3.2 DETERMINING THE FORMULA OF AN UNKNOWN COMPOUND

In Sample Problem 3.3, we knew the formula and used it to find the mass percent (or mass fraction) of an element in a compound *and* the mass of the element in a given mass of the compound. In this section, we do the reverse: use the masses of elements in a compound to find its formula. We'll present the mass data in several ways and then look briefly at molecular structures.

Empirical Formulas

An analytical chemist investigating a compound decomposes it into simpler substances, finds the mass of each component element, converts these masses to numbers of moles, and then arithmetically converts the moles to whole-number (integer) subscripts. This procedure yields the empirical formula, the *simplest whole-number ratio* of moles of each element in the compound (see Section 2.8, p. 62). Let's see how to obtain the subscripts from the moles of each element.

Analysis of an unknown compound shows that the sample contains 0.21 mol of zinc, 0.14 mol of phosphorus, and 0.56 mol of oxygen. Because the subscripts in a formula represent individual atoms or moles of atoms, we write a preliminary formula that contains fractional subscripts: $Zn_{0.21}P_{0.14}O_{0.56}$. Next, we convert these fractional subscripts to whole numbers using one or two simple arithmetic steps (rounding when needed):

1. Divide each subscript by the smallest subscript:

$$Zn_{\frac{0.21}{0.14}}P_{\frac{0.14}{0.14}}O_{\frac{0.56}{0.14}} \longrightarrow Zn_{1.5}P_{1.0}O_{4.0}$$

   This step alone often gives integer subscripts.

2. If any of the subscripts is still not an integer, multiply through by the *smallest integer* that will turn all subscripts into integers. Here, we multiply by 2, the smallest integer that will make 1.5 (the subscript for Zn) into an integer:

$$Zn_{(1.5\times2)}P_{(1.0\times2)}O_{(4.0\times2)} \longrightarrow Zn_{3.0}P_{2.0}O_{8.0}, \text{ or } Zn_3P_2O_8$$

   Notice that the *relative* number of moles has not changed because we multiplied *all* the subscripts by 2.

Always check that the subscripts are the smallest set of integers with the same ratio as the original numbers of moles; that is, 3:2:8 is *in the same ratio* as 0.21:0.14:0.56. A more conventional way to write this formula is $Zn_3(PO_4)_2$; the compound is zinc phosphate, a dental cement.

The following three sample problems (3.4, 3.5, and 3.6) demonstrate how other types of compositional data are used to determine chemical formulas. In the first problem, the empirical formula is found from data given as grams of each element rather than as moles.

A Rose by Any Other Name . . . Chemists studying natural substances obtained from animals and plants isolate compounds and determine their formulas. Geraniol ($C_{10}H_{18}O$), the main compound that gives a rose its odor, is used in many perfumes and cosmetics. Geraniol is also in citronella and lemongrass oils and is part of a larger compound in geranium leaves, from which its name is derived.

SAMPLE PROBLEM 3.4 Determining an Empirical Formula from Masses of Elements

Problem Elemental analysis of a sample of an ionic compound showed 2.82 g of Na, 4.35 g of Cl, and 7.83 g of O. What is the empirical formula and name of the compound?

Plan This problem is similar to the one we just discussed, except that we are given element *masses,* so we must convert the masses into integer subscripts. We first divide each mass by the element's molar mass to find *number of moles.* Then we construct a preliminary formula and convert the numbers of moles to integers.

Solution Finding moles of elements:

$$\text{Moles of Na} = 2.82\ \cancel{\text{g Na}} \times \frac{1 \text{ mol Na}}{22.99\ \cancel{\text{g Na}}} = 0.123 \text{ mol Na}$$

$$\text{Moles of Cl} = 4.35\ \cancel{\text{g Cl}} \times \frac{1 \text{ mol Cl}}{35.45\ \cancel{\text{g Cl}}} = 0.123 \text{ mol Cl}$$

$$\text{Moles of O} = 7.83\ \cancel{\text{g O}} \times \frac{1 \text{ mol O}}{16.00\ \cancel{\text{g O}}} = 0.489 \text{ mol O}$$

Constructing a preliminary formula: $Na_{0.123}Cl_{0.123}O_{0.489}$

Converting to integer subscripts (dividing all by the smallest subscript):

$$Na_{\frac{0.123}{0.123}}Cl_{\frac{0.123}{0.123}}O_{\frac{0.489}{0.123}} \longrightarrow Na_{1.00}Cl_{1.00}O_{3.98} \approx Na_1Cl_1O_4, \text{ or } NaClO_4$$

We rounded the subscript of O from 3.98 to 4. The empirical formula is $NaClO_4$; the name is sodium perchlorate.

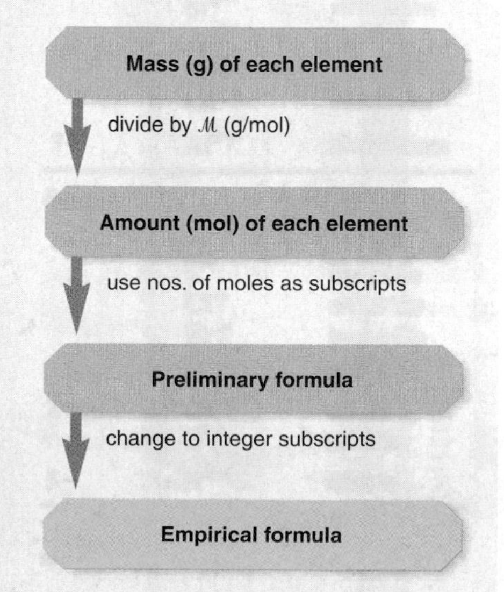

Check The moles seem correct because the masses of Na and Cl are slightly more than 0.1 of their molar masses. The mass of O is greatest and its molar mass is smallest, so it should have the greatest number of moles. The ratio of subscripts, 1:1:4, is the same as the ratio of moles, 0.123:0.123:0.489 (within rounding).

FOLLOW-UP PROBLEM 3.4 An unknown metal M reacts with sulfur to form a compound with the formula M_2S_3. If 3.12 g of M reacts with 2.88 g of S, what are the names of M and M_2S_3? (*Hint:* Determine the number of moles of S and use the formula to find the number of moles of M.)

Molecular Formulas

If we know the molar mass of a compound, we can use the empirical formula to obtain the molecular formula, the *actual* number of moles of each element in 1 mol of compound. In some cases, such as water (H_2O), ammonia (NH_3), and methane (CH_4), the empirical and molecular formulas are identical, but in many others the molecular formula is a *whole-number multiple* of the empirical formula. Hydrogen peroxide, for example, has the empirical formula HO and the molecular formula H_2O_2. Dividing the molar mass of H_2O_2 (34.02 g/mol) by the empirical formula mass (17.01 g/mol) gives the whole-number multiple:

$$\text{Whole-number multiple} = \frac{\text{molar mass (g/mol)}}{\text{empirical formula mass (g/mol)}}$$

$$= \frac{34.02\ \cancel{\text{g/mol}}}{17.01\ \cancel{\text{g/mol}}} = 2.000 = 2$$

Instead of giving compositional data in terms of masses of each element, analytical laboratories provide it as mass percents. From this, we determine the empirical formula by (1) assuming 100.0 g of compound, which allows us to express mass percent directly as mass, (2) converting the mass to number of moles, and (3) constructing the empirical formula. With the molar mass, we can also find the whole-number multiple and then the molecular formula.

SAMPLE PROBLEM 3.5 Determining a Molecular Formula from Elemental Analysis and Molar Mass

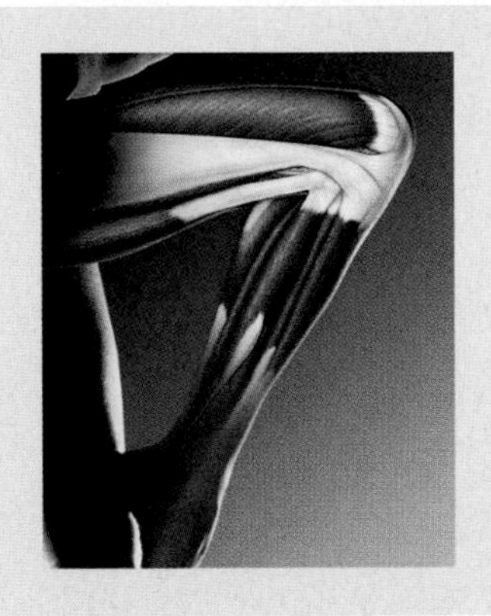

Problem During physical activity, lactic acid ($\mathcal{M}$ = 90.08 g/mol) forms in muscle tissue and is responsible for muscle soreness. Elemental analysis shows that this compound contains 40.0 mass % C, 6.71 mass % H, and 53.3 mass % O.
(a) Determine the empirical formula of lactic acid.
(b) Determine the molecular formula.

(a) Determining the empirical formula
Plan We know the mass % of each element and must convert each to an integer subscript. The mass of lactic acid is not given but, since mass % is the same for any mass of compound, we assume 100.0 g of lactic acid and express each mass % directly as grams. Then, we convert grams to moles and construct the empirical formula as we did in Sample Problem 3.4.
Solution Expressing mass % as grams, assuming 100.0 g of lactic acid:

$$\text{Mass (g) of C} = \frac{\text{40.0 parts C by mass}}{\text{100 parts by mass}} \times 100.0\ \text{g} = 40.0\ \text{g C}$$

Similarly, we have 6.71 g of H and 53.3 g of O.
Converting from grams of each element to moles:

$$\text{Moles of C} = \text{mass of C} \times \frac{1}{\mathcal{M}\ \text{of C}} = 40.0\ \cancel{\text{g C}} \times \frac{1\ \text{mol C}}{12.01\ \cancel{\text{g C}}} = 3.33\ \text{mol C}$$

Similarly, we have 6.66 mol of H and 3.33 mol of O.

Constructing the preliminary formula: $C_{3.33}H_{6.66}O_{3.33}$
Converting to integer subscripts:

$$C_{\frac{3.33}{3.33}}H_{\frac{6.66}{3.33}}O_{\frac{3.33}{3.33}} \longrightarrow C_{1.00}H_{2.00}O_{1.00} \approx C_1H_2O_1\text{; the empirical formula is } \boxed{CH_2O}$$

Check The numbers of moles seem correct: the masses of C and O are each slightly more than 3 times their molar masses (e.g., for C, 40 g/(12 g/mol) > 3 mol), and the mass of H is over 6 times its molar mass.

(b) Determining the molecular formula

Plan The molecular formula subscripts are whole-number multiples of the empirical formula subscripts. To find this whole number, we divide the given molar mass (90.08 g/mol) by the empirical formula mass, which we find from the sum of the elements' molar masses. Then we multiply the whole number by each subscript in the empirical formula.

Solution The empirical-formula molar mass is 30.03 g/mol. Finding the whole-number multiple:

$$\text{Whole-number multiple} = \frac{\mathcal{M}\text{ of lactic acid}}{\mathcal{M}\text{ of empirical formula}} = \frac{90.08\ \cancel{\text{g/mol}}}{30.03\ \cancel{\text{g/mol}}} = 3.000 = 3$$

Determining the molecular formula:

$$C_{(1\times3)}H_{(2\times3)}O_{(1\times3)} = \boxed{C_3H_6O_3}$$

Check The calculated molecular formula has the same ratio of moles of elements (3:6:3) as the empirical formula (1:2:1) and corresponds to the given molar mass:

$$\mathcal{M}\text{ of lactic acid} = (3 \times \mathcal{M}\text{ of C}) + (6 \times \mathcal{M}\text{ of H}) + (3 \times \mathcal{M}\text{ of O})$$
$$= (3 \times 12.01) + (6 \times 1.008) + (3 \times 16.00) = 90.08\text{ g/mol}$$

FOLLOW-UP PROBLEM 3.5 One of the most widespread environmental carcinogens (cancer-causing agents) is benzo[*a*]pyrene ($\mathcal{M}$ = 252.30 g/mol). It is found in coal dust, in cigarette smoke, and even in charcoal-grilled meat. Analysis of this hydrocarbon shows 95.21 mass % C and 4.79 mass % H. What is the molecular formula of benzo[*a*]pyrene?

Combustion Analysis of Organic Compounds Still another type of compositional data is obtained through **combustion analysis,** a method used to measure the amounts of carbon and hydrogen in a combustible organic compound. The unknown is burned in pure O_2 in an apparatus that consists of a combustion chamber and chambers containing compounds that absorb either H_2O or CO_2 (Figure 3.5). All the H in the compound is converted to H_2O, which is absorbed in the first chamber, and all the C is converted to CO_2, which is absorbed in the second. By weighing the contents of the chambers before and after combustion, we find the masses of CO_2 and H_2O and use them to calculate the masses of C and H in the compound, from which we find the empirical formula. As you've seen, many organic compounds also contain oxygen, nitrogen, or halogen. As long as the third element doesn't interfere with the absorption of CO_2 and H_2O, we calculate its mass by subtracting the masses of C and H from the original mass of the compound.

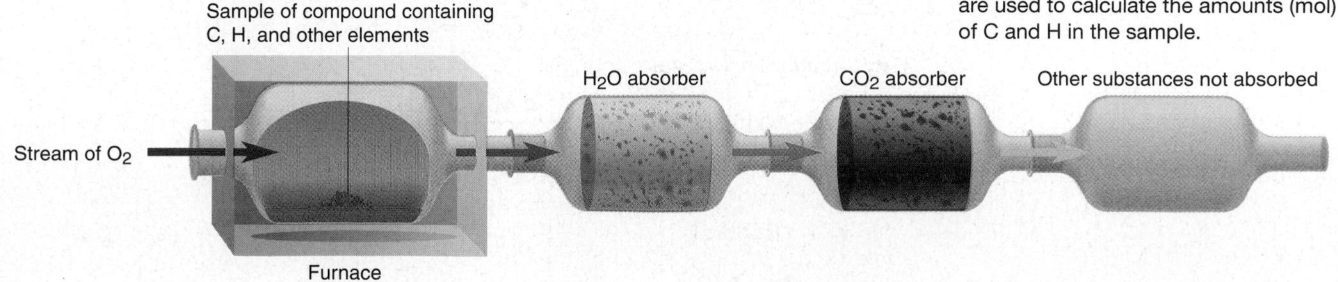

Figure 3.5 Combustion apparatus for determining formulas of organic compounds. A sample of compound that contains C and H (and perhaps other elements) is burned in a stream of O_2 gas. The CO_2 and H_2O formed are absorbed separately, while any other element oxides are carried through by the O_2 gas stream. H_2O is absorbed by $Mg(ClO_4)_2$; CO_2 is absorbed by NaOH on asbestos. The increases in mass of the absorbers are used to calculate the amounts (mol) of C and H in the sample.

SAMPLE PROBLEM 3.6 Determining a Molecular Formula from Combustion Analysis

Problem Vitamin C ($\mathcal{M}$ = 176.12 g/mol) is a compound of C, H, and O found in many natural sources, especially citrus fruits. When a 1.000-g sample of vitamin C is placed in a combustion chamber and burned, the following data are obtained:

$$\begin{aligned}\text{Mass of } CO_2 \text{ absorber after combustion} &= 85.35 \text{ g}\\ \text{Mass of } CO_2 \text{ absorber before combustion} &= 83.85 \text{ g}\\ \text{Mass of } H_2O \text{ absorber after combustion} &= 37.96 \text{ g}\\ \text{Mass of } H_2O \text{ absorber before combustion} &= 37.55 \text{ g}\end{aligned}$$

What is the molecular formula of vitamin C?

Plan We find the masses of CO_2 and H_2O by subtracting the masses of the absorbers before the reaction from the masses after. From the mass of CO_2, we use the mass fraction of C in CO_2 to find the mass of C (see Comment in Sample Problem 3.3). Similarly, we find the mass of H from the mass of H_2O. The mass of vitamin C (1.000 g) minus the sum of the C and H masses gives the mass of O, the third element present. Then, we proceed as in Sample Problem 3.5: calculate numbers of moles using the elements' molar masses, construct the empirical formula, determine the whole-number multiple from the given molar mass, and construct the molecular formula.

Solution Finding the masses of combustion products:

$$\begin{aligned}\text{Mass (g) of } CO_2 &= \text{mass of } CO_2 \text{ absorber after} - \text{mass before}\\ &= 85.35 \text{ g} - 83.85 \text{ g} = 1.50 \text{ g } CO_2\\ \text{Mass (g) of } H_2O &= \text{mass of } H_2O \text{ absorber after} - \text{mass before}\\ &= 37.96 \text{ g} - 37.55 \text{ g} = 0.41 \text{ g } H_2O\end{aligned}$$

Calculating masses of C and H using their mass fractions:

$$\text{Mass of element} = \text{mass of compound} \times \frac{\text{mass of element in compound}}{\text{mass of 1 mol of compound}}$$

$$\text{Mass (g) of C} = \text{mass of } CO_2 \times \frac{1 \text{ mol C} \times \mathcal{M} \text{ of C}}{\text{mass of 1 mol } CO_2} = 1.50 \cancel{\text{g } CO_2} \times \frac{12.01 \text{ g C}}{44.01 \cancel{\text{g } CO_2}}$$

$$= 0.409 \text{ g C}$$

$$\text{Mass (g) of H} = \text{mass of } H_2O \times \frac{2 \text{ mol H} \times \mathcal{M} \text{ of H}}{\text{mass of 1 mol } H_2O} = 0.41 \cancel{\text{g } H_2O} \times \frac{2.016 \text{ g H}}{18.02 \cancel{\text{g } H_2O}}$$

$$= 0.046 \text{ g H}$$

Calculating the mass of O:

$$\begin{aligned}\text{Mass (g) of O} &= \text{mass of vitamin C sample} - (\text{mass of C} + \text{mass of H})\\ &= 1.000 \text{ g} - (0.409 \text{ g} + 0.046 \text{ g}) = 0.545 \text{ g O}\end{aligned}$$

Finding the amounts (mol) of elements: Dividing the mass in grams of each element by its molar mass gives 0.0341 mol of C, 0.046 mol of H, and 0.0341 mol of O.

Constructing the preliminary formula: $C_{0.0341}H_{0.046}O_{0.0341}$

Determining the empirical formula: Dividing through by the smallest subscript gives

$$C_{\frac{0.0341}{0.0341}}H_{\frac{0.046}{0.0341}}O_{\frac{0.0341}{0.0341}} = C_{1.00}H_{1.3}O_{1.00}$$

By trial and error, we find that 3 is the smallest integer that will make all subscripts approximately into integers:

$$C_{(1.00\times3)}H_{(1.3\times3)}O_{(1.00\times3)} = C_{3.00}H_{3.9}O_{3.00} \approx C_3H_4O_3$$

Determining the molecular formula:

$$\text{Whole-number multiple} = \frac{\mathcal{M} \text{ of vitamin C}}{\mathcal{M} \text{ of empirical formula}} = \frac{176.12 \cancel{\text{g/mol}}}{88.06 \cancel{\text{g/mol}}} = 2.000 = 2$$

$$C_{(3\times2)}H_{(4\times2)}O_{(3\times2)} = \boxed{C_6H_8O_6}$$

Check The element masses seem correct: carbon makes up slightly more than 0.25 of the mass of CO_2 (12 g/44 g > 0.25), as do the masses in the problem (0.409 g/1.50 g > 0.25). Hydrogen makes up slightly more than 0.10 of the mass of H_2O (2 g/18 g > 0.10), as do the masses in the problem (0.046 g/0.41 g > 0.10). The molecular formula has the same ratio of subscripts (6:8:6) as the empirical formula (3:4:3) and adds up to the given molar mass:

$$(6 \times \mathcal{M} \text{ of C}) + (8 \times \mathcal{M} \text{ of H}) + (6 \times \mathcal{M} \text{ of O}) = \mathcal{M} \text{ of vitamin C}$$

$$(6 \times 12.01) + (8 \times 1.008) + (6 \times 16.00) = 176.12 \text{ g/mol}$$

Comment In determining the subscript for H, if we string the calculation steps together, we obtain the subscript 4.0, rather than 3.9, and don't need to round:

$$\text{Subscript of H} = 0.41 \text{ g } H_2O \times \frac{2.016 \text{ g H}}{18.02 \text{ g } H_2O} \times \frac{1 \text{ mol H}}{1.008 \text{ g H}} \times \frac{1}{0.0341 \text{ mol}} \times 3 = 4.0$$

FOLLOW-UP PROBLEM 3.6 A dry-cleaning solvent ($\mathcal{M}$ = 146.99 g/mol) that contains C, H, and Cl is suspected to be a cancer-causing agent. When a 0.250-g sample was studied by combustion analysis, 0.451 g of CO_2 and 0.0617 g of H_2O formed. Find the molecular formula.

Chemical Formulas and Molecular Structures

Let's take a short break from calculations to recall that *a formula represents a real three-dimensional object.* How much structural information is contained in the different types of chemical formulas?

1. *Different compounds with the same empirical formula.* The empirical formula tells the *relative* number of each type of atom, but it tells nothing about molecular structure. In fact, *different compounds can have the* ***same*** *empirical formula.* The oxides NO_2 and N_2O_4 are examples among inorganic compounds, but this phenomenon is especially common among organic compounds. While there is no stable hydrocarbon with the formula CH_2, compounds with the general formula C_nH_{2n} are well known, such as ethylene (C_2H_4) and propylene (C_3H_6), starting material for two very common plastics. Table 3.3 shows a few biologically important compounds with a given empirical formula.

Table 3.3 Some Compounds with Empirical Formula CH_2O (Composition by Mass: 40.0% C, 6.71% H, 53.3% O)

Name	Molecular Formula	Whole-Number Multiple	$\mathcal{M}$ (g/mol)	Use or Function
Formaldehyde	CH_2O	1	30.03	Disinfectant; biological preservative
Acetic acid	$C_2H_4O_2$	2	60.05	Acetate polymers; vinegar (5% solution)
Lactic acid	$C_3H_6O_3$	3	90.08	Causes milk to sour; forms in muscle during exercise
Erythrose	$C_4H_8O_4$	4	120.10	Forms during sugar metabolism
Ribose	$C_5H_{10}O_5$	5	150.13	Component of many nucleic acids and vitamin B_2
Glucose	$C_6H_{12}O_6$	6	180.16	Major nutrient for energy in cells

CH_2O $C_2H_4O_2$ $C_3H_6O_3$ $C_4H_8O_4$ $C_5H_{10}O_5$ $C_6H_{12}O_6$

2. *Isomers: different compounds with the same molecular formula.* A molecular formula tells the *actual* number of each type of atom, providing as much information as possible from mass analysis. Yet *different compounds can have the **same** molecular formula* because the atoms can bond to each other in different arrangements to give more than one *structural formula.* **Isomers** are compounds with the same molecular formula but different properties. The simplest type of isomerism, called *constitutional,* or *structural, isomerism,* occurs when the atoms link together in different arrangements. Table 3.4 shows two pairs of examples.

Table 3.4 Two Pairs of Constitutional Isomers

	C_4H_{10}		C_2H_6O	
Property	**Butane**	**2-Methylpropane**	**Ethanol**	**Dimethyl Ether**
$\mathcal{M}$ (g/mol)	58.12	58.12	46.07	46.07
Boiling point	−0.5°C	−11.6°C	78.5°C	−25°C
Density (at 20°C)	0.579 g/mL (gas)	0.549 g/mL (gas)	0.789 g/mL (liquid)	0.00195 g/mL (gas)
Structural formula	H H H H H—C—C—C—C—H H H H H	H H H H—C—C—C—H H H H—C—H H	H H H—C—C—O—H H H	H H H—C—O—C—H H H
Space-filling model				

The left pair is two compounds with the molecular formula C_4H_{10}, butane and 2-methylpropane. One has a four-C chain, and the other has a one-C branch attached to the second C of a three-C chain. Both are small alkanes, so their properties are similar, if not identical. The right pair of constitutional isomers share the molecular formula C_2H_6O but have very different properties because they are different types of compounds—one is an alcohol, and the other is an ether.

As the number and kinds of atoms increase, the number of isomers—that is, the number of structural formulas that can be written for a given molecular formula—also increases: C_2H_6O has two structural formulas, as you've seen, C_3H_8O three, and $C_4H_{10}O$, seven. Imagine how many there are for $C_{16}H_{19}N_3O_4S$! Of all the possible isomers with this formula, only one is the antibiotic ampicillin (Figure 3.6). Only by knowing a molecule's structure—the relative placement of atoms and the distances and angles separating them—can we begin to predict its behavior. (We'll discuss this and other types of isomerism fully later in the text.)

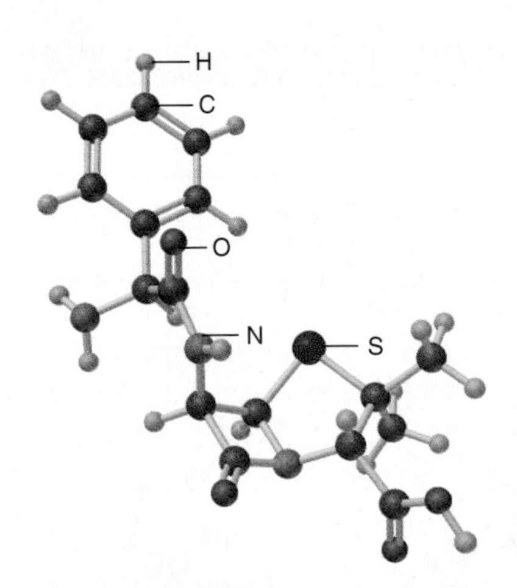

Figure 3.6 Ampicillin. Of the many possible constitutional isomers with the formula $C_{16}H_{19}N_3O_4S$, only this particular arrangement of the atoms is the widely used antibiotic ampicillin.

SECTION SUMMARY

From the masses of elements in an unknown compound, the relative amounts (in moles) can be found and the empirical formula determined. If the molar mass is known, the molecular formula can also be determined. Methods such as combustion analysis provide data on the masses of elements in a compound, which can be used to obtain the formula. Because atoms can bond in different arrangements, more than one compound may have the same molecular formula (constitutional isomers).

3.3 WRITING AND BALANCING CHEMICAL EQUATIONS

Perhaps the most important reason for thinking in terms of moles is because it greatly clarifies the amounts of substances taking part in a reaction. Comparing masses doesn't tell the ratio of substances reacting but comparing numbers of moles does. It allows us to view substances as large populations of interacting particles rather than as grams of material. To clarify this idea, consider the formation of hydrogen fluoride gas from H_2 and F_2, a reaction that occurs explosively at room temperature. If we weigh the gases, we find that

2.016 g of H_2 and 38.00 g of F_2 react to form 40.02 g of HF

This information tells us little except that mass is conserved. However, if we convert these masses (in grams) to amounts (in moles), we find that

1 mol of H_2 and 1 mol of F_2 react to form 2 mol of HF

This information reveals that equal-size populations of H_2 and F_2 molecules combine to form twice as large a population of HF molecules. Dividing through by Avogadro's number shows us the chemical event that occurs between individual molecules:

1 H_2 molecule and 1 F_2 molecule react to form 2 HF molecules

Figure 3.7 shows that when we express the reaction in terms of moles, *the macroscopic (molar) change corresponds to the submicroscopic (molecular) change.* As you'll see, a balanced chemical equation shows both changes.

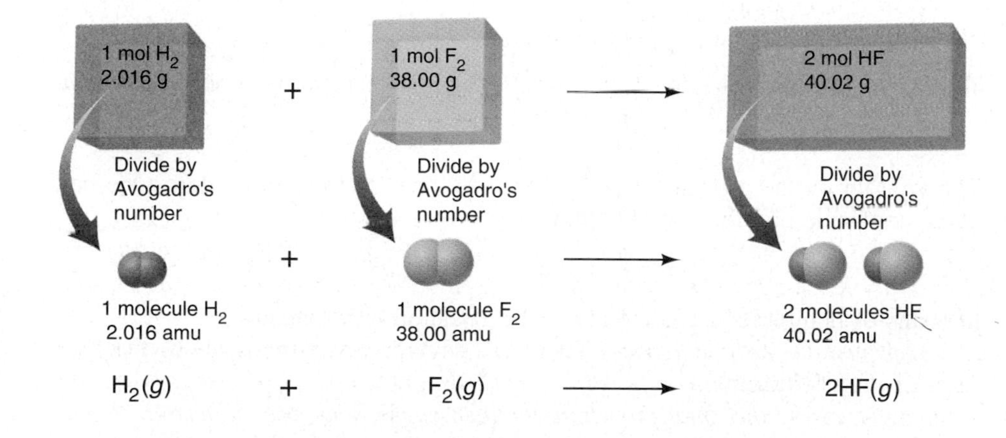

Figure 3.7 The formation of HF gas on the macroscopic and molecular levels. When 1 mol of H_2 (2.016 g) and 1 mol of F_2 (38.00 g) react, 2 mol of HF (40.02 g) forms. Dividing by Avogadro's number shows the change at the molecular level.

A **chemical equation** is a statement in formulas that expresses the identities and quantities of the substances involved in a chemical or physical change. Equations are the "sentences" of chemistry, just as chemical formulas are the "words" and atomic symbols the "letters." The left side of an equation shows the amount of each substance present before the change, and the right side shows the amounts present afterward. *For an equation to depict these amounts accurately, it must be balanced; that is, the same number of each type of atom must appear on both sides of the equation.* This requirement follows directly from the mass laws and the atomic theory:

- In a chemical process, atoms cannot be created, destroyed, or changed, only rearranged into different combinations.
- A formula represents a fixed ratio of the elements in a compound, so a different ratio represents a different compound.

Consider the chemical change that occurs in an old-fashioned photographic flashbulb: magnesium wire and oxygen gas yield powdery magnesium oxide. (Light and heat are produced as well, but here we're concerned only with the substances involved.) Let's convert this chemical statement into a balanced equation through the following steps:

1. *Translating the statement.* We first translate the chemical statement into a "skeleton" equation: chemical formulas arranged in an equation format. All the substances that react during the change, called **reactants,** are placed to the left of a "yield" arrow, which points to all the substances produced, called **products:**

$$\underbrace{\_\_Mg \quad + \quad \_\_O_2}_{\text{reactants}} \xrightarrow{\text{yield}} \underbrace{\_\_MgO}_{\text{product}}$$

magnesium and oxygen yield magnesium oxide

At the beginning of the balancing process, we put a blank in front of each substance to remind us that we have to account for its atoms.

2. *Balancing the atoms.* The next step involves shifting our attention back and forth from right to left in order to *match the number of each type of atom on each side.* At the end of this step, each blank will contain a **balancing (stoichiometric) coefficient,** a numerical multiplier of *all the atoms* in the formula that follows it. In general, balancing is easiest when we

- Start with the most complex substance, the one with the largest number of atoms or different types of atoms.
- End with the least complex substance, such as an element by itself.

In this case, MgO is the most complex, so we place a coefficient 1 *in front of* the compound:

$$\_\_Mg + \_\_O_2 \longrightarrow \underline{1}\,MgO$$

To balance the Mg in MgO on the right, we place a 1 in front of Mg on the left:

$$\underline{1}\,Mg + \_\_O_2 \longrightarrow \underline{1}\,MgO$$

The O atom on the right must be balanced by one O atom on the left. One-half an O_2 molecule provides one O atom:

$$\underline{1}\,Mg + \underline{\tfrac{1}{2}}\,O_2 \longrightarrow \underline{1}\,MgO$$

In terms of number and type of atom, the equation is balanced.

3. *Adjusting the coefficients.* There are several conventions about the final form of the coefficients:

- In most cases, *the smallest whole-number coefficients are preferred.* Whole numbers allow entities such as O_2 molecules to be treated as intact particles. One-half of an O_2 molecule cannot exist, so we multiply the equation by 2:

$$2Mg + 1O_2 \longrightarrow 2MgO$$

- We used the coefficient 1 to remind us to balance each substance. In the final form, a coefficient of 1 is implied just by the presence of the formula of the substance, so we don't need to write it:

$$2Mg + O_2 \longrightarrow 2MgO$$

(This convention is similar to that of not writing a subscript 1 in a formula.)

4. *Checking.* After balancing and adjusting the coefficients, always check that the equation is balanced:

$$\text{Reactants (2 Mg, 2 O)} \longrightarrow \text{products (2 Mg, 2 O)}$$

5. *Specifying the states of matter.* The final equation also indicates the physical state of each substance or whether it is dissolved in water. The abbreviations that are used for these states are solid (*s*), liquid (*l*), gas (*g*), and aqueous solution (*aq*). From the original statement, we know that the Mg "wire" is solid, the O_2 is a gas, and the "powdery" MgO is also solid. The balanced equation, therefore, is

$$2Mg(s) + O_2(g) \longrightarrow 2MgO(s)$$

Of course, the key point to realize is, as was pointed out in Figure 3.7, *the balancing coefficients refer to both individual chemical entities and moles of chemical entities.* Thus, 2 mol of Mg and 1 mol of O_2 yield 2 mol of MgO. Figure 3.8 shows this reaction from three points of view—as you see it on the macroscopic level, as chemists (and you!) can imagine it on the atomic level (darker colored atoms represent the stoichiometry), and on the symbolic level of the chemical equation.

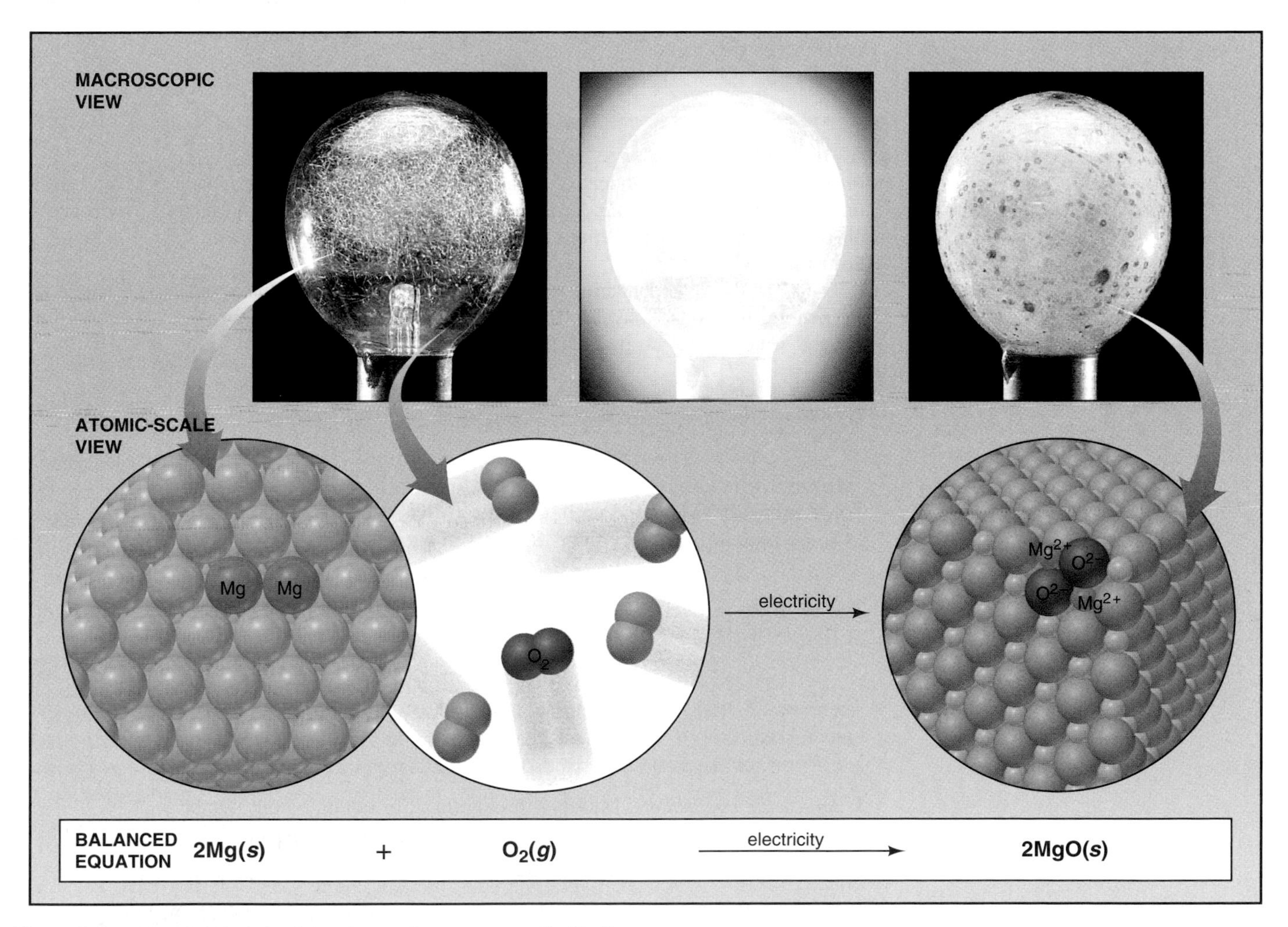

Figure 3.8 A three-level view of the chemical reaction in a flashbulb. The photos present the macroscopic view that you see. Before the reaction occurs, a fine magnesium filament is surrounded by oxygen (*left*). After the reaction, white, powdery magnesium oxide coats the bulb's inner surface (*right*). The blow-up arrows lead to an atomic-scale view, a representation of the chemist's mental picture of the reaction. The darker colored spheres show the stoichiometry. By knowing the substances before and after a reaction, we can write a balanced equation (*bottom*), the chemist's symbolic shorthand for the change.

SAMPLE PROBLEM 3.10 Using Molecular Depictions to Solve a Limiting-Reactant Problem

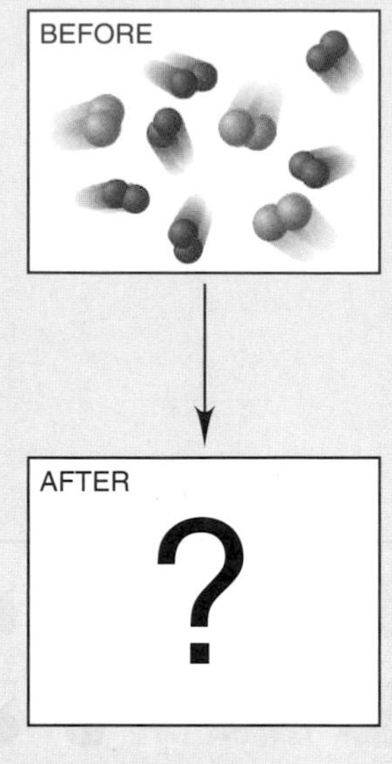

Problem Nuclear engineers use chlorine trifluoride in the processing of uranium fuel for power plants. This extremely reactive substance is formed as a gas in special metal containers by the reaction of elemental chlorine and fluorine.
(a) Suppose the box shown at left represents a container of the reactant mixture before the reaction occurs (with chlorine colored green). Name the limiting reactant and draw the container contents after the reaction is complete.
(b) When the reaction is run again with 0.750 mol of Cl_2 and 3.00 mol of F_2, what mass of chlorine trifluoride will be prepared?

(a) Determining the limiting reactant and drawing the container contents
Plan We first write the balanced equation. From its name, we know that chlorine trifluoride consists of one Cl atom bonded to three F atoms, ClF_3. Elemental chlorine and fluorine refer to the diatomic molecules Cl_2 and F_2. All the substances are gases. To find the limiting reactant, we compare the number of molecules we have of each reactant, with the number we need for the other to react completely. The limiting reactant limits the amount of the other reactant that can react and the amount of product that will form.
Solution The balanced equation is

$$Cl_2(g) + 3F_2(g) \longrightarrow 2ClF_3(g)$$

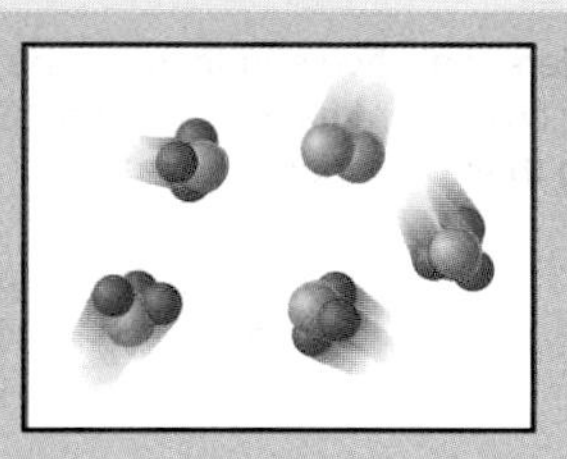

The equation shows that two ClF_3 molecules are formed for every one Cl_2 molecule and three F_2 molecules that react. Before the reaction, there are three Cl_2 molecules (six Cl atoms). For all the Cl_2 to react, we need three times three, or nine, F_2 molecules (18 F atoms). But there are only six F_2 molecules (12 F atoms). Therefore, F_2 is the limiting reactant because it limits the amount of Cl_2 that can react, and thus the amount of ClF_3 that can form. After the reaction, as the box at left depicts, all 12 F atoms and four of the six Cl atoms make four ClF_3 molecules, and one Cl_2 molecule remains in excess.
Check The equation is balanced: reactants (2 Cl, 6 F) $\longrightarrow$ products (2 Cl, 6 F), and, in the boxes, the number of each type of atom before the reaction equals the number after the reaction. You can check the choice of limiting reactant by examining the reaction from the perspective of Cl_2: Two Cl_2 molecules are enough to react with the six F_2 molecules in the container. But there are three Cl_2 molecules, so there is not enough F_2.

(b) Calculating the mass of ClF_3 formed
Plan We first determine the limiting reactant by using the molar ratios from the balanced equation to convert the moles of each reactant to moles of ClF_3 formed, assuming an excess of the other reactant. Whichever reactant forms fewer moles of ClF_3 is limiting. Then we use the molar mass of ClF_3 to convert this lower number of moles to grams.
Solution Determining the limiting reactant:
Finding moles of ClF_3 from moles of Cl_2 (assuming F_2 is in excess):

$$\text{Moles of } ClF_3 = 0.750 \cancel{\text{mol } Cl_2} \times \frac{2 \text{ mol } ClF_3}{1 \cancel{\text{mol } Cl_2}} = 1.50 \text{ mol } ClF_3$$

Finding moles of ClF_3 from moles of F_2 (assuming Cl_2 is in excess):

$$\text{Moles of } ClF_3 = 3.00 \cancel{\text{mol } F_2} \times \frac{2 \text{ mol } ClF_3}{3 \cancel{\text{mol } F_2}} = 2.00 \text{ mol } ClF_3$$

In this experiment, Cl_2 is limiting because it forms fewer moles of ClF_3.
Calculating grams of ClF_3 formed:

$$\text{Mass (g) of } ClF_3 = 1.50 \cancel{\text{mol } ClF_3} \times \frac{92.45 \text{ g } ClF_3}{1 \cancel{\text{mol } ClF_3}} = 139 \text{ g } ClF_3$$

Check Let's check our reasoning that Cl_2 is the limiting reactant by assuming, for the moment, that F_2 is limiting. In that case, all 3.00 mol of F_2 would react to form 2.00 mol of ClF_3. Based on the balanced equation, however, that amount would require that 1.00 mol of Cl_2 reacted. But that result is impossible because only 0.750 mol of Cl_2 is present.
Comment Note that a reactant can be limiting even though it is present in the greater amount. It is the *reactant molar ratio in the balanced equation* that is the determining factor. In both parts (a) and (b), F_2 is present in greater amount than Cl_2. However, in (a), the F_2/Cl_2 ratio is 6/3, or 2/1, which is less than the required molar ratio of 3/1, so F_2 is limiting; in (b), the F_2/Cl_2 ratio is 3.00/0.750, greater than the required 3/1, so F_2 is in excess.

FOLLOW-UP PROBLEM 3.10 B_2 (red spheres) reacts with AB as shown below:

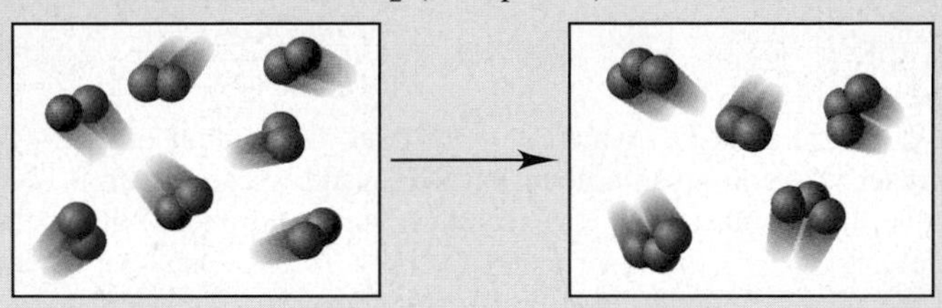

(a) Write a balanced equation for the reaction, and determine the limiting reactant.
(b) How many moles of product can form from the reaction of 1.5 mol of each reactant?

SAMPLE PROBLEM 3.11 Calculating Amounts of Reactant and Product in a Limiting-Reactant Problem

Problem A fuel mixture used in the early days of rocketry is composed of two liquids, hydrazine (N_2H_4) and dinitrogen tetraoxide (N_2O_4), which ignite on contact to form nitrogen gas and water vapor. How many grams of nitrogen gas form when 1.00×10^2 g of N_2H_4 and 2.00×10^2 g of N_2O_4 are mixed?

Plan We first write the balanced equation. *Because the amounts of two reactants are given, we know this is a limiting-reactant problem.* To determine which reactant is limiting, we calculate the mass of N_2 formed from each reactant *assuming an excess of the other.* We convert the grams of each reactant to moles and use the appropriate molar ratio to find the moles of N_2 each forms. Whichever yields *less* N_2 is the limiting reactant. Then, we convert this lower number of moles of N_2 to mass. The roadmap shows the steps.

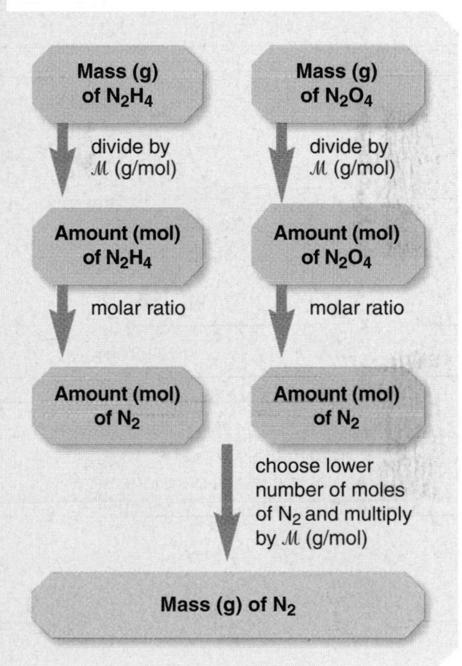

Solution Writing the balanced equation:

$$2N_2H_4(l) + N_2O_4(l) \longrightarrow 3N_2(g) + 4H_2O(g)$$

Finding the moles of N_2 from the moles of N_2H_4 (if N_2H_4 is limiting):

$$\text{Moles of } N_2H_4 = 1.00\times10^2 \cancel{\text{g } N_2H_4} \times \frac{1 \text{ mol } N_2H_4}{32.05 \cancel{\text{g } N_2H_4}} = 3.12 \text{ mol } N_2H_4$$

$$\text{Moles of } N_2 = 3.12 \cancel{\text{mol } N_2H_4} \times \frac{3 \text{ mol } N_2}{2 \cancel{\text{mol } N_2H_4}} = \mathbf{4.68 \text{ mol } N_2}$$

Finding the moles of N_2 from the moles of N_2O_4 (if N_2O_4 is limiting):

$$\text{Moles of } N_2O_4 = 2.00\times10^2 \cancel{\text{g } N_2O_4} \times \frac{1 \text{ mol } N_2O_4}{92.02 \cancel{\text{g } N_2O_4}} = 2.17 \text{ mol } N_2O_4$$

$$\text{Moles of } N_2 = 2.17 \cancel{\text{mol } N_2O_4} \times \frac{3 \text{ mol } N_2}{1 \cancel{\text{mol } N_2O_4}} = \mathbf{6.51 \text{ mol } N_2}$$

Thus, N_2H_4 is the limiting reactant because it yields fewer moles of N_2.
Converting from moles of N_2 to grams:

$$\text{Mass (g) of } N_2 = 4.68 \cancel{\text{mol } N_2} \times \frac{28.02 \text{ g } N_2}{1 \cancel{\text{mol } N_2}} = \boxed{131 \text{ g } N_2}$$

Check The mass of N_2O_4 is greater than that of N_2H_4, but there are fewer moles of N_2O_4 because its $\mathcal{M}$ is much higher. Round off to check the math: for N_2H_4, 100 g N_2H_4 $\times$ 1 mol/32 g $\approx$ 3 mol; ~3 mol $\times \frac{3}{2} \approx$ 4.5 mol N_2; ~4.5 mol $\times$ 30 g/mol $\approx$ 135 g N_2.

Comment 1. Here are two *common mistakes* in solving limiting-reactant problems:

- The limiting reactant is not the *reactant* present in fewer moles (2.17 mol of N_2O_4 vs. 3.12 mol of N_2H_4). Rather, it is the reactant that forms fewer moles of *product.*
- Similarly, the limiting reactant is not the *reactant* present in lower mass. Rather, it is the reactant that forms the lower mass of *product.*

2. Here is an *alternative approach* to finding the limiting reactant. Find the moles of each reactant that would be needed to react with the other reactant. Then see which amount actually given in the problem is sufficient. That substance is in excess, and the other substance is limiting. For example, the balanced equation shows 2 mol of N_2H_4 reacts with 1 mol of N_2O_4. The moles of N_2O_4 needed to react with the given moles of N_2H_4 are

$$\text{Moles of } N_2O_4 \text{ needed} = 3.12 \cancel{\text{mol } N_2H_4} \times \frac{1 \text{ mol } N_2O_4}{2 \cancel{\text{mol } N_2H_4}} = 1.56 \text{ mol } N_2O_4$$

The moles of N_2H_4 needed to react with the given moles of N_2O_4 are

$$\text{Moles of } N_2H_4 \text{ needed} = 2.17 \cancel{\text{mol } N_2O_4} \times \frac{2 \text{ mol } N_2H_4}{1 \cancel{\text{mol } N_2O_4}} = 4.34 \text{ mol } N_2H_4$$

We are given 2.17 mol of N_2O_4, which is *more* than the amount of N_2O_4 that is needed (1.56 mol) to react with the given amount of N_2H_4, and we are given 3.12 mol of N_2H_4, which is *less* than the amount of N_2H_4 needed (4.34 mol) to react with the given amount of N_2O_4. Therefore, N_2H_4 is limiting, and N_2O_4 is in excess. Once we determine this, we continue with the final calculation to find the amount of N_2.

FOLLOW-UP PROBLEM 3.11 How many grams of solid aluminum sulfide can be prepared by the reaction of 10.0 g of aluminum and 15.0 g of sulfur? How much of the nonlimiting reactant is in excess?

Chemical Reactions in Practice: Theoretical, Actual, and Percent Yields

Up until now, we've been optimistic about the amount of product obtained from a reaction. We have assumed that 100% of the limiting reactant becomes product, that ideal separation and purification methods exist for isolating the product, and that we use perfect lab technique to collect all the product formed. In other words, we have assumed that we obtain the **theoretical yield,** the amount indicated by the stoichiometrically equivalent molar ratio in the balanced equation.

It's time to face reality. The theoretical yield is *never* obtained, for reasons that are largely uncontrollable. For one thing, although the major reaction predominates, many reactant mixtures also proceed through one or more **side reactions** that form smaller amounts of different products, as shown in Figure 3.11. In the rocket fuel reaction in Sample Problem 3.11, for example, the reactants might form some NO in the following side reaction:

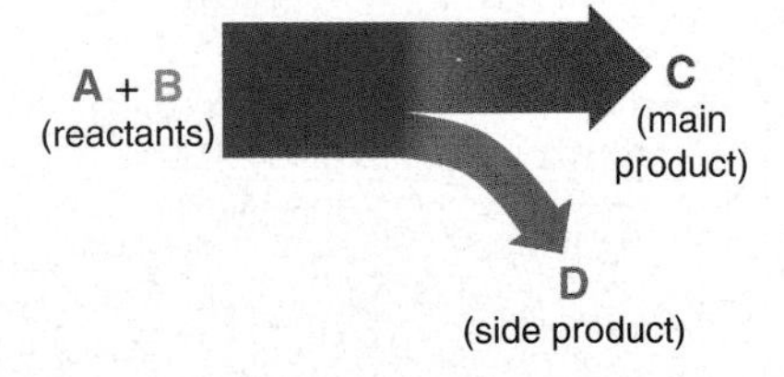

Figure 3.11 The effect of side reactions on yield. One reason the theoretical yield is never obtained is that other reactions lead some of the reactants along side paths to form undesired products.

$$2N_2O_4(l) + N_2H_4(l) \longrightarrow 6NO(g) + 2H_2O(g)$$

This reaction decreases the amounts of reactants available for N_2 production (see Problem 3.116 at the end of the chapter). Even more important, as we'll discuss in Chapter 4, many reactions seem to stop before they are complete, which leaves some limiting reactant unused. But, even when a reaction does go completely to product, losses occur in virtually every step of a separation procedure (see Tools of the Laboratory, Section 2.9): a tiny amount of product clings to filter paper, some distillate evaporates, a small amount of extract remains in the separatory funnel, and so forth. With careful technique, you can minimize these losses but never eliminate them. The amount of product that you actually obtain is the **actual yield.** The **percent yield (% yield)** is the actual yield expressed as a percentage of the theoretical yield:

$$\% \text{ yield} = \frac{\text{actual yield}}{\text{theoretical yield}} \times 100 \qquad \textbf{(3.7)}$$

Since the actual yield must be less than the theoretical yield, the percent yield is always less than 100%. Theoretical and actual yields are expressed in units of amount (moles) or mass (grams).

SAMPLE PROBLEM 3.12 Calculating Percent Yield

Problem Silicon carbide (SiC) is an important ceramic material that is made by allowing sand (silicon dioxide, SiO_2) to react with powdered carbon at high temperature. Carbon monoxide is also formed. When 100.0 kg of sand is processed, 51.4 kg of SiC is recovered. What is the percent yield of SiC from this process?

Plan We are given the actual yield of SiC (51.4 kg), so we need the theoretical yield to calculate the percent yield. After writing the balanced equation, we convert the given mass of SiO_2 (100.0 kg) to amount (mol). We use the molar ratio to find the amount of SiC formed and convert that amount to mass (kg) to obtain the theoretical yield [see Sample Problem 3.8(c)]. Then, we use Equation 3.7 to find the percent yield (see the roadmap).

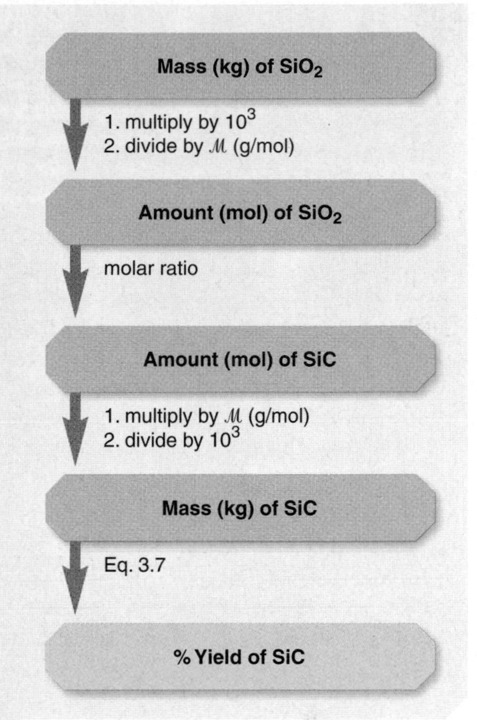

Solution Writing the balanced equation:

$$SiO_2(s) + 3C(s) \longrightarrow SiC(s) + 2CO(g)$$

Converting from kilograms of SiO_2 to moles:

$$\text{Moles of } SiO_2 = 100.0 \text{ kg } SiO_2 \times \frac{1000 \text{ g}}{1 \text{ kg}} \times \frac{1 \text{ mol } SiO_2}{60.09 \text{ g } SiO_2} = 1664 \text{ mol } SiO_2$$

Converting from moles of SiO_2 to moles of SiC: The molar ratio is 1 mol SiC/1 mol SiO_2, so

$$\text{Moles of } SiO_2 = \text{moles of SiC} = 1664 \text{ mol SiC}$$

Converting from moles of SiC to kilograms:

$$\text{Mass (kg) of SiC} = 1664 \text{ mol SiC} \times \frac{40.10 \text{ g SiC}}{1 \text{ mol SiC}} \times \frac{1 \text{ kg}}{1000 \text{ g}} = 66.73 \text{ kg SiC}$$

Calculating the percent yield:

$$\text{\% yield of SiC} = \frac{\text{actual yield}}{\text{theoretical yield}} \times 100 = \frac{51.4 \text{ kg}}{66.73 \text{ kg}} \times 100 = 77.0\%$$

Check Rounding shows that the mass of SiC seems correct: $\sim$1500 mol $\times$ 40 g/mol $\times$ 1 kg/1000 g = 60 kg. The molar ratio of SiC:SiO_2 is 1:1, and the $\mathcal{M}$ of SiC is about two-thirds ($\sim\frac{40}{60}$) the $\mathcal{M}$ of SiO_2, so 100 kg of SiO_2 should form about 66 kg of SiC.

FOLLOW-UP PROBLEM 3.12 Marble (calcium carbonate) reacts with hydrochloric acid solution to form calcium chloride solution, water, and carbon dioxide. What is the percent yield of carbon dioxide if 3.65 g of the gas is collected when 10.0 g of marble reacts?

Yields in Multistep Syntheses In the multistep laboratory synthesis of a complex compound, *the percent yield of each step is expressed as a fraction and multiplied by the others to find the overall yield.* Even when the yield of each step is high, the final result can be surprisingly low. For example, suppose a six-step reaction sequence has a 90.0% yield for each step; that is, you are able to recover 90.0% of the theoretical yield of product in each step. Even so, overall recovery is only slightly more than 50%:

$$\text{Overall \% yield} = (0.900 \times 0.900 \times 0.900 \times 0.900 \times 0.900 \times 0.900) \times 100 = 53.1\%$$

Such multistep sequences are common in the laboratory synthesis of medicines, dyes, pesticides, and many other organic compounds. For example, the antidepressant Sertraline is prepared from a simple starting compound in six steps with yields of 80%, 80%, 50%, 100%, 48%, and 30% respectively. Multiplying these together as fractions gives an overall yield of only 4.6%. Because a typical synthesis begins with large amounts of inexpensive, simple reactants and ends with small amounts of expensive, complex products, the overall yield greatly influences the commercial potential of a product.

Cl
Cl
↓80%
↓80%
↓50%
↓100%
↓48%
↓30%
Cl
CH_3—NH
Cl
Sertraline

A Green Chemistry Perspective on Yield: Atom Economy Reactants are wasted by undesirable side reactions during each step in a synthesis, drastically lowering the overall yield. Moreover, many by-products may be harmful. This fact is one of several concerns addressed by **green chemistry**. A major focus of the academic, industrial, and government chemists who work in this new field is to develop methods that reduce or prevent the release of harmful substances into the environment.

To fully evaluate alternative methods, several green chemistry principles are taken into account, including the quantity of energy needed and the nature of the solvents required. When these factors are similar, the *atom economy,* the proportion of reactant atoms that end up in the desired product, is a useful criterion for choosing the more efficient synthetic route. The efficiency of a synthesis is quantified in terms of the *percent atom economy:*

$$\% \text{ atom economy} = \frac{\text{no. of moles} \times \text{molar mass of desired product}}{\text{sum of no. of moles} \times \text{molar mass for all products}} \times 100$$

Consider two synthetic routes—one starting with benzene (C_6H_6), the other with butane (C_4H_{10})—for the production of maleic anhydride ($C_4H_2O_3$), a key industrial chemical used in the manufacture of polymers, dyes, medicines, pesticides, and other important products:

Route 1. $2C_6H_6(l) + 9O_2(g) \rightarrow \cdots \rightarrow 2C_4H_2O_3(l) + 4H_2O(l) + 4CO_2(g)$

Route 2. $2C_4H_{10}(g) + 7O_2(g) \rightarrow \cdots \rightarrow 2C_4H_2O_3(l) + 8H_2O(l)$

Let's compare the efficiency of these routes in terms of percent atom economy:

Route 1.

$$\% \text{ atom economy} = \frac{2 \times \mathcal{M} \text{ of } C_4H_2O_3}{(2 \times \mathcal{M} \text{ of } C_4H_2O_3) + (4 \times \mathcal{M} \text{ of } H_2O) + (4 \times \mathcal{M} \text{ of } CO_2)} \times 100$$

$$= \frac{2 \times 98.06 \text{ g}}{(2 \times 98.06 \text{ g}) + (4 \times 18.02 \text{ g}) + (4 \times 44.01 \text{ g})} \times 100 = 44.15\%$$

Route 2.

$$\% \text{ atom economy} = \frac{2 \times \mathcal{M} \text{ of } C_4H_2O_3}{(2 \times \mathcal{M} \text{ of } C_4H_2O_3) + (8 \times \mathcal{M} \text{ of } H_2O)} \times 100$$

$$= \frac{2 \times 98.06 \text{ g}}{(2 \times 98.06 \text{ g}) + (8 \times 18.02 \text{ g})} \times 100 = 57.63\%$$

Clearly, from the perspective of atom economy, route 2 is preferable because a larger percentage of reactant atoms end up in the desired product. It is also a "greener" approach than route 1 because it avoids the use of the toxic reactant benzene and does not produce CO_2, a gas that contributes to global warming.

SECTION SUMMARY

The substances in a balanced equation are related to each other by stoichiometrically equivalent molar ratios, which can be used as conversion factors to find the moles of one substance given the moles of another. In limiting-reactant problems, the amounts of two (or more) reactants are given, and one of them limits the amount of product that forms. The limiting reactant is the one that forms the lower amount of product. In practice, side reactions, incomplete reactions, and physical losses result in an actual yield of product that is less than the theoretical yield, the amount based solely on the molar ratio. In multistep reactions, the overall yield is found by multiplying the yields for each step. Atom economy, or the proportion of reactant atoms found in the desired product, is one criterion for choosing a "greener" reaction.

3.5 FUNDAMENTALS OF SOLUTION STOICHIOMETRY

In the popular media, you may have seen a chemist portrayed as a person in a white lab coat, surrounded by odd-shaped glassware, pouring one colored solution into another, which produces frothing bubbles and billowing fumes. Although most reactions in solution are not this dramatic and good technique usually requires safer mixing procedures, the image is true to the extent that aqueous solution chemistry is a central part of laboratory activity. Liquid solutions are more convenient to store and mix than solids or gases, and the amounts of substances in solution can be measured very precisely. Since many environmental reactions and almost all biochemical reactions occur in solution, an understanding of reactions in solution is extremely important in chemistry and related sciences.

We'll discuss solution chemistry at many places in the text, but here we focus on solution stoichiometry. Only one aspect of the stoichiometry of dissolved substances is different from what we've seen so far. We know the amounts of pure substances by converting their masses directly into moles. For dissolved substances, we must know the *concentration*—the number of moles present in a certain volume of solution—to find the volume that contains a given number of moles. Of the various ways to express concentration, the most important is *molarity,* so we discuss it here (and wait until Chapter 13 to discuss the other ways). Then, we see how to prepare a solution of a specific molarity and how to use solutions in stoichiometric calculations.

Expressing Concentration in Terms of Molarity

A typical solution consists of a smaller amount of one substance, the **solute,** dissolved in a larger amount of another substance, the **solvent.** When a solution forms, the solute's individual chemical entities become evenly dispersed throughout the available volume and surrounded by solvent molecules. The **concentration** of a solution is usually expressed as *the amount of solute dissolved in a given amount of solution.* Concentration is an *intensive* quantity (like density or temperature) and thus independent of the volume of solution: a 50-L tank of a given solution has the *same concentration* (solute amount/solution amount) as a 50-mL beaker of the solution. **Molarity *(M)*** expresses the concentration in units of *moles of solute per liter of solution:*

$$\text{Molarity} = \frac{\text{moles of solute}}{\text{liters of solution}} \quad \text{or} \quad M = \frac{\text{mol solute}}{\text{L soln}} \tag{3.8}$$

SAMPLE PROBLEM 3.13 Calculating the Molarity of a Solution

Problem Glycine (H_2NCH_2COOH) is the simplest amino acid. What is the molarity of an aqueous solution that contains 0.715 mol of glycine in 495 mL?

Plan The molarity is the number of moles of solute in each liter of solution. We are given the number of moles (0.715 mol) and the volume (495 mL), so we divide moles by volume and convert the volume to liters to find the molarity.

Amount (mol) of glycine
↓ divide by volume (mL)
Concentration (mol/mL) of glycine
↓ 10^3 mL = 1 L
Molarity (mol/L) of glycine

Solution

$$\text{Molarity} = \frac{0.715\ \text{mol glycine}}{495\ \cancel{\text{mL soln}}} \times \frac{1000\ \cancel{\text{mL}}}{1\ \text{L}} = 1.44\ M\ \text{glycine}$$

Check A quick look at the math shows about 0.7 mol of glycine in about 0.5 L of solution, so the concentration should be about 1.4 mol/L, or 1.4 *M*.

FOLLOW-UP PROBLEM 3.13 How many moles of KI are in 84 mL of 0.50 *M* KI?

Mole-Mass-Number Conversions Involving Solutions

Molarity can be thought of as a conversion factor used to convert between volume of solution and amount (mol) of solute, from which we then find the mass or the number of entities of solute. Figure 3.12 shows this new stoichiometric relationship, and Sample Problem 3.14 applies it.

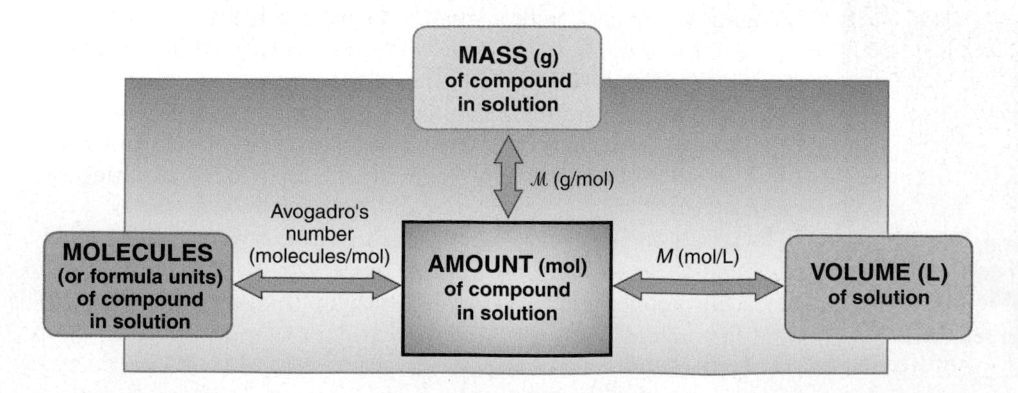

Figure 3.12 Summary of mass-mole-number-volume relationships in solution. The amount (in moles) of a compound in solution is related to the volume of solution in liters through the molarity (*M*) in moles per liter. The other relationships shown are identical to those in Figure 3.4, except that here they refer to the quantities *in solution*. As in previous cases, to find the quantity of substance expressed in one form or another, convert the given information to moles first.

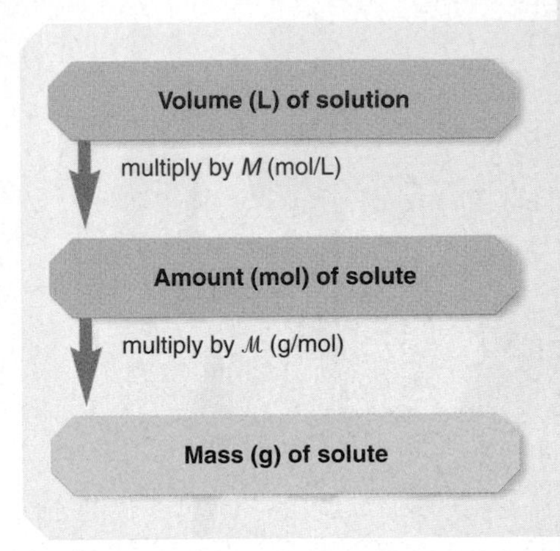

SAMPLE PROBLEM 3.14 Calculating Mass of Solute in a Given Volume of Solution

Problem A *buffered* solution maintains acidity as a reaction occurs. In living cells, phosphate ions play a key buffering role, so biochemists often study reactions in such solutions. How many grams of solute are in 1.75 L of 0.460 *M* sodium monohydrogen phosphate?

Plan We know the solution volume (1.75 L) and molarity (0.460 *M*), and we need the mass of solute. We use the known quantities to find the amount (mol) of solute and then convert moles to grams with the solute molar mass.

Solution Calculating moles of solute in solution:

$$\text{Moles of } Na_2HPO_4 = 1.75 \cancel{\text{L soln}} \times \frac{0.460 \text{ mol } Na_2HPO_4}{1 \cancel{\text{L soln}}} = 0.805 \text{ mol } Na_2HPO_4$$

Converting from moles of solute to grams:

$$\text{Mass (g) } Na_2HPO_4 = 0.805 \cancel{\text{mol } Na_2HPO_4} \times \frac{141.96 \text{ g } Na_2HPO_4}{1 \cancel{\text{mol } Na_2HPO_4}} = 114 \text{ g } Na_2HPO_4$$

Check The answer seems to be correct: ~1.8 L of 0.5 mol/L contains 0.9 mol, and 150 g/mol × 0.9 mol = 135 g, which is close to 114 g of solute.

FOLLOW-UP PROBLEM 3.14 In biochemistry laboratories, solutions of sucrose (table sugar, $C_{12}H_{22}O_{11}$) are used in high-speed centrifuges to separate the parts of a biological cell. How many liters of 3.30 *M* sucrose contain 135 g of solute?

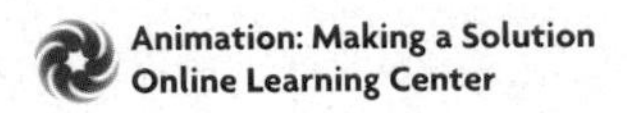

Preparing and Diluting Molar Solutions

Whenever you prepare a solution of specific molarity, remember that the volume term in the denominator of the molarity expression is the *solution* volume, **not** the *solvent* volume. The solution volume includes contributions from solute *and* solvent, so you cannot simply dissolve 1 mol of solute in 1 L of solvent and expect a 1 *M* solution. The solute would increase the solution volume above 1 L, resulting in a lower-than-expected concentration. The correct preparation of a solution containing a solid solute consists of four steps. Let's go through them to prepare 0.500 L of 0.350 *M* nickel(II) nitrate hexahydrate [$Ni(NO_3)_2 \cdot 6H_2O$]:

1. *Weigh the solid needed.* Calculate the mass of solid needed by converting from liters to moles and from moles to grams:

$$\text{Mass (g) of solute} = 0.500 \cancel{\text{L soln}} \times \frac{0.350 \cancel{\text{mol } Ni(NO_3)_2 \cdot 6H_2O}}{1 \cancel{\text{L soln}}} \times \frac{290.82 \text{ g } Ni(NO_3)_2 \cdot 6H_2O}{1 \cancel{\text{mol } Ni(NO_3)_2 \cdot 6H_2O}} = 50.9 \text{ g } Ni(NO_3)_2 \cdot 6H_2O$$

2. *Carefully transfer the solid to a volumetric flask that contains about half the final volume of solvent.* Since we need 0.500 L of solution, we choose a 500-mL volumetric flask. Add about 250 mL of distilled water and then transfer the solid. Wash down any solid clinging to the neck with a small amount of solvent.

3. *Dissolve the solid thoroughly by swirling.* If some solute remains undissolved, the solution will be less concentrated than expected, so be sure the solute is dissolved. If necessary, wait for the solution to reach room temperature. (As we'll discuss in Chapter 13, the solution process is often accompanied by heating or cooling.)

4. *Add solvent until the solution reaches its final volume.* Add distilled water to bring the volume exactly to the line on the flask neck, then cover and mix thoroughly again. Figure 3.13 shows the last three steps.

Figure 3.13 Laboratory preparation of molar solutions.

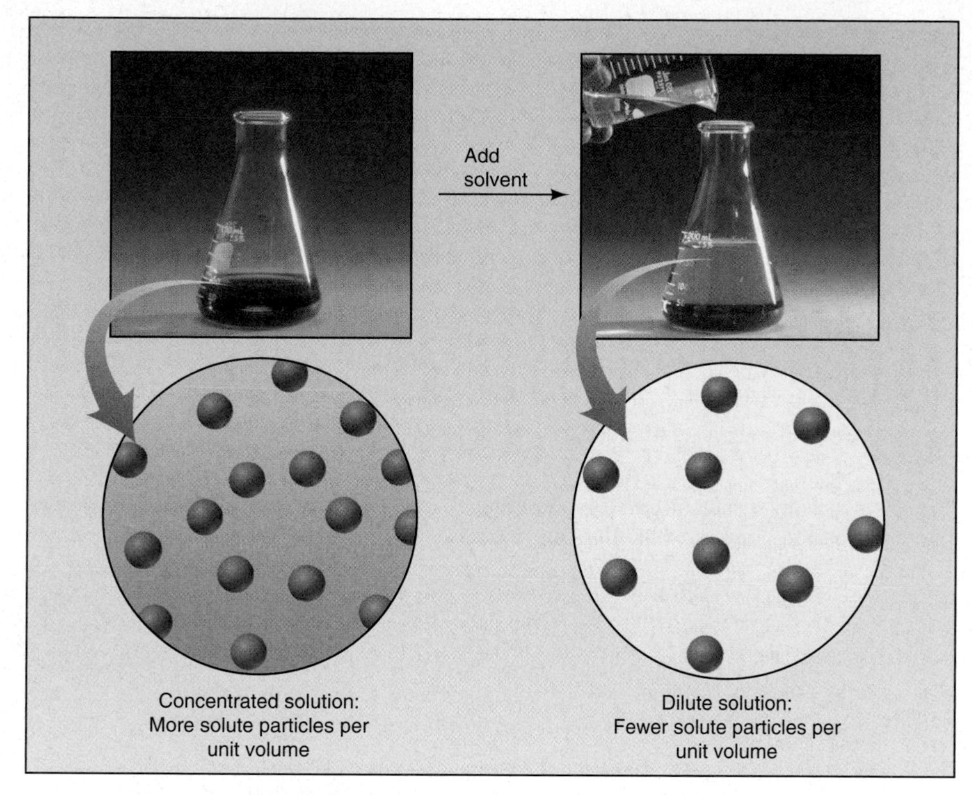

Figure 3.14 Converting a concentrated solution to a dilute solution. When a solution is diluted, only solvent is added. The solution volume increases while the total number of moles of solute remains the same. Therefore, as shown in the blow-up views, a unit volume of concentrated solution contains more solute particles than the same unit volume of dilute solution.

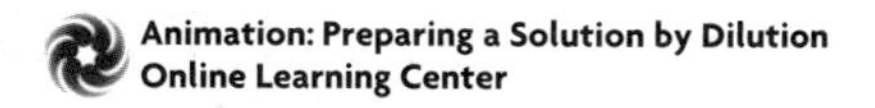

Animation: Preparing a Solution by Dilution
Online Learning Center

As Figure 3.14 shows, *only solvent is added when a solution is diluted,* so the solute is dispersed in a larger final volume. Thus, a given volume of the final solution contains fewer solute particles and has a lower concentration. If various low concentrations of a solution are used frequently, it is common practice to prepare a more concentrated solution (called a *stock solution*), which is stored and diluted as needed.

SAMPLE PROBLEM 3.15 Preparing a Dilute Solution from a Concentrated Solution

Problem Isotonic saline is a 0.15 *M* aqueous solution of NaCl that simulates the total concentration of ions found in many cellular fluids. Its uses range from a cleansing rinse for contact lenses to a washing medium for red blood cells. How would you prepare 0.80 L of isotonic saline from a 6.0 *M* stock solution?

Plan To dilute a concentrated solution, we add only solvent, so the *moles of solute are the same in both solutions*. We know the volume (0.80 L) and molarity (0.15 *M*) of the dilute (dil) NaCl solution we need, so we find the moles of NaCl it contains and then find the volume of concentrated (conc; 6.0 *M*) NaCl solution that contains the same number of moles. Then, we dilute this volume with solvent *up to* the final volume.

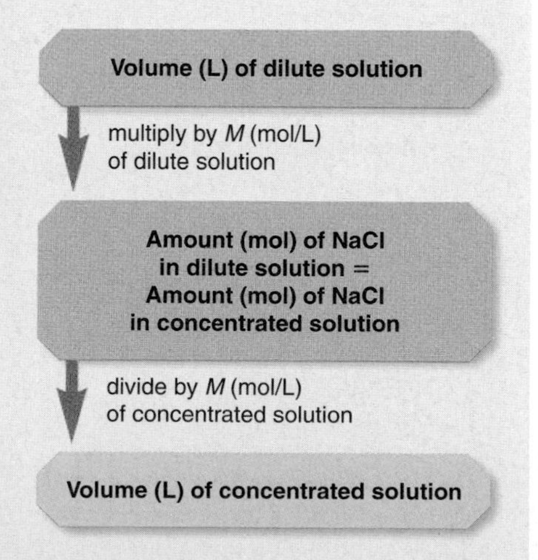

Solution Finding moles of solute in dilute solution:

$$\text{Moles of NaCl in dil soln} = 0.80\ \cancel{\text{L soln}} \times \frac{0.15\ \text{mol NaCl}}{1\ \cancel{\text{L soln}}} = 0.12\ \text{mol NaCl}$$

Finding moles of solute in concentrated solution: Since we add only solvent to dilute the solution,

$$\text{Moles of NaCl in dil soln} = \text{moles of NaCl in conc soln} = 0.12\ \text{mol NaCl}$$

Finding the volume of concentrated solution that contains 0.12 mol of NaCl:

$$\text{Volume (L) of conc NaCl soln} = 0.12\ \cancel{\text{mol NaCl}} \times \frac{1\ \text{L soln}}{6.0\ \cancel{\text{mol NaCl}}} = 0.020\ \text{L soln}$$

To prepare 0.80 L of dilute solution, place 0.020 L of 6.0 *M* NaCl in a 1.0-L cylinder, add distilled water (~780 mL) to the 0.80-L mark, and stir thoroughly.

Check The answer seems reasonable because a small volume of concentrated solution is used to prepare a large volume of dilute solution. Also, the ratio of volumes (0.020 L:0.80 L) is the same as the ratio of concentrations (0.15 *M*:6.0 *M*).

Comment An *alternative approach* to solving dilution problems uses the formula

$$M_{dil} \times V_{dil} = \text{number of moles} = M_{conc} \times V_{conc} \quad \textbf{(3.9)}$$

where the *M* and *V* terms are the molarity and volume of the *dil*ute and *conc*entrated solutions. In this problem, we need the volume of concentrated solution, V_{conc}:

$$V_{conc} = \frac{M_{dil} \times V_{dil}}{M_{conc}} = \frac{0.15\ \cancel{M} \times 0.80\ \text{L}}{6.0\ \cancel{M}} = 0.020\ \text{L}$$

The method worked out in the Solution (above) is actually the same calculation broken into two parts to emphasize the thinking process:

$$V_{conc} = 0.80\ \cancel{\text{L}} \times \frac{0.15\ \cancel{\text{mol NaCl}}}{1\ \cancel{\text{L}}} \times \frac{1\ \text{L}}{6.0\ \cancel{\text{mol NaCl}}} = 0.020\ \text{L}$$

FOLLOW-UP PROBLEM 3.15 To prepare a fertilizer, an engineer dilutes a stock solution of sulfuric acid by adding 25.0 m^3 of 7.50 *M* acid to enough water to make 500. m^3. What is the mass (in g) of sulfuric acid per milliliter of the diluted solution?

Stoichiometry of Chemical Reactions in Solution

Solving stoichiometry problems for reactions in solution requires the same approach as before, with the additional step of converting the volume of reactant or product to moles: (1) balance the equation, (2) find the number of moles of one substance, (3) relate it to the stoichiometrically equivalent number of moles of another substance, and (4) convert to the desired units.

SAMPLE PROBLEM 3.16 Calculating Amounts of Reactants and Products for a Reaction in Solution

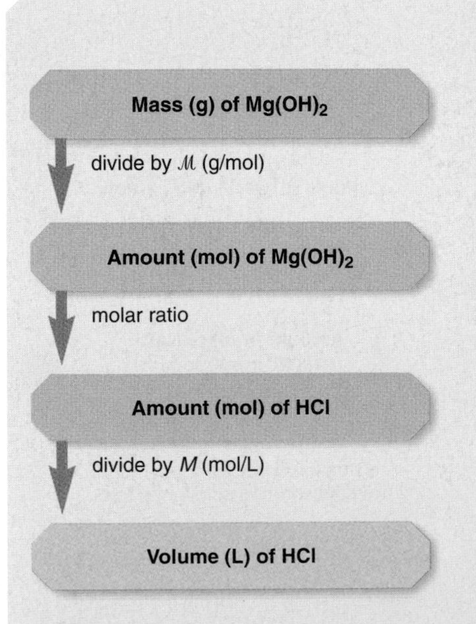

Problem Specialized cells in the stomach release HCl to aid digestion. If they release too much, the excess can be neutralized with an antacid. A common antacid contains magnesium hydroxide, $Mg(OH)_2$, which reacts with the acid to form water and magnesium chloride solution. As a government chemist testing commercial antacids, you use 0.10 *M* HCl to simulate the acid concentration in the stomach. How many liters of "stomach acid" react with a tablet containing 0.10 g of $Mg(OH)_2$?

Plan We know the mass of $Mg(OH)_2$ (0.10 g) that reacts and the acid concentration (0.10 *M*), and we must find the acid volume. After writing the balanced equation, we convert the grams of $Mg(OH)_2$ to moles, use the molar ratio to find the moles of HCl that react with these moles of $Mg(OH)_2$, and then use the molarity of HCl to find the volume that contains this number of moles. The steps appear in the roadmap.

Solution Writing the balanced equation:

$$Mg(OH)_2(s) + 2HCl(aq) \longrightarrow MgCl_2(aq) + 2H_2O(l)$$

Converting from grams of $Mg(OH)_2$ to moles:

$$\text{Moles of } Mg(OH)_2 = 0.10\ \cancel{\text{g } Mg(OH)_2} \times \frac{1\ \text{mol } Mg(OH)_2}{58.33\ \cancel{\text{g } Mg(OH)_2}} = 1.7\times10^{-3}\ \text{mol } Mg(OH)_2$$

Converting from moles of $Mg(OH)_2$ to moles of HCl:

$$\text{Moles of HCl} = 1.7\times10^{-3}\ \cancel{\text{mol } Mg(OH)_2} \times \frac{2\ \text{mol HCl}}{1\ \cancel{\text{mol } Mg(OH)_2}} = 3.4\times10^{-3}\ \text{mol HCl}$$

Converting from moles of HCl to liters:

$$\text{Volume (L) of HCl} = 3.4\times10^{-3}\ \cancel{\text{mol HCl}} \times \frac{1\ \text{L}}{0.10\ \cancel{\text{mol HCl}}} = 3.4\times10^{-2}\ \text{L}$$

Check The size of the answer seems reasonable: a small volume of dilute acid (0.034 L of 0.10 *M*) reacts with a small amount of antacid (0.0017 mol).

Comment The reaction as written is an oversimplification; in reality, HCl and $MgCl_2$ exist as separated ions in solution. This point will be covered in great detail in Chapters 4 and 18.

FOLLOW-UP PROBLEM 3.16 Another active ingredient found in some antacids is aluminum hydroxide. Which is more effective at neutralizing stomach acid, magnesium hydroxide or aluminum hydroxide? [*Hint:* Effectiveness refers to the amount of acid that reacts with a given mass of antacid. You already know the effectiveness of 0.10 g of $Mg(OH)_2$.]

In limiting-reactant problems for reactions in solution, we first determine which reactant is limiting and then determine the yield, as demonstrated in the next sample problem.

SAMPLE PROBLEM 3.17 Solving Limiting-Reactant Problems for Reactions in Solution

Problem Mercury and its compounds have many uses, from fillings for teeth (as an alloy with silver, copper, and tin) to the industrial production of chlorine. Because of their toxicity, however, soluble mercury compounds, such as mercury(II) nitrate, must be removed from industrial wastewater. One removal method reacts the wastewater with sodium sulfide solution to produce solid mercury(II) sulfide and sodium nitrate solution. In a laboratory simulation, 0.050 L of 0.010 *M* mercury(II) nitrate reacts with 0.020 L of 0.10 *M* sodium sulfide. How many grams of mercury(II) sulfide form?

Plan This is a limiting-reactant problem because *the amounts of two reactants are given.* After balancing the equation, we must determine the limiting reactant. The molarity (0.010 *M*) and volume (0.050 L) of the mercury(II) nitrate solution tell us the moles of one reactant, and the molarity (0.10 *M*) and volume (0.020 L) of the sodium sulfide solution tell us the moles of the other. Then, we use the molar ratio to find the moles of HgS that form from each reactant, *assuming the other reactant is present in excess.* The limiting reactant is the one that forms fewer moles of HgS, which we convert to mass using the HgS molar mass. The roadmap shows the process.

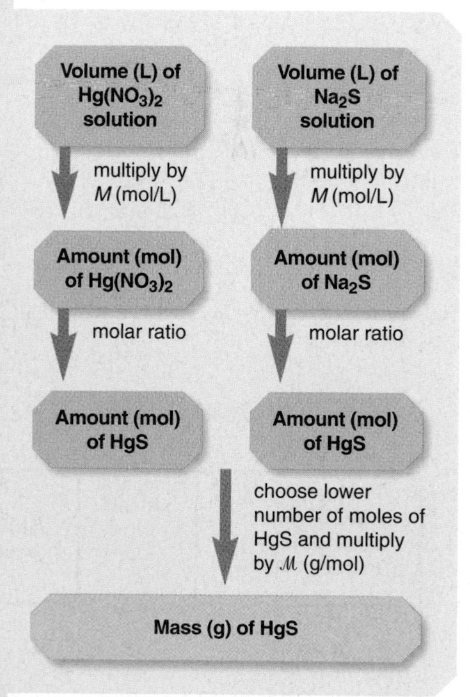

Solution Writing the balanced equation:

$$Hg(NO_3)_2(aq) + Na_2S(aq) \longrightarrow HgS(s) + 2NaNO_3(aq)$$

Finding moles of HgS assuming $Hg(NO_3)_2$ is limiting: Combining the steps gives

$$\text{Moles of HgS} = 0.050\ \cancel{\text{L soln}} \times \frac{0.010\ \cancel{\text{mol Hg(NO}_3)_2}}{1\ \cancel{\text{L soln}}} \times \frac{1\ \text{mol HgS}}{1\ \cancel{\text{mol Hg(NO}_3)_2}} = \mathbf{5.0\times10^{-4}\ mol\ HgS}$$

Finding moles of HgS assuming Na_2S is limiting: Combining the steps gives

$$\text{Moles of HgS} = 0.020\ \cancel{\text{L soln}} \times \frac{0.10\ \cancel{\text{mol Na}_2\text{S}}}{1\ \cancel{\text{L soln}}} \times \frac{1\ \text{mol HgS}}{1\ \cancel{\text{mol Na}_2\text{S}}} = \mathbf{2.0\times10^{-3}\ mol\ HgS}$$

$Hg(NO_3)_2$ is the limiting reactant because it forms fewer moles of HgS. Converting the moles of HgS formed from $Hg(NO_3)_2$ to grams:

$$\text{Mass (g) of HgS} = 5.0\times10^{-4}\ \cancel{\text{mol HgS}} \times \frac{232.7\ \text{g HgS}}{1\ \cancel{\text{mol HgS}}} = 0.12\ \text{g HgS}$$

Check As a check, let's use the alternative method for finding the limiting reactant (see Comment in Sample Problem 3.11, p. 113). Finding moles of reactants available:

$$\text{Moles of Hg(NO}_3)_2 = 0.050\ \cancel{\text{L soln}} \times \frac{0.010\ \text{mol Hg(NO}_3)_2}{1\ \cancel{\text{L soln}}} = 5.0\times10^{-4}\ \text{mol Hg(NO}_3)_2$$

$$\text{Moles of Na}_2\text{S} = 0.020\ \cancel{\text{L soln}} \times \frac{0.10\ \text{mol Na}_2\text{S}}{1\ \cancel{\text{L soln}}} = 2.0\times10^{-3}\ \text{mol Na}_2\text{S}$$

The molar ratio of the reactants is 1 $Hg(NO_3)_2$/1 Na_2S. Therefore, $Hg(NO_3)_2$ is limiting because there are fewer moles of it than are needed to react with the moles of Na_2S. Finding grams of product from moles of limiting reactant and the molar ratio:

$$\text{Mass (g) of HgS} = 5.0\times10^{-4}\ \cancel{\text{mol Hg(NO}_3)_2} \times \frac{1\ \cancel{\text{mol HgS}}}{1\ \cancel{\text{mol Hg(NO}_3)_2}} \times \frac{232.7\ \text{g HgS}}{1\ \cancel{\text{mol HgS}}}$$
$$= 0.12\ \text{g HgS}$$

FOLLOW-UP PROBLEM 3.17 Even though gasoline sold in the United States no longer contains lead, this metal persists in the environment as a poison. Despite their toxicity, many compounds of lead are still used to make pigments.
(a) What volume of 1.50 *M* lead(II) acetate contains 0.400 mol of Pb^{2+} ions?
(b) When this volume reacts with 125 mL of 3.40 *M* sodium chloride, how many grams of solid lead(II) chloride can form? (Sodium acetate solution also forms.)

SECTION SUMMARY

When reactions occur in solution, reactant and product amounts are given in terms of concentration and volume. Molarity is the number of moles of solute dissolved in one liter of solution. Using molarity as a conversion factor, we apply the principles of stoichiometry to all aspects of reactions in solution.

Chapter Perspective

You apply the mole concept every time you weigh a substance, dissolve it, or think about how much of it will react. Figure 3.15 combines the individual stoichiometry summary diagrams into one overall review diagram. Use it for homework, to study for exams, or to obtain an overview of the various ways that the amounts involved in a reaction are interrelated. We apply stoichiometry next to some of the most important types of chemical reactions (Chapter 4), to systems of reacting gases (Chapter 5), and to the heat involved in a reaction (Chapter 6). These concepts and skills appear at many places later in the text as well.

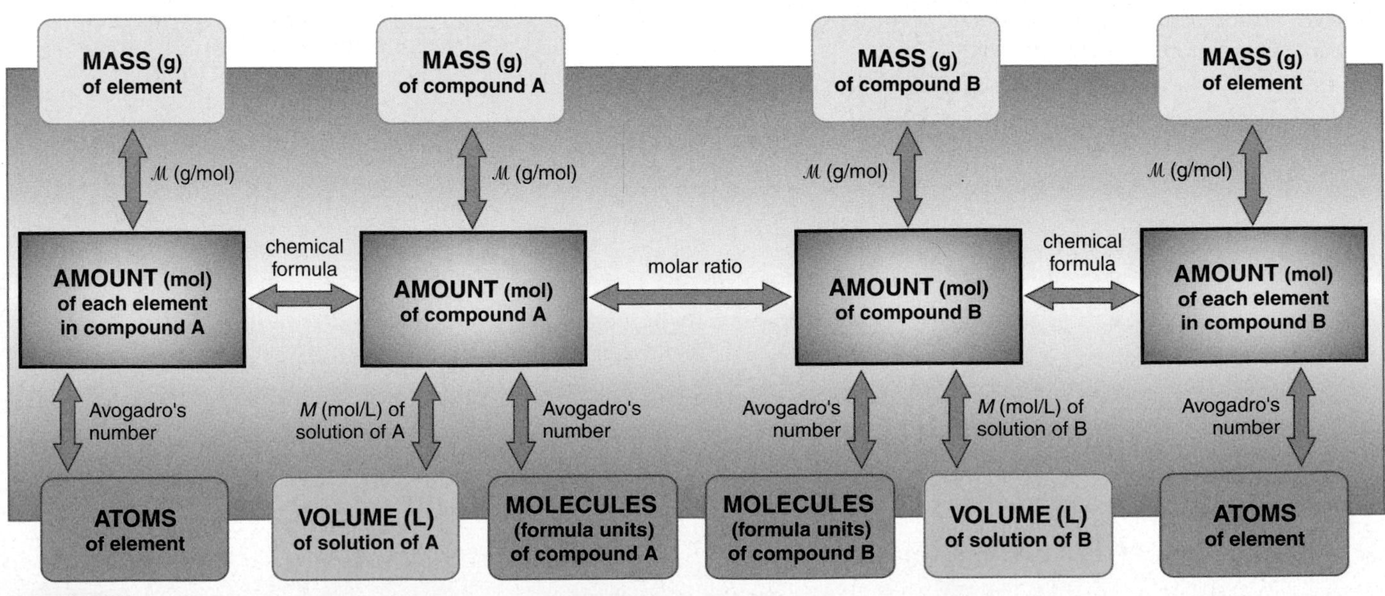

Figure 3.15 **An overview of the key mass-mole-number stoichiometric relationships.**

For Review and Reference

(Numbers in parentheses refer to pages, unless noted otherwise.)

Learning Objectives

Relevant section and/or sample problem (SP) numbers appear in parentheses.

Understand These Concepts

1. The meaning and usefulness of the mole (3.1)
2. The relation between molecular (or formula) mass and molar mass (3.1)
3. The relations among amount of substance (in moles), mass (in grams), and number of chemical entities (3.1)
4. The information in a chemical formula (3.1)
5. The procedure for finding the empirical and molecular formulas of a compound (3.2)
6. How more than one substance can have the same empirical formula and the same molecular formula (isomers) (3.2)
7. The importance of balancing equations for the quantitative study of chemical reactions (3.3)
8. The mole-mass-number information in a balanced equation (3.4)
9. The relation between amounts of reactants and products (3.4)
10. Why one reactant limits the yield of product (3.4)
11. The causes of lower-than-expected yields and the distinction between theoretical and actual yields (3.4)
12. The meanings of concentration and molarity (3.5)
13. The effect of dilution on the concentration of solute (3.5)
14. How reactions in solution differ from those of pure reactants (3.5)

Master These Skills

1. Calculating the molar mass of any substance (3.1; also SPs 3.2, 3.3)
2. Converting between amount of substance (in moles), mass (in grams), and number of chemical entities (SPs 3.1, 3.2)
3. Using mass percent to find the mass of element in a given mass of compound (SP 3.3)
4. Determining empirical and molecular formulas of a compound from mass percent and molar mass of elements (SPs 3.4, 3.5)
5. Determining a molecular formula from combustion analysis (SP 3.6)
6. Converting a chemical statement into a balanced equation (SP 3.7)
7. Using stoichiometrically equivalent molar ratios to calculate amounts of reactants and products in reactions of pure and dissolved substances (SPs 3.8, 3.16)
8. Writing an overall equation from a series of equations (SP 3.9)
9. Solving limiting-reactant problems from molecular depictions and for reactions of pure and dissolved substances (SPs 3.10, 3.11, 3.17)
10. Calculating percent yield (SP 3.12)
11. Calculating molarity and the mass of solute in solution (SPs 3.13, 3.14)
12. Preparing a dilute solution from a concentrated one (SP 3.15)

Key Terms

stoichiometry (87)

Section 3.1

mole (mol) (87)
Avogadro's number (87)
molar mass ($\mathcal{M}$) (89)

Section 3.2

combustion analysis (97)
isomer (100)

Section 3.3

chemical equation (101)
reactant (102)
product (102)
balancing (stoichiometric) coefficient (102)

Section 3.4

overall (net) equation (109)
limiting reactant (110)
theoretical yield (114)
side reaction (114)
actual yield (114)
percent yield (% yield) (114)
green chemistry (115)

Section 3.5

solute (117)
solvent (117)
concentration (117)
molarity (M) (117)

Key Equations and Relationships

3.1 Number of entities in one mole (87):

$$1 \text{ mole contains } 6.022\times10^{23} \text{ entities (to 4 sf)}$$

3.2 Converting amount (mol) to mass using $\mathcal{M}$ (90):

$$\text{Mass (g)} = \text{no. of moles} \times \frac{\text{no. of grams}}{1 \text{ mol}}$$

3.3 Converting mass to amount (mol) using $1/\mathcal{M}$ (90):

$$\text{No. of moles} = \text{mass (g)} \times \frac{1 \text{ mol}}{\text{no. of grams}}$$

3.4 Converting amount (mol) to number of entities (90):

$$\text{No. of entities} = \text{no. of moles} \times \frac{6.022\times10^{23} \text{ entities}}{1 \text{ mol}}$$

3.5 Converting number of entities to amount (mol) (90):

$$\text{No. of moles} = \text{no. of entities} \times \frac{1 \text{ mol}}{6.022\times10^{23} \text{ entities}}$$

3.6 Calculating mass % (93):

Mass % of element X

$$= \frac{\text{moles of X in formula} \times \text{molar mass of X (g/mol)}}{\text{mass (g) of 1 mol of compound}} \times 100$$

3.7 Calculating percent yield (114):

$$\%\text{ yield} = \frac{\text{actual yield}}{\text{theoretical yield}} \times 100$$

3.8 Defining molarity (117):

$$\text{Molarity} = \frac{\text{moles of solute}}{\text{liters of solution}} \quad \text{or} \quad M = \frac{\text{mol solute}}{\text{L soln}}$$

3.9 Diluting a concentrated solution (120):

$$M_{\text{dil}} \times V_{\text{dil}} = \text{number of moles} = M_{\text{conc}} \times V_{\text{conc}}$$

Highlighted Figures and Tables

These figures (F) and tables (T) provide a review of key ideas.

Brief Solutions to Follow-up Problems

3.1 (a) Moles of C $= 315\ \cancel{\text{mg C}} \times \dfrac{1\ \cancel{\text{g}}}{10^3\ \cancel{\text{mg}}} \times \dfrac{1\ \text{mol C}}{12.01\ \cancel{\text{g C}}}$
$= 2.62\times10^{-2}$ mol C

(b) Mass (g) of Mn $= 3.22\times10^{20}\ \cancel{\text{Mn atoms}} \times \dfrac{1\ \cancel{\text{mol Mn}}}{6.022\times10^{23}\ \cancel{\text{Mn atoms}}} \times \dfrac{54.94\ \text{g Mn}}{1\ \cancel{\text{mol Mn}}}$
$= 2.94\times10^{-2}$ g Mn

3.2 (a) Mass (g) of P_4O_{10}
$= 4.65\times10^{22}\ \cancel{\text{molecules } P_4O_{10}} \times \dfrac{1\ \cancel{\text{mol } P_4O_{10}}}{6.022\times10^{23}\ \cancel{\text{molecules } P_4O_{10}}} \times \dfrac{283.88\ \text{g } P_4O_{10}}{1\ \cancel{\text{mol } P_4O_{10}}}$
$= 21.9$ g P_4O_{10}

(b) No. of P atoms $= 4.65\times10^{22}\ \cancel{\text{molecules } P_4O_{10}} \times \dfrac{4\ \text{atoms P}}{1\ \cancel{\text{molecule } P_4O_{10}}}$
$= 1.86\times10^{23}$ P atoms

3.3 (a) Mass % of N $= \dfrac{2\ \cancel{\text{mol N}} \times \dfrac{14.01\ \cancel{\text{g N}}}{1\ \cancel{\text{mol N}}}}{80.05\ \cancel{\text{g } NH_4NO_3}} \times 100$
$= 35.00$ mass % N

(b) Mass (g) of N $= 35.8\ \cancel{\text{kg } NH_4NO_3} \times \dfrac{10^3\ \cancel{\text{g}}}{1\ \cancel{\text{kg}}} \times \dfrac{0.3500\ \text{g N}}{1\ \cancel{\text{g } NH_4NO_3}}$
$= 1.25\times10^4$ g N

3.4 Moles of S $= 2.88\ \cancel{\text{g S}} \times \dfrac{1\ \text{mol S}}{32.07\ \cancel{\text{g S}}} = 0.0898$ mol S

Moles of M $= 0.0898\ \cancel{\text{mol S}} \times \dfrac{2\ \text{mol M}}{3\ \cancel{\text{mol S}}} = 0.0599$ mol M

Molar mass of M $= \dfrac{3.12\ \text{g M}}{0.0599\ \text{mol M}} = 52.1$ g/mol

M is chromium, and M_2S_3 is chromium(III) sulfide.

3.5 Assuming 100.00 g of compound, we have 95.21 g of C and 4.79 g of H:

Moles of C $= 95.21\ \cancel{\text{g C}} \times \dfrac{1\ \text{mol C}}{12.01\ \cancel{\text{g C}}}$
$= 7.928$ mol C

Also, 4.75 mol H
Preliminary formula $= C_{7.928}H_{4.75} \approx C_{1.67}H_{1.00}$
Empirical formula $= C_5H_3$

Whole-number multiple $= \dfrac{252.30\ \cancel{\text{g/mol}}}{63.07\ \cancel{\text{g/mol}}} = 4$

Molecular formula $= C_{20}H_{12}$

3.6 Mass (g) of C $= 0.451\ \cancel{\text{g } CO_2} \times \dfrac{12.01\ \text{g C}}{44.01\ \cancel{\text{g } CO_2}}$
$= 0.123$ g C

Also, 0.00690 g H
Mass (g) of Cl $= 0.250\ \text{g} - (0.123\ \text{g} + 0.00690\ \text{g}) = 0.120$ g Cl
Moles of elements
$= 0.0102$ mol C; 0.00685 mol H; 0.00339 mol Cl
Empirical formula $= C_3H_2Cl$; multiple $= 2$
Molecular formula $= C_6H_4Cl_2$

3.7 (a) $2Na(s) + 2H_2O(l) \longrightarrow H_2(g) + 2NaOH(aq)$
(b) $2HNO_3(aq) + CaCO_3(s) \longrightarrow H_2O(l) + CO_2(g) + Ca(NO_3)_2(aq)$
(c) $PCl_3(g) + 3HF(g) \longrightarrow PF_3(g) + 3HCl(g)$
(d) $4C_3H_5N_3O_9(l) \longrightarrow 12CO_2(g) + 10H_2O(g) + 6N_2(g) + O_2(g)$

3.8 $Fe_2O_3(s) + 2Al(s) \longrightarrow Al_2O_3(s) + 2Fe(l)$
(a) Mass (g) of Fe
$= 135\ \cancel{\text{g Al}} \times \dfrac{1\ \cancel{\text{mol Al}}}{26.98\ \cancel{\text{g Al}}} \times \dfrac{2\ \cancel{\text{mol Fe}}}{2\ \cancel{\text{mol Al}}} \times \dfrac{55.85\ \text{g Fe}}{1\ \cancel{\text{mol Fe}}}$
$= 279$ g Fe

(b) No. of Al atoms $= 1.00\ \cancel{\text{g } Al_2O_3} \times \dfrac{1\ \cancel{\text{mol } Al_2O_3}}{101.96\ \cancel{\text{g } Al_2O_3}} \times \dfrac{2\ \cancel{\text{mol Al}}}{1\ \cancel{\text{mol } Al_2O_3}} \times \dfrac{6.022\times10^{23}\ \text{Al atoms}}{1\ \cancel{\text{mol Al}}}$
$= 1.18\times10^{22}$ Al atoms

3.9
$$2SO_2(g) + O_2(g) \longrightarrow \cancel{2SO_3(g)}$$
$$\cancel{2SO_3(g)} + 2H_2O(l) \longrightarrow 2H_2SO_4(aq)$$
$$\overline{2SO_2(g) + O_2(g) + 2H_2O(l) \longrightarrow 2H_2SO_4(aq)}$$

3.10 (a) $2AB + B_2 \longrightarrow 2AB_2$
In the boxes, the AB/B_2 ratio is 4/3, which is less than the 2/1 ratio in the equation, so there is not enough AB and it is the limiting reactant; note that one B_2 is in excess.

(b) Moles of $AB_2 = 1.5\ \cancel{\text{mol AB}} \times \dfrac{2\ \text{mol } AB_2}{2\ \cancel{\text{mol AB}}} = 1.5$ mol AB_2

Moles of $AB_2 = 1.5\ \cancel{\text{mol } B_2} \times \dfrac{2\ \text{mol } AB_2}{1\ \cancel{\text{mol } B_2}} = 3.0$ mol AB_2

3.11 $2Al(s) + 3S(s) \longrightarrow Al_2S_3(s)$
Mass (g) of Al_2S_3 formed from Al
$= 10.0\ \cancel{\text{g Al}} \times \dfrac{1\ \cancel{\text{mol Al}}}{26.98\ \cancel{\text{g Al}}} \times \dfrac{1\ \cancel{\text{mol } Al_2S_3}}{2\ \cancel{\text{mol Al}}} \times \dfrac{150.17\ \text{g } Al_2S_3}{1\ \cancel{\text{mol } Al_2S_3}}$
$= 27.8$ g Al_2S_3
Similarly, mass (g) of Al_2S_3 formed from S $= 23.4$ g Al_2S_3.
Therefore, S is limiting reactant, and 23.4 g of Al_2S_3 can form.

Mass (g) of Al in excess

$= \text{total mass of Al} - \text{mass of Al used}$

$= 10.0 \text{ g Al}$

$$- \left(15.0 \text{ g S} \times \frac{1 \text{ mol S}}{32.07 \text{ g S}} \times \frac{2 \text{ mol Al}}{3 \text{ mol S}} \times \frac{26.98 \text{ g Al}}{1 \text{ mol Al}}\right)$$

$= 1.6 \text{ g Al}$

(We would obtain the same answer if sulfur were shown more correctly as S_8.)

3.12 $CaCO_3(s) + 2HCl(aq) \longrightarrow CaCl_2(aq) + H_2O(l) + CO_2(g)$

Theoretical yield (g) of CO_2

$$= 10.0 \text{ g } CaCO_3 \times \frac{1 \text{ mol } CaCO_3}{100.09 \text{ g } CaCO_3} \times \frac{1 \text{ mol } CO_2}{1 \text{ mol } CaCO_3}$$

$$\times \frac{44.01 \text{ g } CO_2}{1 \text{ mol } CO_2} = 4.40 \text{ g } CO_2$$

$$\% \text{ yield} = \frac{3.65 \text{ g } CO_2}{4.40 \text{ g } CO_2} \times 100 = 83.0\%$$

3.13 $$\text{Moles of KI} = 84 \text{ mL soln} \times \frac{1 \text{ L}}{10^3 \text{ mL}} \times \frac{0.50 \text{ mol KI}}{1 \text{ L soln}}$$

$= 0.042 \text{ mol KI}$

3.14 Volume (L) of soln

$$= 135 \text{ g sucrose} \times \frac{1 \text{ mol sucrose}}{342.30 \text{ g sucrose}} \times \frac{1 \text{ L soln}}{3.30 \text{ mol sucrose}}$$

$= 0.120 \text{ L soln}$

3.15 $$M_{dil} \text{ of } H_2SO_4 = \frac{7.50\ M \times 25.0 \text{ m}^3}{500. \text{ m}^3} = 0.375\ M\ H_2SO_4$$

Mass (g) of H_2SO_4/mL soln

$$= \frac{0.375 \text{ mol } H_2SO_4}{1 \text{ L soln}} \times \frac{1 \text{ L}}{10^3 \text{ mL}} \times \frac{98.09 \text{ g } H_2SO_4}{1 \text{ mol } H_2SO_4}$$

$= 3.68 \times 10^{-2} \text{ g/mL soln}$

3.16 $Al(OH)_3(s) + 3HCl(aq) \longrightarrow AlCl_3(aq) + 3H_2O(l)$

Volume (L) of HCl consumed

$$= 0.10 \text{ g } Al(OH)_3 \times \frac{1 \text{ mol } Al(OH)_3}{78.00 \text{ g } Al(OH)_3}$$

$$\times \frac{3 \text{ mol HCl}}{1 \text{ mol } Al(OH)_3} \times \frac{1 \text{ L soln}}{0.10 \text{ mol HCl}}$$

$= 3.8 \times 10^{-2} \text{ L soln}$

Therefore, $Al(OH)_3$ is more effective than $Mg(OH)_2$.

3.17 (a) Volume (L) of soln

$= 0.400 \text{ mol } Pb^{2+}$

$$\times \frac{1 \text{ mol } Pb(C_2H_3O_2)_2}{1 \text{ mol } Pb^{2+}} \times \frac{1 \text{ L soln}}{1.50 \text{ mol } Pb(C_2H_3O_2)_2}$$

$= 0.267 \text{ L soln}$

(b) $Pb(C_2H_3O_2)_2(aq) + 2NaCl(aq) \longrightarrow PbCl_2(s) + 2NaC_2H_3O_2(aq)$

Mass (g) of $PbCl_2$ from $Pb(C_2H_3O_2)_2$ soln = 111 g $PbCl_2$

Mass (g) of $PbCl_2$ from NaCl soln = 59.1 g $PbCl_2$

Thus, NaCl is the limiting reactant, and 59.1 g of $PbCl_2$ can form.

Problems

Problems with **colored** numbers are answered in Appendix E. Sections match the text and provide the numbers of relevant sample problems. Most offer Concept Review Questions, Skill-Building Exercises (grouped in pairs covering the same concept), and Problems in Context. Comprehensive Problems are based on material from any section or previous chapter.

The Mole

(Sample Problems 3.1 to 3.3)

Concept Review Questions

3.1 The atomic mass of Cl is 35.45 amu, and the atomic mass of Al is 26.98 amu. What are the masses in grams of 2 mol of Al atoms and of 3 mol of Cl atoms?

3.2 (a) How many moles of C atoms are in 1 mol of sucrose ($C_{12}H_{22}O_{11}$)?
(b) How many C atoms are in 1 mol of sucrose?

3.3 Why might the expression "1 mol of nitrogen" be confusing? What change would remove any uncertainty? For what other elements might a similar confusion exist? Why?

3.4 How is the molecular mass of a compound the same as the molar mass, and how is it different?

3.5 What advantage is there to using a counting unit (the mole) in chemistry rather than a mass unit?

3.6 You need to calculate the number of P_4 molecules that can form from 2.5 g of $Ca_3(PO_4)_2$. Explain how you would proceed. (That is, write a solution "Plan," without actually doing any calculations.)

3.7 Each of the following balances weighs the indicated numbers of atoms of two elements:

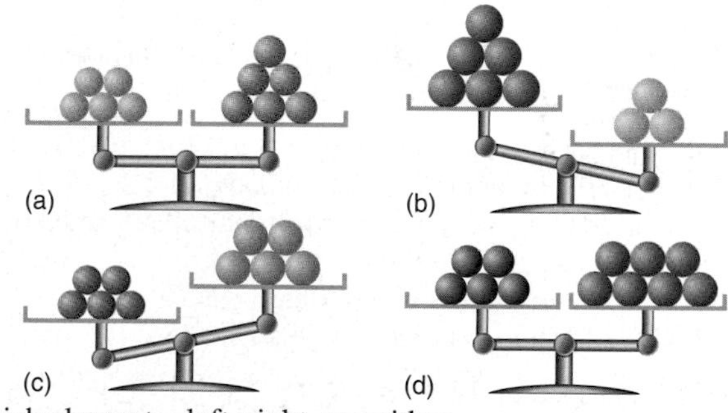

Which element—left, right, or neither,
(a) Has the higher molar mass?
(b) Has more atoms per gram?
(c) Has fewer atoms per gram?
(d) Has more atoms per mole?

Skill-Building Exercises *(grouped in similar pairs)*

3.8 Calculate the molar mass of each of the following:
(a) $Sr(OH)_2$ (b) N_2O (c) $NaClO_3$ (d) Cr_2O_3

3.9 Calculate the molar mass of each of the following:
(a) $(NH_4)_3PO_4$ (b) CH_2Cl_2 (c) $CuSO_4 \cdot 5H_2O$ (d) BrF_5

3.10 Calculate the molar mass of each of the following:
(a) SnO_2 (b) BaF_2 (c) $Al_2(SO_4)_3$ (d) $MnCl_2$

3.11 Calculate the molar mass of each of the following:
(a) N_2O_4 (b) C_8H_{10} (c) $MgSO_4 \cdot 7H_2O$ (d) $Ca(C_2H_3O_2)_2$

3.12 Calculate each of the following quantities:
(a) Mass in grams of 0.57 mol of $KMnO_4$
(b) Moles of O atoms in 8.18 g of $Mg(NO_3)_2$
(c) Number of O atoms in 8.1×10^{-3} g of $CuSO_4{\cdot}5H_2O$

3.13 Calculate each of the following quantities:
(a) Mass in kilograms of 3.8×10^{20} molecules of NO_2
(b) Moles of Cl atoms in 0.0425 g of $C_2H_4Cl_2$
(c) Number of H^- ions in 4.92 g of SrH_2

3.14 Calculate each of the following quantities:
(a) Mass in grams of 0.64 mol of $MnSO_4$
(b) Moles of compound in 15.8 g of $Fe(ClO_4)_3$
(c) Number of N atoms in 92.6 g of NH_4NO_2

3.15 Calculate each of the following quantities:
(a) Total number of ions in 38.1 g of CaF_2
(b) Mass in milligrams of 3.58 mol of $CuCl_2{\cdot}2H_2O$
(c) Mass in kilograms of 2.88×10^{22} formula units of $Bi(NO_3)_3{\cdot}5H_2O$

3.16 Calculate each of the following quantities:
(a) Mass in grams of 8.41 mol of copper(I) carbonate
(b) Mass in grams of 2.04×10^{21} molecules of dinitrogen pentaoxide
(c) Number of moles and formula units in 57.9 g of sodium perchlorate
(d) Number of sodium ions, perchlorate ions, Cl atoms, and O atoms in the mass of compound in part (c)

3.17 Calculate each of the following quantities:
(a) Mass in grams of 3.52 mol of chromium(III) sulfate decahydrate
(b) Mass in grams of 9.64×10^{24} molecules of dichlorine heptaoxide
(c) Number of moles and formula units in 56.2 g of lithium sulfate
(d) Number of lithium ions, sulfate ions, S atoms, and O atoms in the mass of compound in part (c)

3.18 Calculate each of the following:
(a) Mass % of H in ammonium bicarbonate
(b) Mass % of O in sodium dihydrogen phosphate heptahydrate

3.19 Calculate each of the following:
(a) Mass % of I in strontium periodate
(b) Mass % of Mn in potassium permanganate

3.20 Calculate each of the following:
(a) Mass fraction of C in cesium acetate
(b) Mass fraction of O in uranyl sulfate trihydrate (the uranyl ion is UO_2^{2+})

3.21 Calculate each of the following:
(a) Mass fraction of Cl in calcium chlorate
(b) Mass fraction of P in tetraphosphorus hexaoxide

Problems in Context

3.22 Oxygen is required for the metabolic combustion of foods. Calculate the number of atoms in 38.0 g of oxygen gas, the amount absorbed from the lungs at rest in about 15 minutes.

3.23 Cisplatin *(right)*, or Platinol, is a powerful drug used in the treatment of certain cancers. Calculate (a) the moles of compound in 285.3 g of cisplatin; (b) the number of hydrogen atoms in 0.98 mol of cisplatin.

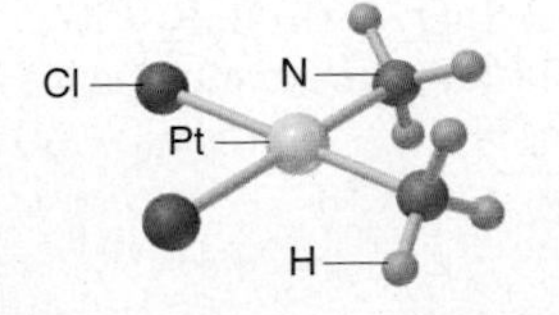

3.24 Allyl sulfide *(below)* gives garlic its characteristic odor.

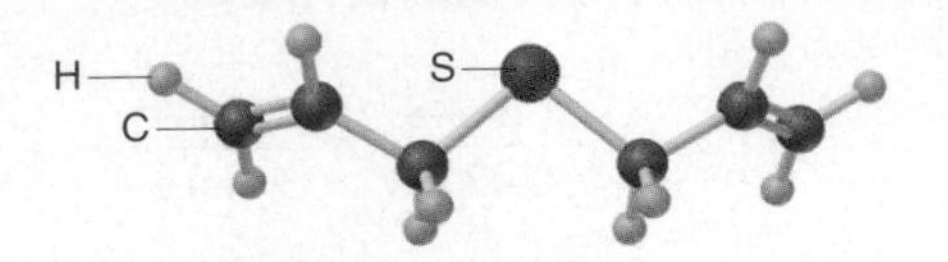

Calculate (a) the mass in grams of 1.63 mol of allyl sulfide; (b) the number of carbon atoms in 4.77 g of allyl sulfide.

3.25 Iron reacts slowly with oxygen and water to form a compound commonly called rust ($Fe_2O_3{\cdot}4H_2O$). For 65.2 kg of rust, calculate (a) the moles of compound; (b) the moles of Fe_2O_3; (c) the grams of iron.

3.26 Propane is widely used in liquid form as a fuel for barbecue grills and camp stoves. For 75.3 g of propane, calculate (a) the moles of compound; (b) the grams of carbon.

3.27 The effectiveness of a nitrogen fertilizer is determined mainly by its mass % N. Rank the following fertilizers in terms of their effectiveness: potassium nitrate; ammonium nitrate; ammonium sulfate; urea, $CO(NH_2)_2$.

3.28 The mineral galena is composed of lead(II) sulfide and has an average density of 7.46 g/cm^3. (a) How many moles of lead(II) sulfide are in 1.00 ft^3 of galena? (b) How many lead atoms are in 1.00 dm^3 of galena?

3.29 Hemoglobin, a protein in red blood cells, carries O_2 from the lungs to the body's cells. Iron (as ferrous ion, Fe^{2+}) makes up 0.33 mass % of hemoglobin. If the molar mass of hemoglobin is 6.8×10^4 g/mol, how many Fe^{2+} ions are in one molecule?

Determining the Formula of an Unknown Compound

(Sample Problems 3.4 to 3.6)

Concept Review Questions

3.30 List three ways compositional data may be given in a problem that involves finding an empirical formula.

3.31 Which of the following sets of information allows you to obtain the molecular formula of a covalent compound? In each case that allows it, explain how you would proceed (write a solution "Plan").
(a) Number of moles of each type of atom in a given sample of the compound
(b) Mass % of each element and the total number of atoms in a molecule of the compound
(c) Mass % of each element and the number of atoms of one element in a molecule of the compound
(d) Empirical formula and the mass % of each element in the compound
(e) Structural formula of the compound

3.32 Is $MgCl_2$ an empirical or a molecular formula for magnesium chloride? Explain.

Skill-Building Exercises *(grouped in similar pairs)*

3.33 What is the empirical formula and empirical formula mass for each of the following compounds?
(a) C_2H_4 (b) $C_2H_6O_2$ (c) N_2O_5 (d) $Ba_3(PO_4)_2$ (e) Te_4I_{16}

3.34 What is the empirical formula and empirical formula mass for each of the following compounds?
(a) C_4H_8 (b) $C_3H_6O_3$ (c) P_4O_{10} (d) $Ga_2(SO_4)_3$ (e) Al_2Br_6

3.35 What is the molecular formula of each compound?
(a) Empirical formula CH_2 ($\mathcal{M}$ = 42.08 g/mol)
(b) Empirical formula NH_2 ($\mathcal{M}$ = 32.05 g/mol)

(c) Empirical formula NO_2 ($\mathcal{M}$ = 92.02 g/mol)
(d) Empirical formula CHN ($\mathcal{M}$ = 135.14 g/mol)

3.36 What is the molecular formula of each compound?
(a) Empirical formula CH ($\mathcal{M}$ = 78.11 g/mol)
(b) Empirical formula $C_3H_6O_2$ ($\mathcal{M}$ = 74.08 g/mol)
(c) Empirical formula HgCl ($\mathcal{M}$ = 472.1 g/mol)
(d) Empirical formula $C_7H_4O_2$ ($\mathcal{M}$ = 240.20 g/mol)

3.37 Determine the empirical formula of each of the following compounds:
(a) 0.063 mol of chlorine atoms combined with 0.22 mol of oxygen atoms
(b) 2.45 g of silicon combined with 12.4 g of chlorine
(c) 27.3 mass % carbon and 72.7 mass % oxygen

3.38 Determine the empirical formula of each of the following compounds:
(a) 0.039 mol of iron atoms combined with 0.052 mol of oxygen atoms
(b) 0.903 g of phosphorus combined with 6.99 g of bromine
(c) A hydrocarbon with 79.9 mass % carbon

3.39 An oxide of nitrogen contains 30.45 mass % N. (a) What is the empirical formula of the oxide? (b) If the molar mass is 90 ± 5 g/mol, what is the molecular formula?

3.40 A chloride of silicon contains 79.1 mass % Cl. (a) What is the empirical formula of the chloride? (b) If the molar mass is 269 g/mol, what is the molecular formula?

3.41 A sample of 0.600 mol of a metal M reacts completely with excess fluorine to form 46.8 g of MF_2.
(a) How many moles of F are in the sample of MF_2 that forms?
(b) How many grams of M are in this sample of MF_2?
(c) What element is represented by the symbol M?

3.42 A sample of 0.370 mol of a metal oxide (M_2O_3) weighs 55.4 g.
(a) How many moles of O are in the sample?
(b) How many grams of M are in the sample?
(c) What element is represented by the symbol M?

Problems in Context

3.43 Nicotine is a poisonous, addictive compound found in tobacco. A sample of nicotine contains 6.16 mmol of C, 8.56 mmol of H, and 1.23 mmol of N [1 mmol (1 millimole) = 10^{-3} mol]. What is the empirical formula?

3.44 Cortisol ($\mathcal{M}$ = 362.47 g/mol), one of the major steroid hormones, is a key factor in the synthesis of protein. Its profound effect on the reduction of inflammation explains its use in the treatment of rheumatoid arthritis. Cortisol is 69.6% C, 8.34% H, and 22.1% O by mass. What is its molecular formula?

3.45 Acetaminophen (*right*) is a popular nonaspirin, "over-the-counter" pain reliever. What is the mass % of each element in acetaminophen?

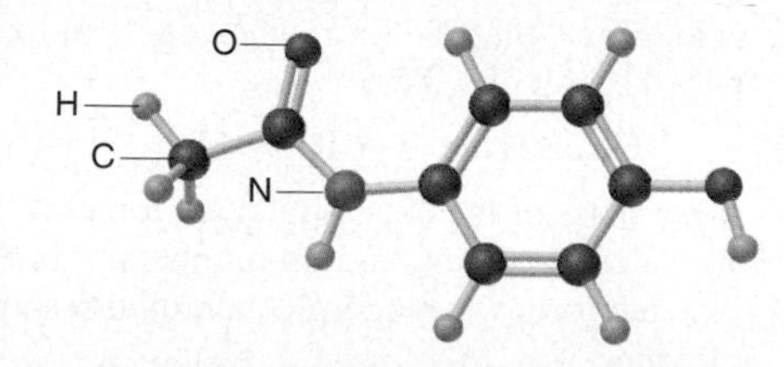

3.46 Menthol ($\mathcal{M}$ = 156.3 g/mol), a strong-smelling substance used in cough drops, is a compound of carbon, hydrogen, and oxygen. When 0.1595 g of menthol was subjected to combustion analysis, it produced 0.449 g of CO_2 and 0.184 g of H_2O. What is menthol's molecular formula?

Writing and Balancing Chemical Equations
(Sample Problem 3.7)

Concept Review Questions

3.47 What three types of information does a balanced chemical equation provide? How?

3.48 How does a balanced chemical equation apply the law of conservation of mass?

3.49 In the process of balancing the equation
$$Al + Cl_2 \longrightarrow AlCl_3$$
Student I writes: $Al + Cl_2 \longrightarrow AlCl_2$
Student II writes: $Al + Cl_2 + Cl \longrightarrow AlCl_3$
Student III writes: $2Al + 3Cl_2 \longrightarrow 2AlCl_3$
Is the approach of Student I valid? Student II? Student III? Explain.

3.50 The boxes below represent a chemical reaction between elements A (red) and B (green):

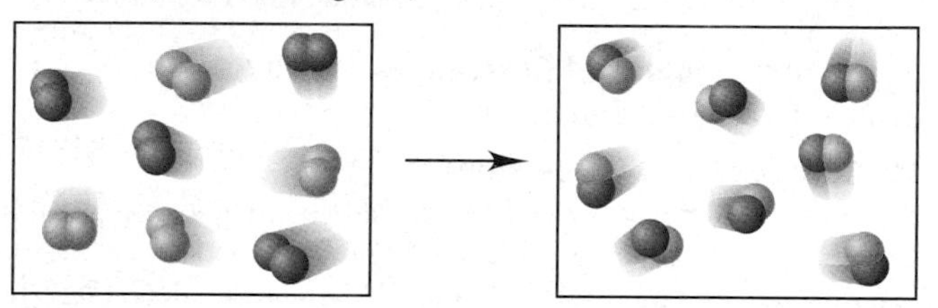

Which of the following best represents the balanced equation for the reaction?
(a) $2A + 2B \longrightarrow A_2 + B_2$ (b) $A_2 + B_2 \longrightarrow 2AB$
(c) $B_2 + 2AB \longrightarrow 2B_2 + A_2$ (d) $4A_2 + 4B_2 \longrightarrow 8AB$

Skill-Building Exercises *(grouped in similar pairs)*

3.51 Write balanced equations for each of the following by inserting the correct coefficients in the blanks:
(a) $\_Cu(s) + \_S_8(s) \longrightarrow \_Cu_2S(s)$
(b) $\_P_4O_{10}(s) + \_H_2O(l) \longrightarrow \_H_3PO_4(l)$
(c) $\_B_2O_3(s) + \_NaOH(aq) \longrightarrow \_Na_3BO_3(aq) + \_H_2O(l)$
(d) $\_CH_3NH_2(g) + \_O_2(g) \longrightarrow \_CO_2(g) + \_H_2O(g) + \_N_2(g)$

3.52 Write balanced equations for each of the following by inserting the correct coefficients in the blanks:
(a) $\_Cu(NO_3)_2(aq) + \_KOH(aq) \longrightarrow \_Cu(OH)_2(s) + \_KNO_3(aq)$
(b) $\_BCl_3(g) + \_H_2O(l) \longrightarrow \_H_3BO_3(s) + \_HCl(g)$
(c) $\_CaSiO_3(s) + \_HF(g) \longrightarrow \_SiF_4(g) + \_CaF_2(s) + \_H_2O(l)$
(d) $\_(CN)_2(g) + \_H_2O(l) \longrightarrow \_H_2C_2O_4(aq) + \_NH_3(g)$

3.53 Write balanced equations for each of the following by inserting the correct coefficients in the blanks:
(a) $\_SO_2(g) + \_O_2(g) \longrightarrow \_SO_3(g)$
(b) $\_Sc_2O_3(s) + \_H_2O(l) \longrightarrow \_Sc(OH)_3(s)$
(c) $\_H_3PO_4(aq) + \_NaOH(aq) \longrightarrow \_Na_2HPO_4(aq) + \_H_2O(l)$
(d) $\_C_6H_{10}O_5(s) + \_O_2(g) \longrightarrow \_CO_2(g) + \_H_2O(g)$

3.54 Write balanced equations for each of the following by inserting the correct coefficients in the blanks:
(a) $\_As_4S_6(s) + \_O_2(g) \longrightarrow \_As_4O_6(s) + \_SO_2(g)$
(b) $\_Ca_3(PO_4)_2(s) + \_SiO_2(s) + \_C(s) \longrightarrow \_P_4(g) + \_CaSiO_3(l) + \_CO(g)$
(c) $\_Fe(s) + \_H_2O(g) \longrightarrow \_Fe_3O_4(s) + \_H_2(g)$
(d) $\_S_2Cl_2(l) + \_NH_3(g) \longrightarrow \_S_4N_4(s) + \_S_8(s) + \_NH_4Cl(s)$

3.55 Convert the following into balanced equations:
(a) When gallium metal is heated in oxygen gas, it melts and forms solid gallium(III) oxide.
(b) Liquid hexane burns in oxygen gas to form carbon dioxide gas and water vapor.
(c) When solutions of calcium chloride and sodium phosphate are mixed, solid calcium phosphate forms and sodium chloride remains in solution.

3.56 Convert the following into balanced equations:
(a) When lead(II) nitrate solution is added to potassium iodide solution, solid lead(II) iodide forms and potassium nitrate solution remains.
(b) Liquid disilicon hexachloride reacts with water to form solid silicon dioxide, hydrogen chloride gas, and hydrogen gas.
(c) When nitrogen dioxide is bubbled into water, a solution of nitric acid forms and gaseous nitrogen monoxide is released.

Calculating Amounts of Reactant and Product

(Sample Problems 3.8 to 3.12)

Concept Review Questions

3.57 What does the term *stoichiometrically equivalent molar ratio* mean, and how is it applied in solving problems?

3.58 Reactants A and B form product C. Write a detailed "Plan" to find the mass of C when 5 g of A reacts with excess B.

3.59 Reactants D and E form product F. Write a detailed "Plan" to find the mass of F when 17 g of D reacts with 11 g of E.

3.60 Percent yields are generally calculated from mass quantities. Would the result be the same if mole quantities were used instead? Why?

Skill-Building Exercises *(grouped in similar pairs)*

3.61 Chlorine gas can be made in the laboratory by the reaction of hydrochloric acid and manganese(IV) oxide:

$$4HCl(aq) + MnO_2(s) \longrightarrow MnCl_2(aq) + 2H_2O(g) + Cl_2(g)$$

When 1.82 mol of HCl reacts with excess MnO_2, (a) how many moles of Cl_2 form? (b) How many grams of Cl_2 form?

3.62 Bismuth oxide reacts with carbon to form bismuth metal:

$$Bi_2O_3(s) + 3C(s) \longrightarrow 2Bi(s) + 3CO(g)$$

When 352 g of Bi_2O_3 reacts with excess carbon, (a) how many moles of Bi_2O_3 react? (b) How many moles of Bi form?

3.63 Potassium nitrate decomposes on heating, producing potassium oxide and gaseous nitrogen and oxygen:

$$4KNO_3(s) \longrightarrow 2K_2O(s) + 2N_2(g) + 5O_2(g)$$

To produce 88.6 kg of oxygen, how many (a) moles of KNO_3 must be heated? (b) Grams of KNO_3 must be heated?

3.64 Chromium(III) oxide reacts with hydrogen sulfide (H_2S) gas to form chromium(III) sulfide and water:

$$Cr_2O_3(s) + 3H_2S(g) \longrightarrow Cr_2S_3(s) + 3H_2O(l)$$

To produce 421 g of Cr_2S_3, (a) how many moles of Cr_2O_3 are required? (b) How many grams of Cr_2O_3 are required?

3.65 Calculate the mass of each product formed when 33.61 g of diborane (B_2H_6) reacts with excess water:

$$B_2H_6(g) + H_2O(l) \longrightarrow H_3BO_3(s) + H_2(g) \text{ [unbalanced]}$$

3.66 Calculate the mass of each product formed when 174 g of silver sulfide reacts with excess hydrochloric acid:

$$Ag_2S(s) + HCl(aq) \longrightarrow AgCl(s) + H_2S(g) \text{ [unbalanced]}$$

3.67 Elemental phosphorus occurs as tetratomic molecules, P_4. What mass of chlorine gas is needed to react completely with 355 g of phosphorus to form phosphorus pentachloride?

3.68 Elemental sulfur occurs as octatomic molecules, S_8. What mass of fluorine gas is needed to react completely with 17.8 g of sulfur to form sulfur hexafluoride?

3.69 Solid iodine trichloride is prepared by reaction between solid iodine and gaseous chlorine to form iodine monochloride crystals, followed by treatment with additional chlorine.
(a) Write a balanced equation for each step.
(b) Write an overall balanced equation for the formation of iodine trichloride.
(c) How many grams of iodine are needed to prepare 31.4 kg of final product?

3.70 Lead can be prepared from galena [lead(II) sulfide] by first roasting the galena in oxygen gas to form lead(II) oxide and sulfur dioxide. Heating the metal oxide with more galena forms the molten metal and more sulfur dioxide.
(a) Write a balanced equation for each step.
(b) Write an overall balanced equation for the process.
(c) How many metric tons of sulfur dioxide form for every metric ton of lead obtained?

3.71 Many metals react with oxygen gas to form the metal oxide. For example, calcium reacts as follows:

$$2Ca(s) + O_2(g) \longrightarrow 2CaO(s)$$

You wish to calculate the mass of calcium oxide that can be prepared from 4.20 g of Ca and 2.80 g of O_2.
(a) How many moles of CaO can be produced from the given mass of Ca?
(b) How many moles of CaO can be produced from the given mass of O_2?
(c) Which is the limiting reactant?
(d) How many grams of CaO can be produced?

3.72 Metal hydrides react with water to form hydrogen gas and the metal hydroxide. For example,

$$SrH_2(s) + 2H_2O(l) \longrightarrow Sr(OH)_2(s) + 2H_2(g)$$

You wish to calculate the mass of hydrogen gas that can be prepared from 5.63 g of SrH_2 and 4.80 g of H_2O.
(a) How many moles of H_2 can be produced from the given mass of SrH_2?
(b) How many moles of H_2 can be produced from the given mass of H_2O?
(c) Which is the limiting reactant?
(d) How many grams of H_2 can be produced?

3.73 Calculate the maximum numbers of moles and grams of iodic acid (HIO_3) that can form when 685 g of iodine trichloride reacts with 117.4 g of water:

$$ICl_3 + H_2O \longrightarrow ICl + HIO_3 + HCl \text{ [unbalanced]}$$

What mass of the excess reactant remains?

3.74 Calculate the maximum numbers of moles and grams of H_2S that can form when 158 g of aluminum sulfide reacts with 131 g of water:

$$Al_2S_3 + H_2O \longrightarrow Al(OH)_3 + H_2S \text{ [unbalanced]}$$

What mass of the excess reactant remains?

3.75 When 0.100 mol of carbon is burned in a closed vessel with 8.00 g of oxygen, how many grams of carbon dioxide can form?

Which reactant is in excess, and how many grams of it remain after the reaction?

3.76 A mixture of 0.0359 g of hydrogen and 0.0175 mol of oxygen in a closed container is sparked to initiate a reaction. How many grams of water can form? Which reactant is in excess, and how many grams of it remain after the reaction?

3.77 Aluminum nitrite and ammonium chloride react to form aluminum chloride, nitrogen, and water. What mass of each substance is present after 62.5 g of aluminum nitrite and 54.6 g of ammonium chloride react completely?

3.78 Calcium nitrate and ammonium fluoride react to form calcium fluoride, dinitrogen monoxide, and water vapor. What mass of each substance is present after 16.8 g of calcium nitrate and 17.50 g of ammonium fluoride react completely?

3.79 Two successive reactions, A $\longrightarrow$ B and B $\longrightarrow$ C, have yields of 82% and 65%, respectively. What is the overall percent yield for conversion of A to C?

3.80 Two successive reactions, D $\longrightarrow$ E and E $\longrightarrow$ F, have yields of 48% and 73%, respectively. What is the overall percent yield for conversion of D to F?

3.81 What is the percent yield of a reaction in which 41.5 g of tungsten(VI) oxide (WO_3) reacts with excess hydrogen gas to produce metallic tungsten and 9.50 mL of water (d = 1.00 g/mL)?

3.82 What is the percent yield of a reaction in which 200. g of phosphorus trichloride reacts with excess water to form 128 g of HCl and aqueous phosphorous acid (H_3PO_3)?

3.83 When 18.5 g of methane and 43.0 g of chlorine gas undergo a reaction that has an 80.0% yield, what mass of chloromethane (CH_3Cl) forms? Hydrogen chloride also forms.

3.84 When 56.6 g of calcium and 30.5 g of nitrogen gas undergo a reaction that has a 93.0% yield, what mass of calcium nitride forms?

Problems in Context

3.85 Cyanogen, $(CN)_2$, has been observed in the atmosphere of Titan, Saturn's largest moon, and in the gases of interstellar nebulas. On Earth, it is used as a welding gas and a fumigant. In its reaction with fluorine gas, carbon tetrafluoride and nitrogen trifluoride gases are produced. What mass of carbon tetrafluoride forms when 80.0 g of each reactant is used?

3.86 An intermediate step in the industrial production of nitric acid involves the reaction of ammonia with oxygen gas to form nitrogen monoxide and water. How many grams of nitrogen monoxide can form by the reaction of 466 g of ammonia with 812 g of oxygen?

3.87 Gaseous butane is compressed and used as a liquid fuel in disposable cigarette lighters and lightweight camping stoves. Suppose a lighter contains 6.50 mL of butane (d = 0.579 g/mL).
(a) How many grams of oxygen are needed to burn the butane completely?
(b) How many moles of CO_2 form when all the butane burns?
(c) How many total molecules of gas form when the butane burns completely?

3.88 Sodium borohydride ($NaBH_4$) is used industrially in many organic syntheses. One way to prepare it is by reacting sodium hydride with gaseous diborane (B_2H_6). Assuming a 95.5% yield, how many grams of $NaBH_4$ can be prepared by reacting 7.88 g of sodium hydride and 8.12 g of diborane?

Fundamentals of Solution Stoichiometry

(Sample Problems 3.13 to 3.17)

Concept Review Questions

3.89 Box A represents a unit volume of a solution. Choose from boxes B and C the one representing the same unit volume of solution that has (a) more solute added; (b) more solvent added; (c) higher molarity; (d) lower concentration.

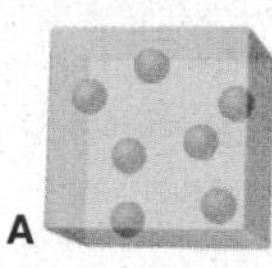

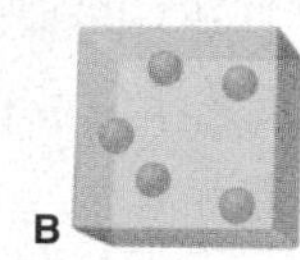

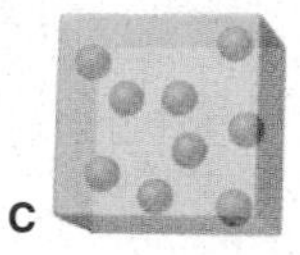

3.90 Given the volume and molarity of a $CaCl_2$ solution, how do you determine the number of moles and the mass of solute?

3.91 Are the following instructions for diluting a 10.0 M solution to a 1.00 M solution correct: "Take 100.0 mL of the 10.0 M solution and add 900.0 mL water"? Explain.

Skill-Building Exercises *(grouped in similar pairs)*

3.92 Calculate each of the following quantities:
(a) Grams of solute in 175.8 mL of 0.207 M calcium acetate
(b) Molarity of 500. mL of solution containing 21.1 g of potassium iodide
(c) Moles of solute in 145.6 L of 0.850 M sodium cyanide

3.93 Calculate each of the following quantities:
(a) Volume in liters of 2.26 M potassium hydroxide that contains 8.42 g of solute
(b) Number of Cu^{2+} ions in 52 L of 2.3 M copper(II) chloride
(c) Molarity of 275 mL of solution containing 135 mmol of glucose

3.94 Calculate each of the following quantities:
(a) Grams of solute needed to make 475 mL of 5.62×10^{-2} M potassium sulfate
(b) Molarity of a solution that contains 6.55 mg of calcium chloride in each milliliter
(c) Number of Mg^{2+} ions in each milliliter of 0.184 M magnesium bromide

3.95 Calculate each of the following quantities:
(a) Molarity of the solution resulting from dissolving 46.0 g of silver nitrate in enough water to give a final volume of 335 mL
(b) Volume in liters of 0.385 M manganese(II) sulfate that contains 57.0 g of solute
(c) Volume in milliliters of 6.44×10^{-2} M adenosine triphosphate (ATP) that contains 1.68 mmol of ATP

3.96 Calculate each of the following quantities:
(a) Molarity of a solution prepared by diluting 37.00 mL of 0.250 M potassium chloride to 150.00 mL
(b) Molarity of a solution prepared by diluting 25.71 mL of 0.0706 M ammonium sulfate to 500.00 mL
(c) Molarity of sodium ion in a solution made by mixing 3.58 mL of 0.288 M sodium chloride with 500. mL of 6.51×10^{-3} M sodium sulfate (assume volumes are additive)

3.97 Calculate each of the following quantities:
(a) Volume of 2.050 M copper(II) nitrate that must be diluted with water to prepare 750.0 mL of a 0.8543 M solution
(b) Volume of 1.03 M calcium chloride that must be diluted with water to prepare 350. mL of a 2.66×10^{-2} M chloride ion solution
(c) Final volume of a 0.0700 M solution prepared by diluting 18.0 mL of 0.155 M lithium carbonate with water

3.98 A sample of concentrated nitric acid has a density of 1.41 g/mL and contains 70.0% HNO_3 by mass.
(a) What mass of HNO_3 is present per liter of solution?
(b) What is the molarity of the solution?

3.99 Concentrated sulfuric acid (18.3 *M*) has a density of 1.84 g/mL.
(a) How many moles of sulfuric acid are present per milliliter of solution?
(b) What is the mass % of H_2SO_4 in the solution?

3.100 How many milliliters of 0.383 *M* HCl are needed to react with 16.2 g of $CaCO_3$?

$$2HCl(aq) + CaCO_3(s) \longrightarrow CaCl_2(aq) + CO_2(g) + H_2O(l)$$

3.101 How many grams of NaH_2PO_4 are needed to react with 38.74 mL of 0.275 *M* NaOH?

$$NaH_2PO_4(s) + 2NaOH(aq) \longrightarrow Na_3PO_4(aq) + 2H_2O(l)$$

3.102 How many grams of solid barium sulfate form when 25.0 mL of 0.160 *M* barium chloride reacts with 68.0 mL of 0.055 *M* sodium sulfate? Aqueous sodium chloride is the other product.

3.103 Which reactant is in excess and by how many moles when 350.0 mL of 0.210 *M* sulfuric acid reacts with 0.500 L of 0.196 *M* sodium hydroxide to form water and aqueous sodium sulfate?

Problems in Context

3.104 Ordinary household bleach is an aqueous solution of sodium hypochlorite. What is the molarity of a bleach solution that contains 20.5 g of sodium hypochlorite in a total volume of 375 mL?

3.105 Muriatic acid, an industrial grade of concentrated HCl, is used to clean masonry and etch cement for painting. Its concentration is 11.7 *M*.
(a) Write instructions for diluting the concentrated acid to make 5.0 gallons of 3.5 *M* acid for routine use (1 gal = 4 qt; 1 qt = 0.946 L).
(b) How many milliliters of the muriatic acid solution contain 9.55 g of HCl?

3.106 A sample of impure magnesium was analyzed by allowing it to react with excess HCl solution:

$$Mg(s) + 2HCl(aq) \longrightarrow MgCl_2(aq) + H_2(g)$$

After 1.32 g of the impure metal was treated with 0.100 L of 0.750 *M* HCl, 0.0125 mol of HCl remained. Assuming the impurities do not react with the acid, what is the mass % of Mg in the sample?

Comprehensive Problems

3.107 The mole is defined in terms of the carbon-12 atom. Use the definition to find (a) the mass in grams equal to 1 atomic mass unit; (b) the ratio of the gram to the atomic mass unit.

3.108 The study of sulfur-nitrogen compounds is an active area of chemical research, made more so by the discovery in the early 1980s of one such compound that conducts electricity like a metal. The first sulfur-nitrogen compound was prepared in 1835 and serves today as a reactant for preparing many of the others. Mass spectrometry of the compound shows a molar mass of 184.27 g/mol, and analysis shows it to contain 2.288 g of S for every 1.000 g of N. What is its molecular formula?

3.109 Narceine is a narcotic in opium. It crystallizes from water solution as a hydrate that contains 10.8 mass % water. If the molar mass of narceine hydrate is 499.52 g/mol, determine *x* in narceine·xH_2O.

3.110 Hydrogen-containing fuels have a "fuel value" based on their mass % H. Rank the following compounds from highest mass % H to lowest: ethane, propane, benzene, ethanol, cetyl palmitate (whale oil, $C_{32}H_{64}O_2$).

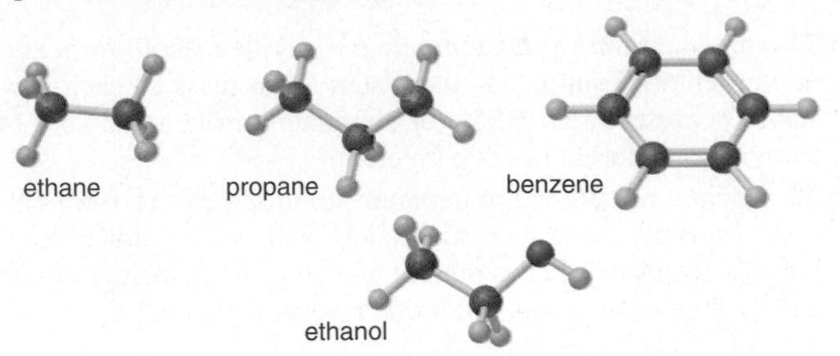

3.111 Serotonin ($\mathcal{M}$ = 176 g/mol) is a compound that transmits nerve impulses between neurons. It contains 68.2 mass % C, 6.86 mass % H, 15.9 mass % N, and 9.08 mass % O. What is its molecular formula?

3.112 Convert the following descriptions of reactions into balanced equations:
(a) In a gaseous reaction, hydrogen sulfide burns in oxygen to form sulfur dioxide and water vapor.
(b) When crystalline potassium chlorate is heated to just above its melting point, it reacts to form two different crystalline compounds, potassium chloride and potassium perchlorate.
(c) When hydrogen gas is passed over powdered iron(III) oxide, iron metal and water vapor form.
(d) The combustion of gaseous ethane in air forms carbon dioxide and water vapor.
(e) Iron(II) chloride is converted to iron(III) fluoride by treatment with chlorine trifluoride gas. Chlorine gas is also formed.

3.113 Isobutylene is a hydrocarbon used in the manufacture of synthetic rubber. When 0.847 g of isobutylene was analyzed by combustion (using an apparatus similar to that in Figure 3.5), the gain in mass of the CO_2 absorber was 2.657 g and that of the H_2O absorber was 1.089 g. What is the empirical formula of isobutylene?

3.114 One of the compounds used to increase the octane rating of gasoline is toluene (*right*). Suppose 15.0 mL of toluene (*d* = 0.867 g/mL) is consumed when a sample of gasoline burns in air.

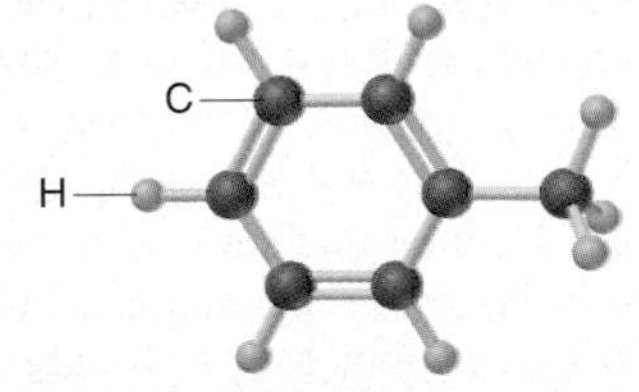

(a) How many grams of oxygen are needed for complete combustion of the toluene?
(b) How many total moles of gaseous products form?
(c) How many molecules of water vapor form?

3.115 The smelting of ferric oxide to form elemental iron occurs at high temperatures in a blast furnace through a reaction sequence with carbon monoxide. In the first step, ferric oxide reacts with carbon monoxide to form Fe_3O_4. This substance reacts with more carbon monoxide to form iron(II) oxide, which reacts with still more carbon monoxide to form molten iron. Carbon dioxide is also produced in each step. (a) Write an overall balanced equation for the iron-smelting process. (b) How many grams of carbon monoxide are required to form 40.0 metric tons of iron from ferric oxide?

3.116 During studies of the reaction in Sample Problem 3.11,

$$N_2O_4(l) + 2N_2H_4(l) \longrightarrow 3N_2(g) + 4H_2O(g)$$

a chemical engineer measured a less-than-expected yield of N_2 and discovered that the following side reaction occurs:

$$2N_2O_4(l) + N_2H_4(l) \longrightarrow 6NO(g) + 2H_2O(g)$$

In one experiment, 10.0 g of NO formed when 100.0 g of each reactant was used. What is the highest percent yield of N_2 that can be expected?

3.117 A mathematical equation useful for dilution calculations is $M_{dil} \times V_{dil} = M_{conc} \times V_{conc}$. What does each symbol mean, and why does the equation work?

3.118 The following boxes represent a chemical reaction between AB_2 and B_2:

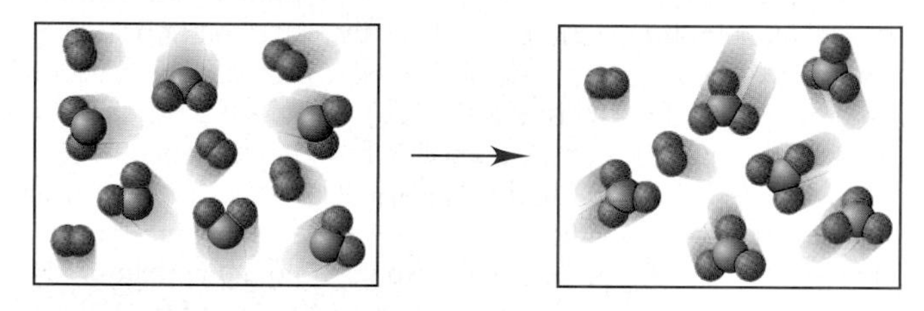

(a) Write a balanced equation for the reaction.
(b) What is the limiting reactant in this reaction?
(c) How many moles of product can be made from 3.0 mol of B_2 and 5.0 mol of AB_2?
(d) How many moles of excess reactant remain after the reaction in part (c)?

3.119 Calculate each of the following quantities:
(a) Volume of 18.0 *M* sulfuric acid that must be added to water to prepare 2.00 L of a 0.309 *M* solution
(b) Molarity of the solution obtained by diluting 80.6 mL of 0.225 *M* ammonium chloride to 0.250 L
(c) Volume of water added to 0.150 L of 0.0262 *M* sodium hydroxide to obtain a 0.0100 *M* solution (assume the volumes are additive at these low concentrations)
(d) Mass of calcium nitrate in each milliliter of a solution prepared by diluting 64.0 mL of 0.745 *M* calcium nitrate to a final volume of 0.100 L

3.120 Agricultural biochemists often use 6-benzylaminopurine ($C_{12}H_{11}N_5$) in trace amounts as a plant growth regulator. In a typical application, 150. mL of a solution contains 0.030 mg of the compound. What is the molarity of the solution?

3.121 One of Germany's largest chemical manufacturers has built a plant to produce 3.74×10^8 lb of vinyl acetate per year. This compound *(right)* is used to make some of the polymers in adhesives, paints, fabric coatings, floppy disks, plastic films, and so on. Assuming production goals are met, how many moles of vinyl acetate will the plant produce each month?

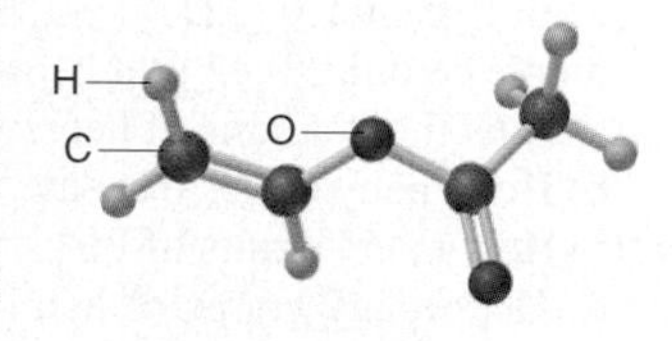

3.122 Seawater is approximately 4.0% by mass dissolved ions. About 85% of the mass of the dissolved ions is from NaCl.
(a) Calculate the mass percent of NaCl in seawater.
(b) Calculate the mass percent of Na^+ ions and of Cl^- ions in seawater.
(c) Calculate the molarity of NaCl in seawater at 15°C (*d* of seawater at 15°C = 1.025 g/mL).

3.123 Is each of the following statements true or false? Correct any that are false:
(a) A mole of one substance has the same number of atoms as a mole of any other substance.
(b) The theoretical yield for a reaction is based on the balanced chemical equation.
(c) A limiting-reactant problem is presented when the quantity of available material is given in moles for one of the reactants.
(d) To prepare 1.00 L of 3.00 *M* NaCl, weigh 175.5 g of NaCl and dissolve it in 1.00 L of distilled water.
(e) The concentration of a solution is an intensive property, but the amount of solute in a solution is an extensive property.

3.124 Box A represents one unit volume of solution A. Which box—B, C, or D—represents one unit volume after adding enough solvent to solution A to (a) triple its volume; (b) double its volume; (c) quadruple its volume?

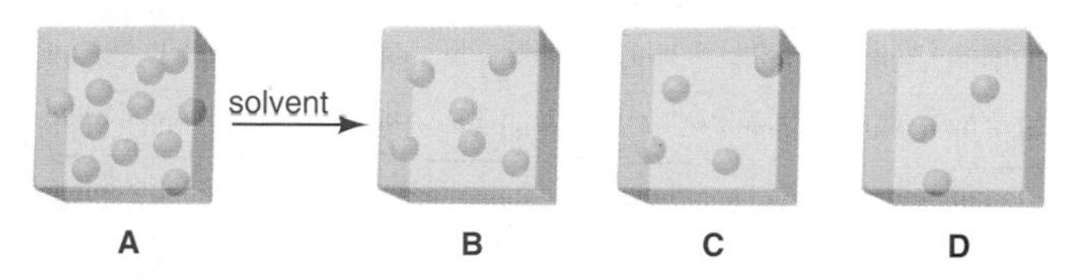

3.125 In each pair, choose the larger of the indicated quantities or state that the samples are equal:
(a) Entities: 0.4 mol of O_3 molecules or 0.4 mol of O atoms
(b) Grams: 0.4 mol of O_3 molecules or 0.4 mol of O atoms
(c) Moles: 4.0 g of N_2O_4 or 3.3 g of SO_2
(d) Grams: 0.6 mol of C_2H_4 or 0.6 mol of F_2
(e) Total ions: 2.3 mol of sodium chlorate or 2.2 mol of magnesium chloride
(f) Molecules: 1.0 g of H_2O or 1.0 g of H_2O_2
(g) Na^+ ions: 0.500 L of 0.500 *M* NaBr or 0.0146 kg of NaCl
(h) Mass: 6.02×10^{23} atoms of ^{235}U or 6.02×10^{23} atoms of ^{238}U

3.126 Balance the equation for the reaction between solid tetraphosphorus trisulfide and oxygen gas to form solid tetraphosphorus decaoxide and gaseous sulfur dioxide. Tabulate the equation (see Table 3.5) in terms of (a) molecules, (b) moles, and (c) grams.

3.127 Hydrogen gas has been suggested as a clean fuel because it produces only water vapor when it burns. If the reaction has a 98.8% yield, what mass of hydrogen forms 85.0 kg of water?

3.128 Assuming that the volumes are additive, what is the concentration of KBr in a solution prepared by mixing 0.200 L of 0.053 *M* KBr with 0.550 L of 0.078 *M* KBr?

3.129 Solar winds composed of free protons, electrons, and α particles bombard Earth constantly, knocking gas molecules out of the atmosphere. In this way, Earth loses about 3.0 kg of matter per second. It is estimated that the atmosphere will be gone in about 50 billion years. Use this estimate to calculate (a) the mass (kg) of Earth's atmosphere and (b) the amount (mol) of nitrogen, which makes up 75.5 mass % of the atmosphere.

3.130 Calculate each of the following quantities:
(a) Moles of compound in 0.588 g of ammonium bromide
(b) Number of potassium ions in 68.5 g of potassium nitrate
(c) Mass in grams of 5.85 mol of glycerol ($C_3H_8O_3$)
(d) Volume of 2.55 mol of chloroform ($CHCl_3$; d = 1.48 g/mL)
(e) Number of sodium ions in 2.11 mol of sodium carbonate
(f) Number of atoms in 10.0 μg of cadmium
(g) Number of atoms in 0.0015 mol of fluorine gas

3.131 Elements X (green) and Y (purple) react according to the following equation: $X_2 + 3Y_2 \longrightarrow 2XY_3$. Which molecular scene represents the product of the reaction?

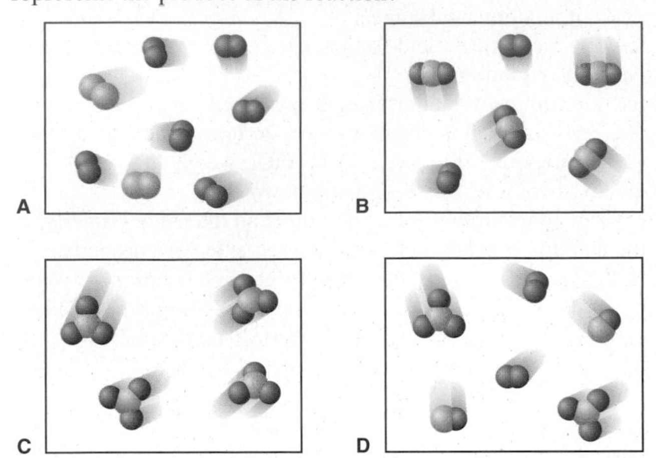

3.132 Hydrocarbon mixtures are used as fuels. How many grams of $CO_2(g)$ are produced by the combustion of 200. g of a mixture that is 25.0% CH_4 and 75.0% C_3H_8 by mass?

3.133 To 1.20 L of 0.325 *M* HCl, you add 3.37 L of a second HCl solution of unknown concentration. The resulting solution is 0.893 *M* HCl. Assuming the volumes are additive, calculate the molarity of the second HCl solution.

3.134 Nitrogen (N), phosphorus (P), and potassium (K) are the main nutrients in plant fertilizers. According to an industry convention, the numbers on the label refer to the mass percents of N, P_2O_5, and K_2O, in that order. Calculate the N:P:K ratio of a 30:10:10 fertilizer in terms of moles of each element, and express it as x:y:1.0.

3.135 A 0.652-g sample of a pure strontium halide reacts with excess sulfuric acid, and the solid strontium sulfate formed is separated, dried, and found to weigh 0.755 g. What is the formula of the original halide?

3.136 Methane and ethane are the two simplest hydrocarbons. What is the mass % C in a mixture that is 40.0% methane and 60.0% ethane by mass?

3.137 When carbon-containing compounds are burned in a limited amount of air, some $CO(g)$ as well as $CO_2(g)$ is produced. A gaseous product mixture is 35.0 mass % CO and 65.0 mass % CO_2. What is the mass % C in the mixture?

3.138 Ferrocene, first synthesized in 1951, was the first organic iron compound with Fe—C bonds. An understanding of the structure of ferrocene gave rise to new ideas about chemical bonding and led to the preparation of many useful compounds. In the combustion analysis of ferrocene, which contains only Fe, C, and H, a 0.9437-g sample produced 2.233 g of CO_2 and 0.457 g of H_2O. What is the empirical formula of ferrocene?

3.139 Citric acid *(right)* is concentrated in citrus fruits and plays a central metabolic role in nearly every animal and plant cell. (a) What are the molar mass and formula of citric acid? (b) How many moles of citric acid are in 1.50 qt of lemon juice (d = 1.09 g/mL) that is 6.82% citric acid by mass?

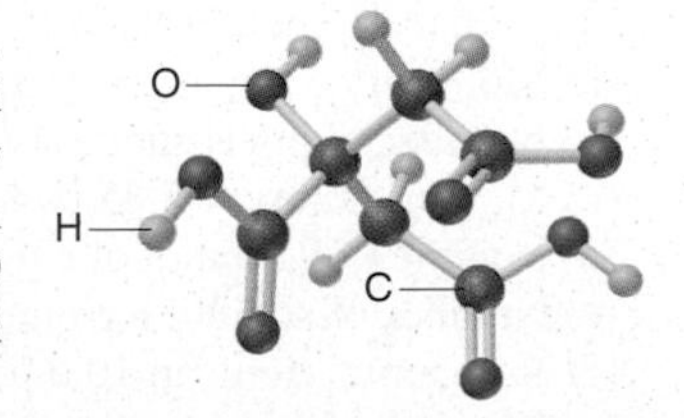

3.140 Various nitrogen oxides, as well as sulfur oxides, contribute to acidic rainfall through complex reaction sequences. Atmospheric nitrogen and oxygen combine to form nitrogen monoxide gas, which reacts with more oxygen to form nitrogen dioxide gas. In contact with water vapor, nitrogen dioxide forms aqueous nitric acid and more nitrogen monoxide. (a) Write balanced equations for these reactions. (b) Use the three equations to write one overall balanced equation that does *not* include nitrogen monoxide and nitrogen dioxide. (c) How many metric tons (t) of nitric acid form when 1.25×10^3 t of atmospheric nitrogen is consumed (1 t = 1000 kg)?

3.141 Alum [$KAl(SO_4)_2 \cdot xH_2O$] is used in food preparation, dye fixation, and water purification. To prepare alum, aluminum is reacted with potassium hydroxide and the product with sulfuric acid. Upon cooling, alum crystallizes from the solution.

(a) A 0.5404-g sample of alum is heated to drive off the waters of hydration, and the resulting $KAl(SO_4)_2$ weighs 0.2941 g. Determine the value of x and the complete formula of alum.

(b) When 0.7500 g of aluminum is used, 8.500 g of alum forms. What is the percent yield?

3.142 When 1.5173 g of an organic iron compound containing Fe, C, H, and O was burned in O_2, 2.838 g of CO_2 and 0.8122 g of H_2O were produced. In a separate experiment to determine the mass % of iron, 0.3355 g of the compound yielded 0.0758 g of Fe_2O_3. What is the empirical formula of the compound?

3.143 Fluorine is so reactive that it forms compounds with materials inert to other treatments.

(a) When 0.327 g of platinum is heated in fluorine, 0.519 g of a dark red, volatile solid forms. What is its empirical formula?

(b) When 0.265 g of this red solid reacts with excess xenon gas, 0.378 g of an orange-yellow solid forms. What is the empirical formula of this compound, the first noble gas compound formed?

(c) Fluorides of xenon can be formed by direct reaction of the elements at high pressure and temperature. Depending on conditions, the product mixture may include the difluoride, the tetrafluoride, and the hexafluoride. Under conditions that produce only the tetra- and hexafluorides, 1.85×10^{-4} mol of xenon reacted with 5.00×10^{-4} mol of fluorine, and 9.00×10^{-6} mol of xenon was found in excess. What are the mass percents of each xenon fluoride in the product mixture?

3.144 Hemoglobin is 6.0% heme ($C_{34}H_{32}FeN_4O_4$) by mass. To remove the heme, hemoglobin is treated with acetic acid and NaCl to form hemin ($C_{34}H_{32}N_4O_4FeCl$). At a crime scene, a blood sample contains 0.45 g of hemoglobin.

(a) How many grams of heme are in the sample?

(b) How many moles of heme?

(c) How many grams of Fe?

(d) How many grams of hemin could be formed for a forensic chemist to measure?

3.145 Manganese is a key component of extremely hard steel. The element occurs naturally in many oxides. A 542.3-g sample of a manganese oxide has an Mn:O ratio of 1.00:1.42 and consists of braunite (Mn_2O_3) and manganosite (MnO).

(a) What masses of braunite and manganosite are in the ore?

(b) What is the ratio Mn^{3+}:Mn^{2+} in the ore?

3.146 Sulfur dioxide is a major industrial gas used primarily for the production of sulfuric acid, but also as a bleach and food preservative. One way to produce it is by roasting iron pyrite (iron disulfide, FeS_2) in oxygen, which yields the gas and solid iron(III) oxide. What mass of each of the other three substances are involved in producing 1.00 kg of sulfur dioxide?

3.147 The human body excretes nitrogen in the form of urea, NH_2CONH_2. The key biochemical step in urea formation is the reaction of water with arginine to produce urea and ornithine:

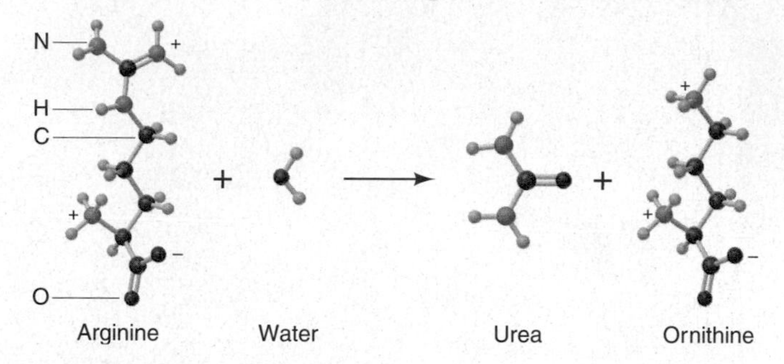

(a) What is the mass percent of nitrogen in urea, arginine, and ornithine? (b) How many grams of nitrogen can be excreted as urea when 143.2 g of ornithine is produced?

3.148 Aspirin (acetylsalicylic acid, $C_9H_8O_4$) can be made by reacting salicylic acid ($C_7H_6O_3$) with acetic anhydride [$(CH_3CO)_2O$]:

$$C_7H_6O_3(s) + (CH_3CO)_2O(l) \longrightarrow C_9H_8O_4(s) + CH_3COOH(l)$$

In one reaction, 3.027 g of salicylic acid and 6.00 mL of acetic anhydride react to form 3.261 g of aspirin.
(a) Which is the limiting reactant (d of acetic anhydride = 1.080 g/mL)?
(b) What is the percent yield of this reaction?
(c) What is the percent atom economy of this reaction?

3.149 The rocket fuel hydrazine (N_2H_4) is manufactured by the Raschig process, a three-step reaction with the following overall equation:

$$NaOCl(aq) + 2NH_3(aq) \longrightarrow N_2H_4(aq) + NaCl(aq) + H_2O(l)$$

What is the percent atom economy of this process?

3.150 Lead(II) chromate ($PbCrO_4$) is used as a yellow pigment to designate traffic lanes, but has been banned from house paint because of the potential for lead poisoning. The compound is produced from chromite ($FeCr_2O_4$), an ore of chromium:

$$4FeCr_2O_4(s) + 8K_2CO_3(aq) + 7O_2(g) \longrightarrow 2Fe_2O_3(s) + 8K_2CrO_4(aq) + 8CO_2(g)$$

Lead(II) ion then replaces the K^+ ion. If a yellow paint is 0.511% $PbCrO_4$ by mass, how many grams of chromite are needed per kilogram of paint?

3.151 Ethanol (CH_3CH_2OH), the intoxicant in alcoholic beverages, is also used to make many other organic compounds. In concentrated sulfuric acid, ethanol forms diethyl ether and water:

$$2CH_3CH_2OH(l) \longrightarrow CH_3CH_2OCH_2CH_3(l) + H_2O(g)$$

In a side reaction, some ethanol forms ethylene and water:

$$CH_3CH_2OH(l) \longrightarrow CH_2{=}CH_2(g) + H_2O(g)$$

(a) If 50.0 g of ethanol yields 33.9 g of diethyl ether, what is the percent yield of diethyl ether?
(b) During the process, 50.0% of the ethanol that did not produce diethyl ether reacts by the side reaction. What mass of ethylene is produced?

3.152 When powdered zinc is heated with sulfur, a violent reaction occurs, and zinc sulfide forms:

$$Zn(s) + S_8(s) \longrightarrow ZnS(s) \text{ [unbalanced]}$$

Some of the reactants also combine with oxygen in air to form zinc oxide and sulfur dioxide. When 85.2 g of Zn reacts with 52.4 g of S_8, 105.4 g of ZnS forms. What is the percent yield of ZnS? (b) If all the remaining reactants combine with oxygen, how many grams of each of the two oxides form?

3.153 Cocaine ($C_{17}H_{21}O_4N$) is a natural substance found in coca leaves, which have been used for centuries as a local anesthetic and stimulant. Illegal cocaine arrives in the United States either as the pure compound or as the hydrochloride salt ($C_{17}H_{21}O_4NHCl$). At 25°C, the salt is very soluble in water (2.50 kg/L), but cocaine is much less so (1.70 g/L). (a) What is the maximum amount (in g) of the salt that can dissolve in 50.0 mL of water? (b) If the solution in part (a) is treated with NaOH, the salt is converted to cocaine. How much additional water (in L) is needed to dissolve it?

3.154 High-temperature superconducting oxides hold great promise in the utility, transportation, and computer industries. (a) One superconductor is $La_{2-x}Sr_xCuO_4$. Calculate the molar mass of this oxide when $x = 0$, $x = 1$, and $x = 0.163$ (the last characterizes the compound with optimum superconducting properties). (b) Another common superconducting oxide is made by heating a mixture of barium carbonate, copper(II) oxide, and yttrium(III) oxide, followed by further heating in O_2:

$$4BaCO_3(s) + 6CuO(s) + Y_2O_3(s) \longrightarrow 2YBa_2Cu_3O_{6.5}(s) + 4CO_2(g)$$

$$2YBa_2Cu_3O_{6.5}(s) + \tfrac{1}{2}O_2(g) \longrightarrow 2YBa_2Cu_3O_7(s)$$

When equal masses of the three reactants are heated, which reactant is limiting?
(c) After the product in part (b) is removed, what is the mass percent of each reactant in the solid mixture remaining?

3.155 The zirconium oxalate $K_2Zr(C_2O_4)_3(H_2C_2O_4){\cdot}H_2O$ was synthesized by mixing 1.60 g of $ZrOCl_2{\cdot}8H_2O$ with 5.20 g of $H_2C_2O_4{\cdot}2H_2O$ and an excess of aqueous KOH. After 2 months, 1.20 g of crystalline product was obtained, as well as aqueous KCl and water. Calculate the percent yield.

3.156 A student weighs a sample of carbon on a balance that is accurate to ± 0.001 g.
(a) How many atoms are in 0.001 g of C?
(b) The carbon is used in the following reaction:

$$Pb_3O_4(s) + C(s) \longrightarrow 3PbO(s) + CO(g)$$

What mass difference in the lead(II) oxide would be caused by an error in the carbon mass of 0.001 g?

3.157 Alcohols are organic compounds that contain an —OH group attached to a hydrocarbon chain. Draw structural formulas like those in Table 3.4 for the four alcohols with the formula $C_4H_{10}O$.

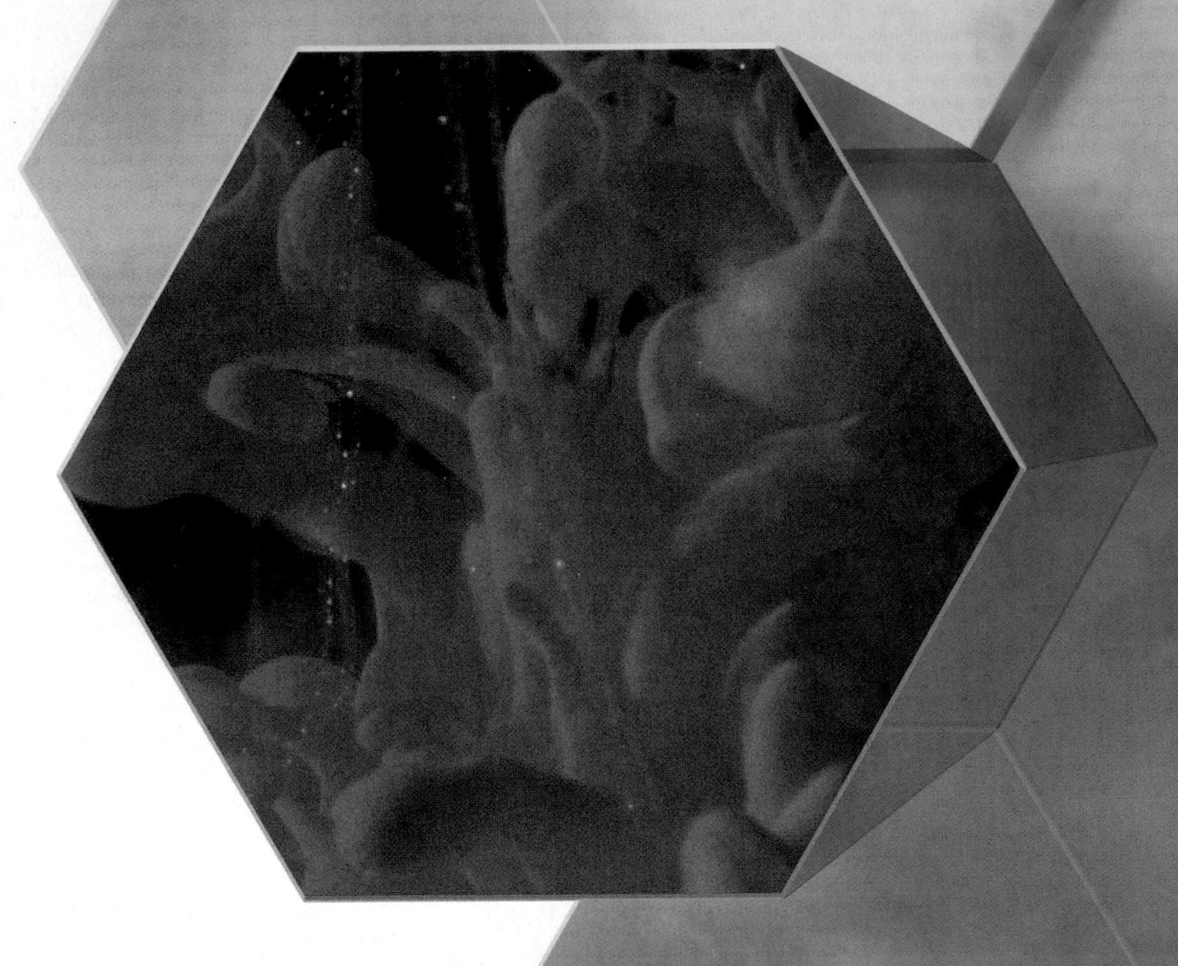

A Select Few Particles of solid lead(II) iodide swirling through a sodium nitrate solution are the products of this precipitation reaction, an example of one of a handful of major reaction classes. Distinguishing the key features of these classes and learning the role of water in many of them form the focus of the chapter.

4

The Major Classes of Chemical Reactions

Rapid chemical changes occur among gas molecules as sunlight bathes the atmosphere or lightning rips through a stormy sky. Aqueous reactions go on unceasingly in the gigantic containers we know as oceans. And, in every cell of your body, thousands of reactions taking place right now allow you to function. Indeed, the amazing variety in nature is largely a consequence of the amazing variety of chemical reactions.

Of the millions of chemical reactions occurring in and around you, we have examined only a tiny fraction so far, and it would be impossible to examine them all. Fortunately, it isn't necessary to catalog every reaction, because when we survey even a small percentage of reactions, a few major patterns emerge.

IN THIS CHAPTER . . . We examine the underlying nature of the three most common reaction processes. One of our main themes is aqueous reaction chemistry, so we first investigate how the molecular structure of water influences its crucial role as a solvent in these reactions. We see how to use ionic equations to describe reactions. We focus in turn on precipitation, acid-base, and oxidation-reduction reactions, examining why they occur and how to quantify them. We classify several important types of oxidation-reduction reactions that include elements as reactants or products. The chapter ends with an introductory look at the reversible nature of all reactions.

Concepts & Skills to Review

before you study this chapter

- names and formulas of compounds (Section 2.8)
- nature of ionic and covalent bonding (Section 2.7)
- mole-mass-number conversions (Section 3.1)
- molarity and mole-volume conversions (Section 3.5)
- balancing chemical equations (Section 3.3)
- calculating amounts of reactants and products (Section 3.4)

4.1 THE ROLE OF WATER AS A SOLVENT

Many reactions take place in aqueous solution, and our first step toward comprehending these reactions is to understand how water acts as a solvent. The role a solvent plays in a reaction depends on its chemical nature. Some solvents play a passive role, dispersing the dissolved substances into individual molecules but doing nothing further. Water plays a much more active role, interacting strongly with the substances and, in some cases, even reacting with them. To understand this active role, we'll examine the structure of water and how it interacts with ionic and covalent solutes.

The Polar Nature of Water

Of the many thousands of reactions that occur in the environment and in organisms, nearly all take place in water. Water's remarkable power as a solvent results from two features of its molecules: *the distribution of the bonding electrons* and *the overall shape*.

Recall from Section 2.7 that the electrons in a covalent bond are shared between the bonded atoms. In a covalent bond between identical atoms (as in H_2, Cl_2, O_2, etc.), the sharing is equal, so no imbalance of charge appears (Figure 4.1A). On the other hand, in covalent bonds between nonidentical atoms, the sharing is unequal: one atom attracts the electron pair more strongly than the other. For reasons discussed in Chapter 9, an O atom attracts electrons more strongly than an H atom. Therefore, in each O—H bond in water, the shared electrons spend more time closer to the O atom (Figure 4.1B).

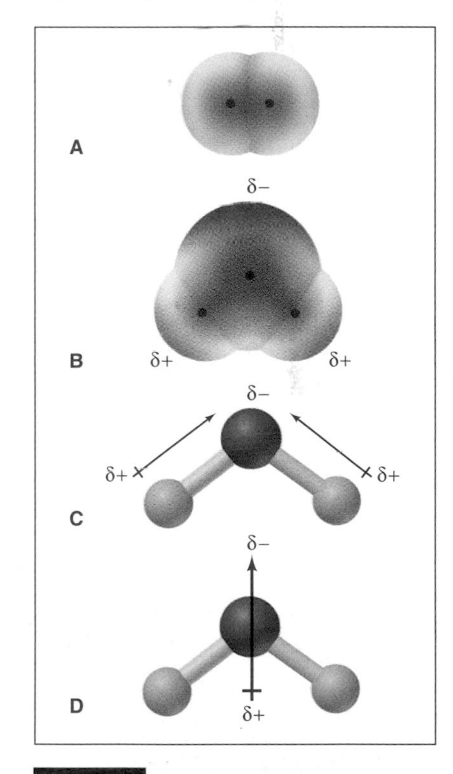

Figure 4.1 Electron distribution in molecules of H_2 and H_2O. A, In H_2, the identical nuclei attract the electrons equally. The central region of higher electron density (red) is balanced by the two outer regions of lower electron density (blue). **B,** In H_2O, the O nucleus attracts the shared electrons more strongly than the H nucleus. **C,** In this ball-and-stick model, a polar arrow points to the negative end of each O—H bond. **D,** The two polar O—H bonds and the bent shape give rise to the polar H_2O molecule.

This unequal distribution of negative charge creates partially charged "poles" at the ends of each O—H bond. The O end acts as a slightly negative pole (represented by the red shading and the $\delta-$), and the H end acts as a slightly positive pole (represented by the blue shading and the $\delta+$). Figure 4.1C indicates the bond's polarity with a *polar arrow* (the arrowhead points to the negative pole and the tail is crossed to make a "plus").

The H—O—H arrangement forms an angle, so the water molecule is bent. The combined effects of its bent shape and its polar bonds make water a **polar molecule:** the O portion of the molecule is the partially negative pole, and the region midway between the H atoms is the partially positive pole (Figure 4.1D).

Ionic Compounds in Water

In an ionic solid, the oppositely charged ions are held next to each other by electrostatic attraction (see Figure 1.3C). *Water separates the ions by replacing that attraction with one between the water molecules and the ions.* Imagine a granule of an ionic compound surrounded by bent, polar water molecules. The negative ends of some water molecules are attracted to the cations, and the positive ends of others are attracted to the anions (Figure 4.2). Gradually, the attraction between each ion and the nearby water molecules outweighs the attraction of the ions for each other. By this process, the ions separate (dissociate) and become **solvated,** surrounded tightly by solvent molecules, as they move randomly in the solution. A similar scene occurs whenever an ionic compound dissolves in water.

Figure 4.2 The dissolution of an ionic compound. When an ionic compound dissolves in water, H_2O molecules separate, surround, and disperse the ions into the liquid. The negative ends of the H_2O molecules face the positive ions and the positive ends face the negative ions.

Although many ionic compounds dissolve in water, many others do not. In the latter cases, the electrostatic attraction among ions in the compound remains greater than the attraction between ions and water molecules, so the solid stays largely intact. Actually, these so-called insoluble substances *do* dissolve to a very small extent, usually several orders of magnitude less than so-called soluble substances. Compare, for example, the solubilities of NaCl (a "soluble" compound) and AgCl (an "insoluble" compound):

$$\text{Solubility of NaCl in } H_2O \text{ at } 20°C = 365 \text{ g/L}$$
$$\text{Solubility of AgCl in } H_2O \text{ at } 20°C = 0.009 \text{ g/L}$$

Actually, the process of dissolving is more complex than just a contest between the relative energies of attraction of the particles for each other and for the solvent. In Chapter 13, we'll see that it also involves the natural tendency of the particles to disperse randomly through the solution.

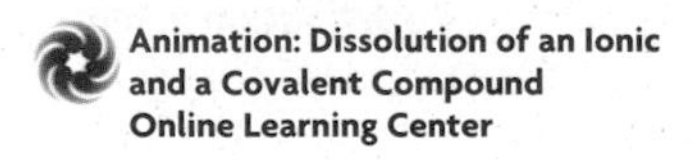

Animation: Dissolution of an Ionic and a Covalent Compound
Online Learning Center

When an ionic compound dissolves, an important change occurs in the solution. Figure 4.3 shows this change with a simple apparatus that demonstrates *electrical conductivity,* the flow of electric current. When the electrodes are immersed in pure water or pushed into an ionic solid, such as potassium bromide (KBr), no current flows. In an aqueous KBr solution, however, a significant current flows, as shown by the brightly lit bulb. This current flow implies the *movement of charged particles:* when KBr dissolves in water, the K^+ and Br^- ions dissociate, become solvated, and move toward the electrode of opposite charge. A substance that conducts a current when dissolved in water is an **electrolyte.** Soluble ionic compounds are called *strong* electrolytes because they dissociate completely into

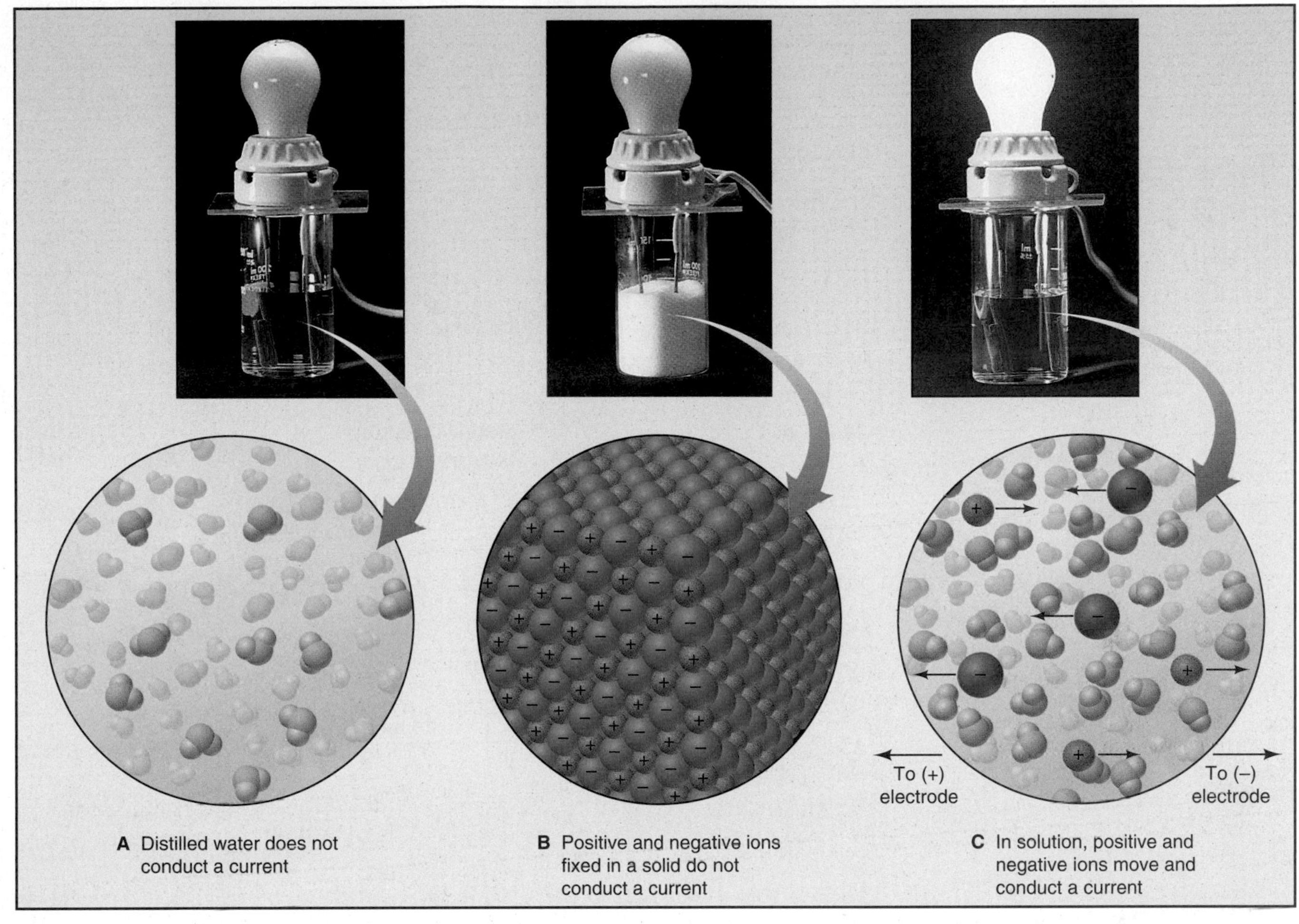
To (+) electrode
To (–) electrode
A Distilled water does not conduct a current
B Positive and negative ions fixed in a solid do not conduct a current
C In solution, positive and negative ions move and conduct a current

Figure 4.3 The electrical conductivity of ionic solutions. A, When electrodes connected to a power source are placed in distilled water, no current flows and the bulb is unlit. **B,** A solid ionic compound, such as KBr, conducts no current because the ions are bound tightly together. **C,** When KBr dissolves in H_2O, the ions separate and move through the solution toward the oppositely charged electrodes, thereby conducting a current.

ions and create a large current. We express the dissociation of KBr into solvated ions in water as follows:

$$\mathrm{KBr}(s) \xrightarrow{\mathrm{H_2O}} \mathrm{K^+}(aq) + \mathrm{Br^-}(aq)$$

The "H_2O" above the arrow indicates that water is required as the solvent but is not a reactant in the usual sense.

The formula of the compound tells the number of moles of different ions that result when the compound dissolves. Thus, 1 mol of KBr dissociates into 2 mol of ions—1 mol of K^+ and 1 mol of Br^-. The upcoming sample problem goes over this idea.

SAMPLE PROBLEM 4.1 Determining Moles of Ions in Aqueous Ionic Solutions

Problem How many moles of each ion are in each solution?

(a) 5.0 mol of ammonium sulfate dissolved in water

(b) 78.5 g of cesium bromide dissolved in water

(c) 7.42×10^{22} formula units of copper(II) nitrate dissolved in water

(d) 35 mL of 0.84 M zinc chloride

Plan We write an equation that shows 1 mol of compound dissociating into ions. In **(a)**, we multiply the moles of ions by 5.0. In **(b)**, we first convert grams to moles. In **(c)**, we first convert formula units to moles. In **(d)**, we first convert molarity and volume to moles.

Solution (a) $(NH_4)_2SO_4(s) \xrightarrow{H_2O} 2NH_4^+(aq) + SO_4^{2-}(aq)$
Remember that, in general, *polyatomic ions remain as intact units in solution.*
Calculating moles of NH_4^+ ions:

$$\text{Moles of } NH_4^+ = 5.0 \cancel{\text{mol } (NH_4)_2SO_4} \times \frac{2 \text{ mol } NH_4^+}{1 \cancel{\text{mol } (NH_4)_2SO_4}} = \boxed{10. \text{ mol } NH_4^+}$$

$\boxed{5.0 \text{ mol of } SO_4^{2-}}$ is also present.

(b) $CsBr(s) \xrightarrow{H_2O} Cs^+(aq) + Br^-(aq)$
Converting from grams to moles:

$$\text{Moles of CsBr} = 78.5 \cancel{\text{g CsBr}} \times \frac{1 \text{ mol CsBr}}{212.8 \cancel{\text{g CsBr}}} = 0.369 \text{ mol CsBr}$$

Thus, $\boxed{0.369 \text{ mol of } Cs^+ \text{ and } 0.369 \text{ mol of } Br^-}$ are present.

Animation: Strong Electrolytes, Weak Electrolytes, and Nonelectrolytes Online Learning Center

(c) $Cu(NO_3)_2(s) \xrightarrow{H_2O} Cu^{2+}(aq) + 2NO_3^-(aq)$
Converting from formula units to moles:

$$\text{Moles of } Cu(NO_3)_2 = 7.42\times10^{22} \cancel{\text{formula units } Cu(NO_3)_2} \times \frac{1 \text{ mol } Cu(NO_3)_2}{6.022\times10^{23} \cancel{\text{formula units } Cu(NO_3)_2}} = 0.123 \text{ mol } Cu(NO_3)_2$$

$$\text{Moles of } NO_3^- = 0.123 \cancel{\text{mol } Cu(NO_3)_2} \times \frac{2 \text{ mol } NO_3^-}{1 \cancel{\text{mol } Cu(NO_3)_2}} = \boxed{0.246 \text{ mol } NO_3^-}$$

$\boxed{0.123 \text{ mol of } Cu^{2+}}$ is also present.

(d) $ZnCl_2(aq) \longrightarrow Zn^{2+}(aq) + 2Cl^-(aq)$
Converting from liters to moles:

$$\text{Moles of } ZnCl_2 = 35 \cancel{\text{mL}} \times \frac{1 \cancel{\text{L}}}{10^3 \cancel{\text{mL}}} \times \frac{0.84 \text{ mol } ZnCl_2}{1 \cancel{\text{L}}} = 2.9\times10^{-2} \text{ mol } ZnCl_2$$

$$\text{Moles of } Cl^- = 2.9\times10^{-2} \cancel{\text{mol } ZnCl_2} \times \frac{2 \text{ mol } Cl^-}{1 \cancel{\text{mol } ZnCl_2}} = \boxed{5.8\times10^{-2} \text{ mol } Cl^-}$$

$\boxed{2.9\times10^{-2} \text{ mol of } Zn^{2+}}$ is also present.

Check After you round off to check the math, see if the relative moles of ions are consistent with the formula. For instance, in **(a)**, 10 mol NH_4^+/5.0 mol SO_4^{2-} = 2 NH_4^+/1 SO_4^{2-}, or $(NH_4)_2SO_4$. In **(d)**, 0.029 mol Zn^{2+}/0.058 mol Cl^- = 1 Zn^{2+}/2 Cl^-, or $ZnCl_2$.

FOLLOW-UP PROBLEM 4.1 How many moles of each ion are in each solution?
(a) 2 mol of potassium perchlorate dissolved in water
(b) 354 g of magnesium acetate dissolved in water
(c) 1.88×10^{24} formula units of ammonium chromate dissolved in water
(d) 1.32 L of 0.55 *M* sodium bisulfate

Covalent Compounds in Water

Water dissolves many covalent compounds also. Table sugar (sucrose, $C_{12}H_{22}O_{11}$), beverage (grain) alcohol (ethanol, CH_3CH_2OH), and automobile antifreeze (ethylene glycol, $HOCH_2CH_2OH$) are some familiar examples. All contain their own polar O—H bonds, which interact with those of water. However, even though these substances dissolve, they do not dissociate into ions but remain as intact molecules. As a result, their aqueous solutions do not conduct an electric current, and these substances are called **nonelectrolytes.** Many other covalent substances, such as benzene (C_6H_6) and octane (C_8H_{18}), do not contain polar bonds, and these substances do not dissolve appreciably in water.

Acids and the Solvated Proton A small, but extremely important, group of H-containing covalent compounds interacts so strongly with water that their molecules *do* dissociate into ions. In aqueous solution, these substances are all *acids,* as you'll see shortly. The molecules contain polar bonds to hydrogen, in which the atom bonded to H pulls more strongly on the shared electron pair. A good example is hydrogen chloride gas. The Cl end of the HCl molecule is partially negative, and the H end is partially positive. When HCl dissolves in water, the partially charged poles of H_2O molecules are attracted to the oppositely charged poles of HCl. The H—Cl bond breaks, with the H becoming the solvated cation $H^+(aq)$ (but see the discussion following the sample problem) and the Cl becoming the solvated anion $Cl^-(aq)$. Hydrogen bromide behaves similarly when it dissolves in water:

$$HBr(g) \xrightarrow{H_2O} H^+(aq) + Br^-(aq)$$

SAMPLE PROBLEM 4.2 Determining the Molarity of H^+ Ions in an Aqueous Solution of an Acid

Problem Nitric acid is a major chemical in the fertilizer and explosives industries. In aqueous solution, each molecule dissociates and the H becomes a solvated H^+ ion. What is the molarity of $H^+(aq)$ in 1.4 *M* nitric acid?

Plan We know the molarity of acid (1.4 *M*), so we just need the formula to find the number of moles of $H^+(aq)$ present in 1 L of solution.

Solution Nitrate ion is NO_3^-, so nitric acid is HNO_3. Thus, 1 mol of $H^+(aq)$ is released per mole of acid:

$$HNO_3(l) \xrightarrow{H_2O} H^+(aq) + NO_3^-(aq)$$

Therefore, 1.4 *M* HNO_3 contains 1.4 mol of $H^+(aq)$ per liter and is 1.4 *M* $H^+(aq)$.

FOLLOW-UP PROBLEM 4.2 How many moles of $H^+(aq)$ are present in 451 mL of 3.20 *M* hydrobromic acid?

Water interacts strongly with many ions, but most strongly with the hydrogen cation, H^+, a very unusual species. The H atom is a proton surrounded by an electron, so the H^+ ion is just a proton. Because its full positive charge is concentrated in such a tiny volume, H^+ attracts the negative pole of surrounding water molecules so strongly that it actually forms a covalent bond to one of them. We usually show this interaction by writing the aqueous H^+ ion as H_3O^+ (hydronium ion). Thus, to show more accurately what takes place when HBr(*g*) dissolves, we should write

$$HBr(g) + H_2O(l) \longrightarrow H_3O^+(aq) + Br^-(aq)$$

To make a point here about the interactions with water, let's write the hydronium ion as $(H_2O)H^+$. The hydronium ion associates with other water molecules to give species such as $H_5O_2^+$ [or $(H_2O)_2H^+$], $H_7O_3^+$ [or $(H_2O)_3H^+$], $H_9O_4^+$ [or $(H_2O)_4H^+$], and still larger aggregates; $H_7O_3^+$ is shown in Figure 4.4. These various species exist together, but we use $H^+(aq)$ as a general, simplified notation. Later in this chapter and much of the rest of the text, we show the solvated proton as $H_3O^+(aq)$ to emphasize water's role.

Water interacts covalently with many metal ions as well. For example, Fe^{3+} exists in water as $Fe(H_2O)_6^{3+}$, an Fe^{3+} ion bound to six H_2O molecules. Similarly, Zn^{2+} exists as $Zn(H_2O)_4^{2+}$ and Ni^{2+} as $Ni(H_2O)_6^{2+}$. We discuss these species fully in later chapters.

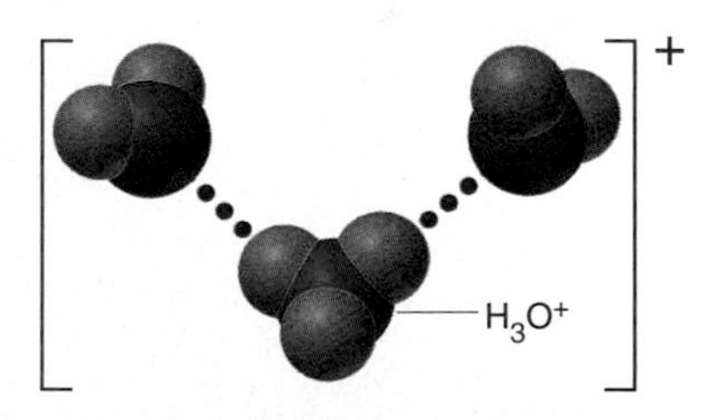

Figure 4.4 The hydrated proton. The charge of the H^+ ion is highly concentrated because the ion is so small. In aqueous solution, it forms a covalent bond to a water molecule, yielding an H_3O^+ ion that associates tightly with other H_2O molecules. Here, the $H_7O_3^+$ ion is shown.

SECTION SUMMARY

Water plays an active role in dissolving ionic compounds because it consists of polar molecules that are attracted to the ions. When an ionic compound dissolves in water, the ions dissociate from each other and become solvated by water molecules. Because the ions are free to move, their solutions conduct electricity. Water also dissolves many covalent substances with polar bonds. It interacts with some H-containing molecules so strongly it breaks their bonds and dissociates them into $H^+(aq)$ ions and anions. In water, the H^+ ion is bonded to an H_2O, forming an H_3O^+.

4.2 WRITING EQUATIONS FOR AQUEOUS IONIC REACTIONS

Chemists use three types of equations to represent aqueous ionic reactions: molecular, total ionic, and net ionic equations. As you'll see in the two types of ionic equations, by balancing the atoms, we also balance the charges.

Let's examine a reaction to see what each of these equations shows. When solutions of silver nitrate and sodium chromate are mixed, the brick-red solid silver chromate (Ag_2CrO_4) forms. Figure 4.5 depicts three views of this reaction: the change you would see if you mixed these solutions in the lab, how you might imagine the change at the atomic level among the ions, and how you can symbolize the change with the three types of equations.

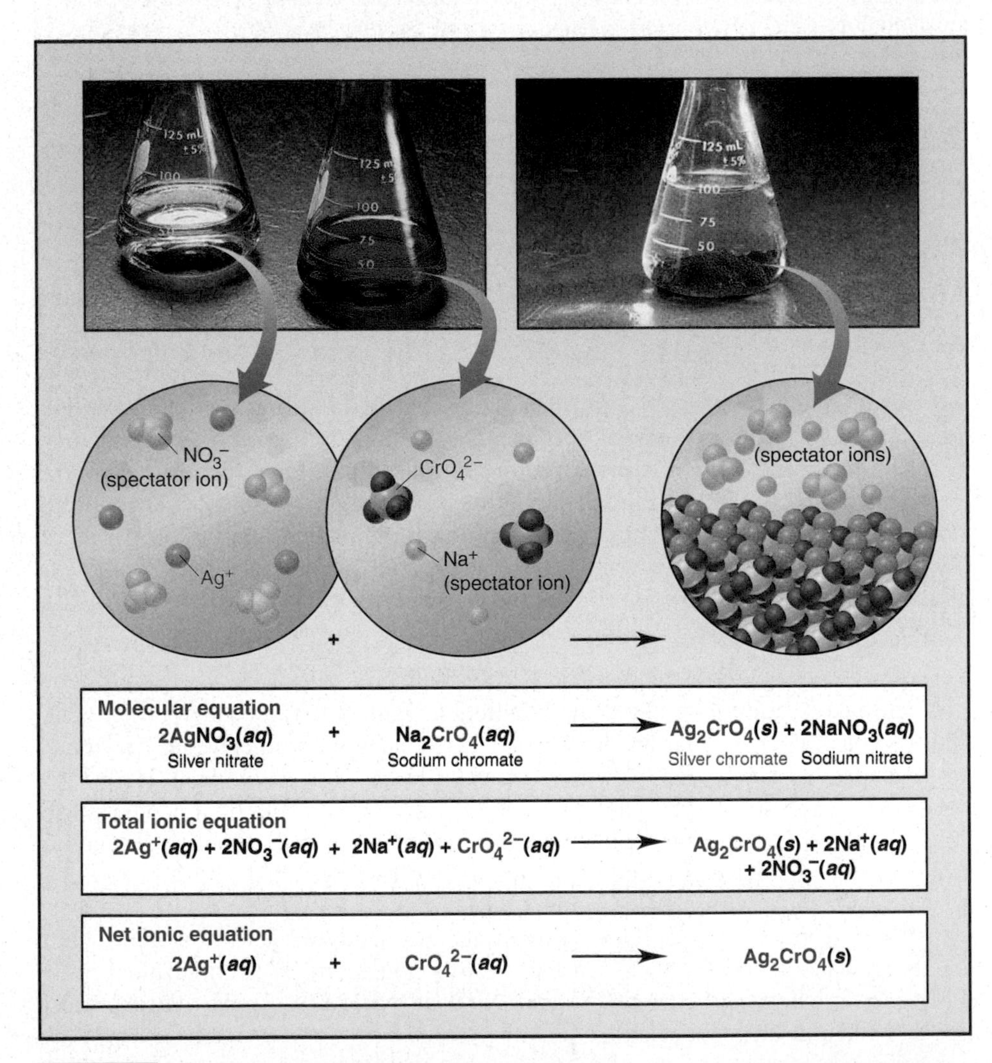

Figure 4.5 An aqueous ionic reaction and its equations. When silver nitrate and sodium chromate solutions are mixed, a reaction occurs that forms solid silver chromate and a solution of sodium nitrate. The photos present the macroscopic view of the reaction, the view the chemist sees in the lab. The blow-up arrows lead to an atomic-scale view, a representation of the chemist's mental picture of the reactants and products. (The pale ions are spectator ions, present for electrical neutrality, but not involved in the reaction.) Three equations represent the reaction in symbols. The *molecular equation* shows all substances intact. The *total ionic equation* shows all *soluble* substances as separate, solvated ions. The *net ionic equation* eliminates the spectator ions to show only the reacting species.

The **molecular equation** (*top*) reveals the least about the species in solution and is actually somewhat misleading, because *it shows all the reactants and products as if they were intact, undissociated compounds*:

$$2AgNO_3(aq) + Na_2CrO_4(aq) \longrightarrow Ag_2CrO_4(s) + 2NaNO_3(aq)$$

Only by examining the state-of-matter designations (s) and (aq), can you tell the change that has occurred.

The **total ionic equation** *(middle)* is a much more accurate representation of the reaction because *it shows all the soluble ionic substances dissociated into ions.* Now the $Ag_2CrO_4(s)$ stands out as the only undissociated substance:

$$2Ag^+(aq) + 2NO_3^-(aq) + 2Na^+(aq) + CrO_4^{2-}(aq) \longrightarrow Ag_2CrO_4(s) + 2Na^+(aq) + 2NO_3^-(aq)$$

Notice that charges balance: there are four positive and four negative charges on the left for a net zero charge, and there are two positive and two negative charges on the right for a net zero charge.

Note that $Na^+(aq)$ and $NO_3^-(aq)$ appear in the same form on both sides of the equation. They are called **spectator ions** because they are not involved in the actual chemical change. These ions are present as part of the reactants to balance the charge. That is, we can't add an Ag^+ ion without also adding an anion, in this case, NO_3^- ion.

The **net ionic equation** *(bottom)* is the most useful because *it eliminates the spectator ions and shows the actual chemical change taking place*:

$$2Ag^+(aq) + CrO_4^{2-}(aq) \longrightarrow Ag_2CrO_4(s)$$

The formation of solid silver chromate from silver ions and chromate ions *is* the only change. In fact, if we had originally mixed solutions of potassium chromate, $K_2CrO_4(aq)$, and silver acetate, $AgC_2H_3O_2(aq)$, instead of sodium chromate and silver nitrate, the same change would have occurred. Only the spectator ions would differ—$K^+(aq)$ and $C_2H_3O_2^-(aq)$ instead of $Na^+(aq)$ and $NO_3^-(aq)$.

Now, let's apply these types of equations to the three most important types of chemical reactions—precipitation, acid-base, and oxidation-reduction.

SECTION SUMMARY

A molecular equation for an aqueous ionic reaction shows undissociated substances. A total ionic equation shows all soluble ionic compounds as separate, solvated ions. Spectator ions appear unchanged on both sides of the equation. By eliminating them, you see the actual chemical change in a net ionic equation.

4.3 PRECIPITATION REACTIONS

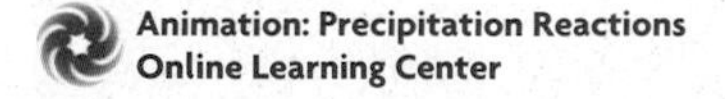

Precipitation reactions are common in both nature and industry. Many geological formations, including coral reefs, some gems and minerals, and deep-sea structures form, in part, through this type of chemical process. And, as you'll see, the chemical industry employs precipitation methods to produce several important inorganic compounds.

The Key Event: Formation of a Solid from Dissolved Ions

In **precipitation reactions,** two soluble ionic compounds react to form an insoluble product, a **precipitate.** The reaction you just saw between silver nitrate and sodium chromate is an example. Precipitates form for the same reason that some ionic compounds do not dissolve: the electrostatic attraction between the ions outweighs the tendency of the ions to remain solvated and move randomly throughout the solution. When solutions of such ions are mixed, the ions collide and stay

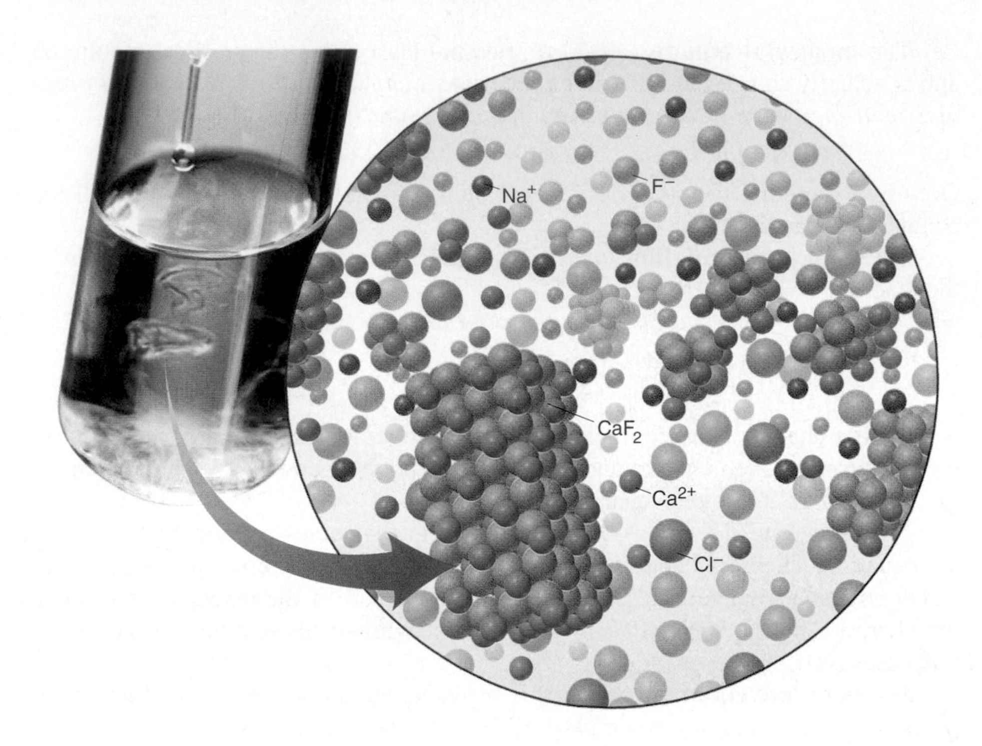

Figure 4.6 The precipitation of calcium fluoride. When an aqueous solution of NaF is added to a solution of $CaCl_2$, Ca^{2+} and F^- ions form particles of solid CaF_2.

together, and the resulting substance "comes out of solution" as a solid, as shown in Figure 4.6 for calcium fluoride.

Thus, the key event in a precipitation reaction is *the formation of an insoluble product through the net removal of solvated ions from solution.* (As you'll see shortly, acid-base reactions have a similar result but the product is water instead of an ionic compound.)

Predicting Whether a Precipitate Will Form

If you mix aqueous solutions of two ionic compounds, can you predict if a precipitate will form? Consider this example. When solid sodium iodide and potassium nitrate are each dissolved in water, each solution consists of separated ions dispersed throughout the solution:

$$NaI(s) \xrightarrow{H_2O} Na^+(aq) + I^-(aq)$$
$$KNO_3(s) \xrightarrow{H_2O} K^+(aq) + NO_3^-(aq)$$

Let's follow three steps to predict whether a precipitate will form:

1. *Note the ions present in the reactants.* The reactant ions are
$$Na^+(aq) + I^-(aq) + K^+(aq) + NO_3^-(aq) \longrightarrow ?$$
2. *Consider the possible cation-anion combinations.* In addition to the two original ones, NaI and KNO_3, which you know are soluble, the other possible cation-anion combinations are $NaNO_3$ and KI.
3. *Decide whether any of the combinations is insoluble.* A reaction does *not* occur when you mix these starting solutions because all the combinations—NaI, KNO_3, $NaNO_3$, and KI—are soluble. All the ions remain in solution. (You'll see shortly a set of rules for deciding if a product is soluble or not.)

Now, what happens if you substitute a solution of lead(II) nitrate, $Pb(NO_3)_2$, for the KNO_3? The reactant ions are Na^+, I^-, Pb^{2+}, and NO_3^-. In addition to the two soluble reactants, NaI and $Pb(NO_3)_2$, the other two possible cation-anion combinations are $NaNO_3$ and PbI_2. Lead(II) iodide is insoluble, so a reaction *does* occur because ions are removed from solution (Figure 4.7):

$$2Na^+(aq) + 2I^-(aq) + Pb^{2+}(aq) + 2NO_3^-(aq) \longrightarrow 2Na^+(aq) + 2NO_3^-(aq) + PbI_2(s)$$

Figure 4.7 The reaction of $Pb(NO_3)_2$ and NaI. When aqueous solutions of these ionic compounds are mixed, the yellow solid PbI_2 forms.

Table 4.1 Solubility Rules for Ionic Compounds in Water

Soluble Ionic Compounds	Insoluble Ionic Compounds
1. All common compounds of Group 1A(1) ions (Li^+, Na^+, K^+, etc.) and ammonium ion (NH_4^+) are soluble.	1. All common metal hydroxides are insoluble, *except* those of Group 1A(1) and the larger members of Group 2A(2) (beginning with Ca^{2+}).
2. All common nitrates (NO_3^-), acetates (CH_3COO^- or $C_2H_3O_2^-$), and most perchlorates (ClO_4^-) are soluble.	2. All common carbonates (CO_3^{2-}) and phosphates (PO_4^{3-}) are insoluble, *except* those of Group 1A(1) and NH_4^+.
3. All common chlorides (Cl^-), bromides (Br^-), and iodides (I^-) are soluble, *except* those of Ag^+, Pb^{2+}, Cu^+, and Hg_2^{2+}. All common fluorides (F^-) are soluble, *except* those of Pb^{2+} and Group 2A(2).	3. All common sulfides are insoluble *except* those of Group 1A(1), Group 2A(2), and NH_4^+.
4. All common sulfates (SO_4^{2-}) are soluble, *except* those of Ca^{2+}, Sr^{2+}, Ba^{2+}, Ag^+, and Pb^{2+}.	

A close look (with color) at the molecular equation shows that *the ions are exchanging partners*:

$$2NaI(aq) + Pb(NO_3)_2(aq) \longrightarrow PbI_2(s) + 2NaNO_3(aq)$$

Such reactions are called double-displacement, or **metathesis** (pronounced *meh-TA-thuh-sis*) **reactions.** Several are important in industry, such as the preparation of silver bromide for the manufacture of black-and-white film:

$$AgNO_3(aq) + KBr(aq) \longrightarrow AgBr(s) + KNO_3(aq)$$

As we said, there is no simple way to decide whether any given ion combination is soluble or not, so Table 4.1 provides a short list of solubility rules to memorize. They allow you to predict the outcome of many precipitation reactions.

SAMPLE PROBLEM 4.3 Predicting Whether a Precipitation Reaction Occurs; Writing Ionic Equations

Problem Predict whether a reaction occurs when each of the following pairs of solutions are mixed. If a reaction does occur, write balanced molecular, total ionic, and net ionic equations, and identify the spectator ions.
(a) Sodium sulfate(*aq*) + strontium nitrate(*aq*) $\longrightarrow$
(b) Ammonium perchlorate(*aq*) + sodium bromide(*aq*) $\longrightarrow$

Plan For each pair of solutions, we note the ions present in the reactants, write the cation-anion combinations, and refer to Table 4.1 to see if any are insoluble. For the molecular equation, we predict the products. For the total ionic equation, we write the soluble compounds as separate ions. For the net ionic equation, we eliminate the spectator ions.

Solution (a) In addition to the reactants, the two other ion combinations are strontium sulfate and sodium nitrate. Table 4.1 shows that strontium sulfate is insoluble, so a reaction *does* occur. Writing the molecular equation:

$$Na_2SO_4(aq) + Sr(NO_3)_2(aq) \longrightarrow SrSO_4(s) + 2NaNO_3(aq)$$

Writing the total ionic equation:

$$2Na^+(aq) + SO_4^{2-}(aq) + Sr^{2+}(aq) + 2NO_3^-(aq) \longrightarrow SrSO_4(s) + 2Na^+(aq) + 2NO_3^-(aq)$$

Writing the net ionic equation:

$$Sr^{2+}(aq) + SO_4^{2-}(aq) \longrightarrow SrSO_4(s)$$

The spectator ions are Na^+ and NO_3^-.

(b) The other ion combinations are ammonium bromide and sodium perchlorate. Table 4.1 shows that all ammonium, sodium, and most perchlorate compounds are soluble, and all bromides are soluble except those of Ag^+, Pb^{2+}, Cu^+, and Hg_2^{2+}. Therefore, ***no*** **reaction occurs.** The compounds remain dissociated in solution as solvated ions.

FOLLOW-UP PROBLEM 4.3 Predict whether a reaction occurs, and write balanced total and net ionic equations:
(a) Iron(III) chloride(*aq*) + cesium phosphate(*aq*) ⟶
(b) Sodium hydroxide(*aq*) + cadmium nitrate(*aq*) ⟶
(c) Magnesium bromide(*aq*) + potassium acetate(*aq*) ⟶
(d) Silver sulfate(*aq*) + barium chloride(*aq*) ⟶

SECTION SUMMARY

Precipitation reactions involve the formation of an insoluble ionic compound from two soluble ones. They occur because electrostatic attractions among certain pairs of solvated ions are strong enough to cause their removal from solution. Such reactions can be predicted by noting whether any possible ion combinations are insoluble, based on a set of solubility rules.

Table 4.2 Strong and Weak Acids and Bases

Acids
Strong
Hydrochloric acid, HCl
Hydrobromic acid, HBr
Hydriodic acid, HI
Nitric acid, HNO_3
Sulfuric acid, H_2SO_4
Perchloric acid, $HClO_4$
Weak
Hydrofluoric acid, HF
Phosphoric acid, H_3PO_4
Acetic acid, CH_3COOH (or $HC_2H_3O_2$)
Bases
Strong
Sodium hydroxide, $NaOH$
Potassium hydroxide, KOH
Calcium hydroxide, $Ca(OH)_2$
Strontium hydroxide, $Sr(OH)_2$
Barium hydroxide, $Ba(OH)_2$
Weak
Ammonia, NH_3

4.4 ACID-BASE REACTIONS

Aqueous acid-base reactions involve water not only as solvent but also in the more active roles of reactant and product. These reactions occur in processes as diverse as the biochemical synthesis of proteins, the production of industrial fertilizer, and some of the methods for revitalizing lakes damaged by acid rain.

Obviously, an **acid-base reaction** (also called a **neutralization reaction**) occurs when an acid reacts with a base, but the definitions of these terms and the scope of this reaction category have changed considerably over the years. For our purposes at this point, we'll use definitions that apply to chemicals you commonly encounter in the lab:

- *An* ***acid*** *is a substance that produces H^+ ions when dissolved in water.*

$$HX \xrightarrow{H_2O} H^+(aq) + X^-(aq)$$

- *A* ***base*** *is a substance that produces OH^- ions when dissolved in water.*

$$MOH \xrightarrow{H_2O} M^+(aq) + OH^-(aq)$$

(Other definitions of *acid* and *base* are presented later in this section and again in Chapter 18, along with a fuller meaning of *neutralization.*)

Acids and bases are electrolytes and are often categorized in terms of their "strength," which refers to the degree to which they dissociate into ions in aqueous solution. Table 4.2 lists some acids and bases in terms of their strength. In water, *strong acids and strong bases dissociate completely* into ions. Therefore, like soluble ionic compounds, they are *strong* electrolytes and conduct a current well (see left photo). In contrast, *weak acids and weak bases dissociate into ions very little, and most of their molecules remain intact.* As a result, they conduct only a small current and are *weak* electrolytes (see right photo).

Strong acids and bases are strong electrolytes.

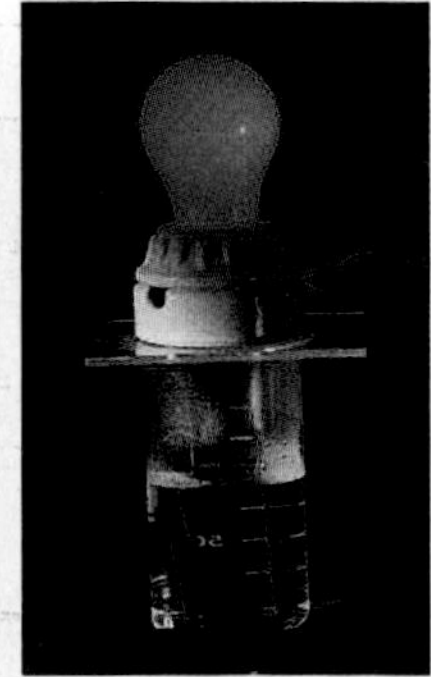

Weak acids and bases are weak electrolytes.

Strong and weak acids have one or more H atoms as part of their structure. Strong bases have either the OH^- or the O^{2-} ion as part of their structure. Soluble ionic oxides, such as K_2O, are strong bases because the oxide ion is not stable in water and reacts immediately to form hydroxide ion:

$$K_2O(s) + H_2O(l) \longrightarrow 2K^+(aq) + 2OH^-(aq)$$

Weak bases, such as ammonia, do not contain OH^- ions, but they produce them in a reaction with water that occurs to a small extent:

$$NH_3(g) + H_2O(l) \rightleftharpoons NH_4^+(aq) + OH^-(aq)$$

(Note the reaction arrow in the preceding equation. This type of arrow indicates that the reaction proceeds in both directions; we'll discuss this important idea further in Section 4.7.)

The Key Event: Formation of H_2O from H^+ and OH^-

Let's use ionic equations to see what occurs in acid-base reactions. We begin with the molecular equation for the reaction between the strong acid HCl and the strong base $Ba(OH)_2$:

$$2HCl(aq) + Ba(OH)_2(aq) \longrightarrow BaCl_2(aq) + 2H_2O(l)$$

Because HCl and $Ba(OH)_2$ dissociate completely and H_2O remains undissociated, the total ionic equation is

$$2H^+(aq) + 2Cl^-(aq) + Ba^{2+}(aq) + 2OH^-(aq) \longrightarrow Ba^{2+}(aq) + 2Cl^-(aq) + 2H_2O(l)$$

In the net ionic equation, we eliminate the spectator ions $Ba^{2+}(aq)$ and $Cl^-(aq)$ and see the actual reaction:

$$2H^+(aq) + 2OH^-(aq) \longrightarrow 2H_2O(l)$$

Or

$$H^+(aq) + OH^-(aq) \longrightarrow H_2O(l)$$

Thus, *the essential change in all aqueous reactions between a strong acid and a strong base is that an H^+ ion from the acid and an OH^- ion from the base form a water molecule. In fact,* only the spectator ions differ from one strong acid–strong base reaction to another.

Now it's easy to understand how these reactions take place: like precipitation reactions, acid-base reactions occur through *the electrostatic attraction of ions and their removal from solution* in the formation of the product. In this case, the ions are H^+ and OH^- and the product is H_2O, which consists almost entirely of undissociated molecules. (Actually, water molecules *do* dissociate, but *very* slightly. As you'll see in Chapter 18, this slight dissociation is very important, but the formation of water in a neutralization reaction nevertheless represents an enormous net removal of H^+ and OH^- ions.)

Evaporate the water from the above reaction mixture, and the ionic solid barium chloride remains. An ionic compound that results from the reaction of an acid and a base is called a **salt.** Thus, in a typical aqueous neutralization reaction, *the reactants are an acid and a base, and the products are a salt solution and water:*

$$\underset{\text{acid}}{HX(aq)} + \underset{\text{base}}{MOH(aq)} \longrightarrow \underset{\text{salt}}{MX(aq)} + \underset{\text{water}}{H_2O(l)}$$

The color shows that *the cation of the salt comes from the base and the anion comes from the acid.*

Note that acid-base reactions, like precipitation reactions, are metathesis (double-displacement) reactions. The molecular equation for the reaction of aluminum hydroxide, the active ingredient in some antacid tablets, with HCl, the major component of stomach acid, shows this clearly:

$$3HCl(aq) + Al(OH)_3(s) \longrightarrow AlCl_3(aq) + 3H_2O(l)$$

Acid-base reactions occur frequently in the synthesis and breakdown of biological macromolecules.

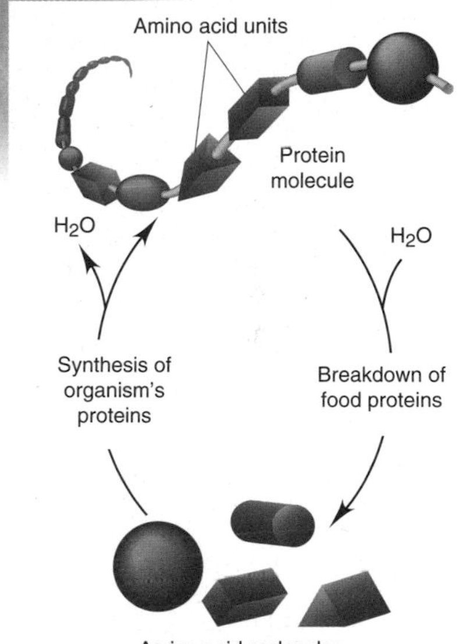

Displacement Reactions Inside You The digestion of food proteins and the formation of an organism's own proteins form a continuous cycle of displacement reactions. A protein consists of hundreds or thousands of smaller molecules, called *amino acids,* linked in a long chain. When you eat proteins, your digestive processes use H_2O to displace one amino acid at a time. These are transported by the blood to your cells, where other metabolic processes link them together, displacing H_2O, to make your own proteins.

SAMPLE PROBLEM 4.4 Writing Ionic Equations for Acid-Base Reactions

Problem Write balanced molecular, total ionic, and net ionic equations for each of the following acid-base reactions and identify the spectator ions:

(a) Strontium hydroxide(*aq*) + perchloric acid(*aq*) ⟶

(b) Barium hydroxide(*aq*) + sulfuric acid(*aq*) ⟶

Plan All are strong acids and bases (see Table 4.2), so the essential reaction is between H^+ and OH^-. The products are H_2O and a salt solution consisting of the spectator ions. Note that in **(b)**, the salt ($BaSO_4$) is insoluble (see Table 4.1), so virtually all ions are removed from solution.

Solution (a) Writing the molecular equation:

$$Sr(OH)_2(aq) + 2HClO_4(aq) \longrightarrow Sr(ClO_4)_2(aq) + 2H_2O(l)$$

Writing the total ionic equation:

$$Sr^{2+}(aq) + 2OH^-(aq) + 2H^+(aq) + 2ClO_4^-(aq) \longrightarrow Sr^{2+}(aq) + 2ClO_4^-(aq) + 2H_2O(l)$$

Writing the net ionic equation:

$$2OH^-(aq) + 2H^+(aq) \longrightarrow 2H_2O(l) \text{ or } OH^-(aq) + H^+(aq) \longrightarrow H_2O(l)$$

$Sr^{2+}(aq)$ and $ClO_4^-(aq)$ are the spectator ions.

(b) Writing the molecular equation:

$$Ba(OH)_2(aq) + H_2SO_4(aq) \longrightarrow BaSO_4(s) + 2H_2O(l)$$

Writing the total ionic equation:

$$Ba^{2+}(aq) + 2OH^-(aq) + 2H^+(aq) + SO_4^{2-}(aq) \longrightarrow BaSO_4(s) + 2H_2O(l)$$

The net ionic equation is the same as the total ionic equation. This is a precipitation *and* a neutralization reaction. There are no spectator ions because all the ions are used to form the two products.

FOLLOW-UP PROBLEM 4.4 Write balanced molecular, total ionic, and net ionic equations for the reaction between aqueous solutions of calcium hydroxide and nitric acid.

Acid-Base Titrations

Chemists study acid-base reactions quantitatively through titrations. In any **titration,** *one solution of known concentration is used to determine the concentration of another solution through a monitored reaction.*

In a typical acid-base titration, a *standardized* solution of base, one whose concentration is *known,* is added slowly to an acid solution of *unknown* concentration (Figure 4.8). A known volume of the acid solution is placed in a flask, and a few drops of indicator solution are added. An *acid-base indicator* is a substance

Figure 4.8 An acid-base titration.
A, In this procedure, a measured volume of the unknown acid solution is placed in a flask beneath a buret containing the known (standardized) base solution. A few drops of indicator are added to the flask; the indicator used here is phenolphthalein, which is colorless in acid and pink in base. After an initial buret reading, base (OH^- ions) is added slowly to the acid (H^+ ions). **B,** Near the end of the titration, the indicator momentarily changes to its base color but reverts to its acid color with swirling. **C,** When the end point is reached, a tiny excess of OH^- is present, shown by the permanent change in color of the indicator. The difference between the final buret reading and the initial buret reading gives the volume of base used.

whose color is different in acid than in base. (We examine indicators in Chapters 18 and 19.) The standardized solution of base is added slowly to the flask from a buret. As the titration is close to its end, indicator molecules near a drop of added base change color due to the temporary excess of OH^- ions there. As soon as the solution is swirled, however, the indicator's acidic color returns. The **equivalence point** in the titration occurs when *all the moles of H^+ ions present in the original volume of acid solution have reacted with an equivalent number of moles of OH^- ions added from the buret:*

$$\text{Moles of H}^+ \text{ (originally in flask)} = \text{moles of OH}^- \text{ (added from buret)}$$

The **end point** of the titration occurs when a tiny excess of OH^- ions changes the indicator permanently to its color in base. In calculations, we assume this tiny excess is insignificant, and therefore *the amount of base needed to reach the end point is the same as the amount needed to reach the equivalence point.*

SAMPLE PROBLEM 4.5 Finding the Concentration of Acid from an Acid-Base Titration

Problem You perform an acid-base titration to standardize an HCl solution by placing 50.00 mL of HCl in a flask with a few drops of indicator solution. You put 0.1524 *M* NaOH into the buret, and the initial reading is 0.55 mL. At the end point, the buret reading is 33.87 mL. What is the concentration of the HCl solution?

Plan We must find the molarity of acid from the volume of acid (50.00 mL), the initial (0.55 mL) and final (33.87 mL) volumes of base, and the molarity of base (0.1524 *M*). First, we balance the equation. We find the volume of base added from the difference in buret readings and use the base's molarity to calculate the amount (mol) of base added. Then, we use the molar ratio from the balanced equation to find the amount (mol) of acid originally present and divide by the acid's original volume to find the molarity.

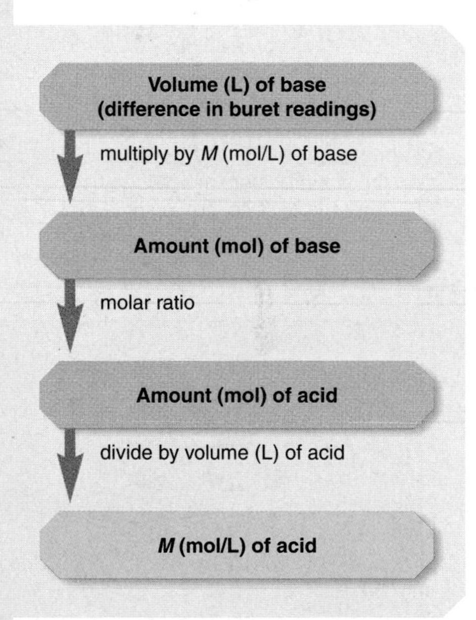

Solution Writing the balanced equation:

$$\text{NaOH}(aq) + \text{HCl}(aq) \longrightarrow \text{NaCl}(aq) + \text{H}_2\text{O}(l)$$

Finding volume (L) of NaOH solution added:

$$\text{Volume (L) of solution} = (33.87 \cancel{\text{mL}} \text{ soln} - 0.55 \cancel{\text{mL}} \text{ soln}) \times \frac{1 \text{ L}}{1000 \cancel{\text{mL}}}$$
$$= 0.03332 \text{ L soln}$$

Finding amount (mol) of NaOH added:

$$\text{Moles of NaOH} = 0.03332 \cancel{\text{L soln}} \times \frac{0.1524 \text{ mol NaOH}}{1 \cancel{\text{L soln}}}$$
$$= 5.078\times10^{-3} \text{ mol NaOH}$$

Finding amount (mol) of HCl originally present: Since the molar ratio is 1:1,

$$\text{Moles of HCl} = 5.078\times10^{-3} \cancel{\text{mol NaOH}} \times \frac{1 \text{ mol HCl}}{1 \cancel{\text{mol NaOH}}} = 5.078\times10^{-3} \text{ mol HCl}$$

Calculating molarity of HCl:

$$\text{Molarity of HCl} = \frac{5.078\times10^{-3} \text{ mol HCl}}{50.00 \cancel{\text{mL}}} \times \frac{1000 \cancel{\text{mL}}}{1 \text{ L}}$$
$$= 0.1016 \; M \text{ HCl}$$

Check The answer makes sense: a larger volume of less concentrated acid neutralized a smaller volume of more concentrated base. Rounding shows that the moles of H^+ and OH^- are about equal: 50 mL × 0.1 *M* H^+ = 0.005 mol = 33 mL × 0.15 *M* OH^-.

FOLLOW-UP PROBLEM 4.5 What volume of 0.1292 *M* $Ba(OH)_2$ would neutralize 50.00 mL of the HCl solution standardized in the preceding sample problem?

Proton Transfer: A Closer Look at Acid-Base Reactions

We gain deeper insight into acid-base reactions if we look closely at the species in solution. Let's see what takes place when HCl gas dissolves in water. Polar water molecules pull apart each HCl molecule and the H^+ ion bonds to a water molecule. In essence, we can say that HCl *transfers its proton* to H_2O:

$$\overbrace{HCl(g) \quad + \quad H_2O(l)}^{H^+ \text{ transfer}} \longrightarrow H_3O^+(aq) + Cl^-(aq)$$

Thus, hydrochloric acid (an aqueous solution of HCl gas) actually consists of solvated H_3O^+ and Cl^- ions.

When sodium hydroxide solution is added, the H_3O^+ ion transfers a proton to the OH^- ion of the base (with the product water shown here as HOH):

$$\overbrace{[H_3O^+(aq) + Cl^-(aq)] + [Na^+(aq) + OH^-(aq)]}^{H^+ \text{ transfer}} \longrightarrow$$
$$H_2O(l) + Cl^-(aq) + Na^+(aq) + HOH(l)$$

Without the spectator ions, the transfer of a proton from H_3O^+ to OH^- is obvious:

$$\overbrace{H_3O^+(aq) \quad + \quad OH^-(aq)}^{H^+ \text{ transfer}} \longrightarrow H_2O(l) + HOH(l) \quad [\text{or } 2H_2O(l)]$$

This net ionic equation is identical with the one we saw earlier (see p. 145),

$$H^+(aq) + OH^-(aq) \longrightarrow H_2O(l)$$

with the additional H_2O molecule coming from the H_3O^+. Thus, *an acid-base reaction is a proton-transfer process.* In this case, the Na^+ and Cl^- ions remain in solution, and if the water is evaporated, they crystallize as the salt NaCl. Figure 4.9 shows this process on the atomic level.

In the early 20th century, the chemists Johannes Brønsted and Thomas Lowry realized the proton-transfer nature of acid-base reactions. They defined *an acid as a molecule (or ion) that donates a proton, and a base as a molecule (or ion) that accepts a proton.* Therefore, in the aqueous reaction between strong acid and

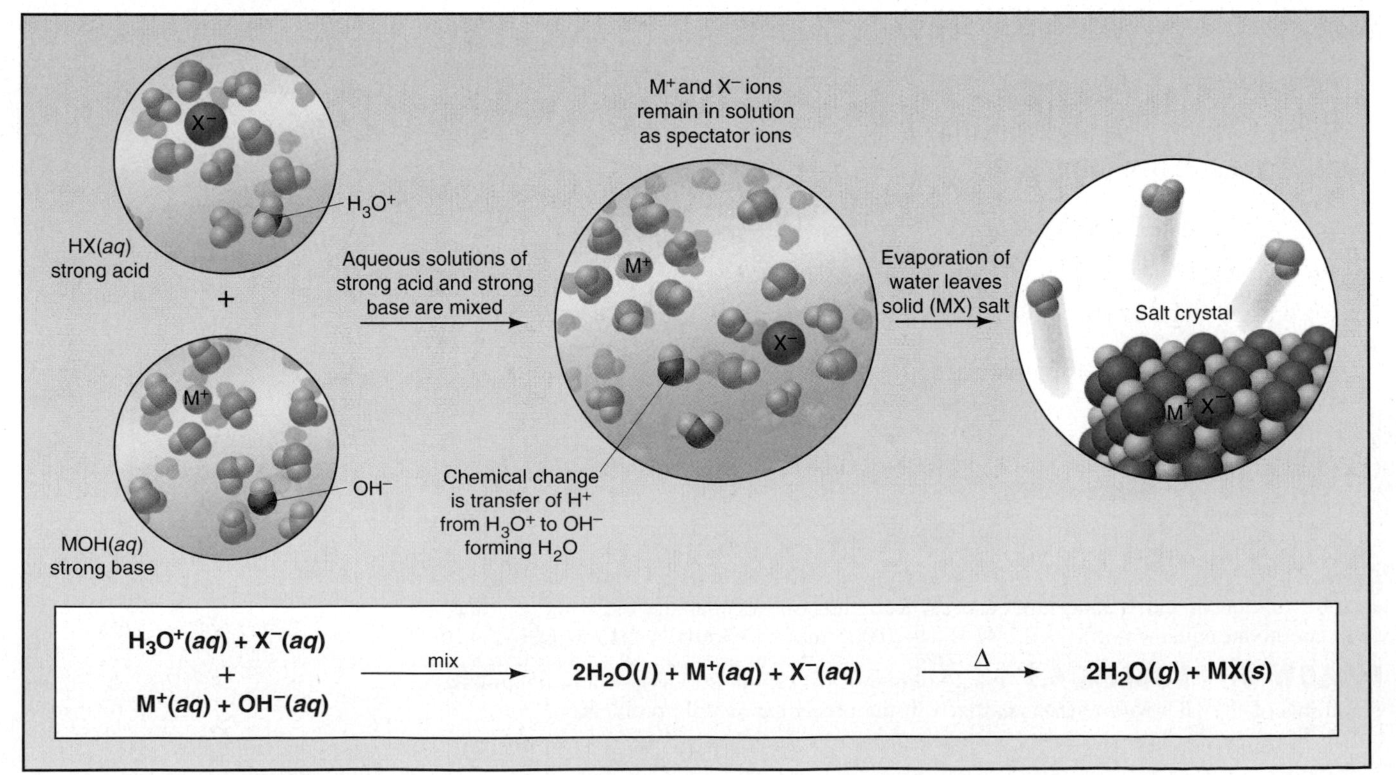

Figure 4.9 An aqueous strong acid–strong base reaction on the atomic scale. When solutions of a strong acid (HX) and a strong base (MOH) are mixed, the H_3O^+ from the acid transfers a proton to the OH^- from the base to form an H_2O molecule. Evaporation of the water leaves the spectator ions, X^- and M^+, as a solid ionic compound called a *salt.*

strong base, *H_3O^+ ion acts as the acid and OH^- ion acts as the base.* Because it ionizes completely, a given amount of strong acid (or strong base) creates an equivalent amount of H_3O^+ (or OH^-) when it dissolves in water. (We discuss the Brønsted-Lowry concept thoroughly in Chapter 18.)

Gas-Forming Reactions Thinking of acid-base reactions as proton-transfer processes helps us understand another common type of aqueous ionic reaction, those that form a gaseous product. For example, when an ionic carbonate, such as K_2CO_3, is treated with an acid, such as HCl, one of the products is carbon dioxide. *Such reactions occur through the formation of a gas **and** water because both products remove reactant ions from solution:*

$$2HCl(aq) + K_2CO_3(aq) \longrightarrow 2KCl(aq) + [H_2CO_3(aq)]$$
$$[H_2CO_3(aq)] \longrightarrow H_2O(l) + CO_2(g)$$

The product H_2CO_3 is shown in square brackets to indicate that it is very unstable. It decomposes immediately into water and carbon dioxide. Combining these two equations gives the overall equation:

$$2HCl(aq) + K_2CO_3(aq) \longrightarrow 2KCl(aq) + H_2O(l) + CO_2(g)$$

When we show H_3O^+ ions from the HCl as the actual species in solution and write the net ionic equation, Cl^- and K^+ ions are eliminated. Note that each of the two H_3O^+ ions transfers a proton to the carbonate ion:

$$\overbrace{2H_3O^+(aq) + CO_3^{2-}}^{2H^+ \text{ transfer}}(aq) \longrightarrow 2H_2O(l) + [H_2CO_3(aq)] \longrightarrow 3H_2O(l) + CO_2(g)$$

In essence, this is an acid-base reaction with carbonate ion accepting the protons and, thus, acting as the base. Several other polyatomic ions react similarly with an acid. In the formation of SO_2 from ionic sulfites, the net ionic equation is

$$\overbrace{2H_3O^+(aq) + SO_3^{2-}}^{2H^+ \text{ transfer}}(aq) \longrightarrow 2H_2O(l) + [H_2SO_3(aq)] \longrightarrow 3H_2O(l) + SO_2(g)$$

Reactions of Weak Acids Ionic equations are written differently for the reactions of weak acids. When solutions of sodium hydroxide and acetic acid (CH_3COOH) are mixed, the total and net ionic equations are

Total ionic equation:

$$CH_3COOH(aq) + Na^+(aq) + OH^-(aq) \longrightarrow CH_3COO^-(aq) + Na^+(aq) + H_2O(l)$$

Net ionic equation:

$$CH_3COOH(aq) + OH^-(aq) \longrightarrow CH_3COO^-(aq) + H_2O(l)$$

Acetic acid dissociates very little because it is a weak acid (see Table 4.2). To show this, *it appears undissociated in both ionic equations.* Note that H_3O^+ does not appear; rather, the proton is transferred from CH_3COOH. Therefore, only $Na^+(aq)$ is a spectator ion; $CH_3COO^-(aq)$ is not. Figure 4.10 shows the gas-forming reaction between vinegar (an aqueous 5% solution of acetic acid) and baking soda (sodium hydrogen carbonate) solution.

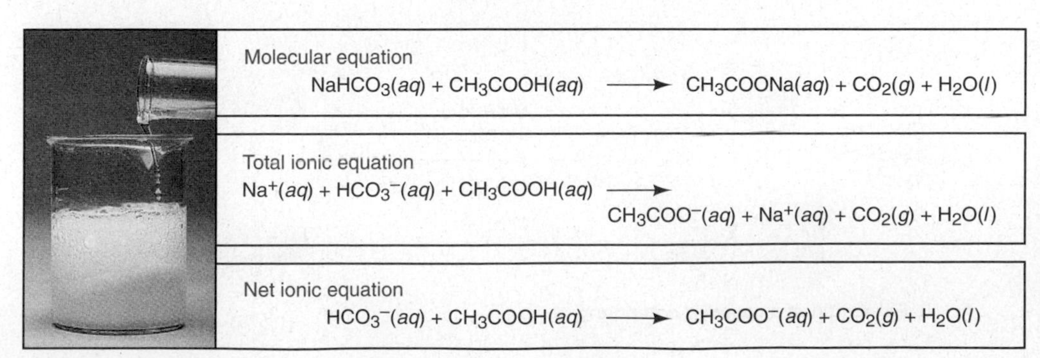
Molecular equation
$NaHCO_3(aq) + CH_3COOH(aq) \longrightarrow CH_3COONa(aq) + CO_2(g) + H_2O(l)$
Total ionic equation
$Na^+(aq) + HCO_3^-(aq) + CH_3COOH(aq) \longrightarrow CH_3COO^-(aq) + Na^+(aq) + CO_2(g) + H_2O(l)$
Net ionic equation
$HCO_3^-(aq) + CH_3COOH(aq) \longrightarrow CH_3COO^-(aq) + CO_2(g) + H_2O(l)$

Figure 4.10 An acid-base reaction that forms a gaseous product. Carbonates and hydrogen carbonates react with acids to form gaseous CO_2 and H_2O. Here, dilute acetic acid solution (vinegar) is added to sodium hydrogen carbonate (baking soda) solution, and bubbles of CO_2 gas form. (Note that the net ionic equation includes acetic acid because it does *not* dissociate into ions to an appreciable extent.)

SECTION SUMMARY

Acid-base (neutralization) reactions occur when an acid (an H^+-yielding substance) and a base (an OH^--yielding substance) react and the H^+ and OH^- ions form a water molecule. Strong acids and bases dissociate completely in water; weak acids and bases dissociate slightly. In a titration, a known concentration of one reactant is used to determine the concentration of the other. An acid-base reaction can also be viewed as the transfer of a proton from an acid to a base. An ionic gas-forming reaction is an acid-base reaction in which an acid transfers a proton to a polyatomic ion (carbonate or sulfite), forming a gas that leaves the reaction mixture. Because weak acids dissociate very little, equations involving them show the acid as an intact molecule.

4.5 OXIDATION-REDUCTION (REDOX) REACTIONS

Redox reactions are the third and, perhaps, most important type of chemical process. They include the formation of a compound from its elements (and vice versa), all combustion reactions, the reactions that generate electricity in batteries, the reactions that produce cellular energy, and many others. In this section, we examine the process, learn some essential terminology, and see one way to balance redox equations and one way to apply them quantitatively.

The Key Event: Movement of Electrons Between Reactants

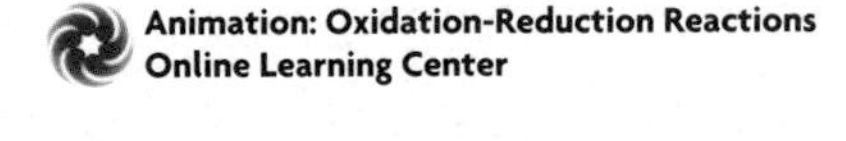

In **oxidation-reduction** (or **redox**) **reactions,** the key chemical event is the *net movement of electrons from one reactant to the other*. This movement of electrons occurs *from the reactant (or atom in the reactant) with less attraction for electrons to the reactant (or atom) with more attraction for electrons.*

Such movement of electron charge occurs in the formation of both ionic and covalent compounds. As an example, let's reconsider the flashbulb reaction (see Figure 3.8, p. 103), in which an ionic compound, MgO, forms from its elements:

$$2Mg(s) + O_2(g) \longrightarrow 2MgO(s)$$

Figure 4.11A shows that during the reaction, each Mg atom loses two electrons and each O atom gains them; that is, two electrons move from each Mg atom to each O atom. This change represents a ***transfer*** *of electron charge* away from

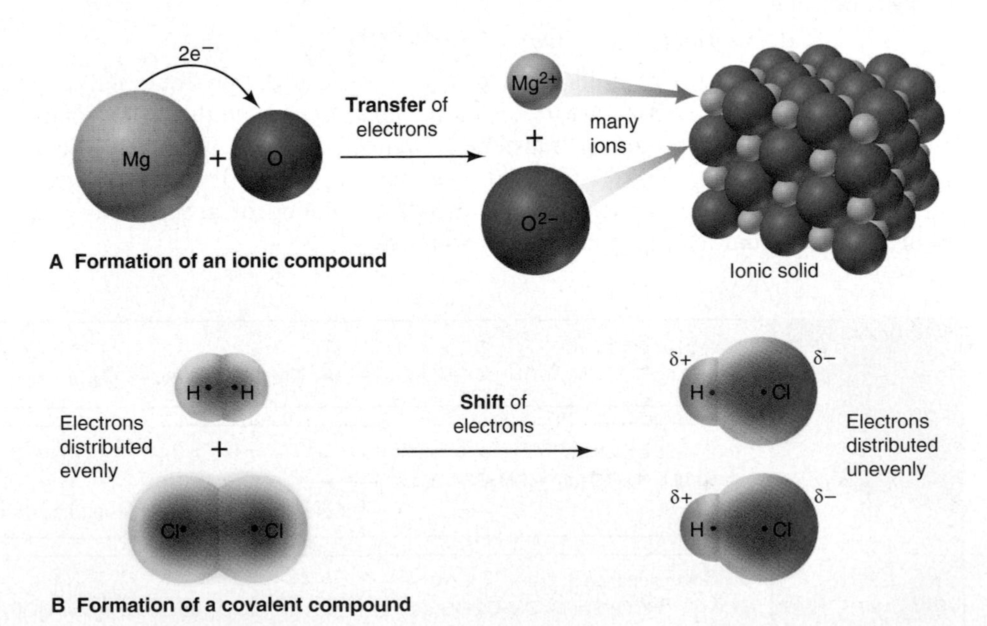

Figure 4.11 The redox process in compound formation. A, In forming the ionic compound MgO, each Mg atom transfers two electrons to each O atom. (Note that atoms become smaller when they lose electrons and larger when they gain electrons.) The resulting Mg^{2+} and O^{2-} ions aggregate with many others to form an ionic solid. **B,** In the reactants H_2 and Cl_2, the electron pairs are shared equally (indicated by even electron density shading). In the covalent product HCl, Cl attracts the shared electrons more strongly than H does. In effect, the H electron shifts toward Cl, as shown by higher electron density *(red)* near the Cl end of the molecule and lower electron density *(blue)* near the H end.

each Mg atom toward each O atom, resulting in the formation of Mg^{2+} and O^{2-} ions. The ions aggregate and form an ionic solid.

During the formation of a covalent compound from its elements, there is again a net movement of electrons, but it is more of a *shift* in electron charge than a full transfer. Thus, *ions do not form.* Consider the formation of HCl gas:

$$H_2(g) + Cl_2(g) \longrightarrow 2HCl(g)$$

To see the electron movement here, compare the electron charge distributions in the reactant bonds and in the product bonds. As Figure 4.11B shows, H_2 and Cl_2 molecules are each held together by covalent bonds in which the electrons are shared equally between the atoms (the tan shading is symmetrical). In the HCl molecule, the electrons are shared unequally because the Cl atom attracts them more strongly than the H atom does. Thus, in HCl, the H has less electron charge *(blue shading)* than it had in H_2, and the Cl has more charge *(red shading)* than it had in Cl_2. In other words, in the formation of HCl, there has been a relative ***shift*** *of electron charge* away from the H atom toward the Cl atom. This electron *shift* is not nearly as extreme as the electron *transfer* during MgO formation. In fact, in some reactions, the net movement of electrons may be very slight, but the reaction is still a redox process.

Some Essential Redox Terminology

Chemists use some important terminology to describe the movement of electrons in oxidation-reduction reactions. **Oxidation** is the *loss* of electrons, and **reduction** is the *gain* of electrons. (The original meaning of *reduction* comes from the process of reducing large amounts of metal ore to smaller amounts of metal, but you'll see shortly why we use the term for the act of gaining.)

For example, during the formation of magnesium oxide, Mg undergoes oxidation (electron loss) and O_2 undergoes reduction (electron gain). The loss and gain are simultaneous, but we can imagine them occurring in separate steps:

Oxidation (electron loss by Mg): $Mg \longrightarrow Mg^{2+} + 2e^-$

Reduction (electron gain by O_2): $\frac{1}{2}O_2 + 2e^- \longrightarrow O^{2-}$

One reactant acts on the other. Thus, we say that *O_2 oxidizes Mg*, and that O_2 is the **oxidizing agent,** the species doing the oxidizing. Similarly, *Mg reduces O_2*, so Mg is the **reducing agent,** the species doing the reducing.

Note especially that O_2 removes the electrons that Mg loses or, put the other way around, Mg gives up the electrons that O_2 gains. This give-and-take of electrons means that *the oxidizing agent is reduced* because it removes the electrons (and thus gains them), and *the reducing agent is oxidized* because it gives up the electrons (and thus loses them). In the formation of HCl, Cl_2 oxidizes H_2 (H loses some electron charge and Cl gains it), which is the same as saying that H_2 reduces Cl_2. The reducing agent, H_2, is oxidized and the oxidizing agent, Cl_2, is reduced.

Using Oxidation Numbers to Monitor the Movement of Electron Charge

Chemists have devised a useful "bookkeeping" system to monitor which atom loses electron charge and which atom gains it. Each atom in a molecule (or ionic compound) is assigned an **oxidation number (O.N.),** or *oxidation state,* the charge the atom would have *if* electrons were not shared but were transferred completely. Oxidation numbers are determined by the set of rules in Table 4.3. [Note that an oxidation number has the sign *before* the number (+2), whereas an ionic charge has the sign *after* the number (2+).]

Table 4.3 Rules for Assigning an Oxidation Number (O.N.)

General Rules

1. For an atom in its elemental form (Na, O_2, Cl_2, etc.): O.N. = 0
2. For a monatomic ion: O.N. = ion charge
3. The sum of O.N. values for the atoms in a molecule or formula unit of a compound equals zero. The sum of O.N. values for the atoms in a polyatomic ion equals the ion's charge.

Rules for Specific Atoms or Periodic Table Groups

1. For Group 1A(1):	O.N. = +1 in all compounds
2. For Group 2A(2):	O.N. = +2 in all compounds
3. For hydrogen:	O.N. = +1 in combination with nonmetals
	O.N. = −1 in combination with metals and boron
4. For fluorine:	O.N. = −1 in all compounds
5. For oxygen:	O.N. = −1 in peroxides
	O.N. = −2 in all other compounds (except with F)
6. For Group 7A(17):	O.N. = −1 in combination with metals, nonmetals (except O), and other halogens lower in the group

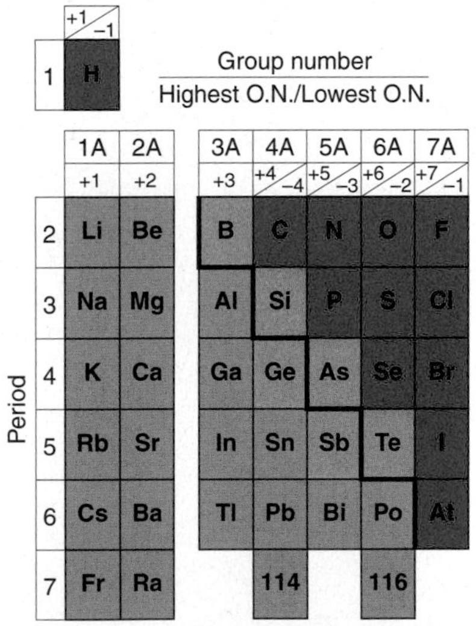

SAMPLE PROBLEM 4.6 Determining the Oxidation Number of an Element

Problem Determine the oxidation number (O.N.) of each element in these compounds:
(a) Zinc chloride **(b)** Sulfur trioxide **(c)** Nitric acid

Plan We apply Table 4.3, noting the general rules that the O.N. values in a compound add up to zero, and the O.N. values in a polyatomic ion add up to the ion's charge.

Solution (a) $ZnCl_2$. The sum of O.N.s for the monatomic ions in the compound must equal zero. The O.N. of the Zn^{2+} ion is +2. The O.N. of each Cl^- ion is −1, for a total of −2. The sum of O.N.s is +2 + (−2), or 0.

(b) SO_3. The O.N. of each oxygen is −2, for a total of −6. Since the O.N.s must add up to zero, the O.N. of S is +6.

(c) HNO_3. The O.N. of H is +1, so the O.N.s of the NO_3 group must add up to −1 to give zero for the compound. The O.N. of each O is −2 for a total of −6. Therefore, the O.N. of N is +5.

FOLLOW-UP PROBLEM 4.6 Determine the O.N. of each element in the following:
(a) Scandium oxide (Sc_2O_3) **(b)** Gallium chloride ($GaCl_3$)
(c) Hydrogen phosphate ion **(d)** Iodine trifluoride

Figure 4.12 Highest and lowest oxidation numbers of reactive main-group elements. The A-group number shows the highest possible oxidation number (O.N.) for a main-group element. (Two important exceptions are O, which never has an O.N. of +6, and F, which never has an O.N. of +7.) For nonmetals *(yellow)* and metalloids *(green)*, the A-group number minus 8 gives the lowest possible oxidation number.

The periodic table is a great help in learning the highest and lowest oxidation numbers of most main-group elements, as Figure 4.12 shows:

- For most main-group elements, the A-group number (1A, 2A, and so on) is the *highest* oxidation number (always positive) of any element in the group. The exceptions are O and F (see Table 4.3).
- For main-group nonmetals and some metalloids, the A-group number minus 8 is the *lowest* oxidation number (always negative) of any element in the group.

For example, the highest oxidation number of S (Group 6A) is +6, as in SF_6, and the lowest is (6 − 8), or −2, as in FeS and other metal sulfides.

As you can see, the oxidation number for an element in a binary *ionic* compound has a realistic value, because it usually equals the ionic charge. On the other hand, the oxidation number for an element in a *covalent* compound (or in a polyatomic ion) has an unrealistic value because the atoms don't have whole charges.

Another way to define a redox reaction is one in which *the oxidation numbers of the species change,* and the most important use of oxidation numbers is to monitor these changes:

- If a given atom has a higher (more positive or less negative) oxidation number in the product than it had in the reactant, the reactant species that contains the atom was oxidized (lost electrons). Thus, *oxidation is represented by an increase in oxidation number.*
- If an atom has a lower (more negative or less positive) oxidation number in the product than it had in the reactant, the reactant species that contains the atom was reduced (gained electrons). Thus, *the gain of electrons is represented by a decrease (a "reduction") in oxidation number.* (Reduction, as mentioned earlier, refers to an ore being "reduced" to the metal. The reducing agents provide electrons that convert the ore's metal ions to atoms.)

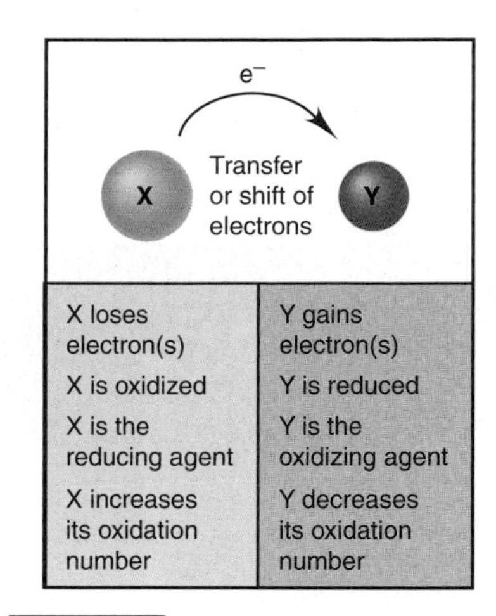

Figure 4.13 **A summary of terminology for oxidation-reduction (redox) reactions.**

Figure 4.13 summarizes redox terminology. Oxidation numbers are assigned according to the relative attraction of an atom for electrons, so they are ultimately based on atomic properties, as you'll see in Chapters 8 and 9. (For the remainder of this section and the next, blue oxidation numbers represent oxidation, and red oxidation numbers indicate reduction.)

SAMPLE PROBLEM 4.7 Recognizing Oxidizing and Reducing Agents

Problem Identify the oxidizing agent and reducing agent in each of the following:

(a) $2Al(s) + 3H_2SO_4(aq) \longrightarrow Al_2(SO_4)_3(aq) + 3H_2(g)$

(b) $PbO(s) + CO(g) \longrightarrow Pb(s) + CO_2(g)$

(c) $2H_2(g) + O_2(g) \longrightarrow 2H_2O(g)$

Plan We first assign an oxidation number (O.N.) to each atom (or ion) based on the rules in Table 4.3. The reactant is the reducing agent if it contains an atom that is oxidized (O.N. increased from left side to right side of the equation). The reactant is the oxidizing agent if it contains an atom that is reduced (O.N. decreased).

Solution (a) Assigning oxidation numbers:

Al: 0; H: +1; S: +6; O: −2 → Al: +3; S: +6; O: −2; H: 0

$$2Al(s) + 3H_2SO_4(aq) \longrightarrow Al_2(SO_4)_3(aq) + 3H_2(g)$$

The O.N. of Al increased from 0 to +3 (Al lost electrons), so Al was oxidized; Al is the reducing agent.

The O.N. of H decreased from +1 to 0 (H gained electrons), so H^+ was reduced; H_2SO_4 is the oxidizing agent.

(b) Assigning oxidation numbers:

Pb: +2; O: −2; C: +2; O: −2 → Pb: 0; C: +4; O: −2

$$PbO(s) + CO(g) \longrightarrow Pb(s) + CO_2(g)$$

Pb decreased its O.N. from +2 to 0, so PbO was reduced; PbO is the oxidizing agent.

C increased its O.N. from +2 to +4, so CO was oxidized; CO is the reducing agent.

In general, when a substance (such as CO) becomes one with more O atoms (as in CO_2), it is oxidized; and when a substance (such as PbO) becomes one with fewer O atoms (as in Pb), it is reduced.

(c) Assigning oxidation numbers:

H: 0; O: 0 → H: +1; O: −2

$$2H_2(g) + O_2(g) \longrightarrow 2H_2O(g)$$

O_2 was reduced (O.N. of O decreased from 0 to −2); O_2 is the oxidizing agent.

H_2 was oxidized (O.N. of H increased from 0 to +1); H_2 is the reducing agent.

Oxygen is always the oxidizing agent in a combustion reaction.

The permanganate ion, MnO_4^-, is a common oxidizing agent in these titrations because it is strongly colored and, thus, also serves as an indicator. In Figure 4.14, MnO_4^- is used to oxidize the oxalate ion, $C_2O_4^{2-}$, to determine its concentration. As long as any $C_2O_4^{2-}$ is present, it reduces the deep purple MnO_4^- to the very faint pink (nearly colorless) Mn^{2+} ion (Figure 4.14, *left*). As soon as all the available $C_2O_4^{2-}$ has been oxidized, the next drop of MnO_4^- turns the solution light purple (Figure 4.14, *right*). This color change indicates the end point, the point at which the electrons lost by the oxidized species ($C_2O_4^{2-}$) equal the electrons gained by the reduced species (MnO_4^-). We then calculate the concentration of the $C_2O_4^{2-}$ solution from its known volume, the known volume and concentration of the MnO_4^- solution, and the balanced equation. Preparing a sample for a redox titration sometimes requires several laboratory steps, as shown in Sample Problem 4.9 for the determination of the Ca^{2+} ion concentration of blood.

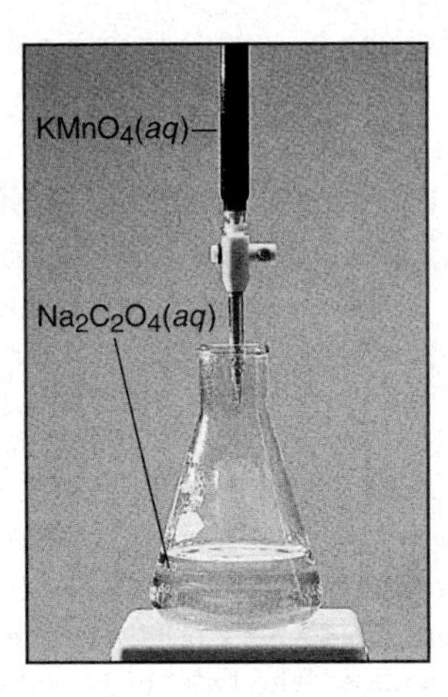

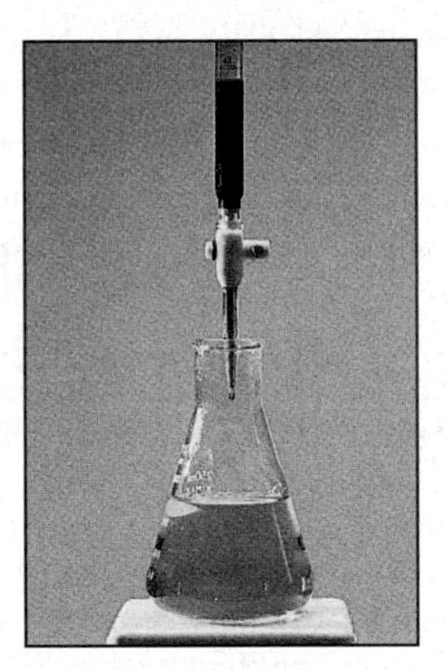

Figure 4.14 A redox titration. The oxidizing agent in the buret, $KMnO_4$, is strongly colored, so it also serves as the indicator. When it reacts with the reducing agent $C_2O_4^{2-}$ in the flask, its color changes from deep purple to very faint pink (nearly colorless) *(left).* When all the $C_2O_4^{2-}$ is oxidized, the next drop of $KMnO_4$ remains unreacted and turns the solution light purple *(right),* signaling the end point of the titration.

Net ionic equation:

$$\overset{+7}{2MnO_4^-}(aq) + \overset{+3}{5C_2O_4^{2-}}(aq) + 16H^+(aq) \longrightarrow \overset{+2}{2Mn^{2+}}(aq) + \overset{+4}{10CO_2}(g) + 8H_2O(l)$$

SAMPLE PROBLEM 4.9 Finding a Concentration by a Redox Titration

Problem Calcium ion (Ca^{2+}) is required for blood to clot and for many other cell processes. An abnormal Ca^{2+} concentration is indicative of disease. To measure the Ca^{2+}concentration, 1.00 mL of human blood is treated with $Na_2C_2O_4$ solution. The resulting CaC_2O_4 precipitate is filtered and dissolved in dilute H_2SO_4 to release $C_2O_4^{2-}$ into solution and allow it to be oxidized. This solution required 2.05 mL of 4.88×10^{-4} *M* $KMnO_4$ to reach the end point. The unbalanced equation is

$$KMnO_4(aq) + CaC_2O_4(s) + H_2SO_4(aq) \longrightarrow MnSO_4(aq) + K_2SO_4(aq) + CaSO_4(s) + CO_2(g) + H_2O(l)$$

(a) Calculate the amount (mol) of Ca^{2+}.
(b) Calculate the Ca^{2+} ion concentration expressed in units of mg Ca^{2+}/100 mL blood.

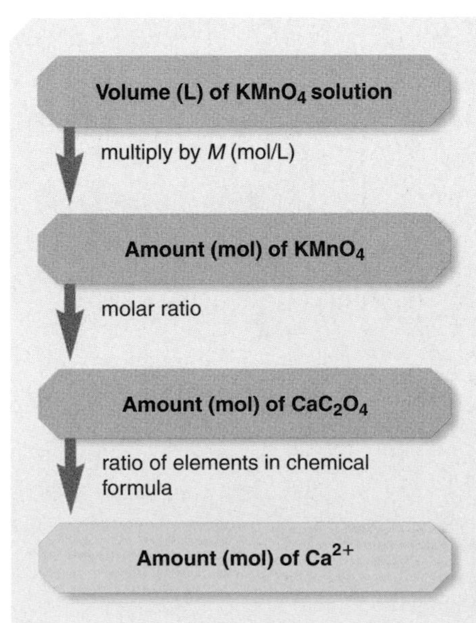

(a) Calculating the moles of Ca^{2+}

Plan As always, we first balance the equation. All the Ca^{2+} ion in the 1.00-mL blood sample is precipitated and then dissolved in the H_2SO_4. We find the number of moles of $KMnO_4$ needed to reach the end point from the volume (2.05 mL) and molarity (4.88×10^{-4} *M*) and use the molar ratio to calculate the number of moles of CaC_2O_4 dissolved in the H_2SO_4. Then, from the chemical formula, we find moles of Ca^{2+} ions.

Solution Balancing the equation:

$$2KMnO_4(aq) + 5CaC_2O_4(s) + 8H_2SO_4(aq) \longrightarrow 2MnSO_4(aq) + K_2SO_4(aq) + 5CaSO_4(s) + 10CO_2(g) + 8H_2O(l)$$

Converting from milliliters and molarity to moles of $KMnO_4$ to reach the end point:

$$\text{Moles of } KMnO_4 = 2.05 \cancel{\text{mL soln}} \times \frac{1 \cancel{\text{L}}}{1000 \cancel{\text{mL}}} \times \frac{4.88\times10^{-4} \text{ mol } KMnO_4}{1 \cancel{\text{L soln}}}$$
$$= 1.00\times10^{-6} \text{ mol } KMnO_4$$

Converting from moles of $KMnO_4$ to moles of CaC_2O_4 titrated:

$$\text{Moles of } CaC_2O_4 = 1.00\times10^{-6} \cancel{\text{mol } KMnO_4} \times \frac{5 \text{ mol } CaC_2O_4}{2 \cancel{\text{mol } KMnO_4}} = 2.50\times10^{-6} \text{ mol } CaC_2O_4$$

Finding moles of Ca^{2+} present:

$$\text{Moles of } Ca^{2+} = 2.50\times10^{-6} \cancel{\text{mol } CaC_2O_4} \times \frac{1 \text{ mol } Ca^{2+}}{1 \cancel{\text{mol } CaC_2O_4}}$$
$$= 2.50\times10^{-6} \text{ mol } Ca^{2+}$$

Check A very small volume of dilute $KMnO_4$ is needed, so 10^{-6} mol of $KMnO_4$ seems reasonable. The molar ratio of 5:2 for CaC_2O_4 to $KMnO_4$ gives 2.5×10^{-6} mol of CaC_2O_4 and thus 2.5×10^{-6} mol of Ca^{2+}.

(b) Expressing the Ca^{2+} concentration as mg/100 mL blood

Plan The amount in part (a) is the moles of Ca^{2+} ion present in 1.00 mL of blood. We multiply by 100 to obtain the moles of Ca^{2+} ion in 100 mL of blood and then use the atomic mass of Ca to convert that amount to grams and then milligrams.

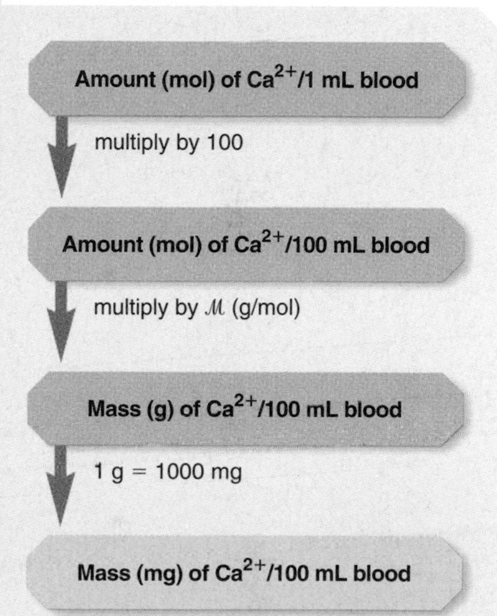

Solution Finding moles of Ca^{2+}/100 mL blood:

$$\text{Moles of } Ca^{2+}/100 \text{ mL blood} = \frac{2.50\times10^{-6} \text{ mol } Ca^{2+}}{1.00 \text{ mL blood}} \times 100$$
$$= 2.50\times10^{-4} \text{ mol } Ca^{2+}/100 \text{ mL blood}$$

Converting from moles of Ca^{2+} to milligrams:

$$\text{Mass (mg) } Ca^{2+}/100 \text{ mL blood} = \frac{2.50\times10^{-4} \cancel{\text{mol } Ca^{2+}}}{100 \text{ mL blood}} \times \frac{40.08 \cancel{\text{g}} \; Ca^{2+}}{1 \cancel{\text{mol } Ca^{2+}}} \times \frac{1000 \text{ mg}}{1 \cancel{\text{g}}}$$
$$= 10.0 \text{ mg } Ca^{2+}/100 \text{ mL blood}$$

Check The relative amounts of Ca^{2+} make sense. If there is 2.5×10^{-6} mol/mL blood, there is 2.5×10^{-4} mol/100 mL blood. A molar mass of about 40 g/mol for Ca^{2+} gives 100×10^{-4} g, or 10×10^{-3} g/100 mL blood. It is easy to make an order-of-magnitude (power of 10) error in this type of calculation, so be sure to include all units.

Comment 1. The normal range for the Ca^{2+} concentration in a human adult is 9.0 to 11.5 mg Ca^{2+}/100 mL blood, so our value seems reasonable.

2. When blood is donated, the receiving bag contains $Na_2C_2O_4$ solution, which precipitates the Ca^{2+} ion and, thus, prevents clotting.

3. A redox titration is analogous to an acid-base titration: in redox processes, electrons are lost and gained, whereas in acid-base processes, H^+ ions are lost and gained.

FOLLOW-UP PROBLEM 4.9 A 2.50-mL sample of low-fat milk was treated with sodium oxalate, and the precipitate was filtered and dissolved in H_2SO_4. This solution required 6.53 mL of 4.56×10^{-3} *M* $KMnO_4$ to reach the end point.

(a) Calculate the molarity of Ca^{2+} in the milk.

(b) What is the concentration of Ca^{2+} in g/L? Is this value consistent with the typical value in milk of about 1.2 g Ca^{2+}/L?

SECTION SUMMARY

When one reactant has a greater attraction for electrons than another, there is a net movement of electron charge, and a redox reaction takes place. Electron gain (reduction) and electron loss (oxidation) occur simultaneously. The redox process is tracked by assigning oxidation numbers to each atom in a reaction. The species that is oxidized (contains an atom that increases in oxidation number) is the reducing agent; the species that is reduced (contains an atom that decreases in oxidation number) is the oxidizing agent. Redox reactions can be balanced by keeping track of the changes in oxidation number. A redox titration is used to determine the concentration of either the oxidizing or the reducing agent from the known concentration of the other.

4.6 ELEMENTS IN REDOX REACTIONS

As we saw in Sample Problem 4.7, if a substance appears as a free element on one side of an equation, it must appear in a different form on the other side, which means there must have been a change in oxidation state. Therefore, *whenever a reaction includes a free element as either reactant or product, it is a redox reaction.* And, while there are many redox reactions that do *not* involve free elements, such as the one between MnO_4^- and $C_2O_4^{2-}$ that we saw earlier, we'll focus here on the many others that do. One way to classify these is by comparing the numbers of reactants and products. By that approach, we have three types:

- *Combination reactions*: two or more reactants form one product:
$$X + Y \longrightarrow Z$$
- *Decomposition reactions:* one reactant forms two or more products:
$$Z \longrightarrow X + Y$$
- *Displacement reactions:* the number of substances is the same but atoms (or ions) exchange places:
$$X + YZ \longrightarrow XZ + Y$$

Combining Two Elements Two elements may react to form binary ionic or covalent compounds. Here are some important examples:

1. *Metal and nonmetal form an ionic compound.* Figure 4.15 shows the reaction between an alkali metal and a halogen on the observable and atomic scales. Note the change in oxidation numbers. As you can see, K is oxidized, so it is the reducing agent; Cl_2 is reduced, so it is the oxidizing agent.

Aluminum reacts with O_2, as does nearly every metal, to form an ionic oxide:

$$4\overset{0}{\text{Al}}(s) + 3\overset{0}{\text{O}}_2(g) \longrightarrow 2\overset{+3}{\text{Al}}_2\overset{-2}{\text{O}}_3(s)$$

2. *Two nonmetals form a covalent compound.* In one of thousands of examples, ammonia forms from nitrogen and hydrogen in a reaction that occurs in industry on an enormous scale:

$$\overset{0}{\text{N}}_2(g) + 3\overset{0}{\text{H}}_2(g) \longrightarrow 2\overset{-3}{\text{N}}\overset{+1}{\text{H}}_3(g)$$

Halogens form many compounds with other nonmetals, as in the formation of phosphorus trichloride, a major reactant in the production of pesticides and other organic compounds:

$$\overset{0}{\text{P}}_4(s) + 6\overset{0}{\text{Cl}}_2(g) \longrightarrow 4\overset{+3}{\text{P}}\overset{-1}{\text{Cl}}_3(l)$$

Nearly every nonmetal reacts with O_2 to form a covalent oxide, as when nitrogen monoxide forms at the very high temperatures created in air by lightning:

$$\overset{0}{\text{N}}_2(g) + \overset{0}{\text{O}}_2(g) \longrightarrow 2\overset{+2}{\text{N}}\overset{-2}{\text{O}}(g)$$

Combining Compound and Element Many binary covalent compounds react with nonmetals to form larger compounds. Many nonmetal oxides react with additional O_2 to form "higher" oxides (those with more O atoms in each molecule). For example,

$$2\overset{+2}{\text{N}}\overset{-2}{\text{O}}(g) + \overset{0}{\text{O}}_2(g) \longrightarrow 2\overset{+4}{\text{N}}\overset{-2}{\text{O}}_2(g)$$

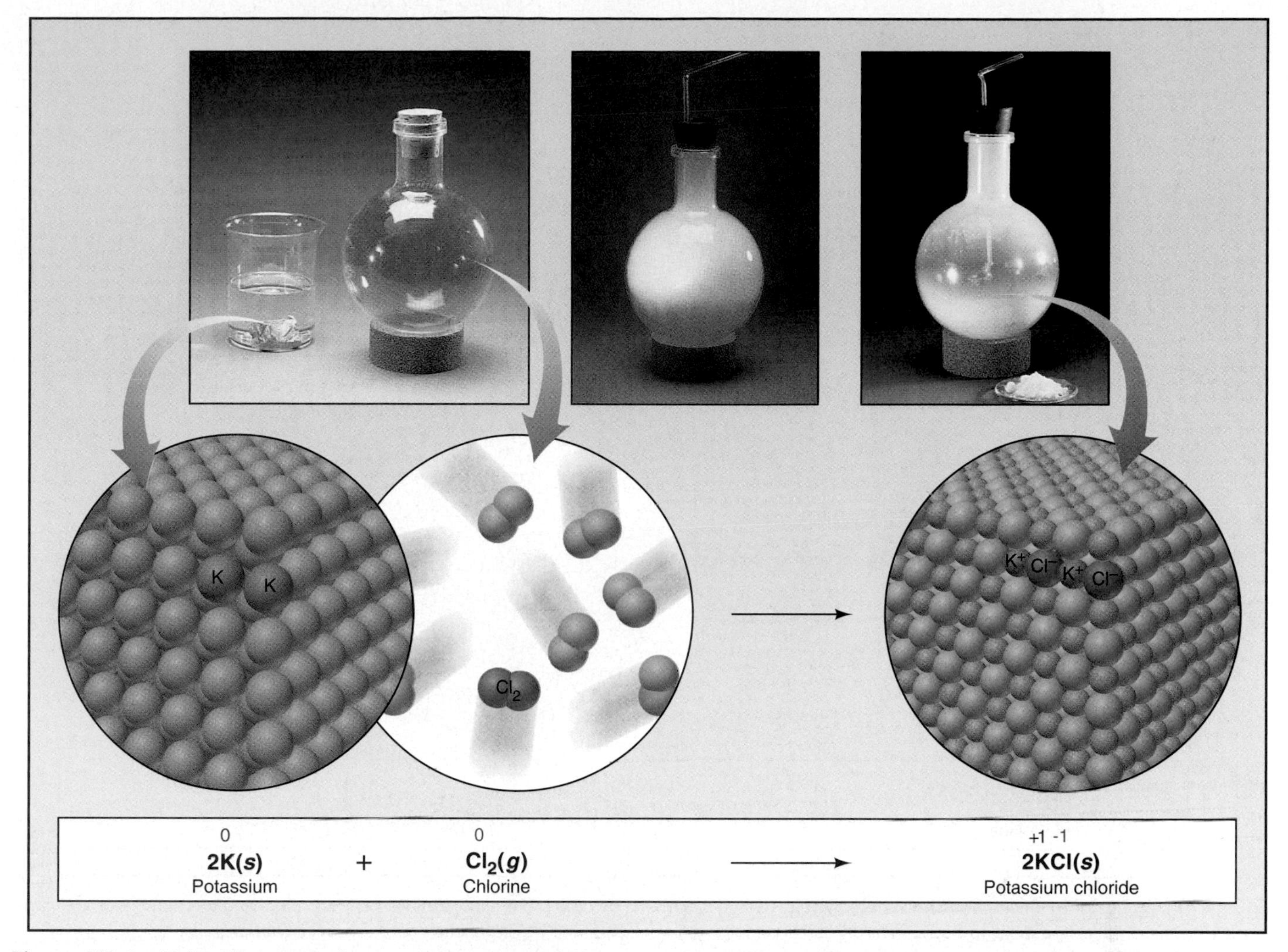

Figure 4.15 Combining elements to form an ionic compound. When the metal potassium and the nonmetal chlorine react, they form the solid ionic compound potassium chloride. The photos *(top)* present the view the chemist sees in the laboratory. The blow-up arrows lead to an atomic-scale view *(middle);* the stoichiometry is indicated by the more darkly colored spheres. The balanced redox equation is shown with oxidation numbers *(bottom).*

Similarly, many nonmetal halides combine with additional halogen:

$$\overset{+3,\,-1}{PCl_3}(l) + \overset{0}{Cl_2}(g) \longrightarrow \overset{+5,\,-1}{PCl_5}(s)$$

Decomposing Compounds into Elements A decomposition reaction occurs when a reactant absorbs enough energy for one or more of its bonds to break. The energy can take many forms—heat, electricity, light, mechanical, and so forth—but we'll focus in this discussion on heat and electricity. The products are either elements or elements and smaller compounds. Following are several common examples:

1. *Thermal decomposition.* When the energy absorbed is heat, the reaction is a thermal decomposition. (A Greek *delta*, Δ, above a reaction arrow indicates that heat is required for the reaction.) Many metal oxides, chlorates, and perchlorates release oxygen when strongly heated. The decomposition of mercury(II) oxide, used by Lavoisier and Priestley in their classic experiments, is shown on the

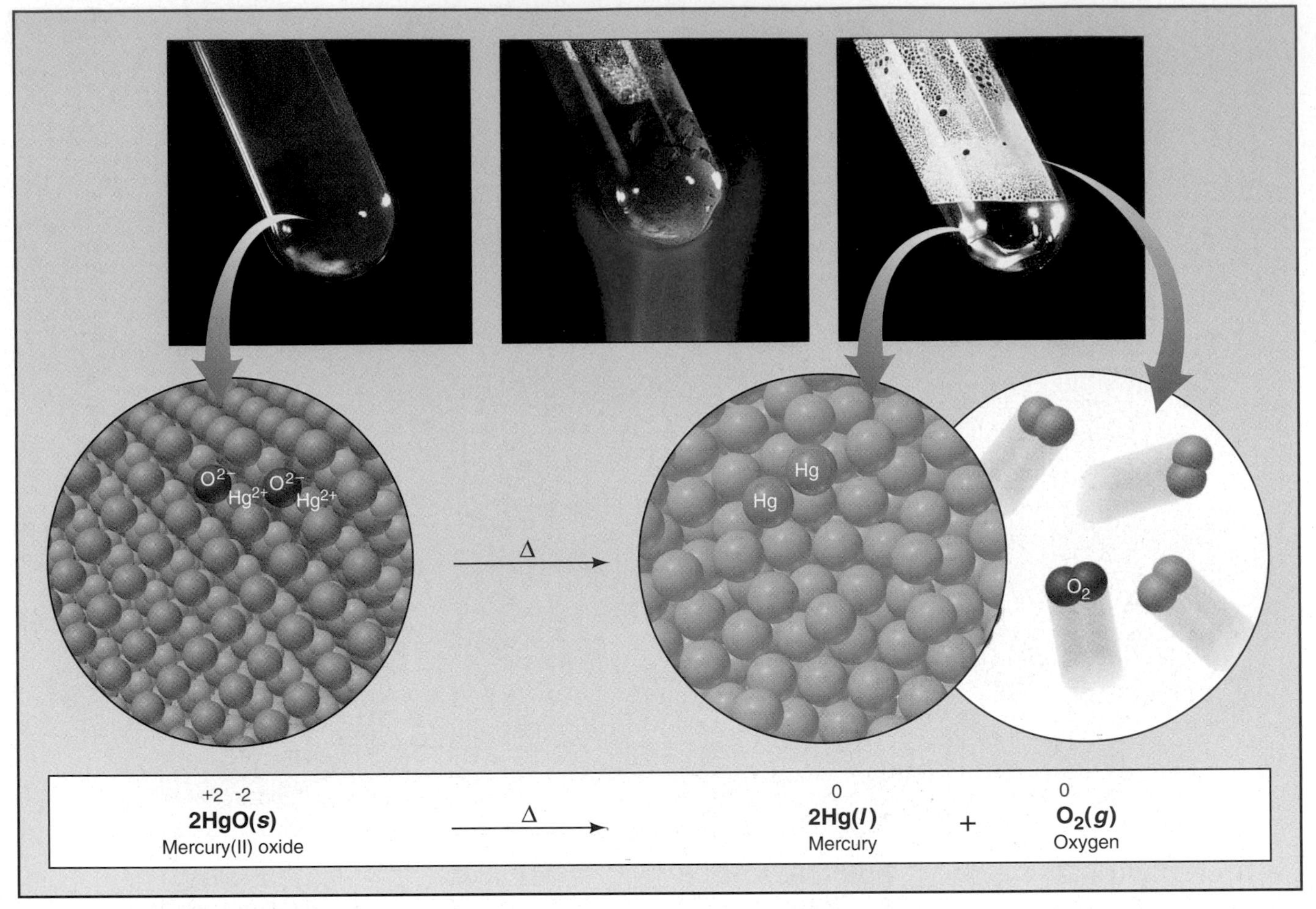

Figure 4.16 Decomposing a compound to its elements. Heating solid mercury(II) oxide decomposes it to liquid mercury and gaseous oxygen: the macroscopic (laboratory) view *(top);* the atomic-scale view, with the more darkly colored spheres showing the stoichiometry *(middle);* and the balanced redox equation *(bottom).*

macroscopic and atomic scales in Figure 4.16. Heating potassium chlorate is a modern method for forming small amounts of oxygen in the laboratory; the same reaction occurs in some explosives and fireworks:

$$\overset{+1\ +5\ -2}{2KClO_3(s)} \xrightarrow{\Delta} \overset{+1\ -1}{2KCl(s)} + \overset{0}{3O_2(g)}$$

Notice that, in these cases, the lone reactant is the oxidizing *and* the reducing agent. For example, with HgO, O^{2-} reduces Hg^{2+} (and Hg^{2+} oxidizes O^{2-}).

2. *Electrolytic decomposition.* In the process of *electrolysis,* a compound absorbs electrical energy and decomposes into its elements. Observing the electrolysis of water was crucial in the establishment of atomic masses:

$$\overset{+1\ -2}{2H_2O(l)} \xrightarrow{\text{electricity}} \overset{0}{2H_2(g)} + \overset{0}{O_2(g)}$$

Many active metals, such as sodium, magnesium, and calcium, are produced industrially by electrolysis of their molten halides:

$$\overset{+2\ -1}{MgCl_2(l)} \xrightarrow{\text{electricity}} \overset{0}{Mg(l)} + \overset{0}{Cl_2(g)}$$

(We'll examine the details of electrolysis in Chapters 21 and 22.)

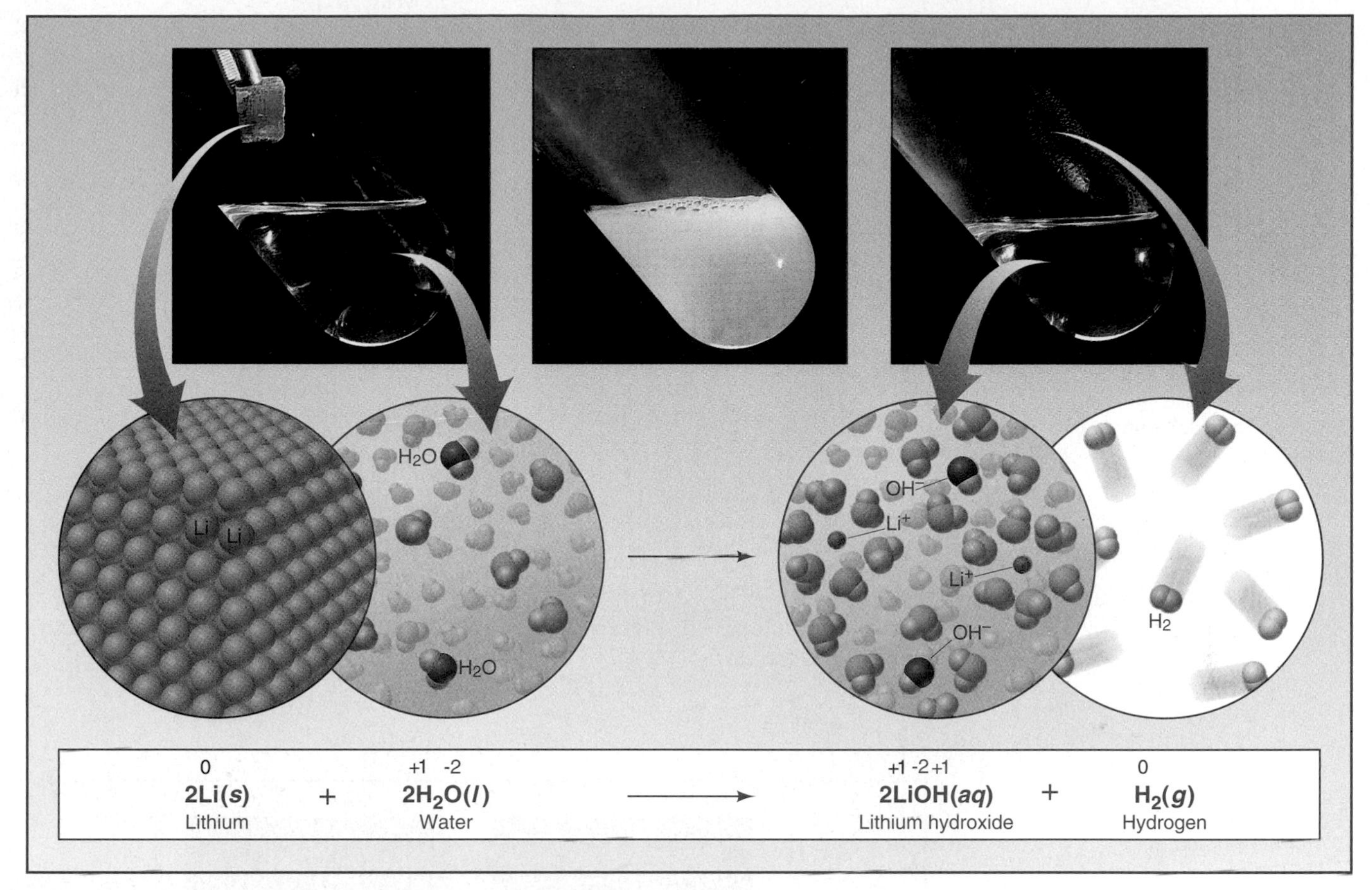

Figure 4.17 An active metal displacing hydrogen from water. Lithium displaces hydrogen from water in a vigorous reaction that yields an aqueous solution of lithium hydroxide and hydrogen gas, as shown on the macroscopic scale *(top)*, at the atomic scale *(middle)*, and as a balanced equation *(bottom)*. (For clarity, the atomic-scale view of water has been greatly simplified, and only water molecules involved in the reaction are colored red and blue.)

Displacing One Element by Another; Activity Series As we said, displacement reactions have the same number of reactants as products. We mentioned double-displacement (metathesis) reactions in discussing precipitation and acid-base reactions. The other type, *single-displacement* reactions, are all oxidation-reduction processes. They occur when one atom displaces the ion of a different atom from solution. When the reaction involves metals, the atom reduces the ion; when it involves nonmetals (specifically halogens), the atom oxidizes the ion. Chemists rank various elements into activity series—one for metals and one for halogens—in order of their ability to displace one another.

1. *The activity series of the metals.* Metals can be ranked by their ability to displace H_2 from various sources or by their ability to displace one another from solution.

- *A metal displaces H_2 from water or acid.* The most reactive metals, such as those from Group 1A(1) and Ca, Sr, and Ba from Group 2A(2), displace H_2 from water, and they do so vigorously. Figure 4.17 shows this reaction for lithium. Heat is needed to speed the reaction of slightly less reactive metals, such as Al and Zn, so these metals displace H_2 from steam:

$$\overset{0}{2\text{Al}}(s) + 6\overset{+1}{\text{H}}_2\overset{-2}{\text{O}}(g) \longrightarrow 2\overset{+3}{\text{Al}}(\overset{-2}{\text{O}}\overset{+1}{\text{H}})_3(s) + 3\overset{0}{\text{H}}_2(g)$$

Figure 4.18 The displacement of H_2 from acid by nickel.

Still less reactive metals, such as nickel and tin, do not react with water but *do* react with acids. Because the concentration of H^+ is higher in acid solutions than in water, H_2 is displaced more easily (Figure 4.18). Here is the net ionic equation:

$$\overset{0}{Ni}(s) + 2\overset{+1}{H^+}(aq) \longrightarrow \overset{+2}{Ni^{2+}}(aq) + \overset{0}{H_2}(g)$$

Notice that in all such reactions, the metal is the reducing agent (O.N. of metal increases), and water or acid is the oxidizing agent (O.N. of H decreases). The least reactive metals, such as silver and gold, cannot displace H_2 from any source.

- *A metal displaces another metal ion from solution.* Direct comparisons of metal reactivity are clearest in these reactions. For example, zinc metal displaces copper(II) ion from copper(II) sulfate solution, as the total ionic equation shows:

$$\overset{+2}{Cu^{2+}}(aq) + \overset{+6\ -2}{SO_4^{2-}}(aq) + \overset{0}{Zn}(s) \longrightarrow \overset{0}{Cu}(s) + \overset{+2}{Zn^{2+}}(aq) + \overset{+6\ -2}{SO_4^{2-}}(aq)$$

Figure 4.19 demonstrates in atomic detail that copper metal can displace silver ion from solution. Thus, zinc is more reactive than copper, which is more reactive than silver.

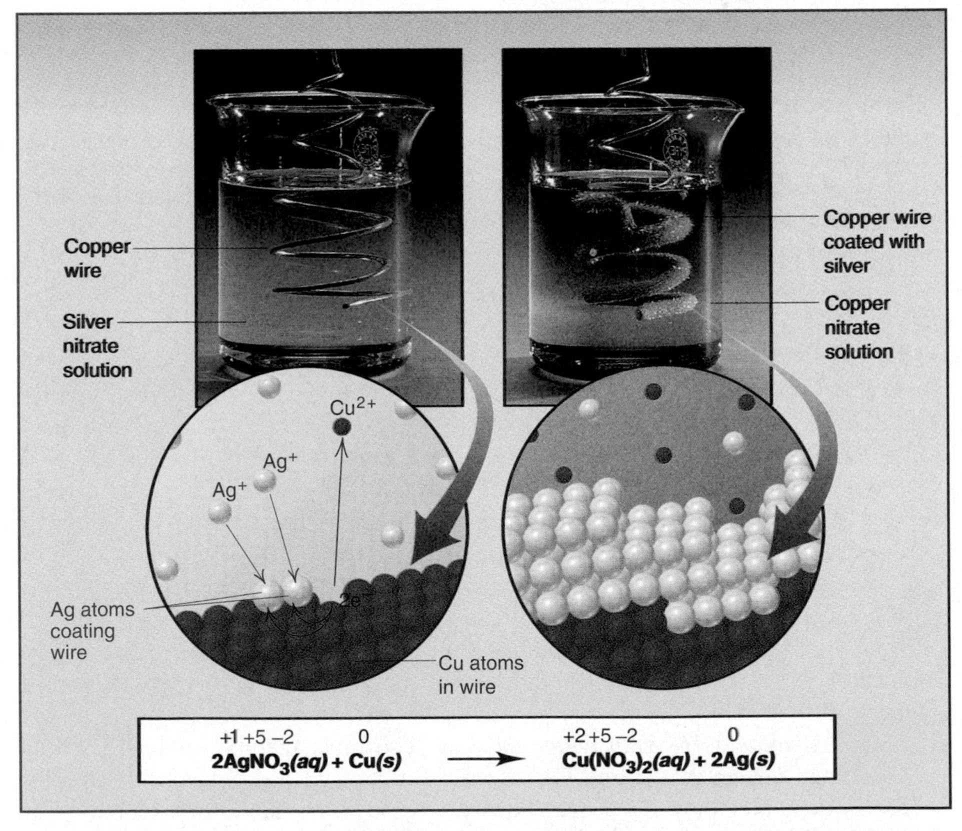

Figure 4.19 Displacing one metal by another. More reactive metals displace less reactive metals from solution. In this reaction, Cu atoms each give up two electrons as they become Cu^{2+} ions and leave the wire. The electrons are transferred to two Ag^+ ions that become Ag atoms and deposit on the wire. With time, a coating of crystalline silver coats the wire. Thus, copper has displaced silver from solution. The reaction is depicted as the laboratory view *(top),* the atomic-scale view *(middle),* and the balanced redox equation *(bottom).*

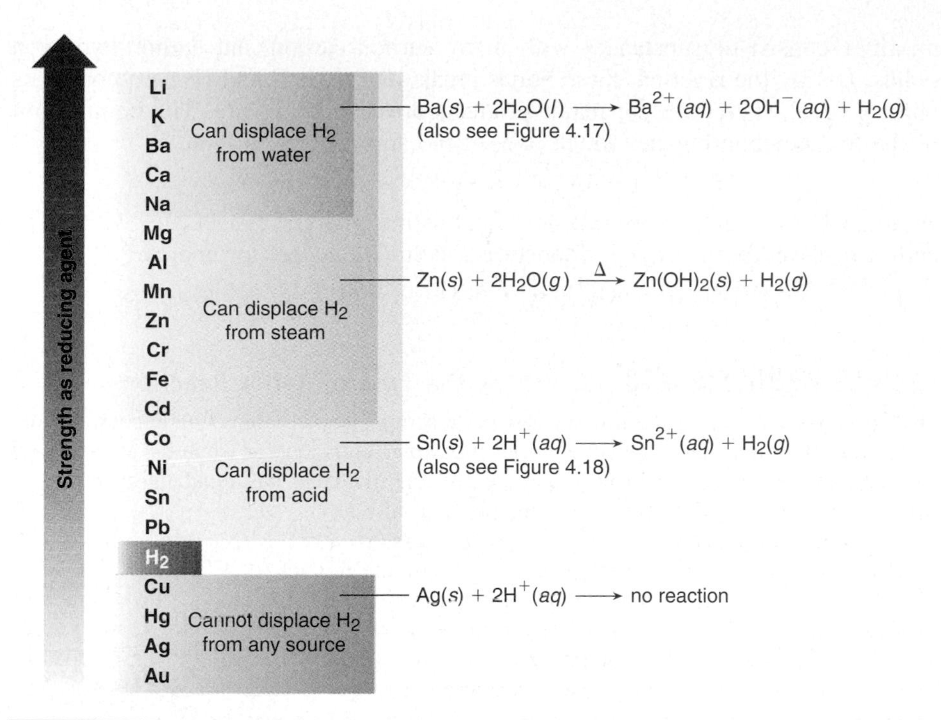

Figure 4.20 The activity series of the metals. This list of metals (and H_2) is arranged with the most active metal (strongest reducing agent) at the top and the least active metal (weakest reducing agent) at the bottom. The four metals below H_2 cannot displace it from any source. An example from each group appears to the right as a net ionic equation. (The ranking refers to behavior of ions in aqueous solution.)

The results of many such reactions between metals and water, aqueous acids, and metal-ion solutions form the basis of the **activity series of the metals.** In Figure 4.20 elements higher on the list are stronger reducing agents than elements lower down; that is, for those that are stable in water, elements higher on the list can reduce aqueous ions of elements lower down. The list also shows whether the metal can displace H_2 and, if so, from which source. Look at the metals in the equations we've just discussed. Note that Li, Al, and Ni lie above H_2, while Ag lies below it; also, Zn lies above Cu, which lies above Ag. The most reactive metals on the list are in Groups 1A(1) and 2A(2) of the periodic table, and the least reactive lie at the right of the transition elements in Groups 1B(11) and 2B(12).

2. *The activity series of the halogens.* Reactivity decreases down Group 7A(17), so we can arrange the halogens into their own activity series:

$$F_2 > Cl_2 > Br_2 > I_2$$

A halogen higher in the periodic table is a stronger oxidizing agent than one lower down. Thus, chlorine can oxidize bromide ions or iodide ions from solution, and bromine can oxidize iodide ions. Here, chlorine displaces bromine:

$$\overset{-1}{2Br^-}(aq) + \overset{0}{Cl_2}(aq) \longrightarrow \overset{0}{Br_2}(aq) + \overset{-1}{2Cl^-}(aq)$$

Combustion Reactions Combustion is the process of combining with oxygen, often with the release of heat and light, as in a flame. Combustion reactions do not fall neatly into classes based on the number of reactants and products, but *all are redox processes* because elemental oxygen is a reactant:

$$2CO(g) + O_2(g) \longrightarrow 2CO_2(g)$$

The combustion reactions that we commonly use to produce energy involve organic mixtures such as coal, gasoline, and natural gas as reactants. These

Space-Age Combustion Without a Flame Combustion reactions are used to generate large amounts of energy. In most applications, the fuel is burned and the energy is released as heat (e.g., in a furnace) or as a combination of work and heat (e.g., in a combustion engine). Aboard the space shuttle, *fuel cells* generate electrical energy from the flameless combustion of hydrogen gas. Here H_2 is the reducing agent, and O_2 is the oxidizing agent in a controlled reaction process that yields water—which the astronauts use for drinking. On Earth, fuel cells based on the reaction of either H_2 or methanol (CH_3OH) with O_2 are being developed for use in car engines.

mixtures consist of substances with many carbon-carbon and carbon-hydrogen bonds. During the reaction, these bonds break, and each C and H atom combines with oxygen. Therefore, the major products are CO_2 and H_2O. The combustion of the hydrocarbon butane, which is used in camp stoves, is typical:

$$2C_4H_{10}(g) + 13O_2(g) \longrightarrow 8CO_2(g) + 10H_2O(g)$$

Biological *respiration* is a multistep combustion process that occurs within our cells when we "burn" organic foodstuffs, such as glucose, for energy:

$$C_6H_{12}O_6(s) + 6O_2(g) \longrightarrow 6CO_2(g) + 6H_2O(g) + \textit{energy}$$

SAMPLE PROBLEM 4.10 Identifying the Type of Redox Reaction

Problem Classify each of the following redox reactions as a combination, decomposition, or displacement reaction, write a balanced molecular equation for each, as well as total and net ionic equations for part **(c)**, and identify the oxidizing and reducing agents:

(a) Magnesium(s) + nitrogen(g) $\longrightarrow$ magnesium nitride(s)
(b) Hydrogen peroxide(l) $\longrightarrow$ water + oxygen gas
(c) Aluminum(s) + lead(II) nitrate(aq) $\longrightarrow$ aluminum nitrate(aq) + lead(s)

Plan To decide on reaction type, recall that combination reactions produce fewer products than reactants, decomposition reactions produce more products, and displacement reactions have the same number of reactants and products. The oxidation number (O.N.) becomes more positive for the reducing agent and less positive for the oxidizing agent.

Solution (a) Combination: two substances form one. This reaction occurs, along with formation of magnesium oxide, when magnesium burns in air:

$$\overset{0}{3\text{Mg}}(s) + \overset{0}{\text{N}_2}(g) \longrightarrow \overset{+2}{\text{Mg}_3}\overset{-3}{\text{N}_2}(s)$$

Mg is the reducing agent; N_2 is the oxidizing agent.

(b) Decomposition: one substance forms two. This reaction occurs within every bottle of this common household antiseptic. Hydrogen peroxide is very unstable and breaks down from heat, light, or just shaking:

$$2\overset{+1}{\text{H}_2}\overset{-1}{\text{O}_2}(l) \longrightarrow 2\overset{+1}{\text{H}_2}\overset{-2}{\text{O}}(l) + \overset{0}{\text{O}_2}(g)$$

H_2O_2 is both the oxidizing *and* the reducing agent. The O.N. of O in peroxides is -1. It increases to 0 in O_2 and decreases to -2 in H_2O.

(c) Displacement: two substances form two others. As Figure 4.20 shows, Al is more active than Pb and, thus, displaces it from aqueous solution:

$$2\overset{0}{\text{Al}}(s) + 3\overset{+2}{\text{Pb}}(\overset{+5}{\text{N}}\overset{-2}{\text{O}_3})_2(aq) \longrightarrow 2\overset{+3}{\text{Al}}(\overset{+5}{\text{N}}\overset{-2}{\text{O}_3})_3(aq) + 3\overset{0}{\text{Pb}}(s)$$

Al is the reducing agent; $Pb(NO_3)_2$ is the oxidizing agent.

The total ionic equation is

$$2Al(s) + 3Pb^{2+}(aq) + 6NO_3^-(aq) \longrightarrow 2Al^{3+}(aq) + 6NO_3^-(aq) + 3Pb(s)$$

The net ionic equation is

$$2Al(s) + 3Pb^{2+}(aq) \longrightarrow 2Al^{3+}(aq) + 3Pb(s)$$

FOLLOW-UP PROBLEM 4.10 Classify each of the following redox reactions as a combination, decomposition, or displacement reaction, write a balanced molecular equation for each, as well as total and net ionic equations for parts (b) and (c), and identify the oxidizing and reducing agents:

(a) $S_8(s) + F_2(g) \longrightarrow SF_4(g)$ **(b)** $CsI(aq) + Cl_2(aq) \longrightarrow CsCl(aq) + I_2(aq)$
(c) $Ni(NO_3)_2(aq) + Cr(s) \longrightarrow Ni(s) + Cr(NO_3)_3(aq)$

SECTION SUMMARY

Any reaction that includes a free element as reactant or product is a redox reaction. In combination reactions, elements combine to form a compound, or a compound and an element combine. Decomposition of compounds by absorption of heat or electricity can form elements or a compound and an element. In displacement reactions, one element displaces another from solution. Activity series rank elements in order of reactivity. The activity series of the metals ranks metals by their ability to displace H_2 from water, steam, acid or to displace one another from solution. Combustion typically releases heat and light energy through reaction of a substance with O_2.

4.7 REVERSIBLE REACTIONS: AN INTRODUCTION TO CHEMICAL EQUILIBRIUM

So far, we have viewed reactions as occurring from "left to right," from reactants to products and continuing until they are complete, that is, until the limiting reactant is used up. However, many reactions seem to stop before this happens. The reason is that two opposing reactions are taking place simultaneously. The forward (left-to-right) reaction has not stopped, but the reverse (right-to-left) reaction is occurring at the same rate. Therefore, *no further changes appear in the amounts of reactants or products.* At this point, the reaction mixture has reached **dynamic equilibrium.** On the macroscopic scale, the reaction is *static,* but it is *dynamic* on the molecular scale. In principle, all reactions are reversible and will eventually reach dynamic equilibrium as long as all products remain available for the reverse reaction.

Let's examine equilibrium with a particular set of substances. Calcium carbonate breaks down when heated to calcium oxide and carbon dioxide:

$$CaCO_3(s) \longrightarrow CaO(s) + CO_2(g) \quad \text{[breakdown]}$$

It also forms when calcium oxide and carbon dioxide react:

$$CaO(s) + CO_2(g) \longrightarrow CaCO_3(s) \quad \text{[formation]}$$

The formation is exactly the reverse of the breakdown. Suppose we place $CaCO_3$ in an *open* steel container and heat it to around 900°C, as shown in Figure 4.21A. The $CaCO_3$ starts breaking down to CaO and CO_2, and the CO_2 escapes from the open container. The reaction goes to completion because the reverse reaction (formation) can occur only if CO_2 is present.

In Figure 4.21B, we perform the same experiment in a *closed* container, so that the CO_2 remains in contact with the CaO. The breakdown (forward reaction) begins, but at first, when very little $CaCO_3$ has broken down, very little CO_2 and CaO are present; thus, the formation (reverse reaction) just barely begins. As the $CaCO_3$ continues to break down, the amounts of CO_2 and CaO increase. They react with each other more frequently, and the formation occurs a bit faster. As the amounts of CaO and CO_2 increase, the formation reaction gradually speeds up. Eventually, the reverse reaction (formation) happens just as fast as the forward reaction (breakdown), and the amounts of $CaCO_3$, CaO, and CO_2 no longer change: the system has reached equilibrium. We indicate this with a pair of arrows pointing in opposite directions:

$$CaCO_3(s) \rightleftharpoons CaO(s) + CO_2(g)$$

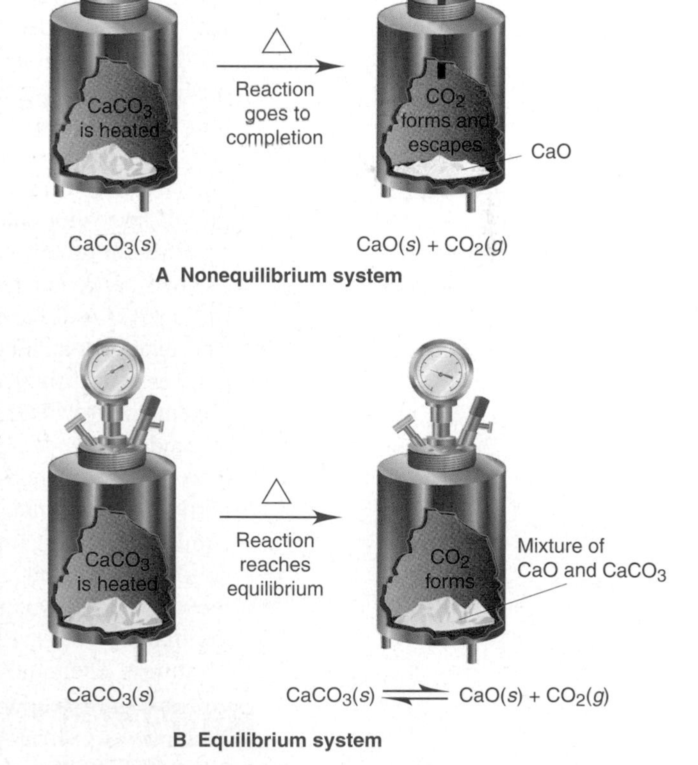

Figure 4.21 The equilibrium state. A, In an *open* steel reaction container, strong heating breaks down $CaCO_3$ completely because the product CO_2 escapes and is not present to react with the other product, CaO. **B,** When $CaCO_3$ breaks down in a *closed* container, the CO_2 *is* present to react with CaO and re-form $CaCO_3$ in a reaction that is the reverse of the breakdown. At a given temperature, no further change in the amounts of products and reactants means that the reaction has reached equilibrium.

Bear in mind that equilibrium can be established only when *all the substances involved are kept in contact with each other.* The breakdown of $CaCO_3$ goes to completion in the open container because the CO_2 escapes.

Reaction reversibility applies equally to the substances and reactions we discussed earlier. Weak acids and bases dissociate into ions only to a small extent because the dissociation quickly becomes balanced by reassociation. For example, when acetic acid dissolves in water, some of the CH_3COOH molecules transfer a proton to H_2O and form H_3O^+ and CH_3COO^- ions. As more of these ions form, they react with each other more often to re-form acetic acid and water:

$$CH_3COOH(aq) + H_2O(l) \rightleftharpoons H_3O^+(aq) + CH_3COO^-(aq)$$

In 0.1 M CH_3COOH at 25°C, only about 1.3% of the acid molecules are dissociated at any given moment. Different weak acids dissociate to different extents. For example, under the same conditions, propanoic acid (CH_3CH_2COOH) is only 1.1% dissociated, but hydrofluoric acid (HF) is 8.6% dissociated.

Similarly, the weak base ammonia reacts with water to form NH_4^+ and OH^- ions. As the ions interact, they re-form ammonia and water, and the rates of the reverse and forward reactions soon balance:

$$NH_3(aq) + H_2O(l) \rightleftharpoons NH_4^+(aq) + OH^-(aq)$$

Another weak base, methylamine (CH_3NH_2), reacts with water to a greater extent before reaching equilibrium, while still another, aniline ($C_6H_5NH_2$), reacts less.

Aqueous acid-base reactions that form a gas go to completion in an open container because the gas escapes. But, if the container were closed and the gas were present for the reverse reaction to take place, the reaction would reach equilibrium. Precipitation and other acid-base reactions seem to "go to completion," even with all the products present, because the *ions* are tied up either as an insoluble solid (precipitation) or as water molecules (acid-base). In truth, however, ionic precipitates and water do dissociate, but to an extremely small extent. Therefore, these reactions also reach equilibrium, but with almost all product formed.

Thus, some reactions proceed very little before they reach equilibrium, while others proceed almost completely, and still others reach equilibrium with a mixture of large amounts of reactants *and* products. So, a fundamental question is why do different processes, even under the same conditions, each reach equilibrium with its own particular ratio of product concentrations to reactant concentrations? The full answer will have to wait until Chapter 20, but we can hint at the factors here. The energy available for a reaction to occur is called its *free energy*. A process reaches equilibrium when the reaction mixture has its lowest free energy. Two components of free energy are the heat a reaction releases (or absorbs) and the change in randomness of the particles—particles are dispersed more randomly in a gas than in a solid, more in a solution than in pure solute and solvent, and so forth. A particular combination of these factors determines the free energy and, thus, the equilibrium point for a given mixture of reactants and products.

Many aspects of dynamic equilibrium are relevant to natural systems, from the cycling of water in the environment to the balance of lion and antelope on the plains of Africa to the nuclear processes occurring in stars. We examine equilibrium in chemical and physical systems in Chapters 12, 13, and 17 through 21.

SECTION SUMMARY

Every reaction is reversible if all the substances are kept in contact with one another. As the amounts of products increase, the reactants begin to re-form. When the reverse reaction happens as rapidly as the forward reaction, the amounts of the substances no longer change, and the reaction mixture has reached dynamic equilibrium. Weak acids and bases reach equilibrium in water with a very small proportion of their molecules dissociated. A reaction "goes to completion" because a product is removed from the system (as a gas) or exists in a form that prevents it from reacting (precipitate or undissociated molecule).

Chapter Perspective

Classifying facts is the first step toward understanding them, and this chapter classified many of the most important facts of reaction chemistry into three major processes—precipitation, acid-base, and oxidation-reduction. We also examined the great influence that water has on reaction chemistry and introduced the state of dynamic equilibrium, which is related to the central question of why physical and chemical changes occur. All these topics appear again at many places in the text.

In the next chapter, our focus changes to the physical behavior of gases. You'll find that your growing appreciation of events on the molecular level has become indispensable for understanding the nature of this physical state.

For Review and Reference

(Numbers in parentheses refer to pages, unless noted otherwise.)

Learning Objectives

Relevant section and/or sample problem (SP) numbers appear in parentheses.

Understand These Concepts

1. Why water is a polar molecule and how it dissolves ionic compounds and dissociates them into ions (4.1)
2. The difference between the species present when ionic and covalent compounds dissolve in water and that between strong and weak electrolytes (4.1)
3. The use of ionic equations to specify the essential nature of an aqueous reaction (4.2)
4. The driving force for aqueous ionic reactions (4.3, 4.4, 4.5)
5. How to decide whether a precipitation reaction occurs (4.3)
6. The main distinction between strong and weak aqueous acids and bases (4.4)
7. The essential character of aqueous acid-base reactions as proton-transfer processes (4.4)
8. The importance of net movement of electrons in the redox process (4.5)
9. The relation between change in oxidation number and identity of oxidizing and reducing agents (4.5)
10. The presence of elements in some important types of redox reactions: combination, decomposition, displacement (4.6)
11. The balance between forward and reverse rates of a chemical reaction that leads to dynamic equilibrium; why some acids and bases are weak (4.7)

Master These Skills

1. Using the formula of a compound to find the number of moles of ions in solution (SP 4.1)
2. Determining the concentration of H^+ ion in an aqueous acid solution (SP 4.2)
3. Predicting whether a precipitation reaction occurs (SP 4.3)
4. Writing ionic equations to describe precipitation and acid-base reactions (SPs 4.3 and 4.4)
5. Calculating an unknown concentration from an acid-base or redox titration (SPs 4.5 and 4.9)
6. Determining the oxidation number of any element in a compound (SP 4.6)
7. Identifying the oxidizing and reducing agents in a redox reaction (SP 4.7)
8. Balancing redox equations (SP 4.8)
9. Identifying combination, decomposition, and displacement redox reactions (SP 4.10)

Key Terms

Section 4.1
polar molecule (135)
solvated (136)
electrolyte (136)
nonelectrolyte (138)

Section 4.2
molecular equation (141)
total ionic equation (141)
spectator ion (141)
net ionic equation (141)

Section 4.3
precipitation reaction (141)
precipitate (141)
metathesis reaction (143)

Section 4.4
acid-base reaction (144)
neutralization reaction (144)
acid (144)
base (144)
salt (145)
titration (146)
equivalence point (147)
end point (147)

Section 4.5
oxidation-reduction (redox) reaction (150)
oxidation (151)
reduction (151)
oxidizing agent (151)
reducing agent (151)
oxidation number (O.N.) (or oxidation state) (151)
oxidation number method (154)

Section 4.6
activity series of the metals (163)

Section 4.7
dynamic equilibrium (165)

Highlighted Figures and Tables

These figures (F) and tables (T) provide a review of key ideas. Entries in **color** contain frequently used data.

Brief Solutions to Follow-up Problems

4.1 (a) $KClO_4(s) \xrightarrow{H_2O} K^+(aq) + ClO_4^-(aq)$;
2 mol of K^+ and 2 mol of ClO_4^-

(b) $Mg(C_2H_3O_2)_2(s) \xrightarrow{H_2O} Mg^{2+}(aq) + 2C_2H_3O_2^-(aq)$;
2.49 mol of Mg^{2+} and 4.97 mol of $C_2H_3O_2^-$

(c) $(NH_4)_2CrO_4(s) \xrightarrow{H_2O} 2NH_4^+(aq) + CrO_4^{2-}(aq)$;
6.24 mol of NH_4^+ and 3.12 mol of CrO_4^{2-}

(d) $NaHSO_4(s) \xrightarrow{H_2O} Na^+(aq) + HSO_4^-(aq)$;
0.73 mol of Na^+ and 0.73 mol of HSO_4^-

4.2 Moles of H^+ $= 451\ \cancel{mL} \times \frac{1\ \cancel{L}}{10^3\ \cancel{mL}} \times \frac{3.20\ \cancel{mol\ HBr}}{1\ \cancel{L\ soln}} \times \frac{1\ mol\ H^+}{1\ \cancel{mol\ HBr}}$
$= 1.44\ mol\ H^+$

4.3 (a) $Fe^{3+}(aq) + 3Cl^-(aq) + 3Cs^+(aq) + PO_4^{3-}(aq) \longrightarrow FePO_4(s) + 3Cl^-(aq) + 3Cs^+(aq)$
$Fe^{3+}(aq) + PO_4^{3-}(aq) \longrightarrow FePO_4(s)$
(b) $2Na^+(aq) + 2OH^-(aq) + Cd^{2+}(aq) + 2NO_3^-(aq) \longrightarrow 2Na^+(aq) + 2NO_3^-(aq) + Cd(OH)_2(s)$
$2OH^-(aq) + Cd^{2+}(aq) \longrightarrow Cd(OH)_2(s)$
(c) No reaction occurs
(d) $2Ag^+(aq) + SO_4^{2-}(aq) + Ba^{2+}(aq) + 2Cl^-(aq) \longrightarrow 2AgCl(s) + BaSO_4(s)$
Total and net ionic equations are identical.

4.4 $Ca(OH)_2(aq) + 2HNO_3(aq) \longrightarrow Ca(NO_3)_2(aq) + 2H_2O(l)$
$Ca^{2+}(aq) + 2OH^-(aq) + 2H^+(aq) + 2NO_3^-(aq) \longrightarrow Ca^{2+}(aq) + 2NO_3^-(aq) + 2H_2O(l)$
$H^+(aq) + OH^-(aq) \longrightarrow H_2O(l)$

4.5 $Ba(OH)_2(aq) + 2HCl(aq) \longrightarrow BaCl_2(aq) + 2H_2O(l)$
Volume (L) of soln
$= 50.00\ \cancel{mL\ HCl\ soln} \times \frac{1\ L}{10^3\ \cancel{mL}} \times \frac{0.1016\ \cancel{mol\ HCl}}{1\ \cancel{L\ soln}} \times \frac{1\ \cancel{mol\ Ba(OH)_2}}{2\ \cancel{mol\ HCl}} \times \frac{1\ \cancel{L\ soln}}{0.1292\ \cancel{mol\ Ba(OH)_2}}$
$= 0.01966\ L$

4.6 (a) O.N. of Sc $= +3$; O.N. of O $= -2$
(b) O.N. of Ga $= +3$; O.N. of Cl $= -1$
(c) O.N. of H $= +1$; O.N. of P $= +5$; O.N. of O $= -2$
(d) O.N. of I $= +3$; O.N. of F $= -1$

4.7 (a) Fe is the reducing agent; Cl_2 is the oxidizing agent.
(b) C_2H_6 is the reducing agent; O_2 is the oxidizing agent.
(c) CO is the reducing agent; I_2O_5 is the oxidizing agent.

4.8 $K_2Cr_2O_7(aq) + 14HI(aq) \longrightarrow 2KI(aq) + 2CrI_3(aq) + 3I_2(s) + 7H_2O(l)$

4.9 (a) Moles of $Ca^{2+} = 6.53\ \cancel{mL\ soln} \times \frac{1\ \cancel{L}}{10^3\ \cancel{mL}} \times \frac{4.56\times10^{-3}\ \cancel{mol\ KMnO_4}}{1\ \cancel{L\ soln}} \times \frac{5\ \cancel{mol\ CaC_2O_4}}{2\ \cancel{mol\ KMnO_4}} \times \frac{1\ mol\ Ca^{2+}}{1\ \cancel{mol\ CaC_2O_4}}$
$= 7.44\times10^{-5}\ mol\ Ca^{2+}$

Molarity of $Ca^{2+} = \frac{7.44\times10^{-2}\ mol\ Ca^{2+}}{2.50\ \cancel{mL\ milk}} \times \frac{10^3\ \cancel{mL}}{1\ L}$
$= 2.98\times10^{-2}\ M\ Ca^{2+}$

(b) Conc. of Ca^{2+} (g/L)
$= \frac{2.98\times10^{-2}\ \cancel{mol\ Ca^{2+}}}{1\ L} \times \frac{40.08\ g\ Ca^{2+}}{1\ \cancel{mol\ Ca^{2+}}}$
$= \frac{1.19\ g\ Ca^{2+}}{1\ L}$

4.10 (a) Combination:
$S_8(s) + 16F_2(g) \longrightarrow 8SF_4(g)$
S_8 is the reducing agent; F_2 is the oxidizing agent.
(b) Displacement:
$2CsI(aq) + Cl_2(aq) \longrightarrow 2CsCl(aq) + I_2(aq)$
Cl_2 is the oxidizing agent; CsI is the reducing agent.
$2Cs^+(aq) + 2I^-(aq) + Cl_2(aq) \longrightarrow 2Cs^+(aq) + 2Cl^-(aq) + I_2(aq)$
$2I^-(aq) + Cl_2(aq) \longrightarrow 2Cl^-(aq) + I_2(aq)$
(c) Displacement:
$3Ni(NO_3)_2(aq) + 2Cr(s) \longrightarrow 3Ni(s) + 2Cr(NO_3)_3(aq)$
$3Ni^{2+}(aq) + 6NO_3^-(aq) + 2Cr(s) \longrightarrow 3Ni(s) + 2Cr^{3+}(aq) + 6NO_3^-(aq)$
$3Ni^{2+}(aq) + 2Cr(s) \longrightarrow 3Ni(s) + 2Cr^{3+}(aq)$
Cr is the reducing agent; $Ni(NO_3)_2$ is the oxidizing agent.

Problems

Problems with **colored** numbers are answered in Appendix E. Sections match the text and provide the numbers of relevant sample problems. Most offer Concept Review Questions, Skill-Building Exercises (grouped in pairs covering the same concept), and Problems in Context. Comprehensive Problems are based on material from any section or previous chapter.

The Role of Water as a Solvent

(Sample Problems 4.1 and 4.2)

Concept Review Questions

4.1 What two factors cause water to be polar?

4.2 What types of substances are most likely to be soluble in water?

4.3 What must be present in an aqueous solution for it to conduct an electric current? What general classes of compounds form solutions that conduct?

4.4 What occurs on the molecular level when an ionic compound dissolves in water?

4.5 Examine each of the following aqueous solutions and determine which represents: (a) $CaCl_2$; (b) Li_2SO_4; (c) NH_4Br.

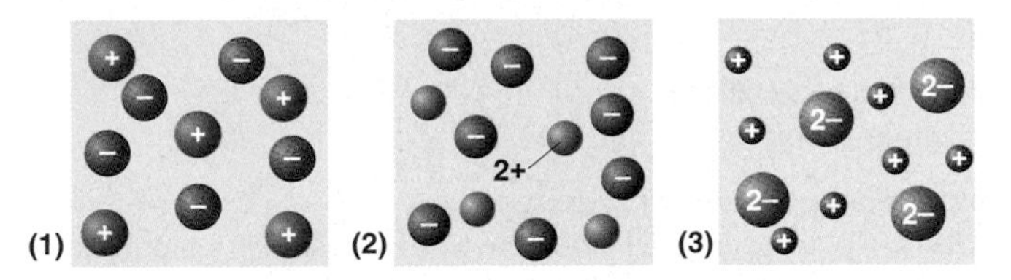

4.6 Which of the following best represents a volume from a solution of magnesium nitrate?

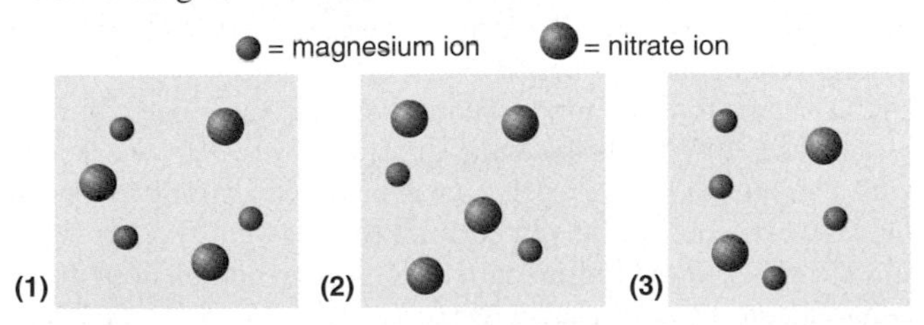

4.7 Why are some ionic compounds soluble in water and others are not?

4.8 Why are some covalent compounds soluble in water and others are not?

4.9 Some covalent compounds dissociate into ions when they dissolve in water. What atom do these compounds have in their structures? What type of aqueous solution do they form? Name three examples of such an aqueous solution.

Skill-Building Exercises *(grouped in similar pairs)*

4.10 State whether each of the following substances is likely to be very soluble in water. Explain.
(a) Benzene, C_6H_6 (b) Sodium hydroxide
(c) Ethanol, CH_3CH_2OH (d) Potassium acetate

4.11 State whether each of the following substances is likely to be very soluble in water. Explain.
(a) Lithium nitrate (b) Glycine, H_2NCH_2COOH
(c) Pentane (d) Ethylene glycol, $HOCH_2CH_2OH$

4.12 State whether an aqueous solution of each of the following substances conducts an electric current. Explain your reasoning.
(a) Cesium iodide (b) Hydrogen bromide

4.13 State whether an aqueous solution of each of the following substances conducts an electric current. Explain your reasoning.
(a) Potassium hydroxide (b) Glucose, $C_6H_{12}O_6$

4.14 How many total moles of ions are released when each of the following samples dissolves completely in water?
(a) 0.37 mol of NH_4Cl (b) 35.4 g of $Ba(OH)_2 \cdot 8H_2O$
(c) 3.55×10^{18} formula units of LiCl

4.15 How many total moles of ions are released when each of the following samples dissolves completely in water?
(a) 0.805 mol of Rb_2SO_4 (b) 3.85×10^{-3} g of $Ca(NO_3)_2$
(c) 4.03×10^{19} formula units of $Sr(HCO_3)_2$

4.16 How many total moles of ions are released when each of the following samples dissolves completely in water?
(a) 0.83 mol of K_3PO_4 (b) 8.11×10^{-3} g of $NiBr_2 \cdot 3H_2O$
(c) 1.23×10^{21} formula units of $FeCl_3$

4.17 How many total moles of ions are released when each of the following samples dissolves completely in water?
(a) 0.734 mol of Na_2HPO_4 (b) 3.86 g of $CuSO_4 \cdot 5H_2O$
(c) 8.66×10^{20} formula units of $NiCl_2$

4.18 How many moles and numbers of ions of each type are present in the following aqueous solutions?
(a) 100. mL of 2.45 *M* aluminum chloride
(b) 1.80 L of a solution containing 2.59 g lithium sulfate/L
(c) 225 mL of a solution containing 1.68×10^{22} formula units of potassium bromide per liter

4.19 How many moles and numbers of ions of each type are present in the following aqueous solutions?
(a) 88 mL of 1.75 *M* magnesium chloride
(b) 321 mL of a solution containing 0.22 g aluminum sulfate/L
(c) 1.65 L of a solution containing 8.83×10^{21} formula units of cesium nitrate per liter

4.20 How many moles of H^+ ions are present in the following aqueous solutions?
(a) 1.40 L of 0.25 *M* perchloric acid
(b) 1.8 mL of 0.72 *M* nitric acid
(c) 7.6 L of 0.056 *M* hydrochloric acid

4.21 How many moles of H^+ ions are present in the following aqueous solutions?
(a) 1.4 mL of 0.75 *M* hydrobromic acid
(b) 2.47 mL of 1.98 *M* hydriodic acid
(c) 395 mL of 0.270 *M* nitric acid

Problems in Context

4.22 To study a marine organism, a biologist prepares a 1.00-kg sample to simulate the ion concentrations in seawater. She mixes 26.5 g of NaCl, 2.40 g of $MgCl_2$, 3.35 g of $MgSO_4$, 1.20 g of $CaCl_2$, 1.05 g of KCl, 0.315 g of $NaHCO_3$, and 0.098 g of NaBr in distilled water. (a) If the density of this solution is 1.04 g/cm^3, what is the molarity of each ion? (b) What is the total molarity of alkali metal ions? (c) What is the total molarity of alkaline earth metal ions? (d) What is the total molarity of anions?

4.23 Water "softeners" remove metal ions such as Ca^{2+} and Fe^{3+} by replacing them with enough Na^+ ions to maintain the same number of positive charges in the solution. If 1.0×10^3 L of "hard" water is 0.015 *M* Ca^{2+} and 0.0010 *M* Fe^{3+}, how many moles of Na^+ are needed to replace these ions?

Writing Equations for Aqueous Ionic Reactions

Concept Review Questions

4.24 Which ions do not appear in a net ionic equation? Why?

4.25 Write two sets of equations (both molecular and total ionic) with different reactants that will give the same net ionic equation, as follows:

$$Ba(NO_3)_2(aq) + Na_2CO_3(aq) \longrightarrow BaCO_3(s) + 2NaNO_3(aq)$$

Precipitation Reactions

(Sample Problem 4.3)

Concept Review Questions

4.26 Why do some pairs of ions precipitate and others do not?

4.27 Use Table 4.1 to determine which of the following combinations leads to a reaction. How can you identify the spectator ions in the reaction?

(a) Calcium nitrate(aq) + sodium chloride(aq) $\longrightarrow$

(b) Potassium chloride(aq) + lead(II) nitrate(aq) $\longrightarrow$

4.28 The beakers represent the aqueous reaction of $AgNO_3$ and NaCl. Silver ions are gray. What colors are used to represent NO_3^-, Na^+, and Cl^-? Write molecular, total ionic, and net ionic equations for the reaction.

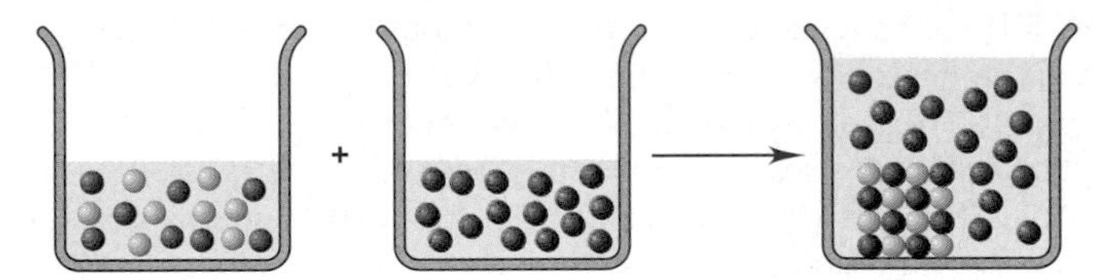

Skill-Building Exercises *(grouped in similar pairs)*

4.29 Complete the following precipitation reactions with balanced molecular, total ionic, and net ionic equations:

(a) $Hg_2(NO_3)_2(aq) + KI(aq) \longrightarrow$

(b) $FeSO_4(aq) + Ba(OH)_2(aq) \longrightarrow$

4.30 Complete the following precipitation reactions with balanced molecular, total ionic, and net ionic equations:

(a) $CaCl_2(aq) + Cs_3PO_4(aq) \longrightarrow$

(b) $Na_2S(aq) + ZnSO_4(aq) \longrightarrow$

4.31 When each of the following pairs of aqueous solutions is mixed, does a precipitation reaction occur? If so, write balanced molecular, total ionic, and net ionic equations:

(a) Sodium nitrate + copper(II) sulfate

(b) Ammonium iodide + silver nitrate

4.32 When each of the following pairs of aqueous solutions is mixed, does a precipitation reaction occur? If so, write balanced molecular, total ionic, and net ionic equations:

(a) Potassium carbonate + barium hydroxide

(b) Aluminum nitrate + sodium phosphate

4.33 When each of the following pairs of aqueous solutions is mixed, does a precipitation reaction occur? If so, write balanced molecular, total ionic, and net ionic equations:

(a) Potassium chloride + iron(II) nitrate

(b) Ammonium sulfate + barium chloride

4.34 When each of the following pairs of aqueous solutions is mixed, does a precipitation reaction occur? If so, write balanced molecular, total ionic, and net ionic equations:

(a) Sodium sulfide + nickel(II) sulfate

(b) Lead(II) nitrate + potassium bromide

4.35 If 35.0 mL of lead(II) nitrate solution reacts completely with excess sodium iodide solution to yield 0.628 g of precipitate, what is the molarity of lead(II) ion in the original solution?

4.36 If 25.0 mL of silver nitrate solution reacts with excess potassium chloride solution to yield 0.842 g of precipitate, what is the molarity of silver ion in the original solution?

Problems in Context

4.37 The mass percent of Cl^- in a seawater sample is determined by titrating 25.00 mL of seawater with $AgNO_3$ solution, causing a precipitation reaction. An indicator is used to detect the end point, which occurs when free Ag^+ ion is present in solution after all the Cl^- has reacted. If 43.63 mL of 0.3020 *M* $AgNO_3$ is required to reach the end point, what is the mass percent of Cl^- in the seawater (d of seawater = 1.04 g/mL)?

4.38 Aluminum sulfate, known as *cake alum*, has a wide range of uses, from dyeing leather and cloth to purifying sewage. In aqueous solution, it reacts with base to form a white precipitate. (a) Write balanced total and net ionic equations for its reaction with aqueous NaOH. (b) What mass of precipitate forms when 185.5 mL of 0.533 *M* NaOH is added to 627 mL of a solution that contains 15.8 g of aluminum sulfate per liter?

Acid-Base Reactions

(Sample Problems 4.4 and 4.5)

Concept Review Questions

4.39 Is the total ionic equation the same as the net ionic equation when $Sr(OH)_2(aq)$ and $H_2SO_4(aq)$ react? Explain.

4.40 State a general equation for a neutralization reaction.

4.41 (a) Name three common strong acids. (b) Name three common strong bases. (c) What is a characteristic behavior of a strong acid or a strong base?

4.42 (a) Name three common weak acids. (b) Name one common weak base. (c) What is the major difference between a weak acid and a strong acid or between a weak base and a strong base, and what experiment would you perform to observe it?

4.43 Do either of the following reactions go to completion? If so, what factor(s) causes each to do so?

(a) $MgSO_3(s) + 2HCl(aq) \longrightarrow MgCl_2(aq) + SO_2(g) + H_2O(l)$

(b) $3Ba(OH)_2(aq) + 2H_3PO_4(aq) \longrightarrow Ba_3(PO_4)_2(s) + 6H_2O(l)$

4.44 The net ionic equation for the aqueous neutralization reaction between acetic acid and sodium hydroxide is different from that for the reaction between hydrochloric acid and sodium hydroxide. Explain by writing balanced net ionic equations.

Skill-Building Exercises *(grouped in similar pairs)*

4.45 Complete the following acid-base reactions with balanced molecular, total ionic, and net ionic equations:

(a) Potassium hydroxide(aq) + hydriodic acid(aq) $\longrightarrow$

(b) Ammonia(aq) + hydrochloric acid(aq) $\longrightarrow$

4.46 Complete the following acid-base reactions with balanced molecular, total ionic, and net ionic equations:

(a) Cesium hydroxide(aq) + nitric acid(aq) $\longrightarrow$

(b) Calcium hydroxide(aq) + acetic acid(aq) $\longrightarrow$

4.47 Limestone (calcium carbonate) is insoluble in water but dissolves when a hydrochloric acid solution is added. Why? Write balanced total ionic and net ionic equations, showing hydrochloric acid as it actually exists in water and the reaction as a proton-transfer process.

4.48 Zinc hydroxide is insoluble in water but dissolves when a nitric acid solution is added. Why? Write balanced total ionic and net ionic equations, showing nitric acid as it actually exists in water and the reaction as a proton-transfer process.

4.49 If 25.98 mL of a standard 0.1180 *M* KOH solution reacts with 52.50 mL of CH_3COOH solution, what is the molarity of the acid solution?

4.50 If 36.25 mL of a standard 0.1750 *M* NaOH solution is required to neutralize 25.00 mL of H_2SO_4, what is the molarity of the acid solution?

Problems in Context

4.51 An auto mechanic spills 78 mL of 2.6 *M* H_2SO_4 solution from a rebuilt auto battery. How many milliliters of 1.5 *M* $NaHCO_3$ must be poured on the spill to react completely with the sulfuric acid?

4.52 Sodium hydroxide is used extensively in acid-base titrations because it is a strong, inexpensive base. A sodium hydroxide solution was standardized by titrating 25.00 mL of 0.1528 *M* standard hydrochloric acid. The initial buret reading of the sodium hydroxide was 2.24 mL, and the final reading was 39.21 mL. What was the molarity of the base solution?

4.53 One of the first steps in the enrichment of uranium for use in nuclear power plants involves a displacement reaction between UO_2 and aqueous HF:

$$UO_2(s) + 4HF(aq) \longrightarrow UF_4(s) + 2H_2O(l)$$

How many liters of 2.50 *M* HF will react with 2.25 kg of UO_2?

Oxidation-Reduction (Redox) Reactions

(Sample Problems 4.6 to 4.9)

Concept Review Questions

4.54 Describe how to determine the oxidation number of sulfur in (a) H_2S and (b) SO_3^{2-}.

4.55 Is the following a redox reaction? Explain.

$$NH_3(aq) + HCl(aq) \longrightarrow NH_4Cl(aq)$$

4.56 Explain why an oxidizing agent undergoes reduction.

4.57 Why must every redox reaction involve an oxidizing agent and a reducing agent?

4.58 In which of the following equations does sulfuric acid act as an oxidizing agent? In which does it act as an acid? Explain.

(a) $4H^+(aq) + SO_4^{2-}(aq) + 2NaI(s) \longrightarrow 2Na^+(aq) + I_2(s) + SO_2(g) + 2H_2O(l)$

(b) $BaF_2(s) + 2H^+(aq) + SO_4^{2-}(aq) \longrightarrow 2HF(aq) + BaSO_4(s)$

4.59 Identify the oxidizing agent and the reducing agent in the following reaction, and explain your answer:

$$8NH_3(g) + 6NO_2(g) \longrightarrow 7N_2(g) + 12H_2O(l)$$

Skill-Building Exercises *(grouped in similar pairs)*

4.60 Give the oxidation number of carbon in the following:
(a) CF_2Cl_2 (b) $Na_2C_2O_4$ (c) HCO_3^- (d) C_2H_6

4.61 Give the oxidation number of bromine in the following:
(a) KBr (b) BrF_3 (c) $HBrO_3$ (d) CBr_4

4.62 Give the oxidation number of nitrogen in the following:
(a) NH_2OH (b) N_2H_4 (c) NH_4^+ (d) HNO_2

4.63 Give the oxidation number of sulfur in the following:
(a) $SOCl_2$ (b) H_2S_2 (c) H_2SO_3 (d) Na_2S

4.64 Give the oxidation number of arsenic in the following:
(a) AsH_3 (b) H_3AsO_4 (c) $AsCl_3$

4.65 Give the oxidation number of phosphorus in the following:
(a) $H_2P_2O_7^{2-}$ (b) PH_4^+ (c) PCl_5

4.66 Give the oxidation number of manganese in the following:
(a) MnO_4^{2-} (b) Mn_2O_3 (c) $KMnO_4$

4.67 Give the oxidation number of chromium in the following:
(a) CrO_3 (b) $Cr_2O_7^{2-}$ (c) $Cr_2(SO_4)_3$

4.68 Identify the oxidizing and reducing agents in the following:

(a) $5H_2C_2O_4(aq) + 2MnO_4^-(aq) + 6H^+(aq) \longrightarrow 2Mn^{2+}(aq) + 10CO_2(g) + 8H_2O(l)$

(b) $3Cu(s) + 8H^+(aq) + 2NO_3^-(aq) \longrightarrow 3Cu^{2+}(aq) + 2NO(g) + 4H_2O(l)$

4.69 Identify the oxidizing and reducing agents in the following:

(a) $Sn(s) + 2H^+(aq) \longrightarrow Sn^{2+}(aq) + H_2(g)$

(b) $2H^+(aq) + H_2O_2(aq) + 2Fe^{2+}(aq) \longrightarrow 2Fe^{3+}(aq) + 2H_2O(l)$

4.70 Identify the oxidizing and reducing agents in the following:

(a) $8H^+(aq) + 6Cl^-(aq) + Sn(s) + 4NO_3^-(aq) \longrightarrow SnCl_6^{2-}(aq) + 4NO_2(g) + 4H_2O(l)$

(b) $2MnO_4^-(aq) + 10Cl^-(aq) + 16H^+(aq) \longrightarrow 5Cl_2(g) + 2Mn^{2+}(aq) + 8H_2O(l)$

4.71 Identify the oxidizing and reducing agents in the following:

(a) $8H^+(aq) + Cr_2O_7^{2-}(aq) + 3SO_3^{2-}(aq) \longrightarrow 2Cr^{3+}(aq) + 3SO_4^{2-}(aq) + 4H_2O(l)$

(b) $NO_3^-(aq) + 4Zn(s) + 7OH^-(aq) + 6H_2O(l) \longrightarrow 4Zn(OH)_4^{2-}(aq) + NH_3(aq)$

4.72 Discuss each conclusion from a study of redox reactions:
(a) The sulfide ion functions only as a reducing agent.
(b) The sulfate ion functions only as an oxidizing agent.
(c) Sulfur dioxide functions as an oxidizing or a reducing agent.

4.73 Discuss each conclusion from a study of redox reactions:
(a) The nitride ion functions only as a reducing agent.
(b) The nitrate ion functions only as an oxidizing agent.
(c) The nitrite ion functions as an oxidizing or a reducing agent.

4.74 Use the oxidation number method to balance the following equations by placing coefficients in the blanks. Identify the reducing and oxidizing agents:

(a) $\_HNO_3(aq) + \_K_2CrO_4(aq) + \_Fe(NO_3)_2(aq) \longrightarrow \_KNO_3(aq) + \_Fe(NO_3)_3(aq) + \_Cr(NO_3)_3(aq) + \_H_2O(l)$

(b) $\_HNO_3(aq) + \_C_2H_6O(l) + \_K_2Cr_2O_7(aq) \longrightarrow \_KNO_3(aq) + \_C_2H_4O(l) + \_H_2O(l) + \_Cr(NO_3)_3(aq)$

(c) $\_HCl(aq) + \_NH_4Cl(aq) + \_K_2Cr_2O_7(aq) \longrightarrow \_KCl(aq) + \_CrCl_3(aq) + \_N_2(g) + \_H_2O(l)$

(d) $\_KClO_3(aq) + \_HBr(aq) \longrightarrow \_Br_2(l) + \_H_2O(l) + \_KCl(aq)$

4.75 Use the oxidation number method to balance the following equations by placing coefficients in the blanks. Identify the reducing and oxidizing agents:

(a) $\_HCl(aq) + \_FeCl_2(aq) + \_H_2O_2(aq) \longrightarrow \_FeCl_3(aq) + \_H_2O(l)$

(b) $\_I_2(s) + \_Na_2S_2O_3(aq) \longrightarrow \_Na_2S_4O_6(aq) + \_NaI(aq)$

4.116 For the following aqueous reactions, complete and balance the molecular equation and write a net ionic equation:
(a) Manganese(II) sulfide + hydrobromic acid
(b) Potassium carbonate + strontium nitrate
(c) Potassium nitrite + hydrochloric acid
(d) Calcium hydroxide + nitric acid
(e) Barium acetate + iron(II) sulfate
(f) Zinc carbonate + sulfuric acid
(g) Copper(II) nitrate + hydrosulfuric acid
(h) Magnesium hydroxide + chloric acid
(i) Potassium chloride + ammonium phosphate
(j) Barium hydroxide + hydrocyanic acid

4.117 Use the oxidation number method to balance the following equations by placing coefficients in the blanks. Identify the reducing and oxidizing agents:
(a) $\_KOH(aq) + \_H_2O_2(aq) + \_Cr(OH)_3(s) \longrightarrow \_K_2CrO_4(aq) + \_H_2O(l)$
(b) $\_MnO_4^-(aq) + \_ClO_2^-(aq) + \_H_2O(l) \longrightarrow \_MnO_2(s) + \_ClO_4^-(aq) + \_OH^-(aq)$
(c) $\_KMnO_4(aq) + \_Na_2SO_3(aq) + \_H_2O(l) \longrightarrow \_MnO_2(s) + \_Na_2SO_4(aq) + \_KOH(aq)$
(d) $\_CrO_4^{2-}(aq) + \_HSnO_2^-(aq) + \_H_2O(l) \longrightarrow \_CrO_2^-(aq) + \_HSnO_3^-(aq) + \_OH^-(aq)$
(e) $\_KMnO_4(aq) + \_NaNO_2(aq) + \_H_2O(l) \longrightarrow \_MnO_2(s) + \_NaNO_3(aq) + \_KOH(aq)$
(f) $\_I^-(aq) + \_O_2(g) + \_H_2O(l) \longrightarrow \_I_2(s) + \_OH^-(aq)$

4.118 In 1995, Mario Molina, Paul Crutzen, and F. Sherwood Rowland shared the Nobel Prize in chemistry for their work on atmospheric chemistry. One of several reaction sequences proposed for the role of chlorine in the decomposition of stratospheric ozone (we'll see another sequence in Chapter 16) is
(1) $Cl(g) + O_3(g) \longrightarrow ClO(g) + O_2(g)$
(2) $ClO(g) + ClO(g) \longrightarrow Cl_2O_2(g)$
(3) $Cl_2O_2(g) \xrightarrow{\text{light}} 2Cl(g) + O_2(g)$
Over the tropics, oxygen atoms are more common in the stratosphere, and ozone is also decomposed by reactions 1 and 4:
(4) $ClO(g) + O_3(g) \longrightarrow Cl(g) + O_2(g)$
(a) Which, if any, of the reactions are oxidation-reduction reactions?
(b) Write an overall equation combining reactions 1-3.

4.119 Sodium peroxide (Na_2O_2) is often used in self-contained breathing devices, such as those used in fire emergencies, because it reacts with exhaled CO_2 to form Na_2CO_3 and O_2. How many liters of respired air can react with 80.0 g of Na_2O_2 if each liter of respired air contains 0.0720 g of CO_2?

4.120 Magnesium is used in airplane bodies and other lightweight alloys. The metal is obtained from seawater in a process that includes precipitation, neutralization, evaporation, and electrolysis. How many kilograms of magnesium can be obtained from 1.00 km^3 of seawater if the initial Mg^{2+} concentration is 0.13% by mass (d of seawater = 1.04 g/mL)?

4.121 A typical formulation for window glass is 75% SiO_2, 15% Na_2O, and 10.% CaO by mass. What masses of sand (SiO_2), sodium carbonate, and calcium carbonate must be combined to produce 1.00 kg of glass after carbon dioxide is driven off by thermal decomposition of the carbonates?

4.122 Physicians who specialize in sports medicine routinely treat athletes and dancers. Ethyl chloride, a local anesthetic commonly used for simple injuries, is the product of the combination of ethylene with hydrogen chloride:

$$C_2H_4(g) + HCl(g) \longrightarrow C_2H_5Cl(g)$$

If 0.100 kg of C_2H_4 and 0.100 kg of HCl react:
(a) How many molecules of gas (reactants plus products) are present when the reaction is complete?
(b) How many moles of gas are present when half the product forms?

4.123 The *salinity* of a solution is defined as the grams of total salts per kilogram of solution. An agricultural chemist uses a solution whose salinity is 35.0 g/kg to test the effect of irrigating farmland with high-salinity river water. The two solutes are NaCl and $MgSO_4$, and there are twice as many moles of NaCl as $MgSO_4$. What masses of NaCl and $MgSO_4$ are contained in 1.00 kg of the solution?

4.124 Thyroxine ($C_{15}H_{11}I_4NO_4$) is a hormone synthesized by the thyroid gland and used to control many metabolic functions in the body. A physiologist determines the mass percent of thyroxine in a thyroid extract by igniting 0.4332 g of extract with sodium carbonate, which converts the iodine to iodide. The iodide is dissolved in water, and bromine and hydrochloric acid are added, which convert the iodide to iodate.
(a) How many moles of iodate form per mole of thyroxine?
(b) Excess bromine is boiled off and more iodide is added, which reacts as shown in the following *unbalanced* equation:

$$IO_3^-(aq) + H^+(aq) + I^-(aq) \longrightarrow I_2(aq) + H_2O(l)$$

How many moles of iodine are produced per mole of thyroxine? (*Hint:* Be sure to balance the charges as well as the atoms.) What are the oxidizing and reducing agents in the reaction?
(c) The iodine reacts completely with 17.23 mL of 0.1000 M thiosulfate as shown in the following *unbalanced* equation:

$$I_2(aq) + S_2O_3^{2-}(aq) \longrightarrow I^-(aq) + S_4O_6^{2-}(aq)$$

What is the mass percent of thyroxine in the thyroid extract?

4.125 Carbon dioxide is removed from the atmosphere of space capsules by reaction with a solid metal hydroxide. The products are water and the metal carbonate.
(a) Calculate the mass of CO_2 that can be removed by reaction with 3.50 kg of lithium hydroxide.
(b) How many grams of CO_2 can be removed by 1.00 g of each of the following: lithium hydroxide, magnesium hydroxide, and aluminum hydroxide?

4.126 Calcium dihydrogen phosphate, $Ca(H_2PO_4)_2$, and sodium hydrogen carbonate, $NaHCO_3$, are ingredients of baking powder that react with each other to produce CO_2, which causes dough or batter to rise:

$$Ca(H_2PO_4)_2(s) + NaHCO_3(s) \longrightarrow CO_2(g) + H_2O(g) + CaHPO_4(s) + Na_2HPO_4(s) \text{ [unbalanced]}$$

If the baking powder contains 31% $NaHCO_3$ and 35% $Ca(H_2PO_4)_2$ by mass:
(a) How many moles of CO_2 are produced from 1.00 g of baking powder?
(b) If 1 mol of CO_2 occupies 37.0 L at 350°F (a typical baking temperature), what volume of CO_2 is produced from 1.00 g of baking powder?

4.127 In a titration of HNO_3, you add a few drops of phenolphthalein indicator to 50.00 mL of acid in a flask. You quickly add 20.00 mL of 0.0502 *M* NaOH but overshoot the end point, and the solution turns deep pink. Instead of starting over, you add 30.00 mL of the acid, and the solution turns colorless. Then, it takes 3.22 mL of the NaOH to reach the end point. (a) What is the concentration of the HNO_3 solution? (b) How many moles of NaOH were in excess after the first addition?

4.128 The active compound in Pepto-Bismol contains C, H, O, and Bi.
(a) When 0.22105 g of it was burned in excess O_2, 0.1422 g of bismuth(III) oxide, 0.1880 g of carbon dioxide, and 0.02750 g of water were formed. What is the empirical formula of this compound?
(b) Given a molar mass of 1086 g/mol, determine the molecular formula.
(c) Complete and balance the acid-base reaction between bismuth(III) hydroxide and salicylic acid ($HC_7H_5O_3$), which is used to form this compound.
(d) A dose of Pepto-Bismol contains 0.600 mg of the active ingredient. If the yield of the reaction in part (c) is 88.0%, what mass (in mg) of bismuth(III) hydroxide is required to prepare one dose?

4.129 Two aqueous solutions contain the ions indicated below.

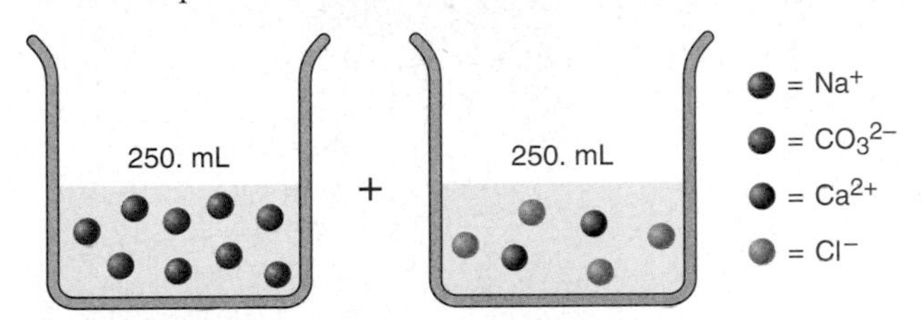

(a) Write balanced molecular, total ionic, and net ionic equations for the reaction that occurs when the solutions are mixed. (b) If each sphere represents 0.050 mol of ion, what mass (in g) of precipitate forms, assuming 100% reaction. (c) What is the concentration of each ion in solution after reaction?

4.130 In 1997, at the United Nations Conference on Climate Change, the major industrial nations agreed to expand their research efforts to develop renewable sources of carbon-based fuels. For more than a decade, Brazil has been engaged in a program to replace gasoline with ethanol derived from the root crop manioc (cassava).
(a) Write separate balanced equations for the complete combustion of ethanol (C_2H_5OH) and of gasoline (represented by the formula C_8H_{18}).
(b) What mass of oxygen is required to burn completely 1.00 L of a mixture that is 90.0% gasoline (d = 0.742 g/mL) and 10.0% ethanol (d = 0.789 g/mL) by volume?
(c) If 1.00 mol of O_2 occupies 22.4 L, what volume of O_2 is needed to burn 1.00 L of the mixture?
(d) Air is 20.9% O_2 by volume. What volume of air is needed to burn 1.00 L of the mixture?

4.131 In a car engine, gasoline (represented by C_8H_{18}) does not burn completely, and some CO, a toxic pollutant, forms along with CO_2 and H_2O. If 5.0% of the gasoline forms CO:
(a) What is the ratio of CO_2 to CO molecules in the exhaust?
(b) What is the mass ratio of CO_2 to CO?
(c) What percentage of the gasoline must form CO for the mass ratio of CO_2 to CO to be exactly 1:1?

4.132 The amount of ascorbic acid (vitamin C; $C_6H_8O_6$) in tablets is determined by reaction with bromine and then titration of the hydrobromic acid with standard base:

$$C_6H_8O_6 + Br_2 \longrightarrow C_6H_6O_6 + 2HBr$$
$$HBr + NaOH \longrightarrow NaBr + H_2O$$

A certain tablet is advertised as containing 500 mg of vitamin C. One tablet was dissolved in water and reacted with Br_2. The solution was then titrated with 43.20 mL of 0.1350 *M* NaOH. Did the tablet contain the advertised quantity of vitamin C?

4.133 In the process of *salting-in*, protein solubility in a dilute salt solution is increased by adding more salt. Because the protein solubility depends on the total ion concentration as well as the ion charge, salts yielding divalent ions are often more effective than those yielding monovalent ions. (a) How many grams of $MgCl_2$ must dissolve to equal the ion concentration of 12.4 g of NaCl? (b) How many grams of CaS must dissolve? (c) Which of the three salt solutions would dissolve the most protein?

4.134 In the process of *pickling*, rust is removed from newly produced steel by washing the steel in hydrochloric acid:

$$(1)\ 6HCl(aq) + Fe_2O_3(s) \longrightarrow 2FeCl_3(aq) + 3H_2O(l)$$

During the process, some iron is lost as well:

$$(2)\ 2HCl(aq) + Fe(s) \longrightarrow FeCl_2(aq) + H_2(g)$$

(a) Which reaction, if either, is a redox process? (b) If reaction 2 did not occur and all the HCl were used, how many grams of Fe_2O_3 could be removed and $FeCl_3$ produced in a 2.50×10^3-L bath of 3.00 *M* HCl? (c) If reaction 1 did not occur and all the HCl were used, how many grams of Fe could be lost and $FeCl_2$ produced in a 2.50×10^3-L bath of 3.00 *M* HCl? (d) If 0.280 g of Fe is lost per gram of Fe_2O_3 removed, what is the mass ratio of $FeCl_2$ to $FeCl_3$?

4.135 At liftoff, the space shuttle uses a solid mixture of ammonium perchlorate and aluminum powder to obtain great thrust from the volume change of solid to gas. In the presence of a catalyst, the mixture forms solid aluminum oxide and aluminum trichloride and gaseous water and nitric oxide. (a) Write a balanced equation for the reaction, and identify the reducing and oxidizing agents. (b) How many total moles of gas (water vapor and nitric oxide) are produced when 50.0 kg of ammonium perchlorate reacts with a stoichiometric amount of Al? (c) What is the volume change from this reaction? (d of NH_4ClO_4 = 1.95 g/cc, Al = 2.70 g/cc, Al_2O_3 = 3.97 g/cc, and $AlCl_3$ = 2.44 g/cc; assume 1 mol of gas occupies 22.4 L.)

4.136 A *reaction cycle* for an element is a series of reactions beginning and ending with that element. In the following copper reaction cycle, copper has either a 0 or a +2 oxidation state. Write balanced molecular and net ionic equations for each step in the cycle.
(1) Copper metal reacts with aqueous bromine to produce a green-blue solution.
(2) Adding aqueous sodium hydroxide forms a blue precipitate.
(3) The precipitate is heated and turns black (water is released).
(4) The black solid dissolves in nitric acid to give a blue solution.
(5) Adding aqueous sodium phosphate forms a green precipitate.
(6) The precipitate forms a blue solution in sulfuric acid.
(7) Copper metal is recovered from the blue solution when zinc metal is added.

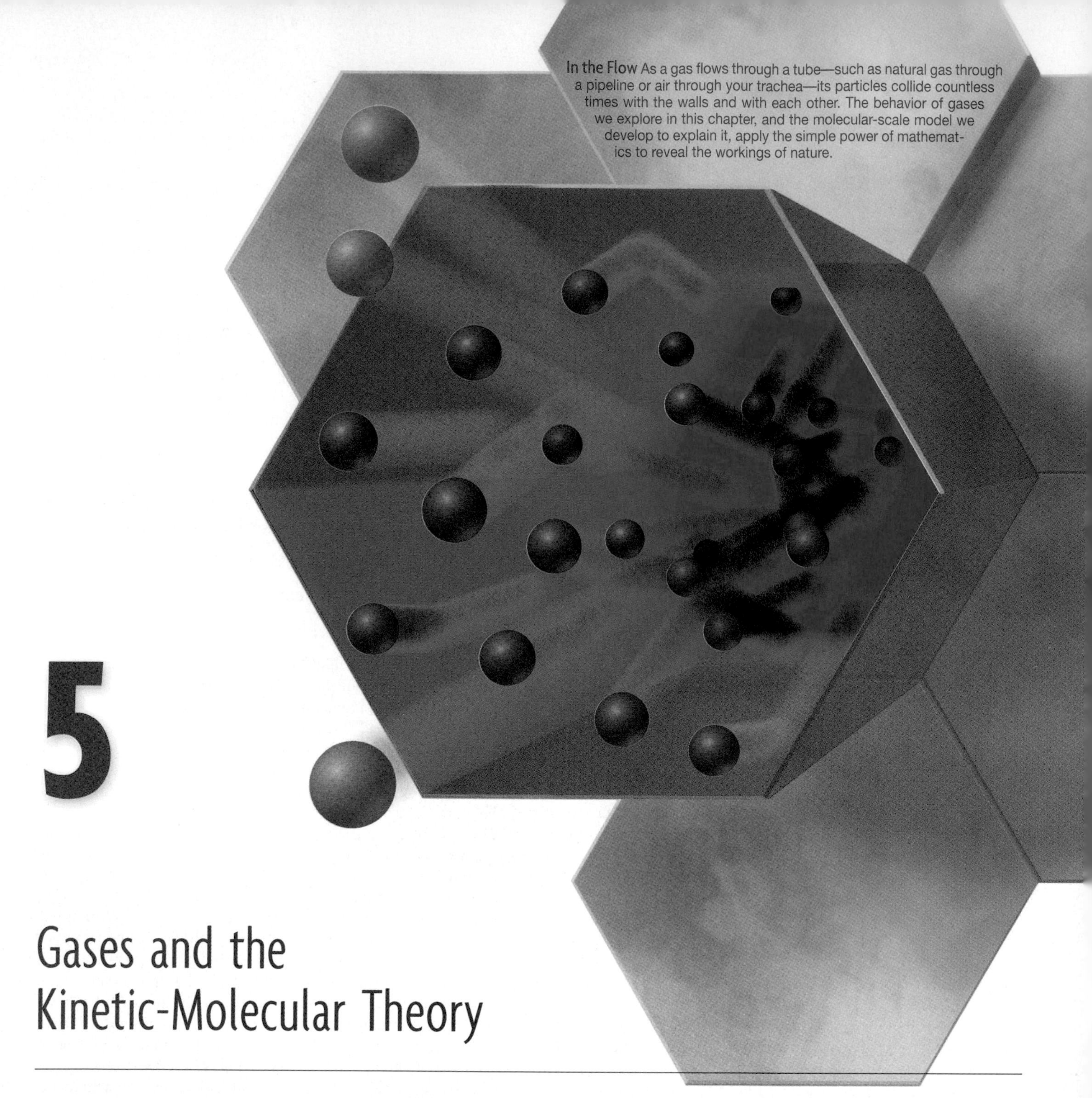
In the Flow As a gas flows through a tube—such as natural gas through a pipeline or air through your trachea—its particles collide countless times with the walls and with each other. The behavior of gases we explore in this chapter, and the molecular-scale model we develop to explain it, apply the simple power of mathematics to reveal the workings of nature.

5

Gases and the Kinetic-Molecular Theory

Gases are everywhere. People have been observing their behavior, and that of matter in other states, throughout history—three of the four "elements" of the ancients were air (gas), water (liquid), and earth (solid). However, many questions remain. In this chapter and its companion, Chapter 12, we examine these states and their interrelations. Here, we highlight the gaseous state, the one we understand best.

Earth's atmosphere—the gaseous envelope that surrounds the planet—is a colorless, odorless mixture of nearly 20 elements and compounds that extends from the surface upward more than 500 km until it merges with outer space. Some components—O_2, N_2, H_2O vapor, and CO_2—take part in complex redox reaction cycles throughout the environment, and you participate in those cycles with every breath you take. Gases also have essential roles in industry (Table 5.1).

Concepts & Skills to Review

before you study this chapter

- physical states of matter (Section 1.1)
- SI unit conversions (Section 1.5)
- mole-mass-number conversions (Section 3.1)

Table 5.1 Some Important Industrial Gases

Name (Formula)	Origin; Use
Methane (CH_4)	Natural deposits; domestic fuel
Ammonia (NH_3)	From N_2 + H_2; fertilizers, explosives
Chlorine (Cl_2)	Electrolysis of seawater; bleaching and disinfecting
Oxygen (O_2)	Liquefied air; steelmaking
Ethylene (C_2H_4)	High-temperature decomposition of natural gas; plastics

Although the *chemical* behavior of a gas depends on its composition, all gases have remarkably similar *physical* behavior, which is the focus of this chapter. For instance, although the particular gases differ, the same physical behavior is at work in the operation of a car and in the baking of bread, in the thrust of a rocket engine and in the explosion of a kernel of popcorn. The process of breathing involves the same physical principles as the creation of thunder.

IN THIS CHAPTER . . . We first contrast gases with liquids and solids and then discuss gas pressure. We consider the mass laws, which describe gas behavior. We then examine the ideal gas law, which encompasses the other mass laws, and apply it to reaction stoichiometry. We explain the observable behavior of gases with the simple kinetic-molecular model. Then, we find that real gas behavior, especially under extreme conditions, requires refinements of the ideal gas law and the model. Finally, we apply these principles to the properties of planetary atmospheres.

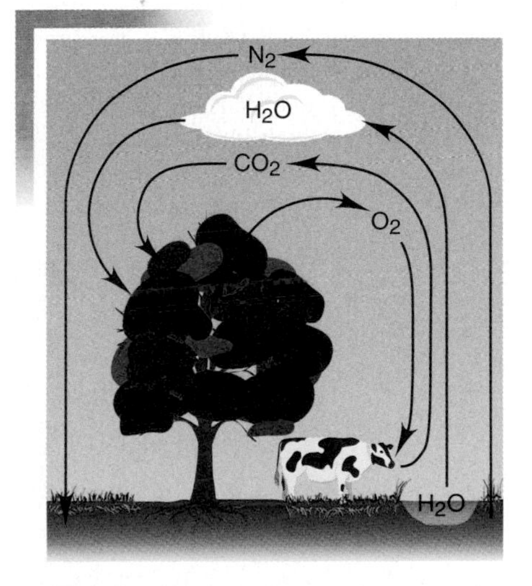

Atmosphere-Biosphere Redox Interconnections The diverse organisms that make up the biosphere interact intimately with the gases of the atmosphere. Powered by solar energy, green plants reduce atmospheric CO_2 and incorporate the C atoms into their own substance. In the process, O atoms in H_2O are oxidized and released to the air as O_2. Certain microbes that live on plant roots reduce N_2 to NH_3 and form compounds that the plant uses to make its proteins. Other microbes that feed on dead plants (and animals) oxidize the proteins and release N_2 again. Animals eat plants and other animals, use O_2 to oxidize their food, and return CO_2 and H_2O to the air.

5.1 AN OVERVIEW OF THE PHYSICAL STATES OF MATTER

Under appropriate conditions of pressure and temperature, most substances can exist as a solid, a liquid, or a gas. In Chapter 1 we described these physical states in terms of how each fills a container and began to develop a molecular view that explains this macroscopic behavior: a solid has a fixed shape regardless of the container shape because its particles are held rigidly in place; a liquid conforms to the container shape but has a definite volume and a surface because its particles are close together but free to move around each other; and a gas fills the container because its particles are far apart and moving randomly. Several other aspects of their behavior distinguish gases from liquids and solids:

1. *Gas volume changes greatly with pressure.* When a sample of gas is confined to a container of variable volume, such as the piston-cylinder assembly of a car engine, an external force can compress the gas. Removing the external force allows the gas volume to increase again. In contrast, a liquid or solid resists significant changes in volume.

2. *Gas volume changes greatly with temperature.* When a gas sample at constant pressure is heated, its volume increases; when it is cooled, its volume decreases. This volume change is 50 to 100 times greater for gases than for liquids or solids.
3. *Gases have relatively low viscosity.* Gases flow much more freely than liquids and solids. Low viscosity allows gases to be transported through pipes over long distances but also to leak rapidly out of small holes.
4. *Most gases have relatively low densities* under normal conditions. Gas density is usually tabulated in units of grams per *liter,* whereas liquid and solid densities are in grams per *milliliter,* about 1000 times as dense (see Table 1.5, p. 21). For example, at 20°C and normal atmospheric pressure, the density of $O_2(g)$ is 1.3 g/**L,** whereas the density of $H_2O(l)$ is 1.0 g/**mL** and that of NaCl(s) is 2.2 g/**mL.** When a gas is cooled, its density increases because its volume decreases: at 0°C, the density of $O_2(g)$ increases to 1.4 g/L.
5. *Gases are miscible. Miscible* substances mix with one another in any proportion to form a solution. Clean dry air, for example, is a solution of about 18 gases. Two liquids, however, may or may not be miscible: water and ethanol are, but water and gasoline are not. Two solids generally do not form a solution unless they are mixed as molten liquids and then allowed to solidify.

Each of these observable properties offers a clue to the molecular properties of gases. For example, consider these density data. At 20°C and normal atmospheric pressure, gaseous N_2 has a density of 1.25 g/L. If cooled below −196°C, it condenses to liquid N_2 and its density becomes 0.808 g/mL. (Note the change in units.) The same amount of nitrogen occupies less than $\frac{1}{600}$ as much space! Further cooling to below −210°C yields solid N_2 (d = 1.03 g/mL), which is only somewhat more dense than the liquid. These values show again that *the molecules are much farther apart in the gas than in either the liquid or the solid.* Moreover, a large amount of space between molecules is consistent with gases' miscibility, low viscosity, and compressibility. Figure 5.1 compares macroscopic and atomic-scale views of the physical states of a real substance.

POW! P-s-s-s-t! POP! A jackhammer uses the force of rapidly expanding compressed air to break through rock and cement. When the nozzle on a can of spray paint is pressed, the pressurized propellant gases expand into the lower pressure of the surroundings and expel droplets of paint. The rapid expansion of heated gases results in such phenomena as the destruction caused by a bomb, the liftoff of a rocket, and the popping of kernels of corn.

A Gas: Molecules are far apart, move freely, and fill the available space
B Liquid: Molecules are close together but move around one another
C Solid: Molecules are close together in a regular array and do not move around one another

Figure 5.1 The three states of matter. Many pure substances, such as bromine (Br_2), can exist under appropriate conditions of pressure and temperature as **A,** a gas; **B,** a liquid; or **C,** a solid. The atomic-scale views show that molecules are much farther apart in a gas than in a liquid or solid.

SECTION SUMMARY

The volume of a gas can be altered significantly by changing the applied external force or the temperature. The corresponding volume changes for liquids and solids are much smaller. Gases flow more freely and have lower densities than liquids and solids, and they mix in any proportion to form solutions. The major reason for these differences in properties is the greater distance between particles in a gas than in a liquid or a solid.

5.2 GAS PRESSURE AND ITS MEASUREMENT

Blowing up a balloon provides clear evidence that a gas exerts pressure on the walls of its container. **Pressure (*P*)** is defined as the force exerted per unit of surface area:

$$\text{Pressure} = \frac{\text{force}}{\text{area}}$$

Earth's gravitational attraction pulls the atmospheric gases toward its surface, where they exert a force on all objects. The force, or weight, of these gases creates a pressure of about 14.7 pounds per square inch (lb/in^2; psi) of surface.

As we'll discuss later, the molecules in a gas are moving in every direction, so the pressure of the atmosphere is exerted uniformly on the floor, walls, ceiling, and every object in a room. The pressure on the outside of your body is equalized by the pressure on the inside, so there is no net pressure on your body's outer surface. What would happen if this were not the case? As an analogy, consider the empty metal can attached to a vacuum pump in Figure 5.2. With the pump off, the can maintains its shape because the pressure on the outside is equal to the pressure on the inside. With the pump on, the internal pressure decreases greatly, and the ever-present external pressure easily crushes the can. A vacuum-filtration flask (and tubing), which you may have used in the lab, has thick walls that can withstand the external pressure when the flask is evacuated.

Snowshoes and the Meaning of Pressure Snowshoes allow you to walk on powdery snow without sinking because they distribute your weight over a much larger area than a boot does, thereby greatly decreasing your weight per square inch. The area of a snowshoe is typically about 10 times as large as that of a boot sole, so the snowshoe exerts only about one-tenth as much pressure as the boot. The wide, padded paws of snow leopards accomplish this, too. For the same reason, high-heeled shoes exert much more pressure than flat shoes.

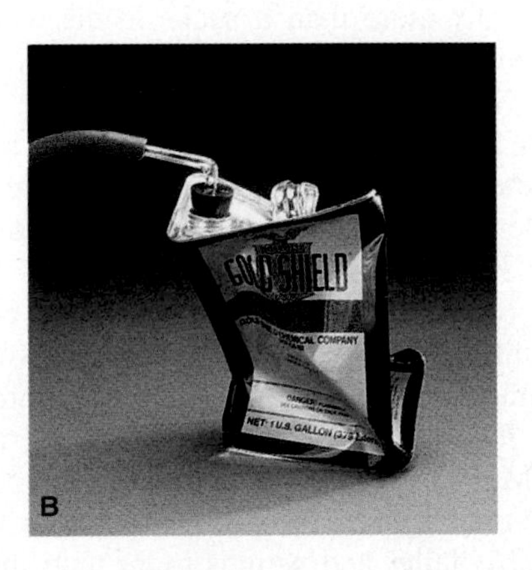

Figure 5.2 Effect of atmospheric pressure on objects at Earth's surface.
A, A metal can filled with air has equal pressure on the inside and outside.
B, When the air inside the can is removed, the atmospheric pressure crushes the can.

Laboratory Devices for Measuring Gas Pressure

The **barometer** is a common device used to measure atmospheric pressure. Invented in 1643 by Evangelista Torricelli, the barometer is still basically just a tube about 1 m long, closed at one end, filled with mercury, and inverted into a dish containing more mercury. When the tube is inverted, some of the mercury flows out into the dish, and a vacuum forms above the mercury remaining in the

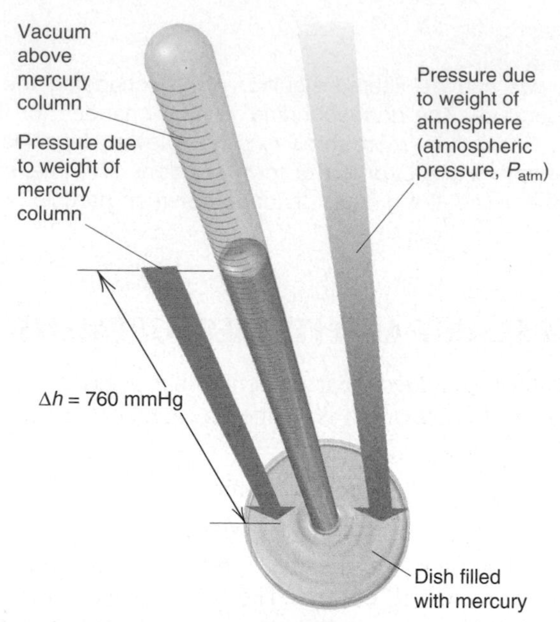

Figure 5.3 A mercury barometer. (See text for explanation.)

tube, as shown in Figure 5.3. At sea level under ordinary atmospheric conditions, the outward flow of mercury stops when the surface of the mercury in the tube is about 760 mm above the surface of the mercury in the dish. It stops at 760 mm because at that point the column of mercury in the tube exerts the same pressure (weight/area) on the mercury surface in the dish as does the column of air that extends from the dish to the outer reaches of the atmosphere. The air pushing down keeps any more of the mercury in the tube from flowing out. Likewise, if you place an evacuated tube into a dish filled with mercury, the mercury rises about 760 mm into the tube because the atmosphere pushes the mercury up to that height.

Several centuries ago, people ascribed mysterious "suction" forces to a vacuum. We know now that a vacuum does not suck up mercury into the barometer tube any more than it sucks in the walls of the crushed can in Figure 5.2. Only matter—in this case, the atmospheric gases—can exert a force.

Notice that we did not specify the diameter of the barometer tube. If the mercury in a 1-cm diameter tube rises to a height of 760 mm, the mercury in a 2-cm diameter tube will rise to that height also. The *weight* of mercury is greater in the wider tube, but the area is larger also; thus the *pressure,* the *ratio* of weight to area, is the same.

Since the pressure of the mercury column is directly proportional to its height, a unit commonly used for pressure is mmHg, the height of the mercury (atomic symbol Hg) column in millimeters (mm). We discuss units of pressure shortly. At sea level and 0°C, normal atmospheric pressure is 760 mmHg, but at the top of Mt. Everest (29,028 ft, or 8848 m), the atmospheric pressure is only about 270 mmHg. Thus, *pressure decreases with altitude:* the column of air above sea level is taller and weighs more than the column of air above Mt. Everest.

Laboratory barometers contain mercury rather than some other liquid because its high density allows the barometer to be a convenient size. For example, the pressure of the atmosphere would equal the pressure of a column of water about 10,300 mm, almost 34 ft, high. Note that, for a given pressure, the ratio of heights (h) of the liquid columns is inversely related to the ratio of the densities (d) of the liquids:

$$\frac{h_{H_2O}}{h_{Hg}} = \frac{d_{Hg}}{d_{H_2O}}$$

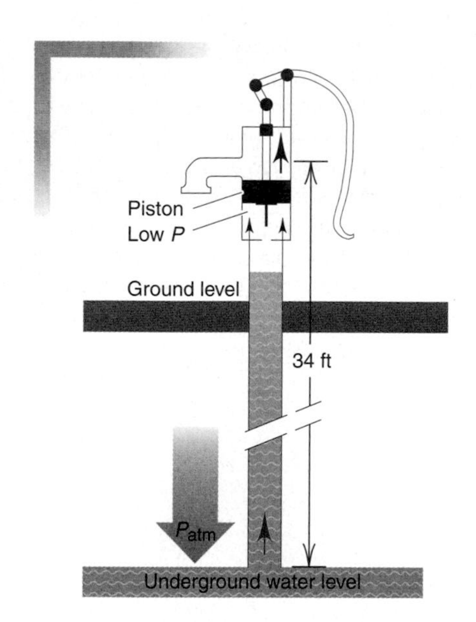

The Mystery of the Suction Pump
When you drink through a straw, you create lower pressure above the liquid, and the atmosphere pushes the liquid up. Similarly, a "suction" pump is a tube dipping into a water source, with a piston and handle that lower the air pressure above the water level. The pump can raise water from a well no deeper than 34 ft. This depth limit was a mystery until the great 17th-century Italian scientist Galileo showed that the atmosphere pushes the water up into the tube and that its pressure can support only a 34-ft column of water. Modern pumps that draw water from deeper sources use compressed air to increase the pressure exerted on the water.

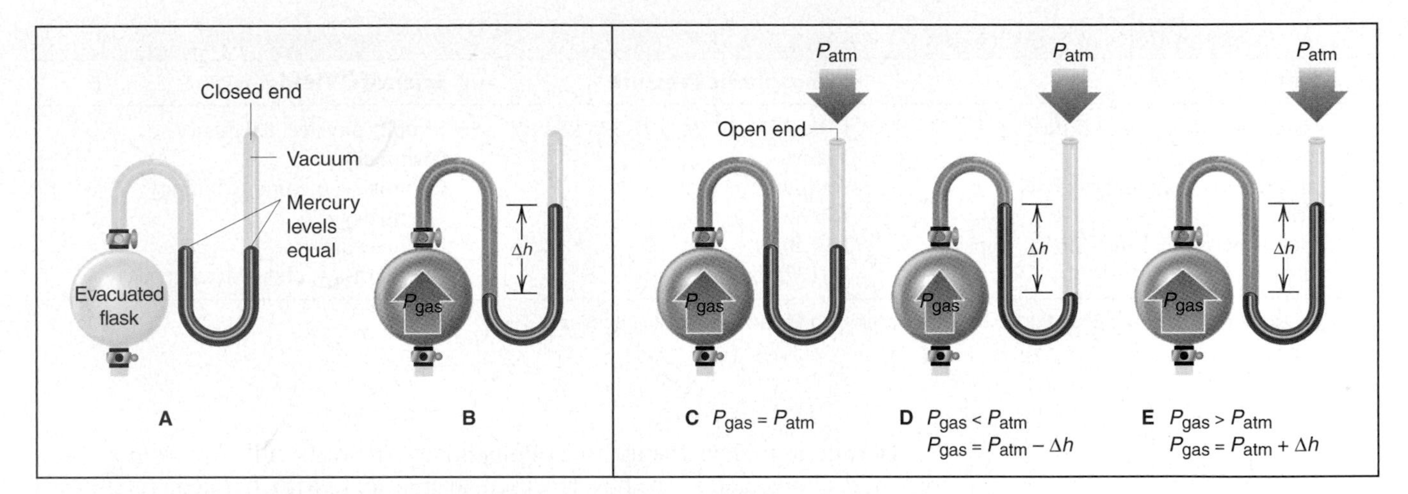

Figure 5.4 Two types of manometer. A, A closed-end manometer with an evacuated flask attached has the mercury levels equal. **B,** A gas exerts pressure on the mercury in the arm closer to the flask. The difference in heights (Δh) equals the gas pressure. **C–E,** An open-end manometer is shown with gas pressure equal to atmospheric pressure, **(C),** gas pressure lower than atmospheric pressure **(D),** and gas pressure higher than atmospheric pressure **(E).**

Manometers are devices used to measure the pressure of a gas in an experiment. Figure 5.4 shows two types of manometer. Part A shows a *closed-end manometer,* a mercury-filled, curved tube, *closed* at one end and attached to a flask at the other. When the flask is evacuated, the mercury levels in the two arms of the tube are the same because no gas exerts pressure on either mercury surface. When a gas is in the flask (part B), it pushes down the mercury level in the near arm, so the level rises in the far arm. The *difference* in column heights (Δh) equals the gas pressure. Note that if we open the lower stopcock of the evacuated flask in part A, air rushes in, Δh equals atmospheric pressure, and the closed-end manometer becomes a barometer.

The *open-end manometer,* shown in parts C–E, also consists of a curved tube filled with mercury, but one end of the tube is *open* to the atmosphere and the other is connected to the gas sample. The atmosphere pushes on one mercury level and the gas pushes on the other. Since Δh equals the difference between two pressures, to calculate the gas pressure with an open-end manometer, we must measure the atmospheric pressure separately with a barometer.

Units of Pressure

Pressure results from a force exerted on an area. The SI unit of force is the newton (N): $1\ \text{N} = 1\ \text{kg·m/s}^2$. The SI unit of pressure is the **pascal (Pa),** which equals a force of one newton exerted on an area of one square meter:

$$1\ \text{Pa} = 1\ \text{N/m}^2$$

A much larger unit is the **standard atmosphere (atm),** the average atmospheric pressure measured at sea level and 0°C. It is defined in terms of the pascal:

$$1\ \text{atm} = 101.325\ \text{kilopascals (kPa)} = 1.01325\times10^5\ \text{Pa}$$

Another common pressure unit is the **millimeter of mercury (mmHg),** which is based on measurement with a barometer or manometer. In honor of Torricelli, this unit has been named the **torr:**

$$1\ \text{torr} = 1\ \text{mmHg} = \frac{1}{760}\ \text{atm} = \frac{101.325}{760}\ \text{kPa} = 133.322\ \text{Pa}$$

The *bar* is coming into more common use in chemistry:

$$1\ \text{bar} = 1\times10^2\ \text{kPa} = 1\times10^5\ \text{Pa}$$

Table 5.2 Common Units of Pressure

Unit	Atmospheric Pressure	Scientific Field
pascal (Pa); kilopascal (kPa)	1.01325×10^5 Pa; 101.325 kPa	SI unit; physics, chemistry
atmosphere (atm)	1 atm*	Chemistry
millimeters of mercury (mmHg)	760 mmHg*	Chemistry, medicine, biology
torr	760 torr*	Chemistry
pounds per square inch (lb/in^2 or psi)	14.7 lb/in^2	Engineering
bar	1.01325 bar	Meteorology, chemistry, physics

*This is an exact quantity; in calculations, we use as many significant figures as necessary.

Despite a gradual change to SI units, many chemists still express pressure in torrs and atmospheres, so they are used in this text, with frequent reference to pascals. Table 5.2 lists some important pressure units used in various scientific fields.

SAMPLE PROBLEM 5.1 Converting Units of Pressure

Problem A geochemist heats a limestone ($CaCO_3$) sample and collects the CO_2 released in an evacuated flask attached to a closed-end manometer (see Figure 5.4B). After the system comes to room temperature, Δh = 291.4 mmHg. Calculate the CO_2 pressure in torrs, atmospheres, and kilopascals.

Plan The CO_2 pressure is given in units of mmHg, so we construct conversion factors from Table 5.2 to find the pressure in the other units.

Solution Converting from mmHg to torr:

$$P_{CO_2}(\text{torr}) = 291.4\ \cancel{\text{mmHg}} \times \frac{1\ \text{torr}}{1\ \cancel{\text{mmHg}}} = \boxed{291.4\ \text{torr}}$$

Converting from torr to atm:

$$P_{CO_2}(\text{atm}) = 291.4\ \cancel{\text{torr}} \times \frac{1\ \text{atm}}{760\ \cancel{\text{torr}}} = \boxed{0.3834\ \text{atm}}$$

Converting from atm to kPa:

$$P_{CO_2}(\text{kPa}) = 0.3834\ \cancel{\text{atm}} \times \frac{101.325\ \text{kPa}}{1\ \cancel{\text{atm}}} = \boxed{38.85\ \text{kPa}}$$

Check There are 760 torr in 1 atm, so ~300 torr should be <0.5 atm. There are ~100 kPa in 1 atm, so <0.5 atm should be <50 kPa.

Comment 1. In the conversion from torr to atm, we retained four significant figures because this unit conversion factor involves *exact* numbers; that is, 760 torr has as many significant figures as the calculation requires.

2. From here on, except in particularly complex situations, *the canceling of units in calculations is no longer shown.*

FOLLOW-UP PROBLEM 5.1 The CO_2 released from another mineral sample was collected in an evacuated flask connected to an open-end manometer (see Figure 5.4D). If the barometer reading is 753.6 mmHg and Δh is 174.0 mmHg, calculate P_{CO_2} in torrs, pascals, and lb/in^2.

SECTION SUMMARY

Gases exert pressure (force/area) on all surfaces with which they make contact. A barometer measures atmospheric pressure in terms of the height of the mercury column that the atmosphere can support (760 mmHg at sea level and 0°C). Both closed-end and open-end manometers are used to measure the pressure of a gas sample. Chemists measure pressure in units of atmospheres (atm), torr (equivalent to mmHg), or pascals (Pa, the SI unit).

5.3 THE GAS LAWS AND THEIR EXPERIMENTAL FOUNDATIONS

The physical behavior of a sample of gas can be described completely by four variables: pressure (P), volume (V), temperature (T), and amount (number of moles, n). The variables are interdependent: *any one of them can be determined by measuring the other three*. We know now that this quantitatively predictable behavior is a direct outcome of the structure of gases on the molecular level. Yet, it was discovered, for the most part, before Dalton's atomic theory was published!

Three key relationships exist among the four gas variables—Boyle's, Charles's, and Avogadro's laws. Each of these gas laws expresses the effect of one variable on another, with the remaining two variables held constant. Since the volume occupied by a gas is so easy to measure, the laws are traditionally expressed as the effect on gas volume of changing the pressure, temperature, or amount of gas.

These three laws are special cases of an all-encompassing relationship among gas variables called the *ideal gas law*. This unifying observation quantitatively describes the state of a so-called **ideal gas,** one that exhibits simple linear relationships among volume, pressure, temperature, and amount. Although no ideal gas actually exists, most simple gases, such as N_2, O_2, H_2, and the noble gases, show nearly ideal behavior at ordinary temperatures and pressures. We discuss the ideal gas law after the three special cases.

The Relationship Between Volume and Pressure: Boyle's Law

Following Torricelli's invention of the barometer, the great English chemist Robert Boyle performed a series of experiments that led him to conclude that at a given temperature, *the volume occupied by a gas is inversely related to its pressure*. Figure 5.5 shows Boyle's experiment and some typical data he might have collected. Boyle fashioned a J-shaped glass tube, sealed the shorter end, and poured mercury into the longer end, thereby trapping some air, the gas in the experiment. From the height of the trapped air column and the diameter of the tube, he calculated the air volume. The total pressure applied to the trapped air was the pressure of the atmosphere (measured with a barometer) plus that of the mercury column (part A). By adding mercury, Boyle increased the total pressure exerted on the air, and the air volume decreased (part B). Since the temperature and amount of air were constant, Boyle could directly measure the effect of the applied pressure on the volume of air.

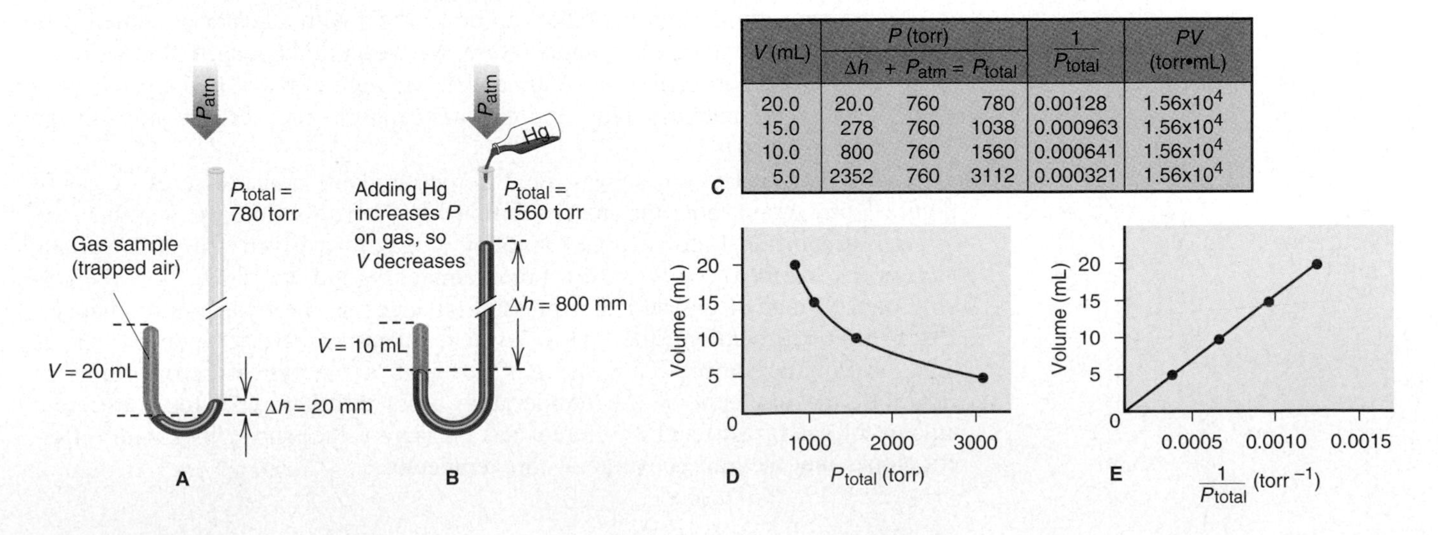

V (mL)	P (torr) Δh	+ P_{atm}	= P_{total}	$\frac{1}{P_{total}}$	PV (torr•mL)
20.0	20.0	760	780	0.00128	1.56×10^4
15.0	278	760	1038	0.000963	1.56×10^4
10.0	800	760	1560	0.000641	1.56×10^4
5.0	2352	760	3112	0.000321	1.56×10^4

Figure 5.5 The relationship between the volume and pressure of a gas. A, A small amount of air (the gas) is trapped in the short arm of a J tube; n and T are fixed. The total pressure on the gas (P_{total}) is the sum of the pressure due to the difference in heights of the mercury columns (Δh) and the pressure of the atmosphere (P_{atm}). If P_{atm} = 760 torr, P_{total} = 780 torr. **B,** As mercury is added, the total pressure on the gas increases and its volume (V) decreases. Note that if P_{total} is doubled (to 1560 torr), V is halved (not drawn to scale). **C,** Some typical pressure-volume data from the experiment. **D,** A plot of V vs. P_{total} shows that V is inversely proportional to P. **E,** A plot of V vs. $1/P_{total}$ is a straight line whose slope is a constant characteristic of *any* gas that behaves ideally.

Note the following results (Figure 5.5):

- The product of corresponding P and V values is a constant (part C, rightmost column)
- V is *inversely* proportional to P (part D)
- V is *directly* proportional to $1/P$ (part E) and generates a linear plot of V against $1/P$. This *linear relationship between two gas variables* is a hallmark of ideal gas behavior.

The generalization of Boyle's observations is known as **Boyle's law:** *at constant temperature, the volume occupied by a fixed amount of gas is inversely proportional to the applied (external) pressure,* or

$$V \propto \frac{1}{P} \qquad [T \text{ and } n \text{ fixed}] \tag{5.1}$$

This relationship can also be expressed as

$$PV = \text{constant} \qquad \text{or} \qquad V = \frac{\text{constant}}{P} \qquad [T \text{ and } n \text{ fixed}]$$

The constant is the same for the great majority of gases. Thus, tripling the external pressure reduces the volume to one-third its initial value; halving the external pressure doubles the volume; and so forth.

The wording of Boyle's law focuses on *external* pressure. In his experiment, however, adding more mercury caused the mercury level to rise until the pressure of the trapped air stopped the rise at some new level. At that point, the pressure exerted *on* the gas equaled the pressure exerted *by* the gas. In other words, by measuring the applied pressure, Boyle was also measuring the gas pressure. Thus, when gas volume doubles, gas pressure is halved. In general, if V_{gas} increases, P_{gas} decreases, and vice versa.

The Relationship Between Volume and Temperature: Charles's Law

One question raised by Boyle's work was why the pressure-volume relationship holds only at constant temperature. Experience had shown that gas volume depends on temperature, but it was not until the early 19th century, through the separate work of the French scientists J. A. C. Charles and J. L. Gay-Lussac, that the relationship was clearly understood.

Let's examine this relationship by measuring the volume of a fixed amount of a gas under constant pressure but at different temperatures. A straight tube, closed at one end, traps a fixed amount of air under a small mercury plug. The tube is immersed in a water bath that can be warmed with a heater or cooled with ice. After each change of water temperature, we measure the length of the air column, which is proportional to its volume. The pressure exerted on the gas is constant because the mercury plug and the atmospheric pressure do not change (Figure 5.6A and B).

Some typical data are shown for different amounts and pressures of gas in Figure 5.6C. Again, note the linear relationships, but this time the variables are *directly* proportional: for a given amount of gas at a given pressure, *volume increases as temperature increases*. For example, the red line shows how the volume of 0.04 mol of gas at 1 atm pressure changes as the temperature changes. Extending (*extrapolating*) the line to lower temperatures (dashed portion) shows that the volume shrinks until the gas occupies a theoretical zero volume at $-273°C$ (the intercept on the temperature axis). Similar plots for a different amount of gas (green) and a different gas pressure (blue) show lines with different slopes, but they all converge at this temperature.

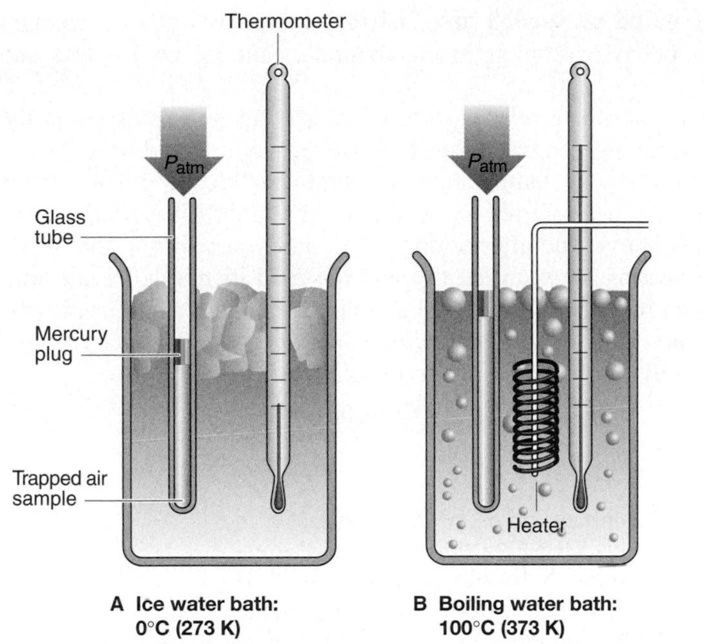

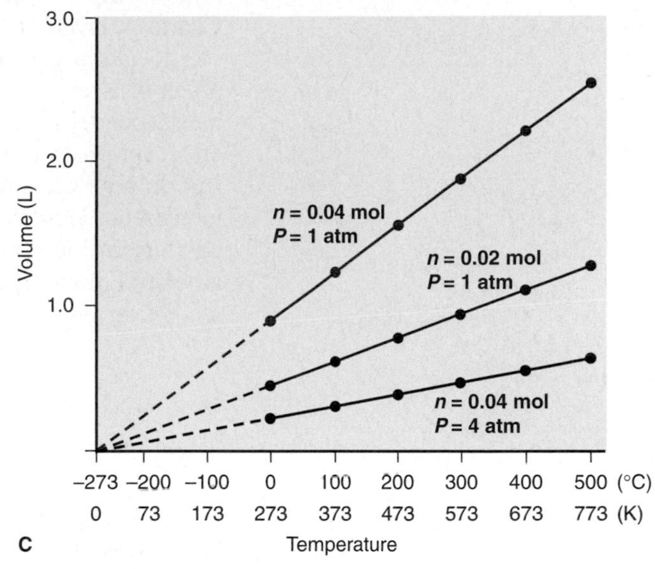

Figure 5.6 The relationship between the volume and temperature of a gas. At constant P, the volume of a given amount of gas is *directly* proportional to the absolute temperature. A fixed amount of gas (air) is trapped under a small plug of mercury at a fixed pressure. **A,** The sample is in an ice water bath. **B,** The sample is in a boiling water bath. As the temperature increases, the volume of the gas increases. **C,** The three lines show the effect of amount (n) of gas (compare red and green) and pressure (P) of gas (compare red and blue). The dashed lines extrapolate the data to lower temperatures. For any amount of an ideal gas at any pressure, the volume is theoretically zero at −273.15°C (0 K).

A half-century after Charles's and Gay-Lussac's work, William Thomson (Lord Kelvin) used this linear relation between gas volume and temperature to devise the absolute temperature scale (Section 1.5). In this scale, absolute zero (0 K or −273.15°C) is the temperature at which an ideal gas would have zero volume. (Absolute zero has never been reached, but physicists have attained temperatures as low as 10^{-9} K.) Of course, no sample of matter can have zero volume, and every real gas condenses to a liquid at some temperature higher than 0 K. Nevertheless, *this linear dependence of volume on absolute temperature holds for most common gases over a wide temperature range.*

The modern statement of the volume-temperature relationship is known as **Charles's law:** *at constant pressure, the volume occupied by a fixed amount of gas is directly proportional to its absolute temperature,* or

$$V \propto T \qquad [P \text{ and } n \text{ fixed}] \tag{5.2}$$

This relationship can also be expressed as

$$\frac{V}{T} = \text{constant} \qquad \text{or} \qquad V = \text{constant} \times T \qquad [P \text{ and } n \text{ fixed}]$$

If T increases, V increases, and vice versa. Once again, for any given P and n, the constant is the same for the great majority of gases.

The dependence of gas volume on *absolute* temperature means that you must *use the Kelvin scale in gas law calculations.* For instance, if the temperature changes from 200 K to 400 K, the volume of 1 mol of gas doubles. But, if the temperature changes from 200°C to 400°C, the volume increases by a factor of 1.42; that is, $\left(\frac{400°\text{C} + 273.15}{200°\text{C} + 273.15}\right) = \frac{673}{473} = 1.42$.

Other Relationships Based on Boyle's and Charles's Laws Two other important relationships in gas behavior emerge from an understanding of Boyle's and Charles's laws:

1. *The pressure-temperature relationship.* Charles's law is expressed as the effect of a temperature change on gas *volume*. However, volume and pressure are interdependent, so the effect of temperature on volume is closely related to its effect on pressure (sometimes referred to as *Amontons's law*). Measure the pressure in your car's tires before and after a long drive, and you will find that it has increased. Frictional heating between the tire and the road increases the air temperature inside the tire, but since the tire volume doesn't change appreciably, the air exerts more pressure. Thus, *at constant volume, the pressure exerted by a fixed amount of gas is directly proportional to the absolute temperature:*

$$P \propto T \qquad [V \text{ and } n \text{ fixed}] \tag{5.3}$$

or

$$\frac{P}{T} = \text{constant} \qquad \text{or} \qquad P = \text{constant} \times T$$

2. *The combined gas law.* A simple combination of Boyle's and Charles's laws gives the *combined gas law,* which applies to situations when two of the three variables (V, P, T) change and you must find the effect on the third:

$$V \propto \frac{T}{P} \qquad \text{or} \qquad V = \text{constant} \times \frac{T}{P} \qquad \text{or} \qquad \frac{PV}{T} = \text{constant}$$

The Relationship Between Volume and Amount: Avogadro's Law

Boyle's and Charles's laws both specify a fixed amount of gas. Let's see why. Figure 5.7 shows an experiment that involves two small test tubes, each fitted with a piston-cylinder assembly. We add 0.10 mol (4.4 g) of dry ice (frozen CO_2) to the first (tube A) and 0.20 mol (8.8 g) to the second (tube B). As the solid warms, it changes directly to gaseous CO_2, which expands into the cylinder and pushes up the piston. When all the solid has changed to gas and the temperature is constant, we find that cylinder A has half the volume of cylinder B. (We can neglect the volume of the tube because it is so much smaller than the volume of the cylinder.)

This experimental result shows that twice the amount (mol) of gas occupies twice the volume. Notice that, for both cylinders, the T of the gas equals room temperature and the P of the gas equals atmospheric pressure. Thus, *at fixed tem-*

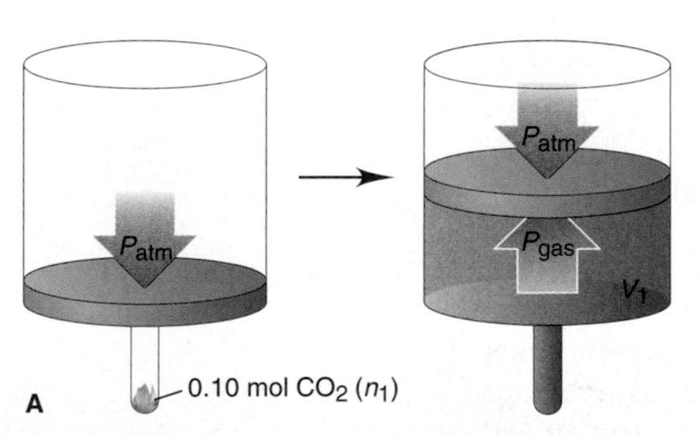

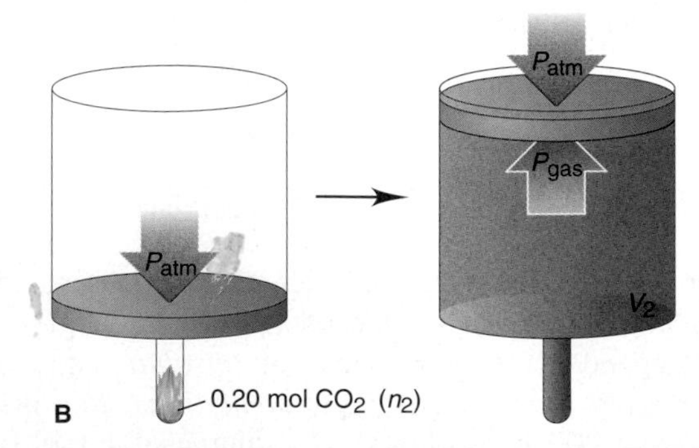

Figure 5.7 An experiment to study the relationship between the volume and amount of a gas. A, At a given external P and T, a given amount (n_1) of $CO_2(s)$ is put into the tube. When the CO_2 changes from solid to gas, it pushes up the piston until $P_{gas} = P_{atm}$, at which point it occupies a given volume of the cylinder. **B,** When twice the amount (n_2) of $CO_2(s)$ is used, twice the volume of the cylinder becomes occupied. Thus, at fixed P and T, the volume (V) of a gas is directly proportional to the amount of gas (n).

perature and pressure, the volume occupied by a gas is directly proportional to the amount (mol) of gas:

$$V \propto n \quad [P \text{ and } T \text{ fixed}] \tag{5.4}$$

As n increases, V increases, and vice versa. This relationship is also expressed as

$$\frac{V}{n} = \text{constant} \qquad \text{or} \qquad V = \text{constant} \times n$$

The constant is the same for all gases at a given temperature and pressure. This relationship is another way of expressing **Avogadro's law,** which states that *at fixed temperature and pressure, equal volumes of* ***any*** *ideal gas contain equal numbers of particles (or moles).*

Many familiar phenomena are based on the relationships among volume, temperature, and amount of gas. For example, in a car engine, a reaction occurs in which fewer moles of gasoline and O_2 form more moles of CO_2 and H_2O vapor, which expand as a result of the released heat and push back the piston. Dynamite explodes because a solid decomposes rapidly to form hot gases. Dough rises in a warm room because yeast forms CO_2 bubbles in the dough, which expand during baking to give the bread a still larger volume.

Breathing and the Gas Laws Taking a deep breath is a combined application of the gas laws. When you inhale, muscles are coordinated such that your diaphragm moves down and your rib cage moves out. This movement increases the volume of the lungs, which decreases the pressure of the air inside them relative to that outside, so air rushes in (Boyle's). The greater amount of air stretches the elastic tissue of the lungs and expands the volume further (Avogadro's). The air also expands slightly as it warms to body temperature (Charles's). When you exhale, the diaphragm relaxes and moves up, the rib cage moves in, and the lung volume decreases. The inside air pressure becomes greater than the outside pressure, and air rushes out.

Gas Behavior at Standard Conditions

To better understand the factors that influence gas behavior, chemists use a set of *standard conditions* called **standard temperature and pressure (STP):**

$$\text{STP:} \quad 0^\circ\text{C (273.15 K) and 1 atm (760 torr)} \tag{5.5}$$

Under these conditions, the volume of 1 mol of an ideal gas is called the **standard molar volume:**

$$\text{Standard molar volume} = 22.4141 \text{ L} \quad \text{or} \quad 22.4 \text{ L} \quad [\text{to 3 sf}] \tag{5.6}$$

Figure 5.8 compares the properties of three simple gases at STP.

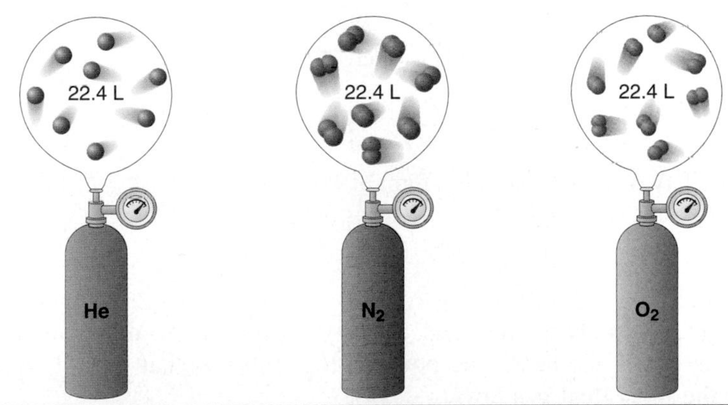

He	N_2	O_2
n = 1 mol	n = 1 mol	n = 1 mol
P = 1 atm (760 torr)	P = 1 atm (760 torr)	P = 1 atm (760 torr)
T = 0°C (273 K)	T = 0°C (273 K)	T = 0°C (273 K)
V = 22.4 L	V = 22.4 L	V = 22.4 L
Number of gas particles = 6.022×10^{23}	Number of gas particles = 6.022×10^{23}	Number of gas particles = 6.022×10^{23}
Mass = 4.003 g	Mass = 28.02 g	Mass = 32.00 g
d = 0.179 g/L	d = 1.25 g/L	d = 1.43 g/L

Figure 5.8 Standard molar volume. One mole of an ideal gas occupies 22.4 L at STP (0°C and 1 atm). At STP, helium, nitrogen, oxygen, and most other simple gases behave ideally. Note that the *mass* of a gas, and thus its density (d), depends on its molar mass.

Figure 5.9 The volumes of 1 mol of an ideal gas and some familiar objects. A basketball (7.5 L), 5-gal fish tank (18.9 L), 13-in television (21.6 L), and 22.4 L of He gas in a balloon.

Figure 5.9 compares the volumes of some familiar objects with the standard molar volume of an ideal gas.

The Ideal Gas Law

Each of the gas laws focuses on the effect that changes in one variable have on gas volume:

- Boyle's law focuses on pressure ($V \propto 1/P$).
- Charles's law focuses on temperature ($V \propto T$).
- Avogadro's law focuses on amount (mol) of gas ($V \propto n$).

We can combine these individual effects into one relationship, called the **ideal gas law** (or *ideal gas equation*):

$$V \propto \frac{nT}{P} \quad \text{or} \quad PV \propto nT \quad \text{or} \quad \frac{PV}{nT} = R$$

where R is a proportionality constant known as the **universal gas constant.** Rearranging gives the most common form of the ideal gas law:

$$PV = nRT \tag{5.7}$$

We can obtain a value of R by measuring the volume, temperature, and pressure of a given amount of gas and substituting the values into the ideal gas law. For example, using standard conditions for the gas variables, we have

$$R = \frac{PV}{nT} = \frac{1\ \text{atm} \times 22.4141\ \text{L}}{1\ \text{mol} \times 273.15\ \text{K}} = 0.082058\ \frac{\text{atm}\cdot\text{L}}{\text{mol}\cdot\text{K}} = 0.0821\ \frac{\text{atm}\cdot\text{L}}{\text{mol}\cdot\text{K}} \quad [3\ \text{sf}] \tag{5.8}$$

This numerical value of R corresponds to the gas variables P, V, and T expressed *in these units*. R has a different numerical value when different units are used. For example, later in this chapter, R has the value 8.314 J/mol·K (J stands for joule, the SI unit of energy).

Figure 5.10 makes a central point: the ideal gas law *becomes* one of the individual gas laws when two of the four variables are kept constant. When initial conditions (subscript $_1$) change to final conditions (subscript $_2$), we have

$$P_1V_1 = n_1RT_1 \quad \text{and} \quad P_2V_2 = n_2RT_2$$

Thus,
$$\frac{P_1V_1}{n_1T_1} = R \quad \text{and} \quad \frac{P_2V_2}{n_2T_2} = R, \quad \text{so} \quad \frac{P_1V_1}{n_1T_1} = \frac{P_2V_2}{n_2T_2}$$

Notice that if two of the variables remain constant, say P and T, then $P_1 = P_2$ and $T_1 = T_2$, and we obtain an expression for Avogadro's law:

$$\frac{\cancel{P_1}V_1}{n_1\cancel{T_1}} = \frac{\cancel{P_2}V_2}{n_2\cancel{T_2}} \quad \text{or} \quad \frac{V_1}{n_1} = \frac{V_2}{n_2}$$

We use rearrangements of the ideal gas law such as this one to solve gas law problems, as you'll see next. The point to remember is that there is no need to memorize the individual gas laws.

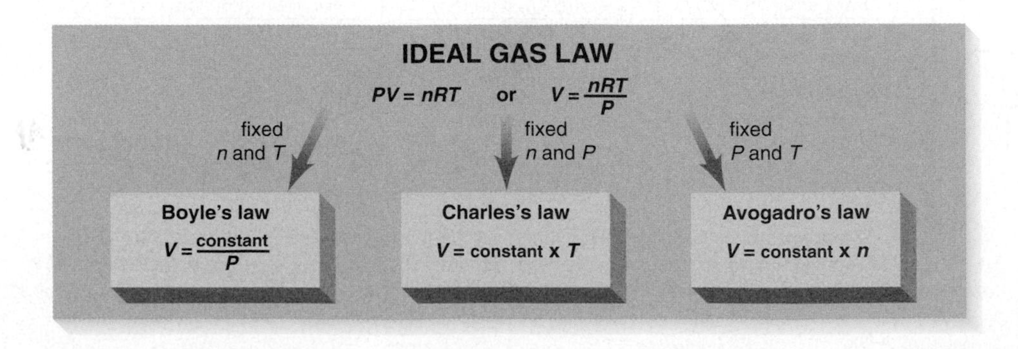

Figure 5.10 Relationship between the ideal gas law and the individual gas laws. Boyle's, Charles's, and Avogadro's laws are contained within the ideal gas law.

Solving Gas Law Problems

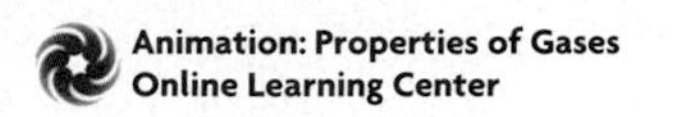

Gas law problems are phrased in many ways, but they can usually be grouped into two main types:

1. *A change in one of the four variables causes a change in another, while the two remaining variables remain constant.* In this type, the ideal gas law reduces to one of the individual gas laws, and you solve for the new value of the variable. Units must be consistent, T must always be in kelvins, but R is not involved. Sample Problems 5.2 to 5.4 and 5.6 are of this type. [A variation on this type involves the combined gas law (p. 186) for simultaneous changes in two of the variables that cause a change in a third.]

2. *One variable is unknown, but the other three are known and no change occurs.* In this type, exemplified by Sample Problem 5.5, the ideal gas law is applied directly to find the unknown, and the units must conform to those in R.

These problems are far easier to solve if you follow a systematic approach:

- Summarize the information: identify the changing gas variables—knowns and unknown—and those held constant.
- Predict the direction of the change, and later check your answer against the prediction.
- Perform any necessary unit conversions.
- Rearrange the ideal gas law to obtain the appropriate relationship of gas variables, and solve for the unknown variable.

The following series of sample problems applies the various gas behaviors.

SAMPLE PROBLEM 5.2 Applying the Volume-Pressure Relationship

Problem Boyle's apprentice finds that the air trapped in a J tube occupies 24.8 cm^3 at 1.12 atm. By adding mercury to the tube, he increases the pressure on the trapped air to 2.64 atm. Assuming constant temperature, what is the new volume of air (in L)?

Plan We must find the final volume (V_2) in liters, given the initial volume (V_1), initial pressure (P_1), and final pressure (P_2). The temperature and amount of gas are fixed. We convert the units of V_1 from cm^3 to mL and then to L, rearrange the ideal gas law to the appropriate form, and solve for V_2. We can predict the direction of the change: since P increases, V will decrease; thus, $V_2 < V_1$. (Note that the roadmap has two parts.)

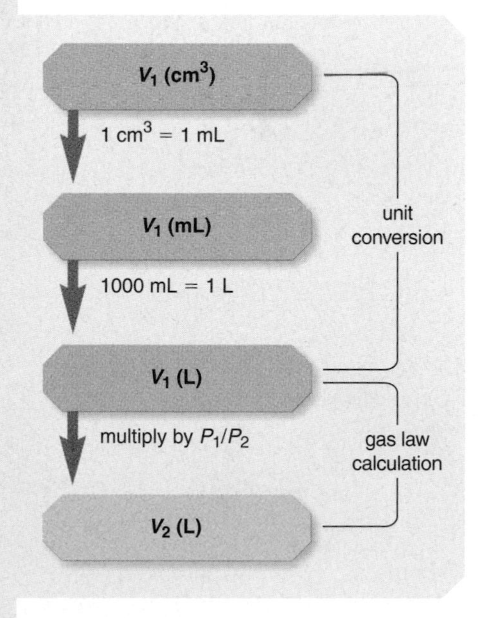

Solution Summarizing the gas variables:

$P_1 = 1.12$ atm $P_2 = 2.64$ atm

$V_1 = 24.8$ cm^3 (convert to L) $V_2 =$ unknown T and n remain constant

Converting V_1 from cm^3 to L:

$$V_1 = 24.8\ \text{cm}^3 \times \frac{1\ \text{mL}}{1\ \text{cm}^3} \times \frac{1\ \text{L}}{1000\ \text{mL}} = 0.0248\ \text{L}$$

Arranging the ideal gas law and solving for V_2: At fixed n and T, we have

$$\frac{P_1V_1}{\cancel{n_1T_1}} = \frac{P_2V_2}{\cancel{n_2T_2}} \quad \text{or} \quad P_1V_1 = P_2V_2$$

$$V_2 = V_1 \times \frac{P_1}{P_2} = 0.0248\ \text{L} \times \frac{1.12\ \text{atm}}{2.64\ \text{atm}} = \boxed{0.0105\ \text{L}}$$

Check As we predicted, $V_2 < V_1$. Let's think about the relative values of P and V as we check the math. P more than doubled, so V_2 should be less than $\frac{1}{2}V_1$ ($0.0105/0.0248 < \frac{1}{2}$).

Comment Predicting the direction of the change provides another check on the problem setup: To make $V_2 < V_1$, we must multiply V_1 by a number *less than* 1. This means the ratio of pressures must be *less than* 1, so the larger pressure (P_2) must be in the denominator, P_1/P_2.

FOLLOW-UP PROBLEM 5.2 A sample of argon gas occupies 105 mL at 0.871 atm. If the temperature remains constant, what is the volume (in L) at 26.3 kPa?

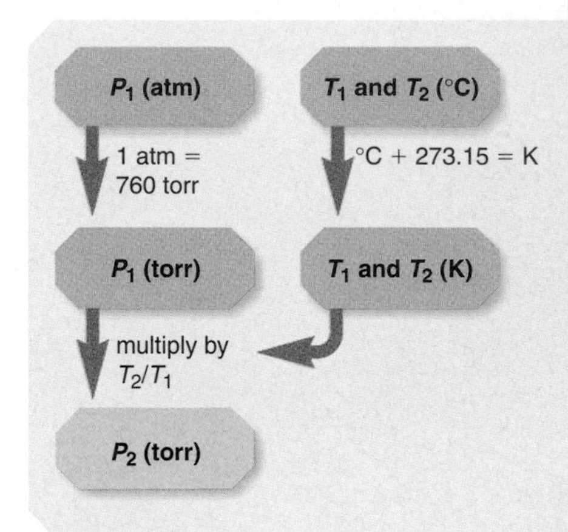

SAMPLE PROBLEM 5.3 Applying the Pressure-Temperature Relationship

Problem A steel tank used for fuel delivery is fitted with a safety valve that opens if the internal pressure exceeds 1.00×10^3 torr. It is filled with methane at 23°C and 0.991 atm and placed in boiling water at exactly 100°C. Will the safety valve open?

Plan The question "Will the safety valve open?" translates into "Is P_2 greater than 1.00×10^3 torr at T_2?" Thus, P_2 is the unknown, and T_1, T_2, and P_1 are given, with V (steel tank) and n fixed. We convert both T values to kelvins and P_1 to torrs in order to compare P_2 with the safety-limit pressure. We rearrange the ideal gas law to the appropriate form and solve for P_2. Since $T_2 > T_1$, we predict that $P_2 > P_1$.

Solution Summary of gas variables:

$P_1 = 0.991$ atm (convert to torr) $\qquad P_2 =$ unknown

$T_1 = 23°\text{C}$ (convert to K) $\qquad T_2 = 100°\text{C}$ (convert to K)

V and n remain constant

Converting T from °C to K:

$$T_1\ (\text{K}) = 23°\text{C} + 273.15 = 296\ \text{K} \qquad T_2\ (\text{K}) = 100°\text{C} + 273.15 = 373\ \text{K}$$

Converting P from atm to torr:

$$P_1\ (\text{torr}) = 0.991\ \text{atm} \times \frac{760\ \text{torr}}{1\ \text{atm}} = 753\ \text{torr}$$

Arranging the ideal gas law and solving for P_2: At fixed n and V, we have

$$\frac{P_1\cancel{V_1}}{\cancel{n_1}T_1} = \frac{P_2\cancel{V_2}}{\cancel{n_2}T_2} \quad \text{or} \quad \frac{P_1}{T_1} = \frac{P_2}{T_2}$$

$$P_2 = P_1 \times \frac{T_2}{T_1} = 753\ \text{torr} \times \frac{373\ \text{K}}{296\ \text{K}} = 949\ \text{torr}$$

P_2 is less than 1.00×10^3 torr, so the valve will *not* open.

Check Our prediction is correct: because $T_2 > T_1$, we have $P_2 > P_1$. Thus, the temperature ratio should be >1 (T_2 in the numerator). The T ratio is about 1.25 (373/296), so the P ratio should also be about 1.25 ($950/750 \approx 1.25$).

FOLLOW-UP PROBLEM 5.3 An engineer pumps air at 0°C into a newly designed piston-cylinder assembly. The volume measures 6.83 cm^3. At what temperature (in K) will the volume be 9.75 cm^3?

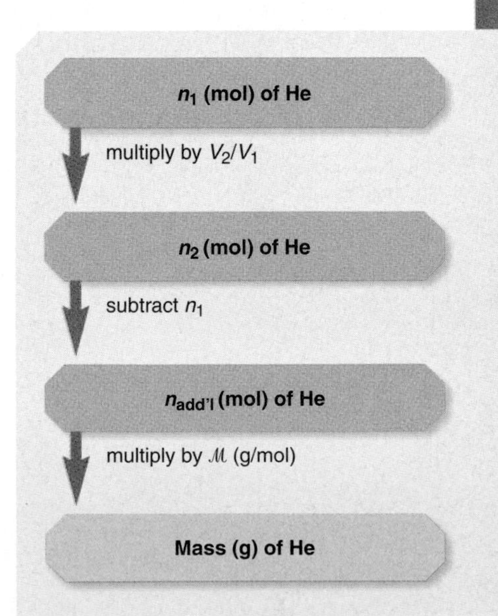

SAMPLE PROBLEM 5.4 Applying the Volume-Amount Relationship

Problem A scale model of a blimp rises when it is filled with helium to a volume of 55.0 dm^3. When 1.10 mol of He is added to the blimp, the volume is 26.2 dm^3. How many more grams of He must be added to make it rise? Assume constant T and P.

Plan We are given the initial amount of helium (n_1), the initial volume of the blimp (V_1), and the volume needed for it to rise (V_2), and we need the additional mass of helium to make it rise. So we first need to find n_2. We rearrange the ideal gas law to the appropriate form, solve for n_2, subtract n_1 to find the additional amount ($n_{add'l}$), and then convert moles to grams. We predict that $n_2 > n_1$ because $V_2 > V_1$.

Solution Summary of gas variables:

$n_1 = 1.10$ mol $\qquad n_2 =$ unknown (find, and then subtract n_1)

$V_1 = 26.2\ \text{dm}^3 \qquad V_2 = 55.0\ \text{dm}^3$

P and T remain constant

Arranging the ideal gas law and solving for n_2: At fixed P and T, we have

$$\frac{\cancel{P_1}V_1}{n_1\cancel{T_1}} = \frac{\cancel{P_2}V_2}{n_2\cancel{T_2}} \quad \text{or} \quad \frac{V_1}{n_1} = \frac{V_2}{n_2}$$

$$n_2 = n_1 \times \frac{V_2}{V_1} = 1.10\ \text{mol He} \times \frac{55.0\ \text{dm}^3}{26.2\ \text{dm}^3} = 2.31\ \text{mol He}$$

Finding the additional amount of He:

$$n_{add'l} = n_2 - n_1 = 2.31 \text{ mol He} - 1.10 \text{ mol He} = 1.21 \text{ mol He}$$

Converting moles of He to grams:

$$\text{Mass (g) of He} = 1.21 \text{ mol He} \times \frac{4.003 \text{ g He}}{1 \text{ mol He}}$$
$$= 4.84 \text{ g He}$$

Check Since V_2 is about twice V_1 ($55/26 \approx 2$), n_2 should be about twice n_1 ($2.3/1.1 \approx 2$). Since $n_2 > n_1$, we were right to multiply n_1 by a number >1 (that is, V_2/V_1). About $1.2 \text{ mol} \times 4 \text{ g/mol} \approx 4.8 \text{ g}$.

Comment 1. A different sequence of steps will give you the same answer: first find the additional volume ($V_{add'l} = V_2 - V_1$), and then solve directly for $n_{add'l}$. Try it for yourself.

2. You saw that Charles's law ($V \propto T$ at fixed P and n) translates into a similar relationship between P and T at fixed V and n. The follow-up problem demonstrates that Avogadro's law ($V \propto n$ at fixed P and T) translates into an analogous relationship at fixed V and T.

FOLLOW-UP PROBLEM 5.4 A rigid plastic container holds 35.0 g of ethylene gas (C_2H_4) at a pressure of 793 torr. What is the pressure if 5.0 g of ethylene is removed at constant temperature?

SAMPLE PROBLEM 5.5 Solving for an Unknown Gas Variable at Fixed Conditions

Problem A steel tank has a volume of 438 L and is filled with 0.885 kg of O_2. Calculate the pressure of O_2 at 21°C.

Plan We are given V, T, and the mass of O_2, and we must find P. Since conditions are not changing, we apply the ideal gas law without rearranging it. We use the given V in liters, convert T to kelvins and mass of O_2 to moles, and solve for P.

Solution Summary of gas variables:

$V = 438 \text{ L}$ $\quad$ $T = 21°\text{C}$ (convert to K)

$n = 0.885 \text{ kg } O_2$ (convert to mol) $\quad$ $P =$ unknown

Converting T from °C to K:

$$T \text{ (K)} = 21°\text{C} + 273.15 = 294 \text{ K}$$

Converting from mass of O_2 to moles:

$$n = \text{mol of } O_2 = 0.885 \text{ kg } O_2 \times \frac{1000 \text{ g}}{1 \text{ kg}} \times \frac{1 \text{ mol } O_2}{32.00 \text{ g } O_2} = 27.7 \text{ mol } O_2$$

Solving for P (note the unit canceling here):

$$P = \frac{nRT}{V} = \frac{27.7 \cancel{\text{mol}} \times 0.0821 \frac{\text{atm}\cdot\cancel{\text{L}}}{\cancel{\text{mol}}\cdot\cancel{\text{K}}} \times 294 \cancel{\text{K}}}{438 \cancel{\text{L}}}$$
$$= 1.53 \text{ atm}$$

Check The amount of O_2 seems correct: ~900 g/(30 g/mol) = 30 mol. To check the approximate size of the final calculation, round off the values, including that for R:

$$P = \frac{30 \text{ mol } O_2 \times 0.1 \frac{\text{atm}\cdot\text{L}}{\text{mol}\cdot\text{K}} \times 300 \text{ K}}{450 \text{ L}} = 2 \text{ atm}$$

which is reasonably close to 1.53 atm.

FOLLOW-UP PROBLEM 5.5 The tank in the sample problem develops a slow leak that is discovered and sealed. The new measured pressure is 1.37 atm. How many grams of O_2 remain?

Finally, in a slightly different type of problem that depicts a simple laboratory scene, we apply the gas laws to determine the correct balanced equation for a process.

SAMPLE PROBLEM 5.6 Using Gas Laws to Determine a Balanced Equation

Problem The piston-cylinders below depict a gaseous reaction carried out at constant pressure. Before the reaction, the temperature is 150 K; when it is complete, the temperature is 300 K.

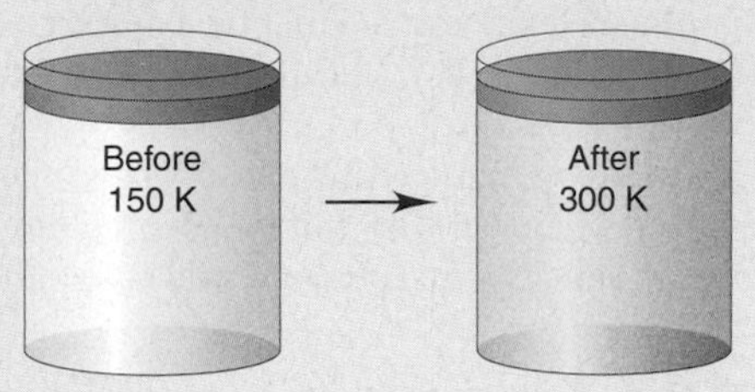

Which of the following balanced equations describes the reaction?

(1) $A_2 + B_2 \longrightarrow 2AB$ (2) $2AB + B_2 \longrightarrow 2AB_2$

(3) $A + B_2 \longrightarrow AB_2$ (4) $2AB_2 \longrightarrow A_2 + 2B_2$

Plan We are shown a depiction of a gaseous reaction and must choose the balanced equation. The problem says that P is constant, and the pictures show that T doubles and V stays the same. If n were also constant, the gas laws tell us that V should double when T doubles. Therefore, n cannot be constant, and the only way to maintain V with P constant and T doubling is for n to be halved. So we examine the four balanced equations and count the number of moles on each side to see in which equation n is halved.

Solution In equation (1), n does not change, so doubling T would double V.

In equation (2), n decreases from 3 mol to 2 mol, so doubling T would increase V by one-third.

In equation (3), n decreases from 2 mol to 1 mol. Doubling T would exactly balance the decrease from halving n, so V would stay the same.

In equation (4), n increases, so doubling T would more than double V.

Equation (3) is correct:

$$A + B_2 \longrightarrow AB_2$$

FOLLOW-UP PROBLEM 5.6 The gaseous reaction depicted below is carried out at constant pressure and an initial temperature of −73°C:

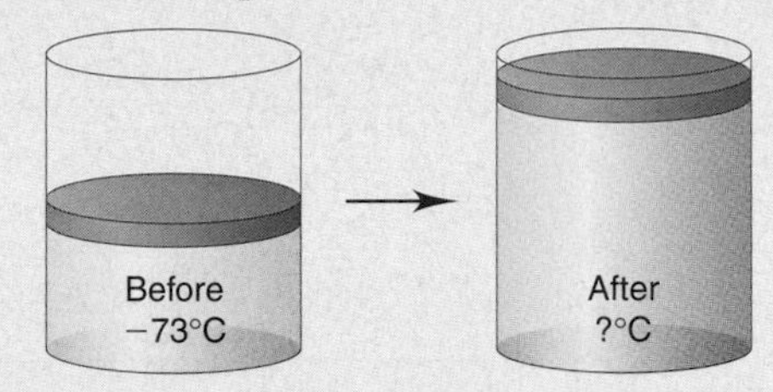

The *unbalanced* equation is $CD \longrightarrow C_2 + D_2$. What is the final temperature (in °C)?

SECTION SUMMARY

Four variables define the physical behavior of an ideal gas: volume (V), pressure (P), temperature (T), and amount (number of moles, n). Most simple gases display nearly ideal behavior at ordinary temperatures and pressures. Boyle's, Charles's, and Avogadro's laws relate volume to pressure, to temperature, and to amount of gas, respectively. At STP (0°C and 1 atm), 1 mol of an ideal gas occupies 22.4 L. The ideal gas law incorporates the individual gas laws into one equation: $PV = nRT$, where R is the universal gas constant.

5.4 FURTHER APPLICATIONS OF THE IDEAL GAS LAW

The ideal gas law can be recast in additional ways to determine other properties of gases. In this section, we use it to find gas density, molar mass, and the partial pressure of each gas in a mixture.

The Density of a Gas

One mole of any gas occupies nearly the same volume at a given temperature and pressure, so differences in gas density ($d = m/V$) depend on differences in molar mass (see Figure 5.8). For example, at STP, 1 mol of O_2 occupies the same volume as 1 mol of N_2, but since each O_2 molecule has a greater mass than each N_2 molecule, O_2 is denser.

All gases are miscible when thoroughly mixed, but in the absence of mixing, a less dense gas will lie above a more dense one. There are many familiar examples of this phenomenon. Some types of fire extinguishers release CO_2 because it is denser than air and will sink onto the fire, preventing more O_2 from reaching the burning material. Enormous air masses of different densities and temperatures moving past each other around the globe give rise to much of our weather.

Gas Density and Human Disasters Many gases that are denser than air have been involved in natural and human-caused disasters. The dense gases in smog that blanket urban centers, such as Los Angeles (see photo), contribute greatly to respiratory illness. On a far more horrific scale, in World War I, phosgene ($COCl_2$) was used against ground troops as they lay in trenches. More recently, the unintentional release of methylisocyanate from a Union Carbide India Ltd. chemical plant in Bhopal, India, killed thousands of people as vapors spread from the outskirts into the city. In 1986 in Cameroon, CO_2 released naturally from Lake Nyos suffocated thousands as it flowed down valleys into villages. Some paleontologists suggest that a similar process in volcanic lakes may have contributed to dinosaur kills.

We can rearrange the ideal gas law to calculate the density of a gas from its molar mass. Recall that the number of moles (n) is the mass (m) divided by the molar mass ($\mathcal{M}$), $n = m/\mathcal{M}$. Substituting for n in the ideal gas law gives

$$PV = \frac{m}{\mathcal{M}}RT$$

Rearranging to isolate m/V gives

$$\frac{m}{V} = d = \frac{\mathcal{M} \times P}{RT} \quad (5.9)$$

Two important ideas are expressed by Equation 5.9:

- *The density of a gas is directly proportional to its molar mass* because a given amount of a heavier gas occupies the same volume as that amount of a lighter gas (Avogadro's law).
- *The density of a gas is inversely proportional to the temperature.* As the volume of a gas increases with temperature (Charles's law), the same mass occupies more space; thus, the density is lower.

Architectural designers and heating engineers apply the second idea when they place heating ducts near the floor of a room: the less dense warm air from the ducts rises and heats the room air. Safety experts recommend staying near the floor when escaping from a fire to avoid the hot, and therefore less dense, noxious gases. We use Equation 5.9 to find the density of a gas at any temperature and pressure near standard conditions.

SAMPLE PROBLEM 5.7 Calculating Gas Density

Problem To apply a green chemistry approach, a chemical engineer uses waste CO_2 from a manufacturing process, instead of chlorofluorocarbons, as a "blowing agent" in the production of polystyrene containers. Find the density (in g/L) of CO_2 and the number of molecules per liter **(a)** at STP (0°C and 1 atm) and **(b)** at room conditions (20.°C and 1.00 atm).

Plan We must find the density (d) and number of molecules of CO_2, given the two sets of P and T data. We find $\mathcal{M}$, convert T to kelvins, and calculate d with Equation 5.9. Then we convert the mass per liter to molecules per liter with Avogadro's number.

Solution (a) Density and molecules per liter of CO_2 at STP. Summary of gas properties:

$$T = 0°C + 273.15 = 273\ K \qquad P = 1\ atm \qquad \mathcal{M}\ of\ CO_2 = 44.01\ g/mol$$

Calculating density (note the unit canceling here):

$$d = \frac{\mathcal{M} \times P}{RT} = \frac{44.01\ g/\cancel{mol} \times 1.00\ \cancel{atm}}{0.0821 \frac{\cancel{atm}\cdot L}{\cancel{mol}\cdot \cancel{K}} \times 273\ \cancel{K}} = \boxed{1.96\ g/L}$$

Converting from mass/L to molecules/L:

$$\text{Molecules } CO_2/L = \frac{1.96\ g\ CO_2}{1\ L} \times \frac{1\ mol\ CO_2}{44.01\ g\ CO_2} \times \frac{6.022\times10^{23}\ molecules\ CO_2}{1\ mol\ CO_2}$$
$$= \boxed{2.68\times10^{22}\ molecules\ CO_2/L}$$

(b) Density and molecules of CO_2 per liter at room conditions. Summary of gas properties:

$$T = 20.°C + 273.15 = 293\ K \qquad P = 1.00\ atm \qquad \mathcal{M}\ of\ CO_2 = 44.01\ g/mol$$

Calculating density:

$$d = \frac{\mathcal{M} \times P}{RT} = \frac{44.01\ g/mol \times 1.00\ atm}{0.0821 \frac{atm\cdot L}{mol\cdot K} \times 293\ K} = \boxed{1.83\ g/L}$$

Converting from mass/L to molecules/L:

$$\text{Molecules } CO_2/L = \frac{1.83\ g\ CO_2}{1\ L} \times \frac{1\ mol\ CO_2}{44.01\ g\ CO_2} \times \frac{6.022\times10^{23}\ molecules\ CO_2}{1\ mol\ CO_2}$$
$$= \boxed{2.50\times10^{22}\ molecules\ CO_2/L}$$

Check Round off to check the density values; for example, in (a), at STP:

$$\frac{50\ g/mol \times 1\ atm}{0.1 \frac{atm\cdot L}{mol\cdot K} \times 250\ K} = 2\ g/L \approx 1.96\ g/L$$

At the higher temperature in (b), the density should decrease, which can happen only if there are fewer molecules per liter, so the answer is reasonable.

Comment 1. An *alternative approach* for finding the density of most simple gases, but *at STP only,* is to divide the molar mass by the standard molar volume, 22.4 L:

$$d = \frac{\mathcal{M}}{V} = \frac{44.01\ g/mol}{22.4\ L/mol} = 1.96\ g/L$$

Then, since you know the density at one temperature (0°C), you can find it at any other temperature with the following relationship: $d_1/d_2 = T_2/T_1$.

2. Note that we have different numbers of significant figures for the pressure values. In (a), "1 atm" is part of the definition of STP, so it is an exact number. In (b), we specified "1.00 atm" to allow three significant figures in the answer.

3. Hot-air balloonists have always applied the change in density with temperature.

FOLLOW-UP PROBLEM 5.7 Compare the density of CO_2 at 0°C and 380 torr with its density at STP.

Up, Up, and Away! When the gas in a hot-air balloon is heated, its volume increases and the balloon inflates. Further heating causes some of the gas to escape. By these means, the gas density decreases and the balloon rises. Two pioneering hot-air balloonists used their knowledge of gas behavior to excel at their hobby. Jacques Charles (of Charles's law) made one of the first balloon flights in 1783. Twenty years later, Joseph Gay-Lussac (who studied the pressure-temperature relationship) set a solo altitude record that held for 50 years.

The Molar Mass of a Gas

Through another simple rearrangement of the ideal gas law, we can determine the molar mass of an unknown gas or volatile liquid (one that is easily vaporized):

$$n = \frac{m}{\mathcal{M}} = \frac{PV}{RT} \quad \text{so} \quad \mathcal{M} = \frac{mRT}{PV} \quad \text{or} \quad \mathcal{M} = \frac{dRT}{P} \qquad \textbf{(5.10)}$$

Notice that this equation is just a rearrangement of Equation 5.9.

The French chemist J. B. A. Dumas (1800–1884) pioneered an ingenious method for finding the molar mass of a volatile liquid. Figure 5.11 shows the apparatus. Place a small volume of the liquid in a preweighed flask of known volume. Close the flask with a stopper that contains a narrow tube and immerse it in a water bath whose fixed temperature exceeds the liquid's boiling point. As the liquid vaporizes, the gas fills the flask and some flows out the tube. When the liquid is gone, the pressure of the gas filling the flask equals the atmospheric pressure. Remove the flask from the water bath and cool it, and the gas condenses to a liquid. Reweigh the flask to obtain the mass of the liquid, *which equals the mass of gas* that remained in the flask.

By this procedure, you have directly measured all the variables needed to calculate the molar mass of the gas: the mass of gas (m) occupies the flask volume (V) at a pressure (P) equal to the barometric pressure and at the temperature (T) of the water bath.

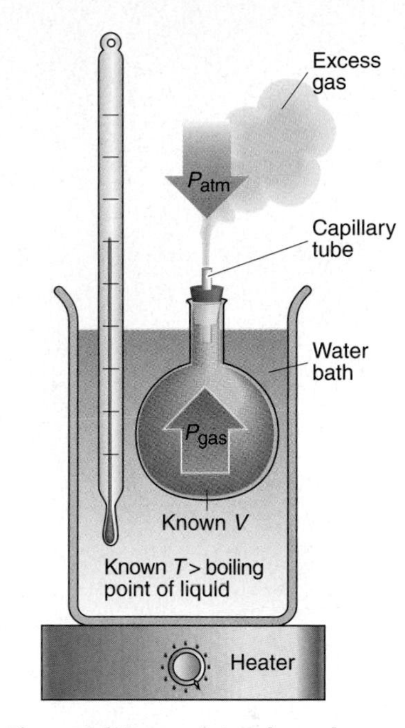

Figure 5.11 Determining the molar mass of an unknown volatile liquid. A small amount of unknown liquid is vaporized, and the gas fills the flask of known volume at the known temperature of the bath. Excess gas escapes through the capillary tube until $P_{gas} = P_{atm}$. When the flask is cooled, the gas condenses, the liquid is weighed, and the ideal gas law is used to calculate $\mathcal{M}$ (see text).

SAMPLE PROBLEM 5.8 Finding the Molar Mass of a Volatile Liquid

Problem An organic chemist isolates a colorless liquid from a petroleum sample. She uses the Dumas method and obtains the following data:

Volume (V) of flask = 213 mL T = 100.0°C P = 754 torr
Mass of flask + gas = 78.416 g Mass of flask = 77.834 g

Calculate the molar mass of the liquid.

Plan We are given V, T, P, and mass data and must find the molar mass ($\mathcal{M}$) of the liquid. We convert V to liters, T to kelvins, and P to atmospheres, find the mass of gas by subtracting the mass of the empty flask, and use Equation 5.10 to solve for $\mathcal{M}$.

Solution Summary of gas variables:

$$m = 78.416\text{ g} - 77.834\text{ g} = 0.582\text{ g} \qquad P\text{ (atm)} = 754\text{ torr} \times \frac{1\text{ atm}}{760\text{ torr}} = 0.992\text{ atm}$$

$$V\text{ (L)} = 213\text{ mL} \times \frac{1\text{ L}}{1000\text{ mL}} = 0.213\text{ L} \qquad T\text{ (K)} = 100.0°\text{C} + 273.15 = 373.2\text{ K}$$

Solving for $\mathcal{M}$:

$$\mathcal{M} = \frac{mRT}{PV} = \frac{0.582\text{ g} \times 0.0821 \dfrac{\text{atm}\cdot\text{L}}{\text{mol}\cdot\text{K}} \times 373.2\text{ K}}{0.992\text{ atm} \times 0.213\text{ L}} = 84.4\text{ g/mol}$$

Check Rounding to check the arithmetic, we have

$$\frac{0.6\text{ g} \times 0.08 \dfrac{\text{atm}\cdot\text{L}}{\text{mol}\cdot\text{K}} \times 375\text{ K}}{1\text{ atm} \times 0.2\text{ L}} = 90\text{ g/mol} \qquad \text{which is close to 84.4 g/mol}$$

FOLLOW-UP PROBLEM 5.8 At 10.0°C and 102.5 kPa, the density of dry air is 1.26 g/L. What is the average "molar mass" of dry air at these conditions?

The Partial Pressure of a Gas in a Mixture of Gases

All of the behaviors we've discussed so far were observed from experiments with air, which is a complex mixture of gases. The ideal gas law holds for virtually any gas, whether pure or a mixture, at ordinary conditions for two reasons:

- Gases mix homogeneously (form a solution) in any proportions.
- Each gas in a mixture behaves as if it were the only gas present (assuming no chemical interactions).

Dalton's Law of Partial Pressures The second point above was discovered by John Dalton in his studies of humidity. He observed that when water vapor is added to dry air, the total air pressure increases by an increment equal to the pressure of the water vapor:

$$P_{\text{humid air}} = P_{\text{dry air}} + P_{\text{added water vapor}}$$

In other words, each gas in the mixture exerts a **partial pressure,** a portion of the total pressure of the mixture, that is the same as the pressure it would exert *by itself.* This observation is formulated as **Dalton's law of partial pressures:** *in a mixture of unreacting gases, the total pressure is the sum of the partial pressures of the individual gases:*

$$P_{\text{total}} = P_1 + P_2 + P_3 + \cdots \qquad (5.11)$$

As an example, suppose you have a tank of fixed volume that contains nitrogen gas at a certain pressure, and you introduce a sample of hydrogen gas into the tank. Each gas behaves independently, so we can write an ideal gas law expression for each:

$$P_{N_2} = \frac{n_{N_2}RT}{V} \quad \text{and} \quad P_{H_2} = \frac{n_{H_2}RT}{V}$$

Because each gas occupies the same total volume and is at the same temperature, the pressure of each gas depends only on its amount, n. Thus, the total pressure is

$$P_{\text{total}} = P_{N_2} + P_{H_2} = \frac{n_{N_2}RT}{V} + \frac{n_{H_2}RT}{V} = \frac{(n_{N_2} + n_{H_2})RT}{V} = \frac{n_{\text{total}}RT}{V}$$

where $n_{\text{total}} = n_{N_2} + n_{H_2}$.

Each component in a mixture contributes a fraction of the total number of moles in the mixture, which is the **mole fraction (X)** of that component. Multiplying X by 100 gives the mole percent. Keep in mind that the sum of the mole fractions of all components in any mixture must be 1, and the sum of the mole percents must be 100%. For N_2, the mole fraction is

$$X_{N_2} = \frac{n_{N_2}}{n_{\text{total}}} = \frac{n_{N_2}}{n_{N_2} + n_{H_2}}$$

Since the total pressure is due to the total number of moles, the partial pressure of gas A is the total pressure multiplied by the mole fraction of A, X_A:

$$P_A = X_A \times P_{\text{total}} \qquad (5.12)$$

Equation 5.12 is a very important result. To see that it is valid for the mixture of N_2 and H_2, we recall that $X_{N_2} + X_{H_2} = 1$ and obtain

$$P_{\text{total}} = P_{N_2} + P_{H_2} = (X_{N_2} \times P_{\text{total}}) + (X_{H_2} \times P_{\text{total}}) = (X_{N_2} + X_{H_2})P_{\text{total}} = 1 \times P_{\text{total}}$$

SAMPLE PROBLEM 5.9 Applying Dalton's Law of Partial Pressures

Problem In a study of O_2 uptake by muscle at high altitude, a physiologist prepares an atmosphere consisting of 79 mole % N_2, 17 mole % $^{16}O_2$, and 4.0 mole % $^{18}O_2$. (The isotope ^{18}O will be measured to determine O_2 uptake.) The total pressure is 0.75 atm to simulate high altitude. Calculate the mole fraction and partial pressure of $^{18}O_2$ in the mixture.

Plan We must find $X_{^{18}O_2}$ and $P_{^{18}O_2}$ from P_{total} (0.75 atm) and the mole % of $^{18}O_2$ (4.0). Dividing the mole % by 100 gives the mole fraction, $X_{^{18}O_2}$. Then, using Equation 5.12, we multiply $X_{^{18}O_2}$ by P_{total} to find $P_{^{18}O_2}$.

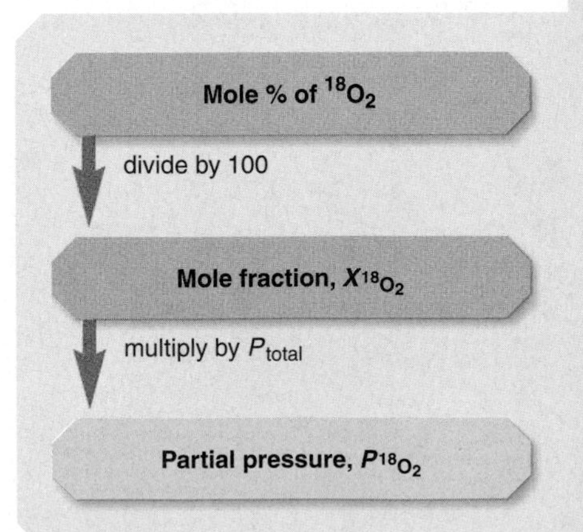

Solution Calculating the mole fraction of $^{18}O_2$:

$$X_{^{18}O_2} = \frac{4.0 \text{ mol \% } ^{18}O_2}{100} = 0.040$$

Solving for the partial pressure of $^{18}O_2$:

$$P_{^{18}O_2} = X_{^{18}O_2} \times P_{\text{total}} = 0.040 \times 0.75 \text{ atm} = 0.030 \text{ atm}$$

Check $X_{^{18}O_2}$ is small because the mole % is small, so $P_{^{18}O_2}$ should be small also.

Comment At high altitudes, specialized brain cells that are sensitive to O_2 and CO_2 levels in the blood trigger an increase in rate and depth of breathing for several days, until a person becomes acclimated.

FOLLOW-UP PROBLEM 5.9 To prevent the presence of air, noble gases are placed over highly reactive chemicals to act as inert "blanketing" gases. A chemical engineer places a mixture of noble gases consisting of 5.50 g of He, 15.0 g of Ne, and 35.0 g of Kr in a piston-cylinder assembly at STP. Calculate the partial pressure of each gas.

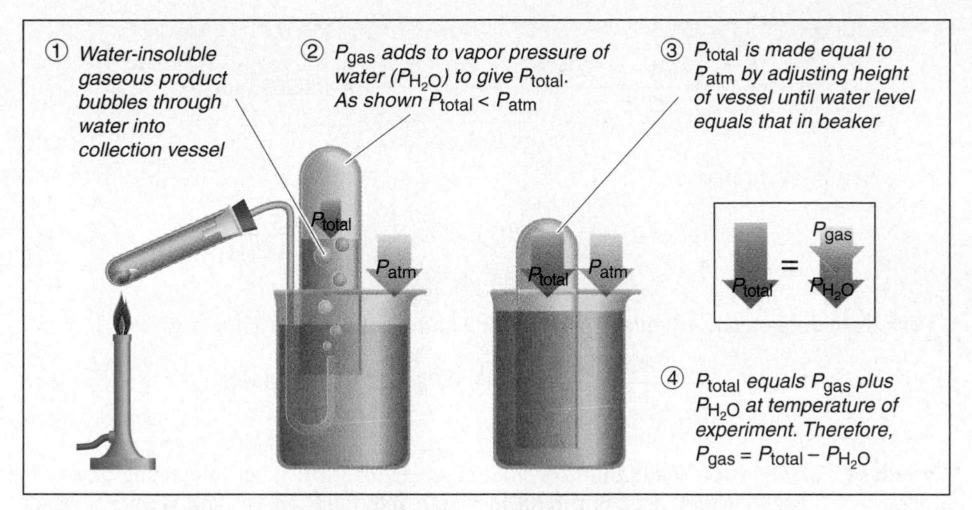
① Water-insoluble gaseous product bubbles through water into collection vessel
② P_{gas} adds to vapor pressure of water (P_{H_2O}) to give P_{total}. As shown $P_{total} < P_{atm}$
③ P_{total} is made equal to P_{atm} by adjusting height of vessel until water level equals that in beaker
④ P_{total} equals P_{gas} plus P_{H_2O} at temperature of experiment. Therefore, $P_{gas} = P_{total} - P_{H_2O}$
P_{total}
P_{atm}
P_{gas}
P_{H_2O}

Figure 5.12 Collecting a water-insoluble gaseous product and determining its pressure.

Table 5.3 Vapor Pressure of Water (P_{H_2O}) at Different T

T (°C)	P (torr)
0	4.6
5	6.5
10	9.2
12	10.5
14	12.0
16	13.6
18	15.5
20	17.5
22	19.8
24	22.4
26	25.2
28	28.3
30	31.8
35	42.2
40	55.3
45	71.9
50	92.5
55	118.0
60	149.4
65	187.5
70	233.7
75	289.1
80	355.1
85	433.6
90	525.8
95	633.9
100	760.0

Collecting a Gas over Water The law of partial pressures is frequently used to determine the yield of a water-insoluble gas formed in a reaction. The gaseous product bubbles through water and is collected into an inverted container, as shown in Figure 5.12. The water vapor that mixes with the gas contributes a portion of the total pressure, called the *vapor pressure,* which depends only on the water temperature.

In order to determine the yield of gaseous product, we find the appropriate vapor pressure value from a list, such as the one in Table 5.3, and subtract it from the total gas pressure (corrected to barometric pressure) to get the partial pressure of the gaseous product. With V and T known, we can calculate the amount of product.

SAMPLE PROBLEM 5.10 Calculating the Amount of Gas Collected over Water

Problem Acetylene (C_2H_2), an important fuel in welding, is produced in the laboratory when calcium carbide (CaC_2) reacts with water:

$$CaC_2(s) + 2H_2O(l) \longrightarrow C_2H_2(g) + Ca(OH)_2(aq)$$

For a sample of acetylene collected over water, total gas pressure (adjusted to barometric pressure) is 738 torr and the volume is 523 mL. At the temperature of the gas (23°C), the vapor pressure of water is 21 torr. How many grams of acetylene are collected?

Plan In order to find the mass of C_2H_2, we first need to find the number of moles of C_2H_2, $n_{C_2H_2}$, which we can obtain from the ideal gas law by calculating $P_{C_2H_2}$. The barometer reading gives us P_{total}, which is the sum of $P_{C_2H_2}$ and P_{H_2O}, and we are given P_{H_2O}, so we subtract to find $P_{C_2H_2}$. We are also given V and T, so we convert to consistent units, and find $n_{C_2H_2}$ from the ideal gas law. Then we convert moles to grams using the molar mass from the formula.

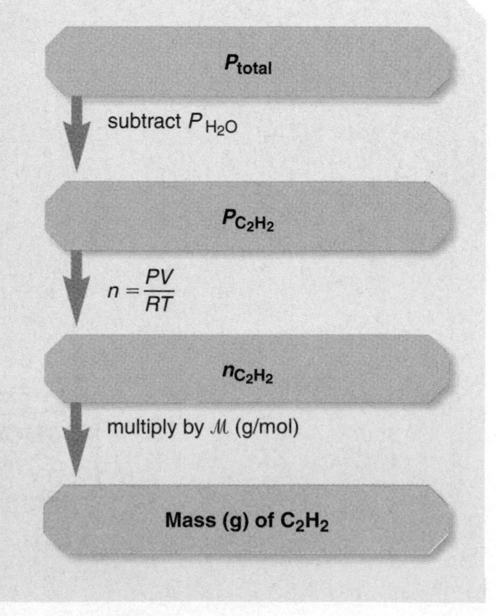
P_{total}
subtract P_{H_2O}
$P_{C_2H_2}$
$n = \frac{PV}{RT}$
$n_{C_2H_2}$
multiply by $\mathcal{M}$ (g/mol)
Mass (g) of C_2H_2

Solution Summary of gas variables:

$$P_{C_2H_2}\text{ (torr)} = P_{total} - P_{H_2O} = 738\text{ torr} - 21\text{ torr} = 717\text{ torr}$$

$$P_{C_2H_2}\text{ (atm)} = 717\text{ torr} \times \frac{1\text{ atm}}{760\text{ torr}} = 0.943\text{ atm}$$

$$V\text{ (L)} = 523\text{ mL} \times \frac{1\text{ L}}{1000\text{ mL}} = 0.523\text{ L}$$

$$T\text{ (K)} = 23°\text{C} + 273.15 = 296\text{ K}$$

$$n_{C_2H_2} = \text{unknown}$$

Solving for $n_{C_2H_2}$:

$$n_{C_2H_2} = \frac{PV}{RT} = \frac{0.943\ \text{atm} \times 0.523\ \text{L}}{0.0821\ \frac{\text{atm·L}}{\text{mol·K}} \times 296\ \text{K}} = 0.0203\ \text{mol}$$

Converting $n_{C_2H_2}$ to mass:

$$\text{Mass (g) of } C_2H_2 = 0.0203\ \text{mol } C_2H_2 \times \frac{26.04\ \text{g } C_2H_2}{1\ \text{mol } C_2H_2}$$

$$= 0.529\ \text{g } C_2H_2$$

Check Rounding to one significant figure, a quick arithmetic check for n gives

$$n \approx \frac{1\ \text{atm} \times 0.5\ \text{L}}{0.08\ \frac{\text{atm·L}}{\text{mol·K}} \times 300\ \text{K}} = 0.02\ \text{mol} \approx 0.0203\ \text{mol}$$

Comment The C_2^{2-} ion (called the *carbide,* or *acetylide, ion*) is an interesting anion. It is simply $^-C{\equiv}C^-$, which acts as a base in water, removing an H^+ ion from two H_2O molecules to form acetylene, $H{-}C{\equiv}C{-}H$.

FOLLOW-UP PROBLEM 5.10 A small piece of zinc reacts with dilute HCl to form H_2, which is collected over water at 16°C into a large flask. The total pressure is adjusted to barometric pressure (752 torr), and the volume is 1495 mL. Use Table 5.3 to help calculate the partial pressure and mass of H_2.

SECTION SUMMARY

Animation: Collecting a Gas over Water
Online Learning Center

The ideal gas law can be rearranged to calculate the density and molar mass of a gas. In a mixture of gases, each component contributes its own partial pressure to the total pressure (Dalton's law of partial pressures). The mole fraction of each component is the ratio of its partial pressure to the total pressure. When a gas is in contact with water, the total pressure is the sum of the gas pressure and the vapor pressure of water at the given temperature.

5.5 THE IDEAL GAS LAW AND REACTION STOICHIOMETRY

In Chapters 3 and 4, we encountered many reactions that involved gases as reactants (e.g., combustion with O_2) or as products (e.g., acid treatment of a carbonate). From the balanced equation, we used stoichiometrically equivalent molar ratios to calculate the amounts (moles) of reactants and products and converted these quantities into masses, numbers of molecules, or solution volumes (see Figures 3.12, p. 117, and 3.15, p. 122). Figure 5.13 shows how you can expand your problem-solving repertoire by using the ideal gas law to convert between gas variables (P, T, and V) and amounts (moles) of gaseous reactants and products. In effect, you combine a gas law problem with a stoichiometry problem; it is more realistic to measure the volume, pressure, and temperature of a gas than its mass.

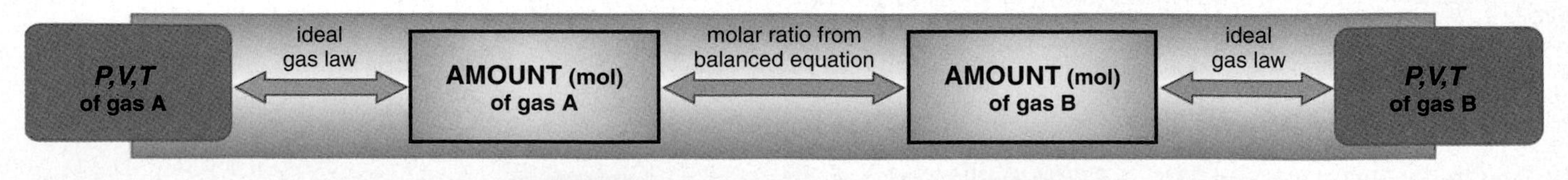

Figure 5.13 Summary of the stoichiometric relationships among the amount (mol, *n*) of gaseous reactant or product and the gas variables pressure (*P*), volume (*V*), and temperature (*T*).

SAMPLE PROBLEM 5.11 Using Gas Variables to Find Amounts of Reactants or Products

Problem Copper dispersed in absorbent beds is used to react with oxygen impurities in the ethylene used for producing polyethylene. The beds are regenerated when hot H_2 reduces the metal oxide, forming the pure metal and H_2O. On a laboratory scale, what volume of H_2 at 765 torr and 225°C is needed to reduce 35.5 g of copper(II) oxide?

Plan This is a stoichiometry *and* gas law problem. To find V_{H_2}, we first need n_{H_2}. We write and balance the equation. Next, we convert the given mass of CuO (35.5 g) to amount (mol) and use the molar ratio to find moles of H_2 needed (stoichiometry portion). Then, we use the ideal gas law to convert moles of H_2 to liters (gas law portion). A roadmap is shown, but you are familiar with all the steps.

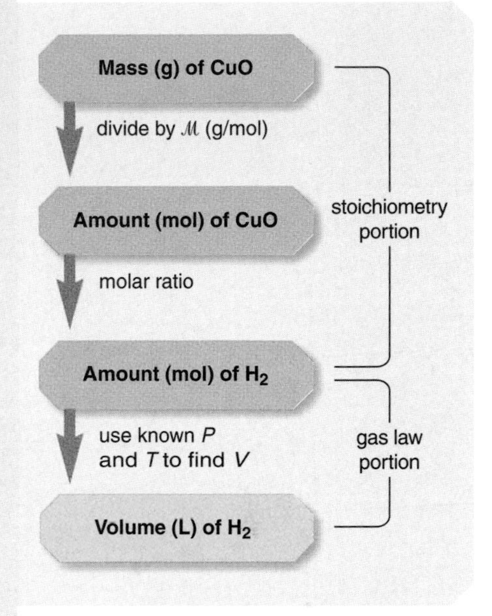

Solution Writing the balanced equation:

$$CuO(s) + H_2(g) \longrightarrow Cu(s) + H_2O(g)$$

Calculating n_{H_2}:

$$n_{H_2} = 35.5 \text{ g CuO} \times \frac{1 \text{ mol CuO}}{79.55 \text{ g CuO}} \times \frac{1 \text{ mol } H_2}{1 \text{ mol CuO}} = 0.446 \text{ mol } H_2$$

Summary of other gas variables:

$$V = \text{unknown} \qquad P \text{ (atm)} = 765 \text{ torr} \times \frac{1 \text{ atm}}{760 \text{ torr}} = 1.01 \text{ atm}$$

$$T \text{ (K)} = 225°\text{C} + 273.15 = 498 \text{ K}$$

Solving for V_{H_2}:

$$V = \frac{nRT}{P} = \frac{0.446 \text{ mol} \times 0.0821 \frac{\text{atm·L}}{\text{mol·K}} \times 498 \text{ K}}{1.01 \text{ atm}} = 18.1 \text{ L}$$

Check One way to check the answer is to compare it with the molar volume of an ideal gas at STP (22.4 L at 273.15 K and 1 atm). One mole of H_2 at STP occupies about 22 L, so less than 0.5 mol occupies less than 11 L. T is less than twice 273 K, so V should be less than twice 11 L. In this case, the decrease in n and the increase in T nearly cancel each other.

Comment The main point here is that the stoichiometry provides one gas variable (n), two more are given, and the ideal gas law is used to find the fourth.

FOLLOW-UP PROBLEM 5.11 Sulfuric acid reacts with sodium chloride to form aqueous sodium sulfate and hydrogen chloride gas. How many milliliters of gas form at STP when 0.117 kg of sodium chloride reacts with excess sulfuric acid?

SAMPLE PROBLEM 5.12 Using the Ideal Gas Law in a Limiting-Reactant Problem

Problem The alkali metals [Group 1A(1)] react with the halogens [Group 7A(17)] to form ionic metal halides. What mass of potassium chloride forms when 5.25 L of chlorine gas at 0.950 atm and 293 K reacts with 17.0 g of potassium (see photo)?

Chlorine gas reacting with potassium.

Plan The only difference between this and previous limiting-reactant problems (see Sample Problem 3.11, p. 113) is that here we use the ideal gas law to find the amount (n) of gaseous reactant from the known V, P, and T. We first write the balanced equation and then use it to find the limiting reactant and the amount of product.

Solution Writing the balanced equation:

$$2K(s) + Cl_2(g) \longrightarrow 2KCl(s)$$

Summary of gas variables:

$$P = 0.950 \text{ atm} \qquad V = 5.25 \text{ L}$$

$$T = 293 \text{ K} \qquad n = \text{unknown}$$

Solving for n_{Cl_2}:

$$n_{Cl_2} = \frac{PV}{RT} = \frac{0.950 \text{ atm} \times 5.25 \text{ L}}{0.0821 \frac{\text{atm·L}}{\text{mol·K}} \times 293 \text{ K}} = 0.207 \text{ mol}$$

2. *Boyle's law* ($V \propto 1/P$). Gas molecules are points of mass with empty space between them (postulate 1), so as the pressure exerted *on* the sample increases at constant temperature, the distance between molecules decreases, and the sample volume decreases. The pressure exerted *by* the gas increases simultaneously because in a smaller volume of gas, there are shorter distances between gas molecules and the walls and between the walls themselves; thus, collisions are more frequent (Figure 5.15). The fact that liquids and solids cannot be compressed means there is little, if any, free space between the molecules.

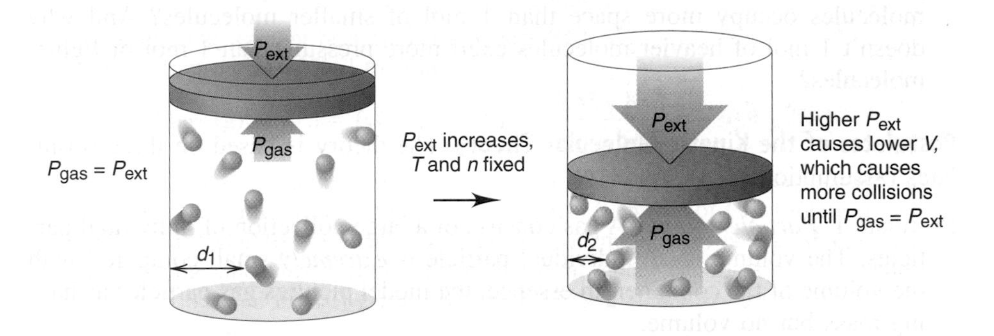

Figure 5.15 A molecular description of Boyle's law. At a given T, gas molecules collide with the walls across an average distance (d_1) and give rise to a pressure (P_{gas}) that equals the external pressure (P_{ext}). If P_{ext} increases, V decreases, and so the average distance between a molecule and the walls is shorter ($d_2 < d_1$). Molecules strike the walls more often, and P_{gas} increases until it again equals P_{ext}. Thus, V decreases when P increases.

3. *Dalton's law of partial pressures* ($P_{total} = P_A + P_B$). Adding a given amount of gas A to a given amount of gas B causes an increase in the total number of molecules in proportion to the amount of A that is added. This increase causes a corresponding increase in the number of collisions per second with the walls (postulate 2), which causes a corresponding increase in the pressure (Figure 5.16). Thus, each gas exerts a fraction of the total pressure based on the fraction of molecules (or fraction of moles; that is, the mole fraction) of that gas in the mixture.

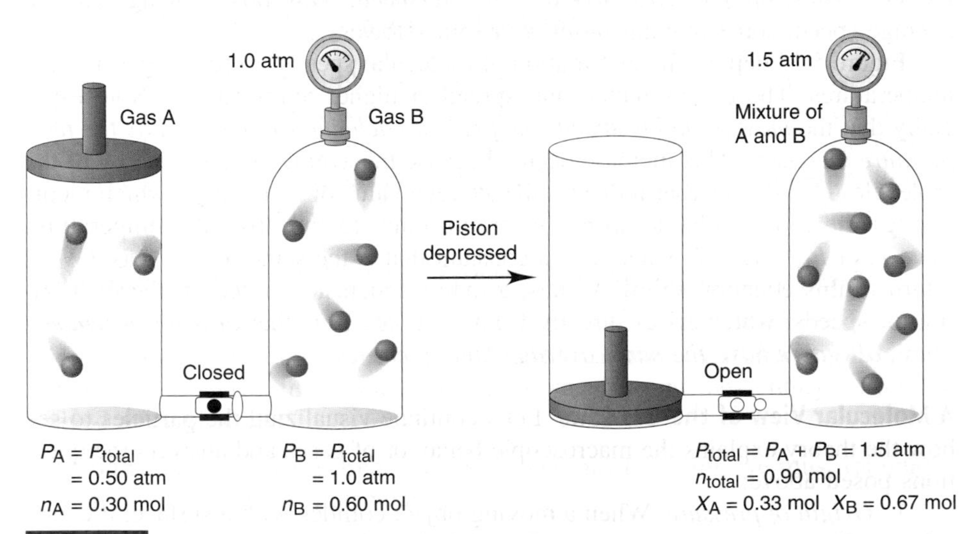

Figure 5.16 A molecular description of Dalton's law of partial pressures. A piston-cylinder assembly containing 0.30 mol of gas A at 0.50 atm is connected to a tank of fixed volume containing 0.60 mol of gas B at 1.0 atm. When the piston is depressed at fixed temperature, gas A is forced into the tank of gas B and the gases mix. The new total pressure, 1.5 atm, equals the sum of the partial pressures, which is related to the new total amount of gas, 0.90 mol. Thus, each gas undergoes a fraction of the total collisions related to its fraction of the total number of molecules (moles), which is equal to its mole fraction.

4. *Charles's law* ($V \propto T$). As the temperature increases, the most probable molecular speed and the average kinetic energy increase (postulate 3). Thus, the molecules hit the walls more frequently *and* more energetically. A higher frequency of collisions causes higher internal pressure. As a result, the walls move outward, which increases the volume and restores the starting pressure (Figure 5.17).

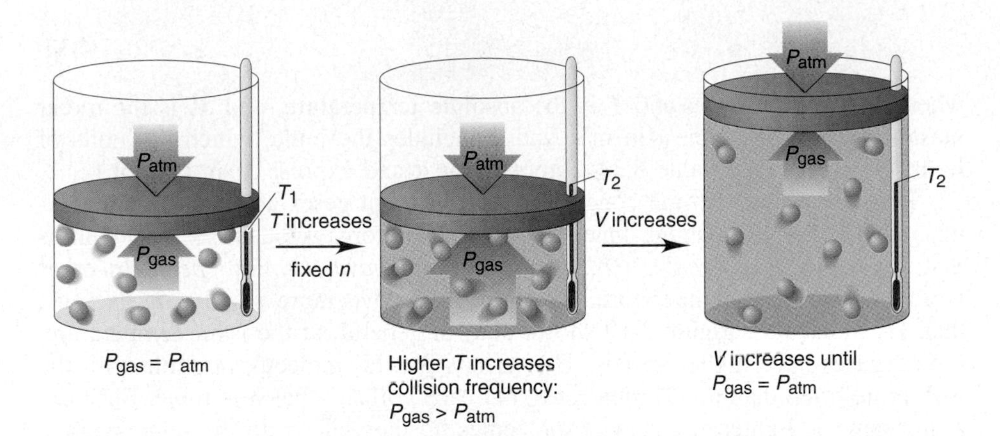

Figure 5.17 A molecular description of Charles's law. At a given temperature (T_1), $P_{gas} = P_{atm}$. When the gas is heated to T_2, the molecules move faster and collide with the walls more often, which increases P_{gas}. This increases V, and so the molecules collide less often until P_{gas} again equals P_{atm}. Thus, V increases when T increases.

5. *Avogadro's law* ($V \propto n$). Adding more molecules to a container increases the total number of collisions with the walls and, therefore, the internal pressure. As a result, the volume expands until the number of collisions per unit of wall area is the same as it was before the addition (Figure 5.18).

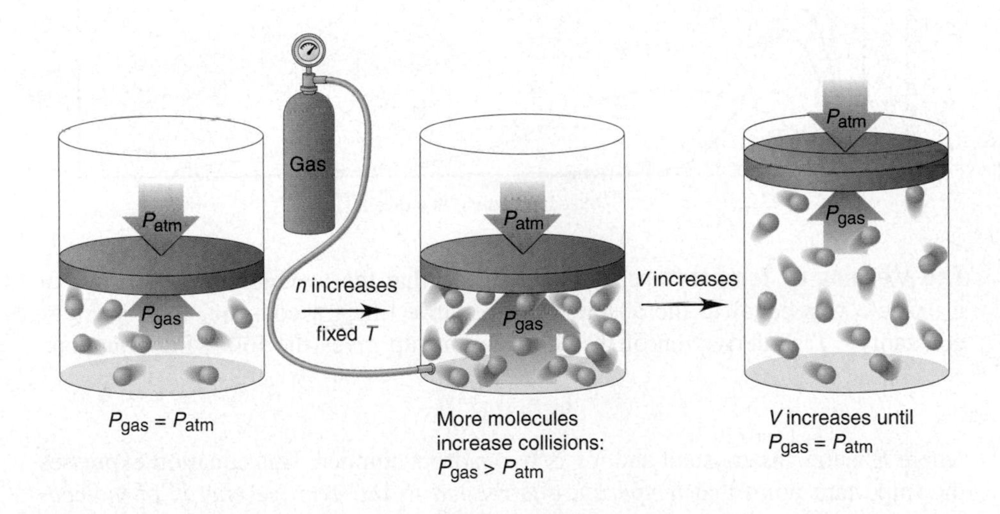

Figure 5.18 A molecular description of Avogadro's law. At a given T, a certain amount (n) of gas gives rise to a pressure (P_{gas}) equal to P_{atm}. When more gas is added, n increases. Collisions with the walls become more frequent, and P_{gas} increases. This leads to an increase in V until $P_{gas} = P_{atm}$ again. Thus, V increases when n increases.

Relation Between Molecular Speed and Mass We still need to explain why equal numbers of molecules of two different gases, such as O_2 and H_2, occupy the same volume. Let's first see why heavier O_2 particles *do not* hit the container walls with more energy than lighter H_2 particles. Recall that the kinetic energy of an object is the energy associated with its motion (Chapter 1). This energy is related to the mass *and* the speed of the object:

$$E_k = \tfrac{1}{2} \text{ mass} \times \text{speed}^2$$

This equation shows that if a heavy object and a light object have the same kinetic energy, *the heavy object must be moving more slowly.* For a large population of molecules, the average kinetic energy is

$$\overline{E_k} = \tfrac{1}{2}m\overline{u^2}$$

where m is the molecular mass and $\overline{u^2}$ is the average of the squares of the molecular speeds. The square root of $\overline{u^2}$ is called the root-mean-square speed, or **rms speed (u_{rms})**. *A molecule moving at this speed has the average kinetic energy.* The rms speed is somewhat higher than the most probable speed, but the speeds are proportional to each other and we will use them interchangeably. The rms speed is related to the temperature and the molar mass as follows:

$$u_{rms} = \sqrt{\frac{3RT}{\mathcal{M}}} \qquad (5.13)$$

where R is the gas constant, T is the absolute temperature, and $\mathcal{M}$ is the molar mass. (Because we want u in m/s, and R includes the joule, which has units of $kg{\cdot}m^2/s^2$, we use the value 8.314 J/mol·K for R and express $\mathcal{M}$ in kg/mol.)

Postulate 3 leads to the conclusion that different gases at the same temperature have the same average kinetic energy. Therefore, Avogadro's law requires that, on average, *molecules with a higher mass have a lower speed.* In other words, at the same temperature, O_2 molecules move more slowly, on average, than H_2 molecules. Figure 5.19 shows that, in general, at the same temperature, lighter gases have higher speeds. This means that H_2 molecules collide with the wall more often than do O_2 molecules, but each collision has less force. Because, at the same T, lighter and heavier molecules hit the walls with the same average kinetic energy, lighter and heavier gases have the same pressure and, thus, the same volume.

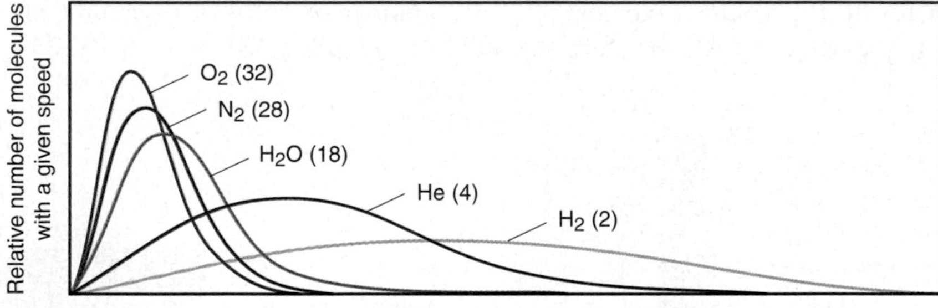

Figure 5.19 Relationship between molar mass and molecular speed. At a given temperature, gases with lower molar masses (numbers in parentheses) have higher most probable speeds (peak of each curve).

The Meaning of Temperature Earlier we said that the average kinetic energy of a particle was equal to the absolute temperature times a constant, that is, $\overline{E_k}$ = constant × T. A derivation of the full relationship gives the following equation:

$$\overline{E_k} = \frac{3}{2}\left(\frac{R}{N_A}\right)T$$

where R is the gas constant and N_A is Avogadro's number. This equation expresses the important point that *temperature is related to the average energy of molecular motion.* Note that it is not related to the *total* energy, which depends on the size of the sample, but to the *average* energy: as T increases, $\overline{E_k}$ increases. What occurs when, say, a beaker of water containing a thermometer is heated over a flame? The rapidly moving gas particles in the flame collide more energetically with the atoms of the beaker, transferring some of their kinetic energy. The atoms vibrate faster and transfer kinetic energy to the molecules of water, which move faster and transfer kinetic energy to the atoms in the thermometer bulb. These, in turn, transfer kinetic energy to the atoms of mercury. They collide with each other more frequently and more energetically, which causes the mercury volume to expand and the level to rise. In the macroscopic world, we see this as a temperature increase; in the molecular world, it is a sequence of kinetic energy transfers from higher energy particles to lower energy particles, such that each group of particles increases its average kinetic energy.

Effusion and Diffusion

The movement of gases, either through one another or into regions of very low pressure, has many important applications.

The Process of Effusion One of the early triumphs of the kinetic-molecular theory was an explanation of **effusion,** the process by which a gas escapes from its container through a tiny hole into an evacuated space. In 1846, Thomas Graham studied this process and concluded that the effusion rate was inversely proportional to the square root of the gas density. The effusion rate is the number of moles (or molecules) of gas effusing per unit time. Since density is directly proportional to molar mass, we state **Graham's law of effusion** as follows: *the rate of effusion of a gas is inversely proportional to the square root of its molar mass,*

$$\text{Rate of effusion} \propto \frac{1}{\sqrt{\mathcal{M}}}$$

Argon (Ar) is lighter than krypton (Kr), so it effuses faster, assuming equal pressures of the two gases. Thus, the ratio of the rates is

$$\frac{\text{Rate}_{\text{Ar}}}{\text{Rate}_{\text{Kr}}} = \frac{\sqrt{\mathcal{M}_{\text{Kr}}}}{\sqrt{\mathcal{M}_{\text{Ar}}}} \quad \text{or, in general,} \quad \frac{\text{rate}_{\text{A}}}{\text{rate}_{\text{B}}} = \frac{\sqrt{\mathcal{M}_{\text{B}}}}{\sqrt{\mathcal{M}_{\text{A}}}} = \sqrt{\frac{\mathcal{M}_{\text{B}}}{\mathcal{M}_{\text{A}}}} \qquad (5.14)$$

The kinetic-molecular theory explains that, at a given temperature and pressure, *the gas with the lower molar mass effuses faster because the most probable speed of its molecules is higher; therefore, more molecules escape per unit time.*

Preparing Nuclear Fuel One of the most important applications of Graham's law is the *enrichment* of nuclear reactor fuel: separating nonfissionable, more abundant ^{238}U from fissionable ^{235}U to increase the proportion of ^{235}U in the mixture. Since the two isotopes have identical chemical properties, they are separated by differences in the effusion rates of their gaseous compounds. Uranium ore is converted to gaseous UF_6 (a mixture of $^{238}UF_6$ and $^{235}UF_6$), which is pumped through a series of chambers with porous barriers. Because they move very slightly faster, molecules of $^{235}UF_6$ ($\mathcal{M} = 349.03$) effuse through each barrier 1.0043 times faster than do molecules of $^{238}UF_6$ ($\mathcal{M} = 352.04$). Many passes are made, each increasing the fraction of $^{235}UF_6$ until a mixture is obtained that contains enough $^{235}UF_6$. This isotope-enrichment process was developed during the latter years of World War II and produced enough ^{235}U for two of the world's first three atomic bombs.

SAMPLE PROBLEM 5.13 Applying Graham's Law of Effusion

Problem Calculate the ratio of the effusion rates of helium and methane (CH_4).

Plan The effusion rate is inversely proportional to $\sqrt{\mathcal{M}}$, so we find the molar mass of each substance from the formula and take its square root. The inverse of the ratio of the square roots is the ratio of the effusion rates.

Solution

$$\mathcal{M} \text{ of } CH_4 = 16.04 \text{ g/mol} \qquad \mathcal{M} \text{ of He} = 4.003 \text{ g/mol}$$

Calculating the ratio of the effusion rates:

$$\frac{\text{Rate}_{\text{He}}}{\text{Rate}_{CH_4}} = \sqrt{\frac{\mathcal{M}_{CH_4}}{\mathcal{M}_{\text{He}}}} = \sqrt{\frac{16.04 \text{ g/mol}}{4.003 \text{ g/mol}}} = \sqrt{4.007} = 2.002$$

Check A ratio >1 makes sense because the lighter He should effuse faster than the heavier CH_4. Because the molar mass of He is about one-fourth that of CH_4, He should effuse about twice as fast (the inverse of $\sqrt{\frac{1}{4}}$).

FOLLOW-UP PROBLEM 5.13 If it takes 1.25 min for 0.010 mol of He to effuse, how long will it take for the same amount of ethane (C_2H_6) to effuse?

Graham's law is also used to determine the molar mass of an unknown gas, X, by comparing its effusion rate with that of a known gas, such as He:

$$\frac{\text{Rate}_{\text{X}}}{\text{Rate}_{\text{He}}} = \sqrt{\frac{\mathcal{M}_{\text{He}}}{\mathcal{M}_{\text{X}}}}$$

Squaring both sides and solving for the molar mass of X gives

$$\mathcal{M}_{\text{X}} = \mathcal{M}_{\text{He}} \times \left(\frac{\text{rate}_{\text{He}}}{\text{rate}_{\text{X}}}\right)^2$$

The Process of Diffusion Closely related to effusion is the process of gaseous **diffusion,** the movement of one gas through another. Diffusion rates are also described generally by Graham's law:

$$\text{Rate of diffusion} \propto \frac{1}{\sqrt{\mathcal{M}}}$$

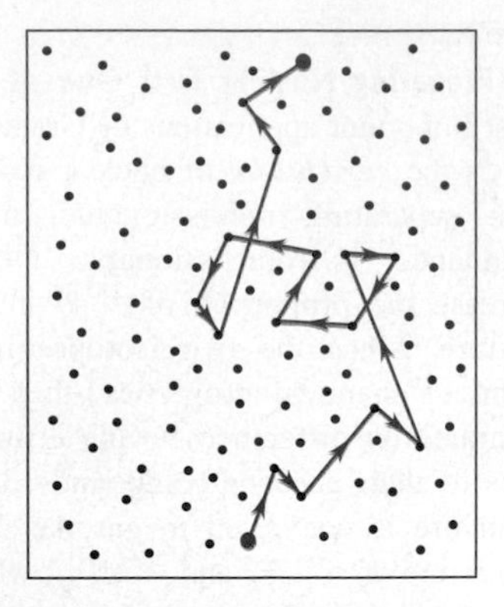

Figure 5.20 Diffusion of a gas particle through a space filled with other particles. In traversing a space, a gas molecule collides with many other molecules, which gives it a tortuous path. For clarity, the path of only one particle *(red dot)* is shown *(red lines)*.

For two gases at equal pressures, such as NH_3 and HCl, moving through another gas or a mixture of gases, such as air, we find

$$\frac{\text{Rate}_{NH_3}}{\text{Rate}_{HCl}} = \sqrt{\frac{\mathcal{M}_{HCl}}{\mathcal{M}_{NH_3}}}$$

The reason for this dependence on molar mass is the same as for effusion rates: lighter molecules have higher molecular speeds than heavier molecules, so they move farther in a given amount of time.

If gas molecules move at hundreds of meters per second at ordinary temperatures (see Figure 5.14), why does it take a second or two after you open a bottle of perfume to smell the fragrance? Although convection plays an important role, a molecule moving by diffusion does not travel very far before it collides with a molecule in the air. As you can see from Figure 5.20, the path of each molecule is tortuous. Imagine walking through an empty room, and then imagine walking through a room crowded with other moving people.

Diffusion also occurs in liquids (and even to a small extent in solids). However, because the distances between molecules are much shorter in a liquid than in a gas, collisions are much more frequent; thus, diffusion is *much* slower. Diffusion of a gas through a liquid is a vital process in biological systems. For example, it plays a key part in the movement of O_2 from lungs to blood. Many organisms have evolved elaborate ways to speed the diffusion of nutrients (for example, sugar and metal ions) through their cell membranes and to slow, or even stop, the diffusion of toxins.

The Chaotic World of Gases: Mean Free Path and Collision Frequency

Refinements of the basic kinetic-molecular theory provide a view into the amazing, chaotic world of gas molecules. Let's follow an "average" N_2 molecule in a room at 20°C and 1 atm pressure—perhaps the room you are in now.

Distribution of Molecular Speeds Our molecule is hurtling through the room at an average speed of 0.47 km/s, or nearly 1100 mi/h (rms speed = 0.51 km/s), and continually changing speed as it collides with other molecules. At any instant, it may be traveling at 2500 mi/h or standing still as it collides head on, but these extreme speeds are *much* less likely than the most probable one (see Figure 5.19).

Mean Free Path From a molecule's diameter, we can use the kinetic-molecular theory to obtain the **mean free path,** the average distance the molecule travels between collisions at a given temperature and pressure. Our average N_2 molecule (3.7×10^{-10} m in diameter) travels 6.6×10^{-8} m before smashing into a fellow traveler, which means it travels about 180 molecular diameters between collisions. (An analogy in the macroscopic world would be an N_2 molecule the size of a billiard ball traveling about 30 ft before colliding with another.) Therefore, even though gas molecules are *not* points of mass, a gas sample *is* mostly empty space. Mean free path is a key factor in the rate of diffusion and the rate of heat flow through a gas.

Collision Frequency Divide the most probable speed (distance per second) by the mean free path (distance per collision) and you obtain the **collision frequency,** the average number of collisions per second that each molecule undergoes:

$$\text{Collision frequency} = \frac{4.7\times10^{2}\ \text{m/s}}{6.6\times10^{-8}\ \text{m/collision}} = 7.1\times10^{9}\ \text{collisions/s}$$

Distribution of speed (and kinetic energy) and collision frequency are essential ideas for understanding the speed of a chemical reaction, as you'll see in Chapter 16. As the upcoming Chemical Connections essay shows, the kinetic-molecular theory applies directly to the behavior of our planet's atmosphere.

Danger on Molecular Highways To give you some idea of how astounding events are in the molecular world, we can express the collision frequency of a molecule in terms of a common experience in the macroscopic world: driving a compact car on the highway. Since a car is much larger than an N_2 molecule, to match the collision frequency of the molecule you would have to travel at 2.8 billion mi/s (an impossibility, given that it is much faster than the speed of light) and would smash into another car every 700 yd!

Structure and Composition of Earth's Atmosphere

An **atmosphere** is the envelope of gases that extends continuously from a planet's surface outward, eventually thinning to a point at which it is indistinguishable from interplanetary space. On Earth, complex changes in pressure, temperature, and composition occur within this mixture of gases, and the present atmosphere is very different from the one that existed during our planet's early history.

Variation in Pressure

Since gases are compressible (Boyle's law), the pressure of the atmosphere *decreases* smoothly with distance from the surface, with a more rapid decrease at lower altitudes (Figure B5.1). Although no specific boundary delineates the outermost fringe of the atmosphere, the density and composition at around 10,000 km from the surface are identical with those of outer space. About 99% of the atmosphere's mass lies within 30 km of the surface, and 75% lies within the lowest 11 km.

Variation in Temperature

Unlike the change in pressure, temperature does *not* decrease smoothly with altitude, and the atmosphere is usually classified into regions based on the direction of temperature change (Figure B5.1). In the *troposphere,* which includes the region from the surface to around 11 km, temperatures *drop* 7°C per kilometer to −55°C (218 K). All our weather occurs in the troposphere, and all but a few aircraft fly there. Temperatures then *rise* through the *stratosphere* from −55°C to about 7°C (280 K) at 50 km; we'll discuss the reason shortly.

In the *mesosphere,* temperatures *drop* smoothly again to −93°C (180 K) at around 80 km. Within the *thermosphere,* which extends to around 500 km, temperatures *rise* again, but vary between 700 and 2000 K, depending on the intensity of solar radiation and sunspot activity.

The *exosphere,* the outermost region, maintains these temperatures and merges with outer space.

(continued)

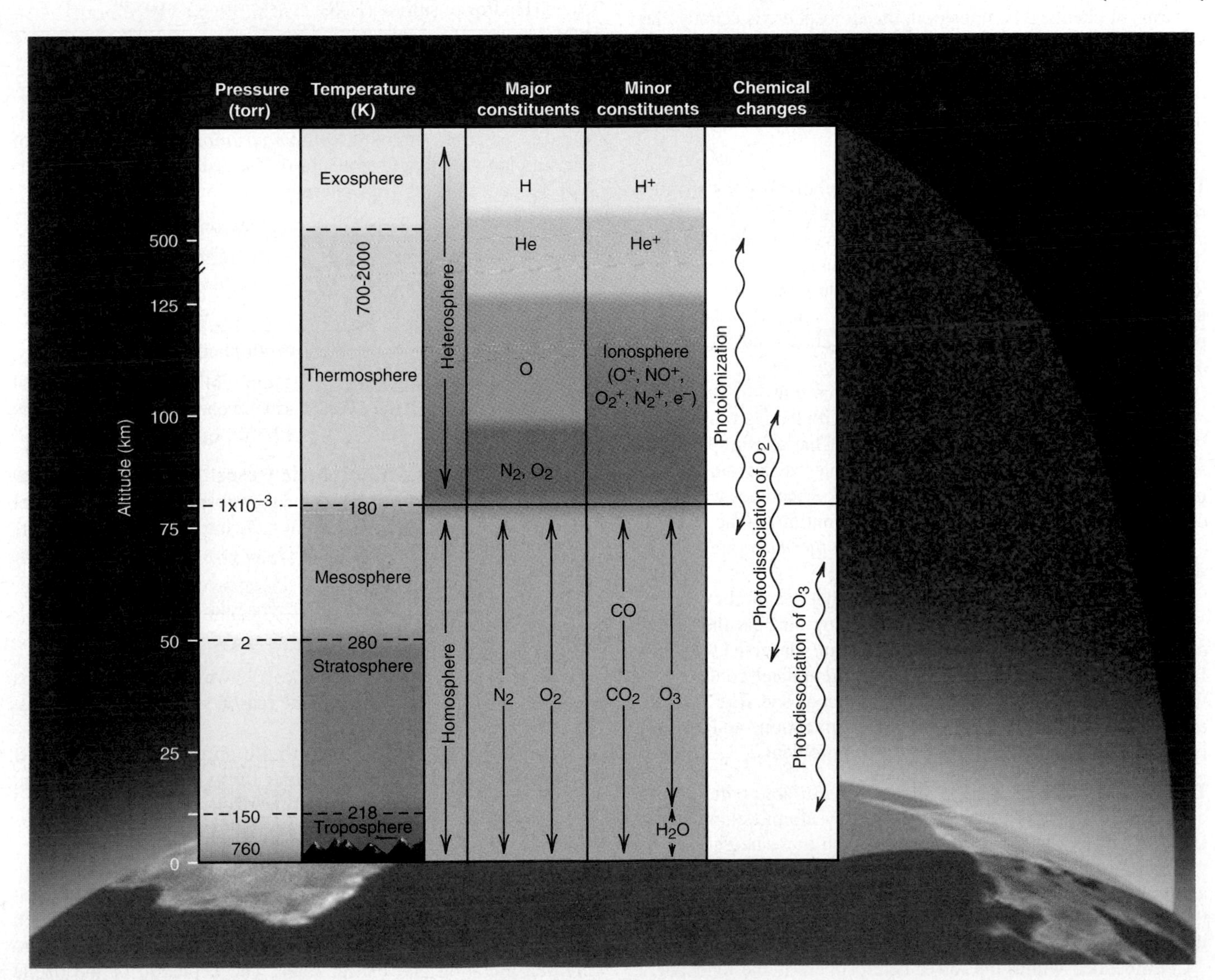

Figure B5.1 **Variations in pressure, temperature, and composition of Earth's atmosphere.**

What does it actually mean to have a temperature of 2000 K at 500 km (300 mi) above Earth's surface? Would a piece of iron (melting point ≈ 1700 K) glow red-hot and melt in the thermosphere? Our everyday use of the words "hot" and "cold" refers to measurements near the surface, where the density of the atmosphere is 10^6 times greater than in the thermosphere. At an altitude of 500 km, where collision frequency is extremely low, a thermometer, or any other object, experiences *very little transfer of kinetic energy.* Thus, the object does not become "hot" in the usual sense; in fact, it is very "cold." Recall that absolute temperature is proportional to the average kinetic energy of the particles. The high-energy solar radiation reaching these outer regions is transferred to relatively few particles, so their average kinetic energy becomes extremely high, as indicated by the high temperature. For this reason, supersonic aircraft do not reach maximum speed until they reach maximum altitude, where the air is less dense, so that collisions with gas molecules are less frequent and the aircraft material becomes less hot.

Variation in Composition

In terms of chemical composition, the atmosphere is usually classified into two major regions, *homosphere* and *heterosphere*. Superimposing the regions defined by temperature on these shows that the homosphere includes the troposphere, stratosphere, and mesosphere, and the heterosphere includes the thermosphere and exosphere (Figure B5.1).

The Homosphere The homosphere has a relatively constant composition, containing, by volume, approximately 78% N_2, 21% O_2, and 1% a mixture of other gases (mostly argon). Under the conditions that occur in the homosphere, the atmospheric gases behave ideally, so volume percent is equal to mole percent (Avogadro's law), and the mole fraction of a component is directly related to its partial pressure (Dalton's law). Table B5.1 shows the components of a sample of clean, dry air at sea level.

The composition of the homosphere is uniform because of *convective mixing*. Air directly in contact with land is warmer than the air above it. The warmer air expands (Charles's law), its density decreases, and it rises through the cooler, denser air, thereby mixing the components. The cooler air sinks, becomes warmer by contact with the land, and the convection continues. The warm air currents that rise from the ground, called *thermals,* are used by soaring birds and glider pilots to stay aloft.

An important effect of convection is that the air above industrialized areas becomes cleaner as the rising air near the surface carries up ground-level pollutants, which are dispersed by winds. However, under certain weather and geographical conditions, a warm air mass remains stationary over a cool one. The resulting *temperature inversion* blocks normal convection, and harmful pollutants build up, causing severe health problems.

The Heterosphere The heterosphere has variable composition, consisting of regions dominated by a few atomic or molecular species. Convective heating does not reach these heights, so the gas particles become layered according to molar mass: nitrogen and oxygen molecules in the lower levels, oxygen atoms (O) in the next, then helium atoms (He), and free hydrogen atoms (H) in the highest level.

Embedded within the lower heterosphere is the ionosphere, containing ionic species such as O^+, NO^+, O_2^+, N_2^+, and free electrons (Figure B5.1). Ionospheric chemistry involves numerous light-induced bond-breaking *(photodissociation)* and light-induced electron-removing *(photoionization)* processes. One of the simpler ways that O atoms form, for instance, involves a four-step sequence that absorbs energy:

Table B5.1 Composition of Clean, Dry Air at Sea Level

Component	Mole Fraction
Nitrogen (N_2)	0.78084
Oxygen (O_2)	0.20946
Argon (Ar)	0.00934
Carbon dioxide (CO_2)	0.00033
Neon (Ne)	1.818×10^{-5}
Helium (He)	5.24×10^{-6}
Methane (CH_4)	2×10^{-6}
Krypton (Kr)	1.14×10^{-6}
Hydrogen (H_2)	5×10^{-7}
Dinitrogen monoxide (N_2O)	5×10^{-7}
Carbon monoxide (CO)	1×10^{-7}
Xenon (Xe)	8×10^{-8}
Ozone (O_3)	2×10^{-8}
Ammonia (NH_3)	6×10^{-9}
Nitrogen dioxide (NO_2)	6×10^{-9}
Nitrogen monoxide (NO)	6×10^{-10}
Sulfur dioxide (SO_2)	2×10^{-10}
Hydrogen sulfide (H_2S)	2×10^{-10}

$$N_2 \longrightarrow N_2^+ + e^- \text{ [photoionization]}$$
$$N_2^+ + e^- \longrightarrow N + N$$
$$N + O_2 \longrightarrow NO + O$$
$$N + NO \longrightarrow N_2 + O$$
$$O_2 \longrightarrow O + O \text{ [overall photodissociation]}$$

When the resulting high-energy O atoms collide with other neutral or ionic components, the average kinetic energy of thermospheric particles increases.

The Importance of Stratospheric Ozone Although most high-energy radiation is absorbed by the thermosphere, a small amount reaches the stratosphere and breaks O_2 into O atoms. The energetic O atoms collide with more O_2 to form ozone (O_3), another molecular form of oxygen:

$$O_2(g) \xrightarrow{\text{high-energy radiation}} 2O(g)$$
$$M + O(g) + O_2(g) \longrightarrow O_3(g) + M$$

where M is any particle that can carry away excess energy. This reaction releases heat, which is the reason stratospheric temperatures increase with altitude.

Stratospheric ozone is vital to life on the surface because it absorbs a great proportion of solar ultraviolet (UV) radiation, which results in decomposition of the ozone:

$$O_3(g) \xrightarrow{\text{UV light}} O_2(g) + O(g)$$

UV radiation is extremely harmful because it is strong enough to break chemical bonds and, thus, interrupt normal biological processes. Without the presence of stratospheric ozone, much more of this radiation would reach the surface, resulting in increased mutation and cancer rates. The depletion of the ozone layer as a result of industrial gases is discussed in Chapter 16.

Earth's Primitive Atmosphere

The composition of the present atmosphere bears little resemblance to that covering the young Earth, but scientists disagree about what that primitive composition actually was: Did the carbon and nitrogen have low oxidation numbers, as in CH_4 (O.N. of C = −4) and NH_3 (O.N. of N = −3)? Or did these atoms have higher oxidation numbers, as in CO_2 (O.N. of C = +4) and N_2 (O.N. of N = 0)? One point generally accepted is that the primitive mixture did not contain free O_2.

Origin-of-life models propose that about 1 billion years after the earliest organisms appeared, blue-green algae evolved. These one-celled plants used solar energy to produce glucose by photosynthesis:

$$6CO_2(g) + 6H_2O(l) \xrightarrow{\text{light}} C_6H_{12}O_6(\text{glucose}) + 6O_2(g)$$

As a result of this reaction, the O_2 content of the atmosphere increased and the CO_2 content decreased. More O_2 allowed more oxidation to occur, which changed the geological and biological makeup of the early Earth. Iron(II) minerals changed to iron(III) minerals, sulfites changed to sulfates, and eventually organisms evolved that could use O_2 to oxidize other organisms to obtain energy. For these organisms to have survived exposure to the more energetic forms of solar radiation (particularly UV radiation), enough O_2 must have formed to create a protective ozone layer. Estimates indicate that the level of O_2 increased to the current level of about 20 mol % approximately 1.5 billion years ago.

A Survey of Planetary Atmospheres

Earth's combination of pressure and temperature and its oxygen-rich atmosphere and watery surface are unique in the Solar System. (Indeed, if similar conditions and composition were discovered on a planet circling any other star, excitement about the possibility of life there would be enormous.) Atmospheres on the Sun's other planets are strikingly different from Earth's. Some, especially those on the outer planets, exist under conditions that cause extreme deviations from ideal gas behavior (Section 5.7). Based on current data from NASA spacecraft and Earth-based observations, Table B5.2 lists conditions and composition of the atmospheres on the planets within the Solar System and on some of their moons.

Table B5.2 Planetary Atmospheres

Planet (Satellite)	Pressure* (atm)	Temperature† (K)	Composition (mol %)
Mercury	$<10^{-12}$	~700 (day) ~100 (night)	He, H_2, O_2, Ar, Ne (Na and K from solar wind)
Venus	~90	~730	CO_2 (96), N_2 (3), He, SO_2, H_2O, Ar, Ne
Earth	1.0	avg. range 250–310	N_2 (78), O_2 (21), Ar (0.9), H_2O, CO_2, Ne, He, CH_4, Kr
(Moon)	$\sim 2\times10^{-14}$	370 (day) 120 (night)	Ne, Ar, He
Mars	7×10^{-3}	300 (summer day) 140 (pole in winter) 218 average	CO_2 (95), N_2(3), Ar (1.6), O_2, H_2O, Ne, CO, Kr
Jupiter	$(\sim 4\times10^{6})$	(~140)	H_2 (89), He (11), CH_4, NH_3, C_2H_6, C_2H_2, PH_3
(Io)	$\sim 10^{-10}$	~110	SO_2, S vapor
Saturn	$(\sim 4\times10^{6})$	(~130)	H_2 (93), He (7), CH_4, NH_3, H_2O, C_2H_6, PH_3
(Titan)	1.6	~94	N_2 (90), Ar (<6), CH_4 (3?), C_2H_6, C_2H_2, C_2H_4, HCN, H_2
Uranus	$(>10^{6})$	(~60)	H_2 (83), He (15), CH_4 (2)
Neptune	$(>10^{6})$	(~60)	H_2 (<90), He (~10), CH_4
Pluto	$\sim 10^{-6}$	~50	N_2, CO, CH_4

*Values in parentheses refer to interior pressures.
†Values in parentheses refer to cloud-top temperatures.

SECTION SUMMARY

The kinetic-molecular theory postulates that gas molecules take up a negligible portion of the gas volume, move in straight-line paths between elastic collisions, and have average kinetic energies proportional to the absolute temperature of the gas. This theory explains the gas laws in terms of changes in distances between molecules and the container walls and changes in molecular speed. Temperature is a measure of the average kinetic energy of molecules. Effusion and diffusion rates are inversely proportional to the square root of the molar mass (Graham's law) because they are directly proportional to molecular speed. Molecular motion is characterized by a temperature-dependent most probable speed within a range of speeds, a mean free path, and a collision frequency. The atmosphere is a complex mixture of gases that exhibits variations in pressure, temperature, and composition with altitude.

5.7 REAL GASES: DEVIATIONS FROM IDEAL BEHAVIOR

A fundamental principle of science is that simpler models are more useful than complex ones—as long as they explain the data. You can certainly appreciate the usefulness of the kinetic-molecular theory. With simple postulates, it explains ideal gas behavior in terms of particles acting like infinitesimal "billiard balls," moving at speeds governed by the absolute temperature, and experiencing only perfectly elastic collisions.

In reality, however, you know that *molecules are not points of mass.* They have volumes determined by the sizes of their atoms and the lengths and directions of their bonds. You also know that atoms contain charged particles, which give rise to *attractive and repulsive forces among molecules.* (In fact, such forces cause substances to undergo changes of states; we'll discuss these forces in great detail in Chapter 12.) Therefore, we expect these real properties of molecules to cause deviations from ideal behavior under some conditions, and this is indeed the case. We must alter the simple model and the ideal gas law to predict gas behavior at low temperatures and very high pressures.

Effects of Extreme Conditions on Gas Behavior

At ordinary conditions—relatively high temperatures and low pressures—most simple gases exhibit nearly ideal behavior. Even at STP (0°C and 1 atm), however, gases deviate *slightly* from ideal behavior. Table 5.4 shows the standard molar volumes of several gases to five significant figures. Note that they do not quite equal the ideal value. The phenomena that cause these slight deviations under standard conditions exert more influence as the temperature decreases toward the condensation point of the gas, the temperature at which it liquefies. As you can see, the largest deviations from ideal behavior in Table 5.4 are for Cl_2 and NH_3 because, at the standard temperature of 0°C, they are already close to their condensation points.

Table 5.4 Molar Volume of Some Common Gases at STP (0°C and 1 atm)

Gas	Molar Volume (L/mol)	Condensation Point (°C)
He	22.435	−268.9
H_2	22.432	−252.8
Ne	22.422	−246.1
Ideal gas	**22.414**	—
Ar	22.397	−185.9
N_2	22.396	−195.8
O_2	22.390	−183.0
CO	22.388	−191.5
Cl_2	22.184	−34.0
NH_3	22.079	−33.4

At pressures greater than 10 atm, we begin to see significant deviations from ideal behavior in many gases. Figure 5.21 shows a plot of PV/RT versus P_{ext} for *1 mol* of several real gases and an ideal gas. For 1 mol of an *ideal* gas, the ratio PV/RT is equal to 1 at any pressure. The values on the horizontal axis are the external pressures at which the PV/RT ratios are calculated. The pressures range from normal (at 1 atm, $PV/RT = 1$) to very high (at ~1000 atm, $PV/RT \approx 1.6$ to 2.3).

The PV/RT curve shown in Figure 5.21 for 1 mol of methane (CH_4) is typical of that for most real gases: it decreases below the ideal value at moderately high pressures and then rises above it as pressure increases further. This shape arises from two overlapping effects of the two characteristics of real molecules just mentioned:

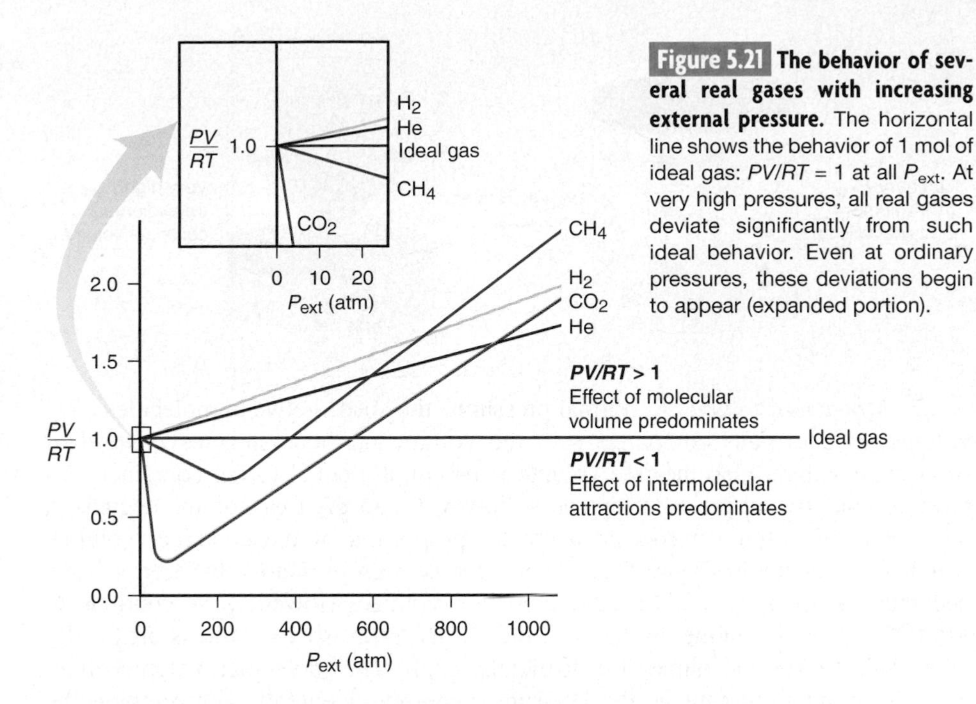

Figure 5.21 The behavior of several real gases with increasing external pressure. The horizontal line shows the behavior of 1 mol of ideal gas: $PV/RT = 1$ at all P_{ext}. At very high pressures, all real gases deviate significantly from such ideal behavior. Even at ordinary pressures, these deviations begin to appear (expanded portion).

1. At moderately high pressure, values of PV/RT lower than ideal (less than 1) are due predominantly to *intermolecular attractions.*
2. At very high pressure, values of PV/RT greater than ideal (more than 1) are due predominantly to *molecular volume.*

Let's examine these effects on the molecular level:

1. *Intermolecular attractions.* Attractive forces between molecules are *much* weaker than the covalent bonding forces that hold a molecule together. Most intermolecular attractions are caused by slight imbalances in electron distributions and are important only over relatively short distances. At normal pressures, the spaces between gas molecules are so large that attractions are negligible, and the gas behaves nearly ideally. As the pressure rises and the volume of the sample decreases, however, the average intermolecular distance becomes smaller and attractions have a greater effect.

Picture a molecule at these higher pressures (Figure 5.22). As it approaches the container wall, nearby molecules attract it, which lessens the force of its impact. *Repeated throughout the sample, this effect results in decreased gas pressure and, thus, a smaller numerator in the PV/RT ratio.* Lowering the temperature has the same effect because it slows the molecules, so attractive forces exert an influence for a longer time. At a low enough temperature, the attractions among molecules become overwhelming, and the gas condenses to a liquid.

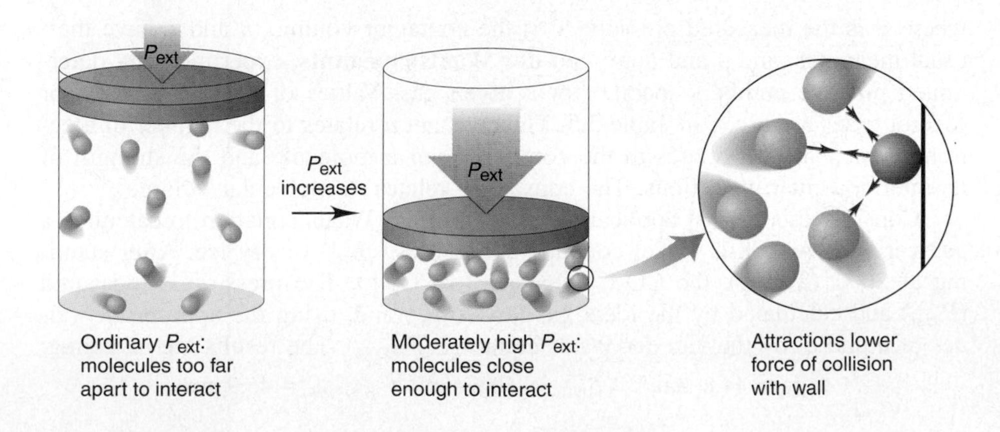

Figure 5.22 The effect of intermolecular attractions on measured gas pressure. At ordinary pressures, the volume is large and gas molecules are too far apart to experience significant attractions. At moderately high external pressures, the volume decreases enough for the molecules to influence each other. As the close-up shows, a gas molecule approaching the container wall experiences intermolecular attractions from neighboring molecules that reduce the force of its impact. As a result, gases exert *less* pressure than the ideal gas law predicts.

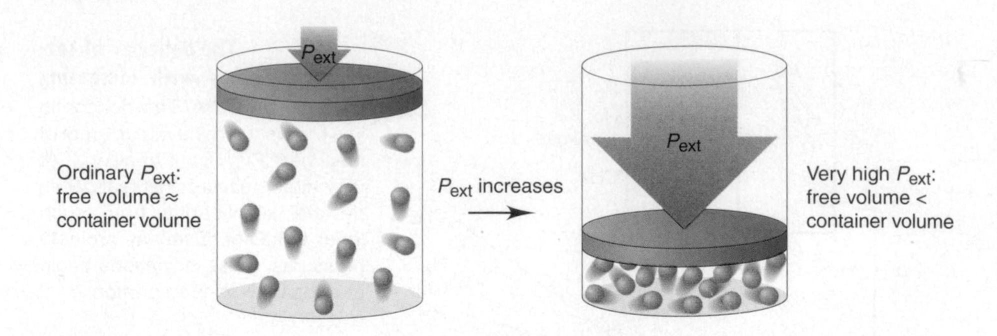

Figure 5.23 The effect of molecular volume on measured gas volume. At ordinary pressures, the volume *between* molecules (free volume) is essentially equal to the container volume because the molecules occupy only a tiny fraction of the available space. At very high external pressures, however, the free volume is significantly *less* than the container volume because of the volume of the molecules themselves.

2. *Molecular volume.* At normal pressures, the space between molecules (free volume) is enormous compared with the volume of the molecules *themselves* (molecular volume), so the free volume is essentially equal to the container volume. As the applied pressure increases, however, and the free volume decreases, the molecular volume makes up a greater proportion of the container volume, which you can see in Figure 5.23. Thus, at very high pressures, the free volume becomes significantly *less* than the container volume. However, we continue to use the container volume as the V in the PV/RT ratio, so the ratio is artificially high. This makes the numerator artificially high. The molecular volume effect becomes more important as the pressure increases, eventually outweighing the effect of the intermolecular attractions and causing PV/RT to rise above the ideal value.

In Figure 5.21, the H_2 and He curves do not show the typical dip at moderate pressures. These gases consist of particles with such weak intermolecular attractions that the molecular volume effect predominates at all pressures.

The van der Waals Equation: The Ideal Gas Law Redesigned

To describe real gas behavior more accurately, we need to "redesign" the ideal gas equation to do two things:

1. Adjust the measured pressure *up* by adding a factor that accounts for intermolecular attractions, and
2. Adjust the measured volume *down* by subtracting a factor from the entire container volume that accounts for the molecular volume.

In 1873, Johannes van der Waals realized the limitations of the ideal gas law and proposed an equation that accounts for the behavior of real gases. The **van der Waals equation** for n moles of a real gas is

$$\underbrace{\left(P + \frac{n^2a}{V^2}\right)}_{\substack{\text{adjusts} \\ P\text{ up}}}\underbrace{(V - nb)}_{\substack{\text{adjusts} \\ V\text{ down}}} = nRT$$

where P is the measured pressure, V is the container volume, n and T have their usual meanings, and a and b are **van der Waals constants,** experimentally determined positive numbers specific for a given gas. Values of these constants for several gases are given in Table 5.5. The constant a relates to the number of electrons, which in turn relates to the complexity of a molecule and the strength of its intermolecular attractions. The constant b relates to molecular volume.

Table 5.5 Van der Waals Constants for Some Common Gases

Gas	$a\left(\frac{\text{atm}\cdot\text{L}^2}{\text{mol}^2}\right)$	$b\left(\frac{\text{L}}{\text{mol}}\right)$
He	0.034	0.0237
Ne	0.211	0.0171
Ar	1.35	0.0322
Kr	2.32	0.0398
Xe	4.19	0.0511
H_2	0.244	0.0266
N_2	1.39	0.0391
O_2	1.36	0.0318
Cl_2	6.49	0.0562
CH_4	2.25	0.0428
CO	1.45	0.0395
CO_2	3.59	0.0427
NH_3	4.17	0.0371
H_2O	5.46	0.0305

Consider this typical application of the van der Waals equation to calculate a gas variable. A 1.98-L vessel contains 215 g (4.89 mol) of dry ice. After standing at 26°C (299 K), the $CO_2(s)$ changes to $CO_2(g)$. The pressure is measured (P_{real}) and calculated by the ideal gas law (P_{IGL}) and, using the appropriate values of a and b, by the van der Waals equation (P_{VDW}). The results are revealing:

$$P_{real} = 44.8 \text{ atm} \qquad P_{IGL} = 60.6 \text{ atm} \qquad P_{VDW} = 45.9 \text{ atm}$$

Comparing the real with each calculated value shows that P_{IGL} is 35.3% greater than P_{real}, but P_{VDW} is only 2.5% greater than P_{real}. At these conditions, CO_2 deviates so much from ideal behavior that the ideal gas law is not very useful.

Here is one final point to realize: According to kinetic-molecular theory, the constants a and b are zero for an ideal gas because the particles do not attract each other and have no volume. Yet, even for a real gas at ordinary pressures, the molecules are very far apart. Thus,

- Attractive forces are miniscule, so $P + \frac{n^2a}{V^2} \approx P$
- The molecular volume is a miniscule fraction of the container volume, so $V - nb \approx V$

Therefore, *at ordinary conditions, the van der Waals equation becomes the ideal gas equation.*

SECTION SUMMARY

At very high pressures or low temperatures, all gases deviate greatly from ideal behavior. As pressure increases, most real gases exhibit first a lower and then a higher *PV/RT* ratio than the value for the same amount (1 mol) of an ideal gas. These deviations are due to attractions between molecules, which lower the pressure (and the ratio), and to the larger fraction of the container volume occupied by the molecules, which increases the ratio. By including parameters characteristic of each gas, the van der Waals equation corrects for these deviations.

Chapter Perspective

As with the atomic model (Chapter 2), we have seen in this chapter how a simple molecular-scale model can explain macroscopic observations and how it often must be revised to predict a wider range of chemical behavior. Gas behavior is relatively easy to understand because gas structure is so randomized—very different from the structures of liquids and solids with their complex molecular interactions, as you'll see in Chapter 12. In Chapter 6, we return to chemical reactions, but from the standpoint of the heat involved in such changes. The meaning of kinetic energy, which we discussed in this chapter, bears directly on this central topic.

For Review and Reference

(Numbers in parentheses refer to pages, unless noted otherwise.)

Learning Objectives

Relevant section and/or sample problem (SP) numbers appear in parentheses.

Understand These Concepts

1. How gases differ in their macroscopic properties from liquids and solids (5.1)
2. The meaning of pressure and the operation of a barometer and a manometer (5.2)
3. The relations among gas variables expressed by Boyle's, Charles's, and Avogadro's laws (5.3)
4. How the individual gas laws are incorporated into the ideal gas law (5.3)
5. How the ideal gas law can be used to study gas density and molar mass (5.4)
6. The relation between the density and the temperature of a gas (5.4)
7. The meaning of Dalton's law and the relation between partial pressure and mole fraction of a gas; how Dalton's law applies to collecting a gas over water (5.4)
8. How the postulates of the kinetic-molecular theory are applied to explain the origin of pressure and the gas laws (5.6)
9. The relations among molecular speed, average kinetic energy, and temperature (5.6)
10. The meanings of *effusion* and *diffusion* and how their rates are related to molar mass (5.6)
11. The relations among mean free path, molecular speed, and collision frequency (5.6)
12. Why intermolecular attractions and molecular volume cause gases to deviate from ideal behavior at low temperatures and high pressures (5.7)
13. How the van der Waals equation corrects the ideal gas law for extreme conditions (5.7)

Problems

Problems with **colored** numbers are answered in Appendix E. Sections match the text and provide the numbers of relevant sample problems. Most offer Concept Review Questions, Skill-Building Exercises (grouped in pairs covering the same concept), and Problems in Context. Comprehensive Problems are based on material from any section or previous chapter.

An Overview of the Physical States of Matter

Concept Review Questions

5.1 How does a sample of gas differ in its behavior from a sample of liquid in each of the following situations?
(a) The sample is transferred from one container to a larger one.
(b) The sample is heated in an expandable container, but no change of state occurs.
(c) The sample is placed in a cylinder with a piston, and an external force is applied.

5.2 Are the particles in a gas farther apart or closer together than the particles in a liquid? Use your answer to this question in order to explain each of the following general observations:
(a) Gases are more compressible than liquids.
(b) Gases have lower viscosities than liquids.
(c) After thorough stirring, all gas mixtures are solutions.
(d) The density of a substance in the gas state is lower than in the liquid state.

Gas Pressure and Its Measurement

(Sample Problem 5.1)

Concept Review Questions

5.3 How does a barometer work? Is the column of mercury in a barometer shorter when it is on a mountaintop or at sea level? Explain.

5.4 How can a unit of length such as millimeter of mercury (mmHg) be used as a unit of pressure, which has the dimensions of force per unit area?

5.5 In a closed-end manometer, the mercury level in the arm attached to the flask can never be higher than the mercury level in the other arm, whereas in an open-end manometer, it *can* be higher. Explain.

Skill-Building Exercises *(grouped in similar pairs)*

5.6 On a cool, rainy day, the barometric pressure is 725 mmHg. Calculate the barometric pressure in centimeters of water (cmH_2O) (*d* of Hg = 13.5 g/mL; *d* of H_2O = 1.00 g/mL).

5.7 A long glass tube, sealed at one end, has an inner diameter of 10.0 mm. The tube is filled with water and inverted into a pail of water. If the atmospheric pressure is 755 mmHg, how high (in mmH_2O) is the column of water in the tube (*d* of Hg = 13.5 g/mL; *d* of H_2O = 1.00 g/mL)?

5.8 Convert the following:
(a) 0.745 atm to mmHg
(b) 992 torr to bar
(c) 365 kPa to atm
(d) 804 mmHg to kPa

5.9 Convert the following:
(a) 74.8 cmHg to atm
(b) 27.0 atm to kPa
(c) 8.50 atm to bar
(d) 0.907 kPa to torr

5.10 In Figure P5.10, what is the pressure of the gas in the flask (in atm) if the barometer reads 738.5 torr?

5.11 In Figure P5.11, what is the pressure of the gas in the flask (in kPa) if the barometer reads 765.2 mmHg?

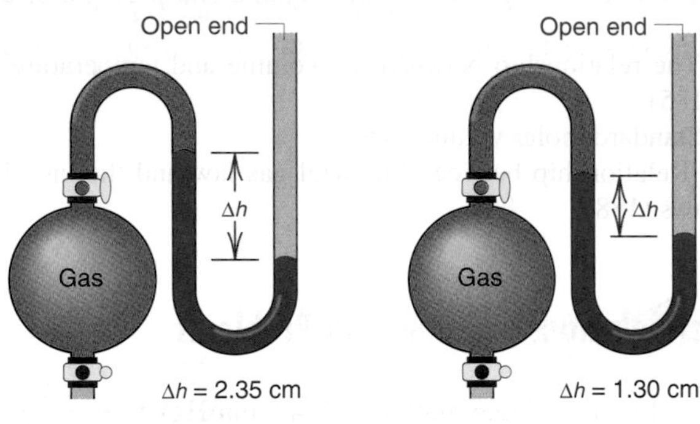

Figure P5.10 Figure P5.11

5.12 If the sample flask in Figure P5.12 is open to the air, what is the atmospheric pressure in atmospheres?

5.13 What is the pressure in pascals of the gas in the flask in Figure P5.13?

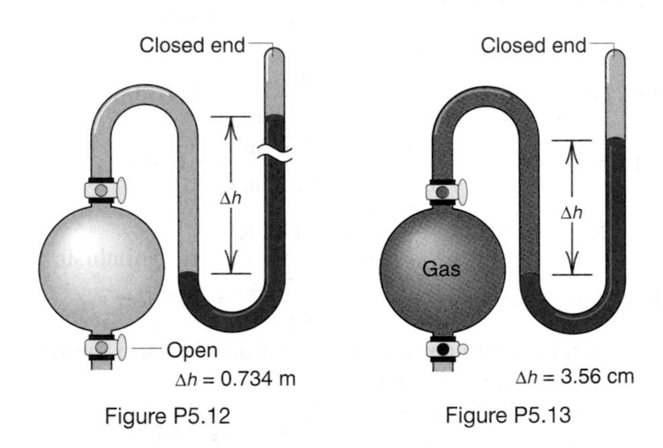

Figure P5.12 Figure P5.13

Problems in Context

5.14 Convert each of the pressures described below into atmospheres:
(a) At the peak of Mt. Everest, atmospheric pressure is only 2.75×10^2 mmHg.
(b) A cyclist fills her bike tires to 91 psi.
(c) The surface of Venus has an atmospheric pressure of 9.15×10^6 Pa.
(d) At 100 ft below sea level, a scuba diver experiences a pressure of 2.44×10^4 torr.

5.15 The gravitational force exerted by an object is given by $F = mg$, where F is the force in newtons, m is the mass in kilograms, and g is the acceleration due to gravity (9.81 m/s^2).
(a) Use the definition of the pascal to calculate the mass (in kg) of the atmosphere on 1 m^2 of ocean.
(b) Osmium ($Z = 76$) has the highest density of any element (22.6 g/mL). If an osmium column is 1 m^2 in area, how high must it be for its pressure to equal atmospheric pressure? [Use the answer from part (a) in your calculation.]

The Gas Laws and Their Experimental Foundations
(Sample Problems 5.2 to 5.6)

Concept Review Questions

5.16 When asked to state Boyle's law, a student replies, "The volume of a gas is inversely proportional to its pressure." How is this statement incomplete? Give a correct statement of Boyle's law.

5.17 Which quantities are variables and which are fixed in each of the following laws: (a) Charles's law; (b) Avogadro's law; (c) Amontons's law?

5.18 Boyle's law relates gas volume to pressure, and Avogadro's law relates gas volume to number of moles. State a relationship between gas pressure and number of moles.

5.19 Each of the following processes caused the gas volume to double, as shown. For each process, state how the remaining gas variable changed or that it remained fixed:
(a) T doubles at fixed P
(b) T and n are fixed
(c) At fixed T, the reaction is $CD_2 \longrightarrow C + D_2$
(d) At fixed P, the reaction is $A_2 + B_2 \longrightarrow 2AB$

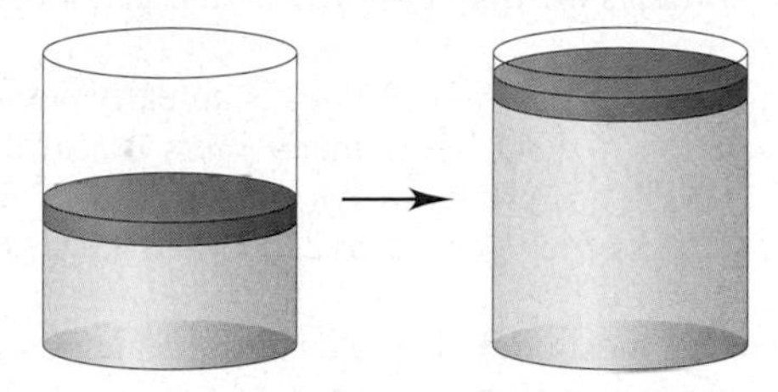

Skill-Building Exercises *(grouped in similar pairs)*

5.20 What is the effect of the following on the volume of 1 mol of an ideal gas?
(a) The pressure is tripled (at constant T).
(b) The absolute temperature is increased by a factor of 2.5 (at constant P).
(c) Two more moles of the gas are added (at constant P and T).

5.21 What is the effect of the following on the volume of 1 mol of an ideal gas?
(a) The pressure is reduced by a factor of 4 (at constant T).
(b) The pressure changes from 760 torr to 202 kPa, and the temperature changes from 37°C to 155 K.
(c) The temperature changes from 305 K to 32°C, and the pressure changes from 2 atm to 101 kPa.

5.22 What is the effect of the following on the volume of 1 mol of an ideal gas?
(a) The temperature is decreased from 700 K to 350 K (at constant P).
(b) The temperature is increased from 350°C to 700°C (at constant P).
(c) The pressure is increased from 2 atm to 8 atm (at constant T).

5.23 What is the effect of the following on the volume of 1 mol of an ideal gas?
(a) Half the gas escapes through a stopcock (at constant P and T).
(b) The initial pressure is 722 torr, and the final pressure is 0.950 atm; the initial temperature is 32°F, and the final temperature is 273 K.
(c) Both the pressure and temperature are reduced by one-fourth of their initial values.

5.24 A sample of sulfur hexafluoride gas occupies a volume of 5.10 L at 198°C. Assuming that the pressure remains constant, what temperature (in °C) is needed to reduce the volume to 2.50 L?

5.25 A 93-L sample of dry air is cooled from 145°C to −22°C while the pressure is maintained at 2.85 atm. What is the final volume?

5.26 A sample of Freon-12 (CF_2Cl_2) occupies 25.5 L at 298 K and 153.3 kPa. Find its volume at STP.

5.27 Calculate the volume of a sample of carbon monoxide that is at −14°C and 367 torr if it occupies 3.65 L at 298 K and 745 torr.

5.28 A sample of chlorine gas is confined in a 5.0-L container at 228 torr and 27°C. How many moles of gas are in the sample?

5.29 If 1.47×10^{-3} mol of argon occupies a 75.0-mL container at 26°C, what is the pressure (in torr)?

5.30 You have 207 mL of chlorine trifluoride gas at 699 mmHg and 45°C. What is the mass (in g) of the sample?

5.31 A 75.0-g sample of dinitrogen monoxide is confined in a 3.1-L vessel. What is the pressure (in atm) at 115°C?

Problems in Context

5.32 In preparation for a demonstration, your professor brings a 1.5-L bottle of sulfur dioxide into the lecture hall before class to allow the gas to reach room temperature. If the pressure gauge reads 85 psi and the lecture hall is 23°C, how many moles of sulfur dioxide are in the bottle? (*Hint:* The gauge reads zero when 14.7 psi of gas remains.)

5.33 A gas-filled weather balloon with a volume of 55.0 L is released at sea-level conditions of 755 torr and 23°C. The balloon can expand to a maximum volume of 835 L. When the balloon rises to an altitude at which the temperature is −5°C and the pressure is 0.066 atm, will it reach its maximum volume?

Further Applications of the Ideal Gas Law
(Sample Problems 5.7 to 5.10)

Concept Review Questions

5.34 Why is moist air less dense than dry air?

5.35 To collect a beaker of H_2 gas by displacing the air already in the beaker, would you hold the beaker upright or inverted? Why? How would you hold the beaker to collect CO_2?

5.36 Why can we use a gas mixture, such as air, to study the general behavior of an ideal gas under ordinary conditions?

5.37 How does the partial pressure of gas A in a mixture compare to its mole fraction in the mixture? Explain.

Skill-Building Exercises *(grouped in similar pairs)*

5.38 What is the density of Xe gas at STP?

5.39 What is the density of Freon-11 ($CFCl_3$) at 120°C and 1.5 atm?

5.40 How many moles of gaseous arsine (AsH_3) will occupy 0.0400 L at STP? What is the density of gaseous arsine?

5.41 The density of a noble gas is 2.71 g/L at 3.00 atm and 0°C. Identify the gas.

5.42 Calculate the molar mass of a gas at 388 torr and 45°C if 206 ng occupies 0.206 μL.

5.43 When an evacuated 63.8-mL glass bulb is filled with a gas at 22°C and 747 mmHg, the bulb gains 0.103 g in mass. Is the gas N_2, Ne, or Ar?

5.44 When 0.600 L of Ar at 1.20 atm and 227°C is mixed with 0.200 L of O_2 at 501 torr and 127°C in a 400-mL flask at 27°C, what is the pressure in the flask?

5.45 A 355-mL container holds 0.146 g of Ne and an unknown amount of Ar at 35°C and a total pressure of 626 mmHg. Calculate the moles of Ar present.

Problems in Context

5.46 The air in a hot-air balloon at 744 torr is heated from 17°C to 60.0°C. Assuming that the moles of air and the pressure remain constant, what is the density of the air at each temperature? (The average molar mass of air is 28.8 g/mol.)

5.47 On a certain winter day in Utah, the average atmospheric pressure is 650. torr. What is the molar density (in mol/L) of air if the temperature is −25°C?

5.48 A sample of a liquid hydrocarbon known to consist of molecules with five carbon atoms is vaporized in a 0.204-L flask by immersion in a water bath at 101°C. The barometric pressure is 767 torr, and the remaining gas condenses to 0.482 g of liquid. What is the molecular formula of the hydrocarbon?

5.49 A sample of air contains 78.08% nitrogen, 20.94% oxygen, 0.05% carbon dioxide, and 0.93% argon, by volume. How many molecules of each gas are present in 1.00 L of the sample at 25°C and 1.00 atm?

5.50 An environmental chemist is sampling industrial exhaust gases from a coal-burning plant. He collects a CO_2-SO_2-H_2O mixture in a 21-L steel tank until the pressure reaches 850. torr at 45°C.
(a) How many moles of gas are collected?
(b) If the SO_2 concentration in the mixture is 7.95×10^3 parts per million by volume (ppmv), what is its partial pressure? [*Hint:* ppmv = (volume of component/volume of mixture) $\times 10^6$.]

The Ideal Gas Law and Reaction Stoichiometry
(Sample Problems 5.11 and 5.12)

Skill-Building Exercises *(grouped in similar pairs)*

5.51 How many grams of phosphorus react with 35.5 L of O_2 at STP to form tetraphosphorus decaoxide?

$$P_4(s) + 5O_2(g) \longrightarrow P_4O_{10}(s)$$

5.52 How many grams of potassium chlorate decompose to potassium chloride and 638 mL of O_2 at 128°C and 752 torr?

$$2KClO_3(s) \longrightarrow 2KCl(s) + 3O_2(g)$$

5.53 How many grams of phosphine (PH_3) can form when 37.5 g of phosphorus and 83.0 L of hydrogen gas react at STP?

$$P_4(s) + H_2(g) \longrightarrow PH_3(g) \quad \text{[unbalanced]}$$

5.54 When 35.6 L of ammonia and 40.5 L of oxygen gas at STP burn, nitrogen monoxide and water are produced. After the products return to STP, how many grams of nitrogen monoxide are present?

$$NH_3(g) + O_2(g) \longrightarrow NO(g) + H_2O(l) \quad \text{[unbalanced]}$$

5.55 Aluminum reacts with excess hydrochloric acid to form aqueous aluminum chloride and 35.8 mL of hydrogen gas over water at 27°C and 751 mmHg. How many grams of aluminum reacted?

5.56 How many liters of hydrogen gas are collected over water at 18°C and 725 mmHg when 0.84 g of lithium reacts with water? Aqueous lithium hydroxide also forms.

Problems in Context

5.57 "Strike anywhere" matches contain the compound tetraphosphorus trisulfide, which burns to form tetraphosphorus decaoxide and sulfur dioxide gas. How many milliliters of sulfur dioxide, measured at 725 torr and 32°C, can be produced from burning 0.800 g of tetraphosphorus trisulfide?

5.58 Freon-12 (CF_2Cl_2), which has been widely used as a refrigerant and aerosol propellant, is a dangerous air pollutant. In the troposphere, it traps heat 25 times as effectively as CO_2, and in the stratosphere, it participates in the breakdown of ozone. Freon-12 is prepared industrially by reaction of gaseous carbon tetrachloride with hydrogen fluoride. Hydrogen chloride gas also forms. How many grams of carbon tetrachloride are required for the production of 16.0 dm^3 of Freon-12 at 27°C and 1.20 atm?

5.59 Xenon hexafluoride was one of the first noble gas compounds synthesized. The solid reacts rapidly with the silicon dioxide in glass or quartz containers to form liquid $XeOF_4$ and gaseous silicon tetrafluoride. What is the pressure in a 1.00-L container at 25°C after 2.00 g of xenon hexafluoride reacts? (Assume that silicon tetrafluoride is the only gas present and that it occupies the entire volume.)

5.60 Roasting galena [lead(II) sulfide] is an early step in the industrial isolation of lead. How many liters of sulfur dioxide, measured at STP, are produced by the reaction of 3.75 kg of galena with 228 L of oxygen gas at 220°C and 2.0 atm? Lead(II) oxide also forms.

5.61 In one of his most critical studies into the nature of combustion, Lavoisier heated mercury(II) oxide and isolated elemental mercury and oxygen gas. If 40.0 g of mercury(II) oxide is heated in a 502-mL vessel and 20.0% (by mass) decomposes, what is the pressure (in atm) of the oxygen that forms at 25.0°C? (Assume that the gas occupies the entire volume.)

The Kinetic-Molecular Theory: A Model for Gas Behavior
(Sample Problem 5.13)

Concept Review Questions

5.62 Use the kinetic-molecular theory to explain the change in gas pressure that results from warming a sample of gas.

5.63 How does the kinetic-molecular theory explain why 1 mol of krypton and 1 mol of helium have the same volume at STP?

5.64 Is the rate of effusion of a gas higher than, lower than, or equal to its rate of diffusion? Explain. For two gases with molecules of approximately the same size, is the ratio of their effusion rates higher than, lower than, or equal to the ratio of their diffusion rates? Explain.

5.65 Consider two 1-L samples of gas: one is H_2 and the other is O_2. Both are at 1 atm and 25°C. How do the samples compare in terms of (a) mass, (b) density, (c) mean free path, (d) average molecular kinetic energy, (e) average molecular speed, and (f) time for a given fraction of molecules to effuse?

5.66 Three 5-L flasks, fixed with pressure gauges and small valves, each contain 4 g of gas at 273 K. Flask A contains H_2, flask B contains He, and flask C contains CH_4. Rank the flask contents in terms of (a) pressure, (b) average molecular kinetic energy, (c) diffusion rate after the valve is opened, (d) total kinetic energy of the molecules, (e) density, and (f) collision frequency.

Skill-Building Exercises *(grouped in similar pairs)*

5.67 What is the ratio of effusion rates for the lightest gas, H_2, and the heaviest known gas, UF_6?

5.68 What is the ratio of effusion rates for O_2 and Kr?

5.69 The graph below shows the distribution of molecular speeds for argon and helium at the same temperature.

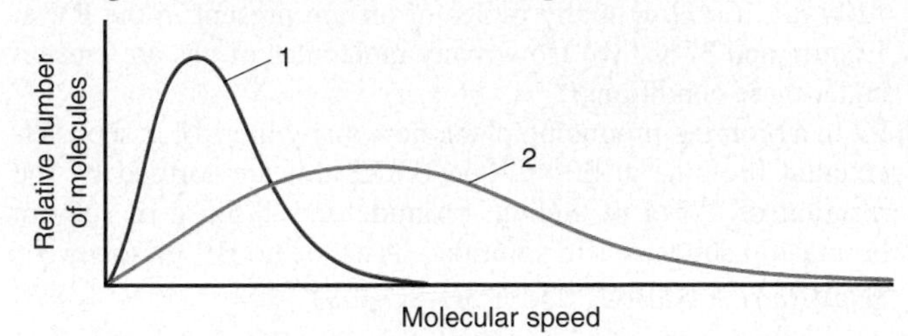

(a) Does curve 1 or 2 better represent the behavior of argon?
(b) Which curve represents the gas that effuses more slowly?
(c) Which curve more closely represents the behavior of fluorine gas? Explain.

5.70 The graph below shows the distribution of molecular speeds for a gas at two different temperatures.

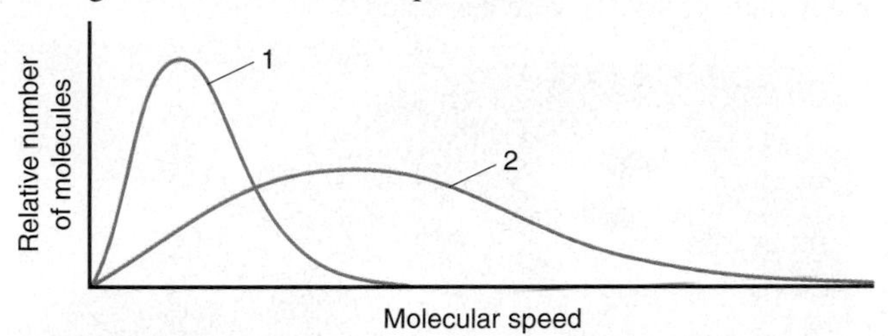

(a) Does curve 1 or 2 better represent the behavior of the gas at the lower temperature?
(b) Which curve represents the sample with the higher $\overline{E}_k$?
(c) Which curve represents the sample that diffuses more quickly?

5.71 At a given pressure and temperature, it takes 4.55 min for a 1.5-L sample of He to effuse through a membrane. How long does it take for 1.5 L of F_2 to effuse under the same conditions?

5.72 A sample of an unknown gas effuses in 11.1 min. An equal volume of H_2 in the same apparatus at the same temperature and pressure effuses in 2.42 min. What is the molar mass of the unknown gas?

Problems in Context

5.73 Solid white phosphorus melts and then vaporizes at high temperature. Gaseous white phosphorus effuses at a rate that is 0.404 times that of neon in the same apparatus under the same conditions. How many atoms are in a molecule of gaseous white phosphorus?

5.74 Helium is the lightest noble gas component of air, and xenon is the heaviest. [For this problem, use $R = 8.314$ J/(mol·K) and $\mathcal{M}$ in kg/mol.]
(a) Calculate the rms speed of helium in winter (0.°C) and in summer (30.°C).
(b) Compare the rms speed of helium with that of xenon at 30.°C.
(c) Calculate the average kinetic energy per mole of helium and of xenon at 30.°C.
(d) Calculate the average kinetic energy per molecule of helium at 30.°C.

Real Gases: Deviations from Ideal Behavior

Skill-Building Exercises *(grouped in similar pairs)*

5.75 Do intermolecular attractions cause negative or positive deviations from the *PV/RT* ratio of an ideal gas? Use data from Table 5.5 to rank Kr, CO_2, and N_2 in order of increasing magnitude of these deviations.

5.76 Does molecular size cause negative or positive deviations from the *PV/RT* ratio of an ideal gas? Use data from Table 5.5 to rank Cl_2, H_2, and O_2 in order of increasing magnitude of these deviations.

5.77 Does N_2 behave more ideally at 1 atm or at 500 atm? Explain.

5.78 Does SF_6 (boiling point = 16°C at 1 atm) behave more ideally at 150°C or at 20°C? Explain.

Comprehensive Problems

5.79 An "empty" gasoline can with dimensions 15.0 cm by 40.0 cm by 12.5 cm is attached to a vacuum pump and evacuated. If the atmospheric pressure is 14.7 lb/in^2, what is the total force (in pounds) on the outside of the can?

5.80 Hemoglobin is the protein that transports O_2 through the blood from the lungs to the rest of the body. In doing so, each molecule of hemoglobin combines with four molecules of O_2. If 1.00 g of hemoglobin combines with 1.53 mL of O_2 at 37°C and 743 torr, what is the molar mass of hemoglobin?

5.81 A baker uses sodium hydrogen carbonate (baking soda) as the leavening agent in a banana-nut quickbread. The baking soda decomposes according to two possible reactions:

Reaction 1. $2NaHCO_3(s) \longrightarrow Na_2CO_3(s) + H_2O(l) + CO_2(g)$

Reaction 2. $NaHCO_3(s) + H^+(aq) \longrightarrow H_2O(l) + CO_2(g) + Na^+(aq)$

Calculate the volume (in mL) of CO_2 that forms at 200.°C and 0.975 atm per gram of $NaHCO_3$ by each of the reaction processes.

5.82 A weather balloon containing 600. L of He is released near the equator at 1.01 atm and 305 K. It rises to a point where conditions are 0.489 atm and 218 K and eventually lands in the northern hemisphere under conditions of 1.01 atm and 250 K. If one-fourth of the helium leaked out during this journey, what is the volume (in L) of the balloon at landing?

5.83 Chlorine is produced from concentrated seawater by the electrochemical chlor-alkali process. During the process, the chlorine is collected in a container that is isolated from the other products to prevent unwanted (and explosive) reactions. If a 15.00-L container holds 0.5850 kg of Cl_2 gas at 225°C, calculate

(a) P_{IGL} (b) P_{VDW} $\left(\text{use } R = 0.08206 \frac{\text{atm·L}}{\text{mol·K}}\right)$

5.84 In a certain experiment, magnesium boride (Mg_3B_2) reacted with acid to form a mixture of four boron hydrides (B_xH_y), three as liquids (labeled I, II, and III) and one as a gas (IV).
(a) When a 0.1000-g sample of each liquid was transferred to an evacuated 750.0-mL container and volatilized at 70.00°C, sample I had a pressure of 0.05951 atm, sample II 0.07045 atm, and sample III 0.05767 atm. What is the molar mass of each liquid?
(b) The mass of boron was found to be 85.63% in sample I, 81.10% in II, and 82.98% in III. What is the molecular formula of each sample?
(c) Sample IV was found to be 78.14% boron. Its rate of effusion was compared to that of sulfur dioxide and under identical

conditions, 350.0 mL of sample IV effused in 12.00 min and 250.0 mL of sulfur dioxide effused in 13.04 min. What is the molecular formula of sample IV?

5.85 Will the volume of a gas increase, decrease, or remain unchanged for each of the following sets of changes?
(a) The pressure is decreased from 2 atm to 1 atm, while the temperature is decreased from 200°C to 100°C.
(b) The pressure is increased from 1 atm to 3 atm, while the temperature is increased from 100°C to 300°C.
(c) The pressure is increased from 3 atm to 6 atm, while the temperature is increased from −73°C to 127°C.
(d) The pressure is increased from 0.2 atm to 0.4 atm, while the temperature is decreased from 300°C to 150°C.

5.86 When air is inhaled, it enters the alveoli of the lungs, and varying amounts of the component gases exchange with dissolved gases in the blood. The resulting alveolar gas mixture is quite different from the atmospheric mixture. The following table presents selected data on the composition and partial pressure of four gases in the atmosphere and in the alveoli:

	Atmosphere (sea level)		Alveoli	
Gas	Mol %	Partial Pressure (torr)	Mol %	Partial Pressure (torr)
N_2	78.6	—	—	569
O_2	20.9	—	—	104
CO_2	0.04	—	—	40
H_2O	0.46	—	—	47

If the total pressure of each gas mixture is 1.00 atm, calculate the following:
(a) The partial pressure (in torr) of each gas in the atmosphere
(b) The mol % of each gas in the alveoli
(c) The number of O_2 molecules in 0.50 L of alveolar air (volume of an average breath at rest) at 37°C

5.87 Radon (Rn) is the heaviest, and only radioactive, member of Group 8A(18) (noble gases). It is a product of the disintegration of heavier radioactive nuclei found in minute concentrations in many common rocks used for building and construction. In recent years, health concerns about the cancers caused from inhaled residential radon have grown. If 1.0×10^{15} atoms of radium (Ra) produce an average of 1.373×10^{4} atoms of Rn per second, how many liters of Rn, measured at STP, are produced per day by 1.0 g of Ra?

5.88 At 1400. mmHg and 286 K, a skin diver exhales a 208-mL bubble of air that is 77% N_2, 17% O_2, and 6.0% CO_2 by volume.
(a) How many milliliters would the volume of the bubble be if it were exhaled at the surface at 1 atm and 298 K?
(b) How many moles of N_2 are in the bubble?

5.89 The mass of Earth's atmosphere is estimated as 5.14×10^{15} t (1 t = 1000 kg).
(a) The average molar mass of air is 28.8 g/mol. How many moles of gas are in the atmosphere?
(b) How many liters would the atmosphere occupy at 25°C and 1 atm?
(c) If the surface area of Earth is 5.100×10^{8} km^2, how high should this volume of air extend? Why does the atmosphere actually extend much higher?

5.90 Nitrogen dioxide is used industrially to produce nitric acid, but it contributes to acid rain and photochemical smog. What volume of nitrogen dioxide is formed at 735 torr and 28.2°C by reacting 4.95 cm^3 of copper (d = 8.95 g/cm^3) with 230.0 mL of nitric acid (d = 1.42 g/cm^3, 68.0% HNO_3 by mass):

$$Cu(s) + 4HNO_3(aq) \longrightarrow Cu(NO_3)_2(aq) + 2NO_2(g) + 2H_2O(l)$$

5.91 In the average adult male, the residual volume (RV) of the lungs, the volume of air remaining after a forced exhalation, is 1200 mL. (a) How many moles of air are present in the RV at 1.0 atm and 37°C? (b) How many molecules of gas are present under these conditions?

5.92 In a bromine-producing plant, how many liters of gaseous elemental bromine at 300°C and 0.855 atm are formed by the reaction of 275 g of sodium bromide and 175.6 g of sodium bromate in aqueous acid solution? (Assume no Br_2 dissolves.)

$$5NaBr(aq) + NaBrO_3(aq) + 3H_2SO_4(aq) \longrightarrow 3Br_2(g) + 3Na_2SO_4(aq) + 3H_2O(g)$$

5.93 In a collision of sufficient force, automobile air bags respond by electrically triggering the explosive decomposition of sodium azide (NaN_3) to its elements. A 50.0-g sample of sodium azide was decomposed, and the nitrogen gas generated was collected over water at 26°C. The total pressure was 745.5 mmHg. How many liters of dry N_2 were generated?

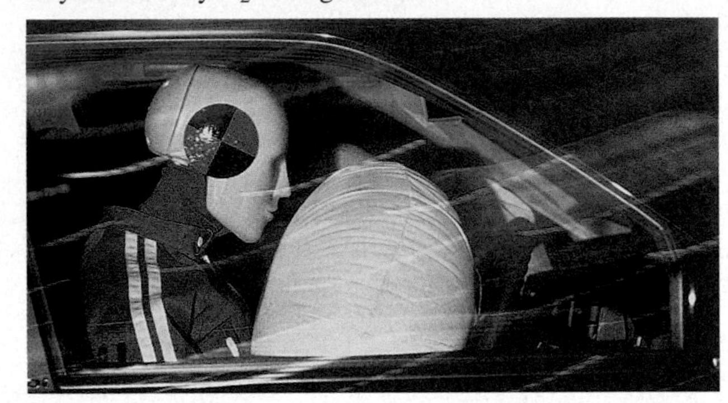

5.94 An anesthetic gas contains 64.81% carbon, 13.60% hydrogen, and 21.59% oxygen, by mass. If 2.00 L of the gas at 25°C and 0.420 atm weighs 2.57 g, what is the molecular formula of the anesthetic?

5.95 Aluminum chloride is easily vaporized at temperatures above 180°C. The gas escapes through a pinhole 0.122 times as fast as helium at the same conditions of temperature and pressure in the same apparatus. What is the molecular formula of gaseous aluminum chloride?

5.96 (a) What is the total volume of gaseous *products*, measured at 350°C and 735 torr, when an automobile engine burns 100. g of C_8H_{18} (a typical component of gasoline)?
(b) For part (a), the source of O_2 is air, which is about 78% N_2, 21% O_2, and 1.0% Ar by volume. Assuming all the O_2 reacts, but none of the N_2 or Ar does, what is the total volume of gaseous *exhaust*?

5.97 An atmospheric chemist studying the reactions of the pollutant SO_2 places a mixture of SO_2 and O_2 in a 2.00-L container at 900. K and an initial pressure of 1.95 atm. When the reaction occurs, gaseous SO_3 forms, and the pressure eventually falls to 1.65 atm. How many moles of SO_3 form?

5.98 Liquid nitrogen trichloride is heated in a 2.50-L closed reaction vessel until it decomposes completely to gaseous elements. The resulting mixture exerts a pressure of 754 mmHg at 95°C.
(a) What is the partial pressure of each gas in the container?
(b) What is the mass of the original sample?

5.99 Ammonium nitrate, a common fertilizer, is used as an explosive in fireworks and by terrorists. It was the material used in the tragic explosion of the Oklahoma City federal building in 1995. How many liters of gas at 307°C and 1.00 atm are formed by the explosive decomposition of 15.0 kg of ammonium nitrate to nitrogen, oxygen, and water vapor?

5.100 An environmental engineer analyzes a sample of air contaminated with sulfur dioxide. To a 500.-mL sample at 700. torr and 38°C, she adds 20.00 mL of 0.01017 *M* aqueous iodine, which reacts as follows:

$$SO_2(g) + I_2(aq) + H_2O(l) \longrightarrow HSO_4^-(aq) + I^-(aq) + H^+(aq) \quad \text{[unbalanced]}$$

Excess I_2 reacts with 11.37 mL of 0.0105 *M* sodium thiosulfate:

$$I_2(aq) + S_2O_3^{2-}(aq) \longrightarrow I^-(aq) + S_4O_6^{2-}(aq) \quad \text{[unbalanced]}$$

What is the volume % of SO_2 in the air sample?

5.101 Canadian chemists have developed a modern variation of the 1899 Mond process for preparing extremely pure metallic nickel. A sample of impure nickel reacts with carbon monoxide at 50°C to form gaseous nickel carbonyl, $Ni(CO)_4$.
(a) How many grams of nickel can be converted to the carbonyl with 3.55 m^3 of CO at 100.7 kPa?
(b) The carbonyl is then decomposed at 21 atm and 155°C to pure (>99.95%) nickel. How many grams of nickel are obtained per cubic meter of the carbonyl?
(c) The released carbon monoxide is cooled and collected for reuse by passing it through water at 35°C. If the barometric pressure is 769 torr, what volume (in m^3) of CO is formed per cubic meter of carbonyl?

5.102 Analysis of a newly discovered gaseous silicon-fluorine compound shows that it contains 33.01 mass % silicon. At 27°C, 2.60 g of the compound exerts a pressure of 1.50 atm in a 0.250-L vessel. What is the molecular formula of the compound?

5.103 A gaseous organic compound containing only carbon, hydrogen, and nitrogen is burned in oxygen gas, and the individual volume of each reactant and product is measured under the same conditions of temperature and pressure. Reaction of four volumes of the compound produces four volumes of CO_2, two volumes of N_2, and ten volumes of water vapor. (a) What volume of oxygen gas was required? (b) What is the empirical formula of the compound?

5.104 A piece of dry ice (solid CO_2, d = 0.900 g/mL) weighing 10.0 g is placed in a 0.800-L bottle filled with air at 0.980 atm and 550.0°C. The bottle is capped, and the dry ice changes to gas. What is the final pressure inside the bottle?

5.105 Containers A, B, and C are attached by closed stopcocks of negligible volume.

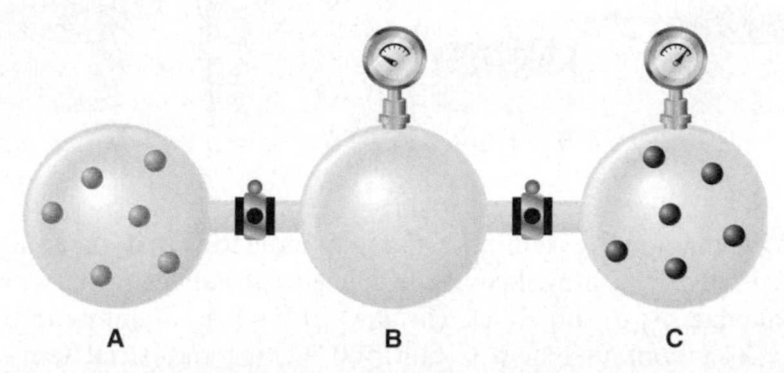

If each particle shown in the picture represents 10^6 particles, (a) How many blue particles and black particles are in B after the stopcocks are opened and the system reaches equilibrium? (b) How many blue particles and black particles are in A after the stopcocks are opened and the system reaches equilibrium? (c) If the pressure in C, P_C, is 750 torr before the stopcocks are opened, what is P_C afterward? (d) What is P_B afterward?

5.106 At the temperatures that exist in the thermosphere (see Figure B5.1), instruments would break down and astronauts would be killed. Yet satellites function in orbit there for many years and astronauts routinely repair equipment on space walks. Explain.

5.107 By what factor would a scuba diver's lungs expand if she ascended rapidly to the surface from a depth of 125 ft without inhaling or exhaling? If an expansion factor greater than 1.5 causes lung rupture, how far could she safely ascend from 125 ft without breathing? Assume constant temperature (d of seawater = 1.04 g/mL; d of Hg = 13.5 g/mL).

5.108 When 15.0 g of fluorite (CaF_2) reacts with excess sulfuric acid, hydrogen fluoride gas is collected at 744 torr and 25.5°C. Solid calcium sulfate is the other product. What gas temperature is required to store the gas in an 8.63-L container at 875 torr?

5.109 Dilute aqueous hydrogen peroxide is used as a bleaching agent and for disinfecting surfaces and small cuts. Its concentration is sometimes given as a certain number of "volumes hydrogen peroxide," which refers to the number of volumes of O_2 gas, measured at STP, that a given volume of hydrogen peroxide solution will release when it decomposes to O_2 and liquid H_2O. How many grams of hydrogen peroxide are in 0.100 L of "20 volumes hydrogen peroxide" solution?

5.110 At a height of 300 km above Earth's surface, an astronaut finds that the atmospheric pressure is about 10^{-8} mmHg and the temperature is 500 K. How many molecules of gas are there per milliliter at this altitude?

5.111 (a) What is the rms speed of O_2 molecules at STP? (b) If the mean free path of O_2 molecules at STP is 6.33×10^{-8} m, what is their collision frequency? [Use R = 8.314 J/(mol·K) and $\mathcal{M}$ in kg/mol.]

5.112 Standard conditions are based on relevant environmental conditions. If normal average surface temperature and pressure on Venus are 730. K and 90 atm, respectively, what is the standard molar volume of an ideal gas on Venus?

5.113 A barometer tube is 1.00×10^2 cm long and has a cross-sectional area of 1.20 cm^2. The height of the mercury column is 74.0 cm, and the temperature is 24°C. A small amount of N_2 is introduced into the evacuated space above the mercury, which causes the mercury level to drop to a height of 66.0 cm. How many grams of N_2 were introduced?

5.114 What is the molar concentration of the cleaning solution formed when 10.0 L of ammonia gas at 31°C and 735 torr dissolves in enough water to give a final volume of 0.750 L?

5.115 The Hawaiian volcano Kilauea emits an average of 1.5×10^3 m^3 of gas each day, when corrected to 298 K and 1.00 atm. The mixture contains gases that contribute to global warming and acid rain, and some are toxic. An atmospheric chemist analyzes a sample and finds the following mole fractions: 0.4896 CO_2, 0.0146 CO, 0.3710 H_2O, 0.1185 SO_2, 0.0003 S_2, 0.0047 H_2, 0.0008 HCl, and 0.0003 H_2S. How many metric tons (t) of each gas is emitted per year (1 t = 1000 kg)?

5.116 To study a key fuel-cell reaction, a chemical engineer has 20.0-L tanks of H_2 and of O_2 and wants to use up both tanks to form 28.0 mol of water at 23.8°C. (a) Use the ideal gas law to find the pressure needed in each tank. (b) Use the van der Waals equation to find the pressure needed in each tank. (c) Compare the results from the two equations.

5.117 For each of the following, which shows the greater deviation from ideal behavior at the same set of conditions? Explain your choice.
(a) Argon or xenon (b) Water vapor or neon
(c) Mercury vapor or radon (d) Water vapor or methane

5.118 How many liters of gaseous hydrogen bromide at 27°C and 0.975 atm will a chemist need if she wishes to prepare 3.50 L of 1.20 *M* hydrobromic acid?

5.119 A mixture consisting of 7.0 g of CO and 10.0 g of SO_2, two atmospheric pollutants, has a pressure of 0.33 atm when placed in a sealed container. What is the partial pressure of CO?

5.120 Sulfur dioxide is used primarily to make sulfuric acid. One method of producing it is by roasting mineral sulfides, for example,

$$FeS_2(s) + O_2(g) \longrightarrow SO_2(g) + Fe_2O_3(s) \qquad \text{[unbalanced]}$$

A production error leads to the sulfide being placed in a 950-L vessel with insufficient oxygen. The partial pressure of O_2 is 0.64 atm and the total pressure is initially 1.05 atm, with the balance N_2. The reaction is run until 85% of the O_2 is consumed, and the vessel is then cooled to its initial temperature. What is the total pressure and partial pressure of each gas in the vessel?

5.121 A mixture of CO_2 and Kr weighs 35.0 g and exerts a pressure of 0.708 atm in its container. Since Kr is expensive, you wish to recover it from the mixture. After the CO_2 is completely removed by absorption with NaOH(*s*), the pressure in the container is 0.250 atm. How many grams of CO_2 were originally present? How many grams of Kr can you recover?

5.122 When a car accelerates quickly, the passengers feel a force that presses them back into their seats, but a balloon filled with helium floats forward. Why?

5.123 Gases such as CO are gradually oxidized in the atmosphere, not by O_2 but by the hydroxyl radical, ·OH, a hydroxide ion with one fewer electron. At night, the radical's concentration is nearly zero, but during the day, it increases to 8.0×10^{12} molecules/m^3. At daytime conditions of 1.00 atm and 22°C, what is the partial pressure and mole percent of the hydroxyl radical?

5.124 Aqueous sulfurous acid (H_2SO_3) was made by dissolving 0.200 L of sulfur dioxide gas at 20.°C and 740. mmHg in water to yield 500.0 mL of solution. The acid solution required 10.0 mL of sodium hydroxide solution to reach the titration end point. What was the molarity of the sodium hydroxide solution?

5.125 During World War II, a portable source of hydrogen gas was needed for weather balloons, and solid metal hydrides were the most convenient form. Many metal hydrides react with water to generate the metal hydroxide and hydrogen. Two candidates were lithium hydride and magnesium hydride. What volume of gas is formed from 1.00 lb of each hydride at 750. torr and 27°C?

5.126 The lunar surface reaches 370 K at midday. The atmosphere consists of neon, argon, and traces of helium at a total pressure of only 2×10^{-14} atm. Calculate the rms speed of each component in the lunar atmosphere. [Use R = 8.314 J/(mol·K) and $\mathcal{M}$ in kg/mol.]

5.127 A person inhales air richer in O_2 and exhales air richer in CO_2 and water vapor. During each hour of sleep, a person exhales a total of about 300 L of this CO_2-enriched and H_2O-enriched air. (a) If the partial pressures of CO_2 and H_2O in exhaled air are each 30.0 torr at 37.0°C, calculate the masses of CO_2 and of H_2O exhaled in 1 h of sleep. (b) How many grams of body mass does the person lose in an 8-h sleep if all the CO_2 and H_2O exhaled come from the metabolism of glucose?

$$C_6H_{12}O_6(s) + 6O_2(g) \longrightarrow 6CO_2(g) + 6H_2O(g)$$

5.128 Popcorn pops because the horny endosperm, a tough, elastic material, resists gas pressure within the heated kernel until it reaches explosive force. A 0.25-mL kernel has a water content of 1.5% by mass, and the water vapor reaches 175°C and 9.0 atm before the kernel ruptures. Assume the water vapor can occupy 75% of the kernel's volume. (a) What is the mass of the kernel? (b) How many milliliters would this amount of water vapor occupy at 25°C and 1.00 atm?

5.129 Sulfur dioxide emissions from coal-based power plants are removed by *flue-gas desulfurization*. The flue gas passes through a scrubber, and a slurry of wet calcium carbonate reacts with it to form carbon dioxide and calcium sulfite. The calcium sulfite then reacts with oxygen to form calcium sulfate, which is sold as gypsum. (a) If the sulfur dioxide concentration is 1000 times higher than its mole fraction in clean dry air (2×10^{-10}), how much calcium sulfate (kg) can be made from scrubbing 4 GL of flue gas (1 GL $= 1\times10^9$ L)? A state-of-the-art scrubber removes at least 95% of the sulfur dioxide. (b) If the mole fraction of oxygen in air is 0.209, what volume (L) of air at 1.00 atm and 25°C is needed to react with all the calcium sulfite?

5.130 Given these relationships for average kinetic energy,

$$\overline{E_k} = \tfrac{1}{2}m\overline{u^2} \qquad \text{and} \qquad \overline{E_k} = \tfrac{3}{2}\left(\frac{R}{N_A}\right)T$$

where m is molecular mass, u is rms speed, R is the gas constant [in J/(mol·K)], N_A is Avogadro's number, and T is absolute temperature: (a) derive Equation 5.13; (b) derive Equation 5.14.

5.131 What would you observe if you tilted a barometer 30° from the vertical? Explain.

5.132 Cylinder A in the picture below contains 0.1 mol of a gas that behaves ideally. Choose the cylinder (B, C, or D) that correctly represents the volume of the gas after each of the following changes. If none of the cylinders is correct, specify "none":
(a) P is doubled at fixed n and T.
(b) T is reduced from 400 K to 200 K at fixed n and P.
(c) T is increased from 100°C to 200°C at fixed n and P.
(d) 0.1 mol of gas is added at fixed P and T.
(e) 0.1 mol of gas is added and P is doubled at fixed T.

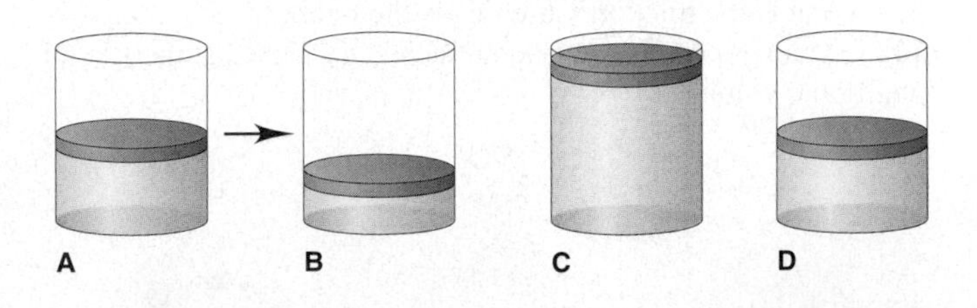

5.133 Ammonia is essential to so many industries that, on a molar basis, it is the most heavily produced substance in the world. Calculate P_{IGL} and P_{VDW} (in atm) of 51.1 g of ammonia in a 3.000-L container at 0°C and 400.°C, the industrial temperature. (From Table 5.5, for NH_3, a = 4.17 atm·L^2/mol^2 and b = 0.0371 L/mol.)

5.134 A 6.0-L flask contains a mixture of methane (CH_4), argon, and helium at 45°C and 1.75 atm. If the mole fractions of helium and argon are 0.25 and 0.35, respectively, how many molecules of methane are present?

5.135 A large portion of metabolic energy arises from the biological combustion of glucose:

$$C_6H_{12}O_6(s) + 6O_2(g) \longrightarrow 6CO_2(g) + 6H_2O(g)$$

(a) If this reaction is carried out in an expandable container at 35°C and 780. torr, what volume of CO_2 is produced from 18.0 g of glucose and excess O_2? (b) If the reaction is carried out at the same conditions with the stoichiometric amount of O_2, what is the partial pressure of each gas when the reaction is 50% complete (9.0 g of glucose remains)?

5.136 What is the average kinetic energy and rms speed of N_2 molecules at STP? Compare these values with those of H_2 molecules at STP. [Use $R = 8.314$ J/(mol·K) and $\mathcal{M}$ in kg/mol.]

5.137 According to the American Conference of Governmental Industrial Hygienists, the *8-h threshold limit value* is 5000 ppmv for CO_2 and 0.1 ppmv for Br_2 (1 ppmv is 1 part by volume in 10^6 parts by volume). Exposure to either gas for 8 h above these limits is unsafe. At STP, which of the following would be unsafe for 8 h of exposure?
(a) Air with a partial pressure of 0.2 torr of Br_2
(b) Air with a partial pressure of 0.2 torr of CO_2
(c) 1000 L of air containing 0.0004 g of Br_2 gas
(d) 1000 L of air containing 2.8×10^{22} molecules of CO_2

5.138 One way to prevent emission of the pollutant NO from industrial plants is by reaction with NH_3:

$$4NH_3(g) + 4NO(g) + O_2(g) \longrightarrow 4N_2(g) + 6H_2O(g)$$

(a) If the NO has a partial pressure of 4.5×10^{-5} atm in the flue gas, how many liters of NH_3 are needed per liter of flue gas at 1.00 atm? (b) If the reaction takes place at 1.00 atm and 150°C, how many grams of NH_3 are needed per kL of flue gas?

5.139 An equimolar mixture of Ne and Xe is accidentally placed in a container that has a tiny leak. After a short while, a very small proportion of the mixture has escaped. What is the mole fraction of Ne in the effusing gas?

5.140 One way to utilize naturally occurring uranium (0.72% ^{235}U and 99.27% ^{238}U) as a nuclear fuel is to enrich it (increase its ^{235}U content) by allowing gaseous UF_6 to effuse through a porous membrane (see the margin note, p. 205). From the relative rates of effusion of $^{235}UF_6$ and $^{238}UF_6$, find the number of steps needed to produce uranium that is 3.0 mole % ^{235}U, the enriched fuel used in many nuclear reactors.

5.141 A slight deviation from ideal behavior exists even at normal conditions. If it behaved ideally, 1 mol of CO would occupy 22.414 L and exert 1 atm pressure at 273.15 K. Calculate P_{VDW} for 1.000 mol of CO at 273.15 K. $\left(\text{Use } R = 0.08206 \frac{\text{atm·L}}{\text{mol·K}}.\right)$

5.142 In preparation for a combustion demonstration, a professor's assistant fills a balloon with equal molar amounts of H_2 and O_2, but the demonstration has to be postponed until the next day. During the night, both gases leak through pores in the balloon. If 45% of the H_2 leaks, what is the O_2/H_2 ratio in the balloon the next day?

5.143 Phosphorus trichloride is important in the manufacture of insecticides, fuel additives, and flame retardants. Phosphorus has only one naturally occurring isotope, ^{31}P, whereas chlorine has two, 75% ^{35}Cl and 25% ^{37}Cl. (a) What different molecular masses (amu) can be found for PCl_3? (b) Which is the most abundant? (c) What is the ratio of the effusion rates for heaviest/lightest PCl_3 molecules?

5.144 A truck tire has a volume of 208 L and is filled with air to 35.0 psi at 295 K. After a drive, the air heats up to 319 K. (a) If the tire volume is constant, what is the pressure? (b) If the tire volume increases 2.0%, what is the pressure? (c) If the tire leaks 1.5 g of air per minute and the temperature is constant, how many minutes will it take for the tire to reach the original pressure of 35.0 psi ($\mathcal{M}$ of air = 28.8 g/mol)?

5.145 Allotropes are different molecular forms of an element. Dioxygen (O_2) and ozone (O_3) are the allotropes of oxygen. (a) What is the density of each allotrope at 0°C and 760 torr? (b) Calculate the ratio of densities of O_3/O_2, and explain the significance of this number.

5.146 When gaseous F_2 and solid I_2 are heated to high temperatures, the I_2 sublimes and gaseous iodine heptafluoride forms. If 350. torr of F_2 and 2.50 g of solid I_2 are put into a 2.50-L container at 250. K and the container is heated to 550. K, what is the final pressure? What is the partial pressure of I_2 gas?

5.147 Many water treatment plants use chlorine gas to kill microorganisms before the water is released for residential use. A plant engineer has to maintain the chlorine pressure in a tank below the 85.0-atm rating and, to be safe, decides to fill the tank to 80.0% of this maximum pressure. (a) How many moles of Cl_2 gas can be kept in the 850.-L tank at 298 K if she uses the ideal gas law in the calculation? (b) What is the tank pressure if she uses the van der Waals equation for this amount of gas? (c) Did the engineer fill the tank to the desired pressure?

Primeval Thermochemistry A campfire in a cave is surely one of the earliest forms of controlled combustion. We focus in this chapter on the release or absorption of heat during a process. Understanding and measuring changes in the energy content of matter is crucial to the fate of Earth's atmosphere and modern society.

6

Thermochemistry: Energy Flow and Chemical Change

Whenever matter changes, whether chemically or physically, the energy content of the matter changes also. In the inferno of a forest fire, the wood and oxygen reactants contain more energy than the ash and gas products, and this difference in energy is *released* as heat and light. In contrast, some of the energy in a flash of lightning is *absorbed* when lower energy N_2 and O_2 in the air react to form higher energy NO. Energy is *absorbed* when snow melts and is *released* when water vapor condenses.

The production and utilization of energy in its many forms have an enormous impact on society. Some of the largest industries manufacture products that release, absorb, or limit the flow of energy. Common fuels—oil, wood, coal, and natural gas—release energy for heating and for powering combustion engines and steam turbines. Fertilizers help crops absorb solar energy and convert it to the chemical energy of food, which our bodies convert into other forms. Numerous plastic, fiberglass, and ceramic materials serve as insulators that limit the flow of energy.

IN THIS CHAPTER . . . We investigate the heat, or *thermal energy,* associated with changes in matter. First, we examine some basic ideas of **thermodynamics,** the study of heat and its transformations. (The discussion begins here and is continued in Chapter 20.) Our focus is on **thermochemistry,** the branch of thermodynamics that deals with the heat involved in chemical and physical change and especially with the concept of *enthalpy.* We describe how heat is measured in a calorimeter and how the quantity of heat released or absorbed is related to the amounts of substances involved in a reaction. You'll learn how to combine equations to obtain the heat change for another equation and see the importance of standard conditions in studying the heat change. The chapter ends with an overview of current and future energy sources and the conflicts between energy demand and environmental quality.

Concepts & Skills to Review

before you study this chapter

- energy and its interconversion (Section 1.1)
- distinction between heat and temperature (Section 1.5)
- nature of chemical bonding (Section 2.7)
- calculations of reaction stoichiometry (Section 3.4)
- properties of the gaseous state (Section 5.1)
- relation between kinetic energy and temperature (Section 5.6)

6.1 FORMS OF ENERGY AND THEIR INTERCONVERSION

As we discussed in Chapter 1, all energy is either potential or kinetic, and these forms are convertible from one to the other. An object has potential energy by virtue of its position and kinetic energy by virtue of its motion. The potential energy of a weight raised above the ground is converted to kinetic energy as it falls (see Figure 1.3, p. 7). When the weight hits the ground, it transfers some of that kinetic energy to the soil and pebbles, causing them to move, and thereby doing *work*. In addition, some of the transferred kinetic energy appears as *heat*, as it slightly warms the soil and pebbles. Thus, the potential energy of the weight is converted to kinetic energy, which is transferred to the ground as work and as heat.

Modern atomic theory allows us to consider other forms of energy—solar, electrical, nuclear, and chemical—as examples of potential and kinetic energy on the atomic and molecular scales. No matter what the details of the situation, *when energy is transferred from one object to another, it appears as work and/or as heat.* In this section, we examine this idea in terms of the loss or gain of energy that takes place during a chemical or physical change.

The System and Its Surroundings

In order to observe and measure a change in energy, we must first define the **system**—the part of the universe that we are going to focus on. The moment that we define the system, everything else relevant to the change is defined as the **surroundings.**

Wherever You Look, There Is a System In the example of the weight hitting the ground, if we define the falling weight as the system, the soil and pebbles that are moved and warmed are the surroundings. An astronomer may define a galaxy as the system and nearby galaxies as the surroundings. An ecologist studying African wildlife can define a zebra herd as the system and other animals, plants, and water supplies as the surroundings. A microbiologist may define a certain cell as the system and the extracellular solution as the surroundings. Thus, in general, it is the experiment and the experimenter that define the system and the surroundings.

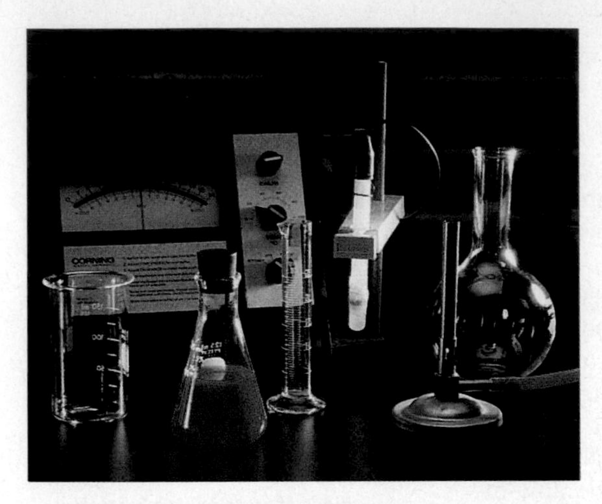

Figure 6.1 A chemical system and its surroundings. Once the contents of the flask (the orange solution) are defined as the system, the flask and the laboratory become defined as the surroundings.

Figure 6.1 shows a typical chemical system and its surroundings: the system is the contents of the flask; the flask itself, the other equipment, and perhaps the rest of the laboratory are the surroundings. In principle, the rest of the universe is the surroundings, but in practice, we need to consider only the portions of the universe relevant to the system. That is, it's not likely that a thunderstorm in central Asia or a methane blizzard on Neptune will affect the contents of the flask, but the temperature, pressure, and humidity of the lab might.

Energy Flow to and from a System

Each particle in a system has potential and kinetic energy, and the sum of these energies for all the particles in the system is the **internal energy, *E*** (some texts use the symbol *U*). When a chemical system, such as the contents of the flask in Figure 6.1, changes from reactants to products and the products return to the starting temperature, the internal energy has changed. To determine this change, ΔE, we measure the difference between the system's internal energy *after* the change (E_{final}) and *before* the change ($E_{initial}$):

$$\Delta E = E_{final} - E_{initial} = E_{products} - E_{reactants} \quad (6.1)$$

where Δ (Greek *delta*) means "change (or difference) in." Note especially that Δ refers to the *final state of the system* ***minus*** *the initial state.*

Because energy must be conserved, *a change in the energy of the system is always accompanied by an* ***opposite*** *change in the energy of the surroundings.* We often represent this change with an *energy diagram* in which the final and initial states are horizontal lines on a vertical energy axis. The change in internal energy, ΔE, is the difference between the heights of the two lines. A system can change its internal energy in one of two ways:

1. By losing some energy *to* the surroundings, as shown in Figure 6.2A:

$$E_{final} < E_{initial} \qquad \Delta E < 0$$

2. By gaining some energy *from* the surroundings, as shown in Figure 6.2B:

$$E_{final} > E_{initial} \qquad \Delta E > 0$$

Note that the change in energy is always a *transfer* of energy from system to surroundings, or vice versa.

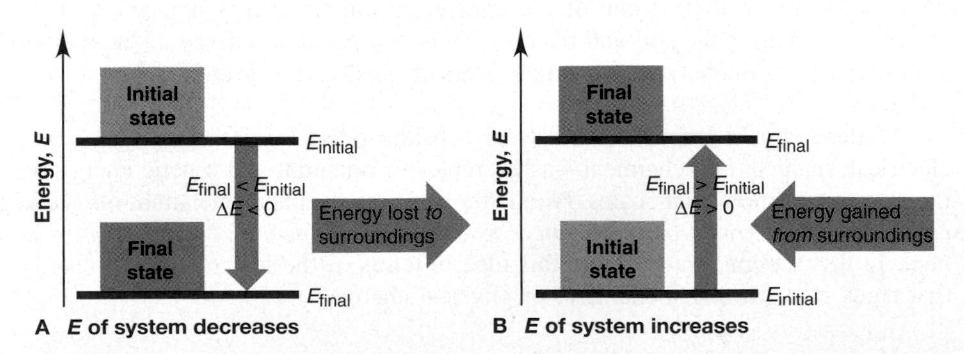

Figure 6.2 Energy diagrams for the transfer of internal energy (*E*) between a system and its surroundings. A, When the internal energy of a system *decreases,* the change in energy (ΔE) is lost *to* the surroundings; therefore, ΔE of the system ($E_{final} - E_{initial}$) is negative. **B,** When the system's internal energy *increases,* ΔE is gained *from* the surroundings and is positive. Note that the vertical yellow arrow, which signifies the direction of the change in energy, *always* has its tail at the initial state and its head at the final state.

Heat and Work: Two Forms of Energy Transfer

Just as we saw when a weight hits the ground, energy transfer outward from the system or inward from the surroundings can appear in two forms, heat and work. **Heat** (or *thermal energy,* symbol ***q***) is the energy transferred between a system and its surroundings as a result of a difference in their temperatures only. Energy in the form of heat is transferred from hot soup (system) to the bowl, air, and table (surroundings) because the surroundings have a lower temperature. All other forms of energy transfer (mechanical, electrical, and so on) involve some type of **work (*w*),** the energy transferred when an object is moved by a force. When you (system) kick a football (surroundings), energy is transferred as work to move the ball. When you inflate the ball, the inside air (system) exerts a force on the inner wall of the ball and nearby air (surroundings) and does work to move it outward.

The total change in a system's internal energy is the sum of the energy transferred as heat and/or work:

$$\Delta E = q + w \tag{6.2}$$

The numerical values of q and w can be either positive or negative, depending on the change the *system* undergoes. In other words, *we define the sign of the energy transfer from the system's perspective.* Energy coming *into* the system is *positive.* Energy going *out from* the system is *negative.* Of the innumerable changes possible in the system's internal energy, we'll examine four simple cases—two that involve only heat and two that involve only work.

Energy Transfer as Heat Only For a system that does no work but transfers energy only as heat (q), we know that $w = 0$. Therefore, from Equation 6.2, we have $\Delta E = q + 0 = q$.

1. *Heat flowing* ***out from*** *a system.* Suppose a sample of hot water is the system; then, the beaker containing it and the rest of the lab are the surroundings. The water transfers energy as heat to the surroundings until the temperature of the water equals that of the surroundings. The system's energy decreases as heat flows *out from* the system, so the final energy of the system is less than its initial energy. Heat was lost by the system, so *q is negative,* and therefore *ΔE is negative.* Figure 6.3A shows this situation.

2. *Heat flowing* ***into*** *a system.* If the system consists of ice water, it gains energy as heat from the surroundings until the temperature of the water equals that of the surroundings. In this case, energy is transferred *into* the system, so the final energy of the system is higher than its initial energy. Heat was gained by the system, so *q is positive,* and therefore *ΔE is positive* (Figure 6.3B).

Thermodynamics in the Kitchen The air in a refrigerator (surroundings) has a lower temperature than a newly added piece of food (system), so the food loses energy as heat to the refrigerator air, $q < 0$. The air in a hot oven (surroundings) has a higher temperature than a newly added piece of food (system), so the food gains energy as heat from the oven air, $q > 0$.

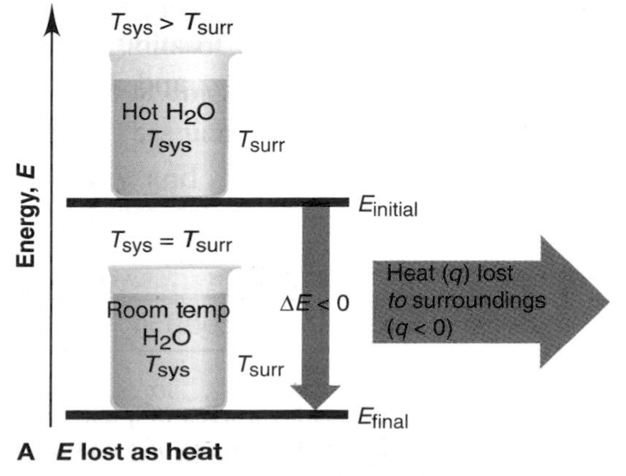

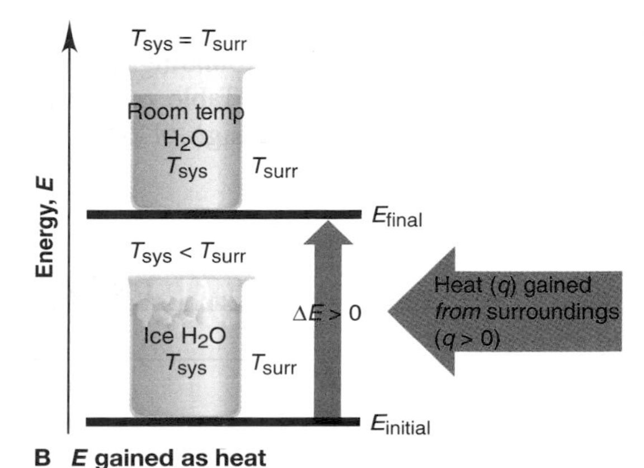

Figure 6.3 A system transferring energy as heat only. A, Hot water (the system, sys) transfers energy as heat (q) *to* the surroundings (surr) until $T_{sys} = T_{surr}$. Since $E_{initial} > E_{final}$ and $w = 0$, $\Delta E < 0$ and the sign of q is negative. **B,** Ice water gains energy as heat (q) *from* the surroundings until $T_{sys} = T_{surr}$. Since $E_{initial} < E_{final}$ and $w = 0$, $\Delta E > 0$ and the sign of q is positive.

6.88 Pure liquid octane (C_8H_{18}; $d = 0.702$ g/mL) is used as the fuel in a test of a new automobile drive train.
(a) How much energy (in kJ) is released by complete combustion of the octane in a 20.4-gal fuel tank to gases ($\Delta H^0_{comb} = -5.45\times10^3$ kJ/mol)?
(b) The energy delivered to the wheels at 65 mph is 5.5×10^4 kJ/h. Assuming all the energy is transferred to the wheels, what is the cruising range (in km) of the car on a full tank?
(c) If the actual cruising range is 455 miles, explain your answer to part (b).

6.89 When simple sugars, called *monosaccharides*, link together, they form a variety of complex sugars and, ultimately, *polysaccharides*, such as starch, glycogen, and cellulose. Glucose and fructose have the same formula, $C_6H_{12}O_6$, but different arrangements of atoms. They link together to form a molecule of sucrose (table sugar) and a molecule of liquid water. The ΔH^0_f values of glucose, fructose, and sucrose are -1273 kJ/mol, -1266 kJ/mol, and -2226 kJ/mol, respectively. Write a balanced equation for this reaction and calculate ΔH^0_{rxn}.

6.90 Physicians and nutritional biochemists recommend eating vegetable oils rather than animal fats to lower risks of heart disease. In olive oil, one of the healthier choices, the main fatty acid is oleic acid ($C_{18}H_{34}O_2$), whose $\Delta H^0_{comb} = -1.11\times10^4$ kJ/mol. Calculate ΔH^0_f of oleic acid. [Assume $H_2O(g)$.]

6.91 Oxidation of gaseous ClF by F_2 yields liquid ClF_3, an important fluorinating agent. Use the following thermochemical equations to calculate ΔH^0_{rxn} for the production of ClF_3:

(1) $2ClF(g) + O_2(g) \longrightarrow Cl_2O(g) + OF_2(g)$ $\quad \Delta H^0 = 167.5$ kJ
(2) $2F_2(g) + O_2(g) \longrightarrow 2OF_2(g)$ $\quad \Delta H^0 = -43.5$ kJ
(3) $2ClF_3(l) + 2O_2(g) \longrightarrow Cl_2O(g) + 3OF_2(g)$
$\Delta H^0 = 394.1$ kJ

6.92 Silver bromide is used to coat ordinary black-and-white photographic film, while high-speed film uses silver iodide.
(a) When 50.0 mL of 5.0 g/L $AgNO_3$ is added to a coffee-cup calorimeter containing 50.0 mL of 5.0 g/L NaI, with both solutions at 25°C, what mass of AgI forms?
(b) Use Appendix B to find ΔH^0_{rxn}.
(c) What is ΔT_{soln} (assume the volumes are additive and the solution has the density and specific heat capacity of water)?

6.93 The calorie (4.184 J) was originally defined as the quantity of energy required to raise the temperature of 1.00 g of liquid water 1.00°C. The British thermal unit (Btu) is defined as the quantity of energy required to raise the temperature of 1.00 lb of liquid water 1.00°F.
(a) How many joules are in 1.00 British thermal unit (1 lb = 453.6 g; a change of 1.0°C = 1.8°F)?
(b) The "therm" is a unit of energy consumption that is used by natural gas companies in the United States and is defined as 100,000 Btu. How many joules are in 1.00 therm?
(c) How many moles of methane must be burned to give 1.00 therm of energy? (Assume water forms as a gas.)
(d) If natural gas costs $0.46 per therm, what is the cost per mole of methane? (Assume natural gas is pure methane.)
(e) How much would it cost to warm 308 gal of water in a hot tub from 15.0°C to 40.0°C (1 gal = 3.78 L)?

6.94 Whenever organic matter is decomposed under oxygen-free (anaerobic) conditions, methane is one of the products. Thus, enormous deposits of natural gas, which is almost entirely methane, exist as a major source of fuel for home and industry.
(a) It is estimated that known sources of natural gas can produce 5,600 EJ of energy (1 EJ = 10^{18} J). Current total global energy usage is 4.0×10^2 EJ per year. Find the mass (in kg) of known sources of natural gas (ΔH^0_{comb} of $CH_4 = -802$ kJ/mol).
(b) For how many years could these sources supply the world's total energy needs?
(c) What volume (in ft^3) of natural gas is required to heat 1.00 qt of water from 20.0°C to 100.0°C (d of H_2O = 1.00 g/mL; d of CH_4 at STP = 0.72 g/L)?
(d) The fission of 1 mol of uranium (about 4×10^{-4} ft^3) in a nuclear reactor produces 2×10^{13} J. What volume (in ft^3) of natural gas would produce the same amount of energy?

6.95 The heat of atomization (ΔH^0_{atom}) is the heat needed to form separated gaseous atoms from a substance in its standard state. The equation for the atomization of graphite is

$$C(graphite) \longrightarrow C(g)$$

Use Hess's law to calculate ΔH^0_{atom} of graphite from these data:
(1) ΔH^0_f of $CH_4 = -74.9$ kJ/mol
(2) ΔH^0_{atom} of CH_4 = 1660 kJ/mol
(3) ΔH^0_{atom} of H_2 = 432 kJ/mol

6.96 A reaction is carried out in a steel vessel within a chamber filled with argon gas. Below are molecular views of the argon adjacent to the surface of the reaction vessel before and after the reaction. Was the reaction exothermic or endothermic? Explain.

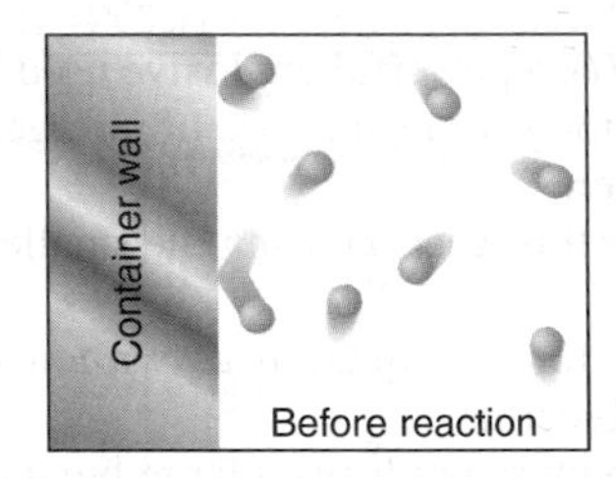

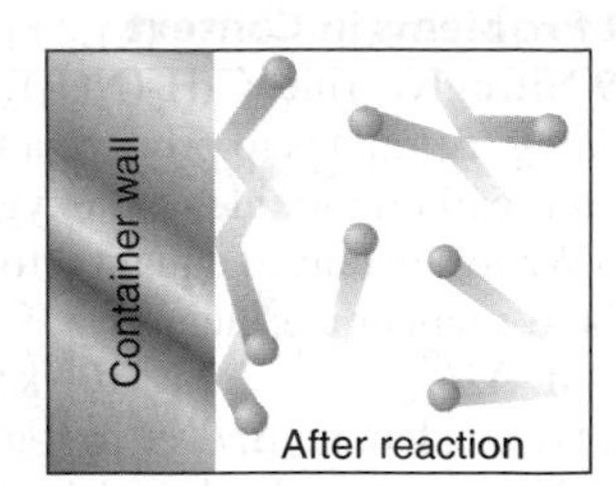

6.97 For each of the following events, the system is in *italics*. State whether heat, work, or both is (are) transferred and specify the direction of each transfer.
(a) You pump air into an automobile *tire*.
(b) A *tree* rots in a forest.
(c) You strike a *match*.
(d) You cool juice in an *ice chest*.
(e) You cook *food* on a kitchen range.

6.98 To make use of ionic hydrates for storing solar energy (see Chemical Connections, p. 247), you use 500.0 kg of sodium sulfate decahydrate on your house roof. Assuming complete reaction and 100% efficiency of heat transfer, how much heat (in kJ) is released to your house at night?

6.99 Benzene (C_6H_6) and acetylene (C_2H_2) have the same empirical formula, CH. Which releases more energy per mole of CH (ΔH^0_f of gaseous C_6H_6 = 82.9 kJ/mol)?

6.100 Kerosene, a common space-heater fuel, is a mixture of hydrocarbons whose "average" formula is $C_{12}H_{26}$.
(a) Write a balanced equation, using the simplest whole-number coefficients, for the complete combustion of kerosene to gases.
(b) If $\Delta H^0_{comb} = -1.50\times10^4$ kJ for the equation as written in part (a), determine ΔH^0_f of kerosene.
(c) Calculate the heat produced by combustion of 0.50 gal of kerosene (d of kerosene = 0.749 g/mL).
(d) How many gallons of kerosene must be burned for a kerosene furnace to produce 1250. Btu (1 Btu = 1.055 kJ)?

6.101 Coal gasification is a multistep process to convert coal into gaseous fuels (see Chemical Connections, p. 245). In one step, a certain coal sample reacts with superheated steam:

$C(coal) + H_2O(g) \longrightarrow CO(g) + H_2(g) \quad \Delta H^0_{rxn} = 129.7\ kJ$

(a) Combine this reaction with the following two to write an overall reaction for the production of methane:

$$CO(g) + H_2O(g) \longrightarrow CO_2(g) + H_2(g)$$
$$CO(g) + 3H_2(g) \longrightarrow CH_4(g) + H_2O(g)$$

(b) Calculate ΔH^0_{rxn} for this overall change.
(c) Using the value in (b) and calculating the ΔH^0_{comb} of methane, find the total heat for gasifying 1.00 kg of coal and burning the methane formed (assume water forms as a gas and $\mathcal{M}$ of coal = 12.00 g/mol).

6.102 Phosphorus pentachloride is used in the industrial preparation of organic phosphorus compounds. Equation 1 shows its preparation from PCl_3 and Cl_2:

(1) $PCl_3(l) + Cl_2(g) \longrightarrow PCl_5(s)$

Use equations 2 and 3 to calculate ΔH_{rxn} of equation 1:

(2) $P_4(s) + 6Cl_2(g) \longrightarrow 4PCl_3(l) \quad \Delta H = -1280\ kJ$
(3) $P_4(s) + 10Cl_2(g) \longrightarrow 4PCl_5(s) \quad \Delta H = -1774\ kJ$

6.103 Consider the following hydrocarbon fuels:

(I) $CH_4(g)$ (II) $C_2H_4(g)$ (III) $C_2H_6(g)$

(a) Rank them in terms of heat released (a) per mole and (b) per gram. [Assume $H_2O(g)$ forms and combustion is complete.]

6.104 A typical candy bar weighs about 2 oz (1.00 oz = 28.4 g).
(a) Assuming that a candy bar is 100% sugar and that 1.0 g of sugar is equivalent to about 4.0 Calories of energy, calculate the energy (in kJ) contained in a typical candy bar.
(b) Assuming that your mass is 58 kg and you convert chemical potential energy to work with 100% efficiency, how high would you have to climb to work off the energy in a candy bar? (Potential energy = mass × g × height, where g = 9.8 m/s^2.)
(c) Why is your actual conversion of potential energy to work less than 100% efficient?

6.105 Silicon tetrachloride is produced annually on the multikiloton scale for making transistor-grade silicon. It can be made directly from the elements (reaction 1) or, more cheaply, by heating sand and graphite with chlorine gas (reaction 2). If water is present in reaction 2, some tetrachloride may be lost in an unwanted side reaction (reaction 3):

(1) $Si(s) + 2Cl_2(g) \longrightarrow SiCl_4(g)$
(2) $SiO_2(s) + 2C(graphite) + 2Cl_2(g) \longrightarrow SiCl_4(g) + 2CO(g)$
(3) $SiCl_4(g) + 2H_2O(g) \longrightarrow SiO_2(s) + 4HCl(g)$
$\Delta H^0_{rxn} = -139.5\ kJ$

(a) Use reaction 3 to calculate the heats of reaction of reactions 1 and 2. (b) What is the heat of reaction for the new reaction that is the sum of reactions 2 and 3?

6.106 Use the following information to find ΔH^0_f of gaseous HCl:

$$N_2(g) + 3H_2(g) \longrightarrow 2NH_3(g) \quad \Delta H^0_{rxn} = -91.8\ kJ$$
$$N_2(g) + 4H_2(g) + Cl_2(g) \longrightarrow 2NH_4Cl(s) \quad \Delta H^0_{rxn} = -628.8\ kJ$$
$$NH_3(g) + HCl(g) \longrightarrow NH_4Cl(s) \quad \Delta H^0_{rxn} = -176.2\ kJ$$

6.107 You want to determine ΔH^0 for the reaction

$$Zn(s) + 2HCl(aq) \longrightarrow ZnCl_2(aq) + H_2(g)$$

(a) To do so, you first determine the heat capacity of a calorimeter using the following reaction, whose ΔH is known:

$NaOH(aq) + HCl(aq) \longrightarrow NaCl(aq) + H_2O(l)$
$\Delta H^0 = -57.32\ kJ$

Calculate the heat capacity of the calorimeter from these data:
Amounts used: 50.0 mL of 2.00 M HCl and 50.0 mL of 2.00 M NaOH
Initial T of both solutions: 16.9°C
Maximum T recorded during reaction: 30.4°C
Density of resulting NaCl solution: 1.04 g/mL
c of 1.00 M NaCl(aq) = 3.93 J/g·K
(b) Use the result from part (a) and the following data to determine ΔH^0_{rxn} for the reaction between zinc and HCl(aq):
Amounts used: 100.0 mL of 1.00 M HCl and 1.3078 g of Zn
Initial T of HCl solution and Zn: 16.8°C
Maximum T recorded during reaction: 24.1°C
Density of 1.0 M HCl solution = 1.015 g/mL
c of resulting $ZnCl_2(aq)$ = 3.95 J/g·K
(c) Given the values below, what is the error in your experiment?

$$\Delta H^0_f \text{ of } HCl(aq) = -1.652\times10^2\ kJ/mol$$
$$\Delta H^0_f \text{ of } ZnCl_2(aq) = -4.822\times10^2\ kJ/mol$$

6.108 One mole of nitrogen gas confined within a cylinder by a piston is heated from 0°C to 819°C at 1.00 atm.
(a) Calculate the work of expansion of the gas in joules (1 J = 9.87×10^{-3} atm·L). Assume all the energy is used to do work.
(b) What would be the temperature change if the gas were heated with the same amount of energy in a container of fixed volume? (Assume the specific heat capacity of N_2 is 1.00 J/g·K.)

6.109 The chemistry of nitrogen oxides is very versatile. Given the following reactions and their standard enthalpy changes,

(1) $NO(g) + NO_2(g) \longrightarrow N_2O_3(g) \quad \Delta H^0_{rxn} = -39.8\ kJ$
(2) $NO(g) + NO_2(g) + O_2(g) \longrightarrow N_2O_5(g) \quad \Delta H^0_{rxn} = -112.5\ kJ$
(3) $2NO_2(g) \longrightarrow N_2O_4(g) \quad \Delta H^0_{rxn} = -57.2\ kJ$
(4) $2NO(g) + O_2(g) \longrightarrow 2NO_2(g) \quad \Delta H^0_{rxn} = -114.2\ kJ$
(5) $N_2O_5(g) \longrightarrow N_2O_5(s) \quad \Delta H^0_{rxn} = -54.1\ kJ$

calculate the heat of reaction for

$$N_2O_3(g) + N_2O_5(s) \longrightarrow 2N_2O_4(g)$$

6.110 Electric generating plants transport large amounts of hot water through metal pipes, and oxygen dissolved in the water can cause a major corrosion problem. Hydrazine (N_2H_4) added to the water avoids the problem by reacting with the oxygen:

$$N_2H_4(aq) + O_2(g) \longrightarrow N_2(g) + 2H_2O(l)$$

About 4×10^7 kg of hydrazine is produced every year by reacting ammonia with sodium hypochlorite in the *Raschig process*:

$2NH_3(aq) + NaOCl(aq) \longrightarrow N_2H_4(aq) + NaCl(aq) + H_2O(l)$
$\Delta H^0_{rxn} = -151\ kJ$

(a) If ΔH^0_f of NaOCl(aq) = −346 kJ/mol, find ΔH^0_f of $N_2H_4(aq)$.
(b) What is the heat released when aqueous N_2H_4 is added to 5.00×10^3 L of plant water that is 2.50×10^{-4} M O_2?

6.111 Liquid methanol (CH_3OH) is used as an alternative fuel in truck engines. An industrial method for preparing it uses the catalytic hydrogenation of carbon monoxide:

$$CO(g) + 2H_2(g) \xrightarrow{\text{catalyst}} CH_3OH(l)$$

How much heat (in kJ) is released when 15.0 L of CO at 85°C and 112 kPa reacts with 18.5 L of H_2 at 75°C and 744 torr?

6.112 (a) How much heat is released when 25.0 g of methane burns in excess O_2 to form gaseous products?
(b) Calculate the temperature of the product mixture if the methane and air are both at an initial temperature of 0.0°C. Assume a stoichiometric ratio of methane to oxygen from the air, with air being 21% O_2 by volume (c of CO_2 = 57.2 J/mol·K; c of $H_2O(g)$ = 36.0 J/mol·K; c of N_2 = 30.5 J/mol·K).

A Spectral Spectacular Why do the exploding compounds in a fireworks display emit only certain colors of light? As you'll learn in this chapter, the answer has revolutionized our conceptions of matter and energy and of the amazing atoms that make up the world around us and the universe beyond.

7 Quantum Theory and Atomic Structure

Over a few remarkable decades—from around 1890 to 1930—a revolution took place in how we view the makeup of the universe. But revolutions in science are not the violent upheavals of political overthrow. Rather, flaws appear in an established model as conflicting evidence mounts, a startling discovery or two widens the flaws into cracks, and the conceptual structure crumbles gradually from its inconsistencies. New insight, verified by experiment, then guides the building of a model more consistent with reality. So it was when Lavoisier's theory of combustion superseded the phlogiston model, when Dalton's atomic theory established the idea of individual units of matter, and when Rutherford's nuclear model substituted atoms with rich internal structure for "billiard balls" or "plum puddings." In this chapter, you will see this process unfold again with the development of modern atomic theory.

Almost as soon as Rutherford proposed his nuclear model, a major problem arose. A nucleus and an electron attract each other, so if they are to remain apart, the energy of the electron's motion (kinetic energy) must balance the energy of attraction (potential energy). However, the laws of classical physics had established that a negative particle moving in a curved path around a positive one *must* emit radiation and thus lose energy. If this requirement applied to atoms, why didn't the orbiting electron lose energy continuously and spiral into the nucleus? Clearly, if electrons behaved the way classical physics predicted, all atoms would have collapsed eons ago! The behavior of subatomic matter seemed to violate real-world experience and accepted principles.

The breakthroughs that soon followed Rutherford's model forced a complete rethinking of the classical picture of matter and energy. In the macroscopic world, the two are distinct. Matter occurs in chunks you can hold and weigh, and you can change the amount of matter in a sample piece by piece. In contrast, energy is "massless," and its quantity changes in a continuous manner. Matter moves in specific paths, whereas light and other types of energy travel in diffuse waves. As soon as 20th-century scientists probed the subatomic world, however, these clear distinctions between particulate matter and wavelike energy began to fade.

IN THIS CHAPTER . . . We discuss *quantum mechanics,* the theory that explains our current picture of atomic structure. We consider the wave properties of energy and then examine the theories and experiments that led to a quantized, or particulate, model of light. We see why the light emitted by excited hydrogen (H) atoms—the *atomic spectrum*—suggests an atom with distinct energy levels, and we look briefly at how atomic spectra are applied to chemical analysis. Wave-particle duality, which reveals two faces of matter and of energy, leads us to the current model of the H atom and the quantum numbers that identify the regions of space an electron occupies in an atom. In Chapter 8, we'll consider atoms that have more than one electron and relate electron number and distribution to chemical behavior.

Concepts & Skills to Review

before you study this chapter

- discovery of the electron and atomic nucleus (Section 2.4)
- major features of atomic structure (Section 2.5)
- changes in energy state of a system (Section 6.1)

Hooray for the Human Mind The invention of the car, radio, and airplane fostered a feeling of unlimited human ability, and the discovery of x-rays, radioactivity, the electron, and the atomic nucleus led to the sense that the human mind would soon unravel all of nature's mysteries. Indeed, some people were convinced that few, if any, mysteries remained.

1895 Röntgen discovers x-rays.
1896 Becquerel discovers radioactivity.
1897 Thomson discovers the electron.
1898 Curie discovers radium.
1900 Freud proposes theory of the unconscious mind.
1900 Planck develops quantum theory.
1901 Marconi invents the radio.
1903 Wright brothers fly an airplane.
1905 Ford uses assembly line to build cars.
1905 Rutherford explains radioactivity.
1905 Einstein publishes relativity and photon theories.
1906 St. Denis develops modern dance.
1908 Matisse and Picasso develop modern art.
1909 Schoenberg and Berg develop modern music.
1911 Rutherford presents nuclear model.
1913 Bohr proposes atomic model.
1914 to 1918 World War I is fought.
1923 Compton demonstrates photon momentum.
1924 De Broglie publishes wave theory of matter.
1926 Schrödinger develops wave equation.
1927 Heisenberg presents uncertainty principle.
1932 Chadwick discovers the neutron.

7.1 THE NATURE OF LIGHT

Visible light is one type of **electromagnetic radiation** (also called *electromagnetic energy* or *radiant energy*). Other familiar types include x-rays, microwaves, and radio waves. All electromagnetic radiation consists of energy propagated by means of electric and magnetic fields that alternately increase and decrease in intensity as they move through space. This classical wave model distinguishes clearly between waves and particles; it is essential for understanding why rainbows form, how magnifying glasses work, why objects look distorted under water, and many other everyday observations. But, it cannot explain observations on the atomic scale because, in that unfamiliar realm, energy behaves as though it consists of particles!

The Wave Nature of Light

The wave properties of electromagnetic radiation are described by two interdependent variables, as Figure 7.1 shows:

- **Frequency** (ν, Greek *nu*) is the number of cycles the wave undergoes per second and is expressed in units of 1/second [s^{-1}; also called *hertz* (Hz)].
- **Wavelength** (λ, Greek *lambda*) is the distance between any point on a wave and the corresponding point on the next crest (or trough) of the wave, that is, the distance the wave travels during one cycle. Wavelength is expressed in meters and often, for very short wavelengths, in nanometers (nm, 10^{-9} m), picometers (pm, 10^{-12} m), or the non-SI unit angstroms (Å, 10^{-10} m).

The speed of the wave, the distance traveled per unit time (in units of meters per second), is the product of its frequency (cycles per second) and its wavelength (meters per cycle):

$$\text{Units for speed of wave:} \quad \frac{\cancel{\text{cycles}}}{\text{s}} \times \frac{\text{m}}{\cancel{\text{cycle}}} = \frac{\text{m}}{\text{s}}$$

In a vacuum, all types of electromagnetic radiation travel at 2.99792458×10^8 m/s (3.00×10^8 m/s to three significant figures), a constant called the **speed of light (*c*)**:

$$c = \nu \times \lambda \tag{7.1}$$

As Equation 7.1 shows, the product of ν and λ is a constant. Thus, the individual terms have a reciprocal relationship to each other: *radiation with a high frequency has a short wavelength, and vice versa.*

Another characteristic of a wave is its **amplitude,** the height of the crest (or depth of the trough) of each wave (Figure 7.2). The amplitude of an electromagnetic wave is a measure of the strength of its electric and magnetic fields. Thus, amplitude is related to the *intensity* of the radiation, which we perceive as brightness in the case of visible light. Light of a particular color, fire-engine red for instance, has a specific frequency and wavelength, but it can be dimmer (lower amplitude) or brighter (higher amplitude).

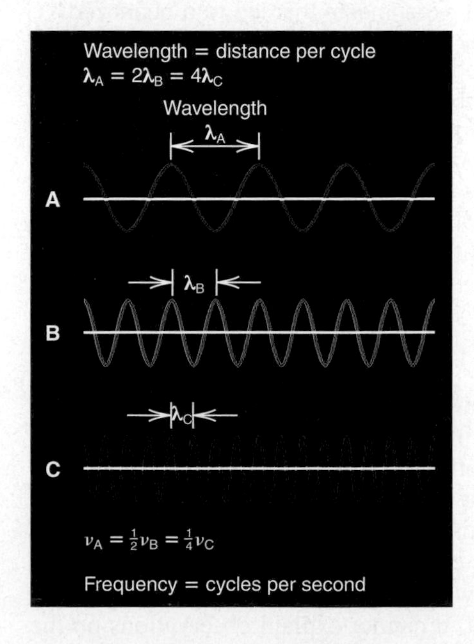

Figure 7.1 Frequency and wavelength. Three waves with different wavelengths (λ) and thus different frequencies (ν) are shown. Note that *as the wavelength decreases, the frequency increases, and vice versa.*

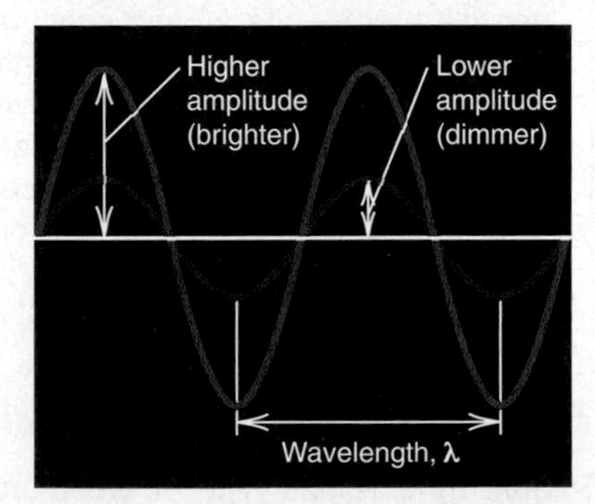

Figure 7.2 Amplitude (intensity) of a wave. Amplitude is represented by the height of the crest (or depth of the trough) of the wave. The two waves shown have the same wavelength (color) but different amplitudes and, therefore, different brightnesses (intensities).

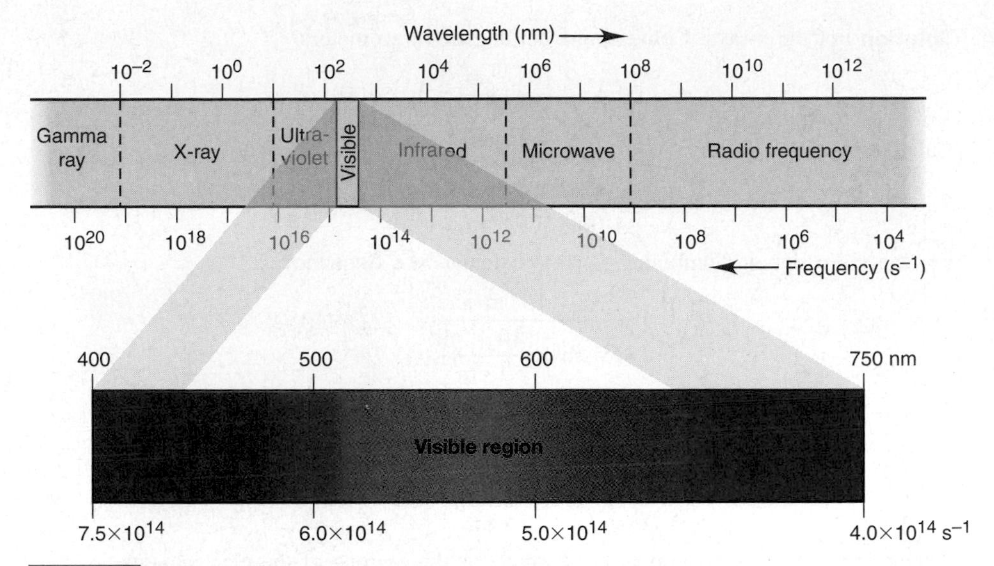

Figure 7.3 Regions of the electromagnetic spectrum. The electromagnetic spectrum extends from the very short wavelengths (very high frequencies) of gamma rays through the very long wavelengths (very low frequencies) of radio waves. The relatively narrow visible region is expanded (and the scale made linear) to show the component colors.

The Electromagnetic Spectrum Visible light represents a small portion of the continuum of radiant energy known as the **electromagnetic spectrum** (Figure 7.3). *All the waves in the spectrum travel at the same speed through a vacuum but differ in frequency and, therefore, wavelength.* Some regions of the spectrum are utilized by particular devices; for example, the long-wavelength, low-frequency radiation is used by microwave ovens and radios. Note that each region meets the next. For instance, the **infrared (IR)** region meets the microwave region on one end and the visible region on the other.

We perceive different wavelengths (or frequencies) of *visible* light as different colors, from red ($\lambda \approx 750$ nm) to violet ($\lambda \approx 400$ nm). Light of a single wavelength is called *monochromatic* (Greek, "one color"), whereas light of many wavelengths is *polychromatic*. White light is polychromatic. The region adjacent to visible light on the short-wavelength end consists of **ultraviolet (UV)** radiation (also called *ultraviolet light*). Still shorter wavelengths (higher frequencies) make up the x-ray and gamma (γ) ray regions. Thus, a TV signal, the green light from a traffic signal, and a gamma ray emitted by a radioactive element all travel at the same speed but differ in their frequency (and wavelength).

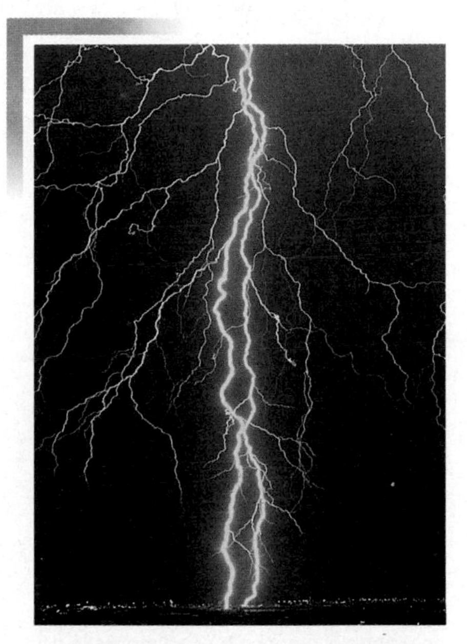

Electromagnetic Emissions Everywhere We are bathed in electromagnetic radiation from the Sun. Radiation from human activities bombards us as well: radio and TV signals; microwaves from traffic monitors and telephone relay stations; from lightbulbs, x-ray equipment, car motors, and so forth. Natural sources on Earth bombard us also: lightning, radioactivity, and even the glow of fireflies! And our knowledge of the distant universe comes from radiation entering our light, x-ray, and radio telescopes.

SAMPLE PROBLEM 7.1 Interconverting Wavelength and Frequency

Problem A dental hygienist uses x-rays ($\lambda = 1.00$ Å) to take a series of dental radiographs while the patient listens to a radio station ($\lambda = 325$ cm) and looks out the window at the blue sky ($\lambda = 473$ nm). What is the frequency (in s^{-1}) of the electromagnetic radiation from each source? (Assume that the radiation travels at the speed of light, 3.00×10^8 m/s.)

Plan We are given the wavelengths, so we use Equation 7.1 to find the frequencies. However, we must first convert the wavelengths to meters because c has units of m/s.

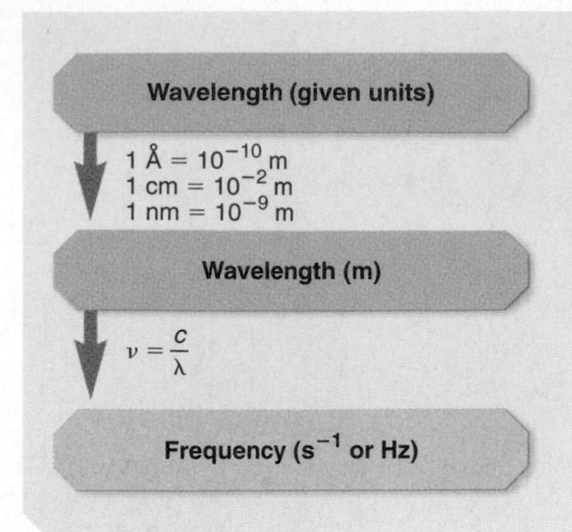

Solution For the x-rays: Converting from angstroms to meters,

$$\lambda = 1.00\ \text{Å} \times \frac{10^{-10}\ \text{m}}{1\ \text{Å}} = 1.00\times10^{-10}\ \text{m}$$

Calculating the frequency:

$$\nu = \frac{c}{\lambda} = \frac{3.00\times10^{8}\ \text{m/s}}{1.00\times10^{-10}\ \text{m}} = 3.00\times10^{18}\ \text{s}^{-1}$$

For the radio signal: Combining steps to calculate the frequency,

$$\nu = \frac{c}{\lambda} = \frac{3.00\times10^{8}\ \text{m/s}}{325\ \text{cm} \times \dfrac{10^{-2}\ \text{m}}{1\ \text{cm}}} = 9.23\times10^{7}\ \text{s}^{-1}$$

For the blue sky: Combining steps to calculate the frequency,

$$\nu = \frac{c}{\lambda} = \frac{3.00\times10^{8}\ \text{m/s}}{473\ \text{nm} \times \dfrac{10^{-9}\ \text{m}}{1\ \text{nm}}} = 6.34\times10^{14}\ \text{s}^{-1}$$

Check The orders of magnitude are correct for the regions of the electromagnetic spectrum (see Figure 7.3): x-rays (10^{19} to 10^{16} s^{-1}), radio waves (10^{9} to 10^{4} s^{-1}), and visible light (7.5×10^{14} to 4.0×10^{14} s^{-1}).

Comment The radio station here is broadcasting at 92.3×10^{6} s^{-1}, or 92.3 million Hz (92.3 MHz), about midway in the FM range.

FOLLOW-UP PROBLEM 7.1 Some diamonds appear yellow because they contain nitrogen compounds that absorb purple light of frequency 7.23×10^{14} Hz. Calculate the wavelength (in nm and Å) of the absorbed light.

The Distinction Between Energy and Matter In the everyday world around us, energy and matter behave very differently. Let's examine some important observations about light and see how they contrast with the behavior of particles. Light of a given wavelength travels at different speeds through different transparent media—vacuum, air, water, quartz, and so forth. Therefore, when a light wave passes from one medium into another, say, from air to water, the speed of the wave changes. Figure 7.4A shows the phenomenon known as **refraction.** If the

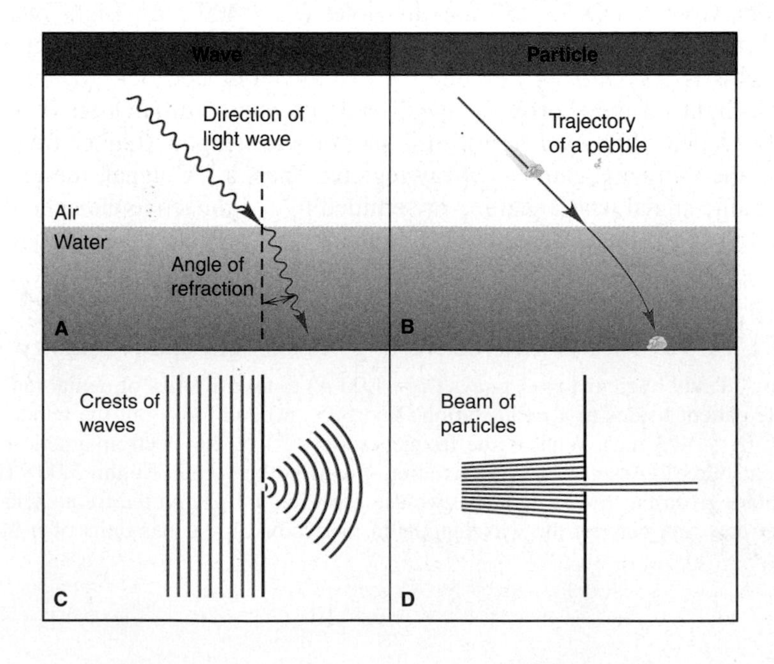

Figure 7.4 Different behaviors of waves and particles. A, A wave passing from air into water is *refracted* (bent at an angle). **B,** In contrast, a particle of matter (such as a pebble) entering a pond moves in a curved path, because gravity and the greater resistance (drag) of the water slow it down gradually. **C,** A wave is *diffracted* through a small opening, which gives rise to a circular wave on the other side. (The lines represent the crests of water waves as seen from above.) **D,** In contrast, when a collection of moving particles encounters a small opening, as when a handful of sand is thrown at a hole in a fence, some particles move through the opening and continue along their individual paths.

wave strikes the boundary, say, between air and water, at an angle other than 90°, the change in speed causes a change in direction, and the wave continues at a different angle. The new angle (angle of refraction) depends on the materials on either side of the boundary and the wavelength of the light. In the process of *dispersion,* white light separates (disperses) into its component colors, as when it passes through a prism, because each incoming wavelength is refracted at a slightly different angle.

In contrast, a particle, like a pebble, does not undergo refraction when passing from one medium to another. Figure 7.4B shows that when a pebble thrown through the air enters a pond, its speed changes abruptly and then it continues to slow down gradually in a curved path.

When a wave strikes the edge of an object, it bends around it in a phenomenon called **diffraction.** If the wave passes through a slit about as wide as its wavelength, it bends around both edges of the slit and forms a semicircular wave on the other side of the opening, as shown in Figure 7.4C.

Once again, particles act very differently. Figure 7.4D shows that if you throw a collection of particles, like a handful of sand, at a small opening, some particles hit the edge, while others go through the opening and continue linearly in a narrower group.

If waves of light pass through two adjacent slits, the emerging circular waves interact with each other through the process of *interference*. If the crests of the waves coincide (*in phase*), they interfere *constructively* and the amplitudes add together. If the crests coincide with troughs (*out of phase*), they interfere *destructively* and the amplitudes cancel. The result is a diffraction pattern of brighter and darker regions (Figure 7.5). In contrast, particles passing through adjacent openings continue in straight paths, some colliding with each other and moving at different angles.

At the end of the 19th century, all everyday and laboratory experience seemed to confirm these classical distinctions between the wave nature of energy and the particle nature of matter.

Rainbows and Diamonds You can see a rainbow only when the Sun is at your back. Light entering the near surface of a water droplet is dispersed and reflected off the far surface. Because red light is bent least, it reaches your eye from droplets higher in the sky, whereas violet appears from lower droplets. The colors in a diamond's sparkle are due to its facets, which are cleaved at angles that disperse and reflect the incoming light, lengthening its path enough for the different wavelengths to separate.

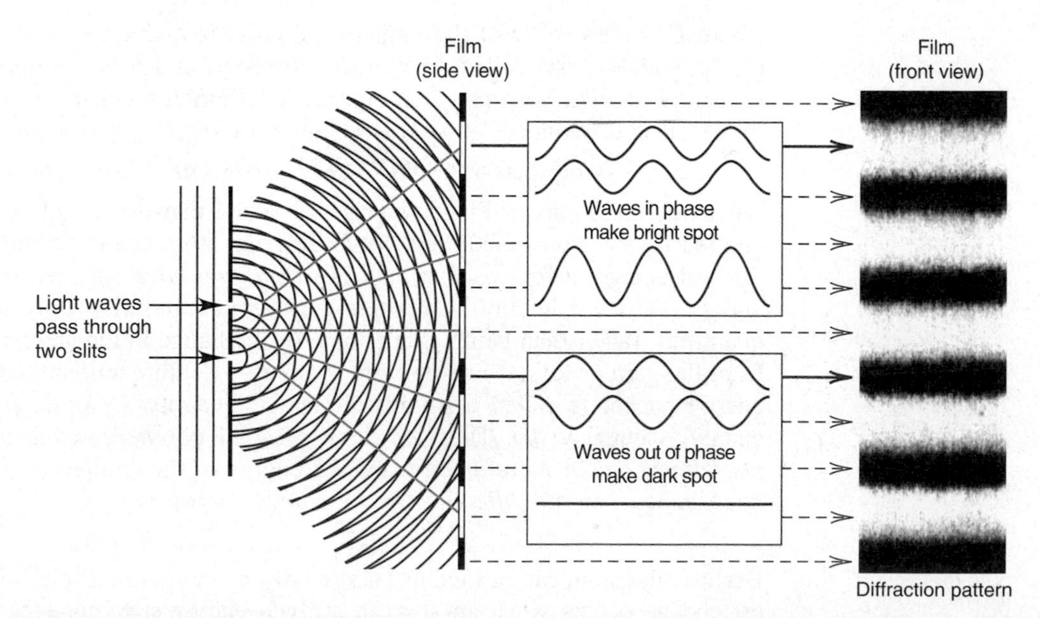

Figure 7.5 The diffraction pattern caused by light passing through two adjacent slits. As light waves pass through two closely spaced slits, they emerge as circular waves and interfere with each other. They create a diffraction (interference) pattern of bright and dark regions on a sheet of film. Bright regions appear where crests coincide and the amplitudes combine with each other (in phase); dark regions appear where crests meet troughs and the amplitudes cancel each other (out of phase).

The Particle Nature of Light

Three phenomena involving matter and light confounded physicists at the turn of the 20th century: (1) blackbody radiation, (2) the photoelectric effect, and (3) atomic spectra. Explaining these phenomena required a radically new picture of energy. We discuss the first two here and the third in the next section.

Blackbody Radiation and the Quantization of Energy When a solid object is heated to about 1000 K, it begins to emit visible light, as you can see in the soft red glow of smoldering coal (Figure 7.6, *left*). At about 1500 K, the light is brighter and more orange, like that from an electric heating coil (Figure 7.6, *center*). At temperatures greater than 2000 K, the light is still brighter and whiter, as seen in the filament of a lightbulb (Figure 7.6, *right*). These changes in intensity and wavelength of emitted light as an object is heated are characteristic of *blackbody radiation,* light given off by a hot *blackbody.** All attempts to account for these observed changes by applying classical electromagnetic theory failed.

Figure 7.6 Blackbody radiation. Familiar examples of the change in intensity and wavelength of light emitted by heated objects.

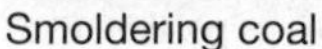

Smoldering coal

Electric heating element

Lightbulb filament

In 1900, the German physicist Max Planck (1858–1947) developed a formula that fit the data perfectly; to find a physical explanation for his formula, however, Planck was forced to make a radical assumption, which eventually led to an entirely new view of energy. He proposed that the hot, glowing object could emit (or absorb) only *certain quantities* of energy:

$$E = nh\nu$$

where E is the energy of the radiation, ν is its frequency, n is a positive integer (1, 2, 3, and so on) called a **quantum number,** and h is a proportionality constant now known very precisely and called **Planck's constant.** With energy in joules (J) and frequency in s^{-1}, h has units of J·s:

$$h = 6.62606876\times10^{-34}\ \text{J·s} = 6.626\times10^{-34}\ \text{J·s} \qquad \text{(4 sf)}$$

Later interpretations of Planck's proposal stated that the hot object's radiation is emitted by the atoms contained within it. If an atom can *emit* only certain quantities of energy, it follows that *the atom itself can have only certain quantities of energy.* Thus, the energy of an atom is *quantized:* it exists only in certain fixed quantities, rather than being continuous. Each change in the atom's energy results from the gain or loss of one or more "packets," definite amounts, of energy. Each energy packet is called a **quantum** ("fixed quantity"; plural, *quanta*), and its energy is equal to $h\nu$. Thus, *an atom changes its energy state by emitting (or absorbing) one or more quanta,* and the energy of the emitted (or absorbed) radiation is equal to the *difference in the atom's energy states:*

$$\Delta E_{\text{atom}} = E_{\text{emitted (or absorbed) radiation}} = \Delta nh\nu$$

Because the atom can change its energy only by integer multiples of $h\nu$, the smallest change occurs when an atom in a given energy state changes to an adjacent state, that is, when $\Delta n = 1$:

$$\Delta E = h\nu \qquad \textbf{(7.2)}$$

*A blackbody is an idealized object that absorbs all the radiation incident on it. A hollow cube with a small hole in one wall approximates a blackbody.

The Photoelectric Effect and the Photon Theory of Light Despite the idea that energy is quantized, Planck and other physicists continued to picture the emitted energy as traveling in waves. However, the wave model could not explain the **photoelectric effect,** the flow of current when monochromatic light of sufficient frequency shines on a metal plate (Figure 7.7). The existence of the current was not puzzling: it could be understood as arising when the light transfers energy to the electrons at the metal surface, which break free and are collected by the positive electrode. However, the photoelectric effect had certain confusing features, in particular, the *presence of a threshold frequency* and the *absence of a time lag:*

1. *Presence of a threshold frequency.* Light shining on the metal must have a minimum *frequency,* or no current flows. (Different metals have different minimum frequencies.) The wave theory, however, associates the energy of the light with the *amplitude* (intensity) of the wave, not with its frequency (color). Thus, the wave theory predicts that an electron would break free when it absorbed enough energy from light of *any* color.
2. *Absence of a time lag.* Current flows the moment light of this minimum frequency shines on the metal, regardless of the light's intensity. The wave theory, however, predicts that in dim light there would be a time lag before the current flowed, because the electrons had to absorb enough energy to break free.

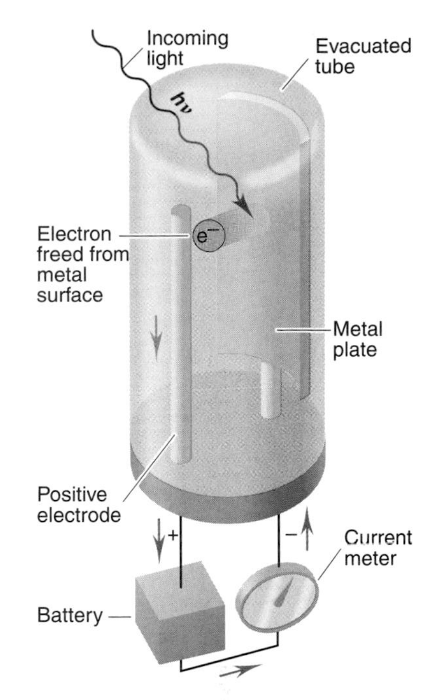

Figure 7.7 Demonstration of the photoelectric effect. When monochromatic light of high enough frequency strikes the metal plate, electrons are freed from the plate and travel to the positive electrode, creating a current.

Carrying Planck's idea of quantized energy further, Einstein proposed that light itself is particulate, that is, quantized into small "bundles" of electromagnetic energy, which were later called **photons.** In terms of Planck's work, we can say that each atom changes its energy whenever it absorbs or emits one photon, one "particle" of light, whose energy is fixed by its *frequency:*

$$E_{\text{photon}} = h\nu = \Delta E_{\text{atom}}$$

Let's see how Einstein's photon theory explains the photoelectric effect:

1. *Explanation of the threshold frequency.* According to the photon theory, a beam of light consists of an enormous number of photons. Light intensity (brightness) is related to the number of photons striking the surface per unit time, but *not* to their energy. Therefore, a photon of a certain minimum *energy* must be absorbed for an electron to be freed. Since energy depends on frequency ($h\nu$), the theory predicts a threshold frequency.
2. *Explanation of the time lag.* An electron cannot "save up" energy from several photons below the minimum energy until it has enough to break free. Rather, one electron breaks free the moment it absorbs one photon of *enough* energy. The current is weaker in dim light than in bright light because fewer photons of enough energy are present, so fewer electrons break free per unit time. But some current flows the moment photons reach the metal plate.

Ping-Pong Photons Consider this analogy for the fact that light of insufficient energy can't free an electron from the metal surface. If one Ping-Pong ball doesn't have enough energy to knock a book off a shelf, neither does a series of Ping-Pong balls, because the book can't save up the energy from the individual impacts. But one baseball traveling at the same speed does have enough energy.

SAMPLE PROBLEM 7.2 Calculating the Energy of Radiation from Its Wavelength

Problem A cook uses a microwave oven to heat a meal. The wavelength of the radiation is 1.20 cm. What is the energy of one photon of this microwave radiation?

Plan We know λ in centimeters (1.20 cm) so we convert to meters, find the frequency with Equation 7.1, and then find the energy of one photon with Equation 7.2.

Solution Combining steps to find the energy:

$$E = h\nu = \frac{hc}{\lambda} = \frac{(6.626\times10^{-34}\ \text{J}\cdot\text{s})(3.00\times10^{8}\ \text{m/s})}{(1.20\ \text{cm})\left(\dfrac{10^{-2}\ \text{m}}{1\ \text{cm}}\right)} = 1.66\times10^{-23}\ \text{J}$$

Check Checking the order of magnitude gives $\dfrac{10^{-33}\ \text{J}\cdot\text{s}\times10^{8}\ \text{m/s}}{10^{-2}\ \text{m}} = 10^{-23}\ \text{J}.$

FOLLOW-UP PROBLEM 7.2 Calculate the energies of one photon of ultraviolet ($\lambda = 1\times10^{-8}$ m), visible ($\lambda = 5\times10^{-7}$ m), and infrared ($\lambda = 1\times10^{-4}$ m) light. What do the answers indicate about the relationship between the wavelength and energy of light?

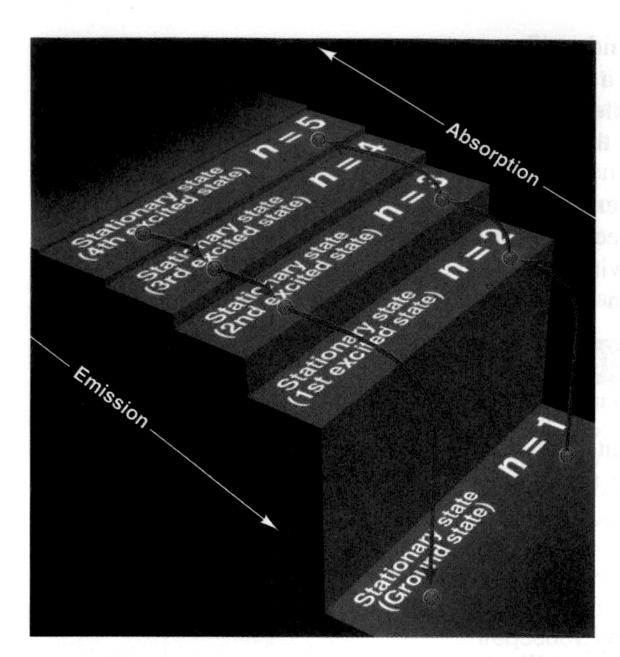

Figure 7.10 Quantum staircase. In this analogy for the energy levels of the hydrogen atom, an electron can absorb a photon and jump up to a higher "step" (stationary state) or emit a photon and jump down to a lower one. But the electron cannot lie between two steps.

3. *The atom changes to another stationary state* (the electron moves to another orbit) *only by absorbing or emitting a photon whose energy equals the difference in energy between the two states:*

$$E_{\text{photon}} = E_{\text{state A}} - E_{\text{state B}} = h\nu$$

where the energy of state A is higher than that of state B. A spectral line results when a photon of specific energy (and thus specific frequency) is *emitted* as the electron moves from a higher energy state to a lower one. Therefore, Bohr's model explains that an atomic spectrum is not continuous because *the atom's energy has only certain discrete levels, or states.*

In Bohr's model, the quantum number n (1, 2, 3, . . .) is associated with the radius of an electron orbit, which is directly related to the electron's energy: *the lower the n value, the smaller the radius of the orbit, and the lower the energy level.* When the electron is in the first orbit ($n = 1$), the orbit closest to the nucleus, the H atom is in its lowest (first) energy level, called the **ground state.** If the H atom absorbs a photon whose energy equals the *difference* between the first and second energy levels, the electron moves to the second orbit ($n = 2$), the next orbit out from the nucleus. When the electron is in the second or any higher orbit, the atom is said to be in an **excited state.** If the H atom in the first excited state (the electron in the second orbit) emits a photon of that same energy, it returns to the ground state. Figure 7.10 shows an analogy for this behavior.

Figure 7.11A shows how Bohr's model accounts for the three line spectra of hydrogen. When a sample of gaseous H atoms is excited, different atoms absorb different quantities of energy. Each atom has one electron, but so many atoms are present that all the energy levels (orbits) are populated by electrons. When the electrons drop from outer orbits to the $n = 3$ orbit (second excited state), the emitted photons create the infrared series of lines. The visible series arises when electrons drop to the $n = 2$ orbit (first excited state). Figure 7.11B shows that the ultraviolet series arises when electrons drop to the $n = 1$ orbit (ground state).

Limitations of the Bohr Model Despite its great success in accounting for the spectral lines of the H atom, the Bohr model failed to predict the spectrum of any other atom, even that of helium, the next simplest element. In essence, the Bohr model is a one-electron model. It works beautifully for the H atom and for other one-electron species, such as several created in the lab or seen in the spectra of stars: He^+ ($Z = 2$), Li^{2+} ($Z = 3$), Be^{3+} ($Z = 4$), B^{4+} ($Z = 5$), C^{5+} ($Z = 6$), N^{6+} ($Z = 7$), and O^{7+} ($Z = 8$). But it does not work for atoms with more than one electron because in these systems, additional nucleus-electron attractions and electron-electron repulsions are present. Moreover, as you'll soon see, electron movement in any real atom is far less clearly defined than the fixed orbits of the model. As a picture of the atom, the Bohr model is incorrect, but we still use the terms "ground state" and "excited state" and retain one of Bohr's central ideas in our current model: *the energy of an atom occurs in discrete levels.*

The Energy States of the Hydrogen Atom

A very useful result from Bohr's work is an equation for calculating the energy levels of an atom, which he derived from the classical principles of electrostatic attraction and circular motion:

$$E = -2.18\times10^{-18}\ \text{J}\left(\frac{Z^2}{n^2}\right)$$

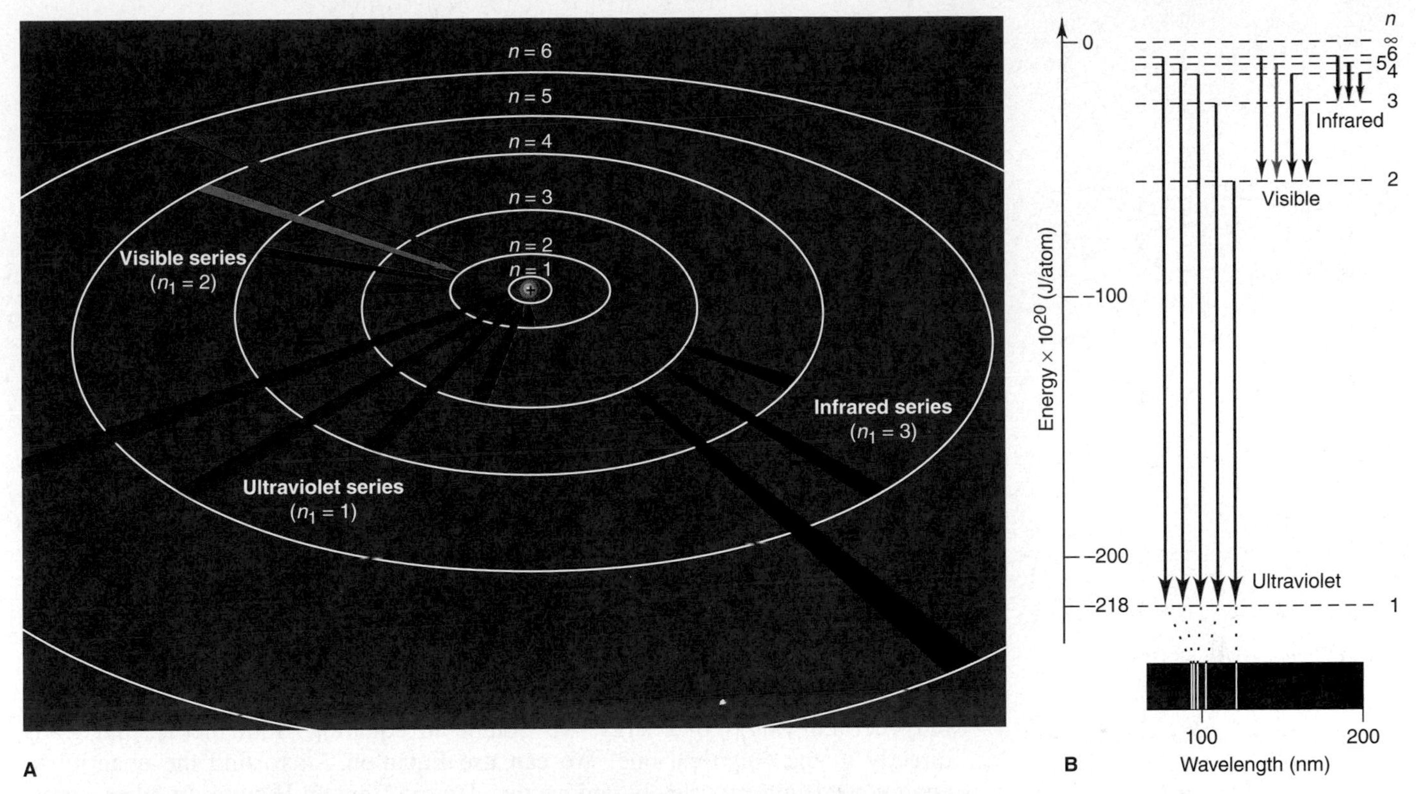

Figure 7.11 The Bohr explanation of three series of spectral lines. **A,** According to the Bohr model, when an electron drops from an outer orbit to an inner one, the atom emits a photon of specific energy that gives rise to a spectral line. In a given series, each electron drop, and thus each emission, has the same inner orbit, that is, the same value of n_1 in the Rydberg equation (see Equation 7.3). (The orbit radius is proportional to n^2. Only the first six orbits are shown.) **B,** An energy diagram shows how the ultraviolet series arises. Within each series, the greater the *difference* in orbit radii, the greater the difference in energy levels (depicted as a downward arrow), and the higher the energy of the photon emitted. For example, in the ultraviolet series, in which $n_1 = 1$, a drop from $n = 5$ to $n = 1$ emits a photon with more energy (shorter λ, higher ν) than a drop from $n = 2$ to $n = 1$. [The axis shows negative values because $n = \infty$ (the electron completely separated from the nucleus) is *defined* as the atom with zero energy.]

where Z is the charge of the nucleus. For the H atom, $Z = 1$, so we have

$$E = -2.18\times10^{-18}\ \text{J}\left(\frac{1^2}{n^2}\right) = -2.18\times10^{-18}\ \text{J}\left(\frac{1}{n^2}\right)$$

Therefore, the energy of the ground state ($n = 1$) is

$$E = -2.18\times10^{-18}\ \text{J}\left(\frac{1}{1^2}\right) = -2.18\times10^{-18}\ \text{J}$$

Don't be confused by the negative sign for the energy values (see the axis in Figure 7.11B). It appears because we *define* the zero point of the atom's energy when *the electron is completely removed from the nucleus.* Thus, $E = 0$ when $n = \infty$, so $E < 0$ for any smaller n. As an analogy, consider a book resting on the floor. You can define the zero point of the book's potential energy in many ways. If you define zero when the book is on the floor, the energy is positive when the book is on a tabletop. But, if you define zero when the book is on a tabletop, the energy is negative when the book lies on the floor; the latter case is analogous to the energy of the H atom (Figure 7.12).

Since n is in the denominator of the energy equation, as the electron moves closer to the nucleus (n decreases), the atom becomes more stable (less energetic) and its energy becomes a *larger negative number.* As the electron moves away from the nucleus (n increases), the atom's energy increases (becomes a smaller negative number).

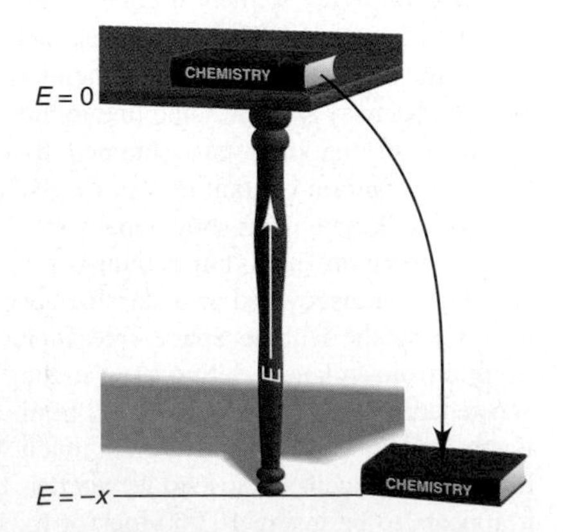

Figure 7.12 A tabletop analogy for the H atom's energy.

This equation is easily adapted to find the energy difference between any two levels:

$$\Delta E = E_{\text{final}} - E_{\text{initial}} = -2.18\times10^{-18}\,\text{J}\left(\frac{1}{n^2_{\text{final}}} - \frac{1}{n^2_{\text{initial}}}\right) \tag{7.4}$$

With it, we can predict the wavelengths of the spectral lines of the H atom. (In fact, Bohr obtained a value for the Rydberg constant that differed from the spectroscopists' value by only 0.05%!) Note that if we combine Equation 7.4 with Planck's expression for the change in an atom's energy (Equation 7.2), we obtain the Rydberg equation (Equation 7.3):

$$\Delta E = h\nu = \frac{hc}{\lambda} = -2.18\times10^{-18}\,\text{J}\left(\frac{1}{n^2_{\text{final}}} - \frac{1}{n^2_{\text{initial}}}\right)$$

Therefore,

$$\frac{1}{\lambda} = -\frac{2.18\times10^{-18}\,\text{J}}{hc}\left(\frac{1}{n^2_{\text{final}}} - \frac{1}{n^2_{\text{initial}}}\right)$$

$$= -\frac{2.18\times10^{-18}\,\text{J}}{(6.626\times10^{-34}\,\text{J·s})(3.00\times10^{8}\,\text{m/s})}\left(\frac{1}{n^2_{\text{final}}} - \frac{1}{n^2_{\text{initial}}}\right)$$

$$= -1.10\times10^{7}\,\text{m}^{-1}\left(\frac{1}{n^2_{\text{final}}} - \frac{1}{n^2_{\text{initial}}}\right)$$

where $n_{\text{final}} = n_2$, $n_{\text{initial}} = n_1$, and $1.10\times10^7\ \text{m}^{-1}$ is the Rydberg constant ($1.096776\times10^7\ \text{m}^{-1}$) to three significant figures. Thus, from classical relationships of charge and of motion combined with the idea that the H atom can have only certain values of energy, we obtain an equation from theory that leads directly to the empirical one! We can use Equation 7.4 to find the quantity of energy needed to completely remove the electron from an H atom. In other words, what is ΔE for the following change?

$$\text{H}(g) \longrightarrow \text{H}^+(g) + \text{e}^-$$

We substitute $n_{\text{final}} = \infty$ and $n_{\text{initial}} = 1$ and obtain

$$\Delta E = E_{\text{final}} - E_{\text{initial}} = -2.18\times10^{-18}\,\text{J}\left(\frac{1}{\infty^2} - \frac{1}{1^2}\right)$$

$$= -2.18\times10^{-18}\,\text{J}(0 - 1) = 2.18\times10^{-18}\,\text{J}$$

ΔE is positive because energy is *absorbed* to remove the electron from the vicinity of the nucleus. For 1 mol of H atoms,

$$\Delta E = \left(2.18\times10^{-18}\,\frac{\text{J}}{\text{atom}}\right)\left(6.022\times10^{23}\,\frac{\text{atoms}}{\text{mol}}\right)\left(\frac{1\ \text{kJ}}{10^3\ \text{J}}\right) = 1.31\times10^3\ \text{kJ/mol}$$

This is the *ionization energy* of the H atom, the quantity of energy required to form 1 mol of gaseous H^+ ions from 1 mol of gaseous H atoms. We return to this idea in Chapter 8.

Spectroscopic analysis of the H atom led to the Bohr model, the first step toward our current model of the atom. From its use by 19th-century chemists as a means of identifying elements and compounds, spectrometry has developed into a major tool of modern chemistry (see Tools of the Laboratory).

What Are Stars and Planets Made Of?
In 1868, the French astronomer Pierre Janssen noted a bright yellow line in the solar emission spectrum and thought it was emitted by an element unique to the Sun, which he named helium (Greek *helios*, "sun"). In 1888, the British chemist William Ramsay saw the same line in the spectrum of the inert gas obtained by heating uranium-containing minerals. Analysis of starlight has shown many elements known on Earth, but helium is the only element discovered on a star. Recent analysis by the Hubble Space Telescope of light from a planet orbiting a star in the constellation Pegasus, 150 light-years from Earth, reveals a hydrogen-gas giant much like Jupiter, except so close to its sun that it appears to be losing 10,000 tons of its atmosphere every second!

SECTION SUMMARY

To explain the line spectrum of atomic hydrogen, Bohr proposed that the atom's energy is quantized because the electron's motion is restricted to fixed orbits. The electron can move from one orbit to another only if the atom absorbs or emits a photon whose energy equals the difference in energy levels (orbits). Line spectra are produced because these energy changes correspond to photons of specific wavelength. Bohr's model predicted the hydrogen atomic spectrum but could not predict that of any other atom because electrons influence one another. Despite this, Bohr's idea that atoms have quantized energy levels is a cornerstone of our current atomic model. Spectrophotometry is an instrumental technique in which emission and absorption spectra are used to identify and measure concentrations of substances.

Spectrophotometry in Chemical Analysis

The use of spectral data to identify and quantify substances is essential to modern chemical analysis. The terms *spectroscopy, spectrometry,* and **spectrophotometry** denote a large group of instrumental techniques that obtain spectra corresponding to a substance's atomic and molecular energy levels.

The two types of spectra most often obtained are emission and absorption spectra. An **emission spectrum,** such as the H atom line spectrum, is produced when atoms in an excited state *emit* photons characteristic of the element as they return to lower energy states. Some elements produce a very intense spectral line (or several closely spaced ones) that serves as a marker of their presence. Such an intense line is the basis of **flame tests,** rapid qualitative procedures performed by placing a granule of an ionic compound or a drop of its solution in a flame (Figure B7.1, A). Some of the colors of fireworks and flares are due to emissions from the same elements shown in the flame tests: crimson from strontium salts and blue-green from copper salts (Figure B7.1, B). The characteristic colors of sodium-vapor and mercury-vapor streetlamps, seen in many towns and cities, are due to one or a few prominent lines in their emission spectra.

An **absorption spectrum** is produced when atoms *absorb* photons of certain wavelengths and become excited from lower to higher energy states. Therefore, the absorption spectrum of an element appears as dark lines against a bright background. When white light passes through sodium vapor, for example, it gives rise to a sodium absorption spectrum, and the dark lines appear at the same wavelengths as those for the yellow-orange lines in the sodium emission spectrum (Figure B7.2).

(continued)

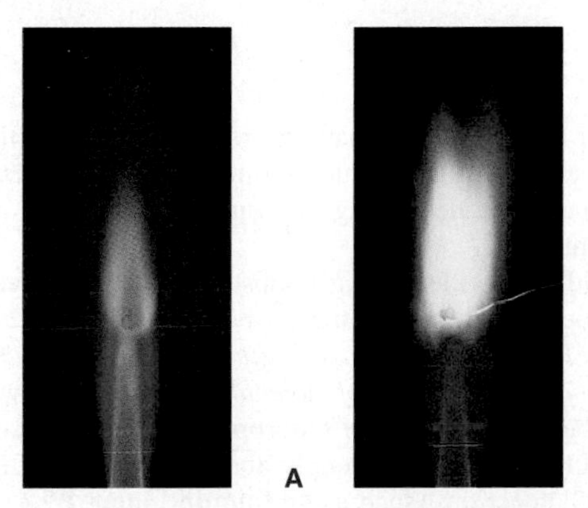

Figure B7.1 Flame tests and fireworks. A, In general, the color of the flame is created by a strong emission in the line spectrum of the element and therefore is often taken as preliminary evidence of the presence of the element in a sample. Shown here are the crimson of strontium and the blue-green of copper. **B,** The same emissions from compounds that contain these elements often appear in the brilliant displays of fireworks.

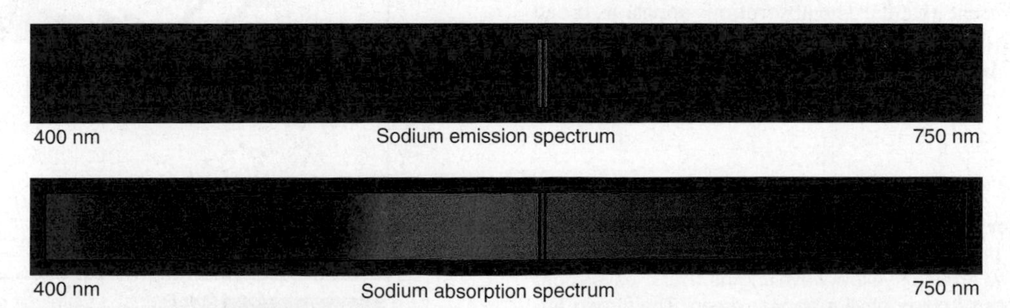

Figure B7.2 Emission and absorption spectra of sodium atoms. The wavelengths of the bright emission lines correspond to those of the dark absorption lines because both are created by the same energy change: $\Delta E_{emission} = -\Delta E_{absorption}$. (Only the two most intense lines in the Na spectra are shown.)

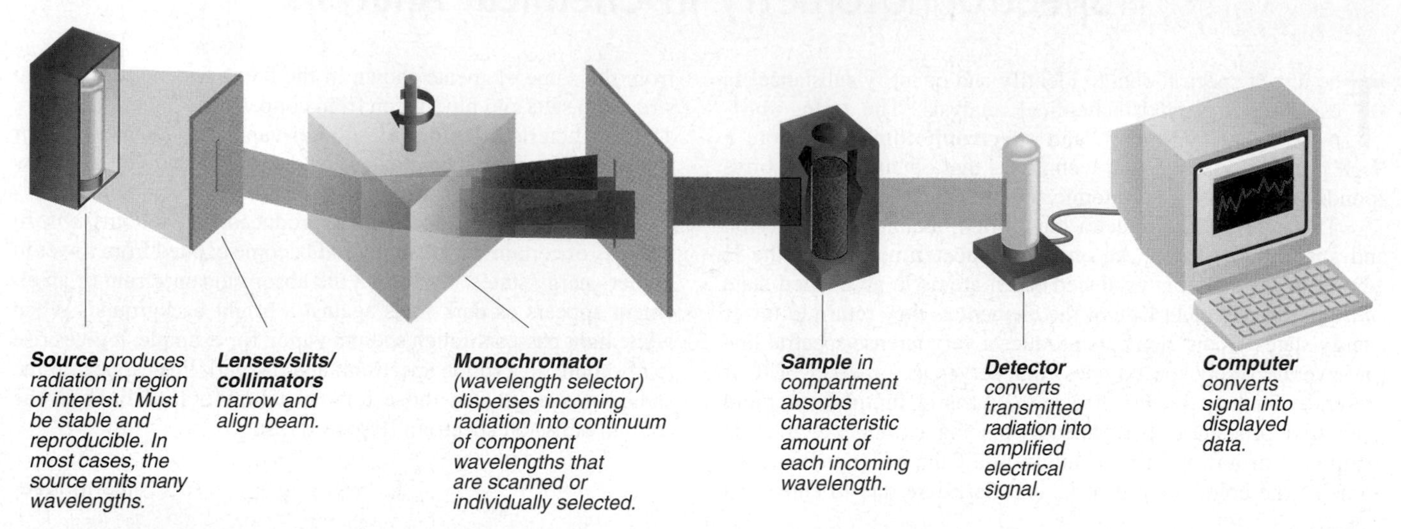
Source produces radiation in region of interest. Must be stable and reproducible. In most cases, the source emits many wavelengths.
Lenses/slits/ collimators narrow and align beam.
Monochromator (wavelength selector) disperses incoming radiation into continuum of component wavelengths that are scanned or individually selected.
Sample in compartment absorbs characteristic amount of each incoming wavelength.
Detector converts transmitted radiation into amplified electrical signal.
Computer converts signal into displayed data.

Figure B7.3 The main components of a typical spectrometer.

Instruments based on absorption spectra are much more common than those based on emission spectra, for several reasons. When a solid, liquid, or dense gas is excited, it *emits* so many lines that the spectrum is a continuum (recall the continuum of colors in sunlight). Absorption is also less destructive of fragile organic and biological molecules.

Despite differences that depend on the region of the electromagnetic spectrum used to irradiate the sample, all modern spectrometers have components that perform the same basic functions (Figure B7.3). (We discuss infrared spectroscopy and nuclear magnetic resonance spectroscopy in later chapters.)

Visible light is often used to study colored substances, which absorb only some of the wavelengths from white light. A leaf looks green, for example, because its chlorophyll absorbs red and blue wavelengths strongly and green weakly, so most of the green light is reflected. The absorption spectrum of chlorophyll *a* in ether solution appears in Figure B7.4.

The overall shape of the curve and the wavelengths of the major peaks are characteristic of chlorophyll *a,* so its spectrum serves as a means of identifying it from an unknown source. The curve varies in height because chlorophyll *a* absorbs incoming wavelengths to different extents. The absorptions appear as broad bands, rather than as the distinct lines we saw earlier for individual gaseous atoms, because dissolved substances, as well as pure solids and liquids, absorb many more wavelengths. This broader absorbance is due to the greater numbers and types of energy levels within a molecule, among molecules, and between molecules and solvent.

In addition to identifying a substance, a spectrometer can be used to measure its concentration because *the absorbance, the amount of light of a given wavelength absorbed by a substance, is proportional to the number of molecules.* Suppose you want to determine the concentration of chlorophyll in an ether solution of leaf extract. You select a strongly absorbed wavelength from the chlorophyll spectrum (such as 663 nm in Figure B7.4), measure the absorbance of the leaf-extract solution, and compare it with the absorbances of a series of ether solutions with known chlorophyll concentrations.

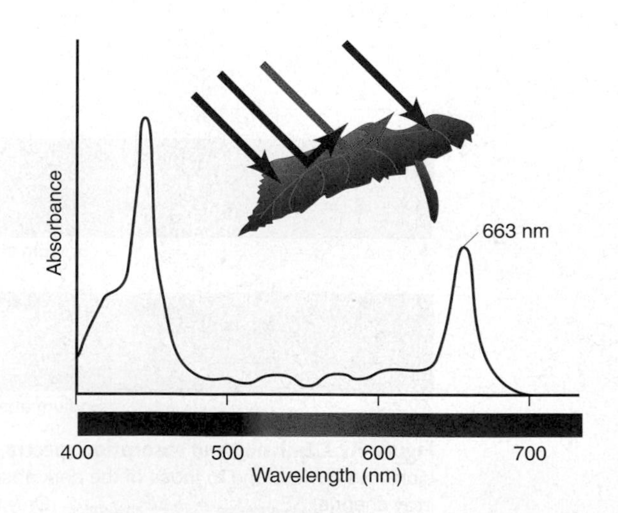
Absorbance
663 nm
400
500
600
700
Wavelength (nm)

Figure B7.4 The absorption spectrum of chlorophyll *a*. Chlorophyll *a* is one of several leaf pigments. It absorbs red and blue wavelengths strongly but almost no green or yellow wavelengths. Thus, leaves containing large amounts of chlorophyll *a* appear green. The strong absorption at 663 nm can be used to quantify the amount of chlorophyll *a* present in a plant extract.

7.3 THE WAVE-PARTICLE DUALITY OF MATTER AND ENERGY

The year 1905 was a busy one for Albert Einstein. In addition to presenting the photon theory of light and explaining the photoelectric effect, he found time to explain Brownian motion (Chapter 13), which helped establish the molecular view of matter, and to introduce a new branch of physics with his theory of relativity. One of its many startling revelations was that matter and energy are alternate forms of the same entity. This idea is embodied in his famous equation $E = mc^2$, which relates the quantity of energy equivalent to a given mass, and vice versa. Relativity theory does not depend on quantum theory, but together they have completely blurred the sharp divisions we normally perceive between matter (chunky and massive) and energy (diffuse and massless).

The early proponents of quantum theory demonstrated that *energy is particle-like*. Physicists who developed the theory turned this proposition upside down and showed that *matter is wavelike*. Strange as this idea may seem, it is the key to our modern atomic model.

"He'll Never Make a Success of Anything" This comment, attributed to the principal of young Albert Einstein's primary school, remains a classic of misperception. Contrary to myth, the greatest physicist of the 20th century (some say of all time) was not a poor student but an independent one, preferring his own path to that prescribed by authority—a trait that gave him the intense focus characteristic of all his work. A friend recalls finding him in his small apartment, rocking his baby in its carriage with one hand while holding a pencil stub and scribbling on a pad with the other. At age 26, he was working on one of the four papers he published in 1905 that would revolutionize the way the universe is perceived and lead to his 1921 Nobel Prize.

The Wave Nature of Electrons and the Particle Nature of Photons

Bohr's efforts were a perfect case of fitting theory to data: he *assumed* that an atom has only certain allowable energy levels in order to *explain* the observed line spectrum. However, his assumption had no basis in physical theory. Then, in the early 1920s, a young French physics student named Louis de Broglie proposed a startling reason for fixed energy levels: *if energy is particle-like, perhaps matter is wavelike*. De Broglie had been thinking of other systems that display only certain allowed motions, such as the wave of a plucked guitar string. Figure 7.13 shows that, because the ends of the string are fixed, only certain vibrational frequencies (and wavelengths) are possible. De Broglie reasoned that *if electrons have wavelike motion* and are restricted to orbits of fixed radii, that would explain why they have only certain possible frequencies and energies.

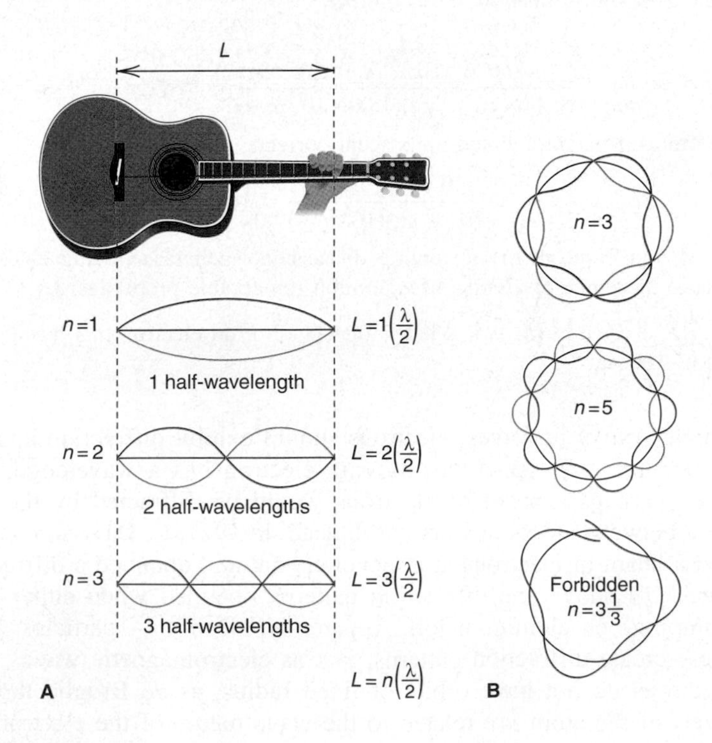

Figure 7.13 Wave motion in restricted systems. A, In a musical analogy to electron waves, one half-wavelength ($\lambda/2$) is the "quantum" of the guitar string's vibration. The string length L is fixed, so the only allowed vibrations occur when L is a whole-number multiple (n) of $\lambda/2$. **B,** If an electron occupies a circular orbit, only whole numbers of wavelengths are allowed ($n = 3$ and $n = 5$ are shown). A wave with a fractional number of wavelengths (such as $n = 3\frac{1}{3}$) is "forbidden" because it rapidly dies out through overlap of crests and troughs.

Combining the equation for mass-energy equivalence ($E = mc^2$) with that for the energy of a photon ($E = h\nu = hc/\lambda$), de Broglie derived an equation for the wavelength of any particle of mass m—whether planet, baseball, or electron—moving at speed u:

$$\lambda = \frac{h}{mu} \quad (7.5)$$

According to this equation for the **de Broglie wavelength,** *matter behaves as though it moves in a wave*. Note also that an object's wavelength is *inversely* proportional to its mass, so heavy objects such as planets and baseballs have wavelengths that are *many* orders of magnitude smaller than the object itself, as you can see in Table 7.1.

Table 7.1 The de Broglie Wavelengths of Several Objects

Substance	Mass (g)	Speed (m/s)	λ (m)
Slow electron	9×10^{-28}	1.0	7×10^{-4}
Fast electron	9×10^{-28}	5.9×10^{6}	1×10^{-10}
Alpha particle	6.6×10^{-24}	1.5×10^{7}	7×10^{-15}
One-gram mass	1.0	0.01	7×10^{-29}
Baseball	142	25.0	2×10^{-34}
Earth	6.0×10^{27}	3.0×10^{4}	4×10^{-63}

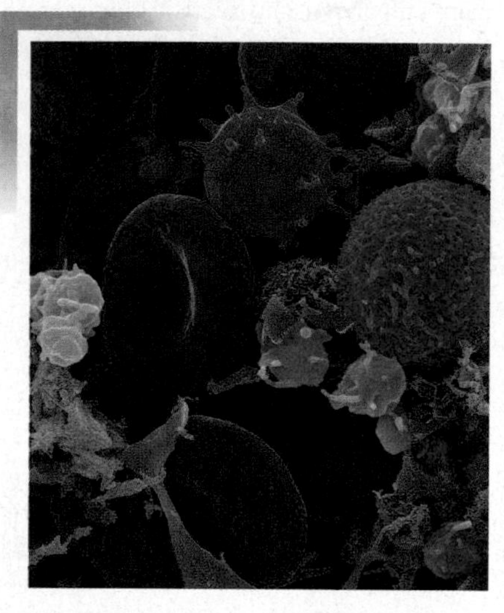

The Electron Microscope In a transmission electron microscope, a beam is focused by a lens and passes through a thin section of the specimen to a second lens. The resulting image is then magnified by a third lens to a final image. The differences between this and a light microscope are that the "beam" consists of high-speed electrons and the "lenses" are electromagnetic fields, which can be adjusted to give up to 200,000-fold magnification and 0.5-nm resolution. In a scanning electron microscope, the electron beam scans the specimen, knocking electrons from it, which creates a current that varies with surface irregularities. The current generates an image that looks like the object's surface, as in this false-color micrograph of various types of blood cells ($\times1200$). The great advantage of electron microscopes is that high-speed electrons have wavelengths much smaller than those of visible light and, thus, allow much higher image resolution.

SAMPLE PROBLEM 7.3 Calculating the de Broglie Wavelength of an Electron

Problem Find the de Broglie wavelength of an electron with a speed of 1.00×10^{6} m/s (electron mass $= 9.11\times10^{-31}$ kg; $h = 6.626\times10^{-34}$ kg·m²/s).

Plan We know the speed (1.00×10^{6} m/s) and mass (9.11×10^{-31} kg) of the electron, so we substitute these into Equation 7.5 to find λ.

Solution

$$\lambda = \frac{h}{mu} = \frac{6.626\times10^{-34}\ \text{kg·m}^2\text{/s}}{(9.11\times10^{-31}\ \text{kg})(1.00\times10^{6}\ \text{m/s})} = 7.27\times10^{-10}\ \text{m}$$

Check The order of magnitude and units seem correct:

$$\lambda \approx \frac{10^{-33}\ \text{kg·m}^2\text{/s}}{(10^{-30}\ \text{kg})(10^{6}\ \text{m/s})} = 10^{-9}\ \text{m}$$

Comment As you'll see in the upcoming discussion, such fast-moving electrons, with wavelengths in the range of atomic sizes, exhibit remarkable properties.

FOLLOW-UP PROBLEM 7.3 What is the speed of an electron that has a de Broglie wavelength of 100. nm?

If particles travel in waves, electrons should exhibit diffraction and interference (see Section 7.1). A fast-moving electron has a wavelength of about 10^{-10} m, so perhaps a beam of electrons would be diffracted by the similarly sized spaces between atoms in a crystal. Indeed, in 1927, C. Davisson and L. Germer guided a beam of electrons at a nickel crystal and obtained a diffraction pattern. Figure 7.14 shows the diffraction patterns obtained when either x-rays or electrons impinge on aluminum foil. Apparently, electrons—particles with mass and charge—create diffraction patterns, just as electromagnetic waves do! Even though electrons do not have orbits of fixed radius, as de Broglie thought, the energy levels of the atom *are* related to the wave nature of the electron.

If electrons have properties of energy, do photons have properties of matter? The de Broglie equation suggests that we can calculate the momentum (p), the product of mass and speed, for a photon of a given wavelength. Substituting the speed of light (c) for speed u in Equation 7.5 and solving for p gives

$$\lambda = \frac{h}{mc} = \frac{h}{p} \qquad \text{and} \qquad p = \frac{h}{\lambda}$$

Notice the inverse relationship between p and λ. This means that shorter wavelength (higher energy) photons have greater momentum. Thus, a decrease in a photon's momentum should appear as an increase in its wavelength. In 1923, Arthur Compton directed a beam of x-ray photons at a sample of graphite and observed that the wavelength of the reflected photons increased. This result means that the photons transferred some of their momentum to the electrons in the carbon atoms of the graphite, just as colliding billiard balls transfer momentum to one another. In this experiment, photons behave as particles with momentum!

To scientists of the time, these results were very unsettling. Classical experiments had shown matter to be particle-like and energy to be wavelike, but these new studies showed that every characteristic trait used to define the one now also defined the other. Figure 7.15 summarizes the conceptual and experimental breakthroughs that led to this juncture.

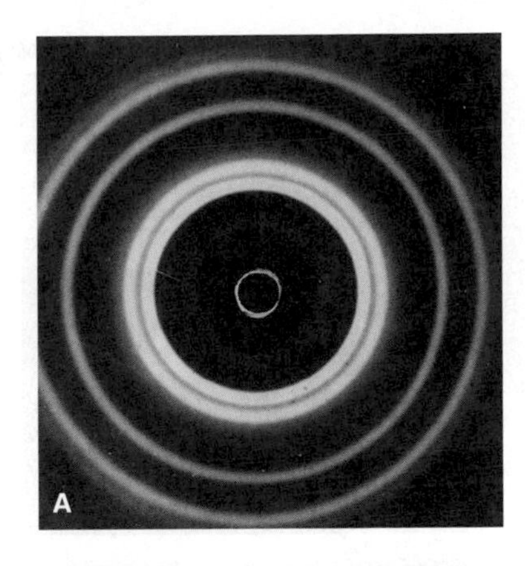

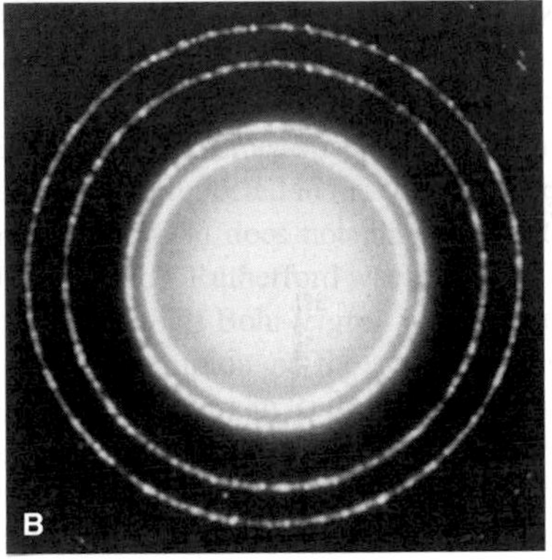

Figure 7.14 Comparing diffraction patterns of x-rays and electrons. A, X-ray diffraction pattern of aluminum. **B,** Electron diffraction pattern of aluminum. This behavior implies that both x-rays, which are electromagnetic radiation, and electrons, which are particles, travel in waves.

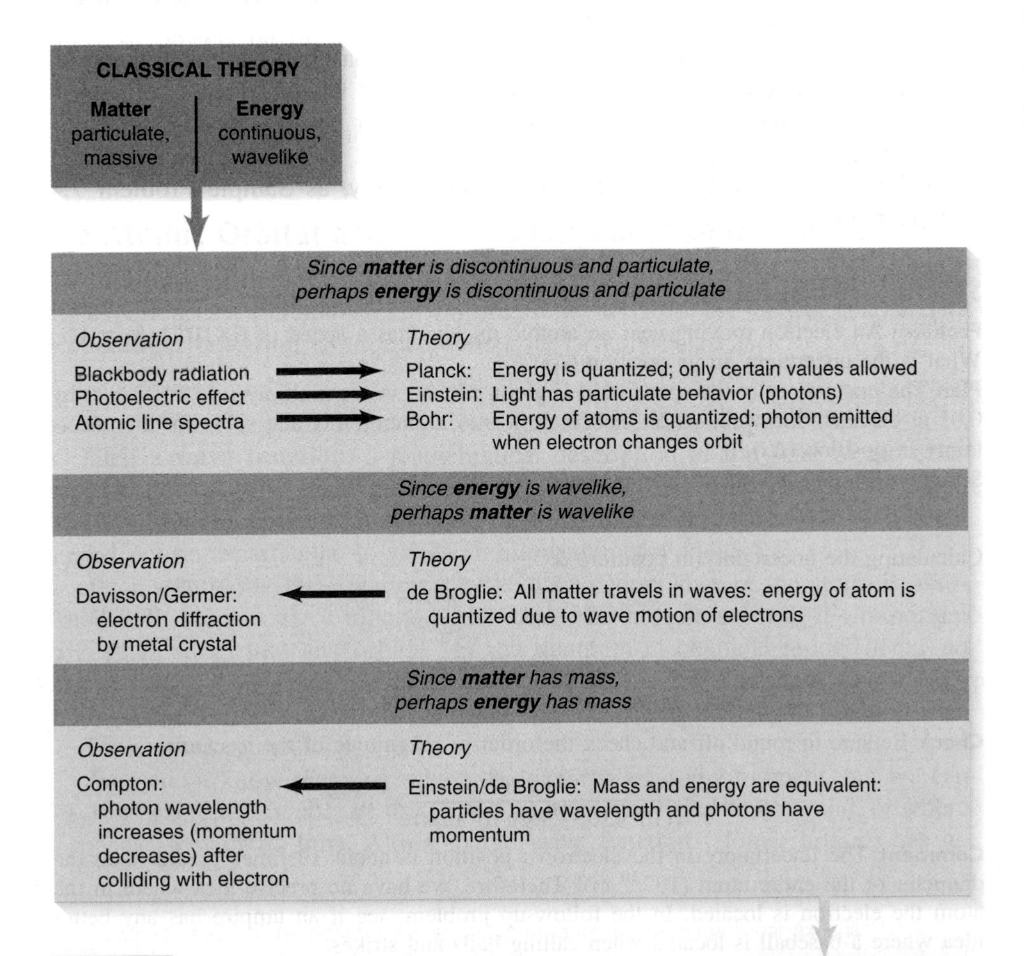

CLASSICAL THEORY
Matter
particulate, massive
Energy
continuous, wavelike
Since matter is discontinuous and particulate, perhaps energy is discontinuous and particulate
Observation
Theory
Blackbody radiation
Photoelectric effect
Atomic line spectra
Planck: Energy is quantized; only certain values allowed
Einstein: Light has particulate behavior (photons)
Bohr: Energy of atoms is quantized; photon emitted when electron changes orbit
Since energy is wavelike, perhaps matter is wavelike
Observation
Theory
Davisson/Germer: electron diffraction by metal crystal
de Broglie: All matter travels in waves: energy of atom is quantized due to wave motion of electrons
Since matter has mass, perhaps energy has mass
Observation
Theory
Compton: photon wavelength increases (momentum decreases) after colliding with electron
Einstein/de Broglie: Mass and energy are equivalent: particles have wavelength and photons have momentum

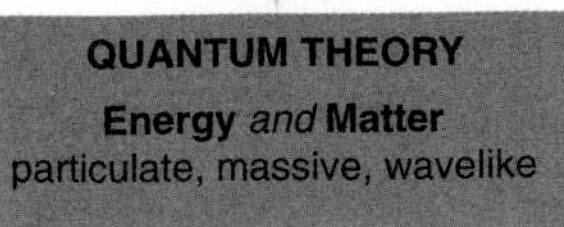

QUANTUM THEORY
Energy and Matter
particulate, massive, wavelike

Figure 7.15 Summary of the major observations and theories leading from classical theory to quantum theory. As often happens in science, an observation (experiment) stimulates the need for an explanation (theory), and/or a theoretical insight provides the impetus for an experimental test.

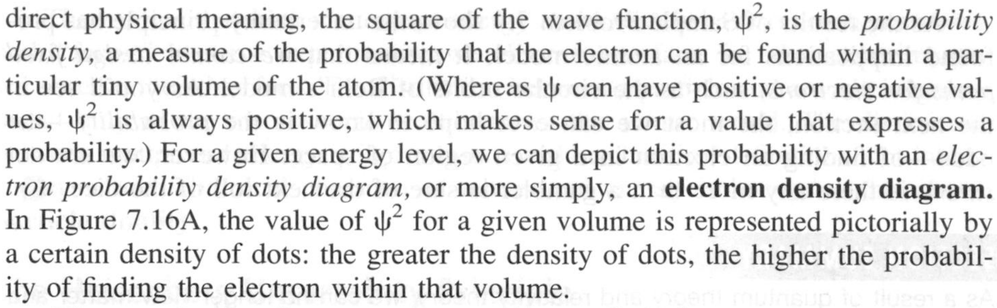

direct physical meaning, the square of the wave function, ψ^2, is the *probability density,* a measure of the probability that the electron can be found within a particular tiny volume of the atom. (Whereas ψ can have positive or negative values, ψ^2 is always positive, which makes sense for a value that expresses a probability.) For a given energy level, we can depict this probability with an *electron probability density diagram,* or more simply, an **electron density diagram.** In Figure 7.16A, the value of ψ^2 for a given volume is represented pictorially by a certain density of dots: the greater the density of dots, the higher the probability of finding the electron within that volume.

Electron density diagrams are sometimes called **electron cloud** representations. If we *could* take a time-exposure photograph of the electron in wavelike motion around the nucleus, it would appear as a "cloud" of electron positions. The electron cloud is an *imaginary* picture of the electron changing its position rapidly over time; it does *not* mean that an electron is a diffuse cloud of charge. Note that *the electron probability density decreases with distance from the nucleus* along a line, r. The same concept is shown graphically in the plot of ψ^2 vs. r in Figure 7.16B. Note that due to the thickness of the printed line, the curve touches the axis; nevertheless, *the probability of the electron being far from the nucleus is very small, but not zero.*

The *total* probability of finding the electron at any distance r from the nucleus is also important. To find this, we mentally divide the volume around the nucleus into thin, concentric, spherical layers, like the layers of an onion (shown in cross section in Figure 7.16C), and ask in which *spherical layer* we are most likely to find the electron. This is the same as asking for the *sum of* ψ^2 *values* within each spherical layer. The steep falloff in probability density with distance (see Figure 7.16B) has an important effect. Near the nucleus, the volume of each layer increases faster than its probability density decreases. As a result, the *total* probability of finding the electron in the second layer is higher than in the first. Electron density drops off so quickly, however, that this effect soon diminishes with greater distance. Thus, even though the volume of each layer continues to increase, the total probability for a given layer gradually decreases. Because of

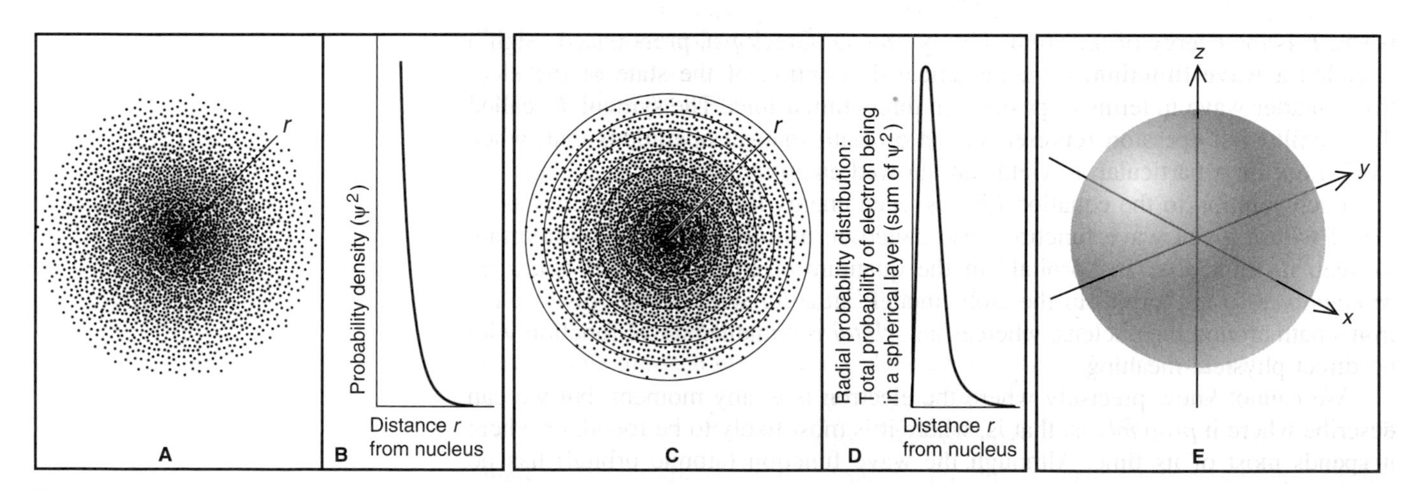

Figure 7.16 Electron probability density in the ground-state H atom. **A,** An electron density diagram shows a cross section of the H atom. The dots, each representing the probability of the electron being within a tiny volume, decrease along a line outward from the nucleus. **B,** A plot of the data in **A** shows that the probability density (ψ^2) decreases with distance from the nucleus but does not reach zero (the thickness of the line makes it appear to do so). **C,** Dividing the atom's volume into thin, concentric, spherical layers (shown in cross section) and counting the dots within each layer gives the total probability of finding the electron within that layer. **D,** A radial probability distribution plot shows total electron density in each spherical layer vs. r. Since electron density decreases more slowly than the volume of each concentric layer increases, the plot shows a peak. **E,** A 90% probability contour shows the ground state of the H atom (orbital of lowest energy) and represents the volume in which the electron spends 90% of its time.

these opposing effects of decreasing probability density and increasing layer volume, the total probability peaks in a layer some distance from the nucleus. Figure 7.16D shows this as a **radial probability distribution plot.**

The peak of the radial probability distribution for the ground-state H atom appears at the same distance from the nucleus (0.529Å, or 5.29×10^{-11} m) as Bohr postulated for the closest orbit. Thus, at least for the ground state, the Schrödinger model predicts that the electron spends *most* of its time at the same distance that the Bohr model predicted it spent *all* of its time. The difference between "most" and "all" reflects the uncertainty of the electron's location in the Schrödinger model.

How far away from the nucleus can we find the electron? This is the same as asking "How large is the atom?" Recall from Figure 7.16B that the probability of finding the electron far from the nucleus is not zero. Therefore, we *cannot* assign a definite volume to an atom. However, we often visualize atoms with a 90% **probability contour,** such as in Figure 7.16E, which shows the volume within which the electron of the hydrogen atom spends 90% of its time.

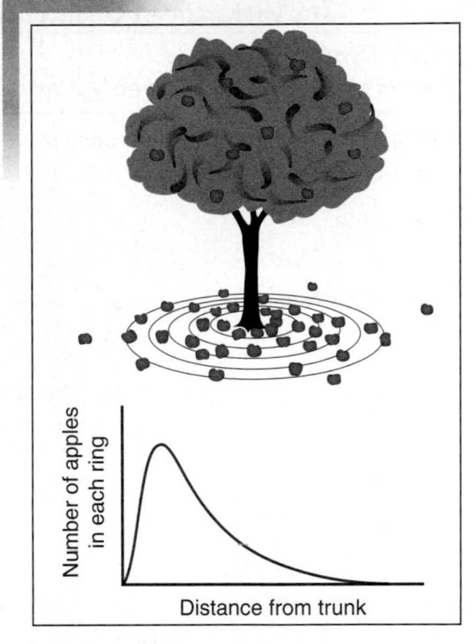

A Radial Probability Distribution of Apples An analogy might clarify why the curve in the radial probability distribution plot peaks and then falls off. Picture fallen apples around the base of an apple tree: the density of apples is greatest near the trunk and decreases with distance. Divide the ground under the tree into foot-wide concentric rings and collect the apples within each ring. Apple density is greatest in the first ring, but the area of the second ring is larger, and so it contains a greater *total* number of apples. Farther out near the edge of the tree, rings have more area but lower apple "density," so the total number of apples decreases. A plot of "number of apples within a ring" vs. "distance of ring from trunk" shows a peak at some distance from the trunk, as in Figure 7.16D.

Quantum Numbers of an Atomic Orbital

So far we have discussed the electron density for the *ground* state of the H atom. When the atom absorbs energy, it exists in an *excited* state and the region of space occupied by the electron is described by a different atomic orbital (wave function). As you'll see, each atomic orbital has a distinctive radial probability distribution and 90% probability contour.

An atomic orbital is specified by three quantum numbers. One is related to the orbital's size, another to its shape, and the third to its orientation in space.* The quantum numbers have a hierarchical relationship: the size-related number limits the shape-related number, which limits the orientation-related number. Let's examine this hierarchy and then look at the shapes and orientations.

1. The **principal quantum number (n)** *is a positive integer* (1, 2, 3, and so forth). It indicates the relative *size* of the orbital and therefore the relative *distance from the nucleus* of the peak in the radial probability distribution plot. The principal quantum number specifies the *energy level* of the H atom: *the higher the n value, the higher the energy level.* When the electron occupies an orbital with $n = 1$, the H atom is in its ground state and has lower energy than when the electron occupies the $n = 2$ orbital (first excited state).

2. The **angular momentum quantum number (l)** *is an integer from* 0 *to* $n - 1$. It is related to the *shape* of the orbital and is sometimes called the *orbital-shape quantum number.* Note that the principal quantum number sets a limit on the values for the angular momentum quantum number; that is, n limits l. For an orbital with $n = 1$, l can have a value of only 0. For orbitals with $n = 2$, l can have a value of 0 or 1; for those with $n = 3$, l can be 0, 1, or 2; and so forth. Note that the number of possible l values equals the value of n.

3. The **magnetic quantum number (m_l)** *is an integer from* $-l$ *through* 0 *to* $+l$. It prescribes the *orientation* of the orbital in the space around the nucleus and is sometimes called the *orbital-orientation quantum number.* The possible values of an orbital's magnetic quantum number are set by its angular momentum quantum number; that is, l sets the possible values of m_l. An orbital with $l = 0$ can have only $m_l = 0$. However, an orbital with $l = 1$ can have any one of three m_l values, -1, 0, or $+1$; thus, there are three possible orbitals with $l = 1$, each with its own orientation. Note that the number of possible m_l values *equals* the number of orbitals, which is $2l + 1$ for a given l value.

*For ease in discussion, we refer to the size, shape, and orientation of an "atomic orbital," although we really mean the size, shape, and orientation of an "atomic orbital's radial probability distribution." This usage is common in both introductory and advanced texts.

Table 7.2 The Hierarchy of Quantum Numbers for Atomic Orbitals

Name, Symbol (Property)	Allowed Values	Quantum Numbers		
Principal, n (size, energy)	Positive integer (1, 2, 3, . . .)	1	2	3
Angular momentum, l (shape)	0 to $n - 1$	0	0 ; 1	0 ; 1 ; 2
Magnetic, m_l (orientation)	$-l, \ldots, 0, \ldots, +l$	0	0 ; −1 0 +1	0 ; −1 0 +1 ; −2 −1 0 +1 +2

Table 7.2 summarizes the relationships among the three quantum numbers. (In Chapter 8, we'll discuss a fourth quantum number that relates to a property of the electron itself.) The total number of orbitals for a given n value is n^2.

SAMPLE PROBLEM 7.5 Determining Quantum Numbers for an Energy Level

Problem What values of the angular momentum (l) and magnetic (m_l) quantum numbers are allowed for a principal quantum number (n) of 3? How many orbitals exist for $n = 3$?

Plan We determine allowable quantum numbers with the rules from the text: l values are integers from 0 to $n - 1$, and m_l values are integers from $-l$ to 0 to $+l$. One m_l value is assigned to each orbital, so the number of m_l values gives the number of orbitals.

Solution Determining l values: for $n = 3$, $l = 0, 1, 2$

Determining m_l for each l value:

For $l = 0$, $m_l = 0$

For $l = 1$, $m_l = -1, 0, +1$

For $l = 2$, $m_l = -2, -1, 0, +1, +2$

There are nine m_l values, so there are nine orbitals with $n = 3$.

Check Table 7.2 shows that we are correct. The total number of orbitals for a given n value is n^2, and for $n = 3$, $n^2 = 9$.

FOLLOW-UP PROBLEM 7.5 Specify the l and m_l values for $n = 4$.

The energy states and orbitals of the atom are described with specific terms and associated with one or more quantum numbers:

1. *Level.* The atom's energy **levels,** or *shells,* are given by the n value: the smaller the n value, the lower the energy level and the greater the probability of the electron being closer to the nucleus.

2. *Sublevel.* The atom's levels contain **sublevels,** or *subshells,* which designate the orbital shape. Each sublevel has a letter designation:

$l = 0$ is an *s* sublevel.

$l = 1$ is a *p* sublevel.

$l = 2$ is a *d* sublevel.

$l = 3$ is an *f* sublevel.

(The letters derive from the names of spectroscopic lines: *s*harp, *p*rincipal, *d*iffuse, and *f*undamental. Sublevels with l values greater than 3 are designated alphabetically: *g* sublevel, *h* sublevel, etc.) Sublevels are named by joining the n value and the letter designation. For example, the sublevel (subshell) with $n = 2$ and $l = 0$ is called the 2*s* sublevel.

3. *Orbital.* Each allowed combination of n, l, and m_l values specifies one of the atom's *orbitals.* Thus, the three quantum numbers that describe an orbital express its size (energy), shape, and spatial orientation. You can easily give the quantum numbers of the orbitals in any sublevel if you know the sublevel letter designation and the quantum number hierarchy. For example, the hierarchy prescribes that the 2s sublevel has only one orbital, and its quantum numbers are $n = 2$, $l = 0$, and $m_l = 0$. The 3p sublevel has three orbitals: one with $n = 3$, $l = 1$, and $m_l = -1$; another with $n = 3$, $l = 1$, and $m_l = 0$; and a third with $n = 3$, $l = 1$, and $m_l = +1$.

SAMPLE PROBLEM 7.6 Determining Sublevel Names and Orbital Quantum Numbers

Problem Give the name, magnetic quantum numbers, and number of orbitals for each sublevel with the given quantum numbers:
(a) $n = 3$, $l = 2$ **(b)** $n = 2$, $l = 0$ **(c)** $n = 5$, $l = 1$ **(d)** $n = 4$, $l = 3$
Plan To name the sublevel (subshell), we combine the n value and l letter designation. Since we know l, we can find the possible m_l values, whose total number equals the number of orbitals.
Solution

n	l	Sublevel Name	Possible m_l Values	No. of Orbitals
(a) 3	2	3d	−2, −1, 0, +1, +2	5
(b) 2	0	2s	0	1
(c) 5	1	5p	−1, 0, +1	3
(d) 4	3	4f	−3, −2, −1, 0, +1, +2, +3	7

Check Check the number of orbitals in each sublevel using

$$\text{No. of orbitals} = \text{no. of } m_l \text{ values} = 2l + 1$$

FOLLOW-UP PROBLEM 7.6 What are the n, l, and possible m_l values for the 2p and 5f sublevels?

SAMPLE PROBLEM 7.7 Identifying Incorrect Quantum Numbers

Problem What is wrong with each of the following quantum number designations and/or sublevel names?

n	l	m_l	Name
(a) 1	1	0	1p
(b) 4	3	+1	4d
(c) 3	1	−2	3p

Solution (a) A sublevel with $n = 1$ can have only $l = 0$, not $l = 1$. The only possible sublevel name is 1s.
(b) A sublevel with $l = 3$ is an f sublevel, not a d sublevel. The name should be 4f.
(c) A sublevel with $l = 1$ can have only m_l of −1, 0, +1, not −2.
Check Check that l is always less than n, and m_l is always $\geq -l$ and $\leq +l$.

FOLLOW-UP PROBLEM 7.7 Supply the missing quantum numbers and sublevel names.

n	l	m_l	Name
(a) ?	?	0	4p
(b) 2	1	0	?
(c) 3	2	−2	?
(d) ?	?	?	2s

Shapes of Atomic Orbitals

Each sublevel of the H atom consists of a set of orbitals with characteristic shapes. As you'll see in Chapter 8, orbitals for the other atoms have similar shapes.

The *s* Orbital An orbital with $l = 0$ has a *spherical* shape with the nucleus at its center and is called an ***s* orbital.** The H atom's ground state, for example, has the electron in the 1*s* orbital, and *the electron probability density is highest at the nucleus*. Figure 7.17A shows this fact graphically *(top),* and an electron density *relief map (inset)* depicts this curve in three dimensions. The quarter-section of an electron cloud representation *(middle)* has the darkest shading at the nucleus. On the other hand, the radial probability distribution plot *(bottom),* which represents the probability of finding the electron (that is, where the electron spends

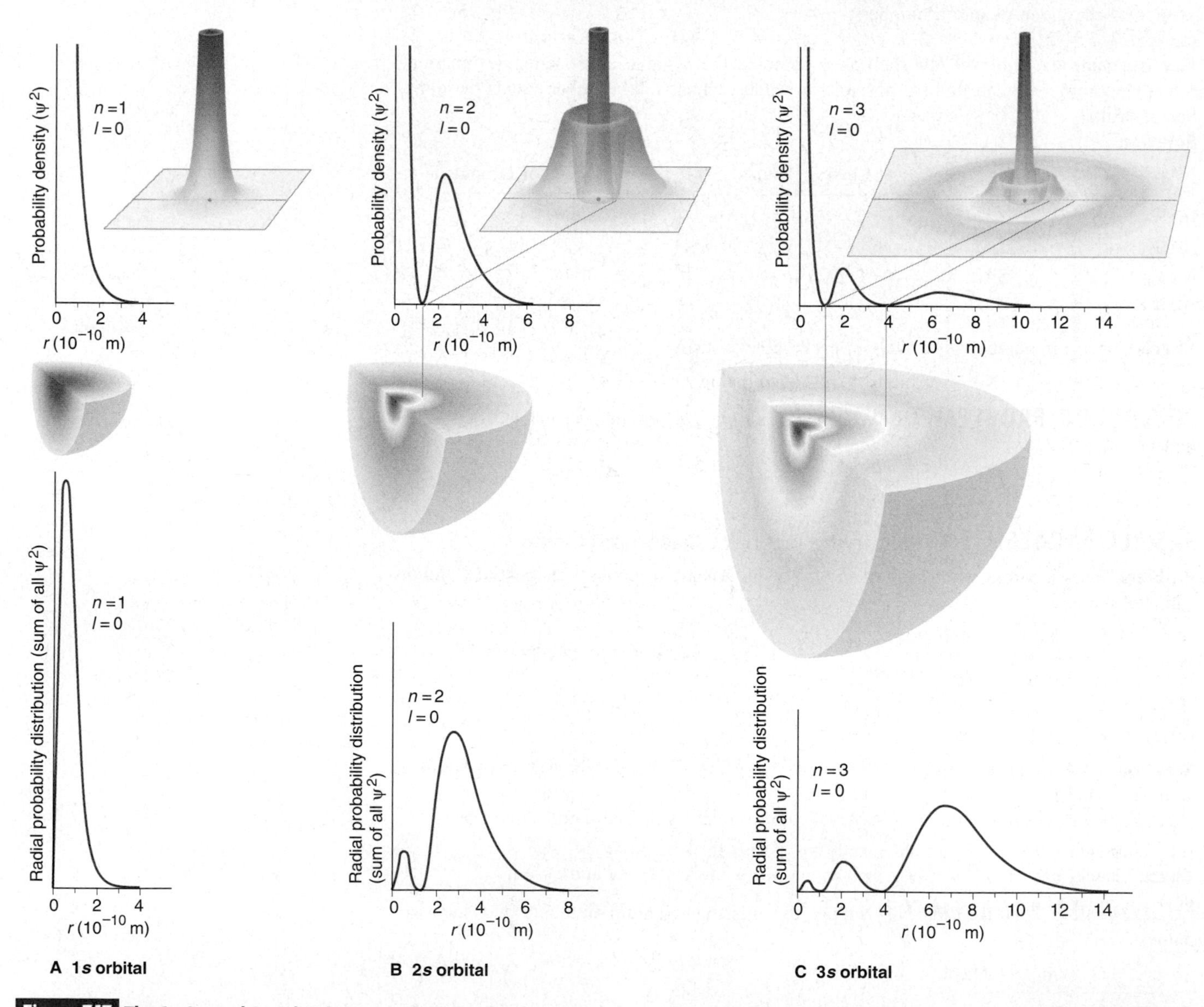

Figure 7.17 The 1*s*, 2*s*, and 3*s* orbitals. Information for each of the *s* orbitals is shown as a plot of probability density vs. distance *(top,* with the relief map, *inset,* showing the plot in three dimensions); as an electron cloud representation *(middle),* in which shading coincides with peaks in the plot above; and as a radial probability distribution *(bottom)* that shows where the electron spends its time. **A,** The 1*s* orbital. **B,** The 2*s* orbital. **C,** The 3*s* orbital. Nodes (regions of zero probability) appear in the 2*s* and 3*s* orbitals.

most of its time), is highest slightly out from the nucleus. Both plots fall off smoothly with distance.

The $2s$ orbital (Figure 7.17B) has two regions of higher electron density. The radial probability distribution (Figure 7.17B, *bottom*) of the more distant region is *higher* than that of the closer one because the sum of its ψ^2 is taken over a much larger volume. Between the two regions is a spherical **node,** a shell-like region where the probability drops to zero ($\psi^2 = 0$ at the node, analogous to zero amplitude of a wave). Because the $2s$ orbital is larger than the $1s$, an electron in the $2s$ spends more time *farther* from the nucleus than when it occupies the $1s$.

The $3s$ orbital, shown in Figure 7.17C, has three regions of high electron density and two nodes. Here again, the highest radial probability is at the greatest distance from the nucleus because the sum of all ψ^2 is taken over a larger volume. This pattern of more nodes and higher probability with distance continues for s orbitals of higher n value. An s orbital has a spherical shape, so it can have only one orientation and, thus, only one value for the magnetic quantum number: for any s orbital, $m_l = 0$.

The *p* Orbital An orbital with $l = 1$ has two regions (lobes) of high probability, one on *either side* of the nucleus, and is called a ***p* orbital.** Thus, as you can see in Figure 7.18, the *nucleus lies at the nodal plane* of this dumbbell-shaped orbital. The maximum value of l is $n - 1$, so only levels with $n = 2$ or higher can have a p orbital. Therefore, the lowest energy p orbital (the one closest to the nucleus) is the $2p$. Keep in mind that *one p orbital consists of both lobes* and that the electron spends *equal* time in both. As we would expect from the pattern of s orbitals, a $3p$ orbital is larger than a $2p$ orbital, a $4p$ orbital is larger than a $3p$ orbital, and so forth.

Unlike an s orbital, each p orbital *does* have a specific orientation in space. The $l = 1$ value has three possible m_l values: -1, 0, and $+1$, which refer to three *mutually perpendicular* p orbitals. They are identical in size, shape, and energy, differing only in orientation. For convenience, we associate p orbitals with the x, y, and z axes (but there is no necessary relation between a spatial axis and a given m_l value): the p_x orbital lies along the x axis, the p_y along the y axis, and the p_z along the z axis.

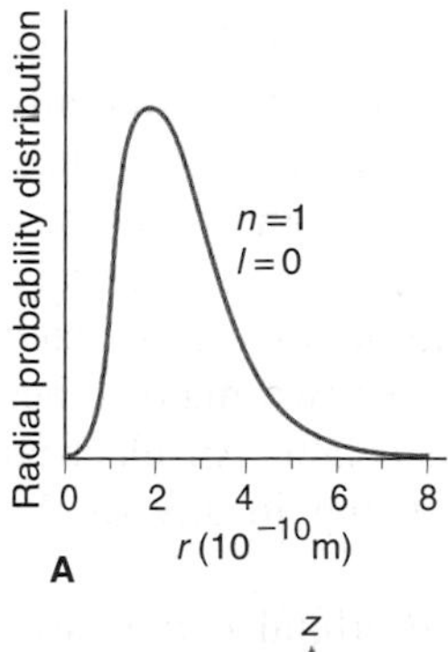

Figure 7.18 The 2*p* orbitals. A, A radial probability distribution plot of the $2p$ orbital shows a single peak. It lies at nearly the same distance from the nucleus as the larger peak in the $2s$ plot (shown in Figure 7.17B). **B,** A cross section shows an electron cloud representation of the 90% probability contour of the $2p_z$ orbital. An electron occupies both regions of a $2p$ orbital equally and spends 90% of its time within this volume. Note the nodal plane at the nucleus. **C,** An accurate representation of the $2p_z$ probability contour. The $2p_x$ and $2p_y$ orbitals have identical shapes but lie along the x and y axes, respectively. **D,** The stylized depiction of the $2p$ probability contour used throughout the text. **E,** In an atom, the three $2p$ orbitals occupy mutually perpendicular regions of space, contributing to the atom's overall spherical shape.

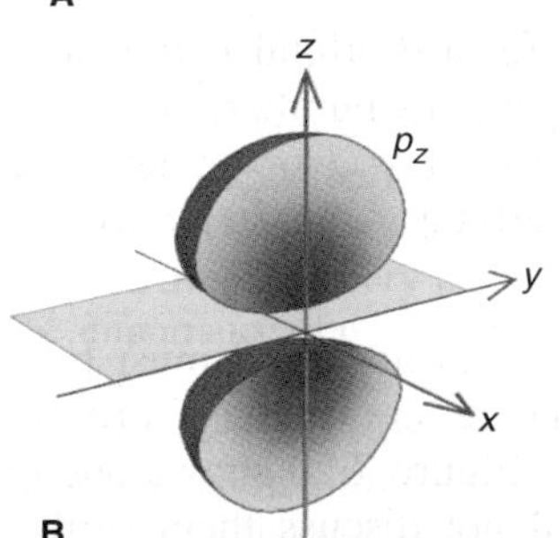

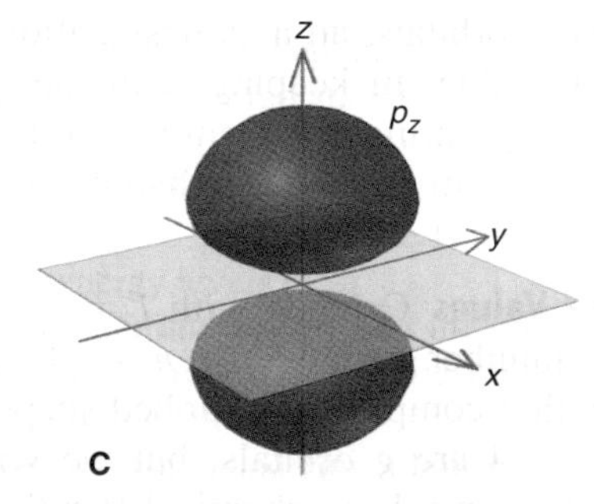

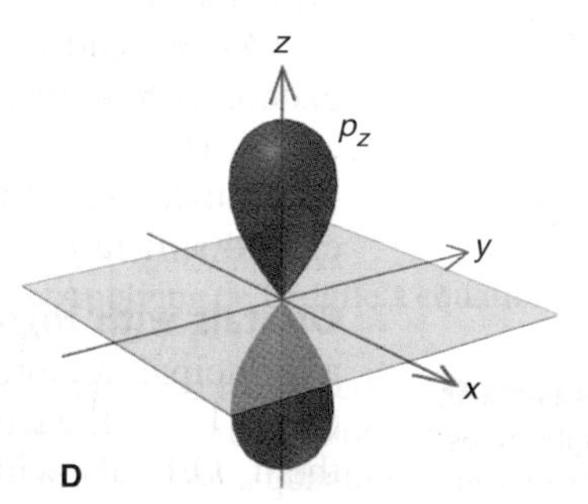

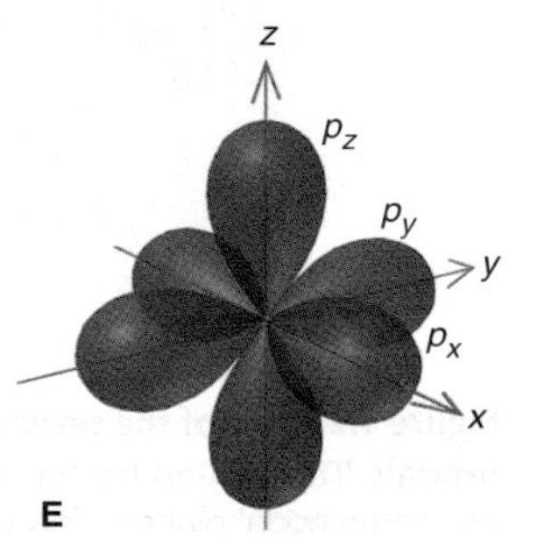

7.30 Arrange the following H atom electron transitions in order of *decreasing* wavelength of the photon absorbed or emitted:
(a) $n = 2$ to $n = \infty$ (b) $n = 4$ to $n = 20$
(c) $n = 3$ to $n = 10$ (d) $n = 2$ to $n = 1$

7.31 The electron in a ground-state H atom absorbs a photon of wavelength 97.20 nm. To what energy level does the electron move?

7.32 An electron in the $n = 5$ level of an H atom emits a photon of wavelength 1281 nm. To what energy level does the electron move?

Problems in Context

7.33 In addition to continuous radiation, fluorescent lamps emit sharp lines in the visible region from a mercury discharge within the tube. Much of this light has a wavelength of 436 nm. What is the energy (in J) of one photon of this light?

7.34 The oxidizing agents that are used in most fireworks consist of potassium salts, such as $KClO_4$ or $KClO_3$, rather than the corresponding sodium salts. One of the problems with using sodium salts is their extremely intense yellow-orange emission at 589 nm, which obscures the other colors in the display. What is the energy (in J) of one photon of this light? What is the energy (in kJ) of 1 einstein of this light (1 einstein = 1 mol of photons)?

The Wave-Particle Duality of Matter and Energy
(Sample Problems 7.3 and 7.4)

Concept Review Questions

7.35 In what sense is the wave motion of a guitar string analogous to the motion of an electron in an atom?

7.36 What experimental support did de Broglie's concept receive?

7.37 If particles have wavelike motion, why don't we observe that motion in the macroscopic world?

7.38 Why can't we overcome the uncertainty predicted by Heisenberg's principle by building more precise devices to reduce the error in measurements below the $h/4\pi$ limit?

Skill-Building Exercises *(grouped in similar pairs)*

7.39 A 220-lb fullback runs the 40-yd dash at a speed of 19.6 ± 0.1 mi/h.
(a) What is his de Broglie wavelength (in meters)?
(b) What is the uncertainty in his position?

7.40 An alpha particle (mass $= 6.6\times10^{-24}$ g) emitted by radium travels at $3.4\times10^7 \pm 0.1\times10^7$ mi/h.
(a) What is its de Broglie wavelength (in meters)?
(b) What is the uncertainty in its position?

7.41 How fast must a 56.5-g tennis ball travel in order to have a de Broglie wavelength that is equal to that of a photon of green light (5400 Å)?

7.42 How fast must a 142-g baseball travel in order to have a de Broglie wavelength that is equal to that of an x-ray photon with $\lambda = 100.$ pm?

7.43 A sodium flame has a characteristic yellow color due to emissions of wavelength 589 nm. What is the mass equivalence of one photon of this wavelength (1 J = 1 kg·m^2/s^2)?

7.44 A lithium flame has a characteristic red color due to emissions of wavelength 671 nm. What is the mass equivalence of 1 mol of photons of this wavelength (1 J = 1 kg·m^2/s^2)?

The Quantum-Mechanical Model of the Atom
(Sample Problems 7.5 to 7.7)

Concept Review Questions

7.45 What physical meaning is attributed to the square of the wave function, ψ^2?

7.46 Explain in your own words what the "electron density" in a particular tiny volume of space means.

7.47 Explain in your own words what it means for the peak in the radial probability distribution plot for the $n = 1$ level of a hydrogen atom to be at 0.529 Å. Is the probability of finding an electron at 0.529 Å from the nucleus greater for the 1*s* or the 2*s* orbital?

7.48 What feature of an orbital is related to each of the following quantum numbers?
(a) Principal quantum number (n)
(b) Angular momentum quantum number (l)
(c) Magnetic quantum number (m_l)

Skill-Building Exercises *(grouped in similar pairs)*

7.49 How many orbitals in an atom can have each of the following designations: (a) 1*s*; (b) 4*d*; (c) 3*p*; (d) $n = 3$?

7.50 How many orbitals in an atom can have each of the following designations: (a) 5*f*; (b) 4*p*; (c) 5*d*; (d) $n = 2$?

7.51 Give all possible m_l values for orbitals that have each of the following: (a) $l = 2$; (b) $n = 1$; (c) $n = 4, l = 3$.

7.52 Give all possible m_l values for orbitals that have each of the following: (a) $l = 3$; (b) $n = 2$; (c) $n = 6, l = 1$.

7.53 Draw 90% probability contours (with axes) for each of the following orbitals: (a) *s*; (b) p_x.

7.54 Draw 90% probability contours (with axes) for each of the following orbitals: (a) p_z; (b) d_{xy}.

7.55 For each of the following, give the sublevel designation, the allowable m_l values, and the number of orbitals:
(a) $n = 4, l = 2$ (b) $n = 5, l = 1$ (c) $n = 6, l = 3$

7.56 For each of the following, give the sublevel designation, the allowable m_l values, and the number of orbitals:
(a) $n = 2, l = 0$ (b) $n = 3, l = 2$ (c) $n = 5, l = 1$

7.57 For each of the following sublevels, give the n and l values and the number of orbitals: (a) 5*s*; (b) 3*p*; (c) 4*f*.

7.58 For each of the following sublevels, give the n and l values and the number of orbitals: (a) 6*g*; (b) 4*s*; (c) 3*d*.

7.59 Are the following quantum number combinations allowed? If not, show two ways to correct them:
(a) $n = 2$; $l = 0$; $m_l = -1$ (b) $n = 4$; $l = 3$; $m_l = -1$
(c) $n = 3$; $l = 1$; $m_l = 0$ (d) $n = 5$; $l = 2$; $m_l = +3$

7.60 Are the following quantum number combinations allowed? If not, show two ways to correct them:
(a) $n = 1$; $l = 0$; $m_l = 0$ (b) $n = 2$; $l = 2$; $m_l = +1$
(c) $n = 7$; $l = 1$; $m_l = +2$ (d) $n = 3$; $l = 1$; $m_l = -2$

Comprehensive Problems

7.61 The orange pigment in carrots and orange peel is mostly β-carotene, an organic compound that is insoluble in water but very soluble in benzene and chloroform. Describe an experimental method for determining the concentration of β-carotene in the oil expressed from orange peel.

7.62 The quantum-mechanical treatment of the hydrogen atom gives the energy, E, of the electron as a function of the principal quantum number, n:

$$E = -\frac{h^2}{8\pi^2 m_e a_0^2 n^2} \qquad (n = 1, 2, 3, \ldots)$$

where h is Planck's constant, m_e is the electron mass, and a_0 is 52.92×10^{-12} m.

(a) Write the expression in the form $E = -(\text{constant})\frac{1}{n^2}$, evaluate the constant (in J), and compare it with the corresponding expression from Bohr's theory.
(b) Use the expression to find ΔE between $n = 2$ and $n = 3$.
(c) Calculate the wavelength of the photon that corresponds to this energy change. Is this photon seen in the hydrogen spectrum obtained from experiment (see Figure 7.8, p. 264)?

7.63 The photoelectric effect is illustrated in a plot of the kinetic energies of electrons ejected from the surface of potassium metal or silver metal at different frequencies of incident light.

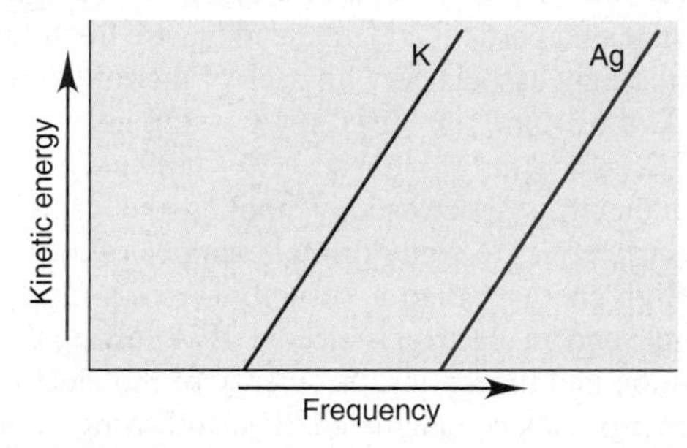

(a) Why don't the lines begin at the origin?
(b) Why don't the lines begin at the same point?
(c) From which metal will light of a shorter wavelength eject an electron?
(d) Why are the slopes of the lines equal?

7.64 The human eye is a complex sensing device for visible light. The optic nerve needs a minimum of 2.0×10^{-17} J of energy to trigger a series of impulses that eventually reach the brain.
(a) How many photons of red light (700. nm) are needed?
(b) How many photons of blue light (475 nm)?

7.65 One reason carbon monoxide (CO) is toxic is that it binds to the blood protein hemoglobin more strongly than oxygen does. The bond between hemoglobin and CO absorbs radiation of 1953 cm^{-1}. (The units are the reciprocal of the wavelength in centimeters.) Calculate the wavelength (in nm and Å) and the frequency (in Hz) of the absorbed radiation.

7.66 A metal ion M^{n+} has a single electron. The highest energy line in its emission spectrum occurs at a frequency of 2.961×10^{16} Hz. Identify the ion.

7.67 TV and radio stations transmit in specific frequency bands of the radio region of the electromagnetic spectrum.
(a) TV channels 2 to 13 (VHF) broadcast signals between the frequencies of 59.5 and 215.8 MHz, whereas FM radio stations broadcast signals with wavelengths between 2.78 and 3.41 m. Do these bands of signals overlap?
(b) AM radio signals have frequencies between 550 and 1600 kHz. Which has a broader transmission band, AM or FM?

7.68 Compare the wavelengths of an electron (mass = 9.11×10^{-31} kg) and a proton (mass = 1.67×10^{-27} kg), each having (a) a speed of 3.0×10^{6} m/s; (b) a kinetic energy of 2.5×10^{-15} J.

7.69 Five lines in the H atom spectrum have wavelengths (in Å): (a) 1212.7; (b) 4340.5; (c) 4861.3; (d) 6562.8; (e) 10938. Three lines result from transitions to $n_{\text{final}} = 2$ (visible series). The other two result from transitions in different series, one with $n_{\text{final}} = 1$ and the other with $n_{\text{final}} = 3$. Identify n_{initial} for each line.

7.70 In his explanation of the threshold frequency in the photoelectric effect, Einstein reasoned that the absorbed photon must have the minimum energy required to dislodge an electron from the metal surface. This energy is called the *work function* (ϕ) of that metal. What is the longest wavelength of radiation (in nm) that could cause the photoelectric effect in each of these metals?
(a) Calcium, $\phi = 4.60\times10^{-19}$ J
(b) Titanium, $\phi = 6.94\times10^{-19}$ J
(c) Sodium, $\phi = 4.41\times10^{-19}$ J

7.71 You have three metal samples—A, B, and C—that are tantalum (Ta), barium (Ba), and tungsten (W), but you don't know which is which. Metal A emits electrons in response to visible light; metals B and C require UV light. (a) Identify metal A, and find the longest wavelength that removes an electron. (b) What range of wavelengths would distinguish B and C? [The work functions are Ta (6.81×10^{-19} J), Ba (4.30×10^{-19} J), and W (7.16×10^{-19} J); work function is explained in Problem 7.70.]

7.72 Refractometry is an analytical method based on the difference between the speed of light as it travels through a substance (v) and its speed in a vacuum (c). In the procedure, light of known wavelength passes through a fixed thickness of the substance at a known temperature. The index of refraction equals c/v. Using yellow light (λ = 589 nm) at 20°C, for example, the index of refraction of water is 1.33 and that of diamond is 2.42. Calculate the speed of light in (a) water and (b) diamond.

7.73 A laser (*l*ight *a*mplification by *s*timulated *e*mission of *r*adiation) provides a coherent (in-phase) nearly monochromatic source of high-intensity light. Lasers are used in eye surgery, CD/DVD players, basic research, etc. Some modern dye lasers can be "tuned" to emit a desired wavelength. Fill in the blanks in the following table of the properties of some common lasers:

Type	λ (nm)	ν (s^{-1})	E (J)	Color
He-Ne	632.8	?	?	?
Ar	?	6.148×10^{14}	?	?
Ar-Kr	?	?	3.499×10^{-19}	?
Dye	663.7	?	?	?

7.74 As space exploration increases, means of communication with humans and probes on other planets are being developed. (a) How much time (in s) does it take for a radio wave of frequency 8.93×10^{7} s^{-1} to reach Mars, which is 8.1×10^{7} km from Earth? (b) If it takes this radiation 1.2 s to reach the Moon, how far (in m) is the Moon from Earth?

7.75 The following quantum number combinations are not allowed. Assuming the n and m_l values are correct, change the l value to create an allowable combination:
(a) $n = 3$; $l = 0$; $m_l = -1$ (b) $n = 3$; $l = 3$; $m_l = +1$
(c) $n = 7$; $l = 2$; $m_l = +3$ (d) $n = 4$; $l = 1$; $m_l = -2$

7.76 A ground-state H atom absorbs a photon of wavelength 94.91 nm, and its electron attains a higher energy level. The atom then emits two photons: one of wavelength 1281 nm to reach an intermediate level, and a second to reach the ground state.
(a) What higher level did the electron reach?
(b) What intermediate level did the electron reach?
(c) What was the wavelength of the second photon emitted?

7.77 Consider these ground-state ionization energies of one-electron species:

$$H = 1.31\times10^3 \text{ kJ/mol}$$
$$He^+ = 5.24\times10^3 \text{ kJ/mol}$$
$$Li^{2+} = 1.18\times10^4 \text{ kJ/mol}$$

(a) Write a general expression for the ionization energy of any one-electron species. (b) Use your expression to calculate the ionization energy of B^{4+}. (c) What is the minimum wavelength required to remove the electron from the $n = 3$ level of He^+? (d) What is the minimum wavelength required to remove the electron from the $n = 2$ level of Be^{3+}?

7.78 Why do the spaces between spectral lines within a series decrease as the wavelength becomes shorter?

7.79 In the course of developing his model, Bohr arrived at the following formula for the radius of the electron's orbit: $r_n = n^2h^2\epsilon_0/\pi m_e e^2$, where m_e is the electron mass, e is its charge, and ϵ_0 is a constant related to charge attraction in a vacuum. Given that $m_e = 9.109\times10^{-31}$ kg, $e = 1.602\times10^{-19}$ C, and $\epsilon_0 = 8.854\times10^{-12}$ C^2/J·m, calculate the following:
(a) The radius of the 1st ($n = 1$) orbit in the H atom
(b) The radius of the 10th ($n = 10$) orbit in the H atom

7.80 (a) Calculate the Bohr radius of an electron in the $n = 3$ orbit of a hydrogen atom. (See Problem 7.79.)
(b) What is the energy (in J) of the atom in part (a)?
(c) What is the energy of an Li^{2+} ion when its electron is in the $n = 3$ orbit?
(d) Why are the answers to parts (b) and (c) different?

7.81 Enormous numbers of microwave photons are needed to warm macroscopic samples of matter. A portion of soup containing 252 g of water is heated in a microwave oven from 20.°C to 98°C, with radiation of wavelength 1.55×10^{-2} m. How many photons are absorbed by the water in the soup?

7.82 The quantum-mechanical treatment of the hydrogen atom gives an expression for the wave function, ψ, of the 1s orbital:

$$\psi = \frac{1}{\sqrt{\pi}}\left(\frac{1}{a_0}\right)^{3/2} e^{-r/a_0}$$

where r is the distance from the nucleus and a_0 is 52.92 pm. The electron probability density is the probability of finding the electron in a tiny volume at distance r from the nucleus and is proportional to ψ^2. The radial probability distribution is the total probability of finding the electron at all points at distance r from the nucleus and is proportional to $4\pi r^2\psi^2$. Calculate the values (to three significant figures) of ψ, ψ^2, and $4\pi r^2\psi^2$ to fill in the following table, and sketch plots of these quantities versus r. Compare the latter two plots with those in Figure 7.17A, p. 280:

r (pm)	ψ ($pm^{-3/2}$)	ψ^2 (pm^{-3})	$4\pi r^2\psi^2$ (pm^{-1})
0			
50			
100			
200			

7.83 Lines in one spectral series can overlap lines in another.
(a) Use the Rydberg equation to show whether the range of wavelengths in the $n_1 = 1$ series overlaps with the range in the $n_1 = 2$ series.
(b) Use the Rydberg equation to show whether the range of wavelengths in the $n_1 = 3$ series overlaps with the range in the $n_1 = 4$ series.
(c) How many lines in the $n_1 = 4$ series lie in the range of the $n_1 = 5$ series?
(d) What does this overlap imply about the hydrogen spectrum at longer wavelengths?

7.84 The following values are the only allowable energy levels of a hypothetical one-electron atom:

$$E_6 = -2\times10^{-19}\text{ J} \qquad E_5 = -7\times10^{-19}\text{ J}$$
$$E_4 = -11\times10^{-19}\text{ J} \qquad E_3 = -15\times10^{-19}\text{ J}$$
$$E_2 = -17\times10^{-19}\text{ J} \qquad E_1 = -20\times10^{-19}\text{ J}$$

(a) If the electron were in the $n = 3$ level, what would be the highest frequency (and minimum wavelength) of radiation that could be emitted?
(b) What is the ionization energy (in kJ/mol) of the atom in its ground state?
(c) If the electron were in the $n = 4$ level, what would be the shortest wavelength (in nm) of radiation that could be absorbed without causing ionization?

7.85 In fireworks displays, light of a given wavelength indicates the presence of a particular element. What are the frequency and color of the light associated with each of the following?
(a) Li^+, $\lambda = 671$ nm (b) Cs^+, $\lambda = 456$ nm
(c) Ca^{2+}, $\lambda = 649$ nm (d) Na^+, $\lambda = 589$ nm

7.86 Photoelectron spectroscopy applies the principle of the photoelectric effect to study orbital energies of atoms and molecules. High-energy radiation (usually UV or x-ray) is absorbed by a sample and an electron is ejected. By knowing the energy of the radiation and measuring the energy of the electron lost, the orbital energy can be calculated. The following energy differences were determined for several electron transitions:

$$\Delta E_{2\longrightarrow1} = 4.088\times10^{-17}\text{ J}$$
$$\Delta E_{3\longrightarrow1} = 4.844\times10^{-17}\text{ J}$$
$$\Delta E_{5\longrightarrow1} = 5.232\times10^{-17}\text{ J}$$
$$\Delta E_{4\longrightarrow2} = 1.022\times10^{-17}\text{ J}$$

Calculate the energy change and the wavelength of a photon emitted in the following transitions.
(a) Level 3 $\longrightarrow$ 2 (b) Level 4 $\longrightarrow$ 1 (c) Level 5 $\longrightarrow$ 4

7.87 Visible light provides green plants with the energy needed to drive photosynthesis. Horticulturists know that, for many plants, leaf color depends on how brightly lit the growing area is: dark green leaves are associated with low light levels, and pale green with high levels. (a) Use the photon theory of light to explain why a plant adapts this way. (b) What change in leaf composition might account for this behavior?

7.88 In compliance with conservation of energy, Einstein explained that in the photoelectric effect, the energy of a photon ($h\nu$) absorbed by a metal is the sum of the work function (ϕ), the minimum energy needed to dislodge an electron from the metal's surface, and the kinetic energy (E_k) of the electron: $h\nu = \phi + E_k$. When light of wavelength 358.1 nm falls on the surface of potassium metal, the speed (u) of the dislodged electron is 6.40×10^5 m/s.
(a) What is E_k ($\frac{1}{2}mu^2$) of the dislodged electron?
(b) What is ϕ (in J) of potassium?

7.89 An electron microscope focuses electrons through magnetic lenses to observe objects at higher magnification than is possible with a light microscope. For any microscope, the smallest object that can be observed is one-half the wavelength of the light used. Thus, for example, the smallest object that can be observed with light of 400 nm is 2×10^{-7} m. (a) What is the smallest object observable with an electron microscope using electrons moving at 5.5×10^4 m/s? (b) At 3.0×10^7 m/s?

7.90 In a typical fireworks device, the heat of the reaction between a strong oxidizing agent, such as $KClO_4$, and an organic compound excites certain salts, which emit specific colors. Strontium salts have an intense emission at 641 nm, and barium salts have one at 493 nm. (a) What colors do these emissions produce? (b) What is the energy (in kJ) of these emissions for 1.00 g each of the chloride salts of Sr and Ba? (Assume that all the heat released is converted to light emitted.)

7.91 Atomic hydrogen produces well-known series of spectral lines in several regions of the electromagnetic spectrum. Each series fits the Rydberg equation with its own particular n_1 value. Calculate the value of n_1 (by trial and error if necessary) that would produce a series of lines in which:
(a) The *highest* energy line has a wavelength of 3282 nm.
(b) The *lowest* energy line has a wavelength of 7460 nm.

7.92 Fish-liver oil is a good source of vitamin A, which is measured spectrophotometrically at a wavelength of 329 nm.
(a) Suggest a reason for using this wavelength.
(b) In what region of the spectrum does this wavelength lie?
(c) When 0.1232 g of fish-liver oil is dissolved in 500. mL of solvent, the absorbance is 0.724 units. When 1.67×10^{-3} g of vitamin A is dissolved in 250. mL of solvent, the absorbance is 1.018 units. Calculate the vitamin A concentration in the fish-liver oil.

7.93 Many calculators use photocells to provide their energy. Find the maximum wavelength needed to remove an electron from silver ($\phi = 7.59\times10^{-19}$ J). Is silver a good choice for a photocell that uses visible light?

7.94 As the uses of aluminum have become widespread, many methods have been developed to measure concentrations of its complexes in solution. In one method, the sodium salt of 2-quinizarinsulfonic acid forms a complex with Al^{3+} that absorbs strongly at 560 nm.
(a) Use the data below to draw a plot of absorbance vs. concentration of a complex in solution and find the slope and *y*-intercept:

Concentration (M)	Absorbance (560 nm)
1.0×10^{-5}	0.131
1.5×10^{-5}	0.201
2.0×10^{-5}	0.265
2.5×10^{-5}	0.329
3.0×10^{-5}	0.396

(b) When 20.0 mL of this complex solution is diluted with water to 150. mL, its absorbance is 0.236. Find the concentrations of the diluted solution and of the original solution.

7.95 In a game of "Clue," Ms. White is killed in the conservatory. You have a device in each room to help you find the murderer—a spectrometer that emits the entire visible spectrum to indicate who is in that room. For example, if someone wearing yellow is in a room, light at 580 nm is reflected. The suspects are Col. Mustard, Prof. Plum, Mr. Green, Ms. Peacock (blue), and Ms. Scarlet. At the time of the murder, the spectrometer in the dining room recorded a reflection at 520 nm, those in the lounge and study recorded reflections of lower frequencies, and the one in the library recorded a reflection of the shortest possible wavelength. Who killed Ms. White? Explain.

7.96 Technetium (Tc; $Z = 43$) is a synthetic element used as a radioactive tracer in medical studies. A Tc atom emits a beta particle (electron) with a kinetic energy (E_k) of 4.71×10^{-15} J. What is the de Broglie wavelength of this electron ($E_k = \frac{1}{2}mv^2$)?

7.97 Electric power is typically stated in units of watts (1 W = 1 J/s). About 95% of the power output of an incandescent bulb is converted to heat and 5% to light. If 10% of that light shines on your chemistry text, how many photons per second shine on the book from a 75-W bulb? (Assume the photons have a wavelength of 550 nm.)

7.98 The flame test for sodium is based on its intense emission at 589 nm, and the test for potassium is based on its emission at 404 nm. When both elements are present, the Na^+ emission is so strong that the K^+ emission can't be seen, except by looking through a cobalt-glass filter. (a) What are the colors of these Na^+ and K^+ emissions? (b) What does the cobalt-glass filter do?

7.99 The net change in the multistep biochemical process of photosynthesis is that CO_2 and H_2O form glucose ($C_6H_{12}O_6$) and O_2. Chlorophyll absorbs light in the 600 to 700 nm region. (a) Write a balanced thermochemical equation for formation of 1.00 mol of glucose. (b) What is the minimum number of photons with $\lambda = 680.$ nm needed to prepare 1.00 mol of glucose?

7.100 Only certain electron transitions are allowed from one energy level to another. In one-electron species, the change in the quantum number l of an allowed transition must be ±1. For example, a $3p$ electron can drop directly to a $2s$ orbital but not to a $2p$. Thus, in the UV series, where $n_{final} = 1$, allowed electron transitions can start in a p orbital ($l = 1$) of $n = 2$ or higher, not in an s ($l = 0$) or d ($l = 2$) orbital of $n = 2$ or higher. From what orbital do each of the allowed electron transitions start for the first four emission lines in the visible series ($n_{final} = 2$)?

7.101 The discharge of phosphate compounds in detergents into the environment has led to serious imbalances in the natural life cycle of freshwater lakes. A chemist studying water pollution used a spectrophotometric method to measure total phosphate and obtained the following data for known standards:

Absorbance (400 nm)	Concentration (mol/L)
0	0.0
0.10	2.5×10^{-5}
0.16	3.2×10^{-5}
0.20	4.4×10^{-5}
0.25	5.6×10^{-5}
0.38	8.4×10^{-5}
0.48	10.5×10^{-5}
0.62	13.8×10^{-5}
0.76	17.0×10^{-5}
0.88	19.4×10^{-5}

(a) Draw a curve of absorbance vs. phosphate concentration.
(b) If a sample of lake water has an absorbance of 0.55, what is its phosphate concentration?

Remarkable Regularity In the regularity of an eclipse, a heartbeat, or the plumage of a peacock, nature exhibits predictably recurring patterns. The arrangement of electrons in atoms recurs periodically, too. As you'll see, this periodicity allows us to predict many properties of the elements, including much of their physical and chemical behavior.

8

Electron Configuration and Chemical Periodicity

In Chapter 7, you saw how an outpouring of scientific creativity by early 20th-century physicists led to a new understanding of matter and energy, which in turn led to the quantum-mechanical model of the atom. But you can be sure that late 19th-century chemists were not sitting idly by, waiting for their colleagues in physics to develop that model. They were exploring the nature of electrolytes, establishing the kinetic-molecular theory, and developing chemical thermodynamics. The fields of organic chemistry and biochemistry were born, as were the fertilizer, explosives, glassmaking, soapmaking, bleaching, and dyestuff industries. And, for the first time, chemistry became a university subject in Europe and America. Superimposed on this activity was the accumulation of an enormous body of facts about the elements, which became organized into the periodic table.

The goal of this chapter is to show how the organization of the table, condensed from countless hours of laboratory work, was explained perfectly by the new quantum-mechanical atomic model. This model answers one of the central questions in chemistry: why do the elements behave as they do? Or, rephrasing the question to fit the main topic of this chapter: how does the **electron configuration** of an element—*the distribution of electrons within the orbitals of its atoms*—relate to its chemical and physical properties?

IN THIS CHAPTER . . . We first discuss the origin of the periodic table. Then we extend the quantum-mechanical model (Chapter 7) to many-electron atoms (those with more than one electron) to define a unique set of quantum numbers for each electron in the atoms of every element. Electrostatic effects lead to the order in which orbitals fill with electrons, and we'll see how that order correlates with the order of elements in the periodic table. We discuss how electron configuration and nuclear charge lead to periodic trends in atomic properties and how these trends account for the patterns of chemical reactivity. Finally, we apply these ideas to the properties of metals, nonmetals, and their ions.

Concepts & Skills to Review

before you study this chapter

- format of the periodic table (Section 2.6)
- characteristics of metals and nonmetals (Section 2.6)
- application of Coulomb's law to electrostatic attraction (Section 2.7)
- characteristics of acids and bases (Section 4.4)
- rules for assigning quantum numbers (Section 7.4)

8.1 DEVELOPMENT OF THE PERIODIC TABLE

An essential requirement for the amazing growth in theoretical and practical chemistry in the second half of the 19th century was the ability to organize the facts known about element behavior. The earliest organizing attempt was made by Johann Döbereiner, who placed groups of three elements with similar properties, such as calcium, strontium, and barium, into "triads." Later, John Newlands noted similarities between every eighth element (arranged by atomic mass), like the similarity between every eighth note in the musical scale, and placed elements into "octaves." As more elements were discovered, however, these early numerical schemes lost much of their validity.

In Chapter 2, you saw that the most successful organizing scheme was made by the Russian chemist Dmitri Mendeleev. In 1870, he arranged the 65 elements then known into a *periodic table* and summarized their behavior in the **periodic law:** when arranged by atomic mass, the elements exhibit a periodic recurrence of similar properties. It is a curious quirk of history that Mendeleev and the German chemist Julius Lothar Meyer arrived at virtually the same organization simultaneously, yet independently. Mendeleev focused on chemical properties and Meyer on physical properties. The greater credit has gone to Mendeleev because he was able to *predict* the properties of several as-yet-undiscovered elements, for which he had left blank spaces in his table. Table 8.1 (on the next page) compares the actual properties of germanium, which Mendeleev gave the provisional name "eka silicon" ("first under silicon"), with his predictions for it.

Today's periodic table, which appears on the inside front cover of the text, resembles Mendeleev's in most details, although it includes 49 elements that were unknown in 1870. The only substantive change is that the elements are now arranged in order of *atomic number* (number of protons) rather than atomic mass.

Mendeleev's Great Contribution Born in a small Siberian town, Dmitri Ivanovich Mendeleev was the youngest of 17 children of the local schoolteacher. He showed early talent for mathematics and science, so his mother took him to St. Petersburg, where he remained to study and work for much of his life. Early research interests centered on the physical properties of gases and liquids, and later he was consulted frequently on the industrial processing of petroleum. In developing his periodic table, he prepared a note card on the properties of each element and arranged and rearranged them until he realized that properties repeated when the elements were placed in order of increasing atomic mass.

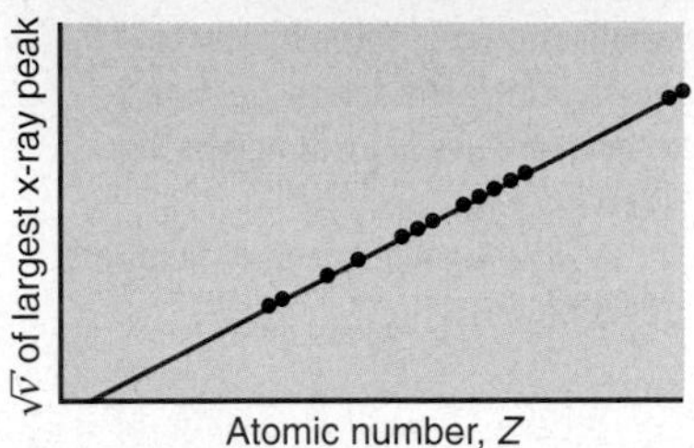

Moseley and Atomic Number When a metal is bombarded with high-energy electrons, an inner electron is knocked from the atom, an outer electron moves down to fill in the space, and x-rays are emitted. Bohr proposed that the x-ray spectrum of an element had wavelengths proportional to the nuclear charge. In 1913, Henry Moseley studied the x-ray spectra of a series of metals and correlated the largest peak in a metal's x-ray spectrum with its order in the periodic table (its *atomic number*). This correlation is known as *Moseley's law,* which may be expressed as $v^{1/2} = K(Z - \sigma)$, where Z represents the atomic number, σ is 1 for electrons nearest the nucleus, v is the frequency of the x-rays, and K is $5.0 \times 10^7\ s^{-1}$. He showed that the nuclear charge increased by 1 for each element (see the graph). Among other results, the findings confirmed the placement of Co ($Z = 27$) *before* Ni ($Z = 28$), despite cobalt's higher atomic mass, and also confirmed that the gap between Cl ($Z = 17$) and K ($Z = 19$) is the place for Ar ($Z = 18$). Tragically, Moseley died in 1915 at the age of 26 while serving as a pilot in the British army during World War I.

Table 8.1 Mendeleev's Predicted Properties of Germanium ("eka Silicon") and Its Actual Properties

Property	Predicted Properties of eka Silicon (E)	Actual Properties of Germanium (Ge)
Atomic mass	72 amu	72.61 amu
Appearance	Gray metal	Gray metal
Density	5.5 g/cm^3	5.32 g/cm^3
Molar volume	13 cm^3/mol	13.65 cm^3/mol
Specific heat capacity	0.31 J/g·K	0.32 J/g·K
Oxide formula	EO_2	GeO_2
Oxide density	4.7 g/cm^3	4.23 g/cm^3
Sulfide formula and solubility	ES_2; insoluble in H_2O; soluble in aqueous $(NH_4)_2S$	GeS_2; insoluble in H_2O; soluble in aqueous $(NH_4)_2S$
Chloride formula (boiling point)	ECl_4 (<100°C)	$GeCl_4$ (84°C)
Chloride density	1.9 g/cm^3	1.844 g/cm^3
Element preparation	Reduction of K_2EF_6 with sodium	Reduction of K_2GeF_6 with sodium

This change was based on the work of the British physicist Henry G. J. Moseley, who found a direct dependence between an element's nuclear charge and its position in the periodic table.

8.2 CHARACTERISTICS OF MANY-ELECTRON ATOMS

Like the Bohr model, the Schrödinger equation does not give *exact* solutions for many-electron atoms. However, unlike the Bohr model, the Schrödinger equation gives very good *approximate* solutions. These solutions show that the atomic orbitals of many-electron atoms are *hydrogen-like;* that is, they resemble those of the H atom. This conclusion means we can use the same quantum numbers that we used for the H atom to describe the orbitals of other atoms.

Nevertheless, the existence of more than one electron in an atom requires us to consider three features that were not relevant in the case of hydrogen: (1) the need for a fourth quantum number, (2) a limit on the number of electrons allowed in a given orbital, and (3) a more complex set of orbital energy levels. Let's examine these new features and then go on to determine the electron configuration for each element.

The Electron-Spin Quantum Number

Recall from Chapter 7 that the three quantum numbers n, l, and m_l describe the size (energy), shape, and orientation, respectively, of an atomic orbital. However, an additional quantum number is needed to describe a property of the electron itself, called *spin,* which is not a property of the orbital. Electron spin becomes important when more than one electron is present.

When a beam of H atoms passes through a nonuniform magnetic field, as shown in Figure 8.1, it splits into two beams that bend away from each other. The explanation of the split beam is that the electron generates a tiny magnetic field, as though it were a spinning charge. The single electron in each H atom can have one of two possible values of *spin,* each of which generates a tiny magnetic field. These two fields have opposing directions, so half of the electrons are attracted into the large external magnetic field and the other half are repelled by it. As a result, the beam of H atoms splits.

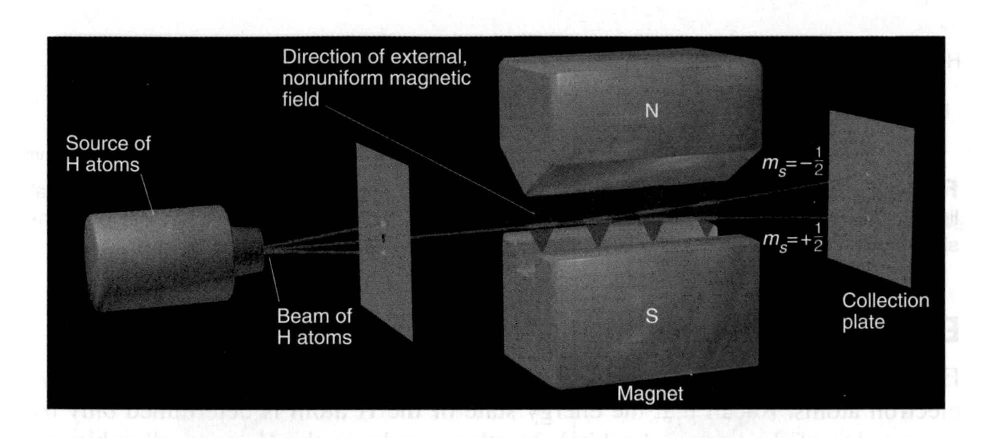

Figure 8.1 Observing the effect of electron spin. A nonuniform magnetic field, created by magnet faces with different shapes, splits a beam of hydrogen atoms in two. The split beam results from the two possible values of electron spin within each atom.

Like its charge, spin is an intrinsic property of the electron, and the **spin quantum number (m_s)** has values of either $+\frac{1}{2}$ or $-\frac{1}{2}$. Thus, *each electron in an atom is described completely by a set of* ***four*** *quantum numbers: the first three describe its orbital, and the fourth describes its spin.* The quantum numbers are summarized in Table 8.2.

Table 8.2 Summary of Quantum Numbers of Electrons in Atoms

Name	Symbol	Permitted Values	Property
Principal	n	Positive integers (1, 2, 3, ...)	Orbital energy (size)
Angular momentum	l	Integers from 0 to $n - 1$	Orbital shape (The l values 0, 1, 2, and 3 correspond to *s*, *p*, *d*, and *f* orbitals, respectively.)
Magnetic	m_l	Integers from $-l$ to 0 to $+l$	Orbital orientation
Spin	m_s	$+\frac{1}{2}$ or $-\frac{1}{2}$	Direction of e^- spin

Now we can write a set of four quantum numbers for any electron in the ground state of any atom. For example, the set of quantum numbers for the lone electron in hydrogen (H; $Z = 1$) is $n = 1$, $l = 0$, $m_l = 0$, and $m_s = +\frac{1}{2}$. (The spin quantum number for this electron could just as well have been $-\frac{1}{2}$, but by convention, we assign $+\frac{1}{2}$ for the first electron in an orbital.)

The Exclusion Principle

The element after hydrogen is helium (He; $Z = 2$), the first with atoms having more than one electron. The first electron in the He ground state has the same set of quantum numbers as the electron in the H atom, but the second He electron does not. Based on observations of the excited states of atoms, the Austrian physicist Wolfgang Pauli formulated the **exclusion principle:** *no two electrons in the same atom can have the same four quantum numbers.* That is, each electron must have a unique "identity" as expressed by its set of quantum numbers. Therefore, the second He electron occupies the same orbital as the first but has an opposite spin: $n = 1$, $l = 0$, $m_l = 0$, and $m_s = -\frac{1}{2}$.

Baseball Quantum Numbers The unique set of quantum numbers that describes an electron is analogous to the unique location of a box seat at a baseball game. The stadium (atom) is divided into section (n, level), box (l, sublevel), row (m_l, orbital), and seat (m_s, spin). Only one person (electron) can have this particular set of stadium "quantum numbers."

Because the spin quantum number (m_s) can have only two values, the major consequence of the exclusion principle is that *an atomic orbital can hold a maximum of two electrons and they must have opposing spins.* We say that the 1*s* orbital in He is *filled* and that the electrons have *paired spins.* Thus, a beam of He atoms is not split in an experiment like that in Figure 8.1.

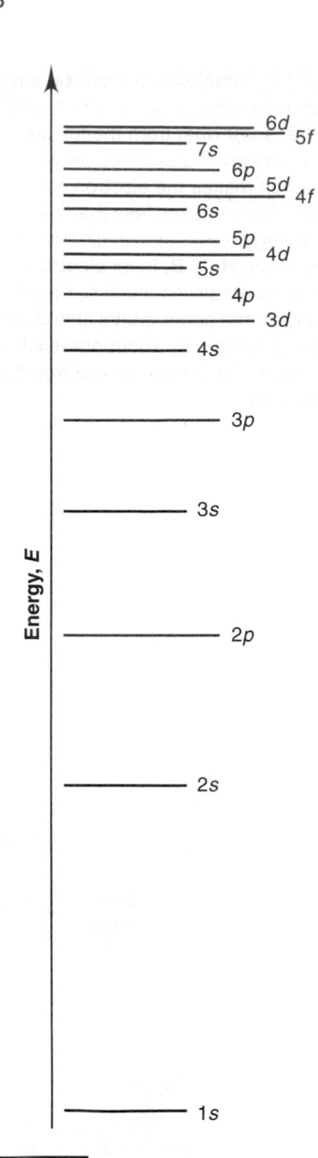

Figure 8.6 Order for filling energy sublevels with electrons. In many-electron atoms, energy levels split into sublevels. The relative energies of sublevels increase with principal quantum number n ($1 < 2 < 3$, etc.) and angular momentum quantum number l ($s < p < d < f$). As n increases, the energies become closer together. The penetration effect, together with this narrowing of energy differences, results in the overlap of some sublevels; for example, the 4s sublevel is slightly lower in energy than the 3d, so it is filled first. (Line color is by sublevel type; line lengths differ for ease in labeling.)

Figure 8.6 shows the general energy order of levels (n value) and how they are split into sublevels (l values) of differing energies. (Compare this with the H atom energy levels in Figure 7.21, p. 283.) Next, we use this energy order to construct a periodic table of ground-state atoms.

SECTION SUMMARY

Identifying electrons in many-electron atoms requires four quantum numbers: three (n, l, m_l) describe the orbital, and a fourth (m_s) describes electron spin. The Pauli exclusion principle requires each electron to have a unique set of four quantum numbers; therefore, an orbital can hold no more than two electrons, and their spins must be paired (opposite). Electrostatic interactions determine orbital energies as follows:

1. Greater nuclear charge lowers orbital energy and makes electrons harder to remove.
2. Electron-electron repulsions raise orbital energy and make electrons easier to remove. Repulsions have the effect of *shielding* electrons from the full nuclear charge, reducing it to an effective nuclear charge, Z_{eff}. Inner electrons shield outer electrons most effectively.
3. Greater radial probability distribution near the nucleus (greater *penetration*) makes an electron harder to remove because it is attracted more strongly and shielded less effectively. As a result, an energy level (shell) is split into sublevels (subshells) with the energy order $s < p < d < f$.

8.3 THE QUANTUM-MECHANICAL MODEL AND THE PERIODIC TABLE

Quantum mechanics provides the theoretical foundation for the experimentally based periodic table. In this section, we fill the table with elements and determine their electron configurations—the distributions of electrons within their atoms' orbitals. Note especially the *recurring pattern in electron configurations, which is the basis for the recurring pattern in chemical behavior.*

Building Up Periods 1 and 2

A useful way to determine the electron configurations of the elements is to start at the beginning of the periodic table and add one electron per element to the *lowest energy orbital available*. (Of course, one proton and one or more neutrons are also added to the nucleus.) This approach is based on the **aufbau principle** (German *aufbauen,* "to build up"), and it results in *ground-state* electron configurations. Let's assign sets of quantum numbers to the electrons in the ground state of the first 10 elements, those in the first two periods (horizontal rows).

For the electron in H, as you've seen, the set of quantum numbers is

$$\text{H } (Z = 1)\text{:}\quad n = 1,\ l = 0,\ m_l = 0,\ m_s = +\tfrac{1}{2}$$

You also saw that the first electron in He has the same set as the electron in H, but the second He electron has opposing spin (exclusion principle):

$$\text{He } (Z = 2)\text{:}\quad n = 1,\ l = 0,\ m_l = 0,\ m_s = -\tfrac{1}{2}$$

(As we go through each element in this discussion, the quantum numbers that follow refer to the element's *last added* electron.)

Here are two common ways to designate the orbital and its electrons:

1. *The electron configuration.* This shorthand notation consists of the principal energy level (n value), the letter designation of the sublevel (l value), and the number of electrons (#) in the sublevel, written as a superscript: $\boldsymbol{nl^{\#}}$. The electron configuration of H is $1s^1$ (spoken "one-ess-one"); that of He is $1s^2$ (spoken "one-ess-two," *not* "one-ess-squared"). This notation does *not* indicate electron spin but assumes you know that the two 1s electrons have paired (opposite) spins.

2. *The orbital diagram.* An **orbital diagram** consists of a box (or circle, or just a line) for each orbital in a given energy level, grouped by sublevel, with an

arrow indicating an electron *and* its spin. (Traditionally, ↑ is $+\frac{1}{2}$ and ↓ is $-\frac{1}{2}$, but these are arbitrary; it is necessary only to be consistent. Throughout the text, orbital occupancy is also indicated by color intensity: an orbital with no color is empty, pale color means half-filled, and full color means filled.) The electron configurations and orbital diagrams for the first two elements are

H ($Z = 1$) $1s^1$ [↑] (1s)   He ($Z = 2$) $1s^2$ [↑↓] (1s)

The exclusion principle tells us that an orbital can hold only two electrons, so the 1*s* orbital in He is filled, and the $n = 1$ level is also filled. The $n = 2$ level is filled next, beginning with the 2*s* orbital, the next lowest in energy. As we said earlier, the first two electrons in Li fill the 1*s* orbital, and the last added Li electron has quantum numbers $n = 2$, $l = 0$, $m_l = 0$, $m_s = +\frac{1}{2}$. The electron configuration for Li is $1s^22s^1$. Note that the orbital diagram shows all the orbitals for $n = 2$, whether or not they are occupied:

Energy, *E* →

Li ($Z = 3$) $1s^22s^1$ [↑↓] (1s) [↑] (2s) [| |] (2p)

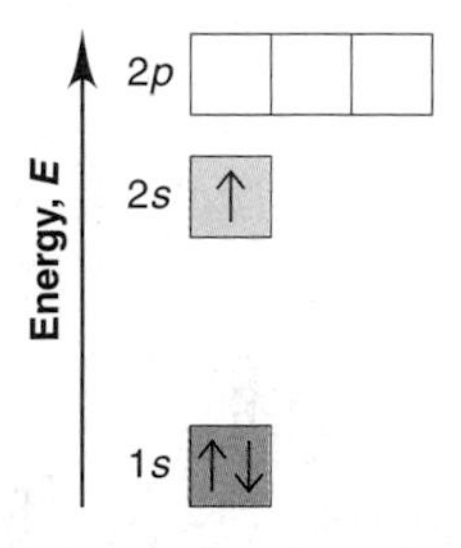

Figure 8.7 A vertical orbital diagram for the Li ground state. Although orbital diagrams are usually written horizontally, this vertical arrangement emphasizes that sublevel energy increases with increasing *n* and *l*.

To save space on a page, orbital diagrams are often written horizontally, but note that *the energy of the sublevels increases from left to right.* Figure 8.7 emphasizes this point by arranging the orbital diagram of lithium vertically.

With the 2*s* orbital only half-filled in Li, the fourth electron of beryllium fills it with the electron's spin paired: $n = 2$, $l = 0$, $m_l = 0$, $m_s = -\frac{1}{2}$.

Be ($Z = 4$) $1s^22s^2$ [↑↓] (1s) [↑↓] (2s) [| |] (2p)

The next lowest energy sublevel is the 2*p*. A *p* sublevel has $l = 1$, so the m_l (orientation) values can be -1, 0, or $+1$. The three orbitals in the 2*p* sublevel have *equal energy* (same *n* and *l* values), which means that the fifth electron of boron can go into *any one of the 2p orbitals.* For convenience, let's label the boxes from left to right, -1, 0, $+1$. By convention, we place the electron in the $m_l = -1$ orbital: $n = 2$, $l = 1$, $m_l = -1$, $m_s = +\frac{1}{2}$.

B ($Z = 5$) $1s^22s^22p^1$ [↑↓] (1s) [↑↓] (2s) [↑ | |] (2p; −1 0 +1)

To minimize electron-electron repulsions, the last added (sixth) electron of carbon enters one of the *unoccupied* 2*p* orbitals; by convention, we place it in the $m_l = 0$ orbital. Experiment shows that the spin of this electron is *parallel* to (the same as) the spin of the other 2*p* electron: $n = 2$, $l = 1$, $m_l = 0$, $m_s = +\frac{1}{2}$.

C ($Z = 6$) $1s^22s^22p^2$ [↑↓] (1s) [↑↓] (2s) [↑ | ↑ |] (2p)

This placement of electrons for carbon exemplifies **Hund's rule:** *when orbitals of equal energy are available, the electron configuration of lowest energy has the maximum number of unpaired electrons with parallel spins.* Based on Hund's rule, nitrogen's seventh electron enters the last empty 2*p* orbital, with its spin parallel to the two other 2*p* electrons: $n = 2$, $l = 1$, $m_l = +1$, $m_s = +\frac{1}{2}$.

N ($Z = 7$) $1s^22s^22p^3$ [↑↓] (1s) [↑↓] (2s) [↑ | ↑ | ↑] (2p)

The eighth electron in oxygen must enter one of these three half-filled 2*p* orbitals and "pair up" with (have opposing spin to) the electron already present.

Since the $2p$ orbitals all have the same energy, we proceed as before and place the electron in the orbital previously designated $m_l = -1$. The quantum numbers are $n = 2$, $l = 1$, $m_l = -1$, $m_s = -\frac{1}{2}$.

O (Z = 8) $1s^22s^22p^4$ [↑↓] 1s [↑↓] 2s [↑↓|↑|↑] 2p

Fluorine's ninth electron enters either of the two remaining half-filled $2p$ orbitals: $n = 2$, $l = 1$, $m_l = 0$, $m_s = -\frac{1}{2}$.

F (Z = 9) $1s^22s^22p^5$ [↑↓] 1s [↑↓] 2s [↑↓|↑↓|↑] 2p

Only one unfilled orbital remains in the $2p$ sublevel, so the tenth electron of neon occupies it: $n = 2$, $l = 1$, $m_l = +1$, $m_s = -\frac{1}{2}$. With neon, the $n = 2$ level is filled.

Ne (Z = 10) $1s^22s^22p^6$ [↑↓] 1s [↑↓] 2s [↑↓|↑↓|↑↓] 2p

SAMPLE PROBLEM 8.1 Determining Quantum Numbers from Orbital Diagrams

Problem Write a set of quantum numbers for the third electron and a set for the eighth electron of the F atom.

Plan Referring to the orbital diagram, we count to the electron of interest and note its level (n), sublevel (l), orbital (m_l), and spin (m_s).

Solution The third electron is in the $2s$ orbital. The upward arrow indicates a spin of $+\frac{1}{2}$:

$$n = 2, l = 0, m_l = 0, m_s = +\tfrac{1}{2}$$

The eighth electron is in the first $2p$ orbital, which is designated $m_l = -1$, and has a downward arrow:

$$n = 2, l = 1, m_l = -1, m_s = -\tfrac{1}{2}$$

FOLLOW-UP PROBLEM 8.1 Use the periodic table to identify the element with the electron configuration $1s^22s^22p^4$. Write its orbital diagram, and give the quantum numbers of its sixth electron.

With so much attention paid to these notations, it's easy to forget that atoms are real spherical objects and that the electrons occupy volumes with specific shapes and orientations. Figure 8.8 shows ground-state electron configurations and orbital contours for the first 10 elements arranged in periodic table format.

Even at this early stage of filling the table, we can make an important correlation between chemical behavior and electron configuration: elements in the same group have similar outer electron configurations. As an example, helium (He) and neon (Ne) in Group 8A(18) both have filled outer sublevels—$1s^2$ for helium and $2p^6$ for neon—and neither element forms compounds. As we'll see often, *filled outer sublevels make elements unreactive*.

Building Up Period 3

The Period 3 elements, sodium through argon, lie directly under the Period 2 elements, lithium through neon. The sublevels of the $n = 3$ level are filled in the order $3s$, $3p$, $3d$. Table 8.3 presents *partial* orbital diagrams ($3s$ and $3p$ sublevels only) and electron configurations for the eight elements in Period 3 (with *filled inner levels* in brackets and the sublevel to which the last electron is added in colored type). Note the *group similarities in outer electron configuration* with the elements in Period 2 (refer to Figure 8.9, p. 300).

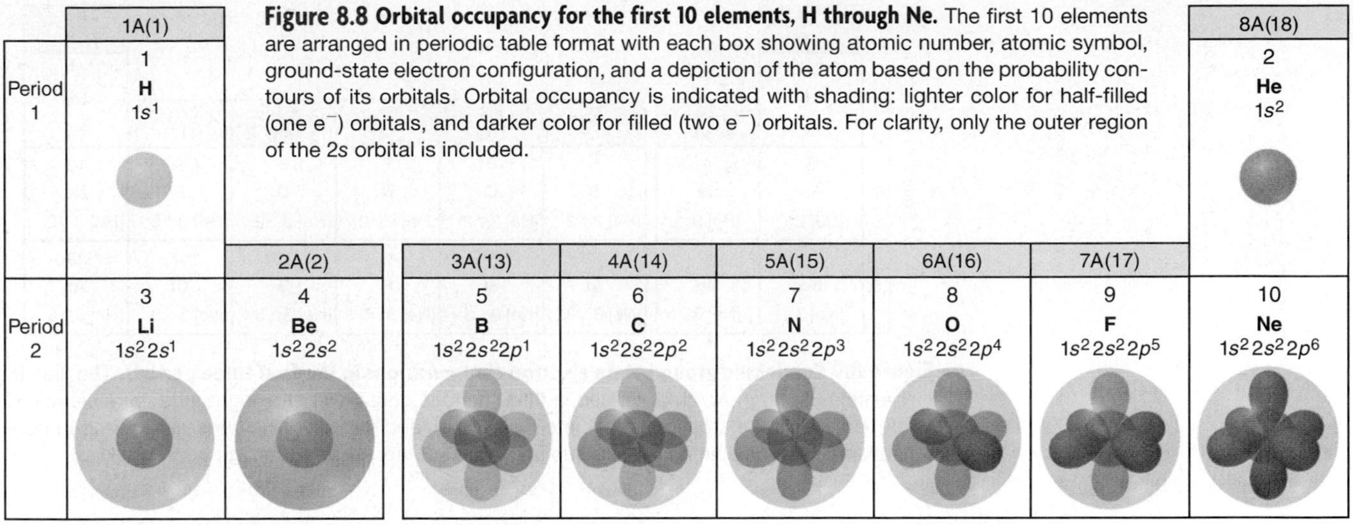

	1A(1)	2A(2)	3A(13)	4A(14)	5A(15)	6A(16)	7A(17)	8A(18)
Period 1	1 **H** $1s^1$							2 **He** $1s^2$
Period 2	3 **Li** $1s^22s^1$	4 **Be** $1s^22s^2$	5 **B** $1s^22s^22p^1$	6 **C** $1s^22s^22p^2$	7 **N** $1s^22s^22p^3$	8 **O** $1s^22s^22p^4$	9 **F** $1s^22s^22p^5$	10 **Ne** $1s^22s^22p^6$

Figure 8.8 Orbital occupancy for the first 10 elements, H through Ne. The first 10 elements are arranged in periodic table format with each box showing atomic number, atomic symbol, ground-state electron configuration, and a depiction of the atom based on the probability contours of its orbitals. Orbital occupancy is indicated with shading: lighter color for half-filled (one e^-) orbitals, and darker color for filled (two e^-) orbitals. For clarity, only the outer region of the 2s orbital is included.

In sodium (the second alkali metal) and magnesium (the second alkaline earth metal), electrons are added to the 3*s* sublevel, which contains the 3*s* orbital only, just as they filled the 2*s* sublevel in lithium and beryllium in Period 2. Then, just as for boron, carbon, and nitrogen in Period 2, the last electrons added to aluminum, silicon, and phosphorus in Period 3 half-fill the three 3*p* orbitals with spins parallel (Hund's rule). The last electrons added to sulfur, chlorine, and argon then successively enter the three half-filled 3*p* orbitals, thereby filling the 3*p* sublevel. With argon, the next noble gas after helium and neon, we arrive at the end of Period 3. (As you'll see shortly, the 3*d* orbitals are filled in Period 4.)

The rightmost column of Table 8.3 shows the *condensed electron configuration*. In this simplified notation, the electron configuration of the previous noble gas is shown by its element symbol in brackets, and it is followed by the electron configuration of the energy level being filled. The condensed electron configuration of sulfur, for example, is [Ne] $3s^23p^4$, where [Ne] stands for $1s^22s^22p^6$.

Table 8.3 Partial Orbital Diagrams and Electron Configurations* for the Elements in Period 3

Atomic Number	Element	Partial Orbital Diagram (3s and 3p Sublevels Only): 3s	3p	Full Electron Configuration	Condensed Electron Configuration
11	Na	↑	_ _ _	$[1s^22s^22p^6]\ 3s^1$	[Ne] $3s^1$
12	Mg	↑↓	_ _ _	$[1s^22s^22p^6]\ 3s^2$	[Ne] $3s^2$
13	Al	↑↓	↑ _ _	$[1s^22s^22p^6]\ 3s^23p^1$	[Ne] $3s^23p^1$
14	Si	↑↓	↑ ↑ _	$[1s^22s^22p^6]\ 3s^23p^2$	[Ne] $3s^23p^2$
15	P	↑↓	↑ ↑ ↑	$[1s^22s^22p^6]\ 3s^23p^3$	[Ne] $3s^23p^3$
16	S	↑↓	↑↓ ↑ ↑	$[1s^22s^22p^6]\ 3s^23p^4$	[Ne] $3s^23p^4$
17	Cl	↑↓	↑↓ ↑↓ ↑	$[1s^22s^22p^6]\ 3s^23p^5$	[Ne] $3s^23p^5$
18	Ar	↑↓	↑↓ ↑↓ ↑↓	$[1s^22s^22p^6]\ 3s^23p^6$	[Ne] $3s^23p^6$

*Colored type indicates the sublevel to which the last electron is added.

Period	1A (1)	2A (2)	3A (13)	4A (14)	5A (15)	6A (16)	7A (17)	8A (18)
1	1 **H** $1s^1$							2 **He** $1s^2$
2	3 **Li** [He] $2s^1$	4 **Be** [He] $2s^2$	5 **B** [He] $2s^22p^1$	6 **C** [He] $2s^22p^2$	7 **N** [He] $2s^22p^3$	8 **O** [He] $2s^22p^4$	9 **F** [He] $2s^22p^5$	10 **Ne** [He] $2s^22p^6$
3	11 **Na** [Ne] $3s^1$	12 **Mg** [Ne] $3s^2$	13 **Al** [Ne] $3s^23p^1$	14 **Si** [Ne] $3s^23p^2$	15 **P** [Ne] $3s^23p^3$	16 **S** [Ne] $3s^23p^4$	17 **Cl** [Ne] $3s^23p^5$	18 **Ar** [Ne] $3s^23p^6$

Figure 8.9 Condensed ground-state electron configurations in the first three periods. The first 18 elements, H through Ar, are arranged in three periods containing two, eight, and eight elements. Each box shows the atomic number, atomic symbol, and condensed ground-state electron configuration. Note that elements in a group have similar outer electron configurations *(color)*.

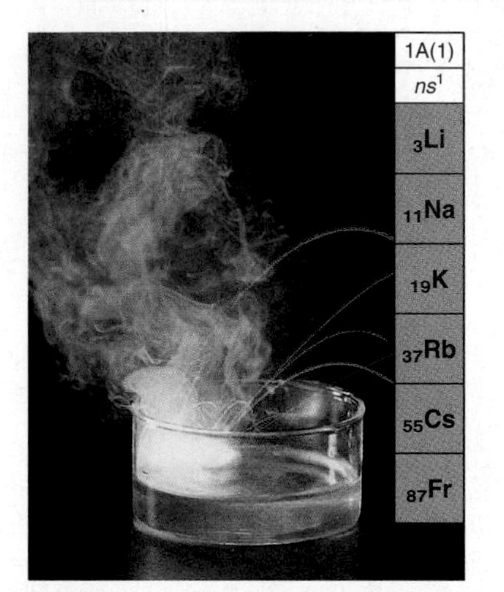

A

B

Figure 8.10 Similar reactivities within a group. A, Potassium metal and water reacting. All alkali metals [Group 1A(1)] react vigorously with water and displace H_2. **B,** Chlorine and potassium metal reacting. All halogens [Group 7A(17)] react with metals to form ionic halides.

Electron Configurations Within Groups

One of the central points in all chemistry is that *similar outer electron configurations correlate with similar chemical behavior.* Figure 8.9 shows the condensed electron configurations of the first 18 elements. Note the similarities within each group. Here are some examples from just two groups:

- In Group 1A(1), lithium and sodium have the condensed electron configuration [noble gas] ns^1 (where n is the quantum number of the outermost energy level), as do all the other alkali metals (K, Rb, Cs, Fr). All are highly reactive metals that form ionic compounds with nonmetals with formulas such as MCl, M_2O, and M_2S (where M represents the alkali metal), and all react vigorously with water to displace H_2 (Figure 8.10A).
- In Group 7A(17), fluorine and chlorine have the condensed electron configuration [noble gas] ns^2np^5, as do the other halogens (Br, I, At). Little is known about rare, radioactive astatine (At), but all the others are reactive nonmetals that occur as diatomic molecules, X_2 (where X represents the halogen). All form ionic compounds with metals (KX, MgX_2) (Figure 8.10B), covalent compounds with hydrogen (HX) that yield acidic solutions in water, and covalent compounds with carbon (CX_4).

To summarize, here is the major connection between quantum mechanics and chemical periodicity: *orbitals are filled in order of increasing energy, which leads to outer electron configurations that recur periodically, which leads to chemical properties that recur periodically.*

The First *d*-Orbital Transition Series: Building Up Period 4

The 3*d* orbitals are filled in Period 4. Note, however, that *the 4s orbital is filled before the 3d.* This switch in filling order is due to the shielding and penetration effects that we discussed in Section 8.2. The radial probability distribution of the 3*d* orbital is greater outside the filled, inner $n = 1$ and $n = 2$ levels, so a 3*d* electron is shielded very effectively from the nuclear charge. In contrast, penetration by the 4*s* electron means that it spends a significant part of its time near the nucleus and feels a greater nuclear attraction. Thus, the 4*s* orbital is slightly *lower* in energy than the 3*d*, and so fills first. Similarly, the 5*s* orbital fills before the 4*d*, and the 6*s* fills before the 5*d*. In general, *the ns sublevel fills before the $(n - 1)d$ sublevel.* As we proceed through the transition series, however, you'll see several exceptions to this pattern because the energies of the *ns* and $(n - 1)d$ sublevels become extremely close at higher values of *n*.

Table 8.4 shows the partial orbital diagrams and ground-state electron configurations for the 18 elements in Period 4 (again with filled inner levels in brackets and the sublevel to which the last electron has been added in colored type). The first two elements of the period, potassium and calcium, are the next alkali and alkaline earth metals, respectively, and their electrons fill the 4*s* sublevel. The third element, scandium ($Z = 21$), is the first of the **transition elements,** those in which *d* orbitals are being filled. The last electron in scandium occupies any one of the five 3*d* orbitals because they are equal in energy. Scandium has the electron configuration [Ar] $4s^23d^1$.

The filling of 3*d* orbitals proceeds one at a time, as with *p* orbitals, except in two cases: chromium ($Z = 24$) and copper ($Z = 29$). Vanadium ($Z = 23$), the element before chromium, has three half-filled *d* orbitals ([Ar] $4s^23d^3$). Rather than having its last electron enter a fourth empty *d* orbital to give [Ar] $4s^23d^4$, chromium has one electron in the 4*s* sublevel and five in the 3*d* sublevel. Thus, both the 4*s* and the 3*d* sublevels are *half-filled* (see margin).

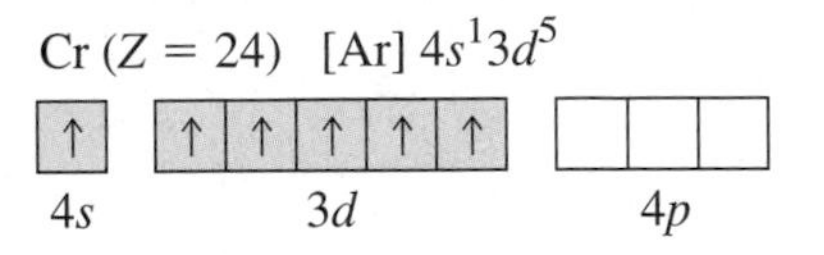

The other anomalous filling pattern occurs with copper. Following nickel ([Ar] $4s^23d^8$), copper would be expected to have the [Ar] $4s^23d^9$ configuration.

Table 8.4 Partial Orbital Diagrams and Electron Configurations* for the Elements in Period 4

Atomic Number	Element	Partial Orbital Diagram (4*s*, 3*d*, and 4*p* Sublevels Only): 4*s*	3*d*	4*p*	Full Electron Configuration	Condensed Electron Configuration
19	K	↑			$[1s^22s^22p^63s^23p^6]\ 4s^1$	$[Ar]\ 4s^1$
20	Ca	↑↓			$[1s^22s^22p^63s^23p^6]\ 4s^2$	$[Ar]\ 4s^2$
21	Sc	↑↓	↑		$[1s^22s^22p^63s^23p^6]\ 4s^23d^1$	$[Ar]\ 4s^23d^1$
22	Ti	↑↓	↑ ↑		$[1s^22s^22p^63s^23p^6]\ 4s^23d^2$	$[Ar]\ 4s^23d^2$
23	V	↑↓	↑ ↑ ↑		$[1s^22s^22p^63s^23p^6]\ 4s^23d^3$	$[Ar]\ 4s^23d^3$
24	Cr	↑	↑ ↑ ↑ ↑ ↑		$[1s^22s^22p^63s^23p^6]\ 4s^13d^5$	$[Ar]\ 4s^13d^5$
25	Mn	↑↓	↑ ↑ ↑ ↑ ↑		$[1s^22s^22p^63s^23p^6]\ 4s^23d^5$	$[Ar]\ 4s^23d^5$
26	Fe	↑↓	↑↓ ↑ ↑ ↑ ↑		$[1s^22s^22p^63s^23p^6]\ 4s^23d^6$	$[Ar]\ 4s^23d^6$
27	Co	↑↓	↑↓ ↑↓ ↑ ↑ ↑		$[1s^22s^22p^63s^23p^6]\ 4s^23d^7$	$[Ar]\ 4s^23d^7$
28	Ni	↑↓	↑↓ ↑↓ ↑↓ ↑ ↑		$[1s^22s^22p^63s^23p^6]\ 4s^23d^8$	$[Ar]\ 4s^23d^8$
29	Cu	↑	↑↓ ↑↓ ↑↓ ↑↓ ↑↓		$[1s^22s^22p^63s^23p^6]\ 4s^13d^{10}$	$[Ar]\ 4s^13d^{10}$
30	Zn	↑↓	↑↓ ↑↓ ↑↓ ↑↓ ↑↓		$[1s^22s^22p^63s^23p^6]\ 4s^23d^{10}$	$[Ar]\ 4s^23d^{10}$
31	Ga	↑↓	↑↓ ↑↓ ↑↓ ↑↓ ↑↓	↑	$[1s^22s^22p^63s^23p^6]\ 4s^23d^{10}4p^1$	$[Ar]\ 4s^23d^{10}4p^1$
32	Ge	↑↓	↑↓ ↑↓ ↑↓ ↑↓ ↑↓	↑ ↑	$[1s^22s^22p^63s^23p^6]\ 4s^23d^{10}4p^2$	$[Ar]\ 4s^23d^{10}4p^2$
33	As	↑↓	↑↓ ↑↓ ↑↓ ↑↓ ↑↓	↑ ↑ ↑	$[1s^22s^22p^63s^23p^6]\ 4s^23d^{10}4p^3$	$[Ar]\ 4s^23d^{10}4p^3$
34	Se	↑↓	↑↓ ↑↓ ↑↓ ↑↓ ↑↓	↑↓ ↑ ↑	$[1s^22s^22p^63s^23p^6]\ 4s^23d^{10}4p^4$	$[Ar]\ 4s^23d^{10}4p^4$
35	Br	↑↓	↑↓ ↑↓ ↑↓ ↑↓ ↑↓	↑↓ ↑↓ ↑	$[1s^22s^22p^63s^23p^6]\ 4s^23d^{10}4p^5$	$[Ar]\ 4s^23d^{10}4p^5$
36	Kr	↑↓	↑↓ ↑↓ ↑↓ ↑↓ ↑↓	↑↓ ↑↓ ↑↓	$[1s^22s^22p^63s^23p^6]\ 4s^23d^{10}4p^6$	$[Ar]\ 4s^23d^{10}4p^6$

*Colored type indicates sublevel(s) whose occupancy changes when the last electron is added.

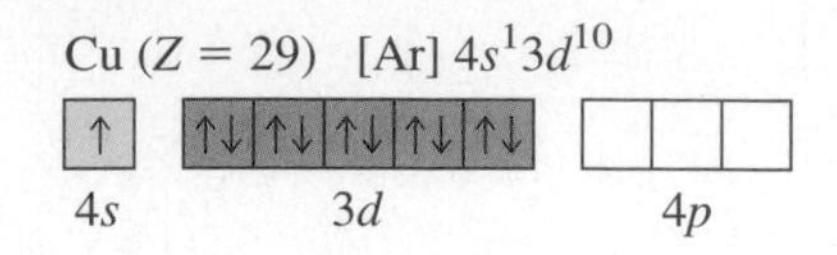

Instead, the 4*s* orbital of copper is half-filled (1 electron), and the 3*d* orbitals are *filled* with 10 electrons (see margin). The anomalous filling patterns in Cr and Cu lead us to conclude that *half-filled and filled sublevels are unexpectedly stable*. These are the first two cases of a pattern seen with many other elements.

In zinc, both the 4*s* and 3*d* sublevels are completely filled, and the first transition series ends. As Table 8.4 shows, the 4*p* sublevel is then filled by the next six elements. Period 4 ends with krypton, the next noble gas.

General Principles of Electron Configurations

There are 77 known elements beyond the 36 we have considered. Let's survey the ground-state electron configurations to highlight some key ideas.

Similar Outer Electron Configurations Within a Group To repeat one of chemistry's central themes and the key to the usefulness of the periodic table, *elements in a group have similar chemical properties because they have similar outer electron configurations* (Figure 8.11). Among the main-group elements (A groups)—the *s*-block and *p*-block elements—outer electron configurations within a group are essentially identical, as shown by the group headings in Figure 8.11. Some variations in the transition elements (B groups, *d* block) and inner transition elements (*f* block) occur, as you'll see.

Figure 8.11 A periodic table of partial ground-state electron configurations. These ground-state electron configurations show the electrons beyond the previous noble gas in the sublevel block being filled (excluding filled inner sublevels). For main-group elements, the group heading identifies the general outer configuration. Anomalous electron configurations occur often among the *d*-block and *f*-block elements, with the first two appearing for Cr (*Z* = 24) and Cu (*Z* = 29). Helium is colored as an *s*-block element but placed with the other members of Group 8A(18). Configurations for elements 112, 114, and 116 have not yet been confirmed.

Main-Group Elements (*s* block) | Transition Elements (*d* block) | Main-Group Elements (*p* block)

Period number: highest occupied energy level

Period	1A (1) ns^1	2A (2) ns^2	3B (3)	4B (4)	5B (5)	6B (6)	7B (7)	8B (8)	8B (9)	8B (10)	1B (11)	2B (12)	3A (13) ns^2np^1	4A (14) ns^2np^2	5A (15) ns^2np^3	6A (16) ns^2np^4	7A (17) ns^2np^5	8A (18) ns^2np^6
1	1 **H** $1s^1$																	2 **He** $1s^2$
2	3 **Li** $2s^1$	4 **Be** $2s^2$											5 **B** $2s^22p^1$	6 **C** $2s^22p^2$	7 **N** $2s^22p^3$	8 **O** $2s^22p^4$	9 **F** $2s^22p^5$	10 **Ne** $2s^22p^6$
3	11 **Na** $3s^1$	12 **Mg** $3s^2$											13 **Al** $3s^23p^1$	14 **Si** $3s^23p^2$	15 **P** $3s^23p^3$	16 **S** $3s^23p^4$	17 **Cl** $3s^23p^5$	18 **Ar** $3s^23p^6$
4	19 **K** $4s^1$	20 **Ca** $4s^2$	21 **Sc** $4s^23d^1$	22 **Ti** $4s^23d^2$	23 **V** $4s^23d^3$	24 **Cr** $4s^13d^5$	25 **Mn** $4s^23d^5$	26 **Fe** $4s^23d^6$	27 **Co** $4s^23d^7$	28 **Ni** $4s^23d^8$	29 **Cu** $4s^13d^{10}$	30 **Zn** $4s^23d^{10}$	31 **Ga** $4s^24p^1$	32 **Ge** $4s^24p^2$	33 **As** $4s^24p^3$	34 **Se** $4s^24p^4$	35 **Br** $4s^24p^5$	36 **Kr** $4s^24p^6$
5	37 **Rb** $5s^1$	38 **Sr** $5s^2$	39 **Y** $5s^24d^1$	40 **Zr** $5s^24d^2$	41 **Nb** $5s^14d^4$	42 **Mo** $5s^14d^5$	43 **Tc** $5s^24d^5$	44 **Ru** $5s^14d^7$	45 **Rh** $5s^14d^8$	46 **Pd** $4d^{10}$	47 **Ag** $5s^14d^{10}$	48 **Cd** $5s^24d^{10}$	49 **In** $5s^25p^1$	50 **Sn** $5s^25p^2$	51 **Sb** $5s^25p^3$	52 **Te** $5s^25p^4$	53 **I** $5s^25p^5$	54 **Xe** $5s^25p^6$
6	55 **Cs** $6s^1$	56 **Ba** $6s^2$	57 **La*** $6s^25d^1$	72 **Hf** $6s^25d^2$	73 **Ta** $6s^25d^3$	74 **W** $6s^25d^4$	75 **Re** $6s^25d^5$	76 **Os** $6s^25d^6$	77 **Ir** $6s^25d^7$	78 **Pt** $6s^15d^9$	79 **Au** $6s^15d^{10}$	80 **Hg** $6s^25d^{10}$	81 **Tl** $6s^26p^1$	82 **Pb** $6s^26p^2$	83 **Bi** $6s^26p^3$	84 **Po** $6s^26p^4$	85 **At** $6s^26p^5$	86 **Rn** $6s^26p^6$
7	87 **Fr** $7s^1$	88 **Ra** $7s^2$	89 **Ac**** $7s^26d^1$	104 **Rf** $7s^26d^2$	105 **Db** $7s^26d^3$	106 **Sg** $7s^26d^4$	107 **Bh** $7s^26d^5$	108 **Hs** $7s^26d^6$	109 **Mt** $7s^26d^7$	110 **Ds** $7s^26d^8$	111 **Rg** $7s^26d^9$	112 $7s^26d^{10}$		114 $7s^27p^2$		116 $7s^27p^4$		

Inner Transition Elements (*f* block)

Period															
6	*Lanthanides	58 **Ce** $6s^24f^15d^1$	59 **Pr** $6s^24f^3$	60 **Nd** $6s^24f^4$	61 **Pm** $6s^24f^5$	62 **Sm** $6s^24f^6$	63 **Eu** $6s^24f^7$	64 **Gd** $6s^24f^75d^1$	65 **Tb** $6s^24f^9$	66 **Dy** $6s^24f^{10}$	67 **Ho** $6s^24f^{11}$	68 **Er** $6s^24f^{12}$	69 **Tm** $6s^24f^{13}$	70 **Yb** $6s^24f^{14}$	71 **Lu** $6s^24f^{14}5d^1$
7	**Actinides	90 **Th** $7s^26d^2$	91 **Pa** $7s^25f^26d^1$	92 **U** $7s^25f^36d^1$	93 **Np** $7s^25f^46d^1$	94 **Pu** $7s^25f^6$	95 **Am** $7s^25f^7$	96 **Cm** $7s^25f^76d^1$	97 **Bk** $7s^25f^9$	98 **Cf** $7s^25f^{10}$	99 **Es** $7s^25f^{11}$	100 **Fm** $7s^25f^{12}$	101 **Md** $7s^25f^{13}$	102 **No** $7s^25f^{14}$	103 **Lr** $7s^25f^{14}6d^1$

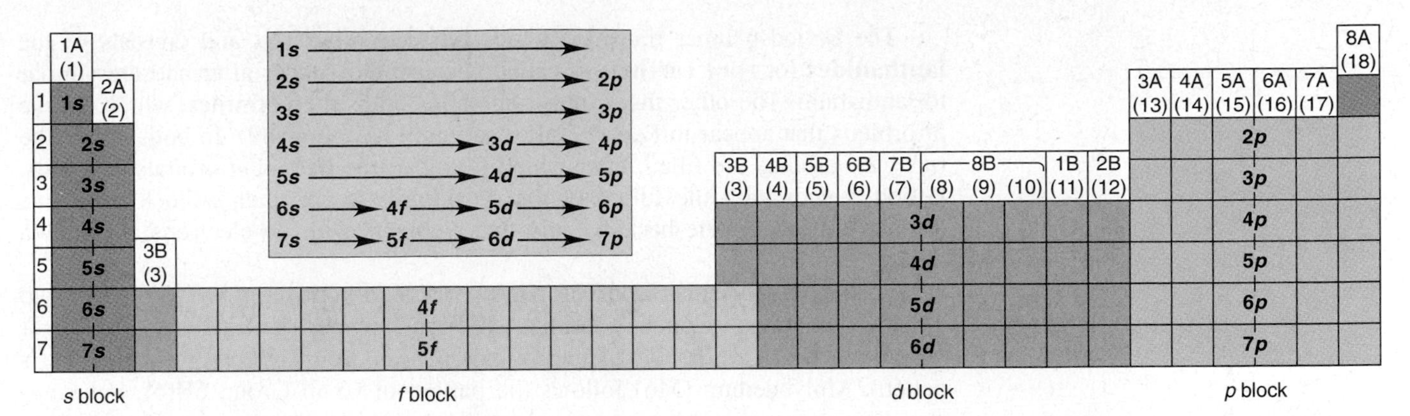

Figure 8.12 The relation between orbital filling and the periodic table. If we "read" the periods like the words on a page, the elements are arranged into sublevel blocks that occur in the order of increasing energy. This form of the periodic table shows the sublevel blocks. (The *f* blocks fit between the first and second elements of the *d* blocks in Periods 6 and 7.) *Inset:* A simple version of sublevel order.

Orbital Filling Order When the elements are "built up" by filling their levels and sublevels in order of increasing energy, we obtain the actual sequence of elements in the periodic table. Thus, reading the table from left to right, as you read words on a page, gives the energy order of levels and sublevels, which is shown in Figure 8.12. The arrangement of the periodic table is the best way to learn the orbital filling order of the elements, but a useful memory aid is shown in Figure 8.13.

Categories of Electrons The elements have three categories of electrons:

1. **Inner (core) electrons** are those seen in the previous noble gas and any completed transition series. They fill all the *lower energy levels* of an atom.
2. **Outer electrons** are those in the *highest energy level* (highest n value). They spend most of their time farthest from the nucleus.
3. **Valence electrons** are those involved in forming compounds. *Among the main-group elements, the valence electrons* ***are*** *the outer electrons.* Among the transition elements, the $(n-1)d$ electrons are counted among the valence electrons because some or all of them are often involved in bonding.

Group and Period Numbers Key information is embedded in the periodic table:

1. Among the main-group elements (A groups), *the group number equals the number of outer electrons* (those with the highest n): chlorine (Cl; Group **7**A) has 7 outer electrons, tellurium (Te; Group **6**A) has 6, and so forth.
2. *The period number is the n value of the highest energy level.* Thus, in Period 2, the $n = 2$ level has the highest energy; in Period 5, it is the $n = 5$ level.
3. The n value squared (n^2) gives the total number of *orbitals* in that energy level. Because an orbital can hold no more than two electrons (exclusion principle), $2n^2$ gives the maximum number of *electrons* (or elements) in the energy level. For example, for the $n = 3$ level, the number of orbitals is $n^2 = 9$: one $3s$, three $3p$, and five $3d$. The number of electrons is $2n^2$, or 18: two $3s$ and six $3p$ electrons occur in the eight elements of Period 3, and ten $3d$ electrons are added in the ten transition elements of Period 4.

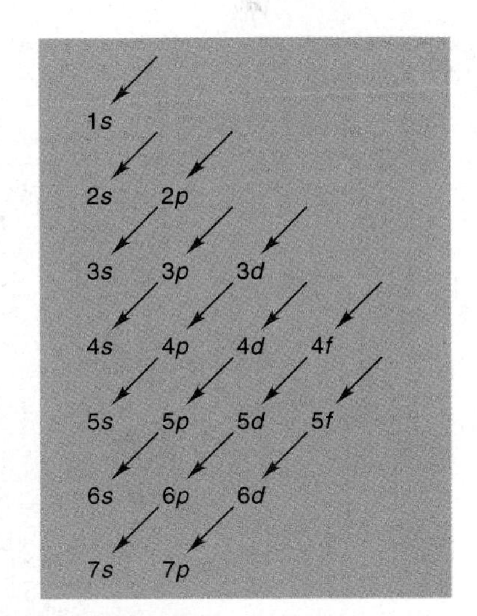

Figure 8.13 Aid to memorizing sublevel filling order. List the sublevels as shown, and read from 1s, following the direction of the arrows. Note that the

- n value is constant horizontally
- l value is constant vertically
- $n + l$ sum is constant diagonally.

Unusual Configurations: Transition and Inner Transition Elements

Periods 4, 5, 6, and 7 incorporate the *d*-block transition elements. The general pattern, as you've seen, is that the $(n-1)d$ orbitals are filled between the ns and np orbitals. Thus, Period 5 follows the same general pattern as Period 4. In Period 6, the $6s$ sublevel is filled in cesium (Cs) and barium (Ba), and then lanthanum (La; $Z = 57$), the first member of the $5d$ transition series, occurs. At this point, the first series of **inner transition elements,** those in which f orbitals are being filled, intervenes (Figure 8.12). The f orbitals have $l = 3$, so the possible m_l values are $-3, -2, -1, 0, +1, +2,$ and $+3$; that is, there are seven f orbitals, for a total of 14 elements in *each* of the two inner transition series.

The Period 6 inner transition series fills the $4f$ orbitals and consists of the **lanthanides** (or **rare earths**), so called because they occur after and are similar to lanthanum. The other inner transition series holds the **actinides,** which fill the $5f$ orbitals that appear in Period 7 after actinium (Ac; $Z = 89$). In both series, the $(n - 2)f$ orbitals are filled, after which filling of the $(n - 1)d$ orbitals proceeds. Period 6 ends with the filling of the $6p$ orbitals as in other p-block elements. Period 7 is incomplete because only two elements with $7p$ electrons are known at this time.

Several irregularities in filling pattern occur in both the d and f blocks. Two already mentioned occur in chromium (Cr) and copper (Cu) in Period 4. Silver (Ag) and gold (Au), the two elements under Cu in Group 1B(11), follow copper's pattern. Molybdenum (Mo) follows the pattern of Cr in Group 6B(6), but tungsten (W) does not. Other anomalous configurations appear among the transition elements in Periods 5 and 6. Note, however, that even though minor variations from the expected configurations occur, the sum of ns electrons and $(n - 1)d$ electrons always equals the *new* group number. For instance, despite variations in the electron configurations in Group 6B(**6**)—Cr, Mo, W, and Sg—the sum of ns and $(n - 1)d$ electrons is 6; in Group 8B(**10**)—Ni, Pd, and Pt—the sum is 10.

Whenever our observations differ from our expectations, remember that the fact always takes precedence over the model; in other words, the electrons don't "care" what orbitals *we* think they should occupy. As the atomic orbitals in larger atoms fill with electrons, sublevel energies differ very little, which results in these variations from the expected pattern.

SAMPLE PROBLEM 8.2 Determining Electron Configurations

Problem Using the periodic table on the inside cover of the text (not Figure 8.11 or Table 8.4), give the full and condensed electron configurations, partial orbital diagrams showing valence electrons, and number of inner electrons for the following elements:
(a) Potassium (K; $Z = 19$) **(b)** Molybdenum (Mo; $Z = 42$) **(c)** Lead (Pb; $Z = 82$)

Plan The atomic number tells us the number of electrons, and the periodic table shows the order for filling sublevels. In the partial orbital diagrams, we include all electrons after those of the previous noble gas *except* those in *filled* inner sublevels. The number of inner electrons is the sum of those in the previous noble gas and in filled d and f sublevels.

Animation: Electron Configurations/Orbital Diagrams
Online Learning Center

Solution (a) For K ($Z = 19$), the full electron configuration is $1s^2 2s^2 2p^6 3s^2 3p^6 4s^1$.
The condensed configuration is [Ar] $4s^1$.
The partial orbital diagram for valence electrons is

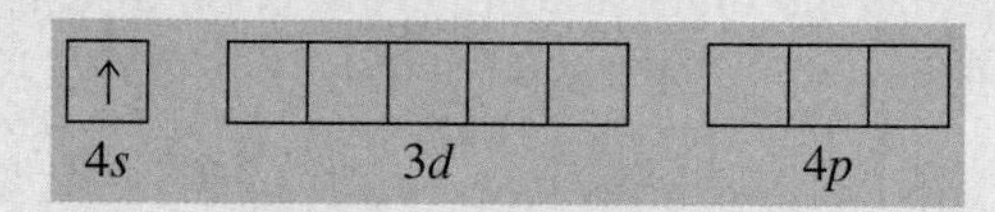

K is a main-group element in Group 1A(1) of Period 4, so there are 18 inner electrons.

(b) For Mo ($Z = 42$), we would expect the full electron configuration to be $1s^2 2s^2 2p^6 3s^2 3p^6 4s^2 3d^{10} 4p^6 5s^2 4d^4$. However, Mo lies under Cr in Group 6B(6) and exhibits the same variation in filling pattern in the ns and $(n - 1)d$ sublevels:
$1s^2 2s^2 2p^6 3s^2 3p^6 4s^2 3d^{10} 4p^6 5s^1 4d^5$.
The condensed electron configuration is [Kr] $5s^1 4d^5$.
The partial orbital diagram for valence electrons is

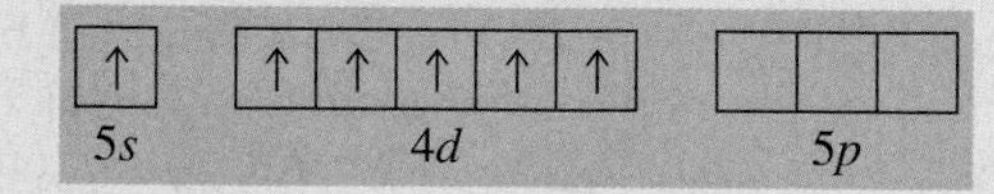

Mo is a transition element in Group 6B(6) of Period 5, so there are 36 inner electrons.

(c) For Pb ($Z = 82$), the full electron configuration is
$1s^2 2s^2 2p^6 3s^2 3p^6 4s^2 3d^{10} 4p^6 5s^2 4d^{10} 5p^6 6s^2 4f^{14} 5d^{10} 6p^2$.
The condensed electron configuration is [Xe] $6s^2 4f^{14} 5d^{10} 6p^2$.

The partial orbital diagram for valence electrons (no filled inner sublevels) is

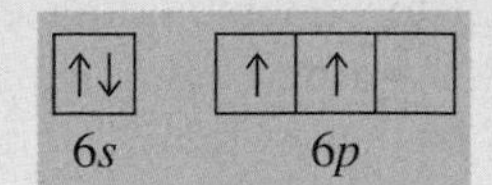

Pb is a main-group element in Group 4A(14) of Period 6, so there are 54 (in Xe) + 14 (in 4*f* series) + 10 (in 5*d* series) = 78 inner electrons.

Check Be sure the sum of the superscripts (electrons) in the full electron configuration equals the atomic number, and that the number of *valence* electrons in the condensed configuration equals the number of electrons in the partial orbital diagram.

FOLLOW-UP PROBLEM 8.2 Without referring to Table 8.4 or Figure 8.11, give full and condensed electron configurations, partial orbital diagrams showing valence electrons, and the number of inner electrons for the following elements:
(a) Ni (Z = 28) **(b)** Sr (Z = 38) **(c)** Po (Z = 84)

SECTION SUMMARY

In the aufbau method, one electron is added to an atom of each successive element in accord with Pauli's exclusion principle (no two electrons can have the same set of quantum numbers) and Hund's rule (orbitals of equal energy become half-filled, with electron spins parallel, before any pairing occurs). The elements of a group have similar outer electron configurations and similar chemical behavior. For the main-group elements, valence electrons (those involved in reactions) are in the outer (highest energy) level only. For transition elements, $(n-1)d$ electrons are also involved in reactions. In general, $(n-1)d$ orbitals fill after ns and before np orbitals. In Periods 6 and 7, $(n-2)f$ orbitals fill between the first and second $(n-1)d$ orbitals.

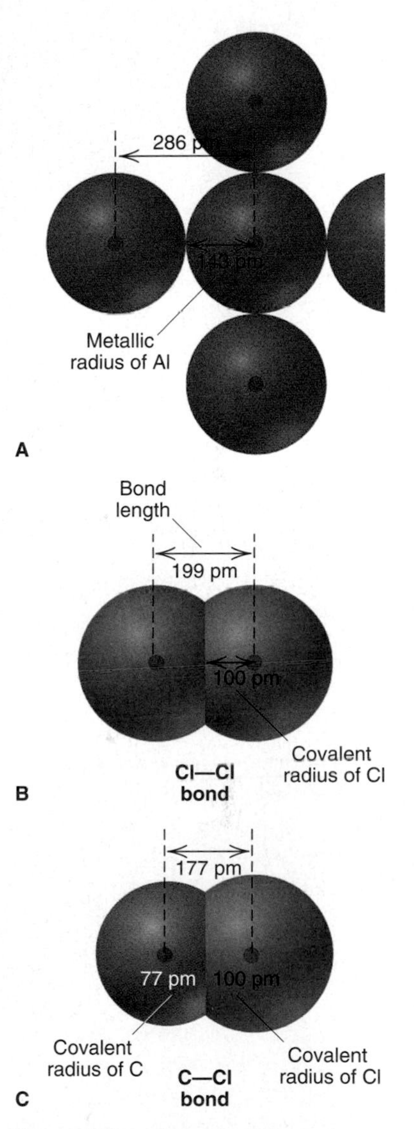

Figure 8.14 Defining metallic and covalent radii. A, The metallic radius is one-half the distance between nuclei of adjacent atoms in a crystal of the element, as shown here for aluminum. **B,** The covalent radius is one-half the distance between bonded nuclei in a molecule of the element, as shown here for chlorine. In effect, it is one-half the bond length. **C,** In a covalent compound, the bond length and known covalent radii are used to determine other radii. Here the C—Cl bond length (177 pm) and the covalent radius of Cl (100 pm) are used to find a value for the covalent radius of C (177 pm − 100 pm = 77 pm).

8.4 TRENDS IN THREE KEY ATOMIC PROPERTIES

All physical and chemical behavior of the elements is based ultimately on the electron configurations of their atoms. In this section, we focus on three properties of atoms that are directly influenced by electron configuration and, thus, effective nuclear charge: atomic size, ionization energy (the energy required to remove an electron from a gaseous atom), and electron affinity (the energy change involved in adding an electron to a gaseous atom). These properties are *periodic:* they generally increase and decrease in a recurring manner throughout the periodic table. As a result, their relative magnitudes can often be predicted, and they often exhibit consistent changes, or *trends,* within a group or period that correlate with element behavior.

Trends in Atomic Size

In Chapter 7, we noted that an electron in an atom can lie relatively far from the nucleus, so we commonly represent atoms as spheres in which the electrons spend 90% of their time. However, we often *define* atomic size in terms of how closely one atom lies next to another. In practice, we measure the distance between identical, adjacent atomic nuclei in a sample of an element and divide that distance in half. (The technique is discussed in Chapter 12.) Because atoms do not have hard surfaces, the size of an atom in a compound depends somewhat on the atoms near it. In other words, *atomic size varies slightly from substance to substance.*

Figure 8.14 shows two common definitions of atomic size. The **metallic radius** is one-half the distance between nuclei of adjacent atoms in a crystal of the element; we typically use this definition for metals. For elements commonly occurring as molecules, mostly nonmetals, we define atomic size by the **covalent radius,** one-half the distance between nuclei of identical covalently bonded atoms.

Trends Among the Main-Group Elements Atomic size greatly influences other atomic properties and is critical to understanding element behavior. Figure 8.15 shows the atomic radii of the main-group elements and most of the transition elements. Among the main-group elements, note that atomic size varies within both a group and a period. These variations in atomic size are the result of two opposing influences:

1. *Changes in n.* As the principal quantum number (n) increases, the probability that the outer electrons will spend more time farther from the nucleus increases as well; thus, the atoms are larger.
2. *Changes in Z_{eff}.* As the effective nuclear charge (Z_{eff})—the positive charge "felt" by an electron—increases, outer electrons are pulled closer to the nucleus; thus, the atoms are smaller.

The net effect of these influences depends on shielding of the increasing nuclear charge by inner electrons:

1. *Down a group, n dominates.* As we move down a main group, each member has one more level of *inner* electrons that shield the *outer* electrons very effectively. Even though calculations show Z_{eff} on the outer electrons rising moderately for each element in the group, the atoms get larger as a result of the increasing n value. *Atomic radius generally* ***increases*** *in a group from top to bottom.*
2. *Across a period, Z_{eff} dominates.* As we move across a period of main-group elements, electrons are added to the same outer level, so the shielding by inner electrons does not change. Because outer electrons shield each other poorly, Z_{eff} on the outer electrons rises significantly, and so they are pulled closer to the nucleus. *Atomic radius generally* ***decreases*** *in a period from left to right.*

Trends Among the Transition Elements As Figure 8.15 shows, these trends hold well for the main-group elements but *not* as consistently for the transition elements. As we move from left to right, size shrinks through the first two or three transition elements because of the increasing nuclear charge. But, from then on, *the size remains relatively constant* because shielding by the inner d electrons counteracts the usual increase in Z_{eff}. For instance, vanadium (V; $Z = 23$), the third Period 4 transition metal, has the same radius as zinc (Zn; $Z = 30$), the last Period 4 transition metal. This pattern of atomic size shrinking also appears in Periods 5 and 6 in the d-block transition series and in both series of inner transition elements.

This shielding by d electrons causes a major size decrease from Group 2A(2) to Group 3A(13), the two main groups that flank the transition series. The size decrease in Periods 4, 5, and 6 (*with* a transition series) is much greater than in Period 3 (*without* a transition series). Because electrons in the np orbitals penetrate more than those in the $(n - 1)d$ orbitals, the first np electron [Group 3A(13)] "feels" a Z_{eff} that has been increased by the protons added to all the intervening transition elements. The greatest change in size occurs in Period 4, in which calcium (Ca; $Z = 20$) is nearly 50% larger than gallium (Ga; $Z = 31$). In fact, filling the d orbitals in the transition series causes such a major size contraction that gallium is slightly *smaller* than aluminum (Al; $Z = 13$), even though Ga is below Al in the same group!

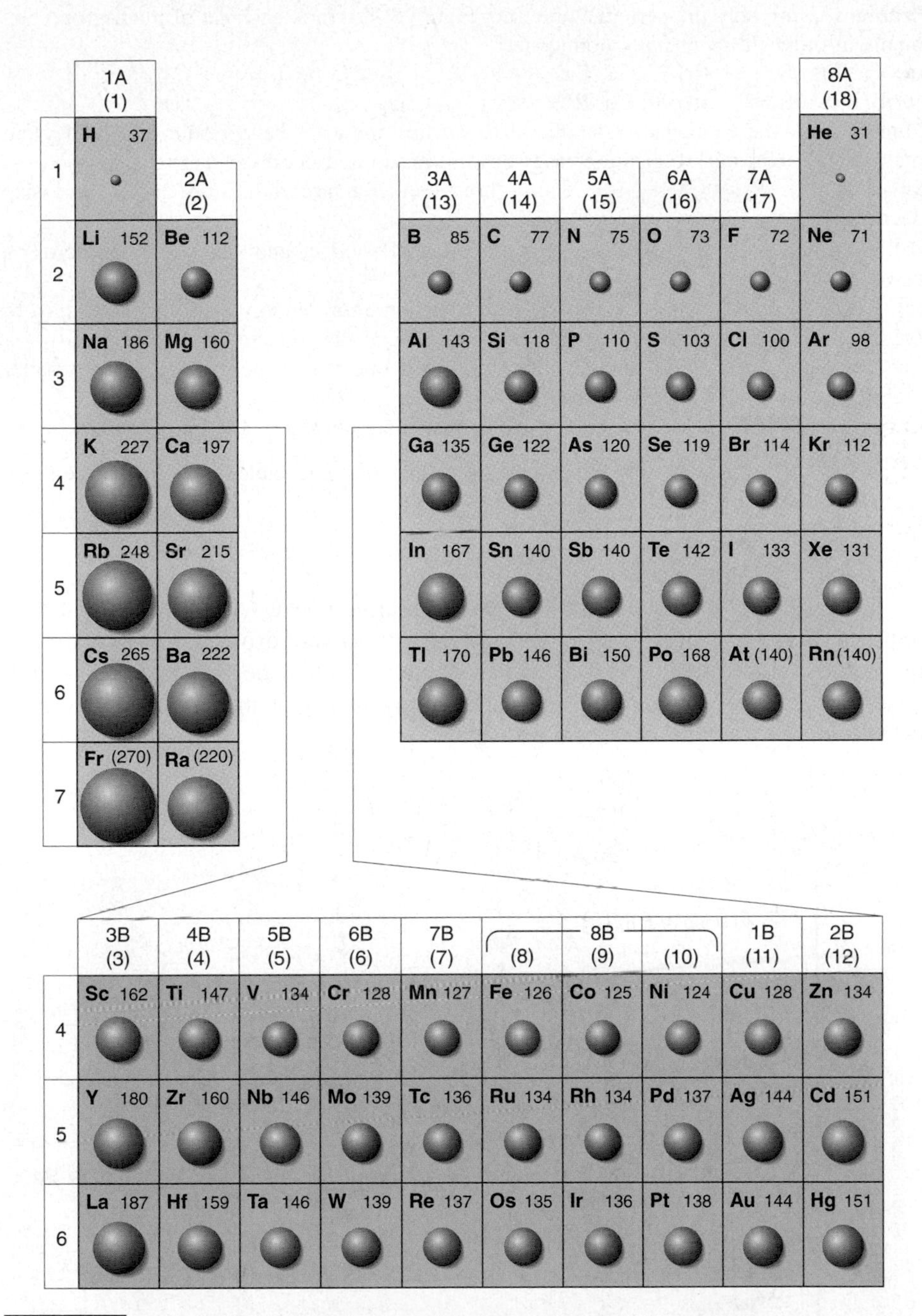

Figure 8.15 **Atomic radii of the main-group and transition elements.** Atomic radii (in picometers) are shown as half-spheres of proportional size for the main-group elements (*tan*) and the transition elements (*blue*). Among the main-group elements, atomic radius generally increases from top to bottom and decreases from left to right. The transition elements do not exhibit this consistent decrease in size. (Values in parentheses have only two significant figures; values for the noble gases are based on quantum-mechanical calculations.)

SAMPLE PROBLEM 8.3 Ranking Elements by Atomic Size

Problem Using only the periodic table (not Figure 8.15), rank each set of main-group elements in order of *decreasing* atomic size:
(a) Ca, Mg, Sr **(b)** K, Ga, Ca
(c) Br, Rb, Kr **(d)** Sr, Ca, Rb

Plan To rank the elements by atomic size, we find them in the periodic table. They are main-group elements, so size increases down a group and decreases across a period.

Solution **(a)** Sr > Ca > Mg. These three elements are in Group 2A(2), and size decreases up the group.

(b) K > Ca > Ga. These three elements are in Period 4, and size decreases across a period.

(c) Rb > Br > Kr. Rb is largest because it has one more energy level and is farthest to the left. Kr is smaller than Br because Kr is farther to the right in Period 4.

(d) Rb > Sr > Ca. Ca is smallest because it has one fewer energy level. Sr is smaller than Rb because it is farther to the right.

Check From Figure 8.15, we see that the rankings are correct.

FOLLOW-UP PROBLEM 8.3 Using only the periodic table, rank the elements in each set in order of *increasing* size:
(a) Se, Br, Cl **(b)** I, Xe, Ba

Packing 'Em In The increasing nuclear charge shrinks the space in which each electron can move. For example, in Group 1A(1), the atomic radius of cesium (Cs; $Z = 55$) is only 1.7 times that of lithium (Li; $Z = 3$); so the volume of Cs is about five times that of Li, even though Cs has 18 times as many electrons. At the opposite ends of Period 2, neon (Ne; $Z = 10$) has about one-tenth the volume of Li ($Z = 3$), but Ne has three times as many electrons. Thus, whether size increases down a group or decreases across a period, the attraction caused by the larger number of protons in the nucleus greatly crowds the electrons.

Figure 8.16 shows the overall variation in atomic size with increasing atomic number. Note the recurring up-and-down pattern as size drops across a period to the noble gas and then leaps up to the alkali metal that begins the next period. Also note how each transition series, beginning with that in Period 4 (K to Kr), throws off the smooth size decrease.

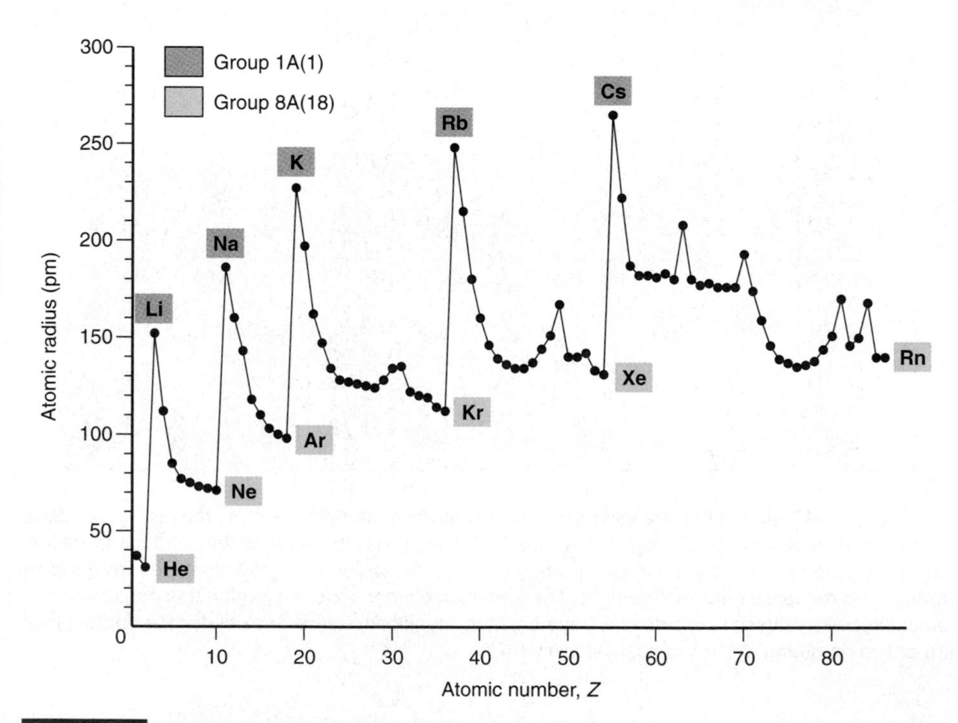

Figure 8.16 Periodicity of atomic radius. A plot of atomic radius vs. atomic number for the elements in Periods 1 through 6 shows a periodic change: the radius generally decreases through a period to the noble gas [Group 8A(18); *purple*] and then increases suddenly to the next alkali metal [Group 1A(1); *brown*]. Deviation from the general decrease occurs among the transition elements.

Trends in Ionization Energy

The **ionization energy (IE)** is the energy (in kJ) required for the *complete removal* of 1 mol of electrons from 1 mol of gaseous atoms or ions. Pulling an electron away from a nucleus *requires* energy to overcome the attraction. Because energy flows *into* the system, the ionization energy is always positive (like ΔH of an endothermic reaction).

In Chapter 7, you saw that the ionization energy of the H atom is the energy difference between $n = 1$ and $n = \infty$, the point at which the electron is completely removed. Many-electron atoms can lose more than one electron. The first ionization energy (IE_1) removes an outermost electron (highest energy sublevel) from the gaseous atom:

$$\text{Atom}(g) \longrightarrow \text{ion}^+(g) + e^- \qquad \Delta E = IE_1 > 0 \qquad (8.2)$$

The second ionization energy (IE_2) removes a second electron. This electron is pulled away from a positively charged ion, so IE_2 is always larger than IE_1:

$$\text{Ion}^+(g) \longrightarrow \text{ion}^{2+}(g) + e^- \qquad \Delta E = IE_2 \text{ (always} > IE_1)$$

The first ionization energy is a key factor in an element's chemical reactivity because, as you'll see, *atoms with a low IE_1 tend to form cations during reactions, whereas those with a high IE_1 (except the noble gases) often form anions.*

Variations in First Ionization Energy The elements exhibit a periodic pattern in first ionization energy, as shown in Figure 8.17. By comparing this figure with Figure 8.16, you can see a roughly *inverse* relationship between IE_1 and atomic size: *as size decreases, it takes more energy to remove an electron.* This inverse relationship appears throughout the groups and periods of the table.

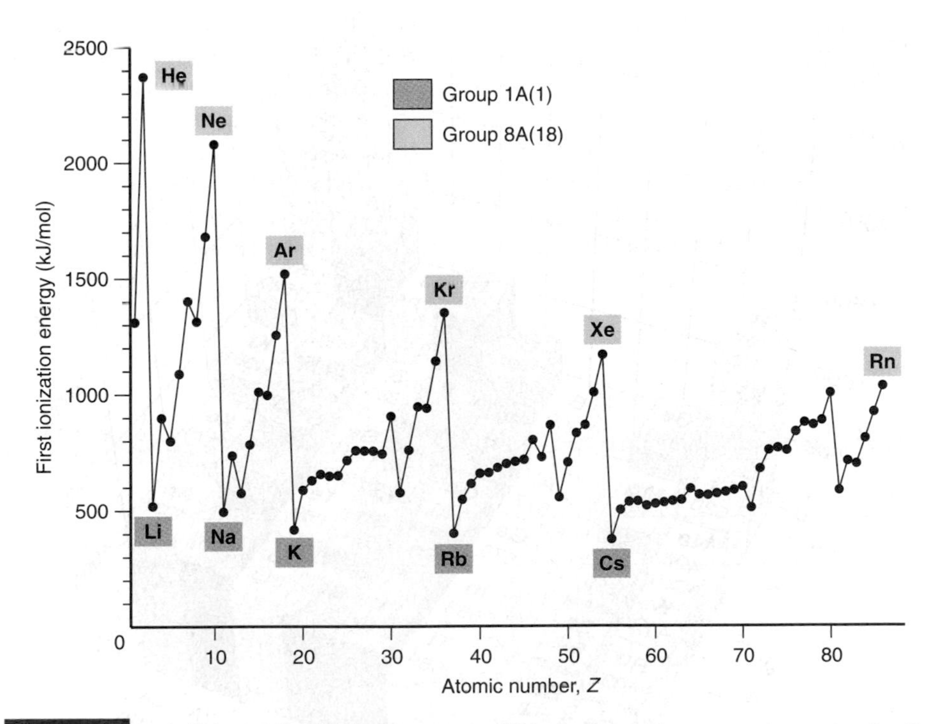

Figure 8.17 Periodicity of first ionization energy (IE_1). A plot of IE_1 vs. atomic number for the elements in Periods 1 through 6 shows a periodic pattern: the lowest values occur for the alkali metals (*brown*) and the highest for the noble gases (*purple*). This is the *inverse* of the trend in atomic size (see Figure 8.16).

SAMPLE PROBLEM 8.5 Identifying an Element from Successive Ionization Energies

Problem Name the Period 3 element with the following ionization energies (in kJ/mol), and write its electron configuration:

IE_1	IE_2	IE_3	IE_4	IE_5	IE_6
1012	1903	2910	4956	6278	22,230

Plan We look for a large jump in the IE values, which occurs after all valence electrons have been removed. Then we refer to the periodic table to find the Period 3 element with this number of valence electrons and write its electron configuration.

Solution The exceptionally large jump occurs after IE_5, indicating that the element has five valence electrons and, thus, is in Group 5A(15). This Period 3 element is phosphorus (P; $Z = 15$). Its electron configuration is $1s^22s^22p^63s^23p^3$.

FOLLOW-UP PROBLEM 8.5 Element Q is in Period 3 and has the following ionization energies (in kJ/mol):

IE_1	IE_2	IE_3	IE_4	IE_5	IE_6
577	1816	2744	11,576	14,829	18,375

Name element Q and write its electron configuration.

Trends in Electron Affinity

The **electron affinity (EA)** is the energy change (in kJ) accompanying the *addition* of 1 mol of electrons to 1 mol of gaseous atoms or ions. As with ionization energy, there is a first electron affinity, a second, and so forth. The *first electron affinity* (EA_1) refers to the formation of 1 mol of monovalent (1−) gaseous anions:

$$\text{Atom}(g) + e^- \longrightarrow \text{ion}^-(g) \qquad \Delta E = EA_1$$

In most cases, *energy is released when the first electron is added* because it is attracted to the atom's nuclear charge. Thus, EA_1 is usually negative (just as ΔH for an exothermic reaction is negative).* The second electron affinity (EA_2), on the other hand, is always positive because energy must be *absorbed* in order to overcome electrostatic repulsions and add another electron to a negative ion.

Factors other than Z_{eff} and atomic size affect electron affinities, so trends are not as regular as those for the previous two properties. For instance, we might expect electron affinities to decrease smoothly down a group (smaller negative number) because the nucleus is farther away from an electron being added. But, as Figure 8.20 shows, only Group 1A(1) exhibits this behavior. We might also expect a regular increase in electron affinities across a period (larger negative number) because size decreases and the increasing Z_{eff} should attract the electron being added more strongly. An overall left-to-right increase in magnitude is there, but we certainly cannot say that it is a regular increase. These exceptions arise from changes in sublevel energy and in electron-electron repulsion.

1A (1)	2A (2)	3A (13)	4A (14)	5A (15)	6A (16)	7A (17)	8A (18)
H −72.8							He (0.0)
Li −59.6	Be ≤0	B −26.7	C −122	N +7	O −141	F −328	Ne (+29)
Na −52.9	Mg ≤0	Al −42.5	Si −134	P −72.0	S −200	Cl −349	Ar (+35)
K −48.4	Ca −2.37	Ga −28.9	Ge −119	As −78.2	Se −195	Br −325	Kr (+39)
Rb −46.9	Sr −5.03	In −28.9	Sn −107	Sb −103	Te −190	I −295	Xe (+41)
Cs −45.5	Ba −13.95	Tl −19.3	Pb −35.1	Bi −91.3	Po −183	At −270	Rn (+41)

Figure 8.20 Electron affinities of the main-group elements. The electron affinities (in kJ/mol) of the main-group elements are shown. Negative values indicate that energy is released when the anion forms. Positive values, which occur in Group 8A(18), indicate that energy is absorbed to form the anion; in fact, these anions are unstable and the values are estimated.

*Tables of first electron affinity often list them as positive if energy is *absorbed* to remove an electron from the anion. Keep this convention in mind when researching these values in reference texts. Electron affinities are difficult to measure, so values are frequently updated with more accurate data. Values for Group 2A(2) reflect recent changes.

Despite the irregularities, three key points emerge when we examine the relative values of ionization energy and electron affinity:

1. *Reactive nonmetals.* The elements in Groups 6A(16) and especially those in Group 7A(17) (halogens) have high ionization energies and highly negative (exothermic) electron affinities. These elements lose electrons with difficulty but attract them strongly. Therefore, *in their ionic compounds, they form negative ions.*
2. *Reactive metals.* The elements in Groups 1A(1) and 2A(2) have low ionization energies and slightly negative (exothermic) electron affinities. Both groups lose electrons readily but attract them only weakly, if at all. Therefore, *in their ionic compounds, they form positive ions.*
3. *Noble gases.* The elements in Group 8A(18) have very high ionization energies and slightly positive (endothermic) electron affinities. Therefore, *these elements tend **not** to lose or gain electrons.* In fact, only the larger members of the group (Kr, Xe, Rn) form any compounds at all.

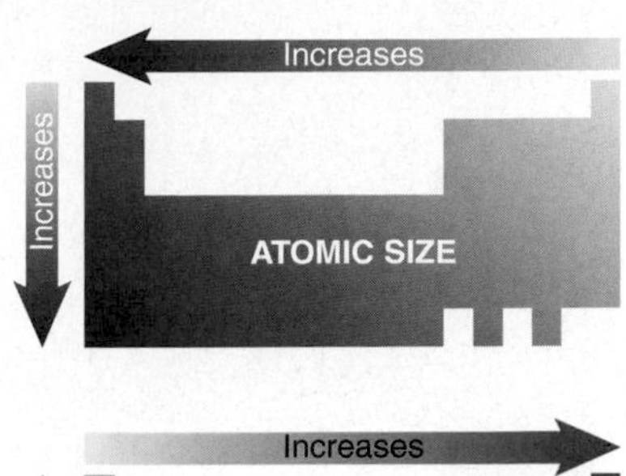

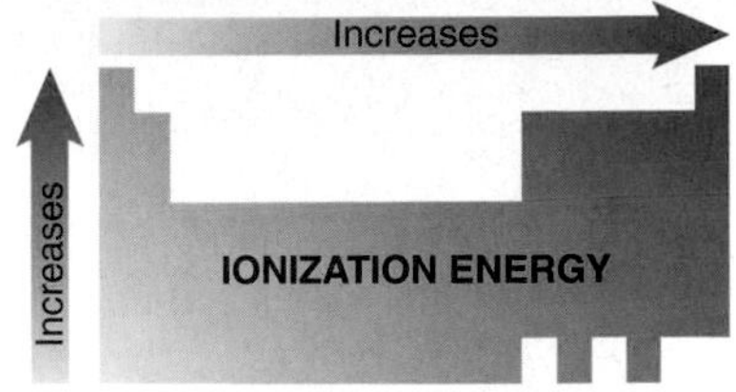

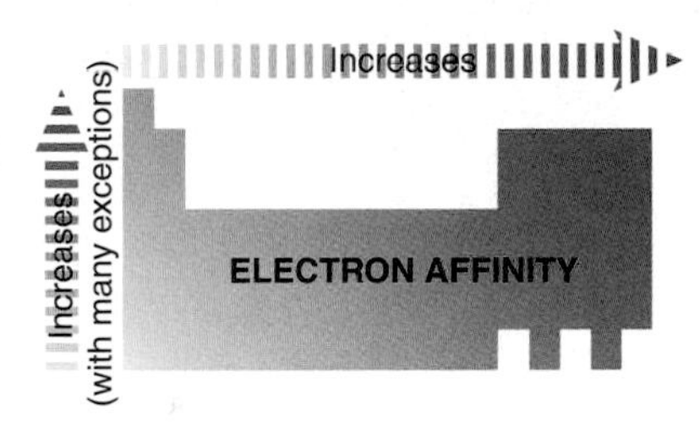

Figure 8.21 Trends in three atomic properties. Periodic trends are depicted as gradations in shading on miniature periodic tables, with arrows indicating the direction of general *increase* in a group or period. For electron affinity, Group 8A(18) is not shown, and the dashed arrows indicate the numerous exceptions to expected trends.

SECTION SUMMARY

Trends in three atomic properties are summarized in Figure 8.21. Atomic size increases down a main group and decreases across a period. Across a transition series, size remains relatively constant. First ionization energy (the energy required to remove the outermost electron from a mole of gaseous atoms) is inversely related to atomic size: IE_1 decreases down a main group and increases across a period. An element's successive ionization energies show a very large increase when the first inner (core) electron is removed. Electron affinity (the energy involved in adding an electron to a mole of gaseous atoms) shows many variations from expected trends. Based on the relative sizes of IEs and EAs, in their ionic compounds, the Group 1A(1) and 2A(2) elements tend to form cations, and the Group 6A(16) and 7A(17) elements tend to form anions.

8.5 ATOMIC STRUCTURE AND CHEMICAL REACTIVITY

Our main purpose for discussing atomic properties is, of course, to see how they affect element behavior. In this section, you'll see how the properties we just examined influence metallic behavior and determine the type of ion an element can form, as well as how electron configuration relates to magnetic properties.

Trends in Metallic Behavior

Metals are located in the left and lower three-quarters of the periodic table. They are typically shiny solids with moderate to high melting points, are good thermal and electrical conductors, can be drawn into wires and rolled into sheets, and tend to lose electrons to nonmetals. *Nonmetals* are located in the upper right quarter of the table. They are typically not shiny, have relatively low melting points, are poor thermal and electrical conductors, are crumbly or gaseous, and tend to gain electrons from metals. *Metalloids* are located in the region between the other two classes and have properties between them as well. Thus, *metallic behavior decreases left to right and increases top to bottom* in the periodic table (Figure 8.22).

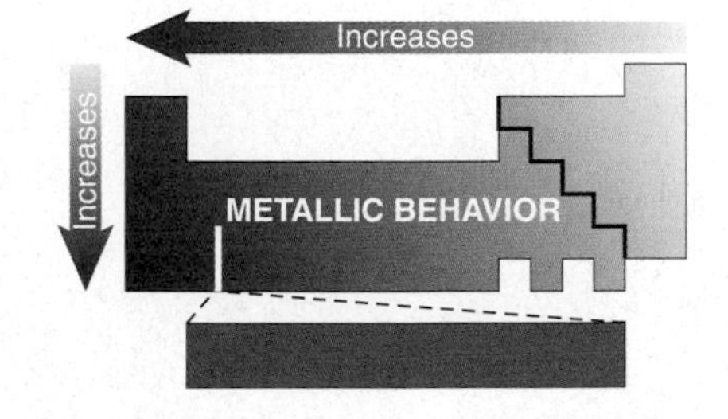

Figure 8.22 Trends in metallic behavior. The gradation in metallic behavior among the elements is depicted as a gradation in shading from bottom left to top right, with arrows showing the direction of increase. Elements that *behave* as metals appear in the left and lower three-quarters of the table. (Hydrogen appears next to helium in this periodic table.)

It's important to realize, however, that an element's properties may not fall neatly into our categories. For instance, the nonmetal carbon in the form of graphite is a good electrical conductor. Iodine, another nonmetal, is a shiny solid. Gallium and cesium are metals that melt at temperatures below body temperature, and mercury is a liquid at room temperature. And iron is quite brittle. Despite such exceptions, we can make several generalizations about metallic behavior.

Relative Tendency to Lose Electrons Metals tend to lose electrons during chemical reactions because they have low ionization energies compared to nonmetals. The increase in metallic behavior down a group is most obvious in the physical and chemical behavior of the elements in Groups 3A(13) through 6A(16), which contain more than one class of element. For example, consider the elements in Group 5A(15), which appear vertically in Figure 8.23. Here, the change is so great that, with regard to monatomic ions, *elements at the top tend to form anions and those at the bottom tend to form cations.* Nitrogen (N) is a gaseous nonmetal, and phosphorus (P) is a solid nonmetal. Both occur occasionally as 3− anions in their compounds. Arsenic (As) and antimony (Sb) are metalloids, with Sb the more metallic of the two; neither forms ions readily. Bismuth (Bi), the largest member, is a typical metal, forming mostly ionic compounds in which it appears as a 3+ cation. Even in Group 2A(2), which consists entirely of metals, the tendency to form cations increases down the group. Beryllium (Be), for example, forms covalent compounds with nonmetals, whereas the compounds of barium (Ba) are ionic.

As we move across a period, it becomes more difficult to lose an electron (IE increases) and easier to gain one (EA becomes more negative). Therefore, with regard to monatomic ions, *elements at the left tend to form cations and those at the right tend to form anions.* The typical decrease in metallic behavior across a period is clear among the elements in Period 3, which appear horizontally in Figure 8.23. Sodium and magnesium are metals. Sodium is shiny when freshly cut under mineral oil, but it loses an electron so readily to O_2 that, if cut in air, its surface is coated immediately with a dull oxide. These metals exist naturally as Na^+ and Mg^{2+} ions in oceans, minerals, and organisms. Aluminum is metallic in its physical properties and forms the Al^{3+} ion in some compounds, but it bonds covalently in most others. Silicon (Si) is a shiny metalloid that does not occur as a monatomic ion. The most common form of phosphorus is a white, waxy nonmetal that, as noted above, forms the P^{3-} ion in a few compounds. Sulfur is a crumbly yellow nonmetal that forms the sulfide ion (S^{2-}) in many compounds. Diatomic chlorine (Cl_2) is a yellow-green, gaseous nonmetal that attracts electrons avidly and exists in nature as the Cl^- ion.

Acid-Base Behavior of the Element Oxides Metals are also distinguished from nonmetals by the acid-base behavior of their oxides in water:

- Most main-group metals *transfer* electrons to oxygen, so their *oxides are ionic. In water, these oxides act as bases,* producing OH^- ions and reacting with acids. Calcium oxide is an example (turns indicator pink; see photo below, *left*).
- Nonmetals *share* electrons with oxygen, so *nonmetal oxides are covalent. In water, they act as acids,* producing H^+ ions and reacting with bases. Tetraphosphorus decaoxide is an example (turns indicator yellow; see photo below, *right*).

Figure 8.23 The change in metallic behavior in Group 5A(15) and Period 3. Moving down from nitrogen to bismuth shows an *increase* in metallic behavior (and thus a decrease in ionization energy). Moving left to right from sodium to chlorine shows a *decrease* in metallic behavior (and thus a general increase in ionization energy). Each box shows the element and its atomic number, symbol, and first ionization energy (in kJ/mol).

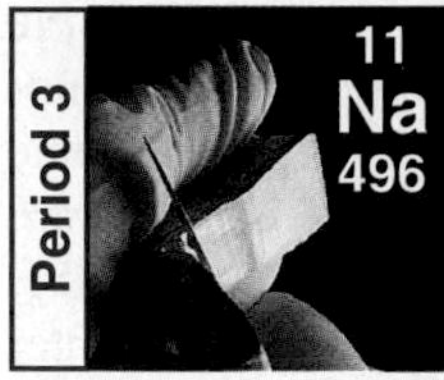

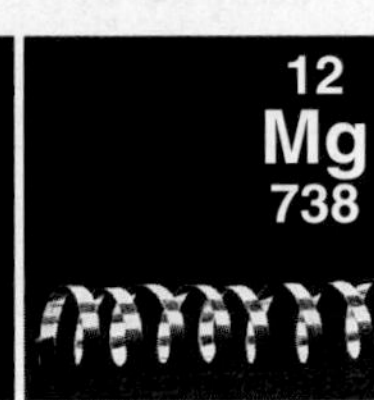

Some metals and many metalloids form oxides that are **amphoteric:** they can act as acids *or* as bases in water.

Figure 8.24 classifies the acid-base behavior of some common oxides, focusing once again on the elements in Group 5A(15) and Period 3. Note that *as the elements become more metallic down a group, their oxides become more basic.* In Group 5A, dinitrogen pentaoxide, N_2O_5, forms nitric acid:

$$N_2O_5(s) + H_2O(l) \longrightarrow 2HNO_3(aq)$$

Tetraphosphorus decaoxide, P_4O_{10}, forms the weaker acid H_3PO_4:

$$P_4O_{10}(s) + 6H_2O(l) \longrightarrow 4H_3PO_4(aq)$$

The oxide of the metalloid arsenic is weakly acidic, whereas that of the metalloid antimony is weakly basic. Bismuth, the most metallic of the group, forms a basic oxide that is insoluble in water but that forms a salt and water with acid:

$$Bi_2O_3(s) + 6HNO_3(aq) \longrightarrow 2Bi(NO_3)_3(aq) + 3H_2O(l)$$

Note that *as the elements become less metallic across a period, their oxides become more acidic.* In Period 3, sodium and magnesium form the strongly basic oxides Na_2O and MgO. Metallic aluminum forms amphoteric aluminum oxide (Al_2O_3), which reacts with acid or with base:

$$Al_2O_3(s) + 6HCl(aq) \longrightarrow 2AlCl_3(aq) + 3H_2O(l)$$

$$Al_2O_3(s) + 2NaOH(aq) + 3H_2O(l) \longrightarrow 2NaAl(OH)_4(aq)$$

Silicon dioxide is weakly acidic, forming a salt and water with base:

$$SiO_2(s) + 2NaOH(aq) \longrightarrow Na_2SiO_3(aq) + H_2O(l)$$

The common oxides of phosphorus, sulfur, and chlorine form acids of increasing strength: H_3PO_4, H_2SO_4, and $HClO_4$.

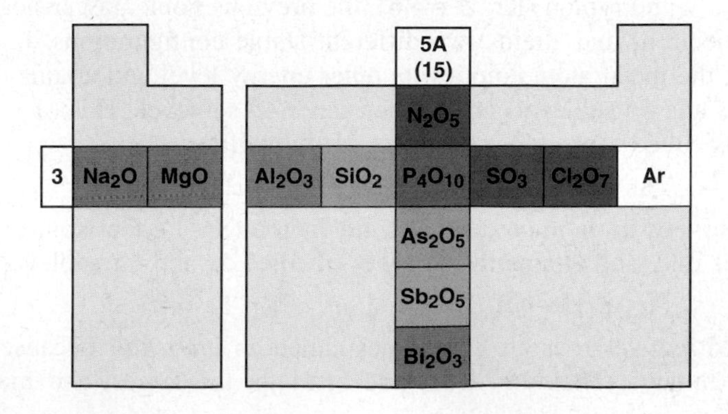

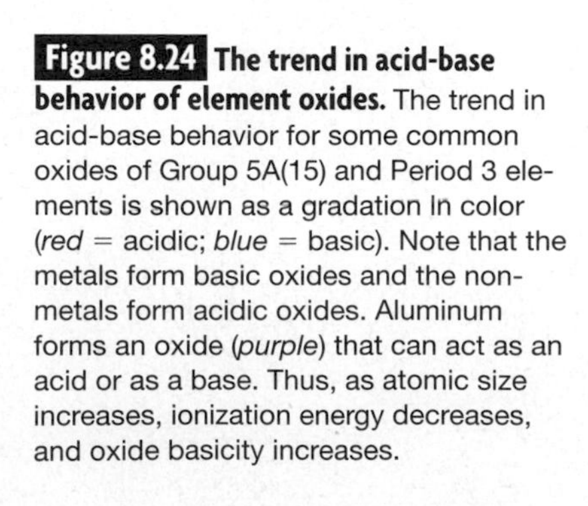
Figure 8.24 The trend in acid-base behavior of element oxides. The trend in acid-base behavior for some common oxides of Group 5A(15) and Period 3 elements is shown as a gradation In color (*red* = acidic; *blue* = basic). Note that the metals form basic oxides and the nonmetals form acidic oxides. Aluminum forms an oxide (*purple*) that can act as an acid or as a base. Thus, as atomic size increases, ionization energy decreases, and oxide basicity increases.

Properties of Monatomic Ions

So far our discussion has focused on the reactants—the atoms—in the process of electron loss and gain. Now we focus on the products—the ions. We examine electron configurations, magnetic properties, and ionic radius relative to atomic radius.

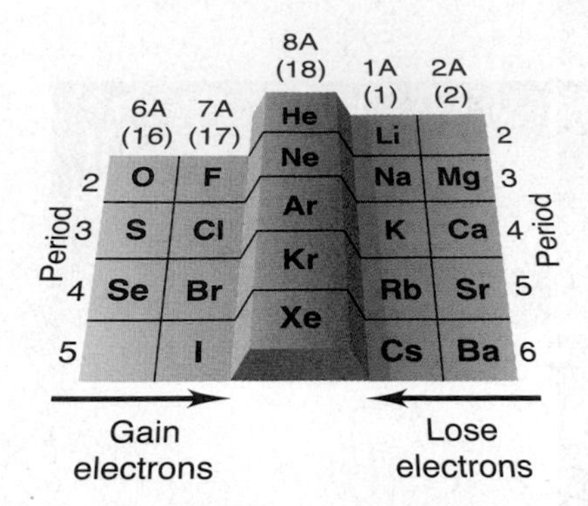

Figure 8.25 Main-group ions and the noble gas electron configurations. Most of the elements that form monatomic ions that are isoelectronic with a noble gas lie in the four groups that flank Group 8A(18), two on either side.

Electron Configurations of Main-Group Ions In Chapter 2, you learned the symbols and charges of many monatomic ions. But *why* does an ion have that charge in its compounds? Why is a sodium ion Na^+ and not Na^{2+}, and why is a fluoride ion F^- and not F^{2-}? For elements at the left and right ends of the periodic table, the explanation concerns the very low reactivity of the noble gases. As we said earlier, because they have high IEs and positive (endothermic) EAs, the noble gases typically do not form ions and remain chemically stable with a *filled* outer energy level (ns^2np^6). *Elements in Groups 1A(1), 2A(2), 6A(16), and 7A(17) that readily form ions either lose or gain electrons to attain a filled outer level and thus a noble gas configuration.* Their ions are said to be **isoelectronic** (Greek *iso,* "same") with the nearest noble gas. Figure 8.25 shows this relationship (also see Figure 2.14, p. 59).

When an alkali metal atom [Group 1A(1)] loses its single valence electron, it becomes isoelectronic with the *previous* noble gas. The Na^+ ion, for example, is isoelectronic with neon (Ne):

$$\text{Na } (1s^22s^22p^63s^1) \longrightarrow \text{Na}^+ \ (1s^22s^22p^6) \text{ [isoelectronic with Ne } (1s^22s^22p^6)] + e^-$$

When a halogen atom [Group 7A(17)] adds a single electron to the five in its *np* sublevel, it becomes isoelectronic with the *next* noble gas. Bromide ion, for example, is isoelectronic with krypton (Kr):

$$\text{Br ([Ar] } 4s^23d^{10}4p^5) + e^- \longrightarrow \text{Br}^- \text{ ([Ar] } 4s^23d^{10}4p^6) \text{ [isoelectronic with Kr ([Ar] } 4s^23d^{10}4p^6)]$$

The energy needed to remove the electrons from metals to attain the previous noble gas configuration is supplied during their exothermic reactions with nonmetals. Removing more than one electron from Na to form Na^{2+} or more than two from Mg to form Mg^{3+} means removing core electrons, which have a much higher Z_{eff}, and overcoming the nucleus-electron attractions requires more energy than is available in a reaction. This is the reason that $NaCl_2$ and MgF_3 do *not* exist. Similarly, adding two electrons to F to form F^{2-} or three to O to form O^{3-} means placing the extra electron into the next energy level. With 18 electrons acting as inner electrons and shielding the nuclear charge very effectively, adding an electron to the negative ion, F^- or O^{2-}, requires too much energy. Thus, we never see Na_2F or Mg_3O_2.

The larger metals of Groups 3A(13), 4A(14), and 5A(15) form cations through a different process, because it would be energetically impossible for them to lose enough electrons to attain a noble gas configuration. For example, tin (Sn; $Z = 50$) would have to lose 14 electrons—two 5*p*, ten 4*d*, and two 5*s*—to be isoelectronic with krypton (Kr; $Z = 36$), the previous noble gas. Instead, tin loses far fewer electrons and attains two different stable configurations. In the tin(IV) ion (Sn^{4+}), the metal atom empties its outer energy level and attains the stability of empty 5*s* and 5*p* sublevels and a filled inner 4*d* sublevel. This $(n - 1)d^{10}$ configuration is called a **pseudo–noble gas configuration:**

$$\text{Sn ([Kr] } 5s^24d^{10}5p^2) \longrightarrow \text{Sn}^{4+} \text{ ([Kr] } 4d^{10}) + 4e^-$$

Alternatively, in the more common tin(II) ion (Sn^{2+}), the atom loses the two 5*p* electrons only and attains the stability of filled 5*s* and 4*d* sublevels:

$$\text{Sn ([Kr] } 5s^24d^{10}5p^2) \longrightarrow \text{Sn}^{2+} \text{ ([Kr] } 5s^24d^{10}) + 2e^-$$

The retained ns^2 electrons are sometimes called an *inert pair* because they seem difficult to remove. Thallium, lead, and bismuth, the largest and most metallic members of Groups 3A(13) to 5A(15), commonly form ions that retain the ns^2 pair of electrons: Tl^+, Pb^{2+}, and Bi^{3+}.

Excessively high energy cost is also the reason that some elements do not form monatomic ions in any of their reactions. For instance, carbon would have to lose four electrons to form C^{4+} and attain the He configuration, or gain four to form C^{4-} and attain the Ne configuration, but neither ion forms. (Such multivalent ions *are* observed in the spectra of stars, however, where temperatures exceed 10^6 K.) As you'll see in Chapter 9, carbon and other atoms that do not form ions attain a filled shell by *sharing* electrons through covalent bonding.

SAMPLE PROBLEM 8.6 Writing Electron Configurations of Main-Group Ions

Problem Using condensed electron configurations, write reactions for the formation of the common ions of the following elements:
(a) Iodine (Z = 53) (b) Potassium (Z = 19) (c) Indium (Z = 49)

Plan We identify the element's position in the periodic table and recall two general points:
- Ions of elements in Groups 1A(1), 2A(2), 6A(16), and 7A(17) are typically isoelectronic with the nearest noble gas.
- Metals in Groups 3A(13) to 5A(15) can lose the *ns* and *np* electrons or just the *np* electrons.

Solution (a) Iodine is in Group 7A(17), so it gains one electron and is isoelectronic with xenon:

$$\text{I ([Kr] } 5s^24d^{10}5p^5) + e^- \longrightarrow \text{I}^- \text{ ([Kr] } 5s^24d^{10}5p^6) \quad \text{(same as Xe)}$$

(b) Potassium is in Group 1A(1), so it loses one electron and is isoelectronic with argon:

$$\text{K ([Ar] } 4s^1) \longrightarrow \text{K}^+ \text{ ([Ar])} + e^-$$

(c) Indium is in Group 3A(13), so it loses either three electrons to form In^{3+} (pseudo–noble gas configuration) or one to form In^+ (inert pair):

$$\text{In ([Kr] } 5s^24d^{10}5p^1) \longrightarrow \text{In}^{3+} \text{ ([Kr] } 4d^{10}) + 3e^-$$

$$\text{In ([Kr] } 5s^24d^{10}5p^1) \longrightarrow \text{In}^{+} \text{ ([Kr] } 5s^24d^{10}) + e^-$$

Check Be sure that the number of electrons in the ion's electron configuration, plus those gained or lost to form the ion, equals Z.

FOLLOW-UP PROBLEM 8.6 Using condensed electron configurations, write reactions showing the formation of the common ions of the following elements:
(a) Ba (Z = 56) (b) O (Z = 8) (c) Pb (Z = 82)

Electron Configurations of Transition Metal Ions In contrast to most main-group ions, *transition metal ions rarely attain a noble gas configuration,* and the reason, once again, is that energy costs are too high. The exceptions in Period 4 are scandium, which forms Sc^{3+}, and titanium, which occasionally forms Ti^{4+} in some compounds. The typical behavior of a transition element is to *form more than one cation by losing all of its ns and some of its (n − 1)d electrons.* (We focus here on the Period 4 series, but these points hold for Periods 5 and 6 also.)

In the aufbau process of building up the ground-state atoms, Period 3 ends with the noble gas argon. At the beginning of Period 4, the radial probability distribution of the 4*s* orbital near the nucleus makes it more stable than the empty 3*d*. Therefore, the first and second electrons added in the period enter the 4*s* in K and Ca. But, as soon as we reach the transition elements and the 3*d* orbitals begin to fill, the increasing nuclear charge attracts their electrons more and more strongly. Moreover, the added 3*d* electrons fill inner orbitals, so they are not very well shielded from the increasing nuclear charge by the 4*s* electrons. As a result,

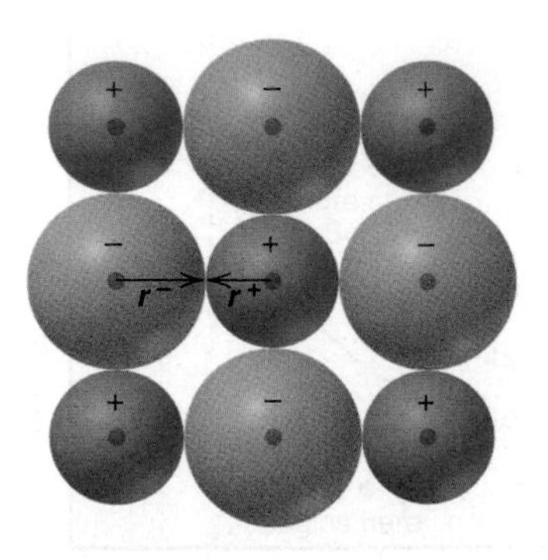

Figure 8.28 Depicting ionic radius. The cation radius (r^+) and the anion radius (r^-) each make up a portion of the total distance between the nuclei of adjacent ions in a crystalline ionic compound.

Ionic Size vs. Atomic Size The **ionic radius** is an estimate of the size of an ion in a crystalline ionic compound. You can picture it as one ion's portion of the distance between the nuclei of neighboring ions in the solid (Figure 8.28). From the relation between effective nuclear charge and atomic size, we can predict the size of an ion relative to its parent atom:

- *Cations are smaller than their parent atoms.* When a cation forms, electrons are *removed from* the outer level. The resulting decrease in electron repulsions allows the nuclear charge to pull the remaining electrons closer.
- *Anions are larger than their parent atoms.* When an anion forms, electrons are *added to* the outer level. The increase in repulsions causes the electrons to occupy more space.

Figure 8.29 shows the radii of some common main-group monatomic ions relative to their parent atoms. As you can see, *ionic size increases down a group* because the number of energy levels increases. Across a period, however, the pattern is more complex. Size decreases among the cations, then increases tremendously with the first of the anions, and finally decreases again among the anions.

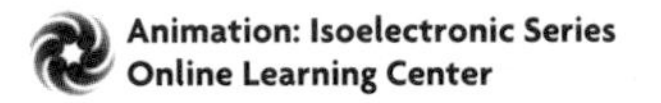

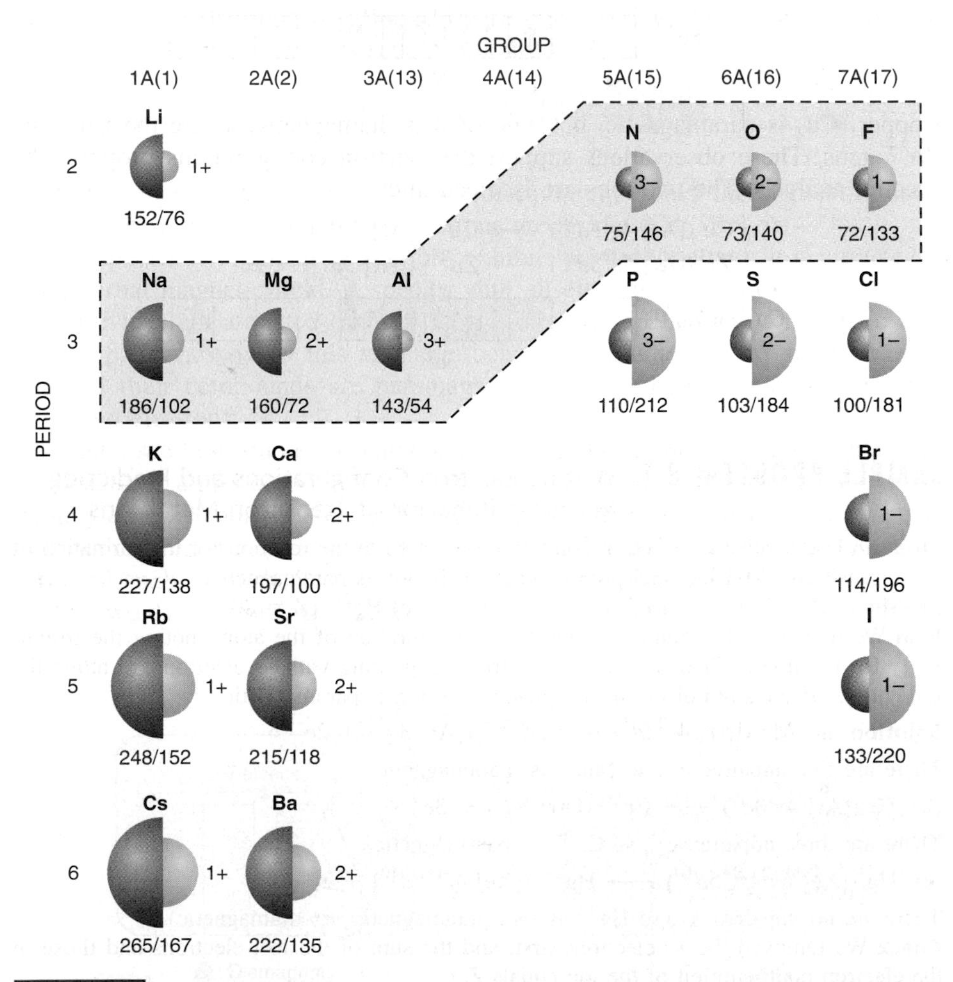

Figure 8.29 Ionic vs. atomic radii. The atomic radii (*colored half-spheres*) and ionic radii (*gray half-spheres*) of some main-group elements are arranged in periodic table format (with all radii values in picometers). Note that metal atoms (*blue*) form *smaller* positive ions, whereas nonmetal atoms (*red*) form *larger* negative ions. The dashed outline sets off ions of Period 2 nonmetals and Period 3 metals that are *isoelectronic* with neon. Note the size decrease from anions to cations.

This pattern results from changes in effective nuclear charge and electron-electron repulsions. In Period 3 (Na through Cl), for example, increasing Z_{eff} from left to right makes Na^+ larger than Mg^{2+}, which in turn is larger than Al^{3+}. The great jump in size from cations to anions occurs because we are *adding* electrons rather than removing them, so repulsions increase sharply. For instance, P^{3-} has eight more electrons than Al^{3+}. Then, the ongoing rise in Z_{eff} makes P^{3-} larger than S^{2-}, which is larger than Cl^-. These factors lead to some striking effects even among ions with the same number of electrons. Look at the ions within the dashed outline in Figure 8.29, which are all isoelectronic with neon. Even though the cations form from elements in the next period, the anions are still much larger. The pattern is

$$3- > 2- > 1- > 1+ > 2+ > 3+$$

When an element forms more than one cation, *the greater the ionic charge, the smaller the ionic radius.* Consider Fe^{2+} and Fe^{3+}. The number of protons is the same, but Fe^{3+} has one fewer electron, so electron repulsions are reduced somewhat. As a result, Z_{eff} increases, which pulls all the electrons closer, so Fe^{3+} is smaller than Fe^{2+}.

To summarize the main points,

- Ionic size increases down a group.
- Ionic size decreases across a period but increases from cations to anions.
- Ionic size decreases with increasing positive (or decreasing negative) charge in an isoelectronic series.
- Ionic size decreases as charge increases for different cations of a given element.

SAMPLE PROBLEM 8.8 Ranking Ions by Size

Problem Rank each set of ions in order of *decreasing* size, and explain your ranking:
(a) Ca^{2+}, Sr^{2+}, Mg^{2+} **(b)** K^+, S^{2-}, Cl^- **(c)** Au^+, Au^{3+}

Plan We find the position of each element in the periodic table and apply the ideas presented in the text.

Solution **(a)** Since Mg^{2+}, Ca^{2+}, and Sr^{2+} are all from Group 2A(2), they decrease in size up the group: $Sr^{2+} > Ca^{2+} > Mg^{2+}$.

(b) The ions K^+, S^{2-}, and Cl^- are isoelectronic. S^{2-} has a lower Z_{eff} than Cl^-, so it is larger. K^+ is a cation, and has the highest Z_{eff}, so it is smallest: $S^{2-} > Cl^- > K^+$.

(c) Au^+ has a lower charge than Au^{3+}, so it is larger: $Au^+ > Au^{3+}$.

FOLLOW-UP PROBLEM 8.8 Rank the ions in each set in order of *increasing* size:
(a) Cl^-, Br^-, F^- **(b)** Na^+, Mg^{2+}, F^- **(c)** Cr^{2+}, Cr^{3+}

SECTION SUMMARY

Highly metallic behavior correlates with large atomic size and low ionization energy. Thus, metallic behavior increases down a group and decreases across a period. Within the main groups, metal oxides are basic and nonmetal oxides acidic. Thus, oxides become more acidic across a period and more basic down a group. Many main-group elements form ions that are isoelectronic with the nearest noble gas. Removing (or adding) more electrons than needed to attain the previous noble gas configuration requires a prohibitive amount of energy. Metals in Groups 3A(13) to 5A(15) lose either their *np* electrons or both their *ns* and *np* electrons. Transition metals lose *ns* electrons before $(n - 1)d$ electrons and commonly form more than one ion. Many transition metals and their compounds are paramagnetic because their atoms (or ions) have unpaired electrons. Cations are smaller and anions larger than their parent atoms. Ionic radius increases down a group. Across a period, cationic and anionic radii decrease, but a large increase occurs from cations to anions.

Chapter Perspective

This chapter is our springboard to understanding the chemistry of the elements. We have begun to see that recurring electron configurations lead to trends in atomic properties, which in turn lead to trends in chemical behavior. With this insight, we can go on to investigate how atoms bond (Chapter 9), how molecular shapes arise (Chapter 10), how molecular shapes and other properties can be explained using certain models (Chapter 11), how the physical properties of liquids and solids emerge from atomic properties (Chapter 12), and how those properties influence the solution process (Chapter 13). We will briefly review these ideas and gain a perspective on where they lead (Interchapter) before we can survey elemental behavior in greater detail (Chapter 14) and see how it applies to the remarkable diversity of organic compounds (Chapter 15). Within this group of chapters, you will see chemical models come alive in chemical facts.

For Review and Reference (Numbers in parentheses refer to pages, unless noted otherwise.)

Learning Objectives

Relevant section and/or sample problem (SP) numbers appear in parentheses.

Understand These Concepts

1. The meaning of the periodic law and the arrangement of elements by atomic number (8.1)
2. The reason for the spin quantum number and its two possible values (8.2)
3. How the exclusion principle applies to orbital filling (8.2)
4. The effects of nuclear charge, shielding, and penetration on the splitting of orbital energies; the meaning of effective nuclear charge (8.2)
5. How the order in the periodic table is based on the order of orbital energies (8.3)
6. How orbitals are filled in main-group and transition elements; the importance of Hund's rule (8.3)
7. How outer electron configuration within a group is related to chemical behavior (8.3)
8. The distinction among inner, outer, and valence electrons (8.3)
9. The meaning of atomic radius, ionization energy, and electron affinity (8.4)
10. How n value and effective nuclear charge give rise to the periodic trends of atomic size and ionization energy (8.4)
11. The importance of core electrons to the pattern of successive ionization energies (8.4)
12. How atomic properties are related to the tendency to form ions (8.4)
13. The general properties of metals and nonmetals (8.5)
14. How the vertical and horizontal trends in metallic behavior are related to ion formation and oxide acidity (8.5)
15. Why main-group ions are either isoelectronic with the nearest noble gas or have a pseudo–noble gas electron configuration (8.5)
16. Why transition elements lose ns electrons first (8.5)
17. The origin of paramagnetic and diamagnetic behavior (8.5)
18. The relation between ionic and atomic size and the trends in ionic size (8.5)

Master These Skills

1. Using orbital diagrams to determine the set of quantum numbers for any electron in an atom (SP 8.1)
2. Writing full and condensed electron configurations of an element (SP 8.2)
3. Using periodic trends to rank elements by atomic size and first ionization energy (SPs 8.3 and 8.4)
4. Identifying an element from its successive ionization energies (SP 8.5)
5. Writing electron configurations of main-group and transition metal ions (SPs 8.6 and 8.7)
6. Using periodic trends to rank ions by relative size (SP 8.8)

Key Terms

electron configuration (291)

Section 8.1

periodic law (291)

Section 8.2

spin quantum number (m_s) (293)
exclusion principle (293)
shielding (294)
effective nuclear charge (Z_{eff}) (294)
penetration (295)

Section 8.3

aufbau principle (296)
orbital diagram (296)
Hund's rule (297)
transition elements (301)
inner (core) electrons (303)
outer electrons (303)
valence electrons (303)
inner transition elements (303)
lanthanides (rare earths) (304)
actinides (304)

Section 8.4

metallic radius (305)
covalent radius (305)
ionization energy (IE) (309)
electron affinity (EA) (312)

Section 8.5

amphoteric (315)
isoelectronic (316)
pseudo–noble gas configuration (316)
paramagnetism (318)
diamagnetism (318)
ionic radius (320)

Key Equations and Relationships

8.1 Defining the energy order of sublevels in terms of the angular momentum quantum number (l value) (295):

Order of sublevel energies: $s < p < d < f$

8.2 Meaning of the first ionization energy (309):

$$\text{Atom}(g) \longrightarrow \text{ion}^+(g) + e^- \qquad \Delta E = \text{IE}_1 > 0$$

Highlighted Figures and Tables

These figures (F) and tables (T) provide a review of key ideas. Entries in **color** contain frequently used data.

T8.2 Summary of quantum numbers of electrons in atoms (293)
F8.3 Lower nuclear charge makes H less stable than He^+ (294)
F8.4 Shielding (295)
F8.5 Penetration by the $2s$ electron in Li makes the $2s$ orbital more stable than the $2p$ (295)
F8.6 Order for filling energy sublevels with electrons (296)
F8.11 Periodic table of partial ground-state electron configurations (302)
F8.12 Relation between orbital filling and the periodic table (303)
F8.15 Atomic radii of the main-group and transition elements (307)
F8.16 Periodicity of atomic radius (308)
F8.17 Periodicity of first ionization energy (IE_1) (309)
F8.18 First ionization energies of the main-group elements (310)
F8.21 Trends in three atomic properties (313)
F8.22 Trends in metallic behavior (313)
F8.24 Trend in acid-base behavior of element oxides (315)
F8.25 Main-group ions and the noble gas electron configurations (316)
F8.26 The Period 4 crossover in orbital energies (318)
F8.29 Ionic vs. atomic radii (320)

Brief Solutions to Follow-up Problems

8.1 The element has eight electrons, so $Z = 8$: oxygen.
Sixth electron: $n = 2, l = 1, m_l = 0, m_s = +\frac{1}{2}$

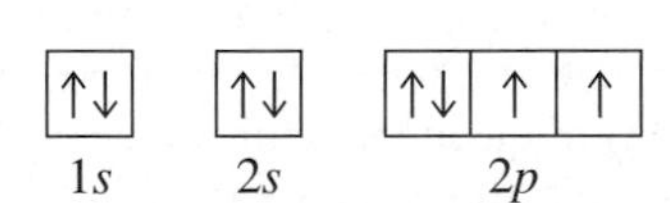

8.2 (a) For Ni, $1s^22s^22p^63s^23p^64s^23d^8$; [Ar] $4s^23d^8$

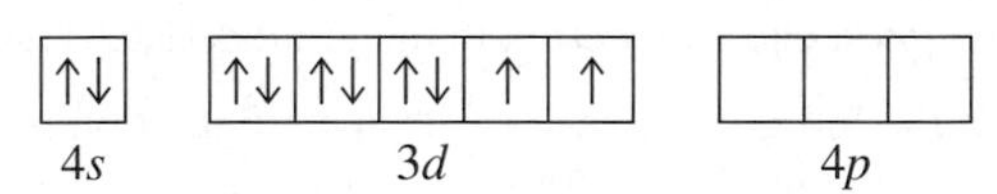

Ni has 18 inner electrons.
(b) For Sr, $1s^22s^22p^63s^23p^64s^23d^{10}4p^65s^2$; [Kr] $5s^2$

5s 4d 5p

Sr has 36 inner electrons.
(c) For Po, $1s^22s^22p^63s^23p^64s^23d^{10}4p^65s^24d^{10}5p^66s^24f^{14}5d^{10}6p^4$; [Xe] $6s^24f^{14}5d^{10}6p^4$

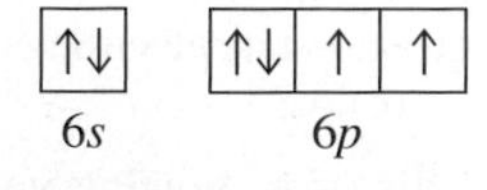

Po has 78 inner electrons.

8.3 (a) Cl < Br < Se; (b) Xe < I < Ba
8.4 (a) Sn < Sb < I; (b) Ba < Sr < Ca
8.5 Q is aluminum: $1s^22s^22p^63s^23p^1$
8.6 (a) Ba ([Xe] $6s^2$) $\longrightarrow$ Ba^{2+} ([Xe]) + $2e^-$
(b) O ([He] $2s^22p^4$) + $2e^-$ $\longrightarrow$ O^{2-} ([He] $2s^22p^6$) (same as Ne)
(c) Pb ([Xe] $6s^24f^{14}5d^{10}6p^2$) $\longrightarrow$ Pb^{2+} ([Xe] $6s^24f^{14}5d^{10}$) + $2e^-$
Pb ([Xe] $6s^24f^{14}5d^{10}6p^2$) $\longrightarrow$ Pb^{4+} ([Xe] $4f^{14}5d^{10}$) + $4e^-$
8.7 (a) V^{3+}: [Ar] $3d^2$; paramagnetic
(b) Ni^{2+}: [Ar] $3d^8$; paramagnetic
(c) La^{3+}: [Xe]; not paramagnetic (diamagnetic)
8.8 (a) $F^- < Cl^- < Br^-$; (b) $Mg^{2+} < Na^+ < F^-$;
(c) $Cr^{3+} < Cr^{2+}$

Problems

Problems with **colored** numbers are answered in Appendix E. Sections match the text and provide the numbers of relevant sample problems. Most offer Concept Review Questions, Skill-Building Exercises (grouped in pairs covering the same concept), and Problems in Context. Comprehensive Problems are based on material from any section or previous chapter.

Development of the Periodic Table

Concept Review Questions

8.1 What would be your reaction to a claim that a new element had been discovered and it fit between tin (Sn) and antimony (Sb) in the periodic table?

8.2 Based on results of his study of atomic x-ray spectra, Moseley discovered a relationship that replaced atomic mass as the criterion for ordering the elements. By what criterion are the elements now ordered in the periodic table? Give an example of a sequence of element order that was confirmed by Moseley's findings.

Skill-Building Exercises *(grouped in similar pairs)*

8.3 Before Mendeleev published his periodic table, Döbereiner grouped similar elements into triads, in which the unknown properties of one member could be predicted by averaging known values of the properties of the others. To test this idea, predict the values of the following quantities:
(a) The atomic mass of K from the atomic masses of Na and Rb
(b) The melting point of Br_2 from the melting points of Cl_2 (−101.0°C) and I_2 (113.6°C) (actual value = −7.2°C)

8.4 To test Döbereiner's idea (Problem 8.3), predict:
(a) The boiling point of HBr from the boiling points of HCl (−84.9°C) and HI (−35.4°C) (actual value = −67.0°C)
(b) The boiling point of AsH_3 from the boiling points of PH_3 (−87.4°C) and SbH_3 (−17.1°C) (actual value = −55°C)

Characteristics of Many-Electron Atoms

Concept Review Questions

8.5 Summarize the rules for the allowable values of the four quantum numbers of an electron in an atom.

8.6 Which of the quantum numbers relate(s) to the electron only? Which relate(s) to the orbital?

8.7 State the exclusion principle. What does it imply about the number and spin of electrons in an atomic orbital?

8.8 What is the key distinction between sublevel energies in one-electron species, such as the H atom, and those in many-electron species, such as the C atom? What factors lead to this distinction? Would you expect the pattern of sublevel energies in Be^{3+} to be more like that in H or that in C? Explain.

8.9 Define *shielding* and *effective nuclear charge*. What is the connection between the two?

8.10 What is the penetration effect? How is it related to shielding? Use the penetration effect to explain the difference in relative orbital energies of a 3*p* and a 3*d* electron in the same atom.

Skill-Building Exercises *(grouped in similar pairs)*

8.11 How many electrons in an atom can have each of the following quantum number or sublevel designations?
(a) $n = 2, l = 1$ (b) 3*d* (c) 4*s*

8.12 How many electrons in an atom can have each of the following quantum number or sublevel designations?
(a) $n = 2, l = 1, m_l = 0$ (b) 5*p* (c) $n = 4, l = 3$

8.13 How many electrons in an atom can have each of the following quantum number or sublevel designations?
(a) 4*p* (b) $n = 3, l = 1, m_l = +1$ (c) $n = 5, l = 3$

8.14 How many electrons in an atom can have each of the following quantum number or sublevel designations?
(a) 2*s* (b) $n = 3, l = 2$ (c) 6*d*

The Quantum-Mechanical Model and the Periodic Table

(Sample Problems 8.1 and 8.2)

Concept Review Questions

8.15 State the periodic law, and explain its relation to electron configuration. (Use Na and K in your explanation.)

8.16 State Hund's rule in your own words, and show its application in the orbital diagram of the nitrogen atom.

8.17 How does the aufbau principle, in connection with the periodic law, lead to the format of the periodic table?

8.18 For main-group elements, are outer electron configurations similar or different within a group? Within a period? Explain.

8.19 For which blocks of elements are outer electrons the same as valence electrons? For which are *d* electrons often included among valence electrons?

8.20 What is the electron capacity of the *n*th energy level? What is the capacity of the fourth energy level?

Skill-Building Exercises *(grouped in similar pairs)*

8.21 Write a full set of quantum numbers for the following:
(a) The outermost electron in an Rb atom
(b) The electron gained when an S^- ion becomes an S^{2-} ion
(c) The electron lost when an Ag atom ionizes
(d) The electron gained when an F^- ion forms from an F atom

8.22 Write a full set of quantum numbers for the following:
(a) The outermost electron in an Li atom
(b) The electron gained when a Br atom becomes a Br^- ion
(c) The electron lost when a Cs atom ionizes
(d) The highest energy electron in the ground-state B atom

8.23 Write the full ground-state electron configuration for each:
(a) Rb (b) Ge (c) Ar

8.24 Write the full ground-state electron configuration for each:
(a) Br (b) Mg (c) Se

8.25 Write the full ground-state electron configuration for each:
(a) Cl (b) Si (c) Sr

8.26 Write the full ground-state electron configuration for each:
(a) S (b) Kr (c) Cs

8.27 Draw an orbital diagram showing valence electrons, and write the condensed ground-state electron configuration for each:
(a) Ti (b) Cl (c) V

8.28 Draw an orbital diagram showing valence electrons, and write the condensed ground-state electron configuration for each:
(a) Ba (b) Co (c) Ag

8.29 Draw an orbital diagram showing valence electrons, and write the condensed ground-state electron configuration for each:
(a) Mn (b) P (c) Fe

8.30 Draw an orbital diagram showing valence electrons, and write the condensed ground-state electron configuration for each:
(a) Ga (b) Zn (c) Sc

8.31 Draw the partial (valence-level) orbital diagram, and write the symbol, group number, and period number of the element:
(a) [He] $2s^22p^4$ (b) [Ne] $3s^23p^3$

8.32 Draw the partial (valence-level) orbital diagram, and write the symbol, group number, and period number of the element:
(a) [Kr] $5s^24d^{10}$ (b) [Ar] $4s^23d^8$

8.33 Draw the partial (valence-level) orbital diagram, and write the symbol, group number, and period number of the element:
(a) [Ne] $3s^23p^5$ (b) [Ar] $4s^23d^{10}4p^3$

8.34 Draw the partial (valence-level) orbital diagram, and write the symbol, group number, and period number of the element:
(a) [Ar] $4s^23d^5$ (b) [Kr] $5s^24d^2$

8.35 From each partial (valence-level) orbital diagram, write the ground-state electron configuration and group number:

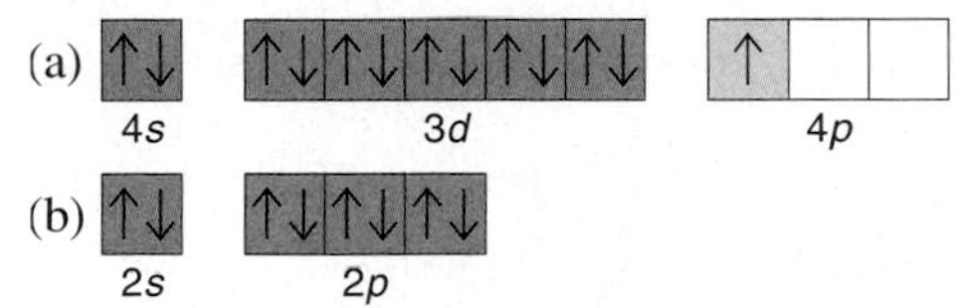

8.36 From each partial (valence-level) orbital diagram, write the ground-state electron configuration and group number:

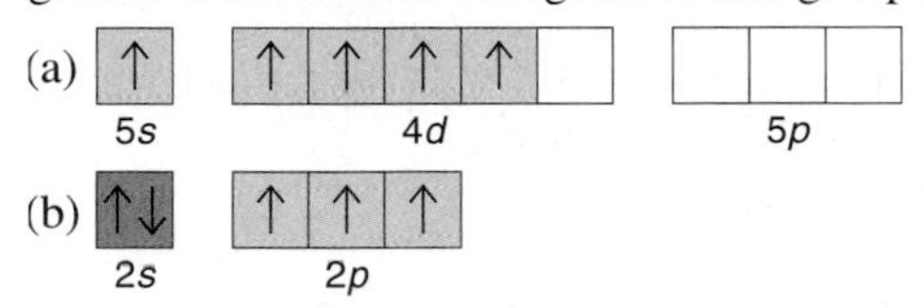

8.37 How many inner, outer, and valence electrons are present in an atom of each of the following elements?
(a) O (b) Sn (c) Ca (d) Fe (e) Se

8.38 How many inner, outer, and valence electrons are present in an atom of each of the following elements?
(a) Br (b) Cs (c) Cr (d) Sr (e) F

8.39 Identify each element below, and give the symbols of the other elements in its group:
(a) [He] $2s^22p^1$ (b) [Ne] $3s^23p^4$ (c) [Xe] $6s^25d^1$

8.40 Identify each element below, and give the symbols of the other elements in its group:
(a) [Ar] $4s^23d^{10}4p^4$ (b) [Xe] $6s^24f^{14}5d^2$ (c) [Ar] $4s^23d^5$

8.41 Identify each element below, and give the symbols of the other elements in its group:
(a) [He] $2s^22p^2$ (b) [Ar] $4s^23d^3$ (c) [Ne] $3s^23p^3$

8.42 Identify each element below, and give the symbols of the other elements in its group:
(a) [Ar] $4s^23d^{10}4p^2$ (b) [Ar] $4s^23d^7$ (c) [Kr] $5s^24d^5$

Problems in Context

8.43 After an atom in its ground state absorbs energy, it exists in an excited state. Spectral lines are produced when the atom returns to its ground state. The yellow-orange line in the sodium spectrum, for example, is produced by the emission of energy when excited sodium atoms return to their ground state. Write the electron configuration and the orbital diagram of the first excited state of sodium. (*Hint:* The outermost electron is excited.)

8.44 One reason spectroscopists study excited states is to gain information about the energies of orbitals that are unoccupied in an atom's ground state. Each of the following electron configurations represents an atom in an excited state. Identify the element, and write its condensed ground-state configuration:
(a) $1s^22s^22p^63s^13p^1$ (b) $1s^22s^22p^63s^23p^44s^1$
(c) $1s^22s^22p^63s^23p^64s^23d^44p^1$ (d) $1s^22s^22p^53s^1$

Trends in Three Key Atomic Properties

(Sample Problems 8.3 to 8.5)

Concept Review Questions

8.45 If the exact outer limit of an isolated atom cannot be measured, what criterion can we use to determine atomic radii? What is the difference between a covalent radius and a metallic radius?

8.46 Explain the relationship between the trends in atomic size and in ionization energy within the main groups.

8.47 In what region of the periodic table will you find elements with relatively high IEs? With relatively low IEs?

8.48 Why do successive IEs of a given element always increase? When the difference between successive IEs of a given element is exceptionally large (for example, between IE_1 and IE_2 of K), what do we learn about its electron configuration?

8.49 In a plot of IE_1 for the Period 3 elements (see Figure 8.17, p. 309), why do the values for elements in Groups 3A(13) and 6A(16) drop slightly below the generally increasing trend?

8.50 Which group in the periodic table has elements with high (endothermic) IE_1 and very negative (exothermic) first electron affinities (EA_1)? Give the charge on the ions these atoms form.

8.51 The EA_2 of an oxygen atom is positive, even though its EA_1 is negative. Why does this change of sign occur? Which other elements exhibit a positive EA_2? Explain.

8.52 How does *d*-electron shielding influence atomic size among the Period 4 transition elements?

Skill-Building Exercises *(grouped in similar pairs)*

8.53 Arrange each set in order of *increasing* atomic size:
(a) Rb, K, Cs (b) C, O, Be (c) Cl, K, S (d) Mg, K, Ca

8.54 Arrange each set in order of *decreasing* atomic size:
(a) Ge, Pb, Sn (b) Sn, Te, Sr (c) F, Ne, Na (d) Be, Mg, Na

8.55 Arrange each set of atoms in order of *increasing* IE_1:
(a) Sr, Ca, Ba (b) N, B, Ne (c) Br, Rb, Se (d) As, Sb, Sn

8.56 Arrange each set of atoms in order of *decreasing* IE_1:
(a) Na, Li, K (b) Be, F, C (c) Cl, Ar, Na (d) Cl, Br, Se

8.57 Write the full electron configuration of the Period 2 element with the following successive IEs (in kJ/mol):
$IE_1 = 801$ $IE_2 = 2427$ $IE_3 = 3659$
$IE_4 = 25{,}022$ $IE_5 = 32{,}822$

8.58 Write the full electron configuration of the Period 3 element with the following successive IEs (in kJ/mol):
$IE_1 = 738$ $IE_2 = 1450$ $IE_3 = 7732$
$IE_4 = 10{,}539$ $IE_5 = 13{,}628$

8.59 Which element in each of the following sets would you expect to have the *highest* IE_2?
(a) Na, Mg, Al (b) Na, K, Fe (c) Sc, Be, Mg

8.60 Which element in each of the following sets would you expect to have the *lowest* IE_3?
(a) Na, Mg, Al (b) K, Ca, Sc (c) Li, Al, B

Atomic Structure and Chemical Reactivity

(Sample Problems 8.6 to 8.8)

Concept Review Questions

8.61 List three ways in which metals and nonmetals differ.

8.62 Summarize the trend in metallic character as a function of position in the periodic table. Is it the same as the trend in atomic size? Ionization energy?

8.63 Summarize the acid-base behavior of the main-group metal and nonmetal oxides in water. How does oxide acidity in water change down a group and across a period?

8.64 What ions are possible for the two largest elements in Group 4A(14)? How does each arise?

8.65 What is a pseudo–noble gas configuration? Give an example of one ion from Group 3A(13) that has it.

8.66 How are measurements of paramagnetism used to support electron configurations derived spectroscopically? Use Cu(I) and Cu(II) chlorides as examples.

8.67 The charges of a set of isoelectronic ions vary from 3+ to 3−. Place the ions in order of increasing size.

Skill-Building Exercises *(grouped in similar pairs)*

8.68 Which element would you expect to be *more* metallic?
(a) Ca or Rb (b) Mg or Ra (c) Br or I

8.69 Which element would you expect to be *more* metallic?
(a) S or Cl (b) In or Al (c) As or Br

8.70 Which element would you expect to be *less* metallic?
(a) Sb or As (b) Si or P (c) Be or Na

8.71 Which element would you expect to be *less* metallic?
(a) Cs or Rn (b) Sn or Te (c) Se or Ge

8.72 Does the reaction of a main-group nonmetal oxide in water produce an acidic or a basic solution? Write a balanced equation for the reaction of a Group 6A(16) nonmetal oxide with water.

8.73 Does the reaction of a main-group metal oxide in water produce an acidic solution or a basic solution? Write a balanced equation for the reaction of a Group 2A(2) oxide with water.

8.74 Write the charge and full ground-state electron configuration of the monatomic ion most likely to be formed by each:
(a) Cl (b) Na (c) Ca

8.75 Write the charge and full ground-state electron configuration of the monatomic ion most likely to be formed by each:
(a) Rb (b) N (c) Br

8.76 Write the charge and full ground-state electron configuration of the monatomic ion most likely to be formed by each:
(a) Al (b) S (c) Sr

8.77 Write the charge and full ground-state electron configuration of the monatomic ion most likely to be formed by each:
(a) P (b) Mg (c) Se

8.78 How many unpaired electrons are present in the ground state of an atom from each of the following groups?
(a) 2A(2) (b) 5A(15) (c) 8A(18) (d) 3A(13)

8.79 How many unpaired electrons are present in the ground state of an atom from each of the following groups?
(a) 4A(14) (b) 7A(17) (c) 1A(1) (d) 6A(16)

8.80 Which of these are paramagnetic in their ground state?
(a) Ga (b) Si (c) Be (d) Te

8.81 Are compounds of these ground-state ions paramagnetic?
(a) Ti^{2+} (b) Zn^{2+} (c) Ca^{2+} (d) Sn^{2+}

8.82 Write the condensed ground-state electron configurations of these transition metal ions, and state which are paramagnetic:
(a) V^{3+} (b) Cd^{2+} (c) Co^{3+} (d) Ag^{+}

8.83 Write the condensed ground-state electron configurations of these transition metal ions, and state which are paramagnetic:
(a) Mo^{3+} (b) Au^{+} (c) Mn^{2+} (d) Hf^{2+}

8.84 Palladium (Pd; $Z = 46$) is diamagnetic. Draw partial orbital diagrams to show which of the following electron configurations is consistent with this fact:
(a) [Kr] $5s^2 4d^8$ (b) [Kr] $4d^{10}$ (c) [Kr] $5s^1 4d^9$

8.85 Niobium (Nb; $Z = 41$) has an anomalous ground-state electron configuration for a Group 5B(5) element: [Kr] $5s^1 4d^4$. What is the expected electron configuration for elements in this group? Draw partial orbital diagrams to show how paramagnetic measurements could support niobium's actual configuration.

8.86 Rank the ions in each set in order of *increasing* size, and explain your ranking:
(a) Li^{+}, K^{+}, Na^{+} (b) Se^{2-}, Rb^{+}, Br^{-} (c) O^{2-}, F^{-}, N^{3-}

8.87 Rank the ions in each set in order of *decreasing* size, and explain your ranking:
(a) Se^{2-}, S^{2-}, O^{2-} (b) Te^{2-}, Cs^{+}, I^{-} (c) Sr^{2+}, Ba^{2+}, Cs^{+}

Comprehensive Problems

8.88 Some versions of the periodic table show hydrogen at the top of Group 1A(1) *and* at the top of Group 7A(17). What properties of hydrogen justify each of these placements?

8.89 Name the element described in each of the following:
(a) Smallest atomic radius in Group 6A
(b) Largest atomic radius in Period 6
(c) Smallest metal in Period 3
(d) Highest IE_1 in Group 14
(e) Lowest IE_1 in Period 5
(f) Most metallic in Group 15
(g) Group 3A element that forms the most basic oxide
(h) Period 4 element with filled outer level
(i) Condensed ground-state electron configuration is [Ne] $3s^2 3p^2$
(j) Condensed ground-state electron configuration is [Kr] $5s^2 4d^6$
(k) Forms 2+ ion with electron configuration [Ar] $3d^3$
(l) Period 5 element that forms 3+ ion with pseudo–noble gas configuration
(m) Period 4 transition element that forms 3+ diamagnetic ion
(n) Period 4 transition element that forms 2+ ion with a half-filled *d* sublevel
(o) Heaviest lanthanide
(p) Period 3 element whose 2− ion is isoelectronic with Ar
(q) Alkaline earth metal whose cation is isoelectronic with Kr
(r) Group 5A(15) metalloid with the most acidic oxide

8.90 Use electron configurations to account for the stability of the lanthanide ions Ce^{4+} and Eu^{2+}.

8.91 The NaCl crystal structure consists of alternating Na^{+} and Cl^{-} ions lying next to each other in three dimensions. If the Na^{+} radius is 56.4% of the Cl^{-} radius and the distance between Na^{+} nuclei is 566 pm, what are the radii of the two ions?

8.92 When a nonmetal oxide reacts with water, it forms an oxoacid with the same nonmetal oxidation state. Give the name and formula of the oxide used to prepare each of these oxoacids: (a) hypochlorous acid; (b) chlorous acid; (c) chloric acid; (d) perchloric acid; (e) sulfuric acid; (f) sulfurous acid; (g) nitric acid; (h) nitrous acid; (i) carbonic acid; (j) phosphoric acid.

8.93 A fundamental relationship of electrostatics states that the energy required to separate opposite charges of magnitudes Q_1 and Q_2 that are the distance d apart is proportional to $\frac{Q_1 \times Q_2}{d}$. Use this relationship and any other factors to explain these observations: (a) the IE_2 of He ($Z = 2$) is *more* than twice the IE_1 of H ($Z = 1$); (b) the IE_1 of He is *less* than twice the IE_1 of H.

8.94 The energy difference between the 5*d* and 6*s* sublevels in gold accounts for its color. Assuming this energy difference is about 2.7 eV (electron volt; 1 eV $= 1.602 \times 10^{-19}$ J), explain why gold has a warm yellow color.

8.95 Write the formula and name of the compound formed from the following ionic interactions: (a) The 2+ ion and the 1− ion are both isoelectronic with the atoms of a chemically unreactive Period 4 element. (b) The 2+ ion and the 2− ion are both isoelectronic with the Period 3 noble gas. (c) The 2+ ion is the smallest with a filled *d* subshell; the anion forms from the smallest halogen. (d) The ions form from the largest and smallest ionizable atoms in Period 2.

8.96 For over a century, chemists have noted several "diagonal relationships" in the periodic table: exceptional similarities between an element in Period 2 and another in Period 3 in the next higher group. Suggest a reason based on atomic and ionic properties for each of the following: (a) the chemical behavior of Li is very similar to that of Mg; (b) the acidity of $Be(OH)_2$ is very similar to that of $Al(OH)_3$.

8.97 The energy changes for many unusual reactions can be determined using Hess's law (Section 6.5).
(a) Calculate ΔE for the conversion of $F^-(g)$ into $F^+(g)$.
(b) Calculate ΔE for the conversion of $Na^+(g)$ into $Na^-(g)$.

8.98 Use Table 8.5, p. 311, to explain the major differences in the relative values of IE_1 through IE_5 for carbon and oxygen.

8.99 The hot glowing gases around the Sun, the *corona*, can reach millions of degrees Celsius, high enough to remove many electrons from gaseous atoms. Iron ions with charges as high as 14+ have been observed in the corona. Which ions from Fe^+ to Fe^{14+} are paramagnetic? Which would be most attracted to a magnetic field?

8.100 There are some exceptions to the trends of first and successive ionization energies. For each of the following pairs, explain which ionization energy would be higher:
(a) IE_1 of Ga or IE_1 of Ge (b) IE_2 of Ga or IE_2 of Ge
(c) IE_3 of Ga or IE_3 of Ge (d) IE_4 of Ga or IE_4 of Ge

8.101 The photoelectric effect is based on the ability of light to ionize metal atoms. Use Figure 8.18, p. 310, to answer: (a) Calculate the longest wavelength of electromagnetic (EM) radiation that can ionize any alkali metal. (b) Calculate the longest wavelength of EM radiation that can ionize any alkaline earth metal. (c) Which elements, other than the alkali and alkaline earth metals, could also be ionized by the radiation from part (b)? (d) In which region of the EM spectrum are these photons found?

8.102 Half of the first 18 elements have an odd number of electrons, and half have an even number. Show why these elements aren't half paramagnetic and half diamagnetic.

8.103 Scientists have speculated that as yet unknown superheavy elements might be moderately stable. In fact, in 1976, it was mistakenly believed that Element 126 had been discovered in a sample of mica. Predict the *n* and *l* quantum numbers for the outermost electron in an atom of this element. What is the sublevel designation of the outermost orbital? How many orbitals would be in this sublevel?

8.104 Two members of the boron family owe their names to bright lines in their emission spectra. Indium has a bright indigo-blue line (451.1 nm), and thallium has a bright green line (535.0 nm) (*thallus* means "green stalk"). What are the energies of these two spectral lines?

8.105 Like main-group metal oxides, transition metal oxides are basic, but this is not obvious in water because of their low solubility. They do, however, react with acids. Complete and write balanced molecular equations for the following reactions:
(a) Iron(II) oxide and hydrochloric acid
(b) Chromium(III) oxide and nitric acid
(c) Cobalt(III) oxide and sulfuric acid
(d) Copper(I) oxide and hydrobromic acid
(e) Manganese(II) oxide and phosphoric acid

8.106 Draw the partial (valence-level) orbital diagram and write the electron configuration of the atom and monatomic ion of the element with the following ionization energies (in kJ/mol):

IE_1	IE_2	IE_3	IE_4	IE_5	IE_6	IE_7	IE_8
999	2251	3361	4564	7013	8495	27,106	31,669

8.107 As a science major, you are assigned to the "Atomic Dorm" at college. The first floor contains one dorm room with one bedroom that has a bunk bed for two students. The second floor contains two dorm rooms, one like the room on the first floor, and the other with three bedrooms, each with a bunk bed for two students. The third floor contains three dorm rooms, two like the rooms on the second floor, and a larger third room that contains five bedrooms, each with a bunk bed for two students. Entering students choose room and bunk on a first-come, first-serve basis by criteria in the following order of importance:
(1) They want to be on the lowest available floor.
(2) They want to be in a smallest available dorm room.
(3) They want to be in a lower bunk if available.
(a) Which bunk does the 1st student choose? (b) How many students are in top bunks when the 17th student chooses? (c) Which bunk does the 21st student choose? (d) How many students are in bottom bunks when the 25th student chooses?

8.108 On the planet Zog in the Andromeda galaxy, all of the stable elements have been studied. Data for some main-group elements are shown below (Zoggian units are unknown on Earth and, therefore, not shown). Limited communications with the Zoggians have indicated that balloonium is a monatomic gas with two positive charges in its nucleus. Use the data to deduce the names that Earthlings give to these elements:

Name	Atomic Radius	IE_1	EA_1
Balloonium	10	339	0
Inertium	24	297	+4.1
Allotropium	34	143	−28.6
Brinium	63	70.9	−7.6
Canium	47	101	−15.3
Fertilium	25	200	0
Liquidium	38	163	−46.4
Utilium	48	82.4	−6.1
Crimsonium	72	78.4	−2.9

Holding parts together The physical links among the metal loops in this piece of medieval chain mail are an analogy for the chemical links among the metal atoms in each loop. In this chapter, we examine the three types of chemical bonds that connect atoms and consider how each type gives rise to the behavior of the substance.

9

Models of Chemical Bonding

Why is table salt (or any other ionic substance) a hard, brittle, high-melting solid that conducts a current only when molten or dissolved in water? Why is candle wax (along with most covalent substances) low melting, soft, and nonconducting, while diamond (and some other exceptions) is high melting and extremely hard? And why is copper (and most other metallic substances) shiny, malleable, and able to conduct a current whether molten or solid? The answers lie in the *type of bonding within the substance*. In Chapter 8, we examined the properties of individual atoms and ions. Yet, in virtually all the substances in and around you, these particles are bonded to one another. As you'll see, deeper insight comes with discovering how the properties of atoms influence the types of chemical bonds they form, because these are ultimately responsible for the behavior of substances.

IN THIS CHAPTER . . . We examine how atomic properties give rise to the three major types of bonding—ionic, covalent, and metallic—and how each model of bonding explains the properties of substances. For ionic bonding, we detail the steps and calculate the energy involved in the formation of an ionic solid from its elements. Covalent bonding occurs in the vast majority of compounds. We examine the formation and characteristics—order, energy, and length—of a covalent bond. We explore the range of bonding, from pure covalent to ionic, in terms of electronegativity and bond polarity and see that most bonds fall somewhere between these extremes. Finally, we consider a simple model that explains metallic bonding.

Concepts & Skills to Review

before you study this chapter

- characteristics of ionic and covalent bonding (Section 2.7)
- polar covalent bonds and the polarity of water (Section 4.1)
- Hess's law, ΔH^0_{rxn}, and ΔH^0_f (Sections 6.5 and 6.6)
- atomic and ionic electron configurations (Sections 8.3 and 8.5)
- trends in atomic properties and metallic behavior (Sections 8.4 and 8.5)

9.1 ATOMIC PROPERTIES AND CHEMICAL BONDS

Before we examine the types of chemical bonding, we should ask why atoms bond at all. In general terms, they do so for one overriding reason: *bonding lowers the potential energy between positive and negative particles*, whether those particles are oppositely charged ions or atomic nuclei and the electrons between them. Just as the electron configuration and the strength of the nucleus-electron attraction determine the properties of an atom, the type and strength of chemical bonds determine the properties of a substance.

The Three Types of Chemical Bonding

On the atomic level, we distinguish a metal from a nonmetal on the basis of several properties that correlate with position in the periodic table (Figure 9.1 and

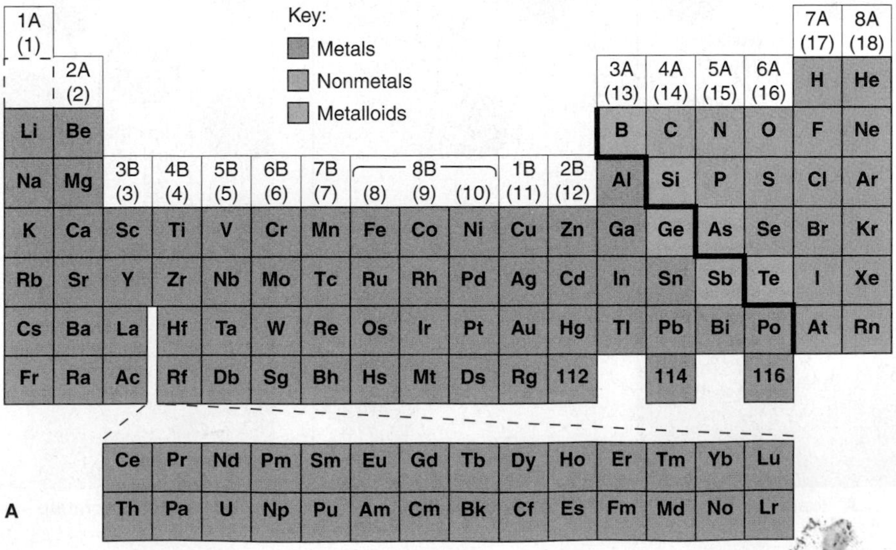

Key: Metals; Nonmetals; Metalloids

1A (1)	2A (2)	3B (3)	4B (4)	5B (5)	6B (6)	7B (7)	8B (8)	(9)	(10)	1B (11)	2B (12)	3A (13)	4A (14)	5A (15)	6A (16)	7A (17)	8A (18)
																H	He
Li	Be											B	C	N	O	F	Ne
Na	Mg											Al	Si	P	S	Cl	Ar
K	Ca	Sc	Ti	V	Cr	Mn	Fe	Co	Ni	Cu	Zn	Ga	Ge	As	Se	Br	Kr
Rb	Sr	Y	Zr	Nb	Mo	Tc	Ru	Rh	Pd	Ag	Cd	In	Sn	Sb	Te	I	Xe
Cs	Ba	La	Hf	Ta	W	Re	Os	Ir	Pt	Au	Hg	Tl	Pb	Bi	Po	At	Rn
Fr	Ra	Ac	Rf	Db	Sg	Bh	Hs	Mt	Ds	Rg	112		114		116		

Ce	Pr	Nd	Pm	Sm	Eu	Gd	Tb	Dy	Ho	Er	Tm	Yb	Lu
Th	Pa	U	Np	Pu	Am	Cm	Bk	Cf	Es	Fm	Md	No	Lr

A

PROPERTY	METAL ATOM	NONMETAL ATOM
Atomic size	Larger	Smaller
Z_{eff}	Lower	Higher
IE	Lower	Higher
EA	Less negative	More negative

B Relative magnitudes of atomic properties within a period

Figure 9.1 A general comparison of metals and nonmetals. A, The key indicates the positions of metals, nonmetals, and metalloids within the periodic table. **B,** The relative magnitudes of some key atomic properties vary from left to right within a period and correlate with whether an element is metallic or nonmetallic.

inside the front cover). Recall from Chapter 8 that, in general, there is a gradation from more metal-like to more nonmetal-like behavior from left to right across a period and from bottom to top within most groups. Three types of bonding result from the three ways these two types of atoms can combine—metal with nonmetal, nonmetal with nonmetal, and metal with metal:

1. *Metal with nonmetal: electron transfer and ionic bonding* (Figure 9.2A). We typically observe **ionic bonding** between atoms with large differences in their tendencies to lose or gain electrons. Such differences occur between reactive metals [Groups 1A(1) and 2A(2)] and nonmetals [Group 7A(17) and the top of Group 6A(16)]. The metal atom (low IE) loses its one or two valence electrons, whereas the nonmetal atom (highly negative EA) gains the electron(s). *Electron transfer* from metal to nonmetal occurs, and each atom forms an ion with a noble gas electron configuration. The electrostatic attraction between these positive and negative ions draws them into the three-dimensional array of an ionic solid, whose chemical formula represents the cation-to-anion ratio (empirical formula).

2. *Nonmetal with nonmetal: electron sharing and covalent bonding* (Figure 9.2B). When two atoms have a small difference in their tendencies to lose or gain electrons, we observe *electron sharing* and **covalent bonding.** This type of bonding most commonly occurs between nonmetal atoms (although a pair of metal atoms can sometimes form a covalent bond also). Each nonmetal atom holds onto its own electrons tightly (high IE) and tends to attract other electrons as well (highly negative EA). The attraction of each nucleus for the valence electrons of the other draws the atoms together. A shared electron pair is considered to be *localized* between the two atoms because it spends most of its time there, linking them in a covalent bond of a particular length and strength. In most cases, separate molecules form when covalent bonding occurs, and the chemical formula reflects the actual numbers of atoms in the molecule (molecular formula).

3. *Metal with metal: electron pooling and metallic bonding* (Figure 9.2C). In general, metal atoms are relatively large, and their few outer electrons are well shielded by filled inner levels. Thus, they lose outer electrons comparatively easily (low IE) but do not gain them very readily (slightly negative or positive EA). These properties lead large numbers of metal atoms to share their valence electrons, but in a way that differs from covalent bonding. In the simplest model of

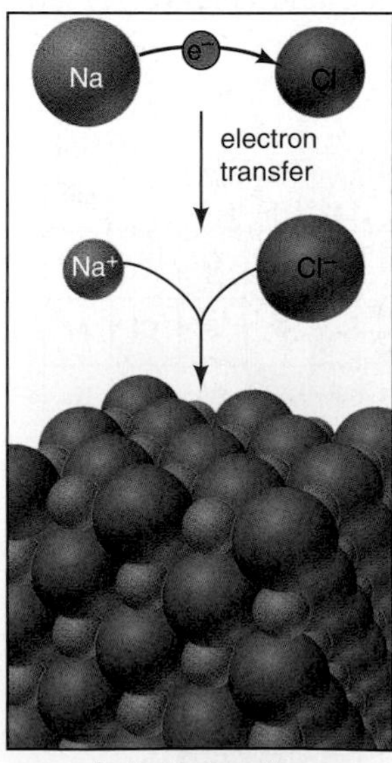

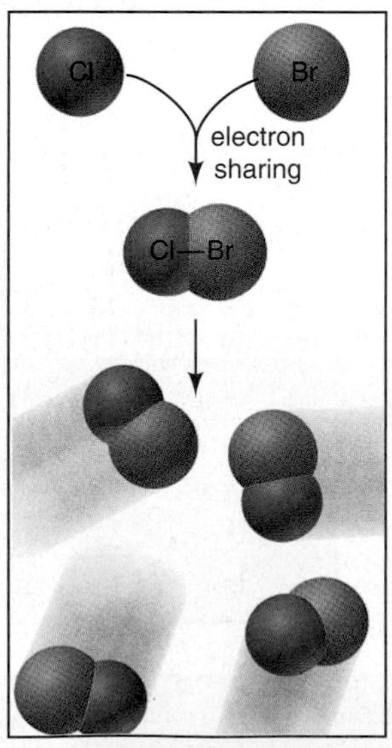

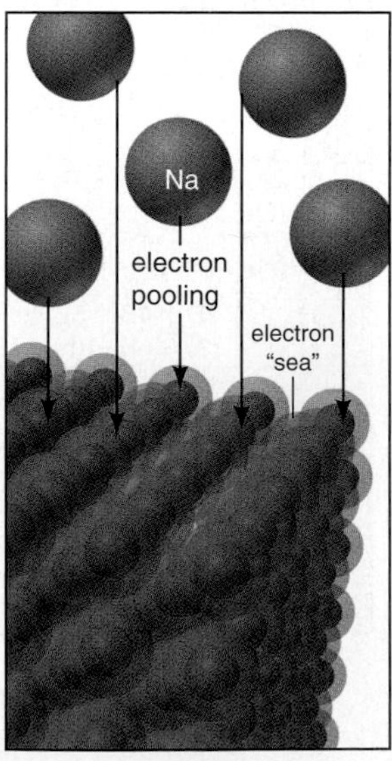

Figure 9.2 The three models of chemical bonding. A, In ionic bonding, metal atoms transfer electron(s) to nonmetal atoms, forming oppositely charged ions that attract each other to form a solid. **B,** In covalent bonding, two atoms share an electron pair localized between their nuclei (shown here as a bond line). Most covalent substances consist of individual molecules, each made from two or more atoms. **C,** In metallic bonding, many metal atoms pool their valence electrons to form a delocalized electron "sea" that holds the metal-ion cores together.

metallic bonding, all the metal atoms in a sample *pool* their valence electrons into an evenly distributed "sea" of electrons that "flows" between and around the metal-ion cores (nucleus plus inner electrons) and attracts them, thereby holding them together. Unlike the localized electrons in covalent bonding, electrons in metallic bonding are *delocalized,* moving freely throughout the piece of metal.

It's important to remember that there are exceptions to these idealized bonding models in the world of real substances. You cannot always predict the bond type solely from the elements' positions in the periodic table. For instance, all binary ionic compounds contain a metal and a nonmetal, but all metals do not form binary ionic compounds with all nonmetals. As just one example, when the metal beryllium [Group 2A(2)] combines with the nonmetal chlorine [Group 7A(17)], the bonding fits the covalent model better than the ionic model. In other words, just as we see a gradation in metallic behavior within groups and periods, we also see a gradation in bonding from one type to another.

Lewis Electron-Dot Symbols: Depicting Atoms in Chemical Bonding

Before turning to the individual models, let's discuss a method for depicting the valence electrons of interacting atoms. In the **Lewis electron-dot symbol** (named for the American chemist G. N. Lewis), the element symbol represents the nucleus *and* inner electrons, and the surrounding dots represent the valence electrons (Figure 9.3). Note that the pattern of dots is the same for elements within a group.

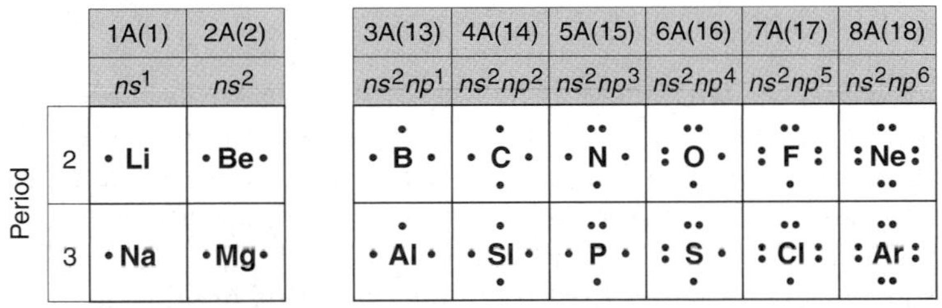

Period	1A(1)	2A(2)	3A(13)	4A(14)	5A(15)	6A(16)	7A(17)	8A(18)
	ns^1	ns^2	ns^2np^1	ns^2np^2	ns^2np^3	ns^2np^4	ns^2np^5	ns^2np^6
2	•Li	•Be•	•Ḃ•	•Ċ• (1 dot below)	•N̈• (1 dot below)	:Ö• (1 dot below)	:F̈: (1 dot below)	:N̈e: (2 dots below)
3	•Na	•Mg•	•Al̇•	•Ṡi• (1 dot below)	•P̈• (1 dot below)	:S̈• (1 dot below)	:C̈l: (1 dot below)	:Är: (2 dots below)

Figure 9.3 Lewis electron-dot symbols for elements in Periods 2 and 3. The element symbol represents the nucleus and inner electrons, and the dots around it represent valence electrons, either paired or unpaired. The number of unpaired dots indicates the number of electrons a metal atom loses, or the number a nonmetal atom gains, or the number of covalent bonds a nonmetal atom usually forms.

It's easy to write the Lewis symbol for any main-group element:

1. Note its A-group number (1A to 8A), which gives the number of valence electrons.
2. Place one dot at a time on the four sides (top, right, bottom, left) of the element symbol.
3. Keep adding dots, pairing the dots until all are used up.

The specific placement of dots is not important; that is, in addition to the one shown in Figure 9.3, the Lewis symbol for nitrogen can *also* be written as

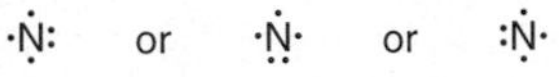

·Ṅ: or ·Ṅ· or :Ṅ·

The number and pairing of dots provide information about an element's bonding behavior:

- For a metal, the *total* number of dots is the maximum number of electrons an atom loses to form a cation.
- For a nonmetal, the number of *unpaired* dots is the number of electrons that become paired either through electron gain or through electron sharing. Thus, the number of unpaired dots equals either the number of electrons a nonmetal atom gains in becoming an anion or the number of covalent bonds it usually forms.

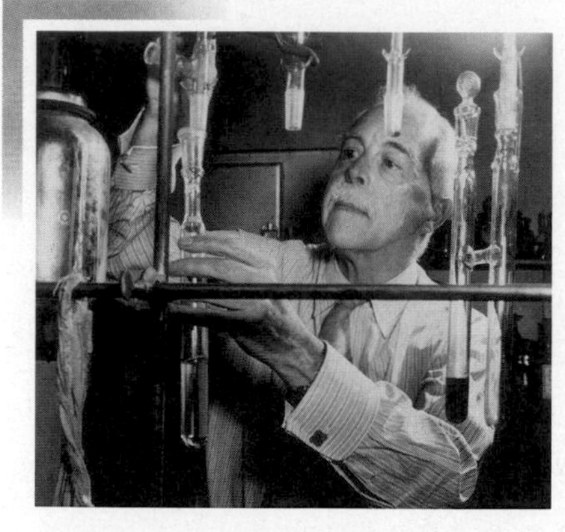

The Remarkable Insights of G. N. Lewis Many of the ideas in this text emerged from the mind of Gilbert Newton Lewis (1875–1946). As early as 1902, nearly a decade *before* Rutherford proposed the nuclear model of the atom, Lewis's notebooks show a scheme that involves the filling of outer electron "shells" to explain the way elements combine. His electron-dot symbols and associated structural formulas (discussed in Chapter 10) have become standards for representing bonding. Among his many other contributions is a more general model for the behavior of acids and bases (which we discuss in Chapter 18).

To illustrate the last point, look at the Lewis symbol for carbon. Rather than one pair of dots and two unpaired dots, as its electron configuration ([He] $2s^22p^2$) would indicate, carbon has four unpaired dots because it forms four bonds. That is, in its compounds, carbon's four electrons are paired with four more electrons from its bonding partners for a total of eight electrons around carbon. (In Chapter 10, we'll see that larger nonmetals can form as many bonds as they have dots in the Lewis symbol.)

In his studies of bonding, Lewis generalized much of bonding behavior into the **octet rule:** *when atoms bond, they lose, gain, or share electrons to attain a filled outer level of eight (or two) electrons.* The octet rule holds for nearly all of the compounds of Period 2 elements and a large number of others as well.

SECTION SUMMARY

Nearly all naturally occurring substances consist of atoms or ions bonded to others. Chemical bonding allows atoms to lower their energy. Ionic bonding occurs when metal atoms transfer electrons to nonmetal atoms, and the resulting ions attract each other and form an ionic solid. Covalent bonding most commonly occurs between nonmetal atoms and usually results in molecules. The bonded atoms share a pair of electrons, which remain localized between them. Metallic bonding occurs when many metal atoms pool their valence electrons in a delocalized electron sea that holds all the atoms together. The Lewis electron-dot symbol of an atom depicts the number of valence electrons for a main-group element. In bonding, many atoms lose, gain, or share electrons to attain a filled outer level of eight (or two).

9.2 THE IONIC BONDING MODEL

The central idea of the ionic bonding model is the *transfer of electrons from metal atoms to nonmetal atoms to form ions that come together in a solid ionic compound.* For nearly every monatomic ion of a main-group element, the electron configuration has a filled outer level: either two or eight electrons, the same number as in the nearest noble gas (octet rule).

The transfer of an electron from a lithium atom to a fluorine atom is depicted in three ways in Figure 9.4. In each, Li loses its single outer electron and is left with a filled $n = 1$ level, while F gains a single electron to fill its $n = 2$ level. In this case, each atom is one electron away from its nearest noble gas—He for Li and Ne for F—so the number of electrons lost by each Li equals the number gained by each F. Therefore, equal numbers of Li^+ and F^- ions form, as the formula LiF indicates. That is, in ionic bonding, *the total number of electrons lost by the metal atoms equals the total number of electrons gained by the nonmetal atoms.*

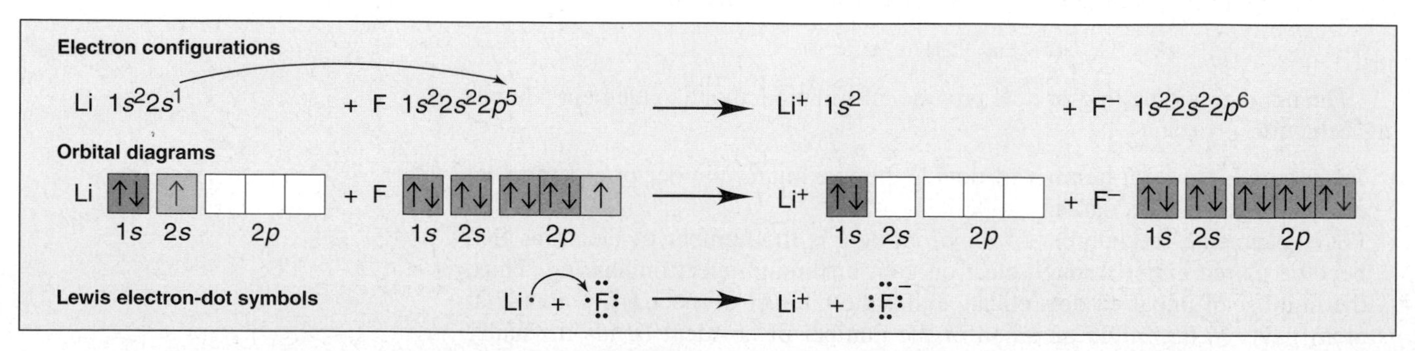

Figure 9.4 Three ways to represent the formation of Li^+ and F^- through electron transfer. The electron being transferred is indicated in red.

SAMPLE PROBLEM 9.1 Depicting Ion Formation

Problem Use partial orbital diagrams and Lewis symbols to depict the formation of Na^+ and O^{2-} ions from the atoms, and determine the formula of the compound the ions form.

Plan First we draw the orbital diagrams and Lewis symbols for the Na and O atoms. To attain filled outer levels, Na loses one electron and O gains two. Thus, to make the number of electrons lost equal the number gained, two Na atoms are needed for each O atom.

Solution

Na [↑] 3s [ ][ ][ ] 3p

Na [↑] 3s [ ][ ][ ] 3p

+ O [↑↓] 2s [↑↓][↑][↑] 2p ⟶ 2Na⁺ + O²⁻ [↑↓] 2s [↑↓][↑↓][↑↓] 2p

$2\,Na\cdot + \cdot\ddot{O}: \longrightarrow 2Na^+ + :\ddot{O}:^{2-}$

The formula is Na_2O.

FOLLOW-UP PROBLEM 9.1 Use condensed electron configurations and Lewis symbols to depict the formation of Mg^{2+} and Cl^- ions from the atoms, and write the formula of the ionic compound.

Energy Considerations in Ionic Bonding: The Importance of Lattice Energy

You may be surprised to learn that the electron-transfer process by itself actually *absorbs* energy! In fact, the reason ionic compounds occur at all is because of the enormous *release* of energy that occurs when the ions come together and form the solid. Consider just the electron-transfer process for the formation of lithium fluoride, which involves two steps—a gaseous Li atom loses an electron, and a gaseous F atom gains it:

- The first ionization energy (IE_1) of Li is the energy change that occurs when 1 mol of gaseous Li atoms loses 1 mol of outer electrons:

$$Li(g) \longrightarrow Li^+(g) + e^- \qquad IE_1 = 520\ kJ$$

- The electron affinity (EA) of F is the energy change that occurs when 1 mol of gaseous F atoms gains 1 mol of electrons:

$$F(g) + e^- \longrightarrow F^-(g) \qquad EA = -328\ kJ$$

Note that the two-step electron-transfer process *by itself* requires energy:

$$Li(g) + F(g) \longrightarrow Li^+(g) + F^-(g) \qquad IE_1 + EA = 192\ kJ$$

The total energy needed for ion formation is even greater than this because metallic lithium and diatomic fluorine must first be converted to separate gaseous atoms, which also requires energy. Despite this, the standard heat of formation (ΔH^0_f) of solid LiF is −617 kJ/mol; that is, 617 kJ is *released* per mole of LiF(*s*) formed. The case of LiF is typical: despite the endothermic electron transfer, ionic solids form readily, often vigorously. Figure 9.5 shows the formation of NaBr.

Clearly, if the overall reaction of Li(*s*) and $F_2(g)$ to form LiF(*s*) releases energy, there must be some exothermic energy component large enough to overcome the endothermic steps. This component arises from the strong *attraction between oppositely charged ions.* When 1 mol of $Li^+(g)$ and 1 mol of $F^-(g)$ form 1 mol of gaseous LiF molecules, a large quantity of heat is released:

$$Li^+(g) + F^-(g) \longrightarrow LiF(g) \qquad \Delta H^0 = -755\ kJ$$

Of course, under ordinary conditions, LiF does not consist of gaseous molecules because *much more energy is released when the gaseous ions coalesce into a crystalline solid.* That occurs because each ion attracts others of opposite charge:

$$Li^+(g) + F^-(g) \longrightarrow LiF(s) \qquad \Delta H^0 = -1050\ kJ$$

A

B

Figure 9.5 The reaction between sodium and bromine. A, Despite the endothermic electron-transfer process, all the Group 1A(1) metals react exothermically with any of the Group 7A(17) nonmetals to form solid alkali-metal halides. The reactants in the example shown are sodium (in beaker under mineral oil) and bromine (dark orange-brown liquid in flask). **B,** The reaction is usually rapid and vigorous.

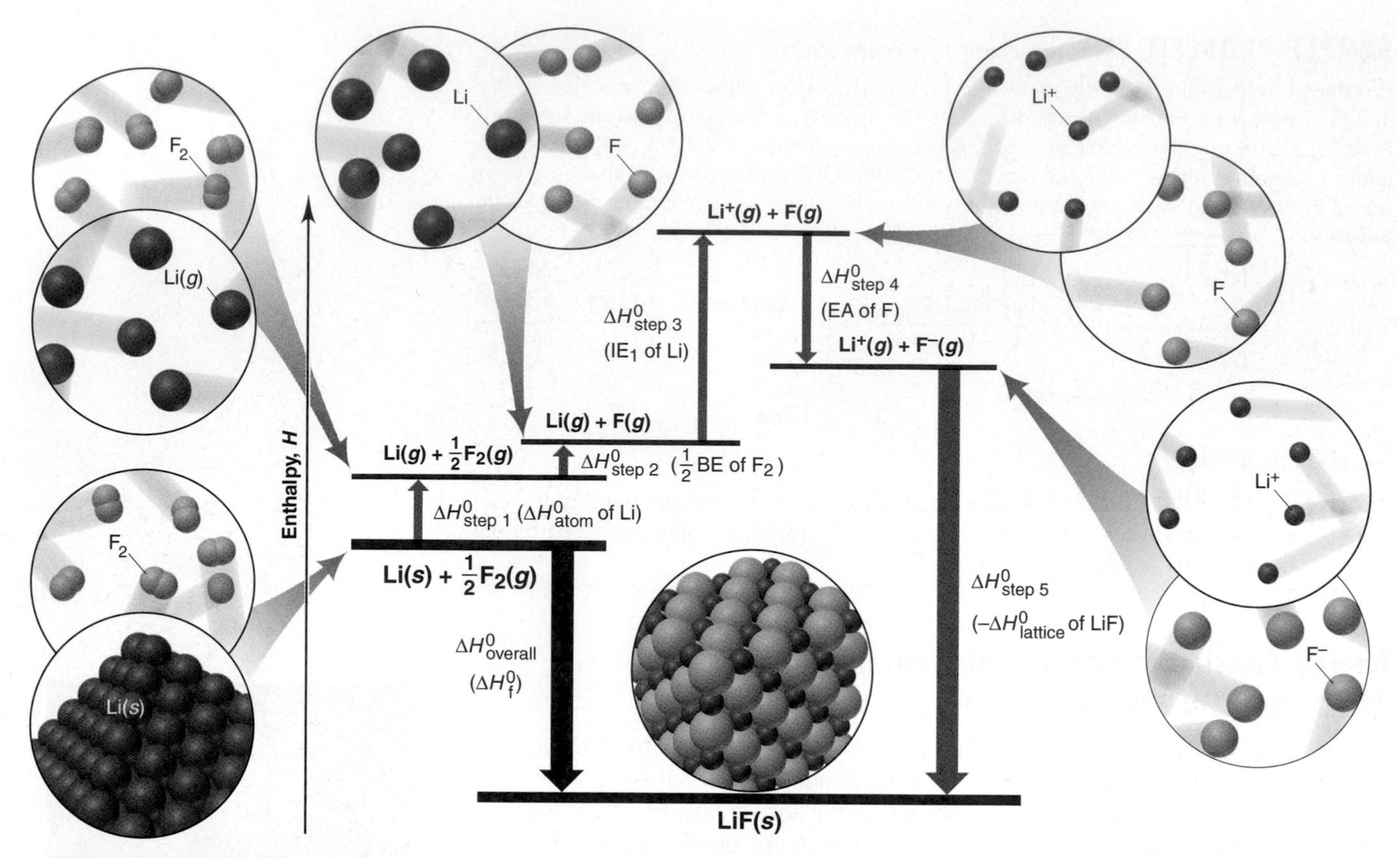

Figure 9.6 The Born-Haber cycle for lithium fluoride. The formation of LiF(*s*) from its elements is shown happening either in one overall reaction *(black arrow)* or in five hypothetical steps, each with its own enthalpy change *(orange arrows).* The overall enthalpy change for the process (ΔH_f^0) is known, as are $\Delta H_{step\ 1}^0$ through $\Delta H_{step\ 4}^0$. Therefore, $\Delta H_{step\ 5}^0$ ($-\Delta H_{lattice}^0$ of LiF) can be calculated to find the lattice energy. (ΔH_{atom}^0 is the heat of atomization; BE is the bond energy.)

Animation: Formation of an Ionic Compound
Online Learning Center

The negative of this value, 1050 kJ, is the lattice energy of LiF. The **lattice energy** ($\boldsymbol{\Delta H_{lattice}^0}$) is the enthalpy change that accompanies the separation of 1 mol of ionic solid into gaseous ions. It indicates the strength of ionic interactions and influences melting point, hardness, solubility, and other properties. (You'll see next how we calculate the lattice energy.)

Lattice energy plays the critical role in the formation of ionic compounds, but it cannot be measured directly. One way to determine lattice energy applies Hess's law (see Section 6.5), which states that the enthalpy change of an overall reaction is the sum of the enthalpy changes for the individual reactions that make it up: $\Delta H_{total} = \Delta H_1 + \Delta H_2 + \cdots$. Lattice energies are calculated by means of a **Born-Haber cycle,** a series of chosen steps from elements to ionic compound for which all the enthalpies* are known except the lattice energy.

Let's go through the Born-Haber cycle for lithium fluoride. Figure 9.6 shows two possible paths, either the direct combination reaction (black arrow) or the multistep path (orange arrows), one step of which involves the unknown lattice energy. Hess's law tells us both paths involve the same overall enthalpy change:

$$\Delta H_f^0 \text{ of LiF}(s) = \text{sum of } \Delta H^0 \text{ values for multistep path}$$

It's important to realize that we *choose hypothetical steps whose enthalpy changes we can measure* to depict the energy components of LiF formation, even though *these are* ***not*** *the actual steps that occur when lithium reacts with fluorine.*

We begin with the elements in their standard states, metallic lithium and gaseous diatomic fluorine. In the multistep process, the elements are converted to individual gaseous atoms (steps 1 and 2), gaseous ions form by electron transfers

*Strictly speaking, ionization energy (IE) and electron affinity (EA) are internal energy changes (ΔE), not enthalpy changes (ΔH), but in these steps, $\Delta H = \Delta E$ because $\Delta V = 0$ (see Section 6.2).

(steps 3 and 4), and the ions come together into a solid (step 5). We identify each ΔH^0 by its step number:

Step 1. Converting 1 mol of solid Li to separate gaseous Li atoms involves breaking metallic bonds, so this step requires energy:

$$\mathrm{Li}(s) \longrightarrow \mathrm{Li}(g) \qquad \Delta H^0_{\text{step 1}} = 161 \text{ kJ}$$

(This process is called *atomization*, and the enthalpy change is ΔH^0_{atom}.)

Step 2. Converting an F_2 molecule to F atoms involves breaking a covalent bond, so it requires energy, which, as we discuss later, is called the *bond energy* (BE). We need 1 mol of F atoms for 1 mol of LiF, so we start with $\frac{1}{2}$ mol of F_2:

$$\tfrac{1}{2}\mathrm{F}_2(g) \longrightarrow \mathrm{F}(g) \qquad \Delta H^0_{\text{step 2}} = \tfrac{1}{2}(\text{BE of F}_2) = \tfrac{1}{2}(159 \text{ kJ}) = 79.5 \text{ kJ}$$

Step 3. Removing the 2*s* electron from 1 mol of Li to form 1 mol of Li^+ requires energy:

$$\mathrm{Li}(g) \longrightarrow \mathrm{Li}^+(g) + \mathrm{e}^- \qquad \Delta H^0_{\text{step 3}} = \text{IE}_1 = 520 \text{ kJ}$$

Step 4. Adding an electron to each atom in 1 mol of F to form 1 mol of F^- releases energy:

$$\mathrm{F}(g) + \mathrm{e}^- \longrightarrow \mathrm{F}^-(g) \qquad \Delta H^0_{\text{step 4}} = \text{EA} = -328 \text{ kJ}$$

Step 5. Forming 1 mol of the crystalline ionic solid from the gaseous ions is the step whose enthalpy change (negative of the lattice energy) is unknown:

$$\mathrm{Li}^+(g) + \mathrm{F}^-(g) \longrightarrow \mathrm{LiF}(s) \qquad \Delta H^0_{\text{step 5}} = -\Delta H^0_{\text{lattice}} \text{ of LiF} = ?$$

We know the enthalpy change of the formation reaction,

$$\mathrm{Li}(s) + \tfrac{1}{2}\mathrm{F}_2(g) \longrightarrow \mathrm{LiF}(s) \qquad \Delta H^0_{\text{overall}} = \Delta H^0_{\text{f}} = -617 \text{ kJ}$$

Using Hess's law, we set the known ΔH^0_f equal to the sum of the ΔH^0 values for the steps and calculate the lattice energy:

$$\Delta H^0_{\text{f}} = \Delta H^0_{\text{step 1}} + \Delta H^0_{\text{step 2}} + \Delta H^0_{\text{step 3}} + \Delta H^0_{\text{step 4}} + (-\Delta H^0_{\text{lattice}} \text{ of LiF})$$

Solving for $-\Delta H^0_{\text{lattice}}$ of LiF gives

$$\begin{aligned} -\Delta H^0_{\text{lattice}} \text{ of LiF} &= \Delta H^0_{\text{f}} - (\Delta H^0_{\text{step 1}} + \Delta H^0_{\text{step 2}} + \Delta H^0_{\text{step 3}} + \Delta H^0_{\text{step 4}}) \\ &= -617 \text{ kJ} - [161 \text{ kJ} + 79.5 \text{ kJ} + 520 \text{ kJ} + (-328 \text{ kJ})] \\ &= -1050 \text{ kJ} \end{aligned}$$

So, $\Delta H^0_{\text{lattice}}$ of LiF $= -(-1050 \text{ kJ}) = 1050$ kJ

Note that *the magnitude of the lattice energy dominates the multistep process.*

The Born-Haber cycle reveals a central point: *ionic solids exist* ***only*** *because the lattice energy exceeds the energetically unfavorable electron transfer.* In other words, the energy *required* for elements to lose or gain electrons is *supplied* by the attraction between the ions they form: energy is expended to form the ions, but it is more than regained when they attract each other and form a solid.

Periodic Trends in Lattice Energy

Because the lattice energy is the result of electrostatic interactions among ions, we expect its magnitude to depend on several factors, including ionic size, ionic charge, and ionic arrangement in the solid. In Chapter 2, you were introduced to Coulomb's law as a fundamental aspect of ionic bonding; in fact, it applies to any situation in which charges attract each other. **Coulomb's law** states that the electrostatic *force* associated with two charges (A and B) is directly proportional to the product of their magnitudes and inversely proportional to the square of the distance between them:

$$\text{Electrostatic force} \propto \frac{\text{charge A} \times \text{charge B}}{\text{distance}^2}$$

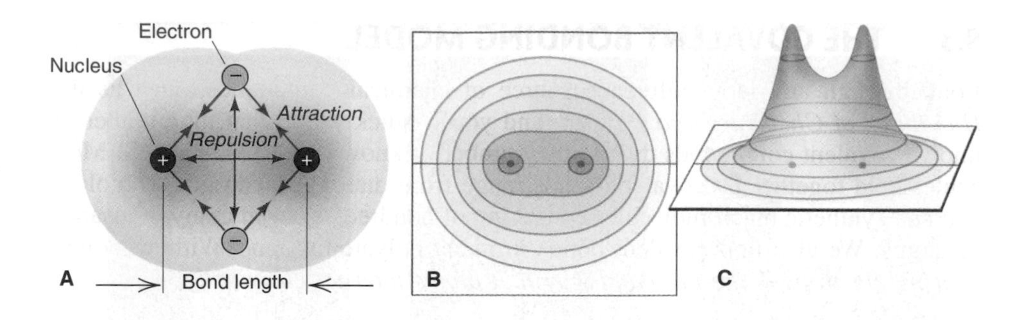

Figure 9.12 Distribution of electron density in H_2. A, At some optimum distance (bond length), attractions balance repulsions. Electron density *(blue shading)* is highest around and between the nuclei. **B,** This *contour map* shows a doubling of electron density with each contour line; the dots represent the nuclei. **C,** This *relief map* depicts the varying electron density of the contour map as peaks. The densest regions, by far, are around the nuclei, but the region between the nuclei—the bonding region—also has higher electron density.

attractions and electron-electron and nucleus-nucleus repulsions. Formation of a bond always results in *greater electron density between the nuclei.* Figure 9.12 depicts this fact in three ways: a cross-section of a space-filling model; an *electron density contour map,* with lines representing regular increments in electron density; and an *electron density relief map* that portrays the contour map three-dimensionally as peaks of electron density.

Bonding Pairs and Lone Pairs In covalent bonding, as in ionic bonding, each atom achieves a full outer (valence) level of electrons, but this is accomplished by different means. *Each atom in a covalent bond "counts" the shared electrons as belonging entirely to itself.* Thus, the two electrons in the shared electron pair of H_2 simultaneously fill the outer level of *both* H atoms. The **shared pair,** or **bonding pair,** is represented by either a pair of dots or a line, H:H or H—H.

An outer-level electron pair that is *not* involved in bonding is called a **lone pair,** or **unshared pair.** The bonding pair in HF fills the outer level of the H atom *and,* together with three lone pairs, fills the outer level of the F atom as well:

bonding pair — H:F: — lone pairs   or   H—F:

Animation: Formation of a Covalent Bond
Online Learning Center

In F_2 the bonding pair and three lone pairs fill the outer level of *each* F atom:

:F:F: or   :F—F:

(This text generally shows bonding pairs as lines and lone pairs as dots.)

Types of Bonds and Bond Order The **bond order** is the number of electron pairs being shared by any pair of bonded atoms. The covalent bond in H_2, HF, or F_2 is a **single bond,** one that consists of a single bonding pair of electrons. A *single bond has a bond order of 1.*

Single bonds are the most common type of bond, but many molecules (and ions) contain multiple bonds. Multiple bonds most frequently involve C, O, N, and/or S atoms. A **double bond** consists of two bonding electron pairs, four electrons shared between two atoms, so *the bond order is 2.* Ethylene (C_2H_4) is a simple hydrocarbon that contains a carbon-carbon double bond and four carbon-hydrogen single bonds:

H H C::C H H   or   H₂C=CH₂ structure (H\C=C/H, H/ \H)

Each carbon "counts" the four electrons in the double bond and the four in its two single bonds to hydrogens to attain an octet.

A **triple bond** consists of three bonding pairs; two atoms share six electrons, so *the bond order is 3.* In the N_2 molecule, the atoms are held together by a triple bond, and each N atom also has a lone pair:

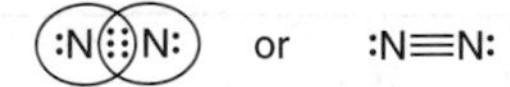

Six shared and two unshared electrons give *each* N atom an octet.

Properties of a Covalent Bond: Bond Energy and Bond Length

The strength of a covalent bond depends on the magnitude of the mutual attraction between bonded nuclei and shared electrons. The **bond energy (BE)** (also called *bond enthalpy* or *bond strength*) is the energy required to overcome this attraction and is defined as the standard enthalpy change for breaking the bond in 1 mol of gaseous molecules. *Bond breakage is an endothermic process, so the bond energy is always positive:*

$$A{-}B(g) \longrightarrow A(g) + B(g) \qquad \Delta H^0_{\text{bond breaking}} = BE_{A-B} \text{ (always} > 0)$$

Stated in another way, the bond energy is the difference in energy between the separated atoms and the bonded atoms (the potential energy difference between points 1 and 3 in Figure 9.11; the depth of the energy well). The same amount of energy that is absorbed to break the bond is released when it forms. *Bond formation is an exothermic process, so the sign of the enthalpy change is negative:*

$$A(g) + B(g) \longrightarrow A{-}B(g) \qquad \Delta H^0_{\text{bond forming}} = -BE_{A-B} \text{ (always} < 0)$$

Because bond energies depend on characteristics of the bonded atoms—their electron configurations, nuclear charges, and atomic radii—each type of bond has its own bond energy. Bond energies for some common bonds are listed in Table 9.2. *Stronger bonds are lower in energy (have a deeper energy well); weaker bonds are higher in energy (have a shallower energy well).* The energy of a given type of bond varies slightly from molecule to molecule (except for symmetrical diatomic molecules such as H_2), and even within the same molecule, so the tabulated value is an *average* bond energy.

Table 9.2 Average Bond Energies (kJ/mol)

	Bond	Energy	Bond	Energy	Bond	Energy	Bond	Energy
Single Bonds								
	H—H	432	N—H	391	Si—H	323	S—H	347
	H—F	565	N—N	160	Si—Si	226	S—S	266
	H—Cl	427	N—P	209	Si—O	368	S—F	327
	H—Br	363	N—O	201	Si—S	226	S—Cl	271
	H—I	295	N—F	272	Si—F	565	S—Br	218
			N—Cl	200	Si—Cl	381	S—I	~170
	C—H	413	N—Br	243	Si—Br	310		
	C—C	347	N—I	159	Si—I	234	F—F	159
	C—Si	301					F—Cl	193
	C—N	305	O—H	467	P—H	320	F—Br	212
	C—O	358	O—P	351	P—Si	213	F—I	263
	C—P	264	O—O	204	P—P	200	Cl—Cl	243
	C—S	259	O—S	265	P—F	490	Cl—Br	215
	C—F	453	O—F	190	P—Cl	331	Cl—I	208
	C—Cl	339	O—Cl	203	P—Br	272	Br—Br	193
	C—Br	276	O—Br	234	P—I	184	Br—I	175
	C—I	216	O—I	234			I—I	151
Multiple Bonds								
	C=C	614	N=N	418	C≡C	839	N≡N	945
	C=N	615	N=O	607	C≡N	891		
	C=O	745 (799 in CO_2)	O_2	498	C≡O	1070		

Table 9.3 Average Bond Lengths (pm)

Bond	Length	Bond	Length	Bond	Length	Bond	Length
Single Bonds							
H—H	74	N—H	101	Si—H	148	S—H	134
H—F	92	N—N	146	Si—Si	234	S—P	210
H—Cl	127	N—P	177	Si—O	161	S—S	204
H—Br	141	N—O	144	Si—S	210	S—F	158
H—I	161	N—S	168	Si—N	172	S—Cl	201
		N—F	139	Si—F	156	S—Br	225
		N—Cl	191	Si—Cl	204	S—I	234
C—H	109	N—Br	214	Si—Br	216		
C—C	154	N—I	222	Si—I	240	F—F	143
C—Si	186					F—Cl	166
C—N	147	O—H	96	P—H	142	F—Br	178
C—O	143	O—P	160	P—Si	227	F—I	187
C—P	187	O—O	148	P—P	221	Cl—Cl	199
C—S	181	O—S	151	P—F	156	Cl—Br	214
C—F	133	O—F	142	P—Cl	204	Cl—I	243
C—Cl	177	O—Cl	164	P—Br	222	Br—Br	228
C—Br	194	O—Br	172	P—I	243	Br—I	248
C—I	213	O—I	194			I—I	266
Multiple Bonds							
C=C	134	N=N	122	C≡C	121	N≡N	110
C=N	127	N=O	120	C≡N	115	N≡O	106
C=O	123	O_2	121	C≡O	113		

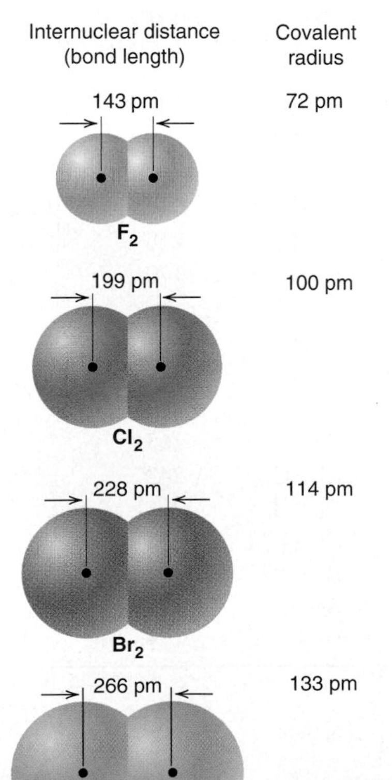

Figure 9.13 Bond length and covalent radius. Within a series of similar molecules, such as the diatomic halogen molecules, bond length increases as covalent radius increases.

A covalent bond has a **bond length,** the distance between the nuclei of two bonded atoms. In Figure 9.11, bond length is shown as the distance between the nuclei at the point of minimum energy. Table 9.3 shows the lengths of some covalent bonds. Here, too, the values represent *average* bond lengths for the given bond in different substances. Bond length is related to the sum of the radii of the bonded atoms. In fact, most atomic radii are calculated from measured bond lengths (see Figure 8.14C, p. 305). Bond lengths for a series of similar bonds increase with atomic size, as shown in Figure 9.13 for the halogens.

A close relationship exists among bond order, bond length, and bond energy. Two nuclei are more strongly attracted to two shared electron pairs than to one: the atoms are drawn closer together *and* are more difficult to pull apart. Therefore, *for a given pair of atoms, a higher bond order results in a shorter bond length and a higher bond energy.* So, as Table 9.4 shows, for a given pair of atoms, *a shorter bond is a stronger bond.*

In some cases, we can extend this relationship among atomic size, bond length, and bond strength by holding one atom in the bond constant and varying the other atom within a group or period. For example, the trend in carbon-halogen single bond lengths, C—I > C—Br > C—Cl, parallels the trend in atomic size, I > Br > Cl, and is opposite to the trend in bond energy, C—Cl > C—Br > C—I. Thus, for *single bonds,* longer bonds are usually weaker.

Table 9.4 The Relation of Bond Order, Bond Length, and Bond Energy

Bond	Bond Order	Average Bond Length (pm)	Average Bond Energy (kJ/mol)
C—O	1	143	358
C=O	2	123	745
C≡O	3	113	1070
C—C	1	154	347
C=C	2	134	614
C≡C	3	121	839
N—N	1	146	160
N=N	2	122	418
N≡N	3	110	945

SAMPLE PROBLEM 9.2 Comparing Bond Length and Bond Strength

Problem Using the periodic table, but not Tables 9.2 and 9.3, rank the bonds in each set in order of *decreasing* bond length and bond strength:
(a) S—F, S—Br, S—Cl **(b)** C=O, C—O, C≡O

Plan In part (a), S is singly bonded to three different halogen atoms, so all members of the set have a bond order of 1. Bond length increases and bond strength decreases as the halogen's atomic radius increases, and that size trend is clear from the periodic table. In all the bonds in part (b), the same two atoms are involved, but the bond orders differ. In this case, bond strength increases and bond length decreases as bond order increases.

Solution (a) Atomic size increases down a group, so F < Cl < Br.

Bond length: S—Br > S—Cl > S—F

Bond strength: S—F > S—Cl > S—Br

(b) By ranking the bond orders, C≡O > C=O > C—O, we obtain

Bond length: C—O > C=O > C≡O

Bond strength: C≡O > C=O > C—O

Check From Tables 9.2 and 9.3, we see that the rankings are correct.

Comment Remember that for bonds involving pairs of different atoms, as in part (a), *the relationship between length and strength holds* only *for single bonds* and not in every case, so apply it carefully.

FOLLOW-UP PROBLEM 9.2 Rank the bonds in each set in order of *increasing* bond length and bond strength: **(a)** Si—F, Si—C, Si—O; **(b)** N=N, N—N, N≡N.

How the Model Explains the Properties of Covalent Substances

The covalent bonding model proposes that electron sharing between pairs of atoms leads to strong, localized bonds, usually within individual molecules. At first glance, however, it seems that the model is inconsistent with some of the familiar physical properties of covalent substances. After all, most are gases (such as methane and ammonia), liquids (such as benzene and water), or low-melting solids (such as sulfur and paraffin wax). Covalent bonds are strong (~200 to 500 kJ/mol), so why do covalent substances melt and boil at such low temperatures?

To answer this question, we must distinguish between two different sets of forces: (1) the *strong covalent bonding forces* holding the atoms together within the molecule (those we have been discussing), and (2) the *weak intermolecular forces* holding the molecules near each other in the macroscopic sample. It is these weak forces *between* the molecules, not the strong covalent bonds *within* each molecule, that are responsible for the physical properties of covalent substances.

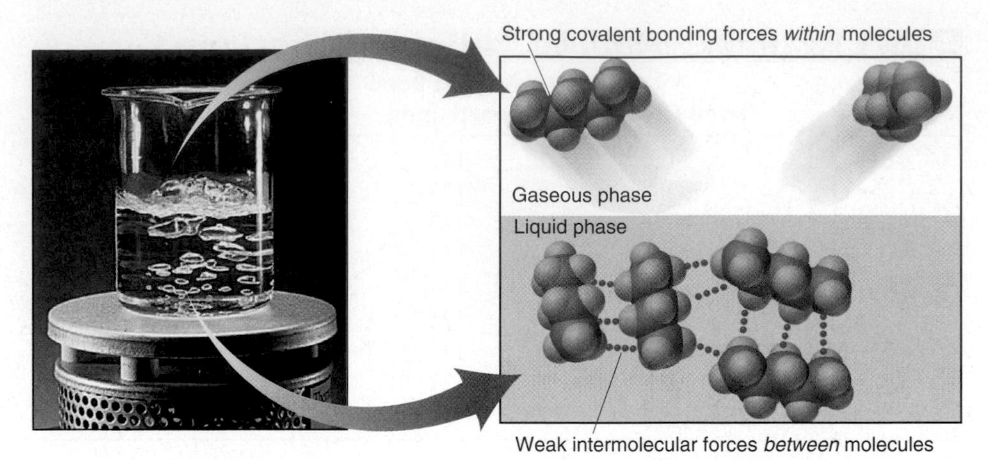

Figure 9.14 Strong forces within molecules and weak forces between them. When pentane boils, weak forces *between* molecules (intermolecular forces) are overcome, but the strong covalent bonds holding the atoms together within each molecule remain unaffected. Thus, the pentane molecules leave the liquid phase as intact units.

Consider, for example, what happens when pentane (C_5H_{12}) boils. As Figure 9.14 shows, the weak interactions *between* the pentane molecules are affected, not the strong C—C and C—H covalent bonds *within* each molecule.

Some covalent substances, called *network covalent solids,* do not consist of separate molecules. Rather, they are held together by covalent bonds that extend in three dimensions *throughout* the sample. If the model is correct, the properties of these substances *should* reflect the strength of their covalent bonds, and this is indeed the case. Two examples, quartz and diamond, are shown in Figure 9.15. Quartz (SiO_2) is very hard and melts at 1550°C. It is composed of silicon and oxygen atoms connected by covalent bonds that extend throughout the sample; no separate SiO_2 molecules exist. Diamond has covalent bonds connecting each of its carbon atoms to four others throughout the sample. It is the hardest substance known and melts at around 3550°C. Clearly, covalent bonds *are* strong, but because most covalent substances consist of separate molecules with weak forces between them, their physical properties do not reflect this bond strength. (We discuss intermolecular forces in detail in Chapter 12.)

Unlike ionic compounds, most covalent substances are poor electrical conductors, even when melted or when dissolved in water. An electric current is carried by either mobile electrons or mobile ions. In covalent substances the electrons are localized as either shared or unshared pairs, so they are not free to move, and no ions are present. The Tools of the Laboratory essay describes a tool used widely for studying the types of bonds in covalent substances.

A Quartz Silicon Oxygen

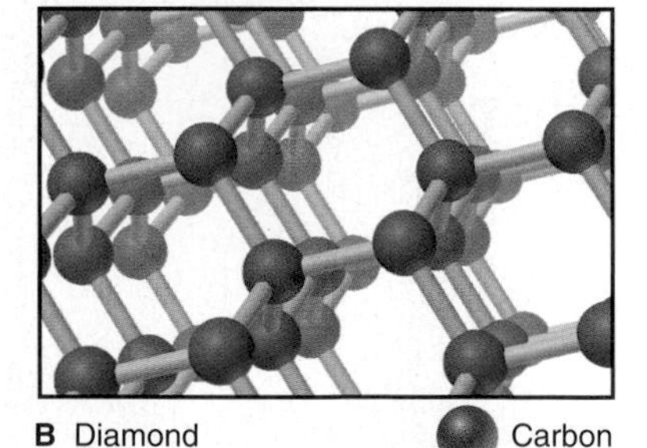

Figure 9.15 Covalent bonds of network covalent solids. A, In quartz (SiO_2), each Si atom is bonded covalently to four O atoms and each O atom is bonded to two Si atoms in a pattern that extends throughout the sample. Because no separate SiO_2 molecules are present, the melting point of quartz is very high, and it is very hard. **B,** In diamond, each C atom is covalently bonded to four other C atoms throughout the crystal. Diamond is the hardest natural substance known and has an extremely high melting point.

SECTION SUMMARY

A shared pair of valence electrons attracts the nuclei of two atoms and holds them together in a covalent bond while filling each atom's outer level. The number of shared pairs between the two atoms is the bond order. For a given type of bond, the bond energy is the average energy required to completely separate the bonded atoms; the bond length is the average distance between their nuclei. For a given pair of bonded atoms, bond order is directly related to bond energy and inversely related to bond length. Substances that consist of separate molecules are generally soft and low melting because of the weak forces between molecules. Solids held together by covalent bonds extending in three dimensions throughout the sample are extremely hard and high melting. Most covalent substances have low electrical conductivity because electrons are localized and ions are absent. The atoms in a covalent bond vibrate, and the energy of these vibrations can be studied with IR spectroscopy.

Infrared Spectroscopy

Infrared (IR) spectroscopy is an instrumental technique used primarily to study the molecular structure of covalently bonded molecules. Found in most research laboratories, the IR spectrometer is an essential part of the chemist's instrumental toolbox—along with the UV-visible spectrophotometer, mass spectrometer, and NMR spectrometer (described in Chapter 15)—for investigating and identifying organic and biological compounds.

The key components of an IR spectrometer are the same as those of similar instruments (see Figure B7.3, p. 270). The source emits radiation of many wavelengths, and those in the IR region are selected and directed at the sample. For organic compounds, the sample is typically either a pure liquid or a solid mixed with an inorganic salt such as KBr. The sample absorbs certain wavelengths of the IR radiation more than others, and an IR spectrum is generated.

What property of a molecule is displayed in its IR spectrum? All molecules, whether occurring in a gas, a liquid, or a solid, undergo continual rotations and vibrations. Consider, for instance, a sample of ethane gas. The $H_3C—CH_3$ molecules zoom throughout the container, colliding with the walls and each other. If we could look closely at one molecule, however, and disregard its motion through space, we would see the whole molecule rotating and its two CH_3 groups rotating relative to each other about the C—C bond. More important to IR spectroscopy, we would also see each of the bonded atoms vibrating, that is, moving closer to and farther from each other, as though the bonds were flexible springs: stretching and compressing, twisting, bending, rocking, and wagging (Figure B9.1). (Thus, the length of a given bond within a particular kind of molecule is actually the *average* distance between nuclei, analogous to the average length of a spring stretching and compressing.)

The energies of IR photons fall in the same range as the energies of these molecular vibrations. Each vibrational motion has its own natural frequency, which is based on the type of motion, the masses of the atoms, and the strength of the bond between them. These frequencies correspond to wavelengths between 2.5 and 25 μm, a part of the IR region of the electromagnetic spectrum (see Figure 7.3, p. 259). The energy of each of these vibrations is quantized. Just as an atom can absorb a photon whose energy corresponds to the difference between two quantized *electron* energy levels, a molecule can absorb an IR photon whose energy corresponds to the difference between two of its quantized *vibrational* energy levels.

The IR spectrum is particularly useful for compound identification because of two related factors. First, *each kind of bond has a characteristic range of IR wavelengths it can absorb.* For example, a C—C bond absorbs IR photons in a different wavelength range from those absorbed by a C=C bond, a C—H bond, a C=O bond, and so forth. Furthermore, groups of atoms that characterize particular types of organic compounds—alcohol, carboxylic acid, ether, and so forth—absorb in slightly different wavelength regions.

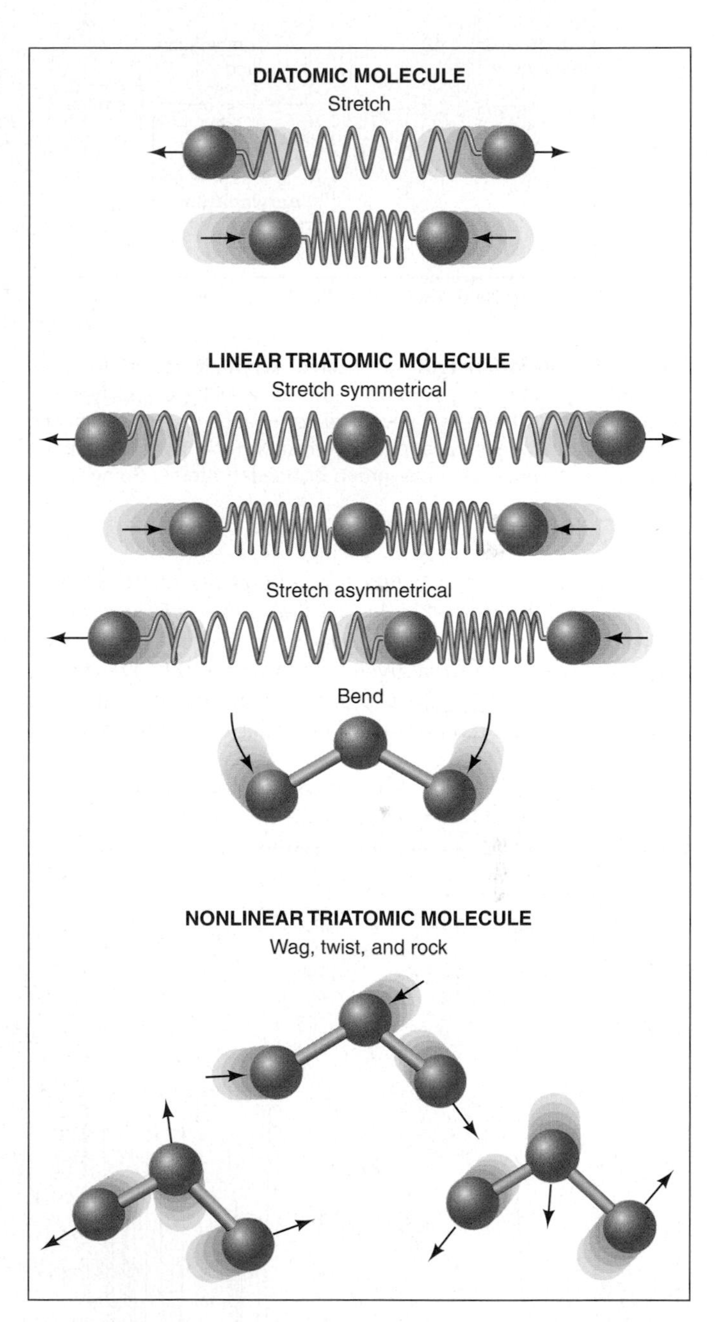

Figure B9.1 Some vibrational motions in general diatomic and triatomic molecules.

(continued)

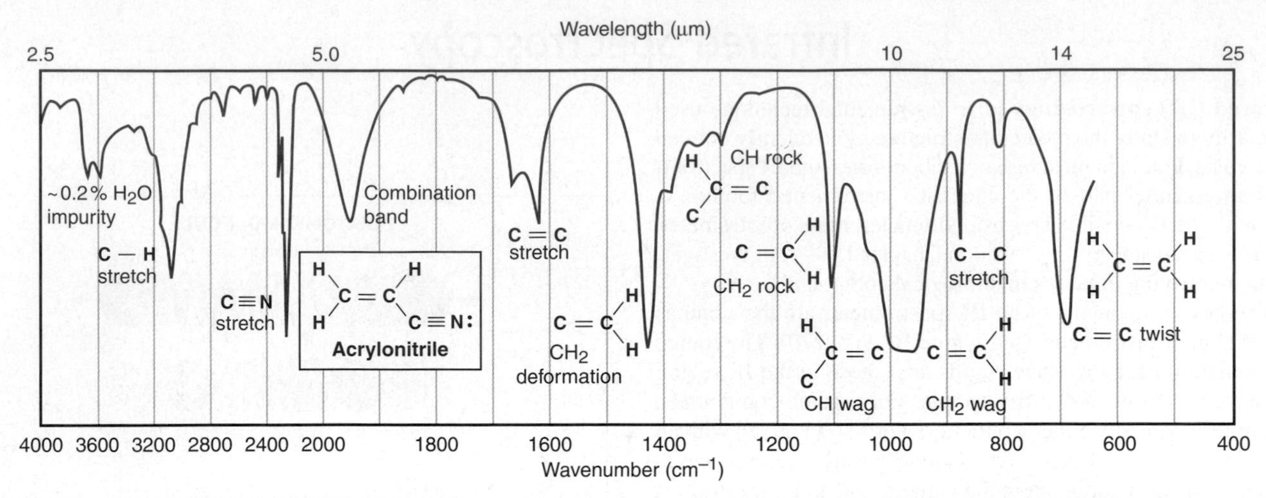

Figure B9.2 The infrared (IR) spectrum of acrylonitrile. The IR spectrum of acrylonitrile is typical of a molecule with several types of covalent bonds. There are many absorption bands (peaks) of differing depths and sharpness. Most peaks correspond to a particular type of vibration involving a particular group of bonded atoms. Some broad peaks (for example, "combination band") represent several overlapping types of vibrations. The spectrum is reproducible and unique for acrylonitrile. (The bottom axis shows wavenumbers, the inverse of wavelength, so its units are those of length^{-1}. The scale expands to the right of 2000 cm^{-1}.)

Second, the exact wavelengths and quantity of IR radiation that a molecule absorbs depend on the *overall structure* of the molecule. Combinations of absorptions overlap to create a very characteristic pattern for a given type of compound. This means that *each compound has a characteristic IR spectrum* that can be used to identify it, much as a fingerprint is used to identify a person.

The spectrum appears as a series of downward pointing peaks of varying depth and sharpness. Figure B9.2 shows the IR spectrum of acrylonitrile, a compound used to manufacture synthetic rubber and plastics. No other compound has exactly the same IR spectrum.

Constitutional (structural) isomers are easily distinguished by their IR spectra. These compounds have the same molecular formula but different structural formulas. We might expect very different isomers such as diethyl ether and 2-butanol to have very different IR spectra because their molecular structures are so dissimilar (Figure B9.3). However, even relatively similar compounds, such as 1,3-dimethylbenzene and 1,4-dimethylbenzene, have clearly different spectra (Figure B9.4).

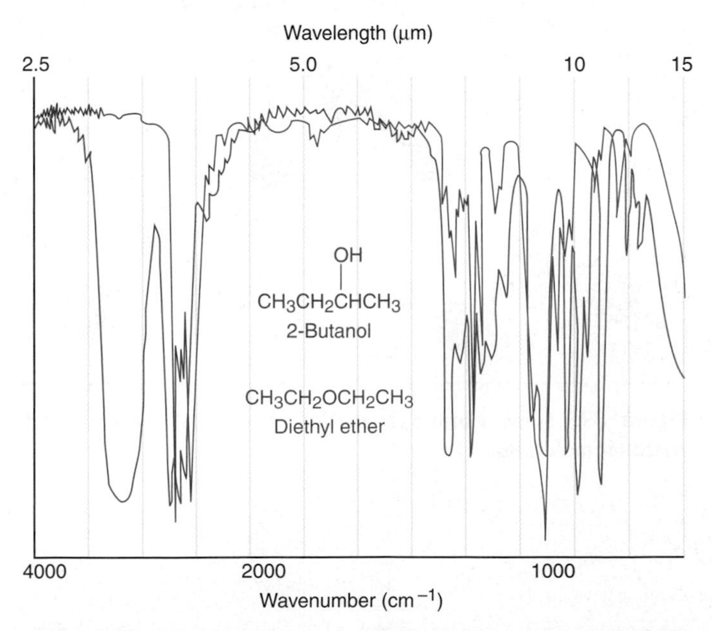

Figure B9.3 The infrared spectra of 2-butanol (*green*) and diethyl ether (*red*).

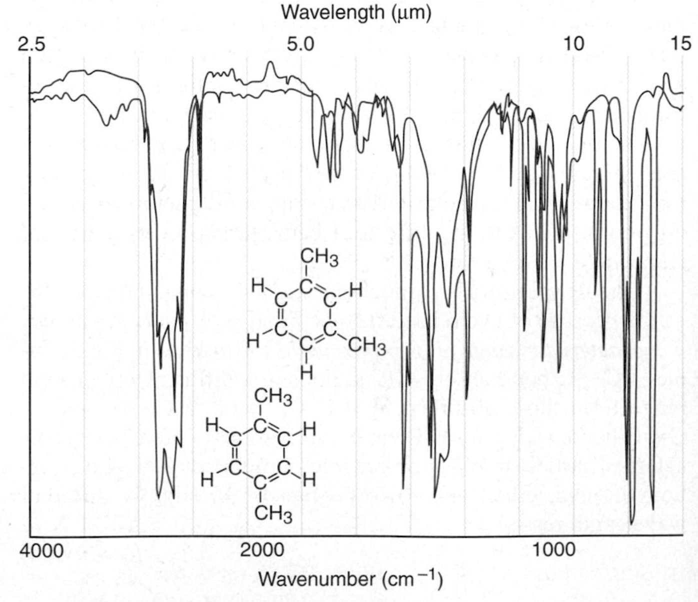

Figure B9.4 The infrared spectra of 1,3-dimethylbenzene (*green*) and 1,4-dimethylbenzene (*red*).

9.4 BOND ENERGY AND CHEMICAL CHANGE

The relative strengths of the bonds in reactants and products of a chemical change determine whether heat is released or absorbed. In fact, as you'll see in Chapter 20, it is one of two essential factors determining whether the change occurs at all. In this section, we'll discuss the importance of bond energy in chemical change, especially in the combustion of fuels and foods.

Changes in Bond Strengths: Where Does ΔH^0_{rxn} Come From?

In Chapter 6, we discussed the heat involved in a chemical change (ΔH^0_{rxn}), but we never stopped to ask a central question. When, for example, 1 mol of H_2 and 1 mol of F_2 react at 298 K, 2 mol of HF form and 546 kJ of heat is released:

$$H_2(g) + F_2(g) \longrightarrow 2HF(g) + 546\ \text{kJ}$$

Where does this heat come from? We find the answer through a very close-up view of the molecules and their energy components.

A system's internal energy has kinetic energy (E_k) and potential energy (E_p) components. Let's examine the contributions to these components to see which one changes during the reaction of H_2 and F_2 to form HF.

Of the various contributions to the kinetic energy, the most important come from the molecules moving through space, rotating, and vibrating and, of course, from the electrons moving within the atoms. Of the various contributions to the potential energy, the most important are electrostatic forces between the vibrating atoms, between nucleus and electrons (and between electrons) in each atom, between protons and neutrons in each nucleus, and, of course, between nuclei and shared electron pair in each bond.

The kinetic energy doesn't change during the reaction because the first three contributions—moving through space, rotating, and vibrating—are proportional to the temperature, which is constant at 298 K; and electron motion is not affected by a reaction. Of the potential energy contributions, those within the atoms and nuclei don't change, and vibrational forces vary only slightly as the bonded atoms change. The only significant change in energy in a chemical reaction is in the strength of attraction of the nuclei for the shared electron pair, that is, in the bond energy.

In other words, the answer to "Where does the heat come from?" is that it doesn't really "come from" anywhere: *the energy released or absorbed during a chemical change is due to differences between the reactant bond energies and the product bond energies.*

Using Bond Energies to Calculate ΔH^0_{rxn}

We can think of a reaction as a two-step process in which *heat is absorbed (ΔH^0 is positive) to break reactant bonds to form separate atoms and is released (ΔH^0 is negative) when the atoms rearrange to form product bonds.* The sum (symbolized by Σ) of these enthalpy changes is the heat of reaction, ΔH^0_{rxn}:

$$\Delta H^0_{rxn} = \Sigma\Delta H^0_{\text{reactant bonds broken}} + \Sigma\Delta H^0_{\text{product bonds formed}} \quad (9.2)$$

- In an exothermic reaction, the total ΔH^0 for product bonds formed is *greater* than that for reactant bonds broken, so the sum, ΔH^0_{rxn}, is *negative.*
- In an endothermic reaction, the total ΔH^0 for product bonds formed is *less* than that for reactant bonds broken, so the sum, ΔH^0_{rxn}, is *positive.*

An equivalent form of Equation 9.2 uses bond energies:

$$\Delta H^0_{rxn} = \Sigma\text{BE}_{\text{reactant bonds broken}} - \Sigma\text{BE}_{\text{product bonds formed}}$$

The minus sign is needed because all bond energies are positive values.

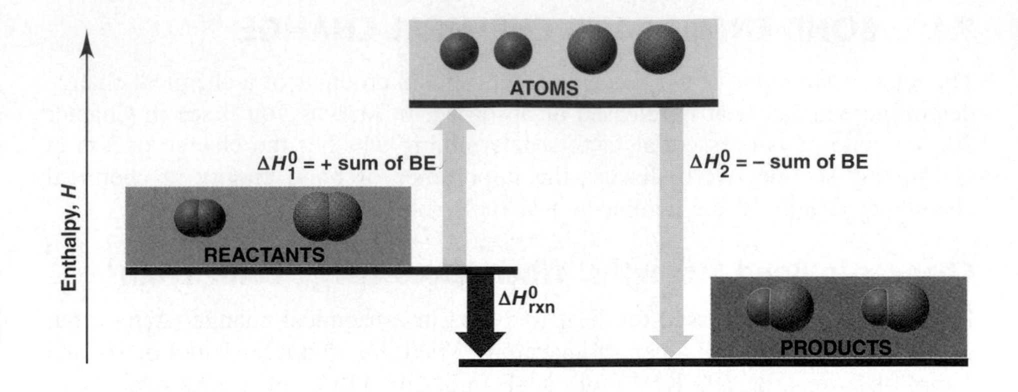

Figure 9.16 Using bond energies to calculate ΔH^0_{rxn}. Any chemical reaction can be divided conceptually into two hypothetical steps: (1) reactant bonds break to yield separate atoms in a step that absorbs heat (+ sum of BE), and (2) the atoms combine to form product bonds in a step that releases heat (– sum of BE). When the total bond energy of the products is greater than that of the reactants, more energy is released than is absorbed, and the reaction is exothermic (as shown); ΔH^0_{rxn} is negative. When the total bond energy of the products is less than that of the reactants, the reaction is endothermic; ΔH^0_{rxn} is positive.

When 1 mol of H—H bonds and 1 mol of F—F bonds absorb energy and break, the 2 mol each of H and F atoms form 2 mol of H—F bonds, which releases energy (Figure 9.16). Recall that *weaker bonds (less stable, more reactive) are easier to break than stronger bonds (more stable, less reactive) because they are higher in energy.* Heat is released when HF forms because the bonds in H_2 and F_2 are weaker (less stable) than the bonds in HF (more stable). Put another way, the sum of the bond energies of 1 mol each of H_2 and F_2 is *smaller* than the sum of the bond energy in 2 mol of HF.

We use bond energies to calculate ΔH^0_{rxn} by assuming that *all the reactant bonds break to give individual atoms, from which all the product bonds form.* Even though the actual reaction may not occur this way—typically, only certain bonds break and form—Hess's law allows us to sum the bond energies (with their appropriate signs) to arrive at the overall heat of reaction. (This method assumes that ΔH^0_{rxn} is due entirely to changes in bond energy, which requires that all reactants and products be in the same physical state. When phase changes occur, additional heat must be taken into account. We address this topic in Chapter 12.)

Let's use bond energies to calculate ΔH^0_{rxn} for the combustion of methane and compare it with the value obtained by calorimetry (Section 6.3), which is

$$CH_4(g) + 2O_2(g) \longrightarrow CO_2(g) + 2H_2O(g) \qquad \Delta H^0_{rxn} = -802\ \text{kJ}$$

Figure 9.17 shows that all the bonds in CH_4 and O_2 break, and the atoms form the bonds in CO_2 and H_2O. We find the bond energy values in Table 9.2 (p. 341), and use a positive sign for bonds broken and a negative sign for bonds formed:

Bonds broken

$$\begin{aligned} 4 \times \text{C—H} &= (4\ \text{mol})(413\ \text{kJ/mol}) = 1652\ \text{kJ} \\ 2 \times O_2 &= (2\ \text{mol})(498\ \text{kJ/mol}) = 996\ \text{kJ} \\ \hline \Sigma\Delta H^0_{\text{reactant bonds broken}} &= 2648\ \text{kJ} \end{aligned}$$

Bonds formed

$$\begin{aligned} 2 \times \text{C=O} &= (2\ \text{mol})(-799\ \text{kJ/mol}) = -1598\ \text{kJ} \\ 4 \times \text{O—H} &= (4\ \text{mol})(-467\ \text{kJ/mol}) = -1868\ \text{kJ} \\ \hline \Sigma\Delta H^0_{\text{product bonds formed}} &= -3466\ \text{kJ} \end{aligned}$$

Applying Equation 9.2 gives

$$\begin{aligned} \Delta H^0_{rxn} &= \Sigma\Delta H^0_{\text{reactant bonds broken}} + \Sigma\Delta H^0_{\text{product bonds formed}} \\ &= 2648\ \text{kJ} + (-3466\ \text{kJ}) = -818\ \text{kJ} \end{aligned}$$

Why is there a discrepancy between the bond energy value (−818 kJ) and the calorimetric value (−802 kJ)? Variations in experimental method always introduce small discrepancies, but there is a more basic reason in this case. Because bond energies are *average* values obtained from many different compounds, the energy of the bond *in a particular substance* is usually close, but not equal, to this average. For example, the tabulated C—H bond energy of 413 kJ/mol is the

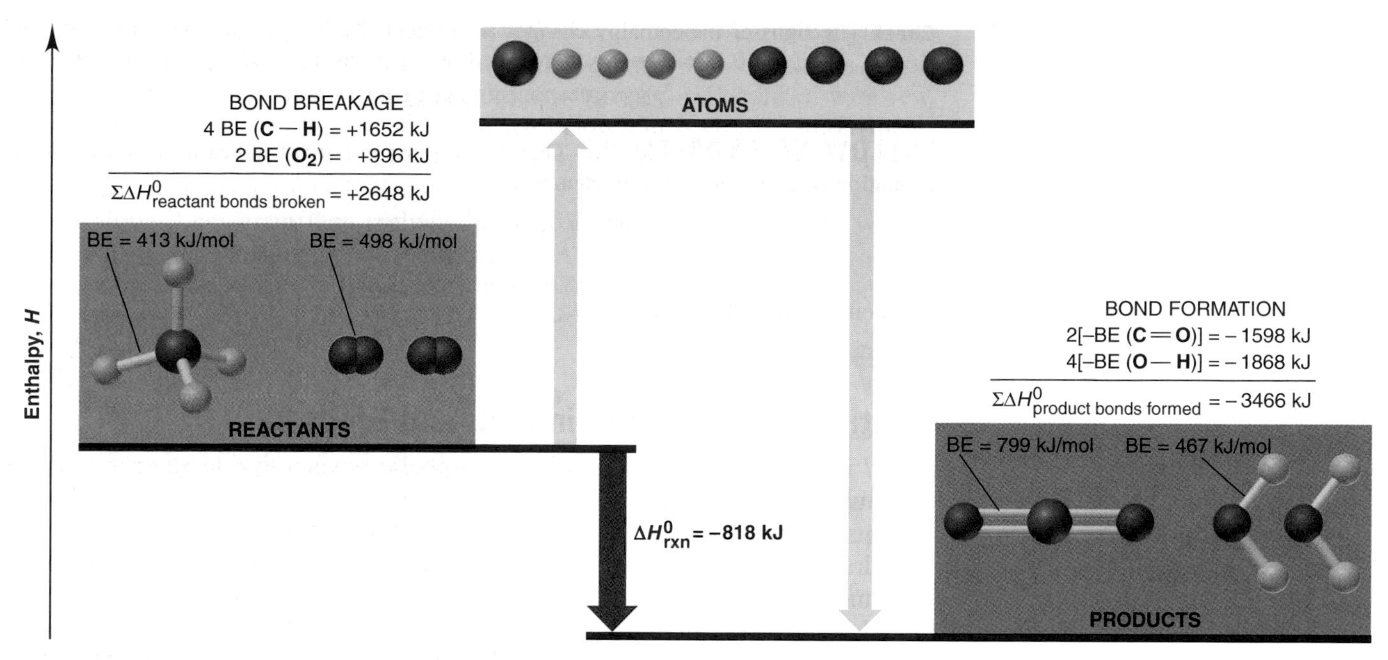

Figure 9.17 Using bond energies to calculate ΔH^0_{rxn} of methane. Treating the combustion of methane as a hypothetical two-step process (see Figure 9.16) means breaking all the bonds in the reactants and forming all the bonds in the products.

average value of C—H bonds in many different molecules. In fact, 415 kJ is actually required to break 1 mol of C—H bonds in methane, or 1660 kJ for 4 mol of these bonds, which gives a ΔH^0_{rxn} even closer to the calorimetric value. Thus, it isn't surprising to find a discrepancy between the two ΔH^0_{rxn} values. What is surprising—and satisfying in its confirmation of bond theory—is that the values are so close.

SAMPLE PROBLEM 9.3 Using Bond Energies to Calculate ΔH^0_{rxn}

Problem Calculate ΔH^0_{rxn} for the chlorination of methane to form chloroform:

$$CH_4 + 3\,Cl\text{—}Cl \longrightarrow CHCl_3 + 3\,H\text{—}Cl$$

(structural formulas: H—C(H)(H)—H + 3 Cl—Cl ⟶ Cl—C(H)(Cl)—Cl + 3 H—Cl)

Plan We assume that, in the reaction, all the reactant bonds break and all the product bonds form. We find the bond energies in Table 9.2 (p. 341) and substitute the two sums, with correct signs, into Equation 9.2.

Solution Finding the standard enthalpy changes for bonds broken and for bonds formed: For bonds broken, the bond energy values are

$$4 \times \text{C—H} = (4\ \text{mol})(413\ \text{kJ/mol}) = 1652\ \text{kJ}$$
$$3 \times \text{Cl—Cl} = (3\ \text{mol})(243\ \text{kJ/mol}) = 729\ \text{kJ}$$
$$\Sigma\Delta H^0_{\text{bonds broken}} = 2381\ \text{kJ}$$

For bonds formed, the values are

$$3 \times \text{C—Cl} = (3\ \text{mol})(-339\ \text{kJ/mol}) = -1017\ \text{kJ}$$
$$1 \times \text{C—H} = (1\ \text{mol})(-413\ \text{kJ/mol}) = -413\ \text{kJ}$$
$$3 \times \text{H—Cl} = (3\ \text{mol})(-427\ \text{kJ/mol}) = -1281\ \text{kJ}$$
$$\Sigma\Delta H^0_{\text{bonds formed}} = -2711\ \text{kJ}$$

Calculating ΔH^0_{rxn}:

$$\Delta H^0_{rxn} = \Sigma\Delta H^0_{\text{bonds broken}} + \Sigma\Delta H^0_{\text{bonds formed}} = 2381\ \text{kJ} + (-2711\ \text{kJ}) = -330\ \text{kJ}$$

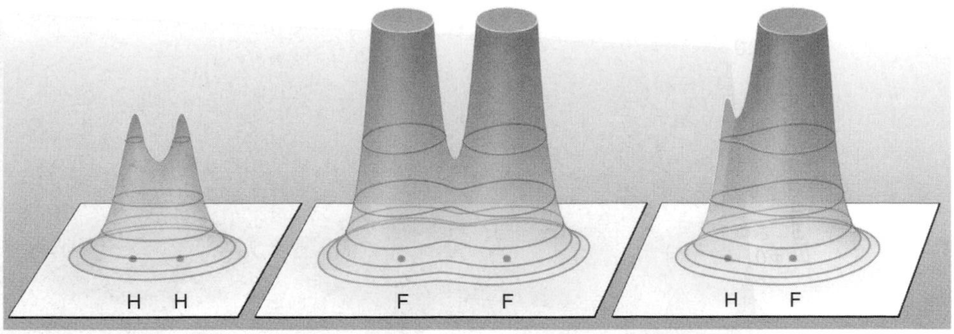

Figure 9.21 Electron density distributions in H_2, F_2, and HF. As these relief maps show, electron density is distributed equally in the nonpolar covalent molecules H_2 and F_2. (The electron density around the F nuclei is so great that the peaks must be "cut off" to fit within the figure.) But, in polar covalent HF, the electron density is shifted away from H and toward F.

SAMPLE PROBLEM 9.4 Determining Bond Polarity from EN Values

Problem (a) Use a polar arrow to indicate the polarity of each bond: N—H, F—N, I—Cl.
(b) Rank the following bonds in order of increasing polarity: H—N, H—O, H—C.

Plan (a) We use Figure 9.19 to find the EN values of the bonded atoms and point the polar arrow toward the negative end. (b) Each choice has H bonded to an atom from Period 2. Since EN increases across a period, the polarity is greatest for the bond whose Period 2 atom is farthest to the right.

Solution (a) The EN of N = 3.0 and the EN of H = 2.1, so N is more electronegative than H: $\overleftarrow{\text{N—H}}$

The EN of F = 4.0 and the EN of N = 3.0, so F is more electronegative: $\overleftarrow{\text{F—N}}$

The EN of I = 2.5 and the EN of Cl = 3.0, so I is less electronegative: $\overrightarrow{\text{I—Cl}}$

(b) The order of increasing EN is C < N < O, and each has a higher EN than H. Therefore, O pulls most on the electron pair shared with H, and C pulls least; so the order of bond polarity is H—C < H—N < H—O.

Comment In Chapter 10, you'll see that the polarity of the bonds in a molecule contributes to the overall polarity of the molecule, which is a major factor determining the magnitudes of several physical properties.

FOLLOW-UP PROBLEM 9.4 Arrange each set of bonds in order of increasing polarity, and indicate bond polarity with δ+ and δ− symbols:
(a) Cl—F, Br—Cl, Cl—Cl (b) Si—Cl, P—Cl, S—Cl, Si—Si

ΔEN	IONIC CHARACTER
>1.7	Mostly ionic
0.4-1.7	Polar covalent
<0.4	Mostly covalent
0	Nonpolar covalent

A

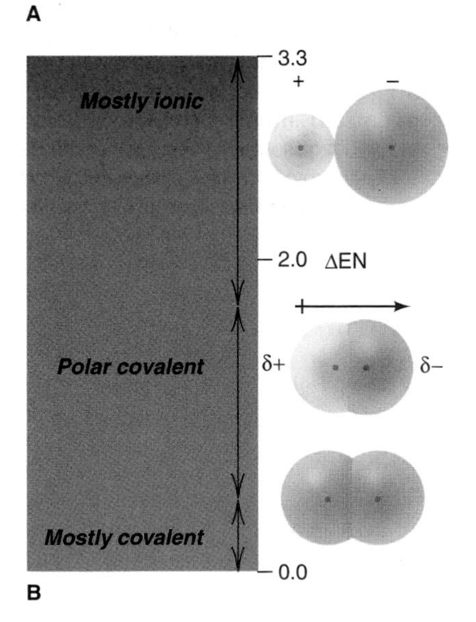

Figure 9.22 Boundary ranges for classifying ionic character of chemical bonds. **A,** The electronegativity difference (ΔEN) between bonded atoms shows cutoff values that act as a general guide to a bond's relative ionic character. **B,** The gradation in ionic character across the entire bonding range is shown as shading from ionic *(green)* to covalent *(yellow)*.

The Partial Ionic Character of Polar Covalent Bonds

If you ask "Is an X—Y bond ionic or covalent?" the answer in almost every case is "Both, partially!" A better question is "*To what extent* is the bond ionic or covalent?" The existence of partial charges means that a polar covalent bond behaves as if it were partially ionic. The **partial ionic character** of a bond is related directly to the **electronegativity difference (ΔEN),** the difference between the EN values of the bonded atoms: *a greater ΔEN results in larger partial charges and a higher partial ionic character.* Consider these three fluorine-containing molecules: ΔEN for LiF(*g*) is 4.0 − 1.0 = 3.0; for HF(*g*), it is 4.0 − 2.1 = 1.9; and for $F_2(g)$, it is 4.0 − 4.0 = 0. Thus, the bond in LiF has more ionic character than the H—F bond, which has more than the F—F bond.

Various attempts have been made to classify the ionic character of bonds, but they all use arbitrary cutoff values, which is inconsistent with the gradation of ionic character observed experimentally. One approach uses ΔEN values to divide bonds into ionic, polar covalent, and nonpolar covalent. Based on a range of ΔEN values from 0 (completely nonpolar) to 3.3 (highly ionic), some approximate guidelines appear in Figure 9.22.

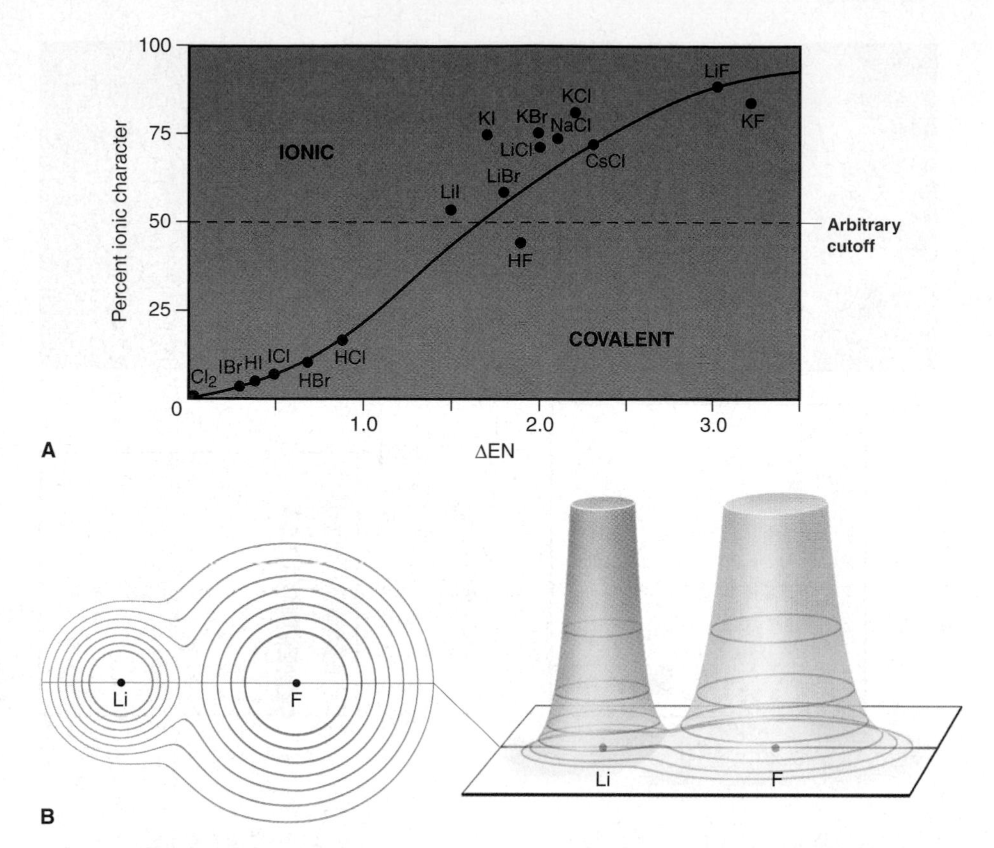

Figure 9.23 Percent ionic character as a function of electronegativity difference (ΔEN). **A,** The percent ionic character is plotted against ΔEN for some gaseous diatomic molecules. Note that, in general, ΔEN correlates with ionic character. (The arbitrary cutoff for an ionic compound is >50% ionic character.) **B,** Even in highly ionic LiF, significant overlap occurs between the ions, indicating some covalent character. The contour map *(left)* and relief map *(right)* depict this overlap.

Another approach calculates the *percent ionic character* of a bond by comparing the actual behavior of a polar molecule in an electric field with the behavior it would show if the electron were transferred completely (a pure ionic bond). A value of 50% ionic character divides substances we call "ionic" from those we call "covalent." Such methods show 43% ionic character for the H—F bond and expected decreases for the other hydrogen halides: H—Cl is 19% ionic, H—Br 11%, and H—I 4%. A plot of percent ionic character vs. ΔEN for a variety of gaseous diatomic molecules is shown in Figure 9.23A. The specific values are not important, but note that *percent ionic character generally increases with ΔEN.* Another point to note is that whereas some molecules, such as $Cl_2(g)$, have 0% ionic character, none has 100% ionic character. Thus, *electron sharing occurs to some extent in every bond,* even one between an alkali metal and a halogen, as electron density contour and relief maps show in Figure 9.23B.

The Continuum of Bonding Across a Period

A metal and a nonmetal—elements from opposite sides of the periodic table—have a relatively large ΔEN and typically interact by electron transfer to form an ionic compound. Two nonmetals—elements from the same side of the table—have a small ΔEN and interact by electron sharing to form a covalent compound. When we combine the nonmetal chlorine with each of the other elements in Period 3, starting with sodium, we should observe a steady decrease in ΔEN and a gradation in bond type from ionic through polar covalent to nonpolar covalent.

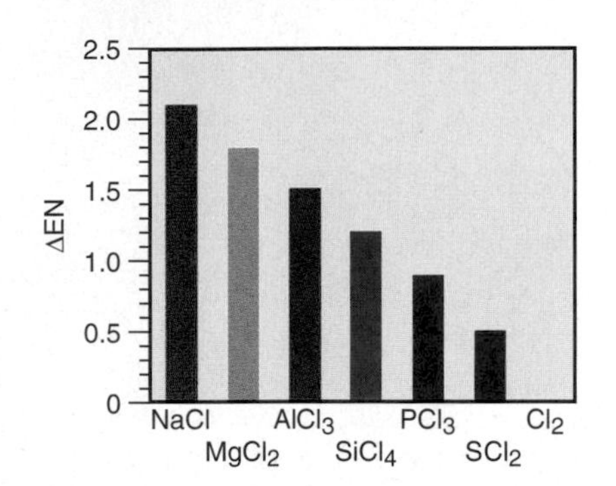

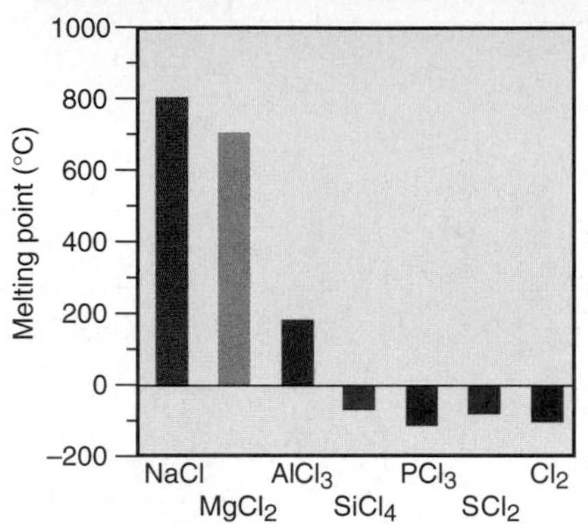

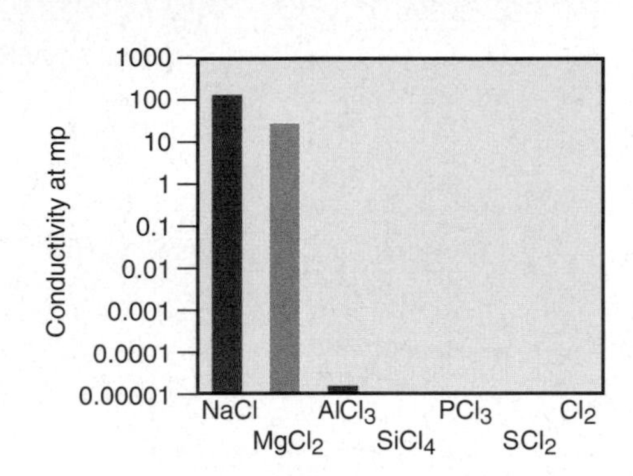

Figure 9.24 Properties of the Period 3 chlorides. Samples of the compounds formed from each of the Period 3 elements with chlorine are shown in periodic table sequence in the photo. Note the trend in properties displayed in the bar graphs: as ΔEN decreases, both melting point and electrical conductivity (at the melting point) decrease. These trends are consistent with a change in bond type from ionic through polar covalent to nonpolar covalent.

Figure 9.24 shows samples of the common Period 3 chlorides—NaCl, $MgCl_2$, $AlCl_3$, $SiCl_4$, PCl_3, and SCl_2, as well as Cl_2—and some key macroscopic properties, while Figure 9.25 shows an electron density relief map of a bond in each of these substances. Note the steady increase in covalent character, as shown by the height of the electron density between the peaks—the bonding region—as we move from ionic NaCl to nonpolar covalent Cl_2.

The first compound in the series is sodium chloride, a white (colorless) crystalline solid with typical ionic properties—high melting point and high electrical conductivity when molten. Having, in addition, a ΔEN of 2.1, NaCl is ionic by any criterion. However, note that, as with LiF (Figure 9.23B), a small but significant covalent region appears in the relief map for NaCl. Magnesium chloride is still considered ionic, with a ΔEN of 1.8, but it has a lower melting point and lower conductivity, as well as a slightly higher bonding region in the relief map.

Aluminum chloride is still less ionic. Rather than having a three-dimensional lattice of Al^{3+} and Cl^- ions, it consists of extended layers of highly polar Al—Cl covalent bonds, as shown by its ΔEN value of 1.5. Weak forces between layers of bonded atoms result in a much lower melting point, and the low electrical conductivity of molten $AlCl_3$ is consistent with a scarcity of ions. And, once again, we see a higher electron density between the nuclei.

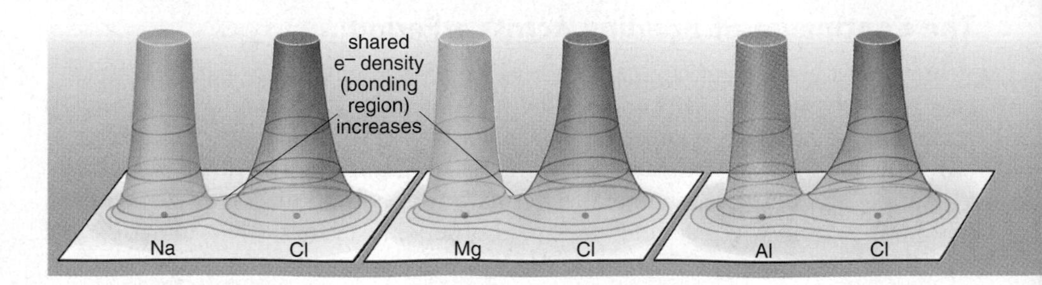

Figure 9.25 Electron density distributions in bonds of the Period 3 chlorides. The relief maps show the electron density for one bond of each chloride. Note the small, but steady, increase in the height of the bonding region between the nuclei, which indicates greater covalent character.

Silicon tetrachloride's melting point is very low because of weak forces *between* its separate molecules, and it has no measurable electrical conductivity. Each molecule has strong Si—Cl bonds with a ΔEN value of 1.2, near the middle of the polar covalent region. With a ΔEN value of 0.9, phosphorus trichloride continues the trend toward lower bond polarity, as evidenced by its very low melting point. The bonds in sulfur dichloride are even less polar (ΔEN = 0.5). Finally, nonpolar, diatomic chlorine is the only one of these substances that is a gas at room temperature. The relief maps show the expected increasing height of the electron density in the bonding region.

Thus, *as ΔEN becomes smaller, the bond becomes more covalent,* with greater electron density between the nuclei, and the macroscopic properties of the Period 3 chlorides and Cl_2 change from those of a solid consisting of ions to those of a gas consisting of individual molecules.

SECTION SUMMARY

An atom's electronegativity refers to its ability to pull bonded electrons toward it, which generates partial charges at the ends of the bond. Electronegativity increases across a period and decreases down a group, the reverse of the trends in atomic size. The greater the ΔEN for the two atoms in a bond, the more polar the bond is and the greater its ionic character. For chlorides of Period 3 elements, there is a gradation of bond type from ionic to polar covalent to nonpolar covalent.

9.6 AN INTRODUCTION TO METALLIC BONDING

The third type of chemical bonding we consider is metallic bonding, which occurs among large numbers of metal atoms. In this section, you'll see a simple, qualitative model; a more detailed one is presented in Chapter 12.

The Electron-Sea Model

In reactions with nonmetals, reactive metals (such as Na) transfer their outer electrons and form ionic solids (such as NaCl). Two metal atoms can also share their valence electrons in a covalent bond and form gaseous, diatomic molecules (such as Na_2). But what holds the atoms together in a piece of sodium metal? The **electron-sea model** of metallic bonding proposes that all the metal atoms in the sample contribute their valence electrons to form an electron "sea" that is delocalized throughout the piece. The metal ions (the nuclei with their core electrons) are submerged within this electron sea in an orderly array (see Figure 9.2C).

In contrast to ionic bonding, the metal ions are not held in place as rigidly as in an ionic solid. In contrast to covalent bonding, no particular pair of metal atoms is bonded through a localized pair of electrons. Rather, *the valence electrons are shared among all the atoms in the sample.* The piece of metal is held together by the mutual attraction of the metal cations for the mobile, highly delocalized valence electrons.

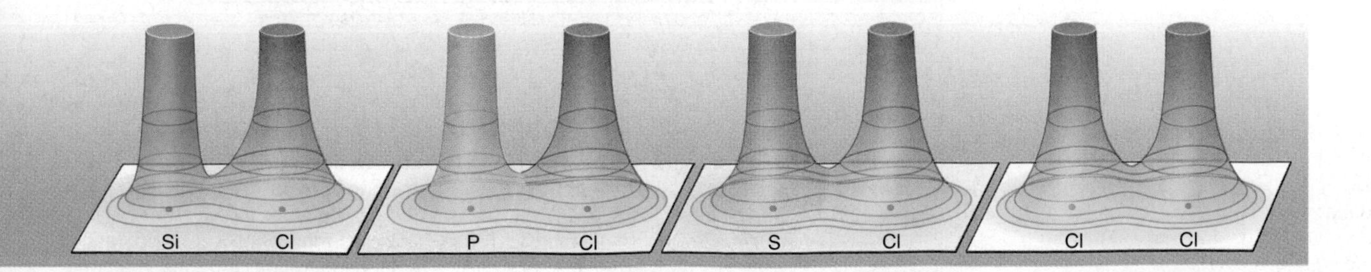

Table 9.7 Melting and Boiling Points of Some Metals

Element	mp (°C)	bp (°C)
Lithium (Li)	180	1347
Tin (Sn)	232	2623
Aluminum (Al)	660	2467
Barium (Ba)	727	1850
Silver (Ag)	961	2155
Copper (Cu)	1083	2570
Uranium (U)	1130	3930

Although there are metallic compounds, two or more metals typically form **alloys,** solid mixtures with variable composition. Many familiar metallic materials are alloys, such as those used for car parts, airplane bodies, building and bridge supports, coins, jewelry, and dental work.

How the Model Explains the Properties of Metals

Although the physical properties of metals vary over a wide range, most are solids with moderate to high melting points and much higher boiling points (Table 9.7). Metals typically bend or dent rather than crack or shatter. Many can be flattened into sheets (malleable) and pulled into wires (ductile). Unlike typical ionic and covalent substances, metals conduct heat and electricity well in *both* the solid and liquid states.

Two features of the electron-sea model that account for these properties are the *regularity,* but not rigidity, of the metal-ion array and the *mobility* of the valence electrons. The melting and boiling points of metals are related to the energy of the metallic bonding. Melting points are only moderately high because the attractions between moveable cations and electrons need not be broken during melting. Boiling a metal requires each cation and its electron(s) to break away from the others, so the boiling points are quite high. Gallium provides a striking example: it melts in your hand (mp 29.8°C) but doesn't boil until the temperature reaches over 2400°C.

Periodic trends are also consistent with the electron-sea model. As Figure 9.26 shows, the alkaline earth metals [Group 2A(2)] have higher melting points than the alkali metals [Group 1A(1)]. The 2A metal atoms have two valence electrons and form 2+ cations. Greater attraction between these cations and twice as many valence electrons means stronger metallic bonding than for the 1A metal atoms, so higher temperatures are needed to melt the 2A solids.

Mechanical and conducting properties are also explained by the model. When a piece of metal is deformed by a hammer, the metal ions slide past each other through the electron sea and end up in new positions. Thus, the metal-ion cores do not repel each other (Figure 9.27). Compare this behavior with the repulsions that occur when an ionic solid is struck (see Figure 9.8, p. 337).

The Amazing Malleability of Gold All the Group 1B(11) metals—copper, silver, and gold—are soft enough to be machined easily, but gold is in a class by itself. One gram of gold forms a cube 0.37 cm on a side or a sphere the size of a small ball-bearing. It is so ductile that it can be drawn into a wire 20 μm thick and 165 m long, and so malleable that it can be hammered into a 1.0-m^2 sheet that is only 230 atoms (about 70 nm) thick!

Metals are good conductors of electricity because they have mobile electrons. When a piece of metal is attached to a battery, electrons flow from one terminal into the metal and replace electrons flowing from the metal into the other terminal. Irregularities in the array of metal atoms reduce this conductivity. Ordinary copper wire used to carry an electric current, for example, is more than 99.99% pure because traces of other atoms can drastically restrict the flow of electrons.

Mobile electrons also make metals good conductors of heat. Place your hand on a piece of metal and a piece of wood that are both at room temperature. The metal feels colder because it conducts body heat away from your hand much faster than the wood. The delocalized electrons in the metal disperse the heat from your hand more quickly than the localized electron pairs in the covalent bonds of wood.

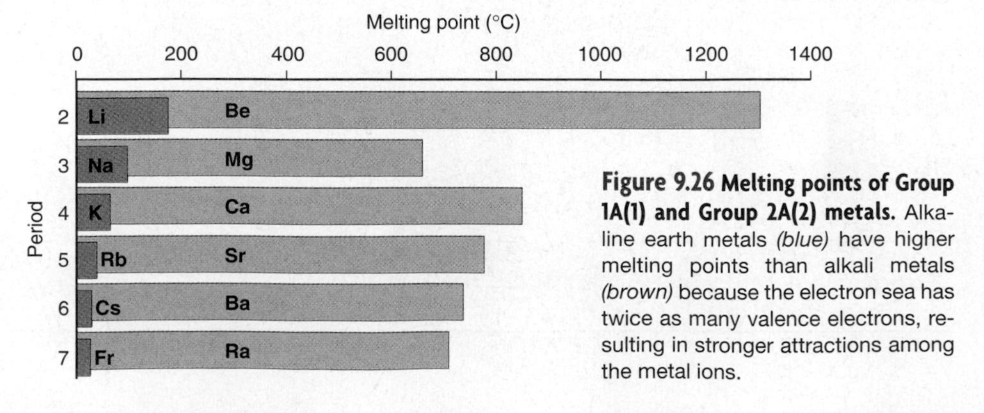

Figure 9.26 Melting points of Group 1A(1) and Group 2A(2) metals. Alkaline earth metals *(blue)* have higher melting points than alkali metals *(brown)* because the electron sea has twice as many valence electrons, resulting in stronger attractions among the metal ions.

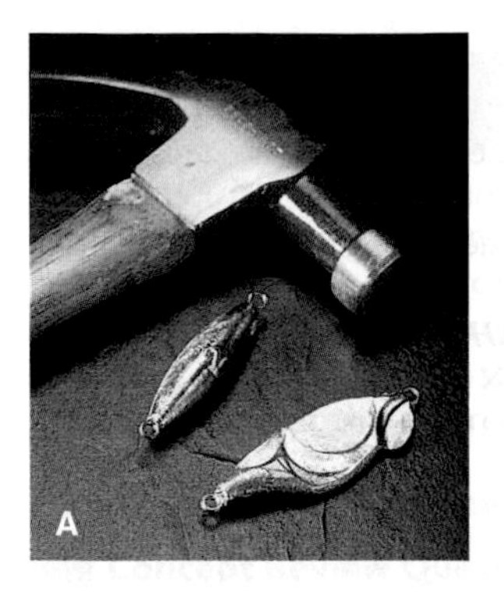

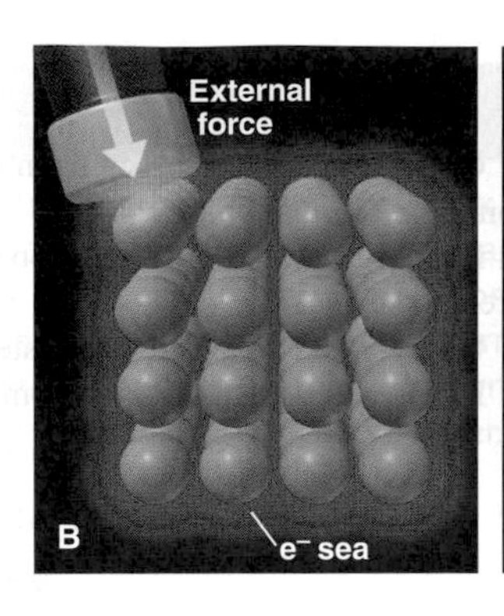

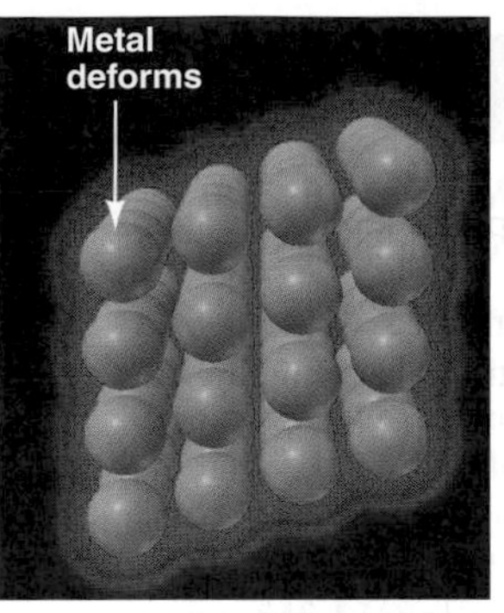

Figure 9.27 The reason metals deform. **A,** An external force applied to a piece of metal deforms but doesn't break it. **B,** The external force merely moves metal ions past each other through the electron sea.

SECTION SUMMARY

According to the electron-sea model, the valence electrons of the metal atoms in a sample are highly delocalized and attract all the metal cations, holding them together. Metals have only moderately high melting points because the metal ions remain attracted to the electron sea even if their relative positions change. Boiling requires completely overcoming these bonding attractions, so metals have very high boiling points. Metals can be deformed because the electron sea prevents repulsions among the cations. Metals conduct electricity and heat because their electrons are mobile.

Chapter Perspective

Our theme in this chapter has been that the type of chemical bonding—ionic, covalent, metallic, or some blend of these—is governed by the properties of the bonding atoms. This fundamental idea reappears as we investigate the forces that give rise to the properties of liquids, solids, and solutions (Chapters 12 and 13), the behavior of the main-group elements (Chapter 14), and the organic chemistry of carbon (Chapter 15). But first, you'll see in Chapter 10 how the relative placement of atoms and the arrangement of bonding and lone pairs gives a molecule its characteristic shape and how that shape influences the compound's properties. Then, in Chapter 11, you'll see how covalent bonding theory explains the nature of the bond itself and the properties of compounds.

For Review and Reference

(Numbers in parentheses refer to pages, unless noted otherwise.)

Learning Objectives

Relevant section and/or sample problem (SP) numbers appear in parentheses.

Understand These Concepts

1. How differences in atomic properties lead to differences in bond type; the basic distinctions among the three types of bonding (9.1)
2. The essential features of ionic bonding: electron transfer to form ions, and their electrostatic attraction to form a solid (9.2)
3. How lattice energy is ultimately responsible for formation of ionic compounds (9.2)
4. How ionic compound formation is conceptualized as occurring in hypothetical steps (Born-Haber cycle) to calculate lattice energy (9.2)
5. How Coulomb's law explains the periodic trends in lattice energy (9.2)
6. Why ionic compounds are brittle and high melting and conduct electricity only when molten or dissolved in water (9.2)
7. How nonmetal atoms form a covalent bond (9.3)
8. How bonding and lone electron pairs fill the outer (valence) level of each atom in a molecule (9.3)
9. The interrelationships among bond order, bond length, and bond energy (9.3)
10. How the distinction between bonding and nonbonding forces explains the properties of covalent molecules and network covalent solids (9.3)
11. How changes in bond strength account for the heat of reaction (9.4)
12. How a reaction can be divided conceptually into bond-breaking and bond-forming steps (9.4)
13. The periodic trends in electronegativity and the inverse relation of EN values to atomic sizes (9.5)
14. How bond polarity arises from differences in electronegativity of bonded atoms; the direction of bond polarity (9.5)
15. The change in partial ionic character with ΔEN and the change in bonding from ionic to polar covalent to nonpolar covalent across a period (9.5)
16. The role of delocalized electrons in metallic bonding (9.6)
17. How the electron-sea model explains why metals bend, have very high boiling points, and conduct electricity in solid or molten form (9.6)

9.78 Gases react explosively with one another if the heat released as some product forms is sufficient to cause more reactant to react, and so on. The rapid heat release expands the gases, and a thermal explosion results. Use bond energies to calculate ΔH^0 of the following reactions, and predict which occurs explosively:
(a) $H_2(g) + Cl_2(g) \longrightarrow 2HCl(g)$
(b) $H_2(g) + I_2(g) \longrightarrow 2HI(g)$
(c) $2H_2(g) + O_2(g) \longrightarrow 2H_2O(g)$

9.79 By using photons of specific wavelengths, chemists can dissociate gaseous HI to produce H atoms with certain speeds. When HI dissociates, the H atoms move away rapidly, whereas the relatively heavy I atoms move more slowly.
(a) What is the longest wavelength (in nm) that can dissociate a molecule of HI?
(b) If a photon of 254 nm is used, what is the excess energy (in J) over that needed for the dissociation?
(c) If all this excess energy is carried away by the H atom as kinetic energy, what is its speed (in m/s)?

9.80 Carbon dioxide is a linear molecule. Its vibrational motions include symmetrical stretching, bending, and asymmetrical stretching (see Figure B9.1, p. 345), and their frequencies are $4.02\times10^{13}\ s^{-1}$, $2.00\times10^{13}\ s^{-1}$, and $7.05\times10^{13}\ s^{-1}$, respectively. (a) In what region of the electromagnetic spectrum are these frequencies? (b) Calculate the energy (in J) of each vibration. Which occurs most readily (takes the least energy)?

9.81 In developing the concept of electronegativity, Pauling used the term *excess bond energy* for the difference between the actual bond energy of X—Y and the average bond energies of X—X and Y—Y (see text discussion for the case of HF). Based on the values in Figure 9.19, p. 352, which of the following substances contains bonds with *no* excess bond energy?
(a) PH_3 (b) CS_2 (c) BrCl (d) BH_3 (e) Se_8

9.82 Use condensed electron configurations to predict the relative hardnesses and melting points of rubidium ($Z = 37$), vanadium ($Z = 23$), and cadmium ($Z = 48$).

9.83 Without stratospheric ozone (O_3), harmful solar radiation would cause gene alterations. Ozone forms when O_2 breaks and each O atom reacts with another O_2 molecule. It is destroyed by reaction with Cl atoms formed when the C—Cl bond in synthetic chemicals breaks. Find the wavelengths of light that can break the C—Cl bond and the bond in O_2.

9.84 "Inert" xenon actually forms many compounds, especially with highly electronegative oxygen and fluorine. The ΔH_f^0 values for xenon difluoride, tetrafluoride, and hexafluoride are −105, −284, and −402 kJ/mol, respectively. Find the average bond energy of the Xe—F bonds in each fluoride.

9.85 The HF bond length is 92 pm, 16% shorter than the sum of the covalent radii of H (37 pm) and F (72 pm). Suggest a reason for this difference. Similar calculations show that the difference becomes smaller down the group from HF to HI. Explain.

9.86 There are two main types of covalent bond breakage. In homolytic breakage (as in Table 9.2, p. 341), each atom in the bond gets one of the shared electrons. In some cases, the electronegativity of adjacent atoms affects the bond energy. In heterolytic breakage, one atom gets both electrons and the other gets none; thus, a cation and an anion form.
(a) Why is the C—C bond in $H_3C—CF_3$ (423 kJ/mol) stronger than the bond in $H_3C—CH_3$ (376 kJ/mol)?
(b) Use bond energy and any other data to calculate the heat of reaction for the heterolytic cleavage of O_2.

9.87 Find the longest wavelengths of light that can cleave the bonds in elemental nitrogen, oxygen, and fluorine.

9.88 The work function (ϕ) of a metal is the minimum energy needed to remove an electron from its surface. (a) Is it easier to remove an electron from a gaseous silver atom or from the surface of solid silver ($\phi = 7.59\times10^{-19}$ J; IE = 731 kJ/mol)? (b) Explain the results in terms of the electron-sea model of metallic bonding.

9.89 Lattice energies can also be calculated for covalent solids using a Born-Haber cycle, and the network solid silicon dioxide has one of the highest. Silicon dioxide is found in pure crystalline form as transparent rock quartz. Much harder than glass, this material was once prized for making lenses for optical devices and expensive spectacles. Use Appendix B and the following data to calculate $\Delta H^0_{lattice}$ of SiO_2:

$Si(s) \longrightarrow Si(g)$ $\quad \Delta H^0 = 454$ kJ
$Si(g) \longrightarrow Si^{4+}(g) + 4e^-$ $\quad \Delta H^0 = 9949$ kJ
$O_2(g) \longrightarrow 2O(g)$ $\quad \Delta H^0 = 498$ kJ
$O(g) + 2e^- \longrightarrow O^{2-}(g)$ $\quad \Delta H^0 = 737$ kJ

9.90 We can write equations for the formation of methane from ethane (C_2H_6) with its C—C bond, from ethene (C_2H_4) with its C=C bond, and from ethyne (C_2H_2) with its C≡C bond:

$C_2H_6(g) + H_2(g) \longrightarrow 2CH_4(g)$ $\quad \Delta H^0_{rxn} = -65.07$ kJ/mol
$C_2H_4(g) + 2H_2(g) \longrightarrow 2CH_4(g)$ $\quad \Delta H^0_{rxn} = -202.21$ kJ/mol
$C_2H_2(g) + 3H_2(g) \longrightarrow 2CH_4(g)$ $\quad \Delta H^0_{rxn} = -376.74$ kJ/mol

Given that the average C—H bond energy in CH_4 is 415 kJ/mol, use Table 9.2 (p. 341) values to calculate the average C—H bond energy in ethane, in ethene, and in ethyne.

9.91 Carbon-carbon bonds form the "backbone" of nearly every organic and biological molecule. The average bond energy of the C—C bond is 347 kJ/mol. Calculate the frequency and wavelength of the least energetic photon that can break an average C—C bond. In what region of the electromagnetic spectrum is this radiation?

9.92 In a future hydrogen-fuel economy, the cheapest source of H_2 will certainly be water. It takes 467 kJ to produce 1 mol of H atoms from water. What is the frequency, wavelength, and minimum energy of a photon that can free an H atom from water?

9.93 The Sun's emissions include infrared, visible, and ultraviolet radiation. Some of this radiation can damage skin tissue, and sunscreens are made to block it. (a) What is the ratio of the energies of yellow light of 575 nm to UV light of 300 nm? (b) What is the ratio of the energies of yellow light of 575 nm to IR radiation of 1000 nm? (c) Can any of these three types of radiation break a C—C bond? (d) A C—H bond?

9.94 Dimethyl ether (CH_3OCH_3) and ethanol (CH_3CH_2OH) are constitutional isomers (see Table 3.4, p. 100). (a) Use Table 9.2, p. 341, to calculate ΔH^0_{rxn} for the formation of each compound as a gas from methane and oxygen; water vapor also forms. (b) State which reaction is more exothermic. (c) Calculate ΔH^0_{rxn} for the conversion of ethanol to dimethyl ether.

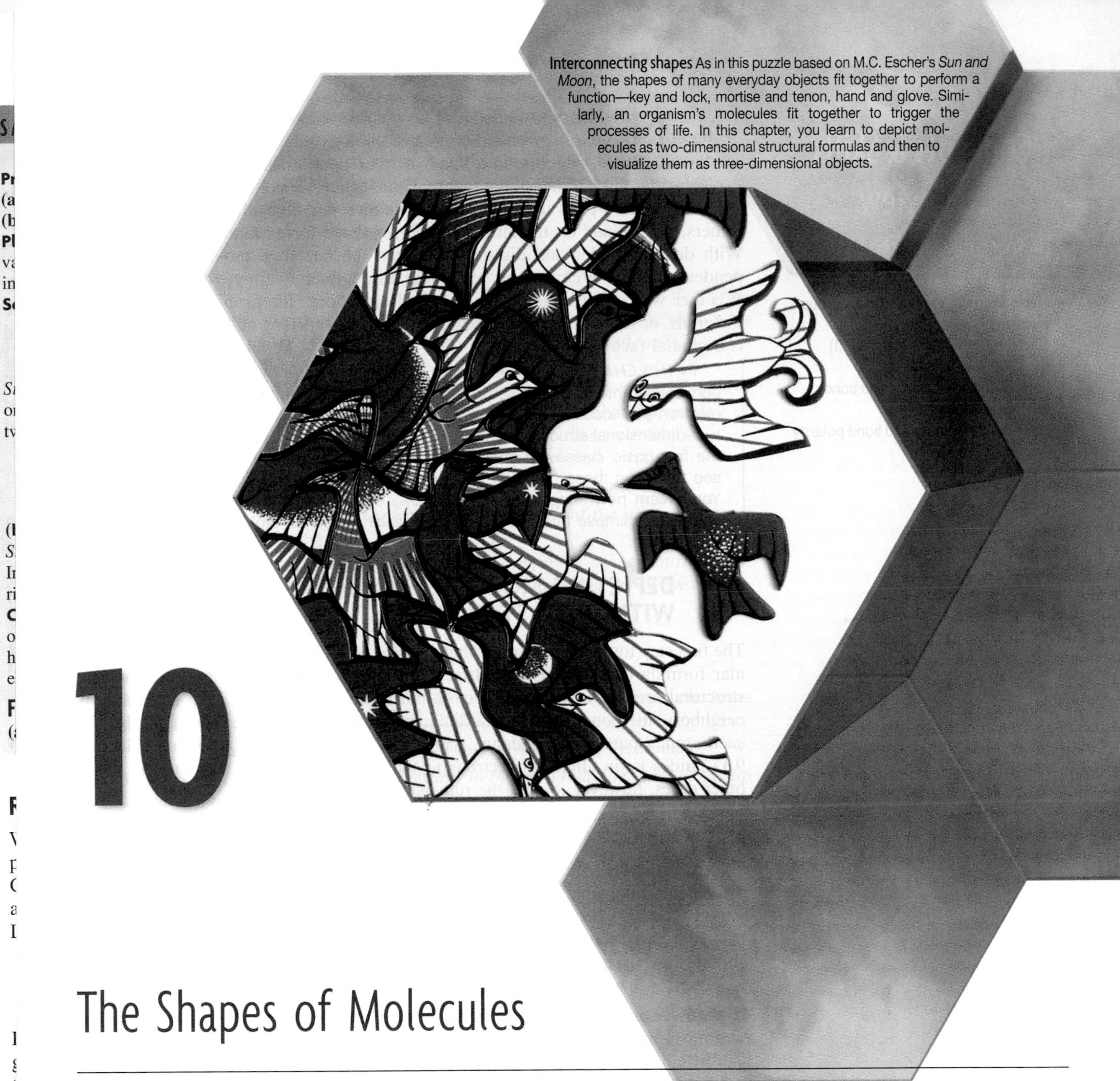

Interconnecting shapes As in this puzzle based on M.C. Escher's *Sun and Moon*, the shapes of many everyday objects fit together to perform a function—key and lock, mortise and tenon, hand and glove. Similarly, an organism's molecules fit together to trigger the processes of life. In this chapter, you learn to depict molecules as two-dimensional structural formulas and then to visualize them as three-dimensional objects.

10 The Shapes of Molecules

Another example is phosphorus pentachloride, PCl_5, a fuming yellow-white solid used in the manufacture of lacquers and films. PCl_5 is formed when phosphorus trichloride, PCl_3, reacts with chlorine gas. The P in PCl_3 has an octet, but it uses the lone pair to form two more bonds to chlorine and expands its valence shell in PCl_5 to a total of 10 electrons. Note that when PCl_5 forms, *one* Cl—Cl bond breaks (left side of the equation), and *two* P—Cl bonds form (right side), for a net increase of one bond:

$$PCl_3 + Cl{-}Cl \longrightarrow PCl_5$$

In SF_6 and PCl_5, the central atom forms bonds to *more than four* atoms. But there are many cases of expanded valence shells in which the central atom bonds to *four or fewer* atoms. Consider sulfuric acid, the industrial chemical produced in the greatest quantity. Two of the resonance forms for H_2SO_4, with formal charges, are

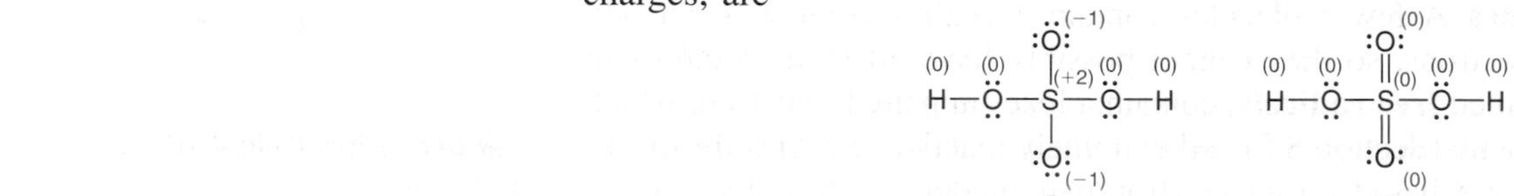

I: H(0)—O(0)—S(+2)(—O(−1))(—O(−1))—O(0)—H(0)

II: H(0)—O(0)—S(0)(=O(0))(=O(0))—O(0)—H(0)

In form II, sulfur has an expanded valence shell of 12 electrons. Based on the formal charge rules, II contributes more than I to the resonance hybrid. More importantly, form II is consistent with observed bond lengths. In gaseous H_2SO_4, the two sulfur-oxygen bonds *with* H atoms attached to O are 157 pm long, whereas the two sulfur-oxygen bonds *without* H atoms attached to O are 142 pm long. This shorter bond indicates double-bond character, which is shown in form II.

When sulfuric acid loses two H^+ ions, it forms the sulfate ion, SO_4^{2-}. All sulfur-oxygen bonds in SO_4^{2-} are 149 pm long, a length that is intermediate between that of the two S═O bonds (~142 pm) and the two S—O bonds (~157 pm) in the parent acid. Two of six resonance forms consistent with these data are

$$\left[\,O{-}S({=}O)_2{-}O\,\right]^{2-} \longleftrightarrow \left[\,O{=}S({-}O)({=}O){-}O\,\right]^{2-}$$

Thus, the SO_4^{2-} ion is a resonance hybrid with four S—O bonds and two more bonding pairs delocalized over the structure; so each sulfur-oxygen bond has a bond order of $1\frac{1}{2}$.

Measurements show that the sulfur-oxygen bonds in SO_2 and SO_3 are all approximately 142 pm long, indicating S═O bonds. Lewis structures consistent with these data have zero formal charges:

$$O{=}S{=}O \qquad\qquad SO_3\ (\text{three } S{=}O)$$

(all atoms (0))

Phosphorus often accommodates 10 electrons in its compounds, sulfur 12, and iodine as many as 14.

It's important to realize that formal charge is a useful, but not perfect, tool for assessing the importance of contributions to a resonance hybrid. You've already seen that it does not predict the most important resonance form of NO_2. In fact, recent theoretical calculations indicate that, for many species with central atoms from Period 3 or higher, forms with expanded valence shells and zero formal charges may be less important than forms with higher formal charges. But we will continue to apply the formal charge rules because it is usually the simplest approach consistent with experimental data.

SAMPLE PROBLEM 10.5 Writing Lewis Structures for Octet Rule Exceptions

Problem Write Lewis structures for (a) H_3PO_4 (pick the most likely structure); (b) $BFCl_2$.

Plan We write each Lewis structure and examine it for exceptions to the octet rule. In (a), the central atom is P, which is in Period 3, so it can use *d* orbitals to have more than an octet. Therefore, we can write more than one Lewis structure. We use formal charges to decide if one resonance form is more important. In (b), the central atom is B, which can have fewer than an octet of electrons.

Solution (a) For H_3PO_4, two possible Lewis structures, with formal charges, are

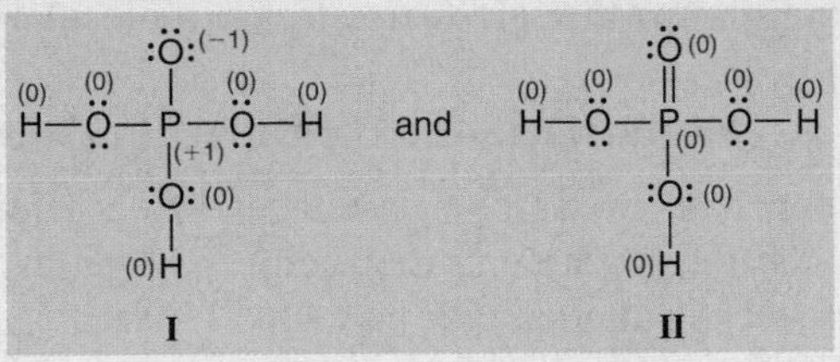

I and II

Structure II has lower formal charges, so it is the more important resonance form.

(b) For $BFCl_2$, the Lewis structure leaves B with only six electrons surrounding it:

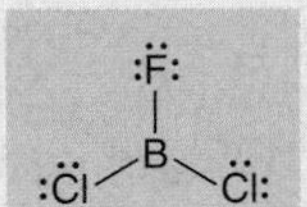

Comment In (a), structure II is also consistent with bond length measurements, which show one shorter (152 pm) phosphorus-oxygen bond and three longer (157 pm) ones.

FOLLOW-UP PROBLEM 10.5 Write the most likely Lewis structure for **(a)** $POCl_3$; **(b)** ClO_2; **(c)** XeF_4.

SECTION SUMMARY

A stepwise process is used to convert a molecular formula into a Lewis structure, a *two-dimensional* representation of a molecule (or ion) that shows the relative placement of atoms and distribution of valence electrons among bonding and lone pairs. When two or more Lewis structures can be drawn for the same relative placement of atoms, the actual structure is a hybrid of those resonance forms. Formal charges are often useful for determining the most important contributor to the hybrid. Electron-deficient molecules (central Be or B) and odd-electron species (free radicals) have less than an octet around the central atom but often attain an octet in reactions. In a molecule (or ion) with a central atom from Period 3 or higher, the atom can hold more than eight electrons by using *d* orbitals to expand its valence shell.

10.2 VALENCE-SHELL ELECTRON-PAIR REPULSION (VSEPR) THEORY AND MOLECULAR SHAPE

Virtually every biochemical process hinges to a great extent on the shapes of interacting molecules. Every medicine you take, odor you smell, or flavor you taste depends on part or all of one molecule fitting physically together with another. This universal importance of molecular shape in the functioning of each organism carries over to the ecosystem. Biologists have learned of complex interactions regulating behaviors, such as mating, defense, navigation, and feeding, that depend on one molecule's shape matching up with that of another. In this section, we discuss a model for understanding and predicting molecular shape.

The Lewis structure of a molecule is something like the blueprint of a building: a flat drawing showing the relative placement of parts (atom cores), structural connections (groups of bonding valence electrons), and various attachments (nonbonding lone pairs of valence electrons). To construct the molecular shape

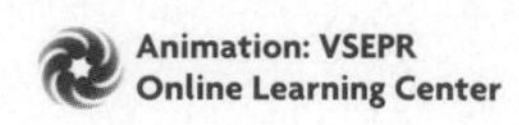

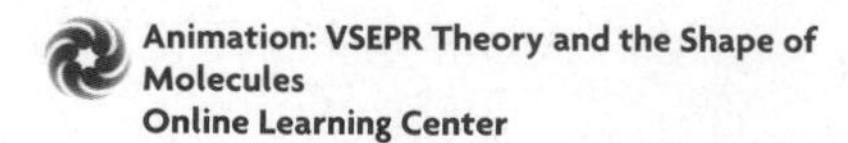

from the Lewis structure, chemists employ **valence-shell electron-pair repulsion (VSEPR) theory.** Its basic principle is that *each group of valence electrons around a central atom is located as far away as possible from the others in order to minimize repulsions.* We define a "group" of electrons as any number of electrons that occupy a localized region around an atom. Thus, an electron group may consist of a single bond, a double bond, a triple bond, a lone pair, or even a lone electron.* Each of these groups of valence electrons repels the other groups to maximize the angles between them. It is the three-dimensional arrangement of nuclei joined by these groups that gives rise to the molecular shape.

Electron-Group Arrangements and Molecular Shapes

When two, three, four, five, or six objects attached to a central point maximize the space that each can occupy around that point, five geometric patterns result. Figure 10.2A depicts these patterns with balloons. If the objects are the valence-electron groups of a central atom, their repulsions maximize the space each occupies and give rise to the five *electron-group arrangements* of minimum energy seen in the great majority of molecules and polyatomic ions.

The electron-group arrangement is defined by the valence-electron groups, both bonding and nonbonding, around the central atom. On the other hand, the **molecular shape** is defined by the relative positions of the atomic nuclei. Figure 10.2B shows the molecular shapes that occur when *all* the surrounding electron groups are *bonding* groups. When some are *nonbonding* groups, different molecular shapes occur. Thus, *the same electron-group arrangement can give rise to different molecular shapes:* some with all bonding groups (as in Figure 10.2B) and others with bonding and nonbonding groups. To classify molecular shapes, we assign each a specific AX_mE_n designation, where m and n are integers, A is the central atom, X is a surrounding atom, and E is a nonbonding valence-electron group (usually a lone pair).

The **bond angle** is the angle formed by the nuclei of two surrounding atoms with the nucleus of the central atom at the vertex. The angles shown for the shapes

*The two electron pairs in a double bond (or the three pairs in a triple bond) occupy separate orbitals, so they remain near each other and act as one electron group (see Chapter 11).

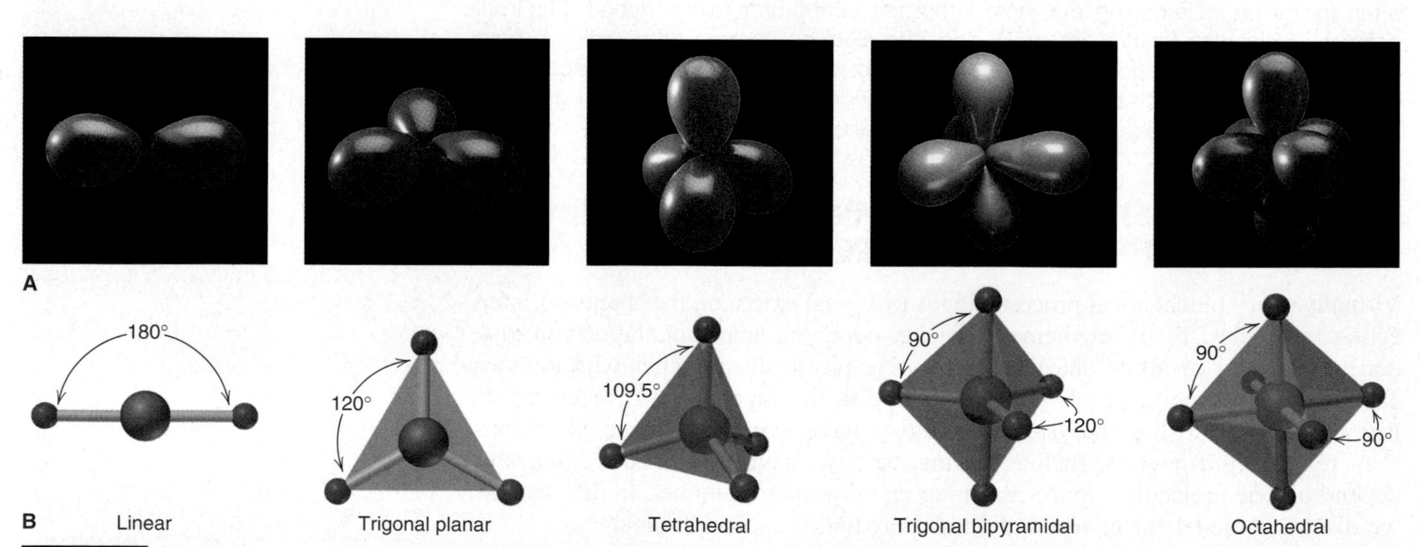

Figure 10.2 Electron-group repulsions and the five basic molecular shapes. A, As an analogy for electron-group arrangements, two to six attached balloons form five geometric orientations such that each balloon occupies as much space as possible. **B,** Mutually repelling electron groups attached to a central atom (*red*) occupy as much space as possible. If each is a bonding group to a surrounding atom (*dark gray*), these molecular shapes and bond angles are observed. The shape has the same name as the electron-group arrangement.

in Figure 10.2B are *ideal* bond angles, those predicted by simple geometry alone. These are observed when all the bonding electron groups around a central atom are identical and are connected to atoms of the same element. When this is not the case, the bond angles deviate from the ideal angles, as you'll see shortly.

It's important to realize that we use the VSEPR model to account for the molecular shapes observed by means of various laboratory instruments. In almost every case, VSEPR predictions are in accord with actual observations. (We discuss some of these observational methods in Chapter 12.)

The Molecular Shape with Two Electron Groups (Linear Arrangement)

When two electron groups attached to a central atom are oriented as far apart as possible, they point in opposite directions. The **linear arrangement** of electron groups results in a molecule with a **linear shape** and a bond angle of 180°. Figure 10.3 shows the general form (top) and shape (middle) with VSEPR shape class (AX_2), and the formulas of some linear molecules.

Gaseous beryllium chloride ($BeCl_2$) is a linear molecule (AX_2). Gaseous Be compounds are electron deficient, with only two electron pairs around the central Be atom:

In carbon dioxide, the central C atom forms two double bonds with the O atoms:

Each double bond acts as a separate electron group and is oriented 180° away from the other, so CO_2 is linear. Notice that the lone pairs on the O atoms of CO_2 or on the Cl atoms of $BeCl_2$ are not involved in the molecular shape: only electron groups around the *central* atom influence shape.

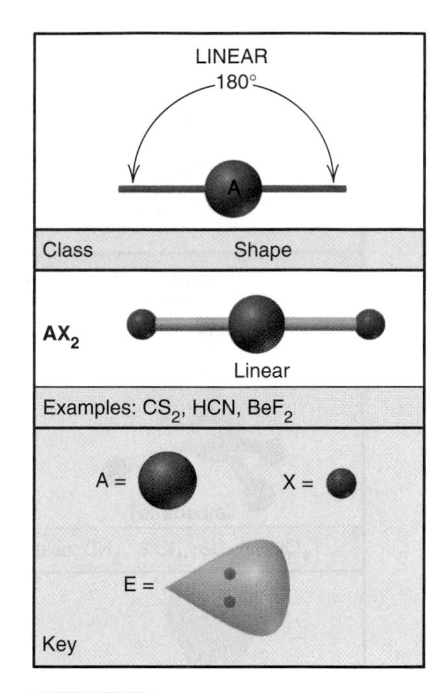

Figure 10.3 The single molecular shape of the linear electron-group arrangement. The key (*bottom*) for A, X, and E also refers to Figures 10.4, 10.5, 10.7, and 10.8.

Molecular Shapes with Three Electron Groups (Trigonal Planar Arrangement)

Three electron groups around the central atom repel each other to the corners of an equilateral triangle, which gives the **trigonal planar arrangement,** shown in Figure 10.4, and an ideal bond angle of 120°. This arrangement has two possible molecular shapes, one with three surrounding atoms and the other with two atoms and one lone pair. It provides our first opportunity to see the effects of double bonds and lone pairs on bond angles.

When the three electron groups are bonding groups, the molecular shape is *trigonal planar* (AX_3). Boron trifluoride (BF_3), another electron-deficient molecule, is an example. It has six electrons around the central B atom in three single bonds to F atoms. The nuclei lie in a plane, and each F—B—F angle is 120°:

The nitrate ion (NO_3^-) is one of several polyatomic ions with the trigonal planar shape. One of three resonance forms of the nitrate ion (Sample Problem 10.4) is

The resonance hybrid has three identical bonds of bond order $1\frac{1}{3}$, so the ideal bond angle is observed.

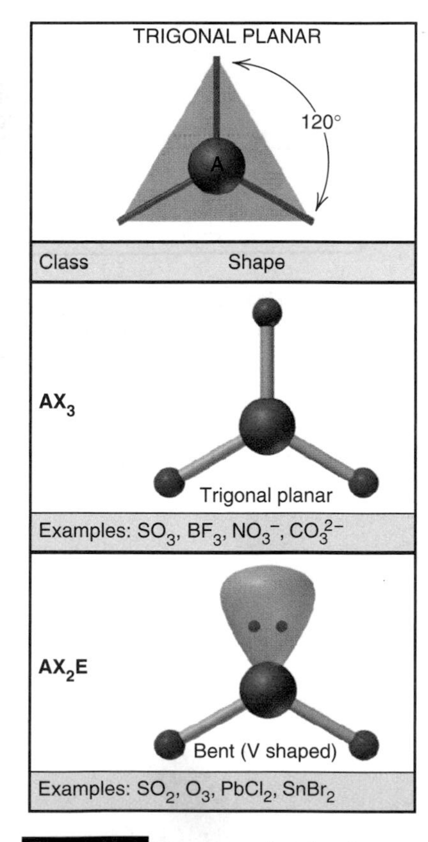

Figure 10.4 The two molecular shapes of the trigonal planar electron-group arrangement.

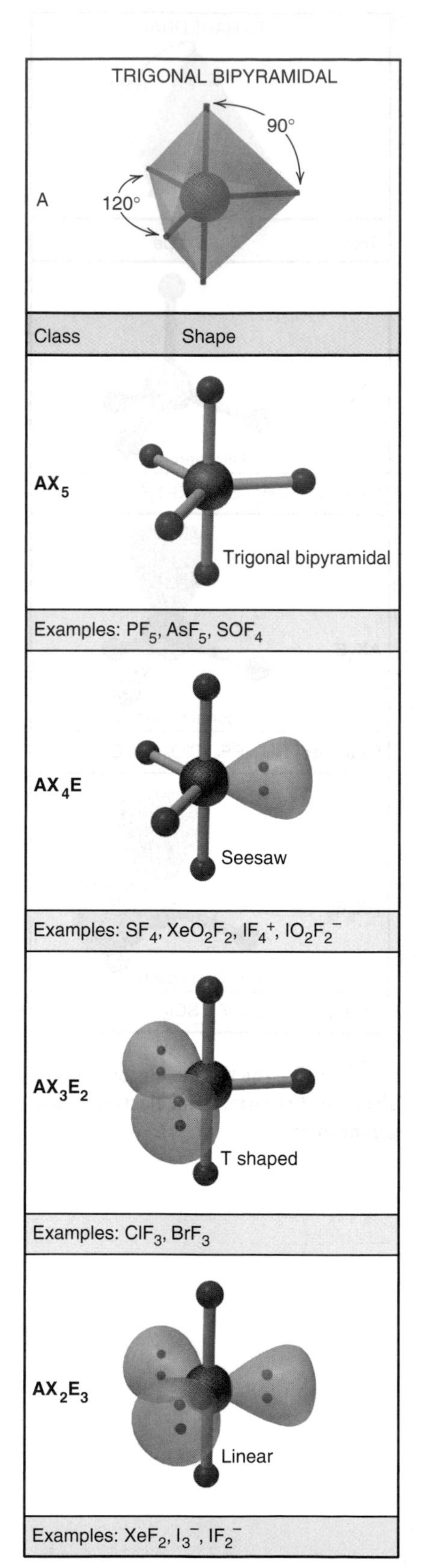

Figure 10.7 The four molecular shapes of the trigonal bipyramidal electron-group arrangement.

on the bond angle than the repulsions from the single lone pair in NH_3. Indeed, the H—O—H bond angle is 104.5°, even less than the H—N—H angle in NH_3:

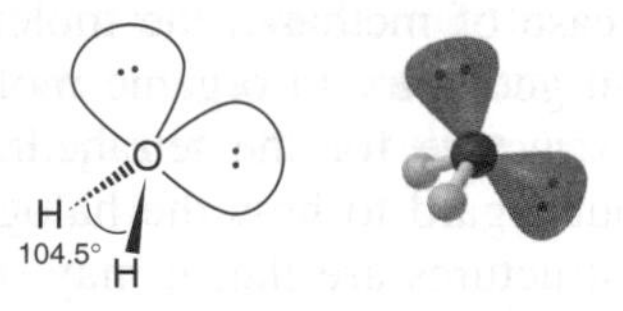

Thus, for similar molecules within a given electron-group arrangement, electron-pair repulsions cause deviations from ideal bond angles in the following order:

Lone pair–lone pair > lone pair–bonding pair > bonding pair–bonding pair **(10.2)**

Molecular Shapes with Five Electron Groups (Trigonal Bipyramidal Arrangement)

All molecules with five or six electron groups have a central atom from Period 3 or higher because only these atoms have the *d* orbitals available to expand the valence shell beyond eight electrons.

When five electron groups maximize their separation, they form the **trigonal bipyramidal arrangement.** In a trigonal bipyramid, two trigonal pyramids share a common base, as shown in Figure 10.7. Note that, in a molecule with this arrangement, *there are two types of positions for surrounding electron groups and two ideal bond angles.* Three **equatorial groups** lie in a trigonal plane that includes the central atom, and two **axial groups** lie above and below this plane. Therefore, a 120° bond angle separates equatorial groups, and a 90° angle separates axial from equatorial groups. In general, the greater the bond angle, the weaker the repulsions, so *equatorial-equatorial (120°) repulsions are weaker than axial-equatorial (90°) repulsions.* The tendency of the electron groups to occupy *equatorial* positions, and thus minimize the stronger axial-equatorial repulsions, governs the four shapes of the trigonal bipyramidal arrangement.

With all five positions occupied by bonded atoms, the molecule has the *trigonal bipyramidal* shape (AX_5), as in phosphorus pentachloride (PCl_5):

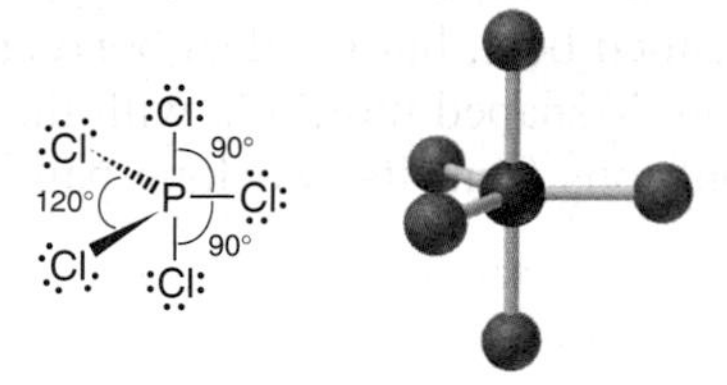

Three other shapes arise for molecules with lone pairs. Since lone pairs exert stronger repulsions than bonding pairs, we find that *lone pairs occupy equatorial positions.* With one lone pair present at an equatorial position, the molecule has a **seesaw shape** (AX_4E). Sulfur tetrafluoride (SF_4), a powerful fluorinating agent, has this shape, shown here and in Figure 10.7 with the "seesaw" tipped up on an end. Note how the equatorial lone pair repels all four bonding pairs to reduce the bond angles:

The tendency of lone pairs to occupy equatorial positions causes molecules with three bonding groups and two lone pairs to have a **T shape** (AX_3E_2). Bromine trifluoride (BrF_3), one of many compounds with fluorine bonded to a larger halogen, has this shape. Note the predicted decrease from the ideal 90° F—Br—F bond angle:

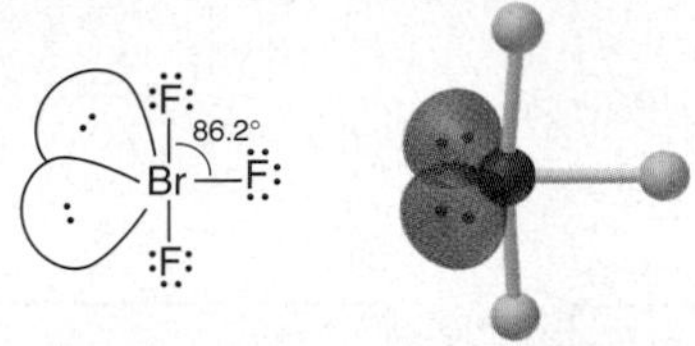

Molecules with three lone pairs in equatorial positions must have the two bonding groups in axial positions, which gives the molecule a *linear* shape (AX_2E_3) and a 180° axial-to-central-to-axial (X—A—X) bond angle. For example, the triiodide ion (I_3^-), which forms when I_2 dissolves in aqueous I^- solution, is linear:

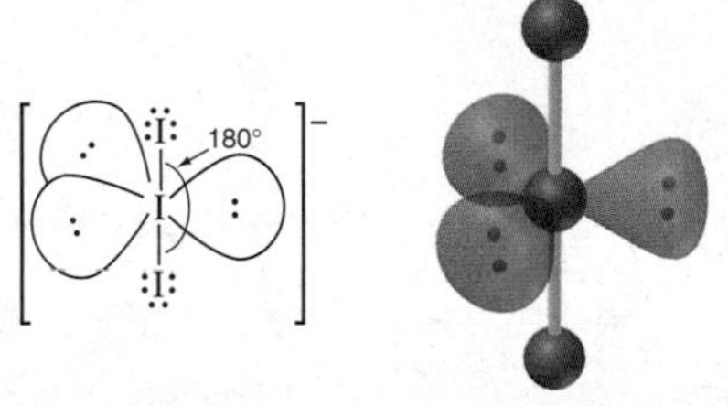

Molecular Shapes with Six Electron Groups (Octahedral Arrangement)

The last of the five major electron-group arrangements is the **octahedral arrangement.** An octahedron is a polyhedron with eight faces made of identical equilateral triangles and six identical vertices, as shown in Figure 10.8. In a molecule (or ion) with this arrangement, six electron groups surround the central atom and each points to one of the six vertices, which gives all the groups a 90° ideal bond angle. Three important molecular shapes occur with this arrangement.

With six bonding groups, the molecular shape is *octahedral* (AX_6). When the seesaw-shaped SF_4 reacts with additional F_2, the central S atom expands its valence shell further to form octahedral sulfur hexafluoride (SF_6):

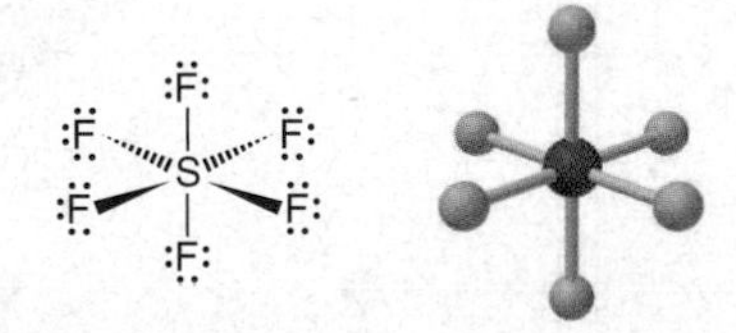

Because all six electron groups have the same ideal bond angle, it makes no difference which position one lone pair occupies. Five bonded atoms and one lone pair define the **square pyramidal shape** (AX_5E), as in iodine pentafluoride (IF_5):

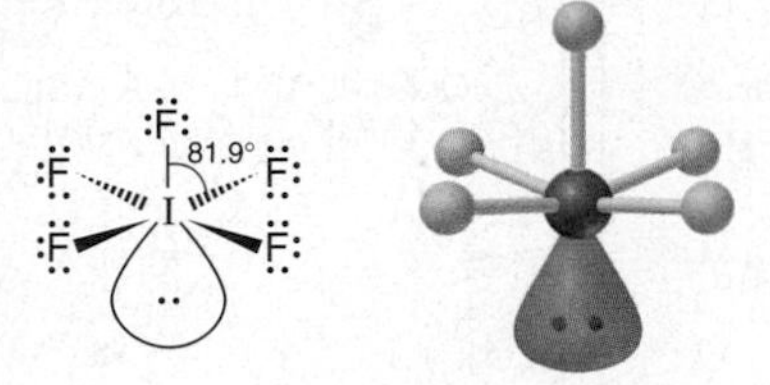

When a molecule has two lone pairs, however, they always lie at *opposite vertices* to avoid the stronger 90° lone pair–lone pair repulsions. This positioning gives the **square planar shape** (AX_4E_2), as in xenon tetrafluoride (XeF_4):

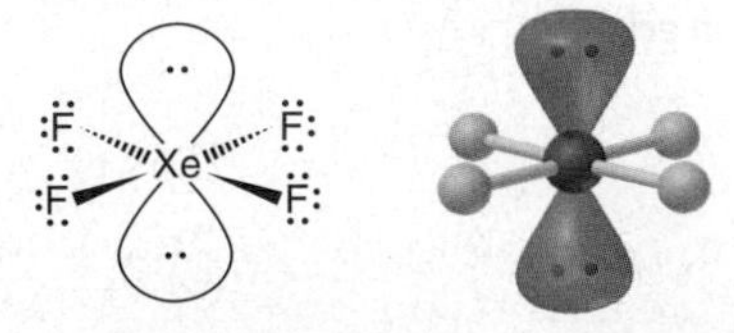

Figure 10.9 (next page) is a summary of the molecular shapes we've discussed.

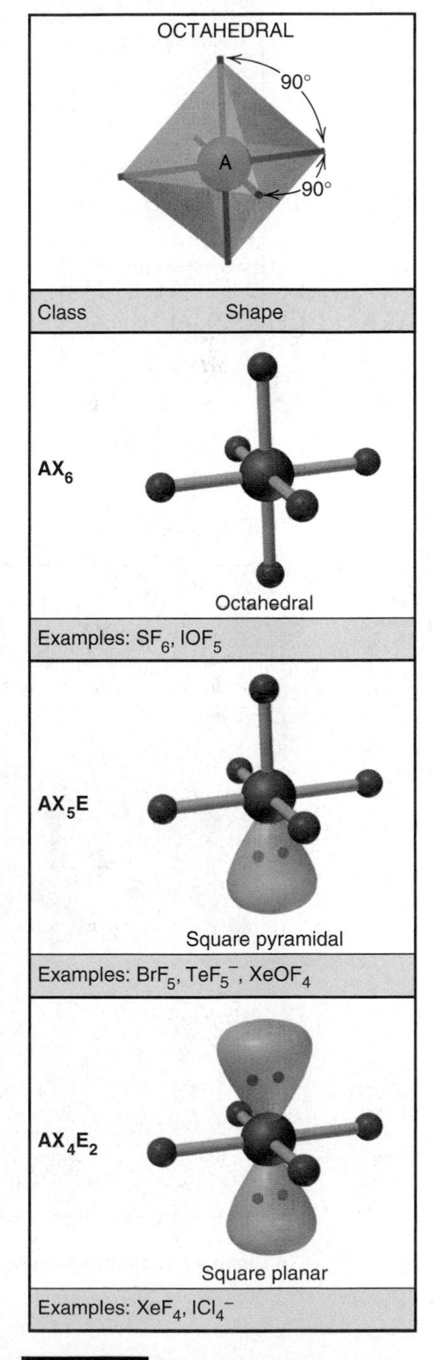

Figure 10.8 The three molecular shapes of the octahedral electron-group arrangement.

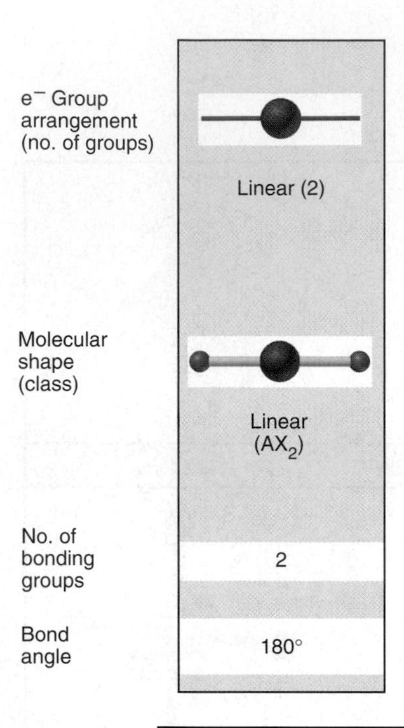

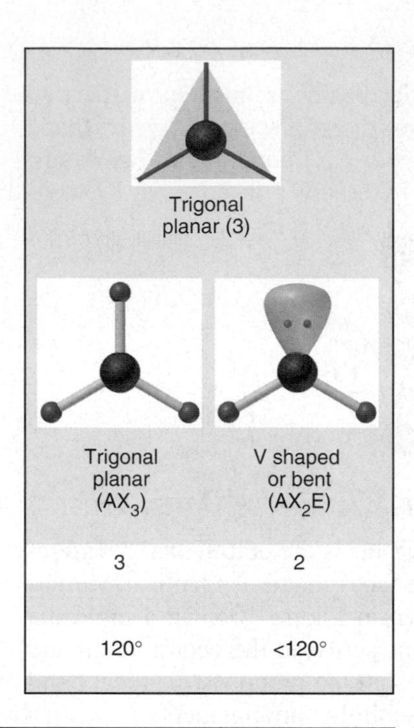

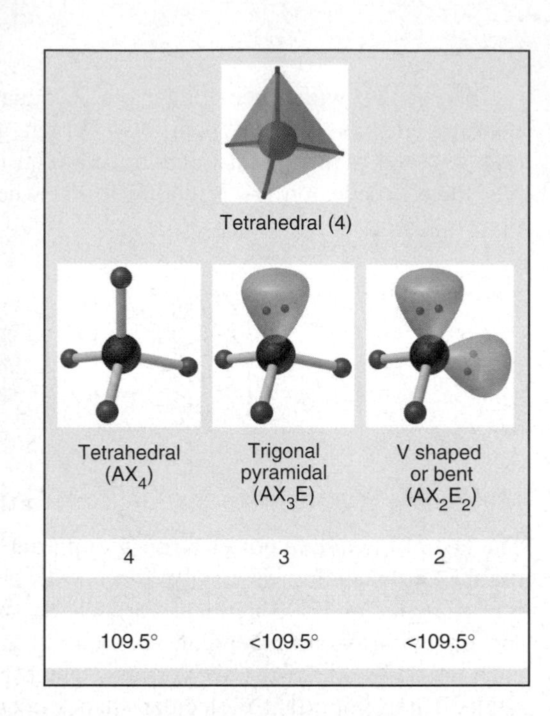

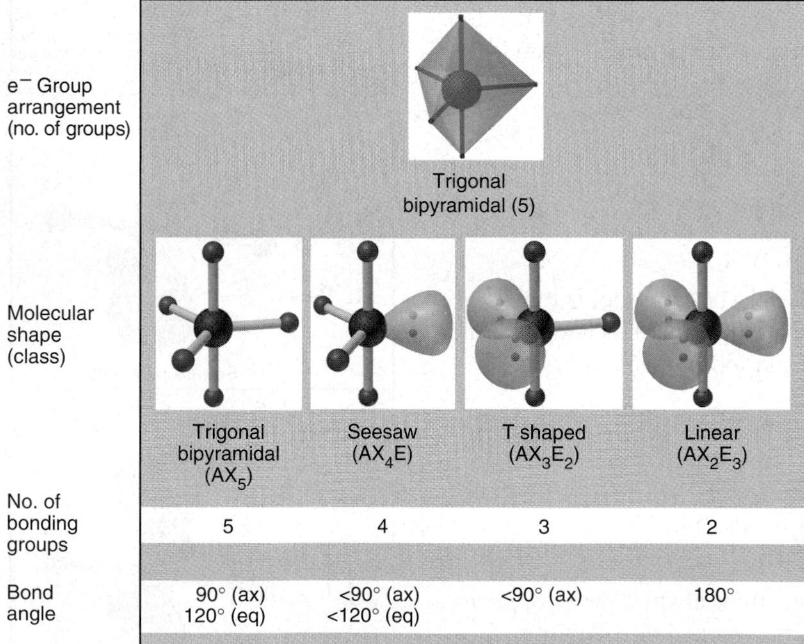

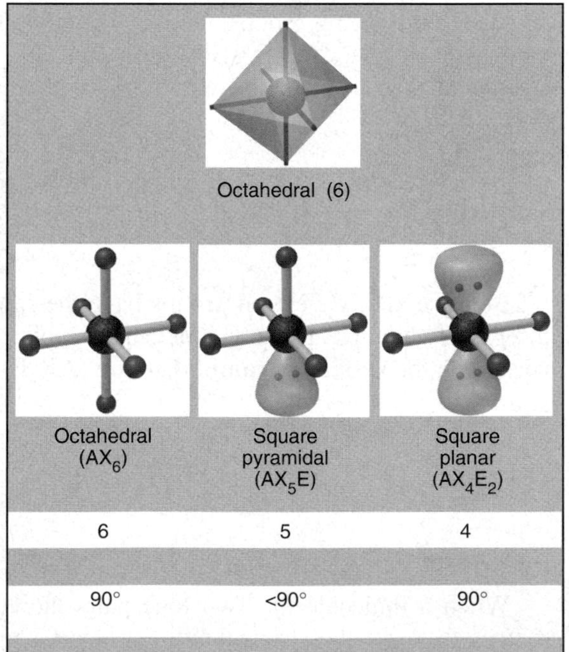

Figure 10.9 **A summary of common molecular shapes with two to six electron groups.**

Using VSEPR Theory to Determine Molecular Shape

Let's apply a stepwise method for using the VSEPR theory to determine a molecular shape from a molecular formula:

Step 1. Write the Lewis structure from the molecular formula (Figure 10.1, p. 366) to see the relative placement of atoms and the number of electron groups.

Step 2. Assign an electron-group arrangement by counting *all* electron groups around the central atom, bonding *plus* nonbonding.

Step 3. Predict the ideal bond angle from the electron-group arrangement and *the direction of any deviation* caused by lone pairs or double bonds.

Step 4. Draw and name the molecular shape by counting bonding groups and nonbonding groups separately.

Figure 10.10 summarizes these steps, and the next two sample problems apply them.

Figure 10.10 The steps in determining a molecular shape. Four steps are needed to convert a molecular formula to a molecular shape.

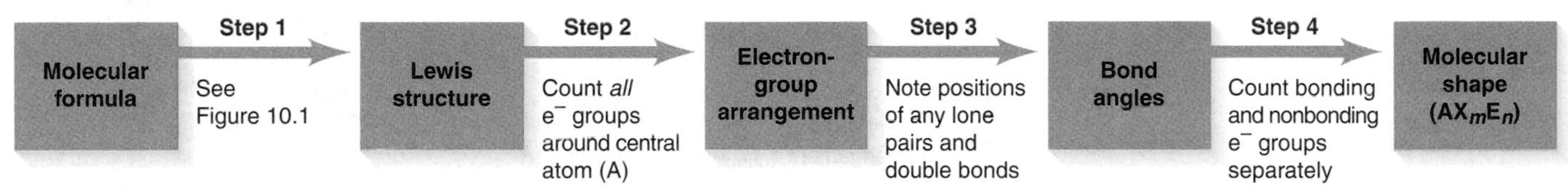

SAMPLE PROBLEM 10.6 Predicting Molecular Shapes with Two, Three, or Four Electron Groups

Problem Draw the molecular shapes and predict the bond angles (relative to the ideal angles) of **(a)** PF_3 and **(b)** $COCl_2$.

Solution (a) For PF_3.

Step 1. Write the Lewis structure from the formula (see below left).

Step 2. Assign the electron-group arrangement: Three bonding groups plus one lone pair give four electron groups around P and the *tetrahedral arrangement.*

Step 3. Predict the bond angle: For the tetrahedral electron-group arrangement, the ideal bond angle is 109.5°. There is one lone pair, so the actual bond angle should be less than 109.5°.

Step 4. Draw and name the molecular shape: With four electron groups, one of them a lone pair, PF_3 has a trigonal pyramidal shape (AX_3E):

4 e^- groups ⟹ Tetrahedral arrangement ⟹ 1 lone pair ⟹ <109.5° ⟹ 3 bonding groups ⟹ 96.3° AX_3E

(b) For $COCl_2$.

Step 1. Write the Lewis structure from the formula (see below left).

Step 2. Assign the electron-group arrangement: Two single bonds plus one double bond give three electron groups around C and the *trigonal planar arrangement.*

Step 3. Predict the bond angles: The ideal bond angle is 120°, but the double bond between C and O should compress the Cl—C—Cl angle to less than 120°.

Step 4. Draw and name the molecular shape: With three electron groups and no lone pairs, $COCl_2$ has a trigonal planar shape (AX_3):

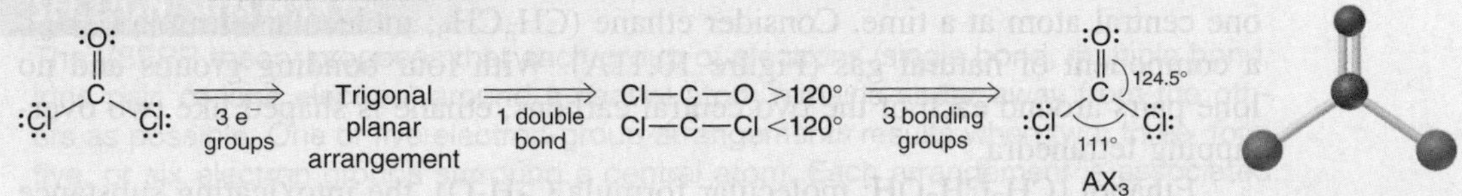

Check We compare the answers with the general information in Figure 10.9.

Comment Be sure the Lewis structure is correct because it determines the other steps.

FOLLOW-UP PROBLEM 10.6 Draw the molecular shapes and predict the bond angles (relative to the ideal angles) of **(a)** CS_2; **(b)** $PbCl_2$; **(c)** CBr_4; **(d)** SF_2.

GALLERY

Molecular Beauty: Odd Shapes with Useful Functions

Chemists see a strange beauty in the geometric intricacy of nature's most minute objects. The simplicity of spherical atoms disappears when they combine to form pentagons, helices, and countless other shapes. Moreover, many "beautiful" molecules have marvelous practical uses.

Fullerenes Buckminsterfullerene ("bucky ball," named for R. Buckminster Fuller, the architect and philosopher who designed domes based on this shape) is a truncated icosahedron of C atoms, a molecular "soccer ball" with 60 vertices and 32 faces (12 pentagons and 20 hexagons). This C_{60} structure, discovered in soot in 1985 and prepared in bulk in 1990, represents a third crystalline form of carbon (graphite and diamond are the other two). It is the parent of a new family of structures called *fullerenes* and has spawned a new field of synthesis. By inserting atoms, such as potassium, inside the ball (see models) and bonding countless types of chemical groups to the outside, researchers have discovered a whole new range of possible applications, including lubricants, superconductors, rocket fuels, lasers, batteries, magnetic films, cancer and AIDS drugs—the list grows daily!

Nanotubes
These younger cousins of fullerenes consist of extremely long, thin, graphite-like cylinders with fullerene ends. They are often nested within one another (as in the photo behind the model). Despite thicknesses of a few nanometers, these structures are highly conductive along their length and about 40 times stronger than steel! Dreams abound of nanoscale electronic components made with nanotubes or atom-thick wires formed by inserting metal atoms within their interiors.

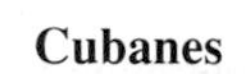

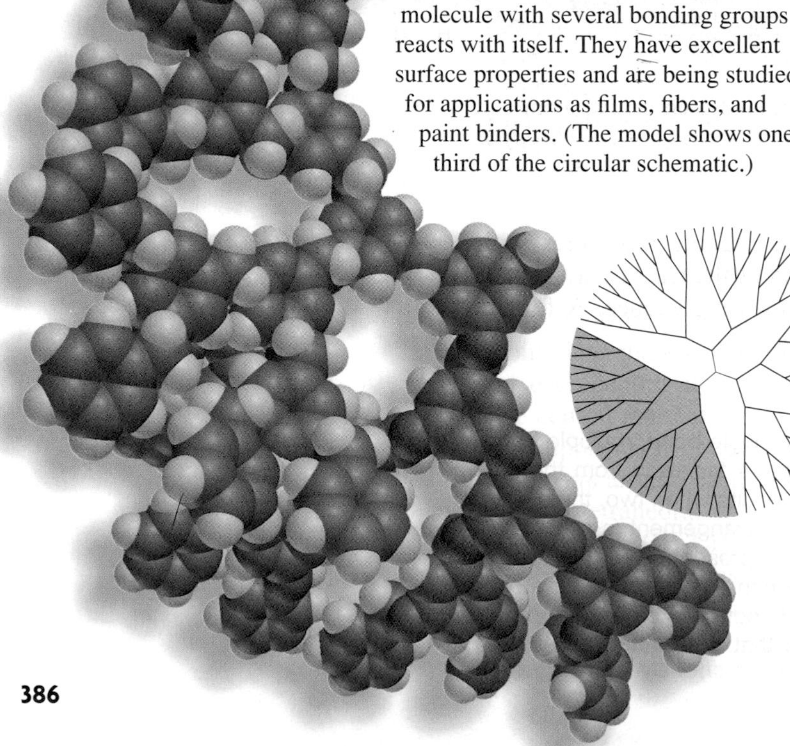

Dendrimers Named for their structural similarity to the branching of trees, dendrimers form when one molecule with several bonding groups reacts with itself. They have excellent surface properties and are being studied for applications as films, fibers, and paint binders. (The model shows one-third of the circular schematic.)

Cubanes
Compressing carbon's ideal tetrahedral bond angle of 109.5° to 90° requires energy, which becomes stored in the bond. Thus, cyclobutane, a square of C atoms with two H atoms at each corner, is unstable above 500 K. Much more energy is stored in the bonds of cubane, a cube of C atoms with one H at each corner, synthesized in 1964. The bond-strain energy of cubane can yield enormous explosive power. Tetranitrocubane (shown here) is a very powerful explosive, and octanitrocubane, which was synthesized in 2000, is considered the most powerful non-nuclear explosive known. Other properties of cubanes, which are apparently unrelated to the bond energy, include attacking cancer cells and inactivating an enzyme involved in Parkinson's disease.

10.3 MOLECULAR SHAPE AND MOLECULAR POLARITY

Knowing the shape of a substance's molecules is a key to understanding its physical and chemical behavior. One of the most important and far-reaching effects of molecular shape is molecular polarity, which can influence melting and boiling points, solubility, chemical reactivity, and even biological function.

In Chapter 9, you learned that a covalent bond is *polar* when it joins atoms of different electronegativities because the atoms share the electrons unequally. In diatomic molecules, such as HF, where there is only one bond, the bond polarity causes the molecule itself to be polar. Molecules with a net imbalance of charge have a **molecular polarity.** In molecules with more than two atoms, *both shape and bond polarity determine molecular polarity.* In an electric field, polar molecules become oriented, on average, with their partial charges pointing toward the oppositely charged electric plates, as shown for HF in Figure 10.12. The **dipole moment (μ)** is the product of these partial charges and the distance between them. It is typically measured in *debye* (D) units; using the SI units of charge (coulomb, C) and length (meter, m), 1 D $= 3.34 \times 10^{-30}$ C·m. [The unit is named for Peter Debye (1884–1966), the Dutch American chemist and physicist who won the Nobel Prize in 1936 for his major contributions to our understanding of molecular structure and solution behavior.]

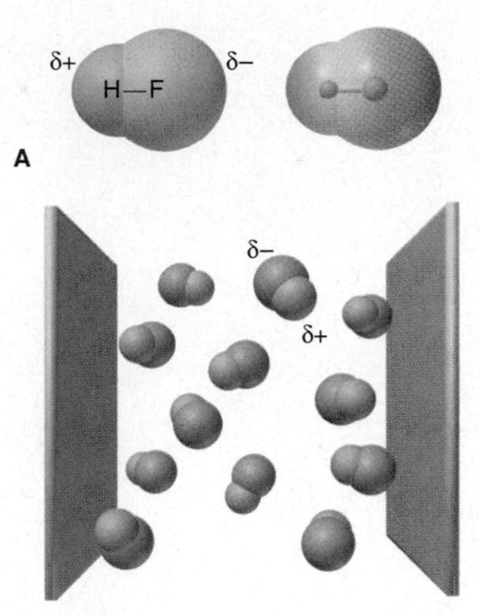

B Electric field off

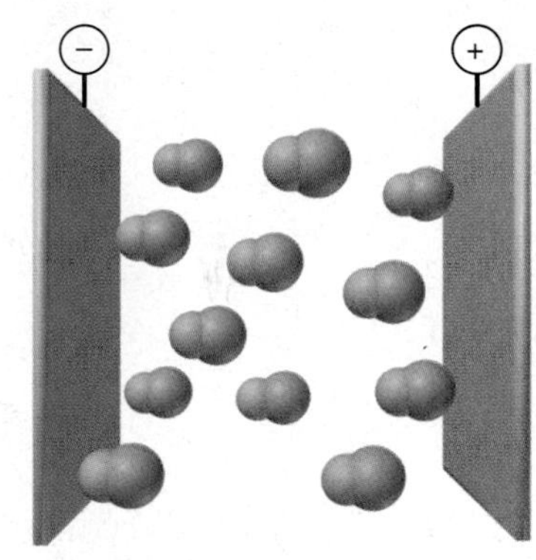

C Electric field on

Figure 10.12 The orientation of polar molecules in an electric field. A, A space-filling model of HF (*left*) shows the partial charges of this polar molecule. The electron density model (*right*) shows high electron density (*red*) associated with the F end and low electron density (*blue*) with the H end. **B,** In the absence of an external electric field, HF molecules are oriented randomly. **C,** In the presence of the field, the molecules, on average, become oriented with their partial charges pointing toward the oppositely charged plates.

Bond Polarity, Bond Angle, and Dipole Moment

When determining molecular polarity, we must take shape into account because the presence of polar bonds does not *always* lead to a polar molecule. In carbon dioxide, for example, the large electronegativity difference between C (EN = 2.5) and O (EN = 3.5) makes each C═O bond quite polar. However, CO_2 is linear, so its bonds point 180° from each other. As a result, the two identical bond polarities are counterbalanced and give the molecule *no net dipole moment* (μ = 0 D). Note that the electron density model shows regions of high negative charge (*red*) distributed equally on either side of the central region of high positive charge (*blue*):

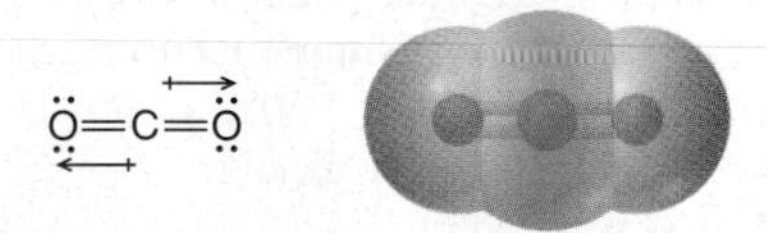

Water also has identical atoms bonded to the central atom, but it *does* have a significant dipole moment (μ = 1.85 D). In each O—H bond, electron density is pulled toward the more electronegative O atom. Here, the bond polarities are *not* counterbalanced, because the water molecule is V shaped (also see Figure 4.1, p. 135). Instead, the bond polarities are partially reinforced, and the O end of the molecule is more negative than the other end (the region between the H atoms), which the electron density model shows clearly:

(The molecular polarity of water has some amazing effects, from determining the composition of the oceans to supporting life itself, as you'll see in Chapter 12.)

In the two previous examples, molecular shape influences polarity. When different molecules have the same shape, the nature of the atoms surrounding the central atom can have a major effect on polarity. Consider carbon tetrachloride (CCl_4) and chloroform ($CHCl_3$), two tetrahedral molecules with very different polarities. In CCl_4, the surrounding atoms are all Cl atoms. Although each C—Cl bond is polar (ΔEN = 0.5), the molecule is nonpolar (μ = 0 D) because the

individual bond polarities counterbalance each other. In $CHCl_3$, H substitutes for one Cl atom, disrupting the balance and giving chloroform a significant dipole moment (μ = 1.01 D):

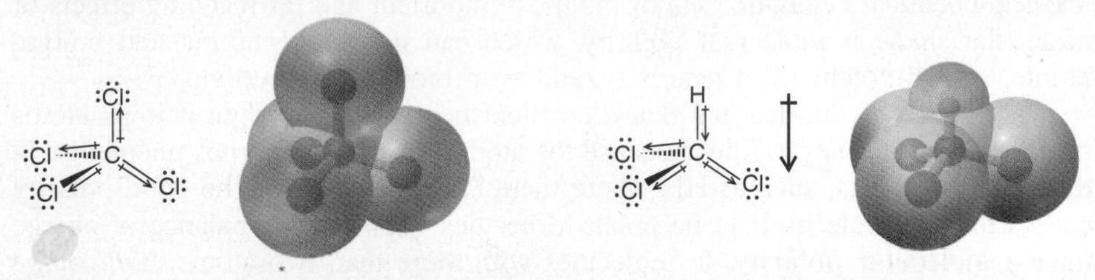

SAMPLE PROBLEM 10.9 Predicting the Polarity of Molecules

Problem From electronegativity (EN) values and their periodic trends (see Figure 9.19, p. 352), predict whether each of the following molecules is polar and show the direction of bond dipoles and the overall molecular dipole when applicable:
(a) Ammonia, NH_3 **(b)** Boron trifluoride, BF_3
(c) Carbonyl sulfide, COS (atom sequence SCO)

Plan First, we draw and name the molecular shape. Then, using relative EN values, we decide on the direction of each bond dipole. Finally, we see if the bond dipoles balance or reinforce each other in the molecule as a whole.

Solution (a) For NH_3. The molecular shape is trigonal pyramidal. From Figure 9.19, we see that N (EN = 3.0) is more electronegative than H (EN = 2.1), so the bond dipoles point toward N. The bond dipoles partially reinforce each other, and thus the molecular dipole points toward N:

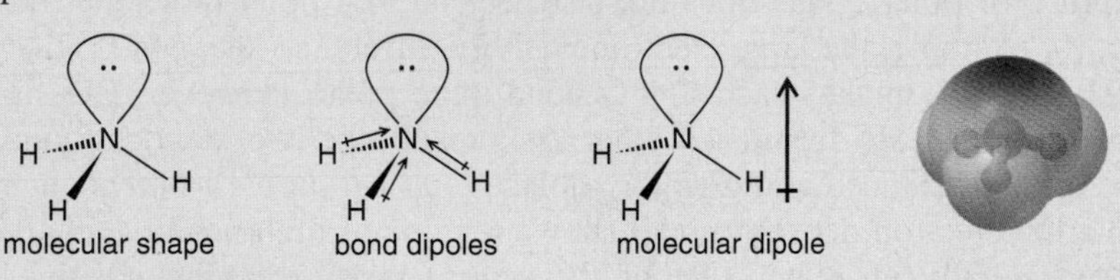

Therefore, ammonia is polar.

(b) For BF_3. The molecular shape is trigonal planar. Because F (EN = 4.0) is farther to the right in Period 2 than B (EN = 2.0), it is more electronegative; thus, each bond dipole points toward F. However, the bond angle is 120°, so the three bond dipoles counterbalance each other, and BF_3 has no molecular dipole:

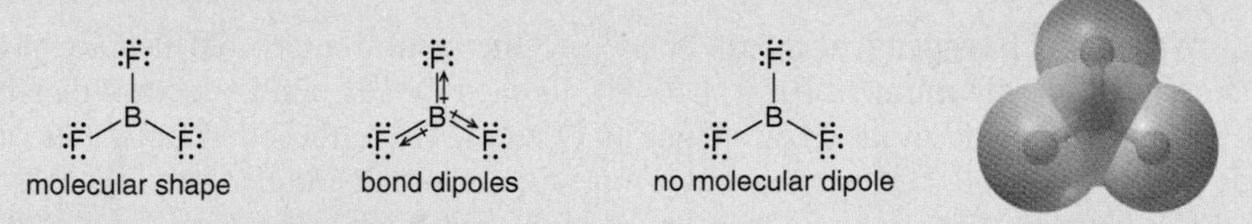

Therefore, boron trifluoride is nonpolar.

(c) For COS. The molecular shape is linear. With C and S having the same EN, the C=S bond is nonpolar, but the C=O bond is quite polar (ΔEN = 1.0), so there is a net molecular dipole toward the O:

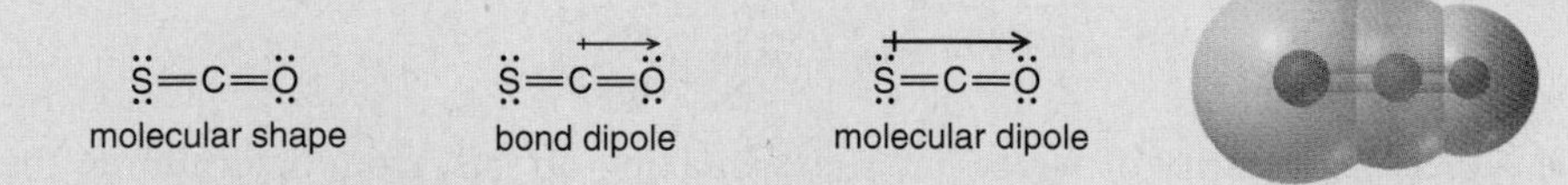

Therefore, carbonyl sulfide is polar.

FOLLOW-UP PROBLEM 10.9 Show the bond dipoles and molecular dipole, if any, for **(a)** dichloromethane (CH_2Cl_2); **(b)** iodine oxide pentafluoride (IOF_5); **(c)** nitrogen tribromide (NBr_3).

The Effect of Molecular Polarity on Behavior

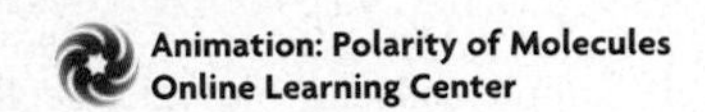

To get a sense of the influence of molecular polarity on physical behavior, consider what effect a molecular dipole might have when many polar molecules lie near each other, as they do in a liquid. How does a molecular property such as dipole moment affect a macroscopic property such as boiling point? A liquid boils when its molecules have enough energy to form bubbles of gas. To enter the bubble, the molecules in the liquid must overcome the weak attractive forces *between* them. A molecular dipole influences the strength of these attractions.

Consider the two dichloroethylenes shown below. These compounds have the *same* molecular formula ($C_2H_2Cl_2$), but they have *different* physical and chemical properties; they are constitutional isomer in particular, *cis*-1,2-dichloroethylene boils 13°C higher than *trans*-1,2-dichloroethylene. VSEPR theory predicts, and laboratory observation confirms, that all the nuclei lie in the same plane, with a trigonal planar shape around each C atom. The *trans* isomer has no dipole moment (μ = 0 D) because the C—Cl bond polarities balance each other. In contrast, the *cis* isomer *is* polar (μ = 1.90 D) because the bond dipoles partially reinforce each other, with the molecular dipole pointing between the Cl atoms. In the liquid state, the polar *cis* molecules attract each other more strongly than the nonpolar *trans* molecules, so more energy is needed to overcome these stronger forces. Therefore, the *cis* isomer has a higher boiling point. We extend these ideas in Chapter 12. The upcoming Chemical Connections essay discusses some fundamental effects of molecular shape and polarity on biological behavior.

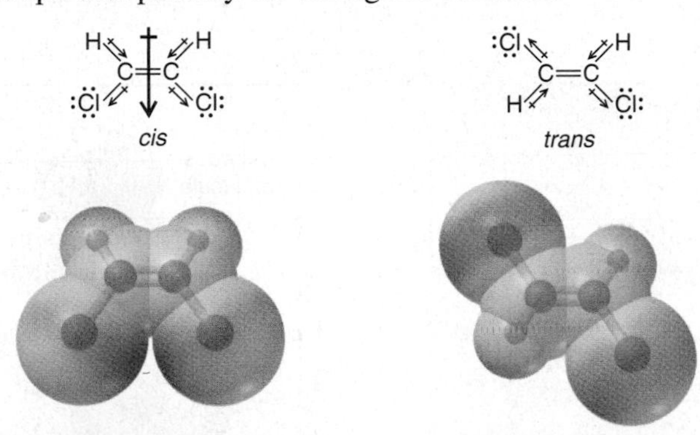

SECTION SUMMARY

Bond polarity and molecular shape determine molecular polarity, which is measured as a dipole moment. When bond polarities counterbalance each other, the molecule is nonpolar; when they reinforce each other, the molecule is polar. Molecular shape and polarity can affect physical properties, such as boiling point, and they play a central role in many aspects of biological function.

Chapter Perspective

In this chapter, you saw how chemists use the simple idea of electrostatic repulsion between electron groups to account for the shapes of molecules and, ultimately, their properties. In the next chapter, you'll learn how chemists explain molecular shape and behavior through the nature of the orbitals that electrons occupy. The landscape of a molecule—its geometric contours and regions of charge—gives rise to all its other properties. In upcoming chapters, we'll explore the physical behavior of liquids and solids (Chapter 12), the properties of solutions (Chapter 13), and the behavior of the main-group elements (Chapter 14) and organic molecules (Chapter 15), all of which emerge from atomic properties and their resulting molecular properties.

Molecular Shape, Biological Receptors, and the Sense of Smell

In a very simple sense, a biological cell is a membrane-bound sack filled with molecular shapes interacting in an aqueous fluid. As a result of the magnificent internal organization of cells, many complex processes in an organism begin when a molecular "key" fits into a correspondingly shaped molecular "lock." The key can be a relatively small molecule circulating in a body fluid, whereas the lock is usually a large molecule, known as a *biological receptor,* that is often found embedded in a cell membrane. The receptor contains a precisely shaped cavity, or *receptor site,* that is exposed to the passing fluid. Thousands of molecules collide with this site, but when one with the correct shape (that is, the molecular key) lands on it, the receptor "grabs" it through intermolecular attractions, and the biological response begins.

Let's see how this fitting together of molecular shapes operates in the sense of smell (olfaction). A substance must have certain properties to have an odor. An odorous molecule travels through the air, so it must come from a gas or a volatile liquid or solid. To reach the receptor, it must be soluble, at least to a small extent, in the thin film of aqueous solution that lines the nasal passages. Most important, the odorous molecule, or a portion of it, must have a shape that fits into one of the olfactory receptor sites that cover the nerve endings deep within the nasal passage (Figure B10.1). When this happens, the resulting nerve impulses travel from these endings to the brain, which interprets the impulses as a specific odor.

In the 1950s, a stereochemical theory of odor (*stereo* means "three-dimensional") was introduced to explain the relationship between odor and molecular shape. Its basic premise is that a molecule's shape (and sometimes its polarity), but *not* its composition, determines its odor. The original theory proposed seven primary odors—camphor-like, musky, floral, minty, ethereal, pungent, and putrid—which correspond to seven different shapes of olfactory receptor sites. Figure B10.2 shows three of the seven receptor sites occupied by a molecule having that odor.

Several predictions of this original theory have been verified. If two substances fit the same receptors, they should have the same odor. The four very different molecules in Figure B10.3 all fit the bowl-shaped camphor-like receptor and all smell like moth repellent. If different portions of one molecule fit different receptors, the molecule should have a mixed odor. Portions of the

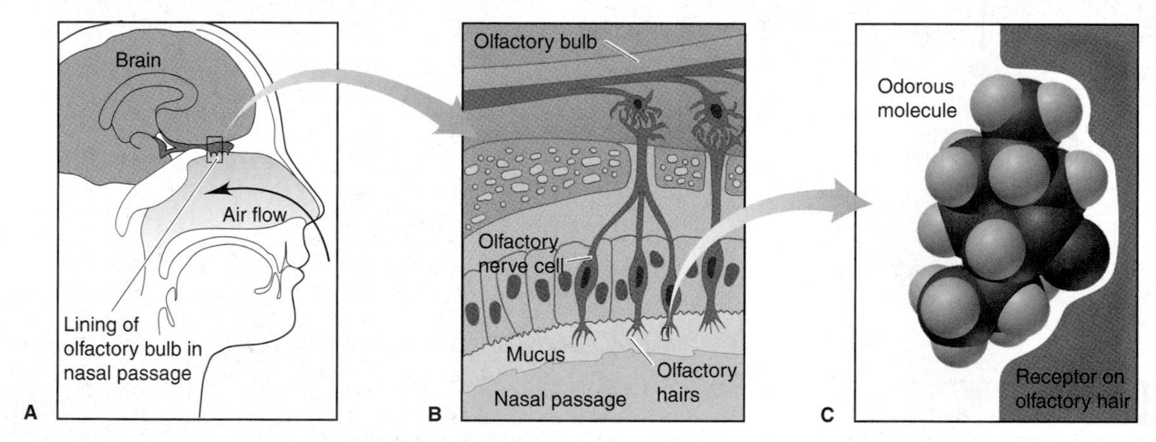

Figure B10.1 The location of olfactory receptors within the nose. A, The olfactory area lies at the top of the nasal passage very close to the brain. Air containing odorous molecules is sniffed in, warmed, moistened, and channeled toward this region. **B,** A blow-up of the region shows olfactory nerve cells with their hairlike endings protruding into the liquid-coated nasal passage. **C,** A greater blow-up shows a receptor site on one of the endings containing an odorous molecule that matches its shape. This particular molecule has a minty odor.

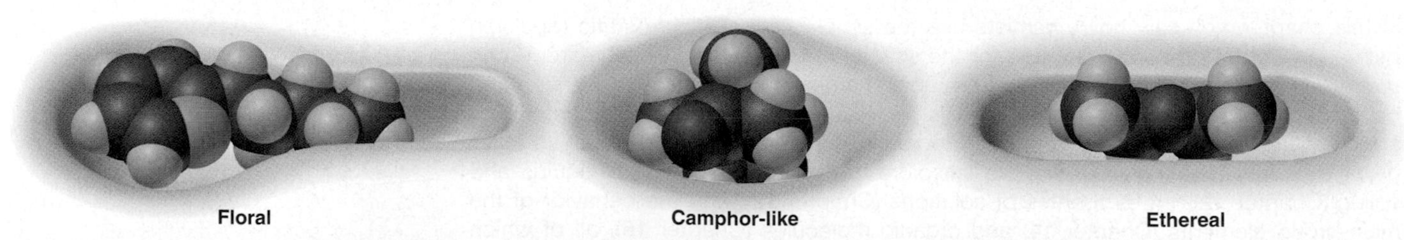

Figure B10.2 Shapes of some olfactory receptor sites. Three of the seven proposed olfactory receptors are shown with a molecule having that odor occupying the site.

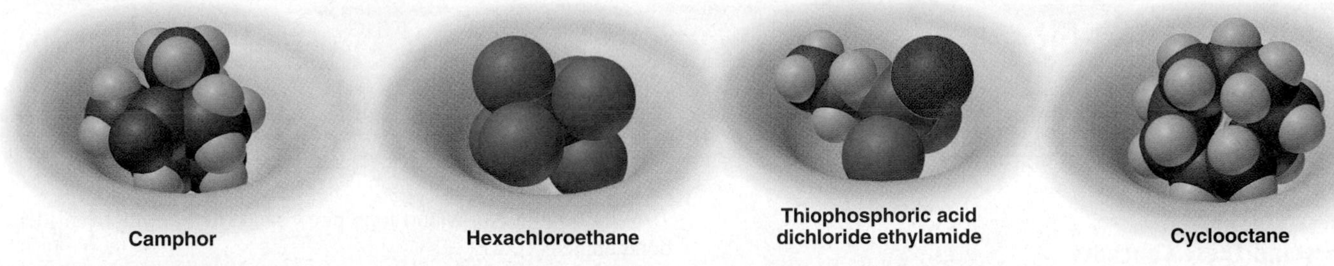

Figure B10.3 Different molecules with the same odor. The theory stating that odor is based on shape, not composition, is supported by these four different substances: all have a camphor-like (moth repellent) odor.

benzaldehyde molecule fit the camphor-like, floral, and minty receptors, which gives it an almond odor; other molecules that smell like almonds fit the same three receptors.

Recent evidence suggests, however, that the original model is far too simple. In many cases, the odor predicted from the shape turns out to be wrong when the substance is smelled. One reason for this discrepancy is that a molecule may have a very different shape in the gas phase than it does when it is in solution at the receptor. By the early 1990s, evidence had indicated that there are about 1000 different receptors and that various combinations of stimulation produce the more than 10,000 odors humans can distinguish. In fact, the 2004 Nobel Prize in medicine or physiology was awarded to Richard Axel and Linda B. Buck for this work. Thus, although a key premise of the theory is still accepted—odor depends on molecular shape—the nature of the dependence is complex and is the focus of active research in the food, cosmetics, and insecticide industries. The sense of smell is so vital for survival that these topics are important areas of study for biologists as well.

Many other biochemical processes are controlled by one molecule fitting into a receptor site on another. Enzymes are proteins that bind cellular reactants and speed their reaction (Figure B10.4). Nerve impulses are transmitted when small molecules released from one nerve fit into receptors on the next. Mind-altering drugs act by chemically disrupting the molecular fit at such nerve receptors in the brain. One type of immune response is triggered when a molecule on a bacterial surface binds to the receptors on "killer" cells in the bloodstream. Hormones regulate energy production and growth by fitting into and activating key receptors. Genes function when certain nucleic acid molecules fit into specific regions of other nucleic acids. Indeed, *no molecular property is more crucial to living systems than shape.*

Figure B10.4 Molecular shape and enzyme action. A small sugar molecule (*bottom*) is about to move up and fit into the receptor site of an enzyme molecule and undergo reaction.

For Review and Reference

(Numbers in parentheses refer to pages, unless noted otherwise.)

Learning Objectives

Relevant section and/or sample problem (SP) numbers appear in parentheses.

Understand These Concepts

1. How Lewis structures depict the atoms and their bonding and lone electron pairs in a molecule or ion (10.1)
2. How resonance and electron delocalization explain bond properties in many compounds with double bonds adjacent to single bonds (10.1)
3. The meaning of formal charge and how it is used to select the most important resonance structure; the difference between formal charge and oxidation number (10.1)
4. The octet rule and its three major exceptions—molecules with a central atom that has an electron deficiency, an odd number of electrons, or an expanded valence shell (10.1)
5. How electron-group repulsions lead to molecular shapes (10.2)
6. The five electron-group arrangements and their associated molecular shapes (10.2)
7. Why double bonds and lone pairs cause deviations from ideal bond angles (10.2)
8. How bond polarities and molecular shape combine to give a molecule polarity (10.3)

Master These Skills

1. Using a stepwise method for writing a Lewis structure from a molecular formula (SPs 10.1 to 10.3 and 10.5)
2. Writing resonance structures for molecules and ions (SP 10.4)
3. Calculating the formal charge of any atom in a molecule or ion (10.5)
4. Predicting molecular shapes from Lewis structures (SPs 10.6 to 10.8)
5. Using molecular shape and electronegativity values to predict the direction of a molecular dipole (SP 10.9)

Key Terms

Section 10.1
Lewis structure (Lewis formula) (366)
resonance structure (resonance form) (369)
resonance hybrid (370)
electron-pair delocalization (370)
formal charge (371)
electron deficient (372)
free radical (373)
expanded valence shell (373)

Section 10.2
valence-shell electron-pair repulsion (VSEPR) theory (376)
molecular shape (376)
bond angle (376)
linear arrangement (377)
linear shape (377)
trigonal planar arrangement (377)
bent shape (V shape) (378)
tetrahedral arrangement (379)
trigonal pyramidal shape (379)
trigonal bipyramidal arrangement (380)
equatorial group (380)
axial group (380)
seesaw shape (380)
T shape (380)
octahedral arrangement (381)
square pyramidal shape (381)
square planar shape (381)

Section 10.3
molecular polarity (387)
dipole moment (μ) (387)

Key Equations and Relationships

10.1 Calculating the formal charge on an atom (371):

$$\text{Formal charge of atom} = \text{no. of valence } e^- - (\text{no. of unshared valence } e^- + \tfrac{1}{2}\text{ no. of shared valence } e^-)$$

10.2 Ranking the effect of electron-pair repulsions on bond angle (380):
Lone pair–lone pair > lone pair–bonding pair > bonding pair–bonding pair

Highlighted Figures and Tables

These figures (F) and tables (T) provide a review of key ideas.

F10.1 The steps in writing a Lewis structure (366)
F10.2 Electron-group repulsions and the five basic molecular shapes (376)
F10.3 The single molecular shape of the linear electron-group arrangement (377)
F10.4 The two molecular shapes of the trigonal planar electron-group arrangement (377)
F10.5 The three molecular shapes of the tetrahedral electron-group arrangement (379)
F10.7 The four molecular shapes of the trigonal bipyramidal electron-group arrangement (380)
F10.8 The three molecular shapes of the octahedral electron-group arrangement (381)
F10.9 A summary of molecular shapes (382)
F10.10 The steps in determining a molecular shape (383)

Brief Solutions to Follow-up Problems

10.1 (a) (b) (c)

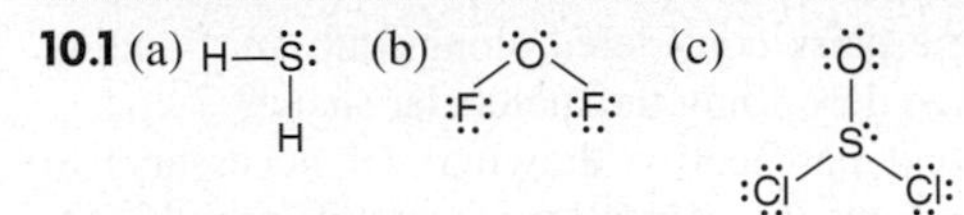

10.2 (a) (b)

10.3 (a) (b) (c)

10.4

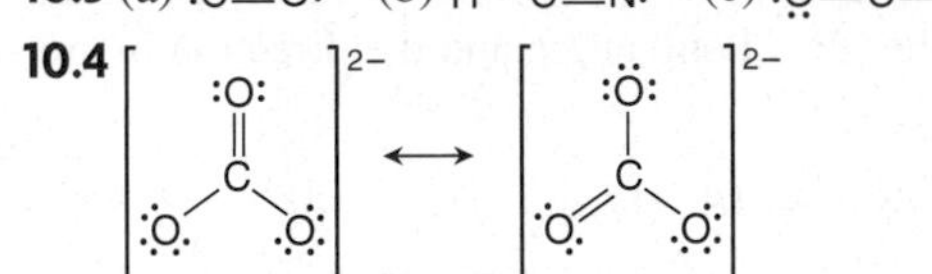

10.5 (a) (b) (c)

10.6 (a) Linear, 180° (b) V shaped, <120°

(c) Tetrahedral, 109.5° (d) V shaped, <109.5°

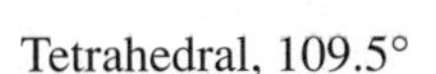

10.7 (a) Linear, 180° (b) T shaped, <90° (c) Trigonal bipyramidal F_{eq}—S—F_{eq} angle <120° F_{ax}—S—F_{eq} angle <90°

10.8 (a) S is tetrahedral; double bonds compress O—S—O angle to <109.5°. Each central O is V shaped; lone pairs compress H—O—S angle to <109.5°.

(b) C in CH_3— is tetrahedral ~109.5°; other C atoms are linear, 180°.

(c) Each S is V shaped; F—S—S angle <109.5°.

10.9 (a) (b) (c)

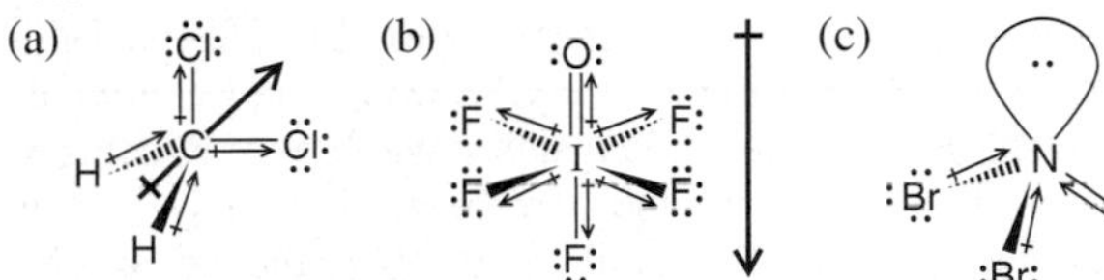

Problems

Problems with **colored** numbers are answered in Appendix E. Sections match the text and provide the number(s) of relevant sample problems. Most offer Concept Review Questions, Skill-Building Exercises (grouped in pairs covering the same concept), and Problems in Context. Comprehensive Problems are based on material from any section or previous chapter.

Depicting Molecules and Ions with Lewis Structures

(Sample Problems 10.1 to 10.5)

Concept Review Questions

10.1 Which of these atoms *cannot* serve as a central atom in a Lewis structure: (a) O; (b) He; (c) F; (d) H; (e) P? Explain.

10.2 When is a resonance hybrid needed to adequately depict the bonding in a molecule? Using NO_2 as an example, explain how a resonance hybrid is consistent with the actual bond length, bond strength, and bond order.

10.3 In which of these bonding patterns does X obey the octet rule?

(a) (b) (c) (d) (e) (f) (g) (h)

10.4 What is required for an atom to expand its valence shell? Which of the following atoms can expand its valence shell: F, S, H, Al, Se, Cl?

Skill-Building Exercises *(grouped in similar pairs)*

10.5 Draw a Lewis structure for (a) SiF_4; (b) $SeCl_2$; (c) COF_2 (C is central).

10.6 Draw a Lewis structure for (a) PH_4^+; (b) C_2F_4; (c) SbH_3.

10.7 Draw a Lewis structure for (a) PF_3; (b) H_2CO_3 (both H atoms are attached to O atoms); (c) CS_2.

10.8 Draw a Lewis structure for (a) CH_4S; (b) S_2Cl_2; (c) $CHCl_3$.

10.9 Draw Lewis structures of all the important resonance forms of (a) NO_2^+; (b) NO_2F (N is central).

10.10 Draw Lewis structures of all the important resonance forms of (a) HNO_3 ($HONO_2$); (b) $HAsO_4^{2-}$ ($HOAsO_3^{2-}$).

10.11 Draw Lewis structures of all the important resonance forms of (a) N_3^-; (b) NO_2^-.

10.12 Draw Lewis structures of all the important resonance forms of (a) HCO_2^- (H is attached to C); (b) $HBrO_4$ ($HOBrO_3$).

10.13 Draw a Lewis structure and calculate the formal charge of each atom in (a) IF_5; (b) AlH_4^-.

10.14 Draw a Lewis structure and calculate the formal charge of each atom in (a) COS (C is central); (b) NO.

10.15 Draw a Lewis structure and calculate the formal charge of each atom in (a) CN^-; (b) ClO^-.

10.16 Draw a Lewis structure and calculate the formal charge of each atom in (a) BF_4^-; (b) ClNO.

10.17 Draw a Lewis structure for the most important resonance form of each ion, showing formal charges and oxidation numbers of the atoms: (a) BrO_3^-; (b) SO_3^{2-}.

10.18 Draw a Lewis structure for the most important resonance form of each ion, showing formal charges and oxidation numbers of the atoms: (a) AsO_4^{3-}; (b) ClO_2^-.

10.19 These species do not obey the octet rule. Draw a Lewis structure for each, and state the type of octet-rule exception:
(a) BH_3 (b) AsF_4^- (c) $SeCl_4$

10.20 These species do not obey the octet rule. Draw a Lewis structure for each, and state the type of octet-rule exception:
(a) PF_6^- (b) ClO_3 (c) H_3PO_3

10.21 These species do not obey the octet rule. Draw a Lewis structure for each, and state the type of octet-rule exception:
(a) BrF_3 (b) ICl_2^- (c) BeF_2

10.22 These species do not obey the octet rule. Draw a Lewis structure for each, and state the type of octet-rule exception:
(a) O_3^- (b) XeF_2 (c) SbF_4^-

Problems in Context

10.23 Molten beryllium chloride reacts with chloride ion from molten NaCl to form the $BeCl_4^{2-}$ ion, in which the Be atom attains an octet. Show the net ionic reaction with Lewis structures.

10.24 Despite many attempts, the perbromate ion (BrO_4^-) was not prepared in the laboratory until about 1970. (Indeed, articles were published explaining theoretically why it could never be prepared!) Draw a Lewis structure for BrO_4^- in which all atoms have lowest formal charges.

10.25 Cryolite (Na_3AlF_6) is an indispensable component in the electrochemical manufacture of aluminum. Draw a Lewis structure for the AlF_6^{3-} ion.

10.26 Phosgene is a colorless, highly toxic gas employed against troops in World War I and used today as a key reactant in organic syntheses. Use formal charges to select the most important of the following resonance structures:

A: :O: =C with :Cl: and :Cl: (C=O double bond, C–Cl single bonds)
B: :Ö: –C, =Cl: and –Cl: (C–O single bond, one C=Cl double bond)
C: :Ö: –C, –Cl: and =Cl: (C–O single bond, one C=Cl double bond)

Valence-Shell Electron-Pair Repulsion (VSEPR) Theory and Molecular Shape

(Sample Problems 10.6 to 10.8)

Concept Review Questions

10.27 If you know the formula of a molecule or ion, what is the first step in predicting its shape?

10.28 In what situation is the name of the molecular shape the same as the name of the electron-group arrangement?

10.29 Which of the following numbers of electron groups can give rise to a bent (V-shaped) molecule: two, three, four, five, six? Draw an example for each case, showing the shape classification (AX_mE_n) and the ideal bond angle.

10.30 Name all the molecular shapes that have a tetrahedral electron-group arrangement.

10.31 Why aren't lone pairs considered along with surrounding bonding groups when describing the molecular shape?

10.32 Use wedge-bond perspective drawings (if necessary) to sketch the atom positions in a general molecule of formula (not shape class) AX_n that has each of the following shapes:
(a) V shaped (b) trigonal planar (c) trigonal bipyramidal
(d) T shaped (e) trigonal pyramidal (f) square pyramidal

10.33 What would you expect to be the electron-group arrangement around atom A in each of the following cases? For each arrangement, give the ideal bond angle and the direction of any expected deviation:

(a) X–A: with X above and X below A (b) X—A≡X (c) X=A—X with X above A (d) X—Ä—X
(e) X=A=X (f) :A with X above, X below, and two X to the right

Skill-Building Exercises *(grouped in similar pairs)*

10.34 Determine the electron-group arrangement, molecular shape, and ideal bond angle(s) for each of the following:
(a) O_3 (b) H_3O^+ (c) NF_3

10.35 Determine the electron-group arrangement, molecular shape, and ideal bond angle(s) for each of the following:
(a) SO_4^{2-} (b) NO_2^- (c) PH_3

10.36 Determine the electron-group arrangement, molecular shape, and ideal bond angle(s) for each of the following:
(a) CO_3^{2-} (b) SO_2 (c) CF_4

10.37 Determine the electron-group arrangement, molecular shape, and ideal bond angle(s) for each of the following:
(a) SO_3 (b) N_2O (N is central) (c) CH_2Cl_2

10.38 Name the shape and give the AX_mE_n classification and ideal bond angle(s) for each of the following general molecules:

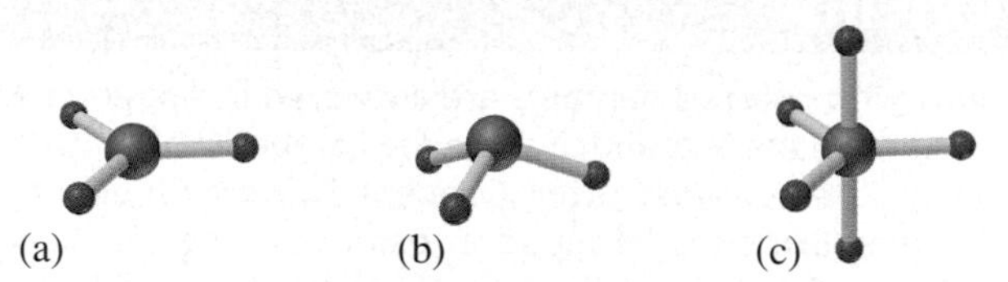

10.39 Name the shape and give the AX_mE_n classification and ideal bond angle(s) for each of the following general molecules:

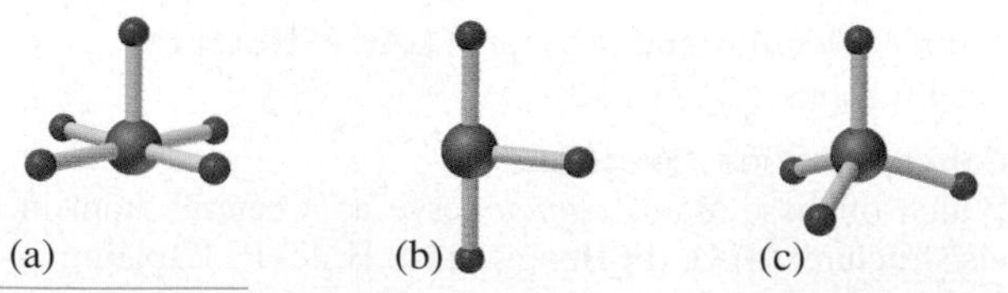

10.40 Determine the shape, ideal bond angle(s), and the direction of any deviation from these angles for each of the following:
(a) ClO_2^- (b) PF_5 (c) SeF_4 (d) KrF_2

10.41 Determine the shape, ideal bond angle(s), and the direction of any deviation from these angles for each of the following:
(a) ClO_3^- (b) IF_4^- (c) $SeOF_2$ (d) TeF_5^-

10.42 Determine the shape around each central atom in each molecule, and explain any deviation from ideal bond angles:
(a) CH_3OH (b) N_2O_4 (O_2NNO_2)

10.43 Determine the shape around each central atom in each molecule, and explain any deviation from ideal bond angles:
(a) H_3PO_4 (no H—P bond) (b) $CH_3—O—CH_2CH_3$

10.44 Determine the shape around each central atom in each molecule, and explain any deviation from ideal bond angles:
(a) CH_3COOH (b) H_2O_2

10.45 Determine the shape around each central atom in each molecule, and explain any deviation from ideal bond angles:
(a) H_2SO_3 (no H—S bond) (b) N_2O_3 ($ONNO_2$)

10.46 Arrange the following AF_n species in order of *increasing* F—A—F bond angles: BF_3, BeF_2, CF_4, NF_3, OF_2.

10.47 Arrange the following ACl_n species in order of *decreasing* Cl—A—Cl bond angles: SCl_2, OCl_2, PCl_3, $SiCl_4$, $SiCl_6^{2-}$.

10.48 State an ideal value for each of the bond angles in each molecule, and note where you expect deviations:

(a)

(b)

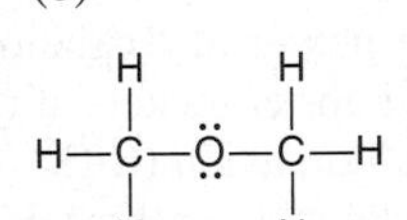

(c) H—Ö—B—Ö—H (with :Ö—H on B)

10.49 State an ideal value for each of the bond angles in each molecule, and note where you expect deviations:

(a) Ö=N—Ö—H (with :Ö: on N)

(b) H—C—C=Ö (H's on each C)

(c) H—C—Ö—H (with :O: double-bonded to C)

Problems in Context

10.50 Because both tin and carbon are members of Group 4A(14), they form structurally similar compounds. However, tin exhibits a greater variety of structures because it forms several ionic species. Predict the shapes and ideal bond angles, including any deviations, for the following:
(a) $Sn(CH_3)_2$ (b) $SnCl_3^-$ (c) $Sn(CH_3)_4$
(d) SnF_5^- (e) SnF_6^{2-}

10.51 In the gas phase, phosphorus pentachloride exists as separate molecules. In the solid phase, however, the compound is composed of alternating PCl_4^+ and PCl_6^- ions. What change(s) in molecular shape occur(s) as PCl_5 solidifies? How does the Cl—P—Cl angle change?

Molecular Shape and Molecular Polarity

(Sample Problem 10.9)

Concept Review Questions

10.52 For molecules of general formula AX_n (where $n > 2$), how do you determine if a molecule is polar?

10.53 How can a molecule with polar covalent bonds fail to be a polar molecule? Give an example.

10.54 Explain in general why the shape of a biomolecule is important to its function.

Skill-Building Exercises *(grouped in similar pairs)*

10.55 Consider the molecules SCl_2, F_2, CS_2, CF_4, and BrCl.
(a) Which has bonds that are the most polar?
(b) Which have a molecular dipole moment?

10.56 Consider the molecules BF_3, PF_3, BrF_3, SF_4, and SF_6.
(a) Which has bonds that are the most polar?
(b) Which have a molecular dipole moment?

10.57 Which molecule in each pair has the greater dipole moment? Give the reason for your choice.
(a) SO_2 or SO_3 (b) ICl or IF
(c) SiF_4 or SF_4 (d) H_2O or H_2S

10.58 Which molecule in each pair has the greater dipole moment? Give the reason for your choice.
(a) ClO_2 or SO_2 (b) HBr or HCl
(c) $BeCl_2$ or SCl_2 (d) AsF_3 or AsF_5

Problems in Context

10.59 There are three different dichloroethylenes (molecular formula $C_2H_2Cl_2$), which we can designate X, Y, and Z. Compound X has no dipole moment, but compound Z does. Compounds X and Z each combine with hydrogen to give the same product:

$$C_2H_2Cl_2 \text{ (X or Z)} + H_2 \longrightarrow ClCH_2—CH_2Cl$$

What are the structures of X, Y, and Z? Would you expect compound Y to have a dipole moment?

10.60 Dinitrogen difluoride, N_2F_2, is the only stable, simple inorganic molecule with an N=N bond. The compound occurs in *cis* and *trans* forms.
(a) Draw the molecular shapes of the two forms of N_2F_2.
(b) Predict the polarity, if any, of each form.

Comprehensive Problems

10.61 In addition to ammonia, nitrogen forms three other hydrides: hydrazine (N_2H_4), diazene (N_2H_2), and tetrazene (N_4H_4).
(a) Use Lewis structures to compare the strength, length, and order of nitrogen-nitrogen bonds in hydrazine, diazene, and N_2.
(b) Tetrazene (atom sequence H_2NNNNH_2) decomposes above 0°C to hydrazine and nitrogen gas. Draw a Lewis structure for tetrazene, and calculate ΔH^0_{rxn} for this decomposition.

10.62 Draw a Lewis structure for each species: (a) PF_5; (b) CCl_4; (c) H_3O^+; (d) ICl_3; (e) BeH_2; (f) PH_2^-; (g) $GeBr_4$; (h) CH_3^-; (i) BCl_3; (j) BrF_4^+; (k) XeO_3; (l) TeF_4.

10.63 Determine the molecular shape of each of the species in Problem 10.62.

10.64 Both aluminum and iodine form chlorides, Al_2Cl_6 and I_2Cl_6, with "bridging" Cl atoms. The Lewis structures are

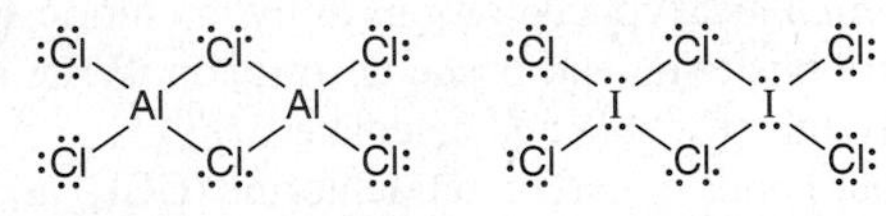

(a) What is the formal charge on each atom? (b) Which of these molecules has a planar shape? Explain.

10.65 Nitrosyl fluoride (NOF) has an atom sequence in which all atoms have formal charges of zero. Write the Lewis structure consistent with this fact.

10.66 The VSEPR model was developed in the 1950s, before any xenon compounds had been prepared. Thus, in the early 1960s, these compounds provided an excellent test of the model's predictive power. What would you have predicted for the shapes of XeF_2, XeF_4, and XeF_6?

10.67 Boron trifluoride is an important reactant in organic syntheses but is a very reactive gas at room temperature and thus difficult to handle. Boron's ability to accept another pair of electrons and attain an octet is used to prepare an easily stored liquid form: a compound of BF_3 and diethyl ether ($CH_3—CH_2—O—CH_2—CH_3$). Draw Lewis structures for BF_3, diethyl ether, and the product, and indicate how the molecular shapes around B and O change during the reaction.

10.68 The actual bond angle in NO_2 is 134.3°, and in NO_2^- it is 115.4°, although the ideal bond angle is 120° in both. Explain.

10.69 "Inert" xenon actually forms several compounds, especially with the highly electronegative elements oxygen and fluorine. The simple fluorides XeF_2, XeF_4, and XeF_6 are all formed by direct reaction of the elements. As you might expect from the size of the xenon atom, the Xe—F bond is not a strong one. Calculate the Xe—F bond energy in XeF_6, given that the heat of formation is −402 kJ/mol.

10.70 Propylene oxide is used to make many products, including plastics such as polyurethane. One method for synthesizing it involves oxidizing propene with hydrogen peroxide:

$$CH_3-CH{=}CH_2 + H_2O_2 \longrightarrow CH_3-\underset{\diagdown O \diagup}{CH-CH_2} + H_2O$$

(a) What is the molecular shape and ideal bond angle around each carbon atom in propylene oxide?
(b) Predict any deviation from the ideal for the actual bond angle (assume the three atoms in the ring form an equilateral triangle).

10.71 Chloral, $Cl_3C-CH{=}O$, reacts with water to form the sedative and hypnotic agent chloral hydrate, $Cl_3C-CH(OH)_2$. Draw Lewis structures for these substances, and describe the change in molecular shape, if any, that occurs around each of the carbon atoms during the reaction.

10.72 Dichlorine heptaoxide, Cl_2O_7, can be viewed as two ClO_4 groups sharing an O atom. Draw a Lewis structure for Cl_2O_7 with the lowest formal charges, and predict any deviation from the ideal for the Cl—O—Cl bond angle.

10.73 Like several other bonds, carbon-oxygen bonds have lengths and strengths that depend on the bond order. Draw Lewis structures for the following species, and arrange them in order of increasing carbon-oxygen bond length and then by increasing carbon-oxygen bond strength: (a) CO; (b) CO_3^{2-}; (c) H_2CO; (d) CH_4O; (e) HCO_3^- (H attached to O).

10.74 In the 1980s, there was an international agreement to destroy all stockpiles of mustard gas, $ClCH_2CH_2SCH_2CH_2Cl$. When this substance contacts the moisture in eyes, nasal passages, and skin, the —OH groups of water replace the Cl atoms and create high local concentrations of hydrochloric acid, which cause severe blistering and tissue destruction. Write a balanced equation for this reaction, and calculate ΔH^0_{rxn}.

10.75 The four bonds of carbon tetrachloride (CCl_4) are polar, but the molecule is nonpolar because the bond polarity is canceled by the symmetric tetrahedral shape. When other atoms substitute for some of the Cl atoms, the symmetry is broken and the molecule becomes polar. Use Figure 9.19 (p. 352) to rank the following molecules from the least polar to the most polar: CH_2Br_2, CF_2Cl_2, CH_2F_2, CH_2Cl_2, CBr_4, CF_2Br_2.

10.76 Ethanol (CH_3CH_2OH) is being used as a gasoline additive or alternative in many parts of the world.
(a) Use bond energies to find the ΔH^0_{comb} of gaseous ethanol. (Assume H_2O forms as a gas.)
(b) In its standard state at 25°C, ethanol is a liquid. Its vaporization requires 40.5 kJ/mol. Correct the value from part (a) to find the heat of combustion of liquid ethanol.
(c) How does the value from part (b) compare with the value you calculate from standard heats of formation (Appendix B)?
(d) "Greener" methods produce ethanol from corn and other plant material, but the main industrial method involves hydrating ethylene from petroleum. Use Lewis structures and bond energies to calculate ΔH^0_{rxn} for the formation of gaseous ethanol from ethylene gas with water vapor.

10.77 In the following compounds, the C atoms form a single ring. Draw a Lewis structure for each compound, identify cases for which resonance exists, and determine the carbon-carbon bond order(s): (a) C_3H_4; (b) C_3H_6; (c) C_4H_6; (d) C_4H_4; (e) C_6H_6.

10.78 An oxide of nitrogen is 25.9% N by mass, has a molar mass of 108 g/mol, and contains no nitrogen-nitrogen or oxygen-oxygen bonds. Draw its Lewis structure, and name it.

10.79 An experiment requires 50.0 mL of 0.040 *M* NaOH for the titration of 1.00 mmol of acid. Mass analysis of the acid shows 2.24% hydrogen, 26.7% carbon, and 71.1% oxygen. Draw the Lewis structure of the acid.

10.80 A gaseous compound has a composition by mass of 24.8% carbon, 2.08% hydrogen, and 73.1% chlorine. At STP, the gas has a density of 4.3 g/L. Draw a Lewis structure that satisfies these facts. Would another structure also satisfy them? Explain.

10.81 Perchlorates are powerful oxidizing agents used in fireworks, flares, and the booster rockets of the space shuttle. Lewis structures for the perchlorate ion (ClO_4^-) can be drawn with all single bonds or with one, two, or three double bonds. Draw each of these possible resonance forms, use formal charges to determine the most important, and calculate its average bond order.

10.82 Methane reacts with oxygen to form carbon dioxide and water vapor. Hydrogen sulfide reacts with oxygen to form sulfur dioxide and water vapor. Use bond energies (Table 9.2, p. 341) to determine the heat of combustion per mole of O_2 for each reaction (assume a Lewis structure for sulfur dioxide with zero formal charge; BE of S=O is 552 kJ/mol).

10.83 Use Lewis structures to determine which *two* of the following are unstable: (a) SF_2; (b) SF_3; (c) SF_4; (d) SF_5; (e) SF_6.

10.84 A major short-lived, neutral species in flames is OH.
(a) What is unusual about the electronic structure of OH?
(b) Use the standard heat of formation of OH(*g*) and bond energies to calculate the O—H bond energy in OH(*g*) (ΔH^0_f of OH(*g*) = 39.0 kJ/mol).
(c) From the average value for the O—H bond energy in Table 9.2 (p. 341) and your value for the O—H bond energy in OH(*g*), find the energy needed to break the first O—H bond in water.

10.85 Pure HN_3 (atom sequence HNNN) is a very explosive compound. In aqueous solution, it is a weak acid (comparable to acetic acid) that yields the azide ion, N_3^-. Draw resonance structures to explain why the nitrogen-nitrogen bond lengths are equal in N_3^- but unequal in HN_3.

10.86 Except for nitrogen, the elements of Group 5A(15) all form pentafluorides, and most form pentachlorides. The chlorine atoms of PCl_5 can be replaced with fluorine atoms one at a time to give, successively, PCl_4F, PCl_3F_2, . . . , PF_5.
(a) Given the sizes of F and Cl, would you expect the first two F substitutions to be at axial or equatorial positions? Explain.
(b) Which of the five fluorine-containing molecules have no dipole moment?

10.87 Dinitrogen monoxide (N_2O), used as the anesthetic "laughing gas" in dental surgery, supports combustion in a manner similar to oxygen, with the nitrogen atoms forming N_2. Draw three resonance structures for N_2O (one N is central), and use formal charges to decide the relative importance of each. What correlation can you suggest between the most important structure and the observation that N_2O supports combustion?

10.88 Oxalic acid ($H_2C_2O_4$) is found in toxic concentrations in rhubarb leaves. The acid forms two ions, $HC_2O_4^-$ and $C_2O_4^{2-}$, by the sequential loss of H^+ ions. Draw Lewis structures for the three species, and comment on the relative lengths and strengths of their carbon-oxygen bonds. The connections among the atoms are shown below with single bonds only.

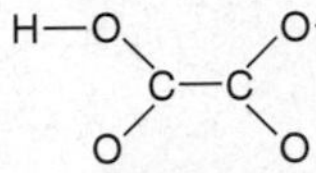

```
H—O         O—H
    \       /
     C — C
    /       \
   O         O
```

10.89 Some scientists speculate that many organic molecules required for life on the young Earth arrived on meteorites. The Murchison meteorite that landed in Australia in 1969 contained 92 different amino acids, including 21 found in Earth organisms. A skeleton structure (single bonds only) of one of these extraterrestrial amino acids is

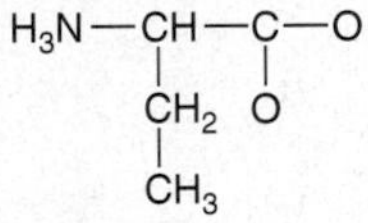

```
H3N—CH—C—O
     |  |
    CH2 O
     |
    CH3
```

Draw a Lewis structure, and identify any atoms with a nonzero formal charge.

10.90 Hydrazine (N_2H_4) is used as a rocket fuel because it reacts very exothermically with oxygen to form nitrogen gas and water vapor. The heat released and the increase in number of moles of gas provide thrust. Calculate the heat of reaction.

10.91 When gaseous sulfur trioxide is dissolved in concentrated sulfuric acid, disulfuric acid forms:

$$SO_3(g) + H_2SO_4(l) \longrightarrow H_2S_2O_7(l)$$

Use bond energies (Table 9.2, p. 341) to determine ΔH^0_{rxn}. (The S atoms in $H_2S_2O_7$ are bonded through an O atom. Assume Lewis structures with zero formal charges; BE of S=O is 552 kJ/mol.)

10.92 In addition to propyne (see Follow-up Problem 10.8), there are two other structures with molecular formula C_3H_4. Draw a Lewis structure for each, determine the shape around each carbon, and predict any deviations from ideal bond angles.

10.93 A molecule of formula AY_3 is found experimentally to be polar. Which molecular shapes are possible and which impossible for AY_3?

10.94 In contrast to the cyanate ion (NCO^-), which is stable and found in many compounds, the fulminate ion (CNO^-), with its different atom sequence, is unstable and forms compounds with heavy metal ions, such as Ag^+ and Hg^{2+}, that are explosive. Like the cyanate ion, the fulminate ion has three resonance structures. Which is the most important contributor to the resonance hybrid? Suggest a reason for the instability of fulminate.

10.95 Hydrogen cyanide (and organic nitriles, which contain the cyano group) can be catalytically reduced with hydrogen to form amines. Use Lewis structures and bond energies to determine ΔH^0_{rxn} for

$$HCN(g) + 2H_2(g) \longrightarrow CH_3NH_2(g)$$

10.96 Ethylene, C_2H_4, and tetrafluoroethylene, C_2F_4, are used to make the polymers polyethylene and polytetrafluoroethylene (Teflon), respectively.
(a) Draw the Lewis structures for C_2H_4 and C_2F_4, and give the ideal H—C—H and F—C—F bond angles.
(b) The actual H—C—H and F—C—F bond angles are 117.4° and 112.4°, respectively. Explain these deviations.

10.97 Lewis structures of mescaline, a hallucinogenic compound in peyote cactus, and dopamine, a neurotransmitter in the mammalian brain, appear below. Suggest a reason for mescaline's ability to disrupt nerve impulses.

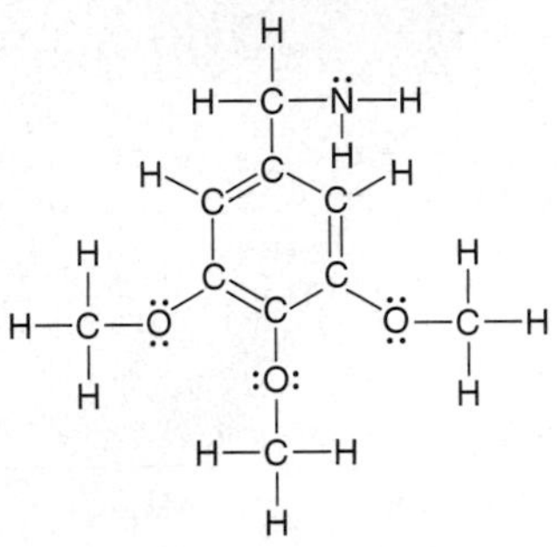

mescaline 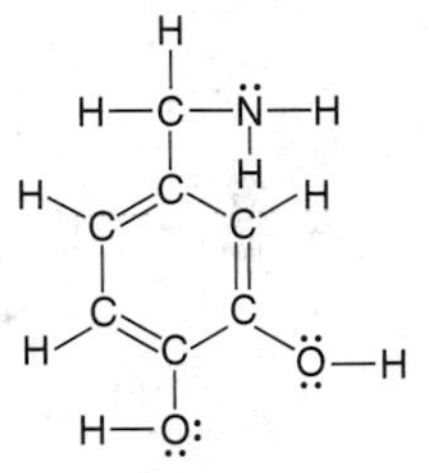   dopamine

10.98 Using bond lengths in Table 9.3 (p. 342) and assuming ideal geometry, calculate each of the following distances:
(a) Between H atoms in C_2H_2
(b) Between F atoms in SF_6 (two answers)
(c) Between equatorial F atoms in PF_5

10.99 Phosphorus pentachloride, a key industrial compound with annual world production of about 2×10^7 kg, is used to make other compounds. It reacts with sulfur dioxide to produce phosphorus oxychloride ($POCl_3$) and thionyl chloride ($SOCl_2$). Draw a Lewis structure and name the molecular shape of each product.

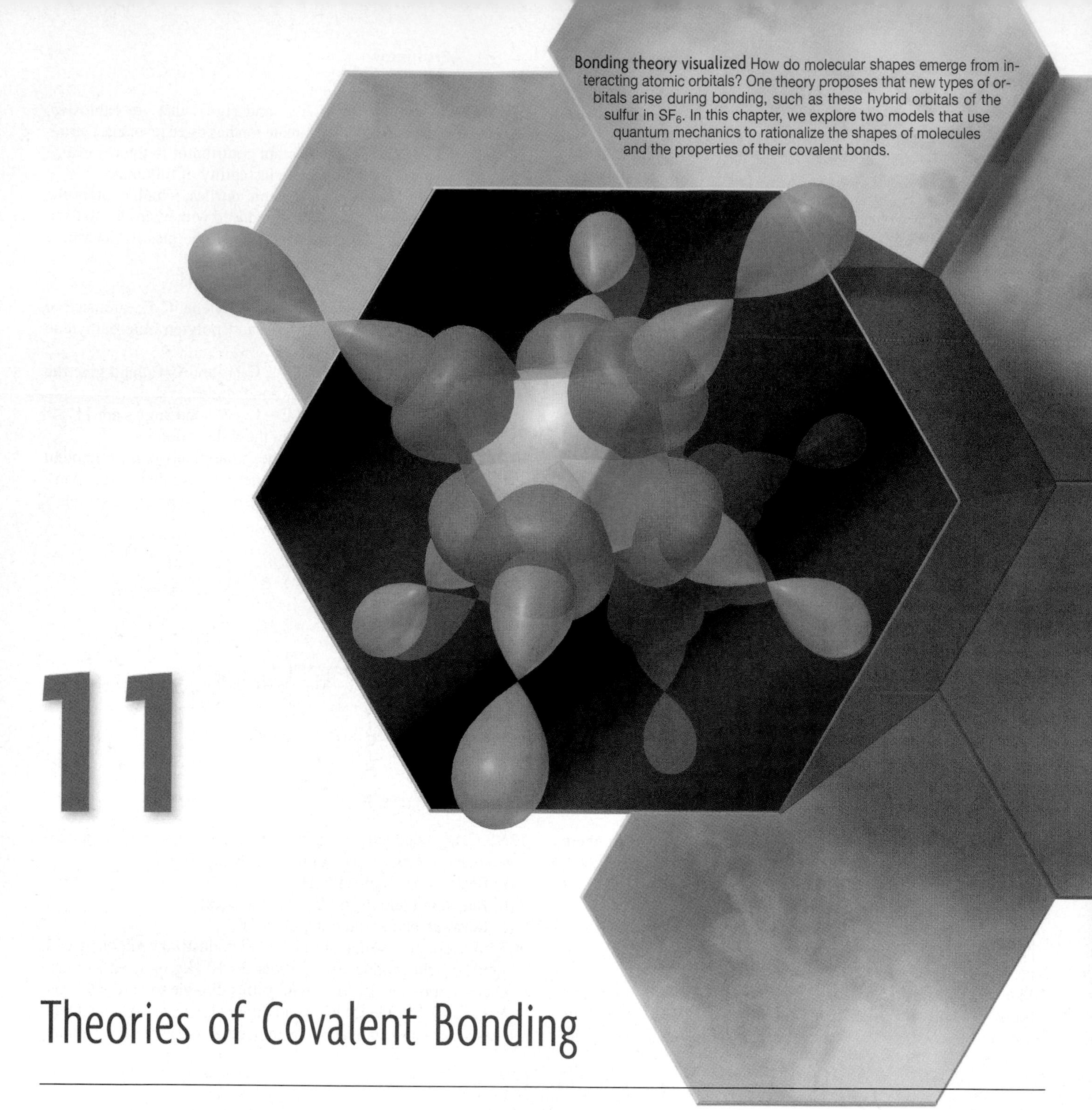

Bonding theory visualized How do molecular shapes emerge from interacting atomic orbitals? One theory proposes that new types of orbitals arise during bonding, such as these hybrid orbitals of the sulfur in SF_6. In this chapter, we explore two models that use quantum mechanics to rationalize the shapes of molecules and the properties of their covalent bonds.

11

Theories of Covalent Bonding

All scientific models have limitations because they are simplifications of reality. The VSEPR model accounts for molecular shapes by assuming that electron groups minimize their repulsions, and thus occupy as much space as possible around a central atom. But it does *not* explain how the shapes arise from interactions of atomic orbitals. After all, the orbitals we examined in Chapter 7 aren't oriented toward the corners of a tetrahedron or a trigonal bipyramid, to mention just two of the common molecular shapes. Moreover, knowing the shape doesn't help us explain the magnetic and spectral properties of molecules; only an understanding of their orbitals and energy levels can do that.

IN THIS CHAPTER . . . We use quantum mechanics, with which we explained the properties of atoms, to discuss two theories of bonding in molecules. Valence bond (VB) theory rationalizes observed molecular shapes through interactions of atomic orbitals that create new "hybrid" orbitals. We apply VB theory to sigma and pi bonding, the two types of covalent bonds. Molecular orbital (MO) theory explains molecular energy levels and properties using orbitals associated with the whole molecule. Simple diatomic and polyatomic molecules are covered. Each theory complements the other, so don't be concerned if we need more than one model to explain a complex phenomenon like bonding. In every science, one model often accounts for some aspect of a topic better than another, and several models are called into service to explain a wider range of phenomena.

Concepts & Skills to Review

before you study this chapter

- atomic orbital shapes (Section 7.4)
- the exclusion principle (Section 8.2)
- Hund's rule (Section 8.3)
- writing Lewis structures (Section 10.1)
- resonance in covalent bonding (Section 10.1)
- molecular shapes (Section 10.2)
- molecular polarity (Section 10.3)

11.1 VALENCE BOND (VB) THEORY AND ORBITAL HYBRIDIZATION

What *is* a covalent bond, and what characteristic gives it strength? And how can we explain *molecular* shapes from the interactions of *atomic* orbitals? The most useful approach for answering these questions is **valence bond (VB) theory.**

The Central Themes of VB Theory

The basic principle of VB theory is that *a covalent bond forms when orbitals of two atoms overlap and the overlap region, which is between the nuclei, is occupied by a pair of electrons.* ("Orbital overlap" is another way of saying that the two wave functions are *in phase,* so the amplitude increases between the nuclei.) The central themes of VB theory derive from this principle:

1. *Opposing spins of the electron pair.* As the exclusion principle (Section 8.2) prescribes, the space formed by the overlapping orbitals *has a maximum capacity of two electrons that must have opposite spins.* When a molecule of H_2 forms, for instance, the two 1*s* electrons of two H atoms occupy the overlapping 1*s* orbitals and have opposite spins (Figure 11.1A).

2. *Maximum overlap of bonding orbitals.* The bond strength depends on the attraction of the nuclei for the shared electrons, so *the greater the orbital overlap, the stronger (more stable) the bond.* The extent of overlap depends on the shapes and directions of the orbitals. An *s* orbital is spherical, but *p* and *d* orbitals have more electron density in one direction than in another. Thus, whenever possible, a bond involving *p* or *d* orbitals will be oriented in the direction that maximizes overlap. In the HF bond, for example, the 1*s* orbital of H overlaps the half-filled 2*p* orbital of F *along the long axis* of that orbital (Figure 11.1B). Any other direction would result in less overlap and, thus, a weaker bond. Similarly, in the F—F bond of F_2, the two half-filled 2*p* orbitals interact end to end, that is, *along the long axes* of the orbitals, to maximize overlap (Figure 11.1C).

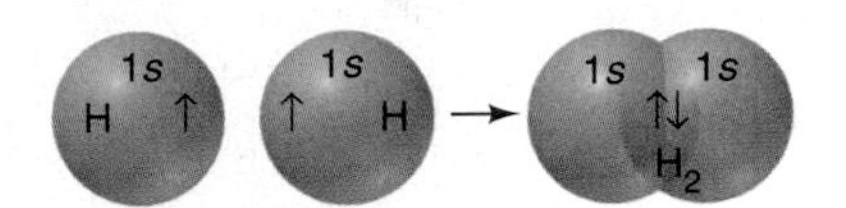

A Hydrogen, H_2

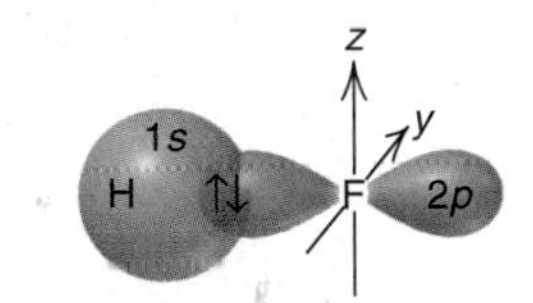

B Hydrogen fluoride, HF

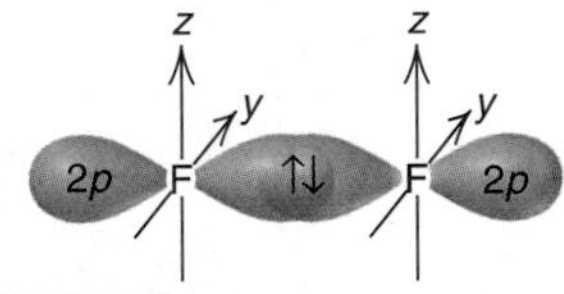

C Fluorine, F_2

Figure 11.1 Orbital overlap and spin pairing in three diatomic molecules. A, In the H_2 molecule, the two overlapping 1*s* orbitals are occupied by the two 1*s* electrons with opposite spins. (The electrons, shown as arrows, spend the most time between the nuclei but move throughout the overlapping orbitals.) **B,** To maximize overlap in HF, half-filled H 1*s* and F 2*p* orbitals overlap along the long axis of the 2*p* orbital involved in bonding. (The $2p_x$ orbital is shown bonding; the other two 2*p* orbitals of F are not shown.) **C,** In F_2, the half-filled $2p_x$ orbital on one F points end to end toward the similar orbital on the other F to maximize overlap.

3. *Hybridization of atomic orbitals.* To account for the bonding in simple diatomic molecules like HF, we picture the direct overlap of *s* and *p* orbitals of isolated atoms. But how can we account for the shapes of so many molecules and polyatomic ions through the overlap of spherical *s* orbitals, dumbbell-shaped *p* orbitals, and cloverleaf-shaped *d* orbitals?

Consider a methane molecule. It has four H atoms bonded to a central C atom. An isolated ground-state C atom ([He] $2s^22p^2$) has four valence electrons: two in the $2s$ orbital and one each in two of the three $2p$ orbitals. We might easily see how the two half-filled *p* orbitals of C could overlap with the $1s$ orbitals of two H atoms to form *two* C—H bonds with a 90° H—C—H bond angle. But methane is not CH_2, and it's not easy to see how the orbitals overlap to form the *four* C—H bonds with the 109.5° bond angle that occurs in methane.

To explain such facts, Linus Pauling proposed that *the valence atomic orbitals in the molecule are* ***different*** *from those in the isolated atoms.* Indeed, quantum-mechanical calculations show that if we "mix" specific combinations of orbitals mathematically, we obtain *new atomic orbitals.* The spatial orientations of these new orbitals lead to more stable bonds and are consistent with observed molecular shapes. The process of orbital mixing is called **hybridization,** and the new atomic orbitals are called **hybrid orbitals.** Two key points about the number and type of hybrid orbitals are that

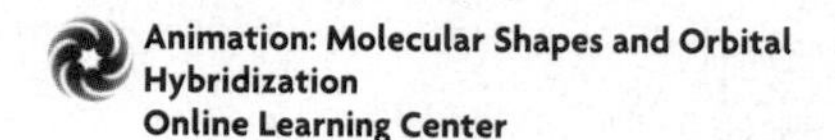

- The *number* of hybrid orbitals obtained *equals* the number of atomic orbitals mixed.
- The *type* of hybrid orbitals obtained *varies* with the types of atomic orbitals mixed.

You can imagine hybridization as a process in which atomic orbitals mix, hybrid orbitals form, and electrons enter them with spins parallel (Hund's rule) to create stable bonds. In truth, though, hybridization is a mathematically derived result from quantum mechanics that accounts for the molecular shapes we observe.

Types of Hybrid Orbitals

We postulate the presence of a certain type of hybrid orbital *after* we observe the molecular shape. As we discuss the five common types of hybridization, notice that the spatial orientation of each type of hybrid orbital corresponds with one of the five common electron-group arrangements predicted by VSEPR theory.

sp* Hybridization** When two electron groups surround the central atom, we observe a linear shape, which means that the bonding orbitals must have a linear orientation. VB theory explains this by proposing that mixing two *nonequivalent* orbitals of a central atom, one *s* and one *p*, gives rise to two *equivalent* ***sp **hybrid orbitals** that lie 180° apart, as shown in Figure 11.2A. Note the shape of the hybrid orbital: with one large and one small lobe, it differs markedly from the shapes of the atomic orbitals that were mixed. The orientations of hybrid orbitals extend electron density in the bonding direction and minimize repulsions between the electrons that occupy them. Thus, *both shape and orientation maximize overlap with the orbital of the other atom in the bond.*

In gaseous $BeCl_2$, for example, the Be atom is said to be *sp* hybridized. The rest of Figure 11.2 depicts the hybridization of Be in $BeCl_2$ in terms of a traditional orbital diagram of the valence level (part B), a similar diagram with orbital contours (C), and the bond formation with Cl (D). The $2s$ and one of the three $2p$ orbitals of Be mix and form two *sp* orbitals. These overlap $3p$ orbitals of two Cl atoms, and the four valence electrons—two from Be and one from each Cl—occupy the overlapped orbitals in pairs with opposite spins. The two unhybridized $2p$ orbitals of Be lie perpendicular to each other and to the bond axes. Thus, through hybridization, the paired $2s$ electrons in the isolated Be atom are distributed into two *sp* orbitals, which form two Be—Cl bonds.

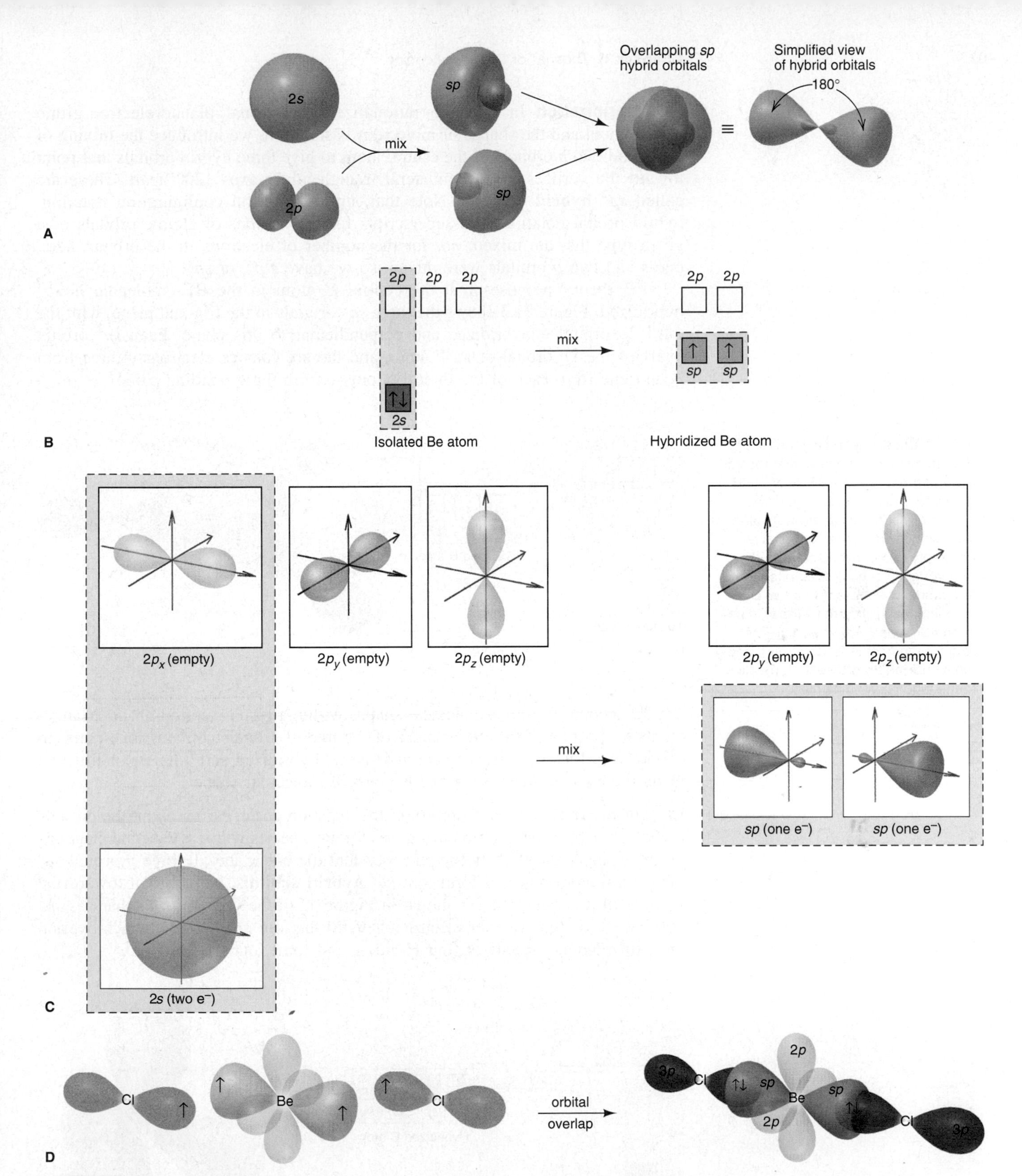

Figure 11.2 **The *sp* hybrid orbitals in gaseous $BeCl_2$. A,** One 2*s* and one 2*p* atomic orbital mix to form two *sp* hybrid orbitals (shown above and below one another and slightly to the side for ease of viewing). Note the large and small lobes of the hybrid orbitals. In the molecule, the two *sp* orbitals of Be are oriented in opposite directions. For clarity, the simplified hybrid orbitals will be used throughout the text, usually without the small lobe. **B,** The orbital diagram for hybridization in Be is drawn vertically and shows that the 2*s* and one of the three 2*p* orbitals form two *sp* hybrid orbitals, and the two other 2*p* orbitals remain unhybridized. Electrons half-fill the *sp* hybrid orbitals. During bonding, each *sp* orbital fills by sharing an electron from Cl (not shown). **C,** The orbital diagram is shown with orbital contours instead of electron arrows. **D,** $BeCl_2$ forms by overlap of the two *sp* hybrids with the 3*p* orbitals of two Cl atoms; the two unhybridized Be 2*p* orbitals lie perpendicular to the *sp* hybrids. (For clarity, only the 3*p* orbital involved in bonding is shown for each Cl.)

sp^2 Hybridization In order to rationalize the trigonal planar electron-group arrangement and the shapes of molecules based on it, we introduce the mixing of one *s* and two *p* orbitals of the central atom to give three hybrid orbitals that point toward the vertices of an equilateral triangle, their axes 120° apart. These are called **sp^2 hybrid orbitals.** Note that, unlike electron configuration notation, hybrid orbital notation uses superscripts for the *number* of atomic orbitals of a given type that are mixed, *not* for the number of electrons in the orbital: here, one *s* and two *p* orbitals were mixed, so we have s^1p^2, or sp^2.

VB theory proposes that the central B atom in the BF_3 molecule is sp^2 hybridized. Figure 11.3 shows the three sp^2 orbitals in the trigonal plane, with the third 2*p* orbital unhybridized and perpendicular to this plane. Each sp^2 orbital overlaps the 2*p* orbital of an F atom, and the six valence electrons—three from B and one from each of the three F atoms—form three bonding pairs.

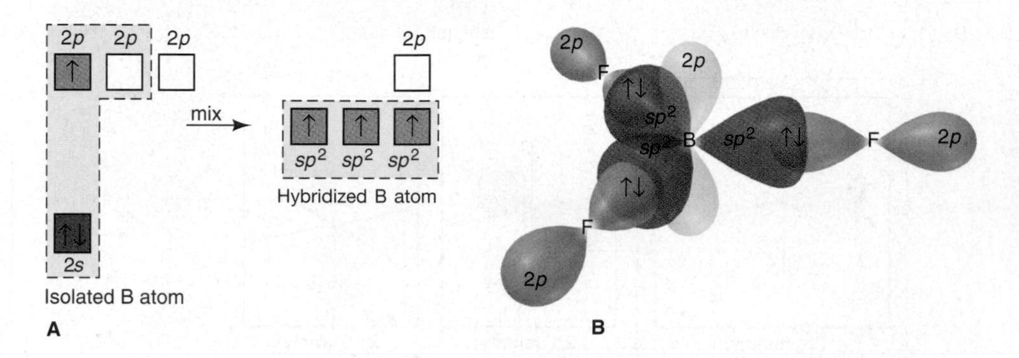

Figure 11.3 The sp^2 hybrid orbitals in BF_3. **A,** The orbital diagram shows that the 2*s* and two of the three 2*p* orbitals of the B atom mix to make three sp^2 hybrid orbitals. Three electrons (*up arrows*) half-fill the sp^2 hybrids. The third 2*p* orbital remains empty and unhybridized. **B,** BF_3 forms through overlap of 2*p* orbitals on three F atoms with the sp^2 hybrids. During bonding, each sp^2 orbital fills with an electron from one F (*down arrow*). The three sp^2 hybrids of B lie 120° apart, and the unhybridized 2*p* orbital is perpendicular to the trigonal bonding plane.

To account for other molecular shapes within a given electron-group arrangement, we postulate that one or more of the hybrid orbitals contains lone pairs. In ozone (O_3), for example, the central O is sp^2 hybridized and a lone pair fills one of its three sp^2 orbitals, so ozone has a bent molecular shape.

sp^3 Hybridization Now let's return to the question posed earlier about the orbitals in methane, the same question that arises for any species with a tetrahedral electron-group arrangement. VB theory proposes that the one *s* and all three *p* orbitals of the central atom mix and form four **sp^3 hybrid orbitals,** which point toward the vertices of a tetrahedron. As shown in Figure 11.4, the C atom in methane is sp^3 hybridized. Its four valence electrons half-fill the four sp^3 hybrids, which overlap the half-filled 1*s* orbitals of four H atoms and form four C—H bonds.

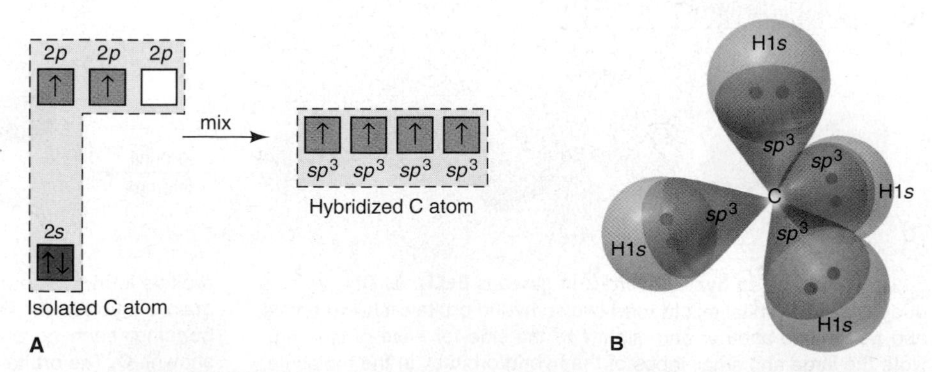

Figure 11.4 The sp^3 hybrid orbitals in CH_4. **A,** The 2*s* and all three 2*p* orbitals of C are mixed to form four sp^3 hybrids. Carbon's four valence electrons half-fill the sp^3 hybrids. **B,** In methane, the four sp^3 orbitals of C point toward the corners of a tetrahedron and overlap the 1*s* orbitals of four H atoms. Each sp^3 orbital fills by addition of an electron from one H (electrons are shown as dots).

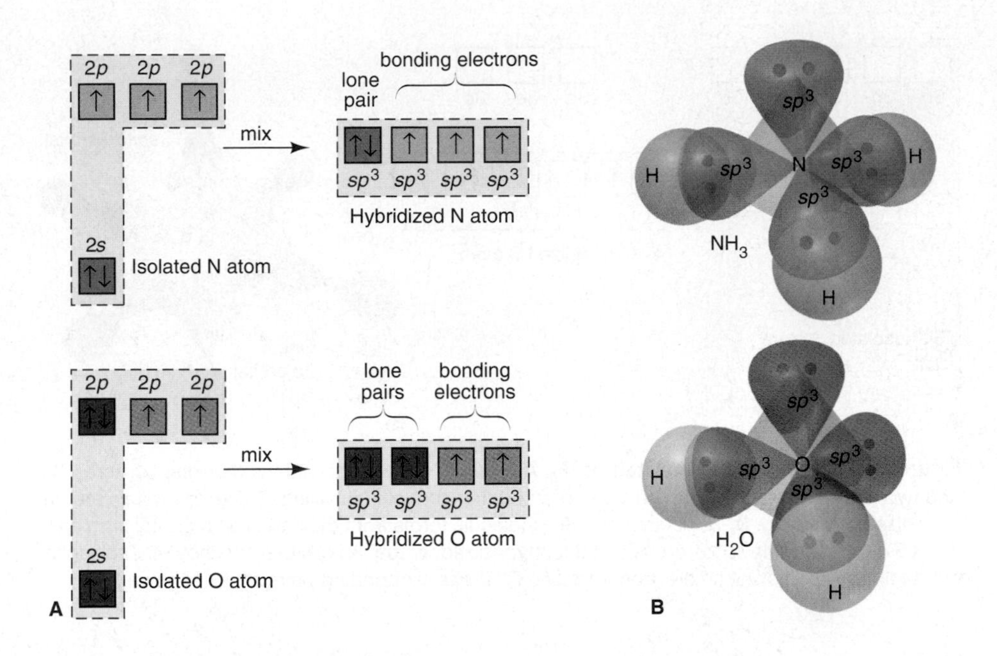

Figure 11.5 The sp^3 hybrid orbitals in NH_3 and H_2O. A, The orbital diagrams show sp^3 hybridization, as in CH_4. In NH_3 (*top*), one sp^3 orbital is filled with a lone pair. In H_2O (*bottom*), two sp^3 orbitals are filled with lone pairs. **B,** Contour diagrams show the tetrahedral orientation of the sp^3 orbitals and the overlap of the bonded H atoms. Each half-filled sp^3 orbital fills by addition of an electron from one H. (Shared pairs and lone pairs are shown as dots.)

Figure 11.5 shows the bonding in molecules of the other shapes with the tetrahedral arrangement. The trigonal pyramidal shape of NH_3 arises when a lone pair fills one of the four sp^3 orbitals of N, and the bent shape of H_2O arises when lone pairs fill two of the sp^3 orbitals of O.

sp^3d Hybridization The shapes of molecules with trigonal bipyramidal or octahedral electron-group arrangements are rationalized with VB theory through similar arguments. The only new point is that such molecules have central atoms from Period 3 or higher, so atomic *d* orbitals, as well as *s* and *p* orbitals, are mixed to form the hybrid orbitals.

To rationalize the trigonal bipyramidal shape of the PCl_5 molecule, for example, the VB model proposes that the one 3*s*, the three 3*p*, and one of the five 3*d* orbitals of the central P atom mix and form five **sp^3d hybrid orbitals,** which point to the vertices of a trigonal bipyramid. Each hybrid orbital overlaps a 3*p* orbital of a Cl atom, and the five valence electrons of P, together with one from each of the five Cl atoms, pair up to form five P—Cl bonds, as shown in Figure 11.6. Other shapes in this electron-group arrangement have lone pairs in one or more of the central atom's sp^3d orbitals.

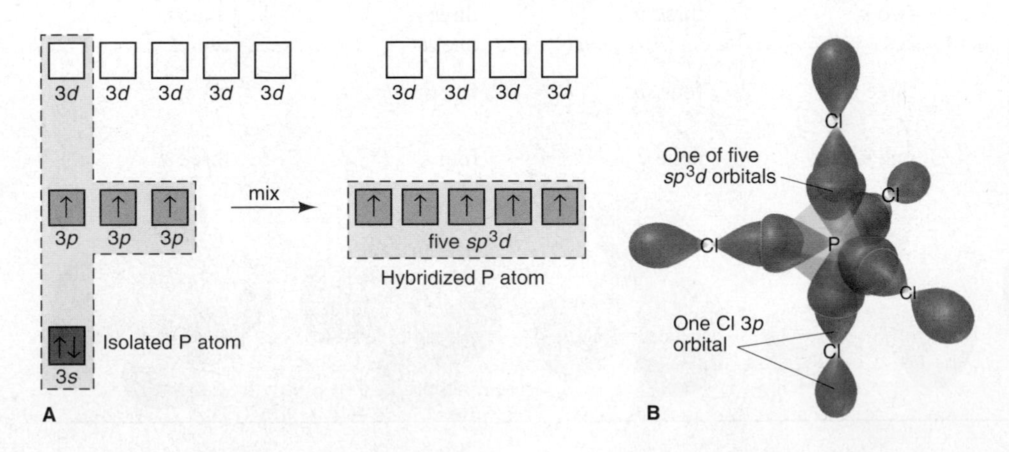

Figure 11.6 The sp^3d hybrid orbitals in PCl_5. A, The orbital diagram shows that one 3*s*, three 3*p*, and one of the five 3*d* orbitals of P mix to form five sp^3d orbitals that are half-filled. Four 3*d* orbitals are unhybridized and empty. **B,** The trigonal bipyramidal PCl_5 molecule forms by the overlap of a 3*p* orbital from each of the five Cl atoms with the sp^3d hybrid orbitals of P (unhybridized, empty 3*d* orbitals not shown). Each sp^3d orbital fills by addition of an electron from one Cl. (These five bonding pairs are not shown.)

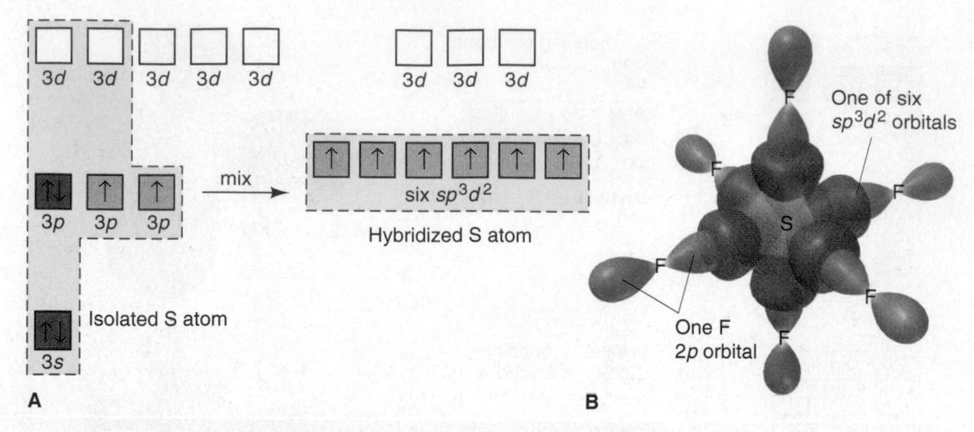

Figure 11.7 The sp^3d^2 hybrid orbitals in SF_6. A, The orbital diagram shows that one 3*s*, three 3*p*, and two 3*d* orbitals of S mix to form six sp^3d^2 orbitals that are half-filled. Three 3*d* orbitals are unhybridized and empty. **B,** The octahedral SF_6 molecule forms from overlap of a 2*p* orbital from each of six F atoms with the sp^3d^2 orbitals of S (unhybridized, empty 3*d* orbitals not shown). Each sp^3d^2 orbital fills by addition of an electron from one F. (These six bonding pairs are not shown.)

sp^3d^2 Hybridization To rationalize the shape of SF_6, the VB model proposes that the one 3*s*, the three 3*p*, and two of the five 3*d* orbitals of the central S atom mix and form six **sp^3d^2 hybrid orbitals,** which point to the vertices of an octahedron. Each hybrid orbital overlaps a 2*p* orbital of an F atom, and the six valence electrons of S, together with one from each of the six F atoms, pair up to form six S—F bonds (Figure 11.7). Square pyramidal and square planar molecules have lone pairs in one and two of the central atom's sp^3d^2 orbitals, respectively.

Table 11.1 summarizes the numbers and types of atomic orbitals that are mixed to obtain the five types of hybrid orbitals. Once again, note the similarities between the orientations of the hybrid orbitals proposed by VB theory and the shapes predicted by VSEPR theory (see Figure 10.2, p. 376). Figure 11.8 shows the three conceptual steps from molecular formula to postulating the hybrid orbitals in the molecule, and Sample Problem 11.1 details the end of that process.

Table 11.1 Composition and Orientation of Hybrid Orbitals

	Linear	Trigonal Planar	Tetrahedral	Trigonal Bipyramidal	Octahedral
Atomic orbitals mixed	one *s* one *p*	one *s* two *p*	one *s* three *p*	one *s* three *p* one *d*	one *s* three *p* two *d*
Hybrid orbitals formed	two *sp*	three sp^2	four sp^3	five sp^3d	six sp^3d^2
Unhybridized orbitals remaining	two *p*	one *p*	none	four *d*	three *d*
Orientation					

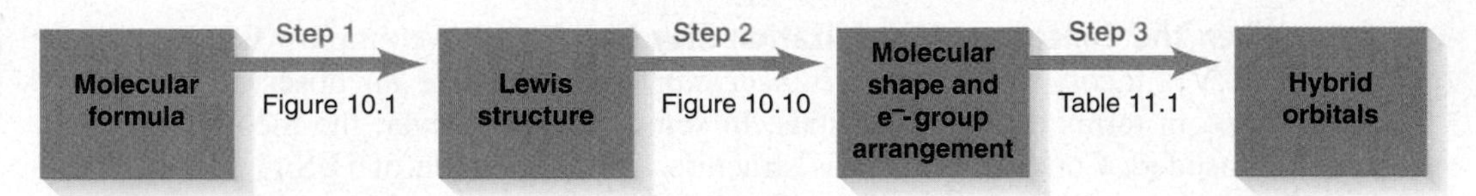

Figure 11.8 The conceptual steps from molecular formula to the hybrid orbitals used in bonding. (See Figures 10.1, page 366, and 10.10, page 383.)

SAMPLE PROBLEM 11.1 Postulating Hybrid Orbitals in a Molecule

Problem Use partial orbital diagrams to describe how mixing of the atomic orbitals of the central atom leads to the hybrid orbitals in each of the following:
(a) Methanol, CH_3OH (b) Sulfur tetrafluoride, SF_4

Plan From the Lewis structure, we determine the number and arrangement of electron groups around the central atoms, along with the molecular shape. From that, we postulate the type of hybrid orbitals involved. Then, we write the partial orbital diagram for each central atom before and after the orbitals are hybridized.

Solution (a) For CH_3OH. The electron-group arrangement is tetrahedral around both C and O atoms. Therefore, each central atom is sp^3 hybridized. The C atom has four half-filled sp^3 orbitals:

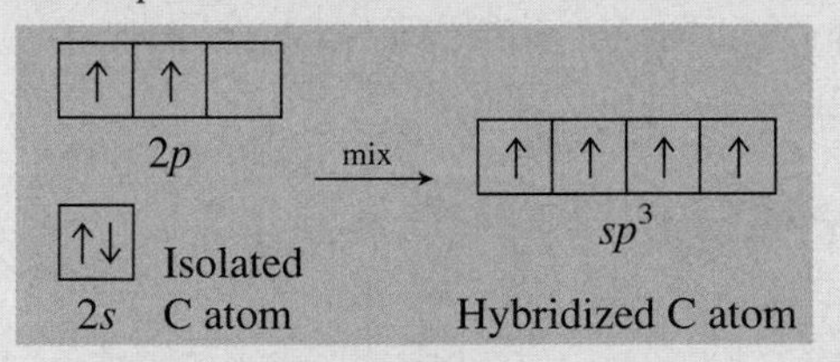

The O atom has two half-filled sp^3 orbitals and two filled with lone pairs:

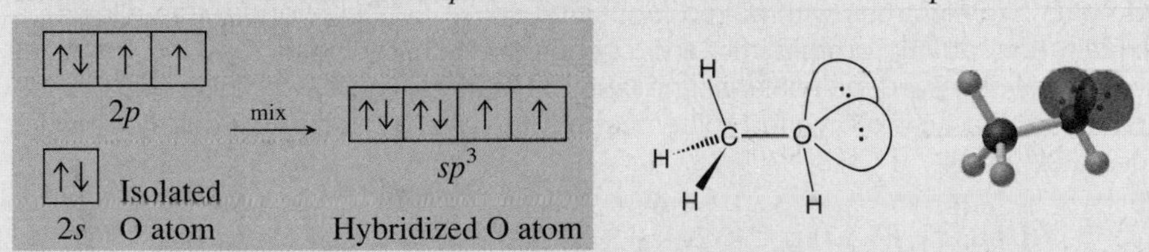

During bonding, each half-filled C or O orbital becomes filled. The second electron comes from an H atom, or, in the C—O bond, from the overlapping C and O orbitals.

(b) For SF_4. The molecular shape is seesaw, which is based on the trigonal bipyramidal electron-group arrangement. Thus, the central S atom is surrounded by five electron groups, which implies sp^3d hybridization. One 3*s* orbital, three 3*p* orbitals, and one 3*d* orbital are mixed. One hybrid orbital is filled with a lone pair, and four are half-filled. Four unhybridized 3*d* orbitals remain empty:

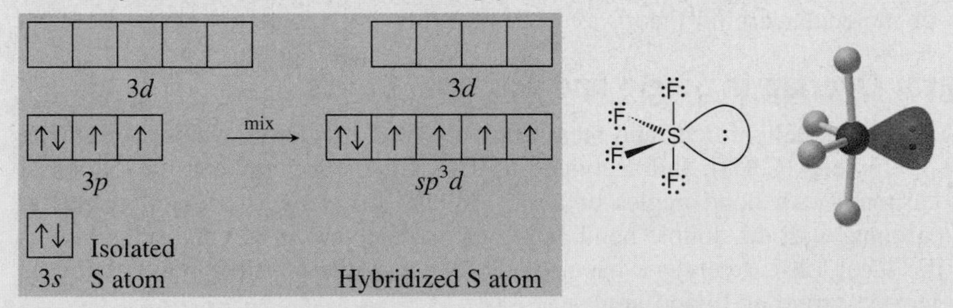

During bonding, each half-filled S orbital becomes filled, with the second electrons coming from F atoms.

FOLLOW-UP PROBLEM 11.1 Use partial orbital diagrams to show how the atomic orbitals of the central atoms mix to form hybrid orbitals in (a) beryllium fluoride, BeF_2; (b) silicon tetrachloride, $SiCl_4$; (c) xenon tetrafluoride, XeF_4.

When the Concept of Hybridization May Not Apply We employ VSEPR theory and VB theory *whenever it is necessary* to rationalize an observed molecular shape in terms of atomic orbitals. In some cases, however, the theories may not be needed. Consider the Lewis structure and bond angle of H_2S:

S
H 92° H

Based on VSEPR theory, we would predict that, as in H_2O, the four electron groups around H_2S point to the vertices of a tetrahedron, and the two lone pairs compress the H—S—H bond angle somewhat below the ideal 109.5°. Based on VB theory, we would propose that the 3*s* and 3*p* orbitals of the central S atom mix and form four sp^3 hybrids, two of which are filled with lone pairs, while the other two overlap 1*s* orbitals of two H atoms and are filled with bonding pairs.

The problem is that observation does *not* support these arguments. In fact, the H_2S molecule has a bond angle of 92°, close to the 90° angle expected between *unhybridized,* perpendicular, atomic *p* orbitals. Similar angles occur in the hydrides of other Group 6A(16) elements and also in the hydrides of larger Group 5A(15) elements. It makes no sense to apply a theory when the facts don't warrant it. In the case of H_2S and these other hydrides, neither VSEPR theory nor the concept of hybridization applies. It's important to remember that real factors, such as bond length, atomic size, and electron-electron repulsions, influence molecular shape. Apparently, with these larger central atoms and their longer bonds to H, crowding and electron-electron repulsions decrease, and the simple overlap of unhybridized atomic orbitals rationalizes the observed shapes perfectly well.

SECTION SUMMARY

VB theory explains that a covalent bond forms when two atomic orbitals overlap. The bond holds two electrons with paired (opposite) spins. Orbital hybridization allows us to explain how atomic orbitals mix and change their characteristics during bonding. Based on the observed molecular shape (and the related electron-group arrangement), we postulate the type of hybrid orbital needed. In certain cases, we need not invoke hybridization at all.

11.2 THE MODE OF ORBITAL OVERLAP AND THE TYPES OF COVALENT BONDS

In this section, we focus on the *mode* by which orbitals overlap—end to end or side to side—to see the detailed makeup of covalent bonds. These two modes give rise to the two types of covalent bonds—*sigma* bonds and *pi* bonds. We'll use valence bond theory to describe the two types here, but they are essential features of molecular orbital theory as well.

Orbital Overlap in Single and Multiple Bonds

The VSEPR model predicts, and measurements verify, different shapes for ethane (C_2H_6), ethylene (C_2H_4), and acetylene (C_2H_2). Ethane is tetrahedrally shaped at both carbons, with bond angles of about 109.5°. Ethylene is trigonal planar at both carbons, with the double bond acting as one electron group and bond angles near the ideal 120°. Acetylene has a linear shape, with the triple bond acting as one electron group and bond angles of 180°:

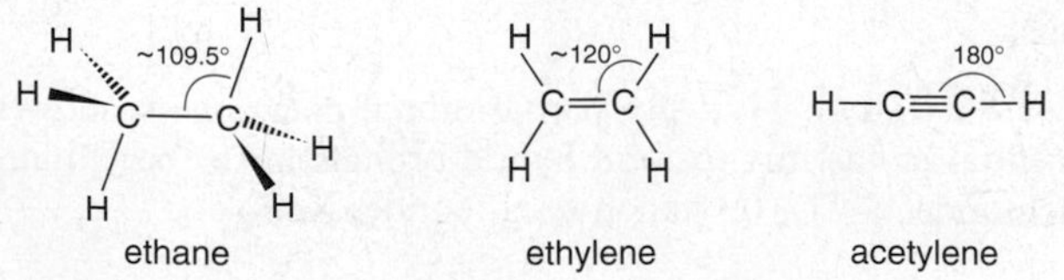

~109.5° ~120° 180°
H—C≡C—H
ethane ethylene acetylene

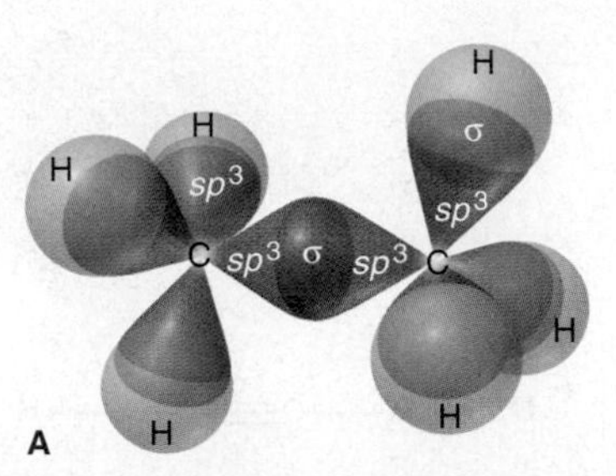

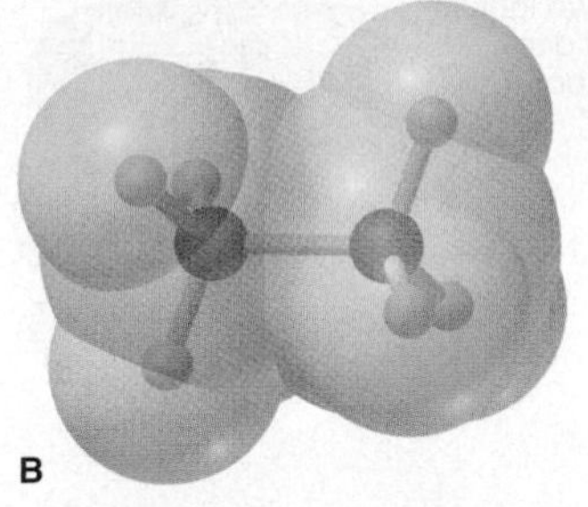

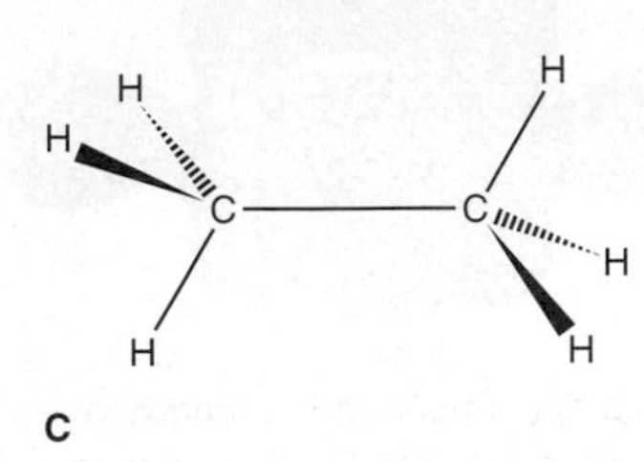

Figure 11.9 The σ bonds in ethane (C_2H_6). **A,** Two sp^3 hybridized C atoms and six H atoms in ethane form one C—C σ bond and six C—H σ bonds. **B,** Electron density is distributed relatively evenly among the σ bonds. **C,** Bond-line drawing.

A close look at the bonds shows two modes of orbital overlap, which correspond to the two types of covalent bonds:

1. *End-to-end overlap and sigma (σ) bonding.* Both C atoms of ethane are sp^3 hybridized (Figure 11.9). The C—C bond involves overlap of one sp^3 orbital from each C, and each of the six C—H bonds involves overlap of a C sp^3 orbital with an H 1*s* orbital. The bonds in ethane are like all the others described so far in this chapter. Look closely at the C—C bond, for example. It involves the overlap of the end of one orbital with the end of the other. The bond resulting from such *end-to-end* overlap is called a **sigma (σ) bond.** It has its *highest electron density along the bond axis* (an imaginary line joining the nuclei) and is shaped like an ellipse rotated about its long axis (the shape resembles a football). All single bonds, formed by any combination of overlapping hybrid, *s*, or *p* orbitals, have their electron density concentrated along the bond axis, and thus are σ bonds.

2. *Side-to-side overlap and pi (π) bonding.* A close look at Figure 11.10 reveals the double nature of the carbon-carbon bond in ethylene. Here, each C atom is sp^2 hybridized. Each C atom's four valence electrons half-fill its three sp^2 orbitals and its unhybridized 2*p* orbital, which lies perpendicular to the sp^2 plane. Two sp^2 orbitals of each C form C—H σ bonds by overlapping the 1*s* orbitals of two H atoms. The third sp^2 orbital forms a C—C σ bond with an sp^2 orbital of the other C because their orientation allows end-to-end overlap. With the σ-bonded C atoms near each other, their half-filled unhybridized 2*p* orbitals are close enough to overlap *side to side*. Such overlap forms another type of covalent bond called a **pi (π) bond.** It has *two regions of electron density,* one above and one below the σ-bond axis. *One π bond holds two electrons that occupy both regions of the bond. A double bond always consists of one σ bond and one π bond.* As shown in Figure 11.10C, the double bond increases electron density between the C atoms.

Now we can answer a question brought up in our discussion of VSEPR theory in Chapter 10: why do the two electron pairs in a double bond act as one electron group; that is, why don't they push each other apart? The answer is that each electron pair occupies a distinct orbital, a specific region of electron density, so repulsions are reduced.

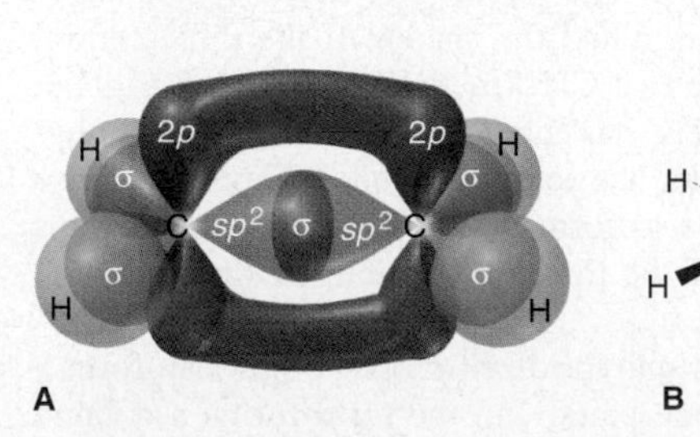

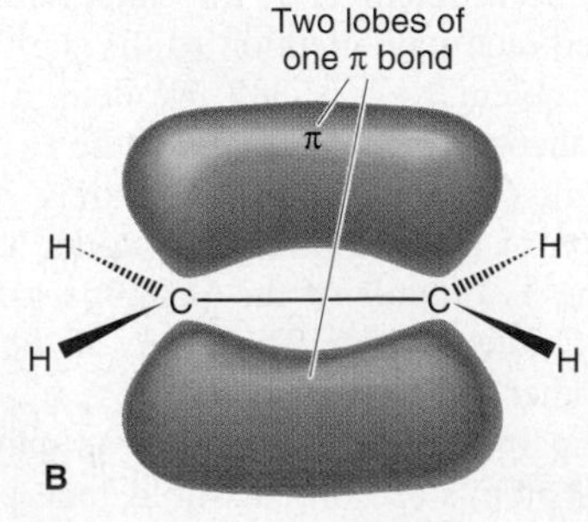

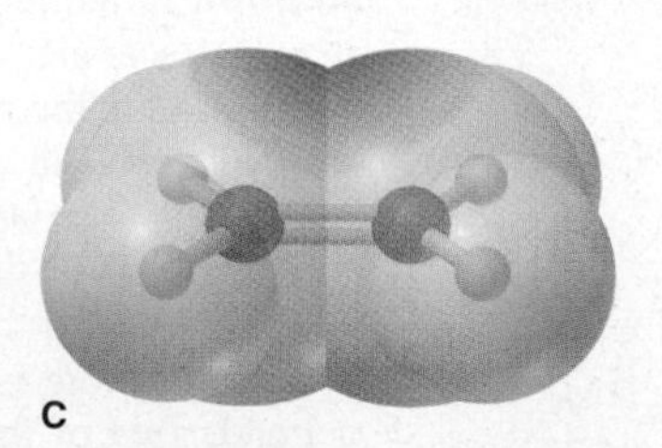

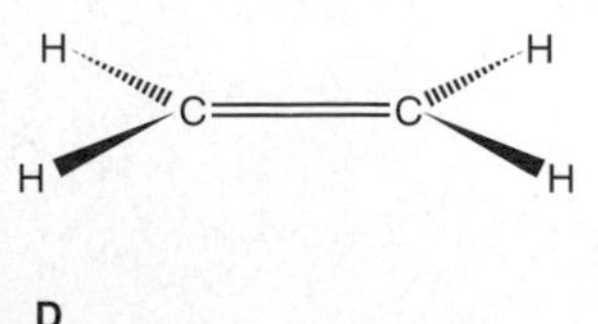

Figure 11.10 The σ and π bonds in ethylene (C_2H_4). **A,** Two sp^2 hybridized C atoms form one C—C σ bond and four C—H σ bonds. Half-filled unhybridized 2*p* orbitals lie perpendicular to the σ-bond axis and overlap side to side. **B,** The two overlapping regions comprise *one* π bond, which is occupied by two electrons. The σ bonds are shown as a line and wedges. **C,** With four electrons (one σ bond and one π bond) between the C atoms, electron density (*red shading*) is high. **D,** Bond-line drawing.

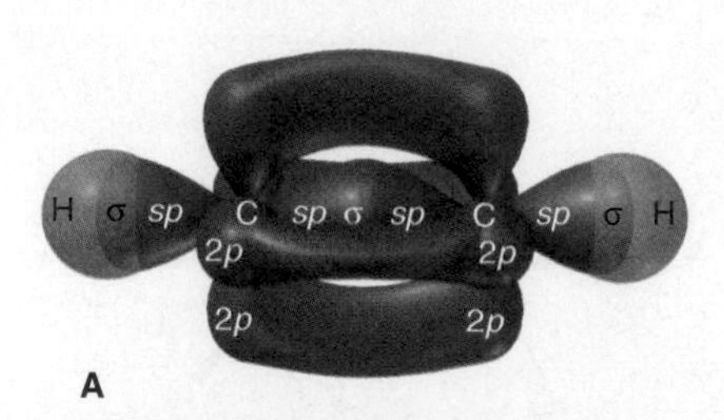

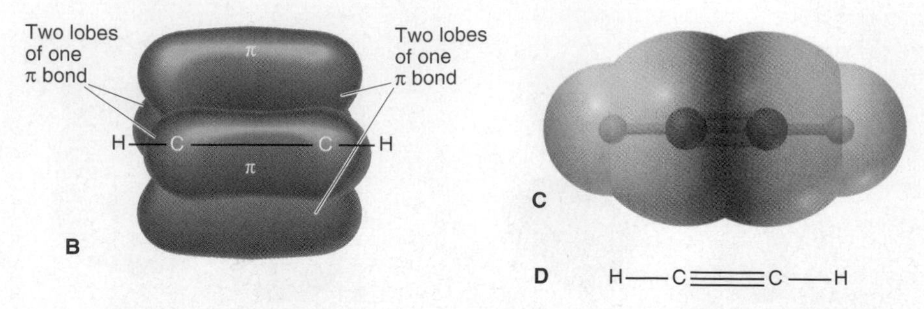

Figure 11.11 The σ bonds and π bonds in acetylene (C_2H_2). **A,** The C—C σ bond in acetylene forms when an *sp* orbital of each C overlaps. Two C—H σ bonds form from overlap of the other *sp* orbital of each C and an *s* orbital of an H. Two unhybridized 2*p* orbitals on each C are shown overlapping. **B,** Side-to-side overlap of pairs of 2*p* orbitals results in two π bonds, one with its lobes above and below the C—C σ bond (shown as a line), and the other with its lobes in front of and behind the σ bond. **C,** The two π bonds give the molecule cylindrical symmetry. Six electrons—one σ bond and two π bonds—create greater electron density between the C atoms than in ethylene. **D,** Bond-line drawing.

A triple bond, such as the C≡C bond in acetylene, *consists of one σ and two π bonds* (Figure 11.11). To maximize overlap in a linear shape, one *s* and one *p* orbital in *each* C atom form two *sp* hybrids, and two 2*p* orbitals remain unhybridized. The four valence electrons half-fill all four orbitals. Each C uses one of its *sp* orbitals to form a σ bond with an H atom and uses the other to form the C—C σ bond. Side-to-side overlap of one pair of 2*p* orbitals gives one π bond, with electron density above and below the σ bond. Side-to-side overlap of the other pair of 2*p* orbitals gives the other π bond, 90° away from the first, with electron density in front and back of the σ bond. The result is a *cylindrically symmetrical* H—C≡C—H molecule. Note the greater electron density between the C atoms created by the six bonding electrons. Figure 11.12 shows electron density relief maps of the carbon-carbon bonds in these compounds; note the increasing electron density between the nuclei from single to double to triple bond.

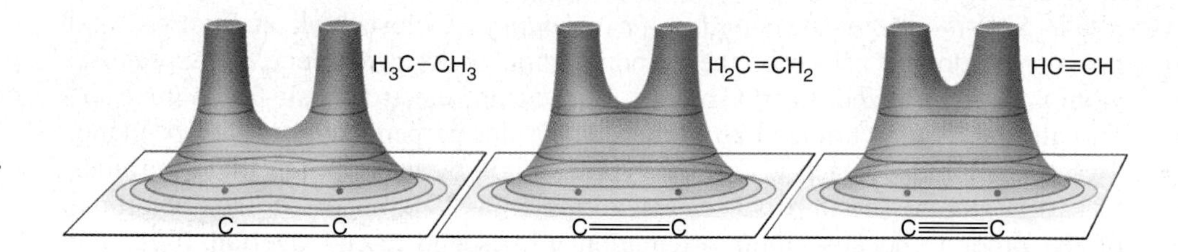

Figure 11.12 Electron density and bond order. Relief maps of the carbon-carbon bonding region in ethane, ethylene, and acetylene show a large increase in electron density as the bond order increases.

The extent of overlap influences bond strength. Because side-to-side overlap is not as extensive as end-to-end overlap, we might expect a π bond to be weaker than a σ bond, and thus a double bond should be less than twice as strong as a single bond. This expectation *is* borne out for carbon-carbon bonds. However, many factors, such as lone-pair repulsions, bond polarities, and other electrostatic contributions, affect overlap and the relative strength of σ and π bonds between other pairs of atoms. Thus, as a rough approximation, a double bond is about twice as strong as a single bond, and a triple bond is about three times as strong.

SAMPLE PROBLEM 11.2 Describing the Types of Bonds in Molecules

Problem Describe the types of bonds and orbitals in acetone, $(CH_3)_2CO$.

Plan We note, in Sample Problem 11.1, the shape around each central atom to postulate the hybrid orbitals used, and pay attention to the multiple bonding of the C=O bond.

Solution In Sample Problem 10.8, p. 385, we determined the shapes of the three central atoms of acetone: tetrahedral around each C of the two CH_3 (methyl) groups and trigonal planar around the middle C atom. Thus, the middle C has three sp^2 orbitals and one unhybridized *p* orbital. Each of the two methyl C atoms has four sp^3 orbitals. Three of these sp^3 orbitals overlap the 1*s* orbitals of the H atoms to form σ bonds; the fourth overlaps an sp^2 orbital of the middle C atom. Thus, two of the three sp^2 orbitals of the middle C form σ bonds to the other two C atoms.

The O atom is also sp^2 hybridized and has an unhybridized *p* orbital that can form a π bond. Two of the O atom's sp^2 orbitals hold lone pairs, and the third forms a σ bond with the third sp^2 orbital of the middle C atom. The unhybridized, half-filled *p* orbitals of C and O form a π bond. The σ and π bonds constitute the C=O bond:

σ bonds

π bond (shown with molecule rotated 90°)

Comment How can we tell if a terminal atom, such as the O atom in acetone, is hybridized? After all, it could use two perpendicular *p* orbitals for the σ and π bonds with C and leave the other *p* and the *s* orbital to hold the two lone pairs. But, having each lone pair in an sp^2 orbital oriented away from the C=O bond, rather than in an *s* orbital, lowers repulsions. Moreover, in some compounds, the O of a C=O group is bonded to another atom, and those bond angles suggest that the O is sp^2 hybridized.

FOLLOW-UP PROBLEM 11.2 Describe the types of bonds and orbitals in **(a)** hydrogen cyanide, HCN, and **(b)** carbon dioxide, CO_2.

Orbital Overlap and Molecular Rotation

The type of bond influences the ability of one part of a molecule to rotate relative to another part. A *σ bond allows free rotation* of the parts of the molecule with respect to each other because the extent of overlap is not affected. If you could hold one CH_3 group of the ethane molecule, the other CH_3 group could spin like a pinwheel without affecting the C—C σ-bond overlap (see Figure 11.9).

However, *p* orbitals must be parallel to engage in side-to-side overlap, so *a π bond restricts rotation* around it. Rotating one CH_2 group in ethylene with respect to the other must decrease the side-to-side overlap and break the π bond. You can see why distinct *cis* and *trans* structures can exist for molecules with double bonds, such as 1,2-dichloroethylene (Section 10.3). As Figure 11.13 shows, the π bond allows two *different* arrangements of atoms around the two C atoms, which can have a major effect on molecular polarity (see p. 389). (Rotation around a triple bond is not meaningful: because each triple-bonded C atom is attached to only one other group, there can be no difference in their relative positions.)

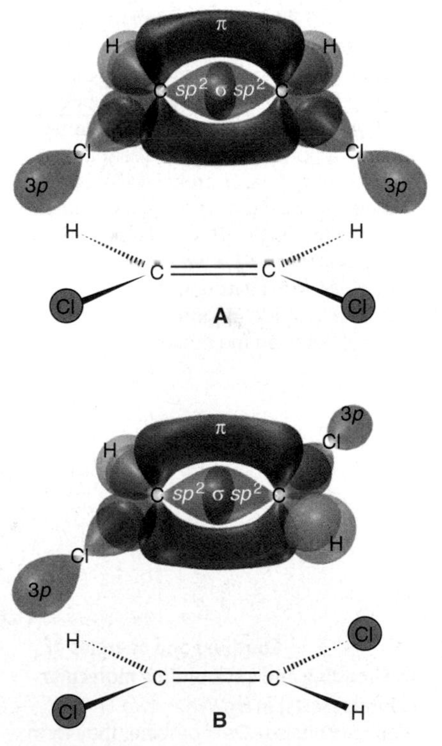

Figure 11.13 Restricted rotation of π-bonded molecules. A, *Cis*- and **B,** *trans*-1,2-dichloroethylene occur as distinct molecules because the π bond between the C atoms restricts rotation and maintains two different relative positions of the H and Cl atoms.

SECTION SUMMARY

End-to-end overlap of atomic orbitals forms a σ bond and allows free rotation of the parts of the molecule. Side-to-side overlap forms a π bond, which restricts rotation. A multiple bond consists of a σ bond and either one π bond (double bond) or two π bonds (triple bond). Multiple bonds have greater electron density than single bonds.

11.3 MOLECULAR ORBITAL (MO) THEORY AND ELECTRON DELOCALIZATION

Scientists choose the model that best helps them answer a particular question. If the question concerns molecular shape, chemists choose the VSEPR model, followed by hybrid-orbital analysis with VB theory. But VB theory does not adequately explain magnetic and spectral properties, and it understates the importance of electron delocalization. In order to deal with these phenomena, which involve molecular energy levels, chemists choose **molecular orbital (MO) theory.**

In VB theory, a molecule is pictured as a group of atoms bound together through *localized* overlap of valence-shell atomic orbitals. In MO theory, a molecule is pictured as a collection of nuclei with the electron orbitals *delocalized* over the entire molecule. The MO model is a quantum-mechanical treatment for molecules similar to the one for atoms in Chapter 8. Just as an atom has atomic orbitals (AOs) with a given energy and shape that are occupied by the atom's electrons, a molecule has **molecular orbitals (MOs)** with a given energy and shape that are occupied by the molecule's electrons. Despite the great usefulness of MO theory, it too has a drawback: MOs are more difficult to visualize than the easily depicted shapes of VSEPR theory or the hybrid orbitals of VB theory.

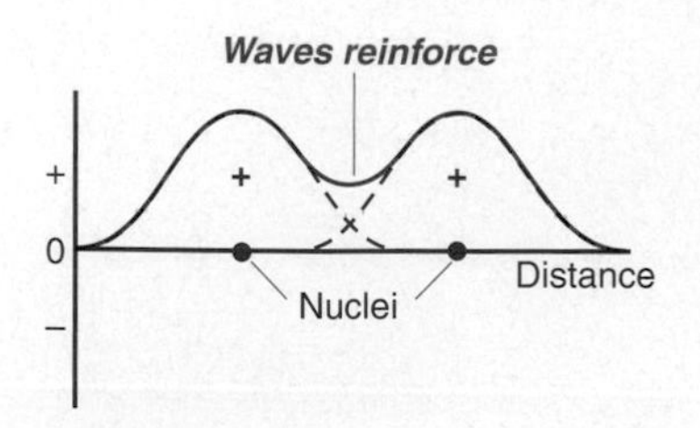

A Amplitudes of wave functions added

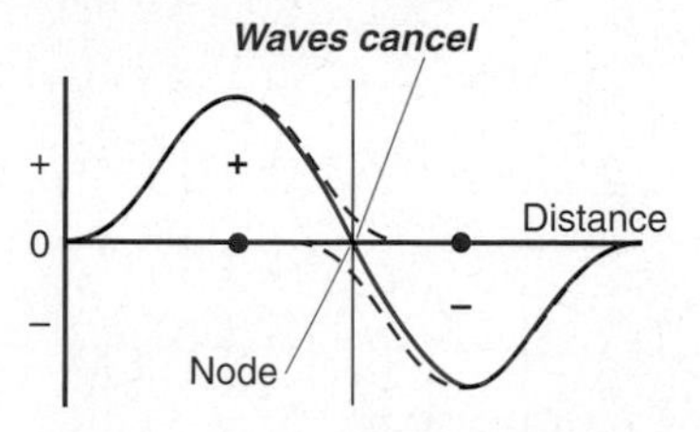

B Amplitudes of wave functions subtracted

Figure 11.14 An analogy between light waves and atomic wave functions. When light waves undergo interference, their amplitudes either add together or subtract. **A,** When the amplitudes of atomic wave functions *(dashed lines)* are added, a bonding molecular orbital (MO) results, and electron density *(red line)* increases between the nuclei. **B,** Conversely, when the amplitudes of the wave functions are subtracted, an antibonding MO results, which has a node (region of zero electron density) between the nuclei.

The Central Themes of MO Theory

Several key ideas of MO theory appear in its description of the hydrogen molecule and other simple species. These ideas include the formation of MOs, their energy and shape, and how they fill with electrons.

Formation of Molecular Orbitals Because electron motion is so complex, we use approximations to solve the Schrödinger equation for an atom with more than one electron. Similar complications arise even with H_2, the simplest molecule, so we use approximations to solve for the properties of MOs. The most common approximation mathematically *combines* (adds or subtracts) the atomic orbitals (atomic wave functions) of nearby atoms to form MOs (molecular wave functions).

When two H nuclei lie near each other, as in H_2, their AOs overlap. The two ways of combining the AOs are as follows:

- *Adding the wave functions together.* This combination forms a **bonding MO,** which has *a region of high electron density between the nuclei.* Additive overlap is analogous to light waves reinforcing each other, making the resulting amplitude higher and the light brighter. For electron waves, the overlap *increases* the probability that the electrons are between the nuclei (Figure 11.14A).
- *Subtracting the wave functions from each other.* This combination forms an **antibonding MO,** which has *a region of zero electron density (a node) between the nuclei* (Figure 11.14B). Subtractive overlap is analogous to light waves canceling each other, so that the light disappears. With electron waves, the probability that the electrons lie between the nuclei *decreases* to zero.

The two possible combinations for hydrogen atoms H_A and H_B are

AO of H_A + AO of H_B = bonding MO of H_2 (more e^- density between nuclei)

AO of H_A − AO of H_B = antibonding MO of H_2 (less e^- density between nuclei)

Notice that *the number of AOs combined always equals the number of MOs formed:* two H atomic orbitals combine to form two H_2 molecular orbitals.

Energy and Shape of H_2 Molecular Orbitals *The bonding MO is lower in energy and the antibonding MO higher in energy than the AOs that combined to form them.* Let's examine Figure 11.15 to see why this is so.

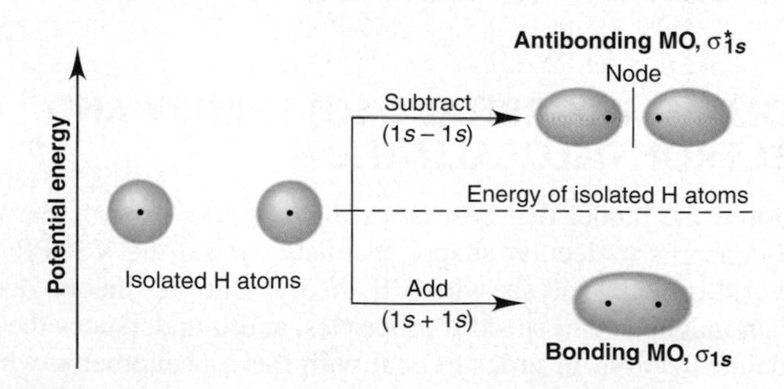

Figure 11.15 Contours and energies of the bonding and antibonding molecular orbitals (MOs) in H_2. When two H 1*s* atomic orbitals (AOs) combine, they form two H_2 MOs. The bonding MO (σ_{1s}) forms from addition of the AOs and is lower in energy than those AOs because most of its electron density lies *between* the nuclei (shown as dots). The antibonding MO (σ^*_{1s}) forms from subtraction of the AOs and is higher in energy because there is a *node* between the nuclei and most of the electron density lies outside the internuclear region.

The bonding MO in H_2 is spread mostly *between* the nuclei, with the nuclei attracted to the intervening electrons. An electron in this MO can delocalize its charge over a much larger volume than is possible in an individual AO in H. Because the electron-electron repulsions are reduced, the bonding MO is *lower* in energy than the isolated AOs. Therefore, when electrons occupy this orbital, the molecule is *more stable* than the separate atoms. In contrast, the antibonding MO has a node between the nuclei and most of its electron density *outside* the internuclear region. The electrons do not shield one nucleus from the other, which increases the nucleus-nucleus repulsion and makes the antibonding MO *higher* in energy than the isolated AOs. Therefore, when the antibonding orbital is occupied, the molecule is *less stable* than when this orbital is empty.

Both the bonding and antibonding MOs of H_2 are **sigma (σ) MOs** because they are cylindrically symmetrical about an imaginary line that runs through the two nuclei. The bonding MO is denoted by σ_{1s}, that is, a σ MO formed by combination of 1*s* AOs. Antibonding orbitals are denoted with a superscript star, so the one derived from the 1*s* AOs is σ^*_{1s} (spoken "sigma, one ess, star").

To interact effectively and form MOs, *atomic orbitals must have similar energy and orientation.* The 1*s* orbitals on two H atoms have identical energy and orientation, so they interact strongly. This point will be important when we consider molecules composed of atoms with many sublevels.

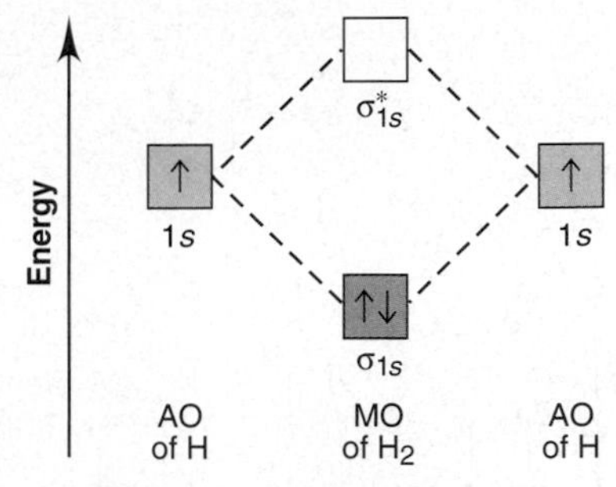

Figure 11.16 The MO diagram for H_2. The positions of the boxes indicate the relative energies and the arrows show the electron occupancy of the MOs and the AOs from which they formed. Two electrons, one from each H atom, fill the lower energy σ_{1s} MO, while the higher energy σ^*_{1s} MO remains empty. Orbital occupancy is also shown by color (darker = full, paler = half-filled, no color = empty).

Filling Molecular Orbitals with Electrons Electrons fill MOs just as they fill AOs:

- Orbitals are filled in order of increasing energy (aufbau principle).
- An orbital has a maximum capacity of two electrons with opposite spins (exclusion principle).
- Orbitals of equal energy are half-filled, with spins parallel, before any is filled (Hund's rule).

A **molecular orbital (MO) diagram** shows the relative energy and number of electrons in each MO, as well as the AOs from which they formed. Figure 11.16 is the MO diagram for H_2.

MO theory redefines bond order. In a Lewis structure, bond order is the number of electron pairs per linkage. The **MO bond order** is the number of electrons in bonding MOs minus the number in antibonding MOs, divided by two:

$$\text{Bond order} = \tfrac{1}{2}[(\text{no. of e}^- \text{ in bonding MO}) - (\text{no. of e}^- \text{ in antibonding MO})] \quad \textbf{(11.1)}$$

Thus, for H_2, the bond order is $\frac{1}{2}(2 - 0) = 1$. A bond order greater than zero indicates that the molecular species is stable relative to the separate atoms, whereas a bond order of zero implies no net stability and, thus, no likelihood that the species will form. In general, *the higher the bond order, the stronger the bond.*

Another similarity of MO theory to the quantum-mechanical model for atoms is that we can write electron configurations for a molecule. The symbol of each occupied MO is shown in parentheses, and the number of electrons in it is written outside as a superscript. Thus, the electron configuration of H_2 is $(\sigma_{1s})^2$.

One of the early triumphs of MO theory was its ability to *predict* the existence of He_2^+, the dihelium molecule-ion, which is composed of two He nuclei and three electrons. Let's use MO theory to see why He_2^+ exists and, at the same time, why He_2 does not. In He_2^+, the 1*s* atomic orbitals form the molecular orbitals, so the MO diagram, shown in Figure 11.17A, is similar to that for H_2. The three electrons enter the MOs to give a pair in the σ_{1s} MO and a lone electron in the σ^*_{1s} MO. The bond order is $\frac{1}{2}(2 - 1) = \frac{1}{2}$. Thus, He_2^+ has a relatively weak bond, but it should exist. Indeed, this molecular ionic species has been observed frequently when He atoms collide with He^+ ions. Its electron configuration is $(\sigma_{1s})^2(\sigma^*_{1s})^1$.

On the other hand, He_2 has four electrons to place in its σ_{1s} and σ^*_{1s} MOs. As Figure 11.17B shows, both the bonding and antibonding orbitals are filled. The

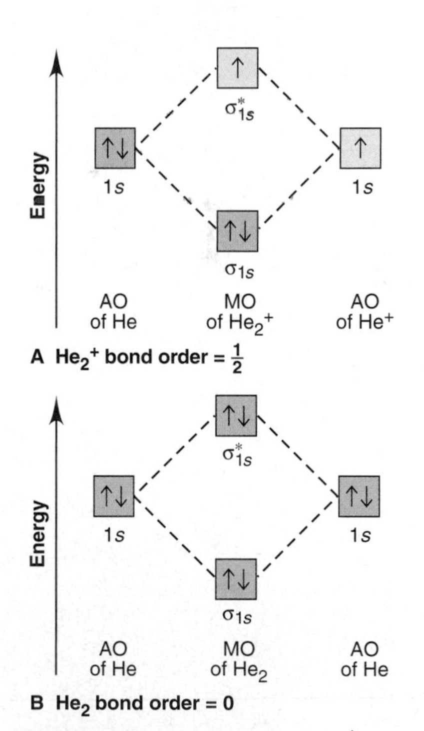

Figure 11.17 MO diagrams for He_2^+ and He_2. A, In He_2^+, three electrons enter MOs in order of increasing energy to give a filled σ_{1s} MO and a half-filled σ^*_{1s} MO. The bond order of $\frac{1}{2}$ implies that He_2^+ exists. **B,** In He_2, the four electrons fill both the σ_{1s} and the σ^*_{1s} MOs, so there is no net stabilization (bond order = 0).

stabilization arising from the electron pair in the bonding MO is canceled by the destabilization due to the electron pair in the antibonding MO. From its zero bond order [$\frac{1}{2}(2 - 2) = 0$], we predict, and experiment has so far confirmed, that a covalent He_2 molecule does not exist.

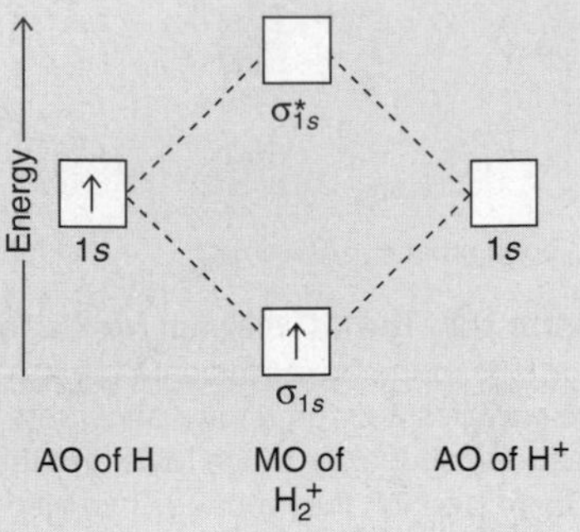

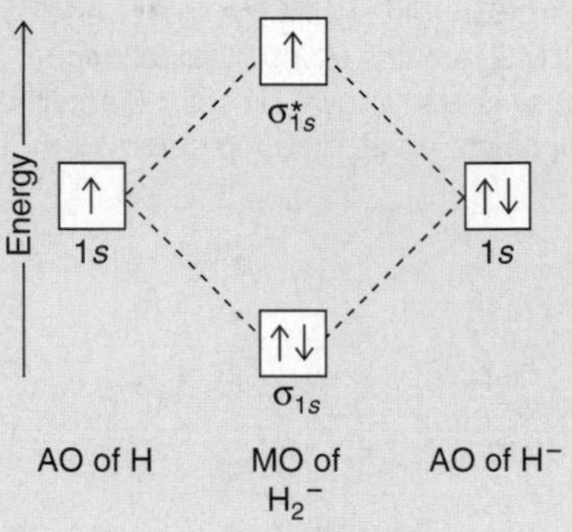

Energy
σ^*_{1s}
1s 1s
σ_{1s}
AO of H MO of H_2^+ AO of H^+

Energy
σ^*_{1s}
1s 1s
σ_{1s}
AO of H MO of H_2^- AO of H^-

SAMPLE PROBLEM 11.3 Predicting Stability of Species Using MO Diagrams

Problem Use MO diagrams to predict whether H_2^+ and H_2^- exist. Determine their bond orders and electron configurations.

Plan In these species, the 1*s* orbitals form MOs, so the MO diagrams are similar to that for H_2. We determine the number of electrons in each species and distribute the electrons in pairs to the bonding and antibonding MOs in order of increasing energy. We obtain the bond order with Equation 11.1 and write the electron configuration as described in the text.

Solution For H_2^+. H_2 has two e^-, so H_2^+ has only one, as shown in the margin (top diagram). The bond order is $\frac{1}{2}(1 - 0) = \frac{1}{2}$, so we predict that H_2^+ does exist. The electron configuration is $(\sigma_{1s})^1$.

For H_2^-. Since H_2 has two e^-, H_2^- has three, as shown in the margin (bottom diagram). The bond order is $\frac{1}{2}(2 - 1) = \frac{1}{2}$, so we predict that H_2^- does exist. The electron configuration is $(\sigma_{1s})^2(\sigma^*_{1s})^1$.

Check The number of electrons in the MOs equals the number of electrons in the AOs, as it should.

Comment Both these species have been detected spectroscopically: H_2^+ occurs in the hydrogen-containing material around stars; H_2^- has been formed in the laboratory.

FOLLOW-UP PROBLEM 11.3 Use an MO diagram to predict whether two hydride ions (H^-) will form H_2^{2-}. Calculate the bond order of H_2^{2-} and write its electron configuration.

Homonuclear Diatomic Molecules of the Period 2 Elements

Homonuclear diatomic molecules are those composed of two identical atoms. You're already familiar with several from Period 2—N_2, O_2, and F_2—as the elemental forms under standard conditions. Others in Period 2—Li_2, Be_2, B_2, C_2, and Ne_2—are observed, if at all, only in high-temperature gas-phase experiments. Molecular orbital descriptions of these species provide some interesting tests of the model. Let's look first at the molecules from the *s* block, Groups 1A(1) and 2A(2), and then at those from the *p* block, Groups 3A(13) through 8A(18).

Bonding in the *s*-Block Homonuclear Diatomic Molecules Both Li and Be occur as metals under normal conditions, but let's see what MO theory predicts for their stability as the diatomic gases dilithium (Li_2) and diberyllium (Be_2).

These atoms have both inner (1*s*) and outer (2*s*) electrons, but the 1*s* orbitals interact negligibly. In general, *only outer (valence) orbitals interact enough to form molecular orbitals.* Like the MOs formed from 1*s* AOs, those formed from 2*s* AOs are σ orbitals, cylindrically symmetrical around the internuclear axis. Bonding (σ_{2s}) and antibonding (σ^*_{2s}) MOs form, and the two valence electrons fill the bonding MO, with opposing spins (Figure 11.18A). Dilithium has two electrons in bonding MOs and none in antibonding MOs; therefore, its bond order is $\frac{1}{2}(2 - 0) = 1$. In fact, Li_2 *has* been observed, and the MO electron configuration is $(\sigma_{2s})^2$.

With its two additional electrons, the MO diagram for Be_2 has filled σ_{2s} and σ^*_{2s} MOs (Figure 11.18B). This is similar to the case of He_2. The bond order is $\frac{1}{2}(2 - 2) = 0$. In keeping with a zero bond order, the ground state of Be_2 has never been observed.

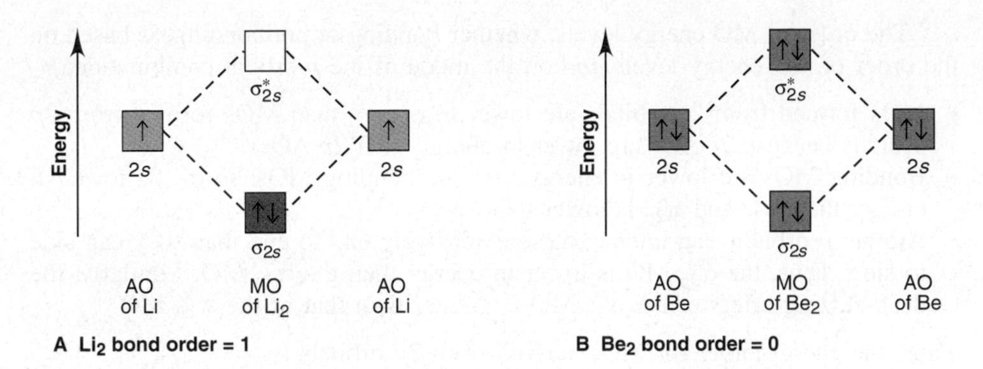

Figure 11.18 Bonding in *s*-block homonuclear diatomic molecules. Only outer (valence) AOs interact enough to form MOs. **A,** Li_2. The two valence electrons from two Li atoms fill the bonding (σ_{2s}) MO, and the antibonding (σ^*_{2s}) remains empty. With a bond order of 1, Li_2 does form. **B,** Be_2. The four valence electrons from two Be atoms fill both MOs to give no net stabilization. Ground-state Be_2 has a zero bond order and has never been observed.

Molecular Orbitals from Atomic *p*-Orbital Combinations As we move to the *p* block, atomic 2*p* orbitals become involved, so we first consider the shapes and energies of the MOs that result from their combinations. Recall that *p* orbitals can overlap with each other in two different modes, as shown in Figure 11.19. End-to-end combination gives a pair of σ MOs, the σ_{2p} and σ^*_{2p}. Side-to-side combination gives a pair of **pi (π) MOs,** π_{2p} and π^*_{2p}. Similar to MOs formed from *s* orbitals, bonding MOs from *p*-orbital combinations have their greatest electron density *between* the nuclei, whereas antibonding MOs from *p*-orbital combinations have a node between the nuclei and most of their electron density *outside* the internuclear region.

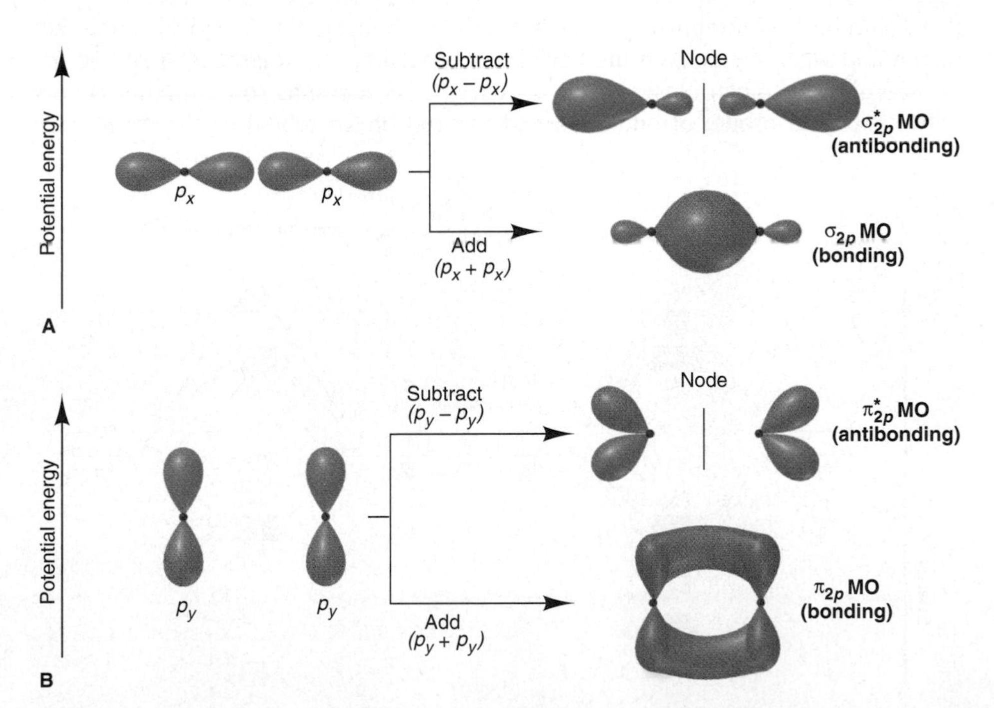

Figure 11.19 Contours and energies of σ and π MOs from combinations of 2*p* atomic orbitals. **A,** The *p* orbitals lying along the line between the atoms (usually designated p_x) undergo end-to-end overlap and form σ_{2p} and σ^*_{2p} MOs. Note the greater electron density between the nuclei for the bonding orbital and the node between the nuclei for the antibonding orbital. **B,** The *p* orbitals that lie perpendicular to the internuclear axis (p_y and p_z) undergo side-to-side overlap to form two π MOs. The p_z interactions are the same as those shown here for the p_y orbitals, so a total of four π MOs form. A π_{2p} is a bonding MO with its greatest density above and below the internuclear axis; a π^*_{2p} is an antibonding MO with a node between the nuclei and its electron density outside the internuclear region.

The order of MO energy levels, whether bonding or antibonding, is based on the order of AO energy levels *and* on the mode of the *p*-orbital combination:

- MOs formed from 2*s* orbitals are lower in energy than MOs formed from 2*p* orbitals because 2*s* AOs are lower in energy than 2*p* AOs.
- Bonding MOs are lower in energy than antibonding MOs, so σ_{2p} is lower in energy than σ^*_{2p} and π_{2p} is lower than π^*_{2p}.
- Atomic *p* orbitals can interact more extensively end to end than they can side to side. Thus, the σ_{2p} MO is lower in energy than the π_{2p} MO. Similarly, the destabilizing effect of the σ^*_{2p} MO is greater than that of the π^*_{2p} MO.

Thus, the energy order for MOs derived from 2*p* orbitals is

$$\sigma_{2p} < \pi_{2p} < \pi^*_{2p} < \sigma^*_{2p}$$

There are three mutually perpendicular 2*p* orbitals in each atom. When the six *p* orbitals in two atoms combine, the two orbitals that interact end to end form a σ and a σ^* MO, and the two pairs of orbitals that interact side to side form two π MOs of the same energy and two π^* MOs of the same energy. Combining these orientations with the energy order gives the *expected* MO diagram for the *p*-block Period 2 homonuclear diatomic molecules (Figure 11.20A).

One other factor influences the MO energy order. Recall that only AOs of similar energy interact to form MOs. The order in Figure 11.20A assumes that the *s* and *p* AOs are so different in energy that they do not interact with each other: the orbitals do not *mix*. This is true for O, F, and Ne. These atoms are small, so relatively strong repulsions occur as the 2*p* electrons pair up; these repulsions raise the energy of the 2*p* orbitals high enough above the energy of the 2*s* orbitals to minimize orbital mixing. In contrast, B, C, and N atoms are larger, and when the 2*p* AOs are half-filled, repulsions are relatively small; so the 2*p* energies are much closer to the 2*s* energy. As a result, some mixing occurs between the 2*s* orbital of one atom and the end-on 2*p* orbital of the other. This

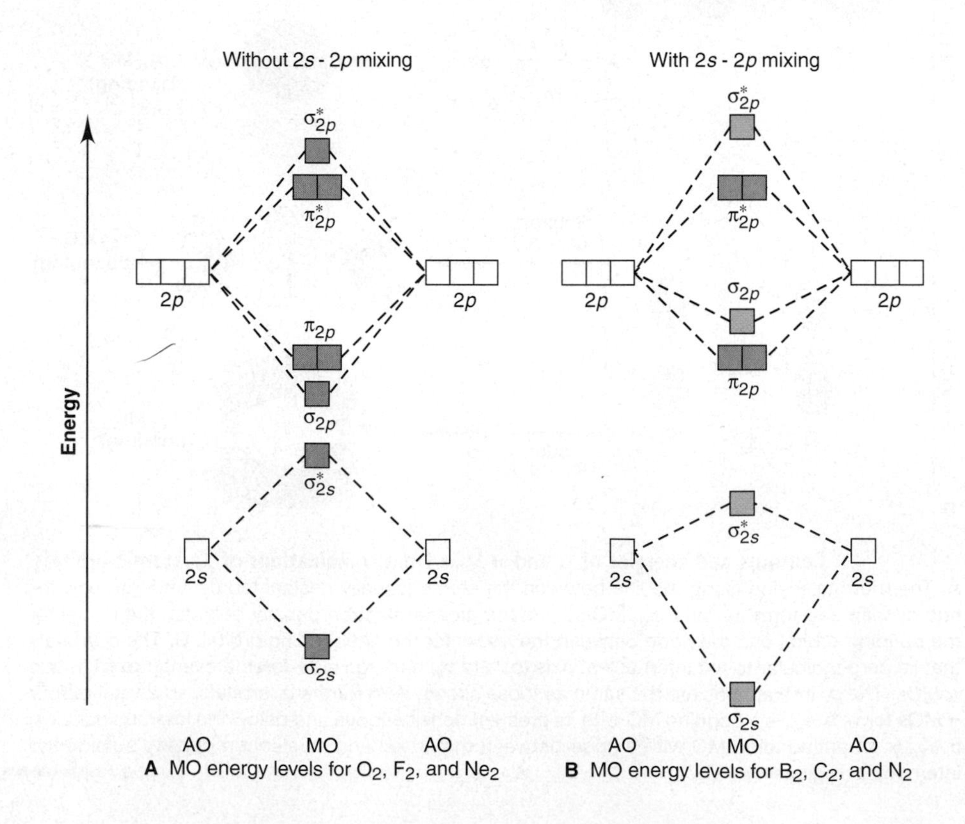

Figure 11.20 Relative MO energy levels for Period 2 homonuclear diatomic molecules. A, MO energy levels for O_2, F_2, and Ne_2. The six 2*p* orbitals of the two atoms form six MOs that are higher in energy than the two MOs formed from the two 2s orbitals. The AOs forming π orbitals give rise to two bonding MOs (π_{2p}) of equal energy and two antibonding MOs (π^*_{2p}) of equal energy. This sequence of energy levels arises from minimal 2*s*-2*p* orbital mixing. **B,** MO energy levels for B_2, C_2, and N_2. Because of significant 2*s*-2*p* orbital mixing, the energies of σ MOs formed from 2*p* orbitals increase and those formed from 2s orbitals decrease. The major effect of this orbital mixing on the MO sequence is that the σ_{2p} is higher in energy than the π_{2p}. (For clarity, those MOs affected by mixing are shown in purple.)

orbital mixing *lowers* the energy of the σ_{2s} and σ^*_{2s} MOs and raises the energy of the σ_{2p} and σ^*_{2p} MOs; the π MOs are not affected. The MO diagram for B_2 through N_2 (Figure 11.20B) reflects this AO mixing. The only change that affects this discussion is the *reverse in energy order* of the σ_{2p} and π_{2p} MOs.

Bonding in the *p*-Block Homonuclear Diatomic Molecules Figure 11.21 shows the MOs, their electron occupancy, and some properties of B_2 through Ne_2. Note how *a higher bond order correlates with a greater bond energy and shorter bond length.* Also note how orbital occupancy correlates with magnetic properties. Recall from Chapter 8 that the spins of unpaired electrons in an atom (or ion) cause the substance to be *paramagnetic*, attracted to an external magnetic field. If all the electron spins are paired, the substance is *diamagnetic*, unaffected (or weakly repelled) by the magnetic field. The same observations apply to molecules. These properties are not addressed directly in VSEPR or VB theory.

The B_2 molecule has six outer electrons to place in its MOs. Four of these fill the σ_{2s} and σ^*_{2s} MOs. The remaining two electrons occupy the two π_{2p} MOs, one in each orbital, in keeping with Hund's rule. With four electrons in bonding MOs and two electrons in antibonding MOs, the bond order of B_2 is $\frac{1}{2}(4 - 2) = 1$. As expected from its MO diagram, B_2 is paramagnetic.

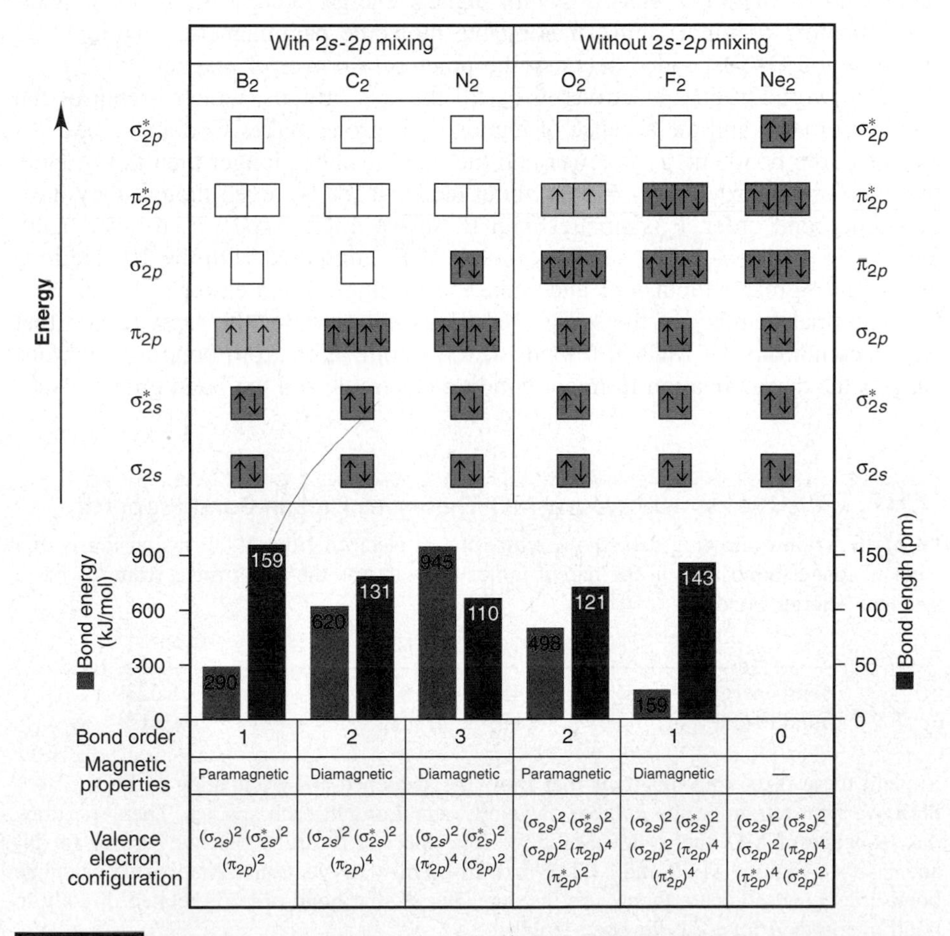

Figure 11.21 MO occupancy and molecular properties for B_2 through Ne_2. The sequence of MOs and their electron populations are shown for the homonuclear diatomic molecules in the *p* block of Period 2 [Groups 3A(13) to 8A(18)]. The bond energy, bond length, bond order, magnetic properties, and outer (valence) electron configuration appear below the orbital diagrams. Note the correlation between bond order and bond energy, both of which are inversely related to bond length.

The two additional electrons present in C_2 fill the two π_{2p} MOs. Since C_2 has two more bonding electrons than B_2, it has a bond order of 2 and the expected stronger, shorter bond. All the electrons are paired, and, as the model predicts, C_2 is diamagnetic.

In N_2, the two additional electrons fill the σ_{2p} MO. The resulting bond order is 3, which is consistent with the triple bond in the Lewis structure. As the model predicts, the bond energy is higher and the bond length shorter than for C_2, and N_2 is diamagnetic.

With O_2, we really see the power of MO theory compared to theories based on localized orbitals. For years, it seemed impossible to reconcile bonding theories with the bond strength and magnetic behavior of O_2. On the one hand, the data show a double-bonded molecule that is paramagnetic. On the other hand, we can write two possible Lewis structures for O_2, but neither gives such a molecule. One has a double bond and all electrons paired, the other a single bond and two electrons unpaired:

$$\ddot{\underset{..}{O}}=\ddot{\underset{..}{O}} \quad \text{or} \quad :\dot{\underset{..}{O}}-\dot{\underset{..}{O}}:$$

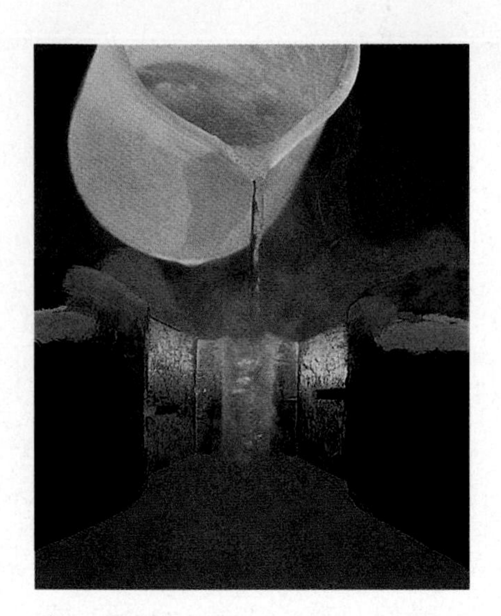

Figure 11.22 The paramagnetic properties of O_2. Liquid O_2 is attracted to the poles of a magnet because it is paramagnetic, as MO theory predicts. A diamagnetic substance would fall between the poles.

MO theory resolves this paradox beautifully. As Figure 11.21 shows, the bond order of O_2 is 2: eight electrons occupy bonding MOs and four occupy antibonding MOs [$\frac{1}{2}(8 - 4) = 2$]. Note the lower bond energy and greater bond length relative to N_2. The *two* electrons with highest energy occupy the *two* π^*_{2p} MOs with unpaired (parallel) spins, making the molecule paramagnetic. Figure 11.22 shows liquid O_2 suspended between the poles of a powerful magnet.

The two additional electrons in F_2 fill the π^*_{2p} orbitals, which decreases the bond order to 1, and the absence of unpaired electrons makes F_2 diamagnetic. As expected, the bond energy is lower and the bond distance longer than in O_2. Note that the bond energy for F_2 is only about half that for B_2, even though they have the same bond order. F is smaller than B, so we might expect a stronger bond. But the 18 electrons in the smaller volume of F_2 compared with the 10 electrons in B_2 cause greater repulsions and make the F_2 single bond easier to break.

The final member of the series, Ne_2, does not exist for the same reason that He_2 does not: all the MOs are filled, so the stabilization from bonding electrons cancels the destabilization from antibonding electrons, and the bond order is zero.

SAMPLE PROBLEM 11.4 Using MO Theory to Explain Bond Properties

Problem As the following data show, removing an electron from N_2 forms an ion with a weaker, longer bond than in the parent molecule, whereas the ion formed from O_2 has a stronger, shorter bond:

	N_2	N_2^+	O_2	O_2^+
Bond energy (kJ/mol)	945	841	498	623
Bond length (pm)	110	112	121	112

Explain these facts with diagrams that show the sequence and occupancy of MOs.

Plan We first determine the number of valence electrons in each species. Then, we draw the sequence of MO energy levels for the four species, recalling that they differ for N_2 and O_2 (see Figures 11.20 and 11.21), and fill them with electrons. Finally, we calculate bond orders and compare them with the data. Recall that bond order is related directly to bond energy and inversely to bond length.

Solution Determining the valence electrons:

N has 5 valence e^-, so N_2 has 10 and N_2^+ has 9

O has 6 valence e^-, so O_2 has 12 and O_2^+ has 11

Drawing and filling the MO diagrams:

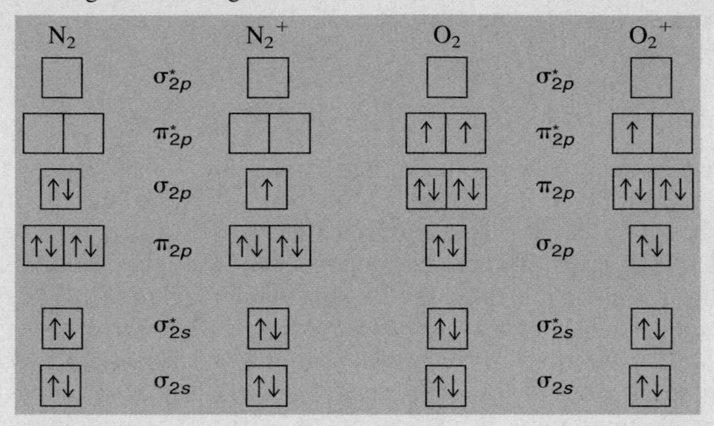

Calculating bond orders:

$\frac{1}{2}(8-2)=3$ $\quad\frac{1}{2}(7-2)=2.5$ $\quad\frac{1}{2}(8-4)=2$ $\quad\frac{1}{2}(8-3)=2.5$

When N_2 forms N_2^+, a *bonding* electron is removed, so the bond order decreases. Thus, N_2^+ has a weaker, longer bond than N_2. When O_2 forms O_2^+, an *antibonding* electron is removed, so the bond order increases. Thus, O_2^+ has a stronger, shorter bond than O_2.

Check The answer makes sense in terms of the relationships among bond order, bond energy, and bond length. Check that the total number of bonding and antibonding electrons equals the number of valence electrons calculated.

FOLLOW-UP PROBLEM 11.4 Determine the bond orders for the following species: F_2^{2-}, F_2^-, F_2, F_2^+, F_2^{2+}. List the species in order of increasing bond energy and in order of increasing bond length.

MO Description of Some Heteronuclear Diatomic Molecules

Heteronuclear diatomic molecules, those composed of two *different* atoms, have asymmetric MO diagrams because the atomic orbitals of the two atoms have unequal energies. Atoms with greater effective nuclear charge (Z_{eff}) draw their electrons closer to the nucleus and thus have atomic orbitals of lower energy (Section 8.2). Greater Z_{eff} also gives these atoms higher electronegativity values. Let's apply MO theory to the bonding in HF and NO.

Bonding in HF To form the MOs in HF, we combine appropriate AOs from isolated H and F atoms. The high effective nuclear charge of F holds all its electrons more tightly than the H nucleus holds its electron. As a result, *all* occupied atomic orbitals of F have lower energy than the 1*s* orbital of H. Therefore, the H 1*s* orbital interacts only with the F 2*p* orbitals, and only one of the three 2*p* orbitals, say the $2p_z$, leads to end-on overlap. This results in a σ MO and a σ* MO. The two other *p* orbitals of F ($2p_x$ and $2p_y$) are not involved in bonding and are called **nonbonding MOs;** they have the same energy as the isolated AOs (Figure 11.23).

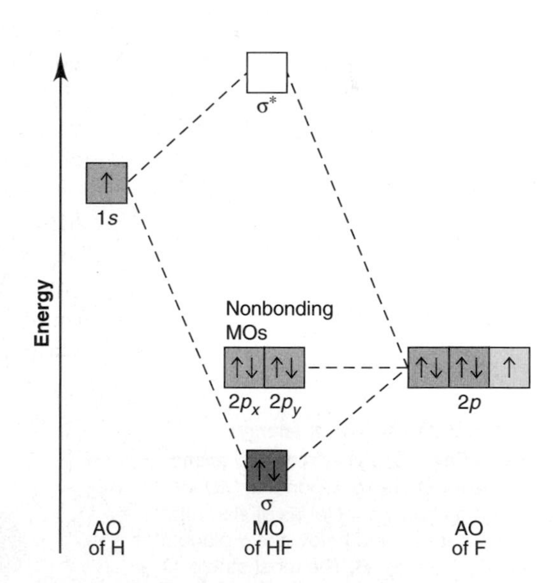

Figure 11.23 The MO diagram for HF. For a polar covalent molecule, the MO diagram is asymmetric because the more electronegative atom has AOs of lower energy. In HF, the bonding MO is closer in energy to the 2*p* orbital of F. Electrons that are not involved in bonding occupy nonbonding MOs. (The 2*s* AO of F is not shown.)

Because the occupied bonding MO is closer in energy to the AOs of F, we say that the F 2*p* orbital contributes more to the bonding in HF than the H 1*s* orbital does. Generally, in polar covalent molecules, *bonding MOs are closer in energy to the AOs of the more electronegative atom.* In effect, fluorine's greater electronegativity lowers the energy of the bonding MO and draws the bonding electrons closer to its nucleus.

Bonding in NO Nitrogen monoxide (nitric oxide) is a highly reactive molecule because it has a lone electron. Two possible Lewis structures for NO, with formal charges (Section 10.1), are

$$\underset{\text{I}}{\overset{0\quad 0}{:\dot{\text{N}}{=}\ddot{\text{O}}:}} \quad \text{or} \quad \underset{\text{II}}{\overset{-1\quad +1}{:\ddot{\text{N}}{=}\dot{\text{O}}:}}$$

Both structures show a double bond, but the *measured* bond energy suggests a bond order *higher* than 2. Moreover, it is not clear where the lone electron resides, although the lower formal charges for structure I suggest that it is on the N atom.

MO theory predicts the bond order and indicates lone-electron placement with no difficulty. The MO diagram in Figure 11.24 is asymmetric, with the AOs of the more electronegative O lower in energy. The 11 valence electrons of NO fill MOs in order of increasing energy, leaving the lone electron in one of the π^*_{2p} orbitals. The eight bonding electrons and three antibonding electrons give a bond order of $\frac{1}{2}(8 - 3) = 2.5$, more in keeping with experiment than either Lewis structure. The bonding electrons lie in MOs closer in energy to the AOs of the O atom. The lone electron occupies an antibonding orbital. Because this orbital receives a greater contribution from the 2*p* orbitals of N, the lone electron resides closer to the N atom.

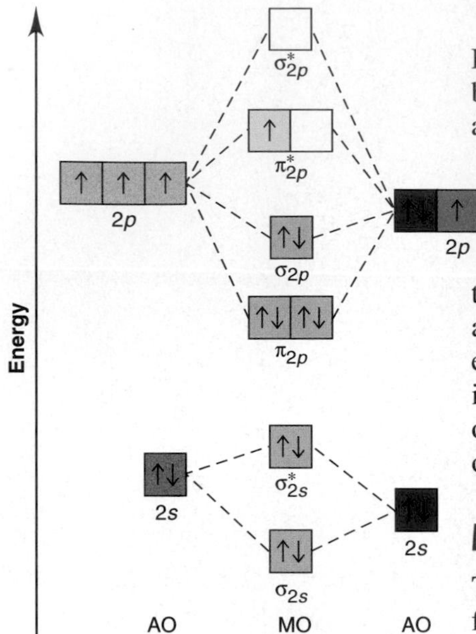

Figure 11.24 The MO diagram for NO. Eleven electrons occupy the MOs in nitric oxide. Note that the lone electron occupies an antibonding MO whose energy is closer to that of the AO of the (less electronegative) N atom.

MO Descriptions of Benzene and Ozone

The orbital shapes and MO diagrams for polyatomic molecules are too complex for a detailed treatment here. However, let's briefly discuss how the model eliminates the need for resonance forms and helps explain the effects of the absorption of energy. Recall that we cannot draw a single Lewis structure that adequately depicts bonding in molecules such as ozone and benzene because the adjacent single and double bonds actually have identical properties. Instead, we draw more than one structure and mentally combine them into a resonance hybrid. The VB model also relies on resonance because it depicts *localized* electron-pair bonds.

In contrast, MO theory pictures a structure of delocalized σ and π bonding and antibonding MOs. Figure 11.25 shows the lowest energy π-bonding MOs in benzene and ozone. Each holds one pair of electrons. The extended electron densities allow delocalization of this π-bonding electron pair over the entire molecule, thus eliminating the need for separate resonance forms. In benzene, the upper and lower hexagonal lobes of this π-bonding MO lie above and below the σ plane of all six carbon nuclei. In ozone, the two lobes of the lowest energy π-bonding MO extend over and under all three oxygen nuclei. Another advantage of MO theory is that it can explain excited states and spectra of molecules: for instance, why O_3 decomposes when it absorbs ultraviolet radiation in the stratosphere (bonding electrons are excited and enter empty antibonding orbitals), and why the ultraviolet spectrum of benzene has its characteristic absorption bands.

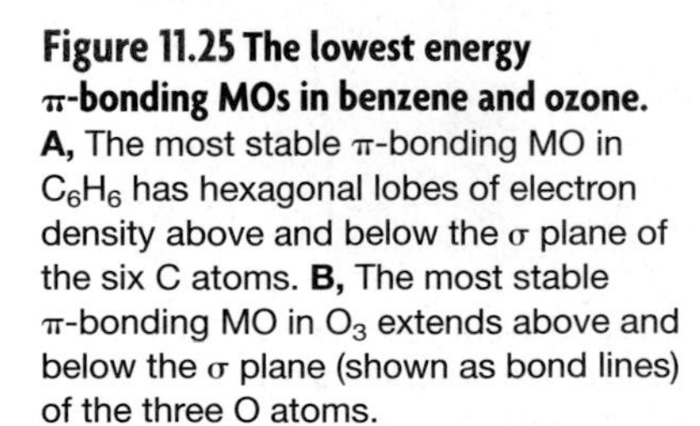

Figure 11.25 The lowest energy π-bonding MOs in benzene and ozone. **A,** The most stable π-bonding MO in C_6H_6 has hexagonal lobes of electron density above and below the σ plane of the six C atoms. **B,** The most stable π-bonding MO in O_3 extends above and below the σ plane (shown as bond lines) of the three O atoms.

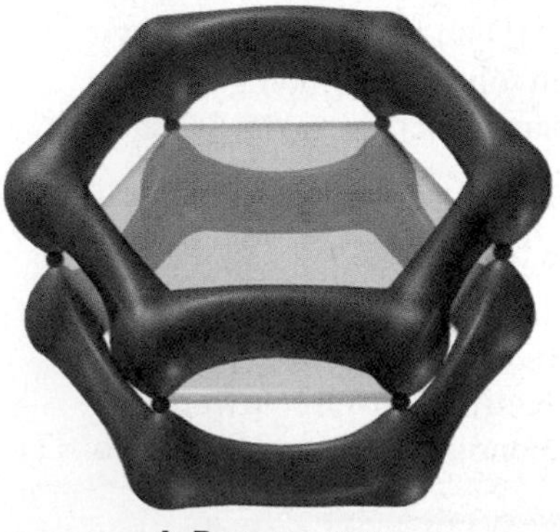

A Benzene, C_6H_6

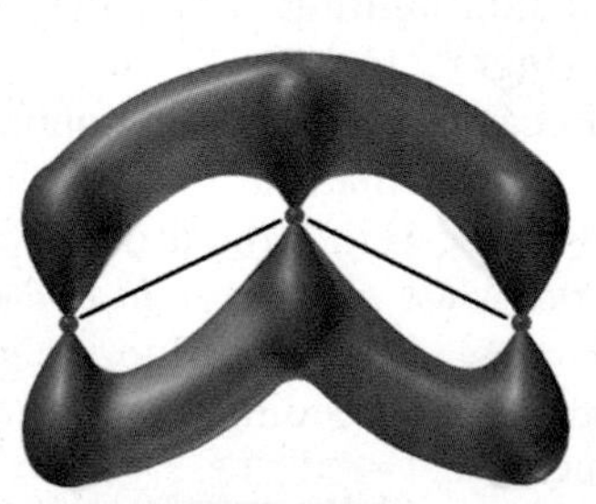

B Ozone, O_3

SECTION SUMMARY

MO theory treats a molecule as a collection of nuclei with molecular orbitals delocalized over the entire structure. Atomic orbitals of comparable energy can be added and subtracted to obtain bonding and antibonding MOs, respectively. Bonding MOs have most of their electron density between the nuclei and are lower in energy than the atomic orbitals; most of the electron density of antibonding MOs does not lie between the nuclei, so these MOs are higher in energy. MOs are filled in order of their energy with paired electrons having opposing spins. MO diagrams show energy levels and orbital occupancy. The diagrams for homonuclear diatomic molecules of Period 2 explain observed bond energy, bond length, and magnetic behavior. In heteronuclear diatomic molecules, the more electronegative atom contributes more to the bonding MOs. MO theory eliminates the need for resonance forms to depict larger molecules.

Chapter Perspective

In this chapter, you've seen that the two most important orbital-based models of covalent bonding, each with its own benefits and limitations, complement each other. Next, you'll see how the electron distribution in the bonds of molecules gives rise to the physical behavior of liquids and solids. MO theory will come into play for a deeper understanding of metallic solids. Then, following Chapter 12 are three chapters that apply many concepts you've learned from Chapter 7 onward to the behavior of solutions and inorganic, organic, and biochemical substances.

For Review and Reference (Numbers in parentheses refer to pages, unless noted otherwise.)

Learning Objectives

Relevant section and/or sample problem (SP) numbers appear in parentheses.

Understand These Concepts

1. The main ideas of valence bond theory—orbital overlap, opposing electron spins, and hybridization—as a means of rationalizing molecular shapes (11.1)
2. How orbitals mix to form hybrid orbitals with different spatial orientations (11.1)
3. The distinction between end-to-end and side-to-side overlap and the origin of sigma (σ) and pi (π) bonds in simple molecules (11.2)
4. How two modes of orbital overlap lead to single, double, and triple bonds (11.2)
5. Why π bonding restricts rotation around double bonds (11.2)
6. The distinction between the localized bonding of VB theory and the delocalized bonding of MO theory (11.3)
7. How addition or subtraction of AOs forms bonding or antibonding MOs (11.3)
8. The shapes of MOs formed from combinations of two *s* orbitals and combinations of two *p* orbitals (11.3)
9. How MO bond order predicts the stability of molecular species (11.3)
10. How MO theory explains the bonding and properties of homonuclear and heteronuclear diatomic molecules of Period 2 (11.3)

Master These Skills

1. Using molecular shape to postulate the hybrid orbitals used by a central atom (SP 11.1)
2. Describing the types of orbitals and bonds in a molecule (SP 11.2)
3. Drawing MO diagrams, writing electron configurations, and calculating bond orders of molecular species (SP 11.3)
4. Explaining bond properties with MO theory (SP 11.4)

Key Terms

Section 11.1
valence bond (VB) theory (399)
hybridization (400)
hybrid orbital (400)
sp hybrid orbital (400)
sp^2 hybrid orbital (402)
sp^3 hybrid orbital (402)
sp^3d hybrid orbital (403)
sp^3d^2 hybrid orbital (404)

Section 11.2
sigma (σ) bond (407)
pi (π) bond (407)

Section 11.3
molecular orbital (MO) theory (409)
molecular orbital (MO) (410)
bonding MO (410)
antibonding MO (410)
sigma (σ) MO (411)
molecular orbital (MO) diagram (411)
MO bond order (411)
homonuclear diatomic molecule (412)
pi (π) MO (413)
nonbonding MO (417)

Key Equations and Relationships

11.1 Calculating the MO bond order (411): Bond order $= \frac{1}{2}$[(no. of e^- in bonding MO) − (no. of e^- in antibonding MO)]

Highlighted Figures and Tables

These figures (F) and tables (T) provide a review of key ideas.

Brief Solutions to Follow-up Problems

11.1 (a) Shape is linear, so Be is *sp* hybridized.

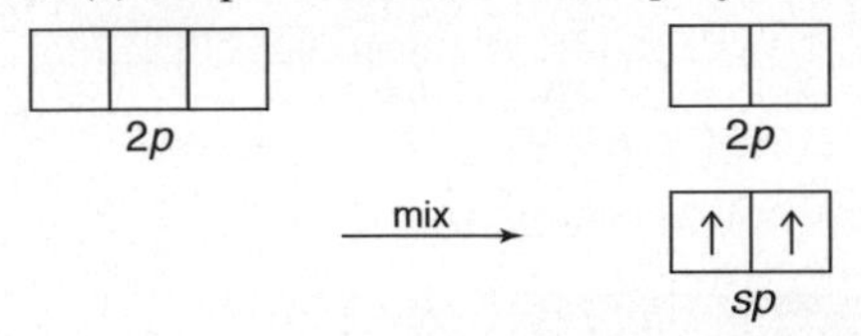

(b) Shape is tetrahedral, so Si is sp^3 hybridized.

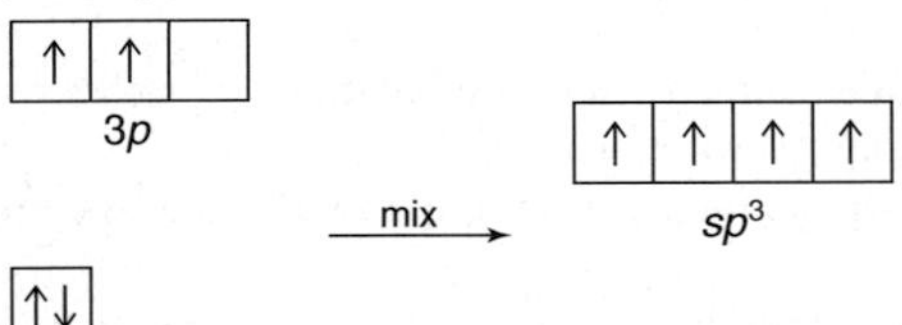

(c) Shape is square planar, so Xe is sp^3d^2 hybridized.

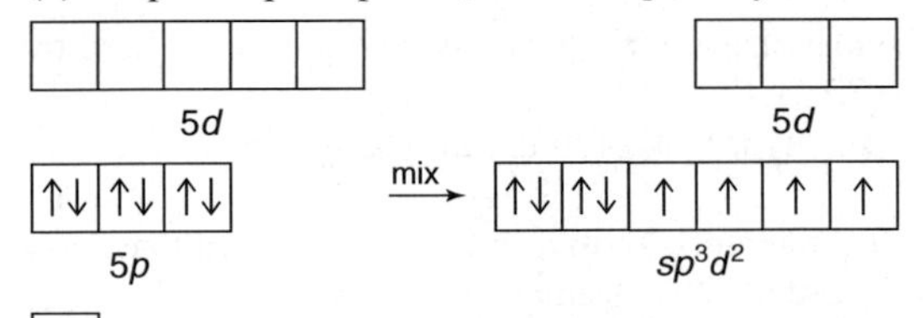

11.2 (a) H—C≡N:

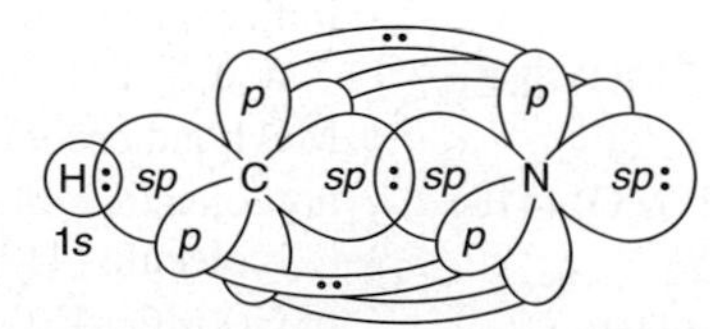

HCN is linear, so C is *sp* hybridized. N is also *sp* hybridized. One *sp* of C overlaps the 1*s* of H to form a σ bond. The other *sp* of C overlaps one *sp* of N to form a σ bond. The other *sp* of N holds a lone pair. Two unhybridized *p* orbitals of N and two of C overlap to form two π bonds.

(b) $\ddot{O}$=C=$\ddot{O}$

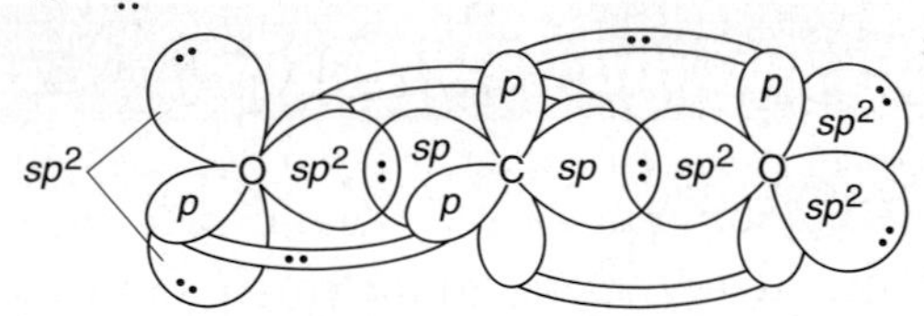

CO_2 is linear, so C is *sp* hybridized. Both O atoms are sp^2 hybridized. Each *sp* of C overlaps one sp^2 of an O to form two σ bonds. Each of the two unhybridized *p* orbitals of C forms a π bond with the unhybridized *p* of one of the two O atoms. Two sp^2 of each O hold lone pairs.

11.3

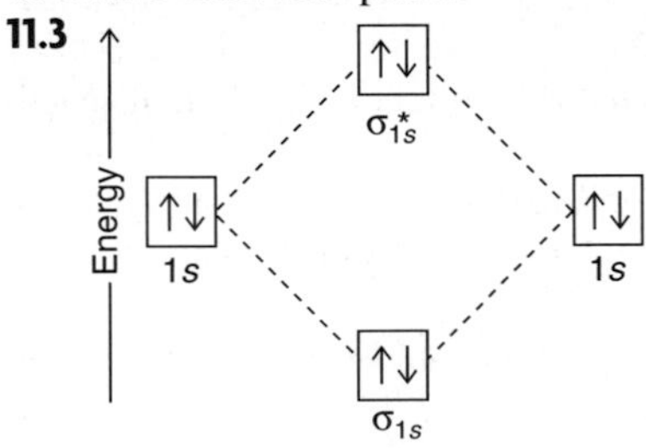

Does not exist: bond order $= \frac{1}{2}(2 - 2) = 0$; $(\sigma_{1s})^2(\sigma^*_{1s})^2$

11.4 Bond orders: $F_2^{2-} = 0$; $F_2^- = \frac{1}{2}$; $F_2 = 1$; $F_2^+ = 1\frac{1}{2}$
$F_2^{2+} = 2$

Bond energy: $F_2^{2-} < F_2^- < F_2 < F_2^+ < F_2^{2+}$

Bond length: $F_2^{2+} < F_2^+ < F_2 < F_2^-$; F_2^{2-} does not exist

Problems

Problems with **colored** numbers are answered in Appendix E. Sections match the text and provide the numbers of relevant sample problems. Most offer Concept Review Questions, Skill-Building Exercises (grouped in pairs covering the same concept), and Problems in Context. Comprehensive Problems are based on material from any section or previous chapter.

Valence Bond (VB) Theory and Orbital Hybridization

(Sample Problem 11.1)

Concept Review Questions

11.1 What type of central-atom orbital hybridization corresponds to each electron-group arrangement: (a) trigonal planar; (b) octahedral; (c) linear; (d) tetrahedral; (e) trigonal bipyramidal?

11.2 What is the orbital hybridization of a central atom that has one lone pair and bonds to: (a) two other atoms; (b) three other atoms; (c) four other atoms; (d) five other atoms?

11.3 How do carbon and silicon differ with regard to the *types* of orbitals available for hybridization? Explain.

11.4 How many hybrid orbitals form when four atomic orbitals of a central atom mix? Explain.

Skill-Building Exercises *(grouped in similar pairs)*

11.5 Give the number and type of hybrid orbital that forms when each set of atomic orbitals mixes:
(a) two *d*, one *s*, and three *p* (b) three *p* and one *s*

11.6 Give the number and type of hybrid orbital that forms when each set of atomic orbitals mixes:
(a) one *p* and one *s* (b) three *p*, one *d*, and one *s*

11.7 What is the hybridization of nitrogen in each of the following: (a) NO; (b) NO_2; (c) NO_2^-?

11.8 What is the hybridization of carbon in each of the following: (a) CO_3^{2-}; (b) $C_2O_4^{2-}$; (c) NCO^-?

11.9 What is the hybridization of chlorine in each of the following: (a) ClO_2; (b) ClO_3^-; (c) ClO_4^-?

11.10 What is the hybridization of bromine in each of the following: (a) BrF_3; (b) BrO_2^-; (c) BrF_5?

11.11 Which types of atomic orbitals of the central atom mix to form hybrid orbitals in (a) $SiClH_3$; (b) CS_2?

11.12 Which types of atomic orbitals of the central atom mix to form hybrid orbitals in (a) Cl_2O; (b) $BrCl_3$?

11.13 Which types of atomic orbitals of the central atom mix to form hybrid orbitals in (a) SCl_3F; (b) NF_3?

11.14 Which types of atomic orbitals of the central atom mix to form hybrid orbitals in (a) PF_5; (b) SO_3^{2-}?

11.15 Use partial orbital diagrams to show how the atomic orbitals of the central atom lead to hybrid orbitals in (a) $GeCl_4$; (b) BCl_3; (c) CH_3^+.

11.16 Use partial orbital diagrams to show how the atomic orbitals of the central atom lead to hybrid orbitals in (a) BF_4^-; (b) PO_4^{3-}; (c) SO_3.

11.17 Use partial orbital diagrams to show how the atomic orbitals of the central atom lead to hybrid orbitals in (a) $SeCl_2$; (b) H_3O^+; (c) IF_4^-.

11.18 Use partial orbital diagrams to show how the atomic orbitals of the central atom lead to hybrid orbitals in (a) $AsCl_3$; (b) $SnCl_2$; (c) PF_6^-.

Problems in Context

11.19 Methyl isocyanate, $CH_3—\ddot{N}=C=\ddot{O}:$, is an intermediate in the manufacture of many pesticides. It received notoriety in 1984 when a leak from a manufacturing plant resulted in the death of more than 2000 people in Bhopal, India. What are the hybridizations of the N atom and the two C atoms in methyl isocyanate? Sketch the molecular shape.

The Mode of Orbital Overlap and the Types of Covalent Bonds

(Sample Problem 11.2)

Concept Review Questions

11.20 Are these statements true or false? Correct any that are false.
(a) Two σ bonds comprise a double bond.
(b) A triple bond consists of one π bond and two σ bonds.
(c) Bonds formed from atomic *s* orbitals are always σ bonds.
(d) A π bond restricts rotation about the σ-bond axis.
(e) A π bond consists of two pairs of electrons.
(f) End-to-end overlap results in a bond with electron density above and below the bond axis.

Skill-Building Exercises *(grouped in similar pairs)*

11.21 Describe the hybrid orbitals used by the central atom and the type(s) of bonds formed in (a) NO_3^-; (b) CS_2; (c) CH_2O.

11.22 Describe the hybrid orbitals used by the central atom and the type(s) of bonds formed in (a) O_3; (b) I_3^-; (c) $COCl_2$ (C is central).

11.23 Describe the hybrid orbitals used by the central atom(s) and the type(s) of bonds formed in (a) FNO; (b) C_2F_4; (c) $(CN)_2$.

11.24 Describe the hybrid orbitals used by the central atom(s) and the type(s) of bonds formed in (a) BrF_3; (b) $CH_3C\equiv CH$; (c) SO_2.

Problems in Context

11.25 2-Butene ($CH_3CH=CHCH_3$) is a starting material in the manufacture of lubricating oils and many other compounds. Draw two different structures for 2-butene, indicating the σ and π bonds in each.

Molecular Orbital (MO) Theory and Electron Delocalization

(Sample Problems 11.3 and 11.4)

Concept Review Questions

11.26 Two *p* orbitals from one atom and two *p* orbitals from another atom are combined to form molecular orbitals for the joined atoms. How many MOs will result from this combination? Explain.

11.27 Certain atomic orbitals on two atoms were combined to form the following MOs. Name the atomic orbitals used and the MOs formed, and explain which MO has higher energy:

11.28 How do the bonding and antibonding MOs formed from a given pair of AOs compare to each other with respect to (a) energy; (b) presence of nodes; (c) internuclear electron density?

11.29 Antibonding MOs always have at least one node. Can a bonding MO have a node? If so, draw an example.

Skill-Building Exercises *(grouped in similar pairs)*

11.30 How many electrons does it take to fill (a) a σ bonding MO; (b) a π antibonding MO; (c) the MOs formed from combination of the 1*s* orbitals of two atoms?

11.31 How many electrons does it take to fill (a) the MOs formed from combination of the 2*p* orbitals of two atoms; (b) a σ^*_{2p} MO; (c) the MOs formed from combination of the 2*s* orbitals of two atoms?

11.32 Show the shapes of bonding and antibonding MOs formed by combination of (a) an *s* orbital and a *p* orbital; (b) two *p* orbitals (end to end).

11.33 Show the shapes of bonding and antibonding MOs formed by combination of (a) two *s* orbitals; (b) two *p* orbitals (side to side).

11.34 Use MO diagrams and the bond orders you obtain from them to answer: (a) Is Be_2^+ stable? (b) Is Be_2^+ diamagnetic? (c) What is the outer (valence) electron configuration of Be_2^+?

11.35 Use MO diagrams and the bond orders you obtain from them to answer: (a) Is O_2^- stable? (b) Is O_2^- paramagnetic? (c) What is the outer (valence) electron configuration of O_2^-?

11.36 Use MO diagrams to place C_2^-, C_2, and C_2^+ in order of (a) increasing bond energy; (b) increasing bond length.

11.37 Use MO diagrams to place B_2^+, B_2, and B_2^- in order of (a) decreasing bond energy; (b) decreasing bond length.

Comprehensive Problems

11.38 Predict the shape, state the hybridization of the central atom, and give the ideal bond angle(s) and any expected deviations for
(a) BrO_3^- (b) $AsCl_4^-$ (c) SeO_4^{2-} (d) BiF_5^{2-}
(e) SbF_4^+ (f) AlF_6^{3-} (g) IF_4^+

11.39 Butadiene (shown below) is a colorless gas used to make synthetic rubber and many other compounds:

(a) How many σ bonds and π bonds does the molecule have?
(b) Are *cis-trans* structural arrangements about the double bonds possible? Explain.

11.40 Epinephrine (or adrenaline) is a naturally occurring hormone that is also manufactured commercially for use as a heart stimulant, a nasal decongestant, and to treat glaucoma. A valid Lewis structure is

(a) What is the hybridization of each C, O, and N atom?
(b) How many σ bonds does the molecule have?
(c) How many π electrons are delocalized in the ring?

11.41 Use partial orbital diagrams to show how the atomic orbitals of the central atom lead to the hybrid orbitals in
(a) IF_2^- (b) ICl_3 (c) $XeOF_4$ (d) BHF_2

11.42 Isoniazid is an antibacterial agent that is very useful against many common strains of tuberculosis. A valid Lewis structure is

(a) How many σ bonds are in the molecule?
(b) What is the hybridization of each C and N atom?

11.43 Hydrazine, N_2H_4, and carbon disulfide, CS_2, form a cyclic molecule with the following Lewis structure:

(a) Draw Lewis structures for N_2H_4 and CS_2.
(b) How do electron-group arrangement, molecular shape, and hybridization of N change when N_2H_4 reacts to form the product?
(c) How do electron-group arrangement, molecular shape, and hybridization of C change when CS_2 reacts to form the product?

11.44 In each of the following equations, what hybridization change, if any, occurs for the underlined atom?
(a) $\underline{B}F_3 + NaF \longrightarrow Na^+BF_4^-$
(b) $\underline{P}Cl_3 + Cl_2 \longrightarrow PCl_5$
(c) $H\underline{C}{\equiv}CH + H_2 \longrightarrow H_2C{=}CH_2$
(d) $\underline{Si}F_4 + 2F^- \longrightarrow SiF_6^{2-}$
(e) $\underline{S}O_2 + \frac{1}{2}O_2 \longrightarrow SO_3$

11.45 The ionosphere lies about 100 km above Earth's surface. This layer consists mostly of NO, O_2, and N_2, and photoionization creates NO^+, O_2^+, and N_2^+. (a) Use MO theory to compare the bond orders of the molecules and ions. (b) Does the magnetic behavior of each species change when its ion forms?

11.46 Glyphosate (*below*) is a common herbicide that is relatively harmless to animals but deadly to most plants. Describe the shape around and the hybridization of the P, N, and three numbered C atoms.

11.47 The sulfate ion can be represented with four S—O bonds or with two S—O and two S=O bonds. (a) Which representation is better from the standpoint of formal charges? (b) What is the shape of the sulfate ion, and what hybrid orbitals of S are postulated for the σ bonding? (c) In view of the answer to part (b), what orbitals of S must be used for the π bonds? What orbitals of O? (d) Draw a diagram to show how one atomic orbital from S and one from O overlap to form a π bond.

11.48 Tryptophan is one of the amino acids found in proteins:

(a) What is the hybridization of each of the numbered C, N, and O atoms?
(b) How many σ bonds are present in tryptophan?
(c) Predict the bond angles at points a, b, and c.

11.49 Sulfur forms oxides, oxoanions, and halides. What is the hybridization of the central S in SO_2, SO_3, SO_3^{2-}, SCl_4, SCl_6, and S_2Cl_2 (atom sequence Cl—S—S—Cl)?

11.50 The hydrocarbon allene, $H_2C{=}C{=}CH_2$, is obtained indirectly from petroleum and used as a precursor for several types of plastics. What is the hybridization of each C atom in allene? Draw a bonding picture for allene with lines for σ bonds, and show the arrangement of the π bonds. Be sure to represent the geometry of the molecule in three dimensions.

11.51 Some species with two oxygen atoms only are the oxygen molecule, O_2, the peroxide ion, O_2^{2-}, the superoxide ion, O_2^-, and the dioxygenyl ion, O_2^+. Draw an MO diagram for each, rank them in order of increasing bond length, and find the number of unpaired electrons in each.

11.52 Linoleic acid is an essential fatty acid found in many vegetable oils, such as soy, peanut, and cottonseed. A key structural feature of the molecule is the *cis* orientation around its two double bonds, where R_1 and R_2 represent two different groups that form the rest of the molecule.

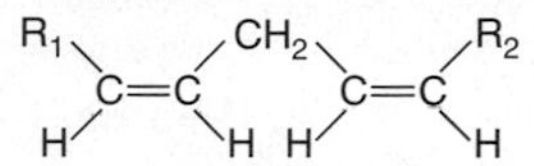

(a) How many different compounds are possible, changing only the *cis-trans* arrangements around these two double bonds?
(b) How many are possible for a similar compound with three double bonds?

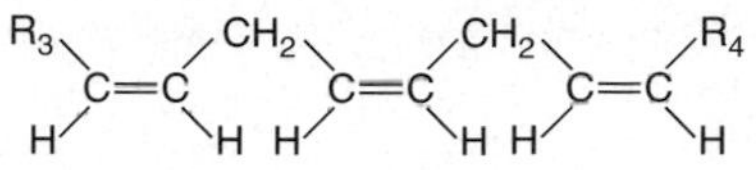

11.53 There is concern in health-related government agencies that the American diet contains too much meat, and numerous recommendations have been made urging people to consume more fruit and vegetables. One of the richest sources of vegetable protein is soy, available in many forms. Among these is soybean curd, or tofu, which is a staple of many Asian diets. Chemists have isolated an anticancer agent called *genistein* from tofu, which may explain the much lower incidence of cancer among people in the Far East. A valid Lewis structure for genistein is

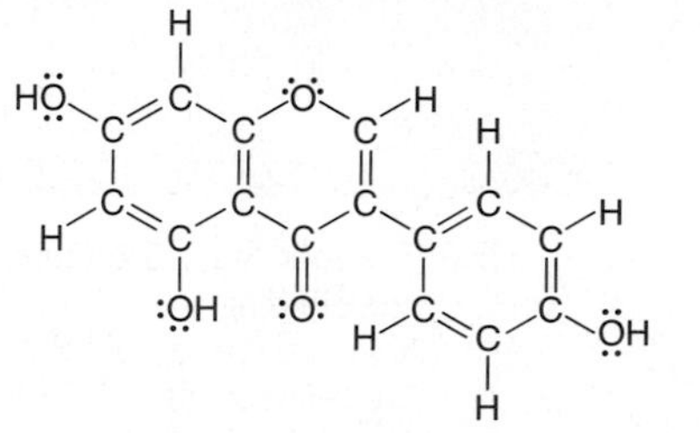

(a) Is the hybridization of each C in the right-hand ring the same? Explain.
(b) Is the hybridization of the O atom in the center ring the same as that of the O atoms in OH groups? Explain.
(c) How many carbon-oxygen σ bonds are there? How many carbon-oxygen π bonds?
(d) Do all the lone pairs on oxygens occupy the same type of hybrid orbital? Explain.

11.54 Molecular nitrogen, carbon monoxide, and cyanide ion are isoelectronic. (a) Draw an MO diagram for each. (b) CO and CN^- are toxic. What property may explain why N_2 isn't?

11.55 Silicon tetrafluoride reacts with F^- to produce the hexafluorosilicate ion, SiF_6^{2-}; GeF_4 behaves similarly, but CF_4 does not. (a) Draw Lewis structures for SiF_4, GeF_6^{2-}, and CF_4. (b) What is the hybridization of the central atom in each species? (c) Why doesn't CF_4 react with F^- to form CF_6^{2-}?

11.56 Simple proteins consist of amino acids linked together in a long chain; a small portion of such a chain is

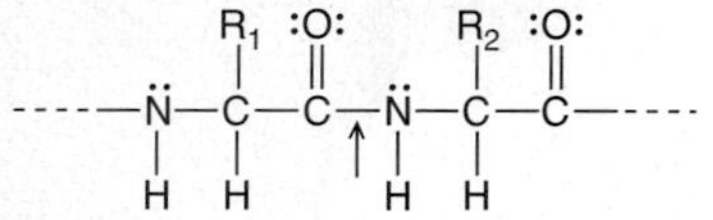

Experiment shows that rotation about the C—N bond (indicated by the arrow) is somewhat restricted. Explain with resonance structures, and show the types of bonding involved.

11.57 The compound 2,6-dimethylpyrazine gives chocolate its odor and is used in flavorings. A valid Lewis structure is

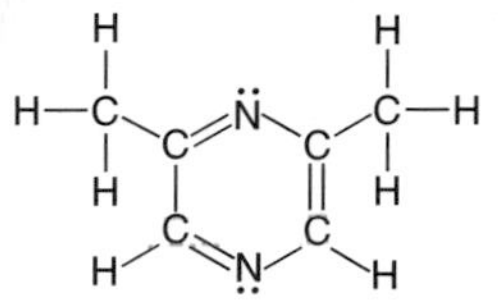

(a) Which atomic orbitals mix to form the hybrid orbitals of N?
(b) In what type of hybrid orbital do the lone pairs of N reside?
(c) Is the hybridization of each C in a CH_3 group the same as that of each C in the ring? Explain.

11.58 Acetylsalicylic acid (aspirin), the most widely used medicine in the world, has the following Lewis structure:

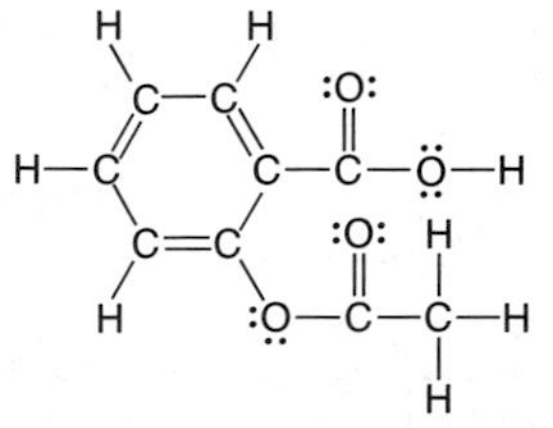

(a) What is the hybridization of each C and each O atom?
(b) How many localized π bonds are present?
(c) How many C atoms have a trigonal planar shape around them? A tetrahedral shape?

Solid water This molecular-scale view reveals the hexagonal order of crystalline water. When the solid melts, this order becomes chaos. In this chapter, we focus on the forces among the molecules that give rise to the liquid and solid states and highlight the unique physical behavior of each state and its applications to everyday life.

12

Intermolecular Forces: Liquids, Solids, and Phase Changes

All the matter around you occurs in one of three physical states—gas, liquid, or solid. Under specific conditions, many pure substances can exist in any of the states. We're all familiar with the three states of water: we inhale and exhale gaseous water; drink, excrete, and wash with liquid water; and shovel, slide on, or cool our drinks with solid water. Look around and you'll find other examples of solids (jewelry, table salt), liquids (gasoline, antifreeze), and gases (air, CO_2 bubbles) that can exist in one or both of the other states, even though the conditions needed may be unusual.

The three states were introduced in Chapter 1 and their properties compared when we examined gases in Chapter 5; now we turn our attention to liquids and solids. A physical state is one type of **phase,** any physically distinct, homogeneous part of a system. The water in a glass constitutes a single phase. Add some ice and you have two phases; or, if there are bubbles in the ice, you have three.

Liquids and solids are called *condensed* phases (or condensed states) because their particles are extremely close together. Electrostatic forces among the particles, called *interparticle forces* or, more commonly, **intermolecular forces,** combine with the particles' kinetic energy to create the properties of each phase as well as **phase changes,** the changes from one phase to another.

IN THIS CHAPTER . . . We begin with a kinetic-molecular overview of the states and then take a similarly close-up view of phase changes. We calculate the energy involved in phase changes and highlight the effects of temperature and pressure with *phase diagrams.* We consider the types of intermolecular forces and apply them first to the properties of liquids. We see how the unique physical properties of water arise inevitably from the electron configurations of its atoms. Then we discuss the properties of solids, emphasizing the relationship between type of bonding and intermolecular force. You will learn how the atomic-scale structures of solids are determined and, in the final section, how some exciting modern materials—semiconductors, liquid crystals, ceramics, polymers, and nanostructures—are produced and used.

Concepts & Skills to Review

before you study this chapter

- properties of gases, liquids, and solids (Section 5.1)
- kinetic-molecular theory of gases (Section 5.6)
- kinetic and potential energy (Section 6.1)
- enthalpy change, heat capacity, and Hess's law (Sections 6.2, 6.3, and 6.5)
- Coulomb's law (Section 9.2)
- chemical bonding models (Chapter 9)
- molecular polarity (Section 10.3)
- molecular orbital treatment of diatomic molecules (Section 11.3)

12.1 AN OVERVIEW OF PHYSICAL STATES AND PHASE CHANGES

Imagine yourself among the particles of a molecular substance, HF, and you'll discover two types of electrostatic forces at work:

- *Intra*molecular forces (bonding forces) exist *within* each molecule and influence the *chemical* properties of the substance.
- *Inter*molecular forces (nonbonding forces) exist *between* the molecules and influence the *physical* properties of the substance.

Now imagine a molecular view of the three states of water, as an example, and focus on just one molecule from each state. They look identical—bent, polar H—O—H molecules. In fact, the *chemical* behavior of the three states *is* identical because their molecules are held together by the same *intra*molecular bonding forces. However, the *physical* behavior of the states differs greatly because the strengths of the *inter*molecular nonbonding forces differ greatly.

A Kinetic-Molecular View of the Three States Whether a substance is a gas, liquid, or solid depends on the interplay of the potential energy of the intermolecular attractions, which tends to draw the molecules together, and the kinetic energy of the molecules, which tends to disperse them. According to Coulomb's law, the potential energy depends on the charges of the particles and the distances between them (see Section 9.2). The average kinetic energy, which is related to the particles' average speed, is proportional to the absolute temperature.

Table 12.1 A Macroscopic Comparison of Gases, Liquids, and Solids

State	Shape and Volume	Compressibility	Ability to Flow
Gas	Conforms to shape and volume of container	High	High
Liquid	Conforms to shape of container; volume limited by surface	Very low	Moderate
Solid	Maintains its own shape and volume	Almost none	Almost none

In Table 12.1, we distinguish among the three states by focusing on three properties—shape, compressibility, and ability to flow. The following kinetic-molecular view of the three states is an extension of the model we used to understand gases (see Section 5.6):

- *In a gas,* the energy of attraction is small relative to the energy of motion; so, on average, the particles are far apart. This large interparticle distance has several macroscopic consequences. A gas moves randomly throughout its container and fills it. Gases are highly compressible, and they flow and diffuse easily through one another.
- *In a liquid,* the attractions are stronger because the particles are in virtual contact. But their kinetic energy still allows them to tumble randomly over and around each other. Therefore, a liquid conforms to the shape of its container but has a surface. With very little free space between the particles, liquids resist an applied external force and thus compress only very slightly. They flow and diffuse but *much* more slowly than gases.
- *In a solid,* the attractions dominate the motion so much that the particles remain in position relative to one another, jiggling in place. With the positions of the particles fixed, a solid has a specific shape and does not flow significantly. The particles are usually slightly closer together than in a liquid, so solids compress even less than liquids.

Environmental Flow The environment demonstrates beautifully the differences in the abilities of the three states to flow and diffuse. Atmospheric gases mix so well that the 80 km of air closest to Earth's surface has a uniform composition. Much less mixing occurs in the oceans, and differences in composition at various depths support different species. Rocky solids (see photo) intermingle so little that adjacent strata remain separated for millions of years.

Types of Phase Changes Phase changes are also determined by the interplay between kinetic energy and intermolecular forces. As the temperature increases, the average kinetic energy increases as well, so the faster moving particles can overcome attractions more easily; conversely, lower temperatures allow the forces to draw the slower moving particles together.

What happens when gaseous water is cooled? A mist appears as the particles form tiny microdroplets that then collect into a bulk sample of liquid with a single surface. The process by which a gas changes into a liquid is called **condensation;** the opposite process, changing from a liquid into a gas, is called **vaporization.** With further cooling, the particles move even more slowly and become fixed in position as the liquid solidifies in the process of **freezing;** the opposite change is called **melting,** or **fusion.** In common speech, the term *freezing* implies low temperature because we typically think of water. But, most metals, for example, freeze (solidify) at much higher temperatures, and this change has many medical, industrial, and artistic applications: gold dental crowns, steel auto bodies, bronze statues, and so forth.

Enthalpy changes accompany phase changes. As the molecules of a gas attract each other and come closer together in the liquid, and then become fixed in the solid, the system of particles loses energy, which is released as heat. Thus, *condensing and freezing are exothermic changes.* On the other hand, energy must be absorbed to overcome the attractive forces that keep the particles in a liquid together and those that keep them fixed in place in a solid. Thus, *melting and vaporizing are endothermic changes.*

For a pure substance, each phase change has a specific enthalpy change *per mole* (measured at 1 atm and the temperature of the change). For vaporization, it is called the **heat of vaporization (ΔH^0_{vap})**, and for fusion, it is the **heat of fusion (ΔH^0_{fus})**. In the case of water, we have

$$H_2O(l) \longrightarrow H_2O(g) \qquad \Delta H = \Delta H^0_{vap} = 40.7 \text{ kJ/mol (at } 100°C)$$
$$H_2O(s) \longrightarrow H_2O(l) \qquad \Delta H = \Delta H^0_{fus} = 6.02 \text{ kJ/mol (at } 0°C)$$

The reverse processes, condensing and freezing, have enthalpy changes of the *same magnitude but opposite sign:*

$$H_2O(g) \longrightarrow H_2O(l) \qquad \Delta H = -\Delta H^0_{vap} = -40.7 \text{ kJ/mol}$$
$$H_2O(l) \longrightarrow H_2O(s) \qquad \Delta H = -\Delta H^0_{fus} = -\ 6.02 \text{ kJ/mol}$$

Water is typical of most pure substances in that it takes less energy to melt 1 mol of solid water than to vaporize 1 mol of liquid water: $\Delta H^0_{fus} < \Delta H^0_{vap}$. Figure 12.1 shows several examples. The reason is that a phase change is essentially a change in intermolecular distance and freedom of motion. Less energy is needed to overcome the forces holding the molecules in fixed positions (melt a solid) than to separate them completely from each other (vaporize a liquid).

The three states of water are so common because they all are stable under ordinary conditions. Carbon dioxide, on the other hand, is familiar as a gas and a solid (dry ice), but liquid CO_2 occurs only at external pressures greater than 5 atm. At ordinary conditions, solid CO_2 becomes a gas without first becoming a liquid. This process is called **sublimation.** On a clear wintry day, you can dry clothes outside on a line, even though it may be too cold for ice to melt, because the ice sublimes. Freeze-dried foods are prepared by sublimation. The opposite process, changing from a gas directly into a solid, is called **deposition**—you may have seen ice crystals form on a cold window from the deposition of water vapor. The **heat of sublimation (ΔH^0_{subl})** is the enthalpy change when 1 mol of a substance sublimes. From Hess's law, it equals the sum of the heats of fusion and vaporization:

$$\begin{array}{lll} \text{Solid} \longrightarrow \text{liquid} & \Delta H^0_{fus} \\ \text{Liquid} \longrightarrow \text{gas} & \Delta H^0_{vap} \\ \hline \text{Solid} \longrightarrow \text{gas} & \Delta H^0_{subl} \end{array}$$

A Cooling Phase Change The evaporation of sweat has a cooling effect because heat from your body is used to vaporize the water. Many fur-covered animals achieve this cooling effect, too. Cats (and many rodents) lick themselves, transferring water from inside their bodies to the surface where it can evaporate, and dogs just open their mouths, hang out their tongues, and pant!

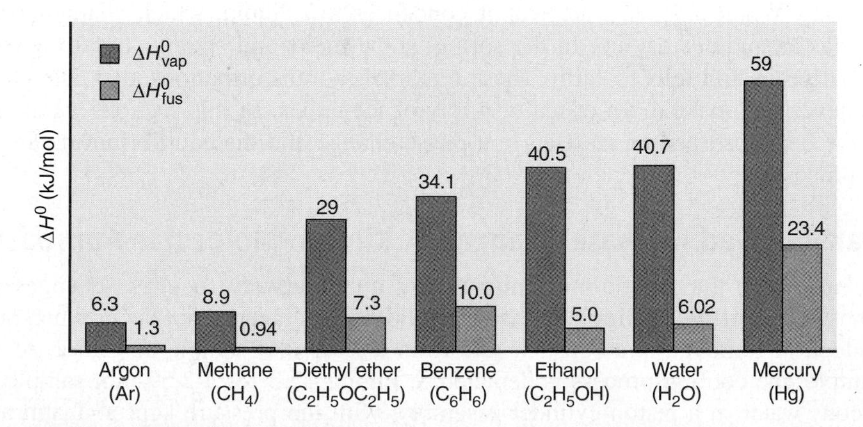

Figure 12.1 Heats of vaporization and fusion for several common substances. ΔH^0_{vap} is always larger than ΔH^0_{fus} because it takes more energy to separate particles completely than just to free them from their fixed positions in the solid.

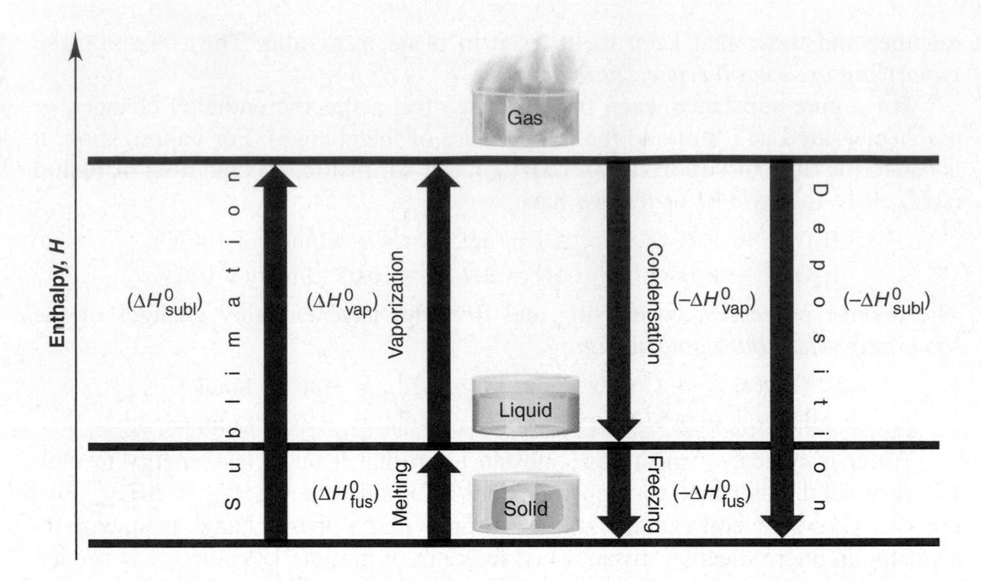

Figure 12.2 Phase changes and their enthalpy changes. Each type of phase change is shown with its associated enthalpy change. Fusion (or melting), vaporization, and sublimation are endothermic changes (positive ΔH^0), whereas freezing, condensation, and deposition are exothermic changes (negative ΔH^0).

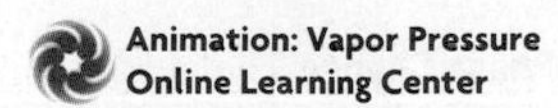
Animation: Vapor Pressure Online Learning Center

Figure 12.2 summarizes the terminology of the various phase changes and shows the enthalpy changes associated with them.

SECTION SUMMARY

Because of the relative magnitudes of intermolecular forces and kinetic energy, the particles in a gas are far apart and moving randomly, those in a liquid are in contact but still moving relative to each other, and those in a solid are in contact and fixed relative to one another in a rigid structure. These molecular-level differences in the states of matter account for macroscopic differences in shape, compressibility, and ability to flow. When a solid becomes a liquid (melting, or fusion) or a liquid becomes a gas (vaporization), energy is absorbed to overcome intermolecular forces and increase the average distance between particles. When particles come closer together in the reverse changes (freezing and condensation), energy is released. Sublimation is the changing of a solid directly into a gas. Each phase change is associated with a given enthalpy change under specified conditions.

12.2 QUANTITATIVE ASPECTS OF PHASE CHANGES

Some of the most spectacular, and familiar, phase changes occur in the weather each day. When it rains, water vapor condenses to a liquid, which changes back to a gas as puddles dry up. In the spring, snow melts and streams fill; in winter, water freezes and falls to earth. These remarkable transformations also take place whenever you make a pot of tea or a tray of ice cubes. In this section, we examine the heat absorbed or released in a phase change and the equilibrium nature of the process.

Heat Involved in Phase Changes: A Kinetic-Molecular Approach

We can apply the kinetic-molecular theory quantitatively to phase changes by means of a **heating-cooling curve,** which shows the changes that occur when heat is added or removed at a constant rate from a particular sample of matter. As an example, the cooling process is depicted in Figure 12.3 for a 2.50-mol sample of gaseous water in a piston-cylinder assembly, with the pressure kept at 1 atm and the temperature changing from 130°C to −40°C. To an observer, the process is continuous, but we can divide it into five heat-releasing (exothermic) stages that correspond to the five portions of the curve—the gas cools, it condenses to a liquid, the liquid cools, it freezes to a solid, and the solid cools a bit further:

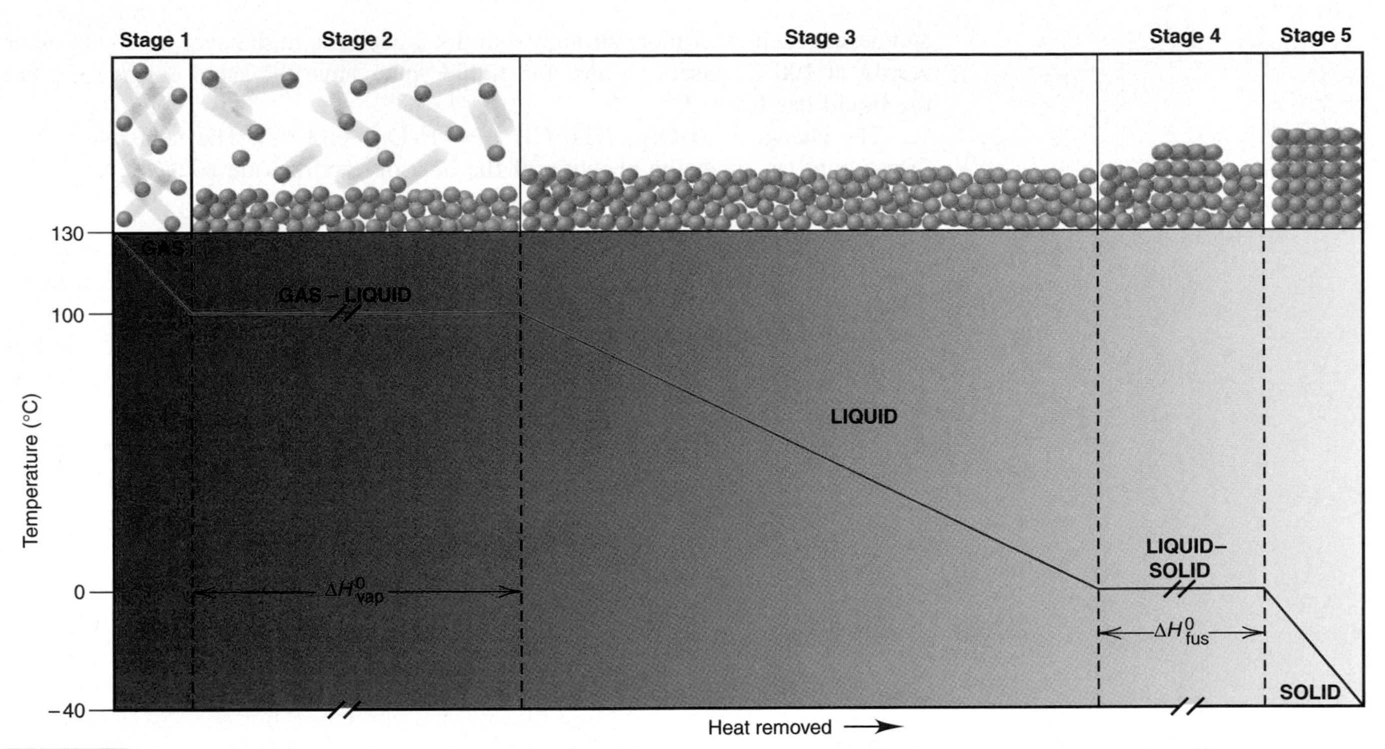

Figure 12.3 A cooling curve for the conversion of gaseous water to ice. A plot is shown of temperature vs. heat removed as gaseous water at 130°C changes to ice at −40°C. This process occurs in five stages, with a molecular-level depiction shown for each stage. Stage 1: Gaseous water cools. Stage 2: Gaseous water condenses. Stage 3: Liquid water cools. Stage 4: Liquid water freezes. Stage 5: Solid water cools. The slopes of the lines in stages 1, 3, and 5 reflect the magnitudes of the molar heat capacities of the phases. Although not drawn to scale, the line in stage 2 is longer than the line in stage 4 because ΔH^0_{vap} of water is greater than ΔH^0_{fus}. A plot of temperature vs. heat added would have the same steps but in reverse order.

Stage 1. Gaseous water cools. Picture a collection of water molecules behaving like a typical gas: zooming around chaotically at a range of speeds, smashing into each other and the container walls. At a high enough temperature, the most probable speed, and thus the average kinetic energy (E_k), of the molecules is high enough to overcome the potential energy (E_p) of attractions among them. As the temperature falls, the average E_k decreases, so the attractions become increasingly important.

The change is $H_2O(g)$ [130°C] $\longrightarrow$ $H_2O(g)$ [100°C]. The heat (q) is the product of the amount (number of moles, n) of water, the molar heat capacity of *gaseous water*, $C_{water(g)}$, and the temperature change, ΔT ($T_{final} - T_{initial}$):

$$q = n \times C_{water(g)} \times \Delta T = (2.50\ \text{mol})\ (33.1\ \text{J/mol·°C})\ (100°\text{C} - 130°\text{C})$$
$$= -2482\ \text{J} = -2.48\ \text{kJ}$$

The minus sign indicates that heat is released. (For purposes of canceling, the units for molar heat capacity, C, include °C, rather than K. The magnitude of C is not affected because the kelvin and the Celsius degree represent the same temperature increment.)

Stage 2. Gaseous water condenses. At the condensation point, the slowest of the molecules are near each other long enough for intermolecular attractions to form groups of molecules, which aggregate into microdroplets and then a bulk liquid. Note that while the state is changing from gas to liquid, *the temperature remains constant,* so the average E_k is constant. This means that, even though molecules in a gas certainly move *farther* between collisions than those in a liquid, their average *speed* is the same at a given temperature. Thus, removing heat from the system involves a decrease in the average E_p, as the molecules approach and attract each other more strongly, but not a decrease in the average E_k. In other words, at 100°C, gaseous water and liquid water have the same average E_k, but the liquid has lower E_p.

The change is $H_2O(g)$ [100°C] $\longrightarrow$ $H_2O(l)$ [100°C]. The heat released is the amount (n) times the negative of the heat of vaporization ($-\Delta H^0_{vap}$):

$$q = n(-\Delta H^0_{vap}) = (2.50 \text{ mol})(-40.7 \text{ kJ/mol}) = -102 \text{ kJ}$$

This step contributes the *greatest portion of the total heat released* because of the decrease in potential energy that occurs with the enormous decrease in distance between molecules in a gas and those in a liquid.

Stage 3. Liquid water cools. The molecules have now condensed to the liquid state. The continued loss of heat appears as a decrease in temperature, that is, as a decrease in the most probable molecular speed and, thus, the average E_k. The temperature decreases as long as the sample remains liquid.

The change is $H_2O(l)$ [100°C] $\longrightarrow$ $H_2O(l)$ [0°C]. The heat depends on amount (n), the molar heat capacity of *liquid water,* and ΔT:

$$q = n \times C_{water(l)} \times \Delta T = (2.50 \text{ mol})(75.4 \text{ J/mol·°C})(0\text{°C} - 100\text{°C})$$
$$= -18{,}850 \text{ J} = -18.8 \text{ kJ}$$

Stage 4. Liquid water freezes. At the freezing temperature of water, 0°C, intermolecular attractions overcome the motion of the molecules around one another. Beginning with the slowest, the molecules lose E_p and align themselves into the crystalline structure of ice. Molecular motion continues, but only as vibration of atoms about their fixed positions. As during condensation, the temperature and average E_k remain constant during freezing.

The change is $H_2O(l)$ [0°C] $\longrightarrow$ $H_2O(s)$ [0°C]. The heat released is n times the negative of the heat of fusion ($-\Delta H^0_{fus}$):

$$q = n(-\Delta H^0_{fus}) = (2.50 \text{ mol})(-6.02 \text{ kJ/mol}) = -15.0 \text{ kJ}$$

Stage 5. Solid water cools. With motion restricted to jiggling in place, further cooling merely reduces the average speed of this jiggling.

The change is $H_2O(s)$ [0°C] $\longrightarrow$ $H_2O(s)$ [−40°C]. The heat released depends on n, the molar heat capacity of *solid water,* and ΔT:

$$q = n \times C_{water(s)} \times \Delta T = (2.50 \text{ mol})(37.6 \text{ J/mol·°C})(-40\text{°C} - 0\text{°C})$$
$$= -3760 \text{ J} = -3.76 \text{ kJ}$$

According to Hess's law, the total heat released is the sum of the heats released for the individual stages. The sum of q for stages 1 to 5 is −142 kJ.

Two key points stand out in this or any similar process (at constant pressure), whether exothermic or endothermic:

- *Within a phase,* a change in heat is accompanied by *a change in temperature,* which is associated with a change in average E_k as *the most probable speed of the molecules changes.* The heat lost or gained depends on the amount of substance, the molar heat capacity for that phase, and the change in temperature.
- *During a phase change,* a change in heat occurs at a *constant temperature,* which is associated with a change in E_p, as *the average distance between molecules changes.* Both physical states are present during a phase change. The heat lost or gained depends on the amount of substance and the enthalpy of the phase change (ΔH^0_{vap} or ΔH^0_{fus}).

The Equilibrium Nature of Phase Changes

In everyday experience, phase changes take place in open containers—the outdoors, a pot on a stove, the freezer compartment of a refrigerator—so such a change is not reversible. In a closed container under controlled conditions, however, *phase changes of many substances are reversible and reach equilibrium,* just as chemical changes do.

Liquid-Gas Equilibria Picture an *open* flask containing a pure liquid at constant temperature and focus on the molecules at the surface. Within their range of molecular speeds, some are moving fast enough and in the right direction to overcome attractions, so they vaporize. Nearby molecules immediately fill the gap, and with energy supplied by the constant-temperature surroundings, the process continues until the entire liquid phase is gone.

Now picture starting with a *closed* flask at constant temperature, as in Figure 12.4A, and assume that a vacuum exists above the liquid. As before, some of the molecules at the surface have a high enough E_k to vaporize. As the number of molecules in the vapor phase increases, the pressure of the vapor increases. At the same time, some of the molecules in the vapor that collide with the surface have a low enough E_k to become attracted too strongly to leave the liquid and they condense. For a given surface area, the number of molecules that make up the surface is constant; therefore, the rate of vaporization—the number of molecules leaving the surface per unit time—is also constant. On the other hand, as the vapor becomes more populated, molecules collide with the surface more often, so the rate of condensation slowly increases. As condensation continues to offset vaporization, the increase in the pressure of the vapor slows. Eventually, the rate of condensation equals the rate of vaporization, as depicted in Figure 12.4B. From this time onward, *the pressure of the vapor is constant at that temperature.* Macroscopically, the situation seems static, but at the molecular level, molecules are entering and leaving the liquid surface at equal rates. The system has reached a state of *dynamic equilibrium:*

$$\text{Liquid} \rightleftharpoons \text{gas}$$

Figure 12.4C depicts the entire process graphically.

The pressure exerted by the vapor at equilibrium is called the *equilibrium vapor pressure*, or just the **vapor pressure,** of the liquid at that temperature. If we use a larger flask, more molecules are present in the gas phase at equilibrium; as long as some liquid remains, however, the vapor pressure does not change. Suppose we disturb this equilibrium system by pumping out some of the vapor at constant temperature, thereby lowering the pressure. (If this were a cylinder with piston, we could lower the pressure by moving the piston outward to increase the volume.) The rate of condensation temporarily falls below the rate of vaporization (the forward process is faster). Fewer molecules re-enter the liquid than leave it, so the pressure of the vapor rises until, after a short period, the condensation rate increases enough for equilibrium to be reached again. Similarly, if we disturb the system by pumping in additional vapor (or moving the piston inward to decrease the volume) at constant temperature, thereby raising the pressure, the rate of condensation temporarily exceeds the rate of vaporization. More molecules enter the liquid than leave it (the backward process is faster), but soon the condensation rate decreases, and the pressure again reaches the equilibrium value.

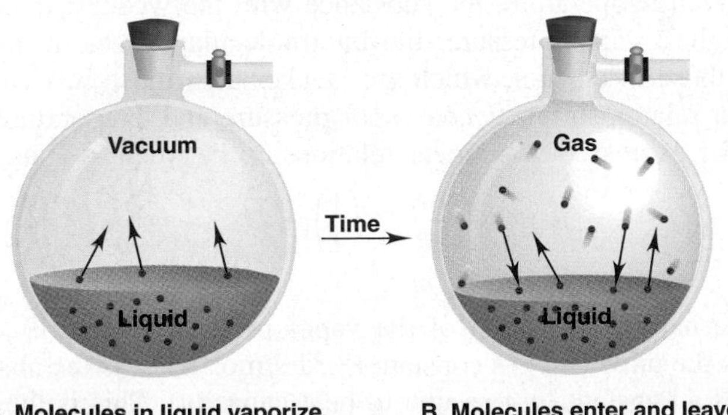

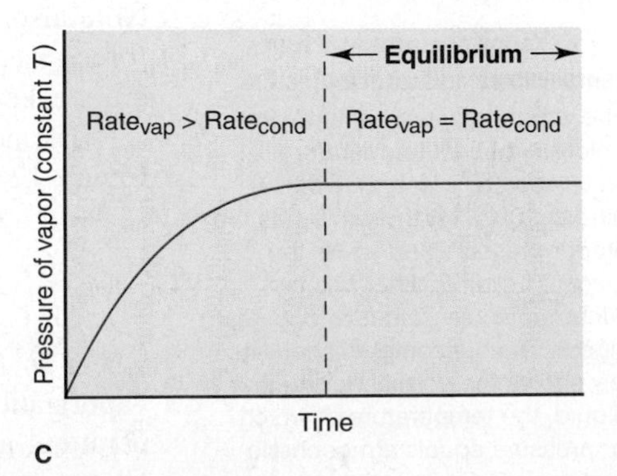

Figure 12.4 Liquid-gas equilibrium. **A,** In a closed flask at constant temperature with the air removed, the initial pressure is zero. As molecules leave the surface and enter the space above the liquid, the pressure of the vapor rises. **B,** At equilibrium, the same number of molecules leave as enter the liquid within a given time, so the pressure of the vapor reaches a constant value. **C,** A plot of pressure vs. time shows that the pressure of the vapor increases as long as the rate of vaporization is greater than the rate of condensation. At equilibrium, the rates are equal, so the pressure is constant. The pressure at this point is the *vapor pressure* of the liquid at that temperature.

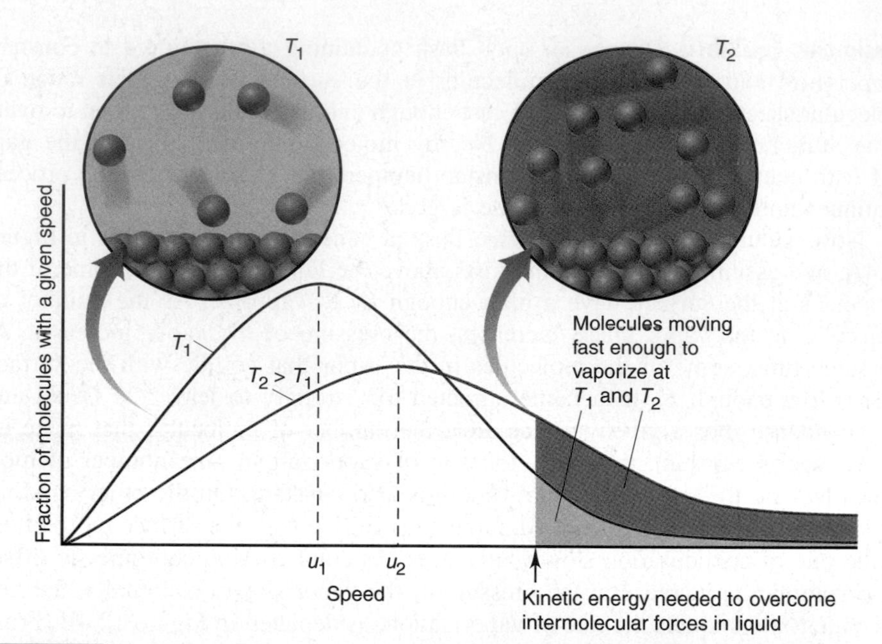

Figure 12.5 The effect of temperature on the distribution of molecular speeds in a liquid. With T_1 lower than T_2, the most probable molecular speed u_1 is less than u_2. (Note the similarity to Figure 5.14, p. 201.) The fraction of molecules with enough energy to escape the liquid *(shaded area)* is *greater at the higher temperature*. The molecular views show that at the higher T, equilibrium is reached with more gas molecules in the same volume and thus at a higher vapor pressure.

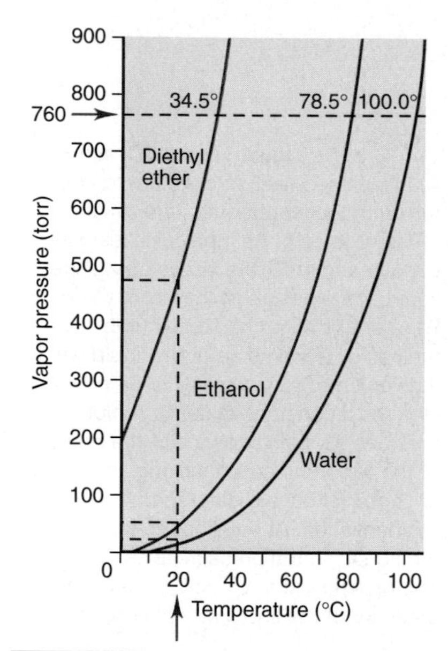

Figure 12.6 Vapor pressure as a function of temperature and intermolecular forces. The vapor pressures of three liquids are plotted against temperature. At any given temperature (see the vertical dashed line at 20°C), diethyl ether has the highest vapor pressure and water the lowest, because diethyl ether has the weakest intermolecular forces and water the strongest. The horizontal dashed line at 760 torr shows the normal boiling point of each liquid, the temperature at which the vapor pressure equals atmospheric pressure at sea level.

This behavior of a pure liquid in contact with its vapor is a general one for any system: *when a system at equilibrium is disturbed, it counteracts the disturbance and eventually re-establishes a state of equilibrium.* We'll return to this key idea often in later chapters.

The Effects of Temperature and Intermolecular Forces on Vapor Pressure The vapor pressure of a substance depends on the temperature. Raising the temperature of a liquid *increases* the fraction of molecules moving fast enough to escape the liquid and *decreases* the fraction moving slowly enough to be recaptured. This important idea is shown in Figure 12.5. In general, *the higher the temperature is, the higher the vapor pressure.*

The vapor pressure also depends on the intermolecular forces present. The average E_k is the same for different substances at a given temperature. Therefore, molecules with weaker intermolecular forces vaporize more easily. In general, *the weaker the intermolecular forces are, the higher the vapor pressure.*

Figure 12.6 shows the vapor pressure of three liquids as a function of temperature. Notice that each curve rises more steeply as the temperature increases. Note also, at a given temperature, the substance with the weakest intermolecular forces has the highest vapor pressure: the intermolecular forces in diethyl ether are weaker than those in ethanol, which are weaker than those in water.

The nonlinear relationship between vapor pressure and temperature shown in Figure 12.6 can be expressed as a linear relationship between ln P and $1/T$:

$$\ln P = \frac{-\Delta H_{\text{vap}}}{R}\left(\frac{1}{T}\right) + C$$
$$y = m \quad x + b$$

where ln P is the natural logarithm of the vapor pressure, ΔH_{vap} is the heat of vaporization, R is the universal gas constant (8.31 J/mol·K), T is the absolute temperature, and C is a constant (not related to heat capacity). This is the **Clausius-Clapeyron equation,** which gives us a way of finding the heat of vaporization, the energy needed to vaporize 1 mol of molecules in the liquid state. The blue equation beneath the Clausius-Clapeyron equation is the equation for a straight line, where $y = \ln P$, $x = 1/T$, m (the slope) $= -\Delta H_{vap}/R$, and b (the y-axis intercept) $= C$. Figure 12.7 shows plots for diethyl ether and water. A two-point version of the equation allows a nongraphical determination of ΔH_{vap}:

$$\ln \frac{P_2}{P_1} = \frac{-\Delta H_{vap}}{R}\left(\frac{1}{T_2} - \frac{1}{T_1}\right) \quad (12.1)$$

If ΔH_{vap} and P_1 at T_1 are known, we can calculate the vapor pressure (P_2) at any other temperature (T_2) or the temperature at any other pressure.

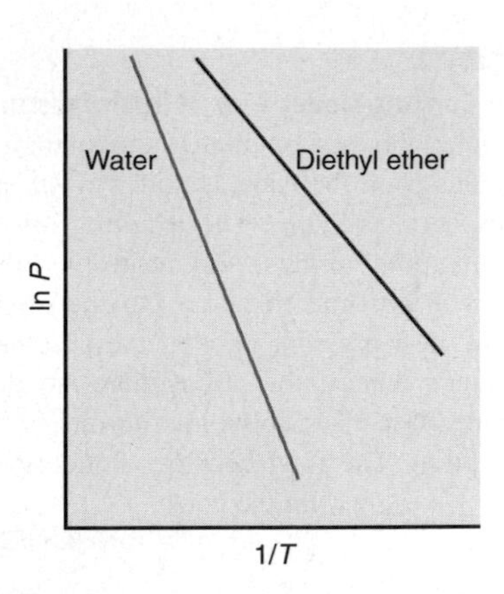

Figure 12.7 A linear plot of the relationship between vapor pressure and temperature. The Clausius-Clapeyron equation gives a straight line when the natural logarithm of the vapor pressure (ln *P*) is plotted against the inverse of the absolute temperature (1/*T*). The slopes ($-\Delta H_{vap}/R$) allow determination of the heats of vaporization of the two liquids. Note that the slope is steeper for water because its ΔH_{vap} is greater.

SAMPLE PROBLEM 12.1 Using the Clausius-Clapeyron Equation

Problem The vapor pressure of ethanol is 115 torr at 34.9°C. If ΔH_{vap} of ethanol is 40.5 kJ/mol, calculate the temperature (in °C) when the vapor pressure is 760 torr.

Plan We are given ΔH_{vap}, P_1, P_2, and T_1 and substitute them into Equation 12.1 to solve for T_2. The value of R here is 8.31 J/mol·K, so we must convert T_1 to K to obtain T_2, and then convert T_2 to °C.

Solution Substituting the values into Equation 12.1 and solving for T_2:

$$\ln \frac{P_2}{P_1} = \frac{-\Delta H_{vap}}{R}\left(\frac{1}{T_2} - \frac{1}{T_1}\right)$$

$$T_1 = 34.9°C + 273.15 = 308.0\ K$$

$$\ln \frac{760\ torr}{115\ torr} = \left(-\frac{40.5\times10^3\ J/mol}{8.314\ J/mol{\cdot}K}\right)\left(\frac{1}{T_2} - \frac{1}{308.0\ K}\right)$$

$$1.888 = (-4.87\times10^3)\left[\frac{1}{T_2} - (3.247\times10^{-3})\right]$$

$$T_2 = 350.\ K$$

Converting T_2 from K to °C:

$$T_2 = 350.\ K - 273.15 = 77°C$$

Check Round off to check the math. The change is in the right direction: higher P should occur at higher T. As we discuss next, a substance has a vapor pressure of 760 torr at its normal boiling point. Checking the *CRC Handbook of Chemistry and Physics* shows that the boiling point of ethanol is 78.5°C, very close to our answer.

FOLLOW-UP PROBLEM 12.1 At 34.1°C, the vapor pressure of water is 40.1 torr. What is the vapor pressure at 85.5°C? The ΔH_{vap} of water is 40.7 kJ/mol.

Vapor Pressure and Boiling Point In an *open* container, the atmosphere bears down on the liquid surface. As the temperature rises, molecules leave the surface more often, and they also move more quickly throughout the liquid. At some temperature, the average E_k of the molecules in the liquid is great enough for bubbles of vapor to form *in the interior,* and the liquid boils. At any lower temperature, the bubbles collapse as soon as they start to form because the external pressure is greater than the vapor pressure inside the bubbles. Thus, the **boiling point** is *the temperature at which the vapor pressure equals the external pressure,* usually that of the atmosphere. Once boiling begins, the temperature remains constant until the liquid phase is gone because the applied heat is used by the molecules to overcome attractions and enter the gas phase.

The boiling point varies with elevation, because the atmospheric pressure does. At high elevations, a lower pressure is exerted on the liquid surface, so molecules in the interior need less kinetic energy to form bubbles. At low elevations, the opposite is true. Thus, *the boiling point depends on the applied pressure*. The *normal boiling point* is observed at standard atmospheric pressure (760 torr, or 101.3 kPa; see the horizontal dashed line in Figure 12.6).

Cooking Under Low or High Pressure People who live or hike in mountainous regions cook their meals under lower atmospheric pressure. The resulting lower boiling point of the liquid means the food takes *more* time to cook. On the other hand, in a pressure cooker, the pressure exceeds that of the atmosphere, so the temperature rises above the normal boiling point. The food becomes hotter and, thus, takes *less* time to cook.

Since the boiling point is the temperature at which the vapor pressure equals the external pressure, we can also interpret the curves in Figure 12.6 as a plot of external pressure vs. boiling point. For instance, the H_2O curve shows that water boils at 100°C at 760 torr (sea level), at 94°C at 610 torr (Boulder, Colorado), and at about 72°C at 270 torr (top of Mt. Everest).

Solid-Liquid Equilibria At the molecular level, the particles in a crystal are continually vibrating about their fixed positions. As the temperature rises, the particles vibrate more violently, until some have enough kinetic energy to break free of their positions, and melting begins. As more molecules enter the liquid (molten) phase, some collide with the solid and become fixed again. Because the phases remain in contact, a dynamic equilibrium is established when the melting rate equals the freezing rate. The temperature at which this occurs is the **melting point;** it is the same temperature as the freezing point, differing only in the direction of the energy flow. As with the boiling point, the temperature remains at the melting point as long as both phases are present.

Because liquids and solids are nearly incompressible, a change in pressure has little effect on the rate of movement to or from the solid. Therefore, in contrast to the boiling point, the melting point is affected by pressure only very slightly, and a plot of pressure (y axis) vs. temperature (x axis) for a solid-liquid phase change is typically a straight, *nearly* vertical line.

Figure 12.8 Iodine subliming. At ordinary atmospheric pressure, solid iodine sublimes (changes directly from a solid into a gas). When the I_2 vapor comes in contact with a cold surface, such as the water-filled inner test tube, it deposits I_2 crystals. Sublimation is a means of purification that, along with distillation, was probably discovered by alchemists.

Solid-Gas Equilibria Solids have much lower vapor pressures than liquids. Sublimation, the process of a solid changing directly into a gas, is much less familiar than vaporization because the necessary conditions of pressure and temperature are uncommon for most substances. Some solids *do* have high enough vapor pressures to sublime at ordinary conditions, including dry ice (carbon dioxide), iodine (Figure 12.8), and solid room deodorizers. A substance sublimes rather than melts because the combination of intermolecular attractions and atmospheric pressure is not great enough to keep the particles near one another when they leave the solid state. The pressure vs. temperature plot for the solid-gas transition shows a large effect of temperature on the pressure of the vapor; thus, this curve resembles the liquid-gas line in curving upward at higher temperatures.

Phase Diagrams: Effect of Pressure and Temperature on Physical State

To describe the phase changes of a substance at various conditions of temperature and pressure, we construct a **phase diagram,** which combines the liquid-gas, solid-liquid, and solid-gas curves. The shape of the phase diagram for CO_2 is typical for most substances (Figure 12.9A). A phase diagram has these four features:

1. *Regions of the diagram.* Each region corresponds to one phase of the substance. A particular phase is stable for any combination of pressure and temperature within its region. If any of the other phases is placed under those conditions, it will change to the stable phase. In general, the solid is stable at low temperature and high pressure, the gas at high temperature and low pressure, and the liquid at intermediate conditions.

2. *Lines between regions.* The lines separating the regions represent the phase-transition curves discussed earlier. Any point along a line shows the pressure and temperature at which the two phases exist in equilibrium. Note that the solid-liquid line has a *positive* slope (slants to the *right* with increasing pressure) because, for most substances, the solid is more dense than the liquid. Because the liquid occupies slightly more space than the solid, an increase in pressure favors the solid phase. (Water is the major exception, as you'll soon see.)

3. *The critical point.* The liquid-gas line ends at the **critical point.** Picture a liquid in a closed container. As it is heated, it expands, so its density decreases. At the same time, more liquid vaporizes, so the density of the vapor increases. The liquid and vapor densities become closer and closer to each other until, at

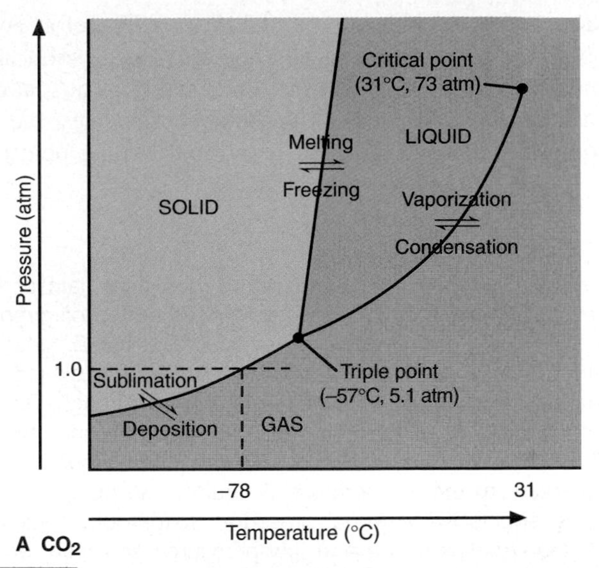

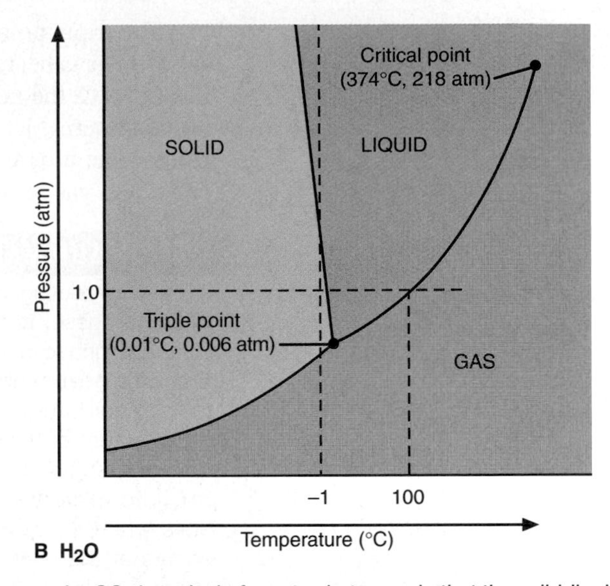

Figure 12.9 Phase diagrams for CO_2 and H_2O. Each region depicts the temperatures and pressures under which the phase is stable. Lines between two regions show conditions at which the two phases exist in equilibrium. The critical point shows conditions beyond which separate liquid and gas phases no longer exist. At the triple point, the three phases exist in equilibrium. (The axes are not linear.) **A,** The phase diagram for CO_2 is typical of most substances in that the solid-liquid line slopes to the right with increasing pressure: the solid is *more* dense than the liquid. **B,** Water is one of the few substances whose solid-liquid line slopes to the left with increasing pressure: the solid is *less* dense than the liquid. (The slopes of the solid-liquid lines in both diagrams are exaggerated.)

Animation: Phase Diagrams of the States of Matter
Online Learning Center

the *critical temperature* (T_c), the two densities are equal and the phase boundary disappears. The pressure at this temperature is the *critical pressure* (P_c). At this point, the average E_k of the molecules is so high that the vapor cannot be condensed no matter what pressure is applied. The two most common gases in air have critical temperatures far below room temperature: no matter what the pressure, O_2 will not condense above $-119°C$, and N_2 will not condense above $-147°C$. Beyond the critical temperature, a *supercritical fluid* exists rather than separate liquid and gaseous phases.

4. *The triple point.* The three phase-transition curves meet at the **triple point:** the pressure and temperature at which three phases are in equilibrium. As strange as it sounds, at the triple point in Figure 12.9A, CO_2 is subliming and depositing, melting and freezing, and vaporizing and condensing simultaneously! Phase diagrams for substances with several solid forms, such as sulfur, have more than one triple point.

The CO_2 phase diagram explains why dry ice (solid CO_2) doesn't melt under ordinary conditions. The triple-point pressure for CO_2 is 5.1 atm; therefore, at around 1 atm, liquid CO_2 does not occur. By following the horizontal dashed line in Figure 12.9A, you can see that when solid CO_2 is heated at 1.0 atm, it sublimes at $-78°C$ to gaseous CO_2 rather than melting. If normal atmospheric pressure were 5.2 atm, liquid CO_2 would be common.

The phase diagram for water differs in one key respect from the general case and reveals an extremely important property (Figure 12.9B). Unlike almost every other substance, solid water is *less dense* than liquid water. Because the solid occupies more space than the liquid, *water expands on freezing.* This behavior results from the unique open crystal structure of ice, which we discuss in a later section. As always, an increase in pressure favors the phase that occupies less space, but in the case of water, this is the *liquid* phase. Therefore, the solid-liquid line for water has a *negative* slope (slants to the *left* with increasing pressure): the higher the pressure, the lower the temperature at which water freezes. In Figure 12.9B, the vertical dashed line at $-1°C$ crosses the solid-liquid line, which means that ice melts at that temperature with only an increase in pressure.

The Remarkable Behavior of a Supercritical Fluid (SCF) What material lies beyond the familiar regions of liquid and gas? An SCF expands and contracts like a gas but has the solvent properties of a liquid, which chemists can alter by controlling the density. Supercritical CO_2 has received the most attention so far. It extracts nonpolar ingredients from complex mixtures, such as caffeine from coffee beans, nicotine from tobacco, and fats from potato chips, while leaving taste and aroma ingredients behind, for healthier consumer products. Lower the pressure and the SCF disperses immediately as a harmless gas. Supercritical CO_2 dissolves the fat from meat, along with pesticide and drug residues, which can then be quantified and monitored. It is being studied as an environmentally friendly dry cleaning agent. In an unexpected finding of great potential use, supercritical H_2O was shown to dissolve nonpolar substances even though liquid water does not! Studies are under way to carry out the large-scale removal of nonpolar organic toxins, such as PCBs, from industrial waste by extracting them into supercritical H_2O; after this step, O_2 gas is added to oxidize the toxins to small, harmless molecules.

The triple point of water occurs at low pressure (0.006 atm). Therefore, when solid water is heated at 1.0 atm, the horizontal dashed line crosses the solid-liquid line (at 0°C, the normal melting point) and enters the liquid region. Thus, at ordinary pressures, ice melts rather than sublimes. As the temperature rises, the horizontal line crosses the liquid-gas curve (at 100°C, the normal boiling point) and enters the gas region.

SECTION SUMMARY

A heating-cooling curve depicts the change in temperature with heat gain or loss. Within a phase, temperature (and average E_k) changes as heat is added or removed. During a phase change, temperature (and average E_k) is constant, but E_p changes. The total heat change for the curve is calculated using Hess's law. In a closed container, equilibrium is established between the liquid and gas phases. Vapor pressure, the pressure of the gas at equilibrium, is related directly to temperature and inversely to the strength of the intermolecular forces. The Clausius-Clapeyron equation uses ΔH_{vap} to relate the vapor pressure to the temperature. A liquid in an open container boils when its vapor pressure equals the external pressure. Solid-liquid equilibrium occurs at the melting point. Many solids sublime at low pressures and high temperatures. A phase diagram shows the phase that exists at a given pressure and temperature and the conditions at the critical point and the triple point of a substance. Water differs from most substances in that its solid phase is less dense than its liquid phase, so its solid-liquid line has a negative slope.

12.3 TYPES OF INTERMOLECULAR FORCES

As we said, the nature of the phases and their changes are due primarily to forces among the molecules. Both bonding (intramolecular) forces and intermolecular forces arise from electrostatic attractions between opposite charges. Bonding forces are due to the attraction between cations and anions (ionic bonding), nuclei and electron pairs (covalent bonding), or metal cations and delocalized valence electrons (metallic bonding). Intermolecular forces, on the other hand, are due to the attraction between molecules as a result of partial charges, or the attraction between ions and molecules. The two types of forces differ in magnitude, and Coulomb's law explains why:

- *Bonding forces are relatively strong* because they involve larger charges that are closer together.
- *Intermolecular forces are relatively weak* because they typically involve smaller charges that are farther apart.

How far apart are the charges between molecules that give rise to intermolecular forces? Consider Cl_2 as an example. When we measure the distances between two Cl nuclei in a sample of solid Cl_2, we obtain two different values, as shown in Figure 12.10. The shorter distance is between *two **bonded** Cl atoms in the same molecule*. It is, as you know, called the *bond length,* and one-half this distance is the *covalent radius*. The longer distance is between *two **nonbonded** Cl atoms in adjacent molecules*. It is called the *van der Waals distance* (named after the Dutch physicist Johannes van der Waals, who studied the effects of intermolecular forces on the behavior of real gases). This distance is the closest one Cl_2 molecule can approach another, the point at which intermolecular attractions balance electron-

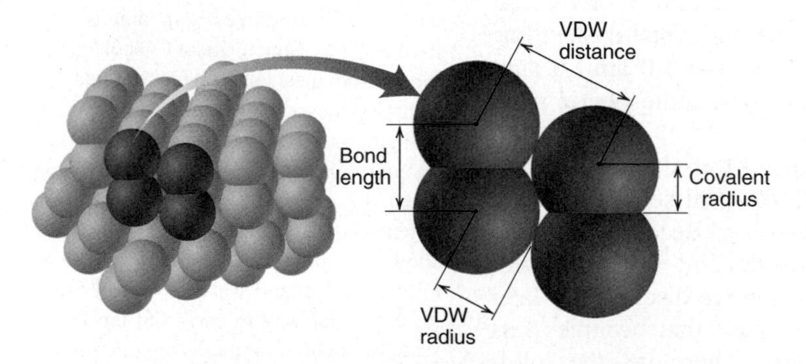

Figure 12.10 Covalent and van der Waals radii. As shown here for solid chlorine, the van der Waals (VDW) radius is one-half the distance between adjacent *nonbonded* atoms ($\frac{1}{2}$ × VDW distance), and the covalent radius is one-half the distance between *bonded* atoms ($\frac{1}{2}$ × bond length).

Table 12.2 Comparison of Bonding and Nonbonding (Intermolecular) Forces

Force	Model	Basis of Attraction	Energy (kJ/mol)	Example
Bonding				
Ionic		Cation–anion	400–4000	NaCl
Covalent		Nuclei–shared e^- pair	150–1100	H—H
Metallic		Cations–delocalized electrons	75–1000	Fe
Nonbonding (Intermolecular)				
Ion-dipole		Ion charge–dipole charge	40–600	Na^+····OH_2
H bond	δ− δ+ δ− —A—H·······:B—	Polar bond to H–dipole charge (high EN of N, O, F)	10–40	:Ö(H)—H····:Ö(H)—H
Dipole-dipole		Dipole charges	5–25	I—Cl···· I—Cl
Ion–induced dipole		Ion charge–polarizable e^- cloud	3–15	Fe^{2+}····O_2
Dipole–induced dipole		Dipole charge–polarizable e^- cloud	2–10	H—Cl····Cl—Cl
Dispersion (London)		Polarizable e^- clouds	0.05–40	F—F····F—F

cloud repulsions. One-half this distance is the **van der Waals radius,** one-half the closest distance between the nuclei of identical *nonbonded* Cl atoms. *The van der Waals radius of an atom is always larger than its covalent radius,* but van der Waals radii decrease across a period and increase down a group, just as covalent radii do. Figure 12.11 shows these relationships for many of the nonmetals.

There are several types of intermolecular forces: ion-dipole, dipole-dipole, hydrogen bonding, dipole–induced dipole, and dispersion forces. As we discuss these *intermolecular* forces (also called *van der Waals forces*), look at Table 12.2, which compares them with the stronger *intramolecular* (bonding) forces.

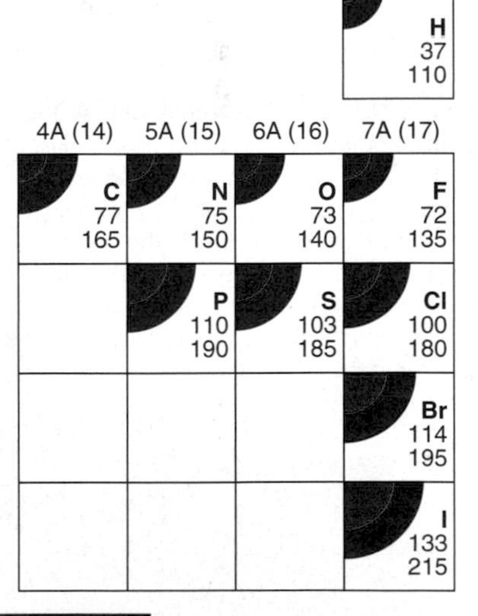

Figure 12.11 Periodic trends in covalent and van der Waals radii (in pm). Like covalent radii (*blue quarter-circles and top numbers*), van der Waals radii (*red quarter-circles and bottom numbers*) increase down a group and decrease across a period. The covalent radius of an element is *always* less than its van der Waals radius.

Ion-Dipole Forces

When an ion and a nearby polar molecule (dipole) attract each other, an **ion-dipole force** results. The most important example takes place when an ionic compound dissolves in water. The ions become separated because the attractions between the ions and the oppositely charged poles of the H_2O molecules overcome the attractions between the ions themselves. Ion-dipole forces in solutions and their associated energy are discussed fully in Chapter 13.

Dipole-Dipole Forces

In Figure 10.12 (p. 387), you saw how an external electric field can orient gaseous polar molecules. When polar molecules lie near one another, as in liquids and solids, their partial charges act as tiny electric fields that orient them and give rise to **dipole-dipole forces:** the positive pole of one molecule attracts the negative pole of another (Figure 12.12).

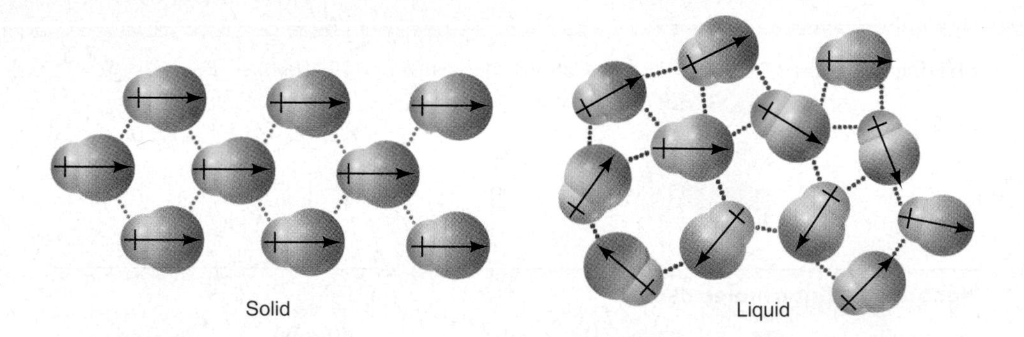

Figure 12.12 Polar molecules and dipole-dipole forces. In a solid or a liquid, the polar molecules are close enough for the partially positive pole of one molecule to attract the partially negative pole of a nearby molecule. The orientation is more orderly in the solid (*left*) than in the liquid (*right*) because, at the lower temperatures required for freezing, the average kinetic energy of the particles is lower. (Interparticle spaces are increased for clarity.)

These are the forces that give polar *cis*-1,2-dichloroethylene a higher boiling point than the nonpolar *trans* compound (see p. 389). In fact, for molecular compounds of approximately the same size and molar mass, the greater the dipole moment, the greater the dipole-dipole forces between the molecules are, and so the more energy it takes to separate them. Consider the boiling points of the compounds in Figure 12.13. Methyl chloride, for instance, has a smaller dipole moment than acetaldehyde, so less energy is needed to overcome the dipole-dipole forces between its molecules and it boils at a lower temperature.

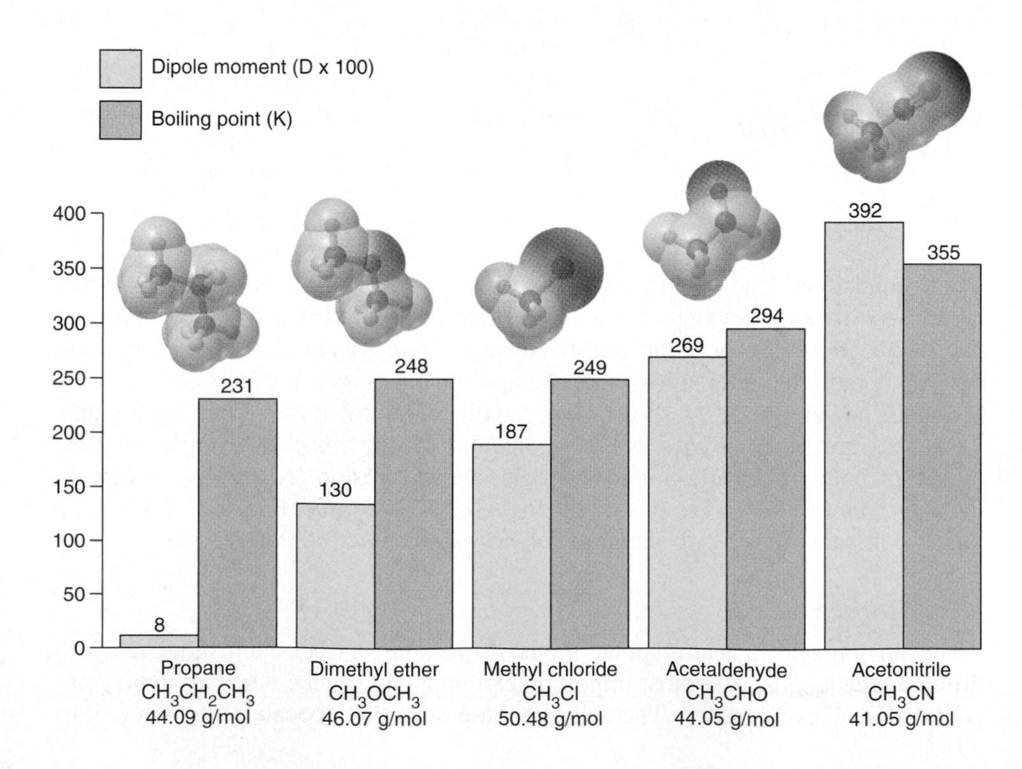

Figure 12.13 Dipole moment and boiling point. For compounds of similar molar mass, the boiling point increases with increasing dipole moment. (Note the increasing color intensities in the electron-density models.) The greater dipole moment creates stronger dipole-dipole forces, which require higher temperatures to overcome.

The Hydrogen Bond

A special type of dipole-dipole force arises between molecules that have *an H atom bonded to a small, highly electronegative atom with lone electron pairs*. The most important atoms that fit this description are N, O, and F. The H—N, H—O, and H—F bonds are very polar, so electron density is withdrawn from H. As a result, the partially positive H of one molecule is attracted to the partially negative lone pair on the N, O, or F of another molecule, and a **hydrogen bond (H bond)** forms. Thus, the atom sequence that allows an H bond (dotted line) to form is —B:····H—A—, where *both* A and B are N, O, or F. Three examples are

—F:····H—O— —O:····H—N— —N:····H—F:

The small sizes of N, O, and F* are essential to H bonding for two reasons:

1. It makes these atoms so electronegative that their covalently bonded H is highly positive.
2. It allows the lone pair on the other N, O, or F to come close to the H.

SAMPLE PROBLEM 12.2 Drawing Hydrogen Bonds Between Molecules of a Substance

Problem Which of the following substances exhibits H bonding? For those that do, draw two molecules of the substance with the H bond(s) between them.

(a) C_2H_6 (b) CH_3OH (c) $CH_3C(=O)—NH_2$

Plan We draw each structure to see if it contains N, O, or F covalently bonded to H. If it does, we draw two molecules of the substance in the —B:····H—A— pattern.

Solution (a) For C_2H_6. No H bonds are formed.

(b) For CH_3OH. The H covalently bonded to the O in one molecule forms an H bond to the lone pair on the O of an adjacent molecule:

H—O:····H—O—C—H (with CH_3 on the first O)

(c) For $CH_3C(=O)—NH_2$. Two of these molecules can form one H bond between an H bonded to N and the O, or they can form two such H bonds:

$CH_3C(=O)$—N(H)—H····:O=C(CH_3)—N(H_2) or two molecules joined by two H bonds (:O:····H—N and N—H····:O)

A third possibility (not shown) could be between an H attached to N in one molecule and the lone pair of N in another molecule.

Check The —B:····H—A— sequence (with A and B either N, O, or F) is present.

Comment Note that *H covalently bonded to C does not form H bonds* because carbon is not electronegative enough to make the C—H bond very polar.

FOLLOW-UP PROBLEM 12.2 Which of these substances exhibits H bonding? Draw the H bond(s) between two molecules of the substance where appropriate.

(a) $CH_3C(=O)—OH$ (b) CH_3CH_2OH (c) $CH_3C(=O)CH_3$

*H-bond–type interactions occur with the larger atoms P, S, and Cl, but those are so much weaker than the interactions with N, O, and F that we will not consider them.

The Significance of Hydrogen Bonding Hydrogen bonding has a profound impact in many systems. Here we'll examine one major effect on physical properties and preview its enormous importance in biological systems, which we address in Chapter 13.

Figure 12.14 shows the effect of H bonding on the boiling points of the binary hydrides of Groups 4A(14) through 7A(17). For reasons we'll discuss shortly, boiling points typically rise as molar mass increases, as you can see in the Group 4A hydrides, CH_4 through SnH_4. In the other groups, however, the first member in each series—NH_3, H_2O, and HF—deviates enormously from this expected increase. The H bonds between the molecules in these substances require additional energy to break before the molecules can separate and enter the gas phase. For example, on the basis of molar mass alone, we would expect water to boil about 200°C lower than it actually does (dashed line). (In Section 12.5 we'll discuss the effects that H bonds in water have in nature.)

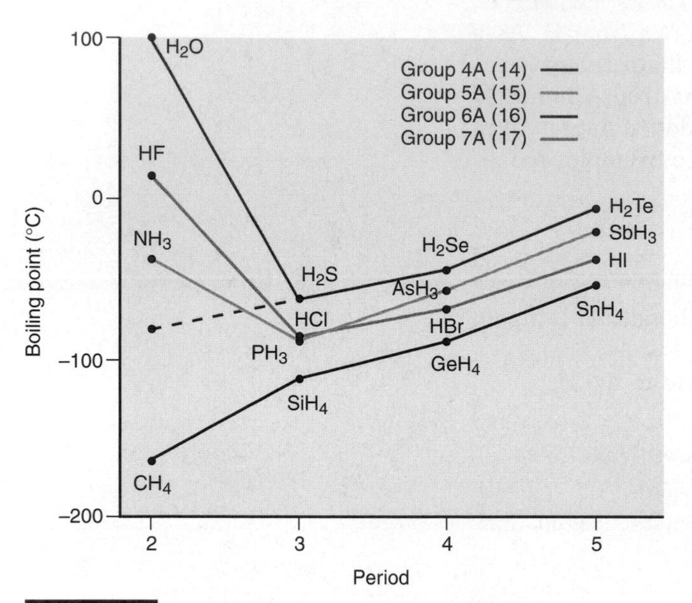

Figure 12.14 Hydrogen bonding and boiling point. Boiling points of the binary hydrides of Groups 4A(14) to 7A(17) are plotted against period number. H bonds in NH_3, H_2O, and HF give them much higher boiling points than if the trend were based on molar mass, as it is for Group 4A. The red dashed line extrapolates to the boiling point H_2O would have if it formed no H bonds.

The significance of hydrogen bonding in biological systems cannot be emphasized too strongly. It is a principal feature in the structures of the three major biopolymers—proteins, nucleic acids, and carbohydrates. In addition, it is responsible for the action of many *enzymes*, the specialized proteins that speed all metabolic reactions. In Chapter 13, after we discuss intermolecular forces in solution, you'll be amazed at the essential roles H bonds and the other intermolecular forces play in the structure and function of the molecules of life.

Polarizability and Charge-Induced Dipole Forces

Even though electrons are localized in bonding or lone pairs, they are in constant motion, so that we often picture them as "clouds" of negative charge. A nearby electric field can distort a cloud, pulling electron density toward a positive charge or pushing it away from a negative charge. In effect, the field *induces* a distortion in the electron cloud. For a nonpolar molecule, this distortion creates a temporary, induced dipole moment; for a polar molecule, it enhances the dipole moment already present. The source of the electric field can be the electrodes of a battery, the charge of an ion, or the partial charges of a polar molecule.

The ease with which the electron cloud of a particle can be distorted is called its **polarizability.** Smaller atoms (or ions) are less polarizable than larger ones because their electrons are closer to the nucleus and therefore are held more tightly. Thus, we observe several trends:

- *Polarizability increases down a group* because size increases, so the larger electron clouds are farther from the nucleus and, thus, more easily distorted.
- *Polarizability decreases from left to right across a period* because the increasing Z_{eff} shrinks atomic size and holds the electrons more tightly.
- Cations are *less* polarizable than their parent atoms because they are smaller; anions are *more* polarizable because they are larger.

Ion–induced dipole and dipole–induced dipole forces are the two types of charge-induced dipole forces; they are most important in solution, so we'll focus on them in Chapter 13. Nevertheless, polarizability affects all intermolecular forces.

Dispersion (London) Forces

Polarizability plays the central role in the most universal intermolecular force. Up to this point, we've discussed forces that depend on an existing charge, of either an ion or a polar molecule. But what forces cause nonpolar substances like octane, chlorine, and argon to condense and solidify? Some force must be acting between the particles, or these substances would be gases under any conditions. *The intermolecular force primarily responsible for the condensed states of nonpolar substances is the* ***dispersion force*** (or ***London force,*** named for Fritz London, the physicist who explained the quantum-mechanical basis of the attraction).

Dispersion forces are caused by *momentary oscillations of electron charge in atoms and, therefore, are present between all particles (atoms, ions, and molecules).* Picture one atom in a sample of argon gas. Averaged over time, the 18 electrons are distributed uniformly around the nucleus, so the atom is nonpolar. But at any instant, there may be more electrons on one side of the nucleus than on the other, so the atom has an *instantaneous dipole*. When far apart, a pair of argon atoms do not influence each other. But when close together, *the instantaneous dipole in one atom induces a dipole in its neighbor.* The result is a synchronized motion of the electrons in the two atoms, which causes an attraction between them. This process occurs with other nearby atoms and, thus, throughout the sample. At low enough temperatures, the attractions among the dipoles keep all the atoms together. Thus, dispersion forces are *instantaneous dipole–induced dipole forces*. Figure 12.15 depicts the dispersion forces among nonpolar particles.

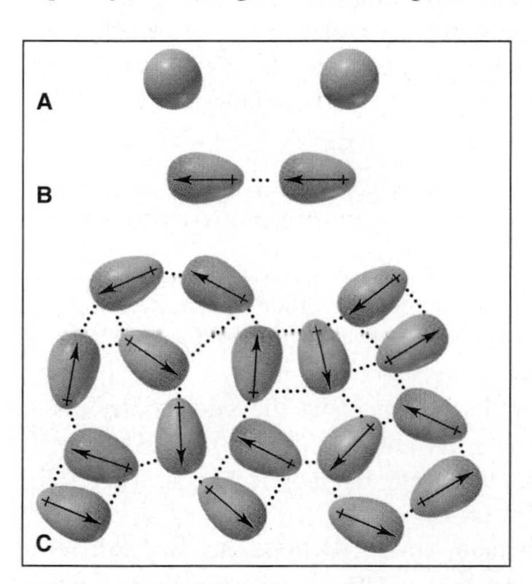

Figure 12.15 Dispersion forces among nonpolar particles. The dispersion force is responsible for the condensed states of noble gases and nonpolar molecules. **A,** Separated Ar atoms are nonpolar. **B,** An instantaneous dipole in one atom induces a dipole in its neighbor. These partial charges attract the atoms together. **C,** This process takes place among atoms throughout the sample.

As we noted, the dispersion force is the *only* force existing between nonpolar particles, but because they exist between *all* particles, *dispersion forces contribute to the overall energy of attraction of all substances.* In fact, except in cases involving small, polar molecules with large dipole moments or those forming strong H bonds, the *dispersion force is the dominant intermolecular force.* Calculations show, for example, that 85% of the total energy of attraction between HCl molecules is due to dispersion forces and only 15% is from dipole-dipole forces. Even for water, estimates indicate that 75% of the total energy of attraction comes from the strong H bonds and nearly 25% from dispersion forces!

Dispersion forces are very weak for small particles, like H_2 and He, but much stronger for larger particles, like I_2 and Xe. The relative strength of the dispersion force depends on the polarizability of the particle. *Polarizability depends on the number of electrons, which correlates closely with molar mass* because heavier particles are either larger atoms or composed of more atoms and thus have more electrons. For example, molar mass increases down the halogens and the noble gases, so dispersion forces increase and so do boiling points (Figure 12.16).

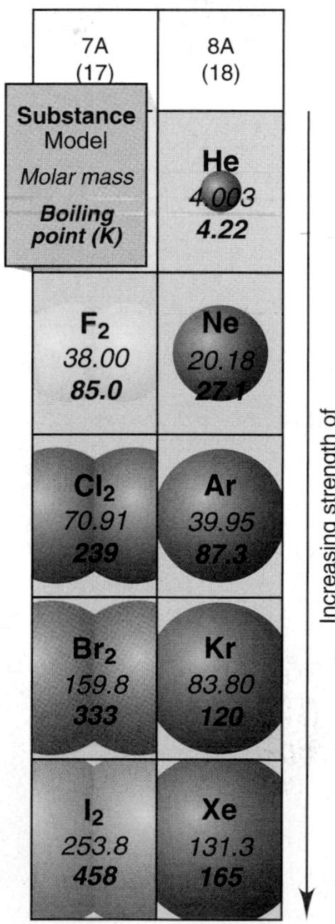

Figure 12.16 Molar mass and boiling point. The strength of dispersion forces increases with number of electrons, which usually correlates with molar mass. As a result, boiling points increase down the halogens and the noble gases.

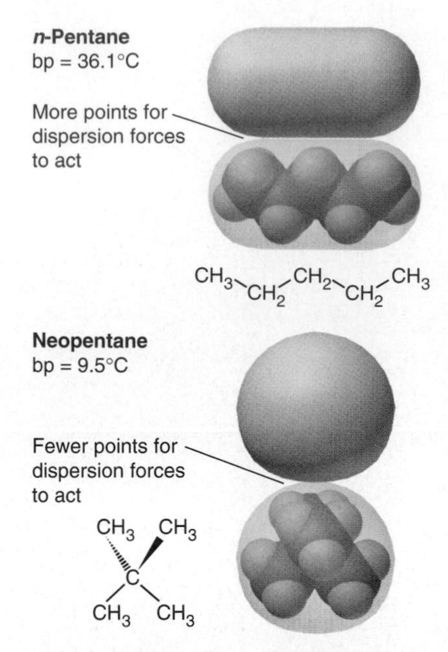

Figure 12.17 Molecular shape and boiling point. Spherical neopentane molecules make less contact with each other than do cylindrical *n*-pentane molecules, so neopentane has a lower boiling point.

For nonpolar substances with the same molar mass, the strength of the dispersion forces is influenced by molecular shape. Shapes that allow more points of contact have more area over which electron clouds can be distorted, so stronger attractions result. Consider the two five-carbon alkanes pentane (also called *n-pentane*) and 2,2-dimethylpropane (also called *neopentane*). These isomers have the same molecular formula (C_5H_{12}) but different shapes. *n*-Pentane is shaped like a cylinder, whereas neopentane has a more compact, spherical shape, as shown in Figure 12.17. Thus, two *n*-pentane molecules make more contact than do two neopentane molecules. Greater contact allows the dispersion forces to act at more points, so *n*-pentane has a higher boiling point. Figure 12.18 summarizes how to analyze the intermolecular forces in a sample.

SAMPLE PROBLEM 12.3 Predicting the Types of Intermolecular Force

Problem For each pair of substances, identify the key intermolecular force(s) in each substance, and select the substance with the higher boiling point:
(a) $MgCl_2$ or PCl_3
(b) CH_3NH_2 or CH_3F
(c) CH_3OH or CH_3CH_2OH
(d) Hexane ($CH_3CH_2CH_2CH_2CH_2CH_3$) or 2,2-dimethylbutane ($CH_3C(CH_3)_2CH_2CH_3$)

Plan We examine the formulas and picture (or draw) the structures to identify key differences between members of each pair: Are ions present? Are molecules polar or nonpolar? Is N, O, or F bonded to H? Do the molecules have different masses or shapes? To rank the boiling points, we consult Table 12.2 and Figure 12.18 and remember that

- Bonding forces are stronger than intermolecular forces.
- Hydrogen bonding is a strong type of dipole-dipole force.
- Dispersion forces are always present, but they are decisive when the difference is molar mass or molecular shape.

Solution (a) $MgCl_2$ consists of Mg^{2+} and Cl^- ions held together by ionic bonding forces; PCl_3 consists of polar molecules, so intermolecular dipole-dipole forces are present. The forces in $MgCl_2$ are stronger, so it has a higher boiling point.
(b) CH_3NH_2 and CH_3F both consist of polar molecules of about the same molar mass. CH_3NH_2 has N—H bonds, so it can form H bonds (see margin). CH_3F contains a C—F bond but no H—F bond, so dipole-dipole forces occur but not H bonds. Therefore, CH_3NH_2 has the higher boiling point.
(c) CH_3OH and CH_3CH_2OH molecules both contain an O—H bond, so they can form H bonds (see margin). CH_3CH_2OH has an additional $—CH_2—$ group and thus a larger molar mass, which correlates with stronger dispersion forces; therefore, it has a higher boiling point.
(d) Hexane and 2,2-dimethylbutane are nonpolar molecules of the same molar mass but different molecular shapes (see margin). Cylindrical hexane molecules can make more extensive intermolecular contact than the more spherical 2,2-dimethylbutane molecules can, so hexane should have greater dispersion forces and a higher boiling point.

(b) $CH_3—\ddot{N}H—H\cdots:NH_2—CH_3$

(c) $CH_3—\ddot{O}—H\cdots:\ddot{O}H—CH_3$

$CH_3CH_2—\ddot{O}—H\cdots:\ddot{O}H—CH_2CH_3$

(d) Hexane 2,2-Dimethylbutane

Check The actual boiling points show that our predictions are correct:
(a) $MgCl_2$ (1412°C) and PCl_3 (76°C)
(b) CH_3NH_2 (−6.3°C) and CH_3F (−78.4°C)
(c) CH_3OH (64.7°C) and CH_3CH_2OH (78.5°C)
(d) Hexane (69°C) and 2,2-dimethylbutane (49.7°C)

Comment Dispersion forces are *always* present, but in parts (a) and (b), they are much less significant than the other forces involved.

FOLLOW-UP PROBLEM 12.3
In each pair, identify all the intermolecular forces present for each substance, and select the substance with the higher boiling point:
(a) CH_3Br or CH_3F (b) $CH_3CH_2CH_2OH$ or $CH_3CH_2OCH_3$ (c) C_2H_6 or C_3H_8

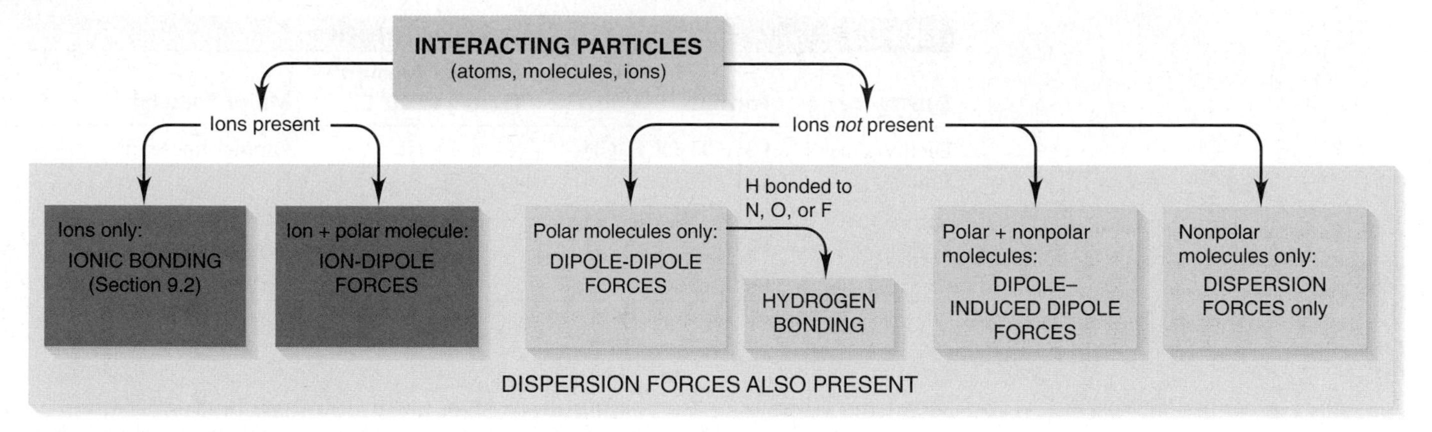

Figure 12.18 **Summary diagram for analyzing the intermolecular forces in a sample.**

SECTION SUMMARY

The van der Waals radius determines the shortest distance over which intermolecular forces operate; it is always larger than the covalent radius. Intermolecular forces are much weaker than bonding (intramolecular) forces. Ion-dipole forces occur between ions and polar molecules. Dipole-dipole forces occur between oppositely charged poles on polar molecules. Hydrogen bonding, a special type of dipole-dipole force, occurs when H bonded to N, O, or F is attracted to the lone pair of N, O, or F in another molecule. Electron clouds can be distorted (polarized) in an electric field. Ion– and dipole–induced dipole forces arise between a charge and the dipole it induces in another molecule. Dispersion (London) forces are instantaneous dipole–induced dipole forces that occur among all particles and increase with number of electrons (molar mass). Molecular shape determines the extent of contact between molecules and can be a factor in the strength of dispersion forces.

12.4 PROPERTIES OF THE LIQUID STATE

Of the three states of matter, the liquid is the least understood at the molecular level. Because of the *randomness* of the particles in a gas, any region of the sample is virtually identical to any other. As you'll see in Section 12.6, different regions of a crystalline solid are identical because of the *orderliness* of the particles. Liquids, however, have a combination of these attributes that changes continually: a region that is orderly one moment becomes random the next, and vice versa. Despite this complexity at the molecular level, the macroscopic properties of liquids are well understood. In this section, we discuss three liquid properties —surface tension, capillarity, and viscosity.

Surface Tension

In a liquid sample, intermolecular forces exert different effects on a molecule at the surface than on one in the interior (Figure 12.19). Interior molecules are attracted by others on all sides, whereas molecules at the surface have others only below and to the sides. As a result, molecules at the surface experience a *net attraction downward* and move toward the interior to increase attractions and become more stable. Therefore, *a liquid surface tends to have the smallest possible area,* that of a sphere, and behaves like a "taut skin" covering the interior.

Figure 12.19 The molecular basis of surface tension. Molecules in the interior of a liquid experience intermolecular attractions in all directions. Molecules at the surface experience a net attraction downward (*red arrow*) and move toward the interior. Thus, a liquid tends to minimize the number of molecules at the surface, which results in surface tension.

To increase the surface area, molecules must move to the surface by breaking some attractions in the interior, which requires energy. The **surface tension** is the energy required to increase the surface area by a unit amount; units of J/m^2 are

Table 12.3 Surface Tension and Forces Between Particles

Substance	Formula	Surface Tension (J/m^2) at 20°C	Major Force(s)
Diethyl ether	$CH_3CH_2OCH_2CH_3$	1.7×10^{-2}	Dipole-dipole; dispersion
Ethanol	CH_3CH_2OH	2.3×10^{-2}	H bonding
Butanol	$CH_3CH_2CH_2CH_2OH$	2.5×10^{-2}	H bonding; dispersion
Water	H_2O	7.3×10^{-2}	H bonding
Mercury	Hg	48×10^{-2}	Metallic bonding

shown in Table 12.3. Comparing these values with those in Table 12.2 (p. 437) shows that, in general, *the stronger the forces are between the particles in a liquid, the greater the surface tension.* Water has a high surface tension because its molecules form multiple H bonds. *Surfactants* (*surf*ace-*act*ive age*nts*), such as soaps, petroleum recovery agents, and biological fat emulsifiers, decrease the surface tension of water by congregating at the surface and disrupting the H bonds.

Capillarity

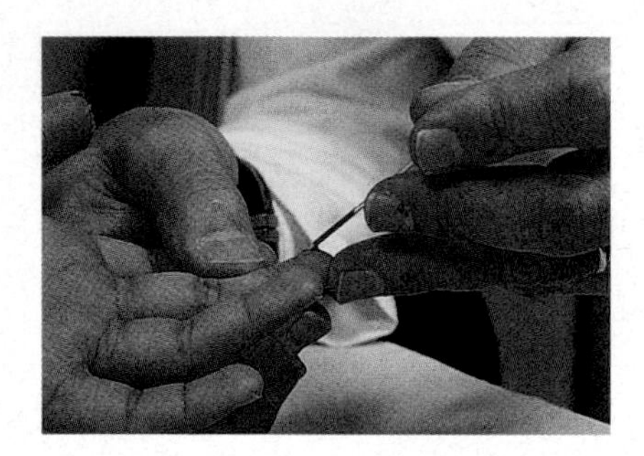

Blood being drawn into a capillary tube for a medical test.

The rising of a liquid through a narrow space against the pull of gravity is called *capillary action,* or **capillarity.** In blood-screening tests, a narrow *capillary tube* is held against the skin opening made by pricking a finger (see photo). Capillarity results from a competition between the intermolecular forces within the liquid (cohesive forces) and those between the liquid and the tube walls (adhesive forces).

Picture what occurs at the molecular level when you place a glass capillary tube in water. Glass is mostly silicon dioxide (SiO_2), so the water molecules form H bonds to the oxygen atoms of the tube's inner wall. Because the adhesive forces (H bonding) between the water and the wall are stronger than the cohesive forces (H bonding) within the water, a thin film of water creeps up the wall. At the same time, the cohesive forces that give rise to surface tension pull the liquid surface taut. These adhesive and cohesive forces combine to raise the water level and produce the familiar concave meniscus (Figure 12.20A). The liquid rises until gravity pulling down is balanced by the adhesive forces pulling up.

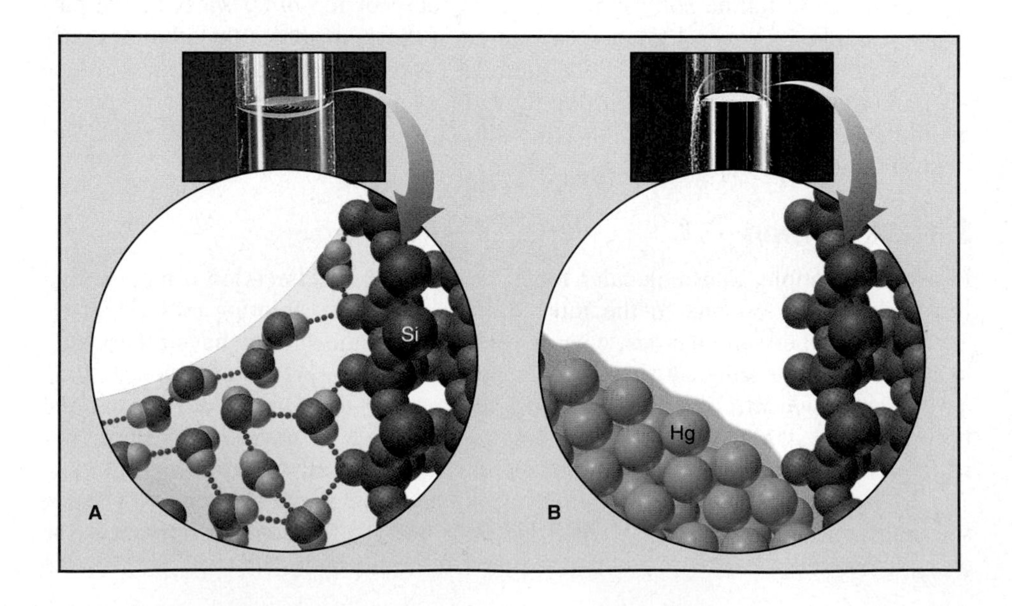

Figure 12.20 Shape of water or mercury meniscus in glass. A, Water displays a concave meniscus in a glass tube because the adhesive (H-bond) forces between the H_2O molecules and the O—Si—O groups of the glass are *stronger* than the cohesive (H-bond) forces within the water. **B,** Mercury displays a convex meniscus in a glass tube because the cohesive (metallic bonding) forces within the mercury are *stronger* than the adhesive (dispersion) forces between the mercury and the glass.

On the other hand, if you place a glass capillary tube in a dish of mercury, the mercury level in the tube drops below that in the dish. Mercury has a higher surface tension than water (see Table 12.3), which means it has stronger cohesive forces (metallic bonding). The cohesive forces among the mercury atoms are much stronger than the adhesive forces (mostly dispersion) between mercury and glass, so the liquid tends to pull away from the walls. At the same time, the surface atoms are being pulled toward the interior of the mercury by its high surface tension, so the level drops. These combined forces produce a convex meniscus (Figure 12.20B, seen in a laboratory barometer or manometer).

Viscosity

When a liquid flows, the molecules slide around and past each other. A liquid's **viscosity,** its resistance to flow, results from intermolecular attractions that impede this movement. Both gases and liquids flow, but liquid viscosities are *much* higher because the intermolecular forces operate over much shorter distances.

Viscosity decreases with heating, as Table 12.4 shows for water. When molecules move faster at higher temperatures, they can overcome intermolecular forces more easily, so the resistance to flow decreases. Next time you put cooking oil in a pan and heat it, watch the oil flow more easily and spread out in a thin layer.

Table 12.4 Viscosity of Water at Several Temperatures

Temperature (°C)	Viscosity ($N \cdot s/m^2$)*
20	1.00×10^{-3}
40	0.65×10^{-3}
60	0.47×10^{-3}
80	0.35×10^{-3}

*The units of viscosity are newton-seconds per square meter.

Molecular shape plays a key role in a liquid's viscosity. Small, spherical molecules make little contact and, like buckshot in a glass, pour easily. Long molecules make more contact and, like spaghetti in a glass, become entangled and pour slowly. Thus, given the same types of forces, liquids containing longer molecules have higher viscosities. A striking example of a change in viscosity occurs during the making of syrup. Even at room temperature, a concentrated aqueous sugar solution has a higher viscosity than water because of H bonding among the many hydroxyl (—OH) groups on the ring-shaped sugar molecules. When the solution is slowly heated to boiling, the sugar molecules react with each other and link covalently, gradually forming long chains. Hydrogen bonds and dispersion forces occur at many points along the chains, and the resulting syrup is a viscous liquid that pours slowly and clings to a spoon. When a viscous syrup is cooled, it may become stiff enough to be picked up and stretched—into taffy candy. The Gallery on the next page shows other familiar examples of these three liquid properties.

SECTION SUMMARY

Surface tension is a measure of the energy required to increase a liquid's surface area. Greater intermolecular forces within a liquid create higher surface tension. Capillary action, the rising of a liquid through a narrow space, occurs when the forces between a liquid and a solid surface (adhesive) are greater than those within the liquid itself (cohesive). Viscosity, the resistance to flow, depends on molecular shape and decreases with temperature. Stronger intermolecular forces create higher viscosity.

12.5 THE UNIQUENESS OF WATER

Water is absolutely amazing stuff, but it is so familiar that we take it for granted. In fact, water has some of the most unusual properties of any substance. These properties, which are vital to our very existence, arise inevitably from the nature of the H and O atoms that make up the molecule. Each atom attains a filled outer level by sharing electrons in single covalent bonds. With two bonding pairs and two lone pairs around the O atom and a large electronegativity difference in each O—H bond, the H_2O molecule is bent and highly polar. This arrangement is crucial because it allows each water molecule to engage in four H bonds with its neighbors (Figure 12.21). From these fundamental atomic and molecular properties emerges unique and remarkable macroscopic behavior.

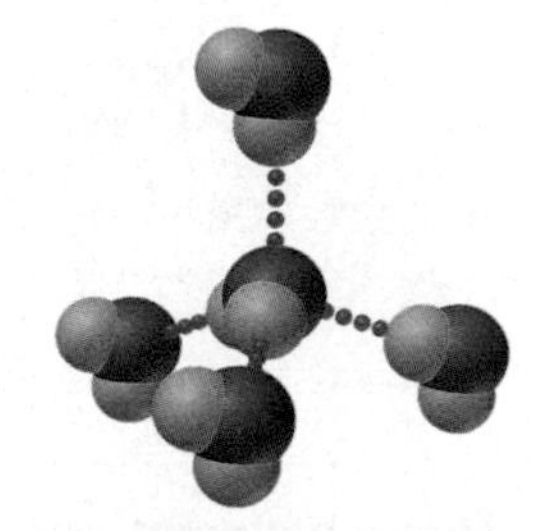

Figure 12.21 The H-bonding ability of the water molecule. Because it has two O—H bonds and two lone pairs, one H_2O molecule can engage in as many as four H bonds to surrounding H_2O molecules, which are arranged tetrahedrally.

Properties of a Liquid

Of the three states of matter, only liquids combine the ability to flow with the strength that comes from intermolecular contact, and this combination appears in numerous applications.

Beaded droplets on waxy surfaces
The adhesive (dipole–induced dipole) forces between water and a nonpolar surface are much weaker than the cohesive (H-bond) forces within water. As a result, water pulls away from a nonpolar surface and forms beaded droplets. You have seen this effect when water beads on a flower petal or a freshly waxed car after a rainfall.

Minimizing a surface
In the low-gravity environment of an orbiting space shuttle, the tendency of a liquid to minimize its surface creates perfectly spherical droplets, unlike the flattened drops we see on Earth. For the same reason, bubbles in a soft drink are spherical because the liquid uses the minimum number of molecules needed to surround the gas. A water strider flits across a pond on widespread legs that do not exert enough pressure to exceed the surface tension.

Maintaining motor oil viscosity
To protect engine parts during long drives or in hot weather, when an oil would ordinarily become too thin, motor oils contain additives, called *polymeric viscosity index improvers,* that act as thickeners. As the oil heats up, the additive molecules change shape from compact spheres to spaghetti-like strands and become tangled with the hydrocarbon oil molecules. As a result of the greater dispersion forces, there is an increase in viscosity that compensates for the decrease due to heating.

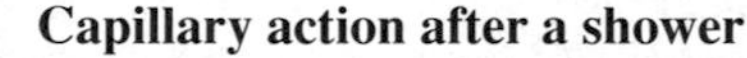

Capillary action after a shower
Paper and cotton consist of fibers of cellulose, long carbon-containing molecules with many attached hydroxyl (—OH) groups. A towel dries you in two ways:
First, capillary action draws the water molecules away from your body between the closely spaced cellulose molecules.
Second, the water molecules themselves form adhesive H bonds to the —OH groups of cellulose.

How a ballpoint pen works
The essential parts of a ballpoint pen are the moving ball and the viscous ink. The material of the ball is chosen for the strong adhesive forces between it and the ink. Cohesive forces within the ink are replaced by those adhesive forces when the ink "wets" the ball. As the ball rolls along the paper, the adhesive forces between ball and ink are overcome by those between ink and paper. The rest of the ink stays in the pen because of its high viscosity.

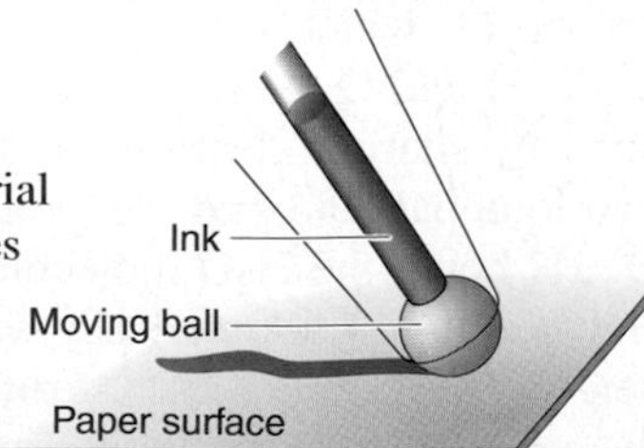

Solvent Properties of Water

The *great solvent power* of water is the result of its polarity and exceptional H-bonding ability. It dissolves ionic compounds through ion-dipole forces that separate the ions from the solid and keep them in solution (see Figure 4.2, p. 136). Water dissolves many polar nonionic substances, such as ethanol (CH_3CH_2OH) and glucose ($C_6H_{12}O_6$), by forming H bonds to them. It even dissolves nonpolar gases, such as those in the atmosphere, to a limited extent, through dipole–induced dipole and dispersion forces. Chapter 13 highlights these solvent properties.

Because it can dissolve so many substances, water is the environmental and biological solvent, forming the complex solutions we know as oceans, rivers, lakes, and cellular fluids. Aquatic animals could not survive without dissolved O_2, nor could aquatic plants survive without dissolved CO_2. The coral reefs that girdle the sea bottom in tropical latitudes are composed of carbonates made from dissolved CO_2 and HCO_3^- by tiny marine animals. Life is thought to have begun in a "primordial soup," an aqueous mixture of simple biomolecules from which emerged the larger molecules whose self-sustaining reactions are characteristic of living things. From a chemical point of view, all organisms, from bacteria to humans, can be thought of as highly organized systems of membranes enclosing and compartmentalizing complex aqueous solutions.

Thermal Properties of Water

Water has an exceptionally *high specific heat capacity,* higher than almost any other liquid. Recall from Section 6.3 that heat capacity is a measure of the heat absorbed by a substance for a given temperature rise. When a substance is heated, some of the energy increases the average molecular speed, some increases molecular vibration and rotation, and some is used to overcome intermolecular forces. Because water has so many strong H bonds, it has a high specific heat capacity. With oceans and seas covering 70% of Earth's surface, daytime heat from the Sun causes relatively small changes in temperature, allowing our planet to support life. On the waterless, airless Moon, temperatures can range from 100°C (212°F) to −150°C (−238°F) in the same lunar day. Even in Earth's deserts, where a dense atmosphere tempers these extremes, day-night temperature differences of 40°C (72°F) are common.

Numerous strong H bonds also give water an exceptionally *high heat of vaporization.* A quick calculation shows how essential this property is to our existence. When 1000 g of water absorbs about 4 kJ of heat, its temperature rises 1°C. With a ΔH^0_{vap} of 2.3 kJ/g, however, less than 2 g of water must evaporate to keep the temperature constant for the remaining 998 g. The average human adult has 40 kg of body water and generates about 10,000 kJ of heat each day from metabolism. If this heat were used only to increase the average E_k of body water, the temperature rise would mean immediate death. Instead, the heat is converted to E_p as it breaks H bonds and evaporates sweat, resulting in a stable body temperature and minimal loss of body fluid. On a planetary scale, the Sun's energy supplies the heat of vaporization for ocean water. Water vapor, formed in warm latitudes, moves through the atmosphere, and its potential energy is released as heat to warm cooler regions when the vapor condenses to rain. The enormous energy involved in this cycling of water powers storms all over the planet.

Surface Properties of Water

The H bonds that give water its remarkable thermal properties are also responsible for its *high surface tension* and *high capillarity.* Except for some metals and molten salts, water has the highest surface tension of any liquid. This property is vital for surface aquatic life because it keeps plant debris resting on a pond surface, which provides shelter and nutrients for many fish, microorganisms, and

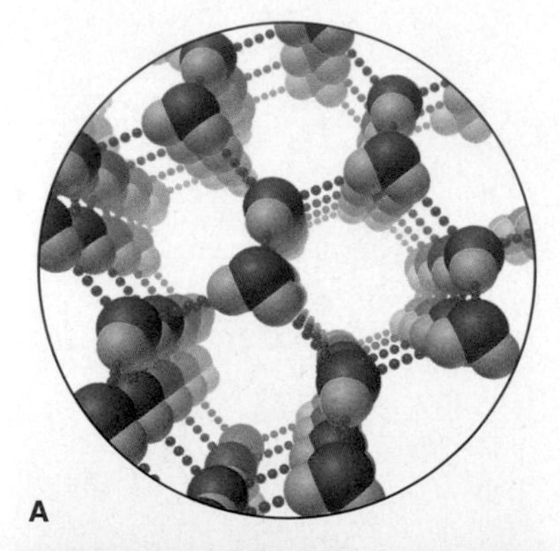

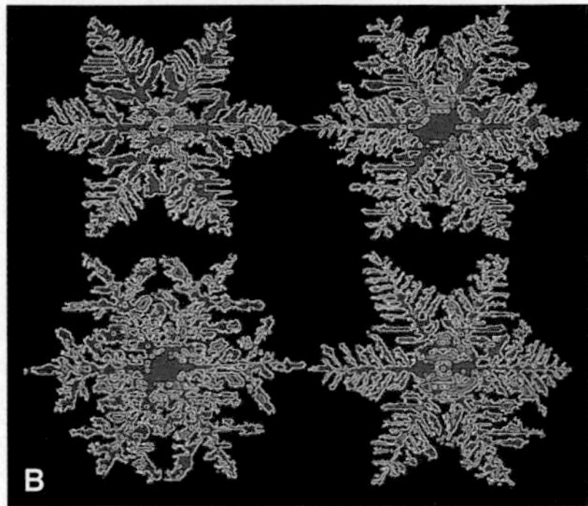

Figure 12.22 The hexagonal structure of ice. A, The geometric arrangement of the H bonds in H_2O leads to the open, hexagonally shaped crystal structure of ice. Thus, when liquid water freezes, the volume *increases.* **B,** The delicate six-pointed beauty of snowflakes reflects the hexagonal crystal structure of ice.

insects. Water's high capillarity, a result of its high surface tension, is crucial to land plants. During dry periods, plant roots absorb deep groundwater, which rises by capillary action through the tiny spaces between soil particles.

The Density of Solid and Liquid Water

As you saw in Figure 12.21, through the H bonds of water, each O atom becomes connected to as many as four other O atoms via four H atoms. Continuing this tetrahedral pattern through many molecules in a fixed array leads to the hexagonal, *open structure* of ice shown in Figure 12.22A. The symmetrical beauty of snowflakes, as shown in Figure 12.22B, reflects this hexagonal organization.

This organization explains the negative slope of the solid-liquid line in the phase diagram for water (see Figure 12.9B, p. 435): As pressure is applied, some H bonds break; as a result, some water molecules enter the spaces. The crystal structure breaks down, and the sample liquefies.

The large spaces within ice give *the solid state a lower density than the liquid state*. When the surface of a lake freezes in winter, the ice floats on the liquid water below. If the solid were denser than the liquid, as is true for nearly every other substance, the surface of a lake would freeze and sink repeatedly until the entire lake was solid. Aquatic life would not survive from year to year.

The density of water changes in a complex way. When ice melts at 0°C, the tetrahedral arrangement around each O atom breaks down, and the loosened molecules pack much more closely, filling spaces in the collapsing solid structure. As a result, water is most dense (1.000 g/mL) at around 4°C (3.98°C). With more heating, the density decreases through normal thermal expansion.

This change in density is vital for freshwater life. As lake water becomes colder in the fall and early winter, it becomes more dense *before* it freezes. Similarly, in spring, less dense ice thaws to form more dense water *before* the water expands. During both of these seasonal density changes, the top layer of water reaches the high-density point first and sinks. The next layer of water rises because it is slightly less dense, reaches 4°C, and likewise sinks. This sinking and rising distribute nutrients and dissolved oxygen.

The expansion and contraction of water dramatically affect the land as well. When rain fills the crevices in rocks and then freezes, great outward stress is applied, to be relieved only when the ice melts. In time, this repeated freeze-thaw stress cracks the rocks apart (Figure 12.23). Over eons of geologic change, this effect has helped to produce the sand and soil of the planet.

The properties of water and their far-reaching consequences illustrate the theme that runs throughout this book: the world we know is the stepwise, macroscopic outgrowth of the atomic world we seek to know. Figure 12.24 offers a fanciful summary of water's unique properties in the form of a tree whose "roots" (atomic properties) give foundation to the base of the "trunk" (molecular properties), which strengthens the next portion (polarity), from which grows another portion (H bonding). These two portions of the trunk "branch" out to the other properties of water and their global effects.

Figure 12.23 The expansion and contraction of water. As water freezes and thaws repeatedly, the stress due to expanding and contracting can break rocks. Over eons of geologic time, this process creates sand and soil.

SECTION SUMMARY

The atomic properties of hydrogen and oxygen atoms result in the water molecule's bent shape, polarity, and H-bonding ability. These properties of water enable it to dissolve many ionic and polar compounds. Water's H bonding results in a high specific heat capacity and a high heat of vaporization, which combine to give Earth and its organisms a narrow temperature range. H bonds also confer high surface tension and capillarity, which are essential to plant and animal life. Water expands on freezing because of its H bonds, which lead to an open crystal structure in ice. Seasonal density changes foster nutrient mixing in lakes. Seasonal freeze-thaw stress on rocks results eventually in soil formation.

Figure 12.24 The macroscopic properties of water and their atomic and molecular "roots."

12.6 THE SOLID STATE: STRUCTURE, PROPERTIES, AND BONDING

Stroll through the mineral collection of any school or museum, and you'll be struck by the extraordinary variety and beauty of these solids. In this section, we first discuss the general structural features of crystalline solids and then examine a laboratory method for studying them. We survey the properties of the major types of solids and find the whole range of intermolecular forces at work. We then present a model for bonding in solids that explains many of their properties.

Structural Features of Solids

We can divide solids into two broad categories based on the orderliness of their shapes, which in turn is based on the orderliness of their particles. **Crystalline solids** generally have a well-defined shape, as shown in Figure 12.25, because their particles—atoms, molecules, or ions—occur in an orderly arrangement. **Amorphous solids** have poorly defined shapes because their particles lack long-range ordering throughout the sample. In this discussion, we focus, for the most part, on crystalline solids.

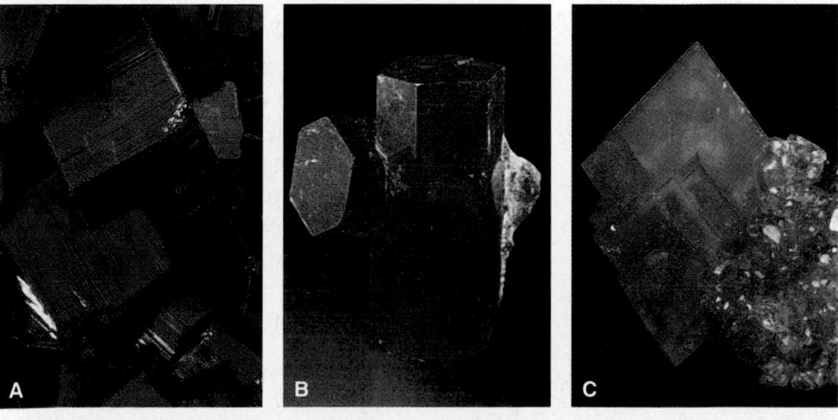

Figure 12.25 The striking beauty of crystalline solids. A, Pyrite. **B,** Beryl. **C,** Barite (*left*) on calcite (*right*).

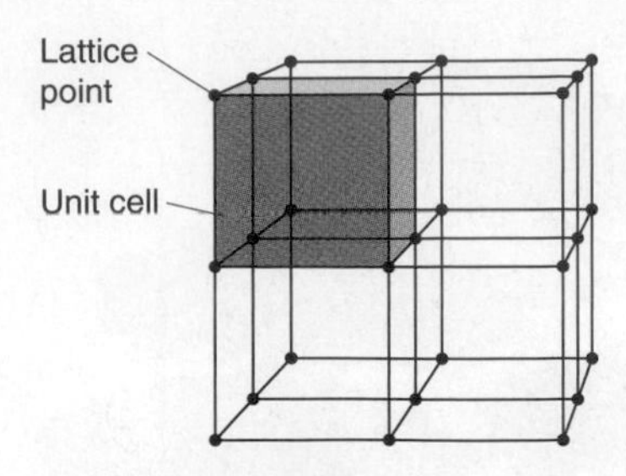

A Portion of 3-D lattice

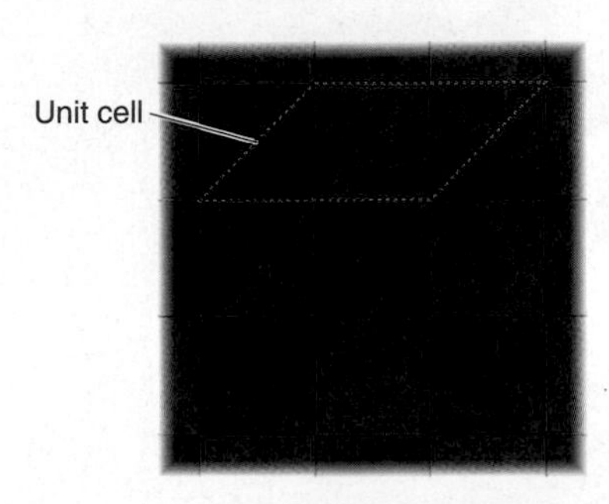

B 2-D analogy for unit cell and lattice

Figure 12.26 The crystal lattice and the unit cell. A, The lattice is an array of points that defines the positions of the particles in a crystal structure. It is shown here as points connected by lines. A unit cell *(colored)* is the simplest array of points that, when repeated in all directions, produces the lattice. A simple cubic unit cell, one of 14 types in nature, is shown. **B,** A checkerboard is a two-dimensional analogy for a lattice.

The Crystal Lattice and the Unit Cell If you could see the particles within a crystal, you would find them packed tightly together in an orderly, three-dimensional array. Consider the simplest case, in which all the particles are *identical* spheres, and imagine a point at the same location within each particle in this array, say, at the center. The points form a regular pattern throughout the crystal that is called the crystal **lattice.** Thus, *the lattice consists of all points with identical surroundings.* Put another way, if the rest of each particle were removed, leaving only the lattice point, and you were transported from one point to another, you would not be able to tell that you had moved. Keep in mind that there is no pre-existing array of lattice points; rather, *the arrangement of the particle points defines the lattice.*

Figure 12.26A shows a portion of a lattice and the **unit cell**, the *smallest* portion of the crystal that, if repeated in all three directions, gives the crystal. A two-dimensional analogy for a unit cell and a crystal lattice can be seen in a checkerboard (as shown in Figure 12.26B), a section of tiled floor, a strip of wallpaper, or any other pattern that is constructed from a repeating unit. The **coordination number** of a particle in a crystal is the number of nearest neighbors surrounding it.

There are 7 crystal systems and 14 types of unit cells that occur in nature, but we will be concerned primarily with the *cubic system,* which gives rise to the cubic lattice. The solid states of a majority of metallic elements, some covalent compounds, and many ionic compounds occur as cubic lattices. (We also describe the hexagonal unit cell a bit later.) There are three types of cubic unit cells within the cubic system:

1. In the **simple cubic unit cell,** shown in Figure 12.27A, the centers of eight identical particles define the corners of a cube. Attractions pull the particles together, so they touch along the cube's edges; but they do not touch diagonally along the cube's faces or through its center. The coordination number of each particle is 6: four in its own layer, one in the layer above, and one in the layer below.
2. In the **body-centered cubic unit cell,** shown in Figure 12.27B, identical particles lie at each corner *and* in the center of the cube. Those at the corners do not touch each other, but they all touch the one in the center. Each particle is surrounded by eight nearest neighbors, four above and four below, so the coordination number is 8.
3. In the **face-centered cubic unit cell,** shown in Figure 12.27C, identical particles lie at each corner *and* in the center of each face but not in the center of the cube. Those at the corners touch those in the faces but not each other. The coordination number is 12.

One unit cell lies adjacent to another throughout the crystal, with no gaps, so a particle at a corner or face is *shared* by adjacent unit cells. As you can see from Figure 12.27 (third row from the top), in the three cubic unit cells, the particle at each corner is part of eight adjacent cells, so one-eighth of each of these particles belongs to each unit cell (bottom row). There are eight corners in a cube, so each simple cubic unit cell contains $8 \times \frac{1}{8}$ particle $= 1$ particle. The body-centered cubic unit cell contains one particle from the eight corners and one in the center, for a total of two particles; and the face-centered cubic unit cell contains four particles, one from the eight corners and three from the half-particles in each of the six faces.

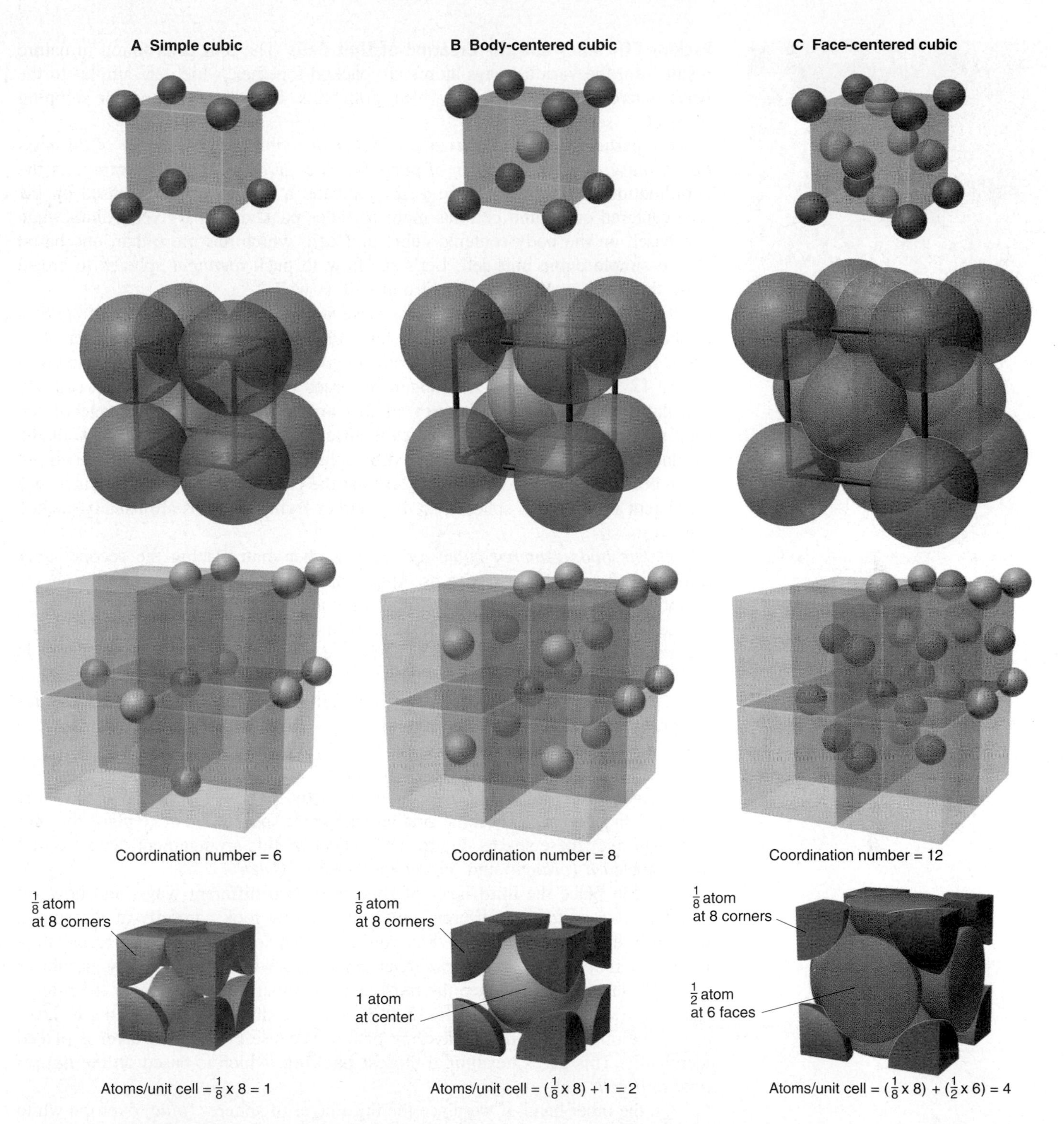

Figure 12.27 The three cubic unit cells. **A,** Simple cubic unit cell. **B,** Body-centered cubic unit cell. **C,** Face-centered cubic unit cell. *Top row:* Cubic arrangements of atoms in expanded view. *Second row:* Space-filling view of these cubic arrangements. All atoms are identical but, for clarity, corner atoms are blue, body-centered atoms pink, and face-centered atoms yellow. *Third row:* A unit cell (*shaded blue*) in a portion of the crystal. The number of nearest neighbors around one particle (*dark blue in center*) is the coordination number. *Bottom row:* The total numbers of atoms in the actual unit cell. The simple cubic has one atom; the body-centered has two; and the face-centered has four.

Animation: Cubic Unit Cells and Their Origins
Online Learning Center

The efficient packing of fruit.

Packing Efficiency and the Creation of Unit Cells The unit cells found in nature result from the various ways atoms are packed together, which are similar to the ways macroscopic spheres—marbles, golf balls, fruit—are packed for shipping or display (see photo).

For particles of the same size, *the higher the coordination number of the crystal is, the greater the number of particles in a given volume*. Therefore, as the coordination numbers in Figure 12.27 indicate, a crystal structure based on the face-centered cubic unit cell has more particles packed into a given volume than one based on the body-centered cubic unit cell, which has more than one based on the simple cubic unit cell. Let's see how to pack *identical* spheres to create these unit cells and the hexagonal unit cell as well:

1. *The simple cubic unit cell.* Suppose we arrange the first layer of spheres as shown in Figure 12.28A. Note the large diamond-shaped spaces (cutaway portion). If we place the next layer of spheres *directly above* the first, as shown in Figure 12.28B, we obtain an arrangement based on the *simple* cubic unit cell. By calculating the **packing efficiency** of this arrangement—the percentage of the total volume occupied by the spheres themselves—we find that only 52% of the available unit-cell volume is occupied by spheres and 48% consists of the empty space between them (see Problem 12.128 at the end of the chapter). This is a very inefficient way to pack spheres, so that neither fruit nor atoms are usually packed this way.

2. *The body-centered cubic unit cell.* Rather than placing the second layer directly above the first, we can use space more efficiently by placing the spheres (colored differently for clarity) on the diamond-shaped spaces in the first layer, as shown in Figure 12.28C. Then we pack the third layer onto the spaces in the second so that the first and third layers line up vertically. This arrangement is based on the *body-centered* cubic unit cell, and its packing efficiency is 68%—much higher than for the simple cubic unit cell. Several metallic elements, including chromium, iron, and all the Group 1A(1) elements, have a crystal structure based on the body-centered cubic unit cell.

3. *The hexagonal and face-centered cubic unit cells.* Spheres can be packed even more efficiently. First, we shift rows in the bottom layer so that the large diamond-shaped spaces become smaller triangular spaces. Then we place the second layer over these spaces. Figure 12.28D shows this arrangement, with the first layer labeled *a* (*orange*) and the second layer *b* (*green*).

We can place the third layer of spheres in two different ways, and how we do so gives rise to two different unit cells. If you look carefully at the spaces formed in layer *b* of Figure 12.28D, you'll see that some are orange because they lie above *spheres* in layer *a*, whereas others are white because they lie above *spaces* in layer *a*. If we place the third layer of spheres (*orange*) over the orange spaces (down and left to Figure 12.28E), they lie directly over spheres in layer *a*, and we obtain an *abab*. . . layering pattern because every other layer is placed identically. This gives **hexagonal closest packing,** which is based on the *hexagonal unit cell*.

On the other hand, if we place the third layer of spheres (*blue*) over the white spaces (down and right to Figure 12.28F), the spheres lie over spaces in layer *a*. This placement is different from both layers *a* and *b*, so we obtain an *abcabc*. . . pattern. This gives **cubic closest packing,** which is based on the *face-centered* cubic unit cell.

The packing efficiency of both hexagonal and cubic closest packing is 74%, and the coordination number of both is 12. There is no way to pack identical spheres more efficiently. Most metallic elements crystallize in either of these arrangements. Magnesium, titanium, and zinc are some elements that adopt the

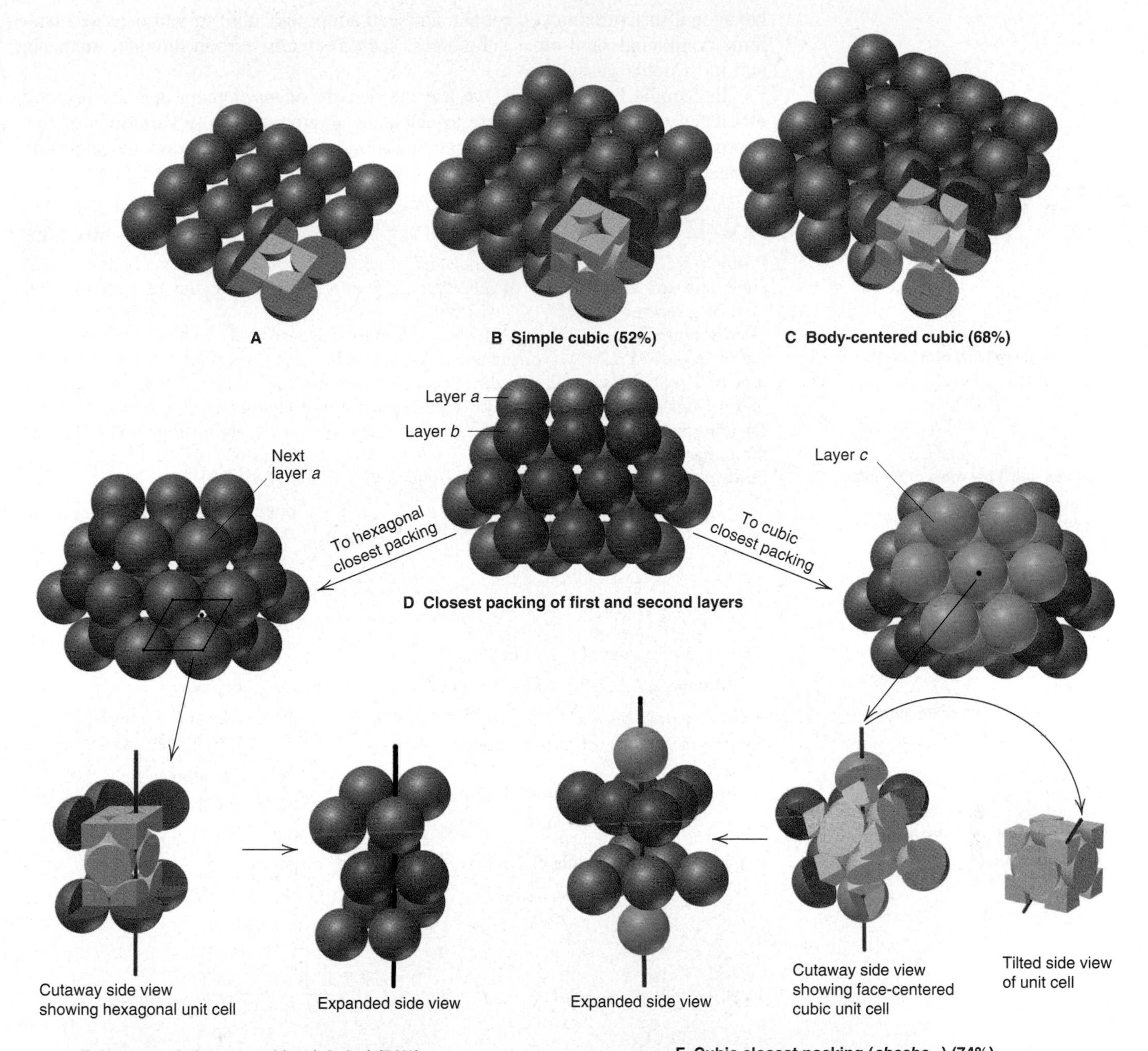

Figure 12.28 Packing identical spheres. A, In the first layer, each sphere lies next to another horizontally and vertically; note the large diamond-shaped spaces (*see cutaway*). **B,** If the spheres in the next layer lie *directly* over those in the first, the packing is based on the *simple cubic* unit cell (*pale orange cube, lower right corner*). **C,** If the spheres in the next layer lie in the diamond-shaped spaces of the first layer, the packing is based on the *body-centered cubic* unit cell (*lower right corner*). **D,** The closest possible packing of the first layer (*layer a, orange*) is obtained by shifting every other row in part A, thus reducing the diamond-shaped spaces to smaller triangular spaces. The spheres of the second layer (*layer b, green*) are placed above these spaces; note the orange and white spaces that result. **E,** Follow the left arrow from part D to obtain hexagonal closest packing. When the third layer (*next layer a, orange*) is placed directly over the first, that is, over the orange spaces, we obtain an *abab. . .* pattern. Rotating the layers 90° produces the side view, with the hexagonal unit cell shown as a cutaway segment, and the expanded side view. **F,** Follow the right arrow from part D to obtain cubic closest packing. When the third layer (*layer c, blue*) covers the white spaces, it lies in a different position from the first *and* second layers to give an *abcabc. . .* pattern. Rotating the layers 90° shows the side view, with the face-centered cubic unit cell as a cutaway, and a further tilt shows the unit cell clearly; finally, we see the expanded view. The packing efficiency for each type of unit cell is given in parentheses.

hexagonal structure; nickel, copper, and lead adopt the cubic structure, as do many ionic compounds and other substances, such as frozen carbon dioxide, methane, and most noble gases.

In Sample Problem 12.4, we use the density of an element and the packing efficiency of its crystal structure to calculate its atomic radius. Variations of this approach are used to find the molar mass and as one of the ways to determine Avogadro's number.

SAMPLE PROBLEM 12.4 Determining Atomic Radius from Crystal Structure

Problem Barium is the largest nonradioactive alkaline earth metal. It has a body-centered cubic unit cell and a density of 3.62 g/cm^3. What is the atomic radius of barium? (Volume of a sphere: $V = \frac{4}{3}\pi r^3$.)

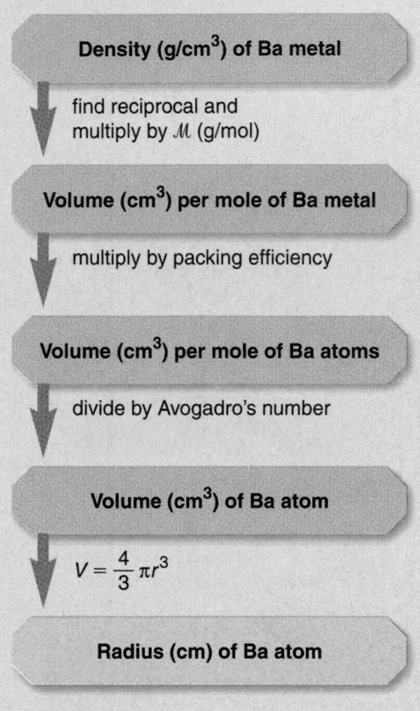

Plan Because an atom is spherical, we can find its radius from its volume. If we multiply the reciprocal of density (volume/mass) by the molar mass (mass/mole), we find the volume of 1 mol of Ba metal. The metal crystallizes in the body-centered cubic structure, so 68% of this volume is occupied by 1 mol of the atoms themselves (see Figure 12.28C). Dividing by Avogadro's number gives the volume of one Ba atom, from which we find the radius.

Density (g/cm^3) of Ba metal

↓ find reciprocal and multiply by $\mathcal{M}$ (g/mol)

Volume (cm^3) per mole of Ba metal

↓ multiply by packing efficiency

Volume (cm^3) per mole of Ba atoms

↓ divide by Avogadro's number

Volume (cm^3) of Ba atom

↓ $V = \frac{4}{3}\pi r^3$

Radius (cm) of Ba atom

Solution Combining steps to find the volume of 1 mol of Ba metal:

$$\text{Volume/mole of Ba metal} = \frac{1}{\text{density}} \times \mathcal{M}$$
$$= \frac{1 \text{ cm}^3}{3.62 \text{ g Ba}} \times \frac{137.3 \text{ g Ba}}{1 \text{ mol Ba}}$$
$$= 37.9 \text{ cm}^3/\text{mol Ba}$$

Finding the volume of 1 mol of Ba *atoms:*

$$\text{Volume/mole of Ba atoms} = \text{volume/mol Ba} \times \text{packing efficiency}$$
$$= 37.9 \text{ cm}^3/\text{mol Ba} \times 0.68 = 26 \text{ cm}^3/\text{mol Ba atoms}$$

Finding the volume of one Ba atom:

$$\text{Volume of Ba atom} = \frac{26 \text{ cm}^3}{1 \text{ mol Ba atoms}} \times \frac{1 \text{ mol Ba atoms}}{6.022\times10^{23} \text{ Ba atoms}}$$
$$= 4.3\times10^{-23} \text{ cm}^3/\text{Ba atom}$$

Finding the atomic radius of Ba from the volume of a sphere:

$$V \text{ of Ba atom} = \tfrac{4}{3}\pi r^3$$

So,
$$r^3 = \frac{3V}{4\pi}$$

Thus,
$$r = \sqrt[3]{\frac{3V}{4\pi}} = \sqrt[3]{\frac{3(4.3\times10^{-23} \text{ cm}^3)}{4 \times 3.14}}$$
$$= 2.2\times10^{-8} \text{ cm}$$

Check The order of magnitude is correct for an atom ($\sim10^{-8}$ cm $\approx 10^{-10}$ m). The actual value for barium is, in fact, 2.22×10^{-8} cm (see Figure 8.15, p. 307).

FOLLOW-UP PROBLEM 12.4 Iron crystallizes in a body-centered cubic structure. The volume of one Fe atom is 8.38×10^{-24} cm^3, and the density of Fe is 7.874 g/cm^3. Calculate an approximate value for Avogadro's number.

Our understanding of solids is based on the ability to "see" their crystal structures. Two techniques for doing this are described in the upcoming Tools of the Laboratory essay.

X-Ray Diffraction Analysis and Scanning Tunneling Microscopy

In this chapter and in Chapter 10, we have discussed crystal structures and molecular shapes as if they have actually been seen. You may have been wondering how chemists know atomic radii, bond lengths, and bond angles when the objects exhibiting them are so incredibly minute. Various tools exist for peering into the molecular world and measuring its dimensions, but two of the most powerful are **x-ray diffraction analysis** and **scanning tunneling microscopy.**

X-Ray Diffraction Analysis

This technique has been used for decades to determine crystal structures. In Chapter 7, we discussed wave diffraction and saw how interference patterns of bright and dark regions appear when light passes through slits that are spaced at the distance of the light's wavelength (see Figure 7.5, p. 261). In 1912, the Swiss physicist Max von Laue suggested that, since x-ray wavelengths are about the same size as the spaces between layers of particles in many solids, the layers might diffract x-rays. (Actually, the suggestion was made to test whether x-rays were particulate or wavelike.) X-ray diffraction was soon recognized as a powerful tool for determining the structure of a solid.

Let's see how this technique is used to measure a key parameter in a crystal structure: the distance (d) between layers of atoms. Figure B12.1 depicts a side view of two layers in a simplified lattice. Two waves impinge on the crystal at an angle θ and are diffracted at the same angle by adjacent layers. When the first wave strikes the top layer and the second strikes the next layer, the waves are *in phase* (peaks aligned with peaks and troughs with troughs). If they are still in phase after being diffracted, a spot appears on a nearby photographic plate. Note that this will occur only if the additional distance traveled by the second wave ($DE + EF$ in the figure) is a whole number of wavelengths, $n\lambda$, where n is an integer (1, 2, 3, and so on). From trigonometry, we find that

$$n\lambda = 2d \sin \theta$$

where θ is the known angle of incoming light, λ is its known wavelength, and d is the unknown distance between the layers in the crystal. This relationship is the *Bragg equation,* named for W. H. Bragg and his son W. L. Bragg, who shared the Nobel Prize in physics in 1915 for their work on crystal structure analysis.

Rotating the crystal changes the angle of incoming radiation and produces a different set of spots, eventually yielding a complete diffraction pattern that is used to determine the distances and angles within the lattice (Figure B12.2). The diffraction pattern is not an actual picture of the structure; the pattern must be analyzed mathematically to obtain the dimensions of the crystal. Modern x-ray diffraction equipment automatically rotates the crystal and measures thousands of diffractions, and a computer calculates the parameters of interest.

(continued)

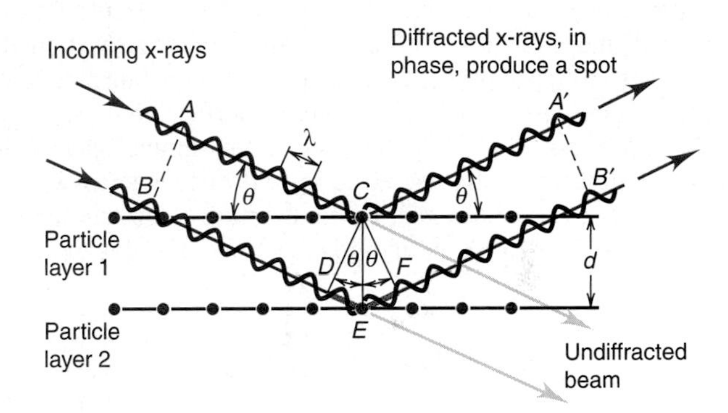

Figure B12.1 Diffraction of x-rays by crystal planes. As in-phase x-ray beams A and B pass into a crystal at angle θ, they are diffracted by interaction with the particles. Beam B travels the distance $DE + EF$ farther than beam A. If this additional distance is equal to a whole number of wavelengths, the beams remain in phase and create a spot on a screen or photographic plate. From the pattern of spots and the Bragg equation, $n\lambda = 2d \sin \theta$, the distance d between layers of particles can be calculated.

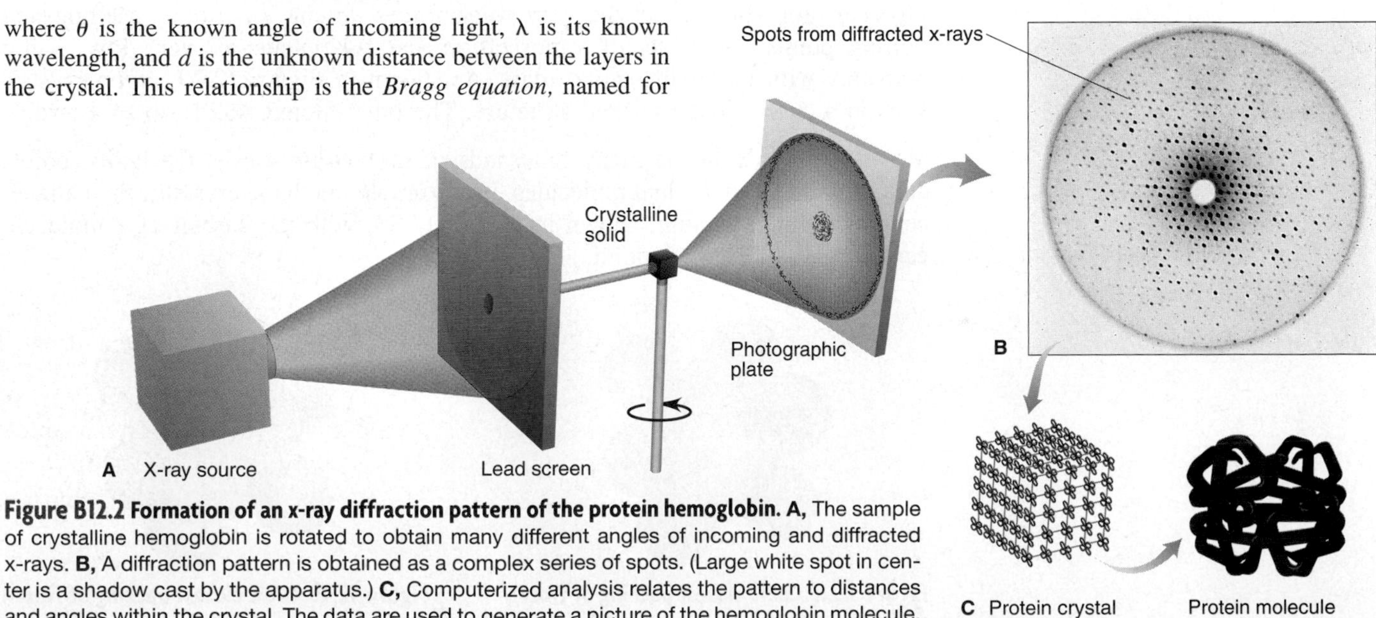

Figure B12.2 Formation of an x-ray diffraction pattern of the protein hemoglobin. A, The sample of crystalline hemoglobin is rotated to obtain many different angles of incoming and diffracted x-rays. **B,** A diffraction pattern is obtained as a complex series of spots. (Large white spot in center is a shadow cast by the apparatus.) **C,** Computerized analysis relates the pattern to distances and angles within the crystal. The data are used to generate a picture of the hemoglobin molecule.

X-ray diffraction analysis is used to answer questions in many branches of chemistry, but its greatest impact has been in biochemistry. It has shown that DNA exists as a double helix, and it is currently helping biochemists learn how a protein's three-dimensional structure is related to its function.

Scanning Tunneling Microscopy

This technique, a much newer method than x-ray diffraction analysis, is used to observe surfaces on the atomic scale. It was invented in the early 1980s by Gerd Binnig and Heinrich Rohrer, two Swiss physicists who won the Nobel Prize in physics in 1986 for their work. The technique is based on the idea that an electron in an atom has a small probability of existing far from the nucleus, so given the right conditions, it can move ("tunnel") to end up closer to another atom.

In practice, the tunneling electrons create a current that can be used to image the atoms of an adjacent surface. An extremely sharp tungsten-tipped probe, the source of the tunneling electrons, is placed very close (about 0.5 nm) to the surface under study. A small electric potential is applied across this minute gap to increase the probability that the electrons will tunnel across it. The size of the gap is kept constant by maintaining a constant tunneling current generated by the moving electrons. For this to occur, the probe must move tiny distances up and down, thus following the atomic contour of the surface. This movement is electronically monitored, and after many scans, a three-dimensional map of the surface is obtained. The method has revealed magnificent images of atoms and molecules coated on surfaces and is being used to study many aspects of surfaces, such as the nature of defects and the adhesion of films (Figure B12.3).

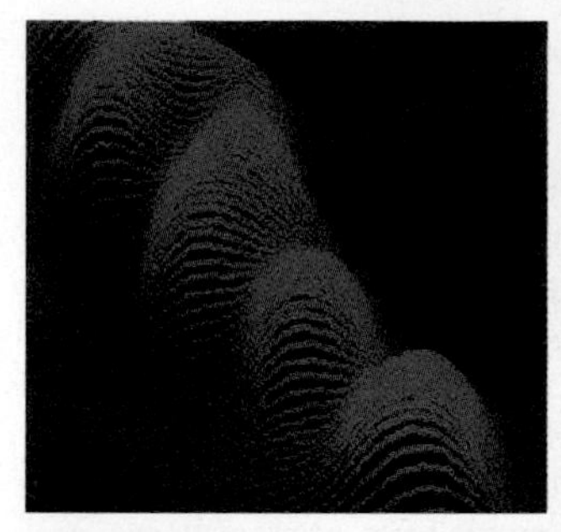
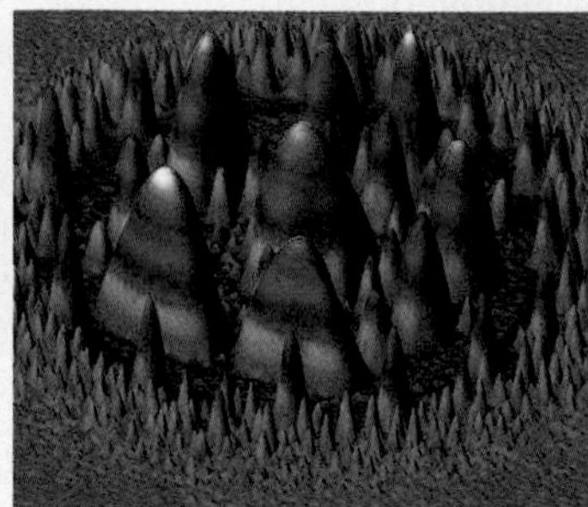

Figure B12.3 Scanning tunneling micrographs. A view of the DNA double helix (*left;* magnification ×1,000,000). Yttrium(II) oxide on an yttrium surface (*right*).

Types and Properties of Crystalline Solids

Now we can turn to the five most important types of solids, which are summarized in Table 12.5. Each is defined by the type of particle in the crystal, which determines the forces between them. You may want to review the bonding models (Chapter 9) to clarify how they relate to the properties of different solids.

Atomic Solids Individual atoms held together by dispersion forces form an **atomic solid.** The noble gases [Group 8A(18)] are the only examples, and their physical properties reflect the very weak forces among the atoms. Melting and boiling points and heats of vaporization and fusion are all very low, rising smoothly with increasing molar mass. As shown in Figure 12.29, argon crystallizes in a cubic closest packed structure. The other atomic solids do so as well.

Molecular Solids In the many thousands of **molecular solids,** the lattice points are occupied by individual molecules. For example, methane crystallizes in a face-centered cubic structure, shown in Figure 12.30, with the carbon of a molecule centered on each lattice point.

Figure 12.29 Cubic closest packing of frozen argon (face-centered cubic unit cell).

Figure 12.30 Cubic closest packing of frozen methane. Only one CH_4 molecule is shown.

Table 12.5 Characteristics of the Major Types of Crystalline Solids

Type	Particle(s)	Interparticle Forces	Physical Properties	Examples [mp, °C]
Atomic	Atoms	Dispersion	Soft, very low mp, poor thermal and electrical conductors	Group 8A(18) [Ne (−249) to Rn (−71)]
Molecular	Molecules	Dispersion, dipole-dipole, H bonds	Fairly soft, low to moderate mp, poor thermal and electrical conductors	*Nonpolar** O_2 [−219], C_4H_{10} [−138] Cl_2 [−101], C_6H_{14} [−95], P_4 [44.1] *Polar* SO_2 [−73], $CHCl_3$ [−64], HNO_3 [−42], H_2O [0.0], CH_3COOH [17]
Ionic	Positive and negative ions	Ion-ion attraction	Hard and brittle, high mp, good thermal and electrical conductors when molten	NaCl [801] CaF_2 [1423] MgO [2852]
Metallic	Atoms	Metallic bond	Soft to hard, low to very high mp, excellent thermal and electrical conductors, malleable and ductile	Na [97.8] Zn [420] Fe [1535]
Network covalent	Atoms	Covalent bond	Very hard, very high mp, usually poor thermal and electrical conductors	SiO_2 (quartz) [1610] C (diamond) [~4000]

*Nonpolar molecular solids are arranged in order of increasing molar mass. Note the correlation with increasing melting point (mp).

Various combinations of dipole-dipole, dispersion, and H-bonding forces are at work in molecular solids, which accounts for their wide range of physical properties. Dispersion forces are the principal force acting in nonpolar substances, so melting points generally increase with molar mass (Table 12.5). Among polar molecules, dipole-dipole forces and, where possible, H bonding dominate. Except for those substances consisting of the simplest molecules, molecular solids have higher melting points than the atomic solids (noble gases). Nevertheless, intermolecular forces are still relatively weak, so the melting points are much lower than those of ionic, metallic, and network covalent solids.

Ionic Solids In crystalline **ionic solids,** the unit cell contains particles with whole, rather than partial, charges. As a result, the interparticle forces (ionic bonds) are *much* stronger than the van der Waals forces in atomic or molecular solids. To maximize attractions, cations are surrounded by as many anions as possible, and vice versa, with *the smaller of the two ions lying in the spaces (holes) formed by the packing of the larger.* Since the unit cell is the smallest portion of the crystal that maintains the overall spatial arrangement, it is also the smallest portion that maintains the overall chemical composition. In other words, *the unit cell has the same cation:anion ratio as the empirical formula.*

Ionic compounds adopt several different crystal structures, but many use cubic closest packing. Let's first consider two structures that have a 1:1 ratio of ions. The *sodium chloride structure* is found in many compounds, including most of the alkali metal [Group 1A(1)] halides and hydrides, the alkaline earth metal [Group 2A(2)] oxides and sulfides, several transition-metal oxides and sulfides, and most of the silver halides. To visualize this structure, first imagine Cl^- anions

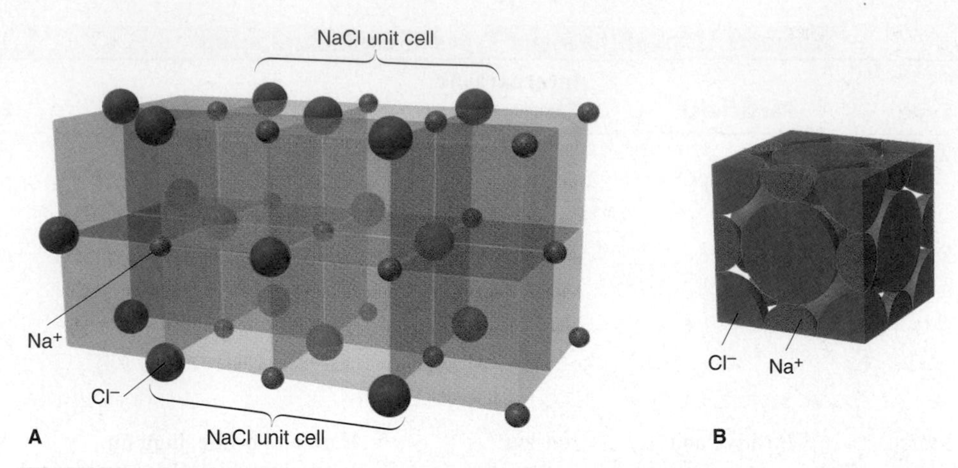

Figure 12.31 The sodium chloride structure. A, In an expanded view, the sodium chloride structure is pictured as resulting from the interpenetration of two face-centered cubic arrangements, one of Na^+ ions (*brown*) and the other of Cl^- ions (*green*). **B,** A space-filling view of the NaCl unit cell (central overlapped portion in part A), which consists of four Cl^- ions and four Na^+ ions.

and Na^+ cations organized separately in face-centered cubic (cubic closest packing) arrays. The crystal structure arises when these two arrays penetrate each other such that the smaller Na^+ ions end up in the holes between the larger Cl^- ions, as shown in Figure 12.31A. Thus, each Na^+ is surrounded by six Cl^-, and vice versa (coordination number = 6). Figure 12.31B is a space-filling depiction of the unit cell showing a face-centered cube of Cl^- ions with Na^+ ions between them. Note the four Cl^- $[(8 \times \frac{1}{8}) + (6 \times \frac{1}{2}) = 4\ Cl^-]$ and four Na^+ $[(12 \times \frac{1}{4}) + 1$ in the center $= 4\ Na^+]$, giving a 1:1 ion ratio.

Another structure with a 1:1 ion ratio is the *zinc blende (ZnS) structure*. It can be pictured as two face-centered cubic arrays, one of Zn^{2+} ions and the other of S^{2-} ions, interpenetrating such that each ion is tetrahedrally surrounded by four ions of opposite charge (coordination number = 4). Note the 1:1 ratio of ions in the unit cell shown in Figure 12.32. Many other compounds, including AgI, CdS, and the Cu(I) halides adopt the zinc blende structure.

The *fluorite (CaF_2) structure* is common among salts with a 1:2 cation:anion ratio, especially those having relatively large cations and relatively small anions.

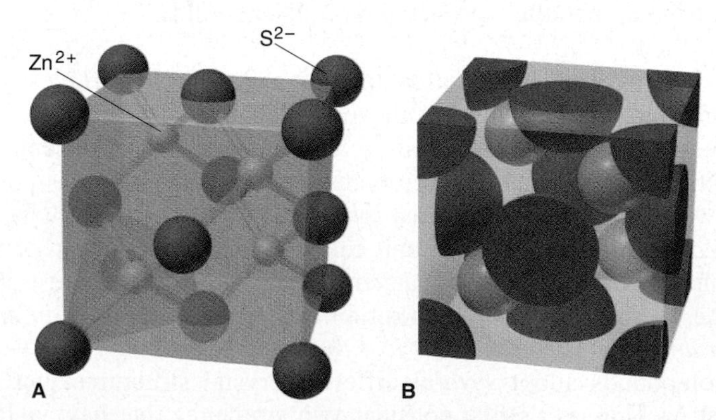

Figure 12.32 The zinc blende structure. Zinc sulfide adopts the zinc blende structure. **A,** The translucent cube shows a face-centered cubic array of four $[(8 \times \frac{1}{8}) + (6 \times \frac{1}{2}) = 4]\ S^{2-}$ ions (*yellow*) tetrahedrally surrounding each of four Zn^{2+} ions (*gray*) to give the 1:1 empirical formula. (Bonds are shown only for clarity.) **B,** The actual unit cell slightly expanded to show the interior ions.

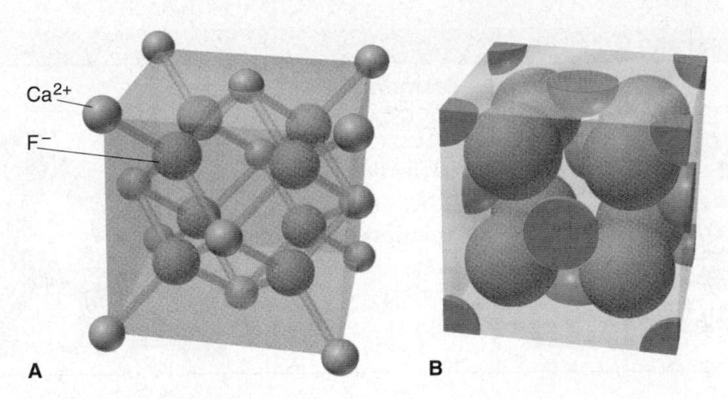

Figure 12.33 The fluorite structure. Calcium fluoride adopts the fluorite structure. **A,** The translucent cube shows a face-centered cubic array of four Ca^{2+} ions (*blue*) tetrahedrally surrounding each of eight F^- ions *(yellow)* to give the 4:8, or 1:2, ratio. **B,** The actual unit cell (slightly expanded).

In the case of CaF_2, the unit cell is a face-centered cubic array of Ca^{2+} ions with F^- ions occupying *all* eight available holes (Figure 12.33). This results in a Ca^{2+}:F^- ratio of 4:8, or 1:2. SrF_2 and $BaCl_2$ have the fluorite structure also. The *antifluorite structure* is often seen in compounds having a cation:anion ratio of 2:1 and a relatively large anion (for example, K_2S). In this structure, the cations occupy all eight holes formed by the cubic closest packing of the anions, just the opposite of the fluorite structure.

The properties of ionic solids are a direct consequence of the *fixed ion positions* and *very strong interionic forces,* which create a high lattice energy. Thus, ionic solids typically have high melting points and low electrical conductivities. When a large amount of heat is supplied and the ions gain enough kinetic energy to break free of their positions, the solid melts and the mobile ions conduct a current. Ionic compounds are hard because only a strong external force can change the relative positions of many trillions of interacting ions. If enough force *is* applied to move them, ions of like charge are brought near each other, and their repulsions crack the crystal (see Figure 9.8, p. 337).

Metallic Solids In contrast to the weak dispersion forces between the atoms in atomic solids, powerful metallic bonding forces hold individual atoms together in **metallic solids.** Most metallic elements crystallize in one of the two closest packed structures (Figure 12.34).

The properties of metals—high electrical and thermal conductivity, luster, and malleability—result from the presence of delocalized electrons, the essential feature of metallic bonding (introduced in Section 9.6). Metals have a wide range of melting points and hardnesses, which are related to the packing efficiency of the crystal structure and the number of valence electrons available for bonding. For example, Group 2A metals are harder and higher melting than Group 1A metals (see Figure 9.26, p. 358), because the 2A metals have closest packed structures (except Ba) and twice as many delocalized valence electrons.

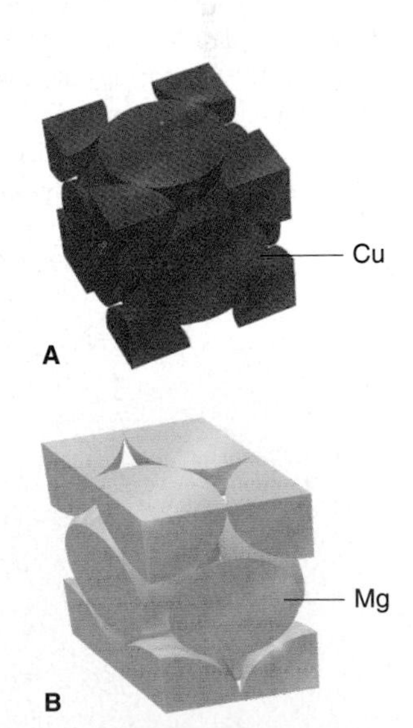

Figure 12.34 Crystal structures of metals. Most metals crystallize in one of the two closest packed arrangements. **A,** Copper adopts cubic closest packing. **B,** Magnesium adopts hexagonal closest packing.

Network Covalent Solids In the final type of crystalline solid, separate particles are not present. Instead, strong covalent bonds link the atoms together throughout a **network covalent solid.** As a consequence of the strong bonding, all these substances have extremely high melting and boiling points, but their conductivity and hardness depend on the details of their bonding.

The two common crystalline forms of elemental carbon are examples of network covalent solids. Although graphite and diamond have the same composition,

Table 12.6 Comparison of the Properties of Diamond and Graphite

Property	Graphite	Diamond
Density (g/cm^3)	2.27	3.51
Hardness	<1 (very soft)	10 (hardest)
Melting point (K)	4100	4100
Color	Shiny black	Colorless transparent
Electrical conductivity	High (along sheet)	None
ΔH^0_{comb} (kJ/mol)	−393.5	−395.4
ΔH^0_f (kJ/mol)	0 (standard state)	1.90

A Diamond Film on Every Pot In the 1950s, synthetic diamonds were manufactured slowly and expensively by exposing graphite to extreme temperatures (~1400°C) and pressures (50,000 atm). In the 1960s, a much cheaper method was discovered; by the 1990s, it was being used to deposit thin films of diamond onto virtually any surface. In the process of chemical vapor deposition (CVD), a stream of methane molecules break down at moderate temperatures (~600°C) and low pressures (~0.001 atm), and the carbon atoms deposit on the surface at rates that build up a thickness greater than 100 μm per hour. Because diamond is incredibly hard and has high thermal conductivity, applications for diamond films are myriad: scratch-proof cookware, watch crystals, hard disks, and eyeglasses; lifetime drill bits, ball bearings, and razor blades; high-temperature semiconductors, and so on. And now the CVD process is even used to make gem-quality synthetic diamonds.

their properties are strikingly different, as Table 12.6 shows. Graphite occurs as stacked flat sheets of hexagonal carbon rings with a strong σ-bond framework and delocalized π bonds, reminiscent of benzene. The arrangement of hexagons looks like chicken wire or honeycomb. Whereas the π-bonding electrons of benzene are delocalized over one ring, those of graphite are delocalized over the entire sheet. These mobile electrons allow graphite to conduct electricity, but only in the plane of the sheets. Graphite is a common electrode material and was once used for lightbulb filaments. The sheets interact via dispersion forces. Common impurities, such as O_2, that lodge between the sheets allow them to slide past each other easily, which explains why graphite is so soft. Diamond crystallizes in a face-centered cubic unit cell, with each carbon atom tetrahedrally surrounded by four others in one virtually endless array. Strong, single bonds throughout the crystal make diamond the hardest substance known. Because of its localized bonding electrons, diamond (like most network covalent solids) is unable to conduct electricity.

By far the most important network covalent solids are the *silicates*. They utilize a variety of bonding patterns, but nearly all consist of extended arrays of covalently bonded silicon and oxygen atoms. Quartz (SiO_2) is a common example. We'll discuss silicates, which form the structure of clays, rocks, and many minerals, when we consider the chemistry of silicon in Chapter 14.

Amorphous Solids

Amorphous solids are noncrystalline. Many have small, somewhat ordered regions connected by large disordered regions. Charcoal, rubber, and glass are some familiar examples of amorphous solids.

The process that forms quartz glass is typical of that for many amorphous solids. Crystalline quartz (SiO_2) has a cubic closest packed structure. The crystalline form is melted, and the viscous liquid is cooled rapidly to prevent it from recrystallizing. The chains of silicon and oxygen atoms cannot orient themselves quickly enough into an orderly structure, so they solidify in a distorted jumble containing many gaps and misaligned rows (Figure 12.35). The absence of regularity in the structure confers some properties of a liquid; in fact, glasses are sometimes referred to as *supercooled liquids*.

Bonding in Solids: Molecular Orbital Band Theory

Chapter 9 introduced a qualitative model of metallic bonding that pictures metal ions submerged in a "sea" of mobile, delocalized valence electrons. Quantum mechanics offers another model, an extension of molecular orbital (MO) theory, called **band theory.** It is more quantitative than the electron-sea model, and therefore more useful. We'll pay special attention to bonding in metals and differences in electrical conductivity of metals, metalloids, and nonmetals.

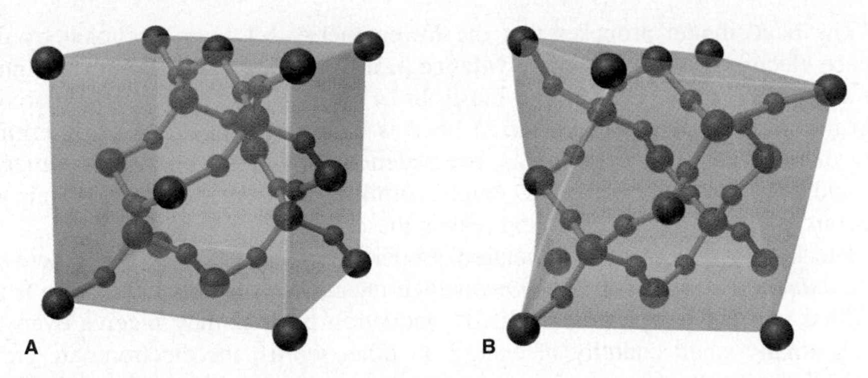

Figure 12.35 Crystalline and amorphous silicon dioxide. A, The atomic arrangement of cristobalite, one of the many crystalline forms of silica (SiO_2), shows the regularity of cubic closest packing. **B,** The atomic arrangement of a quartz glass is amorphous with a generally disordered structure.

Recall that when two atoms form a diatomic molecule, their atomic orbitals (AOs) combine to form an equal number of molecular orbitals (MOs). Let's consider lithium as an example. Figure 12.36 shows the formation of MOs in lithium. In dilithium, Li_2, each atom has four valence orbitals (one 2*s* and three 2*p*). (Recall that in Section 11.3, we focused primarily on the 2*s* orbitals.) They combine to form eight MOs, four bonding and four antibonding, spread over both atoms. If two more Li atoms combine, they form Li_4, a slightly larger aggregate, with 16 delocalized MOs. As more Li atoms join the cluster, more MOs are created, their energy levels lying closer and closer together. Extending this process to a 7-g sample of lithium metal (the molar mass) results in 1 mol of Li atoms (Li_{N_A}) combining to form an extremely large number (4 × Avogadro's number) of delocalized MOs, with *energies so closely spaced that they form a continuum, or band, of MOs*. It is almost as though the entire piece of metal were one enormous Li molecule.

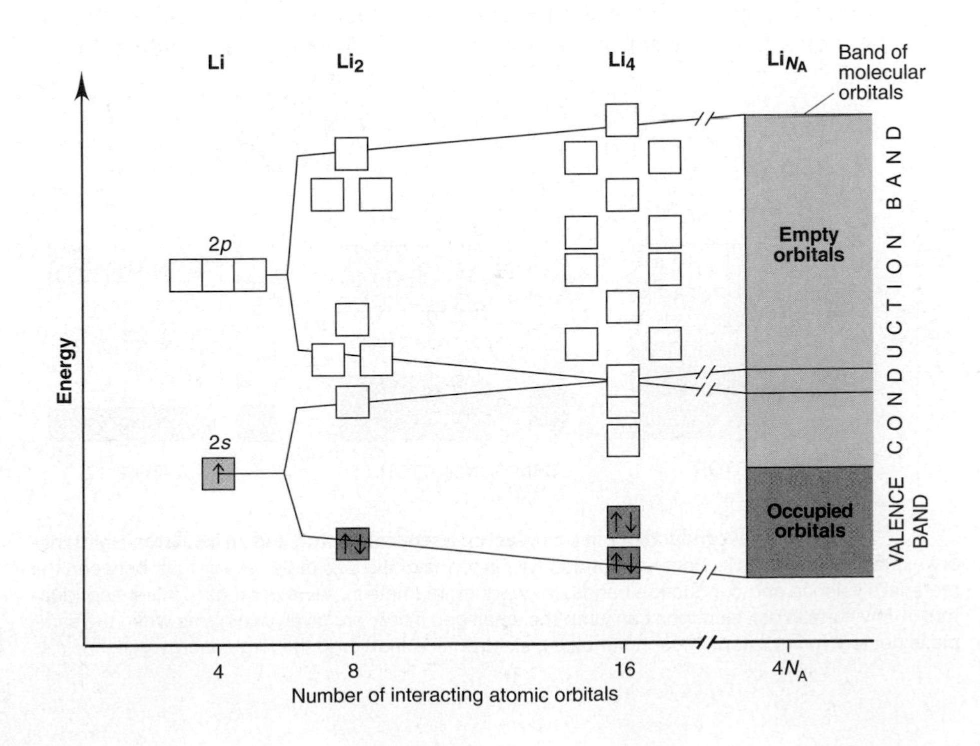

Figure 12.36 The band of molecular orbitals in lithium metal. Lithium atoms contain four valence orbitals, one 2*s* and three 2*p* *(left)*. When two lithium atoms combine (Li_2), their AOs form eight MOs within a certain range of energy. Four Li atoms (Li_4) form 16 MOs. A mole of Li atoms forms $4N_A$ MOs (N_A = Avogadro's number). The orbital energies are so close together that they form a continuous band. The valence electrons enter the lower energy portion (valence band), while the higher energy portion (conduction band) remains empty. In lithium (and other metals), the valence and conduction bands have no gap between them.

The band model proposes that the lower energy MOs are occupied by the valence electrons and make up the **valence band.** The empty MOs that are higher in energy make up the **conduction band.** In Li metal, the valence band is derived from the 2*s* AOs, and the conduction band is derived mostly from an intermingling of the 2*s* and 2*p* AOs. In Li_2, two valence electrons fill the lowest energy MO and leave the antibonding MO empty. Similarly, in Li metal, 1 mol of valence electrons fills the valence band and leaves the conduction band empty.

The key to understanding metallic properties is that *in metals, the valence and conduction bands are contiguous,* which means that electrons can jump from the filled valence band to the unfilled conduction band if they receive even an infinitesimally small quantity of energy. In other words, the electrons are completely delocalized: *they are free to move throughout the piece of metal.* Thus, metals conduct electricity so well because an applied electric field easily excites the highest energy electrons into empty orbitals, and they move through the sample.

Metallic luster (shininess) is another effect of the continuous band of MO energy levels. With so many closely spaced levels available, electrons can absorb and release photons of many frequencies as they move between the valence and conduction bands. Malleability and thermal conductivity also result from the completely delocalized electrons. Under an externally applied force, layers of positive metal ions simply move past each other, always protected from mutual repulsions by the presence of the delocalized electrons (see Figure 9.27B, p. 359). When a metal wire is heated, the highest energy electrons are excited and their extra energy is transferred as kinetic energy along the wire's length.

Large numbers of nonmetal or metalloid atoms can also combine to form bands of MOs. Metals conduct a current well (conductors), whereas most nonmetals do not (insulators), and the conductivity of metalloids lies somewhere in between (semiconductors). Band theory explains these differences in terms of the size of the energy gaps between the valence and conduction bands, as shown in Figure 12.37:

1. *Conductors (metals).* The valence and conduction bands of a **conductor** have no gap between them, so electrons flow when even a tiny electrical potential difference is applied. When the temperature is raised, greater random motion of the atoms hinders electron movement, which *decreases* the conductivity of a metal.

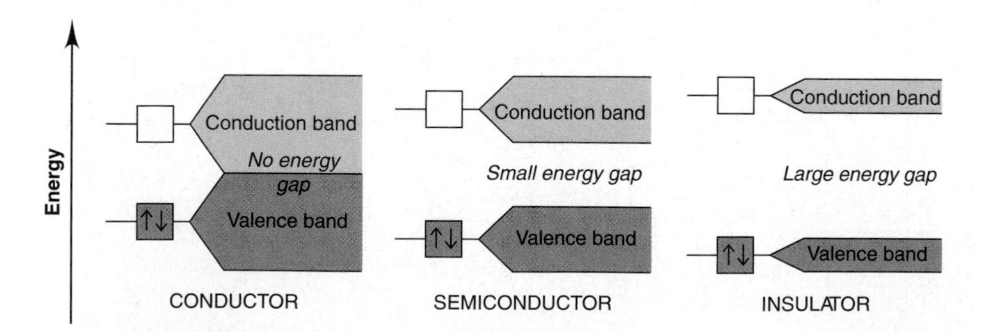

Figure 12.37 Electrical conductivity in a conductor, a semiconductor, and an insulator. Band theory explains differences in electrical conductivity in terms of the size of the energy gap between the material's valence and conduction bands. In conductors (metals), there is no gap. In semiconductors (many metalloids), electrons can jump the small gap if they are given energy, as when the sample is heated. In insulators (most nonmetals), electrons cannot jump the large energy gap.

2. *Semiconductors (metalloids).* In a **semiconductor,** a relatively small energy gap exists between the valence and conduction bands. Thermally excited electrons can cross the gap, allowing a small current to flow. Thus, in contrast to a conductor, the conductivity of a semiconductor *increases* when it is heated.
3. *Insulators (nonmetals).* In an **insulator,** the gap between the bands is too large for electrons to jump even when the substance is heated, so no current is observed.

Another type of electrical conductivity, called **superconductivity,** has been generating intense interest for more than two decades. When metals conduct at ordinary temperatures, electron flow is restricted by collisions with atoms vibrating in their lattice sites. Such restricted flow appears as resistive heating and represents a loss of energy. To conduct with no energy loss—to superconduct—requires extreme cooling to minimize atom movement. This remarkable phenomenon had been observed in metals only by cooling them to near absolute zero, which can be done only with liquid helium (bp = 4 K; price = $11/L).

In 1986, all this changed with the synthesis of certain ionic oxides that superconduct near the boiling point of liquid nitrogen (bp = 77 K; price = $0.25/L). Like metal conductors, oxide superconductors have no band gap. In the case of $YBa_2Cu_3O_7$, x-ray analysis shows that the Cu ions in the oxide lattice are aligned, which may be related to the superconducting property. In 1989, oxides with Bi and Tl instead of Y and Ba were synthesized and found to superconduct at 125 K; recently, an oxide with Hg, Ba, and Ca, in addition to Cu and O, was shown to superconduct at 133 K. Engineering dreams for these materials include storage and transmission of electricity with no loss of energy (allowing power plants to be located far from cities), ultrasmall microchips for ultrafast computers, electromagnets to levitate superfast railway trains (Figure 12.38), and inexpensive medical diagnostic equipment with remarkable image clarity.

Figure 12.38 The levitating power of a superconducting oxide. A magnet is suspended above a cooled high-temperature superconductor. Someday, this phenomenon may be used to levitate trains above their tracks for quiet, fast travel.

However, manufacturing problems must be overcome. The oxides are very brittle, but some recent developments may remedy this drawback. A more fundamental concern is that when the oxide is warmed or placed in a strong magnetic field, the superconductivity may disappear and not return again on cooling. Clearly, research into superconductivity will involve chemists, physicists, and engineers for many years to come. We consider other remarkable materials in the next section.

SECTION SUMMARY

The particles in crystalline solids lie at points that form a structure of repeating unit cells. The three types of unit cells in the cubic system are simple, body-centered, and face-centered. The most efficient packing arrangements are cubic closest packing and hexagonal closest packing. Bond angles and distances in a crystal structure can be determined with x-ray diffraction analysis and scanning tunneling microscopy. Atomic solids have a closest packed structure, with atoms held together by very weak dispersion forces. Molecular solids have molecules at the lattice points, often in a cubic closest packed structure. Their intermolecular forces (dispersion, dipole-dipole, H bonding) and resulting physical properties vary greatly. Ionic solids often crystallize with one type of ion filling holes in a cubic closest packed structure of the other. The high melting points, hardness, and low conductivity of these solids arise from strong ionic attractions. Most metals have a closest packed structure. The atoms of network covalent solids are covalently bonded throughout the sample. Amorphous solids have very little regularity among their particles. Band theory proposes that orbitals in the atoms of solids combine to form a continuum, or band, of molecular orbitals. Metals are electrical conductors because electrons move freely from the filled (valence band) to the empty (conduction band) portions of this energy continuum. Insulators have a large energy gap between the two portions; semiconductors have a small gap.

12.7 ADVANCED MATERIALS

In the last few decades, the exciting field of materials science has grown from solid-state chemistry, physics, and engineering, and it is changing our lives in astonishing ways. Objects that were once considered futuristic fantasies of science-fiction writers are realities or soon will be: powerful, ultrafast computers no bigger than this book, connected electronically to millions of others throughout the world; cars powered by sunlight and made of nonmetallic parts stronger than steel and lighter than aluminum; sporting goods made of the materials of space vehicles; ultrasmall machines constructed by manipulating individual atoms and molecules. In this section, we briefly discuss some of these remarkable materials.

Electronic Materials

The ideal of a perfectly ordered crystal is attainable only if the crystal is grown very slowly under carefully controlled conditions. When crystals form more rapidly, **crystal defects** inevitably form. Planes of particles are misaligned, particles are out of place or missing entirely, and foreign particles are lodged in the lattice.

Although crystal defects usually weaken a substance, they are sometimes introduced intentionally to create materials with improved properties, such as increased strength or hardness or, as you'll see in a moment, to increase a material's conductivity for use in electronic devices. In the process of welding two metals together, for example, *vacancies* form near the surface when atoms vaporize, and then these vacancies move deeper as atoms from lower rows rise to fill the gaps. Welding causes the two types of metal atoms to intermingle and fill each other's vacancies. Metal alloying introduces several kinds of crystal defects, as when some atoms of a second metal occupy lattice sites of the first. Often, the alloy is harder than the pure metal; an example is brass, an alloy of copper with zinc. One reason the welded metals are stronger and the alloy is harder is that the second metal contributes additional valence electrons for metallic bonding.

Doped Semiconductors The manufacture and miniaturization of doped semiconductors have revolutionized the communications, home-entertainment, and information industries. By controlling the number of valence electrons through the creation of specific types of defects, chemists and engineers can greatly increase the conductivity of a semiconductor.

Pure silicon (Si), which lies below carbon in Group 4A(14), conducts poorly at room temperature because an energy gap separates its filled valence band from its conduction band (Figure 12.39A). Its conductivity can be greatly enhanced by **doping,** adding small amounts of other elements to increase or decrease the number of valence electrons in the bands. When Si is doped with phosphorus [or another Group 5A(15) element], P atoms occupy some of the lattice sites. Since P has one more valence electron than Si, this additional electron must enter an empty orbital in the conduction band, thus bridging the energy gap and increasing conductivity. Such doping creates an *n-type semiconductor,* so called because extra *n*egative charges (electrons) are present (Figure 12.39B).

When Si is doped with gallium [or another Group 3A(13) element], Ga atoms occupy some sites (Figure 12.39C). Since Ga has one fewer valence electron than Si, some of the orbitals in the valence band are empty, which creates a positive site. Si electrons can migrate to these empty orbitals, thereby increasing conductivity. Such doping creates a *p-type semiconductor,* so called because the empty orbitals act as *p*ositive holes.

In contact with each other, an n-type and a p-type semiconductor form a *p-n junction.* When the negative terminal of a battery is connected to the

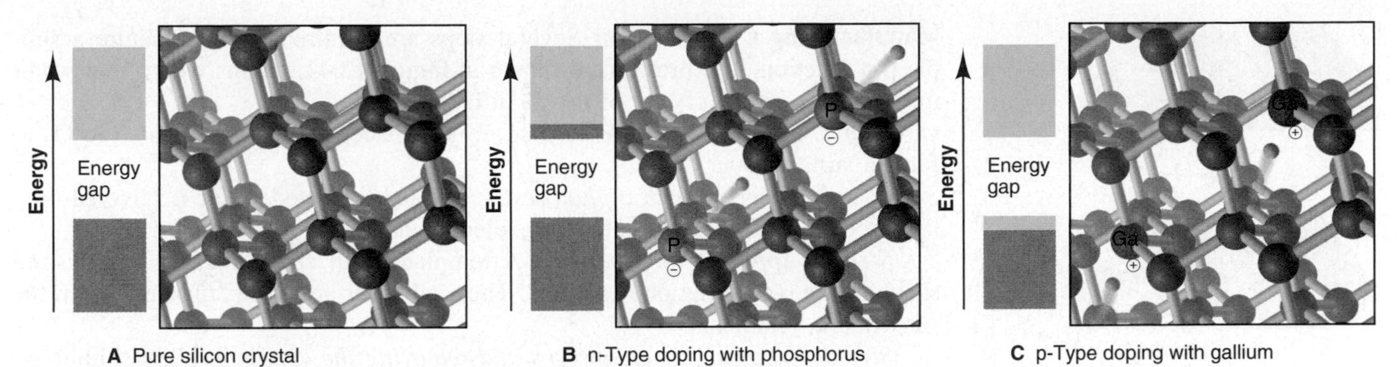

Figure 12.39 Crystal structures and band representations of doped semiconductors. **A,** Pure silicon has the same crystal structure as diamond but acts as a semiconductor; the energy gap between its valence and conduction bands keeps conductivity low at room temperature. **B,** Doping silicon with phosphorus *(purple)* adds additional valence electrons, which are free to move through the crystal. They enter the lower portion of the conduction band, which is adjacent to higher energy empty orbitals, thereby increasing conductivity. **C,** Doping silicon with gallium *(orange)* removes electrons from the valence band and introduces positive holes. Nearby Si electrons can enter these empty orbitals, thereby increasing conductivity. The orbitals from which the Si electrons move become vacant; in effect, the holes move.

n-type portion and the positive terminal to the p-type portion, electrons flow freely in the n-to-p direction, which has the simultaneous effect of moving holes in the p-to-n direction (Figure 12.40). No current flows if the terminals are reversed. Such unidirectional current flow makes a p-n junction act as a *rectifier,* a device that converts alternating current into direct current. A p-n junction in a modern integrated circuit can be made smaller than a square 10 μm on a side. Before the p-n junction was created, rectifiers were bulky, expensive vacuum tubes.

A modern computer chip the size of a nickel may incorporate millions of p-n junctions in the form of *transistors*. One of the most common types, an n-p-n transistor, is made by sandwiching a p-type portion between two n-type portions to form adjacent p-n junctions. The current flowing through one junction controls the current flowing through the other and results in an amplified signal. Once again, to accomplish signal amplification before the advent of doped semiconductors required large, and often unreliable, vacuum tubes. Today, minute transistors are found in every radio, TV, and computer and have made possible the multibillion-dollar electronics industry.

Solar Cells One of the more common types of solar cell is, in essence, a p-n junction with an n-type surface exposed to a light source. The light provides the energy to free electrons from the n-type region and accelerate them through an external circuit into the p-type region, thus producing a current to power a calculator, light a bulb, and so on. Arrays of solar cells supply electric power to many residences and businesses, as well as to the space shuttle and most communications satellites (see the photo).

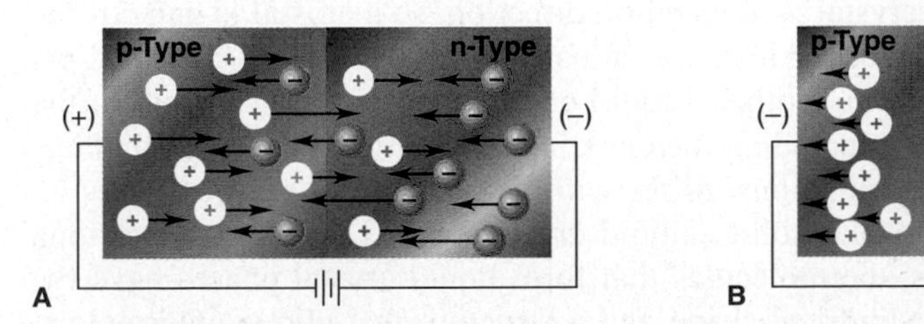

Figure 12.40 The p-n junction. Placing a p-type semiconductor adjacent to an n-type creates a p-n junction. **A,** If the negative battery terminal is connected to the n-type portion, electrons *(yellow sphere with minus sign)* flow toward the p-type portion, which, in effect, moves holes *(white circle with plus sign)* toward the n-type portion. The small arrows indicate the net direction of movement. Note that any hole in the n-type portion lies opposite an electron that moved to the p-type portion. **B,** If the positive battery terminal is connected to the n-type portion, electrons are attracted to it, so no flow takes place across the junction.

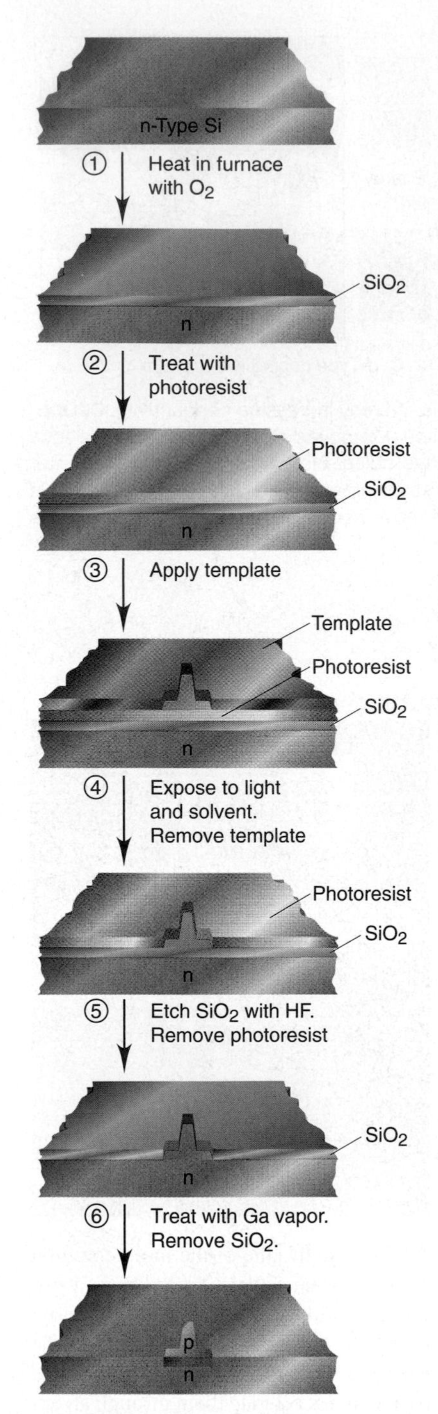

Figure 12.41 Steps in manufacturing a p-n junction. Steps 1 to 5 prepare the desired shape of the p-type portion on the n-type wafer. Step 6 dopes that shape with the Group 3A(13) element.

Manufacturing a p-n Junction Several steps are required to manufacture a simple p-n junction. The process, as shown in Figure 12.41, begins with a thin wafer made from a single crystal of n-type silicon:

Step 1. Forming an oxide surface. In a furnace, the wafer is heated in O_2 to form a surface layer of SiO_2.

Step 2. Covering with a photoresist. The oxide-coated wafer is covered by a light-sensitive wax or polymer film called a *photoresist.*

Step 3. Applying the template. A template with spaces having the desired shape of the p-type region is applied. Thus, all other areas are "masked" from the subsequent treatments.

Step 4. Exposing the photoresist and removing the template. Light shining on the template alters the exposed areas of photoresist so that it can be dissolved away in specific solvents, revealing the oxide surface. The template is removed.

Step 5. Etching the oxide surface and removing the photoresist. Treatment with hydrofluoric acid etches the template-shaped area of oxide coating to expose the n-type Si surface. Following this, the remaining photoresist is also removed.

Step 6. Creating the p-n junction. The wafer is exposed at high temperature to vapor of a Group 3A(13) element, which diffuses into the bare areas to create a template-shaped region of p-type Si adjacent to the n-type Si. The remaining SiO_2 is removed at this point also.

Typically, these steps are repeated to create an n-type region adjacent to the newly formed p-type region, thus forming an n-p-n transistor.

Liquid Crystals

Incorporated in the membrane of every cell in your body and in the display of just about every digital watch, pocket calculator, and laptop computer are unique substances known as **liquid crystals.** These materials flow like liquids but, like crystalline solids, pack at the molecular level with a high degree of order.

Properties, Preparation, and Types of Liquid Crystals To understand the properties of liquid crystals, let's first examine how particles are ordered in the three common physical states and how this affects their properties. The extent of order among particles distinguishes crystalline solids clearly from gases and liquids. Gases have no order, and liquids have little more. Both are considered *isotropic,* which means that their physical properties are the same in every direction within the phase. For example, the viscosity of a gas or a liquid is the same regardless of direction. Glasses and other amorphous solids are also isotropic because they have no regular lattice structure.

In contrast, crystalline solids have a high degree of order among their particles. The properties of a crystal *do* depend on direction, so a crystal is *anisotropic.* The facets in a cut diamond, for instance, arise because the crystal cracks in one direction more easily than in another. Liquid crystals are also anisotropic in that several physical properties, including the electrical and optical properties that lead to their most important applications, differ with direction through the phase.

Like crystalline molecular solids, liquid crystal phases consist of individual molecules. In most cases, the molecules that form liquid crystal phases have two characteristics: a long, cylindrical shape and a structure that allows intermolecular attractions through dispersion and dipole-dipole or H-bonding forces, but that inhibits perfect crystalline packing. Figure 12.42 shows the structures of two molecules that form liquid crystal phases. Note the rodlike shapes and the presence of certain groups—in these cases, flat, benzene-like ring systems—that keep the molecules extended. Many of these types of molecules also have a molecular dipole associated with the long molecular axis. A sufficiently strong electric field

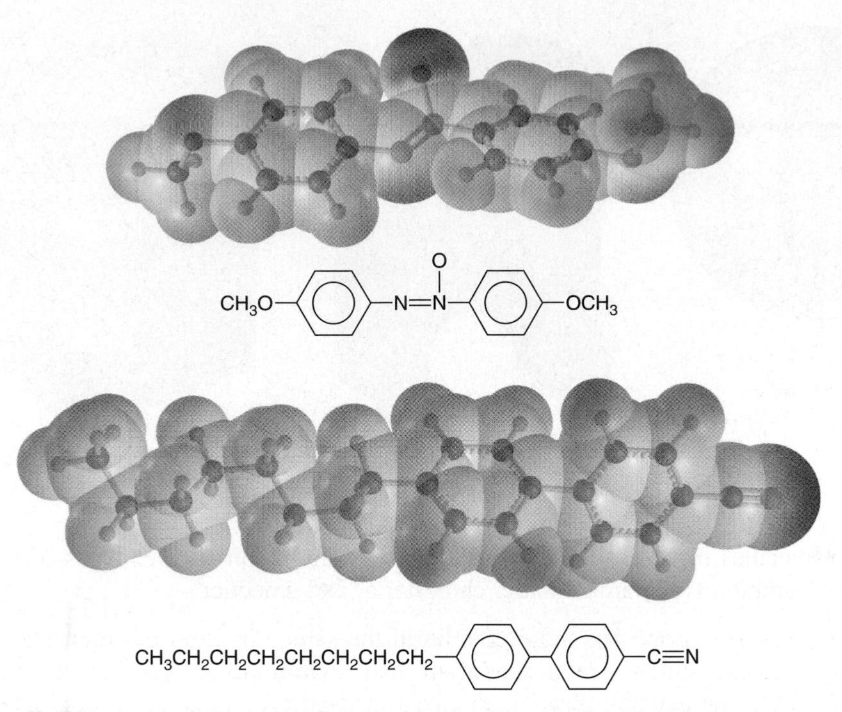

Figure 12.42 Structures of two typical molecules that form liquid crystal phases. Note the long, extended shapes and the regions of high (*red*) and low (*blue*) electron density.

can orient large numbers of these polar molecules in approximately the same direction, like compass needles in a magnetic field.

The viscosity of a liquid crystal phase is lowest in the direction parallel to the long axis. Like moistened microscope slides, it is easier for the molecules to slide along each other (because the total attractive force remains the same), than it is for them to pull apart from each other sideways. As a result, the molecules tend to align while the phase flows.

Liquid crystal phases can arise in two general ways, and, sometimes, either way can occur in the same substance. A *thermotropic* phase develops as a result of a change in temperature. As a crystalline solid is heated, the molecules leave their lattice sites, but the intermolecular interactions are still strong enough to keep the molecules aligned with each other along their long axes. Like any other phase, the liquid crystal phase has sharp transition temperatures; however, it exists over a relatively small temperature range. Further heating provides enough kinetic energy for the molecules to become disordered, as in a normal liquid. The typical range for liquid crystal phases of pure substances is from $<1°C$ to around 10°C, but mixing phases of two or more substances can greatly extend this range. For this reason, the liquid crystal phases used within display devices, as well as those within cell membranes, consist of mixtures of molecules.

A *lyotropic* phase occurs in solution as the result of changes in concentration, but the conditions for forming such a phase vary for different substances. For example, when purified, some biomolecules that exist naturally in mammalian cell membranes form lyotropic phases in water at the moderate temperature that occurs within the organism. At the other extreme, Kevlar, a fiber used in bullet-proof vests and high-performance sports equipment, forms a lyotropic phase at high temperatures in concentrated H_2SO_4 solution.

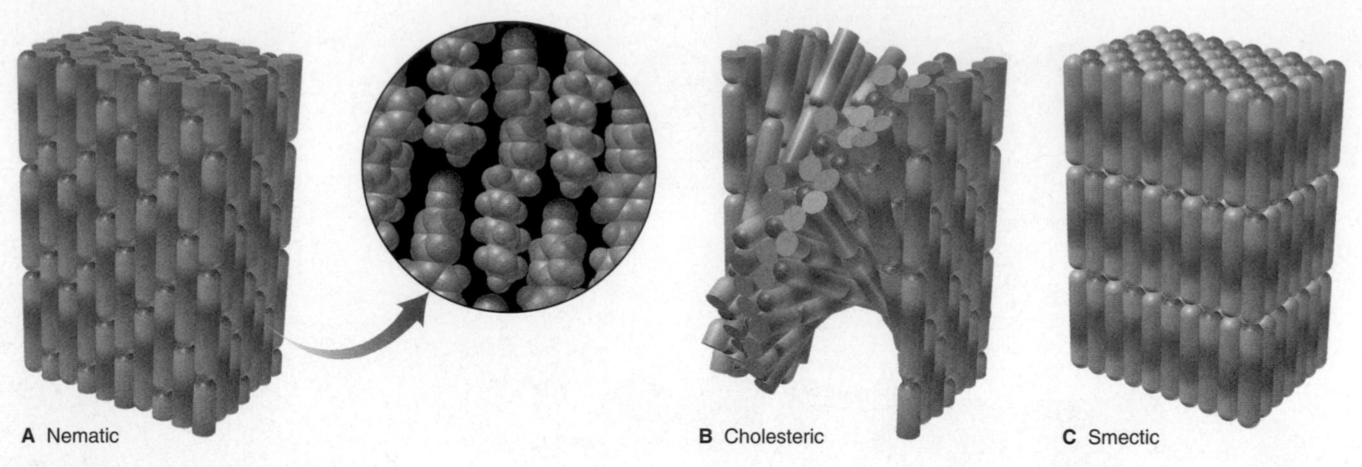

Figure 12.43 The three common types of liquid crystal phases. A, Nematic phase. A rectangular volume of the phase, with expanded view, shows a close-up of the arrangement of molecules. **B,** Cholesteric phase. Note the corkscrew-like arrangement of the layers. **C,** Smectic phase. A rectangular volume of the phase shows the more orderly stacking of layers.

Molecules that form liquid crystal phases can exhibit various types of order. Three common types are nematic, cholesteric, and smectic:

- In a *nematic* phase, the molecules lie in the same direction but their ends are not aligned, much like a school of fish swimming in synchrony (Figure 12.43A). The nematic phase is the least ordered type of liquid crystal phases.
- In a *cholesteric* phase, which is somewhat more ordered, the molecules lie in layers that each exhibit nematic-type ordering. Rather than lying in parallel fashion, however, each layer is rotated by a fixed angle with respect to the next layer. The result is a helical (corkscrew) arrangement, so a cholesteric phase is often called a *twisted nematic phase* (Figure 12.43B).
- In a *smectic* phase, which is the most ordered, the molecules lie parallel to each other, *with* their ends aligned, in layers that are stacked directly over each other, much like a supermarket display of shelves filled with identical bottles (Figure 12.43C). The long molecular axis has a well-defined angle (shown in the figure as 90°) with respect to the plane of the layer. The molecules in Figure 12.42 are typical of those that form nematic or smectic phases. Liquid crystal–type phases appear in many biological systems (Figure 12.44).

In some cases, a substance that forms a given liquid crystal phase under one set of conditions forms other phases under different conditions. Thus, a given thermotropic liquid crystal substance can pass from disordered liquid through a series of distinct liquid crystal phases to an ordered crystal through a decrease in temperature. A lyotropic substance can undergo similar changes through an increase in concentration.

Applications of Liquid Crystals The ability to control the orientation of the molecules in a liquid crystal allows us to produce materials with high strength or unique optical properties.

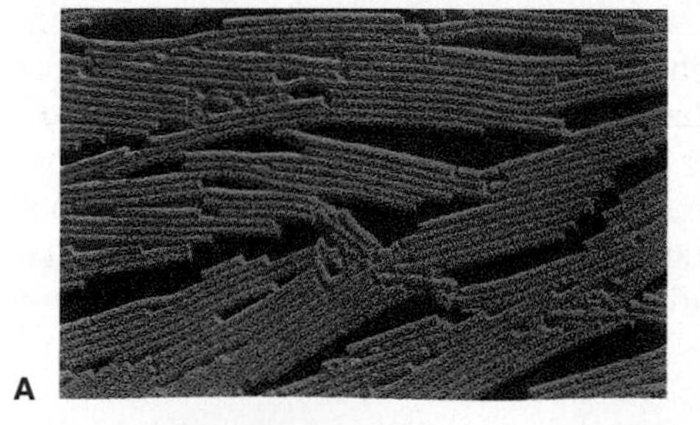

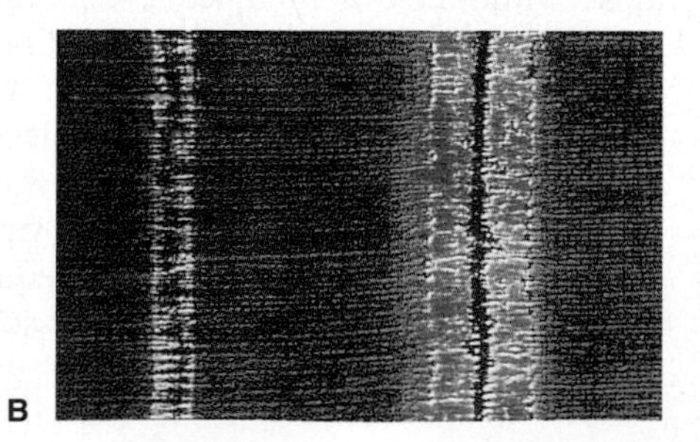

Figure 12.44 Liquid crystal–type phases in biological systems. A, Nematic arrays of tobacco mosaic virus particles within the fluid of a tobacco leaf. **B,** The orderly arrangement of actin and myosin protein filaments in voluntary muscle cells.

High-strength applications involve the use of extremely long molecules called *polymers*. While in a thermotropic liquid crystal phase and during their flow through the processing equipment, these molecules become highly aligned, like the fibers in wood. Cooling solidifies them into fibers, rods, and sheets that can be shaped into materials with superior mechanical properties in the direction of the long molecular axis. Sporting equipment, supersonic aircraft parts, and the sails used in the America's Cup races are fabricated from these polymeric materials. (We discuss the structure and physical behavior of polymers later in this section and their synthesis in Chapter 15.)

Far more important in today's consumer market are the liquid crystal displays (LCDs) used in watches, calculators, cell phones, and computers. All depend on *changes in molecular orientation in an electric field*. The most common type, called a *twisted nematic* display, is shown schematically in Figure 12.45 as a small portion of a wristwatch LCD. The liquid crystal phase consists of layers of nematic phases sandwiched between thin glass plates that incorporate transparent electrodes. As a result of a special coating process, the long molecular axis lies parallel to the plane of the glass plates. The distance between the plates (6–8 μm) is chosen so that the molecular axis within each succeeding layer twists just enough for the molecular orientation at the bottom plate to be 90° from that at the top plate. Above and below this "sandwich" are thin polarizing filters (like those used in Polaroid lenses or camera sunlight filters), which allow light waves oriented in only one direction to pass through. The filters are placed in a "crossed" arrangement, so that light passing through the top filter must twist 90° to pass through the bottom filter. The orientation and optical properties of the molecules twist the orientation of the light by exactly this amount. This whole grouping of filters, plates, and liquid crystal phase lies on a mirror.

A current generated by the watch battery controls the orientation of the molecules within the phase. With the current on in one region of the display, the molecules become oriented *toward* the field, and thus block the light from passing through to the bottom filter, so that region appears dark. With the current off in another region, light passes through the molecules and bottom filter to the mirror and back again, so that region appears bright.

Cholesteric liquid crystals are used in applications that involve color changes with temperature. The twisted molecular orientation in these phases "unwinds" with heating, and the extent of the unwinding determines the color. Liquid crystal thermometers that include a mixture of substances to widen their range of temperatures use this effect. Newer and more important uses include "mapping" the area of a tumor, detecting faulty connections in electronic circuit boards, and nondestructive testing of materials under stress.

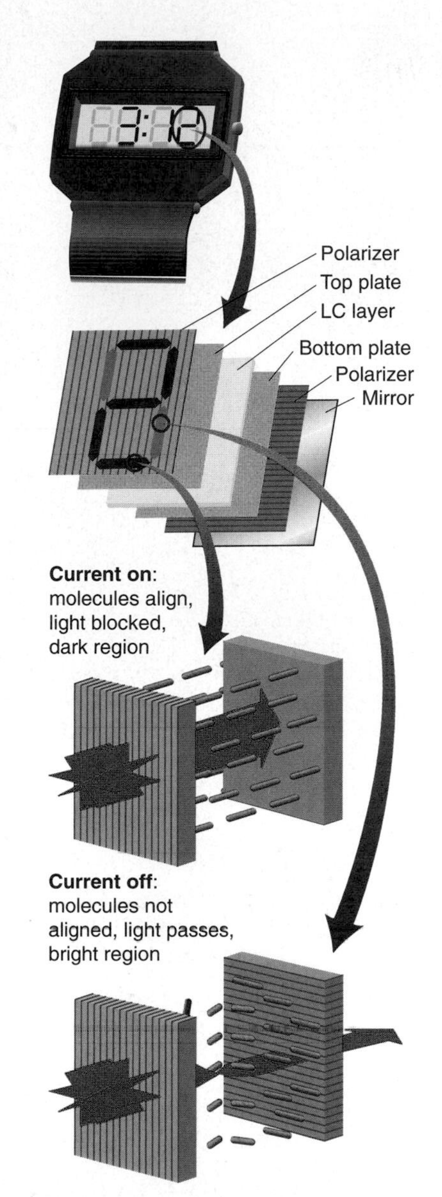

Figure 12.45 Schematic of a liquid crystal display (LCD). A close-up of the "2" on a wristwatch LCD reveals two polarizers sandwiching two glass plates, which sandwich a liquid crystal (LC) layer, all lying on a mirror. When light waves oriented in all directions enter the first polarizer, only waves oriented in one direction emerge to enter the LC layer. Enlarging a dark region of the numeral *(top blow-up)* shows that with the current on, the LC molecules are aligned and the light cannot pass through the other polarizer to the mirror; thus, the viewer sees no light. Enlarging a bright region *(bottom blow-up)* shows that with the current off, the LC molecules lie in a twisted nematic arrangement, which rotates the plane of the light waves and allows them to pass through the other polarizer to the mirror. Reflecting and retracing this path (not shown), the light reaches the viewer.

Ceramic Materials

First developed by Stone Age people, **ceramics** are defined as nonmetallic, nonpolymeric solids that are hardened by heating to high temperatures. Clay ceramics consist of silicate microcrystals suspended in a glassy cementing medium. In "firing" a ceramic pot, for example, a kiln heats the object made of an aluminosilicate clay, such as kaolinite, to 1500°C and the clay loses water:

$$Si_2Al_2O_5(OH)_4(s) \longrightarrow Si_2Al_2O_7(s) + 2H_2O(g)$$

During the heating process, the structure rearranges to an extended network of Si-centered and Al-centered tetrahedra of O atoms (Section 14.6).

Bricks, porcelain, glazes, and other clay ceramics are useful because of their hardness and resistance to heat and chemicals. Today's high-tech ceramics incorporate these traditional characteristics in addition to superior electrical and magnetic properties (Table 12.7, on the next page). As just one example, consider the unusual electrical behavior of certain zinc oxide (ZnO) composites. Ordinarily a

Table 12.7 Some Uses of Modern Ceramics and Ceramic Mixtures

Ceramic	Applications
SiC, Si_3N_4, TiB_2, Al_2O_3	Whiskers (fibers) to strengthen Al and other ceramics
Si_3N_4	Car engine parts; turbine rotors for "turbo" cars; electronic sensor units
Si_3N_4, BN, Al_2O_3	Supports or layering materials (as insulators) in electronic microchips
SiC, Si_3N_4, TiB_2, ZrO_2, Al_2O_3, BN	Cutting tools, edge sharpeners (as coatings and whole devices), scissors, surgical tools, industrial "diamond"
BN, SiC	Armor-plating reinforcement fibers (as in Kevlar composites)
ZrO_2, Al_2O_3	Surgical implants (hip and knee joints)

semiconductor, ZnO can be doped so that it becomes a conductor. Imbedding particles of the doped oxide into an insulating ceramic produces a variable resistor: at low voltage, the material conducts poorly, but at high voltage, it conducts well. Best of all, the changeover voltage can be "preset" by controlling the size of the ZnO particles and the thickness of the insulating medium.

Preparing Modern Ceramics Among the important modern ceramics are silicon carbide (SiC) and nitride (Si_3N_4), boron nitride (BN), and the superconducting oxides. They are prepared by standard chemical methods that involve driving off a volatile component during the reaction.

The SiC ceramics are made from compounds used in silicone polymer manufacture (we'll discuss their structures and uses in Section 14.6):

$$n(CH_3)_2SiCl_2(l) + 2nNa(s) \longrightarrow 2nNaCl(s) + [(CH_3)_2Si]_n(s)$$

This product is heated to 800°C to form the ceramic:

$$[(CH_3)_2Si]_n \longrightarrow nCH_4(g) + nH_2(g) + nSiC(s)$$

SiC can also be prepared by direct reaction of Si and graphite under vacuum:

$$Si(s) + C(graphite) \xrightarrow{\sim 1500°C} SiC(s)$$

The nitride is also prepared by reaction of the elements:

$$3Si(s) + 2N_2(g) \xrightarrow{>1300°C} Si_3N_4(s)$$

The formation of a BN ceramic begins with the reaction of boron trichloride or boric acid with ammonia:

$$B(OH)_3(s) + 3NH_3(g) \longrightarrow B(NH_2)_3(s) + 3H_2O(g)$$

Heat drives off some of the bound nitrogen as NH_3 to yield the ceramic:

$$B(NH_2)_3(s) \xrightarrow{\Delta} 2NH_3(g) + BN(s)$$

One of the common high-temperature superconducting oxides is a ceramic made by heating a mixture of barium carbonate and copper and yttrium oxides, followed by further heating in the presence of O_2:

$$4BaCO_3(s) + 6CuO(s) + Y_2O_3(s) \xrightarrow{\Delta} 2YBa_2Cu_3O_{6.5}(s) + 4CO_2(g)$$

$$YBa_2Cu_3O_{6.5}(s) + \tfrac{1}{4}O_2(g) \xrightarrow{\Delta} YBa_2Cu_3O_7(s)$$

Ceramic Structures and Uses Structures of several ceramic materials are shown in Figure 12.46. Note the diamond-like structure of silicon carbide. Network covalent bonding gives this material great strength. Silicon carbide is made into thin fibers, called *whiskers*, to reinforce other ceramics in a composite structure and prevent cracking, much like steel rods reinforcing concrete. Silicon nitride is

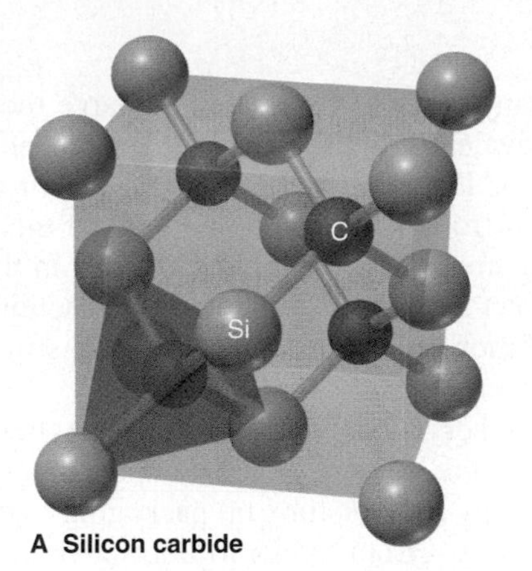

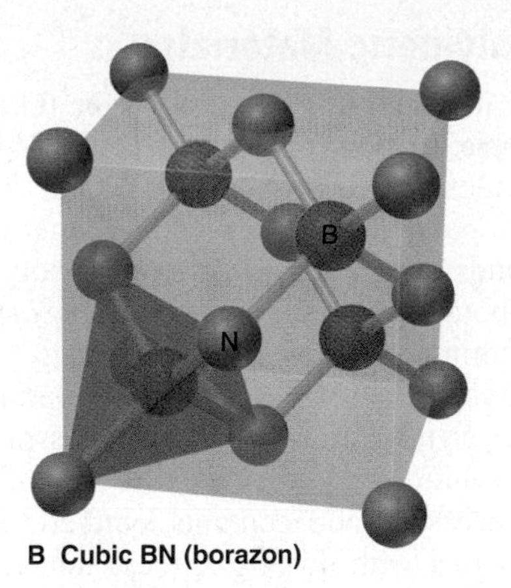

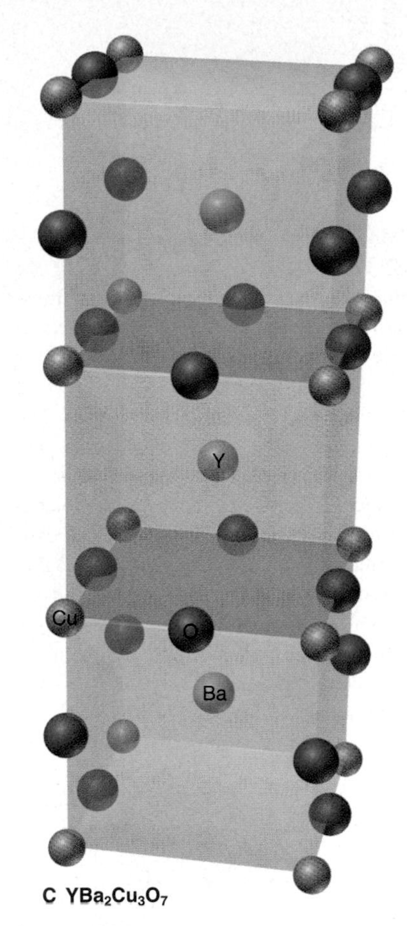

Figure 12.46 Unit cells of some modern ceramic materials. SiC **(A)** and the high-pressure form of BN **(B)** both have a crystal structure similar to that of diamond and are extremely hard. $YBa_2Cu_3O_7$ **(C)** is one of the high-temperature superconducting oxides.

virtually inert chemically, retains its strength and wear resistance for extended periods above 1000°C, is dense and hard, and acts as an electrical insulator. Japanese and American automakers are testing it in high-efficiency car and truck engines because it allows an ideal combination of low weight, high operating temperature, and little need for lubrication.

The BN ceramics exist in two structures (Section 14.5), analogous to the common crystalline forms of carbon. In the graphite-like form, BN has extraordinary properties as an electrical insulator. At high temperature and very high pressure (1800°C and 8.5×10^4 atm), it converts to a diamond-like structure, which is extremely hard and durable. Both forms are virtually invisible to radar.

Earlier, we mentioned some potential uses of the superconducting oxides. In nearly every one of these ceramic materials, copper occurs in an unusual oxidation state. In $YBa_2Cu_3O_7$, for instance, assuming oxidation states of +3 for Y, +2 for Ba, and −2 for O, the three Cu atoms have a total oxidation state of +7. This is allocated as $Cu(II)_2Cu(III)$, with one Cu in the unusual +3 state. X-ray diffraction analysis indicates that a distortion in the structure makes four of the oxide ions unusually close to the Y^{3+} ion, which aligns the Cu ions into chains within the crystal. It is suspected that a specific half-filled 3*d* orbital in Cu oriented toward a neighboring O^{2-} ion may be associated with superconductivity, although the process is still poorly understood. Because of their brittleness, it has been difficult to fashion these ceramics into wires, but methods for making superconducting films and ribbons have recently been developed.

Research in ceramic processing is beginning to overcome the inherent brittleness of this entire class of materials. This brittleness arises from the strength of the ionic-covalent bonding in these solids and their resulting inability to deform. Under stress, a microfine crystal defect widens and lengthens until the material cracks. One new method forms defect-free ceramics using controlled packing and heat-treating of extremely small, uniform oxide particles coated with organic polymers. Another method is aimed at arresting a widening crack. These ceramics are embedded with zirconia (ZrO_2), whose crystal structure expands up to 5% under the mechanical stress of a crack tip: the moment the advancing crack reaches them, the zirconia particles effectively pinch it shut. Very recently, a third method was reported. Japanese researchers prepared a ceramic material made of precisely grown single crystals of Al_2O_3 and $GdAlO_3$, which become entangled during the solidification process. The material bends without cracking at temperatures above 1800 K.

Despite the technical difficulties, chemical ingenuity will continue to develop new ceramic materials and apply their amazing and useful properties well into the 21st century.

Polymeric Materials

In its simplest form, a **polymer** (Greek, "many parts") is an extremely large molecule, or **macromolecule**, consisting of a covalently linked chain of smaller molecules, called **monomers** (Greek, "one part"). The monomer is the *repeat unit* of the polymer, and a typical polymer may have from hundreds to hundreds of thousands of repeat units. *Synthetic* polymers are created by chemical reactions in the laboratory; *natural* polymers (or *biopolymers*) are created by chemical reactions within organisms. There are many types of monomers, and their chemical structures allow for the complete repertoire of intermolecular forces.

One Strand or Many Pieces? By the mid-19th century, entrepreneurs had transformed cheap natural polymers into valuable materials, such as rubber and the cellulose nitrate ("celluloid") film used in the young movie industry. Studies had shown the presence of repeat units, and most believed that polymers were small molecules held together by intermolecular forces. But the young German chemist Hermann Staudinger was convinced that polymers were large molecules held together by covalent bonds, and, despite ridicule from many prominent chemists, his covalent-linkage hypothesis was eventually confirmed. For this work, he was awarded the Nobel Prize in chemistry in 1953.

Synthetic polymers, such as plastics, rubbers, and crosslinked glasses have revolutionized everyday life. Virtually every home, car, electronic component, and processed food contains synthetic polymers in its structure or packaging. You interact with dozens of these materials each day—from paints to floor coverings to clothing to the additives and adhesives in this textbook. Some of these materials, like those used in food containers, are very long-lived in the environment and have created a serious waste-disposal problem. Others are being actively recycled into the same or other useful products, such as garbage bags, outdoor furniture, roofing tiles, and even marine pilings and roadside curbs. Still others, such as artificial skin, heart valve components, and hip joints, are designed to have as long a life as possible.

In this section, we'll examine the physical nature of synthetic polymers and explore the role intermolecular forces play in their properties and uses. In Chapter 15, we'll examine the types of monomers, look at the preparation of synthetic polymers, and then focus on the structures and vital functions of the biopolymers.

Dimensions of a Polymer Chain: Mass, Size, and Shape Because of their great lengths, polymers are unlike smaller molecules in several important ways. Let's see how chemists describe the mass, size, and shape of a polymer chain and how the chains exist in a sample. We'll focus throughout on polyethylene, by far the most common synthetic polymer.

1. *Polymer mass*. The molar mass of a polymer chain ($\mathcal{M}_{polymer}$, in g/mol, often referred to as the *molecular weight*) depends on two parameters—the molar mass of the repeat unit ($\mathcal{M}_{repeat}$) and the **degree of polymerization (n)**, or the number of repeat units in the chain:

$$\mathcal{M}_{polymer} = \mathcal{M}_{repeat} \times n$$

For example, the molar mass of the ethylene repeat unit is 28 g/mol. If an individual polyethylene chain in a plastic grocery bag has a degree of polymerization of 7100, the molar mass of that particular chain is

$$\mathcal{M}_{polymer} = \mathcal{M}_{repeat} \times n = (28 \text{ g/mol})(7.1\times10^3) = 2.0\times10^5 \text{ g/mol}$$

Table 12.8 shows some other examples.

Table 12.8 Molar Masses of Some Common Polymers

Name	$\mathcal{M}_{polymer}$ (g/mol)	n	Uses
Acrylates	2×10^5	2×10^3	Rugs, carpets
Polyamide (nylons)	1.5×10^4	1.2×10^2	Tires, fishing line
Polycarbonate	1×10^5	4×10^2	Compact discs
Polyethylene	3×10^5	1×10^4	Grocery bags
Polyethylene (ultrahigh molecular weight)	5×10^6	2×10^5	Hip joints
Poly(ethylene terephthalate)	2×10^4	1×10^2	Soda bottles
Polystyrene	3×10^5	3×10^3	Packing, coffee cups
Poly(vinyl chloride)	1×10^5	1.5×10^3	Plumbing

However, even though any *given* chain within a sample of a polymer has a fixed molar mass, the degree of polymerization often varies considerably from chain to chain. As a result, *all samples of synthetic polymers have a distribution of chain lengths*. For this reason, polymer chemists use various definitions of *average* molar mass, and a common one is the *number-average molar mass*, $\mathcal{M}_n$:

$$\mathcal{M}_n = \frac{\text{total mass of all chains}}{\text{number of moles of chains}}$$

Thus, even though the number-average molar mass of the polyethylene in grocery bags is, say, 1.6×10^5 g/mol, the chains may vary in molar mass from about 7.0×10^4 to 3.0×10^5 g/mol.

2. *Polymer size and shape*. The long axis of a polymer chain is called its *backbone*. The length of an *extended* backbone is simply the number of repeat units (degree of polymerization, n) times the length of each repeat unit (l_0). For instance, the length of an ethylene repeat unit is about 250 pm, so the extended length of our particular grocery-bag polyethylene chain is

$$\text{Length of extended chain} = n \times l_0 = (7.1\times10^3)\ (2.5\times10^2 \text{ pm}) = 1.8\times10^6 \text{ pm}$$

Comparing this length with the thickness of the chain, which is only about 40 pm, gives a good picture of the threadlike dimensions of the extended chain.

It's very important to understand, however, that a polymer molecule, whether pure or in solution, doesn't exist as an extended chain but, in fact, is far more compact. To picture the actual shape, polymer chemists assume, as a first approximation, that the shape of the chain arises as a result of free rotation around all of its single bonds. Thus, as each repeat unit rotates randomly, the chain continuously changes direction, turning back on itself many times and eventually arriving at the **random coil** shape that most polymers adopt (Figure 12.47). In reality, of course, rotation is not completely free because, as one portion of a chain bends and twists near other portions of the same chain or of other nearby chains, they attract each other. Thus, the nature of intermolecular forces between chain

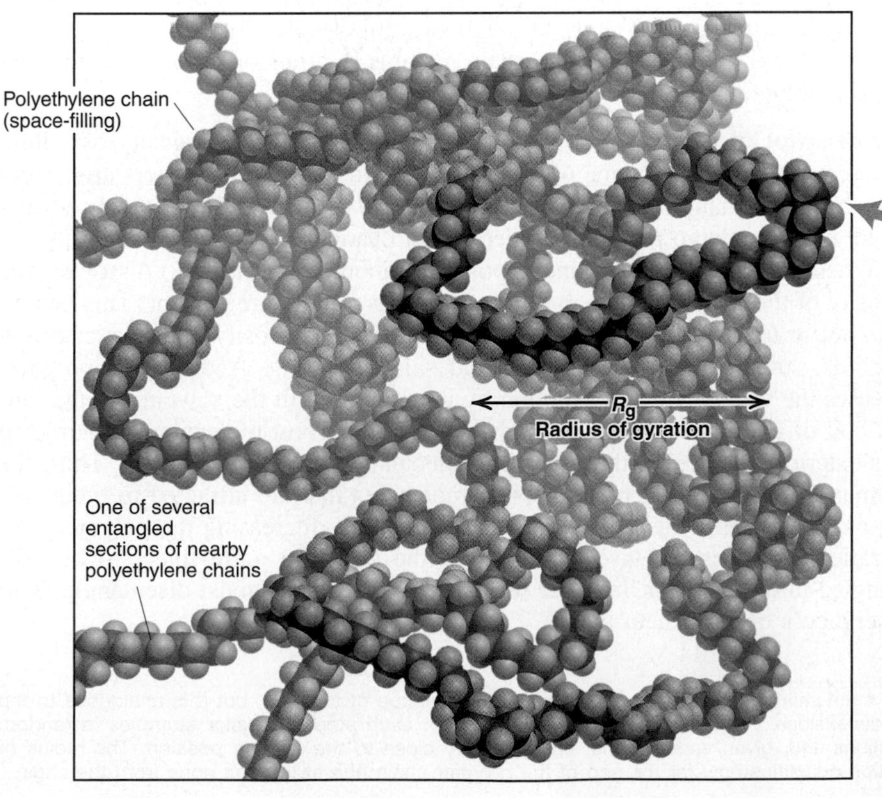

Figure 12.47 The random-coil shape of a polymer chain. Note the random coiling of the chain's carbon atoms *(black)*. Sections of several nearby chains *(red, green,* and *yellow)* are entangled with this chain, kept near one another by dispersion forces. In reality, entangling chains fill any gaps shown here. The radius of gyration (R_g) represents the average distance from the center of mass of the coiled molecule to its outer edge.

portions, between different chains, and/or between chain and solvent becomes a key factor in establishing the actual shape of a polymer chain.

The size of the coiled polymer chain is expressed by its **radius of gyration,** $\boldsymbol{R_g}$, the average distance from the center of mass to the outer edge of the coil (Figure 12.47). Even though R_g is reported as a single value for a given polymer, it represents an average value for many chains. The mathematical expression for the radius of gyration considers the length of each repeat unit and its randomized direction in space, as well as the bond angles between atoms in a unit and between adjacent units:*

$$R_g = \sqrt{\frac{nl_0^2}{6}}$$

As we would expect, the radius of gyration increases with the degree of polymerization, and thus with the molar mass as well. Most importantly, light-scattering experiments and other laboratory measurements correlate with the calculated results, so for many polymers the radius of gyration can be determined experimentally.

For our grocery-bag polyethylene chain, we have

$$R_g = \sqrt{\frac{nl_0^2}{6}} = \sqrt{\frac{(7.1\times10^3)\,(2.5\times10^2\ \text{pm})^2}{6}} = 8.6\times10^3\ \text{pm}$$

Doubling the radius gives a diameter of 1.7×10^4 pm, less than one-hundredth the length of the extended chain!

3. *Polymer crystallinity*. You may get the impression from the discussion so far that a sample of a given polymer is just a disorderly jumble of chains, but this is often not the case. If the molecular structure allows neighboring chains to pack together and if the chemical groups lead to favorable dipole-dipole, H-bonding, or dispersion forces, portions of the chains can align regularly and exhibit crystallinity.

However, the crystallinity of a polymer is very different from the crystal structures of the simple compounds we discussed earlier. There, the orderly array extends over many molecules, and the unit cell includes at least one molecule. In contrast, the orderly regions of a polymer rarely, if ever, involve even one whole molecule (Figure 12.48). At best, a polymer is *semicrystalline,* because only parts of the molecule align with parts of neighboring molecules (or with other parts of the same chain), while most of the chain remains as a random coil. Thus, the unit cell of a polymer includes only a small part of the molecule.

Flow Behavior of Polymers. In Section 12.4, we defined the viscosity of a fluid as its resistance to flow. Some of the most important uses of polymers arise from their ability to change the viscosity of a solvent in which they have been dissolved and to undergo temperature-dependent changes in their own viscosity.

When an appreciable amount of polymer (about 5–15 mass %) dissolves, the viscosity of the solution is much higher than that of the pure solvent. This behavior is put to use by adding polymers to increase the viscosity of many common materials, such as motor oil, paint, and salad dressing. A dissolved polymer increases the viscosity of the solution by interacting with the solvent. As the random coil of a polymer moves through a solution, solvent molecules are attracted to its exterior and interior through intermolecular forces (Figure 12.49). Thus, the polymer coil drags along many solvent molecules that are attracted to other solvent molecules and other coils, and flow is lessened. Increasing the polymer concentration increases the viscosity because the coils are more likely to become entangled in one another. In order for each coil to flow, it must disentangle from its neighbors or drag them along.

*The mathematical derivation of R_g is beyond the scope of this text, but it is analogous to the two-dimensional "walk of the drunken sailor." With each step, the sailor stumbles in random directions and, given enough time, ends up very close to the starting position. The radius of gyration quantifies how far the end of the polymer chain (the sailor) has gone from the origin.

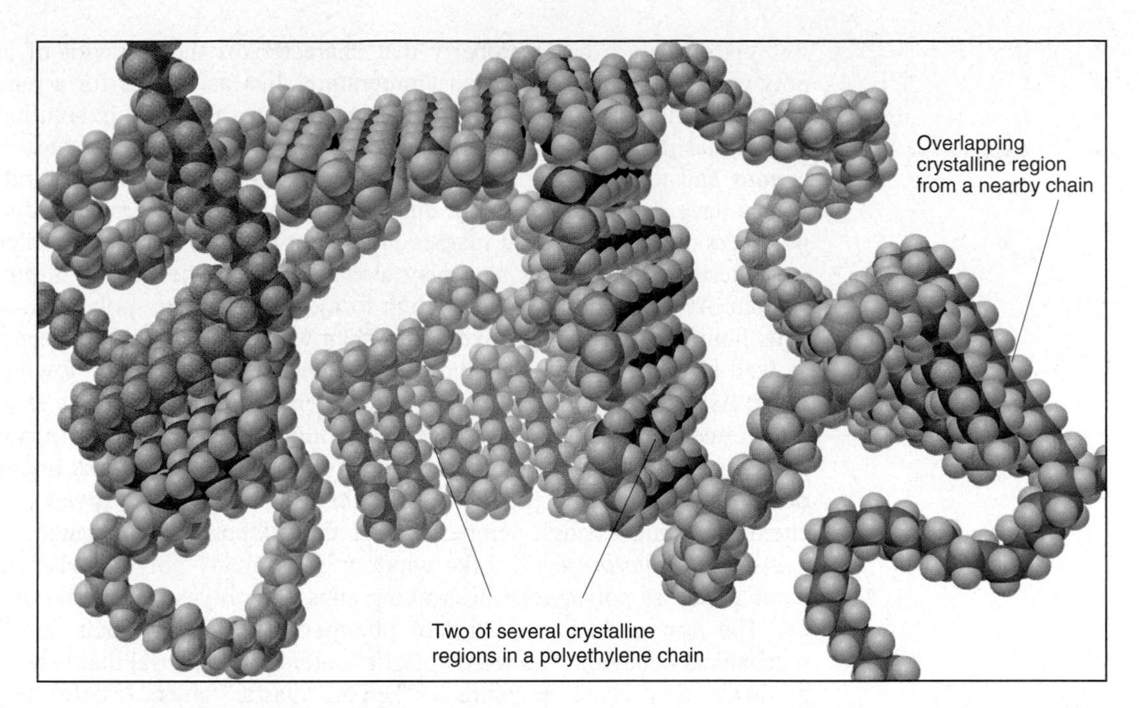

Figure 12.48 The semicrystallinity of a polymer chain. This depiction of polyethylene highlights several ordered regions (*darker color*) with random coils between them. Ordered regions from nearby chains (*red* and *yellow*) are shown overlapping those in the main chain.

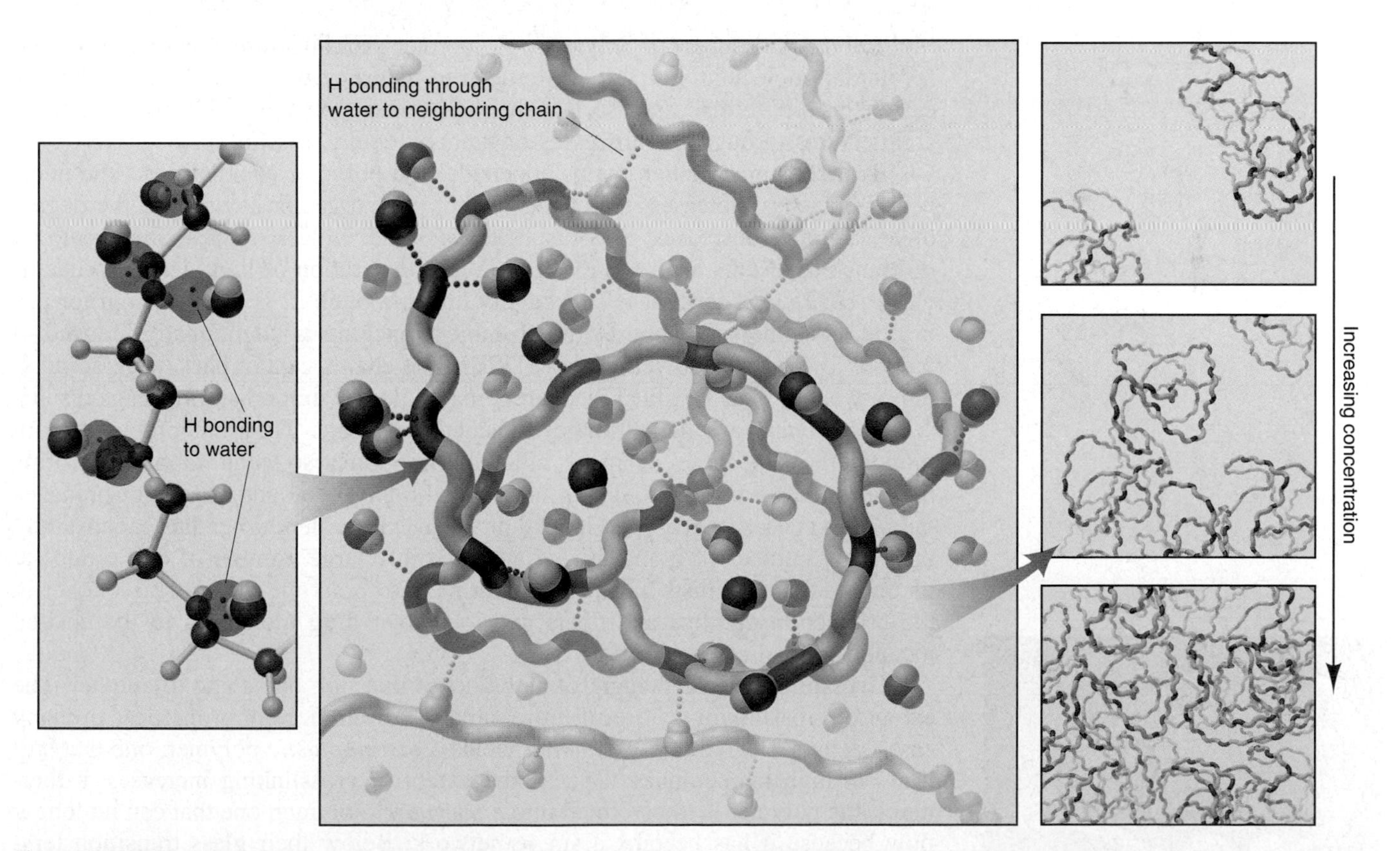

Figure 12.49 The viscosity of a polymer in solution. A ball-and-stick model of a section (*left panel*) of a poly(ethylene oxide) chain in water shows the H bonds that form between the lone pairs of the chain's O atoms and the H atoms of solvent molecules. The polymer chain, depicted as a blue and red coiled rod (*center panel*), forms many H bonds with solvent molecules. Note the H bonds to water molecules that allow one chain to interact with others nearby. As the concentration of polymer increases (*three right panels*), the viscosity of the solution increases because the movement of each chain is restricted through its interactions with solvent and with other chains.

Viscosity is a basic property that characterizes the behavior of a particular polymer-solvent pair at a given temperature. Just as it does for a pure polymer, the size (radius of gyration) of a random coil of a polymer in solution increases with molar mass, and so does the viscosity of the solution. To study basic interactions and to improve polymer manufacture, chemists and other industrial scientists have developed essential quantitative equations to predict the viscosity of polymers of different molar masses in a variety of solvents (see Problem 12.149).

Intermolecular forces also play a major role in the flow of a pure polymer sample. At temperatures high enough to melt them, many polymers exist as viscous liquids, flowing more like honey than water. The forces between the chains, as well as the entangling of chains, hinder the molecules from flowing past each other. As the temperature decreases, the intermolecular attractions exert a greater effect, and the eventual result is a rigid solid. If the chains don't crystallize, the resulting material is called a *polymer glass*. The transition from a liquid to a glass occurs over a narrow (10–20°C) temperature range for a given polymer, but chemists define a single temperature at the midpoint of the range as the *glass transition temperature,* T_g. Like window glass, many polymer glasses are transparent, such as polystyrene in drinking cups and polycarbonate in compact discs.

The flow-related properties of polymers give rise to their familiar **plastic** mechanical behavior. The word "plastic" refers to a material that, when deformed, retains its new shape; in contrast, when an "elastic" object is deformed, it returns to its original shape. Many polymers can be deformed (stretched, bent, twisted) when warm and retain their deformed shape when cooled. In this way, they are made into countless everyday objects—milk bottles, car parts, and so forth.

Molecular Architecture of Polymers A polymer's architecture—its overall spatial layout and molecular structure—is crucial to its properties. In addition to the linear chains we've discussed so far, chemists create polymers with more complex architectures through the processes of branching and crosslinking.

Branches are smaller chains appended to a polymer backbone. As the number of branches increases, the chains cannot pack together as well, so the degree of crystallinity decreases; as a result, the polymer is less rigid. A small amount of branching occurs as a side reaction in the preparation of high-density polyethylene (HDPE). Because it is still largely linear, though, it is rigid enough for use in milk containers. In contrast, much more branching is intentionally induced to prepare low-density polyethylene (LDPE). The chains cannot pack well, so crystallinity is low. The flexible, transparent material used in food storage bags results.

Dendrimers are the ultimate branched polymers. They are prepared from monomers with three or more attachment points, so each monomer forms branches. In essence, then, dendrimers have no backbone and consist of branches only. As you can see in the Gallery on page 386, a dendrimer has a constantly increasing number of branches and an incredibly large number of end groups at its outer edge. Chemists have used dendrimers to bind one polymer to another in the production of films and fibers and to deliver drug molecules to the desired location in medical applications.

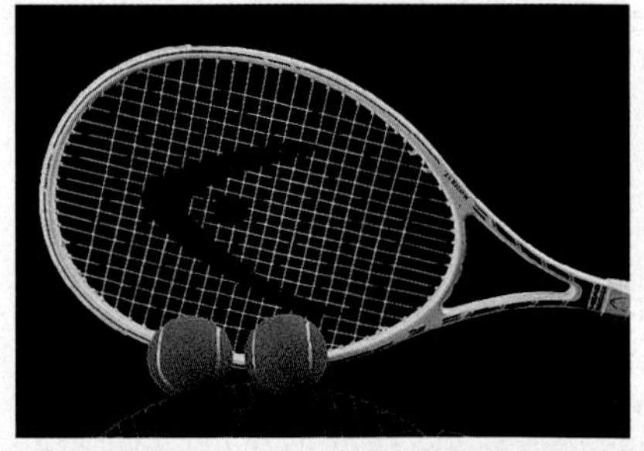

Tennis racket containing a thermoset polymer.

Crosslinks can be thought of as branches that link one chain to another. The extent of crosslinking can result in remarkable differences in properties. In many cases, a small degree of crosslinking yields a *thermoplastic* polymer, one that still flows at high temperatures. But, as the extent of crosslinking increases, a thermoplastic polymer is transformed into a *thermoset* polymer, one that can no longer flow because it has become a single network. Below their glass transition temperatures, some thermosets are extremely rigid and strong, making them ideal as matrix materials in high-strength composites (see photo).

Above their glass transition temperatures, many thermosets become **elastomers**, polymers that can be stretched and immediately spring back to their ini-

Table 12.9 Some Common Elastomers

Name	T_g (°C)	Uses
Poly(dimethyl siloxane)	−123	Breast implants
Polybutadiene	−106	Rubber bands
Polyisoprene	−65	Surgical gloves
Polychloroprene (neoprene)	−43	Footwear, medical tubing

tial shapes when released, like the net under a trapeze artist or a common rubber band. When you stretch a rubber band, individual polymer chains flow for only a short distance before the connectivity of the network returns them to their original positions. Table 12.9 lists some elastomers.

Differences in monomer sequences often influence polymer properties as well. A *homopolymer* consists of one type of monomer (A—A—A—A—A—. . .), whereas a **copolymer** consists of two or more types. The simplest copolymer is called an *AB block copolymer* because a chain (block) of monomer A and a chain of monomer B are linked at one point:

$$\ldots\text{—A—A—A—A—A—B—B—B—B—B—}\ldots$$

If the intermolecular forces between the A and B portions of the chain are weaker than those between different regions within each portion, the A and B portions form their own random coils. This ability makes AB block copolymers ideal adhesives for joining two polymer surfaces covalently. An ABA block copolymer has A chains linked at each end of a B chain:

$$\ldots\text{—A—A—A—B}\text{—}(\text{B})_n\text{—B—A—A—A—}\ldots$$

Some of these block copolymers act as *thermoplastic elastomers*, materials shaped at high temperature that become elastomers at room temperature; not surprisingly, some of these materials have revolutionized the footwear industry.

Polymer chemists tailor polymers to control properties such as viscosity, strength, toughness, and flexibility. The effect of intermolecular forces on their physical properties is only part of the story of these remarkably useful materials. Silicone polymers are described in the Gallery in Chapter 14 (p. 581), and organic reactions that form polymer chains from monomers are examined in Chapter 15.

Nanotechnology: Designing Materials Atom by Atom

At the frontier of interdisciplinary science and expanding at an incredible pace, the exciting new field of *nanotechnology* is joining researchers from physics, materials science, chemistry, biology, environmental science, medicine, and many branches of engineering. International conferences, scientific journals, and an ever-growing list of university and industrial web sites herald the enormous potential impact of nanotechnology on society.

Nanotechnology is the science and engineering of nanoscale systems—those whose sizes range from 1 to 50 nm. Until recently, physical scientists had focused primarily on atoms, which are smaller (around 1×10^{-1} nm), or on crystals, which are larger (around 1×10^{5} nm and up). Working between these size extremes, nanotechnologists examine the chemical and physical properties of nanostructures, manipulating atoms one at a time to synthesize particles, clusters, and layers with properties very different from either individual molecules or their bulk phases. The scanning tunneling microscope (see Tools of the Laboratory, p. 456) and the similar scanning probe and atomic force microscopes are among the precise tools required to place individual atoms, molecules, and clusters into the positions required to build a structure or cause a desired reaction.

The key to this futuristic technology lies in two features of nanoscale construction that occur routinely in nature. The first is *self-assembly*, the ability of smaller, simpler parts to organize themselves into a larger, more complex whole. On the molecular scale, self-assembly refers to atoms or small molecules aggregating through intermolecular forces, especially dipole-dipole, H-bonding, and dispersion forces. Oppositely charged regions on two such particles make contact to form a larger particle, which in turn forms a still larger one. The other feature is *controlled orientation*, the positioning of two molecules near each other long enough for intermolecular forces to take effect. Some industrial catalysts and all biological catalysts (enzymes) function through controlled orientation. (Catalysts, discussed in Chapter 16, are substances that speed reactions.) Common strategies for nanoscale construction include approaches that mimic biological self-assembly, sophisticated precipitation methods, and a variety of physical and chemical aerosol techniques for making nanoclusters and then manipulating them into more consolidated structures.

Because the field is evolving so rapidly, it is possible to provide only a very general overview of some current research directions in nanotechnology. The National Nanotechnology Initiative describes worldwide efforts in many key areas—four of which are dispersions and coatings, high-surface-area materials, functional devices, and consolidated materials.

Dispersions and Coatings Nanostructuring is being used in a wide range of optical, thermal, and electrical applications of dispersions and coatings, including products related to printing, photography, and pharmaceuticals. Some examples are thermal and optical barriers, sunscreens, image enhancers, ink-jet materials, coated abrasive slurries, and information-recording layers. Highly ordered, iron/platinum nanoparticles of extremely uniform size and exceptional magnetic properties may be adapted as coatings for high-density information storage, holding a million times as much data per unit of surface area as current materials. Another exciting application involves highly ordered, one-molecule-thick films—in effect, two-dimensional crystals—that can be coated on a variety of surfaces. One approach is to layer light-sensitive molecules that change reversibly from one form to another. A finely focused laser could trigger the change in the molecular form to create specific patterns, thus storing information at the density of one bit of data per molecule!

High-Surface-Area Materials These applications take advantage of the incredibly large surface areas of nanoscale building blocks. For example, a particle 5 nm in diameter has about half of its atoms on its surface. When such nanoparticles are assembled, the resulting surface area is enormous. Current projects include porous membranes for water purification and batteries, drug-delivery systems, and multilayer films that incorporate photosynthetic molecules for high-efficiency solar cells. There is great potential for molecule-specific sensors. Certain biosensors allow detection of as little as 10^{-14} mol of DNA using color changes in gold nanoparticles. In another development, 50-nm gold particles sense slight differences in two portions of DNA; such sensitivity may allow detection of subtle genetic mutations. An eventual goal is to develop biosensors that could circulate freely in the bloodstream, measure levels of specific disease-related molecules, and deliver drugs to individual cells, even individual genes.

Functional Devices The need for ever smaller machines is the driving force in this area of research, and the nanoscale computer is the dream. The ongoing race for faster, smaller traditional computers will soon hit a fundamental "brick wall." Since the 1960s, computing speeds have doubled about every two years as the light-etched rules between transistors and diodes on the silicon chip were made closer and closer. Currently, UV photolithography makes rules less than 180 nm apart. As this distance approaches 50 nm, however, doping inconsistencies, heat generation, and other inherent limitations arise, and the silicon-based chip reaches its lower size limit. Imagine the potential impact of ongoing work to develop supercomputers consisting of molecule-sized diodes and transistors bound to an organic surface—the ultimate reduction in size and, thus, increase in speed.

To realize this dream, one of the major research efforts focuses on developing the single-electron transistor (SET). SETs will be made into arrays using methods similar to biological self-assembly. Such nanoscale devices must have nanoscale connections, and a related research area involves the fabrication of nanowires. In one method, a metal is electroplated from solution [such as cobalt from a solution of $Co(NO_3)_2$] to fill uniform nanopores created by controlled oxidation of aluminum surfaces. But the greatest research activity is focused on carbon nanotubes (see the Gallery, p. 386). These can be single or nested tubes with insulating, semiconducting, or metallic properties. Produced by high-temperature (800°–1000°C) reduction of a hydrocarbon or by electron-beam irradiation of fullerenes, nanotubes are then physically separated according to size and properties, using an atomic force or scanning tunneling microscope. Adding specific chemical groups to the nanotubes expands their applications further (Figure 12.50).

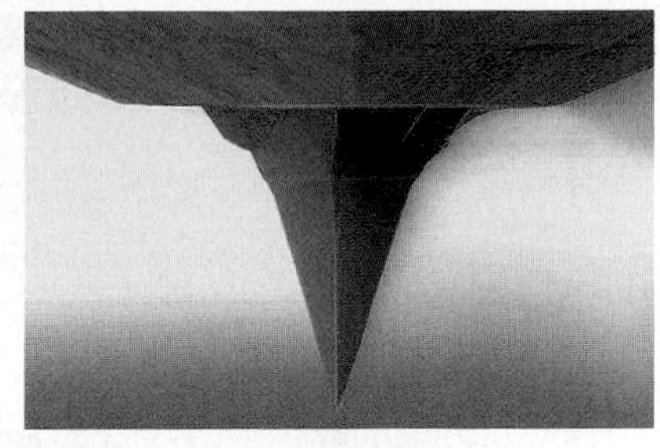

A

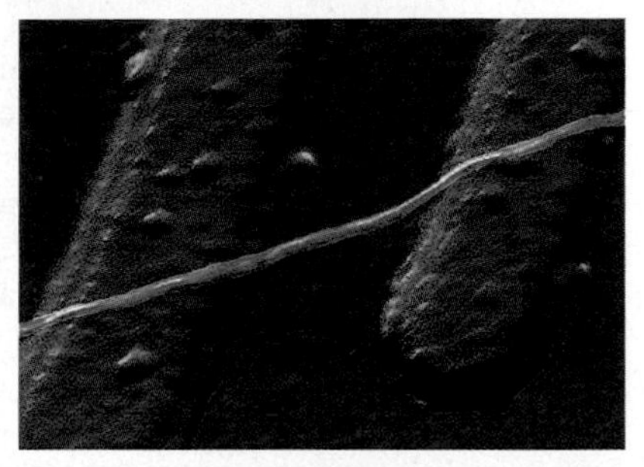

B

Figure 12.50 Manipulating atoms. **A,** The tip of an atomic force microscope, one of the key tools used to build nanodevices. **B,** This atomic force micrograph shows a carbon nanotube "wire" (*blue*) on platinum electrodes (*yellow*). The nanotube is 1.5 nm (10 atoms) across and was made by using a laser to join fullerene molecules into a tube. (Magnification ×120,000.)

Consolidated Materials It is known that mechanical, magnetic, and optical properties change dramatically when bulk materials are consolidated from nanoscale building blocks. Nanostructuring will greatly increase the hardness and strength of metals and the ductility and plasticity of ceramics. Nanoparticle fillers can yield nanocomposites with unique properties—soft magnets, tough cutting tools, ultrastrong ductile cements, magnetic refrigerants, and a wide range of nanoparticle-filled elastomers, thermoplastics, and thermosets.

SECTION SUMMARY

Doping increases the conductivity of semiconductors and is essential to modern electronic materials. Doping silicon with Group 5A atoms introduces negative sites (n-type) by adding valence electrons to the conduction band, whereas doping with Group 3A atoms adds positive holes (p-type) by emptying some orbitals in the valence band. Placing these two types of doped Si next to one another forms a p-n junction. Sandwiching a p-type portion between two n-type portions forms a transistor. Liquid crystal phases flow like liquids but have molecules ordered like crystalline solids. Typically, the molecules have rodlike shapes, and their intermolecular forces keep them aligned. Thermotropic phases are prepared by heating the solid; lyotropic phases form when the solvent concentration is varied. The nematic, cholesteric, and smectic types of liquid crystals differ in their molecular order. Liquid crystal applications depend on controlling the orientation of the molecules. Ceramics are very resistant to heat and chemicals. Most are network covalent solids formed at high temperature from simple reactants. They add lightweight strength to other materials. Polymers are extremely large molecules that adopt the shape of a random coil as a result of intermolecular forces. A polymer sample has an average molar mass because it consists of a range of chain lengths. The high viscosity of a polymer arises from attractions between chains or, in the case of a dissolved polymer, between the chain and the solvent molecules. By varying the degrees of branching, crosslinking, and ordering (crystallinity), chemists tailor polymers with remarkable properties. Nanoscale materials can be made one atom or molecule at a time through construction processes involving self-assembly and controlled orientation of molecules.

Chapter Perspective

Our central focus—macroscopic behavior resulting from molecular behavior—became still clearer in this chapter, as we built on earlier ideas of bonding, molecular shape, and polarity to understand the intermolecular forces that create the properties of liquids and solids. In Chapter 13, we'll find many parallels to the key ideas in this chapter: the same intermolecular forces create solutions, temperature influences solubility just as it does vapor pressure, and the equilibrium state also underlies the properties of solutions.

For Review and Reference (Numbers in parentheses refer to pages, unless noted otherwise.)

Learning Objectives

Relevant section and/or sample problem (SP) numbers appear in parentheses.

Understand These Concepts

1. How the interplay between kinetic and potential energy underlies the properties of the three states of matter and their phase changes (12.1)
2. The processes involved, both within a phase and during a phase change, when heat is added or removed from a pure substance (12.2)
3. The meaning of vapor pressure and how phase changes are dynamic equilibrium processes (12.2)
4. How temperature and intermolecular forces influence vapor pressure (12.2)
5. The relation between vapor pressure and boiling point (12.2)
6. How a phase diagram shows the phases of a substance at differing conditions of pressure and temperature (12.2)
7. The distinction between bonding and intermolecular forces on the basis of Coulomb's law and the meaning of the van der Waals radius of an atom (12.3)
8. The types and relative strengths of intermolecular forces acting in a substance (dipole-dipole, H-bonding, dispersion), the impact of H bonding on physical properties, and the meaning of polarizability (12.3)
9. The meanings of surface tension, capillarity, and viscosity and how intermolecular forces influence their magnitudes (12.4)
10. How the important macroscopic properties of water arise from atomic and molecular properties (12.5)
11. The meaning of *crystal lattice* and the characteristics of the three types of cubic unit cells (12.6)
12. How packing of spheres gives rise to the hexagonal and cubic unit cells (12.6)
13. Types of crystalline solids and how their intermolecular forces give rise to their properties (12.6)
14. How band theory accounts for the properties of metals and the relative conductivities of metals, nonmetals, and metalloids (12.6)
15. The structures, properties, and functions of modern materials (doped semiconductors, liquid crystals, ceramics, polymers, and nanostructures) on the atomic scale (12.7)

Master These Skills

1. Calculating the overall enthalpy change when heat is added to or removed from a pure substance (12.2)
2. Using the Clausius-Clapeyron equation to examine the relationship between vapor pressure and temperature (SP 12.1)
3. Using a phase diagram to predict the physical state and/or phase change of a substance (12.2)
4. Determining whether a substance can form H bonds and drawing the H-bonded structures (SP 12.2)
5. Predicting the types and relative strength of the intermolecular forces acting within a substance from its structure (SP 12.3)
6. Finding the number of particles in a unit cell (12.6)
7. Calculating atomic radius from the density and crystal structure of an element (SP 12.4)

Key Terms

phase (425)
intermolecular forces (425)
phase change (425)

Section 12.1

condensation (426)
vaporization (426)
freezing (426)
melting (fusion) (426)
heat of vaporization (ΔH^0_{vap}) (427)
heat of fusion (ΔH^0_{fus}) (427)
sublimation (427)
deposition (427)
heat of sublimation (ΔH^0_{subl}) (427)

Section 12.2

heating-cooling curve (428)
vapor pressure (431)
Clausius-Clapeyron equation (432)
boiling point (433)
melting point (434)
phase diagram (434)
critical point (434)
triple point (435)

Section 12.3

van der Waals radius (437)
ion-dipole force (437)
dipole-dipole force (438)
hydrogen bond (H bond) (439)
polarizability (440)
dispersion (London) force (441)

Section 12.4

surface tension (443)
capillarity (444)
viscosity (445)

Section 12.6

crystalline solid (449)
amorphous solid (449)
lattice (450)
unit cell (450)
coordination number (450)
simple cubic unit cell (450)
body-centered cubic unit cell (450)
face-centered cubic unit cell (450)
packing efficiency (452)
hexagonal closest packing (452)
cubic closest packing (452)
x-ray diffraction analysis (455)
scanning tunneling microscopy (455)
atomic solid (456)
molecular solid (456)
ionic solid (457)
metallic solid (459)
network covalent solid (459)
band theory (460)
valence band (462)
conduction band (462)
conductor (462)
semiconductor (463)
insulator (463)
superconductivity (463)

Section 12.7

crystal defect (464)
doping (464)
liquid crystal (466)
ceramic (469)
polymer (472)
macromolecule (472)
monomer (472)
degree of polymerization (n) (472)
random coil (473)
radius of gyration (R_g) (474)
plastic (476)
branch (476)
crosslink (476)
elastomer (476)
copolymer (477)
nanotechnology (477)

Key Equations and Relationships

12.1 Using the vapor pressure at one temperature to find the vapor pressure at another temperature (two-point form of the Clausius-Clapeyron equation) (433):

$$\ln \frac{P_2}{P_1} = \frac{-\Delta H_{\text{vap}}}{R}\left(\frac{1}{T_2} - \frac{1}{T_1}\right)$$

Highlighted Figures and Tables

These figures (F) and tables (T) provide a review of key ideas.

Brief Solutions to Follow-up Problems

12.1 $\ln \frac{P_2}{P_1} = \left(\frac{-40.7\times10^3 \text{ J/mol}}{8.314 \text{ J/mol·K}}\right)$

$$\times\left(\frac{1}{273.15 + 85.5 \text{ K}} - \frac{1}{273.15 + 34.1 \text{ K}}\right)$$

$$= (-4.90\times10^3 \text{ K})(-4.66\times10^{-4} \text{ K}^{-1}) = 2.28$$

$\frac{P_2}{P_1} = 9.8$; thus, $P_2 = 40.1 \text{ torr} \times 9.8 = 3.9\times10^2$ torr

12.2 (a)

```
   H        :O:····H—O:        H
   |       //          \       |
H—C—C                   C—C—H
   |       \          //       |
   H        :O—H····:O:        H
```

(b)

```
   H  H
   |  |
H—C—C—O:
   |  |  |
   H  H  H
         :
   H  H  :
   |  |  :
H—C—C—O—H
   |  |
   H  H
```

(c) No H bonding

12.3 (a) Dipole-dipole, dispersion; CH_3Br
(b) H bonds, dipole-dipole, dispersion; $CH_3CH_2CH_2OH$
(c) Dispersion; C_3H_8

12.4 Avogadro's no. $= \frac{1 \text{ cm}^3}{7.874 \text{ g Fe}} \times \frac{55.85 \text{ g Fe}}{1 \text{ mol Fe}} \times 0.68$

$$\times \frac{1 \text{ Fe atom}}{8.38\times10^{-24} \text{ cm}^3}$$

$$= 5.8\times10^{23} \text{ Fe atoms/mol Fe}$$

Problems

Problems with **colored** numbers are answered in Appendix E. Sections match the text and provide the numbers of relevant sample problems. Most offer Concept Review Questions, Skill-Building Exercises (grouped in pairs covering the same concept), and Problems in Context. Comprehensive Problems are based on material from any section or previous chapter.

An Overview of Physical States and Phase Changes

Concept Review Questions

12.1 How does the energy of attraction between particles compare with their energy of motion in a gas and in a solid? As part of your answer, identify two macroscopic properties that differ between a gas and a solid.

12.2 What types of forces, intramolecular or intermolecular,
(a) prevent ice cubes from adopting the shape of their container?
(b) are overcome when ice melts?
(c) are overcome when liquid water is vaporized?
(d) are overcome when gaseous water is converted to hydrogen gas and oxygen gas?

12.3 (a) Why are gases more easily compressed than liquids?
(b) Why do liquids have a greater ability to flow than solids?

12.4 (a) Why is the heat of fusion (ΔH_{fus}) of a substance smaller than its heat of vaporization (ΔH_{vap})?
(b) Why is the heat of sublimation (ΔH_{subl}) of a substance greater than its ΔH_{vap}?
(c) At a given temperature and pressure, how does the magnitude of the heat of vaporization of a substance compare with that of its heat of condensation?

Skill-Building Exercises *(grouped in similar pairs)*

12.5 Which forces are intramolecular and which intermolecular?
(a) Those preventing oil from evaporating at room temperature
(b) Those preventing butter from melting in a refrigerator
(c) Those allowing silver to tarnish
(d) Those preventing O_2 in air from forming O atoms

12.6 Which forces are intramolecular and which intermolecular?
(a) Those allowing fog to form on a cool, humid evening
(b) Those allowing water to form when H_2 is sparked
(c) Those allowing liquid benzene to crystallize when cooled
(d) Those responsible for the low boiling point of hexane

12.7 Name the phase change in each of these events:
(a) Dew appears on a lawn in the morning.
(b) Icicles change into liquid water.
(c) Wet clothes dry on a summer day.

12.8 Name the phase change in each of these events:
(a) A diamond film forms on a surface from gaseous carbon atoms in a vacuum.
(b) Mothballs in a bureau drawer disappear over time.
(c) Molten iron from a blast furnace is cast into ingots ("pigs").

Problems in Context

12.9 Liquid propane, a widely used fuel, is produced by compressing gaseous propane at 20°C. During the process, approximately 15 kJ of energy is released for each mole of gas liquefied. Where does this energy come from?

12.10 Many heat-sensitive and oxygen-sensitive solids, such as camphor, are purified by warming under vacuum. The solid vaporizes directly, and the vapor crystallizes on a cool surface. What phase changes are involved in this method?

Quantitative Aspects of Phase Changes

(Sample Problem 12.1)

Concept Review Questions

12.11 Describe the changes (if any) in potential energy and in kinetic energy among the molecules when gaseous PCl_3 condenses to a liquid at a fixed temperature.

12.12 When benzene is at its melting point, two processes occur simultaneously and balance each other. Describe these processes on the macroscopic and molecular levels.

12.13 Liquid hexane (bp = 69°C) is placed in a closed container at room temperature. At first, the pressure of the vapor phase increases, but after a short time, it stops changing. Why?

12.14 Explain the effect of strong intermolecular forces on each of these parameters: (a) critical temperature; (b) boiling point; (c) vapor pressure; (d) heat of vaporization.

12.15 At 1.1 atm, will water boil at 100.°C? Explain.

12.16 A liquid is in equilibrium with its vapor in a closed vessel at a fixed temperature. The vessel is connected by a stopcock to an evacuated vessel. When the stopcock is opened, will the final pressure of the vapor be different from the original value if (a) some liquid remains; (b) all the liquid is first removed? Explain.

12.17 The phase diagram for substance A has a solid-liquid line with a positive slope, and that for substance B has a solid-liquid line with a negative slope. What macroscopic property can distinguish A from B?

12.18 Why does water vapor at 100°C cause more severe burns than liquid water at 100°C?

Skill-Building Exercises *(grouped in similar pairs)*

12.19 From the data below, calculate the total heat (in J) needed to convert 12.00 g of ice at −5.00°C to liquid water at 0.500°C:
mp at 1 atm: 0.0°C ΔH^0_{fus}: 6.02 kJ/mol
c_{liquid}: 4.21 J/g·°C c_{solid}: 2.09 J/g·°C

12.20 From the data below, calculate the total heat (in J) needed to convert 0.333 mol of gaseous ethanol at 300°C and 1 atm to liquid ethanol at 25.0°C and 1 atm:
bp at 1 atm: 78.5°C ΔH^0_{vap}: 40.5 kJ/mol
c_{gas}: 1.43 J/g·°C c_{liquid}: 2.45 J/g·°C

12.21 A liquid has a ΔH^0_{vap} of 35.5 kJ/mol and a boiling point of 122°C at 1.00 atm. What is its vapor pressure at 109°C?

12.22 Diethyl ether has a ΔH^0_{vap} of 29.1 kJ/mol and a vapor pressure of 0.703 atm at 25.0°C. What is its vapor pressure at 95.0°C?

12.23 What is the ΔH^0_{vap} of a liquid that has a vapor pressure of 641 torr at 85.2°C and a boiling point of 95.6°C at 1 atm?

12.24 Methane (CH_4) has a boiling point of −164°C at 1 atm and a vapor pressure of 42.8 atm at −100°C. What is the heat of vaporization of CH_4?

12.25 Use these data to draw a qualitative phase diagram for ethylene (C_2H_4). Is $C_2H_4(s)$ more or less dense than $C_2H_4(l)$?

bp at 1 atm:	−103.7°C
mp at 1 atm:	−169.16°C
Critical point:	9.9°C and 50.5 atm
Triple point:	−169.17°C and 1.20×10^{-3} atm

12.26 Use these data to draw a qualitative phase diagram for H_2. Does H_2 sublime at 0.05 atm? Explain.

mp at 1 atm:	13.96 K
bp at 1 atm:	20.39 K
Triple point:	13.95 K and 0.07 atm
Critical point:	33.2 K and 13.0 atm
Vapor pressure of solid at 10 K:	0.001 atm

Problems in Context

12.27 Sulfur dioxide is produced in enormous amounts for sulfuric acid production. It melts at −73°C and boils at −10.°C. Its ΔH^0_{fus} is 8.619 kJ/mol and its ΔH^0_{vap} is 25.73 kJ/mol. The specific heat capacities of the liquid and gas are 0.995 J/g·K and 0.622 J/g·K, respectively. How much heat is required to convert 2.500 kg of solid SO_2 at the melting point to a gas at 60.°C?

12.28 Butane is a common fuel used in cigarette lighters and camping stoves. Normally supplied in metal containers under pressure, the fuel exists as a mixture of liquid and gas, so high temperatures may cause the container to explode. At 25.0°C, the vapor pressure of butane is 2.3 atm. What is the pressure in the container at 150.°C (ΔH_{vap} = 24.3 kJ/mol)?

12.29 Use Figure 12.9A, p. 435, to answer the following:
(a) Carbon dioxide is sold in steel cylinders under pressures of approximately 20 atm. Is there liquid CO_2 in the cylinder at room temperature (~20°C)? At 40°C? At −40°C? At −120°C?
(b) Carbon dioxide is also sold as solid chunks, called *dry ice,* in insulated containers. If the chunks are warmed by leaving them in an open container at room temperature, will they melt?
(c) If a container is nearly filled with dry ice and then sealed and warmed to room temperature, will the dry ice melt?
(d) If dry ice is compressed at a temperature below its triple point, will it melt?
(e) Will liquid CO_2 placed in a beaker at room temperature boil?

Types of Intermolecular Forces

(Sample Problems 12.2 and 12.3)

Concept Review Questions

12.30 Why are covalent bonds typically much stronger than intermolecular forces?

12.31 Even though molecules are neutral, the dipole-dipole force is one of the most important interparticle forces that exists among them. Explain.

12.32 Oxygen and selenium are members of Group 6A(16). Water forms H bonds, but H_2Se does not. Explain.

12.33 In solid I_2, is the distance between the two I nuclei of one I_2 molecule longer or shorter than the distance between two I nuclei of adjacent I_2 molecules? Explain.

12.34 Polar molecules exhibit dipole-dipole forces. Do they also exhibit dispersion forces? Explain.

12.35 Distinguish between *polarizability* and *polarity*. How does each influence intermolecular forces?

12.36 How can one nonpolar molecule induce a dipole in a nearby nonpolar molecule?

Skill-Building Exercises *(grouped in similar pairs)*

12.37 What is the strongest interparticle force in each substance?
(a) CH_3OH (b) CCl_4 (c) Cl_2

12.38 What is the strongest interparticle force in each substance?
(a) H_3PO_4 (b) SO_2 (c) $MgCl_2$

12.39 What is the strongest interparticle force in each substance?
(a) CH_3Br (b) CH_3CH_3 (c) NH_3

12.40 What is the strongest interparticle force in each substance?
(a) Kr (b) BrF (c) H_2SO_4

12.41 Which member of each pair of compounds forms intermolecular H bonds? Draw the H-bonded structures in each case:
(a) CH_3CHCH_3 (with OH on the central C) or CH_3SCH_3 (b) HF or HBr

12.42 Which member of each pair of compounds forms intermolecular H bonds? Draw the H-bonded structures in each case:
(a) $(CH_3)_2NH$ or $(CH_3)_3N$ (b) $HOCH_2CH_2OH$ or FCH_2CH_2F

12.43 Which forces oppose vaporization of each substance?
(a) hexane (b) water (c) $SiCl_4$

12.44 Which forces oppose vaporization of each substance?
(a) Br_2 (b) SbH_3 (c) CH_3NH_2

12.45 Which has the greater polarizability? Explain.
(a) Br^- or I^- (b) $CH_2{=}CH_2$ or $CH_3{-}CH_3$ (c) H_2O or H_2Se

12.46 Which has the greater polarizability? Explain.
(a) Ca^{2+} or Ca (b) CH_3CH_3 or $CH_3CH_2CH_3$ (c) CCl_4 or CF_4

12.47 Which member in each pair of liquids has the *higher* vapor pressure at a given temperature? Explain.
(a) C_2H_6 or C_4H_{10} (b) CH_3CH_2OH or CH_3CH_2F
(c) NH_3 or PH_3

12.48 Which member in each pair of liquids has the *lower* vapor pressure at a given temperature? Explain.
(a) $HOCH_2CH_2OH$ or $CH_3CH_2CH_2OH$
(b) CH_3COOH or $(CH_3)_2C{=}O$ (c) HF or HCl

12.49 Which substance has the *higher* boiling point? Explain.
(a) LiCl or HCl (b) NH_3 or PH_3 (c) Xe or I_2

12.50 Which substance has the *higher* boiling point? Explain.
(a) CH_3CH_2OH or $CH_3CH_2CH_3$ (b) NO or N_2 (c) H_2S or H_2Te

12.51 Which substance has the *lower* boiling point? Explain.
(a) $CH_3CH_2CH_2CH_3$ or cyclic $CH_2{-}CH_2$ / $CH_2{-}CH_2$ (four-membered ring) (b) NaBr or PBr_3
(c) H_2O or HBr

12.52 Which substance has the *lower* boiling point? Explain.
(a) CH_3OH or CH_3CH_3 (b) FNO or ClNO
(c) F(H)C=C(F)(H) [F and F on the same side] or H(F)C=C(F)(H) [F and F on opposite sides]

Problems in Context

12.53 For pairs of molecules in the gas phase, average H-bond dissociation energies are 17 kJ/mol for NH_3, 22 kJ/mol for H_2O, and 29 kJ/mol for HF. Explain this increase in H-bond strength.

12.54 Dispersion forces are the only intermolecular forces present in motor oil, yet it has a high boiling point. Explain.

12.55 Why does the antifreeze ingredient ethylene glycol ($HOCH_2CH_2OH$; $\mathcal{M}$ = 62.07 g/mol) have a boiling point of 197.6°C, whereas propanol ($CH_3CH_2CH_2OH$; $\mathcal{M}$ = 60.09 g/mol), a compound with a similar molar mass, has a boiling point of only 97.4°C?

Properties of the Liquid State

Concept Review Questions

12.56 Before the phenomenon of surface tension was understood, physicists described the surface of water as being covered with a "skin." What causes this skinlike phenomenon?

12.57 Small, equal-sized drops of oil, water, and mercury lie on a waxed floor. How does each liquid behave? Explain.

12.58 Why does an aqueous solution of ethanol (CH_3CH_2OH) have a lower surface tension than water?

12.59 Why are units of energy per area (J/m^2) used for surface tension values?

12.60 Does the *strength* of the intermolecular forces in a liquid change as the liquid is heated? Explain. Why does liquid viscosity decrease with rising temperature?

Skill-Building Exercises *(grouped in similar pairs)*

12.61 Rank the following in order of *increasing* surface tension at a given temperature, and explain your ranking:
(a) $CH_3CH_2CH_2OH$ (b) $HOCH_2CH(OH)CH_2OH$
(c) $HOCH_2CH_2OH$

12.62 Rank the following in order of *decreasing* surface tension at a given temperature, and explain your ranking:
(a) CH_3OH (b) CH_3CH_3 (c) $H_2C{=}O$

12.63 Rank the compounds in Problem 12.61 in order of *increasing* viscosity at a given temperature; explain your ranking.

12.64 Rank the compounds in Problem 12.62 in order of *decreasing* viscosity at a given temperature; explain your ranking.

Problems in Context

12.65 Soil vapor extraction (SVE) is used to remove volatile organic pollutants, such as chlorinated solvents, from soil at hazardous waste sites. Vent wells are drilled, and a vacuum pump is applied to the subsurface. (a) How does this remove pollutants? (b) Why does heating combined with SVE speed the process?

12.66 Use Figure 12.1, p. 427 to answer the following: (a) Does it take more heat to melt 12.0 g of CH_4 or 12.0 g of Hg? (b) Does it take more heat to vaporize 12.0 g of CH_4 or 12.0 g of Hg? (c) What is the principal intermolecular force in each sample?

12.67 Pentanol ($C_5H_{11}OH$; $\mathcal{M}$ = 88.15 g/mol) has nearly the same molar mass as hexane (C_6H_{14}; $\mathcal{M}$ = 86.17 g/mol) but is more than 12 times as viscous at 20°C. Explain.

The Uniqueness of Water

Concept Review Questions

12.68 For what types of substances is water a good solvent? For what types is it a poor solvent? Explain.

12.69 A water molecule can engage in as many as four H bonds. Explain.

12.70 Warm-blooded animals have a narrow range of body temperature because their bodies have a high water content. Explain.

12.71 What property of water keeps plant debris on the surface of lakes and ponds? What is the ecological significance of this?

12.72 A drooping plant can be made upright by watering the ground around it. Explain.

12.73 Describe the molecular basis of the property of water responsible for the presence of ice on the surface of a frozen lake.

12.74 Describe in molecular terms what occurs when ice melts.

The Solid State: Structure, Properties, and Bonding

(Sample Problem 12.4)

Concept Review Questions

12.75 What is the difference between an amorphous solid and a crystalline solid on the macroscopic and molecular levels? Give an example of each.

12.76 How are a solid's unit cell and crystal structure related?

12.77 For structures consisting of identical atoms, how many atoms are contained in the simple, body-centered, and face-centered cubic unit cells? Explain how you obtained the values.

12.78 An element has a crystal structure in which the width of the cubic unit cell equals the diameter of an atom. What type of unit cell does it have?

12.79 What specific difference in the positioning of spheres gives a crystal structure based on the face-centered cubic unit cell less empty space than one based on the body-centered cubic unit cell?

12.80 Both solid Kr and solid Cu consist of individual atoms. Why do their physical properties differ so much?

12.81 What is the energy gap in band theory? Compare its size in superconductors, conductors, semiconductors, and insulators.

12.82 Predict the effect (if any) of an increase in temperature on the electrical conductivity of (a) a conductor; (b) a semiconductor; (c) an insulator.

12.83 Besides the type of unit cell, what information is needed to find the density of a solid consisting of identical atoms?

Skill-Building Exercises *(grouped in similar pairs)*

12.84 What type of crystal lattice does each metal form? (The number of atoms per unit cell is given in parentheses.)
(a) Ni (4) (b) Cr (2) (c) Ca (4)

12.85 What is the number of atoms per unit cell for each metal?
(a) Polonium, Po (b) Iron, Fe (c) Silver, Ag

12.86 Of the five major types of crystalline solid, which does each of the following form: (a) Sn; (b) Si; (c) Xe?

12.87 Of the five major types of crystalline solid, which does each of the following form: (a) cholesterol ($C_{27}H_{45}OH$); (b) KCl; (c) BN?

12.88 Of the five major types of crystalline solid, which does each of the following form, and why: (a) Ni; (b) F_2; (c) CH_3OH?

12.89 Of the five major types of crystalline solid, which does each of the following form, and why: (a) SiC; (b) Na_2SO_4; (c) SF_6?

12.90 Zinc oxide adopts the zinc blende crystal structure (Figure P12.90). How many Zn^{2+} ions are in the ZnO unit cell?

12.91 Calcium sulfide adopts the sodium chloride crystal structure (Figure P12.91). How many S^{2-} ions are in the CaS unit cell?

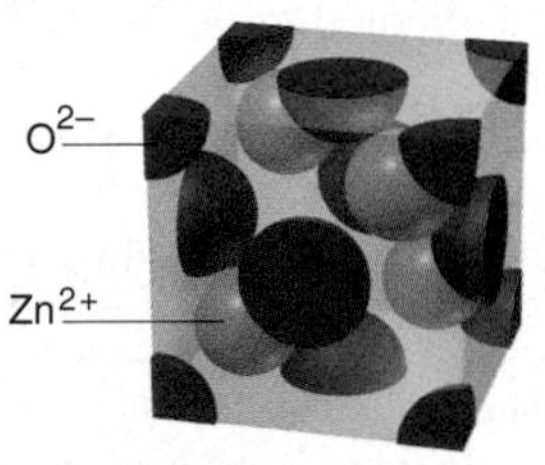

Figure P12.90

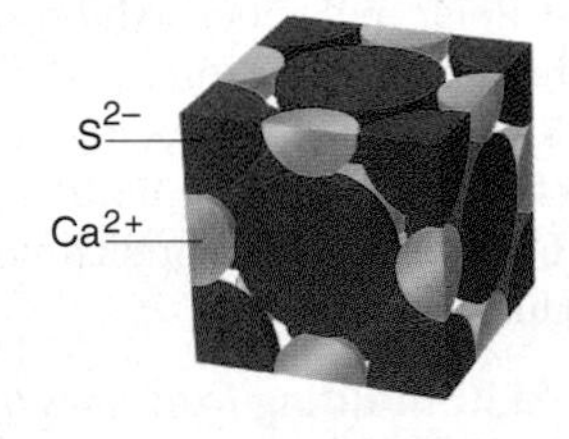

Figure P12.91

12.92 Zinc selenide (ZnSe) crystallizes in the zinc blende structure and has a density of 5.42 g/cm^3.
(a) How many Zn and Se ions are in each unit cell?
(b) What is the mass of a unit cell?
(c) What is the volume of a unit cell?
(d) What is the edge length of a unit cell?

12.93 An element crystallizes in a face-centered cubic lattice and has a density of 1.45 g/cm^3. The edge of its unit cell is 4.52×10^{-8} cm.
(a) How many atoms are in each unit cell?
(b) What is the volume of a unit cell?
(c) What is the mass of a unit cell?
(d) Calculate an approximate atomic mass for the element.

12.94 Classify each of the following as a conductor, insulator, or semiconductor: (a) phosphorus; (b) mercury; (c) germanium.

12.95 Classify each of the following as a conductor, insulator, or semiconductor: (a) carbon (graphite); (b) sulfur; (c) platinum.

12.96 Predict the effect (if any) of an increase in temperature on the electrical conductivity of (a) antimony, Sb; (b) tellurium, Te; (c) bismuth, Bi.

12.97 Predict the effect (if any) of a decrease in temperature on the electrical conductivity of (a) silicon, Si; (b) lead, Pb; (c) germanium, Ge.

Problems in Context

12.98 Polonium, the Period 6 member of Group 6A(16), is a rare radioactive metal that is the only element with a crystal structure based on the simple cubic unit cell. If its density is 9.142 g/cm^3, calculate an approximate atomic radius for polonium.

12.99 The coinage metals—copper, silver, and gold—crystallize in a cubic closest packed structure. Use the density of copper (8.95 g/cm^3) and its molar mass (63.55 g/mol) to calculate an approximate atomic radius for copper.

12.100 One of the most important enzymes in the world—nitrogenase, the plant protein that catalyzes nitrogen fixation—contains active clusters of iron, sulfur, and molydenum atoms. Crystalline molybdenum (Mo) has a body-centered cubic unit cell (d of Mo = 10.28 g/cm^3). (a) Determine the edge length of the unit cell. (b) Calculate the atomic radius of Mo.

12.101 Tantalum (Ta; d = 16.634 g/cm^3 and $\mathcal{M}$ = 180.9479 g/mol) has a body-centered cubic structure with a unit-cell edge length of 3.3058 Å. Use these data to calculate Avogadro's number.

Advanced Materials

Concept Review Questions

12.102 When tin is added to copper, the resulting alloy (bronze) is much harder than copper. Explain.

12.103 In the process of doping a semiconductor, certain impurities are added to increase its electrical conductivity. Explain this process for an n-type and a p-type semiconductor.

12.104 State two molecular characteristics of substances that typically form liquid crystals. How is each of them related to function?

12.105 Distinguish between isotropic and anisotropic substances. To which category do liquid crystals belong?

12.106 How are the properties of high-tech ceramics the same as those of traditional clay ceramics, and how are they different? Refer to specific substances in your answer.

12.107 Why is the average molar mass of a polymer sample different from the molar mass of an individual chain?

12.108 How does the random coil shape relate to the radius of gyration of a polymer chain?

12.109 What factor(s) influence the viscosity of a polymer solution? What factor(s) influence the viscosity of a molten polymer? What is a polymer glass?

12.110 Use an example to show how branching and crosslinking can affect the physical behavior of a polymer.

Skill-Building Exercises (grouped in similar pairs)

12.111 Silicon and germanium are both semiconducting elements from Group 4A(14) that can be doped to improve their conductivity. Would each of the following form an n-type or a p-type semiconductor: (a) Ge doped with P; (b) Si doped with In?

12.112 Would each of the following form an n-type or a p-type semiconductor: (a) Ge doped with As; (b) Si doped with B?

12.113 The repeat unit in a polystyrene coffee cup has the formula $C_6H_5CHCH_2$. If the molar mass of the polymer is 3.5×10^5 g/mol, what is the degree of polymerization?

12.114 The monomer of poly(vinyl chloride) has the formula C_2H_3Cl. If there are 1565 repeat units in a single chain of the polymer, what is the molecular mass (in amu) of that chain?

12.115 The polypropylene (repeat unit CH_3CHCH_2) in a plastic toy has a molar mass of 2.5×10^5 g/mol and a repeat unit length of 0.252 pm. Calculate the radius of gyration.

12.116 The polymer that is used to make 2-L soda bottles [poly(ethylene terephthalate)] has a repeat unit with molecular formula $C_{10}H_8O_4$ and length 1.075 nm. Calculate the radius of gyration of a chain with molar mass of 2.30×10^4 g/mol.

Comprehensive Problems

12.117 A 0.75-L bottle is cleaned, dried, and closed in a room where the air is 22°C and 44% relative humidity (that is, the water vapor in the air is 0.44 of the equilibrium vapor pressure at 22°C). The bottle is brought outside and stored at 0.0°C.
(a) What mass of liquid water condenses inside the bottle?
(b) Would liquid water condense at 10°C? (See Table 5.3, p. 197.)

12.118 In an experiment, 4.00 L of N_2 is saturated with water vapor at 22°C and then compressed to half its volume. (a) What is the partial pressure of H_2O in the compressed gas mixture? (b) What mass of water vapor condenses to liquid?

12.119 Which forces are overcome when the following events occur: (a) NaCl dissolves in water; (b) krypton boils; (c) water boils; (d) CO_2 sublimes?

12.120 Changes in pressure cause phase changes analogous to the changes with temperature shown in Figure 12.3, p. 429.
(a) Refer to Figure 12.9B (p. 435) and draw a curve of pressure vs. time for water similar to Figure 12.3; label the phase changes as the pressure is continuously *increased* at 2°C.
(b) Refer to Figure 12.9A (p. 435) and draw a curve for carbon dioxide as the pressure is continuously *increased* at −50°C.

12.121 Because bismuth has several well-characterized solid, crystalline phases, it is used to calibrate instruments employed in high-pressure studies. The following phase diagram for bismuth shows the liquid phase and five different solid phases stable above 1 katm (1000 atm) and up to 300°C. (a) Which solid phases are stable at 25°C? (b) Which phase is stable at 50 katm and 175°C? (c) Identify the phase transitions that bismuth undergoes at 200°C as the pressure is reduced from 100 to 1 katm. (d) What phases are present at each of the triple points?

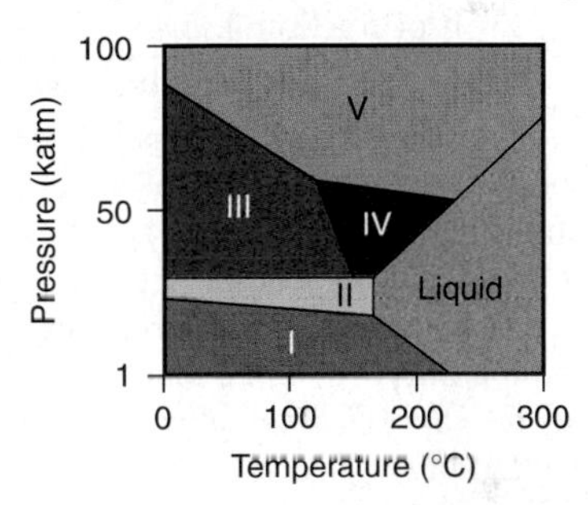

12.122 In the photoelectric effect, the work function (ϕ) is the minimum energy a photon must have to remove an electron from a metal surface (see Problem 7.70). For a given metal, ϕ depends on where the photon strikes the crystal. Copper adopts the face-centered cubic structure. If the photon strikes perpendicular to a face, ϕ is 4.59 eV; if it strikes an edge, ϕ is 4.48 eV; and if it strikes a corner, ϕ is 4.94 eV (1 eV = 1.602×10^{-19} J). Find the wavelength (in nm) of the lowest energy photon that can remove an electron from copper, and state how it strikes the unit cell.

12.123 In making computer chips, a 5.00-kg cylindrical ingot of ultrapure n-type doped silicon that is 5.25 inches in diameter is sliced into wafers 1.02×10^{-4} m thick.
(a) Assuming no waste, how many wafers can be made?
(b) What is the mass of a wafer (d of Si = 2.34 g/cm^3; V of a cylinder = $\pi r^2 h$)?
(c) A key step in making p-n junctions for the chip is chemical removal of the oxide layer on the wafer through treatment with gaseous HF. Write a balanced equation for this reaction.
(d) If 0.750% of the Si atoms are removed during the treatment in part (c), how many moles of HF are required per wafer, assuming 100% reaction yield?

12.124 Methyl salicylate, $C_8H_8O_3$, the odorous constituent of oil of wintergreen, has a vapor pressure of 1.00 torr at 54.3°C and 10.0 torr at 95.3°C. (a) What is its vapor pressure at 25°C? (b) How many liters of air must pass over a sample of the compound at 25°C to vaporize 1.0 mg of it?

12.125 Mercury (Hg) vapor is toxic and readily absorbed from the lungs. At 20.°C, mercury (ΔH_{vap} = 59.1 kJ/mol) has a vapor pressure of 1.20×10^{-3} torr, which is high enough to be hazardous. To reduce the danger to workers in processing plants, Hg is cooled to lower its vapor pressure. At what temperature would the vapor pressure of Hg be at the safer level of 5.0×10^{-5} torr?

12.126 Polytetrafluoroethylene (Teflon) has a repeat unit with the formula F_2C—CF_2. A sample of the polymer consists of fractions with the following distribution of chains:

Fraction	Average no. of repeat units	Amount (mol) of polymer
1	273	0.10
2	330	0.40
3	368	1.00
4	483	0.70
5	525	0.30
6	575	0.10

(a) Determine the molar mass of each fraction.
(b) Determine the number-average molar mass of the sample.
(c) Another type of average molar mass of a polymer sample is called the *weight-average molar mass*, $\mathcal{M}_w$:

$$\mathcal{M}_w = \frac{\Sigma\ (\mathcal{M} \text{ of fraction} \times \text{mass of fraction})}{\text{total mass of all fractions}}$$

Calculate the weight-average molar mass of the above sample.

12.127 A greenhouse contains 216 m^3 of air at a temperature of 26°C, and a humidifier in it vaporizes 4.00 L of water. (a) What is the pressure of water vapor in the greenhouse, assuming that none escapes and that the air was originally completely dry (d of H_2O = 1.00 g/mL)? (b) What total volume of liquid water would have to be vaporized to saturate the air (i.e., achieve 100% relative humidity)? (See Table 5.4, p. 210.)

12.128 The packing efficiency of spheres is the percentage of the total space (volume) of the unit cell occupied by the spheres:

$$\text{Packing efficiency (\%)} = \frac{\text{volume of spheres in unit cell}}{\text{volume of unit cell}} \times 100$$

Using spheres of radius r and unit-cell edge length A, calculate the packing efficiency of (a) a simple cubic unit cell; (b) a face-centered cubic unit cell; (c) a body-centered cubic unit cell (volume of a sphere $= \frac{4}{3}\pi r^3$). (d) A solid consisting of identical atoms has a density of 9.0 g/cm^3 and crystallizes in a cubic closest packed structure. What is its density if the atoms are rearranged to a structure with a body-centered cubic unit cell? A simple cubic unit cell?

12.129 Consider the phase diagram shown for substance X. (a) What phase(s) is (are) present at point A? E? F? H? B? C? (b) Which point corresponds to the critical point? Which point corresponds to the triple point? (c) What curve corresponds to conditions at which the solid and gas are in equilibrium? (d) Describe what happens when you start at point A and increase the temperature at constant pressure. (e) Describe what happens when you start at point H and decrease the pressure at constant temperature. (f) Is liquid X more or less dense than solid X?

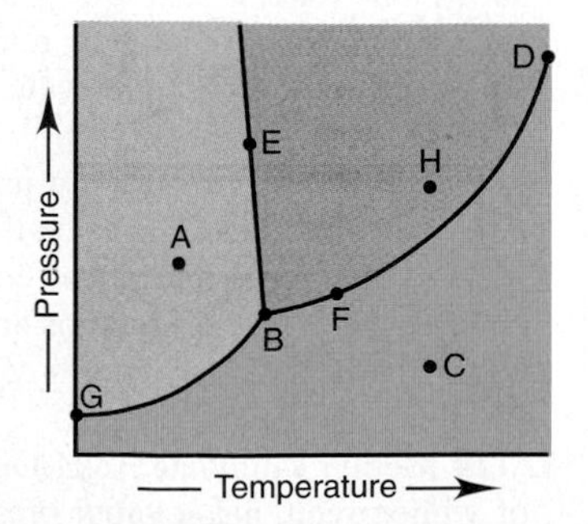

12.130 Some high-temperature superconductors adopt a crystal structure similar to that of *perovskite* ($CaTiO_3$). The unit cell is cubic with a Ti^{4+} ion in each corner, a Ca^{2+} ion in the body center, and O^{2-} ions at the midpoint of each edge. (a) Is this unit cell simple, body-centered, or face-centered? (b) If the unit cell edge length is 3.84 Å, what is the density of perovskite (in g/cm^3)?

12.131 Why do uninsulated water pipes burst in cold weather?

12.132 The only alkali metal halides that do not adopt the NaCl structure are CsCl, CsBr, and CsI, formed from the largest alkali metal cation and the three largest halide ions. These crystallize in the *cesium chloride structure* (shown here for CsCl). This structure has been used as an example of how dispersion forces can dominate in the presence of ionic forces. Use the ideas of coordination number and polarizability to explain why the CsCl structure exists.

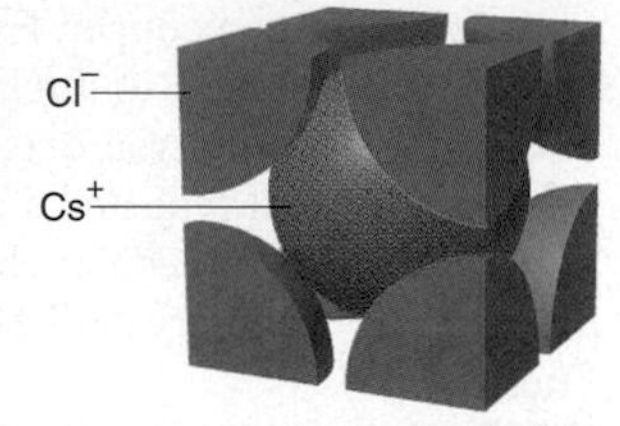

12.133 Hydrogen bonding often causes individual molecules to remain linked together under varying conditions. Each of the following has been identified in either the gas or liquid phase. Draw structures for each showing the H bonds: (a) $(HF)_2$, dimer; (b) $(CH_3OH)_4$, cyclic; (c) $(HF)_6$, cyclic.

12.134 Corn is a valuable source of industrial chemicals. For example, furfural is prepared from corncobs. It is an important reactant in plastics manufacturing and a key solvent for the production of cellulose acetate, which is used to make everything from videotape to waterproof fabric. It can be reduced to furfuryl alcohol or oxidized to 2-furoic acid.

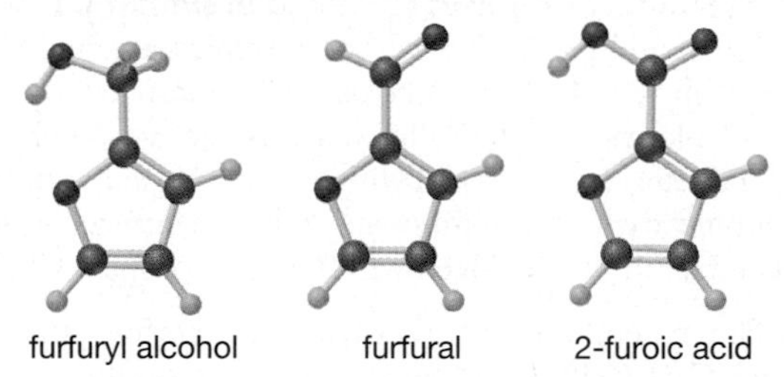

(a) Which of these compounds can form H bonds? Draw structures in each case.
(b) The molecules of some substances can form an "internal" H bond, that is, an H bond *within* a molecule. This takes the form of a polygon with atoms as corners and bonds as sides and an H bond as one of the sides. Which of these molecules is (are) likely to form a stable internal H bond? Draw the structure. (*Hint:* Structures with 5 or 6 atoms as corners are most stable.)

12.135 The density of solid gallium at its melting point is 5.9 g/cm^3, whereas that of liquid gallium is 6.1 g/cm^3. Is the temperature at the triple point higher or lower than the normal melting point? Is the slope of the solid-liquid line for gallium positive or negative?

12.136 A 4.7-L sealed bottle containing 0.33 g of liquid ethanol, C_2H_6O, is placed in a refrigerator and reaches equilibrium with its vapor at −11°C. (a) What mass of ethanol is present in the vapor? (b) When the container is removed and warmed to room temperature, 20.°C, will all the ethanol vaporize? (c) How much liquid ethanol would be present at 0.0°C? The vapor pressure of ethanol is 10. torr at −2.3°C and 40. torr at 19°C.

12.137 On a humid day in New Orleans, the temperature is 22.0°C, and the partial pressure of water vapor in the air is 31.0 torr. The 9000-ton air-conditioning system in the Louisiana Superdome maintains an inside air temperature of 22.0°C also, but a partial pressure of water vapor of 10.0 torr. The volume of air in the dome is 2.4×10^6 m^3, and the total pressure inside and outside the dome are both 1.0 atm. (a) What mass of water (in metric tons) must be removed every time the inside air is completely replaced with outside air? (*Hint:* How many moles of gas are in the dome? How many moles of water vapor? How many moles of dry air? How many moles of outside air must be added to the air in the dome to simulate the composition of outside air?) (b) Find the heat released when this mass of water condenses.

12.138 A cubic unit cell contains atoms of element A at each corner and atoms of element Z on each face. What is the empirical formula of the compound?

12.139 The boiling point of amphetamine, $C_9H_{13}N$, is 201°C at 760 torr and 83°C at 13 torr. What is the concentration (in g/m^3) of amphetamine when it is in contact with 20.°C air?

12.140 Diamond has a face-centered cubic unit cell, with four more C atoms in tetrahedral holes within the cell. Densities of diamonds vary from 3.01 g/cm^3 to 3.52 g/cm^3 because C atoms are missing from some holes. (a) Calculate the unit-cell edge length of the densest diamond. (b) Assuming the cell dimensions are fixed, how many C atoms are in the unit cell of the diamond with the lowest density?

12.141 Is it possible for a salt of formula AB_3 to have a face-centered cubic unit cell of anions with cations in all the eight available holes? Explain.

12.142 Substance A has the following properties.

mp at 1 atm:	−20.°C	bp at 1 atm:	85°C
ΔH_{fus}:	180. J/g	ΔH_{vap}:	500. J/g
c_{solid}:	1.0 J/g·°C	c_{liquid}:	2.5 J/g·°C
c_{gas}:	0.5 J/g·°C		

At 1 atm, a 25-g sample of A is heated from −40.°C to 100.°C at a constant rate of 450. J/min. (a) How many minutes does it take to heat the sample to its melting point? (b) How many minutes does it take to melt the sample? (c) Perform any other necessary calculations, and draw a curve of temperature vs. time for the entire heating process.

12.143 An aerospace manufacturer is building a prototype experimental aircraft that cannot be detected by radar. Boron nitride is chosen for incorporation into the body parts, and the boric acid/ammonia method is used to prepare the ceramic material. Given 85.5% and 86.8% yields for the two reaction steps, how much boron nitride can be prepared from 1.00 metric ton of boric acid and 12.5 m^3 of ammonia at 275 K and 3.07×10^3 kPa? Assume that ammonia does *not* behave ideally under these conditions and is recycled completely in the reaction process.

12.144 A sample of polystyrene (repeat unit $C_6H_5CHCH_2$) has a molar mass of 104,160 g/mol. A section of the backbone is shown with the average C—C bond length and angle. (a) Calculate the extended length of the polymer chain. (b) Calculate the radius of gyration of the polymer chain.

C
154 pm
109.5°
C C

12.145 In a body-centered cubic unit cell, the central atom lies on an internal diagonal of the cell and touches the corner atoms. (a) Find the length of the diagonal in terms of *r*, the atomic radius. (b) If the edge length of the cube is *a*, what is the length of a *face* diagonal? (c) Derive an expression for *a* in terms of *r*. (d) How many atoms are in this unit cell? (e) What fraction of the unit cell volume is filled with spheres?

12.146 The ΔH_f^0 of gaseous dimethyl ether (CH_3OCH_3) is –185.4 kJ/mol; the vapor pressure is 1.00 atm at –23.7°C and 0.526 atm at –37.8°C. (a) Calculate ΔH_{vap}^0 of dimethyl ether. (b) Calculate ΔH_f^0 of liquid dimethyl ether.

12.147 The crystal structure of sodium is based on the body-centered cubic unit cell. What is the mass of one unit cell of sodium?

12.148 KF has the same type of crystal structure as NaCl. The unit cell of KF has an edge length of 5.39 Å. Find the density of KF.

12.149 The *intrinsic viscosity* of a polymer chain in a given solvent, $[\eta]_{solvent}$, is a measure of the intermolecular interactions between solvent and polymer. Generally, a larger $[\eta]_{solvent}$ indicates a stronger interaction. Intrinsic viscosities of polymer chains in solution are given by the Mark-Houwink equation, $[\eta]_{solvent} = K\mathcal{M}^a$, where $\mathcal{M}$ is the molar mass of the polymer, and *K* and *a* are constants specific to the polymer and solvent. Use the data below for 25°C to answer the following questions:

Polymer	Solvent	*K* (mL/g)	*a*
Polystyrene	Benzene	9.5×10^{-3}	0.74
	Cyclohexane	8.1×10^{-2}	0.50
Polyisobutylene	Benzene	8.3×10^{-2}	0.50
	Cyclohexane	2.6×10^{-1}	0.70

(a) A polystyrene sample has a molar mass of 104,160 g/mol. Calculate the intrinsic viscosity in benzene and in cyclohexane. Which solvent has stronger interactions with the polymer?
(b) A different polystyrene sample has a molar mass of 52,000 g/mol. Calculate its $[\eta]_{benzene}$. Given a polymer standard of known $\mathcal{M}$, how could you use its measured $[\eta]$ in a given solvent to determine the molar mass of any sample of that polymer?
(c) Compare $[\eta]$ of a polyisobutylene sample [repeat unit $(CH_3)_2CCH_2$] with a molar mass of 104,160 g/mol with that of the polystyrene in part (a). What does this suggest about the solvent-polymer interactions of the two samples?

12.150 At a local convenience store, you purchase a cup of coffee but, at 98.4°C, it is too hot to drink. You add 23.0 g of ice that is –2.2°C to 248 mL of the coffee. What is the final temperature of the coffee? (Assume the heat capacity and density of the coffee are the same as water and the coffee cup is well insulated.)

12.151 One way of purifying gaseous H_2 is to pass it under high pressure through the holes of a metal's crystal structure. Palladium, which adopts a cubic closest packed structure, absorbs more H_2 than any other element and is one of the metals currently used for this purpose. Although the metal-hydrogen interaction is unclear, it is estimated that the density of absorbed H_2 approaches that of liquid hydrogen (70.8 g/L). What volume (in L) of gaseous H_2, measured at STP, can be packed into the spaces of 1 dm^3 of palladium metal?

Everyday mixtures From mixed paint to brewed tea, mixtures surround us and influence nearly every aspect of daily life. In this chapter, you'll see why mixtures form, how we quantify their composition, and some of the remarkably useful ways their physical behavior differs from that of their components.

13

The Properties of Mixtures: Solutions and Colloids

Nearly all the gases, liquids, and solids that make up our world are *mixtures*—two or more substances physically mixed together but not chemically combined. Synthetic mixtures, such as glass and soap, usually contain relatively few components, whereas natural mixtures, such as seawater and soil, are more complex, often containing more than 50 different substances. Living mixtures, such as trees and students, are the most complex—even a simple bacterial cell contains well over 5000 different compounds (Table 13.1).

Concepts & Skills to Review

before you study this chapter

- classification and separation of mixtures (Section 2.9)
- calculations involving mass percent (Section 3.1) and molarity (Section 3.5)
- electrolytes; water as a solvent (Sections 4.1 and 12.5)
- mole fraction and Dalton's law of partial pressures (Section 5.4)
- types of intermolecular forces and the concept of polarizability (Section 12.3)
- distinction between monomer and polymer (Section 12.7)
- vapor pressure of liquids (Section 12.2)

Table 13.1 Approximate Composition of a Bacterium

Substance	Mass % of Cell	Number of Types	Number of Molecules
Water	~70	1	5×10^{10}
Ions	1	20	?
Sugars*	3	200	3×10^{8}
Amino acids*	0.4	100	5×10^{7}
Lipids*	2	50	3×10^{7}
Nucleotides*	0.4	200	1×10^{7}
Other small molecules	0.2	~200	?
Macromolecules (proteins, nucleic acids, polysaccharides)	23	~5000	6×10^{6}

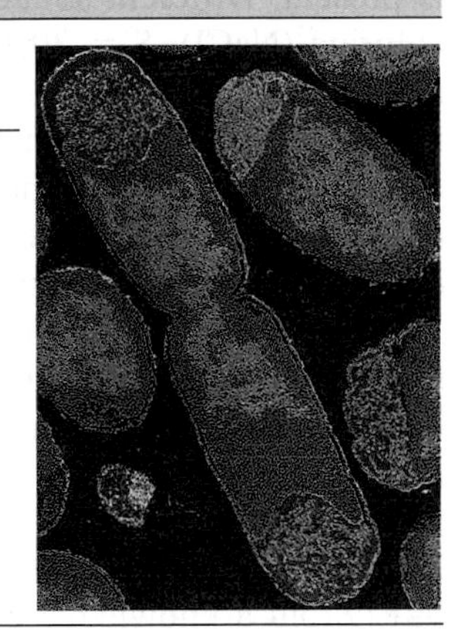

*Includes precursors and metabolites.

Recall from Chapter 2 that a mixture has two defining characteristics: *its composition is variable*, and *it retains some properties of its components*. In this chapter, we focus on two common types of mixtures: solutions and colloids. *A solution is a homogeneous mixture,* one with no boundaries separating its components; thus, a solution exists as one phase. A *heterogeneous mixture* has two or more phases. The pebbles in concrete or the bubbles in champagne are visible indications that these are heterogeneous mixtures. A *colloid* is a heterogeneous mixture in which one component is dispersed as very fine particles in another component, so distinct phases are not easy to see. Smoke and milk are colloids. The essential difference between a solution and a colloid is one of particle *size:*

- In a solution, the particles are individual atoms, ions, or small molecules.
- In a colloid, the particles are typically either macromolecules or aggregations of small molecules that are not large enough to settle out.

As you'll see later in the chapter, these molecular-scale differences result in many observable differences.

IN THIS CHAPTER . . . We examine intermolecular forces between solute and solvent and their role in forming solutions. To predict solubility in different solvents, we focus on the dual polarity of simple organic compounds and then apply this idea to biological macromolecules and the cell membrane. We investigate why a substance dissolves in terms of the enthalpy of solution and the dispersal of matter that occurs during the solution process. To understand the second factor, the concept of entropy is introduced. We examine the equilibrium nature of solubility and see how temperature and pressure affect it. Various concentration units are discussed and interconverted. You'll see why the physical properties of solutions are different from those of pure substances and how we employ those differences. Then we investigate colloids and apply solution and colloid chemistry to the purification of water.

To examine this idea further, let's compare the solubilities of a series of alcohols in two solvents that act through very different intermolecular forces—water and hexane. Alcohols are organic molecules with a hydroxyl (—OH) group bound to a hydrocarbon group. The simplest type of alcohol has the general formula $CH_3(CH_2)_nOH$; we'll consider alcohols with $n = 0$ to 5. We can view an alcohol molecule as consisting of two portions: the polar —OH group and the nonpolar hydrocarbon chain. The —OH portion forms strong H bonds with water and weak dipole–induced dipole forces with hexane. The hydrocarbon portion interacts through dispersion forces with hexane and through dipole–induced dipole forces with water.

In Table 13.2, the models show the relative change in size of the polar and nonpolar portions of the alcohol molecules. In the smaller alcohols (one to three carbons), the hydroxyl group is a relatively large portion, so the molecules interact with each other through H bonding, just as water molecules do. When they mix with water, H bonding within solute and within solvent is replaced by H bonding *between* solute and solvent (Figure 13.3). As a result, these smaller alcohols are miscible with water.

Water solubility decreases dramatically for alcohols larger than three carbons, and those with chains longer than six carbons are insoluble in water. For these larger alcohols to dissolve, the nonpolar chains have to move between the water molecules, substituting their weak attractions with those water molecules for strong H bonds among the water molecules themselves. The —OH portion of the alcohol does form H bonds to water, but these don't outweigh the H bonds between water molecules that have to break to make room for the hydrocarbon portion.

Table 13.2 shows that the opposite trend occurs with hexane. Now the major solute-solvent *and* solvent-solvent interactions are dispersion forces. The weak

Table 13.2 Solubility* of a Series of Alcohols in Water and Hexane

Alcohol	Model	Solubility in Water	Solubility in Hexane
CH_3OH (methanol)		∞	1.2
CH_3CH_2OH (ethanol)		∞	∞
$CH_3(CH_2)_2OH$ (1-propanol)		∞	∞
$CH_3(CH_2)_3OH$ (1-butanol)		1.1	∞
$CH_3(CH_2)_4OH$ (1-pentanol)		0.30	∞
$CH_3(CH_2)_5OH$ (1-hexanol)		0.058	∞

*Expressed in mol alcohol/1000 g solvent at 20°C.

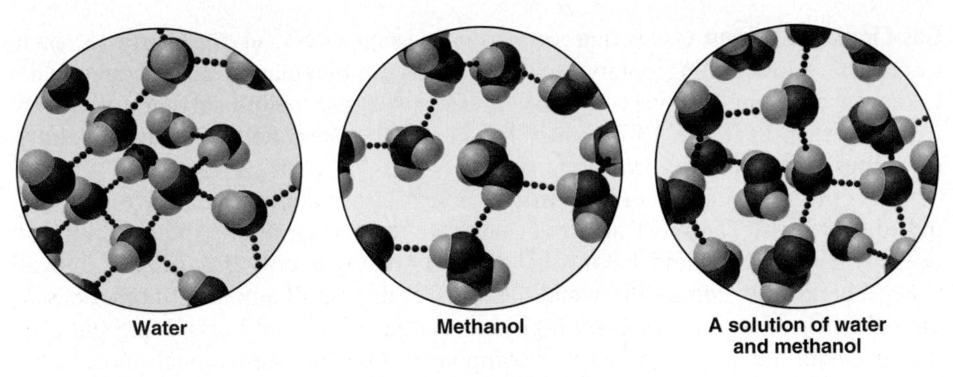

Figure 13.3 Like dissolves like: solubility of methanol in water. The H bonds in water and in methanol are similar in type and strength, so they can substitute for one another. Thus, methanol is soluble in water; in fact, the two substances are miscible.

forces between the —OH group of methanol (CH_3OH) and hexane cannot substitute for the strong H bonding among CH_3OH molecules, so the solubility of methanol in hexane is relatively low. In any larger alcohol, however, dispersion forces become increasingly more important, and these *can* substitute for dispersion forces in pure hexane; thus, solubility increases. With no strong solvent-solvent forces to be replaced by weak solute-solvent forces, even the two-carbon chain of ethanol has strong enough solute-solvent attractions to be miscible in hexane.

Many other organic molecules have polar and nonpolar portions, and the predominance of one portion or the other determines their solubility in different solvents. For example, carboxylic acids and amines behave very much like alcohols. Thus, methanoic acid (HCOOH, formic acid) and methanamine (CH_3NH_2) are miscible in water and much less soluble in hexane, whereas hexanoic acid [$CH_3(CH_2)_4COOH$] and 1-hexanamine [$CH_3(CH_2)_5NH_2$] are slightly soluble in water and very soluble in hexane.

A "Greener" Approach to Working with Organic Solvents Volatile organic solvents have many uses in paints and other coatings, in cleaning agents, and in propellants for hairspray, cooking oils, lubricants, and so forth. But they also play a role in smog formation and may have adverse health effects. These concerns have led to the development of "greener" processes, such as the use of nonvolatile organic solvents, nonaqueous ionic liquid solvents, supercritical fluids, and solvent-free and water-based methods. With the implementation of these approaches, atmospheric emissions and worker exposure are greatly reduced.

SAMPLE PROBLEM 13.1 Predicting Relative Solubilities of Substances

Problem Predict which solvent will dissolve more of the given solute:
(a) Sodium chloride in methanol (CH_3OH) or in 1-propanol ($CH_3CH_2CH_2OH$)
(b) Ethylene glycol ($HOCH_2CH_2OH$) in hexane ($CH_3CH_2CH_2CH_2CH_2CH_3$) or in water
(c) Diethyl ether ($CH_3CH_2OCH_2CH_3$) in water or in ethanol (CH_3CH_2OH)

Plan We examine the formulas of the solute and each solvent to determine the types of forces that could occur. A solute tends to be more soluble in a solvent whose intermolecular forces are similar to, and therefore can substitute for, its own.

Solution (a) Methanol. NaCl is an ionic solid that dissolves through ion-dipole forces. Both methanol and 1-propanol contain a polar —OH group, but 1-propanol's longer hydrocarbon chain can form only weak forces with the ions, so it is less effective at substituting for the ionic attractions in the solute.

(b) Water. Ethylene glycol molecules have two —OH groups, so the molecules interact with each other through H bonding. They are more soluble in H_2O, whose H bonds can substitute for their own H bonds better than the dispersion forces in hexane can.

(c) Ethanol. Diethyl ether molecules interact with each other through dipole-dipole and dispersion forces and can form H bonds to both H_2O and ethanol. The ether is more soluble in ethanol because that solvent can form H bonds *and* substitute for the ether's dispersion forces. Water, on the other hand, can form H bonds with the ether, but it lacks any hydrocarbon portion, so it forms much weaker dispersion forces with that solute.

FOLLOW-UP PROBLEM 13.1 Which solute is more soluble in the given solvent?
(a) 1-Butanol ($CH_3CH_2CH_2CH_2OH$) or 1,4-butanediol ($HOCH_2CH_2CH_2CH_2OH$) in water
(b) Chloroform ($CHCl_3$) or carbon tetrachloride (CCl_4) in water

Table 13.3 Correlation Between Boiling Point and Solubility in Water

Gas	Solubility (*M*)*	bp (K)
He	4.2×10^{-4}	4.2
Ne	6.6×10^{-4}	27.1
N_2	10.4×10^{-4}	77.4
CO	15.6×10^{-4}	81.6
O_2	21.8×10^{-4}	90.2
NO	32.7×10^{-4}	121.4

*At 273 K and 1 atm.

Gas-Liquid Solutions Gases that are nonpolar, such as N_2, or are nearly so, such as NO, have low boiling points because their intermolecular attractions are weak. Likewise, they are not very soluble in water because solute-solvent forces are weak. In fact, as Table 13.3 shows, for nonpolar gases, boiling point generally correlates with solubility in water.

In some cases, the small amount of a nonpolar gas that does dissolve is essential to a process. The most important environmental example is the solubility of O_2 in water. At 25°C and 1 atm, the solubility of O_2 is only 3.2 mL/100. mL of water, but aquatic animal life would die without this small amount. In other cases, the solubility of a gas may *seem* high, but the gas is not only dissolving but also reacting with the solvent or another component. Oxygen seems much more soluble in blood than in water because O_2 molecules are continually bonding with hemoglobin molecules in red blood cells. Similarly, carbon dioxide, which is essential for aquatic plants and coral-reef growth, seems very soluble in water (~81 mL of CO_2/100. mL of H_2O at 25°C and 1 atm) because it is reacting, in addition to simply dissolving:

$$CO_2(g) + H_2O(l) \longrightarrow H^+(aq) + HCO_3^-(aq)$$

Gas Solutions and Solid Solutions

Despite the central place of liquid solutions in chemistry, gaseous solutions and solid solutions also have vital importance and numerous applications.

Gas-Gas Solutions *All gases are infinitely soluble in one another.* Air is the classic example of a gaseous solution, consisting of about 18 gases in widely differing proportions. Anesthetic gas proportions are finely adjusted to the needs of the patient and the length of the surgical procedure. The ratios of components in many industrial gas mixtures, such as $CO:H_2$ in syngas production or $N_2:H_2$ in ammonia production, are controlled to optimize product yield under varying conditions of temperature and pressure.

Gas-Solid Solutions *When a gas dissolves in a solid, it occupies the spaces between the closely packed particles.* Hydrogen gas can be purified by passing an impure sample through a solid metal such as palladium. Only the H_2 molecules are small enough to enter the spaces between the Pd atoms, where they form Pd—H covalent bonds. Under high H_2 pressure, the H atoms are passed along the Pd crystal structure and emerge from the solid as H_2 molecules.

The ability of gases to penetrate a solid also has disadvantages. The electrical conductivity of copper, for example, is drastically reduced by the presence of O_2, which dissolves into the crystal structure and reacts to form copper(I) oxide. High-conductivity copper is prepared by melting and recasting the copper in an O_2-free atmosphere.

Solid-Solid Solutions Because solids diffuse so little, their mixtures are usually heterogeneous, as in gravel mixed with sand. Some solid-solid solutions can be formed by melting the solids and then mixing them and allowing them to freeze. Many **alloys,** mixtures of elements that have an overall metallic character, are examples of solid-solid solutions (although several common alloys have microscopic heterogeneous regions). In *substitutional* alloys, such as brass (Figure 13.4A) and sterling silver, atoms of another element substitute for some atoms of the main element in the structure. In *interstitial* alloys, atoms of another element (often a nonmetal) fill some of the spaces *(interstices)* between atoms of the main element. For example, some forms of carbon steel (Figure 13.4B) are interstitial alloys of iron with a small amount of carbon.

Waxes are another familiar type of solid-solid solution. Most waxes are amorphous solids that may contain small regions of crystalline regularity. A natural *wax* is a solid of biological origin that is insoluble in water but dissolves in non-

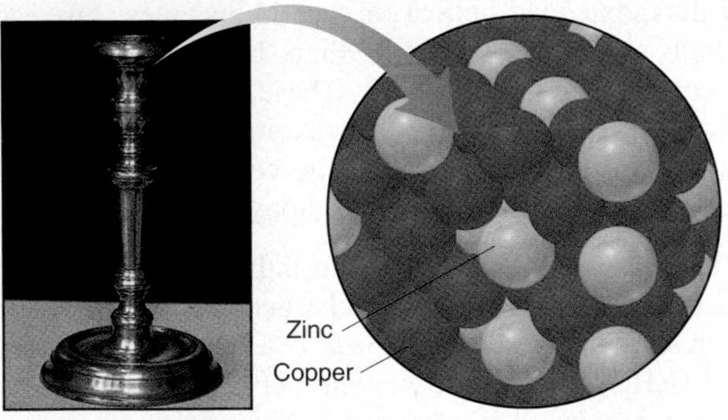

A Brass, a substitutional alloy

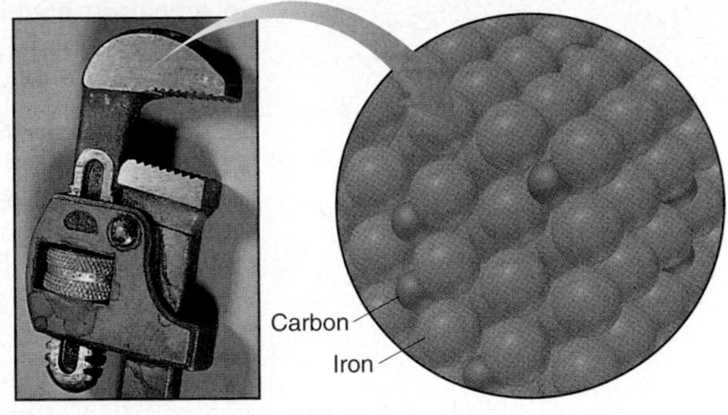

B Carbon steel, an interstitial alloy

Figure 13.4 The arrangement of atoms in two types of alloys. A, Brass is a substitutional alloy in which zinc atoms substitute for copper atoms at many of copper's face-centered cubic packing sites. **B,** Some forms of carbon steel are interstitial alloys in which carbon atoms lie in the holes (interstices) of the body-centered cubic crystal structure of iron.

polar solvents. Many plants have waxy coatings on their leaves and fruit to prevent evaporation of interior water.

Waxes for Home and Auto Beeswax, the remarkable structural material that bees secrete to build their combs, is a complex mixture of fatty acids, long-chain carboxylic acids, and hydrocarbons in which some of the molecules contain chains more than 40 carbon atoms long. Carnauba wax, from a South American palm, is a mixture of compounds, each consisting of a fatty acid bound to a long-chain alcohol. It is hard but forms a thick gel in nonpolar solvents, perfect for car waxes.

SECTION SUMMARY

Solutions are homogeneous mixtures consisting of a solute dissolved in a solvent through the action of intermolecular forces. The solubility of a solute in a given amount of solvent is the maximum amount, with excess solute present, that can dissolve at a specified temperature. (For gaseous solutes, the pressure must be specified also.) In addition to the intermolecular forces that exist in pure substances, ion-dipole, ion–induced dipole, and dipole–induced dipole forces occur in solution. If similar intermolecular forces occur in solute and solvent, a solution will likely form ("like dissolves like"). When ionic compounds dissolve in water, the ions become surrounded by hydration shells of H-bonded water molecules. Solubility of organic molecules in various solvents depends on their polarity and the extent of their polar and nonpolar portions. The solubility of nonpolar gases in water is low because of weak intermolecular forces. Gases are miscible with one another and dissolve in solids by fitting into spaces in the crystal structure. Solid-solid solutions, such as alloys and waxes, form when the components are mixed while molten.

13.2 INTERMOLECULAR FORCES AND BIOLOGICAL MACROMOLECULES

The forces responsible for solutes dissolving in solvents also play a fundamental role in maintaining the complex three-dimensional shapes of cellular biopolymers: proteins, nucleic acids, and polysaccharides. The structures of these molecules, and that of the cell membrane itself, can be understood through two simple ideas involving intermolecular forces: (1) polar and ionic groups interact with the surrounding cellular water, while nonpolar groups do not, and (2) distant groups on a large molecule are often close enough to interact through intermolecular forces.

In this section, we'll see how these forces stabilize the three types of biological macromolecules and the cell membrane; in Chapter 15, we'll see the reactions that form these molecules and how their structure is crucial to their function.

The Structures of Proteins

Proteins are unbranched polymers formed from monomers called **amino acids,** organic compounds with amine ($—NH_2$) and carboxyl ($—COOH$) groups in the same molecule. Proteins range from about 50 amino acids ($\mathcal{M} \approx 5\times10^3$ g/mol) to several thousand ($\mathcal{M} \approx 5\times10^5$ g/mol). *Protein shapes are determined completely by the sequence of amino acids in the chain.* Some proteins have few amino acids in repeating patterns, and thus, extended helical or sheetlike shapes. These function to give structure to hair, skin, and so forth. Others have many more types of amino acids and more complex, globular shapes. These function as antibodies that defend against infection, enzymes that speed metabolic reactions, membrane gates and pumps that control ion concentrations in the cell, and so forth. Let's see how the amino acid composition influences the shapes of globular proteins.

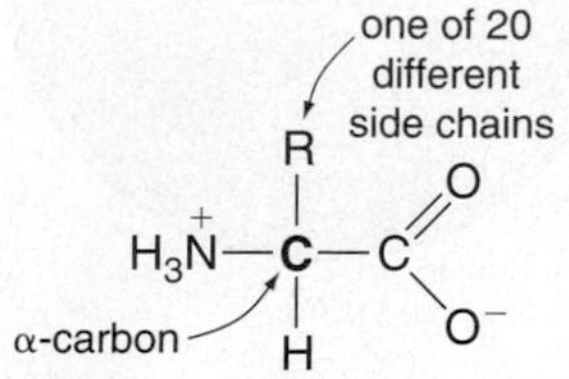

one of 20 different side chains
R
$H_3\overset{+}{N}$—C—COO^-
α-carbon
H

Figure 13.5 The physiological form of an amino acid. Under physiological conditions, amino acids have charged groups. There are about 20 different side chains.

The Polarity of Amino-Acid Side Chains In the cell, an individual amino acid has four groups bonded to its α-carbon (Figure 13.5): charged carboxyl ($-COO^-$) and amine ($-NH_3^+$) groups, an H atom, and a *side chain,* which ranges from an H atom to a two-ringed C_9H_8N group. The *peptide* bond is the covalent linkage between amino acids in a protein and is formed between the carboxyl group of one amino acid and the amine group of the next. (We discuss amino acid structures and details of peptide bond formation in Chapter 15.)

In essence, then, the backbone of a protein consists of an *α-carbon bonded to a peptide bond, which is bonded to the next α-carbon bonded to the next peptide bond,* and so forth, with the various side chains dangling off the α-carbons on alternate sides of the chain (Figure 13.6). Globular proteins are made up of about 20 different types of amino acids, each characterized by its particular side chain. The simple classification of amino acids we use here focuses on their polarity and/or charge: nonpolar, polar, and ionic. A few examples are

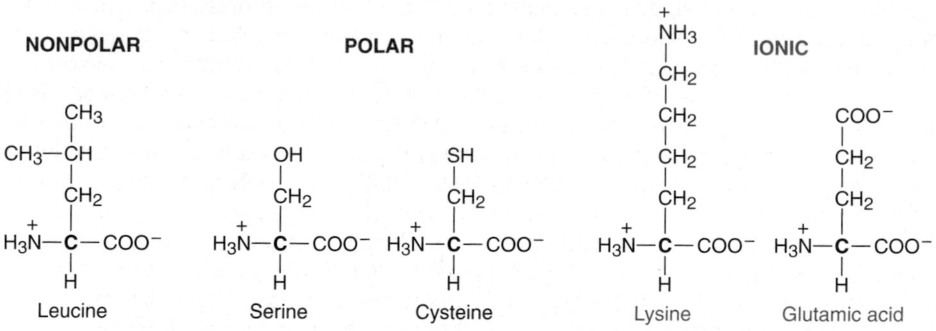

NONPOLAR	POLAR		IONIC	
Leucine	Serine	Cysteine	Lysine	Glutamic acid
side chain: $CH_3-CH(CH_3)-CH_2-$	side chain: $HO-CH_2-$	side chain: $HS-CH_2-$	side chain: $\overset{+}{N}H_3-CH_2-CH_2-CH_2-CH_2-$	side chain: $^-OOC-CH_2-CH_2-$
$H_3\overset{+}{N}-C(H)-COO^-$	$H_3\overset{+}{N}-C(H)-COO^-$	$H_3\overset{+}{N}-C(H)-COO^-$	$H_3\overset{+}{N}-C(H)-COO^-$	$H_3\overset{+}{N}-C(H)-COO^-$

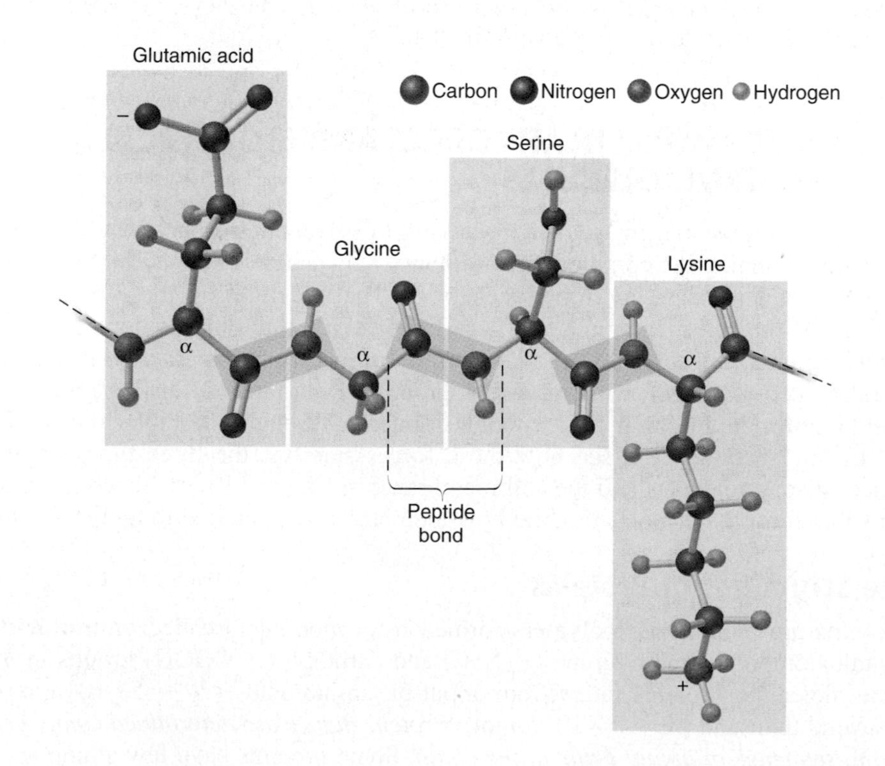

Figure 13.6 A portion of a polypeptide chain. The peptide bond holds the monomers together in a protein. Three peptide bonds (*orange screens*) joining four amino acids (*gray screens*) occur in this portion of a polypeptide chain. Note the repeating pattern of the chain: peptide bond—α-carbon—peptide bond—α-carbon—and so on. Also note that the side chains dangle off the main chain.

Intermolecular Forces and Protein Shape The forces responsible for protein shapes are *the same bonding and intermolecular forces that operate for all molecules.* A long protein chain has many bends, and distant groups of the same molecule end up near enough to interact.

Figure 13.7 depicts the forces operating in proteins dissolved in the aqueous cell medium. (Membrane-bound proteins have an important structural difference, as we'll see shortly.) The covalent peptide bonds create the chain. The —SH ends of two cysteine side chains often form an —S—S— bond, a covalent *disulfide bridge* that brings together distant parts of the chain. Polar and ionic side chains usually protrude into the surrounding cell fluid, interacting with water through ion-dipole forces and H bonds. But sometimes oppositely charged ends of ionic side chains lie near each other, and the $—COO^-$ and $—NH_3^+$ groups form an electrostatic *salt link* (or *ion pair*), which secures the chain's bends. Helical and sheetlike segments arise from *H bonds between the C=O of one peptide bond and the N—H of another.* Other H bonds involving atoms in peptide bonds or in side chains act to keep distant portions of the chain near each other. Nonpolar side chains interact through dispersion forces within the nonaqueous protein interior. Thus, through the influence of intermolecular forces among the protein's side chains, *soluble proteins have polar-ionic exteriors and nonpolar interiors.*

A protein's shape is so important because, as we emphasize in Chapter 15, *the amino acid sequence determines the protein's shape, and the shape determines its function.*

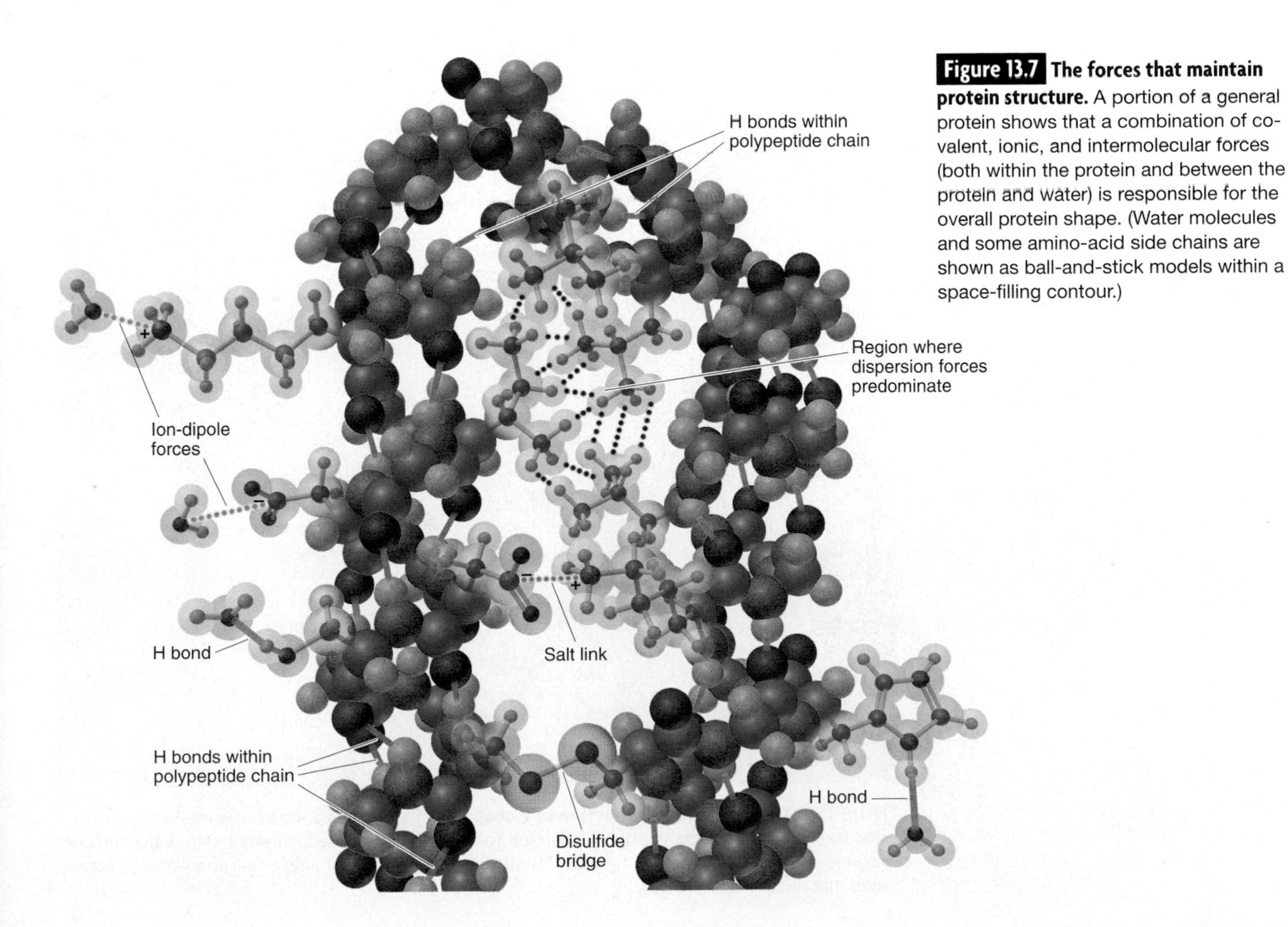

Figure 13.7 The forces that maintain protein structure. A portion of a general protein shows that a combination of covalent, ionic, and intermolecular forces (both within the protein and between the protein and water) is responsible for the overall protein shape. (Water molecules and some amino-acid side chains are shown as ball-and-stick models within a space-filling contour.)

The Structure of the Cell Membrane

The dual polarity that we discussed earlier with reference to solubility can help us understand the function of systems as diverse as soaps, antibiotics, and cell membranes.

The Dual Polarity of Soaps and Detergents A **soap** is the salt of a strong base (metal hydroxide) and a *fatty acid*, a carboxylic acid with a long hydrocarbon chain. A typical soap molecule has an even number of carbon atoms, and is usually made up of a nonpolar "tail" 15–19 carbons long and a polar-ionic "head" consisting of the $—COO^-$ group and a cation. For example, sodium stearate, $CH_3(CH_2)_{16}COONa$, is a major component of many bar soaps:

The Hardness and Softness of Soaps The cation of a soap influences its properties greatly. Lithium soaps are hard and high melting and are used in car lubricants. Softer, more water-soluble sodium soaps are used as common bar soap. Potassium soaps are low melting and used as liquid soaps. The insolubility of calcium soaps explains why it is difficult to clean clothes in hard water, which typically contains Ca^{2+} ions (see Chemical Connections, p. 529).

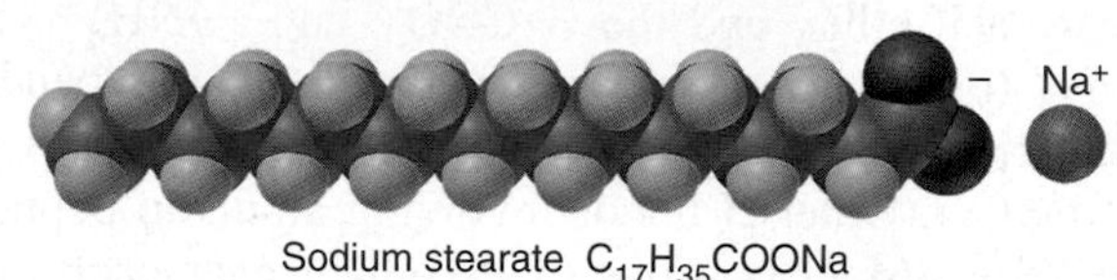

Sodium stearate $C_{17}H_{35}COONa$

When grease on your hands or clothes is immersed in soapy water, the nonpolar tails of the soap molecules interact with the nonpolar grease molecules through dispersion forces, while the polar-ionic heads interact with the water through ion-dipole forces and H bonds. Agitation creates tiny aggregates of grease molecules with embedded soap molecules, whose polar-ionic heads stick into the water. These aggregates are flushed away by added water (Figure 13.8). Detergents have a similar structure of nonpolar tail and polar-ionic head, but they typically consist of an $—SO_3^-$ group (and an inorganic cation) attached to a long hydrocarbon group.

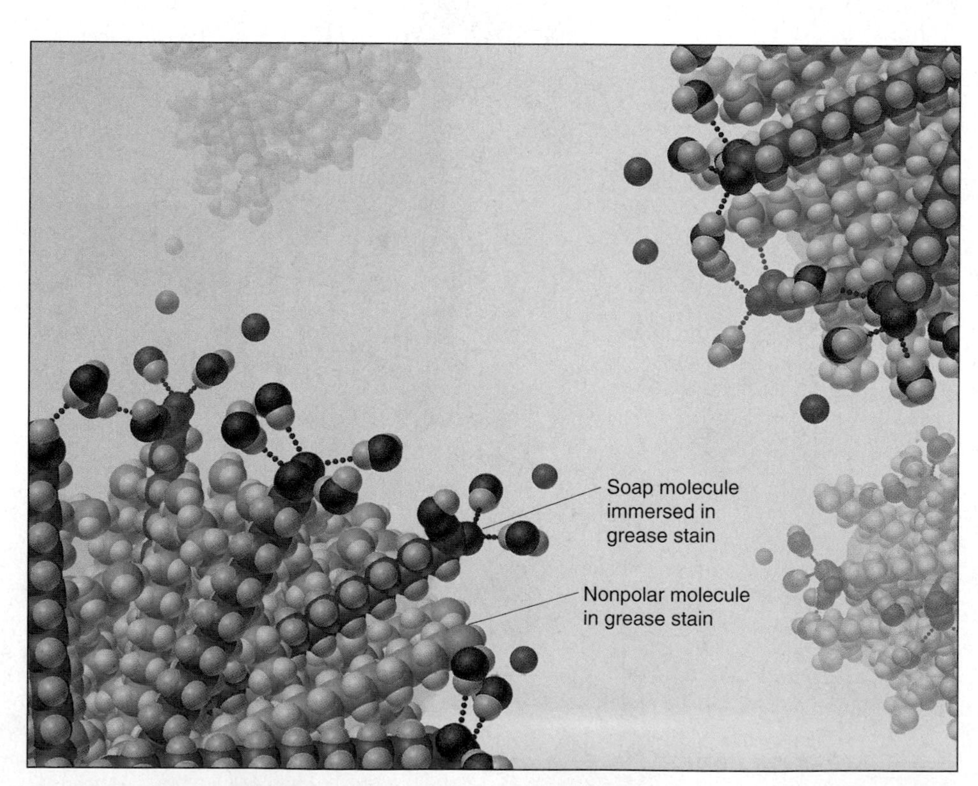

Figure 13.8 The structure and function of a soap. The tails of soap molecules interact with nonpolar molecules in grease through dispersion forces, while the heads interact with water through H bonds and ion-dipole forces. Agitation breaks up the grease stain into small soap-grease aggregates that rinsing flushes away.

The Lipid Bilayer The most abundant molecules in the cell membranes of most species are *phospholipids.* They consist of two long fatty acid chains and a charged organophosphate group all linked to glycerol, a three-carbon trialcohol. Thus, as Figure 13.9 shows, phospholipids, like soaps, have a polar-ionic head and a nonpolar tail.

In water, phospholipids undergo a remarkable self-assembly into an extended sheetlike double layer of molecules, called a **lipid bilayer.** On each surface of the bilayer, the charged heads face the water, and within the interior of the bilayer, the nonpolar tails face each other. So that the tails at the ends of the bilayer are not exposed to the aqueous surroundings, the bilayer tends to close on itself. In the laboratory, bilayers can be made to form spherical vesicles that trap water inside. The result is an energetically favorable structure that optimizes the various intermolecular forces: ion-dipole forces between polar heads and water inside and out, dispersion forces between nonpolar tails within the bilayer, and minimum contact between nonpolar tails and water. Experiments show that the lipid bilayer is impermeable to ions and most polar molecules, including water.

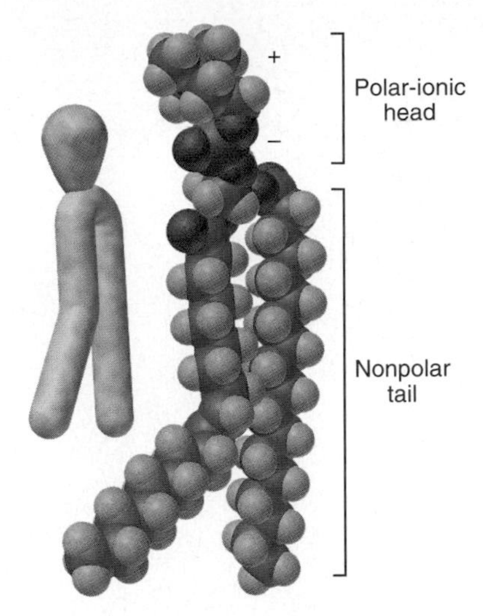

Figure 13.9 Membrane phospholipids. The phospholipids found in cell membranes all have similar shapes, with a polar-ionic head and a nonpolar tail; lecithin (phosphatidylcholine) is shown here. The purple-and-gray shape is a simplified representation of the structure.

Structure of the Cell Membrane The cell membrane is a perfect example of the effect of intermolecular forces in a complex biological assembly. A typical animal cell membrane is about 8 nm thick and consists of the phospholipid bilayer embedded with various proteins that may comprise as much as 50% by mass of the membrane. Depending on the type of cell, membrane proteins function as pumps that actively move ions and specific molecules in or out, channels or gates that passively allow ions or molecules in or out, or receptors for small biomolecules (see Chemical Connections, pp. 390–391), or they may be involved in energy production or metabolic reactions.

The structures of membrane proteins differ in a fundamental way from those of the soluble proteins we discussed previously. In soluble proteins, in most cases, *polar groups are found on the exterior* and *nonpolar groups on the interior,* away from water. Studies in many systems, from red blood cells to photosynthetic bacteria, show that membrane proteins have this arrangement partly inverted: *the exterior of the protein region that lies within the lipid bilayer consists of nonpolar amino-acid side chains.* Dispersion forces between these groups and the hydrocarbon tails of the membrane lipids stabilize this orientation (Figure 13.10).

Figure 13.10 Intermolecular forces and membrane structure.

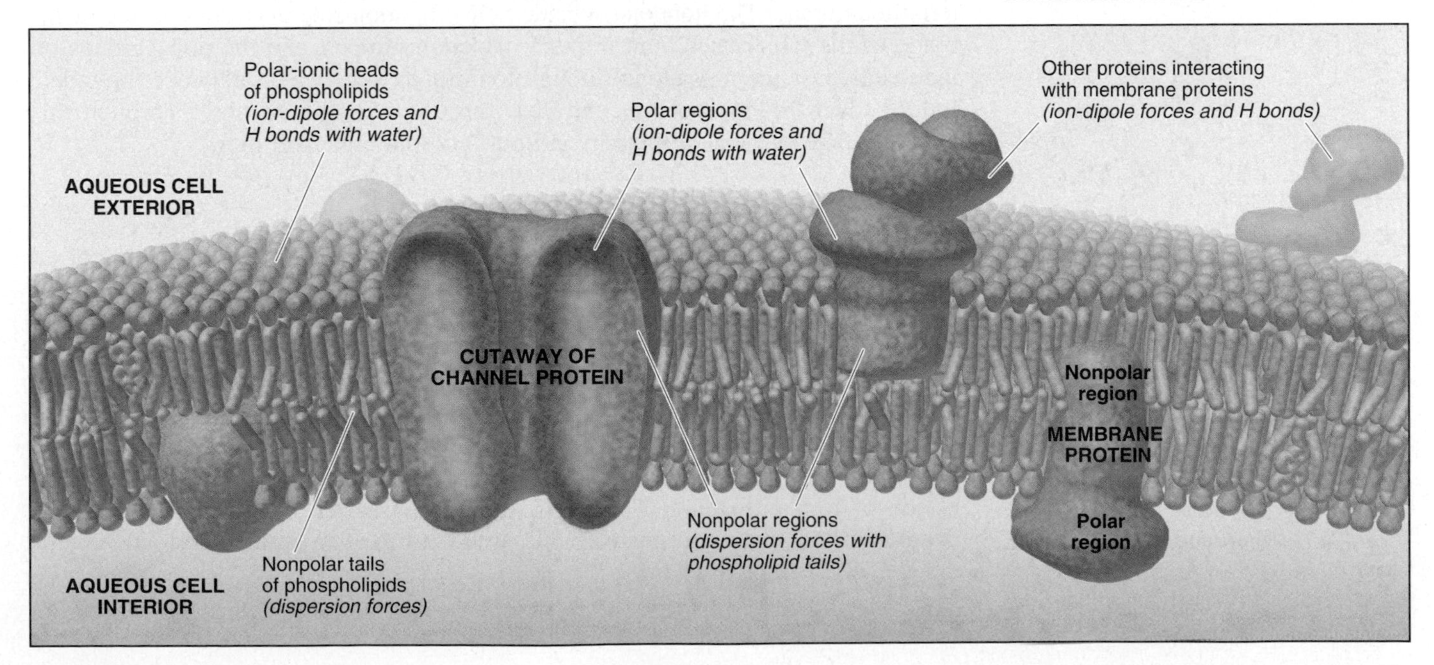

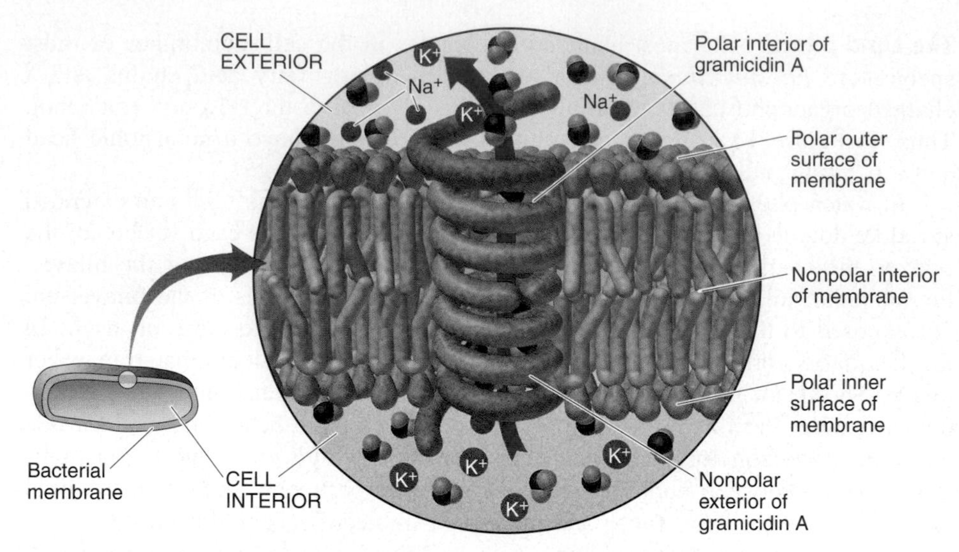

Figure 13.11 The mode of action of an antibiotic. Gramicidin A is a tubelike molecule, nonpolar on the outside and polar on the inside. After dissolving into the nonpolar interior of a bacterial cell membrane via dispersion forces, pairs of gramicidin molecules aligned end to end span the membrane's thickness. Ions diffuse through the polar interior of these molecules, which destroys the cell's ion balance and its segregation of Na^+ outside and K^+ inside.

The Dual Polarity of Antibiotics A clear example of how this orientation of molecules in the cell membrane functions is seen in channel-forming antibiotics. Antibiotics are produced by one microorganism to destroy another, and many have been adapted by medical science to fight bacterial infection.

One function of the membrane is to balance internal and external ion concentrations: Na^+ is excluded from the cell, and K^+ is kept inside. If this ionic segregation is disrupted, the organism dies. Gramicidin A and similar antibiotics disrupt the ion balance by forming a channel in the bacterial membrane through which ions can flow (Figure 13.11). Two helical gramicidin A molecules—with nonpolar groups outside and polar groups inside—lie end to end and span the bacterial membrane. The nonpolar outside helps the molecule dissolve into the membrane of the target bacterium through dispersion forces, and the polar groups on the inside pass the ions along through ion-dipole forces, like a "bucket brigade." Indeed, over 10^7 ions per second pass through a single channel! Through this polar passageway ions diffuse in and out, and the organism dies.

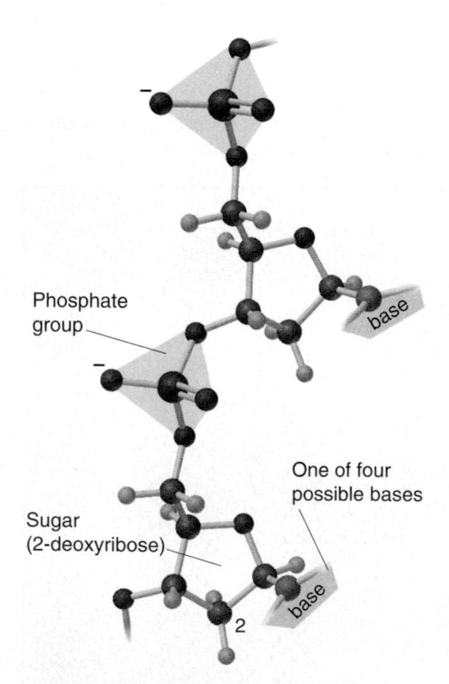

Figure 13.12 A short portion of a DNA chain.

The Structure of DNA

The chemical information that guides the design, construction, and function of all proteins is contained in **nucleic acids,** unbranched polymers that consist of monomers called **mononucleotides.** Each mononucleotide consists of an N-containing base, a sugar, and a phosphate group (Figure 13.12). In DNA (***d**eoxyri-bo**n**ucleic **a**cid*), the sugar is *2-deoxyribose,* in which —H substitutes for —OH on the second C atom of the five-carbon sugar ribose.

The repeating pattern of the backbone of a DNA chain is *sugar linked to phosphate linked to sugar linked to phosphate,* and so on. Attached to each sugar is one of four nitrogen-containing bases. The bases are flat ring structures that dangle off the sugar-phosphate chain, similar to the way amino-acid side chains dangle off the protein chain.

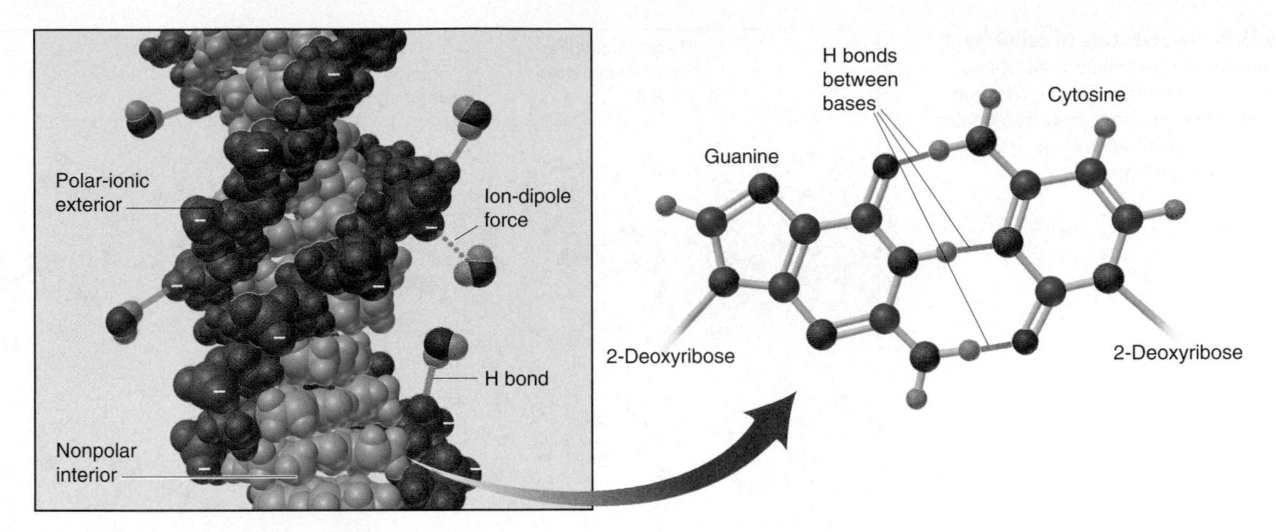

Figure 13.13 The double helix of DNA. In this space-filling segment of DNA (*left*), the polar, negatively charged, sugar-phosphate portion (*pink*) lies on the exterior of the structure and interacts with the aqueous cell fluid through ion-dipole forces and H bonds. Dispersion forces stabilize the stacked, nonpolar bases (*gray*) in the interior. Specific pairs (*right*) are attached by H bonds: a guanine-cytosine pair is shown.

Intermolecular Forces and the Double Helix In the cell nucleus, DNA exists as two chains wrapped around each other in a **double helix** (Figure 13.13). Intermolecular forces play a central role in stabilizing this structure. On the exterior, negatively charged sugar-phosphate chains form ion-dipole and H bonds with the aqueous surroundings. In the interior, the flat, nitrogen-containing bases stack above each other, which allows extensive interaction through dispersion forces. Most important, specific pairs of bases form interchain H bonds; that is, each base in one chain is always H bonded with the same partner base in the other chain. Thus, *the base sequence of one chain is the H-bonded complement of the base sequence of the other.*

A DNA molecule contains millions of H bonds linking bases in these prescribed pairs. The total energy of the H bonds keeps the chains together, but each H bond is weak enough (around 5% of a typical covalent single bond) that a few at a time can break as the chains separate during crucial cellular processes. Linear segments of DNA act as *genes,* chemical blueprints for synthesizing the organism's proteins. In Chapter 15, we'll see how base pairing through H bonding is essential to protein synthesis (the process by which the cell makes its proteins) and DNA replication (the process by which the genes copy themselves during cell division).

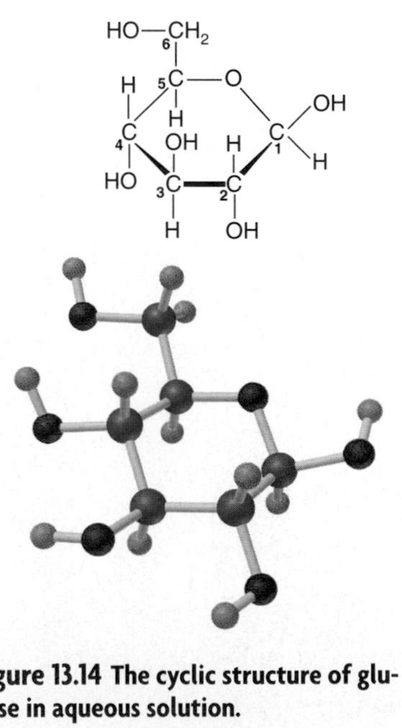

The Structure of Cellulose

Strength and energy storage are the two main functions of **polysaccharides,** polymers consisting of monomers called **monosaccharides,** or simple sugars. The three major natural polysaccharides—cellulose, starch, and glycogen—consist entirely of glucose units, but they differ in the way the units are linked and in the extent of crosslinking. In aqueous systems, glucose occurs as a polyhydroxy six-membered cyclic molecule (Figure 13.14).

The same intermolecular forces operate in all three of these polysaccharides, but we'll focus here on cellulose, the most abundant organic chemical on Earth. Cellulose is the only unbranched polysaccharide of the three and consists entirely of long chains of glucose monomers. Picture parallel chains of linked glucose

Figure 13.14 The cyclic structure of glucose in aqueous solution.

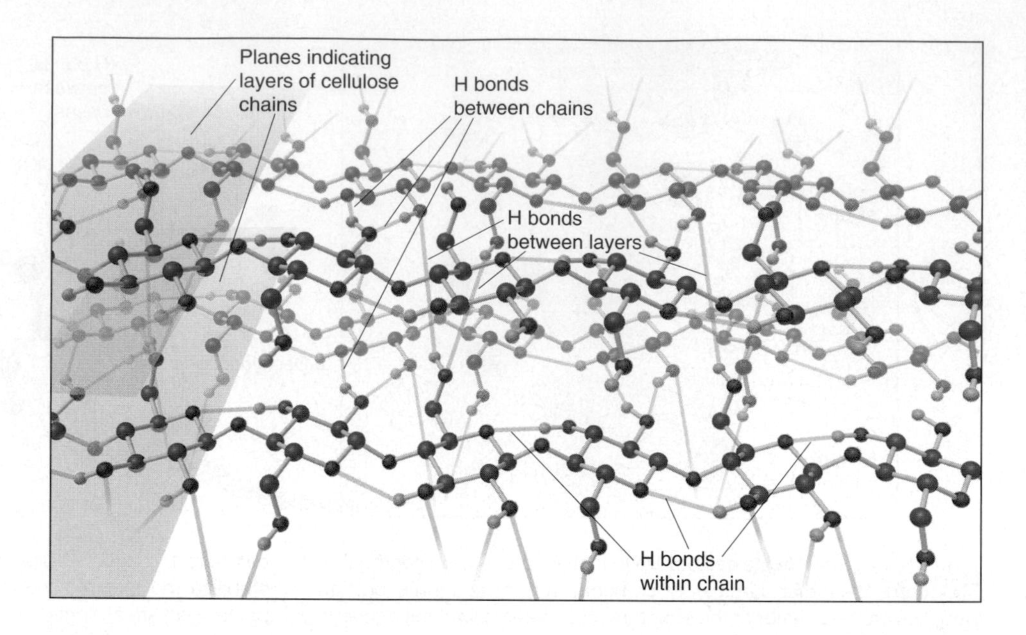

Figure 13.15 The structure of cellulose. The polysaccharide cellulose exists as chains of glucose molecules. Hydrogen bonds within each chain, between chains to form layers, and between layers give cellulose great strength.

units lying next to each other in three dimensions, their —OH groups everywhere (Figure 13.15). The opportunities for dispersion forces and extensive interchain H bonding are enormous. In fact, the great strength of wood is due largely to the countless H bonds among cellulose chains. (We compare the structures of the three polysaccharides in Chapter 15.)

SECTION SUMMARY

Intermolecular forces stabilize the structures of the three biological macromolecules—proteins, nucleic acids, and polysaccharides—and the cell membrane. In soluble proteins, polar and ionic amino-acid side chains face the water, while nonpolar side chains interact with each other in the molecular interior. Soaps have a polar-ionic head and a nonpolar tail that allow them to dissolve grease while interacting with water. Membranes consist of phospholipids that also have two polarities. They assemble in a self-sealing, water-impermeable lipid bilayer. In a cell membrane, the lipid bilayer has embedded proteins whose nonpolar side chains face the surrounding bilayer. Channel-forming antibiotics also have nonpolar exteriors and polar interiors. DNA forms a double helix with a sugar-phosphate, polar-ionic exterior. In the interior, bases H bond in specific pairs and stack through dispersion forces. Cellulose consists of long chains of linked glucose molecules, which interact through extensive H bonding.

13.3 WHY SUBSTANCES DISSOLVE: UNDERSTANDING THE SOLUTION PROCESS

As a qualitative predictive tool, "like dissolves like" is helpful in many cases. As you might expect, this handy *macroscopic* rule is based on the *molecular* interactions that occur between solute and solvent particles. To see *why* like dissolves like, let's break down the solution process conceptually into steps and examine them in terms of changes in *enthalpy* and *entropy* of the system. We discussed enthalpy in Chapter 6 and focus on it first here. The concept of entropy is introduced at the end of this section and treated quantitatively in Chapter 20.

Heats of Solution and Solution Cycles

Picture a general solute and solvent about to form a solution. Both consist of particles attracting each other. For one substance to dissolve in another, three events must occur: (1) solute particles must separate from each other, (2) some solvent

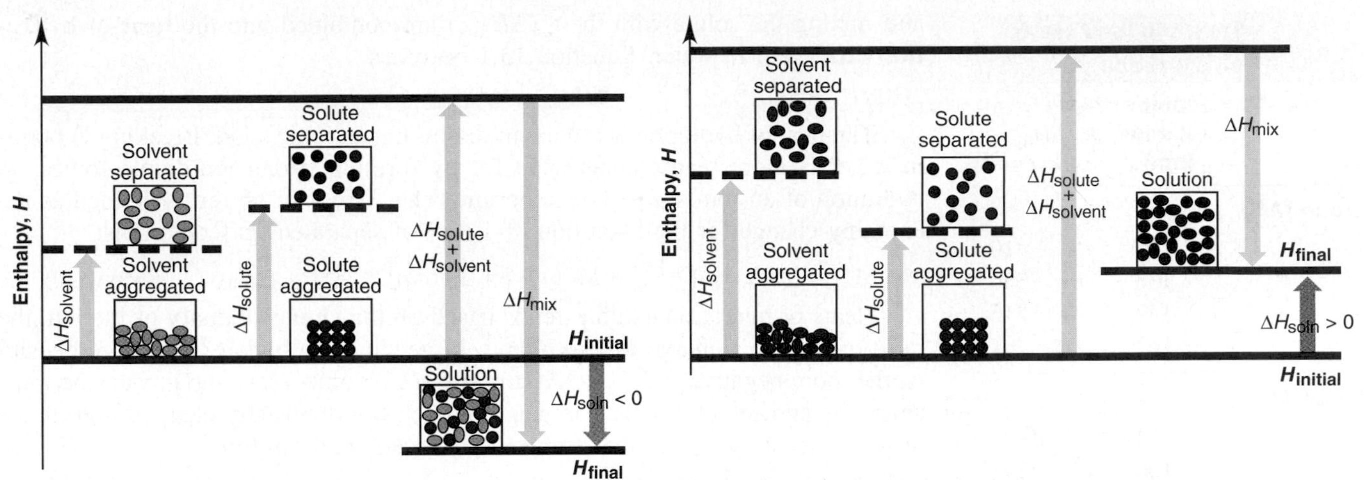

Figure 13.16 Solution cycles and the enthalpy components of the heat of solution. ΔH_{soln} can be thought of as the sum of three enthalpy changes: $\Delta H_{solvent}$ (separating the solvent; always >0), ΔH_{solute} (separating the solute; always >0), and ΔH_{mix} (mixing solute and solvent; always <0). **A,** ΔH_{mix} is larger than the sum of ΔH_{solute} and $\Delta H_{solvent}$, so ΔH_{soln} is negative (exothermic process). **B,** ΔH_{mix} is smaller than the sum of the others, so ΔH_{soln} is positive (endothermic process).

particles must separate to make room for the solute particles, and (3) solute and solvent particles must mix together. No matter what the nature of the attractions within the solute and within the solvent, some energy must be *absorbed* for particles to separate, and some energy is *released* when they mix and attract each other. As a result of these changes, the solution process is typically accompanied by a change in enthalpy. We can divide the process into these three steps, each with its own enthalpy term:

Step 1. Solute particles separate from each other. This step involves overcoming intermolecular attractions, so it is *endothermic:*

$$\text{Solute (aggregated)} + heat \longrightarrow \text{solute (separated)} \qquad \Delta H_{solute} > 0$$

Step 2. Solvent particles separate from each other. This step also involves overcoming attractions, so it is *endothermic*, too:

$$\text{Solvent (aggregated)} + heat \longrightarrow \text{solvent (separated)} \qquad \Delta H_{solvent} > 0$$

Step 3. Solute and solvent particles mix. The particles attract each other, so this step is *exothermic:*

$$\text{Solute (separated)} + \text{solvent (separated)} \longrightarrow \text{solution} + heat \qquad \Delta H_{mix} < 0$$

The total enthalpy change that occurs when a solution forms from solute and solvent is the **heat of solution (ΔH_{soln}),** and we combine the three individual enthalpy changes to find it. The overall process is called a *thermochemical solution cycle* and is yet another application of Hess's law:

$$\Delta H_{soln} = \Delta H_{solute} + \Delta H_{solvent} + \Delta H_{mix} \tag{13.1}$$

(In Section 9.2 we conceptually broke down the heat of formation of an ionic compound into component enthalpies in a similar way using a Born-Haber cycle.) If the sum of the endothermic terms ($\Delta H_{solute} + \Delta H_{solvent}$) is smaller than the exothermic term (ΔH_{mix}), ΔH_{soln} is negative; that is, the process is exothermic. Figure 13.16A is an enthalpy diagram for the formation of such a solution. If the sum of the endothermic terms is larger than the exothermic term, ΔH_{soln} is positive; that is, the process is endothermic (Figure 13.16B). However, if ΔH_{soln} is highly positive, the solute may not dissolve to any significant extent in that solvent.

Heats of Hydration: Ionic Solids in Water

We can simplify the solution cycle for aqueous systems. The $\Delta H_{solvent}$ and ΔH_{mix} components of the heat of solution are difficult to measure individually. Combined, these terms represent the enthalpy change during **solvation,** the process of surrounding a solute particle with solvent particles. Solvation in water is called **hydration.** Thus, enthalpy changes for separating the water molecules ($\Delta H_{solvent}$)

Table 13.4 Trends in Ionic Heats of Hydration

Ion	Ionic Radius (pm)	ΔH_{hydr} (kJ/mol)
Group 1A(1)		
Na^+	102	−410
K^+	138	−336
Rb^+	152	−315
Cs^+	167	−282
Group 2A(2)		
Mg^{2+}	72	−1903
Ca^{2+}	100	−1591
Sr^{2+}	118	−1424
Ba^{2+}	135	−1317
Group 7A(17)		
F^-	133	−431
Cl^-	181	−313
Br^-	196	−284
I^-	220	−247

and mixing the solute with them (ΔH_{mix}) are combined into the **heat of hydration (ΔH_{hydr})**. In water, Equation 13.1 becomes

$$\Delta H_{soln} = \Delta H_{solute} + \Delta H_{hydr}$$

The heat of hydration is crucial in dissolving an ionic solid. Breaking H bonds in water is more than compensated for by forming strong ion-dipole forces, so hydration of an ion is *always* exothermic. The ΔH_{hydr} of an ion is defined as the enthalpy change for the hydration of 1 mol of separated (gaseous) ions:

$$M^+(g) \text{ [or } X^-(g)] \xrightarrow{H_2O} M^+(aq) \text{ [or } X^-(aq)] \qquad \Delta H_{hydr\ of\ the\ ion} \text{ (always } <0)$$

Heats of hydration exhibit trends based on the **charge density** of the ion, the ratio of the ion's charge to its volume. In general, the higher the charge density is, the more negative ΔH_{hydr} is. According to Coulomb's law, the greater the ion's charge is and the closer the ion can approach the oppositely charged end of the water molecule's dipole, the stronger the attraction. Therefore,

- A 2+ ion attracts H_2O molecules more strongly than a 1+ ion of similar size.
- A small 1+ ion attracts H_2O molecules more strongly than a large 1+ ion.

Down a group of ions, for example, from Na^+ to Cs^+ in Group 1A(1), the charge stays the same and the size increases; thus, the charge densities decrease, as do the heats of hydration. Going across the periodic table from Group 1A(1) to Group 2A(2), the 2A ion has a smaller radius *and* a greater charge, so its charge density and ΔH_{hydr} are greater. Table 13.4 shows these relationships for some ions.

The energy required to separate an ionic solute (ΔH_{solute}) into gaseous ions is its lattice energy ($\Delta H_{lattice}$), so ΔH_{solute} is highly positive:

$$M^+X^-(s) \longrightarrow M^+(g) + X^-(g) \qquad \Delta H_{solute} \text{ (always } >0) = \Delta H_{lattice}$$

Thus, the heat of solution for ionic compounds in water combines the lattice energy (always positive) and the combined heats of hydration of cation and anion (always negative),

$$\Delta H_{soln} = \Delta H_{lattice} + \Delta H_{hydr\ of\ the\ ions} \qquad \textbf{(13.2)}$$

The sizes of the individual terms determine the sign of the heat of solution.

Figure 13.17 shows enthalpy diagrams for dissolving three ionic solutes in water. The first, NaCl, has a small positive heat of solution (ΔH_{soln} = 3.9 kJ/mol).

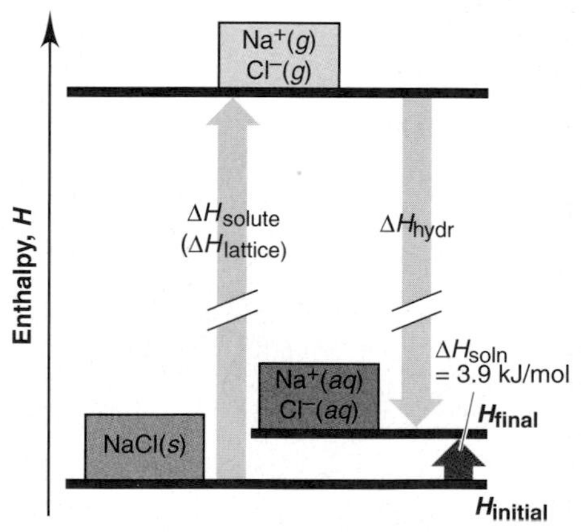

A NaCl. $\Delta H_{lattice}$ is slightly larger than ΔH_{hydr}: ΔH_{soln} is small and positive.

Figure 13.17 Dissolving ionic compounds in water. The enthalpy diagram for an ionic compound in water includes $\Delta H_{lattice}$ (ΔH_{solute}; always positive) and the combined ionic heats of hydration (ΔH_{hydr}; always negative).

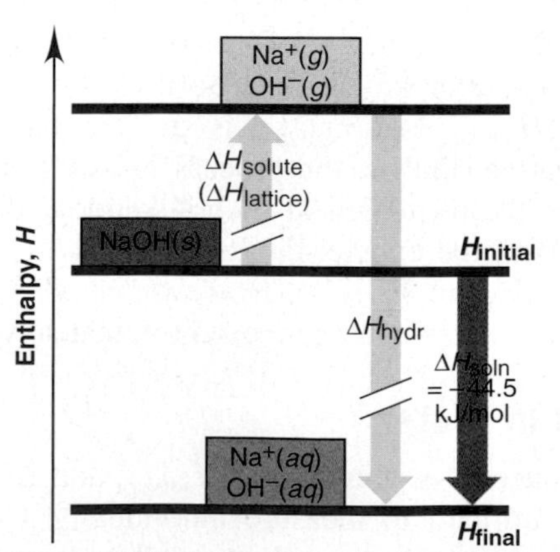

B NaOH. ΔH_{hydr} dominates: ΔH_{soln} is large and negative.

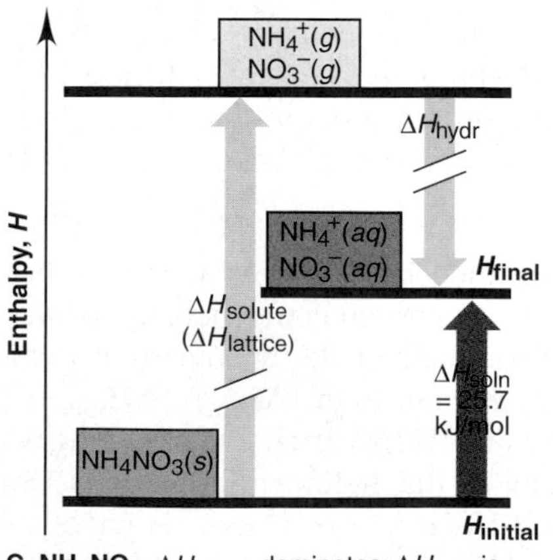

C NH_4NO_3. $\Delta H_{lattice}$ dominates: ΔH_{soln} is large and positive.

Its lattice energy is only slightly greater than the combined ionic heats of hydration, so if you dissolve NaCl in water in a flask, you do not notice any temperature change. However, if you dissolve NaOH in water, the flask feels hot. The lattice energy for NaOH is much smaller than the combined ionic heats of hydration, so dissolving NaOH is highly exothermic ($\Delta H_{soln} = -44.5$ kJ/mol). Finally, if you dissolve NH_4NO_3 in water, the flask feels cold. In this case, the lattice energy is much larger than the combined ionic heats of hydration, so the process is quite endothermic ($\Delta H_{soln} = 25.7$ kJ/mol).

Hot Packs, Cold Packs, and Self-Heating Soup Strain a muscle or sprain a joint and you may obtain temporary relief by applying the appropriate heat of solution. Hot packs and cold packs consist of a thick outer pouch containing water and a thin inner pouch containing a salt. A squeeze on the outer pouch breaks the inner pouch, and the salt dissolves. Most hot packs use anhydrous $CaCl_2$ ($\Delta H_{soln} = -82.8$ kJ/mol), whereas cold packs use NH_4NO_3 ($\Delta H_{soln} = 25.7$ kJ/mol). The change in temperature can be large—a cold pack, for instance, can bring the solution from room temperature down to 0°C—but a pack's usable time is limited to around half an hour. In Japan, some cans of soup have double walls, with an ionic compound in a packet and water between the walls. When you open the can, the packet breaks, and the exothermic solution process, which can quickly reach about 90°C, warms the soup.

The Solution Process and the Change in Entropy

The heat of solution (ΔH_{soln}) is one of two factors that determine whether a solute dissolves in a solvent. The other factor concerns the natural tendency of a system to distribute, or disperse, its energy in as many ways as possible. A thermodynamic variable called **entropy *(S)*** is directly related to the number of ways that a system can distribute its energy, which in turn is closely related to the freedom of motion of the particles and the number of ways they can be arranged.

Let's see what it means for a system to "distribute its energy." We'll first compare the three physical states and then compare solute and solvent with solution. In a solid, the particles are relatively fixed in their positions, but in the liquid, they are free to move around each other. This greater freedom of motion means the particles can distribute their kinetic energy in more ways; thus, the liquid has higher entropy than the solid ($S_{liquid} > S_{solid}$). The gas, in turn, has higher entropy than the liquid because the particles have much more freedom of motion ($S_{gas} > S_{liquid}$). Another way to state this is that the change in entropy when the liquid vaporizes to a gas (ΔS_{vap}) is positive; that is, $\Delta S_{vap} > 0$.

Similarly, *a solution usually has higher entropy than the pure solute and pure solvent.* Here, the number of ways to distribute the energy and the freedom of motion of the particles is related to the number of different interactions between the molecules, and there are far more interactions possible when solute and solvent are mixed than when they are separate; thus, $S_{soln} > (S_{solute} + S_{solvent})$, or $\Delta S_{soln} > 0$.

From everyday experience, we know that solutions form naturally, whereas pure solutes and solvents do not. We'll see in Chapter 20 that energy must be expended to reverse this tendency of systems to distribute their energy in more ways—to get "mixed up." Water treatment facilities, oil refineries, metal foundries, and many other industrial processes expend enormous quantities of energy in order to reverse this natural tendency and separate mixtures into pure components.

The solution process involves the interplay of two factors—the change in enthalpy and the change in entropy. *Systems tend toward a state of lower enthalpy and higher entropy.* In many cases, the relative magnitudes of ΔH_{soln} and ΔS_{soln} determine whether a solution forms.

To see this interplay of enthalpy and entropy, let's consider three solute-solvent pairs in which different influences dominate. In our first example, sodium chloride does not dissolve in hexane (C_6H_{14}), as you would predict from the dissimilar intermolecular forces. As Figure 13.18A on the next page shows, separating the solvent is easy because of the relatively weak dispersion forces, but separating the ionic solute requires supplying $\Delta H_{lattice}$. Mixing releases very little heat because the ion–induced dipole attractions between Na^+ (or Cl^-) ions and hexane molecules are weak. The sum of the endothermic terms is much larger than the exothermic term, so ΔH_{soln} is highly positive. In this case, *a solution does not form because the entropy increase that would accompany the mixing of solute and solvent is much smaller than the enthalpy increase required to separate the ions.*

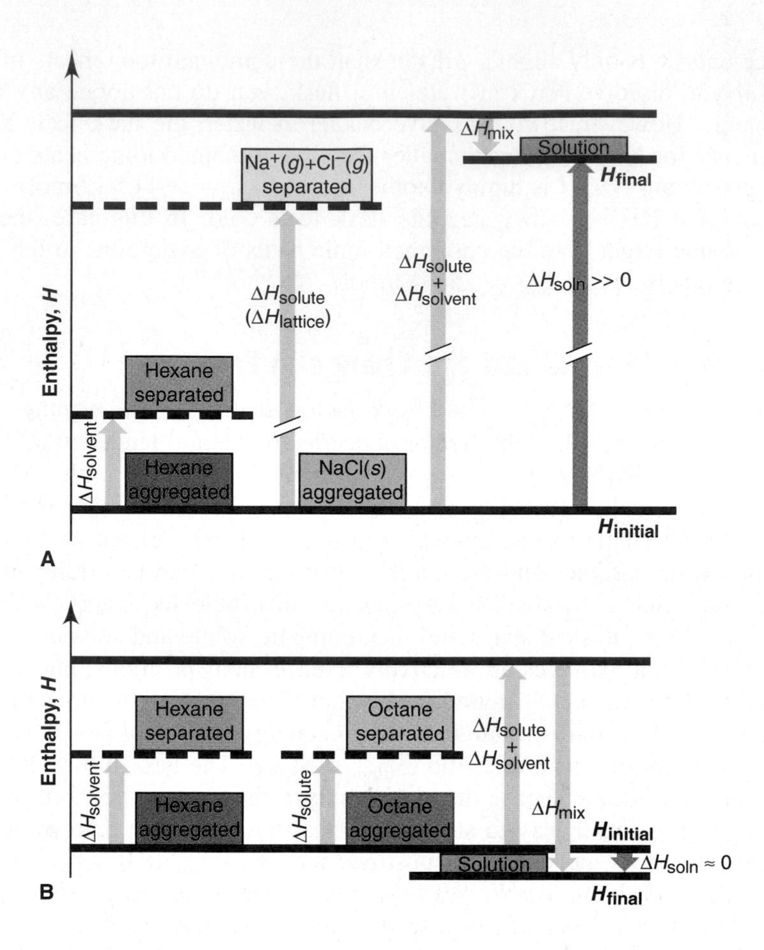

Figure 13.18 Enthalpy diagrams for dissolving NaCl and octane in hexane. **A,** Because attractions between Na^+ (or Cl^-) ions and hexane molecules are weak, ΔH_{mix} is *much* smaller than ΔH_{solute}. Thus, ΔH_{soln} is so positive that NaCl does *not* dissolve in hexane. **B,** Intermolecular forces in octane and in hexane are so similar that ΔH_{soln} is very small. Octane dissolves in hexane because the solution has greater entropy than the pure components.

The second solute-solvent pair is octane (C_8H_{18}) and hexane. Both consist of nonpolar molecules held together by dispersion forces of comparable strength. We therefore predict that octane is soluble in hexane; in fact, they are infinitely soluble (miscible). But this does not necessarily mean that a great deal of heat is released; in fact, ΔH_{soln} is around zero (Figure 13.18B). *With no enthalpy change driving the process, octane dissolves in hexane because the entropy increases greatly when the pure substances mix.* Thus, the large increase in entropy drives formation of this solution.

In some cases, a large enough increase in entropy can cause a solution to form even when the enthalpy increases significantly ($\Delta H_{soln} > 0$). As Figure 13.17C showed, when ammonium nitrate (NH_4NO_3) dissolves in water the process is very endothermic. Nevertheless, in that case, *the increase in entropy that occurs when the crystal breaks down and the ions mix with water molecules more than compensates for the increase in enthalpy.* The enthalpy/entropy interplay in physical and chemical systems is covered in depth in Chapter 20.

SECTION SUMMARY

An overall heat of solution can be obtained from a thermochemical solution cycle as the sum of two endothermic steps (solute separation and solvent separation) and one exothermic step (solute-solvent mixing). In aqueous solutions, the combination of solvent separation and mixing is called *hydration.* Ionic heats of hydration are always negative because of strong ion-dipole forces. Most systems have a natural tendency to increase their entropy (distribute their energy in more ways), and a solution has greater entropy than the pure solute and solvent. The combination of enthalpy and entropy changes determines whether a solution forms. A substance with a positive ΔH_{soln} dissolves *only* if the entropy increase is large enough to outweigh it.

13.4 SOLUBILITY AS AN EQUILIBRIUM PROCESS

When an ionic solid dissolves, ions leave the solid and become dispersed in the solvent. Some dissolved ions collide occasionally with the undissolved solute and recrystallize. As long as the rate of dissolving is greater than the rate of recrystallizing, the concentration of ions rises. Eventually, ions from the solid are dissolving at the same rate as ions in the solution are recrystallizing (Figure 13.19).

Figure 13.19 Equilibrium in a saturated solution. In a saturated solution, equilibrium exists between excess solid solute and dissolved solute. At a particular temperature, the number of solute particles dissolving per unit time equals the number recrystallizing.

At this point, even though dissolving and recrystallizing continue, there is no further change in the concentration with time. The system has reached equilibrium; that is, *excess undissolved solute is in equilibrium with the dissolved solute:*

$$\text{Solute (undissolved)} \rightleftharpoons \text{solute (dissolved)}$$

This solution is called **saturated:** it contains the maximum amount of dissolved solute at a given temperature in the presence of undissolved solute. Filter off the saturated solution and add more solute to it, and the added solute will not dissolve. A solution that contains less than this amount of dissolved solute is called **unsaturated:** add more solute, and more will dissolve until the solution becomes saturated.

In some cases, we can prepare a solution that contains more than the equilibrium amount of dissolved solute. Such a solution is called **supersaturated.** It is unstable relative to the saturated solution: if you add a "seed" crystal of solute, or just tap the container, the excess solute crystallizes immediately, leaving a saturated solution (Figure 13.20). You can often prepare a supersaturated solution of a solute if it has greater solubility at a higher temperature. While warming the contents of the flask, you dissolve more than the amount of solute required to prepare a saturated solution at some lower temperature and then slowly cool the solution. If the excess solute remains dissolved, the solution is supersaturated.

A Saturated Solution Is Like a Pure Liquid and Its Vapor Equilibrium processes in a saturated solution are analogous to those for a pure liquid and its vapor in a closed flask (Section 12.2). For the liquid, rates of vaporizing and condensing are equal; in the solution, rates of dissolving and recrystallizing are equal. In the liquid-vapor system, particles leave the liquid to enter the vapor, and their concentration (pressure) increases until, at equilibrium, the available space above the liquid is "saturated" with vapor at a given temperature. In the solution, solute particles leave the crystal to enter the solvent, and their concentration increases until, at equilibrium, the available solvent is saturated with solute at a given temperature.

Figure 13.20 Sodium acetate crystallizing from a supersaturated solution. When a seed crystal of sodium acetate is added to a supersaturated solution of the compound (**A**), solute begins to crystallize out of solution (**B**) and continues to do so until the remaining solution is saturated (**C**).

Effect of Temperature on Solubility

Temperature affects the solubility of most substances. You may have noticed, for example, that not only does sugar dissolve more quickly in hot tea than in iced tea, but *more* sugar dissolves; in other words, the solubility of sugar in tea is greater at higher temperatures. Let's examine the effects of temperature on the solubility of solids and of gases.

Temperature and the Solubility of Solids Like sugar, *most solids are more soluble at higher temperatures.* Figure 13.21 shows the solubility of several ionic compounds in water as a function of temperature. Note that most of the graphed lines curve upward. Cerium sulfate is the only exception shown in the figure, but several other salts, mostly other sulfates, behave similarly. Some salts exhibit increasing solubility up to a certain temperature and then decreasing solubility at still higher temperatures.

We might think that the sign of ΔH_{soln} would indicate the effect of temperature. Most ionic solids have a positive ΔH_{soln} because their lattice energies are greater than their heats of hydration. Thus, heat is *absorbed* to form the solution from solute and solvent, and if we think of heat as a reactant, a rise in temperature should increase the rate of the forward process:

$$\text{Solute} + \text{solvent} + \textit{heat} \rightleftharpoons \text{saturated solution}$$

Tabulated ΔH_{soln} values refer to the enthalpy change for a solution to reach the standard state of 1 M, but in order to understand the effect of temperature, we need to know the sign of the enthalpy change very close to the point of saturation, which may not be the same. To use earlier examples, tabulated values give a negative ΔH_{soln} for NaOH and a positive one for NH_4NO_3, yet both compounds are more soluble at higher temperatures. The point is that solubility is a complex behavior, and, although the effect of temperature ultimately reflects the equilibrium nature of solubility, no single measure can help us predict the effect for a given solute.

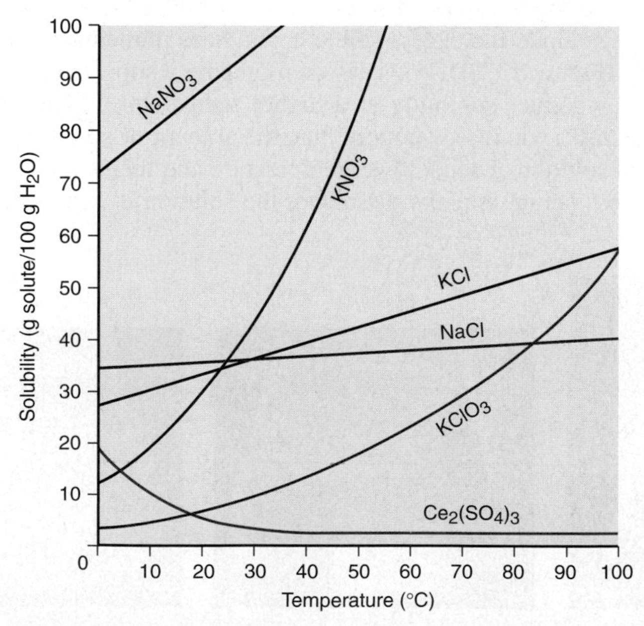

Figure 13.21 The relation between solubility and temperature for several ionic compounds. Most ionic compounds have higher solubilities at higher temperatures. Cerium sulfate is one of several exceptions.

Temperature and Gas Solubility in Water The effect of temperature on gas solubility is much more predictable. When a solid dissolves in a liquid, the solute particles must separate, so energy must be added; thus, for a solid, $\Delta H_{solute} > 0$. In contrast, gas particles are already separated, so $\Delta H_{solute} \approx 0$. Because the hydration step is exothermic ($\Delta H_{hydr} < 0$), the sum of these two terms must be negative. Thus, for all gases in water, $\Delta H_{soln} < 0$:

$$\text{Solute}(g) + \text{water}(l) \rightleftharpoons \text{saturated solution}(aq) + \text{heat}$$

This equation means that *gas solubility in water* ***decreases*** *with rising temperature*. Gases have weak intermolecular forces, so there are relatively weak intermolecular forces between a gas and water. When the temperature rises, the average kinetic energy of the particles in solution increases, allowing the gas particles to easily overcome these weak forces and re-enter the gas phase.

This behavior can lead to an environmental problem known as *thermal pollution*. During many industrial processes, large amounts of water are taken from a nearby river or lake, pumped through the system to cool liquids, gases, and equipment, and then returned to the body of water at a higher temperature. The metabolic rates of fish and other aquatic animals increase in the warmer water released near the plant outlet; thus, their need for O_2 increases, but the concentration of dissolved O_2 is lower at the higher temperature. This oxygen depletion can harm these aquatic populations. Moreover, the warmer water is less dense, so it floats on the cooler water below and prevents O_2 from reaching it. Thus, creatures living at lower levels become oxygen deprived as well. Farther from the plant, the water temperature returns to ambient levels, the O_2 solubility increases, and the temperature layering disappears. One way to lessen the problem is with cooling towers, which cool the water before it exits the plant (Figure 13.22).

Figure 13.22 Fighting thermal pollution. Steam billowing from cooling towers is common at large electric power plants. Hot water resulting from the cooling of gases and equipment enters near the top of the tower and cools in contact with air. The cooler water is then released back into the water source.

Effect of Pressure on Solubility

Because liquids and solids are almost incompressible, pressure has little effect on their solubility, but it has a major effect on gas solubility. Consider the piston-cylinder assembly in Figure 13.23, with a gas above a saturated aqueous solution of the gas. At a given pressure, the same number of gas molecules enter and leave the solution per unit time; that is, the system is at equilibrium:

$$\text{Gas} + \text{solvent} \rightleftharpoons \text{saturated solution}$$

If you push down on the piston, the gas volume decreases, its pressure increases, and gas particles collide with the liquid surface more often. Thus, more gas particles enter than leave the solution per unit time. Higher gas pressure disturbs the

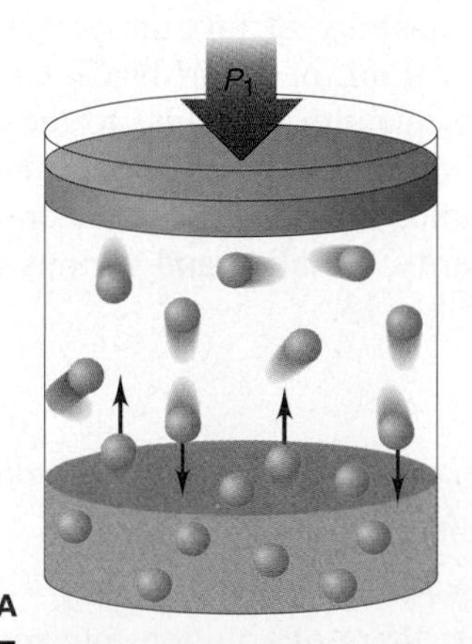

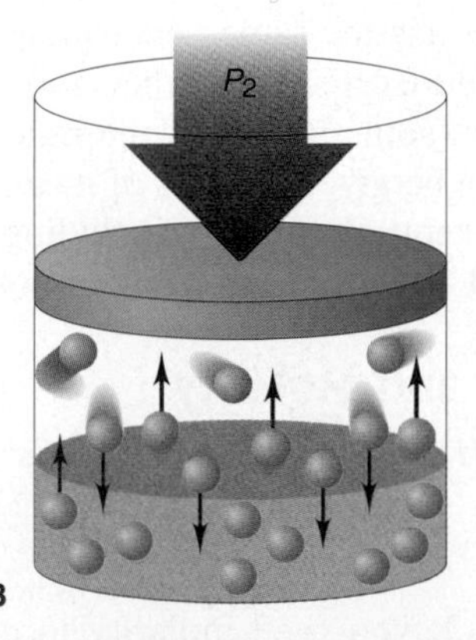

Figure 13.23 The effect of pressure on gas solubility. A, A saturated solution of a gas is in equilibrium at pressure P_1. **B,** If the pressure is increased to P_2, the volume of the gas decreases. As a result, the frequency of collisions with the surface increases, and more gas is in solution when equilibrium is re-established.

Scuba Diving and Soda Pop Although N_2 is barely soluble in water or blood at sea level, high external pressures significantly increase its solubility. Scuba divers who breathe compressed air and dive below about 50 ft have much more N_2 dissolved in their blood. If they ascend quickly to lower pressures, they may be afflicted with decompression sickness (the "bends"), as the dissolved N_2 bubbles out of their blood and causes painful, sometimes fatal, blockage of capillaries. This condition can be avoided by using less soluble gases, such as He, in the breathing mixtures. In principle, the same thing happens when you open a can of a carbonated drink. In the closed can, dissolved CO_2 is at equilibrium with 4 atm of CO_2 gas in the space above it. In the open can, the dissolved CO_2 is under 1 atm of air ($P_{CO_2} = 3\times10^{-4}$ atm), so CO_2 bubbles out of solution. With time, the system reaches equilibrium with the P_{CO_2} in air: the drink goes "flat."

balance at equilibrium, so more gas dissolves to reduce this disturbance (a shift to the right in the preceding equation) until the system re-establishes equilibrium.

Henry's law expresses the quantitative relationship between gas pressure and solubility: *the solubility of a gas (S_{gas}) is directly proportional to the partial pressure of the gas (P_{gas}) above the solution:*

$$S_{gas} = k_H \times P_{gas} \qquad (13.3)$$

where k_H is the Henry's law constant and is specific for a given gas-solvent combination at a given temperature. With S_{gas} in mol/L and P_{gas} in atm, the units of k_H are mol/L·atm.

SAMPLE PROBLEM 13.2 Using Henry's Law to Calculate Gas Solubility

Problem The partial pressure of carbon dioxide gas inside a bottle of cola is 4 atm at 25°C. What is the solubility of CO_2? The Henry's law constant for CO_2 dissolved in water is 3.3×10^{-2} mol/L·atm at 25°C.

Plan We know P_{CO_2} (4 atm) and the value of k_H (3.3×10^{-2} mol/L·atm), so we substitute them into Equation 13.3 to find S_{CO_2}.

Solution $S_{CO_2} = k_H \times P_{CO_2} = (3.3\times10^{-2}$ mol/L·atm$)(4$ atm$) =$ 0.1 mol/L

Check The units are correct for solubility. We rounded the answer to one significant figure because there is only one in the pressure value.

FOLLOW-UP PROBLEM 13.2 If air contains 78% N_2 by volume, what is the solubility of N_2 in water at 25°C and 1 atm (k_H for N_2 in H_2O at 25°C $= 7\times10^{-4}$ mol/L·atm)?

SECTION SUMMARY

A solution that contains the maximum amount of dissolved solute in the presence of excess solute is saturated. A state of equilibrium exists when a saturated solution is in contact with excess solute, because solute particles are entering and leaving the solution at the same rate. Most solids are more soluble at higher temperatures. All gases have a negative ΔH_{soln} in water, so heating lowers gas solubility in water. Henry's law says that the solubility of a gas is directly proportional to its partial pressure above the solution.

13.5 QUANTITATIVE WAYS OF EXPRESSING CONCENTRATION

Concentration is the *proportion* of a substance in a mixture, so it is an intensive property, one that does not depend on the quantity of mixture present: 1.0 L of 0.1 *M* NaCl has the same concentration as 1.0 mL of 0.1 *M* NaCl. Concentration is most often expressed as the ratio of the quantity of solute to the quantity of *solution,* but sometimes it is the ratio of solute to *solvent*. Because both parts of the ratio can be given in units of mass, volume, or amount (mol), chemists employ several concentration terms, including molarity, molality, and various expressions of "parts of solute per part of solution" (Table 13.5).

Molarity and Molality

Molarity (*M*) is the *number of moles of solute dissolved in 1 L of solution:*

$$\text{Molarity } (M) = \frac{\text{amount (mol) of solute}}{\text{volume (L) of solution}} \qquad (13.4)$$

In Chapter 3, you used molarity to convert liters of solution into moles of dissolved solute. Expressing concentration in terms of molarity may have drawbacks, however. Because volume is affected by temperature, so is molarity. A solution expands when heated, so a unit volume of hot solution contains slightly less solute than a unit volume of cold solution. This can be a source of error in very precise

Table 13.5 Concentration Definitions

Concentration Term	Ratio
Molarity (M)	$\frac{\text{amount (mol) of solute}}{\text{volume (L) of solution}}$
Molality (m)	$\frac{\text{amount (mol) of solute}}{\text{mass (kg) of solvent}}$
Parts by mass	$\frac{\text{mass of solute}}{\text{mass of solution}}$
Parts by volume	$\frac{\text{volume of solute}}{\text{volume of solution}}$
Mole fraction (X)	$\frac{\text{amount (mol) of solute}}{\text{amount (mol) of solute + amount (mol) of solvent}}$

work. More importantly, because of solute-solvent interactions that are difficult to predict, *solution volumes may not be additive;* that is, adding 500. mL of one solution to 500. mL of another may not give 1000. mL. Therefore, in precise work, a solution with a desired molarity may not be easy to prepare.

A concentration term that does not contain volume in its ratio is **molality (m),** *the number of moles of solute dissolved in 1000 g (1 kg) of solvent:*

$$\text{Molality } (m) = \frac{\text{amount (mol) of solute}}{\text{mass (kg) of solvent}} \qquad \textbf{(13.5)}$$

Note that molality includes the quantity of *solvent,* not solution. Molal solutions are prepared by measuring *masses* of solute and solvent, not solvent or solution volume. Mass does not change with temperature, so neither does molality. Moreover, unlike volumes, masses *are* additive: adding 500. g of one solution to 500. g of another *does* give 1000. g of final solution. For these reasons, molality is a preferred unit when temperature, and hence density, may change, as in the examination of solutions' physical properties. It's important to realize that, because 1 L of water has a mass of 1 kg, *molality and molarity are nearly the same for dilute aqueous solutions.*

SAMPLE PROBLEM 13.3 Calculating Molality

Problem What is the molality of a solution prepared by dissolving 32.0 g of $CaCl_2$ in 271 g of water?

Plan To use Equation 13.5, we convert mass of $CaCl_2$ (32.0 g) to amount (mol) with the molar mass (g/mol) and then divide by the mass of water (271 g), being sure to convert from grams to kilograms.

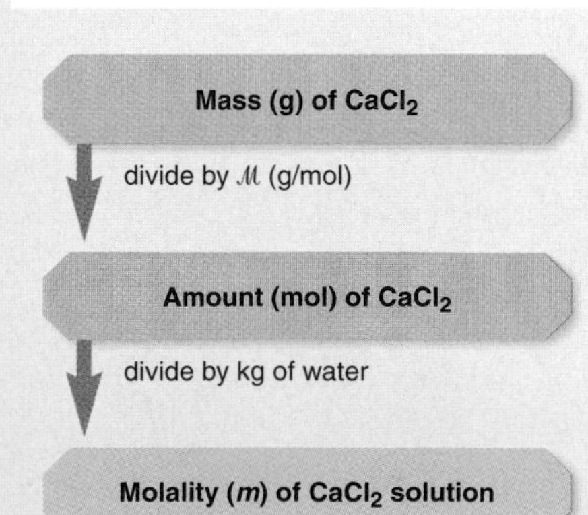

Solution Converting from grams of solute to moles:

$$\text{Moles of } CaCl_2 = 32.0 \text{ g } CaCl_2 \times \frac{1 \text{ mol } CaCl_2}{110.98 \text{ g } CaCl_2} = 0.288 \text{ mol } CaCl_2$$

Finding molality:

$$\text{Molality} = \frac{\text{mol solute}}{\text{kg solvent}} = \frac{0.288 \text{ mol } CaCl_2}{271 \text{ g} \times \frac{1 \text{ kg}}{10^3 \text{ g}}} = 1.06\ m\ CaCl_2$$

Check The answer seems reasonable: the given numbers of moles of $CaCl_2$ and kilograms of H_2O are about the same, so their ratio is about 1.

FOLLOW-UP PROBLEM 13.3 How many grams of glucose ($C_6H_{12}O_6$) must be dissolved in 563 g of ethanol (C_2H_5OH) to prepare a 2.40×10^{-2} m solution?

Parts of Solute by Parts of Solution

Several concentration terms are based on the number of solute (or solvent) parts present in a specific number of *solution* parts. The solution parts can be expressed in terms of mass, volume, or amount (mol).

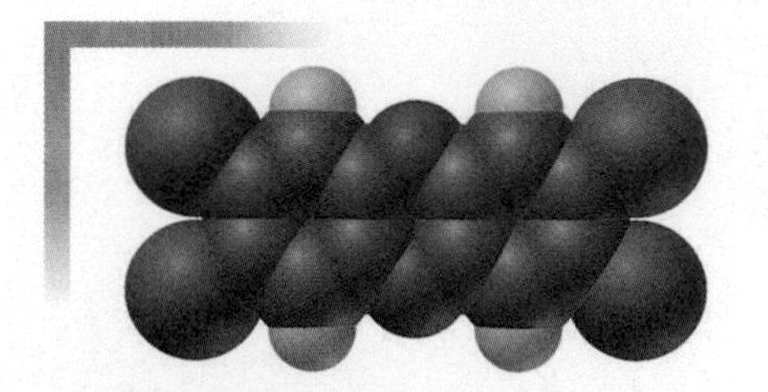

Unhealthy Ultralow Concentrations Environmental toxicologists often measure extremely low concentrations of pollutants. The Centers for Disease Control and Prevention considers TCDD (*t*etra*c*hlorod*ibenzo*d*ioxin, *above*) one of the most toxic substances known. It is a member of the dioxin family of chlorinated hydrocarbons, a by-product of bleaching paper. TCDD is considered unsafe at soil levels above 1 ppb. From contact with air, water, and soil, most North Americans have an average of 0.01 ppb TCDD in their fatty tissues. The Environmental Protection Agency has assessed TCDD as unsafe at *all* levels measured.

Parts by Mass The most common of the parts-by-mass terms is **mass percent,** which you encountered in Chapter 3. The word *percent* means "per hundred," so mass percent of solute means the mass of solute dissolved in every 100. parts by mass of solution, or the mass fraction times 100:

$$\text{Mass percent} = \frac{\text{mass of solute}}{\text{mass of solute} + \text{mass of solvent}} \times 100 = \frac{\text{mass of solute}}{\text{mass of solution}} \times 100 \quad \textbf{(13.6)}$$

Sometimes mass percent is symbolized as **% (w/w),** indicating that the percentage is a ratio of weights (more accurately, masses). You may have seen mass percent values on jars of solid chemicals to indicate the amounts of impurities present. Two very similar terms are parts per million (ppm) by mass and parts per billion (ppb) by mass: grams of solute per million or per billion grams of solution. For these quantities, in Equation 13.6 you multiply by 10^6 or by 10^9, respectively, instead of by 100.

Parts by Volume The most common of the parts-by-volume terms is **volume percent,** the volume of solute in 100. volumes of solution:

$$\text{Volume percent} = \frac{\text{volume of solute}}{\text{volume of solution}} \times 100 \quad \textbf{(13.7)}$$

A common symbol for volume percent is **% (v/v).** Commercial rubbing alcohol, for example, is an aqueous solution of isopropyl alcohol (a three-carbon alcohol) that contains 70 volumes of isopropyl alcohol per 100. volumes of solution, and the label indicates this as "70% (v/v)." Parts-by-volume concentrations are most often used for liquids and gases. Minor atmospheric components occur in parts per million by volume (ppmv). For example, there are about 0.05 ppmv of the toxic gas carbon monoxide (CO) in clean air, 1000 times as much (about 50 ppmv of CO) in air over urban traffic, and 10,000 times as much (about 500 ppmv of CO) in the smoke of one cigarette. Some extremely tiny concentrations are found in nature. *Pheromones* are organic compounds secreted by one member of a species to signal other members about food, danger, sexual readiness, and so forth. Many organisms, including dogs and monkeys, release pheromones, and researchers suspect that humans do as well. Some insect pheromones are active at only a few hundred molecules per milliliter of air, about 100 parts per quadrillion by volume (Figure 13.24).

A measure of concentration frequently used for aqueous solutions is % (w/v), a ratio of solute *weight* (actually mass) to solution *volume*. Thus, a 1.5% (w/v)

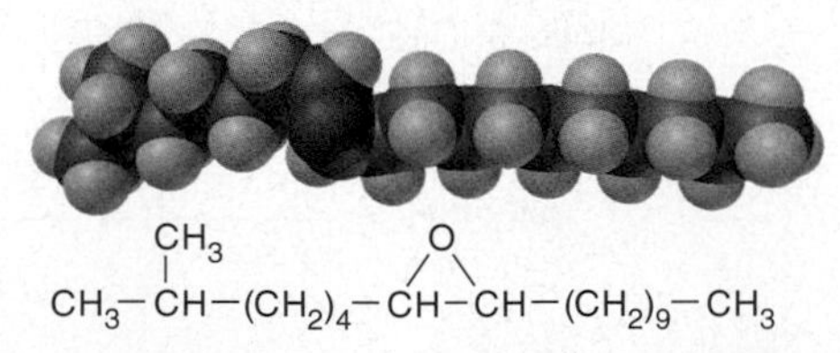

$$\begin{array}{c}\quad CH_3 \qquad\qquad\qquad\quad O \\ CH_3-CH-(CH_2)_4-CH-CH-(CH_2)_9-CH_3\end{array}$$

Figure 13.24 The sex attractant of the gypsy moth. The pheromone secreted by the female gypsy moth (called *disparlure*) can attract a male over 0.5 mi away. Chemists synthesize pheromones for insect pest control.

NaCl solution contains 1.5 g of NaCl per 100. mL of *solution*. This way of expressing concentrations is particularly common in medical labs and other health-related facilities.

Mole Fraction The **mole fraction (*X*)** of a solute is the ratio of number of solute moles to the total number of moles (solute plus solvent), that is, parts by mole. The *mole percent* is the mole fraction expressed as a percentage:

$$\text{Mole fraction } (X) = \frac{\text{amount (mol) of solute}}{\text{amount (mol) of solute} + \text{amount (mol) of solvent}} \quad \textbf{(13.8)}$$

$$\text{Mole percent (mol \%)} = \text{mole fraction} \times 100$$

We discussed these terms in Chapter 5 in relation to Dalton's law of partial pressures for mixtures of gases, but they apply to liquids and solids as well. Concentrations given as mole fractions provide the clearest picture of the actual proportion of solute (or solvent) particles among all the particles in the solution.

SAMPLE PROBLEM 13.4 Expressing Concentrations in Parts by Mass, Parts by Volume, and Mole Fraction

Problem **(a)** Find the concentration of calcium (in ppm) in a 3.50-g pill that contains 40.5 mg of Ca.
(b) The label on a 0.750-L bottle of Italian chianti indicates "11.5% alcohol by volume." How many liters of alcohol does the wine contain?
(c) A sample of rubbing alcohol contains 142 g of isopropyl alcohol (C_3H_7OH) and 58.0 g of water. What are the mole fractions of alcohol and water?

Plan **(a)** We are given the masses of Ca (40.5 mg) and the pill (3.50 g). We convert the mass of Ca from mg to g, find the ratio of mass of Ca to mass of pill, and multiply by 10^6 to obtain ppm. **(b)** We know the volume % (11.5%, or 11.5 parts by volume of alcohol to 100 parts of chianti) and the total volume (0.750 mL), so we use Equation 13.7 to find the volume of alcohol. **(c)** We know the mass and formula of each component, so we convert masses to amounts (mol) and apply Equation 13.8 to find the mole fractions.

Solution **(a)** Finding parts per million by mass of Ca. Combining the steps, we have

$$\text{ppm Ca} = \frac{\text{mass of Ca}}{\text{mass of pill}} \times 10^6 = \frac{40.5 \text{ mg Ca} \times \dfrac{1 \text{ g}}{10^3 \text{ mg}}}{3.50 \text{ g}} \times 10^6 = \boxed{1.16\times10^4 \text{ ppm Ca}}$$

(b) Finding volume (L) of alcohol:

$$\text{Volume (L) of alcohol} = 0.750 \text{ L chianti} \times \frac{11.5 \text{ L alcohol}}{100. \text{ L chianti}} = \boxed{0.0862 \text{ L}}$$

(c) Finding mole fractions. Converting from grams to moles:

$$\text{Moles of } C_3H_7OH = 142 \text{ g } C_3H_7OH \times \frac{1 \text{ mol } C_3H_7OH}{60.09 \text{ g } C_3H_7OH} = 2.36 \text{ mol } C_3H_7OH$$

$$\text{Moles of } H_2O = 58.0 \text{ g } H_2O \times \frac{1 \text{ mol } H_2O}{18.02 \text{ g } H_2O} = 3.22 \text{ mol } H_2O$$

Calculating mole fractions:

$$X_{C_3H_7OH} = \frac{\text{moles of } C_3H_7OH}{\text{total moles}} = \frac{2.36 \text{ mol}}{2.36 \text{ mol} + 3.22 \text{ mol}} = \boxed{0.423}$$

$$X_{H_2O} = \frac{\text{moles of } H_2O}{\text{total moles}} = \frac{3.22 \text{ mol}}{2.36 \text{ mol} + 3.22 \text{ mol}} = \boxed{0.577}$$

Check **(a)** The mass ratio is about 0.04 g/4 g = 10^{-2}, and $10^{-2}\times10^6 = 10^4$ ppm, so it seems correct. **(b)** The volume % is a bit more than 10%, so the volume of alcohol should be a bit more than 75 mL (0.075 L). **(c)** Always check that the *mole fractions add up to 1:* 0.423 + 0.577 = 1.000.

FOLLOW-UP PROBLEM 13.4 An alcohol solution contains 35.0 g of 1-propanol (C_3H_7OH) and 150. g of ethanol (C_2H_5OH). Calculate the mass percent and the mole fraction of each alcohol.

Interconverting Concentration Terms

All the terms we just discussed represent different ways of expressing concentration, so they are interconvertible. Keep these points in mind:

- To convert a term based on amount (mol) to one based on mass, you need the molar mass. These conversions are similar to the mass-mole conversions you've done earlier.
- To convert a term based on mass to one based on volume, you need the solution *density.* Given the mass of a solution, the density (mass/volume) gives you the volume, or vice versa.
- Molality involves quantity of *solvent,* whereas the other concentration terms involve quantity of *solution.*

SAMPLE PROBLEM 13.5 Converting Concentration Terms

Problem Hydrogen peroxide is a powerful oxidizing agent used in concentrated solution in rocket fuels and in dilute solution as a hair bleach. An aqueous solution of H_2O_2 is 30.0% by mass and has a density of 1.11 g/mL. Calculate its
(a) Molality
(b) Mole fraction of H_2O_2
(c) Molarity

Plan We know the mass % (30.0) and the density (1.11 g/mL). **(a)** For molality, we need the amount (mol) of solute and the mass (kg) of *solvent.* Assuming 100.0 g of H_2O_2 solution allows us to express the mass % directly as grams of substance. We subtract the grams of H_2O_2 to obtain the grams of solvent. To find molality, we convert grams of H_2O_2 to moles and divide by mass of solvent (converting g to kg). **(b)** To find the mole fraction, we use the number of moles of H_2O_2 [from part (a)] and convert the grams of H_2O to moles. Then we divide the moles of H_2O_2 by the total moles. **(c)** To find molarity, we assume 100.0 g of solution and use the given solution density to find the volume. Then we divide the amount (mol) of H_2O_2 [from part (a)] by *solution* volume (in L).

Solution (a) From mass % to molality. Finding mass of solvent (assuming 100.0 g of solution):

$$\text{Mass (g) of } H_2O = 100.0 \text{ g solution} - 30.0 \text{ g } H_2O_2 = 70.0 \text{ g } H_2O$$

Converting from grams of H_2O_2 to moles:

$$\text{Moles of } H_2O_2 = 30.0 \text{ g } H_2O_2 \times \frac{1 \text{ mol } H_2O_2}{34.02 \text{ g } H_2O_2} = 0.882 \text{ mol } H_2O_2$$

Calculating molality:

$$\text{Molality of } H_2O_2 = \frac{0.882 \text{ mol } H_2O_2}{70.0 \text{ g} \times \dfrac{1 \text{ kg}}{10^3 \text{ g}}} = 12.6\ m\ H_2O_2$$

(b) From mass % to mole fraction:

$$\text{Moles of } H_2O_2 = 0.882 \text{ mol } H_2O_2 \text{ [from part (a)]}$$

$$\text{Moles of } H_2O = 70.0 \text{ g } H_2O \times \frac{1 \text{ mol } H_2O}{18.02 \text{ g } H_2O} = 3.88 \text{ mol } H_2O$$

$$X_{H_2O_2} = \frac{0.882 \text{ mol}}{0.882 \text{ mol} + 3.88 \text{ mol}} = 0.185$$

(c) From mass % and density to molarity. Converting from solution mass to volume:

$$\text{Volume (mL) of solution} = 100.0 \text{ g} \times \frac{1 \text{ mL}}{1.11 \text{ g}} = 90.1 \text{ mL}$$

Calculating molarity:

$$\text{Molarity} = \frac{\text{mol } H_2O_2}{\text{L soln}} = \frac{0.882 \text{ mol } H_2O_2}{90.1 \text{ mL} \times \dfrac{1 \text{ L soln}}{10^3 \text{ mL}}} = 9.79\ M\ H_2O_2$$

Check Rounding shows the answers seem reasonable: **(a)** The ratio of ~0.9 mol/0.07 kg is greater than 10. **(b)** ~0.9 mol H_2O_2/(1 mol + 4 mol) ≈ 0.2. **(c)** The ratio of moles to liters (0.9/0.09) is around 10.

FOLLOW-UP PROBLEM 13.5 A sample of commercial concentrated hydrochloric acid is 11.8 *M* HCl and has a density of 1.190 g/mL. Calculate the mass %, molality, and mole fraction of HCl.

SECTION SUMMARY

The concentration of a solution is independent of the amount of solution and can be expressed as molarity (mol solute/L solution), molality (mol solute/kg solvent), parts by mass (mass solute/mass solution), parts by volume (volume solute/volume solution), or mole fraction [mol solute/(mol solute + mol solvent)]. The choice of units depends on convenience or the nature of the solution. If, in addition to the quantities of solute and solution, the solution density is also known, all ways of expressing concentration are interconvertible.

13.6 COLLIGATIVE PROPERTIES OF SOLUTIONS

We might expect the presence of solute particles to make the physical properties of a solution different from those of the pure solvent. However, what we might *not* expect is that, in the case of four important solution properties, the *number* of solute particles makes the difference, not their chemical identity. These properties, known as **colligative properties** (*colligative* means "collective"), are vapor pressure lowering, boiling point elevation, freezing point depression, and osmotic pressure. Even though most of these effects are small, they have many practical applications, including some that are vital to biological systems.

Historically, colligative properties were measured to explore the nature of a solute in aqueous solution and its extent of dissociation into ions. In Chapter 4, we classified solutes by their ability to conduct an electric current, which requires moving ions to be present. Recall that an aqueous solution of an **electrolyte** conducts a current because the solute separates into ions as it dissolves. Soluble salts, strong acids, and strong bases dissociate completely. Their solutions conduct a large current, so these solutes are *strong electrolytes*. Weak acids and weak bases dissociate very little. They are *weak electrolytes* because their solutions conduct little current. Many compounds, such as sugar and alcohol, do not dissociate into ions at all. They are **nonelectrolytes** because their solutions do not conduct a current. Figure 13.25 shows these behaviors.

A Strong electrolyte

B Weak electrolyte

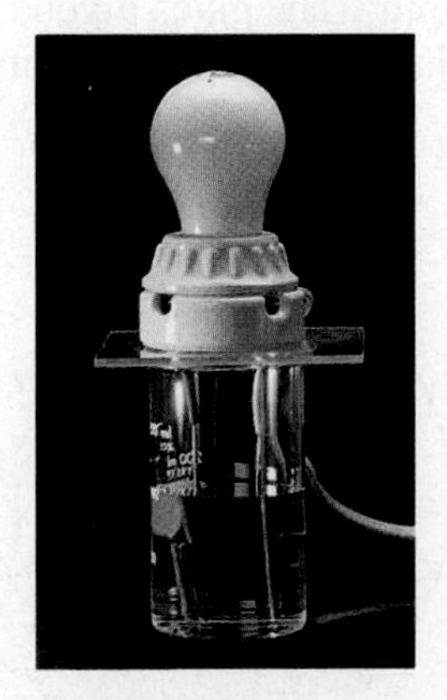

C Nonelectrolyte

Figure 13.25 The three types of electrolytes. A, Strong electrolytes conduct a large current because they dissociate completely into ions. **B,** Weak electrolytes conduct a small current because they dissociate very little. **C,** Nonelectrolytes do not conduct a current because they do not dissociate.

To predict the magnitude of a colligative property, we refer to the solute formula to find the number of particles in solution. Each mole of nonelectrolyte yields 1 mol of particles in the solution. For example, 0.35 *M* glucose contains 0.35 mol of solute particles per liter. In principle, each mole of strong electrolyte dissociates into the number of moles of ions in the formula unit: 0.4 *M* Na_2SO_4 contains 0.8 mol of Na^+ ions and 0.4 mol of SO_4^{2-} ions, or 1.2 mol of particles, per liter (see Sample Problem 4.1, p. 137). (We examine the equilibrium nature of the dissociation of weak electrolytes in Chapters 18 and 19.)

Colligative Properties of Nonvolatile Nonelectrolyte Solutions

In this section, we focus most of our attention on the simplest case, the colligative properties of solutes that do not dissociate into ions and have negligible vapor pressure even at the boiling point of the solvent. Such solutes are called *nonvolatile nonelectrolytes;* sucrose (table sugar) is an example. Later, we briefly explore the properties of volatile nonelectrolytes and of strong electrolytes.

Vapor Pressure Lowering The vapor pressure of a solution of a nonvolatile nonelectrolyte is always *lower* than the vapor pressure of the pure solvent. We can understand this **vapor pressure lowering (ΔP)** in terms of opposing rates and in terms of changes in entropy. Consider a pure solvent and the opposing rates of vaporization (molecules leaving the liquid) and of condensation (molecules reentering the liquid). At equilibrium, the two rates are equal. When we add some nonvolatile solute, the number of solvent molecules on the surface is lower, so fewer vaporize per unit time. To maintain equilibrium, fewer gas molecules can enter the liquid, which occurs only if the concentration of gas, that is, the vapor pressure, is lowered. In terms of the entropy change, recall that natural processes occur in a direction of increasing entropy. A pure solvent vaporizes because the vapor has a greater entropy than the liquid. However, the solvent in a solution already has a greater entropy than it does as pure solvent, so it has less tendency to vaporize in order to gain entropy. Thus, by either argument, equilibrium is reached at a lower vapor pressure for the solution. Figure 13.26 illustrates this point.

In quantitative terms, we find that the vapor pressure of solvent above the solution ($P_{solvent}$) equals the mole fraction of solvent in the solution ($X_{solvent}$) times the vapor pressure of the pure solvent ($P^0_{solvent}$). This relationship is expressed by **Raoult's law:**

$$P_{solvent} = X_{solvent} \times P^0_{solvent} \quad (13.9)$$

In a solution, $X_{solvent}$ is always less than 1, so $P_{solvent}$ is always less than $P^0_{solvent}$. An **ideal solution** is one that follows Raoult's law at any concentration. However, just as most gases deviate from ideality, so do most solutions. In practice, Raoult's law gives a good approximation of the behavior of *dilute* solutions only, and it becomes exact at infinite dilution.

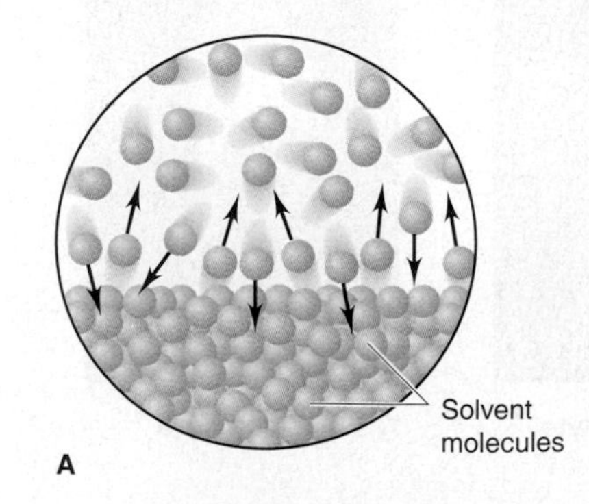

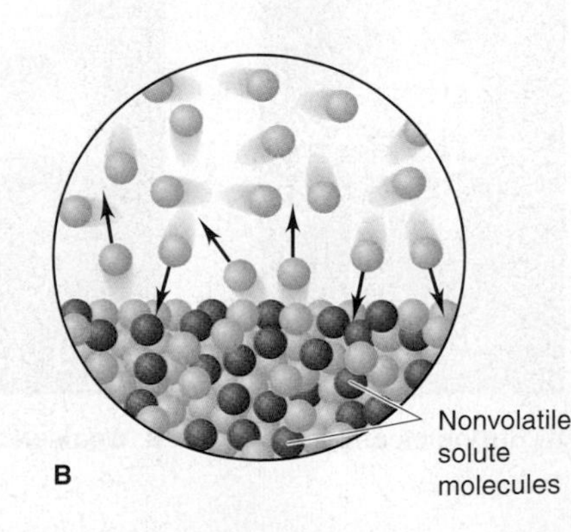

Figure 13.26 The effect of the solute on the vapor pressure of a solution. Vaporization occurs because of the tendency of a system to increase its entropy. **A,** Equilibrium is established between a pure liquid and its vapor when the numbers of molecules vaporizing and condensing in a given time are equal. **B,** The presence of a dissolved solute decreases the number of solvent molecules at the surface and increases the entropy of the system, so fewer solvent molecules vaporize in a given time. Therefore, fewer molecules need to condense to balance them, and equilibrium is established at a lower vapor pressure.

How does the *amount* of solute affect the *magnitude* of the vapor pressure lowering, ΔP? The solution consists of solvent and solute, so the sum of their mole fractions equals 1:

$$X_{solvent} + X_{solute} = 1; \quad \text{thus,} \quad X_{solvent} = 1 - X_{solute}$$

From Raoult's law, we have

$$P_{solvent} = X_{solvent} \times P^0_{solvent} = (1 - X_{solute}) \times P^0_{solvent}$$

Multiplying through on the right side gives

$$P_{solvent} = P^0_{solvent} - (X_{solute} \times P^0_{solvent})$$

Rearranging and introducing ΔP gives

$$P^0_{solvent} - P_{solvent} = \Delta P = X_{solute} \times P^0_{solvent} \quad (13.10)$$

Thus, the *magnitude* of ΔP equals the mole fraction of solute times the vapor pressure of the pure solvent—a relationship applied in the next sample problem.

SAMPLE PROBLEM 13.6 Using Raoult's Law to Find Vapor Pressure Lowering

Problem Calculate the vapor pressure lowering, ΔP, when 10.0 mL of glycerol ($C_3H_8O_3$) is added to 500. mL of water at 50.°C. At this temperature, the vapor pressure of pure water is 92.5 torr and its density is 0.988 g/mL. The density of glycerol is 1.26 g/mL.

Plan To calculate ΔP, we use Equation 13.10. We are given the vapor pressure of pure water ($P^0_{H_2O}$ = 92.5 torr), so we just need the mole fraction of glycerol, $X_{glycerol}$. We convert the given volume of glycerol (10.0 mL) to mass using the given density (1.26 g/L), find the molar mass from the formula, and convert mass (g) to amount (mol). The same procedure gives amount of H_2O. From these, we find $X_{glycerol}$ and ΔP.

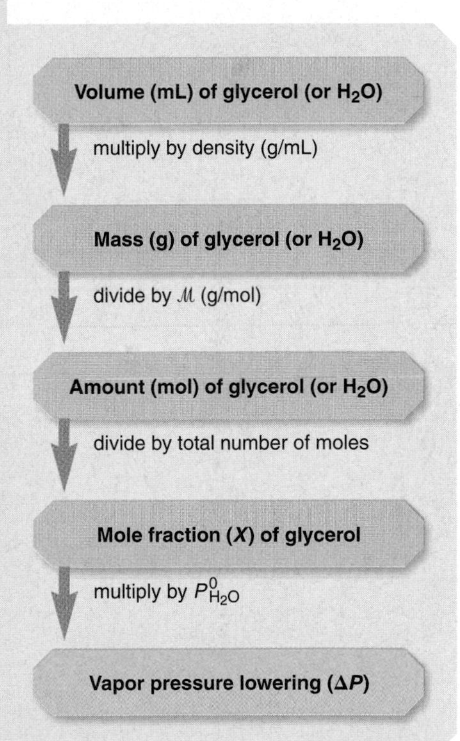

Solution Calculating the amount (mol) of glycerol and of water:

$$\text{Moles of glycerol} = 10.0 \text{ mL glycerol} \times \frac{1.26 \text{ g glycerol}}{1 \text{ mL glycerol}} \times \frac{1 \text{ mol glycerol}}{92.09 \text{ g glycerol}}$$

$$= 0.137 \text{ mol glycerol}$$

$$\text{Moles of } H_2O = 500. \text{ mL } H_2O \times \frac{0.988 \text{ g } H_2O}{1 \text{ mL } H_2O} \times \frac{1 \text{ mol } H_2O}{18.02 \text{ g } H_2O} = 27.4 \text{ mol } H_2O$$

Calculating the mole fraction of glycerol:

$$X_{glycerol} = \frac{0.137 \text{ mol}}{0.137 \text{ mol} + 27.4 \text{ mol}} = 0.00498$$

Finding the vapor pressure lowering:

$$\Delta P = X_{glycerol} \times P^0_{H_2O} = 0.00498 \times 92.5 \text{ torr} = 0.461 \text{ torr}$$

Check The amount of each component seems correct: for glycerol, ~10 mL × 1.25 g/mL ÷ 100 g/mol = 0.125 mol; for H_2O, ~500 mL × 1 g/mL ÷ 20 g/mol = 25 mol. The small ΔP is reasonable because the mole fraction of solute is small.

Comment The calculation assumes that glycerol is nonvolatile. At 1 atm, glycerol boils at 290.0°C, so the vapor pressure of glycerol at 50°C is so low it can be neglected.

FOLLOW-UP PROBLEM 13.6 Calculate the vapor pressure lowering of a solution of 2.00 g of aspirin ($\mathcal{M}$ = 180.15 g/mol) in 50.0 g of methanol (CH_3OH) at 21.2°C. Pure methanol has a vapor pressure of 101 torr at this temperature.

Boiling Point Elevation *A solution boils at a higher temperature than the pure solvent.* Let's see why. The boiling point (boiling temperature, T_b) of a liquid is the temperature at which its vapor pressure equals the external pressure. The vapor pressure of a solution is lower than the external pressure at the solvent's boiling point because the vapor pressure of a solution is lower than that of the pure solvent at any temperature. Therefore, the solution does not yet boil. A higher temperature is needed to raise the solution's vapor pressure to equal the external pressure. We can see this **boiling point elevation (ΔT_b)** by superimposing a phase diagram for the solution on a phase diagram for the pure solvent, as shown in

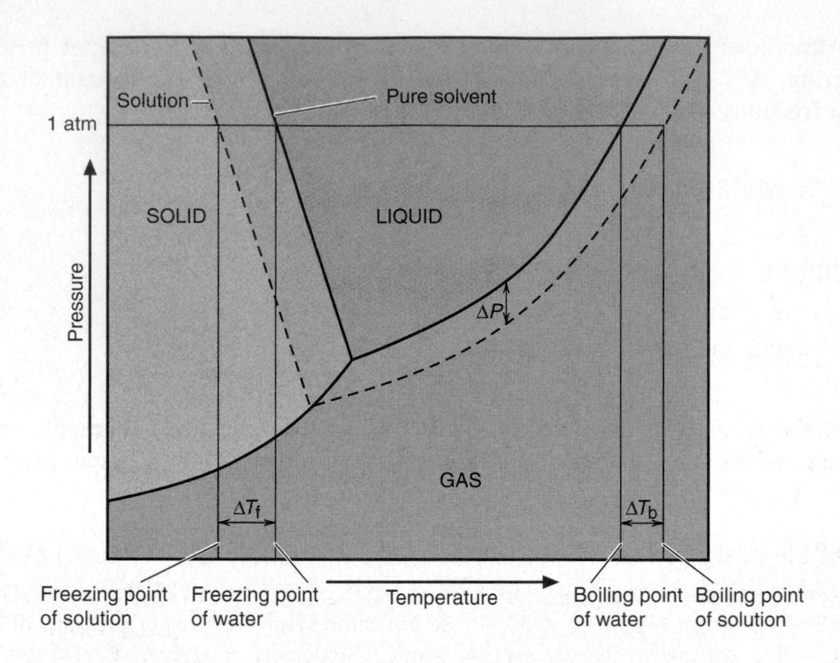

Figure 13.27 Phase diagrams of solvent and solution. Phase diagrams of an aqueous solution (*dashed lines*) and of pure water (*solid lines*) show that, by lowering the vapor pressure (ΔP), a dissolved solute elevates the boiling point (ΔT_b) and depresses the freezing point (ΔT_f).

Figure 13.27. Note that the gas-liquid line for the solution lies *below* that for the pure solvent at any temperature and to the right of it at any pressure.

Like the vapor pressure lowering, the magnitude of the boiling point elevation is proportional to the concentration of solute particles:

$$\Delta T_b \propto m \quad \text{or} \quad \Delta T_b = K_b m \tag{13.11}$$

where m is the solution molality, K_b is the *molal boiling point elevation constant,* and ΔT_b is the boiling point elevation. We typically speak of ΔT_b as a positive value, so we subtract the lower temperature from the higher; that is, we subtract the solvent T_b from the solution T_b:

$$\Delta T_b = T_{b(\text{solution})} - T_{b(\text{solvent})}$$

Molality is the concentration term used because it is related to mole fraction and thus to particles of solute. It also involves mass rather than volume of solvent, so it is not affected by temperature changes. The constant K_b has units of degrees Celsius per molal unit (°C/m) and is specific for a given solvent (Table 13.6).

Notice that the K_b for water is only 0.512°C/m, so the changes in boiling point are quite small: if you dissolved 1.00 mol of glucose (180. g; 1.00 mol of particles) in 1.00 kg of water, or 0.500 mol of NaCl (29.2 g; a strong electrolyte, so also 1.00 mol of particles) in 1.00 kg of water, the boiling points of the resulting solutions at 1 atm would be only 100.512°C instead of 100.000°C.

Table 13.6 Molal Boiling Point Elevation and Freezing Point Depression Constants of Several Solvents

Solvent	Boiling Point (°C)*	K_b (°C/m)	Melting Point (°C)	K_f (°C/m)
Acetic acid	117.9	3.07	16.6	3.90
Benzene	80.1	2.53	5.5	4.90
Carbon disulfide	46.2	2.34	−111.5	3.83
Carbon tetrachloride	76.5	5.03	−23	30.
Chloroform	61.7	3.63	−63.5	4.70
Diethyl ether	34.5	2.02	−116.2	1.79
Ethanol	78.5	1.22	−117.3	1.99
Water	100.0	0.512	0.0	1.86

*At 1 atm.

Freezing Point Depression As you just saw, only solvent molecules can vaporize from the solution, so molecules of the nonvolatile solute are left behind. Similarly, in many cases, *only solvent molecules can solidify,* again leaving solute molecules behind to form a slightly more concentrated solution. The freezing point of a solution is that temperature at which its vapor pressure equals that of the pure solvent. At this temperature, the two phases—solid solvent and liquid solution—are in equilibrium. Because the vapor pressure of the solution is lower than that of the solvent at any temperature, the solution freezes at a lower temperature than the solvent. In other words, the numbers of solvent particles leaving and entering the solid per unit time become equal at a lower temperature. The **freezing point depression (ΔT_f)** is shown in Figure 13.27; note that the solid-liquid line for the solution lies to the left of that for the pure solvent at any pressure.

Like ΔT_b, the freezing point depression has a magnitude proportional to the molal concentration of solute:

$$\Delta T_f \propto m \quad \text{or} \quad \Delta T_f = K_f m \qquad (13.12)$$

where K_f is the *molal freezing point depression constant,* which also has units of °C/m (see Table 13.6). Also like ΔT_b, ΔT_f is considered a positive value, so we subtract the lower temperature from the higher; in this case, however, it is the solution T_f from the solvent T_f:

$$\Delta T_f = T_{f(\text{solvent})} - T_{f(\text{solution})}$$

Here, too, the overall effect in aqueous solution is quite small because the K_f value for water is small—only 1.86°C/m. Thus, 1 m glucose, 0.5 m NaCl, and 0.33 m K_2SO_4, all solutions with 1 mol of particles per kilogram of water, freeze at −1.86°C at 1 atm instead of at 0.00°C.

SAMPLE PROBLEM 13.7 Determining the Boiling Point Elevation and Freezing Point Depression of a Solution

Problem You add 1.00 kg of ethylene glycol ($C_2H_6O_2$) antifreeze to your car radiator, which contains 4450 g of water. What are the boiling and freezing points of the solution?

Plan To find the boiling and freezing points of the solution, we first find the molality by converting the given mass of solute (1.00 kg) to amount (mol) and dividing by mass of solvent (4450 g). Then we calculate ΔT_b and ΔT_f from Equations 13.11 and 13.12 (using constants from Table 13.6). We add ΔT_b to the solvent boiling point and subtract ΔT_f from the solvent freezing point. The roadmap shows the steps.

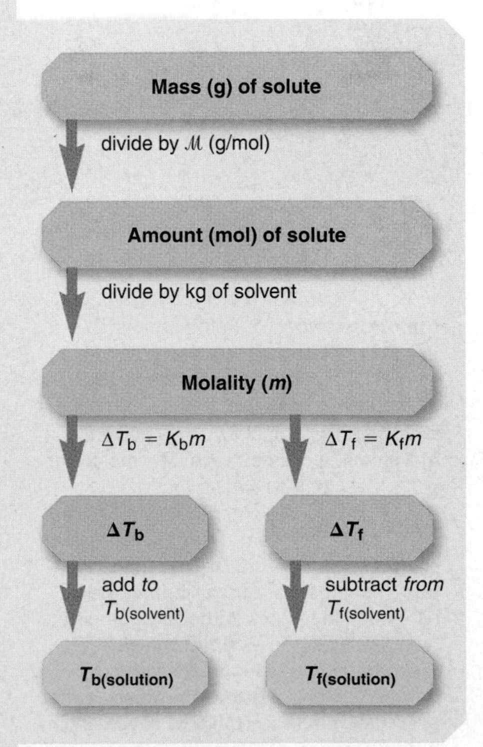

Solution Calculating the molality:

$$\text{Moles of } C_2H_6O_2 = 1.00 \text{ kg } C_2H_6O_2 \times \frac{10^3 \text{ g}}{1 \text{ kg}} \times \frac{1 \text{ mol } C_2H_6O_2}{62.07 \text{ g } C_2H_6O_2} = 16.1 \text{ mol } C_2H_6O_2$$

$$\text{Molality} = \frac{\text{mol solute}}{\text{kg solvent}} = \frac{16.1 \text{ mol } C_2H_6O_2}{4450 \text{ g } H_2O \times \dfrac{1 \text{ kg}}{10^3 \text{ g}}} = 3.62\ m\ C_2H_6O_2$$

Finding the boiling point elevation and $T_{b(\text{solution})}$, with $K_b = 0.512$°C/m:

$$\Delta T_b = \frac{0.512°\text{C}}{m} \times 3.62\ m = 1.85°\text{C}$$

$$T_{b(\text{solution})} = T_{b(\text{solvent})} + \Delta T_b = 100.00°\text{C} + 1.85°\text{C} = 101.85°\text{C}$$

Finding the freezing point depression and $T_{f(\text{solution})}$, with $K_f = 1.86$°C/m:

$$\Delta T_f = \frac{1.86°\text{C}}{m} \times 3.62\ m = 6.73°\text{C}$$

$$T_{f(\text{solution})} = T_{f(\text{solvent})} - \Delta T_f = 0.00°\text{C} - 6.73°\text{C} = -6.73°\text{C}$$

Check The changes in boiling and freezing points should be in the same proportion as the constants used. That is, $\Delta T_b/\Delta T_f$ should equal K_b/K_f: 1.85/6.73 = 0.275 = 0.512/1.86.

Comment These answers are only approximate because the concentration far exceeds that of a *dilute* solution, for which Raoult's law is most useful.

FOLLOW-UP PROBLEM 13.7 What is the minimum concentration (molality) of ethylene glycol solution that will protect the car's cooling system from freezing at 0.00°F? (Assume the solution is ideal.)

Osmotic Pressure The fourth colligative property appears when two solutions of different concentrations are separated by a **semipermeable membrane,** one that allows solvent, but *not* solute, molecules to pass through. This process is called **osmosis.** Many organisms have semipermeable membranes that regulate internal concentrations by osmosis, and synthetic polymers are studied by osmosis.

Consider a simple apparatus in which a semipermeable membrane lies at the curve of a U tube and separates an aqueous sugar solution from pure water. The membrane allows water molecules to pass in *either* direction, but not the larger sugar molecules. Because the solute molecules are present, fewer water molecules touch the membrane on the solution side, so fewer of them leave the solution in a given time than enter it (Figure 13.28A). This *net flow of water into the solution* increases the volume of the solution and thus decreases its concentration.

As the height of the solution rises and that of the solvent falls, the resulting pressure difference pushes some water molecules *from* the solution back through the membrane. At equilibrium, water is pushed out of the solution at the same rate it enters (Figure 13.28B). The pressure difference at equilibrium is the **osmotic pressure (Π),** which is defined as the applied pressure required to *prevent* the net movement of water from solvent to solution (Figure 13.28C).

The osmotic pressure is proportional to the number of solute particles in a given *volume* of solution, that is, to the molarity (M):

$$\Pi \propto \frac{n_{\text{solute}}}{V_{\text{soln}}} \qquad \text{or} \qquad \Pi \propto M$$

The proportionality constant is R times the absolute temperature T. Thus,

$$\Pi = \frac{n_{\text{solute}}}{V_{\text{soln}}} RT = MRT \tag{13.13}$$

The similarity of Equation 13.13 to the ideal gas law ($P = nRT/V$) is not surprising, because both relate the pressure of a system to its concentration and temperature. The Gallery shows some important applications of colligative properties.

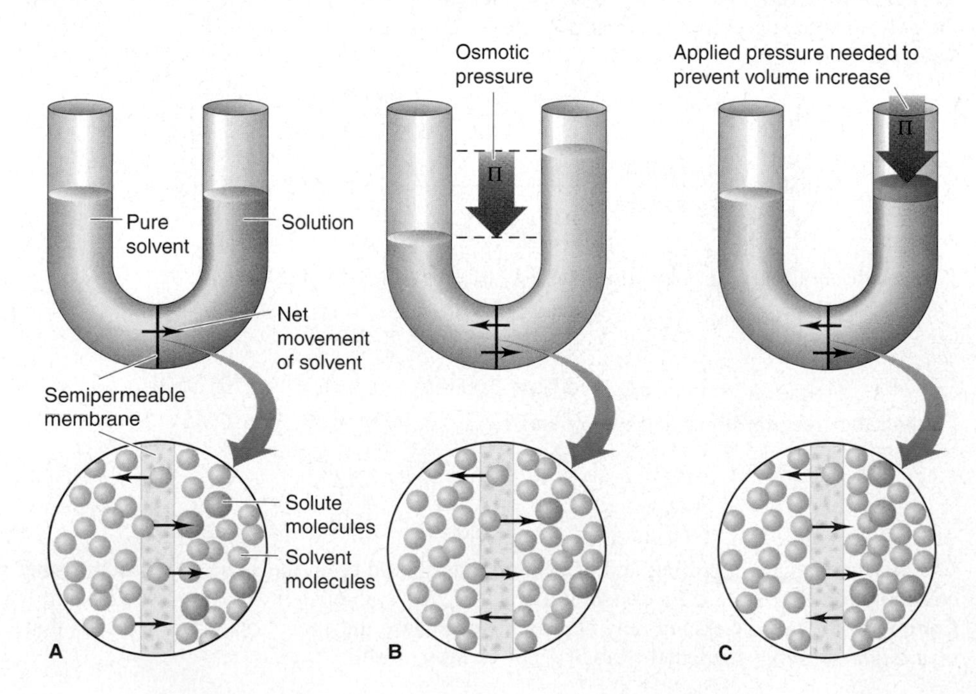

Figure 13.28 The development of osmotic pressure. A, In the process of osmosis, a solution and a solvent (or solutions of different concentrations) are separated by a semipermeable membrane, which allows only solvent molecules to pass through. The molecular-scale view (*below*) shows that more solvent molecules enter the solution than leave it in a given time. **B,** As a result, the solution volume increases, so its concentration decreases. At equilibrium, the difference in heights in the two compartments reflects the *osmotic pressure* (Π). The greater height in the solution compartment exerts a backward pressure that eventually equalizes the flow of solvent in both directions. **C,** Osmotic pressure is defined as the applied pressure required to *prevent* this volume change.

Colligative Properties in Industry and Biology

Colligative properties—especially freezing point depression and osmotic pressure—have much practical relevance to everyday life. Some common applications of freezing point depression make life safer during cold winter months. Others are essential to the electronics industry. Applications of osmotic pressure are found throughout nature and in the health and biological sciences—without question, the most important semipermeable membranes surround living cells.

GALLERY

Applications of Freezing Point Depression

Plane de-icing and car antifreezing Ethylene glycol ($C_2H_6O_2$) is miscible with water through extensive hydrogen bonding and has a high enough boiling point for it to be essentially nonvolatile at 100°C. It is the major ingredient in airplane "de-icers." In "year-round" automobile antifreeze, it lowers the freezing point of water in the radiator in the winter and raises its boiling point in the summer.

Biological antifreeze To survive in the Arctic and in northern winters, many fish and insects, including the common housefly, produce large amounts of glycerol ($C_3H_8O_3$)—a substance with a structure very similar to that of ethylene glycol and also miscible with water—which lowers the freezing point of their blood.

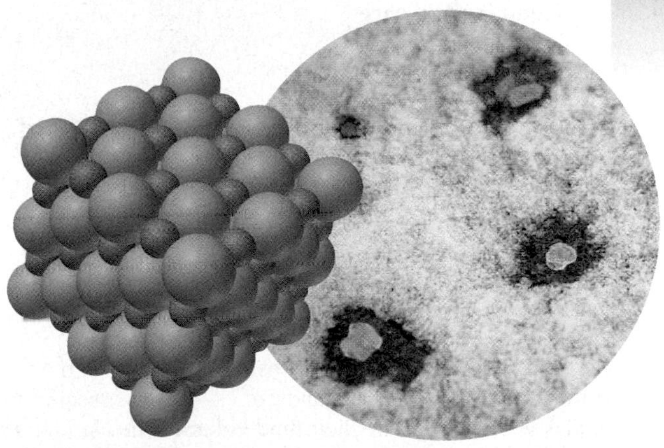

Salts for slippery streets Highway crews use salts, such as mixtures of NaCl and $CaCl_2$, to melt ice on streets. A small amount of salt dissolves in the ice by lowering its freezing point and melting it. More salt dissolves, more ice melts, and so forth. An advantage of $CaCl_2$ is that it has a highly negative ΔH_{soln}, so heat is released when it dissolves, which melts more ice.

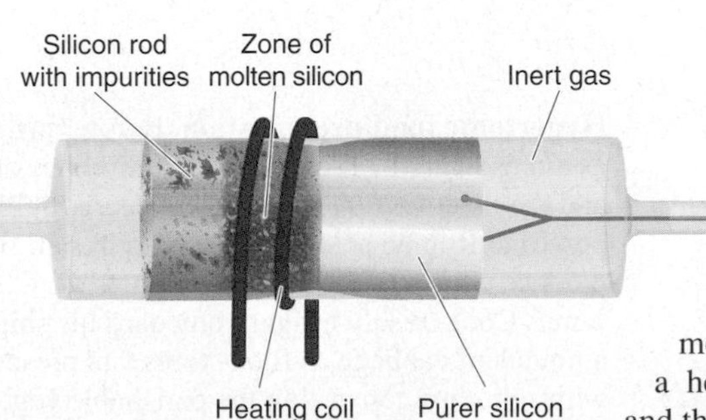

Zone refining Manufacturing computer chips requires extremely pure starting materials. In the process of *zone refining,* a rod of impure silicon (or other metal or metalloid) is passed slowly through a heating coil in an inert atmosphere, and the first narrow zone of impure solid melts. As the next zone melts, the dissolved impurities from the first zone lower the freezing point of the solution, while the purer solvent refreezes. The process continues, zone by zone, throughout the entire rod. After several passes through the coil, in which each zone's impurities move into the adjacent zone, the refrozen silicon at the end of the rod becomes extremely pure—more than 99.999999%. The ultrapure silicon is then sliced into wafers for the production of computer chips (Section 12.7).

(continued)

Applications of Osmotic Pressure

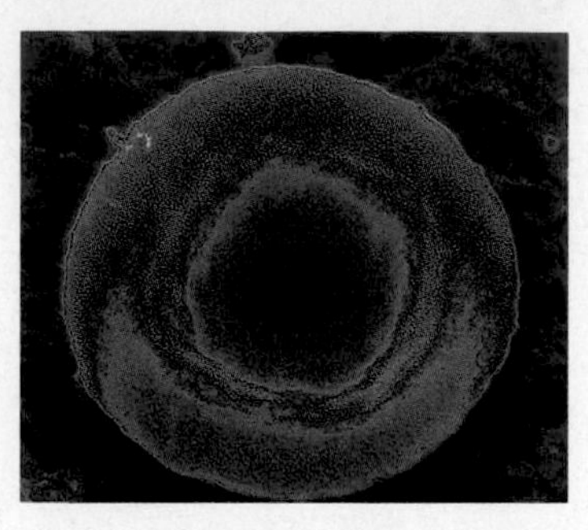

Controlling cell shape The word *tonicity* refers to the tone, or firmness, of a biological cell. An *isotonic* solution has the same concentration of particles as in the cell fluid, so water enters the cell at the same rate that it leaves, thereby maintaining the cell's normal shape.

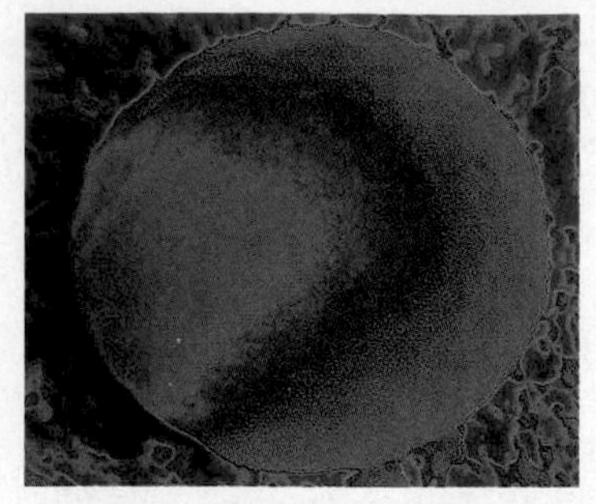

To study cell contents, biochemists rupture membranes by placing the cell in a *hypotonic* solution, one that has a lower concentration of solute particles than the cell fluid. As a result, water enters the cell faster than it leaves, causing the cell to expand and burst.

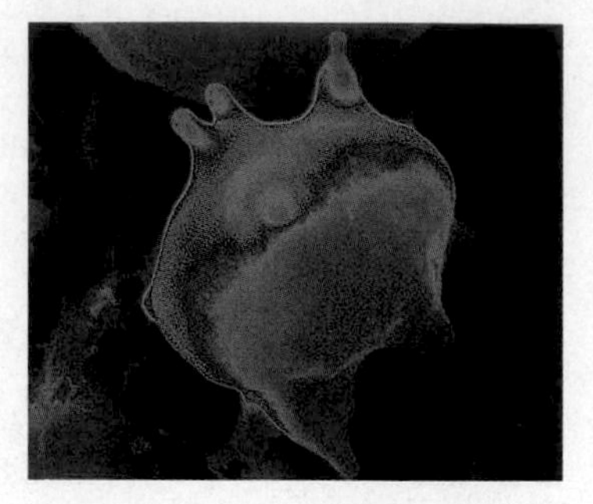

In a *hypertonic* solution, one that contains a higher concentration of solute particles than the cell fluid, the cell shrinks from the net outward flow of water.

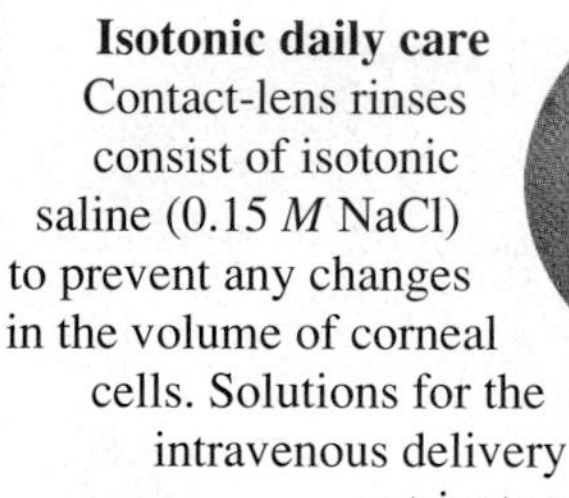

Isotonic daily care Contact-lens rinses consist of isotonic saline (0.15 *M* NaCl) to prevent any changes in the volume of corneal cells. Solutions for the intravenous delivery of nutrients or drugs are always isotonic.

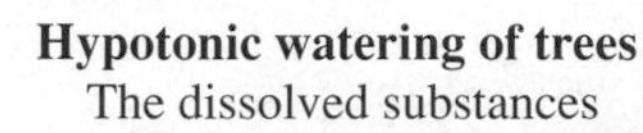

Hypotonic watering of trees The dissolved substances in tree sap create a more concentrated solution than the surrounding groundwater. Water enters membranes in the roots and rises into the tree, creating an osmotic pressure that can exceed 20 atm in the tallest trees!

Sodium ion: the extracellular osmoregulator Of the four major biological cations—Na^+, K^+, Mg^{2+}, and Ca^{2+}—Na^+ is essential for all animals (which includes you!) to regulate their fluid volume. The Na^+ ion accounts for more than 90 mol % of all cations *outside* a cell. A high Na^+ concentration draws water out of the cell by osmosis; a low Na^+ concentration leaves more inside. The primary role of Na^+ is to regulate the water volume of the body, and the primary role of the kidneys is to regulate the concentration of Na^+. Changes in blood pressure (volume) activate nerves and hormones to adjust blood flow and alter kidney function.

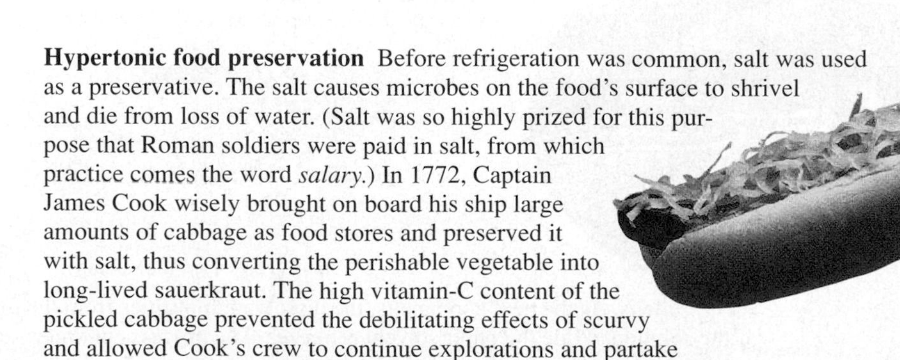

Hypertonic food preservation Before refrigeration was common, salt was used as a preservative. The salt causes microbes on the food's surface to shrivel and die from loss of water. (Salt was so highly prized for this purpose that Roman soldiers were paid in salt, from which practice comes the word *salary.*) In 1772, Captain James Cook wisely brought on board his ship large amounts of cabbage as food stores and preserved it with salt, thus converting the perishable vegetable into long-lived sauerkraut. The high vitamin-C content of the pickled cabbage prevented the debilitating effects of scurvy and allowed Cook's crew to continue explorations and partake in experiments that would revolutionize longitude measurements and, thus, oceangoing navigation.

The Underlying Theme of Colligative Properties A common thread runs through our explanations of the four colligative properties of nonvolatile solutes. Each property rests on the inability of solute particles to cross between two phases. Solute particles cannot enter the gas phase, which leads to vapor pressure lowering and boiling point elevation. They cannot enter the solid phase, which leads to freezing point depression. They cannot cross a semipermeable membrane, which leads to the development of osmotic pressure. The presence of solute decreases the mole fraction of solvent, which lowers the number of solvent particles leaving the solution per unit time; this lowering requires an adjustment to reach equilibrium again. This adjustment to reach the new balance in numbers of particles crossing between two phases per unit time results in the measured colligative property.

Using Colligative Properties to Find Solute Molar Mass

Each colligative property relates concentration to some measurable quantity—the number of degrees the freezing point is lowered, the magnitude of osmotic pressure created, and so forth. From these measurements, we can determine the amount (mol) of solute particles and, for a known mass of solute, the molar mass of the solute as well.

In principle, any of the colligative properties can be used to find the solute's molar mass, but in practice, some systems provide more accurate data than others. For example, to determine the molar mass of an unknown solute by freezing point depression, you would select a solvent with as large a molal freezing point depression constant as possible (see Table 13.6). If the solute is soluble in acetic acid, for instance, a 1 m concentration of it depresses the freezing point of acetic acid by 3.90°C, more than twice the change in water (1.86°C).

Of the four colligative properties, osmotic pressure creates the largest changes and therefore the most precise measurements. Biological and polymer chemists estimate molar masses as great as 10^5 g/mol by measuring osmotic pressure. Because only a tiny fraction of a mole of a macromolecular solute dissolves, it would create too small a change in the other colligative properties.

SAMPLE PROBLEM 13.8 Determining Molar Mass from Osmotic Pressure

Problem Biochemists have discovered more than 400 mutant varieties of hemoglobin, the blood protein that carries oxygen throughout the body. A physician studying a variety associated with a fatal disease first finds its molar mass ($\mathcal{M}$). She dissolves 21.5 mg of the protein in water at 5.0°C to make 1.50 mL of solution and measures an osmotic pressure of 3.61 torr. What is the molar mass of this variety of hemoglobin?

Plan We know the osmotic pressure (Π = 3.61 torr), R, and T (5.0°C). We convert Π from torr to atm, and T from °C to K, and then use Equation 13.13 to solve for molarity (M). Then we calculate the amount (mol) of hemoglobin from the known volume (1.50 mL) and use the known mass (21.5 mg) to find $\mathcal{M}$.

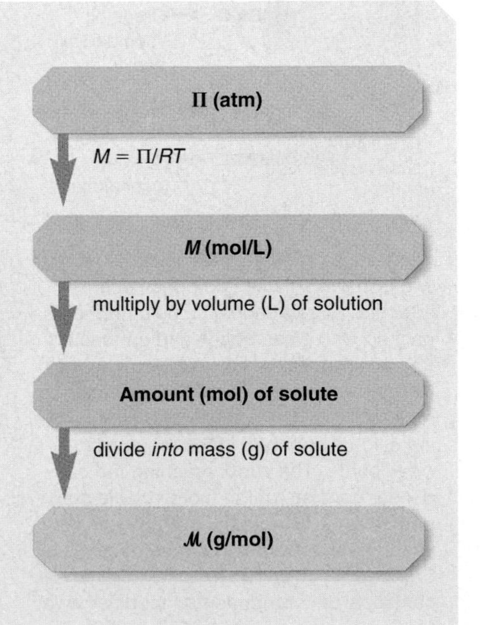

Solution Combining unit conversion steps and solving for molarity from Equation 13.13:

$$M = \frac{\Pi}{RT} = \frac{\dfrac{3.61 \text{ torr}}{760 \text{ torr/1 atm}}}{\left(0.0821 \dfrac{\text{atm·L}}{\text{mol·K}}\right)(273.15 \text{ K} + 5.0)} = 2.08\times10^{-4}\ M$$

Finding amount (mol) of solute (after changing mL to L):

$$\text{Moles of solute} = M \times V = \frac{2.08\times10^{-4} \text{ mol}}{1 \text{ L soln}} \times 0.00150 \text{ L soln} = 3.12\times10^{-7} \text{ mol}$$

Calculating molar mass of hemoglobin (after changing mg to g):

$$\mathcal{M} = \frac{0.0215\ \text{g}}{3.12\times10^{-7}\ \text{mol}} = 6.89\times10^{4}\ \text{g/mol}$$

Check The answers seem reasonable: The small osmotic pressure implies a very low molarity. Hemoglobin is a protein, a biological macromolecule, so we expect a small number of moles [(~2×10^{-4} mol/L) (1.5×10^{-3} L) = 3×10^{-7} mol] and a high molar mass (~21×10^{-3} g/3×10^{-7} mol = 7×10^{4} g/mol).

FOLLOW-UP PROBLEM 13.8 A 0.30 *M* solution of sucrose that is at 37°C has approximately the same osmotic pressure as blood does. What is the osmotic pressure of blood?

Colligative Properties of Volatile Nonelectrolyte Solutions

What is the effect on vapor pressure when the solute *is* volatile, that is, when the vapor consists of solute *and* solvent molecules? From Raoult's law (Equation 13.9), we know that

$$P_{\text{solvent}} = X_{\text{solvent}} \times P^{0}_{\text{solvent}} \quad \text{and} \quad P_{\text{solute}} = X_{\text{solute}} \times P^{0}_{\text{solute}}$$

where X_{solvent} and X_{solute} refer to the mole fractions in the liquid phase. According to Dalton's law of partial pressures, the total vapor pressure is the sum of the partial vapor pressures:

$$P_{\text{total}} = P_{\text{solvent}} + P_{\text{solute}} = (X_{\text{solvent}} \times P^{0}_{\text{solvent}}) + (X_{\text{solute}} \times P^{0}_{\text{solute}})$$

Just as a nonvolatile solute lowers the vapor pressure of the solvent by making the mole fraction of the solvent less than 1, *the presence of each volatile component lowers the vapor pressure of the other* by making each mole fraction less than 1.

Let's examine this effect in a solution that contains equal amounts (mol) of benzene (C_6H_6) and toluene (C_7H_8): $X_{\text{ben}} = X_{\text{tol}} = 0.500$. At 25°C, the vapor pressure of pure benzene (P^{0}_{ben}) is 95.1 torr and that of pure toluene (P^{0}_{tol}) is 28.4 torr; note that benzene is more volatile than toluene. We find the partial pressures from Raoult's law:

$$P_{\text{ben}} = X_{\text{ben}} \times P^{0}_{\text{ben}} = 0.500 \times 95.1\ \text{torr} = 47.6\ \text{torr}$$

$$P_{\text{tol}} = X_{\text{tol}} \times P^{0}_{\text{tol}} = 0.500 \times 28.4\ \text{torr} = 14.2\ \text{torr}$$

As you can see, the presence of benzene lowers the vapor pressure of toluene, and vice versa.

Does the composition of the vapor differ from that of the solution? To see, let's calculate the mole fraction of each substance *in the vapor* by applying Dalton's law. Recall from Section 5.4 that $X_A = P_A/P_{\text{total}}$. Therefore, for benzene and toluene in the vapor,

$$X_{\text{ben}} = \frac{P_{\text{ben}}}{P_{\text{total}}} = \frac{47.6\ \text{torr}}{47.6\ \text{torr} + 14.2\ \text{torr}} = 0.770$$

$$X_{\text{tol}} = \frac{P_{\text{tol}}}{P_{\text{total}}} = \frac{14.2\ \text{torr}}{47.6\ \text{torr} + 14.2\ \text{torr}} = 0.230$$

The vapor composition is very different from the solution composition. The essential point to notice is that *the vapor has a higher mole fraction of the* ***more*** *volatile solution component*. The 50:50 ratio of benzene:toluene in the liquid created a 77:23 ratio of benzene:toluene in the vapor. Condense this vapor into a separate container, and that new *solution* would have this 77:23 composition, and the new *vapor* above it would be enriched still further in benzene.

In the process of **fractional distillation,** this phenomenon is used to separate a mixture of volatile components. Numerous vaporization-condensation steps continually enrich the vapor, until the vapor reaching the top of the fractionating column consists solely of the most volatile component. Figure 13.29 shows how

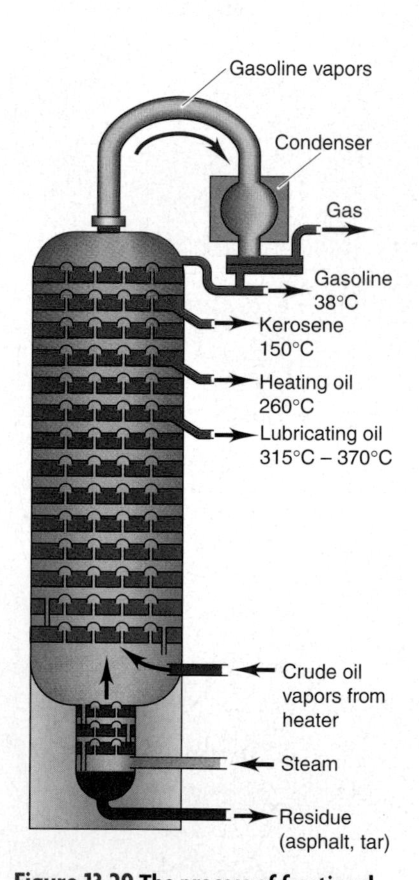

Figure 13.29 The process of fractional distillation. In the laboratory, a solution of two or more volatile components is attached to a *fractionating column* that is packed with glass beads and connected to a condenser. As the solution is heated, the vapor mixture rises and condenses repeatedly on the beads, each time forming a liquid enriched in the more volatile component. The vapor reaching the condenser consists of the more volatile component only. In industry, this process is used to separate petroleum into many products. A 30-m high fractionating tower can separate components that differ by a few tenths of a degree in their boiling points. (The illustration is a simplified version of a multipart process.)

fractional distillation is used in the industrial process of petroleum refining to separate the hundreds of individual compounds in crude oil into a small number of "fractions" based on boiling point range.

Colligative Properties of Strong Electrolyte Solutions

When we consider colligative properties of strong electrolyte solutions, the solute formula tells us the number of particles. For instance, the boiling point elevation (ΔT_b) of 0.050 *m* NaCl should be twice that of 0.050 *m* glucose ($C_6H_{12}O_6$), because NaCl dissociates into two particles per formula unit. Thus, we include a multiplying factor in the equations for the colligative properties of electrolyte solutions. The *van't Hoff factor (i)*, named after the Dutch chemist Jacobus van't Hoff (1852–1911), is the ratio of the *measured* value of the colligative property in the electrolyte solution to the *expected* value for a nonelectrolyte solution:

$$i = \frac{\text{measured value for electrolyte solution}}{\text{expected value for nonelectrolyte solution}}$$

To calculate the colligative properties of strong electrolyte solutions, we incorporate the van't Hoff factor into the equation:

For vapor pressure lowering: $\Delta P = i(X_{\text{solute}} \times P^0_{\text{solvent}})$
For boiling point elevation: $\Delta T_b = i(K_b m)$
For freezing point depression: $\Delta T_f = i(K_f m)$
For osmotic pressure: $\Pi = i(MRT)$

If strong electrolyte solutions behaved ideally, the factor i would be the amount (mol) of particles in solution divided by the amount (mol) of dissolved solute; that is, i would be 2 for NaCl, 3 for $Mg(NO_3)_2$, and so forth. Careful experiment shows, however, that *most strong electrolyte solutions are **not** ideal.* For example, comparing the boiling point elevation for 0.050 *m* NaCl solution with that for 0.050 *m* glucose solution gives a factor i of 1.9, not 2.0:

$$i = \frac{\Delta T_b \text{ of } 0.050\ m \text{ NaCl}}{\Delta T_b \text{ of } 0.050\ m \text{ glucose}} = \frac{0.049°C}{0.026°C} = 1.9$$

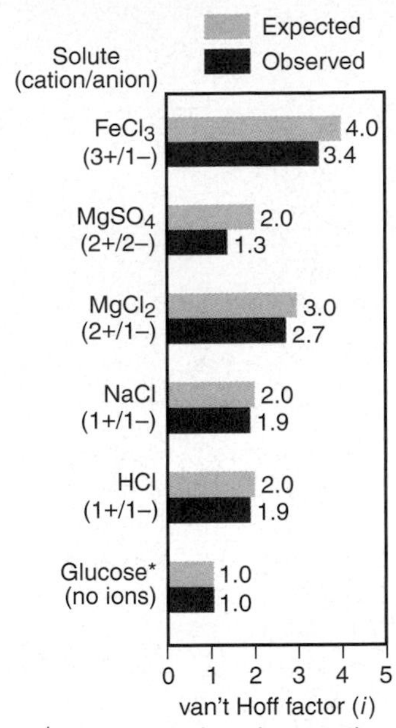

Figure 13.30 Nonideal behavior of strong electrolyte solutions. The van't Hoff factors *(i)* for various ionic solutes in dilute (0.05 *m*) aqueous solution show that the observed value (*dark blue*) is always *lower* than the expected value (*light blue*). This deviation is due to ionic interactions that, in effect, reduce the number of free ions in solution. The deviation is greatest for multivalent ions.

The measured value of the van't Hoff factor is typically *lower* than that expected from the formula. This deviation implies that the ions are not behaving as independent particles. However, we know from other evidence that soluble salts dissociate completely into ions. The fact that the deviation is greater with divalent and trivalent ions is a strong indication that the ionic charge is somehow involved (Figure 13.30).

To explain this nonideal behavior, we picture ions as separate but near each other. Clustered near a positive ion are, on average, more negative ions, and vice versa. Figure 13.31 shows each ion surrounded by an **ionic atmosphere** of net opposite charge. Through these electrostatic associations, each type of ion behaves as if it were "tied up," so its concentration seems *lower* than it actually is. Thus, we often speak of an *effective* concentration, obtained by multiplying i by the *stoichiometric* concentration based on the formula. The greater the charge, the stronger the electrostatic associations, so the deviation from ideal behavior is greater for compounds that dissociate into multivalent ions.

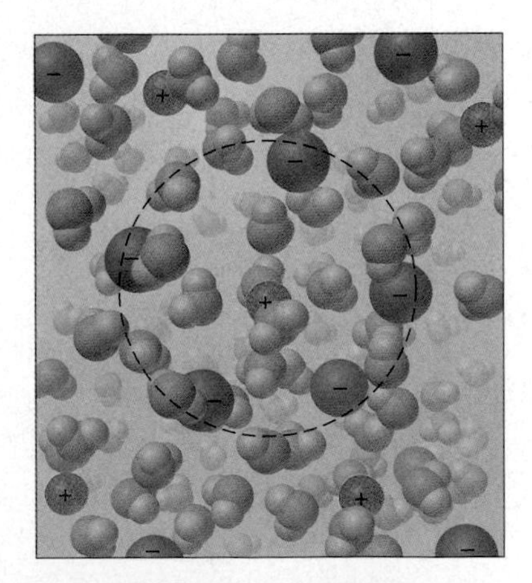

Figure 13.31 An ionic atmosphere model for nonideal behavior of electrolyte solutions. Hydrated anions cluster near cations, and vice versa, to form ionic atmospheres of net opposite charge. Because the ions do not act independently, their concentrations are effectively *less* than expected. Such interactions cause deviations from ideal behavior.

At ordinary conditions and concentrations, nonideal behavior of solutions is much more common (and the deviations much larger) than nonideal behavior of gases, because the particles in solutions are so much closer together. Nevertheless, the two systems exhibit some interesting similarities. Gases display nearly ideal behavior at low pressures because the distances between particles are large. Similarly, van't Hoff factors (i) approach their ideal values as the solution becomes more dilute, that is, as the distance between ions increases. In gases, attractions between particles cause deviations from the expected pressure. In solutions, attractions between particles cause deviations from the expected size of a colligative property. Finally, for both real gases and real solutions, we use empirically determined numbers (van der Waals constants or van't Hoff factors) to transform theories (the ideal gas law or Raoult's law) into more useful relations.

SAMPLE PROBLEM 13.9 Depicting a Solution to Find Its Colligative Properties

Problem A 0.952-g sample of magnesium chloride is dissolved in 100. g of water in a flask. **(a)** Which scene depicts the solution best?

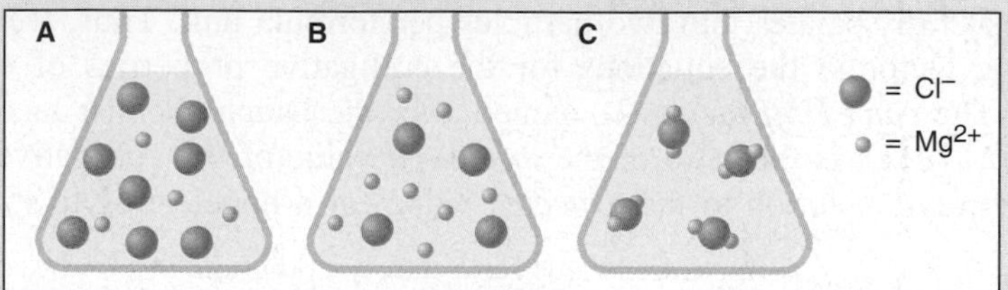

(b) What is the amount (mol) represented by each green sphere?
(c) Assuming the solution is ideal, what is its freezing point (at 1 atm)?

Plan (a) From the name, we recognize an ionic compound, so we determine the formula to find the numbers of cations and anions per formula unit and compare this result with the three scenes: there is 1 magnesium ion for every 2 chloride ions. **(b)** From the given mass of solute, we find the amount (mol); from part (a), there are twice as many moles of chloride ions (green spheres). Dividing by the total number of green spheres gives the moles/sphere. **(c)** From the moles of solute and the given mass (kg) of water, we find the molality (m). We use K_f for water from Table 13.6 and multiply by m to get ΔT_f, and then subtract that from 0.000°C to get the solution freezing point.

Solution (a) The formula is $MgCl_2$; only scene A has 1 Mg^{2+} for every 2 Cl^-.

(b)
$$\text{Moles of } MgCl_2 = \frac{0.952 \text{ g } MgCl_2}{95.21 \text{ g/mol } MgCl_2} = 0.0100 \text{ mol } MgCl_2$$

Therefore,
$$\text{Moles of } Cl^- = 0.0100 \text{ mol } MgCl_2 \times \frac{2\ Cl^-}{1\ MgCl_2} = 0.0200 \text{ mol } Cl^-$$

$$\text{Moles/sphere} = \frac{0.0200 \text{ mol } Cl^-}{8 \text{ spheres}} = 2.50\times10^{-3} \text{ mol/sphere}$$

(c)
$$\text{Molality } (m) = \frac{\text{mol of solute}}{\text{kg of solvent}} = \frac{0.0100 \text{ mol } MgCl_2}{100. \text{ g} \times \frac{1 \text{ kg}}{1000 \text{ g}}} = 0.100\ m\ MgCl_2$$

Assuming an ideal solution, the van't Hoff factor, i, is 3 for $MgCl_2$ because there are 3 ions per formula unit, so we have

$$\Delta T_f = i(K_f m) = 3(1.86°C/m \times 0.100\ m) = 0.558°C$$

And
$$T_f = 0.000°C - 0.558°C = -0.558°C$$

Check Let's quickly check part (c): We have 0.01 mol dissolved in 0.1 kg, which gives 0.1 m. Then, rounding K_f, we have about $3(2°C/m \times 0.1\ m) = 0.6°C$.

FOLLOW-UP PROBLEM 13.9 The $MgCl_2$ solution in the sample problem has a density of 1.006 g/mL at 20.0°C. **(a)** What is the osmotic pressure of the solution? **(b)** A U-tube fitted with a semipermeable membrane is filled with this $MgCl_2$ solution in the left arm and a glucose solution of equal concentration in the right arm. After time, which scene depicts the U-tube best?

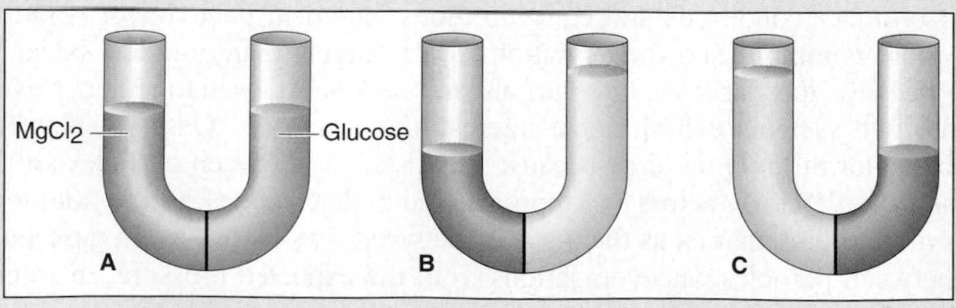

SECTION SUMMARY

Colligative properties are related to the number of dissolved solute particles, not their chemical nature. Compared with the pure solvent, a solution of a nonvolatile nonelectrolyte has a lower vapor pressure (Raoult's law), an elevated boiling point, a depressed freezing point, and an osmotic pressure. Colligative properties can be used to determine the solute molar mass. When solute *and* solvent are volatile, the vapor pressure of each is lowered by the presence of the other. The vapor pressure of the more volatile component is always higher. Electrolyte solutions exhibit nonideal behavior because ionic interactions reduce the effective concentration of the ions.

13.7 THE STRUCTURE AND PROPERTIES OF COLLOIDS

Stir a handful of fine sand into a glass of water, and you'll see the sand particles are suspended at first but gradually settle to the bottom. Sand in water is a **suspension,** a *heterogeneous* mixture containing particles large enough to be seen by the naked eye and clearly distinct from the surrounding fluid. In contrast, stirring sugar into water forms a solution, a *homogeneous* mixture in which the particles are individual molecules distributed evenly throughout the surrounding fluid.

Between the extremes of suspensions and solutions is a large group of mixtures called *colloidal dispersions,* or **colloids,** in which a dispersed (solute-like) substance is distributed throughout a dispersing (solvent-like) substance. Colloidal particles are larger than simple molecules but small enough to remain distributed and not settle out. They range in diameter from 1 to 1000 nm (10^{-9} to 10^{-6} m). A colloidal particle may consist of a single macromolecule (such as a protein or synthetic polymer) or an aggregate of many atoms, ions, or molecules. Whatever their composition, colloidal particles have an enormous total surface area as a result of their small size. A cube with 1-cm sides has a total surface area of 6 cm^2. If it's divided equally into 10^{12} cubes, the cubes are the size of large colloidal particles and have a total surface area of 60,000 cm^2, or 6 m^2. This enormous surface area allows many more interactions to exert a great total adhesive force, which attracts other particles and leads to the practical uses of colloids.

Colloids are classified in Table 13.7 according to whether the dispersed and dispersing substances are gases, liquids, or solids. Many familiar commercial products and natural objects are colloids. Whipped cream is a *foam,* a gas dispersed in a liquid. Firefighting foams are liquid mixtures of proteins in water that are made frothy with fine jets of air. Most biological fluids are aqueous *sols,* solids dispersed in water. Within a typical cell, proteins and nucleic acids are colloidal-size particles dispersed in an aqueous solution of ions and small molecules. The action of soaps and detergents occurs by the formation of an *emulsion,* a liquid (soap dissolved in grease) dispersed in another liquid (water).

"Soaps" in Your Small Intestine In the small intestine, fats are digested by *bile salts,* which are soaplike molecules with smaller polar-ionic and larger nonpolar portions. Secreted by the liver, stored in the gallbladder, and released in the intestine, bile salts emulsify fats just as soap emulsifies grease: fatty aggregates are broken down into particles of colloidal size and dispersed in the watery fluid. The fats are then broken down further and transported into the blood for cellular metabolism.

Table 13.7 Types of Colloids

Colloid Type	Dispersed Substance	Dispersing Medium	Example(s)
Aerosol	Liquid	Gas	Fog
Aerosol	Solid	Gas	Smoke
Foam	Gas	Liquid	Whipped cream
Solid foam	Gas	Solid	Marshmallow
Emulsion	Liquid	Liquid	Milk
Solid emulsion	Liquid	Solid	Butter
Sol	Solid	Liquid	Paint, cell fluid
Solid sol	Solid	Solid	Opal

A

B

Figure 13.32 Light scattering and the Tyndall effect. A, When a beam of light passes through a solution (*left jar*), its path remains narrow and barely visible. When it passes through a colloid (*right jar*), it is scattered and broadened by the particles and thus easily visible. **B,** Sunlight is scattered as it shines through misty air in a forest.

Most colloids are cloudy or opaque, but some are transparent to the naked eye. When light passes through a colloid, it is scattered randomly by the dispersed particles because their sizes are similar to the wavelengths of visible light (400 to 750 nm). Viewed from the side, the scattered light beam is visibly broader than one passing through a solution. This light-scattering phenomenon is called the **Tyndall effect** (Figure 13.32). Dust in air displays this effect when sunlight shines through it, as does mist pierced by headlights at night.

Under low magnification, you can see colloidal particles exhibit *Brownian motion,* a characteristic movement in which the particles change speed and direction erratically. This motion results because the colloidal particles are being pushed this way and that by molecules of the dispersing medium. These collisions are primarily responsible for keeping colloidal particles from settling out. (Einstein's explanation of Brownian motion in 1905 was a principal factor in the acceptance of the molecular nature of matter.)

When colloidal particles collide, why don't they aggregate into larger particles and settle out? Interparticle forces provide the explanation. Water-dispersed colloidal particles remain dispersed because they have charged surfaces that interact strongly with the water molecules through ion-dipole forces. Soap molecules form spherical *micelles,* with the charged heads forming the micelle exterior and the nonpolar tails interacting via dispersion forces in the interior. Aqueous proteins are typically spherical and mimic this micellar arrangement, with charged amino acid groups facing the water and uncharged groups buried within the molecule. Nonpolar oily particles can be dispersed in water by introducing ions, which are adsorbed onto their surfaces by dispersion forces. Charge repulsions between the adsorbed ions prevent the particles from aggregating.

Despite these forces, various methods can coagulate the particles and "destroy" the colloid. Heating a colloid makes the particles move faster and collide more often and with enough force to coalesce into heavier particles that settle out. Adding an electrolyte solution introduces oppositely charged ions that neutralize the particles' surface charges, which allows the particles to coagulate and settle. Uncharged colloidal particles in smokestack gases are removed by creating ions that become adsorbed on the particles, which are then attracted to the charged plates of a Cottrell precipitator. Such precipitators are installed in the smokestacks of coal-burning power plants (Figure 13.33) and collect about 90% of colloidal smoke particulates, thus preventing their release into the air.

The upcoming Chemical Connections essay applies solution and colloid chemistry to the purification of water for residential and industrial use.

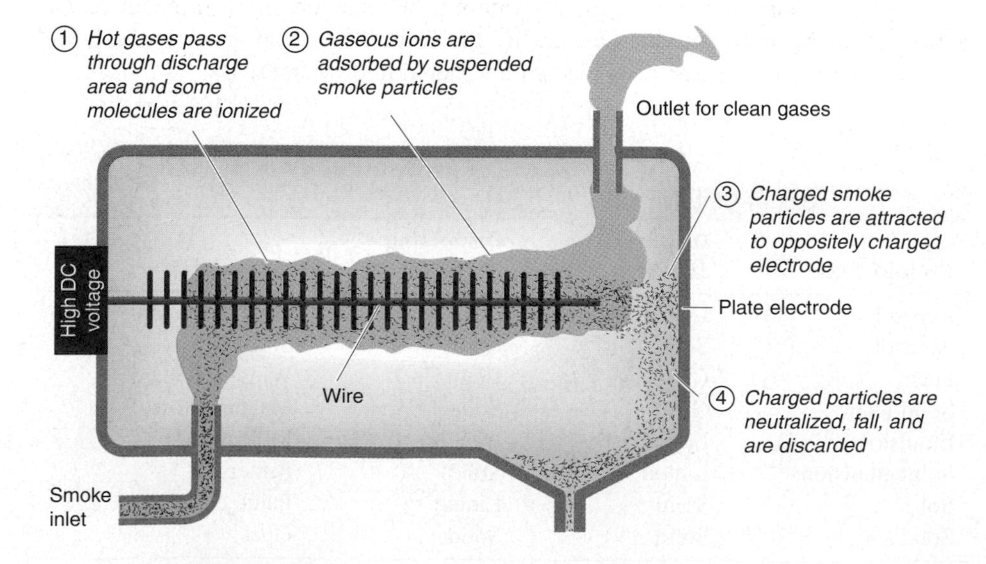
① Hot gases pass through discharge area and some molecules are ionized
② Gaseous ions are adsorbed by suspended smoke particles
Outlet for clean gases
③ Charged smoke particles are attracted to oppositely charged electrode
Plate electrode
High DC voltage
Wire
④ Charged particles are neutralized, fall, and are discarded
Smoke inlet

Figure 13.33 A Cottrell precipitator for removing particulates from industrial smokestack gases.

CHEMICAL CONNECTIONS to Sanitary Engineering

Solutions and Colloids in Water Purification

Clean water is a priceless and limited resource that we've begun to treasure only recently, after decades of pollution and waste. Because of the natural tendency of pure solutes and solvents to form solutions and colloids, it requires energy to remove dissolved, dispersed, and suspended particles from water to make it clean enough for humans to use.

Most water destined for human use comes from lakes, rivers, or reservoirs that may serve also as the final sink after the water is used. Many mineral ions, such as NO_3^- and Fe^{3+}, may be present in high concentrations. Dissolved organic compounds, some of which are toxic, may be present as well. Fine clay particles and a whole spectrum of microorganisms are dispersed in colloidal form. Larger particles and debris of every variety may be present in suspension.

Water Treatment Plants

As a sample of water moves from the natural source into a water treatment facility, the largest particles are physically removed at the intake site by screens (Figure B13.1, step 1). Finer particles, including microorganisms, are removed in large settling tanks by treatment with lime (CaO) and cake alum [$Al_2(SO_4)_3$] (step 2), which react to form a fluffy, gel-like mass of $Al(OH)_3$:

$$3CaO(s) + 3H_2O(l) + Al_2(SO_4)_3(s) \longrightarrow 2Al(OH)_3(\mathit{colloidal\ gel}) + 3CaSO_4(aq)$$

The fine particles are trapped within or adsorbed onto the enormous surface area of the gel, which coagulates, settles out, and is filtered through a sand bed (step 3).

With suspended and colloidal particles removed, the water is aerated in large sprayers to saturate it with oxygen, which speeds the oxidation of dissolved organic compounds (step 4). The water is then disinfected, usually by treatment with Cl_2 gas and/or aqueous solutions of hypochlorite ion, ClO^- (step 5), which may give the water an unpleasant odor. Chlorine can also form toxic chlorinated hydrocarbons, but these can be removed by adsorption onto activated charcoal particles. These steps within the water treatment plant dispose of debris and grit, colloidal clay, microorganisms, and much of the oxidizable organic matter, but dissolved ions remain. Many of these can be removed by water softening and reverse osmosis.

Water Softening via Ion Exchange

Water that contains large amounts of divalent cations, such as Ca^{2+}, Mg^{2+}, and Fe^{2+}, is called **hard water.** These cations cause several problems. During cleaning, they combine with the anions of fatty acids in soaps to produce insoluble deposits on clothes, washing machine parts, and sinks:

$$Ca^{2+}(aq) + \underset{\text{soap}}{2C_{17}H_{35}COONa(aq)} \longrightarrow \underset{\text{insoluble deposit}}{(C_{17}H_{35}COO)_2Ca(s)} + 2Na^+(aq)$$

When a large amount of bicarbonate ion (HCO_3^-) is present in the water, the hard-water cations cause a buildup of *scale*, insoluble carbonate deposits within boilers and hot-water pipes that interfere with the transfer of heat and damage plumbing:

$$Ca^{2+}(aq) + 2HCO_3^-(aq) \xrightarrow{\Delta} CaCO_3(s) + CO_2(g) + H_2O(l)$$

The removal of hard-water ions, called **water softening,** solves these problems. It is accomplished by exchanging "soft-water" Na^+ ions for the hard-water cations. A typical domestic system for **ion exchange** contains an *ion-exchange resin,* an insoluble polymer that has covalently bonded anion groups, such as $—SO_3^-$ or $—COO^-$, to which Na^+ ions are attached to balance

(continued)

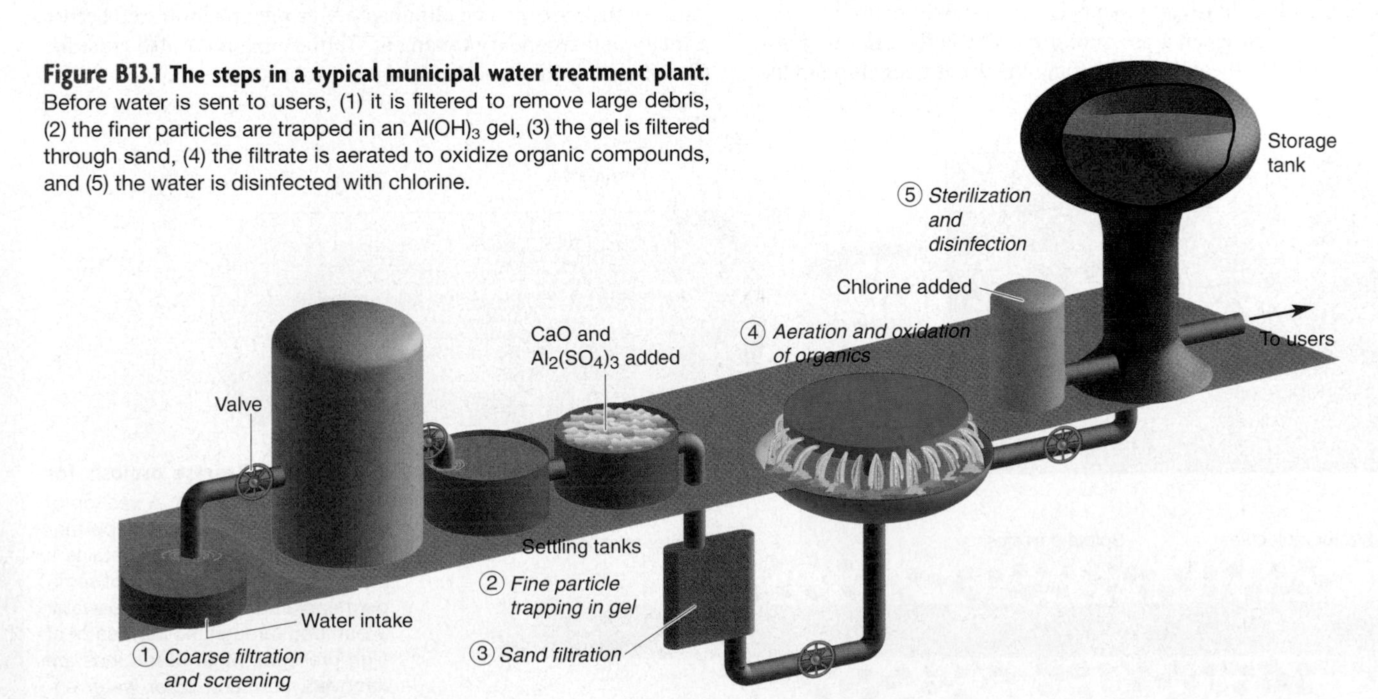

Figure B13.1 The steps in a typical municipal water treatment plant. Before water is sent to users, (1) it is filtered to remove large debris, (2) the finer particles are trapped in an $Al(OH)_3$ gel, (3) the gel is filtered through sand, (4) the filtrate is aerated to oxidize organic compounds, and (5) the water is disinfected with chlorine.

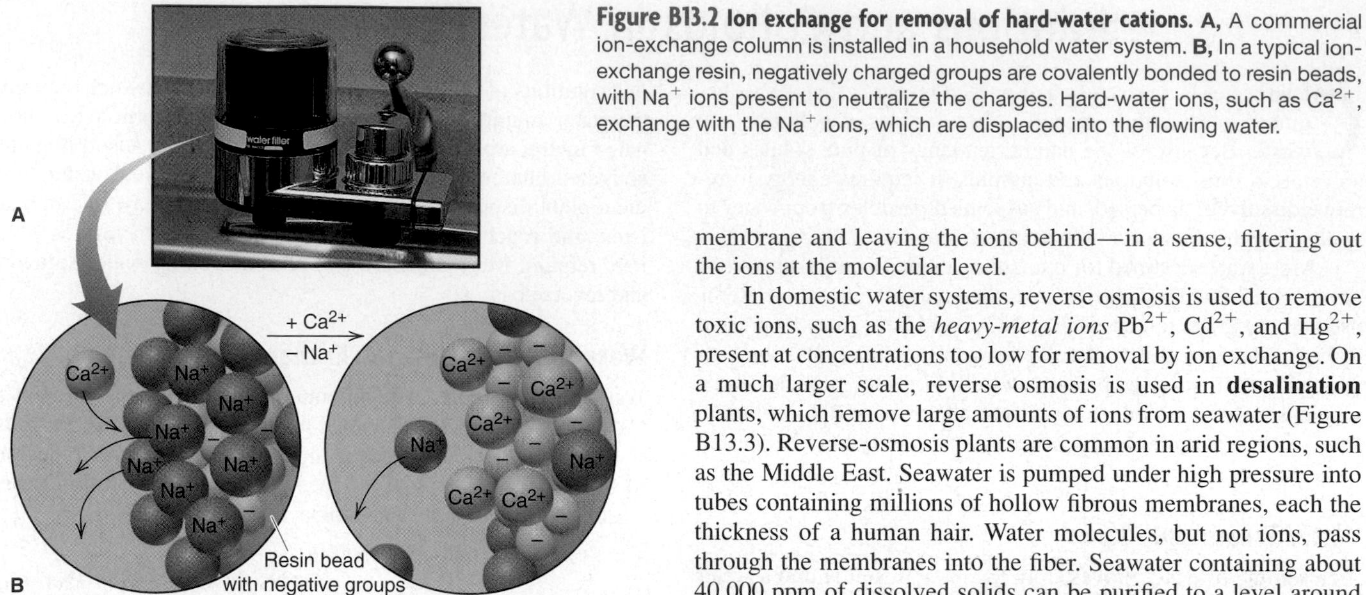

Figure B13.2 Ion exchange for removal of hard-water cations. **A,** A commercial ion-exchange column is installed in a household water system. **B,** In a typical ion-exchange resin, negatively charged groups are covalently bonded to resin beads, with Na^+ ions present to neutralize the charges. Hard-water ions, such as Ca^{2+}, exchange with the Na^+ ions, which are displaced into the flowing water.

the charges (Figure B13.2). The divalent cations in hard water are attracted to the resin's anionic groups and displace the Na^+ ions into the water: one type of ion is exchanged for another. The resin is replaced when all the resin sites are occupied, or it can be "regenerated" by treating it with a very concentrated Na^+ solution, which exchanges Na^+ ions for the bound Ca^{2+}.

Reverse Osmosis

Another way to remove ions and other dissolved substances from water is by **reverse osmosis.** In osmosis, water moves from a dilute to a concentrated solution through a semipermeable membrane. The resulting difference in water volumes creates an osmotic pressure. In reverse osmosis, water moves out of the concentrated solution when a pressure *greater* than the osmotic pressure is *applied* to the solution, forcing the water back through the membrane and leaving the ions behind—in a sense, filtering out the ions at the molecular level.

In domestic water systems, reverse osmosis is used to remove toxic ions, such as the *heavy-metal ions* Pb^{2+}, Cd^{2+}, and Hg^{2+}, present at concentrations too low for removal by ion exchange. On a much larger scale, reverse osmosis is used in **desalination** plants, which remove large amounts of ions from seawater (Figure B13.3). Reverse-osmosis plants are common in arid regions, such as the Middle East. Seawater is pumped under high pressure into tubes containing millions of hollow fibrous membranes, each the thickness of a human hair. Water molecules, but not ions, pass through the membranes into the fiber. Seawater containing about 40,000 ppm of dissolved solids can be purified to a level around 400 ppm (suitable for drinking) in one pass through such a system.

Water Treatment after Use

Used water is called **wastewater,** or *sewage,* and it must be treated before being returned to the groundwater, river, or lake. Sewage treatment is especially important for industrial wastewater, which may contain toxic components. In *primary* sewage treatment, wastewater undergoes the same steps as are used to treat water coming into the system. Most municipalities also include *secondary* sewage treatment. In this stage, bacteria biologically degrade the organic compounds and some microorganisms still present in solution or in the solids from the settling tanks. Secondary treatment is sometimes supplemented by *tertiary* treatment in a process tailored to the specific pollutant involved. Heavy-metal ions, for instance, can be eliminated by a precipitation step before primary and secondary treatment. Tertiary methods also exist for phosphates, nitrate, and toxic organic substances.

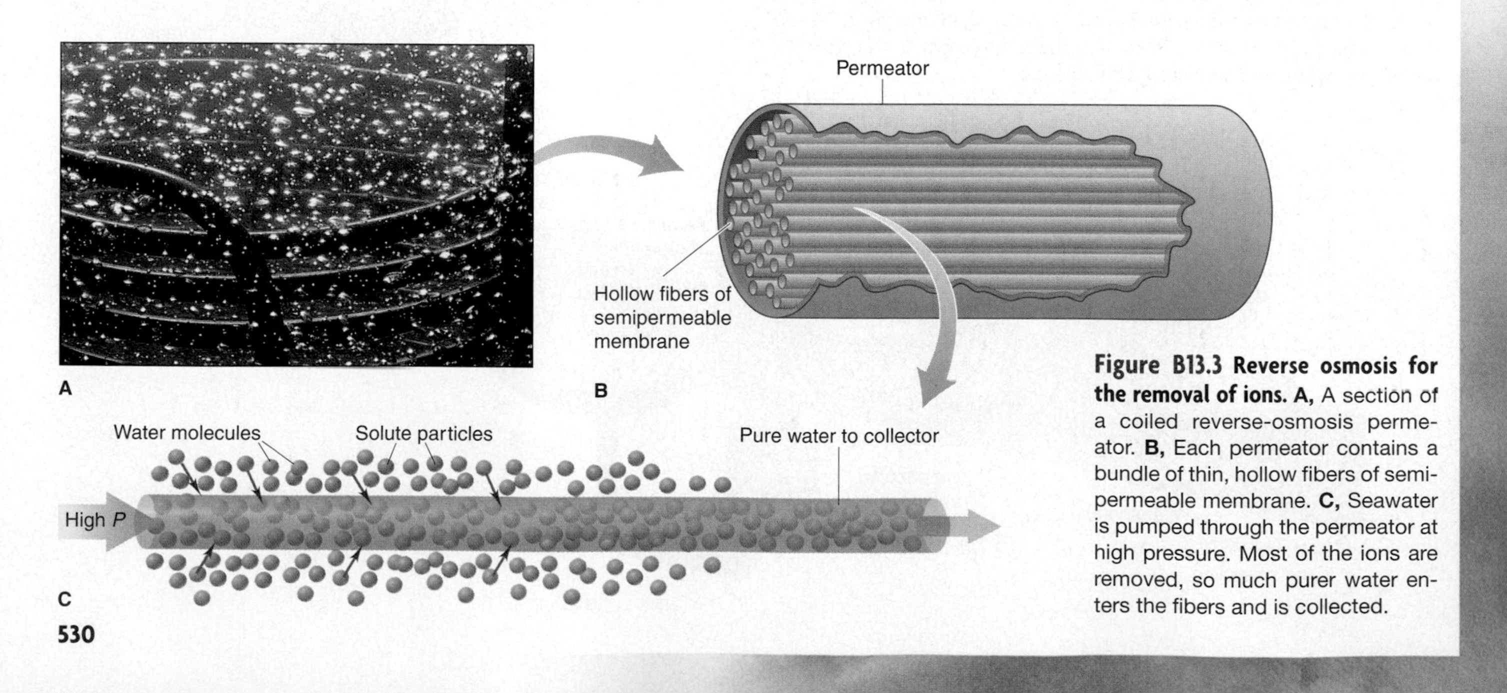

Figure B13.3 Reverse osmosis for the removal of ions. **A,** A section of a coiled reverse-osmosis permeator. **B,** Each permeator contains a bundle of thin, hollow fibers of semipermeable membrane. **C,** Seawater is pumped through the permeator at high pressure. Most of the ions are removed, so much purer water enters the fibers and is collected.

SECTION SUMMARY

Colloidal particles are smaller than those in a suspension and larger than those in a solution. Colloids are classified by the physical states of the dispersed and dispersing substances and are formed from many combinations of gas, liquid, and solid. Colloids have extremely large surface areas, scatter incoming light (Tyndall effect), and exhibit random (Brownian) motion. Colloidal particles in water have charged surfaces that keep them dispersed, but they can be coagulated by heating or by the addition of ions.

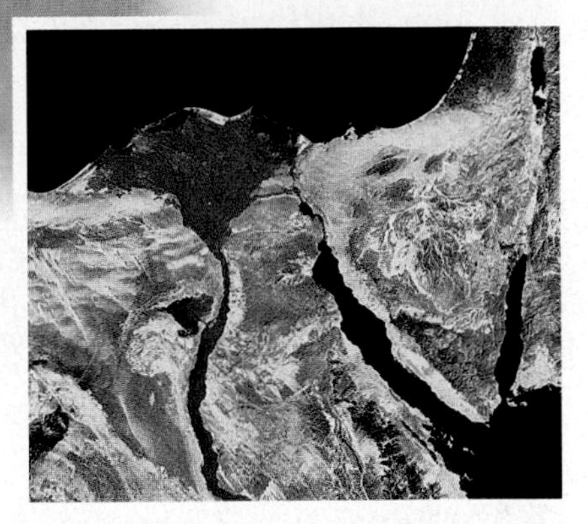

From Colloid to Civilization At times, civilizations have been born where colloids were being coagulated by electrolyte solutions. At the mouths of rivers, where salt concentrations increase near an ocean or sea, the clay particles dispersed in the river water come together to form muddy deltas, such as those of the Nile (*reddish area in photo*) and the Mississippi. The ancient Egyptian empire and the city of New Orleans are the results of this global colloid chemistry.

Chapter Perspective

In Chapter 12, you saw how pure liquids and solids behave, and here you've seen how their behaviors change when they are mixed together. Two features of these systems reappear in later chapters: their equilibrium nature in Chapters 17 to 19 and the importance of entropy in Chapter 20.

This chapter completes our general discussion of atomic and molecular properties and their influence on the properties of matter. Immediately following is an Interchapter feature that reviews these properties in a pictorial format and helps preview their application in the next two chapters. In Chapter 14, we travel through the main groups of the periodic table, applying the ideas you've learned so far to the chemical and physical behavior of the elements. Then, in Chapter 15, we apply them to carbon and its nearest neighbors to see how their properties are responsible for the fascinating world of organic substances, including synthetic polymers and biomolecules.

For Review and Reference (Numbers in parentheses refer to pages, unless noted otherwise.)

Learning Objectives

Relevant section and/or sample problem (SP) numbers appear in parentheses.

Understand These Concepts

1. The quantitative meaning of solubility (13.1)
2. The major types of intermolecular forces in solution and their relative strengths (13.1)
3. How the like-dissolves-like rule depends on intermolecular forces (13.1)
4. Why gases have relatively low solubilities in water (13.1)
5. General characteristics of solutions formed by various combinations of gases, liquids, and solids (13.1)
6. How intermolecular forces stabilize the structures of proteins, the cell membrane, DNA, and cellulose (13.2)
7. The enthalpy components of a solution cycle and their effect on ΔH_{soln} (13.3)
8. The dependence of ΔH_{hydr} on ionic charge density and the factors that determine whether ionic solution processes are exothermic or endothermic (13.3)
9. The meaning of entropy and how the balance between the change in enthalpy and the change in entropy governs the solution process (13.3)
10. The distinctions among saturated, unsaturated, and supersaturated solutions, and the equilibrium nature of a saturated solution (13.4)
11. The relation between temperature and the solubility of solids (13.4)
12. Why the solubility of gases in water decreases with a rise in temperature (13.4)
13. The effect of gas pressure on solubility and its quantitative expression as Henry's law (13.4)
14. The meaning of molarity, molality, mole fraction, and parts by mass or by volume of a solution, and how to convert among them (13.5)
15. The distinction between electrolytes and nonelectrolytes in solution (13.6)
16. The four colligative properties and their dependence on number of dissolved particles (13.6)
17. Ideal solutions and the importance of Raoult's law (13.6)
18. How the phase diagram of a solution differs from that of the pure solvent (13.6)
19. Why the vapor over a solution of volatile nonelectrolyte is richer in the more volatile component (13.6)
20. Why electrolyte solutions are not ideal and the meanings of the van't Hoff factor and ionic atmosphere (13.6)
21. How particle size distinguishes suspensions, colloids, and solutions (13.7)
22. How colloidal behavior is demonstrated by the Tyndall effect and Brownian motion (13.7)

Learning Objectives *(continued)*

Master These Skills

1. Predicting relative solubilities from intermolecular forces (SP 13.1)
2. Using Henry's law to calculate the solubility of a gas (SP 13.2)
3. Expressing concentration in terms of molality, parts by mass, parts by volume, and mole fraction (SPs 13.3 and 13.4)
4. Interconverting among the various terms for expressing concentration (SP 13.5)
5. Using Raoult's law to calculate the vapor pressure lowering of a solution (SP 13.6)
6. Determining boiling point elevation and freezing point depression of a solution (SP 13.7)
7. Using a colligative property to calculate the molar mass of a solute (SP 13.8)
8. Using a solution depiction to determine colligative properties (SP 13.9)
9. Calculating the composition of vapor over a solution of volatile nonelectrolyte (13.6)
10. Calculating the van't Hoff factor i from the magnitude of a colligative property (13.6)

Key Terms

Section 13.1
solute (490)
solvent (490)
miscible (490)
solubility (S) (490)
like-dissolves-like rule (490)
hydration shell (490)
ion–induced dipole force (490)
dipole–induced dipole force (491)
alloy (494)

Section 13.2
protein (495)
amino acid (495)
soap (498)
lipid bilayer (499)
nucleic acid (500)
mononucleotide (500)
double helix (501)
polysaccharide (501)
monosaccharide (501)

Section 13.3
heat of solution (ΔH_{soln}) (503)
solvation (503)
hydration (503)
heat of hydration (ΔH_{hydr}) (504)
charge density (504)
entropy (S) (505)

Section 13.4
saturated solution (507)
unsaturated solution (507)
supersaturated solution (507)
Henry's law (510)

Section 13.5
molality (m) (511)
mass percent [% (w/w)] (512)
volume percent [% (v/v)] (512)
mole fraction (X) (513)

Section 13.6
colligative property (515)
electrolyte (515)
nonelectrolyte (515)
vapor pressure lowering (ΔP) (516)
Raoult's law (516)
ideal solution (516)
boiling point elevation (ΔT_b) (517)
freezing point depression (ΔT_f) (519)
semipermeable membrane (520)
osmosis (520)
osmotic pressure (Π) (520)
fractional distillation (524)
ionic atmosphere (525)

Section 13.7
suspension (527)
colloid (527)
Tyndall effect (528)
hard water (529)
water softening (529)
ion exchange (529)
reverse osmosis (530)
desalination (530)
wastewater (530)

Key Equations and Relationships

13.1 Dividing the general heat of solution into component enthalpies (503):

$$\Delta H_{soln} = \Delta H_{solute} + \Delta H_{solvent} + \Delta H_{mix}$$

13.2 Dividing the heat of solution of an ionic compound in water into component enthalpies (504):

$$\Delta H_{soln} = \Delta H_{lattice} + \Delta H_{hydr\ of\ the\ ions}$$

13.3 Relating gas solubility to its partial pressure (Henry's law) (510):

$$S_{gas} = k_H \times P_{gas}$$

13.4 Defining concentration in terms of molarity (510):

$$\text{Molarity } (M) = \frac{\text{amount (mol) of solute}}{\text{volume (L) of solution}}$$

13.5 Defining concentration in terms of molality (511):

$$\text{Molality } (m) = \frac{\text{amount (mol) of solute}}{\text{mass (kg) of solvent}}$$

13.6 Defining concentration in terms of mass percent (512):

$$\text{Mass percent [\% (w/w)]} = \frac{\text{mass of solute}}{\text{mass of solution}} \times 100$$

13.7 Defining concentration in terms of volume percent (512):

$$\text{Volume percent [\% (v/v)]} = \frac{\text{volume of solute}}{\text{volume of solution}} \times 100$$

13.8 Defining concentration in terms of mole fraction (513):

Mole fraction (X)

$$= \frac{\text{amount (mol) of solute}}{\text{amount (mol) of solute} + \text{amount (mol) of solvent}}$$

13.9 Expressing the relationship between the vapor pressure of solvent above a solution and its mole fraction in the solution (Raoult's law) (516):

$$P_{solvent} = X_{solvent} \times P^0_{solvent}$$

13.10 Calculating the vapor pressure lowering due to solute (517):

$$\Delta P = X_{solute} \times P^0_{solvent}$$

13.11 Calculating the boiling point elevation of a solution (518):

$$\Delta T_b = K_b m$$

13.12 Calculating the freezing point depression of a solution (519):

$$\Delta T_f = K_f m$$

13.13 Calculating the osmotic pressure of a solution (520):

$$\Pi = \frac{n_{solute}}{V_{soln}} RT = MRT$$

Highlighted Figures and Tables

These figures (F) and tables (T) provide a review of key ideas.

Brief Solutions to Follow-up Problems

13.1 (a) 1,4-Butanediol is more soluble in water because it can form more H bonds.
(b) Chloroform is more soluble in water because of dipole-dipole forces.

13.2 $S_{N_2} = (7\times10^{-4}\text{ mol/L·atm})(0.78\text{ atm}) = 5\times10^{-4}\text{ mol/L}$

13.3 Mass (g) of glucose $= 563\text{ g ethanol} \times \dfrac{1\text{ kg}}{10^3\text{ g}} \times \dfrac{2.40\times10^{-2}\text{ mol glucose}}{1\text{ kg ethanol}} \times \dfrac{180.16\text{ g glucose}}{1\text{ mol glucose}} = 2.43\text{ g glucose}$

13.4 Mass % $C_3H_7OH = \dfrac{35.0\text{ g}}{35.0\text{ g} + 150.\text{ g}} \times 100 = 18.9$ mass %

Mass % $C_2H_5OH = 100.0 - 18.9 = 81.1$ mass %

$$X_{C_3H_7OH} = \frac{35.0\text{ g }C_3H_7OH \times \dfrac{1\text{ mol }C_3H_7OH}{60.09\text{ g }C_3H_7OH}}{\left(35.0\text{ g }C_3H_7OH \times \dfrac{1\text{ mol }C_3H_7OH}{60.09\text{ g }C_3H_7OH}\right) + \left(150.\text{ g }C_2H_5OH \times \dfrac{1\text{ mol }C_2H_5OH}{46.07\text{ g }C_2H_5OH}\right)} = 0.152$$

$X_{C_2H_5OH} = 1.000 - 0.152 = 0.848$

13.5 Mass % HCl $= \dfrac{\text{mass of HCl}}{\text{mass of soln}} \times 100 = \dfrac{\dfrac{11.8\text{ mol HCl}}{1\text{ L soln}} \times \dfrac{36.46\text{ g HCl}}{1\text{ mol HCl}}}{\dfrac{1.190\text{ g}}{1\text{ mL soln}} \times \dfrac{10^3\text{ mL}}{1\text{ L}}} \times 100 = 36.2$ mass % HCl

Mass (kg) of soln $= 1\text{ L soln} \times \dfrac{1.190\times10^{-3}\text{ kg soln}}{1\times10^{-3}\text{ L soln}} = 1.190\text{ kg soln}$

Mass (kg) of HCl $= 11.8\text{ mol HCl} \times \dfrac{36.46\text{ g HCl}}{1\text{ mol HCl}} \times \dfrac{1\text{ kg}}{10^3\text{ g}} = 0.430\text{ kg HCl}$

Molality of HCl $= \dfrac{\text{mol HCl}}{\text{kg water}} = \dfrac{\text{mol HCl}}{\text{kg soln} - \text{kg HCl}} = \dfrac{11.8\text{ mol HCl}}{0.760\text{ kg }H_2O} = 15.5\ m$ HCl

$$X_{HCl} = \frac{\text{mol HCl}}{\text{mol HCl} + \text{mol }H_2O} = \frac{11.8\text{ mol}}{11.8\text{ mol} + \left(760\text{ g }H_2O \times \dfrac{1\text{ mol}}{18.02\text{ g }H_2O}\right)} = 0.219$$

13.6 $\Delta P = X_{aspirin} \times P^0_{methanol} = \dfrac{\dfrac{2.00\text{ g}}{180.15\text{ g/mol}}}{\dfrac{2.00\text{ g}}{180.15\text{ g/mol}} + \dfrac{50.0\text{ g}}{32.04\text{ g/mol}}} \times 101\text{ torr} = 0.713\text{ torr}$

13.7 Molality of $C_2H_6O_2 = \dfrac{(0.00°\text{F} - 32°\text{F})\left(\dfrac{5°\text{C}}{9°\text{F}}\right)}{1.86°\text{C}/m} = 9.56\ m$

13.8 $\Pi = MRT = (0.30\text{ mol/L})\left(0.0821\ \dfrac{\text{atm·L}}{\text{mol·K}}\right)(37°\text{C} + 273.15) = 7.6\text{ atm}$

13.9 (a) Mass of 0.100 m solution $= 1\text{ kg water} + 0.100\text{ mol }MgCl_2 = 1000\text{ g} + 9.52\text{ g} = 1009.52\text{ g}$

Volume of solution $= 1009.52\text{ g} \times \dfrac{1\text{ mL}}{1.006\text{ g}} = 1003\text{ mL}$

Molarity $= \dfrac{9.52\text{ g }MgCl_2}{1003\text{ mL soln}} \times \dfrac{1\text{ mol}}{95.21\text{ g }MgCl_2} \times \dfrac{10^3\text{ mL}}{1\text{ L}} = 9.97\times10^{-2}\ M$

Osmotic pressure (Π) $= i(MRT) = 3(9.97\times10^{-2}\text{ mol/L})\left(0.0821\ \dfrac{\text{atm·L}}{\text{mol·K}}\right)(293\text{K}) = 7.19\text{ atm}$

(b) Scene C

Problems

Problems with **colored** numbers are answered in Appendix E. Sections match the text and provide the numbers of relevant sample problems. Most offer Concept Review Questions, Skill-Building Exercises (grouped in pairs covering the same concept), and Problems in Context. Comprehensive Problems are based on material from any section or previous chapter.

Types of Solutions: Intermolecular Forces and Predicting Solubility

(Sample Problem 13.1)

Concept Review Questions

13.1 Describe how properties of seawater illustrate the two characteristics that define mixtures.

13.2 What types of intermolecular forces give rise to hydration shells in an aqueous solution of sodium chloride?

13.3 Acetic acid is miscible with water. Would you expect carboxylic acids, general formula $CH_3(CH_2)_nCOOH$, to become more or less water soluble as *n* increases? Explain.

13.4 Which would you expect to be more effective as a soap, sodium acetate or sodium stearate? Explain.

13.5 Hexane and methanol are miscible as gases but only slightly soluble in each other as liquids. Explain.

13.6 Hydrogen chloride (HCl) gas is much more soluble than propane gas (C_3H_8) in water, even though HCl has a lower boiling point. Explain.

Skill-Building Exercises *(grouped in similar pairs)*

13.7 Which gives the more concentrated solution, (a) KNO_3 in H_2O or (b) KNO_3 in carbon tetrachloride (CCl_4)? Explain.

13.8 Which gives the more concentrated solution, stearic acid [$CH_3(CH_2)_{16}COOH$] in (a) H_2O or (b) CCl_4? Explain.

13.9 What is the strongest type of intermolecular force between solute and solvent in each solution?

(a) $CsCl(s)$ in $H_2O(l)$ (b) $CH_3\overset{\overset{O}{\|}}{C}CH_3(l)$ in $H_2O(l)$
(c) $CH_3OH(l)$ in $CCl_4(l)$

13.10 What is the strongest type of intermolecular force between solute and solvent in each solution?

(a) $Cu(s)$ in $Ag(s)$ (b) $CH_3Cl(g)$ in $CH_3OCH_3(g)$
(c) $CH_3CH_3(g)$ in $CH_3CH_2CH_2NH_2(l)$

13.11 What is the strongest type of intermolecular force between solute and solvent in each solution?

(a) $CH_3OCH_3(g)$ in $H_2O(l)$ (b) $Ne(g)$ in $H_2O(l)$
(c) $N_2(g)$ in $C_4H_{10}(g)$

13.12 What is the strongest type of intermolecular force between solute and solvent in each solution?

(a) $C_6H_{14}(l)$ in $C_8H_{18}(l)$ (b) $H_2C{=}O(g)$ in $CH_3OH(l)$
(c) $Br_2(l)$ in $CCl_4(l)$

13.13 Which member of each pair is more soluble in diethyl ether? Why?

(a) $NaCl(s)$ or $HCl(g)$ (b) $H_2O(l)$ or $CH_3\overset{\overset{O}{\|}}{C}H(l)$
(c) $MgBr_2(s)$ or $CH_3CH_2MgBr(s)$

13.14 Which member of each pair is more soluble in water? Why?

(a) $CH_3CH_2OCH_2CH_3(l)$ or $CH_3CH_2OCH_3(g)$
(b) $CH_2Cl_2(l)$ or $CCl_4(l)$

(c) cyclohexane (H_2C, CH_2 ring of six CH_2 groups) or tetrahydropyran (ring of five CH_2 groups and $\ddot{O}$)

cyclohexane tetrahydropyran

Problems in Context

13.15 The dictionary defines *homogeneous* as "uniform in composition throughout." River water is a mixture of dissolved compounds, such as calcium bicarbonate, and suspended soil particles. Is river water homogeneous? Explain.

13.16 Gluconic acid is a derivative of glucose used in cleaners and in the dairy and brewing industries. Caproic acid is a carboxylic acid used in the flavoring industry. Although both are six-carbon acids (see structures below), gluconic acid is soluble in water and nearly insoluble in hexane, whereas caproic acid has the opposite solubility behavior. Explain.

$CH_2(OH)—CH(OH)—CH(OH)—CH(OH)—CH(OH)—COOH$

gluconic acid

$CH_3—CH_2—CH_2—CH_2—CH_2—COOH$

caproic acid

Intermolecular Forces and Biological Macromolecules

Concept Review Questions

13.17 Name three intermolecular forces that stabilize the shape of a soluble, globular protein, and explain how they act.

13.18 Name three intermolecular forces that stabilize the shape of DNA, and explain how they act.

13.19 Is the sodium salt of propanoic acid as effective a soap as sodium stearate? Explain.

13.20 What intermolecular forces stabilize a lipid bilayer?

13.21 In what way do proteins embedded in a membrane differ structurally from soluble proteins?

13.22 How can wood be so strong if it consists of cellulose chains held together by relatively weak H bonds?

13.23 Histones are proteins that control gene function by attaching through salt links to exterior regions of DNA. Name an amino acid whose side chain is often found on the exterior of histones.

Why Substances Dissolve: Understanding the Solution Process

Concept Review Questions

13.24 What is the relationship between solvation and hydration?

13.25 For a general solvent, which enthalpy terms in the thermochemical solution cycle are combined to obtain $\Delta H_{solvation}$?

13.26 (a) What is the charge density of an ion, and what two properties of an ion affect it?

(b) Arrange the following in order of increasing charge density:

(c) How do the two properties in part (a) affect the ionic heat of hydration, ΔH_{hydr}?

13.27 For ΔH_{soln} to be very small, what quantities must be nearly equal in magnitude? Will their signs be the same or opposite?

13.28 Water is added to a flask containing solid NH_4Cl. As the salt dissolves, the solution becomes colder.
(a) Is the dissolving of NH_4Cl exothermic or endothermic?
(b) Is the magnitude of $\Delta H_{lattice}$ of NH_4Cl larger or smaller than the combined ΔH_{hydr} of the ions? Explain.
(c) Given the answer to (a), why does NH_4Cl dissolve in water?

13.29 An ionic compound has a highly negative ΔH_{soln} in water. Would you expect it to be very soluble or nearly insoluble in water? Explain in terms of enthalpy and entropy changes.

Skill-Building Exercises *(grouped in similar pairs)*

13.30 Sketch a qualitative enthalpy diagram for the process of dissolving KCl(*s*) in H_2O (endothermic).

13.31 Sketch a qualitative enthalpy diagram for the process of dissolving NaI(*s*) in H_2O (exothermic).

13.32 Which ion in each pair has greater charge density? Explain.
(a) Na^+ or Cs^+ (b) Sr^{2+} or Rb^+ (c) Na^+ or Cl^-
(d) O^{2-} or F^- (e) OH^- or SH^-

13.33 Which ion has the lower ratio of charge to volume? Explain.
(a) Br^- or I^- (b) Sc^{3+} or Ca^{2+} (c) Br^- or K^+
(d) S^{2-} or Cl^- (e) Sc^{3+} or Al^{3+}

13.34 Which has the *larger* ΔH_{hydr} in each pair of Problem 13.32?

13.35 Which has the *smaller* ΔH_{hydr} in each pair of Problem 13.33?

13.36 (a) Use the following data to calculate the combined heat of hydration for the ions in potassium bromate ($KBrO_3$):

$$\Delta H_{lattice} = 745 \text{ kJ/mol} \qquad \Delta H_{soln} = 41.1 \text{ kJ/mol}$$

(b) Which ion contributes more to the answer to part (a)? Why?

13.37 (a) Use the following data to calculate the combined heat of hydration for the ions in sodium acetate ($NaC_2H_3O_2$):

$$\Delta H_{lattice} = 763 \text{ kJ/mol} \qquad \Delta H_{soln} = 17.3 \text{ kJ/mol}$$

(b) Which ion contributes more to the answer to part (a)? Why?

13.38 State whether the entropy of the system increases or decreases in each of the following processes:
(a) Gasoline is burned in a car engine.
(b) Gold is extracted and purified from its ore.
(c) Ethanol (CH_3CH_2OH) dissolves in 1-propanol ($CH_3CH_2CH_2OH$).

13.39 State whether the entropy of the system increases or decreases in each of the following processes:
(a) Pure gases are mixed to prepare an anesthetic.
(b) Electronic-grade silicon is prepared from sand.
(c) Dry ice (solid CO_2) sublimes.

Problems in Context

13.40 Besides its use in making black-and-white film, silver nitrate ($AgNO_3$) is used similarly in forensic science. The NaCl left behind in the sweat of a fingerprint is treated with $AgNO_3$ solution to form AgCl. This precipitate is developed to show the black-and-white fingerprint pattern. Given $\Delta H_{lattice}$ of $AgNO_3$ = 822 kJ/mol and $\Delta H_{hydr} = -799$ kJ/mol, calculate its ΔH_{soln}.

Solubility as an Equilibrium Process

(Sample Problem 13.2)

Concept Review Questions

13.41 You are given a bottle of solid X and three aqueous solutions of X—one saturated, one unsaturated, and one supersaturated. How would you determine which solution is which?

13.42 Potassium permanganate ($KMnO_4$) has a solubility of 6.4 g/100 g of H_2O at 20°C and a curve of solubility vs. temperature that slopes upward to the right. How would you prepare a supersaturated solution of $KMnO_4$?

13.43 Why does the solubility of any gas in water decrease with rising temperature?

Skill-Building Exercises *(grouped in similar pairs)*

13.44 For a saturated aqueous solution of each of the following at 20°C and 1 atm, will the solubility increase, decrease, or stay the same when the indicated change occurs?
(a) $O_2(g)$, increase *P* (b) $N_2(g)$, increase *V*

13.45 For a saturated aqueous solution of each of the following at 20°C and 1 atm, will the solubility increase, decrease, or stay the same when the indicated change occurs?
(a) He(*g*), decrease *T* (b) RbI(*s*), increase *P*

13.46 The Henry's law constant (k_H) for O_2 in water at 20°C is 1.28×10^{-3} mol/L·atm. (a) How many grams of O_2 will dissolve in 2.00 L of H_2O that is in contact with pure O_2 at 1.00 atm? (b) How many grams of O_2 will dissolve in 2.00 L of H_2O that is in contact with air, where the partial pressure of O_2 is 0.209 atm?

13.47 Argon makes up 0.93% by volume of air. Calculate its solubility (mol/L) in water at 20°C and 1.0 atm. The Henry's law constant for Ar under these conditions is 1.5×10^{-3} mol/L·atm.

Problems in Context

13.48 Caffeine is about 10 times as soluble in hot water as in cold water. A chemist puts a hot-water extract of caffeine into an ice bath, and some caffeine crystallizes. Is the remaining solution saturated, unsaturated, or supersaturated?

13.49 The partial pressure of CO_2 gas above the liquid in a bottle of champagne at 20°C is 5.5 atm. What is the solubility of CO_2 in champagne? Assume Henry's law constant is the same for champagne as for water: at 20°C, $k_H = 3.7\times10^{-2}$ mol/L·atm.

13.50 Respiratory problems are treated with devices that deliver air with a higher partial pressure of O_2 than normal air. Why?

Quantitative Ways of Expressing Concentration

(Sample Problems 13.3 to 13.5)

Concept Review Questions

13.51 Explain the difference between molarity and molality. Under what circumstances would molality be a more accurate measure of the concentration of a prepared solution than molarity? Why?

13.52 Which way of expressing concentration includes (a) volume of solution; (b) mass of solution; (c) mass of solvent?

13.53 A solute has a solubility in water of 21 g/kg solvent. Is this value the same as 21 g/kg solution? Explain.

13.54 You want to convert among molarity, molality, and mole fraction of a solution. You know the masses of solute and solvent and the volume of solution. Is this enough information to carry out all the conversions? Explain.

13.55 When a solution is heated, which ways of expressing concentration change in value? Which remain unchanged? Explain.

Skill-Building Exercises *(grouped in similar pairs)*

13.56 Calculate the molarity of each aqueous solution:
(a) 42.3 g of table sugar ($C_{12}H_{22}O_{11}$) in 100. mL of solution
(b) 5.50 g of $LiNO_3$ in 505 mL of solution

13.57 Calculate the molarity of each aqueous solution:
(a) 0.82 g of ethanol (C_2H_5OH) in 10.5 mL of solution
(b) 1.22 g of gaseous NH_3 in 33.5 mL of solution

13.58 Calculate the molarity of each aqueous solution:
(a) 75.0 mL of 0.250 *M* NaOH diluted to 0.250 L with water
(b) 35.5 mL of 1.3 *M* HNO_3 diluted to 0.150 L with water

13.59 Calculate the molarity of each aqueous solution:
(a) 25.0 mL of 6.15 *M* HCl diluted to 0.500 L with water
(b) 8.55 mL of 2.00×10^{-2} *M* KI diluted to 10.0 mL with water

13.60 How would you prepare the following aqueous solutions?
(a) 355 mL of 8.74×10^{-2} *M* KH_2PO_4 from solid KH_2PO_4
(b) 425 mL of 0.315 *M* NaOH from 1.25 *M* NaOH

13.61 How would you prepare the following aqueous solutions?
(a) 3.5 L of 0.55 *M* NaCl from solid NaCl
(b) 17.5 L of 0.3 *M* urea [$(NH_2)_2C{=}O$] from 2.2 *M* urea

13.62 How would you prepare the following aqueous solutions?
(a) 1.50 L of 0.257 *M* KBr from solid KBr
(b) 355 mL of 0.0956 *M* $LiNO_3$ from 0.244 *M* $LiNO_3$

13.63 How would you prepare the following aqueous solutions?
(a) 67.5 mL of 1.33×10^{-3} *M* $Cr(NO_3)_3$ from solid $Cr(NO_3)_3$
(b) 6.8×10^3 m^3 of 1.55 *M* NH_4NO_3 from 3.00 *M* NH_4NO_3

13.64 Calculate the molality of the following:
(a) A solution containing 88.4 g of glycine (NH_2CH_2COOH) dissolved in 1.250 kg of H_2O
(b) A solution containing 8.89 g of glycerol ($C_3H_8O_3$) in 75.0 g of ethanol (C_2H_6O)

13.65 Calculate the molality of the following:
(a) A solution containing 164 g of HCl in 753 g of H_2O
(b) A solution containing 16.5 g of naphthalene ($C_{10}H_8$) in 53.3 g of benzene (C_6H_6)

13.66 What is the molality of a solution consisting of 34.0 mL of benzene (C_6H_6; d = 0.877 g/mL) in 187 mL of hexane (C_6H_{14}; d = 0.660 g/mL)?

13.67 What is the molality of a solution consisting of 2.77 mL of carbon tetrachloride (CCl_4; d = 1.59 g/mL) in 79.5 mL of methylene chloride (CH_2Cl_2; d = 1.33 g/mL)?

13.68 How would you prepare the following aqueous solutions?
(a) 3.00×10^2 g of 0.115 *m* ethylene glycol ($C_2H_6O_2$) from ethylene glycol and water
(b) 1.00 kg of 2.00 mass % HNO_3 from 62.0 mass % HNO_3

13.69 How would you prepare the following aqueous solutions?
(a) 1.00 kg of 0.0555 *m* ethanol (C_2H_5OH) from ethanol and water
(b) 475 g of 15.0 mass % HCl from 37.1 mass % HCl

13.70 A solution is made by dissolving 0.30 mol of isopropanol (C_3H_7OH) in 0.80 mol of water. (a) What is the mole fraction of isopropanol? (b) What is the mass percent of isopropanol? (c) What is the molality of isopropanol?

13.71 A solution is made by dissolving 0.100 mol of NaCl in 8.60 mol of water. (a) What is the mole fraction of NaCl? (b) What is the mass percent of NaCl? (c) What is the molality of NaCl?

13.72 What mass of cesium chloride must be added to 0.500 L of water (d = 1.00 g/mL) to produce a 0.400 *m* solution? What are the mole fraction and the mass percent of CsCl?

13.73 What are the mole fraction and the mass percent of a solution made by dissolving 0.30 g of KBr in 0.400 L of water (d = 1.00 g/mL)?

13.74 Calculate the molality, molarity, and mole fraction of NH_3 in an 8.00 mass % aqueous solution (d = 0.9651 g/mL).

13.75 Calculate the molality, molarity, and mole fraction of $FeCl_3$ in a 28.8 mass % aqueous solution (d = 1.280 g/mL).

Problems in Context

13.76 Wastewater from a cement factory contains 0.22 g of Ca^{2+} ion and 0.066 g of Mg^{2+} ion per 100.0 L of solution. The solution density is 1.001 g/mL. Calculate the Ca^{2+} and Mg^{2+} concentrations in ppm (by mass).

13.77 An automobile antifreeze mixture is made by mixing equal volumes of ethylene glycol (d = 1.114 g/mL; $\mathcal{M}$ = 62.07 g/mol) and water (d = 1.00 g/mL) at 20°C. The density of the mixture is 1.070 g/mL. Express the concentration of ethylene glycol as
(a) volume percent (b) mass percent (c) molarity
(d) molality (e) mole fraction

Colligative Properties of Solutions

(Sample Problems 13.6 to 13.9)

Concept Review Questions

13.78 The chemical formula of a solute does *not* affect the extent of the solution's colligative properties. What characteristic of a solute *does* affect these properties? Name a physical property of a solution that *is* affected by the chemical formula of the solute.

13.79 What is a nonvolatile nonelectrolyte? Why is this type of solute the simplest case for examining colligative properties?

13.80 In what sense is a strong electrolyte "strong"? What property of the substance makes it a strong electrolyte?

13.81 Express Raoult's law in words. Is Raoult's law valid for a solution of a volatile solute? Explain.

13.82 What are the most important differences between the phase diagram of a pure solvent and the phase diagram of a solution of that solvent?

13.83 Is the composition of the vapor at the top of a fractionating column different from the composition at the bottom? Explain.

13.84 Is the boiling point of 0.01 *m* KF(*aq*) higher or lower than that of 0.01 *m* glucose(*aq*)? Explain.

13.85 Which aqueous solution has a boiling point closer to its predicted value, 0.050 *m* NaF or 0.50 *m* KCl? Explain.

13.86 Which aqueous solution has a freezing point closer to its predicted value, 0.01 *m* NaBr or 0.01 *m* $MgCl_2$? Explain.

13.87 The freezing point depression constants of the solvents cyclohexane and naphthalene are 20.1°C/*m* and 6.94°C/*m*, respectively. Which solvent would give a more accurate result if you are using freezing point depression to determine the molar mass of a substance that is soluble in either one? Why?

Skill-Building Exercises *(grouped in similar pairs)*

13.88 Classify the following substances as strong electrolytes, weak electrolytes, or nonelectrolytes:
(a) hydrogen chloride (HCl) (b) potassium nitrate (KNO_3)
(c) glucose ($C_6H_{12}O_6$) (d) ammonia (NH_3)

13.89 Classify the following substances as strong electrolytes, weak electrolytes, or nonelectrolytes:
(a) sodium permanganate ($NaMnO_4$)
(b) acetic acid (CH_3COOH)
(c) methanol (CH_3OH) (d) calcium acetate [$Ca(C_2H_3O_2)_2$]

13.90 How many moles of solute particles are present in 1 L of each of the following aqueous solutions?
(a) 0.2 *M* KI (b) 0.070 *M* HNO_3
(c) 10^{-4} *M* K_2SO_4 (d) 0.07 *M* ethanol (C_2H_5OH)

13.91 How many moles of solute particles are present in 1 mL of each of the following aqueous solutions?

(a) 0.01 M $CuSO_4$ (b) 0.005 M $Ba(OH)_2$
(c) 0.06 M pyridine (C_5H_5N) (d) 0.05 M $(NH_4)_2CO_3$

13.92 Which solution has the lower freezing point?
(a) 10.0 g of CH_3OH in 100. g of H_2O *or*
20.0 g of CH_3CH_2OH in 200. g of H_2O
(b) 10.0 g of H_2O in 1.00 kg of CH_3OH *or*
10.0 g of CH_3CH_2OH in 1.00 kg of CH_3OH

13.93 Which solution has the higher boiling point?
(a) 35.0 g of $C_3H_8O_3$ in 250. g of ethanol *or*
35.0 g of $C_2H_6O_2$ in 250. g of ethanol
(b) 20. g of $C_2H_6O_2$ in 0.50 kg of H_2O *or*
20. g of NaCl in 0.50 kg of H_2O

13.94 Rank the following aqueous solutions in order of increasing (a) osmotic pressure; (b) boiling point; (c) freezing point; (d) vapor pressure at 50°C:
(I) 0.100 m $NaNO_3$ (II) 0.200 m glucose (III) 0.100 m $CaCl_2$

13.95 Rank the following aqueous solutions in order of decreasing (a) osmotic pressure; (b) boiling point; (c) freezing point; (d) vapor pressure at 298 K:
(I) 0.04 m urea $[(NH_2)_2C{=}O]$ (II) 0.02 m $AgNO_3$
(III) 0.02 m $CuSO_4$

13.96 Calculate the vapor pressure of a solution of 44.0 g of glycerol ($C_3H_8O_3$) in 500.0 g of water at 25°C. The vapor pressure of water at 25°C is 23.76 torr. (Assume ideal behavior.)

13.97 Calculate the vapor pressure of a solution of 0.39 mol of cholesterol in 5.4 mol of toluene at 32°C. Pure toluene has a vapor pressure of 41 torr at 32°C. (Assume ideal behavior.)

13.98 What is the freezing point of 0.111 m urea in water?

13.99 What is the boiling point of 0.200 m lactose in water?

13.100 The boiling point of ethanol (C_2H_5OH) is 78.5°C. What is the boiling point of a solution of 3.4 g of vanillin ($\mathcal{M}$ = 152.14 g/mol) in 50.0 g of ethanol (K_b of ethanol = 1.22°C/m)?

13.101 The freezing point of benzene is 5.5°C. What is the freezing point of a solution of 5.00 g of naphthalene ($C_{10}H_8$) in 444 g of benzene (K_f of benzene = 4.90°C/m)?

13.102 What is the minimum mass of ethylene glycol ($C_2H_6O_2$) that must be dissolved in 14.5 kg of water to prevent the solution from freezing at −10.0°F? (Assume ideal behavior.)

13.103 What is the minimum mass of glycerol ($C_3H_8O_3$) that must be dissolved in 11.0 mg of water to prevent the solution from freezing at −25°C? (Assume ideal behavior.)

13.104 Calculate the molality and van't Hoff factor (i) for the following aqueous solutions:
(a) 1.00 mass % NaCl, freezing point = −0.593°C
(b) 0.500 mass % CH_3COOH, freezing point = −0.159°C

13.105 Calculate the molality and van't Hoff factor (i) for the following aqueous solutions:
(a) 0.500 mass % KCl, freezing point = −0.234°C
(b) 1.00 mass % H_2SO_4, freezing point = −0.423°C

Problems in Context

13.106 Wastewater discharged into a stream by a sugar refinery contains sucrose ($C_{12}H_{22}O_{11}$) as its main impurity. The solution contains 3.42 g of sucrose/L. A government-industry project is designed to test the feasibility of removing the sugar by reverse osmosis. What pressure must be applied to the apparatus at 20.°C to produce pure water?

13.107 In a study designed to prepare new gasoline-resistant coatings, a polymer chemist dissolves 6.053 g of poly(vinyl alcohol) in enough water to make 100.0 mL of solution. At 25°C, the osmotic pressure of this solution is 0.272 atm. What is the molar mass of the polymer sample?

13.108 The U.S. Food and Drug Administration lists dichloromethane (CH_2Cl_2) and carbon tetrachloride (CCl_4) among the many chlorinated organic compounds that are carcinogenic (cancer-causing). What are the partial pressures of these substances in the vapor above a solution of 1.50 mol of CH_2Cl_2 and 1.00 mol of CCl_4 at 23.5°C? The vapor pressures of pure CH_2Cl_2 and CCl_4 at this temperature are 352 torr and 118 torr, respectively. (Assume ideal behavior.)

The Structure and Properties of Colloids

Concept Review Questions

13.109 Is the fluid inside a bacterial cell considered a solution, a colloid, or both? Explain.

13.110 What type of colloid is each of the following?
(a) milk (b) fog (c) shaving cream

13.111 What is Brownian motion, and what causes it?

13.112 In a movie theater, you can see the beam of projected light. What phenomenon does this exemplify? Why does it occur?

13.113 Why don't soap micelles coagulate and form large globules? Is soap more effective in freshwater or in seawater? Why?

Comprehensive Problems

13.114 Nitrous oxide (N_2O) is used in whipped cream containers as the gas that makes the cream foam. Some experiments to use carbon dioxide (CO_2) instead have proven unsuccessful. What does this suggest about the relative sizes of the Henry's law constant for N_2O and CO_2 in cream? Explain.

13.115 An aqueous solution is 10.% glucose by mass (d = 1.039 g/mL at 20°C). Calculate its freezing point, boiling point at 1 atm, and osmotic pressure.

13.116 Because zinc has nearly the same atomic radius as copper (d = 8.95 g/cm^3), zinc atoms substitute for some copper atoms in the many types of brass. Calculate the density of the brass with (a) 10.0 atom % Zn and (b) 38.0 atom % Zn.

13.117 Which of the following best represents a molecular-scale view of an ionic compound in aqueous solution? Explain.

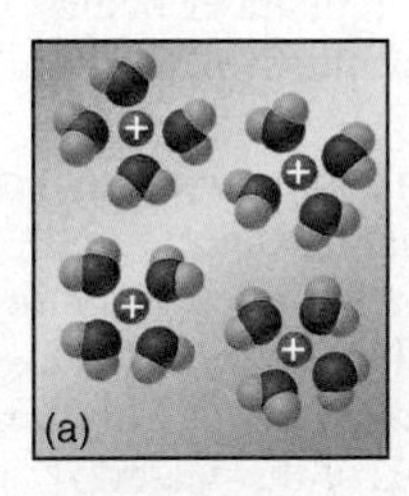

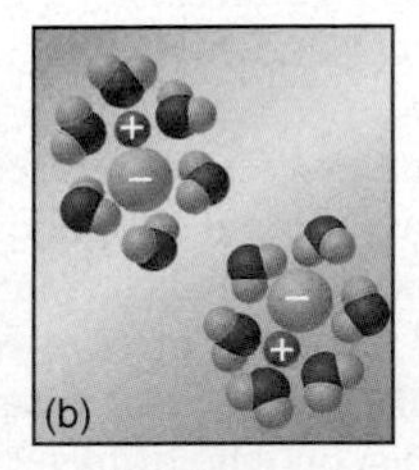

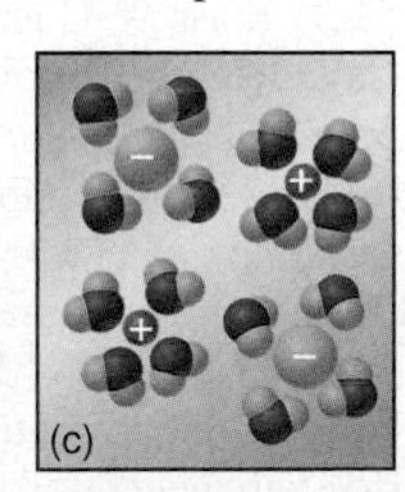

13.118 A car's gas tank is essentially a closed system containing gasoline and its vapor. If the mole fraction of octane is 0.14 and its vapor pressure is 11 torr at 20.°C, what is the partial pressure of octane at 20.°C? (Assume ideal behavior.)

13.119 Gold occurs in seawater at an average concentration of $1.1{\times}10^{-2}$ ppb. How many liters of seawater must be processed to recover 1 troy ounce of gold, assuming 79.5% efficiency (d of seawater = 1.025 g/mL; 1 troy ounce = 31.1 g)?

13.120 Use atomic properties to explain why xenon is more than 25 times as soluble as helium in water at 0°C.

13.121 Thermal pollution from industrial wastewater causes the temperature of river or lake water to increase, which can affect fish survival as the concentration of dissolved O_2 decreases. Use the following data to find the molarity of O_2 at each temperature (assume the solution density is the same as water):

Temperature (°C)	Solubility of O_2 (mg/kg H_2O)	Density of H_2O (g/mL)
0.0	14.5	0.99987
20.0	9.07	0.99823
40.0	6.44	0.99224

13.122 Pyridine (see structure below) is an essential portion of many biologically active compounds, such as nicotine and vitamin B_6. Like ammonia, it has a nitrogen with a lone pair, which makes it act as a weak base. Because it is miscible in a wide range of solvents, from water to benzene, pyridine is one of the most important bases and solvents in organic syntheses. Account for its solubility behavior in terms of intermolecular forces.

pyridine

13.123 "De-icing salt" is used to melt snow and ice on streets. The highway department of a small town is deciding whether to buy NaCl or $CaCl_2$ for the job. The town can obtain NaCl for $0.22/kg. What is the maximum the town should pay for $CaCl_2$ to be cost effective?

13.124 Air in a smoky bar contains 4.0×10^{-6} mol/L of CO. What mass of CO is inhaled by a bartender who respires at a rate of 12 L/min during an 8.0-h shift?

13.125 Is 50% by mass of methanol dissolved in ethanol different from 50% by mass of ethanol dissolved in methanol? Explain.

13.126 An industrial chemist is studying small organic compounds for their potential use as an automobile antifreeze. When 0.243 g of a compound is dissolved in 25.0 mL of water, the freezing point of the solution is −0.201°C.
(a) Calculate the molar mass of the compound (*d* of water = 1.00 g/mL at the temperature of the experiment).
(b) The compositional analysis of the compound shows that it is 53.31 mass % C and 11.18 mass % H, the remainder being O. Calculate the empirical and molecular formulas of the compound.
(c) Draw two possible Lewis structures for a compound with this formula, one that forms H bonds and one that does not.

13.127 A water treatment plant needs to attain a fluoride concentration of 5.00×10^{-5} *M*. (a) What mass of NaF must be added to a 5000.- L blending tank of water? (b) What mass per day of fluoride is ingested by a person who drinks 2.0 L of this water?

13.128 Give brief answers for each of the following:
(a) Why are lime (CaO) and cake alum [$Al_2(SO_4)_3$] added during water purification?
(b) Why is water that contains large amounts of Ca^{2+}, Mg^{2+}, or Fe^{2+} difficult to use for cleaning?
(c) What is the meaning of "reverse" in reverse osmosis?
(d) Why might a water treatment plant use ozone as the final disinfectant instead of chlorine, even though ozone costs more?
(e) How does passing a saturated NaCl solution through a "spent" ion-exchange resin regenerate the resin?

13.129 Which ion in each pair has the *larger* ΔH_{hydr}?
(a) Mg^{2+} or Ba^{2+} (b) Mg^{2+} or Na^+ (c) NO_3^- or CO_3^{2-}
(d) SO_4^{2-} or ClO_4^- (e) Fe^{3+} or Fe^{2+} (f) Ca^{2+} or K^+

13.130 β-Pinene ($C_{10}H_{16}$) and α-terpineol ($C_{10}H_{18}O$) are two of the many compounds used in perfumes and cosmetics to provide a "fresh pine" scent. At 367 K, the pure substances have vapor pressures of 100.3 torr and 9.8 torr, respectively. What is the composition of the vapor (in terms of mole fractions) above a solution containing equal masses of these compounds at 367 K? (Assume ideal behavior.)

13.131 A solution made by dissolving 1.50 g of solute in 25.0 mL of H_2O at 25°C has a boiling point of 100.45°C.
(a) What is the molar mass of the solute if it is a nonvolatile nonelectrolyte and the solution behaves ideally (*d* of H_2O at 25°C = 0.997 g/mL)?
(b) Conductivity measurements indicate that the solute is actually ionic with general formula AB_2 or A_2B. What is the molar mass of the compound *if* the solution behaves ideally?
(c) Analysis indicates an empirical formula of CaN_2O_6. Explain the difference between the actual formula mass and that calculated from the boiling point elevation experiment.
(d) Calculate the van't Hoff factor (*i*) for this solution.

13.132 A pharmaceutical preparation made with ethanol (C_2H_5OH) is contaminated with methanol (CH_3OH). A sample of vapor above the liquid mixture contains a 97:1 mass ratio of C_2H_5OH:CH_3OH. What is the mass ratio of these alcohols in the liquid? At the temperature of the liquid, the vapor pressures of C_2H_5OH and CH_3OH are 60.5 torr and 126.0 torr, respectively.

13.133 Water-treatment plants commonly use chlorination to destroy bacteria. A by-product is chloroform ($CHCl_3$), a suspected carcinogen, produced when HOCl, formed by reaction of Cl_2 and water, reacts with dissolved organic matter. The U.S., Canada, and the World Health Organization have set a limit of 100. ppb of $CHCl_3$ in drinking water. Convert this concentration into molarity, molality, mole fraction, and mass percent.

13.134 A saturated Na_2CO_3 solution is prepared, and a small excess of solid is present. A seed crystal of $Na_2{}^{14}CO_3$ (^{14}C is a radioactive isotope of ^{12}C) is introduced (see the figure), and the radioactivity is measured over time.

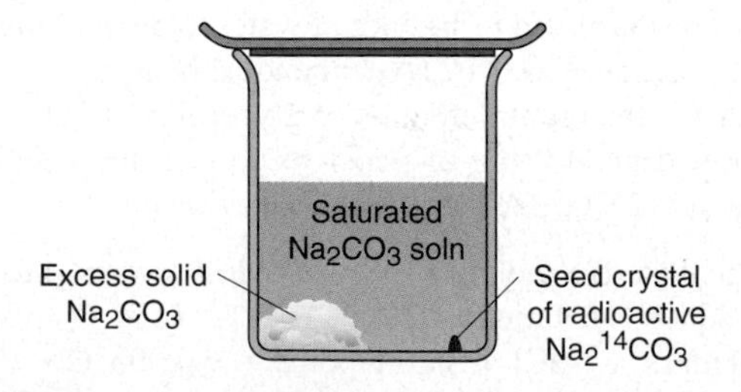

(a) Would you expect radioactivity in the solution? Explain.
(b) Would you expect radioactivity in all the solid or just in the seed crystal? Explain.

13.135 A biochemical engineer isolates a bacterial gene fragment and dissolves a 10.0-mg sample of the material in enough water to make 30.0 mL of solution. The osmotic pressure of the solution is 0.340 torr at 25°C.
(a) What is the molar mass of the gene fragment?
(b) If the solution density is 0.997 g/mL, how large is the freezing point depression for this solution (K_f of water = 1.86°C/*m*)?

13.136 A river is contaminated with 0.75 mg/L of dichloroethylene ($C_2H_2Cl_2$). What is the concentration (in ng/L) of dichloroethylene at 21°C in the air breathed by the people sitting along the riverbank (k_H for $C_2H_2Cl_2$ in water is 0.033 mol/L·atm)?

13.137 At an air-water interface, fatty acids such as oleic acid lie in a one-molecule-thick layer (*monolayer*), with the heads in the water and the tails perpendicular in the air. When 2.50 mg of oleic acid is placed on a water surface, it forms a circular monolayer 38.6 cm in diameter. Find the surface area (in cm^2) occupied by one molecule ($\mathcal{M}$ of oleic acid = 283 g/mol).

13.138 A simple device used for estimating the concentration of total dissolved solids in an aqueous solution works by measuring the electrical conductivity of the solution. The method assumes that equal concentrations of different solids give approximately the same conductivity, and that the conductivity is proportional to concentration. The table below gives some actual electrical conductivities (in arbitrary units) for solutions of selected solids at the indicated concentrations (in ppm by mass):

Sample	Conductivity		
	0 ppm	5.00×10^3 ppm	10.00×10^3 ppm
$CaCl_2$	0.0	8.0	16.0
K_2CO_3	0.0	7.0	14.0
Na_2SO_4	0.0	6.0	11.0
Seawater (dil)	0.0	8.0	15.0
Sucrose ($C_{12}H_{22}O_{11}$)	0.0	0.0	0.0
Urea [$(NH_2)_2C{=}O$]	0.0	0.0	0.0

(a) Comment on the reliability of these measurements for estimating concentrations of dissolved solids.
(b) For what types of substances is this method likely to be seriously in error? Why?
(c) Based on this method, an aqueous $CaCl_2$ solution has a conductivity of 14.0 units. Calculate its mole fraction and molality.

13.139 Two beakers are placed in a closed container (*below, left*). One beaker contains water, the other a concentrated aqueous sugar solution. With time, the solution volume increases and the water volume decreases (*right*). Explain on the molecular level.

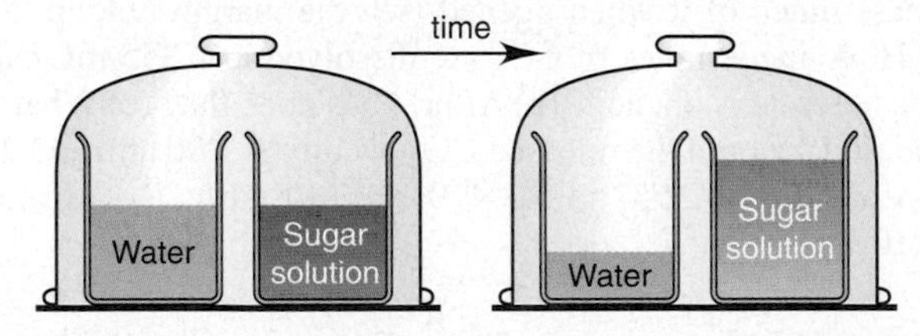

13.140 Glyphosate is the active ingredient in a common weed and grass killer. It is sold as an 18.0% by mass solution with a density of 8.94 lb/gal. (a) How many grams of Glyphosate are in a 16.0 fl oz container (1 gal = 128 fl oz)? (b) To treat a patio area of 300. ft^2, it is recommended that 3.00 fl oz be diluted with water to 1.00 gal. What is the mass percent of Glyphosate in the diluted solution (1 gal = 3.785 L)?

13.141 Although other solvents are available, dichloromethane (CH_2Cl_2) is still often used to "decaffeinate" foods because the solubility of caffeine in CH_2Cl_2 is 8.35 times that in water.
(a) A 100.0-mL sample of cola containing 10.0 mg of caffeine is extracted with 60.0 mL of CH_2Cl_2. What mass of caffeine remains in the aqueous phase?
(b) A second identical cola sample is extracted with two successive 30.0-mL portions of CH_2Cl_2. What mass of caffeine remains in the aqueous phase after each extraction?
(c) Which approach extracts more caffeine?

13.142 How do you prepare 250. g of 0.150 *m* aqueous $NaHCO_3$?

13.143 Tartaric acid can be produced from crystalline residues found in wine vats. It is used in baking powders and as an additive in foods. Analysis shows that it contains 32.3% by mass carbon and 3.97% by mass hydrogen; the balance is oxygen. When 0.981 g of tartaric acid is dissolved in 11.23 g of water, the solution freezes at −1.26°C. Use these data to find the empirical and molecular formulas of tartaric acid.

13.144 Methanol (CH_3OH) and ethanol (C_2H_5OH) are miscible because the strongest intermolecular force for both compounds is hydrogen bonding. In some methanol-ethanol solutions, the mole fraction of methanol is higher, but the mass percent of ethanol is higher. What is the range of mole fraction of methanol for these solutions?

13.145 A solution of 5.0 g of benzoic acid (C_6H_5COOH) in 100.0 g of carbon tetrachloride has a boiling point of 77.5°C.
(a) Calculate the molar mass of benzoic acid in the solution.
(b) Suggest a reason for the difference between the molar mass based on the formula and that found in part (a).

13.146 Derive a general equation that expresses the relationship between the molarity and the molality of a solution, and use it to explain why the numerical values of these two terms are approximately equal for very dilute aqueous solutions.

13.147 A florist prepares a solution of nitrogen-phosphorus fertilizer by dissolving 5.66 g of NH_4NO_3 and 4.42 g of $(NH_4)_3PO_4$ in enough water to make 20.0 L of solution. What are the molarities of NH_4^+ and of PO_4^{3-} in the solution?

13.148 Suppose coal-fired power plants used water in scrubbers to remove SO_2 from smokestack gases (see Chemical Connections, p. 245).
(a) If the partial pressure of SO_2 in the stack gases is 2.0×10^{-3} atm, what is the solubility of SO_2 in the scrubber liquid (k_H for SO_2 in water is 1.23 mol/L·atm at 200.°C)?
(b) From your answer to part (a), why are basic solutions, such as lime-water slurries [$Ca(OH)_2$], used in scrubbers?

13.149 Urea is a white crystalline solid used as a fertilizer, in the pharmaceutical industry, and in the manufacture of certain polymer resins. Analysis of urea reveals that, by mass, it is 20.1% carbon, 6.7% hydrogen, 46.5% nitrogen and the balance oxygen.
(a) Calculate the empirical formula of urea.
(b) A 5.0 g/L solution of urea in water has an osmotic pressure of 2.04 atm, measured at 25°C. What is the molar mass and molecular formula of urea?

13.150 The total concentration of dissolved particles in blood is 0.30 *M*. An intravenous (IV) solution must be isotonic with blood, which means it must have the same concentration.
(a) To relieve dehydration, a patient is given 100. mL/h of IV glucose ($C_6H_{12}O_6$) for 2.5 h. What mass (g) of glucose did she receive?
(b) If isotonic saline (NaCl) were used, what is the molarity of the solution?
(c) If the patient is given 150. mL/h of IV saline for 1.5 h, how many grams of NaCl did she receive?

13.151 To avoid the "bends," divers breathe mixtures of He and O_2, because He has very low solubility in blood. At a pressure of 4.00 atm, what is the molarity of He needed to maintain the molarities that O_2 and N_2 have at 1.00 atm (k_H in water at 37°C is

1.1×10^{-3} mol/L·atm for O_2, 6.2×10^{-4} mol/L·atm for N_2, and 3.7×10^{-4} mol/L·atm for He)?

13.152 The survival of fish depends on the solubility of air in water, which is 0.147 cm^3 of air (measured at STP) per gram of water at 20.°C and 0.10 MPa (megapascal) pressure.
(a) What is the solubility in a mountain stream where the pressure is 0.060 MPa?
(b) Calculate k_H of air in water at 20.°C.
(c) At 20.°C and 0.50 MPa, the solubility is 0.825 cm^3 of air (at STP) per gram of water. What solubility is calculated from Henry's law? Calculate the error, as a percent of the measured solubility, from relying on Henry's law at this high pressure.

13.153 Iodine (I_2) dissolves in a variety of nonpolar or slightly polar solvents: 2.7 g of I_2 dissolves in 100.0 g of chloroform ($CHCl_3$), 2.5 g dissolves in 100.0 g of carbon tetrachloride, and 16 g dissolves in 100.0 g of carbon disulfide. Calculate the mass percent, mole fraction, and molality of I_2 in each solution.

13.154 Volatile organic solvents have been implicated in smog formation and in adverse health effects of industrial workers. Greener methods are phasing them out (see the margin note, p. 493). Rank the solvents in Table 13.6 (p. 518) in terms of increasing volatility.

13.155 At ordinary temperatures, water is a poor solvent for organic substances. But at high pressure and above 200°C, water develops many properties of organic solvents. Find the minimum pressure needed to maintain water as a liquid at 200.°C (ΔH_{vap} = 40.7 kJ/mol at 100°C and 1.00 atm; assume it remains constant with temperature).

13.156 In ice-cream making, the temperature of the ingredients is kept below 0.0°C in an ice-salt bath.
(a) Assuming that NaCl dissolves completely and forms an ideal solution, what mass of it is needed to lower the melting point of 5.5 kg of ice to −5.0°C?
(b) Given the same assumptions as in part (a), what mass of $CaCl_2$ is needed?

13.157 Several different ionic compounds are each being recrystallized by the following procedure:
Step 1. A saturated aqueous solution of the compound is prepared at 50°C.
Step 2. The mixture is filtered to remove undissolved compound.
Step 3. The filtrate is cooled to 0°C.
Step 4. The crystals that form are filtered, dried, and weighed.
(a) Using Figure 13.21 (p. 508), which of the following compounds would have the highest percent recovery and which the lowest: KNO_3, $KClO_3$, KCl, NaCl? Explain.
(b) Starting with 100. g of each compound in your answer to part (a), how many grams of each can be recovered?

13.158 The solubility of N_2 in blood can be a serious problem (the "bends") for divers breathing compressed air (78% N_2 by volume) at depths greater than 50 ft (see the margin note, p. 510).
(a) What is the molarity of N_2 in blood at 1.00 atm?
(b) What is the molarity of N_2 in blood at a depth of 50. ft?
(c) Find the volume (in mL) of N_2, measured at 25°C and 1.00 atm, released per liter of blood when a diver at a depth of 50. ft rises to the surface (k_H for N_2 in water at 25°C is 7.0×10^{-4} mol/L·atm and at 37°C is 6.2×10^{-4} mol/L·atm; assume d of water is 1.00 g/mL).

13.159 Figure 12.9B (p. 435) describes the phase changes when only water is present. Consider how the diagram would change if air were present at 1 atm and dissolved in the water.
(a) Would the three phases of water still attain equilibrium at some temperature? Explain.
(b) Would that temperature be higher, lower, or the same as the triple point for pure water? Explain.
(c) Would ice sublime at a few degrees below the freezing point under this pressure? Explain.
(d) Would the liquid have the same vapor pressure as that shown in Figure 12.9B at 100°C? At 120°C?

13.160 On top of a 1.00×10^4-L pool of water is fitted an airtight covering filled with N_2 and maintained at 1.20 atm.
(a) What volume (in L) of N_2, measured at 25°C and 1.20 atm, is equal to the moles of N_2 that dissolve in the water (k_H for N_2 at 25°C is 7.0×10^{-4} mol/L·atm)?
(b) Repeat the calculation in part (a) for CO_2 (k_H for CO_2 at 25°C is 2.3×10^{-2} mol/L·atm).
(c) Why are the volumes of CO_2 and N_2 so different?

13.161 Eighty proof whiskey is 40% ethanol (C_2H_5OH) by volume. A man has 7.0 L of blood and drinks a 58-mL shot of the whiskey, of which 22% of the ethanol goes into his blood. (a) What concentration (in g/mL) of ethanol is in his blood (d of ethanol = 0.789 g/mL)? (b) What volume (in mL) of whiskey would raise his blood alcohol level to 0.0030 g/mL, at which level a person is considered intoxicated?

13.162 Carbonated soft drinks are canned under 4 atm of CO_2 and release much of it when opened (see the margin note, p. 510). (a) How many moles of CO_2 are dissolved in a 355-mL can of soda before it is opened? (b) After it has gone flat? (c) What volume (in L) would the released CO_2 occupy at 1.00 atm and 25°C (k_H for CO_2 at 25°C is 3.3×10^{-2} mol/L·atm; P_{CO_2} in air is 3×10^{-4} atm)?

The Basis of Element Behavior The dramatic chemical change that occurs when sodium reacts with water is just one example of the countless reactions of the main-group elements. As you have seen, this behavior is based on atomic and molecular properties, which we revisit in the upcoming pages.

INTERCHAPTER
A Perspective on the Properties of the Elements

Chemistry has a central, underlying principle: *The behavior of a sample of matter emerges from the properties of its component atoms.* This illustrated Interchapter reviews many ideas about these properties from earlier chapters so that you can see in upcoming chapters how they lead to the behavior of the main-group elements. Keep in mind that, despite our categories, clear dividing lines rarely appear in nature; instead, the periodic properties of matter display gradual changes from one substance to another.

TOPIC 1

The Key Atomic Properties

Four atomic properties are critical to the behavior of an element: *electron configuration, atomic size, ionization energy,* and *electronegativity.*

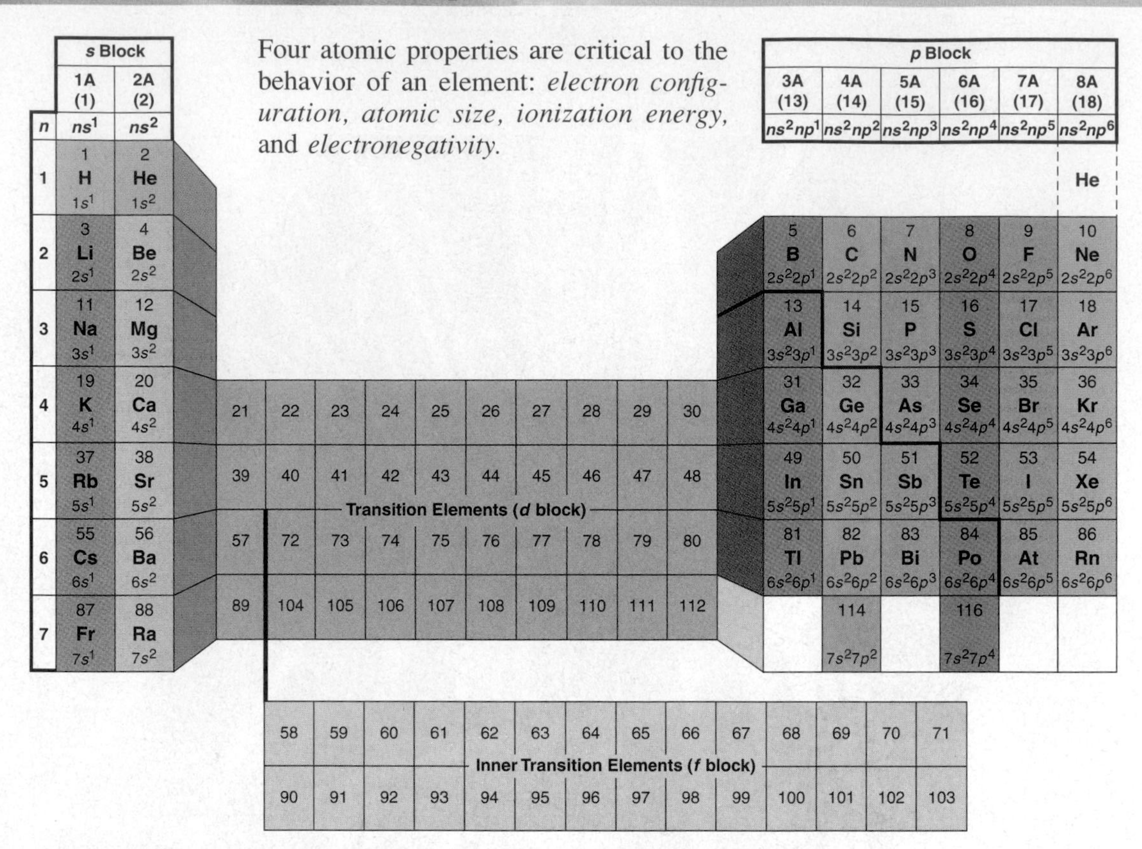

Electron configuration ($nl^{\#}$) is the distribution of electrons in the energy levels and sublevels of an atom *(Sections 7.4 and 8.3)*:

- The *n* value (positive integer) indicates the *energy* and *relative distance* from the nucleus of orbitals in a level.
- The *l* value (or its more commonly used letter designation *s, p, d, f*) indicates the *shape* of orbitals in the sublevel *(right)*.
- The superscript (#) tells the *number* of electrons in the sublevel.

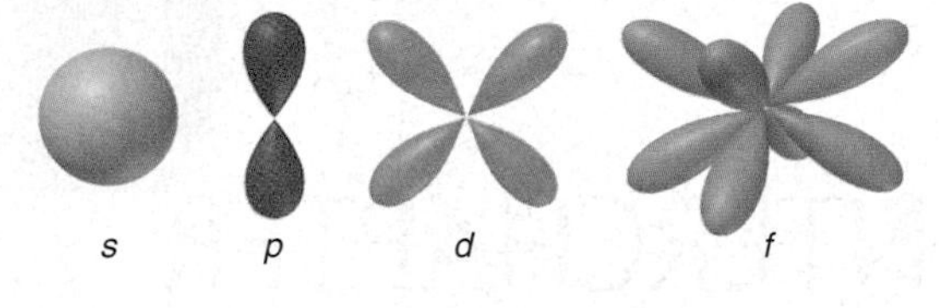

The periodic table above shows the sublevel blocks of elements and the ground-state (lowest energy), outer (valence) electron configuration of the main-group (*s*- and *p*-block) elements. Note that

- The ***s* block** lies to the left of the transition elements and the ***p* block** lies to the right. [Although H is not an alkali metal, like the elements in Group 1A(1), and He belongs in Group 8A(18), both are part of the *s* block.]
- Outer electron configurations are *similar within a group.*
- Outer electron configurations are *different within a period.*
- Outer electrons occupy the *ns* and *np* sublevels (n = period number).
- Four valence-level orbitals (one *ns* + three *np*) occur among the main-group elements.
- Outer electrons *are* the valence electrons of the main-group elements.
- The A-group number (**1**A to **8**A) equals the number of valence electrons.

Outer electrons are *shielded* from the full nuclear charge by electrons in the same level and, especially, by those in lower energy levels. Shielding reduces the attraction they experience to a much lower *effective nuclear charge* (Z_{eff}) *(Section 8.2)*. Within a level, electrons that *penetrate* more (spend more time near the nucleus) shield others more effectively. The extent of penetration is in the order $s > p > d > f$. The radial probability distribution curves *(right)* show that

- Inner ($n = 1$) electrons shield outer ($n = 2$) electrons very effectively.
- 2*s* electrons shield 2*p* electrons somewhat because 2*s* electrons spend more time near the nucleus.

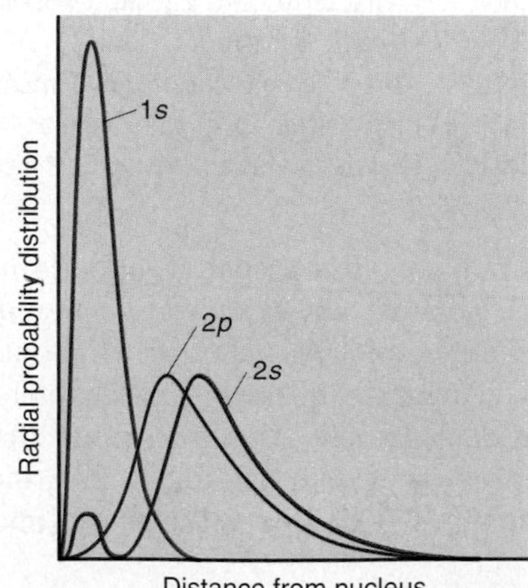

Z_{eff} greatly influences atomic properties. In general,

- Z_{eff} increases significantly left to right *across* a period.
- Z_{eff} increases slightly *down* a group.

Atomic size is based on atomic radius, one-half the distance between nuclei of identical, bonded atoms *(Section 8.4)*. The small red periodic table shows the trends in atomic size among the main-group elements. Note that

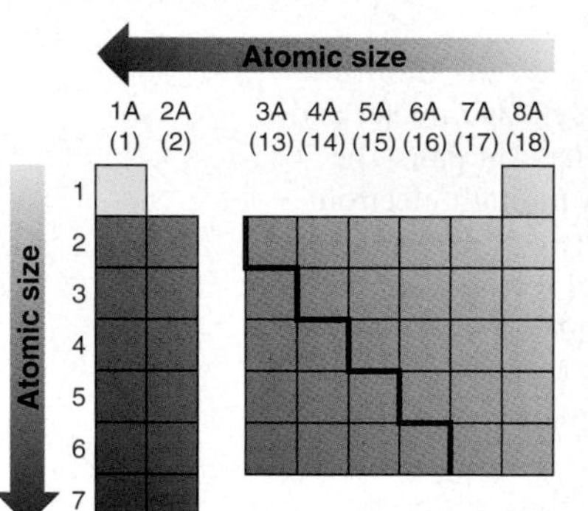

- Atomic size generally *decreases left to right across a period:* increasing Z_{eff} pulls outer electrons closer.
- Atomic size generally *increases down a group:* outer electrons in higher periods lie farther from the nucleus.

Ionization energy (IE) is the energy required to remove the highest energy electron from 1 mol of gaseous atoms *(Section 8.4)*. The relative magnitude of the IE influences the types of bonds an atom forms: an element with a *low IE* is more likely to *lose* electrons, and one with a *high IE* is more likely to *share (or gain)* electrons (excluding the noble gases). From the small yellow periodic table, note that

- IE generally *increases left to right across a period:* higher Z_{eff} holds electrons tighter.
- IE generally *decreases down a group:* greater distance from the nucleus lowers the attraction for electrons.

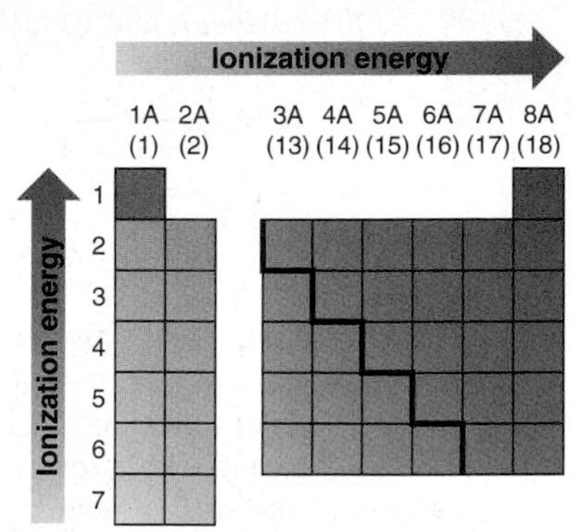

Thus, the trends in IE are *opposite* those in atomic size: it is easier to remove an electron (lower IE) that is farther from the nucleus (larger atomic size).

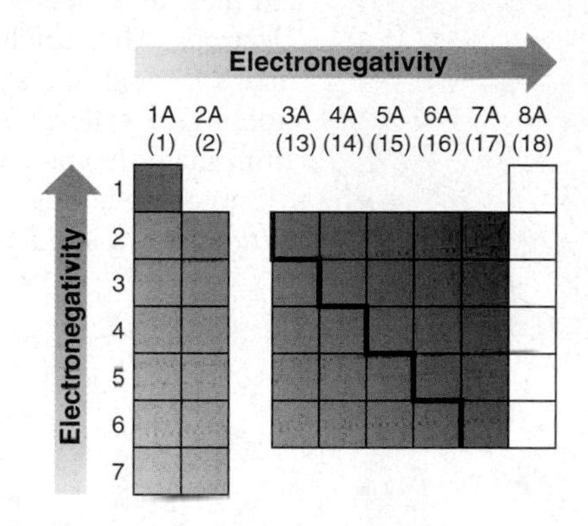

Electronegativity (EN) is a number that refers to the relative ability of an atom in a covalent bond to attract shared electrons *(Section 9.5)*. From the small green periodic table, note that

- EN generally *increases left to right across a period:* higher Z_{eff} and shorter distance from the nucleus strengthen the attraction for the shared pair.
- EN generally *decreases down a group:* greater distance from the nucleus weakens the attraction for the shared pair. (Group 8A is not shaded because the noble gases form few compounds.)

Thus, the trends in EN are *opposite* those in atomic size and the *same* as those in IE.

The graph shows simplified plots of atomic size *(red)*, ionization energy *(yellow)*, and electronegativity *(green)* versus atomic number. Note that

- Atomic size decreases gradually left to right within a period *(vertical gray band)* and then increases suddenly at the beginning of the next period.
- IE and EN display the opposite pattern, increasing left to right within a period and decreasing at the beginning of the next period.

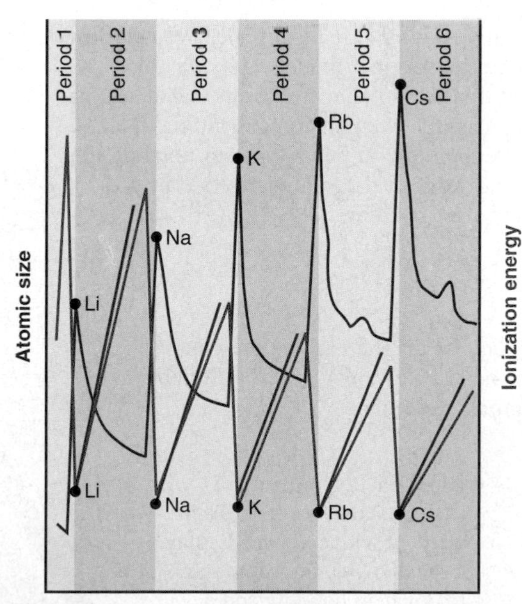

The difference in electronegativity (ΔEN) between the atoms in a bond greatly influences physical and chemical behavior of the compound, as the block diagram shows.

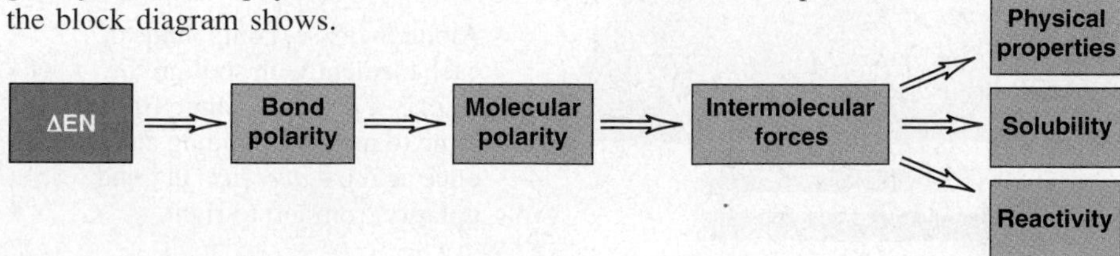

Characteristics of Chemical Bonding

Chemical bonds are the forces that hold atoms (or ions) together in an element or compound. The type of bonding, bond properties, nature of orbital overlap, and number of bonds determine physical and chemical behavior.

Types of Bonding

There are three idealized bonding models: *ionic, covalent,* and *metallic.*

Covalent bonding results from the attraction between two nuclei and a localized electron pair. The bond arises through electron sharing between atoms with a small ΔEN (usually two nonmetals) and leads to discrete molecules with specific shapes or to extended networks *(Section 9.3).*

Ionic bonding results from the attraction between positive and negative ions. The ions arise through electron transfer between atoms with a large ΔEN (from metal to nonmetal). This bonding leads to crystalline solids with ions packed tightly in regular arrays *(Section 9.2).*

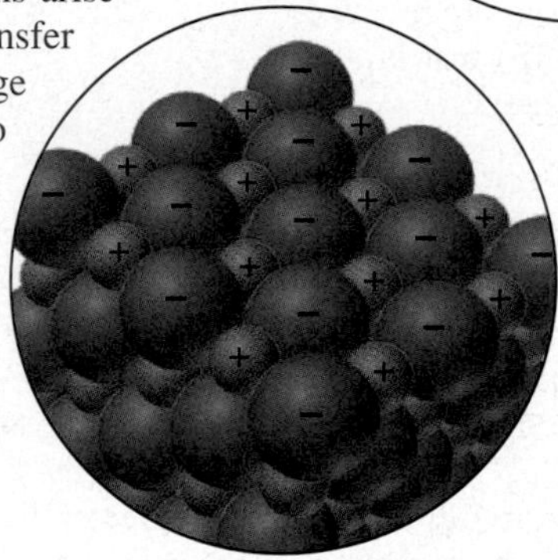

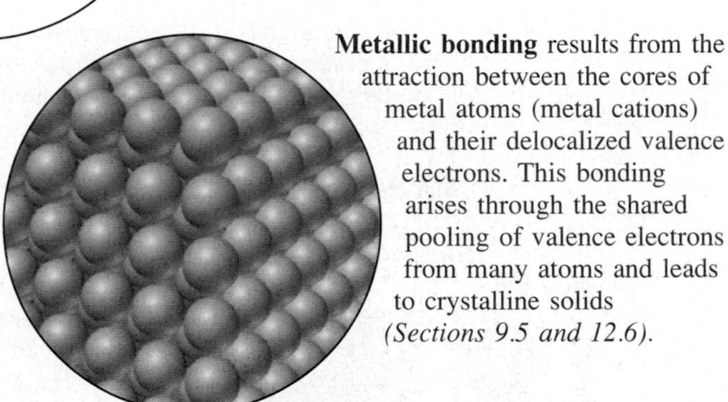

Metallic bonding results from the attraction between the cores of metal atoms (metal cations) and their delocalized valence electrons. This bonding arises through the shared pooling of valence electrons from many atoms and leads to crystalline solids *(Sections 9.5 and 12.6).*

The actual bonding in real substances usually lies between these distinct models *(Section 9.5).* The electron density relief maps show that densities overlap slightly even in ionic bonding (NaCl). They overlap more in polar covalent bonding (an SiCl bond from $SiCl_4$) and even more in nonpolar covalent bonding (Cl_2).

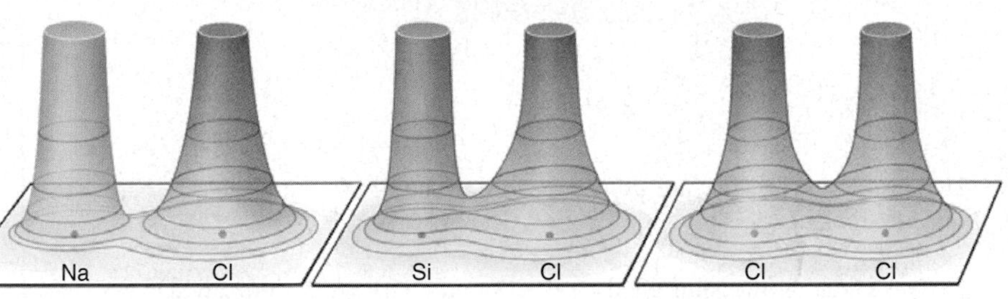

The triangular diagram shows the continuum of bond types among all the Period 3 main-group elements:

- Along the left side of the triangle, compounds of each element with chlorine display a gradual change from ionic to covalent bonding and a *decrease* in bond polarity from bottom to top.

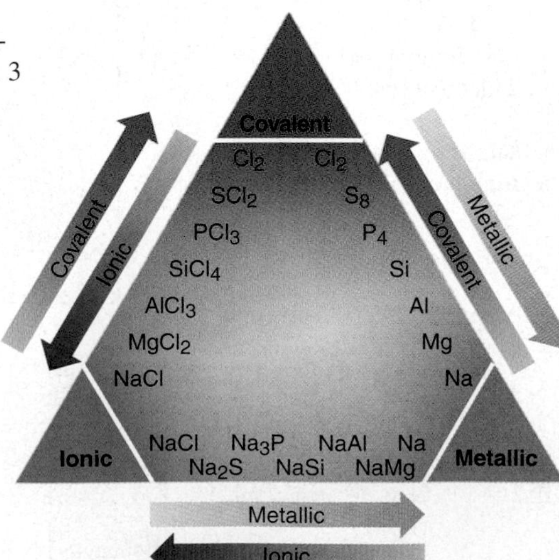

- Along the right side, the elements themselves display a gradual change from covalent to metallic bonding.

- Along the base, compounds of each element with sodium display a gradual change from ionic to metallic bonding and, once again, a *decrease* in bond polarity from left to right.

Bond Properties

There are two important properties of a covalent bond *(Section 9.3):*
Bond length is the distance between the nuclei of bonded atoms.
Bond energy (bond strength) is the enthalpy change required to break a given bond in 1 mol of gaseous molecules.

Among similar compounds, these bond properties are related to each other and to reactivity, as shown in the graph for the carbon tetrahalides (CX_4). Note that

- *As bond length increases, bond energy decreases:* shorter bonds are stronger bonds.
- *As bond energy decreases, reactivity increases.*

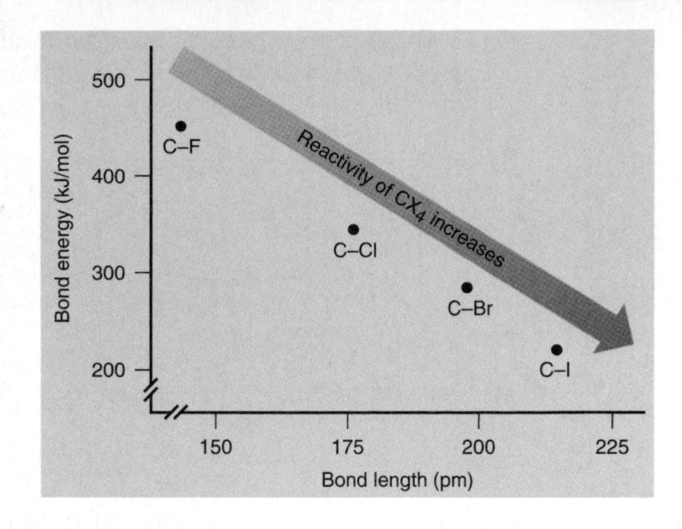

Nature of Orbital Overlap

In a covalent bond, the shared electrons reside in the entire region composed of the overlapping orbitals of the two atoms. The diagram depicts the bonding in ethylene (C_2H_4).

Bond order is one-half the number of electrons shared. Bond orders of 1 (single bond) and 2 (double bond) are common; a bond order of 3 (triple bond) is much less common. Fractional bond orders occur when there are resonance structures for species with adjacent single and double bonds *(Sections 9.3 and 10.1).*

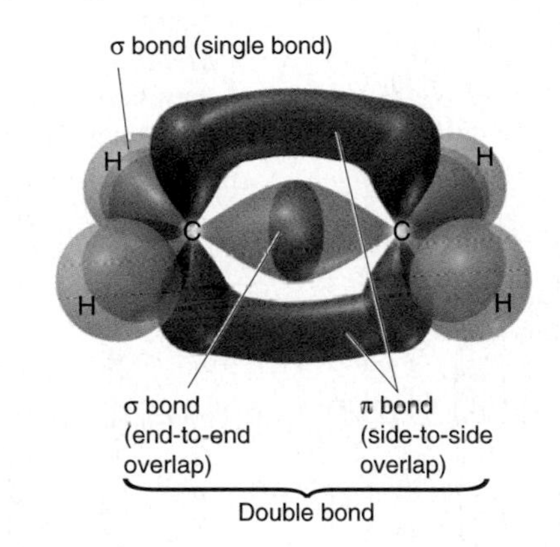

Orbitals overlap in two ways, which leads to two types of bonds *(Section 11.2):*

- **End-to-end overlap** (of *s*, *p*, and hybrid atomic orbitals) leads to a sigma (σ) bond, one with electron density distributed symmetrically along the bond axis. A single bond is a σ bond.
- **Side-to-side overlap** (of *p* with *p*, or sometimes *d*, orbitals) leads to a pi (π) bond, one with electron density distributed above and below the bond axis. A double bond consists of one σ bond and one π bond. A π bond restricts rotation around the bond axis, allowing for different spatial arrangements of the atoms and, therefore, different molecules. Pi bonds are often sites of reactivity; for example,

$$CH_2{=}CH_2(g) + H{-}Cl(g) \longrightarrow CH_3{-}CH_2{-}Cl(g)$$
$$CH_3{-}CH_3(g) + H{-}Cl(g) \longrightarrow \text{no reaction}$$

Number of Bonds and Molecular Shape

The shape of a molecule is defined by the positions of the nuclei of the bonded atoms. According to VSEPR theory *(Section 10.2),* the number of electron groups in the valence level of a central atom, which is based on the number of bonding and lone pairs, is the key factor that determines molecular shape. The small periodic table shows that

- *The elements in Period 2 cannot form more than four bonds* because they have a maximum of four (one *s* and three *p*) valence orbitals. (Only carbon forms four bonds routinely.) Molecular shapes *(small circle)* are based on linear, trigonal planar, and tetrahedral electron-group arrangements.
- *Many elements in Period 3 or higher can form more than four bonds* by using empty *d* orbitals and, thus, expanding their valence levels. Shapes include those above and others based on trigonal bipyramidal and octahedral electron-group arrangements *(large circle).*

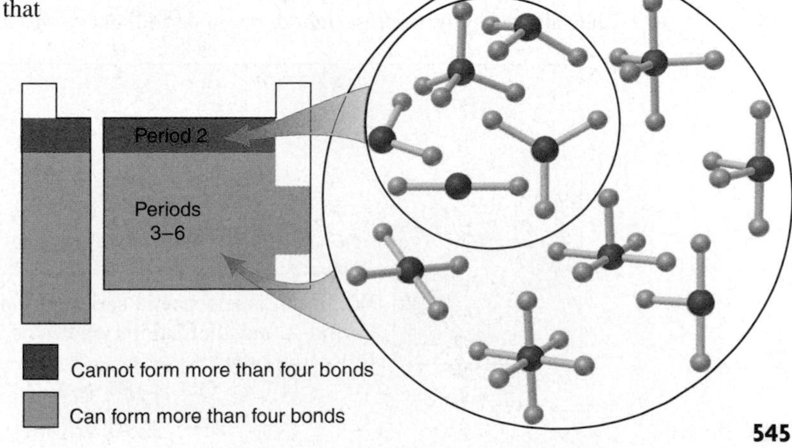

Metallic Behavior

The elements are often classified as metals, metalloids, or nonmetals. Although exceptions exist, this table contrasts the general atomic, physical, and chemical properties of metals with those of nonmetals *(Section 8.5)*.

	Metals	Nonmetals
Atomic Properties	Have fewer valence electrons (in period) Have larger atomic size Have lower ionization energies Have lower electronegativities	Have more valence electrons (in period) Have smaller atomic size Have higher ionization energies Have higher electronegativities
Physical Properties	Occur as solids at room temperature Conduct electricity and heat well Are malleable and ductile	Occur in all three physical states Conduct electricity and heat poorly Are not malleable or ductile
Chemical Properties	Lose electron(s) to become cations React with nonmetals to form ionic compounds Mix with other metals to form solid solutions (alloys)	Gain electron(s) to become anions React with metals to form ionic compounds React with other nonmetals to form covalent compounds

This small periodic table shows the location of metals, metalloids, and nonmetals among the main-group elements. Note that

- Metals lie in the lower-left portion of the table.
- Nonmetals lie in the upper-right portion of the table.
- Metalloids lie between the metals and the nonmetals. These elements have intermediate values of atomic size, IE, and EN and display an intermediate metallic character: shiny solids with low conductivity; cation-like with nonmetals (e.g., AsF_3); and anion-like with metals (e.g., Na_3As).

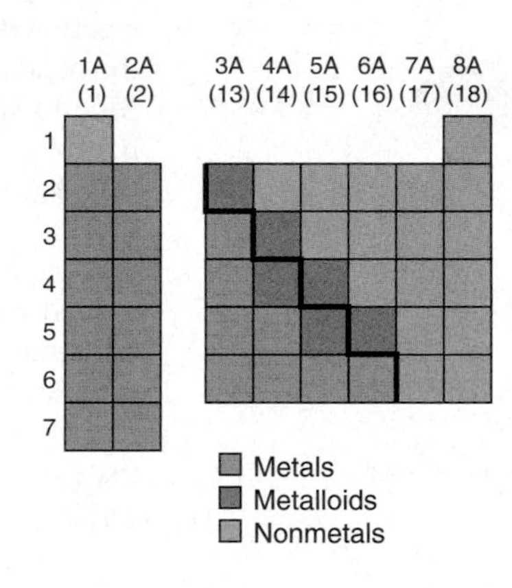

From the blue, shaded periodic table, note that

- *Metallic behavior* changes gradually among the elements.
- *Metallic behavior parallels atomic size:* Larger members of a group *(bottom)* or period *(left)* are more metallic; smaller members are less metallic.

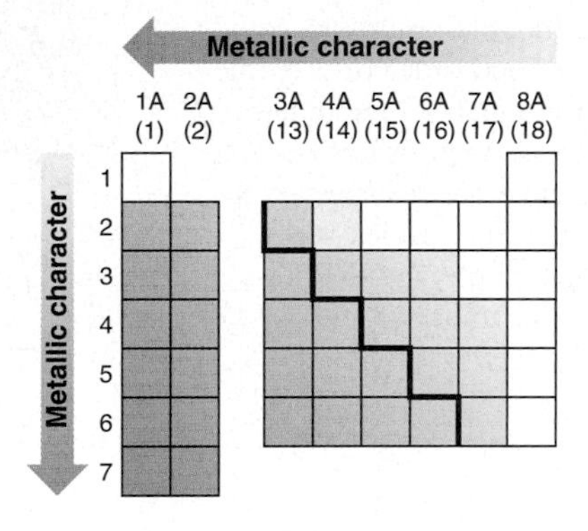

Metals and nonmetals typically form crystalline ionic compounds when they react with each other, and ionic size and charge determine the packing in these solids. Monatomic ions exhibit clear size trends:

- Cations are smaller than their parent atoms, and anions are larger.
- Ionic size *increases down* a group.
- Ionic size *decreases left to right across* a period, but anions are much larger than cations.

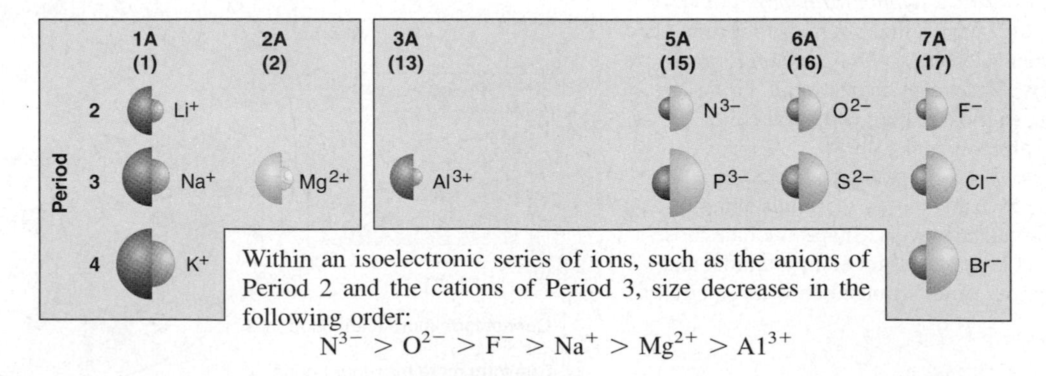

Within an isoelectronic series of ions, such as the anions of Period 2 and the cations of Period 3, size decreases in the following order:

$$N^{3-} > O^{2-} > F^- > Na^+ > Mg^{2+} > Al^{3+}$$

Acid-Base Behavior of the Element Oxides

Oxides are known for almost every element. The metallic behavior of an element correlates with the acid-base behavior of its oxide in water *(Section 8.5)*. Recall that

- An acid produces H^+ ions when dissolved in water and reacts with a base to form a salt and water.
- A base produces OH^- ions when dissolved in water and reacts with an acid to form a salt and water.

This chart shows that the electronegativity and metallic behavior of an element (E) determine the type of bonding (E-to-O) in its oxide and, therefore, the acid-base behavior of the oxide dissolved in water.

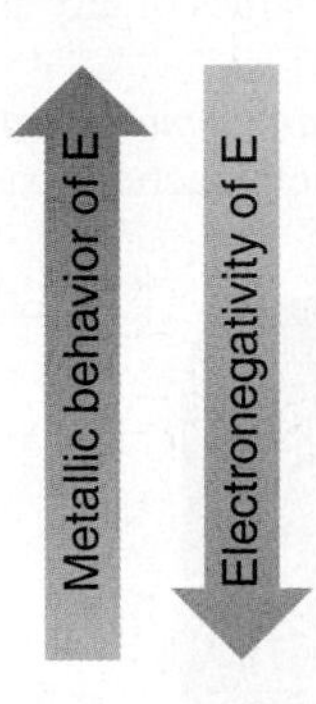

Basic Oxide (Ionic E-to-O)	$+ H_2O \longrightarrow$	more OH^-
	$+$ acid $\longrightarrow$	salt (E cation) + H_2O
Amphoteric Oxide	$+$ acid $\longrightarrow$	salt (E cation) + H_2O
	$+$ base $\longrightarrow$	salt (E oxoanion) + H_2O
Acidic Oxide (Covalent E-to-O)	$+$ base $\longrightarrow$	salt (E oxoanion) + H_2O
	$+ H_2O \longrightarrow$	more H^+

Note that, among the main groups,

- *Elements with low EN (metals) form basic oxides.* The E-to-O bonding is ionic. The O^{2-} ion is the basic species:

  $$O^{2-}(s) + H_2O(l) \longrightarrow 2OH^-(aq)$$

  For example, barium forms the basic oxide BaO:

  Reacts with water: $BaO(s) + H_2O(l) \longrightarrow Ba^{2+}(aq) + 2OH^-(aq)$

  Reacts with acid: $BaO(s) + 2H^+(aq) \longrightarrow Ba^{2+}(aq) + H_2O(l)$

- *Elements with high EN (nonmetals) form acidic oxides.* The E-to-O bonding is covalent. Water bonds to E of the oxide to form an acid, which releases H^+. For example, sulfur forms the acidic oxide SO_2:

  Reacts with water: $SO_2(g) + H_2O(l) \rightleftharpoons H_2SO_3(aq) \rightleftharpoons H^+(aq) + HSO_3^-(aq)$

  Reacts with base: $SO_2(g) + 2OH^-(aq) \longrightarrow SO_3^{2-}(aq) + H_2O(l)$

- *Elements with intermediate EN (some metalloids and metals) form amphoteric oxides, which react with acid and base.* For example, Al_2O_3 is an amphoteric oxide:

  Reacts with acid: $Al_2O_3(s) + 6H^+(aq) \longrightarrow 2Al^{3+}(aq) + 3H_2O(l)$

  Reacts with base: $Al_2O_3(s) + 2OH^-(aq) + 3H_2O(l) \longrightarrow 2Al(OH)_4^-(aq)$

Period	1A (1)	2A (2)	3A (13)	4A (14)	5A (15)	6A (16)	7A (17)	8A (18)
1								
2	Li_2O	BeO	B_2O_3	CO_2	[illegible] N_2O_3			
3	Na_2O	MgO	Al_2O_3	SiO_2	P_4O_{10} P_4O_6	[illegible] SO_2	[illegible] Cl_2O	
4	K_2O	CaO	Ga_2O_3	GeO_2	As_2O_5 As_4O_6	SeO_3 SeO_2	Br_2O	
5	Rb_2O	SrO	In_2O_3 In_2O	SnO_2 SnO	Sb_2O_5 Sb_4O_6	TeO_3 TeO_2	I_2O_5	
6	Cs_2O	BaO	Tl_2O	PbO_2 PbO	Bi_2O_3	PoO_2 PoO		
7	Fr_2O	RaO						

Legend: Strongly basic; Weakly basic; Amphoteric; Strongly acidic; Moderately acidic; Weakly acidic

The periodic table shows the acid-base behavior of many main-group oxides in water. Note that

- *Oxide acidity increases left to right across a period and decreases down a group,* which is *opposite* the trend in metallic behavior (and atomic size).
- When an element forms two oxides, *the element has a higher oxidation number in the more acidic oxide* (see also Topic 5). Thus, for example, SO_2 forms the weak acid H_2SO_3, whereas SO_3 forms the strong acid H_2SO_4.

Redox Behavior of the Elements

The relative ability of an element to lose or gain electrons (or electron charge) when reacting with other elements defines its redox (oxidation-reduction) behavior *(Section 4.5):*

- *The oxidation number, O.N. (or oxidation state)* of an atom in an element is zero. In a compound, the O.N. is the number of electrons that have shifted away from the atom (positive O.N.) or toward it (negative O.N.). O.N. values are determined by a set of rules *(Section 4.5).*
- *An oxidation-reduction (redox) reaction* occurs when the O.N. values of any atoms in the reactants are different from those in the products.
- Among the countless redox reactions are all those that involve *an elemental substance as reactant or product.* These include all combustion reactions and all formation reactions, such as this one between potassium and chlorine: $2K + Cl_2 \longrightarrow 2KCl$

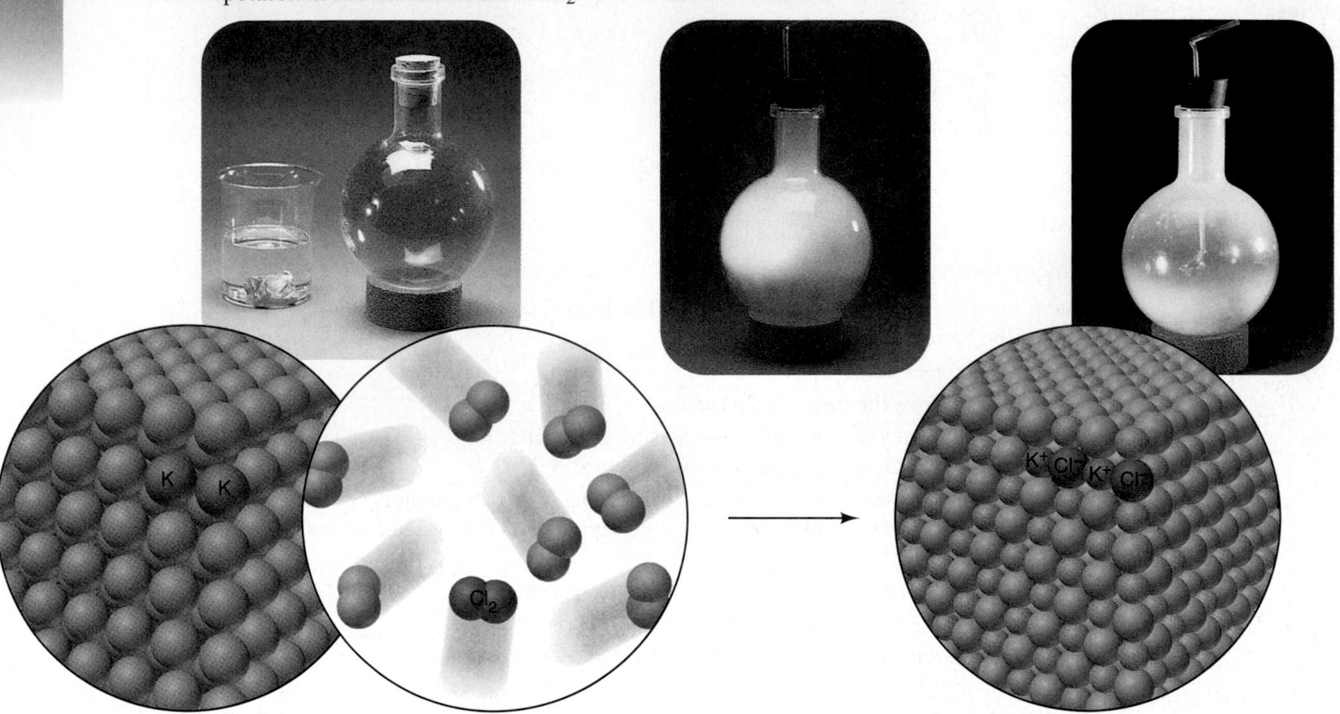

Reducing and Oxidizing Agents

All redox reactions include a reducing *and* an oxidizing agent among the reactants. This chart shows that

- The reducing agent gives electrons *to* (reduces) the oxidizing agent and becomes oxidized (more positive O.N.).
- The oxidizing agent removes electrons *from* (oxidizes) the reducing agent and becomes reduced (more negative O.N.).

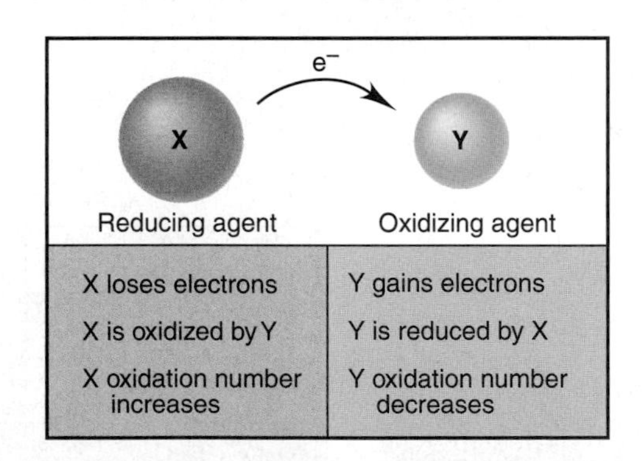

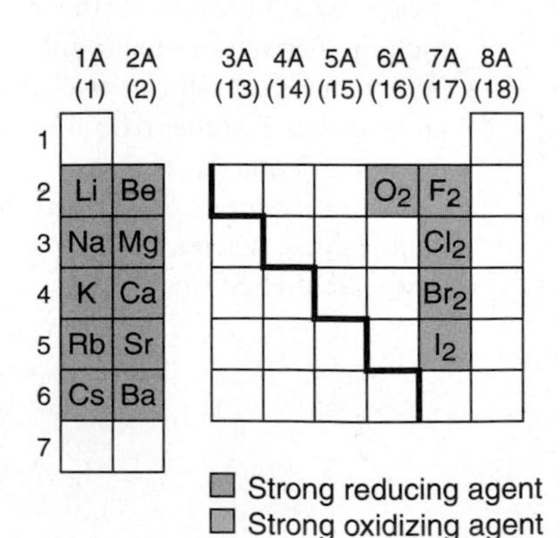

The small periodic table shows that oxidizing and reducing ability are related to atomic properties:

- Elements with low IE and low EN [Groups 1A(1) and 2A(2)] are *strong reducing agents.*
- Elements with high IE and high EN [Group 7A(17) and oxygen in Group 6A(16)] are *strong oxidizing agents.*

Oxidation States of the Main-Group Elements

The periodic table below shows some oxidation states (most common states in **bold** type) of the elements in their compounds. Note that

- *The highest (most positive) state in a group equals the A-group number.* It occurs when all of an atom's outer (valence) electrons shift toward a *more* electronegative atom.
- *Among nonmetals, the lowest (most negative) state equals the A-group number minus eight.* It occurs when a nonmetal atom fills its outer level with electrons that shift away from a *less* electronegative atom.
- *Nonmetals have more oxidation states than metals* in the same group or period. (Oxygen and fluorine are exceptions.)
- Odd-numbered oxidation states are the most common ones in odd-numbered groups, and even-numbered states are the most common ones in even-numbered groups. *Oxidation states differ by units of two* because electrons are lost (or gained) in pairs *(Section 14.9)*.
- For many metals and metalloids with more than one oxidation state [Groups 3A(13) to 5A(15)], *the lower state becomes more common down the group*. This state results when *np* electrons only are lost *(Section 8.5)*.
- An element with more than one oxidation state exhibits *greater metallic behavior in its lower state*. For example, arsenic(III) oxide is more basic, more like a metal oxide, than is arsenic(V) oxide (see Topic 4).

Metals
Metalloids
Nonmetals

Period	1A (1)	2A (2)	3A (13)	4A (14)	5A (15)	6A (16)	7A (17)	8A (18)
1	H −1, +1							He
2	Li **+1**	Be **+2**	B **+3**	C **−4**, **+4**, +2	N **−3**, **+5**, +4, +3, +2, +1	O −1, **−2**	F **−1**	Ne
3	Na **+1**	Mg **+2**	Al **+3**	Si −4, **+4**, +2	P −3, **+5**, +3	S **−2**, **+6**, +4, +2	Cl **−1**, +7, +5, +3, +1	Ar
4	K **+1**	Ca **+2**	Ga **+3**, +1	Ge **+4**, +2	As −3, +5, **+3**	Se −2, +6, **+4**, +2	Br **−1**, +7, +5, +3, +1	Kr +2
5	Rb **+1**	Sr **+2**	In **+3**, +1	Sn **+4**, **+2**	Sb −3, +5, **+3**	Te −2, +6, **+4**, +2	I **−1**, +7, +5, +3, +1	Xe **+8**, **+6**, **+4**, +2
6	Cs **+1**	Ba **+2**	Tl **+1**	Pb +4, **+2**	Bi **+3**	Po **+4**, +2	At −1	Rn +2
7	Fr **+1**	Ra **+2**						

Physical States and Phase Changes

Physical state and the heat of a phase change reflect the relative strengths of the bonding and/or intermolecular forces between the atoms, ions, or molecules that make up an element or compound *(Sections 12.2, 12.3, and 12.6).*

Physical States of the Elements

The large periodic table shows the type of particle or solid structure *(shape in box),* physical state *(color of shape),* and dominant interparticle force *(background color of box)* in the most common form of each of the main-group elements at room temperature. Note that

- *Metals (left and bottom) are solids:* strong metallic bonding holds the atoms in their crystal structures.
- *Metalloids (along staircase line) and carbon are solids:* strong covalent bonding holds the atoms together in extensive networks.
- *Lighter nonmetals and Group 8A(18) (right and top) are gases:* dispersion forces are weak between molecules (H_2, N_2, O_2, F_2, Cl_2) or atoms with smaller, less polarizable electron clouds.
- *Heavier nonmetals are liquid (Br_2) or soft solids (P_4, S_8, and I_2):* dispersion forces are stronger between molecules with larger, more polarizable electron clouds.

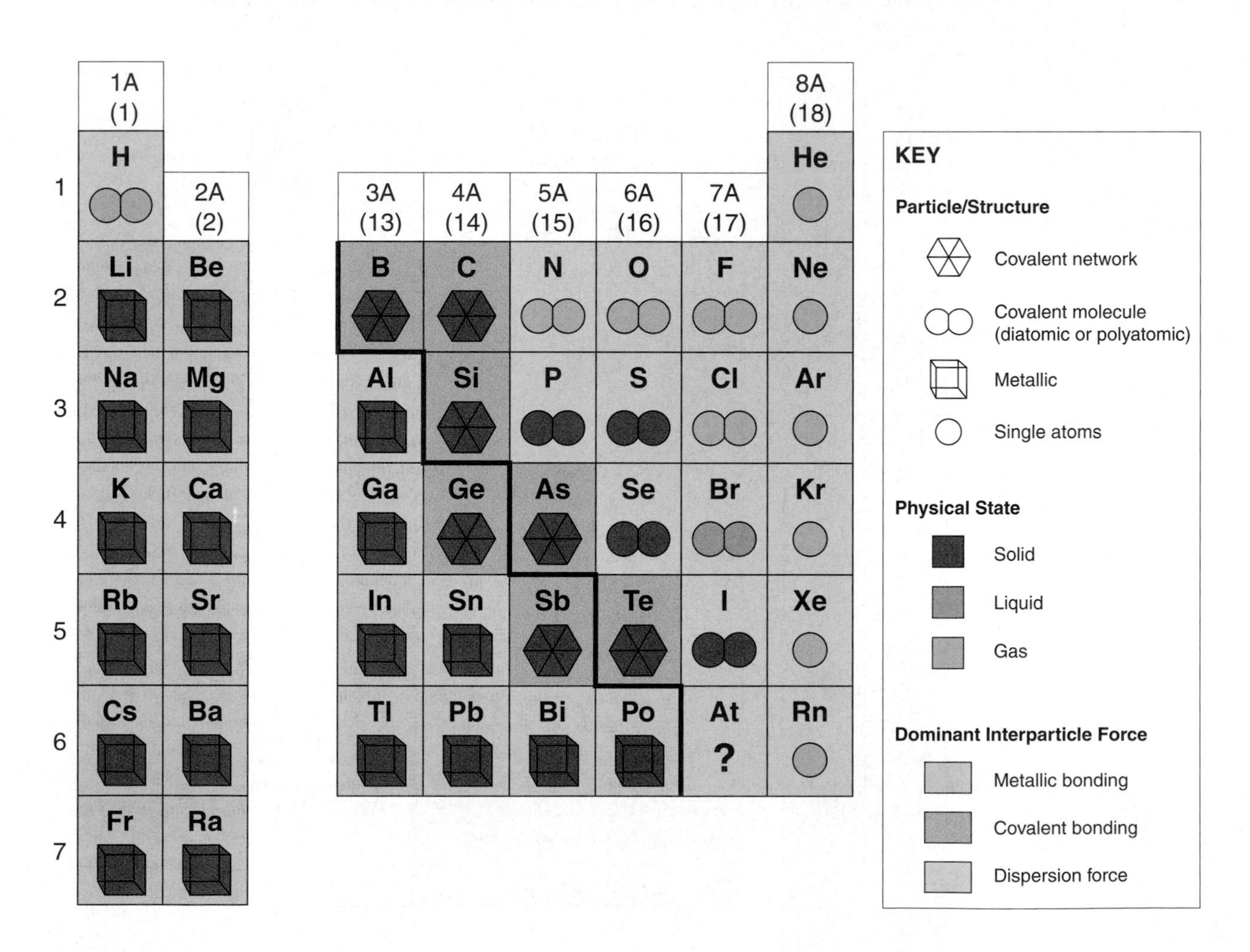

Phase Changes of the Elements

The small periodic table shows trends in melting point, boiling point, ΔH_{fus}, and ΔH_{vap} for Groups 1A(1), 7A(17), and 8A(18):

- In Group 1A(1), these properties generally *increase up* the group: smaller atom cores attract delocalized electrons more strongly, which leads to stronger metallic bonding.
- In Groups 7A(17) and 8A(18), these properties generally *increase down* the group: dispersion forces become stronger with larger, more polarizable atoms.
- In Groups 3A(13) to 6A(16), these properties reflect changes in interparticle forces down the group: lower values for molecular nonmetals, higher values for covalent networks of metalloids (and carbon), and intermediate values for metals.

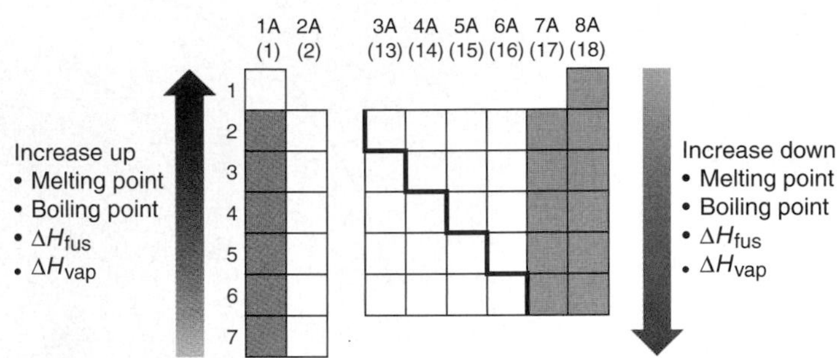

Physical Properties of Compounds

Compounds have physical properties based on the types of bonding and intermolecular forces.

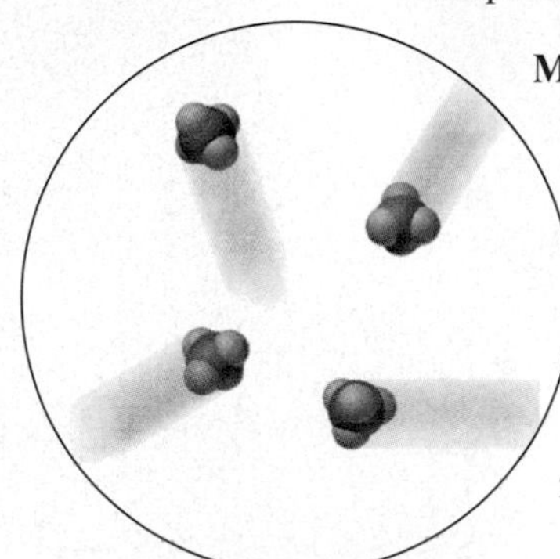

Molecular compounds, such as methane, have a physical state that depends on intermolecular forces. For polar compounds, dipole-dipole forces dominate; for nonpolar compounds, dispersion forces are most important. Most of these substances are gases, liquids, or low-melting solids at room temperature.

Hydrogen bonding arises when H is bonded to N, O, or F. It has a major effect on physical properties. For example, despite similar molar masses, H bonding in H_2O (18.02 g/mol) gives it a much higher melting point, boiling point, ΔH_{fus}, and ΔH_{vap} than CH_4 (16.04 g/mol). Water's unusually high specific heat capacity, surface tension, and viscosity also result from H bonding *(Section 12.5)*.

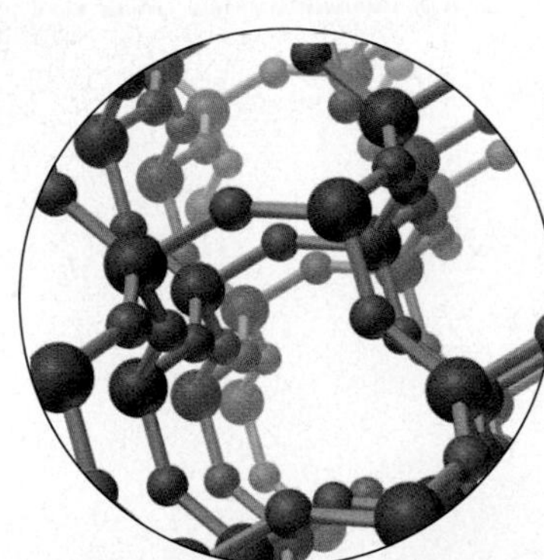

Network covalent compounds, such as silica, have extremely high melting points, boiling points, ΔH_{fus}, and ΔH_{vap}.

Ionic compounds, such as sodium chloride, have very high melting points, boiling points, ΔH_{fus}, and ΔH_{vap}.

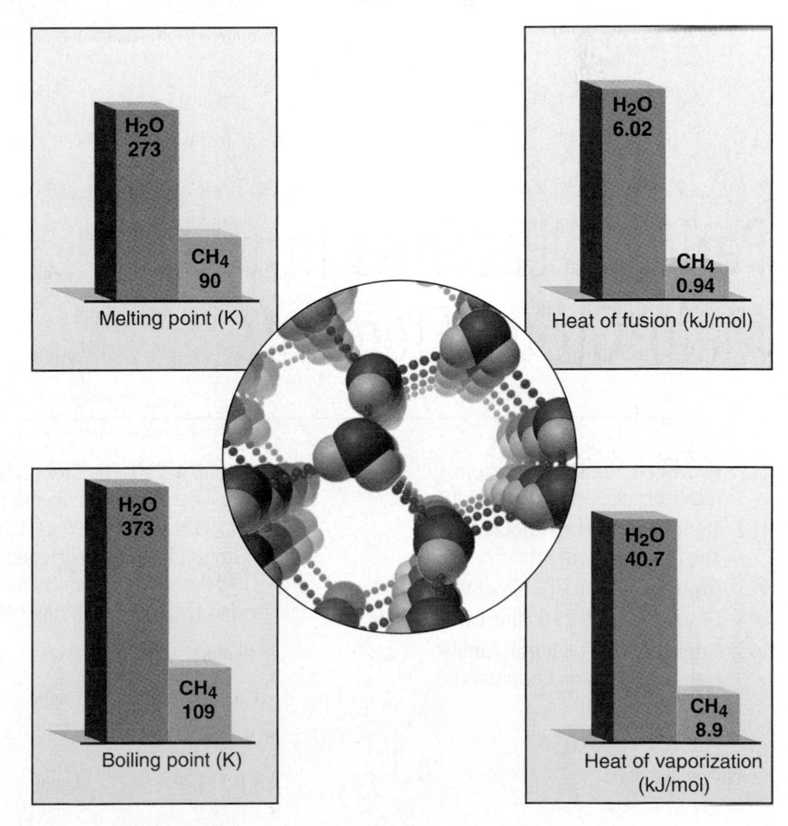

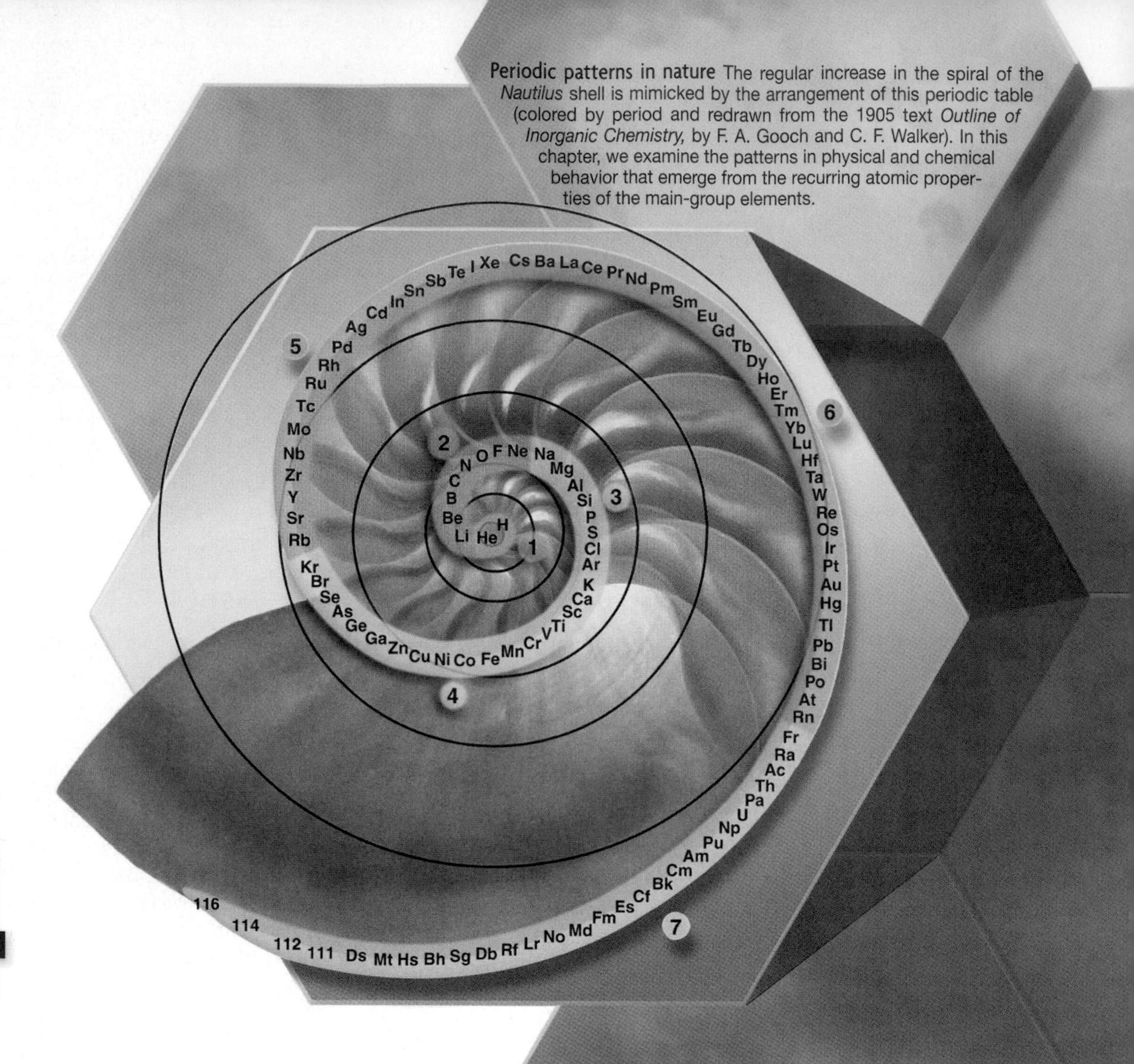

Periodic patterns in nature The regular increase in the spiral of the *Nautilus* shell is mimicked by the arrangement of this periodic table (colored by period and redrawn from the 1905 text *Outline of Inorganic Chemistry,* by F. A. Gooch and C. F. Walker). In this chapter, we examine the patterns in physical and chemical behavior that emerge from the recurring atomic properties of the main-group elements.

14

Periodic Patterns in the Main-Group Elements

In your study of chemistry so far, you've learned how to name compounds, balance equations, and calculate reaction yields. You've seen how heat is related to chemical and physical change, how electron configuration influences atomic properties, how elements bond to form compounds, and how the arrangement of bonding and lone pairs accounts for molecular shapes. You've learned modern theories of bonding and, most recently, seen how atomic and molecular properties give rise to the macroscopic properties of gases, liquids, solids, and solutions.

The purpose of this knowledge, of course, is to make sense of the magnificent diversity of chemical and physical behavior around you. The periodic table, which organizes much of this diversity, was derived from chemical facts observed in countless hours of 18th- and 19th-century research. One of the greatest achievements in science is 20th-century quantum theory, which provides a theoretical foundation for the periodic table's arrangement. But bear in mind that one theory can rarely account for *all* the facts. Therefore, be amazed by the predictive power of the patterns in the table, but do not be concerned by the occasional exceptions to these patterns. After all, our models are simple, but nature is complex.

IN THIS CHAPTER . . . We apply general ideas of bonding, structure, and reactivity to the main-group elements and see how their behavior correlates with their position in the periodic table. We begin with hydrogen, the simplest and, in some ways, most important of the elements, and then consider Period 2 to survey the general changes across the periodic table. In each of the remaining sections, we deal with one of the eight families of main-group elements. We explore vertical trends in each group and also keep an eye on horizontal trends involving neighboring groups. Especially important elements—boron, carbon, silicon, nitrogen, phosphorus, oxygen, sulfur, and the halogens—receive greater attention. (If you haven't already read the Interchapter preceding this chapter, now is the perfect time.)

Concepts & Skills to Review

before you study this chapter

- see the Interchapter

14.1 HYDROGEN, THE SIMPLEST ATOM

A hydrogen atom consists of a nucleus with a single positive charge, surrounded by a single electron. Despite this simple structure, or perhaps because of it, hydrogen may be the most important element of all. In the Sun, hydrogen (H) nuclei combine to form helium (He) nuclei in a process that provides nearly all Earth's energy. About 90% of all the atoms in the universe are H atoms, making it the most abundant element by far. On Earth, only tiny amounts of the free element occur naturally (as H_2), but hydrogen is abundant in combination with oxygen in water. Because of its simple structure and low molar mass, nonpolar gaseous H_2 is colorless and odorless, and its extremely weak dispersion forces result in very low melting (−259°C) and boiling points (−253°C).

Where Does Hydrogen Fit in the Periodic Table?

Hydrogen has no perfectly suitable position in the periodic table (Figure 14.1). Because of its single valence electron and common +1 oxidation state, some think it almost fits in Group 1A(1). However, unlike the alkali metals, hydrogen *shares* its electron with nonmetals rather than losing it to them. Moreover, like a nonmetal, it has a relatively high ionization energy (IE = 1311 kJ/mol)—much higher than that of lithium (IE = 520 kJ/mol), whose IE is the highest in Group 1A—and a much higher electronegativity as well (EN of H = 2.1; EN of Li = 1.0).

Others think hydrogen almost fits in Group 7A(17). Like the halogens, it occurs as diatomic molecules and fills its outer (valence) level either by electron sharing or by gaining one electron from a metal to form a monatomic anion, the hydride ion (H^-; O.N. = −1). However, hydrogen has a lower electronegativity

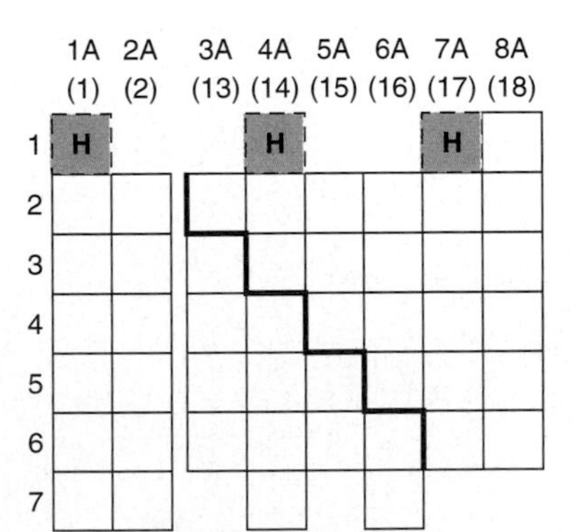

Figure 14.1 Where does hydrogen belong? Hydrogen's small size and single electron confer properties that are not fully consistent with the members of any one periodic group. Depending on the property, hydrogen may fit better in Group 1A(1), 4A(14), or 7A(17).

(EN = 2.1) than any of the halogens (ENs range from 2.2 to 4.0) and lacks their three valence electron pairs. Moreover, the H^- ion is rare and reactive, whereas halide ions are common and stable.

And, still others think that, because of a half-filled valence level and similarities in ionization energy, electron affinity, electronegativity, and bond energies, hydrogen fits best in Group 4A(14).

Hydrogen's unique behavior is attributable to its tiny size. Hydrogen has a high IE because its electron is very close to the nucleus, with no inner electrons to shield it from the positive charge. It has a low EN (for a nonmetal) because it has only one proton to attract bonding electrons. In this chapter, hydrogen will appear in either Group 1A(1) or 7A(17) depending on the property being considered.

Highlights of Hydrogen Chemistry

In Chapters 12 and 13, we discussed hydrogen bonding and its impact on physical properties (melting and boiling points, heats of fusion and vaporization, and specific heat capacities) and solubilities. You learned that H bonding plays a critical part in stabilizing Earth's climate as well as your body temperature, and you glimpsed its role in the functional structures of biomolecules. In Chapter 6 (see Chemical Connections, p. 247) we considered the possible future use of hydrogen as a fuel, and in Chapter 22 we'll see how hydrogen is used to recover some metals from their ores. Hydrogen is very reactive, combining with nearly every element, so we focus here on the three types of hydrides.

Ionic (Saltlike) Hydrides With very reactive metals, such as those in Group 1A(1) and the larger members of Group 2A(2) (Ca, Sr, and Ba), hydrogen forms *saltlike hydrides*—white, crystalline solids composed of the metal cation and the hydride ion:

$$2Li(s) + H_2(g) \longrightarrow 2LiH(s)$$
$$Ca(s) + H_2(g) \longrightarrow CaH_2(s)$$

In water, H^- is a strong base that pulls H^+ from surrounding H_2O molecules to form H_2 and OH^-:

$$NaH(s) + H_2O(l) \longrightarrow Na^+(aq) + OH^-(aq) + H_2(g)$$

The hydride ion is also a powerful reducing agent; for example, it reduces Ti(IV) to the free metal:

$$TiCl_4(l) + 4LiH(s) \longrightarrow Ti(s) + 4LiCl(s) + 2H_2(g)$$

Covalent (Molecular) Hydrides Hydrogen reacts with nonmetals to form many *covalent hydrides,* such as CH_4, NH_3, H_2O, and HF. Most are gases consisting of small molecules, but many hydrides of boron and carbon are liquids or solids that consist of much larger molecules. In most covalent hydrides, hydrogen has an oxidation number of +1 because the other nonmetal has a higher electronegativity than hydrogen.

Conditions for preparing the covalent hydrides depend on the reactivity of the other nonmetal. For example, with stable, triple-bonded N_2, hydrogen reacts at high temperatures (~400°C) and pressures (~250 atm), and the reaction needs a catalyst to proceed at any practical speed:

$$N_2(g) + 3H_2(g) \xrightarrow{\text{catalyst}} 2NH_3(g) \qquad \Delta H^0_{rxn} = -91.8 \text{ kJ}$$

Industrial facilities throughout the world use this reaction to produce millions of tons of ammonia each year for fertilizers, explosives, and synthetic fibers. On the other hand, hydrogen combines rapidly with reactive, single-bonded F_2, even at extremely low temperatures (−196°C):

$$F_2(g) + H_2(g) \longrightarrow 2HF(g) \quad \Delta H^0_{rxn} = -546 \text{ kJ}$$

Metallic (Interstitial) Hydrides Many transition elements form *metallic (interstitial) hydrides,* in which H_2 molecules (and H atoms) occupy the holes in the metal's crystal structure. Thus, such hydrides are *not* compounds but gas-solid solutions. Also, unlike ionic and covalent hydrides, interstitial hydrides, such as $TiH_{1.7}$, typically lack a single, stoichiometric formula because the metal can incorporate variable amounts of hydrogen, depending on the pressure and temperature of the gas.

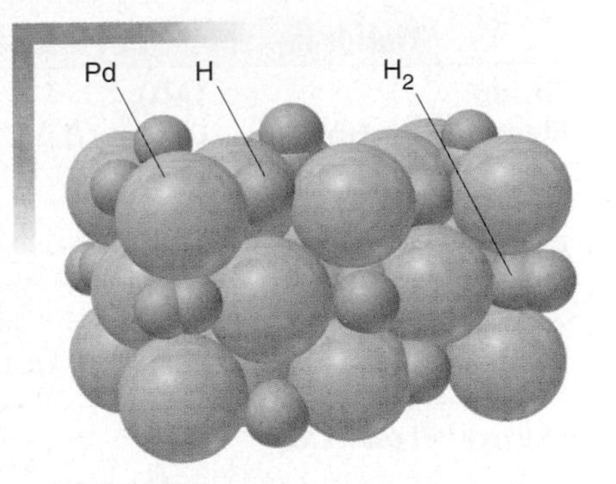

Fill 'Er Up with Hydrogen? Not Soon Dreams of a future hydrogen economy had included using metallic hydrides as "storage containers" for hydrogen fuel in cars. Just drive up to your local refueling station, insert a cylinder of H-packed metal in your car, and zip away again! It is true that some metals, such as palladium (see figure) and niobium, and some alloys, such as $LaNi_5$, store large amounts of hydrogen. However, research uncovered some serious problems, especially the facts that the metals that store the most hydrogen are expensive and heavy and they hold the gas tightest, releasing it only at very high temperatures. As a result, in 1996, U.S. government funding for the research stopped. Some researchers are keeping the dream alive, however, through their studies of carbon nanotubes, which store large amounts of H_2, and of an Na/Al alloy that is lightweight, stores 5% H_2 by mass, and releases it below 200°C. Only time and careful testing will tell!

14.2 TRENDS ACROSS THE PERIODIC TABLE: THE PERIOD 2 ELEMENTS

Table 14.1 on the next two pages presents the trends in atomic properties of the Period 2 elements, lithium through neon, and the physical and chemical properties that emerge from them. In general, these trends apply to the other periods as well. Note the following points:

- Electrons fill the one *ns* and the three *np* orbitals according to Pauli's exclusion principle and Hund's rule.
- As a result of increasing nuclear charge and the addition of electrons to orbitals of the same energy level (same *n* value), atomic size generally decreases, whereas first ionization energy and electronegativity generally increase (see bar graphs on p. 557).
- Metallic character decreases with increasing nuclear charge as period members change from metals to metalloids to nonmetals.
- Reactivity is highest at the left and right ends of the period, except for the inert noble gas.
- Bonding between atoms of an element changes from metallic, to covalent in networks, to covalent in individual molecules, to none (noble gases exist as separate atoms). As expected, physical properties, such as melting point, change abruptly at the network/molecule boundary, which occurs in Period 2 between carbon (solid) and nitrogen (gas).
- Bonding between each element and an active nonmetal changes from ionic, to polar covalent, to covalent. Bonding between each element and an active metal changes from metallic to polar covalent to ionic.
- The acid-base behavior of the common element oxide in water changes from basic to amphoteric to acidic as the bond between the element and oxygen becomes more covalent.
- Reducing strength decreases through the metals, and oxidizing strength increases through the nonmetals. In Period 2, common oxidation numbers (O.N.s) equal the A-group number for Li and Be and the A-group number minus eight for O and F. Boron has several O.N.s, Ne has none, and C and N show all possible O.N.s for their groups.

One point that will be coming up frequently but is not shown in Table 14.1 is the *anomalous behavior of the Period 2 elements within their groups.* Later, we'll highlight some particularly clear examples of such *un*representative behavior, which arises from their relatively *small size* and *small number of orbitals* in the outer energy level.

Table 14.1 Trends in Atomic, Physical, and Chemical Properties of the Period 2 Elements

Group: Element/At. No.:	1A(1) Lithium (Li) $Z = 3$	2A(2) Beryllium (Be) $Z = 4$	3A(13) Boron (B) $Z = 5$	4A(14) Carbon (C) $Z = 6$
Atomic Properties				
Condensed electron configuration; partial orbital diagram	[He] $2s^1$ 2s: ↑ 2p: (empty)	[He] $2s^2$ 2s: ↑↓ 2p: (empty)	[He] $2s^2 2p^1$ 2s: ↑↓ 2p: ↑	[He] $2s^2 2p^2$ 2s: ↑↓ 2p: ↑ ↑
Physical Properties				
Appearance				
Metallic character	Metal	Metal	Metalloid	Nonmetal
Hardness	Soft	Hard	Very hard	Graphite: soft Diamond: extremely hard
Melting point/ boiling point	Low mp for a metal	High mp	Extremely high mp	Extremely high mp
Chemical Properties				
General reactivity	Reactive	Low reactivity at room temperature	Low reactivity at room temperature	Low reactivity at room temperature; graphite more reactive
Bonding among atoms of element	Metallic	Metallic	Network covalent	Network covalent
Bonding with nonmetals	Ionic	Polar covalent	Polar covalent	Covalent (π bonds common)
Bonding with metals	Metallic	Metallic	Polar covalent	Polar covalent
Acid-base behavior of common oxide	Strongly basic	Amphoteric	Very weakly acidic	Very weakly acidic
Redox behavior (O.N.)	Strong reducing agent (+1)	Moderately strong reducing agent (+2)	Complex hydrides good reducing agents (+3, −3)	Every oxidation state from +4 to −4
Relevance/Uses of Element and Compounds	Li soaps for auto grease; thermonuclear bombs; high-voltage, low-weight batteries; treatment of bipolar disorders (Li_2CO_3)	Rocket nose cones; alloys for springs and gears; nuclear reactor parts; x-ray tubes	Cleaning agent (borax); eyewash, antiseptic (boric acid); armor (B_4C); borosilicate glass; plant nutrient	Graphite: lubricant, structural fiber Diamond: jewelry, cutting tools, protective films Limestone ($CaCO_3$) Organic compounds: drugs, fuels, textiles, etc.

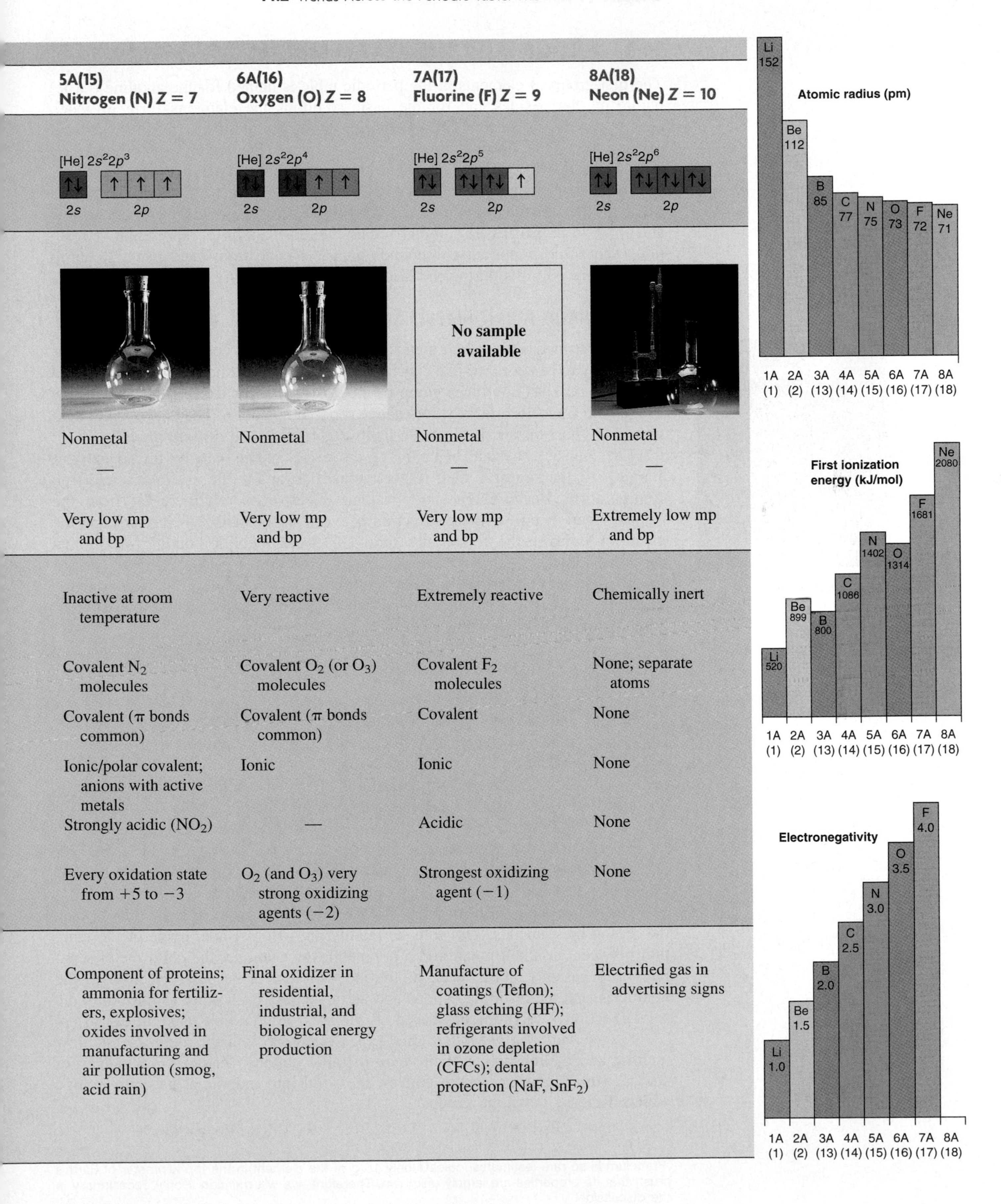

5A(15) Nitrogen (N) $Z = 7$	6A(16) Oxygen (O) $Z = 8$	7A(17) Fluorine (F) $Z = 9$	8A(18) Neon (Ne) $Z = 10$
[He] $2s^22p^3$ ↑↓ / ↑ ↑ ↑ (2s, 2p)	[He] $2s^22p^4$ ↑↓ / ↑↓ ↑ ↑ (2s, 2p)	[He] $2s^22p^5$ ↑↓ / ↑↓ ↑↓ ↑ (2s, 2p)	[He] $2s^22p^6$ ↑↓ / ↑↓ ↑↓ ↑↓ (2s, 2p)
		No sample available	
Nonmetal	Nonmetal	Nonmetal	Nonmetal
—	—	—	—
Very low mp and bp	Very low mp and bp	Very low mp and bp	Extremely low mp and bp
Inactive at room temperature	Very reactive	Extremely reactive	Chemically inert
Covalent N_2 molecules	Covalent O_2 (or O_3) molecules	Covalent F_2 molecules	None; separate atoms
Covalent (π bonds common)	Covalent (π bonds common)	Covalent	None
Ionic/polar covalent; anions with active metals	Ionic	Ionic	None
Strongly acidic (NO_2)	—	Acidic	None
Every oxidation state from +5 to −3	O_2 (and O_3) very strong oxidizing agents (−2)	Strongest oxidizing agent (−1)	None
Component of proteins; ammonia for fertilizers, explosives; oxides involved in manufacturing and air pollution (smog, acid rain)	Final oxidizer in residential, industrial, and biological energy production	Manufacture of coatings (Teflon); glass etching (HF); refrigerants involved in ozone depletion (CFCs); dental protection (NaF, SnF_2)	Electrified gas in advertising signs

14.3 GROUP 1A(1): THE ALKALI METALS

The first group of elements in the periodic table is named for the alkaline (basic) nature of their oxides and for the basic solutions the elements form in water. Group 1A(1) provides the best example of regular trends with no significant exceptions. All the elements in the group—lithium (Li), sodium (Na), potassium (K), rubidium (Rb), cesium (Cs), and rare, radioactive francium (Fr)*—are very reactive metals. The Family Portrait of Group 1A(1), on pages 560 and 561, is the first in a series that provides an overview of each of the main groups. Atomic and physical properties are summarized on the left-hand page; representative reactions and some important compounds appear on the right-hand page. Refer to these family portraits during the discussions for supporting information.

Why Are the Alkali Metals Soft, Low Melting, and Lightweight?

Lithium floating in oil floating on water.

Unlike most metals, the alkali metals are soft: Na has the consistency of cold butter, and K can be squeezed like clay. The alkali metals also have lower melting and boiling points than any other group of metals. Except for Li, they all melt below 100°C, and Cs melts a few degrees above room temperature. They have lower densities than most metals: Li floats on lightweight household oil (see photo).

The unusual physical behavior of the alkali metals can be traced to their atomic size, the largest in their respective periods, and to the ns^1 valence electron configuration. Because the single valence electron is relatively far from the nucleus, there is only a weak attraction between the delocalized electrons and the metal-ion cores in the crystal structure. This weak metallic bonding means that the alkali metal crystal structure can be easily deformed or broken down, which results in a soft consistency and low melting point. These elements also have the lowest molar masses in their periods; with their large atomic radii, they therefore have relatively low densities.

Why Are the Alkali Metals So Reactive?

The alkali metals are extremely reactive elements. They are *powerful reducing agents* and, thus, always occur in nature as 1+ cations rather than as free metals. (As we discuss in Section 22.4, highly endothermic reduction processes are required to prepare the free metals industrially from their molten salts.) The alkali metals reduce the halogens and form ionic solids in reactions that release large quantities of heat. They reduce the hydrogen in water, reacting vigorously (Rb and Cs explosively) to form H_2 and a metal hydroxide solution. They reduce O_2 in the air, and thus tarnish rapidly. Because of this reactivity, Na and K are usually kept under mineral oil (an unreactive liquid) in the laboratory, and Rb and Cs are handled with gloves under an inert argon atmosphere.

The ns^1 configuration, which is the basis for their physical properties, is also the reason these metals form salts so readily. In a Born-Haber cycle of the reaction between an alkali metal and a nonmetal, for example, the solid metal separates into gaseous atoms, each of the atoms transfers its outer electron to the nonmetal, and the resulting cations and anions attract each other to form an ionic solid (Section 9.2). Properties based on the ns^1 configuration relate to each step.

1. *Low heat of atomization* (ΔH^0_{atom}). Consistent with the alkali metals' low melting and boiling points, their weak metallic bonding leads to low values for ΔH^0_{atom} (the energy needed to convert the solid into individual gaseous atoms), which decrease down the group:†

$$E(s) \longrightarrow E(g) \qquad \Delta H_{atom}\ (Li > Na > K > Rb > Cs)$$

*Francium is so rare (estimates indicate only 15 g of the element in the top kilometer of Earth's crust) that its properties are largely unknown. Therefore, we will mention it only occasionally in the discussion.

†Throughout the chapter, we use E to represent any element in a group.

2. *Low IE and small ionic radius.* Each alkali metal has the largest size and the lowest IE in its period. A great *decrease* in size occurs when the outer electron is lost: the volume of the Li^+ ion is less than 13% that of the Li atom! Thus, Group 1A(1) ions are small spheres with considerable charge density.

3. *High lattice energy.* When Group 1A salts crystallize, large amounts of energy are released because the small cations lie close to the anions. Thus, the endothermic atomization and ionization steps are easily outweighed by the highly exothermic formation of the solid. For a given anion, the trend in lattice energy is the inverse of the trend in cation size: *as the cation becomes larger, the lattice energy becomes smaller.* This steady decrease in lattice energy within the Group 1A(1) and 2A(2) chlorides is shown in Figure 14.2.

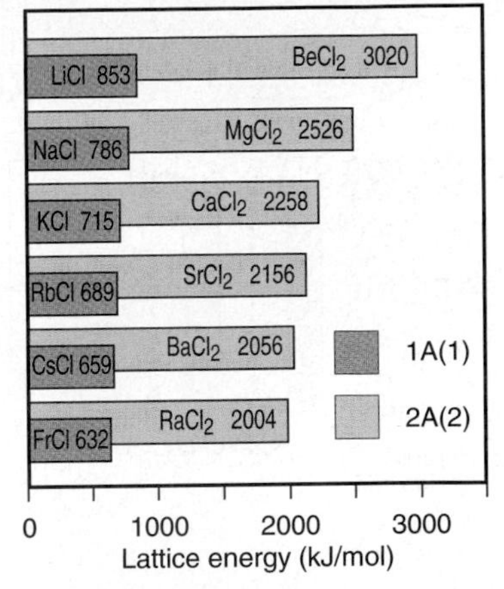

Figure 14.2 Lattice energies of the Group 1A(1) and 2A(2) chlorides. The lattice energy decreases regularly in both groups of metal chlorides as the cations become larger. Lattice energies for the Group 2A chlorides are greater because the 2A cations have higher charge and smaller size.

Despite the strong ionic attractions in the solid, *nearly all Group 1A salts are water soluble.* The attraction between the ions and water molecules creates a highly exothermic heat of hydration (ΔH_{hydr}), and a large increase in entropy occurs when ions in the organized crystal become dispersed and hydrated in solution; together, these factors outweigh the high lattice energy.

The magnitude of the hydration energy *decreases* as ionic size increases:

$$E^+(g) \longrightarrow E^+(aq) \qquad \Delta H = -\Delta H_{hydr} \ (Li^+ > Na^+ > K^+ > Rb^+ > Cs^+)$$

Interestingly, the *smaller* ions attract water molecules strongly enough to form *larger hydrated ions.* This size trend has a major effect on the function of nerves, kidneys, and cell membranes because the *sizes* of $Na^+(aq)$ and $K^+(aq)$, the most common cations in cell fluids, influence their movement in and out of cells.

The Anomalous Behavior of Lithium

As we noted in discussing Table 14.1, all the Period 2 elements display some anomalous (unrepresentative) behavior within their groups. Even within the regular, predictable trends of Group 1A(1), Li has some atypical properties. It is the only member that forms a simple oxide and nitride, Li_2O and Li_3N, on reaction with O_2 and N_2 in air. Only Li forms molecular compounds with hydrocarbon groups from organic halides:

$$2Li(s) + CH_3CH_2Cl(g) \longrightarrow CH_3CH_2Li(s) + LiCl(s)$$

Organolithium compounds, such as CH_3CH_2Li, are liquids or low-melting solids that dissolve in nonpolar solvents and contain polar covalent $^{\delta-}C—Li^{\delta+}$ bonds. They are important reactants in the synthesis of organic compounds.

Because of its small size, Li^+ has a relatively high charge density. Therefore, it can deform nearby electron clouds to a much greater extent than the other 1A ions can, which increases orbital overlap and gives many lithium salts significant covalent character (Figure 14.3). Thus, LiCl, LiBr, and LiI are much more soluble in polar organic solvents, such as ethanol and acetone, than are the halides of Na and K, because the lithium halide dipole interacts with these solvents through dipole-dipole forces. The small, highly positive Li^+ makes dissociation of Li salts into ions more difficult in water; thus, the fluoride, carbonate, hydroxide, and phosphate of Li are much less soluble in water than those of Na and K.

Figure 14.3 The effect of Li^+ charge density on a nearby electron cloud. The ability of the Li^+ ion to deform nearby electron clouds gives rise to many anomalous properties. In this case, Li^+ polarizes an I^- ion, which gives some covalent character to lithium iodide, LiI, as depicted by a space-filling model **(A),** an electron density contour **(B),** and an electron density relief map **(C).**

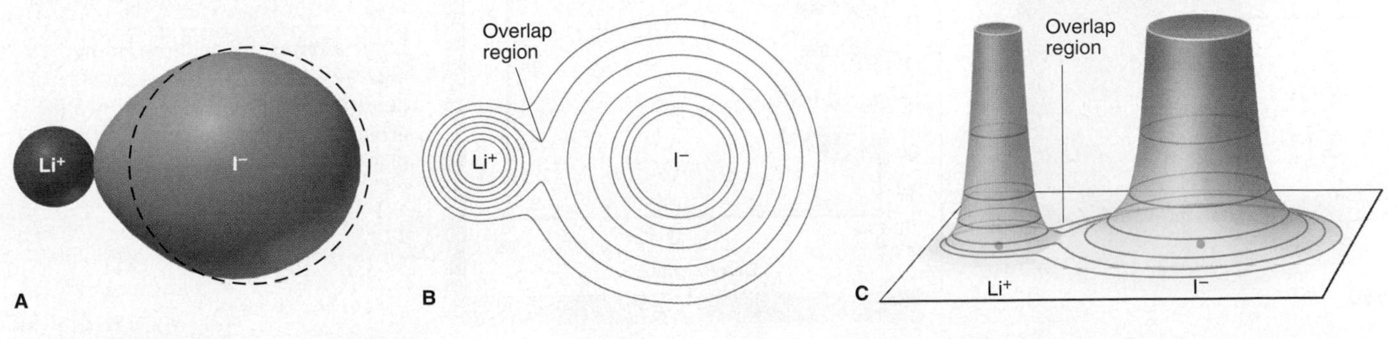

Key Atomic and Physical Properties

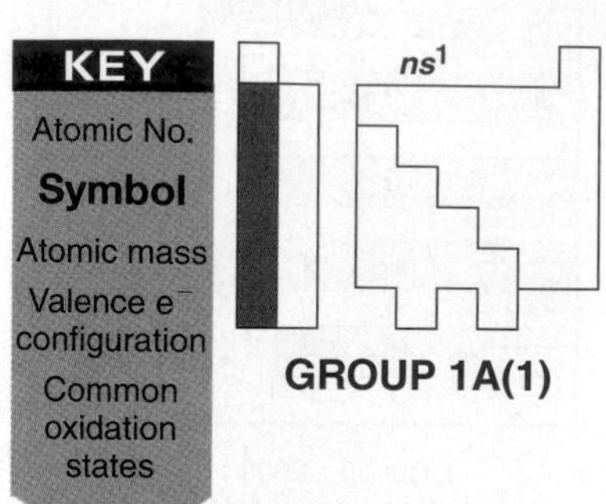

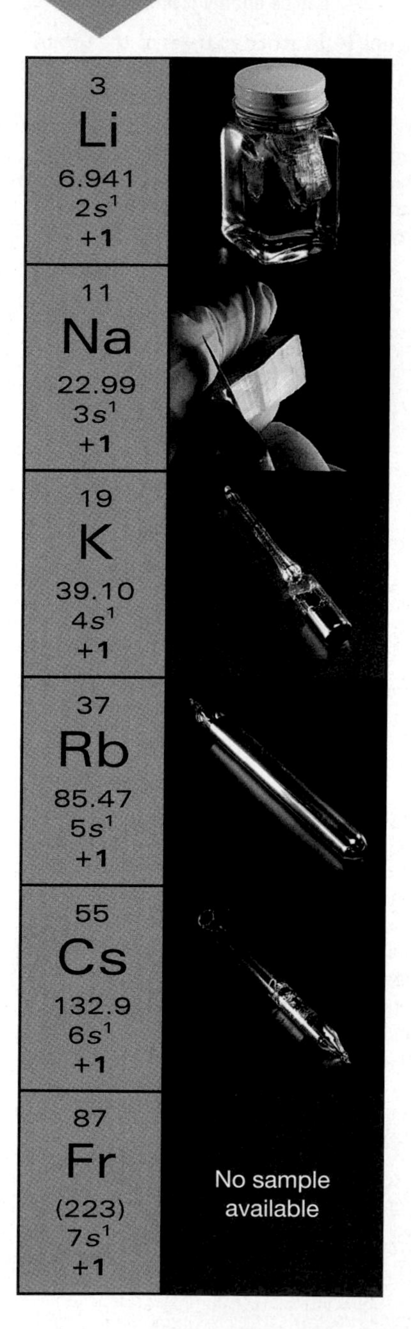

Atomic Properties

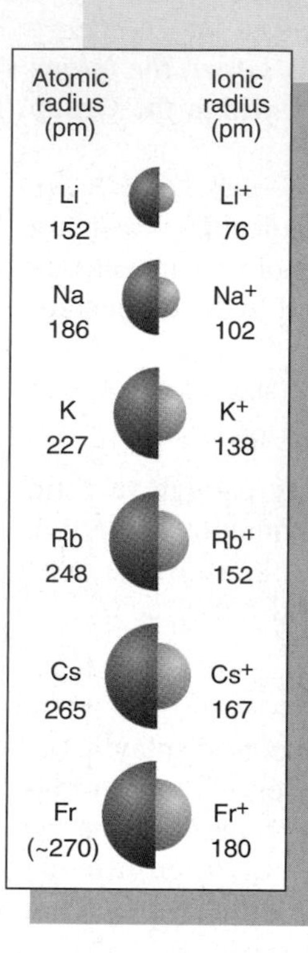

Group electron configuration is ns^1. All members have the +1 oxidation state and form an E^+ ion.

Atoms have the largest size and lowest IE and EN in their periods.

Down the group, atomic and ionic size increase, while IE and EN decrease.

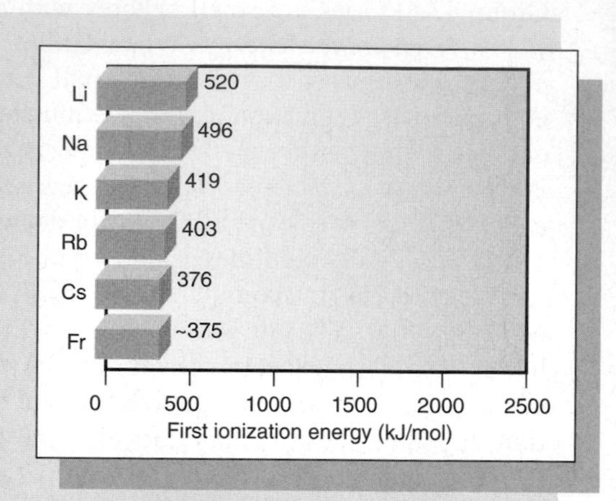

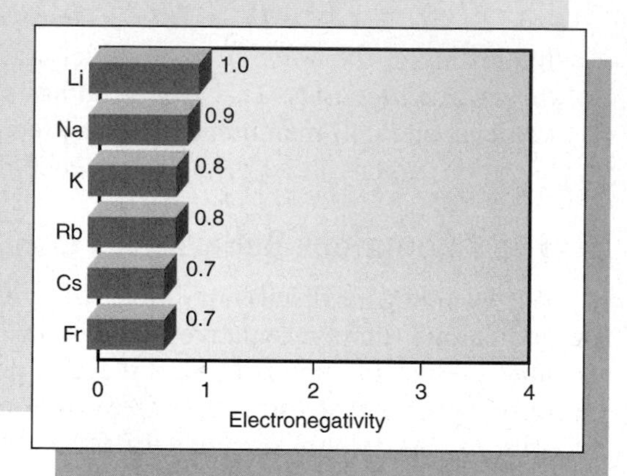

Physical Properties

Metallic bonding is relatively weak because there is only one valence electron. Therefore, these metals are soft with relatively low melting and boiling points. These values decrease down the group because larger atom cores attract delocalized electrons less strongly.

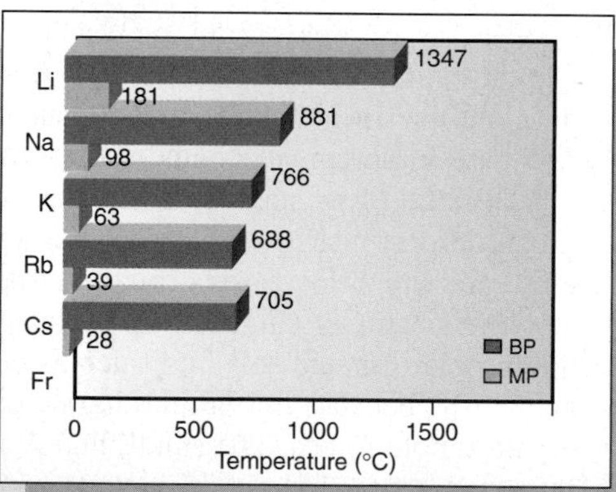

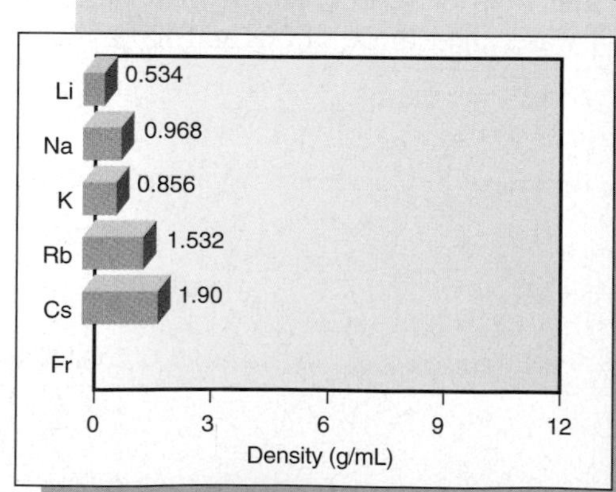

Large atomic size and low atomic mass result in low density; thus density generally increases down the group because mass increases more than size.

Some Reactions and Compounds

Important Reactions

The reducing power of the alkali metals (E) is shown in reactions 1 to 4. Some industrial applications of Group 1A(1) compounds are shown in reactions 5 to 7.

1. The alkali metals reduce H in H_2O from the +1 to the 0 oxidation state:

$$2E(s) + 2H_2O(l) \longrightarrow 2E^+(aq) + 2OH^-(aq) + H_2(g)$$

The reaction becomes more vigorous down the group (*see photo*).

Potassium reacting with water

2. The alkali metals reduce oxygen, but the product depends on the metal. Li forms the oxide, Li_2O; Na forms the peroxide, Na_2O_2; K, Rb, and Cs form the superoxide, EO_2:

$$4Li(s) + O_2(g) \longrightarrow 2Li_2O(s)$$

$$K(s) + O_2(g) \longrightarrow KO_2(s)$$

In emergency breathing units, KO_2 reacts with H_2O and CO_2 in exhaled air to release O_2.

3. The alkali metals reduce hydrogen to form ionic (saltlike) hydrides:

$$2E(s) + H_2(g) \longrightarrow 2EH(s)$$

NaH is an industrial base and reducing agent that is used to prepare other reducing agents, such as $NaBH_4$.

4. The alkali metals reduce halogens to form ionic halides:

$$2E(s) + X_2 \longrightarrow 2EX(s) \quad (X = F, Cl, Br, I)$$

5. Sodium chloride is the most important alkali metal halide.

(a) In the Downs process for the production of sodium metal (*Section 22.4*), reaction 4 is reversed by supplying electricity to molten NaCl:

$$2NaCl(l) \xrightarrow{\text{electricity}} 2Na(l) + Cl_2(g)$$

(b) In the chlor-alkali process (*Section 22.5*), NaCl(*aq*) is electrolyzed to form several key industrial chemicals:

$$2NaCl(aq) + 2H_2O(l) \longrightarrow 2NaOH(aq) + H_2(g) + Cl_2(g)$$

(c) In its reaction with sulfuric acid, NaCl forms two major products:

$$2NaCl(s) + H_2SO_4(aq) \longrightarrow Na_2SO_4(aq) + 2HCl(g)$$

Sodium sulfate is important in the paper industry; HCl is essential in steel, plastics, textile, and food production.

6. Sodium hydroxide is used in the formation of bleaching solutions:

$$2NaOH(aq) + Cl_2(g) \longrightarrow NaClO(aq) + NaCl(aq) + H_2O(l)$$

7. In an ion-exchange process (*Chemical Connections, pp. 529–530*), water is "softened" by removal of dissolved hard-water cations, which displace Na^+ from a resin:

$$M^{2+}(aq) + Na_2Z(s) \longrightarrow MZ(s) + 2Na^+(aq)$$

$$(M = Mg, Ca; Z = \text{resin})$$

Important Compounds

1. Lithium chloride and lithium bromide, LiCl and LiBr. Because the Li^+ ion is so small, Li salts have a high affinity for H_2O and yet a positive heat of solution. Thus, they are used in dehumidifiers and air-cooling units.

2. Lithium carbonate, Li_2CO_3. Used to make porcelain enamels and toughened glasses and as a drug in the treatment of bipolar disorders.

3. Sodium chloride, NaCl. Millions of tons used in the industrial production of Na, NaOH, $Na_2CO_3/NaHCO_3$, Na_2SO_4, and HCl or purified for use as table salt.

4. Sodium carbonate and sodium hydrogen carbonate, Na_2CO_3 and $NaHCO_3$. Carbonate used as an industrial base and to make glass. Hydrogen carbonate, which releases CO_2 at low temperatures (50° to 100°C), used in baking powder and in fire extinguishers.

5. Sodium hydroxide, NaOH. Most important industrial base; used to make bleach, sodium phosphates, and alcohols.

6. Potassium nitrate, KNO_3. Powerful oxidizing agent used in gunpowder and fireworks (*see photo*).

14.4 GROUP 2A(2): THE ALKALINE EARTH METALS

The Group 2A(2) elements are called *alkaline earth metals* because their oxides give basic (alkaline) solutions and melt at such high temperatures that they remained as solids ("earths") in the alchemists' fires. The group includes a fascinating collection of elements: rare beryllium (Be), common magnesium (Mg) and calcium (Ca), less familiar strontium (Sr) and barium (Ba), and radioactive radium (Ra). The Group 2A(2) Family Portrait (pp. 564 and 565) presents an overview of these elements.

Versatile Magnesium Magnesium, second of the alkaline earths, forms a tough, adherent oxide coating that prevents further reaction of the metal in air and confers stability for many uses. The metal can be rolled, forged, welded, and riveted into virtually any shape, and it forms strong, low-density alloys: some weigh 25% as much as steel but are just as strong. Alloyed with aluminum and zinc, magnesium is used for everyday objects, such as camera bodies, luggage, and the "mag" wheels of bicycles and sports cars. Alloyed with the lanthanides (first series of inner transition elements), it is used in objects that require great strength at high temperatures, such as auto engine blocks and missile parts.

How Do the Physical Properties of the Alkaline Earth and Alkali Metals Compare?

In general terms, the elements in Groups 1A(1) and 2A(2) behave as close cousins. Whatever differences occur between the groups are those of degree, not kind, and are due to the change in outer electron configuration: ns^2 vs. ns^1. Two electrons are available from each 2A atom for metallic bonding, and the nucleus contains one additional positive charge. These factors make the attraction between delocalized electrons and atom cores greater. Consequently, melting and boiling points are much higher for 2A metals than for the corresponding 1A metals; in fact, the 2A elements melt at around the same temperatures as the 1A elements boil! Compared with transition metals, such as iron and chromium, the alkaline earths are soft and lightweight, but they are much harder and more dense than the alkali metals.

How Do the Chemical Properties of the Alkaline Earth and Alkali Metals Compare?

The alkaline earth metals display a wider range of chemical behavior than the alkali metals, largely because of the behavior of beryllium, as you'll see shortly. Because the second valence electron lies in the same sublevel as the first, it is not shielded from the additional nuclear charge very well, and so Z_{eff} is greater. Therefore, Group 2A(2) elements have smaller atomic radii and higher ionization energies than Group 1A(1) elements. Despite the higher IEs, *all the alkaline earths (except Be) form ionic compounds* as 2+ cations. Beryllium behaves differently because so much energy is needed to remove two electrons from this tiny atom that it never forms discrete Be^{2+} ions, and its bonds are polar covalent.

Lime: The Most Useful Metal Oxide Calcium oxide (lime) is a slightly soluble, basic oxide that is among the five most heavily produced industrial compounds in the world; it is made by roasting limestone to high temperatures [Group 2A(2) Family Portrait, reaction 7]. It has essential roles in steelmaking, water treatment, and smokestack "scrubbing." Lime reacts with carbon to form calcium carbide, which is used to make acetylene, and with arsenic acid (H_3AsO_4) to make insecticides for treating cotton, tobacco, and potato plants. Most glass contains about 12% lime by mass. The paper industry uses lime to prepare the bleach [$Ca(ClO)_2$] that whitens paper. Farmers "dust" fields with lime to "sweeten" acidic soil. The food industry uses it to neutralize compounds in milk to make cream and butter and to neutralize impurities in crude sugar.

Like the alkali metals, the alkaline earth metals are *strong reducing agents*. Each element reduces O_2 in air to form the oxide (Ba also forms the peroxide, BaO_2). Except for Be and Mg, which form adherent oxide coatings, the alkaline earths reduce H_2O at room temperature to form H_2. And, except for Be, they reduce the halogens, N_2, and H_2 to form ionic compounds. The Group 2A oxides are strongly basic (except for amphoteric BeO) and react with acidic oxides to form salts, such as sulfites and carbonates; for example,

$$SrO(s) + CO_2(g) \longrightarrow SrCO_3(s)$$

Natural carbonates, such as limestone and marble, are major structural materials and the commercial sources for most 2A compounds.

The alkaline earth metals are reactive because the high lattice energies of their compounds more than compensate for the large total IE needed to form the 2+ cation (Section 9.2). Group 2A salts have much higher lattice energies than Group 1A salts (see Figure 14.2) because the 2A cations are smaller and doubly charged.

One of the main differences between the two groups is the lower solubility of 2A salts in water. Their ions are smaller and more highly charged than 1A ions, resulting in much higher charge densities. Even though this factor increases heats of hydration, it increases lattice energies even more. In fact, most 2A fluorides, carbonates, phosphates, and sulfates are considered insoluble, unlike the corresponding 1A compounds. Nevertheless, the ion-dipole attraction of 2+ ions for water is so strong that many slightly soluble 2A salts crystallize as hydrates; two examples are Epsom salt, $MgSO_4 \cdot 7H_2O$, used as a soak for inflammations, and gypsum, $CaSO_4 \cdot 2H_2O$, used as the bonding material between the paper sheets in wallboard and as the cement in surgical casts.

The Anomalous Behavior of Beryllium

The Period 2 element Be displays far more anomalous behavior than Li in Group 1A(1). If the Be^{2+} ion did exist, its high charge density would polarize nearby electron clouds very strongly and cause extensive orbital overlap. In fact, *all Be compounds exhibit covalent bonding*. Even BeF_2, the most ionic Be compound, has a relatively low melting point and, when melted, a low electrical conductivity.

With only two valence electrons, Be does not attain an octet in its simple gaseous compounds (Section 10.1). When it bonds to an electron-rich atom, however, this electron deficiency is overcome as the gas condenses. Consider beryllium chloride ($BeCl_2$) (Figure 14.4). At temperatures greater than 900°C, it consists of linear molecules in which two *sp* hybrid orbitals hold four electrons around the central Be. As it cools, the molecules bond together, solidifying in long chains, with each Be sp^3 hybridized to finally attain an octet.

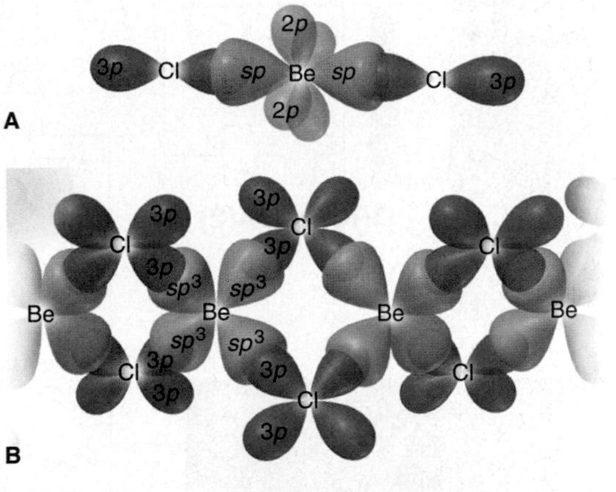

Figure 14.4 Overcoming electron deficiency in beryllium chloride. A, At high temperatures, $BeCl_2$ is a gas with only four electrons around each Be. **B,** Solid $BeCl_2$ occurs in long chains with each Cl bridging two Be atoms, which gives each Be an octet.

Diagonal Relationships: Lithium and Magnesium

One of the clearest ways in which atomic properties influence chemical behavior appears in the **diagonal relationships,** similarities between a Period 2 element and one diagonally down and to the right in Period 3 (Figure 14.5).

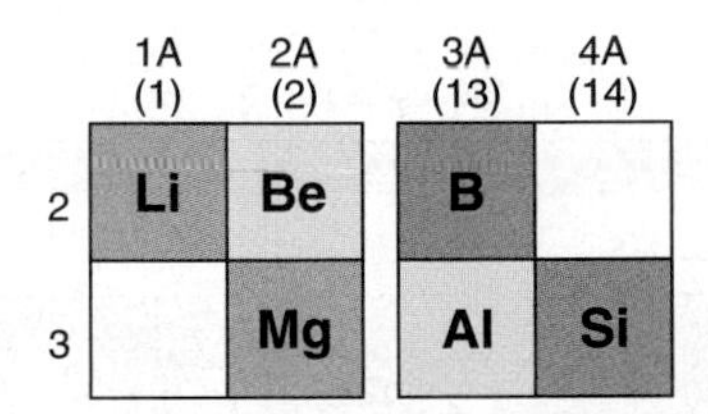

Figure 14.5 Three diagonal relationships in the periodic table. Certain Period 2 elements exhibit behaviors that are very similar to those of the Period 3 elements immediately below and to the right. Three such diagonal relationships exist: Li and Mg, Be and Al, and B and Si.

The first of three such relationships occurs between Li and Mg and reflects similarities in atomic and ionic size. Note that *one period down increases atomic (or ionic) size and one group to the right decreases it.* Thus, Li has a radius of 152 pm and Mg has a radius of 160 pm; the Li^+ radius is 76 pm and that of Mg^{2+} is 72 pm. From similar atomic properties emerge similar chemical properties. Both elements form nitrides with N_2, hydroxides and carbonates that decompose easily with heat, organic compounds with a polar covalent metal-carbon bond, and salts with similar solubilities. We'll discuss the diagonal relationships between Be and Al and between B and Si in later sections.

Looking Backward and Forward: Groups 1A(1), 2A(2), and 3A(13)

Throughout this chapter, comparing the previous, current, and upcoming groups (Figure 14.6) will help you to keep horizontal trends in mind while examining vertical groups. Not much changes from 1A to 2A, and the elements behave as metals both physically and chemically. With smaller atomic sizes and stronger metallic bonding, 2A elements are harder, higher melting, and denser than those in 1A. Nearly all 1A and most 2A compounds are ionic. The higher ionic charge in Group 2A (2+ vs. 1+) leads to higher lattice energies and less soluble salts. The range of behavior in 2A is wider than that in 1A because of Be, and the range widens much further in Group 3A, from metalloid boron to metallic thallium.

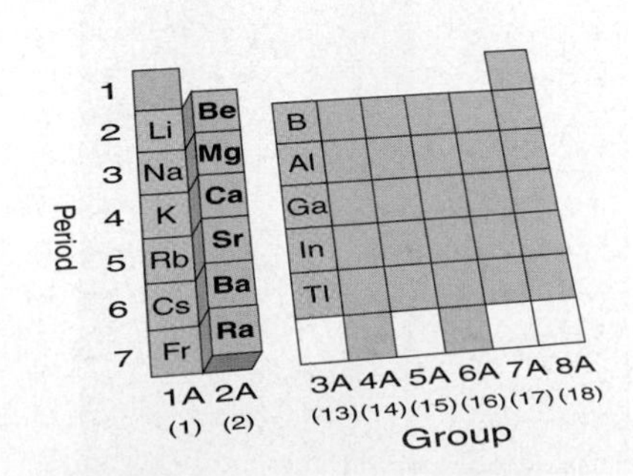

Figure 14.6 Standing in Group 2A(2), looking backward to 1A(1) and forward to 3A(13).

Key Atomic and Physical Properties

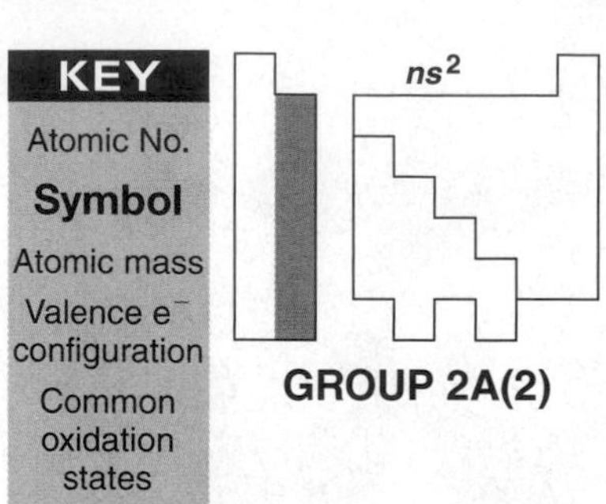

Atomic Properties

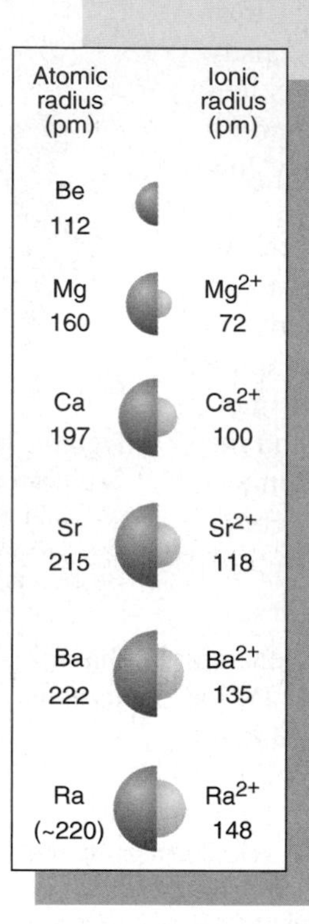

Group electron configuration is ns^2 (filled *ns* sublevel). All members have the +2 oxidation state and, except for Be, form compounds with an E^{2+} ion.

Atomic and ionic sizes increase down the group but are smaller than for the corresponding 1A(1) element.

IE and EN decrease down the group but are higher than for the corresponding 1A(1) element.

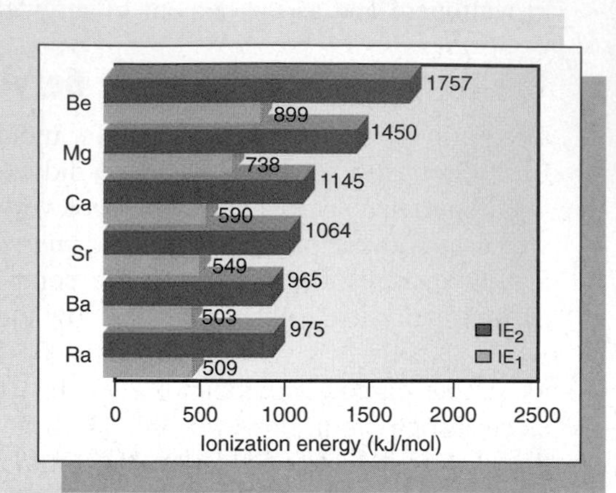

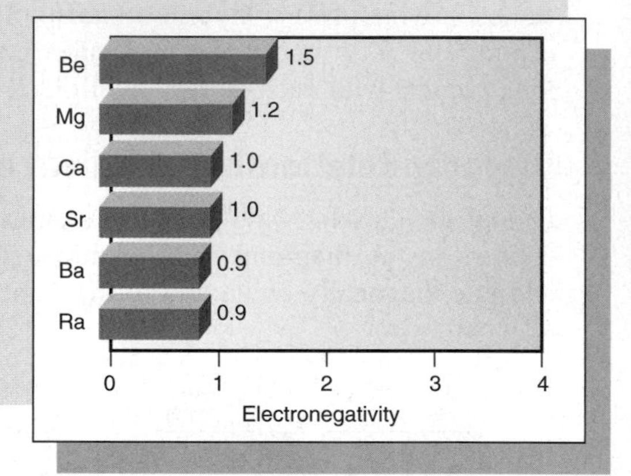

Physical Properties

Metallic bonding involves two valence electrons. These metals are still relatively soft but are much harder than the 1A(1) metals.

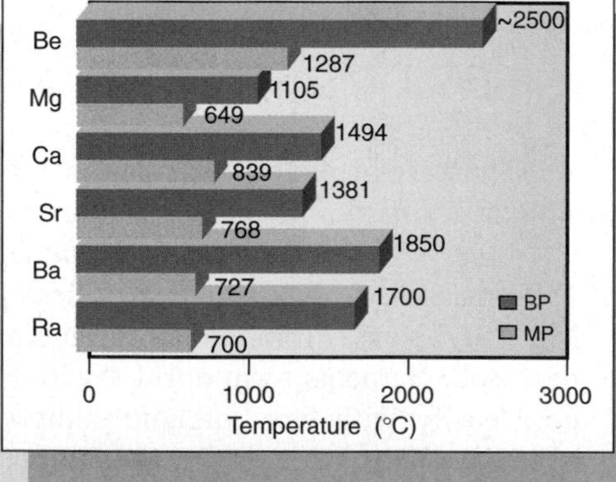

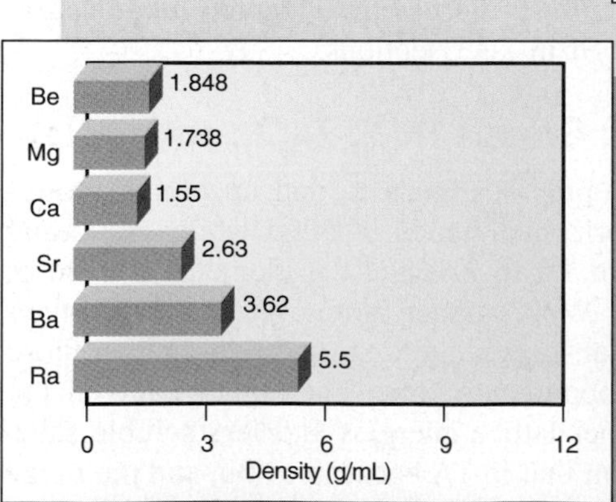

Melting and boiling points generally decrease, and densities generally increase down the group. These values are much higher than for 1A(1) elements, and the trend is not as regular.

Some Reactions and Compounds

Important Reactions

The elements (E) act as reducing agents in reactions 1 to 5; note the similarity to reactions of Group 1A(1). Reaction 6 shows the general basicity of the 2A(2) oxides; reaction 7 shows the general instability of their carbonates at high temperature.

1. The metals reduce O_2 to form the oxides:

$$2E(s) + O_2(g) \longrightarrow 2EO(s)$$

Ba also forms the peroxide, BaO_2.

Magnesium ribbon burning

2. The larger metals reduce water to form hydrogen gas:

$$E(s) + 2H_2O(l) \longrightarrow E^{2+}(aq) + 2OH^-(aq) + H_2(g) \quad (E = Ca, Sr, Ba)$$

Be and Mg form an adherent oxide coating that allows only slight reaction.

3. The metals reduce halogens to form ionic halides:

$$E(s) + X_2 \longrightarrow EX_2(s) \quad (X = F, Cl, Br, I)$$

4. Most of the elements reduce hydrogen to form ionic hydrides:

$$E(s) + H_2(g) \longrightarrow EH_2(s) \quad (E = \text{all except Be})$$

5. Most of the elements reduce nitrogen to form ionic nitrides:

$$3E(s) + N_2(g) \longrightarrow E_3N_2(s) \quad (E = \text{all except Be})$$

6. Except for amphoteric BeO, the element oxides are basic:

$$EO(s) + H_2O(l) \longrightarrow E^{2+}(aq) + 2OH^-(aq)$$

$Ca(OH)_2$ is a component of cement and mortar.

7. All carbonates undergo thermal decomposition to the oxide:

$$ECO_3(s) \xrightarrow{\Delta} EO(s) + CO_2(g)$$

This reaction is used to produce CaO (lime) in huge amounts from naturally occurring limestone (*margin note, p. 562*).

Important Compounds

1. Beryl, $Be_3Al_2Si_6O_{18}$. Beryl, the industrial source of Be metal, also occurs as a gemstone with a variety of colors (*see photo*). It is chemically identical to emerald, except for the trace of Cr^{3+} that gives emerald its green color.

2. Magnesium oxide, MgO. Because of its high melting point (2852°C), MgO is used as a refractory material for furnace brick (*see photo*) and wire insulation.

Industrial kiln

3. Alkylmagnesium halides, RMgX (R = hydrocarbon group; X = halogen). These compounds are used to synthesize many organic compounds. Organotin agricultural fungicides are made by treating RMgX with $SnCl_4$:

$$3RMgCl + SnCl_4 \longrightarrow 3MgCl_2 + R_3SnCl$$

4. Calcium carbonate, $CaCO_3$. Occurs as enormous natural deposits of limestone, marble, chalk, and coral. Used as a building material, to make lime, and, in high purity, as a toothpaste abrasive and an antacid (*see photo*).

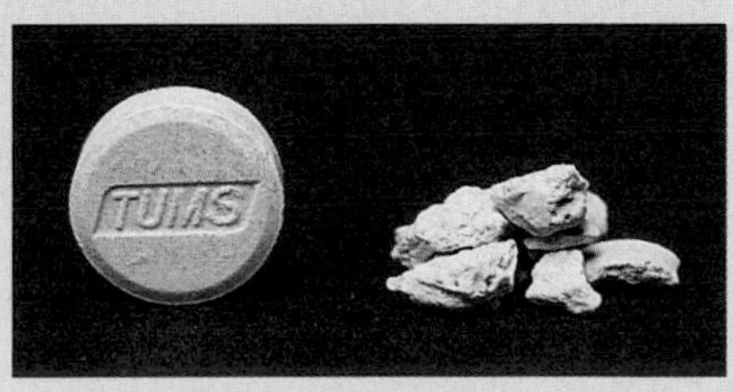

Antacid tablet Limestone

14.5 GROUP 3A(13): THE BORON FAMILY

The third family of main-group elements contains some unusual members and some familiar ones, some exotic bonding, and some strange physical properties. Boron (B) heads the family, but, as you'll see, its properties certainly do not represent the other members. Metallic aluminum (Al) has properties that are more typical of the group, but its great abundance and importance contrast with the rareness of gallium (Ga), indium (In), and thallium (Tl). The atomic, physical, and chemical properties of these elements are summarized in the Group 3A(13) Family Portrait (pp. 568 and 569).

How Do Transition Elements Influence Group 3A(13) Properties?

Group 3A(13) is the first of the *p* block. If you look at the main groups only, the elements of this group seem to be just one group away from those of Group 2A(2). In Period 4 and higher, however, a large gap separates the two groups (see Figure 8.11, p. 302). The gap holds 10 transition elements (*d* block) each in Periods 4, 5, and 6 and an additional 14 inner transition elements (*f* block) in Period 6. Recall from Section 8.2 that *d* and *f* electrons penetrate very little, and so spend very little time near the nucleus. Thus, the heavier 3A members—Ga, In, and Tl—have nuclei with many more protons, but their outer (*s* and *p*) electrons are only partially shielded from the much higher positive charge; as a result, these elements have greater Z_{eff} values than the two lighter members of the group.

Many properties of these heavier 3A elements are influenced by this stronger nuclear attraction. Figure 14.7 compares several properties of Group 3A(13) with those of Group 3B(3) (the first group of transition elements), in which the additional protons of the *d* block and *f* block have *not* been added. Note the regular changes exhibited by the 3B elements in contrast to the irregular patterns for those in 3A. The deviations for Ga reflect the *d*-block *contraction in size* and can be explained by limited shielding by the *d* electrons of the 10 additional protons of the first transition series. Similarly, the deviations for Tl reflect the *f*-block (lanthanide) contraction and can be explained by limited shielding by the *f* electrons of the 14 additional protons of the first inner transition series.

Physical properties are influenced by the type of bonding that occurs in the element. Boron is a network covalent metalloid—black, hard, and very high melting. The other group members are metals—shiny and relatively soft and low melting. Aluminum's low density and three valence electrons make it an exceptional conductor: for a given mass, aluminum conducts a current twice as effectively as copper. Gallium has the largest liquid temperature range of any element: it melts in your hand (see photo, p. 568) but does not boil until 2403°C. Its metallic bonding is too weak to keep the Ga atoms fixed when the solid is warmed, but strong enough to keep them from escaping the molten metal until it is very hot.

Gallium Arsenide: The New Wave of Semiconductors An important modern use of gallium is in the production of gallium arsenide (GaAs) semiconductors. Electrons move 10 times faster through GaAs than through Si-based chips. GaAs chips also have novel optical properties: a current is created when they absorb light, and, conversely, they emit light when a current is supplied. GaAs devices are already used in light-powered calculators, wristwatches, and solar panels. In addition, GaAs-based lasers are much smaller and more powerful than other types.

What New Features Appear in the Chemical Properties of Group 3A(13)?

Looking down Group 3A(13), we see a wide range of chemical behavior. Boron, the anomalous member from Period 2, is the first metalloid we've encountered so far and the only one in the group. It is much less reactive at room temperature than the other members and forms covalent bonds exclusively. Although aluminum acts like a metal physically, its halides exist in the gas phase as covalent *dimers*—molecules formed by joining two identical smaller molecules (Figure 14.8)—and its oxide is amphoteric rather than basic. Most of the other 3A compounds are ionic, but they have more covalent character than similar 2A compounds. Because the 3A cations are smaller and more highly charged than the 2A cations, they polarize an anion's electron cloud more effectively.

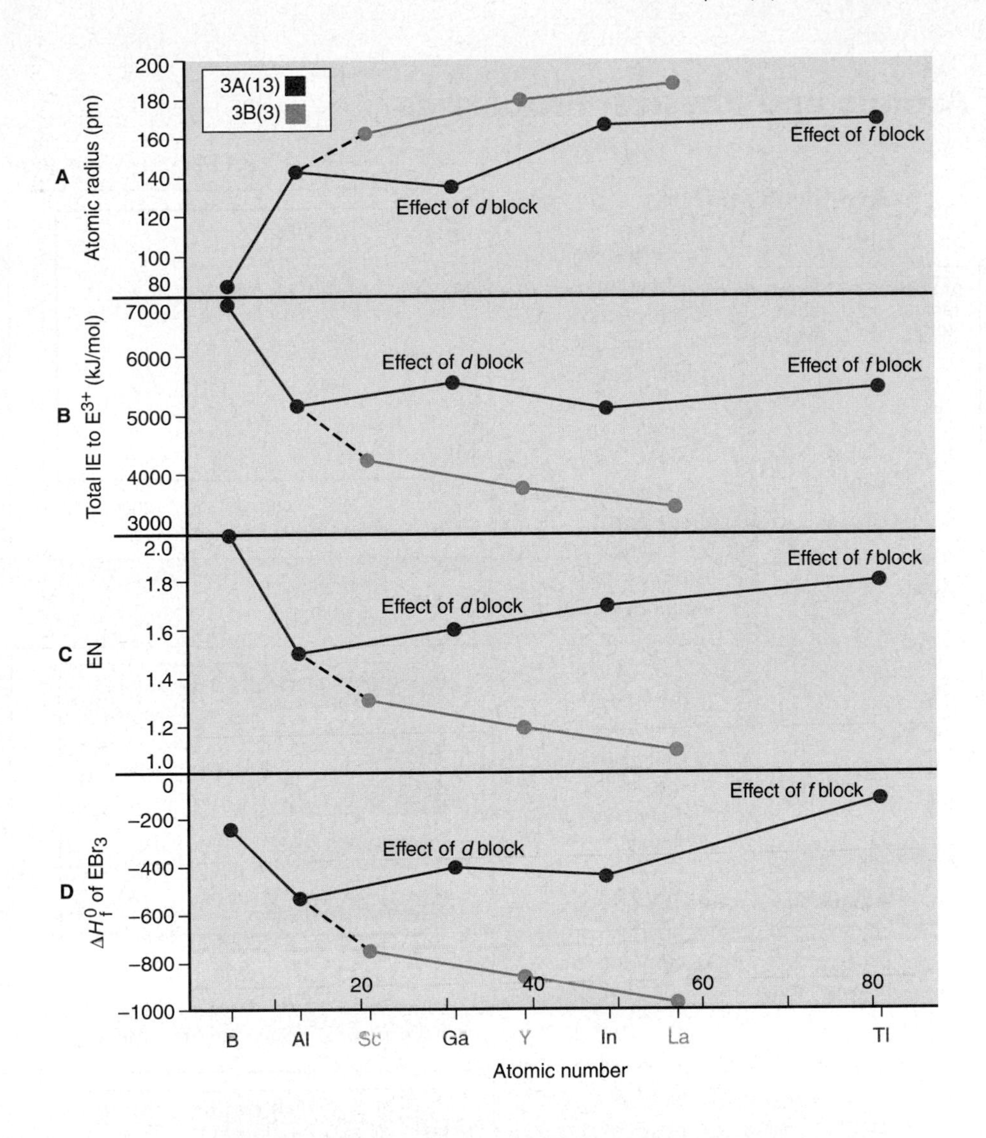

Figure 14.7 The effect of transition elements on properties: Group 3B(3) vs. Group 3A(13). The additional protons in the nuclei of transition elements exert an exceptionally strong attraction because *d* and *f* electrons shield the nuclear charge poorly. This greater effective nuclear charge affects the *p*-block elements in Periods 4 to 6, as can be seen by comparing properties of Group 3B(3), the first group after the *s* block, with those of Group 3A(13), the first group after the *d* block. **A,** Atomic size. In Group 3B, size increases smoothly, whereas in 3A, Ga and Tl are smaller than expected. **B,** Total ionization energy for E^{3+}. The deviations in size lead to deviations in total IE ($IE_1 + IE_2 + IE_3$). Note the regular decrease for Group 3B and the unexpectedly higher values for Ga and Tl in 3A. **C,** Electronegativity. The deviations in size also make Ga and Tl more electronegative than expected. **D,** Heat of formation of EBr_3. The higher IE values for Ga and Tl mean less heat is released upon formation of the ionic compound; thus, the magnitudes of the ΔH^0_f values for $GaBr_3$ and $TlBr_3$ are smaller than expected.

The redox behavior of the elements in this group provides a chance to note three general principles that appear in Groups 3A(13) to 6A(16):

1. *Presence of multiple oxidation states.* Many of the larger elements in these groups also have an important oxidation state *two lower than the A-group number.* The lower state occurs when the atoms lose their *np* electrons only, not their two *ns* electrons. This fact is often called the *inert-pair effect* (Section 8.5), but it has nothing to do with any inertness of *ns* electrons; for example, the 6*s* electrons in Tl are lost *more* easily than the 4*s* electrons in Ga. The reason for the common appearance of the lower state involves bond energies of the compounds. Bond energy decreases as atomic size, and therefore bond length, increases. Consider the Tl—Cl bond. It is relatively long and weak, and it takes more energy to remove the two 6*s* electrons from Tl^+ to make Tl^{3+} than is released when two more weak Tl—Cl bonds form in $TlCl_3$. In fact, $TlCl_3$ is so unstable that it decomposes readily to Cl_2 gas and TlCl.

2. *Increasing prominence of the lower oxidation state.* When a group exhibits more than one oxidation state, *the lower state becomes more prominent going down the group.* In Group 3A(13), for instance, all members exhibit the +3 state, but the +1 state first appears with some compounds of gallium and becomes the only important state of thallium.

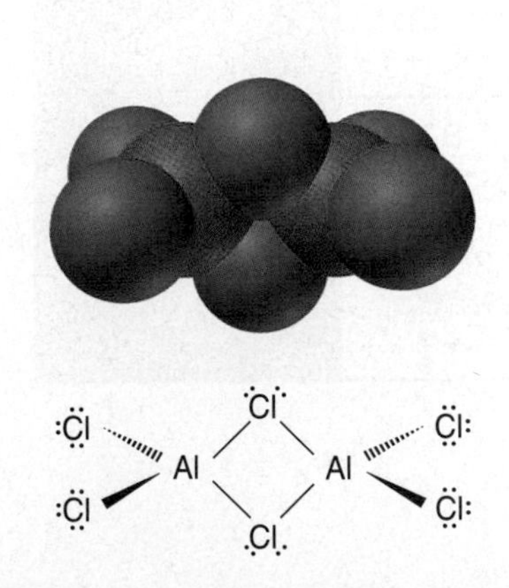

Figure 14.8 The dimeric structure of gaseous aluminum chloride. Despite its name, aluminum trichloride exists in the gas phase as the dimer, Al_2Cl_6.

Key Atomic and Physical Properties

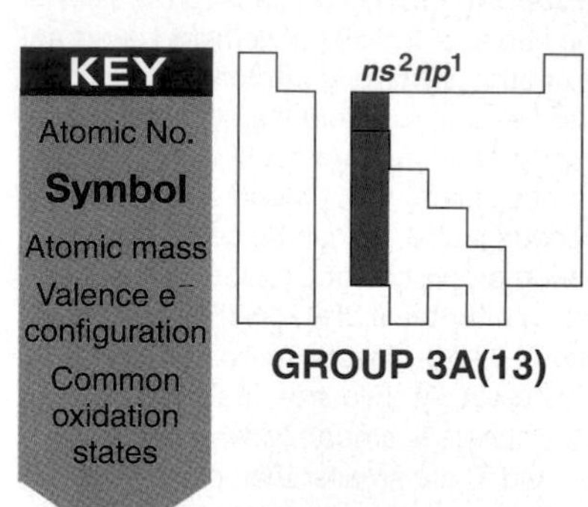

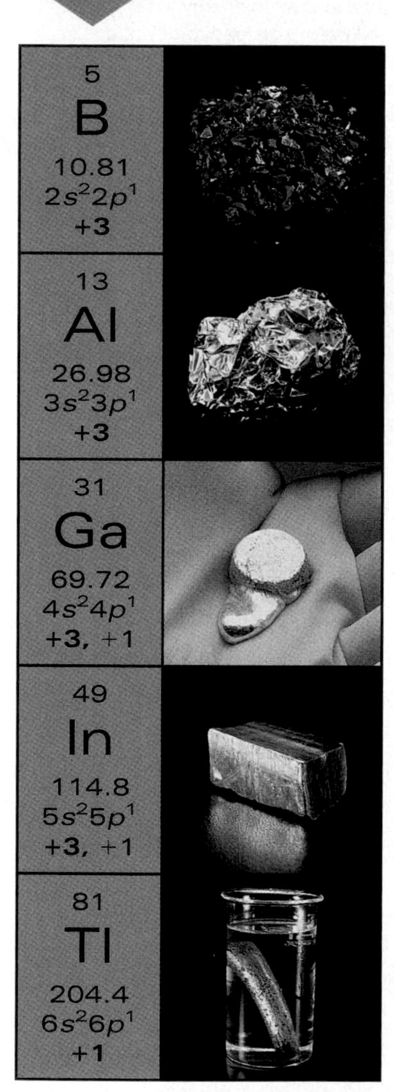

Atomic Properties

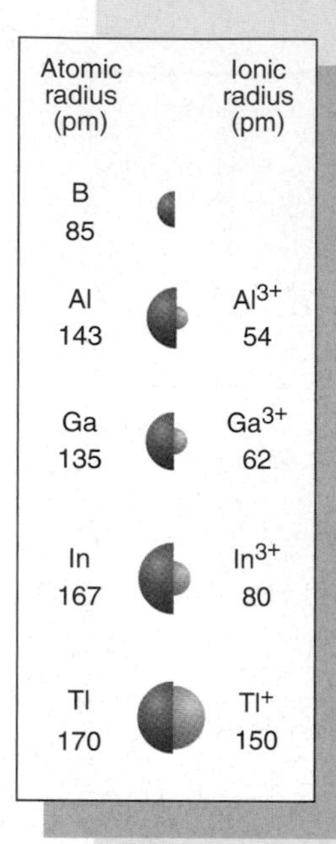

Group electron configuration is ns^2np^1. All except Tl commonly display the +3 oxidation state. The +1 state becomes more common down the group.

Atomic size is smaller and EN is higher than for 2A(2) elements; IE is lower, however, because it is easier to remove an electron from the higher energy *p* sublevel.

Atomic size, IE, and EN do not change as expected down the group because there are intervening transition and inner transition elements.

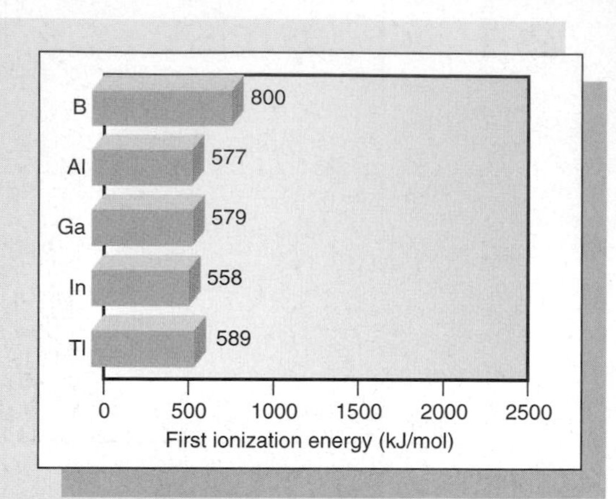

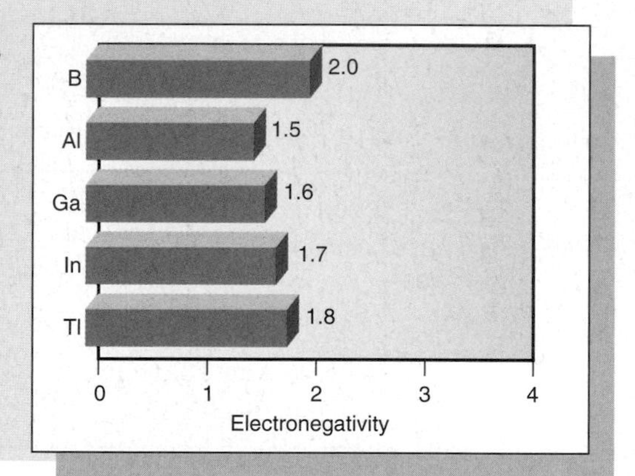

Physical Properties

Bonding changes from network covalent in B to metallic in the rest of the group. Thus, B has a much higher melting point than the others, but there is no overall trend. Boiling points decrease down the group.

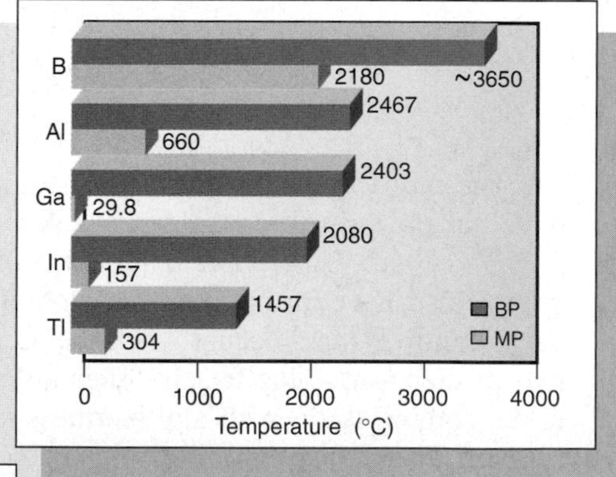

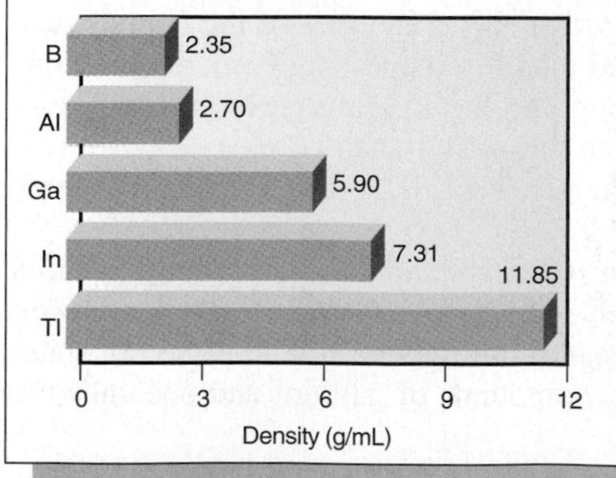

Densities increase down the group.

Some Reactions and Compounds

Important Reactions

In reactions 1 to 3, the elements (E) usually require higher temperatures to react than those of Groups 1A(1) and 2A(2); note the lower oxidation state of Tl. Reactions 4 to 6 show some compounds in key industrial processes.

1. The elements react sluggishly, if at all, with water:

$$2Ga(s) + 6H_2O(hot) \longrightarrow 2Ga^{3+}(aq) + 6OH^-(aq) + 3H_2(g)$$
$$2Tl(s) + 2H_2O(steam) \longrightarrow 2Tl^+(aq) + 2OH^-(aq) + H_2(g)$$

Al becomes covered with a layer of Al_2O_3 that prevents further reaction (*see photo*).

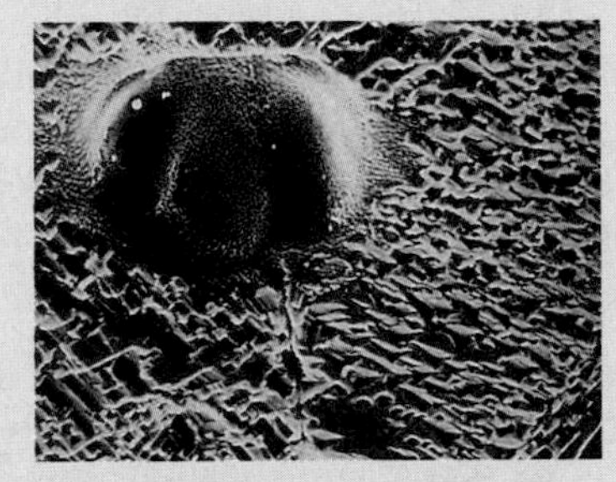

Electron micrograph of oxide coating on Al surface

2. When strongly heated in pure O_2, all members form oxides:

$$4E(s) + 3O_2(g) \xrightarrow{\Delta} 2E_2O_3(s) \qquad (E = B, Al, Ga, In)$$
$$4Tl(s) + O_2(g) \xrightarrow{\Delta} 2Tl_2O(s)$$

Oxide acidity decreases down the group: B_2O_3 (weakly acidic) > Al_2O_3 > Ga_2O_3 > In_2O_3 > Tl_2O (strongly basic), and the +1 oxide is more basic than the +3 oxide.

3. All members reduce halogens (X_2):

$$2E(s) + 3X_2 \longrightarrow 2EX_3 \qquad (E = B, Al, Ga, In)$$
$$2Tl(s) + X_2 \longrightarrow 2TlX(s)$$

The BX_3 compounds are volatile and covalent. Trihalides of Al, Ga, and In are (mostly) ionic solids but occur as covalent dimers in the gas phase; in this way, the 3A atom attains a filled outer level.

4. Acid treatment of Al_2O_3 is important in water purification:

$$Al_2O_3(s) + 3H_2SO_4(l) \longrightarrow Al_2(SO_4)_3(s) + 3H_2O(l)$$

In water, $Al_2(SO_4)_3$ and CaO form a colloid that aids in removing suspended particles.

5. The overall reaction in the production of aluminum metal is a redox process:

$$2Al_2O_3(s) + 3C(s) \longrightarrow 4Al(s) + 3CO_2(g)$$

This process is carried out electrochemically in the presence of cryolite (Na_3AlF_6), which lowers the melting point of the reactant mixture and takes part in the change (*Section 22.4*).

6. A displacement reaction produces gallium arsenide, GaAs (*margin note, p. 566*):

$$(CH_3)_3Ga(g) + AsH_3(g) \longrightarrow 3CH_4(g) + GaAs(s)$$

Important Compounds

1. Boron oxide, B_2O_3. Used in the production of borosilicate glass (*see margin note, p. 570*).

2. Borax, $Na_2[B_4O_5(OH)_4]\cdot 8H_2O$. Major mineral source of boron compounds and B_2O_3. Used as a fireproof insulation material and as a washing powder (20-Mule Team Borax).

3. Boric acid, H_3BO_3 [or $B(OH)_3$]. Used as external disinfectant, eyewash, and insecticide.

4. Diborane, B_2H_6. A powerful reducing agent for possible use as a rocket fuel. Used to synthesize higher boranes, compounds that led to new theories of chemical bonding.

5. Aluminum sulfate (cake alum), $Al_2(SO_4)_3\cdot 18H_2O$. Used in purifying water, tanning leather, and sizing paper; as a fixative for dyeing cloth (*see photo*); and as an antiperspirant.

6. Aluminum oxide, Al_2O_3. Major compound in natural source (bauxite) of Al metal. Used as abrasive in sandpaper, sanding and cutting tools, and toothpaste. Large crystals with metal ion impurities often of gemstone quality (*see photo*). Inert support for chromatography. In fibrous forms, woven into heat-resistant fabrics; also used to strengthen ceramics and metals.

Ruby

7. $Tl_2Ba_2Ca_2Cu_3O_{10}$. Becomes a high-temperature superconductor at 125 K, which is readily attained with liquid N_2 (*see photo*).

Magnet levitated by superconductor

3. *Relative basicity of oxides.* In general, *oxides with the element in a lower oxidation state are more basic than oxides with the element in a higher oxidation state.* A good example in Group 3A is that In_2O is more basic than In_2O_3. In general, when an element has more than one oxidation state, *it acts more like a metal in its lower state*, and this, too, is related to ionic charge density. In this example, the lower charge of In^+ does not polarize the O^{2-} ion as much as the higher charge of In^{3+} does. Thus, in compounds of general formula E_2O, the E-to-O bonding is more ionic than it is in E_2O_3 compounds, so the O^{2-} ion is more available to act as a base (Interchapter, Topic 4).

Highlights of Boron Chemistry

Like the other Period 2 elements, the chemical behavior of boron is strikingly different from that of the other members of its group. As was pointed out earlier, *all boron compounds are covalent,* and unlike the other Group 3A(13) members, boron forms network covalent compounds or large molecules with metals, H, O, N, and C. The unifying feature of many boron compounds is the element's *electron deficiency,* but boron adopts two strategies to fill its outer level: accepting a bonding pair from an electron-rich atom and forming bridge bonds with electron-poor atoms.

Accepting a Bonding Pair from an Electron-Rich Atom In gaseous boron trihalides (BX_3), the B atom is electron deficient, with only six electrons around it (Section 10.1). To attain an octet, the B atom accepts a lone pair of electrons from an electron-rich atom and forms a covalent bond:

$$BF_3(g) + :NH_3(g) \longrightarrow F_3B{-}NH_3(g)$$

(Such reactions, in which one reactant accepts an electron pair from another to form a covalent bond, are very widespread in inorganic, organic, and biochemical processes. They are known as *Lewis acid-base reactions,* and we'll discuss them in Chapters 18 and 23 and see examples of them throughout the second half of the text.)

Similarly, B has only six electrons in boric acid, $B(OH)_3$ (sometimes written as H_3BO_3). In water, the acid itself does not release a proton. Rather, it accepts an electron pair from the O in H_2O, forming a fourth bond and releasing an H^+ ion:

$$B(OH)_3(s) + H_2O(l) \rightleftharpoons B(OH)_4^-(aq) + H^+(aq)$$

Boron's outer shell is filled in the wide variety of borate salts, such as the mineral borax (sodium borate), $Na_2[B_4O_5(OH)_4]{\cdot}8H_2O$, used for decades as a household cleaning agent.

Borates in Your Labware Strong heating of boric acid (or borate salts) drives off water molecules and gives molten boron oxide:

$$2B(OH)_3(s) \xrightarrow{\Delta} B_2O_3(l) + 3H_2O(g)$$

The molten oxide dissolves metal oxides to form borate glasses. When mixed with silica (SiO_2), it forms borosilicate glass. Its high transparency and small change in size when heated or cooled make borosilicate glass useful in cookware and in the glassware you use in the lab.

Boron also attains an octet in several boron-nitrogen compounds whose crystal or molecular structures are amazingly similar to those of elemental carbon and some of its organic compounds (Figure 14.9). These similarities are due to the fact that the size, IE, and EN of C are between those of B and N (see Table 14.1). Moreover, C has four valence electrons, whereas B has three and N has five, so the $\cdot\dot{\underset{\cdot}{C}}{-}\dot{\underset{\cdot}{C}}\cdot$ and $\cdot\dot{B}{-}\underset{\cdot}{\dot{N}}\cdot$ groupings have the same number of valence electrons. Therefore, $H_3C{-}CH_3$ (ethane) and $H_3B{-}NH_3$ (amine-borane) are isoelectronic, as are benzene and borazine (Figure 14.9A and B).

Boron nitride (Figure 14.9C) has a structure consisting of hexagons fused into sheets, very similar to that of graphite. Moreover, like graphite, its π electrons are highly delocalized through resonance. However, molecular-orbital calculations show that there is a large energy gap between the filled valence band and the empty conductance band in boron nitride, but no such gap in graphite (see Figure 12.37, p. 462). As a result, boron nitride is a white electrical insulator, whereas graphite is a black electrical conductor. At high pressure and temperature, boron nitride forms borazon (Figure 14.9D), which has a crystal structure like that of diamond and is extremely hard and abrasive.

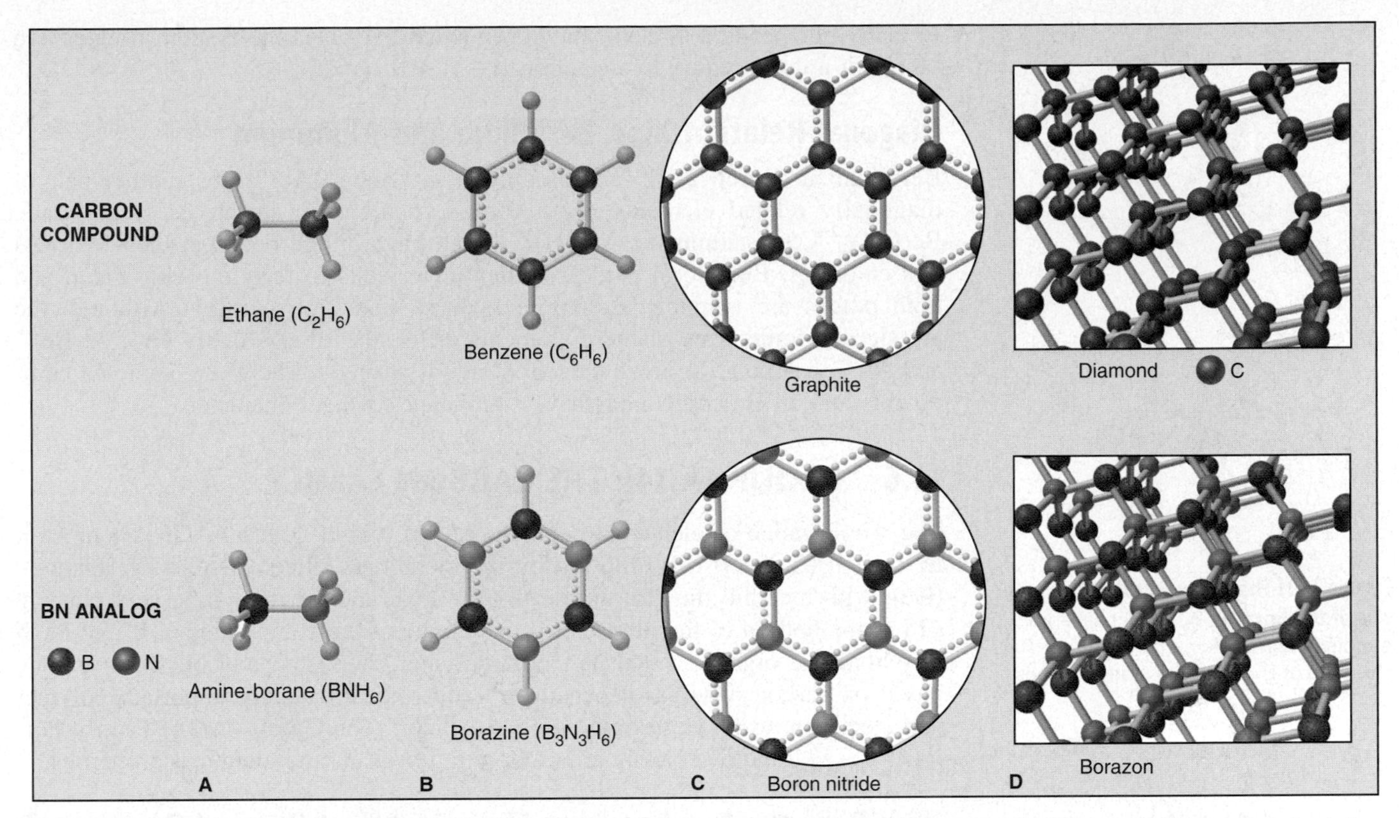

Figure 14.9 Similarities between substances with C—C bonds and those with B—N bonds. Note that in each case, B attains an octet of electrons by bonding with electron-rich N. **A,** Ethane and its BN analog. **B,** Benzene and borazine, which is often referred to as "inorganic" benzene. **C,** Graphite and the similar extended hexagonal sheet structure of boron nitride. **D,** Diamond and borazon have the same crystal structure and are among the hardest substances known.

Forming Bridge Bonds with Electron-Poor Atoms In elemental boron and its many hydrides (boranes), there is no electron-rich atom to supply boron with electrons. In these substances, boron attains an octet through some unusual bonding. In diborane (B_2H_6) and many larger boranes, for example, two types of B—H bonds exist. The first type is a typical electron-pair bond. The valence bond picture in Figure 14.10 shows an sp^3 orbital of B overlapping a 1*s* orbital of H in each of the four terminal B—H bonds, using two of the three electrons in the valence level of each B atom.

The other type of bond is a hydride **bridge bond** (or three-center, two-electron bond), in which *each B—H—B grouping is held together by only two electrons.* Two sp^3 orbitals, one from *each* B, overlap an H 1*s* orbital between them. Two electrons move through this extended bonding orbital—one from one of the B atoms and the other from the H atom—and join the two B atoms via the H atom bridge. Notice that *each B atom is surrounded by eight electrons:* four from the two normal B—H bonds and four from the two B—H—B bridge bonds.

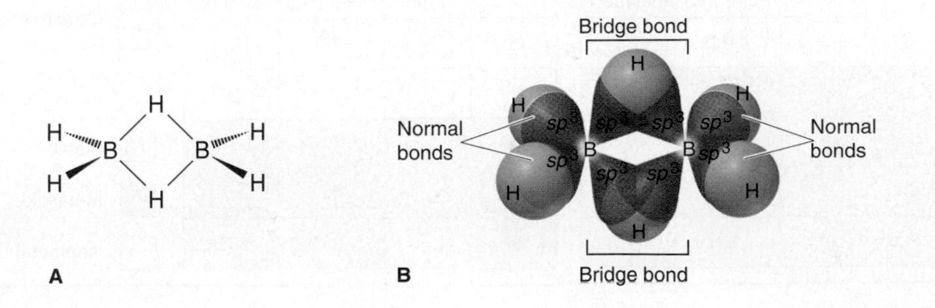

Figure 14.10 The two types of covalent bonding in diborane. A, A perspective diagram of B_2H_6 shows the unusual B—H—B bridge bond and the tetrahedral arrangement around each B atom. **B,** A valence bond depiction shows each sp^3-hybridized B forming normal covalent bonds with two hydrogens and two bridge bonds, in which two electrons bind three atoms, at the two central B—H—B groupings.

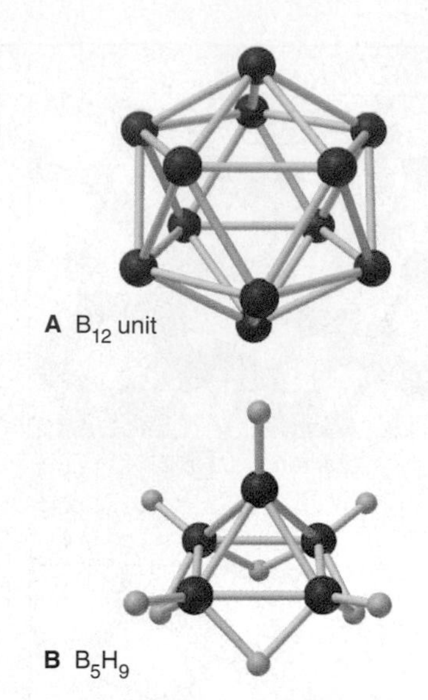

Figure 14.11 The boron icosahedron and one of the boranes. A, The icosahedral structural unit of elemental boron. **B,** The structure of B_5H_9, one of many boranes.

In many boranes and in elemental boron (Figure 14.11), one B atom bridges two others in a three-center, two-electron B—B—B bond.

Diagonal Relationships: Beryllium and Aluminum

Beryllium in Group 2A(2) and aluminum in Group 3A(13) are another pair of diagonally related elements. Both form oxoanions in strong base: beryllate, $Be(OH)_4^{2-}$, and aluminate, $Al(OH)_4^-$. Both have bridge bonds in their hydrides and chlorides. Both form oxide coatings impervious to reaction with water, and both oxides are amphoteric, extremely hard, and high melting. Although the atomic and ionic sizes of these elements differ, the small, highly charged Be^{2+} and Al^{3+} ions polarize nearby electron clouds strongly. Therefore, some Al compounds and all Be compounds have significant covalent character.

14.6 GROUP 4A(14): THE CARBON FAMILY

The whole range of elemental behavior occurs within Group 4A(14): nonmetallic carbon (C) leads off, followed by the metalloids silicon (Si) and germanium (Ge), with metallic tin (Sn) and lead (Pb) next, and newly synthesized element 114 at the bottom of the group. Information about the compounds of C and of Si fills libraries: organic chemistry, most polymer chemistry, and biochemistry are based on carbon, whereas geochemistry and some extremely important polymer and electronic technologies are based on silicon. The Group 4A(14) Family Portrait (pp. 574 and 575) summarizes atomic, physical, and chemical properties.

How Does the Bonding in an Element Affect Physical Properties?

The elements of Group 4A(14) and their neighbors in Groups 3A(13) and 5A(15) illustrate how physical properties, such as melting point and heat of fusion (ΔH_{fus}), depend on the type of bonding in an element (Table 14.2). Within Group 4A, the large decrease in melting point between the network covalent solids C and Si is due to longer, weaker bonds in the Si structure; the large decrease between Ge and Sn is due to the change from covalent network to metallic bonding. Similarly, considering horizontal trends, the large increases in melting point and ΔH_{fus} across a period between Al and Si and between Ga and Ge reflect the change from metallic to covalent network bonding. Note the abrupt rises in the values for these properties from metallic Al, Ga, and Sn to the network covalent metalloids Si, Ge, and Sb, and note the abrupt drops from the covalent networks of C and Si to the individual molecules of N and P in Group 5A.

Table 14.2 Bond Type and the Melting Process in Groups 3A(13) to 5A(15)

Period	Group 3A(13)				Group 4A(14)				Group 5A(15)			
	Element	Bond Type	Melting Point (°C)	ΔH_{fus} (kJ/mol)	Element	Bond Type	Melting Point (°C)	ΔH_{fus} (kJ/mol)	Element	Bond Type	Melting Point (°C)	ΔH_{fus} (kJ/mol)
2	**B**	Covalent network	2180	23.6	**C**	Covalent network	4100	Very high	**N**	Covalent molecule	−210	0.7
3	**Al**	Metallic	660	10.5	**Si**	Covalent network	1420	50.6	**P**	Covalent molecule	44.1	2.5
4	**Ga**	Metallic	30	5.6	**Ge**	Covalent network	945	36.8	**As**	Covalent network	816	27.7
5	**In**	Metallic	157	3.3	**Sn**	Metallic	232	7.1	**Sb**	Covalent network	631	20.0
6	**Tl**	Metallic	304	4.3	**Pb**	Metallic	327	4.8	**Bi**	Metallic	271	10.5

Key: Metallic; Covalent network; Covalent molecule; Metal; Metalloid; Nonmetal

Allotropism: Different Forms of an Element Striking variations in physical properties often appear among **allotropes,** different crystalline or molecular forms of a substance. One allotrope is usually more stable than another at a particular pressure and temperature. Group 4A(14) provides the first dramatic example of allotropism, in the forms of carbon. It is difficult to imagine two substances made entirely of the same atom that are more different than graphite and diamond. Graphite is a black electrical conductor that is soft and "greasy," whereas diamond is a colorless electrical insulator that is extremely hard. Graphite is the standard state of carbon, the more stable form at ordinary temperature and pressure, as the phase diagram in Figure 14.12 shows. Fortunately for jewelry owners, diamond changes to graphite at a negligible rate under normal conditions.

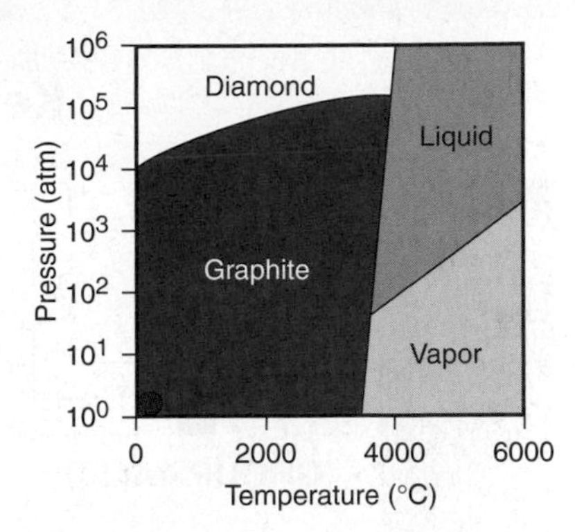

Figure 14.12 Phase diagram of carbon. Graphite is the more stable form of carbon at ordinary conditions (*small red circle at extreme lower left*). Diamond is more stable at very high pressure.

In the mid-1980s, a newly discovered allotrope of carbon began generating great interest. Mass spectrometric analysis of soot had shown evidence for a soccer ball–shaped molecule of formula C_{60} (Figure 14.13A; see also the Gallery, p. 386). More recently, the molecule has also been found in geological samples formed by meteorite impacts, even the one that occurred around the time the dinosaurs became extinct. The molecule has been dubbed *buckminsterfullerene* (informally called a "buckyball") after the architect-engineer R. Buckminster Fuller, who designed structures with similar shapes. Excitement rose in 1990, when scientists learned how to prepare multigram quantities of C_{60} and related fullerenes, enough to study macroscopic behavior and possible applications. Since then, metal atoms have been incorporated into the structure and many different atoms and groups (fluorine, hydroxyl groups, sugars, etc.) have been attached, resulting in compounds with a range of useful properties.

Then, in 1991, scientists passed an electric discharge through graphite rods sealed with helium gas in a container and obtained extremely thin (~1 nm in diameter) graphite-like tubes with fullerene ends (Figure 14.13B). These *nanotubes* are rigid and, on a mass basis, much stronger than steel along their long axis. They also conduct electricity along this axis because of the delocalized electrons. Reports about fullerenes and nanotubes appear almost daily in scientific journals. With potential applications in nanoscale electronics, catalysis, polymers, and medicine, fullerene and nanotube chemistry will receive increasing attention well into the 21st century.

Tin has two allotropes. White β-tin is stable at room temperature and above, whereas gray α-tin is the more stable form below 13°C (56°F). When white tin is kept for long periods at a low temperature, some converts to microcrystals of gray tin. The random formation and growth of these regions of gray tin, which has a different crystal structure, weaken the metal and make it crumble. In the unheated cathedrals of medieval northern Europe, tin pipes of magnificent organs sometimes crumbled as a result of "tin disease" caused by this allotropic transition from white to gray tin.

Figure 14.13 Buckyballs and nanotubes. **A,** Crystals of buckminsterfullerene (C_{60}) are shown leading to a ball-and-stick model. The parent of the fullerenes, the "buckyball," is a soccer ball–shaped molecule of 60 carbon atoms. **B,** Nanotubes are single or, as shown in this colorized transmission electron micrograph, concentric graphite-like tubes with fullerene ends (see the Gallery, p. 386).

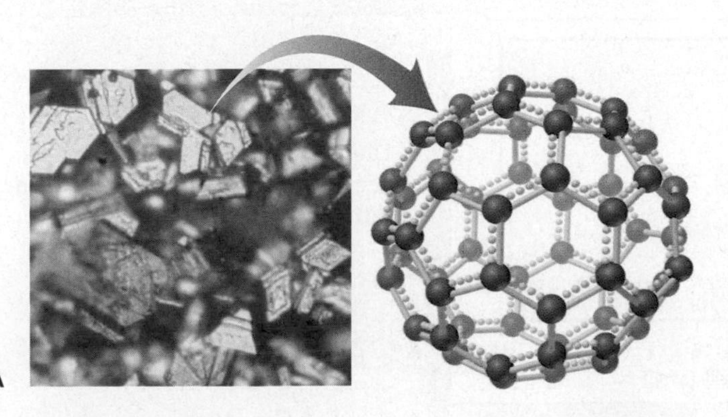

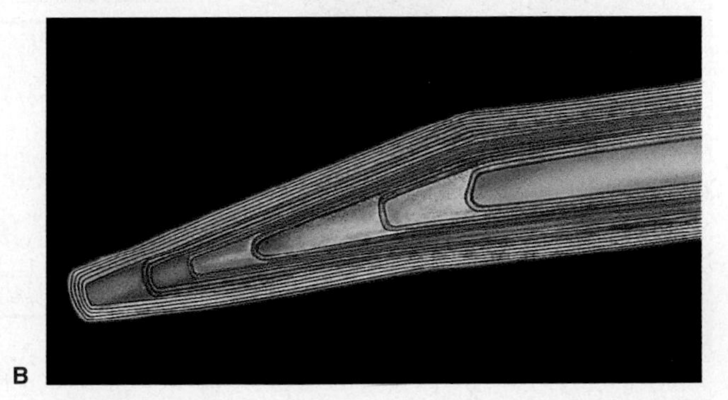

Group 4A(14): The Carbon Family

Key Atomic and Physical Properties

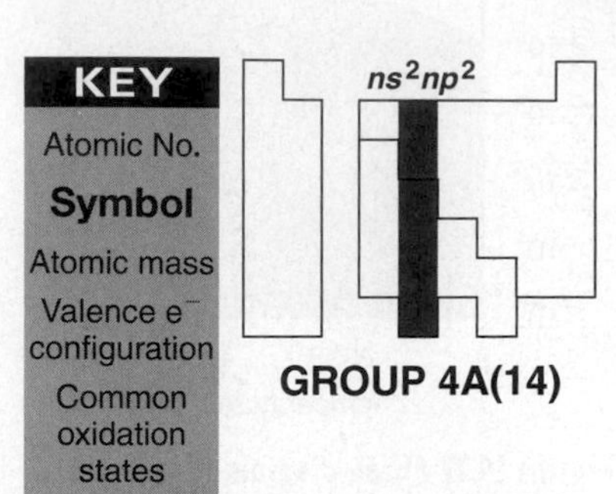

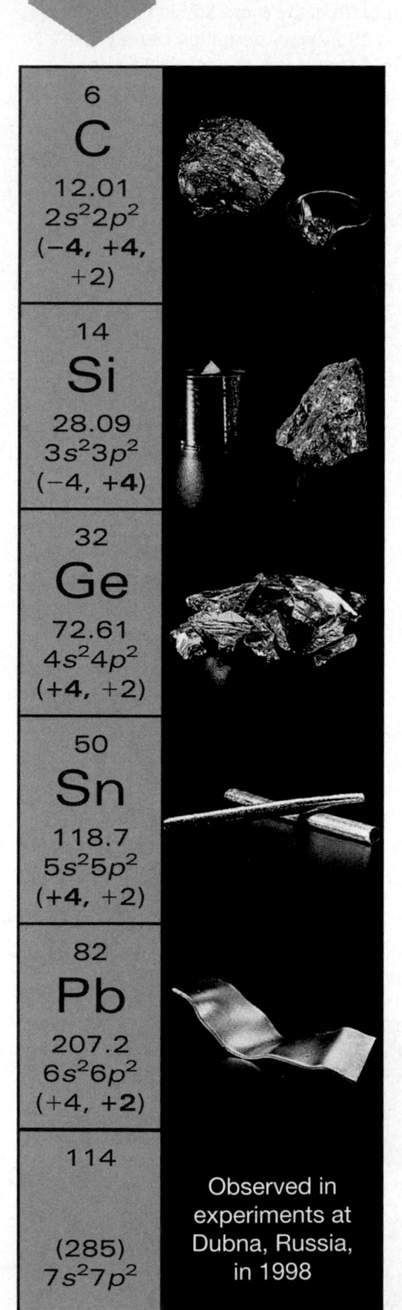

6
C
12.01
2s²2p²
(−4, +4, +2)
14
Si
28.09
3s²3p²
(−4, +4)
32
Ge
72.61
4s²4p²
(+4, +2)
50
Sn
118.7
5s²5p²
(+4, +2)
82
Pb
207.2
6s²6p²
(+4, +2)
114
(285)
7s²7p²
Observed in experiments at Dubna, Russia, in 1998

Atomic Properties

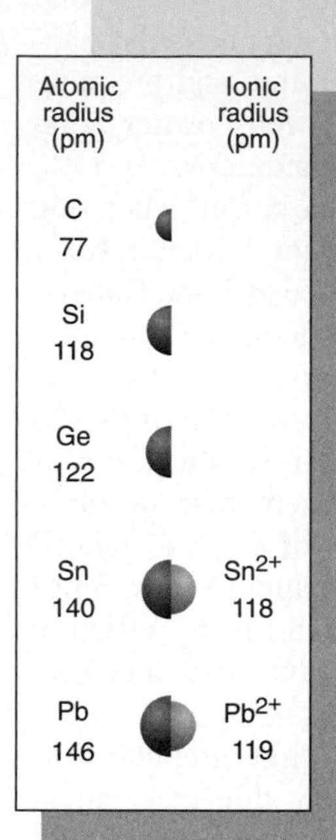

Group electron configuration is ns^2np^2. Down the group, the number of oxidation states decreases, and the lower (+2) state becomes more common.

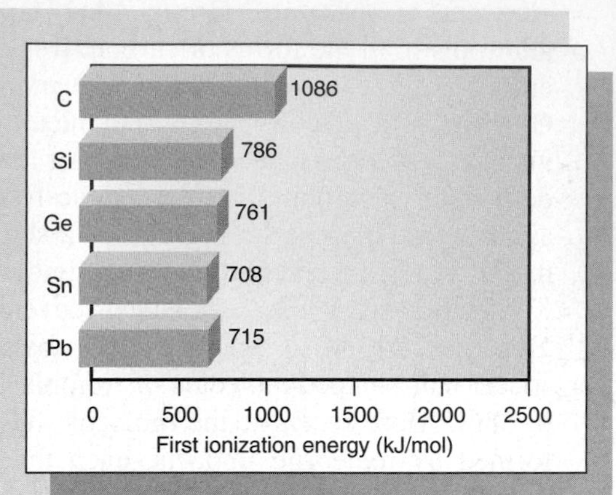

Down the group, size increases. Because transition and inner transition elements intervene, IE and EN do not decrease smoothly.

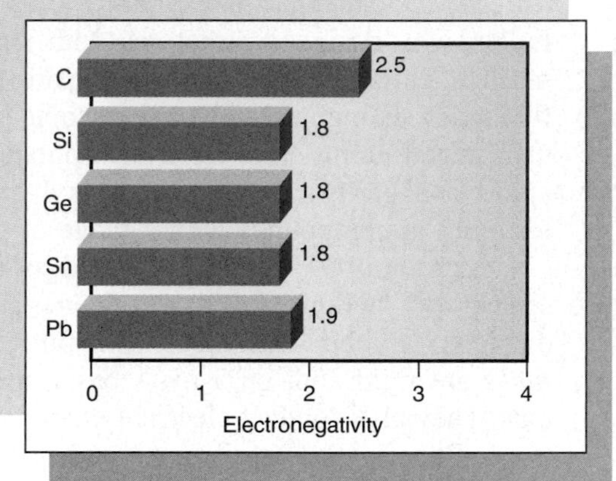

Physical Properties

Trends in properties, such as decreasing hardness and melting point, are due to changes in types of bonding within the solid: network covalent in C, Si, and Ge; metallic in Sn and Pb (*see text*).

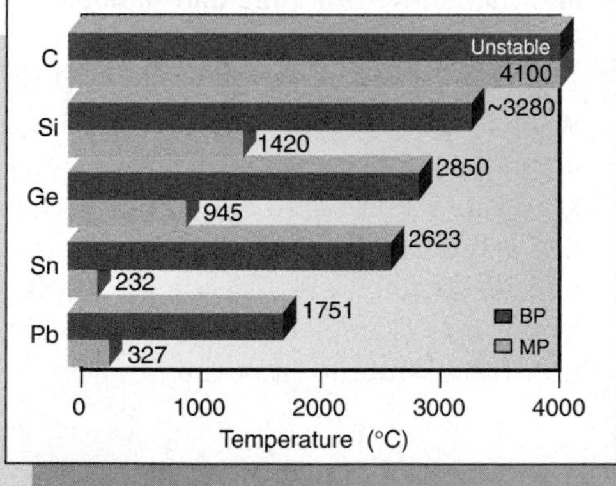

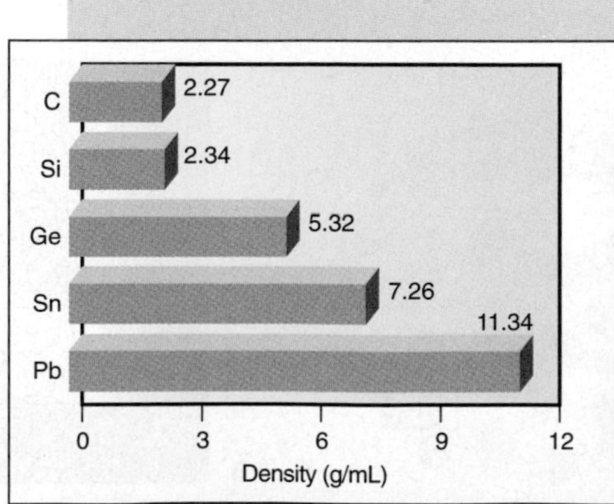

Down the group, density increases because of several factors, including differences in crystal packing.

Some Reactions and Compounds

Important Reactions

Reactions 1 and 2 pertain to all the elements (E); reactions 3 to 7 concern industrial uses for compounds of C and Si.

1. The elements are oxidized by halogens:

$$E(s) + 2X_2 \longrightarrow EX_4 \quad (E = C, Si, Ge)$$

The +2 halides are more stable for tin and lead, SnX_2 and PbX_2.

2. The elements are oxidized by O_2:

$$E(s) + O_2(g) \longrightarrow EO_2 \quad (E = C, Si, Ge, Sn)$$

Pb forms the +2 oxide, PbO. Oxides become more basic down the group. The reaction of CO_2 and H_2O provides the weak acidity of natural unpolluted waters:

$$CO_2(g) + H_2O(l) \rightleftharpoons [H_2CO_3(aq)] \rightleftharpoons H^+(aq) + HCO_3^-(aq)$$

3. Air and steam passed over hot coke produce gaseous fuel mixtures (producer gas and water gas):

$$C(s) + air(g) + H_2O(g) \longrightarrow CO(g) + CO_2(g) + N_2(g) + H_2(g)$$
[not balanced]

4. Hydrocarbons react with O_2 to form CO_2 and H_2O. The reaction for methane is adapted to yield heat or electricity:

$$CH_4(g) + 2O_2(g) \longrightarrow CO_2(g) + 2H_2O(g)$$

Oxyacetylene torch

5. Certain metal carbides react with water to produce acetylene:

$$CaC_2(s) + 2H_2O(l) \longrightarrow Ca(OH)_2(aq) + C_2H_2(g)$$

The gas is used to make other organic compounds and as a fuel in welding (*see photo*).

6. Freons (chlorofluorocarbons) are formed by fluorinating carbon tetrachloride:

$$CCl_4(l) + HF(g) \longrightarrow CFCl_3(g) + HCl(g)$$

Production of trichlorofluoromethane (Freon-11), the most widely used refrigerant in the world, is being eliminated because of adverse effects on the environment (*margin note, p. 577*).

7. Silica is reduced to form elemental silicon:

$$SiO_2(s) + 2C(s) \longrightarrow Si(s) + 2CO(g)$$

This crude silicon is made ultrapure through zone refining (*Gallery, p. 521*) for the manufacture of computer chips (*see photo*).

Computer chip

Important Compounds

1. Carbon monoxide, CO. Used as a gaseous fuel, as a precursor for organic compounds, and as a reactant in the purification of nickel. Formed in internal combustion engines and released as a toxic air pollutant.

2. Carbon dioxide, CO_2. Atmospheric component used by photosynthetic plants to make carbohydrates and O_2. The final oxidation product of all carbon-based fuels; its increase in the atmosphere is causing global warming. Used industrially as a refrigerant gas, a blanketing gas in fire extinguishers, and an effervescent gas in beverages (*see photo*). Combined with NH_3 to form urea for fertilizers and plastics manufacture.

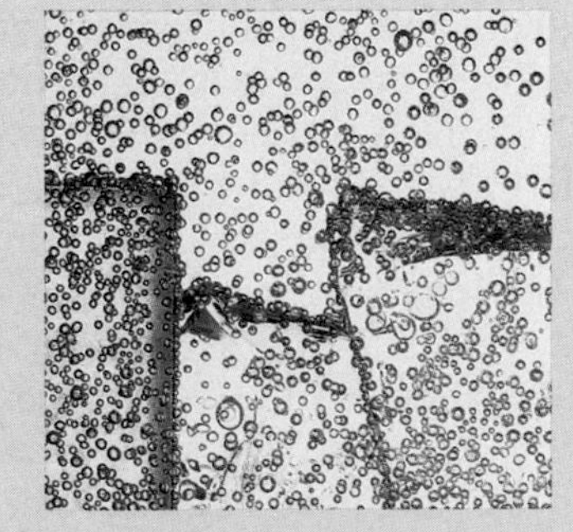

3. Methane, CH_4. Used as a fuel and in the production of many organic compounds. Major component of natural gas. Formed by anaerobic decomposition of plants (swamp gas) and by microbes in termites and certain mammals. May contribute to global warming.

4. Silicon dioxide, SiO_2. Occurs in many amorphous (glassy) and crystalline forms, quartz being the most common. Used to make glass and as an inert chromatography support material.

5. Silicon carbide, SiC. Known as *carborundum*, a major industrial abrasive and a highly refractory ceramic for tough, high-temperature uses. Can be doped to form a high-temperature semiconductor.

6. Organotin compounds, R_4Sn. Used to stabilize PVC, or poly(vinyl chloride), plastics (*see photo*) and to cure silicone rubbers. Agricultural biocide for insects, fungi, and weeds.

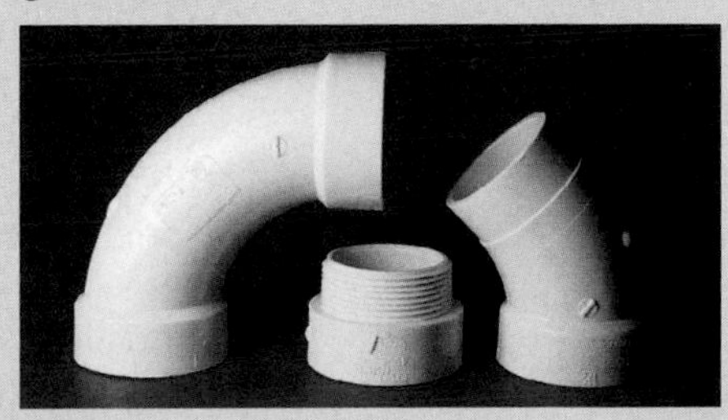

7. Tetraethyl lead, $(C_2H_5)_4Pb$. Once used as a gasoline additive to improve fuel efficiency, but now banned because of its inactivation of auto catalytic converters. Major source of lead as a toxic air pollutant.

How Does the Type of Bonding Change in Group 4A(14) Compounds?

The Group 4A(14) elements display a wide range of chemical behavior, from the covalent compounds of carbon to the ionic compounds of lead. Carbon's intermediate EN of 2.5 ensures that it virtually always forms covalent bonds, but the larger members of the group form bonds with increasing ionic character. With nonmetals, Si and Ge form strong polar covalent bonds, such as the Si—O bond, one of the strongest of any Period 3 element (BE = 368 kJ/mol). This bond is responsible for the physical and chemical stability of Earth's solid surface, as we discuss later in this section. Although individual Sn or Pb ions rarely exist, the bonding of either element with a nonmetal has considerable ionic character.

The pattern of elements with more than one oxidation state that we observed in Group 3A(13) also appears here. After Si, the shift to greater importance of the lower oxidation state occurs: compounds with Si in the +4 state are much more stable than those with Si in the +2 state, whereas compounds with Pb in the +2 state are more stable than those with Pb in the +4 state. The elements also behave more like metals in the lower oxidation state. Consider the chlorides and oxides of Sn and Pb. The +2 chlorides $SnCl_2$ and $PbCl_2$ are white, relatively high-melting, water-soluble crystals—typical properties of a salt (Figure 14.14). In contrast, $SnCl_4$ is a volatile, benzene-soluble liquid, and $PbCl_4$ is a thermally unstable oil. Both likely consist of individual, tetrahedral molecules. The +2 oxides SnO and PbO are more basic than the +4 oxides SnO_2 and PbO_2: because the +2 metals are less able to polarize the O^{2-} ion, the E-to-O bonding is more ionic.

Figure 14.14 The greater metallic character of tin and lead in the lower oxidation state. Metals with more than one oxidation state exhibit more metallic behavior in the lower state. Tin(II) and lead(II) chlorides are white, crystalline, saltlike solids. In contrast, tin(IV) and lead(IV) chlorides are volatile liquids, indicating the presence of individual molecules.

Highlights of Carbon Chemistry

Like the other Period 2 elements, carbon is an anomaly in its group; indeed, it may be an anomaly in the entire periodic table. Carbon forms bonds with the smaller Group 1A(1) and 2A(2) metals, many transition metals, the halogens, and its neighbors B, Si, N, O, P, and S. It also exhibits every oxidation state possible for its group, from +4 in CO_2 and halides like CCl_4 through −4 in CH_4.

Two features stand out in the chemistry of carbon: its ability to bond to itself to form chains, a process known as *catenation*, and its ability to form multiple bonds. As a result of its small size and its capacity for four bonds, carbon can form chains, branches, and rings that lead to myriad structures. Add a lot of H, some O and N, a bit of S, P, halogens, and a few metals, and you have the whole organic world! Figure 14.15 shows three of the several million organic compounds known. Multiple bonds are common in carbon structures because the C—C bond is short enough for side-to-side overlap of two half-filled $2p$ orbitals to form π bonds. (In Chapter 15, we discuss in detail how the atomic properties of carbon give rise to the diverse structures and reactivities of organic compounds.)

Because the other 4A members are larger, E—E bonds become longer and weaker down the group; thus, in terms of bond strength, C—C > Si—Si > Ge—Ge. The empty d orbitals of these larger atoms make the chains much more

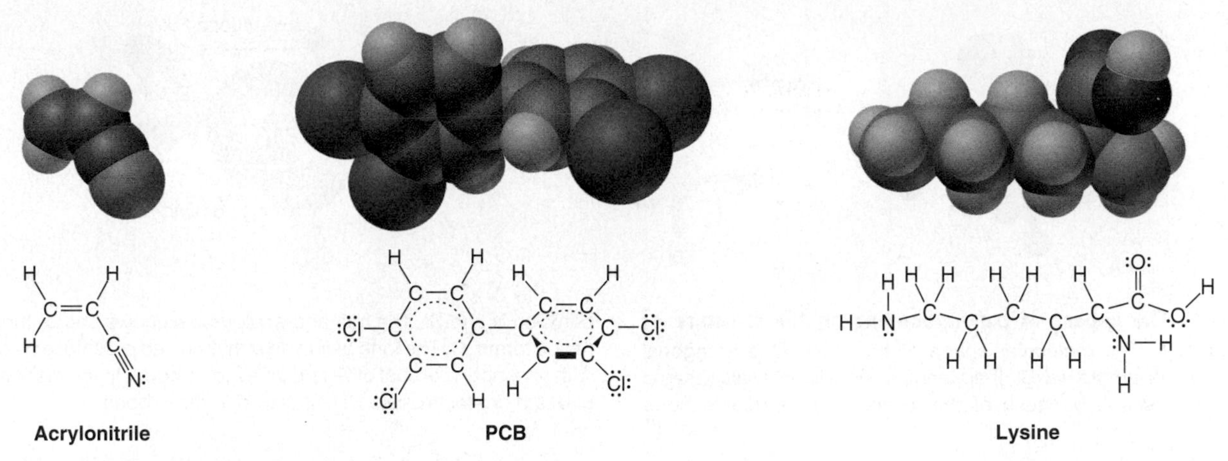

Figure 14.15 Three of the several million known organic compounds of carbon. Acrylonitrile, a precursor of acrylic fibers. PCB (one of the polychlorinated biphenyls). Lysine, one of about 20 amino acids that occur in proteins.

susceptible to chemical attack. Thus, some compounds with long Si chains are known but they are very reactive, and the longest Ge chain has only eight atoms. Moreover, longer bonds do not typically allow sufficient overlap of *p* orbitals for π bonding. Only very recently have compounds with π bonds between Si atoms or between Ge atoms been prepared, and they are also extremely reactive.

In contrast to its organic compounds, carbon's inorganic compounds are simple. Metal carbonates are the main mineral form. Marble, limestone, chalk, coral, and several other types are found in enormous deposits throughout the world. Many of these compounds are remnants of fossilized marine organisms. Carbonates are used in several common antacids because they react with the HCl in stomach acid [see the Group 2A(2) Family Portrait, compound 4, p. 565]:

$$CaCO_3(s) + 2HCl(aq) \longrightarrow CaCl_2(aq) + CO_2(g) + H_2O(l)$$

Identical net ionic reactions with sulfuric and nitric acids protect lakes bounded by limestone deposits from the harmful effects of acid rain.

Unlike the other 4A members, which form only solid network-covalent or ionic oxides, carbon forms two common gaseous oxides, CO_2 and CO. Carbon dioxide is essential to all life: it is the primary source of carbon in plants and animals through photosynthesis. Its aqueous solution is the cause of acidity in natural waters. However, its atmospheric buildup from deforestation and excessive use of fossil fuels may severely affect the global climate. Carbon monoxide forms when carbon or its compounds burn in an inadequate supply of O_2:

$$2C(s) + O_2(g) \longrightarrow 2CO(g)$$

Carbon monoxide is a key component of syngas fuels (see Chemical Connections, p. 245) and is widely used in the production of methanol, formaldehyde, and other major industrial compounds.

CO binds strongly to many transition metals. When inhaled in cigarette smoke or polluted air, it enters the blood and binds strongly to the Fe(II) in hemoglobin, preventing the normal binding of O_2, and to other iron-containing proteins. The cyanide ion (CN^-) is *isoelectronic* with CO:

$$[:C{\equiv}N:]^- \quad \text{same electronic structure as} \quad :C{\equiv}O:$$

Cyanide binds to many of the same iron-containing proteins and is also toxic.

Monocarbon halides (or halomethanes) are tetrahedral molecules whose stability to heat and light decreases as the size of the halogen atom increases and the C—X bond becomes longer (weaker) (see the Interchapter, Topic 2). The chlorofluorocarbons (CFCs, or Freons) have important industrial uses; their short, strong carbon-halogen bonds are exceptionally stable, however, and the environmental persistence of these compounds has created major problems.

CFCs: The Good, the Bad, and the Strong The strengths of the C—F and C—Cl bonds make CFCs, such as Freon-12, shown here, both useful and harmful. Chemically and thermally stable, nontoxic, and nonflammable, they are excellent cleaners for electronic parts, coolants in refrigerators and air conditioners, and propellants in aerosol cans. However, their bond strengths also mean that CFCs decompose *very* slowly near Earth's surface. In the lower atmosphere, they contribute to climate warming, absorbing infrared radiation 16,000 times as effectively as CO_2. Once in the stratosphere, however, they are bombarded by ultraviolet (UV) radiation, which breaks the otherwise stable C—Cl bonds, releasing free Cl atoms that initiate ozone-destroying reactions (Chapter 16). Legal production of CFCs has ended in the United States, but international production and smuggling are widespread.

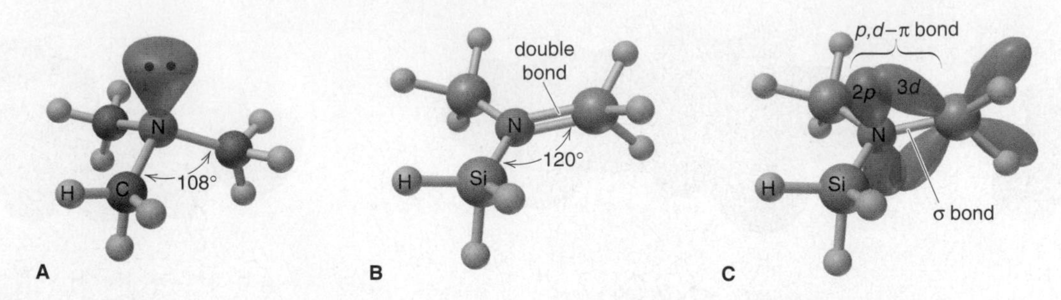

Figure 14.16 The impact of *p,d*–π bonding on the structure of trisilylamine. **A,** The molecular shape of trimethylamine is trigonal *pyramidal*, as for ammonia. **B,** Trisilylamine, the silicon analog, has a trigonal *planar* shape because of the formation of a double bond between N and Si. The ball-and-stick model shows one of three resonance forms. **C,** The lone pair in an unhybridized *p* orbital of N overlaps with an empty *d* orbital of Si to give a *p,d*–π bond. In the resonance hybrid, a *d* orbital on each Si is involved in the π bond.

Highlights of Silicon Chemistry

Silicon halides are more reactive than carbon halides because Si has empty *d* orbitals that are available for bond formation. Surprisingly, the silicon halides have longer yet *stronger* bonds than the corresponding carbon halides (see Tables 9.2 and 9.3, pp. 341 and 342). One explanation for this unusual strength is that the Si—X bond has some double-bond character because of the presence of a σ bond *and* a different type of π bond, called a *p,d–π bond*. This bond arises from side-to-side overlap of an Si *d* orbital and a halogen *p* orbital. A similar type of *p,d*–π bonding leads to the trigonal planar molecular shape of trisilylamine, in contrast to the trigonal pyramidal shape of trimethylamine (Figure 14.16).

To a great extent, the chemistry of silicon is the chemistry of the *silicon-oxygen bond*. Just as carbon forms unending C—C chains, the —Si—O— grouping repeats itself endlessly in a wide variety of **silicates,** the most important minerals on the planet, and in **silicones,** synthetic polymers that have many applications:

1. *Silicate minerals.* From common sand and clay to semiprecious amethyst and carnelian, silicate minerals are the dominant form of matter in the nonliving world. Oxygen, the most abundant element on Earth, and silicon, the next most abundant, compose these minerals and thus account for four of every five atoms on the surface of the planet!

The silicate building unit is the *orthosilicate* grouping, $—SiO_4—$, a tetrahedral arrangement of four oxygens around a central silicon. Several well-known minerals contain SiO_4^{4-} ions or small groups of them linked together. The gemstone zircon ($ZrSiO_4$) contains one unit; hemimorphite [$Zn_4(OH)_2Si_2O_7{\cdot}H_2O$] contains two units linked through an oxygen corner; and beryl ($Be_3Al_2Si_6O_{18}$), the major source of beryllium, contains six units joined into a cyclic ion (Figure 14.17). As you'll see shortly, in addition to these separate ions, SiO_4 units are linked more extensively to create much of the planet's mineral structure.

2. *Silicone polymers.* Unlike the naturally occurring silicates, silicone polymers are manufactured substances, consisting of alternating Si and O atoms with two organic groups also bonded to each Si atom. A key starting material is formed by the reaction of silicon with methyl chloride:

$$Si(s) + 2CH_3Cl(g) \xrightarrow[\sim 300^\circ C]{\text{Cu catalyst}} (CH_3)_2SiCl_2(g)$$

Although the corresponding carbon compound [$(CH_3)_2CCl_2$] is inert in water, the empty *d* orbitals of Si make $(CH_3)_2SiCl_2$ reactive:

$$(CH_3)_2SiCl_2(l) + 2H_2O(l) \longrightarrow (CH_3)_2Si(OH)_2(l) + 2HCl(g)$$

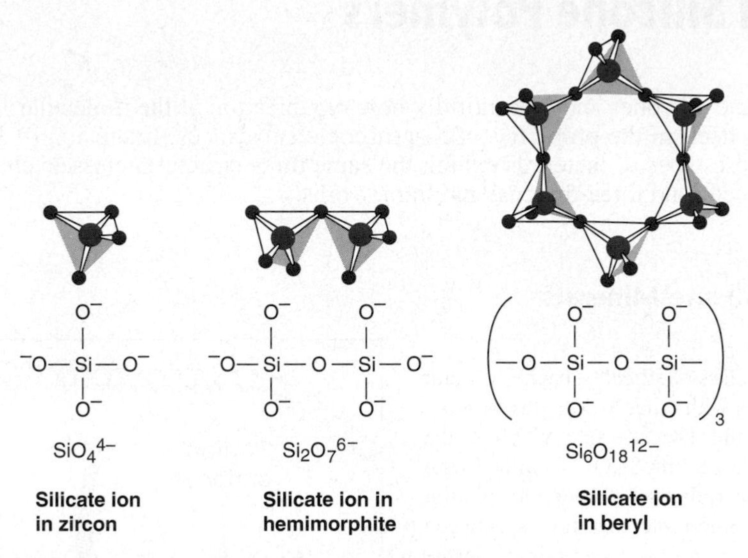

Figure 14.17 Structures of the silicate anions in some minerals.

The product reacts with itself to form water and a silicone chain molecule called a *poly(dimethyl siloxane)*:

$$n(CH_3)_2Si(OH)_2 \longrightarrow \left[-O-\underset{CH_3}{\overset{CH_3}{\mid}}\!\!\!\!\!Si- \right]_n + nH_2O$$

Silicones have properties of both plastics and minerals. The organic groups give them the flexibility and weak intermolecular forces between chains that are characteristic of a plastic, while the O—Si—O backbone confers the thermal stability and nonflammability of a mineral. In the Gallery on the following pages, note the structural parallels between silicates and silicones.

Diagonal Relationships: Boron and Silicon

The final diagonal relationship that we consider occurs between the metalloids boron and silicon. Both of these elements exhibit the electrical properties of a semiconductor (Section 12.6). Both B and Si and their mineral oxoanions—borates and silicates—occur in extended covalent networks. Both boric acid [$B(OH)_3$] and silicic acid [$Si(OH)_4$] are weakly acidic solids that occur as layers held together by widespread H bonding. Their hydrides—the compact boranes and the extended silanes—are both flammable, low-melting compounds that act as reducing agents.

Looking Backward and Forward: Groups 3A(13), 4A(14), and 5A(15)

Standing in Group 4A(14), we look back at Group 3A(13) as the transition from the *s* block of metals to the *p* block of mostly metalloids and nonmetals (Figure 14.18). Major changes in physical behavior occur moving across a period from metals to covalent networks. Changes in chemical behavior occur as well, as cations give way to covalent tetrahedra. Looking ahead to Group 5A(15), we note the more frequent occurrence of covalent bonding and (as electronegativity increases) the expanded valence shells of nonmetals, as well as the first appearance of monatomic anions.

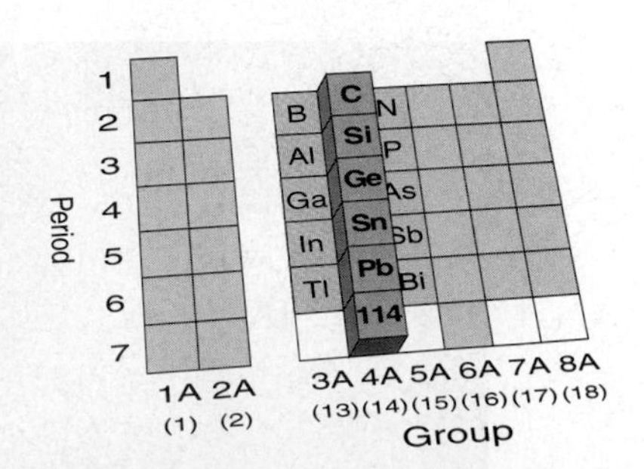

Figure 14.18 Standing in Group 4A(14), looking backward to 3A(13) and forward to 5A(15).

GALLERY

Silicate Minerals and Silicone Polymers

The silicates and silicones show beautifully how organization at the molecular level manifests itself in the properties of macroscopic substances. Interestingly, both of these types of materials exhibit the same three structural classes: chains, sheets, and three-dimensional frameworks.

Silicate Minerals

Chain Silicates

In the simplest structural class of silicates, each SiO_4 unit shares two of its O corners with other SiO_4 units, forming a chain. Two chains can link laterally into a ribbon; the most common ribbon has repeating $Si_4O_{11}^{6-}$ units. Metal ions bind the polyanionic ribbons together into neutral sheaths. With only weak intermolecular forces between sheaths, the material occurs in fibrous strands, as in the family of asbestos minerals.

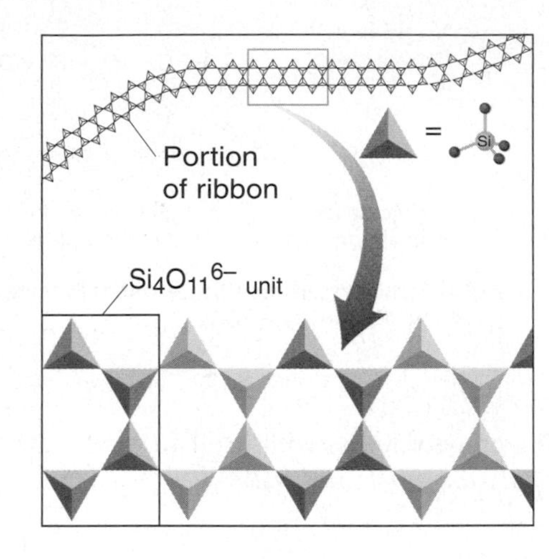

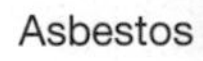
Asbestos

Sheet (Layer) Silicates

In the next structural class of silicates, each SiO_4 unit shares three of its four O corners with other SiO_4 units to form a sheet; double sheets arise when the fourth O is shared with another sheet. In talc, the softest mineral, the sheets interact through weak forces, so talcum powder feels slippery. If Al substitutes for some Si, or if $Al(OH)_3$ layers interleave with silicate sheets, an aluminosilicate results, such as the clay kaolinite. Different substitutions for Si and/or interlayers of other ions give the micas. In muscovite mica, ions lie between aluminosilicate double layers. Mica flakes when the ionic attractions are overcome.

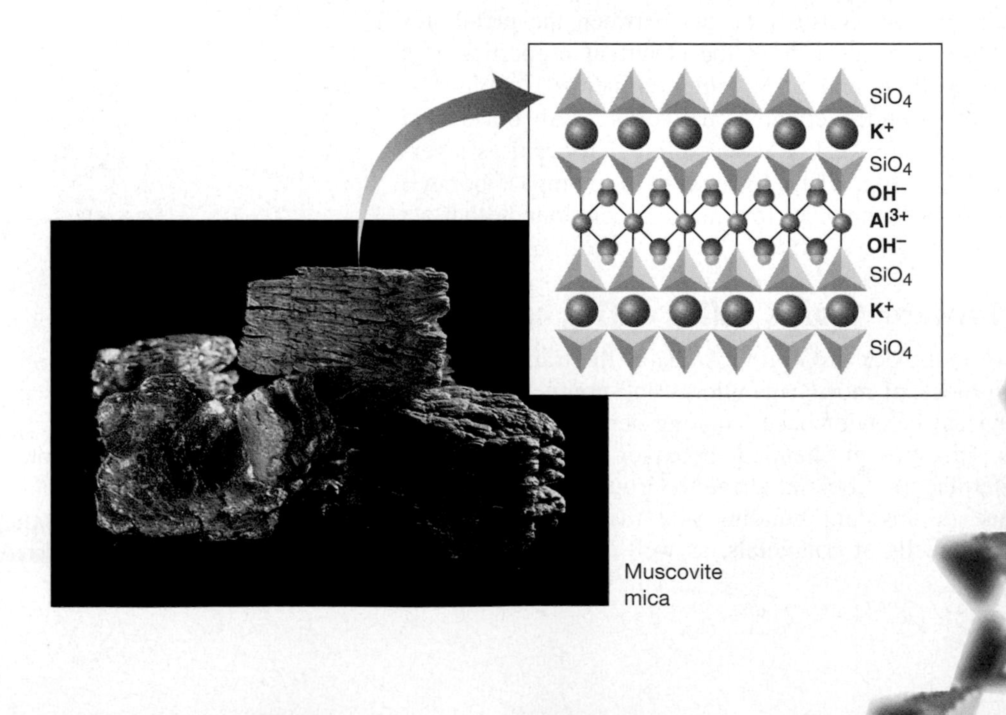

Muscovite mica

Framework Silicates

In the final structural class of silicates, SiO_4 units share all four O corners to give a three-dimensional framework silicate, such as silica (SiO_2), which occurs most often as α-quartz. Some of silica's 12 crystalline forms exist as semiprecious gems. Feldspars, which comprise 60% of Earth's crust, result when some Si in a framework silicate is replaced by Al; granite consists of microcrystals of feldspar, mica, and quartz. Eons of weathering convert feldspars into clays. Zeolites have open frameworks of polyhedra that give rise to minute tunnels. Synthetic zeolites are made with cavities of specific sizes to trap particular molecules; they are used to dry gas mixtures, separate hydrocarbons, and prepare catalysts.

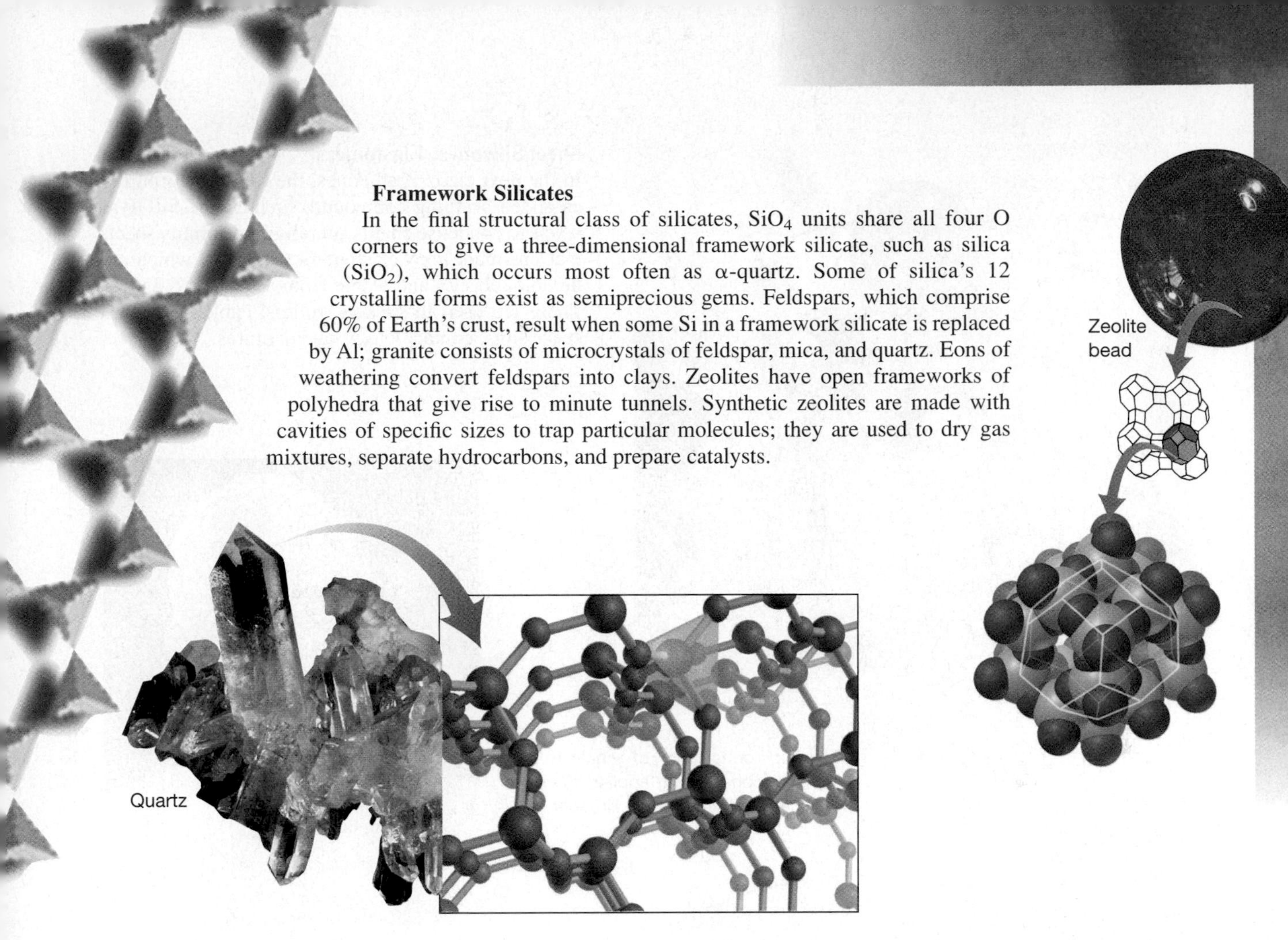

Silicone Polymers

Polymer Chemistry

The design, synthesis, and production of polymers form one of the largest branches of modern chemical science. Nearly half of all industrial chemists and chemical engineers are involved in this pursuit. A major branch of polymer chemistry is devoted to exploring the properties and countless uses of silicones.

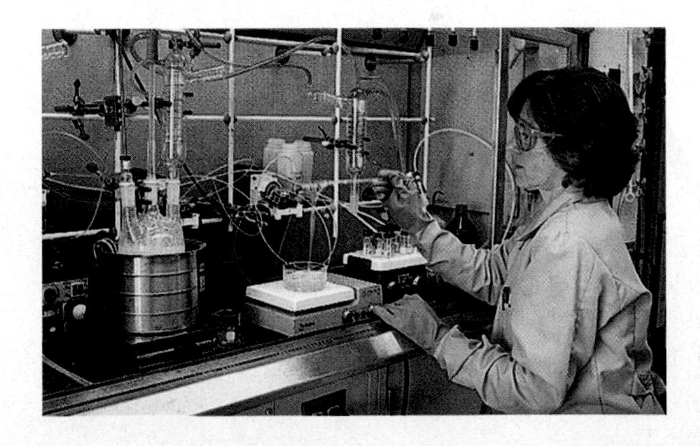

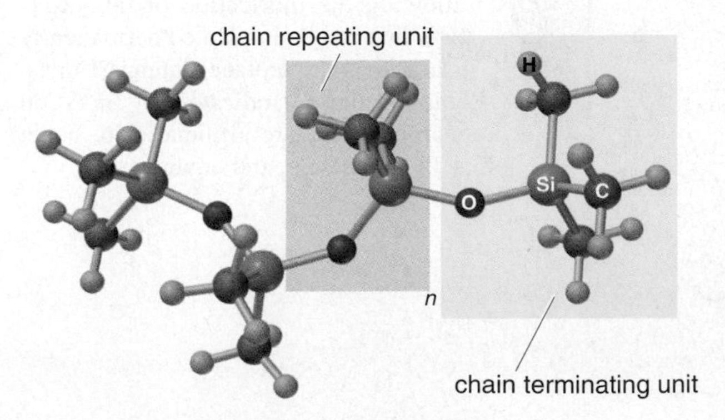

Chain Silicones: Oils and Greases

The simplest structural class of silicones consists of the poly(dimethyl siloxane) chain, in which each $(CH_3)_2Si(OH)_2$ unit uses both OH groups to link to two others. A chain-terminating compound with a third organic group, such as $(CH_3)_3SiOH$, is added to control chain length. These polymers are unreactive, oily liquids with high viscosity and low surface tension. They are used as hydraulic oils and lubricants, as antifoaming agents for frying potato chips, and as components of suntan oil, car polish, digestive aids, and makeup.

(continued)

Sheet Silicones: Elastomers

In the next class of silicones, the third OH group of an added bridging compound, such as $CH_3Si(OH)_3$, reacts to condense chains laterally into gummy sheets that are made into elastomers (rubbers), which are flexible, elastic, and stable from $-100°C$ to $250°C$. These are used in gaskets, rollers, cable insulation, space suits, contact lenses, and dentures.

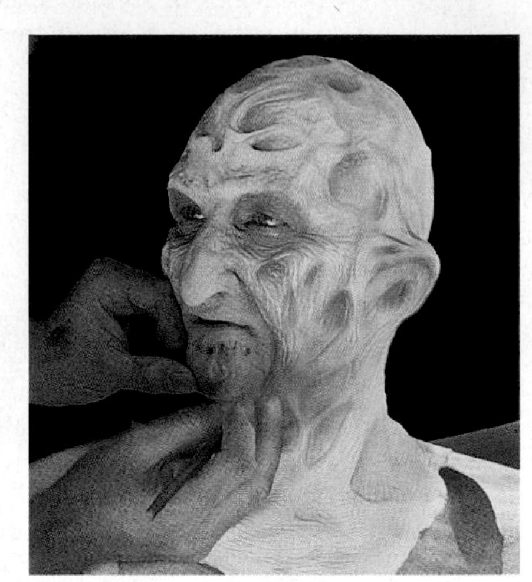

Makeup consisting of chain and sheet silicones being applied to create the character Freddie Krueger

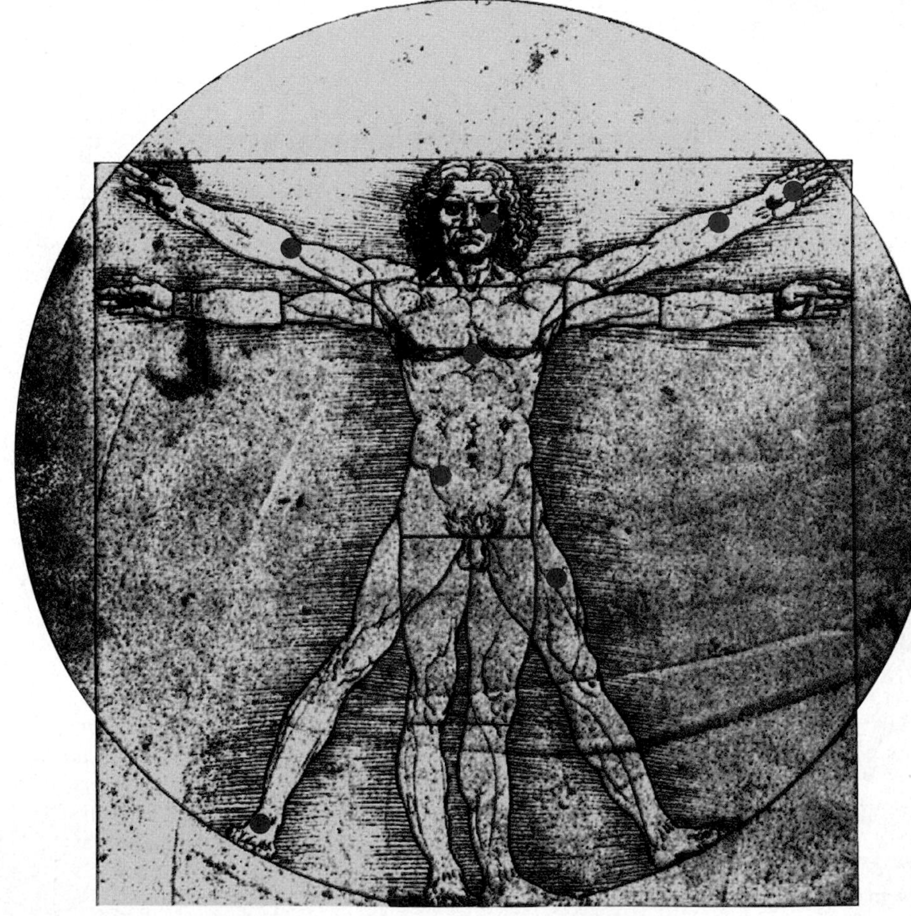

Framework Silicones: Resins

In the final structural class of silicones, reactions that free OH groups, while substituting larger organic groups for some CH_3 groups, interlink sheets to produce strong, thermally stable resins. These are used as insulating laminates on printed circuit boards and as nonstick coatings on cookware. Sheet and framework silicones have revolutionized modern surgical practice by allowing the fabrication of numerous parts that can be implanted permanently in a patient to replace damaged ones. Some of these (indicated by dots on Leonardo's man) are artificial skin, bone, joints, blood vessels, and organ parts.

14.7 GROUP 5A(15): THE NITROGEN FAMILY

The first two elements of Group 5A(15), gaseous nonmetallic nitrogen (N) and solid nonmetallic phosphorus (P), will occupy most of our attention. The industrial and environmental significance of their compounds is matched only by their importance in the structures and functions of biomolecules. Below these nonmetals are two metalloids, arsenic (As) and antimony (Sb), followed by the sole metal, bismuth (Bi), the last nonradioactive element in the periodic table. The Group 5A(15) Family Portrait (pp. 584 and 585) provides an overview.

What Accounts for the Wide Range of Physical Behavior in Group 5A(15)?

Group 5A(15) displays the widest range of physical behavior we've seen so far because of large changes in bonding and intermolecular forces. Nitrogen occurs as a gas consisting of N_2 molecules, which interact through such weak dispersion forces that the element *boils* more than 200°C below room temperature. Elemental phosphorus exists most commonly as tetrahedral P_4 molecules. However, because P is heavier and more polarizable than N, stronger dispersion forces are present, and the element melts about 25°C above room temperature. Arsenic consists of extended, puckered sheets in which each As atom is covalently bonded to three others and forms nonbonding interactions with three nearest neighbors in adjacent sheets. This arrangement gives As the highest melting point in the group. A similar covalent network for Sb gives it a much higher melting point than the next member of the group, Bi, which has metallic bonding.

Phosphorus has several allotropes. The white and red forms have very different properties because of the way the atoms are linked. The highly reactive white form is prepared as a gas and condensed to a whitish, waxy solid under cold water to prevent it from igniting in air. It consists of tetrahedral molecules of four P atoms bonded *to each other,* with no atom in the center (Figure 14.19A). Each P atom uses its half-filled $3p$ orbitals to bond to the other three. Whereas the $3p$ orbitals of an isolated P atom lie 90° apart, the bond angles in P_4 are 60°. With the smaller angle comes poor orbital overlap and a P—P bond strength only about 80% that of a normal P—P bond (Figure 14.19B). Weaker bonds break more easily, contributing to the white allotrope's high reactivity. Heating the white form in the absence of air breaks one of the P—P bonds in each tetrahedron, and those $3p$ orbitals overlap with others to form the chains of P_4 units that make up the red form (Figure 14.19C). The individual molecules of white P make it highly reactive, low melting (44.1°C), and soluble in nonpolar solvents; the chains of red P make it much less reactive, high melting (~600°C), and insoluble.

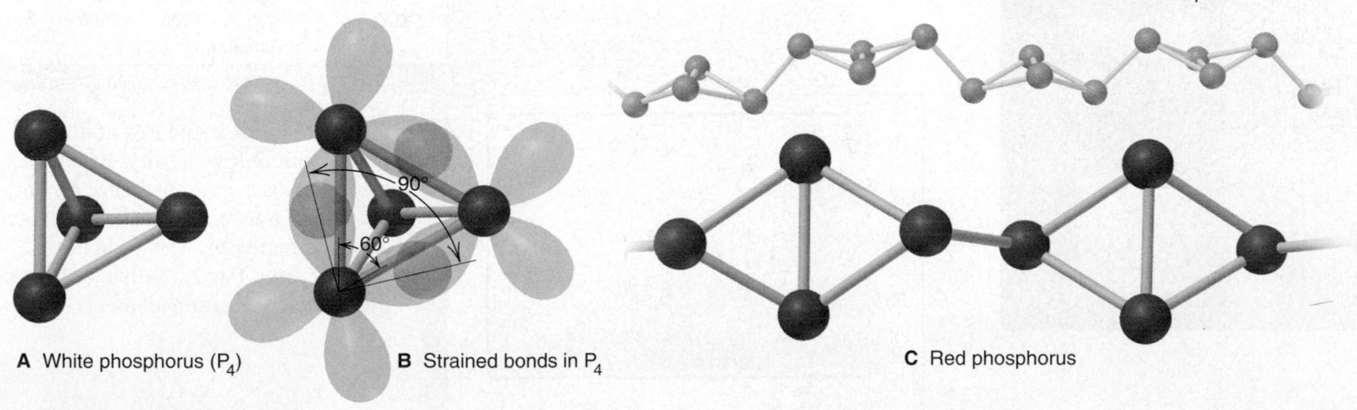

Figure 14.19 Two allotropes of phosphorus. A, White phosphorus exists as individual P_4 molecules, with the P—P bonds forming the edges of a tetrahedron. **B,** The reactivity of P_4 is due in part to the bond strain that arises from the 60° bond angle. Note how overlap of the $3p$ orbitals is decreased because they do not meet directly end to end (overlap is shown here for only three of the P—P bonds), which makes the bonds easier to break. **C,** In red phosphorus, one of the P—P bonds of the white form has broken and links the P_4 units together into long chains. Lone pairs (not shown) reside in *s* orbitals in both allotropes.

Key Atomic and Physical Properties

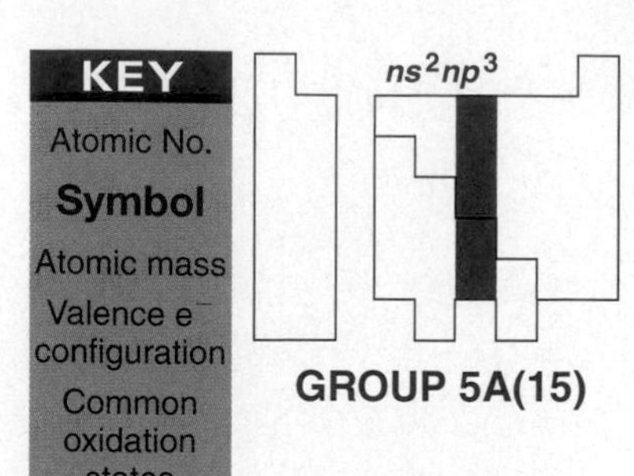

Atomic Properties

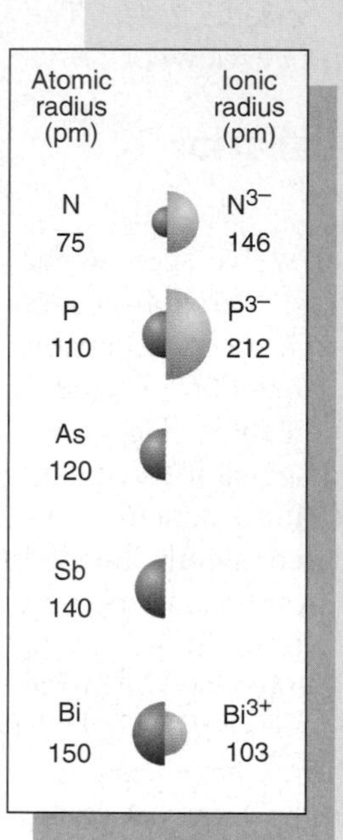

Group electron configuration is ns^2np^3. The np sublevel is half-filled, with each p orbital containing one electron. The number of oxidation states decreases down the group, and the lower (+3) state becomes more common.

Atomic properties follow generally expected trends. The large (~50%) increase in size from N to P correlates with the much lower IE and EN of P.

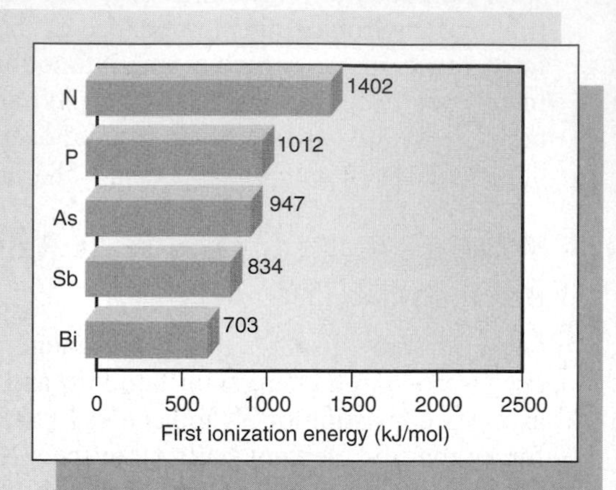

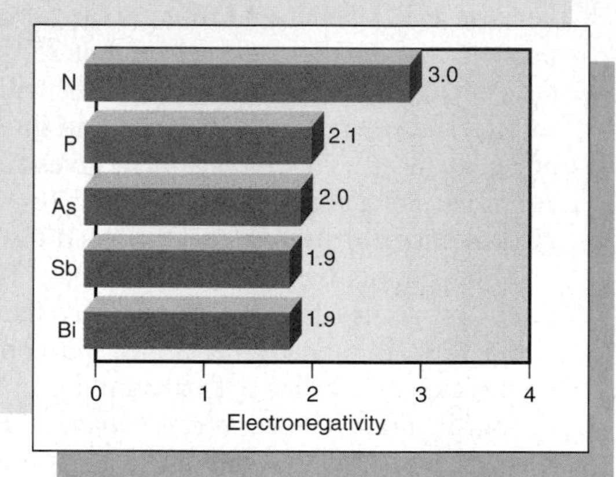

Physical Properties

Physical properties reflect the change from individual molecules (N, P) to network covalent solid (As, Sb) to metal (Bi). Thus, melting points increase and then decrease.

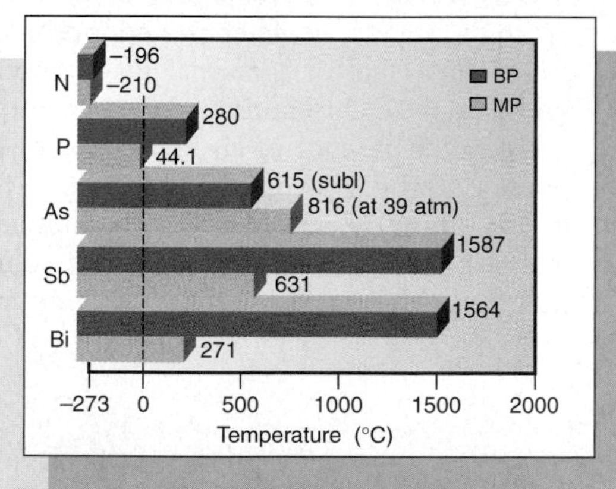

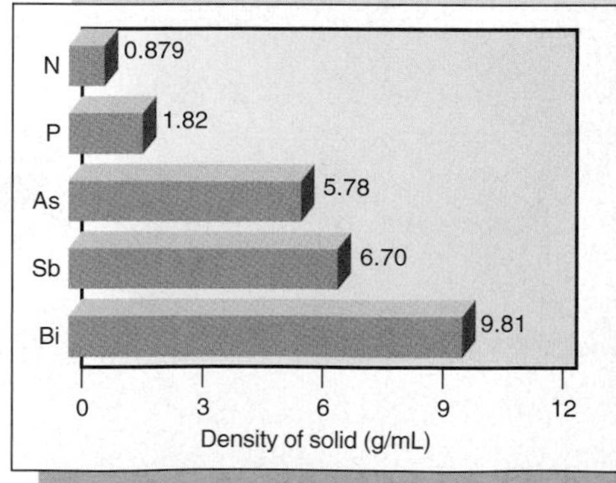

Large atomic size and low atomic mass result in low density. Because mass increases more than size down the group, the density of the elements as solids increases. The dramatic increase from P to As is due to the intervening transition elements.

Some Reactions and Compounds

Important Reactions

General group behavior is shown in reactions 1 to 3, whereas phosphorus chemistry is the theme in reactions 4 and 5.

1. Nitrogen is "fixed" industrially in the Haber process:

$$N_2(g) + 3H_2(g) \rightleftharpoons 2NH_3(g)$$

Further reactions convert NH_3 to NO, NO_2, and HNO_3 (*Highlights of Nitrogen Chemistry*). Hydrides of some other group members are formed from reaction in water (or with H_3O^+) of a metal phosphide, arsenide, and so forth:

$$Ca_3P_2(s) + 6H_2O(l) \longrightarrow 2PH_3(g) + 3Ca(OH)_2(aq)$$

2. Halides are formed by direct combination of the elements:

$$2E(s) + 3X_2 \longrightarrow 2EX_3 \quad (E = \text{all except N})$$

$$EX_3 + X_2 \longrightarrow EX_5 \quad (E = \text{all except N and Bi with X = F and Cl, but no } BiCl_5\text{; E = P for X = Br})$$

3. Oxoacids are formed from the halides in a reaction with water that is common to many nonmetal halides:

$$EX_3 + 3H_2O(l) \longrightarrow H_3EO_3(aq) + 3HX(aq) \quad (E = \text{all except N})$$

$$EX_5 + 4H_2O(l) \longrightarrow H_3EO_4(aq) + 5HX(aq) \quad (E = \text{all except N and Bi})$$

Note that the oxidation number of E does *not* change.

4. Phosphate ions are dehydrated to form polyphosphates:

$$3NaH_2PO_4(s) \longrightarrow Na_3P_3O_9(s) + 3H_2O(g)$$

5. When P_4 reacts in basic solution, its oxidation number both decreases *and* increases (disproportionation):

$$P_4(s) + 3OH^-(aq) + 3H_2O(l) \longrightarrow PH_3(g) + 3H_2PO_2^-(aq)$$

Analogous reactions are typical of many nonmetals, such as S_8 and X_2 (halogens).

Important Compounds

1. Ammonia, NH_3. First substance formed when atmospheric N_2 is used to make N-containing compounds. Annual multimillion-ton production for use in fertilizers, explosives, rayon, and polymers such as nylon, urea-formaldehyde resins, and acrylics.

2. Hydrazine, N_2H_4 (*margin note, p. 586*).

3. Nitric oxide (NO), nitrogen dioxide (NO_2), and nitric acid (HNO_3). Oxides are intermediates in HNO_3 production. This acid is used in fertilizer manufacture, nylon production, metal etching, and the explosives industry (*Highlights of Nitrogen Chemistry*).

Detonation of explosives

4. Amino acids, $H_3N^+—CH(R)—COO^-$ (R = one of 20 different organic groups). Occur in every organism, both free and linked together into proteins. Essential to the growth and function of all cells. Synthetic amino acids are used as dietary supplements.

5. Phosphorus trichloride, PCl_3. Used to form many organic phosphorus compounds, including oil and fuel additives, plasticizers, flame retardants, and insecticides. Also used to make PCl_5, $POCl_3$, and other important P-containing compounds.

6. Tetraphosphorus decaoxide (P_4O_{10}) and phosphoric acid (H_3PO_4) (*Highlights of Phosphorus Chemistry*).

7. Sodium tripolyphosphate, $Na_5P_3O_{10}$. As a water-softening agent (Calgon), it combines with hard-water Mg^{2+} and Ca^{2+} ions, preventing them from reacting with soap anions, and thus improving cleaning action. Its use has been curtailed in the United States because it pollutes lakes and streams by causing excessive algal growth (*see photo*).

Algal growth in polluted lake

8. Adenosine triphosphate (ATP) and other biophosphates. ATP acts to transfer chemical energy in the cell; it is necessary for all biological processes requiring energy. Phosphate groups also occur in sugars, fats, proteins, and nucleic acids.

9. Bismuth subsalicylate, $BiO(C_7H_5O_3)$. The active ingredient in Pepto-Bismol (*see photo*), a widely used remedy for diarrhea and nausea. (The pink color is not due to this white compound.)

What Patterns Appear in the Chemical Behavior of Group 5A(15)?

The same general pattern of chemical behavior that we discussed for Group 4A(14) appears again in this group, reflected in the change from nonmetallic N to metallic Bi. The overwhelming majority of Group 5A(15) compounds have *covalent bonds.* Whereas N can form no more than four bonds, the next three members can expand their valence shells by using empty *d* orbitals.

For a 5A element to form an ion with a noble gas electron configuration, it must *gain* three electrons, the last two in endothermic steps. Nevertheless, the enormous lattice energy released when such highly charged anions attract cations drives their formation. However, the 3− anion of N occurs only in compounds with active metals, such as Li_3N and Mg_3N_2 (and that of P may occur in Na_3P). Metallic Bi forms mostly covalent compounds but exists as a cation in a few compounds, such as BiF_3 and $Bi(NO_3)_3 \cdot 5H_2O$, through *loss* of its three valence *p* electrons.

Again, as we move down this group, we see the patterns first seen in Groups 3A and 4A. Fewer oxidation states occur, with the lower state becoming more prominent: N exhibits every state possible for a 5A element, from +5 to −3; only the +5 and +3 states are common for P, As, and Sb; and +3 is the only common state of Bi. The oxides change from acidic to amphoteric to basic, reflecting the increase in the metallic character of the elements. In addition, the lower oxide of an element is more basic than the higher oxide, reflecting the greater ionic character of the E-to-O bonding in the lower oxide.

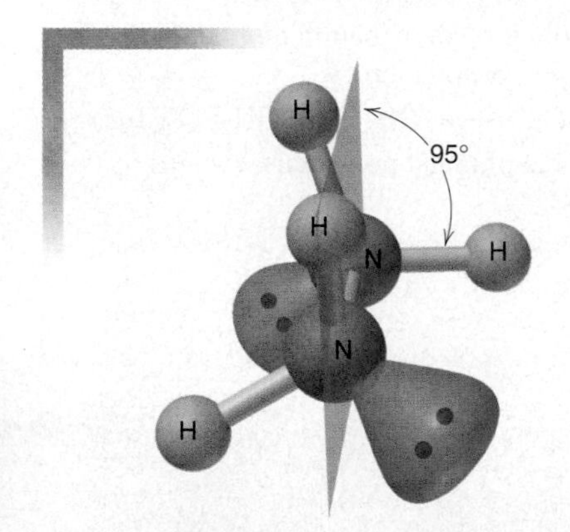

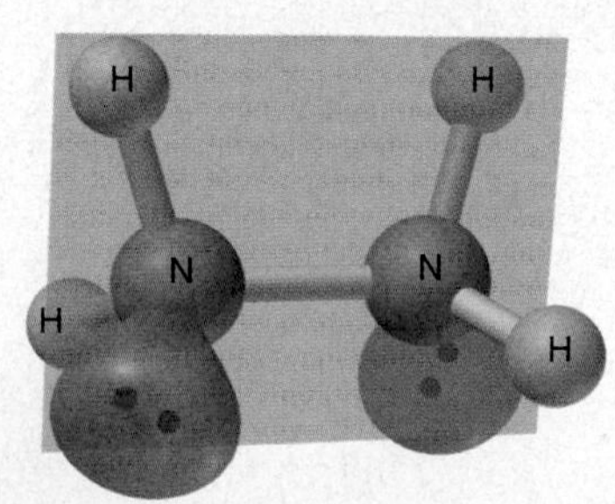

Hydrazine, Nitrogen's Other Hydride Aside from NH_3, the most important Group 5A(15) hydride is hydrazine, N_2H_4. Like NH_3, it is a weak base, forming $N_2H_5^+$ and $N_2H_6^{2+}$ ions with acids. Hydrazine is also used to make antituberculin drugs, plant growth regulators, and fungicides, and organic hydrazine derivatives are used in rocket and missile fuels. We might expect the lone pairs in the molecule to lie opposite each other and give a symmetrical structure with no dipole moment. However, repulsions between each lone pair and the bonding pairs on the other atom force the lone pairs to lie below the N—N bond (see model), resulting in a large dipole moment ($\mu = 1.85$ D in the gas phase). A similar arrangement occurs in hydrogen peroxide (H_2O_2), the analogous hydride of oxygen.

All the Group 5A(15) elements form gaseous hydrides of formula EH_3. Except for NH_3, these are extremely reactive and poisonous and are synthesized by reaction of a metal phosphide, arsenide, and so forth, which acts as a strong base in water or aqueous acid. For example,

$$Ca_3As_2(s) + 6H_2O(l) \longrightarrow 2AsH_3(g) + 3Ca(OH)_2(aq)$$

Ammonia is made industrially by direct combination of the elements at high pressure and moderately high temperature:

$$N_2(g) + 3H_2(g) \rightleftharpoons 2NH_3(g)$$

Molecular properties of the Group 5A(15) hydrides reveal some interesting bonding and structural patterns:

- Despite its much lower molar mass, NH_3 melts and boils at higher temperatures than the other 5A hydrides as a result of *H bonding.*
- Bond angles decrease from 107.3° for NH_3 to around 90° for the other hydrides, which suggests that the larger atoms use unhybridized *p* orbitals.
- E—H bond lengths increase down the group, so bond strength and thermal stability decrease: AsH_3 decomposes at 250°C, SbH_3 at 20°C, and BiH_3 at −45°C.

We'll see these features—H bonding for the smallest member, change in bond angles, change in bond energies—in the hydrides of Group 6A(16) as well.

The Group 5A(15) elements all form trihalides (EX_3). All except nitrogen form pentafluorides (EF_5), but only a few other pentahalides (PCl_5, PBr_5, $AsCl_5$, and $SbCl_5$) are known. Nitrogen cannot form pentahalides because it cannot expand its valence shell. Most trihalides are prepared by direct combination:

$$P_4(s) + 6Cl_2(g) \longrightarrow 4PCl_3(l)$$

The pentahalides form with excess halogen:

$$PCl_3(l) + Cl_2(g) \longrightarrow PCl_5(s)$$

As with the hydrides, the thermal stability of the halides decreases as the E—X bond becomes longer. Among the nitrogen halides, for example, NF_3 is a stable, rather unreactive gas. NCl_3 is explosive and reacts rapidly with water. (The chemist who first prepared it lost three fingers and an eye!) NBr_3 can only be made below −87°C. NI_3 has never been prepared, but an ammoniated product ($NI_3 \cdot NH_3$) explodes at the slightest touch.

In an aqueous reaction pattern *typical of many nonmetal halides,* each 5A halide reacts with water to yield the hydrogen halide and the oxoacid, in which E has the *same* oxidation number as it had in the original halide. For example, PX_5 (O.N. of P = +5) produces phosphoric acid (O.N. of P = +5) and HX:

$$PCl_5(s) + 4H_2O(l) \longrightarrow H_3PO_4(l) + 5HCl(g)$$

Highlights of Nitrogen Chemistry

Surely the most striking highlight of nitrogen chemistry is the inertness of N_2 itself. Nearly four-fifths of the atmosphere consists of N_2, and the other fifth is nearly all O_2, a very strong oxidizing agent. Nevertheless, the searing temperature of a lightning bolt is required for significant amounts of atmospheric nitrogen oxides to form. Thus, even though N_2 is inert at moderate temperatures, it reacts at high temperatures with H_2, Li, Group 2A(2) members, B, Al, C, Si, Ge, O_2, and many transition elements. In fact, nearly every element in the periodic table forms bonds to N. Here we focus on the oxides and the oxoacids and their salts.

Nitrogen Oxides Nitrogen is remarkable for having six stable oxides, each with a *positive* heat of formation because of the great strength of the N≡N bond (BE = 945 kJ/mol). Their structures and some properties are shown in Table 14.3. Unlike the hydrides and halides of nitrogen, the oxides are planar. Nitrogen displays all its positive oxidation states in these compounds, and in N_2O and N_2O_3, the two N atoms have different states.

Table 14.3 Structures and Properties of the Nitrogen Oxides

Formula	Name	Space-filling Model	Lewis Structure	Oxidation State of N	ΔH_f^0 (kJ/mol) at 298 K	Comment
N_2O	Dinitrogen monoxide (dinitrogen oxide; nitrous oxide)		:N≡N—Ö:	+1 (0, +2)	82.0	Colorless gas; used as dental anesthetic ("laughing gas") and aerosol propellant
NO	Nitrogen monoxide (nitrogen oxide; nitric oxide)		:Ṅ=Ö:	+2	90.3	Colorless, paramagnetic gas; biochemical messenger; air pollutant
N_2O_3	Dinitrogen trioxide		O=N—N(=O)—O	+3 (+2, +4)	83.7	Reddish brown gas (reversibly dissociates to NO and NO_2)
NO_2	Nitrogen dioxide		O—Ṅ=O	+4	33.2	Orange-brown, paramagnetic gas formed during HNO_3 manufacture; poisonous air pollutant
N_2O_4	Dinitrogen tetraoxide		O_2N—NO_2	+4	9.16	Colorless to yellow liquid (reversibly dissociates to NO_2)
N_2O_5	Dinitrogen pentaoxide		O_2N—Ö—NO_2	+5	11.3	Colorless, volatile solid consisting of NO_2^+ and NO_3^-; gas consists of N_2O_5 molecules

Dinitrogen monoxide (N_2O; also called dinitrogen oxide or nitrous oxide) is the dental anesthetic "laughing gas" and the propellant in canned whipped cream. It is a linear molecule with an electronic structure best described by three resonance forms (note formal charges; see Problem 10.87):

$$\overset{0}{:\text{N}}\equiv\overset{+1}{\text{N}}-\overset{-1}{\ddot{\underset{\cdot\cdot}{\text{O}}}}: \longleftrightarrow \overset{-1}{:\underset{\cdot\cdot}{\text{N}}}=\overset{+1}{\text{N}}=\overset{0}{\underset{\cdot\cdot}{\text{O}}}: \longleftrightarrow \overset{-2}{:\ddot{\underset{\cdot\cdot}{\text{N}}}}-\overset{+1}{\text{N}}\equiv\overset{+1}{\text{O}}:$$

most important ... least important

> **Nitric Oxide: A Biochemical Surprise**
> The tiny inorganic free radical NO is a remarkable player in an enormous range of biochemical systems. NO is biosynthesized in animal species from barnacles to humans. In the late 1980s and early 1990s, it was shown to play various roles in neurotransmission, blood clotting, blood pressure, and the destruction of cancer cells. The list of effects is now much longer and includes the harmful and beneficial ones of the NO-derived species NO_2, N_2O_3, $ONOO^-$, and HNO. Living systems are full of surprises!

Nitrogen monoxide (NO; also called nitrogen oxide or nitric oxide) is an odd-electron molecule with a vital biochemical function. In Section 11.3, we used MO theory to explain its bonding. The commercial preparation of NO through the oxidation of ammonia occurs as a first step in the production of nitric acid:

$$4NH_3(g) + 5O_2(g) \longrightarrow 4NO(g) + 6H_2O(g)$$

Nitrogen monoxide is also produced whenever air is heated to high temperatures, as in a car engine or a lightning storm:

$$N_2(g) + O_2(g) \xrightarrow{\text{high } T} 2NO(g)$$

Heating converts NO to two other oxides:

$$3NO(g) \xrightarrow{\Delta} N_2O(g) + NO_2(g)$$

This type of redox reaction is called a **disproportionation.** It occurs when a substance *acts as both an oxidizing and a reducing agent in a reaction.* In the process, an atom with an intermediate oxidation state in the reactant occurs in the products in both lower and higher states: the oxidation state of N in NO (+2) is intermediate between that in N_2O (+1) and that in NO_2 (+4).

Nitrogen dioxide (NO_2), a brown poisonous gas, forms to a small extent when NO reacts with additional oxygen:

$$2NO(g) + O_2(g) \rightleftharpoons 2NO_2(g)$$

Like NO, NO_2 is also an odd-electron molecule, but the electron is more localized on the N atom. Thus, NO_2 dimerizes reversibly to dinitrogen tetraoxide:

$$O_2N\cdot(g) + \cdot NO_2(g) \rightleftharpoons O_2N-NO_2(g) \quad (\text{or } N_2O_4)$$

Thunderstorms form NO and NO_2 and carry them down to the soil, where they act as natural fertilizers. In urban traffic, however, their formation leads to *photochemical smog* (Figure 14.20). During the morning commute, NO forms in car and truck (especially diesel) engines. It is then oxidized to NO_2 by free radicals ($RO_2\cdot$), which form by oxidation of unburned hydrocarbons in gasoline fumes. As the Sun climbs later in the day, radiant energy breaks down some NO_2:

$$NO_2(g) \xrightarrow{\text{sunlight}} NO(g) + O(g)$$

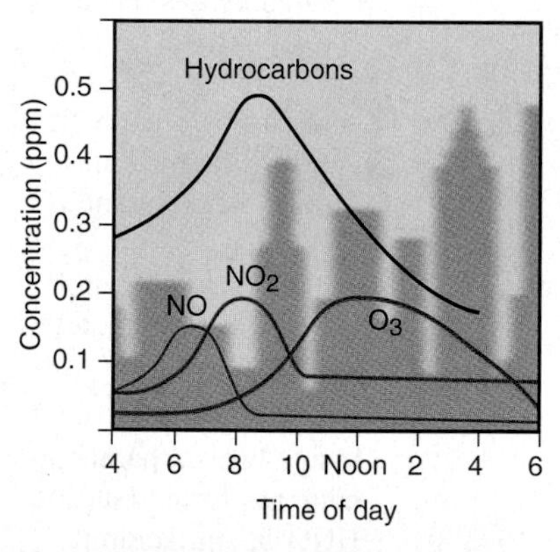

Figure 14.20 The formation of photochemical smog. Localized atmospheric concentrations of the precursor components of smog change during the day. In early morning traffic, exhaust from motor vehicles raises NO and hydrocarbon levels, followed by increases in NO_2 as NO is oxidized by free radicals. Ozone peaks later, as NO_2 breaks down in stronger sunlight and releases O atoms that react with O_2. By midafternoon, all the components for the production of peroxyacylnitrates (PANs) are present, and photochemical smog results.

The O atoms collide with O_2 molecules and form ozone, O_3, a powerful oxidizing agent that damages synthetic rubber, plastics, and plant and animal tissue:

$$O_2(g) + O(g) \longrightarrow O_3(g)$$

A complex series of reactions involving NO, NO_2, O_3, unburned gasoline, and various free-radical species forms ***peroxyacylnitrates*** (PANs), a group of potent nose and eye irritants. The outcome of this fascinating atmospheric chemistry is choking, brown smog.

Nitrogen Oxoacids and Oxoanions The two common nitrogen oxoacids are nitric acid and nitrous acid (Figure 14.21). The first two steps in the *Ostwald process* for the production of nitric acid—the oxidations of NH_3 to NO and of NO to NO_2—have been shown. The final step is a disproportionation, as the oxidation numbers show:

$$3\overset{+4}{N}O_2(g) + H_2O(l) \longrightarrow 2H\overset{+5}{N}O_3(aq) + \overset{+2}{N}O(g)$$

The NO is recycled to make more NO_2.

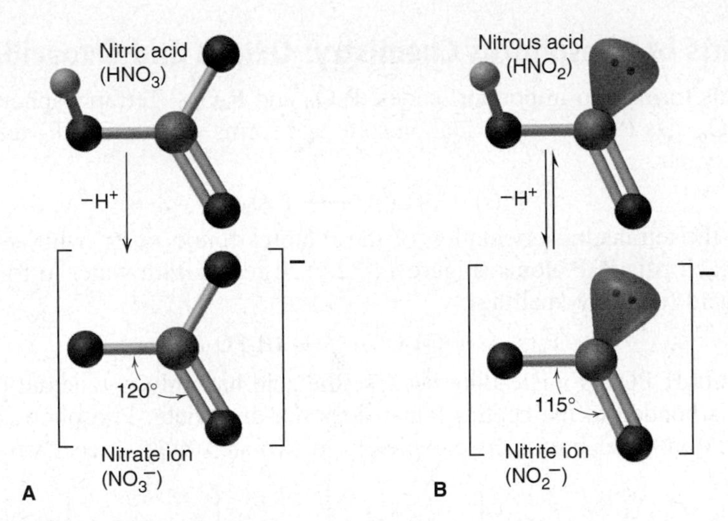

Figure 14.21 The structures of nitric and nitrous acids and their oxoanions. A, Nitric acid loses a proton (H^+) to form the trigonal planar nitrate ion (one of three resonance forms is shown). **B,** Nitrous acid, a much weaker acid, forms the planar nitrite ion (one of two resonance forms is shown). Note the effect of nitrogen's lone pair (the lone pairs of the oxygens are not shown) in reducing the ideal 120° bond angle to 115°.

In nitric acid, as in all oxoacids, *the acidic H is attached to one of the O atoms*. In the laboratory, nitric acid is used as a strong oxidizing acid. The products of its reactions with metals vary with the metal's reactivity and the acid's concentration. In the following examples, notice from the net ionic equations that *the NO_3^- ion is the oxidizing agent*. Nitrate ion that is not reduced is a spectator ion and does not appear in the net ionic equations.

- With an active metal, such as Al, and dilute acid, N is reduced from the +5 state all the way to the −3 state in the ammonium ion, NH_4^+:

$$8Al(s) + 30HNO_3(aq;\ 1\ M) \longrightarrow 8Al(NO_3)_3(aq) + 3NH_4NO_3(aq) + 9H_2O(l)$$

$$8Al(s) + 30H^+(aq) + 3NO_3^-(aq) \longrightarrow 8Al^{3+}(aq) + 3NH_4^+(aq) + 9H_2O(l)$$

- With a less reactive metal, such as Cu, and more concentrated acid, N is reduced to the +2 state in NO:

$$3Cu(s) + 8HNO_3(aq;\ 3\ \text{to}\ 6\ M) \longrightarrow 3Cu(NO_3)_2(aq) + 4H_2O(l) + 2NO(g)$$

$$3Cu(s) + 8H^+(aq) + 2NO_3^-(aq) \longrightarrow 3Cu^{2+}(aq) + 4H_2O(l) + 2NO(g)$$

- With still more concentrated acid, N is reduced only to the +4 state in NO_2:

$$Cu(s) + 4HNO_3(aq;\ 12\ M) \longrightarrow Cu(NO_3)_2(aq) + 2H_2O(l) + 2NO_2(g)$$

$$Cu(s) + 4H^+(aq) + 2NO_3^-(aq) \longrightarrow Cu^{2+}(aq) + 2H_2O(l) + 2NO_2(g)$$

Nitrates form when HNO_3 reacts with metals or with their hydroxides, oxides, or carbonates. *All nitrates are soluble in water.*

Nitrous acid, HNO_2, a much weaker acid than HNO_3, forms when metal nitrites are treated with a strong acid:

$$NaNO_2(aq) + HCl(aq) \longrightarrow HNO_2(aq) + NaCl(aq)$$

These two acids reveal a *general pattern in relative acid strength among oxoacids:* the more O atoms bonded to the central nonmetal, the stronger the acid. Thus, HNO_3 is stronger than HNO_2. The O atoms pull electron density from the N atom, which in turn pulls electron density from the O of the O—H bond, facilitating the release of the H^+ ion. The O atoms also act to stabilize the resulting oxoanion by delocalizing its negative charge. The same pattern occurs in the oxoacids of sulfur and the halogens; we'll discuss the pattern quantitatively in Chapter 18.

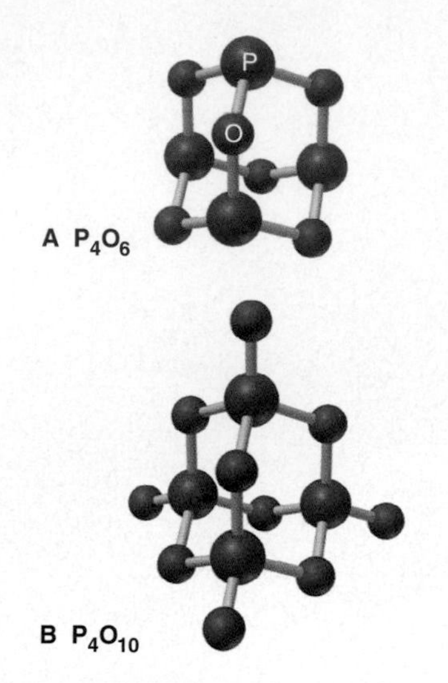

Figure 14.22 Important oxides of phosphorus. A, P_4O_6. **B,** P_4O_{10}.

Highlights of Phosphorus Chemistry: Oxides and Oxoacids

Phosphorus forms two important oxides, P_4O_6 and P_4O_{10}. Tetraphosphorus hexaoxide, P_4O_6, has P in its +3 oxidation state and forms when white P_4 reacts with limited oxygen:

$$P_4(s) + 3O_2(g) \longrightarrow P_4O_6(s)$$

P_4O_6 has the tetrahedral orientation of the P atoms found in P_4, with an O atom between each pair of P atoms (Figure 14.22A). It reacts with water to form phosphor*ous* acid (note the spelling):

$$P_4O_6(s) + 6H_2O(l) \longrightarrow 4H_3PO_3(l)$$

The formula H_3PO_3 is misleading because the acid has only two acidic H atoms; the third is bonded to the central P and does not dissociate. Phosphorous acid is a weak acid in water but reacts completely in two steps with excess strong base:

$$\text{HO–P(=O)(OH)–H} \xrightarrow[-H^+]{OH^-} [\text{HO–P(=O)(O)–H}]^- \xrightarrow[-H^+]{OH^-} [\text{O–P(=O)(O)–H}]^{2-}$$

Salts of phosphorous acid contain the phosphite ion, $HPO_3^{\,2-}$.

In tetraphosphorus decaoxide, P_4O_{10}, P is in the +5 oxidation state. Commonly known as "phosphorus pentoxide" from the empirical formula (P_2O_5), it forms when P_4 burns in excess O_2:

$$P_4(s) + 5O_2(g) \longrightarrow P_4O_{10}(s)$$

Its structure can be viewed as that of P_4O_6 with another O atom bonded to each of the four corner P atoms (Figure 14.22B). P_4O_{10} is a powerful drying agent and, in a vigorous exothermic reaction with water, forms phosphoric acid (H_3PO_4), one of the "top-10" most important compounds in chemical manufacturing:

$$P_4O_{10}(s) + 6H_2O(l) \longrightarrow 4H_3PO_4(l)$$

The presence of many H bonds makes pure H_3PO_4 syrupy, more than 75 times as viscous as water. The laboratory-grade concentrated acid is an 85% by mass aqueous solution. H_3PO_4 is a weak triprotic acid; in water, it loses one proton in the following equilibrium reaction:

$$H_3PO_4(l) + H_2O(l) \rightleftharpoons H_2PO_4^-(aq) + H_3O^+(aq)$$

In excess strong base, however, the three protons dissociate completely in three steps to give the three phosphate oxoanions:

$$\text{HO–P(=O)(OH)–OH} \xrightarrow[-H^+]{OH^-} [\text{HO–P(=O)(OH)–O}]^- \xrightarrow[-H^+]{OH^-} [\text{HO–P(=O)(O)–O}]^{2-} \xrightarrow[-H^+]{OH^-} [\text{O–P(=O)(O)–O}]^{3-}$$

dihydrogen phosphate ion; hydrogen phosphate ion; phosphate ion

Phosphoric acid has a central role in fertilizer production, and it is also used as a polishing agent for aluminum car trim and as an additive in soft drinks to give a touch of tartness. The various phosphate salts have numerous essential applications.

Polyphosphates are formed by heating hydrogen phosphates, which lose water as they form P—O—P linkages. This type of reaction, in which an H_2O molecule is lost for every pair of OH groups that join, is called a **dehydration-condensation;** it occurs frequently in the formation of polyoxoanion chains and other polymeric structures, both synthetic and natural. For example, sodium diphosphate, $Na_4P_2O_7$, is prepared by heating sodium hydrogen phosphate:

$$2Na_2HPO_4(s) \xrightarrow{\Delta} Na_4P_2O_7(s) + H_2O(g)$$

The Countless Uses of Phosphates

Phosphates have an amazing array of applications in home and industry. Na_3PO_4 is a paint stripper and grease remover and is still used in cleaning powders in other countries. Na_2HPO_4 is a laxative ingredient and is used to adjust the acidity of boiler water. The potassium salt K_3PO_4 is used to stabilize latex for synthetic rubber, and K_2HPO_4 is a radiator corrosion inhibitor. Ammonium phosphates are used as fertilizers and as flame retardants on curtains and paper costumes. Calcium phosphates are used in baking powders and toothpastes, as mineral supplements in livestock feed, and (on the hundred-million-ton scale) as fertilizers throughout the world.

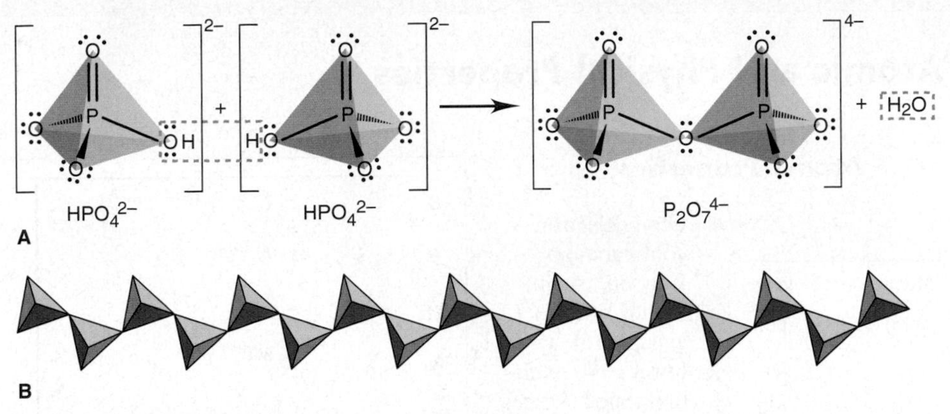

Figure 14.23 The diphosphate ion and polyphosphates. A, When two hydrogen phosphate ions undergo a dehydration-condensation reaction, they lose a water molecule and join through a shared O atom to form a diphosphate ion. **B,** Polyphosphates are chains of many such tetrahedral PO_4 units. Chain shapes depend on the orientation of the individual units. Note the similarity to silicate chains (see the Gallery, p. 580).

The diphosphate ion, $P_2O_7^{4-}$, the smallest of the polyphosphates, consists of two PO_4 units linked through a common oxygen corner (Figure 14.23A). Its reaction with water, the reverse of the previous reaction, generates heat:

$$P_2O_7^{4-}(aq) + H_2O(l) \longrightarrow 2HPO_4^{-}(aq) + heat$$

A similar process is put to vital use by organisms, when a third PO_4 unit linked to diphosphate creates the triphosphate grouping, part of the all-important high-energy biomolecule adenosine triphosphate (ATP). In Chapters 20 and 21, we discuss the central role of ATP in biological energy production. Extended polyphosphate chains consist of many tetrahedral PO_4 units (Figure 14.23B) and are structurally similar to silicate chains. As a final look at the chemical versatility of phosphorus, consider some of its numerous important compounds with sulfur and with nitrogen.

Match Heads, Bug Sprays, and O-Rings
Phosphorus forms many sulfides and nitrides. P_4S_3 is used in "strike-anywhere" match heads, and P_4S_{10} is used in the manufacture of organophosphorus pesticides, such as malathion. Polyphosphazenes have properties similar to those of silicones. Indeed, the $-(R_2)P{=}N-$ unit is isoelectronic with the silicone unit, $-(R_2)Si-O-$. Sheets, films, fibers, and foams of polyphosphazene are water repellent, flame resistant, solvent resistant, and flexible at low temperatures—perfect for the gaskets and O-rings in spacecraft and polar vehicles.

14.8 GROUP 6A(16): THE OXYGEN FAMILY

The first two members of this family—gaseous nonmetallic oxygen (O) and solid nonmetallic sulfur (S)—are among the most important elements in industry, the environment, and living things. Two metalloids, selenium (Se) and tellurium (Te), appear below them, then comes the lone metal, radioactive polonium (Po), and newly synthesized element 116 ends the group. The Group 6A(16) Family Portrait (pp. 592 and 593) displays the features of these elements.

How Do the Oxygen and Nitrogen Families Compare Physically?

Group 6A(16) resembles Group 5A(15) in many respects, so let's look at some common themes. The pattern of physical properties we saw in Group 5A appears again in this group. Like nitrogen, oxygen occurs as a low-boiling diatomic gas. Like phosphorus, sulfur occurs as a polyatomic molecular solid. Like arsenic, selenium commonly occurs as a gray metalloid. Like antimony, tellurium is slightly more metallic than the preceding group member but still displays network covalent bonding. Finally, like bismuth, polonium has a metallic crystal structure. As we expect by comparison with the Group 5A elements, electrical conductivities increase steadily down Group 6A as bonding changes from individual molecules (insulators) to metalloid networks (semiconductors) to a metallic solid (conductor).

Key Atomic and Physical Properties

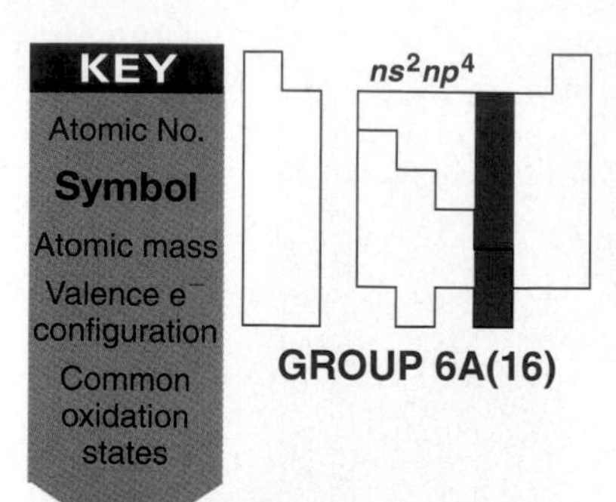

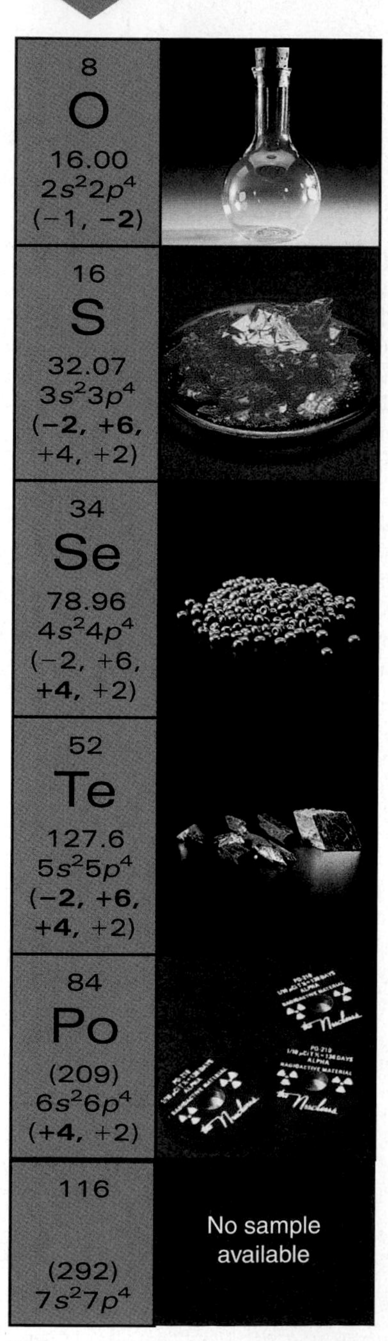

Atomic Properties

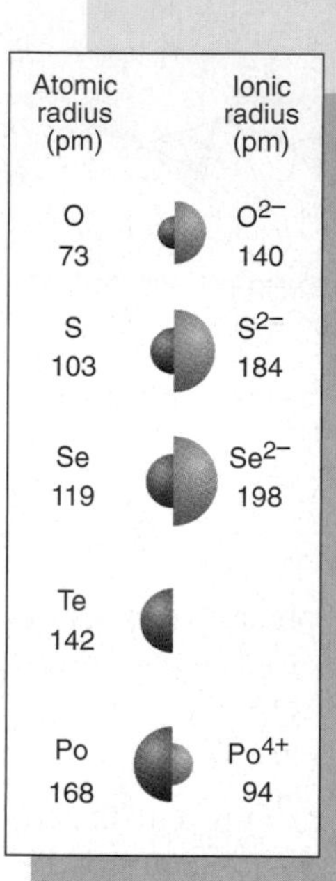

Group electron configuration is ns^2np^4. As in Groups 3A(13) and 5A(15), a lower (+4) oxidation state becomes more common down the group.

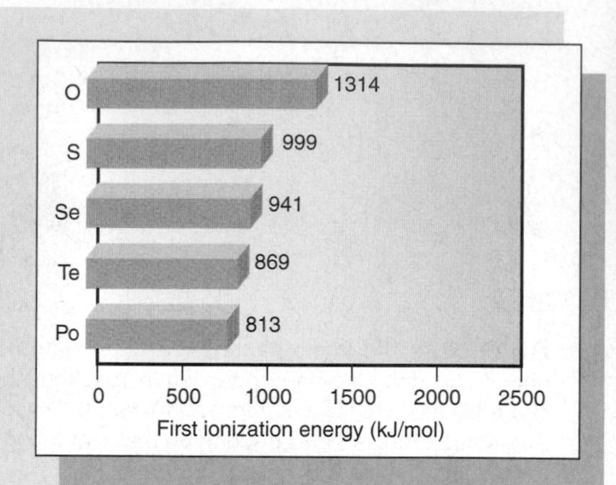

Down the group, atomic and ionic size increase, and IE and EN decrease.

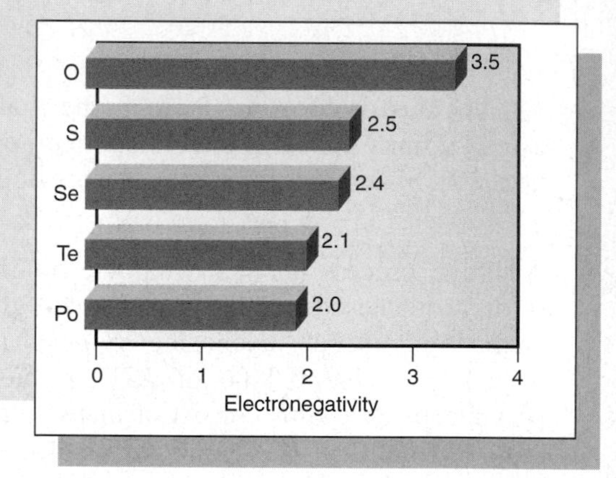

Physical Properties

Melting points increase through Te, which has covalent bonding, and then decrease for Po, which has metallic bonding.

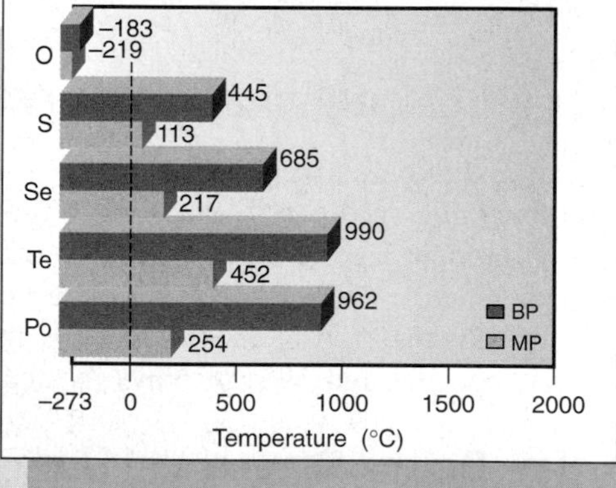

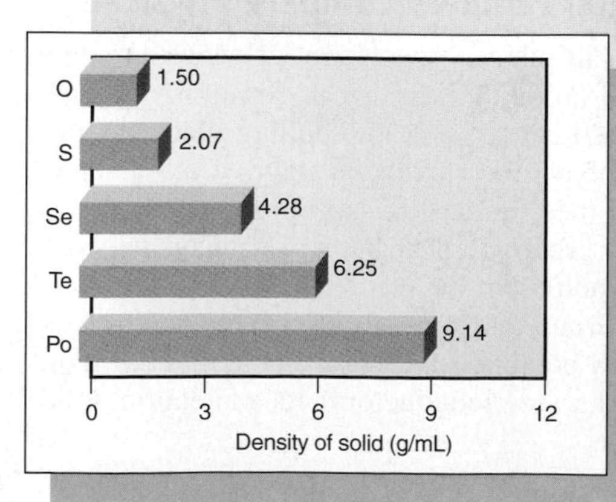

Densities of the elements as solids increase steadily.

Some Reactions and Compounds

Important Reactions

Halogenation and oxidation of the elements (E) appear in reactions 1 and 2, and sulfur chemistry in reactions 3 and 4.

1. Halides are formed by direct combination:

$$E(s) + X_2(g) \longrightarrow \text{various halides} \quad (E = S, Se, Te; X = F, Cl)$$

2. The other elements in the group are oxidized by O_2:

$$E(s) + O_2(g) \longrightarrow EO_2 \quad (E = S, Se, Te, Po)$$

SO_2 is oxidized further, and the product is used in the final step of H_2SO_4 manufacture (*Highlights of Sulfur Chemistry*):

$$2SO_2(g) + O_2(g) \longrightarrow 2SO_3(g)$$

3. Sulfur is recovered when hydrogen sulfide is oxidized:

$$8H_2S(g) + 4O_2(g) \longrightarrow S_8(s) + 8H_2O(g)$$

This reaction is used to obtain sulfur when natural deposits are not available.

4. The thiosulfate ion is formed when an alkali metal sulfite reacts with sulfur, as in the preparation of "hypo," photographer's developing solution:

$$S_8(s) + 8Na_2SO_3(aq) \longrightarrow 8Na_2S_2O_3(aq)$$

Important Compounds

1. Water, H_2O. The single most important compound on Earth (*Section 12.5*).

2. Hydrogen peroxide, H_2O_2. Used as an oxidizing agent, disinfectant, and bleach, and in the production of peroxy compounds for polymerization (*margin note, p. 595*).

3. Hydrogen sulfide, H_2S. Vile-smelling toxic gas formed during anaerobic decomposition of plant and animal matter, in volcanoes, and in deep-sea thermal vents. Used as a source of sulfur and in the manufacture of paper. Atmospheric traces cause silver to tarnish through formation of black Ag_2S (*see photo*).

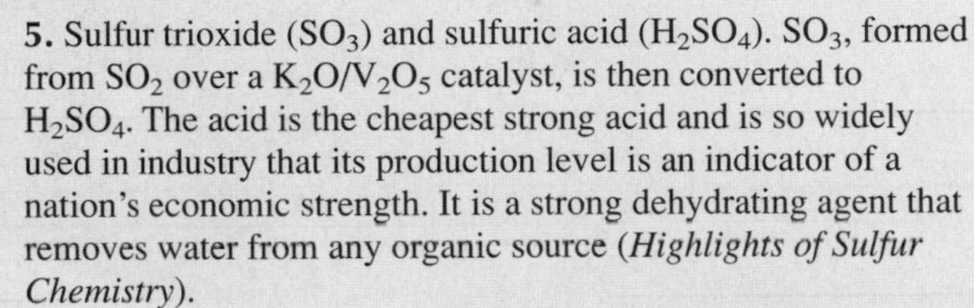

4. Sulfur dioxide, SO_2. Colorless, choking gas formed in volcanoes (*see photo*) or whenever an S-containing material (coal, oil, metal sulfide ores, and so on) is burned. More than 90% of SO_2 produced is used to make sulfuric acid. Also used as a fumigant and a preservative of fruit, syrups, and wine. As a reducing agent, removes excess Cl_2 from industrial wastewater, removes O_2 from petroleum handling tanks, and prepares ClO_2 for bleaching paper. Atmospheric pollutant in acid rain.

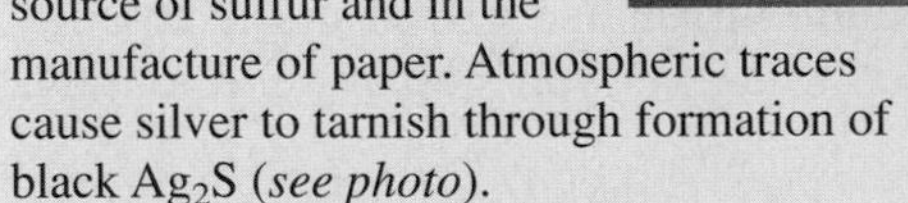

5. Sulfur trioxide (SO_3) and sulfuric acid (H_2SO_4). SO_3, formed from SO_2 over a K_2O/V_2O_5 catalyst, is then converted to H_2SO_4. The acid is the cheapest strong acid and is so widely used in industry that its production level is an indicator of a nation's economic strength. It is a strong dehydrating agent that removes water from any organic source (*Highlights of Sulfur Chemistry*).

6. Sulfur hexafluoride, SF_6. Extremely inert gas used as an electrical insulator.

Allotropism is more common in Group 6A(16) than in Group 5A(15). Oxygen has two allotropes: life-giving dioxygen (O_2), and poisonous triatomic ozone (O_3). Oxygen gas is colorless, odorless, paramagnetic, and thermally stable. In contrast, ozone gas is bluish, has a pungent odor, is diamagnetic, and decomposes in heat and especially in ultraviolet (UV) light:

$$2O_3(g) \xrightarrow{\text{UV}} 3O_2(g)$$

This ability to absorb high-energy photons makes stratospheric ozone vital to life. A thinning of the ozone layer, observed above the North Pole and especially the South Pole, means that more UV light will reach Earth's surface, with potentially hazardous effects. (We'll discuss the chemical causes of ozone depletion in Chapter 16.)

Sulfur is the allotrope "champion" of the periodic table, with more than 10 forms. The S atom's ability to bond to other S atoms creates numerous rings and chains, with S—S bond lengths that range from 180 pm to 260 pm and bond angles from 90° to 180°. At room temperature, the sulfur molecule is a crown-shaped ring of eight atoms, called *cyclo-*S_8 (Figure 14.24). The most stable allotrope is orthorhombic α-S_8, which consists entirely of these molecules; all other S allotropes eventually revert to this one.

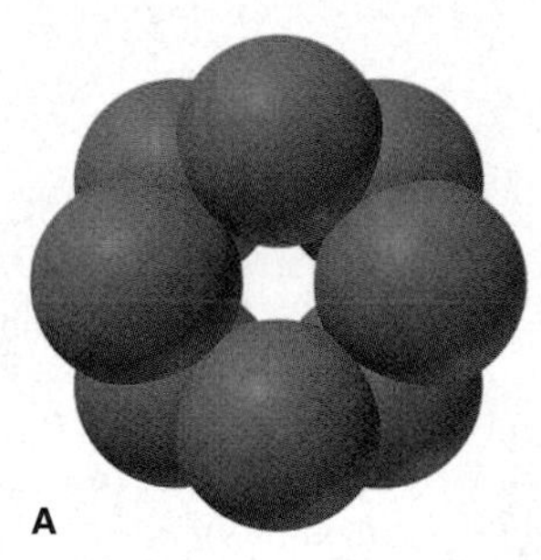

Figure 14.24 The cyclo-S_8 molecule. A, Top view of a space-filling model of the cyclo-S_8 molecule. **B,** Side view of a ball-and-stick model of the molecule; note the crownlike shape.

Selenium also has several allotropes, some consisting of crown-shaped Se_8 molecules. Gray Se is composed of layers of helical chains. Its electrical conductivity in visible light revolutionized the photocopying industry. When molten glass, cadmium sulfide, and gray Se are mixed and heated in the absence of air, a ruby-red glass forms, which you see whenever you stop at a traffic light.

Selenium and Xerography Photocopying was invented in the 1940s as a rapid, inexpensive, dry means of copying documents (*xerography,* Greek "dry writing"). It is based on the ability of Se to conduct a current when illuminated. A film of amorphous Se is deposited on an aluminum drum and electrostatically charged. Exposure to a document produces an "image" of low and high positive charges corresponding to the document's bright and dark areas. Negatively charged black, dry ink (toner) particles are attracted to the regions of high charge more than to those of low charge. This pattern of black particles is transferred electrostatically to paper, and the particles are fused to the paper's surface by heat or solvent. Excess toner is removed from the Se film, the charges are "erased" by exposure to light, and the film is ready for the next page.

How Do the Oxygen and Nitrogen Families Compare Chemically?

Changes in chemical behavior in Group 6A(16) are also similar to those in the previous group. Even though O and S occur as anions much more often than do N and P, like N and P, they bond covalently with almost every other nonmetal. Covalent bonds appear in the compounds of Se and Te (as in those of As and Sb), whereas Po behaves like a metal (as does Bi) in some of its saltlike compounds. In contrast to nitrogen, oxygen has few common oxidation states, but the earlier pattern returns with the other members: the +6, +4, and −2 states occur most often, with the lower positive (+4) state becoming more common in Te and Po [as the lower positive (+3) state does in Sb and Bi].

The range in atomic properties is wider in this group than in Group 5A(15) because of oxygen's high EN (3.5) and great oxidizing strength, second only to that of fluorine. As in 5A and earlier groups, the behavior of the Period 2 element stands out. In fact, aside from a similar outer electron configuration, the larger members of Group 6A behave very little like oxygen: they are much less electronegative, form anions much less often (S^{2-} occurs with active metals), and their hydrides exhibit no H bonding.

Except for O, all the 6A elements form foul-smelling, poisonous, gaseous hydrides (H_2E) on treatment with acid of the metal sulfide, selenide, and so forth. For example,

$$FeSe(s) + 2HCl(aq) \longrightarrow H_2Se(g) + FeCl_2(aq)$$

Hydrogen sulfide also forms naturally in swamps from the breakdown of organic matter. It is as toxic as HCN, and even worse, it anesthetizes your olfactory nerves, so that as its concentration increases, you smell it less! The other hydrides are about 100 times *more* toxic.

In their bonding and thermal stability, these Group 6A hydrides have several features in common with those of Group 5A:

- Only water can form H bonds, so it melts and boils much higher than the other H_2E compounds (see Figure 12.14, p. 440). (Oxygen's other hydride, H_2O_2, is also extensively H bonded.)
- Bond angles drop from the nearly tetrahedral value for H_2O (104.5°) to around 90° for the larger 6A hydrides, suggesting that the central atom uses unhybridized *p* orbitals.
- E—H bond length increases (bond energy decreases) down the group. Thus, H_2Te decomposes above 0°C, and H_2Po can be made only in extreme cold because thermal energy from the radioactive Po decomposes it. Another result of longer (weaker) bonds is that the 6A hydrides are acids in water, and their acidity increases from H_2S to H_2Po.

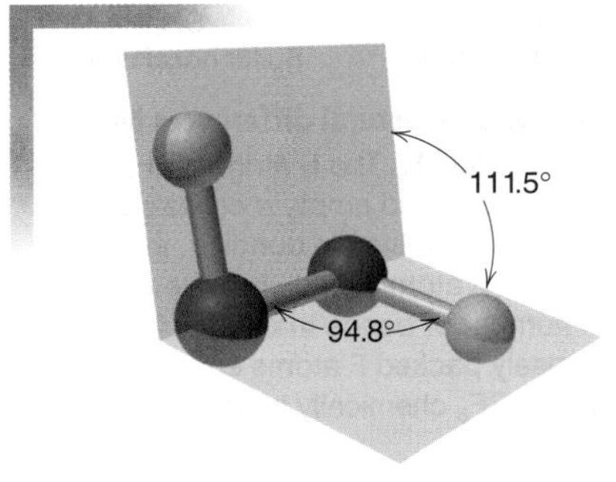

Hydrogen Peroxide: Hydrazine's Cousin In addition to water, oxygen forms another hydride, called hydrogen peroxide, H_2O_2 (HO—OH). Like hydrazine ($H_2N—NH_2$), the related hydride of nitrogen, hydrogen peroxide has a skewed molecular shape. It is a colorless liquid with a high density, viscosity, and boiling point because of extensive H bonding. In peroxides, O has the -1 oxidation state, midway between that in O_2 (zero) and that in oxides (-2); thus, H_2O_2 readily disproportionates:

$$H_2O_2(l) \longrightarrow H_2O(l) + \tfrac{1}{2}O_2(g)$$

Aside from the familiar use of H_2O_2 as a hair bleach and disinfectant, more than 70% of the half-million tons produced each year are used to bleach paper pulp, textiles, straw, and leather and to make other chemicals. H_2O_2 is also used in tertiary sewage treatment (Chemical Connections, p. 530) to oxidize foul-smelling effluents and restore O_2 to wastewater.

Except for O, the Group 6A elements form a wide range of halides, whose structure and reactivity patterns depend on the *sizes of the central atom and the surrounding halogens:*

- Sulfur forms many fluorides, a few chlorides, one bromide, but no stable iodides.
- As the central atom becomes larger, the halides become more stable. Thus, tetrachlorides and tetrabromides of Se, Te, and Po are known, as are tetraiodides of Te and Po. Hexafluorides are known only for S, Se, and Te.

The inverse relationship between bond length and bond strength that we've seen previously does not account for this pattern. Rather, it is based on the effect of electron repulsions due to crowding of lone pairs and halogen (X) atoms around the central Group 6A atom. With S, the larger X atoms would be too crowded, which explains why sulfur iodides do not occur. With increasing size of E, and therefore increasing length of E—X bonds, however, lone pairs and X atoms do not crowd each other as much, and a greater number of stable halides form.

Two sulfur fluorides illustrate clearly how crowding and orbital availability affect reactivity. Sulfur tetrafluoride (SF_4) is extremely reactive. It forms SO_2 and HF when exposed to moisture and is commonly used to fluorinate many compounds:

$$3SF_4(g) + 4BCl_3(g) \longrightarrow 4BF_3(g) + 3SCl_2(l) + 3Cl_2(g)$$

In contrast, sulfur hexafluoride (SF_6) is almost as inert as a noble gas! It is odorless, tasteless, nonflammable, nontoxic, and insoluble. Hot metals, boiling HCl, molten KOH, and high-pressure steam have no effect on it. It is used as an insulating gas in high-voltage generators, withstanding over 10^6 volts across electrodes only 50 mm apart. A look at the structures of these two fluorides provides

Group 8A(18): The Noble Gases

Key Atomic and Physical Properties

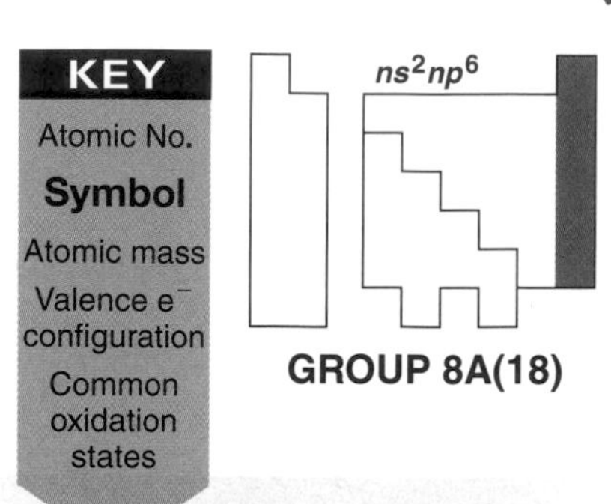

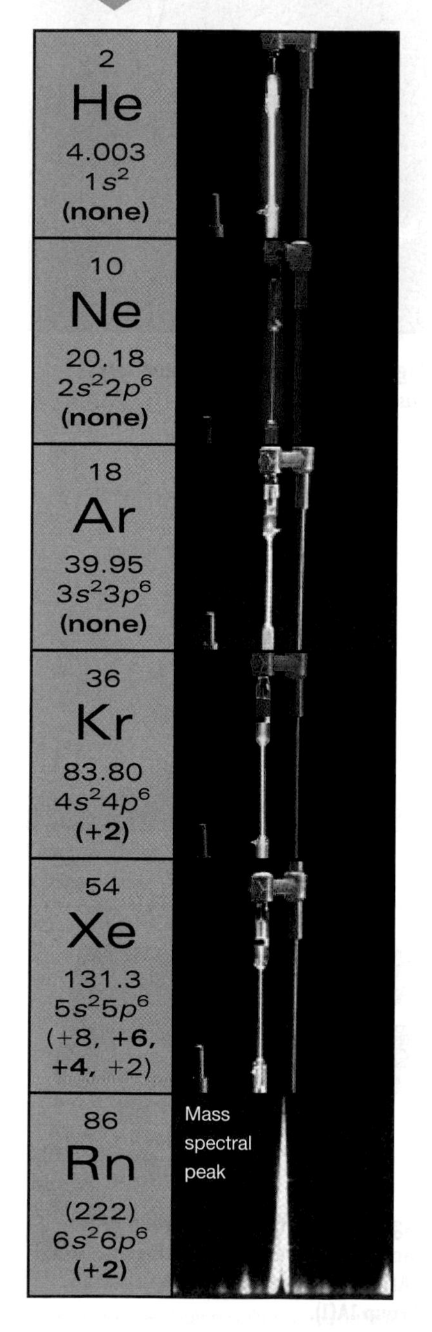

Atomic Properties

Atomic radius (pm)

He	31
Ne	71
Ar	98
Kr	112
Xe	131
Rn	(140)

Group electron configuration is $1s^2$ for He and ns^2np^6 for the others. The valence shell is filled. Only Kr and Xe (and perhaps Rn) are known to form compounds. The more reactive Xe exhibits all even oxidation states (+2 to +8).

This group contains the smallest atoms with the highest IEs in their periods. Down the group, atomic size increases and IE decreases steadily. (EN values are given only for Kr and Xe.)

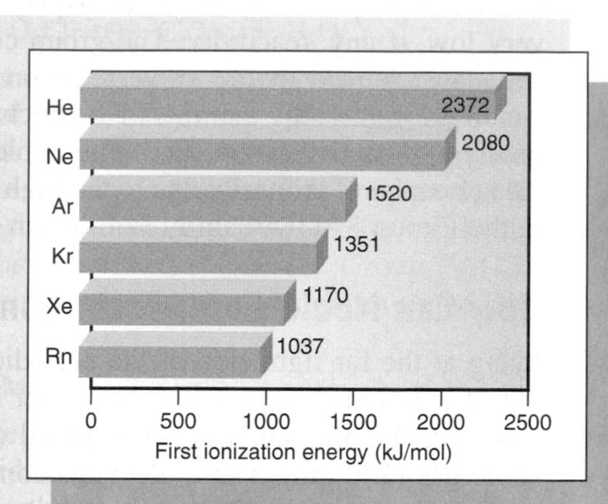

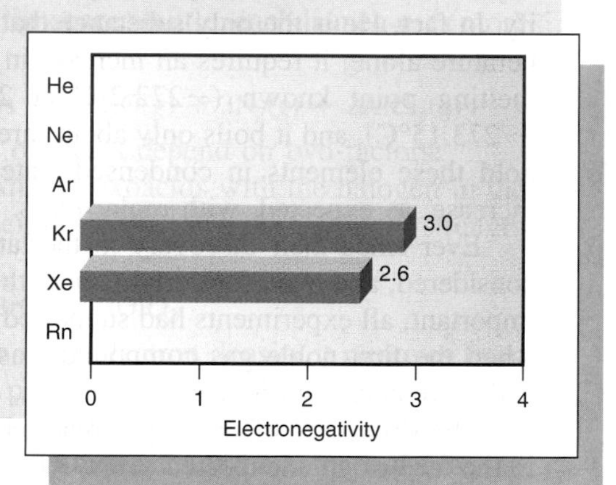

Physical Properties

Melting and boiling points of these gaseous elements are extremely low but increase down the group because of stronger dispersion forces. Note the extremely small liquid ranges.

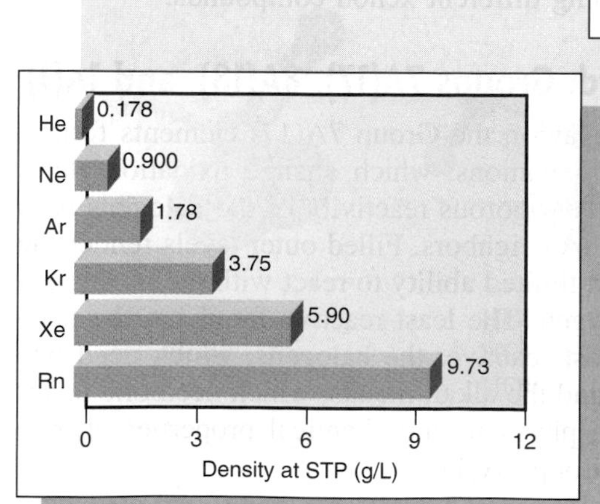

Densities (at STP) increase steadily, as expected.

Chapter Perspective

Our excursion through the bonding and reactivity patterns of the elements has come full circle, but we've only been able to touch on some of the most important features. Clearly, the properties of the atoms and the resultant physical and chemical behaviors of the elements are magnificent testimony to nature's diversity. In Chapter 15, we continue with our theme of macroscopic behavior emerging from atomic properties in an investigation of the marvelously complex organic compounds of carbon. In Chapter 22, we revisit the most important main-group elements to see how they occur in nature and how we isolate and use them.

For Review and Reference (Numbers in parentheses refer to pages, unless noted otherwise.)

Learning Objectives

Relevant section numbers appear in parentheses.

Understand These Concepts

Note: Many characteristic reactions appear in the "Important Reactions" section of each group's Family Portrait.

1. How hydrogen is similar to, yet different from, alkali metals and halogens; the differences between ionic, covalent, and metallic hydrides (Section 14.1)
2. Key horizontal trends in atomic properties, types of bonding, oxide acid-base properties, and redox behavior as the elements change from metals to nonmetals (Section 14.2)
3. How the ns^1 configuration accounts for the physical and chemical properties of the alkali metals (Section 14.3)
4. How small atomic size and limited number of valence orbitals account for the anomalous behavior of the Period 2 member of each group (see each section for details)
5. How the ns^2 configuration accounts for the key differences between Groups 1A(1) and 2A(2) (Section 14.4)
6. The basis of the three important diagonal relationships (Li/Mg, Be/Al, B/Si) (Sections 14.4 to 14.6)
7. How the presence of inner $(n-1)d$ electrons affects properties in Group 3A(13) (Section 14.5)
8. Patterns among larger members of Groups 3A(13) to 6A(16): two common oxidation states (inert-pair effect), lower state more important down the group, and more basic lower oxide (Sections 14.5 to 14.8)
9. How boron attains an octet of electrons (Section 14.5)
10. The effect of bonding on the physical behavior of Groups 4A(14) to 6A(16) (Sections 14.6 to 14.8)
11. Allotropism in carbon, phosphorus, and sulfur (Sections 14.6 to 14.8)
12. How atomic properties lead to catenation and multiple bonds in organic compounds (Section 14.6)
13. Structures and properties of the silicates and silicones (Section 14.6)
14. Patterns of behavior among hydrides and halides of Groups 5A(15) and 6A(16) (Sections 14.7 and 14.8)
15. The meaning of disproportionation (Section 14.7) and its importance in halogen aqueous chemistry (Section 14.9)
16. Structure and chemistry of the nitrogen oxides and oxoacids (Section 14.7)
17. Structure and chemistry of the phosphorus oxides and oxoacids (Section 14.7)
18. Dehydration-condensation reactions and polyphosphate structures (Section 14.7)
19. Structure and chemistry of the sulfur oxides and oxoacids (Section 14.8)
20. How the ns^2np^5 configuration accounts for halogen reactivity with metals (Section 14.9)
21. Why the oxidation states of an element change by two units (Section 14.9)
22. Structure and chemistry of the halogen oxides and oxoacids (Section 14.9)
23. How the ns^2np^6 configuration accounts for the relative inertness of the noble gases (Section 14.10)

Key Terms

Section 14.4
diagonal relationship (563)

Section 14.5
bridge bond (571)

Section 14.6
allotrope (573)
silicate (578)
silicone (578)

Section 14.7
disproportionation reaction (588)
dehydration-condensation reaction (590)

Section 14.9
interhalogen compound (602)

Highlighted Figures and Tables

These figures (F) and tables (T) provide a review of key ideas.

Problems

Problems with **colored** numbers are answered in Appendix E. Sections match the text and provide the numbers of relevant sample problems. Most offer Concept Review Questions, Skill-Building Exercises (grouped in pairs covering the same concept), and Problems in Context. Comprehensive Problems are based on material from any section or previous chapter.

Note: This set begins with problems based on material from the Interchapter.

Interchapter Review

Concept Review Questions

14.1 (a) Define the four atomic properties that are reviewed in the Interchapter.
(b) To what do the three parts of the electron configuration ($nl^{\#}$) correlate?
(c) Which part of the electron configuration is primarily associated with the size of the atom?
(d) What is the major distinction between the outer electron configurations within a group and those within a period?
(e) What correlation, if any, exists between the group number and the number of valence electrons?

14.2 (a) What trends, if any, exist for Z_{eff} across a period and down a group?
(b) How does Z_{eff} influence atomic size, IE_1, and EN across a period?

14.3 Iodine monochloride and elemental bromine have nearly the same molar mass and liquid density but very different boiling points.
(a) What molecular property is primarily responsible for this difference in boiling point? What atomic property gives rise to it? Explain.
(b) Which substance has a higher boiling point? Why?

14.4 How does the trend in atomic size differ from the trend in ionization energy? Explain.

14.5 How are bond energy, bond length, and reactivity related for similar compounds?

14.6 How are covalent and metallic bonding similar? How are they different?

14.7 If the leftmost element in a period combined with each of the others in the period, how would the type of bonding change from left to right? Explain in terms of atomic properties.

14.8 Why is rotation about the bond axis possible for single-bonded atoms but not double-bonded atoms?

14.9 Explain the horizontal irregularity in size of the most common ions of Period 3 elements.

14.10 Would you expect S^{2-} to be larger or smaller than Cl^-? Would you expect Mg^{2+} to be larger or smaller than Na^+? Explain.

14.11 (a) How does the type of bonding in element oxides correlate with the electronegativity of the elements?
(b) How does the acid-base behavior of element oxides correlate with the electronegativity of the elements?

14.12 (a) How does the metallic character of an element correlate with the *acidity* of its oxide?
(b) What trends, if any, exist in oxide *basicity* across a period and down a group?

14.13 How are atomic size, IE_1, and EN related to redox behavior of the elements in Groups 1A(1), 2A(2), 6A(16), and 7A(17)?

14.14 How do the physical properties of a network covalent solid and a molecular covalent solid differ? Why?

Skill-Building Exercises *(grouped in similar pairs)*

14.15 Rank the following elements in order of *increasing*
(a) atomic size: Ba, Mg, Sr
(b) IE_1: P, Na, Al
(c) EN: Br, Cl, Se
(d) number of valence electrons: Bi, Ga, Sn

14.16 Rank the following elements in order of *decreasing*
(a) atomic size: N, Si, P
(b) IE_1: Kr, K, Ar
(c) EN: In, Rb, I
(d) number of valence electrons: Sb, S, Cs

14.17 Which of the following pairs react to form *ionic* compounds: (a) Cl and Br; (b) Na and Br; (c) P and Se; (d) H and Ba?

14.18 Which of the following pairs react to form *covalent* compounds: (a) Be and C; (b) Sr and O; (c) Ca and Cl; (d) P and F?

14.19 Draw a Lewis structure for a compound with a bond order of 2 throughout the molecule.

14.20 Draw a Lewis structure for a polyatomic ion with a fractional bond order throughout the molecule.

14.21 Rank the following in order of *increasing* bond length: $SiCl_4$, CF_4, $GeBr_4$.

14.22 Rank the following in order of *decreasing* bond energy: NF_3, NI_3, NCl_3.

14.23 Rank the following in order of *increasing* radius:
(a) O^{2-}, F^-, Na^+ (b) S^{2-}, P^{3-}, Cl^-

14.24 Rank the following in order of *decreasing* radius: (a) Ca^{2+}, K^+, Ga^{3+} (b) Br^-, Sr^{2+}, Rb^+

14.25 Rank O^-, O^{2-}, and O in order of *increasing* size.

14.26 Rank Tl^{3+}, Tl, and Tl^+ in order of *decreasing* size.

14.27 Which member of each pair gives the more *basic* solution in water: (a) CaO or SO_3; (b) BeO or BaO; (c) CO_2 or SO_2; (d) P_4O_{10} or K_2O? For the pair in part (a), write an equation for the dissolving of each oxide to support your answer.

14.28 Which member of each pair gives the more *acidic* solution in water: (a) CO_2 or SrO; (b) SnO or SnO_2; (c) Cl_2O or Na_2O; (d) SO_2 or MgO? For the pair in part (a), write an equation for the dissolving of each oxide to support your answer.

14.29 Which member of each pair has more *covalent* character in its bonds: (a) LiCl or KCl; (b) $AlCl_3$ or PCl_3; (c) NCl_3 or $AsCl_3$?

14.30 Which member of each pair has more *ionic* character in its bonds: (a) BeF_2 or CaF_2; (b) PbF_2 or PbF_4; (c) GeF_4 or PF_3?

14.31 Rank the following in order of *increasing*
(a) Melting point: Na, Si, Ar (b) ΔH_{fus}: Rb, Cs, Li

14.32 Rank the following in order of *decreasing*
(a) Boiling point: O_2, Br_2, $As(s)$ (b) ΔH_{vap}: Cl_2, Ar, I_2

Hydrogen, the Simplest Atom

Concept Review Questions

14.33 Hydrogen has only one proton, but its IE_1 is much greater than that of lithium, which has three protons. Explain.

14.34 Sketch a periodic table, and label the areas containing elements that give rise to the three types of hydrides.

Skill-Building Exercises *(grouped in similar pairs)*

14.35 Draw Lewis structures for the following compounds, and predict which member of each pair will form hydrogen bonds:
(a) NF_3 or NH_3 (b) CH_3OCH_3 or CH_3CH_2OH

14.36 Draw Lewis structures for the following compounds, and predict which member of each pair will form hydrogen bonds:
(a) NH_3 or AsH_3 (b) CH_4 or H_2O

14.37 Complete and balance the following equations:
(a) An active metal reacting with acid,
$$Al(s) + HCl(aq) \longrightarrow$$
(b) A saltlike (alkali metal) hydride reacting with water,
$$LiH(s) + H_2O(l) \longrightarrow$$

14.38 Complete and balance the following equations:
(a) A saltlike (alkaline earth metal) hydride reacting with water,
$$CaH_2(s) + H_2O(l) \longrightarrow$$
(b) Reduction of a metal halide by hydrogen to form a metal,
$$PdCl_2(aq) + H_2(g) \longrightarrow$$

Problems in Context

14.39 Compounds such as $NaBH_4$, $Al(BH_4)_3$, and $LiAlH_4$ are complex hydrides used as reducing agents in many syntheses.
(a) Give the oxidation state of each element in these compounds.
(b) Write a Lewis structure for the polyatomic anion in $NaBH_4$, and predict its shape.

14.40 Unlike the F^- ion, which has an ionic radius close to 133 pm in all alkali metal fluorides, the ionic radius of H^- varies from 137 pm in LiH to 152 pm in CsH. Suggest an explanation for the large variability in the size of H^- but not F^-.

Trends Across the Periodic Table

Concept Review Questions

14.41 How does the maximum oxidation number vary across a period in the main groups? Is the pattern in Period 2 different?

14.42 What correlation, if any, exists for the Period 2 elements between group number and the number of covalent bonds the element typically forms? How is the correlation different for elements in Periods 3 to 6?

14.43 Each of the chemically active Period 2 elements forms stable compounds that have bonds to fluorine.
(a) What are the names and formulas of these compounds?
(b) Does ΔEN increase or decrease left to right?
(c) Does percent ionic character increase or decrease left to right?
(d) Draw Lewis structures for these compounds.

14.44 Period 6 is unusual in several ways.
(a) It is the longest period in the table. How many elements belong to Period 6? How many metals?
(b) It contains no metalloids. Where is the metal/nonmetal boundary in Period 6?

14.45 An element forms an oxide, E_2O_3, and a fluoride, EF_3.
(a) Of which two groups might E be a member?
(b) How does the group to which E belongs affect the properties of the oxide and the fluoride?

14.46 Fluorine lies between oxygen and neon in Period 2. Whereas atomic sizes and ionization energies of these three elements change smoothly, their electronegativities display a dramatic change. What is this change, and how do their electron configurations explain it?

Group 1A(1): The Alkali Metals

Concept Review Questions

14.47 Lithium salts are often much less soluble in water than the corresponding salts of other alkali metals. For example, at 18°C, the concentration of a saturated LiF solution is 1.0×10^{-2} *M*, whereas that of a saturated KF solution is 1.6 *M*. How would you explain this behavior?

14.48 The alkali metals play virtually the same general chemical role in all their reactions.
(a) What is this role?
(b) How is it based on atomic properties?
(c) Using sodium, write two balanced equations that illustrate this role.

14.49 How do atomic properties account for the low densities of the Group 1A(1) elements?

Skill-Building Exercises *(grouped in similar pairs)*

14.50 Each of the following properties shows regular trends in Group 1A(1). Predict whether each increases or decreases *down* the group: (a) density; (b) ionic size; (c) E—E bond energy; (d) IE_1; (e) magnitude of ΔH_{hydr} of E^+ ion.

14.51 Each of the following properties shows regular trends in Group 1A(1). Predict whether each increases or decreases *up* the group: (a) melting point; (b) E—E bond length; (c) hardness; (d) molar volume; (e) lattice energy of EBr.

14.52 Write a balanced equation for the formation from its elements of sodium peroxide, an industrial bleach.

14.53 Write a balanced equation for the formation of rubidium bromide through a reaction of a strong acid and a strong base.

Problems in Context

14.54 Although the alkali metal halides can be prepared directly from the elements, the far less expensive industrial route is treatment of the carbonate or hydroxide with aqueous hydrohalic acid (HX) followed by recrystallization. Balance the reaction between potassium carbonate and aqueous hydriodic acid.

14.55 Lithium forms several useful organolithium compounds. Calculate the mass percent of Li in the following:
(a) Lithium stearate ($C_{17}H_{35}COOLi$), a water-resistant grease used in cars because it does not harden at cold temperatures
(b) Butyllithium (LiC_4H_9), a reagent in organic syntheses

Group 2A(2): The Alkaline Earth Metals

Concept Review Questions

14.56 How do Groups 1A(1) and 2A(2) compare with respect to reaction of the metals with water?

14.57 Alkaline earth metals are involved in two key diagonal relationships in the periodic table. (a) Give the two pairs of elements in these diagonal relationships. (b) For each pair, cite two similarities that demonstrate the relationship. (c) Why are the members of each pair so similar in behavior?

14.58 The melting points of alkaline earth metals are many times higher than those of the alkali metals. Explain this difference on the basis of atomic properties. Name three other physical properties for which Group 2A(2) metals have higher values than the corresponding 1A(1) metals.

Skill-Building Exercises *(grouped in similar pairs)*

14.59 Write a balanced equation for each reaction:
(a) "Slaking" of lime (treatment with water)
(b) Combustion of calcium in air

14.60 Write a balanced equation for each reaction:
(a) Thermal decomposition of witherite (barium carbonate)
(b) Neutralization of stomach acid (HCl) by milk of magnesia (magnesium hydroxide)

Problems in Context

14.61 Lime (CaO) is one of the most abundantly produced chemicals in the world. Write balanced equations for
(a) The preparation of lime from natural sources
(b) The use of slaked lime to remove SO_2 from flue gases
(c) The reaction of lime with arsenic acid (H_3AsO_4) to manufacture the insecticide calcium arsenate
(d) The regeneration of NaOH in the paper industry by reaction of lime with aqueous sodium carbonate

14.62 In some reactions, Be behaves like a typical alkaline earth metal; in others, it does not. Complete and balance the following equations:
(a) $BeO(s) + H_2O(l) \longrightarrow$
(b) $BeCl_2(l) + Cl^-(l;\text{ from molten NaCl}) \longrightarrow$
In which reaction does Be behave like the other Group 2A(2) members?

Group 3A(13): The Boron Family

Concept Review Questions

14.63 How do the transition metals in Period 4 affect the pattern of ionization energies in Group 3A(13)? How does this pattern compare with that in Group 3B(3)?

14.64 How do the acidities of aqueous solutions of Tl_2O and Tl_2O_3 compare with each other? Explain.

14.65 Despite the expected decrease in atomic size, there is an unexpected drop in the first ionization energy between Groups 2A(2) and 3A(13) in Periods 2 through 4. Explain this pattern in terms of electron configurations and orbital energies.

14.66 Many compounds of Group 3A(13) elements have chemical behavior that reflects an electron deficiency.
(a) What is the meaning of *electron deficiency?*
(b) Give two reactions that illustrate this behavior.

14.67 Boron's chemistry is not typical of its group.
(a) Cite three ways in which boron and its compounds differ significantly from the other 3A(13) members and their compounds.
(b) What is the reason for these differences?

Skill-Building Exercises *(grouped in similar pairs)*

14.68 Rank the following oxides in order of increasing aqueous *acidity:* Ga_2O_3, Al_2O_3, In_2O_3.

14.69 Rank the following hydroxides in order of increasing aqueous *basicity:* $Al(OH)_3$, $B(OH)_3$, $In(OH)_3$.

14.70 Thallium forms the compound TlI_3. What is the apparent oxidation state of Tl in this compound? Given that the anion is I_3^-, what is the actual oxidation state of Tl? Draw the shape of the anion, giving its VSEPR class and bond angles. Propose a reason why the compound does not exist as $(Tl^{3+})(I^-)_3$.

14.71 Very stable dihalides of the Group 3A(13) metals are known. What is the apparent oxidation state of Ga in $GaCl_2$? Given that $GaCl_2$ consists of a Ga^+ cation and a $GaCl_4^-$ anion, what are the actual oxidation states of Ga? Draw the shape of the anion, giving its VSEPR class and bond angles.

Problems in Context

14.72 Give the name and symbol or formula of a Group 3A(13) element or compound that fits each description or use:
(a) Component of heat-resistant (Pyrex-type) glass
(b) Manufacture of high-speed computer chips
(c) Largest temperature range for liquid state of an element
(d) Elementary substance with three-center, two-electron bonds
(e) Metal protected from oxidation by adherent oxide coat
(f) Mild antibacterial agent (e.g., for eye infections)
(g) Toxic metal that lies between two other toxic metals

14.73 Indium (In) reacts with HCl to form a diamagnetic solid with the formula $InCl_2$.
(a) Write condensed electron configurations for In, In^+, In^{2+}, and In^{3+}.
(b) Which of these species is (are) diamagnetic and which paramagnetic?
(c) What is the apparent oxidation state of In in $InCl_2$?
(d) Given your answers to parts (b) and (c), explain how $InCl_2$ can be diamagnetic.

14.74 Use VSEPR theory to draw structures, with ideal bond angles, for boric acid and the anion it forms in reaction with water.

14.75 Boron nitride (BN) has a structure similar to graphite, but is a white insulator rather than a black conductor. It is synthesized by heating diboron trioxide with ammonia at about 1000°C.
(a) Write a balanced equation for the formation of BN; water forms also.
(b) Calculate ΔH^0_{rxn} for the production of BN (ΔH^0_f of BN is −254 kJ/mol).
(c) Boron is obtained from the mineral borax, $Na_2B_4O_7 \cdot 10H_2O$. How much borax is needed to produce 1.0 kg of BN, assuming 72% yield?

Group 4A(14): The Carbon Family

Concept Review Questions

14.76 How does the basicity of SnO_2 in water compare with that of CO_2? Explain.

14.77 Nearly every compound of silicon has the element in the +4 oxidation state. In contrast, most compounds of lead have the element in the +2 state.
(a) What general observation do these facts illustrate?
(b) Explain in terms of atomic and molecular properties.
(c) Give an analogous example from Group 3A(13).

14.78 The sum of IE_1 through IE_4 for Group 4A(14) elements shows a decrease from C to Si, a slight increase from Si to Ge, a decrease from Ge to Sn, and an increase from Sn to Pb.
(a) What is the expected trend for ionization energy down a group?
(b) Suggest a reason for the deviations from the expected trend.
(c) Which group might show even greater deviations?

14.79 Give explanations for the large drops in melting point from C to Si and from Ge to Sn.

14.80 What is an allotrope? Name two Group 4A(14) elements that exhibit allotropism, and name two of their allotropes.

14.81 Even though EN values vary relatively little down Group 4A(14), the elements change from nonmetal to metal. Explain.

14.82 How do atomic properties account for the enormous number of carbon compounds? Why don't other Group 4A(14) elements behave similarly?

Skill-Building Exercises *(grouped in similar pairs)*

14.83 Draw a Lewis structure for
(a) The cyclic silicate ion $Si_4O_{12}{}^{8-}$
(b) A cyclic hydrocarbon with formula C_4H_8

14.84 Draw a Lewis structure for
(a) The cyclic silicate ion $Si_6O_{18}{}^{12-}$
(b) A cyclic hydrocarbon with formula C_6H_{12}

14.85 Show three units of a linear silicone polymer made from $(CH_3)_2Si(OH)_2$ with some $(CH_3)_3SiOH$ added to end the chain.

14.86 Show two chains of three units each of a sheet silicone polymer made from $(CH_3)_2Si(OH)_2$ with $CH_3Si(OH)_3$ added to crosslink the chains.

Problems in Context

14.87 Zeolite A, $Na_{12}[(AlO_2)_{12}(SiO_2)_{12}]\cdot 27H_2O$, is used to soften water by replacing Ca^{2+} and Mg^{2+} with Na^+. A source of hard water is 4.5×10^{-3} *M* Ca^{2+} and 9.2×10^{-4} *M* Mg^{2+}, and a pipe delivers 25,000 L of this hard water per day. What mass (in kg) of zeolite A is needed to soften a week's supply of the water? (Assume zeolite A loses its capacity to exchange ions when 85 mol % of its Na^+ has been lost.)

14.88 Give the name and symbol or formula of a Group 4A(14) element or compound that fits each description or use:
(a) Hardest known substance
(b) Medicinal antacid
(c) Atmospheric gas implicated in the greenhouse effect
(d) Waterproofing polymer based on silicon
(e) Synthetic abrasive composed entirely of Group 4A elements
(f) Product formed when coke burns in a limited supply of air
(g) Toxic metal found in plumbing and paints

14.89 One similarity between B and Si is the explosive combustion of their hydrides in air. Write balanced equations for the combustion of B_2H_6 and of Si_4H_{10}.

Group 5A(15): The Nitrogen Family

Concept Review Questions

14.90 Which Group 5A(15) elements form trihalides? Pentahalides? Explain.

14.91 As you move down Group 5A(15), the melting points of the elements increase and then decrease. Explain.

14.92 (a) What is the range of oxidation states shown by the elements of Group 5A(15) as you move down the group? (b) How does this range illustrate the general rule for the range of oxidation states in groups on the right side of the periodic table?

14.93 Bismuth(V) compounds are such powerful oxidizing agents that they have not been prepared in pure form. How is this fact consistent with the location of Bi in the periodic table?

14.94 Rank the following oxides in order of increasing acidity in water: Sb_2O_3, Bi_2O_3, P_4O_{10}, Sb_2O_5.

Skill-Building Exercises *(grouped in similar pairs)*

14.95 Assuming acid strength relates directly to electronegativity of the central atom, rank H_3PO_4, HNO_3, and H_3AsO_4 in order of *increasing* acid strength.

14.96 Assuming acid strength relates directly to number of O atoms bonded to the central atom, rank $H_2N_2O_2$ [or $(HON)_2$]; HNO_3 (or $HONO_2$); and HNO_2 (or HONO) in order of *decreasing* acid strength.

14.97 Complete and balance the following:
(a) $As(s) + \text{excess } O_2(g) \longrightarrow$
(b) $Bi(s) + \text{excess } F_2(g) \longrightarrow$
(c) $Ca_3As_2(s) + H_2O(l) \longrightarrow$

14.98 Complete and balance the following:
(a) $\text{Excess } Sb(s) + Br_2(l) \longrightarrow$
(b) $HNO_3(aq) + MgCO_3(s) \longrightarrow$
(c) $K_2HPO_4(s) \xrightarrow{\Delta}$

14.99 Complete and balance the following:
(a) $N_2(g) + Al(s) \xrightarrow{\Delta}$
(b) $PF_5(g) + H_2O(l) \longrightarrow$

14.100 Complete and balance the following:
(a) $AsCl_3(l) + H_2O(l) \longrightarrow$
(b) $Sb_2O_3(s) + NaOH(aq) \longrightarrow$

14.101 Based on the relative sizes of F and Cl, predict the structure of PF_2Cl_3.

14.102 Use the VSEPR model to predict the structure of the cyclic ion $P_3O_9{}^{3-}$.

Problems in Context

14.103 The pentafluorides of the larger members of Group 5A(15) have been prepared, but N can have only eight electrons. A claim has been made that, at low temperatures, a compound with the empirical formula NF_5 forms. Draw a possible Lewis structure for this compound. (*Hint:* NF_5 is ionic.)

14.104 Give the name and symbol or formula of a Group 5A(15) element or compound that fits each description or use:
(a) Hydride produced at multimillion-ton level
(b) Element(s) essential in plant nutrition
(c) Hydride used to manufacture rocket propellants and drugs
(d) Odd-electron molecule (two examples)
(e) Amphoteric hydroxide with a 5A element in its +3 state
(f) Phosphorus-containing water softener
(g) Element that is an electrical conductor

14.105 In addition to those in Table 14.3, other less stable nitrogen oxides exist. Draw a Lewis structure for each of the following:
(a) N_2O_2, a dimer of nitrogen monoxide with an N—N bond
(b) N_2O_2, a dimer of nitrogen monoxide with no N—N bond
(c) N_2O_3 with no N—N bond
(d) NO^+ and NO_3^-, products of the ionization of liquid N_2O_4

14.106 Nitrous oxide (N_2O), the "laughing gas" used as an anesthetic by dentists, is made by thermal decomposition of solid NH_4NO_3. Write a balanced equation for this reaction. What are the oxidation states of N in NH_4NO_3 and in N_2O?

14.107 Write balanced equations for the thermal decomposition of potassium nitrate (O_2 is also formed in both cases): (a) at low temperature to the nitrite; (b) at high temperature to the metal oxide and nitrogen.

Group 6A(16): The Oxygen Family

Concept Review Questions

14.108 Rank the following in order of increasing electrical conductivity, and explain your ranking: Po, S, Se.

14.109 The oxygen and nitrogen families have some obvious similarities and differences.
(a) State two general physical similarities between Group 5A(15) and 6A(16) elements.
(b) State two general chemical similarities between Group 5A(15) and 6A(16) elements.
(c) State two chemical similarities between P and S.
(d) State two physical similarities between N and O.
(e) State two chemical differences between N and O.

14.110 A molecular property of the Group 6A(16) hydrides changes abruptly down the group. This change has been explained in terms of a change in orbital hybridization.
(a) Between what periods does the change occur?
(b) What is the change in the molecular property?
(c) What is the change in hybridization?
(d) What other group displays a similar change?

Skill-Building Exercises *(grouped in similar pairs)*

14.111 Complete and balance the following:
(a) $NaHSO_4(aq) + NaOH(aq) \longrightarrow$
(b) $S_8(s) + \text{excess } F_2(g) \longrightarrow$
(c) $FeS(s) + HCl(aq) \longrightarrow$
(d) $Te(s) + I_2(s) \longrightarrow$

14.112 Complete and balance the following:
(a) $H_2S(g) + O_2(g) \longrightarrow$
(b) $SO_3(g) + H_2O(l) \longrightarrow$
(c) $SF_4(g) + H_2O(l) \longrightarrow$
(d) $Al_2Se_3(s) + H_2O(l) \longrightarrow$

14.113 Is each oxide basic, acidic, or amphoteric in water: (a) SeO_2; (b) N_2O_3; (c) K_2O; (d) BeO; (e) BaO?

14.114 Is each oxide basic, acidic, or amphoteric in water: (a) MgO; (b) N_2O_5; (c) CaO; (d) CO_2; (e) TeO_2?

14.115 Rank the following hydrides in order of *increasing* acid strength: H_2S, H_2O, H_2Te.

14.116 Rank the following species in order of *decreasing* acid strength: H_2SO_4, H_2SO_3, HSO_3^-.

Problems in Context

14.117 Describe the physical changes observed when solid sulfur is heated from room temperature to 440°C and then poured quickly into cold water. Explain the molecular changes responsible for the macroscopic changes.

14.118 Give the name and symbol or formula of a Group 6A(16) element or compound that fits each description or use:
(a) Unreactive gas used as an electrical insulator
(b) Unstable allotrope of oxygen
(c) Oxide having sulfur in the same oxidation state as in sulfuric acid
(d) Air pollutant produced by burning sulfur-containing coal
(e) Powerful dehydrating agent
(f) Compound used in solution in the photographic process
(g) Gas in trace amounts in air that tarnishes silver

14.119 Give the oxidation state of sulfur in (a) S_8; (b) SF_4; (c) SF_6; (d) H_2S; (e) FeS_2; (f) H_2SO_4; (g) $Na_2S_2O_3{\cdot}5H_2O$.

14.120 Disulfur decafluoride is intermediate in reactivity between SF_4 and SF_6. It disproportionates at 150°C to these monosulfur fluorides. Write a balanced equation for this reaction, and give the oxidation state of S in each compound.

Group 7A(17): The Halogens

Concept Review Questions

14.121 (a) What is the physical state and color of each of the halogens at STP?
(b) Explain the change in physical state down Group 7A(17) in terms of molecular properties.

14.122 (a) What are the common oxidation states of the halogens?
(b) Give an explanation based on electron configuration for the range and values of the oxidation states of chlorine.
(c) Why is fluorine an exception to the pattern of oxidation states found for the other group members?

14.123 How many electrons does a halogen atom need to complete its octet? Give examples of the ways a Cl atom can do so.

14.124 Select the stronger bond in each pair:
(a) Cl—Cl or Br—Br
(b) Br—Br or I—I
(c) F—F or Cl—Cl. Why doesn't the F—F bond strength follow the group trend?

14.125 In addition to interhalogen compounds, many interhalogen ions exist. Would you expect interhalogen ions with a 1+ or a 1− charge to have an even or odd number of atoms? Explain.

14.126 (a) A halogen (X_2) disproportionates in base in several steps to X^- and XO_3^-. Write the overall equation for the disproportionation of Br_2 to Br^- and BrO_3^-.
(b) Write a balanced equation for the reaction of ClF_5 with aqueous base (see the reaction of BrF_5 shown on p. 602).

Skill-Building Exercises *(grouped in similar pairs)*

14.127 Complete and balance the following equations. If no reaction occurs, write NR:
(a) $Rb(s) + Br_2(l) \longrightarrow$
(b) $I_2(s) + H_2O(l) \longrightarrow$
(c) $Br_2(l) + I^-(aq) \longrightarrow$
(d) $CaF_2(s) + H_2SO_4(l) \longrightarrow$

14.128 Complete and balance the following equations. If no reaction occurs, write NR:
(a) $H_3PO_4(l) + NaI(s) \longrightarrow$
(b) $Cl_2(g) + I^-(aq) \longrightarrow$
(c) $Br_2(l) + Cl^-(aq) \longrightarrow$
(d) $ClF(g) + F_2(g) \longrightarrow$

14.129 Rank the following acids in order of *increasing* acid strength: HClO, $HClO_2$, HBrO, HIO.

14.130 Rank the following acids in order of *decreasing* acid strength: $HBrO_3$, $HBrO_4$, HIO_3, $HClO_4$.

Problems in Context

14.131 Give the name and symbol or formula of a Group 7A(17) element or compound that fits each description or use:
(a) Used in etching glass
(b) Naturally occurring source (ore) of fluorine
(c) Oxide used in bleaching paper pulp and textiles
(d) Weakest hydrohalic acid
(e) Compound used as food additive to prevent goiter (thyroid disorder)
(f) Element that is produced in the largest quantity
(g) Organic chloride used to make plastics

14.132 An industrial chemist treats solid NaCl with concentrated H_2SO_4 and obtains gaseous HCl and $NaHSO_4$. When she substitutes solid NaI for NaCl, gaseous H_2S, solid I_2, and S_8 are obtained but no HI.
(a) What type of reaction did the H_2SO_4 undergo with NaI?
(b) Why does NaI, but not NaCl, cause this type of reaction?
(c) To produce HI(*g*) by the reaction of NaI with an acid, how does the acid have to differ from sulfuric acid?

14.133 Rank the halogens Cl_2, Br_2, and I_2 in order of increasing oxidizing strength based on their products with metallic Re: $ReCl_6$, $ReBr_5$, ReI_4. Explain your ranking.

Group 8A(18): The Noble Gases

14.134 Which noble gas is the most abundant in the universe? In Earth's atmosphere?

14.135 What oxidation states does Xe show in its compounds?

14.136 Why do the noble gases have such low boiling points?

14.137 Explain why Xe, and to a limited extent Kr, form compounds, whereas He, Ne, and Ar do not.

14.138 (a) Why do stable xenon fluorides have an even number of F atoms? (b) Why do the ionic species XeF_3^+ and XeF_7^- have odd numbers of F atoms? (c) Predict the shape of XeF_3^+.

Comprehensive Problems

14.139 Sodium hydride is used to prepare sodium dithionate ($Na_2S_2O_4$), used in bleaching paper pulp:

$$NaH(s) + SO_2(l) \longrightarrow Na_2S_2O_4(s) + H_2(g)$$

(a) Balance the equation, and give the O.N. of each element.
(b) How many liters of hydrogen gas measured at 758 torr and −45°C can form when exactly 1 metric ton of NaH reacts with 1 metric ton of SO_2?

14.140 Given the following information,

$$H^+(g) + H_2O(g) \longrightarrow H_3O^+(g) \qquad \Delta H = -720 \text{ kJ}$$
$$H^+(g) + H_2O(l) \longrightarrow H_3O^+(aq) \qquad \Delta H = -1090 \text{ kJ}$$
$$H_2O(l) \longrightarrow H_2O(g) \qquad \Delta H = 40.7 \text{kJ}$$

calculate the heat of solution of the hydronium ion:

$$H_3O^+(g) \xrightarrow{H_2O} H_3O^+(aq)$$

14.141 The electronic transition in Na from $3p^1$ to $3s^1$ gives rise to a bright yellow-orange emission at 589.2 nm. What is the energy of this transition?

14.142 Unlike other Group 2A(2) metals, beryllium reacts like aluminum and zinc with concentrated aqueous base to release hydrogen gas and form oxoanions of formula $M(OH)_4^{n-}$. Write equations for the reactions of these three metals with NaOH.

14.143 Just as boron nitride is isoelectronic with carbon, other compounds of Groups 3A(13) and 5A(15) are isoelectronic with elements in Group 4A(14). What element is isoelectronic with (a) aluminum phosphide; (b) gallium arsenide?

14.144 Cyclopropane (C_3H_6), the smallest cyclic hydrocarbon, is reactive for the same reason white phosphorus is. Use bond properties and valence bond theory to explain its high reactivity.

14.145 Thionyl chloride ($SOCl_2$) is a sulfur oxohalide used industrially to dehydrate metal halide hydrates.
(a) Write a balanced equation for its reaction with magnesium chloride hexahydrate, in which SO_2 and HCl form along with the metal halide.
(b) Draw a Lewis structure of $SOCl_2$ with minimal formal charges.

14.146 The main reason alkali metal dihalides (MX_2) do *not* form is the high IE_2 of the metal.
(a) Why is IE_2 so high for alkali metals?
(b) The IE_2 for Cs is 2255 kJ/mol, low enough for CsF_2 to form exothermically ($\Delta H_f^0 = -125$ kJ/mol). This compound cannot be synthesized, however, because CsF forms with a much greater release of heat ($\Delta H_f^0 = -530$ kJ/mol). Thus, the breakdown of CsF_2 to CsF happens readily. Write the equation for this breakdown, and calculate the heat of reaction per mole of CsF.

14.147 Semiconductors made from elements in Groups 3A(13) and 5A(15) are typically prepared by direct reaction of the elements at high temperature. An engineer treats 32.5 g of molten gallium with 20.4 L of white phosphorus vapor at 515 K and 195 kPa. If purification losses are 7.2% by mass, how many grams of gallium phosphide will be prepared?

14.148 Two substances with empirical formula HNO are hyponitrous acid ($\mathcal{M}$ = 62.04 g/mol) and nitroxyl ($\mathcal{M}$ = 31.02 g/mol).
(a) What is the molecular formula of each species?
(b) For each species, draw the Lewis structure having the lowest formal charges. (*Hint:* Hyponitrous acid has an N=N bond.)
(c) Predict the shape around the N atoms of each species.
(d) When hyponitrous acid loses two protons, it forms the hyponitrite ion. Draw *cis* and *trans* forms of this ion.

14.149 The species CO, CN^-, and C_2^{2-} are isoelectronic.
(a) Draw their Lewis structures.
(b) Draw their MO diagrams (assume 2*s*-2*p* mixing, as in N_2), and give the bond order and electron configuration for each.

14.150 The Ostwald process is a series of three reactions used for the industrial production of nitric acid from ammonia.
(a) Write a series of balanced equations for the Ostwald process.
(b) If NO is *not* recycled, how many moles of NH_3 are consumed per mole of HNO_3 produced?
(c) In a typical industrial unit, the process is very efficient, with a 96% yield for the first step. Assuming 100% yields for the subsequent steps, what volume of nitric acid (60.% by mass; d = 1.37 g/mL) can be prepared for each cubic meter of a gas mixture that is 90.% air and 10.% NH_3 by volume at the industrial conditions of 5.0 atm and 850.°C?

14.151 Buckminsterfullerene (C_{60}) is a soccer-ball–shaped molecule with a C atom at each vertex. It has a face-centered cubic unit cell that contains 240 C atoms and a density of only 1.693 g/cm^3, less than half that of diamond. Calculate the volume and edge length of the unit cell, and suggest a reason for the low density despite the closest packing.

14.152 Perhaps surprisingly, some pure liquid interhalogens that contain fluorine have high electrical conductivity. The explanation is that one molecule transfers a fluoride ion to another molecule. Write the equation that would explain the high electrical conductivity of bromine trifluoride.

14.153 All common plant fertilizers contain nitrogen compounds. Determine the mass % of N in (a) ammonia; (b) ammonium nitrate; (c) ammonium hydrogen phosphate.

14.154 Producer gas is a fuel formed by passing air over red-hot coke (amorphous carbon). It consists of approximately 25% CO, 5.0% CO_2, and 70.% N_2 by mass. What mass of producer gas can be formed from 1.75 metric tons of coke, assuming an 87% yield?

14.155 Gaseous F_2 reacts with water to form HF and O_2. In NaOH solution, F_2 forms F^-, water, and oxygen difluoride (OF_2), a highly toxic gas and powerful oxidizing agent. The OF_2 reacts with excess OH^-, forming O_2, water, and F^-.
(a) For each reaction, write a balanced equation, give the oxidation state of O in all compounds, and identify the oxidizing and reducing agents.
(b) Draw a Lewis structure for OF_2 and predict the molecule's shape.

14.156 What is a disproportionation reaction, and which of the following fit the description?
(a) $I_2(s) + KI(aq) \longrightarrow KI_3(aq)$
(b) $2ClO_2(g) + H_2O(l) \longrightarrow HClO_3(aq) + HClO_2(aq)$
(c) $Cl_2(g) + 2NaOH(aq) \longrightarrow NaCl(aq) + NaClO(aq) + H_2O(l)$
(d) $NH_4NO_2(s) \longrightarrow N_2(g) + 2H_2O(g)$
(e) $3MnO_4^{2-}(aq) + 2H_2O(l) \longrightarrow 2MnO_4^-(aq) + MnO_2(s) + 4OH^-(aq)$
(f) $3AuCl(s) \longrightarrow AuCl_3(s) + 2Au(s)$

14.157 Explain the following observations:
(a) In reactions with Cl_2, phosphorus forms PCl_5 in addition to the expected PCl_3, but nitrogen forms only NCl_3.
(b) Carbon tetrachloride is unreactive toward water, but silicon tetrachloride reacts rapidly and completely. (To give what?)
(c) The sulfur-oxygen bond in SO_4^{2-} is shorter than expected for an S—O single bond.
(d) Chlorine forms ClF_3 and ClF_5, but ClF_4 is unknown.

14.158 Which group(s) of the periodic table is (are) described by each of the following general statements?
(a) The elements form neutral compounds of VSEPR class AX_3E.
(b) The free elements are strong oxidizing agents and form monatomic ions and oxoanions.
(c) The valence electron configuration allows the atoms to form compounds by combining with two atoms that donate one electron each.
(d) The free elements are strong reducing agents, show only one nonzero oxidation state, and form mainly ionic compounds.
(e) The elements can form stable compounds with only three bonds, but as a central atom, they can accept a pair of electrons from a fourth atom without expanding their valence shell.
(f) Only larger members of the group are chemically active.

14.159 Diiodine pentaoxide (I_2O_5) was discovered by Joseph Gay-Lussac in 1813, but its structure was unknown until 1970! Like Cl_2O_7, it can be prepared by the dehydration-condensation of the corresponding oxoacid.
(a) Name the precursor oxoacid, write a reaction for formation of the oxide, and draw a likely Lewis structure.
(b) Data show that the bonds to the terminal O are shorter than the bonds to the bridging O. Why?
(c) I_2O_5 is one of the few chemicals that can oxidize CO rapidly and completely; elemental iodine forms in the process. Write a balanced equation for this reaction.

14.160 An important starting material for the manufacture of polyphosphazenes is the cyclic molecule $(NPCl_2)_3$. The molecule has a symmetrical six-membered ring of alternating N and P atoms, with the Cl atoms bonded to the P atoms. The nitrogen-phosphorus bond length is significantly less than that expected for an N—P single bond.
(a) Draw a likely Lewis structure for the molecule.
(b) How many lone pairs of electrons do the ring atoms have?
(c) What is the order of the nitrogen-phosphorus bond?

14.161 Potassium fluotitanate (K_2TiF_6), a starting material for producing other titanium salts, is made by dissolving titanium(IV) oxide in concentrated HF, adding aqueous KF, and then heating to dryness to drive off excess HF. The impure solid is recrystallized by dissolving in hot water and then cooling to 0°C, which leaves excess KF in solution. For the reaction of 5.00 g of TiO_2, 250. mL of hot water is used to recrystallize the product. How many grams of purified K_2TiF_6 are obtained? (Solubility of K_2TiF_6 in water at 0°C is 0.60 g/100 mL.)

14.162 Bromine monofluoride (BrF) disproportionates to bromine gas and bromine tri- and pentafluorides. Use the following to find ΔH^0_{rxn} for the decomposition of BrF to its elements:

$$3BrF(g) \longrightarrow Br_2(g) + BrF_3(l) \qquad \Delta H_{rxn} = -125.3 \text{ kJ}$$
$$5BrF(g) \longrightarrow 2Br_2(g) + BrF_5(l) \qquad \Delta H_{rxn} = -166.1 \text{ kJ}$$
$$BrF_3(l) + F_2(g) \longrightarrow BrF_5(l) \qquad \Delta H_{rxn} = -158.0 \text{ kJ}$$

14.163 Account for the following facts:
(a) Ca^{2+} and Na^+ have very nearly the same radii.
(b) CaF_2 is insoluble in water, but NaF is quite soluble.
(c) Molten $BeCl_2$ is a poor electrical conductor, whereas molten $CaCl_2$ is an excellent one.

14.164 Cake alum (aluminum sulfate) is used as a *flocculating agent* in water purification. The "floc" is a gelatinous precipitate of aluminum hydroxide that carries out of solution small suspended particles and bacteria for removal by filtration.
(a) The hydroxide is formed by the reaction of the aluminum ion with water. Write the equation for this reaction.
(b) The acidity from this reaction is partially neutralized by the sulfate ion. Write an equation for this reaction.

14.165 Carbon tetrachloride is made by passing Cl_2 in the presence of a catalyst through liquid CS_2 near its boiling point. Disulfur dichloride (S_2Cl_2) also forms, in addition to the by-products SCl_2 and CCl_3SCl. How many grams of CS_2 reacted if excess Cl_2 yielded 50.0 g of CCl_4 and 0.500 g of CCl_3SCl? Is it possible to determine how much S_2Cl_2 and SCl_2 also formed? Explain.

14.166 H_2 may act as a reducing agent or an oxidizing agent, depending on the substance reacting with it. Using, in turn, sodium or chlorine as the other reactant, write balanced equations for the two reactions, and characterize the redox role of hydrogen.

14.167 P_4 is prepared by heating phosphate rock [principally $Ca_3(PO_4)_2$] with sand and coke:

$$Ca_3(PO_4)_2(s) + SiO_2(s) + C(s) \longrightarrow CaSiO_3(s) + CO(g) + P_4(g) \text{ [unbalanced]}$$

How many kilograms of phosphate rock are needed to produce 315 mol of P_4, assuming that the conversion is 90.% efficient?

14.168 Assume you can use either NH_4NO_3 or N_2H_4 as fertilizer. Which contains more moles of N per mole of compound? Which contains the higher mass of N per gram of compound?

14.169 From its formula, one might expect CO to be quite polar, but its dipole moment is actually low (0.11 D).
(a) Draw the Lewis structure for CO.
(b) Calculate the formal charges.
(c) Based on your answers to parts (a) and (b), explain why the dipole moment is so low.

14.170 In addition to Al_2Cl_6, aluminum forms other species with bridging halide ions to two aluminum atoms. One such species is the ion $Al_2Cl_7^-$. The ion is symmetrical, with a 180° Al—Cl—Al bond angle.
(a) What orbitals does Al use to bond with the Cl atoms?
(b) What is the shape around each Al?
(c) What is the hybridization of the central Cl?
(d) What do the shape and hybridization suggest about the presence of lone pairs of electrons on the central Cl?

14.171 When an alkaline earth carbonate is heated, it releases CO_2, leaving the metal oxide. The temperature at which each Group 2A carbonate yields a CO_2 partial pressure of 1 atm is

Carbonate	Temperature (°C)
$MgCO_3$	542
$CaCO_3$	882
$SrCO_3$	1155
$BaCO_3$	1360

(a) Suggest a reason for this trend.
(b) Mixtures of $CaCO_3$ and MgO are used to absorb dissolved silicates from boiler water. How would you prepare a mixture of $CaCO_3$ and MgO from dolomite, which contains $CaCO_3$ and $MgCO_3$?

14.172 The bond angles in the nitrite ion, nitrogen dioxide, and the nitronium ion (NO_2^+) are 115°, 134°, and 180°, respectively. Explain these values using Lewis structures and VSEPR theory.

14.173 A common method for producing a gaseous hydride is to treat a salt containing the anion of the volatile hydride with a strong acid.
(a) Write an equation for each of the following examples: (1) the production of HF from CaF_2; (2) the production of HCl from NaCl; (3) the production of H_2S from FeS.
(b) In some cases even a weak acid such as water will suffice if the anion of the salt has a sufficiently strong attraction for protons. An example is the production of PH_3 from Ca_3P_2 and water. Write the equation for this reaction.
(c) By analogy, predict the products and write the equation for the reaction of Al_4C_3 with water.

14.174 Chlorine trifluoride was formerly used in the production of uranium hexafluoride for the U.S. nuclear industry:

$$U(s) + 3ClF_3(l) \longrightarrow UF_6(l) + 3ClF(g)$$

How many grams of UF_6 can form from 1.00 metric ton of uranium ore that is 1.55% by mass uranium and 12.75 L of chlorine trifluoride (d = 1.88 g/mL)?

14.175 Chlorine is used to make household bleach solutions containing 5.25% (by mass) NaClO. Assuming 100% yield in the reaction producing NaClO from Cl_2, how many liters of $Cl_2(g)$ at STP will be needed to make 1000. L of bleach solution (d = 1.07 g/mL)?

14.176 The triatomic molecular ion H_3^+ was first detected and characterized by J. J. Thomson using mass spectrometry. Use the bond energy of H_2 (432 kJ/mol) and the proton affinity of H_2 ($H_2 + H^+ \longrightarrow H_3^+$; $\Delta H = -337$ kJ/mol) to calculate the heat of reaction for

$$H + H + H^+ \longrightarrow H_3^+$$

14.177 An atomic hydrogen torch is used for cutting and welding thick sheets of metal. When H_2 passes through an electric arc, the molecules decompose into atoms, which react with O_2 at the end of the torch. Temperatures in excess of 5000°C are reached, which can melt all metals. Write equations for H_2 breakdown to H atoms and for the subsequent overall reaction of the H atoms with oxygen. Use Appendix B to find the standard heat of each reaction per mole of product.

14.178 In aqueous HF solution, an important species is the ion HF_2^-, which has the bonding arrangement FHF^-. Draw the Lewis structure for this ion, and explain how it arises.

14.179 Which of the following oxygen ions are paramagnetic: O^+, O^-, O^{2-}, O^{2+}?

14.180 Copper(II) hydrogen arsenite ($CuHAsO_3$) is a green pigment once used in wallpaper; in fact, forensic evidence suggests that Napoleon may have been poisoned by arsenic from his wallpaper. In damp conditions, mold metabolizes this compound to trimethylarsenic [$(CH_3)_3As$], a highly toxic gas.
(a) Calculate the mass percent of As in each compound.
(b) How much $CuHAsO_3$ must react to reach a toxic level in a room that measures 12.35 m × 7.52 m × 2.98 m (arsenic is toxic at 0.50 mg/m^3)?

14.181 Hydrogen peroxide can act as either an oxidizing agent or a reducing agent.
(a) When H_2O_2 is treated with aqueous KI, I_2 forms. In which role is H_2O_2 acting? What is the oxygen-containing product formed?
(b) When H_2O_2 is treated with aqueous $KMnO_4$, the purple color of MnO_4^- disappears and a gas forms. In which role is H_2O_2 acting? What is the oxygen-containing product formed?

Precursor extraordinaire Benzene, almost an icon for organic compounds, is used to synthesize countless medicines, polymers, dyes, and other chemicals. Ultimately, our biochemical nature derives from our organochemical nature, which, as you'll see in this chapter, emerges from the amazing properties of the carbon atom.

15

Organic Compounds and the Atomic Properties of Carbon

Is there any chemical system more remarkable than a living cell? Through delicately controlled mechanisms, it oxidizes food for energy, maintains the concentrations of thousands of aqueous components, interacts continuously with its environment, synthesizes both simple and complex molecules, and even reproduces itself! For all our technological prowess, no human-made system even approaches the cell for sheer elegance of function.

This amazing chemical machine consumes, creates, and consists largely of *organic compounds*. Except for a few inorganic salts and ever-present water, nearly everything you put into or on your body—food, medicine, cosmetics, and clothing—consists of organic compounds. Organic fuels warm our homes, cook our meals, and power our society. Major industries are devoted to producing organic compounds, such as polymers, pharmaceuticals, and insecticides.

What *is* an organic compound? Dictionaries define it as "a compound of carbon," but that definition is too general because it includes carbonates, cyanides, carbides, cyanates, and other carbon-containing ionic compounds that most chemists classify as inorganic. Here is a more specific definition: all **organic compounds** contain carbon, nearly always bonded to other carbons and hydrogen, and often to other elements.

The word *organic* has a biological connotation arising from a major misconception that stifled research into the chemistry of living systems for many decades. In the early 19th century, many prominent thinkers believed that an unobservable spiritual energy, a "vital force," existed *within* the compounds of living things, making them impossible to synthesize and fundamentally different from compounds of the mineral world. This idea of *vitalism* was challenged in 1828, when the young German chemist Friedrich Wöhler heated ammonium cyanate, a "mineral-world" compound, and produced urea, a "living-world" compound:

$$NH_4OCN \xrightarrow{\Delta} H_2N\text{—}\overset{\overset{\displaystyle O}{\|}}{C}\text{—}NH_2$$

Although Wöhler did not appreciate the significance of this reaction—he was more interested in the fact that two compounds can have the same molecular formula—his experiment is considered a key event in the origin of organic chemistry. Chemists soon synthesized methane, acetic acid, acetylene, and many other organic compounds from inorganic sources. Today, we know that *the same chemical principles govern organic and inorganic systems* because the behavior of a compound arises from the properties of its elements, no matter how marvelous that behavior may be.

IN THIS CHAPTER ... We apply the text's central theme to the enormous field of organic chemistry—how the structure and reactivity of organic molecules emerge naturally from the properties of their component atoms. First, we review the special atomic properties of carbon and see how they lead to the complex structure and reactivity of organic molecules. Next, we focus on writing and naming hydrocarbons as a prelude to naming other types of organic compounds. We classify the main types of organic reactions and apply them to various families of organic compounds. Finally, we extend these ideas to the giant molecules of commerce and life—synthetic and natural polymers.

15.1 THE SPECIAL NATURE OF CARBON AND THE CHARACTERISTICS OF ORGANIC MOLECULES

Although there is nothing mystical about organic molecules, their indispensable role in biology and industry leads us to ask if carbon has some extraordinary attributes that give it a special chemical "personality." Of course, each element has its own specific properties, and carbon is no more unique than sodium,

Concepts & Skills to Review

before you study this chapter

- naming simple organic compounds; the functional-group concept (Section 2.8)
- constitutional isomerism (Section 3.2)
- ΔEN and bond polarity (Section 9.5)
- resonance structures (Section 10.1)
- VSEPR theory (Section 10.2)
- orbital hybridization (Section 11.1)
- σ and π bonding (Section 11.2)
- types of intermolecular forces and the shapes of biological macromolecules (Sections 12.3, 12.7, and 13.2)
- properties of the Period 2 elements (Section 14.2)
- properties of the Group 4A(14) elements (Section 14.6)

Organic Chemistry Is Enough to Drive One Mad The vitalists did not change their beliefs overnight. Indeed, organic compounds *do* seem different from inorganic compounds because of their complex structures and compositions. Imagine the consternation of the 19th-century chemist who was accustomed to studying compounds with formulas such as $CaCO_3$, $Ba(NO_3)_2$, and $CuSO_4{\cdot}5H_2O$ and then isolated white crystals from a gallstone and found their empirical formula to be $C_{27}H_{46}O$ (cholesterol). Even Wöhler later said, "Organic chemistry . . . is enough to drive one mad. It gives me the impression of a primeval tropical forest, full of the most remarkable things, a monstrous and boundless thicket, with no way of escape, into which one may well dread to enter."

hafnium, or any other element. But the atomic properties of carbon do give it bonding capabilities beyond those of any other element, which in turn lead to the two obvious characteristics of organic molecules—structural complexity and chemical diversity.

The Structural Complexity of Organic Molecules

Most organic molecules have much more complex structures than most inorganic molecules, and a quick review of carbon's atomic properties and bonding behavior shows why:

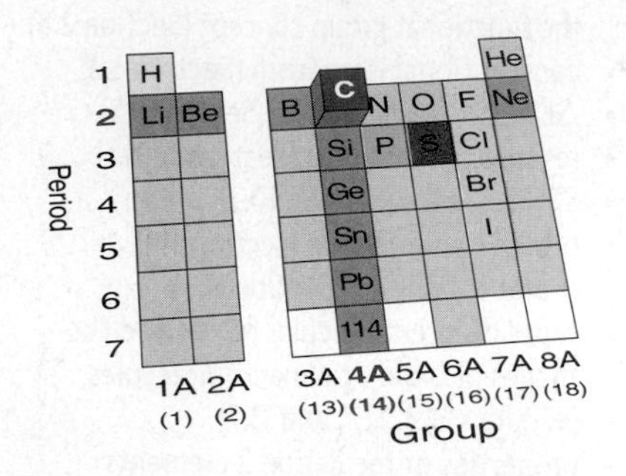

Figure 15.1 The position of carbon in the periodic table. Lying at the center of Period 2, carbon has an intermediate electronegativity (EN), and its position at the top of Group 4A(14) means it is relatively small. Other elements common in organic compounds are H, N, O, P, S, and the halogens.

1. *Electron configuration, electronegativity, and covalent bonding.* Carbon's ground-state electron configuration of [He] $2s^22p^2$—four electrons more than He and four fewer than Ne—means that the formation of carbon ions is energetically impossible under ordinary conditions: to form a C^{4+} cation requires energy equal to the sum of IE_1 through IE_4, and a C^{4-} anion requires the sum of EA_1 through EA_4, the last three steps being endothermic. Carbon's position in the periodic table (Figure 15.1) and its electronegativity are midway between the most metallic and nonmetallic elements of Period 2: Li = 1.0, C = 2.5, F = 4.0. Therefore, *carbon shares electrons to attain a filled outer (valence) level,* bonding covalently in all its elemental forms and compounds.

2. *Bond properties, catenation, and molecular shape.* The *number* and *strength* of carbon's bonds lead to its outstanding ability to *catenate* (form chains of atoms), which allows it to form a multitude of chemically and thermally stable chain, ring, and branched compounds. Through the process of orbital hybridization (Section 11.1), *carbon forms four bonds in virtually all its compounds,* and they point in as many as four different directions. The small size of carbon allows close approach to another atom and thus greater orbital overlap, meaning that *carbon forms relatively short, strong bonds.* The C—C bond is short enough to allow side-to-side overlap of half-filled, unhybridized *p* orbitals and the formation of *multiple bonds,* which restrict rotation of attached groups (see Figure 11.13, p. 409). These features add more possibilities for the shapes of carbon compounds.

3. *Molecular stability.* Although silicon and several other elements also catenate, none can compete with carbon. Atomic and bonding properties confer three crucial differences between C and Si chains that explain why C chains are so stable and, therefore, so common:

- *Atomic size and bond strength.* As atomic size increases down Group 4A(14), bonds between identical atoms become longer and weaker. Thus, a C—C bond (347 kJ/mol) is much stronger than an Si—Si bond (226 kJ/mol).
- *Relative heats of reaction.* A C—C bond (347 kJ/mol) and a C—O bond (358 kJ/mol) have nearly the same energy, so relatively little heat is released when a C chain reacts and one bond replaces the other. In contrast, an Si—O bond (368 kJ/mol) is much stronger than an Si—Si bond (226 kJ/mol), so a large quantity of heat is released when an Si chain reacts.
- *Orbitals available for reaction.* Unlike C, Si has low-energy *d* orbitals that can be attacked (occupied) by the lone pairs of incoming reactants. Thus, for example, ethane ($CH_3—CH_3$) is stable in water and does not react in air unless sparked, whereas disilane ($SiH_3—SiH_3$) breaks down in water and ignites spontaneously in air.

The Chemical Diversity of Organic Molecules

In addition to their elaborate geometries, organic compounds are noted for their sheer number and diverse chemical behavior. Look in any compendium of known compounds, such as the *CRC Handbook of Chemistry and Physics,* and you'll find that the number of organic compounds dwarfs the number of inorganic compounds of all the other elements combined! Several million organic compounds are known, and thousands more are discovered or synthesized each year.

This incredible diversity is also founded on atomic and bonding behavior and is due to three interrelated factors:

1. *Bonding to heteroatoms.* Many organic compounds contain **heteroatoms,** atoms other than C or H. The most common heteroatoms are N and O, but S, P, and the halogens often occur, and organic compounds with other elements are known as well. To get an idea of the diversity arising from the presence of a heteroatom, look at Figure 15.2, which shows that 23 different molecular structures are produced by various arrangements of four C atoms singly bonded to each other, the necessary number of H atoms, and just one O atom (either singly or doubly bonded).

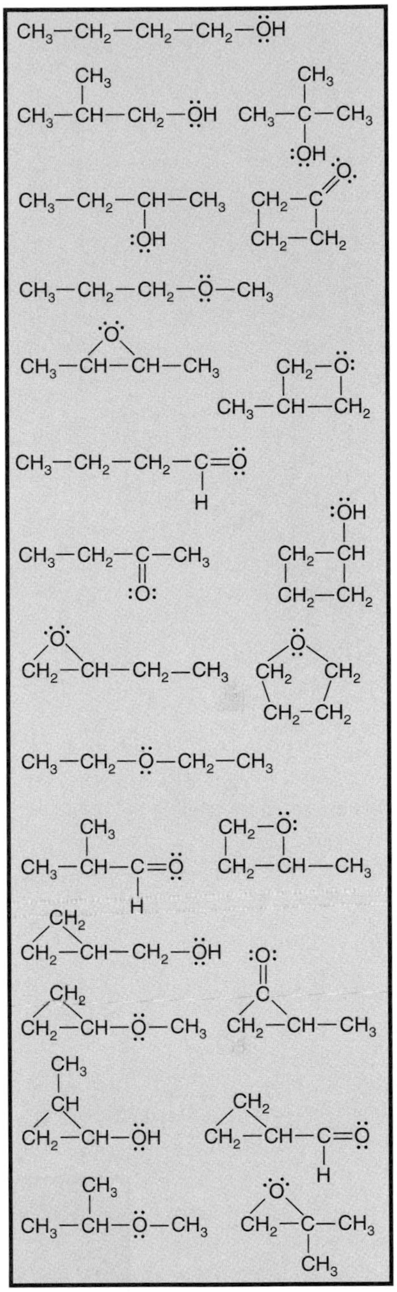

2. *Electron density and reactivity.* Most reactions start—that is, a new bond begins to form—*when a region of high electron density on one molecule meets a region of low electron density on another.* These regions may be due to the presence of a multiple bond or to the partial charges that occur in carbon-heteroatom bonds. For example, consider four bonds that are commonly found in organic molecules:

- *The C—C bond.* When C is singly bonded to itself, as it is in portions of nearly every organic molecule, the EN values are equal and the bond is nonpolar. Therefore, in general, *C—C bonds are unreactive.*
- *The C—H bond.* This bond, which also occurs in nearly every organic molecule, is very nearly nonpolar because it is short (109 pm) and the EN values of H (2.1) and C (2.5) are close. Thus, *C—H bonds are largely unreactive* as well.
- *The C—O bond.* In contrast to the previous two bonds, this bond, which occurs in many types of organic molecules, is highly polar ($\Delta EN = 1.0$), with the O end of the bond electron rich and the C end electron poor. As a result of this imbalance in electron density, *the C—O bond is reactive* and, given appropriate conditions, a reaction will occur there.
- *Bonds to other heteroatoms.* Even when a carbon-heteroatom bond has a small ΔEN, such as that for C—Br ($\Delta EN = 0.3$), or none at all, as for C—S ($\Delta EN = 0$), heteroatoms like these are large, and so their bonds to carbon are long, weak, and thus reactive.

3. *Nature of functional groups.* One of the most important ideas in organic chemistry is that of the **functional group,** a specific combination of bonded atoms that reacts in a *characteristic* way, no matter what molecule it occurs in (we mentioned this idea briefly in Section 2.8). In nearly every case, *the reaction of an organic compound takes place at the functional group.* In fact, as you'll see, we often substitute a general symbol for the remainder of the molecule because it usually stays the same while the functional group reacts. Functional groups vary from carbon-carbon multiple bonds to several combinations of carbon-heteroatom bonds, and each has its own pattern of reactivity. A particular bond may be a functional group itself or *part* of one or more functional groups. For example, the C—O bond occurs in four functional groups. We will discuss the reactivity of three of these groups in this chapter:

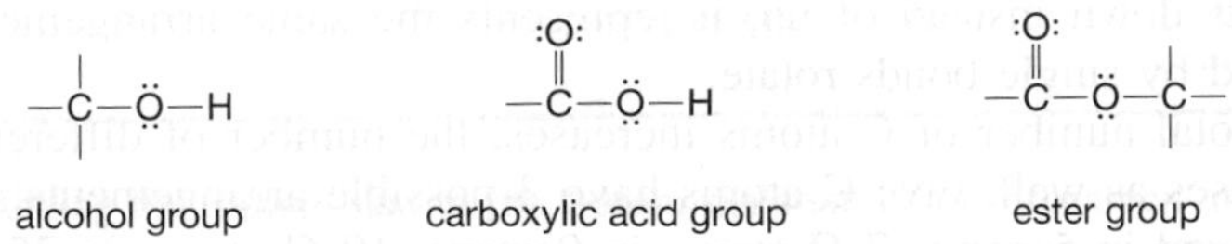

—C—O—H (alcohol group)

—C(=O)—O—H (carboxylic acid group)

—C(=O)—O—C— (ester group)

$CH_3—CH_2—CH_2—CH_2—OH$

$CH_3—CH(CH_3)—CH_2—OH$ $\quad$ $CH_3—C(CH_3)(OH)—CH_3$

$CH_3—CH_2—CH(OH)—CH_3$ $\quad$ cyclic $CH_2—C(=O)$ / $CH_2—CH_2$

$CH_3—CH_2—CH_2—O—CH_3$

$CH_3—CH—CH—CH_3$ (O bridging) $\quad$ $CH_2—O$ / $CH_3—CH—CH_2$

$CH_3—CH_2—CH_2—C(H)=O$

$CH_3—CH_2—C(=O)—CH_3$ $\quad$ $CH_2—CH(OH)$ / $CH_2—CH_2$

$CH_2—CH—CH_2—CH_3$ (O bridging) $\quad$ CH_2—O—CH_2 / $CH_2—CH_2$

$CH_3—CH_2—O—CH_2—CH_3$

$CH_3—CH(CH_3)—C(H)=O$ $\quad$ $CH_2—O$ / $CH_2—CH—CH_3$

$CH_2(CH_2)CH—CH_2—OH$ $\quad$ $CH_2(C=O)CH—CH_3$

$CH_2(CH_2)CH—O—CH_3$

$CH_2(CH(CH_3))CH—OH$ $\quad$ $CH_2(CH_2)CH—C(H)=O$

$CH_3—CH(CH_3)—O—CH_3$ $\quad$ $CH_2(O)C(CH_3)—CH_3$

Figure 15.2 The chemical diversity of organic compounds. Different arrangements of chains, branches, rings, and heteroatoms give rise to many structures. There are 23 different compounds possible from just four C atoms joined by single bonds, one O atom, and the necessary H atoms.

SAMPLE PROBLEM 15.1 Drawing Hydrocarbons

Problem Draw structures that have different atom arrangements for hydrocarbons with:
(a) Six C atoms, no multiple bonds, and no rings
(b) Four C atoms, one double bond, and no rings
(c) Four C atoms, no multiple bonds, and one ring

Plan In each case, we draw the longest carbon chain first and then work down to smaller chains with branches at different points along them. The process typically involves trial and error. Then, we add H atoms to give each C a total of four bonds.

Solution (a) Compounds with six C atoms:

```
             H   H   H   H   H   H
             |   |   |   |   |   |
6-C chain: H-C---C---C---C---C---C-H
             |   |   |   |   |   |
             H   H   H   H   H   H

                                                 H
                                                 |
                                               H-C-H
             H   H   H       H             H   H     H   H
             |   |   |       |             |   |     |   |
5-C chains: H-C---C---C---C---C-H        H-C---C---C---C---C-H
             |   |   |       |             |   |   |   |   |
             H   H   H     H-C-H           H   H   H   H   H
                             |
                             H

                 H                             H
                 |                             |
               H-C-H                         H-C-H
             H   |   H   H                 H   H     H
             |   |   |   |                 |   |     |
4-C chains: H-C---C---C---C-H            H-C---C---C---C-H
             |   |   |   |                 |     |   |
             H   |   H   H                 H     H   H
               H-C-H                         H-C-H
                 |                             |
                 H                             H
```

(b) Compounds with four C atoms and one double bond:

```
             H  H  H  H          H  H   H  H                    H     H
             |  |  |  |          |  |   |  |                    |     |
4-C chains: H-C--C--C=C-H      H-C--C--C=C-H     3-C chain: H-C---C=C-H
             |  |                |  |                          |  |
             H  H                H  H                          H  |
                                                             H-C-H
                                                               |
                                                               H
```

(c) Compounds with four C atoms and one ring:

```
              H   H                            H   H
              |   |                             \ /
            H-C---C-H                            C      H
4-C ring:     |   |              3-C ring:      / \     |
            H-C---C-H                      H-C-----C---C-H
              |   |                          |     |   |
              H   H                          H     H   H
```

Check Be sure each skeleton has the correct number of C atoms, multiple bonds, and/or rings, and no arrangements are repeated or omitted; remember a double bond counts as two bonds.

Comment Avoid some *common mistakes:*

```
                                                        C
                                                        |
In (a): C-C-C-C-C    is the same skeleton as   C-C-C-C-C
          |
          C

            C                                      C  C
            |                                      |  |
        C-C-C-C      is the same skeleton as   C-C-C-C
            |
            C
```

In (b): C—C—C=C is the same skeleton as C=C—C—C

The double bond restricts rotation, so, in addition to the *cis* form shown in part (b), another possibility is the *trans* form:

```
      H   H       H
      |   |       |
    H-C---C=C-----C-H
      |     |     |
      H     H     H
```

(We discuss *cis-trans* isomers fully later in this section.)

Too many bonds to one C in

```
          5 bonds
   H  H  /H
   |  | ↙|
H—C—C=C—H
   |  |
   H  |
    H—C—H
      |
      H
```

In (c): Too many bonds to one C in

```
     H   H
      \ /   5 bonds
       C   H /H
      / \ /↙ |
H—C———C—C—H
   |    |  |
   H    H  H
```

FOLLOW-UP PROBLEM 15.1 Draw all hydrocarbons that have different atom arrangements with:
(a) Seven C atoms, no multiple bonds, and no rings (nine arrangements)
(b) Five C atoms, one triple bond, and no rings (three arrangements)

Hydrocarbons can be classified into four main groups. In the remainder of this section, we examine some structural features and physical properties of each group. Later, we discuss the chemical behavior of the hydrocarbons.

Alkanes: Hydrocarbons with Only Single Bonds

A hydrocarbon that contains only single bonds is an **alkane** (general formula $\mathbf{C_nH_{2n+2}}$, where n is a positive integer). For example, if $n = 5$, the formula is $C_5H_{[(2\times5)+2]}$, or C_5H_{12}. The alkanes comprise a **homologous series,** one in which each member differs from the next by a $—CH_2—$ (methylene) group. In an alkane, each C is sp^3 hybridized. Because each C is bonded to the *maximum number of other atoms* (C or H), alkanes are referred to as **saturated hydrocarbons.**

Naming Alkanes You learned how to name simple alkanes in Section 2.8. Here we discuss general rules for naming any alkane and, by extension, other organic compounds as well. The key point is that *each chain, branch, or ring has a name based on the* ***number*** *of C atoms.* The name of a compound has three portions:

PREFIX + ROOT + SUFFIX

- *Root:* The root tells the number of C atoms in the longest *continuous* chain in the molecule. The roots for the ten smallest alkanes are shown in Table 15.1. Recall that there are special roots for compounds of one to four C atoms; roots of longer chains are based on Greek numbers.
- *Prefix:* Each prefix identifies a *group attached to the main chain* and the number of the carbon to which it is attached. Prefixes identifying hydrocarbon branches are the same as root names (Table 15.1) but have *-yl* as their ending. Each prefix is placed *before* the root.
- *Suffix:* The suffix tells the *type of organic compound* the molecule represents; that is, it identifies the key functional group the molecule possesses. The suffix is placed *after* the root.

For example, in the name 2-methylbutane, *2-methyl-* is the prefix (a one-carbon branch is attached to C-2 of the main chain), *-but-* is the root (the main chain has four C atoms), and *-ane* is the suffix (the compound is an alkane).

To obtain the systematic name of a compound,

1. Name the longest chain (root).
2. Add the compound type (suffix).
3. Name any branches (prefix).

Table 15.1 Numerical Roots for Carbon Chains and Branches

Roots	Number of C Atoms
meth-	1
eth-	2
prop-	3
but-	4
pent-	5
hex-	6
hept-	7
oct-	8
non-	9
dec-	10

Table 15.2 Rules for Naming an Organic Compound

Rule	Example
1. Naming the longest chain (root) (a) Find the longest *continuous* chain of C atoms. (b) Select the root that corresponds to the number of C atoms in this chain.	CH_3 (branch on C2) $CH_3—CH—CH—CH_2—CH_2—CH_3$ $CH_2—CH_3$ (branch on C3) 6 carbons ⟹ hex-
2. Naming the compound type (suffix) (a) For alkanes, add the suffix *-ane* to the chain root. (Other suffixes appear in Table 15.5 with their functional group and compound type.) (b) If the chain forms a ring, the name is preceded by *cyclo-*.	hex- + -ane ⟹ hexane
3. Naming the branches (prefixes) (If the compound has no branches, the name consists of the root and suffix.) (a) Each branch name consists of a subroot (number of C atoms) and the ending *-yl* to signify that it is not part of the main chain. (b) Branch names precede the chain name. When two or more branches are present, their names appear in *alphabetical* order.	CH_3 methyl $CH_3—CH—CH—CH_2—CH_2—CH_3$ $CH_2—CH_3$ ethyl ethylmethylhexane
(c) To specify where the branch occurs along the chain, number the main-chain C atoms consecutively, starting at the end *closer* to a branch, to achieve the *lowest* numbers for the branches. Precede each branch name with the number of the main-chain C to which that branch is attached.	CH_3 (on C2) $\overset{1}{C}H_3—\overset{2}{C}H—\overset{3}{C}H—\overset{4}{C}H_2—\overset{5}{C}H_2—\overset{6}{C}H_3$ $CH_2—CH_3$ (on C3) 3-ethyl-2-methylhexane

Table 15.2 presents the rules for naming any organic compound and applies them to an alkane component of gasoline. Other organic compounds are named with a variety of other prefixes and suffixes (see Table 15.5, p. 639). In addition to *systematic* names, we'll also note important *common* names still in use.

Depicting Alkanes with Formulas and Models Chemists have several ways to depict organic compounds. Expanded, condensed, and carbon-skeleton formulas are easy to draw; ball-and-stick and space-filling models show the actual shapes.

The *expanded formula* shows each atom and bond. One type of *condensed formula* groups each C atom with its H atoms. *Carbon-skeleton formulas* show only carbon-carbon bonds and appear as zig-zag lines, often with branches. *Each end or bend of a zig-zag line or branch represents a C atom attached to the number of H atoms that gives it a total of four bonds*:

propane $CH_3—CH_2—CH_3$

2,3-dimethylbutane $CH_3—\underset{}{CH}(CH_3)—CH(CH_3)—CH_3$

Figure 15.5 shows these types of formulas, together with ball-and-stick and space-filling models, of the compound named in Table 15.2.

Figure 15.5 Ways of depicting an alkane.

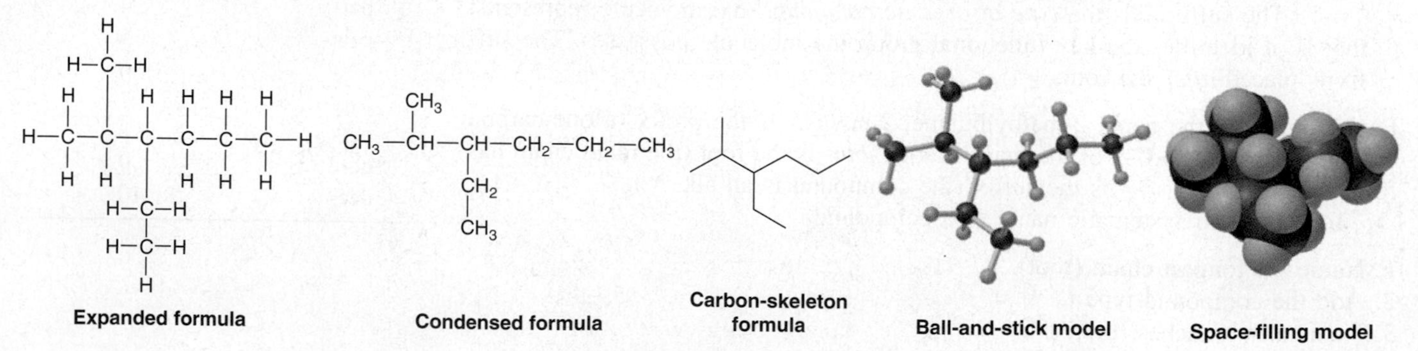

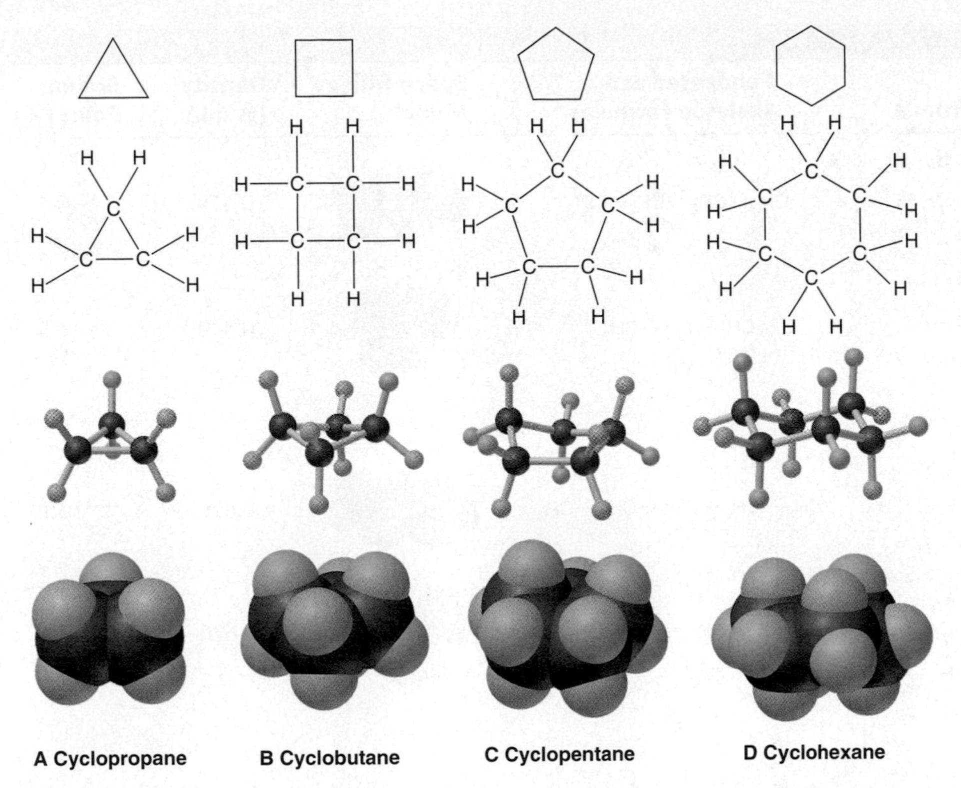

Figure 15.6 Depicting cycloalkanes. Cycloalkanes are usually drawn as regular polygons. Each side is a C—C bond, and each corner represents a C atom with its required number of H atoms. The expanded formulas show each bond in the molecule. The ball-and-stick and space-filling models show that, except for cyclopropane, the rings are not planar. These conformations minimize electron repulsions between adjacent H atoms. Cyclohexane **(D)** is shown in its more stable chair conformation.

Cyclic Hydrocarbons A **cyclic hydrocarbon** contains one or more rings in its structure. When a straight-chain alkane (C_nH_{2n+2}) forms a ring, two H atoms are lost as the C—C bond forms to join the two ends of the chain. Thus, *cycloalkanes* have the general formula C_nH_{2n}. Cyclic hydrocarbons are often drawn with carbon-skeleton formulas, as shown at the top of Figure 15.6. Except for three-carbon rings, *cycloalkanes are nonplanar.* This structural feature arises from the tetrahedral shape around each C atom and the need to minimize electron repulsions between adjacent H atoms. As a result, orbital overlap of adjacent C atoms is maximized. The most stable form of cyclohexane, called the *chair conformation,* is shown in Figure 15.6D.

Constitutional Isomerism and the Physical Properties of Alkanes

Recall from Section 3.2 that two or more compounds with the same molecular formula but different properties are called **isomers.** Those with *different arrangements of bonded atoms* are **constitutional** (or **structural) isomers;** alkanes with the same number of C atoms but different skeletons are examples. The smallest alkane to exhibit constitutional isomerism has four C atoms: two different compounds have the formula C_4H_{10}, as shown in Table 15.3, on the next page. The unbranched one is butane (common name, *n*-butane; *n*- stands for "normal," or having a straight chain), and the other is 2-methylpropane (common name, *iso*butane). Similarly, three compounds have the formula C_5H_{12} (shown in Table 15.3). The unbranched isomer is pentane (common name, *n*-pentane); the one with a

Table 15.3 The Constitutional Isomers of C_4H_{10} and C_5H_{12}

Systematic Name (Common Name)	Expanded Formula	Condensed and Skeleton Formulas	Space-filling Model	Density (g/mL)	Boiling Point (°C)
Butane (*n*-butane)	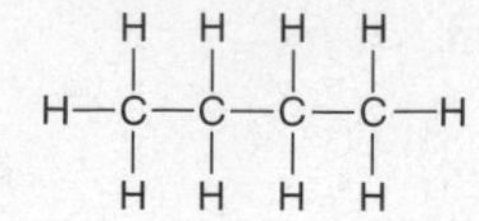	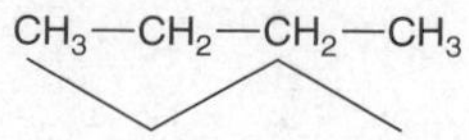 $CH_3—CH_2—CH_2—CH_3$	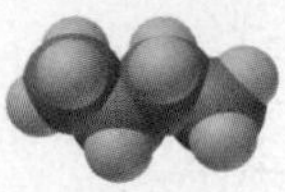	0.579	−0.5
2-Methylpropane (isobutane)	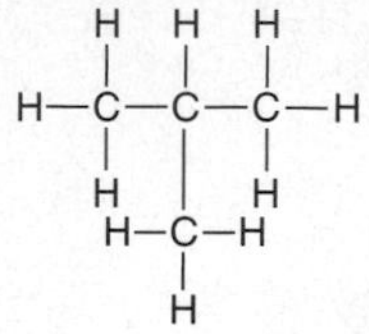	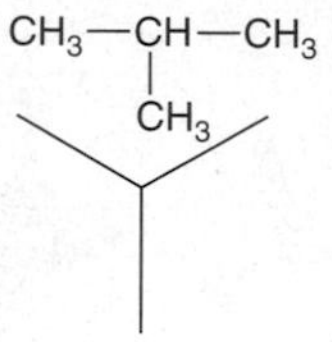 $CH_3—CH(CH_3)—CH_3$		0.549	−11.6
Pentane (*n*-pentane)	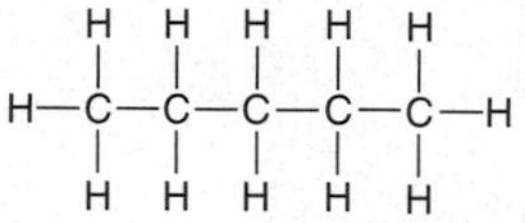	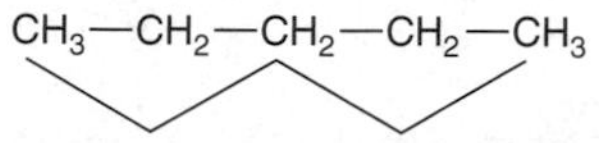 $CH_3—CH_2—CH_2—CH_2—CH_3$		0.626	36.1
2-Methylbutane (isopentane)	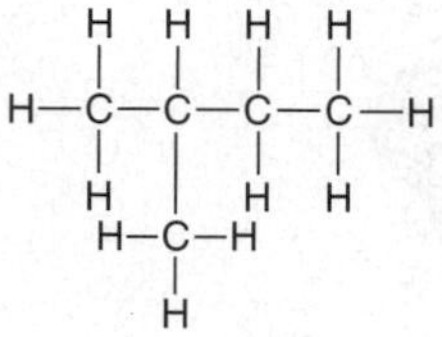	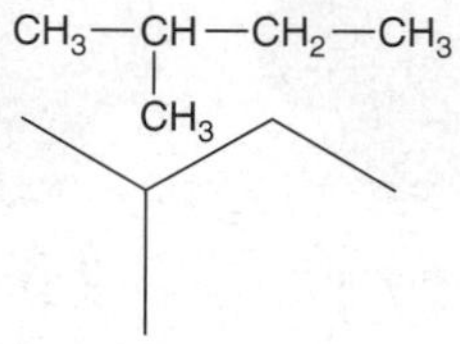 $CH_3—CH(CH_3)—CH_2—CH_3$		0.620	27.8
2,2-Dimethylpropane (neopentane)	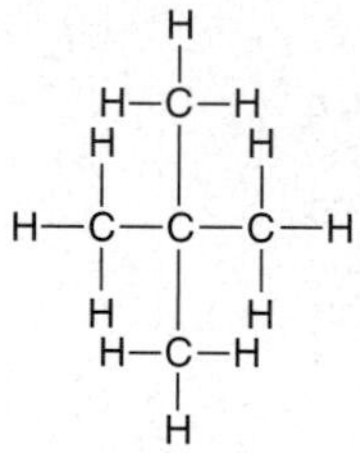	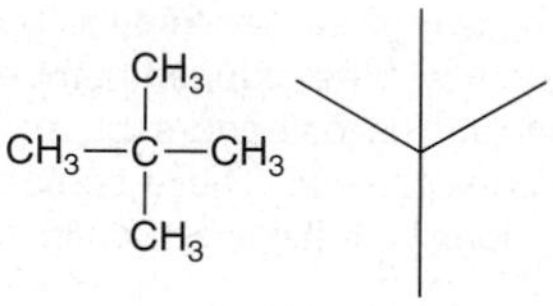 $CH_3—C(CH_3)_2—CH_3$		0.614	9.5

methyl group at C-2 of a four-C chain is 2-methylbutane (common name, *iso*pentane). The third isomer has two methyl branches on C-2 of a three-C chain, so its name is 2,2-dimethylpropane (common name, *neo*pentane).

Because alkanes are nearly nonpolar, we expect their physical properties to be determined by dispersion forces, and the boiling points in Table 15.3 certainly bear this out. The four-C alkanes boil lower than the five-C compounds. Moreover, within each group of isomers, the more spherical member (isobutane or neopentane) boils lower than the more elongated one (*n*-butane or *n*-pentane). As you saw in Chapter 12, this trend occurs because a spherical shape leads to less intermolecular contact, and thus lower total dispersion forces, than does an elongated shape.

A particularly clear example of the effect of dispersion forces on physical properties occurs among the unbranched alkanes (*n*-alkanes). Among these compounds, boiling points increase steadily with chain length: the longer the chain, the greater the intermolecular contact, the stronger the dispersion forces, and the higher the boiling point (Figure 15.7). Pentane (five C atoms) is the smallest *n*-alkane that exists as a liquid at room temperature. The solubility of alkanes, and of all hydrocarbons, is easy to predict from the like-dissolves-like rule (Section 13.1). Alkanes are miscible in each other and in other nonpolar solvents, such as benzene, but are nearly insoluble in water. The solubility of pentane in water, for example, is only 0.36 g/L at room temperature.

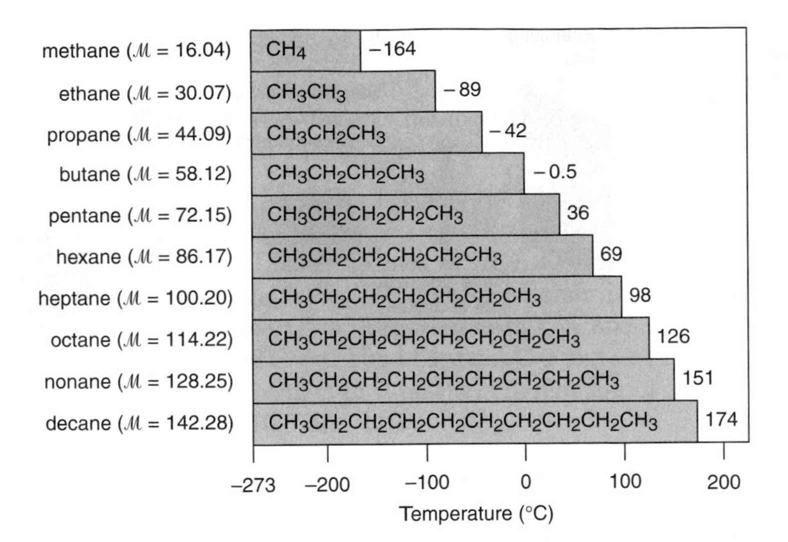

Figure 15.7 Boiling points of the first 10 unbranched alkanes. Boiling point increases smoothly with chain length because dispersion forces increase. Each entry includes the name, molar mass (ℳ, in g/mol), formula, and boiling point at 1 atm pressure.

Chiral Molecules and Optical Isomerism

Another type of isomerism exhibited by some alkanes and many other organic (as well as some inorganic) compounds is called *stereoisomerism*. **Stereoisomers** are molecules with the same arrangement of atoms *but different orientations of groups in space*. *Optical isomerism* is one type of stereoisomerism: *when two objects are mirror images of each other and cannot be superimposed, they are* ***optical isomers,*** also called *enantiomers*. To use a familiar example, your right hand is an optical isomer of your left. Look at your right hand in a mirror, and you will see that the *image* is identical to your left hand (Figure 15.8). No matter how you twist your arms around, however, your hands cannot lie on top of each other with all parts superimposed. They are not superimposable because each is *asymmetric:* there is no plane of symmetry that divides your hand into two identical parts.

An asymmetric molecule is called **chiral** (Greek *cheir,* "hand"). Typically, an organic molecule *is chiral if it contains a carbon atom that is bonded to four* ***different*** *groups*. This C atom is called a *chiral center* or an asymmetric carbon. In 3-methylhexane, for example, C-3 is a chiral center because it is bonded to four different groups: H—, CH_3—, CH_3—CH_2—, and CH_3—CH_2—CH_2— (Figure 15.9A). Like your two hands, the two forms are mirror images and cannot be superimposed on each other: when two of the groups are superimposed, the other two are opposite each other. Thus, the two forms are optical isomers. The central C atom in the amino acid alanine is also a chiral center (Figure 15.9B).

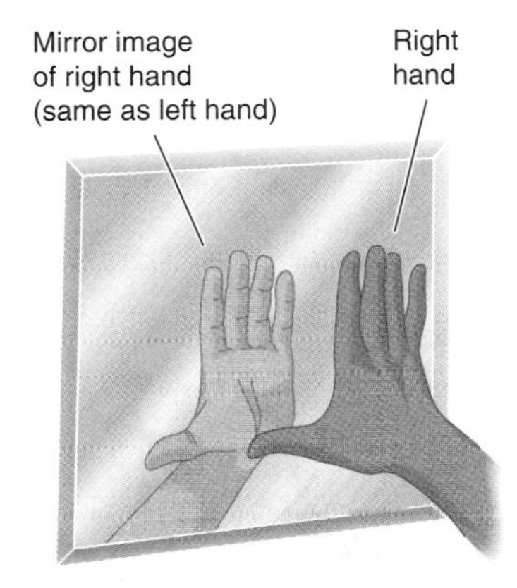

Figure 15.8 An analogy for optical isomers. The reflection of your right hand looks like your left hand. Each hand is asymmetric, so you cannot superimpose them with your palms facing in the same direction.

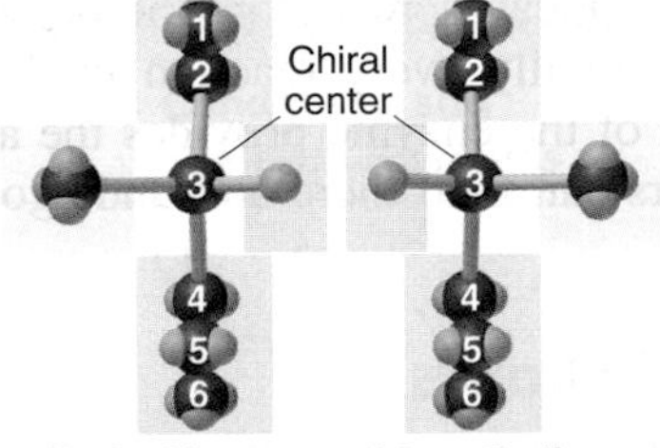

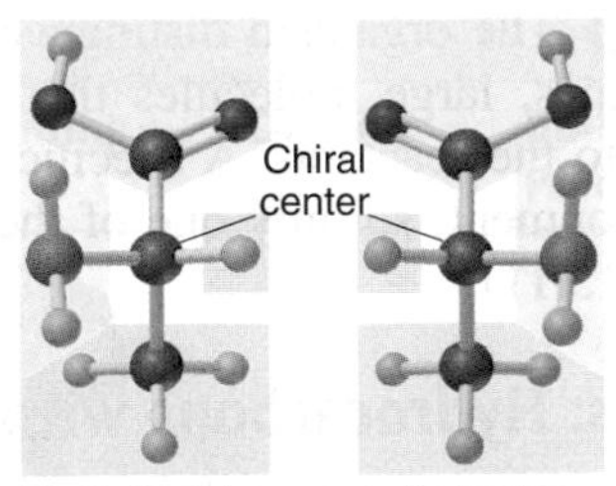

Figure 15.9 Two chiral molecules. A, 3-Methylhexane is chiral because C-3 is bonded to four different groups. These two models are optical isomers (enantiomers). **B,** The central C in the amino acid alanine is also bonded to four different groups.

Elimination of two H atoms requires inorganic oxidizing agents, such as $K_2Cr_2O_7$ in aqueous H_2SO_4. As we saw in Section 15.3, the reaction is an oxidation and produces the C=O group (shown with condensed and carbon-skeleton formulas):

$$\mathrm{CH_3{-}CH_2{-}\underset{}{\overset{OH}{\overset{|}{C}}}H{-}CH_3} \xrightarrow[H_2SO_4]{K_2Cr_2O_7} \mathrm{CH_3{-}CH_2{-}\overset{O}{\overset{\|}{C}}{-}CH_3}$$

2-butanol → 2-butanone

For alcohols with an OH group at the end of the chain (R—CH$_2$—OH), another oxidation occurs. Wine turns sour, for example, when the ethanol in contact with air is oxidized to acetic acid:

$$\mathrm{CH_3{-}\overset{OH}{\overset{|}{C}}H_2} \xrightarrow[-H_2O]{\frac{1}{2}O_2} \mathrm{CH_3{-}\overset{O}{\overset{\|}{C}}H} \xrightarrow{\frac{1}{2}O_2} \mathrm{CH_3{-}\overset{O}{\overset{\|}{C}}{-}OH}$$

Substitution yields products with other single-bonded functional groups. With hydrohalic acids, many alcohols give haloalkanes:

$$\mathrm{R_2CH{-}OH + HBr \longrightarrow R_2CH{-}Br + HOH}$$

As you'll see below, *the C atom undergoing the change in a substitution is bonded to a more electronegative element,* which makes it partially positive and, thus, a target for a negatively charged or electron-rich group of an incoming reactant.

Haloalkanes A *halogen* atom (X) bonded to C gives the **haloalkane** functional group, $-\overset{|}{\underset{|}{C}}-\ddot{\underset{..}{X}}:$, and compounds with the general formula R—X. Haloalkanes (common name, **alkyl halides**) are named by adding the halogen as a prefix to the hydrocarbon name and numbering the C atom to which the halogen is attached, as in bromomethane, 2-chloropropane, or 1,3-diiodohexane.

Just as many alcohols undergo substitution to alkyl halides when treated with halide ions in acid, many halides undergo substitution to alcohols in base. For example, OH^- attacks the positive C end of the C—X bond and displaces X^-:

$$\mathrm{CH_3{-}CH_2{-}CH_2{-}CH_2{-}Br + OH^- \longrightarrow CH_3{-}CH_2{-}CH_2{-}CH_2{-}OH + Br^-}$$

1-bromobutane → 1-butanol

Substitution by groups such as —CN, —SH, —OR, and —NH$_2$ allows chemists to convert alkyl halides to a host of other compounds.

Just as addition of HX *to* an alkene produces haloalkanes, elimination of HX *from* a haloalkane by reaction with a strong base, such as potassium ethoxide, produces an alkene:

$$\mathrm{CH_3{-}\underset{Cl}{\underset{|}{\overset{CH_3}{\overset{|}{C}}}}{-}CH_3 + CH_3{-}CH_2{-}OK \longrightarrow CH_3{-}\overset{CH_3}{\overset{|}{C}}{=}CH_2 + KCl + CH_3{-}CH_2{-}OH}$$

2-chloro-2-methylpropane, potassium ethoxide → 2-methylpropene

Haloalkanes have many important uses, but many are carcinogenic in mammals and have severe neurological effects in humans. Their stability has made them notorious environmental contaminants.

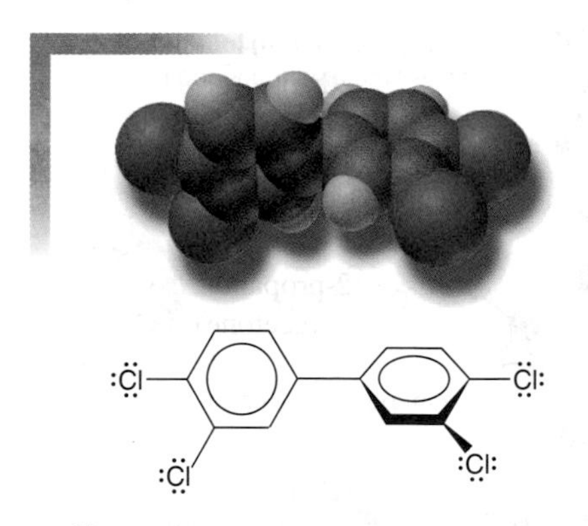

Pollutants in the Food Chain Until recently, halogenated aromatic hydrocarbons, such as the *polychlorinated biphenyls* (*PCBs;* one of 209 different compounds is shown above), were used as insulating fluids in electrical transformers and then discharged in wastewater. Because of their low solubility and high stability, they accumulate for decades in river and lake sediment and are eaten by microbes and invertebrates. Fish eat the invertebrates, and birds and mammals, including humans, eat the fish. PCBs become increasingly concentrated in body fat at each stage. As a result of their health risks, PCBs in natural waters present an enormous cleanup problem.

Amines The **amine** functional group is $-\overset{|}{\underset{|}{C}}-\overset{|}{\underset{|}{N}}:$. Chemists classify amines as derivatives of ammonia, with R groups in place of one or more H atoms. *Primary* (1°) amines are RNH$_2$, *secondary* (2°) amines are R$_2$NH, and *tertiary* (3°) amines are R$_3$N. Like ammonia, amines have trigonal pyramidal shapes and a lone pair

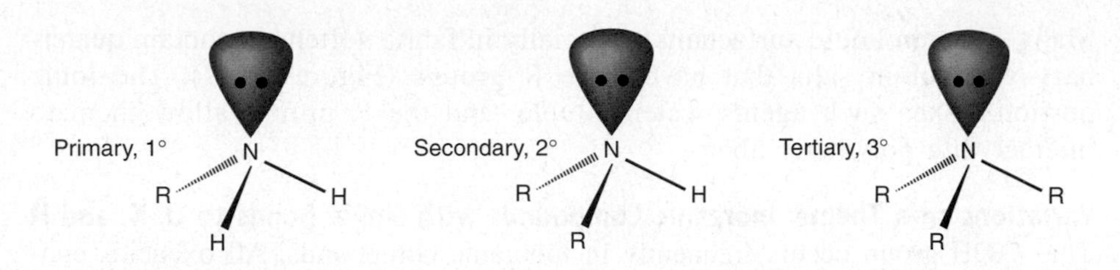

Figure 15.15 General structures of amines. Amines have a trigonal pyramidal shape and are classified by the number of R groups bonded to N. The lone pair on the nitrogen atom is the key to amine reactivity.

of electrons on a partially negative N atom (Figure 15.15). Systematic names drop the final *-e* of the alkane and add the suffix *-amine,* as in ethanamine. However, there is still wide usage of common names, in which the suffix *-amine* follows the name of the alkyl group; thus, methylamine has one methyl group attached to N, diethylamine has two ethyl groups attached, and so forth. Figure 15.16 shows that the amine functional group occurs in many biomolecules.

Primary and secondary amines can form H bonds, so they have higher melting and boiling points than hydrocarbons and alkyl halides of similar molar mass. For example, dimethylamine ($\mathcal{M}$ = 45.09 g/mol) boils 45°C higher than ethyl fluoride ($\mathcal{M}$ = 48.06 g/mol). Trimethylamine has a greater molar mass than dimethylamine, but it melts more than 20°C *lower* because trimethylamine molecules are not H bonded.

Amines of low molar mass are fishy smelling, water soluble, and weakly basic. The reaction with water proceeds only slightly to the right to reach equilibrium:

$$CH_3-\ddot{N}H_2 + H_2O \rightleftharpoons CH_3-\overset{+}{N}H_3 + OH^-$$

Amines undergo substitution reactions in which the lone pair on N attacks the partially positive C in alkyl halides to displace X^- and form a larger amine:

$$2CH_3-CH_2-\ddot{N}H_2 + CH_3-CH_2-Cl \longrightarrow CH_3-CH_2-\ddot{N}H(CH_2-CH_3) + CH_3-CH_2-\overset{+}{N}H_3Cl^-$$

ethylamine, chloroethane, diethylamine, ethylammonium chloride

(One molecule of ethylamine participates in the substitution, while the other binds the released H^+ and prevents it from remaining on the diethylamine product.)

Four R groups bonded to an N atom give an ionic compound called a *quaternary (4°) ammonium salt:*

$$4NH_3 + 4RCl \longrightarrow 3NH_4Cl + R_4N^+Cl^- \quad \text{(a quaternary ammonium chloride)}$$

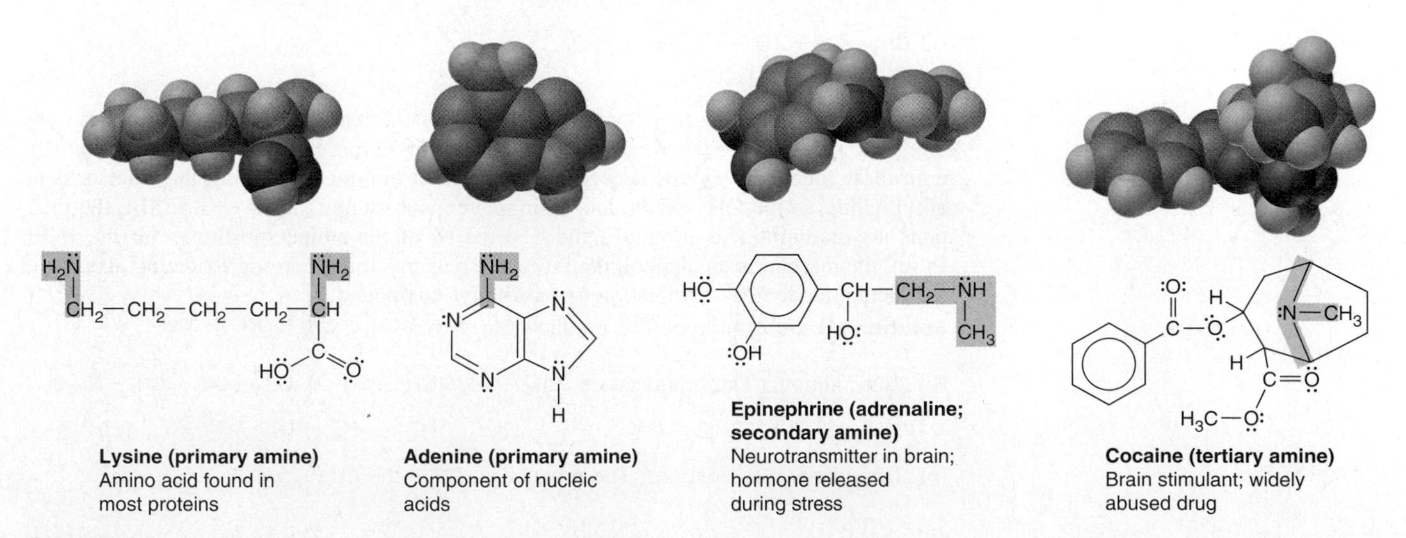

Figure 15.16 Some biomolecules with the amine functional group.

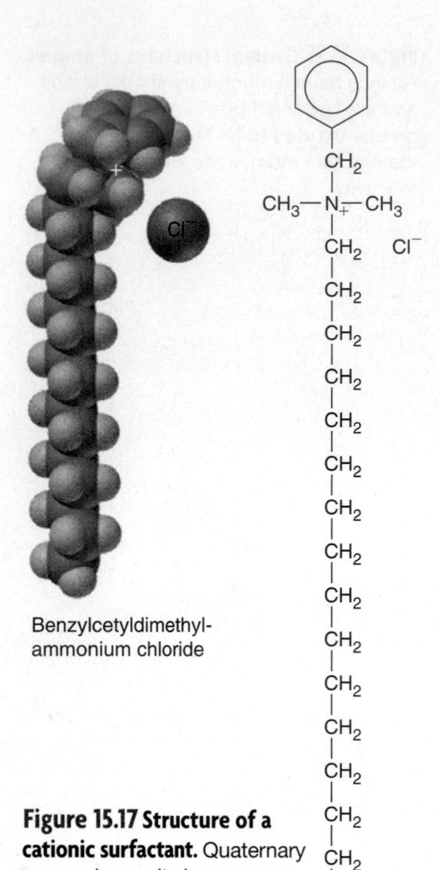

Figure 15.17 Structure of a cationic surfactant. Quaternary ammonium salts have many industrial uses.

Many common liquid surfactants, especially in fabric softeners, contain quaternary ammonium salts that have large R groups (Figure 15.17). The ionic portion makes such agents water soluble, and the R groups allow them to interact with polymeric fibers.

Variations on a Theme: Inorganic Compounds with Single Bonds to O, X, and N The —OH group occurs frequently in inorganic compounds. All oxoacids contain at least one —OH, usually bonded to a relatively electronegative nonmetal atom, which in most cases is bonded to other O atoms. Oxoacids are acidic in water because these additional O atoms pull electron density from the central nonmetal, which pulls electron density from the O—H bond, releasing an H^+ ion and stabilizing the oxoanion through resonance. Alcohols are *not* acidic in water because they lack the additional O atoms and the electronegative nonmetal.

Halides of nearly every nonmetal are known, and many undergo substitution reactions in base. As in the case of an alkyl halide, the process involves an attack on the partially positive central atom by OH^-:

$$H{-}O^- + \overset{\delta+}{C}{-}\overset{\delta-}{Cl} \longrightarrow C{-}OH + Cl^-$$

$$H{-}O^- + \overset{\delta+}{B}{-}\overset{\delta-}{Cl} \longrightarrow B{-}OH + Cl^-$$

Thus, alkyl halides undergo the same general reaction as other nonmetal halides, such as BCl_3, SiF_4, and PCl_5.

The bonds between nitrogen and larger nonmetals, such as Si, P, and S, have significant double-bond character, which affects structure and reactivity. For example, trisilylamine, the Si analog of trimethylamine (see Figure 14.16, p. 578), is planar, rather than trigonal pyramidal, partially as a result of p,d-π bonding. The lone pair on N is delocalized in this π bond, so trisilylamine is not basic.

SAMPLE PROBLEM 15.4 Predicting the Reactions of Alcohols, Alkyl Halides, and Amines

Problem Determine the reaction type and predict the product(s) for each of the following reactions:

(a) $CH_3{-}CH_2{-}CH_2{-}I + NaOH \longrightarrow$

(b) $CH_3{-}CH_2{-}Br + 2CH_3{-}CH_2{-}CH_2{-}NH_2 \longrightarrow$

(c) $CH_3{-}\underset{OH}{\underset{|}{CH}}{-}CH_3 \xrightarrow[H_2SO_4]{Cr_2O_7^{2-}}$

Plan We first determine the functional group(s) of the reactant(s) and then examine any inorganic reagent(s) to decide on the possible reaction type, keeping in mind that, in general, these functional groups undergo substitution or elimination. In **(a)**, the reactant is an alkyl halide, so the OH^- of the inorganic reagent substitutes for the —I. In **(b)**, the reactants are an amine and an alkyl halide, so the N: of the amine substitutes for the —Br. In **(c)**, the reactant is an alcohol, the inorganic reagents form a strong oxidizing agent, and the alcohol undergoes elimination to a carbonyl compound.

Solution (a) Substitution: The products are $CH_3{-}CH_2{-}CH_2{-}OH + NaI$

(b) Substitution: The products are $CH_3{-}CH_2{-}CH_2{-}\underset{CH_2-CH_3}{\underset{|}{NH}} + CH_3{-}CH_2{-}CH_2{-}\overset{+}{N}H_3Br^-$

(c) Elimination: (oxidation): The product is $CH_3{-}\underset{O}{\underset{\|}{C}}{-}CH_3$

Check The only changes should be at the functional group.

FOLLOW-UP PROBLEM 15.4 Fill in the blank in each reaction. (*Hint:* Examine any inorganic compounds and the organic product to determine the organic reactant.)

(a) ____ + $CH_3{-}ONa \longrightarrow CH_3{-}CH{=}C(CH_3){-}CH_3 + NaCl + CH_3{-}OH$

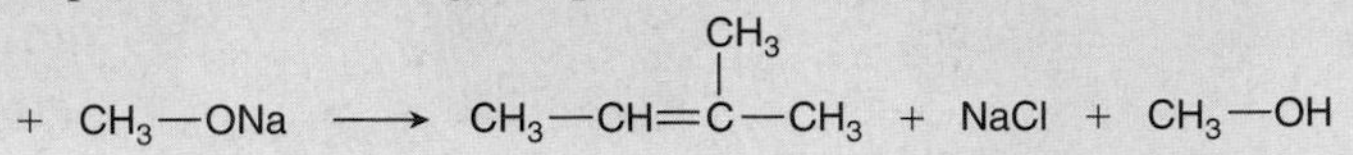

(b) ____ $\xrightarrow[H_2SO_4]{Cr_2O_7^{2-}} CH_3{-}CH_2{-}C({=}O){-}OH$

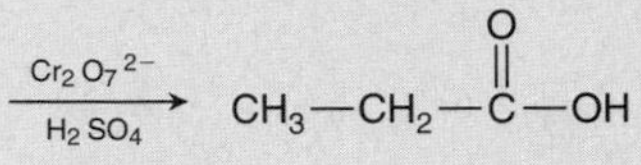

Functional Groups with Double Bonds

The most important functional groups with double bonds are the C=C of alkenes and the C=O of aldehydes and ketones. Both appear in many organic and biological molecules. *Their most common reaction type is addition.*

Comparing the Reactivity of Alkenes and Aromatic Compounds The C=C bond is the essential portion of the alkene functional group, >C=C<. Although they can be further unsaturated to alkynes, *alkenes typically undergo addition.* The electron-rich double bond is readily attracted to the partially positive H atoms of hydronium ions and hydrohalic acids, yielding alcohols and alkyl halides, respectively:

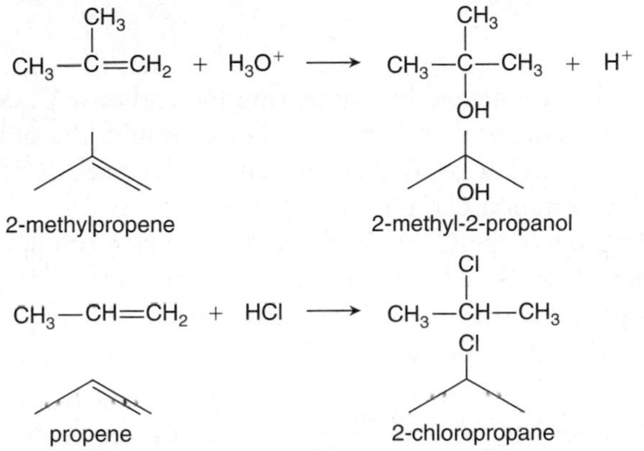

$$CH_3{-}C(CH_3){=}CH_2 + H_3O^+ \longrightarrow CH_3{-}C(CH_3)(OH){-}CH_3 + H^+$$

2-methylpropene → 2-methyl-2-propanol

$$CH_3{-}CH{=}CH_2 + HCl \longrightarrow CH_3{-}CH(Cl){-}CH_3$$

propene → 2-chloropropane

The *localized* unsaturation of alkenes is very different from the *delocalized* unsaturation of benzene. Since benzene does *not* have double bonds, despite the way we depict its resonance forms, it does *not* behave like an alkene. Benzene, for example, does not decolorize bromine because there are no isolated π-electron pairs to bond with Br_2.

In general, aromatic rings are much *less* reactive than alkenes because delocalized π electrons stabilize such a ring, making it *lower* in energy than one with localized π electrons. Data indicate that benzene is about 150 kJ/mol more stable than a six-C ring with three C=C bonds would be; in other words, benzene requires about 150 kJ/mol more energy to react.

An *addition* reaction with benzene, therefore, requires additional energy to break up the delocalized π system. Benzene does undergo many *substitution* reactions, however, in which the delocalization is retained when an H atom attached to a ring C is replaced by another group:

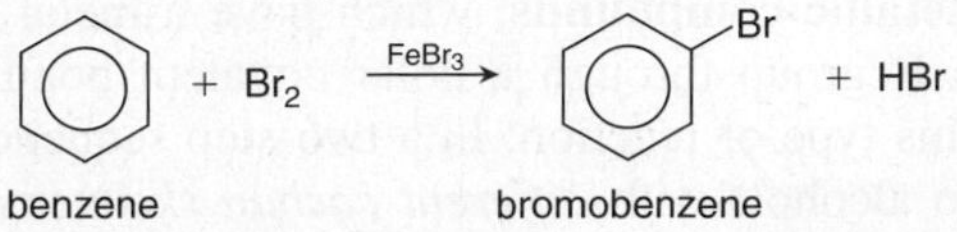

$$C_6H_6 + Br_2 \xrightarrow{FeBr_3} C_6H_5Br + HBr$$

benzene → bromobenzene

Aldehydes and Ketones The C=O bond, or **carbonyl group,** is one of the most chemically versatile. In the **aldehyde** functional group, the carbonyl C is bonded to H (and often to another C), so it occurs *at the end of a chain,* R—C(H)=O.

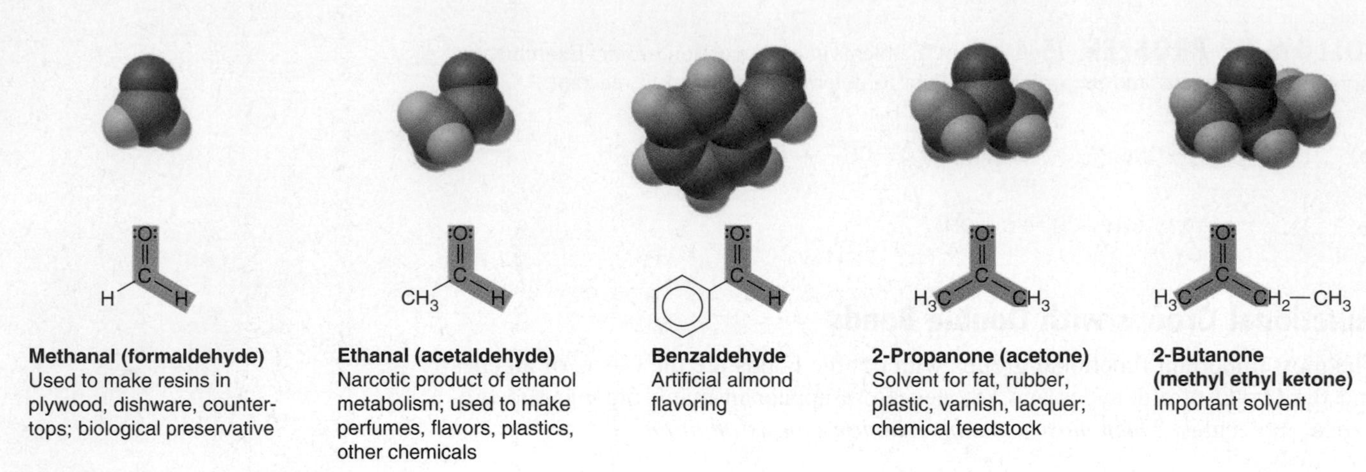

Methanal (formaldehyde) Used to make resins in plywood, dishware, countertops; biological preservative

Ethanal (acetaldehyde) Narcotic product of ethanol metabolism; used to make perfumes, flavors, plastics, other chemicals

Benzaldehyde Artificial almond flavoring

2-Propanone (acetone) Solvent for fat, rubber, plastic, varnish, lacquer; chemical feedstock

2-Butanone (methyl ethyl ketone) Important solvent

Figure 15.18 Some common aldehydes and ketones.

Aldehyde names drop the final *-e* from the alkane name and add *-al;* thus, the three-C aldehyde is propanal. In the **ketone** functional group, the carbonyl C is bonded to two other C atoms, $-\overset{|}{\underset{|}{C}}-\overset{:O:}{\overset{\|}{C}}-\overset{|}{\underset{|}{C}}-$, so it occurs *within the chain.* Ketones, $R-\overset{O}{\overset{\|}{C}}-R'$, are named by numbering the carbonyl C, dropping the final *-e* from the alkane name, and adding *-one*. For example, the unbranched, five-C ketone with the carbonyl C as C-2 in the chain is named 2-pentanone. Figure 15.18 shows some common carbonyl compounds.

Like the C=C bond, the C=O bond is *electron rich;* unlike the C=C bond, it is *highly polar* ($\Delta EN = 1.0$). Figure 15.19 emphasizes this polarity with an electron density model and a charged resonance form. Aldehydes and ketones are formed by the oxidation of alcohols:

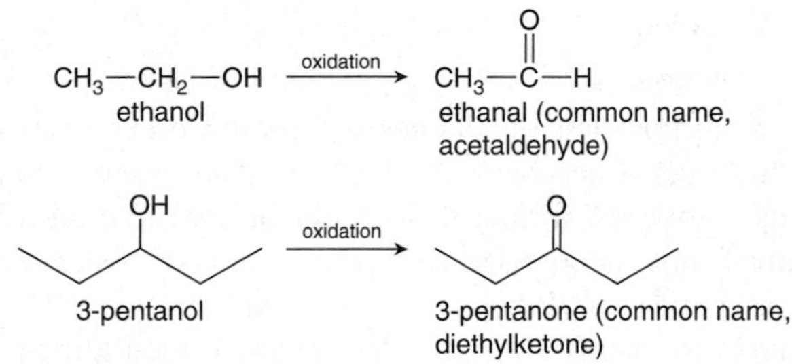

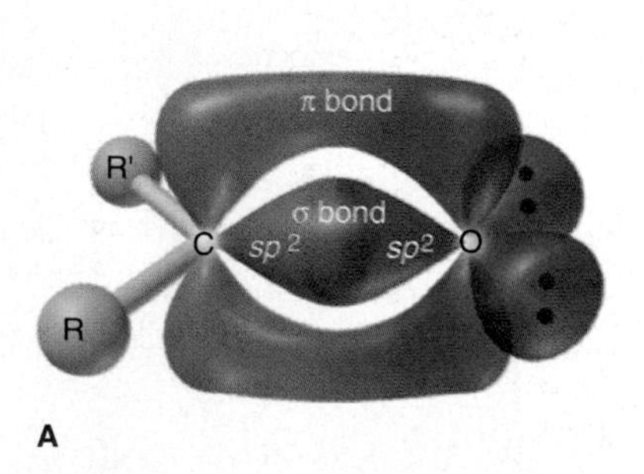

A

$$R'R\,C=\ddot{O} \longleftrightarrow R'R\,C^{+}-\ddot{O}:^{-}$$

B

Figure 15.19 The carbonyl group. A, The σ and π bonds that make up the C=O bond of the carbonyl group. **B,** The charged resonance form shows that the C=O bond is polar (ΔEN = 1.0).

Conversely, as a result of their unsaturation, carbonyl compounds can undergo *addition* and be reduced to alcohols:

$$\text{cyclobutanone} \xrightarrow{\text{reduction}} \text{cyclobutanol}$$

As a result of the bond polarity, addition often occurs with an electron-rich group bonding to the carbonyl C and an electron-poor group bonding to the carbonyl O. **Organometallic compounds,** which have a metal atom (usually Li or Mg) attached to an R group through a polar covalent bond (Sections 14.3 and 14.4), take part in this type of reaction. In a two-step sequence, they convert carbonyl compounds to alcohols with *different carbon skeletons:*

$$R-\overset{\delta+}{C}H=\overset{\delta-}{O} + \overset{\delta-}{R'}-\overset{\delta+}{Li} \longrightarrow \xrightarrow{H_2O} R-\overset{OH}{\overset{|}{C}}H-R' + LiOH$$

In the following reaction steps, for example, the electron-rich C bonded to Li in ethyllithium, CH_3CH_2—Li, attacks the electron-poor carbonyl C of 2-propanone, adding its ethyl group; at the same time, the Li adds to the carbonyl O. Treating the mixture with water forms the C—OH group:

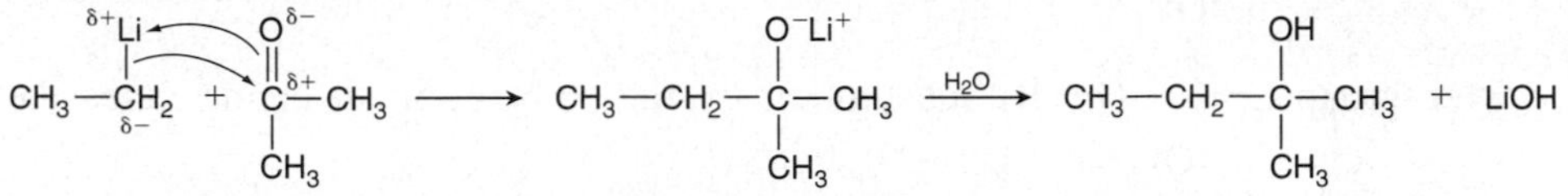

$$\overset{\delta+}{\text{Li}}\text{—}\underset{\delta-}{\text{CH}_2}\text{—CH}_3 + \overset{\delta+}{\text{C}}(=\overset{\delta-}{\text{O}})(\text{CH}_3)\text{—CH}_3 \longrightarrow \text{CH}_3\text{—CH}_2\text{—C}(\text{O}^-\text{Li}^+)(\text{CH}_3)\text{—CH}_3 \xrightarrow{H_2O} \text{CH}_3\text{—CH}_2\text{—C}(\text{OH})(\text{CH}_3)\text{—CH}_3 + \text{LiOH}$$

Note that the product skeleton combines the two reactant skeletons. The field of *organic synthesis* often employs organometallic compounds to create molecules with different skeletons and, thus, synthesize new compounds.

SAMPLE PROBLEM 15.5 Predicting the Steps in a Reaction Sequence

Problem Fill in the blanks in the following reaction sequence:

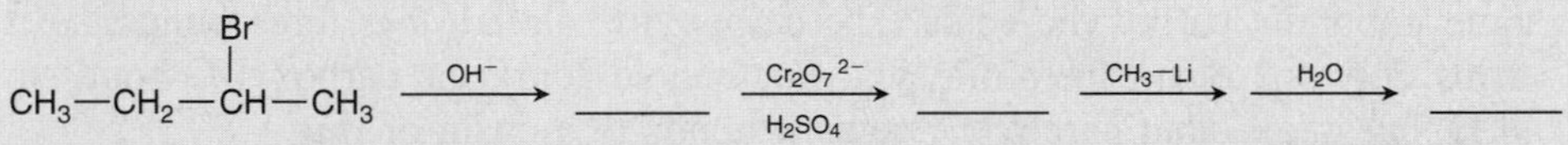

$$\text{CH}_3\text{—CH}_2\text{—CH(Br)—CH}_3 \xrightarrow{OH^-} \underline{\qquad} \xrightarrow[H_2SO_4]{Cr_2O_7^{2-}} \underline{\qquad} \xrightarrow{CH_3\text{—}Li} \xrightarrow{H_2O} \underline{\qquad}$$

Plan For each step, we examine the functional group of the reactant and the reagent above the yield arrow to decide on the most likely product.

Solution The sequence starts with an alkyl halide reacting with OH^-. Substitution gives an alcohol. Oxidation of this alcohol with acidic dichromate gives a ketone. Finally, a two-step reaction of a ketone with CH_3—Li and then water forms an alcohol with a carbon skeleton that has the CH_3— group attached to the carbonyl C:

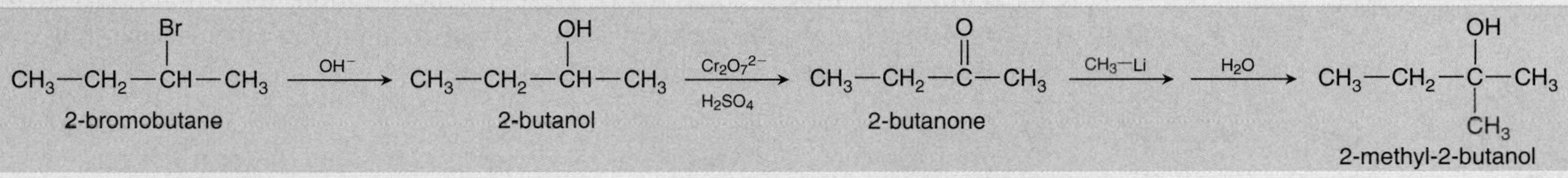

$$\underset{\text{2-bromobutane}}{\text{CH}_3\text{—CH}_2\text{—CH(Br)—CH}_3} \xrightarrow{OH^-} \underset{\text{2-butanol}}{\text{CH}_3\text{—CH}_2\text{—CH(OH)—CH}_3} \xrightarrow[H_2SO_4]{Cr_2O_7^{2-}} \underset{\text{2-butanone}}{\text{CH}_3\text{—CH}_2\text{—C(=O)—CH}_3} \xrightarrow{CH_3\text{—}Li} \xrightarrow{H_2O} \underset{\text{2-methyl-2-butanol}}{\text{CH}_3\text{—CH}_2\text{—C(OH)(CH}_3\text{)—CH}_3}$$

Check In this case, make sure that the first two reactions alter the functional group only and that the final steps change the C skeleton.

FOLLOW-UP PROBLEM 15.5 Choose reactants to obtain the following products:

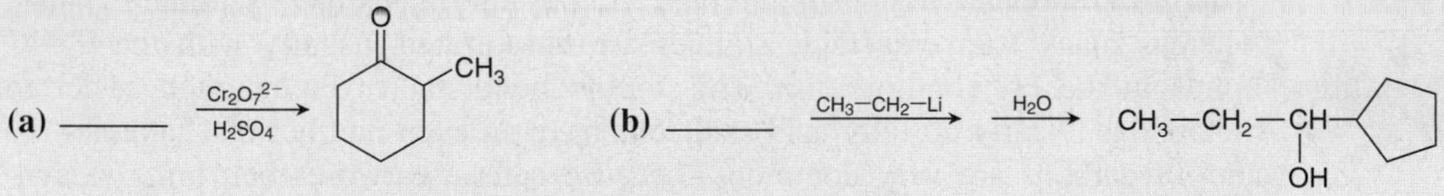

(a) $\underline{\qquad} \xrightarrow[H_2SO_4]{Cr_2O_7^{2-}}$ 2-methylcyclohexanone (cyclohexane ring with C=O and adjacent CH_3)

(b) $\underline{\qquad} \xrightarrow{CH_3\text{—}CH_2\text{—}Li} \xrightarrow{H_2O} \text{CH}_3\text{—CH}_2\text{—CH(OH)—}$cyclopentyl

Variations on a Theme: Inorganic Compounds with Double Bonds Homonuclear (same kind of atom) double bonds are rare for atoms other than C. However, we've seen many double bonds between O and other nonmetals, for example, in the oxides of S, N, and the halogens. Like carbonyl compounds, these substances undergo addition reactions. For example, the partially negative O of water attacks the partially positive S of SO_3 to form sulfuric acid:

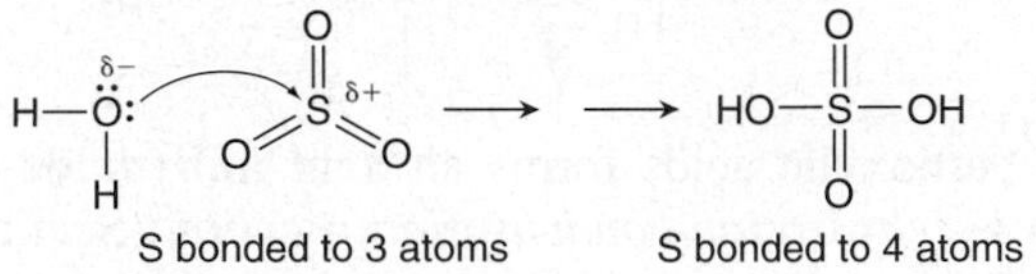

$$\text{H—}\overset{\delta-}{\ddot{\text{O}}}\text{:(—H)} + \overset{\delta+}{\text{S}}\text{O}_3 \longrightarrow \longrightarrow \text{HO—S(=O)}_2\text{—OH}$$

S bonded to 3 atoms S bonded to 4 atoms

Functional Groups with Both Single and Double Bonds

A family of three functional groups contains C double bonded to O (a carbonyl group) *and* single bonded to O or N. The parent of the family is the **carboxylic acid** group, —C(=Ö:)—ÖH, also called the *carboxyl group* and written —COOH.

The most important reaction type of this family is substitution from one member to another. Substitution for the —OH by the —OR of an alcohol gives the **ester** group, $-\overset{\overset{\Large :O:}{\|}}{C}-\ddot{\underset{..}{O}}-R$; substitution by the $-\ddot{N}-$ of an amine gives the **amide** group, $-\overset{\overset{\Large :O:}{\|}}{C}-\underset{..}{N}-$.

Carboxylic Acids Carboxylic acids, $R-\overset{\overset{O}{\|}}{C}-OH$, are named by dropping the *-e* from the alkane name and adding *-oic acid;* however, many common names are used. For example, the four-C acid is butanoic acid (the carboxyl C is counted when choosing the root); its common name is butyric acid. Figure 15.20 shows some important carboxylic acids. The carboxyl C already has three bonds, so it forms only one other. In formic acid (methanoic acid), the carboxyl C bonds to an H, but in all other carboxylic acids it bonds to a chain or ring.

Carboxylic acids are weak acids in water:

$$CH_3-\overset{\overset{O}{\|}}{C}-OH(l) + H_2O(l) \rightleftharpoons CH_3-\overset{\overset{O}{\|}}{C}-O^-(aq) + H_3O^+(aq)$$

ethanoic acid
(acetic acid)

At equilibrium in acid solutions of typical concentration, more than 99% of the acid molecules are undissociated at any given moment. In strong base, however, they react completely to form a salt and water:

$$CH_3-\overset{\overset{O}{\|}}{C}-OH(l) + NaOH(aq) \longrightarrow CH_3-\overset{\overset{O}{\|}}{C}-O^-(aq) + Na^+(aq) + H_2O(l)$$

The anion is the *carboxylate ion,* named by dropping *-oic acid* and adding *-oate;* the sodium salt of butanoic acid, for instance, is sodium butanoate.

Carboxylic acids with long hydrocarbon chains are **fatty acids,** an essential group of compounds found in all cells. Animal fatty acids have saturated chains, whereas many from vegetable sources are unsaturated, usually with the C=C bonds in the *cis* configuration. The double bond makes them much easier to metabolize. Nearly all fatty acid skeletons have an even number of C atoms—16 and 18 carbons are very common—because cells use two-carbon units in synthesizing them. Fatty acid salts are soaps, with the cation usually from Group 1A(1) or 2A(2) (Section 13.2).

Substitution of carboxylic acids occurs through a two-step sequence: *addition plus elimination equals substitution.* Addition to the trigonal planar shape of the carbonyl group gives an unstable tetrahedral intermediate, which immediately undergoes elimination to revert to a trigonal planar product:

$$R-\overset{\overset{O}{\|}}{C}-X + Z-Y \underset{}{\overset{\text{addition}}{\rightleftharpoons}} \left[R-\overset{\overset{O-Z}{|}}{\underset{\underset{Y}{|}}{C}}-X \right] \overset{\text{elimination}}{\rightleftharpoons} R-\overset{\overset{O}{\|}}{C}-Y + Z-X$$

Strong heating of carboxylic acids forms an **acid anhydride** through a type of substitution called a *dehydration-condensation reaction* (Section 14.7), in which two molecules condense into one with loss of water:

$$R-\overset{\overset{O}{\|}}{C}-\boxed{OH + H}O-\overset{\overset{O}{\|}}{C}-R \xrightarrow{\Delta} R-\overset{\overset{O}{\|}}{C}-O-\overset{\overset{O}{\|}}{C}-R + HOH$$

Esters *An alcohol and a carboxylic acid form an ester.* The first part of an ester name designates the alcohol portion and the second the acid portion (named in the same way as the carboxylate ion). For example, the ester formed between

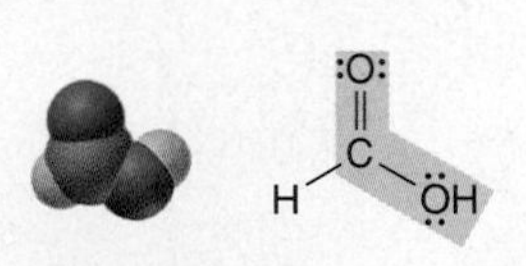

Methanoic acid (formic acid)
An irritating component of ant and bee stings

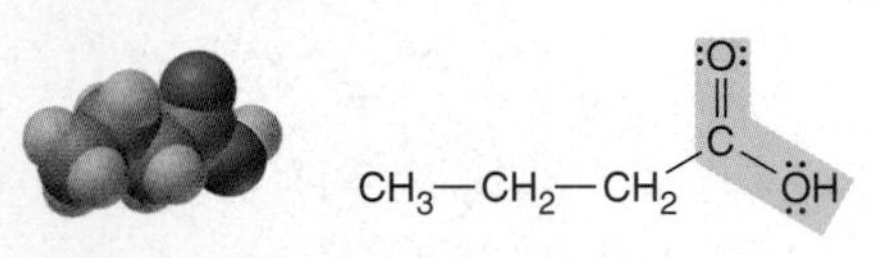

Butanoic acid (butyric acid)
Odor of rancid butter; suspected component of monkey sex attractant

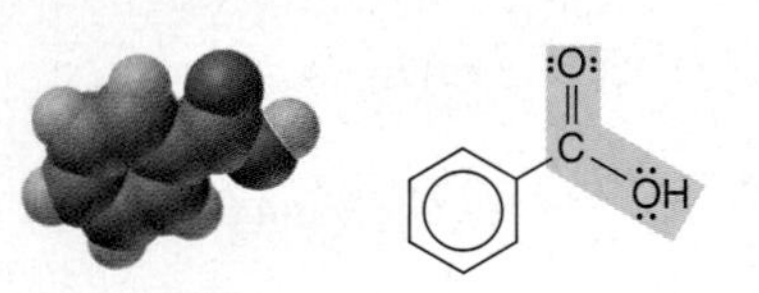

Benzoic acid
Calorimetric standard; used in preserving food, dyeing fabric, curing tobacco

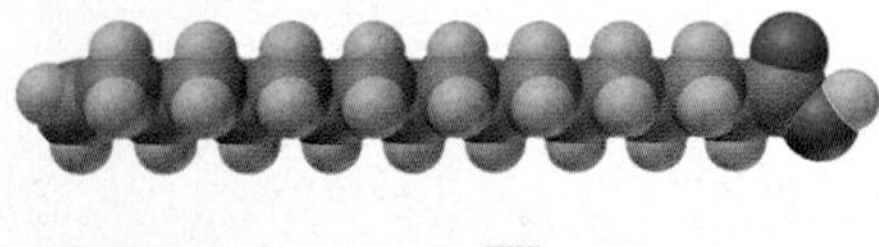

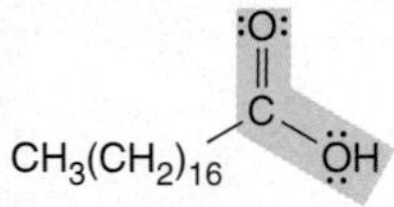

Octadecanoic acid (stearic acid)
Found in animal fats; used in making candles and soaps

Figure 15.20 Some molecules with the carboxylic acid functional group.

$CH_3(CH_2)_{14}C(=O)-O-CH_2(CH_2)_{14}CH_3$

Cetyl palmitate The most common lipid in whale blubber

$CH_3(CH_2)_{14}C(=O)-O-CH_2$
$CH_3(CH_2)_{16}C(=O)-O-CH$
$CH_2-O-P(=O)(O^-)-O-CH_2-CH_2-N^+(CH_3)_3$

Lecithin Phospholipid found in all cell membranes

$CH_3(CH_2)_{16}C(=O)-O-CH_2$
$CH_3(CH_2)_{16}C(=O)-O-CH$
$CH_3(CH_2)_{16}C(=O)-O-CH_2$

Tristearin Typical dietary fat used as an energy store in animals

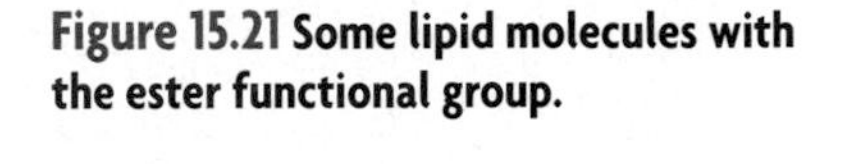

Figure 15.21 Some lipid molecules with the ester functional group.

ethanol and ethanoic acid is ethyl ethanoate (common name, ethyl acetate), a solvent for nail polish and model glue.

The ester group occurs commonly in **lipids,** a large group of fatty biological substances. Most dietary fats are *triglycerides,* esters composed of three fatty acids linked to the alcohol 1,2,3-trihydroxypropane (common name, glycerol) that function as energy stores. Some important lipids are shown in Figure 15.21; lecithin is one of several phospholipids that make up the lipid bilayer in all cell membranes (Section 13.2).

Esters, like acid anhydrides, form through a dehydration-condensation reaction; in this case, it is called an *esterification*:

$$R-\overset{O}{\overset{\|}{C}}-OH + H-O-R' \underset{}{\overset{H^+}{\rightleftharpoons}} R-\overset{O}{\overset{\|}{C}}-O-R' + HOH$$

In Chapter 16, we will discuss the role of H^+ in increasing the rate of this multistep reaction.

Note that the esterification reaction is reversible. The opposite of dehydration-condensation is called **hydrolysis,** in which the O atom of water is attracted to the partially positive C atom of the ester, cleaving (lysing) the molecule into two parts. One part receives water's —OH, and the other part receives water's other H. In the process of soap manufacture, or *saponification* (Latin *sapon,* "soap"), begun in ancient times, ester bonds in animal or vegetable fats are hydrolyzed with strong base:

$$\begin{matrix} R-\overset{O}{\overset{\|}{C}}-O-CH_2 \\ R'-\overset{O}{\overset{\|}{C}}-O-CH \\ R''-\overset{O}{\overset{\|}{C}}-O-CH_2 \end{matrix} + 3NaOH \xrightarrow{\Delta} \begin{matrix} R-\overset{O}{\overset{\|}{C}}-O^-Na^+ \\ + \\ R'-\overset{O}{\overset{\|}{C}}-O^-Na^+ \\ + \\ R''-\overset{O}{\overset{\|}{C}}-O^-Na^+ \end{matrix} + \begin{matrix} HO-CH_2 \\ HO-CH \\ HO-CH_2 \end{matrix}$$

a triglyceride — 3 soaps — glycerol

Amides The product of a substitution between an amine (or NH_3) and an ester is an amide. The partially negative N of the amine is attracted to the partially positive C of the ester, an alcohol (ROH) is lost, and an amide forms:

$$CH_3-\overset{O}{\overset{\|}{C}}-O-CH_3 + H\ddot{N}(H)-CH_2-CH_3 \longrightarrow CH_3-\overset{O}{\overset{\|}{C}}-\overset{H}{\overset{|}{N}}-CH_2-CH_3 + CH_3-OH$$

methyl ethanoate (methyl acetate) — ethanamine (ethylamine) — *N*-ethylethanamide (*N*-ethylacetamide) — methanol

Amides are named by denoting the amine portion with *N*- and replacing *-oic acid* from the parent carboxylic acid with *-amide*. In the amide from the previous

A Pungent, Pleasant Banquet Many organic compounds have strong odors, but none can rival the diverse odors of carboxylic acids and esters. From the pungent, vinegary odors of the one-C and two-C acids to the cheesy stench of slightly larger ones, carboxylic acids possess some awful odors. When butanoic acid reacts with ethanol, however, its rancid-butter smell becomes the peachy pineapple scent of ethyl butanoate, and when pentanoic acid reacts with pentanol, its Limburger cheese odor becomes the fresh apple aroma of pentyl pentanoate. Naturally occurring and synthetic esters are used to add fruity, floral, and herbal odors to foods, cosmetics, household deodorizers, and medicines.

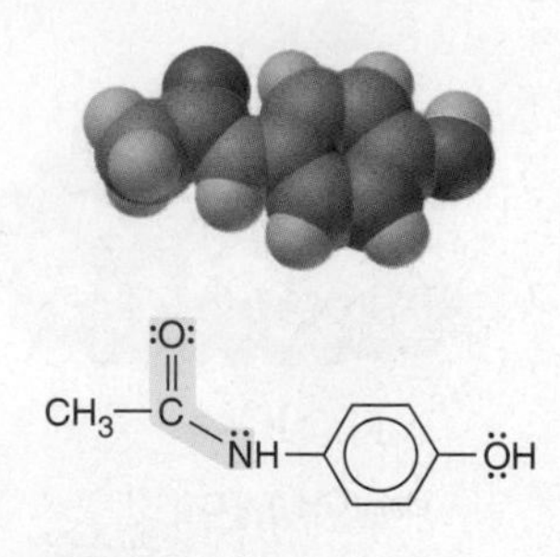

Acetaminophen
Active ingredient in nonaspirin pain relievers; used to make dyes and photographic chemicals

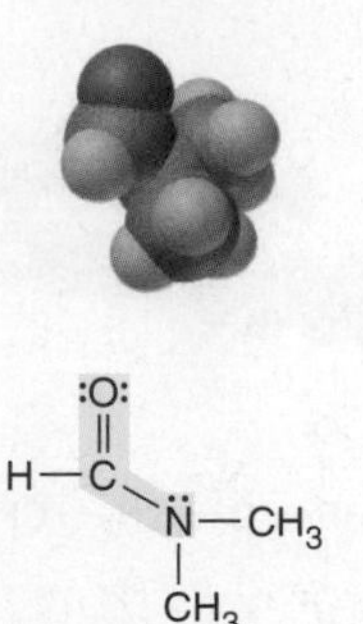

***N,N*-Dimethylmethanamide (dimethylformamide)**
Major organic solvent; used in production of synthetic fibers

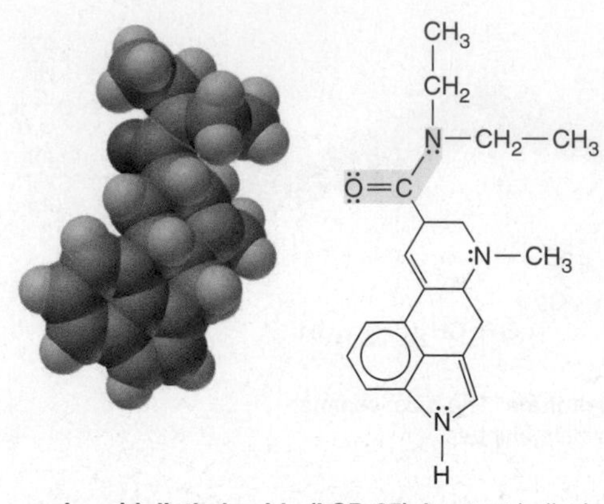

Lysergic acid diethylamide (LSD-25) A potent hallucinogen

Figure 15.22 Some molecules with the amide functional group.

reaction, the ethyl group comes from the amine, and the acid portion comes from ethanoic acid (acetic acid). Some amides are shown in Figure 15.22.

Amides are hydrolyzed in hot water (or base) to a carboxylic acid and an amine. Thus, even though amides are not normally formed in the following way, they can be viewed as the result of a reversible dehydration-condensation:

$$R{-}\overset{O}{\overset{\|}{C}}\boxed{-OH + H}{-}\overset{H}{\overset{|}{N}}{-}R' \rightleftharpoons R{-}\overset{O}{\overset{\|}{C}}{-}\overset{H}{\overset{|}{N}}{-}R' + HOH$$

The most important example of the amide group is the *peptide bond* (discussed in Sections 13.2 and 15.6), which links amino acids in a protein.

The carboxylic acid family also undergoes reduction to form other functional groups. For example, certain inorganic reducing agents convert acids or esters to alcohols and convert amides to amines:

$$R{-}\overset{O}{\overset{\|}{C}}{-}OH \text{ (or } R{-}\overset{O}{\overset{\|}{C}}{-}O{-}R'\text{)} \xrightarrow{\text{reduction}} R{-}CH_2{-}OH + HOH \text{ (or } R'{-}OH\text{)}$$

$$R{-}\overset{O}{\overset{\|}{C}}{-}NH{-}R' \xrightarrow{\text{reduction}} R{-}CH_2{-}NH{-}R' + H_2O$$

SAMPLE PROBLEM 15.6 Predicting Reactions of the Carboxylic Acid Family

Problem Predict the product(s) of the following reactions:

(a) $CH_3{-}CH_2{-}CH_2{-}\overset{O}{\overset{\|}{C}}{-}OH + CH_3{-}\overset{OH}{\overset{|}{C}H}{-}CH_3 \overset{H^+}{\rightleftharpoons}$

(b) $CH_3{-}\overset{CH_3}{\overset{|}{C}H}{-}CH_2{-}CH_2{-}\overset{O}{\overset{\|}{C}}{-}NH{-}CH_2{-}CH_3 \xrightarrow[H_2O]{NaOH}$

Plan We discussed substitution reactions (including addition-elimination and dehydration-condensation) and hydrolysis. In **(a)**, a carboxylic acid and an alcohol react, so the reaction must be a substitution to form an ester and water. In **(b)**, an amide reacts with OH^-, so it is hydrolyzed to an amine and a sodium carboxylate.

Solution (a) Formation of an ester:

$$CH_3{-}CH_2{-}CH_2{-}\overset{O}{\overset{\|}{C}}{-}O{-}\overset{CH_3}{\overset{|}{C}H}{-}CH_3 + H_2O$$

(b) Basic hydrolysis of an amide:

$$CH_3{-}\overset{CH_3}{\overset{|}{C}H}{-}CH_2{-}CH_2{-}\overset{O}{\overset{\|}{C}}{-}O^- + Na^+ + H_2N{-}CH_2{-}CH_3$$

Check Note that in part (b), the carboxylate ion forms, rather than the acid, because the aqueous NaOH that is present reacts with the carboxylic acid.

FOLLOW-UP PROBLEM 15.6 Fill in the blanks in the following reactions:

(a) ____ + $CH_3—OH$ $\overset{H^+}{\rightleftharpoons}$ $C_6H_5—CH_2—C(=O)—O—CH_3$ + H_2O

(b) ____ + ____ ⟶ $CH_3—CH_2—CH_2—C(=O)—NH—CH_2—CH_3$ + $CH_3—OH$

Variations on a Theme: Oxoacids, Esters, and Amides of Other Nonmetals A nonmetal that is both double and single bonded to O occurs in most inorganic oxoacids, such as phosphoric, sulfuric, and chlorous acids. Those with additional O atoms are stronger acids than carboxylic acids.

Diphosphoric and disulfuric acids are acid anhydrides formed by dehydration-condensation reactions, just as a carboxylic acid anhydride is formed (Figure 15.23). Inorganic oxoacids form esters and amides that are part of many biological molecules. We already saw that certain lipids are phosphate esters (see Figure 15.21). The first compound formed when glucose is digested is a phosphate ester (Figure 15.24A); a similar phosphate ester is a major structural feature of nucleic acids, as we'll see shortly. Amides of organic sulfur-containing oxoacids, called *sulfonamides*, are potent antibiotics; the simplest of these is depicted in Figure 15.24B. More than 10,000 different sulfonamides have been synthesized.

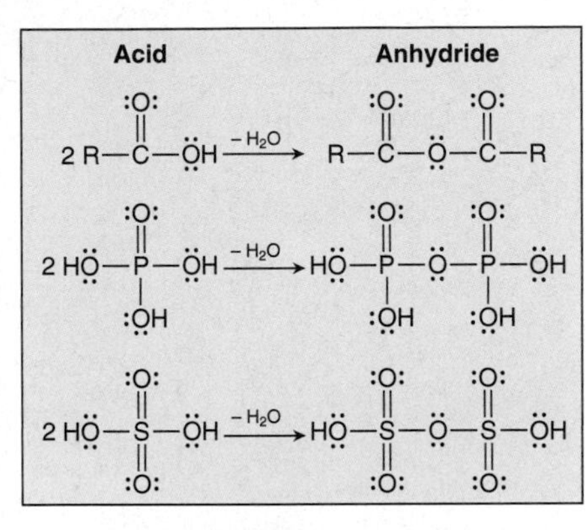

Figure 15.23 The formation of carboxylic, phosphoric, and sulfuric acid anhydrides.

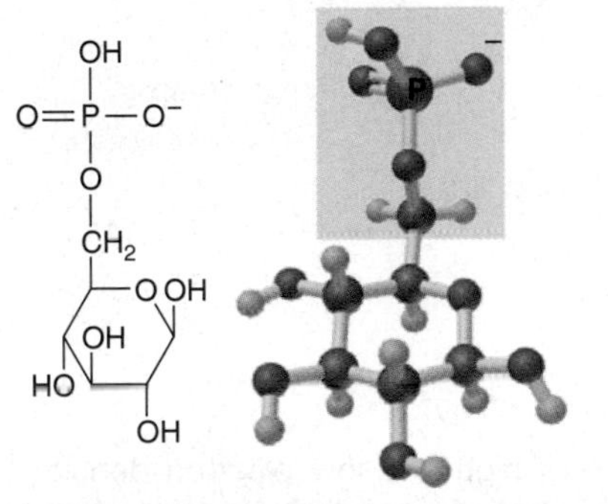

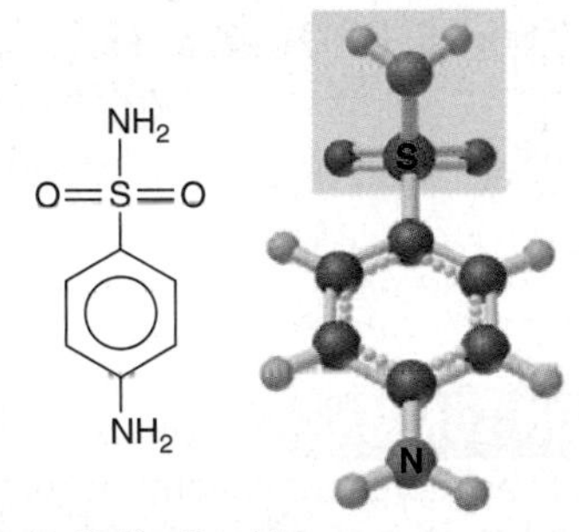

A Glucose-6-phosphate B Sulfanilamide

Figure 15.24 An ester and an amide with another nonmetal. A, Glucose-6-phosphate contains a phosphate ester group. **B,** Sulfanilamide contains a sulfonamide group.

Functional Groups with Triple Bonds

There are only two important functional groups with triple bonds. *Alkynes,* with their electron-rich —C≡C— group, undergo addition (by H_2O, H_2, HX, X_2, and so forth) to form double-bonded or saturated compounds:

$$\underset{\text{propyne}}{CH_3—C{\equiv}CH} \xrightarrow{H_2} \underset{\text{propene}}{CH_3—CH{=}CH_2} \xrightarrow{H_2} \underset{\text{propane}}{CH_3—CH_2—CH_3}$$

Nitriles (R—C≡N) contain the **nitrile** group (—C≡N:) and are made by substituting a CN^- (cyanide) ion for X^- in a reaction with an alkyl halide:

$$CH_3—CH_2—Cl + NaCN \longrightarrow CH_3—CH_2—C{\equiv}N + NaCl$$

This reaction is useful because it *increases the hydrocarbon chain by one C atom.* Nitriles are versatile because once they are formed, they can be reduced to amines or hydrolyzed to carboxylic acids:

$$CH_3—CH_2—CH_2—NH_2 \xleftarrow{\text{reduction}} CH_3—CH_2—C{\equiv}N \xrightarrow[\text{hydrolysis}]{H_3O^+,\ H_2O} CH_3—CH_2—C(=O)—OH + NH_4^+$$

Variations on a Theme: Inorganic Compounds with Triple Bonds Triple bonds are as scarce in the inorganic world as in the organic world. Carbon monoxide (:C≡O:), elemental nitrogen (:N≡N:), and the cyanide ion ([:C≡N:]$^-$) are the only common examples.

You've seen quite a few functional groups by this time, and it is especially important that you can recognize them in a complex organic molecule. Sample Problem 15.7 provides some practice.

SAMPLE PROBLEM 15.7 Recognizing Functional Groups

Problem Circle and name the functional groups in the following molecules:

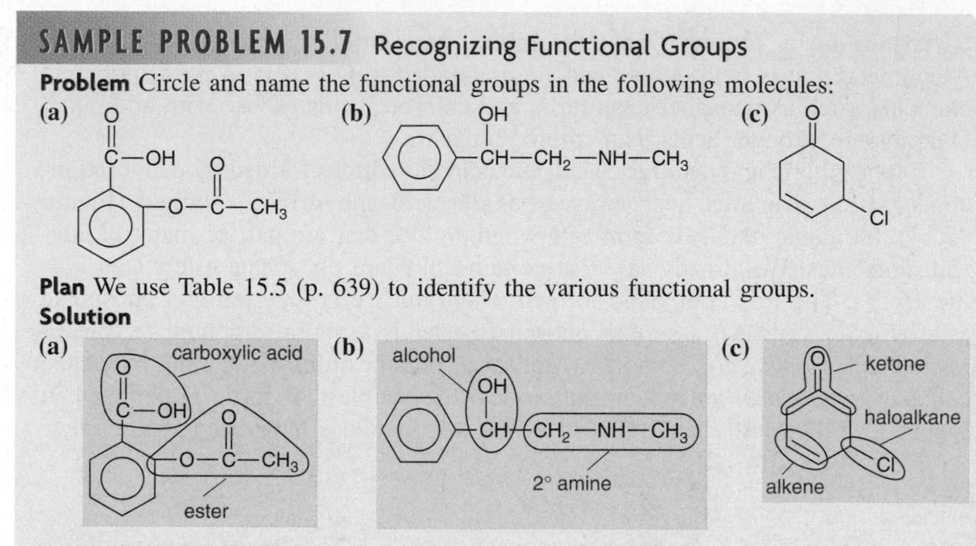

Plan We use Table 15.5 (p. 639) to identify the various functional groups.

Solution

FOLLOW-UP PROBLEM 15.7 Circle and name the functional groups:

(a) [benzene ring]—CH=CH—C(=O)—H

(b) $H_2N—C(=O)—CH_2—CH(Br)—CH_3$

SECTION SUMMARY

Organic reactions are initiated when regions of high and low electron density of different reactant molecules attract each other. Groups containing only single bonds—alcohols, amines, and alkyl halides—take part in substitution and elimination reactions. Groups with double or triple bonds—alkenes, aldehydes, ketones, alkynes, and nitriles—generally take part in addition reactions. Aromatic compounds typically undergo substitution, rather than addition, because delocalization of the π electrons stabilizes the ring. Groups with both double and single bonds—carboxylic acids, esters, and amides—generally take part in substitution reactions. Many reactions change one functional group to another, but some, especially reactions with organometallic compounds and with the cyanide ion, change the C skeleton.

15.5 THE MONOMER-POLYMER THEME I: SYNTHETIC MACROMOLECULES

In our survey of advanced materials in Chapter 12, you saw that polymers are extremely large molecules that consist of many monomeric repeat units. In that chapter, we focused on the mass, shape, and physical properties of polymers. Now we'll see how polymers are named and discuss the two types of reactions that link monomers covalently into a chain. To name a polymer, just add the prefix *poly-* to the monomer name, as in *polyethylene* or *polystyrene*. When the monomer has a two-word name, parentheses are used, as in *poly(vinyl chloride)*.

Because synthetic polymers consist of petroleum-based monomers, a major shortage of raw materials for their manufacture is expected by 2150. To avoid this, green chemists are devising polymers based on corn, sugarcane, and even cashew-nut shells. Before being embraced as a "green" technology, however, the total impact of growing, transporting, and manufacturing must be assessed.

The two major types of reaction processes that form synthetic polymers lend their names to the resulting classes of polymer—addition and condensation.

Addition Polymers

Addition polymers form when monomers undergo an addition reaction with one another. These are also called *chain-reaction (or chain-growth) polymers* because as each monomer adds to the chain, it forms a new reactive site to continue the process. The monomers of most addition polymers have the $>C=C<$ grouping. As you can see from Table 15.6, the essential differences between an acrylic sweater, a plastic grocery bag, and a bowling ball are due to the different chemical groups that are attached to the double-bonded C atoms of the monomer.

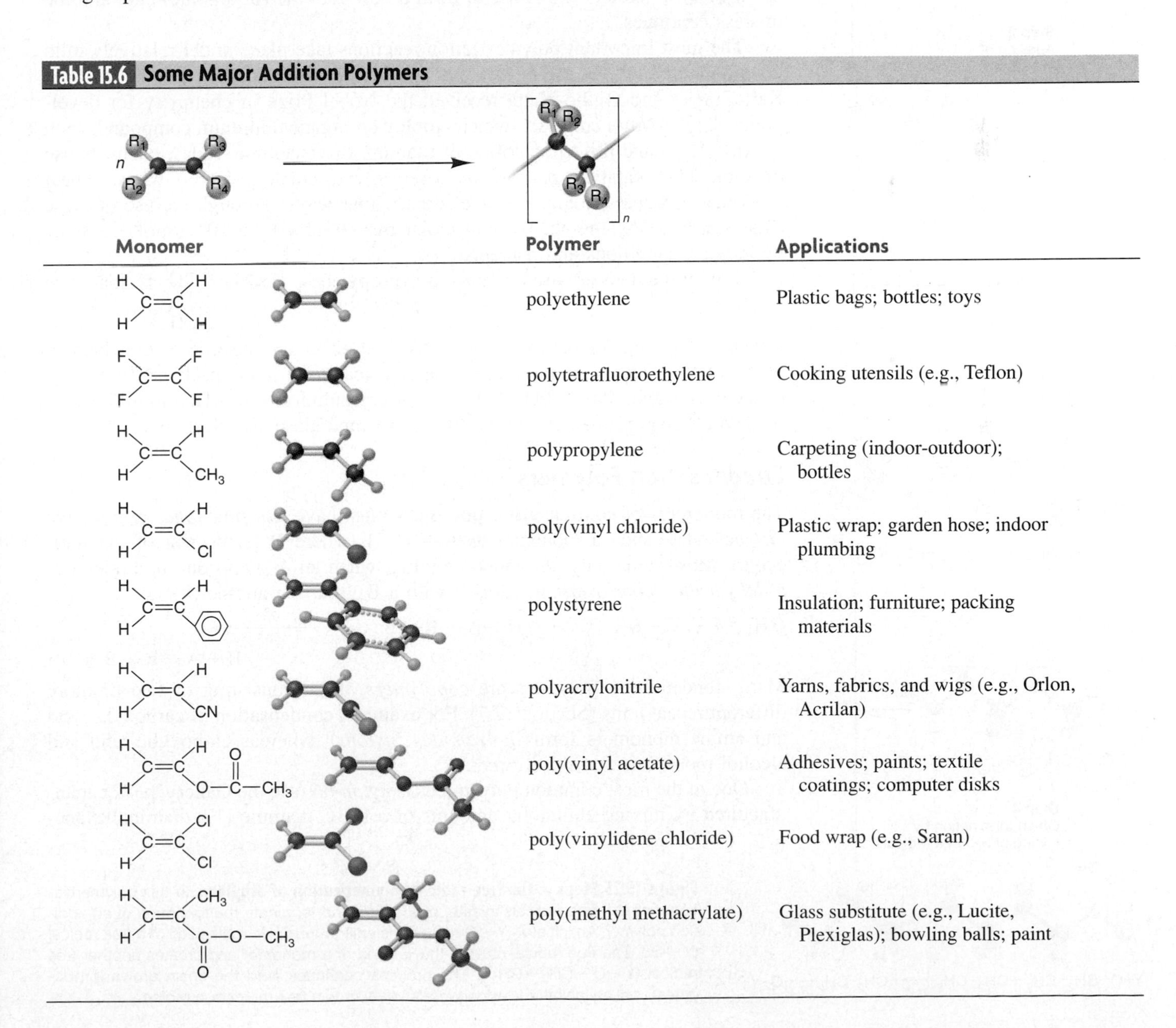

Table 15.6 Some Major Addition Polymers

Monomer	Polymer	Applications
$H_2C{=}CH_2$	polyethylene	Plastic bags; bottles; toys
$F_2C{=}CF_2$	polytetrafluoroethylene	Cooking utensils (e.g., Teflon)
$H_2C{=}CH{-}CH_3$	polypropylene	Carpeting (indoor-outdoor); bottles
$H_2C{=}CH{-}Cl$	poly(vinyl chloride)	Plastic wrap; garden hose; indoor plumbing
$H_2C{=}CH{-}C_6H_5$	polystyrene	Insulation; furniture; packing materials
$H_2C{=}CH{-}CN$	polyacrylonitrile	Yarns, fabrics, and wigs (e.g., Orlon, Acrilan)
$H_2C{=}CH{-}O{-}C({=}O){-}CH_3$	poly(vinyl acetate)	Adhesives; paints; textile coatings; computer disks
$H_2C{=}CCl_2$	poly(vinylidene chloride)	Food wrap (e.g., Saran)
$H_2C{=}C(CH_3){-}C({=}O){-}O{-}CH_3$	poly(methyl methacrylate)	Glass substitute (e.g., Lucite, Plexiglas); bowling balls; paint

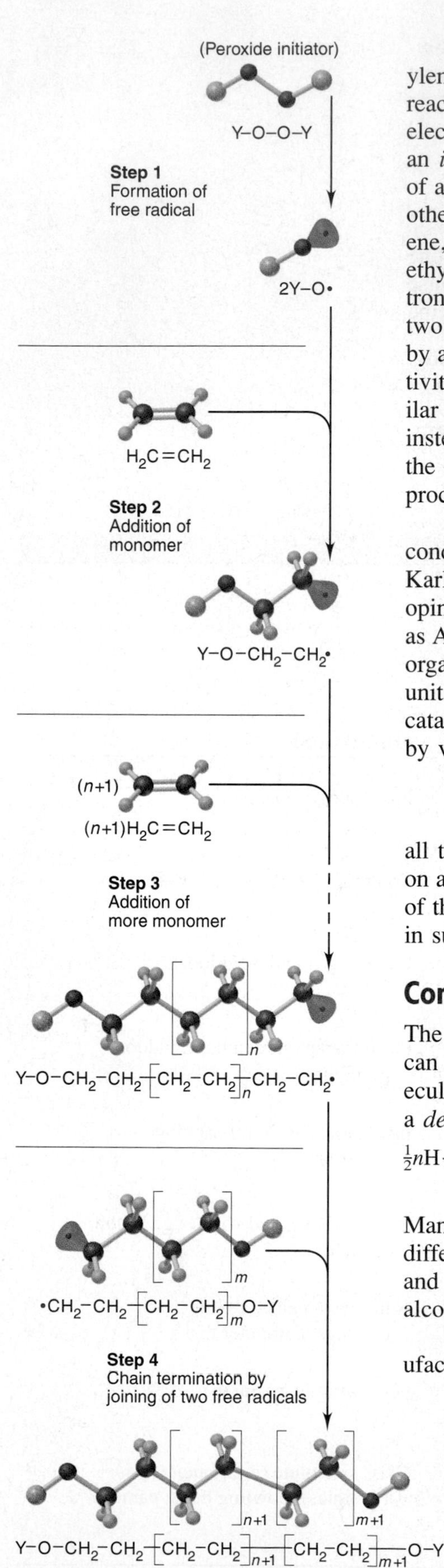

The *free-radical polymerization* of ethene (ethylene, $CH_2{=}CH_2$) to polyethylene is a simple example of the addition process. In Figure 15.25, the monomer reacts to form a *free radical,* a species with an unpaired electron, that seeks an electron from another monomer to form a covalent bond. The process begins when an *initiator,* usually a peroxide, generates a free radical that attacks the π bond of an ethylene unit, forming a σ bond with one of the p electrons and leaving the other unpaired. This new free radical then attacks the π bond of another ethylene, joining it to the chain end, and the backbone of the polymer grows. As each ethylene adds, it leaves an unpaired electron on the growing end to find an electron "mate" and make the chain one repeat unit longer. This process stops when two free radicals form a covalent bond or when a very stable free radical is formed by addition of an *inhibitor* molecule. Recent progress in controlling the high reactivity of free-radical species promises an even wider range of polymers. In a similar method, polymerization is initiated by the formation of a cation (or anion) instead of a free radical. The cationic (or anionic) reactive end of the chain attacks the π bond of another monomer to form a new cationic (or anionic) end, and the process continues.

The most important polymerization reactions take place under relatively mild conditions through the use of catalysts that incorporate transition metals. In 1963, Karl Ziegler and Giulio Natta received the Nobel Prize in chemistry for developing *Ziegler-Natta catalysts,* which employ an organoaluminum compound, such as $Al(C_2H_5)_3$, and the tetrachloride of titanium or vanadium. Today, chemists use organometallic catalysts that are *stereoselective* to create polymers whose repeat units have groups spatially oriented in particular ways. Through the use of these catalysts, polyethylene chains with molar masses of 10^4 to 10^5 g/mol are made by varying conditions and reagents.

Similar methods are used to make polypropylenes, $\left[CH_2{-}\underset{\displaystyle CH_3}{\underset{|}{CH}}\right]_n$, that have all the CH_3 groups of the repeat units oriented either on one side of the chain or on alternating sides. The different orientations lead to different packing efficiencies of the chains and, thus, different degrees of crystallinity, which lead to differences in such physical properties as density, rigidity, and elasticity (Section 12.7).

Condensation Polymers

The monomers of **condensation polymers** must have *two functional groups;* we can designate such a monomer as A—**R**—B (where **R** is the rest of the molecule). Most commonly, the monomers link when an A group on one undergoes a *dehydration-condensation reaction* with a B group on another:

$$\tfrac{1}{2}n\text{H—A—}\mathbf{R}\text{—B—OH} + \tfrac{1}{2}n\text{H—A—}\mathbf{R}\text{—B—OH} \xrightarrow{-(n-1)\text{HOH}} \text{H}\left[\text{A—}\mathbf{R}\text{—B}\right]_n\text{OH}$$

Many condensation polymers are *copolymers,* those consisting of two or more different repeat units (Section 12.7). For example, condensation of carboxylic acid and amine monomers forms *polyamides (nylons),* whereas carboxylic acid and alcohol monomers form *polyesters.*

One of the most common polyamides is *nylon-66* (see the Gallery, p. 72), manufactured by mixing equimolar amounts of a six-C diamine (1,6-diaminohexane)

Figure 15.25 Steps in the free-radical polymerization of ethylene. In this polymerization method, free radicals initiate, propagate, and terminate the formation of an addition polymer. An initiator (Y—O—O—Y) is split to form two molecules of a free radical (Y—O·). The free radical attacks the π bond of a monomer and creates another free radical (Y—O—CH_2—CH_2·). The process continues, and the chain grows (propagates) until an inhibitor is added (not shown) or two free radicals combine.

and a six-C diacid (1,6-hexanedioic acid). The basic amine reacts with the acid to form a "nylon salt." Heating drives off water and forms the amide bonds:

$$n\text{HO}-\overset{\text{O}}{\overset{\|}{\text{C}}}-(\text{CH}_2)_4-\overset{\text{O}}{\overset{\|}{\text{C}}}-\text{OH} + n\text{H}_2\text{N}-(\text{CH}_2)_6-\text{NH}_2 \xrightarrow[-(2n-1)\text{H}_2\text{O}]{\Delta} \text{HO}\left[\overset{\text{O}}{\overset{\|}{\text{C}}}-(\text{CH}_2)_4-\overset{\text{O}}{\overset{\|}{\text{C}}}-\text{NH}-(\text{CH}_2)_6-\text{NH}\right]_n\text{H}$$

In the laboratory, this nylon is made without heating by using a more reactive acid component (Figure 15.26). Covalent bonds within the chains give nylons great strength, and H bonds between chains give them great flexibility. About half of all nylons are made to reinforce automobile tires, the remainder being used for rugs, clothing, fishing line, and so forth.

Dacron, a popular polyester fiber, is woven from polymer strands formed when equimolar amounts of 1,4-benzenedicarboxylic acid and 1,2-ethanediol react. Blending these polyester fibers with various amounts of cotton gives fabrics that are durable, easily dyed, and crease resistant. Extremely thin Mylar films, used for recording tape and food packaging, are also made from this polymer.

Variations on a Theme: Inorganic Polymers You already know that some synthetic polymers have inorganic backbones. In Chapter 14, we discussed the silicones, polymers with the repeat unit $-(\text{R}_2)\text{Si}-\text{O}-$. Depending on the chain crosslinks and the R groups, silicones range from oily liquids to elastic sheets to rigid solids and have applications that include artificial limbs and spacesuits. We also mentioned the polyphosphazenes in Chapter 14, which exist as flexible chains even at low temperatures, and have the repeat unit $-(\text{R}_2)\text{P}{=}\text{N}-$.

Figure 15.26 The formation of nylon-66. Nylons are formed industrially by the reaction of diamine and diacid monomers. In this laboratory demonstration, the six-C diacid chloride, which is more reactive than the diacid, is used as one monomer; the polyamide forms between the two liquid phases.

SECTION SUMMARY

Polymers are extremely large molecules that are made of repeat units called monomers. Addition polymers are formed from unsaturated monomers that commonly link through free-radical reactions. Most condensation polymers are formed by linking two types of monomer through dehydration-condensation reactions. Reaction conditions, catalysts, and monomers can be varied to produce polymers with different properties.

15.6 THE MONOMER-POLYMER THEME II: BIOLOGICAL MACROMOLECULES

The monomer-polymer theme was being played out in nature eons before humans employed it to such great advantage. Biological macromolecules are nothing more than condensation polymers created by nature's reaction chemistry and improved through evolution. These remarkable molecules are the greatest proof of the versatility of carbon and its handful of atomic partners.

Natural polymers are the "stuff of life"—polysaccharides, proteins, and nucleic acids. Some have structures that make wood strong, hair curly, fingernails hard, and wool flexible. Others speed up the myriad reactions that occur in every cell or defend the body against infection. Still others possess the genetic information needed to forge other biomolecules. Remarkable as these giant molecules are, the functional groups of their monomers and the reactions that link them are identical to those of other, smaller organic molecules. Moreover, as you saw in Section 13.2, the same intermolecular forces that dissolve smaller molecules stabilize these giant molecules in the aqueous medium of the cell.

Sugars and Polysaccharides

In essence, the same chemical change occurs when you burn a piece of wood or eat a piece of bread. Wood and bread are mixtures of carbohydrates, substances that provide energy through oxidation.

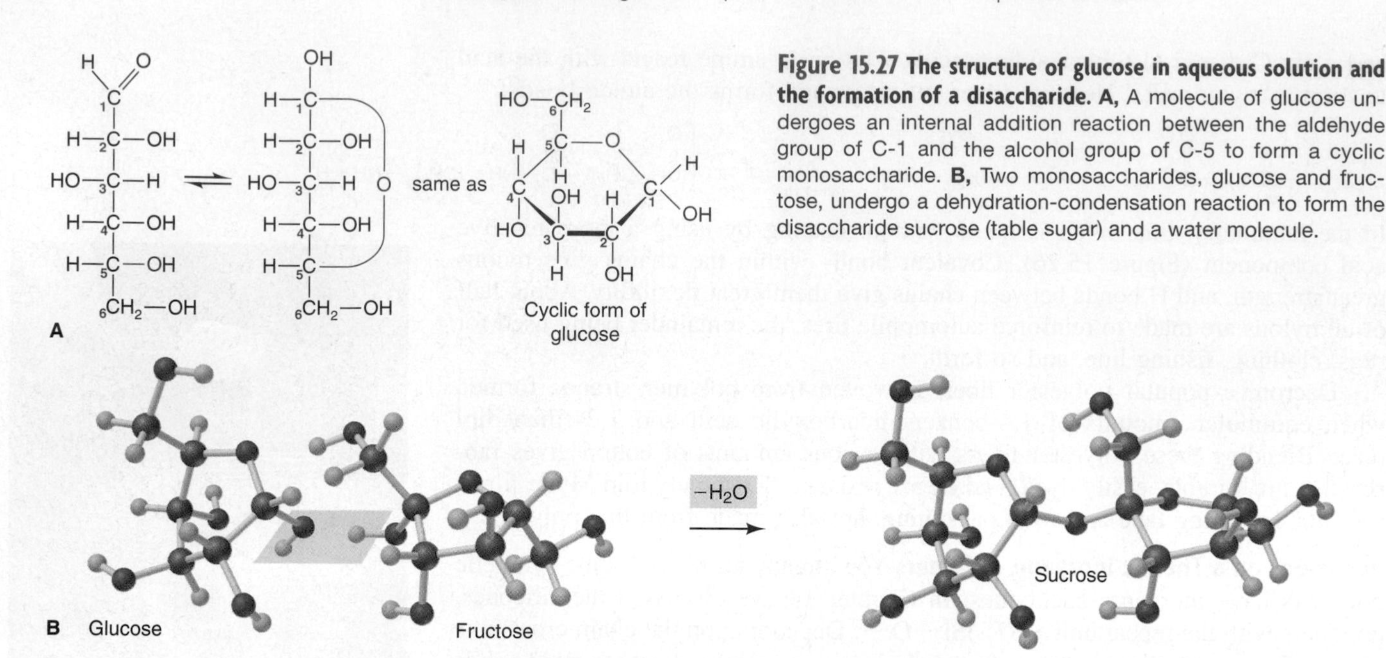

Figure 15.27 The structure of glucose in aqueous solution and the formation of a disaccharide. A, A molecule of glucose undergoes an internal addition reaction between the aldehyde group of C-1 and the alcohol group of C-5 to form a cyclic monosaccharide. **B,** Two monosaccharides, glucose and fructose, undergo a dehydration-condensation reaction to form the disaccharide sucrose (table sugar) and a water molecule.

Monomer Structure and Linkage Glucose and other simple sugars, from the three-C *trioses* to the seven-C *heptoses,* are called **monosaccharides** and consist of carbon chains with attached hydroxyl and carbonyl groups. In addition to their roles as individual molecules engaged in energy metabolism, they serve as the monomer units of **polysaccharides.** Most natural polysaccharides are formed from five- and six-C units. In aqueous solution, *an alcohol group and the aldehyde (or ketone) group of a given monosaccharide react with each other to form a cyclic molecule with either a five- or six-membered ring* (Figure 15.27A).

When two monosaccharides undergo a dehydration-condensation reaction, a **disaccharide** forms. For example, sucrose (table sugar) is a disaccharide of glucose (linked at C-1) and fructose (linked at C-2) (Figure 15.27B); lactose (milk sugar) is a disaccharide of glucose (C-1) and galactose (C-4); and maltose, used in brewing and as a sweetener, is a disaccharide of two glucose units (C-1 to C-4).

Types of Polysaccharides A polysaccharide consists of *many* monosaccharide units linked together. The three major natural polysaccharides—cellulose, starch, and glycogen—consist entirely of glucose units, but they differ in the ring positions of the links, in the orientation of certain bonds, and in the extent of crosslinking. Some other polysaccharides contain nitrogen in their attached groups.

Polysaccharide Skeletons of Lobsters and Roaches The variety of polysaccharide properties that arises from simple changes in the monomers is amazing. For example, substituting an $-NH_2$ group for the $-OH$ group at C-2 in glucose produces *glucosamine.* The amide that is formed from glucosamine and acetic acid (*N*-acetylglucosamine) is the monomer of *chitin* (pronounced "KY-tin"), a polysaccharide that is the main component of the tough, brittle, external skeletons of insects and crustaceans.

Cellulose is the most abundant organic chemical on Earth. More than 50% of the carbon in plants occurs in the cellulose of stems and leaves; wood is largely cellulose, and cotton is more than 90% cellulose (see Section 13.2). Cellulose consists of long chains of glucose. The great strength of wood is due largely to the H bonds between cellulose chains. The monomers are linked in a particular way from C-1 in one unit to C-4 in the next. Humans lack the enzymes to break this link, so we cannot digest cellulose (unfortunately!); however, microorganisms in the digestive tracts of some animals, such as cows, sheep, and termites, can.

Starch is a mixture of polysaccharides of glucose and serves as an *energy store* in plants. When a plant needs energy, some starch is broken down by hydrolysis of the bonds between units, and the released glucose is oxidized through a multistep metabolic pathway. Starch occurs in plant cells as insoluble granules of amylose, a helical molecule of several thousand glucose units, and amylopectin, a highly branched, bushlike molecule of up to a million glucose units. Most of the glucose units are linked by C-1 to C-4 bonds, as in cellulose, but a different orientation around the chiral C-1 allows our digestive enzymes to break starch down into monomers. A C-6 to C-1 crosslink joins chains every 24 to 30 units.

Glycogen functions as the energy storage molecule in animals. It occurs in liver and muscle cells as large, insoluble granules consisting of glycogen molecules made from 1000 to more than 500,000 glucose units. In glycogen, the units are also linked by C-1 to C-4 bonds, but the molecule is more highly crosslinked than starch, with C-6 to C-1 crosslinks every 8 to 12 units.

Amino Acids and Proteins

As you saw in Section 15.5, synthetic polyamides (such as nylon-66) are formed from two monomers, one with a carboxyl group at each end and the other with an amine group at each end. **Proteins,** the polyamides of nature, are unbranched polymers formed from monomers called **amino acids.** As we discussed in Section 13.2, *each amino acid has a carboxyl group and an amine group.*

Monomer Structure and Linkage An amino acid has both its carboxyl group and its amine group attached to the α-*carbon,* the second C atom in the chain. Proteins are made up of about 20 different types of amino acids, each with its own particular R group, ranging from an H atom to a polycyclic N-containing aromatic structure (Figure 15.28).

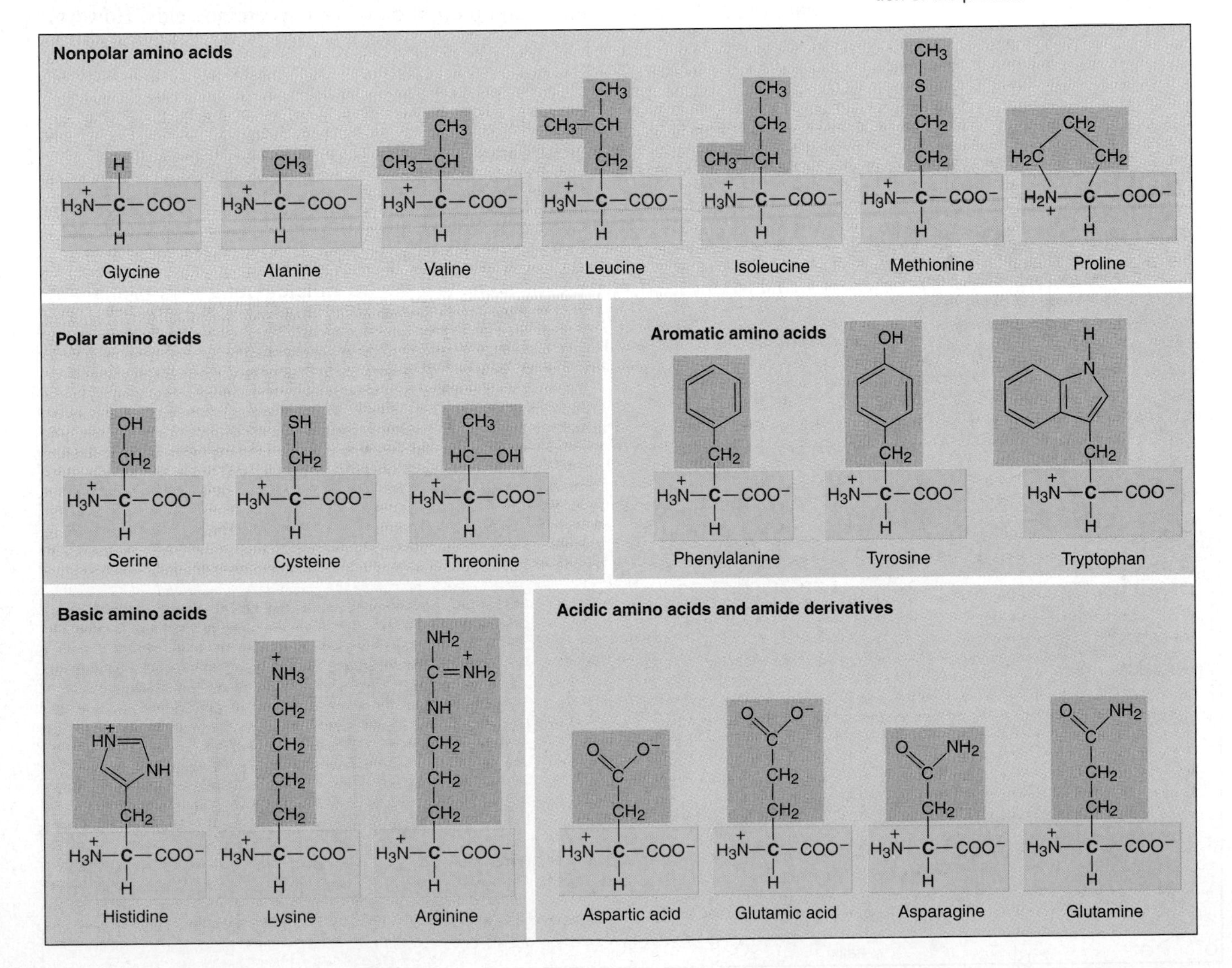

Figure 15.28 The common amino acids. About 20 different amino acids occur in proteins. The R groups are screened yellow, and the α-carbons (**boldface**), with carboxyl and amino groups, are screened grey. Here the amino acids are shown with the charges they have under physiological conditions. They are grouped by polarity, acid-base character, and presence of an aromatic ring. The R groups play a major role in the shape and function of the protein.

In the aqueous cell fluid, the amino-acid NH_2 and COOH groups are charged because the carboxyl group transfers an H^+ ion to H_2O to form H_3O^+, which transfers the H^+ to the amine group. The overall process is, in effect, an intramolecular acid-base reaction:

$$H_2N-\underset{H}{\overset{R}{C}}-COOH \xrightarrow{-H_2O,\ H_3O^+,\ +H_2O} H_3\overset{+}{N}-\underset{H}{\overset{R}{C}}-COO^-$$

(α-carbon; gives; R = one of 20 different groups)

An H atom is the third group bonded to the α-carbon, and the fourth is the R group (also called the *side chain*).

Each amino acid is linked to the next one through a *peptide* (amide) bond formed by a dehydration-condensation reaction in which the carboxyl group of one monomer reacts with the amine group of the next. Therefore, as was pointed out in Section 13.2, the polypeptide chain, the backbone of the protein, consists of an *α-carbon bonded to an amide group bonded to the next α-carbon bonded to the next amide group,* and so forth (see Figure 13.6, p. 496). The various R groups dangle from the α-carbons on alternate sides of the chain.

The Hierarchy of Protein Structure *Each type of protein has its own amino acid composition,* a specific number and proportion of various amino acids. However, it is not the composition that defines the protein's role in the cell; rather, *the sequence of amino acids determines the protein's shape and function.* Proteins range from about 50 to several thousand amino acids, yet from a purely mathematical point of view, even a small protein of 100 amino acids has a virtually limitless number of possible sequences of the 20 types of amino acids ($20^{100} \approx 10^{130}$). In fact, though, only a tiny fraction of these possibilities occur in actual proteins. For example, even in an organism as complex as a human being, there are only about 10^5 different types of protein.

As we discuss later in the section, a protein folds into its *native* shape as it is being synthesized in the cell. Some shapes are simple—long rods or undulating sheets. Others are far more complex—baskets, Y shapes, spheroid blobs, and countless other globular forms. Biochemists identify a hierarchy for the overall structure of a protein (Figure 15.29):

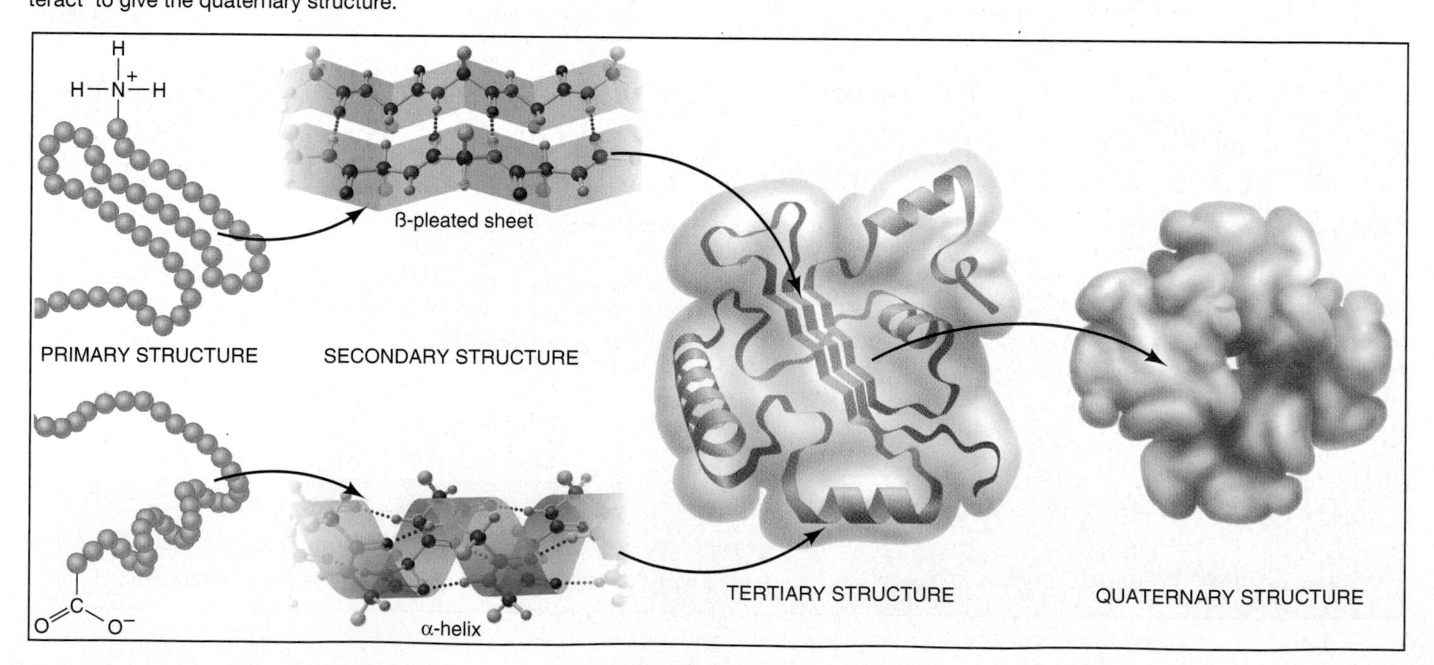

Figure 15.29 The structural hierarchy of proteins. A typical protein's structure can be viewed at different levels. Primary structure (shown as a long string of balls leaving and returning to the picture frame) is the sequence of amino acids. Secondary structure consists of highly ordered regions that occur as an α-helix or a β-sheet. Tertiary structure combines these regions with random coil sections. In many proteins, several tertiary units interact to give the quaternary structure.

1. *Primary (1°) structure*, the most basic level, refers to the sequence of covalently bonded amino acids in the polypeptide chain.
2. *Secondary (2°) structure* refers to sections of the chain that, as a result of H bonding between nearby peptide groupings, adopt shapes called α-helices and β-pleated sheets.
3. *Tertiary (3°) structure* refers to the three-dimensional folding of the whole polypeptide chain. In some proteins, certain folding patterns form characteristic regions that play a role in the protein's function—binding a hormone, attaching to a membrane, forming a polar channel, and so forth.
4. *Quaternary (4°) structure*, the most complex level, occurs in proteins made up of several polypeptide chains (subunits) and refers to the way the chains assemble into the overall protein.

Note that *only the 1° structure involves covalent bonds; the 2°, 3°, and 4° structures rely on intermolecular forces*, as we discussed in Section 13.2.

The Relation Between Structure and Function Two broad classes of proteins differ in the complexity of their amino acid compositions and sequences and, therefore, in their structure and function:

1. *Fibrous proteins* have relatively simple amino acid compositions and correspondingly simple structures. They are key components of hair, wool, skin, and connective tissue—materials that require strength and flexibility. Like synthetic polymers, these proteins have a small number of different amino acids in a repeating sequence. Consider collagen, the most common animal protein, which makes up as much as 40% of human body weight. More than 30% of its amino acids are glycine (G) and another 20% are proline (P). It exists as three chains, each an extended helix, that wind tightly around each other as the peptide C=O groups in one chain form H bonds to the peptide N—H groups in another. The result is a long, triple-helical cable with the sequence —G—X—P—G—X—P— and so on (where X is another amino acid), as shown in Figure 15.30A. As the main component of tendons, skin, and blood vessels, collagen has a high tensile strength; in fact, a 1-mm thick strand can support a 10-kg weight!

In silk fibroin, secreted by the silk moth caterpillar, more than 85% of the amino acids are glycine, alanine, and serine (Figure 15.30B). Fibroin chain segments bend back and forth, running alongside each other, and form interchain H bonds to create a *pleated sheet*. Stacks of sheets interact through dispersion forces, which make fibroin strong and flexible but not very extendable—perfect for a silkworm's cocoon.

2. *Globular proteins* have much more complex compositions, often containing varying proportions of all 20 different amino acids. As the name implies, globular proteins are typically more rounded and compact, with a wide variety of shapes and a correspondingly wide range of functions: defenders against bacterial invasion, messengers that trigger cell actions, catalysts of chemical change, membrane gatekeepers that maintain aqueous concentrations, and many others. The locations of particular amino-acid R groups are crucial to the protein's function. For example, in catalytic proteins, a few R groups form a crevice that closely matches the shapes of reactant molecules. These groups typically hold the reactants through intermolecular forces and speed their reaction to products by bringing them together and twisting and stretching their bonds. Experiment has shown repeatedly that a slight change in one of these critical R groups decreases function dramatically. This fact supports the essential idea that *the protein's amino acid sequence determines its structure, which in turn determines its function*:

$$\text{SEQUENCE} \Longrightarrow \text{STRUCTURE} \Longrightarrow \text{FUNCTION}$$

Next, we'll see how the amino acid sequence of every protein in every organism is prescribed by the genetic information that is held within the organism's nucleic acids.

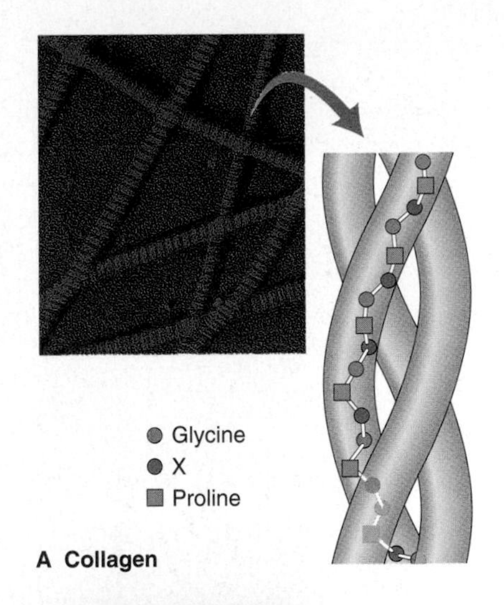

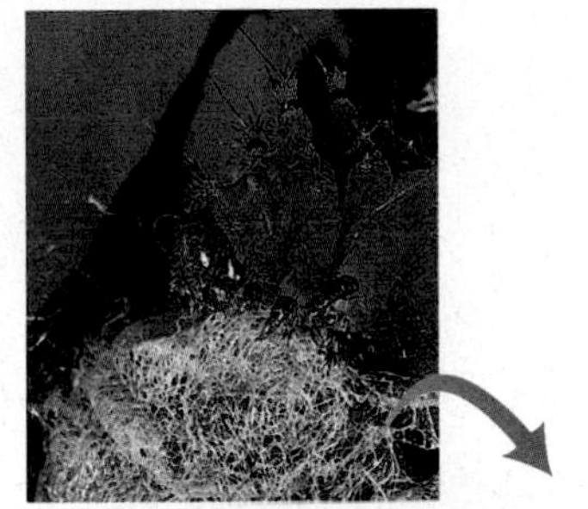

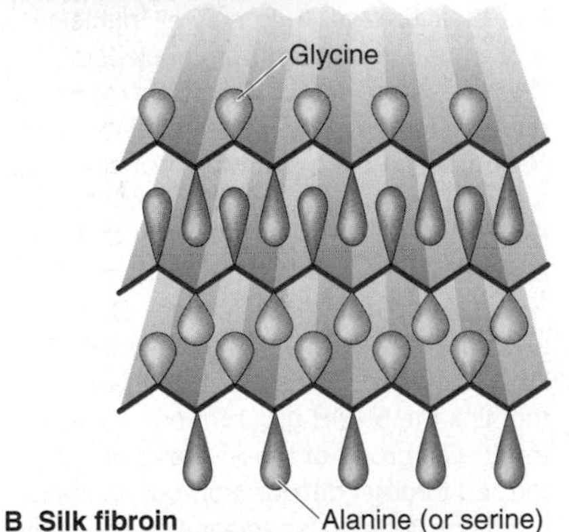

Figure 15.30 The shapes of fibrous proteins. Fibrous proteins have largely structural roles in an organism, and their amino acid sequence is relatively simple. **A,** The triple helix of a collagen molecule has a glycine as every third amino acid along the chain. **B,** The pleated sheet structure of silk fibroin has a repeating sequence of glycine alternating with either alanine or serine. The sheets interact through dispersion forces.

Nucleotides and Nucleic Acids

An organism's nucleic acids construct its proteins. And, given that the proteins determine how the organism looks and behaves, no job could be more essential.

Monomer Structure and Linkage **Nucleic acids** are unbranched polymers that consist of linked monomer units called **mononucleotides,** which consist of an N-containing base, a sugar, and a phosphate group. The two types of nucleic acid, *ribo**nu**cleic **a**cid* (RNA) and *deoxyribo**nu**cleic **a**cid* (DNA), differ in the sugar portions of their mononucleotides. RNA contains *ribose,* a five-C sugar, and DNA contains *deoxyribose,* in which —H substitutes for —OH on the 2′ position of ribose. (Carbon atoms in the sugar portion are given numbers with a prime to distinguish them from carbon atoms in the base.)

The cellular precursors that form a nucleic acid are *nucleoside triphosphates* (Figure 15.31A). Dehydration-condensation reactions between them release inorganic diphosphate ($H_2P_2O_7^{2-}$) and create *phosphodiester* bonds to form a polynucleotide chain. Therefore, the repeating pattern of the nucleic acid backbone is *—sugar—phosphate—sugar—phosphate—,* and so on (Figure 15.31B). Attached to each sugar is one of four N-containing bases, either a pyrimidine (six-membered ring) or a purine (six- and five-membered rings sharing a side). The pyrimidines are thymine (T) and cytosine (C); the purines are guanine (G) and adenine (A); in RNA, uracil (U) substitutes for thymine. The bases dangle off the sugar-phosphate chain, much as R groups dangle off the polypeptide chain of a protein.

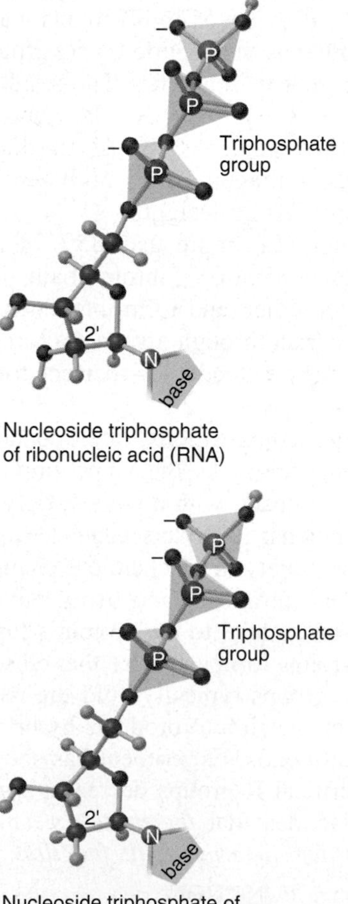

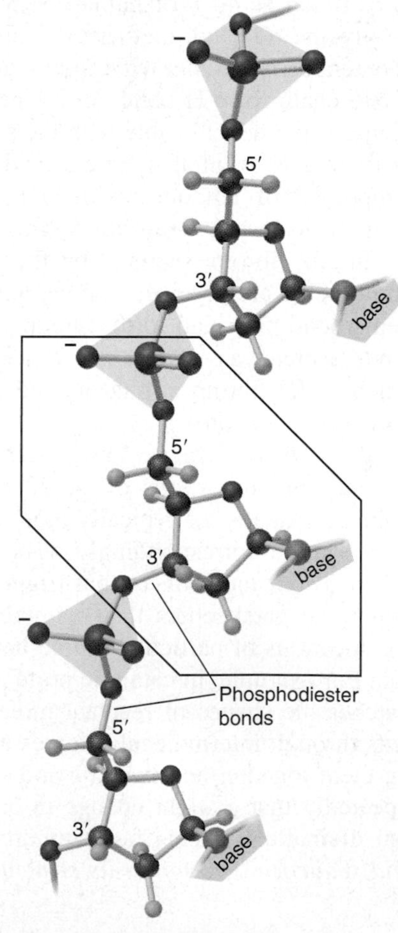

Figure 15.31 Nucleic acid precursors and their linkage. A, In the cell, nucleic acids are constructed from nucleoside triphosphates, precursors of the mononucleotide units. Each one consists of an N-containing base (structure not shown), a sugar, and a triphosphate group. In RNA (*top*), the sugar is ribose; in DNA, it is 2′-deoxyribose (note the absence of an —OH group on C-2 of the ring). **B,** A tiny segment of the polynucleotide chain of DNA shows the phosphodiester bonds that link the 5′-OH group of one sugar to the 3′-OH group of the next and are formed through dehydration-condensation reactions (which also release diphosphate ion). The bases dangle off the chain.

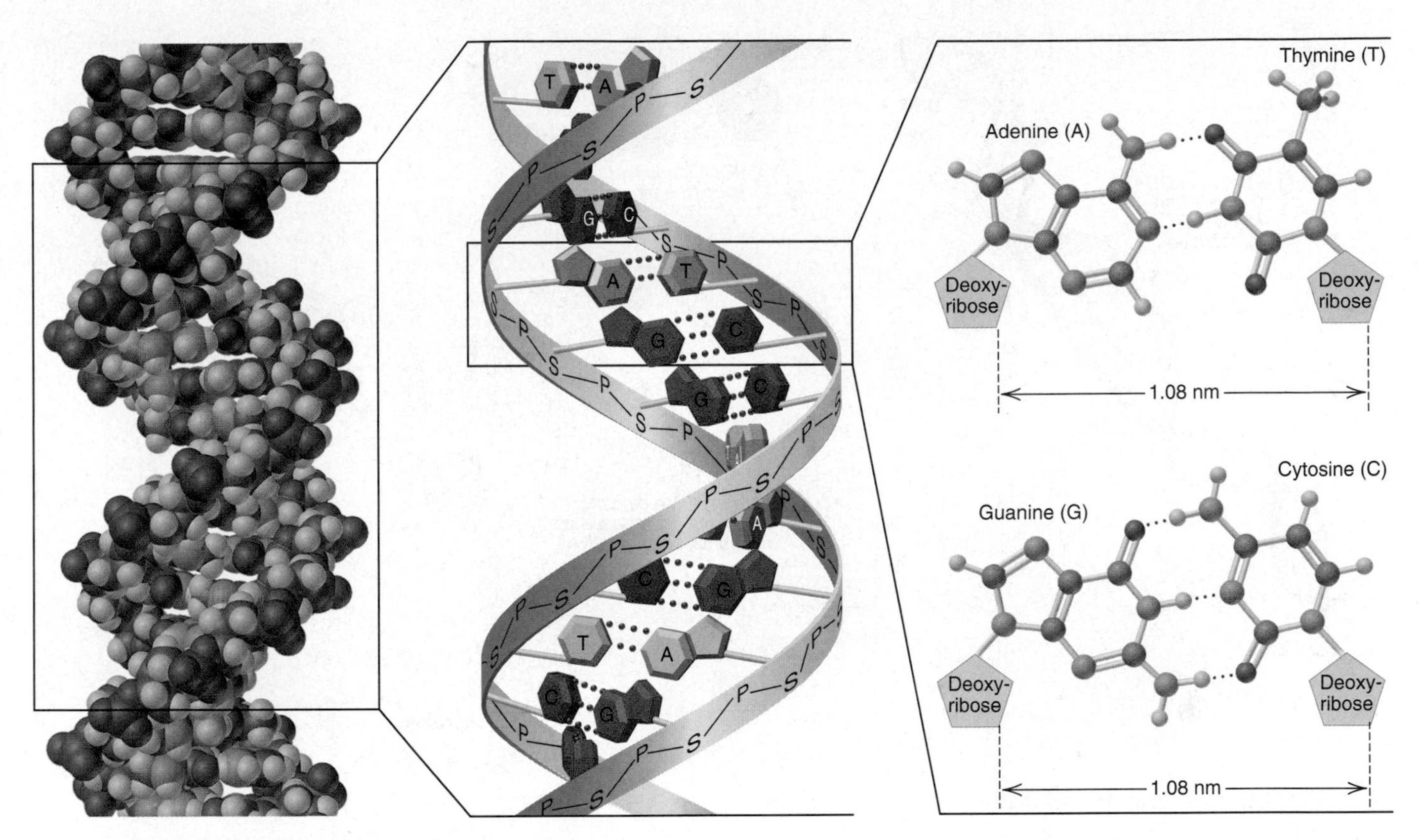

Figure 15.32 The double helix of DNA. A segment of DNA is shown as a space-filling model (*left*). The boxed area is expanded (*center*) to show how the polar sugar(S)-phosphate(P) backbone faces the watery outside, and the nonpolar bases form H bonds to each other in the DNA core. The boxed area is expanded (*right*) to show how a pyrimidine and a purine always form H-bonded base pairs to maintain the double helix width. The members of the pairs are always the same: A pairs with T, and G pairs with C.

The Central Importance of Base Pairing In the nucleus of the cell, DNA exists as two chains wrapped around each other in a **double helix** (Figure 15.32). Each base in one chain "pairs" with a base in the other through H bonding. A double-helical DNA molecule may contain many millions of H-bonded bases in specified pairs. Two features of these **base pairs** are crucial to the structure and function of DNA:

- A pyrimidine and a purine are always paired, which gives the double helix a constant diameter.
- Each base is always paired with the same partner: A with T and G with C. Thus, *the base sequence on one chain is the complement of the sequence on the other.* For example, the sequence A—C—T on one chain is *always* paired with T—G—A on the other: A with T, C with G, and T with A.

Each DNA molecule is folded into a tangled mass that forms one of the cell's *chromosomes*. The DNA molecule is amazingly long and thin: if the largest human chromosome were stretched out, it would be 4 cm long; in the cell nucleus, however, it is wound into a structure only 5 nm long—8 million times shorter! Segments of the DNA chains are the *genes* that contain the chemical information for synthesizing the organism's proteins.

An Outline of Protein Synthesis *The information content of a gene resides in its base sequence.* In the **genetic code,** each base acts as a "letter," each three-base sequence as a "word," and *each word codes for a specific amino acid.* For example, the sequence C—A—C codes for the amino acid histidine, A—A—G codes for lysine, and so on. Through a complex series of interactions, greatly simplified

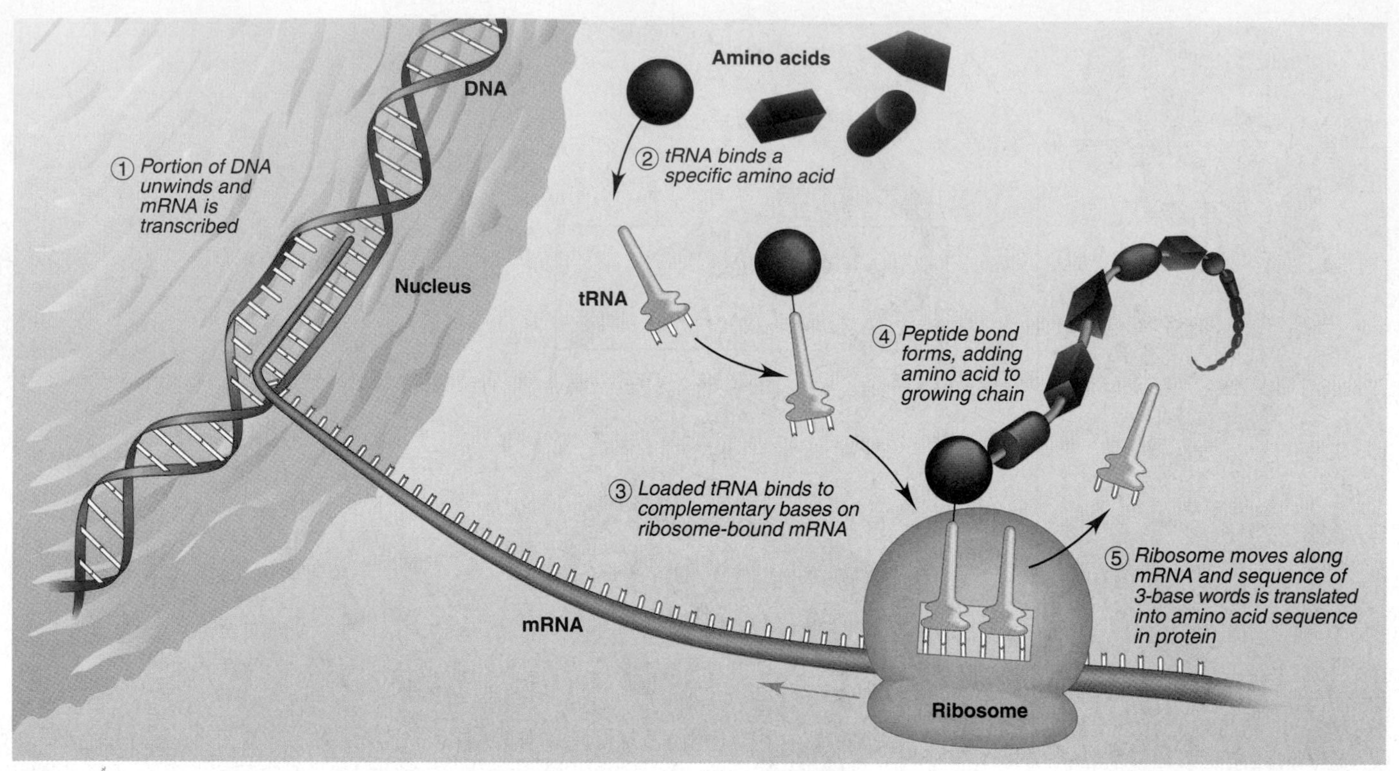

Amino acids
DNA
① Portion of DNA unwinds and mRNA is transcribed
② tRNA binds a specific amino acid
Nucleus
tRNA
④ Peptide bond forms, adding amino acid to growing chain
③ Loaded tRNA binds to complementary bases on ribosome-bound mRNA
⑤ Ribosome moves along mRNA and sequence of 3-base words is translated into amino acid sequence in protein
mRNA
Ribosome

Figure 15.33 Key stages in protein synthesis.

in the overview in Figure 15.33, one amino acid at a time is positioned and linked to the next in the process of protein synthesis. To fully appreciate this aspect of the chemical basis of biology, keep in mind that *this amazingly complex process occurs largely through H bonding between base pairs.*

Here is an outline of *protein synthesis*. DNA occurs in the cell nucleus, but the genetic message is decoded outside it, so the information must be sent to the synthesis site. RNA serves in this messenger role, as well as in several others. A portion of the DNA is temporarily unwound and one chain segment acts as a *template* for the formation of a complementary chain of *messenger RNA* (mRNA) made by individual mononucleoside triphosphates linking together; thus, *the DNA code words are transcribed into RNA code words* through base pairing. The mRNA leaves the nucleus and binds, again through base pairing, to an RNA-rich particle in the cell called a *ribosome*. The words (three-base sequences) in the mRNA are then decoded by molecules of *transfer RNA* (tRNA). These smaller nucleic acid "shuttles" have two key portions on opposite ends of their structures:

- A three-base sequence that is the complement of a word on the mRNA
- A binding site for the amino acid coded by that word

The ribosome moves along the bound mRNA, one word at a time, while tRNAs bind to the mRNA and position their amino acids near one another in preparation for peptide bond formation and synthesis of the protein.

In essence, then, protein synthesis involves the DNA message of three-base words being transcribed into the RNA message of three-base words, which is then translated into a sequence of amino acids that are linked to make a protein:

$$\text{DNA BASE SEQUENCE} \Longrightarrow \text{RNA BASE SEQUENCE} \Longrightarrow \text{PROTEIN AMINO ACID SEQUENCE}$$

An Outline of DNA Replication Another complex series of interactions allows *DNA replication,* the ability of DNA to copy itself. When a cell divides, its chromosomes are replicated, or reproduced, ensuring that the new cells have the same number and types of chromosomes. In this process, a small portion of the dou-

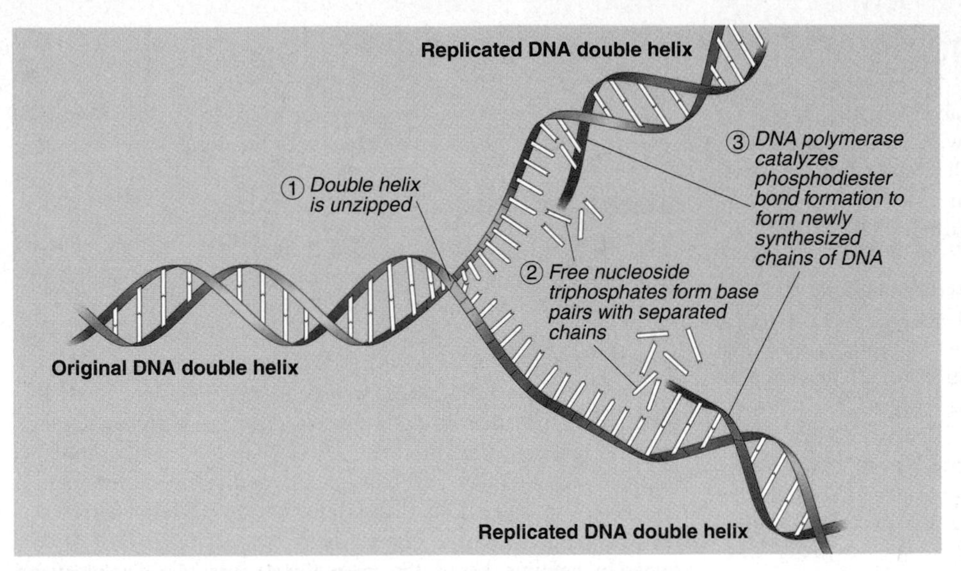
Replicated DNA double helix
① Double helix is unzipped
③ DNA polymerase catalyzes phosphodiester bond formation to form newly synthesized chains of DNA
② Free nucleoside triphosphates form base pairs with separated chains
Original DNA double helix
Replicated DNA double helix

Figure 15.34 Key stages in DNA replication.

ble helix is "unzipped," and each DNA chain acts as a template for the base pairing of its mononucleotide monomers with free mononucleoside triphosphates (Figure 15.34). These new units, H-bonded to their complements, are linked through phosphodiester bonds into a chain by an enzyme called *DNA polymerase*. Gradually, each of the unzipped chains forms the complementary half of a double helix, leading to *two* double helices. Because each base always pairs with its complement, the original double helix is copied and the genetic makeup of the cell preserved. (Base pairing is central to methods for learning the sequence of nucleotides in our genes, as the upcoming Chemical Connections essay shows.)

The biopolymers provide striking evidence for the folly of vitalism—that some sort of vital force exists only in substances from living systems. No unknowable force is required to explain the marvels of the living world. The same atomic properties that give rise to covalent bonds, molecular shape, and intermolecular forces provide the means for all life forms to flourish.

SECTION SUMMARY

The three types of natural polymers—polysaccharides, proteins, and nucleic acids—are formed by dehydration-condensation reactions. Polysaccharides are formed from cyclic monosaccharides, such as glucose. Cellulose, starch, and glycogen have structural or energy-storage roles. Proteins are polyamides formed from as many as 20 different types of amino acids. Fibrous proteins have extended shapes and play structural roles. Globular proteins have compact shapes and play metabolic, immunologic, and hormonal roles. The amino acid sequence of a protein determines its shape and function. Nucleic acids (DNA and RNA) are polynucleotides consisting of four different mononucleotides. The base sequence of the DNA chain determines the sequence of amino acids in an organism's proteins. Hydrogen bonding between specific base pairs is the key to protein synthesis and DNA replication.

Chapter Perspective

The amazing diversity of organic compounds highlights the importance of atomic properties. While gaining an appreciation for the special properties of carbon, you've seen similar types of bonding, molecular shapes, intermolecular forces, and reactivities in organic and inorganic compounds. Upcoming chapters focus on the dynamic aspects of all reactions—their speed, extent, and direction—as we explore the kinetics, equilibrium, and thermodynamics of chemical and physical change. We'll have many opportunities to exemplify these topics with organic compounds and biochemical systems.

DNA Sequencing and the Human Genome Project

As a result of one of the most remarkable scientific achievements in modern times, we now know our complete genetic makeup, the sequence of the 3 billion nucleotide base pairs in the DNA of the entire human genome! The potential benefits for medical science and for an understanding of our relationship with other creatures are profound.

As is true of so many large scientific endeavors, the Human Genome Project (www.genome.gov) began as a vision of forward-thinking scientists meeting at a conference—in 1986, at the Cold Spring Harbor Symposium in New York. In 1988, after much debate over feasibility—and even advisability—the U.S. Congress funded the $3-billion, 15-year program. Government scientists worked independently of, but in parallel with, private researchers at Celera Genomics, and in early 2003, the project was completed.

Central to this accomplishment is *DNA sequencing,* the process used to determine the identity and order of bases. Sequencing is an indispensable tool in molecular biology and biochemical genetics research. The process usually begins with a series of enzymatic and cloning steps to prepare the sample, but we will focus on the chemical steps that yield the final result.

An Overview of DNA Sequencing

The genome consists of DNA molecules; segments of their chains are the genes, the functional units of heredity. The DNA exists in each cell's nucleus as tightly coiled threads arranged, in the human, into 23 pairs of chromosomes (Figure B15.6). A given chromosome may have 100 million nucleotide bases, but the sequencing process can handle, at one time, DNA fragments only about 2000 bases long. Therefore, the chromosome is first broken into pieces by enzymes that cleave at specific sites. Then, through a variety of "amplification" methods, many copies of the individual DNA "target" fragments are made. After each target fragment has been sequenced, the data are interpreted to give a continuous sequence that represents the structure of the entire chromosome.

Steps in the Sequencing

Once a DNA target fragment is prepared, either of two methods are used to sequence it. We'll introduce the first and detail the second because it is much more common today.

1. *The Maxam-Gilbert chemical cleavage method* uses chemical reactions to break the DNA at specific locations. The total amount of the DNA target fragment is divided into four samples, and each is labeled with phosphorus-32, a radioactive "tag" that will make the final product easy to visualize. Each of the four samples is treated with reactants under conditions of temperature, concentration, and acidity that will cleave the DNA in characteristic ways. For instance, dimethylsulfate ($CH_3OSO_2OCH_3$) followed by heating cleaves the chain at a guanine (G). The cleavage products in each tube are then separated by size (length) through gel electrophoresis (discussed below). Photographic film placed on the gel is exposed by the radioactive tags in the cleavage products. The result is a series of dark bands, from which the sequence of the original DNA target fragment is determined.

2. *The Sanger chain-termination method* uses chemically altered bases to stop the growth of a complementary DNA chain at specific locations. As you've seen, the chain consists of linked 2′-deoxyribonucleoside monophosphate units (dNMP, where N represents A, T, G, or C). The link is a phosphodiester bond from the 3′-OH of one unit to the 5′-OH of the next. The free monomers used to construct the chain are 2′-deoxyribonucleoside triphosphates (dNTP). The Sanger method uses a modified monomer, called a *di*deoxyribonucleoside triphosphate (ddNTP), in which the 3′-OH group is also missing from the ribose unit (Figure B15.7). As soon as the ddNTP is incorporated into the growing chain,

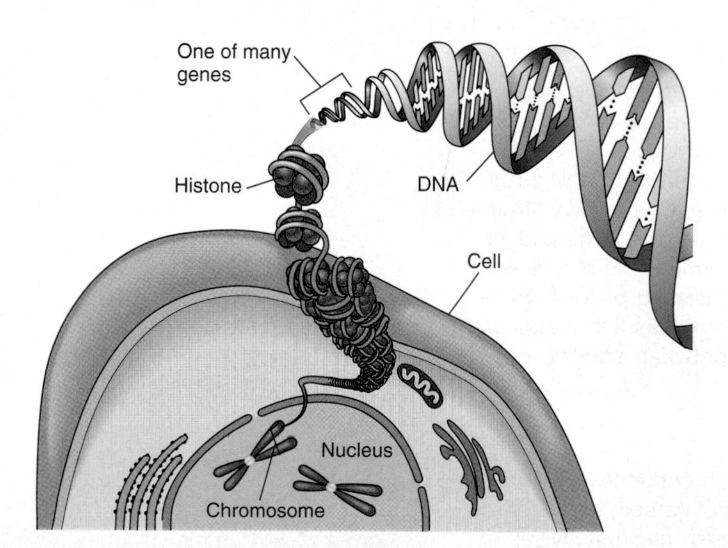

Figure B15.6 DNA, the genetic material. In the cell nucleus, each chromosome consists of a DNA molecule wrapped around globular proteins called histones. Segments of the DNA chains are genes.

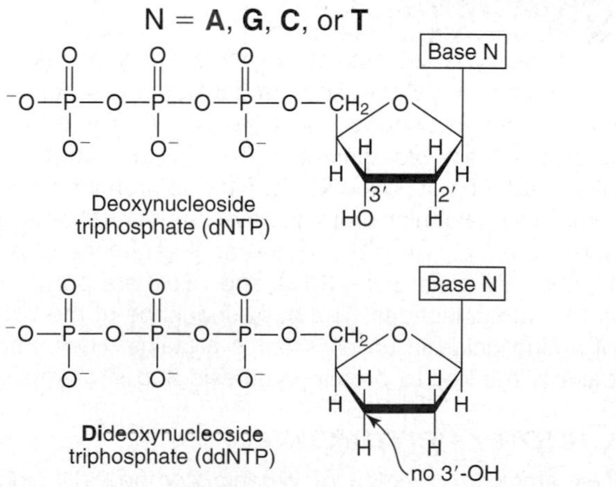

N = **A**, **G**, **C**, or **T**

Base N

Deoxynucleoside triphosphate (dNTP)

Base N

Dideoxynucleoside triphosphate (ddNTP)

no 3′-OH

Figure B15.7 Nucleoside triphosphate monomers. The normal *deoxy* monomer (dNTP; *top*) has no 2′-OH group but *does* have a 3′-OH group to continue growth of the polynucleotide chain. The modified *dideoxy* monomer in the Sanger method (ddNTP; *bottom*) also lacks the 3′-OH group.

polymerization stops, because there is no —OH on the 3′ position to form a phosphodiester bond.

Here is a simplified description of the procedure. After several preparation steps, the sample to be sequenced consists of a single-stranded DNA target fragment, which is attached to one strand of a double-stranded segment of DNA (Figure B15.8A). This sample is divided into four tubes, and to each tube is added a mixture of DNA polymerase, large amounts of all four dNTP's, and a small amount of one of the four ddNTP's. Thus, tube 1 contains polymerase, dATP, dGTP, dCTP, and dTTP, and, say, ddATP; tube 2 contains the same, except ddGTP instead of ddATP; and so forth. After the polymerization reaction is complete, each tube contains the original target fragment paired to complementary chains of varying length (Figure B15.8B). The chain lengths vary because in tube 1, each complementary chain ends in ddA (designated **A** in the figure); in tube 2, each ends in ddG (**G**); in tube 3, each ends in ddC (**C**); and in tube 4, each ends in ddT (**T**).

Each double-stranded product is divided into single strands, and then the complementary chains are separated by means of *high-resolution polyacrylamide-gel electrophoresis*. This technique, which is also used in the Maxam-Gilbert method, separates charged species by size through differences in their rate of migration through pores in the gel in the presence of an electric field: smaller species move faster than larger species. Because they have charged phosphate groups, polynucleotide fragments are commonly separated by electrophoresis. Polyacrylamide gels can be made with pores of varying size, and high-resolution gels can separate fragments that differ by a single nucleotide.

Each sample is applied to its own "lane" on a gel and, after electrophoresis, the gel is scanned to locate the chains, which appear in bands. Because all four ddNTP's were used, all possible chain fragments are formed, so the sequence of the original DNA fragment can be determined (Figure B15.8C).

An automated approach begins with each ddNTP tagged with a different fluorescent dye that emits light of a distinct color. In this method, the entire mixture of complementary chains is introduced onto the gel and separated. The gel then passes a laser, which activates the dyes. The fluorescent intensity versus distance along the chain is detected and plotted by a computer (Figure B15.8D).

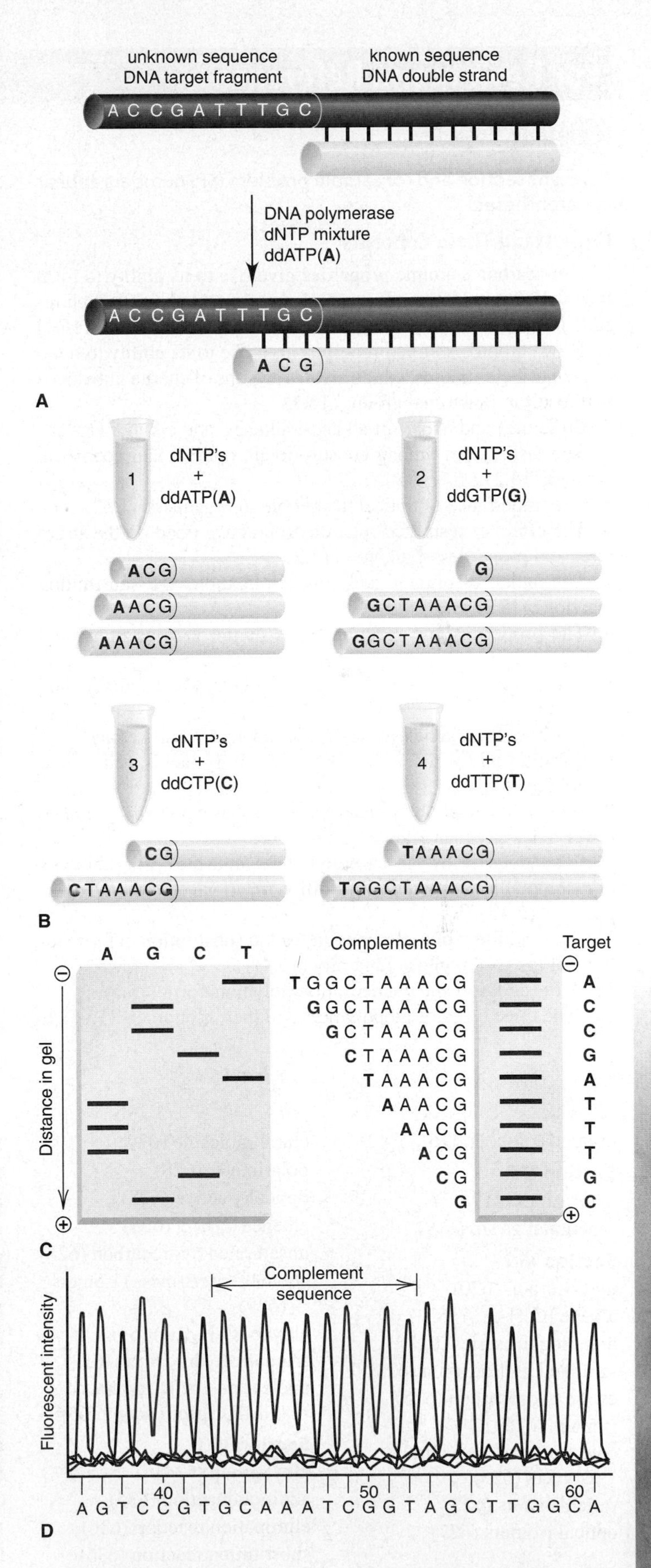

Figure B15.8 Steps in the Sanger method of DNA sequencing. A, The sample is the DNA target fragment, which is bonded to one strand (*black*) of a double-stranded segment. After polymerization, shown here in the presence of fluorescent-tagged ddATP, a complementary chain forms that ends with the base A. **B,** The sample is divided into four tubes, and each is treated with the dNTP mixture and a different ddNTP. Thus, every possible complementary chain forms. **C,** Electrophoresis of each mixture on its own lane of a polyacrylamide gel (*left*) separates the complements by length. The complement and target sequences are shown next to the bands. In the automated approach, all four ddNTP's are present in one tube, and electrophoresis gives bands on a gel (*right*). The gel is then scanned by a laser, which activates the fluorescent tags to emit their specific colors. **D,** A computer plot of light intensity against position in the chain shows the target sequence within the sequences of neighboring fragments.

For Review and Reference

(Numbers in parentheses refer to pages, unless noted otherwise.)

Learning Objectives

Relevant section and/or sample problem (SP) numbers appear in parentheses.

Understand These Concepts

1. How carbon's atomic properties give rise to its ability to form four strong covalent bonds, multiple bonds, and chains, which results in the great structural diversity of organic compounds (15.1)
2. How carbon's atomic properties give rise to its ability to bond to various heteroatoms, which creates regions of charge imbalance that result in functional groups (15.1)
3. Structures and names of alkanes, alkenes, and alkynes (15.2)
4. The distinctions among constitutional, optical, and geometric isomers (15.2)
5. The importance of optical isomerism in organisms (15.2)
6. The effect of restricted rotation around a π bond on the structures and properties of alkenes (15.2)
7. The nature of organic addition, elimination, and substitution reactions (15.3)
8. The properties and reaction types of the various functional groups (15.4):
   - Substitution and elimination for alcohols, alkyl halides, and amines
   - Addition for alkenes, alkynes, and aldehydes and ketones
   - Substitution for the carboxylic acid family (acids, esters, and amides)
9. Why delocalization of electrons causes aromatic rings to have lower reactivity than alkenes (15.4)
10. The polarity of the carbonyl bond and the importance of organometallic compounds in addition reactions of carbonyl compounds (15.4)
11. How addition plus elimination lead to substitution in the reactions of the carboxylic acid family (15.4)
12. How addition and condensation polymers form (15.5)
13. The three types of biopolymers and their monomers (15.6)
14. How amino acid sequence determines protein shape, which determines function (15.6)
15. How complementary base pairing controls the processes of protein synthesis and DNA replication (15.6)
16. How DNA base sequence determines RNA base sequence, which determines amino acid sequence (15.6)

Master These Skills

1. Drawing hydrocarbon structures given the number of C atoms, multiple bonds, and rings (SP 15.1)
2. Naming hydrocarbons and drawing expanded, condensed, and carbon skeleton formulas (15.2 and SP 15.2)
3. Drawing geometric isomers and identifying chiral centers of molecules (SP 15.2)
4. Recognizing the type of reaction from the structures of reactants and products (SP 15.3)
5. Recognizing a reaction as an oxidation or reduction from the structures of reactants and products (15.3)
6. Determining the reactants and products of the reactions of alcohols, alkyl halides, and amines (SP 15.4 and Follow-up)
7. Determining the products in a stepwise reaction sequence (SP 15.5)
8. Determining the reactants of the reactions of aldehydes and ketones (Follow-up Problem 15.5)
9. Determining the reactants and products of the reactions of the carboxylic acid family (SP 15.6 and Follow-up)
10. Recognizing and naming the functional groups in an organic molecule (SP 15.7)
11. Drawing an abbreviated synthetic polymer structure based on monomer structures (15.5)
12. Drawing small peptides from amino acid structures (15.6)
13. Using the base-paired sequence of one DNA strand to predict the sequence of the other (15.6)

Key Terms

organic compound (617)

Section 15.1

heteroatom (619)
functional group (619)

Section 15.2

hydrocarbon (620)
alkane (C_nH_{2n+2}) (623)
homologous series (623)
saturated hydrocarbon (623)
cyclic hydrocarbon (625)
isomers (625)
constitutional (structural) isomers (625)
stereoisomers (627)
optical isomers (627)
chiral molecule (627)
polarimeter (628)
optically active (628)
alkene (C_nH_{2n}) (628)
unsaturated hydrocarbon (628)
geometric (*cis-trans*) isomers (629)
alkyne (C_nH_{2n-2}) (631)
aromatic hydrocarbon (632)
nuclear magnetic resonance (NMR) spectroscopy (634)

Section 15.3

alkyl group (634)
addition reaction (635)
elimination reaction (636)
substitution reaction (636)

Section 15.4

alcohol (638)
haloalkane (alkyl halide) (640)
amine (640)
carbonyl group (643)
aldehyde (643)
ketone (644)
organometallic compound (644)
carboxylic acid (645)
ester (646)
amide (646)
fatty acid (646)
acid anhydride (646)
lipid (647)
hydrolysis (647)
nitrile (649)

Section 15.5

addition polymer (651)
condensation polymer (652)

Section 15.6

monosaccharide (654)
polysaccharide (654)
disaccharide (654)
protein (655)
amino acid (655)
nucleic acid (658)
mononucleotide (658)
double helix (659)
base pair (659)
genetic code (659)

Highlighted Figures and Tables

These figures (F) and tables (T) provide a review of key ideas.

Brief Solutions to Follow-up Problems

15.1 (a)

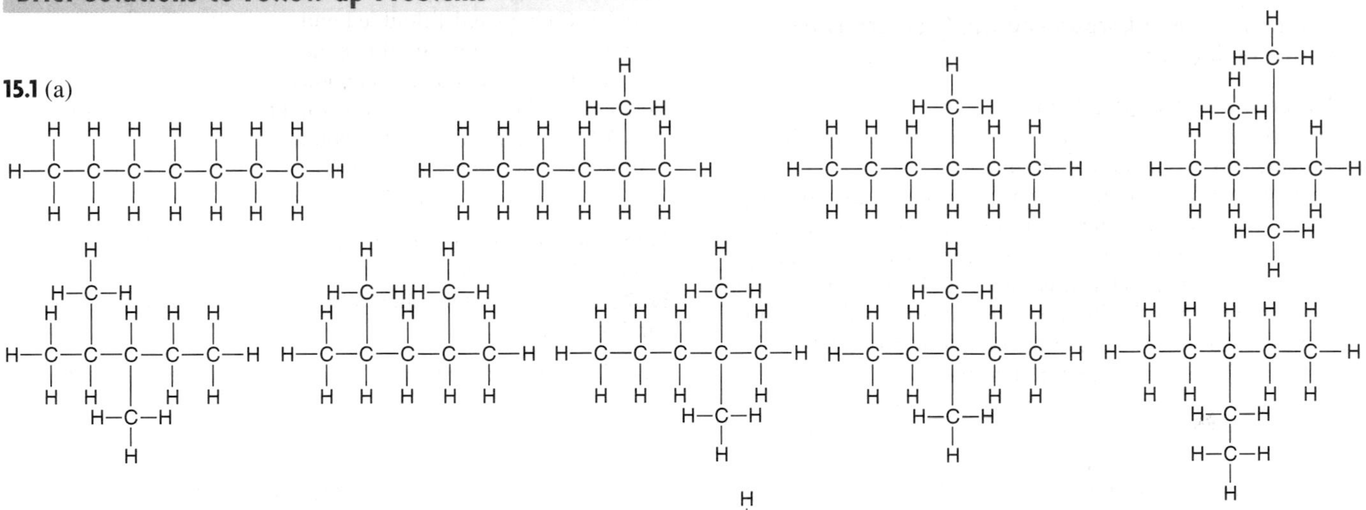

```
   H  H  H  H  H  H  H
   |  |  |  |  |  |  |
H—C——C——C——C——C——C——C—H
   |  |  |  |  |  |  |
   H  H  H  H  H  H  H

               H
               |
             H-C-H
   H  H  H  H  |  H
   |  |  |  |  |  |
H—C——C——C——C——C——C—H
   |  |  |  |  |  |
   H  H  H  H  H  H

            H
            |
          H-C-H
   H  H  H  |  H  H
   |  |  |  |  |  |
H—C——C——C——C——C——C—H
   |  |  |  |  |  |
   H  H  H  H  H  H

            H
            |
          H-C-H
      H     |
      |     |
    H-C-H   |
   H  |     |  H
   |  |     |  |
H—C——C——C——C—H
   |  |  |  |
   H  H  |  H
       H-C-H
         |
         H

      H
      |
    H-C-H
   H  |  H  H  H
   |  |  |  |  |
H—C——C——C——C——C—H
   |  |  |  |  |
   H  H  |  H  H
       H-C-H
         |
         H

      H     H
      |     |
    H-C-HH-C-H
   H  |  H  |  H
   |  |  |  |  |
H—C——C——C——C——C—H
   |  |  |  |  |
   H  H  H  H  H

            H
            |
          H-C-H
   H  H  H  |  H
   |  |  |  |  |
H—C——C——C——C——C—H
   |  |  |  |  |
   H  H  H  |  H
          H-C-H
            |
            H

         H
         |
       H-C-H
   H  H  |  H  H
   |  |  |  |  |
H—C——C——C——C——C—H
   |  |  |  |  |
   H  H  |  H  H
       H-C-H
         |
         H

   H  H  H  H  H
   |  |  |  |  |
H—C——C——C——C——C—H
   |  |  |  |  |
   H  H  |  H  H
       H-C-H
         |
       H-C-H
         |
         H
```

(b)

```
   H  H  H
   |  |  |
H—C——C——C——C≡C—H
   |  |  |
   H  H  H

   H  H        H
   |  |        |
H—C——C——C≡C——C—H
   |  |        |
   H  H        H

      H
      |
    H-C-H
   H  |
   |  |
H—C——C——C≡C—H
   |  |
   H  H
```

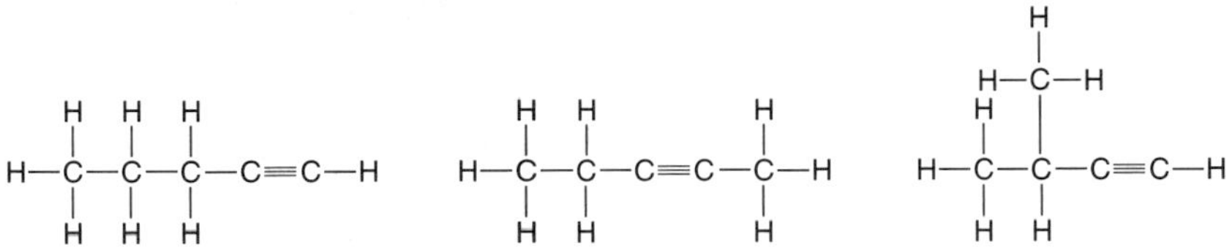

15.2 (a) $CH_3—CH_2—CH_2—CH_2—CH_2—C(CH_2—CH_3)(CH_3)—CH_2—CH_3$

(b) $CH_2—CH_2—CH_3$; $CH_2—CH_3$ and

(c) $HC{\equiv}C—C(CH_2—CH_3)_2—CH_2—CH_2—CH_3$

(d) $CH_3—CH_2$, CH_3 $C{=}C$ H, $CH_2—CH_2—CH_3$

15.3 (a) $CH_3—CH{=}CH—CH_3 + Cl_2 \longrightarrow CH_3—CH(Cl)—CH(Cl)—CH_3$

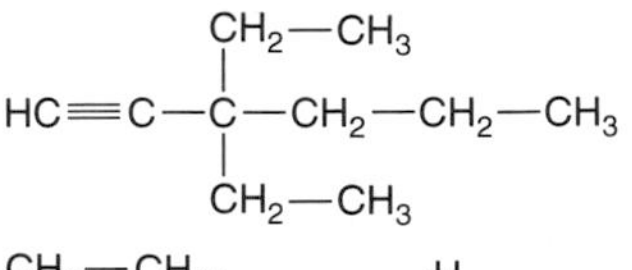

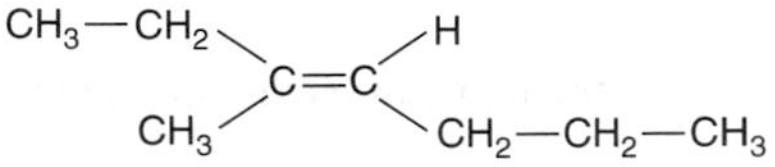

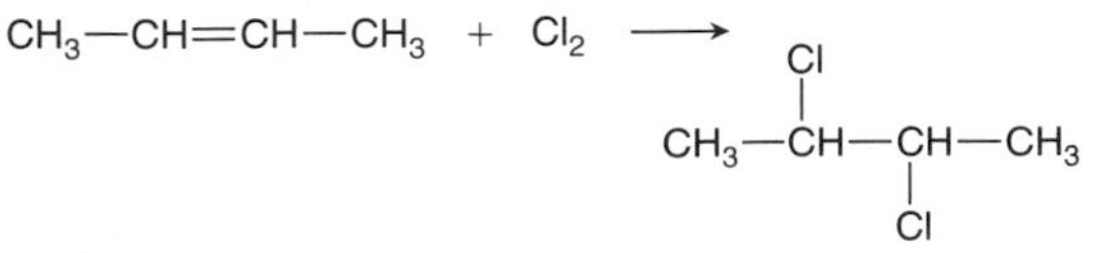

(b) $CH_3—CH_2—CH_2—Br + OH^- \longrightarrow CH_3—CH_2—CH_2—OH + Br^-$

(c) $CH_3—C(CH_3)(OH)—CH_3 \xrightarrow{-H_2O} CH_3—C(CH_3){=}CH_2$

15.4 (a) $CH_3—CH(CH_3)—CH(Cl)—CH_3$ or $CH_3—CH_2—C(CH_3)(Cl)—CH_3$

(b) $CH_3—CH_2—CH_2—OH$

15.5 (a) OH, CH_3 (b) $—C(=O)—H$

15.6 (a) $—CH_2—C(=O)—OH$

(b) $CH_3—CH_2—CH_2—C(=O)—O—CH_3 + CH_3—CH_2—NH_2$

15.7 (a) O, H; alkene, aldehyde

(b) $H_2N—C(=O)—CH_2—CH(Br)—CH_3$; amide, haloalkane

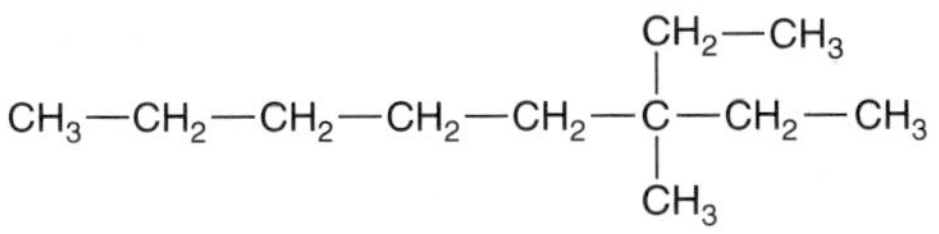

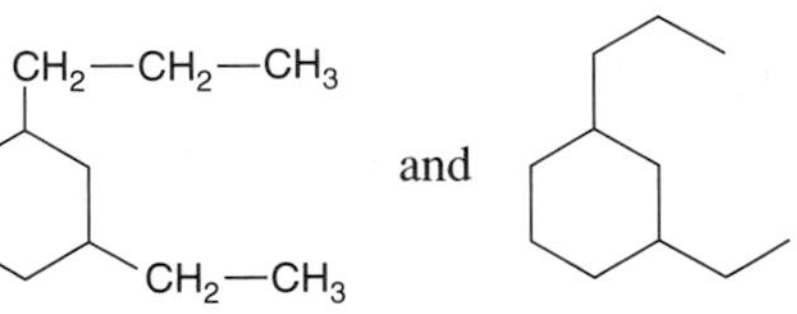

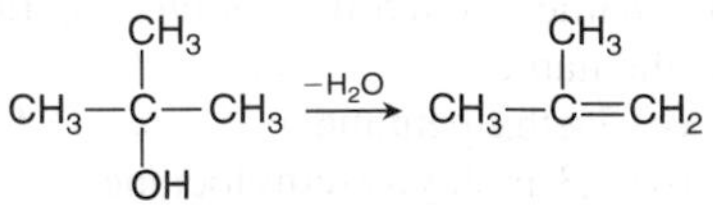

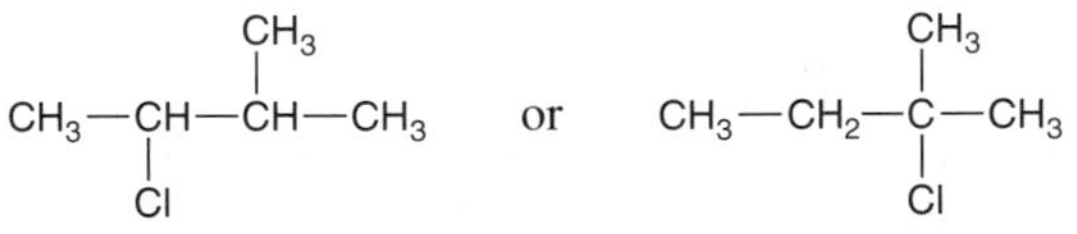

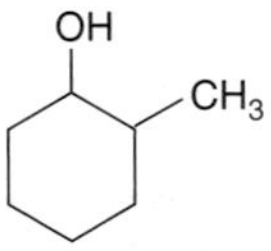

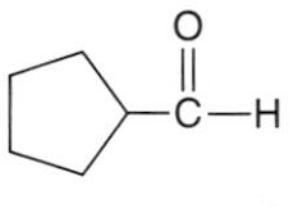

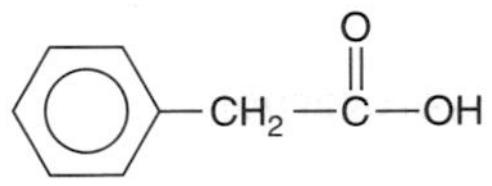

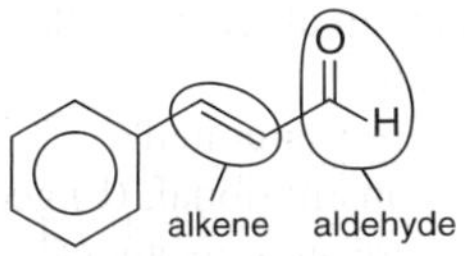

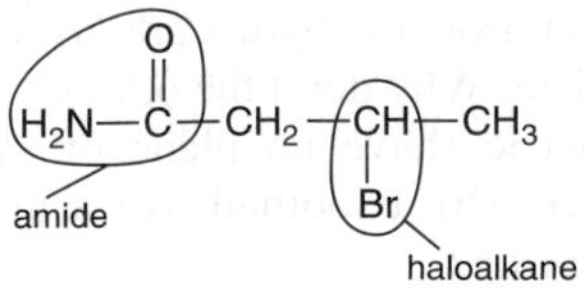

Temperature and biological activity. The metabolic processes of cold-blooded animals like this thorny devil (*Moloch horridus*) speed up as temperatures rise toward midday. In this chapter, you'll see how the speed of a reaction is influenced by several factors, including temperature, and how we can control them.

16

Kinetics: Rates and Mechanisms of Chemical Reactions

Until now we've taken a rather simple approach to chemical change: reactants mix and products form. A balanced equation is an essential quantitative tool for calculating product yields from reactant amounts, but it tells us nothing about three dynamic aspects of the reaction, which are essential to understanding chemical change and which we examine in the next several chapters:

- How fast is the reaction proceeding at a given moment?
- What will the reactant and product concentrations be when the reaction is complete?
- Will the reaction proceed by itself and release energy, or will it require energy to proceed?

This chapter addresses the first of these questions and focuses on the field of *kinetics,* which deals with the speed of a reaction and its mechanism, the stepwise changes that reactants undergo in their conversion to products. Chapters 17 through 19 are concerned with *equilibrium,* the dynamic balance between forward and reverse reactions and how external influences alter reactant and product concentrations. Chapters 20 and 21 present *thermodynamics* and its application to electrochemistry; there, we investigate why reactions occur and how we can put them to use. These central subdisciplines of chemistry apply to all physical and chemical change and, thus, are crucial to our understanding of modern technology, the environment, and the reactions in living things.

Chemical kinetics is the study of *reaction rates,* the changes in concentrations of reactants (or products) as a function of time (Figure 16.1).

Concepts & Skills to Review

before you study this chapter

- influence of temperature on molecular speed and collision frequency (Section 5.6)

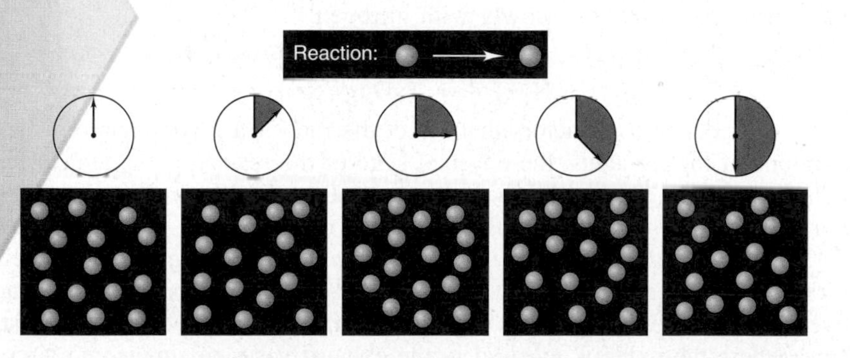

Figure 16.1 Reaction rate: the central focus of chemical kinetics. The rate at which reactant becomes product is the underlying theme of chemical kinetics. As time elapses, reactant *(purple) decreases* and product *(green) increases.*

Reactions occur at a wide range of rates (Figure 16.2). Some, like a neutralization, a precipitation, or an explosive redox process, seem to be over as soon as the reactants make contact—in a fraction of a second. Others, such as the reactions involved in cooking or rusting, take a moderate length of time, from minutes to months. Still others take much longer: the reactions that make up the

A

B

C

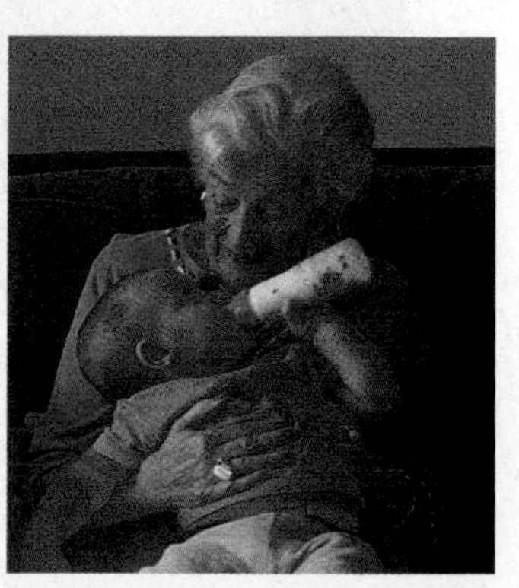

D

Figure 16.2 The wide range of reaction rates. Reactions proceed at a wide range of rates. An explosion **(A)** is much faster than the process of ripening **(B),** which is much faster than the process of rusting **(C),** which is much faster than the process of human aging **(D).**

human aging process continue for decades, and those involved in the formation of coal from dead plants take hundreds of millions of years.

Knowing how fast a chemical change occurs can be essential. How quickly a medicine acts or blood clots can make the difference between life and death. How long it takes for cement to harden or polyethylene to form can make the difference between profit and loss. In general, the rates of these diverse processes depend on the same variables, most of which chemists can manipulate to maximize yields within a given time or to slow down an unwanted reaction.

IN THIS CHAPTER . . . We first overview three key factors that affect reaction rate. Then we see how to express rate in the form of a *rate law* and how the components of a rate law are determined experimentally. We focus on the effects of concentration and temperature on rate and examine the models that explain those effects. Only then can we develop a reaction mechanism by noting the steps the reaction goes through and picturing the structure that exists as reactant bonds are breaking and product bonds are forming. Finally, we discuss how catalysts increase reaction rates, highlighting two vital areas—the reactions of a living cell and the depletion of atmospheric ozone.

16.1 FACTORS THAT INFLUENCE REACTION RATE

Let's begin our study of kinetics with a qualitative look at the key factors that affect how fast a reaction proceeds. Under any given set of conditions, *each reaction has its own characteristic rate,* which is determined by the chemical nature of the reactants. At room temperature, for example, hydrogen reacts explosively with fluorine but extremely slowly with nitrogen:

$$H_2(g) + F_2(g) \longrightarrow 2HF(g) \qquad \text{[very fast]}$$
$$3H_2(g) + N_2(g) \longrightarrow 2NH_3(g) \qquad \text{[very slow]}$$

We can control four factors that affect the rate of a given reaction: the concentrations of the reactants, the physical state of the reactants, the temperature at which the reaction occurs, and the use of a catalyst. We'll consider the first three factors here and discuss the fourth later in the chapter.

1. *Concentration: molecules must collide to react.* A major factor influencing the rate of a given reaction is reactant concentration. Consider the reaction between ozone and nitric oxide (nitrogen monoxide) that occurs in the stratosphere, where the oxide is released in the exhaust gases of supersonic aircraft:

$$NO(g) + O_3(g) \longrightarrow NO_2(g) + O_2(g)$$

Imagine what this might look like at the molecular level with the reactants confined in a reaction vessel. Nitric oxide and ozone molecules zoom every which way, crashing into each other and the vessel walls. A reaction between NO and O_3 can occur only when the molecules collide. The more molecules present in the container, the more frequently they collide, and the more often a reaction occurs. Thus, *reaction rate is proportional to the concentration of reactants:*

$$\text{Rate} \propto \text{collision frequency} \propto \text{concentration}$$

In this case, we're looking at a very simple reaction, one in which reactant molecules collide and form product molecules in one step, but even the rates of complex reactions depend on reactant concentration.

2. *Physical state: molecules must mix to collide.* The frequency of collisions between molecules also depends on the physical states of the reactants. When the reactants are in the same phase, as in an aqueous solution, thermal motion brings them into contact. When they are in different phases, contact occurs only at the interface, so vigorous stirring and grinding may be needed. In these cases, *the more finely divided a solid or liquid reactant, the greater its surface area per unit volume, the more contact it makes with the other reactant, and the faster the reaction occurs.* Figure 16.3A shows a steel nail heated in oxygen glowing feebly; in

A

B

Figure 16.3 The effect of surface area on reaction rate. A, A hot nail glows in O_2. **B,** The same mass of hot steel wool bursts into flame in O_2. The greater surface area per unit volume of the steel wool means that more metal makes contact with O_2, so the reaction is faster.

Figure 16.3B, the same mass of steel wool bursts into flame. For the same reason, you start a campfire with wood chips and thin branches, not logs.

3. *Temperature: molecules must collide with enough energy to react.* Temperature usually has a major effect on the speed of a reaction. Two familiar kitchen appliances employ this effect: a refrigerator slows down chemical processes that spoil food, whereas an oven speeds up other chemical processes that cook it.

Recall that molecules in a sample of gas have a range of speeds, with the most probable speed dependent on the temperature (see Figure 5.14, p. 201). Thus, *at a higher temperature, more collisions occur in a given time.* Even more important, however, is the fact that temperature affects the kinetic energy of the molecules, and thus the *energy* of the collisions. In the jumble of molecules in the reaction of NO and O_3, mentioned previously, most collisions result in the molecules simply recoiling, like billiard balls, with no reaction taking place. However, some collisions occur with so much energy that the molecules react (Figure 16.4). And, at a higher temperature, more of these sufficiently energetic collisions occur. Thus, *raising the temperature increases the reaction rate by increasing the number and, especially, the energy of the collisions:*

$$\text{Rate} \propto \text{collision energy} \propto \text{temperature}$$

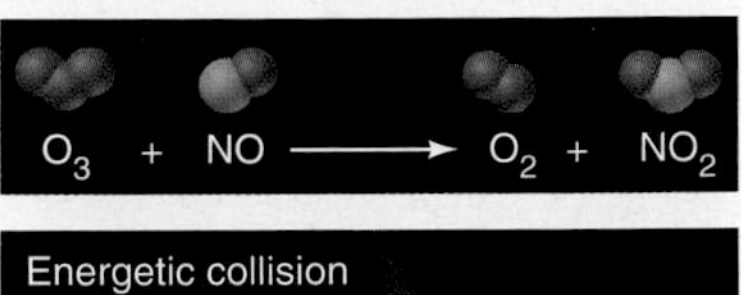

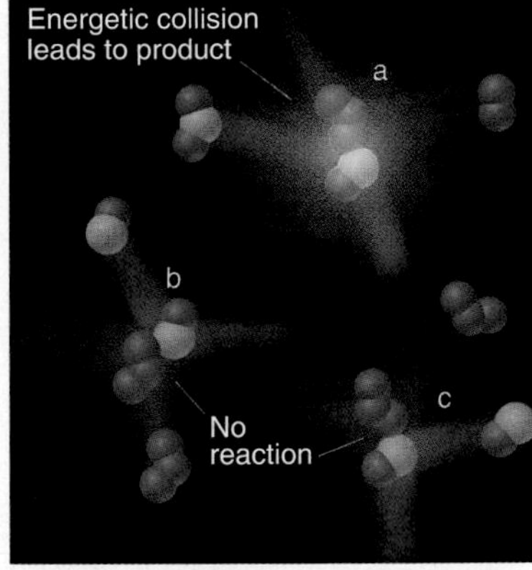

Figure 16.4 Collision energy and reaction rate. The reaction equation is shown in the panel. Although many collisions between NO and O_3 molecules occur, relatively few have enough energy to cause reaction. At this temperature, only collision *a* is energetic enough to lead to product; the reactant molecules in collisions *b* and *c* just bounce off each other.

The qualitative idea that reaction rate is influenced by the frequency and energy of reactant collisions leads to several quantitative questions: How can we describe the dependence of rate on reactant concentration mathematically? Do all changes in concentration affect the rate to the same extent? Do all rates increase to the same extent with a given rise in temperature? How do reactant molecules use the energy of collision to form product molecules, and is there a way to determine this energy? What do the reactants look like as they are turning into products? We address these questions in the following sections.

SECTION SUMMARY

Chemical kinetics deals with reaction rates and the stepwise molecular events by which a reaction occurs. Under a given set of conditions, each reaction has its own rate. Concentration affects rate by influencing the frequency of collisions between reactant molecules. Physical state affects rate by determining the surface area per unit volume of reactant(s). Temperature affects rate by influencing the frequency and, even more importantly, the energy of the reactant collisions.

16.2 EXPRESSING THE REACTION RATE

Before we can deal quantitatively with the effects of concentration and temperature on reaction rate, we must express the rate mathematically. A *rate* is a change in some variable per unit of time. The most common examples relate to the rate of motion (speed) of an object, which is the change in its position divided by the change in time. Suppose, for instance, we measure a runner's starting position, x_1, at time t_1 and final position, x_2, at time t_2. The runner's average speed is

$$\text{Rate of motion} = \frac{\text{change in position}}{\text{change in time}} = \frac{x_2 - x_1}{t_2 - t_1} = \frac{\Delta x}{\Delta t}$$

A runner changing position with time

In the case of a chemical change, we are concerned with the **reaction rate,** the changes in concentrations of reactants or products per unit time: *reactant concentrations decrease while product concentrations increase.* Consider a general reaction, A $\longrightarrow$ B. We quickly measure the starting reactant concentration (conc A_1) at t_1, allow the reaction to proceed, and then quickly measure the reactant concentration again (conc A_2) at t_2. The change in concentration divided by the change in time gives the *average* rate:

$$\text{Rate of reaction} = -\frac{\text{change in concentration of A}}{\text{change in time}} = -\frac{\text{conc A}_2 - \text{conc A}_1}{t_2 - t_1} = -\frac{\Delta(\text{conc A})}{\Delta t}$$

Note the minus sign. By convention, reaction rate is a *positive* number, but conc A_2 will always be *lower* than conc A_1, so the *change in (final − initial) concentration of reactant A is always negative.* We use the minus sign simply to convert the negative change in reactant concentration to a positive value for the rate. Suppose the concentration of A changes from 1.2 mol/L (conc A_1) to 0.75 mol/L (conc A_2) over a 125-s period. The average rate is

$$\text{Rate} = -\frac{0.75 \text{ mol/L} - 1.2 \text{ mol/L}}{125 \text{ s} - 0 \text{ s}} = 3.6\times10^{-3} \text{ mol/L·s}$$

We use *square brackets,* [ ], *to express concentration in moles per liter.* That is, [A] is the concentration of A in mol/L, so the rate expressed in terms of A is

$$\text{Rate} = -\frac{\Delta[\text{A}]}{\Delta t} \quad \textbf{(16.1)}$$

The rate has units of moles per liter per second (mol L^{-1} s^{-1}, or mol/L·s), or any time unit convenient for the particular reaction (minutes, years, and so on).

If instead we measure the *product* to determine the reaction rate, we find its concentration *increasing* over time. That is, conc B_2 is always *higher* than conc B_1. Thus, the *change* in product concentration, Δ[B], is *positive,* and the reaction rate for A $\longrightarrow$ B expressed in terms of B is

$$\text{Rate} = \frac{\Delta[\text{B}]}{\Delta t}$$

Average, Instantaneous, and Initial Reaction Rates

Examining the rate of a real reaction reveals an important point: *the rate itself varies with time as the reaction proceeds.* Consider the reversible gas-phase reaction between ethylene and ozone, one of many reactions that can be involved in the formation of photochemical smog:

$$C_2H_4(g) + O_3(g) \rightleftharpoons C_2H_4O(g) + O_2(g)$$

For now, we consider only reactant concentrations. You can see from the equation coefficients that for every molecule of C_2H_4 that reacts, a molecule of O_3 reacts with it. In other words, the concentrations of both reactants decrease at the same rate in this particular reaction:

$$\text{Rate} = -\frac{\Delta[C_2H_4]}{\Delta t} = -\frac{\Delta[O_3]}{\Delta t}$$

By measuring the concentration of either reactant, we can follow the reaction rate.

Suppose we have a known concentration of O_3 in a closed reaction vessel kept at 30°C (303 K). Table 16.1 shows the concentration of O_3 at various times during the first minute after we introduce C_2H_4 gas. The rate over the entire 60.0 s is the total change in concentration divided by the change in time:

$$\text{Rate} = -\frac{\Delta[O_3]}{\Delta t} = -\frac{(1.10\times10^{-5} \text{ mol/L}) - (3.20\times10^{-5} \text{ mol/L})}{60.0 \text{ s} - 0.0 \text{ s}} = 3.50\times10^{-7} \text{ mol/L·s}$$

This calculation gives us the **average rate** over that period; that is, during the first 60.0 s of the reaction, ozone concentration decreases an *average* of 3.50×10^{-7} mol/L each second. However, the average rate does not show that the rate is changing, and it tells us nothing about how fast the ozone concentration is decreasing *at any given instant.*

We can see the rate change during the reaction by calculating the average rate over two shorter periods—one earlier and one later. Between the starting time 0.0 s and 10.0 s, the average rate is

$$\text{Rate} = -\frac{\Delta[O_3]}{\Delta t} = -\frac{(2.42\times10^{-5} \text{ mol/L}) - (3.20\times10^{-5} \text{ mol/L})}{10.0 \text{ s} - 0.0 \text{ s}} = 7.80\times10^{-7} \text{ mol/L·s}$$

Table 16.1 Concentration of O_3 at Various Times in Its Reaction with C_2H_4 at 303 K

Time (s)	Concentration of O_3 (mol/L)
0.0	3.20×10^{-5}
10.0	2.42×10^{-5}
20.0	1.95×10^{-5}
30.0	1.63×10^{-5}
40.0	1.40×10^{-5}
50.0	1.23×10^{-5}
60.0	1.10×10^{-5}

During the last 10.0 s, between 50.0 s and 60.0 s, the average rate is

$$\text{Rate} = -\frac{\Delta[O_3]}{\Delta t} = -\frac{(1.10\times10^{-5}\ \text{mol/L}) - (1.23\times10^{-5}\ \text{mol/L})}{60.0\ \text{s} - 50.0\ \text{s}} = 1.30\times10^{-7}\ \text{mol/L·s}$$

The earlier rate is six times as fast as the later rate. Thus, *the rate decreases during the course of the reaction*. This makes perfect sense from a molecular point of view: as O_3 molecules are used up, fewer of them are present to collide with C_2H_4 molecules, so the rate decreases.

The change in rate can also be seen by plotting the concentrations vs. the times at which they were measured (Figure 16.5). A curve is obtained, which means that the rate changes. *The slope of the straight line ($\Delta y/\Delta x$, that is, $\Delta[O_3]/\Delta t$) joining any two points gives the average rate over that period.*

The shorter the time period we choose, the closer we come to the **instantaneous rate,** the rate at a particular instant during the reaction. *The slope of a line tangent to the curve at a particular point gives the instantaneous rate at that time.* For example, the rate of the reaction of C_2H_4 and O_3 35.0 s after it began is 2.50×10^{-7} mol/L·s, the slope of the line drawn tangent to the curve through the point at which $t = 35.0$ s (line *d* in Figure 16.5). In general, we use the term *reaction rate* to mean the *instantaneous* reaction rate.

As a reaction continues, the product concentrations increase, and so the reverse reaction proceeds more quickly. To find the overall (net) rate, we would have to take both forward and reverse reactions into account and calculate the difference between their rates. A common way to avoid this complication for many reactions is to measure the **initial rate,** the instantaneous rate at the moment the reactants are mixed. Under these conditions, the product concentrations are negligible, so the reverse rate is negligible. Moreover, we know the reactant concentrations from the concentrations and volumes of the solutions we mix together. The initial rate is measured by determining the slope of the line tangent to the curve at $t = 0$ s. In Figure 16.5, the initial rate is 10.0×10^{-7} mol/L·s (line *a*). Unless stated otherwise, we will use initial rate data to determine other kinetic parameters.

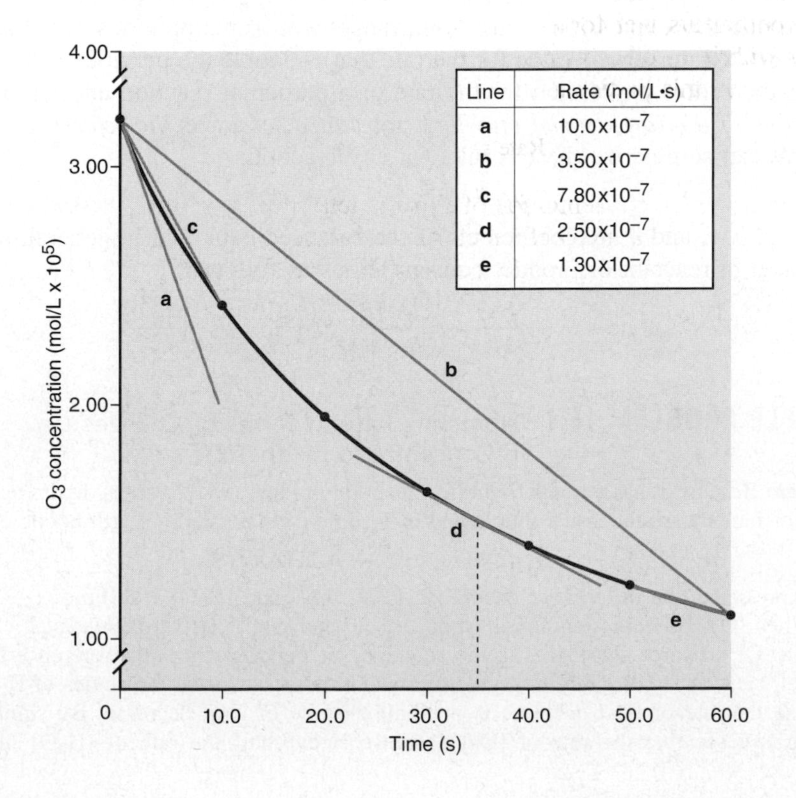

Figure 16.5 The concentration of O_3 vs. time during its reaction with C_2H_4. Plotting the data in Table 16.1 gives a curve because the rate changes during the reaction. The *average* rate over a given period is the slope of a line joining two points along the curve. The slope of line *b* is the average rate over the first 60.0 s of the reaction. The slopes of lines *c* and *e* give the average rate over the first and last 10.0-s intervals, respectively. Line *c* is steeper than line *e* because the average rate over the earlier period is higher. The *instantaneous* rate at 35.0 s is the slope of line *d*, the tangent to the curve at $t = 35.0$ s. The *initial* rate is the slope of line *a*, the tangent to the curve at $t = 0$ s.

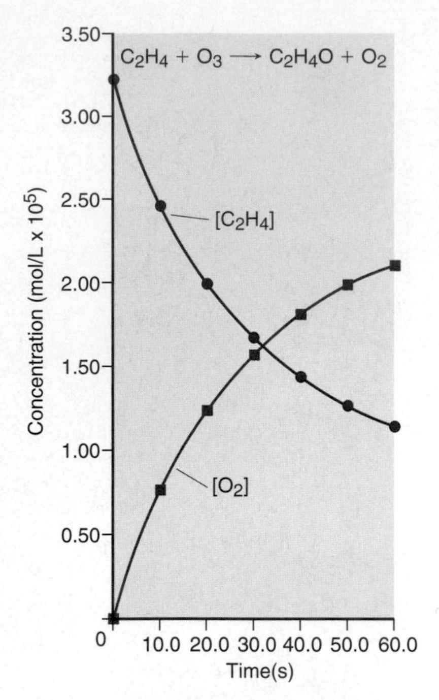

Figure 16.6 Plots of $[C_2H_4]$ and $[O_2]$ vs. time. Measuring reactant concentration, $[C_2H_4]$, and product concentration, $[O_2]$, gives curves of identical shapes but changing in opposite directions. The steep upward (positive) slope of $[O_2]$ early in the reaction mirrors the steep downward (negative) slope of $[C_2H_4]$ because the faster C_2H_4 is used up, the faster O_2 is formed. The curve shapes are identical in this case because the equation coefficients are identical.

Expressing Rate in Terms of Reactant and Product Concentrations

So far, in our discussion of the reaction of C_2H_4 and O_3, we've expressed the rate in terms of the decreasing concentration of O_3. The rate is the same in terms of C_2H_4, but it is exactly the opposite in terms of the products because their concentrations are *increasing*. From the balanced equation, we see that one molecule of C_2H_4O and one of O_2 appear for every molecule of C_2H_4 and of O_3 that disappear. We can express the rate in terms of any of the four substances involved:

$$\text{Rate} = -\frac{\Delta[C_2H_4]}{\Delta t} = -\frac{\Delta[O_3]}{\Delta t} = +\frac{\Delta[C_2H_4O]}{\Delta t} = +\frac{\Delta[O_2]}{\Delta t}$$

Again, note the negative values for the reactants and the positive values for the products (usually written without the plus sign). Figure 16.6 shows a plot of the simultaneous monitoring of one reactant and one product. Because, in this case, product concentration increases at the same rate that reactant concentration decreases, the curves have the same shapes but are inverted.

In the reaction between ethylene and ozone, the reactants disappear and the products appear at the same rate because all the coefficients in the balanced equation are equal. Consider next the reaction between hydrogen and iodine to form hydrogen iodide:

$$H_2(g) + I_2(g) \longrightarrow 2HI(g)$$

For every molecule of H_2 that disappears, one molecule of I_2 disappears and *two* molecules of HI appear. In other words, the rate of $[H_2]$ decrease is the same as the rate of $[I_2]$ decrease, but both are only half the rate of [HI] increase. By referring the change in $[I_2]$ and [HI] to the change in $[H_2]$, we have

$$\text{Rate} = -\frac{\Delta[H_2]}{\Delta t} = -\frac{\Delta[I_2]}{\Delta t} = \frac{1}{2}\frac{\Delta[HI]}{\Delta t}$$

If we refer the change in $[H_2]$ and $[I_2]$ to the change in [HI] instead, we obtain

$$\text{Rate} = \frac{\Delta[HI]}{\Delta t} = -2\frac{\Delta[H_2]}{\Delta t} = -2\frac{\Delta[I_2]}{\Delta t}$$

Notice that this expression is just a rearrangement of the previous one; also note that it gives a numerical value for the rate that is double the previous value. Thus, the mathematical expression for the rate of a particular reaction and *the numerical value of the rate depend on which substance serves as the reference.*

We can summarize these results for any reaction:

$$aA + bB \longrightarrow cC + dD$$

where a, b, c, and d are coefficients of the balanced equation. In general, the rate is related to reactant or product concentrations as follows:

$$\text{Rate} = -\frac{1}{a}\frac{\Delta[A]}{\Delta t} = -\frac{1}{b}\frac{\Delta[B]}{\Delta t} = \frac{1}{c}\frac{\Delta[C]}{\Delta t} = \frac{1}{d}\frac{\Delta[D]}{\Delta t} \quad (16.2)$$

SAMPLE PROBLEM 16.1 Expressing Rate in Terms of Changes in Concentration with Time

Problem Because it has a nonpolluting combustion product (water vapor), hydrogen gas is used for fuel aboard the space shuttle and may some day be used by Earth-bound engines:

$$2H_2(g) + O_2(g) \longrightarrow 2H_2O(g)$$

(a) Express the rate in terms of changes in $[H_2]$, $[O_2]$, and $[H_2O]$ with time.
(b) When $[O_2]$ is decreasing at 0.23 mol/L·s, at what rate is $[H_2O]$ increasing?

Plan (a) Of the three substances in the equation, let's choose O_2 as the reference because its coefficient is 1. For every molecule of O_2 that disappears, two molecules of H_2 disappear, so the rate of $[O_2]$ decrease is one-half the rate of $[H_2]$ decrease. By similar reasoning, we see that the rate of $[O_2]$ decrease is one-half the rate of $[H_2O]$ increase.

(b) Because $[O_2]$ is decreasing, the change in its concentration must be negative. We substitute the negative value into the expression and solve for $\Delta[H_2O]/\Delta t$.

Solution (a) Expressing the rate in terms of each component:

$$\text{Rate} = -\frac{1}{2}\frac{\Delta[H_2]}{\Delta t} = -\frac{\Delta[O_2]}{\Delta t} = \frac{1}{2}\frac{\Delta[H_2O]}{\Delta t}$$

(b) Calculating the rate of change of $[H_2O]$:

$$\frac{1}{2}\frac{\Delta[H_2O]}{\Delta t} = -\frac{\Delta[O_2]}{\Delta t} = -(-0.23 \text{ mol/L·s})$$

$$\frac{\Delta[H_2O]}{\Delta t} = 2(0.23 \text{ mol/L·s}) = 0.46 \text{ mol/L·s}$$

Check (a) A good check is to use the rate expression to obtain the balanced equation: $[H_2]$ changes twice as fast as $[O_2]$, so two H_2 molecules react for each O_2. $[H_2O]$ changes twice as fast as $[O_2]$, so two H_2O molecules form from each O_2. From this reasoning, we get $2H_2 + O_2 \longrightarrow 2H_2O$. The $[H_2]$ and $[O_2]$ decrease, so they take minus signs; $[H_2O]$ increases, so it takes a plus sign. Another check is to use Equation 16.2, with $A = H_2$, $a = 2$; $B = O_2$, $b = 1$; $C = H_2O$, $c = 2$. Thus,

$$\text{Rate} = -\frac{1}{a}\frac{\Delta[A]}{\Delta t} = -\frac{1}{b}\frac{\Delta[B]}{\Delta t} = \frac{1}{c}\frac{\Delta[C]}{\Delta t}$$

or

$$\text{Rate} = -\frac{1}{2}\frac{\Delta[H_2]}{\Delta t} = -\frac{\Delta[O_2]}{\Delta t} = \frac{1}{2}\frac{\Delta[H_2O]}{\Delta t}$$

(b) Given the rate expression, it makes sense that the numerical value of the rate of $[H_2O]$ increase is twice that of $[O_2]$ decrease.

Comment Thinking through this type of problem at the molecular level is the best approach, but use Equation 16.2 to confirm your answer.

FOLLOW-UP PROBLEM 16.1 **(a)** Balance the following equation and express the rate in terms of the change in concentration with time for each substance:

$$NO(g) + O_2(g) \longrightarrow N_2O_3(g)$$

(b) How fast is $[O_2]$ decreasing when [NO] is decreasing at a rate of 1.60×10^{-4} mol/L·s?

SECTION SUMMARY

The average reaction rate is the change in reactant (or product) concentration over a change in time, Δt. The rate slows as reactants are used up. The instantaneous rate at time t is obtained from the slope of the tangent to a concentration vs. time curve at time t. The initial rate, the instantaneous rate at $t = 0$, occurs when reactants are just mixed and before any product accumulates. The expression for a reaction rate and its numerical value depend on which reaction component is being monitored.

16.3 THE RATE LAW AND ITS COMPONENTS

The centerpiece of any kinetic study is the **rate law** (or **rate equation**) for the reaction in question. The rate law expresses the rate as a function of reactant concentrations, product concentrations, and temperature. Any hypothesis we make about how the reaction occurs on the molecular level must conform to the rate law because it is based on experimental fact.

In this discussion, we generally consider reactions for which the products do not appear in the rate law. In these cases, *the reaction rate depends only on reactant concentrations and temperature*. First, we look at the effect of concentration on rate for reactions occurring at a fixed temperature. For a general reaction,

$$aA + bB + \cdots \longrightarrow cC + dD + \cdots$$

the rate law has the form

$$\text{Rate} = k[A]^m[B]^n \cdots \qquad (16.3)$$

Aside from the concentration terms, [A] and [B], the other parameters in Equation 16.3 require some definition. The proportionality constant k, called the **rate constant,** is specific for a given reaction at a given temperature; it does *not* change as the reaction proceeds. (As you'll see in Section 16.5, k *does* change with temperature and therefore determines how temperature affects the rate.) The exponents m and n, called the **reaction orders,** define how the rate is affected by reactant concentration. Thus, if the rate doubles when [A] doubles, the rate depends on [A] raised to the first power, $[A]^1$, so $m = 1$. Similarly, if the rate quadruples when [B] doubles, the rate depends on [B] raised to the second power, $[B]^2$, so $n = 2$. In another reaction, the rate may not change at all when [A] doubles; in that case, the rate does *not* depend on [A] or, to put it another way, the rate depends on [A] raised to the zero power, $[A]^0$, so $m = 0$. Keep in mind that the coefficients a and b in the general balanced equation are *not* necessarily related in any way to the reaction orders m and n.

A key point to remember is that *the components of the rate law—rate, reaction orders, and rate constant—must be found by experiment;* they cannot be

TOOLS OF THE LABORATORY

Measuring Reaction Rates

Speculation about how a reaction occurs at the molecular level must be based on measurements of reaction rates. There are many experimental approaches, but all must obtain the results quickly and reproducibly. We consider four common methods, with specific examples.

Spectrometric Methods

These methods are used to measure the concentration of a reactant or product that absorbs (or emits) light of a narrow range of wavelengths. The reaction is typically performed *within* the sample compartment of a spectrometer that has been set to measure a wavelength characteristic of one of the species (Figure B16.1; see also Tools of the Laboratory, pp. 269–270). For example, in the reaction of NO and O_3, only NO_2 has a color:

$$NO(g, \text{colorless}) + O_3(g, \text{colorless}) \longrightarrow O_2(g, \text{colorless}) + NO_2(g, \text{brown})$$

Known amounts of the reactants are injected into a gas sample tube of known volume, and the rate of NO_2 formation is measured by monitoring the color over time. Reactions in aqueous solution are studied similarly.

Conductometric Methods

When nonionic reactants form ionic products, or vice versa, the change in conductivity of the solution over time can be used to measure the rate. Electrodes are immersed in the reaction mixture, and the increase (or decrease) in conductivity correlates with the formation of product (Figure B16.2). Consider the reaction between an organic halide, such as 2-bromo-2-methylpropane, and water:

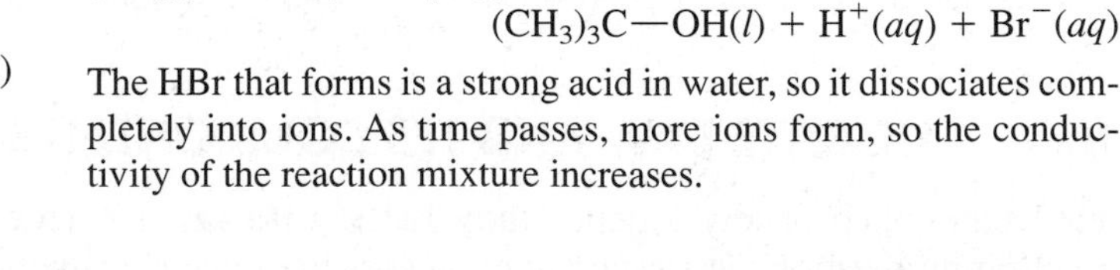

$$(CH_3)_3C{-}Br(l) + H_2O(l) \longrightarrow (CH_3)_3C{-}OH(l) + H^+(aq) + Br^-(aq)$$

The HBr that forms is a strong acid in water, so it dissociates completely into ions. As time passes, more ions form, so the conductivity of the reaction mixture increases.

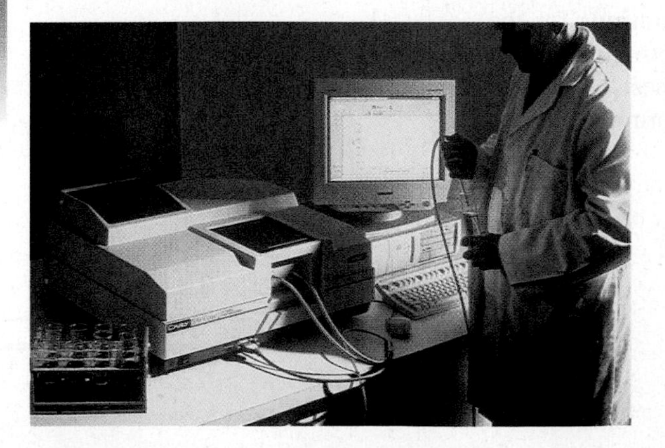

Absorbance: 1.50, 1.25, 1.00, 0.75, 0.50, 0.25; Time (s): 0, 30, 60, 90, 120, 150, 180

Figure B16.1 Spectrometric monitoring of a reaction. The investigator adds the reactant(s) to the sample tube and immediately places it in the spectrometer. For a reactant (or product) that is colored, rate data are determined from a plot of absorbance (of light of particular wavelength) vs. time. The rate of disappearance of a blue dye is being studied here.

deduced from the reaction stoichiometry. So let's take an experimental approach to finding the components by

1. Using concentration measurements to find the *initial rate*
2. Using initial rates from several experiments to find the *reaction orders*
3. Using these values to calculate the *rate constant*

Once we know the rate law, we can use it to predict the rate for any initial reactant concentrations.

Determining the Initial Rate

In the last section, we showed how initial rates are determined from a plot of concentration vs. time. Because we use initial rate data to determine the reaction orders and rate constant, an accurate experimental method for measuring concentration at various times during the reaction is essential to constructing the rate law. A few of the many techniques used for measuring changes in concentration over time are presented in the Tools of the Laboratory essay.

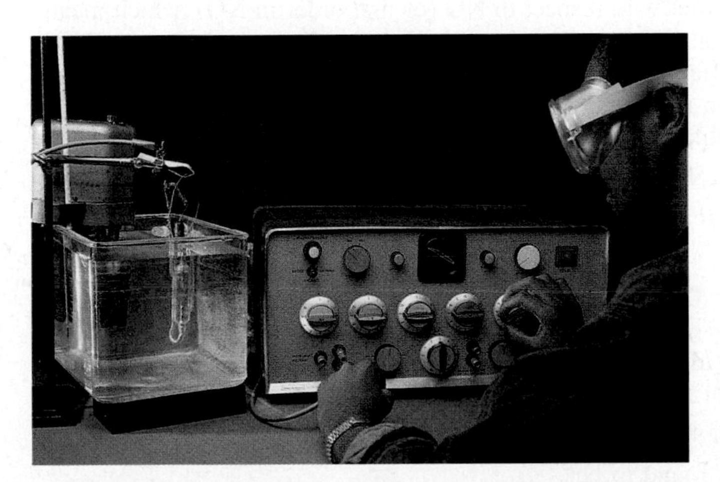

Figure B16.2 Conductometric monitoring of a reaction. When a reactant mixture differs in conductivity from the product mixture, the change in conductivity is proportional to the reaction rate. This method is usually used when nonionic reactants form ionic products.

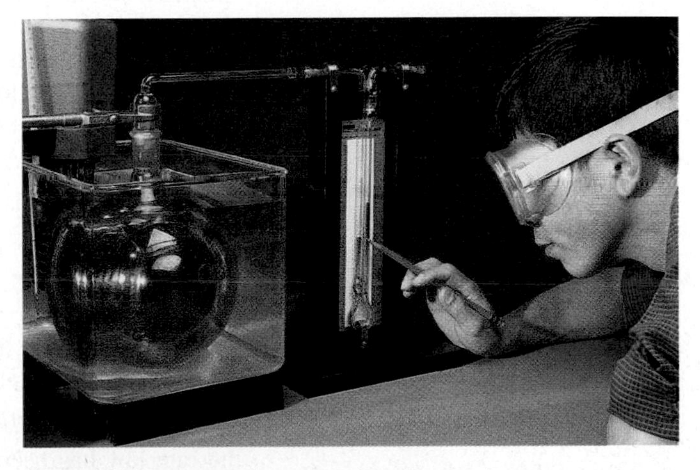

Figure B16.3 Manometric monitoring of a reaction. When a reaction results in a change in the number of moles of gas, the change in pressure with time corresponds to a change in reaction rate. The rate of formation of NO_2 from N_2O_4 is being studied here.

Manometric Methods

If a reaction involves a change in the number of moles of gas, the rate can be determined from the change in pressure (at constant volume and temperature) over time. In practice, a manometer is attached to a reaction vessel of known volume that is immersed in a constant-temperature bath. For example, the reaction between zinc and acetic acid can be monitored by this method:

$$Zn(s) + 2CH_3COOH(aq) \longrightarrow Zn^{2+}(aq) + 2CH_3COO^-(aq) + H_2(g)$$

As H_2 forms, the gas pressure increases. Thus, the reaction rate is directly proportional to the rate of increase of the H_2 gas pressure. In Figure B16.3, this method is being used to study a reaction involving nitrogen dioxide.

Direct Chemical Methods

Rates of slow reactions, or of those that can be easily slowed, are often studied by direct chemical methods. A small, measured portion (called an *aliquot*) of the reaction mixture is removed, and the reaction in this portion is stopped, usually by rapid cooling. The concentration of reactant or product in the aliquot is measured, while the bulk of the reaction mixture continues to react and is sampled later. For example, the earlier reaction between an organic halide and water can also be studied by titration. The reaction rate in an aliquot is slowed by quickly transferring it to a chilled flask in an ice bath. HBr concentration in the aliquot is determined by titrating with standardized NaOH solution. To determine the change in HBr concentration with time, the procedure is repeated at regular intervals during the reaction.

Reaction Order Terminology

Before we see how reaction orders are determined from initial rate data, let's discuss the meaning of reaction order and some important terminology. We speak of a reaction as having an *individual* order "with respect to" or "in" each reactant as well as an *overall* order, which is simply the sum of the individual orders.

In the simplest case, a reaction with a single reactant A, the reaction is *first order* overall if the rate is directly proportional to [A]:

$$\text{Rate} = k[\text{A}]$$

It is *second order* overall if the rate is directly proportional to the square of [A]:

$$\text{Rate} = k[\text{A}]^2$$

And it is *zero order* overall if the rate is *not* dependent on [A] at all, a situation that is quite common in metal-catalyzed and biochemical processes:

$$\text{Rate} = k[\text{A}]^0 = k(1) = k$$

Here are some real examples. For the reaction between nitrogen monoxide and ozone,

$$\text{NO}(g) + \text{O}_3(g) \longrightarrow \text{NO}_2(g) + \text{O}_2(g)$$

the rate law has been experimentally determined to be

$$\text{Rate} = k[\text{NO}][\text{O}_3]$$

This reaction is first order with respect to NO (or first order in NO), which means that the rate depends on NO concentration raised to the first power, that is, $[\text{NO}]^1$ (an exponent of 1 is generally omitted). It is also first order with respect to O_3, or $[\text{O}_3]^1$. This reaction is second order overall (1 + 1 = 2).

Now consider a different gas-phase reaction:

$$2\text{NO}(g) + 2\text{H}_2(g) \longrightarrow \text{N}_2(g) + 2\text{H}_2\text{O}(g)$$

The rate law for this reaction has been determined to be

$$\text{Rate} = k[\text{NO}]^2[\text{H}_2]$$

The reaction is second order in NO and first order in H_2, so it is third order overall.

Finally, for the hydrolysis of 2-bromo-2-methylpropane, which we just considered in the Tools of the Laboratory essay,

$$(\text{CH}_3)_3\text{C—Br}(l) + \text{H}_2\text{O}(l) \longrightarrow (\text{CH}_3)_3\text{C—OH}(l) + \text{H}^+(aq) + \text{Br}^-(aq)$$

the rate law has been found to be

$$\text{Rate} = k[(\text{CH}_3)_3\text{CBr}]$$

This reaction is first order in 2-bromo-2-methylpropane. Note that the concentration of H_2O does not even appear in the rate law. Thus, the reaction is zero order with respect to H_2O ($[\text{H}_2\text{O}]^0$). This means that the rate does not depend on the concentration of H_2O. We can also write the rate law as

$$\text{Rate} = k[(\text{CH}_3)_3\text{CBr}][\text{H}_2\text{O}]^0$$

Overall, this is a first-order reaction.

These examples demonstrate a major point: *reaction orders* ***cannot*** *be deduced from the balanced equation.* For the reaction between NO and H_2 and for the hydrolysis of 2-bromo-2-methylpropane, the reaction orders in the rate laws do *not* correspond to the coefficients of the balanced equations. Reaction orders *must* be determined from rate data.

Reaction orders are usually positive integers or zero, but they can also be fractional or negative. For the reaction

$$\text{CHCl}_3(g) + \text{Cl}_2(g) \longrightarrow \text{CCl}_4(g) + \text{HCl}(g)$$

a fractional order appears in the rate law:

$$\text{Rate} = k[\text{CHCl}_3][\text{Cl}_2]^{1/2}$$

This reaction order means that the rate depends on the square root of the Cl_2 concentration. For example, if the initial Cl_2 concentration is increased by a factor of 4, while the initial $CHCl_3$ concentration is kept the same, the rate increases by a factor of 2, the square root of the change in $[Cl_2]$. A negative exponent means that the rate *decreases* when the concentration of that component increases. Negative orders are often seen for reactions whose rate laws include products. For example, for the atmospheric reaction

$$2\text{O}_3(g) \rightleftharpoons 3\text{O}_2(g)$$

the rate law has been shown to be

$$\text{Rate} = k[\text{O}_3]^2[\text{O}_2]^{-1} = k\frac{[\text{O}_3]^2}{[\text{O}_2]}$$

If the O_2 concentration doubles, the reaction proceeds half as fast.

SAMPLE PROBLEM 16.2 Determining Reaction Order from Rate Laws

Problem For each of the following reactions, use the given rate law to determine the reaction order with respect to each reactant and the overall order:
(a) $2\text{NO}(g) + \text{O}_2(g) \longrightarrow 2\text{NO}_2(g)$; rate $= k[\text{NO}]^2[\text{O}_2]$
(b) $\text{CH}_3\text{CHO}(g) \longrightarrow \text{CH}_4(g) + \text{CO}(g)$; rate $= k[\text{CH}_3\text{CHO}]^{3/2}$
(c) $\text{H}_2\text{O}_2(aq) + 3\text{I}^-(aq) + 2\text{H}^+(aq) \longrightarrow \text{I}_3^-(aq) + 2\text{H}_2\text{O}(l)$; rate $= k[\text{H}_2\text{O}_2][\text{I}^-]$
Plan We inspect the exponents in the rate law, *not* the coefficients of the balanced equation, to find the individual orders, and then take their sum to find the overall reaction order.
Solution (a) The exponent of [NO] is 2, so the reaction is second order with respect to NO, first order with respect to O_2, and third order overall.
(b) The reaction is $\frac{3}{2}$ order in CH_3CHO and $\frac{3}{2}$ order overall.
(c) The reaction is first order in H_2O_2, first order in I^-, and second order overall.
The reactant H^+ does not appear in the rate law, so the reaction is zero order in H^+.
Check Be sure that each reactant has an order and that the sum of the individual orders gives the overall order.

FOLLOW-UP PROBLEM 16.2 Experiment shows that the reaction

$$5\text{Br}^-(aq) + \text{BrO}_3^-(aq) + 6\text{H}^+(aq) \longrightarrow 3\text{Br}_2(l) + 3\text{H}_2\text{O}(l)$$

obeys this rate law: rate $= k[\text{Br}^-][\text{BrO}_3^-][\text{H}^+]^2$. What are the reaction orders in each reactant and the overall reaction order?

Determining Reaction Orders

Sample Problem 16.2 shows how to find the reaction orders from a known rate law. Now let's see how they are found from data *before* the rate law is known. Consider the reaction between oxygen and nitrogen monoxide, a key step in the formation of acid rain and in the industrial production of nitric acid:

$$\text{O}_2(g) + 2\text{NO}(g) \longrightarrow 2\text{NO}_2(g)$$

The rate law, expressed in general form, is

$$\text{Rate} = k[\text{O}_2]^m[\text{NO}]^n$$

To find the reaction orders, *we run a series of experiments, starting each one with a different set of reactant concentrations and obtaining an initial rate in each case.*

Table 16.2 Initial Rates for a Series of Experiments with the Reaction Between O_2 and NO

Experiment	Initial Reactant Concentrations (mol/L) O_2	NO	Initial Rate (mol/L·s)
1	1.10×10^{-2}	1.30×10^{-2}	3.21×10^{-3}
2	2.20×10^{-2}	1.30×10^{-2}	6.40×10^{-3}
3	1.10×10^{-2}	2.60×10^{-2}	12.8×10^{-3}
4	3.30×10^{-2}	1.30×10^{-2}	9.60×10^{-3}
5	1.10×10^{-2}	3.90×10^{-2}	28.8×10^{-3}

Table 16.2 shows experiments that change one reactant concentration while keeping the other constant. If we compare experiments 1 and 2, we see the effect of doubling $[O_2]$ on the rate. First, we take the ratio of their rate laws:

$$\frac{\text{Rate 2}}{\text{Rate 1}} = \frac{k[O_2]_2^m[NO]_2^n}{k[O_2]_1^m[NO]_1^n}$$

where $[O_2]_2$ is the O_2 concentration for experiment 2, $[NO]_1$ is the NO concentration for experiment 1, and so forth. Because k is a constant and [NO] does not change between these two experiments, these quantities cancel:

$$\frac{\text{Rate 2}}{\text{Rate 1}} = \frac{[O_2]_2^m}{[O_2]_1^m} = \left(\frac{[O_2]_2}{[O_2]_1}\right)^m$$

Substituting the values from Table 16.2, we obtain

$$\frac{6.40\times10^{-3}\text{ mol/L·s}}{3.21\times10^{-3}\text{ mol/L·s}} = \left(\frac{2.20\times10^{-2}\text{ mol/L}}{1.10\times10^{-2}\text{ mol/L}}\right)^m$$

Dividing, we obtain

$$1.99 = (2.00)^m$$

Rounding to one significant figure gives

$$2 = 2^m; \quad \text{therefore,} \quad m = 1$$

The reaction is first order in O_2: when $[O_2]$ doubles, the rate doubles.

To find the order with respect to NO, we compare experiments 3 and 1, in which $[O_2]$ is held constant and [NO] is doubled:

$$\frac{\text{Rate 3}}{\text{Rate 1}} = \frac{k[O_2]_3^m[NO]_3^n}{k[O_2]_1^m[NO]_1^n}$$

As before, k is constant, and in this pair of experiments $[O_2]$ does not change, so these quantities cancel:

$$\frac{\text{Rate 3}}{\text{Rate 1}} = \left(\frac{[NO]_3}{[NO]_1}\right)^n$$

The actual values give

$$\frac{12.8\times10^{-3}\text{ mol/L·s}}{3.21\times10^{-3}\text{ mol/L·s}} = \left(\frac{2.60\times10^{-2}\text{ mol/L}}{1.30\times10^{-2}\text{ mol/L}}\right)^n$$

Dividing, we obtain

$$3.99 = (2.00)^n$$

Rounding gives

$$4 = 2^n; \quad \text{therefore,} \quad n = 2$$

The reaction is second order in NO: when [NO] doubles, the rate quadruples. Thus, the rate law is

$$\text{Rate} = k[O_2][NO]^2$$

You may want to use experiment 1 in combination with experiments 4 and 5 to check this result.

SAMPLE PROBLEM 16.3 Determining Reaction Orders from Initial Rate Data

Problem Many gaseous reactions occur in car engines and exhaust systems. One of these is

$$NO_2(g) + CO(g) \longrightarrow NO(g) + CO_2(g) \qquad \text{rate} = k[NO_2]^m[CO]^n$$

Use the following data to determine the individual and overall reaction orders:

Experiment	Initial Rate (mol/L·s)	Initial $[NO_2]$ (mol/L)	Initial [CO] (mol/L)
1	0.0050	0.10	0.10
2	0.080	0.40	0.10
3	0.0050	0.10	0.20

Plan We need to solve the general rate law for the reaction orders *m* and *n*. To solve for each exponent, we proceed as in the text, taking the ratio of the rate laws for two experiments in which only the reactant in question changes.

Solution Calculating *m* in $[NO_2]^m$: We take the ratio of the rate laws for experiments 1 and 2, in which $[NO_2]$ varies but [CO] is constant:

$$\frac{\text{Rate 2}}{\text{Rate 1}} = \frac{k[NO_2]_2^m[CO]_2^n}{k[NO_2]_1^m[CO]_1^n} = \left(\frac{[NO_2]_2}{[NO_2]_1}\right)^m \quad \text{or} \quad \frac{0.080 \text{ mol/L·s}}{0.0050 \text{ mol/L·s}} = \left(\frac{0.40 \text{ mol/L}}{0.10 \text{ mol/L}}\right)^m$$

This gives $16 = 4.0^m$, so $m = 2.0$. The reaction is second order in NO_2.

Calculating *n* in $[CO]^n$: We take the ratio of the rate laws for experiments 1 and 3, in which [CO] varies but $[NO_2]$ is constant:

$$\frac{\text{Rate 3}}{\text{Rate 1}} = \frac{k[NO_2]_3^2[CO]_3^n}{k[NO_2]_1^2[CO]_1^n} = \left(\frac{[CO]_3}{[CO]_1}\right)^n \quad \text{or} \quad \frac{0.0050 \text{ mol/L·s}}{0.0050 \text{ mol/L·s}} = \left(\frac{0.20 \text{ mol/L}}{0.10 \text{ mol/L}}\right)^n$$

We have $1.0 = (2.0)^n$, so $n = 0$. The rate does not change when [CO] varies, so the reaction is zero order in CO.

Therefore, the rate law is

$$\text{Rate} = k[NO_2]^2[CO]^0 = k[NO_2]^2(1) = k[NO_2]^2$$

The reaction is second order overall.

Check A good check is to reason through the orders. If $m = 1$, quadrupling $[NO_2]$ would quadruple the rate; but the rate *more* than quadruples; so $m > 1$. If $m = 2$, quadrupling $[NO_2]$ would increase the rate by a factor of 16 (4^2). The ratio of rates is $0.080/0.005 = 16$, so $m = 2$. In contrast, increasing [CO] has no effect on the rate, which can happen only if $[CO]^n = 1$, so $n = 0$.

FOLLOW-UP PROBLEM 16.3 Find the rate law and the overall reaction order for the reaction $H_2 + I_2 \longrightarrow 2HI$ from the following data at 450°C:

Experiment	Initial Rate (mol/L·s)	Initial $[H_2]$ (mol/L)	Initial $[I_2]$ (mol/L)
1	1.9×10^{-23}	0.0113	0.0011
2	1.1×10^{-22}	0.0220	0.0033
3	9.3×10^{-23}	0.0550	0.0011
4	1.9×10^{-22}	0.0220	0.0056

Determining the Rate Constant

With the rate, reactant concentrations, and reaction orders known, the sole remaining unknown in the rate law is the rate constant, *k*. The rate constant is specific for a particular reaction *at a particular temperature*. The experiments with the reaction of O_2 and NO were run at the same temperature, so we can use data from any to solve for *k*. From experiment 1 in Table 16.2, for instance, we obtain

$$k = \frac{\text{rate 1}}{[O_2]_1[NO]_1^2} = \frac{3.21\times10^{-3} \text{ mol/L·s}}{(1.10\times10^{-2} \text{ mol/L})(1.30\times10^{-2} \text{ mol/L})^2}$$

$$= \frac{3.21\times10^{-3} \text{ mol/L·s}}{1.86\times10^{-6} \text{ mol}^3/\text{L}^3} = 1.73\times10^3 \text{ L}^2/\text{mol}^2\text{·s}$$

Table 16.3 Units of the Rate Constant k for Several Overall Reaction Orders

Overall Reaction Order	Units of k (t in seconds)
0	mol/L·s (or mol L^{-1} s^{-1})
1	1/s (or s^{-1})
2	L/mol·s (or L mol^{-1} s^{-1})
3	L^2/mol^2·s (or L^2 mol^{-2} s^{-1})

General formula:

$$\text{Units of } k = \frac{\left(\dfrac{\text{L}}{\text{mol}}\right)^{\text{order}-1}}{\text{unit of } t}$$

Always check that the values of k for a series are constant within experimental error. To three significant figures, the average value of k for the five experiments in Table 16.2 is 1.72×10^3 L^2/mol^2·s.

Note the units for the rate constant. With concentrations in mol/L and the reaction rate in units of mol/L·time, the units for k depend on the order of the reaction and, of course, the time unit. The units for k in our example, L^2/mol^2·s, are required to give a rate with units of mol/L·s:

$$\frac{\text{mol}}{\text{L}\cdot\text{s}} = \frac{\text{L}^2}{\text{mol}^2\cdot\text{s}} \times \frac{\text{mol}}{\text{L}} \times \left(\frac{\text{mol}}{\text{L}}\right)^2$$

The rate constant will *always* have these units for an overall third-order reaction with the time unit in seconds. Table 16.3 shows the units of k for some common overall reaction orders, but you can always determine the units mathematically.

SECTION SUMMARY

An experimentally determined rate law shows how the rate of a reaction depends on concentration. If we consider only initial rates, the rate law often takes this form: rate $= k[A]^m[B]^n \cdots$. With an accurate method for obtaining initial rates, reaction orders are determined experimentally by comparing rates for different initial concentrations, that is, by performing several experiments and varying the concentration of one reactant at a time to see its effect on the rate. With rate, concentrations, and reaction orders known, the rate constant is the only remaining unknown in the rate law, so it can be calculated.

16.4 INTEGRATED RATE LAWS: CONCENTRATION CHANGES OVER TIME

Notice that the rate laws we've developed so far do not include time as a variable. They tell us the rate or concentration at a given instant, allowing us to answer a critical question, "How fast is the reaction proceeding at the moment when y moles per liter of A are reacting with z moles per liter of B?" However, by employing different forms of the rate laws, called **integrated rate laws,** we can consider the time factor and answer other questions, such as "How long will it take for x moles per liter of A to be used up?" and "What is the concentration of A after y minutes of reaction?"

Integrated Rate Laws for First-, Second-, and Zero-Order Reactions

Consider a simple first-order reaction, A $\longrightarrow$ B. (Because first- and second-order reactions are more common, we'll discuss them before zero-order reactions.) As we discussed previously, the rate can be expressed as the change in the concentration of A divided by the change in time:

$$\text{Rate} = -\frac{\Delta[\text{A}]}{\Delta t}$$

It can also be expressed in terms of the rate law:

$$\text{Rate} = k[\text{A}]$$

Setting these different expressions equal to each other gives

$$-\frac{\Delta[\text{A}]}{\Delta t} = k[\text{A}]$$

Through the methods of calculus, this expression is integrated over time to obtain the integrated rate law for a first-order reaction:

$$\ln\frac{[\text{A}]_0}{[\text{A}]_t} = kt \quad \text{(first-order reaction; rate} = k[\text{A}]) \qquad \textbf{(16.4)}$$

where ln is the natural logarithm, $[A]_0$ is the concentration of A at $t = 0$, and $[A]_t$ is the concentration of A at any time t during an experiment. In mathematical terms, $\ln \frac{a}{b} = \ln a - \ln b$, so we have

$$\ln [A]_0 - \ln [A]_t = kt$$

For a general second-order reaction, the expression including time is quite complex, so let's consider the case in which the rate law contains only one reactant. Setting the rate expressions equal to each other gives

$$\text{Rate} = -\frac{\Delta[A]}{\Delta t} = k[A]^2$$

Integrating over time gives the integrated rate law for a second-order reaction involving one reactant:

$$\frac{1}{[A]_t} - \frac{1}{[A]_0} = kt \quad \text{(second-order reaction; rate} = k[A]^2) \qquad \textbf{(16.5)}$$

For a zero-order reaction, we have

$$\text{Rate} = -\frac{\Delta[A]}{\Delta t} = k[A]^0$$

Integrating over time gives the integrated rate law for a zero-order reaction:

$$[A]_t - [A]_0 = -kt \quad \text{(zero-order reaction; rate} = k[A]^0 = k) \qquad \textbf{(16.6)}$$

Sample Problem 16.4 shows one way integrated rate laws are applied.

SAMPLE PROBLEM 16.4 Determining the Reactant Concentration at a Given Time

Problem At 1000°C, cyclobutane (C_4H_8) decomposes in a first-order reaction, with the very high rate constant of 87 s^{-1}, to two molecules of ethylene (C_2H_4).
(a) If the initial C_4H_8 concentration is 2.00 M, what is the concentration after 0.010 s?
(b) What fraction of C_4H_8 has decomposed in this time?

Plan (a) We must find the concentration of cyclobutane at time t, $[C_4H_8]_t$. The problem tells us this is a first-order reaction, so we use the integrated first-order rate law:

$$\ln \frac{[C_4H_8]_0}{[C_4H_8]_t} = kt$$

We know k (87 s^{-1}), t (0.010 s), and $[C_4H_8]_0$ (2.00 M), so we can solve for $[C_4H_8]_t$.
(b) The fraction decomposed is the concentration that has decomposed divided by the initial concentration:

$$\text{Fraction decomposed} = \frac{[C_4H_8]_0 - [C_4H_8]_t}{[C_4H_8]_0}$$

Solution (a) Substituting the data into the integrated rate law:

$$\ln \frac{2.00 \text{ mol/L}}{[C_4H_8]_t} = (87 \text{ s}^{-1})(0.010 \text{ s}) = 0.87$$

Taking the antilog of both sides:

$$\frac{2.00 \text{ mol/L}}{[C_4H_8]_t} = e^{0.87} = 2.4$$

Solving for $[C_4H_8]_t$:

$$[C_4H_8]_t = \frac{2.00 \text{ mol/L}}{2.4} = 0.83 \text{ mol/L}$$

(b) Finding the fraction that has decomposed after 0.010 s:

$$\frac{[C_4H_8]_0 - [C_4H_8]_t}{[C_4H_8]_0} = \frac{2.00 \text{ mol/L} - 0.83 \text{ mol/L}}{2.00 \text{ mol/L}} = 0.58$$

Check The concentration remaining after 0.010 s (0.83 mol/L) is less than the starting concentration (2.00 mol/L), which makes sense. Raising e to an exponent slightly less than 1 should give a number (2.4) slightly less than the value of e (2.718). Moreover, the final result makes sense: a high rate constant indicates a fast reaction, so it's not surprising that so much decomposes in such a short time.

Comment Integrated rate laws are also used to solve for the time it takes to reach a certain reactant concentration, as in the follow-up problem.

FOLLOW-UP PROBLEM 16.4 At 25°C, hydrogen iodide breaks down very slowly to hydrogen and iodine: rate $= k[\text{HI}]^2$. The rate constant at 25°C is 2.4×10^{-21} L/mol·s. If 0.0100 mol of HI(g) is placed in a 1.0-L container, how long will it take for the concentration of HI to reach 0.00900 mol/L (10.0% reacted)?

Determining the Reaction Order from the Integrated Rate Law

Suppose you don't know the rate law for a reaction and don't have the initial rate data needed to determine the reaction orders (which we did have in Sample Problem 16.3). Another method for finding reaction orders is a graphical technique that uses concentration and time data directly.

An integrated rate law can be rearranged into the form of an equation for a straight line, $y = mx + b$, where m is the slope and b is the y-axis intercept. For a first-order reaction, we have

$$\ln\,[\text{A}]_0 - \ln\,[\text{A}]_t = kt$$

Rearranging and changing signs gives

$$\begin{aligned} \ln\,[\text{A}]_t &= -kt + \ln\,[\text{A}]_0 \\ y &= mx + b \end{aligned}$$

Therefore, a plot of $\ln\,[\text{A}]_t$ vs. time gives a straight line with slope $= -k$ and y intercept $= \ln\,[\text{A}]_0$ (Figure 16.7A).

For a simple second-order reaction, we have

$$\frac{1}{[\text{A}]_t} - \frac{1}{[\text{A}]_0} = kt$$

Rearranging gives

$$\begin{aligned} \frac{1}{[\text{A}]_t} &= kt + \frac{1}{[\text{A}]_0} \\ y &= mx + b \end{aligned}$$

In this case, a plot of $1/[\text{A}]_t$ vs. time gives a straight line with slope $= k$ and y intercept $= 1/[\text{A}]_0$ (Figure 16.7B).

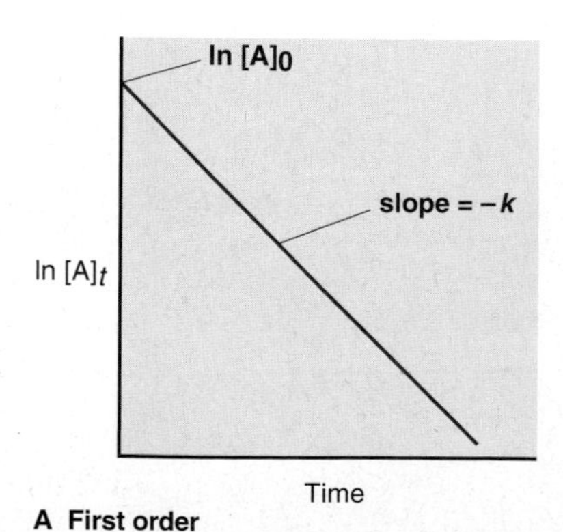

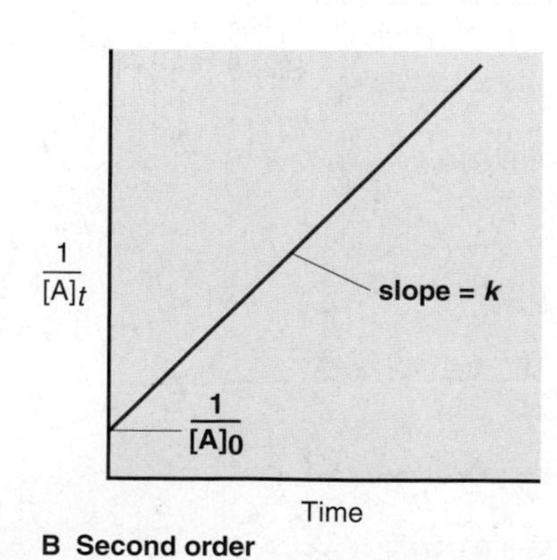

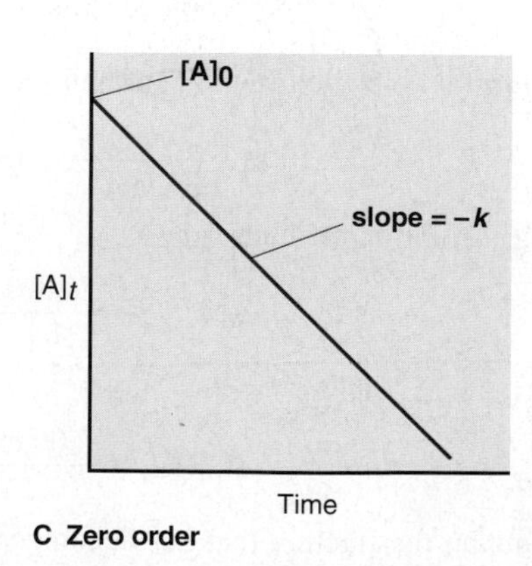

Figure 16.7 Integrated rate laws and reaction orders. A, Plot of $\ln\,[\text{A}]_t$ vs. time gives a straight line for a reaction that is first order in A. **B,** Plot of $1/[\text{A}]_t$ vs. time gives a straight line for a reaction that is second order in A. **C,** Plot of $[\text{A}]_t$ vs. time gives a straight line for a reaction that is zero order in A.

For a zero-order reaction, we have

$$[A]_t - [A]_0 = -kt$$

Rearranging gives

$$[A]_t = -kt + [A]_0$$
$$y = mx + b$$

Thus, a plot of $[A]_t$ vs. time gives a straight line with slope $= -k$ and y intercept $= [A]_0$ (Figure 16.7C).

Therefore, some trial-and-error graphical plotting is required to find the reaction order from the concentration and time data:

- If you obtain a straight line when you plot ln [reactant] vs. time, the reaction is *first order* with respect to that reactant.
- If you obtain a straight line when you plot 1/[reactant] vs. time, the reaction is *second order* with respect to that reactant.
- If you obtain a straight line when you plot [reactant] vs. time, the reaction is *zero order* with respect to that reactant.

Figure 16.8 shows how this approach is used to determine the order for the decomposition of N_2O_5. Since the plot of ln $[N_2O_5]$ *is* linear and the plot of $1/[N_2O_5]$ *is not,* the decomposition of N_2O_5 must be first order in N_2O_5.

Reaction Half-Life

The **half-life** ($t_{1/2}$) of a reaction is the time required for the reactant concentration to reach half its initial value. A half-life is expressed in time units appropriate for a given reaction and is characteristic of that reaction at a given temperature.

At fixed conditions, *the half-life of a first-order reaction is a constant, independent of reactant concentration.* For example, the half-life for the first-order decomposition of N_2O_5 at 45°C is 24.0 min. The meaning of this value is that if

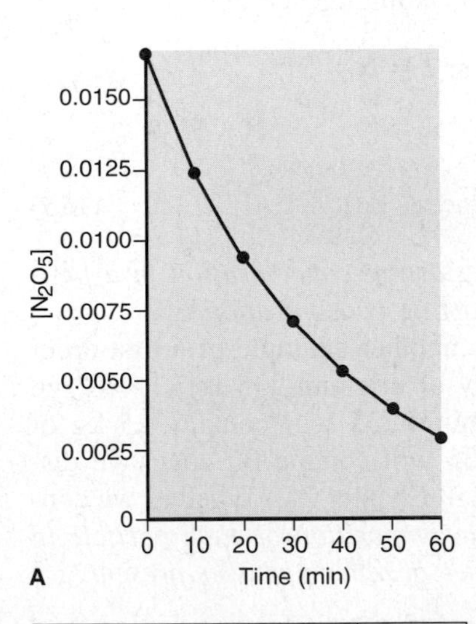

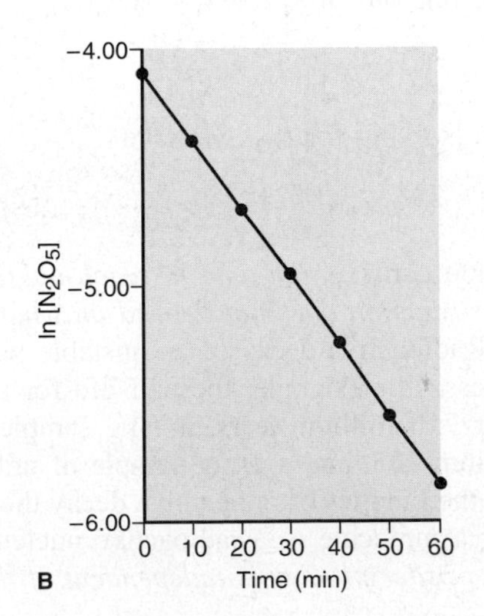

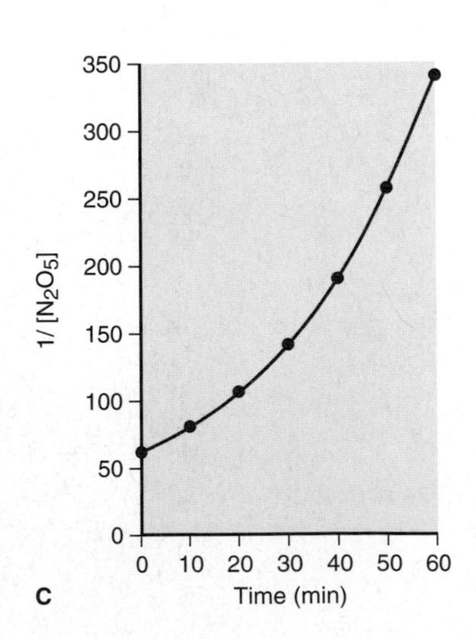

Time (min)	$[N_2O_5]$	ln $[N_2O_5]$	$1/[N_2O_5]$
0	0.0165	−4.104	60.6
10	0.0124	−4.390	80.6
20	0.0093	−4.68	1.1×10^2
30	0.0071	−4.95	1.4×10^2
40	0.0053	−5.24	1.9×10^2
50	0.0039	−5.55	2.6×10^2
60	0.0029	−5.84	3.4×10^2

Figure 16.8 Graphical determination of the reaction order for the decomposition of N_2O_5. A table of time and concentration data for determining reaction order appears below the graphs. **A,** A plot of $[N_2O_5]$ vs. time is curved, indicating that the reaction is *not* zero order in N_2O_5. **B,** A plot of ln $[N_2O_5]$ vs. time gives a straight line, indicating that the reaction *is* first order in N_2O_5. **C,** A plot of $1/[N_2O_5]$ vs. time is curved, indicating that the reaction is *not* second order in N_2O_5. Plots A and C support the conclusion from plot B.

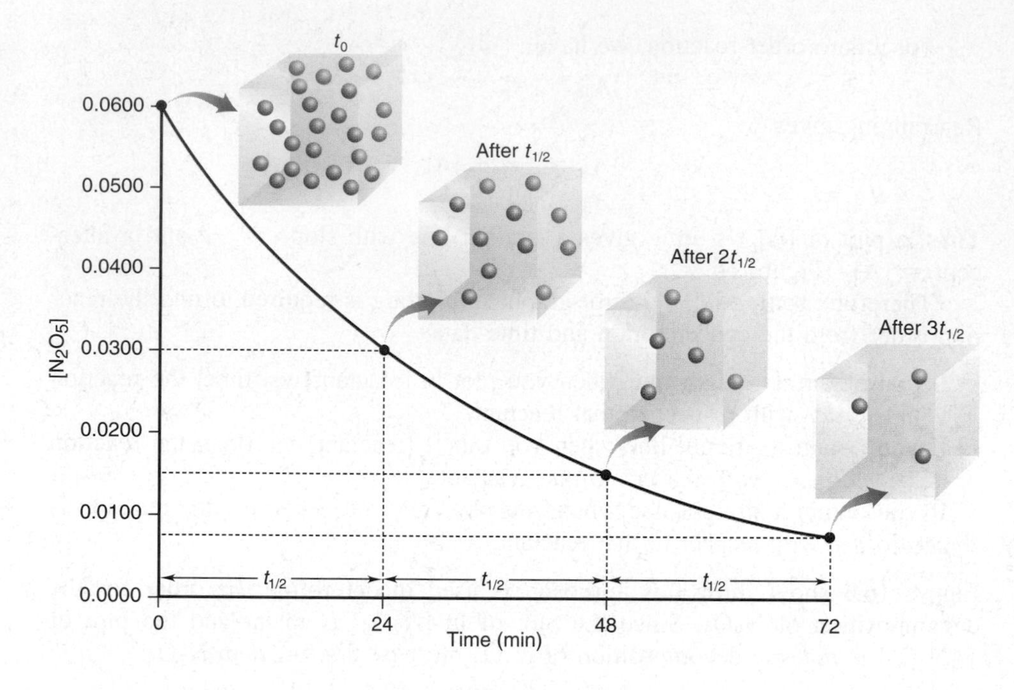

Figure 16.9 A plot of $[N_2O_5]$ vs. time for three half-lives. During each half-life, the concentration is halved (T = 45°C and $[N_2O_5]_0$ = 0.0600 mol/L). The blow-up volumes, with N_2O_5 molecules as colored spheres, show that after three half-lives, $\frac{1}{2} \times \frac{1}{2} \times \frac{1}{2} = \frac{1}{8}$ of the original concentration remains.

we start with, say, 0.0600 mol/L of N_2O_5 at 45°C, after 24 min (one half-life), 0.0300 mol/L has been consumed and 0.0300 mol/L remains; after 48 min (two half-lives), 0.0150 mol/L remains; after 72 min (three half-lives), 0.0075 mol/L remains, and so forth (Figure 16.9).

We can see from the integrated rate law why the half-life of a first-order reaction is independent of concentration:

$$\ln \frac{[A]_0}{[A]_t} = kt$$

After one half-life, $t = t_{1/2}$, and $[A]_t = \frac{1}{2}[A]_0$. Substituting, we obtain

$$\ln \frac{\cancel{[A]_0}}{\frac{1}{2}\cancel{[A]_0}} = kt_{1/2} \qquad \text{or} \qquad \ln 2 = kt_{1/2}$$

Then, solving for $t_{1/2}$, we have

$$t_{1/2} = \frac{\ln 2}{k} = \frac{0.693}{k} \qquad \text{(first-order process; rate} = k[A]) \tag{16.7}$$

As you can see, *the time to reach one-half the starting concentration in a first-order reaction does not depend on what that starting concentration is.*

Radioactive decay of an unstable nucleus is another example of a first-order process. For example, the half-life for the decay of uranium-235 is 7.1×10^8 yr. After 710 million years, a 1-kg sample of uranium-235 will contain 0.5 kg of uranium-235, and a 1-mg sample of uranium-235 will contain 0.5 mg. (We discuss the kinetics of radioactive decay thoroughly in Chapter 24.) Whether we consider a molecule or a radioactive nucleus, the *decomposition of each particle in a first-order process is independent of the number of other particles present.*

SAMPLE PROBLEM 16.5 Determining the Half-Life of a First-Order Reaction

Problem Cyclopropane is the smallest cyclic hydrocarbon. Because its 60° bond angles reduce orbital overlap, its bonds are weak. As a result, it is thermally unstable and rearranges to propene at 1000°C via the following first-order reaction:

$$\text{H}_2\text{C}\overset{\text{CH}_2}{\overset{/\ \backslash}{—}}\text{CH}_2(g) \xrightarrow{\Delta} \text{CH}_3\text{—CH=CH}_2(g)$$

The rate constant is 9.2 s^{-1}. **(a)** What is the half-life of the reaction? **(b)** How long does it take for the concentration of cyclopropane to reach one-quarter of the initial value?

Plan **(a)** The cyclopropane rearrangement is first order, so to find $t_{1/2}$ we use Equation 16.7 and substitute for k (9.2 s^{-1}). **(b)** Each half-life decreases the concentration to one-half of its initial value, so two half-lives decrease it to one-quarter.

Solution **(a)** Solving for $t_{1/2}$:

$$t_{1/2} = \frac{\ln 2}{k} = \frac{0.693}{9.2\ s^{-1}} = 0.075\ s$$

It takes 0.075 s for half the cyclopropane to form propene at this temperature.

(b) Finding the time to reach one-quarter of the initial concentration:

$$\text{Time} = 2(t_{1/2}) = 2(0.075\ s) = 0.15\ s$$

Check For (a), rounding gives 0.7/9 s^{-1} = 0.08 s, so the answer seems correct.

FOLLOW-UP PROBLEM 16.5 Iodine-123 is used to study thyroid gland function. This radioactive isotope breaks down in a first-order process with a half-life of 13.1 h. What is the rate constant for the process?

In contrast to the half-life of a first-order reaction, the half-life of a second-order reaction *does* depend on reactant concentration:

$$t_{1/2} = \frac{1}{k[A]_0} \quad \text{(second-order process; rate} = k[A]^2)$$

Note that here *the half-life is* ***inversely*** *proportional to the initial reactant concentration.* This relationship means that a second-order reaction with a high initial reactant concentration has a shorter half-life, and one with a low initial reactant concentration has a longer half-life. Therefore, *as a second-order reaction proceeds, the half-life increases.*

In contrast to the half-life of a second-order reaction, *the half-life of a zero-order reaction is* ***directly*** *proportional to the initial reactant concentration:*

$$t_{1/2} = \frac{[A]_0}{2k} \quad \text{(zero-order process; rate} = k)$$

Thus, if a zero-order reaction begins with a high reactant concentration, it has a longer half-life than if it begins with a low reactant concentration. Table 16.4 summarizes the essential features of zero-, first-, and second-order reactions.

Table 16.4 An Overview of Zero-Order, First-Order, and Simple Second-Order Reactions

	Zero Order	First Order	Second Order
Rate law	rate = k	rate = k[A]	rate = $k[A]^2$
Units for k	mol/L·s	1/s	L/mol·s
Integrated rate law in straight-line form	$[A]_t = -kt + [A]_0$	$\ln [A]_t = -kt + \ln [A]_0$	$1/[A]_t = kt + 1/[A]_0$
Plot for straight line	$[A]_t$ vs. time	$\ln [A]_t$ vs. time	$1/[A]_t$ vs. time
Slope, y intercept	$-k$, $[A]_0$	$-k$, $\ln [A]_0$	k, $1/[A]_0$
Half-life	$[A]_0/2k$	$(\ln 2)/k$	$1/k[A]_0$

SECTION SUMMARY

Integrated rate laws are used to find either the time needed to reach a certain concentration of reactant or the concentration present after a given time. Rearrangements of the integrated rate laws allow us to determine reaction orders and rate constants graphically. The half-life is the time needed for the reaction to consume half the reactant; for first-order reactions, it is independent of concentration.

Figure 16.10 Dependence of the rate constant on temperature. A, In the hydrolysis of an ester, when reactant concentrations are held constant and temperature increases, the rate and rate constant increase. Note the near doubling of k with each rise of 10 K (10°C). **B,** A plot of rate constant vs. temperature for this reaction shows an exponentially increasing curve.

Exp't	[Ester]	[H_2O]	T (K)	Rate (mol/L·s)	k (L/mol·s)
1	0.100	0.200	288	1.04×10^{-3}	0.0521
2	0.100	0.200	298	2.02×10^{-3}	0.101
3	0.100	0.200	308	3.68×10^{-3}	0.184
4	0.100	0.200	318	6.64×10^{-3}	0.332

A

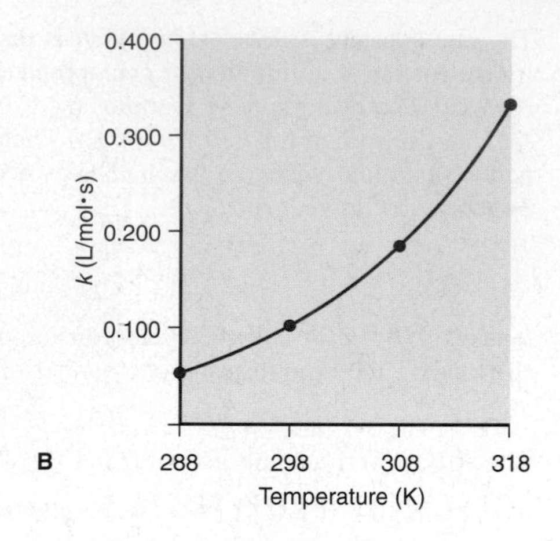

16.5 THE EFFECT OF TEMPERATURE ON REACTION RATE

Temperature often has a major effect on reaction rate. As Figure 16.10A shows for a common organic reaction (hydrolysis, or reaction with water, of an ester), when reactant concentrations are held constant, the rate nearly doubles with each rise in temperature of 10 K (or 10°C). In fact, for many reactions near room temperature, an increase of 10°C causes a doubling or tripling of the rate.

Aside from mammals and birds, most animals—all reptiles and many species of amphibians, insects, and fish—absorb heat from the external environment to maintain metabolic and physical activity. For example, lizards and crocodiles bask in the sun or lie in the shade to regulate their internal reaction rates.

How does the rate law express this effect of temperature? If we collect concentration and time data for the same reaction run at *different* temperatures (T), and then solve each rate expression for k, we find that k increases as T increases. In other words, *temperature affects the rate by affecting the rate constant.* A plot of k vs. T gives a curve that increases exponentially (Figure 16.10B).

These results are consistent with studies made in 1889 by the Swedish chemist Svante Arrhenius, who discovered a key relationship between T and k. In its modern form, the **Arrhenius equation** is

$$k = Ae^{-E_a/RT} \tag{16.8}$$

where k is the rate constant, e is the base of natural logarithms, T is the absolute temperature, and R is the universal gas constant. We'll discuss the meaning of A, which is related to the orientation of the colliding molecules, in the next section. The $\mathbf{E_a}$ term is the **activation energy** of the reaction, which Arrhenius considered the *minimum energy* the molecules must have to react; we'll explore its meaning in the next section as well. This negative exponential relationship between T and k means that *as T increases, the negative exponent becomes smaller, so the value of k becomes larger, which means that the rate increases:*

$$\text{Higher } T \Longrightarrow \text{larger } k \Longrightarrow \text{increased rate}$$

We can calculate E_a from the Arrhenius equation by taking the natural logarithm of both sides and recasting the equation into an equation for a straight line:

$$\ln k = \ln A - \frac{E_a}{R}\left(\frac{1}{T}\right)$$
$$y = b + mx$$

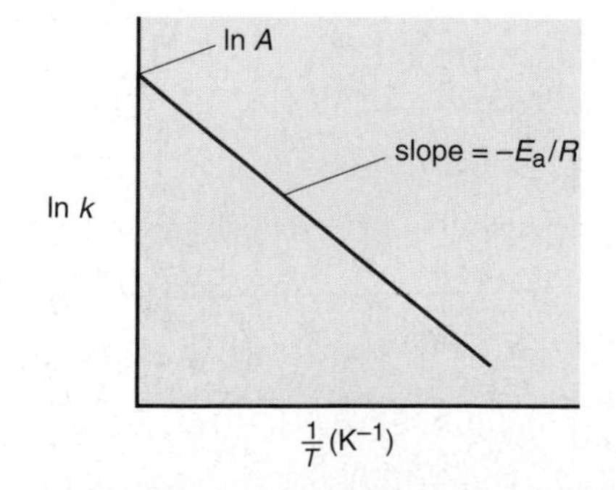

Figure 16.11 Graphical determination of the activation energy. A plot of ln k vs. $1/T$ gives a straight line with slope = $-E_a/R$.

A plot of ln k vs. $1/T$ gives a straight line whose slope is $-E_a/R$ and whose y intercept is ln A (Figure 16.11). Therefore, with the constant R known, we can determine E_a graphically from a series of k values at different temperatures.

Because the relationship between ln k and $1/T$ is linear, we can use a simpler method to find E_a if we know the rate constants at two temperatures, T_2 and T_1:

$$\ln k_2 = \ln A - \frac{E_a}{R}\left(\frac{1}{T_2}\right) \qquad \ln k_1 = \ln A - \frac{E_a}{R}\left(\frac{1}{T_1}\right)$$

When we subtract ln k_1 from ln k_2, the term ln A drops out and the other terms can be rearranged to give

$$\ln \frac{k_2}{k_1} = -\frac{E_a}{R}\left(\frac{1}{T_2} - \frac{1}{T_1}\right) \tag{16.9}$$

From this, we can solve for E_a.

The Significance of R The importance of R extends beyond the study of gases (or osmotic pressure). By expressing pressure and volume in more fundamental quantities, we obtain dimensions for energy, E:

$$P = \frac{\text{force}}{\text{area}} = \frac{\text{force}}{(\text{length})^2} \qquad V = (\text{length})^3$$

Thus,

$$PV = \frac{\text{force}}{(\text{length})^2} \times (\text{length})^3 = \text{force} \times \text{length}$$

Energy is used when a force moves an object over a distance. Thus, E = force × distance (or length), so

$$PV = \text{force} \times \text{length} = E$$

Solving for R in the ideal gas law and substituting for PV, we obtain

$$R = \frac{PV}{nT} = \frac{E}{\text{amount} \times T}$$

Thus, R is the proportionality constant that relates the energy, amount (mol), and temperature of any chemical system.

SAMPLE PROBLEM 16.6 Determining the Energy of Activation

Problem The decomposition of hydrogen iodide,

$$2HI(g) \longrightarrow H_2(g) + I_2(g)$$

has rate constants of 9.51×10^{-9} L/mol·s at 500. K and 1.10×10^{-5} L/mol·s at 600. K. Find E_a.

Plan We are given the rate constants, k_1 and k_2, at two temperatures, T_1 and T_2, so we substitute into Equation 16.9 and solve for E_a.

Solution Rearranging Equation 16.9 to solve for E_a:

$$\ln \frac{k_2}{k_1} = -\frac{E_a}{R}\left(\frac{1}{T_2} - \frac{1}{T_1}\right)$$

$$E_a = -R\left(\ln \frac{k_2}{k_1}\right)\left(\frac{1}{T_2} - \frac{1}{T_1}\right)^{-1}$$

$$= -(8.314\ \text{J/mol·K})\left(\ln \frac{1.10\times10^{-5}\ \text{L/mol·s}}{9.51\times10^{-9}\ \text{L/mol·s}}\right)\left(\frac{1}{600.\ \text{K}} - \frac{1}{500.\ \text{K}}\right)^{-1}$$

$$= 1.76\times10^5\ \text{J/mol} = 1.76\times10^2\ \text{kJ/mol}$$

Comment Be sure to retain the same number of significant figures in $1/T$ as you have in T, or a significant error could be introduced. Round to the correct number of significant figures only at the final answer. On most pocket calculators, the expression $(1/T_2 - 1/T_1)$ is entered as follows: $(T_2)(1/x) - (T_1)(1/x) =$.

FOLLOW-UP PROBLEM 16.6

The reaction $2NOCl(g) \longrightarrow 2NO(g) + Cl_2(g)$ has an E_a of 1.00×10^2 kJ/mol and a rate constant of 0.286 L/mol·s at 500. K. What is the rate constant at 490. K?

In this section and the previous two, we discussed a series of experimental and mathematical methods for the study of reaction kinetics. Figure 16.12 is a useful summary of this information. Note that the integrated rate law provides an alternative method for obtaining reaction orders and the rate constant.

Series of plots of concentration vs. time
Determine slope of tangent at t_0 for each plot
Initial rates
Compare initial rates when [A] changes and [B] is held constant (and vice versa)
Reaction orders
Substitute initial rates, orders, and concentrations into general rate law: rate = $k[A]^m[B]^n$
Rate constant (k) and actual rate law
Determine k at different temperatures
Activation energy, E_a
Integrated rate law (half-life, $t_{1/2}$)
Rearrange to linear form and graph
Rate constant and reaction order

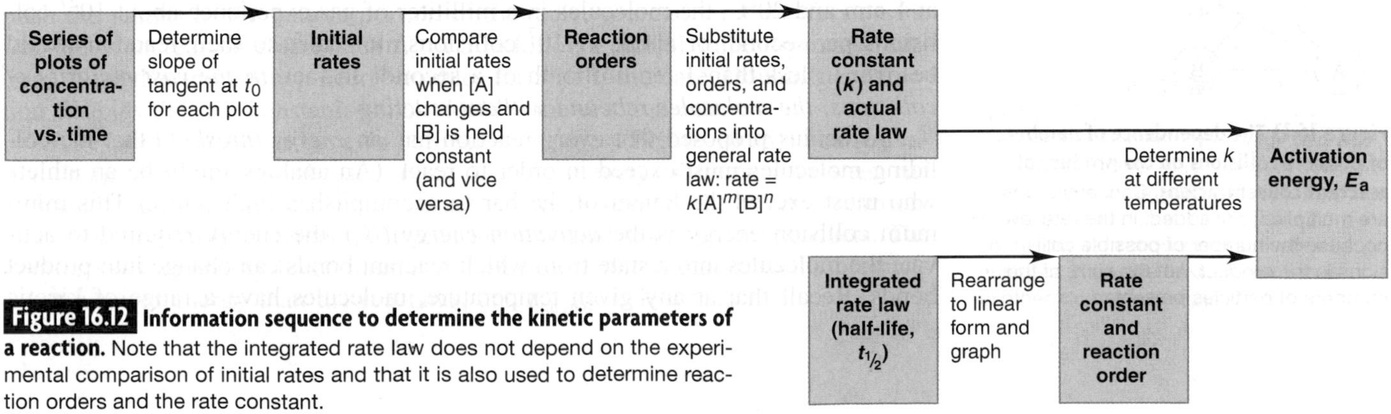

Figure 16.12 Information sequence to determine the kinetic parameters of a reaction. Note that the integrated rate law does not depend on the experimental comparison of initial rates and that it is also used to determine reaction orders and the rate constant.

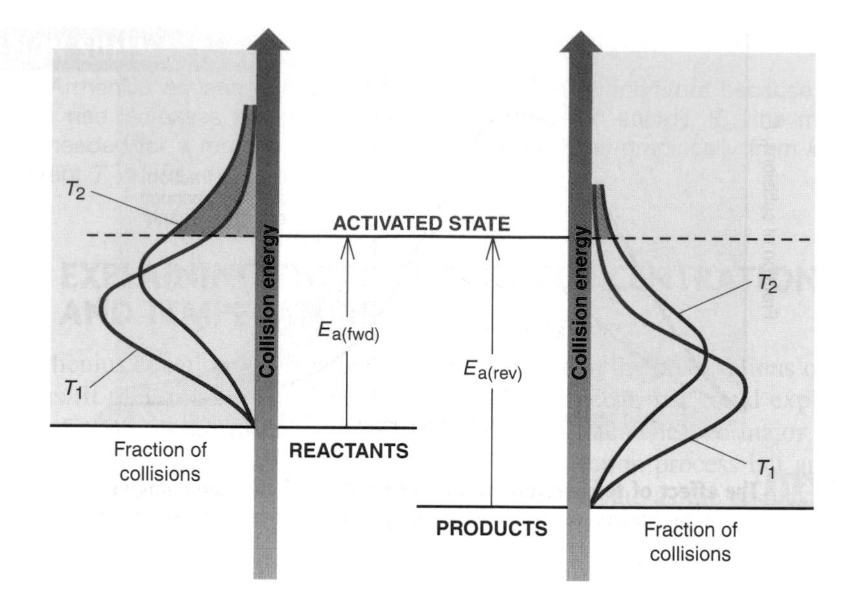

Figure 16.16 An energy-level diagram of the fraction of collisions whose energy exceeds E_a. When Figure 16.14 is aligned vertically on the left and right axes of Figure 16.15, we see that

- In either direction, the fraction of collisions with energy exceeding E_a is larger at the higher T.
- In an *exothermic* reaction at any temperature, the fraction of collisions with energy exceeding $E_{a(fwd)}$ is larger than the fraction exceeding $E_{a(rev)}$.

By turning the collision-energy distribution curve of Figure 16.14 vertically and aligning it on both sides of Figure 16.15, we obtain Figure 16.16, which shows how temperature affects the fraction of collisions that have energy exceeding the activation energy for both the forward and reverse reactions. Several ideas are illustrated in this composite figure:

- In both reaction directions, a larger fraction of collisions exceeds the activation energy at the higher temperature, T_2: higher T increases reaction rate.
- For an *exothermic* process (forward reaction here) at any temperature, the fraction of reactant collisions with energy exceeding $E_{a(fwd)}$ is *larger* than the fraction of product collisions with energy exceeding $E_{a(rev)}$; thus, the forward reaction is faster. On the other hand, in an *endothermic* process (reverse reaction here), $E_{a(fwd)}$ is greater than $E_{a(rev)}$, so the fraction of product collisions with energy exceeding $E_{a(rev)}$ is larger, and the reverse reaction is faster.

These conclusions are consistent with the Arrhenius equation; that is, *the larger the E_a, the smaller the value of k, and the slower the reaction:*

$$\text{Larger } E_a \Longrightarrow \text{smaller } k \Longrightarrow \text{decreased rate}$$

How Molecular Structure Affects Rate You've seen that the enormous number of collisions per second is greatly reduced when we count only those with enough energy to react. However, even this tiny fraction of the total collisions does not reveal the true number of **effective collisions,** those that actually lead to product. In addition to colliding with enough energy, *the molecules must collide so that the reacting atoms make contact.* In other words, a collision must have enough energy *and* a particular *molecular orientation* to be an effective collision.

In the Arrhenius equation, the effect of molecular orientation is contained in the term A:

$$k = Ae^{-E_a/RT}$$

This term is called the **frequency factor,** the product of the collision frequency Z and an *orientation probability factor, p,* which is specific for each reaction: $A = pZ$. The factor p is related to the structural complexity of the colliding particles. You can think of it as the ratio of effectively oriented collisions to all possible collisions. For example, Figure 16.17 shows a few of the possible collision orientations for the following simple gaseous reaction:

$$NO(g) + NO_3(g) \longrightarrow 2NO_2(g)$$

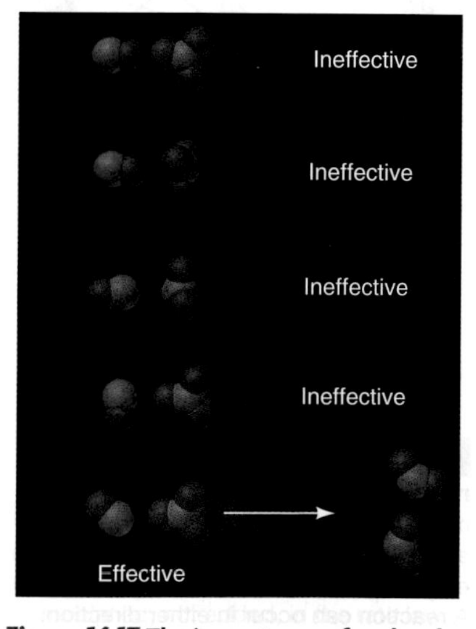

Figure 16.17 The importance of molecular orientation to an effective collision. Only one of the five orientations shown for the collision between NO and NO_3 has the correct orientation to lead to product. In the effective orientation, contact occurs between the atoms that will become bonded in the product.

Of the five collisions shown, only one has an orientation in which the N of NO makes contact with an O of NO_3. Actually, the orientation probability factor (*p* value) for this reaction is 0.006: only 6 collisions in every 1000 (1 in 167) have an orientation that can lead to reaction.

Collisions between individual atoms have *p* values near 1: almost no matter how they hit, as long as the collision has enough energy, the particles react. In such cases, the rate constant depends only on the frequency and energy of the collisions. At the other extreme are biochemical reactions, in which the reactants are often two small molecules that can react only when they collide with a specific tiny region of a giant molecule—a protein or nucleic acid. The orientation probability factor for these reactions is often less than 10^{-6}: fewer than one in a million sufficiently energetic collisions leads to product. The fact that countless such biochemical reactions are occurring right now, as you read this sentence, helps make the point that the number of collisions per second is truly astounding.

Transition State Theory: Molecular Nature of the Activated Complex

Collision theory is a simple, easy to visualize model, but it provides no insight about why the activation energy is crucial and how the activated molecules look. To understand these aspects of the process, we turn to **transition state theory,** which focuses on the high-energy species that forms through an effective collision.

Visualizing the Transition State Recall from our discussion of energy changes (Chapter 6) that the internal energy of a system is the sum of its kinetic and potential energies. Two molecules that are far apart but speeding toward each other have high kinetic energy and low potential energy. As the molecules get closer, some kinetic energy is converted to potential energy as the electron clouds repel each other. At the moment of a head-on collision, the molecules stop, and their kinetic energy is converted to the potential energy of the collision. *If this potential energy is less than the activation energy, the molecules recoil,* bouncing off each other like billiard balls. Repulsions decrease, speeds increase, and the molecules zoom apart without reacting.

The tiny fraction of molecules that are oriented effectively *and* moving at the highest speed behave differently. *Their kinetic energy pushes them together with enough force to overcome repulsions and react.* Nuclei in one atom attract electrons in another; atomic orbitals overlap and electron density shifts; some bonds lengthen and weaken while others start to form. If we could watch this process in slow motion, we would see the reactant molecules gradually change their bonds and shapes as they turn into product molecules. At some point during this smooth transformation, what exists is *neither reactant nor product but a transitional species with partial bonds.* This species is extremely unstable (has very high potential energy) and exists only at the instant when the reacting system is highest in energy. It is called the **transition state,** or **activated complex,** and it forms only if the molecules collide in an effective orientation *and* with energy equal to or greater than the activation energy. Thus, *the activation energy is the quantity needed to stretch and deform bonds in order to reach the transition state.*

Because transition states *cannot* be isolated, our knowledge of them, until recently, came from reasoning and studies of analogous, more stable species. This situation is changing rapidly due to the pioneering work of Ahmed H. Zewail, who received the 1999 Nobel Prize in chemistry. He used lasers pulsing on the same femtosecond (10^{-15}) time scale as bond vibrations to observe the detailed changes occurring when transition states form and decompose.

Consider the reaction between methyl bromide and hydroxide ion:

$$CH_3Br + OH^- \longrightarrow CH_3OH + Br^-$$

The electronegative bromine makes the carbon of methyl bromide partially positive. If the reactants are moving toward each other fast enough and are oriented

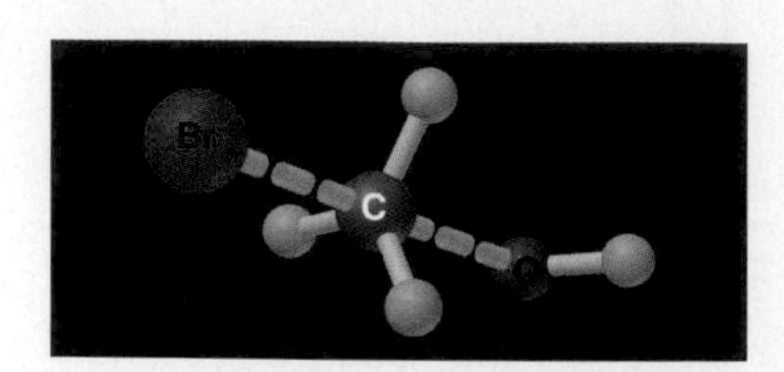

Figure 16.18 Nature of the transition state in the reaction between CH_3Br and OH^-. Note the partial (elongated) C—O and C—Br bonds and the trigonal bipyramidal shape of the transition state of this reaction.

effectively when they collide, the negatively charged oxygen in OH^- approaches the carbon with enough energy to begin forming a C—O bond, which causes the C—Br bond to weaken. In the transition state (Figure 16.18), C is surrounded by five atoms (trigonal bipyramidal), which never occurs in its stable compounds. This high-energy species has three normal C—H bonds and two partial bonds, one from C to O and the other from C to Br. Reaching this transition state is no guarantee that the reaction will proceed to products. A transition state can change in either direction: if the C—O bond continues to shorten and strengthen, products form; however, if the C—Br bond becomes shorter and stronger again, the transition state reverts to reactants.

Depicting the Change with Reaction Energy Diagrams A useful way to depict the events we just described is with a **reaction energy diagram,** which shows the potential energy of the system during the reaction as a smooth curve. Figure 16.19 shows the reaction energy diagram for the reaction of CH_3Br and OH^-, with an electron density relief map, structural formula, and molecular-scale view at various points during the change.

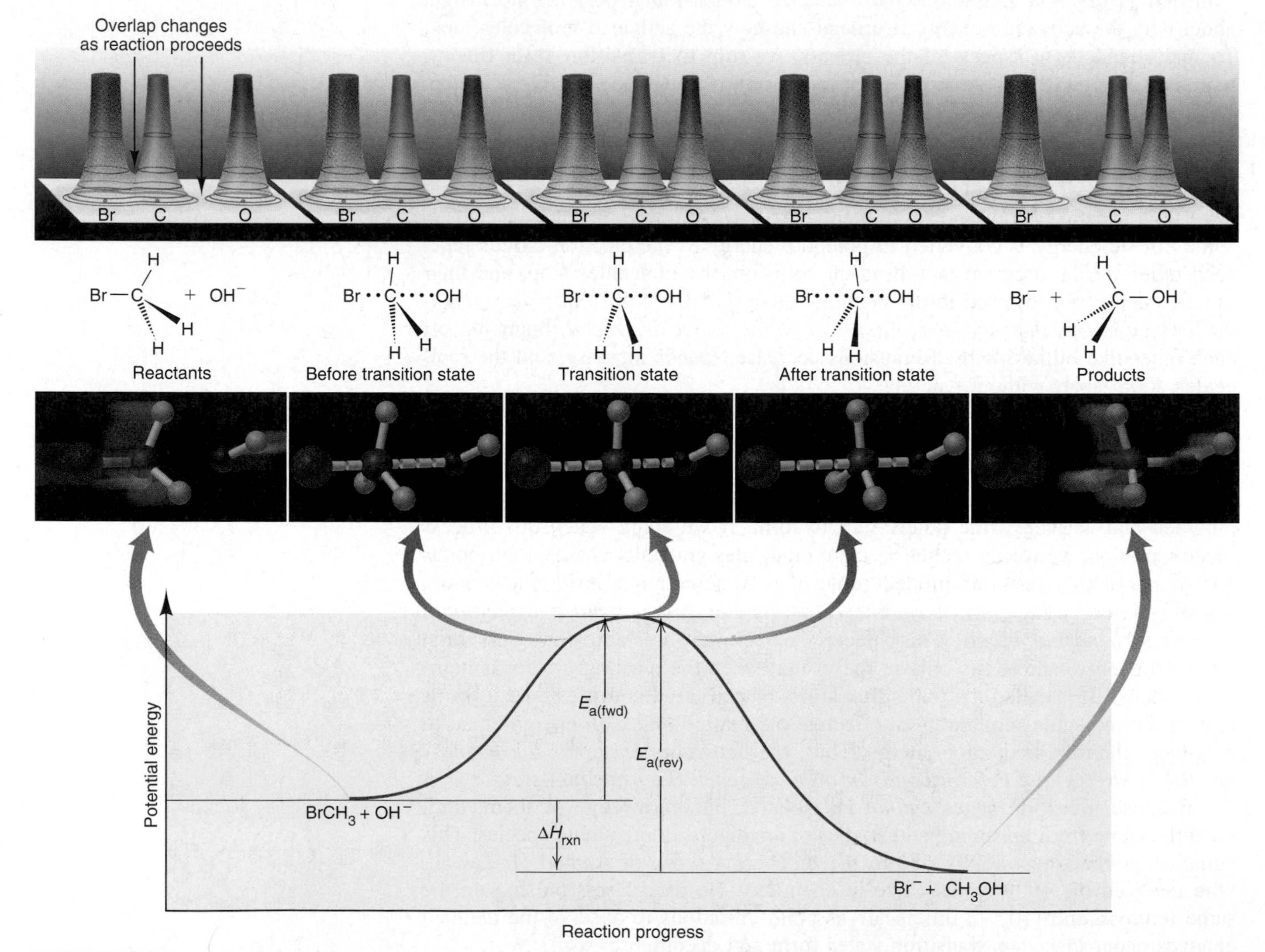

Figure 16.19 Reaction energy diagram for the reaction between CH_3Br and OH^-. A plot of potential energy vs. reaction progress shows the relative energy levels of reactants, products, and transition state joined by a curved line, as well as the activation energies of the forward and reverse reactions and the heat of reaction. The electron density relief maps, structural formulas, and molecular-scale views depict the change at five points. Note the gradual bond forming and bond breaking as the system goes through the transition state.

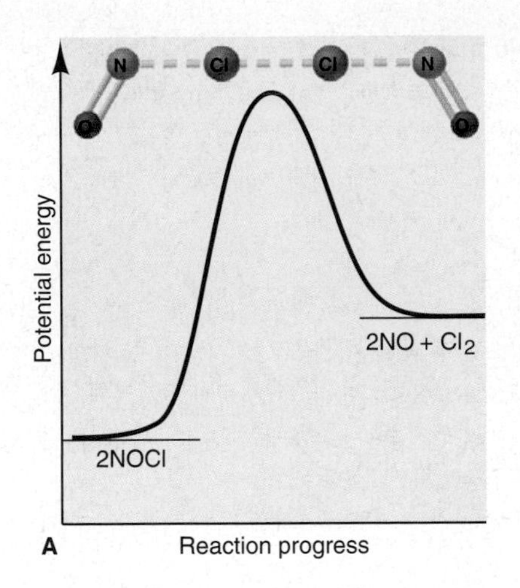

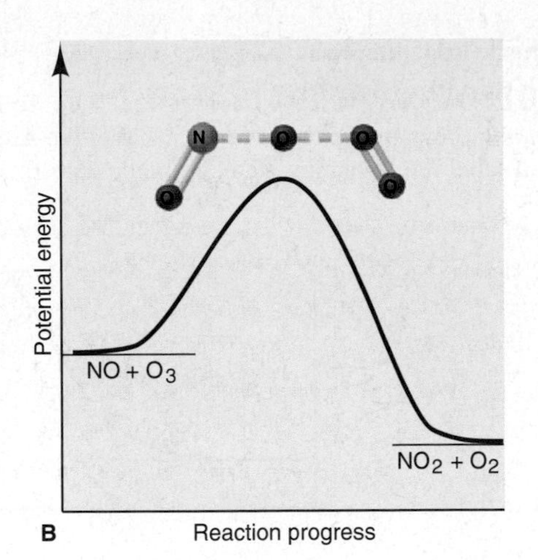

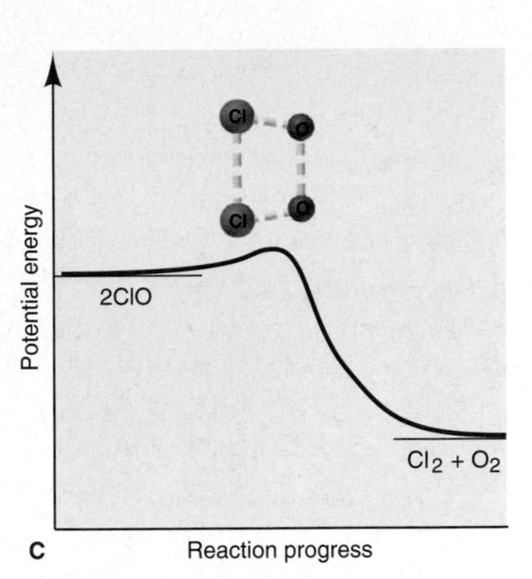

Figure 16.20 Reaction energy diagrams and possible transition states for three reactions.
A, $2NOCl(g) \longrightarrow 2NO(g) + Cl_2(g)$ (Despite the formula NOCl, the atom sequence is ClNO.)
B, $NO(g) + O_3(g) \longrightarrow NO_2(g) + O_2(g)$
C, $2ClO(g) \longrightarrow Cl_2(g) + O_2(g)$
Note that reaction A is endothermic, B and C are exothermic, and C has a very small $E_{a(fwd)}$.

The horizontal axis, labeled "Reaction progress," means reactants change to products from left to right. This reaction is exothermic, so reactants are higher in energy than products. The diagram also shows activation energies for the forward and reverse reactions; in this case, $E_{a(fwd)}$ is less than $E_{a(rev)}$. This difference, which reflects the change in bond energies, equals the heat of reaction, ΔH_{rxn}:

$$\Delta H_{rxn} = E_{a(fwd)} - E_{a(rev)} \qquad (16.10)$$

Transition state theory proposes that *every reaction (and every step in an overall reaction) goes through its own transition state,* from which it can continue in either direction. We imagine how a transition state might look by examining the reactant and product bonds that change. Figure 16.20 depicts reaction energy diagrams for three simple reactions. Note that the shape of the postulated transition state in each case is based on a specific collision orientation between the atoms that become bonded to form the product.

SAMPLE PROBLEM 16.7 Drawing Reaction Energy Diagrams and Transition States

Problem A key reaction in the upper atmosphere is

$$O_3(g) + O(g) \longrightarrow 2O_2(g)$$

The $E_{a(fwd)}$ is 19 kJ, and the ΔH_{rxn} for the reaction as written is -392 kJ. Draw a reaction energy diagram for this reaction, postulate a transition state, and calculate $E_{a(rev)}$.

Plan The reaction is highly exothermic ($\Delta H_{rxn} = -392$ kJ), so the products are much lower in energy than the reactants. The small $E_{a(fwd)}$ (19 kJ) means the energy of the reactants lies slightly below that of the transition state. We use Equation 16.10 to calculate $E_{a(rev)}$. To postulate the transition state, we sketch the species and note that one of the bonds in O_3 weakens, and this partially bonded O begins forming a bond to the separate O atom.

Solution Solving for $E_{a(rev)}$:

$$\Delta H_{rxn} = E_{a(fwd)} - E_{a(rev)}$$

So, $$E_{a(rev)} = E_{a(fwd)} - \Delta H_{rxn} = 19 \text{ kJ} - (-392 \text{ kJ}) = 411 \text{ kJ}$$

The reaction energy diagram (not drawn to scale), with transition state, is

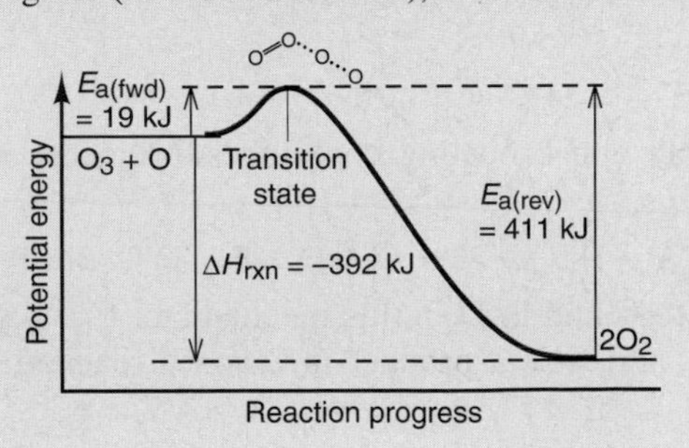

Check Rounding to find $E_{a(rev)}$ gives ~20 + 390 = 410.

FOLLOW-UP PROBLEM 16.7 The following reaction energy diagram depicts another key atmospheric reaction. Label the axes, identify $E_{a(fwd)}$, $E_{a(rev)}$, and ΔH_{rxn}, draw and label the transition state, and calculate $E_{a(rev)}$ for the reaction.

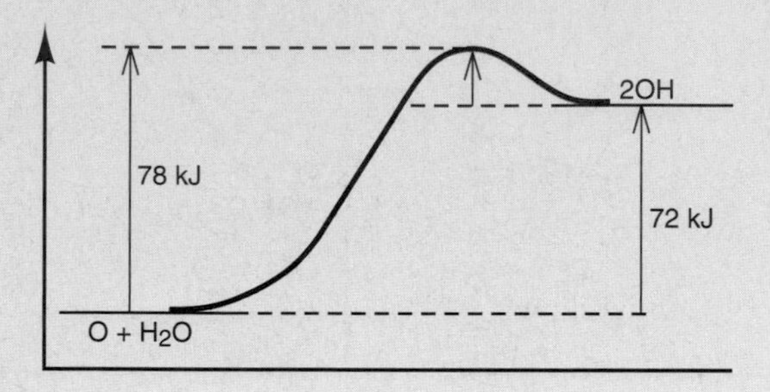

SECTION SUMMARY

According to collision theory, reactant particles must collide to react, and the number of collisions depends on the product of the reactant concentrations. At higher temperatures, more collisions have enough energy to exceed the activation energy (E_a). The relative E_a values for the forward and the reverse reactions depend on whether the overall reaction is exothermic or endothermic. Molecules must collide with an effective orientation for reaction to occur, so structural complexity decreases rate.

Transition state theory pictures the kinetic energy of the particles changing to potential energy during a collision. Given a sufficiently energetic collision and an effective molecular orientation, the reactant species become an unstable transition state, which either forms product(s) or reverts to reactant(s). Reaction energy diagrams depict the changing energy of the chemical system as it progresses from reactants through transition state(s) to products.

16.7 REACTION MECHANISMS: STEPS IN THE OVERALL REACTION

Imagine trying to figure out how a car works just by examining the body, wheels, and dashboard. It can't be done—you need to look under the hood and inside the engine to see how the parts fit together and function. Similarly, because our main purpose is to know how a reaction works *at the molecular level,* examining the overall balanced equation is not much help—we must "look under the yield arrow and inside the reaction" to see how reactants change into products.

When we do so, we find that most reactions occur through a **reaction mechanism,** a sequence of single reaction steps that sum to the overall reaction. For example, a possible mechanism for the overall reaction

$$2A + B \longrightarrow E + F$$

might involve these three simpler steps:

(1) $A + B \longrightarrow C$
(2) $C + A \longrightarrow D$
(3) $D \longrightarrow E + F$

Adding them together and canceling common substances, we obtain the overall equation:

$$A + B + \cancel{C} + A + \cancel{D} \longrightarrow \cancel{C} + \cancel{D} + E + F \quad \text{or} \quad 2A + B \longrightarrow E + F$$

Note what happens to C and to D in this mechanism. C is a product in step 1 and a reactant in step 2, and D is a product in 2 and a reactant in 3. Each functions as a **reaction intermediate,** a substance that is formed and used up during the overall reaction. Reaction intermediates do not appear in the overall balanced equation but are absolutely essential for the reaction to occur. They are usually unstable relative to the reactants and products but are far more stable than transition states (activated complexes). Reaction intermediates are molecules with normal bonds and are sometimes stable enough to be isolated.

Chemists *propose* a reaction mechanism to explain how a particular reaction might occur, and then they *test* the mechanism. This section focuses on the nature of the individual steps and how they fit together to give a rate law consistent with experimental results.

Elementary Reactions and Molecularity

The individual steps, which together make up the proposed reaction mechanism, are called **elementary reactions** (or **elementary steps**). Each describes a *single molecular event,* such as one particle decomposing or two particles colliding and combining. *An elementary step is* ***not*** *made up of simpler steps.*

An elementary step is characterized by its **molecularity,** the number of *reactant* particles involved in the step. Consider the mechanism for the breakdown of ozone in the stratosphere. The overall reaction is

$$2O_3(g) \longrightarrow 3O_2(g)$$

A two-step mechanism has been proposed for this reaction. Notice that the two steps sum to the overall reaction. The first elementary step is a **unimolecular reaction,** one that involves the decomposition or rearrangement of a single particle:

(1) $O_3(g) \longrightarrow O_2(g) + O(g)$

The second step is a **bimolecular reaction,** one in which two particles react:

(2) $O_3(g) + O(g) \longrightarrow 2O_2(g)$

Some *termolecular* elementary steps occur, but they are extremely rare because the probability of three particles colliding simultaneously with enough energy and with an effective orientation is very small. Higher molecularities are not known. Unless evidence exists to the contrary, it makes good chemical sense to propose only unimolecular or bimolecular reactions as the elementary steps in a reaction mechanism.

The rate law for an elementary reaction, unlike that for an overall reaction, *can* be deduced from the reaction stoichiometry. An elementary reaction occurs in one step, so its rate must be proportional to the product of the reactant concentrations. Therefore, *we use the equation coefficients as the reaction orders in the rate law for an elementary step; that is, reaction order equals molecularity* (Table 16.6). Remember that this statement holds *only* when we know that the reaction is elementary; you've already seen that for an overall reaction, the reaction orders must be determined experimentally.

Table 16.6 Rate Laws for General Elementary Steps

Elementary Step	Molecularity	Rate Law
$A \longrightarrow$ product	Unimolecular	Rate = $k[A]$
$2A \longrightarrow$ product	Bimolecular	Rate = $k[A]^2$
$A + B \longrightarrow$ product	Bimolecular	Rate = $k[A][B]$
$2A + B \longrightarrow$ product	Termolecular	Rate = $k[A]^2[B]$

SAMPLE PROBLEM 16.8 Determining Molecularity and Rate Laws for Elementary Steps

Problem The following two reactions are proposed as elementary steps in the mechanism for an overall reaction:

(1) $NO_2Cl(g) \longrightarrow NO_2(g) + Cl(g)$

(2) $NO_2Cl(g) + Cl(g) \longrightarrow NO_2(g) + Cl_2(g)$

(a) Write the overall balanced equation.

(b) Determine the molecularity of each step.

(c) Write the rate law for each step.

Plan We find the overall equation from the sum of the elementary steps. The molecularity of each step equals the total number of reactant particles. We write the rate law for each step using the molecularities as reaction orders.

Solution (a) Writing the overall balanced equation:

$$\begin{array}{r l}
NO_2Cl(g) \longrightarrow & NO_2(g) + Cl(g) \\
NO_2Cl(g) + Cl(g) \longrightarrow & NO_2(g) + Cl_2(g) \\
\hline
NO_2Cl(g) + NO_2Cl(g) + \cancel{Cl(g)} \longrightarrow & NO_2(g) + \cancel{Cl(g)} + NO_2(g) + Cl_2(g) \\
2NO_2Cl(g) \longrightarrow & 2NO_2(g) + Cl_2(g)
\end{array}$$

(b) Determining the molecularity of each step: The first elementary step has only one reactant, NO_2Cl, so it is unimolecular. The second elementary step has two reactants, NO_2Cl and Cl, so it is bimolecular.

(c) Writing rate laws for the elementary reactions:

(1) $\text{Rate}_1 = k_1[NO_2Cl]$

(2) $\text{Rate}_2 = k_2[NO_2Cl][Cl]$

Check In part (a), be sure the equation is balanced; in part (c), be sure the substances in brackets are the reactants of each elementary step.

FOLLOW-UP PROBLEM 16.8 The following elementary steps constitute a proposed mechanism for a reaction:

(1) $2NO(g) \longrightarrow N_2O_2(g)$

(2) $2H_2(g) \longrightarrow 4H(g)$

(3) $N_2O_2(g) + H(g) \longrightarrow N_2O(g) + HO(g)$

(4) $2HO(g) + 2H(g) \longrightarrow 2H_2O(g)$

(5) $H(g) + N_2O(g) \longrightarrow HO(g) + N_2(g)$

(a) Write the balanced equation for the overall reaction.

(b) Determine the molecularity of each step.

(c) Write the rate law for each step.

Sleeping Through the Rate-Determining Step Baking bread provides a macroscopic example of a rate-limiting step. The five steps for making a loaf of tasty French bread are (1) mixing the ingredients (15 min), (2) kneading the dough (10 min), (3) letting the dough rise in a refrigerator (400 min), (4) shaping the loaf (2 min), and (5) baking the loaf (25 min). Obviously, the dough-rising step limits how fast a baker can produce a loaf because it is so much slower than the other steps. Keenly aware of the baking "reaction mechanism," bakers let the dough rise overnight, which allows them to sleep through the rate-determining step.

The Rate-Determining Step of a Reaction Mechanism

All the elementary steps in a mechanism do not have the same rate. Usually, one of the elementary steps is much slower than the others, so it limits how fast the overall reaction proceeds. This step is called the **rate-determining step** (or **rate-limiting step**).

Because the rate-determining step limits the rate of the overall reaction, its rate law represents the rate law for the overall reaction. Consider the reaction between nitrogen dioxide and carbon monoxide:

$$NO_2(g) + CO(g) \longrightarrow NO(g) + CO_2(g)$$

If the overall reaction were an elementary reaction—that is, if the mechanism consisted of only one step—we could immediately write the overall rate law as

$$\text{Rate} = k[NO_2][CO]$$

However, as you saw in Sample Problem 16.3, experiment shows that the actual rate law is

$$\text{Rate} = k[NO_2]^2$$

From this, we know immediately that the reaction shown cannot be elementary.

A proposed two-step mechanism is

(1) $NO_2(g) + NO_2(g) \longrightarrow NO_3(g) + NO(g)$ [slow; rate determining]

(2) $NO_3(g) + CO(g) \longrightarrow NO_2(g) + CO_2(g)$ [fast]

Note that NO_3 functions as a reaction intermediate in the mechanism. Rate laws for these elementary steps are

(1) $\text{Rate}_1 = k_1[NO_2][NO_2] = k_1[NO_2]^2$

(2) $\text{Rate}_2 = k_2[NO_3][CO]$

Note that if $k_1 = k$, *the rate law for the rate-determining step (step 1) is identical to the experimental rate law.* The first step is so slow compared with the second that the overall reaction takes essentially as long as the first step. Here you can see that one reason that a reactant (in this case, CO) has a reaction order of zero is that it takes part in the reaction only *after* the rate-determining step.

Correlating the Mechanism with the Rate Law

Conjuring up a reasonable reaction mechanism is one of the most exciting aspects of chemical kinetics and can be a classic example of the use of the scientific method. We use observations and data from rate experiments to hypothesize what the individual steps might be and then test our hypothesis by gathering further evidence. If the evidence supports it, we continue to apply that mechanism; if not, we propose a new one. However, *we can never prove, just from data, that a particular mechanism represents the actual chemical change.*

Regardless of the elementary steps proposed for a mechanism, they must meet three criteria:

1. *The elementary steps must add up to the overall balanced equation.* We cannot wind up with more (or fewer) reactants or products than are present in the balanced equation.
2. *The elementary steps must be physically reasonable.* As we noted, most steps should involve one reactant particle (unimolecular) or two (bimolecular). Steps with three reactant particles (termolecular) are very unlikely.
3. *The mechanism must correlate with the rate law.* Most importantly, a mechanism must support the experimental facts shown by the rate law, not the other way around.

Let's see how the mechanisms of several reactions conform to these criteria and how the elementary steps fit together.

Mechanisms with a Slow Initial Step We've already seen one mechanism with a rate-determining first step—that for the reaction of NO_2 and CO. Another example is the reaction between nitrogen dioxide and fluorine gas:

$$2NO_2(g) + F_2(g) \longrightarrow 2NO_2F(g)$$

The experimental rate law is first order in NO_2 and in F_2:

$$\text{Rate} = k[NO_2][F_2]$$

The accepted mechanism for the reaction is

(1) $NO_2(g) + F_2(g) \longrightarrow NO_2F(g) + F(g)$ [slow; rate determining]

(2) $NO_2(g) + F(g) \longrightarrow NO_2F(g)$ [fast]

Molecules of reactant and product appear in both elementary steps. The free fluorine atom is a reaction intermediate.

Does this mechanism meet the three crucial criteria?

1. The elementary reactions sum to the balanced equation:

$$NO_2(g) + NO_2(g) + F_2(g) + \cancel{F(g)} \longrightarrow NO_2F(g) + NO_2F(g) + \cancel{F(g)}$$

or

$$2NO_2(g) + F_2(g) \longrightarrow 2NO_2F(g)$$

2. Both steps are bimolecular, so they are chemically reasonable.

3. The mechanism gives the rate law for the overall equation. To show this, we write the rate laws for the elementary steps:

   (1) $\text{Rate}_1 = k_1[NO_2][F_2]$

   (2) $\text{Rate}_2 = k_2[NO_2][F]$

Step 1 is the rate-determining step and therefore gives the overall rate law, with $k_1 = k$. Because the second molecule of NO_2 appears in the step that follows the rate-determining step, it does not appear in the overall rate law. Thus, we see that *the overall rate law includes only species active in the reaction up to and including those in the rate-determining step.* This point was also illustrated by the mechanism for NO_2 and CO shown earlier. Carbon monoxide was absent from the overall rate law because it appeared *after* the rate-determining step.

Figure 16.21 is a reaction energy diagram for the reaction of NO_2 and F_2. Note that

- *Each step in the mechanism has its own transition state.* (Note that only one molecule of NO_2 reacts in step 1, and only the first transition state is depicted.)
- The F atom intermediate is a reactive, unstable species (as you know from halogen chemistry), so it is higher in energy than the reactants or product.
- The first step is slower (rate limiting), so its activation energy is *larger* than that of the second step.
- The overall reaction is exothermic, so the product is lower in energy than the reactants.

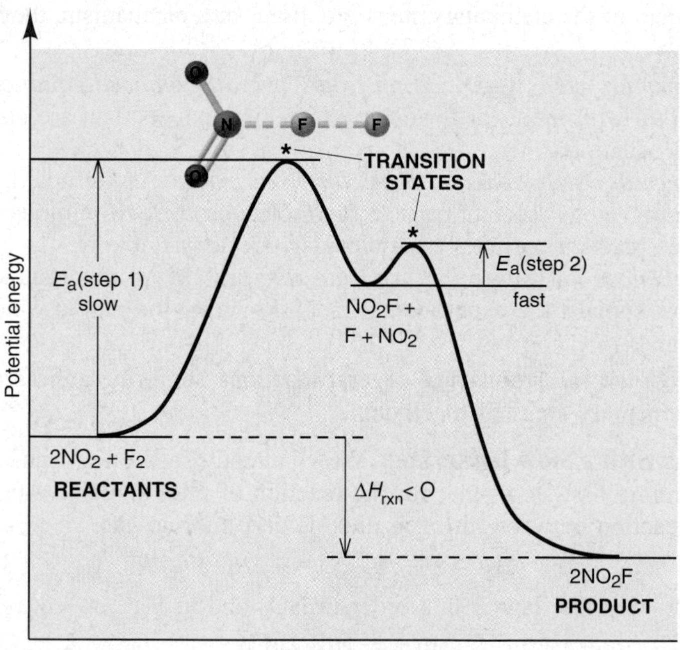

Figure 16.21 Reaction energy diagram for the two-step reaction of NO_2 and F_2. Each step in the mechanism has its own transition state. The proposed transition state is shown for step 1. Reactants for the second step are the F atom intermediate and the second molecule of NO_2. Note that the first step is slower (higher E_a). The overall reaction is exothermic ($\Delta H_{rxn} < 0$).

Mechanisms with a Fast Initial Step If the rate-limiting step in a mechanism is *not* the initial step, it acts as a bottleneck later in the reaction sequence. As a result, the product of a fast initial step builds up and starts reverting to reactant, while waiting for the slow step to remove it. With time, the product of the initial step is changing back to reactant as fast as it is forming. In other words, the *fast initial step reaches equilibrium.* As you'll see, this situation allows us to fit the mechanism to the overall rate law.

Consider once again the oxidation of nitric oxide:

$$2NO(g) + O_2(g) \longrightarrow 2NO_2(g)$$

The experimentally determined rate law is

$$\text{Rate} = k[\text{NO}]^2[\text{O}_2]$$

and a proposed mechanism is

(1) $NO(g) + O_2(g) \rightleftharpoons NO_3(g)$ [fast, reversible]

(2) $NO_3(g) + NO(g) \longrightarrow 2NO_2(g)$ [slow; rate determining]

Note that, with cancellation of the reaction intermediate NO_3, the first criterion is met because the sum of the steps gives the overall equation. Also note that the second criterion is met because both steps are bimolecular.

To meet the third criterion (that the mechanism conforms to the overall rate law), we first write rate laws for the elementary steps:

(1) $\text{Rate}_{1(\text{fwd})} = k_1[\text{NO}][\text{O}_2]$

$\text{Rate}_{1(\text{rev})} = k_{-1}[\text{NO}_3]$

where k_{-1} is the rate constant for the reverse reaction.

(2) $\text{Rate}_2 = k_2[\text{NO}_3][\text{NO}]$

Now we must show that the rate law for the rate-determining step (step 2) gives the overall rate law. As written, it does not, because it contains the intermediate NO_3, and *an overall rate law can include only reactants (and products).* Therefore, we must eliminate $[NO_3]$ from the step 2 rate law. To do so, we express $[NO_3]$ in terms of reactants. Step 1 reaches equilibrium when the forward and reverse rates are equal:

$$\text{Rate}_{1(\text{fwd})} = \text{Rate}_{1(\text{rev})} \quad \text{or} \quad k_1[\text{NO}][\text{O}_2] = k_{-1}[\text{NO}_3]$$

To express $[NO_3]$ in terms of reactants, we isolate it algebraically:

$$[\text{NO}_3] = \frac{k_1}{k_{-1}}[\text{NO}][\text{O}_2]$$

Then, substituting for $[NO_3]$ in the rate law for step 2, we obtain

$$\text{Rate}_2 = k_2[\text{NO}_3][\text{NO}] = k_2\left(\frac{k_1}{k_{-1}}[\text{NO}][\text{O}_2]\right)[\text{NO}] = \frac{k_2k_1}{k_{-1}}[\text{NO}]^2[\text{O}_2]$$

This rate law is identical to the overall rate law, with $k = \dfrac{k_2k_1}{k_{-1}}$.

Thus, to test the validity of a mechanism with a fast initial, reversible step:

1. Write rate laws for both directions of the fast step and for the slow step.
2. Show the slow step's rate law is equivalent to the overall rate law, by *expressing [intermediate] in terms of [reactant]:* set the forward rate law of the fast, reversible step equal to the reverse rate law, and solve for [intermediate].
3. Substitute the expression for [intermediate] into the rate law for the slow step to obtain the overall rate law.

Several end-of-chapter problems, including 16.72 and 16.73, provide additional examples of this approach.

SECTION SUMMARY

The mechanisms of most common reactions consist of two or more elementary steps, reactions that occur in one step and depict a single chemical change. The molecularity of an elementary step equals the number of reactant particles and is the same as the reaction order of its rate law. Unimolecular and bimolecular steps are common. The rate-determining, or rate-limiting (slowest), step determines how fast the overall reaction occurs, and its rate law represents the overall rate law. Reaction intermediates are species that form in one step and react in a later one. The steps in a proposed mechanism must add up to the overall reaction, be physically reasonable, and conform to the overall rate law. If a fast step precedes a slow step, the fast step reaches equilibrium, and the concentrations of intermediates in the rate law of the slow step must be expressed in terms of reactants.

16.8 CATALYSIS: SPEEDING UP A CHEMICAL REACTION

There are many situations in which the rate of a reaction must be increased for it to be useful. In an industrial process, for example, a higher rate often determines whether a new product can be made economically. Sometimes, we can speed up a reaction sufficiently with a higher temperature, but energy is costly and many substances are heat sensitive and easily decomposed. Alternatively, we can often employ a **catalyst,** a substance that increases the rate *without* being consumed in the reaction. Because catalysts are not consumed, only very small, nonstoichiometric quantities are generally required. Nevertheless, these substances are employed in so many important processes that several million tons of industrial catalysts are produced annually in the United States alone! Nature is the master designer and user of catalysts. Even the simplest bacterium employs thousands of biological catalysts, known as *enzymes,* to speed up its cellular reactions. Every organism relies on enzymes to sustain life.

Each catalyst has its own specific way of functioning, but in general, *a catalyst effects a lower activation energy, which in turn makes the rate constant larger and the rate higher*. Two important points stand out in Figure 16.22:

- A catalyst speeds up the forward *and* reverse reactions. A reaction with a catalyst *does not yield more product* than one without a catalyst, but it yields the product *more quickly*.
- A catalyst effects a *lower activation energy* by providing a *different mechanism* for the reaction, a new, lower energy pathway.

Consider a general *uncatalyzed* reaction that proceeds by a one-step mechanism involving a bimolecular collision:

$$A + B \longrightarrow \text{product} \qquad \text{[slower]}$$

In the *catalyzed* reaction, a reactant molecule interacts with the catalyst, so the mechanism might involve a two-step pathway:

$$A + \text{catalyst} \longrightarrow C \qquad \text{[faster]}$$
$$C + B \longrightarrow \text{product} + \text{catalyst} \qquad \text{[faster]}$$

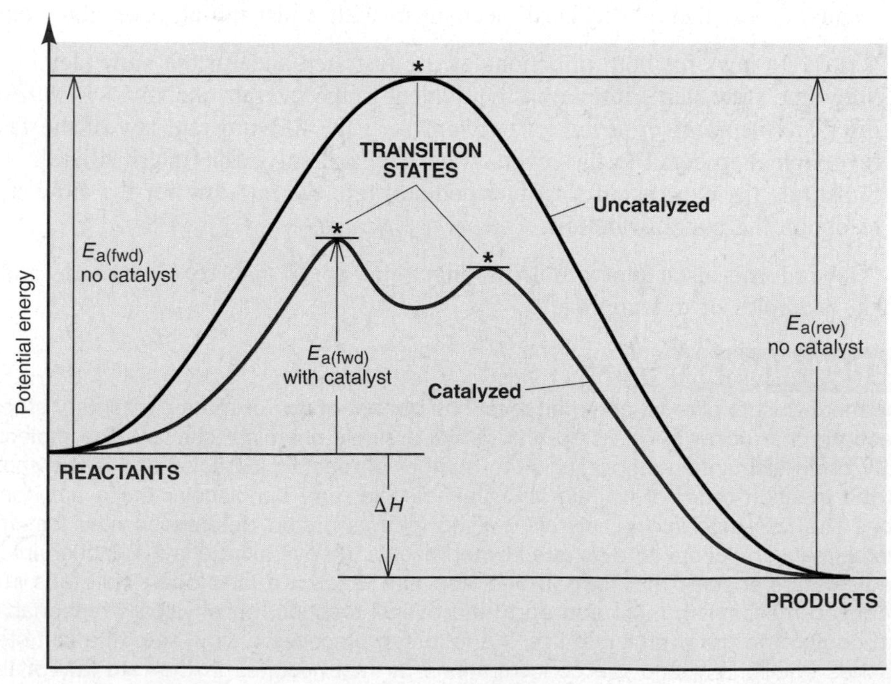

Figure 16.22 Reaction energy diagram for a catalyzed and an uncatalyzed process. A catalyst speeds a reaction by providing a new, lower energy pathway, in this case by replacing the one-step mechanism with a two-step mechanism. Both forward and reverse rates are increased to the same extent, so *a catalyst does not affect the overall reaction yield.* (The only activation energy shown for the catalyzed reaction is the larger one for the forward direction.)

Note that *the catalyst is not consumed,* as its definition requires. Rather, it is used and then regenerated, and the activation energies of both steps are lower than the activation energy of the uncatalyzed pathway.

There are two general categories of catalyst—homogeneous catalysts and heterogeneous catalysts—based on whether the catalyst is in the same phase as the reactant and product.

Homogeneous Catalysis

A **homogeneous catalyst** exists in solution with the reaction mixture. All homogeneous catalysts are gases, liquids, or soluble solids. Some industrial processes that employ these catalysts are shown in Table 16.7 *(top).*

A thoroughly studied example of homogeneous catalysis is the hydrolysis of an organic ester (RCOOR′), a reaction we examined in Section 15.4:

$$R{-}\overset{\displaystyle O}{\overset{\|}{C}}{-}O{-}R' + H_2O \rightleftharpoons R{-}\overset{\displaystyle O}{\overset{\|}{C}}{-}OH + R'{-}OH$$

Here R and R′ are hydrocarbon groups, $R{-}\overset{O}{\overset{\|}{C}}{-}OH$ is a carboxylic acid, and R′—OH is an alcohol. The reaction rate is low at room temperature but can be increased greatly by adding a small amount of strong acid, which provides H^+ ion, the catalyst in the reaction; strong bases, which supply OH^- ions, also speed ester hydrolysis, but by a slightly different mechanism.

In the first step of the acid-catalyzed reaction (Figure 16.23), the H^+ of a hydronium ion forms a bond to the double-bonded O atom. From the resonance forms, we see that the bonding of H^+ then makes the C atom more positive, which *increases its attraction* for the partially negative O atom of water. In effect, H^+ increases the likelihood that the bonding of water, which is the rate-determining step, will take place. Several steps later, a water molecule, acting as a base, removes the H^+ and returns it to solution. Thus, H^+ acts as a catalyst because it speeds up the reaction but is not itself consumed: it is used up in one step and re-formed in another.

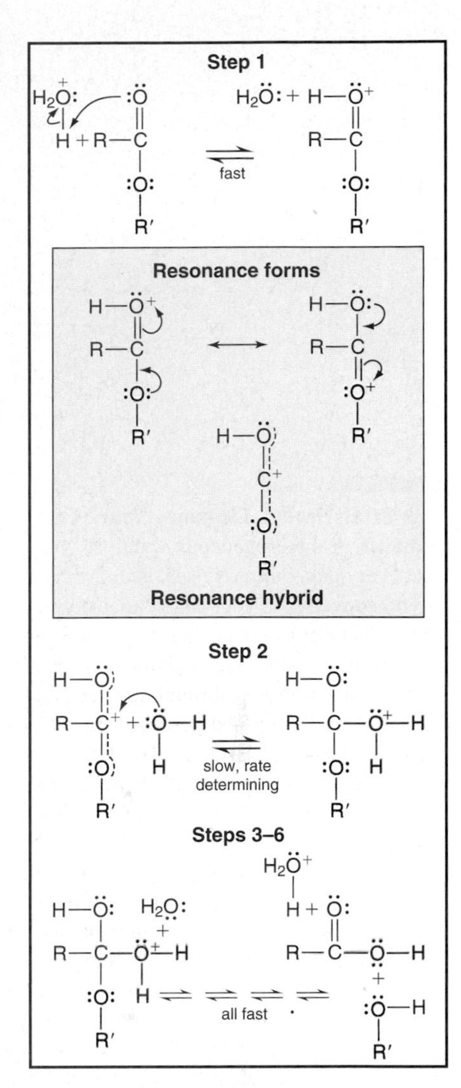

Figure 16.23 Mechanism for the catalyzed hydrolysis of an organic ester. In step 1, the catalytic H^+ ion binds to the electron-rich oxygen. The resonance hybrid of this product *(see gray panel)* shows the C atom is more positive than it would ordinarily be. The enhanced charge on C attracts the partially negative O of water more strongly, increasing the fraction of effective collisions and thus speeding up step 2, the rate-determining step. Loss of R′OH and removal of H^+ by water occur in a final series of fast steps.

Table 16.7 Some Modern Processes That Employ Catalysts

Reactants	Catalyst	Product	Use
Homogeneous			
Propylene, oxidizer	Mo(VI) complexes	Propylene oxide	Polyurethane foams; polyesters
Methanol, CO	$[Rh(CO)_2I_2]^-$	Acetic acid	Poly(vinyl acetate) coatings; poly(vinyl alcohol)
Butadiene, HCN	Ni/P compounds	Adiponitrile	Nylons (fibers, plastics)
α-Olefins, CO, H_2	Rh/P compounds	Aldehydes	Plasticizers, lubricants
Heterogeneous			
Ethylene, O_2	Silver, cesium chloride on alumina	Ethylene oxide	Polyesters, ethylene glycol, lubricants
Propylene, NH_3, O_2	Bismuth molybdates	Acrylonitrile	Plastics, fibers, resins
Ethylene	Organochromium and titanium halides on silica	High-density polyethylene	Molded products

Many digestive enzymes, which catalyze the hydrolysis of proteins, fats, and carbohydrates during the digestion of foods, employ very similar mechanisms. The difference is that the acids or bases that speed these reactions are not the strong inorganic reagents used in the lab, but rather specific amino-acid side chains of the enzymes that release or abstract H^+ ions.

Heterogeneous Catalysis

A **heterogeneous catalyst** speeds up a reaction that occurs in a separate phase. The catalyst is most often a solid interacting with gaseous or liquid reactants. Because reaction occurs on the solid's surface, heterogeneous catalysts usually have enormous surface areas for contact, between 1 and 500 m^2/g. Interestingly, many reactions that occur on a metal surface, such as the decomposition of HI on gold and the decomposition of N_2O on platinum, are zero order because the rate-determining step occurs on the surface itself. Thus, despite an enormous surface area, once the reactant gas covers the surface, increasing the reactant concentration cannot increase the rate. Table 16.7 *(bottom)* lists some polymer manufacturing processes that employ heterogeneous catalysts.

One of the most important examples of heterogeneous catalysis is the addition of H_2 to the C=C bonds of organic compounds to form C—C bonds. The petroleum, plastics, and food industries frequently use catalytic **hydrogenation.** The conversion of vegetable oil into margarine is one example.

The simplest hydrogenation converts ethylene to ethane:

$$H_2C{=}CH_2(g) + H_2(g) \longrightarrow H_3C{-}CH_3(g)$$

In the absence of a catalyst, the reaction occurs very slowly. At high H_2 pressure in the presence of finely divided Ni, Pd, or Pt, the reaction becomes rapid even at ordinary temperatures. These Group 8B(10) metals catalyze by *chemically adsorbing the reactants onto their surface* (Figure 16.24). The H_2 lands and splits into separate H atoms chemically bound to the solid catalyst's metal atoms (catM):

$$H{-}H(g) + 2\text{catM}(s) \longrightarrow 2\text{catM}{-}H \text{ (H atoms bound to metal surface)}$$

Then, C_2H_4 adsorbs and reacts with two H atoms, one at a time, to form C_2H_6. The H—H bond breakage is the rate-determining step in the overall process, and interaction with the catalyst's surface provides the low-E_a step as part of an alternative reaction mechanism.

Two Chemical Connections essays follow. The first introduces the remarkable abilities of the catalytic enzymes inside you, and the second discusses how catalysis operates in the depletion of stratospheric ozone.

Catalytically Cleaning Your Car's Exhaust A heterogeneous catalyst you use every day, but rarely see, is in the catalytic converter in your car's exhaust system. This device is designed to convert polluting exhaust gases into nontoxic ones. In a single pass through the catalyst bed, CO and unburned gasoline are oxidized to CO_2 and H_2O, while NO is reduced to N_2. As in the mechanism for catalytic hydrogenation of an alkene, the catalyst lowers the activation energy of the rate-determining step by adsorbing the molecules, thereby weakening their bonds. Mixtures of transition metals and their oxides embedded in inert supports convert as much exhaust gas as possible in the shortest time. It is estimated, for example, that an NO molecule is adsorbed and split into catalyst-bound N and O atoms in less than 2×10^{-12} s!

Figure 16.24 The metal-catalyzed hydrogenation of ethylene.

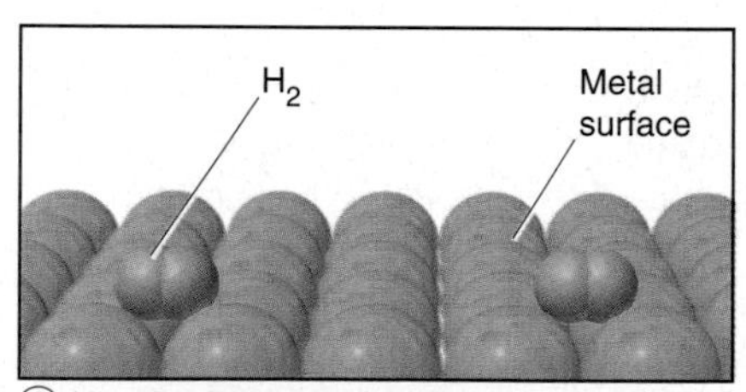

① H_2 adsorbs to metal surface

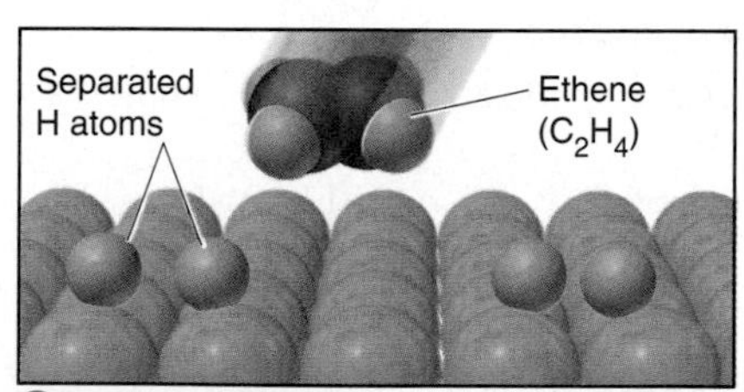

② Rate-limiting step is H—H bond breakage.

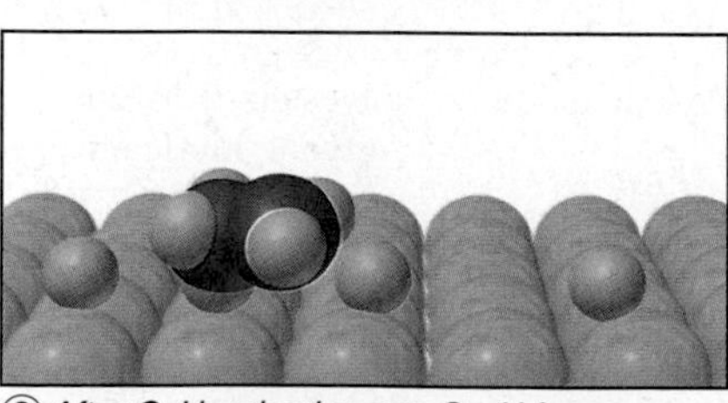
③ After C_2H_4 adsorbs, one C—H forms.

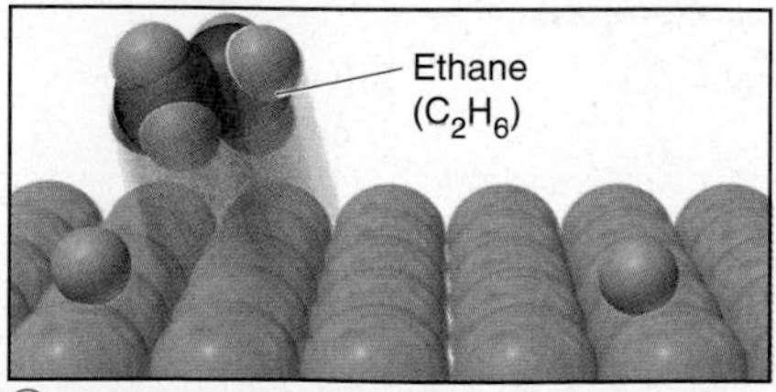

④ Another C—H bond forms; C_2H_6 leaves surface.

CHEMICAL CONNECTIONS to Enzymology

Kinetics and Function of Biological Catalysts

Within every living cell, thousands of individual reactions occur. Many involve complex chemical changes, yet they take place in dilute solution at ordinary temperatures and pressures. The rate of each reaction responds smoothly to momentary or permanent changes in other reaction rates, various signals from other cells, and environmental stresses. Virtually every reaction in this marvelous chemical harmony is catalyzed by its own specific **enzyme,** a protein catalyst whose function has been perfected through evolution.

Enzymes have complex three-dimensional shapes and molar masses ranging from about 15,000 to 1,000,000 g/mol (Section 15.6). At a specific region on an enzyme's surface is its **active site,** a crevice whose shape results from the shapes of the amino acid side chains (R groups) directly involved in catalyzing the reaction (also see Figure B10.4, p. 391). When the reactant molecules, called the **substrates** of the reaction, collide with the active site, the chemical change is initiated. The active site makes up only a small part of the enzyme's surface—like a tiny hollow carved into a mountainside—and often includes amino acid R groups from distant regions of the molecular chain that lie near each other because of the folding of the protein's backbone, as depicted in Figure B16.4. In most cases, *substrates bind to the active site through intermolecular forces:* H bonds, dipole forces, and other weak attractions.

(continued)

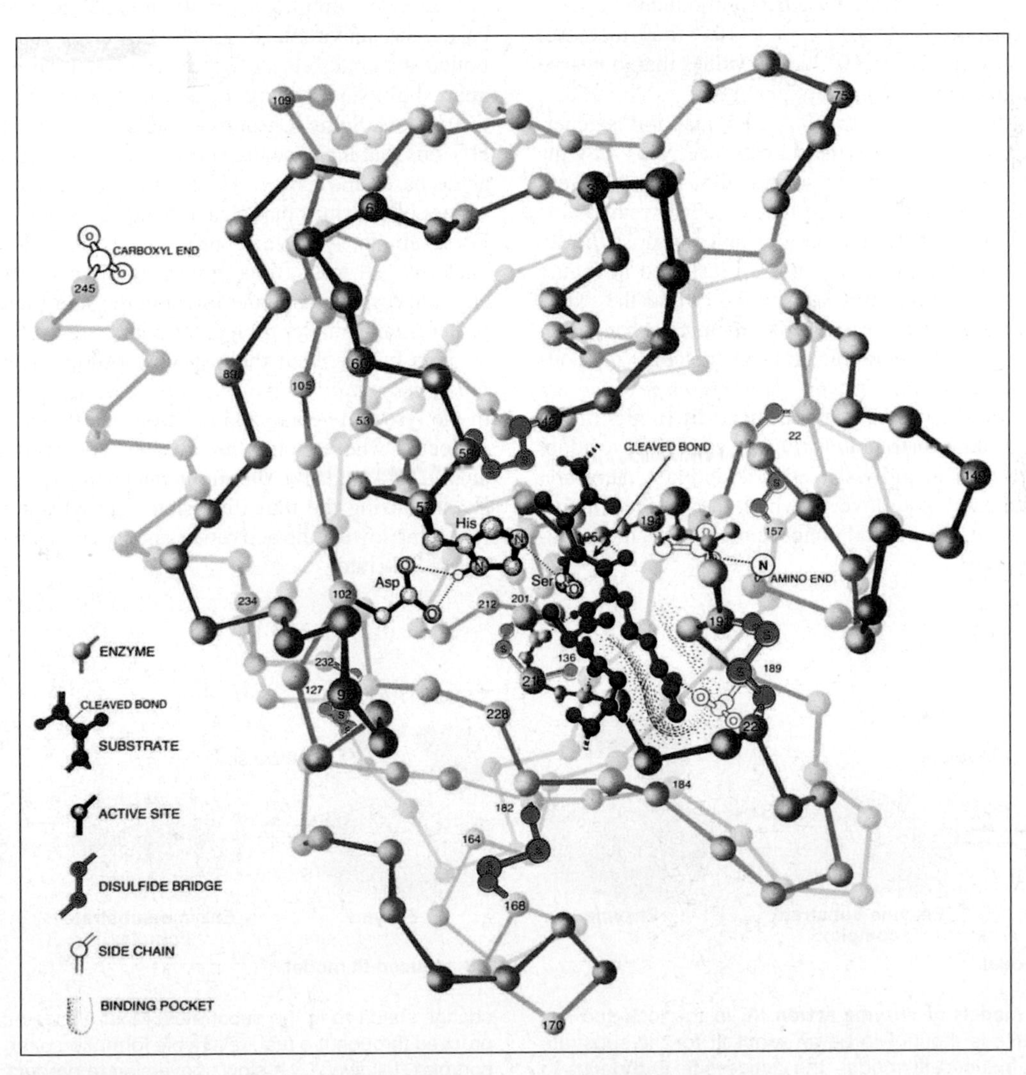

Figure B16.4 The widely separated amino acid groups that form an active site. The amino acids in chymotrypsin, a digestive enzyme, are shown as linked spheres and numbered consecutively from the beginning of the chain. The R groups of the active-site amino acids 57 (histidine, His), 102 (aspartic acid, Asp), and 195 (serine, Ser) are shown because they play a crucial role in the enzyme-catalyzed reaction. Part of the substrate molecule lies within the active site's crevice (binding pocket) to anchor it during reaction. (Illustration by Irving Geis. Rights owned by Howard Hughes Medical Institute. Not to be used without permission.)

An enzyme has features of both a homogeneous and a heterogeneous catalyst. Most enzymes are enormous compared with their substrates, and they are often found embedded within membranes of larger cell parts. Thus, like a heterogeneous catalyst, an enzyme provides an active surface on which a reactant is immobilized temporarily, waiting for its reaction partner to land nearby. Like a homogeneous catalyst, the enzyme's amino acid R groups interact actively with the substrates in multistep sequences involving intermediates.

Enzymes are incredibly *efficient* catalysts. Consider the hydrolysis of urea as an example:

$$(NH_2)_2C{=}O(aq) + 2H_2O(l) + H^+(aq) \longrightarrow 2NH_4^+(aq) + HCO_3^-(aq)$$

In water at room temperature, the rate constant for the uncatalyzed reaction is approximately 3×10^{-10} s^{-1}. Under the same conditions in the presence of the enzyme *urease* (pronounced "*yur*-ee-ase"), the rate constant is 3×10^4 s^{-1}, a 10^{14}-fold increase! Enzymes increase rates by 10^8 to 10^{20} times, values that industrial chemists who design catalysts can only dream of.

Enzymes are also extremely *specific:* each reaction is generally catalyzed by a particular enzyme. Urease catalyzes *only* the hydrolysis of urea, and none of the several thousand other enzymes present in the cell catalyzes that reaction. This remarkable specificity results from the particular groups that comprise the active site. Two models of enzyme action are illustrated in Figure B16.5. According to the **lock-and-key model,** when the "key" (substrate) fits the "lock" (active site), the chemical change begins. However, x-ray crystallographic and spectroscopic methods show that, in many cases, *the enzyme changes shape when the substrate lands at the active site.* This **induced-fit model** of enzyme action pictures the substrate inducing the active site to adopt a perfect fit. Rather than a rigidly shaped lock and key, therefore, we should picture a hand in a glove, in which the "glove" (active site) does not attain its functional shape until the "hand" (substrate) moves into place.

The kinetics of enzyme catalysis has many features in common with that of ordinary catalysis. In an uncatalyzed reaction, the rate is affected by the concentrations of the reactants; in a catalyzed reaction, the rate is affected by the *concentration of reactant bound to catalyst.* In the enzyme-catalyzed case, substrate (S) and enzyme (E) form an intermediate **enzyme-substrate complex (ES),** whose concentration determines the rate of product (P) formation. The steps common to virtually all enzyme-catalyzed reactions are

(1) $E + S \rightleftharpoons ES$ [fast, reversible]

(2) $ES \longrightarrow E + P$ [slow; rate determining]

Thus, rate $= k[ES]$. When all the available enzyme molecules are bound to substrate and exist as ES, increasing [S] has no effect on rate and the process is zero order. To cite one very common example, the breakdown of ethanol in the body is zero order.

Enzymes employ a variety of catalytic mechanisms. In some cases, the active site R groups bring the reacting atoms of the bound substrates closer together. In other cases, the groups move apart slightly, stretching the substrate bond that is to be broken in the process. Some R groups are acidic and thus are able to provide H^+ ions that increase the speed of a rate-determining step; others act as bases and remove H^+ ions at a critical step. *Hydrolases* are a class of enzymes that cleave bonds by such *acid-base catalysis.* For example, lysozyme, an enzyme found in tears, hydrolyzes bacterial cell walls, thus protecting the eyes from microbes, and chymotrypsin, an enzyme found in the small intestine, hydrolyzes proteins into smaller molecules during digestion.

No matter what their specific mode of action, *all enzymes function by stabilizing the reaction's transition state.* For instance, in the lysozyme-catalyzed reaction, the transition state is a sugar molecule whose bonds are twisted and stretched into unusual lengths and angles so that it fits the lysozyme active site perfectly. By stabilizing the transition state through binding it effectively, lysozyme lowers the activation energy of the reaction and thus increases the rate.

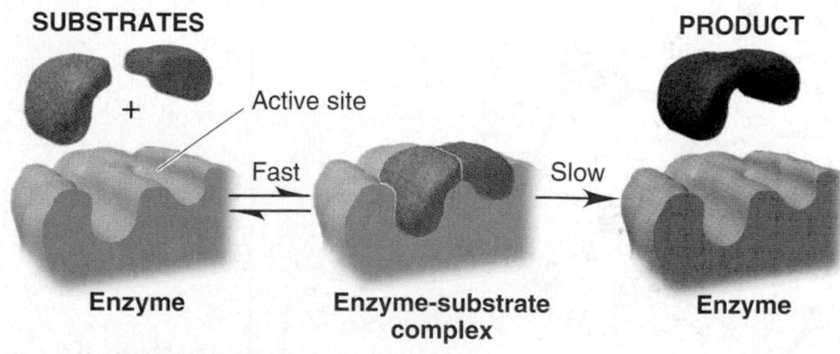

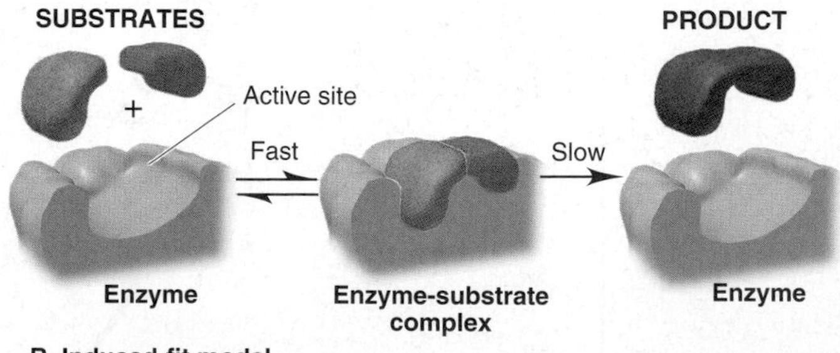

Figure B16.5 Two models of enzyme action. A, In the lock-and-key model, the active site is thought to be an exact fit for the substrate shapes. **B,** In the induced-fit model, the active site is thought to change shape to fit the substrates. Most enzyme-catalyzed reactions proceed through the fast, reversible formation of an enzyme-substrate complex, followed by a slow conversion to product and free enzyme.

Depletion of the Earth's Ozone Layer

Both homogeneous and heterogeneous catalysts play key roles in one of today's most serious environmental concerns—the depletion of ozone from the stratosphere. The stratospheric ozone layer absorbs UV radiation with wavelengths between 280 and 320 nm emitted by the Sun that would otherwise reach Earth's surface. This radiation (called UV-B) has enough energy to break bonds in deoxyribonucleic acid (DNA) and thereby damage genes. Depletion of stratospheric ozone could significantly increase risks to human health from UV-B, particularly a higher incidence of skin cancer and cataracts (loss of transparency of the lens of the eye). Plant life, especially the simpler forms at the base of the food chain, may also be damaged.

Historically, stratospheric ozone concentration varied seasonally but remained nearly constant from year to year through a series of atmospheric reactions. Oxygen atoms formed from dissociation of O_2 by UV radiation with wavelengths less than 242 nm react with O_2 to form O_3, and with O_3 to regenerate O_2:

$$O_2 \xrightarrow{UV} 2O$$
$$O + O_2 \longrightarrow O_3 \qquad \text{[ozone formation]}$$
$$O + O_3 \longrightarrow 2O_2 \qquad \text{[ozone breakdown]}$$

However, research by Paul J. Crutzen, Mario J. Molina, and F. Sherwood Rowland, for which they received the Nobel Prize in chemistry in 1995, revealed that industrially produced chlorofluorocarbons (CFCs) shifted this balance by catalyzing the breakdown reaction.

CFCs were widely used as aerosol propellants, foam blowing agents, and air-conditioning coolants, so large quantities were released into the atmosphere. Unreactive in the troposphere, CFC molecules gradually reach the stratosphere, where they encounter UV radiation that can split them and release chlorine atoms:

$$CF_2Cl_2 \xrightarrow{UV} CF_2Cl\cdot + Cl\cdot$$

(The dots are unpaired electrons resulting from bond cleavage.)

Like many species with unpaired electrons (free radicals), atomic Cl is very reactive. The Cl atoms react with stratospheric ozone to produce an intermediate, chlorine monoxide ($ClO\cdot$), which then reacts with free O atoms to regenerate Cl atoms:

$$O_3 + Cl\cdot \longrightarrow ClO\cdot + O_2$$
$$ClO\cdot + O \longrightarrow \cdot Cl + O_2$$

The sum of these steps is the ozone breakdown reaction:

$$O_3 + \cancel{\cdot Cl} + \cancel{\cdot ClO} + O \longrightarrow \cancel{\cdot ClO} + O_2 + \cancel{\cdot Cl} + O_2$$

or

$$O_3 + O \longrightarrow 2O_2$$

Within a region low in ozone, the concentration of O atoms is also low, so Cl atoms are regenerated from ClO:

$$2\,ClO\cdot \longrightarrow ClOOCl$$
$$ClOOCl + \text{light} \longrightarrow Cl\cdot + \cdot OOCl$$
$$ClOO\cdot + \text{light} \longrightarrow Cl\cdot + O_2$$

Thus, the overall process converts two O_3 to three O_2 molecules.

Note that the Cl atom acts as a homogeneous catalyst: it exists in the same phase as the reactants, speeds up the process via a different mechanism, and is regenerated. The problem is that each Cl atom has a stratospheric half-life of about 2 years, during which time it speeds the breakdown of about 100,000 ozone molecules. Bromine is a more effective catalyst, but much less ends up in the stratosphere. Halons, used as fire suppressants, and methyl bromide, an agricultural insecticide, are the main anthropogenic (human-made) sources of stratospheric bromine.

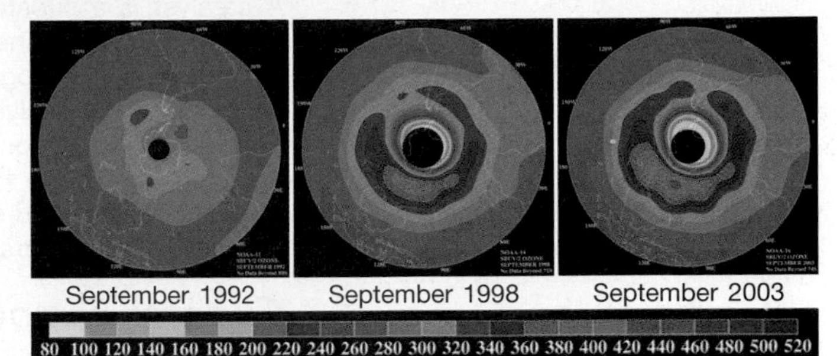

Figure B16.6 The increasing size of the Antarctic ozone hole. Satellite images show changes in ozone concentration over the South Pole for September 1992, 1998, and 2003. Note the increasing size of the "hole" in the ozone layer (indicated by the four leftmost colors in the key; the black circle is an instrument artifact). Since 1995, data show a similar thinning, though not yet as severe, over the North Pole.

A variety of Cl-containing species cause the Antarctic *ozone hole*, the severe reduction of stratospheric ozone over the Antarctic region when the Sun rises after the long winter darkness. Measurements of high [$ClO\cdot$] over Antarctica are consistent with the breakdown mechanism. Atmospheric scientists have documented more than 80% ozone depletion over the South Pole (Figure B16.6). The ozone hole enlarges by heterogeneous catalysis involving polar stratospheric clouds. The clouds provide a surface for reactions that convert inactive chlorine compounds, such as HCl and chlorine nitrate ($ClONO_2$), to substances, such as Cl_2, that are cleaved by UV radiation to Cl atoms. Heterogeneous catalysis also occurs on fine particles in the stratosphere. Dust from the eruption of Mt. Pinatubo in 1991 reduced stratospheric ozone for 2 years. There is ozone thinning over the North Pole, and NASA has also documented the loss of stratospheric ozone at middle latitudes, such as over the United States. So far, both of these losses occur to a lower extent and at a lower rate.

The Montreal Protocol in 1987 and later amendments curtailed growth in production of CFCs and set dates for phasing out these and other chlorinated and brominated compounds, such as CCl_4, CCl_3CH_3, and CH_3Br. Another change was replacement of CFC aerosol propellants by hydrocarbons, such as isobutane, $(CH_3)_3CH$. Replacement of CFCs in refrigeration units is more difficult, because it will require changes in equipment. The first replacements for CFCs were hydrochlorofluorocarbons (HCFCs), such as CHF_2Cl. These deplete ozone less than CFCs do because they have relatively short tropospheric lifetimes due to abstraction of H atoms by hydroxyl radicals ($\cdot OH$):

$$CHF_2Cl + \cdot OH \longrightarrow H_2O + \cdot CF_2Cl$$

After this initial attack, further degradation occurs rapidly.

HCFCs will be phased out by 2040 and replaced by hydrofluorocarbons (HFCs, which contain only C, H, and F) in devices such as car air conditioners. HFCs are preferable to CFCs or HCFCs because fluorine is a poor catalyst of ozone breakdown. Nevertheless, because of the long lifetimes of CFCs and the ongoing emissions of HCFCs until their phase-out, full recovery of the ozone layer may take another century! The good news is that tropospheric halogen levels have already begun to fall.

SECTION SUMMARY

A catalyst is a substance that increases the rate of a reaction without being consumed. It accomplishes this by providing an alternative mechanism with a lower activation energy. Homogeneous catalysts function in the same phase as the reactants. Heterogeneous catalysts act in a different phase from the reactants. The hydrogenation of carbon-carbon double bonds takes place on a solid catalyst, which speeds the breakage of the H—H bond in H_2. Enzymes are biological catalysts with spectacular efficiency and specificity. Chlorine atoms derived from CFC molecules catalyze the breakdown of stratospheric ozone.

Chapter Perspective

With this introduction to chemical kinetics, we have begun to explore the dynamic inner workings of chemical change. Variations in reaction rate are observed with concentration and temperature changes, which operate on the molecular level through the frequency and energetics of particle collisions and the details of reactant structure. Kinetics allows us to speculate about the molecular pathway of a reaction. Modern industry and biochemistry depend on its principles. However, speed and yield are very different aspects of a reaction. In Chapter 17, we'll see how opposing reaction rates give rise to the equilibrium state and examine how much product has formed once the net reaction stops.

For Review and Reference

(Numbers in parentheses refer to pages, unless noted otherwise.)

Learning Objectives

Relevant section and/or sample problem (SP) numbers appear in parentheses.

Understand These Concepts

1. How reaction rate depends on concentration, physical state, and temperature (Section 16.1)
2. The meaning of reaction rate in terms of changing concentrations over time (Section 16.2)
3. How the rate can be expressed in terms of reactant or product concentrations (Section 16.2)
4. The distinction between average and instantaneous rate and why the instantaneous rate changes during the reaction (Section 16.2)
5. The interpretation of reaction rate in terms of reactant and product concentrations (Section 16.2)
6. The experimental basis of the rate law and the information needed to determine it—initial rate data, reaction orders, and rate constant (Section 16.3)
7. The importance of reaction order in determining the rate (Section 16.3)
8. How reaction order is determined from initial rates at different concentrations (Section 16.3)
9. How integrated rate laws show the dependence of concentration on time (Section 16.4)
10. What reaction half-life means and why it is constant for a first-order reaction (Section 16.4)
11. Activation energy and the effect of temperature on the rate constant (Arrhenius equation) (Section 16.5)
12. Why concentrations are multiplied in the rate law (Section 16.6)
13. How temperature affects rate by influencing collision energy and, thus, the fraction of collisions with energy exceeding the activation energy (Section 16.6)
14. Why molecular orientation and complexity influence the number of effective collisions and the rate (Section 16.6)
15. The transition state as the momentary species between reactants and products whose formation requires the activation energy (Section 16.6)
16. How an elementary step represents a single molecular event and its molecularity equals the number of colliding particles (Section 16.7)
17. How a reaction mechanism consists of several elementary steps, with the slowest step determining the overall rate (Section 16.7)
18. The criteria for a valid reaction mechanism (Section 16.7)
19. How a catalyst speeds a reaction by lowering the activation energy (Section 16.8)
20. The distinction between homogeneous and heterogeneous catalysis (Section 16.8)

Master These Skills

1. Calculating instantaneous rate from the slope of a tangent to a concentration vs. time plot (Section 16.2)
2. Expressing reaction rate in terms of changes in concentration over time (SP 16.1)
3. Determining reaction order from a known rate law (SP 16.2)
4. Determining reaction order from changes in initial rate with concentration (SP 16.3)
5. Calculating the rate constant and its units (Section 16.3)
6. Using an integrated rate law to find concentration at a given time or the time to reach a given concentration (SP 16.4)
7. Determining reaction order graphically with a rearranged integrated rate law (Section 16.4)
8. Determining the half-life of a first-order reaction (SP 16.5)
9. Using a form of the Arrhenius equation to calculate the activation energy (SP 16.6)

Learning Objectives *(continued)*

10. Using reaction energy diagrams to depict the energy changes during a reaction (SP 16.7)
11. Postulating a transition state for a simple reaction (SP 16.7)
12. Determining the molecularity and rate law for an elementary step (SP 16.8)
13. Constructing a mechanism with either a slow or a fast initial step (Section 16.7)

Key Terms

chemical kinetics (673)

Section 16.2
reaction rate (675)
average rate (676)
instantaneous rate (677)
initial rate (677)

Section 16.3
rate law (rate equation) (679)
rate constant (680)
reaction orders (680)

Section 16.4
integrated rate law (686)
half-life ($t_{1/2}$) (689)

Section 16.5
Arrhenius equation (692)
activation energy (E_a) (692)

Section 16.6
collision theory (694)
effective collision (696)
frequency factor (696)
transition state theory (697)
transition state (activated complex) (697)
reaction energy diagram (698)

Section 16.7
reaction mechanism (700)
reaction intermediate (701)
elementary reaction (elementary step) (701)
molecularity (701)
unimolecular reaction (701)
bimolecular reaction (701)
rate-determining (rate-limiting) step (702)

Section 16.8
catalyst (706)
homogeneous catalyst (707)
heterogeneous catalyst (708)
hydrogenation (708)
enzyme (709)
active site (709)
substrate (709)
lock-and-key model (710)
induced-fit model (710)
enzyme-substrate complex (ES) (710)

Key Equations and Relationships

16.1 Expressing reaction rate in terms of reactant A (676):

$$\text{Rate} = -\frac{\Delta[\text{A}]}{\Delta t}$$

16.2 Expressing the rate of a general reaction (678):

$$a\text{A} + b\text{B} \longrightarrow c\text{C} + d\text{D}$$

$$\text{Rate} = -\frac{1}{a}\frac{\Delta[\text{A}]}{\Delta t} = -\frac{1}{b}\frac{\Delta[\text{B}]}{\Delta t} = \frac{1}{c}\frac{\Delta[\text{C}]}{\Delta t} = \frac{1}{d}\frac{\Delta[\text{D}]}{\Delta t}$$

16.3 Writing a general rate law (for a case not involving products) (679):

$$\text{Rate} = k[\text{A}]^m[\text{B}]^n \cdots$$

16.4 Calculating the time to reach a given [A] in a first-order reaction (rate $= k[\text{A}]$) (686):

$$\ln\frac{[\text{A}]_0}{[\text{A}]_t} = kt$$

16.5 Calculating the time to reach a given [A] in a simple second-order reaction (rate $= k[\text{A}]^2$) (687):

$$\frac{1}{[\text{A}]_t} - \frac{1}{[\text{A}]_0} = kt$$

16.6 Calculating the time to reach a given [A] in a zero-order reaction (rate $= k$) (687):

$$[\text{A}]_t - [\text{A}]_0 = -kt$$

16.7 Finding the half-life of a first-order process (690):

$$t_{1/2} = \frac{\ln 2}{k} = \frac{0.693}{k}$$

16.8 Relating the rate constant to the temperature (Arrhenius equation) (692):

$$k = Ae^{-E_a/RT}$$

16.9 Calculating the activation energy (rearranged form of Arrhenius equation) (693):

$$\ln\frac{k_2}{k_1} = -\frac{E_a}{R}\left(\frac{1}{T_2} - \frac{1}{T_1}\right)$$

Highlighted Figures and Tables

These figures (F) and tables (T) provide a review of key ideas.

Brief Solutions to Follow-up Problems

16.1 (a) $4NO(g) + O_2(g) \longrightarrow 2N_2O_3(g)$;

$$\text{rate} = -\frac{\Delta[O_2]}{\Delta t} = -\frac{1}{4}\frac{\Delta[NO]}{\Delta t} = \frac{1}{2}\frac{\Delta[N_2O_3]}{\Delta t}$$

(b) $$-\frac{\Delta[O_2]}{\Delta t} = -\frac{1}{4}\frac{\Delta[NO]}{\Delta t}$$

$$= -\frac{1}{4}(-1.60\times10^{-4}\ \text{mol/L·s}) = 4.00\times10^{-5}\ \text{mol/L·s}$$

16.2 First order in Br^-, first order in BrO_3^-, second order in H^+, fourth order overall.

16.3 Rate $= k[H_2]^m[I_2]^n$. From experiments 1 and 3, $m = 1$. From experiments 2 and 4, $n = 1$.
Therefore, rate $= k[H_2][I_2]$; second order overall.

16.4 $1/[HI]_1 - 1/[HI]_0 = kt$;
111 L/mol − 100 L/mol = $(2.4\times10^{-21}$ L/mol·s$)(t)$
$t = 4.6\times10^{21}$ s (or 1.5×10^{14} yr)

16.5 $t_{1/2} = (\ln 2)/k$; $k = 0.693/13.1\ \text{h} = 5.29\times10^{-2}\ \text{h}^{-1}$

16.6 $$\ln\frac{0.286\ \text{L/mol·s}}{k_1} = -\frac{1.00\times10^5\ \text{J/mol}}{8.314\ \text{J/mol·K}}\times\left(\frac{1}{500.\ \text{K}} - \frac{1}{490.\ \text{K}}\right) = 0.491$$

$k_1 = 0.175$ L/mol·s

16.7

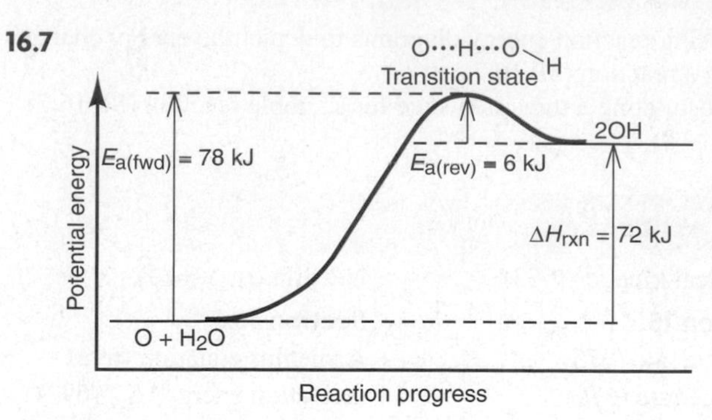

16.8 (a) Balanced equation:
$2NO(g) + 2H_2(g) \longrightarrow N_2(g) + 2H_2O(g)$
(b) Step 2 is unimolecular. All others are bimolecular.
(c) Rate$_1$ = $k_1[NO]^2$; rate$_2$ = $k_2[H_2]$; rate$_3$ = $k_3[N_2O_2][H]$; rate$_4$ = $k_4[HO][H]$; rate$_5$ = $k_5[H][N_2O]$.

Problems

Problems with **colored** numbers are answered in Appendix E. Sections match the text and provide the numbers of relevant sample problems. Most offer Concept Review Questions, Skill-Building Exercises (grouped in pairs covering the same concept), and Problems in Context. Comprehensive Problems are based on material from any section or previous chapter.

Factors That Influence Reaction Rate

Concept Review Questions

16.1 What variable of a chemical reaction is measured over time to obtain the reaction rate?

16.2 How does an increase in pressure affect the rate of a gas-phase reaction? Explain.

16.3 A reaction is carried out with water as the solvent. How does the addition of more water to the reaction vessel affect the rate of the reaction? Explain.

16.4 A gas reacts with a solid that is present in large chunks. Then the reaction is run again with the solid pulverized. How does the increase in the surface area of the solid affect the rate of its reaction with the gas? Explain.

16.5 How does an increase in temperature affect the rate of a reaction? Explain the two factors involved.

16.6 In a kinetics experiment, a chemist places crystals of iodine in a closed reaction vessel, introduces a given quantity of hydrogen gas, and obtains data to calculate the rate of hydrogen iodide formation. In a second experiment, she uses the same amounts of iodine and hydrogen, but she first warms the flask to 130°C, a temperature above the sublimation point of iodine. In which of these two experiments does the reaction proceed at a higher rate? Explain.

Expressing the Reaction Rate
(Sample Problem 16.1)

Concept Review Questions

16.7 Define *reaction rate*. Assuming constant temperature and a closed reaction vessel, why does the rate change with time?

16.8 (a) What is the difference between an average rate and an instantaneous rate? (b) What is the difference between an initial rate and an instantaneous rate?

16.9 Give two reasons to measure initial rates in a kinetics study.

16.10 For the reaction $A(g) \longrightarrow B(g)$, sketch two curves on the same set of axes that show
(a) The formation of product as a function of time
(b) The consumption of reactant as a function of time

16.11 For the reaction $C(g) \longrightarrow D(g)$, [C] vs. time is plotted:

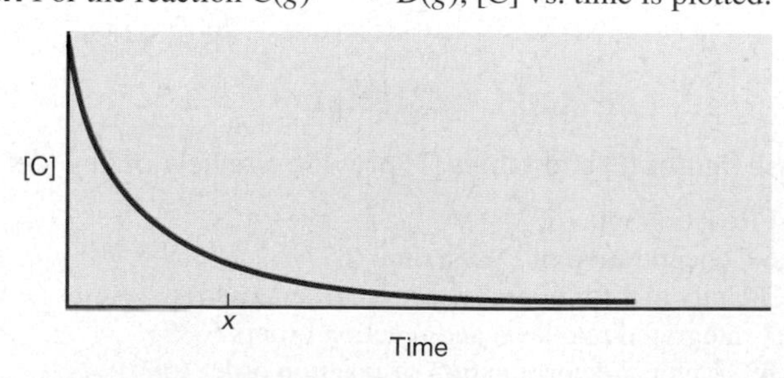

How do you determine each of the following?
(a) The average rate over the entire experiment
(b) The reaction rate at time x
(c) The initial reaction rate
(d) Would the values in parts (a), (b), and (c) be different if you plotted [D] vs. time? Explain.

Skill-Building Exercises *(grouped in similar pairs)*

16.12 The compound AX_2 decomposes according to the equation $2AX_2(g) \longrightarrow 2AX(g) + X_2(g)$. In one experiment, $[AX_2]$ was measured at various times and these data were obtained:

Time (s)	$[AX_2]$ (mol/L)
0.0	0.0500
2.0	0.0448
6.0	0.0300
8.0	0.0249
10.0	0.0209
20.0	0.0088

(a) Find the average rate over the entire experiment.
(b) Is the initial rate higher or lower than the rate in part (a)? Use graphical methods to estimate the initial rate.

16.13 (a) Use the data from Problem 16.12 to calculate the average rate from 8.0 to 20.0 s.
(b) Is the rate at exactly 5.0 s higher or lower than the rate in part (a)? Use graphical methods to estimate the rate at 5.0 s.

16.14 Express the rate of reaction in terms of the change in concentration of each of the reactants and products:

$$2A(g) \longrightarrow B(g) + C(g)$$

When [C] is increasing at 2 mol/L·s, how fast is [A] decreasing?

16.15 Express the rate of reaction in terms of the change in concentration of each of the reactants and products:

$$D(g) \longrightarrow \tfrac{3}{2}E(g) + \tfrac{5}{2}F(g)$$

When [E] is increasing at 0.25 mol/L·s, how fast is [F] increasing?

16.16 Express the rate of reaction in terms of the change in concentration of each of the reactants and products:

$$A(g) + 2B(g) \longrightarrow C(g)$$

When [B] is decreasing at 0.5 mol/L·s, how fast is [A] decreasing?

16.17 Express the rate of reaction in terms of the change in concentration of each of the reactants and products:

$$2D(g) + 3E(g) + F(g) \longrightarrow 2G(g) + H(g)$$

When [D] is decreasing at 0.1 mol/L·s, how fast is [H] increasing?

16.18 Reaction rate is expressed in terms of changes in concentration of reactants and products. Write a balanced equation for

$$\text{Rate} = -\frac{1}{2}\frac{\Delta[N_2O_5]}{\Delta t} = \frac{1}{4}\frac{\Delta[NO_2]}{\Delta t} = \frac{\Delta[O_2]}{\Delta t}$$

16.19 Reaction rate is expressed in terms of changes in concentration of reactants and products. Write a balanced equation for

$$\text{Rate} = -\frac{\Delta[CH_4]}{\Delta t} = -\frac{1}{2}\frac{\Delta[O_2]}{\Delta t} = \frac{1}{2}\frac{\Delta[H_2O]}{\Delta t} = \frac{\Delta[CO_2]}{\Delta t}$$

Problems in Context

16.20 The decomposition of nitrosyl bromide is followed manometrically because the number of moles of gas changes; it cannot be followed colorimetrically because both NOBr and Br_2 are reddish brown:

$$2NOBr(g) \longrightarrow 2NO(g) + Br_2(g)$$

Use the data in the next column to answer the following:
(a) Determine the average rate over the entire experiment.
(b) Determine the average rate between 2.00 and 4.00 s.
(c) Use graphical methods to estimate the initial reaction rate.
(d) Use graphical methods to estimate the rate at 7.00 s.
(e) At what time does the instantaneous rate equal the average rate over the entire experiment?

Time (s)	[NOBr] (mol/L)
0.00	0.0100
2.00	0.0071
4.00	0.0055
6.00	0.0045
8.00	0.0038
10.00	0.0033

16.21 The formation of ammonia is one of the most important processes in the chemical industry:

$$N_2(g) + 3H_2(g) \longrightarrow 2NH_3(g)$$

Express the rate in terms of changes in $[N_2]$, $[H_2]$, and $[NH_3]$.

16.22 Just as the depletion of stratospheric ozone threatens life on Earth today, its accumulation was one of the crucial processes that allowed life to develop in prehistoric times:

$$3O_2(g) \longrightarrow 2O_3(g)$$

(a) Express the reaction rate in terms of $[O_2]$ and $[O_3]$.
(b) At a given instant, the reaction rate in terms of $[O_2]$ is 2.17×10^{-5} mol/L·s. What is it in terms of $[O_3]$?

The Rate Law and Its Components

(Sample Problems 16.2 and 16.3)

Concept Review Questions

16.23 The rate law for the general reaction

$$aA + bB + \cdots \longrightarrow cC + dD + \cdots$$

is rate $= k[A]^m[B]^n \cdots$.
(a) Explain the meaning of k.
(b) Explain the meanings of m and n. Does $m = a$ and $n = b$? Explain.
(c) If the reaction is first order in A and second order in B, and time is measured in minutes (min), what are the units for k?

16.24 You are studying the reaction

$$A_2(g) + B_2(g) \longrightarrow 2AB(g)$$

to determine its rate law. Assuming that you have a valid experimental procedure for obtaining $[A_2]$ and $[B_2]$ at various times, explain how you determine (a) the initial rate, (b) the reaction orders, and (c) the rate constant.

16.25 By what factor does the rate change in each of the following cases (assuming constant temperature)?
(a) A reaction is first order in reactant A, and [A] is doubled.
(b) A reaction is second order in reactant B, and [B] is halved.
(c) A reaction is second order in reactant C, and [C] is tripled.

Skill-Building Exercises *(grouped in similar pairs)*

16.26 Give the individual reaction orders for all substances and the overall reaction order from the following rate law:

$$\text{Rate} = k[BrO_3^-][Br^-][H^+]^2$$

16.27 Give the individual reaction orders for all substances and the overall reaction order from the following rate law:

$$\text{Rate} = k\frac{[O_3]^2}{[O_2]}$$

16.28 By what factor does the rate in Problem 16.26 change if each of the following changes occurs: (a) $[BrO_3^-]$ is doubled; (b) $[Br^-]$ is halved; (c) $[H^+]$ is quadrupled?

16.29 By what factor does the rate in Problem 16.27 change if each of the following changes occurs: (a) $[O_3]$ is doubled; (b) $[O_2]$ is doubled; (c) $[O_2]$ is halved?

16.30 Give the individual reaction orders for all substances and the overall reaction order from this rate law:

$$\text{Rate} = k[NO_2]^2[Cl_2]$$

16.31 Give the individual reaction orders for all substances and the overall reaction order from this rate law:

$$\text{Rate} = k\frac{[HNO_2]^4}{[NO]^2}$$

16.32 By what factor does the rate in Problem 16.30 change if each of the following changes occurs: (a) $[NO_2]$ is tripled; (b) $[NO_2]$ and $[Cl_2]$ are doubled; (c) $[Cl_2]$ is halved?

16.33 By what factor does the rate in Problem 16.31 change if each of the following changes occurs: (a) $[HNO_2]$ is doubled; (b) [NO] is doubled; (c) $[HNO_2]$ is halved?

16.34 For the reaction

$$4A(g) + 3B(g) \longrightarrow 2C(g)$$

the following data were obtained at constant temperature:

Experiment	Initial [A] (mol/L)	Initial [B] (mol/L)	Initial Rate (mol/L·min)
1	0.100	0.100	5.00
2	0.300	0.100	45.0
3	0.100	0.200	10.0
4	0.300	0.200	90.0

(a) What is the order with respect to each reactant? (b) Write the rate law. (c) Calculate k (using the data from experiment 1).

16.35 For the reaction

$$A(g) + B(g) + C(g) \longrightarrow D(g)$$

the following data were obtained at constant temperature:

Exp't	Initial [A] (mol/L)	Initial [B] (mol/L)	Initial [C] (mol/L)	Initial Rate (mol/L·s)
1	0.0500	0.0500	0.0100	6.25×10^{-3}
2	0.1000	0.0500	0.0100	1.25×10^{-2}
3	0.1000	0.1000	0.0100	5.00×10^{-2}
4	0.0500	0.0500	0.0200	6.25×10^{-3}

(a) What is the order with respect to each reactant? (b) Write the rate law. (c) Calculate k (using the data from experiment 1).

16.36 Without consulting Table 16.3, give the units of the rate constants for reactions with the following overall orders: (a) first order; (b) second order; (c) third order; (d) $\frac{5}{2}$ order.

16.37 Give the overall reaction order that corresponds to a rate constant with each of the following units: (a) mol/L·s; (b) yr^{-1}; (c) $(mol/L)^{1/2}\cdot s^{-1}$; (d) $(mol/L)^{-5/2}\cdot min^{-1}$.

Problems in Context

16.38 Phosgene is a toxic gas prepared by the reaction of carbon monoxide with chlorine:

$$CO(g) + Cl_2(g) \longrightarrow COCl_2(g)$$

These data were obtained in a kinetics study of its formation:

Experiment	Initial [CO] (mol/L)	Initial $[Cl_2]$ (mol/L)	Initial Rate (mol/L·s)
1	1.00	0.100	1.29×10^{-29}
2	0.100	0.100	1.33×10^{-30}
3	0.100	1.00	1.30×10^{-29}
4	0.100	0.0100	1.32×10^{-31}

(a) Write the rate law for the formation of phosgene.
(b) Calculate the average value of the rate constant.

Integrated Rate Laws: Concentration Changes over Time

(Sample Problems 16.4 and 16.5)

Concept Review Questions

16.39 How are integrated rate laws used to determine reaction order? What is the order in reactant if a plot of
(a) The natural logarithm of [reactant] vs. time is linear?
(b) The inverse of [reactant] vs. time is linear?
(c) [Reactant] vs. time is linear?

16.40 Define the *half-life* of a reaction. Explain on the molecular level why the half-life of a first-order reaction is constant.

Skill-Building Exercises *(grouped in similar pairs)*

16.41 For the simple decomposition reaction

$$AB(g) \longrightarrow A(g) + B(g)$$

rate $= k[AB]^2$ and $k = 0.2$ L/mol·s. How long will it take for [AB] to reach $\frac{1}{3}$ of its initial concentration of 1.50 M?

16.42 For the reaction in Problem 16.41, what is [AB] after 10.0 s?

16.43 In a first-order decomposition reaction, 50.0% of a compound decomposes in 10.5 min. (a) What is the rate constant of the reaction? (b) How long does it take for 75.0% of the compound to decompose?

16.44 A decomposition reaction has a rate constant of 0.0012 yr^{-1}. (a) What is the half-life of the reaction? (b) How long does it take for [reactant] to reach 12.5% of its original value?

Problems in Context

16.45 In a study of ammonia production, an industrial chemist discovers that the compound decomposes to its elements N_2 and H_2 in a first-order process. She collects the following data:

Time (s)	0	1.000	2.000
$[NH_3]$ (mol/L)	4.000	3.986	3.974

(a) Use graphical methods to determine the rate constant.
(b) What is the half-life for ammonia decomposition?

The Effect of Temperature on Reaction Rate

(Sample Problem 16.6)

Concept Review Questions

16.46 Use the exponential term in the Arrhenius equation to explain how temperature affects reaction rate.

16.47 How is the activation energy determined from the Arrhenius equation?

16.48 (a) Graph the relationship between k (y axis) and T (x axis). (b) Graph the relationship between ln k (y axis) and $1/T$ (x axis). How is the activation energy determined from this graph?

Skill-Building Exercises *(grouped in similar pairs)*

16.49 The rate constant of a reaction is 4.7×10^{-3} s^{-1} at 25°C, and the activation energy is 33.6 kJ/mol. What is k at 75°C?

16.50 The rate constant of a reaction is 4.50×10^{-5} L/mol·s at 195°C and 3.20×10^{-3} L/mol·s at 258°C. What is the activation energy of the reaction?

Problems in Context

16.51 Understanding the high-temperature formation and breakdown of the nitrogen oxides is essential for controlling the pollutants generated from power plants and cars. The first-order breakdown of dinitrogen monoxide to its elements has rate constants of 0.76/s at 727°C and 0.87/s at 757°C. What is the activation energy of this reaction?

Explaining the Effects of Concentration and Temperature
(Sample Problem 16.7)

Concept Review Questions

16.52 What is the central idea of collision theory? How does this idea explain the effect of concentration on reaction rate?

16.53 Is collision frequency the only factor affecting rate? Explain.

16.54 Arrhenius proposed that each reaction has an energy threshold that must be reached for the particles to react. The kinetic theory of gases proposes that the average kinetic energy of the particles is proportional to the absolute temperature. How do these concepts relate to the effect of temperature on rate?

16.55 (a) For a reaction with a given E_a, how does an increase in T affect the rate? (b) For a reaction at a given T, how does a decrease in E_a affect the rate?

16.56 In the reaction $AB + CD \rightleftharpoons EF$, 4×10^{-5} mol of AB molecules collide with 4×10^{-5} mol of CD molecules. Will 4×10^{-5} mol of EF form? Explain.

16.57 Assuming the activation energies are equal, which of the following reactions will occur at a higher rate at 50°C? Explain:

$$NH_3(g) + HCl(g) \longrightarrow NH_4Cl(s)$$
$$N(CH_3)_3(g) + HCl(g) \longrightarrow (CH_3)_3NHCl(s)$$

Skill-Building Exercises *(grouped in similar pairs)*

16.58 For the reaction $A(g) + B(g) \longrightarrow AB(g)$, how many unique collisions between A and B are possible if there are four particles of A and three particles of B present in the vessel?

16.59 For the reaction $A(g) + B(g) \longrightarrow AB(g)$, how many unique collisions between A and B are possible if 1.01 mol of $A(g)$ and 2.12 mol of $B(g)$ are present in the vessel?

16.60 At 25°C, what is the fraction of collisions with energy equal to or greater than an activation energy of 100. kJ/mol?

16.61 If the temperature in Problem 16.60 is increased to 50.°C, by what factor does the fraction of collisions with energy equal to or greater than the activation energy change?

16.62 For the reaction $ABC + D \rightleftharpoons AB + CD$, $\Delta H^0_{rxn} = -55$ kJ/mol and $E_{a(fwd)} = 215$ kJ/mol. Assuming a one-step reaction, (a) draw a reaction energy diagram; (b) calculate $E_{a(rev)}$; and (c) sketch a possible transition state if ABC is V-shaped.

16.63 For the reaction $A_2 + B_2 \longrightarrow 2AB$, $E_{a(fwd)} = 125$ kJ/mol and $E_{a(rev)} = 85$ kJ/mol. Assuming the reaction occurs in one step, (a) draw a reaction energy diagram; (b) calculate ΔH^0_{rxn}; and (c) sketch a possible transition state.

Problems in Context

16.64 Aqua regia, a mixture of HCl and HNO_3, has been used since alchemical times to dissolve many metals, including gold. Its orange color is due to the presence of nitrosyl chloride. Consider this one-step reaction for the formation of this compound:

$$NO(g) + Cl_2(g) \longrightarrow NOCl(g) + Cl(g) \qquad \Delta H^0 = 83 \text{ kJ}$$

(a) Draw a reaction energy diagram, given $E_{a(fwd)}$ is 86 kJ/mol.
(b) Calculate $E_{a(rev)}$.
(c) Sketch a possible transition state for the reaction. (*Note:* The atom sequence of nitrosyl chloride is Cl—N—O.)

Reaction Mechanisms: Steps in the Overall Reaction
(Sample Problem 16.8)

Concept Review Questions

16.65 Is the rate of an overall reaction lower, higher, or equal to the average rate of the individual steps? Explain.

16.66 Explain why the coefficients of an elementary step equal the reaction orders of its rate law but those of an overall reaction do not.

16.67 Is it possible for more than one mechanism to be consistent with the rate law of a given reaction? Explain.

16.68 What is the difference between a reaction intermediate and a transition state?

16.69 Why is a bimolecular step more reasonable physically than a termolecular step?

16.70 If a slow step precedes a fast step in a two-step mechanism, do the substances in the fast step appear in the rate law? Explain.

16.71 If a fast step precedes a slow step in a two-step mechanism, how is the fast step affected? How is this effect used to determine the validity of the mechanism?

Skill-Building Exercises *(grouped in similar pairs)*

16.72 The proposed mechanism for a reaction is

(1) $A(g) + B(g) \rightleftharpoons X(g)$ [fast]
(2) $X(g) + C(g) \longrightarrow Y(g)$ [slow]
(3) $Y(g) \longrightarrow D(g)$ [fast]

(a) What is the overall equation?
(b) Identify the intermediate(s), if any.
(c) What are the molecularity and the rate law for each step?
(d) Is the mechanism consistent with the actual rate law: rate = $k[A][B][C]$?
(e) Is the following one-step mechanism equally valid: $A(g) + B(g) + C(g) \longrightarrow D(g)$?

16.73 Consider the following mechanism:

(1) $ClO^-(aq) + H_2O(l) \rightleftharpoons HClO(aq) + OH^-(aq)$ [fast]
(2) $I^-(aq) + HClO(aq) \longrightarrow HIO(aq) + Cl^-(aq)$ [slow]
(3) $OH^-(aq) + HIO(aq) \longrightarrow H_2O(l) + IO^-(aq)$ [fast]

(a) What is the overall equation?
(b) Identify the intermediate(s), if any.
(c) What are the molecularity and the rate law for each step?
(d) Is the mechanism consistent with the actual rate law: rate = $k[ClO^-][I^-]$?

Problems in Context

16.74 In a study of nitrosyl halides, a chemist proposes the following mechanism for the synthesis of nitrosyl bromide:

$$NO(g) + Br_2(g) \rightleftharpoons NOBr_2(g) \quad \text{[fast]}$$
$$NOBr_2(g) + NO(g) \longrightarrow 2NOBr(g) \quad \text{[slow]}$$

If the rate law is rate = $k[NO]^2[Br_2]$, is the proposed mechanism valid? If so, show that it satisfies the three criteria for validity.

16.75 The rate law for $2NO(g) + O_2(g) \longrightarrow 2NO_2(g)$ is rate = $k[NO]^2[O_2]$. In addition to the mechanism in the text, the following ones have been proposed:

I $2NO(g) + O_2(g) \longrightarrow 2NO_2(g)$
II $2NO(g) \rightleftharpoons N_2O_2(g)$ [fast]
$N_2O_2(g) + O_2(g) \longrightarrow 2NO_2(g)$ [slow]
III $2NO(g) \rightleftharpoons N_2(g) + O_2(g)$ [fast]
$N_2(g) + 2O_2(g) \longrightarrow 2NO_2(g)$ [slow]

(a) Which of these mechanisms is consistent with the rate law?
(b) Which is most reasonable chemically? Why?

Catalysis: Speeding Up a Chemical Reaction

Concept Review Questions

16.76 Consider the reaction $N_2O(g) \xrightarrow{Au} N_2(g) + \frac{1}{2}O_2(g)$.
(a) Is the gold a homogeneous or a heterogeneous catalyst?
(b) On the same set of axes, sketch the reaction energy diagrams for the catalyzed and the uncatalyzed reactions.

16.77 Does a catalyst increase reaction rate by the same means as a rise in temperature does? Explain.

16.78 In a classroom demonstration, hydrogen gas and oxygen gas are mixed in a balloon. The mixture is stable under normal conditions, but if a spark is applied to it or some powdered metal is added, the mixture explodes. (a) Is the spark acting as a catalyst? Explain. (b) Is the metal acting as a catalyst? Explain.

16.79 A principle of green chemistry is that the energy requirements of industrial processes should have minimal environmental impact. How can the use of catalysts lead to "greener" technologies?

Comprehensive Problems

16.80 Experiments show that each of the following redox reactions is second order overall:

Reaction 1: $NO_2(g) + CO(g) \longrightarrow NO(g) + CO_2(g)$

Reaction 2: $NO(g) + O_3(g) \longrightarrow NO_2(g) + O_2(g)$

(a) When $[NO_2]$ in reaction 1 is doubled, the rate quadruples. Write the rate law for this reaction.

(b) When [NO] in reaction 2 is doubled, the rate doubles. Write the rate law for this reaction.

(c) In each reaction, the initial concentrations of the reactants are equal. For each reaction, what is the ratio of the initial rate to the rate when the reaction is 50% complete?

(d) In reaction 1, the initial $[NO_2]$ is twice the initial [CO]. What is the ratio of the initial rate to the rate at 50% completion?

(e) In reaction 2, the initial [NO] is twice the initial $[O_3]$. What is the ratio of the initial rate to the rate at 50% completion?

16.81 Consider the following reaction energy diagram:

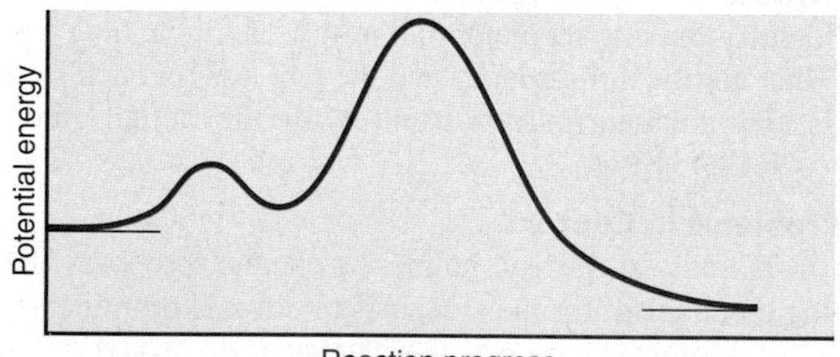

(a) How many elementary steps are in the reaction mechanism?

(b) Which step is rate limiting?

(c) Is the overall reaction exothermic or endothermic?

16.82 Reactions between certain organic (alkyl) halides and water produce alcohols. Consider the overall reaction for *t*-butyl bromide (2-bromo-2-methylpropane):

$$(CH_3)_3CBr(aq) + H_2O(l) \longrightarrow (CH_3)_3COH(aq) + H^+(aq) + Br^-(aq)$$

The experimental rate law is rate $= k[(CH_3)_3CBr]$. The accepted mechanism for the reaction is

(1) $(CH_3)_3C{-}Br(aq) \longrightarrow (CH_3)_3C^+(aq) + Br^-(aq)$ [slow]

(2) $(CH_3)_3C^+(aq) + H_2O(l) \longrightarrow (CH_3)_3C{-}OH_2{}^+(aq)$ [fast]

(3) $(CH_3)_3C{-}OH_2{}^+(aq) \longrightarrow H^+(aq) + (CH_3)_3C{-}OH(aq)$ [fast]

(a) Why doesn't H_2O appear in the rate law?

(b) Write rate laws for the elementary steps.

(c) What reaction intermediates appear in the mechanism?

(d) Show that the mechanism is consistent with the experimental rate law.

16.83 The catalytic destruction of ozone occurs via a two-step mechanism, where X can be any of several species:

(1) $X + O_3 \longrightarrow XO + O_2$ [slow]

(2) $XO + O \longrightarrow X + O_2$ [fast]

(a) Write the overall reaction.

(b) Write the rate law for each step.

(c) X acts as ______, and XO acts as ______.

(d) High-flying aircraft release NO into the stratosphere, which catalyzes this process. When O_3 and NO concentrations are 5×10^{12} molecule/cm^3 and 1.0×10^9 molecule/cm^3, respectively, what is the rate of O_3 depletion (*k* for the rate-determining step is 6×10^{-15} cm^3/molecule·s)?

(e) Is the O_3 concentration in part (d) reasonable for this reaction, given that stratospheric O_3 never exceeds 10 mg/L?

16.84 Archeologists can determine the age of artifacts made of wood or bone by measuring the amount of the radioactive isotope ^{14}C present in the object. The amount of isotope decreases in a first-order process. If 15.5% of the original amount of ^{14}C is present in a wooden tool at the time of analysis, what is the age of the tool? The half-life of ^{14}C is 5730 yr.

16.85 A slightly bruised apple will rot extensively in about 4 days at room temperature (20°C). If it is kept in the refrigerator at 0°C, the same extent of rotting takes about 16 days. What is the activation energy for the rotting reaction?

16.86 Benzoyl peroxide, the substance most widely used against acne, has a half-life of 9.8×10^3 days when refrigerated. How long will it take to lose 5% of its potency (95% remaining)?

16.87 The rate law for the reaction

$$NO_2(g) + CO(g) \longrightarrow NO(g) + CO_2(g)$$

is rate $= k[NO_2]^2$; one possible mechanism is shown on p. 703.

(a) Draw a reaction energy diagram for that mechanism, given that $\Delta H^0_{overall} = -226$ kJ/mol.

(b) The following alternative mechanism has been proposed:

(1) $2NO_2(g) \longrightarrow N_2(g) + 2O_2(g)$ [slow]

(2) $2CO(g) + O_2(g) \longrightarrow 2CO_2(g)$ [fast]

(3) $N_2(g) + O_2(g) \longrightarrow 2NO(g)$ [fast]

Is the alternative mechanism consistent with the rate law? Is one mechanism more reasonable physically? Explain.

16.88 Consider the following general reaction and data:

$$2A + 2B + C \longrightarrow D + 3E$$

Exp't	Initial [A] (mol/L)	Initial [B] (mol/L)	Initial [C] (mol/L)	Initial Rate (mol/L·s)
1	0.024	0.085	0.032	6.0×10^{-6}
2	0.096	0.085	0.032	9.6×10^{-5}
3	0.024	0.034	0.080	1.5×10^{-5}
4	0.012	0.170	0.032	1.5×10^{-6}

(a) What is the reaction order with respect to each reactant?

(b) Calculate the rate constant.

(c) Write the rate law for this reaction.

(d) Express the rate in terms of changes in concentration with time for each of the components.

16.89 In acidic solution, the breakdown of sucrose into glucose and fructose has this rate law: rate $= k[H^+][\text{sucrose}]$. The initial rate of sucrose breakdown is measured in a solution that is 0.01 *M* H^+, 1.0 *M* sucrose, 0.1 *M* fructose, and 0.1 *M* glucose. How does the rate change if

(a) [Sucrose] is changed to 2.5 *M*?
(b) [Sucrose], [fructose], and [glucose] are all changed to 0.5 *M*?
(c) $[H^+]$ is changed to 0.0001 *M*?
(d) [Sucrose] and $[H^+]$ are both changed to 0.1 *M*?

16.90 Enzymes are remarkably efficient catalysts that can increase reaction rates by as many as 20 orders of magnitude.
(a) How does an enzyme affect the transition state of a reaction, and how does this effect increase the reaction rate?
(b) What characteristics of enzymes give them this tremendous effectiveness as catalysts?

16.91 Biacetyl, the flavoring that makes margarine taste "just like butter," is extremely stable at room temperature, but at 200°C it undergoes a first-order breakdown with a half-life of 9.0 min. An industrial flavor-enhancing process requires that a biacetyl-flavored food be heated briefly at 200°C. How long can the food be heated and retain 85% of its buttery flavor?

16.92 Biochemists consider the citric acid cycle to be the central reaction sequence in metabolism. One of the key steps is an oxidation catalyzed by the enzyme isocitrate dehydrogenase and the oxidizing agent NAD^+. Under certain conditions, the reaction in yeast obeys 11th-order kinetics:

$$\text{Rate} = k[\text{enzyme}][\text{isocitrate}]^4[\text{AMP}]^2[\text{NAD}^+]^m[\text{Mg}^{2+}]^2$$

What is the order with respect to NAD^+?

16.93 At body temperature (37°C), *k* of an enzyme-catalyzed reaction is 2.3×10^{14} times greater than *k* of the uncatalyzed reaction. Assuming that the frequency factor *A* is the same for both reactions, by how much does the enzyme lower the E_a?

16.94 Enzymes in human liver catalyze a large number of reactions that degrade ingested toxic chemicals. By what factor is the rate of a detoxification reaction changed if a liver enzyme lowers the activation energy by 5 kJ/mol at 37°C?

16.95 Experiment shows that the rate of formation of carbon tetrachloride from chloroform,

$$CHCl_3(g) + Cl_2(g) \longrightarrow CCl_4(g) + HCl(g)$$

is first order in $CHCl_3$, $\frac{1}{2}$ order in Cl_2, and $\frac{3}{2}$ order overall. Show that the following mechanism is consistent with the overall rate law:

(1) $Cl_2(g) \rightleftharpoons 2Cl(g)$ [fast]
(2) $Cl(g) + CHCl_3(g) \longrightarrow HCl(g) + CCl_3(g)$ [slow]
(3) $CCl_3(g) + Cl(g) \longrightarrow CCl_4(g)$ [fast]

16.96 A biochemist studying breakdown of the insecticide DDT finds that it decomposes by a first-order reaction with a half-life of 12 yr. How long does it take DDT in a soil sample to decompose from 275 ppbm to 10. ppbm (parts per billion by mass)?

16.97 Proteins in the body undergo continual breakdown and synthesis. Insulin is a polypeptide hormone that stimulates fat and muscle to take up glucose. Once released from the pancreas, it has a first-order half-life in the blood of 8.0 min. To maintain an adequate blood concentration of insulin, it must be replenished in a time interval equal to $1/k$. How long is this interval?

16.98 For the reaction $A(g) + B(g) \longrightarrow AB(g)$, the rate is 0.20 mol/L·s, when $[A]_0 = [B]_0 = 1.0$ mol/L. If the reaction is first order in B and second order in A, what is the rate when $[A]_0 = 2.0$ mol/L and $[B]_0 = 3.0$ mol/L?

16.99 The hydrolysis of sucrose occurs by this overall reaction:

$$\underset{\text{sucrose}}{C_{12}H_{22}O_{11}(s)} + H_2O(l) \longrightarrow \underset{\text{glucose}}{C_6H_{12}O_6(aq)} + \underset{\text{fructose}}{C_6H_{12}O_6(aq)}$$

A nutritional biochemist obtains the following kinetic data:

[Sucrose] (mol/L)	Time (h)
0.501	0.00
0.451	0.50
0.404	1.00
0.363	1.50
0.267	3.00

(a) Determine the rate constant and the half-life of the reaction.
(b) How long does it take to hydrolyze 75% of the sucrose?
(c) Other studies have shown that this reaction is actually second order overall but appears to follow first-order kinetics. (Such a reaction is called a *pseudo–first-order reaction.*) Suggest a reason for this apparent first-order behavior.

16.100 Is each of these statements true? If not, explain why.
(a) At a given *T*, all molecules have the same kinetic energy.
(b) Halving the *P* of a gaseous reaction doubles the rate.
(c) A higher activation energy gives a lower reaction rate.
(d) A temperature rise of 10°C doubles the rate of any reaction.
(e) If reactant molecules collide with greater energy than the activation energy, they change into product molecules.
(f) The activation energy of a reaction depends on temperature.
(g) The rate of a reaction increases as the reaction proceeds.
(h) Activation energy depends on collision frequency.
(i) A catalyst increases the rate by increasing collision frequency.
(j) Exothermic reactions are faster than endothermic reactions.
(k) Temperature has no effect on the frequency factor (*A*).
(l) The activation energy of a reaction is lowered by a catalyst.
(m) For most reactions, ΔH_{rxn} is lowered by a catalyst.
(n) The orientation probability factor (*p*) is near unity for reactions between single atoms.
(o) The initial rate of a reaction is its maximum rate.
(p) A bimolecular reaction is generally twice as fast as a unimolecular reaction.
(q) The molecularity of an elementary reaction is proportional to the molecular complexity of the reactant(s).

16.101 For the decomposition of gaseous dinitrogen pentaoxide, $2N_2O_5(g) \longrightarrow 4NO_2(g) + O_2(g)$, the rate constant is $k = 2.8\times10^{-3}\ s^{-1}$ at 60°C. The initial concentration of N_2O_5 is 1.58 mol/L. (a) What is $[N_2O_5]$ after 5.00 min? (b) What fraction of the N_2O_5 has decomposed after 5.00 min?

16.102 Even when a mechanism is consistent with the rate law, later experimentation may show it to be incorrect or only one of several alternatives. As an example, the reaction between hydrogen and iodine has the following rate law: rate $= k[H_2][I_2]$. The long-accepted mechanism proposed a single bimolecular step; that is, the overall reaction was thought to be elementary:

$$H_2(g) + I_2(g) \longrightarrow 2HI(g)$$

In the 1960s, however, spectroscopic evidence showed the presence of free I atoms during the reaction. Kineticists have since proposed a three-step mechanism:

(1) $I_2(g) \rightleftharpoons 2I(g)$ [fast]
(2) $H_2(g) + I(g) \rightleftharpoons H_2I(g)$ [fast]
(3) $H_2I(g) + I(g) \longrightarrow 2HI(g)$ [slow]

Show that this mechanism is consistent with the rate law.

16.103 Suggest an experimental method for measuring the change in concentration with time for each of the following reactions:

(a) $CH_3CH_2Br(l) + H_2O(l) \longrightarrow CH_3CH_2OH(l) + HBr(aq)$
(b) $2NO(g) + Cl_2(g) \longrightarrow 2NOCl(g)$

16.104 An atmospheric chemist fills a container with gaseous N_2O_5 to a pressure of 125 kPa, and the gas decomposes to NO_2 and O_2. What is the partial pressure of NO_2, P_{NO_2} (in kPa), when the total pressure is 178 kPa?

16.105 Many drugs decompose in blood by a first-order process.
(a) Two tablets of aspirin supply 0.60 g of the active compound. After 30 min, this compound reaches a maximum concentration of 2 mg/100 mL of blood. If the half-life for its breakdown is 90 min, what is its concentration (in mg/100 mL) 2.5 h after it reaches its maximum concentration?
(b) In 8.0 h, secobarbital sodium, a common sedative, reaches a blood level that is 18% of its maximum. What is $t_{1/2}$ of the decomposition of secobarbital sodium in blood?
(c) The blood level of the sedative phenobarbital sodium drops to 59% of its maximum after 20. h. What is $t_{1/2}$ for its breakdown in blood?
(d) For the decomposition of an antibiotic in a person with a normal temperature (98.6°F), $k = 3.1\times10^{-5}\ s^{-1}$; for a person with a fever at 101.9°F, $k = 3.9\times10^{-5}\ s^{-1}$. If the sick person must take another pill when $\frac{2}{3}$ of the first pill has decomposed, how many hours should she wait to take a second pill? A third pill?
(e) Calculate E_a for decomposition of the antibiotic in part (d).

16.106 Iodide ion reacts with chloroform to displace chloride ion in a common organic substitution reaction:

$$I^- + CH_3Cl \longrightarrow CH_3I + Cl^-$$

(a) Draw a wedge-bond structural formula of chloroform and indicate the most effective direction of I^- attack.
(b) The analogous reaction with 2-chlorobutane [Figure P16.106(b)] results in a major change in specific rotation as measured by polarimetry. Explain, showing a wedge-bond structural formula of the product.
(c) Under different conditions, 2-chlorobutane loses Cl^- in a rate-determining step to form a planar intermediate [Figure P16.106(c)]. This cationic species reacts with HI and then loses H^+ to form a product that exhibits no optical activity. Explain, showing a wedge-bond structural formula.

Figure P16.106(b)

Figure P16.106(c)

16.107 In Houston (near sea level), water boils at 100.0°C. In Cripple Creek, Colorado (near 9500 ft), it boils at 90.0°C. If it takes 4.8 min to cook an egg in Cripple Creek and 4.5 min in Houston, what is E_a for this process?

16.108 Sulfonation of benzene has the following mechanism:
(1) $2H_2SO_4 \longrightarrow H_3O^+ + HSO_4^- + SO_3$ [fast]
(2) $SO_3 + C_6H_6 \longrightarrow H(C_6H_5^+)SO_3^-$ [slow]
(3) $H(C_6H_5^+)SO_3^- + HSO_4^- \longrightarrow C_6H_5SO_3^- + H_2SO_4$ [fast]
(4) $C_6H_5SO_3^- + H_3O^+ \longrightarrow C_6H_5SO_3H + H_2O$ [fast]
(a) Write an overall equation for the reaction.
(b) Write the overall rate law for the initial rate of the reaction.

16.109 Acetone is one of the most important solvents in organic chemistry, used to dissolve everything from fats and waxes to airplane glue and nail polish. At high temperatures, it decomposes in a first-order process to methane and ketene ($CH_2{=}C{=}O$). At 600°C, the rate constant is $8.7\times10^{-3}\ s^{-1}$.
(a) What is the half-life of the reaction?
(b) How much time is required for 40.% of a sample of acetone to decompose?
(c) How much time is required for 90.% of a sample of acetone to decompose?

16.110 In the lower troposphere, ozone is one of the components of photochemical smog. It is generated in air when nitrogen dioxide, formed by the oxidation of nitrogen monoxide from car exhaust, reacts by the following mechanism:
(1) $NO_2(g) \xrightarrow{k_1} NO(g) + O(g)$
(2) $O(g) + O_2(g) \xrightarrow{k_2} O_3(g)$
Assuming the rate of formation of atomic oxygen in step 1 equals the rate of its consumption in step 2, use the data below to calculate (a) the concentration of atomic oxygen [O]; (b) the rate of ozone formation.

$k_1 = 6.0\times10^{-3}\ s^{-1}$ $\quad [NO_2] = 4.0\times10^{-9}\ M$
$k_2 = 1.0\times10^{6}$ L/mol·s $\quad [O_2] = 1.0\times10^{-2}\ M$

16.111 Chlorine is commonly used to disinfect drinking water, and inactivation of pathogens by chlorine follows first-order kinetics. The following data show *E. coli* inactivation:

Contact time (min)	Percent (%) inactivation
0.00	0.0
0.50	68.3
1.00	90.0
1.50	96.8
2.00	99.0
2.50	99.7
3.00	99.9

(a) Determine the first-order inactivation constant, k. [*Hint:* % inactivation $= 100 \times (1 - [A]_t/[A]_0)$.]
(b) How much contact time is required for 95% inactivation?

16.112 The reaction and rate law for the gas-phase decomposition of dinitrogen pentaoxide are

$$2N_2O_5(g) \longrightarrow 4NO_2(g) + O_2(g) \qquad \text{rate} = k[N_2O_5]$$

Which of the following can be considered valid mechanisms for the reaction?

I One-step collision
II $2N_2O_5(g) \longrightarrow 2NO_3(g) + 2NO_2(g)$ [slow]
$2NO_3(g) \longrightarrow 2NO_2(g) + 2O(g)$ [fast]
$2O(g) \longrightarrow O_2(g)$ [fast]
III $N_2O_5(g) \rightleftharpoons NO_3(g) + NO_2(g)$ [fast]
$NO_2(g) + N_2O_5(g) \longrightarrow 3NO_2(g) + O(g)$ [slow]
$NO_3(g) + O(g) \longrightarrow NO_2(g) + O_2(g)$ [fast]
IV $2N_2O_5(g) \rightleftharpoons 2NO_2(g) + N_2O_3(g) + 3O(g)$ [fast]
$N_2O_3(g) + O(g) \longrightarrow 2NO_2(g)$ [slow]
$2O(g) \longrightarrow O_2(g)$ [fast]
V $2N_2O_5(g) \longrightarrow N_4O_{10}(g)$ [slow]
$N_4O_{10}(g) \longrightarrow 4NO_2(g) + O_2(g)$ [fast]

16.113 Nitrification is a biological process for removing NH_3 from wastewater as NH_4^+:

$$NH_4^+ + 2O_2 \longrightarrow NO_3^- + 2H^+ + H_2O$$

The first-order rate constant is given as

$$k_1 = 0.47e^{0.095(T - 15°C)}$$

where k_1 is in day^{-1} and T is in °C.
(a) If the initial concentration of NH_3 is 3.0 mol/m^3, how long will it take to reduce the concentration to 0.35 mol/m^3 in the spring ($T = 20°C$)? (b) In the winter ($T = 10°C$)? (c) Using your answer to part (a), what is the rate of O_2 consumption?

16.114 Carbon disulfide, a poisonous, flammable liquid, is an excellent solvent for phosphorus, sulfur, and some other nonmetals. A kinetic study of its gaseous decomposition reveals these data:

Experiment	Initial [CS_2] (mol/L)	Initial Rate (mol/L·s)
1	0.100	2.7×10^{-7}
2	0.080	2.2×10^{-7}
3	0.055	1.5×10^{-7}
4	0.044	1.2×10^{-7}

(a) Write the rate law for the decomposition of CS_2.
(b) Calculate the average value of the rate constant.

16.115 Water exchanges between the human body and surroundings in a first-order process with an average $t_{1/2}$ of 11 days. (a) What is the rate constant in day^{-1}? (b) A person receives ^3H_2O, and the radioactive ^3H distributes uniformly throughout the body very rapidly. How long does it take for the radioactivity to be reduced by 90% ($t_{1/2}$ of ^3H is 12.3 yr)?

16.116 In a *clock reaction,* a dramatic color change occurs at a time determined by concentration and temperature. One of the most famous is the iodine clock reaction. The overall equation is

$$2I^-(aq) + S_2O_8^{2-}(aq) \longrightarrow I_2(aq) + 2SO_4^{2-}(aq)$$

As I_2 forms, it is immediately consumed by its reaction with a fixed amount of added $S_2O_3^{2-}$:

$$I_2(aq) + 2S_2O_3^{2-}(aq) \longrightarrow 2I^-(aq) + S_4O_6^{2-}(aq)$$

Once the $S_2O_3^{2-}$ is consumed, the excess I_2 forms a blue-black product with starch present in solution:

$$I_2 + \text{starch} \longrightarrow \text{starch}\cdot I_2 \text{ (blue-black)}$$

The rate of the reaction is also influenced by the total concentration of ions, so KCl and $(NH_4)_2SO_4$ are added to maintain a constant value. Use the data below to determine the following:
(a) The average rate for each trial
(b) The order with respect to each reactant
(c) The rate constant at 23°C
(d) The rate law for the overall reaction

	Exp't 1	Exp't 2	Exp't 3
0.200 *M* KI (mL)	10.0	20.0	20.0
0.100 *M* $Na_2S_2O_8$ (mL)	20.0	20.0	10.0
0.0050 *M* $Na_2S_2O_3$ (mL)	10.0	10.0	10.0
0.200 *M* KCl (mL)	10.0	0.0	0.0
0.100 *M* $(NH_4)_2SO_4$ (mL)	0.0	0.0	10.0
Time to color (s)	29.0	14.5	14.5

16.117 Two oxidations start with O_2 at high pressure and a large excess of the solid being oxidized. One reaction is first order in O_2, and the other is second order in O_2. Both rate constants have the same numerical value, 1.7×10^{-3}, and each reaction starts with $[O_2] = 2.045$ *M*.
(a) How much time does it take each reaction to consume $\frac{1}{2}$ of the O_2? Which takes longer?
(b) How much time does it take each reaction to consume $\frac{3}{4}$ of the O_2? Which takes longer?
(c) From graphs of concentration vs. time for each reaction, find the time when the O_2 concentrations are equal.

16.118 The mathematics of the first-order rate law can be applied to any situation in which a quantity decreases by a constant fraction per unit of time (or any other variable).
(a) As light moves through a solution, its intensity decreases per unit distance traveled in the solution. Show that

$$\ln\left(\frac{\text{intensity of light leaving the solution}}{\text{intensity of light entering the solution}}\right) = -\text{fraction of light removed per unit of length} \times \text{distance traveled in solution}$$

(b) The value of your savings declines under conditions of constant inflation. Show that

$$\ln(\text{value remaining}) = -\text{fraction lost per unit of time} \times \text{savings time interval}$$

16.119 The growth of *Pseudomonas* bacteria is modeled as a first-order process with $k = 0.035\ min^{-1}$ at 37°C. The initial *Pseudomonas* population density is 1.0×10^3 cells/L. (a) What is the population density after 2 h? (b) What is the time required for the population to go from 1.0×10^3 to 2.0×10^3 cells/ L?

16.120 Consider the following organic reaction, in which one halogen replaces another in an alkyl halide:

$$CH_3CH_2Br + KI \longrightarrow CH_3CH_2I + KBr$$

In acetone, this particular reaction goes to completion because KI is soluble in acetone but KBr is not. In the mechanism, I^- approaches the carbon *opposite* to the Br (see Figure 16.19, with I^- instead of OH^-). After Br^- has been replaced by I^- and precipitates as KBr, other I^- ions react with the ethyl iodide by the same mechanism.
(a) If we designate the carbon bonded to the halogen as C-1, what is the shape around C-1 and the hybridization of C-1 in ethyl iodide?
(b) In the transition state, one of the two lobes of the unhybridized 2*p* orbital of C-1 overlaps a *p* orbital of I, while the other lobe overlaps a *p* orbital of Br. What is the shape around C-1 and the hybridization of C-1 in the transition state?
(c) The deuterated reactant, CH_3CHDBr (where D is ^2H), has two optical isomers because C-1 is chiral. If the reaction is run with one of the isomers, the ethyl iodide is *not* optically active. Explain.

16.121 Another radioisotope of iodine, ^{131}I, is also used to study thyroid function (see Follow-up Problem 16.5). A patient is given a sample that is 1.7×10^{-4} *M* ^{131}I. If the half-life is 8.04 days, what fraction of the radioactivity remains after 30. days?

16.122 The effect of substrate concentration on the first-order microbial growth rate follows the Monod equation:

$$\mu = \frac{\mu_{max} S}{K_s + S}$$

where μ is the first-order growth rate (s^{-1}), μ_{max} is the maximum growth rate (s^{-1}), S is the substrate concentration (kg/m^3), and K_s is the half-saturation coefficient (kg/m^3). For $\mu_{max} = 1.5\times10^{-4}\ s^{-1}$ and $K_s = 0.03$ kg/m^3:
(a) Prepare a plot of μ vs. S for S between 0.0 and 1.0 kg/m^3.
(b) The initial population density is 5.0×10^3 cells/m^3. What is the density after 1.0 h, if the initial S is 0.30 kg/m^3?
(c) What is it if the initial S is 0.70 kg/m^3?

Escalator equilibrium. Like the continual up-and-down of people on an escalator, the forward and reverse steps of a chemical reaction can maintain constant concentrations by moving constant numbers of entities in both directions. In this chapter, you'll see how a system attains equilibrium, how we determine the extent of the process, and how these systems adapt to external forces.

17

Equilibrium: The Extent of Chemical Reactions

Our study of kinetics in the previous chapter addressed one of three central questions in reaction chemistry, that of reaction rate, which also concerns mechanism. Now we turn to the second question, previewed in Chapter 4: how much product forms under a given set of starting concentrations and conditions? The principles of kinetics and equilibrium apply to different aspects of a reaction:

- Kinetics applies to the *speed* of a reaction, the concentration of product that appears (or of reactant that disappears) per unit time.
- Equilibrium applies to the *extent* of a reaction, the concentration of product that has appeared after an unlimited time, or once no further change occurs.

Just as reactions vary greatly in their speed, they also vary in their extent. A fast reaction may go almost completely or barely at all toward products. Consider the dissociation of an acid in water. In 1 M HCl, virtually all the hydrogen chloride molecules are dissociated into ions. In contrast, in 1 M CH_3COOH, fewer than 1% of the acetic acid molecules are dissociated at any given time. Yet both reactions take less than a second to reach completion. Similarly, some slow reactions eventually yield a large amount of product, whereas others yield very little. After a few years at ordinary temperatures, a steel water-storage tank will rust, and it will do so completely given enough time; but no matter how long you wait, the water inside will not decompose to hydrogen and oxygen.

Knowing the extent of a given reaction is crucial. How much product—medicine, polymer, or fuel—can you obtain from a particular reaction mixture? How can you adjust conditions to obtain more? If a reaction is slow but has a good yield, will a catalyst speed it up enough to make it useful?

IN THIS CHAPTER . . . We consider equilibrium principles in systems of gases and pure liquids and solids; we'll discuss various solution equilibria in the next two chapters. Throughout these chapters, you'll learn how to solve the major kinds of equilibrium problems. We first examine the equilibrium state at the macroscopic and molecular levels. Then we focus on the relation between the reaction quotient, which changes as the reaction proceeds, and the equilibrium constant, which applies to the system at equilibrium. We express the equilibrium condition in terms of concentrations or pressures. Then we see how to determine whether a system is proceeding toward products or reactants as it approaches equilibrium. We examine how reaction conditions affect the equilibrium state and end with applications of equilibrium in metabolism and in industrial production.

Concepts & Skills to Review

before you study this chapter

- reversibility of reactions (Section 4.7)
- equilibrium vapor pressure (Section 12.2)
- equilibrium nature of a saturated solution (Section 13.4)
- dependence of rate on concentration (Sections 16.2 and 16.6)
- rate laws for elementary reactions (Section 16.7)
- function of a catalyst (Section 16.8)

17.1 THE EQUILIBRIUM STATE AND THE EQUILIBRIUM CONSTANT

Countless experiments with chemical systems have shown that, in a state of equilibrium, *the concentrations of reactants and products no longer change with time.* This apparent cessation of chemical activity occurs because *all reactions are reversible.* Let's examine a chemical system at the macroscopic and molecular levels to see how the equilibrium state arises. The system consists of two gases, colorless dinitrogen tetraoxide and brown nitrogen dioxide:

$$N_2O_4(g;\ \text{colorless}) \rightleftharpoons 2NO_2(g;\ \text{brown})$$

When we introduce some $N_2O_4(l)$ into a sealed flask kept at 200°C, a change occurs immediately. The liquid vaporizes (bp = 21°C) and the gas begins to turn pale brown. The color slowly darkens, but after a few moments, the color stops changing, as shown in Figure 17.1 on the next page.

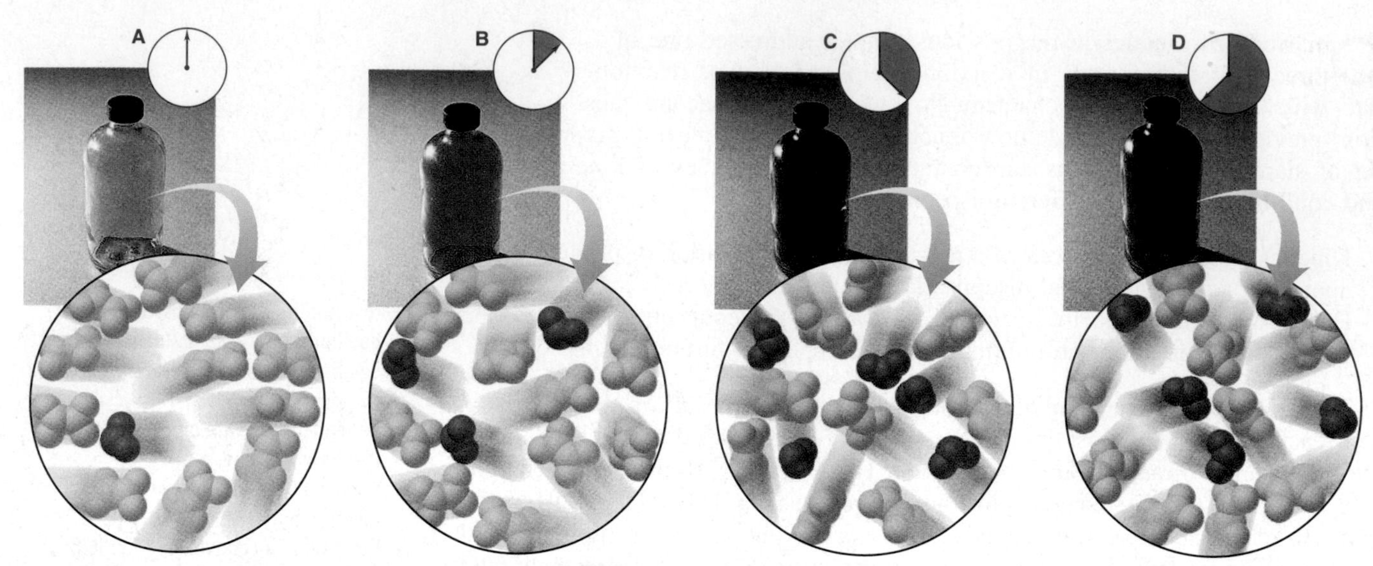

Figure 17.1 Reaching equilibrium on the macroscopic and molecular levels. A, When the experiment begins, the reaction mixture consists mostly of colorless N_2O_4. **B,** As N_2O_4 decomposes to reddish brown NO_2, the color of the mixture becomes pale brown. **C,** When equilibrium is reached, the concentrations of NO_2 and N_2O_4 are constant, and the color reaches its final intensity. **D,** Because the reaction continues in the forward and reverse directions at equal rates, the concentrations (and color) remain constant.

Now we repeat the process, and as we close in on the molecular level, a much more dynamic scene unfolds. The N_2O_4 molecules fly wildly throughout the flask, a few splitting into two NO_2 molecules. As time passes, more N_2O_4 molecules decompose and the concentration of NO_2 rises. As observers in the macroscopic world, we see the flask contents darken, because NO_2 is reddish brown. As the number of N_2O_4 molecules decreases, N_2O_4 decomposition slows. At the same time, increasing numbers of NO_2 molecules collide and combine, so re-formation of N_2O_4 speeds up. Eventually, N_2O_4 molecules decompose into NO_2 molecules as fast as the NO_2 molecules combine into N_2O_4. The system has reached equilibrium: *reactant and product concentrations stop changing because the forward and reverse rates have become equal:*

$$\text{At equilibrium:}\quad \text{rate}_{\text{fwd}} = \text{rate}_{\text{rev}} \tag{17.1}$$

Thus, a system at equilibrium continues to be dynamic at the molecular level, but we observe *no further* ***net*** *change because changes in one direction are balanced by changes in the other.*

At a particular temperature, when the system reaches equilibrium, product and reactant concentrations are constant. Therefore, their ratio must be a constant. We'll use the N_2O_4-NO_2 system to derive this constant. At equilibrium, we have

$$\text{rate}_{\text{fwd}} = \text{rate}_{\text{rev}}$$

In this case, both forward and reverse reactions are elementary steps (Section 16.7), so we can write their rate laws directly from the balanced equation:

$$k_{\text{fwd}}[N_2O_4]_{\text{eq}} = k_{\text{rev}}[NO_2]^2_{\text{eq}}$$

where k_{fwd} and k_{rev} are the forward and reverse rate constants, respectively, and the subscript "eq" refers to concentrations at equilibrium. By rearranging, we set the ratio of the rate constants equal to the ratio of the concentration terms:

$$\frac{k_{\text{fwd}}}{k_{\text{rev}}} = \frac{[NO_2]^2_{\text{eq}}}{[N_2O_4]_{\text{eq}}}$$

The ratio of constants gives rise to a new overall constant called the **equilibrium constant (*K*)**:

$$K = \frac{k_{fwd}}{k_{rev}} = \frac{[NO_2]^2_{eq}}{[N_2O_4]_{eq}} \quad (17.2)$$

The equilibrium constant K is a number equal to a particular ratio of equilibrium concentrations of product and reactant at a particular temperature. We examine this idea closely in the next section and show that it holds as well for overall reactions made up of several elementary steps.

The magnitude of K is an indication of how far a reaction proceeds toward product at a given temperature. Remember, it is the opposing *rates* that are equal at equilibrium, not necessarily the concentrations. Indeed, different reactions, even at the same temperature, have a wide range of concentrations at equilibrium—from almost all reactant to almost all product—and, therefore, they have a wide range of equilibrium constants (Figure 17.2). Here are three examples of different magnitudes of K:

1. *Small K.* If a reaction yields very little product before reaching equilibrium, it has a small K, and we may even say there is "no reaction." For example, the oxidation of nitrogen barely proceeds at 1000 K:*

   $$N_2(g) + O_2(g) \rightleftharpoons 2NO(g) \qquad K = 1\times10^{-30}$$

2. *Large K.* Conversely, if a reaction reaches equilibrium with very little reactant remaining, it has a large K, and we say it "goes to completion." The oxidation of carbon monoxide goes to completion at 1000 K:

   $$2CO(g) + O_2(g) \rightleftharpoons 2CO_2(g) \qquad K = 2.2\times10^{22}$$

3. *Intermediate K.* When significant amounts of both reactant and product are present at equilibrium, K has an intermediate value, as when bromine monochloride breaks down to its elements at 1000 K:

   $$2BrCl(g) \rightleftharpoons Br_2(g) + Cl_2(g) \qquad K = 5$$

*To distinguish the equilibrium constant from the Kelvin temperature unit, the equilibrium constant is written with an uppercase italic *K*, whereas the kelvin is an uppercase roman K. Also, since the kelvin is a unit, it always follows a number.

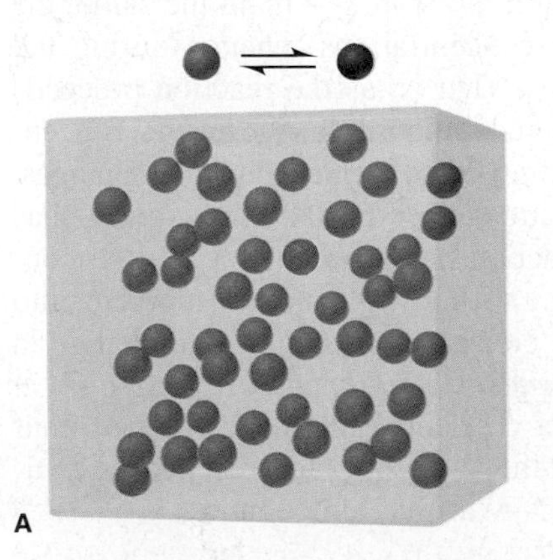

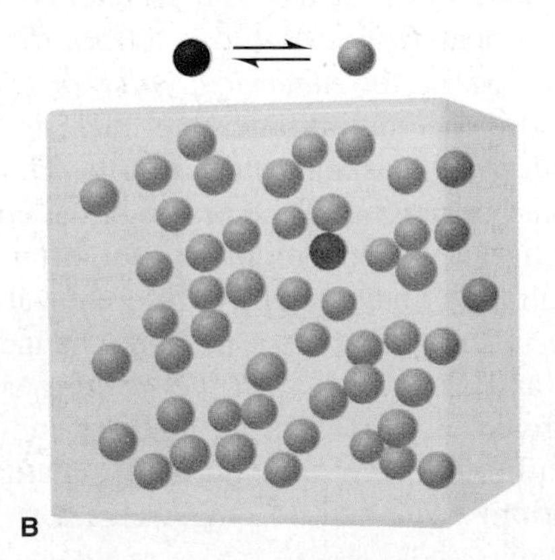

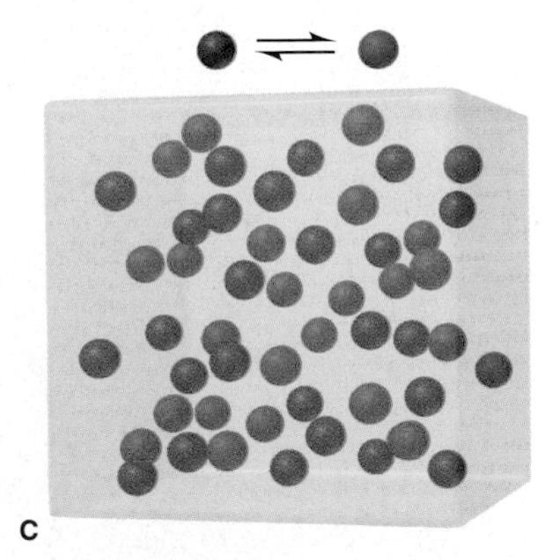

Figure 17.2 The range of equilibrium constants. A, A system that reaches equilibrium with very little product has a small K. For this reaction, $K = 1/49 = 0.020$. **B,** A system that reaches equilibrium with nearly all product has a large K. For this reaction, $K = 49/1 = 49$. **C,** A system that reaches equilibrium with significant concentrations of reactant and product has an intermediate K. For this reaction, $K = 25/25 = 1.0$.

SECTION SUMMARY

Kinetics and equilibrium are distinct aspects of a reaction system; that is, rate and yield are not necessarily related. When the forward and reverse reactions occur at the same rate, the system has reached dynamic equilibrium and concentrations no longer change. The equilibrium constant (K) is a number based on a particular ratio of product and reactant concentrations. K is small for reactions that reach equilibrium with a high concentration of reactant(s) and large for reactions that reach equilibrium with a low concentration of reactant(s).

17.2 THE REACTION QUOTIENT AND THE EQUILIBRIUM CONSTANT

Our derivation of the equilibrium constant in Section 17.1 was based on kinetics. But the fundamental observation of equilibrium studies was stated many years before the principles of kinetics were developed. In 1864, two Norwegian chemists, Cato Guldberg and Peter Waage, observed that *at a given temperature, a chemical system reaches a state in which a particular ratio of reactant and product concentrations has a constant value*. This is one way of stating the **law of chemical equilibrium,** or the **law of mass action.** No mention of rates appears.

In their discovery of the law of mass action, Guldberg and Waage studied many reactions in which reactant and product concentrations varied widely. They found that, *for a particular system and temperature, the same equilibrium state is attained regardless of how the reaction is run*. For example, in the N_2O_4-NO_2 system at 200°C, we can start with pure reactant (N_2O_4), pure product (NO_2), or any mixture of NO_2 and N_2O_4, and the ratio of concentrations will attain the same equilibrium value (within experimental error).

The particular ratio of concentration terms that we write for a given reaction is called the **reaction quotient (Q, or mass-action expression).** For the breakdown of N_2O_4 to form NO_2, the reaction quotient, which is based directly on the balanced equation, is

$$Q = \frac{[NO_2]^2}{[N_2O_4]}$$

Although the reactant and product concentration *terms* in Q remain the same, the numerical *values* of those terms (the actual concentrations) change during the reaction, so the *numerical value of Q changes*. That is, as the reaction proceeds toward the equilibrium state, there is a continual, smooth change in the concentrations of reactants and products. Thus, the ratio of concentrations also changes: at the beginning of the reaction, the concentrations have initial values, and Q has an initial value; a moment later, after the reaction has proceeded a bit, the concentrations have slightly different values, and so does Q; another moment into the reaction, and there is more change in the concentrations and more change in Q; and on and on, *until the reacting system reaches equilibrium*. At that point, at a given temperature, the reactant and product concentrations have reached their equilibrium levels and no longer change. And the value of Q has reached its equilibrium value and no longer changes. It equals K at that temperature:

$$\text{At equilibrium: } Q = K \qquad \textbf{(17.3)}$$

So, monitoring Q tells whether the system has reached equilibrium, how far away it is if it has not, and, as we discuss later, in which direction it is changing to reach equilibrium. Table 17.1 presents four experiments, each a different run of the N_2O_4-NO_2 reaction at 200°C. The two essential points to note are:

Table 17.1 Initial and Equilibrium Concentration Ratios for the N_2O_4-NO_2 System at 200°C (473 K)

	Initial		Ratio (Q)	Equilibrium		Ratio (K)
Exp't.	$[N_2O_4]$	$[NO_2]$	$[NO_2]^2/[N_2O_4]$	$[N_2O_4]_{eq}$	$[NO_2]_{eq}$	$[NO_2]^2_{eq}/[N_2O_4]_{eq}$
1	0.1000	0.0000	0.0000	0.00357	0.193	10.4
2	0.0000	0.1000	∞	9.24×10^{-4}	9.82×10^{-2}	10.4
3	0.0500	0.0500	0.0500	0.00204	0.146	10.4
4	0.0750	0.0250	0.00833	0.00275	0.170	10.5

- The ratio of *initial* concentrations varies widely but always gives the same ratio of *equilibrium* concentrations.
- The *individual* equilibrium concentrations are different in each case, but the *ratio* of these equilibrium concentrations is constant.

The curves in Figure 17.3 show experiment 1 in Table 17.1. Note that $[N_2O_4]$ and $[NO_2]$ change smoothly during the course of the reaction and, thus, so does the value of Q. Once the system reaches equilibrium, as indicated by the constant brown color, the concentrations no longer change and Q equals K. In other words, for any given chemical system, *K is a special value of Q that occurs when the reactant and product terms have their equilibrium values.*

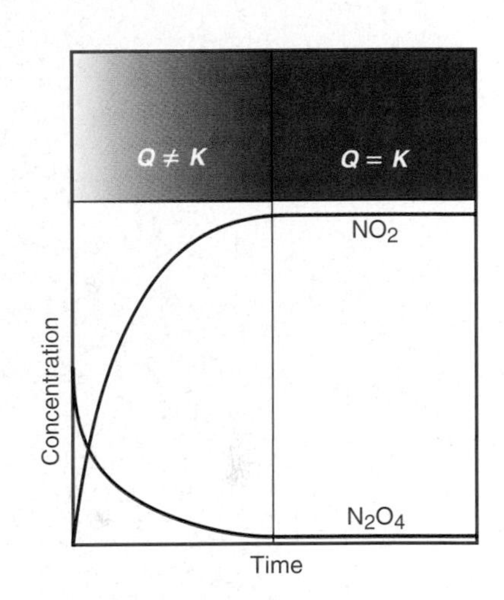

Figure 17.3 The change in Q during the N_2O_4-NO_2 reaction. The curved plots and the darkening brown screen above them show that $[N_2O_4]$ and $[NO_2]$, and therefore the value of Q, change with time. Before equilibrium is reached, the concentrations are changing continuously, so $Q \neq K$. Once equilibrium is reached (*vertical line*) and any time thereafter, $Q = K$.

Writing the Reaction Quotient

In Chapter 16, you saw that the rate law for an overall reaction cannot be written from the balanced equation, but must be determined from rate data. In contrast, the reaction quotient *can* be written directly from the balanced equation: *Q is a ratio made up of product concentration terms multiplied together and divided by reactant concentration terms multiplied together, with each term raised to the power of its stoichiometric coefficient in the balanced equation.*

The most common form of the reaction quotient shows reactant and product terms as molar concentrations, which are designated by square brackets, [ ]. In these cases, which you've seen so far, K is the *equilibrium constant based on concentrations*, designated from now on as K_c. Similarly, we designate the reaction quotient based on concentrations as Q_c. For the general balanced equation

$$a\text{A} + b\text{B} \rightleftharpoons c\text{C} + d\text{D}$$

where a, b, c, and d are the stoichiometric coefficients, the reaction quotient is

$$Q_c = \frac{[\text{C}]^c[\text{D}]^d}{[\text{A}]^a[\text{B}]^b} \quad (17.4)$$

(Another form of the reaction quotient that we discuss later shows gaseous reactant and product terms as pressures.)

To construct the reaction quotient for any reaction, write the balanced equation first. For the formation of ammonia from its elements, for example, the balanced equation (with colored coefficients for easy reference) is

$$N_2(g) + 3H_2(g) \rightleftharpoons 2NH_3(g)$$

To construct the reaction quotient, we place the product term in the numerator and the reactant terms in the denominator, multiplied by each other, and raise each term to the power of its balancing coefficient (colored here as in the equation):

$$Q_c = \frac{[NH_3]^2}{[N_2][H_2]^3}$$

Let's practice this essential skill.

SAMPLE PROBLEM 17.1 Writing the Reaction Quotient from the Balanced Equation

Problem Write the reaction quotient, Q_c, for each of the following reactions:
(a) The decomposition of dinitrogen pentaoxide, $N_2O_5(g) \rightleftharpoons NO_2(g) + O_2(g)$
(b) The combustion of propane gas, $C_3H_8(g) + O_2(g) \rightleftharpoons CO_2(g) + H_2O(g)$
Plan We balance the equations and then construct the reaction quotient (Equation 17.4).

Solution **(a)** $2N_2O_5(g) \rightleftharpoons 4NO_2(g) + O_2(g)$ $\quad Q_c = \frac{[NO_2]^4[O_2]}{[N_2O_5]^2}$

(b) $C_3H_8(g) + 5O_2(g) \rightleftharpoons 3CO_2(g) + 4H_2O(g)$ $\quad Q_c = \frac{[CO_2]^3[H_2O]^4}{[C_3H_8][O_2]^5}$

Check Always be sure that the exponents in Q are the same as the balancing coefficients. A good check is to reverse the process: turn the numerator into products and the denominator into reactants, and change the exponents to coefficients.

FOLLOW-UP PROBLEM 17.1 Write a reaction quotient, Q_c, for each of the following reactions (unbalanced):
(a) The first step in nitric acid production, $NH_3(g) + O_2(g) \rightleftharpoons NO(g) + H_2O(g)$
(b) The disproportionation of nitric oxide, $NO(g) \rightleftharpoons N_2O(g) + NO_2(g)$

Variations in the Form of the Reaction Quotient

As you'll see in the upcoming discussion, the reaction quotient Q is a collection of terms based on the balanced equation *exactly as written* for a given reaction. Therefore, the value of Q, which varies during the reaction, and the value of K, the constant value of Q when the system has reached equilibrium, also depend on how the balanced equation is written.

A Word about Units for *Q* and *K* In this text (and most others), *the values of Q and K are shown as unitless numbers*. This is because each term in the reaction quotient represents the *ratio* of the measured quantity of the substance (molar concentration or pressure) to the thermodynamic standard-state quantity of the substance. Recall from Section 6.6 that these standard states are 1 *M* for a substance in solution, 1 atm for gases, and the pure substance for a liquid or solid. Thus, a concentration of 1.20 *M* becomes $\frac{1.20\ M}{1\ M} = 1.20$; similarly, a pressure of 0.53 atm becomes $\frac{0.53\ \text{atm}}{1\ \text{atm}} = 0.53$. (As you'll see shortly, the molar "concentration" of a pure liquid or solid, that is, the number of moles per liter of the substance, is a constant, and the term does not appear in Q at all.) With these quantity terms unitless, the ratio of terms we use to find the value of Q (or K) is also unitless.

Form of *Q* for an Overall Reaction Notice that we've been writing reaction quotients without knowing whether an equation represents an individual reaction step or an overall multistep reaction. We can do this because we obtain the same expression for the overall reaction as we do when we combine the expressions for the individual steps. That is, *if an overall reaction is the* ***sum*** *of two or more reactions, the overall reaction quotient (or equilibrium constant) is the* ***product*** *of the reaction quotients (or equilibrium constants) for the steps:*

$$Q_{\text{overall}} = Q_1 \times Q_2 \times Q_3 \times \cdots$$

and

$$K_{\text{overall}} = K_1 \times K_2 \times K_3 \times \cdots \qquad (17.5)$$

Sample Problem 17.2 demonstrates this point.

SAMPLE PROBLEM 17.2 Writing the Reaction Quotient for an Overall Reaction

Problem Understanding reactions involving normal components of air is essential for solving problems dealing with atmospheric pollution. Here is a reaction sequence involving N_2 and O_2, the most abundant gases in air. Nitrogen dioxide is a toxic pollutant that contributes to photochemical smog (see photo).

(1) $N_2(g) + O_2(g) \rightleftharpoons 2NO(g) \qquad K_{c1} = 4.3\times10^{-25}$

(2) $2NO(g) + O_2(g) \rightleftharpoons 2NO_2(g) \qquad K_{c2} = 6.4\times10^{9}$

(a) Show that the overall Q_c for this reaction sequence is the same as the product of the Q_c's for the individual reactions.

(b) Given that the K_c's occur at the same temperature, find K_c for the overall reaction.

Plan In **(a),** we first write the overall reaction by adding the individual reactions and then write the overall Q_c. Next, we write the Q_c for each reaction. Because we *add* the individual steps, we *multiply* their Q_c's and cancel common terms to obtain the overall Q_c. In **(b),** we are given the individual K_c values (4.3×10^{-25} and 6.4×10^{9}), so we multiply them to find $K_{c(\text{overall})}$.

Solution (a) Writing the overall reaction and its reaction quotient:

$$\begin{array}{ll} (1) & N_2(g) + O_2(g) \rightleftharpoons \cancel{2NO(g)} \\ (2) & \cancel{2NO(g)} + O_2(g) \rightleftharpoons 2NO_2(g) \\ \hline \text{Overall:} & N_2(g) + 2O_2(g) \rightleftharpoons 2NO_2(g) \end{array}$$

$$Q_{c(\text{overall})} = \frac{[NO_2]^2}{[N_2][O_2]^2}$$

Writing the reaction quotients for the individual steps:

For step 1, $$Q_{c1} = \frac{[NO]^2}{[N_2][O_2]}$$

For step 2, $$Q_{c2} = \frac{[NO_2]^2}{[NO]^2[O_2]}$$

Multiplying the individual reaction quotients and canceling:

$$Q_{c1} \times Q_{c2} = \frac{\cancel{[NO]^2}}{[N_2][O_2]} \times \frac{[NO_2]^2}{\cancel{[NO]^2}[O_2]} = \frac{[NO_2]^2}{[N_2][O_2]^2} = Q_{c(\text{overall})}$$

(b) Calculating the overall K_c:

$$K_{c(\text{overall})} = K_{c1} \times K_{c2} = (4.3\times10^{-25})(6.4\times10^{9}) = 2.8\times10^{-15}$$

Check Round off and check the calculation in part (b):

$$K_c \approx (4\times10^{-25})(6\times10^{9}) = 24\times10^{-16} = 2.4\times10^{-15}$$

FOLLOW-UP PROBLEM 17.2 The following sequence of individual steps has been proposed for the overall reaction between H_2 and Br_2 to form HBr:

(1) $Br_2(g) \rightleftharpoons 2Br(g)$

(2) $Br(g) + H_2(g) \rightleftharpoons HBr(g) + H(g)$

(3) $H(g) + Br(g) \rightleftharpoons HBr(g)$

Write the overall balanced equation and show that the overall Q_c is the product of the Q_c's for the individual steps.

Smog over Los Angeles.

Form of *Q* for a Forward and Reverse Reaction The form of the reaction quotient depends on the *direction* in which the balanced equation is written. Consider, for example, the oxidation of sulfur dioxide to sulfur trioxide. This reaction is a key step in acid rain formation and sulfuric acid production (see photo). The balanced equation is

$$2SO_2(g) + O_2(g) \rightleftharpoons 2SO_3(g)$$

A sulfuric acid manufacturing plant.

The reaction quotient for this equation *as written* is

$$Q_{c(fwd)} = \frac{[SO_3]^2}{[SO_2]^2[O_2]}$$

If we had written the reverse reaction, the decomposition of sulfur trioxide,

$$2SO_3(g) \rightleftharpoons 2SO_2(g) + O_2(g)$$

the reaction quotient would be the *reciprocal* of $Q_{c(fwd)}$:

$$Q_{c(rev)} = \frac{[SO_2]^2[O_2]}{[SO_3]^2} = \frac{1}{Q_{c(fwd)}}$$

Thus, *a reaction quotient (or equilibrium constant) for a forward reaction is the* ***reciprocal*** *of the reaction quotient (or equilibrium constant) for the reverse reaction:*

$$Q_{c(fwd)} = \frac{1}{Q_{c(rev)}} \quad \text{and} \quad K_{c(fwd)} = \frac{1}{K_{c(rev)}} \tag{17.6}$$

The K_c values for the forward and reverse reactions at 1000 K are

$$K_{c(fwd)} = 261 \quad \text{and} \quad K_{c(rev)} = \frac{1}{K_{c(fwd)}} = \frac{1}{261} = 3.83\times10^{-3}$$

These values make sense: if the forward reaction goes far to the right (high K_c), the reverse reaction does not (low K_c).

Form of *Q* for a Reaction with Coefficients Multiplied by a Common Factor

Multiplying all the coefficients of the equation by some factor also changes the form of Q. For example, multiplying all the coefficients in the previous equation for the formation of SO_3 by $\frac{1}{2}$ gives

$$SO_2(g) + \tfrac{1}{2}O_2(g) \rightleftharpoons SO_3(g)$$

For this equation, the reaction quotient is

$$Q'_{c(fwd)} = \frac{[SO_3]}{[SO_2][O_2]^{1/2}}$$

Notice that Q_c for the halved equation equals Q_c for the original equation raised to the $\frac{1}{2}$ power:

$$Q'_{c(fwd)} = Q^{1/2}_{c(fwd)} = \left(\frac{[SO_3]^2}{[SO_2]^2[O_2]}\right)^{1/2} = \frac{[SO_3]}{[SO_2][O_2]^{1/2}}$$

Once again, the same property holds for the equilibrium constants. Relating the halved reaction to the original, we have

$$K'_{c(fwd)} = K^{1/2}_{c(fwd)} = (261)^{1/2} = 16.2$$

Similarly, if you double coefficients, the reaction quotient is the original expression squared; if you triple coefficients, it is the original expression cubed; and so on. It may seem that we have changed the extent of the reaction, as indicated by a change in K, merely by changing the balancing coefficients of the equation, but this clearly cannot be true. *A particular K has meaning only in relation to a particular balanced equation.* In this case, $K_{c(fwd)}$ and $K'_{c(fwd)}$ relate to different equations and thus cannot be compared directly.

In general, *if all the coefficients of the balanced equation are multiplied by some factor, that factor becomes the exponent for relating the reaction quotients and the equilibrium constants.* For a multiplying factor n, which we can write as

$$n(aA + bB \rightleftharpoons cC + dD)$$

the reaction quotient and equilibrium constant are

$$Q' = Q^n = \left(\frac{[C]^c[D]^d}{[A]^a[B]^b}\right)^n \quad \text{and} \quad K' = K^n \tag{17.7}$$

SAMPLE PROBLEM 17.3 Finding the Equilibrium Constant for an Equation Multiplied by a Common Factor

Problem For the ammonia-formation reaction,

$$N_2(g) + 3H_2(g) \rightleftharpoons 2NH_3(g)$$

the equilibrium constant, K_c, is 2.4×10^{-3} at 1000 K. If we change the coefficients of this equation, which we'll call the reference (ref) equation, what are the values of K_c for the following balanced equations?

(a) $\frac{1}{3}N_2(g) + H_2(g) \rightleftharpoons \frac{2}{3}NH_3(g)$ (b) $NH_3(g) \rightleftharpoons \frac{1}{2}N_2(g) + \frac{3}{2}H_2(g)$

Plan We compare each equation with the reference equation to see how the direction and coefficients have changed. In **(a)**, the equation is the reference equation multiplied by $\frac{1}{3}$, so K_c equals $K_{c(ref)}$ (2.4×10^{-3}) raised to the $\frac{1}{3}$ power. In **(b)**, the equation is one-half the *reverse* of the reference equation, so K_c is the reciprocal of $K_{c(ref)}$ raised to the $\frac{1}{2}$ power.

Solution The reaction quotient for the reference equation is $Q_{c(ref)} = \frac{[NH_3]^2}{[N_2][H_2]^3}$.

(a)
$$Q_c = Q_{c(ref)}^{1/3} = \left(\frac{[NH_3]^2}{[N_2][H_2]^3}\right)^{1/3} = \frac{[NH_3]^{2/3}}{[N_2]^{1/3}[H_2]}$$

Thus,
$$K_c = K_{c(ref)}^{1/3} = (2.4\times10^{-3})^{1/3} = 0.13$$

(b)
$$Q_c = \left(\frac{1}{Q_{c(ref)}}\right)^{1/2} = \left(\frac{1}{\frac{[NH_3]^2}{[N_2][H_2]^3}}\right)^{1/2} = \frac{[N_2]^{1/2}[H_2]^{3/2}}{[NH_3]}$$

Thus,
$$K_c = \left(\frac{1}{K_{c(ref)}}\right)^{1/2} = \left(\frac{1}{2.4\times10^{-3}}\right)^{1/2} = 20.$$

Check A good check is to work the math backward. For **(a)**, $(0.13)^3 = 2.2\times10^{-3}$, within rounding of 2.4×10^{-3}. The reaction goes in the same direction, so at equilibrium, there should be mostly reactants, as $K_c < 1$ indicates. For **(b)**, $1/(20.)^2 = 2.5\times10^{-3}$, again within rounding. At equilibrium, the reverse reaction should yield mostly products, as $K_c > 1$ indicates.

FOLLOW-UP PROBLEM 17.3 At 1200 K, the reaction of hydrogen and chlorine to form hydrogen chloride is

$$H_2(g) + Cl_2(g) \rightleftharpoons 2HCl(g) \qquad K_c = 7.6\times10^{8}$$

Calculate K_c for the following reactions:

(a) $\frac{1}{2}H_2(g) + \frac{1}{2}Cl_2(g) \rightleftharpoons HCl(g)$ (b) $\frac{4}{3}HCl(g) \rightleftharpoons \frac{2}{3}H_2(g) + \frac{2}{3}Cl_2(g)$

Form of *Q* for a Reaction Involving Pure Liquids and Solids Until now, we've looked at *homogeneous* equilibria, systems in which all the components of the reaction are in the same phase, such as a system of reacting gases. When the components are in different phases, the system reaches *heterogeneous* equilibrium.

Consider the decomposition of limestone to lime and carbon dioxide, in which a gas and two solids make up the reaction components:

$$CaCO_3(s) \rightleftharpoons CaO(s) + CO_2(g)$$

Based on the rules for writing the reaction quotient, we have

$$Q_c = \frac{[CaO][CO_2]}{[CaCO_3]}$$

A pure solid, however, such as $CaCO_3$ or CaO, always has the same *concentration* at a given temperature, that is, the same number of moles per liter of the solid, just as it has the same density (g/cm^3) at a given temperature. Moreover, a solid's volume changes very little with temperature, so its concentration also changes very little. For these reasons, the concentration of a pure solid is constant, and the same argument applies to the concentration of a pure liquid.

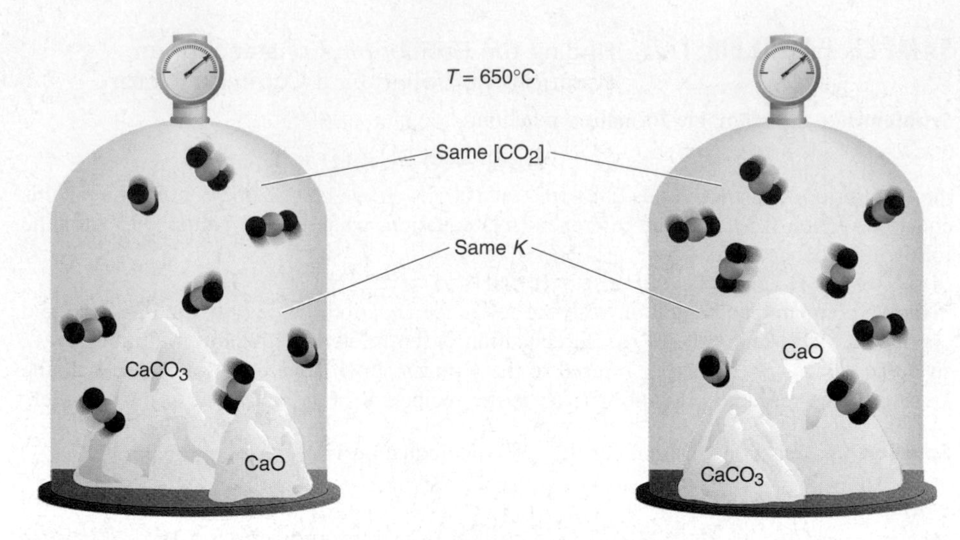

Figure 17.4 The reaction quotient for a heterogeneous system. Even though the two containers have different amounts of the two solids CaO and $CaCO_3$, as long as both solids are present, at a given temperature, the containers have the same $[CO_2]$ at equilibrium.

Because we are concerned only with concentrations that *change* as they approach equilibrium, *we eliminate the terms for pure liquids and solids from the reaction quotient*. We do this by incorporating their constant concentrations into a rearranged reaction quotient, Q'_c. We multiply both sides of the equation by $[CaCO_3]$ and divide both sides by [CaO]. Thus, the only substance whose concentration can change is the gas CO_2:

$$Q'_c = Q_c\frac{[CaCO_3]}{[CaO]} = [CO_2]$$

No matter how much CaO and $CaCO_3$ are in the reaction vessel, *as long as some of each is present*, the reaction quotient for the reaction equals the CO_2 concentration (Figure 17.4).

Table 17.2 summarizes the ways of writing Q and calculating K.

Table 17.2 Ways of Expressing Q and Calculating K

Form of Chemical Equation	Form of Q	Value of K
Reference reaction: $A \rightleftharpoons B$	$Q_{(ref)} = \frac{[B]}{[A]}$	$K_{(ref)} = \frac{[B]_{eq}}{[A]_{eq}}$
Reverse reaction: $B \rightleftharpoons A$	$Q = \frac{1}{Q_{(ref)}} = \frac{[A]}{[B]}$	$K = \frac{1}{K_{(ref)}}$
Reaction as sum of two steps: (1) $A \rightleftharpoons C$	$Q_1 = \frac{[C]}{[A]}; Q_2 = \frac{[B]}{[C]}$	
(2) $C \rightleftharpoons B$	$Q_{overall} = Q_1 \times Q_2 = Q_{(ref)}$ $= \frac{\cancel{[C]}}{[A]} \times \frac{[B]}{\cancel{[C]}} = \frac{[B]}{[A]}$	$K_{overall} = K_1 \times K_2$ $= K_{(ref)}$
Coefficients multiplied by n	$Q = Q^n_{(ref)}$	$K = K^n_{(ref)}$
Reaction with pure solid or liquid component, such as A(*s*)	$Q = Q_{(ref)}[A] = [B]$	$K = K_{(ref)}[A] = [B]$

SECTION SUMMARY

The reaction quotient, Q, is a particular ratio of product to reactant terms. Substituting experimental values into this expression gives the value of Q, which changes as the reaction proceeds. When the system reaches equilibrium at a particular temperature, $Q = K$. If a reaction is the sum of two or more steps, the overall Q (or K) is the product of the individual Q's (or K's). The *form* of Q is based directly on the balanced equation for the reaction exactly as written, so it changes if the equation is reversed or multiplied by some factor, and K changes accordingly. Pure liquids or solids do not appear as terms in the expression for Q because their concentrations are constant.

17.3 EXPRESSING EQUILIBRIA WITH PRESSURE TERMS: RELATION BETWEEN K_c AND K_p

It is easier to measure the pressure of a gas than its concentration and, as long as the gas behaves ideally under the conditions of the experiment, the ideal gas law (Section 5.3) allows us to relate these variables to each other:

$$PV = nRT, \quad \text{so} \quad P = \frac{n}{V}RT \quad \text{or} \quad \frac{P}{RT} = \frac{n}{V}$$

where P is the pressure of a gas and n/V is its molar concentration (M). Thus, with R a constant and T kept constant, *pressure is directly proportional to molar concentration.* When the substances involved in the reaction are gases, we can express the reaction quotient and calculate its value in terms of partial pressures instead of concentrations. For example, in the reaction between gaseous NO and O_2,

$$2NO(g) + O_2(g) \rightleftharpoons 2NO_2(g)$$

the reaction quotient based on partial pressures, Q_p, is

$$Q_p = \frac{P_{NO_2}^2}{P_{NO}^2 \times P_{O_2}}$$

(In later chapters, you'll see cases where some reaction components are expressed as concentrations and others as partial pressures.) The equilibrium constant obtained when all components are present at their equilibrium partial pressures is designated K_p, the *equilibrium constant based on pressures.* In many cases, K_p has a value different from K_c, but the two constants are related; thus, if you know one, you can calculate the other by noting the *change in amount (mol) of gas,* Δn_{gas}, from the balanced equation. Let's see this relationship by converting the terms in Q_c for the reaction of NO and O_2 to those in Q_p:

$$2NO(g) + O_2(g) \rightleftharpoons 2NO_2(g)$$

As the balanced equation shows,

$$\text{3 mol (2 mol + 1 mol) gaseous reactants} \rightleftharpoons \text{2 mol gaseous products}$$

With Δ meaning final minus initial (products minus reactants), we have

$$\Delta n_{gas} = \text{moles of gaseous product} - \text{moles of gaseous reactant} = 2 - 3 = -1$$

Keep this value of Δn_{gas} in mind because it appears in the algebraic conversion that follows. The reaction quotient based on concentrations is

$$Q_c = \frac{[NO_2]^2}{[NO]^2[O_2]}$$

Using the ideal gas law as $n/V = P/RT$, we first express concentrations as n/V and convert them to partial pressures, P; then we collect the RT terms and cancel:

$$Q_c = \frac{\dfrac{n_{NO_2}^2}{V^2}}{\dfrac{n_{NO}^2}{V^2} \times \dfrac{n_{O_2}}{V}} = \frac{\dfrac{P_{NO_2}^2}{(RT)^2}}{\dfrac{P_{NO}^2}{(RT)^2} \times \dfrac{P_{O_2}}{RT}} = \frac{P_{NO_2}^2}{P_{NO}^2 \times P_{O_2}} \times \frac{\dfrac{1}{\cancel{(RT)^2}}}{\dfrac{1}{\cancel{(RT)^2}} \times \dfrac{1}{RT}} = \frac{P_{NO_2}^2}{P_{NO}^2 \times P_{O_2}} \times RT$$

The far right side of the previous expression is Q_p multiplied by RT: $Q_c = Q_p(RT)$. Also, at equilibrium, $K_c = K_p(RT)$; thus, $K_p = \frac{K_c}{RT}$, or $K_c(RT)^{-1}$.

Notice that *the exponent of the RT term equals the change in the amount (mol) of gas (Δn_{gas}) from the balanced equation,* -1. Thus, in general, we have

$$K_p = K_c(RT)^{\Delta n_{gas}} \quad (17.8)$$

The units for the partial pressure terms in K_p are generally atmospheres, pascals, or torr, raised to some power, and the units of R must be consistent with those units. As Equation 17.8 shows, for those reactions in which the amount (mol) of gas does not change, we have $\Delta n_{gas} = 0$, so the RT term drops out and $K_p = K_c$.

SAMPLE PROBLEM 17.4 Converting Between K_c and K_p

Problem A chemical engineer injects limestone ($CaCO_3$) into the hot flue gas of a coal-burning power plant to form lime (CaO), which scrubs SO_2 from the gas and forms gypsum ($CaSO_4 \cdot 2H_2O$). Find K_c for the following reaction, if CO_2 pressure is in atmospheres:

$$CaCO_3(s) \rightleftharpoons CaO(s) + CO_2(g) \qquad K_p = 2.1\times10^{-4} \text{ (at 1000. K)}$$

Plan We know K_p (2.1×10^{-4}), so to convert between K_p and K_c, we must first determine Δn_{gas} from the balanced equation. Then we rearrange Equation 17.8. With gas pressure in atmospheres, R is 0.0821 atm·L/mol·K.

Solution Determining Δn_{gas}: There is 1 mol of gaseous product and no gaseous reactant, so $\Delta n_{gas} = 1 - 0 = 1$.

Rearranging Equation 17.8 and calculating K_c:

$$K_p = K_c(RT)^1 \qquad \text{so} \qquad K_c = K_p(RT)^{-1}$$

$$K_c = (2.1\times10^{-4})(0.0821 \times 1000.)^{-1} = 2.6\times10^{-6}$$

Check Work backward to see whether you obtain the given K_p:

$$K_p = (2.6\times10^{-6})(0.0821 \times 1000.) = 2.1\times10^{-4}$$

FOLLOW-UP PROBLEM 17.4 Calculate K_p for the following reaction:

$$PCl_3(g) + Cl_2(g) \rightleftharpoons PCl_5(g) \qquad K_c = 1.67 \text{ (at 500. K)}$$

SECTION SUMMARY

The reaction quotient and the equilibrium constant can be expressed in terms of concentrations (Q_c and K_c); for gases, they are expressed in terms of partial pressures (Q_p and K_p). The values of K_p and K_c are related by using the ideal gas law: $K_p = K_c(RT)^{\Delta n_{gas}}$.

17.4 REACTION DIRECTION: COMPARING *Q* AND *K*

Suppose you start a reaction with a mixture of reactants and products and you know the equilibrium constant at the temperature of the reaction. How do you know if the reaction has reached equilibrium? And, if it hasn't, how do you know in which direction it is progressing to reach equilibrium? The value of Q can change; thus, at any particular time during the reaction, Q can be smaller than K, larger than K, or, when the system reaches equilibrium, equal to K. By comparing the value of Q at a particular time with the known K, you can tell whether the reaction has attained equilibrium or, if not, in which direction it is progressing. With product terms in the numerator of Q and reactant terms in the denominator, *more product makes the ratio of terms larger, and more reactant makes the ratio smaller.*

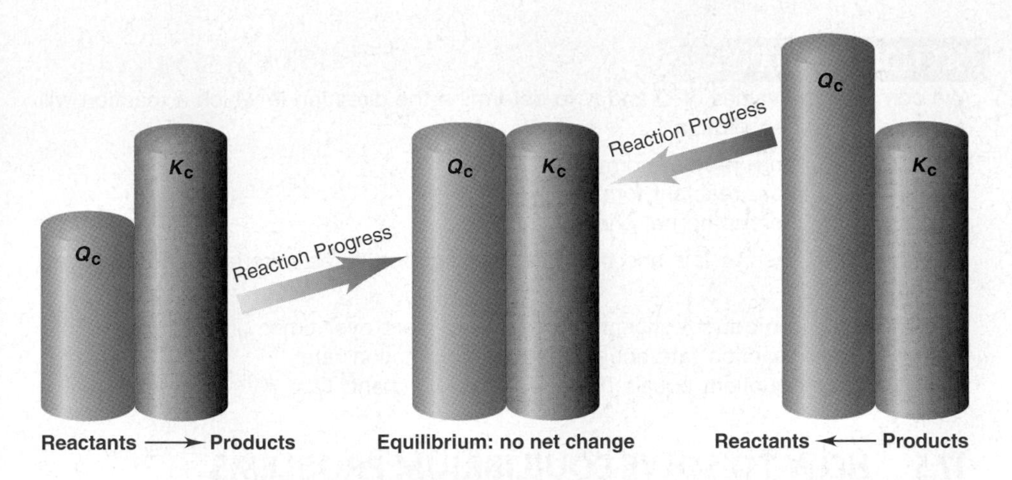

Figure 17.5 Reaction direction and the relative sizes of Q and K. When Q_c is smaller than K_c, the equilibrium of the reaction system shifts to the right, that is, toward products. When Q_c is larger than K_c, the equilibrium of the reaction system shifts to the left. Both shifts continue until $Q_c = K_c$. Note that the size of K_c remains the same throughout.

The three possible relative sizes of Q and K are shown in Figure 17.5.

- $Q < K$. If the value of Q is smaller than K, the denominator (reactants) is large relative to the numerator (products). For Q to become equal to K, the denominator must decrease and the numerator increase. In other words, the reaction will progress to the right, toward products, until equilibrium is reached:

$$\text{If } Q < K, \text{ reactants} \longrightarrow \text{products}$$

- $Q > K$. If Q is larger than K, the numerator (products) will decrease and the denominator (reactants) increase until equilibrium is reached. Therefore, the reaction will progress to the left, toward reactants:

$$\text{If } Q > K, \text{ reactants} \longleftarrow \text{products}$$

- $Q = K$. This situation exists only when the reactant and product concentrations (or pressures) have attained their equilibrium values. Thus, despite the dynamic processes occurring at the molecular level, no further net change occurs:

$$\text{If } Q = K, \text{ reactants} \rightleftharpoons \text{products}$$

SAMPLE PROBLEM 17.5 Comparing Q and K to Determine Reaction Direction

Problem For the reaction $N_2O_4(g) \rightleftharpoons 2NO_2(g)$, $K_c = 0.21$ at 100°C. At a point during the reaction, $[N_2O_4] = 0.12\ M$ and $[NO_2] = 0.55\ M$. Is the reaction at equilibrium? If not, in which direction is it progressing?

Plan We write the expression for Q_c, find its value by substituting the given concentrations, and then compare its value with the given K_c.

Solution Writing the reaction quotient and solving for Q_c:

$$Q_c = \frac{[NO_2]^2}{[N_2O_4]} = \frac{0.55^2}{0.12} = 2.5$$

With $Q_c > K_c$, the reaction is not at equilibrium and will proceed to the left until $Q_c = K_c$.

Check With $[NO_2] > [N_2O_4]$, we expect to obtain a value for Q_c that is greater than 0.21. If $Q_c > K_c$, the numerator will decrease and the denominator will increase until $Q_c = K_c$; that is, this reaction will proceed toward reactants.

FOLLOW-UP PROBLEM 17.5 Chloromethane forms by the reaction

$$CH_4(g) + Cl_2(g) \rightleftharpoons CH_3Cl(g) + HCl(g)$$

At 1500 K, $K_p = 1.6\times10^4$. In the reaction mixture, $P_{CH_4} = 0.13$ atm, $P_{Cl_2} = 0.035$ atm, $P_{CH_3Cl} = 0.24$ atm, and $P_{HCl} = 0.47$ atm. Is CH_3Cl or CH_4 forming?

SECTION SUMMARY

We compare the values of Q and K to determine the direction in which a reaction will proceed toward equilibrium.

- If $Q_c < K_c$, more product forms.
- If $Q_c > K_c$, more reactant forms.
- If $Q_c = K_c$, there is no net change.

As you've seen in this and the previous sections, three criteria define a system at equilibrium:

- Reactant and product concentrations are constant over time.
- The forward reaction rate equals the reverse reaction rate.
- The reaction quotient equals the equilibrium constant: $Q = K$.

17.5 HOW TO SOLVE EQUILIBRIUM PROBLEMS

Many kinds of equilibrium problems arise in the real world, as well as on chemistry exams, but we can group most of them into two types:

1. In one type, we are given equilibrium quantities (concentrations or partial pressures) and solve for K.
2. In the other type, we are given K and initial quantities and solve for the equilibrium quantities.

Using Quantities to Determine the Equilibrium Constant

There are two common variations on the type of equilibrium problem in which we solve for K: one involves a straightforward substitution of quantities, and the other requires first finding some of the quantities.

Substituting Given Equilibrium Quantities into Q to Find K In the straightforward case, we are given the equilibrium quantities and we must calculate K.

Suppose, for example, that equal amounts of gaseous hydrogen and iodine are injected into a 1.50-L reaction flask at a fixed temperature. In time, the following equilibrium is attained:

$$H_2(g) + I_2(g) \rightleftharpoons 2HI(g)$$

At equilibrium, analysis shows that the flask contains 1.80 mol of H_2, 1.80 mol of I_2, and 0.520 mol of HI. We calculate K_c by finding the concentrations and substituting them into the reaction quotient. From the balanced equation, we write the reaction quotient:

$$Q_c = \frac{[HI]^2}{[H_2][I_2]}$$

We first have to convert the amounts (mol) to concentrations (mol/L), using the flask volume of 1.50 L:

$$[H_2] = \frac{1.80\ \text{mol}}{1.50\ \text{L}} = 1.20\ M$$

Similarly, $[I_2] = 1.20\ M$, and $[HI] = 0.347\ M$. Substituting these values into the expression for Q_c gives K_c:

$$K_c = \frac{(0.347)^2}{(1.20)(1.20)} = 8.36\times10^{-2}$$

Using a Reaction Table to Determine Equilibrium Quantities and Find K When some quantities are not given, we determine them first from the reaction stoichiometry and then find K. In the following example, pay close attention to a valuable tool being introduced: *the reaction table*.

In a study of carbon oxidation, an evacuated vessel containing a small amount of powdered graphite is heated to 1080 K, and then CO_2 is added to a pressure of 0.458 atm. Once the CO_2 is added, the system starts to produce CO. After equi-

Using the Equilibrium Constant to Determine Quantities

Like the type of problem that involves finding K, the type that involves finding equilibrium concentrations (or pressures) has several variations. Sample Problem 17.7 is one variation, in which we know K and some of the equilibrium concentrations and must find another equilibrium concentration.

SAMPLE PROBLEM 17.7 Determining Equilibrium Concentrations from K_c

Problem In a study of the conversion of methane to other fuels, a chemical engineer mixes gaseous CH_4 and H_2O in a 0.32-L flask at 1200 K. At equilibrium, the flask contains 0.26 mol of CO, 0.091 mol of H_2, and 0.041 mol of CH_4. What is $[H_2O]$ at equilibrium? $K_c = 0.26$ for the equation

$$CH_4(g) + H_2O(g) \rightleftharpoons CO(g) + 3H_2(g)$$

Plan First, we use the balanced equation to write the reaction quotient. We can calculate the equilibrium concentrations from the given numbers of moles and the flask volume (0.32 L). Substituting these into Q_c and setting it equal to the given K_c (0.26), we solve for the unknown equilibrium concentration, $[H_2O]$.

Solution Writing the reaction quotient:

$$CH_4(g) + H_2O(g) \rightleftharpoons CO(g) + 3H_2(g) \qquad Q_c = \frac{[CO][H_2]^3}{[CH_4][H_2O]}$$

Determining the equilibrium concentrations:

$$[CH_4] = \frac{0.041 \text{ mol}}{0.32 \text{ L}} = 0.13\ M$$

Similarly, $[CO] = 0.81\ M$ and $[H_2] = 0.28\ M$.

Calculating $[H_2O]$ at equilibrium: Since $Q_c = K_c$, rearranging gives

$$[H_2O] = \frac{[CO][H_2]^3}{[CH_4]K_c} = \frac{(0.81)(0.28)^3}{(0.13)(0.26)} = 0.53\ M$$

Check Always check by substituting the concentrations into Q_c to confirm K_c:

$$Q_c = \frac{[CO][H_2]^3}{[CH_4][H_2O]} = \frac{(0.81)(0.28)^3}{(0.13)(0.53)} = 0.26 = K_c$$

FOLLOW-UP PROBLEM 17.7 Nitric oxide, oxygen, and nitrogen react by the following equation: $2NO(g) \rightleftharpoons N_2(g) + O_2(g)$; $K_c = 2.3\times10^{30}$ at 298 K. In the atmosphere, $P_{O_2} = 0.209$ atm and $P_{N_2} = 0.781$ atm. What is the equilibrium partial pressure of NO in the air we breathe? [*Hint*: You need K_p to find the partial pressure.]

In a somewhat more involved variation, we know K and *initial* quantities and must find *equilibrium* quantities, for which we use a reaction table. In Sample Problem 17.8, the amounts were chosen to simplify the math, allowing us to focus more easily on the overall approach.

SAMPLE PROBLEM 17.8 Determining Equilibrium Concentrations from Initial Concentrations and K_c

Problem Fuel engineers use the extent of the change from CO and H_2O to CO_2 and H_2 to regulate the proportions of synthetic fuel mixtures. If 0.250 mol of CO and 0.250 mol of H_2O are placed in a 125-mL flask at 900 K, what is the composition of the equilibrium mixture? At this temperature, K_c is 1.56 for the equation

$$CO(g) + H_2O(g) \rightleftharpoons CO_2(g) + H_2(g)$$

Plan We have to find the "composition" of the equilibrium mixture, in other words, the equilibrium concentrations. As always, we use the balanced equation to write the reaction

quotient. We find the initial [CO] and [H_2O] from the given amounts (0.250 mol of each) and volume (0.125 L), use the balanced equation to define x and set up a reaction table, substitute into Q_c, and solve for x, from which we calculate the concentrations.
Solution Writing the reaction quotient:

$$CO(g) + H_2O(g) \rightleftharpoons CO_2(g) + H_2(g) \qquad Q_c = \frac{[CO_2][H_2]}{[CO][H_2O]}$$

Calculating initial reactant concentrations:

$$[CO] = [H_2O] = \frac{0.250\ \text{mol}}{0.125\ \text{L}} = 2.00\ M$$

Setting up the reaction table, with x = [CO] and [H_2O] that react:

Concentration (M)	$CO(g)$	+	$H_2O(g)$	$\rightleftharpoons$	$CO_2(g)$	+	$H_2(g)$
Initial	2.00		2.00		0		0
Change	$-x$		$-x$		$+x$		$+x$
Equilibrium	$2.00 - x$		$2.00 - x$		x		x

Substituting into the reaction quotient and solving for x:

$$Q_c = \frac{[CO_2][H_2]}{[CO][H_2O]} = \frac{(x)(x)}{(2.00 - x)(2.00 - x)} = \frac{x^2}{(2.00 - x)^2}$$

At equilibrium, we have

$$Q_c = K_c = 1.56 = \frac{x^2}{(2.00 - x)^2}$$

We can apply the following math shortcut in this case *but not in general:* Because the right side of the equation is a perfect square, we take the square root of both sides:

$$\sqrt{1.56} = \frac{x}{2.00 - x} = \pm 1.25$$

A positive number (1.56) has a positive *and* a negative square root, but *only the positive root has any chemical meaning,* so we ignore the negative root:*

$$1.25 = \frac{x}{2.00 - x} \quad \text{or} \quad 2.50 - 1.25x = x$$

So $\quad 2.50 = 2.25x; \quad$ therefore, $\quad x = 1.11\ M$

Calculating equilibrium concentrations:

$$[CO] = [H_2O] = 2.00\ M - x = 2.00\ M - 1.11\ M = 0.89\ M$$
$$[CO_2] = [H_2] = x = 1.11\ M$$

Check From the intermediate size of K_c, it makes sense that the changes in concentration are moderate. It's a good idea to check that the sign of x in the reaction table makes sense—only reactants were initially present, so the change had to proceed to the right: x is the change in concentration, so it has a negative sign for reactants and a positive sign for products. Also check that the equilibrium concentrations give the known K_c: $\frac{(1.11)(1.11)}{(0.89)(0.89)} = 1.56$.

FOLLOW-UP PROBLEM 17.8 The decomposition of HI at low temperature was studied by injecting 2.50 mol of HI into a 10.32-L vessel at 25°C. What is [H_2] at equilibrium for the reaction $2HI(g) \rightleftharpoons H_2(g) + I_2(g)$; $K_c = 1.26\times10^{-3}$?

*The negative root gives $-1.25 = \frac{x}{2.00 - x}$, or $-2.50 + 1.25x = x$.

So $\quad -2.50 = -0.25x$, and $x = 10.\ M$

This value has no chemical meaning because we started with 2.00 M of each reactant, so it is impossible for 10. M to react. Moreover, the square root of an equilibrium constant is another equilibrium constant, which cannot have a negative value.

Using the Quadratic Formula to Solve for the Unknown The shortcut that we used to simplify the math in Sample Problem 17.8 is a special case that occurs when both numerator and denominator of the reaction quotient are perfect squares. It worked out that way because we started with equal concentrations of the two reactants, but that is not ordinarily the case.

Suppose, for example, we instead start the reaction in the sample problem with 2.00 *M* CO and 1.00 *M* H_2O. The reaction table is

Concentration (*M*)	$CO(g)$	+	$H_2O(g)$	$\rightleftharpoons$	$CO_2(g)$	+	$H_2(g)$
Initial	2.00		1.00		0		0
Change	$-x$		$-x$		$+x$		$+x$
Equilibrium	$2.00 - x$		$1.00 - x$		x		x

Substituting these values into Q_c, we obtain

$$Q_c = \frac{[CO_2][H_2]}{[CO][H_2O]} = \frac{(x)(x)}{(2.00 - x)(1.00 - x)} = \frac{x^2}{x^2 - 3.00x + 2.00}$$

At equilibrium, we have

$$1.56 = \frac{x^2}{x^2 - 3.00x + 2.00}$$

To solve for x in this case, we rearrange the previous expression into the form of a *quadratic equation:*

$$a\,x^2 + \;b\,x + \;c \;= 0$$
$$0.56x^2 - 4.68x + 3.12 = 0$$

where $a = 0.56$, $b = -4.68$, and $c = 3.12$. Then we can find x with the quadratic formula (Appendix A):

$$x = \frac{-b \pm \sqrt{b^2 - 4ac}}{2a}$$

The $\pm$ sign means that we obtain two possible values for x:

$$x = \frac{4.68 \pm \sqrt{(-4.68)^2 - 4(0.56)(3.12)}}{2(0.56)}$$

$$x = 7.6\ M \qquad \text{and} \qquad x = 0.73\ M$$

Note that only one of the values for x makes sense chemically. The larger value gives negative concentrations at equilibrium (for example, $2.00\ M - 7.6\ M = -5.6\ M$), which have no meaning. Therefore, $x = 0.73\ M$, and we have

$$[CO] = 2.00\ M - x = 2.00\ M - 0.73\ M = 1.27\ M$$
$$[H_2O] = 1.00\ M - x = 0.27\ M$$
$$[CO_2] = [H_2] = x = 0.73\ M$$

Checking to see if these values give the known K_c, we have

$$K_c = \frac{(0.73)(0.73)}{(1.27)(0.27)} = 1.6 \text{ (within rounding of 1.56)}$$

Simplifying Assumptions for Finding an Unknown Quantity In many cases, we can use chemical "common sense" to make an assumption that avoids the use of the quadratic formula to find x. In general, *if a reaction has a relatively small K and a relatively large initial reactant concentration, the concentration change (x) can often be neglected* without introducing significant error. This assumption does not mean that $x = 0$, because then there would be no reaction. It means that if a reaction proceeds very little (small K) and if there is a high initial reactant concentration, only a very small amount will be used up; therefore, at equilibrium, the reactant concentration will have hardly changed:

$$[\text{reactant}]_{\text{init}} - x = [\text{reactant}]_{\text{eq}} \approx [\text{reactant}]_{\text{init}}$$

You can imagine a similar situation in everyday life. On a bathroom scale, you weigh 158 lb. Take off your wristwatch, and you still weigh 158 lb. Within the precision of the measurement, the weight of the wristwatch is so small compared with your body weight that it can be neglected:

Initial body weight − weight of watch = final body weight ≈ initial body weight

Similarly, if the initial concentration of A is, for example, 0.500 *M* and, because of a small K_c, the concentration of A that reacts is 0.002 *M*, we can assume that

$$0.500\ M - 0.002\ M = 0.498\ M \approx 0.500\ M$$

that is,

$$[A]_{init} - [A]_{reacting} = [A]_{eq} \approx [A]_{init} \quad (17.9)$$

For the assumption that x is negligible to be justified, you must check that the error introduced is not significant. But how much is "significant"? One common, though somewhat arbitrary, criterion is the 5% rule: *if the assumption results in a change (error) in a concentration that is less than 5%, the error is not significant, and the assumption is justified.* Let's go through a sample problem and make this assumption to see how it simplifies the math, and then we'll see if the assumption is justified in the case of two different initial concentrations. We make this assumption often in Chapters 18 and 19.

SAMPLE PROBLEM 17.9 Calculating Equilibrium Concentrations with Simplifying Assumptions

Problem Phosgene is a potent chemical warfare agent that is now outlawed by international agreement. It decomposes by the reaction

$$COCl_2(g) \rightleftharpoons CO(g) + Cl_2(g) \qquad K_c = 8.3\times10^{-4} \text{ (at 360°C)}$$

Calculate [CO], $[Cl_2]$, and $[COCl_2]$ when each of the following amounts of phosgene decomposes and reaches equilibrium in a 10.0-L flask:
(a) 5.00 mol of $COCl_2$ **(b)** 0.100 mol of $COCl_2$

Plan We know from the balanced equation that when x mol of $COCl_2$ decomposes, x mol of CO and x mol of Cl_2 form. We convert amount (5.00 mol or 0.100 mol) to concentration, define x and set up the reaction table, and substitute the values into Q_c. Before using the quadratic formula, we simplify the calculation by assuming that x is negligibly small. After solving for x, we check the assumption and find the concentrations. If the assumption is not justified, we must use the quadratic formula to find x.

Solution (a) For 5.00 mol of $COCl_2$. Writing the reaction quotient:

$$Q_c = \frac{[CO][Cl_2]}{[COCl_2]}$$

Calculating initial $[COCl_2]$:

$$[COCl_2]_{init} = \frac{5.00\text{ mol}}{10.0\text{ L}} = 0.500\ M$$

Setting up the reaction table, with $x = [COCl_2]_{reacting}$:

Concentration (*M*)	$COCl_2(g)$	⇌	$CO(g)$	+	$Cl_2(g)$
Initial	0.500		0		0
Change	$-x$		$+x$		$+x$
Equilibrium	$0.500 - x$		x		x

If we use the equilibrium values in Q_c, we obtain

$$Q_c = \frac{[CO][Cl_2]}{[COCl_2]} = \frac{x^2}{0.500 - x} = K_c = 8.3\times10^{-4}$$

Because K_c is small, the reaction does not proceed very far to the right, so let's assume that x (the $[COCl_2]$ that reacts) is so much smaller than the initial concentration, 0.500 *M*, that the equilibrium concentration is nearly the same. Therefore,

$$0.500\ M - x \approx 0.500\ M$$

Using this assumption, we substitute and solve for x:

$$K_c = 8.3\times10^{-4} \approx \frac{x^2}{0.500}$$

$$x^2 \approx (8.3\times10^{-4})(0.500) \quad \text{so} \quad x \approx 2.0\times10^{-2}$$

Checking the assumption by finding the percent error:

$$\frac{2.0\times10^{-2}}{0.500} \times 100 = 4\% \quad \text{(less than 5\%, so the assumption is justified)}$$

Solving for the equilibrium concentrations:

$$[CO] = [Cl_2] = x = 2.0\times10^{-2}\ M$$

$$[COCl_2] = 0.500\ M - x = 0.480\ M$$

(b) For 0.100 mol of $COCl_2$. The calculation in this case is the same as the calculation in part (a), except that $[COCl_2]_{init} = 0.100\ \text{mol}/10.0\ \text{L} = 0.0100\ M$. Thus, at equilibrium, we have

$$Q_c = \frac{[CO][Cl_2]}{[COCl_2]} = \frac{x^2}{0.0100 - x}$$

$$= K_c = 8.3\times10^{-4}$$

Making the assumption that $0.0100\ M - x \approx 0.0100\ M$ and solving for x:

$$K_c = 8.3\times10^{-4} \approx \frac{x^2}{0.0100}$$

$$x \approx 2.9\times10^{-3}$$

Checking the assumption:

$$\frac{2.9\times10^{-3}}{0.0100} \times 100 = 29\% \quad \text{(more than 5\%, so the assumption is } \textit{not} \text{ justified)}$$

We must solve the quadratic equation, $x^2 + (8.3\times10^{-4})x - (8.3\times10^{-6}) = 0$, for which the only meaningful value of x is 2.5×10^{-3} (see Appendix A).

Solving for the equilibrium concentrations:

$$[CO] = [Cl_2] = 2.5\times10^{-3}\ M$$

$$[COCl_2] = 1.00\times10^{-2}\ M - x = 7.5\times10^{-3}\ M$$

Check Once again, the best check is to use the calculated values to be sure you obtain the given K_c.

Comment Note that the assumption was justified at the high initial concentration, but *not* at the low initial concentration.

FOLLOW-UP PROBLEM 17.9 In a study of the effect of temperature on halogen decomposition, 0.50 mol of I_2 was heated in a 2.5-L vessel, and the following reaction occurred: $I_2(g) \rightleftharpoons 2I(g)$.

(a) Calculate $[I_2]$ and $[I]$ at equilibrium at 600 K; $K_c = 2.94\times10^{-10}$.

(b) Calculate $[I_2]$ and $[I]$ at equilibrium at 2000 K; $K_c = 0.209$.

Mixtures of Reactants and Products: Determining Reaction Direction

In the problems we've worked so far, the direction of the reaction was obvious: with only reactants present at the start, the reaction *had* to go toward products. Thus, in the reaction tables, we knew that the unknown change in reactant concentration had a negative sign ($-x$) and the change in product concentration had a positive sign ($+x$). Suppose, however, we start with a *mixture* of reactants and products. Whenever the reaction direction is not obvious, we first *compare the value of Q with K to find the direction* in which the reaction proceeds to reach equilibrium. This tells us the sign of x, the unknown change in concentration. (In order to focus on this idea, the next sample problem eliminates the need for the quadratic formula.)

SAMPLE PROBLEM 17.10 Predicting Reaction Direction and Calculating Equilibrium Concentrations

Problem The research and development unit of a chemical company is studying the reaction of CH_4 and H_2S, two components of natural gas:

$$CH_4(g) + 2H_2S(g) \rightleftharpoons CS_2(g) + 4H_2(g)$$

In one experiment, 1.00 mol of CH_4, 1.00 mol of CS_2, 2.00 mol of H_2S, and 2.00 mol of H_2 are mixed in a 250-mL vessel at 960°C. At this temperature, $K_c = 0.036$.
(a) In which direction will the reaction proceed to reach equilibrium?
(b) If $[CH_4] = 5.56\ M$ at equilibrium, what are the equilibrium concentrations of the other substances?

Plan **(a)** To find the direction, we convert the given initial amounts and volume (0.250 L) to concentrations, calculate Q_c, and compare it with K_c. **(b)** Based on the results from (a), we determine the sign of each concentration change for the reaction table and then use the known $[CH_4]$ at equilibrium (5.56 M) to determine x and the other equilibrium concentrations.

Solution **(a)** Calculating the initial concentrations:

$$[CH_4] = \frac{1.00\ \text{mol}}{0.250\ \text{L}} = 4.00\ M$$

Similarly, $[H_2S] = 8.00\ M$, $[CS_2] = 4.00\ M$, and $[H_2] = 8.00\ M$.
Calculating the value of Q_c:

$$Q_c = \frac{[CS_2][H_2]^4}{[CH_4][H_2S]^2} = \frac{(4.00)(8.00)^4}{(4.00)(8.00)^2} = 64.0$$

Comparing Q_c and K_c: $Q_c > K_c$ (64.0 > 0.036), so the reaction goes to the left. Therefore, concentrations of reactants increase and those of products decrease.
(b) Setting up a reaction table, with $x = [CS_2]$ that reacts, which equals $[CH_4]$ that forms:

Concentration (M)	$CH_4(g)$	+	$2H_2S(g)$	$\rightleftharpoons$	$CS_2(g)$	+	$4H_2(g)$
Initial	4.00		8.00		4.00		8.00
Change	$+x$		$+2x$		$-x$		$-4x$
Equilibrium	$4.00 + x$		$8.00 + 2x$		$4.00 - x$		$8.00 - 4x$

Solving for x: At equilibrium,

$$[CH_4] = 5.56\ M = 4.00\ M + x$$

So,

$$x = 1.56\ M$$

Thus,

$$[H_2S] = 8.00\ M + 2x = 8.00\ M + 2(1.56\ M) = 11.12\ M$$

$$[CS_2] = 4.00\ M - x = 2.44\ M$$

$$[H_2] = 8.00\ M - 4x = 1.76\ M$$

Check The comparison of Q_c and K_c showed the reaction proceeding to the left. The given data from part (b) confirm this because $[CH_4]$ increases from 4.00 M to 5.56 M during the reaction. Check that the concentrations give the known K_c:

$$\frac{(2.44)(1.76)^4}{(5.56)(11.12)^2} = 0.0341, \text{ which is close to } 0.036$$

FOLLOW-UP PROBLEM 17.10 An inorganic chemist studying the reactions of phosphorus halides mixes 0.1050 mol of PCl_5 with 0.0450 mol of Cl_2 and 0.0450 mol of PCl_3 in a 0.5000-L flask at 250°C: $PCl_5(g) \rightleftharpoons PCl_3(g) + Cl_2(g)$; $K_c = 4.2\times10^{-2}$.
(a) In which direction will the reaction proceed?
(b) If $[PCl_5] = 0.2065\ M$ at equilibrium, what are the equilibrium concentrations of the other components?

By this time, you've seen quite a few variations on the type of equilibrium problem in which you know K and some initial quantities and must find the equilibrium quantities. Figure 17.6 presents a useful summary of the steps involved in solving these types of equilibrium problems. A good way to organize the steps is to group them into three overall parts.

SOLVING EQUILIBRIUM PROBLEMS

PRELIMINARY SETTING UP

1. Write the balanced equation
2. Write the reaction quotient, Q
3. Convert all amounts into the correct units (M or atm)

WORKING ON THE REACTION TABLE

4. When reaction direction is not known, compare Q with K
5. Construct a reaction table

✓Check the sign of x, the change in the quantity

SOLVING FOR x AND EQUILIBRIUM QUANTITIES

6. Substitute the quantities into Q
7. To simplify the math, assume that x is negligible ($[A]_{init} - x = [A]_{eq} \approx [A]_{init}$)
8. Solve for x

✓Check that assumption is justified (< 5% error). If not, solve quadratic equation for x.

9. Find the equilibrium quantities

✓Check to see that calculated values give the known K

Figure 17.6 Steps in solving equilibrium problems. These nine steps, grouped into three tasks, provide a useful approach to calculating equilibrium quantities, given initial quantities and K.

SECTION SUMMARY

In most equilibrium problems, we use quantities (concentrations or pressures) of reactants and products to find K, or we use K to find quantities. We use a reaction table to summarize the initial quantities, how they change, and the equilibrium quantities. When K is small and the initial quantity of reactant is large, we assume the unknown change in the quantity (x) is so much smaller than the initial quantity that it can be neglected. If this assumption is not justified (that is, if the error is greater than 5%), we use the quadratic formula to find x. To determine reaction direction, we compare the values of Q and K.

17.6 REACTION CONDITIONS AND THE EQUILIBRIUM STATE: LE CHÂTELIER'S PRINCIPLE

The most remarkable feature of a system at equilibrium is its ability to return to equilibrium after a change in conditions moves it away from that state. This drive to reattain equilibrium is stated in **Le Châtelier's principle:** when a chemical system at equilibrium is disturbed, it reattains equilibrium by undergoing a net reaction that reduces the effect of the disturbance.

Two phrases in this statement need further explanation. First, what does it mean to "disturb" a system? At equilibrium, Q equals K. When a change in conditions forces the system temporarily out of equilibrium ($Q \neq K$), we say that the system has been stressed, or disturbed. Three common disturbances are a change in concentration of a component (that appears in Q), a change in pressure (caused by a change in volume), or a change in temperature. We'll discuss each of these changes below.

The other phrase, "net reaction," is often referred to as a shift in the *equilibrium position* of the system to the right or left. The equilibrium position is just the specific equilibrium concentrations (or pressures). A shift in the equilibrium position to the right means that there is a net reaction to the right (reactant to product) until equilibrium is reattained; a shift to the left means that there is a net reaction to the left (product to reactant). Thus, when a disturbance occurs, we say that the equilibrium position shifts, which means that *concentrations (or pressures) change in a way that reduces the disturbance, and the system attains a new equilibrium position* ($Q = K$ again).

Le Châtelier's principle allows us to predict the direction of the shift in equilibrium position. Most importantly, it helps research and industrial chemists create conditions that maximize yields. For the remainder of this section, we examine each of the three kinds of disturbances—concentration, pressure (volume), and temperature—to see how a system at equilibrium responds; then, we'll note whether a catalyst has any effect.

In the following discussions, we focus on the reversible gaseous reaction between phosphorus trichloride and chlorine to produce phosphorus pentachloride:

$$PCl_3(g) + Cl_2(g) \rightleftharpoons PCl_5(g)$$

However, the basis of Le Châtelier's principle holds for any system at equilibrium, whether in the natural or social sciences.

The Universality of Le Châtelier's Principle Although Le Châtelier limited the scope of his principle to physical and chemical systems, its application is much wider. Ecologists and economists, for example, often see it at work. On the African savannah, the numbers of herbivores (antelope, wildebeest, zebra) and carnivores (lion, cheetah) are in delicate balance. Any disturbance (drought, disease) causes shifts in the relative numbers until they attain a new balance. The economic principle of supply-and-demand is another example of this equilibrium principle. A given demand-supply balance for a product establishes a given price. If either demand or supply is disturbed, the disturbance causes a shift in the other and, therefore, in the prevailing price until a new balance is attained.

The Effect of a Change in Concentration

When a system at equilibrium is disturbed by a change in concentration of one of the components, the system reacts in the direction that reduces the change:

- If the concentration increases, the system reacts to consume some of it.
- If the concentration decreases, the system reacts to produce some of it.

Of course, the component must be one that appears in Q; thus, pure liquids and solids, which do not appear in Q because their concentrations are constant, are not involved.

At 523 K, the PCl_3-Cl_2-PCl_5 system reaches equilibrium when

$$Q_c = \frac{[PCl_5]}{[PCl_3][Cl_2]} = 24.0 = K_c$$

What happens if we now inject some Cl_2 gas, one of the reactants? The system will always act to reduce the disturbance, so it will reduce the increase in reactant by proceeding toward the product side, thereby consuming some additional Cl_2. In terms of the reaction quotient, when we add Cl_2, the $[Cl_2]$ term increases, so the value of Q_c immediately falls as the denominator becomes larger; thus, the system is no longer at equilibrium. As some of the added Cl_2 reacts with some of the PCl_3 present and produces more PCl_5, the denominator becomes smaller once again and the numerator larger, until eventually Q_c again equals K_c. The concentrations of the components have changed, however: the concentrations of Cl_2 and PCl_5 are higher than in the original equilibrium position, and the concentration of PCl_3 is lower. Nevertheless, the ratio of values gives the same K_c. We describe this change by saying that *the equilibrium position shifts to the right when a component on the left is added:*

$$PCl_3 + Cl_2(\text{added}) \longrightarrow PCl_5$$

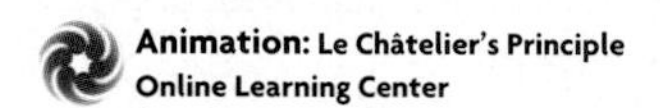

Animation: Le Châtelier's Principle
Online Learning Center

What happens if, instead of adding Cl_2, we remove some PCl_3, the other reactant? In this case, the system reduces the disturbance (the decrease in reactant), by proceeding toward the reactant side, thereby consuming some PCl_5. Once again, thinking in terms of Q_c, when we remove PCl_3, the $[PCl_3]$ term decreases, the denominator becomes smaller, and the value of Q_c rises above K_c. As some PCl_5 decomposes to PCl_3 and Cl_2, the numerator decreases and the denominator increases until Q_c equals K_c again. Here, too, the concentrations are different from those of the original equilibrium position, but K_c is not. We say that *the equilibrium position shifts to the left when a component on the left is removed:*

$$PCl_3(\text{removed}) + Cl_2 \longleftarrow PCl_5$$

The same points we just made for adding or removing a reactant also hold for adding or removing a product. If we add PCl_5, its concentration rises and the equilibrium position shifts to the left, just as it did when we removed some PCl_3; if we remove some PCl_5, the equilibrium position shifts to the right, just as it did when we added some Cl_2. In other words, no matter how the disturbance in concentration comes about, the system responds to make Q_c and K_c equal again. To summarize the effects of concentration changes (see margin):

- The equilibrium position shifts to the *right* if a reactant is added or a product is removed: [reactant] increases or [product] decreases.
- The equilibrium position shifts to the *left* if a reactant is removed or a product is added: [reactant] decreases or [product] increases.

ANY OF THESE CHANGES CAUSES A SHIFT TO THE *RIGHT*

increase increase decrease

$$PCl_3 + Cl_2 \rightleftharpoons PCl_5$$

decrease decrease increase

ANY OF THESE CHANGES CAUSES A SHIFT TO THE *LEFT*

In general, whenever the concentration of a component changes, *the equilibrium system reacts to consume some of the added substance or produce some of the removed substance*. In this way, the system "reduces the effect of the disturbance."

The effect is not completely eliminated, however, as we will see next from a quantitative comparison of original and new equilibrium positions.

Consider the case in which we added Cl_2 to the system at equilibrium. Suppose the original equilibrium position was established with the following concentrations: $[PCl_3] = 0.200\ M$, $[Cl_2] = 0.125\ M$, and $[PCl_5] = 0.600\ M$. Thus,

$$Q_c = \frac{[PCl_5]}{[PCl_3][Cl_2]} = \frac{0.600}{(0.200)(0.125)} = 24.0 = K_c$$

Now we add enough Cl_2 to increase its concentration by 0.075 M. Before any reaction occurs, this addition creates a new set of initial concentrations. Then the system reacts and comes to a new equilibrium position. From Le Châtelier's principle, we predict that adding more reactant will produce more product, that is, shift the equilibrium position to the right. Experiment shows that the new $[PCl_5]$ at equilibrium is 0.637 M.

Table 17.3 shows a reaction table of the entire process: the original equilibrium position, the disturbance, the (new) initial concentrations, the size and direction of the change needed to reattain equilibrium, and the new equilibrium position. Figure 17.7 depicts the process.

Table 17.3 The Effect of Added Cl_2 on the PCl_3-Cl_2-PCl_5 System

Concentration (M)	$PCl_3(g)$	+	$Cl_2(g)$	$\rightleftharpoons$	$PCl_5(g)$
Original equilibrium	0.200		0.125		0.600
Disturbance			+0.075		
New initial	0.200		0.200		0.600
Change	$-x$		$-x$		$+x$
New equilibrium	$0.200 - x$		$0.200 - x$		$0.600 + x$ (0.637)*

*Experimentally determined value.

From Table 17.3,

$$[PCl_5] = 0.600\ M + x = 0.637\ M, \qquad \text{so } x = 0.037\ M$$

Also, $[PCl_3] = [Cl_2] = 0.200\ M - x = 0.163\ M$

Therefore, at equilibrium,

$$K_{c(\text{original})} = \frac{0.600}{(0.200)(0.125)} = 24.0$$

$$K_{c(\text{new})} = \frac{0.637}{(0.163)(0.163)} = 24.0$$

There are several key points to notice about the new equilibrium concentrations that exist after Cl_2 is added:

- As we predicted, $[PCl_5]$ (0.637 M) is higher than its original concentration (0.600 M).
- $[Cl_2]$ (0.163 M) is higher than its original equilibrium concentration (0.125 M), but lower than its initial concentration just after the addition (0.200 M); thus, the disturbance (addition of Cl_2) is *reduced but not eliminated.*
- $[PCl_3]$ (0.163 M), the other left-side component, is lower than its original concentration (0.200 M) because some reacted with the added Cl_2.
- Most importantly, although the position of equilibrium shifted to the right, K_c *remains the same.*

Be sure to note that the system adjusts by changing concentrations, but the value of Q_c *at equilibrium* is the same as in the original system. In other words, *at a given temperature, K_c does **not** change with a change in concentration.*

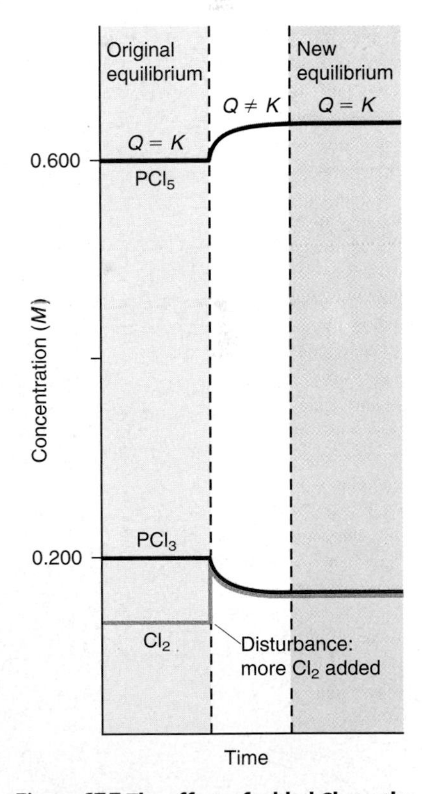

Figure 17.7 The effect of added Cl_2 on the PCl_3-Cl_2-PCl_5 system. In the original equilibrium (*gray region*), all concentrations are constant. When Cl_2 (*yellow curve*) is added, its concentration jumps and then starts to fall as Cl_2 reacts with some PCl_3 to form more PCl_5. After a period of time, equilibrium is re-established at new concentrations (*blue region*) but with the same K.

SAMPLE PROBLEM 17.11 Predicting the Effect of a Change in Concentration on the Equilibrium Position

Problem To improve air quality and obtain a useful product, chemists often remove sulfur from coal and natural gas by treating the fuel contaminant hydrogen sulfide with O_2:

$$2H_2S(g) + O_2(g) \rightleftharpoons 2S(s) + 2H_2O(g)$$

What happens to
(a) $[H_2O]$ if O_2 is added? **(b)** $[H_2S]$ if O_2 is added?
(c) $[O_2]$ if H_2S is removed? **(d)** $[H_2S]$ if sulfur is added?

Plan We write the reaction quotient to see how Q_c is affected by each disturbance, relative to K_c. This effect tells us the direction in which the reaction proceeds for the system to reattain equilibrium and how each concentration changes.

Solution Writing the reaction quotient: $Q_c = \dfrac{[H_2O]^2}{[H_2S]^2[O_2]}$

(a) When O_2 is added, the denominator of Q_c increases, so $Q_c < K_c$. The reaction proceeds to the right until $Q_c = K_c$ again, so $[H_2O]$ increases.

(b) As in part (a), when O_2 is added, $Q_c < K_c$. Some H_2S reacts with the added O_2 as the reaction proceeds to the right, so $[H_2S]$ decreases.

(c) When H_2S is removed, the denominator of Q_c decreases, so $Q_c > K_c$. As the reaction proceeds to the left to re-form H_2S, more O_2 is produced as well, so $[O_2]$ increases.

(d) The concentration of solid S is unchanged as long as some is present, so it does not appear in the reaction quotient. Adding more S has no effect, so $[H_2S]$ is unchanged (but see Comment 2 below).

Check Apply Le Châtelier's principle to see that the reaction proceeds in the direction that lowers the increased concentration or raises the decreased concentration.

Comment 1. As you know, sulfur exists most commonly as S_8. How would this change in formula affect the answers? The balanced equation and Q_c would be

$$8H_2S(g) + 4O_2(g) \rightleftharpoons S_8(s) + 8H_2O(g) \qquad Q_c = \frac{[H_2O]^8}{[H_2S]^8[O_2]^4}$$

The value of K_c is different for this equation, but the changes described in the problem have the same effects. For example, in (a), if O_2 were added, the denominator of Q_c would increase, so $Q_c < K_c$. As above, the reaction would proceed to the right until $Q_c = K_c$ again. In other words, changes predicted by Le Châtelier's principle for a given reaction are not affected by a change in the balancing coefficients.

2. In (d), you saw adding a solid has no effect on the concentrations of other components: because *the **concentration** of the solid cannot change*, it does not appear in Q. But *the **amount** of solid can change*. Adding H_2S shifts the reaction to the right, and more S forms.

FOLLOW-UP PROBLEM 17.11 In a study of the chemistry of glass etching, an inorganic chemist examines the reaction between sand (SiO_2) and hydrogen fluoride at a temperature above the boiling point of water:

$$SiO_2(s) + 4HF(g) \rightleftharpoons SiF_4(g) + 2H_2O(g)$$

Predict the effect on $[SiF_4]$ when **(a)** $H_2O(g)$ is removed; **(b)** some liquid water is added; **(c)** HF is removed; **(d)** some sand is removed.

The Effect of a Change in Pressure (Volume)

Changes in pressure have significant effects only on equilibrium systems with gaseous components. Aside from phase changes, a change in pressure has a negligible effect on liquids and solids because they are nearly incompressible. Pressure changes can occur in three ways:

- Changing the concentration of a gaseous component
- Adding an inert gas (one that does not take part in the reaction)
- Changing the volume of the reaction vessel

We just considered the effect of changing the concentration of a component, and that reasoning holds here. Next, let's see why *adding an inert gas has no effect on the equilibrium position.* Adding an inert gas does not change the volume, so all *reactant and product concentrations remain the same.* In other words, the volume and the number of moles of the reactant and product gases do not change, so their *partial pressures do not change.* Because we use these (unchanged) partial pressures in the reaction quotient, the equilibrium position cannot change. Moreover, the inert gas does not appear in Q, so it cannot have an effect.

On the other hand, changing the pressure by changing the volume often causes a large shift in the equilibrium position. Suppose we let the PCl_3-Cl_2-PCl_5 system come to equilibrium in a cylinder-piston assembly. Then, we press down on the piston to halve the volume: the gas pressure immediately doubles. To reduce this increase in gas pressure, the system responds by reducing the number of gas molecules. And it does so in the only possible way—by shifting the reaction toward the side with *fewer moles of gas,* in this case, toward the product side:

$$PCl_3(g) + Cl_2(g) \longrightarrow PCl_5(g)$$
$$\text{2 mol gas} \longrightarrow \text{1 mol gas}$$

Notice that *a change in volume results in a change in concentration:* a decrease in container volume raises the concentration, and an increase in volume lowers the concentration. Recall that $Q_c = \frac{[PCl_5]}{[PCl_3][Cl_2]}$. When the volume is halved, the concentrations double, but the denominator of Q_c is the product of two concentrations, so it quadruples while the numerator only doubles. Thus, Q_c becomes less than K_c. As a result, the system forms more PCl_5 and a new equilibrium position is reached. Because it is just another way to change the concentration, *a change in pressure due to a change in volume does **not** alter K_c.*

Thus, for a system that contains gases at equilibrium, in which the amount (mol) of gas, n_{gas}, changes during the reaction (Figure 17.8):

- If the volume becomes smaller (pressure is higher), the reaction shifts so that the total number of gas molecules decreases.
- If the volume becomes larger (pressure is lower), the reaction shifts so that the total number of gas molecules increases.

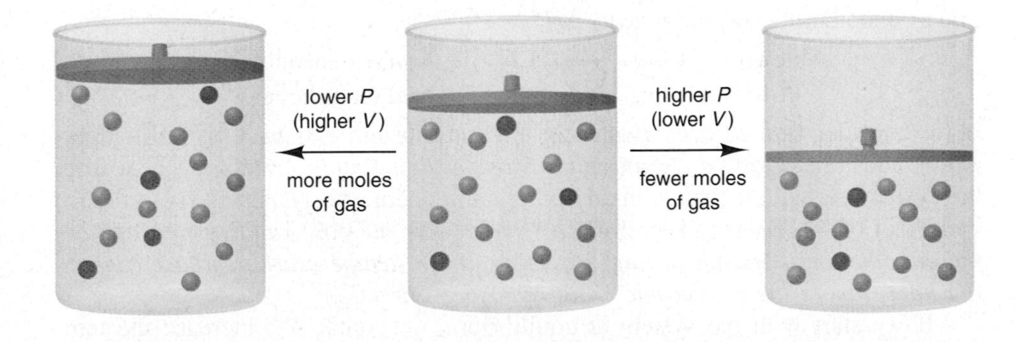

Figure 17.8 The effect of pressure (volume) on a system at equilibrium. The system of gases *(center)* is at equilibrium. For the reaction

● + ● ⇌ ●

an increase in pressure (*right*) decreases the volume, so the reaction shifts to the right to make fewer molecules. A decrease in pressure (*left*) increases the volume, so the reaction shifts to the left to make more molecules.

In many cases, however, n_{gas} does not change ($\Delta n_{gas} = 0$). For example,

$$H_2(g) + I_2(g) \rightleftharpoons 2HI(g)$$
$$\text{2 mol gas} \longrightarrow \text{2 mol gas}$$

Q_c has the same number of terms in the numerator and denominator:

$$Q_c = \frac{[HI]^2}{[H_2][I_2]} = \frac{[HI][HI]}{[H_2][I_2]}$$

Therefore, a change in volume has the same effect on the numerator and denominator. Thus, *if $\Delta n_{gas} = 0$, there is no effect on the equilibrium position.*

SAMPLE PROBLEM 17.12 Predicting the Effect of a Change in Volume (Pressure) on the Equilibrium Position

Problem How would you change the volume of each of the following reactions to *increase* the yield of the products?
(a) $CaCO_3(s) \rightleftharpoons CaO(s) + CO_2(g)$ (b) $S(s) + 3F_2(g) \rightleftharpoons SF_6(g)$
(c) $Cl_2(g) + I_2(g) \rightleftharpoons 2ICl(g)$

Plan Whenever gases are present, a change in volume causes a change in concentration. For reactions in which the number of moles of gas changes, if the volume decreases (pressure increases), the equilibrium position shifts to relieve the pressure by reducing the number of moles of gas. A volume increase (pressure decrease) has the opposite effect.

Solution (a) The only gas is the product CO_2. To make the system produce more CO_2, we increase the volume (decrease the pressure).

(b) With 3 mol of gas on the left and only 1 mol on the right, we decrease the volume (increase the pressure) to form more SF_6.

(c) The number of moles of gas is the same on both sides of the equation, so a change in volume (pressure) will have no effect on the yield of ICl.

Check Let's predict the relative values of Q_c and K_c. In (a), $Q_c = [CO_2]$, so increasing the volume will make $Q_c < K_c$, and the system will make more CO_2. In (b), $Q_c = [SF_6]/[F_2]^3$. Lowering the volume increases $[F_2]$ and $[SF_6]$ proportionately, but Q_c decreases because of the exponent 3 in the denominator. To make $Q_c = K_c$ again, $[SF_6]$ must increase. In (c), $Q_c = [ICl]^2/[Cl_2][I_2]$. A change in volume (pressure) affects the numerator (2 mol) and denominator (2 mol) equally, so it will have no effect.

FOLLOW-UP PROBLEM 17.12 Would you increase or decrease the pressure (via a volume change) of each of the following reaction mixtures to *decrease* the yield of products?
(a) $2SO_2(g) + O_2(g) \rightleftharpoons 2SO_3(g)$ (b) $4NH_3(g) + 5O_2(g) \rightleftharpoons 4NO(g) + 6H_2O(g)$
(c) $CaC_2O_4(s) \rightleftharpoons CaCO_3(s) + CO(g)$

The Effect of a Change in Temperature

Of the three types of disturbances—a change in concentration, in pressure, or in temperature—*only temperature changes alter K*. To see why, we must take the heat of reaction into account:

$$PCl_3(g) + Cl_2(g) \rightleftharpoons PCl_5(g) \quad \Delta H^0_{rxn} = -111 \text{ kJ}$$

The forward reaction is exothermic (releases heat; $\Delta H^0 < 0$), so the reverse reaction is endothermic (absorbs heat; $\Delta H^0 > 0$):

$$PCl_3(g) + Cl_2(g) \longrightarrow PCl_5(g) + \textbf{heat} \text{ (exothermic)}$$
$$PCl_3(g) + Cl_2(g) \longleftarrow PCl_5(g) + \textbf{heat} \text{ (endothermic)}$$

If we consider *heat as a component of the equilibrium system,* a rise in temperature "adds" heat to the system and a drop in temperature "removes" heat from the system. As with a change in any other component, the system shifts to reduce the effect of the change. Therefore, *a temperature increase (adding heat) favors the endothermic (heat-absorbing) direction, and a temperature decrease (removing heat) favors the exothermic (heat-releasing) direction.*

If we start with the system at equilibrium, Q_c equals K_c. Increase the temperature, and the system responds by decomposing some PCl_5 to PCl_3 and Cl_2, which absorbs the added heat. The denominator of Q_c becomes larger and the numerator smaller, so the system reaches a new equilibrium position at a smaller ratio of concentration terms, that is, a lower K_c. Similarly, the system responds to a drop in temperature by forming more PCl_5 from some PCl_3 and Cl_2, which releases more heat. The numerator of Q_c becomes larger, the denominator smaller, and the new equilibrium position has a higher K_c. Thus,

- *A temperature rise will increase K_c for a system with a positive ΔH^0_{rxn}.*
- *A temperature rise will decrease K_c for a system with a negative ΔH^0_{rxn}.*

Let's review these ideas with a sample problem.

SAMPLE PROBLEM 17.13 Predicting the Effect of a Change in Temperature on the Equilibrium Position

Problem How does an *increase* in temperature affect the equilibrium concentration of the underlined substance and K_c for each of the following reactions?

(a) $CaO(s) + H_2O(l) \rightleftharpoons \underline{Ca(OH)_2(aq)}$ $\Delta H^0 = -82$ kJ

(b) $CaCO_3(s) \rightleftharpoons CaO(s) + \underline{CO_2(g)}$ $\Delta H^0 = 178$ kJ

(c) $\underline{SO_2(g)} \rightleftharpoons S(s) + O_2(g)$ $\Delta H^0 = 297$ kJ

Plan We write each equation to show heat as a reactant or product. Increasing the temperature adds heat, so the system shifts to absorb the heat; that is, the endothermic reaction occurs. K_c will increase if the forward reaction is endothermic and decrease if it is exothermic.

Solution (a) $CaO(s) + H_2O(l) \rightleftharpoons Ca(OH)_2(aq) + \textit{heat}$

Adding heat shifts the system to the left: $[Ca(OH)_2]$ and K_c will decrease.

(b) $CaCO_3(s) + \textit{heat} \rightleftharpoons CaO(s) + CO_2(g)$

Adding heat shifts the system to the right: $[CO_2]$ and K_c will increase.

(c) $SO_2(g) + \textit{heat} \rightleftharpoons S(s) + O_2(g)$

Adding heat shifts the system to the right: $[SO_2]$ will decrease and K_c will increase.

Check You can check your answers by going through the reasoning for a *decrease* in temperature: heat is removed and the exothermic direction is favored. All the answers should be opposite.

FOLLOW-UP PROBLEM 17.13 How does a *decrease* in temperature affect the partial pressure of the underlined substance and the value of K_p for each of the following reactions?

(a) $C(graphite) + \underline{2H_2(g)} \rightleftharpoons CH_4(g)$ $\Delta H^0 = -75$ kJ

(b) $\underline{N_2(g)} + O_2(g) \rightleftharpoons 2NO(g)$ $\Delta H^0 = 181$ kJ

(c) $P_4(s) + 10Cl_2(g) \rightleftharpoons \underline{4PCl_5(g)}$ $\Delta H^0 = -1528$ kJ

The van't Hoff Equation: The Effect of *T* on *K* The *van't Hoff equation* shows how the equilibrium constant is affected by changes in temperature:

$$\ln \frac{K_2}{K_1} = -\frac{\Delta H^0_{rxn}}{R}\left(\frac{1}{T_2} - \frac{1}{T_1}\right) \quad (17.10)$$

where K_1 is the equilibrium constant at T_1, K_2 is the equilibrium constant at T_2, and R is the universal gas constant (8.314 J/mol·K). If we know ΔH^0_{rxn} and K at one temperature, the van't Hoff equation allows us to find K at any other temperature (or to find ΔH^0_{rxn}, given the two K's at two T's).

Here's a typical problem that requires the van't Hoff equation. Coal gasification processes usually begin with the formation of syngas from carbon and steam:

$$C(s) + H_2O(g) \rightleftharpoons CO(g) + H_2(g) \qquad \Delta H^0_{rxn} = 131 \text{ kJ/mol}$$

An engineer knows that K_p is only 9.36×10^{-17} at 25°C and therefore wants to find a temperature that allows a much higher yield. Calculate K_p at 700°C.

$$\ln \frac{K_2}{K_1} = -\frac{\Delta H^0_{rxn}}{R}\left(\frac{1}{T_2} - \frac{1}{T_1}\right)$$

The temperatures must be converted, and we make the units of ΔH^0 and R consistent:

$$\ln\left(\frac{K_{p2}}{9.36\times10^{-17}}\right) = -\frac{131\times10^3 \text{ J/mol}}{8.314 \text{ J/mol·K}}\left(\frac{1}{973 \text{ K}} - \frac{1}{298 \text{ K}}\right)$$

$$\frac{K_{p2}}{9.36\times10^{-17}} = 8.51\times10^{15}$$

$$K_{p2} = 0.797$$

Temperature-Dependent System There is a striking similarity among expressions for the temperature dependence of K, k (rate constant), and P (equilibrium vapor pressure):

$$\ln\frac{K_2}{K_1} = -\frac{\Delta H^0_{rxn}}{R}\left(\frac{1}{T_2} - \frac{1}{T_1}\right)$$

$$\ln\frac{k_2}{k_1} = -\frac{E_a}{R}\left(\frac{1}{T_2} - \frac{1}{T_1}\right)$$

$$\ln\frac{P_2}{P_1} = -\frac{\Delta H_{vap}}{R}\left(\frac{1}{T_2} - \frac{1}{T_1}\right)$$

Each concentration-related term (K, k, or P) is dependent on T through an energy term (ΔH^0_{rxn}, E_a, or ΔH_{vap}, respectively) divided by R. The similarity arises because the equations express the same relationship. K equals a ratio of rate constants, and $E_{a(fwd)} - E_{a(rev)} = \Delta H^0_{rxn}$. In the vapor pressure case, for the phase change $A(l) \rightleftharpoons A(g)$, the heat of reaction is equal to the heat of vaporization: $\Delta H_{vap} = \Delta H^0_{rxn}$. Also, the equilibrium constant equals the equilibrium vapor pressure: $K_p = P_A$.

Equation 17.10 confirms the qualitative prediction from Le Châtelier's principle: for a temperature rise, we have

$$T_2 > T_1 \quad \text{and} \quad 1/T_2 < 1/T_1, \quad \text{so } 1/T_2 - 1/T_1 < 0$$

Therefore,

- For an endothermic reaction ($\Delta H^0_{rxn} > 0$), the $-(\Delta H^0_{rxn}/R)$ term is < 0. With $1/T_2 - 1/T_1 < 0$, the right side of the equation is > 0. Thus, $\ln(K_2/K_1) > 0$, so $K_2 > K_1$.
- For an exothermic reaction ($\Delta H^0_{rxn} < 0$), the $-(\Delta H^0_{rxn}/R)$ term is > 0. With $1/T_2 - 1/T_1 < 0$, the right side of the equation is < 0. Thus, $\ln(K_2/K_1) < 0$, so $K_2 < K_1$.

(For further practice with the van't Hoff equation, see Problems 17.73 and 17.74.)

The Lack of Effect of a Catalyst

Let's briefly consider a final external change to the reacting system: adding a catalyst. Recall from Chapter 16 that a catalyst speeds up a reaction by providing an alternative mechanism with a lower activation energy, thereby increasing the forward *and* reverse rates to the same extent. In other words, it shortens the time needed to attain the *final* concentrations. Thus, *a catalyst shortens the time it takes to reach equilibrium but has* ***no*** *effect on the equilibrium position.*

If, for instance, we add a catalyst to a mixture of PCl_3 and Cl_2 at 523 K, the system will attain the *same* equilibrium concentrations of PCl_3, Cl_2, and PCl_5 *more quickly* than it did without the catalyst. As you'll see in a moment, however, catalysts often play key roles in optimizing reaction systems. The upcoming Chemical Connections essays highlight two areas that apply equilibrium principles: the first examines metabolic processes in organisms, and the second describes a major industrial process.

Table 17.4 summarizes the effects of changing conditions on the position of equilibrium. Note that many changes alter the equilibrium position, but only temperature changes alter the value of the equilibrium constant.

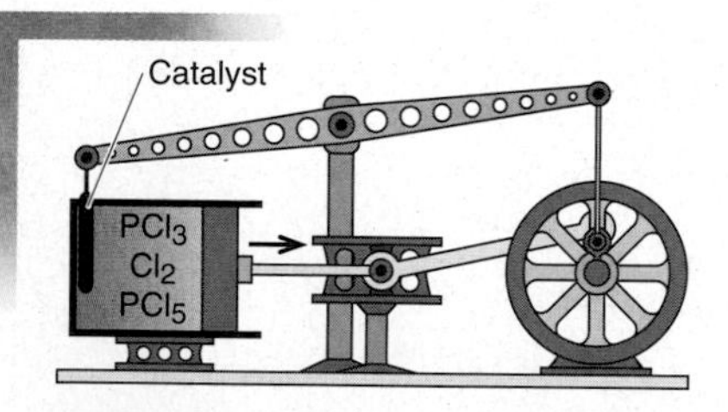

Catalyzed Perpetual Motion? This engine consists of a piston attached to a flywheel, whose rocker arm holds a catalyst and moves in and out of the reaction in the cylinder. What if the catalyst *could* increase the rate of PCl_5 breakdown but not of its formation? In the cylinder, the catalyst would speed up the breakdown of PCl_5 to PCl_3 and Cl_2—1 mol of gas to 2 mol—increasing gas pressure and pushing the piston out. With the catalyst out of the cylinder, PCl_3 and Cl_2 would re-form PCl_5, lowering gas pressure and moving the piston in. The process would supply power with no input of external energy. If a catalyst could change the rate in only one direction, it would change K, and we could design some amazing machines!

Table 17.4 Effect of Various Disturbances on a System at Equilibrium

Disturbance	Net Direction of Reaction	Effect on Value of K
Concentration		
Increase [reactant]	Toward formation of product	None
Decrease [reactant]	Toward formation of reactant	None
Increase [product]	Toward formation of reactant	None
Decrease [product]	Toward formation of product	None
Pressure		
Increase P (decrease V)	Toward formation of fewer moles of gas	None
Decrease P (increase V)	Toward formation of more moles of gas	None
Increase P (add inert gas, no change in V)	None; concentrations unchanged	None
Temperature		
Increase T	Toward absorption of heat	Increases if $\Delta H^0_{rxn} > 0$ Decreases if $\Delta H^0_{rxn} < 0$
Decrease T	Toward release of heat	Increases if $\Delta H^0_{rxn} < 0$ Decreases if $\Delta H^0_{rxn} > 0$
Catalyst added	None; forward and reverse equilibrium attained sooner; rates increase equally	None

SAMPLE PROBLEM 17.14 Determining Equilibrium Parameters from Molecular Scenes

Problem For the reaction,

$$X(g) + Y_2(g) \rightleftharpoons XY(g) + Y(g) \qquad \Delta H > 0$$

the following molecular scenes depict different reaction mixtures (X = green, Y = purple):

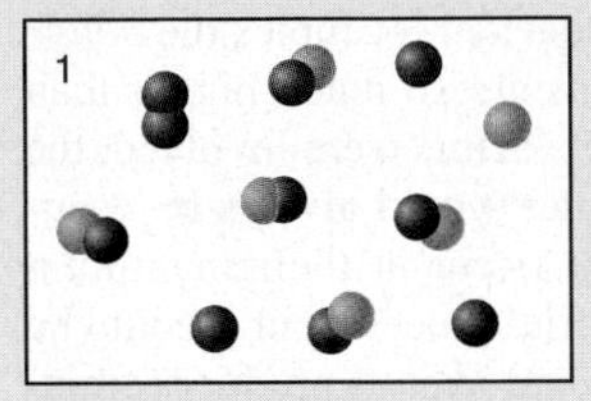

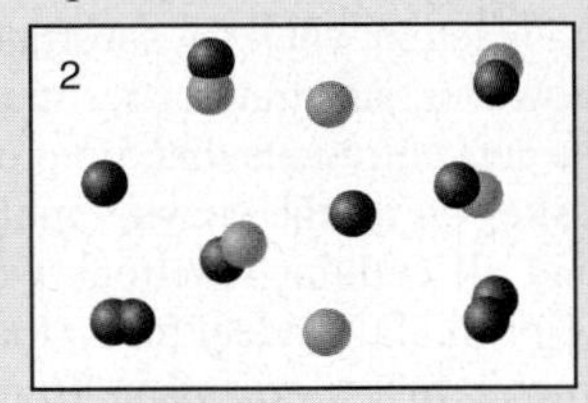

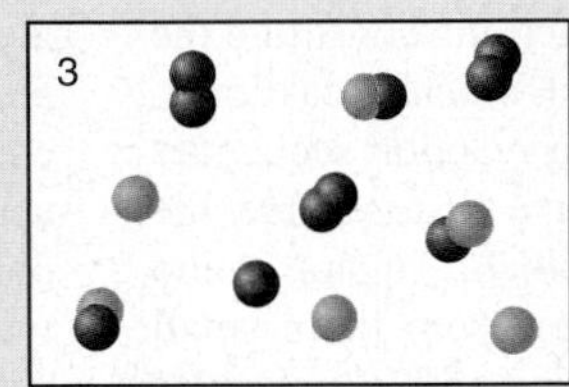

(a) If $K_c = 2$ at the temperature of the reaction, which scene represents the mixture at equilibrium?
(b) Will the reaction mixtures in the other two scenes proceed toward reactants or toward products to reach equilibrium?
(c) For the mixture at equilibrium, how will a rise in temperature affect $[Y_2]$?

Plan (a) We are given the balanced equation and the value of K_c and must choose the scene representing the mixture at equilibrium. We write the expression for Q_c, and for each scene, count particles and plug in the numbers to solve for the value of Q_c. Whichever scene gives a Q_c equal to K_c represents the mixture at equilibrium. **(b)** To determine the direction each reaction proceeds in the other two scenes, we compare the value of Q_c with the given K_c. If $Q_c > K_c$, the numerator (product side) is too high, so the reaction proceeds toward reactants; if $Q_c < K_c$, the reaction proceeds toward products. **(c)** We are given the sign of ΔH and must see whether a rise in T (corresponding to supplying heat) will increase or decrease the amount of the reactant Y_2. We treat heat as a reactant or product and see whether adding heat shifts the reaction right or left.

Solution (a) For the reaction, we have

$$Q_c = \frac{[XY][Y]}{[X][Y_2]}$$

scene 1: $Q_c = \dfrac{5 \times 3}{1 \times 1} = 15$ scene 2: $Q_c = \dfrac{4 \times 2}{2 \times 2} = 2$ scene 3: $Q_c = \dfrac{3 \times 1}{3 \times 3} = \dfrac{1}{3}$

For scene 2, $Q_c = K_c$, so it represents the mixture at equilibrium.
(b) For scene 1, Q_c (15) $> K_c$ (2), so the reaction proceeds toward reactants. For scene 3, Q_c ($\frac{1}{3}$) $< K_c$ (2), so the reaction proceeds toward products.
(c) The reaction is endothermic, so heat acts as a reactant:

$$X(g) + Y_2(g) + \textit{heat} \rightleftharpoons XY(g) + Y(g)$$

Therefore, adding heat to the left shifts the reaction to the right, so $[Y_2]$ decreases.

Check (a) Remember that quantities in the numerator (or denominator) of Q_c are multiplied, not added. For example, the denominator for scene 1 is $1 \times 1 = 1$, not $1 + 1 = 2$.
(c) A good check is to imagine that $\Delta H < 0$ and see if you get the opposite result:

$$X(g) + Y_2(g) \rightleftharpoons XY(g) + Y(g) + \textit{heat}$$

If $\Delta H < 0$, adding heat would shift the reaction to the left and increase $[Y_2]$.

FOLLOW-UP PROBLEM 17.14 For the reaction

$$C_2(g) + D_2(g) \rightleftharpoons 2CD(g) \qquad \Delta H < 0$$

these molecular scenes depict different reaction mixtures (C = red and D = blue):

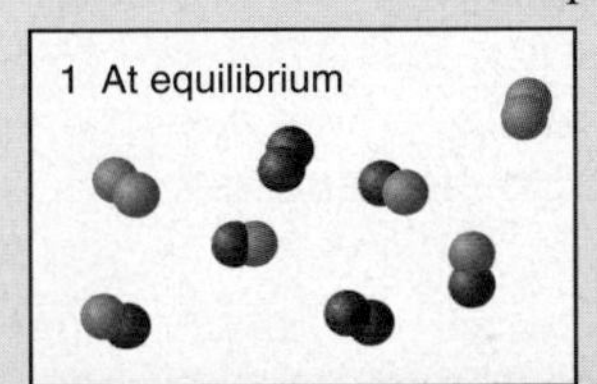

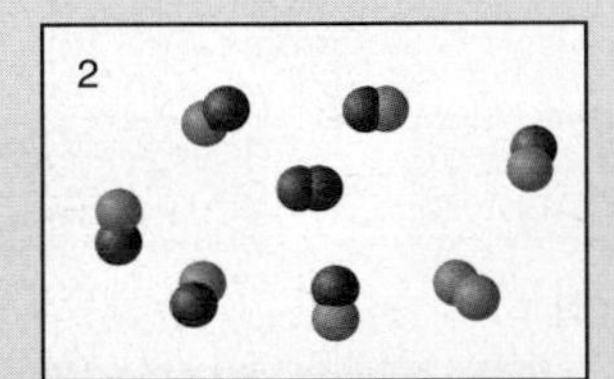

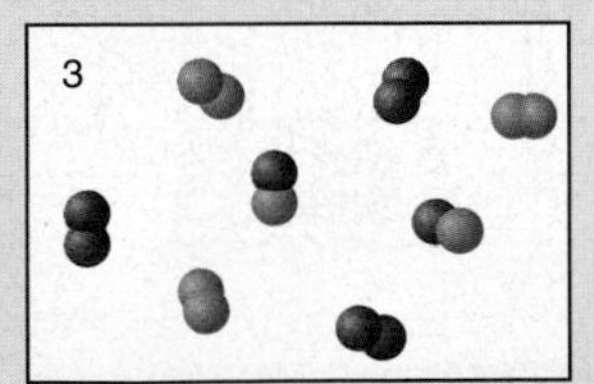

(a) Calculate the value of K_p. **(b)** In which direction will the reaction proceed for the mixtures *not* at equilibrium? **(c)** For the mixture at equilibrium, what effect will a rise in T have on the total moles of gas (increase, decrease, no effect)? Explain.

Design and Control of a Metabolic Pathway

Biological cells are microscopic wizards of chemical change. From the simplest bacterium to the most specialized neuron, every cell performs thousands of individual reactions that allow it to grow and reproduce, feed and excrete, move and communicate. Taken together, these chemical reactions constitute the cell's *metabolism*. These myriad feats of biochemical breakdown, synthesis, and energy flow are organized into reaction sequences called **metabolic pathways.** Some pathways disassemble the biopolymers in food into monomer building blocks: sugars, amino acids, and nucleotides. Others extract energy from these small molecules and then use much of the energy to assemble molecules with other functions, such as hormones, neurotransmitters, defensive toxins, and the cell's own biopolymers.

In principle, *each step in a metabolic pathway is a reversible reaction catalyzed by a specific enzyme* (Section 16.8). However, *equilibrium is never reached* in a pathway. As in other reaction sequences, the product of the first reaction becomes the reactant of the second, the product of the second becomes the reactant of the third, and so on. Consider the five-step pathway shown in Figure B17.1, in which one amino acid, threonine, is converted into another, isoleucine, in the cells of a bread mold. Threonine, supplied from a different region of the cell, forms ketobutyrate through the catalytic action of the first enzyme (reaction 1). The equilibrium position of reaction 1 continuously shifts to the right because the ketobutyrate is continuously removed as the reactant in reaction 2. Similarly, reaction 2 shifts to the right as its product is used in reaction 3. Each subsequent reaction shifts the equilibrium position of the previous reaction in the direction of product. The final product, isoleucine, is typically removed to make proteins elsewhere in the cell. Thus, *the entire pathway operates in one direction.*

This continuous shift in equilibrium position results in two major features of metabolic pathways. First, *each step proceeds with nearly 100% yield:* virtually every molecule of threonine that enters this region of the cell eventually changes to ketobutyrate, every molecule of ketobutyrate to the next product, and so on.

Second, *reactant and product concentrations remain constant* or vary within extremely narrow limits, even though the system never attains equilibrium. This situation, called a *steady state,* is different in a basic way from those we have studied so far. In equilibrium systems, equal reaction rates in *opposing* directions give rise to constant concentrations of reactants and products. In steady-state systems, on the other hand, the rates of reactions in *one* direction—into, through, and out of the system—give rise to nearly constant concentrations of intermediates. Ketobutyrate, for example, is formed via reaction 1 just as fast as it is used via reaction 2, so its concentration is kept constant. (You can establish a steady-state amount of water by filling a sink and then opening the faucet and drain so that water enters just as fast as it leaves.)

It is critical that a cell regulate the concentration of the final product of a pathway. Recall from Chapter 16 that an enzyme catalyzes a reaction when the substrate (reactant) occupies the *active site*. However, substrate concentrations are so much higher than enzyme concentrations that, if no other factors were involved, the active sites on all the enzyme molecules would always be occupied and all cellular reactions would occur at their maximum rates. This might be ideal for an industrial process, but it could be catastrophic for an organism. To regulate overall product formation, the rates of certain key steps are controlled by *regulatory enzymes,* which contain an *inhibitor site* on their surface in addition to an active site. The enzyme's three-dimensional structure is such that when the inhibitor site is occupied, the shape of the active site is deformed and the reaction cannot be catalyzed (Figure B17.2).

In the simplest case of metabolic control (Figure B17.1), *the final product of the pathway is also the inhibitor molecule, and the regulatory enzyme catalyzes the first step.* Suppose, for instance, that a bread mold cell is temporarily making less protein, so the isoleucine synthesized by the pathway in Figure B17.1 is not being removed as quickly. As the concentration of isoleucine rises, the chance increases that an isoleucine molecule will land on the inhibitor site of the first enzyme in the pathway, thereby inhibiting its own production. This process is called *end-product feedback inhibition.* More complex pathways have more elaborate regulatory schemes, but many operate through similar types of inhibitory feedback (see Problem 17.112). The exquisitely detailed regulation of a cell's numerous metabolic pathways arises directly from the principles of chemical kinetics and equilibrium.

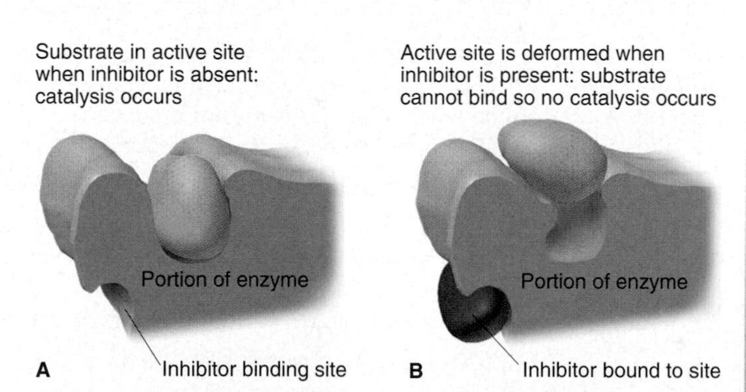

Figure B17.2 Effect of inhibitor binding on shape of active site. **A,** If inhibitor site is not occupied, the enzyme catalyzes the reaction. **B,** If inhibitor site is occupied, the enzyme molecule does not function.

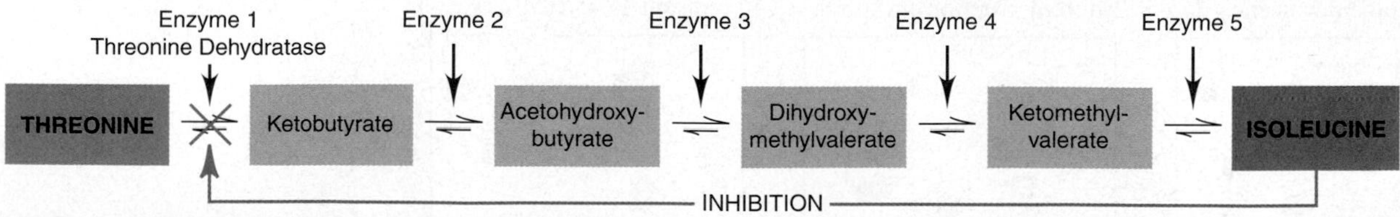

Figure B17.1 The biosynthesis of isoleucine from threonine. Isoleucine is synthesized from threonine in a sequence of five enzyme-catalyzed reactions. The unequal equilibrium arrows show that each step is reversible but is shifted toward product as each product becomes the reactant of the subsequent step. When the cell's need for isoleucine is temporarily satisfied, the concentration of isoleucine, the end product, builds up and inhibits the catalytic activity of threonine dehydratase, the first enzyme in the pathway.

CHEMICAL CONNECTIONS to Industrial Production

The Haber Process for the Synthesis of Ammonia

Nitrogen occurs in many essential natural and synthetic compounds. By far the richest source of nitrogen is the atmosphere, where four of every five molecules are N_2. Despite this abundance, the supply of *usable* nitrogen for biological and manufacturing processes is limited because of the low chemical reactivity of N_2. Because of the strong triple bond holding the two N atoms together, the nitrogen atom is very difficult to "fix," that is, to combine with other atoms.

Natural nitrogen fixation occurs either through the elegant specificity of enzymes found in bacteria that live on plant roots or through the brute force of lightning. Nearly 13% of nitrogen fixation on Earth is accomplished industrially through the **Haber process** for the formation of ammonia from its elements:

$$N_2(g) + 3H_2(g) \rightleftharpoons 2NH_3(g) \qquad \Delta H^0_{rxn} = -91.8 \text{ kJ}$$

The process was developed by the German chemist Fritz Haber and first used in 1913. From its humble beginnings in a plant with a capacity of 12,000 tons a year, world production of ammonia has exploded to a current level of more than 110 million tons a year. On a mole basis, more ammonia is produced industrially than any other compound. Over 80% of this ammonia is used in fertilizer applications. In fact, the most common form of fertilizer is compressed anhydrous, liquid NH_3 injected directly into the soil (Figure B17.3). Other uses of NH_3 include the production of explosives, via the formation of HNO_3, and the making of nylons and other polymers. Smaller amounts are used as refrigerants, rubber stabilizers, and household cleaners and in the synthesis of pharmaceuticals and other organic chemicals.

The Haber process provides an excellent opportunity to apply equilibrium principles and see the compromises needed to make an industrial process economically worthwhile. From inspection of the balanced equation, we can see three ways to maximize the yield of ammonia:

1. *Decrease [NH_3].* Ammonia is the product, so removing it as it forms will make the system produce more in a continual drive to reattain equilibrium.

Figure B17.3 Liquid ammonia used as fertilizer.

Table B17.1 Effect of Temperature on K_c for Ammonia Synthesis

T (K)	K_c
200.	7.17×10^{15}
300.	2.69×10^{8}
400.	3.94×10^{4}
500.	1.72×10^{2}
600.	4.53×10^{0}
700.	2.96×10^{-1}
800.	3.96×10^{-2}

2. *Decrease volume (increase pressure).* Because 4 mol of gas reacts to form 2 mol of gas, decreasing the volume will shift the equilibrium position toward fewer moles of gas, that is, toward ammonia formation.
3. *Decrease temperature.* Because the formation of ammonia is exothermic, decreasing the temperature (removing heat) will shift the equilibrium position toward the product, thereby increasing K_c (Table B17.1).

Therefore, the ideal conditions for maximizing the yield of ammonia are continual removal of NH_3 as it forms, high pressure, and low temperature. Figure B17.4 shows the percent yield of ammonia at various conditions of pressure and temperature. Note the almost complete conversion (98.3%) to ammonia at 1000 atm and the relatively low temperature of 473 K (200.°C).

Unfortunately, a problem arises that highlights the distinction between the principles of equilibrium and kinetics. Although the *yield* is favored by low temperature, the *rate* of formation is not. In fact, ammonia forms so slowly at low temperatures that the process becomes uneconomical. In practice, a compromise is achieved that optimizes yield *and* rate. High pressure and continuous removal are used to increase yield, but the temperature is

(continued)

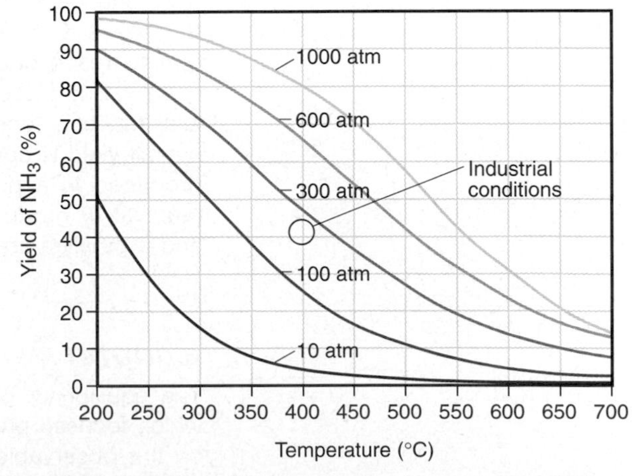

Figure B17.4 Percent yield of ammonia vs. temperature (°C) at five different operating pressures. At very high pressure and low temperature (*top left*), the yield is high, but the rate of formation is low. Industrial conditions (*circle*) are between 200 and 300 atm at about 400°C.

raised to a moderate level and a catalyst is used to increase the rate. Achieving the same rate without employing a catalyst would require much higher temperatures and would result in a much lower yield.

The stages in the industrial production of ammonia are shown schematically in Figure B17.5. To extend equipment life and minimize cost, modern ammonia plants operate at pressures of about 200 to 300 atm and temperatures of around 673 K (400.°C). The catalyst consists of 5-mm to 10-mm chunks of iron crystals embedded in a fused mixture of MgO, Al_2O_3, and SiO_2. The stoichiometric ratio of compressed reactant gases (N_2:H_2 = 1:3 by volume) is injected into the heated, pressurized reaction chamber, where it flows over the catalyst beds. Some of the heat needed is supplied by the enthalpy change of the reaction. The emerging equilibrium mixture, which contains about 35% NH_3 by volume, is cooled by refrigeration coils until the NH_3 (boiling point, −33.4°C) condenses and is removed. Because N_2 and H_2 have much lower boiling points than NH_3, they remain gaseous and are recycled by pumps back into the reaction chamber to continue the process.

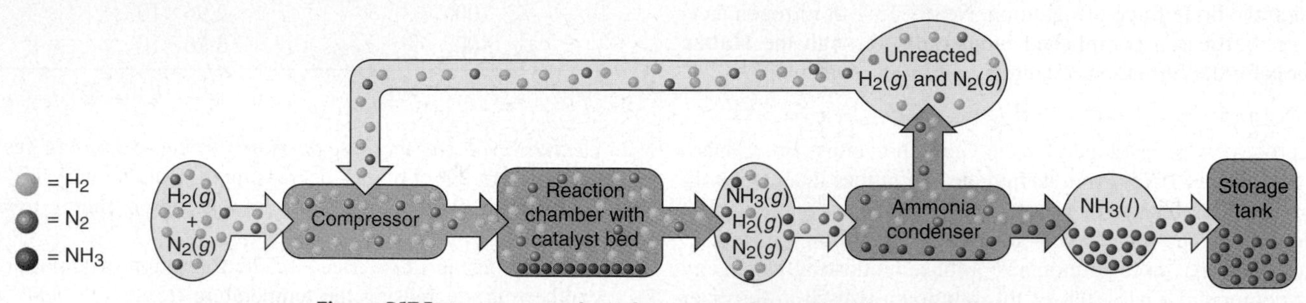

Figure B17.5 Key stages in the Haber process for synthesizing ammonia.

SECTION SUMMARY

Le Châtelier's principle states that if a system at equilibrium is disturbed, the system undergoes a net reaction that reduces the disturbance and allows equilibrium to be reattained. Changes in concentration cause a net reaction away from the added component or toward the removed component. For a reaction that involves a change in number of moles of gas, an increase in pressure (decrease in volume) causes a net reaction toward fewer moles of gas, and a decrease in pressure causes the opposite change. Although the equilibrium concentrations of components change as a result of concentration and volume changes, K does not change. A temperature change *does* change K: higher T increases K for an endothermic reaction (positive ΔH^0_{rxn}) and decreases K for an exothermic reaction (negative ΔH^0_{rxn}). A catalyst causes the system to reach equilibrium more quickly by speeding forward and reverse reactions equally, but it does not affect the equilibrium position. A metabolic pathway is a cellular reaction sequence in which each step is shifted completely toward product. Its overall yield is controlled by feedback inhibition of certain key enzymes. Ammonia is produced in a process favored by high pressure, low temperature, and continual removal of product. To make the process economical, an intermediate temperature and a catalyst are used.

Chapter Perspective

The equilibrium phenomenon is central to all natural systems. In this introduction, which focused primarily on gaseous systems, we discussed the nature of equilibrium on the observable and molecular levels, ways to solve relevant problems, and how conditions affect systems at equilibrium. In the next two chapters, we apply these ideas to acids and bases and to other aqueous ionic systems. Then we examine the fundamental relationship between the equilibrium constant and the energy change that accompanies a reaction to understand why chemical reactions occur.

For Review and Reference

(Numbers in parentheses refer to pages, unless noted otherwise.)

Learning Objectives

Relevant section and/or sample problem (SP) numbers appear in parentheses.

Understand These Concepts

1. The distinction between the rate and the extent of a reaction (Introduction)
2. Why a system attains dynamic equilibrium when forward and reverse reaction rates are equal (Section 17.1)
3. The equilibrium constant as a number that is equal to a particular ratio of rate constants and of concentration terms (Section 17.1)
4. How the magnitude of K is related to the extent of the reaction (Section 17.1)
5. Why the same equilibrium state is reached no matter what the starting concentrations of the reacting system (Section 17.2)
6. How the reaction quotient (Q) changes continuously until the system reaches equilibrium, at which point $Q = K$ (Section 17.2)
7. Why the form of Q is based exactly on the balanced equation *as written* (Section 17.2)
8. How the *sum* of reaction steps gives the overall reaction, and the *product* of Q's (or K's) gives the overall Q (or K) (Section 17.2)
9. Why pure solids and liquids do not appear in Q (Section 17.2)
10. How the interconversion of K_c and K_p is based on the ideal gas law and Δn_{gas} (Section 17.3)
11. How the reaction direction depends on the relative values of Q and K (Section 17.4)
12. How a reaction table is used to find an unknown quantity (concentration or pressure) (Section 17.5)
13. How assuming that the change in [reactant] is relatively small simplifies finding equilibrium quantities (Section 17.5)
14. How Le Châtelier's principle explains the effects of a change in concentration, pressure (volume), or temperature on a system at equilibrium and on K (Section 17.6)
15. Why a change in temperature *does* affect K (Section 17.6)
16. Why the addition of a catalyst does *not* affect K (Section 17.6)

Master These Skills

1. Writing the reaction quotient (Q) from a balanced equation (SP 17.1)
2. Writing Q and calculating K for a reaction consisting of more than one step (SP 17.2)
3. Writing Q and finding K for a reaction multiplied by a common factor (SP 17.3)
4. Writing Q for heterogeneous equilibria (Section 17.2)
5. Converting between K_c and K_p (SP 17.4)
6. Comparing Q and K to determine reaction direction (SP 17.5)
7. Substituting quantities (concentrations or pressures) into Q to find K (Section 17.5)
8. Using a reaction table to determine quantities and find K (SP 17.6)
9. Finding one equilibrium quantity from other equilibrium quantities and K (SP 17.7)
10. Finding an equilibrium quantity from initial quantities and K (SP 17.8)
11. Solving a quadratic equation for an unknown equilibrium quantity (Section 17.5)
12. Assuming that the change in [reactant] is relatively small to find equilibrium quantities and checking the assumption (SP 17.9)
13. Comparing the values of Q and K to find reaction direction and sign of x, the unknown change in a quantity (SP 17.10)
14. Using the relative values of Q and K to predict the effect of a change in concentration on the equilibrium position and on K (SP 17.11)
15. Using Le Châtelier's principle and Δn_{gas} to predict the effect of a change in pressure (volume) on the equilibrium position (SP 17.12)
16. Using Le Châtelier's principle and ΔH^0 to predict the effect of a change in temperature on the equilibrium position and on K (SP 17.13)
17. Using the van't Hoff equation to calculate K at one temperature given K at another temperature (Section 17.6)
18. Using molecular scenes to find equilibrium parameters (SP 17.14)

Key Terms

Section 17.1
equilibrium constant (K) (725)

Section 17.2
law of chemical equilibrium (law of mass action) (726)
reaction quotient (Q) (mass-action expression) (726)

Section 17.6
Le Châtelier's principle (745)
metabolic pathway (754)
Haber process (755)

Key Equations and Relationships

17.1 Defining equilibrium in terms of reaction rates (724):

$$\text{At equilibrium: } \text{rate}_{fwd} = \text{rate}_{rev}$$

17.2 Defining the equilibrium constant for the reaction $A \rightleftharpoons 2B$ (725):

$$K = \frac{k_{fwd}}{k_{rev}} = \frac{[B]^2_{eq}}{[A]_{eq}}$$

17.3 Defining the equilibrium constant in terms of the reaction quotient (726):

$$\text{At equilibrium: } Q = K$$

17.4 Expressing Q_c for the reaction $aA + bB \rightleftharpoons cC + dD$ (727):

$$Q_c = \frac{[C]^c[D]^d}{[A]^a[B]^b}$$

Key Equations and Relationships *(continued)*

17.5 Finding the overall K for a reaction sequence (728):

$$K_{overall} = K_1 \times K_2 \times K_3 \times \cdots$$

17.6 Finding K of a reaction from K of the reverse reaction (730):

$$K_{fwd} = \frac{1}{K_{rev}}$$

17.7 Finding K of a reaction multiplied by a factor n (730):

$$K' = K^n$$

17.8 Relating K based on pressures to K based on concentrations (734):

$$K_p = K_c(RT)^{\Delta n_{gas}}$$

17.9 Assuming that ignoring the concentration that reacts introduces no significant error (742):

$$[A]_{init} - [A]_{reacting} = [A]_{eq} \approx [A]_{init}$$

17.10 Finding K at one temperature given K at another (van't Hoff equation) (751):

$$\ln \frac{K_2}{K_1} = -\frac{\Delta H^0_{rxn}}{R}\left(\frac{1}{T_2} - \frac{1}{T_1}\right)$$

Highlighted Figures and Tables

These figures (F) and tables (T) provide a review of key ideas.

F17.2 The range of equilibrium constants (725)
F17.3 The change in Q during a reaction (727)
T17.2 Ways of expressing Q and calculating K (732)
F17.5 Reaction direction and the relative sizes of Q and K (735)
F17.6 Steps in solving equilibrium problems (745)
T17.3 Effect of added Cl_2 on the PCl_3-Cl_2-PCl_5 system (747)
F17.8 Effect of pressure (volume) on equilibrium (749)
T17.4 Effects of disturbances on equilibrium (752)

Brief Solutions to Follow-up Problems

17.1 (a) $Q_c = \dfrac{[NO]^4[H_2O]^6}{[NH_3]^4[O_2]^5}$ (b) $Q_c = \dfrac{[N_2O][NO_2]}{[NO]^3}$

17.2 $H_2(g) + Br_2(g) \rightleftharpoons 2HBr(g)$;

$$Q_{c(overall)} = \frac{[HBr]^2}{[H_2][Br_2]}$$

$$Q_{c(overall)} = Q_{c1} \times Q_{c2} \times Q_{c3}$$

$$= \frac{[\cancel{Br}]^{\cancel{2}}}{[Br_2]} \times \frac{[HBr]\cancel{[H]}}{\cancel{[Br]}[H_2]} \times \frac{[HBr]}{\cancel{[H]}\cancel{[Br]}} = \frac{[HBr]^2}{[H_2][Br_2]}$$

17.3 (a) $K_c = K_{c(ref)}^{1/2} = 2.8\times10^4$

(b) $K_c = \left(\dfrac{1}{K_{c(ref)}}\right)^{2/3} = 1.2\times10^{-6}$

17.4 $K_p = K_c(RT)^{-1} = 1.67\left(0.0821\,\dfrac{\text{atm·L}}{\text{mol·K}} \times 500.\text{ K}\right)^{-1}$

$= 4.07\times10^{-2}$

17.5 $Q_p = \dfrac{(P_{CH_3Cl})(P_{HCl})}{(P_{CH_4})(P_{Cl_2})} = \dfrac{(0.24)(0.47)}{(0.13)(0.035)} = 25$;

$Q_p < K_p$, so CH_3Cl is forming.

17.6 From the reaction table for $2NO + O_2 \rightleftharpoons 2NO_2$,

$P_{O_2} = 1.000\text{ atm} - x = 0.506\text{ atm}; x = 0.494\text{ atm}$

Also, $P_{NO} = 0.012$ atm and $P_{NO_2} = 0.988$ atm, so

$$K_p = \frac{0.988^2}{0.012^2(0.506)} = 1.3\times10^4$$

17.7 Since $\Delta n_{gas} = 0$, $K_p = K_c = 2.3\times10^{30} = \dfrac{(0.781)(0.209)}{P^2_{NO}}$

Thus, $P_{NO} = 2.7\times10^{-16}$ atm

17.8 From the reaction table, $[H_2] = [I_2] = x$; $[HI] = 0.242 - 2x$.

Thus, $K_c = 1.26\times10^{-3} = \dfrac{x^2}{(0.242 - 2x)^2}$

Taking the square root of both sides, ignoring the negative root, and solving gives $x = [H_2] = 8.02\times10^{-3}\ M$.

17.9 (a) Based on the reaction table, and assuming that $0.20\ M - x \approx 0.20\ M$,

$$K_c = 2.94\times10^{-10} \approx \frac{4x^2}{0.20} \qquad x \approx 3.8\times10^{-6}$$

Error $= 1.9\times10^{-3}\%$, so assumption is justified; therefore, at equilibrium, $[I_2] = 0.20\ M$ and $[I] = 7.6\times10^{-6}\ M$.

(b) Based on the same reaction table and assumption, $x \approx 0.10$; error is 50%, so assumption is *not* justified. Solve equation:

$$4x^2 + 0.209x - 0.042 = 0 \qquad x = 0.080\ M$$

Therefore, at equilibrium, $[I_2] = 0.12\ M$ and $[I] = 0.16\ M$.

17.10 (a) $Q_c = \dfrac{(0.0900)(0.0900)}{0.2100} = 3.86\times10^{-2}$

$Q_c < K_c$, so reaction proceeds to the right.

(b) From the reaction table,

$$[PCl_5] = 0.2100\ M - x = 0.2065\ M \qquad x = 0.0035\ M$$

So, $[Cl_2] = [PCl_3] = 0.0900\ M + x = 0.0935\ M$.

17.11 (a) $[SiF_4]$ increases; (b) decreases; (c) decreases; (d) no effect.

17.12 (a) Decrease P; (b) increase P; (c) increase P.

17.13 (a) P_{H_2} will decrease; K_p will increase; (b) P_{N_2} will increase; K_p will decrease; (c) P_{PCl_5} will increase; K_p will increase.

17.14 (a) Since $P = \dfrac{n}{V}RT$ and, in this case, V, R, and T cancel,

$$K_p = \frac{n^2_{CD}}{n_{C_2}\times n_{D_2}} = \frac{16}{(2)(2)} = 4$$

(b) Scene 2 to the left; scene 3 to the right. (c) There are 2 mol of gas on each side of the balanced equation, so there is no effect on total moles of gas.

Problems

Problems with **colored** numbers are answered in Appendix E. Sections match the text and provide the numbers of relevant sample problems. Most offer Concept Review Questions, Skill-Building Exercises (grouped in pairs covering the same concept), and Problems in Context. Comprehensive Problems are based on material from any section or previous chapter.

The Equilibrium State and the Equilibrium Constant

Concept Review Questions

17.1 A change in reaction conditions increases the rate of a certain forward reaction more than that of the reverse reaction. What is the effect on the equilibrium constant and the concentrations of reactants and products at equilibrium?

17.2 When a chemical company employs a new reaction to manufacture a product, the chemists consider its rate (kinetics) and yield (equilibrium). How do each of these affect the usefulness of a manufacturing process?

17.3 If there is no change in concentrations, why is the equilibrium state considered dynamic?

17.4 (a) Is K very large or very small for a reaction that goes essentially to completion? Explain. (b) White phosphorus, P_4, is produced by the reduction of phosphate rock, $Ca_3(PO_4)_2$. If exposed to oxygen, the waxy, white solid smokes, bursts into flames, and releases a large quantity of heat. Does the reaction

$$P_4(g) + 5O_2(g) \rightleftharpoons P_4O_{10}(s)$$

have a large or small equilibrium constant? Explain.

17.5 Does the following energy diagram depict a reaction in which K is small or large? Explain.

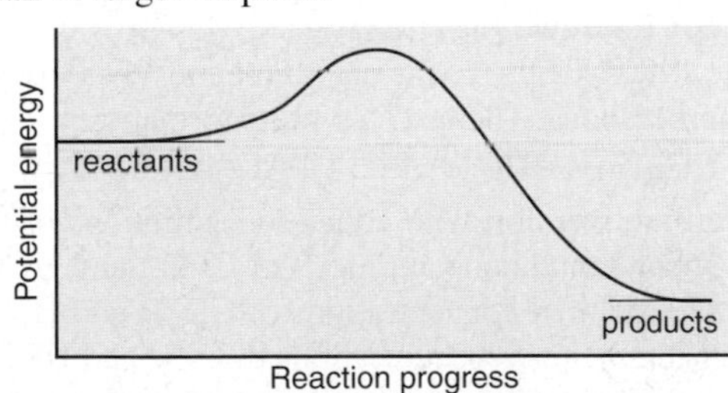

The Reaction Quotient and the Equilibrium Constant
(Sample Problems 17.1 to 17.3)

Concept Review Questions

17.6 For a given reaction at a given temperature, the value of K is constant. Is the value of Q also constant? Explain.

17.7 In a series of experiments on the thermal decomposition of lithium peroxide,

$$2Li_2O_2(s) \rightleftharpoons 2Li_2O(s) + O_2(g)$$

a chemist finds that, as long as some Li_2O_2 is present at the end of the experiment, the amount of O_2 obtained in a given container at a given T is the same. Explain.

17.8 In a study of the formation of HI from its elements,

$$H_2(g) + I_2(g) \rightleftharpoons 2HI(g)$$

equal amounts of H_2 and I_2 were placed in a container, which was then sealed and heated.
(a) On one set of axes, sketch concentration vs. time curves for H_2 and HI, and explain how Q changes as a function of time.
(b) Is the value of Q different if $[I_2]$ is plotted instead of $[H_2]$?

17.9 Explain the difference between a heterogeneous and a homogeneous equilibrium. Give an example of each.

17.10 Does Q for the formation of 1 mol of NO from its elements differ from Q for the decomposition of 1 mol of NO to its elements? Explain and give the relationship between the two Q's.

17.11 Does Q for the formation of 1 mol of NH_3 from H_2 and N_2 differ from Q for the formation of NH_3 from H_2 and 1 mol of N_2? Explain and give the relationship between the two Q's.

Skill-Building Exercises *(grouped in similar pairs)*

17.12 Balance each reaction and write its reaction quotient, Q_c:
(a) $NO(g) + O_2(g) \rightleftharpoons N_2O_3(g)$
(b) $SF_6(g) + SO_3(g) \rightleftharpoons SO_2F_2(g)$
(c) $SClF_5(g) + H_2(g) \rightleftharpoons S_2F_{10}(g) + HCl(g)$

17.13 Balance each reaction and write its reaction quotient, Q_c:
(a) $C_2H_6(g) + O_2(g) \rightleftharpoons CO_2(g) + H_2O(g)$
(b) $CH_4(g) + F_2(g) \rightleftharpoons CF_4(g) + HF(g)$
(c) $SO_3(g) \rightleftharpoons SO_2(g) + O_2(g)$

17.14 Balance each reaction and write its reaction quotient, Q_c:
(a) $NO_2Cl(g) \rightleftharpoons NO_2(g) + Cl_2(g)$
(b) $POCl_3(g) \rightleftharpoons PCl_3(g) + O_2(g)$
(c) $NH_3(g) + O_2(g) \rightleftharpoons N_2(g) + H_2O(g)$

17.15 Balance each reaction and write its reaction quotient, Q_c:
(a) $O_2(g) \rightleftharpoons O_3(g)$
(b) $NO(g) + O_3(g) \rightleftharpoons NO_2(g) + O_2(g)$
(c) $N_2O(g) + H_2(g) \rightleftharpoons NH_3(g) + H_2O(g)$

17.16 At a particular temperature, $K_c = 1.6\times10^{-2}$ for

$$2H_2S(g) \rightleftharpoons 2H_2(g) + S_2(g)$$

Calculate K_c for each of the following reactions:
(a) $\frac{1}{2}S_2(g) + H_2(g) \rightleftharpoons H_2S(g)$
(b) $5H_2S(g) \rightleftharpoons 5H_2(g) + \frac{5}{2}S_2(g)$

17.17 At a particular temperature, $K_c = 6.5\times10^{2}$ for

$$2NO(g) + 2H_2(g) \rightleftharpoons N_2(g) + 2H_2O(g)$$

Calculate K_c for each of the following reactions:
(a) $NO(g) + H_2(g) \rightleftharpoons \frac{1}{2}N_2(g) + H_2O(g)$
(b) $2N_2(g) + 4H_2O(g) \rightleftharpoons 4NO(g) + 4H_2(g)$

17.18 Balance each of the following examples of heterogeneous equilibria and write its reaction quotient, Q_c:
(a) $Na_2O_2(s) + CO_2(g) \rightleftharpoons Na_2CO_3(s) + O_2(g)$
(b) $H_2O(l) \rightleftharpoons H_2O(g)$
(c) $NH_4Cl(s) \rightleftharpoons NH_3(g) + HCl(g)$

17.19 Balance each of the following examples of heterogeneous equilibria and write its reaction quotient, Q_c:
(a) $H_2O(l) + SO_3(g) \rightleftharpoons H_2SO_4(aq)$
(b) $KNO_3(s) \rightleftharpoons KNO_2(s) + O_2(g)$
(c) $S_8(s) + F_2(g) \rightleftharpoons SF_6(g)$

17.20 Balance each of the following examples of heterogeneous equilibria and write its reaction quotient, Q_c:
(a) $NaHCO_3(s) \rightleftharpoons Na_2CO_3(s) + CO_2(g) + H_2O(g)$
(b) $SnO_2(s) + H_2(g) \rightleftharpoons Sn(s) + H_2O(g)$
(c) $H_2SO_4(l) + SO_3(g) \rightleftharpoons H_2S_2O_7(l)$

17.21 Balance each of the following examples of heterogeneous equilibria and write its reaction quotient, Q_c:
(a) $Al(s) + NaOH(aq) + H_2O(l) \rightleftharpoons Na[Al(OH)_4](aq) + H_2(g)$

(b) $CO_2(s) \rightleftharpoons CO_2(g)$
(c) $N_2O_5(s) \rightleftharpoons NO_2(g) + O_2(g)$

Problems in Context

17.22 Write Q_c for each of the following:
(a) Hydrogen chloride gas reacts with oxygen gas to produce chlorine gas and water vapor.
(b) Solid diarsenic trioxide reacts with fluorine gas to produce liquid arsenic pentafluoride and oxygen gas.
(c) Gaseous sulfur tetrafluoride reacts with liquid water to produce gaseous sulfur dioxide and hydrogen fluoride gas.
(d) Solid molybdenum(VI) oxide reacts with gaseous xenon difluoride to form liquid molybdenum(VI) fluoride, xenon gas, and oxygen gas.

17.23 The interhalogen ClF_3 is prepared in a two-step fluorination of chlorine gas:

$$Cl_2(g) + F_2(g) \rightleftharpoons ClF(g)$$
$$ClF(g) + F_2(g) \rightleftharpoons ClF_3(g)$$

(a) Balance each step and write the overall equation.
(b) Show that the overall Q_c equals the product of the Q_c's for the individual steps.

Expressing Equilibria with Pressure Terms: Relation Between K_c and K_p

(Sample Problem 17.4)

Concept Review Questions

17.24 Guldberg and Waage proposed the definition of the equilibrium constant as a certain ratio of *concentrations*. What relationship allows us to use a particular ratio of *partial pressures* (for a gaseous reaction) to express an equilibrium constant? Explain.

17.25 When are K_c and K_p equal, and when are they not?

17.26 A certain reaction at equilibrium has more moles of gaseous products than of gaseous reactants.
(a) Is K_c larger or smaller than K_p?
(b) Write a general statement about the relative sizes of K_c and K_p for any gaseous equilibrium.

Skill-Building Exercises *(grouped in similar pairs)*

17.27 Determine Δn_{gas} for each of the following reactions:
(a) $2KClO_3(s) \rightleftharpoons 2KCl(s) + 3O_2(g)$
(b) $2PbO(s) + O_2(g) \rightleftharpoons 2PbO_2(s)$
(c) $I_2(s) + 3XeF_2(s) \rightleftharpoons 2IF_3(s) + 3Xe(g)$

17.28 Determine Δn_{gas} for each of the following reactions:
(a) $MgCO_3(s) \rightleftharpoons MgO(s) + CO_2(g)$
(b) $2H_2(g) + O_2(g) \rightleftharpoons 2H_2O(l)$
(c) $HNO_3(l) + ClF(g) \rightleftharpoons ClONO_2(g) + HF(g)$

17.29 Calculate K_c for each of the following equilibria:
(a) $CO(g) + Cl_2(g) \rightleftharpoons COCl_2(g)$; $K_p = 3.9\times10^{-2}$ at 1000. K
(b) $S_2(g) + C(s) \rightleftharpoons CS_2(g)$; $K_p = 28.5$ at 500. K

17.30 Calculate K_c for each of the following equilibria:
(a) $H_2(g) + I_2(g) \rightleftharpoons 2HI(g)$; $K_p = 49$ at 730. K
(b) $2SO_2(g) + O_2(g) \rightleftharpoons 2SO_3(g)$; $K_p = 2.5\times10^{10}$ at 500. K

17.31 Calculate K_p for each of the following equilibria:
(a) $N_2O_4(g) \rightleftharpoons 2NO_2(g)$; $K_c = 6.1\times10^{-3}$ at 298 K
(b) $N_2(g) + 3H_2(g) \rightleftharpoons 2NH_3(g)$; $K_c = 2.4\times10^{-3}$ at 1000. K

17.32 Calculate K_p for each of the following equilibria:
(a) $H_2(g) + CO_2(g) \rightleftharpoons H_2O(g) + CO(g)$; $K_c = 0.77$ at 1020. K
(b) $3O_2(g) \rightleftharpoons 2O_3(g)$; $K_c = 1.8\times10^{-56}$ at 570. K

Reaction Direction: Comparing Q and K

(Sample Problem 17.5)

Concept Review Questions

17.33 When the numerical value of Q is less than K, in which direction does the reaction proceed to reach equilibrium? Explain.

17.34 What three criteria characterize a chemical system at equilibrium?

Skill-Building Exercises *(grouped in similar pairs)*

17.35 At 425°C, $K_p = 4.18\times10^{-9}$ for the reaction

$$2HBr(g) \rightleftharpoons H_2(g) + Br_2(g)$$

In one experiment, 0.20 atm of HBr(*g*), 0.010 atm of $H_2(g)$, and 0.010 atm of $Br_2(g)$ are introduced into a container. Is the reaction at equilibrium? If not, in which direction will it proceed?

17.36 At 100°C, $K_p = 60.6$ for the reaction

$$2NOBr(g) \rightleftharpoons 2NO(g) + Br_2(g)$$

In a given experiment, 0.10 atm of each component is placed in a container. Is the system at equilibrium? If not, in which direction will the reaction proceed?

Problems in Context

17.37 The water-gas shift reaction plays a central role in the chemical methods for obtaining cleaner fuels from coal:

$$CO(g) + H_2O(g) \rightleftharpoons CO_2(g) + H_2(g)$$

At a given temperature, $K_p = 2.7$. If 0.13 mol of CO, 0.56 mol of H_2O, 0.62 mol of CO_2, and 0.43 mol of H_2 are introduced into a 2.0-L flask, in which direction must the reaction proceed to reach equilibrium?

How to Solve Equilibrium Problems

(Sample Problems 17.6 to 17.10)

Concept Review Questions

17.38 In the 1980s, CFC-11 was one of the most heavily produced chlorofluorocarbons. The last step in its formation is

$$CCl_4(g) + HF(g) \rightleftharpoons CFCl_3(g) + HCl(g)$$

If you start the reaction with equal concentrations of CCl_4 and HF, you obtain equal concentrations of $CFCl_3$ and HCl at equilibrium. Are the final concentrations of $CFCl_3$ and HCl equal if you start with unequal concentrations of CCl_4 and HF? Explain.

17.39 For a problem involving the catalyzed reaction of methane and steam, the following reaction table was prepared:

Pressure (atm)	$CH_4(g)$	+ $2H_2O(g)$	$\rightleftharpoons$ $CO_2(g)$	+ $4H_2(g)$
Initial	0.30	0.40	0	0
Change	$-x$	$-2x$	$+x$	$+4x$
Equilibrium	$0.30 - x$	$0.40 - 2x$	x	$4x$

Explain the entries in the "Change" and "Equilibrium" rows.

17.40 (a) What is the basis of the approximation that avoids using the quadratic formula to find an equilibrium concentration?
(b) When should this approximation *not* be made?

Skill-Building Exercises *(grouped in similar pairs)*

17.41 In an experiment to study the formation of HI(*g*),

$$H_2(g) + I_2(g) \rightleftharpoons 2HI(g)$$

$H_2(g)$ and $I_2(g)$ were placed in a sealed container at a certain temperature. At equilibrium, $[H_2] = 6.50\times10^{-5}$ *M*, $[I_2] = 1.06\times10^{-3}$ *M*, and $[HI] = 1.87\times10^{-3}$ *M*. Calculate K_c for the reaction at this temperature.

17.42 Gaseous ammonia was introduced into a sealed container and heated to a certain temperature:

$$2NH_3(g) \rightleftharpoons N_2(g) + 3H_2(g)$$

At equilibrium, $[NH_3] = 0.0225\ M$, $[N_2] = 0.114\ M$, and $[H_2] = 0.342\ M$. Calculate K_c for the reaction at this temperature.

17.43 Gaseous PCl_5 decomposes according to the reaction

$$PCl_5(g) \rightleftharpoons PCl_3(g) + Cl_2(g)$$

In one experiment, 0.15 mol of $PCl_5(g)$ was introduced into a 2.0-L container. Construct the reaction table for this process.

17.44 Hydrogen fluoride, HF, can be made from the reaction

$$H_2(g) + F_2(g) \rightleftharpoons 2HF(g)$$

In one experiment, 0.10 mol of $H_2(g)$ and 0.050 mol of $F_2(g)$ are added to a 0.50-L flask. Write a reaction table for this process.

17.45 For the following reaction, $K_p = 6.5\times10^4$ at 308 K:

$$2NO(g) + Cl_2(g) \rightleftharpoons 2NOCl(g)$$

At equilibrium, $P_{NO} = 0.35$ atm and $P_{Cl_2} = 0.10$ atm. What is the equilibrium partial pressure of NOCl(*g*)?

17.46 For the following reaction, $K_p = 0.262$ at 1000°C:

$$C(s) + 2H_2(g) \rightleftharpoons CH_4(g)$$

At equilibrium, P_{H_2} is 1.22 atm. What is the equilibrium partial pressure of $CH_4(g)$?

17.47 Ammonium hydrogen sulfide decomposes according to the following reaction, for which $K_p = 0.11$ at 250°C:

$$NH_4HS(s) \rightleftharpoons H_2S(g) + NH_3(g)$$

If 55.0 g of $NH_4HS(s)$ is placed in a sealed 5.0-L container, what is the partial pressure of $NH_3(g)$ at equilibrium?

17.48 Hydrogen sulfide decomposes according to the following reaction, for which $K_c = 9.30\times10^{-8}$ at 700°C:

$$2H_2S(g) \rightleftharpoons 2H_2(g) + S_2(g)$$

If 0.45 mol of H_2S is placed in a 3.0-L container, what is the equilibrium concentration of $H_2(g)$ at 700°C?

17.49 Even at high *T*, the formation of nitric oxide is not favored:

$$N_2(g) + O_2(g) \rightleftharpoons 2NO(g) \qquad K_c = 4.10\times10^{-4} \text{ at } 2000°C$$

What is [NO] when a mixture of 0.20 mol of $N_2(g)$ and 0.15 mol of $O_2(g)$ reach equilibrium in a 1.0-L container at 2000°C?

17.50 Nitrogen dioxide decomposes according to the reaction

$$2NO_2(g) \rightleftharpoons 2NO(g) + O_2(g)$$

where $K_p = 4.48\times10^{-13}$ at a certain temperature. A pressure of 0.75 atm of NO_2 is introduced into a container and allowed to come to equilibrium. What are the equilibrium partial pressures of NO(*g*) and $O_2(g)$?

17.51 Hydrogen iodide decomposes according to the reaction

$$2HI(g) \rightleftharpoons H_2(g) + I_2(g)$$

A sealed 1.50-L container initially holds 0.00623 mol of H_2, 0.00414 mol of I_2, and 0.0244 mol of HI at 703 K. When equilibrium is reached, the concentration of $H_2(g)$ is 0.00467 *M*. What are the concentrations of HI(*g*) and $I_2(g)$?

17.52 Compound A decomposes according to the equation

$$A(g) \rightleftharpoons 2B(g) + C(g)$$

A sealed 1.00-L container initially contains 1.75×10^{-3} mol of A(*g*), 1.25×10^{-3} mol of B(*g*), and 6.50×10^{-4} mol of C(*g*) at 100°C. At equilibrium, [A] is 2.15×10^{-3} *M*. Find [B] and [C].

Problems in Context

17.53 In an analysis of interhalogen reactivity, 0.500 mol of ICl was placed in a 5.00-L flask, where it decomposed at a high *T*: $2ICl(g) \rightleftharpoons I_2(g) + Cl_2(g)$. Calculate the equilibrium concentrations of I_2, Cl_2, and ICl ($K_c = 0.110$ at this temperature).

17.54 A United Nations toxicologist studying the properties of mustard gas, $S(CH_2CH_2Cl)_2$, a blistering agent used in warfare, prepares a mixture of 0.675 *M* SCl_2 and 0.973 *M* C_2H_4 and allows it to react at room temperature (20.0°C):

$$SCl_2(g) + 2C_2H_4(g) \rightleftharpoons S(CH_2CH_2Cl)_2(g)$$

At equilibrium, $[S(CH_2CH_2Cl)_2] = 0.350\ M$. Calculate K_p.

17.55 The first step in industrial production of nitric acid is the catalyzed oxidation of ammonia. Without a catalyst, a different reaction predominates:

$$4NH_3(g) + 3O_2(g) \rightleftharpoons 2N_2(g) + 6H_2O(g)$$

When 0.0150 mol of $NH_3(g)$ and 0.0150 mol of $O_2(g)$ are placed in a 1.00-L container at a certain temperature, the N_2 concentration at equilibrium is 1.96×10^{-3} *M*. Calculate K_c.

17.56 A key step in the extraction of iron from its ore is

$$FeO(s) + CO(g) \rightleftharpoons Fe(s) + CO_2(g) \quad K_p = 0.403 \text{ at } 1000°C$$

This step occurs in the 700°C to 1200°C zone within a blast furnace. What are the equilibrium partial pressures of CO(*g*) and $CO_2(g)$ when 1.00 atm of CO(*g*) and excess FeO(*s*) react in a sealed container at 1000°C?

Reaction Conditions and the Equilibrium State: Le Châtelier's Principle

(Sample Problems 17.11 to 17.14)

Concept Review Questions

17.57 What is meant by the word "disturbance" in the statement of Le Châtelier's principle?

17.58 What is the difference between the equilibrium position and the equilibrium constant of a reaction? Which changes as a result of a change in reactant concentration?

17.59 Scenes A, B, and C depict this reaction at three temperatures:

$$NH_4Cl(s) \rightleftharpoons NH_3(g) + HCl(g) \qquad \Delta H^0_{rxn} = 176 \text{ kJ}$$

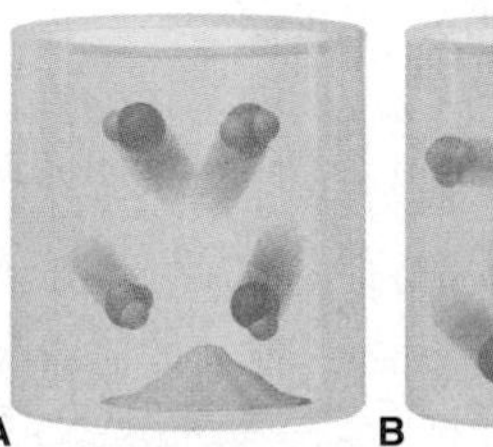
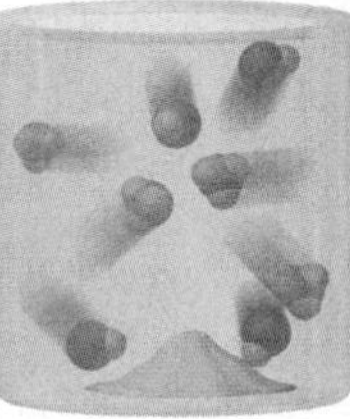
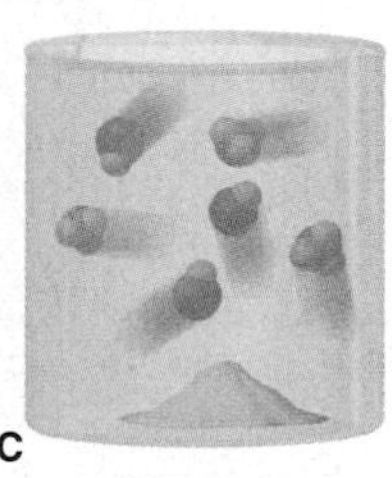

A B C

(a) Which best represents the reaction mixture at the highest temperature? Explain.
(b) Which best represents the reaction mixture at the lowest temperature? Explain.

17.60 What is implied by the word "constant" in the term *equilibrium constant?* Give two reaction parameters that can be changed without changing the value of an equilibrium constant.

17.61 Le Châtelier's principle is related ultimately to the rates of the forward and reverse steps in a reaction. Explain (a) why an increase in reactant concentration shifts the equilibrium position to the right but does not change *K*; (b) why a decrease in *V* shifts the equilibrium position toward fewer moles of gas but does not change *K*; and (c) why a rise in *T* shifts the equilibrium position of an exothermic reaction toward reactants and also changes *K*.

17.62 Le Châtelier's principle predicts that a rise in the temperature of an endothermic reaction from T_1 to T_2 results in K_2 being larger than K_1. Explain.

Skill-Building Exercises *(grouped in similar pairs)*

17.63 Consider this equilibrium system:

$$CO(g) + Fe_3O_4(s) \rightleftharpoons CO_2(g) + 3FeO(s)$$

How does the equilibrium position shift as a result of each of the following disturbances?
(a) CO is added.
(b) CO_2 is removed by adding solid NaOH.
(c) Additional $Fe_3O_4(s)$ is added to the system.
(d) Dry ice is added at constant temperature.

17.64 Sodium bicarbonate undergoes thermal decomposition according to the reaction

$$2NaHCO_3(s) \rightleftharpoons Na_2CO_3(s) + CO_2(g) + H_2O(g)$$

How does the equilibrium position shift as a result of each of the following disturbances?
(a) 0.20 atm of argon gas is added.
(b) $NaHCO_3(s)$ is added.
(c) $Mg(ClO_4)_2(s)$ is added as a drying agent to remove H_2O.
(d) Dry ice is added at constant temperature.

17.65 Predict the effect of *increasing* the container volume on the amounts of each reactant and product in the following reactions:
(a) $F_2(g) \rightleftharpoons 2F(g)$
(b) $2CH_4(g) \rightleftharpoons C_2H_2(g) + 3H_2(g)$

17.66 Predict the effect of *increasing* the container volume on the amounts of each reactant and product in the following reactions:
(a) $CH_3OH(l) \rightleftharpoons CH_3OH(g)$
(b) $CH_4(g) + NH_3(g) \rightleftharpoons HCN(g) + 3H_2(g)$

17.67 Predict the effect of *decreasing* the container volume on the amounts of each reactant and product in the following reactions:
(a) $H_2(g) + Cl_2(g) \rightleftharpoons 2HCl(g)$
(b) $2H_2(g) + O_2(g) \rightleftharpoons 2H_2O(l)$

17.68 Predict the effect of *decreasing* the container volume on the amounts of each reactant and product in the following reactions:
(a) $C_3H_8(g) + 5O_2(g) \rightleftharpoons 3CO_2(g) + 4H_2O(l)$
(b) $4NH_3(g) + 3O_2(g) \rightleftharpoons 2N_2(g) + 6H_2O(g)$

17.69 How would you adjust the *volume* of the reaction vessel in order to maximize product yield in each of the following reactions?
(a) $Fe_3O_4(s) + 4H_2(g) \rightleftharpoons 3Fe(s) + 4H_2O(g)$
(b) $2C(s) + O_2(g) \rightleftharpoons 2CO(g)$

17.70 How would you adjust the *volume* of the reaction vessel in order to maximize product yield in each of the following reactions?
(a) $Na_2O_2(s) \rightleftharpoons 2Na(l) + O_2(g)$
(b) $C_2H_2(g) + 2H_2(g) \rightleftharpoons C_2H_6(g)$

17.71 Predict the effect of *increasing* the temperature on the amounts of products in the following reactions:
(a) $CO(g) + 2H_2(g) \rightleftharpoons CH_3OH(g) \quad \Delta H^0_{rxn} = -90.7$ kJ
(b) $C(s) + H_2O(g) \rightleftharpoons CO(g) + H_2(g) \quad \Delta H^0_{rxn} = 131$ kJ
(c) $2NO_2(g) \rightleftharpoons 2NO(g) + O_2(g)$ (endothermic)
(d) $2C(s) + O_2(g) \rightleftharpoons 2CO(g)$ (exothermic)

17.72 Predict the effect of *decreasing* the temperature on the amounts of reactants in the following reactions:
(a) $C_2H_2(g) + H_2O(g) \rightleftharpoons CH_3CHO(g) \quad \Delta H^0_{rxn} = -151$ kJ
(b) $CH_3CH_2OH(l) + O_2(g) \rightleftharpoons CH_3CO_2H(l) + H_2O(g) \quad \Delta H^0_{rxn} = -451$ kJ
(c) $2C_2H_4(g) + O_2(g) \rightleftharpoons 2CH_3CHO(g)$ (exothermic)
(d) $N_2O_4(g) \rightleftharpoons 2NO_2(g)$ (endothermic)

17.73 Deuterium (D or ^2H) is an isotope of hydrogen. The molecule D_2 undergoes an exchange reaction with ordinary H_2 that leads to isotopic equilibrium:

$$D_2(g) + H_2(g) \rightleftharpoons 2DH(g) \qquad K_p = 1.80 \text{ at } 298 \text{ K}$$

If ΔH^0_{rxn} is 0.32 kJ/mol DH, calculate K_p at 500. K.

17.74 The formation of methanol is important to the processing of new fuels. At 298 K, $K_p = 2.25\times10^4$ for the reaction

$$CO(g) + 2H_2(g) \rightleftharpoons CH_3OH(l)$$

If $\Delta H^0_{rxn} = -128$ kJ/mol CH_3OH, calculate K_p at 0°C.

Problems in Context

17.75 The minerals hematite (Fe_2O_3) and magnetite (Fe_3O_4) exist in equilibrium with atmospheric oxygen:

$$4Fe_3O_4(s) + O_2(g) \rightleftharpoons 6Fe_2O_3(s) \quad K_p = 2.5\times10^{87} \text{ at } 298 \text{ K}$$

(a) Determine P_{O_2} at equilibrium. (b) Given that P_{O_2} in air is 0.21 atm, in which direction will the reaction proceed to reach equilibrium? (c) Calculate K_c at 298 K.

17.76 The oxidation of SO_2 is the key step in H_2SO_4 production:

$$SO_2(g) + \tfrac{1}{2}O_2(g) \rightleftharpoons SO_3(g) \qquad \Delta H^0_{rxn} = -99.2 \text{ kJ}$$

(a) What qualitative combination of T and P maximizes SO_3 yield?
(b) How does addition of O_2 affect Q? K?
(c) Why is catalysis used for this reaction?

Comprehensive Problems

17.77 The "filmstrip" represents five molecular-level scenes of a gaseous mixture as it reaches equilibrium over time:

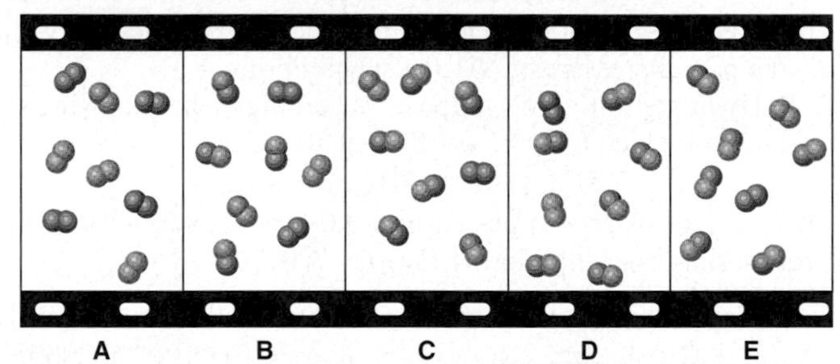

X is purple and Y is orange: $X_2(g) + Y_2(g) \rightleftharpoons 2XY(g)$.
(a) Write the reaction quotient, Q, for this reaction.
(b) If each particle represents 0.1 mol of particles, calculate Q for each scene.
(c) If $K > 1$, is time progressing to the right or to the left? Explain.
(d) Calculate K at this temperature.
(e) If $\Delta H^0_{rxn} < 0$, which scene, if any, best represents the mixture at a higher temperature? Explain.
(f) Which scene, if any, best represents the mixture at a higher pressure (lower volume)? Explain.

17.78 The powerful chlorinating agent sulfuryl dichloride (SO_2Cl_2) can be prepared by the following two-step sequence:

$$H_2S(g) + O_2(g) \rightleftharpoons SO_2(g) + H_2O(g)$$
$$SO_2(g) + Cl_2(g) \rightleftharpoons SO_2Cl_2(g)$$

(a) Balance each step, and write the overall equation.
(b) Show that the overall Q_c equals the product of the Q_c's for the individual steps.

17.79 A mixture of 5.00 volumes of N_2 and 1.00 volume of O_2 passes through a heated furnace and reaches equilibrium at 900. K and 5.00 atm:

$$N_2(g) + O_2(g) \rightleftharpoons 2NO(g) \qquad K_p = 6.70\times10^{-10}$$

(a) What is the partial pressure of NO? (b) What is the concentration in micrograms per liter (μg/L) of NO in the mixture?

17.80 For the following equilibrium system, which of the changes will form more $CaCO_3$?

$$CO_2(g) + Ca(OH)_2(s) \rightleftharpoons CaCO_3(s) + H_2O(l) \quad \Delta H^0 = -113 \text{ kJ}$$

(a) Decrease temperature at constant pressure (no phase change)
(b) Increase volume at constant temperature
(c) Increase partial pressure of CO_2
(d) Remove one-half of the initial $CaCO_3$

17.81 Ammonium carbamate (NH_2COONH_4) is a salt of carbamic acid that is found in the blood and urine of mammals. At 250.°C, $K_c = 1.58\times10^{-8}$ for the following equilibrium:

$$NH_2COONH_4(s) \rightleftharpoons 2NH_3(g) + CO_2(g)$$

If 7.80 g of NH_2COONH_4 is put into a 0.500-L evacuated container, what is the total pressure at equilibrium?

17.82 A study of the water-gas shift reaction (see Problem 17.37) was made in which equilibrium was reached with $[CO] = [H_2O] = [H_2] = 0.10\ M$ and $[CO_2] = 0.40\ M$. After 0.60 mol of H_2 is added to the 2.0-L container and equilibrium is re-established, what are the new concentrations of all the components?

17.83 Isolation of Group 8B(10) elements, used as industrial catalysts, involves a series of steps. For nickel, the sulfide ore is roasted in air: $Ni_3S_2(s) + O_2(g) \rightleftharpoons NiO(s) + SO_2(g)$. The metal oxide is reduced by the H_2 in water gas ($CO + H_2$) to impure Ni: $NiO(s) + H_2(g) \rightleftharpoons Ni(s) + H_2O(g)$. The CO in water gas then reacts with the metal in the Mond process to form gaseous nickel carbonyl, $Ni(s) + CO(g) \rightleftharpoons Ni(CO)_4(g)$, which is subsequently decomposed to the metal.
(a) Balance each of the three steps, and obtain an overall balanced equation for the conversion of Ni_3S_2 to $Ni(CO)_4$.
(b) Show that the overall Q_c is the product of the Q_c's for the individual reactions.

17.84 One of the most important industrial sources of ethanol is the reaction of steam with ethene derived from crude oil:

$$C_2H_4(g) + H_2O(g) \rightleftharpoons C_2H_5OH(g)$$
$$\Delta H^0_{rxn} = -47.8 \text{ kJ} \qquad K_c = 9\times10^3 \text{ at } 600. \text{ K}$$

(a) At equilibrium, $P_{C_2H_5OH} = 200.$ atm and $P_{H_2O} = 400.$ atm. Calculate $P_{C_2H_4}$.
(b) Is the highest yield of ethanol obtained at high or low pressures? High or low temperatures?
(c) Calculate K_c at 450. K.
(d) In manufacturing, the yield of ammonia is increased by condensing it to a liquid and removing it from the vessel. Would condensing the C_2H_5OH work in this process? Explain.

17.85 Which of the following situations represent equilibrium?
(a) Migratory birds fly north in summer and south in winter.
(b) In a grocery store, some carts are kept inside and some outside. Customers bring carts out to their cars, while store clerks bring carts in to replace those taken out.
(c) In a tug o' war, a ribbon tied to the center of the rope moves back and forth until one side loses, after which the ribbon goes all the way to the winner's side.
(d) As a stew is cooking, water in the stew vaporizes, and the vapor condenses on the lid to droplets that drip into the stew.

17.86 An industrial chemist introduces 2.0 atm of H_2 and 2.0 atm of CO_2 into a 1.00-L container at 25.0°C and then raises the temperature to 700.°C, at which $K_c = 0.534$:

$$H_2(g) + CO_2(g) \rightleftharpoons H_2O(g) + CO(g)$$

How many grams of H_2 are present at equilibrium?

17.87 As an EPA scientist studying catalytic converters and urban smog, you want to find K_c for the following reaction:

$$2NO_2(g) \rightleftharpoons N_2(g) + 2O_2(g) \qquad K_c = ?$$

Use the following data to find the unknown K_c:

$$\tfrac{1}{2}N_2(g) + \tfrac{1}{2}O_2(g) \rightleftharpoons NO(g) \qquad K_c = 4.8\times10^{-10}$$
$$2NO_2(g) \rightleftharpoons 2NO(g) + O_2(g) \qquad K_c = 1.1\times10^{-5}$$

17.88 An inorganic chemist places 1 mol of BrCl in container A and 0.5 mol of Br_2 and 0.5 mol of Cl_2 in container B. She seals the containers and heats them to 300°C. With time, both containers hold identical mixtures of BrCl, Br_2, and Cl_2.
(a) Write a balanced equation for the reaction in container A.
(b) Write the reaction quotient, Q, for this reaction.
(c) How do the values of Q in A and in B compare over time?
(d) Explain on the molecular level how it is possible for both containers to end up with identical mixtures.

17.89 An engineer examining the oxidation of SO_2 in the manufacture of sulfuric acid determines that $K_c = 1.7\times10^8$ at 600. K:

$$2SO_2(g) + O_2(g) \rightleftharpoons 2SO_3(g)$$

(a) At equilibrium, $P_{SO_3} = 300.$ atm and $P_{O_2} = 100.$ atm. Calculate P_{SO_2}.
(b) The engineer places a mixture of 0.0040 mol of $SO_2(g)$ and 0.0028 mol of $O_2(g)$ in a 1.0-L container and raises the temperature to 1000 K. At equilibrium, 0.0020 mol of $SO_3(g)$ is present. Calculate K_c and P_{SO_2} for this reaction at 1000. K.

17.90 Phosgene ($COCl_2$) is a toxic substance that forms readily from carbon monoxide and chlorine at elevated temperatures:

$$CO(g) + Cl_2(g) \rightleftharpoons COCl_2(g)$$

If 0.350 mol of each reactant is placed in a 0.500-L flask at 600 K, what are the concentrations of all three substances at equilibrium ($K_c = 4.95$ at this temperature)?

17.91 When 0.100 mol of $CaCO_3(s)$ and 0.100 mol of $CaO(s)$ are placed in an evacuated sealed 10.0-L container and heated to 385 K, $P_{CO_2} = 0.220$ atm after equilibrium is established:

$$CaCO_3(s) \rightleftharpoons CaO(s) + CO_2(g)$$

An additional 0.300 atm of $CO_2(g)$ is then pumped into the container. What is the total mass (in g) of $CaCO_3$ after equilibrium is re-established?

17.92 Use each of the following reaction quotients to write the balanced equation:

(a) $Q = \dfrac{[CO_2]^2[H_2O]^2}{[C_2H_4][O_2]^3}$ (b) $Q = \dfrac{[NH_3]^4[O_2]^7}{[NO_2]^4[H_2O]^6}$

17.93 Hydrogenation of carbon-carbon π bonds is important in the petroleum and food industries. The conversion of acetylene to ethylene is a simple example of the process:

$$C_2H_2(g) + H_2(g) \rightleftharpoons C_2H_4(g) \qquad K_c = 2.9\times10^8 \text{ at } 2000. \text{ K}$$

The process is usually performed at much lower temperatures with the aid of a catalyst. Use ΔH^0_f values from Appendix B to calculate K_c for this reaction at 300. K.

17.94 In combustion studies of H_2 as an alternative fuel, you find evidence that the hydroxyl radical (HO) is formed in flames by the reaction $H(g) + \tfrac{1}{2}O_2(g) \rightleftharpoons HO(g)$. Use the following data to calculate K_c for the reaction:

$$\tfrac{1}{2}H_2(g) + \tfrac{1}{2}O_2(g) \rightleftharpoons HO(g) \qquad K_c = 0.58$$
$$\tfrac{1}{2}H_2(g) \rightleftharpoons H(g) \qquad K_c = 1.6\times10^{-3}$$

17.95 Highly toxic disulfur decafluoride decomposes by a free-radical process: $S_2F_{10}(g) \rightleftharpoons SF_4(g) + SF_6(g)$. In a study of the decomposition, S_2F_{10} was placed in a 2.0-L flask and heated to 100°C; $[S_2F_{10}]$ was 0.50 *M* at equilibrium. More S_2F_{10} was added, and when equilibrium was reattained, $[S_2F_{10}]$ was 2.5 *M*. How did $[SF_4]$ and $[SF_6]$ change from the original to the new equilibrium position after the addition of more S_2F_{10}?

17.96 Aluminum is one of the most versatile metals. It is produced by the Hall-Heroult process, in which molten cryolite, Na_3AlF_6, is used as a solvent for the aluminum ore. Cryolite undergoes very slight decomposition with heat to produce a tiny amount of F_2, which escapes into the atmosphere above the solvent. K_c is 2×10^{-104} at 1300 K for the reaction:

$$Na_3AlF_6(l) \rightleftharpoons 3Na(l) + Al(l) + 3F_2(g)$$

What is the concentration of F_2 over a bath of molten cryolite at this temperature?

17.97 An equilibrium mixture of car exhaust gases consisting of 10.0 volumes of CO_2, 1.00 volume of unreacted O_2, and 50.0 volumes of unreacted N_2 leaves the engine at 4.0 atm and 800. K.

(a) Given this equilibrium, what is the partial pressure of CO?

$$2CO_2(g) \rightleftharpoons 2CO(g) + O_2(g) \qquad K_p = 1.4\times10^{-28} \text{ at } 800.\text{ K}$$

(b) Assuming the mixture has enough time to reach equilibrium, what is the concentration in picograms per liter (pg/L) of CO in the exhaust gas? (The actual concentration of CO in car exhaust is much higher because the gases do *not* reach equilibrium in the short transit time through the engine and exhaust system.)

17.98 Consider the following reaction:

$$3Fe(s) + 4H_2O(g) \rightleftharpoons Fe_3O_4(s) + 4H_2(g)$$

(a) What are the apparent oxidation states of Fe and of O in Fe_3O_4?

(b) Fe_3O_4 is a compound of iron in which Fe occurs in two oxidation states. What are the oxidation states of Fe in Fe_3O_4?

(c) At 900°C, K_c for the reaction is 5.1. If 0.050 mol of $H_2O(g)$ and 0.100 mol of Fe(*s*) are placed in a 1.0-L container at 900°C, how many grams of Fe_3O_4 are present at equilibrium?

Note: The synthesis of ammonia is a major process throughout the industrialized world. Problems 17.99 to 17.106 refer to various aspects of this all-important reaction:

$$N_2(g) + 3H_2(g) \rightleftharpoons 2NH_3(g) \qquad \Delta H^0_{rxn} = -91.8 \text{ kJ}$$

17.99 When ammonia is made industrially, the mixture of N_2, H_2, and NH_3 that emerges from the reaction chamber is far from equilibrium. Why does the plant supervisor use reaction conditions that produce less than the maximum yield of ammonia?

17.100 The following reaction is sometimes used to produce the H_2 needed for the synthesis of ammonia:

$$CH_4(g) + CO_2(g) \rightleftharpoons 2CO(g) + 2H_2(g)$$

(a) What is the percent yield of H_2 when an equimolar mixture of CH_4 and CO_2 with a total pressure of 20.0 atm reaches equilibrium at 1200. K, at which $K_p = 3.548\times10^6$?

(b) What is the percent yield of H_2 for this system at 1300. K, at which $K_p = 2.626\times10^7$?

(c) Use the van't Hoff equation to find ΔH^0_{rxn}.

17.101 The methane used to obtain H_2 for NH_3 manufacture is impure and usually contains other hydrocarbons, such as propane, C_3H_8. Imagine the reaction of propane occurring in two steps:

$$C_3H_8(g) + 3H_2O(g) \rightleftharpoons 3CO(g) + 7H_2(g) \qquad K_p = 8.175\times10^{15} \text{ at } 1200.\text{ K}$$

$$CO(g) + H_2O(g) \rightleftharpoons CO_2(g) + H_2(g) \qquad K_p = 0.6944 \text{ at } 1200.\text{ K}$$

(a) Write the overall equation for the reaction of propane and steam to produce carbon dioxide and hydrogen.

(b) Calculate K_p for the overall process at 1200. K.

(c) When 1.00 volume of C_3H_8 and 4.00 volumes of H_2O, each at 1200. K and 5.0 atm, are mixed in a container, what is the final pressure? Assume the total volume remains constant, that the reaction is essentially complete, and that the gases behave ideally.

(d) What percentage of the C_3H_8 remains unreacted?

17.102 Using CH_4 and steam as a source of H_2 for NH_3 synthesis requires high temperatures. Rather than burning CH_4 separately to heat the mixture, it is more efficient to inject some oxygen into the reaction mixture. All of the H_2 is thus released for the synthesis, and the heat of combustion of CH_4 helps maintain the required temperature. Imagine the reaction occurring in two steps:

$$2CH_4(g) + O_2(g) \rightleftharpoons 2CO(g) + 4H_2(g) \qquad K_p = 9.34\times10^{28} \text{ at } 1000.\text{ K}$$

$$CO(g) + H_2O(g) \rightleftharpoons CO_2(g) + H_2(g) \qquad K_p = 1.374 \text{ at } 1000.\text{ K}$$

(a) Write the overall equation for the reaction of methane, steam, and oxygen to form carbon dioxide and hydrogen.

(b) What is K_p for the overall reaction?

(c) What is K_c for the overall reaction?

(d) A mixture of 2.0 mol of CH_4, 1.0 mol of O_2, and 2.0 mol of steam with a total pressure of 30. atm reacts at 1000. K at constant volume. Assuming that the reaction is complete and the ideal gas law is a valid approximation, what is the final pressure?

17.103 A mixture of 3.00 volumes of H_2 and 1.00 volume of N_2 reacts at 344°C to form ammonia. The equilibrium mixture at 110. atm contains 41.49% NH_3 by volume. Calculate K_p for the reaction, assuming that the gases behave ideally.

17.104 One mechanism for the synthesis of ammonia proposes that N_2 and H_2 molecules catalytically dissociate into atoms:

$$N_2(g) \rightleftharpoons 2N(g) \qquad \log K_p = -43.10$$

$$H_2(g) \rightleftharpoons 2H(g) \qquad \log K_p = -17.30$$

(a) Find the partial pressure of N in N_2 at 1000. K and 200. atm.

(b) Find the partial pressure of H in H_2 at 1000. K and 600. atm.

(c) How many N atoms and H atoms are present per liter?

(d) Based on these answers, which of the following is a more reasonable step to continue the mechanism after the catalytic dissociation? Explain.

$$N(g) + H(g) \longrightarrow NH(g)$$

$$N_2(g) + H(g) \longrightarrow NH(g) + N(g)$$

17.105 Consider the formation of ammonia in two experiments.

(a) To a 1.00-L container at 727°C, 1.30 mol of N_2 and 1.65 mol of H_2 are added. At equilibrium, 0.100 mol of NH_3 is present. Calculate the equilibrium concentrations of N_2 and H_2, and find K_c for the reaction: $2NH_3(g) \rightleftharpoons N_2(g) + 3H_2(g)$.

(b) In a different 1.00-L container at the same temperature, equilibrium is established with 8.34×10^{-2} mol of NH_3, 1.50 mol of N_2, and 1.25 mol of H_2 present. Calculate K_c for the reaction: $NH_3(g) \rightleftharpoons \frac{1}{2}N_2(g) + \frac{3}{2}H_2(g)$.

(c) What is the relationship between the K_c values in parts (a) and (b)? Why aren't these values the same?

17.106 You are a member of a research team of chemists discussing the plans to operate an ammonia processing plant:

$$N_2(g) + 3H_2(g) \rightleftharpoons 2NH_3(g)$$

(a) The plant operates at close to 700 K, at which K_p is 1.00×10^{-4}, and employs the stoichiometric 1:3 ratio of N_2:H_2.

At equilibrium, the partial pressure of NH_3 is 50. atm. Calculate the partial pressures of each reactant and P_{total}.
(b) One member of the team makes the following suggestion: since the partial pressure of H_2 is cubed in the reaction quotient, the plant could produce the same amount of NH_3 if the reactants were in a 1:6 ratio of N_2:H_2 and could do so at a lower pressure, which would cut operating costs. Calculate the partial pressure of each reactant and P_{total} under these conditions, assuming an unchanged partial pressure of 50. atm for NH_3. Is the team member's argument valid?

17.107 The atmospheric breakdown of CO_2, the main greenhouse gas, to CO and O_2 is highly temperature dependent, with K_p varying over many orders of magnitude for a temperature range of 1000 K.
(a) At a low T, a flask initially contains 1.00 atm of CO_2. What is P_{CO} if $K_p = 1.00\times10^{22}$ at this temperature for the *formation* of CO_2 from CO and O_2?
(b) What is P_{total} in the flask at this low T?
(c) The same experiment is run at a high T, and 35% of the CO_2 decomposes. Calculate K_p for CO_2 formation at this T.
(d) What is P_{total} in the flask at this high T?
(e) Is CO_2 breakdown exothermic or endothermic? Explain.

17.108 The two most abundant atmospheric gases react to a tiny extent at 298 K in the presence of a catalyst:

$$N_2(g) + O_2(g) \rightleftharpoons 2NO(g) \qquad K_p = 4.35\times10^{-31}$$

(a) What are the equilibrium pressures of the three components when the atmospheric partial pressures of O_2 (0.210 atm) and of N_2 (0.780 atm) are put into an evacuated 1.00-L flask at 298 K with catalyst?
(b) What is P_{total} in the container?
(c) Find K_c for this reaction at 298 K.

17.109 The oxidation of nitric oxide is favored at 457 K:

$$2NO(g) + O_2(g) \rightleftharpoons 2NO_2(g) \qquad K_p = 1.3\times10^4$$

(a) Calculate K_c at 457 K.
(b) Find ΔH^0_{rxn} from standard heats of formation.
(c) At what temperature does $K_c = 6.4\times10^9$?

17.110 The kinetics and equilibrium of the decomposition of hydrogen iodide have been studied extensively:

$$2HI(g) \rightleftharpoons H_2(g) + I_2(g)$$

(a) At 298 K, $K_c = 1.26\times10^{-3}$ for this reaction. Calculate K_p.
(b) Calculate K_c for the *formation* of HI at 298 K.
(c) Calculate ΔH^0_{rxn} for HI decomposition from ΔH^0_f values.
(d) At 729 K, $K_c = 2.0\times10^{-2}$ for HI decomposition. Calculate ΔH_{rxn} for this reaction from the van't Hoff equation.

17.111 Isopentyl alcohol reacts with pure acetic acid to form isopentyl acetate, the essence of banana oil:

$$C_5H_{11}OH + CH_3COOH \rightleftharpoons CH_3COOC_5H_{11} + H_2O$$

A student adds a drying agent to remove H_2O and thus increase the yield of banana oil. Is this approach reasonable? Explain.

17.112 Many essential metabolites are products in branched pathways, such as the one shown below:

A —1→ B —2→ C; C —3→ D —4→ E —5→ F; C —6→ G —7→ H —8→ I

One method of control of these pathways occurs through inhibition of the first enzyme specific for a branch.
(a) Which enzyme is inhibited by F?
(b) Which enzyme is inhibited by I?
(c) What disadvantage would there be if F inhibited enzyme 1?
(d) What disadvantage would there be if F inhibited enzyme 6?

17.113 For the equilibrium

$$H_2S(g) \rightleftharpoons 2H_2(g) + S_2(g) \qquad K_c = 9.0\times10^{-8} \text{ at } 700°C$$

the initial concentrations of the three gases are 0.300 M H_2S, 0.300 M H_2, and 0.150 M S_2. Determine the equilibrium concentrations of the gases.

17.114 Glauber's salt, $Na_2SO_4\cdot10H_2O$, was used by J. R. Glauber in the 17th century as a medicinal agent. At 25°C, $K_p = 4.08\times10^{-25}$ for the loss of water of hydration from Glauber's salt:

$$Na_2SO_4\cdot10H_2O(s) \rightleftharpoons Na_2SO_4(s) + 10H_2O(g)$$

(a) What is the vapor pressure of water at 25°C in a closed container holding a sample of $Na_2SO_4\cdot10H_2O(s)$?
(b) How do the following changes affect the ratio (higher, lower, same) of hydrated form to anhydrous form for the system above?
(1) Add more $Na_2SO_4(s)$ (2) Reduce the container volume
(3) Add more water vapor (4) Add N_2 gas

17.115 In a study of synthetic fuels, 0.100 mol of CO and 0.100 mol of water vapor are added to a 20.00-L container at 900.°C, and they react to form CO_2 and H_2. At equilibrium, [CO] is 2.24×10^{-3} M. (a) Calculate K_c at this temperature. (b) Calculate P_{total} in the flask at equilibrium. (c) How many moles of CO must be added to double this pressure? (d) After P_{total} is doubled and the system reattains equilibrium, what is $[CO]_{eq}$?

17.116 Synthetic diamonds are made under conditions of high temperature (2000 K) and high pressure (10^{10} Pa; 10^5 atm) in the presence of catalysts. The phase diagram for carbon is useful for finding the conditions for the formation of natural and synthetic diamonds. Along the diamond-graphite line, the two allotropes are in equilibrium. (a) At point A, what is the sign of ΔH for the formation of diamond from graphite? Explain. (b) Which allotrope is denser? Explain.

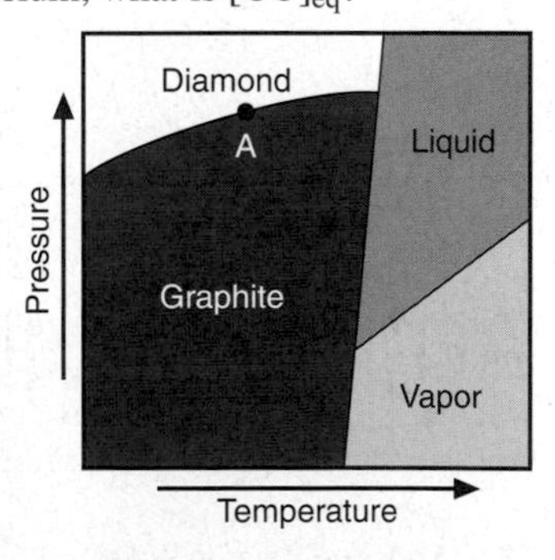

At last! In this titration of HCl with NaOH, the first swirls of phenolphthalein indicator changing from colorless to magenta signal the balance being tipped from acidic to basic solution. In this chapter, we'll examine acids and bases from several vantage points and see how to apply the principles of equilibrium to their reactions.

18

Acid-Base Equilibria

Acids and bases have been used as laboratory chemicals since the time of the alchemists, and they remain indispensable, not only in academic and industrial labs, but in the home as well (Table 18.1).

Notice that some of the acids (e.g., acetic and citric) have a sour taste. In fact, sourness had been a defining property since the 17th century: an acid was any substance that had a sour taste; reacted with active metals, such as aluminum and zinc, to produce hydrogen gas; and turned certain organic compounds characteristic colors. (We discuss *indicators* later and in Chapter 19.) A base was any substance that had a bitter taste and slippery feel and turned the same organic compounds different characteristic colors. (Please remember *NEVER* to taste or touch laboratory chemicals; instead, try some acetic acid in the form of vinegar on your next salad.) Moreover, it was known that *when acids and bases react, each cancels the properties of the other in a process called neutralization.* But definitions in science evolve because, as descriptions become too limited, they must be replaced by broader ones. Although the early definitions of acids and bases described distinctive properties, they inevitably gave way to definitions based on molecular behavior.

IN THIS CHAPTER . . . We develop three definitions of acids and bases that allow us to understand ever-increasing numbers of reactions. In the process, we apply the principles of chemical equilibrium to this essential group of substances. After presenting the classical (Arrhenius) acid-base definition, we examine acid dissociation to see why acids vary in strength. The pH scale is introduced as a means of comparing the acidity or basicity of aqueous solutions. Then, we'll see that the Brønsted-Lowry acid-base definition greatly expands the meaning of "base," along with the scope of chemical changes considered acid-base reactions. We explore the molecular structures of acids and bases to rationalize variations in their strengths and see that the very designations "acid" and "base" depend on relative strengths and on the solvent. Finally, we'll see that the Lewis acid-base definition expands the meaning of "acid" and acid-base behavior even further.

Concepts & Skills to Review

before you study this chapter

- role of water as solvent (Section 4.1)
- writing ionic equations (Section 4.2)
- acids, bases, and acid-base reactions (Section 4.4)
- proton transfer in acid-base reactions (Section 4.4)
- properties of an equilibrium constant (Section 17.2)
- solving equilibrium problems (Section 17.5)

Table 18.1 Some Common Acids and Bases and Their Household Uses

Substance	Use
Acids	
Acetic acid, CH_3COOH	Flavoring, preservative
Citric acid, $H_3C_6H_5O_7$	Flavoring
Phosphoric acid, H_3PO_4	Rust remover
Boric acid, H_3BO_3	Mild antiseptic, insecticide
Aluminum salts, $NaAl(SO_4)_2 \cdot 12H_2O$	In baking powder, with sodium hydrogen carbonate
Hydrochloric acid (muriatic acid), HCl	Brick and ceramic tile cleaner
Bases	
Sodium hydroxide (lye), NaOH	Oven and drain cleaners
Ammonia, NH_3	Household cleaner
Sodium carbonate, Na_2CO_3	Water softener, grease remover
Sodium hydrogen carbonate, $NaHCO_3$	Fire extinguisher, rising agent in cake mixes (baking soda), mild antacid
Sodium phosphate, Na_3PO_4	Cleaner for surfaces before painting or wallpapering

Pioneers of Acid-Base Chemistry
Three 17th-century chemists laid the foundations of acid-base chemistry. Johann Glauber (1604–1668) became renowned for his ability to prepare acids and their salts. Glauber's salt ($Na_2SO_4 \cdot 10H_2O$) is still used today for preparing dyes and printing textiles. Otto Tachenius (c. 1620–1690) is credited with first recognizing that a salt is the product of the reaction between an acid and a base. Robert Boyle (1627–1691), in addition to studying gas behavior, fostered the use of spot tests, flame colors, fume odors, and precipitates to analyze reactions. He was the first to associate the color change in syrup of violets (an organic dye) with the acidic or basic nature of the test solution.

18.1 ACIDS AND BASES IN WATER

Although water is not an essential participant in all modern acid-base definitions, most laboratory work with acids and bases involves water, as do most environmental, biological, and industrial applications. Recall from our discussion in Chapter 4 that *water is a product in all reactions between strong acids and strong bases:*

$$HCl(aq) + NaOH(aq) \longrightarrow NaCl(aq) + H_2O(l)$$

Indeed, as the net ionic equation of this reaction shows, water is *the* product:

$$H^+(aq) + OH^-(aq) \longrightarrow H_2O(l)$$

Furthermore, whenever an acid dissociates in water, solvent molecules participate in the reaction:

$$HA(g \text{ or } l) + H_2O(l) \longrightarrow A^-(aq) + H_3O^+(aq)$$

As you saw in that earlier discussion, water surrounds the proton to form H-bonded species with the general formula $H(H_2O)_n{}^+$. Because the proton is so small, its charge density is very high, so its attraction to water is especially strong. The proton bonds covalently to one of the lone electron pairs of a water molecule's O atom to form a **hydronium ion, H_3O^+,** which forms H bonds to several other water molecules (see Figure 4.4, p. 139). To emphasize the active role of water and the nature of the proton-water interaction, the hydrated proton is usually shown in the text as $H_3O^+(aq)$, although in some cases this hydrated species is shown more simply as $H^+(aq)$.

Release of H^+ or OH^- and the Classical Acid-Base Definition

The earliest and simplest definition of acids and bases that reflects their molecular nature was suggested by Svante Arrhenius, whose work on the rate constant we encountered in Chapter 16. In the **classical** (or **Arrhenius**) **acid-base definition,** acids and bases are classified in terms of their formulas and their behavior *in water:*

- An *acid* is a substance that has H in its formula and dissociates in water to yield H_3O^+.
- A *base* is a substance that has OH in its formula and dissociates in water to yield OH^-.

Some typical Arrhenius acids are HCl, HNO_3, and HCN, and some typical bases are NaOH, KOH, and $Ba(OH)_2$. Although Arrhenius bases contain discrete OH^- ions in their structures, Arrhenius acids *never* contain H^+ ions. On the contrary, these acids contain *covalently bonded H atoms that ionize in water.*

When an acid and a base react, they undergo **neutralization.** The meaning of acid-base reactions has changed along with the definitions of acid and base, but in the Arrhenius sense, neutralization occurs when *the H^+ ion from the acid and the OH^- ion from the base combine to form H_2O.* This description explains an observation that puzzled many of Arrhenius's colleagues. They observed that all neutralization reactions between what we now call strong acids and strong bases (those that dissociate completely in water) had the same heat of reaction. No matter which strong acid and base reacted, and no matter which salt formed, ΔH^0_{rxn} was about -56 kJ per mole of water formed. Arrhenius suggested that the heat of reaction was always the same because the actual reaction was always the same—a hydrogen ion and a hydroxide ion formed water:

$$H^+(aq) + OH^-(aq) \longrightarrow H_2O(l) \qquad \Delta H^0_{rxn} = -55.9 \text{ kJ}$$

Solving for pH:

$$[H_3O^+] = \frac{K_w}{[OH^-]} = \frac{1.0\times10^{-14}}{1.2\times10^{-5}} = 8.3\times10^{-10}\,M$$

$$pH = -\log\,(8.3\times10^{-10}) = 9.08$$

Check The K_b calculation seems reasonable: $\sim10\times10^{-15}/2\times10^{-5} = 5\times10^{-10}$. Because Ac^- is a weak base, $[OH^-] > [H_3O^+]$; thus, pH > 7, which makes sense.

FOLLOW-UP PROBLEM 18.10 Sodium hypochlorite (NaClO) is the active ingredient in household laundry bleach. What is the pH of 0.20 *M* NaClO?

SECTION SUMMARY

The extent to which a weak base accepts a proton from water to form OH^- is expressed by a base-dissociation constant, K_b. Brønsted-Lowry bases include NH_3 and amines and the anions of weak acids. All produce basic solutions by accepting H^+ from water, which yields OH^- and thus makes $[H_3O^+] < [OH^-]$. A solution of HA is acidic because $[HA] >> [A^-]$, so $[H_3O^+] > [OH^-]$. A solution of A^- is basic because $[A^-] >> [HA]$, so $[OH^-] > [H_3O^+]$. By multiplying the expressions for K_a of HA and K_b of A^-, we obtain K_w. This relationship allows us to calculate either K_a of BH^+, the cationic conjugate acid of a molecular weak base B, or K_b of A^-, the anionic conjugate base of a molecular weak acid HA.

18.6 MOLECULAR PROPERTIES AND ACID STRENGTH

The strength of an acid depends on its ability to donate a proton, which depends in turn on the strength of the bond to the acidic proton. In this section, we apply trends in atomic and bond properties to determine the trends in acid strength of nonmetal hydrides and oxoacids and discuss the acidity of hydrated metal ions.

Trends in Acid Strength of Nonmetal Hydrides

Two factors determine how easily a proton is released from a nonmetal hydride: the electronegativity of the central nonmetal (E) and the strength of the E—H bond. Figure 18.11 displays two periodic trends:

1. *Across a period, nonmetal hydride acid strength increases.* Across a period, the electronegativity of the nonmetal E determines the trend. As E becomes more electronegative, electron density around H is withdrawn, and the E—H bond becomes more polar. As a result, an H^+ is released more easily to an O atom of a surrounding water molecule. In aqueous solution, the hydrides of Groups 3A(13) to 5A(15) do not behave as acids, but an increase in acid strength is seen in Groups 6A(16) and 7A(17). Thus, HCl is a stronger acid than H_2S because Cl is more electronegative (EN = 3.0) than S (EN = 2.5). The same relationship holds across each period.

2. *Down a group, nonmetal hydride acid strength increases.* Down a group, E—H bond strength determines the trend. As E becomes larger, the E—H bond becomes longer and weaker, so H^+ comes off more easily.* Thus, the hydrohalic acids increase in strength down the group:

$$HF << HCl < HBr < HI$$

A similar trend in increasing acid strength is seen down Group 6A(16). (The trend in hydrohalic acid strength is not seen in aqueous solution, where HCl, HBr, and HI are all equally strong; we discuss how this trend is observed in Section 18.8.)

*Actually, bond energy refers to bond breakage that forms an H atom, whereas acidity refers to bond breakage that forms an H^+ ion, so the two processes are not the same. Nevertheless, the magnitudes of the two types of bond breakage parallel each other.

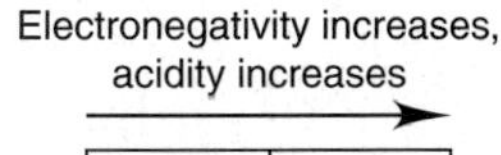

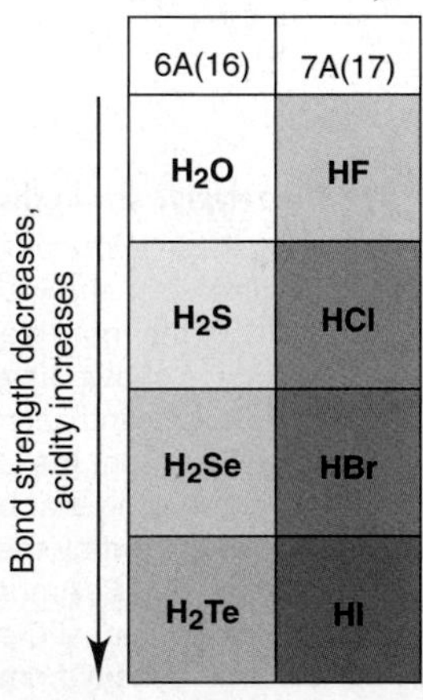

Figure 18.11 The effect of atomic and molecular properties on nonmetal hydride acidity. As the electronegativity of the nonmetal (E) bonded to the ionizable proton increases *(left to right),* the acidity increases. As the length of the E—H bond increases *(top to bottom),* the bond strength decreases, so the acidity increases. (In water, HCl, HBr, and HI are equally strong, for reasons discussed in Section 18.8.)

Trends in Acid Strength of Oxoacids

All oxoacids have the acidic H atom bonded to an O atom, so bond strength (length) is not a factor in their acidity, though it is with the nonmetal hydrides. Rather, as you saw in Section 14.8, two factors determine the acid strength of oxoacids: the electronegativity of the central nonmetal (E) and the number of O atoms.

1. *For oxoacids with the* ***same*** *number of oxygens around E, acid strength increases with the electronegativity of E.* Consider the hypohalous acids (written here as HOE, where E is a halogen atom). The more electronegative E is, the more electron density it pulls from the O—H bond; the more polar the O—H bond becomes, the more easily H^+ is lost (Figure 18.12A). Because electronegativity decreases down the group, we predict that acid strength decreases: HOCl > HOBr > HOI. Our prediction is confirmed by the K_a values:

$$K_a \text{ of HOCl} = 2.9\times10^{-8} \qquad K_a \text{ of HOBr} = 2.3\times10^{-9} \qquad K_a \text{ of HOI} = 2.3\times10^{-11}$$

We also predict (correctly) that in Group 6A(16), H_2SO_4 is stronger than H_2SeO_4; in Group 5A(15), H_3PO_4 is stronger than H_3AsO_4, and so forth.

2. *For oxoacids with* ***different*** *numbers of oxygens around a given E, acid strength increases with number of O atoms.* The electronegative O atoms pull electron density away from E, which makes the O—H bond more polar. The more O atoms present, the greater the shift in electron density, and the more easily the H^+ ion comes off (Figure 18.12B). Therefore, we predict, for instance, that chlorine oxoacids (written here as $HOClO_n$, with n from 0 to 3) increase in strength in the order HOCl < HOClO < $HOClO_2$ < $HOClO_3$. Once again, the K_a values support the prediction:

$$K_a \text{ of HOCl (hypochlorous acid)} = 2.9\times10^{-8}$$
$$K_a \text{ of HOClO (chlorous acid)} = 1.12\times10^{-2}$$
$$K_a \text{ of } HOClO_2 \text{ (chloric acid)} \approx 1$$
$$K_a \text{ of } HOClO_3 \text{ (perchloric acid)} = >10^7$$

It follows from this that HNO_3 is stronger than HNO_2, that H_2SO_4 is stronger than H_2SO_3, and so forth.

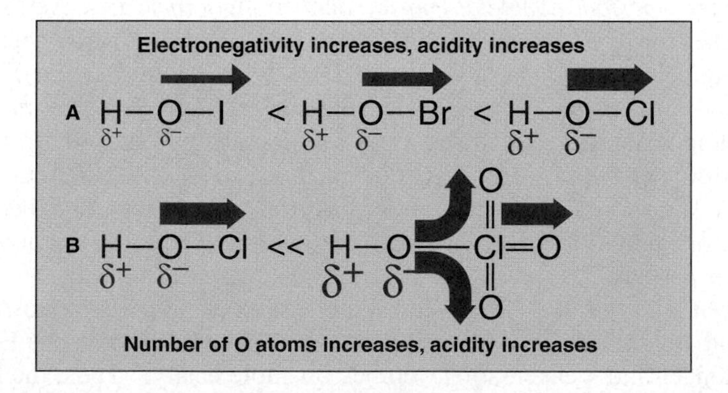

Figure 18.12 The relative strengths of oxoacids. A, Among these hypohalous acids, HOCl is the strongest and HOI the weakest. Because Cl is the most electronegative of the halogens shown here, it withdraws electron density (indicated by thickness of green arrow) from the O—H bond most effectively, making that bond most polar in HOCl (indicated by the relative sizes of the δ symbols). **B,** Among the chlorine oxoacids, the additional O atoms in $HOClO_3$ pull electron density from the O—H bond, making the bond much more polar than that in HOCl.

Acidity of Hydrated Metal Ions

The aqueous solutions of certain metal ions are acidic because the *hydrated* metal ion transfers an H^+ ion to water. Consider a general metal nitrate, $M(NO_3)_n$, as it dissolves in water. The ions separate and become bonded to a specific number of surrounding H_2O molecules. This equation shows the hydration of the cation (M^{n+}) with H_2O molecules; hydration of the anion (NO_3^-) is indicated by (*aq*):

$$M(NO_3)_n(s) + xH_2O(l) \longrightarrow M(H_2O)_x^{n+}(aq) + nNO_3^-(aq)$$

If the metal ion, M^{n+}, is *small and highly charged,* it has a high charge density and withdraws sufficient electron density from the O—H bonds of these bonded water molecules for a proton to be released. That is, the hydrated cation, $M(H_2O)_x^{n+}$, acts as a typical Brønsted-Lowry acid. In the process, the bound H_2O molecule that releases the proton becomes a bound OH^- ion:

$$M(H_2O)_x^{n+}(aq) + H_2O(l) \rightleftharpoons M(H_2O)_{x-1}OH^{(n-1)+}(aq) + H_3O^+(aq)$$

Each type of hydrated metal ion that releases a proton has a characteristic K_a value. Table 18.7 shows some common examples (also see Appendix C).

Aluminum ion, for example, has the small size and high positive charge needed to produce an acidic solution. When an aluminum salt, such as $Al(NO_3)_3$, dissolves in water, the following steps occur:

$$Al(NO_3)_3(s) + 6H_2O(l) \longrightarrow Al(H_2O)_6^{3+}(aq) + 3NO_3^-(aq)$$

[dissolution and hydration]

$$Al(H_2O)_6^{3+}(aq) + H_2O(l) \rightleftharpoons Al(H_2O)_5OH^{2+}(aq) + H_3O^+(aq)$$

[dissociation of weak acid]

Note the formulas of the hydrated metal ions in the last step. When H^+ is released, the number of bound H_2O molecules decreases by 1 (from 6 to 5) and the number of bound OH^- ions increases by 1 (from 0 to 1), which reduces the ion's positive charge by 1 (from 3 to 2) (Figure 18.13).

Through its ability to withdraw electron density from the O—H bonds of the bonded water molecules, a small, highly charged central metal ion behaves like a central electronegative atom in an oxoacid. Salts of most M^{2+} and M^{3+} ions yield acidic aqueous solutions.

Table 18.7 K_a **Values of Some Hydrated Metal Ions at 25°C**

Free Ion	Hydrated Ion	K_a
Fe^{3+}	$Fe(H_2O)_6^{3+}(aq)$	6×10^{-3}
Sn^{2+}	$Sn(H_2O)_6^{2+}(aq)$	4×10^{-4}
Cr^{3+}	$Cr(H_2O)_6^{3+}(aq)$	1×10^{-4}
Al^{3+}	$Al(H_2O)_6^{3+}(aq)$	1×10^{-5}
Cu^{2+}	$Cu(H_2O)_6^{2+}(aq)$	3×10^{-8}
Pb^{2+}	$Pb(H_2O)_6^{2+}(aq)$	3×10^{-8}
Zn^{2+}	$Zn(H_2O)_6^{2+}(aq)$	1×10^{-9}
Co^{2+}	$Co(H_2O)_6^{2+}(aq)$	2×10^{-10}
Ni^{2+}	$Ni(H_2O)_6^{2+}(aq)$	1×10^{-10}

ACID STRENGTH (increasing upward)

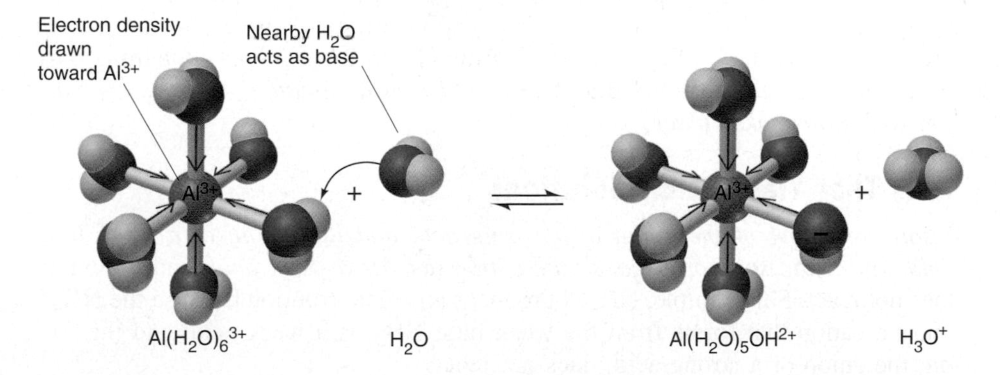

Figure 18.13 The acidic behavior of the hydrated Al^{3+} ion. When a metal ion enters water, it is hydrated as water molecules bond to it. If the ion is small and multiply charged, as is the Al^{3+} ion, it pulls sufficient electron density from the O—H bonds of the attached water molecules to make the bonds more polar, and an H^+ ion is transferred to a nearby water molecule.

SECTION SUMMARY

The strength of an acid depends on the ease with which the ionizable proton is released. For nonmetal hydrides, acid strength increases across a period, with the electronegativity of the nonmetal (E), and down a group, with the length of the E—H bond. For oxoacids with the same number of O atoms, acid strength increases with electronegativity of E; for oxoacids with the same E, acid strength increases with number of O atoms. Small, highly charged metal ions are acidic in water because they withdraw electron density from the O—H bonds of bound H_2O molecules, releasing an H^+ ion to the solution.

18.7 ACID-BASE PROPERTIES OF SALT SOLUTIONS

Up to now you've seen that cations of weak bases (such as NH_4^+) are acidic, anions of weak acids (such as CN^-) are basic, anions of polyprotic acids (such as $H_2PO_4^-$) are often acidic, and small, highly charged metal cations (such as Al^{3+}) are acidic. Therefore, when salts containing these ions dissolve in water, the pH of the solution is affected. You can predict the relative acidity of a salt solution from the relative ability of the cation and/or anion to react with water. Let's examine the ionic makeup of salts that yield neutral, acidic, or basic solutions to see how we make this prediction.

Salts That Yield Neutral Solutions

A salt consisting of the anion of a strong acid and the cation of a strong base yields a neutral solution because the ions do not react with water. In order to see why the ions don't react, let's consider the dissociation of the parent acid and base. When a strong acid such as HNO_3 dissolves, complete dissociation takes place:

$$HNO_3(l) + H_2O(l) \longrightarrow NO_3^-(aq) + H_3O^+(aq)$$

H_2O is a much stronger base than NO_3^-, so the reaction proceeds essentially to completion. The same argument can be made for any strong acid: *the anion of a strong acid is a much weaker base than water.* Therefore, a strong acid anion is hydrated, but nothing further happens.

Now consider the dissociation of a strong base, such as NaOH:

$$NaOH(s) \xrightarrow{H_2O} Na^+(aq) + OH^-(aq)$$

The Na^+ ion has a relatively large size and low charge and therefore does not bond strongly with the water molecules around it. When Na^+ enters water, it becomes hydrated but nothing further happens. *The cations of all strong bases behave this way.*

The anions of strong acids are the halide ions, except F^-, and those of strong oxoacids, such as NO_3^- and ClO_4^-. The cations of strong bases are those from Group 1A(1) and Ca^{2+}, Sr^{2+}, and Ba^{2+} from Group 2A(2). Salts containing only these ions, such as NaCl and $Ba(NO_3)_2$, *yield neutral solutions because no reaction with water takes place.*

Salts That Yield Acidic Solutions

A salt consisting of the anion of a strong acid and the cation of a weak base yields an acidic solution because the cation acts as a weak acid, and the anion does not react. For example, NH_4Cl produces an acidic solution because the NH_4^+ ion, the cation that forms from the weak base NH_3, is a weak acid, and the Cl^- ion, the anion of a strong acid, does not react:

$$NH_4Cl(s) \xrightarrow{H_2O} NH_4^+(aq) + Cl^-(aq) \qquad \text{[dissolution and hydration]}$$

$$NH_4^+(aq) + H_2O(l) \rightleftharpoons NH_3(aq) + H_3O^+(aq) \qquad \text{[dissociation of weak acid]}$$

As you saw earlier, *small, highly charged metal ions* make up another group of cations that yield H_3O^+ in solution. For example, $Fe(NO_3)_3$ produces an acidic solution because the hydrated Fe^{3+} ion acts as a weak acid, whereas the NO_3^- ion, the anion of a strong acid, does not react:

$$Fe(NO_3)_3(s) + 6H_2O(l) \xrightarrow{H_2O} Fe(H_2O)_6^{3+}(aq) + 3NO_3^-(aq) \qquad \text{[dissolution and hydration]}$$

$$Fe(H_2O)_6^{3+}(aq) + H_2O(l) \rightleftharpoons Fe(H_2O)_5OH^{2+}(aq) + H_3O^+(aq) \qquad \text{[dissociation of weak acid]}$$

A third group of salts that yield H_3O^+ ions in solutions consists of *cations of strong bases and anions of polyprotic acids that still have one or more ionizable protons*. For example, NaH_2PO_4 yields an acidic solution because Na^+, the cation of a strong base, does not react, while $H_2PO_4^-$, the first anion of the weak polyprotic acid H_3PO_4, is also a weak acid:

$$NaH_2PO_4(s) \xrightarrow{H_2O} Na^+(aq) + H_2PO_4^-(aq) \quad \text{[dissolution and hydration]}$$

$$H_2PO_4^-(aq) + H_2O(l) \rightleftharpoons HPO_4^{2-}(aq) + H_3O^+(aq) \quad \text{[dissociation of weak acid]}$$

Salts That Yield Basic Solutions

A salt consisting of the anion of a weak acid and the cation of a strong base yields a basic solution in water because the anion acts as a weak base, and the cation does not react. The anion of a weak acid accepts a proton from water to yield OH^- ion. Sodium acetate, for example, yields a basic solution because the Na^+ ion, the cation of a strong base, does not react with water, and the CH_3COO^- ion, the anion of the weak acid CH_3COOH, acts as a weak base:

$$CH_3COONa(s) \xrightarrow{H_2O} Na^+(aq) + CH_3COO^-(aq) \quad \text{[dissolution and hydration]}$$

$$CH_3COO^-(aq) + H_2O(l) \rightleftharpoons CH_3COOH(aq) + OH^-(aq) \quad \text{[reaction of weak base]}$$

Table 18.8 displays the acid-base behavior of the various types of salts in water.

Table 18.8 The Behavior of Salts in Water

Salt Solution (Examples)	pH	Nature of Ions	Ion That Reacts with Water
Neutral [$NaCl$, KBr, $Ba(NO_3)_2$]	7.0	Cation of strong base; Anion of strong acid	None 
Acidic (NH_4Cl, NH_4NO_3, CH_3NH_3Br)	<7.0	Cation of weak base; Anion of strong acid	Cation 
Acidic [$Al(NO_3)_3$, $CrCl_3$, $FeBr_3$]	<7.0	Small, highly charged cation; Anion of strong acid	Cation 
Acidic (NaH_2PO_4, $KHSO_4$, $NaHSO_3$)	<7.0	Cation of strong base; First anion of polyprotic acid	Anion 
Basic (CH_3COONa, KF, Na_2CO_3)	>7.0	Cation of strong base; Anion of weak acid	Anion 

SAMPLE PROBLEM 18.11 Predicting Relative Acidity of Salt Solutions

Problem Predict whether aqueous solutions of the following are acidic, basic, or neutral, and write an equation for the reaction of any ion with water:
(a) Potassium perchlorate, $KClO_4$ (b) Sodium benzoate, C_6H_5COONa
(c) Chromium trichloride, $CrCl_3$ (d) Sodium hydrogen sulfate, $NaHSO_4$

Plan We examine the formulas to determine the cations and anions. Depending on the nature of these ions, the solution will be neutral (strong-acid anion and strong-base cation), acidic (weak-base cation and strong-acid anion, highly charged metal cation, or first anion of a polyprotic acid), or basic (weak-acid anion and strong-base cation).

Solution (a) Neutral. The ions are K^+ and ClO_4^-. The K^+ ion is from the strong base KOH, and the ClO_4^- anion is from the strong acid $HClO_4$. Neither ion reacts with water.
(b) Basic. The ions are Na^+ and $C_6H_5COO^-$. Na^+ is the cation of the strong base NaOH and does not react with water. The benzoate ion, $C_6H_5COO^-$, is from the weak acid benzoic acid, so it reacts with water to produce OH^- ion:

$$C_6H_5COO^-(aq) + H_2O(l) \rightleftharpoons C_6H_5COOH(aq) + OH^-(aq)$$

(c) Acidic. The ions are Cr^{3+} and Cl^-. Cl^- is the anion of the strong acid HCl, so it does not react with water. Cr^{3+} is a small metal ion with a high positive charge, so the hydrated ion, $Cr(H_2O)_6^{3+}$, reacts with water to produce H_3O^+:

$$Cr(H_2O)_6^{3+}(aq) + H_2O(l) \rightleftharpoons Cr(H_2O)_5OH^{2+}(aq) + H_3O^+(aq)$$

(d) Acidic. The ions are Na^+ and HSO_4^-. Na^+ is the cation of the strong base NaOH, so it does not react with water. HSO_4^- is the first anion of the diprotic acid H_2SO_4, and it reacts with water to produce H_3O^+:

$$HSO_4^-(aq) + H_2O(l) \rightleftharpoons SO_4^{2-}(aq) + H_3O^+(aq)$$

FOLLOW-UP PROBLEM 18.11 Write equations to predict whether solutions of the following salts are acidic, basic, or neutral: (a) $KClO_2$; (b) $CH_3NH_3NO_3$; (c) CsI.

Salts of Weakly Acidic Cations and Weakly Basic Anions

The only salts left to consider are those consisting of a cation that acts as a weak acid *and* an anion that acts as a weak base. In these cases, and there are quite a few, both ions react with water. It makes sense, then, that the overall acidity of the solution will depend on the relative acid strength or base strength of the separated ions, which can be determined by comparing their equilibrium constants.

For example, will an aqueous solution of ammonium hydrogen sulfide, NH_4HS, be acidic or basic? First, we write equations for any reactions that occur between the separated ions and water. Ammonium ion is the conjugate acid of a weak base, so it acts as a weak acid:

$$NH_4^+(aq) + H_2O(l) \rightleftharpoons NH_3(aq) + H_3O^+(aq)$$

Hydrogen sulfide ion is the anion of the weak acid H_2S, so it acts as a weak base:

$$HS^-(aq) + H_2O(l) \rightleftharpoons H_2S(aq) + OH^-(aq)$$

The reaction that goes farther to the right will have the greater influence on the pH of the solution, so we must compare the K_a of NH_4^+ with the K_b of HS^-. Recall that only molecular compounds are listed in K_a and K_b tables, so we have to calculate these values for the ions:

$$K_a \text{ of } NH_4^+ = \frac{K_w}{K_b \text{ of } NH_3} = \frac{1.0\times10^{-14}}{1.76\times10^{-5}} = 5.7\times10^{-10}$$

$$K_b \text{ of } HS^- = \frac{K_w}{K_{a1} \text{ of } H_2S} = \frac{1.0\times10^{-14}}{9\times10^{-8}} = 1\times10^{-7}$$

The difference in magnitude of the equilibrium constants ($K_b \approx 200K_a$) tells us that the acceptance of a proton from H_2O by HS^- proceeds further than the donation of a proton to H_2O by NH_4^+. In other words, because K_b of $HS^- > K_a$ of NH_4^+, the NH_4HS solution is basic.

SAMPLE PROBLEM 18.12 Predicting the Relative Acidity of a Salt Solution from K_a and K_b of the Ions

Problem Determine whether an aqueous solution of zinc formate, $Zn(HCOO)_2$, is acidic, basic, or neutral.
Plan The formula consists of the small, highly charged, and therefore weakly acidic, Zn^{2+} cation and the weakly basic $HCOO^-$ anion of the weak acid HCOOH. To determine the relative acidity of the solution, we write equations that show the reactions of the ions with water, and then find K_a of Zn^{2+} (from Appendix C) and calculate K_b of $HCOO^-$ (from K_a of HCOOH in Appendix C) to see which ion reacts to a greater extent.
Solution Writing the reactions with water:

$$Zn(H_2O)_6^{2+}(aq) + H_2O(l) \rightleftharpoons Zn(H_2O)_5OH^+(aq) + H_3O^+(aq)$$
$$HCOO^-(aq) + H_2O(l) \rightleftharpoons HCOOH(aq) + OH^-(aq)$$

Obtaining K_a and K_b of the ions: The K_a of $Zn(H_2O)_6^{2+}(aq)$ is 1×10^{-9}. We obtain K_a of HCOOH and solve for K_b of $HCOO^-$:

$$K_b \text{ of } HCOO^- = \frac{K_w}{K_a \text{ of } HCOOH} = \frac{1.0\times10^{-14}}{1.8\times10^{-4}} = 5.6\times10^{-11}$$

K_a of $Zn(H_2O)_6^{2+} > K_b$ of $HCOO^-$, so the solution is acidic.

FOLLOW-UP PROBLEM 18.12 Determine whether solutions of the following salts are acidic, basic, or neutral: **(a)** $Cu(CH_3COO)_2$; **(b)** NH_4F.

SECTION SUMMARY

Salts that yield a neutral solution consist of ions that do not react with water. Salts that yield an acidic solution contain an unreactive anion and a cation that releases a proton to water. Salts that yield a basic solution contain an unreactive cation and an anion that accepts a proton from water. If both cation and anion react with water, the ion that reacts to the greater extent (higher *K*) determines the acidity or basicity of the salt solution.

18.8 GENERALIZING THE BRØNSTED-LOWRY CONCEPT: THE LEVELING EFFECT

We conclude our focus on the Brønsted-Lowry concept with an important principle that holds for acid-base behavior in any solvent. Notice that, in H_2O, all Brønsted-Lowry acids yield H_3O^+ and all Brønsted-Lowry bases yield OH^-—the ions that form when the solvent autoionizes. In general, *an acid yields the cation and a base yields the anion of solvent autoionization.*

This idea lets us examine a question you may have been wondering about: why are all strong acids and strong bases *equally* strong in water? The answer is that *in water, the strongest acid possible is H_3O^+ and the strongest base possible is OH^-*. The moment we put some gaseous HCl in water, it reacts with the base H_2O and forms H_3O^+. The same holds for HNO_3, H_2SO_4, and any strong acid. All strong acids are equally strong in water because they dissociate *completely* to form H_3O^+. Given that the strong acid is no longer present, we are actually observing the acid strength of H_3O^+.

Similarly, strong bases, such as $Ba(OH)_2$, dissociate completely in water to yield OH^-. Even those that do not contain hydroxide ions in the solid, such as K_2O, do so. The oxide ion, which is a stronger base than OH^-, immediately takes a proton from water to form OH^-:

$$2K^+(aq) + O^{2-}(aq) + H_2O(l) \longrightarrow 2K^+(aq) + 2OH^-(aq)$$

No matter what species we try, any acid stronger than H_3O^+ simply donates its proton to H_2O, and any base stronger than OH^- accepts a proton from H_2O. Thus, water exerts a **leveling effect** on any strong acid or base by reacting with it to form the products of water's autoionization. Acting as a base, water levels the strength of all strong acids by making them appear equally strong, and acting as an acid, it levels the strength of all strong bases as well.

To rank strong acids in terms of relative strength, we must dissolve them in a solvent that is a *weaker* base than water, one that accepts their protons less readily. For example, you saw in Figure 18.11 that the hydrohalic acids increase in strength as the halogen becomes larger, as a result of the longer, weaker H—X bond. In water, HF is weaker than the other hydrogen halides, but HCl, HBr, and HI appear equally strong because water causes them to dissociate completely. When we dissolve them in pure acetic acid, however, *the acetic acid acts as the base* and accepts a proton from the acids:

$$\underset{\text{acid}}{HCl(g)} + \underset{\text{base}}{CH_3COOH(l)} \rightleftharpoons \underset{\text{base}}{Cl^-(acet)} + \underset{\text{acid}}{CH_3COOH_2^+(acet)}$$
$$HBr(g) + CH_3COOH(l) \rightleftharpoons Br^-(acet) + CH_3COOH_2^+(acet)$$
$$HI(g) + CH_3COOH(l) \rightleftharpoons I^-(acet) + CH_3COOH_2^+(acet)$$

[The use of (*acet*) instead of (*aq*) indicates solvation by CH_3COOH.] However, because acetic acid is a *weaker base* than water, the three acids protonate it to *different* extents. Measurements show that HI protonates the solvent to a greater extent than HBr, and HBr does so more than HCl; that is, in pure acetic acid, $K_{HI} > K_{HBr} > K_{HCl}$. Therefore, HCl is a weaker acid than HBr, which is weaker than HI. Similarly, the relative strength of strong bases is determined in a solvent that is a weaker acid than H_2O, such as liquid NH_3.

SECTION SUMMARY

Strong acids (or strong bases) dissociate completely to yield H_3O^+ (or OH^-) in water; in effect, water equalizes (levels) their strengths. Acids that are equally strong in water show differences in strength when dissolved in a solvent that is a weaker base than water, such as acetic acid.

18.9 ELECTRON-PAIR DONATION AND THE LEWIS ACID-BASE DEFINITION

The final acid-base concept we consider was developed by Gilbert N. Lewis, whose contribution to understanding the importance of valence electron pairs in molecular bonding we discussed in Chapter 9. Whereas the Brønsted-Lowry concept focuses on the proton in defining a species as an acid or a base, the Lewis concept highlights the role of the *electron pair*. The **Lewis acid-base definition** holds that

- A *base* is any species that *donates* an electron pair.
- An *acid* is any species that *accepts* an electron pair.

The Lewis definition, like the Brønsted-Lowry definition, requires that a base have an electron pair to donate, so it does not expand the classes of bases. However, *it greatly expands the classes of acids*. Many species, such as CO_2 and Cu^{2+}, that do not contain H in their formula (and thus cannot be Brønsted-Lowry acids) function as Lewis acids by accepting an electron pair in their reactions. Lewis stated his objection to the proton as the defining feature of an acid this way: "To restrict the group of acids to those substances which contain hydrogen interferes as seriously with the systematic understanding of chemistry as would the restriction of the term oxidizing agent to those substances containing oxygen." Moreover, in the Lewis sense, the proton itself functions as an acid because it accepts the electron pair donated by a base:

$$B\!: + H^+ \rightleftharpoons B{-}H^+$$

Thus, *all Brønsted-Lowry acids donate* H^+*, a Lewis acid.*

The product of any Lewis acid-base reaction is called an **adduct,** *a single species that contains a* ***new*** *covalent bond:*

$$A + :B \rightleftharpoons A{-}B \text{ (adduct)}$$

Thus, the Lewis concept radically broadens the idea of acid-base reactions. What to Arrhenius was the formation of H_2O from H^+ and OH^- became, to Brønsted and Lowry, the transfer of a proton from a stronger acid to a stronger base to form a weaker base and weaker acid. To Lewis, the same process became *the donation and acceptance of an electron pair to form a covalent bond in an adduct.*

As we've seen, the key feature of a *Lewis base is a lone pair of electrons to donate*. The key feature of a *Lewis acid is a vacant orbital* (or the ability to rearrange its bonds to form one) to accept that lone pair and form a new bond. There are a variety of neutral molecules and positively charged ions that satisfy this requirement.

Molecules as Lewis Acids

Many neutral molecules function as Lewis acids. In every case, the atom that accepts the electron pair is low in electron density because of either an electron deficiency or a polar multiple bond.

Lewis Acids with Electron-Deficient Atoms Some molecular Lewis acids contain a central atom that is *electron deficient,* one surrounded by fewer than eight valence electrons. The most important are covalent compounds of the Group 3A(13) elements boron and aluminum. As noted in Chapters 10 and 14, these compounds react vigorously to complete their octet. For example, boron trifluoride accepts an electron pair from ammonia to form a covalent bond in a gaseous Lewis acid-base reaction:

$$BF_3 + :NH_3 \rightleftharpoons F_3B{-}NH_3$$

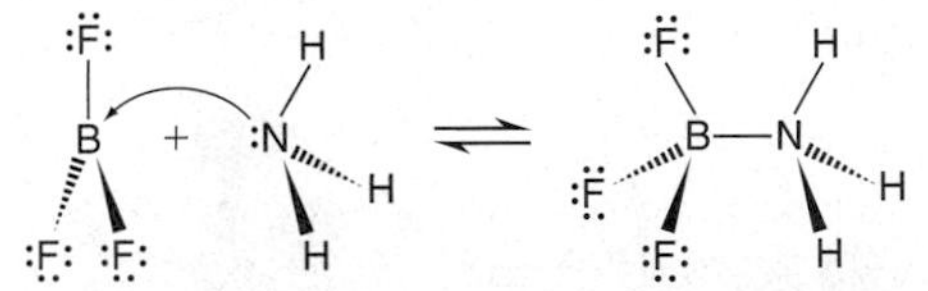

acid base adduct

Unexpected solubility behavior is sometimes due to adduct formation. Aluminum chloride, for instance, dissolves freely in relatively nonpolar diethyl ether because of a Lewis acid-base reaction, in which the ether's O atom donates an electron pair to Al to form a covalent bond:

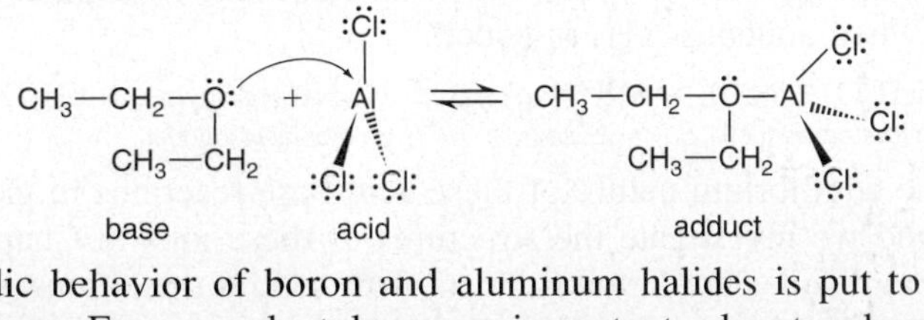

$$(CH_3{-}CH_2)_2\ddot{O}: + AlCl_3 \rightleftharpoons (CH_3{-}CH_2)_2\ddot{O}{-}AlCl_3$$

base acid adduct

This acidic behavior of boron and aluminum halides is put to use in many organic syntheses. For example, toluene, an important solvent and organic reagent, can be made by the action of CH_3Cl on benzene in the presence of $AlCl_3$. The

Lewis acid $AlCl_3$ abstracts the Lewis base Cl^- from CH_3Cl to form an adduct that has a reactive CH_3^+ group, which attacks the benzene ring:

$$\underset{\text{base}}{CH_3Cl} + \underset{\text{acid}}{AlCl_3} \rightleftharpoons \underset{\text{adduct}}{[CH_3]^+[Cl{-}AlCl_3]^-}$$

$$\underset{\text{benzene}}{C_6H_6} + [CH_3]^+[Cl{-}AlCl_3]^- \rightleftharpoons \underset{\text{toluene}}{C_6H_5CH_3} + AlCl_3 + HCl$$

Lewis Acids with Polar Multiple Bonds Molecules that contain a polar double bond also function as Lewis acids. As the electron pair on the Lewis base approaches the partially positive end of the double bond, one of the bonds breaks to form the new bond in the adduct. For example, consider the reaction that occurs when SO_2 dissolves in water. The electronegative O atoms in SO_2 withdraw electron density from the central S, so it is partially positive. The O atom of water donates a lone pair to the S, breaking one of the π bonds and forming an S—O bond, and a proton is transferred from water to that O. The resulting adduct is sulfurous acid, and the overall process is

$$\underset{\text{acid}}{SO_2} + \underset{\text{base}}{H{-}O{-}H} \rightleftharpoons \underset{\text{adduct}}{H_2SO_3}$$

The formation of carbonates from a metal oxide and carbon dioxide is an analogous reaction that occurs in a nonaqueous heterogeneous system. The O^{2-} ion (shown below from CaO) donates an electron pair to the partially positive C in CO_2, a π bond breaks, and the CO_3^{2-} ion forms as the adduct:

$$\underset{\text{base}}{Ca^{2+}\,O^{2-}} + \underset{\text{acid}}{O{=}C{=}O} \rightleftharpoons \underset{\text{adduct}}{Ca^{2+}\,[CO_3]^{2-}}$$

Metal Cations as Lewis Acids

Earlier we saw that certain hydrated metal ions act as Brønsted-Lowry acids. In the Lewis sense, the hydration process itself is an acid-base reaction. The hydrated cation is the adduct, as lone electron pairs on the O atoms of water form covalent bonds to the positively charged ion; thus, *any metal ion acts as a Lewis acid when it dissolves in water:*

$$\underset{\text{acid}}{M^{2+}} + \underset{\text{base}}{6H_2O(l)} \rightleftharpoons \underset{\text{adduct}}{M(H_2O)_6^{2+}(aq)}$$

Ammonia is a stronger Lewis base than water because it displaces H_2O from a hydrated ion when aqueous NH_3 is added:

$$\underset{\text{hydrated adduct}}{Ni(H_2O)_6^{2+}(aq)} + \underset{\text{base}}{6NH_3(aq)} \rightleftharpoons \underset{\text{ammoniated adduct}}{Ni(NH_3)_6^{2+}(aq)} + 6H_2O(l)$$

We discuss the equilibrium nature of these acid-base reactions in greater detail in Chapter 19, and we investigate the structures of these ions in Chapter 23.

Many biomolecules are Lewis adducts with central metal ions. Most often, O and N atoms of organic groups, with their lone pairs, serve as the Lewis bases.

Chlorophyll is a Lewis adduct of a central Mg^{2+} ion and the four N atoms of an organic tetrapyrrole ring system (Figure 18.14). Vitamin B_{12} has a similar structure with a central Co^{3+}, and so does heme, but with a central Fe^{2+}. Several other metal ions, such as Zn^{2+}, Mo^{2+}, and Cu^{2+}, are bound at the active sites of enzymes and function as Lewis acids in the catalytic action.

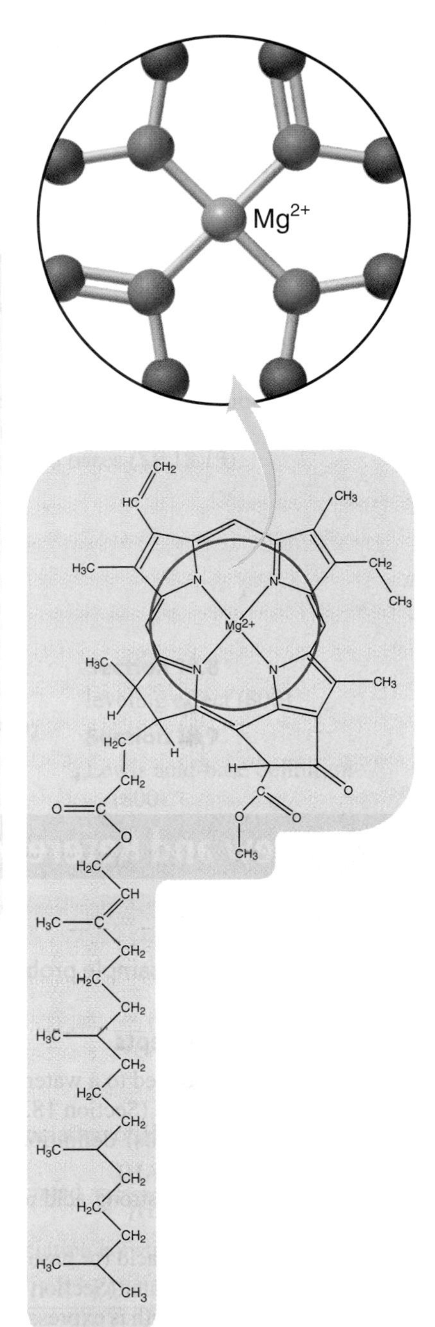

Figure 18.14 The Mg^{2+} ion as a Lewis acid in the chlorophyll molecule. Many biomolecules contain metal ions that act as Lewis acids. In chlorophyll, Mg^{2+} accepts electron pairs from surrounding N atoms that are part of the large organic portion of the molecule.

SAMPLE PROBLEM 18.13 Identifying Lewis Acids and Bases

Problem Identify the Lewis acids and Lewis bases in the following reactions:
(a) $H^+ + OH^- \rightleftharpoons H_2O$
(b) $Cl^- + BCl_3 \rightleftharpoons BCl_4^-$
(c) $K^+ + 6H_2O \rightleftharpoons K(H_2O)_6^+$

Plan We examine the formulas to see which species accepts the electron pair (Lewis acid) and which donates it (Lewis base) in forming the adduct.

Solution (a) The H^+ ion accepts an electron pair from the OH^- ion in forming a bond. H^+ is the acid and OH^- is the base.

(b) The Cl^- ion has four lone pairs and uses one to form a new bond to the central B. Therefore, BCl_3 is the acid and Cl^- is the base.

(c) The K^+ ion does not have any valence electrons to provide, so the bond is formed when electron pairs from O atoms of water enter empty orbitals on K^+. Thus, K^+ is the acid and H_2O is the base.

Check The Lewis acids (H^+, BCl_3, and K^+) each have an unfilled valence shell that can accept an electron pair from the Lewis bases (OH^-, Cl^-, and H_2O).

FOLLOW-UP PROBLEM 18.13 Identify the Lewis acids and Lewis bases in the following reactions:
(a) $OH^- + Al(OH)_3 \rightleftharpoons Al(OH)_4^-$
(b) $SO_3 + H_2O \rightleftharpoons H_2SO_4$
(c) $Co^{3+} + 6NH_3 \rightleftharpoons Co(NH_3)_6^{3+}$

An Overview of Acid-Base Definitions

By looking closely at the essential chemical change involved, chemists can see a common theme in reactions as diverse as a standardized base being used to analyze an unknown fatty acid, baking soda being used in breadmaking, and even oxygen binding to hemoglobin in a blood cell. From this wider perspective, the diversity of acid-base reactions takes on more unity. Let's stand back and survey the scope of the three acid-base definitions and see how they fit together.

The *classical (Arrhenius) definition,* which was the first attempt at describing acids and bases on the molecular level, is the most limited and narrow of the three definitions. It applies only to species whose structures include an H atom or an OH group that is released as an ion when the species dissolves in water. Because relatively few species have these prerequisites, Arrhenius acid-base reactions are relatively few in number, and all such reactions result in the formation of H_2O.

The *Brønsted-Lowry definition* is more general, seeing acid-base reactions as proton-transfer processes and eliminating the requirement that they occur in water. Whereas a Brønsted-Lowry acid, like an Arrhenius acid, still must have an H, a Brønsted-Lowry base is defined as any species with an electron pair available to accept a transferred proton. This definition includes a great many more species as bases. Furthermore, it defines the acid-base reaction in terms of conjugate acid-base pairs, with an acid and a base on both sides of the reaction. The system reaches an equilibrium state based on the relative strengths of the acid, the base, and their conjugates.

18.158 Three beakers contain 100. mL of 0.10 M acid, either HCl, $HClO_2$, or HClO. (a) Find the pH of each. (b) Describe quantitatively how to make the pH equal in the beakers through the addition of water only.

18.159 Human urine has a normal pH of 6.2. If a person eliminates an average of 1250. mL of urine per day, how many H^+ ions are eliminated per week?

18.160 Liquid ammonia autoionizes like water:

$$2NH_3(l) \longrightarrow NH_4^+(am) + NH_2^-(am)$$

where (am) represents solvation by ammonia.
(a) Write the ion-product constant expression, K_{am}.
(b) What are the strongest acid and strongest base that can exist in liquid ammonia?
(c) HNO_3 and HCOOH are leveled in liquid NH_3. Explain with equations.
(d) At the boiling point (−33°C), $K_{am} = 5.1\times10^{-27}$. Calculate $[NH_4^+]$ at this temperature.
(e) Pure sulfuric acid also autoionizes. Write the ion-product constant expression, K_{sulf}, and find the concentration of the conjugate base at 20°C ($K_{sulf} = 2.7\times10^{-4}$ at 20°C).

18.161 Autoionization (see Problem 18.160) occurs in methanol (CH_3OH) and in ethylenediamine ($NH_2CH_2CH_2NH_2$).
(a) The autoionization constant of methanol (K_{met}) is 2×10^{-17}. What is $[CH_3O^-]$ in pure CH_3OH?
(b) The concentration of $NH_2CH_2CH_2NH_3^+$ in pure $NH_2CH_2CH_2NH_2$ is 2×10^{-8} M. What is the autoionization constant of ethylenediamine (K_{en})?

18.162 Thiamine hydrochloride ($C_{12}H_{18}ON_4SCl_2$) is a water-soluble form of thiamine (vitamin B_1; $K_a = 3.37\times10^{-7}$). How many grams of the hydrochloride must be dissolved in 10.00 mL of water to give a pH of 3.50?

18.163 Tris(hydroxymethyl)aminomethane, known as TRIS or THAM, is a water-soluble base that is a reactant in the synthesis of surfactants and pharmaceuticals, an emulsifying agent in cosmetic creams and lotions, and a component of various cleaning and polishing mixtures for textiles and leather. In biomedical research, solutions of TRIS are used to maintain nearly constant pH for the study of enzymes and other cellular components. Given that the pK_b is 5.91, calculate the pH of 0.060 M TRIS.

18.164 When Fe^{3+} salts are dissolved in water, the solution becomes acidic due to formation of $Fe(H_2O)_5OH^{2+}$ and H_3O^+. The overall process involves both Lewis and Brønsted-Lowry acid-base reactions. Write the equations for the process.

18.165 Vinegar is a 5.0% (w/v) solution of acetic acid in water. What is the pH of vinegar?

18.166 How would you differentiate between a strong and a weak monoprotic acid from the results of the following procedures?
(a) Electrical conductivity of an equimolar solution of each acid is measured.
(b) Equal molarities of each are tested with pH paper.
(c) Zinc metal is added to solutions of equal concentration.

18.167 At 50°C and 1 atm, $K_w = 5.19\times10^{-14}$. Calculate parts (a)–(c) under these conditions:
(a) $[H_3O^+]$ in pure water
(b) $[H_3O^+]$ in 0.010 M NaOH
(c) $[OH^-]$ in 0.0010 M $HClO_4$
(d) Calculate $[H_3O^+]$ in 0.0100 M KOH at 100°C and 1000 atm pressure ($K_w = 1.10\times10^{-12}$).
(e) Calculate the pH of pure water at 100°C and 1000 atm.

18.168 The catalytic efficiency of an enzyme is called its *activity* and refers to the rate at which it catalyzes the reaction. Most enzymes have optimum activity over a relatively narrow pH range, which is related to the pH of the local cellular fluid. The pH profiles of three digestive enzymes are shown.

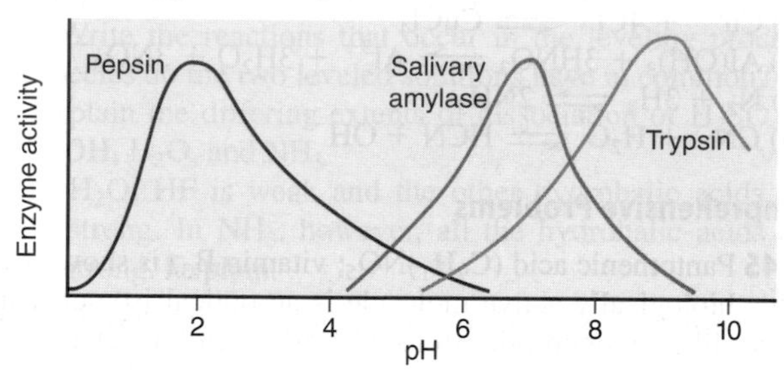

Salivary amylase begins digestion of starches in the mouth and has optimum activity at a pH of 6.8; pepsin begins protein digestion in the stomach and has optimum activity at a pH of 2.0; and trypsin, released in pancreatic juices, continues protein digestion in the small intestine and has optimum activity at a pH of 9.5. Calculate $[H_3O^+]$ in the local cellular fluid for each enzyme.

18.169 Acetic acid has a K_a of 1.8×10^{-5}, and ammonia has a K_b of 1.8×10^{-5}. Find $[H_3O^+]$, $[OH^-]$, pH, and pOH for (a) 0.240 M acetic acid and (b) 0.240 M ammonia.

18.170 Sodium phosphate has industrial uses ranging from clarifying crude sugar to manufacturing paper. Sold as TSP, it is used in solution to remove boiler scale and to wash painted brick and concrete. What is the pH of a solution containing 33 g of Na_3PO_4 per liter? What is $[OH^-]$ of this solution?

18.171 The Group 5A(15) hydrides react with boron trihalides in a reversible Lewis acid-base reaction. When 0.15 mol of $PH_3BCl_3(s)$ is introduced into a 3.0-L container at a certain temperature, 8.4×10^{-3} mol of PH_3 is present at equilibrium: $PH_3BCl_3(s) \rightleftharpoons PH_3(g) + BCl_3(g)$.
(a) Calculate K_c for the reaction at this temperature.
(b) Draw a Lewis structure for the reactant.

18.172 A 1.000 m solution of chloroacetic acid ($ClCH_2COOH$) freezes at −1.93°C. Use these data to find the K_a of chloroacetic acid. (Assume the molarities equal the molalities.)

18.173 Sodium stearate ($C_{17}H_{35}COONa$) is a major component of bar soap (see Chapter 13, p. 498). The K_a of the stearic acid is 1.3×10^{-5}. What is the pH of 10.0 mL of a solution containing 0.42 g of sodium stearate?

18.174 Calcium propionate [$Ca(CH_3CH_2COO)_2$; calcium propanoate] is a mold inhibitor used in food, tobacco, and pharmaceuticals. (a) Use balanced equations to show whether aqueous calcium propionate is acidic, basic, or neutral. (b) Use Appendix C to find the pH of a solution made by dissolving 7.05 g of $Ca(CH_3CH_2COO)_2$ in water to give 0.500 L of solution.

18.175 Carbon dioxide is less soluble in dilute HCl than in dilute NaOH. Explain.

18.176 (a) If $K_w = 1.139\times10^{-15}$ at 0°C and 5.474×10^{-14} at 50°C, find $[H_3O^+]$ and pH of water at 0°C and 50°C.
(b) The autoionization constant for heavy water (deuterium oxide, D_2O) is 3.64×10^{-16} at 0°C and 7.89×10^{-15} at 50°C. Find $[D_3O^+]$ and pD of heavy water at 0°C and 50°C.
(c) Suggest a reason for these differences.

18.177 HX ($\mathcal{M}$ = 150. g/mol) and HY ($\mathcal{M}$ = 50.0 g/mol) are weak acids. A solution of 12.0 g/L of HX has the same pH as one containing 6.00 g/L of HY. Which is the stronger acid? Why?

18.178 In his acid-base studies, Arrhenius discovered an important fact involving reactions like the following:

$$KOH(aq) + HNO_3(aq) \longrightarrow ?$$
$$NaOH(aq) + HCl(aq) \longrightarrow ?$$

(a) Complete the reactions and use the data for the individual ions in Appendix B to calculate each ΔH^0_{rxn}.
(b) Explain your results and use them to predict ΔH^0_{rxn} for

$$KOH(aq) + HCl(aq) \longrightarrow ?$$

18.179 Amines have foul odors. Putrescine [$NH_2(CH_2)_4NH_2$], once thought to be found only in rotting animal tissue, is now known to be a component of all cells and essential for normal and abnormal (cancerous) growth. It also plays a key role in formation of GABA, a neurotransmitter. A 0.10 *M* aqueous solution of putrescine has $[OH^-] = 2.1\times10^{-3}$. What is the K_b?

18.180 Nitrogen is discharged from wastewater treatment facilities into rivers and streams, usually as NH_3 and NH_4^+:

$$NH_3(aq) + H_2O(l) \rightleftharpoons NH_4^+(aq) + OH^-(aq) \quad K_b = 1.76\times10^{-5}$$

One strategy for removing it is to raise the pH and "strip" the NH_3 from solution by bubbling air through the water. (a) At pH 7.00, what fraction of the total nitrogen in solution is NH_3, defined as $[NH_3]/([NH_3] + [NH_4^+])$? (b) What is the fraction at pH 10.00? (c) Explain the basis of ammonia stripping.

18.181 Polymers and other large molecules are not very soluble in water, but their solubility increases if they have charged groups.
(a) Casein is a protein in milk that contains many carboxylic acid groups on its side chains. Explain how the solubility of casein in water varies with pH.
(b) Histones are proteins that are essential to the proper function of DNA. They are weakly basic due to the presence of side chains with $-NH_2$ and $=NH$ groups. Explain how the solubility of histones in water varies with pH.

18.182 Hemoglobin (Hb) transports oxygen in the blood:

$$HbH^+(aq) + O_2(aq) + H_2O(l) \longrightarrow H_3O^+(aq) + HbO_2(aq)$$

In blood, $[H_3O^+]$ is held nearly constant at 4×10^{-8} *M*.
(a) How does the equilibrium position change in the lungs?
(b) How does it change in O_2-deficient cells?
(c) Excessive vomiting may lead to metabolic *alkalosis*, in which $[H_3O^+]$ in blood *decreases*. How does this condition affect the ability of Hb to transport O_2?
(d) Diabetes mellitus may lead to metabolic *acidosis*, in which $[H_3O^+]$ in blood *increases*. How does this condition affect the ability of Hb to transport O_2?

18.183 Vitamin C (ascorbic acid, $H_2C_6H_6O_6$) is a weak diprotic acid. It is essential for the synthesis of collagen, the major protein in connective tissue. (a) If the pH of a 5.0% (w/v) solution of vitamin C in water is 2.77, calculate the K_{a1} of vitamin C. (b) The vitamin is also taken as its sodium salt. What is the pH of a 10.0 g/L solution of sodium ascorbate (NaAsc)?

18.184 Because of the behavior of their R groups in water, lysine is called a basic amino acid and aspartic acid an acidic amino acid. Write balanced equations that demonstrate this behavior.

18.185 A solution of propanoic acid (CH_3CH_2COOH), made by dissolving 7.500 g in sufficient water to make 100.0 mL, has a freezing point of -1.890°C.
(a) Calculate the molarity of the solution.
(b) Calculate the molarity of the propanoate ion. (Assume the molarity of the solution equals the molality.)
(c) Calculate the percent dissociation of propanoic acid.

18.186 Quinine ($C_{20}H_{24}N_2O_2$; *see below*) is a natural product with antimalarial properties that saved thousands during construction of the Panama Canal. It stands as a classic example of the medicinal wealth of tropical forests. Both N atoms are basic, but the N (colored) of the 3° amine group is far more basic ($pK_b = 5.1$) than the N within the aromatic ring system ($pK_b = 9.7$).

$CH_2{=}CH$
N
HO—CH
CH_3O
N

(a) Quinine is not very soluble in water: a saturated solution is only 1.6×10^{-3} *M*. What is the pH of this solution?
(b) Show that the aromatic N contributes negligibly to the pH of the solution.
(c) Because of its low solubility as a free base in water, quinine is given as an amine salt. For instance, quinine hydrochloride ($C_{20}H_{24}N_2O_2{\cdot}HCl$) is about 120 times more soluble in water than quinine. What is the pH of 0.53 *M* quinine hydrochloride?
(d) An antimalarial concentration in water is 1.5% quinine hydrochloride by mass ($d = 1.0$ g/mL). What is the pH?

18.187 Drinking water is often disinfected with chlorine gas, which hydrolyzes to form hypochlorous acid (HClO), a weak acid but powerful disinfectant:

$$Cl_2(aq) + 2H_2O(l) \longrightarrow HClO(aq) + H_3O^+(aq) + Cl^-(aq)$$

The fraction of HClO in solution is defined as

$$\frac{[HClO]}{[HClO] + [ClO^-]}$$

(a) What is the fraction of HClO at pH 7.00 (K_a of HClO $= 2.9\times10^{-8}$)? (b) What is the fraction at pH 10.00?

18.188 The following scenes represent three weak acids HA (where A = X, Y, or Z) dissolved in water (H_2O is not shown):

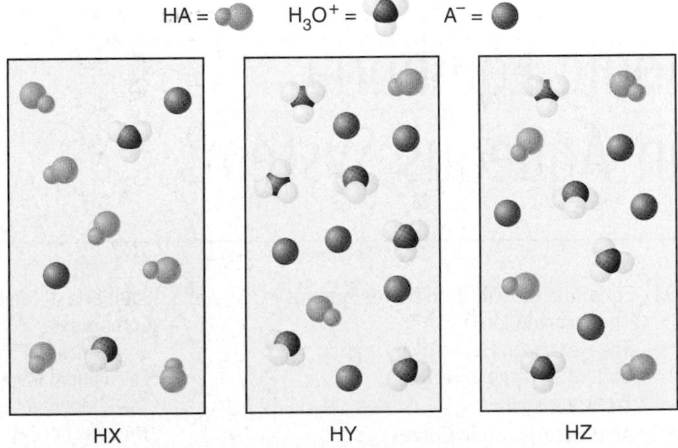

(a) Rank the acids in order of increasing K_a.
(b) Rank the acids in order of increasing pK_a.
(c) Rank the conjugate bases in order of increasing pK_b.
(d) What is the percent dissociation of HX?
(e) If equimolar amounts of the sodium salts of the acids (NaX, NaY, and NaZ) were dissolved in water, which solution would have the highest pOH? The lowest pH?

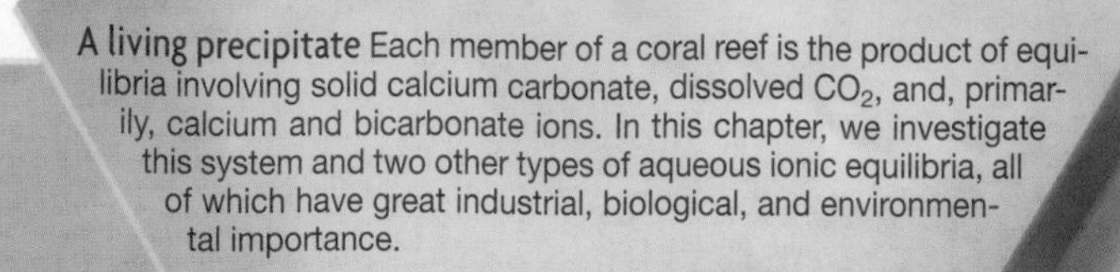

A living precipitate Each member of a coral reef is the product of equilibria involving solid calcium carbonate, dissolved CO_2, and, primarily, calcium and bicarbonate ions. In this chapter, we investigate this system and two other types of aqueous ionic equilibria, all of which have great industrial, biological, and environmental importance.

19

Ionic Equilibria in Aqueous Systems

Europa, one of Jupiter's moons, has an icy surface with hints of vast oceans of liquid water beneath. Is there life on Europa? Perhaps some Europan astronomer viewing Earth is asking a similar question, because liquid water is essential for the aqueous systems that maintain life. Every astronaut has felt awe at seeing our "beautiful blue orb" from space. A biologist peering at the fabulous watery world of a living cell probably feels the same way. A chemist is awed by the principles of equilibrium and their universal application to aqueous solutions wherever they occur.

Consider just a few cases of aqueous equilibria. The magnificent formations in limestone caves and the vast expanses of oceanic coral reefs result from subtle shifts in carbonate solubility equilibria. Carbonates also influence soil pH and prevent acidification of lakes by acid rain. Equilibria involving carbon dioxide and phosphates help organisms maintain cellular pH within narrow limits. Equilibria involving clays in soils control the availability of ionic nutrients for plants. The principles of ionic equilibrium also govern how water is softened, how substances are purified by precipitation of unwanted ions, and even how the weak acids in wine and vinegar influence the delicate taste of a fine French sauce.

IN THIS CHAPTER . . . We explore three aqueous ionic equilibrium systems: acid-base buffers, slightly soluble salts, and complex ions. Our discussion of buffers introduces the common-ion effect, an important phenomenon in many ionic equilibria. We discuss why buffers are important, how they work, and how to prepare them, and we then examine the various types of acid-base titrations and how buffered solutions are involved in them. Slightly soluble salts have major roles in the laboratory and in nature, and we'll see how conditions influence their solubility equilibria. Next, we investigate how complex ions form and change from one type to another. Finally, we see how various aqueous equilibria are employed in chemical analysis.

Concepts & Skills to Review

before you study this chapter

- solubility rules for ionic compounds (Section 4.3)
- effect of concentration on equilibrium position (Section 17.6)
- conjugate acid-base pairs (Section 18.3)
- calculations for weak-acid and weak-base equilibria (Sections 18.4 and 18.5)
- acid-base properties of salt solutions (Section 18.7)
- Lewis acids and bases (Section 18.9)

19.1 EQUILIBRIA OF ACID-BASE BUFFER SYSTEMS

Why do some lakes become acidic when showered by acid rain, while others remain unaffected? How does blood maintain a constant pH in contact with countless cellular acid-base reactions? How can a chemist sustain a nearly constant $[H_3O^+]$ in reactions that consume or produce H_3O^+ or OH^-? The answer in each case depends on the action of a buffer.

In everyday language, a buffer is something that lessens the impact of an external force. An **acid-base buffer** is a solution that *lessens the impact on pH from the addition of acid or base*. Figure 19.1 shows that a small amount of H_3O^+ or OH^- added to an unbuffered solution (or just water) changes the pH by several units. Indeed, because of the logarithmic nature of pH, *this change is several orders of magnitude larger* than the change that results from the same addition

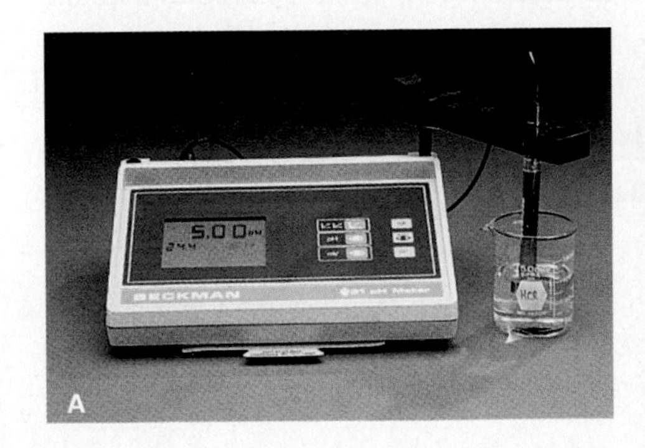

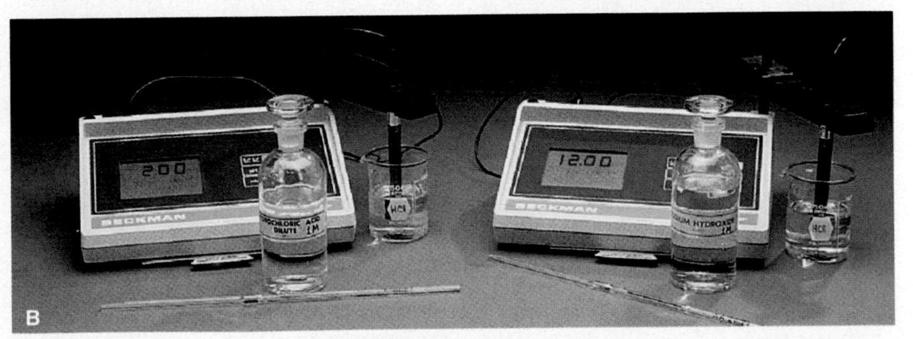

Figure 19.1 The effect of addition of acid or base to an unbuffered solution.
A, A 100-mL sample of dilute HCl is adjusted to pH 5.00. **B,** After the addition of 1 mL of 1 *M* HCl (*left*) or of 1 *M* NaOH (*right*), the pH changes by several units.

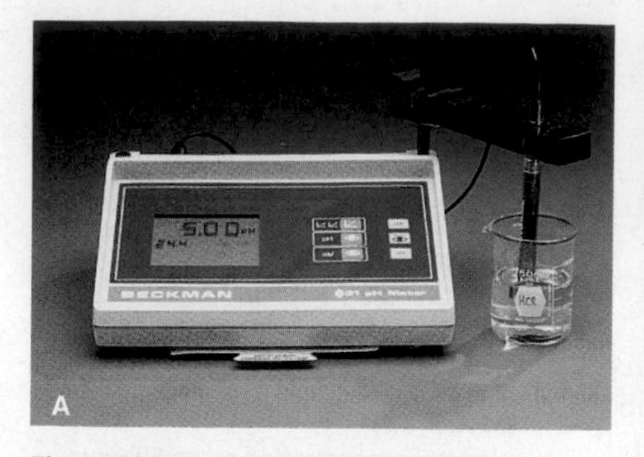

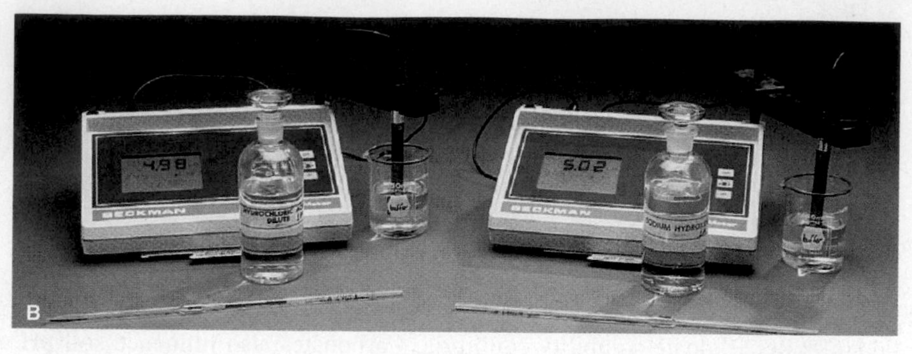

Figure 19.2 The effect of addition of acid or base to a buffered solution.
A, A 100-mL sample of a buffered solution, made by mixing 1 *M* CH_3COOH with 1 *M* CH_3COONa, is adjusted to pH 5.00. **B,** After the addition of 1 mL of 1 *M* HCl (*left*) or of 1 *M* NaOH (*right*), the pH change is negligibly small. Compare these changes with those in Figure 19.1.

to a buffered solution, shown in Figure 19.2. To withstand the addition of strong acid or strong base without significantly changing its pH, a buffer must contain an acidic component that can react with the added OH^- ion *and* a basic component that can react with added H_3O^+ ion. However, these buffer components cannot be just any acid and base because they would neutralize each other. Most commonly, *the components of a buffer are the conjugate acid-base pair of a weak acid.* The buffer used in Figure 19.2, for example, is a mixture of acetic acid (CH_3COOH) and acetate ion (CH_3COO^-).

How a Buffer Works: The Common-Ion Effect

Buffers work through a phenomenon known as the **common-ion effect.** An example of this effect occurs when acetic acid dissociates in water and some sodium acetate is added. As you know, acetic acid dissociates only slightly in water:

$$CH_3COOH(aq) + H_2O(l) \rightleftharpoons H_3O^+(aq) + CH_3COO^-(aq)$$

From Le Châtelier's principle (Section 17.6), we know that if some CH_3COO^- ion is added (from the soluble sodium acetate), the equilibrium position shifts to the left; thus, $[H_3O^+]$ decreases, in effect lowering the extent of acid dissociation:

$$CH_3COOH(aq) + H_2O(l) \xleftarrow{\rightleftharpoons} H_3O^+(aq) + CH_3COO^-(aq;\ \text{added})$$

Similarly, if we dissolve acetic acid in a sodium acetate solution, acetate ion and H_3O^+ ion from the acid enter the solution. The acetate ion already present combines with some of the H_3O^+, which lowers the $[H_3O^+]$. The effect again is to lower the acid dissociation. Acetate ion is called *the common ion* in this case because it is "common" to both the acetic acid and sodium acetate solutions; that is, acetate ion from the acid enters a solution in which it is already present. *The common-ion effect occurs when a reactant containing a given ion is added to an equilibrium mixture that already contains that ion, and the position of equilibrium shifts away from forming more of it.*

Table 19.1 shows the percent dissociation and the pH of an acetic acid solution containing various concentrations of acetate ion (supplied from solid sodium

Table 19.1 The Effect of Added Acetate Ion on the Dissociation of Acetic Acid

$[CH_3COOH]_{init}$	$[CH_3COO^-]_{added}$	% Dissociation*	pH
0.10	0.00	1.3	2.89
0.10	0.050	0.036	4.44
0.10	0.10	0.018	4.74
0.10	0.15	0.012	4.92

$$\text{*\% Dissociation} = \frac{[CH_3COOH]_{dissoc}}{[CH_3COOH]_{init}} \times 100$$

acetate). Note that the *common ion, CH_3COO^-, suppresses the dissociation of CH_3COOH,* which makes the solution less acidic (higher pH).

The Essential Feature of a Buffer In the previous case, we prepared a buffer by mixing a weak acid (CH_3COOH) and its conjugate base (CH_3COO^-). *How* does this solution resist pH changes when H_3O^+ or OH^- is added? The essential feature of a buffer is that *it consists of high concentrations of the acidic (HA) and basic (A^-) components.* When small amounts of H_3O^+ or OH^- ions are added to the buffer, they cause *a small amount of one buffer component to convert into the other,* which changes the relative concentrations of the two components. As long as the amount of H_3O^+ or OH^- added is much smaller than the amounts of HA and A^- present originally, *the added ions have little effect on the pH because they are consumed by one or the other buffer component:* the A^- consumes added H_3O^+, and the HA consumes added OH^-.

Consider what happens to a solution containing high $[CH_3COOH]$ and $[CH_3COO^-]$ when we add small amounts of strong acid or base. The expression for HA dissociation at equilibrium is

$$K_a = \frac{[CH_3COO^-][H_3O^+]}{[CH_3COOH]}$$

Solving for $[H_3O^+]$ gives

$$[H_3O^+] = K_a \times \frac{[CH_3COOH]}{[CH_3COO^-]}$$

Note that because K_a is constant, *the $[H_3O^+]$ of the solution depends directly on the buffer-component concentration ratio,* $\frac{[CH_3COOH]}{[CH_3COO^-]}$:

- If the ratio $[HA]/[A^-]$ goes up, $[H_3O^+]$ goes up.
- If the ratio $[HA]/[A^-]$ goes down, $[H_3O^+]$ goes down.

When we add a small amount of strong acid, the increased amount of H_3O^+ ion reacts with a nearly *stoichiometric amount* of acetate ion from the buffer to form more acetic acid:

$$H_3O^+(aq;\ \text{added}) + CH_3COO^-(aq;\ \text{from buffer}) \longrightarrow CH_3COOH(aq) + H_2O(l)$$

As a result, $[CH_3COO^-]$ goes down by that amount and $[CH_3COOH]$ goes up by that amount, which increases the buffer-component concentration ratio, as you can see in Figure 19.3. The $[H_3O^+]$ also increases but only very slightly.

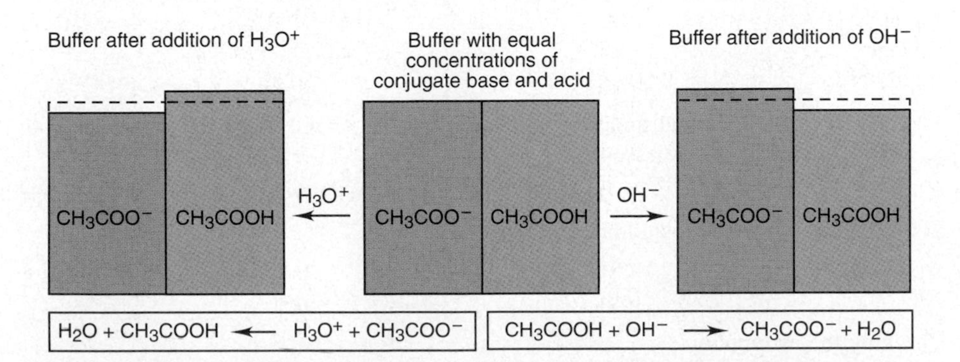

Figure 19.3 How a buffer works. A buffer consists of high concentrations of a conjugate acid-base pair, in this case, acetic acid (CH_3COOH) and acetate ion (CH_3COO^-). When a small amount of H_3O^+ is added (*left*), that same amount of CH_3COO^- combines with it, which increases the amount of CH_3COOH slightly. Similarly, when a small amount of OH^- is added (*right*), that amount of CH_3COOH combines with it, which increases the amount of CH_3COO^- slightly. In both cases, the relative changes in amounts of the buffer components are small, so their concentration ratio, and therefore the pH, changes very little.

Adding a small amount of strong base produces the opposite result. It supplies OH^- ions, which react with a nearly *stoichiometric amount* of CH_3COOH from the buffer, forming that much more CH_3COO^-:

$$CH_3COOH(aq;\ \text{from buffer}) + OH^-(aq;\ \text{added}) \longrightarrow CH_3COO^-(aq) + H_2O(l)$$

The buffer-component concentration ratio decreases, which decreases $[H_3O^+]$, but once again, the change is very slight.

Thus, the buffer components consume virtually all the added H_3O^+ or OH^-. To reiterate, as long as the amount of added H_3O^+ or OH^- is small compared with the amounts of the buffer components, *the conversion of one component into the other produces a small change in the buffer-component concentration ratio and, consequently, a small change in $[H_3O^+]$ and in pH.* Sample Problem 19.1 demonstrates how small these pH changes typically are. Note that the latter two parts of the problem combine a stoichiometry portion, like the problems in Chapter 3, and a weak-acid dissociation portion, like those in Chapter 18.

SAMPLE PROBLEM 19.1 Calculating the Effect of Added H_3O^+ or OH^- on Buffer pH

Problem Calculate the pH:
(a) Of a buffer solution consisting of 0.50 *M* CH_3COOH and 0.50 *M* CH_3COONa
(b) After adding 0.020 mol of solid NaOH to 1.0 L of the buffer solution in part (a)
(c) After adding 0.020 mol of HCl to 1.0 L of the buffer solution in part (a)
K_a of $CH_3COOH = 1.8\times10^{-5}$. (Assume the additions cause negligible volume changes.)

Plan In each case, we know, or can find, $[CH_3COOH]_{init}$ and $[CH_3COO^-]_{init}$ and the K_a of CH_3COOH (1.8×10^{-5}), and we need to find $[H_3O^+]$ at equilibrium and convert it to pH. In **(a)**, we use the given concentrations of buffer components (each 0.50 *M*) as the initial values. As in earlier problems, we assume that x, the $[CH_3COOH]$ that dissociates, which equals $[H_3O^+]$, is so small relative to $[CH_3COOH]_{init}$ that it can be neglected. We set up a reaction table, solve for x, and check the assumption. In **(b)** and **(c)**, we assume that the added OH^- or H_3O^+ reacts completely with the buffer components to yield new $[CH_3COOH]_{init}$ and $[CH_3COO^-]_{init}$, which then dissociate to an unknown extent. We set up two reaction tables. The first summarizes the stoichiometry of adding strong base (0.020 mol) or acid (0.020 mol). The second summarizes the dissociation of the new HA concentrations, so we proceed as in part (a) to find the new $[H_3O^+]$.

Solution **(a)** The original pH: $[H_3O^+]$ in the original buffer.
Setting up a reaction table with $x = [CH_3COOH]_{dissoc} = [H_3O^+]$ (as in Chapter 18, we assume that $[H_3O^+]$ from H_2O is negligible and disregard it):

Concentration (*M*)	$CH_3COOH(aq)$	+	$H_2O(l)$	$\rightleftharpoons$	$CH_3COO^-(aq)$	+	$H_3O^+(aq)$
Initial	0.50		—		0.50		0
Change	$-x$		—		$+x$		$+x$
Equilibrium	$0.50 - x$		—		$0.50 + x$		x

Making the assumption and finding the equilibrium $[CH_3COOH]$ and $[CH_3COO^-]$: With K_a small, x is small, so we assume

$$[CH_3COOH] = 0.50\ M - x \approx 0.50\ M \quad \text{and} \quad [CH_3COO^-] = 0.50\ M + x \approx 0.50\ M$$

Solving for x ($[H_3O^+]$ at equilibrium):

$$x = [H_3O^+] = K_a \times \frac{[CH_3COOH]}{[CH_3COO^-]} \approx (1.8\times10^{-5}) \times \frac{0.50}{0.50} = 1.8\times10^{-5}\ M$$

Checking the assumption:

$$\frac{1.8\times10^{-5}\ M}{0.50\ M} \times 100 = 3.6\times10^{-3}\% < 5\%$$

The assumption is justified, and we will use the same assumption in parts (b) and (c).
Calculating pH:

$$\text{pH} = -\log\ [H_3O^+] = -\log\ (1.8\times10^{-5}) = 4.74$$

(b) The pH after adding base (0.020 mol of NaOH to 1.0 L of buffer). Finding $[OH^-]_{added}$:

$$[OH^-]_{added} = \frac{0.020 \text{ mol } OH^-}{1.0 \text{ L soln}} = 0.020\ M\ OH^-$$

Setting up a reaction table for the *stoichiometry* of adding OH^- to CH_3COOH:

Concentration (*M*)	$CH_3COOH(aq)$	+ $OH^-(aq)$	⟶ $CH_3COO^-(aq)$	+ $H_2O(aq)$
Before addition	0.50	—	0.50	—
Addition	—	0.020	—	—
After addition	0.48	0	0.52	—

Setting up a reaction table for the *acid dissociation,* using these new initial concentrations. As in part (a), $x = [CH_3COOH]_{dissoc} = [H_3O^+]$:

Concentration (*M*)	$CH_3COOH(aq)$	+ $H_2O(l)$	⇌ $CH_3COO^-(aq)$	+ $H_3O^+(aq)$
Initial	0.48	—	0.52	0
Change	$-x$	—	$+x$	$+x$
Equilibrium	$0.48 - x$	—	$0.52 + x$	x

Making the assumption that x is small, and solving for x:

$$[CH_3COOH] = 0.48\ M - x \approx 0.48\ M \quad \text{and} \quad [CH_3COO^-] = 0.52\ M + x \approx 0.52\ M$$

$$x = [H_3O^+] = K_a \times \frac{[CH_3COOH]}{[CH_3COO^-]} \approx (1.8\times10^{-5}) \times \frac{0.48}{0.52} = 1.7\times10^{-5}\ M$$

Calculating the pH:

$$pH = -\log [H_3O^+] = -\log (1.7\times10^{-5}) = 4.77$$

The addition of strong base increased the concentration of the basic buffer component at the expense of the acidic buffer component. Note especially that the pH *increased only slightly,* from 4.74 to 4.77.

(c) The pH after adding acid (0.020 mol of HCl to 1.0 L of buffer). Finding $[H_3O^+]_{added}$:

$$[H_3O^+]_{added} = \frac{0.020 \text{ mol } H_3O^+}{1.0 \text{ L soln}} = 0.020\ M\ H_3O^+$$

Now we proceed as in part (b), by first setting up a reaction table for the *stoichiometry* of adding H_3O^+ to CH_3COO^-:

Concentration (*M*)	$CH_3COO^-(aq)$	+ $H_3O^+(aq)$	⟶ $CH_3COOH(aq)$	+ $H_2O(l)$
Before addition	0.50	—	0.50	—
Addition	—	0.020	—	—
After addition	0.48	0	0.52	—

The reaction table for the acid dissociation, with $x = [CH_3COOH]_{dissoc} = [H_3O^+]$ is

Concentration (*M*)	$CH_3COOH(aq)$	+ $H_2O(l)$	⇌ $CH_3COO^-(aq)$	+ $H_3O^+(aq)$
Initial	0.52	—	0.48	0
Change	$-x$	—	$+x$	$+x$
Equilibrium	$0.52 - x$	—	$0.48 + x$	x

Making the assumption that x is small, and solving for x:

$$[CH_3COOH] = 0.52\ M - x \approx 0.52\ M \quad \text{and} \quad [CH_3COO^-] = 0.48\ M + x \approx 0.48\ M$$

$$x = [H_3O^+] = K_a \times \frac{[CH_3COOH]}{[CH_3COO^-]} \approx (1.8\times10^{-5}) \times \frac{0.52}{0.48} = 2.0\times10^{-5}\ M$$

Calculating the pH:

$$pH = -\log [H_3O^+] = -\log (2.0\times10^{-5}) = 4.70$$

The addition of strong acid increased the concentration of the acidic buffer component at the expense of the basic buffer component and *lowered* the pH only slightly, from 4.74 to 4.70.

Check The changes in $[CH_3COOH]$ and $[CH_3COO^-]$ occur in opposite directions in parts (b) and (c), which makes sense. The additions were of equal amounts, so the pH increase in (b) should equal the pH decrease in (c), within rounding.

Comment In part (a), we justified our assumption that x can be neglected. Therefore, in parts (b) and (c), we could have used the "After addition" values from the last line of the stoichiometry tables directly for the ratio of buffer components; that would have allowed us to dispense with a reaction table for the dissociation. In subsequent problems in this chapter, we will follow this simplified approach.

FOLLOW-UP PROBLEM 19.1 Calculate the pH of a buffer consisting of 0.50 M HF and 0.45 M F^- **(a)** before and **(b)** after addition of 0.40 g of NaOH to 1.0 L of the buffer (K_a of HF = 6.8×10^{-4}).

The Henderson-Hasselbalch Equation

For any weak acid, HA, the equation and K_a expression for the acid dissociation at equilibrium are

$$HA + H_2O \rightleftharpoons H_3O^+ + A^-$$

$$K_a = \frac{[H_3O^+][A^-]}{[HA]}$$

A simple mathematical rearrangement turns this expression into a more useful form for buffer calculations. Remember that the key variable that determines $[H_3O^+]$ is the concentration *ratio* of acid species to base species. Isolating $[H_3O^+]$ gives

$$[H_3O^+] = K_a \times \frac{[HA]}{[A^-]}$$

Taking the negative common logarithm (base 10) of both sides gives

$$-\log\,[H_3O^+] = -\log K_a - \log\left(\frac{[HA]}{[A^-]}\right)$$

from which we obtain

$$pH = pK_a + \log\left(\frac{[A^-]}{[HA]}\right)$$

(Note the inversion of the buffer-component concentration ratio when the sign of the logarithm is changed.) A key point we'll emphasize again later is that when $[A^-] = [HA]$, their ratio becomes 1; the log term then becomes 0, and thus $pH = pK_a$.

Generalizing the previous equation for any conjugate acid-base pair gives the **Henderson-Hasselbalch equation:**

$$pH = pK_a + \log\left(\frac{[\text{base}]}{[\text{acid}]}\right) \tag{19.1}$$

This relationship allows us to solve directly for pH instead of having to calculate $[H_3O^+]$ first. For instance, by applying the Henderson-Hasselbalch equation in part (b) of Sample Problem 19.1, we could have found the pH of the buffer after the addition of NaOH as follows:

$$pH = pK_a + \log\left(\frac{[CH_3COO^-]}{[CH_3COOH]}\right)$$

$$= 4.74 + \log\left(\frac{0.52}{0.48}\right) = 4.77$$

(We just derived the Henderson-Hasselbalch equation from fundamental definitions and simple algebra. It's always a good idea to derive simple relationships this way rather than having to memorize them.)

Buffer Capacity and Buffer Range

As you've seen, a buffer resists a pH change as long as the concentrations of buffer components are *large* compared with the amount of strong acid or base added. **Buffer capacity** is a measure of this ability to resist pH change and depends on both the absolute and relative component concentrations.

In absolute terms, *the more concentrated the components of a buffer, the greater the buffer capacity.* In other words, you must add more H_3O^+ or OH^- to a high-capacity (concentrated) buffer than to a low-capacity (dilute) buffer to obtain the same pH change. Conversely, adding the same amount of H_3O^+ or OH^- to buffers of different capacities produces a smaller pH change in the higher capacity buffer (Figure 19.4). It's important to realize that *the pH of a buffer is distinct from its buffer capacity.* A buffer made of equal volumes of 1.0 *M* CH_3COOH and 1.0 *M* CH_3COO^- has the same pH (4.74) as a buffer made of equal volumes of 0.10 *M* CH_3COOH and 0.10 *M* CH_3COO^-, but the more concentrated buffer has a much larger capacity for resisting a pH change.

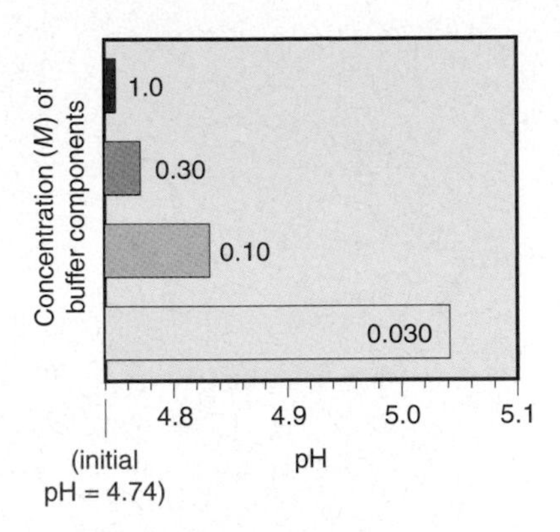

Figure 19.4 The relation between buffer capacity and pH change. The four bars in the graph represent CH_3COOH-CH_3COO^- buffers with the same initial pH (4.74) but different component concentrations (labeled on or near each bar). When a given amount of strong base is added to each buffer, the pH increases. The length of the bar corresponds to the pH increase. Note that the more concentrated the buffer, the greater its capacity, and the smaller the pH change.

Buffer capacity is also affected by the *relative* concentrations of the buffer components. As a buffer functions, the concentration of one component increases relative to the other. Because the ratio of these concentrations determines the pH, the less the ratio changes, the less the pH changes. *For a given addition of acid or base, the concentration ratio changes less when the buffer-component concentrations are similar than when they are different.* Suppose that a buffer has $[HA] = [A^-] = 1.000\ M$. When we add 0.010 mol of OH^- to 1.00 L of buffer, $[A^-]$ becomes 1.010 *M* and [HA] becomes 0.990 *M*:

$$\frac{[A^-]_{init}}{[HA]_{init}} = \frac{1.000\ M}{1.000\ M} = 1.000 \qquad \frac{[A^-]_{final}}{[HA]_{final}} = \frac{1.010\ M}{0.990\ M} = 1.02$$

$$\text{Percent change} = \frac{1.02 - 1.000}{1.000} \times 100 = 2\%$$

Now suppose that the buffer-component concentrations are $[HA] = 0.250\ M$ and $[A^-] = 1.750\ M$. The same addition of 0.010 mol of OH^- to 1.00 L of buffer gives $[HA] = 0.240\ M$ and $[A^-] = 1.760\ M$, so the ratios are

$$\frac{[A^-]_{init}}{[HA]_{init}} = \frac{1.750\ M}{0.250\ M} = 7.00 \qquad \frac{[A^-]_{final}}{[HA]_{final}} = \frac{1.760\ M}{0.240\ M} = 7.33$$

$$\text{Percent change} = \frac{7.33 - 7.00}{7.00} \times 100 = 4.7\%$$

As you can see, the change in the buffer-component concentration ratio is much larger when the initial concentrations of the components are very different.

It follows that *a buffer has the highest capacity when the component concentrations are equal,* that is, when $[A^-]/[HA] = 1$:

$$pH = pK_a + \log\left(\frac{[A^-]}{[HA]}\right) = pK_a + \log 1 = pK_a + 0 = pK_a$$

Note this important result: for a given concentration, *a buffer whose pH is equal to or near the pK_a of its acid component has the highest buffer capacity.*

The **buffer range** is the pH range over which the buffer acts effectively, and it is related to the relative component concentrations. The further the buffer-component concentration ratio is from 1, the less effective the buffering action (that is, the lower the buffer capacity). In practice, if the $[A^-]/[HA]$ ratio is greater than 10 or less than 0.1—that is, if one component concentration is more than 10 times the other—buffering action is poor. Recalling that $\log 10 = +1$ and $\log 0.1 = -1$, we find that *buffers have a usable range within ±1 pH unit of the pK_a of the acid component:*

$$pH = pK_a + \log\left(\frac{10}{1}\right) = pK_a + 1 \qquad \text{and} \qquad pH = pK_a + \log\left(\frac{1}{10}\right) = pK_a - 1$$

Preparing a Buffer

Any large chemical supply-house catalog lists many common buffers in a variety of pH values and concentrations. So, you might ask, why learn how to prepare a buffer? In many cases, a common buffer is simply not available with the desired pH or concentration, and you have to make it yourself. In many environmental or biomedical research applications, a buffer of unusual composition may be required to simulate an ecosystem or stabilize a fragile biological macromolecule. Even the most sophisticated, automated laboratory frequently relies on personnel with a good knowledge of basic chemistry to prepare a buffer. Several steps are required to prepare a buffer of a desired pH:

1. *Choose the conjugate acid-base pair.* First, decide on the chemical composition of the buffer, that is, the conjugate acid-base pair. This choice is determined to a large extent by the desired pH. Remember that a buffer is most effective when the ratio of its component concentrations is close to 1, in which case the pH $\approx$ pK_a of the acid.

Suppose that you need a buffer whose pH is 3.90: the pK_a of the acid component should be as close to 3.90 as possible; or $K_a = 10^{-3.90} = 1.3\times10^{-4}$. Scanning a table of acid-dissociation constants (see Appendix C) shows that lactic acid (pK_a = 3.86), glycolic acid (pK_a = 3.83), and formic acid (pK_a = 3.74) are good choices. To avoid adding common biological species, let's use formic acid. Therefore, the buffer components will be formic acid, HCOOH, and formate ion, $HCOO^-$, supplied by a soluble salt, such as sodium formate, HCOONa.

2. *Calculate the ratio of buffer component concentrations.* Next, find the ratio of $[A^-]/[HA]$ that gives the desired pH. With the Henderson-Hasselbalch equation, we have

$$\text{pH} = \text{p}K_a + \log\left(\frac{[A^-]}{[HA]}\right) \quad \text{or} \quad 3.90 = 3.74 + \log\left(\frac{[HCOO^-]}{[HCOOH]}\right)$$

$$\log\left(\frac{[HCOO^-]}{[HCOOH]}\right) = 0.16 \quad \text{so} \quad \left(\frac{[HCOO^-]}{[HCOOH]}\right) = 10^{0.16} = 1.4$$

Thus, for every 1.0 mol of HCOOH in a given volume of solution, you need 1.4 mol of HCOONa.

3. *Determine the buffer concentration.* Next, decide how concentrated the buffer should be. Remember that the higher the concentrations of components, the greater the buffer capacity. For most laboratory-scale applications, concentrations of about 0.50 *M* are suitable, but the decision is often based on availability of stock solutions.

Suppose you have a large stock of 0.40 *M* HCOOH and you need approximately 1.0 L of final buffer. First, find the moles of sodium formate that will give the needed 1.4:1.0 ratio, and then convert to grams:

$$\text{Moles of HCOOH} = 1.0\ \text{L soln} \times \frac{0.40\ \text{mol HCOOH}}{1.0\ \text{L soln}} = 0.40\ \text{mol HCOOH}$$

$$\text{Moles of HCOONa} = 0.40\ \text{mol HCOOH} \times \frac{1.4\ \text{mol HCOONa}}{1.0\ \text{mol HCOOH}} = 0.56\ \text{mol HCOONa}$$

$$\text{Mass (g) of HCOONa} = 0.56\ \text{mol HCOONa} \times \frac{68.01\ \text{g HCOONa}}{1\ \text{mol HCOONa}} = 38\ \text{g HCOONa}$$

4. *Mix the solution and adjust the pH.* Thoroughly dissolve 38 g of solid sodium formate in 0.40 *M* HCOOH to a total volume of 1.0 L. Finally, note that because of the behavior of nonideal solutions (Section 13.6), a buffer prepared in this way may vary from the desired pH by a few tenths of a pH unit. Therefore, after making up the solution, *adjust the buffer pH* to the desired value by adding strong acid or strong base, while monitoring the solution with a pH meter. (The following sample problem does not refer to the Henderson-Hasselbalch equation.)

SAMPLE PROBLEM 19.2 Preparing a Buffer

Problem An environmental chemist needs a carbonate buffer of pH 10.00 to study the effects of the acid rain on limestone-rich soils. How many grams of Na_2CO_3 must she add to 1.5 L of freshly prepared 0.20 *M* $NaHCO_3$ to make the buffer? K_a of HCO_3^- is 4.7×10^{-11}.

Plan The conjugate pair is already chosen, HCO_3^- (acid) and CO_3^{2-} (base), as are the volume (1.5 L) and concentration (0.20 *M*) of HCO_3^-, so we must find the buffer-component concentration ratio that gives pH 10.00 and the mass of Na_2CO_3 to dissolve. We first convert pH to $[H_3O^+]$ and use the K_a expression from the equation to solve for the required $[CO_3^{2-}]$. Multiplying by the volume of solution gives the amount (mol) of CO_3^{2-} required, and then we use the molar mass to find the mass (g) of Na_2CO_3.

Solution Calculating $[H_3O^+]$: $[H_3O^+] = 10^{-pH} = 10^{-10.00} = 1.0\times10^{-10}\ M$

Solving for $[CO_3^{2-}]$ in the concentration ratio:

$$HCO_3^-(aq) + H_2O(l) \rightleftharpoons H_3O^+(aq) + CO_3^{2-}(aq) \qquad K_a = \frac{[H_3O^+][CO_3^{2-}]}{[HCO_3^-]}$$

So $$[CO_3^{2-}] = K_a\frac{[HCO_3^-]}{[H_3O^+]} = \frac{(4.7\times10^{-11})(0.20)}{1.0\times10^{-10}} = 0.094\ M$$

Calculating the amount (mol) of CO_3^{2-} needed for the given volume:

$$\text{Moles of } CO_3^{2-} = 1.5\text{ L soln} \times \frac{0.094\text{ mol } CO_3^{2-}}{1\text{ L soln}} = 0.14\text{ mol } CO_3^{2-}$$

Calculating the mass (g) of Na_2CO_3 needed:

$$\text{Mass (g) of } Na_2CO_3 = 0.14\text{ mol } Na_2CO_3 \times \frac{105.99\text{ g } Na_2CO_3}{1\text{ mol } Na_2CO_3} = 15\text{ g } Na_2CO_3$$

We dissolve 15 g of Na_2CO_3 into 1.3 L of 0.20 *M* $NaHCO_3$ and add 0.20 *M* $NaHCO_3$ to make 1.5 L. Using a pH meter, we adjust the pH to 10.00 with strong acid or base.

Check For a useful buffer range, the concentration of the acidic component, $[HCO_3^-]$, must be within a factor of 10 of the concentration of the basic component, $[CO_3^{2-}]$. We have (1.5 L)(0.20 *M* HCO_3^-), or 0.30 mol of HCO_3^-, and 0.14 mol of CO_3^{2-}; 0.30:0.14 = 2.1, which seems fine. Make sure the relative amounts of components seem reasonable: since we want a pH lower than the pK_a of HCO_3^- (10.33), it makes sense that we have more of the acidic than the basic species.

FOLLOW-UP PROBLEM 19.2 How would you prepare a benzoic acid–benzoate buffer with pH = 4.25, starting with 5.0 L of 0.050 *M* sodium benzoate (C_6H_5COONa) solution and adding the acidic component? K_a of benzoic acid (C_6H_5COOH) is 6.3×10^{-5}.

Another way to prepare a buffer is to form one of the components during the final mixing step by *partial neutralization* of the other component. For example, you can prepare an $HCOOH$-$HCOO^-$ buffer by mixing appropriate amounts of HCOOH solution and NaOH solution. As the OH^- ions react with the HCOOH molecules, neutralization of part of the total HCOOH present produces the $HCOO^-$ needed:

$$\text{HCOOH (HA total)} + OH^- \text{ (amt added)} \longrightarrow$$
$$\text{HCOOH (HA total} - OH^- \text{ amt added)} + HCOO^- \text{ (}OH^-\text{ amt added)} + H_2O$$

This method is based on the same chemical process that occurs when a weak acid is titrated with a strong base, as you'll see in the next section.

SECTION SUMMARY

A buffered solution exhibits a much smaller change in pH when H_3O^+ or OH^- is added than does an unbuffered solution. A buffer consists of relatively high concentrations of the components of a conjugate weak acid–base pair. The buffer-component concentration ratio determines the pH, and the ratio and pH are related by the Henderson-Hasselbalch equation. As H_3O^+ or OH^- is added, one buffer component reacts with it and is converted into the other component; therefore, the buffer-component concentration ratio, and consequently the free $[H_3O^+]$ (and pH), changes only slightly. A concentrated buffer undergoes smaller changes in pH than a dilute buffer. When the buffer pH equals the pK_a of the acid component, the buffer has its highest capacity. A buffer has an effective range of $pK_a \pm 1$ pH unit. To prepare a buffer, you choose the conjugate acid-base pair, calculate the ratio of buffer components, determine buffer concentration, and adjust the final solution to the desired pH.

19.2 ACID-BASE TITRATION CURVES

In Chapter 4, we discussed the acid-base titration as an analytical method. Let's re-examine it, this time tracking the change in pH with an **acid-base titration curve,** a plot of pH vs. volume of titrant added. The behavior of an acid-base indicator and its role in the titration are described first. To better understand the titration process, we apply the principles of the acid-base behavior of salt solutions (Section 18.7) and, later in the section, the principles of buffer action.

Monitoring pH with Acid-Base Indicators

The two common devices for measuring pH in the laboratory are pH meters and acid-base indicators. (We discuss the operation of pH meters in Chapter 21.) An *acid-base indicator* is a weak organic acid (denoted as HIn) that has a different color than its conjugate base (In^-), with the color change occurring over a specific and relatively narrow pH range. Typically, one or both of the forms are intensely colored, so only a tiny amount of indicator is needed, far too little to affect the pH of the solution being studied.

Indicators are used for estimating the pH of a solution and for monitoring the pH in acid-base titrations and in reactions. Figure 19.5 shows the color changes and their pH ranges for some common acid-base indicators. Selecting an indicator requires that you know the approximate pH of the titration end point, which in turn requires that you know which ionic species are present. You'll see how to identify those in our discussion of acid-base titration curves.

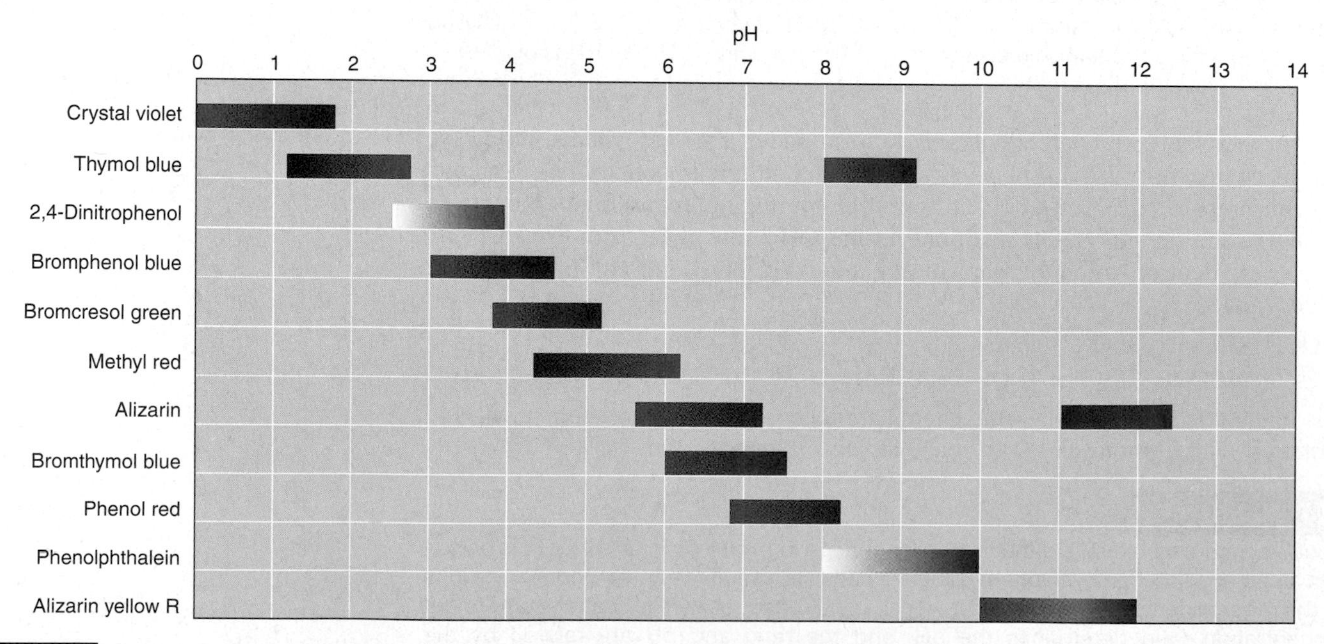

Figure 19.5 Colors and approximate pH range of some common acid-base indicators. Most have a range of about 2 pH units, in keeping with the useful buffer range of 2 pH units ($pK_a \pm 1$). (pH range depends to some extent on the solvent used to prepare the indicator.)

Because the indicator molecule is a weak acid, the ratio of the two forms is governed by the $[H_3O^+]$ of the test solution:

$$HIn(aq) + H_2O(l) \rightleftharpoons H_3O^+(aq) + In^-(aq) \qquad K_a \text{ of HIn} = \frac{[H_3O^+][In^-]}{[HIn]}$$

Therefore,

$$\frac{[HIn]}{[In^-]} = \frac{[H_3O^+]}{K_a}$$

How we perceive colors has a major influence on the use of indicators. Typically, the experimenter will see the HIn color if the $[HIn]/[In^-]$ ratio is 10:1 or greater and the In^- color if the $[HIn]/[In^-]$ ratio is 1:10 or less. Between these extremes, the colors of the two forms merge into an intermediate hue. Therefore, an indicator has a *color range* that reflects a 100-fold range in the [HIn]/[In] ratio, which means that an *indicator changes color over a range of about 2 pH units.* For example, as you can see in Figure 19.5, bromthymol blue has a pH range of about 6.0 to 7.6 and, as Figure 19.6 shows, it is yellow below that range, blue above it, and greenish in between.

Figure 19.6 The color change of the indicator bromthymol blue. The acidic form of bromthymol blue is yellow (*left*) and the basic form is blue (*right*). Over the pH range in which the indicator is changing, both forms are present, so the mixture appears greenish (*center*).

Strong Acid–Strong Base Titration Curves

A typical curve for the titration of a strong acid with a strong base appears in Figure 19.7, along with the data used to construct it.

Features of the Curve There are three distinct regions of the curve, which correspond to three major changes in slope:

1. The pH starts out low, reflecting the high $[H_3O^+]$ of the strong acid, and increases slowly as acid is gradually neutralized by the added base.
2. Suddenly, the pH rises steeply. This rise begins when the moles of OH^- that have been added nearly equal the moles of H_3O^+ originally present in the acid. An additional drop or two of base neutralizes the final tiny excess of acid and introduces a tiny excess of base, so the pH jumps 6 to 8 units.
3. Beyond this steep portion, the pH increases slowly as more base is added.

The **equivalence point,** which occurs within the nearly vertical portion of the curve, is the point at which the *number of moles of added OH^- equals the number of moles of H_3O^+ originally present.* At the equivalence point of a strong

Volume of NaOH added (mL)	pH
00.00	1.00
10.00	1.22
20.00	1.48
30.00	1.85
35.00	2.18
39.00	2.89
39.50	3.20
39.75	3.50
39.90	3.90
39.95	4.20
39.99	4.90
40.00	7.00
40.01	9.10
40.05	9.80
40.10	10.10
40.25	10.50
40.50	10.79
41.00	11.09
45.00	11.76
50.00	12.05
60.00	12.30
70.00	12.43
80.00	12.52

A

Titration of 40.00 mL of 0.1000 *M* HCl with 0.1000 *M* NaOH

pH

Phenolphthalein

pH = 7.00 at equivalence point

Methyl red

Volume of NaOH added (mL)

B

Figure 19.7 Curve for a strong acid–strong base titration. A, Data obtained from the titration of 40.00 mL of 0.1000 *M* HCl with 0.1000 *M* NaOH. **B,** Acid-base titration curve from data in part A. The pH increases gradually at first. When the amount (mol) of OH^- added is slightly less than the amount (mol) of H_3O^+ originally present, a large pH change accompanies a small addition of OH^-. The equivalence point occurs when amount (mol) of OH^- added = amount (mol) of H_3O^+ originally present. Note that, for a strong acid–strong base titration, pH = 7.00 at the equivalence point. Added before the titration begins, either methyl red or phenolphthalein is a suitable indicator in this case because each changes color on the steep portion of the curve, as shown by the color strips. Photos showing the color changes from 1–2 drops of indicator appear nearby. Beyond this point, added OH^- causes a gradual pH increase again.

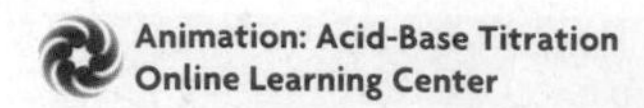

acid–strong base titration, *the solution consists of the anion of the strong acid and the cation of the strong base.* Recall from Chapter 18 that *these ions do not react with water, so the solution is neutral: pH = 7.00.* The volume and concentration of base needed to reach the equivalence point allow us to calculate the amount of acid originally present (see Sample Problem 4.5, p. 147).

Before the titration begins, we add a few drops of an appropriate indicator to the acid solution to signal when we reach the equivalence point. The **end point** of the titration occurs when the indicator changes color. *We choose an indicator with an end point close to the equivalence point,* one that changes color in the pH range on the steep vertical portion of the curve. Figure 19.7 shows the color changes for two indicators that are suitable for a strong acid–strong base titration. Methyl red changes from red at pH 4.2 to yellow at pH 6.3, whereas phenolphthalein changes from colorless at pH 8.3 to pink at pH 10.0. Even though neither color change occurs *at* the equivalence point (pH 7.00), both occur on the vertical portion of the curve, where a single drop of base causes a large pH change: when methyl red turns yellow, or when phenolphthalein turns pink, we know we are within a fraction of a drop from the equivalence point. For example, in going from 39.90 to 39.99 mL, one to two drops, the pH changes one whole unit. For all practical purposes, then, the *visible* change in color of the indicator (end point) signals the *invisible* point at which moles of added base equal the original moles of acid (equivalence point).

Calculating the pH By knowing the chemical species present during the titration, we can calculate the pH at various points along the way:

1. *Original solution of strong HA.* In Figure 19.7, 40.00 mL of 0.1000 *M* HCl is titrated with 0.1000 *M* NaOH. Because a strong acid is completely dissociated, $[HCl] = [H_3O^+] = 0.1000\ M$. Therefore, the initial pH is*

$$pH = -\log [H_3O^+] = -\log (0.1000) = 1.00$$

2. *Before the equivalence point.* As soon as we start adding base, two changes occur that affect the pH calculations: (1) some acid is neutralized, and (2) the volume of solution increases. To find the pH at various points up to the equivalence point, we find the initial *amount (mol)* of H_3O^+ present, subtract the amount reacted, *which equals the amount (mol) of OH^- added*, and then use the change in volume to calculate the *concentration*, $[H_3O^+]$, and convert to pH. For example, after adding 20.00 mL of 0.1000 *M* NaOH:

- *Find the moles of H_3O^+ remaining.* Subtracting the number of moles of H_3O^+ reacted from the number initially present gives the number remaining. Moles of H_3O^+ reacted equals moles of OH^- added, so

$$\begin{array}{lr} \text{Initial moles of } H_3O^+ = 0.04000\ L \times 0.1000\ M = & 0.004000\ \text{mol } H_3O^+ \\ -\text{Moles of } OH^- \text{ added} = 0.02000\ L \times 0.1000\ M = & 0.002000\ \text{mol } OH^- \\ \hline \text{Moles of } H_3O^+ \text{ remaining} = & 0.002000\ \text{mol } H_3O^+ \end{array}$$

- *Calculate $[H_3O^+]$, taking the total volume into account.* To find the ion concentrations, we use the *total volume* because the water of one solution dilutes the ions of the other:

$$[H_3O^+] = \frac{\text{amount (mol) of } H_3O^+ \text{ remaining}}{\text{original volume of acid + volume of added base}}$$

$$= \frac{0.002000\ \text{mol } H_3O^+}{0.04000\ L + 0.02000\ L} = 0.03333\ M \qquad pH = 1.48$$

Given the moles of OH^- added, we are halfway to the equivalence point; but we are still on the initial slow rise of the curve, so the pH is still very low. Similar calculations give values up to the equivalence point.

*In acid-base titrations, volumes and concentrations are usually known to four significant figures, but pH is generally reported to no more than two digits to the right of the decimal point.

3. *At the equivalence point.* After 40.00 mL of 0.1000 M NaOH has been added, the equivalence point is reached. All the H_3O^+ from the acid has been neutralized, and the solution contains Na^+ and Cl^-, neither of which reacts with water. Because of the autoionization of water, however,

$$[H_3O^+] = 1.0\times10^{-7}\ M \qquad pH = 7.00$$

In this example, 0.004000 mol of OH^- reacted with 0.004000 mol of H_3O^+ to reach the equivalence point.

4. *After the equivalence point.* From the equivalence point on, the pH calculation is based on the moles of *excess* OH^- present. For example, after adding 50.00 mL of NaOH, we have

$$\begin{array}{lr} \text{Total moles of } OH^- \text{ added} = 0.05000\ L \times 0.1000\ M = & 0.005000 \text{ mol } OH^- \\ -\text{Moles of } H_3O^+ \text{ consumed} = 0.04000\ L \times 0.1000\ M = & 0.004000 \text{ mol } H_3O^+ \\ \hline \text{Moles of excess } OH^- = & 0.001000 \text{ mol } OH^- \end{array}$$

$$[OH^-] = \frac{0.001000 \text{ mol } OH^-}{0.04000\ L + 0.05000\ L} = 0.01111\ M \qquad pOH = 1.95$$

$$pH = pK_w - pOH = 14.00 - 1.95 = 12.05$$

Weak Acid–Strong Base Titration Curves

Now let's turn to the titration of a weak acid with a strong base. Figure 19.8 shows the curve obtained when we use 0.1000 M NaOH to titrate 40.00 mL of 0.1000 M propanoic acid, a weak organic acid (CH_3CH_2COOH; $K_a = 1.3\times10^{-5}$). (We abbreviate the acid as HPr and the conjugate base, $CH_3CH_2COO^-$, as Pr^-.)

Features of the Curve When we compare this weak acid–strong base titration curve with the strong acid–strong base titration curve (dotted curve portion in Figure 19.8 corresponds to bottom half of curve in Figure 19.7), four key regions appear, and the first three differ from the strong acid case:

1. *The initial pH is higher.* Because the weak acid (HPr) dissociates slightly, less H_3O^+ is present than with the strong acid.
2. *A gradually rising portion of the curve, called the buffer region, appears before the steep rise to the equivalence point.* As HPr reacts with the strong base, more and more conjugate base (Pr^-) forms, which creates an HPr-Pr^- buffer. At the midpoint of the buffer region, *half the original HPr has reacted,* so [HPr] = [Pr^-], or [Pr^-]/[HPr] = 1. Therefore, *the pH equals the* pK_a*:*

$$pH = pK_a + \log\left(\frac{[Pr^-]}{[HPr]}\right) = pK_a + \log 1 = pK_a + 0 = pK_a$$

Observing the pH at the midpoint of the buffer region is a common method for estimating the pK_a of an unknown acid.

3. *The pH at the equivalence point is greater than 7.00.* The solution contains the strong-base cation Na^+, which does not react with water, and the weak-acid anion Pr^-, which acts as a weak base to accept a proton from H_2O and yield OH^-.
4. Beyond the equivalence point, the pH increases slowly as *excess* OH^- is added.

Titration of 40.00 mL of 0.1000 M HPr with 0.1000 M NaOH

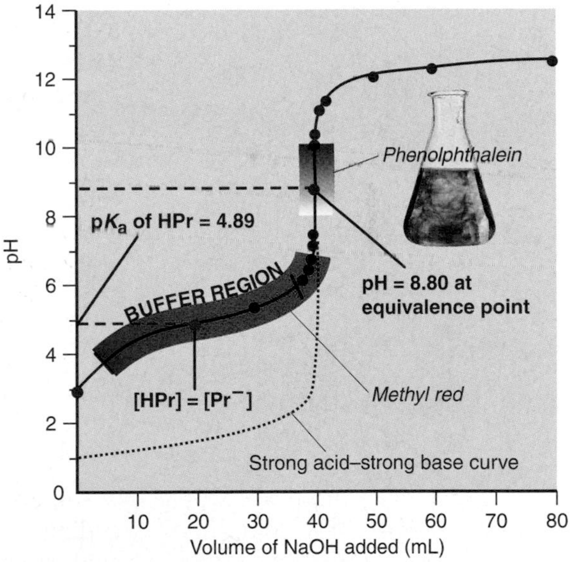

Figure 19.8 Curve for a weak acid–strong base titration. The curve for the titration of 40.00 mL of 0.1000 M CH_3CH_2COOH (HPr) with 0.1000 M NaOH is compared with that for the strong acid HCl (*dotted curve portion*). Phenolphthalein (*photo*) is a suitable indicator here.

Our choice of indicator is more limited here than for a strong acid–strong base titration because the steep rise occurs over a smaller pH range. Phenolphthalein is suitable because its color change lies within this range (Figure 19.8). However, the figure shows that methyl red, our other choice for the strong acid–strong base titration, changes color earlier and slowly over a large volume (~10 mL) of titrant, thereby giving a vague and false indication of the equivalence point.

Calculating the pH The calculation procedure for the weak acid–strong base titration is different from the previous case because we have to consider the partial dissociation of the weak acid and the reaction of the conjugate base with water. There are four key regions of the titration curve, each of which requires a different type of calculation to find $[H_3O^+]$:

1. *Solution of HA.* Before base is added, the $[H_3O^+]$ is that of a weak-acid solution, so we find $[H_3O^+]$ as in Section 18.4: we set up a reaction table with $x = [HPr]_{dissoc}$, assume $[H_3O^+] = [HPr]_{dissoc} << [HPr]_{init}$, and solve for x:

$$K_a = \frac{[H_3O^+][Pr^-]}{[HPr]} \approx \frac{x^2}{[HPr]_{init}} \qquad \text{therefore,} \qquad x = [H_3O^+] \approx \sqrt{K_a \times [HPr]_{init}}$$

2. *Solution of HA and added base.* As soon as we add NaOH, it reacts with HPr to form Pr^-. This means that up to the equivalence point, we have a mixture of acid and conjugate base, and an HPr-Pr^- buffer solution exists over much of that interval. Therefore, we find $[H_3O^+]$ from the relationship

$$[H_3O^+] = K_a \times \frac{[HPr]}{[Pr^-]}$$

(Of course, we can find pH directly with the Henderson-Hasselbalch equation, which is just an alternative form of this relationship.) Note that in this calculation we do *not* have to consider the new total volume because *the volumes cancel in the ratio of concentrations.* That is, $[HPr]/[Pr^-]$ = moles of HPr/moles of Pr^-, so we need not calculate concentrations.

3. *Equivalent amounts of HA and added base.* At the equivalence point, the original amount of HPr has reacted, so the flask contains a solution of Pr^-, a weak base that reacts with water to form OH^-:

$$Pr^-(aq) + H_2O(l) \rightleftharpoons HPr(aq) + OH^-(aq)$$

Therefore, as mentioned previously, in a weak acid–strong base titration, the solution at the equivalence point is slightly basic, pH > 7.00. We calculate $[H_3O^+]$ as in Section 18.5: we first find K_b of Pr^- from K_a of HPr, set up a reaction table (assume $[Pr^-] >> [Pr^-]_{reacting}$), and solve for $[OH^-]$. We need a single concentration, $[Pr^-]$, to solve for $[OH^-]$, so we *do* need the total volume. Then, we convert to $[H_3O^+]$. These two steps are

$$(1)\ [OH^-] \approx \sqrt{K_b \times [Pr^-]}, \quad \text{where } K_b = \frac{K_w}{K_a} \quad \text{and} \quad [Pr^-] = \frac{\text{moles of HPr}_{init}}{\text{total volume}}$$

$$(2)\ [H_3O^+] = \frac{K_w}{[OH^-]}$$

Combining them into one step gives

$$[H_3O^+] \approx \frac{K_w}{\sqrt{K_b \times [Pr^-]}}$$

4. *Solution of excess added base.* Beyond the equivalence point, we are just adding excess OH^- ion, so the calculation is the same as for the strong acid–strong base titration:

$$[H_3O^+] = \frac{K_w}{[OH^-]}, \qquad \text{where } [OH^-] = \frac{\text{moles of excess OH}^-}{\text{total volume}}$$

Sample Problem 19.3 shows the overall approach.

SAMPLE PROBLEM 19.3 Finding the pH During a Weak Acid–Strong Base Titration

Problem Calculate the pH during the titration of 40.00 mL of 0.1000 *M* propanoic acid (HPr; $K_a = 1.3\times10^{-5}$) after adding the following volumes of 0.1000 *M* NaOH:
(a) 0.00 mL **(b)** 30.00 mL **(c)** 40.00 mL **(d)** 50.00 mL

Plan (a) 0.00 mL: No base has been added yet, so this is a weak-acid solution. Thus, we calculate the pH as we did in Section 18.4. **(b)** 30.00 mL: A mixture of Pr^- and HPr is present. We find the amount (mol) of each, substitute into the K_a expression to solve for $[H_3O^+]$, and convert to pH. **(c)** 40.00 mL: The amount (mol) of NaOH added equals the initial amount (mol) of HPr, so a solution of Na^+ and the weak base Pr^- exists. We calculate the pH as we did in Section 18.5, except that we need *total* volume to find $[Pr^-]$. **(d)** 50.00 mL: Excess NaOH is added, so we calculate the amount (mol) of excess OH^- in the total volume and convert to $[H_3O^+]$ and then pH.

Solution (a) 0.00 mL of 0.1000 *M* NaOH added. Following the approach used in Sample Problem 18.7 and just described in the text, we obtain

$$[H_3O^+] \approx \sqrt{K_a \times [HPr]_{init}} = \sqrt{(1.3\times10^{-5})(0.1000)} = 1.1\times10^{-3}\ M$$

$$pH = 2.96$$

(b) 30.00 mL of 0.1000 *M* NaOH added. Calculating the ratio of moles of HPr to Pr^-:

$$\text{Original moles of HPr} = 0.04000\ L \times 0.1000\ M = 0.004000\ \text{mol HPr}$$

$$\text{Moles of NaOH added} = 0.03000\ L \times 0.1000\ M = 0.003000\ \text{mol } OH^-$$

For 1 mol of NaOH that reacts, 1 mol of Pr^- forms, so we construct the following reaction table for the stoichiometry:

Amount (mol)	$HPr(aq)$	+	$OH^-(aq)$	⟶	$Pr^-(aq)$	+	$H_2O(l)$
Before addition	0.004000		—		0		—
Addition	—		0.003000		—		—
After addition	0.001000		0		0.003000		—

The last line of this table shows the new initial amounts of HPr and Pr^- that will react to attain a new equilibrium. However, with *x* very small, we assume that the $[HPr]/[Pr^-]$ ratio at equilibrium is essentially equal to the ratio of these new initial amounts (see Comment in Sample Problem 19.1). Thus,

$$\frac{[HPr]}{[Pr^-]} = \frac{0.001000\ \text{mol}}{0.003000\ \text{mol}} = 0.3333$$

Solving for $[H_3O^+]$: $$[H_3O^+] = K_a \times \frac{[HPr]}{[Pr^-]} = (1.3\times10^{-5})(0.3333) = 4.3\times10^{-6}\ M$$

$$pH = 5.37$$

(c) 40.00 mL of 0.1000 *M* NaOH added. Calculating $[Pr^-]$ after all HPr has reacted:

$$[Pr^-] = \frac{0.004000\ \text{mol}}{0.04000\ L + 0.04000\ L} = 0.05000\ M$$

Calculating K_b: $$K_b = \frac{K_w}{K_a} = \frac{1.0\times10^{-14}}{1.3\times10^{-5}} = 7.7\times10^{-10}$$

Solving for $[H_3O^+]$ as described in the text:

$$[H_3O^+] \approx \frac{K_w}{\sqrt{K_b \times [Pr^-]}} = \frac{1.0\times10^{-14}}{\sqrt{(7.7\times10^{-10})(0.05000)}} = 1.6\times10^{-9}\ M$$

$$pH = 8.80$$

(d) 50.00 mL of 0.1000 *M* NaOH added.

$$\text{Moles of excess } OH^- = (0.1000\ M)(0.05000\ L - 0.04000\ L) = 0.001000\ \text{mol}$$

$$[OH^-] = \frac{\text{moles of excess } OH^-}{\text{total volume}} = \frac{0.001000\ \text{mol}}{0.09000\ L} = 0.01111\ M$$

$$[H_3O^+] = \frac{K_w}{[OH^-]} = \frac{1.0\times10^{-14}}{0.01111} = 9.0\times10^{-13}\ M$$

$$pH = 12.05$$

Check As expected from the continuous addition of base, the pH increases through the four stages. Be sure to round off and check the arithmetic along the way.

FOLLOW-UP PROBLEM 19.3 A chemist titrates 20.00 mL of 0.2000 *M* HBrO ($K_a = 2.3\times10^{-9}$) with 0.1000 *M* NaOH. What is the pH **(a)** before any base is added; **(b)** when $[HBrO] = [BrO^-]$; **(c)** at the equivalence point; **(d)** when the moles of OH^- added are twice the moles of HBrO originally present? **(e)** Sketch the titration curve.

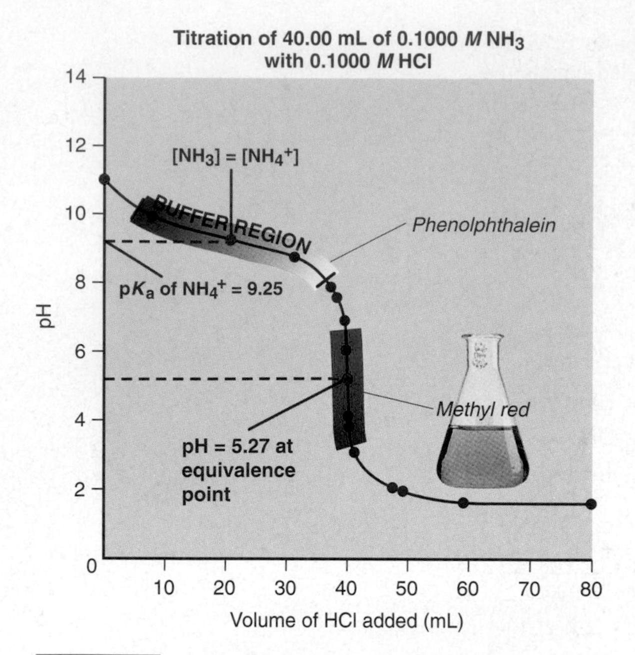

Figure 19.9 Curve for a weak base–strong acid titration. Titrating 40.00 mL of 0.1000 *M* NH_3 with a solution of 0.1000 *M* HCl leads to a curve whose shape is the same as that of the weak acid–strong base curve in Figure 19.8 but inverted. The midpoint of the buffer region occurs when $[NH_3] = [NH_4^+]$; the pH at this point equals the pK_a of NH_4^+. Methyl red (*photo*) is a suitable indicator here.

Weak Base–Strong Acid Titration Curves

In the previous case, we titrated a weak acid with a strong base. The opposite process is the titration of a weak base (NH_3) with a strong acid (HCl), shown in Figure 19.9. Note that *the curve has the same shape as the weak acid–strong base curve* (Figure 19.8, p. 827), *but it is inverted.* Thus, the regions of the curve have the same features, but *the pH decreases* throughout the process:

1. The initial solution is that of a weak base, so *the pH starts out above 7.00.*
2. The pH decreases gradually in the buffer region, where significant amounts of base (NH_3) and conjugate acid (NH_4^+) are present. At the midpoint of the buffer region, *the pH equals the pK_a of the ammonium ion.*
3. After the buffer region, the curve drops vertically to the equivalence point, at which all the NH_3 has reacted and the solution contains only NH_4^+ and Cl^-. Note that *the pH at the equivalence point is below 7.00* because Cl^- does not react with water and NH_4^+ is acidic:
$$NH_4^+(aq) + H_2O(l) \rightleftharpoons NH_3(aq) + H_3O^+(aq)$$
4. Beyond the equivalence point, the pH decreases slowly as *excess H_3O^+* is added.

For this titration also, we must be more careful in choosing the indicator than for a strong acid–strong base titration. Phenolphthalein changes color too soon and too slowly to indicate the equivalence point; but methyl red lies on the steep portion of the curve and straddles the equivalence point, so it is a perfect choice.

Titration Curves for Polyprotic Acids

As we discussed in Section 18.4, polyprotic acids have more than one ionizable proton. Except for sulfuric acid, the common polyprotic acids are all weak. The successive K_a values for a polyprotic acid differ by several orders of magnitude, which means that the first H^+ is lost much more easily than subsequent ones, as these values for sulfurous acid show:

$$H_2SO_3(aq) + H_2O(l) \rightleftharpoons HSO_3^-(aq) + H_3O^+(aq)$$
$$K_{a1} = 1.4\times10^{-2} \quad \text{and} \quad pK_{a1} = 1.85$$

$$HSO_3^-(aq) + H_2O(l) \rightleftharpoons SO_3^{2-}(aq) + H_3O^+(aq)$$
$$K_{a2} = 6.5\times10^{-8} \quad \text{and} \quad pK_{a2} = 7.19$$

In a titration of a diprotic acid such as H_2SO_3, two OH^- ions are required to completely remove both H^+ ions from each molecule of acid. Figure 19.10 shows the titration curve for sulfurous acid with strong base. Because of the large difference in K_a values, we assume that each mole of H^+ is titrated separately; that is, all H_2SO_3 molecules lose one H^+ before any HSO_3^- ions lose one:

$$H_2SO_3 \xrightarrow{\text{1 mol OH}^-} HSO_3^- \xrightarrow{\text{1 mol OH}^-} SO_3^{2-}$$

As you can see from the curve in Figure 19.10, the loss of each mole of H^+ shows up as a separate equivalence point and buffer region. As on the curve for a weak monoprotic acid (Figure 19.8, p. 827), the pH at the midpoint of the buffer region is equal to the pK_a of that acidic species. Note also that the same volume of added base (in this case, 40.00 mL of 0.1000 *M* OH^-) is required to remove each mole of H^+.

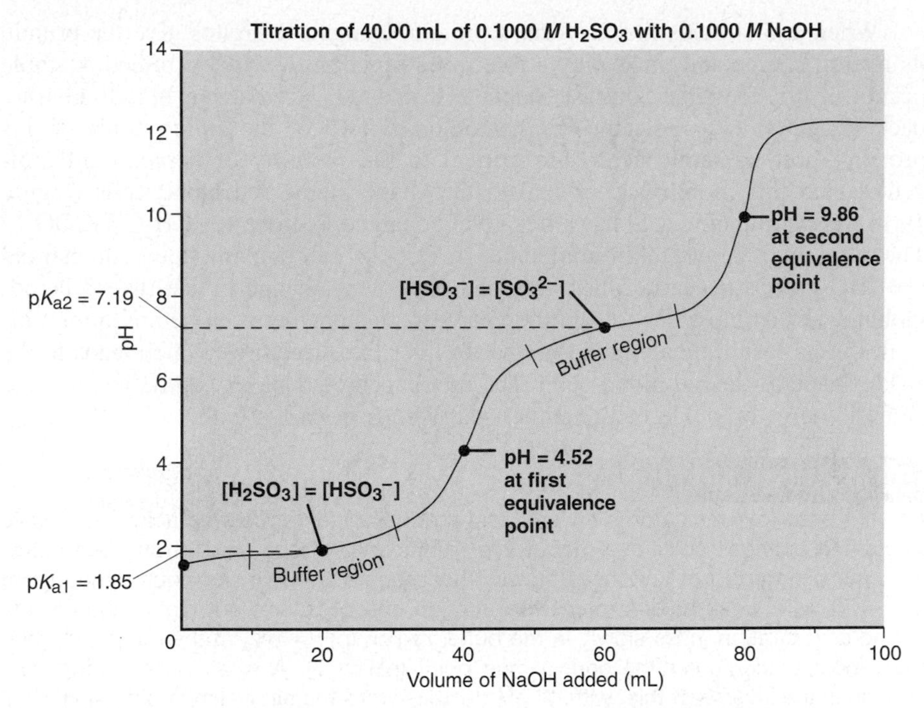

Figure 19.10 Curve for the titration of a weak polyprotic acid. Titrating 40.00 mL of 0.1000 *M* H_2SO_3 with 0.1000 *M* NaOH leads to a curve with two buffer regions and two equivalence points. Because the K_a values are separated by several orders of magnitude, the titration curve looks like two weak acid–strong base curves joined end to end. The pH of the first equivalence point is below 7.00 because the solution contains HSO_3^-, which is a stronger acid than it is a base (K_a of $HSO_3^- = 6.5\times10^{-8}$; K_b of $HSO_3^- = 7.1\times10^{-13}$).

Amino Acids as Biological Polyprotic Acids

Amino acids, the molecules that link together to form proteins (Sections 13.2 and 15.6), have the general formula NH_2—CH(R)—COOH, where R can be one of about 20 different groups. At low pH, both the amino group (—NH_2) and the acid group (—COOH) are protonated: $^+NH_3$—CH(R)—COOH. Thus, in this form the amino acid behaves like a polyprotic acid. In the case of glycine (R = H), the simplest amino acid, the dissociation reactions and pK_a values are

$$^+NH_3CH_2COOH(aq) + H_2O(l) \rightleftharpoons {}^+NH_3CH_2COO^-(aq) + H_3O^+(aq) \quad pK_{a1} = 2.35$$

$$^+NH_3CH_2COO^-(aq) + H_2O(l) \rightleftharpoons NH_2CH_2COO^-(aq) + H_3O^+(aq) \quad pK_{a2} = 9.78$$

The pK_a values show that the —COOH group is much more acidic than the —NH_3^+ group. As we saw with H_2SO_3, the protons are titrated separately, so virtually all the —COOH protons are removed before any —NH_3^+ protons are:

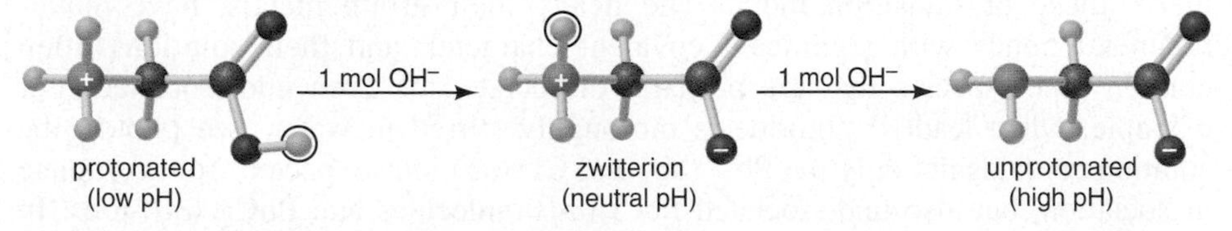

Thus, at physiological pH (~7), *glycine exists predominantly as a zwitterion* (German *zwitter,* "double"), a species with opposite charges on the same molecule: $^+NH_3CH_2COO^-$. Among the 20 different R groups of amino acids in proteins, several have *additional* —COO^- or —NH_3^+ groups at pH 7 (see Figure 15.28, p. 655).

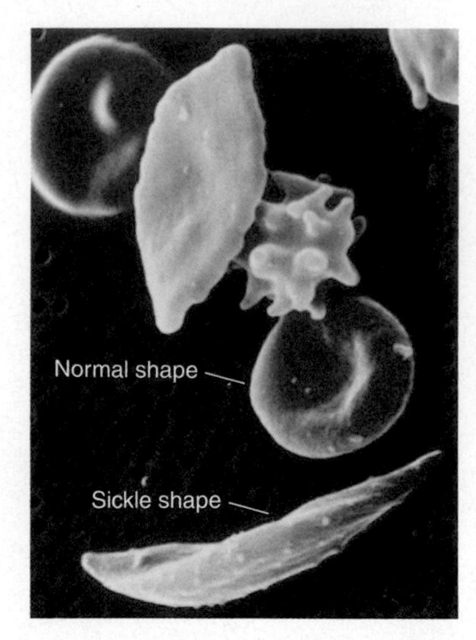

Figure 19.11 Sickle shape of red blood cells in sickle cell anemia.

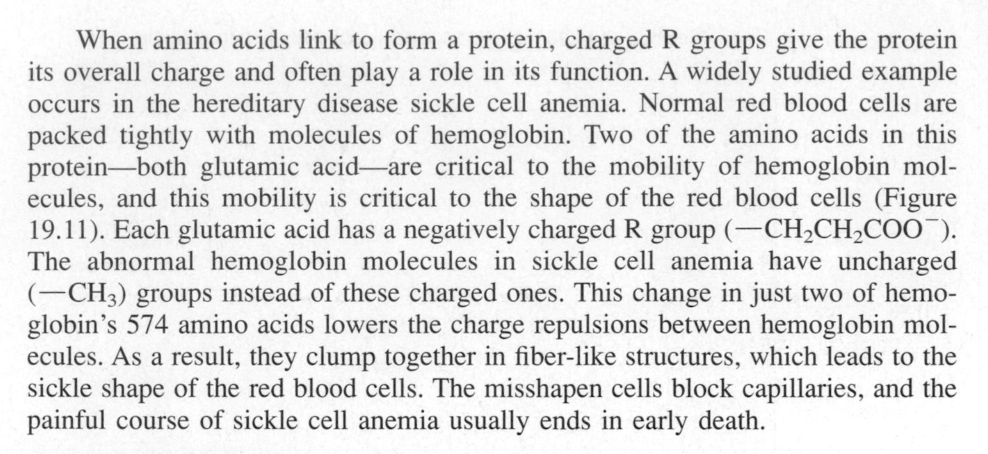

When amino acids link to form a protein, charged R groups give the protein its overall charge and often play a role in its function. A widely studied example occurs in the hereditary disease sickle cell anemia. Normal red blood cells are packed tightly with molecules of hemoglobin. Two of the amino acids in this protein—both glutamic acid—are critical to the mobility of hemoglobin molecules, and this mobility is critical to the shape of the red blood cells (Figure 19.11). Each glutamic acid has a negatively charged R group ($—CH_2CH_2COO^-$). The abnormal hemoglobin molecules in sickle cell anemia have uncharged ($—CH_3$) groups instead of these charged ones. This change in just two of hemoglobin's 574 amino acids lowers the charge repulsions between hemoglobin molecules. As a result, they clump together in fiber-like structures, which leads to the sickle shape of the red blood cells. The misshapen cells block capillaries, and the painful course of sickle cell anemia usually ends in early death.

SECTION SUMMARY

An acid-base (pH) indicator is a weak acid that has different colored acidic and basic forms and changes color over about 2 pH units. In a strong acid–strong base titration, the pH starts out low, rises slowly, then shoots up near the equivalence point (pH = 7). In a weak acid–strong base titration, the pH starts out higher than in the strong acid titration, rises slowly in the buffer region (pH = pK_a at the midpoint), then rises more quickly near the equivalence point (pH $>$ 7). A weak base–strong acid titration is the inverse of this, with its pH decreasing to the equivalence point (pH $<$ 7). Polyprotic acids have two or more acidic protons, each with its own K_a value. Because the K_a's differ by several orders of magnitude, each proton is titrated separately. Amino acids exist in different charged forms that depend on the pH of the solution.

19.3 EQUILIBRIA OF SLIGHTLY SOLUBLE IONIC COMPOUNDS

In this section, we explore the aqueous equilibria of slightly soluble ionic compounds, which up to now we've called "insoluble." In Chapter 13, we found that most solutes, even those called "soluble," have a limited solubility in a particular solvent. Add more than this amount, and some solute remains undissolved. In a saturated solution at a particular temperature, equilibrium exists between the undissolved and dissolved solute. Slightly soluble ionic compounds have a relatively low solubility, so they reach equilibrium with relatively little solute dissolved. At this point, it would be a good idea for you to review the solubility rules listed in Table 4.1 (p. 143).

$PbCl_2$, a slightly soluble ionic compound.

When a soluble ionic compound dissolves in water, it dissociates completely into ions. In this discussion, we will assume that the small amount of a slightly soluble ionic compound that dissolves in water also dissociates completely into ions. In reality, however, this is not the case. Many slightly soluble salts, particularly those of transition metals and heavy main-group metals, have metal-nonmetal bonds with significant covalent character, and their solutions often contain other species that are partially dissociated or even undissociated. For example, when lead(II) chloride is thoroughly stirred in water (see photo), the solution contains not only the $Pb^{2+}(aq)$ and $Cl^-(aq)$ ions expected from complete dissociation, but also undissociated $PbCl_2(aq)$ molecules and $PbCl^+(aq)$ ions. In solutions of some other salts, such as $CaSO_4$, there are no molecules, but pairs of ions exist, such as $Ca^{2+}SO_4^{2-}(aq)$. These species increase the solubility above what we calculate assuming complete dissociation. More advanced courses address these complexities, but we will only hint at some of them in the Comments of several sample problems. For these reasons, it is best to treat our calculations here as first approximations.

The Ion-Product Expression (Q_{sp}) and the Solubility-Product Constant (K_{sp})

If we make the assumption that there is complete dissociation of a slightly soluble ionic compound into its component ions, then *equilibrium exists between solid solute and aqueous ions.* Thus, for example, for a saturated solution of lead(II) sulfate in water, we have

$$PbSO_4(s) \rightleftharpoons Pb^{2+}(aq) + SO_4^{2-}(aq)$$

As with all the other equilibrium systems we've looked at, this one can be expressed by a reaction quotient:

$$Q_c = \frac{[Pb^{2+}][SO_4^{2-}]}{[PbSO_4]}$$

As in previous cases, we combine the constant concentration of the solid, $[PbSO_4]$, with the value of Q_c and eliminate it. This gives the *ion-product expression*, Q_{sp}:

$$Q_{sp} = Q_c[PbSO_4] = [Pb^{2+}][SO_4^{2-}]$$

And, when solid $PbSO_4$ attains equilibrium with Pb^{2+} and SO_4^{2-} ions, that is, when the solution reaches saturation, the numerical value of Q_{sp} attains a constant value. This new equilibrium constant is called the **solubility-product constant, K_{sp}.** The K_{sp} for $PbSO_4$ at 25°C, for example, is 1.6×10^{-8}.

As we've seen with other equilibrium constants, a given K_{sp} value depends only on the temperature, not on the individual ion concentrations. Suppose, for example, you add some lead(II) nitrate, a soluble lead salt, to increase the solution's $[Pb^{2+}]$. The equilibrium position shifts to the left, and $[SO_4^{2-}]$ goes down as more $PbSO_4$ precipitates; so the K_{sp} value is maintained.

The form of Q_{sp} is identical to that of the other reaction quotients we have written: each ion concentration is raised to an exponent equal to the coefficient in the balanced equation, which in this case also *equals the subscript of each ion in the compound's formula.* Thus, in general, for a saturated solution of a slightly soluble ionic compound, M_pX_q, composed of the ions M^{n+} and X^{z-}, the equilibrium condition is

$$Q_{sp} = [M^{n+}]^p[X^{z-}]^q = K_{sp} \qquad \textbf{(19.2)}$$

Of course, at saturation, the concentration terms represent equilibrium concentrations, so from here on, we write the ion-product expression directly with the symbol K_{sp}. For example, the equation and ion-product expression that describe a saturated solution of $Cu(OH)_2$ are

$$Cu(OH)_2(s) \rightleftharpoons Cu^{2+}(aq) + 2OH^-(aq) \qquad K_{sp} = [Cu^{2+}][OH^-]^2$$

Insoluble metal sulfides present a slightly different case. The sulfide ion, S^{2-}, is so basic that it is not stable in water and reacts completely to form the hydrogen sulfide ion (HS^-) and the hydroxide ion (OH^-):

$$S^{2-}(aq) + H_2O(l) \longrightarrow HS^-(aq) + OH^-(aq)$$

For instance, when manganese(II) sulfide is shaken with water, the solution contains Mn^{2+}, HS^-, and OH^- ions. Although the sulfide ion does not exist as such in water, you can imagine the dissolution process as the sum of two steps, with S^{2-} occurring as an intermediate that is consumed immediately:

$$MnS(s) \rightleftharpoons Mn^{2+}(aq) + \cancel{S^{2-}(aq)}$$
$$\cancel{S^{2-}(aq)} + H_2O(l) \longrightarrow HS^-(aq) + OH^-(aq)$$
$$MnS(s) + H_2O(l) \rightleftharpoons Mn^{2+}(aq) + HS^-(aq) + OH^-(aq)$$

Therefore, the ion-product expression is

$$K_{sp} = [Mn^{2+}][HS^-][OH^-]$$

SAMPLE PROBLEM 19.4 Writing Ion-Product Expressions for Slightly Soluble Ionic Compounds

Problem Write the ion-product expression for each of the following compounds:
(a) Magnesium carbonate
(b) Iron(II) hydroxide
(c) Calcium phosphate
(d) Silver sulfide

Plan We write an equation that describes a saturated solution and then write the ion-product expression, K_{sp}, according to Equation 19.2, noting the sulfide in part (d).

Solution (a) Magnesium carbonate:

$$MgCO_3(s) \rightleftharpoons Mg^{2+}(aq) + CO_3^{2-}(aq) \qquad K_{sp} = [Mg^{2+}][CO_3^{2-}]$$

(b) Iron(II) hydroxide:

$$Fe(OH)_2(s) \rightleftharpoons Fe^{2+}(aq) + 2OH^-(aq) \qquad K_{sp} = [Fe^{2+}][OH^-]^2$$

(c) Calcium phosphate:

$$Ca_3(PO_4)_2(s) \rightleftharpoons 3Ca^{2+}(aq) + 2PO_4^{3-}(aq) \qquad K_{sp} = [Ca^{2+}]^3[PO_4^{3-}]^2$$

(d) Silver sulfide:

$$Ag_2S(s) \rightleftharpoons 2Ag^+(aq) + \cancel{S^{2-}(aq)}$$
$$\cancel{S^{2-}(aq)} + H_2O(l) \longrightarrow HS^-(aq) + OH^-(aq)$$
$$Ag_2S(s) + H_2O(l) \rightleftharpoons 2Ag^+(aq) + HS^-(aq) + OH^-(aq) \qquad K_{sp} = [Ag^+]^2[HS^-][OH^-]$$

Check Except for part (d), you can check by reversing the process to see if you obtain the formula of the compound from K_{sp}.

Comment In part (d), we include H_2O as reactant to obtain a balanced equation.

FOLLOW-UP PROBLEM 19.4 Write the ion-product expression for each of the following compounds:
(a) Calcium sulfate
(b) Chromium(III) carbonate
(c) Magnesium hydroxide
(d) Arsenic(III) sulfide

Table 19.2 Solubility-Product Constants (K_{sp}) of Selected Ionic Compounds at 25°C

Name, Formula	K_{sp}
Aluminum hydroxide, $Al(OH)_3$	3×10^{-34}
Cobalt(II) carbonate, $CoCO_3$	1.0×10^{-10}
Iron(II) hydroxide, $Fe(OH)_2$	4.1×10^{-15}
Lead(II) fluoride, PbF_2	3.6×10^{-8}
Lead(II) sulfate, $PbSO_4$	1.6×10^{-8}
Mercury(I) iodide, Hg_2I_2	4.7×10^{-29}
Silver sulfide, Ag_2S	8×10^{-48}
Zinc iodate, $Zn(IO_3)_2$	3.9×10^{-6}

The magnitude of the solubility-product constant, K_{sp}, is a measure of how far to the right the dissolution proceeds at equilibrium (saturation). We'll use K_{sp} values later to compare solubilities. Table 19.2 presents some representative K_{sp} values of slightly soluble ionic compounds. (Appendix C includes a much more extensive list.) Note that, even though the values are all quite low, they range over many orders of magnitude.

Calculations Involving the Solubility-Product Constant

In Chapters 17 and 18, we described two types of equilibrium problems. In one type, we use concentrations to find K, and in the other, we use K to find concentrations. Here we encounter the same two types.

Determining K_{sp} from Solubility The solubilities of ionic compounds are determined experimentally, and several chemical handbooks tabulate them. Most solubility values are given in units of grams of solute dissolved in 100 grams of H_2O. Because the mass of compound in solution is small, a negligible error is introduced if we assume that "100 g of water" is equal to "100 mL of solution." We then convert the solubility from grams of solute per 100 mL of solution to **molar solubility,** the amount (mol) of solute dissolved per liter of solution (that is, the molarity of the solute). Next, we use the equation showing the dissolution of the solute to find the molarity of each ion and substitute into the ion-product expression to find the value of K_{sp}.

SAMPLE PROBLEM 19.5 Determining K_{sp} from Solubility

Problem (a) Lead(II) sulfate ($PbSO_4$) is a key component in lead-acid car batteries. Its solubility in water at 25°C is 4.25×10^{-3} g/100 mL solution. What is the K_{sp} of $PbSO_4$?
(b) When lead(II) fluoride (PbF_2) is shaken with pure water at 25°C, the solubility is found to be 0.64 g/L. Calculate the K_{sp} of PbF_2.

Plan We are given the solubilities in various units and must find K_{sp}. For each compound, we write an equation for its dissolution to see the number of moles of each ion, and then write the ion-product expression. We convert the solubility to molar solubility, find the molarity of each ion, and substitute into the ion-product expression to calculate K_{sp}.

Solution (a) For $PbSO_4$. Writing the equation and ion-product (K_{sp}) expression:

$$PbSO_4(s) \rightleftharpoons Pb^{2+}(aq) + SO_4^{2-}(aq) \qquad K_{sp} = [Pb^{2+}][SO_4^{2-}]$$

Converting solubility to molar solubility:

$$\text{Molar solubility of } PbSO_4 = \frac{0.00425\text{ g } PbSO_4}{100\text{ mL soln}} \times \frac{1000\text{ mL}}{1\text{ L}} \times \frac{1\text{ mol } PbSO_4}{303.3\text{ g } PbSO_4}$$
$$= 1.40\times10^{-4}\ M\ PbSO_4$$

Determining molarities of the ions: Because 1 mol of Pb^{2+} and 1 mol of SO_4^{2-} form when 1 mol of $PbSO_4$ dissolves, $[Pb^{2+}] = [SO_4^{2-}] = 1.40\times10^{-4}\ M$.

Calculating K_{sp}:

$$K_{sp} = [Pb^{2+}][SO_4^{2-}] = (1.40\times10^{-4})^2$$
$$= 1.96\times10^{-8}$$

(b) For PbF_2. Writing the equation and K_{sp} expression:

$$PbF_2(s) \rightleftharpoons Pb^{2+}(aq) + 2F^{-}(aq) \qquad K_{sp} = [Pb^{2+}][F^{-}]^2$$

Converting solubility to molar solubility:

$$\text{Molar solubility of } PbF_2 = \frac{0.64\text{ g } PbF_2}{1\text{ L soln}} \times \frac{1\text{ mol } PbF_2}{245.2\text{ g } PbF_2} = 2.6\times10^{-3}\ M\ PbF_2$$

Determining molarities of the ions: Since 1 mol of Pb^{2+} and 2 mol of F^- form when 1 mol of PbF_2 dissolves,

$$[Pb^{2+}] = 2.6\times10^{-3}\ M \qquad \text{and} \qquad [F^-] = 2(2.6\times10^{-3}\ M) = 5.2\times10^{-3}\ M$$

Calculating K_{sp}:

$$K_{sp} = [Pb^{2+}][F^-]^2 = (2.6\times10^{-3})(5.2\times10^{-3})^2$$
$$= 7.0\times10^{-8}$$

Check The low solubilities are consistent with K_{sp} values being small. In **(a)**, the molar solubility seems about right: $\sim \dfrac{4\times10^{-2}\text{ g/L}}{3\times10^{2}\text{ g/mol}} \approx 1.3\times10^{-4}\ M$. Squaring this number gives 1.7×10^{-8}, close to the calculated K_{sp}. In **(b)**, we check the final step: $\sim(3\times10^{-3})(5\times10^{-3})^2 = 7.5\times10^{-8}$, close to the calculated K_{sp}.

Comment 1. In part (b), the formula PbF_2 means that $[F^-]$ is twice $[Pb^{2+}]$. Then we square this value of $[F^-]$. Always follow the ion-product expression explicitly.

2. The tabulated K_{sp} values for these compounds (Table 19.2) are lower than our calculated values. For PbF_2, for instance, the tabulated value is 3.6×10^{-8}, but we calculated 7.0×10^{-8} from solubility data. The discrepancy arises because we assumed that the PbF_2 in solution dissociates completely to Pb^{2+} and F^-. Here is an example of the complexity pointed out at the beginning of this section. Actually, about a third of the PbF_2 dissolves as $PbF^+(aq)$ and a small amount as undissociated $PbF_2(aq)$. The solubility (0.64 g/L) is determined experimentally and includes these other species, which we did not include in our simple calculation. This is why we treat such calculated K_{sp} values as approximations.

Fluorite

FOLLOW-UP PROBLEM 19.5 When powdered fluorite (CaF_2; see photo) is shaken with pure water at 18°C, 1.5×10^{-4} g dissolves for every 10.0 mL of solution. Calculate the K_{sp} of CaF_2 at 18°C.

Determining Solubility from K_{sp} The reverse of the previous type of problem involves finding the solubility of a compound based on its formula and K_{sp} value. An approach similar to the one we used for weak acids in Sample Problem 18.7 is to define the unknown amount dissolved—molar solubility—as S. Then we define the ion concentrations in terms of this unknown in a reaction table, and solve for S.

SAMPLE PROBLEM 19.6 Determining Solubility from K_{sp}

Problem Calcium hydroxide (slaked lime) is a major component of mortar, plaster, and cement, and solutions of $Ca(OH)_2$ are used in industry as a cheap, strong base. Calculate the solubility of $Ca(OH)_2$ in water if the K_{sp} is 6.5×10^{-6}.

Plan We write the dissolution equation and the ion-product expression. We know K_{sp} (6.5×10^{-6}); to find molar solubility (S), we set up a reaction table that expresses $[Ca^{2+}]$ and $[OH^-]$ in terms of S, substitute into the ion-product expression, and solve for S.

Solution Writing the equation and ion-product expression:

$$Ca(OH)_2(s) \rightleftharpoons Ca^{2+}(aq) + 2OH^-(aq) \qquad K_{sp} = [Ca^{2+}][OH^-]^2 = 6.5\times10^{-6}$$

Setting up a reaction table, with S = molar solubility:

Concentration (M)	$Ca(OH)_2(s)$	$\rightleftharpoons$	$Ca^{2+}(aq)$	+	$2OH^-(aq)$
Initial	—		0		0
Change	—		$+S$		$+2S$
Equilibrium	—		S		$2S$

Substituting into the ion-product expression and solving for S:

$$K_{sp} = [Ca^{2+}][OH^-]^2 = (S)(2S)^2 = (S)(4S^2) = 4S^3 = 6.5\times10^{-6}$$

$$S = \sqrt[3]{\frac{6.5\times10^{-6}}{4}} = 1.2\times10^{-2}\ M$$

Check We expect a low solubility from a slightly soluble salt. If we reverse the calculation, we should obtain the given K_{sp}: $4(1.2\times10^{-2})^3 = 6.9\times10^{-6}$, close to 6.5×10^{-6}.

Comment 1. Note that we did not double and *then* square $[OH^-]$. $2S$ *is* the $[OH^-]$, so we just squared it, as the ion-product expression required.

2. Once again, we assumed that the solid dissociates completely. Actually, the solubility is increased to about $2.0\times10^{-2}\ M$ by the presence of $CaOH^+(aq)$ formed in the reaction $Ca(OH)_2(s) \rightleftharpoons CaOH^+(aq) + OH^-(aq)$. Our calculated answer is only approximate because we did not take this other species into account.

FOLLOW-UP PROBLEM 19.6 A suspension of $Mg(OH)_2$ in water is marketed as "milk of magnesia," which alleviates minor symptoms of indigestion by neutralizing stomach acid. The $[OH^-]$ is too low to harm the mouth and throat, but the suspension dissolves in the acidic stomach juices. What is the molar solubility of $Mg(OH)_2$ ($K_{sp} = 6.3\times10^{-10}$) in pure water?

Using K_{sp} Values to Compare Solubilities The K_{sp} values provide a guide to *relative* solubility, as long as we compare compounds whose formulas contain the *same total number of ions*. In such cases, *the higher the K_{sp}, the greater the solubility.* Table 19.3 shows this point for several compounds. Note that for compounds that form three ions, the relationship holds whether the cation:anion ratio is 1:2 or 2:1, because the mathematical expression containing S is the same ($4S^3$) in the calculation (see Sample Problem 19.6).

Table 19.3 Relationship Between K_{sp} and Solubility at 25°C

No. of Ions	Formula	Cation:Anion	K_{sp}	Solubility (*M*)
2	$MgCO_3$	1:1	3.5×10^{-8}	1.9×10^{-4}
2	$PbSO_4$	1:1	1.6×10^{-8}	1.3×10^{-4}
2	$BaCrO_4$	1:1	2.1×10^{-10}	1.4×10^{-5}
3	$Ca(OH)_2$	1:2	6.5×10^{-6}	1.2×10^{-2}
3	BaF_2	1:2	1.5×10^{-6}	7.2×10^{-3}
3	CaF_2	1:2	3.2×10^{-11}	2.0×10^{-4}
3	Ag_2CrO_4	2:1	2.6×10^{-12}	8.7×10^{-5}

The Effect of a Common Ion on Solubility

The presence of a common ion decreases the solubility of a slightly soluble ionic compound. As we saw in the case of acid-base systems, Le Châtelier's principle helps explain this effect. Let's examine the equilibrium condition for a saturated solution of lead(II) chromate:

$$PbCrO_4(s) \rightleftharpoons Pb^{2+}(aq) + CrO_4^{2-}(aq) \qquad K_{sp} = [Pb^{2+}][CrO_4^{2-}] = 2.3\times10^{-13}$$

At a given temperature, K_{sp} depends only on the product of the ion concentrations. If the concentration of either ion goes up, the other must go down to maintain the constant K_{sp}. Suppose we add Na_2CrO_4, a very soluble salt, to the saturated $PbCrO_4$ solution. The concentration of the common ion, CrO_4^{2-}, increases, and some of it combines with Pb^{2+} ion to form more solid $PbCrO_4$ (Figure 19.12). The overall effect is a shift in the position of equilibrium to the left:

$$PbCrO_4(s) \overset{\longleftarrow}{\rightleftharpoons} Pb^{2+}(aq) + CrO_4^{2-}(aq;\ \text{added})$$

After the addition, $[CrO_4^{2-}]$ is higher, but $[Pb^{2+}]$ is lower. In this case, $[Pb^{2+}]$ represents the amount of $PbCrO_4$ dissolved; thus, in effect, the solubility of $PbCrO_4$ has decreased. The same result is obtained if we dissolve $PbCrO_4$ in a Na_2CrO_4 solution. We also obtain this result by adding a soluble lead(II) salt, such as $Pb(NO_3)_2$. The added Pb^{2+} ion combines with some $CrO_4^{2-}(aq)$, thereby lowering the amount of dissolved $PbCrO_4$.

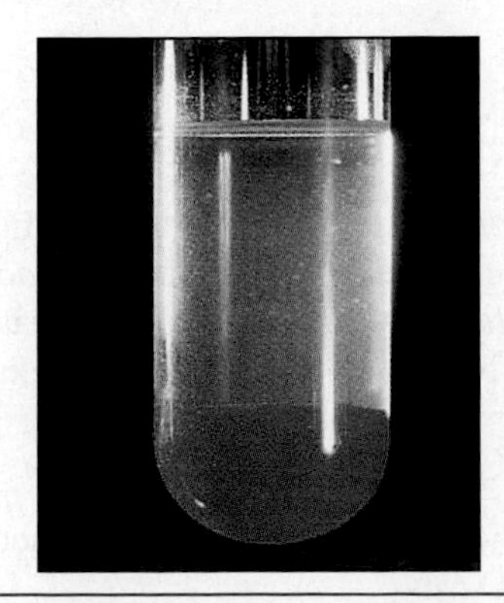

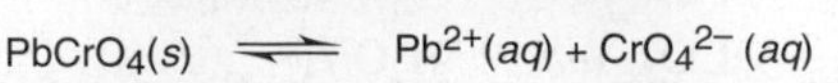

$PbCrO_4(s) \overset{\longleftarrow}{\rightleftharpoons} Pb^{2+}(aq) + CrO_4^{2-}$ (*aq*; added)

A

B

Figure 19.12 The effect of a common ion on solubility. When a common ion is added to a saturated solution of an ionic compound, the solubility is lowered and more of the compound precipitates. **A,** Lead(II) chromate, a slightly soluble salt, forms a saturated aqueous solution. **B,** When Na_2CrO_4 solution is added, the amount of $PbCrO_4(s)$ increases. Thus, $PbCrO_4$ is less soluble in the presence of the common ion CrO_4^{2-}.

SAMPLE PROBLEM 19.7 Calculating the Effect of a Common Ion on Solubility

Problem In Sample Problem 19.6, we calculated the solubility of $Ca(OH)_2$ in water. What is its solubility in 0.10 *M* $Ca(NO_3)_2$? K_{sp} of $Ca(OH)_2$ is 6.5×10^{-6}.

Plan From the equation and the ion-product expression for $Ca(OH)_2$, we predict that the addition of Ca^{2+}, the common ion, will lower the solubility. We set up a reaction table with $[Ca^{2+}]_{init}$ coming from $Ca(NO_3)_2$ and S equal to $[Ca^{2+}]_{from\ Ca(OH)_2}$. To simplify the math, we assume that, because K_{sp} is low, S is so small relative to $[Ca^{2+}]_{init}$ that it can be neglected. Then we solve for S and check the assumption.

Solution Writing the equation and ion-product expression:

$$Ca(OH)_2(s) \rightleftharpoons Ca^{2+}(aq) + 2OH^-(aq) \qquad K_{sp} = [Ca^{2+}][OH^-]^2 = 6.5\times10^{-6}$$

Setting up the reaction table, with $S = [Ca^{2+}]_{from\ Ca(OH)_2}$:

Concentration (*M*)	$Ca(OH)_2(s)$	$\rightleftharpoons$	$Ca^{2+}(aq)$	+	$2OH^-(aq)$
Initial	—		0.10		0
Change	—		$+S$		$+2S$
Equilibrium	—		$0.10 + S$		$2S$

Making the assumption: K_{sp} is small, so $S << 0.10\ M$; thus, $0.10\ M + S \approx 0.10\ M$.
Substituting into the ion-product expression and solving for S:

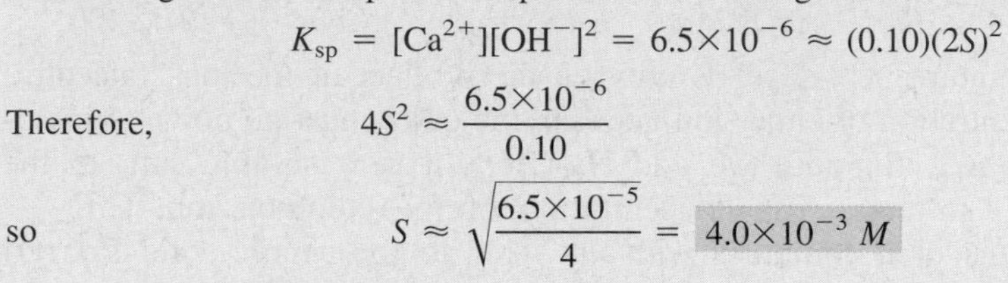

$$K_{sp} = [Ca^{2+}][OH^-]^2 = 6.5\times10^{-6} \approx (0.10)(2S)^2$$

Therefore, $$4S^2 \approx \frac{6.5\times10^{-6}}{0.10}$$

so $$S \approx \sqrt{\frac{6.5\times10^{-5}}{4}} = 4.0\times10^{-3}\ M$$

Checking the assumption:

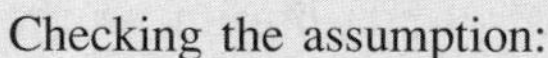

$$\frac{4.0\times10^{-3}\ M}{0.10\ M} \times 100 = 4.0\% < 5\%$$

Check In Sample Problem 19.6, the solubility of $Ca(OH)_2$ was 0.012 *M*, but here, it is 0.0040 *M*, so the solubility *decreased* in the presence of added Ca^{2+}, the common ion, as we predicted.

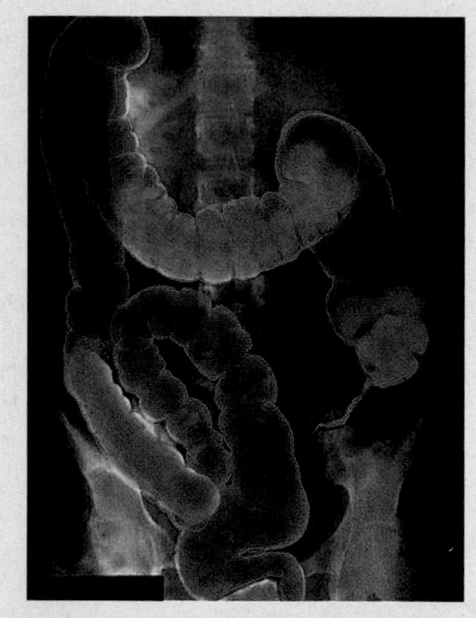

$BaSO_4$ imaging of a human large intestine.

FOLLOW-UP PROBLEM 19.7 To improve the quality of x-ray images used in the diagnosis of intestinal disorders, the patient drinks an aqueous suspension of $BaSO_4$ before the x-ray procedure (see photo). The Ba^{2+} in the suspension is opaque to x-rays, but it is also toxic; thus, the Ba^{2+} concentration is lowered by the addition of dilute Na_2SO_4. What is the solubility of $BaSO_4$ ($K_{sp} = 1.1\times10^{-10}$) in **(a)** pure water and in **(b)** 0.10 *M* Na_2SO_4?

The Effect of pH on Solubility

The hydronium ion concentration can have a profound effect on the solubility of an ionic compound. *If the compound contains the anion of a weak acid, addition of H_3O^+ (from a strong acid) increases its solubility.* Once again, Le Châtelier's principle explains why. In a saturated solution of calcium carbonate, for example, we have

$$CaCO_3(s) \rightleftharpoons Ca^{2+}(aq) + CO_3^{2-}(aq)$$

Adding some strong acid introduces a large amount of H_3O^+, which immediately reacts with CO_3^{2-} to form the weak acid HCO_3^-:

$$CO_3^{2-}(aq) + H_3O^+(aq) \longrightarrow HCO_3^-(aq) + H_2O(l)$$

If enough H_3O^+ is added, further reaction occurs to form carbonic acid, which decomposes immediately to H_2O and CO_2, which escapes the container:

$$HCO_3^-(aq) + H_3O^+(aq) \longrightarrow H_2CO_3(aq) + H_2O(l) \longrightarrow CO_2(g) + 2H_2O(l)$$

Thus, the net effect of added H_3O^+ is a shift in the equilibrium position to the right, and more $CaCO_3$ dissolves:

$$CaCO_3(s) \rightleftharpoons Ca^{2+} + CO_3^{2-} \xrightarrow{H_3O^+} HCO_3^- \xrightarrow{H_3O^+} H_2CO_3 \longrightarrow CO_2(g) + H_2O + Ca^{2+}$$

This particular case illustrates a qualitative field test for carbonate minerals because the CO_2 bubbles vigorously (Figure 19.13).

In contrast, adding H_3O^+ to a saturated solution of a compound with a strong-acid anion, such as silver chloride, has no effect on the equilibrium position:

$$AgCl(s) \rightleftharpoons Ag^+(aq) + Cl^-(aq)$$

Because Cl^- ion is the conjugate base of a strong acid (HCl), it can coexist in solution with high $[H_3O^+]$. The Cl^- does not leave the system, so the equilibrium position is not affected.

Figure 19.13 Test for the presence of a carbonate. When a mineral that contains carbonate ion is treated with strong acid, the added H_3O^+ shifts the equilibrium position of the carbonate solubility. More carbonate dissolves, and the carbonic acid that is formed breaks down to water and gaseous CO_2.

SAMPLE PROBLEM 19.8 Predicting the Effect on Solubility of Adding Strong Acid

Problem Write balanced equations to explain whether addition of H_3O^+ from a strong acid affects the solubility of these ionic compounds:
(a) Lead(II) bromide
(b) Copper(II) hydroxide
(c) Iron(II) sulfide

Plan We write the balanced dissolution equation and note the anion: Weak-acid anions react with H_3O^+ and shift the equilibrium position toward more dissolution. Strong-acid anions do not react, so added H_3O^+ has no effect.

Solution (a) $PbBr_2(s) \rightleftharpoons Pb^{2+}(aq) + 2Br^-(aq)$

No effect. Br^- is the anion of HBr, a strong acid, so it does not react with H_3O^+.

(b) $Cu(OH)_2(s) \rightleftharpoons Cu^{2+}(aq) + 2OH^-(aq)$

Increases solubility. OH^- is the anion of H_2O, a very weak acid, so it reacts with the added H_3O^+:

$$OH^-(aq) + H_3O^+(aq) \longrightarrow 2H_2O(l)$$

(c) $FeS(s) + H_2O(l) \rightleftharpoons Fe^{2+}(aq) + HS^-(aq) + OH^-(aq)$

Increases solubility. We noted earlier that the S^{2-} ion reacts immediately with water to form HS^-. The added H_3O^+ reacts with both weak-acid anions, HS^- and OH^-:

$$HS^-(aq) + H_3O^+(aq) \longrightarrow H_2S(aq) + H_2O(l)$$
$$OH^-(aq) + H_3O^+(aq) \longrightarrow 2H_2O(l)$$

FOLLOW-UP PROBLEM 19.8 Write balanced equations to show how addition of $HNO_3(aq)$ affects the solubility of these ionic compounds:
(a) Calcium fluoride
(b) Zinc sulfide
(c) Silver iodide

Many principles of ionic equilibria are manifested in natural formations. The upcoming Chemical Connections essay illustrates the fascinating carbonate equilibria that give rise to limestone caves.

CHEMICAL CONNECTIONS to Geology

Creation of a Limestone Cave

Limestone caves and the intricate structures within them provide striking evidence of the workings of aqueous ionic equilibria (Figure B19.1). The spires and vaults of these natural cathedrals are the products of reactions between carbonate rocks and the water that has run through them for millennia. Limestone is predominantly calcium carbonate ($CaCO_3$), a slightly soluble ionic compound with a K_{sp} of 3.3×10^{-9}. This rocky material began accumulating in the Earth over 400 million years ago, and a relatively young cave, such as Howe Caverns in eastern New York State, began forming about 800,000 years ago.

Two key facts help us understand how limestone caves form:

1. Gaseous CO_2 is in equilibrium with aqueous CO_2 in natural waters:

$$CO_2(g) \xrightleftharpoons{H_2O(l)} CO_2(aq) \qquad \textbf{(equation 1)}$$

The concentration of CO_2 in the water is proportional to the partial pressure of $CO_2(g)$ in contact with the water (Henry's law; Section 13.4):

$$[CO_2(aq)] \propto P_{CO_2}$$

Because of the continual release of CO_2 from within the Earth (outgassing), the P_{CO_2} in soil-trapped air is *higher* than the P_{CO_2} in the atmosphere.

2. As discussed in the text, the presence of $H_3O^+(aq)$ increases the solubility of ionic compounds that contain the anion of a weak acid. The reaction of CO_2 with water produces H_3O^+:

$$CO_2(aq) + 2H_2O(l) \rightleftharpoons H_3O^+(aq) + HCO_3^-(aq)$$

Thus, the presence of $CO_2(aq)$ leads to the formation of H_3O^+, which increases the solubility of $CaCO_3$:

$$CaCO_3(s) + CO_2(aq) + H_2O(l) \rightleftharpoons Ca^{2+}(aq) + 2HCO_3^-(aq) \qquad \textbf{(equation 2)}$$

Here is an overview of the principal cave-forming process in most limestone caves: As surface water trickles through cracks in the ground, it meets soil-trapped air with its high P_{CO_2}. As a result, $[CO_2(aq)]$ increases (equation 1 shifts to the right), and the solution becomes more acidic. When this CO_2-rich water contacts limestone, more $CaCO_3$ dissolves (equation 2 shifts to the right). As a result, more rock is carved out, more water flows in, and so on. Centuries pass and a cave slowly begins to form.

Eating its way through underground tunnels, some of the aqueous solution, largely dilute $Ca(HCO_3)_2$, passes through the ceiling of the growing cave. As it drips, it meets air, which has a lower P_{CO_2} than the soil, so some $CO_2(aq)$ comes out of solution (equation 1 shifts to the left). This causes some $CaCO_3$ to precipitate on the ceiling and on the floor below, where the drops land (equation 2 shifts to the left). Decades pass, and the ceiling bears an "icicle" of $CaCO_3$, called a *stalactite,* while a spike of $CaCO_3$, called a *stalagmite,* grows upward from the cave floor. With time, they meet to form a column of precipitated limestone.

The same chemical process can lead to different shapes. Standing pools of $Ca(HCO_3)_2$ solution form limestone "lily pads" or "corals." Cascades of solution form delicate limestone "draperies" on a cave wall, with fabulous colors arising from trace metal ions, such as iron (reddish brown) or copper (bluish green).

Recently, vast, spectacularly shaped deposits of gypsum ($CaSO_4{\cdot}2H_2O$; Figure B19.2) found in some limestone caves in Mexico point strongly to another process of cave formation. Certain bacteria found in abundance in these caves can survive at very low pH and consume hydrogen sulfide (H_2S), which leaks into the cave from oil deposits deep below it. The evidence shows that these *acidophilic* microbes use O_2 to oxidize the H_2S to sulfuric acid (H_2SO_4), a strong acid:

$$H_2S(g) + 2O_2(g) + H_2O(l) \xrightarrow{\text{bacteria}} H_3O^+(aq) + HSO_4^-(aq)$$

The sulfuric acid contributes greatly to the formation of the cave.

Figure B19.1 A view inside Carlsbad Caverns, New Mexico. The formations within this limestone cave result from subtle shifts in carbonate equilibria acting over millions of years.

Figure B19.2 Gypsum deposits. Formations of gypsum in the Chandelier Ballroom of Lechuguilla Cave in Mexico.

Predicting the Formation of a Precipitate: Q_{sp} vs. K_{sp}

In Chapter 17, we compared the values of Q and K to see if a reaction had reached equilibrium and, if not, in which net direction it would move until it did. In this discussion, we use the same approach to see if a precipitation reaction has reached equilibrium, that is, whether a precipitate will form and, if not, what changes in the concentrations of the component ions will cause it to do so.

As you know, $Q_{sp} = K_{sp}$ when the solution is saturated. If Q_{sp} is greater than K_{sp}, the solution is momentarily supersaturated, and some solid precipitates until the remaining solution becomes saturated ($Q_{sp} = K_{sp}$). If Q_{sp} is less than K_{sp}, the solution is unsaturated, and no precipitate forms at that temperature (more solid can dissolve). To summarize,

- $Q_{sp} = K_{sp}$: solution is saturated and no change occurs.
- $Q_{sp} > K_{sp}$: precipitate forms until solution is saturated.
- $Q_{sp} < K_{sp}$: solution is unsaturated and no precipitate forms.

SAMPLE PROBLEM 19.9 Predicting Whether a Precipitate Will Form

Problem A common laboratory method for preparing a precipitate is to mix solutions containing the component ions. Does a precipitate form when 0.100 L of 0.30 *M* $Ca(NO_3)_2$ is mixed with 0.200 L of 0.060 *M* NaF?

Plan First, we must decide which slightly soluble salt could form and look up its K_{sp} value in Appendix C. To see whether mixing these solutions will form the precipitate, we find the initial ion concentrations by calculating the amount (mol) of each ion from its concentration and volume, and then dividing by the *total* volume, since one solution dilutes the other. Finally, we write the ion-product expression, calculate Q_{sp}, and compare Q_{sp} with K_{sp}.

Solution The ions present are Ca^{2+}, Na^+, F^-, and NO_3^-. All sodium and all nitrate salts are soluble (Table 4.1, p.143), so the only possibility is CaF_2 ($K_{sp} = 3.2\times10^{-11}$). Calculating the ion concentrations:

$$\text{Moles of } Ca^{2+} = 0.30\ M\ Ca^{2+} \times 0.100\ L = 0.030\ \text{mol } Ca^{2+}$$

$$[Ca^{2+}]_{init} = \frac{0.030\ \text{mol } Ca^{2+}}{0.100\ L + 0.200\ L} = 0.10\ M\ Ca^{2+}$$

$$\text{Moles of } F^- = 0.060\ M\ F^- \times 0.200\ L = 0.012\ \text{mol } F^-$$

$$[F^-]_{init} = \frac{0.012\ \text{mol } F^-}{0.100\ L + 0.200\ L} = 0.040\ M\ F^-$$

Substituting into the ion-product expression and comparing Q_{sp} with K_{sp}:

$$Q_{sp} = [Ca^{2+}]_{init}[F^-]^2_{init} = (0.10)(0.040)^2 = 1.6\times10^{-4}$$

Because $Q_{sp} > K_{sp}$, CaF_2 will precipitate until $Q_{sp} = 3.2\times10^{-11}$.

Check Make sure you remember to round off and quickly check the math. For example, $Q_{sp} = (1\times10^{-1})(4\times10^{-2})^2 = 1.6\times10^{-4}$. With K_{sp} so low, CaF_2 must have a low solubility, and given the sizable concentrations being mixed, we would expect CaF_2 to precipitate.

Precipitation of CaF_2.

FOLLOW-UP PROBLEM 19.9 As a result of mineral erosion and biological activity, phosphate ion is common in natural waters, where it often precipitates as insoluble salts, such as $Ca_3(PO_4)_2$. If $[Ca^{2+}]_{init} = [PO_4^{3-}]_{init} = 1.0\times10^{-9}$ *M* in a given river, will $Ca_3(PO_4)_2$ precipitate? K_{sp} of $Ca_3(PO_4)_2$ is 1.2×10^{-29}.

As the upcoming Chemical Connections essay demonstrates, the principles of ionic equilibria often help us understand the chemical basis of complex environmental problems and may provide ways to solve them.

The Acid-Rain Problem

The conflict between industrial society and the environment is very clear in the problem of *acid rain,* acids resulting from human activity that occur as wet deposition in rain, snow, or fog, and as dry deposition on solid particles. Acidic precipitation has been recorded in all parts of the United States, in Canada, Mexico, the Amazon basin, throughout Europe, Russia, and many parts of Asia, and even at the North and South Poles. We've addressed several aspects of this problem earlier; now, let's examine some effects of acidic precipitation on biological and mineral systems and see how to prevent them. Several chemical culprits are

1. *Sulfurous acid.* Sulfur dioxide (SO_2) formed primarily by the burning of high-sulfur coal, forms sulfurous acid (H_2SO_3) in contact with water. Oxidants, such as hydrogen peroxide and ozone, that are present as pollutants in the atmosphere also dissolve in water and oxidize the sulfurous acid to sulfuric acid:

$$H_2O_2(aq) + H_2SO_3(aq) \longrightarrow H_2SO_4(aq) + H_2O(l)$$

2. *Sulfuric acid.* Sulfur trioxide forms through the atmospheric oxidation of SO_2. It forms sulfuric acid (H_2SO_4) on contact with water, including that in the atmosphere.

3. *Nitric acid.* Nitrogen oxides (collectively known as NO_x) form in the reaction of N_2 and O_2. NO is formed during combustion, primarily in car engines and electric power plants, and it forms NO_2 and HNO_3 in air in the process that creates smog (p. 588). At night, NO_x is converted to N_2O_5, which hydrolyzes to HNO_3 in the presence of water. The strong acids H_2SO_4 and HNO_3 cause the greatest concern (Figure B19.3).

How does the pH of acidic precipitation compare with that of natural waters? Normal rainwater is weakly acidic because it contains dissolved CO_2 from the air:

$$CO_2(g) + 2\,H_2O(l) \rightleftharpoons H_3O^+(aq) + HCO_3^-(aq)$$

Based on the volume percent of CO_2 in air, the solubility of CO_2 in water, and the K_{a1} of H_2CO_3, the pH of normal rainwater is about 5.6 (see Problem 19.148 at the end of the chapter). In stark contrast, the average pH of rainfall in many parts of the United States was 4.2 as recently as 1984, representing 25 times as much H_3O^+. Worldwide, rain in Sweden and Pennsylvania shared second prize with a pH of 2.7, about the same as vinegar. Rain in Wheeling, West Virginia, won first prize with a pH of 1.8, between that of lemon juice and stomach acid! Acidic fog in California sometimes has a pH of 1.6 because evaporation of water from the particles concentrates the acid.

These 10- to 10,000-fold excesses of $[H_3O^+]$ are very destructive to living things. Some fish and shellfish die at pH values between 4.5 and 5.0. The young of most species are generally most sensitive. At pH 5, most fish eggs cannot hatch. With tens of thousands of rivers and lakes around the world becoming acidified, the loss of fish became a major concern long ago. In addition, acres of forest have been harmed by the acid, which removes nutrients and releases toxic substances from the soil. Even if the soil is buffered, acidic fog and clouds can remove essential nutrients from leaf surfaces (Figure B19.4).

Many principles of aqueous equilibria bear directly on the effects of acid rain. The aluminosilicates that make up most soils are extremely insoluble in water. In these materials, the Al^{3+} ion is bonded to OH^- and O^{2-} ions in complex structures (see the Gallery in Chapter 14, p. 580). Continual contact with the H_3O^+ in acid rain causes these ions to react, and some of the bound Al^{3+}, which is toxic to fish, dissolves. Along with dissolved Al^{3+} ions, the acid rain carries away ions that serve as nutrients for plants and animals.

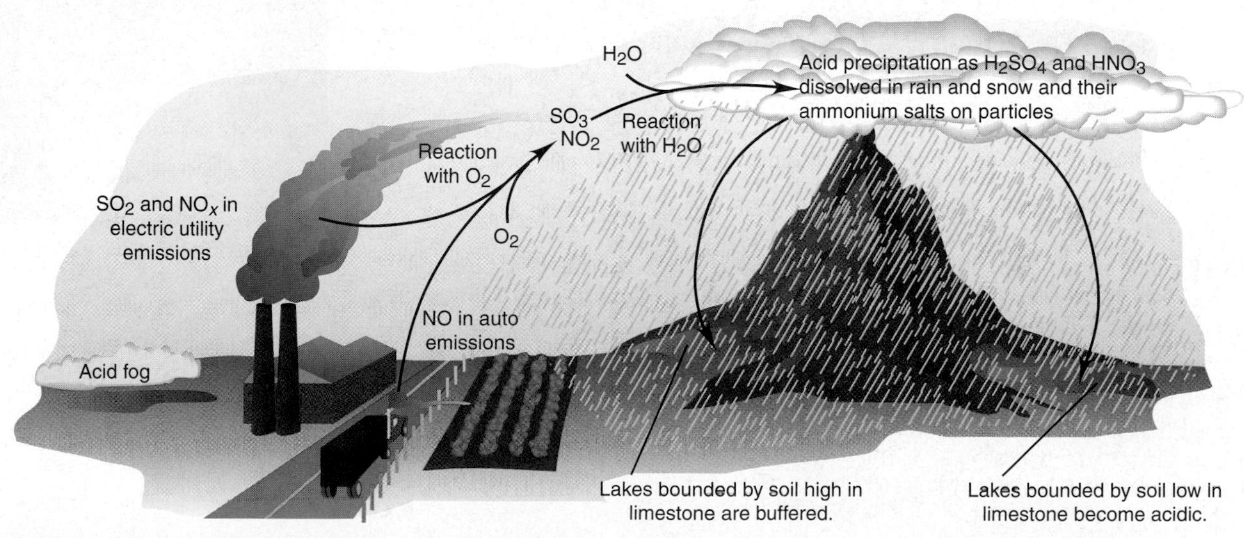

Figure B19.3 Formation of acidic precipitation. A complex interplay of human activities, atmospheric chemistry, and environmental distribution leads to acidic precipitation and its harmful effects. Car exhaust and electrical utility waste gases contain lower oxides of nitrogen and sulfur. These are oxidized in the atmosphere by O_2 (or O_3, not shown) to higher oxides (NO_2, SO_3), which react with moisture to form acidic rain, snow, and fog. In contact with acidic precipitation, many lakes become acidified, whereas limestone-bounded lakes form a carbonate buffer that prevents acidification.

Figure B19.4 A forest damaged by acid rain.

Acid rain also dissolves the calcium carbonate in the marble and limestone of buildings and monuments (Figure B19.5). Ironically, the same chemical process that destroys these structures is responsible for saving those lakes that lie on or are bounded by limestone-rich soil. As acid rain falls, the H_3O^+ reacts with dissolved carbonate ion in the lake to form bicarbonate:

$$CO_3^{2-}(aq) + H_3O^+(aq) \rightleftharpoons HCO_3^-(aq) + H_2O(l)$$

In essence, limestone-bounded lakes function as enormous HCO_3^--CO_3^{2-} buffer solutions, absorbing the additional H_3O^+ and maintaining a relatively stable pH. In fact, lakes, rivers, and groundwater in limestone-rich soils actually remain mildly basic.

For lakes and rivers in contact with limestone-poor soils, expensive remediation methods are needed. A direct attack on the symptoms is carried out by *liming* (treating with limestone) lakes and rivers. Sweden spent tens of millions of dollars during the 1990s to neutralize slightly more than 3000 lakes by adding limestone. This approach is, at best, only a stopgap because the lakes are acidic again within several years.

As we pointed out earlier (see Chemical Connections, Chapter 6, p. 245), the principal means of controlling sulfur dioxide is by "scrubbing" the emissions from power plants with limestone. Both dry and wet scrubbers are used. Another method reduces some of the SO_2 with methane or coal to H_2S, and the mixture is catalytically converted to sulfur, which is sold:

$$16H_2S(g) + 8SO_2(g) \longrightarrow 3S_8(s) + 16H_2O(l)$$

Burning low-sulfur coal to reduce SO_2 formation is sometimes an option, but such coal deposits are rare and expensive to mine. Coal can also be converted into gaseous and liquid low-sulfur fuels (see Chemical Connections, p. 245). The sulfur is removed (as H_2S) in an acid-gas scrubber after gasification.

1944 1994

Figure B19.5 The effect of acid rain on marble statuary. Calcium carbonate, which is the major component of marble, decomposes slowly when exposed to acid rain. These photos of the same statue of George Washington in Washington Square, New York City, were taken 50 years apart.

Through the use of a catalytic converter in an auto exhaust system, NO_x species are reduced to N_2 and NH_3. In power plants, the amount of NO_x is decreased by adjusting combustion conditions, but it also can be removed from the hot stack gases by treatment with ammonia:

$$4NO(g) + 4NH_3(g) + O_2(g) \longrightarrow 4N_2(g) + 6H_2O(g)$$

Several 1990 amendments to the Clean Air Act led to large reductions in SO_2 emissions from coal-burning utilities. Emissions of NO_x must be curbed substantially in the eastern United States under new rules designed to help states meet ozone standards, and HNO_3 will be reduced in the process.

SECTION SUMMARY

As a first approximation, the dissolved portion of a slightly soluble salt dissociates completely into ions. In a saturated solution, the ions are in equilibrium with the solid, and the product of the ion concentrations, each raised to the power of its subscript in the compound's formula, has a constant value ($Q_{sp} = K_{sp}$). The value of K_{sp} can be obtained from the solubility, and vice versa. Adding a common ion lowers an ionic compound's solubility. Adding H_3O^+ (lowering the pH) increases a compound's solubility if the anion of the compound is that of a weak acid. If $Q_{sp} > K_{sp}$ for an ionic compound, a precipitate forms when two solutions, each containing one of the compound's ions, are mixed. Limestone caves result from shifts in the $CaCO_3$-CO_2 equilibrium system. Lakes bounded by limestone-rich soils form buffer systems that prevent harmful acidification.

19.4 EQUILIBRIA INVOLVING COMPLEX IONS

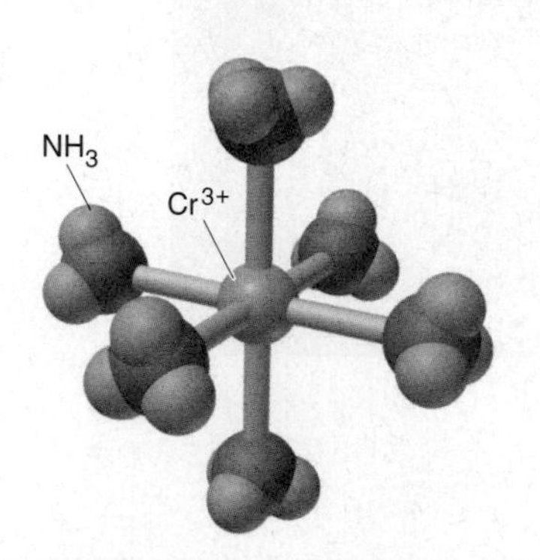

Figure 19.14 $Cr(NH_3)_6^{3+}$, a typical complex ion. A complex ion consists of a central metal ion, such as Cr^{3+}, covalently bonded to a specific number of ligands, such as NH_3.

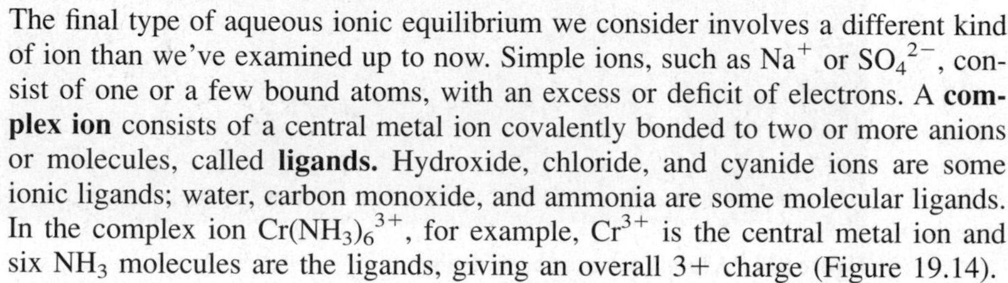

The final type of aqueous ionic equilibrium we consider involves a different kind of ion than we've examined up to now. Simple ions, such as Na^+ or SO_4^{2-}, consist of one or a few bound atoms, with an excess or deficit of electrons. A **complex ion** consists of a central metal ion covalently bonded to two or more anions or molecules, called **ligands.** Hydroxide, chloride, and cyanide ions are some ionic ligands; water, carbon monoxide, and ammonia are some molecular ligands. In the complex ion $Cr(NH_3)_6^{3+}$, for example, Cr^{3+} is the central metal ion and six NH_3 molecules are the ligands, giving an overall 3+ charge (Figure 19.14).

As we discussed in Section 18.9, *all complex ions are Lewis adducts.* The metal ion acts as a Lewis acid (accepts an electron pair) and the ligand acts as a Lewis base (donates an electron pair). The acidic hydrated metal ions that we discussed in Section 18.6 are complex ions with water molecules as ligands. In Chapter 23, we discuss the transition metals and the structures and properties of the numerous complex ions they form. Our focus here is on equilibria of hydrated ions with ligands other than water.

Formation of Complex Ions

Whenever a metal ion enters water, a complex ion forms, with water as the ligand. In many cases, when we treat this hydrated cation with a solution of another ligand, the bound water molecules exchange for the other ligand. For example, a hydrated M^{2+} ion, $M(H_2O)_4^{2+}$, forms the complex ion $M(NH_3)_4^{2+}$ in aqueous NH_3:

$$M(H_2O)_4^{2+}(aq) + 4NH_3(aq) \rightleftharpoons M(NH_3)_4^{2+}(aq) + 4H_2O(l)$$

At equilibrium, this system is expressed by a ratio of concentration terms whose form follows that of any other equilibrium expression:

$$K_c = \frac{[M(NH_3)_4^{2+}][H_2O]^4}{[M(H_2O)_4^{2+}][NH_3]^4}$$

Once again, because the concentration of water is essentially constant in aqueous reactions, we incorporate it into K_c and obtain the expression for a new equilibrium constant, the **formation constant, K_f:**

$$K_f = \frac{K_c}{[H_2O]^4} = \frac{[M(NH_3)_4^{2+}]}{[M(H_2O)_4^{2+}][NH_3]^4}$$

At the molecular level, the actual process is stepwise, with ammonia molecules replacing water molecules one at a time to give a series of intermediate species, each with its own formation constant:

$$M(H_2O)_4^{2+}(aq) + NH_3(aq) \rightleftharpoons M(H_2O)_3(NH_3)^{2+}(aq) + H_2O(l)$$

$$K_{f1} = \frac{[M(H_2O)_3(NH_3)^{2+}]}{[M(H_2O)_4^{2+}][NH_3]}$$

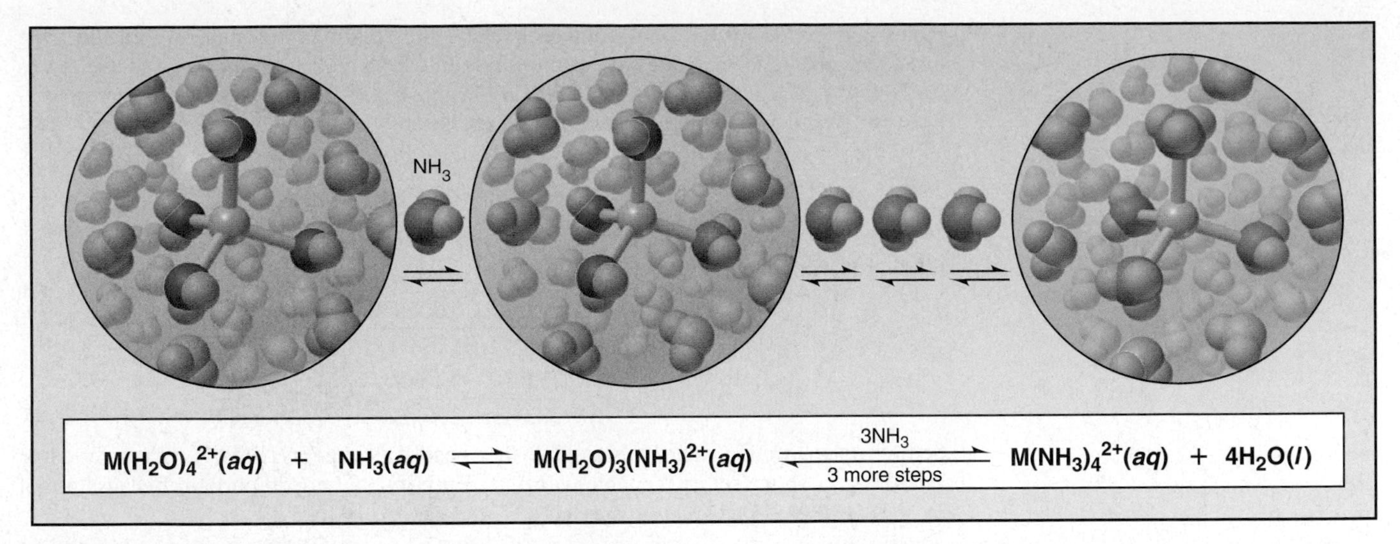

Figure 19.15 The stepwise exchange of NH_3 for H_2O in $M(H_2O)_4^{2+}$. The ligands of a complex ion can exchange for other ligands. When ammonia is added to a solution of the hydrated M^{2+} ion, $M(H_2O)_4^{2+}$, NH_3 molecules replace the bound H_2O molecules one at a time to form the $M(NH_3)_4^{2+}$ ion. The molecular-scale views show the first exchange and the fully ammoniated ion.

$$M(H_2O)_3(NH_3)^{2+}(aq) + NH_3(aq) \rightleftharpoons M(H_2O)_2(NH_3)_2^{2+}(aq) + H_2O(l)$$

$$K_{f2} = \frac{[M(H_2O)_2(NH_3)_2^{2+}]}{[M(H_2O)_3(NH_3)^{2+}][NH_3]}$$

$$M(H_2O)_2(NH_3)_2^{2+}(aq) + NH_3(aq) \rightleftharpoons M(H_2O)(NH_3)_3^{2+}(aq) + H_2O(l)$$

$$K_{f3} = \frac{[M(H_2O)(NH_3)_3^{2+}]}{[M(H_2O)_2(NH_3)_2^{2+}][NH_3]}$$

$$M(H_2O)(NH_3)_3^{2+}(aq) + NH_3(aq) \rightleftharpoons M(NH_3)_4^{2+}(aq) + H_2O(l)$$

$$K_{f4} = \frac{[M(NH_3)_4^{2+}]}{[M(H_2O)(NH_3)_3^{2+}][NH_3]}$$

The *sum* of the equations gives the overall equation, so the *product* of the individual formation constants gives the overall formation constant:

$$K_f = K_{f1} \times K_{f2} \times K_{f3} \times K_{f4}$$

Figure 19.15 summarizes the process on the molecular level. In this case, the K_f for each step is much larger than 1 because ammonia is a stronger Lewis base than water. Therefore, if we add excess ammonia to the $M(H_2O)_4^{2+}$ solution, the H_2O ligands are replaced and essentially all the M^{2+} ion exists as $M(NH_3)_4^{2+}$.

Table 19.4 (and Appendix C) shows the formation constants of some complex ions. Notice that the K_f values are all 10^6 or greater, which means that these ions form readily. Because of this behavior, complex-ion formation is used to retrieve a metal from its ore, eliminate a toxic or unwanted metal ion from a solution, or convert a metal ion to a different form, as Sample Problem 19.10 shows for the zinc ion.

Table 19.4 Formation Constants (K_f) of Some Complex Ions at 25°C

Complex Ion	K_f
$Ag(CN)_2^-$	3.0×10^{20}
$Ag(NH_3)_2^+$	1.7×10^{7}
$Ag(S_2O_3)_2^{3-}$	4.7×10^{13}
AlF_6^{3-}	$4\ \times10^{19}$
$Al(OH)_4^-$	$3\ \times10^{33}$
$Be(OH)_4^{2-}$	$4\ \times10^{18}$
CdI_4^{2-}	$1\ \times10^{6}$
$Co(OH)_4^{2-}$	$5\ \times10^{9}$
$Cr(OH)_4^-$	8.0×10^{29}
$Cu(NH_3)_4^{2+}$	5.6×10^{11}
$Fe(CN)_6^{4-}$	$3\ \times10^{35}$
$Fe(CN)_6^{3-}$	4.0×10^{43}
$Hg(CN)_4^{2-}$	9.3×10^{38}
$Ni(NH_3)_6^{2+}$	2.0×10^{8}
$Pb(OH)_3^-$	$8\ \times10^{13}$
$Sn(OH)_3^-$	$3\ \times10^{25}$
$Zn(CN)_4^{2-}$	4.2×10^{19}
$Zn(NH_3)_4^{2+}$	7.8×10^{8}
$Zn(OH)_4^{2-}$	$3\ \times10^{15}$

SAMPLE PROBLEM 19.10 Calculating the Concentration of a Complex Ion

Problem An industrial chemist converts $Zn(H_2O)_4^{2+}$ to the more stable $Zn(NH_3)_4^{2+}$ by mixing 50.0 L of 0.0020 *M* $Zn(H_2O)_4^{2+}$ and 25.0 L of 0.15 *M* NH_3. What is the final $[Zn(H_2O)_4^{2+}]$? K_f of $Zn(NH_3)_4^{2+}$ is 7.8×10^8.

Plan We write the equation and the K_f expression and use a reaction table to calculate the equilibrium concentrations. To set up the table, we must first find $[Zn(H_2O)_4^{2+}]_{init}$ and $[NH_3]_{init}$. We are given the individual volumes and molar concentrations, so we find the moles and divide by the *total* volume, because the solutions are mixed. With the large excess of NH_3 and high K_f, we assume that almost all the $Zn(H_2O)_4^{2+}$ is converted to $Zn(NH_3)_4^{2+}$. Because $[Zn(H_2O)_4^{2+}]$ at equilibrium is very small, we use x to represent it.

Solution Writing the equation and the K_f expression:

$$Zn(H_2O)_4^{2+}(aq) + 4NH_3(aq) \rightleftharpoons Zn(NH_3)_4^{2+}(aq) + 4H_2O(l)$$

$$K_f = \frac{[Zn(NH_3)_4^{2+}]}{[Zn(H_2O)_4^{2+}][NH_3]^4}$$

Finding the initial reactant concentrations:

$$[Zn(H_2O)_4^{2+}]_{init} = \frac{50.0\ L \times 0.0020\ M}{50.0\ L + 25.0\ L} = 1.3\times10^{-3}\ M$$

$$[NH_3]_{init} = \frac{25.0\ L \times 0.15\ M}{50.0\ L + 25.0\ L} = 5.0\times10^{-2}\ M$$

Setting up a reaction table: We assume that nearly all the $Zn(H_2O)_4^{2+}$ is converted to $Zn(NH_3)_4^{2+}$, so we set up the table with $x = [Zn(H_2O)_4^{2+}]$ at equilibrium. Because 4 mol of NH_3 is needed per mole of $Zn(H_2O)_4^{2+}$, the change in $[NH_3]$ is

$$[NH_3]_{reacted} \approx 4(1.3\times10^{-3}\ M) = 5.2\times10^{-3}\ M$$

and

$$[Zn(NH_3)_4^{2+}] \approx 1.3\times10^{-3}\ M$$

Concentration (M)	$Zn(H_2O)_4^{2+}(aq)$	+	$4NH_3(aq)$	$\rightleftharpoons$	$Zn(NH_3)_4^{2+}(aq)$	+	$4H_2O(l)$
Initial	1.3×10^{-3}		5.0×10^{-2}		0		—
Change	$\sim(-1.3\times10^{-3})$		$\sim(-5.2\times10^{-3})$		$\sim(+1.3\times10^{-3})$		—
Equilibrium	x		4.5×10^{-2}		1.3×10^{-3}		—

Solving for x, the $[Zn(H_2O)_4^{2+}]$ remaining at equilibrium:

$$K_f = \frac{[Zn(NH_3)_4^{2+}]}{[Zn(H_2O)_4^{2+}][NH_3]^4} = 7.8\times10^8 \approx \frac{1.3\times10^{-3}}{x(4.5\times10^{-2})^4}$$

$$x = [Zn(H_2O)_4^{2+}] \approx 4.1\times10^{-7}\ M$$

Check The K_f is large, so we expect the $[Zn(H_2O)_4^{2+}]$ remaining to be very low.

FOLLOW-UP PROBLEM 19.10 Cyanide ion is toxic because it forms stable complex ions with the Fe^{3+} ion in certain iron-containing proteins engaged in energy production. To study this effect, a biochemist mixes 25.5 mL of $3.1\times10^{-2}\ M$ $Fe(H_2O)_6^{3+}$ with 35.0 mL of 1.5 M NaCN. What is the final $[Fe(H_2O)_6^{3+}]$? K_f of $Fe(CN)_6^{3-}$ is 4.0×10^{43}.

Complex Ions and the Solubility of Precipitates

In Section 19.3, you saw that H_3O^+ increases the solubility of a slightly soluble ionic compound if its anion is that of a weak acid. Similarly, *a ligand increases the solubility of a slightly soluble ionic compound if it forms a complex ion with the cation.* For example, zinc sulfide is very slightly soluble:

$$ZnS(s) + H_2O(l) \rightleftharpoons Zn^{2+}(aq) + HS^-(aq) + OH^-(aq) \qquad K_{sp} = 2.0\times10^{-22}$$

When we add some 1.0 M NaCN, the CN^- ions act as ligands and react with the small amount of $Zn^{2+}(aq)$ to form the complex ion $Zn(CN)_4^{2-}$:

$$Zn^{2+}(aq) + 4CN^-(aq) \rightleftharpoons Zn(CN)_4^{2-}(aq) \qquad K_f = 4.2\times10^{19}$$

To see the effect of complex ion formation on the solubility of ZnS, we add the equations and, therefore, multiply their equilibrium constants:

$$ZnS(s) + 4CN^-(aq) + H_2O(l) \rightleftharpoons Zn(CN)_4^{2-}(aq) + HS^-(aq) + OH^-(aq)$$

$$K_{overall} = K_{sp} \times K_f = (2.0\times10^{-22})(4.2\times10^{19}) = 8.4\times10^{-3}$$

The overall equilibrium constant increased by more than a factor of 10^{19} in the presence of the ligand; this reflects the increased amount of ZnS in solution.

SAMPLE PROBLEM 19.11 Calculating the Effect of Complex-Ion Formation on Solubility

Problem In black-and-white film developing (see photo), excess AgBr is removed from the film negative by "hypo," an aqueous solution of sodium thiosulfate ($Na_2S_2O_3$), which forms the complex ion $Ag(S_2O_3)_2^{3-}$. Calculate the solubility of AgBr in **(a)** H_2O; **(b)** 1.0 *M* hypo. K_f of $Ag(S_2O_3)_2^{3-}$ is 4.7×10^{13} and K_{sp} of AgBr is 5.0×10^{-13}.

Developing the image in "hypo."

Plan (a) After writing the equation and the ion-product expression, we use the given K_{sp} to solve for S, the molar solubility of AgBr. **(b)** In hypo, Ag^+ forms a complex ion with $S_2O_3^{2-}$, which shifts the equilibrium and dissolves more AgBr. We write the complex-ion equation and add it to the equation for dissolving AgBr to obtain the overall equation for dissolving AgBr in hypo. We multiply K_{sp} by K_f to find $K_{overall}$. To find the solubility of AgBr in hypo, we set up a reaction table, with $S = [Ag(S_2O_3)_2^{3-}]$, substitute into the expression for $K_{overall}$, and solve for S.

Solution (a) Solubility in water. Writing the equation for the saturated solution and the ion-product expression:

$$AgBr(s) \rightleftharpoons Ag^+(aq) + Br^-(aq) \qquad K_{sp} = [Ag^+][Br^-]$$

Solving for solubility (S) directly from the equation: We know that

$$S = [AgBr]_{dissolved} = [Ag^+] = [Br^-]$$

Thus,

$$K_{sp} = [Ag^+][Br^-] = S^2 = 5.0\times10^{-13}$$

so

$$S = 7.1\times10^{-7}\ M$$

(b) Solubility in 1.0 *M* hypo. Writing the overall equation:

$$AgBr(s) \rightleftharpoons \cancel{Ag^+(aq)} + Br^-(aq)$$
$$\cancel{Ag^+(aq)} + 2S_2O_3^{2-}(aq) \rightleftharpoons Ag(S_2O_3)_2^{3-}(aq)$$
$$AgBr(s) + 2S_2O_3^{2-}(aq) \rightleftharpoons Ag(S_2O_3)_2^{3-}(aq) + Br^-(aq)$$

Calculating $K_{overall}$:

$$K_{overall} = \frac{[Ag(S_2O_3)_2^{3-}][Br^-]}{[S_2O_3^{2-}]^2} = K_{sp} \times K_f = (5.0\times10^{-13})(4.7\times10^{13}) = 24$$

Setting up a reaction table, with $S = [AgBr]_{dissolved} = [Ag(S_2O_3)_2^{3-}]$:

Concentration (*M*)	AgBr(*s*)	+	$2S_2O_3^{2-}(aq)$	$\rightleftharpoons$	$Ag(S_2O_3)_2^{3-}(aq)$	+	$Br^-(aq)$
Initial	—		1.0		0		0
Change	—		$-2S$		$+S$		$+S$
Equilibrium	—		$1.0 - 2S$		S		S

Substituting the values into the expression for $K_{overall}$ and solving for S:

$$K_{overall} = \frac{[Ag(S_2O_3)_2^{3-}][Br^-]}{[S_2O_3^{2-}]^2} = \frac{S^2}{(1.0\ M - 2S)^2} = 24$$

Taking the square root of both sides gives

$$\frac{S}{1.0\ M - 2S} = \sqrt{24} = 4.9 \qquad [Ag(S_2O_3)_2^{3-}] = S = 0.45\ M$$

Check (a) From the number of ions in the formula of AgBr, we know that $S = \sqrt{K_{sp}}$, so the order of magnitude seems right: $\sim\sqrt{10^{-14}} = 10^{-7}$. **(b)** The $K_{overall}$ seems correct: the exponents cancel, and $5 \times 5 = 25$. Most importantly, the answer makes sense because the photographic process requires the remaining AgBr to be washed off the film and the large $K_{overall}$ confirms that. We can check S by rounding and working backward to find $K_{overall}$: from the reaction table, we find that

$$[(S_2O_3)^{2-}] = 1.0\ M - 2S = 1.0\ M - 2(0.45\ M) = 1.0\ M - 0.90\ M = 0.1\ M$$

so $K_{overall} \approx (0.45)^2/(0.1)^2 = 20$, within rounding of the calculated value.

FOLLOW-UP PROBLEM 19.11 How does the solubility of AgBr in 1.0 *M* NH_3 compare with its solubility in hypo? K_f of $Ag(NH_3)_2^+$ is 1.7×10^7.

Complex Ions of Amphoteric Hydroxides

Many of the same metals that form amphoteric oxides (Chapter 8, p. 315, and Interchapter Topic 4, p. 547) also form slightly soluble *amphoteric hydroxides*. These compounds dissolve very little in water, but they dissolve to a much greater extent in both acidic and basic solutions. Aluminum hydroxide is one of several examples:

$$Al(OH)_3(s) \rightleftharpoons Al^{3+}(aq) + 3OH^-(aq)$$

It is insoluble in water ($K_{sp} = 3\times10^{-34}$), but

- It dissolves in acid because H_3O^+ reacts with the OH^- anion (Section 19.3),
$$3H_3O^+(aq) + 3OH^-(aq) \longrightarrow 6H_2O(l)$$
giving the overall equation
$$Al(OH)_3(s) + 3H_3O^+(aq) \longrightarrow Al^{3+}(aq) + 6H_2O(l)$$
- It dissolves in base through the formation of a complex ion:
$$Al(OH)_3(s) + OH^-(aq) \longrightarrow Al(OH)_4^-(aq)$$

Figure 19.16 illustrates this amphoteric behavior but includes some unexpected formulas for the species. Let's see how these species arise. When we dissolve a soluble aluminum salt, such as $Al(NO_3)_3$, in water and then slowly add a strong base, a white precipitate first forms and then dissolves as more base is added.

What reactions are occurring? The formula for the hydrated Al^{3+} ion is $Al(H_2O)_6^{3+}(aq)$. It acts as a weak polyprotic acid and reacts with added OH^- ions in a stepwise manner. In each step, one of the bound H_2O molecules loses a proton and becomes a bound OH^- ion, so the number of bound H_2O molecules is reduced by 1:

$$Al(H_2O)_6^{3+}(aq) + OH^-(aq) \rightleftharpoons Al(H_2O)_5OH^{2+}(aq) + H_2O(l)$$
$$Al(H_2O)_5OH^{2+}(aq) + OH^-(aq) \rightleftharpoons Al(H_2O)_4(OH)_2^+(aq) + H_2O(l)$$
$$Al(H_2O)_4(OH)_2^+(aq) + OH^-(aq) \rightleftharpoons Al(H_2O)_3(OH)_3(s) + H_2O(l)$$

After three protons have been removed from each $Al(H_2O)_6^{3+}$, the white precipitate has formed, which is the insoluble hydroxide $Al(H_2O)_3(OH)_3(s)$, often written more simply as $Al(OH)_3(s)$. Now you can see that the precipitate actually consists of the hydrated Al^{3+} ion with an H^+ removed from each of three bound H_2O molecules. Addition of H_3O^+ protonates the OH^- ions and re-forms the hydrated Al^{3+} ion.

Further addition of OH^- removes a fourth H^+ and the precipitate dissolves as the soluble ion $Al(H_2O)_2(OH)_4^-(aq)$ forms, which we usually write with the formula $Al(OH)_4^-(aq)$:

$$Al(H_2O)_3(OH)_3(s) + OH^-(aq) \rightleftharpoons Al(H_2O)_2(OH)_4^-(aq) + H_2O(l)$$

In other words, this complex ion is not created by ligands substituting for bound water molecules but through an acid-base reaction in which added OH^- ions titrate bound water molecules.

Several other slightly soluble hydroxides, including those of cadmium, chromium(III), cobalt(III), lead(II), tin(II), and zinc, are amphoteric and exhibit similar reactions:

$$Zn(H_2O)_2(OH)_2(s) + OH^-(aq) \rightleftharpoons Zn(H_2O)(OH)_3^-(aq) + H_2O(l)$$

In contrast to the preceding hydroxides, the slightly soluble hydroxides of iron(II), iron(III), and calcium dissolve in acid, but do *not* dissolve in base, because the

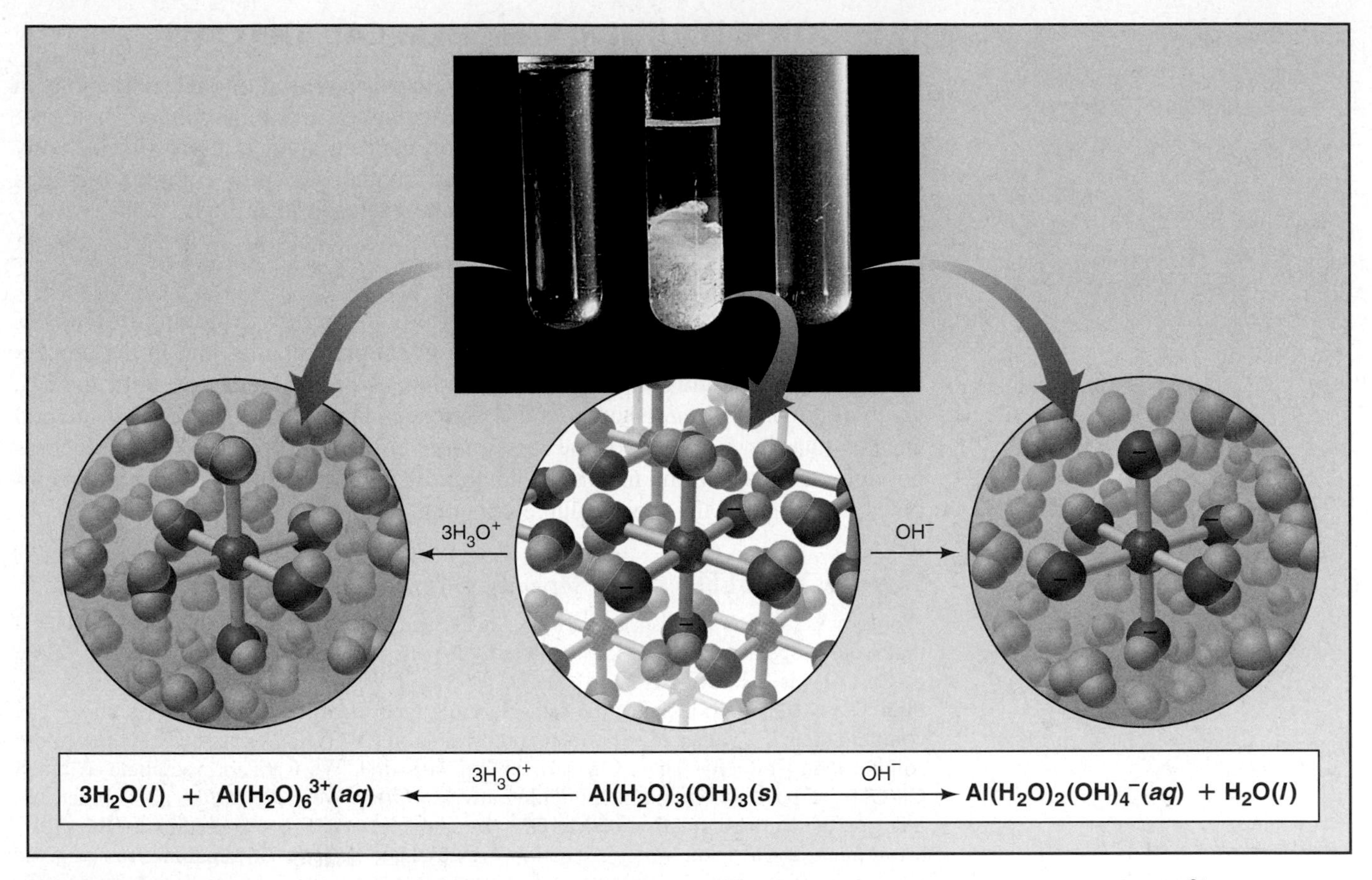

Figure 19.16 The amphoteric behavior of aluminum hydroxide. When solid $Al(OH)_3$ is treated with H_3O^+ (*left*) or with OH^- (*right*), it dissolves as a result of the formation of soluble complex ions. The molecular views show that $Al(OH)_3$ is actually the hydrated species $Al(H_2O)_3(OH)_3$. (The extensive OH bridging that occurs between Al^{3+} ions throughout the solid is not shown.) Addition of OH^- (*right*) forms the soluble $Al(H_2O)_2(OH)_4^-$ ion; addition of H_3O^+ (*left*) forms the soluble $Al(H_2O)_6^{3+}$ ion.

three remaining bound water molecules are not acidic enough to lose any of their protons:

$$Fe(H_2O)_3(OH)_3(s) + 3H_3O^+(aq) \longrightarrow Fe(H_2O)_6^{3+}(aq) + 3H_2O(l)$$
$$Fe(H_2O)_3(OH)_3(s) + OH^-(aq) \longrightarrow \text{no reaction}$$

This difference in solubility in base between $Al(OH)_3$ and $Fe(OH)_3$ is the key to an important separation step in the production of aluminum metal, so we'll consider it again in Section 22.4. It is also employed in analyzing a mixture of ions, a process introduced briefly in the next section.

SECTION SUMMARY

A complex ion consists of a central metal ion covalently bonded to two or more negatively charged or neutral ligands. Its formation is described by a formation constant, K_f. A hydrated metal ion is a complex ion with water molecules as ligands. Other ligands can displace the water in a stepwise process. In most cases, the K_f value of each step is large, so the fully substituted complex ion forms almost completely in the presence of excess ligand. Adding a solution containing a ligand increases the solubility of an ionic precipitate if the cation forms a complex ion with the ligand. Amphoteric metal hydroxides dissolve in acid and base due to acid-base reactions that form soluble complex ions.

19.5 IONIC EQUILIBRIA IN CHEMICAL ANALYSIS

Many of the ideas we've discussed in this chapter are used to analyze the ions in a mixture. In this brief introduction to an extensive and time-honored field, we discuss how control of the precipitating-ion concentration is used to selectively precipitate one metal ion in the presence of another and how complex mixtures of ions are separated into smaller groups and each ion identified.

Selective Precipitation

We can often select one ion in a solution from another by exploiting differences in the solubility of their compounds with a given precipitating ion. In the process of **selective precipitation,** we add a solution of precipitating ion until the Q_{sp} value of the *more soluble* compound is almost equal to its K_{sp} value. This method ensures that the K_{sp} value of the *less soluble* compound is exceeded as much as possible. As a result, the maximum amount of the less soluble compound precipitates, but none of the more soluble compound does.

SAMPLE PROBLEM 19.12 Separating Ions by Selective Precipitation

Problem A solution consists of 0.20 M $MgCl_2$ and 0.10 M $CuCl_2$. Calculate the $[OH^-]$ that would separate the metal ions as their hydroxides. K_{sp} of $Mg(OH)_2$ is 6.3×10^{-10}; K_{sp} of $Cu(OH)_2$ is 2.2×10^{-20}.

Plan The two hydroxides have the same formula type (1:2) (see Section 19.3), so we can compare their K_{sp} values to see which is more soluble: $Mg(OH)_2$ is about 10^{10} times more soluble than $Cu(OH)_2$. Thus, $Cu(OH)_2$ precipitates first. We want to precipitate as much $Cu(OH)_2$ as possible without precipitating any $Mg(OH)_2$. We solve for the $[OH^-]$ that will just give a saturated solution of $Mg(OH)_2$ because this $[OH^-]$ will precipitate the greatest amount of Cu^{2+} ion. As confirmation, we calculate the $[Cu^{2+}]$ remaining to see if the separation was accomplished.

Solution Writing the equations and ion-product expressions:

$$Mg(OH)_2(s) \rightleftharpoons Mg^{2+}(aq) + 2OH^-(aq) \qquad K_{sp} = [Mg^{2+}][OH^-]^2$$
$$Cu(OH)_2(s) \rightleftharpoons Cu^{2+}(aq) + 2OH^-(aq) \qquad K_{sp} = [Cu^{2+}][OH^-]^2$$

Calculating the $[OH^-]$ that gives a saturated $Mg(OH)_2$ solution:

$$[OH^-] = \sqrt{\frac{K_{sp}}{[Mg^{2+}]}} = \sqrt{\frac{6.3\times10^{-10}}{0.20}} = 5.6\times10^{-5}\ M$$

This is the maximum $[OH^-]$ that will *not* precipitate Mg^{2+} ion.
Calculating the $[Cu^{2+}]$ remaining in the solution with this $[OH^-]$:

$$[Cu^{2+}] = \frac{K_{sp}}{[OH^-]^2} = \frac{2.2\times10^{-20}}{(5.6\times10^{-5})^2} = 7.0\times10^{-12}\ M$$

Since the initial $[Cu^{2+}]$ was 0.10 M, virtually all the Cu^{2+} ion is precipitated.

Check Rounding, we find that the $[OH^-]$ seems right: $\sim\sqrt{(6\times10^{-10})/0.2} = 5\times10^{-5}$. The $[Cu^{2+}]$ remaining also seems correct: $(200\times10^{-22})/(5\times10^{-5})^2 = 8\times10^{-12}$.

Comment Often, more than one approach will accomplish the same analytical step. Another possibility in this case is to add excess ammonia to make the solution basic enough to precipitate both hydroxides:

$$Mg^{2+}(aq) + 2NH_3(aq) + 2H_2O(l) \rightleftharpoons Mg(OH)_2(s) + 2NH_4^+(aq)$$
$$Cu^{2+}(aq) + 2NH_3(aq) + 2H_2O(l) \rightleftharpoons Cu(OH)_2(s) + 2NH_4^+(aq)$$

Then, the $Cu(OH)_2$ dissolves in excess NH_3 by forming a soluble complex ion:

$$Cu(OH)_2(s) + 4NH_3(aq) \rightleftharpoons Cu(NH_3)_4^{2+}(aq) + 2OH^-(aq)$$

FOLLOW-UP PROBLEM 19.12 A solution containing two alkaline earth metal ions is made from 0.050 M $BaCl_2$ and 0.025 M $CaCl_2$. What concentration of SO_4^{2-} must be present to leave 99.99% of only one of the cations in solution? K_{sp} of $BaSO_4$ is 1.1×10^{-10}, and K_{sp} of $CaSO_4$ is 2.4×10^{-5}.

Sometimes two or more types of ionic equilibria are controlled simultaneously to precipitate ions selectively. This approach is used commonly to separate ions as their sulfides, so the HS^- ion is the precipitating ion. In the manner used in Sample Problem 19.12, we control the $[HS^-]$ to exceed the K_{sp} value of one metal sulfide but not another. We exert this control on the $[HS^-]$ by controlling H_2S dissociation because H_2S is the source of HS^-; we control H_2S dissociation by adjusting the $[H_3O^+]$:

$$H_2S(aq) + H_2O(l) \rightleftharpoons H_3O^+(aq) + HS^-(aq)$$

These interactions are controlled in the following way: If we add strong acid to the solution, $[H_3O^+]$ is high, so H_2S dissociation shifts to the left, which decreases $[HS^-]$. With a low $[HS^-]$, the *less* soluble sulfide precipitates. Conversely, if we add strong base, $[H_3O^+]$ is low, so H_2S dissociation shifts to the right, which increases $[HS^-]$, and then the *more* soluble sulfide precipitates. In essence, what we have done is shift one equilibrium system (H_2S dissociation) by adjusting a second (H_2O ionization) to control a third (metal sulfide solubility).

Qualitative Analysis: Identifying Ions in Complex Mixtures

The practice of inorganic **qualitative analysis,** the separation and identification of the ions in a mixture, was once an essential part of a chemist's skills. Today, many of these wet chemical techniques have been replaced by instrumental methods. Nevertheless, they provide an excellent means for studying ionic equilibria, including some of the very ones utilized by those modern instruments. In this brief discussion, we apply solubility and complex-ion equilibria to separate and characterize a mixture of cations; similar procedures exist for anions.

Separation into Ion Groups The general approach begins by separating the unknown solution into *ion groups* (Figure 19.17). (Ion groups have *nothing* to do with periodic table groups.) The mixture of metal ions is treated with a solution that precipitates a certain group of them and leaves the others in solution. Filtration or centrifugation (the rapid spinning of the tube to collect the solid in a compact pellet at the bottom) separates the compounds containing the precipitated metal ions. The solution containing the remaining ions is decanted (poured off) and treated with a solution that precipitates a different group of ions. These steps are repeated until the original mixture has been separated into specific ion groups. In an actual analysis, a *known solution* (one that contains all the ions under study) and a *blank* of distilled water are treated in exactly the same way as the unknown solution, making it much easier to judge a positive or negative result.

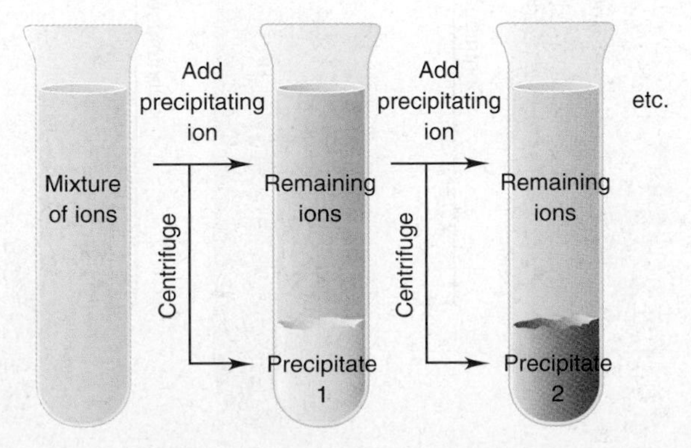

Figure 19.17 The general procedure for separating ions in qualitative analysis. A precipitating ion is added to a mixture of ions, the precipitate is separated by centrifugation, and the remaining dissolved ions are treated with another precipitating ion. The process is repeated until the ions are separated into ion groups.

Figure 19.18 shows a common scheme for separating cations into ion groups.

Ion group 1: Insoluble chlorides. The entire mixture of soluble ions is treated with 6 *M* HCl. Most metal chlorides are soluble, so only those few ions that form insoluble chlorides precipitate in this step. If a precipitate appears, it consists of chlorides of any of the following:

$$Ag^+, Hg_2^{2+}, Pb^{2+}$$

If none appears, no ions from ion group 1 are present in the mixture. If a precipitate forms, the tube is centrifuged, and the solution is carefully decanted.

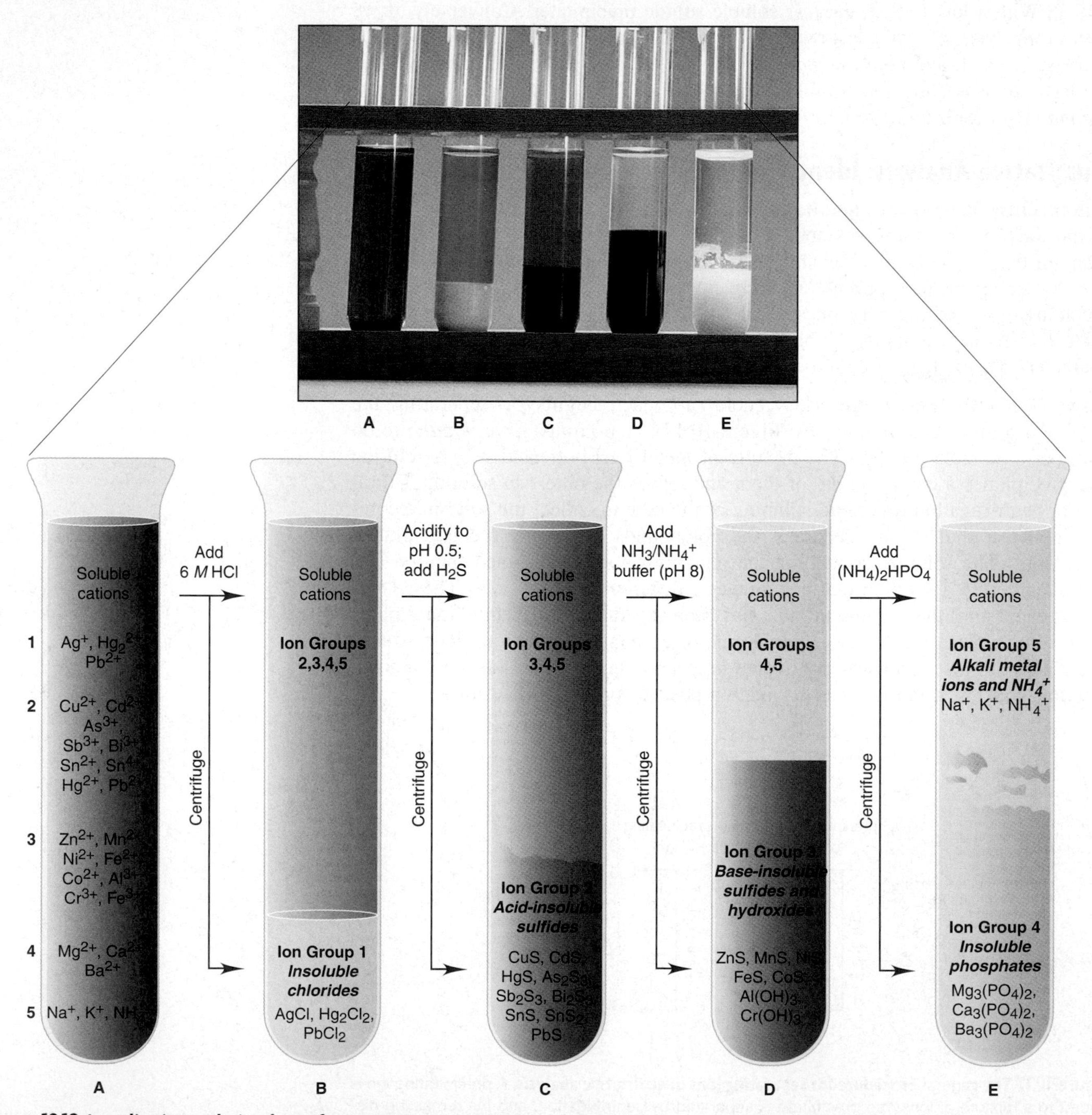

Figure 19.18 A qualitative analysis scheme for separating cations into five ion groups. The first tube contains a solution of ions (listed by ion group). It is treated with the first precipitating solution (6 *M* HCl), and the procedure continues, as shown in Figure 19.17.

Ion group 2: Acid-insoluble sulfides. The decanted solution is already acidic from the previous treatment with HCl. The pH is adjusted to 0.5 and the solution treated with aqueous H_2S. The H_3O^+ present keeps the $[HS^-]$ very low, so any precipitate contains one or more of the least soluble sulfides, which are those of

$$Cu^{2+}, Cd^{2+}, As^{3+}, Sb^{3+}, Bi^{3+}, Sn^{2+}, Sn^{4+}, Hg^{2+}, Pb^{2+}$$

($PbCl_2$ is slightly soluble in water, so a small amount of Pb^{2+} remains in solution after addition of HCl and appears in ion groups 1 *and* 2.) If no precipitate forms under these conditions, no members of ion group 2 are present. The tube is centrifuged and the solution decanted.

Ion group 3: Base-insoluble sulfides and hydroxides. The decanted solution is made slightly basic with an NH_4^+-NH_3 buffer. The OH^- present increases the $[HS^-]$, which causes precipitation of the more soluble sulfides and some hydroxides. The cations included are

$$Zn^{2+}, Mn^{2+}, Ni^{2+}, Fe^{2+}, Co^{2+} \text{ as sulfides, and } Al^{3+}, Cr^{3+} \text{ as hydroxides}$$

(Fe^{3+} is reduced to Fe^{2+} in this step.) If no precipitate appears, none of these ions is present. Centrifuging and decanting gives the next solution.

Ion group 4: Insoluble phosphates. To the slightly basic solution, $(NH_4)_2HPO_4$ is added, which precipitates any alkaline earth ions as phosphates. Alternatively, Na_2CO_3 is added, and the precipitate contains alkaline earth carbonates. In either case, any of the following ions are precipitated:

$$Mg^{2+}, Ca^{2+}, Ba^{2+}$$

If no precipitate appears, none of these ions is present.

Ion group 5: Alkali metal and ammonium ions. After centrifuging and decanting, the final solution contains any of the following ions:

$$Na^+, K^+, NH_4^+$$

With the ions separated into ion groups, the analyst then devises schemes to identify various ions in the groups. For example, to separate a mixture of Ag^+ and Cu^{2+} ions, we note that Cl^- can act as both a ligand and a precipitating ion. We add dilute NaCl to the solution and centrifuge, and the Cl^- ion forms a white precipitate with $Ag^+(aq)$ and a green complex ion with $Cu^{2+}(aq)$:

$$Ag^+(aq) + Cl^-(aq;\text{ added}) \longrightarrow AgCl(s) \text{ [white]}$$
$$Cu^{2+}(aq) + 4Cl^-(aq;\text{ added}) \longrightarrow CuCl_4^{2-}(aq) \text{ [green]}$$

As another example, the ions in ion group 5 are identified through flame and color tests (Figure 19.19). (Because NH_4^+ ion is added during earlier steps and Na^+ is a common contaminant in ammonium salts, the analyst performs tests for these ions on the original ion mixture.) In flame tests, sodium has a characteristic yellow-orange color, and potassium has a violet color. Acid-base behavior is used to identify NH_4^+. The solution is made basic with NaOH and warmed gently; then moist red litmus paper is held over it. If the paper turns blue, NH_4^+ is present because the OH^- reacts to form NH_3, which reacts with the H_2O on the paper:

$$NH_4^+(aq) + OH^-(aq;\text{ added}) \longrightarrow NH_3(g) + H_2O(l)$$
$$NH_3(g) + H_2O\text{ (on paper)} \rightleftharpoons NH_4^+(aq) + OH^- \text{ (turns red litmus blue)}$$

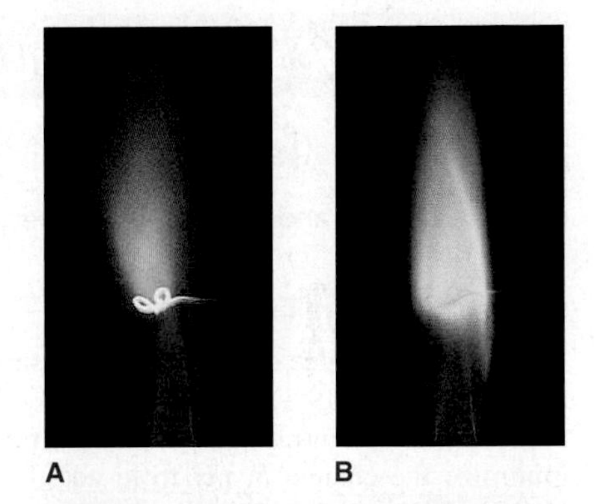

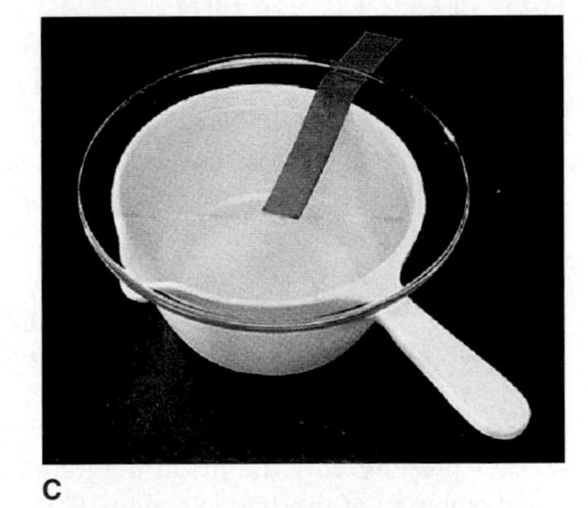

Figure 19.19 Tests to determine the presence of cations in ion group 5. Ion group 5 is soluble through all the steps in Figure 19.18. It consists of some alkali metal ions and NH_4^+ ion. Flame tests give characteristic colors for Na^+ ion **(A)** and K^+ ion **(B)**. Adding OH^- to NH_4^+ and warming the solution forms gaseous NH_3, which turns moistened red litmus paper blue **(C)**.

SECTION SUMMARY

Ions are precipitated selectively by adding a precipitating ion until the K_{sp} of one compound is exceeded as much as possible without exceeding the K_{sp} of the other. An extension of this approach is to control the equilibrium of the slightly soluble compound by simultaneously controlling an equilibrium system that contains the precipitating ion. Qualitative analysis of ion mixtures involves adding precipitating ions to separate the unknown ions into ion groups. The groups are then analyzed further through precipitation and complex-ion formation.

Chapter Perspective

This chapter is the last of three that explore the nature and variety of equilibrium systems. In Chapter 17, we discussed the central ideas of equilibrium in the context of gaseous systems. In Chapter 18, we extended our understanding to acid-base equilibria. In this chapter, we highlighted three types of aqueous ionic systems and examined their role in the laboratory and the environment. The equilibrium constant, in all its forms, is a number that provides a limit to changes in a system, whether a chemical reaction, a physical change, or the dissolution of a substance. You now have the skills to predict whether a change will take place and to calculate its result, but you still do not know why the change occurs in the first place, or why it stops when it does. In Chapter 20, you'll find out.

For Review and Reference (Numbers in parentheses refer to pages, unless noted otherwise.)

Learning Objectives

Relevant section and/or sample problem (SP) numbers appear in parentheses.

Understand These Concepts

1. How the presence of a common ion suppresses a reaction that forms it (Section 19.1)
2. Why the concentrations of buffer components must be high to minimize the change in pH from addition of small amounts of H_3O^+ or OH^- (Section 19.1)
3. How buffer capacity depends on buffer concentration and on the pK_a of the acid component; why buffer range is within ± 1 pH unit of the pK_a (Section 19.1)
4. The nature of an acid-base indicator as a conjugate acid-base pair with differently colored acidic and basic forms (Section 19.2)
5. The distinction between equivalence point and end point in an acid-base titration (Section 19.2)
6. Why the shapes of strong acid–strong base, weak acid–strong base, and strong acid–weak base titration curves differ (Section 19.2)
7. How the pH at the equivalence point is determined by the species present; why the pH at the midpoint of the buffer region equals the pK_a of the acid (Section 19.2)
8. How the titration curve of a polyprotic acid has a buffer region and equivalence point for each ionizable proton (Section 19.2)
9. How a slightly soluble ionic compound reaches equilibrium in water, expressed by an equilibrium (solubility-product) constant, K_{sp} (Section 19.3)
10. Why incomplete dissociation of an ionic compound leads to approximate calculated values for K_{sp} and solubility (Section 19.3)
11. Why a common ion in a solution decreases the solubility of its compounds (Section 19.3)
12. How pH affects the solubility of a compound that contains a weak-acid anion (Section 19.3)
13. How precipitate formation depends on the relative values of Q_{sp} and K_{sp} (Section 19.3)
14. How complex-ion formation is stepwise and expressed by an overall equilibrium (formation) constant, K_f (Section 19.4)
15. Why addition of a ligand increases the solubility of a compound whose metal ion forms a complex ion (Section 19.4)
16. How the aqueous chemistry of amphoteric hydroxides involves precipitation, complex-ion formation, and acid-base equilibria (Section 19.4)
17. How selective precipitation and simultaneous equilibria are used to separate ions (Section 19.5)
18. How qualitative analysis is used to separate and identify the ions in a mixture (Section 19.5)

Master These Skills

1. Using stoichiometry and equilibrium problem-solving techniques to calculate the effect of added H_3O^+ or OH^- on buffer pH (SP 19.1)
2. Using the Henderson-Hasselbalch equation to calculate buffer pH (Section 19.1)
3. Choosing the components of a buffer with a given pH and calculating their quantities (SP 19.2)
4. Calculating the pH at any point in an acid-base titration (Section 19.2 and SP 19.3)
5. Choosing an appropriate indicator based on the pH at various points in a titration (Section 19.2)
6. Writing K_{sp} expressions for slightly soluble ionic compounds (SP 19.4)
7. Calculating a K_{sp} value from solubility data (SP 19.5)
8. Calculating solubility from a K_{sp} value (SP 19.6)
9. Using K_{sp} values to compare solubilities for compounds with the same total number of ions (Section 19.3)
10. Calculating the decrease in solubility caused by the presence of a common ion (SP 19.7)
11. Predicting the effect of added H_3O^+ on solubility (SP 19.8)
12. Using ion concentrations to calculate Q_{sp} and compare it with K_{sp} to predict whether a precipitate forms (SP 19.9)
13. Calculating the concentration of metal ion remaining after addition of excess ligand forms a complex ion (SP 19.10)
14. Using an overall equilibrium constant ($K_{sp} \times K_f$) to calculate the effect of complex-ion formation on solubility (SP 19.11)
15. Comparing K_{sp} values in order to separate ions by selective precipitation (SP 19.12)

Key Terms

Section 19.1
acid-base buffer (815)
common-ion effect (816)
Henderson-Hasselbalch equation (820)
buffer capacity (821)
buffer range (821)

Section 19.2
acid-base titration curve (824)
equivalence point (825)
end point (826)

Section 19.3
solubility-product constant (K_{sp}) (833)
molar solubility (834)

Section 19.4
complex ion (844)
ligand (844)
formation constant (K_f) (844)

Section 19.5
selective precipitation (850)
qualitative analysis (851)

Key Equations and Relationships

19.1 Finding the pH from known concentrations of a conjugate acid-base pair (Henderson-Hasselbalch equation) (820):

$$\text{pH} = \text{p}K_a + \log\left(\frac{[\text{base}]}{[\text{acid}]}\right)$$

19.2 Defining the equilibrium condition for a saturated solution of a slightly soluble compound, M_pX_q, composed of M^{n+} and X^{z-} ions (833):

$$Q_{sp} = [M^{n+}]^p[X^{z-}]^q = K_{sp}$$

Highlighted Figures and Tables

These figures (F) and tables (T) provide a review of key ideas.

F19.3 How a buffer works (817)
F19.4 Buffer capacity and pH change (821)
F19.5 Colors and pH ranges of acid-base indicators (824)
F19.7 A strong acid–strong base titration curve (825)
F19.8 A weak acid–strong base titration curve (827)
F19.9 A weak base–strong acid titration curve (830)

Brief Solutions to Follow-up Problems

19.1 (a) Before addition:
Assuming x is small enough to be neglected,
$[\text{HF}] = 0.50\ M$ and $[\text{F}^-] = 0.45\ M$

$$[\text{H}_3\text{O}^+] = K_a \times \frac{[\text{HF}]}{[\text{F}^-]} \approx (6.8\times10^{-4})\left(\frac{0.50}{0.45}\right) = 7.6\times10^{-4}\ M$$

$\text{pH} = 3.12$

(b) After addition of 0.40 g of NaOH (0.010 mol of NaOH) to 1.0 L of buffer,
$[\text{HF}] = 0.49\ M$ and $[\text{F}^-] = 0.46\ M$

$$[\text{H}_3\text{O}^+] \approx (6.8\times10^{-4})\left(\frac{0.49}{0.46}\right) = 7.2\times10^{-4}\ M;\ \text{pH} = 3.14$$

19.2 $[\text{H}_3\text{O}^+] = 10^{-\text{pH}} = 10^{-4.25} = 5.6\times10^{-5}$

$$[\text{C}_6\text{H}_5\text{COOH}] = \frac{[\text{H}_3\text{O}^+][\text{C}_6\text{H}_5\text{COO}^-]}{K_a} = \frac{(5.6\times10^{-5})(0.050)}{6.3\times10^{-5}} = 0.044\ M$$

Mass (g) of C_6H_5COOH

$$= 5.0\ \text{L soln} \times \frac{0.044\ \text{mol C}_6\text{H}_5\text{COOH}}{1\ \text{L soln}} \times \frac{122.12\ \text{g C}_6\text{H}_5\text{COOH}}{1\ \text{mol C}_6\text{H}_5\text{COOH}} = 27\ \text{g C}_6\text{H}_5\text{COOH}$$

Dissolve 27 g of C_6H_5COOH in 4.9 L of 0.050 M C_6H_5COONa and add solution to make 5.0 L. Adjust pH to 4.25 with strong acid or base.

19.3 (a) $[\text{H}_3\text{O}^+] \approx \sqrt{(2.3\times10^{-9})(0.2000)} = 2.1\times10^{-5}\ M$
$\text{pH} = 4.68$

(b) $[\text{H}_3\text{O}^+] = K_a \times \frac{[\text{HBrO}]}{[\text{BrO}^-]} = (2.3\times10^{-9})(1) = 2.3\times10^{-9}\ M$
$\text{pH} = 8.64$

(c) $$[\text{BrO}^-] = \frac{\text{moles of BrO}^-}{\text{total volume}} = \frac{0.004000\ \text{mol}}{0.06000\ \text{L}} = 0.06667\ M$$

$$K_b \text{ of BrO}^- = \frac{K_w}{K_a \text{ of HBrO}} = 4.3\times10^{-6}$$

$$[\text{H}_3\text{O}^+] = \frac{K_w}{\sqrt{K_b \times [\text{BrO}^-]}} \approx \frac{1.0\times10^{-14}}{\sqrt{(4.3\times10^{-6})(0.06667)}} = 1.9\times10^{-11}\ M$$

$\text{pH} = 10.72$

(d) Moles of OH^- added = 0.008000 mol
Volume (L) of OH^- soln = 0.08000 L

$$[\text{OH}^-] = \frac{\text{moles of OH}^-\ \text{unreacted}}{\text{total volume}} = \frac{0.008000\ \text{mol} - 0.004000\ \text{mol}}{(0.02000 + 0.08000)\ \text{L}} = 0.04000\ M$$

$$[\text{H}_3\text{O}^+] = \frac{K_w}{[\text{OH}^-]} = 2.5\times10^{-13}$$

$\text{pH} = 12.60$

Brief Solutions to Follow-up Problems *(continued)*

(e)

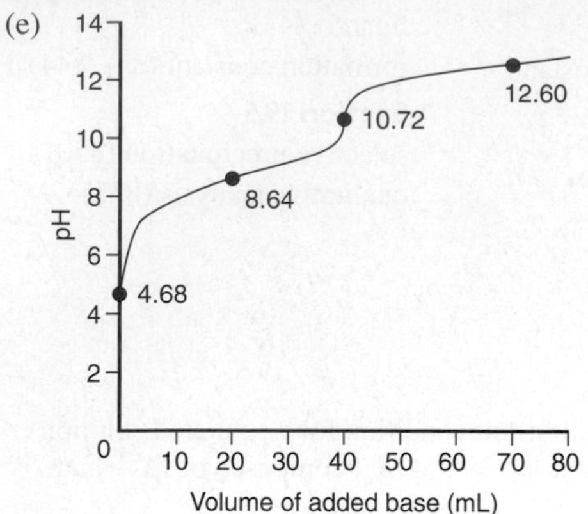

19.4 (a) $K_{sp} = [Ca^{2+}][SO_4^{2-}]$
(b) $K_{sp} = [Cr^{3+}]^2[CO_3^{2-}]^3$
(c) $K_{sp} = [Mg^{2+}][OH^-]^2$
(d) $K_{sp} = [As^{3+}]^2[HS^-]^3[OH^-]^3$

19.5 $[CaF_2] = \frac{1.5\times10^{-4}\text{ g } CaF_2}{10.0\text{ mL soln}} \times \frac{1000\text{ mL}}{1\text{ L}} \times \frac{1\text{ mol } CaF_2}{78.08\text{ g } CaF_2}$
$= 1.9\times10^{-4}\ M$
$CaF_2(s) \rightleftharpoons Ca^{2+}(aq) + 2F^-(aq)$
$[Ca^{2+}] = 1.9\times10^{-4}\ M$ and $[F^-] = 3.8\times10^{-4}\ M$
$K_{sp} = [Ca^{2+}][F^-]^2 = (1.9\times10^{-4})(3.8\times10^{-4})^2 = 2.7\times10^{-11}$

19.6 From the reaction table, $[Mg^{2+}] = S$ and $[OH^-] = 2S$
$K_{sp} = [Mg^{2+}][OH^-]^2 = 4S^3 = 6.3\times10^{-10}$; $S = 5.4\times10^{-4}\ M$

19.7 (a) In water: $K_{sp} = [Ba^{2+}][SO_4^{2-}] = S^2 = 1.1\times10^{-10}$;
$S = 1.0\times10^{-5}$
(b) In 0.10 M Na_2SO_4: $[SO_4^{2-}] = 0.10\ M$
$K_{sp} = 1.1\times10^{-10} \approx S \times 0.10$; $S = 1.1\times10^{-9}\ M$
S decreases in presence of the common ion SO_4^{2-}.

19.8 (a) Increases solubility.
$CaF_2(s) \rightleftharpoons Ca^{2+}(aq) + 2F^-(aq)$
$F^-(aq) + H_3O^+(aq) \longrightarrow HF(aq) + H_2O(l)$
(b) Increases solubility.
$ZnS(s) + H_2O(l) \rightleftharpoons Zn^{2+}(aq) + HS^-(aq) + OH^-(aq)$
$HS^-(aq) + H_3O^+(aq) \longrightarrow H_2S(aq) + H_2O(l)$
$OH^-(aq) + H_3O^+(aq) \longrightarrow 2H_2O(l)$
(c) No effect. $I^-(aq)$ is conjugate base of strong acid, HI.

19.9 $Ca_3(PO_4)_2(s) \rightleftharpoons 3Ca^{2+}(aq) + 2PO_4^{3-}(aq)$
$Q_{sp} = [Ca^{2+}]^3[PO_4^{3-}]^2 = (1.0\times10^{-9})^5 = 1.0\times10^{-45}$
$Q_{sp} < K_{sp}$, so $Ca_3(PO_4)_2$ will not precipitate.

19.10 $[Fe(H_2O)_6^{3+}]_{init} = \frac{(0.0255\text{ L})(3.1\times10^{-2}\ M)}{0.0255\text{ L} + 0.0350\text{ L}}$
$= 1.3\times10^{-2}\ M$
Similarly, $[CN^-]_{init} = 0.87\ M$. From the reaction table,
$K_f = \frac{[Fe(CN)_6^{3-}]}{[Fe(H_2O)_6^{3+}][CN^-]^6} = 4.0\times10^{43} \approx \frac{1.3\times10^{-2}}{x(0.79)^6}$
$x = [Fe(H_2O)_6^{3+}] \approx 1.3\times10^{-45}$

19.11 $AgBr(s) + 2NH_3(aq) \rightleftharpoons Ag(NH_3)_2^+(aq) + Br^-(aq)$
$K_{overall} = K_{sp}$ of AgBr $\times K_f$ of $Ag(NH_3)_2^+$
$= 8.5\times10^{-6}$
From the reaction table,
$\frac{S}{1.0 - 2S} = \sqrt{8.5\times10^{-6}} = 2.9\times10^{-3}$
$S = [Ag(NH_3)_2^+] = 2.9\times10^{-3}\ M$
Solubility is greater in 1 M hypo than in 1 M NH_3.

19.12 Both 1:1 salts, so K_{sp} values show that $CaSO_4$ is more soluble:
$[SO_4^{2-}] = \frac{K_{sp}}{[Ca^{2+}]} = \frac{2.4\times10^{-5}}{(0.025)(0.9999)} = 9.6\times10^{-4}\ M$

Problems

Problems with **colored** numbers are answered in Appendix E. Sections match the text and provide the numbers of relevant sample problems. Most offer Concept Review Questions, Skill-Building Exercises (grouped in pairs covering the same concept), and Problems in Context. Comprehensive Problems are based on material from any section or previous chapter.

Note: Unless stated otherwise, all of the problems for this chapter refer to aqueous solutions at 298 K (25°C).

Equilibria of Acid-Base Buffer Systems
(Sample Problems 19.1 and 19.2)

Concept Review Questions

19.1 What is the purpose of an acid-base buffer?

19.2 How do the acid and base components of a buffer function? Why are they often a conjugate acid-base pair of a weak acid?

19.3 What is the common-ion effect? How is it related to Le Châtelier's principle? Explain with equations that include HF and NaF.

19.4 When a small amount of H_3O^+ is added to a buffer, does the pH remain constant? Explain.

19.5 What is the difference between buffers with high and low capacities? Will adding 0.01 mol of HCl produce a greater pH change in a buffer with a high or a low capacity? Explain.

19.6 Which of these factors influence buffer capacity? How?
(a) Conjugate acid-base pair (b) pH of the buffer
(c) Concentration of buffer components (d) Buffer range
(e) pK_a of the acid component

19.7 What is the relationship between the buffer range and the buffer-component concentration ratio?

19.8 A chemist needs a pH 3.5 buffer. Should she use NaOH with formic acid ($K_a = 1.8\times10^{-4}$) or with acetic acid ($K_a = 1.8\times10^{-5}$)? Why? What is the disadvantage of choosing the other acid? What is the role of the NaOH?

19.9 State and explain the relative change in pH and in buffer-component concentration ratio, [NaA]/[HA], for each addition:
(a) Add 0.1 M NaOH to the buffer
(b) Add 0.1 M HCl to the buffer

(c) Dissolve pure NaA in the buffer
(d) Dissolve pure HA in the buffer

19.10 Does the pH increase or decrease, and does it do so to a large or small extent, with each of the following additions?
(a) 5 drops of 0.1 M NaOH to 100 mL of 0.5 M acetate buffer
(b) 5 drops of 0.1 M HCl to 100 mL of 0.5 M acetate buffer
(c) 5 drops of 0.1 M NaOH to 100 mL of 0.5 M HCl
(d) 5 drops of 0.1 M NaOH to distilled water

Skill-Building Exercises *(grouped in similar pairs)*

19.11 What are the $[H_3O^+]$ and the pH of a propanoic acid–propanoate buffer that consists of 0.25 M CH_3CH_2COONa and 0.15 M CH_3CH_2COOH (K_a of propanoic acid = 1.3×10^{-5})?

19.12 What are the $[H_3O^+]$ and the pH of a benzoic acid–benzoate buffer that consists of 0.33 M C_6H_5COOH and 0.28 M C_6H_5COONa (K_a of benzoic acid = 6.3×10^{-5})?

19.13 What are the $[H_3O^+]$ and the pH of a buffer that consists of 0.50 M HNO_2 and 0.65 M KNO_2 (K_a of HNO_2 = 7.1×10^{-4})?

19.14 What are the $[H_3O^+]$ and the pH of a buffer that consists of 0.20 M HF and 0.25 M KF (K_a of HF = 6.8×10^{-4})?

19.15 Find the pH of a buffer that consists of 0.55 M HCOOH and 0.63 M HCOONa (pK_a of HCOOH = 3.74).

19.16 Find the pH of a buffer that consists of 0.95 M HBrO and 0.68 M KBrO (pK_a of HBrO = 8.64).

19.17 Find the pH of a buffer that consists of 1.0 M sodium phenolate (C_6H_5ONa) and 1.2 M phenol (C_6H_5OH) (pK_a of phenol = 10.00).

19.18 Find the pH of a buffer that consists of 0.12 M boric acid (H_3BO_3) and 0.82 M sodium borate (NaH_2BO_3) (pK_a of boric acid = 9.24).

19.19 Find the pH of a buffer that consists of 0.20 M NH_3 and 0.10 M NH_4Cl (pK_b of NH_3 = 4.75).

19.20 Find the pH of a buffer that consists of 0.50 M methylamine (CH_3NH_2) and 0.60 M CH_3NH_3Cl (pK_b of CH_3NH_2 = 3.35).

19.21 A buffer consists of 0.25 M $KHCO_3$ and 0.32 M K_2CO_3. Carbonic acid is a diprotic acid with $K_{a1} = 4.5\times10^{-7}$ and $K_{a2} = 4.7\times10^{-11}$. (a) Which K_a value is more important to this buffer? (b) What is the buffer pH?

19.22 A buffer consists of 0.50 M NaH_2PO_4 and 0.40 M Na_2HPO_4. Phosphoric acid is a triprotic acid ($K_{a1} = 7.2\times10^{-3}$, $K_{a2} = 6.3\times10^{-8}$, and $K_{a3} = 4.2\times10^{-13}$). (a) Which K_a value is most important to this buffer? (b) What is the buffer pH?

19.23 What is the buffer-component concentration ratio, $[Pr^-]/[HPr]$, of a buffer that has a pH of 5.11 (K_a of HPr = 1.3×10^{-5})?

19.24 What is the buffer-component concentration ratio, $[NO_2^-]/[HNO_2]$, of a buffer that has a pH of 2.95 (K_a of HNO_2 = 7.1×10^{-4})?

19.25 What is the buffer-component concentration ratio, $[BrO^-]/[HBrO]$, of a buffer that has a pH of 7.88 (K_a of HBrO = 2.3×10^{-9})?

19.26 What is the buffer-component concentration ratio, $[CH_3COO^-]/[CH_3COOH]$, of a buffer that has a pH of 4.39 (K_a of CH_3COOH = 1.8×10^{-5})?

19.27 A buffer containing 0.2000 M of acid, HA, and 0.1500 M of its conjugate base, A^-, has a pH of 3.35. What is the pH after 0.0015 mol of NaOH is added to 0.5000 L of this solution?

19.28 A buffer that contains 0.40 M base, B, and 0.25 M of its conjugate acid, BH^+, has a pH of 8.88. What is the pH after 0.0020 mol of HCl is added to 0.25 L of this solution?

19.29 A buffer that contains 0.110 M HY and 0.220 M Y^- has a pH of 8.77. What is the pH after 0.0010 mol of $Ba(OH)_2$ is added to 0.750 L of this solution?

19.30 A buffer that contains 1.05 M B and 0.750 M BH^+ has a pH of 9.50. What is the pH after 0.0050 mol of HCl is added to 0.500 L of this solution?

19.31 A buffer is prepared by mixing 184 mL of 0.442 M HCl and 0.500 L of 0.400 M sodium acetate. (See Appendix C.) (a) What is the pH? (b) How many grams of KOH must be added to 0.500 L of the buffer to change the pH by 0.15 units?

19.32 A buffer is prepared by mixing 50.0 mL of 0.050 M sodium bicarbonate and 10.7 mL of 0.10 M NaOH. (See Appendix C.) (a) What is the pH? (b) How many grams of HCl must be added to 25.0 mL of the buffer to change the pH by 0.07 units?

19.33 Choose specific acid-base conjugate pairs suitable for preparing the following buffers: (a) pH $\approx$ 4.0; (b) pH $\approx$ 7.0. (See Appendix C.)

19.34 Choose specific acid-base conjugate pairs suitable for preparing the following buffers: (a) $[H_3O^+] \approx 1\times10^{-9}$ M; (b) $[OH^-] \approx 3\times10^{-5}$ M. (See Appendix C.)

19.35 Choose specific acid-base conjugate pairs suitable for preparing the following buffers: (a) pH $\approx$ 2.5; (b) pH $\approx$ 5.5. (See Appendix C.)

19.36 Choose specific acid-base conjugate pairs suitable for preparing the following buffers: (a) $[OH^-] \approx 1\times10^{-6}$ M; (b) $[H_3O^+] \approx 4\times10^{-4}$ M. (See Appendix C.)

Problems in Context

19.37 An industrial chemist studying the effect of pH on bleaching and sterilizing processes prepares several hypochlorite buffers. Calculate the pH of the following buffers:
(a) 0.100 M HClO and 0.100 M NaClO
(b) 0.100 M HClO and 0.150 M NaClO
(c) 0.150 M HClO and 0.100 M NaClO
(d) One liter of the solution in part (a) after 0.0050 mol of NaOH has been added

19.38 Oxoanions of phosphorus are buffer components in blood. For a KH_2PO_4-Na_2HPO_4 solution with pH = 7.40 (pH of normal arterial blood), what is the buffer-component concentration ratio?

Acid-Base Titration Curves

(Sample Problem 19.3)

Concept Review Questions

19.39 How can you estimate the pH range of an indicator's color change? Why do some indicators have two separate pH ranges?

19.40 Why does the color change of an indicator take place over a range of about 2 pH units?

19.41 Why doesn't the addition of an acid-base indicator affect the pH of the test solution?

19.42 What is the difference between the end point of a titration and the equivalence point? Is the equivalence point always reached first? Explain.

19.43 Some automatic titrators measure the slope of a titration curve to determine the equivalence point. What happens to the slope that enables the instrument to recognize this point?

19.44 Explain how *strong acid*–strong base, *weak acid*–strong base, and *weak base*–strong acid titrations using the same concentrations differ in terms of (a) the initial pH and (b) the pH at the equivalence point. (The component in italics is in the flask.)

19.45 What species are in the buffer region of a weak acid–strong base titration? How are they different from the species at the equivalence point? How are they different from the species in the buffer region of a weak base–strong acid titration?

19.46 Why is the center of the buffer region of a weak acid–strong base titration significant?

19.47 How does the titration curve of a monoprotic acid differ from that of a diprotic acid?

Skill-Building Exercises *(grouped in similar pairs)*

19.48 The indicator cresol red has $K_a = 5.0\times10^{-9}$. Over what approximate pH range does it change color?

19.49 The indicator thymolphthalein has $K_a = 7.9\times10^{-11}$. Over what approximate pH range does it change color?

19.50 Use Figure 19.5 to find an indicator for these titrations:
(a) 0.10 *M* HCl with 0.10 *M* NaOH
(b) 0.10 *M* HCOOH (Appendix C) with 0.10 *M* NaOH

19.51 Use Figure 19.5 to find an indicator for these titrations:
(a) 0.10 *M* CH_3NH_2 (Appendix C) with 0.10 *M* HCl
(b) 0.50 *M* HI with 0.10 *M* KOH

19.52 Use Figure 19.5 to find an indicator for these titrations:
(a) 0.5 *M* $(CH_3)_2NH$ (Appendix C) with 0.5 *M* HBr
(b) 0.2 *M* KOH with 0.2 *M* HNO_3

19.53 Use Figure 19.5 to find an indicator for these titrations:
(a) 0.25 *M* C_6H_5COOH (Appendix C) with 0.25 *M* KOH
(b) 0.50 *M* NH_4Cl (Appendix C) with 0.50 *M* NaOH

19.54 Calculate the pH during the titration of 50.00 mL of 0.1000 *M* HCl with 0.1000 *M* NaOH solution after the following additions of base: (a) 0 mL; (b) 25.00 mL; (c) 49.00 mL; (d) 49.90 mL; (e) 50.00 mL; (f) 50.10 mL; (g) 60.00 mL.

19.55 Calculate the pH during the titration of 30.00 mL of 0.1000 *M* KOH with 0.1000 *M* HBr solution after the following additions of acid: (a) 0 mL; (b) 15.00 mL; (c) 29.00 mL; (d) 29.90 mL; (e) 30.00 mL; (f) 30.10 mL; (g) 40.00 mL.

19.56 Find the pH during the titration of 20.00 mL of 0.1000 *M* butanoic acid, $CH_3CH_2CH_2COOH$ ($K_a = 1.54\times10^{-5}$), with 0.1000 *M* NaOH solution after the following additions of titrant: (a) 0 mL; (b) 10.00 mL; (c) 15.00 mL; (d) 19.00 mL; (e) 19.95 mL; (f) 20.00 mL; (g) 20.05 mL; (h) 25.00 mL.

19.57 Find the pH during the titration of 20.00 mL of 0.1000 *M* triethylamine, $(CH_3CH_2)_3N$ ($K_b = 5.2\times10^{-4}$), with 0.1000 *M* HCl solution after the following additions of titrant: (a) 0 mL; (b) 10.00 mL; (c) 15.00 mL; (d) 19.00 mL; (e) 19.95 mL; (f) 20.00 mL; (g) 20.05 mL; (h) 25.00 mL.

19.58 Find the pH and volume (mL) of 0.0372 *M* NaOH needed to reach the equivalence point(s) in titrations of
(a) 42.2 mL of 0.0520 *M* CH_3COOH
(b) 18.9 mL of 0.0890 *M* H_2SO_3 (two equivalence points)

19.59 Find the pH and volume (mL) of 0.0588 *M* KOH needed to reach the equivalence point(s) in titrations of
(a) 23.4 mL of 0.0390 *M* HNO_2
(b) 17.3 mL of 0.130 *M* H_2CO_3 (two equivalence points)

19.60 Find the pH and the volume (mL) of 0.135 *M* HCl needed to reach the equivalence point(s) in titrations of the following:
(a) 55.5 mL of 0.234 *M* NH_3
(b) 17.8 mL of 1.11 *M* CH_3NH_2

19.61 Find the pH and volume (mL) of 0.447 *M* HNO_3 needed to reach the equivalence point(s) in titrations of
(a) 2.65 L of 0.0750 *M* pyridine (C_5H_5N)
(b) 0.188 L of 0.250 *M* ethylenediamine ($H_2NCH_2CH_2NH_2$)

Equilibria of Slightly Soluble Ionic Compounds

(Sample Problems 19.4 to 19.9)

Concept Review Questions

19.62 The molar solubility of M_2X is 5×10^{-5} *M*. What is the molarity of each ion? How do you set up the calculation to find K_{sp}? What assumption must you make about the dissociation of M_2X into ions? Why is the calculated K_{sp} higher than the actual value?

19.63 Why does pH affect the solubility of CaF_2 but not of $CaCl_2$?

19.64 A list of K_{sp} values like that in Appendix C can be used to compare the solubility of silver chloride directly with that of silver bromide but not with that of silver chromate. Explain.

19.65 In a gaseous equilibrium, the reverse reaction occurs when $Q_c > K_c$. What occurs in aqueous solution when $Q_{sp} > K_{sp}$?

Skill-Building Exercises *(grouped in similar pairs)*

19.66 Write the ion-product expressions for (a) silver carbonate; (b) barium fluoride; (c) copper(II) sulfide.

19.67 Write the ion-product expressions for (a) iron(III) hydroxide; (b) barium phosphate; (c) tin(II) sulfide.

19.68 Write the ion-product expressions for (a) calcium chromate; (b) silver cyanide; (c) nickel(II) sulfide.

19.69 Write the ion-product expressions for (a) lead(II) iodide; (b) strontium sulfate; (c) cadmium sulfide.

19.70 The solubility of silver carbonate is 0.032 *M* at 20°C. Calculate its K_{sp}.

19.71 The solubility of zinc oxalate is 7.9×10^{-3} *M* at 18°C. Calculate its K_{sp}.

19.72 The solubility of silver dichromate at 15°C is 8.3×10^{-3} g/100 mL solution. Calculate its K_{sp}.

19.73 The solubility of calcium sulfate at 30°C is 0.209 g/100 mL solution. Calculate its K_{sp}.

19.74 Find the molar solubility of $SrCO_3$ ($K_{sp} = 5.4\times10^{-10}$) in (a) pure water and (b) 0.13 *M* $Sr(NO_3)_2$.

19.75 Find the molar solubility of $BaCrO_4$ ($K_{sp} = 2.1\times10^{-10}$) in (a) pure water and (b) 1.5×10^{-3} *M* Na_2CrO_4.

19.76 Calculate the molar solubility of $Ca(IO_3)_2$ in (a) 0.060 *M* $Ca(NO_3)_2$ and (b) 0.060 *M* $NaIO_3$. (See Appendix C.)

19.77 Calculate the molar solubility of Ag_2SO_4 in (a) 0.22 *M* $AgNO_3$ and (b) 0.22 *M* Na_2SO_4. (See Appendix C.)

19.78 Which compound in each pair is more soluble in water?
(a) Magnesium hydroxide or nickel(II) hydroxide
(b) Lead(II) sulfide or copper(II) sulfide
(c) Silver sulfate or magnesium fluoride

19.79 Which compound in each pair is more soluble in water?
(a) Strontium sulfate or barium chromate
(b) Calcium carbonate or copper(II) carbonate
(c) Barium iodate or silver chromate

19.80 Which compound in each pair is more soluble in water?
(a) Barium sulfate or calcium sulfate
(b) Calcium phosphate or magnesium phosphate
(c) Silver chloride or lead(II) sulfate

19.81 Which compound in each pair is more soluble in water?
(a) Manganese(II) hydroxide or calcium iodate
(b) Strontium carbonate or cadmium sulfide
(c) Silver cyanide or copper(I) iodide

19.82 Write equations to show whether the solubility of either of the following is affected by pH: (a) AgCl; (b) $SrCO_3$.

19.83 Write equations to show whether the solubility of either of the following is affected by pH: (a) CuBr; (b) $Ca_3(PO_4)_2$.

19.84 Write equations to show whether the solubility of either of the following is affected by pH: (a) $Fe(OH)_2$; (b) CuS.

19.85 Write equations to show whether the solubility of either of the following is affected by pH: (a) PbI_2; (b) $Hg_2(CN)_2$.

19.86 Does any solid $Cu(OH)_2$ form when 0.075 g of KOH is dissolved in 1.0 L of 1.0×10^{-3} *M* $Cu(NO_3)_2$?

19.87 Does any solid $PbCl_2$ form when 3.5 mg of NaCl is dissolved in 0.250 L of 0.12 *M* $Pb(NO_3)_2$?

19.88 Does any solid $Ba(IO_3)_2$ form when 6.5 mg of $BaCl_2$ is dissolved in 500. mL of 0.033 *M* $NaIO_3$?

19.89 Does any solid Ag_2CrO_4 form when 2.7×10^{-5} g of $AgNO_3$ is dissolved in 15.0 mL of 4.0×10^{-4} *M* K_2CrO_4?

Problems in Context

19.90 When blood is donated, sodium oxalate solution is used to precipitate Ca^{2+}, which triggers clotting. A 104-mL sample of blood contains 9.7×10^{-5} g Ca^{2+}/mL. A technologist treats the sample with 100.0 mL of 0.1550 *M* $Na_2C_2O_4$. Calculate $[Ca^{2+}]$ after the treatment. (See Appendix C for K_{sp} of $CaC_2O_4{\cdot}H_2O$.)

Equilibria Involving Complex Ions

(Sample Problems 19.10 and 19.11)

Concept Review Questions

19.91 How can a positive metal ion be at the center of a negative complex ion?

19.92 Write equations to show the stepwise reaction of $Cd(H_2O)_4^{2+}$ in an aqueous solution of KI to form CdI_4^{2-}. Show that $K_{f(overall)} = K_{f1} \times K_{f2} \times K_{f3} \times K_{f4}$.

19.93 Consider the dissolution of PbS in water:

$$PbS(s) + H_2O(l) \rightleftharpoons Pb^{2+}(aq) + HS^-(aq) + OH^-(aq)$$

Adding aqueous NaOH causes more PbS to dissolve. Does this violate Le Châtelier's principle? Explain.

Skill-Building Exercises *(grouped in similar pairs)*

19.94 Write a balanced equation for the reaction of $Hg(H_2O)_4^{2+}$ in aqueous KCN.

19.95 Write a balanced equation for the reaction of $Zn(H_2O)_4^{2+}$ in aqueous NaCN.

19.96 Write a balanced equation for the reaction of $Ag(H_2O)_2^+$ in aqueous $Na_2S_2O_3$.

19.97 Write a balanced equation for the reaction of $Al(H_2O)_6^{3+}$ in aqueous KF.

19.98 Potassium thiocyanate, KSCN, is often used to detect the presence of Fe^{3+} ions in solution through the formation of the red $Fe(H_2O)_5SCN^{2+}$ (or, more simply, $FeSCN^{2+}$). What is $[Fe^{3+}]$ when 0.50 L each of 0.0015 *M* $Fe(NO_3)_3$ and 0.20 *M* KSCN are mixed? K_f of $FeSCN^{2+} = 8.9\times10^2$.

19.99 What is $[Ag^+]$ when 25.0 mL each of 0.044 *M* $AgNO_3$ and 0.57 *M* $Na_2S_2O_3$ are mixed [K_f of $Ag(S_2O_3)_2^{3+} = 4.7\times10^{13}$]?

19.100 When 0.82 g of $ZnCl_2$ is dissolved in 255 mL of 0.150 *M* NaCN, what are $[Zn^{2+}]$, $[Zn(CN)_4^{2-}]$, and $[CN^-]$ [K_f of $Zn(CN)_4^{2-} = 4.2\times10^{19}$]?

19.101 When 2.4 g of $Co(NO_3)_2$ is dissolved in 0.350 L of 0.22 *M* KOH, what are $[Co^{2+}]$, $[Co(OH)_4^{2-}]$, and $[OH^-]$ [K_f of $Co(OH)_4^{2-} = 5\times10^9$]?

19.102 Find the solubility of AgI in 2.5 *M* NH_3 [K_{sp} of AgI $= 8.3\times10^{-17}$; K_f of $Ag(NH_3)_2^+ = 1.7\times10^7$].

19.103 Find the solubility of $Cr(OH)_3$ in a buffer of pH 13.0 [K_{sp} of $Cr(OH)_3 = 6.3\times10^{-31}$; K_f of $Cr(OH)_4^- = 8.0\times10^{29}$].

Ionic Equilibria in Chemical Analysis

(Sample Problem 19.12)

Problems in Context

19.104 A 50.0-mL volume of 0.50 *M* $Fe(NO_3)_3$ is mixed with 125 mL of 0.25 *M* $Cd(NO_3)_2$.
(a) If aqueous NaOH is added, which ion precipitates first? (See Appendix C.)
(b) Describe how the metal ions can be separated using NaOH.
(c) Calculate the $[OH^-]$ that will accomplish the separation.

19.105 In a qualitative analysis procedure, a chemist adds 0.3 *M* HCl to a group of ions and then saturates the solution with H_2S.
(a) What is the $[HS^-]$ of the solution?
(b) If 0.01 *M* of each of these ions is present, which will form a precipitate: Pb^{2+}, Mn^{2+}, Hg^{2+}, Cu^{2+}, Ni^{2+}, Fe^{2+}, Ag^+, K^+?

Comprehensive Problems

19.106 What volumes of 0.200 *M* HCOOH and 2.00 *M* NaOH would make 500. mL of a buffer with the same pH as one made from 475 mL of 0.200 *M* benzoic acid and 25 mL of 2.00 *M* NaOH?

19.107 A microbiologist is preparing a medium on which to culture *E. coli* bacteria. She buffers the medium at pH 7.00 to minimize the effect of acid-producing fermentation. What volumes of equimolar aqueous solutions of K_2HPO_4 and KH_2PO_4 must she combine to make 100. mL of the pH 7.00 buffer?

19.108 As an FDA physiologist, you need 0.600 L of formic acid–formate buffer with a pH of 3.74.
(a) What is the required buffer-component concentration ratio?
(b) How do you prepare this solution from stock solutions of 1.0 *M* HCOOH and 1.0 *M* NaOH?
(c) What is the final concentration of HCOOH in this solution?

19.109 Tris(hydroxymethyl)aminomethane [$(HOCH_2)_3CNH_2$, known as TRIS or THAM] is a weak base widely used in biochemical experiments to make buffer solutions in the pH range of 7 to 9. A certain TRIS buffer has a pH of 8.10 at 25°C and a pH of 7.80 at 37°C. Why does the pH change with temperature?

19.110 Gout is caused by an error in nucleic acid metabolism that leads to a buildup of uric acid in body fluids, which is deposited as slightly soluble sodium urate ($C_5H_3N_4O_3Na$) in the soft tissues of joints. If the extracellular $[Na^+]$ is 0.15 *M* and the solubility in water of sodium urate is 0.085 g/100. mL, what is the minimum urate ion concentration (abbreviated $[Ur^-]$) that will cause a deposit of sodium urate?

19.111 Cadmium ion in solution is analyzed by precipitation as the sulfide, a yellow compound used as a pigment in everything from artists' oil paints to glass and rubber. Calculate the molar solubility of cadmium sulfide at 25°C.

19.112 In the discussion of cave formation in the Chemical Connections essay (p. 840), the dissolution of CO_2 (equation 1) has a K_{eq} of 3.1×10^{-2}, and the formation of aqueous $Ca(HCO_3)_2$ (equation 2) has a K_{eq} of 1×10^{-12}. The fraction by volume of atmospheric CO_2 is 3×10^{-4}. (a) Find $[CO_2(aq)]$ in equilibrium with atmospheric CO_2. (b) Determine $[Ca^{2+}]$ arising from (equation 2) with atmospheric CO_2. (c) Calculate $[Ca^{2+}]$ if atmospheric CO_2 doubles.

19.113 Phosphate systems form essential buffers in organisms. Calculate the pH of a buffer made by dissolving 0.80 mol of NaOH in 0.50 L of 1.0 *M* H_3PO_4.

19.114 The solubility of KCl is 3.7 *M* at 20°C. Two beakers contain 100. mL of saturated KCl solution: 100. mL of 6.0 *M* HCl is added to the first beaker and 100. mL of 12 *M* HCl to the second. (a) Find the ion-product constant of KCl at 20°C. (b) What mass, if any, of KCl will precipitate from each beaker?

19.115 It is possible to detect NH_3 gas over 10^{-2} *M* NH_3. To what pH must 0.15 *M* NH_4Cl be raised to form detectable NH_3?

19.116 Manganese(II) sulfide is one of the compounds found in the nodules on the ocean floor that may eventually be a primary source of many transition metals. The solubility of MnS is 4.7×10^{-4} g/100 mL solution. Estimate the K_{sp} of MnS.

19.117 The normal pH of blood is 7.40 ± 0.05 and is controlled in part by the H_2CO_3-HCO_3^- buffer system.
(a) Assuming that the K_a value for carbonic acid at 25°C applies to blood, what is the $[H_2CO_3]/[HCO_3^-]$ ratio in normal blood?
(b) In a condition called *acidosis*, the blood is too acidic. What is the $[H_2CO_3]/[HCO_3^-]$ ratio in a patient whose blood pH is 7.20 (severe acidosis)?

19.118 Consider the following reaction:

$$Fe(OH)_3(s) + OH^-(aq) \rightleftharpoons Fe(OH)_4^-(aq) \qquad K_c = 4\times10^{-5}$$

(a) What is the solubility of $Fe(OH)_3$ in 0.010 *M* NaOH?
(b) How might this result be used in qualitative analysis?

19.119 A bioengineer preparing cells for cloning bathes a small piece of rat epithelial tissue in a TRIS buffer (see Problem 19.109). The buffer is made by dissolving 43.0 g of TRIS ($pK_b = 5.91$) in enough 0.095 *M* HCl to make 1.00 L of solution. What is the molarity of TRIS and the pH of the buffer?

19.120 Sketch a qualitative curve for the titration of ethylenediamine, $H_2NCH_2CH_2NH_2$, with 0.1 *M* HCl.

19.121 A solution contains 0.10 *M* $ZnCl_2$ and 0.020 *M* $MnCl_2$. Given the following information, how would you adjust the pH to separate the ions as their sulfides ($[H_2S]$ of a saturated aqueous solution at 25°C = 0.10 *M*; $K_w = 1.0\times10^{-14}$ at 25°C)?

$$MnS + H_2O \rightleftharpoons Mn^{2+} + HS^- + OH^- \qquad K_{sp} = 3\times10^{-11}$$
$$ZnS + H_2O \rightleftharpoons Zn^{2+} + HS^- + OH^- \qquad K_{sp} = 2\times10^{-22}$$
$$H_2S + H_2O \rightleftharpoons H_3O^+ + HS^- \qquad K_{a1} = 9\times10^{-8}$$

19.122 Amino acids [general formula $NH_2CH(R)COOH$] can be considered polyprotic acids. In many cases, the R group contains additional amine and carboxyl groups.
(a) Can an amino acid dissolved in pure water have a protonated COOH group and an unprotonated NH_2 group (K_a of COOH group = 4.47×10^{-3}; K_b of NH_2 group = 6.03×10^{-5})? Use glycine, NH_2CH_2COOH, to explain why.
(b) Calculate $[^+NH_3CH_2COO^-]/[^+NH_3CH_2COOH]$ at pH 5.5.
(c) The R group of lysine is $-CH_2CH_2CH_2CH_2NH_2$ (pK_b = 3.47). Draw the structure of lysine at pH 1, physiological pH (~7), and pH 13.
(d) The R group of glutamic acid is $-CH_2CH_2COOH$ (pK_a = 4.07). Four ionic forms of glutamic acid (A to D) are shown below. Select the form that predominates at pH 1, at physiological pH (~7), and at pH 13.

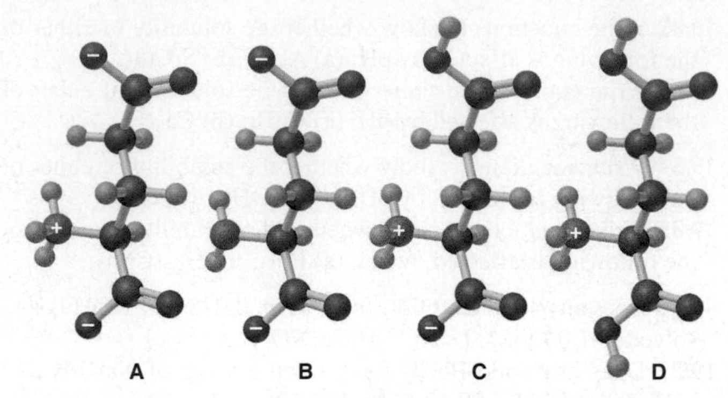

19.123 Tooth enamel consists of hydroxyapatite, $Ca_5(PO_4)_3OH$ ($K_{sp} = 6.8\times10^{-37}$). Fluoride ion added to drinking water reacts with $Ca_5(PO_4)_3OH$ to form the more tooth decay–resistant fluorapatite, $Ca_5(PO_4)_3F$ ($K_{sp} = 1.0\times10^{-60}$). Fluoridated water has dramatically decreased cavities among children. Calculate the solubility of $Ca_5(PO_4)_3OH$ and of $Ca_5(PO_4)_3F$ in water.

19.124 The acid-base indicator ethyl orange turns from red to yellow over the pH range 3.4 to 4.8. Estimate K_a for ethyl orange.

19.125 The titration of a weak acid with a strong base has an end point at pH = 9.0. What indicator is suitable for the titration?

19.126 Use the values obtained in Problem 19.54 to sketch a curve of $[H_3O^+]$ vs. mL of added titrant. Are there advantages or disadvantages to viewing the results in this form? Explain.

19.127 Instrumental acid-base titrations use a pH meter to monitor the changes in pH and volume. The equivalence point is found from the volume at which the curve has the steepest slope.
(a) Use Figure 19.7 to calculate the slope $\Delta pH/\Delta V$ for all pairs of adjacent points and the average volume (V_{avg}) for each interval.
(b) Plot $\Delta pH/\Delta V$ vs. V_{avg} to find the steepest slope, and thus the volume at the equivalence point. (For example, the first pair of points gives $\Delta pH = 0.22$, $\Delta V = 10.00$ mL; hence, $\Delta pH/\Delta V = 0.022\ mL^{-1}$, and $V_{avg} = 5.00$ mL.)

19.128 What is the pH of a solution of 6.5×10^{-9} mol of $Ca(OH)_2$ in 10.0 L of water [K_{sp} of $Ca(OH)_2 = 6.5\times10^{-6}$]?

19.129 Muscle physiologists study the accumulation of lactic acid [$CH_3CH(OH)COOH$] during exercise. Food chemists study its occurrence in sour milk, beer, wine, and fruit. Industrial microbiologists study its formation by various bacterial species from carbohydrates. A biochemist prepares a lactic acid–lactate buffer by mixing 225 mL of 0.85 *M* lactic acid ($K_a = 1.38\times10^{-4}$) with 435 mL of 0.68 *M* sodium lactate. What is the buffer pH?

19.130 A student wants to dissolve the maximum amount of CaF_2 ($K_{sp} = 3.2\times10^{-11}$) to make 1 L of aqueous solution.
(a) Into which of the following should she dissolve the salt?
(I) Pure water (II) 0.01 *M* HF (III) 0.01 *M* NaOH
(IV) 0.01 *M* HCl (V) 0.01 *M* $Ca(OH)_2$
(b) Which would dissolve the least amount of salt?

19.131 A 500.-mL solution consists of 0.050 mol of solid NaOH and 0.13 mol of hypochlorous acid (HClO; $K_a = 3.0\times10^{-8}$) dissolved in water. (a) Aside from water, what is the concentration of each species present? (b) What is the pH of the solution? (c) What is the pH after adding 0.0050 mol of HCl to the flask?

19.132 The Henderson-Hasselbalch equation gives a relationship for obtaining the pH of a buffer solution consisting of HA and A^-. Derive an analogous relationship for obtaining the pOH of a buffer solution consisting of B and BH^+.

19.133 Calculate the molar solubility of $Hg_2C_2O_4$ (K_{sp} = 1.75×10^{-13}) in 0.13 *M* $Hg_2(NO_3)_2$.

19.134 The well water in an area is "hard" because it is in equilibrium with $CaCO_3$ in the surrounding rocks. What is the concentration of Ca^{2+} in the well water (assuming the water's pH is such that the CO_3^{2-} ion is not hydrolyzed)? (See Appendix C for K_{sp} of $CaCO_3$.)

19.135 Four samples of an unknown solution are tested individually by adding HCl, H_2S in acidic solution, H_2S in basic solution, or Na_2CO_3, respectively. No precipitate forms in any of the tests. When NaOH is added to the original solution, moistened litmus paper held above the solution turns blue. A drop of the original solution turns the color of a flame violet. Which ions are present in the unknown solution?

19.136 Human blood contains one buffer system based on phosphate species and one on carbonate species. Assuming that blood has a normal pH of 7.4, what are the principal phosphate and carbonate species present? What is the ratio of the two phosphate species? (In the presence of the dissolved ions and other species in blood, K_{a1} of H_3PO_4 = 1.3×10^{-2}, K_{a2} = 2.3×10^{-7}, and $K_{a3} = 6\times10^{-12}$; K_{a1} of $H_2CO_3 = 8\times10^{-7}$ and $K_{a2} = 1.6\times10^{-10}$.)

19.137 Most qualitative analysis schemes start with the separation of ion group 1, Ag^+, Hg_2^{2+}, and Pb^{2+}, by precipitating AgCl, Hg_2Cl_2, and $PbCl_2$. Develop a procedure for confirming the presence or absence of these ions from the following:

- $PbCl_2$ is soluble in hot water, but AgCl and Hg_2Cl_2 are not.
- $PbCrO_4$ is a yellow solid that is insoluble in hot water.
- AgCl forms colorless $Ag(NH_3)_2^+(aq)$ in aqueous NH_3.
- When a solution containing $Ag(NH_3)_2^+$ is acidified with HCl, AgCl precipitates.
- Hg_2Cl_2 forms an insoluble mixture of Hg(*l*; black) and $HgNH_2Cl(s$; white) in aqueous NH_3.

19.138 An environmental technician collects a sample of rainwater. A light on her portable pH meter indicates low battery power, so she uses indicator solutions to estimate the pH. A piece of litmus paper turns red, indicating acidity, so she divides the sample into thirds and obtains the following results: thymol blue turns yellow; bromphenol blue turns green; and methyl red turns red. Estimate the pH of the rainwater.

19.139 A 0.050 *M* H_2S solution contains 0.15 *M* $NiCl_2$ and 0.35 *M* $Hg(NO_3)_2$. What pH is required to precipitate the maximum amount of HgS but none of the NiS? (See Appendix C.)

19.140 Quantitative analysis of Cl^- ion is often performed by a titration with silver nitrate, using sodium chromate as an indicator. As standardized $AgNO_3$ is added, both white AgCl and red Ag_2CrO_4 precipitate, but so long as some Cl^- remains, the Ag_2CrO_4 redissolves as the mixture is stirred. When the red color is permanent, the equivalence point has been reached.
(a) Calculate the equilibrium constant for the reaction

$$2AgCl(s) + CrO_4^{2-}(aq) \rightleftharpoons Ag_2CrO_4(s) + 2Cl^-(aq)$$

(b) Explain why the silver chromate redissolves.
(c) If 25.00 cm^3 of 0.1000 *M* NaCl is mixed with 25.00 cm^3 of 0.1000 *M* $AgNO_3$, what is the concentration of Ag^+ remaining in solution? Is this sufficient to precipitate any silver chromate?

19.141 An ecobotanist separates the components of a tropical bark extract by chromatography. She discovers a large proportion of quinidine, a dextrorotatory isomer of quinine used for control of arrhythmic heartbeat. Quinidine has two basic nitrogens ($K_{b1} = 4.0\times10^{-6}$ and $K_{b2} = 1.0\times10^{-10}$). To measure the concentration, she carries out a titration. Because of the low solubility of quinidine, she first protonates both nitrogens with excess HCl and titrates the acidified solution with standardized base. A 33.85-mg sample of quinidine ($\mathcal{M}$ = 324.41 g/mol) is acidified with 6.55 mL of 0.150 *M* HCl. (a) How many milliliters of 0.0133 *M* NaOH are needed to titrate the excess HCl? (b) How many additional milliliters of titrant are needed to reach the first equivalence point of quinidine dihydrochloride? (c) What is the pH at the first equivalence point?

19.142 Some kidney stones form by the precipitation of calcium oxalate monohydrate ($CaC_2O_4 \cdot H_2O$, $K_{sp} = 2.3\times10^{-9}$). The pH of urine varies from 5.5 to 7.0, and the average $[Ca^{2+}]$ in urine is 2.6×10^{-3} *M*.
(a) If the concentration of oxalic acid in urine is 3.0×10^{-13} *M*, will kidney stones form at pH = 5.5?
(b) At pH = 7.0?
(c) Vegetarians have a urine pH above 7. Are they more or less likely to form kidney stones?

19.143 A biochemist needs a medium for acid-producing bacteria. The pH of the medium must not change by more than 0.05 pH units for every 0.0010 mol of H_3O^+ generated by the organisms per liter of medium. A buffer consisting of 0.10 *M* HA and 0.10 *M* A^- is included in the medium to control its pH. What volume of this buffer must be included in 1.0 L of medium?

19.144 A 35.00-mL solution of 0.2500 *M* HF is titrated with a standardized 0.1532 *M* solution of NaOH at 25°C.
(a) What is the pH of the HF solution before titrant is added?
(b) How many milliliters of titrant are required to reach the equivalence point?
(c) What is the pH at 0.50 mL before the equivalence point?
(d) What is the pH at the equivalence point?
(e) What is the pH at 0.50 mL after the equivalence point?

19.145 Because of the toxicity of mercury compounds, mercury(I) chloride is used in antibacterial salves. The mercury(I) ion (Hg_2^{2+}) consists of two bound Hg^+ ions.
(a) What is the empirical formula of mercury(I) chloride?
(b) Calculate $[Hg_2^{2+}]$ in a saturated solution of mercury(I) chloride ($K_{sp} = 1.5\times10^{-18}$).
(c) A seawater sample contains 0.20 lb of NaCl per gallon. Find $[Hg_2^{2+}]$ if the seawater is saturated with mercury(I) chloride.
(d) How many grams of mercury(I) chloride are needed to saturate 4900 km^3 of water (the volume of Lake Michigan)?
(e) How many grams of mercury(I) chloride are needed to saturate 4900 km^3 of seawater?

19.146 A lake that has a surface area of 10.0 acres (1 acre = 4.840×10^3 yd^2) receives 1.00 in. of rain of pH 4.20. (Assume the acidity of the rain is due to a strong, monoprotic acid.)
(a) How many moles of H_3O^+ are in the rain falling on the lake?
(b) If the lake is unbuffered (pH = 7.00) and its average depth is 10.0 ft before the rain, find the pH after the rain has been mixed with lake water. (Ignore runoff from the surrounding land.)
(c) If the lake contains hydrogen carbonate ions (HCO_3^-), what mass of HCO_3^- would neutralize the acid in the rain?

19.147 A 35.0-mL solution of 0.075 M $CaCl_2$ is mixed with 25.0 mL of 0.090 M $BaCl_2$. (a) If aqueous KF is added, which fluoride precipitates first? (b) Describe how the metal ions can be separated using KF to form the fluorides. (c) Calculate the fluoride ion concentration that will accomplish the separation.

19.148 Even before the industrial age, rainwater was slightly acidic due to dissolved CO_2. Use the following data to calculate pH of unpolluted rainwater at 25°C: vol % in air of CO_2 = 0.033 vol %; solubility of CO_2 in pure water at 25°C and 1 atm = 88 mL CO_2/100 mL H_2O; K_{a1} of $H_2CO_3 = 4.5\times10^{-7}$.

19.149 Seawater at the surface has a pH of about 8.5.
(a) Which of the following species has the highest concentration at this pH: H_2CO_3; HCO_3^-; CO_3^{2-}? Explain.
(b) What are the concentration ratios $[CO_3^{2-}]/[HCO_3^-]$ and $[HCO_3^-]/[H_2CO_3]$ at this pH?
(c) In the deep sea, light levels are low, and the pH is around 7.5. Suggest a reason for the lower pH at the greater ocean depth. (*Hint:* Consider the presence or absence of plant and animal life, and the effects on carbon dioxide concentrations.)

19.150 Ethylenediaminetetraacetic acid (abbreviated H_4EDTA) is a tetraprotic acid. Its salts are used to treat toxic metal poisoning by forming soluble complex ions that are then excreted. Because $EDTA^{4-}$ also binds essential calcium ions, it is often administered as the calcium disodium salt. For example, when $Na_2Ca(EDTA)$ is given to a patient, the $[Ca(EDTA)]^{2-}$ ions react with circulating Pb^{2+} ions and the metal ions are exchanged:

$$[Ca(EDTA)]^{2-}(aq) + Pb^{2+}(aq) \rightleftharpoons [Pb(EDTA)]^{2-}(aq) + Ca^{2+}(aq) \qquad K_c = 2.5\times10^{7}$$

A child has a dangerous blood lead level of 120 μg/100 mL. If the child is administered 100. mL of 0.10 M $Na_2Ca(EDTA)$, assuming the exchange reaction and excretion process are 100% efficient, what is the final concentration of Pb^{2+} in μg/100 mL blood? (Total blood volume is 1.5 L.)

19.151 EDTA binds metal ions to form complex ions (see Problem 19.150), so it is used to determine the concentrations of metal ions in solution:

$$M^{n+}(aq) + EDTA^{4-}(aq) \longrightarrow MEDTA^{n-4}(aq)$$

A 50.0-mL sample of 0.048 M Co^{2+} is titrated with 0.050 M $EDTA^{4-}$. Find $[Co^{2+}]$ and $[EDTA^{4-}]$ after (a) 25.0 mL and (b) 75.0 mL of $EDTA^{4-}$ are added (log K_f of $CoEDTA^{2-} = 16.31$).

19.152 Sodium chloride is purified for use as table salt by adding HCl to a saturated solution of NaCl (317 g/L). When 25.5 mL of 7.85 M HCl is added to 0.100 L of saturated solution, how many grams of purified NaCl precipitate?

19.153 The solubility of Ag(I) in aqueous solutions containing different concentrations of Cl^- is based on the following equilibria:

$$Ag^+(aq) + Cl^-(aq) \rightleftharpoons AgCl(s) \qquad K_{sp} = 1.8\times10^{-10}$$
$$Ag^+(aq) + 2Cl^-(aq) \rightleftharpoons AgCl_2^-(aq) \qquad K_f = 1.8\times10^{5}$$

When solid AgCl is shaken with a solution containing Cl^-, Ag(I) is present as both Ag^+ and $AgCl_2^-$. The solubility of AgCl is the sum of the concentrations of Ag^+ and $AgCl_2^-$.
(a) Show that $[Ag^+]$ in solution is given by

$$[Ag^+] = 1.8\times10^{-10}/[Cl^-]$$

and that $[AgCl_2^-]$ in solution is given by

$$[AgCl_2^-] = (3.2\times10^{-5})([Cl^-])$$

(b) Find the $[Cl^-]$ at which $[Ag^+] = [AgCl_2^-]$.
(c) Explain the shape of a plot of AgCl solubility vs. $[Cl^-]$.
(d) Find the solubility of AgCl at the $[Cl^-]$ in part (b), which is the minimum solubility of AgCl in the presence of Cl^-.

19.154 Environmental engineers use *alkalinity* as a measure of the capacity of carbonate buffering systems in water samples:

$$\text{Alkalinity (mol/L)} = [HCO_3^-] + 2[CO_3^{2-}] + [OH^-] + [H^+]$$

Find the alkalinity of a water sample that has a pH of 9.5, 26.0 mg/L CO_3^{2-}, and 65.0 mg/L HCO_3^-.

19.155 Calcium ion present in water supplies is easily precipitated as calcite ($CaCO_3$):

$$Ca^{2+}(aq) + CO_3^{2-}(aq) \rightleftharpoons CaCO_3(s)$$

Because the K_{sp} decreases with temperature, heating hard water forms a calcite "scale," which clogs pipes and water heaters. Find the solubility of calcite in water (a) at 10°C ($K_{sp} = 4.4\times10^{-9}$) and (b) at 30°C ($K_{sp} = 3.1\times10^{-9}$).

19.156 Litmus is an organic dye extracted from lichens. It is red below pH 4.5 and blue above pH 8.3. One drop of either 0.1 M HCl (pH 1) or a pH 3 buffer changes blue litmus paper to red, but a drop of 0.001 M HCl (also pH 3) does not. Explain.

19.157 Buffers that are based on 3-morpholinopropanesulfonic acid (MOPS) are often used in RNA analysis. The useful pH range of a MOPS buffer is 6.5 to 7.9. Estimate the K_a of MOPS.

19.158 Scenes A to D represent tiny portions of 0.10 M aqueous solutions of a weak acid HA ($K_a = 4.5\times10^{-5}$), its conjugate base A^-, or a mixture of the two (only these species are shown):

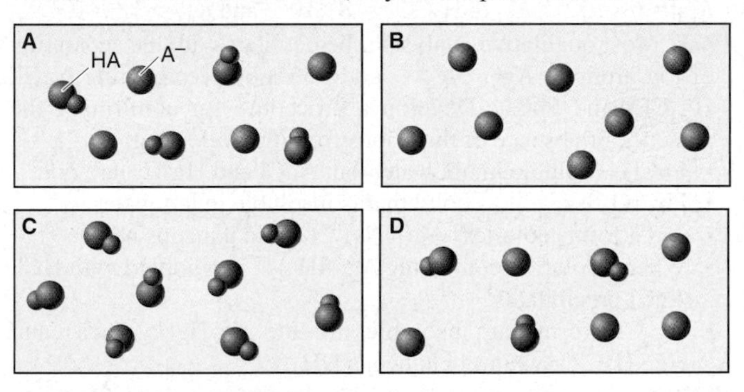

(a) Which scene(s) show(s) a buffer?
(b) What is the pH of each solution?
(c) Arrange the scenes in sequence, assuming that they represent stages in a weak acid–strong base titration.
(d) Which scene represents the titration at its equivalence point?

19.159 Scenes A to C represent aqueous solutions of the slightly soluble salt MZ (only the ions of this salt are shown):

$$MZ(s) \rightleftharpoons M^{2+}(aq) + Z^{2-}(aq)$$

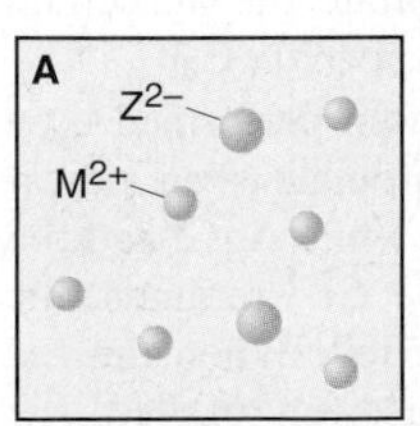

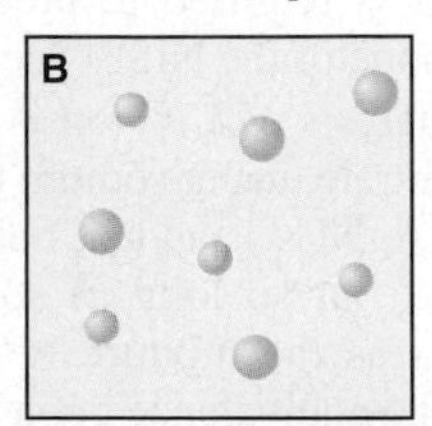

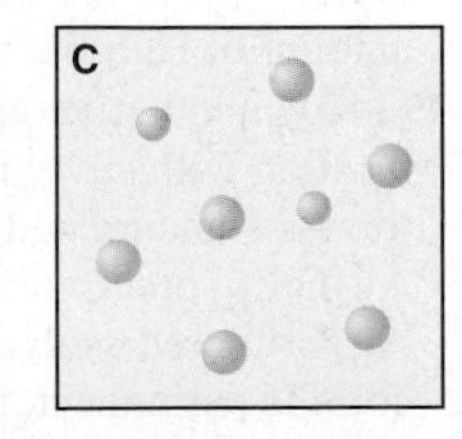

(a) Which scene represents the solution just after solid MZ is stirred thoroughly in distilled water?
(b) If each sphere represents 2.5×10^{-6} M of ions, what is the K_{sp} of MZ?
(c) Which scene represents the solution after $Na_2Z(aq)$ is added?
(d) If Z^{2-} is CO_3^{2-}, which scene represents the solution after the pH has been lowered?

Thermodynamics and the steam engine The relationship between heat and work was discovered in the search to understand and improve the newly invented steam engine. In this chapter we examine this relationship to learn the real-world limit of the energy that can be obtained from any chemical or physical change. In the process, we see *why* reactions occur.

20 Thermodynamics: Entropy, Free Energy, and the Direction of Chemical Reactions

Concepts & Skills to Review

before you study this chapter

- internal energy, heat, and work (Section 6.1)
- state functions (Section 6.1) and standard states (Section 6.6)
- enthalpy, ΔH, and Hess's law (Sections 6.2 and 6.5)
- entropy and solution formation (Section 13.3)
- comparing Q and K to find reaction direction (Section 17.4)

In the last few chapters, we've posed and answered some essential questions about chemical and physical change: How fast does the change occur, and how is this rate affected by concentration and temperature? How much product will be present when the net change ceases, and how is this yield affected by concentration and temperature? We've explored these questions for systems ranging from the stratosphere to a limestone cave and from the cells of your body to the lakes of Sweden.

Now it's time to stand back and ask the most profound question of all: why does a change occur in the first place? From everyday experience, it seems that some changes happen by themselves, almost as if a force were driving them in one direction and not the other. Turn on a gas stove, for example, and the methane mixes with oxygen and burns immediately with a vigorous burst of heat to yield carbon dioxide and water vapor. But those products will not remake methane and oxygen no matter how long they mix. A steel shovel left outside slowly rusts, but put a rusty one outside and it won't become shiny. A cube of sugar dissolves in a cup of coffee after a few seconds of stirring, but stir for another century and the dissolved sugar will never reappear as a cube.

Chemists speak of a process that occurs by itself as being *spontaneous*. Some absorb energy whereas others release it. The principles of thermodynamics, which were developed in the early 19th century to help utilize the newly invented steam engine, allow us to understand the nature of spontaneous change. Despite their narrow historical focus, these principles apply, as far as we know, to every system in the universe!

IN THIS CHAPTER . . . We begin by looking for a criterion to predict the direction of a spontaneous change. A review of the first law of thermodynamics shows that it accounts for the quantity of energy in a change but not the direction of the change. The sign of the enthalpy change is also *not* a criterion for predicting direction. Rather, we find this criterion in the second law of thermodynamics and its quantitative application of entropy (S), the state function that relates to the natural tendency of a system's energy to become more dispersed, and we examine these ideas for both exothermic and endothermic changes. The concept of free energy we develop gives a simplified criterion for spontaneous change, and we see how it relates to the work a system can do. Finally, we consider the key relationship between the free energy change and the equilibrium constant of a reaction.

20.1 THE SECOND LAW OF THERMODYNAMICS: PREDICTING SPONTANEOUS CHANGE

In a formal sense, a **spontaneous change** of a system, whether a chemical or physical change or just a change in location, is one that occurs by itself under specified conditions, without an ongoing input of energy from outside the system. The freezing of water, for example, is spontaneous at 1 atm and $-5°C$. A spontaneous process such as burning or falling may need a little "push" to get started—a spark to ignite gasoline, a shove to knock a book off your desk—but once the process begins, it continues without external aid because the system releases enough energy to keep the process going.

In contrast, for a nonspontaneous change to occur, the surroundings must supply the system with a *continuous* input of energy. A book falls spontaneously, but it rises only if something else, such as a human hand (or a hurricane-force wind), supplies energy in the form of work. Under a given set of conditions, *if a change is spontaneous in one direction, it is **not** spontaneous in the other.*

Note that the term *spontaneous* does not mean *instantaneous* and has nothing to do with how long a process takes to occur; it means that, given enough time, the process will happen by itself. Many processes are spontaneous but slow—ripening, rusting, and (happily!) aging.

A chemical reaction proceeding toward equilibrium is an example of a spontaneous change. As you learned in Chapter 17, we can predict the net direction of the reaction—its spontaneous direction—by comparing the reaction quotient (Q) with the equilibrium constant (K). But *why* is there a drive to attain equilibrium? And what determines the value of the equilibrium constant? Can we tell the direction of a spontaneous change in cases that are not as obvious as burning gasoline or falling books? Because energy changes seem to be involved, let's begin by reviewing the idea of conservation of energy to see whether it can help uncover the criterion for spontaneity.

Limitations of the First Law of Thermodynamics

In Chapter 6, we discussed the first law of thermodynamics (the law of conservation of energy) and its application to chemical and physical systems. The first law states that the internal energy (E) of a system, the sum of the kinetic and potential energy of all its particles, changes when heat (q) and/or work (w) are added or removed:

$$\Delta E = q + w$$

Whatever is not part of the system (sys) is part of the surroundings (surr), so the system and surroundings together constitute the universe (univ):

$$E_{\text{univ}} = E_{\text{sys}} + E_{\text{surr}}$$

Heat and/or work gained by the system is lost by the surroundings, and vice versa:

$$(q + w)_{\text{sys}} = -(q + w)_{\text{surr}}$$

It follows from these ideas that *the total energy of the universe is constant:**

$$\Delta E_{\text{sys}} = -\Delta E_{\text{surr}} \qquad \text{therefore} \qquad \Delta E_{\text{sys}} + \Delta E_{\text{surr}} = 0 = \Delta E_{\text{univ}}$$

Is the first law sufficient to explain why a natural process takes place as it does? It certainly accounts for the energy involved. When a book that was resting on your desk falls to the floor, the first law guides us through the conversion from the potential energy of the resting book to the kinetic energy of the falling book to the heat dispersed in the floor near the point of contact. When gasoline burns in your car's engine, the first law explains that the potential energy difference between the chemical bonds in the fuel mixture and those in the exhaust gases is converted to the kinetic energy of the moving car and its parts plus the heat released to the environment. When an ice cube melts in your hand, the first law tells that energy from your hand was transferred to the ice to change it to a liquid. If you could measure the work and heat involved in each case, you would find that energy is conserved as it is converted from one form to another.

However, the first law does not help us make sense of the *direction* of the change. Why doesn't the heat in the floor near the fallen book change to kinetic energy in the book and move it back onto your desk? Why doesn't the heat released in the car engine convert exhaust fumes back into gasoline and oxygen? Why doesn't the pool of water in your cupped hand transfer the heat back to your hand and refreeze? None of these events would violate the first law—energy would still be conserved—but they never happen. The first law by itself tells nothing about the direction of a spontaneous change, so we must look elsewhere for a way to predict that direction.

---

*Any modern statement of conservation of energy must take into account mass-energy equivalence and the processes in stars, which convert enormous amounts of matter into energy. These can be included by stating that the total *mass-energy* of the universe is constant.

The Sign of ΔH Cannot Predict Spontaneous Change

In the mid-19th century, some thought that the sign of the enthalpy change (ΔH), the heat added or removed at constant pressure (q_P), was the criterion for spontaneity. They thought that exothermic processes ($\Delta H < 0$) were spontaneous and endothermic ones ($\Delta H > 0$) were nonspontaneous. This hypothesis had some experimental support; after all, many spontaneous processes *are* exothermic. All combustion reactions, such as methane burning, are spontaneous and exothermic:

$$CH_4(g) + 2O_2(g) \longrightarrow CO_2(g) + 2H_2O(g) \qquad \Delta H^0_{rxn} = -802 \text{ kJ}$$

Iron metal oxidizes spontaneously and exothermically:

$$2Fe(s) + \tfrac{3}{2}O_2(g) \longrightarrow Fe_2O_3(s) \qquad \Delta H^0_{rxn} = -826 \text{ kJ}$$

Ionic compounds, such as NaCl, form spontaneously and exothermically from their elements:

$$Na(s) + \tfrac{1}{2}Cl_2(g) \longrightarrow NaCl(s) \qquad \Delta H^0_{rxn} = -411 \text{ kJ}$$

However, in many other cases, the sign of ΔH is no help. An exothermic process occurs spontaneously under certain conditions, whereas the opposite, endothermic, process occurs spontaneously under other conditions. Consider the following examples of phase changes, dissolving salts, and chemical changes.

At ordinary pressure, water freezes below 0°C but melts above 0°C. Both changes are spontaneous, but the first is exothermic and the second endothermic:

$$H_2O(l) \longrightarrow H_2O(s) \qquad \Delta H^0_{rxn} = -6.02 \text{ kJ (\textit{exo}thermic; spontaneous at } T < 0°\text{C)}$$
$$H_2O(s) \longrightarrow H_2O(l) \qquad \Delta H^0_{rxn} = +6.02 \text{ kJ (\textit{endo}thermic; spontaneous at } T > 0°\text{C)}$$

At ordinary pressure and room temperature, liquid water vaporizes spontaneously in dry air, another endothermic change:

$$H_2O(l) \longrightarrow H_2O(g) \qquad \Delta H^0_{rxn} = +44.0 \text{ kJ}$$

In fact, all melting and vaporizing are endothermic changes that are spontaneous under proper conditions.

Recall from Chapter 13 that most water-soluble salts have a positive ΔH^0_{soln} but dissolve spontaneously:

$$NaCl(s) \xrightarrow{H_2O} Na^+(aq) + Cl^-(aq) \qquad \Delta H^0_{soln} = +3.9 \text{ kJ}$$
$$RbClO_3(s) \xrightarrow{H_2O} Rb^+(aq) + ClO_3^-(aq) \qquad \Delta H^0_{soln} = +47.7 \text{ kJ}$$
$$NH_4NO_3(s) \xrightarrow{H_2O} NH_4^+(aq) + NO_3^-(aq) \qquad \Delta H^0_{soln} = +25.7 \text{ kJ}$$

Some endothermic chemical changes are also spontaneous:

$$N_2O_5(s) \longrightarrow 2NO_2(g) + \tfrac{1}{2}O_2(g) \qquad \Delta H^0_{rxn} = +109.5 \text{ kJ}$$
$$Ba(OH)_2{\cdot}8H_2O(s) + 2NH_4NO_3(s) \longrightarrow$$
$$Ba^{2+}(aq) + 2NO_3^-(aq) + 2NH_3(aq) + 10H_2O(l) \qquad \Delta H^0_{rxn} = +62.3 \text{ kJ}$$

In the second reaction, the released waters of hydration solvate the ions, but the reaction mixture cannot absorb heat from the surroundings quickly enough, and the container becomes so cold that a wet block of wood freezes to it (Figure 20.1).

A

B

Figure 20.1 A spontaneous endothermic chemical reaction. A, When the crystalline solid reactants barium hydroxide octahydrate and ammonium nitrate are mixed, a slurry soon forms as waters of hydration are released. **B,** The reaction mixture absorbs heat so quickly from the surroundings that the beaker becomes covered with frost, and a moistened block of wood freezes to it.

Freedom of Particle Motion and Dispersal of Particle Energy

What features common to the previous endothermic processes can help us see why they occur spontaneously? In each case, the particles that make up the matter have more freedom of motion after the change occurs. And this means that their energy of motion becomes more dispersed. As we'll see below, "dispersed" means spread over more quantized energy levels.

Phase changes lead from a solid, in which particle motion is restricted, to a liquid, in which the particles have more freedom to move around each other, to a gas, with its much greater freedom of particle motion. Along with this greater freedom of motion, the energy of the particles becomes dispersed over more levels. Dissolving a salt leads from a crystalline solid and pure liquid to ions and solvent molecules moving and interacting throughout the solution; their energy of motion, therefore, is much more dispersed. In the chemical reactions shown, *fewer* moles of crystalline solids produce *more* moles of gases and/or solvated ions. In these cases, there is not only more freedom of motion, but more particles to disperse their energy over more levels.

Thus, in each process, the particles have more freedom of motion and, therefore, their energy of motion has more levels over which to be dispersed:

	less freedom of particle motion $\longrightarrow$	more freedom of particle motion
	localized energy of motion $\longrightarrow$	dispersed energy of motion
Phase change:	solid $\longrightarrow$	liquid $\longrightarrow$ gas
Dissolving of salt:	crystalline solid + liquid $\longrightarrow$	ions in solution
Chemical change:	crystalline solids $\longrightarrow$	gases + ions in solution

In thermodynamic terms, a change in the freedom of motion of particles in a system and in the dispersal of their energy of motion is a key factor determining the direction of a spontaneous process.

Entropy and the Number of Microstates

Let's see how freedom of motion and dispersal of energy relate to spontaneous change. In earlier chapters, we discussed the quantization of energy, not only quantized electronic energy levels of an atom or molecule, but also quantized kinetic energy levels—vibrational, rotational, and translational—of a molecule and its atoms (see Tools of the Laboratory, p. 345).

Picture a system of, say, 1 mol of N_2 gas and focus on one molecule. At any instant, it is moving through space (translating) at a certain speed and rotating at a certain speed, and its atoms are vibrating at a certain speed. In the next instant, the molecule collides with another, and these motional energy states change. The complete quantum state of the molecule at any instant is given by a combination of its particular electronic, translational, rotational, and vibrational states. Clearly, many such combinations are possible for this single molecule, and the number of quantized energy states possible for the system of a mole of molecules is staggering—on the order of $10^{10^{23}}$. Each quantized state of the system is called a *microstate*, and every microstate has the same total energy at a given set of conditions. With each microstate equally possible for the system, the laws of probability say that, over time, all microstates are equally occupied. If we focus only on microstates associated with thermal energy, we say that the number of microstates of a system is the number of ways it can disperse its thermal energy among the various modes of motion of all its molecules.

In 1877, the Austrian mathematician and physicist Ludwig Boltzmann related the number of microstates (W) to the **entropy** (**S**) of the system:

$$S = k \ln W \tag{20.1}$$

where k, the *Boltzmann constant,* is the universal gas constant divided by Avogadro's number, R/N_A, and equals 1.38×10^{-23} J/K. Because W is just a number of microstates and has no units, S has units of joules/kelvin (J/K). From this relationship, we conclude that

- A system with fewer microstates (smaller W) among which to spread its energy has *lower entropy (lower S).*
- A system with more microstates (larger W) among which to spread its energy has *higher entropy (higher S).*

Thus, for our earlier examples,

	lower entropy (fewer microstates) $\longrightarrow$ higher entropy (more microstates)
Phase change:	solid $\longrightarrow$ liquid $\longrightarrow$ gas
Dissolving of salt:	crystalline solid + liquid $\longrightarrow$ ions in solution
Chemical change:	crystalline solids $\longrightarrow$ gases + ions in solution

(In Chapter 13, we used some of these ideas to explain solution behavior.)

Changes in Entropy If the number of microstates increases during a physical or chemical change, there are more ways for the energy of the system to be dispersed among them. Thus, the entropy increases:

$$S_{\text{more microstates}} > S_{\text{fewer microstates}}$$

If the number of microstates decreases, the entropy decreases.

Like internal energy (E) and enthalpy (H), entropy is a state function, which means it depends only on the present state of the system, not on the path it took to arrive at that state (Chapter 6, p. 230). Therefore, the change in entropy of the system (ΔS_{sys}) depends only on the difference between its final and initial values:

$$\Delta S_{\text{sys}} = S_{\text{final}} - S_{\text{initial}}$$

Like any state function, $\Delta S_{\text{sys}} > 0$ when its value increases during a change. For example, when dry ice sublimes to gaseous carbon dioxide, we have

$$CO_2(s) \longrightarrow CO_2(g) \qquad \Delta S_{\text{sys}} = S_{\text{gaseous } CO_2} - S_{\text{solid } CO_2} > 0$$

Similarly, $\Delta S_{\text{sys}} < 0$ when the entropy decreases during a change, as when water vapor condenses:

$$H_2O(g) \longrightarrow H_2O(l) \qquad \Delta S_{\text{sys}} = S_{\text{liquid } H_2O} - S_{\text{gaseous } H_2O} < 0$$

Or consider the decomposition of dinitrogen tetraoxide (written as $O_2N—NO_2$):

$$O_2N—NO_2(g) \longrightarrow 2NO_2(g)$$

When the N—N bond in 1 mol of dinitrogen tetraoxide molecules breaks, the 2 mol of NO_2 molecules have much more freedom of motion; thus, their energy is spread over more microstates:

$$\Delta S_{\text{sys}} = \Delta S_{\text{rxn}} = S_{\text{final}} - S_{\text{initial}} = S_{\text{products}} - S_{\text{reactants}} = 2S_{NO_2} - S_{N_2O_4} > 0$$

Quantitative Meaning of the Entropy Change Two approaches for quantifying an entropy change look different but give the same result. The first is a statistical approach based on the number of microstates possible for the particles in a system. The second is based on the heat absorbed (or released) by a system. We'll explore both in a simple case of 1 mol of an ideal gas, say neon, expanding from 10 L to 20 L at 298 K:

$$\text{1 mol neon (initial: 10 L and 298 K)} \longrightarrow \text{1 mol neon (final: 20 L and 298 K)}$$

We use a statistical approach to find ΔS_{sys} by applying the definition of entropy expressed by Equation 20.1. Figure 20.2 shows a container consisting of

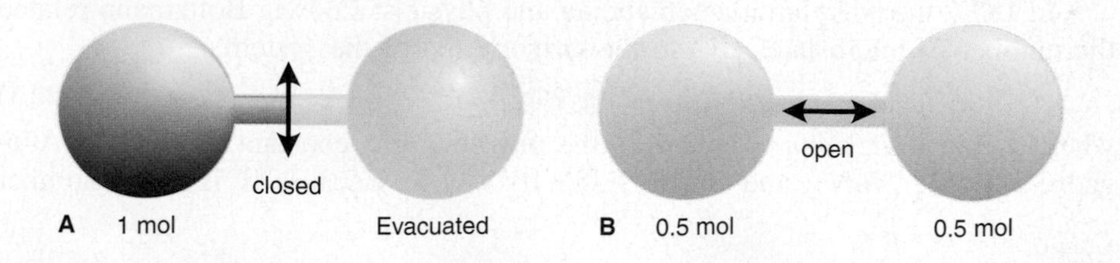

Figure 20.2 Spontaneous expansion of a gas. The container consists of two identical flasks connected by a stopcock. **A,** With the stopcock closed, 1 mol of neon gas occupies one flask, and the other is evacuated. **B,** Open the stopcock, and the gas expands spontaneously until each flask contains 0.5 mol.

two identical flasks connected by a stopcock, with 1 mol of neon in the left flask and an evacuated right flask. We know from experience that when we open the stopcock, the gas will expand to fill both flasks with 0.5 mol each—but *why*?

Let's start with one neon atom and think through what happens as we add more atoms and open the stopcock (Figure 20.3). One atom has some number of microstates (W) possible for it in the left flask and the same number possible in the right flask. Opening the stopcock increases the volume, which increases the number of possible translational energy levels. As a result, the system has 2^1, or 2 times as many microstates possible when the atom moves through both flasks (final state, W_{final}) as when it is confined to one flask (initial state, W_{initial}).

With more atoms, different combinations of atoms can occupy various energy levels, and each combination represents a microstate. With 2 atoms, A and B, moving through both flasks, there are 2^2, or 4 times as many microstates as when they are confined initially to one flask—some number of microstates with A and B in the left, the same number with A in the left and B in the right, that number with B in the left and A in the right, and that number with A and B in the right. Add another atom and there are 2^3 or 8 times as many microstates when the stopcock is open—some number with all three in the left, that number with A and B in the left and C in the right, that number with A and C in the left and B in the right, and so on. With 10 neon atoms, there are 2^{10} or 1024 times as many microstates for the gas in both flasks. Finally, with 1 mol (Avogadro's number, N_A) of neon atoms, there are 2^{N_A} times as many microstates possible for the atoms in both flasks (W_{final}) as in one flask (W_{initial}). In other words, for 1 mol, we have

$$W_{\text{final}}/W_{\text{initial}} = 2^{N_A}$$

Now let's find ΔS_{sys} through the Boltzmann equation. From Appendix A, we know that $\ln A - \ln B = \ln A/B$. Thus,

$$\Delta S_{\text{sys}} = S_{\text{final}} - S_{\text{initial}} = k \ln W_{\text{final}} - k \ln W_{\text{initial}} = k \ln (W_{\text{final}}/W_{\text{initial}})$$

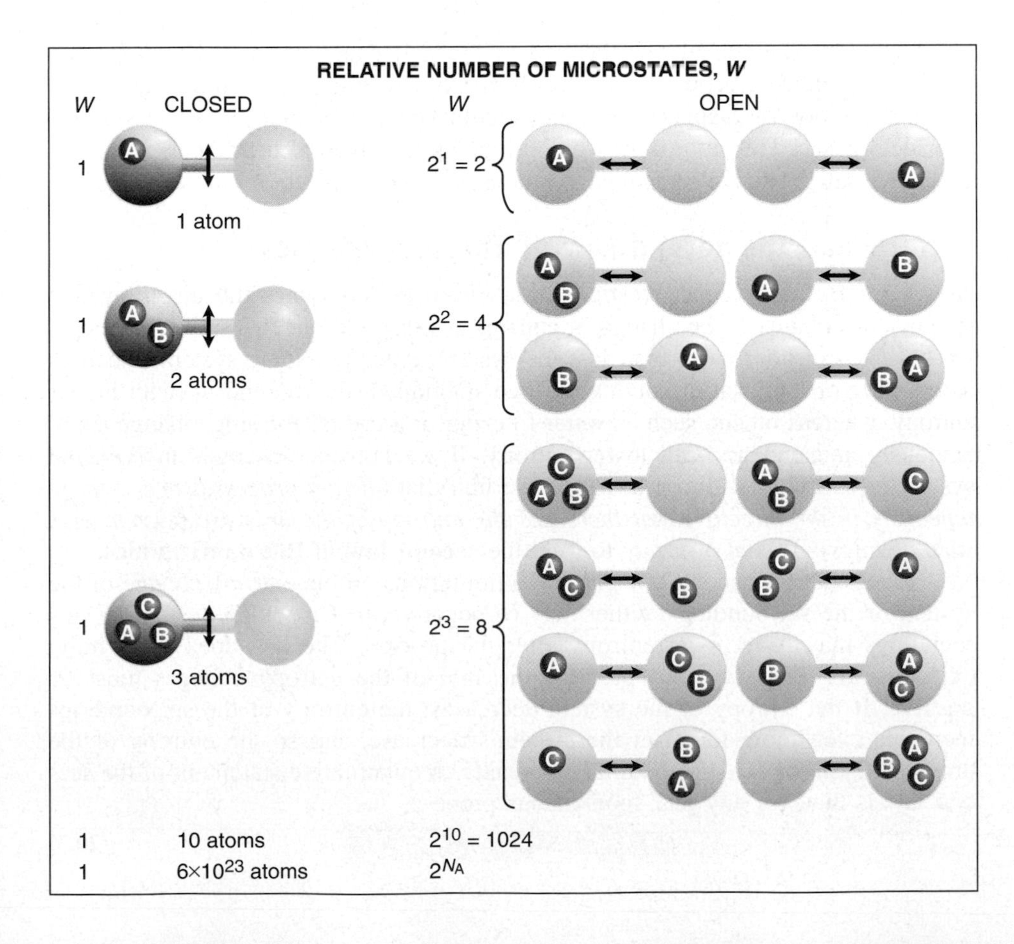

Figure 20.3 Expansion of a gas and the increase in number of microstates. When a gas confined to one flask is allowed to spread through two flasks, the energy of the particles is dispersed over more microstates, and so the entropy is higher. Each combination of particles in the available volume represents a different microstate. The increase in the number of possible microstates that occurs when the volume increases is given by 2^n, where n is the number of particles.

Also from Appendix A, $\ln A^y = y \ln A$; with $k = R/N_A$, we have

$$\Delta S_{sys} = R/N_A \ln 2^{N_A} = (R/N_A)N_A \ln 2 = R \ln 2 = (8.314\ \text{J/mol·K})(0.693)$$
$$= 5.76\ \text{J/mol·K}$$

or, for 1 mol, $\Delta S_{sys} = 5.76\ \text{J/K}$

The second approach for finding ΔS_{sys} is based on heat changes and relates closely to 19th-century attempts to understand the work done by steam engines. In such a process, the entropy change is defined by

$$\Delta S_{sys} = \frac{q_{rev}}{T} \quad \textbf{(20.2)}$$

where T is the temperature at which the heat change occurs and q is the heat absorbed. The subscript "rev" refers to a *reversible process*, one that occurs slowly enough for equilibrium to be maintained continuously, so that the direction of the change can be reversed by an infinitesimal reversal of conditions.

A truly reversible expansion of an ideal gas can only be imagined, but we can approximate it by placing the 10-L neon sample in a piston-cylinder assembly surrounded by a heat reservoir maintained at 298 K, with a beaker of sand on the piston exerting the pressure. We remove one grain of sand (an "infinitesimal" change in pressure) with a pair of tweezers, and the gas expands a tiny amount, raising the piston and doing work on the surroundings, $-w$. If the neon behaves ideally, it absorbs from the reservoir a tiny increment of heat q, equivalent to $-w$. We remove another grain of sand, and the gas expands a tiny bit more and absorbs another tiny increment of heat. This expansion is very close to being reversible because we can reverse it at any point by putting a grain of sand back into the beaker, which causes a tiny compression of the gas. If we continue this expansion process to 20 L and apply calculus to add together all the tiny increments of heat, we find q_{rev} is 1718 J. Thus, applying Equation 20.2, the entropy change is

$$\Delta S_{sys} = q_{rev}/T = 1718\ \text{J}/298\ \text{K} = 5.76\ \text{J/K}$$

This is the same result we obtained by the statistical approach. That approach helps us visualize entropy changes in terms of the number of microstates over which the energy is dispersed, but the calculations are limited to simple systems like ideal gases. This approach, which involves incremental heat changes, is less easy to visualize but can be applied to liquids, solids, and solutions, as well as gases.

Entropy and the Second Law of Thermodynamics

Now back to our original question: what criterion determines the direction of a spontaneous change? The change in entropy is essential, but to apply it correctly, we have to consider more than just the system. After all, some systems, such as ice melting or a crystal dissolving, change spontaneously and end up with higher entropy, whereas others, such as water freezing or a crystal forming, change spontaneously and end up with lower entropy. If we consider changes in *both* the system *and* its surroundings, however, we find that *all real processes occur spontaneously in the direction that increases the entropy of the universe (system plus surroundings)*. This is one way to state the **second law of thermodynamics.**

Notice that the second law places no limitations on the entropy change of the system *or* the surroundings: either may be negative; that is, either system *or* surroundings may have lower entropy after the process. The law does state, however, that for a spontaneous process, the *sum* of the entropy changes must be positive. If the entropy of the system decreases, the entropy of the surroundings increases even more to offset the system's decrease, and so the entropy of the universe (system *plus* surroundings) increases. A quantitative statement of the second law is that, for any real spontaneous process,

$$\Delta S_{univ} = \Delta S_{sys} + \Delta S_{surr} > 0 \quad \textbf{(20.3)}$$

Standard Molar Entropies and the Third Law

Both entropy and enthalpy are state functions, but the nature of their values differs in a fundamental way. Recall that we cannot determine absolute enthalpies because we have no easily measurable starting point, no baseline value for the enthalpy of a substance. Therefore, we measure only enthalpy *changes*.

In contrast, we *can* determine the absolute entropy of a substance. To do so requires application of the **third law of thermodynamics,** which states that *a perfect crystal has zero entropy at a temperature of absolute zero:* $S_{sys} = 0$ at 0 K. "Perfect" means that all the particles are aligned flawlessly in the crystal structure, with no defects of any kind. At absolute zero, all particles in the crystal have the minimum energy, and there is only one way it can be dispersed: thus, in Equation 20.1, $W = 1$, so $S = k \ln 1 = 0$. When we warm the crystal, its total energy increases, so the particles' energy can be dispersed over more microstates (Figure 20.4). Thus, $W > 1$, $\ln W > 0$, and $S > 0$.

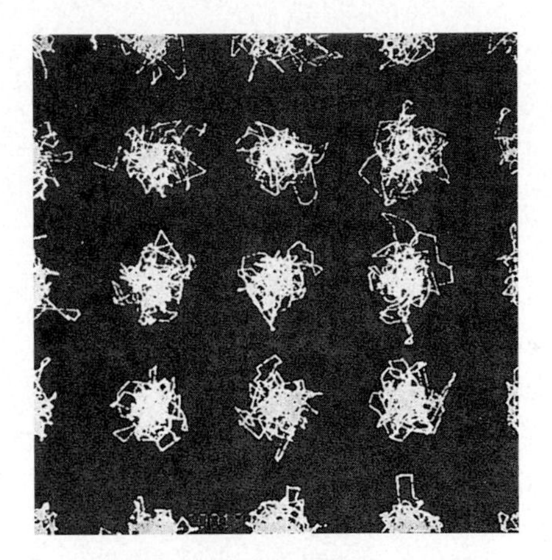

Figure 20.4 Random motion in a crystal. This computer simulation shows the paths of the particle centers in a crystalline solid. At any temperature greater than 0 K, each particle moves about its lattice position. The higher the temperature, the more vigorous the movement. Adding thermal energy increases the total energy, and the particle energies can be distributed over more microstates; thus, the entropy increases.

To obtain a value for S at a given temperature, we first cool a crystalline sample of the substance as close to 0 K as possible. Then we heat it in small increments, dividing q by T to get the increase in S for each increment, and add up all the entropy increases to the temperature of interest, usually 298 K. The entropy of a substance at a given temperature is therefore an *absolute* value that is equal to the entropy increase obtained when the substance is heated from 0 K to that temperature.

As with other thermodynamic variables, we usually compare entropy values for substances in their *standard states* at the temperature of interest: *1 atm for gases, 1 M for solutions, and the pure substance in its most stable form for solids or liquids.* Because entropy is an *extensive* property, that is, one that depends on the amount of substance, we are interested in the **standard molar entropy (S^0)** in units of J/mol·K (or $J{\cdot}mol^{-1}{\cdot}K^{-1}$). The S^0 values at 298 K for many elements, compounds, and ions appear, with other thermodynamic variables, in Appendix B.

Predicting Relative S^0 Values of a System Based on an understanding of systems at the molecular level and the effects of heat absorbed, we can often predict how the entropy of a substance is affected by temperature, physical state, dissolution, and atomic or molecular complexity. (All S^0 values in the following discussion have units of J/mol·K and, unless stated otherwise, refer to the system at 298 K.)

1. *Temperature changes.* For a given substance, S^0 *increases as the temperature rises.* Consider these typical values for copper metal:

T (K):	273	295	298
S^0:	31.0	32.9	33.2

The temperature increases as heat is absorbed ($q > 0$), which represents an increase in the average kinetic energy of the particles. Recall from Figure 5.14 (p. 201) that the kinetic energies of gas particles in a sample are distributed over a range, which becomes wider as the temperature rises. The same general behavior occurs for liquids and solids. With more microstates in which the energy can be dispersed, the entropy of the substance goes up. In other words, raising the temperature populates more microstates. Thus, S^0 *increases for a substance as it is heated.*

2. *Physical states and phase changes.* For a phase change such as melting or vaporizing, heat is absorbed ($q > 0$). The particles have more freedom of motion and their energy is more dispersed, so the entropy change is positive. Thus, S^0 *increases for a substance as it changes from a solid to a liquid to a gas:*

	Na	H_2O	C(graphite)
S^0(s or l):	51.4(s)	69.9(l)	5.7(s)
S^0(g):	153.6	188.7	158.0

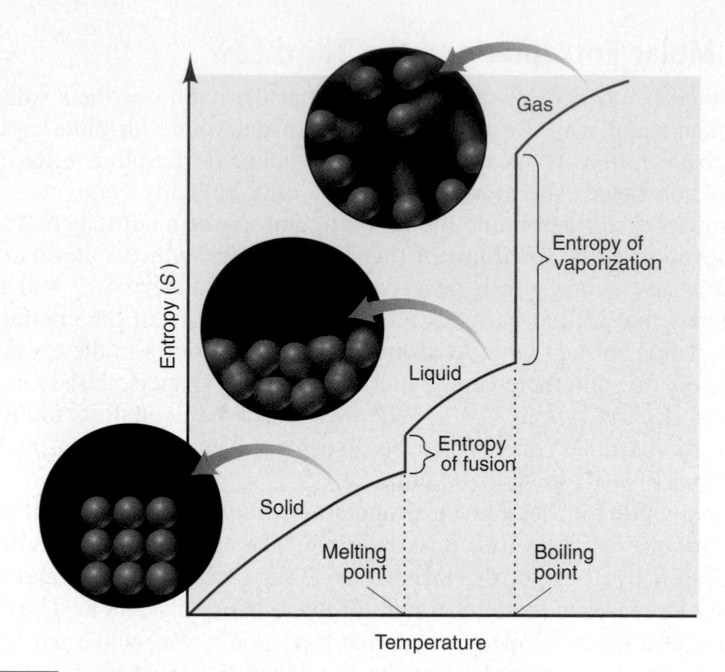

Figure 20.5 The increase in entropy from solid to liquid to gas. A plot of entropy vs. temperature shows the gradual increase in entropy within a phase and the abrupt increase with a phase change. The molecular-scale views depict the increase in freedom of motion of the particles as the solid melts and, even more so, as the liquid vaporizes.

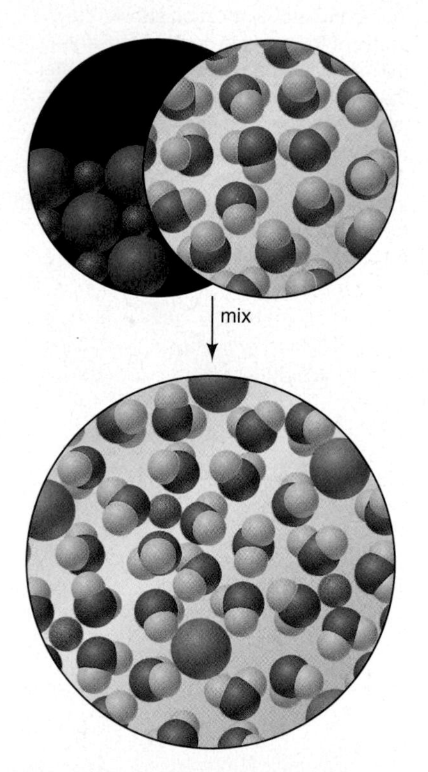

Figure 20.6 The entropy change accompanying the dissolution of a salt. When a crystalline salt and pure liquid water form a solution, the entropy change has two contributions: a positive contribution as the crystal separates into ions and the pure liquid disperses them, and a negative contribution as water molecules become organized around each ion. The relative magnitudes of these contributions determine the overall entropy change. The entropy of a solution is usually *greater* than that of the solid and water.

Figure 20.5 shows the entropy of a typical substance as it is heated and undergoes a phase change. Note the gradual increase within a phase as the temperature rises and the large, sudden increase at the phase change. The solid has the least energy dispersed within it and, thus, the lowest entropy. Its particles vibrate about their positions but, on average, remain fixed. As the temperature rises, the entropy gradually increases with the increase in the particles' kinetic energy. When the solid melts, the particles move much more freely between and around each other, so there is an abrupt increase in entropy. Further heating increases the speed of the particles in the liquid, and the entropy increases gradually. Finally, freed from intermolecular forces, the particles undergo another abrupt entropy increase and move chaotically as a gas. Note that *the increase in entropy from liquid to gas is much larger than that from solid to liquid:* $\Delta S^0_{vap} >> \Delta S^0_{fus}$.

3. *Dissolving a solid or liquid.* The entropy of a dissolved solid or liquid is usually *greater* than the entropy of the pure solute, but the nature of solute *and* solvent and the dissolving process affect the overall entropy change (Figure 20.6):

	NaCl	$AlCl_3$	CH_3OH
S^0(*s* or *l*):	72.1(*s*)	167(*s*)	127(*l*)
S^0(*aq*):	115.1	−148	132

When an ionic solid dissolves in water, the crystal breaks down, and the ions experience a great increase in freedom of motion as they become hydrated and separate, with their energy dispersed over more microstates. We expect the entropy of the ions themselves to be greater in the solution than in the crystal. However, some of the water molecules become organized around the ions (see Figure 13.2, p. 491), which makes a negative contribution to the overall entropy change. In fact, for small, multiply charged ions, the solvent becomes so attracted to the ions, making its energy localized rather than dispersed, that this negative contribution can dominate and lead to negative S^0 values for the ions in solution.

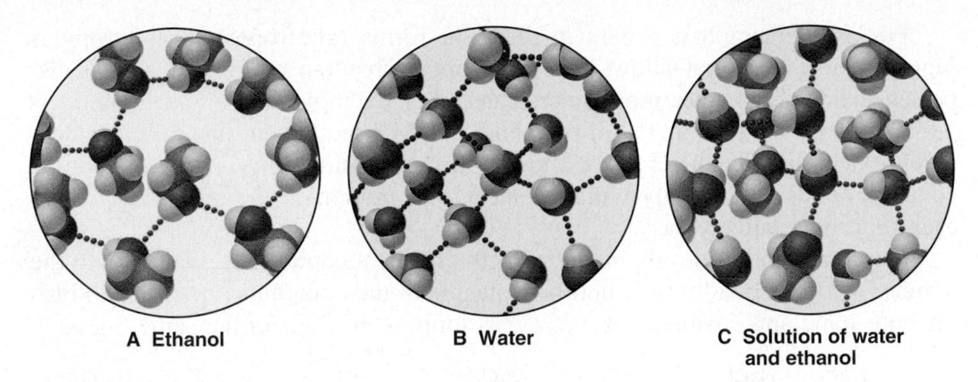

Figure 20.7 The small increase in entropy when ethanol dissolves in water. Pure ethanol **(A)** and pure water **(B)** have many intermolecular H bonds. **C,** In a solution of these two substances, the molecules form H bonds to one another, so their freedom of motion does not change significantly. Thus, the entropy increase is relatively small and is due solely to random mixing.

For example, the $Al^{3+}(aq)$ ion has such a negative S^0 value (-313 J/mol·K) that when $AlCl_3$ dissolves in water, even though S^0 of $Cl^-(aq)$ is positive, the overall entropy of aqueous $AlCl_3$ is lower than that of solid $AlCl_3$.*

For molecular solutes, the increase in entropy upon dissolving is typically much smaller than for ionic solutes. For a solid such as glucose, there is no separation into ions, and for a liquid such as ethanol, the breakdown of a crystal structure is absent as well. Furthermore, in pure ethanol and in pure water, the molecules form many H bonds, so there is relatively little change in their freedom of motion when they are mixed (Figure 20.7). The small increase in the entropy of dissolved ethanol arises from the random mixing of the molecules.

4. *Dissolving a gas.* The particles in a gas already have so much freedom of motion and such highly dispersed energy that they always lose freedom when they dissolve in a liquid or solid. Therefore, the entropy of a solution of a gas in a liquid or a solid is always *less* than the entropy of the gas. For instance, when gaseous O_2 [$S^0(g) = 205.0$ J/mol·K] dissolves in water, its entropy decreases dramatically [$S^0(aq) = 110.9$ J/mol·K] (Figure 20.8). When a gas dissolves in another gas, however, the entropy increases from the mixing of the molecules.

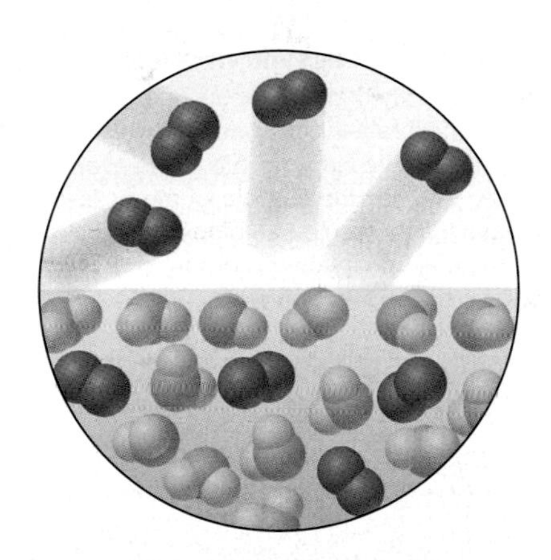

Figure 20.8 The large *decrease* in entropy of a gas when it dissolves in a liquid. The chaotic movement and high entropy of molecules of O_2 are reduced greatly when the gas dissolves in water.

5. *Atomic size or molecular complexity.* In general, differences in entropy values for substances in the same phase are based on atomic size and molecular complexity. For elements within a periodic group, energy levels (microstates) become closer together for heavier atoms, so entropy increases down the group:

	Li	Na	K	Rb	Cs
Atomic radius (pm):	152	186	227	248	265
Molar mass (g/mol):	6.941	22.99	39.10	85.47	132.9
$S^0(s)$:	29.1	51.4	64.7	69.5	85.2

The same trend of increasing entropy down a group holds for similar compounds:

	HF	HCl	HBr	HI
Molar mass (g/mol):	20.01	36.46	80.91	127.9
$S^0(g)$:	173.7	186.8	198.6	206.3

*An S^0 value for a hydrated ion can be negative because it is relative to the S^0 value for the hydrated proton, $H^+(aq)$, which is assigned a value of 0. In other words, $Al^{3+}(aq)$ has a lower entropy than $H^+(aq)$.

For an element that occurs in different forms (allotropes), the entropy is *higher* in the form that allows the atoms more freedom of motion, which disperses their energy over more microstates. For example, the S^0 of graphite is 5.69 J/mol·K, whereas the S^0 of diamond is 2.44 J/mol·K. In diamond, covalent bonds extend in three dimensions, allowing the atoms little movement; in graphite, covalent bonds extend only within a sheet, and motion of the sheets relative to each other is relatively easy.

For compounds, entropy increases with chemical complexity, that is, with the number of atoms in a formula unit or molecule of the compound. This trend holds for both ionic and covalent substances, as long as they are in the same phase:

	NaCl	$AlCl_3$	P_4O_{10}	NO	NO_2	N_2O_4
$S^0(s)$:	72.1	167	229			
$S^0(g)$:				211	240	304

The trend is based on the types of movement, and thus number of microstates, possible for the atoms (or ions) in each compound. For example, as Figure 20.9 shows, among the nitrogen oxides listed above, the two atoms of NO can vibrate only toward and away from each other. The three atoms of NO_2 have more vibrational motions, and the six atoms of N_2O_4 have even more.

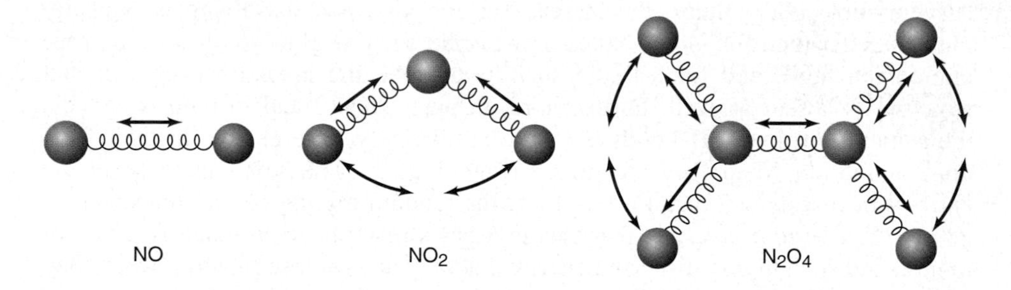

Figure 20.9 Entropy and vibrational motion. A diatomic molecule, such as NO, can vibrate in only one way. NO_2 can vibrate in more ways, and N_2O_4 in even more. Thus, as the number of atoms increases, a molecule can disperse its vibrational energy over more microstates, and so has higher entropy.

For larger molecules, we also consider how parts of the molecule move relative to other parts. A long hydrocarbon chain can rotate and vibrate in more ways than a short one, so entropy increases with chain length. A ring compound, such as cyclopentane (C_5H_{10}), has lower entropy than the corresponding chain compound, pentene (C_5H_{10}), because the ring structure restricts freedom of motion:

	$CH_4(g)$	$C_2H_6(g)$	$C_3H_8(g)$	$C_4H_{10}(g)$	$C_5H_{10}(g)$	C_5H_{10}(cyclo, g)	$C_2H_5OH(l)$
S^0:	186	230	270	310	348	293	161

Remember these trends hold only for *substances in the same physical state*. Gaseous methane (CH_4) has a greater entropy than liquid ethanol (C_2H_5OH), even though ethanol molecules are more complex. When gases are compared with liquids, *the effect of physical state usually dominates that of molecular complexity.*

SAMPLE PROBLEM 20.1 Predicting Relative Entropy Values

Problem Choose the member with the higher entropy in each of the following pairs, and justify your choice [assume constant temperature, except in part (e)]:
(a) 1 mol of $SO_2(g)$ or 1 mol of $SO_3(g)$
(b) 1 mol of $CO_2(s)$ or 1 mol of $CO_2(g)$
(c) 3 mol of $O_2(g)$ or 2 mol of $O_3(g)$
(d) 1 mol of KBr(s) or 1 mol of KBr(aq)
(e) Seawater in midwinter at 2°C or in midsummer at 23°C
(f) 1 mol of $CF_4(g)$ or 1 mol of $CCl_4(g)$

Plan In general, we know that particles with more freedom of motion or more dispersed energy have higher entropy and that raising the temperature increases entropy. We apply the general categories described in the text to choose the member with the higher entropy.

Solution (a) 1 mol of $SO_3(g)$. For equal numbers of moles of substances with the same types of atoms in the same physical state, the more atoms in the molecule, the more types of motion available, and thus the higher the entropy.

(b) 1 mol of $CO_2(g)$. For a given substance, entropy increases in the sequence $s < l < g$.

(c) 3 mol of $O_2(g)$. The two samples contain the same number of oxygen atoms but different numbers of molecules. Despite the greater complexity of O_3, the greater number of molecules dominates in this case because there are many more microstates possible for three moles of particles than for two moles.

(d) 1 mol of $KBr(aq)$. The two samples have the same number of ions, but their motion is more limited and their energy less dispersed in the solid than in the solution.

(e) Seawater in summer. Entropy increases with rising temperature.

(f) 1 mol of $CCl_4(g)$. For similar compounds, entropy increases with molar mass.

FOLLOW-UP PROBLEM 20.1 For 1 mol of substance at a given temperature, select the member in each pair with the higher entropy, and give the reason for your choice:

(a) $PCl_3(g)$ or $PCl_5(g)$

(b) $CaF_2(s)$ or $BaCl_2(s)$

(c) $Br_2(g)$ or $Br_2(l)$

SECTION SUMMARY

A change is spontaneous if it occurs in a given direction under specified conditions without a continuous input of energy. Neither the first law of thermodynamics nor the sign of ΔH predicts the direction. All spontaneous processes involve an increase in the dispersion of energy. Entropy is a state function that measures the extent of energy dispersal through the number of microstates possible for a system, which is related to the freedom of motion of its particles. The second law of thermodynamics states that, in a spontaneous process, the entropy of the universe (system plus surroundings) increases. Absolute entropy values can be found because perfect crystals have zero entropy at 0 K (third law). Standard molar entropy S^0 (J/mol·K) is affected by temperature, phase changes, dissolution, and atomic size or molecular complexity.

20.2 CALCULATING THE CHANGE IN ENTROPY OF A REACTION

In addition to understanding trends in S^0 values for different substances or for the same substance in different phases, chemists are especially interested in learning how to predict the sign *and* calculate the value of the change in entropy as a reaction occurs.

Entropy Changes in the System: Standard Entropy of Reaction (ΔS^0_{rxn})

Based on the ideas we just discussed, we can often predict the sign of the **standard entropy of reaction, ΔS^0_{rxn},** the entropy change that occurs when all reactants and products are in their standard states. A deciding event is usually a change in the number of moles of gas. Because gases have such great freedom of motion and thus high molar entropies, *if the number of moles of gas increases, ΔS^0_{rxn} is usually positive; if the number decreases, ΔS^0_{rxn} is usually negative.*

For example, when $H_2(g)$ and $I_2(s)$ form $HI(g)$, the total number of moles stays the same, but we predict that the entropy increases because the number of moles of *gas* increases:

$$H_2(g) + I_2(s) \longrightarrow 2HI(g) \qquad \Delta S^0_{rxn} = S^0_{products} - S^0_{reactants} > 0$$

When ammonia forms from its elements, 4 mol of gas produces 2 mol of gas, so we predict that the entropy decreases during the reaction:

$$N_2(g) + 3H_2(g) \rightleftharpoons 2NH_3(g) \qquad \Delta S^0_{rxn} = S^0_{products} - S^0_{reactants} < 0$$

Sometimes, even when the amount (mol) of gases stays the same, another factor related to freedom of motion may help predict the sign of the entropy change. For example, when cyclopropane is heated to 500°C, the ring opens and propene forms. The chain has more freedom of motion than the ring, so the standard molar entropy of the product is greater than that of the reactant:

$$\text{(cyclo-}CH_2\text{—}H_2C\text{—}CH_2)(g) \longrightarrow CH_3\text{—}CH\text{=}CH_2(g) \qquad \Delta S^0 = S^0_{products} - S^0_{reactants} > 0$$

Keep in mind, however, that in general *we cannot predict the sign of the entropy change unless the reaction involves a change in number of moles of gas.*

Recall that by applying Hess's law (Chapter 6), we can combine ΔH^0_f values to find the standard heat of reaction, ΔH^0_{rxn}. Similarly, we combine S^0 values to find the standard entropy of reaction, ΔS^0_{rxn}:

$$\Delta S^0_{rxn} = \Sigma m S^0_{products} - \Sigma n S^0_{reactants} \qquad \textbf{(20.4)}$$

where m and n are the amounts of the individual species, represented by their coefficients in the balanced equation. For the formation of ammonia, we have

$$\Delta S^0_{rxn} = [(2 \text{ mol } NH_3)(S^0 \text{ of } NH_3)] - [(1 \text{ mol } N_2)(S^0 \text{ of } N_2) + (3 \text{ mol } H_2)(S^0 \text{ of } H_2)]$$

From Appendix B, we find the appropriate S^0 values:

$$\Delta S^0_{rxn} = [(2 \text{ mol})(193 \text{ J/mol·K})] - [(1 \text{ mol})(191.5 \text{ J/mol·K}) + (3 \text{ mol})(130.6 \text{ J/mol·K})]$$
$$= -197 \text{ J/K}$$

As we predicted, $\Delta S^0_{rxn} < 0$.

SAMPLE PROBLEM 20.2 Calculating the Standard Entropy of Reaction, ΔS^0_{rxn}

Problem Calculate ΔS^0_{rxn} for the combustion of 1 mol of propane at 25°C:

$$C_3H_8(g) + 5O_2(g) \longrightarrow 3CO_2(g) + 4H_2O(l)$$

Plan To determine ΔS^0_{rxn}, we apply Equation 20.4. We predict the sign of ΔS^0_{rxn} from the change in the number of moles of gas: 6 mol of gas yields 3 mol of gas, so the entropy will probably decrease ($\Delta S^0_{rxn} < 0$).

Solution Calculating ΔS^0_{rxn}. Using Appendix B values,

$$\Delta S^0_{rxn} = [(3 \text{ mol } CO_2)(S^0 \text{ of } CO_2) + (4 \text{ mol } H_2O)(S^0 \text{ of } H_2O)]$$
$$- [(1 \text{ mol } C_3H_8)(S^0 \text{ of } C_3H_8) + (5 \text{ mol } O_2)(S^0 \text{ of } O_2)]$$
$$= [(3 \text{ mol})(213.7 \text{ J/mol·K}) + (4 \text{ mol})(69.9 \text{ J/mol·K})]$$
$$- [(1 \text{ mol})(269.9 \text{ J/mol·K}) + (5 \text{ mol})(205.0 \text{ J/mol·K})]$$
$$= -374 \text{ J/K}$$

Check $\Delta S^0 < 0$, so our prediction is correct. Rounding gives $[3(200) + 4(70)] - [270 + 5(200)] = 880 - 1270 = -390$, close to the calculated value.

Comment We based our prediction on the fact that S^0 values of gases are much greater than those of solids or liquids. (This is usually true even when the condensed phases consist of more complex molecules.) Remember that when there is no change in the amount (mol) of gas, you *cannot* confidently predict the sign of ΔS^0_{rxn}.

FOLLOW-UP PROBLEM 20.2 Balance the following equations, predict the sign of ΔS^0_{rxn} if possible, and calculate its value at 25°C:

(a) $NaOH(s) + CO_2(g) \longrightarrow Na_2CO_3(s) + H_2O(l)$

(b) $Fe(s) + H_2O(g) \longrightarrow Fe_2O_3(s) + H_2(g)$

Entropy Changes in the Surroundings: The Other Part of the Total

In many spontaneous reactions, such as the synthesis of ammonia and the combustion of propane, we see that the entropy of the reacting system decreases ($\Delta S^0_{rxn} < 0$). The second law dictates that ***decreases** in the entropy of the system can occur only if **increases** in the entropy of the surroundings outweigh them.* Let's examine the influence of the surroundings—in particular, the addition (or removal) of heat and the temperature at which this heat change occurs—on the *total* entropy change.

The essential role of the surroundings is to *either add heat to the system or remove heat from it.* In essence, the surroundings function as an enormous heat source or heat sink, one so large that its temperature remains constant, even though its entropy changes through the loss or gain of heat. The surroundings participate in the two possible types of enthalpy changes as follows:

1. *Exothermic change.* Heat lost by the system is gained by the surroundings. This heat gain increases the freedom of motion of particles in the surroundings, which disperses their energy; so the entropy of the surroundings increases:

   For an exothermic change: $q_{sys} < 0$, $q_{surr} > 0$, and $\Delta S_{surr} > 0$

2. *Endothermic change.* Heat gained by the system is lost by the surroundings. This heat loss reduces the freedom of motion of particles in the surroundings, which localizes their energy; so the entropy of the surroundings decreases:

   For an endothermic change: $q_{sys} > 0$, $q_{surr} < 0$, and $\Delta S_{surr} < 0$

The *temperature* of the surroundings at which the heat is transferred also affects ΔS_{surr}. Consider the effect of an exothermic reaction at a low and at a high temperature. At a low temperature, such as 20 K, there is very little random motion in the surroundings, that is, relatively little energy is dispersed there. Therefore, transferring heat to the surroundings has a large effect on how much energy is dispersed. At a higher temperature, such as 298 K, the surroundings already have a relatively large quantity of energy dispersed, so transferring the same amount of heat has a smaller effect on the total energy dispersed. In other words, the change in entropy of the surroundings is greater when heat is added at a lower temperature. Putting these ideas together, *the change in entropy of the surroundings is directly related to an opposite change in the heat of the system and inversely related to the temperature at which the heat is transferred.* Combining these relationships gives an equation that is closely related to Equation 20.2:

$$\Delta S_{surr} = -\frac{q_{sys}}{T}$$

Recall that for a process at *constant pressure,* the heat (q_P) is ΔH, so

$$\Delta S_{surr} = -\frac{\Delta H_{sys}}{T} \tag{20.5}$$

This means that we can calculate ΔS_{surr} by measuring ΔH_{sys} and the temperature T at which the change takes place.

To restate the central point, if a spontaneous reaction has a negative ΔS_{sys} (energy dispersed over fewer microstates), ΔS_{surr} must be positive enough (energy dispersed over many more microstates) that ΔS_{univ} is positive (energy dispersed over more microstates). Sample Problem 20.3 illustrates this situation for one of the reactions we considered earlier.

A Checkbook Analogy for Heating the Surroundings A monetary analogy may clarify the relative sizes of the changes that arise from heating the surroundings at different initial temperatures. If you have \$10 in your checking account, a \$10 deposit represents a 100% increase in your net worth; that is, a given change to a low initial state has a large impact. If, however, you have a \$1000 balance, a \$10 deposit represents only a 1% increase. That is, the same change to a high initial state has a smaller impact.

SAMPLE PROBLEM 20.3 Determining Reaction Spontaneity

Problem At 298 K, the formation of ammonia has a negative ΔS^0_{sys}:

$$N_2(g) + 3H_2(g) \longrightarrow 2NH_3(g) \qquad \Delta S^0_{sys} = -197 \text{ J/K}$$

Calculate ΔS_{univ}, and state whether the reaction occurs spontaneously at this temperature.

Plan For the reaction to occur spontaneously, $\Delta S_{univ} > 0$, and so ΔS_{surr} must be greater than +197 J/K. To find ΔS_{surr}, we need ΔH^0_{sys}, which is the same as ΔH^0_{rxn}. We use ΔH^0_f values from Appendix B to find ΔH^0_{rxn}. Then, we use ΔH^0_{rxn} and the given T (298 K) to find ΔS_{surr}. To find ΔS_{univ}, we add the calculated ΔS_{surr} to the given ΔS^0_{sys} (−197 J/K).

Solution Calculating ΔH^0_{sys}:

$$\Delta H^0_{sys} = \Delta H^0_{rxn}$$
$$= [(2 \text{ mol } NH_3)(-45.9 \text{ kJ/mol})] - [(3 \text{ mol } H_2)(0 \text{ kJ/mol}) + (1 \text{ mol } N_2)(0 \text{ kJ/mol})]$$
$$= -91.8 \text{ kJ}$$

Calculating ΔS_{surr}:

$$\Delta S_{surr} = -\frac{\Delta H^0_{sys}}{T} = -\frac{-91.8 \text{ kJ} \times \frac{1000 \text{ J}}{1 \text{ kJ}}}{298 \text{ K}} = 308 \text{ J/K}$$

Determining ΔS_{univ}:

$$\Delta S_{univ} = \Delta S^0_{sys} + \Delta S_{surr} = -197 \text{ J/K} + 308 \text{ J/K} = 111 \text{ J/K}$$

$\Delta S_{univ} > 0$, so the reaction occurs spontaneously at 298 K (see figure in margin).

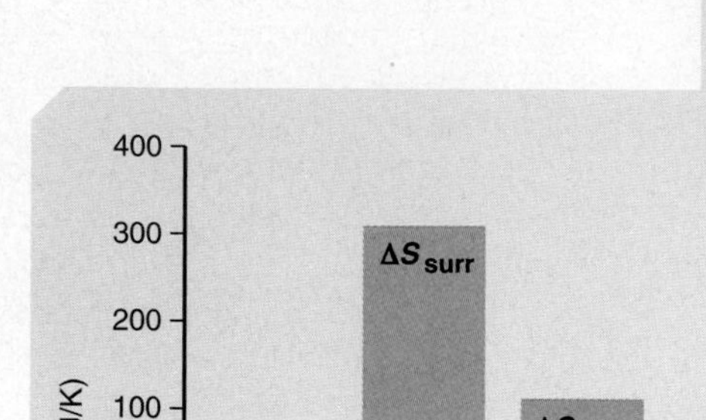

Check Rounding to check the math, we have

$$\Delta H^0_{rxn} \approx 2(-45 \text{ kJ}) = -90 \text{ kJ}$$
$$\Delta S_{surr} \approx -(-90{,}000 \text{ J})/300 \text{ K} = 300 \text{ J/K}$$
$$\Delta S_{univ} \approx -200 \text{ J/K} + 300 \text{ J/K} = 100 \text{ J/K}$$

Given the negative ΔH^0_{rxn}, Le Châtelier's principle predicts that low temperature should favor NH_3 formation, and so the answer is reasonable (see the Chemical Connections essay on the Haber process, pp. 755–756).

Comment 1. Note that ΔH^0 has units of kJ, whereas ΔS has units of J/K. Don't forget to convert kJ to J, or you'll introduce a large error.

2. This example highlights the distinction between thermodynamic and kinetic considerations. Even though NH_3 forms spontaneously, it does so slowly; the chemical industry expends great effort to catalyze its formation at a practical rate.

FOLLOW-UP PROBLEM 20.3 Does the oxidation of FeO(s) to Fe_2O_3(s) occur spontaneously at 298 K?

As Sample Problem 20.3 shows, taking the surroundings into account is crucial to determining reaction spontaneity. Moreover, it clarifies the relevance of thermodynamics to biology, as the Chemical Connections essay shows.

The Entropy Change and the Equilibrium State

A process spontaneously approaches equilibrium, so $\Delta S_{univ} > 0$. When the process reaches equilibrium, there is no longer any force driving it to proceed further and, thus, no net change in either direction; that is, $\Delta S_{univ} = 0$. At that point, any entropy change in the system is exactly balanced by an opposite entropy change in the surroundings:

$$\text{At equilibrium:} \quad \Delta S_{univ} = \Delta S_{sys} + \Delta S_{surr} = 0 \quad \text{or} \quad \Delta S_{sys} = -\Delta S_{surr}$$

For example, let's calculate ΔS_{univ} for a phase change. For the vaporization-condensation of 1 mol of water at 100°C (373 K),

$$H_2O(l; 373 \text{ K}) \rightleftharpoons H_2O(g; 373 \text{ K})$$

Do Living Things Obey the Laws of Thermodynamics?

Organisms can be thought of as chemical machines that evolved by extracting energy as efficiently as possible from the environment. Such a mechanical view holds that all processes, whether they involve living or nonliving systems, are consistent with thermodynamic principles. Let's examine the first and second laws of thermodynamics to see if they apply to living systems.

Organisms certainly comply with the first law. The chemical bond energy in food is converted into the mechanical energy of sprouting, crawling, swimming, and countless other movements; the electrical energy of nerve conduction; the thermal energy of warming the body; and so forth. Many experiments have demonstrated that in all these energy conversions, the total energy is conserved. Some of the earliest studies of this question were performed by Lavoisier, who included animal respiration in his new theory of combustion (Figure B20.1). He was the first to show that "animal heat" was produced by a slow combustion process occurring continually in the body. In experiments with guinea pigs, he measured the intake of food and O_2 and the output of CO_2 and heat, for which he invented a calorimeter based on the melting of ice, and he established the principles and methods for measuring the basal rate of metabolism. Modern experiments using a calorimeter large enough to contain an exercising human continue to confirm the conservation of energy (Figure B20.2).

It may not seem as clear, however, that an organism, and even the whole parade of life, complies with the second law. Mature humans are far more complex than the simple egg and sperm cells from which they develop, and modern organisms are far more complex than the one-celled ancestral specks from which they evolved. Energy must be highly localized and macromolecules constrained to carry out myriad reactions that synthesize biopolymers from their monomers. Are the growth of an organism and the evolution of life exceptions to the spontaneous tendency of natural processes to disperse their energy and increase molecular freedom? Does biology violate the second law? Not at all, if we examine the system *and* its surroundings. For an organism to grow or a species to evolve, countless moles of food molecules—carbohydrates, proteins, and fats—and oxygen molecules undergo exothermic combustion reactions to form many more moles of gaseous CO_2 and H_2O. The formation and discharge of these waste gases represent a tremendous net increase in the entropy of the surroundings, as does the heat released. Thus, the localization of energy and restrictions on molecular freedom apparent in the growth and evolution of organisms occur at the expense of a far greater dispersal of energy and freedom of motion in the Earth-Sun surroundings. When system and surroundings are considered together, the entropy of the universe, as always, increases.

Figure B20.2 A whole-body calorimeter. In this room-sized apparatus, a subject exercises while respiratory gases, energy input and output, and other physiological variables are monitored.

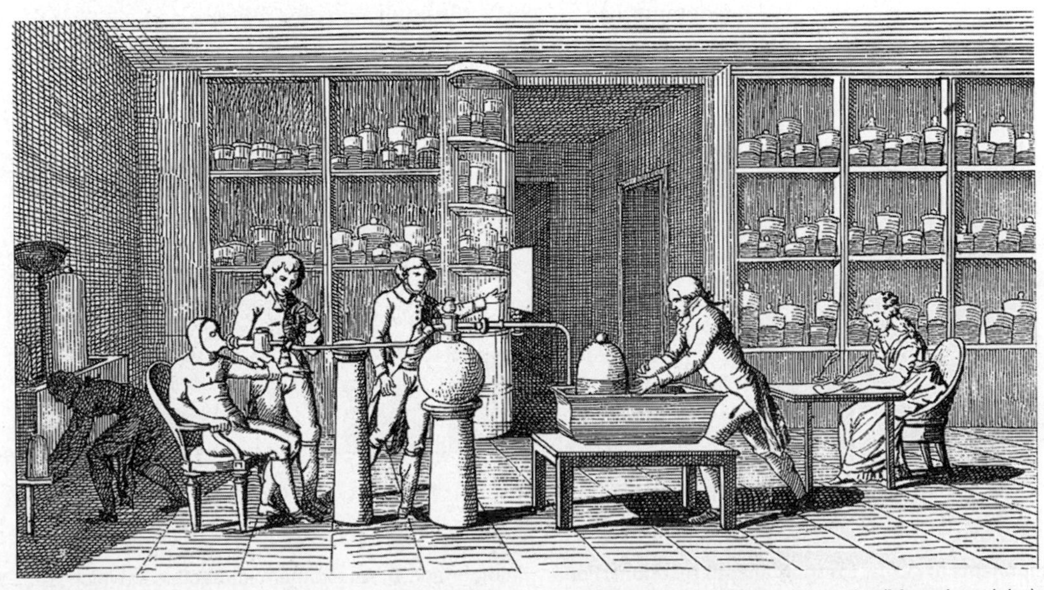

Figure B20.1 Lavoisier studying human respiration as a form of combustion. Lavoisier (standing at right) measured substances consumed and excreted in addition to changes in heat to understand the chemical nature of respiration. The artist (Mme. Lavoisier) shows herself taking notes (seated at right).

First, we find ΔS_{univ} for the forward change (vaporization) by calculating ΔS^0_{sys}:

$$\Delta S^0_{sys} = \Sigma m S^0_{products} - \Sigma n S^0_{reactants} = S^0 \text{ of } H_2O(g;\ 373\text{ K}) - S^0 \text{ of } H_2O(l;\ 373\text{ K})$$
$$= 195.9\text{ J/K} - 86.8\text{ J/K} = 109.1\text{ J/K}$$

As we expect, the entropy of the system increases ($\Delta S^0_{sys} > 0$) as the liquid absorbs heat and changes to a gas.

For ΔS_{surr}, we have

$$\Delta S_{surr} = -\frac{\Delta H^0_{sys}}{T}$$

where $\Delta H^0_{sys} = \Delta H^0_{vap}$ at 373 K $= 40.7$ kJ/mol $= 40.7\times10^3$ J/mol. For 1 mol of water, we have

$$\Delta S_{surr} = -\frac{\Delta H^0_{vap}}{T} = -\frac{40.7\times10^3\text{ J}}{373\text{ K}} = -109\text{ J/K}$$

The surroundings lose heat, and the negative sign means that the entropy of the surroundings decreases. The two entropy changes have the same magnitude but opposite signs, so they cancel:

$$\Delta S_{univ} = 109\text{ J/K} + (-109\text{ J/K}) = 0$$

For the reverse change (condensation), ΔS_{univ} also equals zero, but ΔS^0_{sys} and ΔS_{surr} have signs opposite those for vaporization. A similar treatment of a chemical change shows the same result: the entropy change of the forward reaction is *equal in magnitude but opposite in sign* to the entropy change of the reverse reaction. Thus, *when a system reaches equilibrium, neither the forward nor the reverse reaction is spontaneous,* and so there is no net reaction in either direction.

Spontaneous Exothermic and Endothermic Reactions: A Summary

We can now see why exothermic and endothermic spontaneous reactions occur. No matter what its *enthalpy* change, a reaction occurs because the total *entropy* of the reacting system *and* its surroundings increases. The two possibilities are

1. *For an exothermic reaction* ($\Delta H_{sys} < 0$), heat is released by the system, which increases the freedom of motion and energy dispersed and, thus, the entropy of the surroundings ($\Delta S_{surr} > 0$).

- If the reacting system yields products whose entropy is greater than that of the reactants ($\Delta S_{sys} > 0$), the total entropy change ($\Delta S_{sys} + \Delta S_{surr}$) will be positive (Figure 20.10A). For example, in the oxidation of glucose, an essential reaction for all higher organisms,

$$C_6H_{12}O_6(s) + 6O_2(g) \longrightarrow 6CO_2(g) + 6H_2O(g) + \textbf{\textit{heat}}$$

6 mol of gas yields 12 mol of gas and heat; thus, $\Delta S_{sys} > 0$, $\Delta S_{surr} > 0$, and $\Delta S_{univ} > 0$.

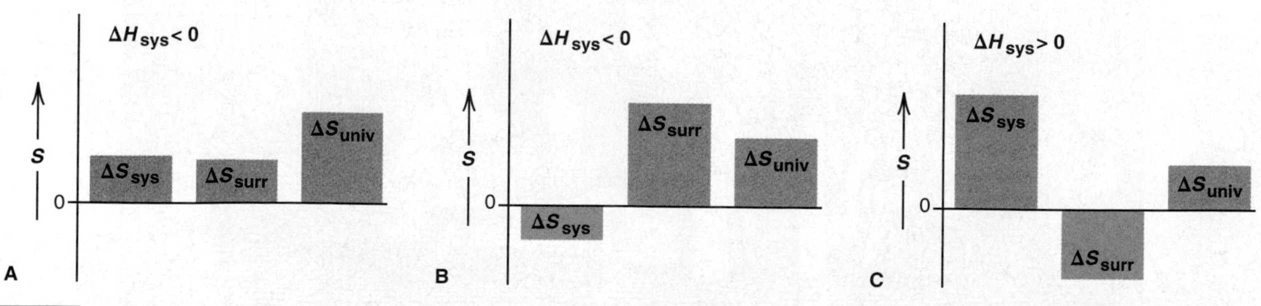

Figure 20.10 Components of ΔS_{univ} for spontaneous reactions. For a reaction to occur spontaneously, ΔS_{univ} must be positive. **A,** An exothermic reaction in which ΔS_{sys} increases; the size of ΔS_{surr} is not important. **B,** An exothermic reaction in which ΔS_{sys} decreases; ΔS_{surr} must be larger than ΔS_{sys}. **C,** An endothermic reaction in which ΔS_{sys} increases; ΔS_{surr} must be smaller than ΔS_{sys}.

- If, on the other hand, the entropy of the system decreases as the reaction occurs ($\Delta S_{sys} < 0$), the entropy of the surroundings must increase even more ($\Delta S_{surr} >> 0$) to make the total ΔS positive (Figure 20.10B). For example, when calcium oxide and carbon dioxide form calcium carbonate,

$$CaO(s) + CO_2(g) \longrightarrow CaCO_3(s) + \textbf{heat}$$

the entropy of the system decreases because the amount (mol) of gas decreases. However, the heat released increases the entropy of the surroundings even more; thus, $\Delta S_{sys} < 0$, but $\Delta S_{surr} >> 0$, so $\Delta S_{univ} > 0$.

2. *For an endothermic reaction* ($\Delta H_{sys} > 0$), the heat lost by the surroundings decreases molecular freedom of motion and dispersal of energy and, thus, decreases the entropy of the surroundings ($\Delta S_{surr} < 0$). Therefore, the only way an endothermic reaction can occur spontaneously is if ΔS_{sys} is positive and large enough to outweigh the negative ΔS_{surr} (Figure 20.10C).

- In the solution process for many ionic compounds, heat is absorbed to form the solution, so the entropy of the surroundings decreases ($\Delta S_{surr} < 0$). However, when the crystalline solid becomes freely moving ions, the entropy increase is so large ($\Delta S_{sys} >> 0$) that it outweighs the negative ΔS_{surr}. Thus, ΔS_{univ} is positive.
- Spontaneous endothermic reactions have similar features. For example, in the reaction between barium hydroxide octahydrate and ammonium nitrate (see Figure 20.1, p. 866),

$$\textbf{heat} + Ba(OH)_2{\cdot}8H_2O(s) + 2NH_4NO_3(s) \longrightarrow Ba^{2+}(aq) + 2NO_3^-(aq) + 2NH_3(aq) + 10H_2O(l)$$

3 mol of crystalline solids absorb heat from the surroundings ($\Delta S_{surr} < 0$) and yield 15 mol of dissolved ions and molecules, which have much more freedom of motion and so much greater entropy ($\Delta S_{sys} >> 0$).

The Greatness and Obscurity of J. Willard Gibbs Even today, one of the greatest minds in science is barely known outside chemistry and physics. In 1878, Josiah Willard Gibbs (1839–1903), a professor of mathematical physics at Yale, completed a 323-page paper that virtually established the science of chemical thermodynamics and included major principles governing chemical equilibria, phase-change equilibria, and the energy changes in electrochemical cells. The European scientists James Clerk Maxwell, Wilhelm Ostwald, and Henri Le Châtelier appreciated the significance of Gibbs's achievements long before most of his American colleagues did. In fact, he was elected to the Hall of Fame of Distinguished Americans only in 1950, because he had not received enough votes until then!

SECTION SUMMARY

The standard entropy of reaction, ΔS^0_{rxn}, is calculated from S^0 values. When the amount (mol) of gas (Δn_{gas}) increases in a reaction, usually $\Delta S^0_{rxn} > 0$. The value of ΔS_{surr} is related directly to ΔH^0_{sys} and inversely to the T at which the change occurs. In a spontaneous change, the entropy of the system can decrease only if the entropy of the surroundings increases even more. For a system at equilibrium, $\Delta S_{univ} = 0$.

20.3 ENTROPY, FREE ENERGY, AND WORK

By making *two* separate measurements, ΔS_{sys} and ΔS_{surr}, we can predict whether a reaction will be spontaneous at a particular temperature. It would be useful, however, to have *one* criterion for spontaneity that applies only to the system. The Gibbs free energy, or simply **free energy (G)**, is a function that combines the system's enthalpy and entropy:

$$G = H - TS$$

Named for Josiah Willard Gibbs, who proposed it and laid much of the foundation for chemical thermodynamics, this function provides the criterion for spontaneity we've been seeking.

Free Energy Change and Reaction Spontaneity

The free energy change (ΔG) is a measure of the spontaneity of a process and of the useful energy available from it. Let's see how the free energy change is derived from the second law of thermodynamics. Recall that by definition, the entropy change of the universe is the sum of the entropy changes of the system and the surroundings:

$$\Delta S_{univ} = \Delta S_{sys} + \Delta S_{surr}$$

At constant pressure,

$$\Delta S_{surr} = -\frac{\Delta H_{sys}}{T}$$

Substituting for ΔS_{surr} gives a relationship that lets us focus solely on the system:

$$\Delta S_{univ} = \Delta S_{sys} - \frac{\Delta H_{sys}}{T}$$

Multiplying both sides by $-T$ gives

$$-T\Delta S_{univ} = \Delta H_{sys} - T\Delta S_{sys}$$

Now we can introduce the new free energy quantity to replace the enthalpy and entropy terms. From $G = H - TS$, the *Gibbs equation* shows us the change in the free energy of the system (ΔG_{sys}) at constant temperature and pressure:

$$\Delta G_{sys} = \Delta H_{sys} - T\Delta S_{sys} \quad \textbf{(20.6)}$$

Combining this equation with the previous one shows that

$$-T\Delta S_{univ} = \Delta H_{sys} - T\Delta S_{sys} = \Delta G_{sys}$$

*The **sign** of ΔG tells if a reaction is spontaneous.* The second law dictates

- $\Delta S_{univ} > 0$ for a spontaneous process
- $\Delta S_{univ} < 0$ for a nonspontaneous process
- $\Delta S_{univ} = 0$ for a process at equilibrium

Of course, absolute temperature is always positive, so

$$T\Delta S_{univ} > 0 \quad \text{or} \quad -T\Delta S_{univ} < 0 \text{ for a spontaneous process}$$

Because $\Delta G = -T\Delta S_{univ}$, we know that

- $\Delta G < 0$ for a spontaneous process
- $\Delta G > 0$ for a nonspontaneous process
- $\Delta G = 0$ for a process at equilibrium

An important point to keep in mind is that if a process is *nonspontaneous* in one direction ($\Delta G > 0$), it is *spontaneous* in the opposite direction ($\Delta G < 0$). By using ΔG, we have not incorporated any new ideas, but we can predict reaction spontaneity from one variable (ΔG_{sys}) rather than two (ΔS_{sys} and ΔS_{surr}).

As we noted at the beginning of the chapter, the degree of spontaneity of a reaction—that is, the sign *and* magnitude of ΔG—tells us nothing about its rate. Remember that some spontaneous reactions are extremely slow. For example, the reaction between $H_2(g)$ and $O_2(g)$ at room temperature is highly spontaneous—ΔG has a large negative value—but in the absence of a catalyst or a flame, the reaction doesn't occur to a measurable extent because its rate is so low.

Calculating Standard Free Energy Changes

Because free energy (G) combines three state functions, H, S, and T, it is also a state function. As with enthalpy, we focus on the free energy *change* (ΔG).

The Standard Free Energy Change As we did with the other thermodynamic variables, to compare the free energy changes of different reactions we calculate the ***standard* free energy change (ΔG^0),** which occurs when all components of the system are in their standard states. Adapting the Gibbs equation (20.6), we have

$$\Delta G^0_{sys} = \Delta H^0_{sys} - T\Delta S^0_{sys} \quad \textbf{(20.7)}$$

This important relationship is used frequently to find any one of these three central thermodynamic variables, given the other two, as in this sample problem.

SAMPLE PROBLEM 20.4 Calculating ΔG^0_{rxn} from Enthalpy and Entropy Values

Problem Potassium chlorate, a common oxidizing agent in fireworks (see photo) and matchheads, undergoes a solid-state disproportionation reaction when heated:

$$4\overset{+5}{KClO_3}(s) \xrightarrow{\Delta} 3\overset{+7}{KClO_4}(s) + \overset{-1}{KCl}(s)$$

Use ΔH^0_f and S^0 values to calculate ΔG^0_{sys} (ΔG^0_{rxn}) at 25°C for this reaction.

Plan To solve for ΔG^0, we need values from Appendix B. We use ΔH^0_f values to calculate ΔH^0_{rxn} (ΔH^0_{sys}), use S^0 values to calculate ΔS^0_{rxn} (ΔS^0_{sys}), and then apply Equation 20.7.

Solution Calculating ΔH^0_{sys} from ΔH^0_f values (with Equation 6.8):

$$\begin{aligned}\Delta H^0_{sys} = \Delta H^0_{rxn} &= \Sigma m\Delta H^0_{f(products)} - \Sigma n\Delta H^0_{f(reactants)}\\ &= [(3\text{ mol }KClO_4)(\Delta H^0_f\text{ of }KClO_4) + (1\text{ mol }KCl)(\Delta H^0_f\text{ of }KCl)]\\ &\quad - [(4\text{ mol }KClO_3)(\Delta H^0_f\text{ of }KClO_3)]\\ &= [(3\text{ mol})(-432.8\text{ kJ/mol}) + (1\text{ mol})(-436.7\text{ kJ/mol})]\\ &\quad - [(4\text{ mol})(-397.7\text{ kJ/mol})]\\ &= -144\text{ kJ}\end{aligned}$$

Calculating ΔS^0_{sys} from S^0 values (with Equation 20.4):

$$\begin{aligned}\Delta S^0_{sys} = \Delta S^0_{rxn} &= [(3\text{ mol }KClO_4)(S^0\text{ of }KClO_4) + (1\text{ mol }KCl)(S^0\text{ of }KCl)]\\ &\quad - [(4\text{ mol }KClO_3)(S^0\text{ of }KClO_3)]\\ &= [(3\text{ mol})(151.0\text{ J/mol·K}) + (1\text{ mol})(82.6\text{ J/mol·K})]\\ &\quad - [(4\text{ mol})(143.1\text{ J/mol·K})]\\ &= -36.8\text{ J/K}\end{aligned}$$

Calculating ΔG^0_{sys} at 298 K:

$$\Delta G^0_{sys} = \Delta H^0_{sys} - T\Delta S^0_{sys} = -144\text{ kJ} - \left[(298\text{ K})(-36.8\text{ J/K})\left(\frac{1\text{ kJ}}{1000\text{ J}}\right)\right] = -133\text{ kJ}$$

Check Rounding to check the math:

$\Delta H^0 \approx [3(-433\text{ kJ}) + (-440\text{ kJ})] - [4(-400\text{ kJ})] = -1740\text{ kJ} + 1600\text{ kJ} = -140\text{ kJ}$

$\Delta S^0 \approx [3(150\text{ J/K}) + 85\text{ J/K}] - [4(145\text{ J/K})] = 535\text{ J/K} - 580\text{ J/K} = -45\text{ J/K}$

$\Delta G^0 \approx -140\text{ kJ} - 300\text{ K}(-0.04\text{ kJ/K}) = -140\text{ kJ} + 12\text{ kJ} = -128\text{ kJ}$

All values are close to the calculated ones. Another way to calculate ΔG^0 for this reaction is presented in Sample Problem 20.5.

Comment 1. For a spontaneous reaction under *any* conditions, the free energy change, ΔG, is negative. Under standard-state conditions, a spontaneous reaction has a negative *standard* free energy change; that is, $\Delta G^0 < 0$.

2. This reaction is spontaneous, but the rate is very low in the solid. When $KClO_3$ is heated slightly above its melting point, the ions are free to move and the reaction occurs readily.

Potassium chlorate is the oxidizing agent in fireworks.

FOLLOW-UP PROBLEM 20.4 Determine the standard free energy change at 298 K for the reaction $2NO(g) + O_2(g) \longrightarrow 2NO_2(g)$.

The Standard Free Energy of Formation Another way to calculate ΔG^0_{rxn} is with values for the **standard free energy of formation (ΔG^0_f)** of the components; ΔG^0_f is the free energy change that occurs when 1 mol of compound is made *from its elements,* with all components in their standard states. Because free energy is a state function, we can combine ΔG^0_f values of reactants and products to calculate ΔG^0_{rxn} no matter how the reaction takes place:

$$\Delta G^0_{rxn} = \Sigma m\Delta G^0_{f(products)} - \Sigma n\Delta G^0_{f(reactants)} \qquad (20.8)$$

ΔG^0_f values have properties similar to ΔH^0_f values:

- ΔG^0_f of an element in its standard state is zero.
- An equation coefficient (m or n above) multiplies ΔG^0_f by that number.
- Reversing a reaction changes the sign of ΔG^0_f.

Many ΔG^0_f values appear along with those for ΔH^0_f and S^0 in Appendix B.

SAMPLE PROBLEM 20.5 Calculating ΔG^0_{rxn} from ΔG^0_f Values

Problem Use ΔG^0_f values to calculate ΔG^0_{rxn} for the reaction in Sample Problem 20.4:

$$4KClO_3(s) \longrightarrow 3KClO_4(s) + KCl(s)$$

Plan We apply Equation 20.8 to calculate ΔG^0_{rxn}.

Solution

$$\begin{aligned}\Delta G^0_{rxn} &= \Sigma m\Delta G^0_{f(products)} - \Sigma n\Delta G^0_{f(reactants)}\\ &= [(3 \text{ mol } KClO_4)(\Delta G^0_f \text{ of } KClO_4) + (1 \text{ mol } KCl)(\Delta G^0_f \text{ of } KCl)]\\ &\quad - [(4 \text{ mol } KClO_3)(\Delta G^0_f \text{ of } KClO_3)]\\ &= [(3 \text{ mol})(-303.2 \text{ kJ/mol}) + (1 \text{ mol})(-409.2 \text{ kJ/mol})]\\ &\quad - [(4 \text{ mol})(-296.3 \text{ kJ/mol})]\\ &= -134 \text{ kJ}\end{aligned}$$

Check Rounding to check the math:

$$\Delta G^0_{rxn} \approx [3(-300 \text{ kJ}) + 1(-400 \text{ kJ})] - 4(-300 \text{ kJ})$$
$$= -1300 \text{ kJ} + 1200 \text{ kJ} = -100 \text{ kJ}$$

Comment The slight discrepancy between this answer and that obtained in Sample Problem 20.4 is within experimental error. As you can see, when ΔG^0_f values are available for a reaction taking place at 25°C, this method is simpler than that in Sample Problem 20.4.

FOLLOW-UP PROBLEM 20.5 Use ΔG^0_f values to calculate the free energy change at 25°C for each of the following reactions:
(a) $2NO(g) + O_2(g) \longrightarrow 2NO_2(g)$ (from Follow-up Problem 20.4)
(b) $2C(graphite) + O_2(g) \longrightarrow 2CO(g)$

ΔG and the Work a System Can Do

Recall that the science of thermodynamics was born soon after the invention of the steam engine, and one of the most practical relationships in the field is that between the free energy change and the work a system can do:

- For a spontaneous process ($\Delta G < 0$) at constant T and P, ΔG is the *maximum useful work obtainable* ***from*** *the system* ($-w$) *as the process takes place:*

$$\Delta G = -w_{max} \qquad (20.9)$$

- For a nonspontaneous process ($\Delta G > 0$) at constant T and P, ΔG is the *minimum work that must be done* ***to*** *the system to make the process take place.*

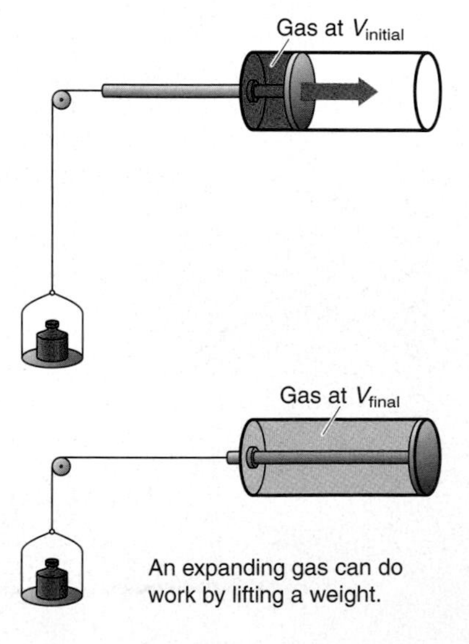

An expanding gas can do work by lifting a weight.

The phrase "maximum useful work obtainable" requires some explanation. First, what do we mean by the "maximum work obtainable," and what determines this limit? The free energy change is the maximum work the system can *possibly* do. But the work the system *actually* does depends on how the free energy is released. To understand this, let's first consider a nonchemical process. Suppose a gas is confined within a cylinder, at $V_{initial}$, by a piston attached to a 1-kg weight (see margin). As the gas expands and lifts the weight, its pressure becomes just balanced by the weight at some final volume, V_{final}. The gas lifted the weight in one step, thus doing a certain quantity of work. The gas can do more work in two steps, by lifting a 2-kg weight to one-half V_{final} and then lifting a 1-kg weight the rest of the way to V_{final}. The gas can do even more work in three steps, by lifting a 3-kg weight to one-third V_{final}, the 2-kg weight to one-half V_{final}, and then the 1-kg weight the rest of the way to V_{final}. As the number of steps increases, the quantity of work done by the gas increases. The gas would do *much* more work, nearly the maximum, if the weights were replaced by a container of sand and the weight adjusted continually grain by grain, as described earlier. In this way, the gas would lift the container in a very high number of steps. Note that, in a limiting hypothetical process, the work would be done in an infinite number of steps, and an infinitesimal increase in the weight would reverse the expansion. In general, *the maximum work is done by a spontaneous process only if the work is carried out reversibly.*

Of course, in any *real* process, work is performed *irreversibly*, that is, in a finite number of steps, so *we can never obtain the maximum work.* The free energy not used for work is lost to the surroundings as heat. This "unharnessed" energy is a consequence of any real process.

Next, what do we mean by "useful work"? As the gas expands and raises the weights, some of the work just pushes back the atmosphere. That work increases the atmosphere's energy, but it doesn't lift the weights, run a car, or do anything else useful. By the same token, if we wanted to compress the gas, the atmosphere would do part of the work by pushing on the piston, but that part of the work doesn't come from our muscles or from gasoline or from the power company. The ΔG function ignores the work done to or by the atmosphere and tells us about the work we can use or have to do ourselves.

Let's consider two examples of systems that use a chemical reaction to do work—a car engine and a battery. When gasoline (represented here by octane) is burned in a car engine,

$$C_8H_{18}(l) + \tfrac{25}{2}O_2(g) \longrightarrow 8CO_2(g) + 9H_2O(g),$$

a large amount of energy is given off as heat ($\Delta H_{sys} < 0$), and because the number of moles of gas increases, the entropy of the system increases ($\Delta S_{sys} > 0$). Therefore, the reaction is spontaneous ($\Delta G_{sys} < 0$). The free energy available does work turning wheels, moving belts, regenerating the battery, and so on. However, only if it is released reversibly, that is, in a series of infinitesimal steps, do we obtain the maximum work available from this reaction. In reality, the process is carried out irreversibly and much of the total free energy just warms the engine and the outside air, which increases the freedom of motion of the particles in the universe, in accord with the second law.

A battery is essentially a packaged spontaneous redox reaction that releases free energy to the surroundings (flashlight, radio, motor, etc.). If we connect the battery terminals to each other through a short piece of wire, the free energy change is released all at once but does no work—it just heats the wire and battery. If we connect the terminals to a motor, the free energy is released more slowly, and a sizeable portion of it runs the motor, but some is still converted to heat in the battery and the motor. If we discharge the battery still more slowly, more of the free energy change does work and less is converted to heat, but only when the battery discharges infinitely slowly can we obtain the maximum work.

This is the compromise that all engineers and machine designers must face—*in the real world, some free energy is always changed to heat and is, thereby, unharnessed: no real process uses all the available free energy to do work.*

The Wide Range of Energy Efficiency One definition for the efficiency of a device is the percentage of the energy input that results in work output. The range of efficiencies is enormous for the common "energy-conversion systems" of society. For instance, a common incandescent lightbulb converts only 5% of incoming electrical energy to light, the rest is given off as heat. At the other extreme, a large electrical generator converts 99% of the incoming mechanical energy to electricity. Although improvements are continually being made, here are some efficiency values for other devices: home oil furnace, 65%; hand tool motor, 63%; liquid fuel rocket, 50%; car engine, <30%; fluorescent lamp, 20%; solar cell, ~15%.

Let's summarize the relationship between the free energy change of a reaction and the work it can actually do:

- A spontaneous reaction ($\Delta G_{sys} < 0$) will occur *and* can do work on the surroundings. In any real machine, however, the work actually obtained from the reaction is *always less than the maximum possible* because some of the ΔG released is lost as heat.
- A nonspontaneous reaction ($\Delta G_{sys} > 0$) will not occur unless the surroundings do work on it. In any real machine, however, the work needed to make the reaction occur is *always more than the minimum* because some of the ΔG added is lost as heat.
- A reaction at equilibrium ($\Delta G_{sys} = 0$) can no longer do any work.

The Effect of Temperature on Reaction Spontaneity

In most cases, the enthalpy contribution (ΔH) to the free energy change (ΔG) is much *larger* than the entropy contribution ($T\Delta S$). For this reason, most exothermic reactions are spontaneous: the negative ΔH helps make ΔG negative. However, the *temperature of a reaction influences the magnitude of the* $T\Delta S$ *term,* so the overall spontaneity of many reactions depends on the temperature.

By scrutinizing the signs of ΔH and ΔS, we can predict the effect of temperature on the sign of ΔG and thus on the spontaneity of a process at any temperature. The values for the thermodynamic variables in this discussion are based on standard state values from Appendix B, but we show them without the superscript zero to emphasize that the relationships among ΔG, ΔH, and ΔS are valid at any conditions. Also, we assume that ΔH and ΔS change little with temperature, which is true as long as no phase changes occur.

Let's examine the four combinations of positive and negative ΔH and ΔS; two combinations do not depend on temperature and two do:

- *Temperature-independent cases.* When ΔH and ΔS have *opposite* signs, the reaction occurs spontaneously either at all temperatures or at none.
  1. *Reaction is spontaneous at all temperatures:* $\Delta H < 0$, $\Delta S > 0$. Both contributions favor the spontaneity of the reaction. ΔH is negative and ΔS is positive, so $-T\Delta S$ is negative; thus, ΔG is always negative. Many combustion reactions are in this category, including those of glucose and octane that we considered earlier. The decomposition of hydrogen peroxide, a common disinfectant, is also spontaneous at all temperatures:
  $$2H_2O_2(l) \longrightarrow 2H_2O(l) + O_2(g) \qquad \Delta H = -196 \text{ kJ and } \Delta S = 125 \text{ J/K}$$
  2. *Reaction is nonspontaneous at all temperatures:* $\Delta H > 0$, $\Delta S < 0$. Both contributions oppose the spontaneity of the reaction. ΔH is positive and ΔS is negative, so $-T\Delta S$ is positive; thus, ΔG is always positive. The formation of ozone from oxygen is not spontaneous at any temperature:
  $$3O_2(g) \longrightarrow 2O_3(g) \qquad \Delta H = 286 \text{ kJ and } \Delta S = -137 \text{ J/K}$$
  This reaction occurs only if enough energy is supplied from the surroundings, as when ozone is synthesized by passing an electrical discharge through pure O_2.

- *Temperature-dependent cases.* When ΔH and ΔS have the *same* sign, the relative magnitudes of the $-T\Delta S$ and ΔH terms determine the sign of ΔG. In these cases, the magnitude of T is crucial to reaction spontaneity.
  3. *Reaction is spontaneous at higher temperatures:* $\Delta H > 0$ *and* $\Delta S > 0$. In these cases, ΔS favors spontaneity ($-T\Delta S < 0$), but ΔH does not. For example,
  $$2N_2O(g) + O_2(g) \longrightarrow 4NO(g) \qquad \Delta H = 197.1 \text{ kJ and } \Delta S = 198.2 \text{ J/K}$$
  With a positive ΔH, the reaction will occur spontaneously only when $-T\Delta S$ is large enough to make ΔG negative, which will happen at higher temperatures. The oxidation of N_2O occurs spontaneously at $T > 994$ K.
  4. *Reaction is spontaneous at lower temperatures:* $\Delta H < 0$ *and* $\Delta S < 0$. In these cases, ΔH favors spontaneity, but ΔS does not ($-T\Delta S > 0$). For example,
  $$4Fe(s) + 3O_2(g) \longrightarrow 2Fe_2O_3(s) \qquad \Delta H = -1651 \text{ kJ and } \Delta S = -549.4 \text{ J/K}$$
  With a negative ΔH, the reaction will occur spontaneously only if the $-T\Delta S$ term is smaller than the ΔH term, and this happens at lower temperatures. The production of iron(III) oxide occurs spontaneously at any $T < 3005$ K. Other examples are the formation of an ammonium halide from ammonia and a hydrogen halide, and the formation of metal oxides, fluorides, and chlorides from their elements.

Table 20.1 Reaction Spontaneity and the Signs of ΔH, ΔS, and ΔG

ΔH	ΔS	$-T\Delta S$	ΔG	Description
−	+	−	−	Spontaneous at all T
+	−	+	+	Nonspontaneous at all T
+	+	−	+ or −	Spontaneous at higher T; nonspontaneous at lower T
−	−	+	+ or −	Spontaneous at lower T; nonspontaneous at higher T

Table 20.1 summarizes these four possible combinations of ΔH and ΔS.

As you saw in Sample Problem 20.4, one way to calculate ΔG is from enthalpy and entropy changes. Because ΔH and ΔS usually change little with temperature, if no phase changes occur, we can use their values at 298 K to examine the effect of temperature on ΔG and, thus, on reaction spontaneity.

SAMPLE PROBLEM 20.6 Determining the Effect of Temperature on ΔG

Problem A key step in the production of sulfuric acid is the oxidation of $SO_2(g)$ to $SO_3(g)$:

$$2SO_2(g) + O_2(g) \longrightarrow 2SO_3(g)$$

At 298 K, $\Delta G = -141.6$ kJ; $\Delta H = -198.4$ kJ; and $\Delta S = -187.9$ J/K.
(a) Use the data to decide if this reaction is spontaneous at 25°C, and predict how ΔG will change with increasing T.
(b) Assuming that ΔH and ΔS are constant with T, is the reaction spontaneous at 900.°C?
Plan (a) We note the sign of ΔG to see if the reaction is spontaneous and the signs of ΔH and ΔS to see the effect of T. **(b)** We use Equation 20.7 to calculate ΔG from the given ΔH and ΔS at the higher T (in K).
Solution (a) $\Delta G < 0$, so the reaction is spontaneous at 298 K: SO_2 and O_2 will form SO_3 spontaneously. With $\Delta S < 0$, the term $-T\Delta S > 0$, and this term will become more positive at higher T. Therefore,

ΔG will become less negative, and the reaction less spontaneous, with increasing T.

(b) Calculating ΔG at 900.°C ($T = 273 + 900. = 1173$ K):

$$\Delta G = \Delta H - T\Delta S = -198.4 \text{ kJ} - [(1173 \text{ K})(-187.9 \text{ J/K})(1 \text{ kJ}/1000 \text{ J})] = 22.0 \text{ kJ}$$

$\Delta G > 0$, so the reaction is nonspontaneous at the higher T.
Check The answer in part (b) seems reasonable based on our prediction in part (a). The arithmetic seems correct, given considerable rounding:

$$\Delta G \approx -200 \text{ kJ} - [(1200 \text{ K})(-200 \text{ J/K})/1000 \text{ J}] = +40 \text{ kJ}$$

FOLLOW-UP PROBLEM 20.6 A reaction is nonspontaneous at room temperature but *is* spontaneous at −40°C. What can you say about the signs and relative magnitudes of ΔH, ΔS, and $-T\Delta S$?

The Temperature at Which a Reaction Becomes Spontaneous As you have just seen, when the signs of ΔH and ΔS are the same, some reactions that are nonspontaneous at one temperature become spontaneous at another, and vice versa. It would certainly be useful to know the temperature at which a reaction becomes spontaneous. This is the temperature at which a positive ΔG switches to a negative ΔG because of the changing magnitude of the $-T\Delta S$ term. We find this crossover temperature by setting ΔG equal to zero and solving for T:

$$\Delta G = \Delta H - T\Delta S = 0$$

Therefore,

$$\Delta H = T\Delta S \quad \text{and} \quad T = \frac{\Delta H}{\Delta S} \qquad \textbf{(20.10)}$$

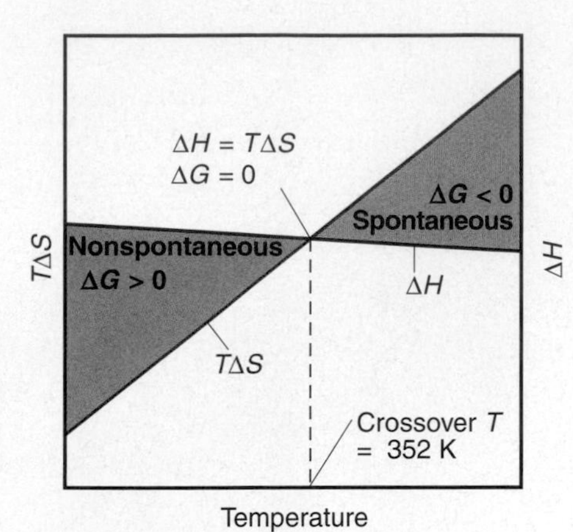

Figure 20.11 The effect of temperature on reaction spontaneity. The two terms that make up ΔG are plotted against T. The figure shows a relatively constant ΔH and a steadily increasing $T\Delta S$ (and thus more negative $-T\Delta S$) for the reaction between Cu_2O and C. At low T, the reaction is nonspontaneous ($\Delta G > 0$) because the positive ΔH term has a greater magnitude than the negative $T\Delta S$ term. At 352 K, $\Delta H = T\Delta S$, so $\Delta G = 0$. At any higher T, the reaction becomes spontaneous ($\Delta G < 0$) because the $-T\Delta S$ term dominates.

Consider the reaction of copper(I) oxide with carbon, which does *not* occur at lower temperatures but is used at higher temperatures in a step during the extraction of copper metal from chalcocite:

$$Cu_2O(s) + C(s) \longrightarrow 2Cu(s) + CO(g)$$

We predict this reaction has a positive entropy change because the number of moles of gas increases; in fact, $\Delta S = 165$ J/K. Furthermore, because the reaction is *non*spontaneous at lower temperatures, it must have a positive ΔH (58.1 kJ). As the $-T\Delta S$ term becomes more negative at higher temperatures, it will eventually outweigh the positive ΔH term, and the reaction will occur spontaneously.

We'll calculate ΔG for this reaction at 25°C and then find the temperature above which the reaction is spontaneous. At 25°C (298 K),

$$\Delta G = \Delta H - T\Delta S = 58.1 \text{ kJ} - \left(298 \text{ K} \times 165 \text{ J/K} \times \frac{1 \text{ kJ}}{1000 \text{ J}}\right) = 8.9 \text{ kJ}$$

Because ΔG is positive, the reaction will not proceed on its own at 25°C. At the crossover temperature, $\Delta G = 0$, so

$$T = \frac{\Delta H}{\Delta S} = \frac{58.1 \text{ kJ} \times \dfrac{1000 \text{ J}}{1 \text{ kJ}}}{165 \text{ J/K}} = 352 \text{ K}$$

At any temperature above 352 K (79°C), a moderate one for recovering a metal from its ore, the reaction occurs spontaneously. Figure 20.11 depicts this result. The line for $T\Delta S$ increases steadily (and thus the $-T\Delta S$ term becomes more negative) with rising temperature. This line crosses the relatively constant ΔH line at 352 K. At any higher temperature, the $-T\Delta S$ term is greater than the ΔH term, so ΔG is negative.

Coupling of Reactions to Drive a Nonspontaneous Change

When considering the spontaneity of a complex, multistep reaction, we often find that a nonspontaneous step is driven by a spontaneous step in a **coupling of reactions.** *One step supplies enough free energy for the other to occur*, just as the combustion of gasoline supplies enough free energy to move a car.

Look again at the reduction of copper(I) oxide by carbon. Previously, we found that the *overall* reaction becomes spontaneous at any temperature above 352 K. Dividing the reaction into two steps, however, we find that even at a higher temperature, such as 375 K, copper(I) oxide does not spontaneously decompose to its elements:

$$Cu_2O(s) \longrightarrow 2Cu(s) + \tfrac{1}{2}O_2(g) \qquad \Delta G_{375} = 140.0 \text{ kJ}$$

However, the oxidation of carbon to CO at 375 K is quite spontaneous:

$$C(s) + \tfrac{1}{2}O_2(g) \longrightarrow CO(g) \qquad \Delta G_{375} = -143.8 \text{ kJ}$$

Coupling these reactions means having the carbon in contact with the Cu_2O, which allows the reaction with the larger negative ΔG to "drive" the one with the smaller positive ΔG. Adding the reactions together and canceling the common substance ($\tfrac{1}{2}O_2$) gives an overall negative ΔG:

$$Cu_2O(s) + C(s) \longrightarrow 2Cu(s) + CO(g) \qquad \Delta G_{375} = -3.8 \text{ kJ}$$

Many biochemical reactions are also nonspontaneous. Key steps in the synthesis of proteins and nucleic acids, the formation of fatty acids, the maintenance of ion balance, and the breakdown of nutrients are among the many essential processes with positive ΔG values. Driving these energetically unfavorable steps by coupling them to a spontaneous one is a life-sustaining strategy common to all organisms—animals, plants, and microbes—as you'll see in the Chemical Connections essay.

The Universal Role of ATP

One of the most remarkable features of organisms, as well as a strong indication of a common ancestry, is the utilization of the same few biomolecules for all the reactions of life. Despite their incredible diversity of appearance and behavior, virtually all organisms use the same amino acids to make their proteins, the same nucleotides to make their nucleic acids, and the same carbohydrate (glucose) to provide energy.

In addition, *all organisms use the same spontaneous reaction to provide the free energy needed to drive a wide variety of nonspontaneous ones*. This reaction is the hydrolysis of a high-energy molecule called **adenosine triphosphate (ATP)** to adenosine diphosphate (ADP):

$$ATP^{4-} + H_2O \rightleftharpoons ADP^{3-} + HPO_4^{2-} + H^+ \qquad \Delta G^{0'} = -30.5 \text{ kJ}$$

(In biochemical systems, the standard-state concentration of H^+ is 10^{-7} *M*, not the usual 1 *M*, and the standard free energy change has the symbol $\Delta G^{0'}$.) In the metabolic breakdown of glucose, for example, the initial step is the addition of a phosphate group to a glucose molecule in a dehydration-condensation reaction:

$$\text{Glucose} + HPO_4^{2-} + H^+ \rightleftharpoons [\text{glucose phosphate}]^- + H_2O \qquad \Delta G^{0'} = 13.8 \text{ kJ}$$

Coupling this nonspontaneous reaction to ATP hydrolysis makes the overall process spontaneous. If we add the two reactions, HPO_4^{2-}, H^+, and H_2O cancel, and we obtain

$$\text{Glucose} + ATP^{4-} \rightleftharpoons [\text{glucose phosphate}]^- + ADP^{3-} \qquad \Delta G^{0'} = -16.7 \text{ kJ}$$

Like the reactions to extract copper that we just discussed, these two reactions cannot affect each other if they are physically separated. Coupling of the reactions is accomplished through an enzyme, a biological catalyst (Section 16.8) that simultaneously binds glucose and ATP such that the phosphate group of ATP to be transferred lies next to the particular —OH group of glucose that will accept it (Figure B20.3). Enzymes play similar catalytic roles in all the reactions driven by ATP hydrolysis.

The ADP formed in these energy-releasing reactions is combined with phosphate to regenerate ATP in energy-absorbing reactions catalyzed by other enzymes. In fact, one major biochemical reason an organism eats and breathes is to make ATP, so that it has the energy to move, grow, reproduce—and study chemistry, of course. Thus, there is a continuous cycling of ATP to ADP and back to ATP again to supply energy to the cells (Figure B20.4).

(continued)

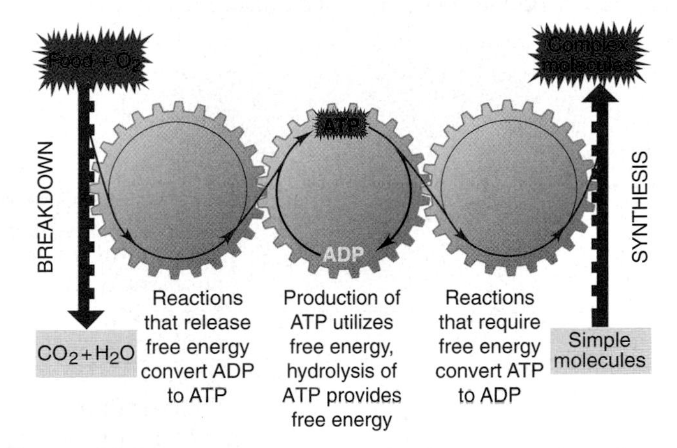

Figure B20.4 The cycling of metabolic free energy through ATP. Processes that release free energy are coupled to the formation of ATP from ADP, whereas those that require free energy are coupled to the hydrolysis of ATP to ADP.

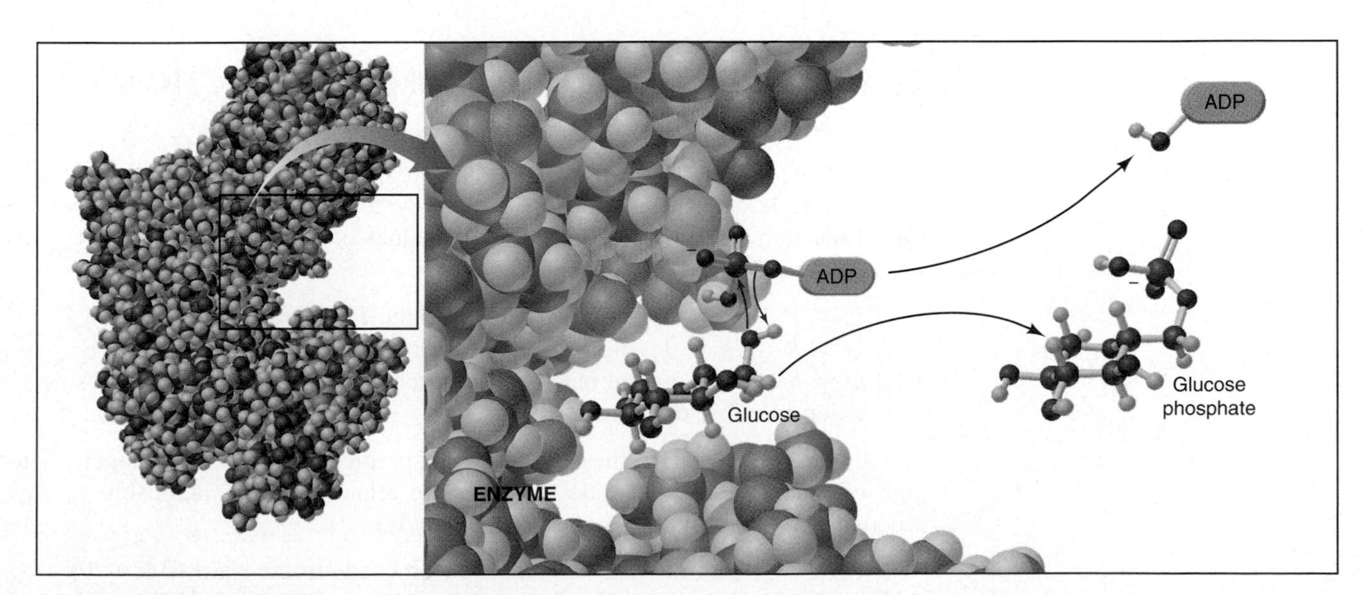

Figure B20.3 The coupling of a nonspontaneous reaction to the hydrolysis of ATP. The glucose molecule must lie next to the ATP molecule (shown as ADP—O—PO_3H) in the enzyme's active site for the correct atoms to form bonds. ADP (shown as ADP—OH) and glucose phosphate are released.

What makes ATP hydrolysis such a good supplier of free energy? By examining the phosphate portions of ATP, ADP, and HPO_4^{2-}, we can see two reasons (Figure B20.5). The first is that, at physiological pH (~7), the triphosphate portion of the ATP molecule has an average of four negative charges grouped closely together. As a result, the ATP molecule has a *high charge repulsion* built into its structure. In ADP, however, this charge repulsion is reduced.

The second reason relates to the greater delocalization of π electrons in the hydrolysis products, which we can see from resonance structures. For example, once ATP is hydrolyzed, the π electrons in HPO_4^{2-} can be more readily delocalized, which stabilizes the ion. Thus, greater charge repulsion and less electron delocalization make ATP higher in energy than the sum of ADP and HPO_4^{2-}. When ATP is hydrolyzed, some of this additional energy is released, to be harnessed by the organism to drive the metabolic reactions that could not otherwise take place.

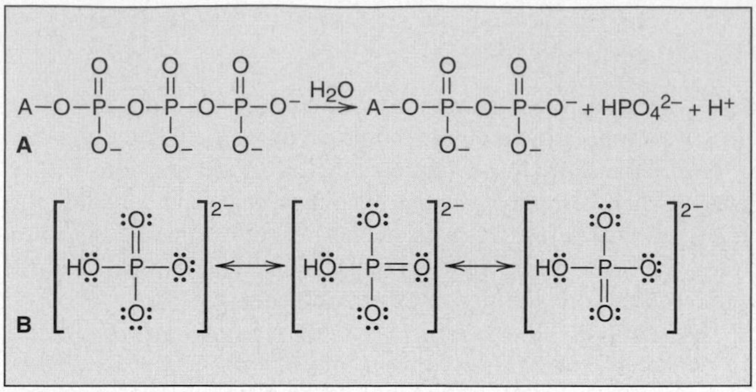

Figure B20.5 Why is ATP a high-energy molecule? The ATP molecule releases a large amount of free energy when it is hydrolyzed because **(A)** the high charge repulsion in the triphosphate portion of ATP is reduced, and **(B)** the free HPO_4^{2-} ion is stabilized by delocalization of its π electrons, as shown by these resonance forms.

SECTION SUMMARY

The sign of the free energy change, $\Delta G = \Delta H - T\Delta S$, is directly related to reaction spontaneity: a negative ΔG corresponds to a positive ΔS_{univ}. We use the standard free energy change (ΔG^0) to evaluate a reaction's spontaneity, and we use the standard free energy of formation (ΔG_f^0) to calculate ΔG_{rxn}^0 at 25°C. The maximum work a system can do is never obtained from a real (irreversible) process because some free energy is always converted to heat. The magnitude of T influences the spontaneity of temperature-dependent reactions (same signs of ΔH and ΔS) by affecting the size of $T\Delta S$. For such reactions, the T at which the reaction becomes spontaneous can be estimated by setting $\Delta G = 0$. A nonspontaneous reaction ($\Delta G > 0$) can be coupled to a more spontaneous one ($\Delta G << 0$) to make it occur. In organisms, the hydrolysis of ATP drives many reactions with a positive ΔG.

20.4 FREE ENERGY, EQUILIBRIUM, AND REACTION DIRECTION

The sign of ΔG allows us to predict reaction spontaneity and thus direction, but you already know that it is not the only way to do so. In Chapter 17, we predicted reaction direction by comparing the values of the reaction quotient (Q) and the equilibrium constant (K). Recall that

- If $Q < K$ ($Q/K < 1$), the reaction as written proceeds to the right.
- If $Q > K$ ($Q/K > 1$), the reaction as written proceeds to the left.
- If $Q = K$ ($Q/K = 1$), the reaction has reached equilibrium, and there is no net reaction in either direction.

As you might expect, these two ways of predicting reaction spontaneity—the sign of ΔG and the magnitude of Q/K—are related. Their relationship emerges when we compare the signs of $\ln Q/K$ with ΔG:

- If $Q/K < 1$, then $\ln Q/K < 0$: reaction proceeds to the right ($\Delta G < 0$).
- If $Q/K > 1$, then $\ln Q/K > 0$: reaction proceeds to the left ($\Delta G > 0$).
- If $Q/K = 1$, then $\ln Q/K = 0$: reaction is at equilibrium ($\Delta G = 0$).

Note that the signs of ΔG and $\ln Q/K$ are identical for a given reaction direction. In fact, ΔG and $\ln Q/K$ are proportional to each other and made equal through the constant RT:

$$\Delta G = RT \ln \frac{Q}{K} = RT \ln Q - RT \ln K \tag{20.11}$$

What does this central relationship mean? As you know, Q represents the concentrations (or pressures) of a system's components at any time during the reaction, whereas K represents them when the reaction has reached equilibrium. Therefore, Equation 20.11 says the free energy change for a reaction, ΔG, depends on how different the ratio of concentrations, Q, is from the equilibrium ratio, K.

The last term in Equation 20.11 is very important. By choosing standard-state values for Q, we obtain the standard free energy change (ΔG^0). When all concentrations are 1 M (or all pressures 1 atm), ΔG equals ΔG^0 and Q equals 1:

$$\Delta G^0 = RT \ln 1 - RT \ln K$$

We know that $\ln 1 = 0$, so the $RT \ln Q$ term drops out, and we have

$$\Delta G^0 = -RT \ln K \tag{20.12}$$

This relationship allows us to calculate the standard free energy change of a reaction (ΔG^0) from its equilibrium constant, or vice versa. Because ΔG^0 is related logarithmically to K, even a small change in the value of ΔG^0 has a large effect on the value of K. Table 20.2 shows the K values that correspond to a range of ΔG^0 values. Note that as ΔG^0 becomes more positive, the equilibrium constant becomes smaller, which means the reaction reaches equilibrium with less product and more reactant. Similarly, as ΔG^0 becomes more negative, the reaction reaches equilibrium with more product and less reactant. For example, if $\Delta G^0 = +10$ kJ, $K \approx 0.02$, which means that the magnitudes of the product terms are about $\frac{1}{50}$ those of the reactant terms; whereas, if $\Delta G^0 = -10$ kJ, they are 50 times larger.

Of course, most reactions do not begin with all components in their standard states. By substituting the relationship between ΔG^0 and K (Equation 20.12) into the expression for ΔG (Equation 20.11), we obtain a relationship that applies to any starting concentrations:

$$\Delta G = \Delta G^0 + RT \ln Q \tag{20.13}$$

Sample Problem 20.7 illustrates how Equations 20.12 and 20.13 are applied.

Table 20.2 The Relationship Between ΔG^0 and K at 298 K

ΔG^0 (kJ)	K	Significance
200	9×10^{-36}	Essentially no forward reaction; reverse reaction goes to completion
100	3×10^{-18}	
50	2×10^{-9}	
10	2×10^{-2}	
1	7×10^{-1}	
0	1	Forward and reverse reactions proceed to same extent
−1	1.5	
−10	5×10^{1}	
−50	6×10^{8}	
−100	3×10^{17}	
−200	1×10^{35}	Forward reaction goes to completion; essentially no reverse reaction

FORWARD REACTION (↓) / REVERSE REACTION (↑)

SAMPLE PROBLEM 20.7 Calculating ΔG at Nonstandard Conditions

Problem The oxidation of SO_2, which we considered in Sample Problem 20.6,

$$2SO_2(g) + O_2(g) \longrightarrow 2SO_3(g)$$

is too slow at 298 K to be useful in the manufacture of sulfuric acid. To overcome this low rate, the process is conducted at an elevated temperature.
(a) Calculate K at 298 K and at 973 K. ($\Delta G^0_{298} = -141.6$ kJ/mol for reaction as written; using ΔH^0 and ΔS^0 values at 973 K, $\Delta G^0_{973} = -12.12$ kJ/mol for reaction as written.)
(b) In experiments to determine the effect of temperature on reaction spontaneity, two sealed containers are filled with 0.500 atm of SO_2, 0.0100 atm of O_2, and 0.100 atm of SO_3 and kept at 25°C and at 700.°C. In which direction, if any, will the reaction proceed to reach equilibrium at each temperature?
(c) Calculate ΔG for the system in part (b) at each temperature.
Plan (a) We know ΔG^0, T, and R, so we can calculate the K's from Equation 20.12. **(b)** To determine if a net reaction will occur at the given pressures, we calculate Q with the given partial pressures and compare it with each K from part (a). **(c)** Because these are not standard-state pressures, we calculate ΔG at each T from Equation 20.13 with the values of ΔG^0 (given) and Q [found in part (b)].
Solution (a) Calculating K at the two temperatures:

$$\Delta G^0 = -RT \ln K \quad \text{so} \quad K = e^{-(\Delta G^0/RT)}$$

At 298 K, the exponent is

$$-(\Delta G^0/RT) = -\left(\frac{-141.6 \text{ kJ/mol} \times \dfrac{1000 \text{ J}}{1 \text{ kJ}}}{8.314 \text{ J/mol·K} \times 298 \text{ K}}\right) = 57.2$$

So $$K = e^{-(\Delta G^0/RT)} = e^{57.2} = 7\times10^{24}$$

At 973 K, the exponent is

$$-(\Delta G^0/RT) = -\left(\frac{-12.12 \text{ kJ/mol} \times \dfrac{1000 \text{ J}}{1 \text{ kJ}}}{8.314 \text{ J/mol·K} \times 973 \text{ K}}\right) = 1.50$$

So $$K = e^{-(\Delta G^0/RT)} = e^{1.50} = 4.5$$

(b) Calculating the value of Q:

$$Q = \frac{P^2_{SO_3}}{P^2_{SO_2} \times P_{O_2}} = \frac{0.100^2}{0.500^2 \times 0.0100} = 4.00$$

Because $Q < K$ at both temperatures, the denominator will decrease and the numerator increase—more SO_3 will form—until Q equals K. However, the reaction will go far to the right at 298 K before reaching equilibrium, whereas it will move only slightly to the right at 973 K.
(c) Calculating ΔG, the nonstandard free energy change, at 298 K:

$$\begin{aligned}\Delta G_{298} &= \Delta G^0 + RT \ln Q \\ &= -141.6 \text{ kJ/mol} + \left(8.314 \text{ J/mol·K} \times \frac{1 \text{ kJ}}{1000 \text{ J}} \times 298 \text{ K} \times \ln 4.00\right) \\ &= -138.2 \text{ kJ/mol}\end{aligned}$$

Calculating ΔG at 973 K:

$$\begin{aligned}\Delta G_{973} &= \Delta G^0 + RT \ln Q \\ &= -12.12 \text{ kJ/mol} + \left(8.314 \text{ J/mol·K} \times \frac{1 \text{ kJ}}{1000 \text{ J}} \times 973 \text{ K} \times \ln 4.00\right) \\ &= -0.9 \text{ kJ/mol}\end{aligned}$$

Check Note that in parts (a) and (c) we made the energy units in free energy changes (kJ) consistent with those in R (J). Based on the rules for significant figures in addition and subtraction, we retain one digit to the right of the decimal place in part (c).
Comment For these starting gas pressures at 973 K, the process is barely spontaneous ($\Delta G = -0.9$ kJ/mol), so why use a higher temperature? Like the synthesis of NH_3 (Chem-

ical Connections, p. 755), this process is carried out at a higher temperature *with a catalyst* to attain a higher *rate*, even though the *yield* is greater at a lower temperature. We discuss the details of the industrial production of sulfuric acid in Chapter 22.

FOLLOW-UP PROBLEM 20.7 At 298 K, hypobromous acid (HBrO) dissociates in water with a K_a of 2.3×10^{-9}.
(a) Calculate ΔG^0 for the dissociation of HBrO.
(b) Calculate ΔG if $[H_3O^+] = 6.0\times10^{-4}\ M$, $[BrO^-] = 0.10\ M$, and $[HBrO] = 0.20\ M$.

Another Look at the Meaning of Spontaneity At this point, let's consider some terminology related to, but distinct from, the terms *spontaneous* and *nonspontaneous*. Consider the general reaction A $\rightleftharpoons$ B, for which $K = [B]/[A] > 1$; therefore, the reaction proceeds largely from left to right (Figure 20.12A). From pure A to the equilibrium point, $Q < K$ and the curved *green* arrow indicates the reaction is spontaneous ($\Delta G < 0$). From there on, the curved *red* arrow shows the reaction is nonspontaneous ($\Delta G > 0$). From pure B to the equilibrium point, $Q > K$ and the reaction is also spontaneous ($\Delta G < 0$), but not thereafter. In either case, *the free energy decreases as the reaction proceeds, until it reaches a minimum at the equilibrium mixture:* $Q = K$ *and* $\Delta G = 0$. For the overall reaction A $\rightleftharpoons$ B (starting with all components in their standard states), G^0_B is smaller than G^0_A, so ΔG^0 is negative, which corresponds to $K > 1$. We call this a *product-favored* reaction because the final state of the system contains mostly product.

Now consider the opposite situation, a general reaction C $\rightleftharpoons$ D, for which $K = [D]/[C] < 1$: the reaction proceeds only slightly from left to right (Figure 20.12B). Here, too, whether we start with pure C or pure D, the reaction is spontaneous ($\Delta G < 0$) until the equilibrium point. But in this case, the equilibrium mixture contains mostly C (the reactant), so we say the reaction is *reactant favored.* In this case, G^0_D is *larger* than G^0_C, so ΔG^0 is *positive*, which corresponds to $K < 1$. The point is that *spontaneous* refers to that portion of a reaction in which the free energy is decreasing, that is, from some starting mixture to the equilibrium mixture, whereas *product-favored* refers to a reaction that goes predominantly, but not necessarily completely, to product (see Table 20.2).

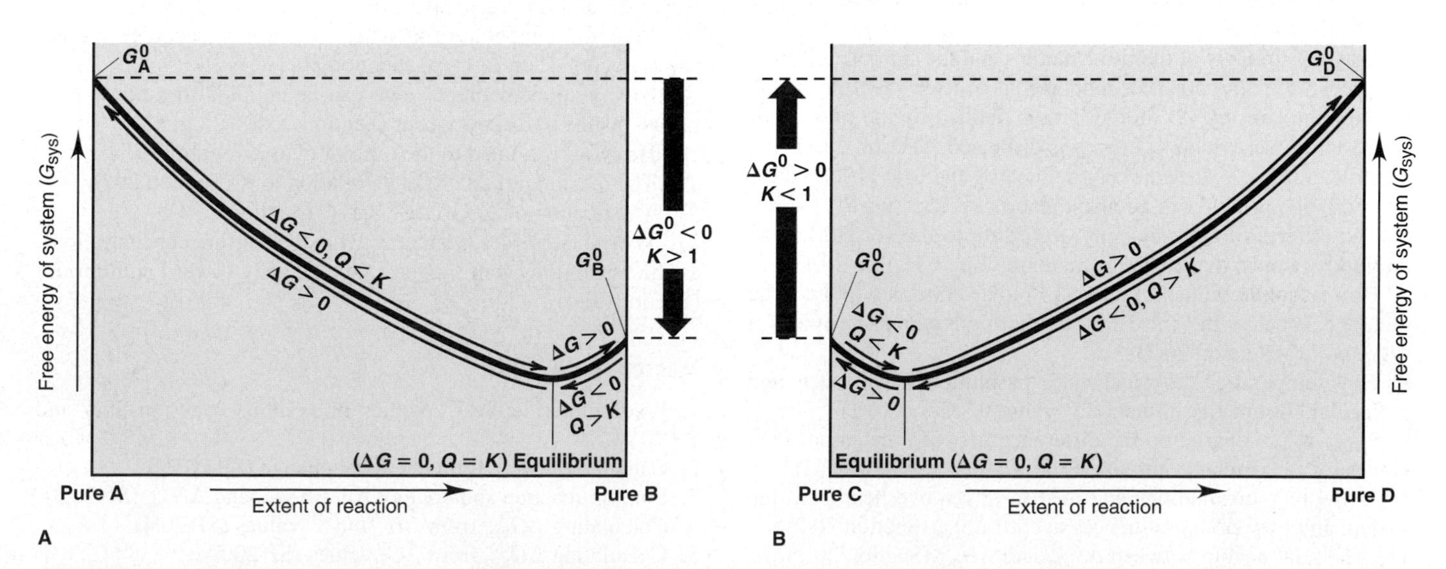

Figure 20.12 The relation between free energy and the extent of reaction. The free energy of the system is plotted against the extent of reaction. Each reaction proceeds spontaneously ($Q \neq K$ and $\Delta G < 0$; *curved green arrows*) from either pure reactants (A or C) or pure products (B or D) *to* the equilibrium mixture, at which point $\Delta G = 0$. The reaction *from* the equilibrium mixture to either pure reactants or products is nonspontaneous ($\Delta G > 0$; *curved red arrows*). **A,** For the product-favored reaction A $\rightleftharpoons$ B, $G^0_A > G^0_B$, so $\Delta G^0 < 0$ and $K > 1$. **B,** For the reactant-favored reaction C $\rightleftharpoons$ D, $G^0_D > G^0_C$, so $\Delta G^0 > 0$ and $K < 1$.

SECTION SUMMARY

Two ways of predicting reaction direction are from the value of ΔG and from the relation of Q to K. These variables represent different aspects of the same phenomenon and are related to each other by $\Delta G = RT \ln Q/K$. When $Q = K$, the system can release no more free energy. Beginning with Q at the standard state, the free energy change is ΔG^0, and it is related to the equilibrium constant by $\Delta G^0 = -RT \ln K$. For nonstandard conditions, ΔG has two components: ΔG^0 and $RT \ln Q$. Any nonequilibrium mixture of reactants and products moves spontaneously ($\Delta G < 0$) toward the equilibrium mixture. A product-favored reaction has $K > 1$ and, thus, $\Delta G^0 < 0$.

Chapter Perspective

As processes move toward equilibrium, some of the energy released becomes degraded to unusable heat, and the entropy of the universe increases. This unalterable necessity built into all natural processes has led to speculation about the "End of Everything," when all possible processes have stopped occurring, no free energy remains in any system, and nothing but waste heat is dispersed evenly throughout the universe—the final equilibrium! Even if this grim future is in store, however, it is many billions of years away, so you have plenty of time to appreciate more hopeful applications of thermodynamics. In Chapter 21, you'll see how spontaneous reactions can generate electricity and how electricity supplies free energy to drive nonspontaneous ones.

For Review and Reference

(Numbers in parentheses refer to pages, unless noted otherwise.)

Learning Objectives

Relevant section and/or sample problem (SP) numbers appear in parentheses.

Understand These Concepts

1. How the tendency of a process to occur by itself is distinct from how long it takes to occur (Introduction)
2. The distinction between a spontaneous and a nonspontaneous change (Section 20.1)
3. Why the first law of thermodynamics and the sign of ΔH^0 cannot predict the direction of a spontaneous process (Section 20.1)
4. How the entropy (S) of a system is defined by the number of microstates over which its energy is dispersed (Section 20.1)
5. How entropy is alternatively defined by the heat absorbed (or released) at constant T in a reversible process (Section 20.1)
6. The criterion for spontaneity according to the second law of thermodynamics: that a change increases S_{univ} (Section 20.1)
7. How absolute values of standard molar entropies (S^0) can be obtained because the third law of thermodynamics provides a "zero point" (Section 20.1)
8. How temperature, physical state, dissolution, atomic size, and molecular complexity influence S^0 values (Section 20.1)
9. How ΔS^0_{rxn} is based on the difference between the summed S^0 values for the reactants and those for products (Section 20.2)
10. How the surroundings add heat to or remove heat from the system and how ΔS_{surr} influences overall ΔS^0_{rxn} (Section 20.2)
11. The relationship between ΔS_{surr} and ΔH_{sys} (Section 20.2)
12. How reactions proceed spontaneously toward equilibrium ($\Delta S_{univ} > 0$) but proceed no further at equilibrium ($\Delta S_{univ} = 0$) (Section 20.2)
13. How the free energy change (ΔG) combines the system's entropy and enthalpy changes (Section 20.3)
14. How the expression for the free energy change is derived from the second law (Section 20.3)
15. The relationship between ΔG and the maximum work a system can perform and why this quantity of work is never performed in a real process (Section 20.3)
16. How temperature determines spontaneity for reactions in which ΔS and ΔH have the same sign (Section 20.3)
17. Why the temperature at which a reaction becomes spontaneous occurs when $\Delta G = 0$ (Section 20.3)
18. How a spontaneous change can be coupled to a nonspontaneous change to make it occur (Section 20.3)
19. How ΔG is related to the ratio of Q to K (Section 20.4)
20. The meaning of ΔG^0 and its relation to K (Section 20.4)
21. The relation of ΔG to ΔG^0 and Q (Section 20.4)
22. Why G decreases, no matter what the starting concentrations, as the reacting system moves spontaneously toward equilibrium (Section 20.4)

Master These Skills

1. Predicting relative S^0 values of systems (Section 20.1 and SP 20.1)
2. Calculating ΔS^0_{rxn} for a chemical change (SP 20.2)
3. Finding reaction spontaneity from ΔS_{surr} and ΔH^0_{sys} (SP 20.3)
4. Calculating ΔG^0_{rxn} from ΔH^0_f and S^0 values (SP 20.4)
5. Calculating ΔG^0_{rxn} from ΔG^0_f values (SP 20.5)
6. Calculating the effect of temperature on ΔG (SP 20.6)
7. Calculating the temperature at which a reaction becomes spontaneous (Section 20.3)
8. Calculating K from ΔG^0 (Section 20.4 and SP 20.7)
9. Using ΔG^0 and Q to calculate ΔG at any conditions (SP 20.7)

Key Terms

Section 20.1
spontaneous change (864)
entropy (S) (867)
second law of thermodynamics (870)
third law of thermodynamics (871)
standard molar entropy (S^0) (871)

Section 20.2
standard entropy of reaction (ΔS^0_{rxn}) (875)

Section 20.3
free energy (G) (881)
standard free energy change (ΔG^0) (882)
standard free energy of formation (ΔG^0_f) (883)
coupling of reactions (888)
adenosine triphosphate (ATP) (889)

Key Equations and Relationships

20.1 Quantifying entropy in terms of the number of microstates (W) over which the energy of a system can be distributed (867):

$$S = k \ln W$$

20.2 Quantifying the entropy change in terms of heat absorbed (or released) in a reversible process (870)

$$\Delta S_{sys} = \frac{q_{rev}}{T}$$

20.3 Stating the second law of thermodynamics, for a spontaneous process (870):

$$\Delta S_{univ} = \Delta S_{sys} + \Delta S_{surr} > 0$$

20.4 Calculating the standard entropy of reaction from the standard molar entropies of reactants and products (876):

$$\Delta S^0_{rxn} = \Sigma m S^0_{products} - \Sigma n S^0_{reactants}$$

20.5 Relating the entropy change in the surroundings to the enthalpy change of the system and the temperature (877):

$$\Delta S_{surr} = -\frac{\Delta H_{sys}}{T}$$

20.6 Expressing the free energy change of the system in terms of its component enthalpy and entropy changes (Gibbs equation) (882):

$$\Delta G_{sys} = \Delta H_{sys} - T\Delta S_{sys}$$

20.7 Calculating the standard free energy change from standard enthalpy and entropy changes (882):

$$\Delta G^0_{sys} = \Delta H^0_{sys} - T\Delta S^0_{sys}$$

20.8 Calculating the standard free energy change from the standard free energies of formation (883):

$$\Delta G^0_{rxn} = \Sigma m \Delta G^0_{f(products)} - \Sigma n \Delta G^0_{f(reactants)}$$

20.9 Relating the free energy change to the maximum work a process can perform (884):

$$\Delta G = -w_{max}$$

20.10 Finding the temperature at which a reaction becomes spontaneous (887):

$$T = \frac{\Delta H}{\Delta S}$$

20.11 Expressing the free energy change in terms of Q and K (891):

$$\Delta G = RT \ln \frac{Q}{K} = RT \ln Q - RT \ln K$$

20.12 Expressing the free energy change when Q is evaluated at the standard state (891):

$$\Delta G^0 = -RT \ln K$$

20.13 Expressing the free energy change for a nonstandard initial state (891):

$$\Delta G = \Delta G^0 + RT \ln Q$$

Highlighted Figures and Tables

These figures (F) and tables (T) provide a review of key ideas.

F20.5 Entropy and phase changes (872)
F20.10 Components of ΔS_{univ} for spontaneous reactions (880)
T20.1 Reaction spontaneity and the signs of ΔH, ΔS, and ΔG (887)
T20.2 The relationship of ΔG^0 and K (891)
F20.12 Free energy and extent of reaction (893)

Brief Solutions to Follow-up Problems

20.1 (a) $PCl_5(g)$: higher molar mass and more complex molecule; (b) $BaCl_2(s)$: higher molar mass; (c) $Br_2(g)$: gases have more freedom of motion and dispersal of energy than liquids.

20.2 (a) $2NaOH(s) + CO_2(g) \longrightarrow Na_2CO_3(s) + H_2O(l)$

$\Delta n_{gas} = -1$, so $\Delta S^0_{rxn} < 0$

$$\begin{aligned}\Delta S^0_{rxn} &= [(1\text{ mol } H_2O)(69.9\text{ J/mol·K}) \\ &\quad + (1\text{ mol } Na_2CO_3)(139\text{ J/mol·K})] \\ &\quad - [(1\text{ mol } CO_2)(213.7\text{ J/mol·K}) \\ &\quad + (2\text{ mol NaOH})(64.5\text{ J/mol·K})] \\ &= -134\text{ J/K}\end{aligned}$$

(b) $2Fe(s) + 3H_2O(g) \longrightarrow Fe_2O_3(s) + 3H_2(g)$

$\Delta n_{gas} = 0$, so cannot predict sign of ΔS^0_{rxn}

$$\begin{aligned}\Delta S^0_{rxn} &= [(1\text{ mol } Fe_2O_3)(87.4\text{ J/mol·K}) \\ &\quad + (3\text{ mol } H_2)(130.6\text{ J/mol·K})] \\ &\quad - [(2\text{ mol Fe})(27.3\text{ J/mol·K}) \\ &\quad + (3\text{ mol } H_2O)(188.7\text{ J/mol·K})] \\ &= -141.5\text{ J/K}\end{aligned}$$

Brief Solutions to Follow-up Problems *(continued)*

20.3 $2FeO(s) + \frac{1}{2}O_2(g) \longrightarrow Fe_2O_3(s)$

$$\Delta S^0_{sys} = (1 \text{ mol } Fe_2O_3)(87.4 \text{ J/mol·K}) - [(2 \text{ mol FeO})(60.75 \text{ J/mol·K}) + (\tfrac{1}{2} \text{ mol } O_2)(205.0 \text{ J/mol·K})] = -136.6 \text{ J/K}$$

$$\Delta H^0_{sys} = (1 \text{ mol } Fe_2O_3)(-825.5 \text{ kJ/mol}) - [(2 \text{ mol FeO})(-272.0 \text{ kJ/mol}) + (\tfrac{1}{2} \text{ mol } O_2)(0 \text{ kJ/mol})] = -281.5 \text{ kJ}$$

$$\Delta S_{surr} = -\frac{\Delta H^0_{sys}}{T} = -\frac{(-281.5 \text{ kJ} \times 1000 \text{ J/kJ})}{298 \text{ K}} = +945 \text{ J/K}$$

$$\Delta S_{univ} = \Delta S^0_{sys} + \Delta S_{surr} = -136.6 \text{ J/K} + 945 \text{ J/K}$$
$= 808$ J/K; reaction is spontaneous at 298 K.

20.4 Using ΔH^0_f and S^0 values from Appendix B,

$$\Delta H^0_{rxn} = -114.2 \text{ kJ and } \Delta S^0_{rxn} = -146.5 \text{ J/K}$$

$$\Delta G^0_{rxn} = \Delta H^0_{rxn} - T\Delta S^0_{rxn} = -114.2 \text{ kJ} - [(298 \text{ K})(-146.5 \text{ J/K})(1 \text{ kJ}/1000 \text{ J})] = -70.5 \text{ kJ}$$

20.5 (a) $\Delta G^0_{rxn} = (2 \text{ mol } NO_2)(51 \text{ kJ/mol}) - [(2 \text{ mol NO})(86.60 \text{ kJ/mol}) + (1 \text{ mol } O_2)(0 \text{ kJ/mol})] = -71 \text{ kJ}$

(b) $\Delta G^0_{rxn} = (2 \text{ mol CO})(-137.2 \text{ kJ/mol}) - [(2 \text{ mol C})(0 \text{ kJ/mol}) + (1 \text{ mol } O_2)(0 \text{ kJ/mol})] = -274.4 \text{ kJ}$

20.6 ΔG becomes negative at lower T, so $\Delta H < 0$, $\Delta S < 0$, and $-T\Delta S > 0$. At lower T, the negative ΔH value becomes larger than the positive $-T\Delta S$ value.

20.7 (a) $\Delta G^0 = -RT \ln K$

$$= -8.314 \text{ J/mol·K} \times \frac{1 \text{ kJ}}{1000 \text{ J}} \times 298 \text{ K} \times \ln (2.3\times10^{-9}) = 49 \text{ kJ/mol}$$

(b) $$Q = \frac{[H_3O^+][BrO^-]}{[HBrO]} = \frac{(6.0\times10^{-4})(0.10)}{0.20} = 3.0\times10^{-4}$$

$$\Delta G = \Delta G^0 + RT \ln Q = 49 \text{ kJ/mol} + \left[8.314 \text{ J/mol·K} \times \frac{1 \text{ kJ}}{1000 \text{ J}} \times 298 \text{ K} \times \ln (3.0\times10^{-4})\right] = 29 \text{ kJ/mol}$$

Problems

Problems with **colored** numbers are answered in Appendix E. Sections match the text and provide the numbers of relevant sample problems. Most offer Concept Review Questions, Skill-Building Exercises (grouped in pairs covering the same concept), and Problems in Context. Comprehensive Problems are based on material from any section or previous chapter.

Note: Unless stated otherwise, problems refer to systems at 298 K (25°C). Solving these problems may require values from Appendix B.

The Second Law of Thermodynamics: Predicting Spontaneous Change

(Sample Problem 20.1)

Concept Review Questions

20.1 Distinguish between the terms *spontaneous* and *instantaneous*. Give an example of a process that is spontaneous but very slow, and one that is very fast but not spontaneous.

20.2 Distinguish between the terms *spontaneous* and *nonspontaneous*. Can a nonspontaneous process occur? Explain.

20.3 State the first law of thermodynamics in terms of (a) the energy of the universe; (b) the creation or destruction of energy; (c) the energy change of system and surroundings. Does the first law reveal the direction of spontaneous change? Explain.

20.4 State qualitatively the relationship between entropy and freedom of particle motion. Use this idea to explain why you will probably never (a) be suffocated because all the air near you has moved to the other side of the room; (b) see half the water in your cup of tea freeze while the other half boils.

20.5 Why is ΔS_{vap} of a substance always larger than ΔS_{fus}?

20.6 How does the entropy of the surroundings change during an exothermic reaction? An endothermic reaction? Other than the examples cited in text, describe a spontaneous endothermic process.

20.7 (a) What is the entropy of a perfect crystal at 0 K?
(b) Does entropy increase or decrease as the temperature rises?
(c) Why is $\Delta H^0_f = 0$ but $S^0 > 0$ for an element?
(d) Why does Appendix B list ΔH^0_f values but not ΔS^0_f values?

Skill-Building Exercises *(grouped in similar pairs)*

20.8 Which of the following processes are spontaneous?
(a) Water evaporating from a puddle in summer
(b) A lion chasing an antelope
(c) An unstable isotope undergoing radioactive disintegration

20.9 Which of the following processes are spontaneous?
(a) Earth moving around the Sun
(b) A boulder rolling up a hill
(c) Sodium metal and chlorine gas reacting to form solid sodium chloride

20.10 Which of the following processes are spontaneous?
(a) Methane burning in air
(b) A teaspoonful of sugar dissolving in a cup of hot coffee
(c) A soft-boiled egg becoming raw

20.11 Which of the following processes are spontaneous?
(a) A satellite falling to Earth's surface
(b) Water decomposing to H_2 and O_2 at 298 K and 1 atm
(c) Average car prices increasing

20.12 Predict the sign of ΔS_{sys} for each process:
(a) A piece of wax melting
(b) Silver chloride precipitating from solution
(c) Dew forming

20.13 Predict the sign of ΔS_{sys} for each process:
(a) Gasoline vapors mixing with air in a car engine
(b) Hot air expanding
(c) Breath condensing in cold air

20.14 Predict the sign of ΔS_{sys} for each process:
(a) Alcohol evaporating
(b) A solid explosive converting to a gas
(c) Perfume vapors diffusing through a room

20.15 Predict the sign of ΔS_{sys} for each process:
(a) A pond freezing in winter
(b) Atmospheric CO_2 dissolving in the ocean
(c) An apple tree bearing fruit

20.16 Without using Appendix B, predict the sign of ΔS^0 for
(a) $2K(s) + F_2(g) \longrightarrow 2KF(s)$
(b) $NH_3(g) + HBr(g) \longrightarrow NH_4Br(s)$
(c) $NaClO_3(s) \longrightarrow Na^+(aq) + ClO_3^-(aq)$

20.17 Without using Appendix B, predict the sign of ΔS^0 for
(a) $H_2S(g) + \frac{1}{2}O_2(g) \longrightarrow \frac{1}{8}S_8(s) + H_2O(g)$
(b) $HCl(aq) + NaOH(aq) \longrightarrow NaCl(aq) + H_2O(l)$
(c) $2NO_2(g) \longrightarrow N_2O_4(g)$

20.18 Without using Appendix B, predict the sign of ΔS^0 for
(a) $CaCO_3(s) + 2HCl(aq) \longrightarrow CaCl_2(aq) + H_2O(l) + CO_2(g)$
(b) $2NO(g) + O_2(g) \longrightarrow 2NO_2(g)$
(c) $2KClO_3(s) \longrightarrow 2KCl(s) + 3O_2(g)$

20.19 Without using Appendix B, predict the sign of ΔS^0 for
(a) $Ag^+(aq) + Cl^-(aq) \longrightarrow AgCl(s)$
(b) $KBr(s) \longrightarrow KBr(aq)$
(c) $CH_3CH{=}CH_2(g) \longrightarrow H_2C{-}CH_2(g)$ (cyclopropane: ring with CH_2 at apex bonded to H_2C and CH_2)

20.20 Predict the sign of ΔS for each process:
(a) $C_2H_5OH(g)$ (350 K and 500 torr) $\longrightarrow$ $C_2H_5OH(g)$ (350 K and 250 torr)
(b) $N_2(g)$ (298 K and 1 atm) $\longrightarrow$ $N_2(aq)$ (298 K and 1 atm)
(c) $O_2(aq)$ (303 K and 1 atm) $\longrightarrow$ $O_2(g)$ (303 K and 1 atm)

20.21 Predict the sign of ΔS for each process:
(a) $O_2(g)$ (1.0 L at 1 atm) $\longrightarrow$ $O_2(g)$ (0.10 L at 10 atm)
(b) $Cu(s)$ (350°C and 2.5 atm) $\longrightarrow$ $Cu(s)$ (450°C and 2.5 atm)
(c) $Cl_2(g)$ (100°C and 1 atm) $\longrightarrow$ $Cl_2(g)$ (10°C and 1 atm)

20.22 Predict which substance has greater molar entropy. Explain.
(a) Butane $CH_3CH_2CH_2CH_3(g)$ or 2-butene $CH_3CH{=}CHCH_3(g)$
(b) $Ne(g)$ or $Xe(g)$ (c) $CH_4(g)$ or $CCl_4(l)$

20.23 Predict which substance has greater molar entropy. Explain.
(a) $NO_2(g)$ or $N_2O_4(g)$ (b) $CH_3OCH_3(l)$ or $CH_3CH_2OH(l)$
(c) $HCl(g)$ or $HBr(g)$

20.24 Predict which substance has greater molar entropy. Explain.
(a) $CH_3OH(l)$ or $C_2H_5OH(l)$ (b) $KClO_3(s)$ or $KClO_3(aq)$
(c) $Na(s)$ or $K(s)$

20.25 Predict which substance has greater molar entropy. Explain.
(a) $P_4(g)$ or $P_2(g)$ (b) $HNO_3(aq)$ or $HNO_3(l)$
(c) $CuSO_4(s)$ or $CuSO_4{\cdot}5H_2O(s)$

20.26 Without consulting Appendix B, arrange each group in order of *increasing* standard molar entropy (S^0). Explain.
(a) Graphite, diamond, charcoal
(b) Ice, water vapor, liquid water (c) O_2, O_3, O atoms

20.27 Without consulting Appendix B, arrange each group in order of *increasing* standard molar entropy (S^0). Explain.
(a) Glucose ($C_6H_{12}O_6$), sucrose ($C_{12}H_{22}O_{11}$), ribose ($C_5H_{10}O_5$)
(b) $CaCO_3$, $Ca + C + \frac{3}{2}O_2$, $CaO + CO_2$
(c) $SF_6(g)$, $SF_4(g)$, $S_2F_{10}(g)$

20.28 Without consulting Appendix B, arrange each group in order of *decreasing* standard molar entropy (S^0). Explain.
(a) $ClO_4^-(aq)$, $ClO_2^-(aq)$, $ClO_3^-(aq)$
(b) $NO_2(g)$, $NO(g)$, $N_2(g)$ (c) $Fe_2O_3(s)$, $Al_2O_3(s)$, $Fe_3O_4(s)$

20.29 Without consulting Appendix B, arrange each group in order of *decreasing* standard molar entropy (S^0). Explain.
(a) Mg metal, Ca metal, Ba metal
(b) Hexane (C_6H_{14}), benzene (C_6H_6), cyclohexane (C_6H_{12})
(c) $PF_2Cl_3(g)$, $PF_5(g)$, $PF_3(g)$

Calculating the Change in Entropy of a Reaction

(Sample Problems 20.2 and 20.3)

Concept Review Questions

20.30 What property of entropy allows Hess's law to be used in the calculation of entropy changes?

20.31 Describe the equilibrium condition in terms of the entropy changes of a system and its surroundings. What does this description mean about the entropy change of the universe?

20.32 For the reaction $H_2O(g) + Cl_2O(g) \longrightarrow 2HClO(g)$, you know ΔS^0_{rxn} and S^0 of $HClO(g)$ and of $H_2O(g)$. Write an expression to determine S^0 of $Cl_2O(g)$.

Skill-Building Exercises *(grouped in similar pairs)*

20.33 For each reaction, predict the sign and find the value of ΔS^0:
(a) $3NO(g) \longrightarrow N_2O(g) + NO_2(g)$
(b) $3H_2(g) + Fe_2O_3(s) \longrightarrow 2Fe(s) + 3H_2O(g)$
(c) $P_4(s) + 5O_2(g) \longrightarrow P_4O_{10}(s)$

20.34 For each reaction, predict the sign and find the value of ΔS^0:
(a) $3NO_2(g) + H_2O(l) \longrightarrow 2HNO_3(l) + NO(g)$
(b) $N_2(g) + 3F_2(g) \longrightarrow 2NF_3(g)$
(c) $C_6H_{12}O_6(s) + 6O_2(g) \longrightarrow 6CO_2(g) + 6H_2O(g)$

20.35 Find ΔS^0 for the combustion of ethane (C_2H_6) to carbon dioxide and gaseous water. Is the sign of ΔS^0 as expected?

20.36 Find ΔS^0 for the combustion of methane to carbon dioxide and liquid water. Is the sign of ΔS^0 as expected?

20.37 Find ΔS^0 for the reaction of nitric oxide with hydrogen to form ammonia and water vapor. Is the sign of ΔS^0 as expected?

20.38 Find ΔS^0 for the combustion of ammonia to nitrogen dioxide and water vapor. Is the sign of ΔS^0 as expected?

20.39 Find ΔS^0 for the formation of $Cu_2O(s)$ from its elements.

20.40 Find ΔS^0 for the formation of $HI(g)$ from its elements.

20.41 Find ΔS^0 for the formation of $CH_3OH(l)$ from its elements.

20.42 Find ΔS^0 for the formation of $PCl_5(g)$ from its elements.

Problems in Context

20.43 Sulfur dioxide is released in the combustion of coal. Scrubbers use lime slurries of calcium hydroxide to remove the SO_2 from flue gases. Write a balanced equation for this reaction and calculate ΔS^0 at 298 K [S^0 of $CaSO_3(s)$ = 101.4 J/mol·K].

20.44 Oxyacetylene welding is used to repair metal structures, including bridges, buildings, and even the Statue of Liberty. Calculate ΔS^0 for the combustion of 1 mol of acetylene (C_2H_2).

Entropy, Free Energy, and Work
(Sample Problems 20.4 to 20.6)

Concept Review Questions

20.45 What is the advantage of calculating free energy changes rather than entropy changes to determine reaction spontaneity?

20.46 Given that $\Delta G_{sys} = -T\Delta S_{univ}$, explain how the sign of ΔG_{sys} correlates with reaction spontaneity.

20.47 Is an endothermic reaction more likely to be spontaneous at higher temperatures or lower temperatures? Explain.

20.48 With its components in their standard states, a certain reaction is spontaneous only at high T. What do you know about the signs of ΔH^0 and ΔS^0? Describe a process for which this is true.

20.49 How can ΔS^0 be relatively independent of T if S^0 of each reactant and product increases with T?

Skill-Building Exercises *(grouped in similar pairs)*

20.50 Calculate ΔG^0 for each reaction using ΔG_f^0 values:
(a) $2Mg(s) + O_2(g) \longrightarrow 2MgO(s)$
(b) $2CH_3OH(g) + 3O_2(g) \longrightarrow 2CO_2(g) + 4H_2O(g)$
(c) $BaO(s) + CO_2(g) \longrightarrow BaCO_3(s)$

20.51 Calculate ΔG^0 for each reaction using ΔG_f^0 values:
(a) $H_2(g) + I_2(s) \longrightarrow 2HI(g)$
(b) $MnO_2(s) + 2CO(g) \longrightarrow Mn(s) + 2CO_2(g)$
(c) $NH_4Cl(s) \longrightarrow NH_3(g) + HCl(g)$

20.52 Find ΔG^0 for the reactions in Problem 20.50 using ΔH_f^0 and S^0 values.

20.53 Find ΔG^0 for the reactions in Problem 20.51 using ΔH_f^0 and S^0 values.

20.54 Consider the oxidation of carbon monoxide:
$$CO(g) + \tfrac{1}{2}O_2(g) \longrightarrow CO_2(g)$$
(a) Predict the signs of ΔS^0 and ΔH^0. Explain.
(b) Calculate ΔG^0 by two different methods.

20.55 Consider the combustion of butane gas:
$$C_4H_{10}(g) + \tfrac{13}{2}O_2(g) \longrightarrow 4CO_2(g) + 5H_2O(g)$$
(a) Predict the signs of ΔS^0 and ΔH^0. Explain.
(b) Calculate ΔG^0 by two different methods.

20.56 For the gaseous reaction of xenon and fluorine to form xenon hexafluoride:
(a) Calculate ΔS^0 at 298 K ($\Delta H^0 = -402$ kJ/mol and $\Delta G^0 = -280.$ kJ/mol).
(b) Assuming that ΔS^0 and ΔH^0 change little with temperature, calculate ΔG^0 at 500. K.

20.57 For the gaseous reaction of carbon monoxide and chlorine to form phosgene ($COCl_2$):
(a) Calculate ΔS^0 at 298 K ($\Delta H^0 = -220.$ kJ/mol and $\Delta G^0 = -206$ kJ/mol).
(b) Assuming that ΔS^0 and ΔH^0 change little with temperature, calculate ΔG^0 at 450. K.

20.58 One reaction used to produce small quantities of pure H_2 is
$$CH_3OH(g) \rightleftharpoons CO(g) + 2H_2(g)$$
(a) Determine ΔH^0 and ΔS^0 for the reaction at 298 K.
(b) Assuming that these values are relatively independent of temperature, calculate ΔG^0 at 38°C, 138°C, and 238°C.
(c) What is the significance of the different values of ΔG^0?

20.59 A reaction that occurs in the internal combustion engine is
$$N_2(g) + O_2(g) \rightleftharpoons 2NO(g)$$
(a) Determine ΔH^0 and ΔS^0 for the reaction at 298 K.
(b) Assuming that these values are relatively independent of temperature, calculate ΔG^0 at 100.°C, 2560.°C, and 3540.°C.
(c) What is the significance of the different values of ΔG^0?

20.60 Use ΔH^0 and ΔS^0 values for the following process at 1 atm to find the normal boiling point of Br_2:
$$Br_2(l) \rightleftharpoons Br_2(g)$$

20.61 Use ΔH^0 and ΔS^0 values to find the temperature at which these sulfur allotropes reach equilibrium at 1 atm:
$$S(\text{rhombic}) \rightleftharpoons S(\text{monoclinic})$$

Problems in Context

20.62 As a fuel, $H_2(g)$ produces only nonpolluting $H_2O(g)$ when it burns. Moreover, it combines with $O_2(g)$ in a fuel cell (Chapter 21) to provide electrical energy.
(a) Calculate ΔH^0, ΔS^0, and ΔG^0 per mol of H_2 at 298 K.
(b) Is the spontaneity of this reaction dependent on T? Explain.
(c) At what temperature does the reaction become spontaneous?

20.63 The U.S. requires automobile fuels to contain a renewable component. The fermentation of glucose from corn produces ethanol, which is added to gasoline to fulfill this requirement:
$$C_6H_{12}O_6(s) \longrightarrow 2C_2H_5OH(l) + 2CO_2(g)$$
Calculate ΔH^0, ΔS^0, and ΔG^0 for the reaction at 25°C. Is the spontaneity of this reaction dependent on T? Explain.

Free Energy, Equilibrium, and Reaction Direction
(Sample Problem 20.7)

Concept Review Questions

20.64 (a) If $K << 1$ for a reaction, what do you know about the sign and magnitude of ΔG^0?
(b) If $\Delta G^0 << 0$ for a reaction, what do you know about the magnitude of K? Of Q?

20.65 How is the free energy change of a process related to the work that can be obtained from the process? Is this quantity of work obtainable in practice? Explain.

20.66 What happens to a portion of the useful energy as it does work?

20.67 What is the difference between ΔG^0 and ΔG? Under what circumstances does $\Delta G = \Delta G^0$?

Skill-Building Exercises *(grouped in similar pairs)*

20.68 Calculate K at 298 K for each reaction:
(a) $NO(g) + \tfrac{1}{2}O_2(g) \rightleftharpoons NO_2(g)$
(b) $2HCl(g) \rightleftharpoons H_2(g) + Cl_2(g)$
(c) $2C(\text{graphite}) + O_2(g) \rightleftharpoons 2CO(g)$

20.69 Calculate K at 298 K for each reaction:
(a) $MgCO_3(s) \rightleftharpoons Mg^{2+}(aq) + CO_3^{2-}(aq)$
(b) $2HCl(g) + Br_2(l) \rightleftharpoons 2HBr(g) + Cl_2(g)$
(c) $H_2(g) + O_2(g) \rightleftharpoons H_2O_2(l)$

20.70 Calculate K at 298 K for each reaction:
(a) $2H_2S(g) + 3O_2(g) \rightleftharpoons 2H_2O(g) + 2SO_2(g)$
(b) $H_2SO_4(l) \rightleftharpoons H_2O(l) + SO_3(g)$
(c) $HCN(aq) + NaOH(aq) \rightleftharpoons NaCN(aq) + H_2O(l)$

20.71 Calculate K at 298 K for each reaction:
(a) $SrSO_4(s) \rightleftharpoons Sr^{2+}(aq) + SO_4^{2-}(aq)$
(b) $2NO(g) + Cl_2(g) \rightleftharpoons 2NOCl(g)$
(c) $Cu_2S(s) + O_2(g) \rightleftharpoons 2Cu(s) + SO_2(g)$

20.72 Use Appendix B to determine the K_{sp} of Ag_2S.

20.73 Use Appendix B to determine the K_{sp} of CaF_2.

20.74 For the reaction $I_2(g) + Cl_2(g) \rightleftharpoons 2ICl(g)$, calculate K_p at 25°C [ΔG^0_f of $ICl(g) = -6.075$ kJ/mol].

20.75 For the reaction $CaCO_3(s) \rightleftharpoons CaO(s) + CO_2(g)$, calculate the equilibrium P_{CO_2} at 25°C.

20.76 The K_{sp} of $PbCl_2$ is 1.7×10^{-5} at 25°C. What is ΔG^0? Is it possible to prepare a solution that contains $Pb^{2+}(aq)$ and $Cl^-(aq)$, at their standard-state concentrations?

20.77 The K_{sp} of ZnF_2 is 3.0×10^{-2} at 25°C. What is ΔG^0? Is it possible to prepare a solution that contains $Zn^{2+}(aq)$ and $F^-(aq)$ at their standard-state concentrations?

20.78 The equilibrium constant for the reaction

$$2Fe^{3+}(aq) + Hg_2^{2+}(aq) \rightleftharpoons 2Fe^{2+}(aq) + 2Hg^{2+}(aq)$$

is $K_c = 9.1\times10^{-6}$ at 298 K.
(a) What is ΔG^0 at this temperature?
(b) If standard-state concentrations of the reactants and products are mixed, in which direction does the reaction proceed?
(c) Calculate ΔG when $[Fe^{3+}] = 0.20\ M$, $[Hg_2^{2+}] = 0.010\ M$, $[Fe^{2+}] = 0.010\ M$, and $[Hg^{2+}] = 0.025\ M$. In which direction will the reaction proceed to achieve equilibrium?

20.79 The formation constant for the reaction

$$Ni^{2+}(aq) + 6NH_3(aq) \rightleftharpoons Ni(NH_3)_6^{2+}(aq)$$

is $K_f = 5.6\times10^8$ at 25°C.
(a) What is ΔG^0 at this temperature?
(b) If standard-state concentrations of the reactants and products are mixed, in which direction does the reaction proceed?
(c) Determine ΔG when $[Ni(NH_3)_6^{2+}] = 0.010\ M$, $[Ni^{2+}] = 0.0010\ M$, and $[NH_3] = 0.0050\ M$. In which direction will the reaction proceed to achieve equilibrium?

Problems in Context

20.80 High levels of ozone (O_3) make rubber deteriorate, green plants turn brown, and many people have difficulty breathing.
(a) Is the formation of O_3 from O_2 favored at all T, no T, high T, or low T?
(b) Calculate ΔG^0 for this reaction at 298 K.
(c) Calculate ΔG at 298 K for this reaction in urban smog where $[O_2] = 0.21$ atm and $[O_3] = 5\times10^{-7}$ atm.

20.81 A $BaSO_4$ slurry is ingested before the gastrointestinal tract is x-rayed because it is opaque to x-rays and defines the contours of the tract. Ba^{2+} ion is toxic, but the compound is nearly insoluble. If ΔG^0 at 37°C (body temperature) is 59.1 kJ/mol for the process

$$BaSO_4(s) \rightleftharpoons Ba^{2+}(aq) + SO_4^{2-}(aq)$$

what is $[Ba^{2+}]$ in the intestinal tract? (Assume that the only source of SO_4^{2-} is the ingested slurry.)

Comprehensive Problems

20.82 According to the advertisement, "a diamond is forever."
(a) Calculate ΔH^0, ΔS^0, and ΔG^0 at 298 K for the phase change

$$\text{Diamond} \longrightarrow \text{graphite}$$

(b) Given the conditions under which diamond jewelry is normally kept, argue for and against the statement in the ad.
(c) Given the answers in part (a), what would need to be done to make synthetic diamonds from graphite?
(d) Assuming ΔH^0 and ΔS^0 do not change with temperature, can graphite be converted to diamond spontaneously at 1 atm?

20.83 Replace each question mark with the correct information:

	ΔS_{rxn}	ΔH_{rxn}	ΔG_{rxn}	Comment
(a)	+	−	−	?
(b)	?	0	−	Spontaneous
(c)	−	+	?	Not spontaneous
(d)	0	?	−	Spontaneous
(e)	?	0	+	?
(f)	+	+	?	$T\Delta S > \Delta H$

20.84 Among the many complex ions of cobalt are the following:

$$Co(NH_3)_6^{3+}(aq) + 3en(aq) \rightleftharpoons Co(en)_3^{3+}(aq) + 6NH_3(aq)$$

where "en" stands for ethylenediamine, $H_2NCH_2CH_2NH_2$. Six Co—N bonds are broken and six Co—N bonds are formed in this reaction, so $\Delta H^0_{rxn} \approx 0$; yet $K > 1$. What are the signs of ΔS^0 and ΔG^0? What drives the reaction?

20.85 What is the change in entropy when 0.200 mol of potassium freezes at 63.7°C ($\Delta H_{fus} = 2.39$ kJ/mol)?

20.86 Is each statement true or false? If false, correct it.
(a) All spontaneous reactions occur quickly.
(b) The reverse of a spontaneous reaction is nonspontaneous.
(c) All spontaneous processes release heat.
(d) The boiling of water at 100°C and 1 atm is spontaneous.
(e) If a process increases the freedom of motion of the particles of a system, the entropy of the system decreases.
(f) The energy of the universe is constant; the entropy of the universe decreases toward a minimum.
(g) All systems disperse their energy spontaneously.
(h) Both ΔS_{sys} and ΔS_{surr} equal zero at equilibrium.

20.87 Hemoglobin carries O_2 from the lungs to tissue cells, where the O_2 is released. The protein is represented as Hb in its unoxygenated form and as $Hb{\cdot}O_2$ in its oxygenated form. One reason CO is toxic is that it competes with O_2 in binding to Hb:

$$Hb{\cdot}O_2(aq) + CO(g) \rightleftharpoons Hb{\cdot}CO(aq) + O_2(g)$$

(a) If $\Delta G^0 \approx -14$ kJ at 37°C (body temperature), what is the ratio of $[Hb{\cdot}CO]$ to $[Hb{\cdot}O_2]$ at 37°C with $[O_2] = [CO]$?
(b) How is Le Châtelier's principle used to treat CO poisoning?

20.88 An important ore of lithium is spodumene, $LiAlSi_2O_6$. To extract the metal, the α form of spodumene is first converted into the less dense β form in preparation for subsequent leaching and washing steps. Use the following data to calculate the lowest temperature at which the α ⟶ β conversion is feasible:

	ΔH^0_f (kJ/mol)	S^0 (J/mol·K)
α-spodumene	−3055	129.3
β-spodumene	−3027	154.4

20.89 Magnesia (MgO) is used for fire brick, crucibles, and furnace linings because of its high melting point. It is produced by decomposing magnesite ($MgCO_3$) at around 1200°C.
(a) Write a balanced equation for magnesite decomposition.
(b) Use ΔH^0 and S^0 values to find ΔG^0 at 298 K.
(c) Assuming ΔH^0 and S^0 do not change with temperature, find the minimum temperature at which the reaction is spontaneous.
(d) Calculate the equilibrium P_{CO_2} above $MgCO_3$ at 298 K.
(e) Calculate the equilibrium P_{CO_2} above $MgCO_3$ at 1200 K.

20.90 To prepare nuclear fuel, U_3O_8 is converted to $UO_2(NO_3)_2$, which is then converted to UO_3 and finally UO_2 ("yellow cake"). The fuel is enriched (the proportion of the ^{235}U is increased) by a two-step conversion of UO_2 into UF_6, a volatile solid, followed by a gaseous-diffusion separation of the ^{235}U and ^{238}U isotopes:

$$UO_2(s) + 4HF(g) \longrightarrow UF_4(s) + 2H_2O(g)$$
$$UF_4(s) + F_2(g) \longrightarrow UF_6(s)$$

Calculate ΔG^0 for the overall process at 85°C:

	ΔH_f^0 (kJ/mol)	S^0 (J/mol·K)	ΔG_f^0 (kJ/mol)
$UO_2(s)$	−1085	77.0	−1032
$UF_4(s)$	−1921	152	−1830.
$UF_6(s)$	−2197	225	−2068

20.91 Methanol, a major industrial feedstock, is made by several catalyzed reactions, such as $CO(g) + 2H_2(g) \longrightarrow CH_3OH(l)$.
(a) Show that this reaction is thermodynamically feasible.
(b) Is it favored at low or at high temperatures?
(c) One concern about using CH_3OH as an auto fuel is its oxidation in air to yield formaldehyde, $CH_2O(g)$, which poses a health hazard. Calculate ΔG^0 at 100.°C for this oxidation.

20.92 (a) Write a balanced equation for the gaseous reaction between N_2O_5 and F_2 to form NF_3 and O_2.
(b) Determine ΔG^0_{rxn}.
(c) Find ΔG_{rxn} at 298 K if $P_{N_2O_5} = P_{F_2} = 0.20$ atm, $P_{NF_3} =$ 0.25 atm, and $P_{O_2} = 0.50$ atm.

20.93 Consider the following reaction:

$$2NOBr(g) \rightleftharpoons 2NO(g) + Br_2(g) \qquad K = 0.42 \text{ at } 373 \text{ K}$$

Given that S^0 of $NOBr(g)$ = 272.6 J/mol·K and that ΔS^0_{rxn} and ΔH^0_{rxn} are constant with temperature, find
(a) ΔS^0_{rxn} at 298 K
(b) ΔG^0_{rxn} at 373 K
(c) ΔH^0_{rxn} at 373 K
(d) ΔH_f^0 of NOBr at 298 K
(e) ΔG^0_{rxn} at 298 K
(f) ΔG_f^0 of NOBr at 298 K

20.94 Calculate the equilibrium constants for decomposition of the hydrogen halides at 298 K:

$$2HX(g) \rightleftharpoons H_2(g) + X_2(g)$$

What do these values indicate about the extent of decomposition of HX at 298 K? Suggest a reason for this trend.

20.95 Hydrogenation is the addition of H_2 to double (or triple) carbon-carbon bonds. Peanut butter and most commercial baked goods include hydrogenated oils. Find ΔH^0, ΔS^0, and ΔG^0 for the hydrogenation of ethene (C_2H_4) to ethane (C_2H_6) at 25°C.

20.96 The key process in a blast furnace during the production of iron is the reaction of Fe_2O_3 and carbon to yield Fe and CO_2.
(a) Calculate ΔH^0 and ΔS^0. [Assume C(graphite).]
(b) Is the reaction spontaneous at low or at high T? Explain.
(c) Is the reaction spontaneous at 298 K?
(d) At what temperature does the reaction become spontaneous?

20.97 Bromine monochloride is formed from the elements:

$$Cl_2(g) + Br_2(g) \longrightarrow 2BrCl(g)$$
$$\Delta H^0_{rxn} = -1.35 \text{ kJ/mol} \qquad \Delta G_f^0 = -0.88 \text{ kJ/mol}$$

Calculate (a) ΔH_f^0 and (b) S^0 of $BrCl(g)$.

20.98 Solid N_2O_5 reacts with water to form liquid HNO_3. Consider the reaction with all substances in their standard states.
(a) Is the reaction spontaneous at 25°C?
(b) The solid decomposes to NO_2 and O_2 at 25°C. Is the decomposition spontaneous at 25°C? At what T is it spontaneous?
(c) At what T does *gaseous* N_2O_5 decompose spontaneously? Explain the difference between this T and that in part (b).

Note: **Problems 20.99 to 20.102 relate to the thermodynamics of adenosine triphosphate (ATP). Refer to the Chemical Connections essay on pp. 889–890.**

20.99 Find K for (a) the hydrolysis of ATP, (b) the dehydration-condensation to form glucose phosphate, and (c) the coupled reaction between ATP and glucose. (d) How does each K change when T changes from 25°C to 37°C?

20.100 The complete oxidation of 1 mol of glucose supplies enough metabolic energy to form 36 mol of ATP. But oxidation of a fat yields much more. For example, oxidation of 1 mol of tristearin ($C_{57}H_{116}O_6$), a typical dietary fat, yields enough energy to form 458 mol of ATP. (a) How many molecules of ATP can form per gram of glucose? (b) Per gram of tristearin?

20.101 Nonspontaneous processes like muscle contraction, protein building, and nerve conduction are "coupled" to the spontaneous hydrolysis of ATP to ADP. ATP is then regenerated by coupling its synthesis to other energy-yielding reactions such as

$$\text{Creatine phosphate} \longrightarrow \text{creatine} + \text{phosphate} \qquad \Delta G^{0'} = -43.1 \text{ kJ/mol}$$
$$\text{ADP} + \text{phosphate} \longrightarrow \text{ATP} \qquad \Delta G^{0'} = +30.5 \text{ kJ/mol}$$

Calculate $\Delta G^{0'}$ for the overall reaction that regenerates ATP.

20.102 Energy from ATP hydrolysis drives many nonspontaneous cell reactions:

$$ATP^{4-}(aq) + H_2O(l) \rightleftharpoons ADP^{3-}(aq) + HPO_4^{\ 2-}(aq) + H^+(aq) \qquad \Delta G^{0'} = -30.5 \text{ kJ}$$

Energy for the reverse process comes ultimately from glucose metabolism:

$$C_6H_{12}O_6(s) + 6O_2(g) \longrightarrow 6CO_2(g) + 6H_2O(l)$$

(a) Find K for the hydrolysis of ATP at 37°C.
(b) Find $\Delta G^{0'}_{rxn}$ for metabolism of 1 mol of glucose.
(c) How many moles of ATP can be produced by metabolism of 1 mol of glucose?
(d) If 36 mol of ATP is formed, what is the actual yield?

20.103 From the following reaction and data, find (a) S^0 of $SOCl_2$ and (b) T at which the reaction becomes nonspontaneous:

$$SO_3(g) + SCl_2(l) \longrightarrow SOCl_2(l) + SO_2(g) \qquad \Delta G^0_{rxn} = -75.2 \text{ kJ}$$

	$SO_3(g)$	$SCl_2(l)$	$SOCl_2(l)$	$SO_2(g)$
ΔH_f^0 (kJ/mol)	−396	−50.0	−245.6	−296.8
S^0 (J/mol·K)	256.7	184	—	248.1

20.104 Oxidation of a metal in air is called *corrosion*. Write equations for the corrosion of iron and aluminum. Use ΔG_f^0 to determine whether either process is spontaneous at 25°C.

20.105 Antimony forms strong alloys with lead that are used in car batteries. Its main ore is stibnite (Sb_2S_3), which can be reduced to Sb with Fe:

$$Sb_2S_3(s) + 3Fe(s) \longrightarrow 2Sb(s) + 3FeS(s) \qquad \Delta H^0_{rxn} = -125 \text{ kJ}$$

Calculate (a) ΔH_f^0 for Sb_2S_3, (b) ΔG^0_{rxn}, and (c) S^0 for Sb(*s*). [For FeS(*s*), $\Delta G_f^0 = -100.$ kJ/mol, $\Delta H_f^0 = -100.$ kJ/mol, and S^0 = 60.3 J/mol·K. For $Sb_2S_3(s)$, ΔG_f^0 = −174 kJ/mol and S^0 = 182 J/mol·K.]

20.106 A key step in the metabolism of glucose for energy is the isomerization of glucose-6-phosphate (G6P) to fructose-6-phosphate (F6P): G6P $\rightleftharpoons$ F6P; K = 0.510 at 298 K.
(a) Calculate ΔG^0 at 298 K.
(b) Calculate ΔG when Q, the [F6P]/[G6P] ratio, equals 10.0.
(c) Calculate ΔG when Q = 0.100.
(d) Calculate Q if ΔG = −2.50 kJ/mol.

20.107 A chemical reaction, such as HI forming from its elements, can reach equilibrium at many temperatures. In contrast, a phase change, such as ice melting, is in equilibrium at a given pressure only at the melting point. (a) Which graph depicts how G_{sys} changes for the formation of HI? Explain. (b) Which graph depicts how G_{sys} changes as ice melts at 1°C and 1 atm? Explain.

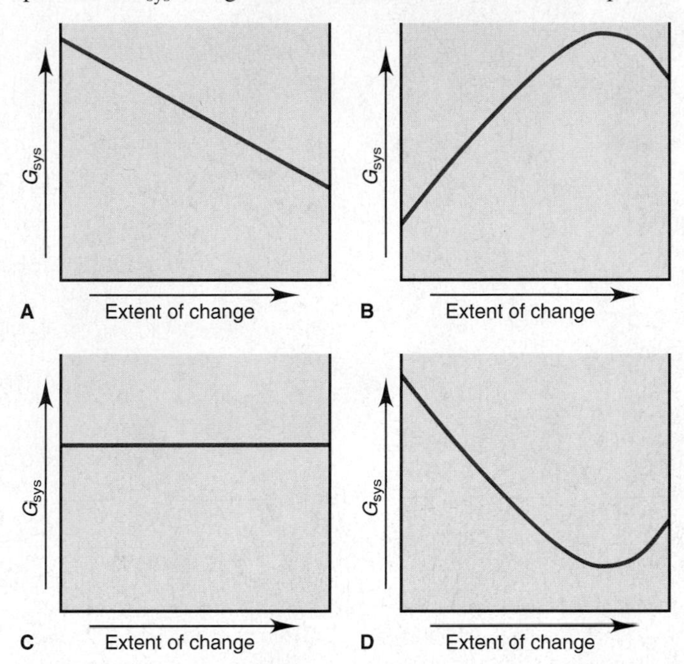

20.108 When heated, the DNA double helix separates into two random-coil single strands. When cooled, the random coils re-form the double helix: double helix $\rightleftharpoons$ 2 random coils.
(a) What is the sign of ΔS for the forward process? Why?
(b) Energy must be added to overcome H bonds and dispersion forces between the strands. What is the sign of ΔG for the forward process when $T\Delta S$ is smaller than ΔH?
(c) Write an expression that shows T in terms of ΔH and ΔS when the reaction is at equilibrium. (This temperature is called the *melting temperature* of the nucleic acid.)

20.109 In the process of respiration, glucose is oxidized completely. In fermentation, O_2 is absent and glucose is broken down to ethanol and CO_2. Ethanol is oxidized to CO_2 and H_2O.
(a) Balance the following equations for these processes:
Respiration: $C_6H_{12}O_6(s) + O_2(g) \longrightarrow CO_2(g) + H_2O(l)$
Fermentation: $C_6H_{12}O_6(s) \longrightarrow C_2H_5OH(l) + CO_2(g)$
Ethanol oxidation: $C_2H_5OH(l) + O_2(g) \longrightarrow CO_2(g) + H_2O(l)$
(b) Calculate ΔG^0_{rxn} for respiration of 1.00 g of glucose.
(c) Calculate ΔG^0_{rxn} for fermentation of 1.00 g of glucose.
(d) Calculate ΔG^0_{rxn} for oxidation of the ethanol in part (c).

20.110 Consider the formation of ammonia:

$$N_2(g) + 3H_2(g) \rightleftharpoons 2NH_3(g)$$

(a) Assuming that ΔH^0 and ΔS^0 are constant with temperature, find the temperature at which $K_p = 1.00$.
(b) Find K_p at 400.°C, a typical temperature for NH_3 production.
(c) Given the lower K_p at the higher temperature, why are these conditions used industrially?

20.111 Kyanite, sillimanite, and andalusite all have the formula Al_2SiO_5. Each is stable under different conditions:

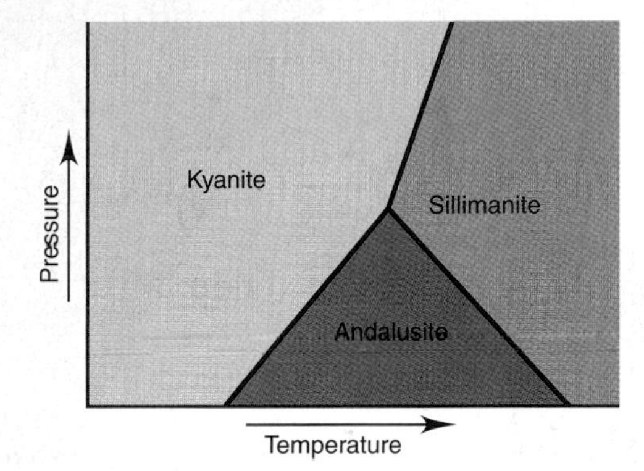

At the point where the three phases intersect:
(a) Which mineral, if any, has the lowest free energy?
(b) Which mineral, if any, has the lowest enthalpy?
(c) Which mineral, if any, has the highest entropy?
(d) Which mineral, if any, has the lowest density?

The bad side of a good thing The process in batteries that supplies electrical energy is, in principle, the same as the one that corrodes iron. In this chapter, we explore the two faces of electrochemistry—processes that create electricity as they occur and those that require electricity in order to occur. Both are indispensable to our way of life.

21

Electrochemistry: Chemical Change and Electrical Work

If you think thermodynamics relates mostly to expanding gases inside steam engines and has few practical, everyday applications, just look around. Some applications are probably within your reach right now, in the form of battery-operated devices—laptop computer, palm organizer, DVD remote, and, of course, wristwatch and calculator—or in the form of metal-plated jewelry or silverware. The operation and creation of these objects, and the many similar ones you use daily, involve the principles of electrochemistry, certainly one of the most important areas of applied thermodynamics.

Electrochemistry is the study of the relationship between chemical change and electrical work. It is typically investigated through the use of **electrochemical cells,** systems that incorporate a redox reaction to produce or utilize electrical energy. The common objects just mentioned highlight the essential difference between the two types of electrochemical cells:

- One type of cell *does work by releasing free energy from a spontaneous reaction to produce electricity*. A battery houses this type of cell.
- The other type of cell *does work by absorbing free energy from a source of electricity to drive a nonspontaneous reaction*. Such cells are used to plate a thin layer of metal on objects and, in major industrial processes, produce some compounds and nonmetals and recover many metals from their ores.

Concepts & Skills to Review

before you study this chapter

- redox terminology (Section 4.5 and Interchapter Topic 5)
- balancing redox reactions (Section 4.5)
- activity series of the metals (Section 4.6)
- free energy, work, and equilibrium (Sections 20.3 and 20.4)
- Q vs. K (Section 17.4) and ΔG vs. ΔG^0 (Section 20.4)

IN THIS CHAPTER . . . We review redox concepts and highlight a method for balancing redox equations (mentioned briefly in Chapter 4) that is particularly useful for electrochemical cells. An overview of the two cell types follows. The first type releases free energy to do electrical work, and we see how and why it operates. The cell's electrical output and its relation to the relative strengths of the redox species concern us next. We then examine the free energy change and equilibrium nature of the cell reaction and how they relate to the cell output. The concentration cells and batteries we consider are useful examples of this type of cell, and corrosion is a destructive electrochemical process that operates by the same principle. Next, we focus on cells that absorb free energy to do electrical work and see how they are used to isolate elements from their compounds and how to determine the identity and amount of product formed. Finally, we examine the redox system that generates energy in living cells.

The Electrochemical Future Is Here
As the combustion products of coal and gasoline continue to threaten our atmosphere, a new generation of electrochemical devices is being developed. Battery-gasoline hybrid cars are already common, achieving double or even triple the gas mileage of traditional gasoline-powered cars. Soon, electric cars, powered by banks of advanced fuel cells, may reduce the need for gasoline to a minimum. Virtually every car company has a fuel-cell prototype in operation. And these devices, once used principally on space missions, are now being produced for everyday residential and industrial applications as well.

21.1 REDOX REACTIONS AND ELECTROCHEMICAL CELLS

Whether an electrochemical process releases or absorbs free energy, it always involves the *movement of electrons from one chemical species to another* in an oxidation-reduction (redox) reaction. In this section, we review the redox process and describe the half-reaction method of balancing redox reactions that was mentioned in Section 4.5. Then we see how such reactions are used in the two types of electrochemical cells.

A Quick Review of Oxidation-Reduction Concepts

In electrochemical reactions, as in any redox process, *oxidation* is the loss of electrons, and *reduction* is the gain of electrons. An *oxidizing agent* is the species that does the oxidizing, taking electrons from the substance being oxidized. A *reducing agent* is the species that does the reducing, giving electrons to the substance being reduced. After the reaction, the oxidized substance has a higher (more positive or less negative) oxidation number, and the reduced substance has a lower (less positive or more negative) one. Keep in mind three key points:

- Oxidation (electron loss) always accompanies reduction (electron gain).
- The oxidizing agent is reduced, and the reducing agent is oxidized.
- The total number of electrons gained by the atoms/ions of the oxidizing agent always equals the total number lost by the atoms/ions of the reducing agent.

Figure 21.1 A summary of redox terminology. In the reaction between zinc and hydrogen ion, Zn is oxidized and H^+ is reduced.

PROCESS	$Zn(s) + 2H^+(aq) \longrightarrow Zn^{2+}(aq) + H_2(g)$
OXIDATION • One reactant loses electrons. • Reducing agent is oxidized. • Oxidation number increases.	Zinc **loses** electrons. Zinc is the reducing agent and becomes **oxidized.** The oxidation number of Zn **increases** from 0 to +2.
REDUCTION • Other reactant gains electrons. • Oxidizing agent is reduced. • Oxidation number decreases.	Hydrogen ion **gains** electrons. Hydrogen ion is the oxidizing agent and becomes **reduced.** The oxidation number of H **decreases** from +1 to 0.

Figure 21.1 presents these ideas for the aqueous reaction between zinc metal and a strong acid. Be sure you can identify the oxidation and reduction parts of a redox process. If you're having trouble, you may want to review the full discussion in Chapter 4 and the summary in Topic 5 of the Interchapter.

Half-Reaction Method for Balancing Redox Reactions

In Chapter 4, two methods for balancing redox reactions were mentioned—the oxidation number method and the half-reaction method—but only the first was discussed in detail (see Sample Problem 4.8, p. 154). The essential difference between the two methods is that the **half-reaction method** *divides the overall redox reaction into oxidation and reduction half-reactions.* Each half-reaction is balanced for atoms and charge. Then, one or both are multiplied by some integer to make electrons gained equal electrons lost, and the half-reactions are recombined to give the balanced redox equation. The half-reaction method offers several advantages for studying electrochemistry:

- It separates the oxidation and reduction steps, which reflects their actual physical separation in electrochemical cells.
- It makes it easier to balance redox reactions that take place in acidic or basic solution, which is common in these cells.
- It (usually) does *not* require assigning O.N.s. (In cases where the half-reactions are not obvious, we assign O.N.s to determine which atoms undergo a change and write half-reactions with the species that contain those atoms.)

In general, we begin with a "skeleton" ionic reaction, which shows only the species that are oxidized and reduced. *If the oxidized form of a species is on the left side of the skeleton reaction, the reduced form of that species is on the right, and vice versa.* Unless H_2O, H^+, and OH^- are being oxidized or reduced, they do not appear in the skeleton reaction. The following steps are used in balancing a redox reaction by the half-reaction method:

Step 1. Divide the skeleton reaction into two half-reactions, each of which contains the oxidized and reduced forms of one of the species. (Which half-reaction is the oxidation and which the reduction becomes clear after we balance the charges in the next step.)

Step 2. Balance the atoms and charges in each half-reaction.

- Atoms are balanced in order: atoms other than O and H, then O, and then H.
- Charge is balanced by *adding electrons* (e^-). They are added *to the left in the reduction half-reaction* because the reactant gains them; they are added *to the right in the oxidation half-reaction* because the reactant loses them.

Step 3. If necessary, multiply one or both half-reactions by an integer to make the number of e^- gained in the reduction equals the number lost in the oxidation.

Step 4. Add the balanced half-reactions, and include states of matter.

Step 5. Check that the atoms and charges are balanced.

We'll balance a redox reaction that occurs in acidic solution first and then go through Sample Problem 21.1 to balance one in basic solution.

Balancing Redox Reactions in Acidic Solution When a redox reaction occurs in acidic solution, H_2O molecules and H^+ ions are available for balancing. Even though we've used H_3O^+ to indicate the proton in water, we use H^+ in this chapter because it makes the balanced equations less complex. However, after using H^+ in this example, we'll balance the same equation with H_3O^+ just to show you that the only difference is in the number of water molecules needed.

Let's balance the redox reaction between dichromate ion and iodide ion to form chromium(III) ion and solid iodine, which occurs in acidic solution (Figure 21.2). The skeleton ionic reaction shows only the oxidized and reduced species:

$$Cr_2O_7^{2-}(aq) + I^-(aq) \longrightarrow Cr^{3+}(aq) + I_2(s) \qquad \text{[acidic solution]}$$

Figure 21.2 The redox reaction between dichromate ion and iodide ion. When $Cr_2O_7^{2-}$ (*left*) and I^- (*center*) are mixed in acid solution, they react to form Cr^{3+} and I_2 (*right*).

Step 1. Divide the reaction into half-reactions, each of which contains the oxidized and reduced forms of one species. The two chromium species make up one half-reaction, and the two iodine species make up the other:

$$Cr_2O_7^{2-} \longrightarrow Cr^{3+}$$
$$I^- \longrightarrow I_2$$

Step 2. Balance atoms and charges in each half-reaction. We use H_2O to balance O atoms, H^+ to balance H atoms, and e^- to balance positive charges.

- For the $Cr_2O_7^{2-}/Cr^{3+}$ half-reaction:

  a. *Balance atoms other than O and H.* We balance the two Cr on the left with a coefficient 2 on the right:

  $$Cr_2O_7^{2-} \longrightarrow 2Cr^{3+}$$

  b. *Balance O atoms by adding H_2O molecules.* Each H_2O has one O atom, so we add seven H_2O on the right to balance the seven O in $Cr_2O_7^{2-}$:

  $$Cr_2O_7^{2-} \longrightarrow 2Cr^{3+} + 7H_2O$$

  c. *Balance H atoms by adding H^+ ions.* Each H_2O contains two H, and we added seven H_2O, so we add 14 H^+ ions on the left:

  $$14H^+ + Cr_2O_7^{2-} \longrightarrow 2Cr^{3+} + 7H_2O$$

  d. *Balance charge by adding electrons.* Each H^+ ion has a 1+ charge, and 14 H^+ plus $Cr_2O_7^{2-}$ gives 12+ on the left. Two Cr^{3+} give 6+ on the right. There is an excess of 6+ on the left, so we add six e^- on the left:

  $$6e^- + 14H^+ + Cr_2O_7^{2-} \longrightarrow 2Cr^{3+} + 7H_2O$$

  This half-reaction is balanced, and we see it is the *reduction* because electrons appear on the *left, as reactants:* the reactant $Cr_2O_7^{2-}$ gains electrons (is reduced), so $Cr_2O_7^{2-}$ is the *oxidizing agent.* (Note that the O.N. of Cr decreases from +6 on the left to +3 on the right.)

- For the I^-/I_2 half-reaction:

  a. *Balance atoms other than O and H.* Two I atoms on the right require a coefficient 2 on the left:

  $$2I^- \longrightarrow I_2$$

  b. *Balance O atoms with H_2O.* Not needed; there are no O atoms.

  c. *Balance H atoms with H^+.* Not needed; there are no H atoms.

  d. *Balance charge with e^-.* To balance the 2− on the left, we add two e^- on the right:

  $$2I^- \longrightarrow I_2 + 2e^-$$

  This half-reaction is balanced, and it is the *oxidation* because electrons appear on the *right, as products:* the reactant I^- loses electrons (is oxidized), so I^- is the *reducing agent.* (Note that the O.N. of I increases from −1 to 0.)

Step 3. Multiply each half-reaction, if necessary, by an integer so that the number of e^- lost in the oxidation equals the number of e^- gained in the reduction. Two e^- are lost in the oxidation and six e^- are gained in the reduction, so we multiply the oxidation by 3:

$$3(2I^- \longrightarrow I_2 + 2e^-)$$
$$6I^- \longrightarrow 3I_2 + 6e^-$$

Step 4. Add the half-reactions together, canceling substances that appear on both sides, and include states of matter. In this example, only the electrons cancel:

$$\cancel{6e^-} + 14H^+ + Cr_2O_7^{2-} \longrightarrow 2Cr^{3+} + 7H_2O$$
$$6I^- \longrightarrow 3I_2 + \cancel{6e^-}$$
$$6I^-(aq) + 14H^+(aq) + Cr_2O_7^{2-}(aq) \longrightarrow 3I_2(s) + 7H_2O(l) + 2Cr^{3+}(aq)$$

Step 5. Check that atoms and charges balance:

Reactants (6I, 14H, 2Cr, 7O; 6+) $\longrightarrow$ products (6I, 14H, 2Cr, 7O; 6+)

Balancing a Redox Reaction Using H_3O^+ Now let's balance the reduction half-reaction with H_3O^+ as the supplier of H atoms. In step 2, balancing Cr atoms and balancing O atoms with H_2O are the same as before, so we start with

$$Cr_2O_7^{2-} \longrightarrow 2Cr^{3+} + 7H_2O$$

Now we use H_3O^+ instead of H^+ to balance H atoms. Because each H_3O^+ is an H^+ bonded to an H_2O, we balance the 14 H on the right with 14 H_3O^+ on the left and immediately balance the H_2O that are part of those 14 H_3O^+ by adding 14 more H_2O on the right:

$$14H_3O^+ + Cr_2O_7^{2-} \longrightarrow 2Cr^{3+} + 7H_2O + 14H_2O$$

Simplifying the right side by taking the sum of the H_2O molecules gives

$$14H_3O^+ + Cr_2O_7^{2-} \longrightarrow 2Cr^{3+} + 21H_2O$$

None of this affects the redox change, so balancing the charge still requires six e^- on the left, and we obtain the balanced reduction half-reaction:

$$6e^- + 14H_3O^+ + Cr_2O_7^{2-} \longrightarrow 2Cr^{3+} + 21H_2O$$

Adding the balanced oxidation half-reaction gives the balanced redox equation:

$$6I^-(aq) + 14H_3O^+(aq) + Cr_2O_7^{2-}(aq) \longrightarrow 3I_2(s) + 21H_2O(l) + 2Cr^{3+}(aq)$$

Note, as mentioned earlier, that the only difference is the number of H_2O molecules: there are 14 more H_2O (21 instead of 7) from the 14 H_3O^+.

To depict the reaction even more accurately, we could show the metal ion in its hydrated form, which would also affect only the total number of H_2O:

$$6I^-(aq) + 14H_3O^+(aq) + Cr_2O_7^{2-}(aq) \longrightarrow 3I_2(s) + 9H_2O(l) + 2Cr(H_2O)_6^{3+}(aq)$$

Although these added steps present the species in solution more accurately, they change only the number of water molecules and make balancing more difficult; thus, they detract somewhat from the key chemical event—the redox change. Therefore, we'll employ the method that uses H^+ to balance H atoms.

Balancing Redox Reactions in Basic Solution As you just saw, in acidic solution, H_2O molecules and H^+ (or H_3O^+) ions are available for balancing. As Sample Problem 21.1 shows, in basic solution, H_2O molecules and OH^- ions are available. Only one additional step is needed to balance a redox equation that takes place in basic solution. It appears after both half-reactions have first been balanced *as if they took place in acidic solution* (steps 1 and 2), the e^- lost have been made equal to the e^- gained (step 3), and the half-reactions have been combined (step 4). At this point, *we add one OH^- ion **to both sides of the equation** for every H^+ ion present.* (We label this step "4 Basic.") The H^+ ions on one side are combined with the added OH^- ions to form H_2O, and OH^- ions appear on the other side of the equation. Excess H_2O molecules are canceled, and states of matter are identified. Finally, we check that atoms and charges balance (step 5).

SAMPLE PROBLEM 21.1 Balancing Redox Reactions by the Half-Reaction Method

Problem Permanganate ion is a strong oxidizing agent, and its deep purple color makes it useful as an indicator in redox titrations (see Figure 4.14, p. 156). It reacts in basic solution with the oxalate ion to form carbonate ion and solid manganese dioxide. Balance the skeleton ionic equation for the reaction between $NaMnO_4$ and $Na_2C_2O_4$ in basic solution:

$$MnO_4^-(aq) + C_2O_4^{2-}(aq) \longrightarrow MnO_2(s) + CO_3^{2-}(aq) \quad \text{[basic solution]}$$

Plan We proceed through step 4 as if this took place in acidic solution. Then, we add the appropriate number of OH^- ions and cancel excess H_2O molecules (step 4 Basic).

Solution

1. Divide into half-reactions.

$MnO_4^- \longrightarrow MnO_2$	$C_2O_4^{2-} \longrightarrow CO_3^{2-}$

2. Balance.

a. Atoms other than O and H,	a. Atoms other than O and H,
Not needed	$C_2O_4^{2-} \longrightarrow 2CO_3^{2-}$
b. O atoms with H_2O,	b. O atoms with H_2O,
$MnO_4^- \longrightarrow MnO_2 + 2H_2O$	$2H_2O + C_2O_4^{2-} \longrightarrow 2CO_3^{2-}$
c. H atoms with H^+,	c. H atoms with H^+,
$4H^+ + MnO_4^- \longrightarrow MnO_2 + 2H_2O$	$2H_2O + C_2O_4^{2-} \longrightarrow 2CO_3^{2-} + 4H^+$
d. Charge with e^-,	d. Charge with e^-,
$3e^- + 4H^+ + MnO_4^- \longrightarrow MnO_2 + 2H_2O$ [reduction]	$2H_2O + C_2O_4^{2-} \longrightarrow 2CO_3^{2-} + 4H^+ + 2e^-$ [oxidation]

3. Multiply each half-reaction, if necessary, by some integer to make e^- lost equal e^- gained.

$2(3e^- + 4H^+ + MnO_4^- \longrightarrow MnO_2 + 2H_2O)$	$3(2H_2O + C_2O_4^{2-} \longrightarrow 2CO_3^{2-} + 4H^+ + 2e^-)$
$6e^- + 8H^+ + 2MnO_4^- \longrightarrow 2MnO_2 + 4H_2O$	$6H_2O + 3C_2O_4^{2-} \longrightarrow 6CO_3^{2-} + 12H^+ + 6e^-$

4. Add half-reactions, and cancel substances appearing on both sides.
The six e^- cancel, eight H^+ cancel to leave four H^+ on the right, and four H_2O cancel to leave two H_2O on the left:

$$\cancel{6e^-} + \cancel{8H^+} + 2MnO_4^- \longrightarrow 2MnO_2 + \cancel{4H_2O}$$
$$2\,\cancel{6}H_2O + 3C_2O_4^{2-} \longrightarrow 6CO_3^{2-} + 4\,\cancel{12}H^+ + \cancel{6e^-}$$
$$2MnO_4^- + 2H_2O + 3C_2O_4^{2-} \longrightarrow 2MnO_2 + 6CO_3^{2-} + 4H^+$$

4 Basic. Add OH^- to both sides to neutralize H^+, and cancel H_2O.
Adding four OH^- to both sides forms four H_2O on the right, two of which cancel the two H_2O on the left, leaving two H_2O on the right:

$$2MnO_4^- + 2H_2O + 3C_2O_4^{2-} + 4OH^- \longrightarrow 2MnO_2 + 6CO_3^{2-} + [4H^+ + 4OH^-]$$
$$2MnO_4^- + \cancel{2H_2O} + 3C_2O_4^{2-} + 4OH^- \longrightarrow 2MnO_2 + 6CO_3^{2-} + 2\,\cancel{4}H_2O$$

Including states of matter gives the final balanced equation:

$$2MnO_4^-(aq) + 3C_2O_4^{2-}(aq) + 4OH^-(aq) \longrightarrow 2MnO_2(s) + 6CO_3^{2-}(aq) + 2H_2O(l)$$

5. Check that atoms and charges balance.

$$(2Mn, 24O, 6C, 4H; 12-) \longrightarrow (2Mn, 24O, 6C, 4H; 12-)$$

Comment As a final step, we can obtain the balanced *molecular* equation for this reaction by noting the number of moles of each anion in the balanced ionic equation and adding the correct number of moles of spectator ions (in this case, Na^+) to obtain neutral compounds. Thus, for instance, balancing the charge of 2 mol of MnO_4^- requires 2 mol of Na^+, so we have $2NaMnO_4$. The balanced molecular equation is

$$2NaMnO_4(aq) + 3Na_2C_2O_4(aq) + 4NaOH(aq) \longrightarrow 2MnO_2(s) + 6Na_2CO_3(aq) + 2H_2O(l)$$

FOLLOW-UP PROBLEM 21.1 Write a balanced molecular equation for the reaction between $KMnO_4$ and KI in basic solution. The skeleton ionic reaction is

$$MnO_4^-(aq) + I^-(aq) \longrightarrow MnO_4^{2-}(aq) + IO_3^-(aq) \quad \text{[basic solution]}$$

The half-reaction method reveals a great deal about redox processes and is essential to understanding electrochemical cells. The major points are

- Any redox reaction can be treated as the sum of a reduction and an oxidation half-reaction.
- Atoms and charge are conserved in each half-reaction.
- Electrons lost in one half-reaction are gained in the other.
- Although the half-reactions are treated separately, electron loss and electron gain occur simultaneously.

An Overview of Electrochemical Cells

We distinguish two types of electrochemical cells based on the general thermodynamic nature of the reaction:

1. A **voltaic cell** (or **galvanic cell**) uses a spontaneous reaction ($\Delta G < 0$) to generate electrical energy. In the cell reaction, the difference in chemical potential energy between higher energy reactants and lower energy products is converted into electrical energy. This energy is used to operate the load—flashlight bulb, CD player, car starter motor, or other electrical device. In other words, *the system does work on the surroundings.* All batteries contain voltaic cells.
2. An **electrolytic cell** uses electrical energy to drive a nonspontaneous reaction ($\Delta G > 0$). In the cell reaction, electrical energy from an external power supply converts lower energy reactants into higher energy products. Thus, *the surroundings do work on the system.* Electroplating and recovering metals from ores involve electrolytic cells.

The two types of cell have certain design features in common (Figure 21.3). Two **electrodes,** which conduct the electricity between cell and surroundings, are dipped into an **electrolyte,** a mixture of ions (usually in aqueous solution) that

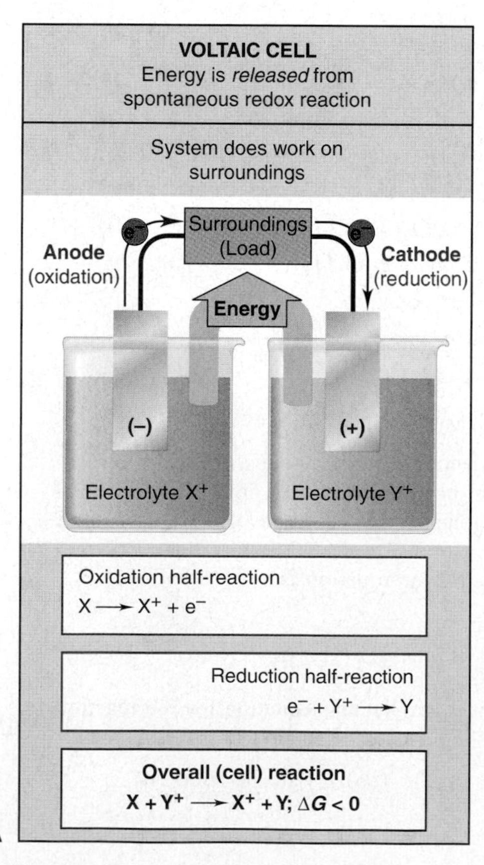

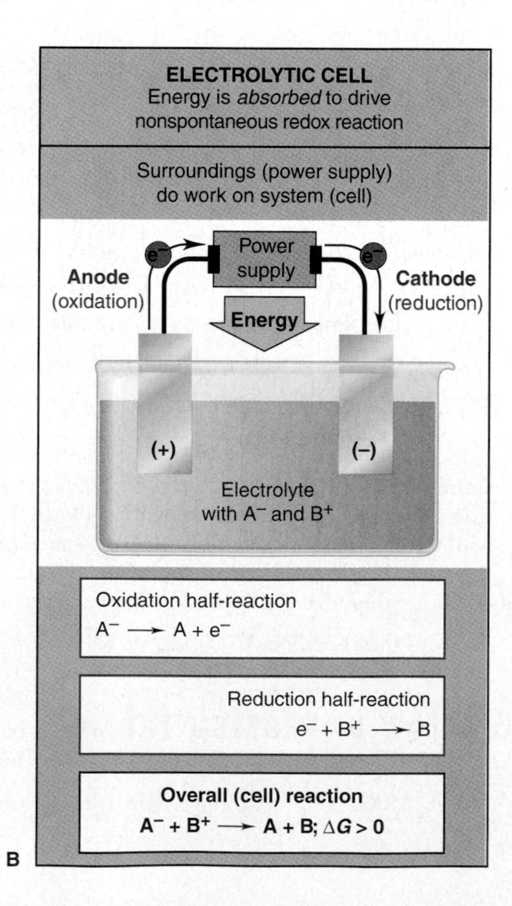

Figure 21.3 General characteristics of voltaic and electrolytic cells. A voltaic cell **(A)** generates energy from a spontaneous reaction ($\Delta G < 0$), whereas an electrolytic cell **(B)** requires energy to drive a nonspontaneous reaction ($\Delta G > 0$). In both types of cell, two electrodes dip into electrolyte solutions, and an external circuit provides the means for electrons to flow between them. Most important, notice that oxidation takes place at the anode and reduction takes place at the cathode, but the relative electrode charges are opposite in the two cells.

are involved in the reaction or that carry the charge. An electrode is identified as either **anode** or **cathode** depending on the half-reaction that takes place there:

- *The oxidation half-reaction occurs at the anode.* Electrons are lost by the substance being oxidized (reducing agent) and *leave the cell* at the anode.
- *The reduction half-reaction occurs at the cathode.* Electrons are gained by the substance being reduced (oxidizing agent) and *enter the cell* at the cathode.

As shown in Figure 21.3, the relative charges of the electrodes are *opposite* in the two types of cell. As you'll see in the following sections, these opposite charges result from the different phenomena that cause the electrons to flow.

Which Half-Reaction Occurs at Which Electrode? Here are some memory aids to help you remember which half-reaction occurs at which electrode:

1. The words *anode* and *oxidation* start with vowels; the words *cathode* and *reduction* start with consonants.
2. Alphabetically, the *A* in anode comes before the *C* in cathode, and the *O* in oxidation comes before the *R* in reduction.
3. Look at the first syllables and use your imagination:

ANode, OXidation; REDuction, CAThode
⇒ AN OX and a RED CAT

SECTION SUMMARY

An oxidation-reduction (redox) reaction involves the transfer of electrons from a reducing agent to an oxidizing agent. The half-reaction method of balancing divides the overall reaction into half-reactions that are balanced separately and then recombined. Both types of electrochemical cells are based on redox reactions. In a voltaic cell, a spontaneous reaction generates electricity and does work on the surroundings; in an electrolytic cell, the surroundings supply electricity that does work to drive a nonspontaneous reaction. In both types, two electrodes dip into electrolyte solutions; oxidation occurs at the anode, and reduction occurs at the cathode.

21.2 VOLTAIC CELLS: USING SPONTANEOUS REACTIONS TO GENERATE ELECTRICAL ENERGY

If you put a strip of zinc metal in a solution of Cu^{2+} ion, the blue color of the solution fades as a brown-black crust of Cu metal forms on the Zn strip (Figure 21.4). Judging from what we see, the reaction involves the reduction of Cu^{2+} ion to Cu metal, which must be accompanied by the oxidation of Zn metal to Zn^{2+} ion. The overall reaction consists of two half-reactions:

$$\begin{array}{rcll} Cu^{2+}(aq) + 2e^- & \longrightarrow & Cu(s) & \text{[reduction]} \\ Zn(s) & \longrightarrow & Zn^{2+}(aq) + 2e^- & \text{[oxidation]} \\ \hline Zn(s) + Cu^{2+}(aq) & \longrightarrow & Zn^{2+}(aq) + Cu(s) & \text{[overall reaction]} \end{array}$$

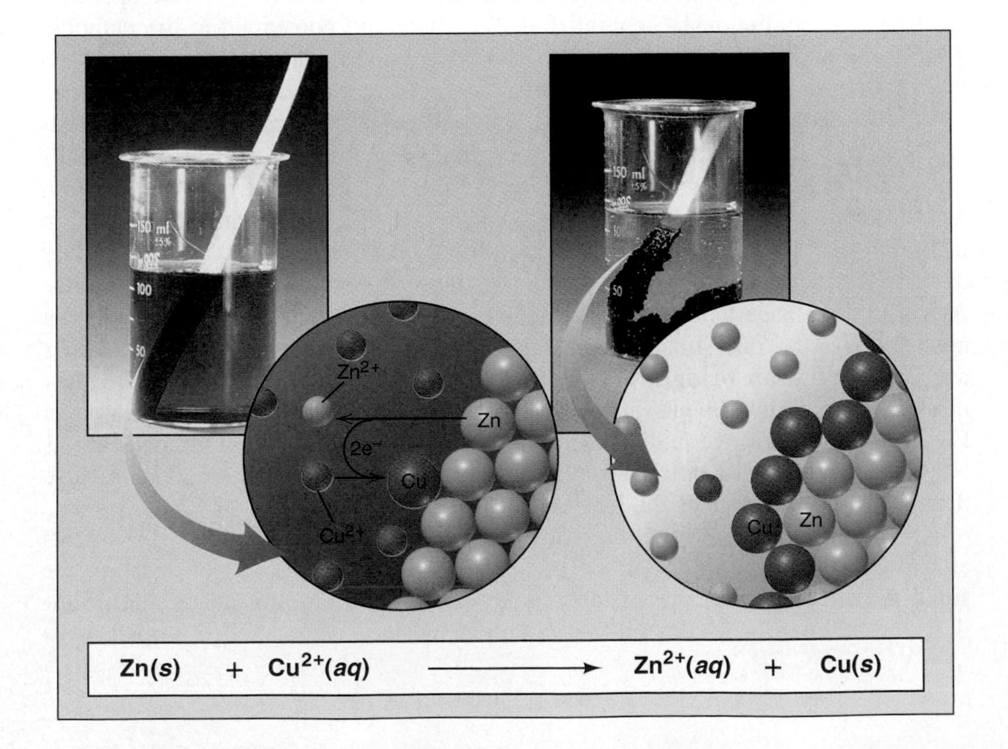

Figure 21.4 The spontaneous reaction between zinc and copper(II) ion. When a strip of zinc metal is placed in a solution of Cu^{2+} ion, a redox reaction begins (*left*), in which the zinc is oxidized to Zn^{2+} and the Cu^{2+} is reduced to copper metal. As the reaction proceeds (*right*), the deep blue color of the solution of hydrated Cu^{2+} ion lightens, and the Cu "plates out" on the Zn and falls off in chunks. (The Cu appears black because it is very finely divided.) At the atomic scale, each Zn atom loses two electrons, which are gained by a Cu^{2+} ion. The process is summarized with symbols in the balanced equation.

In the remainder of this section, we examine this spontaneous reaction as the basis of a voltaic (galvanic) cell.

Construction and Operation of a Voltaic Cell

Animation: Operation of a Voltaic Cell
Online Learning Center

Electrons are being transferred in the Zn/Cu^{2+} reaction (Figure 21.4), but the system does not generate electrical energy because the oxidizing agent (Cu^{2+}) and the reducing agent (Zn) are in the same beaker. If, however, the half-reactions are physically separated and connected by an external circuit, the electrons are transferred by traveling through the circuit, thus producing an electric current.

This separation of half-reactions is the essential idea behind a voltaic cell (Figure 21.5A). The components of each half-reaction are placed in a separate container, or **half-cell,** which consists of one electrode dipping into an electrolyte solution. The two half-cells are joined by the circuit, which consists of a wire and a salt bridge (the inverted U tube in the figure; we will discuss its function shortly). In order to measure the voltage generated by the cell, a voltmeter is inserted in the path of the wire connecting the electrodes. A switch (not shown) closes (completes) or opens (breaks) the circuit. By convention, *the oxidation half-cell (anode compartment) is shown on the left and the reduction half-cell (cathode compartment) on the right.* Here are the key points about the Zn/Cu^{2+} voltaic cell:

1. *The oxidation half-cell.* In this case, the anode compartment consists of a zinc bar (the anode) immersed in a Zn^{2+} electrolyte (such as a solution of zinc sulfate, $ZnSO_4$). The zinc bar is the reactant in the oxidation half-reaction, and it conducts the released electrons *out* of its half-cell.

2. *The reduction half-cell.* In this case, the cathode compartment consists of a copper bar (the cathode) immersed in a Cu^{2+} electrolyte [such as a solution of copper(II) sulfate, $CuSO_4$]. The copper bar is the product in the reduction half-reaction, and it conducts electrons *into* its half-cell.

3. *Relative charges on the electrodes.* The electrode charges are determined by the *source of electrons* and the *direction of electron flow* through the circuit. In this cell, zinc atoms are oxidized at the anode to Zn^{2+} ions and electrons. The Zn^{2+} ions enter the solution, while the electrons enter the bar and then the wire. *The electrons flow left to right* through the wire to the cathode, where Cu^{2+} ions in the solution accept them and are reduced to Cu atoms. As the cell operates, electrons are continuously generated at the anode and consumed at the cathode. Therefore, the anode has an excess of electrons and a negative charge *relative* to the cathode. *In any* ***voltaic*** *cell, the anode is negative and the cathode is positive.*

4. *The purpose of the salt bridge.* The cell cannot operate unless the circuit is complete. The oxidation half-cell originally contains a neutral solution of Zn^{2+} and SO_4^{2-} ions, but as Zn atoms in the bar lose electrons, the solution would develop a net positive charge from the Zn^{2+} ions entering. Similarly, in the reduction half-cell, the neutral solution of Cu^{2+} and SO_4^{2-} ions would develop a net negative charge as Cu^{2+} ions leave the solution to form Cu atoms. A charge imbalance would arise and stop cell operation if the half-cells were not neutral. To avoid this situation and enable the cell to operate, the two half-cells are joined by a **salt bridge,** which acts as a "liquid wire," allowing ions to flow through both compartments and complete the circuit. The salt bridge shown in Figure 21.5A is an inverted U tube containing a solution of the nonreacting ions Na^+ and SO_4^{2-} in a gel. The solution cannot pour out, but ions can diffuse through it into and out of the half-cells.

To maintain neutrality in the reduction half-cell (right; cathode compartment) as Cu^{2+} ions change to Cu atoms, Na^+ ions move from the salt bridge into the solution (and some SO_4^{2-} ions move from the solution into the salt bridge). Similarly, to maintain neutrality in the oxidation half-cell (left; anode compartment)

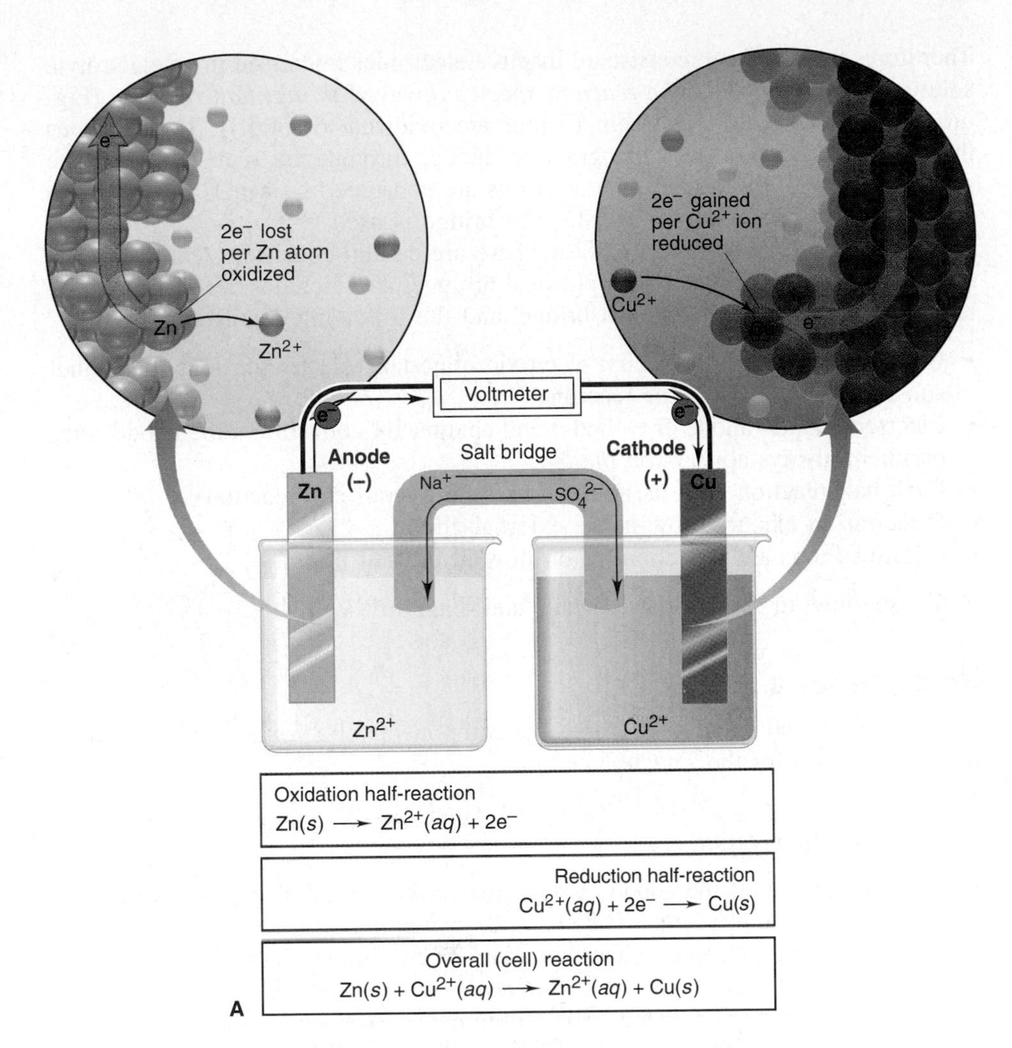

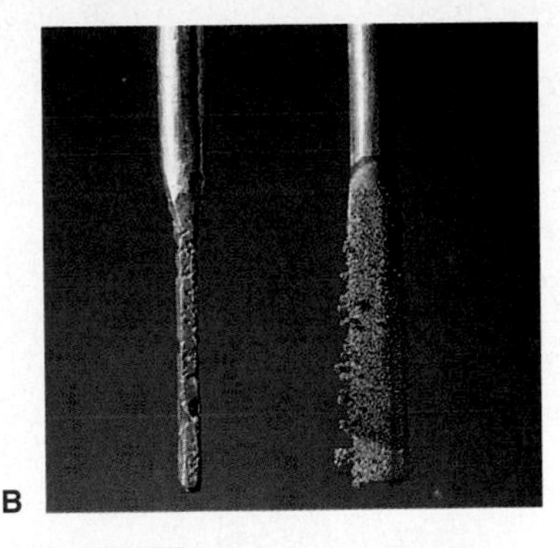

Figure 21.5 A voltaic cell based on the zinc-copper reaction. A, The anode half-cell (oxidation) consists of a Zn electrode dipping into a Zn^{2+} solution. The two electrons generated in the oxidation of each Zn atom move through the Zn bar and the wire, and into the Cu electrode, which dips into a Cu^{2+} solution in the cathode half-cell (reduction). There, the electrons reduce Cu^{2+} ions. Thus, electrons flow left to right through electrodes and wire. A salt bridge contains unreactive Na^+ and SO_4^{2-} ions that maintain neutral charge in the electrolyte solutions: anions in the salt bridge flow to the left, and cations flow to the right. The voltmeter registers the electrical output of the cell. **B,** After the cell runs for several hours, the Zn anode weighs less because Zn atoms have been oxidized to aqueous Zn^{2+} ions, and the Cu cathode weighs more because aqueous Cu^{2+} ions have been reduced to Cu metal.

as Zn atoms change to Zn^{2+} ions, SO_4^{2-} ions move from the salt bridge into that solution (and some Zn^{2+} ions move from the solution into the salt bridge). Thus, as Figure 21.5A shows, the circuit is completed *as electrons move left to right through the wire, while anions move right to left and cations move left to right through the salt bridge.*

5. *Active vs. inactive electrodes.* The electrodes in the Zn/Cu^{2+} cell are *active* because the metal bars themselves are components of the half-reactions. As the cell operates, the mass of the zinc electrode gradually decreases, and the $[Zn^{2+}]$ in the anode half-cell increases. At the same time, the mass of the copper electrode increases, and the $[Cu^{2+}]$ in the cathode half-cell decreases; we say that the Cu^{2+} "plates out" on the electrode. Look at Figure 21.5B to see how the electrodes look, removed from their half-cells, after several hours of operation.

For many redox reactions, there are no reactants or products capable of serving as electrodes, so *inactive* electrodes are used. Most commonly, inactive electrodes are rods of *graphite* or *platinum*: they conduct electrons into or out of the half-cells but cannot take part in the half-reactions. In a voltaic cell based on the following half-reactions, for instance, the reacting species cannot act as electrodes:

$$2I^-(aq) \longrightarrow I_2(s) + 2e^- \qquad \text{[anode; oxidation]}$$

$$MnO_4^-(aq) + 8H^+(aq) + 5e^- \longrightarrow Mn^{2+}(aq) + 4H_2O(l) \qquad \text{[cathode; reduction]}$$

Therefore, each half-cell consists of inactive electrodes immersed in an electrolyte solution that contains *all the reactant species involved in that half-reaction* (Figure 21.6). In the anode half-cell, I^- ions are oxidized to solid I_2. The electrons that are released flow into the graphite anode, through the wire, and into the graphite cathode. From there, the electrons are consumed by MnO_4^- ions, which are reduced to Mn^{2+} ions. (A KNO_3 salt bridge is used.)

As Figures 21.5A and 21.6 show, there are certain consistent features in the *diagram* of any voltaic cell. The physical arrangement includes the half-cell containers, electrodes, wire, and salt bridge, and the following details appear:

- Components of the half-cells: electrode materials, electrolyte ions, and other substances involved in the reaction
- Electrode name (anode or cathode) and charge. By convention, the anode compartment always appears *on the left*.
- Each half-reaction with its half-cell and the overall cell reaction
- Direction of electron flow in the external circuit
- Nature of ions and direction of ion flow in the salt bridge

You'll see how to specify these details and diagram a cell shortly.

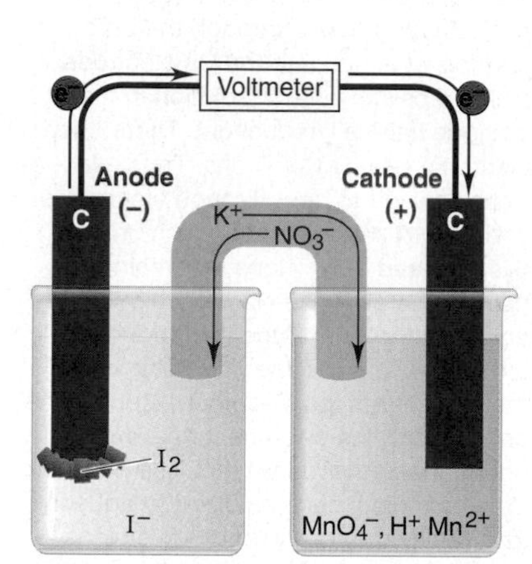

Oxidation half-reaction
$2I^-(aq) \longrightarrow I_2(s) + 2e^-$

Reduction half-reaction
$MnO_4^-(aq) + 8H^+(aq) + 5e^- \longrightarrow Mn^{2+}(aq) + 4H_2O(l)$

Overall (cell) reaction
$2MnO_4^-(aq) + 16H^+(aq) + 10I^-(aq) \longrightarrow 2Mn^{2+}(aq) + 5I_2(s) + 8H_2O(l)$

Figure 21.6 A voltaic cell using inactive electrodes. The reaction between I^- and MnO_4^- in acidic solution does not have species that can be used as electrodes, so inactive graphite (C) electrodes are used.

Notation for a Voltaic Cell

A useful shorthand notation describes the components of a voltaic cell. For example, the notation for the Zn/Cu^{2+} cell is

$$Zn(s) \mid Zn^{2+}(aq) \parallel Cu^{2+}(aq) \mid Cu(s)$$

Key parts of the notation are

- The components of the anode compartment (oxidation half-cell) are written *to the left* of the components of the cathode compartment (reduction half-cell).
- A vertical line represents a phase boundary. For example, $Zn(s) \mid Zn^{2+}(aq)$ indicates that the *solid* Zn is a *different* phase from the *aqueous* Zn^{2+}. A comma separates the half-cell components that are in the *same* phase. For example, the notation for the voltaic cell housing the reaction between I^- and MnO_4^- shown in Figure 21.6 is

$$\text{graphite} \mid I^-(aq) \mid I_2(s) \parallel H^+(aq), MnO_4^-(aq), Mn^{2+}(aq) \mid \text{graphite}$$

That is, in the cathode compartment, H^+, MnO_4^-, and Mn^{2+} ions are all in aqueous solution with solid graphite immersed in it. Often, we specify the concentrations of dissolved components; for example, if the concentrations of Zn^{2+} and Cu^{2+} are 1 *M*, we write

$$Zn(s) \mid Zn^{2+}(1\ M) \parallel Cu^{2+}(1\ M) \mid Cu(s)$$

- Half-cell components usually appear in the same order as in the half-reaction, and electrodes appear at the far left and right of the notation.
- A double vertical line separates the half-cells and represents the phase boundary on either side of the salt bridge (the ions in the salt bridge are omitted because they are not part of the reaction).

SAMPLE PROBLEM 21.2 Diagramming Voltaic Cells

Problem Diagram, show balanced equations, and write the notation for a voltaic cell that consists of one half-cell with a Cr bar in a $Cr(NO_3)_3$ solution, another half-cell with an Ag bar in an $AgNO_3$ solution, and a KNO_3 salt bridge. Measurement indicates that the Cr electrode is negative relative to the Ag electrode.

Plan From the given contents of the half-cells, we can write the half-reactions. We must determine which is the anode compartment (oxidation) and which is the cathode (reduction). To do so, we must find the direction of the spontaneous redox reaction, which is given by the relative electrode charges. Since electrons are released into the anode during oxidation, it has a negative charge. We are told that Cr is negative, so it must be the anode; and, therefore, Ag is the cathode.

Solution Writing the balanced half-reactions. Since the Ag electrode is positive, the half-reaction consumes e^-:

$$Ag^+(aq) + e^- \longrightarrow Ag(s) \quad \text{[reduction; cathode]}$$

Since the Cr electrode is negative, the half-reaction releases e^-:

$$Cr(s) \longrightarrow Cr^{3+}(aq) + 3e^- \quad \text{[oxidation; anode]}$$

Writing the balanced overall cell reaction. We triple the reduction half-reaction in order to balance e^-, and then we combine the half-reactions to obtain the overall spontaneous redox reaction:

$$Cr(s) + 3Ag^+(aq) \longrightarrow Cr^{3+}(aq) + 3Ag(s)$$

Determining direction of electron and ion flow. The released e^- in the Cr electrode (negative) flow through the external circuit to the Ag electrode (positive). As Cr^{3+} ions enter the anode electrolyte, NO_3^- ions enter from the salt bridge to maintain neutrality. As Ag^+ ions leave the cathode electrolyte and plate out on the Ag electrode, K^+ ions enter from the salt bridge to maintain neutrality. The diagram of this cell is shown in the margin.

Writing the cell notation:

$$Cr(s) \mid Cr^{3+}(aq) \parallel Ag^+(aq) \mid Ag(s)$$

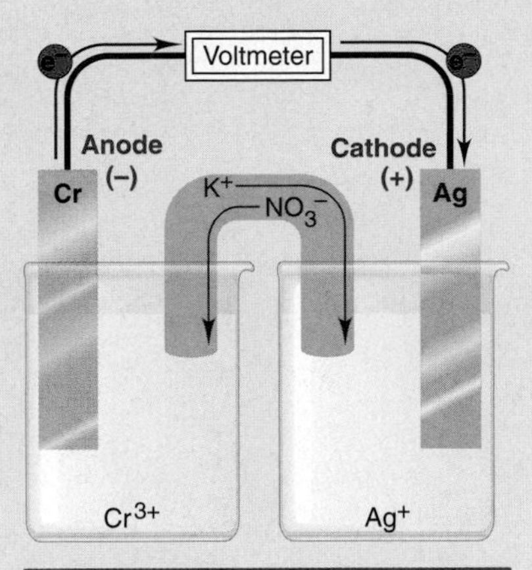

Check Always be sure that the half-reactions and cell reaction are balanced, the half-cells contain *all* components of the half-reactions, and the electron and ion flow are shown. You should be able to write the half-reactions from the cell notation as a check.

Comment The key to diagramming a voltaic cell is to use the direction of the spontaneous reaction to identify the oxidation (anode) and reduction (cathode) half-reactions.

FOLLOW-UP PROBLEM 21.2 In one compartment of a voltaic cell, a graphite rod dips into an acidic solution of $K_2Cr_2O_7$ and $Cr(NO_3)_3$; in the other compartment, a tin bar dips into a $Sn(NO_3)_2$ solution. A KNO_3 salt bridge joins them. The tin electrode is negative relative to the graphite. Diagram the cell, write the balanced equations, and show the cell notation.

Why Does a Voltaic Cell Work?

By placing a lightbulb in the circuit or looking at the voltmeter, we can see that the Zn/Cu^{2+} cell generates electrical energy. But what principle explains *how* the reaction takes place, and *why* do electrons flow in the direction shown?

Let's examine what happens when the switch is open and no reaction is occurring. In each half-cell, we can consider the metal electrode to be in equilibrium with the metal ions in the electrolyte and the electrons residing in the metal:

$$Zn(s) \rightleftharpoons Zn^{2+}(aq) + 2e^-(\text{in Zn metal})$$
$$Cu(s) \rightleftharpoons Cu^{2+}(aq) + 2e^-(\text{in Cu metal})$$

From the direction of the overall spontaneous reaction, we know that Zn gives up its electrons more easily than Cu does; thus, Zn is a stronger reducing agent. Therefore, the equilibrium position of the Zn half-reaction lies farther to the right: Zn produces more electrons than Cu does. You might think of the electrons in the Zn electrode as being subject to a greater electron "pressure" than those in the Cu electrode, a greater potential energy (referred to as *electrical potential*) ready to "push" them through the circuit. Close the switch, and electrons flow from the Zn to the Cu electrode to equalize this difference in electrical potential. The flow disturbs the equilibrium at each electrode. The Zn half-reaction shifts to the right to restore the electrons flowing out, and the Cu half-reaction shifts to the left to remove the electrons flowing in. Thus, *the spontaneous reaction occurs as a result of the different abilities of these metals to give up their electrons and the ability of the electrons to flow through the circuit.*

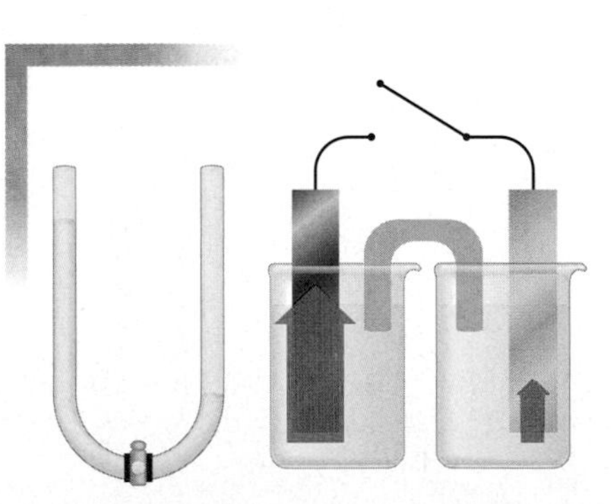

Electron Flow and Water Flow Consider this analogy between electron "pressure" and water pressure. A U tube (cell) is separated into two arms (two half-cells) by a stopcock (switch), and the two arms contain water at different heights (the half-cells contain half-reactions with different electrical potentials). Open the stopcock (close the switch), and the different heights (potential difference) become equal as water flows (electrons flow and current is generated).

SECTION SUMMARY

A voltaic cell consists of oxidation (anode) and reduction (cathode) half-cells, connected by a wire to conduct electrons and a salt bridge to maintain charge neutrality as the cell operates. Electrons move from anode (left) to cathode (right), while cations move from the salt bridge into the cathode half-cell and anions from the salt bridge into the anode half-cell. The cell notation shows the species and their phases in each half-cell, as well as the direction of current flow. A voltaic cell operates because species in the two half-cells differ in their tendency to lose electrons.

21.3 CELL POTENTIAL: OUTPUT OF A VOLTAIC CELL

The purpose of a voltaic cell is to convert the free energy change of a spontaneous reaction into the kinetic energy of electrons moving through an external circuit (electrical energy). This electrical energy can do work and is proportional to the *difference in electrical potential between the two electrodes*. This difference in the electrical potential of the electrodes is the **cell potential (E_{cell})**, also called the **voltage** of the cell or the **electromotive force (emf).**

Electrons flow spontaneously from the negative to the positive electrode, that is, toward the electrode with the more positive electrical potential. Thus, when the cell operates *spontaneously,* the difference in the electrical potential of the electrodes is positive; that is, there is a *positive* cell potential:

$$E_{cell} > 0 \text{ for a spontaneous process} \quad \textbf{(21.1)}$$

The more positive E_{cell} is, the more work the cell can do, and the farther the reaction proceeds to the right as written. A *negative* cell potential, on the other hand, is associated with a *nonspontaneous* cell reaction. If $E_{cell} = 0$, the reaction has reached equilibrium and the cell can do no more work. (There is a clear relationship between E_{cell}, K, and ΔG that we'll discuss in Section 21.4.)

How are the units of cell potential related to those of energy available to do work? As you've seen, work is done when charge moves between electrode compartments that differ in electrical potential. The SI unit of electrical potential is the **volt (V),** and the SI unit of electrical charge is the **coulomb (C).** By definition, for two electrodes that differ by 1 volt of electrical potential, 1 joule of energy is released (that is, 1 joule of work can be done) for each coulomb of charge that moves between the electrodes. Thus,

$$1 \text{ V} = 1 \text{ J/C} \quad \textbf{(21.2)}$$

Table 21.1 lists the voltages of some commercial and natural voltaic cells. Let's see how to measure cell potential.

Table 21.1 Voltages of Some Voltaic Cells

Voltaic Cell	Voltage (V)
Common alkaline flashlight battery	1.5
Lead-acid car battery (6 cells = 12 V)	2.0
Calculator battery (mercury)	1.3
Lithium-ion laptop battery	3.7
Electric eel (~5000 cells in 6-ft eel = 750 V)	0.15
Nerve of giant squid (across cell membrane)	0.070

Standard Cell Potentials

The measured potential of a voltaic cell is affected by changes in concentration as the reaction proceeds and by energy losses due to heating of the cell and the external circuit. Therefore, in order to compare the output of different cells, we obtain a **standard cell potential (E^0_{cell})**, the potential measured at a specified temperature (usually 298 K) with no current flowing* and *all components in their standard states:* 1 atm for gases, 1 M for solutions, the pure solid for electrodes. When the zinc-copper cell that we diagrammed in Figure 21.5 begins operating under standard state conditions, that is, when $[Zn^{2+}] = [Cu^{2+}] = 1\ M$, the cell produces 1.10 V at 298 K (see photo):

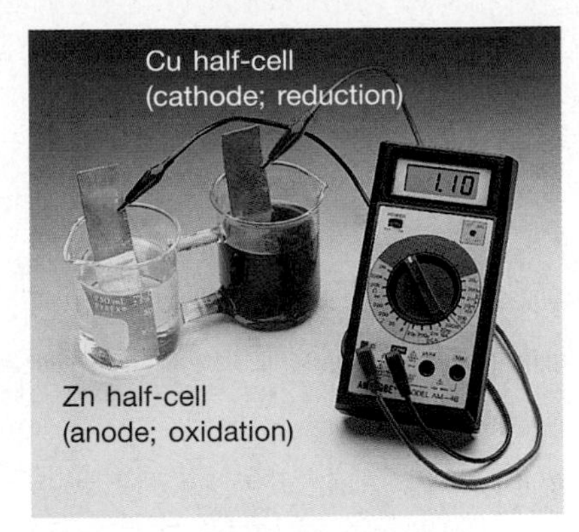

The zinc-copper cell at 298 K under standard-state conditions.

$$Zn(s) + Cu^{2+}(aq;\ 1\ M) \longrightarrow Zn^{2+}(aq;\ 1\ M) + Cu(s) \qquad E^0_{cell} = 1.10\ V$$

Standard Electrode (Half-Cell) Potentials Just as each half-reaction makes up part of the overall reaction, the potential of each half-cell makes up a part of the overall cell potential. The **standard electrode potential ($E^0_{half\text{-}cell}$)** is the potential associated with a given half-reaction (electrode compartment) when all the components are in their standard states.

By convention, *a standard electrode potential always refers to the half-reaction written as a* ***reduction.*** For the zinc-copper reaction, for example, the standard electrode potentials for the zinc half-reaction (E^0_{zinc}, anode compartment) and for the copper half-reaction (E^0_{copper}, cathode compartment) refer to the processes written as reductions:

$$Zn^{2+}(aq) + 2e^- \longrightarrow Zn(s) \qquad E^0_{zinc}\ (E^0_{anode}) \qquad [\text{reduction}]$$
$$Cu^{2+}(aq) + 2e^- \longrightarrow Cu(s) \qquad E^0_{copper}\ (E^0_{cathode}) \qquad [\text{reduction}]$$

The overall cell reaction involves the *oxidation* of zinc at the anode, not the *reduction* of Zn^{2+}, so we reverse the zinc half-reaction:

$$Zn(s) \longrightarrow Zn^{2+}(aq) + 2e^- \qquad [\text{oxidation}]$$
$$Cu^{2+}(aq) + 2e^- \longrightarrow Cu(s) \qquad [\text{reduction}]$$

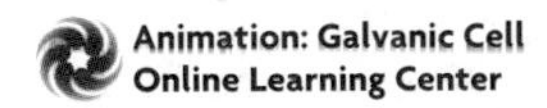
Animation: Galvanic Cell
Online Learning Center

The overall redox reaction is the sum of these half-reactions:

$$Zn(s) + Cu^{2+}(aq) \longrightarrow Zn^{2+}(aq) + Cu(s)$$

Because electrons flow spontaneously toward the copper electrode (cathode), it must have a more positive $E^0_{half\text{-}cell}$ than the zinc electrode (anode). Therefore, to obtain a positive E^0_{cell}, we subtract E^0_{zinc} from E^0_{copper}:

$$E^0_{cell} = E^0_{copper} - E^0_{zinc}$$

We can generalize this result for any voltaic cell: *the standard cell potential is the difference between the standard electrode potential of the cathode (reduction) half-cell and the standard electrode potential of the anode (oxidation) half-cell:*

$$E^0_{cell} = E^0_{cathode\ (reduction)} - E^0_{anode\ (oxidation)} \qquad (21.3)$$

Determining $E^0_{half\text{-}cell}$: The Standard Hydrogen Electrode What portion of E^0_{cell} for the zinc-copper reaction is contributed by the anode half-cell (oxidation of Zn) and what portion by the cathode half-cell (reduction of Cu^{2+})? That is, how can we know half-cell potentials if we can only measure the potential of the complete cell? Half-cell potentials, such as E^0_{zinc} and E^0_{copper}, are not absolute quantities, but rather are values *relative* to that of a standard. *This standard reference half-cell has its standard electrode potential defined as zero ($E^0_{reference} \equiv 0.00$ V).*

*The current required to operate modern digital voltmeters makes a negligible difference in the value of E^0_{cell}.

The **standard reference half-cell** is a **standard hydrogen electrode,** which consists of a specially prepared platinum electrode immersed in a 1 *M* aqueous solution of a strong acid, $H^+(aq)$ [or $H_3O^+(aq)$], through which H_2 gas at 1 atm is bubbled. Thus, the reference half-reaction is

$$2H^+(aq;\ 1\ M) + 2e^- \rightleftharpoons H_2(g;\ 1\ atm) \qquad E^0_{reference} = 0.00\ V$$

Now we can construct a voltaic cell consisting of this reference half-cell and another half-cell whose potential we want to determine. With $E^0_{reference}$ defined as zero, the overall E^0_{cell} allows us to find the unknown standard electrode potential, $E^0_{unknown}$. When H_2 is oxidized, the reference half-cell is the anode, and so reduction occurs at the unknown half-cell:

$$E^0_{cell} = E^0_{cathode} - E^0_{anode} = E^0_{unknown} - E^0_{reference} = E^0_{unknown} - 0.00\ V = E^0_{unknown}$$

When H^+ is reduced, the reference half-cell is the cathode, and so oxidation occurs at the unknown half-cell:

$$E^0_{cell} = E^0_{cathode} - E^0_{anode} = E^0_{reference} - E^0_{unknown} = 0.00\ V - E^0_{unknown} = -E^0_{unknown}$$

Figure 21.7 shows a voltaic cell that has the Zn/Zn^{2+} half-reaction in one compartment and the H^+/H_2 (or H_3O^+/H_2) half-reaction in the other. The zinc electrode is negative relative to the hydrogen electrode, so we know that the zinc is being oxidized and is the anode. The measured E^0_{cell} is +0.76 V, and we use this value to find the unknown standard electrode potential, E^0_{zinc}:

$$2H^+(aq) + 2e^- \longrightarrow H_2(g) \qquad E^0_{reference} = 0.00\ V \qquad \text{[cathode; reduction]}$$
$$Zn(s) \longrightarrow Zn^{2+}(aq) + 2e^- \qquad E^0_{zinc} = ?\ V \qquad \text{[anode; oxidation]}$$
$$Zn(s) + 2H^+(aq) \longrightarrow Zn^{2+}(aq) + H_2(g) \qquad E^0_{cell} = 0.76\ V$$
$$E^0_{cell} = E^0_{cathode} - E^0_{anode} = E^0_{reference} - E^0_{zinc}$$
$$E^0_{zinc} = E^0_{reference} - E^0_{cell} = 0.00\ V - 0.76\ V = -0.76\ V$$

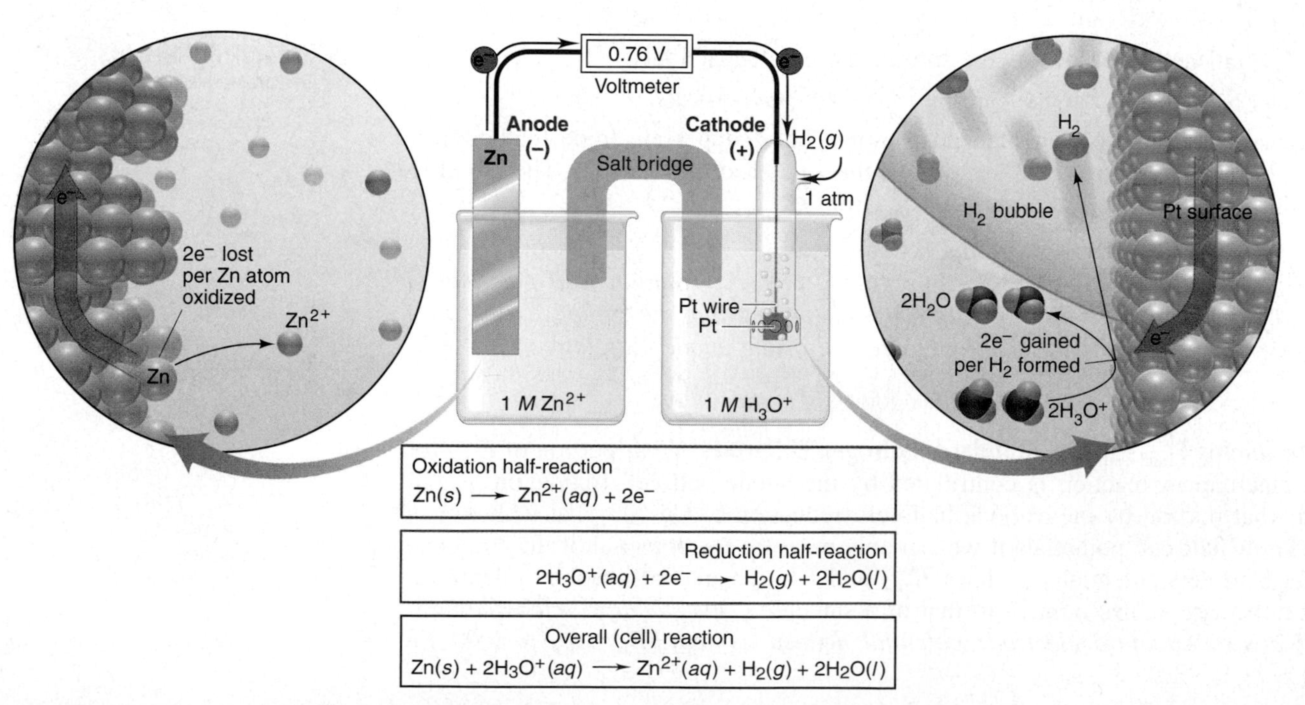

Figure 21.7 Determining an unknown $E^0_{half\text{-}cell}$ with the standard reference (hydrogen) electrode. A voltaic cell has the Zn half-reaction in one half-cell and the hydrogen reference half-reaction in the other. The magnified view of the hydrogen half-reaction shows two H_3O^+ ions being reduced to two H_2O molecules and an H_2 molecule, which enters the H_2 bubble. The Zn/Zn^{2+} half-cell potential is negative (anode), and the cell potential is 0.76 V. The potential of the standard reference electrode is defined as 0.00 V, so the cell potential equals the negative of the anode potential; that is,

$$0.76\ V = 0.00\ V - E^0_{zinc} \qquad \text{so} \qquad E^0_{zinc} = -0.76\ V$$

Now let's return to the zinc-copper cell and use the measured value of E^0_{cell} (1.10 V) and the value we just found for E^0_{zinc} to calculate E^0_{copper}:

$$E^0_{cell} = E^0_{cathode} - E^0_{anode} = E^0_{copper} - E^0_{zinc}$$
$$E^0_{copper} = E^0_{cell} + E^0_{zinc} = 1.10 \text{ V} + (-0.76 \text{ V}) = 0.34 \text{ V}$$

By continuing this process of constructing cells with one known and one unknown electrode potential, we can find many other standard electrode potentials. Let's go over these ideas once more with a sample problem.

SAMPLE PROBLEM 21.3 Calculating an Unknown $E^0_{half\text{-}cell}$ from E^0_{cell}

Problem A voltaic cell houses the reaction between aqueous bromine and zinc metal:

$$Br_2(aq) + Zn(s) \longrightarrow Zn^{2+}(aq) + 2Br^-(aq) \qquad E^0_{cell} = 1.83 \text{ V}$$

Calculate $E^0_{bromine}$, given $E^0_{zinc} = -0.76$ V.

Plan E^0_{cell} is positive, so the reaction is spontaneous as written. By dividing the reaction into half-reactions, we see that Br_2 is reduced and Zn is oxidized; thus, the zinc half-cell contains the anode. We use Equation 21.3 to find $E^0_{unknown}$ ($E^0_{bromine}$).

Solution Dividing the reaction into half-reactions:

$$Br_2(aq) + 2e^- \longrightarrow 2Br^-(aq) \qquad E^0_{unknown} = E^0_{bromine} = ?\ \text{V}$$
$$Zn(s) \longrightarrow Zn^{2+}(aq) + 2e^- \qquad E^0_{zinc} = -0.76 \text{ V}$$

Calculating $E^0_{bromine}$:

$$E^0_{cell} = E^0_{cathode} - E^0_{anode} = E^0_{bromine} - E^0_{zinc}$$
$$E^0_{bromine} = E^0_{cell} + E^0_{zinc} = 1.83 \text{ V} + (-0.76 \text{ V}) = 1.07 \text{ V}$$

Check A good check is to make sure that calculating $E^0_{bromine} - E^0_{zinc}$ gives E^0_{cell}: 1.07 V − (−0.76 V) = 1.83 V.

Comment Keep in mind that, whichever is the unknown half-cell, reduction is the cathode half-reaction and oxidation is the anode half-reaction. Always subtract E^0_{anode} from $E^0_{cathode}$ to get E^0_{cell}.

FOLLOW-UP PROBLEM 21.3 A voltaic cell based on the reaction between aqueous Br_2 and vanadium(III) ions has $E^0_{cell} = 1.39$ V:

$$Br_2(aq) + 2V^{3+}(aq) + 2H_2O(l) \longrightarrow 2VO^{2+}(aq) + 4H^+(aq) + 2Br^-(aq)$$

What is $E^0_{vanadium}$, the standard electrode potential for the reduction of VO^{2+} to V^{3+}?

Relative Strengths of Oxidizing and Reducing Agents

One of the things we can learn from measuring potentials of voltaic cells is the relative strengths of the oxidizing and reducing agents involved. Three oxidizing agents present in the voltaic cell just discussed are Cu^{2+}, H^+, and Zn^{2+}. We can rank their relative oxidizing strengths by writing each half-reaction as a gain of electrons (reduction), with its corresponding standard electrode potential:

$$Cu^{2+}(aq) + 2e^- \longrightarrow Cu(s) \qquad E^0 = 0.34 \text{ V}$$
$$2H^+(aq) + 2e^- \longrightarrow H_2(g) \qquad E^0 = 0.00 \text{ V}$$
$$Zn^{2+}(aq) + 2e^- \longrightarrow Zn(s) \qquad E^0 = -0.76 \text{ V}$$

The more positive the E^0 value, the more readily the reaction (as written) occurs; thus, Cu^{2+} gains two electrons more readily than H^+, which gains them more readily than Zn^{2+}. In terms of strength as an oxidizing agent, therefore, $Cu^{2+} > H^+ > Zn^{2+}$. Moreover, this listing also ranks the strengths of the reducing agents: $Zn > H_2 > Cu$. Notice that this list of half-reactions in order of *decreasing* half-cell potential shows, *from top to bottom*, the oxidizing agents (reactants) *decreasing* in strength and the reducing agents (products) *increasing* in strength; that is, Cu^{2+} (top left) is the strongest oxidizing agent, and Zn (bottom right) is the strongest reducing agent.

By combining many pairs of half-cells into voltaic cells, we can create a list of reduction half-reactions and arrange them in *decreasing* order of standard electrode potential (from most positive to most negative). Such a list, called an *emf series* or a *table of standard electrode potentials,* appears in Appendix D, with a few examples in Table 21.2. There are several key points to keep in mind:

- All values are relative to the standard hydrogen (reference) electrode:
$$2H^+(aq;\ 1\ M) + 2e^- \rightleftharpoons H_2(g;\ 1\ \text{atm}) \qquad E^0_{\text{reference}} = 0.00\ \text{V}$$
- By convention, the half-reactions are written as *reductions,* which means that *only reactants are oxidizing agents and only products are reducing agents.*
- The more positive the $E^0_{\text{half-cell}}$, the more readily the half-reaction occurs.
- Half-reactions are shown with an equilibrium arrow because each can occur as a reduction or an oxidation (that is, take place at the cathode or anode, respectively), depending on the $E^0_{\text{half-cell}}$ of the other half-reaction.
- As Appendix D (and Table 21.2) is arranged, the strength of the oxidizing agent (reactant) *increases going up (bottom to top),* and the strength of the reducing agent (product) *increases going down (top to bottom).*

Table 21.2 Selected Standard Electrode Potentials (298 K)

Strength of oxidizing agent (increases going up); Strength of reducing agent (increases going down)

Half-Reaction	$E^0_{\text{half-cell}}$ (V)
$F_2(g) + 2e^- \rightleftharpoons 2F^-(aq)$	+2.87
$Cl_2(g) + 2e^- \rightleftharpoons 2Cl^-(aq)$	+1.36
$MnO_2(s) + 4H^+(aq) + 2e^- \rightleftharpoons Mn^{2+}(aq) + 2H_2O(l)$	+1.23
$NO_3^-(aq) + 4H^+(aq) + 3e^- \rightleftharpoons NO(g) + 2H_2O(l)$	+0.96
$Ag^+(aq) + e^- \rightleftharpoons Ag(s)$	+0.80
$Fe^{3+}(aq) + e^- \rightleftharpoons Fe^{2+}(aq)$	+0.77
$O_2(g) + 2H_2O(l) + 4e^- \rightleftharpoons 4OH^-(aq)$	+0.40
$Cu^{2+}(aq) + 2e^- \rightleftharpoons Cu(s)$	+0.34
$2H^+(aq) + 2e^- \rightleftharpoons H_2(g)$	**0.00**
$N_2(g) + 5H^+(aq) + 4e^- \rightleftharpoons N_2H_5^+(aq)$	−0.23
$Fe^{2+}(aq) + 2e^- \rightleftharpoons Fe(s)$	−0.44
$Zn^{2+}(aq) + 2e^- \rightleftharpoons Zn(s)$	−0.76
$2H_2O(l) + 2e^- \rightleftharpoons H_2(g) + 2OH^-(aq)$	−0.83
$Na^+(aq) + e^- \rightleftharpoons Na(s)$	−2.71
$Li^+(aq) + e^- \rightleftharpoons Li(s)$	−3.05

Thus, $F_2(g)$ is the strongest oxidizing agent (has the largest positive E^0), which means $F^-(aq)$ is the weakest reducing agent. Similarly, $Li^+(aq)$ is the weakest oxidizing agent (has the most negative E^0), which means $Li(s)$ is the strongest reducing agent. (You may have noticed an analogy to conjugate acid-base pairs: a strong acid forms a weak conjugate base, and vice versa, just as a *strong oxidizing agent forms a weak reducing agent,* and vice versa.)

If you forget the ranking in the table, just rely on your chemical knowledge of the elements. You know that F_2 is very electronegative and typically occurs as F^-. It is easily reduced (gains electrons), so it must be a strong oxidizing agent (high, positive E^0). Similarly, Li metal has a low ionization energy and typically occurs as Li^+. Therefore, it is easily oxidized (loses electrons), so it must be a strong reducing agent (low, negative E^0).

Writing Spontaneous Redox Reactions Appendix D can be used to write spontaneous redox reactions, which is useful for constructing voltaic cells.

Every redox reaction is the sum of two half-reactions, so there is a reducing agent and an oxidizing agent on each side. In the zinc-copper reaction, for instance, Zn and Cu are the reducing agents, and Cu^{2+} and Zn^{2+} are the oxidizing agents. The stronger oxidizing and reducing agents react spontaneously to form the weaker oxidizing and reducing agents:

$$\underset{\text{stronger reducing agent}}{Zn(s)} + \underset{\text{stronger oxidizing agent}}{Cu^{2+}(aq)} \longrightarrow \underset{\text{weaker oxidizing agent}}{Zn^{2+}(aq)} + \underset{\text{weaker reducing agent}}{Cu(s)}$$

Here, too, note the similarity to acid-base chemistry. The stronger acid and base spontaneously form the weaker base and acid, respectively. The members of a conjugate acid-base pair differ by a proton: the acid has the proton and the base does not. The members of a redox pair, or *redox couple,* such as Zn and Zn^{2+}, differ by one or more electrons: the reduced form (Zn) has the electrons and the oxidized form (Zn^{2+}) does not. In acid-base reactions, we compare acid and base strength using K_a and K_b values. In redox reactions, we compare oxidizing and reducing strength using E^0 values.

Based on the order of the E^0 values in Appendix D, *the stronger oxidizing agent (species on the left) has a half-reaction with a larger (more positive or less negative) E^0 value, and the stronger reducing agent (species on the right) has a half-reaction with a smaller (less positive or more negative) E^0 value*. Therefore, a spontaneous reaction ($E^0_{cell} > 0$) will occur between an oxidizing agent and a reducing agent that lies *below* it in the list. For instance, Cu^{2+} (left) and Zn (right) react spontaneously, and Zn lies below Cu^{2+}. In other words, *for a spontaneous reaction to occur, the half-reaction higher in the list proceeds at the cathode as written, and the half-reaction lower in the list proceeds at the anode in reverse.* This pairing ensures that the stronger oxidizing agent (higher on the left) and stronger reducing agent (lower on the right) will be the reactants.

However, if we know the electrode potentials, we can write a spontaneous redox reaction even if Appendix D is not available. Let's choose a pair of half-reactions from the appendix and, without referring to their relative positions in the list, arrange them into a spontaneous redox reaction:

$$Ag^+(aq) + e^- \longrightarrow Ag(s) \qquad E^0_{silver} = 0.80\ V$$
$$Sn^{2+}(aq) + 2e^- \longrightarrow Sn(s) \qquad E^0_{tin} = -0.14\ V$$

There are two steps involved:

1. Reverse one of the half-reactions into an oxidation step such that the difference of the electrode potentials (cathode *minus* anode) gives a *positive* E^0_{cell}. Note that when we reverse the half-reaction, we do *not* reverse the sign of $E^0_{half\text{-}cell}$ because the minus sign in Equation 21.3 ($E^0_{cell} = E^0_{cathode} - E^0_{anode}$) will do that.
2. Add the rearranged half-reactions to obtain a balanced overall equation. Be sure to multiply by coefficients so that e^- lost equals e^- gained and to cancel species common to both sides.

(You may be tempted in this case to add the two half-reactions as written, because you obtain a positive E^0_{cell}, but you would then have two oxidizing agents forming two reducing agents, which cannot occur.)

We want to pair the stronger oxidizing and reducing agents as reactants. The larger (more positive) E^0 value for the silver half-reaction means that Ag^+ is a stronger oxidizing agent (gains electrons more readily) than Sn^{2+}, and the smaller (more negative) E^0 value for the tin half-reaction means that Sn is a stronger reducing agent (loses electrons more readily) than Ag. Therefore, we reverse the tin half-reaction (but *not* the sign of E^0_{tin}):

$$Sn(s) \longrightarrow Sn^{2+}(aq) + 2e^- \qquad E^0_{tin} = -0.14\ V$$

Subtracting $E^0_{half\text{-}cell}$ of the tin half-reaction (anode, oxidation) from $E^0_{half\text{-}cell}$ of the silver half-reaction (cathode, reduction) gives a positive E^0_{cell}; that is, 0.80 V − (−0.14 V) = 0.94 V.

With the half-reactions written in the correct direction, we must next make sure that the *number of electrons lost in the oxidation equals the number gained in the reduction.* In this case, we double the silver (reduction) half-reaction and add the half-reactions to obtain the balanced equation:

$$2Ag^+(aq) + 2e^- \longrightarrow 2Ag(s) \qquad E^0 = 0.80\ V \qquad \text{[reduction]}$$
$$Sn(s) \longrightarrow Sn^{2+}(aq) + 2e^- \qquad E^0 = -0.14\ V \qquad \text{[oxidation]}$$
$$Sn(s) + 2Ag^+(aq) \longrightarrow Sn^{2+}(aq) + 2Ag(s) \qquad E^0_{cell} = E^0_{silver} - E^0_{tin} = 0.94\ V$$

With the reaction spontaneous as written, the stronger oxidizing and reducing agents are reactants, which confirms that Sn is a stronger reducing agent than Ag, and Ag^+ is a stronger oxidizing agent than Sn^{2+}.

A very important point to note is that, when we doubled the coefficients of the silver half-reaction to balance the number of electrons, we did *not* double its E^0 value—it remained 0.80 V. That is, *changing the balancing coefficients of a half-reaction does* ***not*** *change its E^0 value.* The reason is that a standard electrode potential is an *intensive* property, one that does *not* depend on the amount of substance present. The potential is the *ratio* of energy to charge. When we change the coefficients, thus changing the amount of substance, the energy *and* the charge change proportionately, so their ratio stays the same. (Recall that density, which is also an intensive property, does not change with the amount of substance because the mass *and* the volume change proportionately.)

SAMPLE PROBLEM 21.4 Writing Spontaneous Redox Reactions and Ranking Oxidizing and Reducing Agents by Strength

Problem (a) Combine the following three half-reactions into three balanced equations (A, B, and C) for spontaneous reactions, and calculate E^0_{cell} for each. **(b)** Rank the relative strengths of the oxidizing and reducing agents.

$$(1)\ NO_3^-(aq) + 4H^+(aq) + 3e^- \longrightarrow NO(g) + 2H_2O(l) \qquad E^0 = 0.96\ V$$
$$(2)\ N_2(g) + 5H^+(aq) + 4e^- \longrightarrow N_2H_5^+(aq) \qquad E^0 = -0.23\ V$$
$$(3)\ MnO_2(s) + 4H^+(aq) + 2e^- \longrightarrow Mn^{2+}(aq) + 2H_2O(l) \qquad E^0 = 1.23\ V$$

Plan (a) To write the redox equations, we combine the possible pairs of half-reactions: (1) and (2), (1) and (3), and (2) and (3). They are all written as reductions, so the oxidizing agents appear as reactants and the reducing agents appear as products. In each pair, we reverse the reduction half-reaction that has the smaller (less positive or more negative) E^0 value to an oxidation to obtain a positive E^0_{cell}. We make e^- lost equal e^- gained, without changing the magnitude of the E^0 value, add the half-reactions together, and then apply Equation 21.3 to find E^0_{cell}. **(b)** Because each reaction is spontaneous as written, the stronger oxidizing and reducing agents are the reactants. To obtain the overall ranking, we first rank the relative strengths within each equation and then compare them.

Solution (a) Combining half-reactions (1) and (2) gives equation A. The E^0 value for half-reaction (1) is larger (more positive) than that for (2), so we reverse (2) to obtain a positive E^0_{cell}:

$$(1)\ NO_3^-(aq) + 4H^+(aq) + 3e^- \longrightarrow NO(g) + 2H_2O(l) \qquad E^0 = 0.96\ V$$
$$\text{(rev 2)}\quad N_2H_5^+(aq) \longrightarrow N_2(g) + 5H^+(aq) + 4e^- \qquad E^0 = -0.23\ V$$

To make e^- lost equal e^- gained, we multiply (1) by four and the reversed (2) by three; then add half-reactions and cancel appropriate numbers of common species (H^+ and e^-):

$$4NO_3^-(aq) + 16H^+(aq) + 12e^- \longrightarrow 4NO(g) + 8H_2O(l) \qquad E^0 = 0.96\ V$$
$$3N_2H_5^+(aq) \longrightarrow 3N_2(g) + 15H^+(aq) + 12e^- \qquad E^0 = -0.23\ V$$

$$\text{(A)}\ 3N_2H_5^+(aq) + 4NO_3^-(aq) + H^+(aq) \longrightarrow 3N_2(g) + 4NO(g) + 8H_2O(l)$$
$$E^0_{cell} = 0.96\ V - (-0.23\ V) = 1.19\ V$$

Combining half-reaction (1) and half-reaction (3) gives equation B. Half-reaction (1) must be reversed:

$$\text{(rev 1)} \quad NO(g) + 2H_2O(l) \longrightarrow NO_3^-(aq) + 4H^+(aq) + 3e^- \qquad E^0 = 0.96\text{ V}$$
$$\text{(3)} \quad MnO_2(s) + 4H^+(aq) + 2e^- \longrightarrow Mn^{2+}(aq) + 2H_2O(l) \qquad E^0 = 1.23\text{ V}$$

We multiply reversed (1) by two and (3) by three, then add and cancel:

$$2NO(g) + 4H_2O(l) \longrightarrow 2NO_3^-(aq) + 8H^+(aq) + 6e^- \qquad E^0 = 0.96\text{ V}$$
$$3MnO_2(s) + 12H^+(aq) + 6e^- \longrightarrow 3Mn^{2+}(aq) + 6H_2O(l) \qquad E^0 = 1.23\text{ V}$$

$$\text{(B)} \quad 3MnO_2(s) + 4H^+(aq) + 2NO(g) \longrightarrow 3Mn^{2+}(aq) + 2H_2O(l) + 2NO_3^-(aq)$$

$$E^0_{cell} = 1.23\text{ V} - 0.96\text{ V} = 0.27\text{ V}$$

Combining half-reaction (2) and half-reaction (3) gives equation C. Half-reaction (2) must be reversed:

$$\text{(rev 2)} \quad N_2H_5^+(aq) \longrightarrow N_2(g) + 5H^+(aq) + 4e^- \qquad E^0 = -0.23\text{ V}$$
$$\text{(3)} \quad MnO_2(s) + 4H^+(aq) + 2e^- \longrightarrow Mn^{2+}(aq) + 2H_2O(l) \qquad E^0 = 1.23\text{ V}$$

We multiply reaction (3) by two, add the half-reactions, and cancel:

$$N_2H_5^+(aq) \longrightarrow N_2(g) + 5H^+(aq) + 4e^- \qquad E^0 = -0.23\text{ V}$$
$$2MnO_2(s) + 8H^+(aq) + 4e^- \longrightarrow 2Mn^{2+}(aq) + 4H_2O(l) \qquad E^0 = 1.23\text{ V}$$

$$\text{(C)} \quad N_2H_5^+(aq) + 2MnO_2(s) + 3H^+(aq) \longrightarrow N_2(g) + 2Mn^{2+}(aq) + 4H_2O(l)$$

$$E^0_{cell} = 1.23\text{ V} - (-0.23\text{ V}) = 1.46\text{ V}$$

(b) Ranking the oxidizing and reducing agents within each equation:

Equation (A): Oxidizing agents: $NO_3^- > N_2$ Reducing agents: $N_2H_5^+ > NO$
Equation (B): Oxidizing agents: $MnO_2 > NO_3^-$ Reducing agents: $NO > Mn^{2+}$
Equation (C): Oxidizing agents: $MnO_2 > N_2$ Reducing agents: $N_2H_5^+ > Mn^{2+}$

Determining the overall ranking of oxidizing and reducing agents. Comparing the relative strengths from the three balanced equations gives

Oxidizing agents: $MnO_2 > NO_3^- > N_2$

Reducing agents: $N_2H_5^+ > NO > Mn^{2+}$

Check As always, check that atoms and charge balance on each side of the equation. A good way to check the ranking and equations is to list the given half-reactions in order of decreasing E^0 value:

$$MnO_2(s) + 4H^+(aq) + 2e^- \longrightarrow Mn^{2+}(aq) + 2H_2O(l) \qquad E^0 = 1.23\text{ V}$$
$$NO_3^-(aq) + 4H^+(aq) + 3e^- \longrightarrow NO(g) + 2H_2O(l) \qquad E^0 = 0.96\text{ V}$$
$$N_2(g) + 5H^+(aq) + 4e^- \longrightarrow N_2H_5^+(aq) \qquad E^0 = -0.23\text{ V}$$

Then the oxidizing agents (reactants) decrease in strength going down the list, so the reducing agents (products) decrease in strength going up. Moreover, each of the three spontaneous reactions (A, B, and C) should combine a reactant with a product from lower down on this list.

FOLLOW-UP PROBLEM 21.4 Is the following reaction spontaneous as written?

$$3Fe^{2+}(aq) \longrightarrow Fe(s) + 2Fe^{3+}(aq)$$

If not, write the equation for the spontaneous reaction, calculate E^0_{cell}, and rank the three species of iron in order of decreasing reducing strength.

Relative Reactivities of Metals In Chapter 4, we discussed the activity series of the metals (see Figure 4.20, p. 163), which ranks metals by their ability to "displace" one another from aqueous solution. Now you'll see *why* this displacement occurs, as well as why many, but not all, metals react with acid to form H_2, and why a few metals form H_2 even in water.

1. *Metals that can displace H_2 from acid.* The standard hydrogen half-reaction represents the reduction of H^+ ions from an acid to H_2:

$$2H^+(aq) + 2e^- \longrightarrow H_2(g) \qquad E^0 = 0.00 \text{ V}$$

To see which metals reduce H^+ (referred to as "displacing H_2") from acids, choose a metal, write its half-reaction as an oxidation, combine this half-reaction with the hydrogen half-reaction, and see if E^0_{cell} is positive. What you find is that the metals Li through Pb, those that lie *below* the standard hydrogen (reference) half-reaction in Appendix D, give a positive E^0_{cell} when reducing H^+. Iron, for example, reduces H^+ from an acid to H_2:

$$\begin{array}{lll} Fe(s) \longrightarrow Fe^{2+}(aq) + 2e^- & E^0 = -0.44 \text{ V} & \text{[anode; oxidation]} \\ 2H^+(aq) + 2e^- \longrightarrow H_2(g) & E^0 = 0.00 \text{ V} & \text{[cathode; reduction]} \\ \hline Fe(s) + 2H^+(aq) \longrightarrow H_2(g) + Fe^{2+}(aq) & E^0_{cell} = 0.00 \text{ V} - (-0.44 \text{ V}) = 0.44 \text{ V} & \end{array}$$

The lower the metal in the list, the stronger it is as a reducing agent; therefore, the more positive its half-cell potential when the half-reaction is reversed, and the higher the E^0_{cell} for its reduction of H^+ to H_2. *If E^0_{cell} for the reduction of H^+ is more positive for metal A than it is for metal B, metal A is a stronger reducing agent than metal B and a more* ***active*** *metal.*

2. *Metals that cannot displace H_2 from acid.* Metals that are *above* the standard hydrogen (reference) half-reaction *cannot* reduce H^+ from acids. When we reverse the metal half-reaction, the E^0_{cell} is negative, so the reaction does not occur. For example, the coinage metals—copper, silver, and gold, which are in Group 1B(11)—are not strong enough reducing agents to reduce H^+ from acids:

$$\begin{array}{lll} Ag(s) \longrightarrow Ag^+(aq) + e^- & E^0 = 0.80 \text{ V} & \text{[anode; oxidation]} \\ 2H^+(aq) + 2e^- \longrightarrow H_2(g) & E^0 = 0.00 \text{ V} & \text{[cathode; reduction]} \\ \hline 2Ag(s) + 2H^+(aq) \longrightarrow 2Ag^+(aq) + H_2(g) & E^0_{cell} = 0.00 \text{ V} - 0.80 \text{ V} = -0.80 \text{ V} & \end{array}$$

The higher the metal in the list, the more negative is its E^0_{cell} for the reduction of H^+ to H_2, the lower is its reducing strength, and the less active it is. Thus, gold is less active than silver, which is less active than copper.

3. *Metals that can displace H_2 from water.* Metals active enough to displace H_2 from water lie *below the half-reaction for the reduction of water:*

$$2H_2O(l) + 2e^- \longrightarrow H_2(g) + 2OH^-(aq) \qquad E = -0.42 \text{ V}$$

(The value shown here is the *nonstandard* electrode potential because, in pure water, $[OH^-]$ is 1.0×10^{-7} *M*, not the standard-state value of 1 *M*.) For example, consider the reaction of sodium in water (with the Na half-reaction reversed and doubled):

$$\begin{array}{lll} 2Na(s) \longrightarrow 2Na^+(aq) + 2e^- & E^0 = -2.71 \text{ V} & \text{[anode; oxidation]} \\ 2H_2O(l) + 2e^- \longrightarrow H_2(g) + 2OH^-(aq) & E = -0.42 \text{ V} & \text{[cathode; reduction]} \\ \hline 2Na(s) + 2H_2O(l) \longrightarrow 2Na^+(aq) + H_2(g) + 2OH^-(aq) & & \\ & E_{cell} = -0.42 \text{ V} - (-2.71 \text{ V}) = 2.29 \text{ V} & \end{array}$$

Oxidation half-reaction
$Ca(s) \longrightarrow Ca^{2+}(aq) + 2e^-$

Reduction half-reaction
$2H_2O(l) + 2e^- \longrightarrow H_2(g) + 2OH^-(aq)$

Overall (cell) reaction
$Ca(s) + 2H_2O(l) \longrightarrow Ca(OH)_2(aq) + H_2(g)$

Figure 21.8 The reaction of calcium in water. Calcium is one of the metals active enough to displace H_2 from H_2O.

The alkali metals [Group 1A(1)] and the larger alkaline earth metals [Group 2A(2)] can displace H_2 from H_2O (Figure 21.8).

4. *Metals that can displace other metals from solution.* We can also predict whether one metal can reduce the aqueous ion of another metal. Any metal that is lower in the list in Appendix D can reduce the ion of a metal that is higher up, and thus displace that metal from solution. For example, zinc can displace iron from solution:

$$\begin{array}{lll} Zn(s) \longrightarrow Zn^{2+}(aq) + 2e^- & E^0 = -0.76 \text{ V} & \text{[anode; oxidation]} \\ Fe^{2+}(aq) + 2e^- \longrightarrow Fe(s) & E^0 = -0.44 \text{ V} & \text{[cathode; reduction]} \\ \hline Zn(s) + Fe^{2+}(aq) \longrightarrow Zn^{2+}(aq) + Fe(s) & E^0_{cell} = -0.44 \text{ V} - (-0.76 \text{ V}) = 0.32 \text{ V} & \end{array}$$

This particular reaction has tremendous economic importance in protecting iron from rusting, as you'll see shortly. The reducing power of metals has other, more personal, consequences, as the margin note points out.

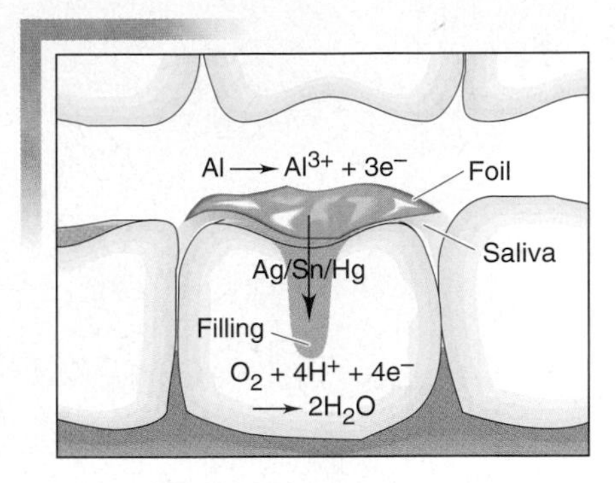

The Pain of a Dental Voltaic Cell Have you ever felt a jolt of pain when biting down with a filled tooth on a scrap of foil left on a piece of food? Here's the reason. The aluminum foil acts as an active anode (E^0 of Al = −1.66 V), saliva as the electrolyte, and the filling (usually a silver/tin/mercury alloy) as an inactive cathode. O_2 is reduced to water, and the short circuit between the foil in contact with the filling creates a current that is sensed by the nerve of the tooth.

SECTION SUMMARY

The output of a cell is called the cell potential (E_{cell}) and is measured in volts (1 V = 1 J/C). When all substances are in their standard states, the output is the standard cell potential (E^0_{cell}). $E^0_{cell} > 0$ for a spontaneous reaction at standard-state conditions. By convention, a standard electrode potential ($E^0_{half\text{-}cell}$) refers to the *reduction* half-reaction. E^0_{cell} equals $E^0_{half\text{-}cell}$ of the cathode *minus* $E^0_{half\text{-}cell}$ of the anode. Using a standard hydrogen (reference) electrode, other $E^0_{half\text{-}cell}$ values can be measured and used to rank oxidizing (or reducing) agents (see Appendix D). Spontaneous redox reactions combine stronger oxidizing and reducing agents to form weaker ones. A metal can reduce another species (H^+, H_2O, or an ion of another metal) if E^0_{cell} for the reaction is positive.

21.4 FREE ENERGY AND ELECTRICAL WORK

In Chapter 20, we discussed the relationship of useful work, free energy, and the equilibrium constant. In this section, we examine this relationship in the context of electrochemical cells and see the effect of concentration on cell potential.

Standard Cell Potential and the Equilibrium Constant

As you know from Section 20.3, a spontaneous reaction has a *negative* free energy change ($\Delta G < 0$), and you've just seen that a spontaneous electrochemical reaction has a *positive* cell potential ($E_{cell} > 0$). Note that *the signs of ΔG and E_{cell} are opposite for a spontaneous reaction*. These two indications of spontaneity are proportional to each other:

$$\Delta G \propto -E_{cell}$$

Let's determine this proportionality constant by focusing on the electrical work done (w, in joules), which is the product of the potential (E_{cell}, in volts) and the amount of charge that flows (in coulombs):

$$w = E_{cell} \times \text{charge}$$

The value used for E_{cell} is measured with no current flowing and, therefore, no energy lost to heating the cell. Thus, E_{cell} is the maximum voltage the cell can generate, that is, the maximum work the system can do *on* the surroundings. Recall from Chapter 20 that *only a reversible process can do maximum work*. For no current to flow and the process to be reversible, E_{cell} must be opposed by an equal potential in the measuring circuit. (In this case, a reversible process means that, if the opposing potential is infinitesimally smaller, the cell reaction goes forward; if it is infinitesimally larger, the reaction goes backward.) Equation 20.9, p. 884, shows that the maximum work done *on* the surroundings is $-\Delta G$:

$$w_{max} = E_{cell} \times \text{charge} = -\Delta G \quad \text{or} \quad \Delta G = -E_{cell} \times \text{charge}$$

The charge that flows through the cell equals the number of moles of electrons (n) transferred times the charge of 1 mol of electrons (symbol F):

$$\text{Charge} = \text{moles of } e^- \times \frac{\text{charge}}{\text{mol } e^-} \quad \text{or} \quad \text{charge} = nF$$

The charge of 1 mol of electrons is the **Faraday constant (*F*)**, named in honor of Michael Faraday, the 19th-century British scientist who pioneered the study of electrochemistry:

$$F = \frac{96{,}485 \text{ C}}{\text{mol } e^-}$$

Because 1 V = 1 J/C, we have 1 C = 1 J/V, and

$$F = 9.65\times10^4 \frac{\text{J}}{\text{V}\cdot\text{mol e}^-} \quad \text{(3 sf)} \tag{21.4}$$

Substituting for charge, the proportionality constant is nF:

$$\Delta G = -nFE_{\text{cell}} \tag{21.5}$$

When all of the components are in their standard states, we have

$$\Delta G^0 = -nFE^0_{\text{cell}} \tag{21.6}$$

Using this relationship, we can relate the standard cell potential to the equilibrium constant of the redox reaction. Recall that

$$\Delta G^0 = -RT \ln K$$

Substituting for ΔG^0 from Equation 21.6 gives

$$-nFE^0_{\text{cell}} = -RT \ln K$$

Solving for E^0_{cell} gives

$$E^0_{\text{cell}} = \frac{RT}{nF} \ln K \tag{21.7}$$

Figure 21.9 summarizes the interconnections among the standard free energy change, the equilibrium constant, and the standard cell potential. The procedures presented in Chapter 20 for determining K required that we know ΔG^0, either from ΔH^0 and ΔS^0 values or from ΔG^0_f values. For redox reactions, we now have a direct experimental method for determining K *and* ΔG^0: measure E^0_{cell}.

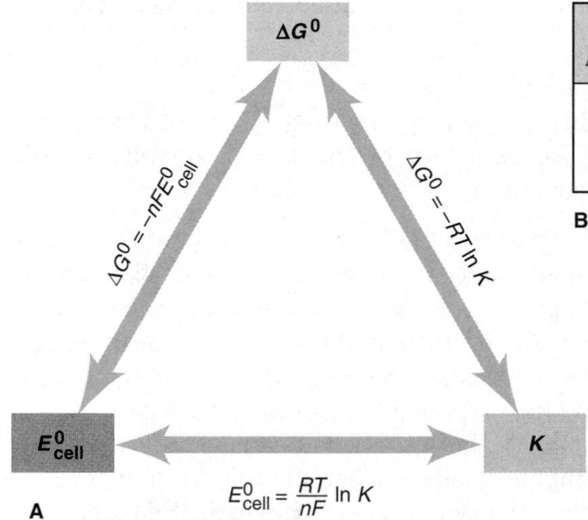

ΔG^0	K	E^0_{cell}	Reaction at standard-state conditions
<0	>1	>0	Spontaneous
0	1	0	At equilibrium
>0	<1	<0	Nonspontaneous

B

Figure 21.9 The interrelationship of ΔG^0, E^0_{cell}, and K. **A,** Any one of these three central thermodynamic parameters can be used to find the other two. **B,** The signs of ΔG^0 and E^0_{cell} determine the reaction direction at standard-state conditions.

It is common practice to simplify Equation 21.7 in calculations by

- Substituting the known value of 8.314 J/(mol rxn·K) for the constant R
- Substituting the known value of 9.65×10^4 J/(V·mol e^-) for the constant F
- Substituting the standard temperature of 298.15 K for T, but keeping in mind that the cell can run at other temperatures.
- Multiplying by 2.303 to convert from natural to common (base-10) logarithms. This conversion shows that a *10-fold change in K makes E^0_{cell} change by 1,* which arises from the close connection between cell voltage and pH.

Thus, when n moles of e^- are transferred per mole of reaction in the balanced equation, this simplified relation between E^0_{cell} and K gives

$$E^0_{cell} = \frac{RT}{nF}\ln K = 2.303\frac{RT}{nF}\log K = 2.303 \times \frac{8.314\,\frac{\cancel{J}}{\cancel{mol\,rxn\cdot K}} \times 298.15\,\cancel{K}}{\frac{n\,\cancel{mol\,e^-}}{\cancel{mol\,rxn}}\left(9.65\times10^4\,\frac{\cancel{J}}{V\cdot\cancel{mol\,e^-}}\right)}\log K$$

And, we have

$$E^0_{cell} = \frac{0.0592\text{ V}}{n}\log K \quad \text{or} \quad \log K = \frac{nE^0_{cell}}{0.0592\text{ V}} \quad \text{(at 298.15 K)} \quad \textbf{(21.8)}$$

SAMPLE PROBLEM 21.5 Calculating K and ΔG^0 from E^0_{cell}

Problem Lead can displace silver from solution:

$$Pb(s) + 2Ag^+(aq) \longrightarrow Pb^{2+}(aq) + 2Ag(s)$$

As a consequence, silver is a valuable by-product in the industrial extraction of lead from its ore. Calculate K and ΔG^0 at 298.15 K for this reaction.

Plan We divide the spontaneous redox equation into the half-reactions and use values from Appendix D to calculate E^0_{cell}. Then, we substitute this result into Equation 21.8 to find K and into Equation 21.6 to find ΔG^0.

Solution Writing the half-reactions and their E^0 values:

(1) $Ag^+(aq) + e^- \longrightarrow Ag(s) \qquad E^0 = 0.80\text{ V}$

(2) $Pb^{2+}(aq) + 2e^- \longrightarrow Pb(s) \qquad E^0 = -0.13\text{ V}$

Calculating E^0_{cell}: We double (1), reverse (2), add the half-reactions, and subtract E^0_{lead} from E^0_{silver}:

$$2Ag^+(aq) + 2e^- \longrightarrow 2Ag(s) \qquad E^0 = 0.80\text{ V}$$
$$Pb(s) \longrightarrow Pb^{2+}(aq) + 2e^- \qquad E^0 = -0.13\text{ V}$$
$$Pb(s) + 2Ag^+(aq) \longrightarrow Pb^{2+}(aq) + 2Ag(s) \qquad E^0_{cell} = 0.80\text{ V} - (-0.13\text{ V}) = 0.93\text{ V}$$

Calculating K with Equations 21.7 and 21.8:

$$E^0_{cell} = \frac{RT}{nF}\ln K = 2.303\frac{RT}{nF}\log K$$

The adjusted half-reactions show that 2 mol of e^- are transferred per mole of reaction as written, so $n = 2$. Then, performing the substitutions for R and F that we just discussed with the cell running at 25°C (298.15 K), we have

$$E^0_{cell} = \frac{0.0592\text{ V}}{2}\log K = 0.93\text{ V}$$

So, $$\log K = \frac{0.93\text{ V} \times 2}{0.0592\text{ V}} = 31.42 \quad \text{and} \quad K = 2.6\times10^{31}$$

Calculating ΔG^0 (Equation 21.6):

$$\Delta G^0 = -nFE^0_{cell} = -\frac{2\text{ mol }e^-}{\text{mol rxn}} \times \frac{96.5\text{ kJ}}{\text{V}\cdot\text{mol }e^-} \times 0.93\text{ V} = -1.8\times10^2\text{ kJ/mol rxn}$$

Check The three variables are consistent with the reaction being spontaneous at standard-state conditions: $E^0_{cell} > 0$, $\Delta G^0 < 0$, and $K > 1$. Be sure to round and check the order of magnitude: in the ΔG^0 calculation, for instance, $\Delta G^0 \approx -2 \times 100 \times 1 = -200$, so the overall math seems right. Another check would be to obtain ΔG^0 directly from its relation with K:

$$\Delta G^0 = -RT\ln K = -8.314\text{ J/mol rxn}\cdot\text{K} \times 298.15\text{ K} \times \ln(2.6\times10^{31})$$
$$= -1.8\times10^5\text{ J/mol rxn} = -1.8\times10^2\text{ kJ/mol rxn}$$

FOLLOW-UP PROBLEM 21.5 When cadmium metal reduces Cu^{2+} in solution, Cd^{2+} forms in addition to copper metal. If $\Delta G^0 = -143$ kJ, calculate K at 25°C. What is E^0_{cell} of a voltaic cell that uses this reaction?

The Effect of Concentration on Cell Potential

So far, we've considered cells with all components in their standard states and found standard cell potential (E^0_{cell}) from standard half-cell potentials ($E^0_{half\text{-}cell}$). However, most cells do not start with all components in their standard states, and even if they did, the concentrations change after a few moments of operation. Moreover, in all practical voltaic cells, such as batteries, reactant concentrations are far from standard-state values. Clearly, we must be able to determine E_{cell}, the cell potential under nonstandard conditions.

To do so, let's derive an expression for the relation between cell potential and concentration based on the relation between free energy and concentration. Recall from Chapter 20 (Equation 20.13) that ΔG equals ΔG^0 (the free energy change when the system moves from standard-state concentrations to equilibrium) *plus* $RT \ln Q$ (the free energy change when the system moves from nonstandard-state to standard-state concentrations):

$$\Delta G = \Delta G^0 + RT \ln Q$$

ΔG is related to E_{cell} and ΔG^0 to E^0_{cell} (Equations 21.5 and 21.6), so we substitute for them and get

$$-nFE_{cell} = -nFE^0_{cell} + RT \ln Q$$

Dividing both sides by $-nF$, we obtain the **Nernst equation,** developed by the German chemist Walther Hermann Nernst in 1889:

Walther Hermann Nernst (1864–1941) This great German physical chemist was only 25 years old when he developed the equation for the relationship between cell voltage and concentration. His career, which culminated in the Nobel Prize in chemistry in 1920, included formulating the third law of thermodynamics, establishing the principle of the solubility product, and contributing key ideas to photochemistry and the Haber process. He is shown here in later years making a point to a few of his celebrated colleagues: from left to right, Nernst, Albert Einstein, Max Planck, Robert Millikan, and Max von Laue.

$$E_{cell} = E^0_{cell} - \frac{RT}{nF} \ln Q \qquad \textbf{(21.9)}$$

The Nernst equation says that a cell potential under any conditions depends on the potential at standard-state concentrations *and* a term for the potential at nonstandard-state concentrations. How do changes in Q affect cell potential? From Equation 21.9, we see that

- When $Q < 1$ and thus [reactant] > [product], $\ln Q < 0$, so $E_{cell} > E^0_{cell}$.
- When $Q = 1$ and thus [reactant] = [product], $\ln Q = 0$, so $E_{cell} = E^0_{cell}$.
- When $Q > 1$ and thus [reactant] < [product], $\ln Q > 0$, so $E_{cell} < E^0_{cell}$.

As before, to obtain a simplified form of the Nernst equation for use in calculations, let's substitute known values of R and F, operate the cell at 298.15 K, and convert to common (base-10) logarithms:

$$E_{cell} = E^0_{cell} - \frac{RT}{nF} \ln Q = E^0_{cell} - 2.303 \frac{RT}{nF} \log Q$$

$$= E^0_{cell} - 2.303 \times \frac{8.314 \frac{\cancel{J}}{\cancel{mol\ rxn \cdot K}} \times 298.15 \cancel{K}}{n \frac{\cancel{mol\ e^-}}{\cancel{mol\ rxn}} \left(9.65 \times 10^4 \frac{\cancel{J}}{V \cdot \cancel{mol\ e^-}}\right)} \log Q$$

And we obtain:

$$E_{cell} = E^0_{cell} - \frac{0.0592\ V}{n} \log Q \quad \text{(at 298.15 K)} \qquad \textbf{(21.10)}$$

Remember that the expression for Q *contains only those species with concentrations (and/or pressures) that can vary;* thus, solids do not appear, even when they are the electrodes. For example, in the reaction between cadmium and silver ion, the Cd and Ag electrodes do not appear in the expression for Q:

$$Cd(s) + 2Ag^+(aq) \longrightarrow Cd^{2+}(aq) + 2Ag(s) \qquad Q = \frac{[Cd^{2+}]}{[Ag^+]^2}$$

SAMPLE PROBLEM 21.6 Using the Nernst Equation to Calculate E_{cell}

Problem In a test of a new reference electrode, a chemist constructs a voltaic cell consisting of a Zn/Zn^{2+} half-cell and an H_2/H^+ half-cell under the following conditions:

$$[Zn^{2+}] = 0.010\ M \qquad [H^+] = 2.5\ M \qquad P_{H_2} = 0.30\ \text{atm}$$

Calculate E_{cell} at 298.15 K.

Plan To apply the Nernst equation and determine E_{cell}, we must know E^0_{cell} and Q. We write the spontaneous reaction, calculate E^0_{cell} from standard electrode potentials (Appendix D), and use the given pressure and concentrations to find Q. (Recall that the ideal gas law allows us to use P at constant T as another way of writing concentration, n/V.) Then we substitute into Equation 21.10.

Solution Determining the cell reaction and E^0_{cell}:

$$2H^+(aq) + 2e^- \longrightarrow H_2(g) \qquad E^0 = 0.00\ \text{V}$$

$$Zn(s) \longrightarrow Zn^{2+}(aq) + 2e^- \qquad E^0 = -0.76\ \text{V}$$

$$2H^+(aq) + Zn(s) \longrightarrow H_2(g) + Zn^{2+}(aq) \qquad E^0_{cell} = 0.00\ \text{V} - (-0.76\ \text{V}) = 0.76\ \text{V}$$

Calculating Q:

$$Q = \frac{P_{H_2} \times [Zn^{2+}]}{[H^+]^2} = \frac{0.30 \times 0.010}{2.5^2} = 4.8\times10^{-4}$$

Solving for E_{cell} at 25°C (298.15 K), with $n = 2$:

$$E_{cell} = E^0_{cell} - 2.303\frac{RT}{nF}\log Q = E^0_{cell} - \frac{0.0592\ \text{V}}{n}\log Q$$

$$= 0.76\ \text{V} - \left[\frac{0.0592\ \text{V}}{2}\log(4.8\times10^{-4})\right] = 0.76\ \text{V} - (-0.0982\ \text{V}) = 0.86\ \text{V}$$

Check After you check the arithmetic, reason through the answer: $E_{cell} > E^0_{cell}$ $(0.86 > 0.76)$ because the log Q term was negative, which is consistent with $Q < 1$; that is, the amounts of products, P_{H_2} and $[Zn^{2+}]$, are smaller than the amount of reactant, $[H^+]$.

FOLLOW-UP PROBLEM 21.6 Consider a voltaic cell based on the following reaction: $Fe(s) + Cu^{2+}(aq) \longrightarrow Fe^{2+}(aq) + Cu(s)$. If $[Cu^{2+}] = 0.30\ M$, what must $[Fe^{2+}]$ be to increase E_{cell} by 0.25 V above E^0_{cell} at 25°C?

Changes in Potential During Cell Operation

Let's see how the potential of the zinc-copper cell changes as concentrations change during cell operation. In this case, the only concentrations that change are [reactant] = $[Cu^{2+}]$ and [product] = $[Zn^{2+}]$:

$$Zn(s) + Cu^{2+}(aq) \longrightarrow Zn^{2+}(aq) + Cu(s) \qquad Q = \frac{[Zn^{2+}]}{[Cu^{2+}]}$$

The positive E^0_{cell} (1.10 V) means that this reaction proceeds *spontaneously* from standard-state conditions, at which $[Zn^{2+}] = [Cu^{2+}] = 1\ M$ $(Q = 1)$, to some point at which $[Zn^{2+}] > [Cu^{2+}]$ $(Q > 1)$. Now, suppose that we start the cell when $[Zn^{2+}] < [Cu^{2+}]$ $(Q < 1)$, for example, when $[Zn^{2+}] = 1.0\times10^{-4}\ M$ and $[Cu^{2+}] = 2.0\ M$. Thus, the cell potential is *higher* than the standard cell potential:

$$E_{cell} = E^0_{cell} - \frac{0.0592\ \text{V}}{2}\log\frac{[Zn^{2+}]}{[Cu^{2+}]} = 1.10\ \text{V} - \left(\frac{0.0592\ \text{V}}{2}\log\frac{1.0\times10^{-4}}{2.0}\right)$$

$$= 1.10\ \text{V} - \left[\frac{0.0592\ \text{V}}{2}(-4.30)\right] = 1.10\ \text{V} + 0.127\ \text{V} = 1.23\ \text{V}$$

As the cell operates, $[Zn^{2+}]$ increases (as the Zn electrode deteriorates) and $[Cu^{2+}]$ decreases (as Cu plates out on the Cu electrode). Although the changes during this process occur smoothly, if we keep Equation 21.10 in mind, we can

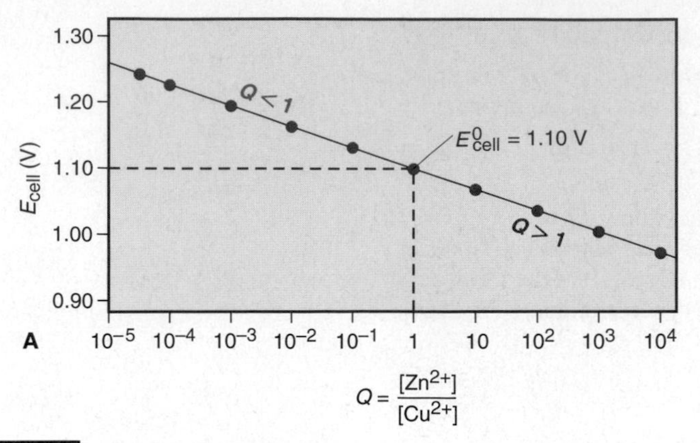

Changes in E_{cell} and Concentration

Stage in cell operation	Q	Relative [P] and [R]	$\frac{0.0592\text{ V}}{n}\log Q$
1. $E > E^0$	<1	[P] < [R]	<0
2. $E = E^0$	=1	[P] = [R]	=0
3. $E < E^0$	>1	[P] > [R]	>0
4. $E = 0$	$=K$	[P] ≫ [R]	$=E^0$

B

Figure 21.10 The relation between E_{cell} and log Q for the zinc-copper cell. A, A plot of E_{cell} vs. Q (on a logarithmic scale) for the zinc-copper cell shows a linear decrease. When $Q < 1$ (*left*), [reactant] is relatively high, and the cell can do relatively more work. When $Q = 1$, $E_{cell} = E^0_{cell}$. When $Q > 1$ (*right*), [reactant] is relatively low, and the cell can do relatively less work. **B,** A summary of the changes in E_{cell} as the cell operates, including the changes in [Zn^{2+}], denoted [P] for [product], and [Cu^{2+}], denoted [R] for [reactant].

identify four general stages of operation. Figure 21.10A shows the first three. The main point to note is *as the cell operates, its potential decreases:*

Stage 1. $E_{cell} > E^0_{cell}$ when $Q < 1$: When the cell begins operation, [Cu^{2+}] > [Zn^{2+}], so the [(0.0592 V/n) log Q] term < 0 and $E_{cell} > E^0_{cell}$.

As cell operation continues, [Zn^{2+}] increases and [Cu^{2+}] decreases; thus, Q becomes larger, the [(0.0592 V/n) log Q] term becomes less negative (more positive), and E_{cell} decreases.

Stage 2. $E_{cell} = E^0_{cell}$ when $Q = 1$: At the point when [Cu^{2+}] = [Zn^{2+}], $Q = 1$, so the [(0.0592 V/n) log Q] term = 0 and $E_{cell} = E^0_{cell}$.

Stage 3. $E_{cell} < E^0_{cell}$ when $Q > 1$: As the [Zn^{2+}]/[Cu^{2+}] ratio continues to increase, the [(0.0592 V/n) log Q] term > 0, so $E_{cell} < E^0_{cell}$.

Stage 4. $E_{cell} = 0$ when $Q = K$: Eventually, the [(0.0592 V/n) log Q] term becomes so large that it equals E^0_{cell}, which means that E_{cell} is zero. This occurs *when the system reaches* ***equilibrium:*** *no more free energy is released, so the cell can do no more work.* At this point, we say that a battery is "dead."

Figure 21.10B summarizes these four key stages in the operation of a voltaic cell.

Let's find K for the zinc-copper cell. At equilibrium, Equation 21.10 becomes

$$0 = E^0_{cell} - \left(\frac{0.0592\text{ V}}{n}\right)\log K, \text{ which rearranges to } E^0_{cell} = \frac{0.0592\text{ V}}{n}\log K$$

Note that this result is identical to Equation 21.8, which we obtained from ΔG^0. Solving for K of the zinc-copper cell (E^0_{cell} = 1.10 V),

$$\log K = \frac{2 \times E^0_{cell}}{0.0592\text{ V}}, \quad \text{so} \quad K = 10^{(2\times1.10\text{ V})/0.0592\text{ V}} = 10^{37.16} = 1.4\times10^{37}$$

Thus, the zinc-copper cell does work until the [Zn^{2+}]/[Cu^{2+}] ratio is *very* high.

To conclude, let's examine cell potential in terms of the *starting* Q/K ratio:

- If $Q/K < 1$, E_{cell} is positive for the reaction *as written*. The smaller the Q/K ratio, the greater the value of E_{cell}, and the more electrical work the cell can do.
- If $Q/K = 1$, $E_{cell} = 0$. The cell is at equilibrium and can no longer do work.
- If $Q/K > 1$, E_{cell} is negative for the reaction as written. The reverse reaction will take place, and the cell will do work until Q/K equals 1 at equilibrium.

Concentration Cells

If you mix a concentrated solution and a dilute solution of a salt, you know that the final concentration equals some intermediate value. A **concentration cell** employs this phenomenon to generate electrical energy. The two solutions are in separate half-cells, so they do not mix; rather, their concentrations become equal as the cell operates.

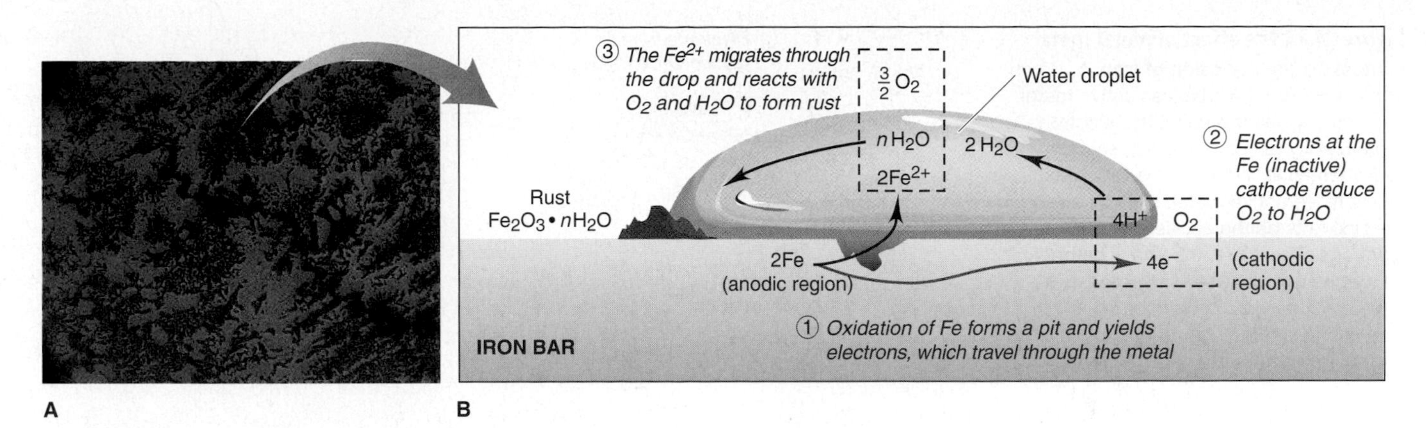

Figure 21.20 The corrosion of iron. A, A close-up view of an iron surface. Corrosion usually occurs at a surface irregularity. **B,** A schematic depiction of a small area of the surface, showing the steps in the corrosion process.

the surface of a surrounding water droplet, for instance. At this *cathodic region,* the electrons released from the iron atoms reduce O_2 molecules:

$$O_2(g) + 4H^+(aq) + 4e^- \longrightarrow 2H_2O(l) \qquad \text{[cathodic region; reduction]}$$

Notice that this overall redox process is complete; thus, the iron loss has occurred without any rust forming:

$$2Fe(s) + O_2(g) + 4H^+(aq) \longrightarrow 2Fe^{2+}(aq) + 2H_2O(l)$$

Rust forms through another redox reaction in which the reactants make direct contact. The Fe^{2+} ions formed originally at the anodic region disperse through the surrounding water and react with O_2, often at some distance from the pit (fact 3). The overall reaction for this step is

$$2Fe^{2+}(aq) + \tfrac{1}{2}O_2(g) + (2+n)H_2O(l) \longrightarrow Fe_2O_3{\cdot}nH_2O(s) + 4H^+(aq)$$

[The inexact coefficient n for H_2O in the above equation appears because rust, $Fe_2O_3{\cdot}nH_2O$, is a form of iron(III) oxide with a variable number of waters of hydration.] The rust deposit is really incidental to the damage caused by loss of iron—a chemical insult added to the original injury.

Adding the previous two equations together shows the overall equation for the rusting of iron:

$$2Fe(s) + \tfrac{3}{2}O_2(g) + nH_2O(l) + \cancel{4H^+(aq)} \longrightarrow Fe_2O_3{\cdot}nH_2O(s) + \cancel{4H^+(aq)}$$

The canceled H^+ ions are shown to emphasize that they act as a catalyst; that is, they speed the process as they are used up in one step of the overall reaction and created in another. As a result of this action, rusting is faster at low pH (high $[H^+]$) (fact 4). Ionic solutions speed rusting by improving the conductivity of the aqueous medium near the anodic and cathodic regions (fact 5). The effect of ions is especially evident on ocean-going vessels (Figure 21.21) and on the underbodies and around the wheel wells of cars driven in cold climates, where salts are used to melt ice on slippery roads.

Figure 21.21 Enhanced corrosion at sea. The high ion concentration of seawater leads to its high conductivity, which enhances the corrosion of iron in the hulls and anchors of ocean-going vessels.

In many ways, the components of the corrosion process resemble those of a voltaic cell:

- Anodic and cathodic regions are separated in space.
- The regions are connected via an external circuit through which the electrons travel.
- In the anodic region, iron behaves like an active electrode, whereas in the cathodic region, it is inactive.
- The moisture surrounding the pit functions somewhat like a salt bridge, a means for ions to ferry back and forth and keep the solution neutral.

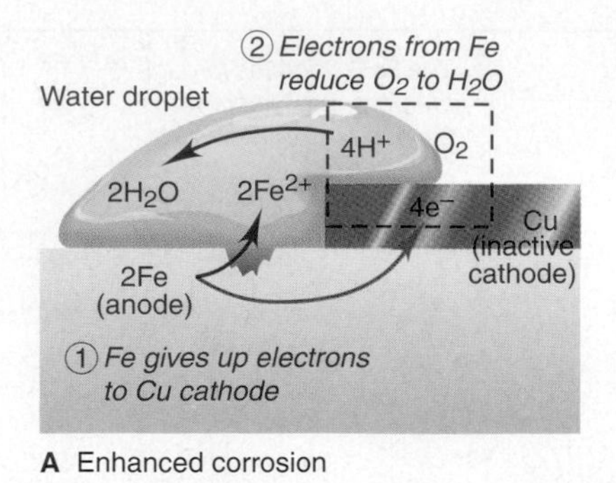

A Enhanced corrosion

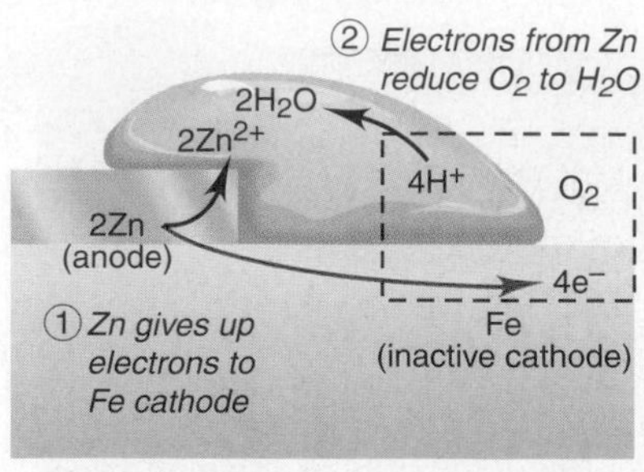

B Cathodic protection

Figure 21.22 The effect of metal-metal contact on the corrosion of iron. A, When iron is in contact with a less active metal, such as copper, the iron loses electrons more readily (is more anodic), so it corrodes faster. **B,** When iron is in contact with a more active metal, such as zinc, the zinc acts as the anode and loses electrons. Therefore, the iron is cathodic, so it does not corrode. The process is known as *cathodic protection.*

Protecting Against the Corrosion of Iron

A common approach to preventing or limiting corrosion is to eliminate contact with the corrosive factors. The simple act of washing off road salt removes the ionic solution from auto bodies. Iron objects are frequently painted to keep out O_2 and moisture, but if the paint layer chips, rusting proceeds. More permanent coatings include chromium plated on plumbing fixtures. In the "blueing" of gun barrels, wood stoves, and other steel objects, an adherent coating of Fe_3O_4 (magnetite) is bonded to the surface.

The only fact regarding corrosion that we have not yet addressed concerns the relative activity of other metals in contact with iron (fact 6), which leads to the most effective way to prevent corrosion. The essential idea is that *iron functions as both anode and cathode in the rusting process, but it is lost only at the anode*. Therefore, it makes sense that anything that makes iron behave more like the anode increases corrosion. As you can see in Figure 21.22A, when iron is in contact with a *less* active metal (weaker reducing agent), such as copper, its anodic function is enhanced. As a result, when iron plumbing is connected directly to copper plumbing with no electrical insulation between them, the iron pipe corrodes rapidly.

On the other hand, anything that makes iron behave more like the cathode prevents corrosion. Application of this principle is called *cathodic protection.* For example, if the iron makes contact with a *more* active metal (stronger reducing agent), such as zinc, the iron becomes cathodic and remains intact, while the zinc acts as the anode and loses electrons (Figure 21.22B). Coating steel with a "sacrificial" layer of zinc is the basis of the *galvanizing* process. In addition to blocking physical contact with H_2O and O_2, the zinc is "sacrificed" (oxidized) instead of the iron.

Sacrificial anodes are employed to protect iron and steel structures (pipes, tanks, oil rigs, and so on) in marine and moist underground environments. The metals most frequently used for this purpose are magnesium and aluminum; because these elements are much more active than iron, they act as the anode while iron acts as the cathode (Figure 21.23). Another advantage of these metals is that they form adherent oxide coatings, which slows their own corrosion.

Mg rod
Iron pipe

Figure 21.23 The use of sacrificial anodes to prevent iron corrosion. In cathodic protection, an active metal, such as magnesium or aluminum, is connected to underground iron pipes to prevent their corrosion. The active metal is sacrificed instead of the iron.

SECTION SUMMARY

Corrosion damages metal structures through a natural electrochemical change. Iron corrosion occurs in the presence of oxygen and moisture and is increased by high $[H^+]$, high [ion], or contact with a less active metal, such as Cu. Fe is oxidized and O_2 is reduced in one redox reaction, while rust (hydrated form of Fe_2O_3) is formed in another reaction that often takes place at a different location. Because Fe functions as both anode and cathode in the process, an iron or steel object can be protected by physically covering its surface or joining it to a more active metal (such as Zn, Mg, or Al), which acts as the anode in place of the Fe.

21.7 ELECTROLYTIC CELLS: USING ELECTRICAL ENERGY TO DRIVE NONSPONTANEOUS REACTIONS

Up to now, we've been considering voltaic cells, those that generate electrical energy from a spontaneous redox reaction. The principle of an electrolytic cell is exactly the opposite: *electrical energy from an external source drives a nonspontaneous reaction.*

Construction and Operation of an Electrolytic Cell

Let's examine the operation of an electrolytic cell by constructing one from a voltaic cell. Consider the tin-copper voltaic cell in Figure 21.24A. The Sn anode will gradually become oxidized to Sn^{2+} ions, and the Cu^{2+} ions will gradually be reduced and plate out on the Cu cathode because the cell reaction is spontaneous in that direction:

For the voltaic cell

$$\begin{array}{lll} Sn(s) \longrightarrow Sn^{2+}(aq) + 2e^- & & \text{[anode; oxidation]} \\ Cu^{2+}(aq) + 2e^- \longrightarrow Cu(s) & & \text{[cathode; reduction]} \\ \hline Sn(s) + Cu^{2+}(aq) \longrightarrow Sn^{2+}(aq) + Cu(s) & & E^0_{cell} = 0.48 \text{ V and } \Delta G^0 = -93 \text{ kJ} \end{array}$$

Therefore, the *reverse* cell reaction is *non*spontaneous and never happens of its own accord, as the negative E^0_{cell} and positive ΔG^0 indicate:

$$Cu(s) + Sn^{2+}(aq) \longrightarrow Cu^{2+}(aq) + Sn(s) \qquad E^0_{cell} = -0.48 \text{ V and } \Delta G^0 = 93 \text{ kJ}$$

However, we can make this process happen by supplying from an external source an electric potential *greater than* E^0_{cell}. In effect, we have converted the voltaic cell into an electrolytic cell and changed the nature of the electrodes—anode is now cathode, and cathode is now anode (Figure 21.24B):

For the electrolytic cell

$$\begin{array}{lll} Cu(s) \longrightarrow Cu^{2+}(aq) + 2e^- & & \text{[anode; oxidation]} \\ Sn^{2+}(aq) + 2e^- \longrightarrow Sn(s) & & \text{[cathode; reduction]} \\ \hline Cu(s) + Sn^{2+}(aq) \longrightarrow Cu^{2+}(aq) + Sn(s) & & \text{[overall (cell) reaction]} \end{array}$$

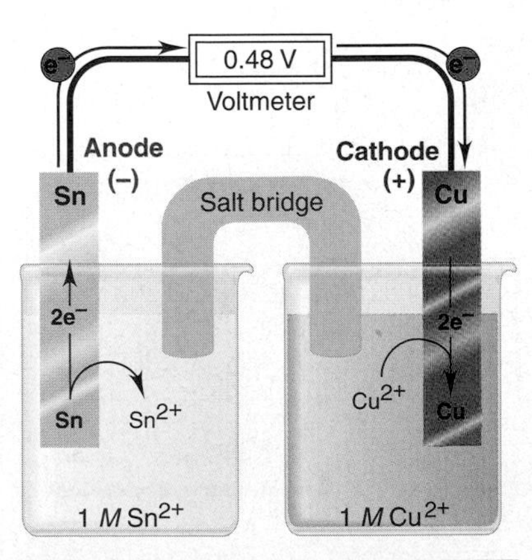

Oxidation half-reaction
$Sn(s) \longrightarrow Sn^{2+}(aq) + 2e^-$

Reduction half-reaction
$Cu^{2+}(aq) + 2e^- \longrightarrow Cu(s)$

Overall (cell) reaction
$Sn(s) + Cu^{2+}(aq) \longrightarrow Sn^{2+}(aq) + Cu(s)$

A Voltaic cell

External source *greater* than 0.48 V
Cathode (–)
Anode (+)
Salt bridge
Sn
Cu
2e⁻
Sn2+
Cu2+
1 M Sn2+
1 M Cu2+

Oxidation half-reaction
$Cu(s) \longrightarrow Cu^{2+}(aq) + 2e^-$

Reduction half-reaction
$Sn^{2+}(aq) + 2e^- \longrightarrow Sn(s)$

Overall (cell) reaction
$Cu(s) + Sn^{2+}(aq) \longrightarrow Cu^{2+}(aq) + Sn(s)$

B Electrolytic cell

Figure 21.24 The tin-copper reaction as the basis of a voltaic and an electrolytic cell. A, The spontaneous reaction between Sn and Cu^{2+} generates 0.48 V in a voltaic cell. **B,** If more than 0.48 V is supplied, the same apparatus becomes an electrolytic cell, and the nonspontaneous reaction between Cu and Sn^{2+} occurs. Note the changes in electrode charges and direction of electron flow.

Note that in an electrolytic cell, as in a voltaic cell, *oxidation takes place at the anode and reduction takes place at the cathode, but the direction of electron flow and the signs of the electrodes are reversed.*

To understand these changes, keep in mind the *cause* of the electron flow:

- In a voltaic cell, electrons are generated at the anode, so it is negative, and electrons are consumed at the cathode, so it is positive.
- In an electrolytic cell, the electrons come from the external power source, which *supplies* them *to* the cathode, so it is negative, and *removes* them *from* the anode, so it is positive.

A rechargeable battery functions as a voltaic cell when it is discharging and as an electrolytic cell when it is recharging, so it provides a good way to compare these two cell types and the changes in the processes and charges of the electrodes. Figure 21.25 shows these changes in the lead-acid battery. In the discharge mode (voltaic cell), oxidation occurs at electrode I, thus making the *negative* electrode the anode. In the recharge mode (electrolytic cell), oxidation occurs at electrode II, thus making the *positive* electrode the anode. Similarly, the cathode is *positive* during discharge (electrode II) and *negative* during recharge (electrode I). To reiterate, regardless of whether the cell is discharging or recharging, oxidation occurs at the anode and reduction occurs at the cathode.

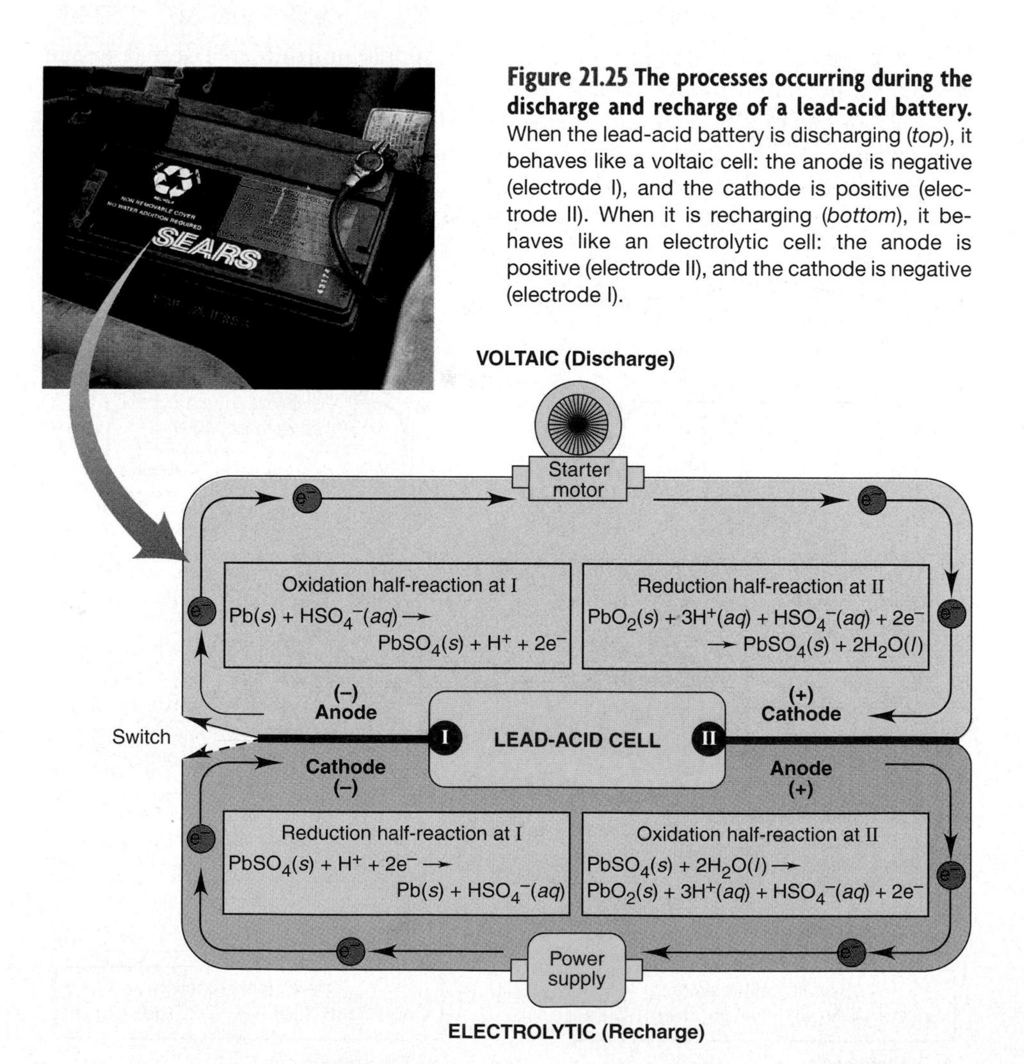

Figure 21.25 The processes occurring during the discharge and recharge of a lead-acid battery. When the lead-acid battery is discharging (*top*), it behaves like a voltaic cell: the anode is negative (electrode I), and the cathode is positive (electrode II). When it is recharging (*bottom*), it behaves like an electrolytic cell: the anode is positive (electrode II), and the cathode is negative (electrode I).

Table 21.4 Comparison of Voltaic and Electrolytic Cells

			Electrode		
Cell Type	**ΔG**	**E_{cell}**	**Name**	**Process**	**Sign**
Voltaic	<0	>0	Anode	Oxidation	$-$
Voltaic	<0	>0	Cathode	Reduction	$+$
Electrolytic	>0	<0	Anode	Oxidation	$+$
Electrolytic	>0	<0	Cathode	Reduction	$-$

Table 21.4 summarizes the processes and signs in the two types of electrochemical cells.

Predicting the Products of Electrolysis

Electrolysis, the splitting (lysing) of a substance by the input of electrical energy, is often used to decompose a compound into its elements. Electrolytic cells are involved in key industrial production steps for some of the most commercially important elements, including chlorine, aluminum, and copper, as you'll see in the next chapter. The first laboratory electrolysis of H_2O to H_2 and O_2 was performed in 1800, and the process is still used to produce these gases in ultrahigh purity. The electrolyte in an electrolytic cell can be the pure compound (such as H_2O or a molten salt), a mixture of molten salts, or an aqueous solution of a salt. The products obtained depend on atomic properties and several other factors, so let's examine some actual cases.

Electrolysis of Pure Molten Salts Many electrolytic applications involve isolating a metal or nonmetal from a molten salt. Predicting the product at each electrode is simple if the salt is pure because *the cation will be reduced and the anion oxidized.* The electrolyte is the molten salt itself, and the ions move through the cell because they are attracted by the oppositely charged electrodes.

Consider the electrolysis of molten (fused) calcium chloride. The two species present are Ca^{2+} and Cl^-, so Ca^{2+} ion is reduced and Cl^- ion is oxidized:

$$2Cl^-(l) \longrightarrow Cl_2(g) + 2e^- \quad \text{[anode; oxidation]}$$
$$Ca^{2+}(l) + 2e^- \longrightarrow Ca(s) \quad \text{[cathode; reduction]}$$
$$Ca^{2+}(l) + 2Cl^-(l) \longrightarrow Ca(s) + Cl_2(g) \quad \text{[overall]}$$

Metallic calcium is prepared industrially this way, as are several other active metals, such as Na and Mg, and the halogens Cl_2 and Br_2. We examine the details of these and several other electrolytic processes in Chapter 22.

Electrolysis of Mixed Molten Salts More typically, the electrolyte is a mixture of molten salts, which is then electrolyzed to obtain a particular metal. When we have a choice of product, how can we tell which species will react at which electrode? The general rule for all electrolytic cells is that *the more easily oxidized species (stronger reducing agent) reacts at the anode, and the more easily reduced species (stronger oxidizing agent) reacts at the cathode.*

It's important to realize that for electrolysis of mixtures of molten salts, we *cannot* use tabulated E^0 values to tell the relative strength of the oxidizing and reducing agents. Those values refer to the *change from aqueous ion to free element,* $M^{n+}(aq) + ne^- \longrightarrow M(s)$, under standard-state conditions, but there are no aqueous ions in a molten salt. Instead, we rely on our knowledge of periodic atomic trends to predict which of the ions present gains or loses electrons more easily (Sections 8.4 and 9.5).

SAMPLE PROBLEM 21.8 Predicting the Electrolysis Products of a Molten Salt Mixture

Problem A chemical engineer melts a naturally occurring mixture of NaBr and $MgCl_2$ and decomposes it in an electrolytic cell. Predict the substance formed at each electrode, and write balanced half-reactions and the overall cell reaction.

Plan We have to determine which metal and nonmetal will form more easily at the electrodes. We first list the ions as oxidizing or reducing agents. If one metal holds its electrons more tightly than another, it has a higher ionization energy (IE). Therefore, as a cation, it gains electrons more easily; it is the stronger oxidizing agent and is reduced at the cathode. Similarly, if one nonmetal holds its electrons less tightly than another, it has a lower electronegativity (EN). Therefore, as an anion, it loses electrons more easily; it is the stronger reducing agent and is oxidized at the anode.

Solution Listing the ions as oxidizing or reducing agents:

The possible oxidizing agents are Na^+ and Mg^{2+}.

The possible reducing agents are Br^- and Cl^-.

Determining the cathode product (more easily reduced cation): Mg is to the right of Na in Period 3. IE increases from left to right, so Mg has a higher IE. It takes more energy to remove an e^- from Mg than from Na, so it follows that Mg^{2+} has a greater attraction for e^- and thus is more easily reduced (stronger oxidizing agent):

$$Mg^{2+}(l) + 2e^- \longrightarrow Mg(l) \qquad \text{[cathode; reduction]}$$

Determining the anode product (more easily oxidized anion): Br is below Cl in Group 7A(17). EN decreases down the group, so Br has a lower EN than Cl. Therefore, it follows that Br^- holds its e^- less tightly than Cl^-, so Br^- is more easily oxidized (stronger reducing agent):

$$2Br^-(l) \longrightarrow Br_2(g) + 2e^- \qquad \text{[anode; oxidation]}$$

Writing the overall cell reaction:

$$Mg^{2+}(l) + 2Br^-(l) \longrightarrow Mg(l) + Br_2(g) \qquad \text{[overall]}$$

Comment The cell temperature must be high enough to keep the salt mixture molten. In this case, the temperature is greater than the melting point of Mg, so it appears as a liquid in the equation, and greater than the boiling point of Br_2, so it appears as a gas.

FOLLOW-UP PROBLEM 21.8 A sample of $AlBr_3$ contaminated with KF is melted and electrolyzed. Determine the electrode products and write the overall cell reaction.

Oxidation half-reaction
$2H_2O(l) \longrightarrow O_2(g) + 4H^+(aq) + 4e^-$

Reduction half-reaction
$2H_2O(l) + 2e^- \longrightarrow H_2(g) + 2OH^-(aq)$

Overall (cell) reaction
$2H_2O(l) \longrightarrow 2H_2(g) + O_2(g)$

Figure 21.26 The electrolysis of water. A certain volume of oxygen forms through the oxidation of H_2O at the anode (*right*), and twice that volume of hydrogen forms through the reduction of H_2O at the cathode (*left*).

Electrolysis of Water and Nonstandard Half-Cell Potentials Before we can analyze the electrolysis products of aqueous salt solutions, we must examine the electrolysis of water itself. Extremely pure water is difficult to electrolyze because very few ions are present to conduct a current. If we add a small amount of a salt that cannot be electrolyzed in water (such as Na_2SO_4), however, electrolysis proceeds rapidly. A glass electrolytic cell with separated gas compartments is used to keep the H_2 and O_2 gases from mixing (Figure 21.26). At the anode, water is oxidized as the O.N. of O changes from -2 to 0:

$$2H_2O(l) \longrightarrow O_2(g) + 4H^+(aq) + 4e^- \qquad E = 0.82\text{ V} \qquad \text{[anode; oxidation]}$$

At the cathode, water is reduced as the O.N. of H changes from $+1$ to 0:

$$2H_2O(l) + 2e^- \longrightarrow H_2(g) + 2OH^-(aq) \qquad E = -0.42\text{ V} \qquad \text{[cathode; reduction]}$$

After doubling the cathode half-reaction to equate e^- loss and gain, adding the half-reactions (which involves combining the H^+ and OH^- into H_2O and canceling e^- and excess H_2O), and calculating E_{cell}, the overall reaction is

$$2H_2O(l) \longrightarrow 2H_2(g) + O_2(g) \qquad E_{cell} = -0.42\text{ V} - 0.82\text{ V} = -1.24\text{ V} \quad \text{[overall]}$$

Notice that these electrode potentials are not written with a superscript zero because they are *not* standard electrode potentials. The $[H^+]$ and $[OH^-]$ are

1.0×10^{-7} M rather than the standard-state value of 1 M. These E values are obtained by applying the Nernst equation. For example, the calculation for the anode potential (with $n = 4$) is

$$E_{cell} = E^0_{cell} - \frac{0.0592\ V}{4} \log (P_{O_2} \times [H^+]^4)$$

The standard potential for the *oxidation* of water is -1.23 V (from Appendix D) and $P_{O_2} \approx 1$ atm in the half-cell, so we have

$$E_{cell} = -1.23\ V - \left\{\frac{0.0592\ V}{4} \times [\log 1 + 4 \log (1.0\times10^{-7})]\right\} = -0.82\ V$$

In aqueous ionic solutions, $[H^+]$ and $[OH^-]$ are approximately 10^{-7} M also, so we use these nonstandard E_{cell} values to predict electrode products.

Electrolysis of Aqueous Ionic Solutions and the Phenomenon of Overvoltage

Aqueous salt solutions are mixtures of ions *and* water, so we have to compare the various electrode potentials to predict the electrode products. When two half-reactions are possible at an electrode,

- *The reduction with the less negative (more positive) electrode potential occurs.*
- *The oxidation with the less positive (more negative) electrode potential occurs.*

What happens, for instance, when a solution of potassium iodide is electrolyzed? The possible oxidizing agents are K^+ and H_2O, and their reduction half-reactions are

$$K^+(aq) + e^- \longrightarrow K(s) \qquad E^0 = -2.93\ V$$
$$2H_2O(l) + 2e^- \longrightarrow H_2(g) + 2OH^-(aq) \qquad E = -0.42\ V \qquad \text{[reduction]}$$

The less *negative* electrode potential for water means that it is much easier to reduce than K^+, so H_2 forms at the cathode. The possible reducing agents are I^- and H_2O, and their oxidation half-reactions are

$$2I^-(aq) \longrightarrow I_2(s) + 2e^- \qquad E^0 = 0.53\ V \qquad \text{[oxidation]}$$
$$2H_2O(l) \longrightarrow O_2(g) + 4H^+(aq) + 4e^- \qquad E = 0.82\ V$$

The less *positive* electrode potential for I^- means that a lower potential is needed to oxidize it than to oxidize H_2O, so I_2 forms at the anode.

However, the products predicted from this type of comparison of electrode potentials are not always the actual products. For gases such as $H_2(g)$ and $O_2(g)$ to be produced at metal electrodes, an additional voltage is required. This increment above the expected voltage is called the **overvoltage,** and it is 0.4 to 0.6 V for these gases. The overvoltage results from kinetic factors, such as the large activation energy (Section 16.5) required for gases to form at the electrode.

Overvoltage has major practical significance. A multibillion-dollar example is the industrial production of chlorine from concentrated NaCl solution. Water is easier to reduce than Na^+, so H_2 forms at the cathode even with an overvoltage of 0.6 V:

$$Na^+(aq) + e^- \longrightarrow Na(s) \qquad E^0 = -2.71\ V$$
$$2H_2O(l) + 2e^- \longrightarrow H_2(g) + 2OH^-(aq) \qquad E = -0.42\ V\ (\approx -1\ V \text{ with overvoltage}) \qquad \text{[reduction]}$$

But Cl_2 forms at the anode, even though the electrode potentials themselves would lead us to predict that O_2 should form:

$$2H_2O(l) \longrightarrow O_2(g) + 4H^+(aq) + 4e^- \qquad E = 0.82\ V\ (\sim 1.4\ V \text{ with overvoltage})$$
$$2Cl^-(aq) \longrightarrow Cl_2(g) + 2e^- \qquad E^0 = 1.36\ V \qquad \text{[oxidation]}$$

An overvoltage of ~0.6 V makes the potential needed to form O_2 slightly above that for Cl_2. Keeping the $[Cl^-]$ high also favors Cl_2 formation. Thus, chlorine, which is one of the 10 most heavily produced industrial chemicals, can be formed from plentiful natural sources of aqueous sodium chloride (see Chapter 22).

From these and other examples, we can determine which elements can be prepared electrolytically from aqueous solutions of their salts:

1. Cations of less active metals *are* reduced to the metal, including gold, silver, copper, chromium, platinum, and cadmium.
2. Cations of more active metals *are not* reduced, including those in Groups 1A(1) and 2A(2), and Al from 3A(13). Water is reduced to H_2 and OH^- instead.
3. Anions that *are* oxidized, because of overvoltage from O_2 formation, include the halides ($[Cl^-]$ must be high), except for F^-.
4. Anions that *are not* oxidized include F^- and common oxoanions, such as SO_4^{2-}, CO_3^{2-}, NO_3^-, and PO_4^{3-}, because the central nonmetal in these oxoanions is already in its highest oxidation state. Water is oxidized to O_2 and H^+ instead.

SAMPLE PROBLEM 21.9 Predicting the Electrolysis Products of Aqueous Ionic Solutions

Problem What products form during electrolysis of aqueous solutions of the following salts: **(a)** KBr; **(b)** $AgNO_3$; **(c)** $MgSO_4$?

Plan We identify the reacting ions and compare their electrode potentials with those of water, taking the 0.4 to 0.6 V overvoltage into consideration. The reduction half-reaction with the less negative electrode potential occurs at the cathode, and the oxidation half-reaction with the less positive electrode potential occurs at the anode.

Solution

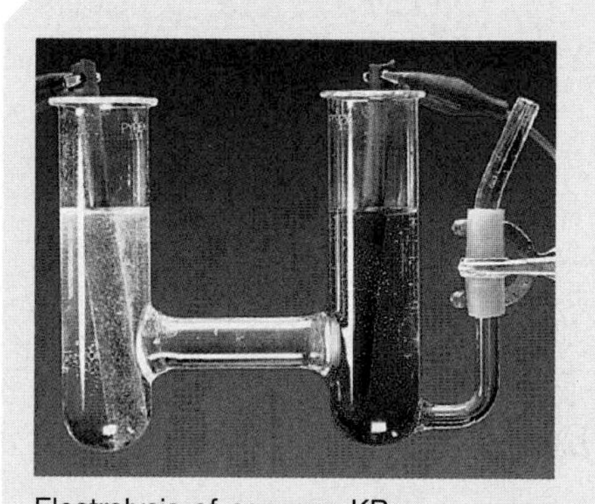

Electrolysis of aqueous KBr.

(a)

$$K^+(aq) + e^- \longrightarrow K(s) \qquad E^0 = -2.93\ V$$
$$2H_2O(l) + 2e^- \longrightarrow H_2(g) + 2OH^-(aq) \qquad E = -0.42\ V$$

Despite the overvoltage, which makes E for the reduction of water between −0.8 and −1.0 V, H_2O is still easier to reduce than K^+, so $H_2(g)$ forms at the cathode.

$$2Br^-(aq) \longrightarrow Br_2(l) + 2e^- \qquad E^0 = 1.07\ V$$
$$2H_2O(l) \longrightarrow O_2(g) + 4H^+(aq) + 4e^- \qquad E = 0.82\ V$$

Because of the overvoltage, which makes E for the oxidation of water between 1.2 and 1.4 V, Br^- is easier to oxidize than water, so $Br_2(l)$ forms at the anode (see photo).

(b)

$$Ag^+(aq) + e^- \longrightarrow Ag(s) \qquad E^0 = 0.80\ V$$
$$2H_2O(l) + 2e^- \longrightarrow H_2(g) + 2OH^-(aq) \qquad E = -0.42\ V$$

As the cation of an inactive metal, Ag^+ is a better oxidizing agent than H_2O, so Ag forms at the cathode. NO_3^- cannot be oxidized, because N is already in its highest (+5) oxidation state. Thus, O_2 forms at the anode:

$$2H_2O(l) \longrightarrow O_2(g) + 4H^+(aq) + 4e^-$$

(c)

$$Mg^{2+}(aq) + 2e^- \longrightarrow Mg(s) \qquad E^0 = -2.37\ V$$

Like K^+ in part (a), Mg^{2+} cannot be reduced in the presence of water, so H_2 forms at the cathode. The SO_4^{2-} ion cannot be oxidized because S is in its highest (+6) oxidation state. Thus, H_2O is oxidized, and O_2 forms at the anode:

$$2H_2O(l) \longrightarrow O_2(g) + 4H^+(aq) + 4e^-$$

FOLLOW-UP PROBLEM 21.9 Write half-reactions showing the products you predict will form in the electrolysis of aqueous $AuBr_3$.

The Stoichiometry of Electrolysis: The Relation Between Amounts of Charge and Product

As you've seen, the charge flowing through an electrolytic cell yields products at the electrodes. In the electrolysis of molten NaCl, for example, the power source supplies electrons to the cathode, where Na^+ ions migrate to pick them up and become Na metal. At the same time, the power source pulls from the anode the electrons that Cl^- ions release as they become Cl_2 gas. It follows that the more electrons picked up by Na^+ ions and released by Cl^- ions, the greater the amounts of Na and Cl_2 that form. This relationship was first determined experimentally by Michael Faraday and is referred to as *Faraday's law of electrolysis: the amount of substance produced at each electrode is directly proportional to the quantity of charge flowing through the cell.*

Each balanced half-reaction shows the amounts (mol) of reactant, electrons, and product involved in the change, so it contains the information we need to answer such questions as "How much material will form as a result of a given quantity of charge?" or, conversely, "How much charge is needed to produce a given amount of material?" To apply Faraday's law,

1. Balance the half-reaction to find the number of moles of electrons needed per mole of product.
2. Use the Faraday constant ($F = 9.65 \times 10^4$ C/mol e^-) to find the corresponding charge.
3. Use the molar mass to find the charge needed for a given mass of product.

In practice, to supply the correct amount of electricity, we need some means of finding the charge flowing through the cell. We cannot measure charge directly, but we *can* measure current, the charge flowing per unit time. The SI unit of current is the **ampere (A),** which is defined as 1 coulomb flowing through a conductor in 1 second:

$$1 \text{ ampere} = 1 \text{ coulomb/second} \quad \text{or} \quad 1 \text{ A} = 1 \text{ C/s} \qquad (21.11)$$

Thus, the current multiplied by the time gives the charge:

$$\text{Current} \times \text{time} = \text{charge} \quad \text{or} \quad \text{A} \times \text{s} = \frac{\text{C}}{\text{s}} \times \text{s} = \text{C}$$

Therefore, we find the charge by measuring the current *and* the time during which the current flows. This, in turn, relates to the amount of product formed. Figure 21.27 summarizes these relationships.

The Father of Electrochemistry and Much More The investigations by Michael Faraday (1791–1867) into the mass of an element that is equivalent to a given amount of charge established electrochemistry as a quantitative science, but his breakthroughs in physics are even more celebrated. In 1821, Faraday developed the precursor of the electric motor, and his later studies of currents induced by electric and magnetic fields eventually led to the electric generator and the transformer. After his death, this self-educated blacksmith's son was rated by Albert Einstein as the peer of Newton, Galileo, and Maxwell. He is shown here delivering one of his famous lectures on the "Chemical History of a Candle."

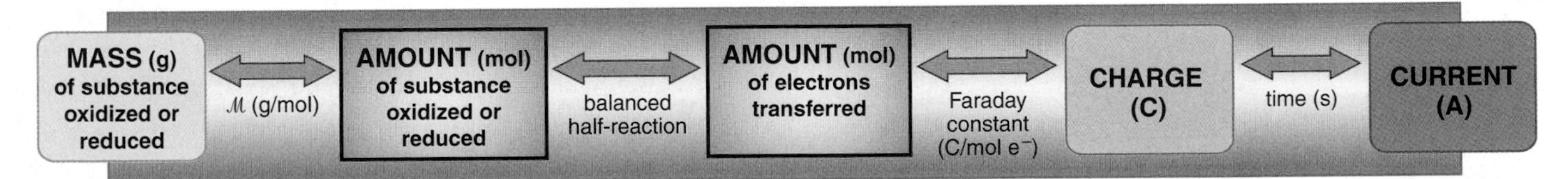

Figure 21.27 **A summary diagram for the stoichiometry of electrolysis.**

Problems based on Faraday's law often ask you to calculate current, mass of material, or time. The electrode half-reaction provides the key to solving these problems because it is related to the mass for a certain quantity of charge.

As an example, let's consider a typical problem in practical electrolysis: how long does it take to produce 3.0 g of $Cl_2(g)$ by electrolysis of aqueous NaCl using a power supply with a current of 12 A? The problem asks for the time needed to produce a certain mass, so let's first relate mass to number of moles of electrons to find the charge needed. Then, we'll relate the charge to the current to find the time.

We know the mass of Cl_2 produced, so we can find the amount (mol) of Cl_2. The half-reaction tells us that the loss of 2 mol of electrons produces 1 mol of chlorine gas:

$$2Cl^-(aq) \longrightarrow Cl_2(g) + 2e^-$$

We use this relationship as a conversion factor, and multiplying by the Faraday constant gives us the total charge:

$$\text{Charge (C)} = 3.0 \text{ g } Cl_2 \times \frac{1 \text{ mol } Cl_2}{70.90 \text{ g } Cl_2} \times \frac{2 \text{ mol } e^-}{1 \text{ mol } Cl_2} \times \frac{9.65\times10^4 \text{ C}}{1 \text{ mol } e^-} = 8.2\times10^3 \text{ C}$$

Now we use the relationship between charge and current to find the time needed:

$$\text{Time (s)} = \frac{\text{charge (C)}}{\text{current (A, or C/s)}} = 8.2\times10^3 \text{ C} \times \frac{1 \text{ s}}{12 \text{ C}} = 6.8\times10^2 \text{ s } (\sim 11 \text{ min})$$

Note that the entire calculation follows Figure 21.27 until the last step:

grams of Cl_2 ⇒ moles of Cl_2 ⇒ moles of e^- ⇒ coulombs ⇒ seconds

Sample Problem 21.10 demonstrates the steps as they appear in Figure 21.27.

SAMPLE PROBLEM 21.10 Applying the Relationship Among Current, Time, and Amount of Substance

Problem A technician is plating a faucet with 0.86 g of chromium from an electrolytic bath containing aqueous $Cr_2(SO_4)_3$. If 12.5 min is allowed for the plating, what current is needed?

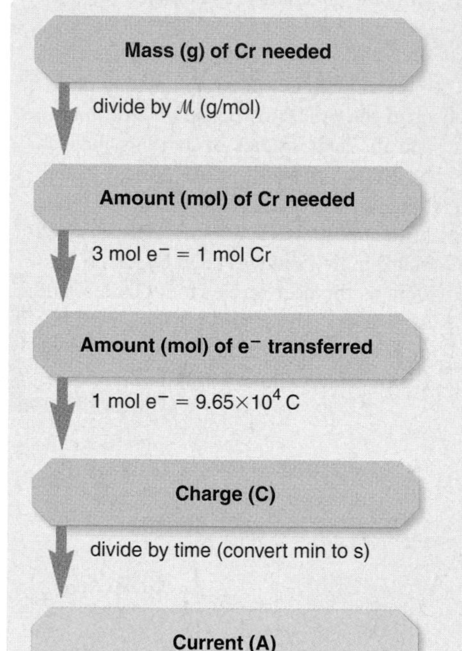

Plan To find the current, we divide the charge by the time; therefore, we need to find the charge. First we write the half-reaction for Cr^{3+} reduction. From it, we know the number of moles of e^- required per mole of Cr. As the roadmap shows, to find the charge, we convert the mass of Cr needed (0.86 g) to amount (mol) of Cr. The balanced half-reaction gives the amount (mol) of e^- transferred. Then, we use the Faraday constant (9.65×10^4 C/mol e^-) to find the charge and divide by the time (12.5 min, converted to seconds) to obtain the current.

Solution Writing the balanced half-reaction:

$$Cr^{3+}(aq) + 3e^- \longrightarrow Cr(s)$$

Combining steps to find amount (mol) of e^- transferred for mass of Cr needed:

$$\text{Moles of } e^- \text{ transferred} = 0.86 \text{ g Cr} \times \frac{1 \text{ mol Cr}}{52.00 \text{ g Cr}} \times \frac{3 \text{ mol } e^-}{1 \text{ mol Cr}} = 0.050 \text{ mol } e^-$$

Calculating the charge:

$$\text{Charge (C)} = 0.050 \text{ mol } e^- \times \frac{9.65\times10^4 \text{ C}}{1 \text{ mol } e^-} = 4.8\times10^3 \text{ C}$$

Calculating the current:

$$\text{Current (A)} = \frac{\text{charge (C)}}{\text{time (s)}} = \frac{4.8\times10^3 \text{ C}}{12.5 \text{ min}} \times \frac{1 \text{ min}}{60 \text{ s}} = 6.4 \text{ C/s} = \boxed{6.4 \text{ A}}$$

Check Rounding gives

$$(\sim 0.9 \text{ g})(1 \text{ mol Cr}/50 \text{ g})(3 \text{ mol } e^-/1 \text{ mol Cr}) = 5\times10^{-2} \text{ mol } e^-$$

then $(5\times10^{-2} \text{ mol } e^-)(\sim 1\times10^5 \text{ C/mol } e^-) = 5\times10^3 \text{ C}$

and $(5\times10^3 \text{ C}/12 \text{ min})(1 \text{ min}/60 \text{ s}) = 7 \text{ A}$

Comment For the sake of introducing Faraday's law, the details of the electroplating process have been simplified here. Actually, electroplating chromium is only 30% to 40% efficient and must be run at a particular temperature range for the plate to appear bright. Nearly 10,000 metric tons (2×10^8 mol) of chromium is used annually for electroplating.

FOLLOW-UP PROBLEM 21.10 Using a current of 4.75 A, how many minutes does it take to plate onto a sculpture 1.50 g of Cu from a $CuSO_4$ solution?

The following Chemical Connections essay links several themes of this chapter in the setting of a living cell.

Cellular Electrochemistry and the Production of ATP

Biological cells apply the principles of electrochemical cells to generate energy. The complex multistep process can be divided into two parts:

1. Bond energy in food is used to generate an electrochemical potential.
2. The potential is used to create the bond energy of the high-energy molecule adenosine triphosphate (ATP; see Chemical Connections, pp. 889–890).

The redox species that accomplish these steps are part of the *electron-transport chain* (ETC), which lies on the inner membranes of *mitochondria,* the subcellular particles that produce the cell's energy (Figure B21.1).

The ETC is a series of large molecules (mostly proteins), each of which contains a *redox couple* (the oxidized and reduced forms of a species), such as Fe^{3+}/Fe^{2+}, that passes electrons down the chain. At three points along the chain, large potential differences supply enough free energy to convert adenosine diphosphate (ADP) into ATP.

Bond Energy to Electrochemical Potential

Cells utilize the energy in food by releasing it in controlled steps rather than all at once. The reaction that ultimately powers the ETC is the oxidation of hydrogen to form water:

$$H_2 + \frac{1}{2}O_2 \longrightarrow H_2O$$

However, instead of H_2 gas, which does not occur in organisms, the hydrogen takes the form of two H^+ ions and two e^-. A biological oxidizing agent called NAD^+ (*n*icotinamide *a*denine *d*inucleotide) acquires these protons and electrons in the process of oxidizing the molecules in food. To show this process, we use the following half-reaction (without canceling an H^+ on both sides):

$$NAD^+(aq) + 2H^+(aq) + 2e^- \longrightarrow NADH(aq) + H^+(aq)$$

At the mitochondrial inner membrane, the NADH and H^+ transfer the two e^- to the first redox couple of the ETC and release the two H^+. The electrons are transported down the chain of redox couples, where they finally reduce O_2 to H_2O. The overall process, with standard electrode potentials,* is

$$NADH(aq) + H^+(aq) \longrightarrow NAD^+(aq) + 2H^+(aq) + 2e^- \qquad E^{0'} = -0.315\text{ V}$$

$$\tfrac{1}{2}O_2(aq) + 2e^- + 2H^+(aq) \longrightarrow H_2O(l) \qquad E^{0'} = 0.815\text{ V}$$

$$NADH(aq) + H^+(aq) + \tfrac{1}{2}O_2(aq) \longrightarrow NAD^+(aq) + H_2O(l)$$

$$E^{0'}_{\text{overall}} = 0.815\text{ V} - (-0.315\text{ V}) = 1.130\text{ V}$$

Thus, for each mole of NADH that enters the ETC, the free-energy equivalent of 1.14 V is available:

$$\begin{aligned}\Delta G^{0'} &= -nFE^{0'}\\ &= -(2\text{ mol e}^-/\text{mol NADH})(96.5\text{ kJ/V·mol e}^-)(1.130\text{ V})\\ &= -218\text{ kJ/mol NADH}\end{aligned}$$

(continued)

*In biological systems, standard potentials are designated $E^{0'}$ and the standard states include a pH of 7.0 ($[H^+] = 1\times10^{-7}$ *M*).

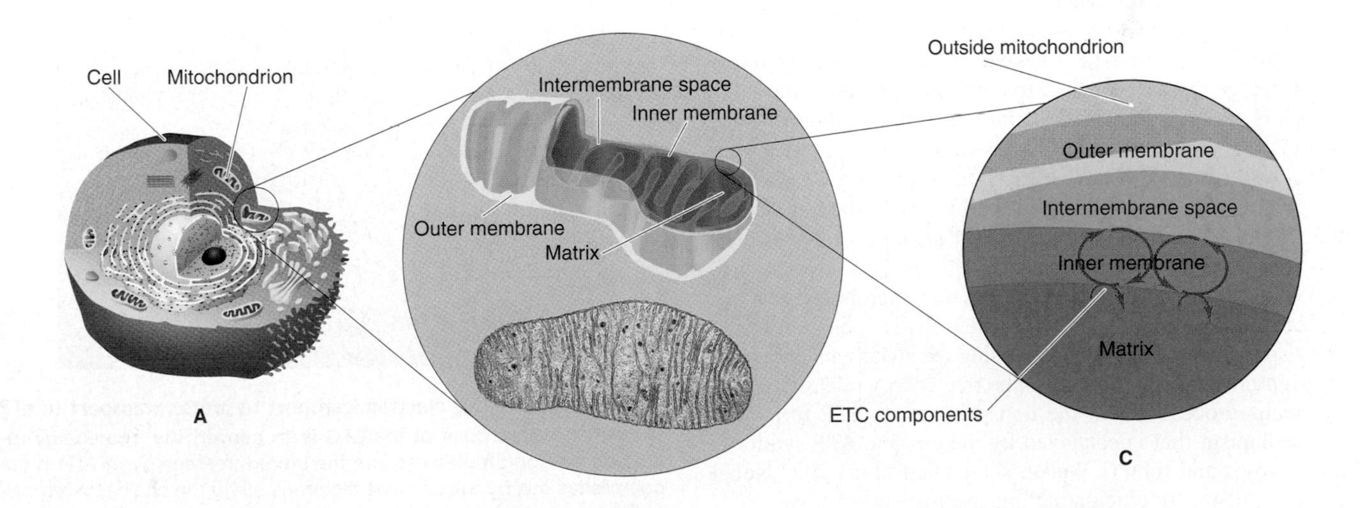

Figure B21.1 The mitochondrion. A, Mitochondria are subcellular particles outside the cell nucleus. **B,** They have a smooth outer membrane and a highly folded inner membrane, shown schematically and in an electron micrograph. **C,** The components of the electron-transport chain are attached to the inner membrane.

Note that this aspect of the process functions like a voltaic cell: a spontaneous reaction, the reduction of O_2 to H_2O, is used to generate a potential. In contrast to a laboratory voltaic cell, in which the overall process occurs in one step, this process occurs in many small steps. Figure B21.2 is a greatly simplified diagram of the three key steps in the ETC that generate the high potential to produce ATP.

In the cell, each of these steps is part of a complex consisting of several components, most of which are proteins. Electrons are passed from one redox couple to the next down the chain, such that the reduced form of the first couple reduces the oxidized form of the second, and so forth. Most of the ETC components are iron-containing proteins, and the redox change consists of the oxidation of Fe^{2+} to Fe^{3+} in one component accompanied by the reduction of Fe^{3+} to Fe^{2+} in another:

$$Fe^{2+} \text{ (in A)} + Fe^{3+} \text{ (in B)} \longrightarrow Fe^{3+} \text{ (in A)} + Fe^{2+} \text{ (in B)}$$

In other words, metal ions within the ETC proteins are the actual species undergoing the redox reactions.

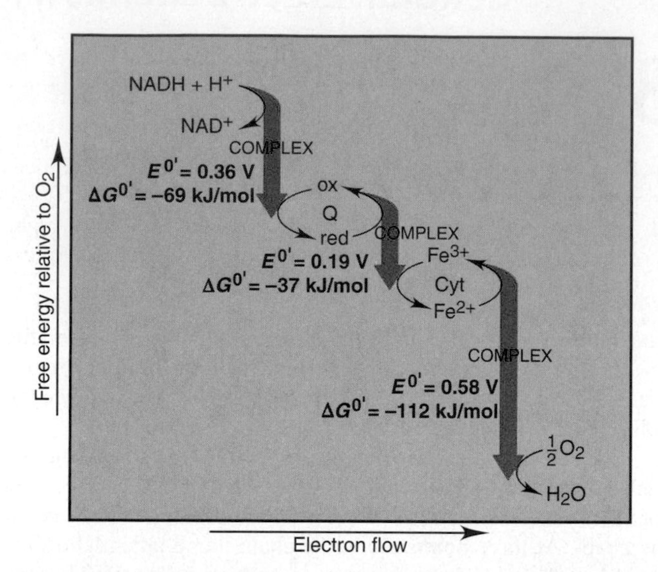

Figure B21.2 The main energy-yielding steps in the electron-transport chain (ETC). The electron carriers in the ETC undergo oxidation and reduction as they pass electrons to one another along the chain. At the three points shown, the difference in potential $E^{0'}$ (or free energy, $\Delta G^{0'}$) is large enough to be used for ATP production. (A complex consists of many components, mostly proteins; Q is a large organic molecule; and Cyt is the abbreviation for a cytochrome, a protein that contains a metal-ion redox couple, such as Fe^{3+}/Fe^{2+}, which is the actual electron carrier.)

Electrochemical Potential to Bond Energy

At the three points shown in Figure B21.2, the large potential difference is used to form ATP:

$$ADP^{3-}(aq) + HPO_4^{2-}(aq) + H^+(aq) \longrightarrow ATP^{4-}(aq) + H_2O(l) \qquad \Delta G^{0'} = 30.5 \text{ kJ/mol}$$

Note that the free energy that is *released* at each of the three ATP-producing points exceeds 30.5 kJ, the free energy that must be *absorbed* to form ATP. Thus, just as in an electrolytic cell, an electrochemical potential is supplied to drive a nonspontaneous reaction.

So far, we've followed the flow of *electrons* through the members of the ETC, but where have the released *protons* gone? The answer is the key to *how* electrochemical potential is converted to bond energy in ATP. As the electrons flow and the redox couples change oxidation states, free energy released at the three key steps is used to force H^+ ions into the intermembrane space, so that the $[H^+]$ of the intermembrane space soon becomes higher than that of the matrix (Figure B21.3). In other words, this aspect of the process acts as an electrolytic cell by using the free energy supplied by the three steps to *create what amounts to an H^+ concentration cell across the membrane.*

When the $[H^+]$ difference across the membrane reaches about 2.5-fold, it triggers the membrane to let H^+ ions flow back through spontaneously (in effect, closing the switch and allowing the concentration cell to operate). The free energy released in this spontaneous process drives the nonspontaneous ATP formation via a mechanism that is catalyzed by the enzyme ATP synthase. (Paul D. Boyer and John E. Walker shared half of the 1997 Nobel Prize in chemistry for elucidating this mechanism.)

Thus, the mitochondrion uses the "electron-motive force" of redox couples on the membrane to generate a "proton-motive force" across the membrane, which converts a potential difference to bond energy.

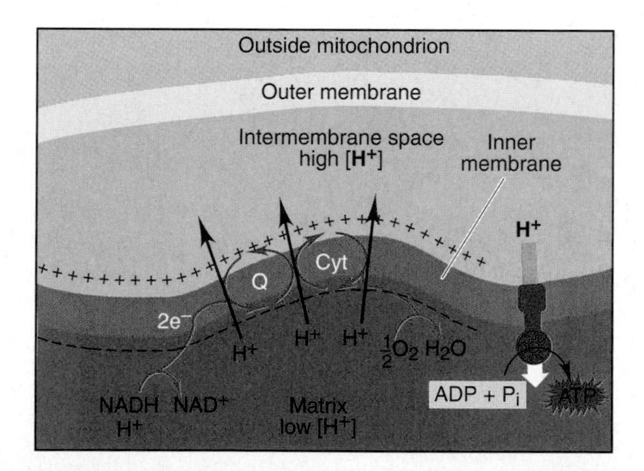

Figure B21.3 Coupling electron transport to proton transport to ATP synthesis. The purpose of the ETC is to convert the free energy released from food molecules into the stored free energy of ATP. It accomplishes this by transporting electrons along the chain (*curved yellow line*), while protons are pumped out of the mitochondrial matrix. This pumping creates an $[H^+]$ difference and generates a potential across the inner membrane (in effect, a concentration cell). When this potential reaches a "trigger" value, H^+ flows back into the inner space, and the free energy released drives the formation of ATP.

SECTION SUMMARY

An electrolytic cell uses electrical energy to drive a nonspontaneous reaction. Oxidation occurs at the anode and reduction at the cathode, but the direction of electron flow and the charges of the electrodes are opposite those in voltaic cells. When two products can form at each electrode, the more easily oxidized substance reacts at the anode and the more easily reduced at the cathode. The reduction or oxidation of water takes place at nonstandard conditions. Overvoltage causes the actual voltage to be unexpectedly high and can affect the electrode product that forms. The amount of product that forms depends on the quantity of charge flowing through the cell, which is related to the magnitude of the current and the time it flows. Cellular redox systems combine aspects of voltaic, concentration, and electrolytic cells to convert bond energy in food into electrochemical potential and then into the bond energy of ATP.

Chapter Perspective

The field of electrochemistry is one of the many areas in which the principles of thermodynamics lead to practical benefits. As you've seen, electrochemical cells can use a reaction to generate energy or use energy to drive a reaction. Such processes are central not only to our mobile way of life, but also to our biological existence. In Chapter 22, we examine the electrochemical (and other) methods used by industry to convert raw natural resources into some of the materials modern society finds indispensable.

For Review and Reference

(Numbers in parentheses refer to pages, unless noted otherwise.)

Learning Objectives

Relevant section and/or sample problem (SP) numbers appear in parentheses.

Understand These Concepts

1. The meanings of *oxidation* and *reduction;* why an oxidizing agent is reduced and a reducing agent is oxidized (Section 21.1; also Section 4.5)
2. How the half-reaction method is used to balance redox reactions in acidic or basic solution (Section 21.1)
3. The distinction between voltaic and electrolytic cells in terms of the sign of ΔG (Section 21.1)
4. How voltaic cells use a spontaneous reaction to release electrical energy (Section 21.2)
5. The physical makeup of a voltaic cell: arrangement and composition of half-cells, relative charges of electrodes, and purpose of a salt bridge (Section 21.2)
6. How the difference in reducing strength of the electrodes determines the direction of electron flow (Section 21.2)
7. The correspondence between a positive E_{cell} and a spontaneous cell reaction (Section 21.3)
8. The usefulness and significance of standard electrode potentials ($E^0_{half\text{-}cell}$) (Section 21.3)
9. How $E^0_{half\text{-}cell}$ values are combined to give E^0_{cell} (Section 21.3)
10. How the standard reference electrode is used to find an unknown $E^0_{half\text{-}cell}$ (Section 21.3)
11. How an emf series (e.g., Table 21.2 or Appendix D) is used to write spontaneous redox reactions (Section 21.3)
12. How the relative reactivity of a metal is determined by its reducing power and is related to the negative of its $E^0_{half\text{-}cell}$ (Section 21.3)
13. How E_{cell} (the nonstandard cell potential) is related to ΔG (maximum work) and the charge (moles of electrons times the Faraday constant) flowing through the cell (Section 21.4)
14. The interrelationship of ΔG^0, E^0_{cell}, and K (Section 21.4)
15. How E_{cell} changes as the cell operates (Q changes) (Section 21.4)
16. Why a voltaic cell can do work until $Q = K$ (Section 21.4)
17. How a concentration cell does work until the half-cell concentrations are equal (Section 21.4)
18. The distinction between primary (nonrechargeable) and secondary (rechargeable) batteries (Section 21.5)
19. How corrosion occurs and is prevented; the similarities between a corroding metal and a voltaic cell (Section 21.6)
20. How electrolytic cells use nonspontaneous redox reactions driven by an external source of electricity (Section 21.7)
21. How atomic properties (ionization energy and electronegativity) determine the products of the electrolysis of molten salt mixtures (Section 21.7)
22. How the electrolysis of water influences the products of aqueous electrolysis; the importance of overvoltage (Section 21.7)
23. The relationship between the quantity of charge flowing through the cell and the amount of product formed (Section 21.7)

Learning Objectives *(continued)*

Master These Skills

1. Balancing redox reactions by the half-reaction method (Section 21.1 and SP 21.1)
2. Diagramming and notating a voltaic cell (Section 21.2 and SP 21.2)
3. Combining $E^0_{\text{half-cell}}$ values to obtain E^0_{cell} (Section 21.3)
4. Using E^0_{cell} and a known $E^0_{\text{half-cell}}$ to find an unknown $E^0_{\text{half-cell}}$ (SP 21.3)
5. Manipulating half-reactions to write a spontaneous redox reaction and calculate its E^0_{cell} (SP 21.4)
6. Ranking the relative strengths of oxidizing and reducing agents in a redox reaction (SP 21.4)
7. Predicting whether a metal can displace hydrogen or another metal from solution (Section 21.3)
8. Using the interrelationship of ΔG^0, E^0_{cell}, and K to calculate one of the three given the other two (Section 21.4 and SP 21.5)
9. Using the Nernst equation to calculate the nonstandard cell potential (E_{cell}) (SP 21.6)
10. Calculating E_{cell} of a concentration cell (SP 21.7)
11. Predicting the products of the electrolysis of a mixture of molten salts (SP 21.8)
12. Predicting the products of the electrolysis of aqueous salt solutions (SP 21.9)
13. Calculating the current (or time) needed to produce a given amount of product by electrolysis (SP 21.10)

Key Terms

electrochemistry (903)
electrochemical cell (903)

Section 21.1
half-reaction method (904)
voltaic (galvanic) cell (908)
electrolytic cell (908)
electrode (908)
electrolyte (908)
anode (909)
cathode (909)

Section 21.2
half-cell (910)
salt bridge (910)

Section 21.3
cell potential (E_{cell}) (914)
voltage (914)
electromotive force (emf) (914)
volt (V) (914)
coulomb (C) (914)
standard cell potential (E^0_{cell}) (915)
standard electrode (half-cell) potential ($E^0_{\text{half-cell}}$) (915)
standard reference half-cell (standard hydrogen electrode) (916)

Section 21.4
Faraday constant (F) (923)
Nernst equation (926)
concentration cell (928)

Section 21.5
battery (932)
fuel cell (935)

Section 21.6
corrosion (936)

Section 21.7
electrolysis (941)
overvoltage (943)
ampere (A) (945)

Key Equations and Relationships

21.1 Relating a spontaneous process to the sign of the cell potential (914):

$$E_{\text{cell}} > 0 \text{ for a spontaneous process}$$

21.2 Relating electric potential to energy and charge in SI units (914):

$$\text{Potential} = \text{energy/charge} \quad \text{or} \quad 1\text{ V} = 1\text{ J/C}$$

21.3 Relating standard cell potential to standard electrode potentials in a voltaic cell (915):

$$E^0_{\text{cell}} = E^0_{\text{cathode (reduction)}} - E^0_{\text{anode (oxidation)}}$$

21.4 Defining the Faraday constant (924):

$$F = 9.65\times10^4 \frac{\text{J}}{\text{V·mol e}^-} \quad (3\text{ sf})$$

21.5 Relating the free energy change to electrical work and cell potential (924):

$$\Delta G = -w_{\text{max}} = -nFE_{\text{cell}}$$

21.6 Finding the standard free energy change from the standard cell potential (924):

$$\Delta G^0 = -nFE^0_{\text{cell}}$$

21.7 Finding the equilibrium constant from the standard cell potential (924):

$$E^0_{\text{cell}} = \frac{RT}{nF}\ln K$$

21.8 Substituting known values of R, F, and T into Equation 21.7 and converting to common logarithms (925):

$$E^0_{\text{cell}} = \frac{0.0592\text{ V}}{n}\log K \quad \text{or} \quad \log K = \frac{nE^0_{\text{cell}}}{0.0592\text{ V}} \quad (\text{at } 298.15\text{ K})$$

21.9 Calculating the nonstandard cell potential (Nernst equation) (926):

$$E_{\text{cell}} = E^0_{\text{cell}} - \frac{RT}{nF}\ln Q$$

21.10 Substituting known values of R, F, and T into the Nernst equation and converting to common logarithms (926):

$$E_{\text{cell}} = E^0_{\text{cell}} - \frac{0.0592\text{ V}}{n}\log Q \quad (\text{at } 298.15\text{ K})$$

21.11 Relating current to charge and time (945):

$$\text{Current} = \text{charge/time} \quad \text{or} \quad 1\text{ A} = 1\text{ C/s}$$

Problems in Context

21.20 In many residential water systems, the aqueous Fe^{3+} concentration is high enough to stain sinks and turn drinking water light brown. The iron content is analyzed by first reducing the Fe^{3+} to Fe^{2+} and then titrating with MnO_4^- in acidic solution. Balance the skeleton reaction of the titration step:

$$Fe^{2+}(aq) + MnO_4^-(aq) \longrightarrow Mn^{2+}(aq) + Fe^{3+}(aq)$$

21.21 *Aqua regia,* a mixture of concentrated HNO_3 and HCl, was developed by alchemists as a means to "dissolve" gold. The process is actually a redox reaction with the following simplified skeleton reaction:

$$Au(s) + NO_3^-(aq) + Cl^-(aq) \longrightarrow AuCl_4^-(aq) + NO_2(g)$$

(a) Balance the reaction by the half-reaction method.
(b) What are the oxidizing and reducing agents?
(c) What is the function of HCl in aqua regia?

Voltaic Cells: Using Spontaneous Reactions to Generate Electrical Energy

(Sample Problem 21.2)

Concept Review Questions

21.22 Consider the following general voltaic cell:

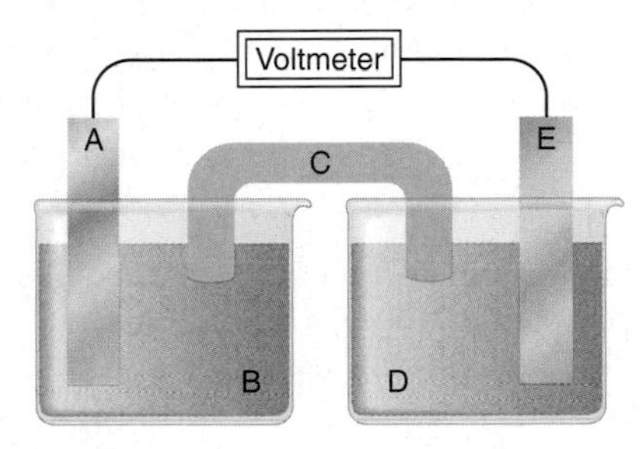

Identify the (a) anode, (b) cathode, (c) salt bridge, (d) electrode at which e^- leave the cell, (e) electrode with a positive charge, and (f) electrode that gains mass as the cell operates (assuming that a metal plates out).

21.23 Why does a voltaic cell not operate unless the two compartments are connected through an external circuit?

21.24 What purpose does the salt bridge serve in a voltaic cell, and how does it accomplish this purpose?

21.25 What is the difference between an active and an inactive electrode? Why are inactive electrodes used? Name two substances commonly used for inactive electrodes.

21.26 When a piece of metal A is placed in a solution containing ions of metal B, metal B plates out on the piece of A.
(a) Which metal is being oxidized?
(b) Which metal is being displaced?
(c) Which metal would you use as the anode in a voltaic cell incorporating these two metals?
(d) If bubbles of H_2 form when B is placed in acid, will they form if A is placed in acid? Explain.

Skill-Building Exercises *(grouped in similar pairs)*

21.27 A voltaic cell is constructed with an Sn/Sn^{2+} half-cell and a Zn/Zn^{2+} half-cell. The zinc electrode is negative.
(a) Write balanced half-reactions and the overall reaction.
(b) Diagram the cell, labeling electrodes with their charges and showing the directions of electron flow in the circuit and of cation and anion flow in the salt bridge.

21.28 A voltaic cell is constructed with an Ag/Ag^+ half-cell and a Pb/Pb^{2+} half-cell. The silver electrode is positive.
(a) Write balanced half-reactions and the overall reaction.
(b) Diagram the cell, labeling electrodes with their charges and showing the directions of electron flow in the circuit and of cation and anion flow in the salt bridge.

21.29 Consider the following voltaic cell:

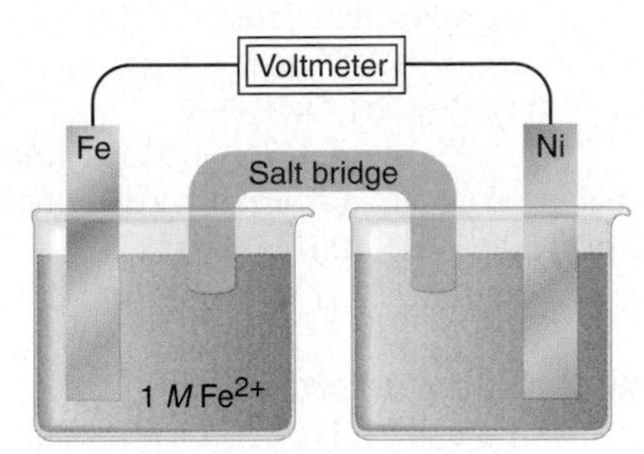

(a) In which direction do electrons flow in the external circuit?
(b) In which half-cell does oxidation occur?
(c) In which half-cell do electrons enter the cell?
(d) At which electrode are electrons consumed?
(e) Which electrode is negatively charged?
(f) Which electrode decreases in mass during cell operation?
(g) Suggest a solution for the cathode electrolyte.
(h) Suggest a pair of ions for the salt bridge.
(i) For which electrode could you use an inactive material?
(j) In which direction do anions within the salt bridge move to maintain charge neutrality?
(k) Write balanced half-reactions and an overall cell reaction.

21.30 Consider the following voltaic cell:

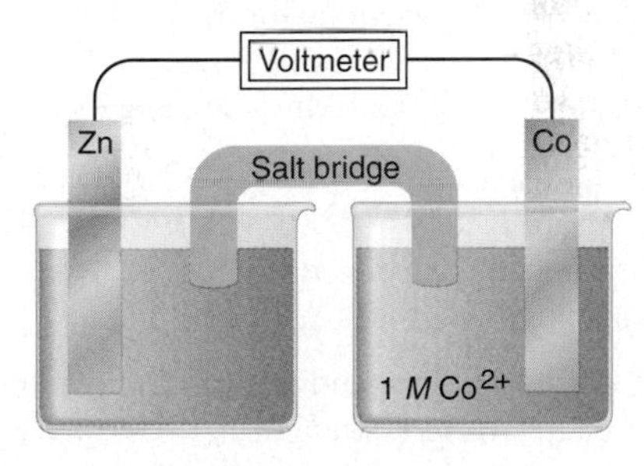

(a) In which direction do electrons flow in the external circuit?
(b) In which half-cell does reduction occur?
(c) In which half-cell do electrons leave the cell?
(d) At which electrode are electrons generated?
(e) Which electrode is positively charged?
(f) Which electrode increases in mass during cell operation?
(g) Suggest a solution for the anode electrolyte.
(h) Suggest a pair of ions for the salt bridge.
(i) For which electrode could you use an inactive material?
(j) In which direction do cations within the salt bridge move to maintain charge neutrality?
(k) Write balanced half-reactions and an overall cell reaction.

21.31 A voltaic cell is constructed with an Fe/Fe^{2+} half-cell and an Mn/Mn^{2+} half-cell. The iron electrode is positive.
(a) Write balanced half-reactions and the overall reaction.
(b) Diagram the cell, labeling electrodes with their charges and showing the directions of electron flow in the circuit and of cation and anion flow in the salt bridge.

21.32 A voltaic cell is constructed with a Cu/Cu^{2+} half-cell and an Ni/Ni^{2+} half-cell. The nickel electrode is negative.
(a) Write balanced half-reactions and the overall reaction.
(b) Diagram the cell, labeling electrodes with their charges and showing the directions of electron flow in the circuit and of cation and anion flow in the salt bridge.

21.33 Write the cell notation for the voltaic cell that incorporates each of the following redox reactions:
(a) $Al(s) + Cr^{3+}(aq) \longrightarrow Al^{3+}(aq) + Cr(s)$
(b) $Cu^{2+}(aq) + SO_2(g) + 2H_2O(l) \longrightarrow Cu(s) + SO_4^{2-}(aq) + 4H^+(aq)$

21.34 Write a balanced equation from each cell notation:
(a) $Mn(s) \mid Mn^{2+}(aq) \parallel Cd^{2+}(aq) \mid Cd(s)$
(b) $Fe(s) \mid Fe^{2+}(aq) \parallel NO_3^-(aq) \mid NO(g) \mid Pt(s)$

Cell Potential: Output of a Voltaic Cell

(Sample Problems 21.3 and 21.4)

Concept Review Questions

21.35 How is a standard reference electrode used to determine unknown $E^0_{half\text{-}cell}$ values?

21.36 What does a negative E^0_{cell} indicate about a redox reaction? What does it indicate about the reverse reaction?

21.37 The standard cell potential is a thermodynamic state function. How are E^0 values treated similarly to ΔH^0, ΔG^0, and S^0 values? How are they treated differently?

Skill-Building Exercises *(grouped in similar pairs)*

21.38 In basic solution, Se^{2-} and SO_3^{2-} ions react spontaneously:

$$2Se^{2-}(aq) + 2SO_3^{2-}(aq) + 3H_2O(l) \longrightarrow 2Se(s) + 6OH^-(aq) + S_2O_3^{2-}(aq) \qquad E^0_{cell} = 0.35\ V$$

(a) Write balanced half-reactions for the process.
(b) If $E^0_{sulfite}$ is -0.57 V, calculate $E^0_{selenium}$.

21.39 In acidic solution, O_3 and Mn^{2+} ion react spontaneously:

$$O_3(g) + Mn^{2+}(aq) + H_2O(l) \longrightarrow O_2(g) + MnO_2(s) + 2H^+(aq) \qquad E^0_{cell} = 0.84\ V$$

(a) Write the balanced half-reactions.
(b) Using Appendix D to find E^0_{ozone}, calculate $E^0_{manganese}$.

21.40 Use the emf series (Appendix D) to arrange the species.
(a) In order of *decreasing* strength as *oxidizing* agents: Fe^{3+}, Br_2, Cu^{2+}
(b) In order of *increasing* strength as *oxidizing* agents: Ca^{2+}, $Cr_2O_7^{2-}$, Ag^+

21.41 Use the emf series (Appendix D) to arrange the species.
(a) In order of *decreasing* strength as *reducing* agents: SO_2, $PbSO_4$, MnO_2
(b) In order of *increasing* strength as *reducing* agents: Hg, Fe, Sn

21.42 Balance each skeleton reaction, calculate E^0_{cell}, and state whether the reaction is spontaneous:
(a) $Co(s) + H^+(aq) \longrightarrow Co^{2+}(aq) + H_2(g)$
(b) $Mn^{2+}(aq) + Br_2(l) \longrightarrow MnO_4^-(aq) + Br^-(aq)$ [acidic]
(c) $Hg_2^{2+}(aq) \longrightarrow Hg^{2+}(aq) + Hg(l)$

21.43 Balance each skeleton reaction, calculate E^0_{cell}, and state whether the reaction is spontaneous:
(a) $Cl_2(g) + Fe^{2+}(aq) \longrightarrow Cl^-(aq) + Fe^{3+}(aq)$
(b) $Mn^{2+}(aq) + Co^{3+}(aq) \longrightarrow MnO_2(s) + Co^{2+}(aq)$ [acidic]
(c) $AgCl(s) + NO(g) \longrightarrow Ag(s) + Cl^-(aq) + NO_3^-(aq)$ [acidic]

21.44 Balance each skeleton reaction, calculate E^0_{cell}, and state whether the reaction is spontaneous:
(a) $Ag(s) + Cu^{2+}(aq) \longrightarrow Ag^+(aq) + Cu(s)$
(b) $Cd(s) + Cr_2O_7^{2-}(aq) \longrightarrow Cd^{2+}(aq) + Cr^{3+}(aq)$
(c) $Ni^{2+}(aq) + Pb(s) \longrightarrow Ni(s) + Pb^{2+}(aq)$

21.45 Balance each skeleton reaction, calculate E^0_{cell}, and state whether the reaction is spontaneous:
(a) $Cu^+(aq) + PbO_2(s) + SO_4^{2-}(aq) \longrightarrow PbSO_4(s) + Cu^{2+}(aq)$ [acidic]
(b) $H_2O_2(aq) + Ni^{2+}(aq) \longrightarrow O_2(g) + Ni(s)$ [acidic]
(c) $MnO_2(s) + Ag^+(aq) \longrightarrow MnO_4^-(aq) + Ag(s)$ [basic]

21.46 Use the following half-reactions to write three spontaneous reactions, calculate E^0_{cell} for each reaction, and rank the strengths of the oxidizing and reducing agents:
(1) $Al^{3+}(aq) + 3e^- \longrightarrow Al(s) \qquad E^0 = -1.66\ V$
(2) $N_2O_4(g) + 2e^- \longrightarrow 2NO_2^-(aq) \qquad E^0 = 0.867\ V$
(3) $SO_4^{2-}(aq) + H_2O(l) + 2e^- \longrightarrow SO_3^{2-}(aq) + 2OH^-(aq) \qquad E^0 = 0.93\ V$

21.47 Use the following half-reactions to write three spontaneous reactions, calculate E^0_{cell} for each reaction, and rank the strengths of the oxidizing and reducing agents:
(1) $Au^+(aq) + e^- \longrightarrow Au(s) \qquad E^0 = 1.69\ V$
(2) $N_2O(g) + 2H^+(aq) + 2e^- \longrightarrow N_2(g) + H_2O(l) \qquad E^0 = 1.77\ V$
(3) $Cr^{3+}(aq) + 3e^- \longrightarrow Cr(s) \qquad E^0 = -0.74\ V$

21.48 Use the following half-reactions to write three spontaneous reactions, calculate E^0_{cell} for each reaction, and rank the strengths of the oxidizing and reducing agents:
(1) $2HClO(aq) + 2H^+(aq) + 2e^- \longrightarrow Cl_2(g) + 2H_2O(l) \qquad E^0 = 1.63\ V$
(2) $Pt^{2+}(aq) + 2e^- \longrightarrow Pt(s) \qquad E^0 = 1.20\ V$
(3) $PbSO_4(s) + 2e^- \longrightarrow Pb(s) + SO_4^{2-}(aq) \qquad E^0 = -0.31\ V$

21.49 Use the following half-reactions to write three spontaneous reactions, calculate E^0_{cell} for each reaction, and rank the strengths of the oxidizing and reducing agents:
(1) $I_2(s) + 2e^- \longrightarrow 2I^-(aq) \qquad E^0 = 0.53\ V$
(2) $S_2O_8^{2-}(aq) + 2e^- \longrightarrow 2SO_4^{2-}(aq) \qquad E^0 = 2.01\ V$
(3) $Cr_2O_7^{2-}(aq) + 14H^+(aq) + 6e^- \longrightarrow 2Cr^{3+}(aq) + 7H_2O(l) \qquad E^0 = 1.33\ V$

Problems in Context

21.50 When metal A is placed in a solution of a salt of metal B, the surface of metal A changes color. When metal B is placed in acid solution, gas bubbles form on the surface of the metal. When metal A is placed in a solution of a salt of metal C, no change is observed in the solution or on the metal A surface. Will metal C cause formation of H_2 when placed in acid solution? Rank metals A, B, and C in order of *decreasing* reducing strength.

21.51 When a clean iron nail is placed in an aqueous solution of copper(II) sulfate, the nail becomes coated with a brownish black material.
(a) What is the material coating the iron?
(b) What are the oxidizing and reducing agents?
(c) Can this reaction be made into a voltaic cell?
(d) Write the balanced equation for the reaction.
(e) Calculate E^0_{cell} for the process.

In this chapter we return to the periodic table, but from a new perspective. In Chapter 14, we considered the atomic, physical, and chemical behavior of the main-group elements. Here, we examine various elements from a practical viewpoint to learn how much of an element is present in nature, in what form it occurs, how organisms affect its distribution, and how we isolate it for our own use.

Remember that chemistry is, above all, a practical science, and its concepts were developed to address real-life problems: How do we isolate and recycle aluminum? How do we make iron stronger? How do we lower energy costs and increase yield in the production of sulfuric acid? How does fossil-fuel combustion affect the environmental cycling of carbon? Here, we apply concepts you learned from the chapters on kinetics, equilibrium, thermodynamics, and electrochemistry to explain chemical processes in nature and industry that influence our lives.

IN THIS CHAPTER . . . We first discuss the abundances and sources of elements on Earth, then consider how three essential elements—carbon, nitrogen, and phosphorus—cycle through the environment. After seeing where and in what forms elements are found, we focus on their isolation and utilization. We discuss general redox and metallurgical procedures for extracting an element from its ore and examine in detail the isolation and uses of certain key elements. The chapter ends with a close look at two of the most important processes in chemical manufacturing: the production of sulfuric acid and the isolation of chlorine.

Concepts & Skills to Review

before you study this chapter

- catalysts and reaction rate (Section 16.8)
- Le Châtelier's principle (Section 17.6)
- acid-base equilibria (Sections 18.3 and 18.9)
- solubility and complex-ion equilibria (Sections 19.3 and 19.4)
- temperature and reaction spontaneity (Section 20.3)
- free energy and equilibrium (Section 20.4)
- standard electrode potentials (Section 21.3)
- electrolysis of molten salts and aqueous solutions (Section 21.7)

22.1 HOW THE ELEMENTS OCCUR IN NATURE

To begin our examination of how we use the elements, let's take inventory of our elemental stock—the distribution and relative amounts of the elements on Earth, especially that thin outer portion of the planet that we can reach.

Earth's Structure and the Abundance of the Elements

Any attempt to isolate an element must begin with a knowledge of its **abundance,** the amount of the element in a particular region of the natural world. The abundances of the elements on Earth and in its various regions are the result of the specific details of our planet's evolution.

Formation and Layering of Planet Earth About 4.5 billion years ago, vast clouds of cold gases and interstellar debris from exploded older stars gradually coalesced into the Sun and planets. At first, Earth was a cold, solid sphere of uniformly distributed elements and simple compounds. In the next billion years or so, heat from radioactive decay and virtually continuous meteor impacts raised the planet's temperature to around 10^4 K, sufficient to form an enormous molten mass. Any remaining gaseous elements, such as the cosmically abundant hydrogen and helium, were ejected into space.

As Earth cooled, chemical and physical processes resulted in its **differentiation,** the formation of regions of different composition and density. Differentiation gave Earth its layered internal structure, which consists of a dense (10–15 g/cm^3) **core,** composed of a molten outer core and a solid Moon-sized inner core. (Recent evidence has revealed remarkable properties of the inner core: it is nearly as hot as the surface of the Sun and spins within the molten outer core slightly faster than does Earth itself!) Around the core lies a thick homogeneous **mantle,** consisting of lower and upper portions, with an overall density of 4–6 g/cm^3. All the comings and goings of life take place on the thin heterogeneous **crust** (average density = 2.8 g/cm^3). Figure 22.1 on the next page shows these layers and compares the abundances of some key elements in the universe, the whole Earth (actually core plus mantle only, which account for more than 99%

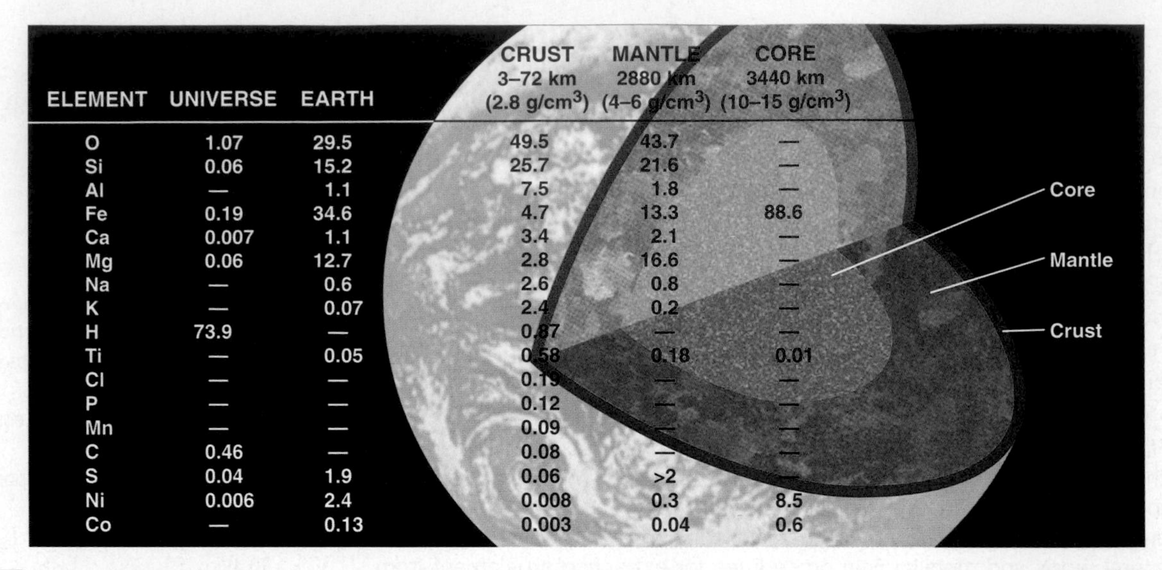

ELEMENT	UNIVERSE	EARTH	CRUST 3–72 km (2.8 g/cm^3)	MANTLE 2880 km (4–6 g/cm^3)	CORE 3440 km (10–15 g/cm^3)
O	1.07	29.5	49.5	43.7	—
Si	0.06	15.2	25.7	21.6	—
Al	—	1.1	7.5	1.8	—
Fe	0.19	34.6	4.7	13.3	88.6
Ca	0.007	1.1	3.4	2.1	—
Mg	0.06	12.7	2.8	16.6	—
Na	—	0.6	2.6	0.8	—
K	—	0.07	2.4	0.2	—
H	73.9	—	0.87	—	—
Ti	—	0.05	0.58	0.18	0.01
Cl	—	—	0.19	—	—
P	—	—	0.12	—	—
Mn	—	—	0.09	—	—
C	0.46	—	0.08	—	—
S	0.04	1.9	0.06	>2	—
Ni	0.006	2.4	0.008	0.3	8.5
Co	—	0.13	0.003	0.04	0.6

Figure 22.1 Cosmic and terrestrial abundances of selected elements (mass percent). The mass percents of the elements are superimposed on a cutaway of Earth. The internal structure consists of a large core, thick mantle, and thin crust. The densities of these regions are indicated in the column heads. Blank abundance values mean either that reliable data are not available or that the value is less than 0.001% by mass. (Helium is not listed because it is not abundant on Earth, but it accounts for 24.0% by mass of the universe.)

of its mass), and Earth's three regions. Because the deepest terrestrial sampling can penetrate only a few kilometers into the crust, some of these data represent extrapolations from meteor samples and from seismic studies of earthquakes. Several points stand out:

1. Cosmic and whole-Earth abundances are very different, particularly for H.
2. The elements O, Si, Fe, and Mg are abundant both cosmically and on Earth. Together, they account for more than 90% of Earth's mass.
3. The core is particularly rich in the dense Group 8B metals: Co, Ni, and especially Fe, the most abundant element in the whole Earth.
4. Crustal abundances are very different from whole-Earth abundances. The crust makes up only 0.4% of Earth's mass, but has the largest share of nonmetals, metalloids, and light, active metals: Al, Ca, Na, and K. The mantle contains smaller proportions of these, and the core has none. Oxygen is the most abundant element in the crust and mantle but is absent from the core.

These compositional differences in Earth's major layers, or *phases,* arose from the effects of thermal energy. When Earth was molten, gravity and convection caused more dense materials to sink and less dense materials to rise. Most of the Fe sank to form the core, or *iron phase.* In the light outer phase, oxygen combined with Si, Al, Mg, and some Fe to form silicates, the material of rocks. This *silicate phase* later separated into the mantle and crust. The *sulfide phase,* intermediate in density and insoluble in the other two, consisted mostly of iron sulfide and mixed with parts of the silicate phase above and the iron phase below. *Outgassing,* the expulsion of trapped gases, produced a thin, primitive atmosphere, probably a mixture of water vapor (which gave rise to the oceans), carbon monoxide, and nitrogen (or ammonia).

The distribution of the remaining elements (discussed here using the new group numbers) was controlled by their chemical affinity for one of the three phases. In general terms, as Figure 22.2 shows, elements with low or high electronegativity—active metals (Groups 1 through 5, Cr, and Mn) and nonmetals (O, lighter members of Groups 13 to 15, and all of Group 17)—tended to congregate in the silicate phase as ionic compounds. Metals with intermediate electronegativities (many from Groups 6 to 10) dissolved in the iron core. Lower-melting transition metals and many metals and metalloids in Groups 11 to 16 became concentrated in the sulfide phase.

Atmosphere (crust)
Silicate phase (crust and mantle)
Sulfide phase (mantle)
Iron phase (core)

1A (1)	2A (2)	3B (3)	4B (4)	5B (5)	6B (6)	7B (7)	8B (8)	8B (9)	8B (10)	1B (11)	2B (12)	3A (13)	4A (14)	5A (15)	6A (16)	7A (17)	8A (18)
H																	He
Li	Be											B	C	N	O	F	Ne
Na	Mg											Al	Si	P	S	Cl	Ar
K	Ca	Sc	Ti	V	Cr	Mn	Fe	Co	Ni	Cu	Zn	Ga	Ge	As	Se	Br	Kr
Rb	Sr	Y	Zr	Nb	Mo		Ru	Rh	Pd	Ag	Cd	In	Sn	Sb	Te	I	Xe
Cs	Ba	La	Hf	Ta	W	Re	Os	Ir	Pt	Au	Hg	Tl	Pb	Bi			

Ce	Pr	Nd	Pm	Sm	Eu	Gd	Tb	Dy	Ho	Er	Tm	Yb	Lu
Th		U											

Figure 22.2 Geochemical differentiation of the elements. During Earth's formation, elements became concentrated primarily in the atmosphere or one of three phases: silicate, sulfide, or iron.

The Impact of Life on Crustal Abundances At present, the crust is the only accessible portion of our planet, so only crustal abundances have practical significance. The crust is divided into solid, liquid, and gaseous portions called the **lithosphere, hydrosphere,** and **atmosphere.** Over billions of years, weathering and volcanic upsurges have dramatically altered the composition of the crust. The **biosphere,** which consists of the living systems that have inhabited the planet, has been another *major influence on crustal element composition and distribution.*

When the earliest rocks were forming and the ocean basins filling with water, binary inorganic molecules in the atmosphere were reacting to form first simple, and then more complex, organic molecules. The energy for these (mostly) endothermic changes was supplied by lightning, solar radiation, geologic heating, and meteoric impact. In an amazingly short period of time, probably no more than 500 million years, the first organisms appeared. It took perhaps another billion years for these to evolve into simple algae that could derive metabolic energy from photosynthesis, converting CO_2 and H_2O into organic molecules and releasing O_2 as a by-product. The importance of this process to crustal chemistry cannot be overstated. Over the next 300 million years, the atmosphere gradually became richer in O_2, and *oxidation became the major source of free energy in the crust and biosphere.* Geologists mark this period by the appearance of Fe(III)-containing minerals in place of the Fe(II)-containing minerals that had predominated (Figure 22.3A). Paleontologists see in it an explosion of O_2-utilizing life-forms, a few of which evolved into the organisms of today (Figure 22.3B).

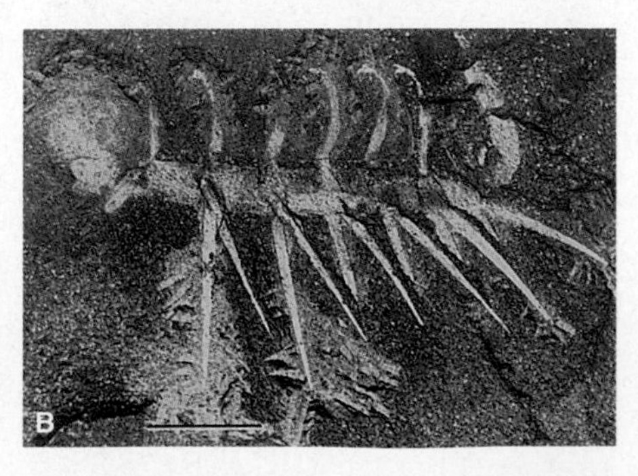

Figure 22.3 Ancient effects of an O_2-rich atmosphere. A, With the appearance of photosynthetic organisms and the resulting oxidizing environment, much of the iron(II) in minerals was oxidized to iron(III). These ancient banded-iron formations (red beds) from Michigan contain hematite, Fe_2O_3. **B,** A fossil of *Hallucigenia sparsa,* one of the multicellular organisms found in the Burgess shale, British Columbia, Canada, whose appearance coincided with the increase in O_2. Many of these organisms have no known modern descendants, having disappeared in a mass extinction at the end of the Cambrian Period about 505 million years ago.

In addition to creating an oxidizing environment, organisms have had profound effects on the *distribution* of specific elements. Why, for instance, is the K^+ concentration of the oceans so much lower than the Na^+ concentration? Forming over eons from outgassed water vapor condensing into rain and streaming over the land, the oceans became complex ionic solutions of dissolved minerals with 30 times as much Na^+ as K^+. There are two principal reasons for this difference. Clays, which make up a large portion of many soils, bind K^+ preferentially through an ion-exchange process. Even more importantly, plants require K^+ for growth, absorbing dissolved K^+ that would otherwise wash down to the sea.

Dolomite Mountains, Italy

As another example, consider the enormous subterranean deposits of organic carbon. Buried deeply and decomposing under high pressure and temperature in the absence of free O_2, ancient plants gradually turned into coal, and animals buried under similar conditions turned into petroleum. Crustal deposits of this organic carbon provide the fuels that move our cars, heat our homes, and electrify our cities. Moreover, carbon *and* calcium, the fifth most abundant element in the crust, occur together in vast sedimentary deposits all over the world as limestone, dolomite, marble, and chalk, the fossilized skeletal remains of early marine organisms (see photo).

Table 22.1 compares selected elemental abundances in the whole crust, the three crustal regions, and the human body, a representative portion of the biosphere. Note the quantities of the four major elements of life—O, C, H, and N. Oxygen is either the first or second most abundant element in all cases. The biosphere contains large amounts of carbon in its biomolecules and large amounts of hydrogen as water. Nitrogen is abundant in the atmosphere as free N_2 and in organisms as part of their proteins. Phosphorus and sulfur occur in exceptionally high amounts in organisms, too.

One striking difference among the lithosphere, hydrosphere, and biosphere (human) is in the abundances of the transition metals vanadium through zinc.

Table 22.1 Abundance of Selected Elements in the Crust, Its Regions, and the Human Body as Representative of the Biosphere (Mass %)

		Crustal Regions			
Element	**Crust**	**Lithosphere**	**Hydrosphere**	**Atmosphere**	**Human**
O	49.5	45.5	85.8	23.0	65.0
C	0.08	0.018	—	0.01	18.0
H	0.87	0.15	10.7	0.02	10.0
N	0.03	0.002	—	75.5	3.0
P	0.12	0.11	—	—	1.0
Mg	1.9	2.76	0.13	—	0.50
K	2.4	1.84	0.04	—	0.34
Ca	3.4	4.66	0.05	—	2.4
S	0.06	0.034	—	—	0.26
Na	2.6	2.27	1.1	—	0.14
Cl	0.19	0.013	2.1	—	0.15
Fe	4.7	6.2	—	—	0.005
Zn	0.013	0.008	—	—	0.003
Cr	0.02	0.012	—	—	3×10^{-6}
Co	0.003	0.003	—	—	3×10^{-6}
Cu	0.007	0.007	—	—	4×10^{-4}
Mn	0.09	0.11	—	—	1×10^{-4}
Ni	0.008	0.010	—	—	3×10^{-6}
V	0.015	0.014	—	—	3×10^{-6}

Their relatively insoluble oxides and sulfides make them much scarcer in water than on land. Yet organisms, which evolved in the seas, developed the ability to concentrate the trace amounts of these elements present in their aqueous environment. In every case, the biological concentration shows an increase of at least 100-fold, with Mn increasing about 1000-fold, and Cu, Zn, and Fe even more. Each of these elements performs an essential role in living systems (see Chemical Connections, pp. 1035–1036).

Sources of the Elements

Given an element's abundance in a region of the crust, we must next determine its **occurrence,** or **source,** the form(s) in which the element exists. Practical considerations often determine the commercial source. Oxygen, for example, is abundant in all three crustal regions, but the atmosphere is its primary industrial source because it occurs there as the free element. Nitrogen and the noble gases (except helium) are also obtained from the atmosphere. Several other elements occur uncombined, formed in large deposits by prehistoric biological action; examples are sulfur in caprock salt domes and nearly pure carbon in coal. The relatively unreactive elements gold and platinum also occur in an uncombined (*native*) state.

The overwhelming majority of elements, however, occur in **ores,** natural compounds or mixtures of compounds from which an element can be extracted by economically feasible means. The phrase "economically feasible" refers to an important point: *the financial costs of mining, isolating, and purifying an element must be considered when choosing a process to obtain it.*

Figure 22.4 shows the most useful sources of the elements. Alkali metal halides are ores for both of their component groups of elements. One example of how cost influences recovery concerns the feasibility of using silicates as ores. Even though many elements occur as silicates, most of these sources are very stable thermodynamically. Thus, the cost of the energy required to process them prohibits their use—aside from silicon, only lithium and beryllium are obtained from their silicates. The Group 2A(2) metals occur as carbonates in the marble and limestone of mountain ranges, although magnesium's great abundance in seawater makes that the preferred source, as we discuss later.

Deposit of borate (tufa), the ore of boron.

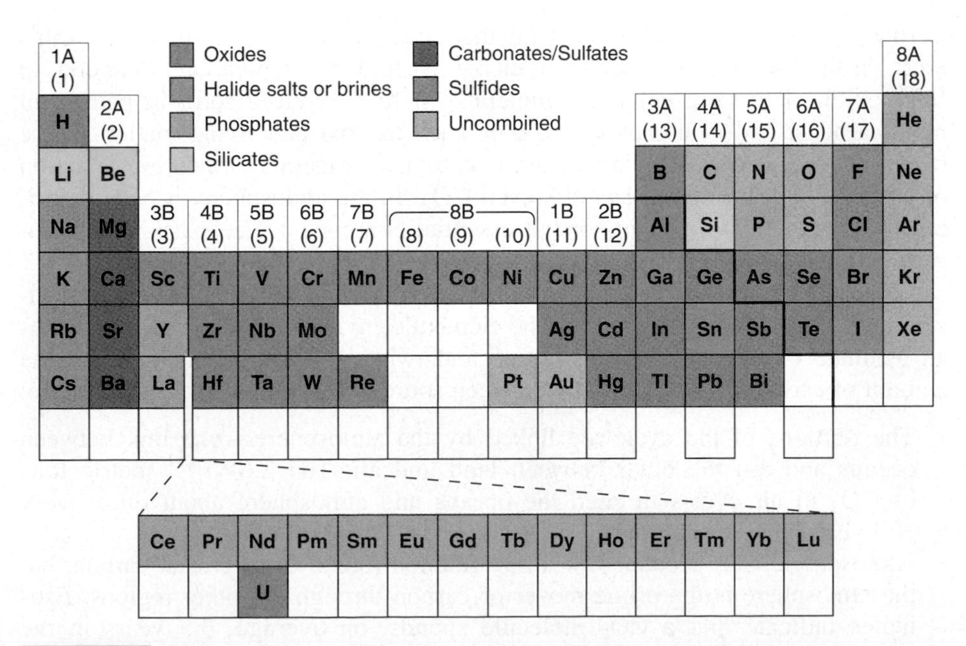

Figure 22.4 Sources of the elements. The major sources of most of the elements are shown. Although a few elements occur uncombined, most are obtained from their oxide or sulfide ores.

The ores of most industrially important metals are either *oxides,* which dominate the left half of the transition series, or *sulfides,* which dominate the right half of the transition series and a few of the main groups beyond. The reasons for the prominence of oxides and sulfides are complex and include processes of weathering, selective precipitation, and relative solubilities. Certain atomic properties are relevant as well. Elements toward the left side of the periodic table have lower ionization energies and electronegativities, so they tend to give up electrons or hold them loosely in bonds. The O^{2-} ion is small enough to approach a metal cation closely, which results in a high lattice energy for the oxide. In contrast, elements toward the right side have higher ionization energies and electronegativities. Thus, they tend to form bonds that are more covalent, which suits the larger, more polarizable S^{2-} ion.

SECTION SUMMARY

As the young Earth cooled, the elements became differentiated into a dense, metallic core, a silicate-rich mantle, and a low-density crust. High abundances of light metals, metalloids, and nonmetals are concentrated in the crust, which consists of the lithosphere (solid), hydrosphere (liquid), and atmosphere (gaseous). The biosphere (living systems) has affected crustal chemistry primarily by producing free O_2, and thus an oxidizing environment. Some elements occur in their native state, but most are combined in ores. The ores of most important metallic elements are oxides or sulfides.

22.2 THE CYCLING OF ELEMENTS THROUGH THE ENVIRONMENT

The distributions of many elements are in flux, changing at widely differing rates. The physical, chemical, and biological paths that the atoms of an element take on their journey through regions of the crust constitute the element's **environmental cycle.** In this section, we consider the cycles for three important elements—carbon, nitrogen, and phosphorus—and highlight the effects of humans on them.

The Carbon Cycle

Carbon is one of a handful of elements that appear in all three portions of Earth's crust. In the lithosphere, it occurs in elemental form as graphite and diamond, in fully oxidized form in carbonate minerals, in fully reduced form in petroleum hydrocarbons, and in complex mixtures, such as coal and living matter. In the hydrosphere, it occurs in living matter, in carbonate minerals formed by the action of coral-reef organisms, and in dissolved CO_2. In the atmosphere, it occurs principally in gaseous CO_2, a minor but essential component that exists in equilibrium with the aqueous fraction.

Figure 22.5 depicts the complex interplay of the carbon sources and the considerable effect of the biosphere on the element's environmental cycle. It provides an estimate of the size of each source and, where reliable numbers exist, the amount of carbon moving annually between sources. Some key points to note are:

- The portions of the cycle are linked by the atmosphere—one link between oceans and air, the other between land and air. The 2.6×10^{12} metric tons of CO_2 in air cycles through the oceans and atmosphere about once every 300 years but spends a much longer time in carbonate minerals.
- Atmospheric CO_2 accounts for a tiny fraction (0.003%) of crustal carbon, but the atmosphere is the prime mover of carbon through the other regions. Estimates indicate that a CO_2 molecule spends, on average, 3.5 years in the atmosphere.

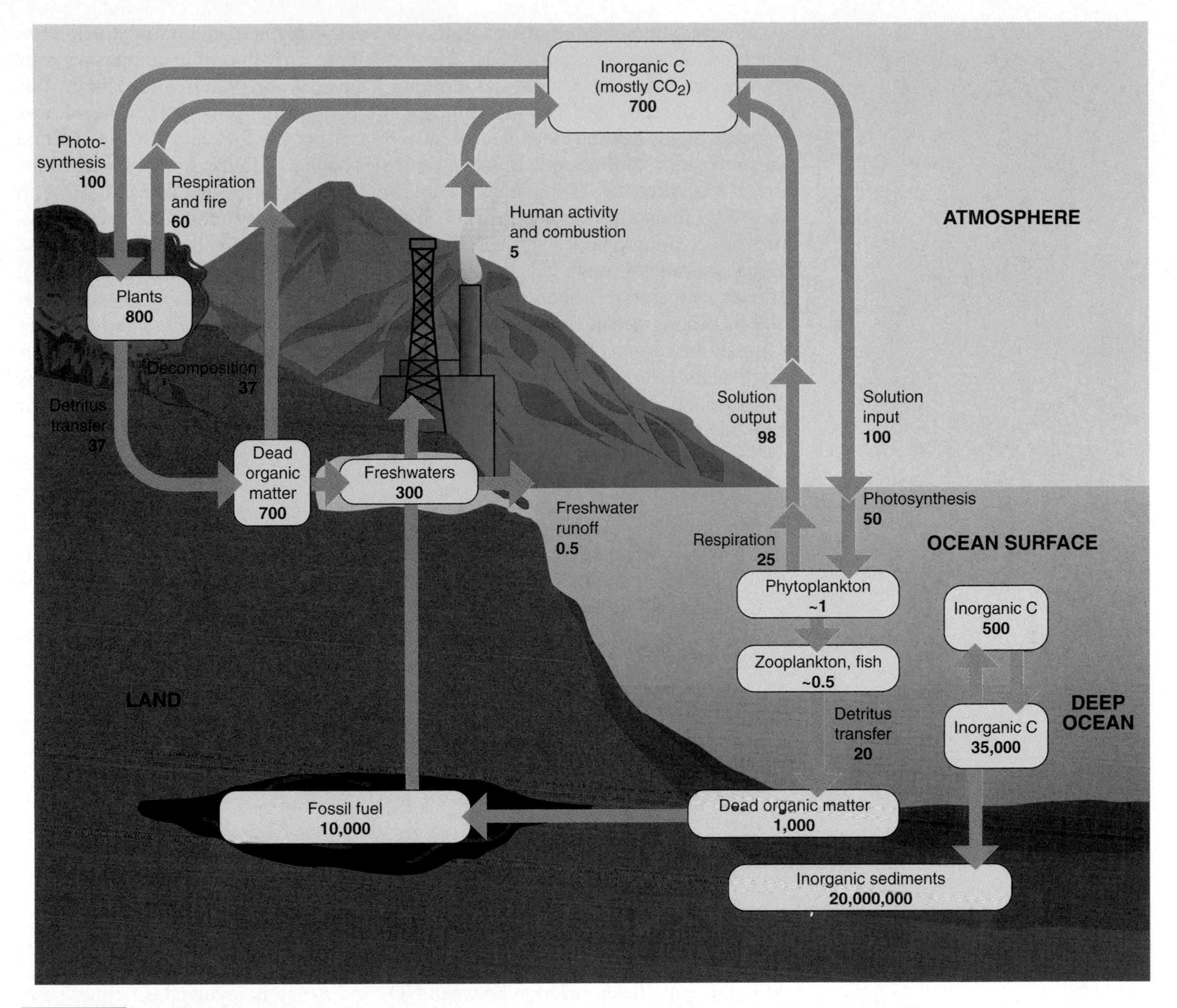

Figure 22.5 The carbon cycle. The major sources and changes in carbon distribution are shown. Note that the atmosphere is the major conduit of carbon between land and sea and that CO_2 becomes fixed through the action of organisms. Numbers in boxes refer to the size of the source; numbers along arrows refer to the annual movement of the element from source to source. Values are in 10^9 metric tons of C (1 metric ton = 1000 kg = 1.1 tons).

- The land and oceans are also in contact with the largest sources of carbon, which are effectively immobilized as carbonates, coal, and oil in the rocky sediment beneath the soil.
- The biological processes of *photosynthesis, respiration,* and *decay* are major factors in the carbon cycle. **Fixation** is the process of converting a gaseous substance into a condensed, more usable form. Via photosynthesis, marine plankton and terrestrial plants use sunlight to fix atmospheric CO_2 into carbohydrates. Plants release CO_2 by respiration at night and, much more slowly, when they decay. Animals eat the plants and release CO_2 by respiration and in their decay.
- Natural fires and volcanoes also release CO_2 into the air.

For hundreds of millions of years, the cycle has maintained a relatively constant amount of atmospheric CO_2. The tremendous growth of human industry over the past century and a half, and especially since World War II, has shifted this natural balance by increasing atmospheric CO_2. The principal cause of this increase is the combustion of coal, wood, and oil for fuel and for the thermal decomposition of limestone to make cement, coupled with the simultaneous clearing of vast stretches of forest and jungle for lumber, paper, and agriculture. With the 1990s the hottest decade ever recorded, evidence is mounting that these activities are creating one of the most dire environmental changes in human history—global warming through the greenhouse effect (see Chemical Connections, p. 246). Higher temperatures, alterations in dry and rainy seasons, melting of polar ice, and increasing ocean acidity, with associated changes in carbonate equilibria, are some of the results currently being assessed. Most environmental and science policy experts are pressing for the immediate adoption of a program that combines conservation of carbon-based fuels, an end to deforestation, extensive planting of trees, and development of alternative energy sources.

The Nitrogen Cycle

In contrast to the carbon cycle, the nitrogen cycle includes a direct interaction of land and sea (Figure 22.6). All nitrites and nitrates are soluble, so rain and runoff contribute huge amounts of nitrogen to lakes, rivers, and oceans. Human activity, in the form of fertilizer production and use, plays a major role in this cycle.

Like carbon dioxide, atmospheric nitrogen must be fixed to be used by organisms. However, whereas CO_2 can be incorporated by plants in either its gaseous or aqueous form, the great stability of N_2 prevents plants from using it directly. Fixation of N_2 requires a great deal of energy and occurs through atmospheric, industrial, and biological processes:

1. *Atmospheric fixation.* Lightning brings about the high-temperature endothermic reaction of N_2 and O_2 to form NO, which is then oxidized exothermically to NO_2:

$$N_2(g) + O_2(g) \longrightarrow 2NO(g) \qquad \Delta H^0 = 180.6 \text{ kJ}$$
$$2NO(g) + O_2(g) \longrightarrow 2NO_2(g) \qquad \Delta H^0 = -114.2 \text{ kJ}$$

During the day, NO_2 reacts with hydroxyl radical to form HNO_3:

$$NO_2(g) + HO\cdot(g) \longrightarrow HNO_3(g)$$

At night, a multistep reaction between NO_2 and ozone is involved. In either case, rain dissociates the nitric acid, and $NO_3^-(aq)$ enters both sea and land to be utilized by plants.

2. *Industrial fixation.* Most human-caused fixation occurs industrially during ammonia synthesis via the Haber process (see Chemical Connections, pp. 755–756). The process takes place on an enormous scale: on a mole basis, NH_3 ranks first among compounds in amount produced. Some of this NH_3 is converted to HNO_3 in the Ostwald process (Chapter 14, p. 588), but most is used as fertilizer, either directly or in the form of urea and ammonium salts (sulfate, phosphate, and nitrate), which enter the biosphere as they are taken up by plants (item 3 below). In recent decades, high-temperature combustion in electric power plants and in car, truck, and plane engines has become an important contributor to total fixed nitrogen, mainly because of the enormous rise in vehicle traffic. High engine operating temperatures mimic lightning to form NO from the air taken in to burn the hydrocarbon fuel. The NO in exhaust gases reacts to form nitric acid in the atmosphere, which adds to the nitrate load that reaches the ground.

The overuse of fertilizers and automobiles presents an increasingly serious water pollution problem in many areas. Leaching of the land by rain causes nitrate from fertilizer use and vehicle operation to enter natural waters (lakes, rivers, and

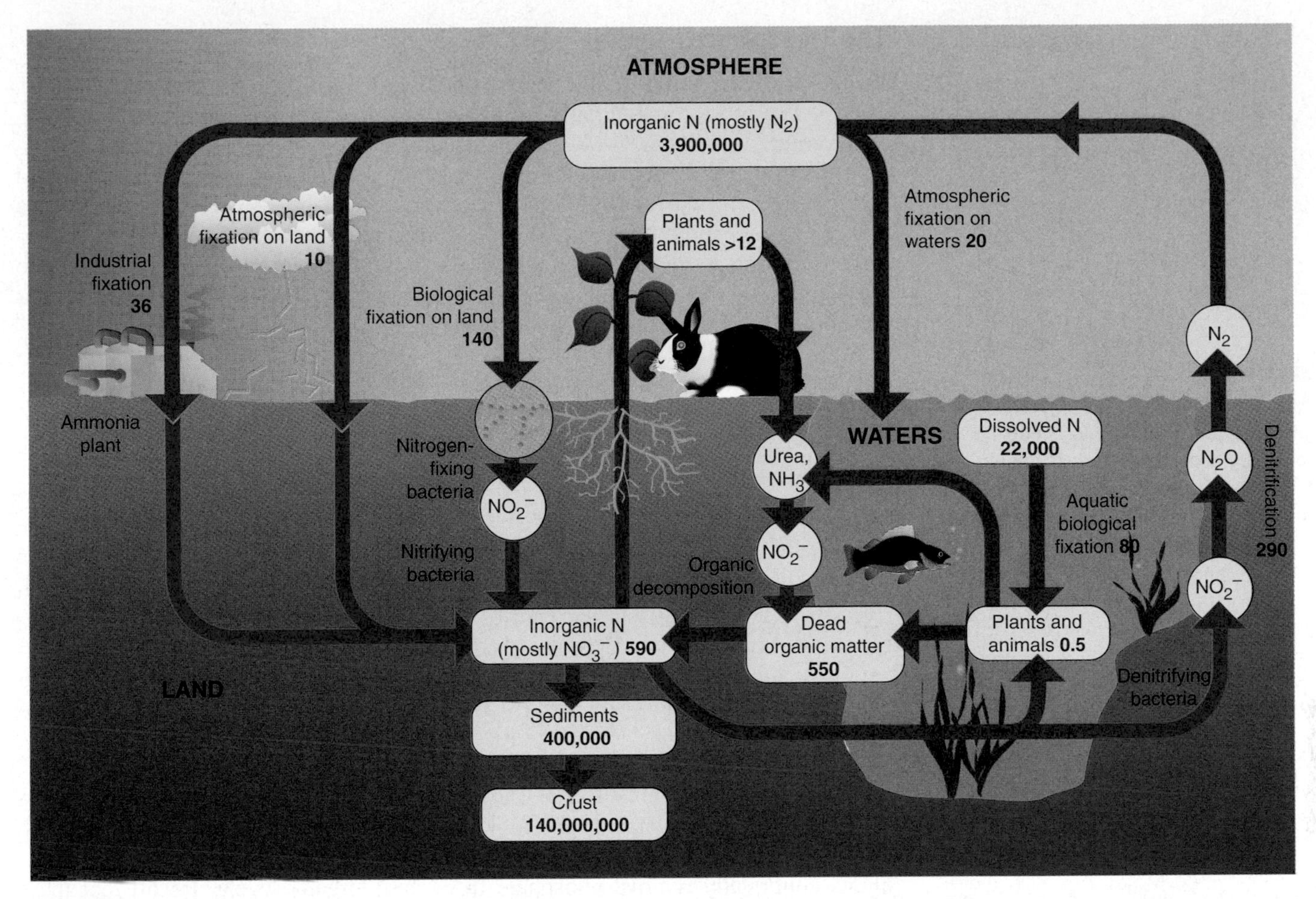

Figure 22.6 The nitrogen cycle. The major source of nitrogen, N_2, is fixed atmospherically, industrially, and biologically. Various types of bacteria are indispensable throughout the cycle. Numbers in boxes are in 10^9 metric tons of N and refer to the size of the source; numbers along arrows are in 10^6 metric tons of N and refer to the annual movement of the element between sources.

coastal estuaries) and cause *eutrophication,* the depletion of O_2 and the death of aquatic animal life from excessive algal and plant growth and decay. Excess nitrate also spoils nearby drinkable water.

3. *Biological fixation.* The biological fixation of atmospheric N_2 occurs in blue-green algae and in nitrogen-fixing bacteria that live on the roots of leguminous plants (such as peas, alfalfa, and clover). On a planetary scale, these microbial processes dwarf the previous two, fixing more than seven times as much nitrogen as the atmosphere and six times as much as industry. Root bacteria fix N_2 by reducing it to NH_3 and NH_4^+ through the action of enzymes that contain the transition metal molybdenum at their active site. Enzymes in other soil bacteria catalyze the multistep oxidation of NH_4^+ to NO_2^- and finally NO_3^-, which the plants reduce again to make their proteins. When the plants die, still other soil bacteria oxidize the proteins to NO_3^-.

Animals eat the plants, use the plant proteins to make their own proteins, and excrete nitrogenous wastes, such as urea $[(H_2N)_2C{=}O]$. The nitrogen in the proteins is released when the animals die and decay, and is converted by soil bacteria to NO_2^- and NO_3^- again. This central pool of inorganic nitrate has three main fates: some enters marine and terrestrial plants, some enters the enormous sediment store of mineral nitrates, and some is reduced by denitrifying bacteria to NO_2^- and then to N_2O and N_2, which re-enter the atmosphere to complete the cycle.

The Phosphorus Cycle

Virtually all the mineral sources of phosphorus contain the phosphate group, PO_4^{3-}. The most commercially important ores are **apatites,** compounds of general formula $Ca_5(PO_4)_3X$, where X is usually F, Cl, or OH. The cycling of phosphorus through the environment involves three interlocking subcycles, illustrated in Figure 22.7. Two rapid biological cycles—a land-based cycle completed in a matter of years and a water-based cycle completed in weeks to years—are superimposed on an inorganic cycle that takes millions of years to complete. Unlike the carbon and nitrogen cycles, the phosphorus cycle has no gaseous component and thus does *not* involve the atmosphere.

Phosphorus from Outer Space Although most phosphate occurs in the rocks formed by geological activity on Earth, a sizeable amount arrived here from outer space and continues to do so. About 100 metric tons of meteorites enter our atmosphere each day. With an average P content of 0.1% by mass, they contribute about 36 metric tons of P per year. Over Earth's 4.6-billion-year lifetime, about 10^{11} metric tons of P has arrived from extraterrestrial sources!

The Inorganic Cycle Most phosphate rock formed when Earth's lithosphere solidified. The most important components of this material are nearly insoluble phosphate salts:

$$K_{sp} \text{ of } Ca_3(PO_4)_2 \approx 10^{-29}$$
$$K_{sp} \text{ of } Ca_5(PO_4)_3OH \approx 10^{-51}$$
$$K_{sp} \text{ of } Ca_5(PO_4)_3F \approx 10^{-60}$$

Weathering slowly leaches phosphates from the soil and carries the ions through rivers to the sea. Plants in the land-based biological cycle speed this process. In the ocean, some phosphate is absorbed by organisms in the water-based biological cycle, but the majority is precipitated again by Ca^{2+} ion and deposited on the continental shelf. Geologic activity lifts the continental shelves, returning the phosphate to the land.

The Land-Based Biological Cycle The biological cycles involve the incorporation of phosphate into organisms (biomolecules, bones, teeth, and so forth) and the release of phosphate through their excretion and decay. In the land-based cycle, plants continually remove phosphate from the inorganic cycle. Recall that there are three phosphate oxoanions, and they exist in equilibrium in the aqueous environment:

$$H_2PO_4^- \rightleftharpoons \underset{+\,H^+}{HPO_4^{2-}} \rightleftharpoons \underset{+\,H^+}{PO_4^{3-}}$$

In topsoil, phosphates occur as insoluble compounds of Ca^{2+}, Fe^{3+}, and Al^{3+}. Because plants can absorb only the water-soluble dihydrogen phosphate, they have evolved the ability to secrete acids near their roots to convert the insoluble salts gradually into the soluble ion:

$$Ca_3(PO_4)_2(s;\text{ soil}) + 4H^+(aq;\text{ secreted by plants}) \longrightarrow 3Ca^{2+}(aq) + 2H_2PO_4^-(aq;\text{ absorbed by plants})$$

Animals that eat the plants excrete soluble phosphate, which is used in turn by newly growing plants. As the plants and animals excrete, die, and decay, some phosphate is washed into rivers and from there to the ocean. Most of the 2 million metric tons of phosphate that washes from land to sea each year comes from biological decay. Thus, the biosphere greatly increases the movement of phosphate from lithosphere to hydrosphere.

The Water-Based Biological Cycle Various phosphate oxoanions continually enter the aquatic environment. Studies that incorporate trace amounts of radioactive phosphorus into $H_2PO_4^-$ and monitor its uptake show that, within 1 minute, 50% of the ion is taken up by photosynthetic algae, which use it to synthesize

Figure 22.7 The phosphorus cycle. The inorganic subcycle *(purple arrows),* water-based biological subcycle *(blue arrows),* and land-based biological subcycle *(yellow arrows)* interact through erosion and through the growth and decay of organisms. The phosphorus cycle has no gaseous component.

their biomolecules (this synthesis is denoted by the top curved arrow in the upcoming equation). The overall process might be represented by the following equation:

$$106CO_2(g) + 16NO_3^-(aq) + H_2PO_4^-(aq) + 122H_2O(l) + 17H^+(aq) \underset{\text{decay}}{\overset{\text{synthesis}}{\rightleftharpoons}} C_{106}H_{263}O_{110}N_{16}P(aq;\ \text{in algal cell fluid}) + 138O_2(g)$$

(The complex formula on the right represents the total composition of algal biomolecules, not some particular compound.)

As on land, animals eat the plants and are eaten by other animals, and they all excrete phosphate, die, and decay (denoted by the bottom curved arrow in the preceding equation). Some of the released phosphate is used by other aquatic organisms, and some returns to the land when fish-eating animals (mostly humans and birds) excrete phosphate, die, and decay. In addition, some aquatic phosphate precipitates with Ca^{2+}, sinks to the seabed, and returns to the long-term mineral deposits of the inorganic cycle.

Human Effects on Phosphorus Movement From prehistoric through preindustrial times, these interlocking phosphorus cycles were balanced. Modern human activity, however, alters the movement of phosphorus considerably. Our influence on the inorganic cycle is minimal, despite our annual removal of more than

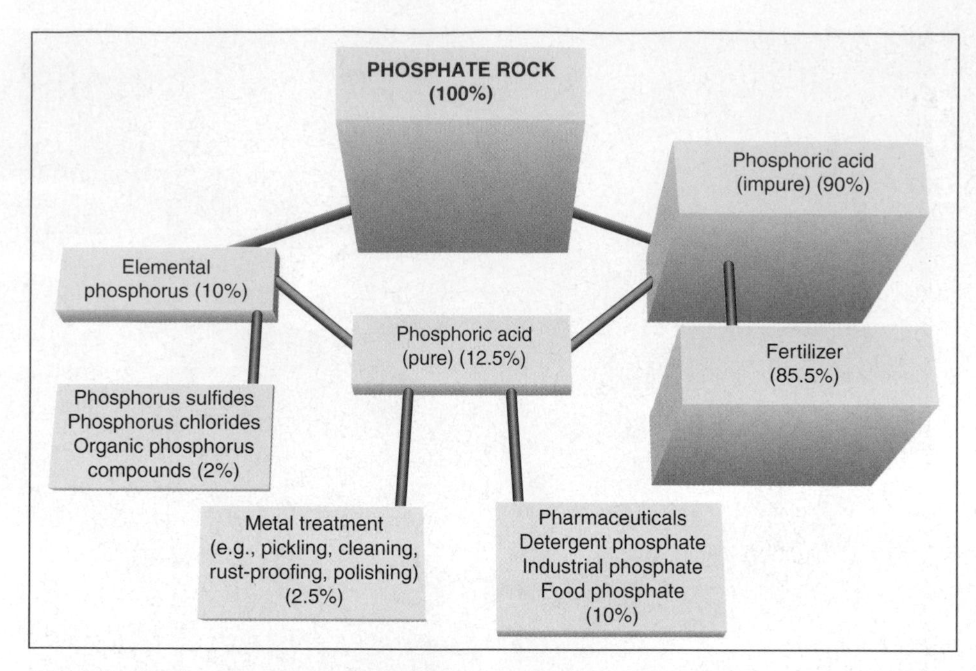

Figure 22.8 Industrial uses of phosphorus. The major end products from phosphate rock are shown, with box heights proportional to the amounts (mass %) of phosphate rock used annually to make the end product. Over 100 million metric tons of phosphate rock is mined each year (100%).

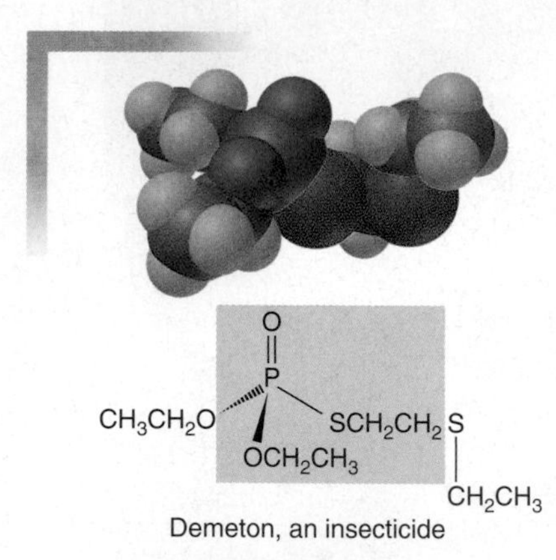

Demeton, an insecticide

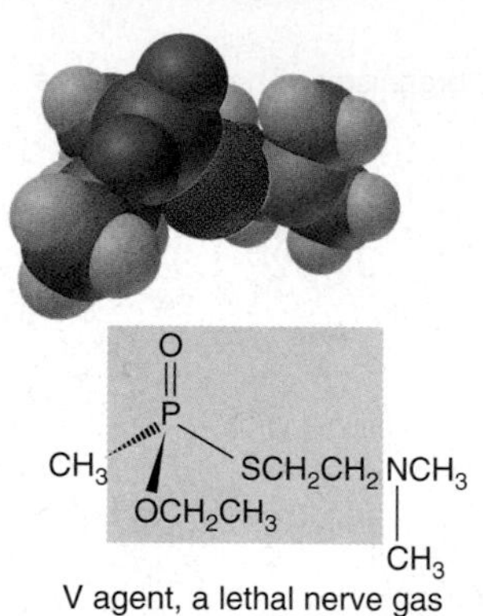

V agent, a lethal nerve gas

Phosphorus Nerve Poisons Because phosphorus is essential for all organisms, its compounds have many biological actions. In addition to their widespread use as fertilizers, phosphorus compounds are also used as pesticides *and* as military nerve gases. In fact, these two types of substances often have similar structures (see above) and modes of action. Both types affect the target organism by inactivating an enzyme involved in muscle contraction. They bind an amino acid critical to the enzyme's function, allowing the muscle to contract but not relax; the effect on the organism—insect or human—is often fatal. Chemists have synthesized a nerve-gas antidote whose shape mimics the binding site of the enzyme. The drug binds the poison, removing it from the enzyme, and is excreted.

100 million metric tons of phosphate rock for a variety of uses (Figure 22.8). The end products of this removal, however, have unbalanced the two biological cycles.

The imbalance arises from our overuse of *soluble phosphate fertilizers,* such as $Ca(H_2PO_4)_2$ and $NH_4H_2PO_4$, the end products of about 85% of phosphate rock mined. Some fertilizer finds its way into rivers, lakes, and oceans to enter the water-based cycle. Much larger pollution sources, however, are the crops grown with the fertilizer and the detergents made from phosphate rock. The great majority of this phosphate arrives eventually in cities as crops, as animals that were fed crops, and as consumer products, such as the tripolyphosphates in detergents. Human garbage, excrement, wash water with detergents, and industrial wastewater containing phosphate return through sewers to the aquatic system. This human contribution equals the natural contribution—another 2 million metric tons of phosphate per year. The increased concentration in rivers and lakes causes eutrophication, which robs the water of O_2 so that it cannot support life. Such "dead" rivers and lakes are no longer usable for fishing, drinking, or recreation.

As early as 1912, a chemical solution to this problem was tried in Switzerland, where the enormous Lake Zurich had been choked with algae and become devoid of fish because of the release of human sewage. Treatment with $FeCl_3$ precipitated enough phosphate to return the lake gradually to its natural state. Soluble aluminum compounds have the same effect:

$$FeCl_3(aq) + PO_4^{3-}(aq) \longrightarrow FePO_4(s) + 3Cl^-(aq)$$

$$KAl(SO_4)_2 \cdot 12H_2O(aq) + PO_4^{3-}(aq) \longrightarrow AlPO_4(s) + K^+(aq) + 2SO_4^{2-}(aq) + 12H_2O(l)$$

After several lakes in the United States became polluted in this way, phosphates in detergents were drastically reduced, and stringent requirements for sewage treatment, including tertiary treatments to remove phosphate (see Chemical Connections, pp. 529–530), were put into effect. Lake Erie, which was severely polluted by the late 1960s and early 1970s, was the target of a program to reduce phosphates and is today relatively clean and restocked with fish.

SECTION SUMMARY

The environmental distribution of many elements changes cyclically with time and is affected in major ways by organisms. Carbon occurs in all three regions of Earth's crust, with atmospheric CO_2 linking the other two regions. Photosynthesis and decay of organisms alter the amount of carbon in the land and oceans. Human activity has increased atmospheric CO_2. Nitrogen is fixed by lightning, by industry, and primarily by microorganisms. When plants decay, bacteria decompose organic nitrogen and eventually return N_2 to the atmosphere. Through extensive use of fertilizers, humans have added excess nitrogen to freshwaters. The phosphorus cycle has no gaseous component. Inorganic phosphates leach slowly into land and water, where biological cycles interact. When plants and animals decay, they release phosphate to natural waters, where it is then absorbed by other plants and animals. Through overuse of phosphate fertilizers and detergents, humans double the amount of phosphorus entering aqueous systems.

22.3 METALLURGY: EXTRACTING A METAL FROM ITS ORE

Metallurgy is the branch of materials science concerned with the extraction and utilization of metals. In this section, we discuss the general procedures for isolating metals and, occasionally, the nonmetals that are found in their ores. The extraction process applies one or more of these three types of metallurgy: *pyrometallurgy* uses heat to obtain the metal, *electrometallurgy* employs an electrochemical step, and *hydrometallurgy* relies on the metal's aqueous solution chemistry.

The extraction of an element begins with *mining the ore*. Most ores consist of mineral and gangue. The **mineral** contains the element; it is defined as a naturally occurring, homogeneous, crystalline inorganic solid, with a well-defined composition. The **gangue** is the portion of the ore with no commercial value, such as sand, rock, and clay attached to the mineral. Table 22.2 lists the common mineral sources of some metals.

Table 22.2 Common Mineral Sources of Some Elements

Element	Mineral, Formula
Al	Gibbsite (in bauxite), $Al(OH)_3$
Ba	Barite, $BaSO_4$
Be	Beryl, $Be_3Al_2Si_6O_{18}$
Ca	Limestone, $CaCO_3$
Fe	Hematite, Fe_2O_3
Hg	Cinnabar, HgS
Na	Halite, NaCl
Pb	Galena, PbS
Sn	Cassiterite, SnO_2
Zn	Sphalerite, ZnS

Humans have been removing metals from Earth's crust for thousands of years, so most of the known concentrated sources are long gone. In many cases, ores containing very low mass percents of a metal are all that remain, and these require elaborate—and ingenious—extraction methods. Nevertheless, the general procedure for extracting most metals (and many nonmetals) involves a few basic steps, described in the upcoming subsections and previewed in Figure 22.9.

Figure 22.9 Steps in metallurgy.

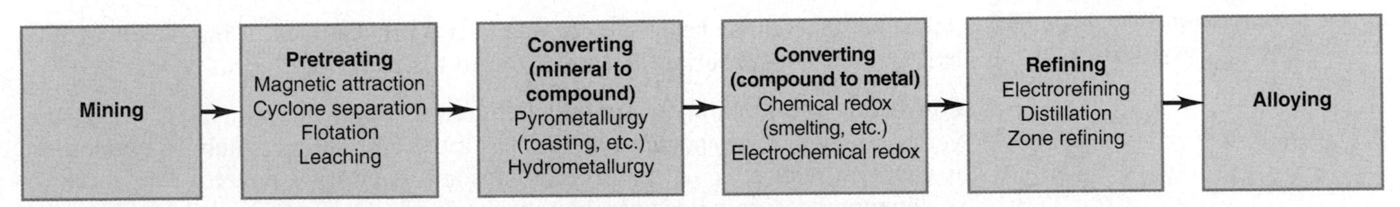

Pretreating the Ore

Following a crushing, grinding, or pulverizing step, which can be very expensive, pretreatment usually takes advantage of some physical or chemical difference to separate mineral from gangue. For magnetic minerals, such as magnetite (Fe_3O_4), a magnet can remove the mineral and leave the gangue behind. Where large density differences exist, a *cyclone separator* is used; it blows high-pressure air through the pulverized mixture to separate the particles. The lighter silicate-rich

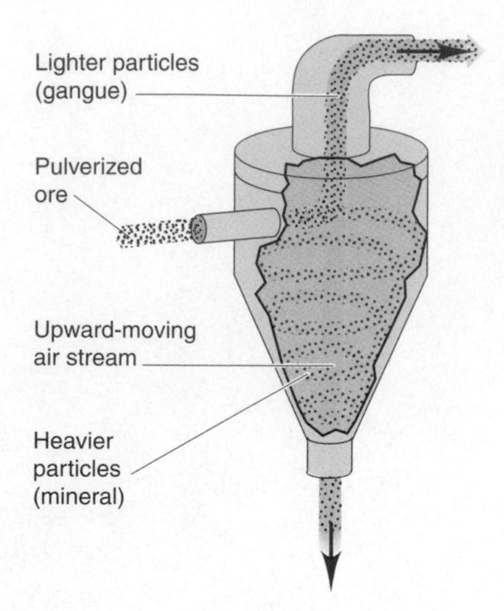

Figure 22.10 The cyclone separator. The pulverized ore enters at an angle, with the more dense particles (mineral) hitting the walls and spiraling downward, while the less dense particles (silicate-rich gangue) are carried upward in a stream of air.

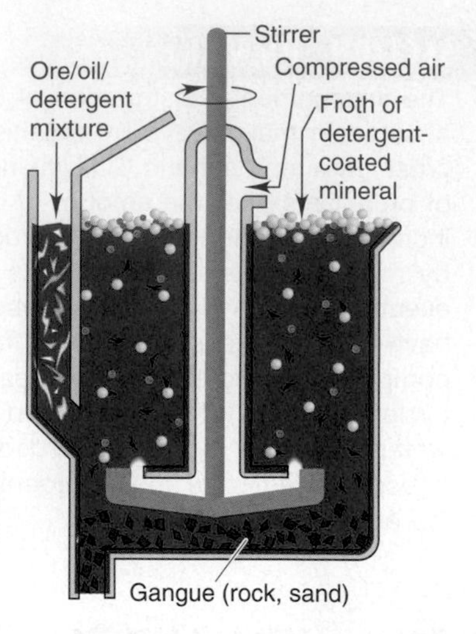

Figure 22.11 The flotation process. The pulverized ore is treated with an oil-detergent mixture, then stirred and aerated to create a froth that separates the detergent-coated mineral from the gangue. The mineral-rich froth is collected for further processing.

gangue is blown away, while the denser mineral-rich particles hit the walls of the separator and fall through the open bottom (Figure 22.10).

In **flotation,** an oil-detergent mixture is stirred with the pulverized ore in water to form a slurry (Figure 22.11). Flotation takes advantage of the different abilities of the surfaces of mineral and of gangue to become wet in contact with water and detergent. Rapid mixing moves air bubbles through the mixture and produces an oily, mineral-rich froth that floats, while the silicate particles sink. Skimming, followed by solvent removal of the oil-detergent mixture, isolates the concentrated mineral fraction. Flotation is a key step in copper recovery.

Leaching is a hydrometallurgical process that selectively extracts the metal, usually by forming a complex ion. The modern extraction of gold is a good example of this technique, because nuggets of the precious metal are found only in museums nowadays. The crushed ore, which often contains as little as 25 ppm of gold, is contained in a plastic-lined pool, treated with a cyanide ion solution, and aerated. In the presence of CN^-, the O_2 in air oxidizes gold metal to gold(I) ion, Au^+, which forms the soluble complex ion, $Au(CN)_2^-$:

$$4Au(s) + O_2(g) + 8CN^-(aq) + 2H_2O(l) \longrightarrow 4Au(CN)_2^-(aq) + 4OH^-(aq)$$

(The method has become controversial in some areas because cyanide lost from the contained area enters streams and lakes, where it poisons fish and birds.)

Panning and Fleecing for Gold The picture of the prospector, alone but for his faithful mule, hunched over the bank of a mountain stream patiently panning for gold, is mostly a relic of old movies. However, the process is valid. It is based on the very large density difference between gold (d = 19.3 g/cm^3) and sand ($d \approx$ 2.5 g/cm^3). In ancient times, river sands were washed over a sheep's fleece to trap the gold grains, a practice historians think represents the origin of the Golden Fleece of Greek mythology.

Converting Mineral to Element

After the mineral has been freed of debris and concentrated, it may undergo several chemical steps during its conversion to the element.

Converting the Mineral to Another Compound First, the mineral is often converted to another compound, one that has more appropriate solubility properties, is easier to reduce, or is free of a troublesome impurity. Conversion to an oxide is common because oxides can be reduced easily. Carbonates are heated to convert them to the oxide:

$$CaCO_3(s) \xrightarrow{\Delta} CaO(s) + CO_2(g)$$

Metal sulfides, such as ZnS, can be converted to oxides by **roasting** in air:

$$2ZnS(s) + 3O_2(g) \xrightarrow{\Delta} 2ZnO(s) + 2SO_2(g)$$

In most countries, hydrometallurgical methods are now used to avoid the atmospheric release of SO_2 during roasting. One way of processing copper, for example, is by bubbling air through an acidic slurry of insoluble Cu_2S, which completes both the pretreatment and conversion steps:

$$2Cu_2S(s) + 5O_2(g) + 4H^+(aq) \longrightarrow 4Cu^{2+}(aq) + 2SO_4^{2-}(aq) + 2H_2O(l)$$

Converting the Compound to the Element Through Chemical Redox The next step converts the new mineral form (usually an oxide) to the free element by either chemical or electrochemical methods. In chemical redox, a reducing agent reacts directly with the compound. The most common reducing agents are carbon, hydrogen, and an active metal.

1. *Reduction with carbon.* Carbon, in the form of coke (a porous residue from incomplete combustion of coal) or charcoal, is a very common reducing agent. Heating an oxide with a reducing agent such as coke to obtain the metal is called **smelting.** Many metal oxides, such as zinc oxide and tin(IV) oxide, are smelted with carbon to free the metal, which may need to be condensed and solidified:

$$ZnO(s) + C(s) \longrightarrow Zn(g) + CO(g)$$
$$SnO_2(s) + 2C(s) \longrightarrow Sn(l) + 2CO(g)$$

Several nonmetals that occur with positive oxidation states in minerals can be reduced with carbon as well. Phosphorus, for example, is produced from calcium phosphate:

$$2Ca_3(PO_4)_2(s) + 10C(s) + 6SiO_2(s) \longrightarrow 6CaSiO_3(s) + 10CO(g) + P_4(s)$$

(Metallic calcium is a much stronger reducing agent than carbon, so it is not formed.)

Low cost and ready availability explain why carbon is such a common reducing agent, and thermodynamic principles explain why it is such an effective one in many cases. Consider the standard free energy change (ΔG^0) for the reduction of tin(IV) oxide:

$$SnO_2(s) + 2C(s) \longrightarrow Sn(s) + 2CO(g) \qquad \Delta G^0 = 245 \text{ kJ at } 25°C \text{ (298 K)}$$

The magnitude and sign of ΔG^0 indicate a highly *non*spontaneous reaction for standard conditions at 25°C. However, because solid C becomes gaseous CO, the standard molar entropy change is very positive ($\Delta S^0 \approx 380$ J/K). Therefore, *the* $-T\Delta S^0$ *term of* ΔG^0 *becomes more negative with higher temperature* (Section 20.3). As the temperature increases, ΔG^0 decreases, and at a high enough temperature, the reaction becomes spontaneous ($\Delta G^0 < 0$). At 1000°C (1273 K), for example, $\Delta G^0 \approx -62.8$ kJ. The actual process is carried out at a temperature of around 1250°C (1523 K), so ΔG^0 is even more negative.

As you know, overall reactions often mask several intermediate steps. For instance, the reduction of tin(IV) oxide may begin with formation of tin(II) oxide:

$$SnO_2(s) + C(s) \longrightarrow \cancel{SnO(s)} + CO(g)$$
$$\cancel{SnO(s)} + C(s) \longrightarrow Sn(l) + CO(g)$$

[Molten tin (mp = 232 K) is obtained at this temperature.] The second step may even be composed of others, in which CO, not C, is the actual reducing agent:

$$SnO(s) + \cancel{CO(g)} \longrightarrow Sn(l) + \cancel{CO_2(g)}$$
$$\cancel{CO_2(g)} + C(s) \rightleftharpoons 2CO(g)$$

Other pyrometallurgical processes are similarly complex. Shortly, you'll see that the overall reaction for the smelting of iron is

$$2Fe_2O_3(s) + 3C(s) \longrightarrow 4Fe(l) + 3CO_2(g)$$

However, the actual process is a multistep one with CO as the reducing agent. The advantage of CO over C is the far greater contact the gaseous reducing agent can make with the other reactant, which speeds the process.

2. *Reduction with hydrogen.* For oxides of some metals, especially some members of Groups 6B(6) and 7B(7), reduction with carbon forms metal carbides. These carbides are difficult to convert further, so other reducing agents are used. Hydrogen gas is used for less active metals, such as tungsten:

$$WO_3(s) + 3H_2(g) \longrightarrow W(s) + 3H_2O(g)$$

One step in the purification of the metalloid germanium uses hydrogen:

$$GeO_2(s) + 2H_2(g) \longrightarrow Ge(s) + 2H_2O(g)$$

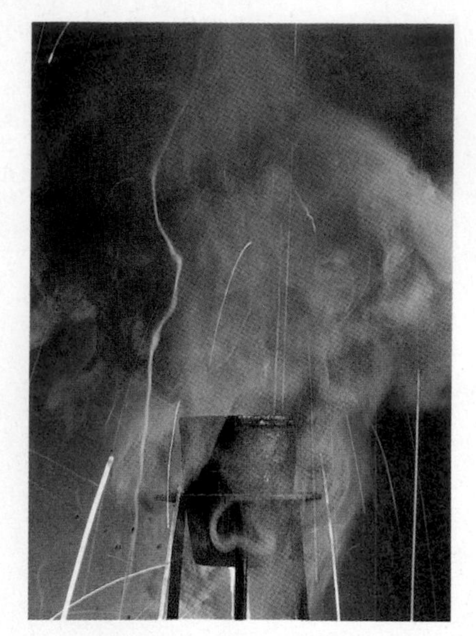

Figure 22.12 The thermite reaction. This highly exothermic redox reaction uses powdered aluminum to reduce metal oxides.

3. *Reduction with an active metal.* When a metal might form an undesirable hydride, its oxide is reduced by a more active metal. In the *thermite reaction,* aluminum powder reduces the metal oxide in a spectacular exothermic reaction to give the molten metal (Figure 22.12). The reaction for chromium is

$$Cr_2O_3(s) + 2Al(s) \longrightarrow 2Cr(l) + Al_2O_3(s) \qquad \Delta H^0 << 0$$

Reduction by a more active metal is also used in situations that do not involve an oxide. In the extraction of gold, after the pretreatment by leaching, the gold(I) complex ion is reduced with zinc to form the metal and a zinc complex ion:

$$2Au(CN)_2^-(aq) + Zn(s) \longrightarrow 2Au(s) + Zn(CN)_4^{2-}(aq)$$

In some cases, an active metal, such as calcium, is used to recover an even more active metal, such as rubidium, from its molten salt:

$$Ca(l) + 2RbCl(l) \longrightarrow CaCl_2(l) + 2Rb(g)$$

A similar process is used to recover sodium, as detailed in the next section.

4. *Oxidation with an active nonmetal.* Just as chemical *reduction* of a mineral is used to obtain the metal, chemical *oxidation* of a mineral is sometimes used to obtain a nonmetal. A stronger oxidizing agent is used to remove electrons from the nonmetal anion to give the free nonmetal, as in the industrial production of iodine from concentrated brines by oxidation with chlorine gas:

$$2I^-(aq) + Cl_2(g) \longrightarrow 2Cl^-(aq) + I_2(s)$$

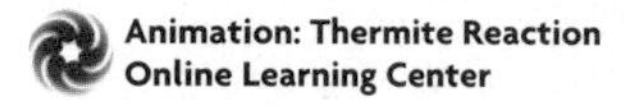

Animation: Thermite Reaction
Online Learning Center

Converting the Compound to the Element Through Electrochemical Redox In these processes, the mineral components are converted to the elements in an electrolytic cell (Section 21.7). Sometimes, the *pure* mineral, in the form of the molten halide or oxide, is used to prevent unwanted side reactions. The cation is reduced to the metal at the cathode, and the anion is oxidized to the nonmetal at the anode:

$$BeCl_2(l) \longrightarrow Be(s) + Cl_2(g)$$

High-purity hydrogen gas is prepared by electrochemical reduction:

$$2H_2O(l) \longrightarrow 2H_2(g) + O_2(g)$$

Specially designed cells separate the products to prevent recombination. Cost is a major factor in the use of electrolysis, and an inexpensive source of electricity is essential for large-scale methods. The current and voltage requirements depend on the electrochemical potential and any overvoltage effects. Figure 22.13 shows the most common step for obtaining each free element from its mineral.

Figure 22.13 The redox step in converting a mineral to the element. Most elements are reduced from the compound with carbon, a more active metal, or H_2.

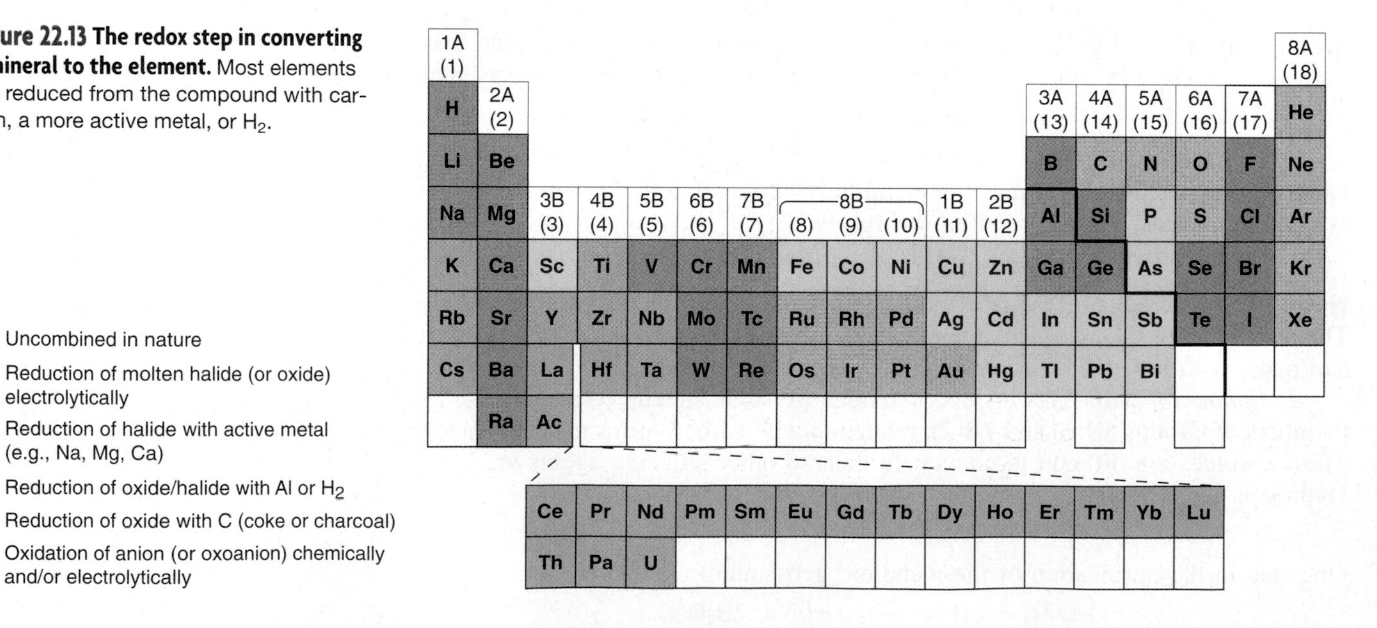

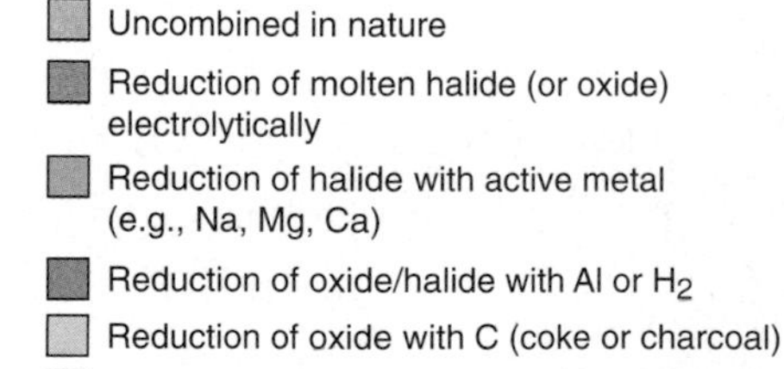

Refining and Alloying the Element

At this point in the isolation process, the element typically contains impurities from previous steps, so it must be refined. Then, once purified, metallic elements are often alloyed to improve their properties.

Refining (Purifying) the Element Refining is a purification procedure, often carried out by one of three common methods. An electrolytic cell is employed in **electrorefining,** in which the impure metal acts as the anode and a sample of the pure metal acts as the cathode. As the reaction proceeds, the anode slowly disintegrates, and the metal ions are reduced to deposit on the cathode. In some cases, the impurities, which fall beneath the disintegrating anode, are the source of several valuable and less abundant elements. (We examine the electrorefining of copper shortly.)

Metals with relatively low boiling points, such as zinc and mercury, are refined by *distillation.*

In the process of **zone refining** (see the Gallery, p. 521), impurities are removed from a bar of the element by concentrating them in a thin molten zone, while the purified element recrystallizes. Metalloids used in electronic semiconductors, such as silicon and germanium, must be zone refined to greater than 99.999999% purity.

Alloying the Purified Element An **alloy** is a metal-like mixture consisting of solid phases of two or more pure elements, a solid solution, or, in some cases, distinct intermediate phases (these alloys are sometimes referred to as *intermetallic compounds*). The separate phases are often so finely divided that they can only be distinguished microscopically. Alloying a metal with other metals (and, in some cases, nonmetals) is done to alter the metal's melting point and to enhance properties such as luster, conductivity, malleability, ductility, and strength.

Iron, probably the most important metal, is used only when it is alloyed. In pure form, it is soft and corrodes easily. However, when alloyed with carbon and other metals, such as Mo for hardness and Cr and Ni for corrosion resistance, it forms the various steels. Copper stiffens when zinc is added to make brass. Mercury solidifies when alloyed with sodium, and vanadium becomes extremely tough with some carbon added. Table 22.3 shows the composition and uses of some common alloys.

The simplest alloys, called *binary alloys,* contain only two elements. In some cases, *the added element enters interstices,* spaces between the parent metal atoms

Table 22.3 Some Familiar Alloys and Their Composition

Name	Composition (Mass %)	Uses
Stainless steel	73–79 Fe, 14–18 Cr, 7–9 Ni	Cutlery, instruments
Nickel steel	96–98 Fe, 2–4 Ni	Cables, gears
High-speed steels	80–94 Fe, 14–20 W (or 6–12 Mo)	Cutting tools
Permalloy	78 Ni, 22 Fe	Ocean cables
Bronzes	70–95 Cu, 1–25 Zn, 1–18 Sn	Statues, castings
Brasses	50–80 Cu, 20–50 Zn	Plating, ornamental objects
Sterling silver	92.5 Ag, 7.5 Cu	Jewelry, tableware
14-Carat gold	58 Au, 4–28 Ag, 14–28 Cu	Jewelry
18-Carat white gold	75 Au, 12.5 Ag, 12.5 Cu	Jewelry
Typical tin solder	67 Pb, 33 Sn	Electrical connections
Dental amalgam	69 Ag, 18 Sn, 12 Cu, 1 Zn (dissolved in Hg)	Dental fillings

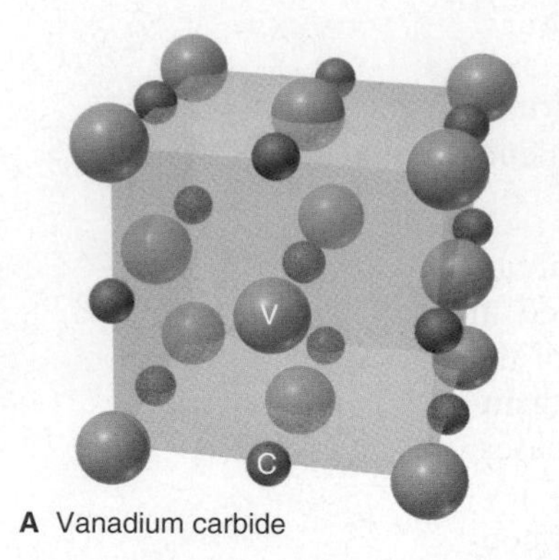

A Vanadium carbide

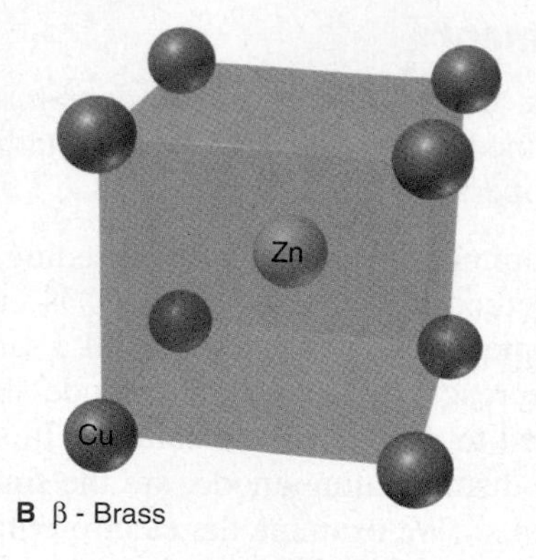

B β - Brass

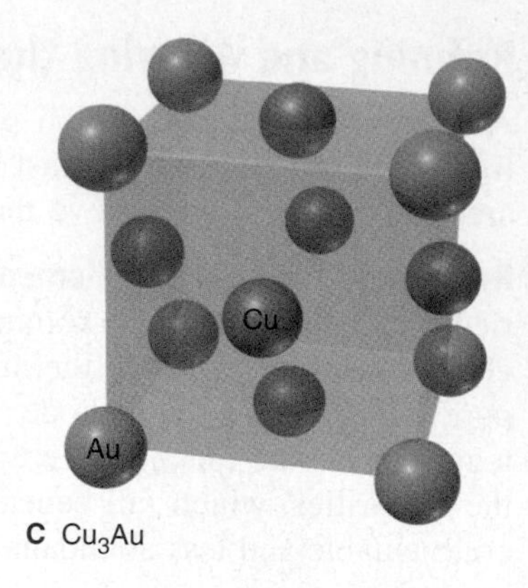

C Cu_3Au

Figure 22.14 Three binary alloys.

in the crystal structure (Section 12.6). In vanadium carbide, for instance, carbon atoms occupy holes of a face-centered cubic vanadium unit cell (Figure 22.14A). In other cases, the *added element substitutes for parent atoms* in the unit cell. In many types of brass, for example, relatively few zinc atoms substitute randomly for copper atoms in the face-centered cubic copper unit cell (see Figure 13.4A). On the other hand, β-brass is an intermediate phase that crystallizes as the Zn:Cu ratio approaches 1:1 in a structure with a Zn atom surrounded by eight Cu atoms (Figure 22.14B). In one alloy of copper and gold, Cu atoms occupy the faces of a face-centered cube, and Au atoms lie at the corners (Figure 22.14C).

Atomic properties, including size, electron configuration, and number of valence electrons, determine which metals form stable alloys. Quantitative predictions indicate that transition metals having few *d* electrons often form stable alloys with transition metals having many *d* electrons. Thus, metals from the left half of the *d* block (Groups 3 to 5) often form alloys with metals from the right half (Groups 9 to 11). For example, electron-poor Zr, Nb, and Ta form strong alloys with electron-rich Ir, Pt, or Au; an example is the very stable $ZrPt_3$.

SECTION SUMMARY

Metallurgy involves mining an ore, separating it from debris, pretreating it to concentrate the mineral source, converting the mineral to another compound that is easier to process further, reducing this compound to the metal, purifying the metal, and in many cases, alloying it to obtain a more useful material.

22.4 TAPPING THE CRUST: ISOLATION AND USES OF THE ELEMENTS

Once an element's abundance and sources are known and the metallurgical methods chosen, the isolation process begins. Obviously, the process depends on the physical and chemical properties of the source: we isolate an element that occurs uncombined in the air differently from one dissolved in the sea or one found in a rocky ore. In this section, we detail methods for recovering some important elements; pay special attention to the application of ideas from earlier chapters to these key industrial processes.

Producing the Alkali Metals: Sodium and Potassium

The alkali metals are among the most reactive elements and thus are always found as ions in nature, either in solid minerals or in aqueous solution. In the laboratory, the free elements must be protected from contact with air (for example, by being stored under mineral oil) to prevent their immediate oxidation and to minimize the possibility of fires. The two most important alkali metals are sodium and potassium. Their abundant, water-soluble compounds are used throughout industry and research, and the Na^+ and K^+ ions are essential to organisms.

Industrial Production of Sodium and Potassium The sodium ore is *halite* (largely NaCl), which is obtained either by evaporation of concentrated salt solutions (brines) or by mining vast salt deposits formed from the evaporation of prehistoric seas. The Cheshire salt field in Britain, for example, is 60 km by 24 km by 400 m thick and contains more than 90% NaCl. Other large deposits occur in New Mexico, Michigan, New York, and Kansas.

A Plentiful Oceanic Supply of NaCl Will we ever run out of sodium chloride? Not likely. The rock-salt (halite) equivalent of NaCl in the world's oceans has been estimated at 1.9×10^7 km^3, about 150% of the volume of North America above sea level. Or, looking at this amount another way: a row of 1-km^3 cubes of NaCl would stretch from Earth to the Moon 47 times! The photo shows evaporative salt beds near San Francisco Bay.

The brine is evaporated to dryness, and the solid is crushed and fused (melted) for use in an electrolytic apparatus called the **Downs cell** (Figure 22.15). To reduce heating costs, the NaCl (mp = 801°C) is mixed with $1\frac{1}{2}$ parts $CaCl_2$ to form a mixture that melts at only 580°C. Reduction of the metal ions to Na and Ca takes place at a cylindrical steel cathode, with the molten metals floating on the denser molten salt mixture. As they rise through a short collecting pipe, the liquid Na is siphoned off, while a higher melting Na/Ca alloy solidifies and falls back into the molten electrolyte. Chloride ions are oxidized to Cl_2 gas at a large anode within an inverted cone-shaped chamber. The cell design separates the metals from the Cl_2 to prevent their explosive recombination. The Cl_2 gas is collected, purified, and sold as a valuable by-product.

Sylvite (mostly KCl) is the major ore of potassium. The metal is too soluble in molten KCl to be prepared by a method similar to that for sodium. Instead, chemical reduction of K^+ ions by liquid Na is used. The method is based on atomic properties and the nature of equilibrium systems. Recall that an Na atom is smaller than a K atom, so it holds its outer electron more tightly, as the first ionization energies indicate: IE_1 of Na = 496 kJ/mol; IE_1 of K = 419 kJ/mol. Thus, if the isolation method for K were based solely on this atomic property, Na would not be very effective at reducing K^+, and the reaction would not move very far toward products. An ingenious approach, which applies Le Châtelier's

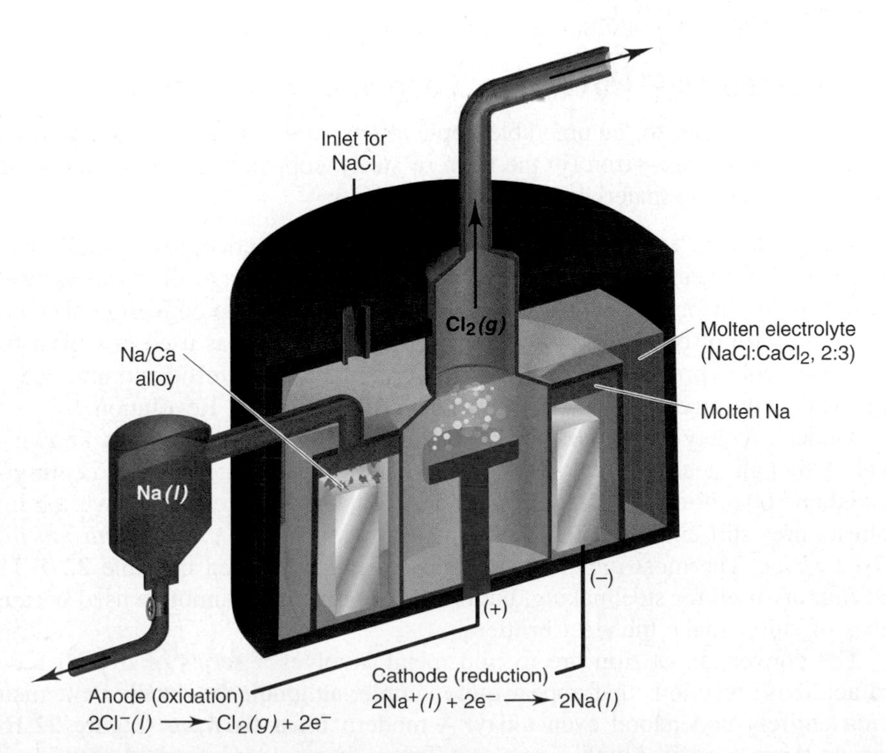

Figure 22.15 The Downs cell for production of sodium. The mixture of solid NaCl and $CaCl_2$ forms the molten electrolyte. Sodium and calcium are formed at the cathode and float, but an Na/Ca alloy solidifies and falls back into the bath. Chlorine gas forms at the anode.

principle, overcomes this drawback. The reduction is carried out at 850°C, which is above the boiling point of K, so the equilibrium mixture contains *gaseous* K:

$$Na(l) + K^+(l) \rightleftharpoons Na^+(l) + K(g)$$

As the K gas is removed, the system shifts to produce more K. The gas is then condensed and purified by fractional distillation. The same general method (with Ca as the reducing agent) is used to produce rubidium and cesium.

Uses of Sodium and Potassium The compounds of Na and K (and indeed, of all the alkali metals) have many more uses than the elements themselves [see the Group 1A(1) Family Portrait, p. 561]. Nevertheless, there are some interesting uses of the metals that take advantage of their strong reducing power. Large amounts of Na were used as an alloy with lead to make gasoline antiknock additives, such as tetraethyllead:

$$4C_2H_5Cl(g) + 4Na(s) + Pb(s) \longrightarrow (C_2H_5)_4Pb(l) + 4NaCl(s)$$

The toxic effects of environmental lead have made this use virtually nonexistent in the United States. Moreover, although leaded gasoline is still used in parts of Europe, recent legislation promises to eliminate this product early in this decade.

If certain types of nuclear reactors, called *breeder reactors* (Chapter 24), become a practical way to generate energy in the United States, Na production would increase enormously. Its low melting point, viscosity, and absorption of neutrons, combined with its high thermal conductivity and heat capacity, make it perfect for cooling the reactor and exchanging heat to the steam generator.

Firefighter with KO_2 breathing mask.

The major use of potassium at present is in an alloy with sodium for use as a heat exchanger in chemical and nuclear reactors. Another application relies on production of its superoxide, which it forms by direct contact with O_2:

$$K(s) + O_2(g) \longrightarrow KO_2(s)$$

This material is employed as an emergency source of O_2 in breathing masks for miners, divers, submarine crews, and firefighters (see photo):

$$4KO_2(s) + 4CO_2(g) + 2H_2O(g) \longrightarrow 4KHCO_3(s) + 3O_2(g)$$

The Indispensable Three: Iron, Copper, and Aluminum

Their familiar presence, innumerable applications, and enormous production levels make three metals—iron (in the form of steel), copper, and aluminum—stand out as indispensable materials of industrial society.

Metallurgy of Iron and Steel Although people have practiced iron smelting for more than 3000 years, it was only a little over 225 years ago that iron assumed its current dominant role. In 1773, an inexpensive process to convert coal to carbon in the form of coke was discovered, and the material was used in a blast furnace. The coke process made iron smelting cheap and efficient and led to large-scale iron production, which ushered in the Industrial Revolution.

Modern society rests, quite literally, on the various alloys of iron known as **steel.** Although steel production has grown enormously since the 18th century—more than 700 million tons are produced annually—the process of recovering iron from its ores still employs the same general approach: *reduction by carbon in a blast furnace*. The most important minerals of iron are listed in Table 22.4. The first four are used for steelmaking, but the sulfide minerals cannot be used because traces of sulfur make the steel brittle.

Table 22.4 Important Minerals of Iron

Mineral Type	Mineral, Formula
Oxide	Hematite, Fe_2O_3
	Magnetite, Fe_3O_4
	Ilmenite, $FeTiO_3$
Carbonate	Siderite, $FeCO_3$
Sulfide	Pyrite, FeS_2
	Pyrrhotite, FeS

The conversion of iron ore to iron metal involves a series of overall redox and acid-base reactions that appear quite simple, although the detailed chemistry is not entirely understood even today. A modern **blast furnace** (Figure 22.16), such as those used in South Korea and Japan, is a tower about 14 m wide by 40 m high, made of a brick that can withstand intense heat. The *charge,* which consists of an iron ore (usually hematite) containing the mineral, coke, and lime-

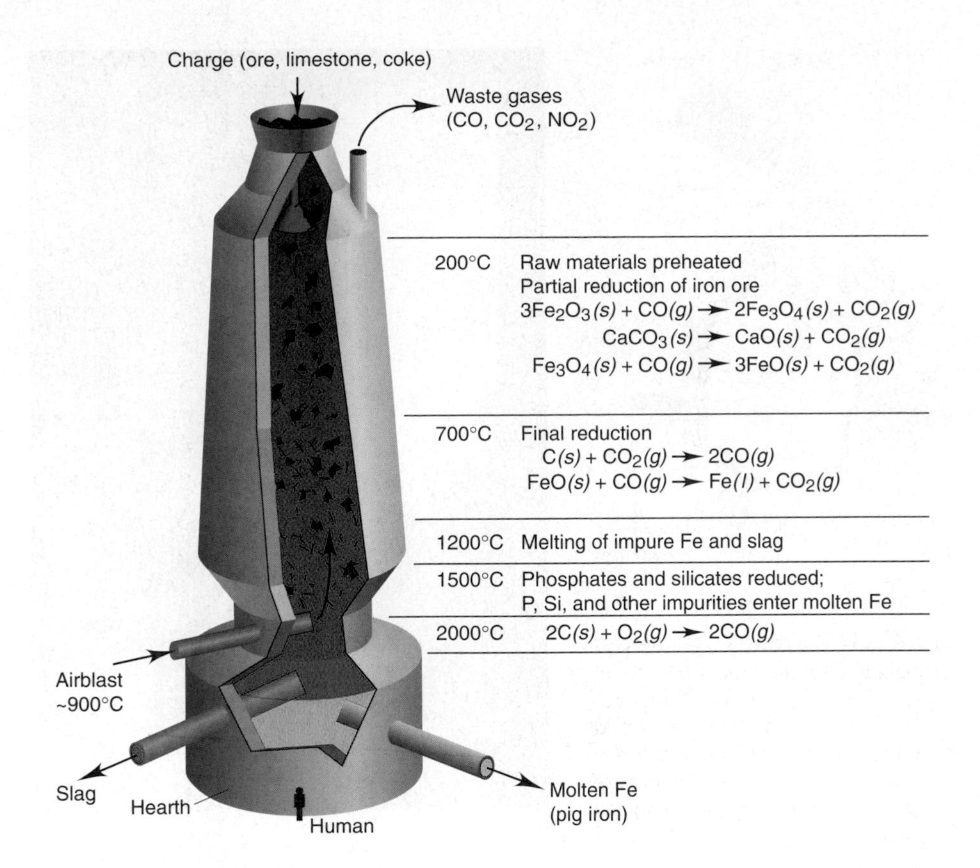

Figure 22.16 The major reactions in a blast furnace. Iron is produced from iron ore, coke, and limestone in a complex process that depends on the temperature at different heights in the furnace. (A human is shown for scale.)

Animation: Iron Smelting
Online Learning Center

stone, is fed through the top. More coke is burned in air at the bottom. The charge falls and meets a *blast* of rapidly rising hot air created by the burning coke:

$$2C(s) + O_2(g) \longrightarrow 2CO(g) + \textbf{\textit{heat}}$$

At the bottom of the furnace, the temperature exceeds 2000°C, while at the top, it reaches only 200°C. As a result, different stages of the overall reaction process occur at different heights as the charge descends:

1. In the upper part of the furnace (200°C to 700°C), the charge is preheated, and a *partial reduction* step occurs. The hematite is reduced to magnetite and then to iron(II) oxide (FeO) by CO, the actual reducing agent for the iron oxides. Carbon dioxide is formed as well. Because the blast of hot air passes through the entire furnace in only 10 s, the various gas-solid reactions do *not* reach equilibrium, and many intermediate products form. Limestone also decomposes to form CO_2 and the basic oxide CaO, which is important later in the process.
2. Lower down, at 700°C to 1200°C, a *final reduction* step occurs, as some of the coke reduces CO_2 to form more CO, which reduces the FeO to Fe.
3. At still higher temperatures, between 1200°C and 1500°C, the iron melts and drips to the bottom of the furnace. Acidic silica particles from the gangue react with basic calcium oxide in a Lewis acid-base reaction to form a molten waste product called **slag:**

$$CaO(s) + SiO_2(s) \longrightarrow CaSiO_3(l)$$

The siliceous slag drips down and floats on the denser iron (much as Earth's silicate-rich mantle and crust float on its molten iron core).

4. Some unwanted reactions occur at this hottest stage, between 1500°C and 2000°C. Any remaining phosphates and silicates are reduced to P and Si, and some Mn and traces of S dissolve into the molten iron along with carbon. The resulting impure product is called *pig iron* and contains about 3% to 4% C. A small amount of pig iron is used to make *cast iron,* but most is purified and alloyed to make various kinds of steel.

Was It Slag That Made the Great Ship Go Down? Direct observation of the sunken luxury liner *Titanic* shows that six small slits caused by collision with an iceberg, rather than an enormous gash, sank the ship. The slits appear between sections of steel plates in the lower hull and suggest weakness in the rivets that connected the plates. Metallurgical analysis of the rivets shows three times as much slag content as modern rivets. Too much of this silicate-rich material makes the wrought iron used in the rivets brittle and easier to break under stress. Moreover, rather than being evenly distributed, the slag particles within the iron, viewed microscopically, appear to be clumped, which causes weak spots. Could low-grade rivets with too much slag have caused this tragedy? Follow-up analysis of more rivets and a historical search of rivet standards in the early 1900s may finally solve the mystery of what brought down the unsinkable *Titanic* and caused the deaths of 1500 passengers.

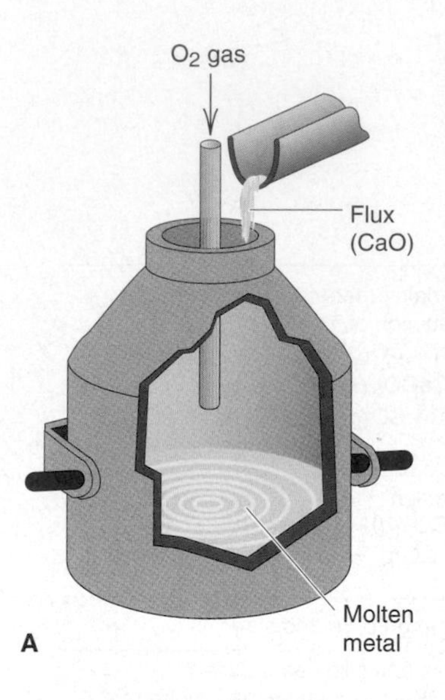

Figure 22.17 The basic-oxygen process for making steel. A, Jets of pure O_2 in combination with a basic flux (CaO) are used to remove impurities and lower the carbon content of pig iron in the manufacture of steel. **B,** Impure molten iron is added to a basic-oxygen furnace.

Pig iron is converted to steel in a separate furnace by means of the **basic-oxygen process,** as shown in Figure 22.17. High-pressure O_2 is blown over and through the molten iron, so that impurities (C, Si, P, Mn, S) are oxidized rapidly. The highly negative standard heats of formation of their oxides (such as ΔH_f^0 of $CO_2 = -394$ kJ/mol and ΔH_f^0 of $SiO_2 = -911$ kJ/mol) raise the temperature, which speeds the reaction. A lime (CaO) flux is added, which converts the oxides to a molten slag [primarily $CaSiO_3$ and $Ca_3(PO_4)_2$] that is decanted from the molten steel. The product is **carbon steel,** which contains 1% to 1.5% C and other impurities. It is alloyed with metals that prevent corrosion and increase its strength or flexibility.

Mining chalcopyrite.

Isolation and Electrorefining of Copper After many centuries of being mined to make bronze and brass articles, copper ores have become less plentiful and less rich in copper, so the metal is more expensive to extract. Despite this, more than 2.5 billion pounds of copper is produced in the United States annually. The most common copper ore is chalcopyrite, $CuFeS_2$, a mixed sulfide of FeS and CuS (see photo). Most remaining deposits contain less than 0.5% Cu by mass. To "win" this small amount of copper from the ore requires several metallurgical steps, including a final refining to achieve the 99.99% purity needed for electrical wiring, copper's most important application.

The low copper content in chalcopyrite must be enriched by removing the iron. The first step in copper extraction is pretreatment by flotation (see Figure 22.11, p. 974), which concentrates the ore to around 15% Cu by mass. The next step in many processing plants is a controlled roasting step, which oxidizes the FeS but not the CuS:

$$2FeCuS_2(s) + 3O_2(g) \longrightarrow 2CuS(s) + 2FeO(s) + 2SO_2(g)$$

To remove the FeO and convert the CuS to a more convenient form, the mixture is heated to 1100°C with sand and more of the concentrated ore. Several reactions occur in this step. The FeO reacts with sand to form a molten slag:

$$FeO(s) + SiO_2(s) \longrightarrow FeSiO_3(l)$$

The CuS is thermodynamically unstable at the elevated temperature and decomposes to yield Cu_2S, which is drawn off as a liquid.

In the final smelting step, the Cu_2S is roasted in air, which converts some of it to Cu_2O:

$$2Cu_2S(s) + 3O_2(g) \longrightarrow 2Cu_2O(s) + 2SO_2(g)$$

The two copper(I) compounds then react, with sulfide ion acting as the reducing agent:

$$Cu_2S(s) + 2Cu_2O(s) \longrightarrow 6Cu(l) + SO_2(g)$$

The copper obtained at this stage is usable for plumbing, but it must be purified further for electrical applications by removing unwanted impurities (Fe and Ni) as well as valuable ones (Ag, Au, and Pt). Purification is accomplished by *electrorefining,* which involves the oxidation of Cu and the formation of Cu^{2+} ions in solution, followed by their reduction and the plating out of Cu metal (Figure 22.18). The impure copper obtained from smelting is cast into plates to be used as anodes, and cathodes are made from already purified copper. The electrodes are immersed in acidified $CuSO_4$ solution, and a controlled voltage is applied that accomplishes two tasks simultaneously:

1. Copper and the more active impurities (Fe, Ni) are oxidized to their cations, while the less active ones (Ag, Au, Pt) are not. As the anode slabs react, these unoxidized metals fall off as a valuable "anode mud" and are purified separately. Sale of the precious metals in the anode mud nearly offsets the cost of electricity to operate the cell, making Cu wire inexpensive.
2. Because Cu is much less active than the Fe and Ni impurities, Cu^{2+} ions are reduced at the cathode, but Fe^{2+} and Ni^{2+} ions remain in solution:

$$Cu^{2+}(aq) + 2e^- \longrightarrow Cu(s) \qquad E^0 = 0.34\ V$$
$$Ni^{2+}(aq) + 2e^- \longrightarrow Ni(s) \qquad E^0 = -0.25\ V$$
$$Fe^{2+}(aq) + 2e^- \longrightarrow Fe(s) \qquad E^0 = -0.44\ V$$

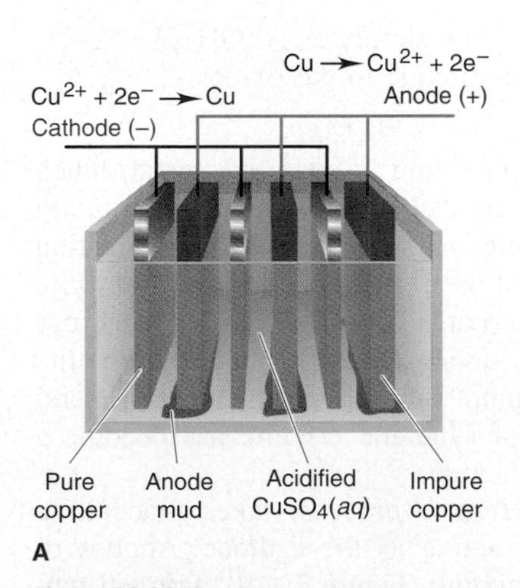

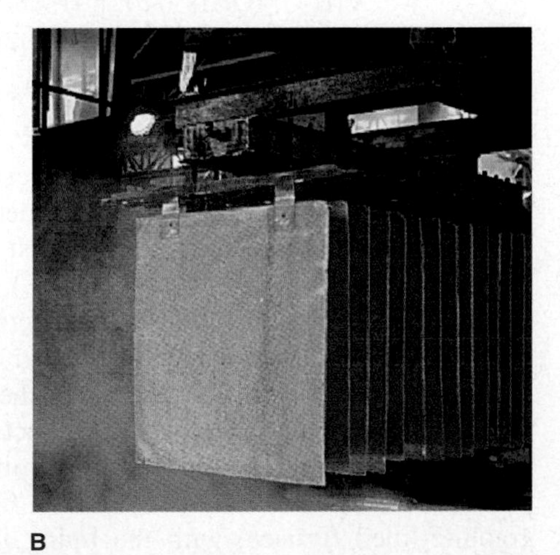

Figure 22.18 The electrorefining of copper. A, Copper is refined electrolytically, using impure slabs of copper as anodes and sheets of pure copper as cathodes. The Cu^{2+} ions released from the anode are reduced to Cu metal and plate out at the cathode. The "anode mud" contains valuable metal by-products. **B,** A small section of an industrial facility for electrorefining copper.

Isolation, Uses, and Recycling of Aluminum Aluminum is the most abundant metal in Earth's crust by mass and the third most abundant element (after O and Si). It is found in numerous aluminosilicate minerals (Gallery, pp. 580–581), such as feldspars, micas, and clays, and in the rare gems garnet, beryl, spinel, and turquoise. Corundum, pure aluminum oxide (Al_2O_3), is extremely hard; mixed with traces of transition metals, it exists as ruby and sapphire. Impure Al_2O_3 is used in sandpaper and other abrasives.

Mining bauxite.

Through eons of weathering, certain clays became *bauxite,* the major ore of aluminum. This mixed oxide-hydroxide occurs in enormous surface deposits in Mediterranean and tropical regions (see photo), but with world aluminum production approaching 100 million tons annually, it may someday be scarce. In addition to hydrated Al_2O_3 (about 75%), industrial-grade bauxite also contains Fe_2O_3, SiO_2, and TiO_2, which are removed during the extraction.

The overall two-step process combines hydro- and electrometallurgical techniques. In the first step, Al_2O_3 is separated from bauxite; in the second, it is converted to the metal.

1. *Isolating Al_2O_3 from bauxite.* After mining, bauxite is pretreated by extended boiling in 30% NaOH in the *Bayer process,* which involves acid-base, solubility, and complex-ion equilibria. The acidic SiO_2 and the amphoteric Al_2O_3 dissolve in the base, but the basic Fe_2O_3 and TiO_2 do not:

$$SiO_2(s) + 2NaOH(aq) + 2H_2O(l) \longrightarrow Na_2Si(OH)_6(aq)$$
$$Al_2O_3(s) + 2NaOH(aq) + 3H_2O(l) \longrightarrow 2NaAl(OH)_4(aq)$$
$$Fe_2O_3(s) + NaOH(aq) \longrightarrow \text{no reaction}$$
$$TiO_2(s) + NaOH(aq) \longrightarrow \text{no reaction}$$

Further heating slowly precipitates the $Na_2Si(OH)_6$ as an aluminosilicate, which is filtered out with the insoluble Fe_2O_3 and TiO_2 ("red mud").

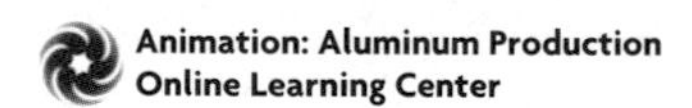
Animation: Aluminum Production
Online Learning Center

Acidifying the filtrate precipitates Al^{3+} as $Al(OH)_3$. Recall from our discussion of complex-ion equilibria that the aluminate ion, $Al(OH)_4^-(aq)$, is actually the complex ion $Al(H_2O)_2(OH)_4^-$, in which four of the six water molecules surrounding Al^{3+} have each lost a proton (see Figure 19.16, p. 849). Weakly acidic CO_2 is added to produce a small amount of H^+ ion, which reacts with this complex ion. Cooling supersaturates the solution, and the solid that forms is filtered out:

$$CO_2(g) + H_2O \rightleftharpoons H^+(aq) + HCO_3^-(aq)$$
$$Al(H_2O)_2(OH)_4^-(aq) + H^+(aq) \longrightarrow Al(H_2O)_3(OH)_3(s)$$

[Recall that we usually write $Al(H_2O)_3(OH)_3$ more simply as $Al(OH)_3$.]

Drying at high temperature converts the hydroxide to the oxide:

$$2Al(H_2O)_3(OH)_3(s) \xrightarrow{\Delta} Al_2O_3(s) + 9H_2O(g)$$

2. *Converting Al_2O_3 to the free metal.* Aluminum is an active metal, much too strong a reducing agent to be formed at the cathode from aqueous solution (Section 21.7), so the oxide itself must be electrolyzed. However, the melting point of Al_2O_3 is very high (2030°C), so it is dissolved in molten *cryolite* (Na_3AlF_6) to give a mixture that is electrolyzed at ~1000°C. Obviously, the use of cryolite provides a major energy (and cost) savings. The only sizeable cryolite mines are in Greenland, however, and they cannot supply enough natural mineral to meet the demand. Therefore, production of synthetic cryolite has become a major subsidiary industry in aluminum manufacture.

The electrolytic step, called the *Hall-Heroult process,* takes place in a graphite-lined furnace, with the lining itself acting as the cathode. Anodes of graphite dip into the molten Al_2O_3-Na_3AlF_6 mixture (Figure 22.19). The cell typically operates at a moderate voltage of 4.5 V, but with an enormous current flow of 1.0×10^5 to 2.5×10^5 A.

The process is complex and its details are still not entirely known. Therefore, the specific reactions shown below are chosen from among several other possibilities. Molten cryolite contains several ions (including AlF_6^{3-}, AlF_4^-, and F^-), which react with Al_2O_3 to form fluoro-oxy ions (including $AlOF_3^{2-}$, $Al_2OF_6^{2-}$, and $Al_2O_2F_4^{2-}$) that dissolve in the mixture. For example,

$$2Al_2O_3(s) + 2AlF_6^{3-}(l) \longrightarrow 3Al_2O_2F_4^{2-}(l)$$

Al forms at the cathode (reduction), shown here with AlF_6^{3-} as reactant:

$$AlF_6^{3-}(l) + 3e^- \longrightarrow Al(l) + 6F^-(l) \qquad \text{[cathode; reduction]}$$

The graphite anodes are oxidized and form carbon dioxide gas. Using one of the fluoro-oxy species as an example, the anode reaction is

$$Al_2O_2F_4^{2-}(l) + 8F^-(l) + C(\text{graphite}) \longrightarrow 2AlF_6^{3-}(l) + CO_2(g) + 4e^- \qquad \text{[anode; oxidation]}$$

Thus, the anodes are consumed in this half-reaction and must be replaced frequently.

Combining the three previous equations and making sure that e^- gained at the cathode equal e^- lost at the anode gives the overall reaction:

$$2Al_2O_3(\text{in } Na_3AlF_6) + 3C(\text{graphite}) \longrightarrow 4Al(l) + 3CO_2(g) \qquad \text{[overall (cell) reaction]}$$

The Hall-Heroult process uses an enormous quantity of energy: aluminum production accounts for more than 5% of total U.S. electrical usage.

Aluminum is a superb decorative, functional, and structural metal. It is lightweight, attractive, easy to work, and forms strong alloys. Although Al is very active, it does not corrode readily because of an adherent oxide layer that forms rapidly in air and prevents more O_2 from penetrating. Nevertheless, when it is in

Energy Received and Returned

Aluminum manufacture in the United States uses more electricity in 1 day than a city of 100,000 uses in 1 year! The reason for the high energy needs of Al manufacture is the electron configuration of Al ([Ne] $3s^23p^1$). Each Al^{3+} ion needs $3e^-$ to form an Al atom, and the atomic mass of Al is so low (~27 g/mol) that 1 mol of e^- produces only 9 g of Al. Compare this with the mass of Mg or Ca (two other lightweight structural metals) that 1 mol of e^- produces:

$$\tfrac{1}{3}Al^{3+} + e^- = \tfrac{1}{3}Al(s), \sim 9\text{ g}$$

$$\tfrac{1}{2}Mg^{2+} + e^- = \tfrac{1}{2}Mg(s), \sim 12\text{ g}$$

$$\tfrac{1}{2}Ca^{2+} + e^- = \tfrac{1}{2}Ca(s), \sim 20\text{ g}$$

An Al battery can turn this disadvantage around. Once produced, Al represents a concentrated form of electrical energy that can deliver 1 mol of e^- (96,500 C) for every 9 g of Al consumed in the battery. Thus, its electrical output per gram of metal is high. In fact, aluminum-air batteries are now being produced.

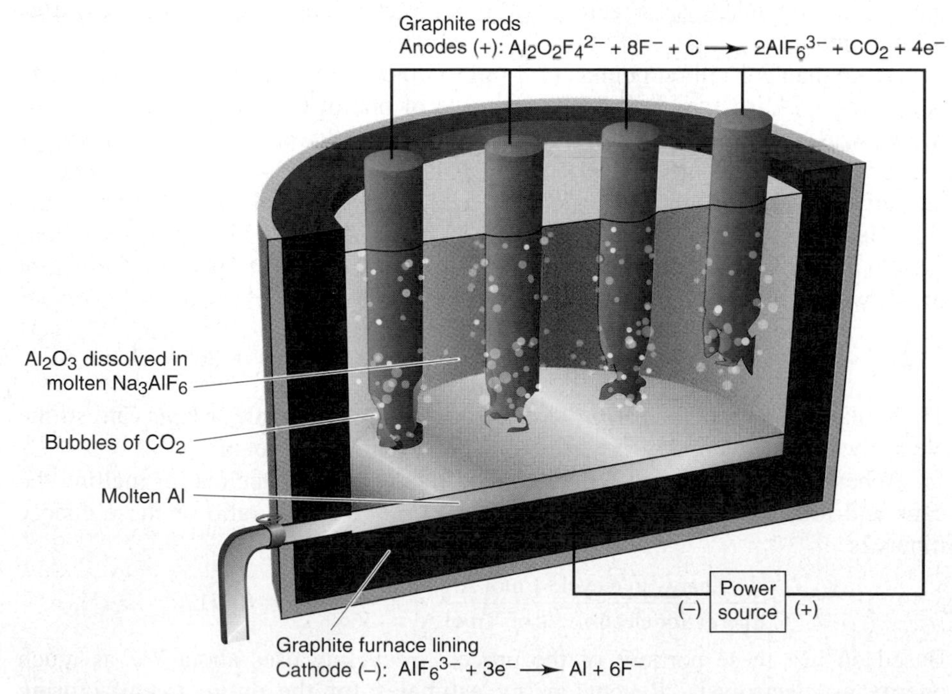

Figure 22.19 The electrolytic cell in the manufacture of aluminum. Purified Al_2O_3 is mixed with cryolite (Na_3AlF_6) and melted. Reduction at the graphite furnace lining (cathode) gives molten Al. Oxidation at the graphite rods (anodes) slowly converts them to CO_2, so they must be replaced periodically.

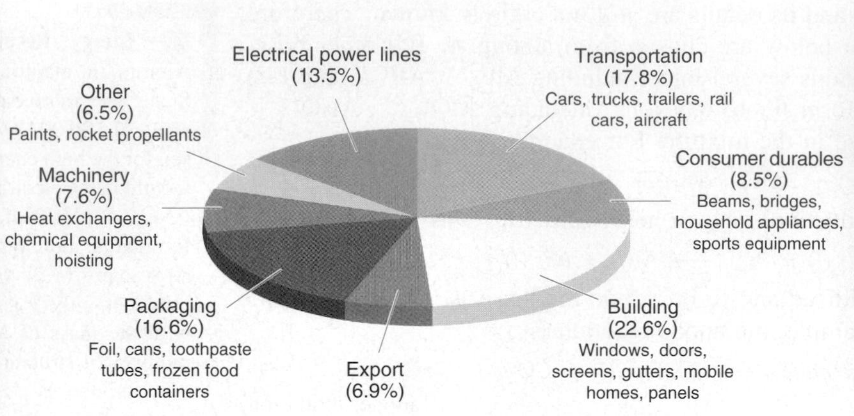

Figure 22.20 The many familiar and essential uses of aluminum.

contact with less active metals such as Fe, Cu, and Pb, aluminum becomes the anode and deteriorates rapidly (Section 21.6). To prevent this, aluminum objects are often *anodized,* that is, made to act as the anode in an electrolysis that coats them with an oxide layer. The object is immersed in a 20% H_2SO_4 bath and connected to a graphite cathode:

$$6H^+(aq) + 6e^- \longrightarrow 3H_2(g) \quad \text{[cathode; reduction]}$$
$$2Al(s) + 3H_2O(l) \longrightarrow Al_2O_3(s) + 6H^+(aq) + 6e^- \quad \text{[anode; oxidation]}$$
$$2Al(s) + 3H_2O(l) \longrightarrow Al_2O_3(s) + 3H_2(g) \quad \text{[overall (cell) reaction]}$$

The Al_2O_3 layer deposited is typically from 10 μ to 100 μ thick, depending on the object's intended use. Figure 22.20 is a pie chart showing the countless uses of aluminum.

More than 3.5 billion pounds (1.5 million metric tons) of aluminum cans and packaging are discarded each year—a waste of one of the most useful materials in the world *and* the energy used to make it. A quick calculation of the energy needed to prepare 1 mol of Al from *purified* Al_2O_3, compared with the energy needed for recycling, conveys a clear message. The overall cell reaction in the Hall-Heroult process has a ΔH^0 of 2272 kJ and a ΔS^0 of 635.4 J/K. If we consider *only* the standard free energy change of the reaction at 1000.°C, for 1 mol of Al, we obtain

$$\Delta G^0 = \Delta H^0 - T\Delta S^0 = \frac{2272\text{ kJ}}{4\text{ mol Al}} - \left(1273\text{ K} \times \frac{0.6354\text{ kJ/K}}{4\text{ mol Al}}\right) = 365.8\text{ kJ/mol Al}$$

The molar mass of Al is nearly twice the mass of a soft-drink or beer can, so the electrolysis step requires nearly 200 kJ of energy for each can!

When aluminum is recycled, the major energy input (which is for melting the cans and foil) has been calculated as ~26 kJ/mol Al. The ratio of these energy inputs is

$$\frac{\text{Energy to recycle 1 mol Al}}{\text{Energy for electrolysis of 1 mol Al}} = \frac{26\text{ kJ}}{365.8\text{ kJ}} = 0.071$$

Based on just these portions of the process, recycling uses about 7% as much energy as electrolysis. Recent energy estimates for the entire manufacturing process (including mining, pretreating, maintaining operating conditions, electrolyzing, and so forth) are about 6000 kJ/mol Al, which means recycling requires less than 1% as much energy as manufacturing! The economic advantages, not to mention the environmental ones, are obvious, and recycling of aluminum has become common in the United States (see photo).

Crushed aluminum cans ready for recycling.

Mining the Sea: Magnesium and Bromine

In the not too distant future, as terrestrial sources of certain elements become scarce or too costly to mine, the oceans will become an important source. Despite the abundant distribution of magnesium on land, it is already being obtained from the sea, and its ocean-based production is a good example of the approach we might one day use for other elements. Bromine, too, is recovered from the sea as well as from inland brines.

Isolation and Uses of Magnesium The *Dow process* for the isolation of magnesium from the sea involves steps that are similar to the procedures used for rocky ores (Figure 22.21):

1. *Mining*. Intake of seawater and straining the debris are the "mining" steps. No pretreatment is needed.
2. *Converting to mineral*. The dissolved Mg^{2+} ion is converted to the mineral $Mg(OH)_2$ with $Ca(OH)_2$, which is generated on-site (at the plant). Seashells ($CaCO_3$) are crushed, decomposed with heat to CaO, and mixed with water to make slaked lime [$Ca(OH)_2$]. This is pumped into the seawater intake tank to precipitate the dissolved Mg^{2+} as the hydroxide ($K_{sp} \approx 10^{-9}$):
$$Ca(OH)_2(aq) + Mg^{2+}(aq) \longrightarrow Mg(OH)_2(s) + Ca^{2+}(aq)$$
3. *Converting to compound*. The solid $Mg(OH)_2$ is filtered and mixed with excess HCl, which is also made on-site, to form aqueous $MgCl_2$:
$$Mg(OH)_2(s) + 2HCl(aq) \longrightarrow MgCl_2(aq) + 2H_2O(l)$$
The water is evaporated in stages to give solid, hydrated $MgCl_2{\cdot}nH_2O$.
4. *Electrochemical redox*. Heating above 700°C drives off the water of hydration and melts the $MgCl_2$. Electrolysis gives chlorine gas and the molten metal, which floats on the denser molten salt:
$$MgCl_2(l) \longrightarrow Mg(l) + Cl_2(g)$$
The Cl_2 that forms is recycled to make the HCl used in step 3.

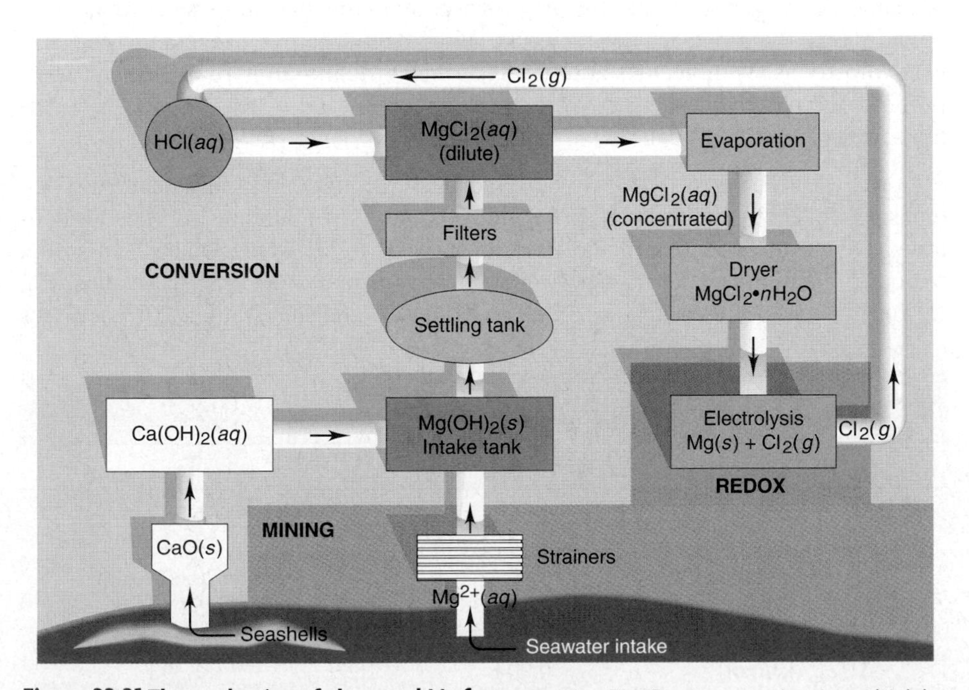

Figure 22.21 The production of elemental Mg from seawater. The Dow process involves obtaining Mg^{2+} ion from seawater; the Mg^{2+} is converted to $Mg(OH)_2$ and then to aqueous $MgCl_2$, which is evaporated to dryness and then electrolytically reduced to Mg metal and Cl_2 gas.

Jet engine assembly.

Magnesium is the lightest structural metal available (about one-half the density of Al and one-fifth that of steel). Although Mg is quite reactive, it forms an extremely adherent, high-melting oxide layer (MgO); thus, it finds many uses in metal alloys and can be machined into any form. Magnesium alloys occur in everything from aircraft bodies to camera bodies, and from luggage to auto engine blocks. The pure metal is a strong reducing agent, which makes it useful for sacrificial anodes (Section 21.6) and in the metallurgical extraction of other metals, such as Be, Ti, Zr, Hf, and U. For example, titanium, an essential component of jet engines (see photo), is made from its major ore, ilmenite, in two steps:

$$(1)\quad 2FeTiO_3(s) + 7Cl_2(g) + 6C(s) \longrightarrow 2TiCl_4(l) + 2FeCl_3(s) + 6CO(g)$$

$$(2)\quad TiCl_4(l) + 2Mg(l) \longrightarrow Ti(s) + 2MgCl_2(l)$$

Isolation and Uses of Bromine The largest source of bromine is the oceans, where it occurs as Br^- at a concentration of 0.065 g/L (65 ppm). However, the cost of concentrating such a dilute solution shifts the choice of source, when available, to much more concentrated salt lakes and natural brines. Arkansas brines, the most important source in the United States, contain about 4.5 g Br^-/L (4500 ppm). At standard-state conditions, the Br^- is readily oxidized to Br_2 with aqueous Cl_2:

$$2Br^-(aq) + Cl_2(aq) \longrightarrow Br_2(l) + 2Cl^-(aq) \qquad \Delta G^0 = -61.5 \text{ kJ}$$

The Br_2 is removed by passing steam through the mixture and then cooling and drying the liquid bromine. Iodide is present in these brines also, and it is oxidized to I_2 by aqueous Cl_2 in a similar way.

World production of Br_2 is about 1% that of Cl_2. Its major use is in the preparation of organic chemicals, such as ethylene dibromide, which was added to leaded gasoline to prevent lead oxides from forming and depositing on engine parts. Newer products requiring bromine include the flame retardants in rugs and textiles. Bromine is also needed in the synthesis of inorganic chemicals, especially silver bromide for photographic emulsions.

The Many Sources and Uses of Hydrogen

Although hydrogen accounts for 90% of the atoms in the universe, it makes up only 15% of the atoms in Earth's crust. Moreover, whereas it occurs in the universe mostly as H_2 molecules and free H atoms, virtually all hydrogen in the crust is combined with other elements, either oxygen in natural waters or carbon in biomass, petroleum, and coal.

Properties of the Hydrogen Isotopes Hydrogen has three naturally occurring isotopes. Ordinary hydrogen (^1H), or protium, is the most abundant and has no neutrons. Deuterium (^2H or D), the next in abundance, has one neutron, and rare, radioactive tritium (^3H or T) has two. All three occur as diatomic molecules, H_2, D_2 (or $^2H_2$), and T_2 (or $^3H_2$), as well as HD, DT, and HT. Table 22.5 compares

Table 22.5 Some Molecular and Physical Properties of Diatomic Protium, Deuterium, and Tritium

Property	H_2	D_2	T_2
Molar mass (g/mol)	2.016	4.028	6.032
Bond length (pm)	74.14	74.14	74.14
Melting point (K)	13.96	18.73	20.62
Boiling point (K)	20.39	23.67	25.04
ΔH^0_{fus} (kJ/mol)	0.117	0.197	0.250
ΔH^0_{vap} (kJ/mol)	0.904	1.226	1.393
Bond energy (kJ/mol at 298 K)	432	443	447

some molecular and physical properties of H_2, D_2, and T_2. As you can see, the heavier the isotope is, the higher the molar mass of the molecule, and the higher its melting point, boiling point, and heats of phase change; it also effuses at a lower rate (Section 5.6).

Because hydrogen is so light, the relative difference in mass of its isotopes is enormous compared to that of isotopes of other common elements. (For example, the mass of ^{13}C is only 8% greater than that of ^{12}C.) Note, especially, that the mass difference leads to different bond energies, which affect reactivity. Because the mass of D is twice the mass of H, H atoms bonded to a given atom vibrate at a higher frequency than do D atoms, so their bonds are higher in energy. As a result, the *bond to H is weaker and thus quicker to break.* Therefore, any reaction that includes breaking a bond to hydrogen in the rate-determining step occurs *faster with H than with D*. This phenomenon is called a *kinetic isotope effect;* we'll see an example shortly. No other element displays a kinetic isotope effect nearly as large.

Industrial Production of Hydrogen Hydrogen gas (H_2) is produced on an industrial scale worldwide: 250,000 metric tons (3×10^{12} L at STP; equivalent to $\sim1\times10^{11}$ mol) are produced annually in the United States alone. All production methods are energy intensive, so the choice is determined by energy costs. In Scandinavia, where hydroelectric power is plentiful, electrolysis is the chosen method; on the other hand, where natural gas from oil refineries is plentiful, as in the United States and Great Britain, thermal methods are used to produce hydrogen.

The most common thermal methods use water and a simple hydrocarbon in two steps. Modern U.S. plants use methane, which has the highest H:C ratio of any hydrocarbon. In the first step, the reactants are heated to around 1000°C over a nickel-based catalyst in the endothermic *steam-reforming process:*

$$CH_4(g) + H_2O(g) \longrightarrow CO(g) + 3H_2(g) \qquad \Delta H = 206 \text{ kJ}$$

Heat is supplied by burning methane at the refinery. To generate more H_2, the product mixture (called *water gas*) is heated with steam at 400°C over an iron or cobalt oxide catalyst in the exothermic *water-gas shift reaction,* and the CO reacts:

$$H_2O(g) + CO(g) \rightleftharpoons CO_2(g) + H_2(g) \qquad \Delta H = -41 \text{ kJ}$$

The reaction mixture is recycled several times, which decreases the CO to around 0.2% by volume. Passing the mixture through liquid water removes the more soluble CO_2 (solubility = 0.034 mol/L) from the H_2 (solubility < 0.001 mol/L). Calcium oxide can also be used to remove CO_2 by formation of $CaCO_3$. By removing CO_2, these steps shift the equilibrium position to the right and produce H_2 that is about 98% pure. To attain greater purity (~99.9%), the gas mixture is passed through a *synthetic zeolite* (see the Gallery, p. 581) selected to filter out nearly all molecules larger than H_2.

In regions where inexpensive electricity is available, very pure H_2 is prepared through electrolysis of water with Pt (or Ni) electrodes:

$$2H_2O(l) + 2e^- \longrightarrow H_2(g) + 2OH^-(aq) \qquad E = -0.42 \text{ V} \qquad \text{[cathode; reduction]}$$
$$H_2O(l) \longrightarrow \tfrac{1}{2}O_2(g) + 2H^+(aq) + 2e^- \qquad E = 0.82 \text{ V} \qquad \text{[anode; oxidation]}$$
$$H_2O(l) \longrightarrow H_2(g) + \tfrac{1}{2}O_2(g) \qquad E_{cell} = -0.42 \text{ V} - 0.82 \text{ V} = -1.24 \text{ V}$$

Overvoltage makes the cell potential about −2 V (Section 21.7). Therefore, under typical operating conditions, it takes about 400 kJ of energy to produce 1 mol of H_2:

$$\Delta G = -nFE = (-2 \text{ mol e}^-)\left(\frac{96.5 \text{ kJ}}{\text{V·mol e}^-}\right)(-2 \text{ V}) = 4\times10^2 \text{ kJ/mol } H_2$$

High-purity O_2 is a valuable by-product of this process that offsets some of the costs of isolating H_2.

Calcium reacts with water to form H_2.

Laboratory Production of Hydrogen You may already have produced small amounts of H_2 in the lab by one of several methods. A characteristic reaction of the very active metals in Groups 1A(1) and 2A(2) is the reduction of water to H_2 and OH^- (see photo):

$$Ca(s) + 2H_2O(l) \longrightarrow Ca^{2+}(aq) + 2OH^-(aq) + H_2(g)$$

Alternatively, the strongly reducing hydride ion can be used:

$$NaH(s) + H_2O(l) \longrightarrow Na^+(aq) + OH^-(aq) + H_2(g)$$

Less active metals reduce H^+ in acids:

$$Zn(s) + 2H^+(aq) \longrightarrow Zn^{2+}(aq) + H_2(g)$$

In view of hydrogen's great potential as a fuel, chemists are seeking ways to decompose water by lowering the overall activation energy of the process. More than 10,000 thermodynamically *and* kinetically favorable *water-splitting* schemes have been devised.

However, at 298 K, ΔG^0 for the decomposition of 1 mol of $H_2O(g)$ is 229 kJ, so all water-splitting schemes require energy. Given the increase in the number of moles of gas, ΔS^0 is positive; therefore, this is an endothermic process ($\Delta H^0 > 0$) and, at standard-state conditions, requires heat to occur spontaneously. One promising source of this heat is focused sunlight.

Industrial Uses of Hydrogen Typically, a plant produces H_2 to make some other end product. In fact, more than 95% of H_2 produced industrially is consumed on-site in ammonia or petrochemical facilities. In a plant that synthesizes NH_3 from N_2 and H_2, the reactant gases are formed through a series of reactions that involve methane, including the steam-reforming and water-gas shift reactions discussed previously. For this reason, the cost of NH_3 is closely correlated with that of CH_4.

Here is a typical example of how the reaction series in an ammonia plant works. The steam-reforming reaction is performed with excess CH_4, which depletes the reaction mixture of H_2O:

$$CH_4(g;\text{ excess}) + H_2O(g) \longrightarrow CO(g) + 3H_2(g)$$

In the next step, an excess of the product mixture (CH_4, CO, and H_2) is burned in an amount of air ($N_2 + O_2$) that is insufficient to effect complete combustion, but is just enough to consume the O_2, heat the mixture to 1100°C, and form additional H_2O:

$$4CH_4(g;\text{ excess}) + 7O_2(g) \longrightarrow 2CO_2(g) + 2CO(g) + 8H_2O(g)$$
$$2H_2(g;\text{ excess}) + O_2(g) \longrightarrow 2H_2O(g)$$
$$2CO(g;\text{ excess}) + O_2(g) \longrightarrow 2CO_2(g)$$

Any remaining CH_4 reacts by the steam-reforming reaction, the remaining CO reacts by the water-gas shift reaction ($H_2O + CO \rightleftharpoons CO_2 + H_2$) to form more H_2, and then the CO_2 is removed with CaO. The amounts are carefully adjusted to produce a final mixture that contains a 1:3 ratio of N_2 (from the added air) to H_2 (with traces of CH_4, Ar, and CO). This mixture is used directly in the synthesis of ammonia, described in Chapter 17 (see the Chemical Connections essay, pp. 755–756).

A second major use of H_2 is in *hydrogenation* of the C=C bonds in liquid oils to form the C—C bonds in solid fats and margarine. The process uses H_2 in contact with transition metal catalysts, such as powdered nickel (see Figure 16.24, p. 708). Solid fats are also used in commercial baked goods. Look at the list of ingredients on most packages of bread, cake, and cookies, and you'll see the "partially hydrogenated vegetable oils" made by this process.

Hydrogen is also essential in the manufacture of numerous "bulk" chemicals, those that are produced in large amounts because they have many further uses. One application that has been gaining great attention is the production of methanol. In this process, carbon monoxide reacts with hydrogen over a copper–zinc oxide catalyst:

$$CO(g) + 2H_2(g) \xrightarrow{\text{Cu-ZnO catalyst}} CH_3OH(l)$$

Many automotive engineers expect methanol to be used as a gasoline additive.

Of course, as less expensive sources of hydrogen become available, it will be used directly in fuel cells (Section 21.5).

Production and Uses of Deuterium Deuterium and its compounds are produced from D_2O (heavy water), which is present as a minor component (0.016 mol % D_2O) in normal water and is isolated on the multiton scale by *electrolytic enrichment*. This process is based on the kinetic isotope effect for hydrogen noted earlier, specifically on the *higher rate of bond breaking for O—H bonds compared to O—D bonds*, and thus on the higher rate of electrolysis of H_2O compared with D_2O.

For example, with Pt electrodes, H_2O is electrolyzed about 14 times faster than D_2O. As some of the liquid decomposes to the elemental gases, the remainder becomes enriched in D_2O. Thus, by the time the volume of water has been reduced to 1/20,000 of its original volume, the remaining water is around 99% D_2O. By combining samples and repeating the electrolysis, more than 99.9% D_2O is obtained.

Deuterium gas is produced by electrolysis of D_2O or by any of the chemical reactions that produce hydrogen gas from water, such as

$$2Na(s) + 2D_2O(l) \longrightarrow 2Na^+(aq) + 2OD^-(aq) + D_2(g)$$

Compounds containing deuterium (or tritium) are produced from reactions that give rise to the corresponding hydrogen-containing compound; for example,

$$SiCl_4(l) + 2D_2O(l) \longrightarrow SiO_2(s) + 4DCl(g)$$

Compounds with acidic protons undergo hydrogen/deuterium exchange:

$$CH_3COOH(l) + D_2O(l;\ \text{excess}) \longrightarrow CH_3COOD(l) + DHO(l;\ \text{small amount})$$

Notice that only the acidic H atom, the one in the COOH group, is exchanged, not any of those attached to carbon. We discuss the natural and synthetic formation of tritium in Chapter 24.

SECTION SUMMARY

Production highlights of key elements are as follows:

- Na is isolated by electrolysis of molten NaCl in the Downs process; Cl_2 is a by-product.
- K is produced by reduction with Na in a thermal process.
- Fe is produced through a multistep high-temperature process in a blast furnace. The crude pig iron is converted to carbon steel in the basic-oxygen process and then alloyed with other metals to make different steels.
- Cu is produced by concentration of the ore through flotation, reduction to the metal by smelting, and purification by electrorefining. The metal has extensive electrical and plumbing uses.
- Al is extracted from bauxite by pretreating the ore with concentrated base, followed by electrolysis of the Al_2O_3 in molten cryolite. Al alloys are widely used in homes and industry. The total energy required to extract Al from its ore is over 100 times that needed for recycling it.
- The Mg^{2+} in seawater is converted to $Mg(OH)_2$ and then to $MgCl_2$, which is electrolyzed to obtain the metal; Mg forms strong, lightweight alloys.
- Br_2 is obtained from brines by oxidizing Br^- with Cl_2.
- H_2 is produced by electrolysis of water or in the formation of gaseous fuels from hydrocarbons. It is used in NH_3 production and in hydrogenation of vegetable oils. The isotopes of hydrogen differ significantly in atomic mass and thus in the rate at which their bonds to other atoms break. This difference is used to obtain D_2O from water.

22.5 CHEMICAL MANUFACTURING: TWO CASE STUDIES

From laboratory-scale endeavors of the first half of the 19th century, today's chemical industries have grown to multinational corporations producing materials that define modern life: polymers, electronics, pharmaceuticals, consumer goods, and fuels. In this final section, we examine the interplay of theory and practice in two of the most important processes in the inorganic chemical industry: (1) the contact process for the production of sulfuric acid, and (2) the chlor-alkali process for the production of chlorine.

Sulfuric Acid, the Most Important Chemical

The manufacture of sulfuric acid began more than 400 years ago, when the acid was known as *oil of vitriol* and was distilled from "green vitriol" ($FeSO_4{\cdot}7H_2O$). Considering its countless uses, it is not surprising that today sulfuric acid is produced throughout the world on a gigantic scale—more than 150 million tons a year. The green vitriol method and the many production methods used since have all been superseded by the modern **contact process,** which is based on the *catalyzed oxidation of SO_2*. Here are the key steps.

1. *Obtaining sulfur.* In most countries today, the production of sulfuric acid starts with the production of elemental sulfur, often by the *Claus process,* in which the H_2S in "sour" natural gas is chemically separated and then oxidized:

$$2H_2S(g) + 2O_2(g) \xrightarrow{\text{low temperature}} \tfrac{1}{8}S_8(g) + SO_2(g) + 2H_2O(g)$$

$$2H_2S(g) + SO_2(g) \xrightarrow{Fe_2O_3\text{ catalyst}} \tfrac{3}{8}S_8(g) + 2H_2O(g)$$

Where natural gas is not abundant but natural underground deposits of the element are, sulfur is obtained by the *Frasch process,* a nonchemical method that taps these deposits. A hole is drilled to the deposit, and superheated water (about 160°C) is pumped down two outer concentric pipes to melt the sulfur (Figure 22.22A). Then, a combination of the hydrostatic pressure in the outermost pipe and the pressure of compressed air sent through a narrow inner pipe forces the sulfur to the surface (Figure 22.22B and C). The costs of drilling, pumping, and supplying water ($5{\times}10^6$ gallons per day) are balanced somewhat by the fact that the product is very pure (~99.7% S).

2. *From sulfur to sulfur dioxide.* Once obtained, the sulfur is burned in air to form SO_2:

$$\tfrac{1}{8}S_8(s) + O_2(g) \longrightarrow SO_2(g) \qquad \Delta H^0 = -297\text{ kJ}$$

Some SO_2 is also obtained from the roasting of metal sulfide ores. About 90% of processed sulfur is used in making sulfur dioxide for production of the all-important sulfuric acid. Indeed, this end product from sulfur is central to so many chemical industries that a nation's level of sulfur production is a reliable indicator of its overall industrial capacity: the United States, Russia, Japan, and Germany are the top four sulfur producers.

3. *From sulfur dioxide to trioxide.* The contact process oxidizes SO_2 with O_2 to SO_3:

$$SO_2(g) + \tfrac{1}{2}O_2(g) \rightleftharpoons SO_3(g) \qquad \Delta H^0 = -99\text{ kJ}$$

The reaction is *exothermic* and very *slow* at room temperature. From Le Châtelier's principle (Section 17.6), we know that the yield of SO_3 can be increased by (1) changing the temperature, (2) increasing the pressure (more moles of gas are on the left than on the right), or (3) adjusting the concentrations (adding excess O_2 and removing SO_3).

First, let's examine the temperature effect. Adding heat (raising the temperature) increases the frequency of SO_2-O_2 collisions and thus increases the *rate* of SO_3 formation. However, because the formation of SO_3 is exothermic, removing heat (lowering the temperature) shifts the equilibrium position to the right and

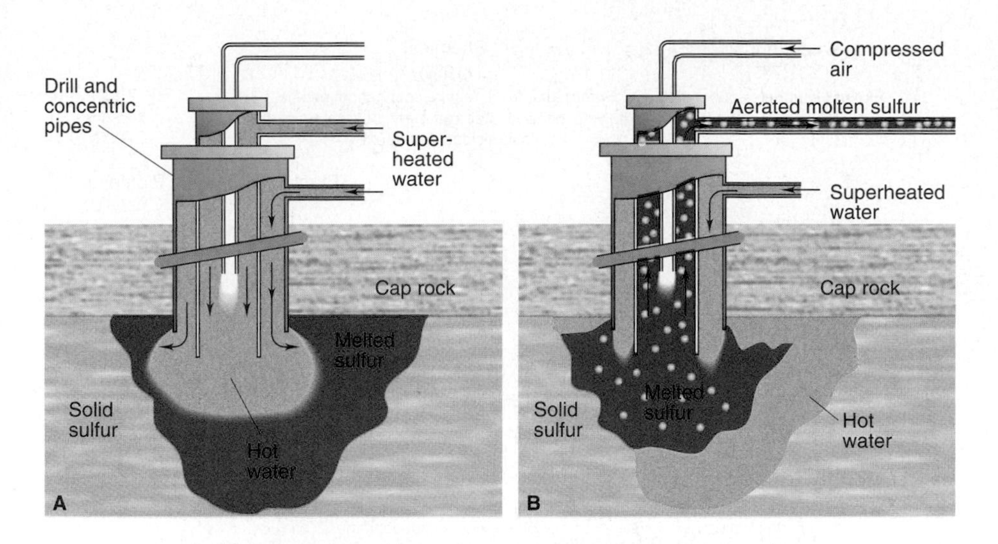

Figure 22.22 The Frasch process for mining elemental sulfur. This nonchemical process uses superheated water and compressed air to melt underground sulfur and bring it to the surface. **A,** Superheated water is forced down a drilled hole into the subterranean sulfur deposit, and the sulfur melts. **B,** Compressed air is sent down an inner pipe in the drilling apparatus to force up the molten sulfur. **C,** Sulfur obtained from the Frasch process.

thus increases the *yield* of SO_3. This is a classic situation that calls for use of a catalyst. By lowering the activation energy, *a catalyst allows equilibrium to be reached more quickly and at a lower temperature;* thus, rate *and* yield are optimized (Section 16.8). The catalyst in the contact process is V_2O_5 on inert silica, which is active between 400°C and 600°C.

The pressure effect is small and economically not worth exploiting. The concentration effects are controlled by providing an excess of O_2 in the form of a 5:1 mixture of air:SO_2, or about 1:1 O_2:SO_2, about twice as much O_2 as is called for by the reaction stoichiometry. The mixture is passed over catalyst beds in four stages, and the SO_3 is removed at several points to favor more SO_3 formation. The overall yield of SO_3 is 99.5%.

4. *From sulfur trioxide to acid.* Sulfur trioxide is the anhydride of sulfuric acid, so a hydration step is next. However, SO_3 cannot be added to water because, at the operating temperature, it would first meet water vapor, which catalyzes its polymerization to $(SO_3)_x$, and results in a smoke of solid particles that makes poor contact with water and yields little acid. To prevent this, previously formed H_2SO_4 absorbs the SO_3 and forms pyrosulfuric acid (or disulfuric acid, $H_2S_2O_7$; see margin), which is then hydrolyzed with sufficient water:

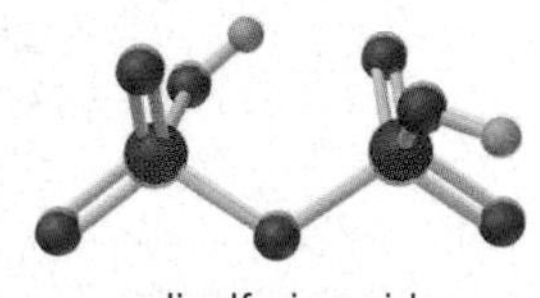

disulfuric acid

$$SO_3(g) + H_2O(l) \longrightarrow H_2SO_4(l)\text{–[low yield]}$$
$$SO_3(g) + H_2SO_4(l) \longrightarrow H_2S_2O_7(l)$$
$$H_2S_2O_7(l) + H_2O(l) \longrightarrow 2H_2SO_4(l)$$

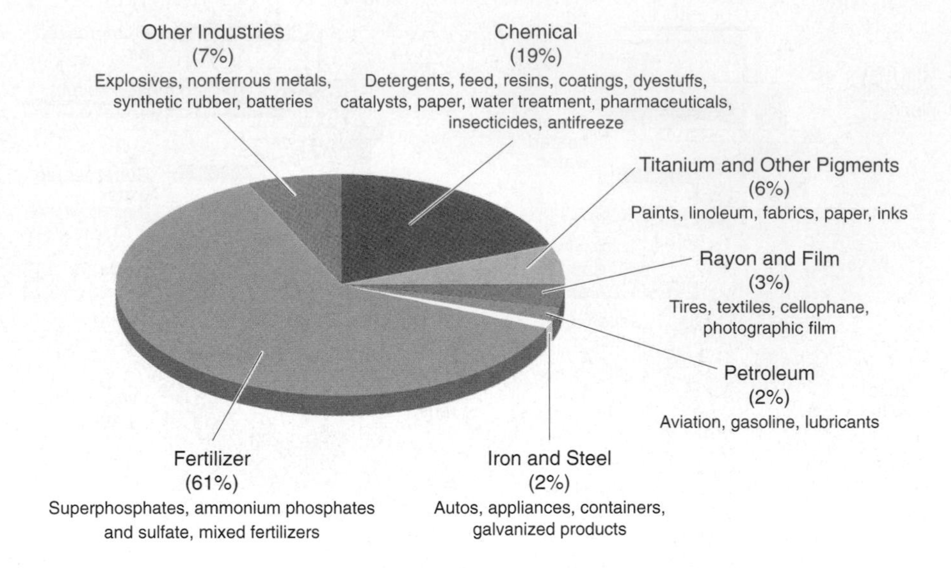

Figure 22.23 The many indispensable applications of sulfuric acid.

The uses of sulfuric acid are legion, as Figure 22.23 indicates.

Sulfuric acid is remarkably inexpensive (about \$150/ton), largely because each step in the process is exothermic—burning S ($\Delta H^0 = -297$ kJ/mol), oxidizing SO_2 ($\Delta H^0 = -99$ kJ/mol), hydrating SO_3 ($\Delta H^0 = -132$ kJ/mol)—and the heat is a valuable by-product. Three-quarters of the heat is sold as steam, and the rest of it is used to pump gases through the plant. A typical plant making 825 tons of H_2SO_4 per day produces enough steam to generate 7×10^6 watts of electric power.

The Chlor-Alkali Process

Chlorine is produced and used in amounts many times greater than all the other halogens combined, ranking among the top 10 chemicals produced in the United States. Its production methods depend on *the oxidation of Cl^- ion from NaCl.*

Earlier, we discussed the Downs process for isolating sodium, which yields Cl_2 gas as the other product (Section 22.4). The **chlor-alkali process,** which forms the basis of one of the largest inorganic chemical industries, electrolyzes concentrated aqueous NaCl to produce Cl_2 and several other important chemicals. As you learned in Section 21.7, the electrolysis of aqueous NaCl does not yield both of the component elements. Because of the effects of overvoltage, Cl^- ions rather than H_2O molecules are oxidized at the anode. However, Na^+ ions are not reduced at the cathode because the half-cell potential (−2.71 V) is much more negative than that for reduction of H_2O (−0.42 V), even with the normal overvoltage (around −0.6 V). Therefore, the half-reactions for electrolysis of aqueous NaCl are

$$2Cl^-(aq) \longrightarrow Cl_2(g) + 2e^- \qquad E^0 = 1.36\text{ V} \quad \text{[anode; oxidation]}$$

$$2H_2O(l) + 2e^- \longrightarrow 2OH^-(aq) + H_2(g) \qquad E \approx -1.0\text{ V} \quad \text{[cathode; reduction]}$$

$$2Cl^-(aq) + 2H_2O(l) \longrightarrow 2OH^-(aq) + H_2(g) + Cl_2(g) \qquad E_{cell} = -1.0\text{ V} - 1.36\text{ V} = -2.4\text{ V}$$

To obtain commercially meaningful amounts of Cl_2, however, a voltage almost twice this value and a current in excess of 3×10^4 A are used.

When we include the spectator ion Na^+, the total ionic equation shows another important product made by the process:

$$2Na^+(aq) + 2Cl^-(aq) + 2H_2O(l) \longrightarrow 2Na^+(aq) + 2OH^-(aq) + H_2(g) + Cl_2(g)$$

As Figure 22.24 shows, the sodium salts in the cathode compartment exist as an aqueous mixture of NaCl and NaOH; the NaCl is removed by fractional crystallization. Thus, in this version of the chlor-alkali process, which uses an *asbestos*

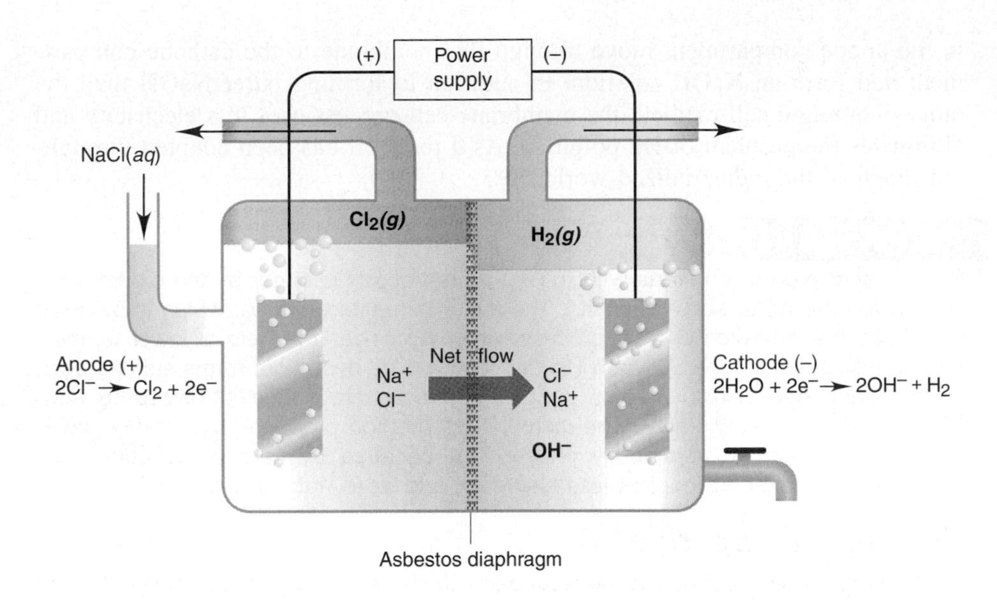

Figure 22.24 A diaphragm cell for the chlor-alkali process. This process uses concentrated aqueous NaCl to make NaOH, Cl_2, and H_2 in an electrolytic cell. The difference in liquid level between compartments keeps a net movement of solution into the cathode compartment, which prevents reaction between OH^- and Cl_2. The cathode electrolyte is concentrated and fractionally crystallized to give industrial-grade NaOH.

diaphragm to separate the anode and cathode compartments, electrolysis of NaCl brines yields Cl_2, H_2, and industrial-grade NaOH, an important base. Like other reactive products, H_2 and Cl_2 are kept apart to prevent explosive recombination. Note the higher liquid level in the anode compartment. This slight hydrostatic pressure difference minimizes backflow of NaOH, which prevents the disproportionation (self–oxidation-reduction) reactions of Cl_2 that occur in the presence of OH^- (Section 14.9), such as

$$Cl_2(g) + 2OH^-(aq) \longrightarrow Cl^-(aq) + ClO^-(aq) + H_2O(l)$$

If high-purity NaOH is desired, a slightly different version, called the chlor-alkali *mercury-cell* process, is employed. Mercury is used as the cathode, which creates such a large overvoltage for reduction of H_2O to H_2 that the process *does* favor reduction of Na^+. The sodium dissolves in the mercury to form sodium amalgam, Na(Hg). In the mercury-cell version, the half-reactions are

$$2Cl^-(aq) \longrightarrow Cl_2(g) + 2e^- \qquad \text{[anode; oxidation]}$$

$$2Na^+(aq) + 2e^- \xrightarrow{\text{Hg}} 2Na(Hg) \qquad \text{[cathode; reduction]}$$

To obtain sodium hydroxide, the sodium amalgam is pumped out of the system and treated with H_2O, which is reduced by the Na:

$$2Na(Hg) + 2H_2O(l) \xrightarrow{-\text{Hg}} 2Na^+(aq) + 2OH^-(aq) + H_2(g)$$

The mercury released in this step is recycled back to the electrolysis bath. Therefore, *the products are the same in both versions,* but the purity of NaOH from the mercury-cell version is much higher.

Despite the formation of purer NaOH, the mercury-cell method is being steadily phased out in the United States and has been eliminated in Japan for more than a decade. Cost is not the major reason for the phase-out, although the method does consume about 15% more electricity than the diaphragm-cell version. Rather, the problem is that as the mercury is recycled, some is lost in the industrial wastewater. On average, 200 g of Hg is lost per ton of Cl_2 produced. In the 1980s, annual U.S. production via the mercury-cell method was 2.75 million tons of Cl_2; thus, during that decade, 550,000 kg of this toxic heavy metal was flowing into U.S. waterways each year!

The newer chlor-alkali *membrane-cell* process replaces the diaphragm with a polymeric membrane to separate the cell compartments. The membrane allows only cations to move through it and only from anode to cathode compartments. Thus, as Cl^- ions are removed at the anode through oxidation to Cl_2, Na^+ ions in the anode compartment move through the membrane to the cathode compartment and form an NaOH solution. In addition to forming purer NaOH than the older diaphragm-cell method, the membrane-cell process uses less electricity and eliminates the problem of Hg pollution. As a result, it has been adopted throughout much of the industrialized world.

SECTION SUMMARY

Sulfuric acid production starts with the extraction of sulfur, either by the oxidation of H_2S or the mining of sulfur deposits. The sulfur is roasted to SO_2, which is oxidized to SO_3 by the catalyzed contact process, which optimizes the yield at lower temperatures. Absorption of the SO_3 into H_2SO_4, followed by hydration, forms sulfuric acid. In the diaphragm-cell chlor-alkali process, aqueous NaCl is electrolyzed to form Cl_2, H_2, and low-purity NaOH. The mercury-cell method produces high-purity NaOH but has been almost completely phased out because of mercury pollution. The membrane-cell method requires less electricity and does not use Hg.

Chapter Perspective

In this chapter, we took a realistic approach to the chemistry of the elements that allowed a glimpse at the intertwining natural cycles in which various elements take part, as well as at the ingenious methods that have been developed to extract them. The impact of living systems in general, and of humans in particular, on the origin and distribution of elements reveals a complex, evolving relationship. Understanding of and respect for that relationship are required for a healthy, productive balance between society's needs and the environment.

In the next chapter, we conclude our exploration of the elements with a close look at the transition metals, a fascinating and extremely useful collection of metals. Then, in the final chapter, we dive down to the atom's nuclear core and investigate its properties and its great potential for energy production and medical science.

For Review and Reference

(Numbers in parentheses refer to pages, unless noted otherwise.)

Learning Objectives

Relevant section numbers appear in parentheses.

Understand These Concepts

1. How gravity, thermal convection, and elemental properties led to the silicate, sulfide, and iron phases and the predominance of certain elements in Earth's crust, mantle, and core (Section 22.1)
2. How organisms affect crustal abundances of elements, especially oxygen, carbon, calcium, and some transition metals; the onset of oxidation as an energy source (Section 22.1)
3. How atomic properties influence which elements have oxide ores and which have sulfide ores (Section 22.1)
4. The central role of CO_2 and the importance of photosynthesis, respiration, and decay in the carbon cycle (Section 22.2)
5. The central role of N_2 and the importance of atmospheric, industrial, and biological fixation in the nitrogen cycle (Section 22.2)
6. The absence of a gaseous component, the interactions of the inorganic and biological cycles, and the impact of humans on the phosphorus cycle (Section 22.2)
7. How pyro-, electro-, and hydrometallurgical processes are employed to extract a metal from its ore; the importance of the reduction step from compound to metal; refining and alloying processes (Section 22.3)
8. Functioning of the Downs cell for Na production and the application of Le Châtelier's principle for K production (Section 22.4)
9. How iron ore is reduced in a blast furnace and how pig iron is purified by the basic-oxygen process (Section 22.4)
10. How Fe is removed from copper ore and impure Cu is electrorefined (Section 22.4)
11. The importance of amphoterism in the Bayer process for isolating Al_2O_3 from bauxite; the significance of cryolite in the electrolytic step; the energy advantage of Al recycling (Section 22.4)
12. The steps in the Dow process for the extraction of Mg from seawater (Section 22.4)
13. How H_2 production, whether by chemical or electrolytic means, is tied to NH_3 production (Section 22.4)
14. How the kinetic isotope effect is applied to produce deuterium (Section 22.4)
15. How the Frasch process is used to obtain sulfur from natural deposits (Section 22.5)
16. The importance of equilibrium and kinetic factors in H_2SO_4 production (Section 22.5)
17. How overvoltage allows electrolysis of aqueous NaCl to form Cl_2 gas in the chlor-alkali process; coproduction of NaOH; comparison of diaphragm-cell, mercury-cell, and membrane-cell methods (Section 22.5)

Key Terms

Highlighted Figures and Tables

These figures (F) and tables (T) provide a review of key ideas.

Problems

Problems with **colored** numbers are answered in Appendix E. Sections match the text and provide the numbers of relevant sample problems. Most offer Concept Review Questions and Problems in Context. Comprehensive Problems are based on material from any section or previous chapter.

How the Elements Occur in Nature

Concept Review Questions

22.1 Hydrogen is by far the most abundant element cosmically. In interstellar space, it exists mainly as H_2. In contrast, on Earth, it exists very rarely as H_2 and is ninth in abundance in the crust. Why is hydrogen so abundant in the universe? Why is hydrogen so rare as a diatomic gas in Earth's atmosphere?

22.2 Metallic elements can be recovered from ores that are oxides, carbonates, halides, or sulfides. Give an example of each type.

22.3 The location of elements in the regions of Earth has enormous practical importance.
(a) Define the term *differentiation* and explain which physical property of a substance is primarily responsible for this process.
(b) What are the four most abundant elements in the crust?
(c) Which element is abundant in the crust and mantle but not the core?

22.4 How does the position of a metal in the periodic table relate to whether it occurs primarily as an oxide or as a sulfide?

Problems in Context

22.5 What material is the source for commercial production of each of the following elements: (a) aluminum; (b) nitrogen; (c) chlorine; (d) calcium; (e) sodium?

22.6 Aluminum is widely distributed throughout the world in the form of aluminosilicates. What property of these minerals prevents them from being a source of aluminum?

22.7 Describe two ways in which the biosphere has influenced the composition of Earth's crust.

The Cycling of Elements Through the Environment

Concept Review Questions

22.8 Use atomic and molecular properties to explain why life is based on the chemistry of carbon, rather than some other element such as silicon.

22.9 Define the term *fixation*. Name two elements that undergo environmental fixation. What natural forms of them are fixed?

22.10 Carbon dioxide enters the atmosphere by natural processes and as a result of human activity. Why is the latter source a cause of great concern?

22.11 Diagrams of environmental cycles are simplified to show overall changes and omit relatively minor contributors. For example, the production of lime from limestone is not explicitly shown in the cycle for carbon (Figure 22.5, p. 967). Which labeled category in the figure includes this process? Name two other processes that contribute to this category.

22.12 Describe three pathways for the utilization of atmospheric nitrogen. Is human activity a significant factor? Explain.

22.13 Why do the nitrogen-containing species shown in Figure 22.6 (p. 969) not include ring compounds or long-chain compounds with N—N bonds?

22.14 (a) Which region of Earth's crust is not involved in the phosphorus cycle?
(b) Describe two roles organisms play in the phosphorus cycle.

Problems in Context

22.15 Nitrogen fixation requires a great deal of energy because the N_2 bond is strong (activation energy for its breakage is high).
(a) How do the processes of atmospheric and industrial fixation reflect this energy requirement?
(b) How do the thermodynamics of the two processes differ? (*Hint:* Examine the respective heats of formation.)
(c) In view of the mild conditions for biological fixation, what must be the source of the "great deal of energy" for this process?

(d) What would be the most obvious environmental result of a low activation energy for N_2 fixation?

22.16 The following steps are *unbalanced* half-reactions involved in the nitrogen cycle. Balance each half-reaction to show the number of electrons lost or gained, and state whether it is an oxidation or a reduction (all occur in acidic conditions):
(a) $N_2(g) \longrightarrow NO(g)$ (b) $N_2O(g) \longrightarrow NO_2(g)$
(c) $NH_3(aq) \longrightarrow NO_2^-(aq)$ (d) $NO_3^-(aq) \longrightarrow NO_2^-(aq)$
(e) $N_2(g) \longrightarrow NO_3^-(aq)$

22.17 The use of silica to form slag in the production of phosphorus from phosphate rock was introduced by Robert Boyle more than 300 years ago. When fluorapatite [$Ca_5(PO_4)_3F$] is used in phosphorus production, most of the fluorine atoms appear in the slag, but some end up in toxic and corrosive $SiF_4(g)$.
(a) If 15% by mass of the fluorine in 100. kg of $Ca_5(PO_4)_3F$ forms SiF_4, what volume of this gas is collected at 1.00 atm and the industrial furnace temperature of 1450.°C?
(b) In some facilities, the SiF_4 is used to produce sodium hexafluorosilicate (Na_2SiF_6) which is sold for water fluoridation:

$$2SiF_4(g) + Na_2CO_3(s) + H_2O(l) \longrightarrow Na_2SiF_6(aq) + SiO_2(s) + CO_2(g) + 2HF(aq)$$

How many cubic meters of drinking water can be fluoridated to a level of 1.0 ppm of F^- using the SiF_4 produced in part (a)?

22.18 An impurity sometimes found in $Ca_3(PO_4)_2$ is Fe_2O_3, which is removed during the production of phosphorus as *ferrophosphorus* (Fe_2P). (a) Why is this impurity troubling from an economic standpoint? (b) If 50. metric tons of crude $Ca_3(PO_4)_2$ contains 2.0% Fe_2O_3 by mass and the overall yield of phosphorus is 90.%, how many metric tons of P_4 can be isolated?

Metallurgy: Extracting a Metal from Its Ore

Concept Review Questions

22.19 Define: (a) ore; (b) mineral; (c) gangue; (d) brine.

22.20 Define: (a) roasting; (b) smelting; (c) flotation; (d) refining.

22.21 What factors determine which reducing agent is selected for producing a specific metal?

22.22 Use atomic properties to explain the reduction of a less active metal by a more active one: (a) in aqueous solution; (b) in the molten state. Give a specific example of each process.

22.23 What class of element is obtained by oxidation of a mineral? What class of element is obtained by reduction of a mineral?

Problems in Context

22.24 Which group of elements gives each of the following alloys: (a) brass; (b) stainless steel; (c) bronze; (d) sterling silver?
1. Cu, Ag 2. Cu, Sn, Zn 3. Ag, Au
4. Fe, Cr, Ni 5. Fe, V 6. Cu, Zn

Tapping the Crust: Isolation and Uses of the Elements

Concept Review Questions

22.25 How are each of the following involved in iron metallurgy: (a) slag; (b) pig iron; (c) steel; (d) basic-oxygen process?

22.26 What are the distinguishing features of each extraction process: pyrometallurgy, electrometallurgy, and hydrometallurgy? Explain briefly how the types of metallurgy are used in the production of (a) Fe; (b) Na; (c) Au; (d) Al.

22.27 What property allows copper to be purified in the presence of iron and nickel impurities? Explain.

22.28 What is the practical reason for using cryolite in the electrolysis of aluminum oxide?

22.29 (a) What is a kinetic isotope effect?
(b) Do compounds of hydrogen exhibit a relatively large or small kinetic isotope effect? Explain.
(c) Carbon compounds also exhibit a kinetic isotope effect. How do you expect it to compare in magnitude with that for hydrogen compounds? Why?

22.30 How is Le Châtelier's principle involved in the production of elemental potassium?

Problems in Context

22.31 Elemental Li and Na are prepared by electrolysis of a molten salt, whereas K, Rb, and Cs are prepared by chemical reduction.
(a) In general terms, explain why the alkali metals cannot be prepared by electrolysis of their aqueous salt solutions.
(b) Use ionization energies (see the Family Portraits, pp. 560 and 564) to explain why calcium should *not* be able to isolate Rb from molten RbX (X = halide).
(c) Use physical properties to explain why calcium *is* used to isolate Rb from molten RbX.
(d) Can Ca be used to isolate Cs from molten CsX? Explain.

22.32 A Downs cell operating at 75.0 A produces 30.0 kg of Na.
(a) What volume of $Cl_2(g)$ is produced at 1.0 atm and 580.°C?
(b) How many coulombs were passed through the cell?
(c) How long did the cell operate?

22.33 (a) In the industrial production of iron, what is the reducing substance loaded into the blast furnace?
(b) In addition to furnishing the reducing power, what other function does this substance serve?
(c) What is the formula of the active reducing agent in the process?
(d) Write equations for the stepwise reduction of Fe_2O_3 to iron in the furnace.

22.34 One of the substances loaded into a blast furnace is limestone, which produces lime in the furnace.
(a) Give the chemical equation for the reaction forming lime.
(b) Explain the purpose of lime in the furnace. The term *flux* is often used as a label for a substance acting as the lime does. What is the derivation of this word, and how does it relate to the function of the lime?
(c) Write a chemical equation describing the action of lime flux.

22.35 The last step in the Dow process for the production of magnesium metal involves electrolysis of molten $MgCl_2$.
(a) Why isn't the electrolysis carried out with aqueous $MgCl_2$? What are the products of this aqueous electrolysis?
(b) Do the high temperatures required to melt $MgCl_2$ favor products or reactants? (*Hint:* Consider the ΔH_f^0 of $MgCl_2$.)

22.36 Iodine is the only halogen that occurs in a positive oxidation state, in $NaIO_3$ impurities within Chile saltpeter, $NaNO_3$.
(a) Is this mode of occurrence consistent with iodine's location in the periodic table? Explain.
(b) In the production of I_2, IO_3^- reacts with HSO_3^-:

$$IO_3^-(aq) + HSO_3^-(aq) \longrightarrow HSO_4^-(aq) + SO_4^-(aq) + H_2O(l) + I_2(s) \quad \text{[unbalanced]}$$

Identify the oxidizing and reducing agents.
(c) If 0.82 mol % of an $NaNO_3$ deposit is $NaIO_3$, how much I_2 (in g) can be obtained from 1.000 ton (2000. lb) of the deposit?

22.37 Selenium is prepared by the reaction of H_2SeO_3 with gaseous SO_2.
(a) What redox process does the sulfur dioxide undergo? What is the oxidation state of sulfur in the product?

(b) Given that the reaction occurs in acidic aqueous solution, what is the formula of the sulfur-containing species?
(c) Write the balanced redox equation for the process.

22.38 The halogens F_2 and Cl_2 are produced by electrolytic oxidation, whereas Br_2 and I_2 are produced by chemical oxidation of the halide ions in a concentrated aqueous solution (brine) by a more electronegative halogen. State two reasons why Cl_2 is not prepared by this method.

22.39 Silicon is prepared by the reduction of K_2SiF_6 with Al. Write the equation for this reaction. (*Hint:* Can F^- be oxidized in this reaction? Can K^+ be reduced?)

22.40 What is the mass percent of iron in each of the following iron ores: Fe_2O_3, Fe_3O_4, FeS_2?

22.41 Phosphorus is one of the impurities present in pig iron that is removed in the basic-oxygen process. Assuming that the phosphorus is present as P atoms, write equations for its oxidation and subsequent reaction in the basic slag.

22.42 The final step in the smelting of $FeCuS_2$ is

$$Cu_2S(s) + 2Cu_2O(s) \longrightarrow 6Cu(l) + SO_2(g)$$

(a) Give the oxidation states of copper in Cu_2S, Cu_2O, and Cu.
(b) What are the oxidizing and reducing agents in this reaction?

22.43 Use balanced equations to explain how acid-base properties are used to separate iron and titanium oxides from aluminum oxide in the Bayer process.

22.44 A piece of Al with a surface area of 2.3 m^2 is anodized to produce a film of Al_2O_3 that is 20. μ ($20.\times10^{-6}$ m) thick.
(a) How many coulombs flow through the cell in this process (assume that the density of the Al_2O_3 layer is 3.97 g/cm^3)?
(b) If it takes 15 min to produce this film, what current must flow through the cell?

22.45 The production of H_2 gas by the electrolysis of water typically requires about 400 kJ of energy per mole.
(a) Use the relationship between work and cell potential (Section 21.4) to calculate the minimum work needed to form 1.0 mol of H_2 gas at a cell potential of 1.24 V.
(b) What is the energy efficiency of the cell operation?
(c) Calculate the cost of producing 500. mol of H_2 if the cost of electrical energy is exactly \$0.05 per kilowatt·hour (1 watt·second = 1 joule).

22.46 (a) What are the components of the reaction mixture following the water-gas shift reaction? (b) Explain how zeolites (Gallery, p. 581) are used to purify the H_2 formed.

22.47 Metal sulfides are often first converted to oxides by roasting in air and then reduced with carbon to produce the metal. Why are the metal sulfides not reduced directly by carbon to yield CS_2? Give a thermodynamic analysis of both processes for a typical case such as ZnS.

Chemical Manufacturing: Two Case Studies

Concept Review Questions

22.48 Explain in detail why a catalyst is used to produce SO_3.

22.49 Among the exothermic steps in the manufacture of sulfuric acid is the process of hydrating SO_3. (a) Write two chemical reactions that show this process. (b) Why is the direct reaction of SO_3 with water not feasible?

22.50 Why is commercial H_2SO_4 so inexpensive?

22.51 (a) What are the three commercial products formed in the chlor-alkali process? (b) State an advantage and a disadvantage of the mercury-cell method for this process.

Problems in Context

22.52 Consider the oxidation of SO_2 to SO_3 at standard conditions.
(a) Calculate ΔG^0 at 25°C. Is the reaction spontaneous?
(b) Why is the reaction not performed at 25°C?
(c) Is the reaction spontaneous at 500.°C? (Assume that ΔH^0 and ΔS^0 are constant with temperature.)
(d) Compare K at 500.°C and at 25°C.
(e) What is the highest T at which the reaction is spontaneous?

22.53 If a chlor-alkali cell used a current of 3×10^4 A, how many pounds of Cl_2 would be produced in a typical 8-h operating day?

22.54 In the chlor-alkali process, the products chlorine and sodium hydroxide are kept separate from each other.
(a) Why is this necessary when Cl_2 is the desired product?
(b) Hypochlorite or chlorate may form by disproportionation of Cl_2 in basic solution. What determines which product forms?
(c) What mole ratio of Cl_2 to OH^- will produce ClO^-? ClO_3^-?

Comprehensive Problems

22.55 The key step in the manufacture of sulfuric acid is the oxidation of sulfur dioxide in the presence of a catalyst, such as V_2O_5. At 727°C, 0.010 mol of SO_2 is injected into an empty 2.00-L container ($K_p = 3.18$). (a) What is the equilibrium pressure of O_2 that is needed to maintain a 1:1 mole ratio of SO_3:SO_2? (b) What is the equilibrium pressure of O_2 that is needed to maintain a 95:5 mole ratio of SO_3:SO_2?

22.56 Many metal oxides are converted to the free metal by reduction with other elements, such as C or Si. For each of the following reactions, calculate the temperature at which the reduction occurs spontaneously:
(a) $MnO_2(s) \longrightarrow Mn(s) + O_2(g)$
(b) $MnO_2(s) + 2C(\text{graphite}) \longrightarrow Mn(s) + 2CO(g)$
(c) $MnO_2(s) + C(\text{graphite}) \longrightarrow Mn(s) + CO_2(g)$
(d) $MnO_2(s) + Si(s) \longrightarrow Mn(s) + SiO_2(s)$

22.57 Tetraphosphorus decaoxide (P_4O_{10}) is made from phosphate rock and used as a drying agent in the laboratory.
(a) Write a balanced equation for its reaction with water.
(b) What is the pH of a solution formed from the addition of 5.0 g of P_4O_{10} in sufficient water to form 0.500 L? (See Table 18.5, p. 786, for additional information.)

22.58 Heavy water (D_2O) is used to make deuterated chemicals.
(a) What major species, aside from the starting compounds, do you expect to find in a solution of CH_3OH and D_2O?
(b) Write equations to explain how these various species arise. (*Hint:* Consider the autoionization of both components.)

22.59 A blast furnace uses Fe_2O_3 to produce 8400. t of Fe per day.
(a) What mass of CO_2 is produced each day?
(b) Compare this amount of CO_2 with that produced by 1.0 million automobiles, each burning 5.0 gal of gasoline a day. Assume that gasoline has the formula C_8H_{18} and a density of 0.74 g/mL, and that it burns completely. (Note that U.S. gasoline consumption is over 4×10^8 gal/day.)

22.60 Several decades ago, various empirical rules were used to predict the formation of alloys. Two of the rules propose that stable alloys form if the ratio of the number of valence electrons (*s* plus *p*) to the number of atoms is 3:2 or 21:13. For example, β-brass (CuZn) is a very stable alloy: Cu has one 4*s* and Zn has two 4*s* electrons, for 3 valence electrons; one Cu atom plus one Zn atom gives 2 atoms; thus, a 3:2 ratio of electrons to atoms. What is the electron-to-atom ratio for each of the following alloys: (a) Cu_5Sn; (b) Cu_5Zn_8; (c) $Cu_{31}Sn_8$; (d) $AgZn_3$; (e) Ag_3Al?

22.61 In the production of magnesium, $Mg(OH)_2$ is precipitated by using $Ca(OH)_2$, which itself is "insoluble." (a) Use K_{sp} values to show that $Mg(OH)_2$ can be precipitated from seawater in which $[Mg^{2+}]$ is initially 0.051 *M*. (b) If the seawater is saturated with $Ca(OH)_2$, what fraction of the Mg^{2+} is precipitated?

22.62 The Ostwald process for the production of HNO_3 is

(1) $4NH_3(g) + 5O_2(g) \xrightarrow{\text{Pt/Rh catalyst}} 4NO(g) + 6H_2O(g)$
(2) $2NO(g) + O_2(g) \longrightarrow 2NO_2(g)$
(3) $3NO_2(g) + H_2O(l) \longrightarrow 2HNO_3(aq) + NO(g)$

(a) Describe the nature of the change that occurs in step 3.
(b) Write an overall equation that includes NH_3 and HNO_3 as the only nitrogen-containing species.
(c) Calculate ΔH^0_{rxn} (in kJ/mol N atoms) for this reaction at 25°C.

22.63 Step 1 of the Ostwald process for nitric acid production is

$$4NH_3(g) + 5O_2(g) \xrightarrow{\text{Pt/Rh catalyst}} 4NO(g) + 6H_2O(g)$$

An unwanted side reaction for this step is

$$4NH_3(g) + 3O_2(g) \longrightarrow 2N_2(g) + 6H_2O(g)$$

(a) Calculate K_p for these two NH_3 oxidations at 25°C.
(b) Calculate K_p for these two NH_3 oxidations at 900.°C.
(c) The Pt/Rh catalyst is one of the most efficient in the chemical industry, achieving 96% yield in 1 millisecond of contact with the reactants. However, at normal operating conditions (5 atm and 850°C), about 175 mg of Pt is lost per metric ton (t) of HNO_3 produced. If the annual U.S. production of HNO_3 is 1.01×10^7 t and the market price of Pt is $807/troy oz, what is the annual cost of the lost Pt (1 kg = 32.15 troy oz)?

22.64 The compounds $(NH_4)_2HPO_4$ and $NH_4(H_2PO_4)$ are water-soluble fertilizers. (a) Which has the higher mass % of P? (b) Why might the one with the lower mass % of P be preferred?

22.65 Several transition metals are prepared by reduction of the metal halide with magnesium. Titanium is prepared by the Kroll method, in which ore (ilmenite) is converted to the gaseous chloride, which is then reduced to Ti metal by molten Mg (see p. 988). Assuming yields of 84% for step 1 and 93% for step 2, and an excess of the other reactants, what mass of Ti metal can be prepared from 17.5 metric tons of ilmenite?

22.66 The production of S_8 from the $H_2S(g)$ found in natural gas deposits occurs through the Claus process (Section 22.5):
(a) Use these two unbalanced steps to write an overall balanced equation for this process:

(1) $H_2S(g) + O_2(g) \longrightarrow S_8(g) + SO_2(g) + H_2O(g)$
(2) $H_2S(g) + SO_2(g) \longrightarrow S_8(g) + H_2O(g)$

(b) Write the overall reaction with Cl_2 as the oxidizing agent instead of O_2. Use thermodynamic data to show whether $Cl_2(g)$ can be used to oxidize $H_2S(g)$.
(c) Why is oxidation by O_2 preferred to oxidation by Cl_2?

22.67 Acid mine drainage (AMD) occurs when geologic deposits containing pyrite (FeS_2) are exposed to oxygen and moisture. AMD is generated in a multistep process catalyzed by acidophilic (acid-loving) bacteria. Balance each step and identify those that increase acidity:

(1) $FeS_2(s) + O_2(g) \longrightarrow Fe^{2+}(aq) + SO_4^{2-}(aq)$
(2) $Fe^{2+}(aq) + O_2(g) \longrightarrow Fe^{3+}(aq) + H_2O(l)$
(3) $Fe^{3+}(aq) + H_2O(l) \longrightarrow Fe(OH)_3(s) + 12H^+(aq)$
(4) $FeS_2(s) + Fe^{3+}(aq) \longrightarrow Fe^{2+}(aq) + SO_4^{2-}(aq)$

22.68 Based on a concentration of 0.065 g Br^-/L, the ocean is estimated to have $244,000,000 of Br_2/mi^3 at recent prices.
(a) What is the price per gram of Br_2 in the ocean?
(b) Find the mass (in g) of $AgNO_3$ needed to precipitate 90.0% of the Br^- in 1.00 L of ocean water.
(c) Use Appendix C to find how low $[Cl^-]$ must be to prevent the unwanted precipitation of AgCl.
(d) Given a $[Cl^-]$ of about 0.3 *M*, is precipitation with silver ion a practical way to obtain bromine from the ocean? Explain.

22.69 Below 912°C, pure iron crystallizes in a body-centered cubic structure (ferrite) with a density of 7.86 g/cm^3; from 912°C to 1394°C, it adopts a face-centered cubic structure (austenite) with a density of 7.40 g/cm^3. Both types of iron form interstitial alloys with carbon. The maximum amount of carbon is 0.0218 mass% in ferrite and 2.08 mass% in austenite. Calculate the density of each alloy.

22.70 Why is nitric acid not produced by oxidizing atmospheric nitrogen directly in the following way?

(1) $N_2(g) + 2O_2(g) \longrightarrow 2NO_2(g)$
(2) $3NO_2(g) + H_2O(l) \longrightarrow 2HNO_3(aq) + NO(g)$
(3) $2NO(g) + O_2(g) \longrightarrow 2NO_2(g)$

$$3N_2(g) + 6O_2(g) + 2H_2O(l) \longrightarrow 4HNO_3(aq) + 2NO(g)$$

(*Hint:* Evaluate the thermodynamics of each step.)

22.71 Before the development of the Downs cell, the Castner cell was used for the industrial production of Na metal. The Castner cell was based on the electrolysis of molten NaOH.
(a) Write balanced cathode and anode half-reactions for this cell.
(b) A major problem with this cell was that the water produced at one electrode diffused to the other and reacted with the Na. If all the water produced reacted with Na, what would be the maximum efficiency of the Castner cell expressed as moles of Na produced per mole of electrons flowing through the cell?

22.72 When gold ores are leached with CN^- solutions, gold forms a complex ion, $Au(CN)_2^-$. (a) Find E_{cell} for the oxidation in air ($P_{O_2} = 0.21$) of Au to Au^+ in basic (pH 13.55) solution with $[Au^+] = 0.50$ *M*. Is the reaction $Au^+(aq) + e^- \longrightarrow Au(s)$, $E^0 = 1.68$ V, spontaneous? (b) How does formation of the complex ion change E^0 so that the oxidation can be accomplished?

22.73 Nitric oxide is an important species in the tropospheric nitrogen cycle, but it destroys ozone in the stratosphere.
(a) Write a balanced equation for its reversible reaction with ozone. (b) Given that the forward and reverse steps are first order in each component, write general rate laws for them. (c) Calculate ΔG^0 for this reaction at 280. K, the average temperature in the stratosphere. (Assume that the ΔH^0 and S^0 values in Appendix B do not change with temperature.) (d) Calculate the ratio of rate constants for the reaction at this temperature.

22.74 A key part of the carbon cycle is the fixation of CO_2 by photosynthesis to produce carbohydrates and oxygen gas.
(a) Using the formula $(CH_2O)_n$ to represent a carbohydrate, write a balanced equation for the photosynthetic reaction.
(b) If a tree fixes 45 g of CO_2 a day, what volume of O_2 gas measured at 1.0 atm and 80.°F does the tree produce per day?
(c) What volume of air (0.033 mol % CO_2) at the same conditions contains this amount of CO_2?

22.75 Farmers use ammonium sulfate as a fertilizer. In the soil, nitrifying bacteria oxidize NH_4^+ to NO_3^-, a groundwater contaminant that causes methemoglobinemia ("blue baby" syndrome). The World Health Organization standard for maximum $[NO_3^-]$ in groundwater is 45 mg/L. A farmer adds 200. kg of $(NH_4)_2SO_4$ to a field and 35% is oxidized to NO_3^-. What is the groundwater $[NO_3^-]$ (in mg/L) if 1000. m^3 of the water is contaminated?

22.76 In the 1980s, U.S. Fish and Wildlife Service researchers found high mortality among newborn coots and ducks in parts of California. U.S. Geological Survey scientists later found the cause to be a high selenium concentration in agricultural drainage water. Balance these half-reactions in the selenium cycle:
(a) $Se^{2-}(aq) \longrightarrow Se(s)$
(b) $SeO_3^{2-}(aq) \longrightarrow SeO_4^{2-}(aq)$
(c) $SeO_4^{2-}(aq) \longrightarrow Se(s)$

22.77 The key reaction (unbalanced) in the manufacture of synthetic cryolite for aluminum electrolysis is

$$HF(g) + Al(OH)_3(s) + NaOH(aq) \longrightarrow Na_3AlF_6(aq) + H_2O(l)$$

Assuming a 96.6% yield of dried, crystallized product, what mass (in kg) of cryolite can be obtained from the reaction of 353 kg of $Al(OH)_3$, 1.10 m^3 of 50.0% by mass aqueous NaOH (d = 1.53 g/mL), and 225 m^3 of gaseous HF at 315 kPa and 89.5°C? (Assume that the ideal gas law holds.)

22.78 Because of their different molar masses, H_2 and D_2 effuse at different rates (Section 5.6). (a) If it takes 14.5 min for 0.10 mol of H_2 to effuse, how long does it take for 0.10 mol of D_2 to do so in the same apparatus at the same T and P? (b) How many effusion steps does it take to separate an equimolar mixture of D_2 and H_2 to 99 mol % purity?

22.79 The disproportionation of carbon monoxide to graphite and carbon dioxide is thermodynamically favored but slow.
(a) What does this mean in terms of the magnitudes of the equilibrium constant (K), rate constant (k), and activation energy (E_a)?
(b) Write a balanced equation for the disproportionation of CO.
(c) Calculate K_c at 298 K. (d) Calculate K_p at 298 K.

22.80 The overall cell reaction for aluminum production is

$$2Al_2O_3(\text{in } Na_3AlF_6) + 3C(\text{graphite}) \longrightarrow 4Al(l) + 3CO_2(g)$$

(a) Assuming 100% efficiency, how many metric tons (t) of Al_2O_3 are consumed per metric ton of Al produced?
(b) Assuming 100% efficiency, how many metric tons of the graphite anode are consumed per metric ton of Al produced?
(c) Actual conditions in an aluminum plant require 1.89 t of Al_2O_3 and 0.45 t of graphite per metric ton of Al. What is the percent yield of Al with respect to Al_2O_3?
(d) What is the percent yield of Al with respect to graphite?
(e) What volume of CO_2 (in m^3) is produced per metric ton of Al at operating conditions of 960.°C and exactly 1 atm?

22.81 World production of chromite ($FeCr_2O_4$), the main ore of chromium, was 1.5×10^7 metric tons in 2003. To isolate chromium, a mixture of chromite and sodium hydroxide is heated in air to form sodium chromate, iron(III) oxide, and water vapor. The sodium chromate is dissolved in water, and this solution is acidified with sulfuric acid to produce the less soluble sodium dichromate. The sodium dichromate is filtered out and reduced with carbon to produce chromium(III) oxide, sodium carbonate, and carbon monoxide. The chromium(III) oxide is then reduced to chromium with aluminum metal. (a) Write balanced equations for each step. (b) What mass of chromium (in kg) could be prepared from the 2003 world production of chromite?

22.82 The following nitrogen-containing species play roles in the nitrogen cycle: NH_3, N_2O, NO, NO_2, NO_2^-, NO_3^-. Draw a Lewis structure for each species, showing minimal formal charges, and indicate the shape, including ideal bond angles.

22.83 Even though most metal sulfides are sparingly soluble in water, their solubilities differ by several orders of magnitude. This difference is sometimes used to separate the metals by controlling the pH. Use the following data to find the pH at which you can separate 0.10 M Cu^{2+} and 0.10 M Ni^{2+}:
Saturated H_2S = 0.10 M
K_{a1} of $H_2S = 9\times10^{-8}$ K_{a2} of $H_2S = 1\times10^{-17}$
K_{sp} of NiS $= 1.1\times10^{-18}$ K_{sp} of CuS $= 8\times10^{-34}$

22.84 Ores with as little as 0.25% by mass of copper are used as sources of the metal. (a) How many kilograms of such an ore would be needed for another Statue of Liberty, which contains 2.0×10^5 lb of copper? (b) If the mineral in the ore is chalcopyrite ($FeCuS_2$), what is the mass % of chalcopyrite in the ore?

22.85 How does acid rain affect the leaching of phosphate into groundwater from terrestrial phosphate rock? Calculate the solubility of $Ca_3(PO_4)_2$ in each of the following: (a) Pure water, pH 7.0 (Assume that PO_4^{3-} does not react with water.) (b) Moderately acidic rainwater, pH 4.5 (*Hint:* Assume that all the phosphate exists in the form that predominates at this pH.)

22.86 The lead(IV) oxide used in car batteries is prepared by coating the electrode plate with PbO and then oxidizing it to PbO_2. Despite its formula, PbO_2 has a nonstoichiometric ratio of lead to oxygen of about 1:1.888. In fact, the holes in the PbO_2 crystal structure due to missing O atoms are responsible for the oxide's conductivity. (a) What is the mole % of O missing from the PbO_2 structure? (b) What is the molar mass of the nonstoichiometric compound?

22.87 Chemosynthetic bacteria reduce CO_2 by "splitting" $H_2S(g)$ instead of the $H_2O(g)$ used by photosynthetic organisms. Compare the free energy change for splitting H_2S with that for splitting H_2O. Is there an advantage to using H_2S instead of H_2O?

22.88 Silver has a face-centered cubic structure with a unit cell edge length of 408.6 pm. Sterling silver is a substitutional alloy that contains 7.5% copper atoms. Assuming the unit cell remains the same, find the density of silver and of sterling silver.

22.89 Earth's mass is estimated to be 5.98×10^{24} kg, and titanium represents 0.05% by mass of this total. (a) How many moles of Ti are present? (b) If half of the Ti is found as ilmenite ($FeTiO_3$), what mass of ilmenite is present? (c) If the airline and auto industries use 1.00×10^5 tons of Ti per year, how many years would it take to use up all the Ti (1 ton = 2000 lb)?

22.90 In 1790, Nicolas Leblanc found a way to form Na_2CO_3 from NaCl. His process, now obsolete, consisted of three steps:

$$2NaCl(s) + H_2SO_4(aq) \longrightarrow Na_2SO_4(aq) + 2HCl(g)$$
$$Na_2SO_4(s) + 2C(s) \longrightarrow Na_2S(s) + 2CO_2(g)$$
$$Na_2S(s) + CaCO_3(s) \longrightarrow Na_2CO_3(s) + CaS(s)$$

(a) Write a balanced overall equation for the process.
(b) Calculate the ΔH_f^0 of CaS if ΔH_{rxn}^0 is 351.8 kJ/mol.
(c) Is the overall process spontaneous at standard-state conditions and 298 K?
(d) How many grams of Na_2CO_3 form from 100. g of NaCl if the process is 78% efficient?

22.91 Limestone ($CaCO_3$) is the second most abundant mineral on Earth after SiO_2. For many of its uses, it is first decomposed thermally to quicklime (CaO). MgO is prepared similarly from $MgCO_3$. (a) Find the temperature at which each decomposition is spontaneous. (b) Quicklime reacts with SiO_2 to form a slag ($CaSiO_3$) used in steelmaking. In 2003, the total steelmaking capacity of the U.S. steel industry was 2,370,000 tons per week, but only 84% of this capacity was utilized. If 50 kg of slag is needed per ton of steel, what mass (in kg) of limestone was used to make slag in 2003?

The colors of rocks Much of the color in the mineral world comes from transition-metal ions. In fact, it is the Cr^{3+} within the crystal structure of otherwise colorless aluminum oxide that makes rubies red. In this chapter, we examine the transition metals and some of their numerous applications as well as the structures and properties of their compounds.

23

The Transition Elements and Their Coordination Compounds

Our exploration of the elements to this point is far from complete; in fact, we have skirted the majority of them and some of the most familiar. Whereas most important uses of the main-group elements involve their compounds, the transition elements are remarkably useful in their uncombined form. Figure 23.1 shows that the **transition elements** *(transition metals)* make up the *d* block (B groups) and *f* block *(inner transition elements)*.

In Chapter 22, you saw how two of the most important—copper and iron—are extracted from their ores. Aside from those two, with their countless essential uses, many other transition elements are indispensable as well: chromium in automobile parts, gold and silver in jewelry, tungsten in lightbulb filaments, platinum in automobile catalytic converters, titanium in bicycle frames and aircraft parts, and zinc in batteries, to mention just a few of the better known elements. You may be less aware of zirconium in nuclear-reactor liners, vanadium in axles and crankshafts, molybdenum in boiler plates, nickel in coins, tantalum in organ-replacement parts, palladium in telephone-relay contacts—the list goes on and on. As ions, many of these elements also play vital roles in living organisms.

IN THIS CHAPTER . . . We first discuss some atomic, physical, and chemical properties of the transition elements and then focus on the chemistry of four familiar ones: chromium, manganese, silver, and mercury. Next, we concentrate on the most distinctive feature of transition element chemistry, the formation of coordination compounds, substances that contain complex ions. We consider two models that explain the striking colors of these compounds, as well as their magnetic properties and structures, and then end with some essential biochemical functions of transition metal ions.

Concepts & Skills to Review

before you study this chapter

- properties of light (Section 7.1)
- electron shielding of nuclear charge (Section 8.2)
- electron configuration, ionic size, and magnetic behavior (Sections 8.3 to 8.5)
- valence bond theory (Section 11.1)
- constitutional, geometric, and optical isomerism (Section 15.2)
- Lewis acid-base concepts (Section 18.9)
- complex-ion formation (Section 19.4)
- redox behavior and standard electrode potentials (Section 21.3)

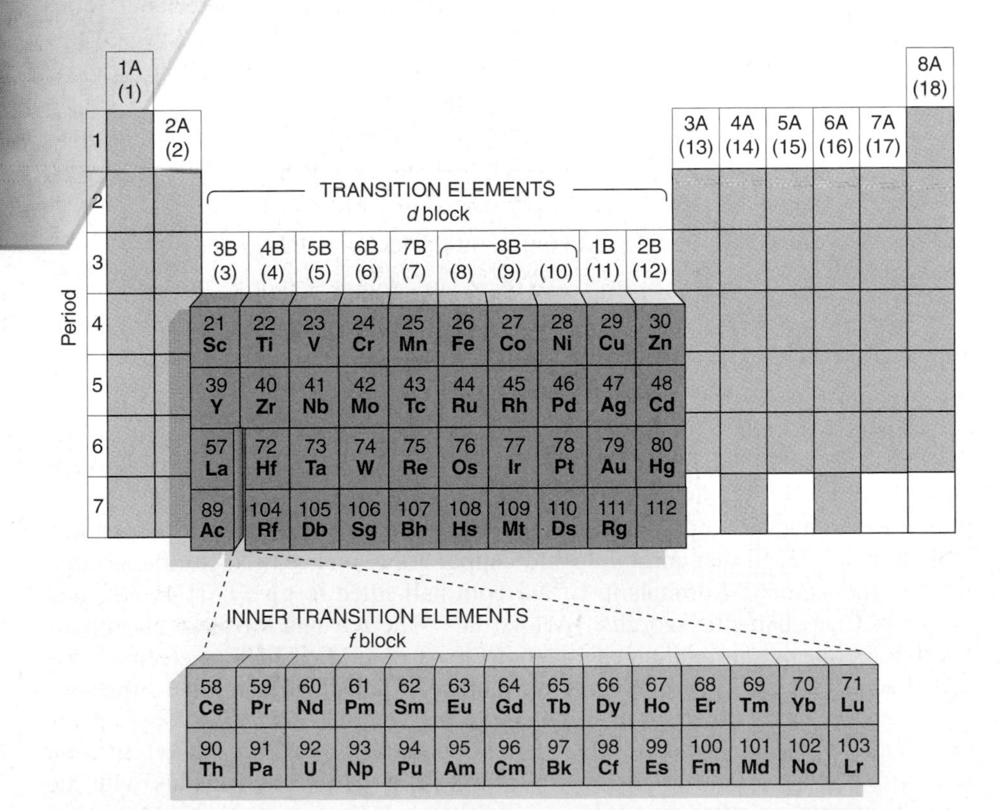

Figure 23.1 The transition elements (*d* block) and inner transition elements (*f* block) in the periodic table.

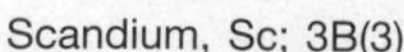

Figure 23.2 The Period 4 transition metals. Samples of all ten elements appear as pure metals, in chunk or powder form, in periodic-table order on this and the facing page.

23.1 PROPERTIES OF THE TRANSITION ELEMENTS

The transition elements differ considerably in physical and chemical behavior from the main-group elements. In some ways, they are more uniform: main-group elements in each period change from metal to nonmetal, but *all transition elements are metals.* In other ways, the transition elements are more diverse: most main-group ionic compounds are colorless and diamagnetic, but *many transition metal compounds are highly colored and paramagnetic.* We first discuss electron configurations of the atoms and ions, and then examine certain key properties of transition elements, with an occasional comparison to the main-group elements.

Electron Configurations of the Transition Metals and Their Ions

As with any of the elements, the properties of the transition elements and their compounds arise largely from the electron configurations of their atoms (Section 8.3) and ions (Section 8.5). The *d*-block (B-group) elements occur in four series that lie within Periods 4 through 7 between the last *ns*-block element [Group 2A(2)] and the first *np*-block element [Group 3A(13)]. Each series represents the filling of five *d* orbitals and, thus, contains ten elements. In 1996 and 1997, elements 110 through 112 were synthesized in particle accelerators, so the Period 7 series is complete; thus, all 40 *d*-block transition elements are known. Lying between the first and second members of the *d*-block transition series in Periods 6 and 7 are the inner transition elements, whose *f* orbitals are being filled.

Even though there are several exceptions, in general, the *condensed* ground-state electron configuration for the elements in each *d*-block series is

$$[\text{noble gas}]\ ns^2(n-1)d^x, \text{ with } n = 4 \text{ to } 7 \text{ and } x = 1 \text{ to } 10$$

In Periods 6 and 7, the condensed configuration includes the *f* sublevel:

$$[\text{noble gas}]\ ns^2(n-2)f^{14}(n-1)d^x, \text{ with } n = 6 \text{ or } 7$$

The *partial* (valence-level) electron configuration for the *d*-block elements excludes the noble gas core and the filled inner *f* sublevel:

$$ns^2(n-1)d^x$$

The first transition series occurs in Period 4 and consists of scandium (Sc) through zinc (Zn) (Figure 23.2 and Table 23.1). Scandium has the electron configuration [Ar] $4s^23d^1$, and the addition of one electron at a time (along with one proton in the nucleus) first half-fills, then fills, the $3d$ orbitals across the periodic table to zinc. Recall that chromium and copper are two exceptions to this general pattern: the $4s$ and $3d$ orbitals in Cr are both half-filled to give [Ar] $4s^13d^5$, and the $4s$ in Cu is half-filled to give [Ar] $4s^13d^{10}$. The reasons for these exceptions involve the change in relative energies of the $4s$ and $3d$ orbitals as electrons are added across the series and the unusual stability of half-filled and filled sublevels.

Transition metal ions form through *the loss of the ns electrons before the (n − 1)d electrons.* Therefore, the electron configuration of Ti^{2+} is [Ar] $3d^2$, *not* [Ar] $4s^2$, and Ti^{2+} is referred to as a d^2 ion. Ions of different metals with the same configuration often have similar properties. For example, both Mn^{2+} and Fe^{3+} are d^5 ions; both have pale colors in aqueous solution and, as we'll discuss later, form complex ions with similar magnetic properties.

Table 23.1 shows a general pattern in number of unpaired electrons (or half-filled orbitals) across the Period 4 transition series. Note that the number increases

Iron, Fe; 8B(8)

Cobalt, Co; 8B(9)

Nickel, Ni; 8B(10)

Copper, Cu; 1B(11)

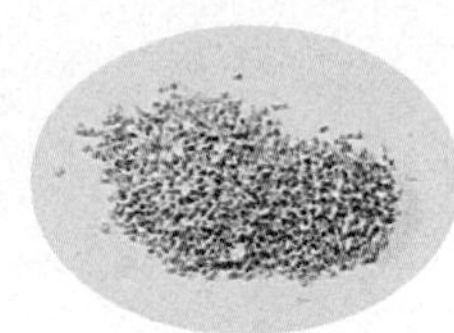
Zinc, Zn; 2B(12)

Table 23.1 Orbital Occupancy of the Period 4 Transition Metals

Element	Partial Orbital Diagram: 4s	3d					4p			Unpaired Electrons
Sc	↑↓	↑								1
Ti	↑↓	↑	↑							2
V	↑↓	↑	↑	↑						3
Cr	↑	↑	↑	↑	↑	↑				6
Mn	↑↓	↑	↑	↑	↑	↑				5
Fe	↑↓	↑↓	↑	↑	↑	↑				4
Co	↑↓	↑↓	↑↓	↑	↑	↑				3
Ni	↑↓	↑↓	↑↓	↑↓	↑	↑				2
Cu	↑	↑↓	↑↓	↑↓	↑↓	↑↓				1
Zn	↑↓	↑↓	↑↓	↑↓	↑↓	↑↓				0

in the first half of the series and, when pairing begins, decreases through the second half. As you'll see, it is the electron configuration of the transition metal *atom* that correlates with physical properties of the *element,* such as density and magnetic behavior, whereas it is the electron configuration of the *ion* that determines the properties of the *compounds.*

SAMPLE PROBLEM 23.1 Writing Electron Configurations of Transition Metal Atoms and Ions

Problem Write *condensed* electron configurations for the following: **(a)** Zr; **(b)** V^{3+}; **(c)** Mo^{3+}. (Assume that elements in higher periods behave like those in Period 4.)

Plan We locate the element in the periodic table and count its position in the respective transition series. These elements are in Periods 4 and 5, so the general configuration is [noble gas] $ns^2(n-1)d^x$. For the ions, we recall that *ns* electrons are lost first.

Solution (a) Zr is the second element in the 4*d* series: [Kr] $5s^24d^2$.

(b) V is the third element in the 3*d* series: [Ar] $4s^23d^3$. In forming V^{3+}, three electrons are lost (two 4*s* and one 3*d*), so V^{3+} is a d^2 ion: [Ar] $3d^2$.

(c) Mo lies below Cr in Group 6B(6), so we expect the same exception as for Cr. Thus, Mo is [Kr] $5s^14d^5$. To form the ion, Mo loses the one 5*s* and two of the 4*d* electrons, so Mo^{3+} is a d^3 ion: [Kr] $4d^3$.

Check Figure 8.11 (p. 302) shows we're correct for the atoms. Be sure that charge plus number of *d* electrons in the ion equals the sum of outer *s* and *d* electrons in the atom.

FOLLOW-UP PROBLEM 23.1 Write *partial* electron configurations for the following: **(a)** Ag^+; **(b)** Cd^{2+}; **(c)** Ir^{3+}.

Atomic and Physical Properties of the Transition Elements

The atomic properties of the transition elements contrast in several ways with those of a comparable set of main-group elements.

Trends Across a Period Consider the variations in atomic size, electronegativity, and ionization energy across Period 4 (Figure 23.3):

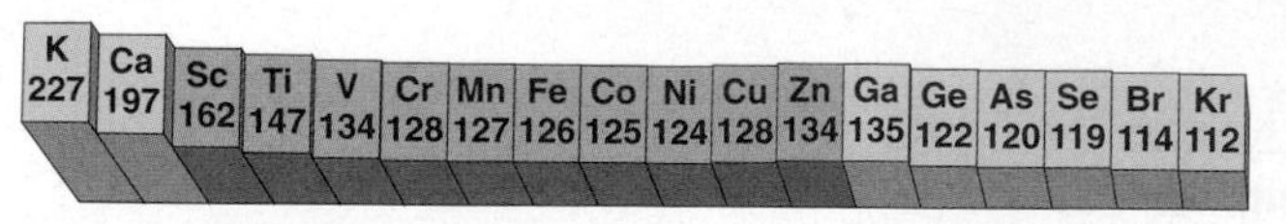

A Atomic radius (pm)

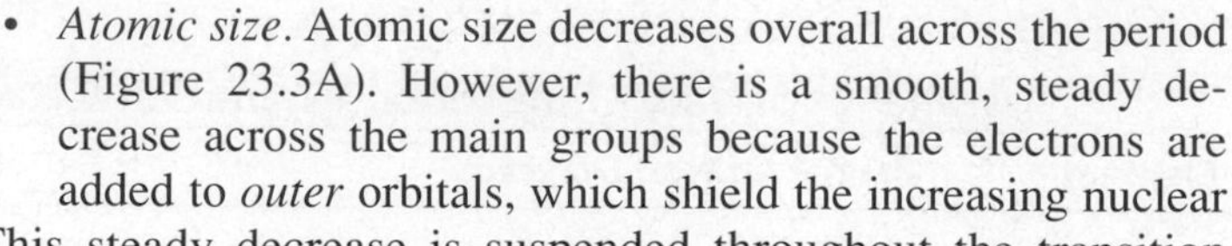

- *Atomic size.* Atomic size decreases overall across the period (Figure 23.3A). However, there is a smooth, steady decrease across the main groups because the electrons are added to *outer* orbitals, which shield the increasing nuclear charge poorly. This steady decrease is suspended throughout the transition series, where *atomic size decreases at first but then remains fairly constant.* Recall that the *d* electrons fill *inner* orbitals, so they shield outer electrons from the increasing nuclear charge very efficiently. As a result, the outer 4*s* electrons are not pulled closer.

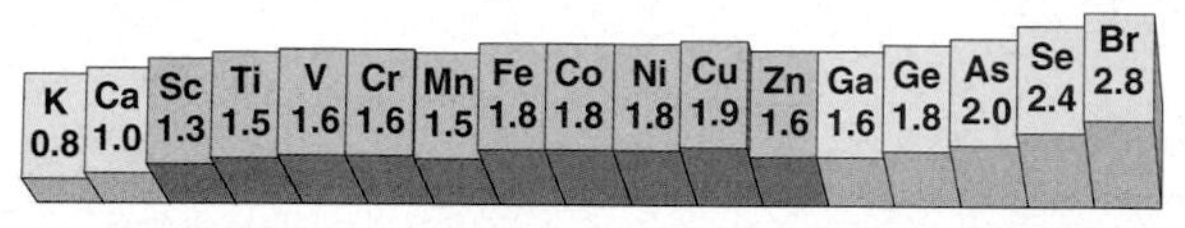

B Electronegativity

- *Electronegativity.* Electronegativity generally increases across the period but, once again, the transition elements exhibit a relatively *small change in electronegativity* (Figure 23.3B), consistent with the relatively small change in size. In contrast, the main groups show a steady, much steeper increase between the metal potassium (0.8) and the nonmetal bromine (2.8). The transition elements all have intermediate electronegativity values, much like the large, metallic members of Groups 3A(13) to 5A(15).

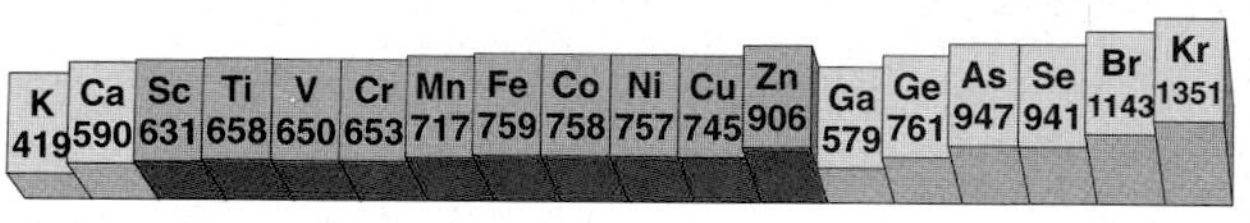

C First ionization energy (kJ/mol)

Figure 23.3 Horizontal trends in key atomic properties of the Period 4 elements. The atomic radius (**A**), electronegativity (**B**), and first ionization energy (**C**) of the elements in Period 4 are shown as posts of different heights, with darker shades for the transition series. The transition elements exhibit smaller, less regular changes for these properties than do the main-group elements.

- *Ionization energy.* The ionization energies of the Period 4 main-group elements rise steeply from left to right, more than tripling from potassium (419 kJ/mol) to krypton (1351 kJ/mol), as electrons become more difficult to remove from the poorly shielded, increasing nuclear charge. In the transition metals, however, the *first ionization energies increase relatively little* because the inner 3*d* electrons shield effectively (Figure 23.3C); thus, the outer 4*s* electron experiences only a slightly higher effective nuclear charge. [Recall from Section 8.4 that the drop at Group 3A(13) occurs because it is relatively easy to remove the first electron from the outer *np* orbital.]

Trends Within a Group Vertical trends for transition elements are also different from those for the main groups.

- *Atomic size.* As expected, atomic size increases from Period 4 to 5, as it does for the main-group elements, but there is virtually *no size increase from Period 5 to 6* (Figure 23.4A). Remember that the lanthanides, with their buried 4*f* sublevel, appear between the 4*d* (Period 5) and 5*d* (Period 6) series. Therefore, an element in Period 6 is separated from the one above it in Period 5 by 32 elements (ten 4*d*, six 5*p*, two 6*s*, and fourteen 4*f* orbitals) instead of just 18. The extra shrinkage that results from the increase in nuclear charge due to the addition of 14 protons is called the **lanthanide contraction.** By coincidence, this *decrease* is about equal to the normal *increase* between periods, so the Periods 5 and 6 transition elements have about the same atomic sizes.
- *Electronegativity.* The vertical trend in electronegativity seen in most transition groups is opposite the trend in main groups. Here, we see an *increase* in electronegativity from Period 4 to Period 5, but then no further increase in Period 6 (Figure 23.4B). The heavier elements, especially gold (EN = 2.4), become quite electronegative, with values exceeding those of most metalloids and even some nonmetals (e.g., EN of Te and of P = 2.1). (In fact, gold forms the saltlike CsAu and the Au^- ion, which exists in liquid ammonia.) Although the

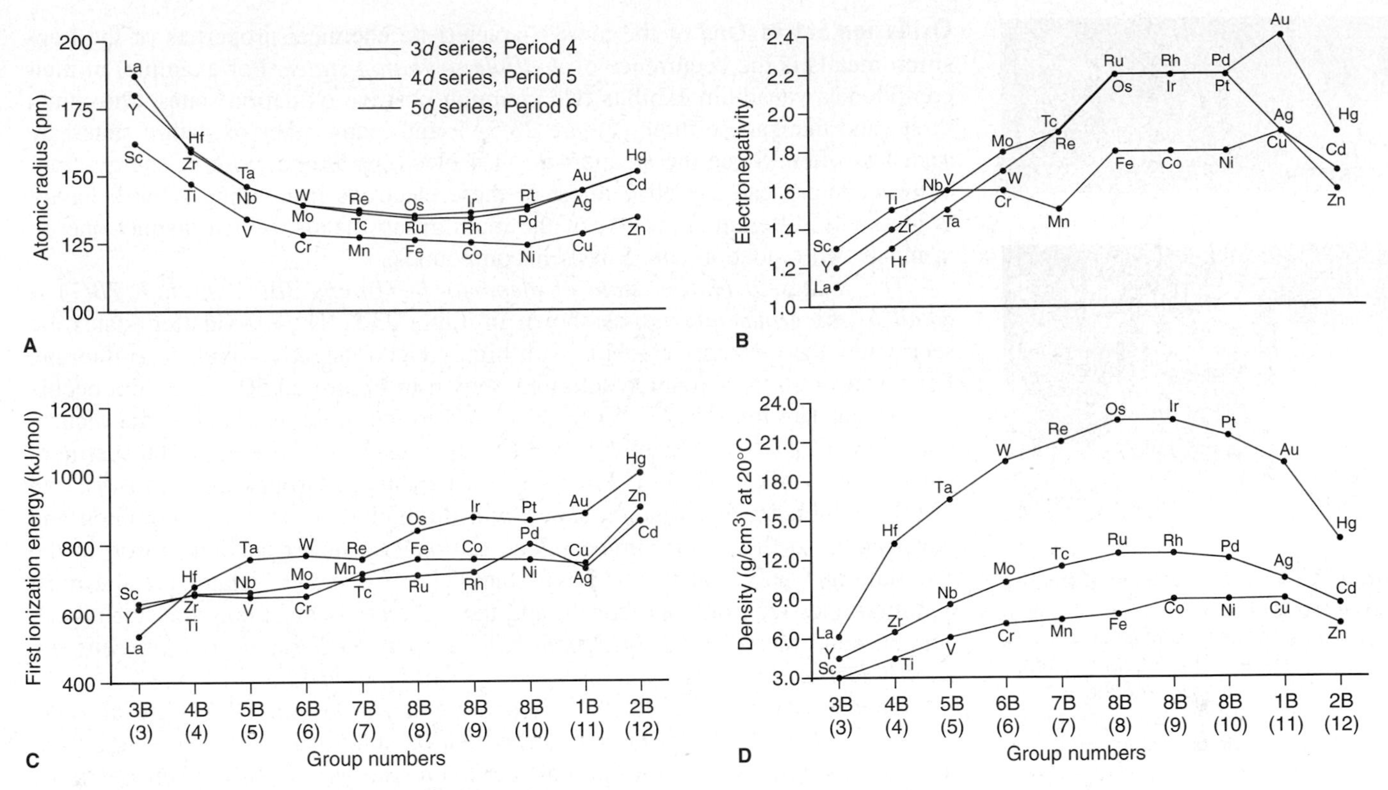

Figure 23.4 Vertical trends in key properties within the transition elements. The trends are unlike those for the main-group elements in several ways: **A,** The second and third members of a transition metal group are nearly the same size. **B,** Electronegativity increases down a transition group. **C,** First ionization energies are highest at the bottom of a transition group. **D,** Densities increase down a transition group because mass increases faster than volume.

atomic size increases slightly from the top to the bottom of a group, the nuclear charge increases much more. Therefore, the heavier transition metals exhibit more covalent character in their bonds and attract electrons more strongly than do main-group metals.

- *Ionization energy.* The relatively small increase in size combined with the relatively large increase in nuclear charge also explains why *the first ionization energy generally increases* down a transition group (Figure 23.4C). This trend also runs counter to the pattern in the main groups, in which heavier members are so much larger that their outer electron is easier to remove.
- *Density.* Atomic size, and therefore volume, is inversely related to density. Across a period, densities increase, then level off, and finally dip a bit at the end of a series (Figure 23.4D). Down a transition group, densities increase dramatically because atomic volumes change little from Period 5 to 6, but atomic masses increase significantly. As a result, the Period 6 series contains some of the densest elements known: tungsten, rhenium, osmium, iridium, platinum, and gold have densities about 20 times that of water and twice that of lead.

Chemical Properties of the Transition Metals

Like their atomic and physical properties, the chemical properties of the transition elements are very different from those of the main-group elements. Let's examine the key properties in the Period 4 transition series and then see how behavior changes within a group.

A

B

Figure 23.5 Aqueous oxoanions of transition elements. A, Often, a given transition element has multiple oxidation states. Here, Mn is shown in the +2 (Mn^{2+}, *left*), the +6 (MnO_4^{2-}, *middle*), and the +7 states (MnO_4^-, *right*). **B,** The highest possible oxidation state equals the group number in these oxoanions: VO_4^{3-} (*left*), $Cr_2O_7^{2-}$ (*middle*), and MnO_4^- (*right*).

Oxidation States One of the most characteristic chemical properties of the transition metals is the occurrence of *multiple oxidation states*. For example, in their compounds, vanadium exhibits two common positive oxidation states, chromium three, and manganese three (Figure 23.5A), and many other oxidation states are seen less often. Since the *ns* and $(n-1)d$ electrons are so close in energy, transition elements can use all or most of these electrons in bonding. This behavior is markedly different from that of the main-group metals, which display one or, at most, two oxidation states in their compounds.

The highest oxidation state of elements in Groups 3B(3) through 7B(7) is equal to the group number, as shown in Table 23.2. These oxidation states are seen when the elements combine with highly electronegative oxygen or fluorine. For instance, in the oxoanion solutions shown in Figure 23.5B, vanadium occurs as the vanadate ion (VO_4^{3-}; O.N. of V = +5), chromium occurs as the dichromate ion ($Cr_2O_7^{2-}$; O.N. of Cr = +6), and manganese occurs as the permanganate ion (MnO_4^-; O.N. of Mn = +7). Elements in Groups 8B(8), 8B(9), and 8B(10) exhibit fewer oxidation states, and the highest state is less common and never equal to the group number. For example, we never encounter iron in the +8 state and only rarely in the +6 state. The +2 and +3 states are the most common ones for iron* and cobalt, and the +2 state is most common for nickel, copper, and zinc. *The +2 oxidation state is common because ns^2 electrons are readily lost.*

Copper, silver, and gold (the coinage metals) in Group 1B(11) are unusual. Although copper has a fairly common oxidation state of +1, which results from loss of its single 4*s* electron, its most common state is +2. Silver behaves more predictably, exhibiting primarily the +1 oxidation state. Gold exhibits the +3 state and, less often, the +1 state. Zinc, cadmium, and mercury in Group 2B(12) exhibit the +2 state, but mercury also exhibits the +1 state in the Hg_2^{2+} ion, with its Hg—Hg bond.†

*Iron may seem to have unusual oxidation states in the common ores magnetite (Fe_3O_4) and pyrite (FeS_2), but this is not really the case. In magnetite, one-third of the metal ions are Fe^{2+} and two-thirds are Fe^{3+}, which is equivalent to a 1:1 ratio of FeO:Fe_2O_3 and gives an overall formula of Fe_3O_4. Pyrite contains Fe^{2+} combined with the disulfide ion, S_2^{2-}.

†Some evidence suggests that the Hg_2^{2+} ion may exist as an Hg atom bonded to an Hg^{2+} ion, with the atom donating its two 6*s* electrons for the covalent bond.

Table 23.2 Oxidation States and *d*-Orbital Occupancy of the Period 4 Transition Metals*

Oxidation State	3B (3) Sc	4B (4) Ti	5B (5) V	6B (6) Cr	7B (7) Mn	8B (8) Fe	8B (9) Co	8B (10) Ni	1B (11) Cu	2B (12) Zn
0	d^1	d^2	d^3	d^5	d^5	d^6	d^7	d^8	d^{10}	d^{10}
+1			d^3	d^5	d^5	d^6	d^7	d^8	d^{10}	
+2		d^2	d^3	d^4	d^5	d^6	d^7	d^8	d^9	d^{10}
+3	d^0	d^1	d^2	d^3	d^4	d^5	d^6	d^7	d^8	
+4		d^0	d^1	d^2	d^3	d^4	d^5	d^6		
+5			d^0	d^1	d^2		d^4			
+6				d^0	d^1	d^2				
+7					d^0					

*The most important orbital occupancies are in color.

Metallic Behavior and Reducing Strength Atomic size and oxidation state have a major effect on the nature of bonding in transition metal compounds. Like the metals in Groups 3A(13), 4A(14), and 5A(15), the transition elements in their *lower* oxidation states behave chemically more like metals. That is, *ionic bonding is more prevalent for the lower oxidation states, and covalent bonding is more prevalent for the higher states.* For example, at room temperature, $TiCl_2$ is an ionic solid, whereas $TiCl_4$ is a molecular liquid. In the higher oxidation states, the atoms have higher charge densities, so they polarize the electron clouds of the nonmetal ions more strongly and the bonding becomes more covalent. For the same reason, the oxides become less basic as the oxidation state increases: TiO is weakly basic in water, whereas TiO_2 is amphoteric (reacts with both acid and base).

Table 23.3 shows the standard electrode potentials of the Period 4 transition metals in their +2 oxidation state in acid solution. Note that, in general, reducing strength decreases across the series. All the Period 4 transition metals, except copper, are active enough to reduce H^+ from aqueous acid to form hydrogen gas. In contrast to the rapid reaction at room temperature of the Group 1A(1) and 2A(2) metals with water, however, the transition metals have an oxide coating that allows rapid reaction only with hot water or steam.

Table 23.3 Standard Electrode Potentials of Period 4 M^{2+} Ions

Half-Reaction	E^o (V)
$Ti^{2+}(aq) + 2e^- \rightleftharpoons Ti(s)$	−1.63
$V^{2+}(aq) + 2e^- \rightleftharpoons V(s)$	−1.19
$Cr^{2+}(aq) + 2e^- \rightleftharpoons Cr(s)$	−0.91
$Mn^{2+}(aq) + 2e^- \rightleftharpoons Mn(s)$	−1.18
$Fe^{2+}(aq) + 2e^- \rightleftharpoons Fe(s)$	−0.44
$Co^{2+}(aq) + 2e^- \rightleftharpoons Co(s)$	−0.28
$Ni^{2+}(aq) + 2e^- \rightleftharpoons Ni(s)$	−0.25
$Cu^{2+}(aq) + 2e^- \rightleftharpoons Cu(s)$	0.34
$Zn^{2+}(aq) + 2e^- \rightleftharpoons Zn(s)$	−0.76

Color and Magnetism of Compounds *Most main-group ionic compounds are colorless* because the metal ion has a filled outer level (noble gas electron configuration). With only much higher energy orbitals available to receive an excited electron, the ion does not absorb visible light. In contrast, electrons in a partially filled *d* sublevel can absorb visible wavelengths and move to slightly higher energy *d* orbitals. As a result, *many transition metal compounds have striking colors.* Exceptions are the compounds of scandium, titanium(IV), and zinc, which are colorless because their metal ions have either an empty *d* sublevel (Sc^{3+} or Ti^{4+}: [Ar] $3d^0$) or a filled one (Zn^{2+}: [Ar] $3d^{10}$) (Figure 23.6).

Magnetic properties are also related to sublevel occupancy (Section 8.5). Recall that a *paramagnetic* substance has atoms or ions with unpaired electrons, which cause it to be attracted to an external magnetic field. A *diamagnetic* substance has only paired electrons, so it is unaffected (or slightly repelled) by a magnetic field. *Most main-group metal ions are diamagnetic* for the same reason they are colorless: all their electrons are paired. In contrast, *many transition metal compounds are paramagnetic because of their unpaired d electrons.* For example, $MnSO_4$ is paramagnetic, but $CaSO_4$ is diamagnetic. The Ca^{2+} ion has the electron configuration of argon, whereas Mn^{2+} has a d^5 configuration. Transition metal ions with a d^0 or d^{10} configuration are also colorless and diamagnetic.

Figure 23.6 Colors of representative compounds of the Period 4 transition metals. Staggered from left to right, the compounds are scandium oxide *(white)*, titanium(IV) oxide *(white)*, vanadyl sulfate dihydrate *(light blue)*, sodium chromate *(yellow)*, manganese(II) chloride tetrahydrate *(light pink)*, potassium ferricyanide *(red-orange)*, cobalt(II) chloride hexahydrate *(violet)*, nickel(II) nitrate hexahydrate *(green)*, copper(II) sulfate pentahydrate *(blue)*, and zinc sulfate heptahydrate *(white)*.

Chemical Behavior Within a Group The *increase* in reactivity going down a group of main-group metals, as shown by the *decrease* in first ionization energy (IE_1), does *not* occur going down a group of transition metals. Consider the chromium (Cr) group [6B(6)], which shows a typical pattern (Table 23.4). IE_1 *increases* down the group, which makes the two heavier metals *less* reactive than the lightest one. Chromium is also a much stronger reducing agent than molybdenum (Mo) or tungsten (W), as shown by the standard electrode potentials.

Table 23.4 Some Properties of Group 6B(6) Elements

Element	Atomic Radius (pm)	IE_1 (kJ/mol)	E^0 (V) for $M^{3+}(aq)\|M(s)$
Cr	128	653	−0.74
Mo	139	685	−0.20
W	139	770	−0.11

The similarity in atomic size of Period 5 and 6 members also leads to similar chemical behavior, a fact that has some important practical consequences. Because Mo and W compounds behave similarly, for example, their ores often occur together in nature, which makes the elements very difficult to separate from each other. The same situation occurs with zirconium and hafnium in Group 4B(4) and with niobium and tantalum in Group 5B(5).

SECTION SUMMARY

All transition elements are metals. Atoms of the *d*-block elements have $(n - 1)d$ orbitals being filled, and their ions have an empty *ns* orbital. Unlike the trends in the main-group elements, atomic size, electronegativity, and first ionization energy change relatively little across a transition series. Because of the lanthanide contraction, atomic size changes little from Period 5 to 6 in a transition metal group; thus, electronegativity, first ionization energy, and density *increase* down a group. Transition metals typically have several oxidation states, with the +2 state most common. The elements exhibit more metallic behavior in their lower states. Most Period 4 transition metals are active enough to reduce hydrogen ion from acid solution. Many transition metal compounds are colored and paramagnetic because the metal ion has unpaired *d* electrons.

23.2 THE INNER TRANSITION ELEMENTS

The 14 **lanthanides**—cerium (Ce; $Z = 58$) through lutetium (Lu; $Z = 71$)—lie between lanthanum ($Z = 57$) and hafnium ($Z = 72$) in the third *d*-block transition series. Below them are the 14 radioactive **actinides,** thorium (Th; $Z = 90$) through lawrencium (Lr; $Z = 103$), which lie between actinium (Ac; $Z = 89$) and rutherfordium (Rf; $Z = 104$). The lanthanides and actinides are called **inner transition elements** because, in most cases, their seven inner 4*f* or 5*f* orbitals are being filled.

The Lanthanides

The lanthanides are sometimes called the *rare earth elements,* a term referring to their presence in unfamiliar oxides, but they are actually not rare at all. Cerium (Ce), for instance, ranks 26th in natural abundance (by mass %) and is five times more abundant than lead. All the lanthanides are silvery, high-melting (800°C to 1600°C) metals. Their chemical properties represent an extreme case of the small variations typical of transition elements in a period or a group, which makes the lanthanides very difficult to separate.

A Remarkable Laboratory Feat
The discovery of the lanthanides is a testimony to the remarkable laboratory skills of 19th-century chemists. The two mineral sources, ceria and yttria, are mixtures of compounds of all 14 lanthanides; yttria also includes oxides of Sc, Y, and La. Purification of these elements was so difficult that many claims of new elements turned out to be mixtures of similar elements. To confuse matters, the periodic table of the time had space for only one element between Ba and Hf, which turned out to be La. Not until 1913, when the atomic-number basis of the table was established, did chemists realize that 14 new elements could fit, and in 1918, Niels Bohr proposed that the $n = 4$ level be expanded to include the *f* sublevel.

The natural co-occurrence of the lanthanides arises because they exist as M^{3+} ions of very similar radii in their common ores. Most lanthanides have the ground-state electron configuration [Xe] $6s^24f^x5d^0$, where x varies across the series. The three exceptions (Ce, Gd, and Lu) have a single electron in one of their $5d$ orbitals: Ce ([Xe] $6s^24f^15d^1$) forms a stable 4+ ion with an empty (f^0) sublevel, and the Gd^{3+} and Lu^{3+} ions have a stable half-filled (f^7) or filled (f^{14}) sublevel.

Lanthanide compounds and their mixtures have many uses. Several oxides are used for tinting sunglasses and welder's goggles and for adding color to the fluorescent powder coatings in TV screens. High-quality camera lenses incorporate La_2O_3 because of its extremely high index of refraction. Samarium forms an alloy with cobalt, $SmCo_5$, that is used in the strongest known permanent magnet (see Sample Problem 23.2). Two industrial uses account for more than 60% of rare earth applications. In gasoline refining, the zeolite catalysts (Gallery, p. 581) used to "crack" hydrocarbon components into smaller molecules contain 5% by mass rare earth oxides. In steelmaking, a mixture of lanthanides, called *misch metal,* is used to remove carbon impurities from molten iron and steel.

SAMPLE PROBLEM 23.2 Finding the Number of Unpaired Electrons

Problem The alloy $SmCo_5$ forms a permanent magnet because both samarium and cobalt have unpaired electrons. How many unpaired electrons are in the Sm atom ($Z = 62$)?

Plan We write the condensed electron configuration of Sm and then, using Hund's rule and the aufbau principle, place the electrons in a partial orbital diagram and count the unpaired electrons.

Solution Samarium is the eighth element after Xe. Two electrons go into the $6s$ sublevel. In general, the $4f$ sublevel fills before the $5d$ (among the lanthanides, only Ce, Gd, and Lu have $5d$ electrons), so the remaining electrons go into the $4f$. Thus, Sm is [Xe] $6s^24f^6$. There are seven f orbitals, so each of the six f electrons enters a separate orbital:

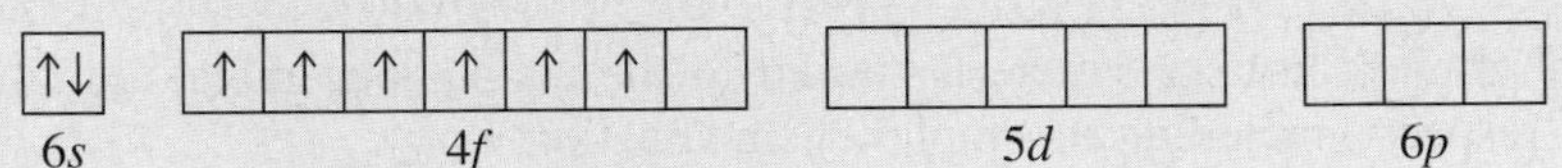

Thus, Sm has six unpaired electrons.

Check Six $4f$ e^- plus two $6s$ e^- plus the 54 e^- in Xe gives 62, the atomic number of Sm.

FOLLOW-UP PROBLEM 23.2 How many unpaired electrons are in the Er^{3+} ion?

The Actinides

All actinides are radioactive. Like the lanthanides, they share very similar physical and chemical properties. Thorium and uranium occur in nature, but the transuranium elements, those with Z greater than 92, have been synthesized in high-energy particle accelerators, some in only tiny amounts (Section 24.3). In fact, macroscopic samples of mendelevium (Md), nobelium (No), and lawrencium (Lr) have never been seen. The actinides that have been isolated are silvery and chemically reactive and, like the lanthanides, form highly colored compounds. The actinides and lanthanides have similar outer-electron configurations. Although the +3 oxidation state is characteristic of the actinides, as it is for the lanthanides, other states also occur. For example, uranium exhibits +3 through +6 states, with the +6 state the most prevalent; thus, the most common oxide of uranium is UO_3.

SECTION SUMMARY

There are two series of inner transition elements. The lanthanides ($4f$ series) have a common +3 oxidation state and exhibit very similar properties. The actinides ($5f$ series) are radioactive. All actinides have a +3 oxidation state; several, including uranium, have higher states as well.

23.3 HIGHLIGHTS OF SELECTED TRANSITION METALS

Let's investigate several transition elements, focusing on the general patterns in the aqueous chemistry of each. We examine chromium and manganese from Period 4, silver from Period 5, and mercury from Period 6.

Chromium

Chromium is a very shiny, silvery metal, whose name (from the Greek *chroma,* "color") refers to its many colorful compounds. A solution of Cr^{3+} is deep violet, for example, and a trace of Cr^{3+} in the Al_2O_3 crystal structure gives ruby its beautiful red hue (see photo p. 1002). Chromium readily forms a thin, adherent, transparent coating of Cr_2O_3 in air, making the metal extremely useful as an attractive protective coating on easily corroded metals, such as iron. "Stainless" steels often contain as much as 18% chromium by mass and are highly resistant to corrosion.

With six valence electrons ([Ar] $4s^13d^5$), chromium occurs in all possible positive oxidation states, but the three most important are +2, +3, and +6 (see Table 23.2). In its three common oxides, chromium exhibits the pattern seen in many elements: *nonmetallic character and oxide acidity increase with metal oxidation state.* Chromium(II) oxide (CrO) is basic and largely ionic. It forms an insoluble hydroxide in neutral or basic solution but dissolves in acidic solution to yield the Cr^{2+} ion:

$$CrO(s) + 2H^+(aq) \longrightarrow Cr^{2+}(aq) + H_2O(l)$$

Chromium(III) oxide (Cr_2O_3) is amphoteric, dissolving in acid to yield the violet Cr^{3+} ion,

$$Cr_2O_3(s) + 6H^+(aq) \longrightarrow 2Cr^{3+}(aq) + 3H_2O(l)$$

and in base to form the green $Cr(OH)_4^-$ ion:

$$Cr_2O_3(s) + 3H_2O(l) + 2OH^-(aq) \longrightarrow 2Cr(OH)_4^-(aq)$$

Thus, chromium in its +3 state is similar to the main-group metal aluminum in several respects, including its amphoterism (Section 19.4).

Deep-red chromium(VI) oxide (CrO_3) is covalent and acidic, forming chromic acid (H_2CrO_4) in water,

$$CrO_3(s) + H_2O(l) \longrightarrow H_2CrO_4(aq)$$

which yields the yellow chromate ion (CrO_4^{2-}) in base:

$$H_2CrO_4(aq) + 2OH^-(aq) \longrightarrow CrO_4^{2-}(aq) + 2H_2O(l)$$

In acidic solution, the chromate ion immediately forms the orange dichromate ion ($Cr_2O_7^{2-}$):

$$2CrO_4^{2-}(aq) + 2H^+(aq) \rightleftharpoons Cr_2O_7^{2-}(aq) + H_2O(l)$$

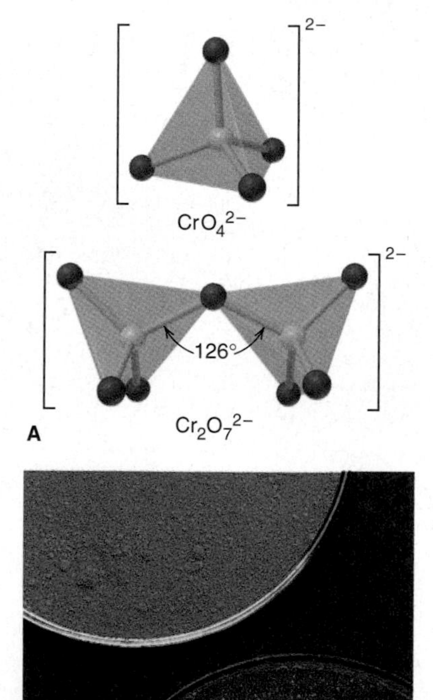

Figure 23.7 The bright colors of chromium(VI) compounds. A, Structures of the chromate (CrO_4^{2-}) and dichromate ($Cr_2O_7^{2-}$) ions. **B,** Samples of K_2CrO_4 (*yellow*) and $K_2Cr_2O_7$ (*orange*).

Because both ions contain chromium(VI), this is not a redox reaction; rather, it is a dehydration-condensation, as you can see from the structures in Figure 23.7. Hydrogen-ion concentration controls the equilibrium position: yellow CrO_4^{2-} predominates at high pH and orange $Cr_2O_7^{2-}$ at low pH. The bright colors of chromium(VI) compounds lead to their wide use in pigments for artist's paints and ceramic glazes. Lead chromate (chrome yellow) is used as an oil-paint color, and also in the yellow stripes that delineate traffic lanes.

Chromium metal and the Cr^{2+} ion are potent reducing agents. The metal displaces hydrogen from dilute acids to form blue $Cr^{2+}(aq)$, which reduces O_2 in air within minutes to form the violet Cr^{3+} ion:

$$4Cr^{2+}(aq) + O_2(g) + 4H^+(aq) \longrightarrow 4Cr^{3+}(aq) + 2H_2O(l) \qquad E^0_{overall} = 1.64 \text{ V}$$

Chromium(VI) compounds in acid solution are strong oxidizing agents (concentrated solutions are *extremely* corrosive!), the chromium(VI) being readily reduced to chromium(III):

$$Cr_2O_7^{2-}(aq) + 14H^+(aq) + 6e^- \longrightarrow 2Cr^{3+}(aq) + 7H_2O(l) \qquad E^0 = 1.33 \text{ V}$$

This reaction is often used to determine the iron content of a water or soil sample by oxidizing Fe^{2+} to Fe^{3+} ion. In basic solution, the CrO_4^{2-} ion, which is a much weaker oxidizing agent, predominates:

$$CrO_4^{2-}(aq) + 4H_2O(l) + 3e^- \longrightarrow Cr(OH)_3(s) + 5OH^-(aq) \qquad E^0 = -0.13\ V$$

The Concept of Valence-State Electronegativity Why does oxide acidity increase with oxidation state? And how can a metal, like chromium, form an oxoanion? To answer such questions, we must apply the concept of electronegativity to the various oxidation states of an element. A metal in a higher oxidation state is more positively charged, which increases its attraction for electrons; in effect, its *electronegativity increases*. This effective electronegativity, called *valence-state electronegativity,* also has numerical values. The electronegativity of chromium metal is 1.6, close to that of aluminum (1.5), another active metal. For chromium(III), the value increases to 1.7, still characteristic of a metal. However, the electronegativity of chromium(VI) is 2.3, close to the values of some nonmetals, such as phosphorus (2.1), selenium (2.4), and carbon (2.5). Thus, like P in PO_4^{3-}, Cr in chromium(VI) compounds often occurs covalently bonded at the center of the oxoanion of a relatively strong acid.

Manganese

Elemental manganese is hard and shiny and, like vanadium and chromium, is used mostly to make steel alloys. A small amount of Mn (<1%) makes steel easier to roll, forge, and weld. Steel made with 12% Mn is tough enough to be used for naval armor, front-end loader buckets (see photo), and other extremely hard steel objects. Small amounts of manganese are added to aluminum beverage cans and bronze alloys to make them stiffer and tougher as well.

Front-end loader parts are made of steel containing manganese.

The chemistry of manganese resembles that of chromium in some respects. The free metal is quite reactive and readily reduces H^+ from acids, forming the pale-pink Mn^{2+} ion:

$$Mn(s) + 2H^+(aq) \longrightarrow Mn^{2+}(aq) + H_2(g) \qquad E^0 = 1.18\ V$$

Like chromium, manganese can use all its valence electrons in its compounds, exhibiting every possible positive oxidation state, with the +2, +4, and +7 states most common (Table 23.5). As the oxidation state of manganese rises, its valence-state electronegativity increases and its oxides change from basic to acidic. Manganese(II) oxide (MnO) is basic, and manganese(III) oxide (Mn_2O_3) is amphoteric. Manganese(IV) oxide (MnO_2) is insoluble and shows no acid-base properties. [It is used in dry cells and alkaline batteries as the oxidizing agent in a redox reaction with zinc (Chapter 21, p. 932).] Manganese(VII) oxide (Mn_2O_7),

Table 23.5 Some Oxidation States of Manganese

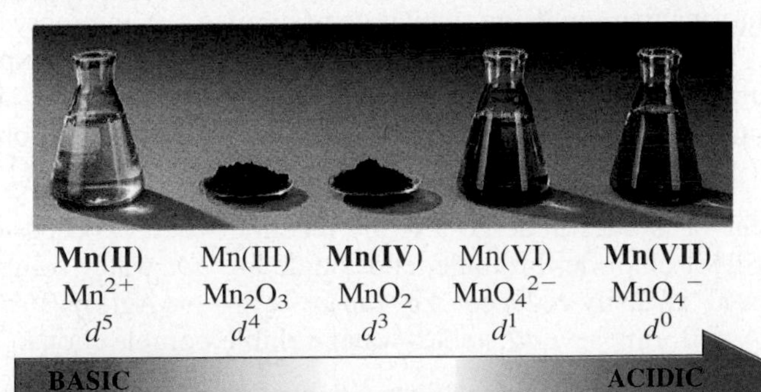

Oxidation state*	**Mn(II)**	Mn(III)	**Mn(IV)**	Mn(VI)	**Mn(VII)**
Example	Mn^{2+}	Mn_2O_3	MnO_2	MnO_4^{2-}	MnO_4^-
Orbital occupancy	d^5	d^4	d^3	d^1	d^0
Oxide acidity	BASIC				ACIDIC

*Most common states in **boldface**.

Sharing the Ocean's Wealth At their current rate of usage, known reserves of many key transition metals will be depleted in less than 50 years. Other sources must be found. One promising source is nodules strewn over large portions of the ocean floor. Varying from a few millimeters to a few meters in diameter, these chunks consist mainly of manganese and iron oxides, with oxides of other elements present in smaller amounts. Billions of tons exist, but mining them presents major technical and political challenges. Global agreements have designated the ocean floor as international property, so cooperation will be required to mine the nodules and share the mineral rewards.

which forms by reaction of Mn with pure O_2, reacts with water to form permanganic acid ($HMnO_4$), which is as strong as perchloric acid ($HClO_4$).

All manganese species with oxidation states greater than +2 act as oxidizing agents, but the purple permanganate (MnO_4^-) ion is particularly powerful. Like ions with chromium in its highest oxidation state, MnO_4^- is a much stronger oxidizing agent in acidic than in basic solution:

$$MnO_4^-(aq) + 4H^+(aq) + 3e^- \longrightarrow MnO_2(s) + 2H_2O(l) \qquad E^0 = 1.68\text{ V}$$

$$MnO_4^-(aq) + 2H_2O(l) + 3e^- \longrightarrow MnO_2(s) + 4OH^-(aq) \qquad E^0 = 0.59\text{ V}$$

Unlike Cr^{2+} and Fe^{2+}, the Mn^{2+} ion resists oxidation in air. The Cr^{2+} ion is a d^4 species and readily loses a 3*d* electron to form the d^3 ion Cr^{3+}, which is more stable. The Fe^{2+} ion is a d^6 species, and removing a 3*d* electron yields the stable, half-filled d^5 configuration of Fe^{3+}. Removing an electron from Mn^{2+} disrupts its stable d^5 configuration.

Silver

Silver, the second member of the coinage metals [Group 1B(11)], has been admired for thousands of years and is still treasured for use in jewelry and fine flatware. Because the pure metal is too soft for these purposes, however, it is alloyed with copper to form the harder sterling silver. In former times, silver was used in coins, but it has been replaced almost universally by copper-nickel alloys. Silver has the *highest electrical conductivity of any element* but is not used in wiring because copper is cheaper and more plentiful. In the past, silver was found in nuggets and veins of rock, often mixed with gold, because both elements are chemically inert enough to exist uncombined. Nearly all of those deposits have been mined, so most silver is now obtained from the anode mud formed during the electrorefining of copper (Section 22.4).

The only important oxidation state of silver is +1. Its most important *soluble* compound is silver nitrate, used for electroplating and in the manufacture of the halides used for photographic film. Although silver forms no oxide in air, it tarnishes to black Ag_2S by reaction with traces of sulfur-containing compounds. Some polishes remove the Ag_2S, along with some silver, by physically abrading the surface. An alternative "home remedy" that removes the tarnish and restores the metal involves heating the object in a solution of table salt or baking soda ($NaHCO_3$) in an aluminum pan. Aluminum, a strong reducing agent, reduces the Ag^+ ions back to the metal:

$$2Al(s) + 3Ag_2S(s) + 6H_2O(l) \longrightarrow 2Al(OH)_3(s) + 6Ag(s) + 3H_2S(g) \qquad E^0 = 0.86\text{ V}$$

The Chemistry of Black-and-White Photography The most widespread use of silver compounds—particularly the three halides AgCl, AgBr, and AgI—is in black-and-white photography, an art that applies transition metal chemistry and solution kinetics. The photographic film itself is simply a flexible plastic support for the light-sensitive emulsion, which consists of AgBr microcrystals dispersed in gelatin. The five steps in obtaining a final photograph are exposing the film, developing the image, fixing the image, washing the negative, and printing the image. Figure 23.8 summarizes the first four of these steps. The process depends on several key chemical properties of silver and its compounds:

- Silver halides undergo a redox reaction when exposed to visible light.
- Silver chloride, bromide, and iodide are *not* water soluble.
- Ag^+ is easily reduced: $Ag^+(aq) + e^- \longrightarrow Ag(s)$; $E^0 = 0.80$ V.
- Ag^+ forms several stable, water-soluble complex ions.

1. *Exposing the film.* Light reflected from the objects in a scene—more light from bright objects than from dark ones—enters the camera lens and strikes the film. Exposed AgBr crystals absorb photons ($h\nu$) in a very localized redox reac-

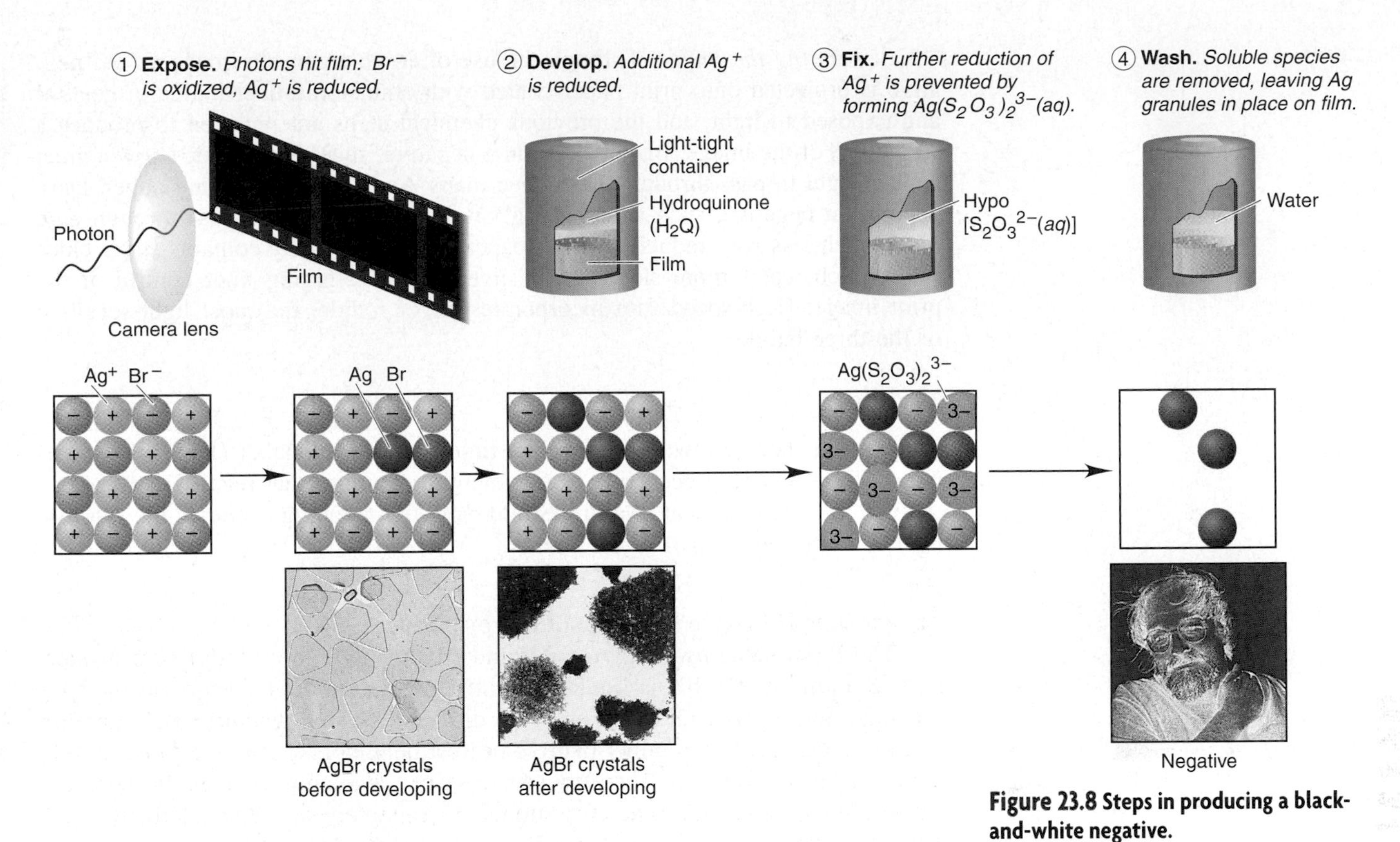

① **Expose.** *Photons hit film: Br⁻ is oxidized, Ag⁺ is reduced.*
② **Develop.** *Additional Ag⁺ is reduced.*
③ **Fix.** *Further reduction of Ag⁺ is prevented by forming Ag(S₂O₃)₂³⁻(aq).*
④ **Wash.** *Soluble species are removed, leaving Ag granules in place on film.*
Light-tight container
Hydroquinone (H₂Q)
Film
Hypo [S₂O₃²⁻(aq)]
Water
Photon
Film
Camera lens
Ag⁺ Br⁻
Ag Br
Ag(S₂O₃)₂³⁻
AgBr crystals before developing
AgBr crystals after developing
Negative

Figure 23.8 Steps in producing a black-and-white negative.

tion. A Br^- ion is excited by a photon and oxidized, and the released electron almost immediately reduces a nearby Ag^+ ion:

$$\begin{array}{rcl} Br^- & \xrightarrow{h\nu} & Br + e^- \\ Ag^+ + e^- & \longrightarrow & Ag \\ \hline Ag^+ + Br^- & \xrightarrow{h\nu} & Ag + Br \end{array}$$

Wherever more light strikes a microcrystal, more Ag atoms form. The exposed crystals are called a *latent image* because the few scattered atoms of photoreduced Ag are not yet visible. Nevertheless, their presence as crystal defects within the AgBr crystal makes it highly susceptible to further reduction.

2. *Developing the image.* The latent image is developed into the actual image by reducing more of the silver ions in the crystal in a controlled manner. Developing is a rate-dependent step: crystals with many photoreduced Ag atoms react more quickly than those with only a few. The developer is a weak reducing agent, such as the organic substance hydroquinone ($H_2C_6H_4O_2$; H_2Q) (see margin):

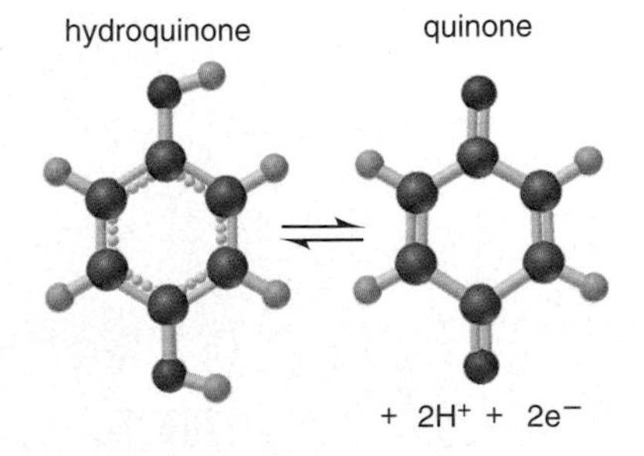

hydroquinone
quinone
+ 2H⁺ + 2e⁻

$$2Ag^+(s) + H_2Q(aq;\ \text{reduced form}) \longrightarrow 2Ag(s) + Q(aq;\ \text{oxidized form}) + 2H^+(aq)$$

The reaction rate depends on H_2Q concentration, solution temperature, and the length of time the emulsion is bathed in the solution. After developing, approximately 10^6 as many Ag atoms are present on the film as there were in the latent image, and they form very small, black clusters of silver.

3. *Fixing the image.* After the image is developed, it must be "fixed"; that is, the reduction of Ag^+ must be stopped, or the entire film will blacken on exposure to more light. Fixing involves removing the remaining Ag^+ chemically by converting it to a soluble complex ion with sodium thiosulfate solution ("hypo"):

$$AgBr(s) + 2S_2O_3^{2-}(aq) \longrightarrow Ag(S_2O_3)_2^{3-}(aq) + Br^-(aq)$$

4. *Washing the negative.* The water-soluble ions are washed away in water. Washing is the final step in producing a photographic *negative,* in which dark objects in the scene appear bright in the image, and vice versa.

5. *Printing the image.* Through the use of an enlarger, the image on the negative is projected onto print paper coated with emulsion (silver halide in gelatin) and exposed to light, and the previous chemical steps are repeated to produce a "positive" of the image. Bright areas on a negative, such as car tires, allow a great deal of light to pass through and reduce many Ag^+ ions on the print paper. Dark areas on a negative, such as clouds, allow much less light to pass through and, thus, much less Ag^+ reduction. Print-paper emulsion usually contains silver chloride, which reacts more slowly than silver bromide, giving finer control of the print image. High-speed film incorporates silver iodide, the most light-sensitive of the three halides.

Mercury

Mercury has been known since ancient times because cinnabar (HgS), its principal ore, is a naturally occurring red pigment (vermilion) that readily undergoes a redox reaction in the heat of a fire. Sulfide ion, the reducing agent for the process, is already present as part of the ore:

$$HgS(s) + O_2(g) \longrightarrow Hg(g) + SO_2(g)$$

The gaseous Hg condenses on cool nearby surfaces.

The Latin name *hydrargyrum* ("liquid silver") is a good description of mercury, the only metal that is liquid at room temperature. Two factors account for this unusual property. First, because of a distorted crystal structure, each mercury atom is surrounded by 6 rather than 12 nearest neighbors. Second, a filled, tightly held *d* sublevel leaves only the two 6*s* electrons available for metallic bonding. Thus, interactions among mercury atoms are relatively few and relatively weak and, as a result, the solid form breaks down at $-38.9°C$.

Many of mercury's uses arise from its unusual physical properties. Its liquid range ($-39°C$ to $357°C$) encompasses most everyday temperatures, so Hg is commonly used in thermometers. As you might expect from its position in Period 6 following the lanthanide contraction, mercury is quite dense (13.5 g/mL), which makes it convenient for use in barometers and manometers. Mercury's fluidity and conductivity make it useful for "silent" switches in thermostats. At high pressures, mercury vapor can be excited electrically to emit the bright white light seen in sports stadium and highway lights.

Mercury is a good solvent for other metals, and many *amalgams* (alloys of mercury) exist. In mercury batteries, a zinc amalgam acts as the anode and mercury(II) oxide acts as the cathode (Chapter 21, p. 933). In the mercury-cell version of the chlor-alkali process (Chapter 22, p. 995), mercury acts as the cathode and as the solvent for the sodium metal that forms. Recall that, when treated with water, the resulting sodium amalgam, Na(Hg), forms two important by-products:

$$2Na(Hg) + 2H_2O(l) \xrightarrow{-Hg} 2NaOH(aq) + H_2(g)$$

The mercury is released in this step and reused in the electrolytic cell. In view of mercury's toxicity, which we discuss shortly, a newer method involving a polymeric membrane is replacing the mercury-cell method. Nevertheless, past contamination of wastewater from the chlor-alkali process and the disposal of old mercury batteries remain serious environmental concerns.

The chemical properties of mercury are unique within Group 2B(12). The other members, zinc and cadmium, occur in the +2 oxidation state as d^{10} ions, but mercury occurs in the +1 state as well, with the condensed electron configuration [Xe] $6s^1 4f^{14} 5d^{10}$. We can picture the unpaired 6*s* electron allowing two Hg(I) species to form the *diatomic ion* $[Hg—Hg]^{2+}$ (written Hg_2^{2+}), one of the first species known with a covalent metal-metal bond (but also see the footnote on p. 1008). Mercury's more common oxidation state is +2. Whereas HgF_2 is largely ionic, many other compounds, such as $HgCl_2$, contain bonds that are predominantly covalent. Most mercury(II) compounds are insoluble in water.

The common ions Zn^{2+}, Cd^{2+}, and Hg^{2+} are biopoisons. Zinc oxide is used as an external antiseptic ointment. The Cd^{2+} and Hg^{2+} ions are two of the so-called toxic heavy-metal ions. The cadmium in solder may be more responsible than the lead for solder's high toxicity. Mercury compounds have been used in agriculture as fungicides and pesticides and in medicine as internal drugs, but these uses have been largely phased out.

Because most mercury(II) compounds are insoluble in water, they were once thought to be harmless in the environment; but we now know otherwise. Microorganisms in sludge and river sediment convert mercury atoms and ions to the methyl mercury ion, $CH_3—Hg^+$, and then to organomercury compounds, such as dimethylmercury, $CH_3—Hg—CH_3$. These toxic compounds are nonpolar and, like chlorinated hydrocarbons, become increasingly concentrated in fatty tissues, as they move up a food chain from microorganisms to worms to fish and, finally, to birds and mammals. Indeed, fish living in a mercury-polluted lake or coastal estuary can have a mercury concentration thousands of times higher than that of the water itself.

The mechanism for the toxicity of mercury and other heavy-metal ions is not fully understood. It is thought that the ions migrate from fatty tissue and bind strongly to thiol (—SH) groups of amino acids in proteins, thereby disrupting the proteins' structure and function. Because the brain has a high fat content, small amounts of lead, cadmium, and mercury ions circulating in the blood become deposited in the brain's fatty tissue, interact with its proteins, and often cause devastating neurological and psychological effects.

Mad as a Hatter The felt top hat and zany manner of the Mad Hatter in *Alice in Wonderland* are literary references to the toxicity of mercury compounds. Mercury(II) nitrate and chloride were used with HNO_3 to make felt for hats from animal hair. Inhalation of the dust generated by the process led to "hatter's shakes," abnormal behavior, and other neurological and psychological symptoms.

SECTION SUMMARY

Chromium and manganese add corrosion resistance and hardness to steels. They are typical of transition metals in having several oxidation states. Valence-state electronegativity refers to the ability of an element in its various oxidation states to attract bonding electrons. It increases with oxidation number, which is the reason elements act more metallic (more ionic compounds, more basic oxides) in lower states and more nonmetallic (more acidic oxides, oxoanions of acids) in higher states. Cr and Mn produce H_2 in acid. Cr(VI) undergoes a pH-sensitive dehydration-condensation reaction. Both Cr(VI) and Mn(VII) are stronger oxidizing agents in acid than in base. The only important oxidation state for silver is +1. The silver halides are light sensitive and are used in photography. Mercury, the only metal that is liquid at room temperature, dissolves many other metals in important applications. The mercury(I) ion is diatomic and has a metal-metal covalent bond. The element and its compounds are toxic and become concentrated as they move up a food chain.

23.4 COORDINATION COMPOUNDS

The most distinctive aspect of transition metal chemistry is the formation of **coordination compounds** (also called *complexes*). These are substances that contain at least one **complex ion,** a species consisting of *a central metal cation (either a transition metal or a main-group metal) that is bonded to molecules and/or anions called* **ligands.** In order to maintain charge neutrality in the coordination compound, the complex ion is typically associated with other ions, called **counter ions.**

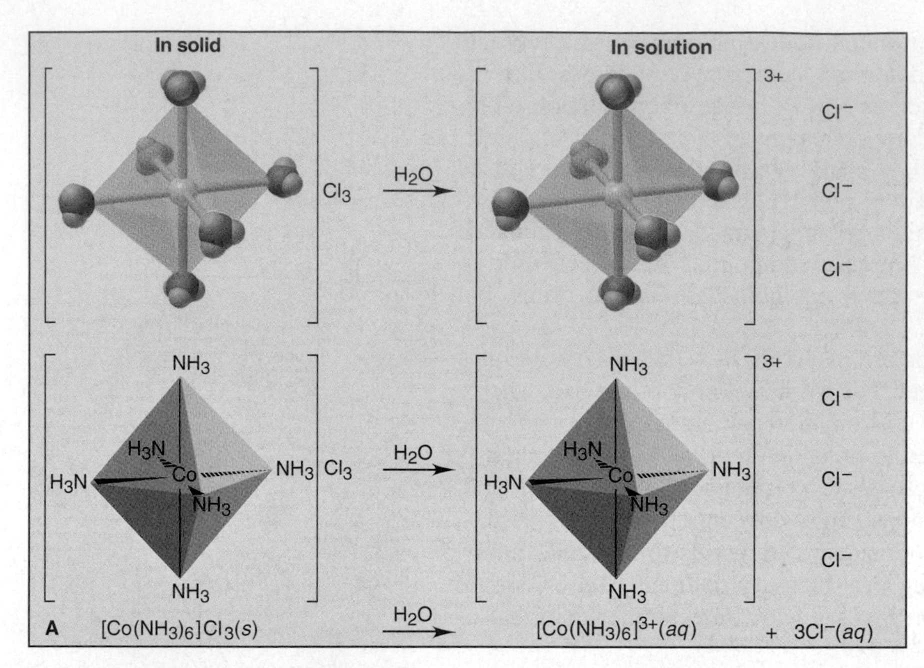

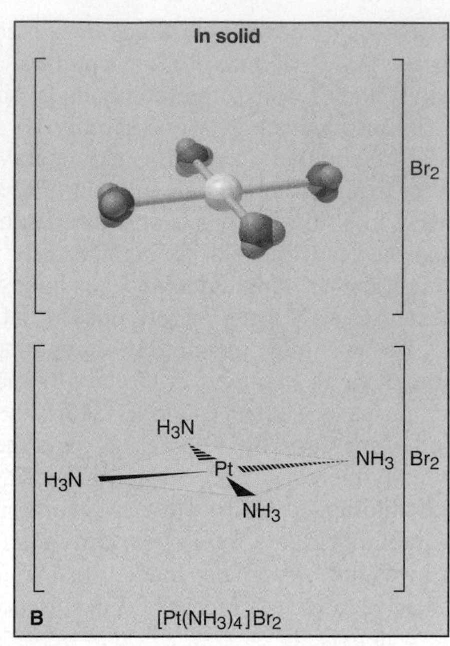

Figure 23.9 Components of a coordination compound. Coordination compounds, shown here as models (*top*), perspective drawings (*middle*), and chemical formulas (*bottom*), typically consist of a complex ion and counter ions to neutralize the charge. The complex ion has a central metal ion surrounded by ligands. **A,** When solid $[Co(NH_3)_6]Cl_3$ dissolves, the complex ion and the counter ions separate, but the ligands remain bound to the metal ion. Six ligands around the metal ion give the complex ion an octahedral geometry. **B,** Complex ions with a central d^8 metal ion have four ligands and a square planar geometry.

A typical coordination compound appears in Figure 23.9A: the coordination compound is $[Co(NH_3)_6]Cl_3$, the complex ion (always enclosed in square brackets) is $[Co(NH_3)_6]^{3+}$, the six NH_3 molecules bonded to the central Co^{3+} are ligands, and the three Cl^- ions are counter ions. *A coordination compound behaves like an electrolyte in water:* the complex ion and counter ions separate from each other. But the complex ion behaves like a polyatomic ion: *the ligands and central metal ion remain attached.* Thus, as Figure 23.9A shows, 1 mol of $[Co(NH_3)_6]Cl_3$ yields 1 mol of $[Co(NH_3)_6]^{3+}$ ions and 3 mol of Cl^- ions.

We discussed the Lewis acid-base properties of hydrated metal ions, which are a type of complex ion, in Section 18.9, and we examined complex-ion equilibria in Section 19.4. In this section, we consider the bonding, structure, and properties of complex ions.

Complex Ions: Coordination Numbers, Geometries, and Ligands

A complex ion is described by the metal ion and the number and types of ligands attached to it. Its structure is related to three characteristics—coordination number, geometry, and number of donor atoms per ligand:

- *Coordination number.* The **coordination number** is the *number of ligand atoms* that are bonded directly to the central metal ion and is *specific* for a given metal ion in a particular oxidation state and compound. The coordination number of the Co^{3+} ion in $[Co(NH_3)_6]^{3+}$ is 6 because six ligand atoms (N from NH_3) are bonded to it. The coordination number of the Pt^{2+} ion in many of its complexes is 4, whereas that of the Pt^{4+} ion in its complexes is 6. Copper(II) may have a coordination number of 2, 4, or 6 in different complex ions. In general, *the most common coordination number in complex ions is 6,* but 2 and 4 are often seen, and some higher ones are also known.
- *Geometry. The geometry (shape) of a complex ion depends on the coordination number and nature of the metal ion.* Table 23.6 shows the geometries associated with the coordination numbers 2, 4, and 6, with some examples of each. A complex ion whose metal ion has a coordination number of 2, such as $[Ag(NH_3)_2]^+$, is *linear.* The coordination number 4 gives rise to either of two geometries—square planar or tetrahedral. Most d^8 metal ions form *square planar* complex ions, depicted in Figure 23.9B. The d^{10} ions are among those that

Table 23.6 Coordination Numbers and Shapes of Some Complex Ions

Coordination Number	Shape	Examples
2	Linear	 $[CuCl_2]^-$, $[Ag(NH_3)_2]^+$, $[AuCl_2]^-$
4	Square planar	 $[Ni(CN)_4]^{2-}$, $[PdCl_4]^{2-}$, $[Pt(NH_3)_4]^{2+}$, $[Cu(NH_3)_4]^{2+}$
4	Tetrahedral	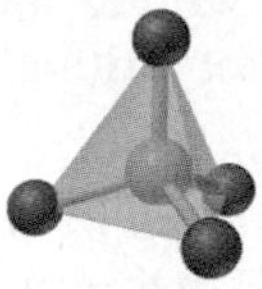 $[Cu(CN)_4]^{3-}$, $[Zn(NH_3)_4]^{2+}$, $[CdCl_4]^{2-}$, $[MnCl_4]^{2-}$
6	Octahedral	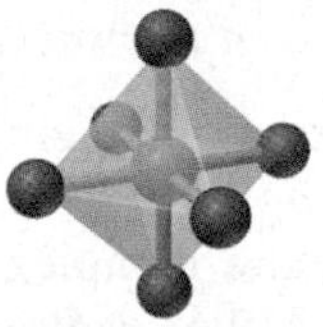 $[Ti(H_2O)_6]^{3+}$, $[V(CN)_6]^{4-}$, $[Cr(NH_3)_4Cl_2]^+$, $[Mn(H_2O)_6]^{2+}$, $[FeCl_6]^{3-}$, $[Co(en)_3]^{3+}$

form *tetrahedral* complex ions. A coordination number of 6 results in an *octahedral* geometry, as shown by $[Co(NH_3)_6]^{3+}$ in Figure 23.9A. Note the similarity with some of the molecular shapes in VSEPR theory (Section 10.2).

- *Donor atoms per ligand.* The ligands of complex ions are *molecules or anions* with one or more **donor atoms** that each *donate a lone pair of electrons* to the metal ion to form a covalent bond. Because they have at least one lone pair, donor atoms often come from Group 5A(15), 6A(16), or 7A(17).

Ligands are classified in terms of the number of donor atoms, or "teeth," that each uses to bond to the central metal ion. *Monodentate* (Latin, "one-toothed") ligands, such as Cl^- and NH_3, use a single donor atom. *Bidentate* ligands have two donor atoms, each of which bonds to the metal ion. *Polydentate* ligands have more than two donor atoms. Table 23.7 shows some common ligands in coordination compounds; note that each ligand has one or more donor atoms (colored type), each with a lone pair of electrons to donate. Bidentate and polydentate

Table 23.7 Some Common Ligands in Coordination Compounds

Ligand Type	Examples
Monodentate	$H_2\ddot{O}$: water; $:NH_3$ ammonia; $:\ddot{F}:^-$ fluoride ion; $:\ddot{Cl}:^-$ chloride ion; $[:C{\equiv}N:]^-$ cyanide ion; $[:\ddot{S}{=}C{=}\ddot{N}:]^-$ thiocyanate ion (S or N); $[:\ddot{O}{-}H]^-$ hydroxide ion; $[:\ddot{O}{-}\ddot{N}{=}\ddot{O}:]^-$ nitrite ion (O or N)
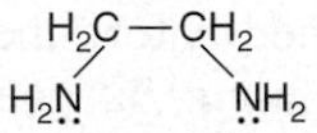 Bidentate	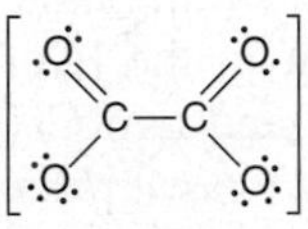 $H_2\ddot{N}{-}CH_2{-}CH_2{-}\ddot{N}H_2$ ethylenediamine (en); $[C_2O_4]^{2-}$ oxalate ion
Polydentate	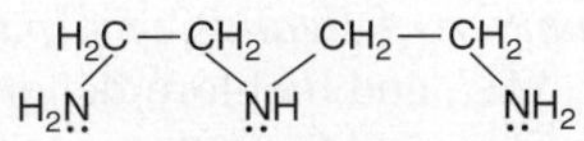 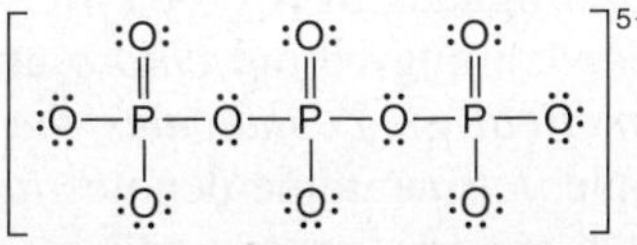 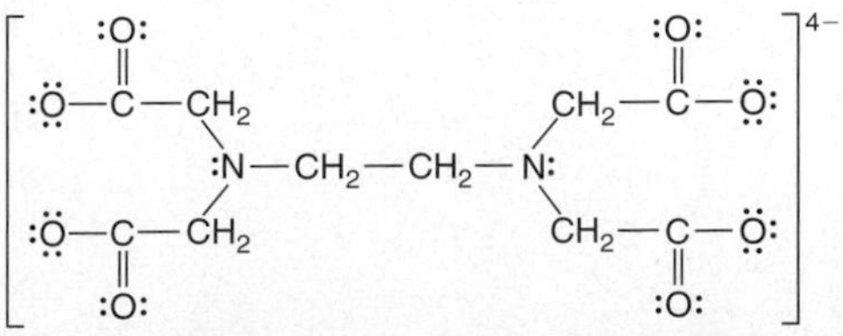 $H_2\ddot{N}{-}CH_2{-}CH_2{-}\ddot{N}H{-}CH_2{-}CH_2{-}\ddot{N}H_2$ diethylenetriamine; $[P_3O_{10}]^{5-}$ triphosphate ion; ethylenediaminetetraacetate ion ($EDTA^{4-}$)

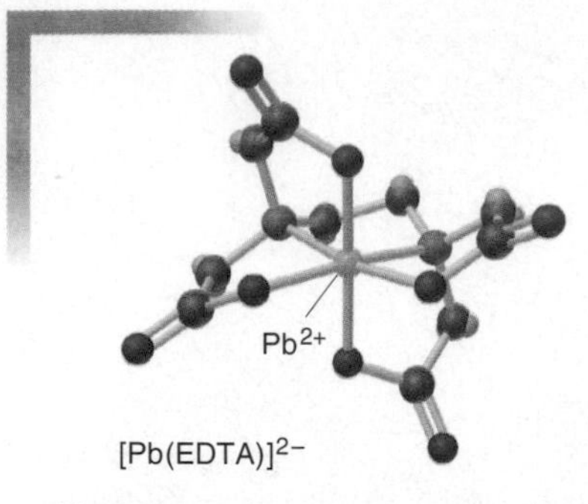

$[Pb(EDTA)]^{2-}$

Grabbing Ions Because it has 6 donor atoms, the ethylenediaminetetraacetate ($EDTA^{4-}$) ion forms very stable complexes with many metal ions. This property makes EDTA useful in treating heavy-metal poisoning. Once ingested by the patient, the ion acts as a scavenger to remove lead and other heavy-metal ions from the blood and other body fluids.

ligands give rise to *rings* in the complex ion. For instance, ethylenediamine (abbreviated *en* in formulas) has a chain of four atoms (:N—C—C—N:), so it forms a five-membered ring, with the two electron-donating N atoms bonding to the metal atom. Such ligands seem to grab the metal ion like claws, so a complex ion that contains them is also called a **chelate** (pronounced "KEY-late"; Greek *chela,* "crab's claw").

Formulas and Names of Coordination Compounds

There are three important rules for writing the formulas of coordination compounds, the first two being the same for writing formulas of any ionic compound:

1. *The cation is written before the anion.*
2. *The charge of the cation(s) is balanced by the charge of the anion(s).*
3. *In the complex ion, neutral ligands are written before anionic ligands, and the formula for the whole ion is placed in brackets.*

Let's apply these rules as we examine the combinations of ions in coordination compounds. *The whole complex ion may be a cation or an anion.* A complex cation has anionic counter ions, and a complex anion has cationic counter ions. It's easy to find the charge of the central metal ion. For example, in $K_2[Co(NH_3)_2Cl_4]$, two K^+ counter ions balance the charge of the complex anion $[Co(NH_3)_2Cl_4]^{2-}$, which contains two NH_3 molecules and four Cl^- ions as ligands. The two NH_3 are neutral, the four Cl^- have a total charge of 4−, and the entire complex ion has a charge of 2−, so the central metal ion must be Co^{2+}:

$$\text{Charge of complex ion} = \text{Charge of metal ion} + \text{total charge of ligands}$$
$$2- = \text{Charge of metal ion} + [(2 \times 0) + (4 \times 1-)]$$

So, $$\text{Charge of metal ion} = (2-) - (4-) = 2+$$

In the compound $[Co(NH_3)_4Cl_2]Cl$, the complex ion is $[Co(NH_3)_4Cl_2]^+$ and one Cl^- is the counter ion. The four NH_3 ligands are neutral, the two Cl^- ligands have a total charge of 2−, and the complex cation has a charge of 1+, so the central metal ion must be Co^{3+} [that is, 1+ = (3+) + (2−)]. Some coordination compounds have a complex cation *and* a complex anion, as in $[Co(NH_3)_5Br]_2[Fe(CN)_6]$. In this compound, the complex cation is $[Co(NH_3)_5Br]^{2+}$, with Co^{3+}, and the complex anion is $[Fe(CN)_6]^{4-}$, with Fe^{2+}.

Coordination compounds were originally named after the person who first prepared them or from their color, and some of these common names are still used, but most coordination compounds are named systematically through a set of rules:

1. *The cation is named before the anion.* In naming $[Co(NH_3)_4Cl_2]Cl$, for example, we name the $[Co(NH_3)_4Cl_2]^+$ ion before the Cl^- ion. Thus, the name is

   tetraamminedichlorocobalt(III) chloride

   The only space in the name appears between the cation and the anion.
2. *Within the complex ion, the ligands are named, in alphabetical order,* ***before*** *the metal ion.* Note that in the $[Co(NH_3)_4Cl_2]^+$ ion of the compound named in rule 1, the four NH_3 and two Cl^- are named before the Co^{3+}.
3. *Neutral ligands generally have the molecule name,* but there are a few exceptions (Table 23.8). *Anionic ligands drop the -ide and add -o after the root name;* thus, the name *fluoride* for the F^- ion becomes the ligand name *fluoro.* The two ligands in $[Co(NH_3)_4Cl_2]^+$ are *ammine* (NH_3) and *chloro* (Cl^-) with *ammine* coming before *chloro* alphabetically.
4. *A numerical prefix indicates the number of ligands of a particular type.* For example, *tetra*ammine denotes *four* NH_3, and *di*chloro denotes *two* Cl^-. Other prefixes are *tri-, penta-,* and *hexa-.* These prefixes do *not* affect the alphabetical order; thus, *tetraammine* comes before *dichloro.* Because some ligand

Table 23.8 Names of Some Neutral and Anionic Ligands

Neutral		Anionic	
Name	**Formula**	**Name**	**Formula**
Aqua	H_2O	Fluoro	F^-
Ammine	NH_3	Chloro	Cl^-
Carbonyl	CO	Bromo	Br^-
Nitrosyl	NO	Iodo	I^-
		Hydroxo	OH^-
		Cyano	CN^-

names already contain a numerical prefix (such as ethylene*di*amine), we use *bis* (2), *tris* (3), or *tetrakis* (4) to indicate the number of such ligands, followed by the ligand name in parentheses. Therefore, a complex ion that has two ethylenediamine ligands has *bis(ethylenediamine)* in its name.

5. *The oxidation state of the central metal ion is given by a Roman numeral (in parentheses) only* if the metal ion can have more than one state, as in the compound named in rule 1.
6. *If the complex ion is an anion, we drop the ending of the metal name and add -ate.* Thus, the name for $K[Pt(NH_3)Cl_5]$ is

   potassium amminepentachloroplatinate(IV)

   (Note that there is one K^+ counter ion, so the complex anion has a charge of 1−. The five Cl^- ligands have a total charge of 5−, so Pt must be in the +4 oxidation state.) For some metals, we use the Latin root with the *-ate* ending, shown in Table 23.9. For example, the name for $Na_4[FeBr_6]$ is

   sodium hexabromoferrate(II)

Table 23.9 Names of Some Metal Ions in Complex Anions

Metal	Name in Anion
Iron	Ferrate
Copper	Cuprate
Lead	Plumbate
Silver	Argentate
Gold	Aurate
Tin	Stannate

SAMPLE PROBLEM 23.3 Writing Names and Formulas of Coordination Compounds

Problem (a) What is the systematic name of $Na_3[AlF_6]$?
(b) What is the systematic name of $[Co(en)_2Cl_2]NO_3$?
(c) What is the formula of tetraamminebromochloroplatinum(IV) chloride?
(d) What is the formula of hexaamminecobalt(III) tetrachloroferrate(III)?

Plan We use the rules that were presented above and refer to Tables 23.8 and 23.9.

Solution (a) The complex ion is $[AlF_6]^{3-}$. There are six *(hexa-)* F^- ions *(fluoro)* as ligands, so we have *hexafluoro*. The complex ion is an anion, so the ending of the metal ion (aluminum) must be changed to *-ate:* hexafluoroaluminate. Aluminum has only the +3 oxidation state, so we do *not* use a Roman numeral. The positive counter ion is named first and separated from the anion by a space: sodium hexafluoroaluminate.

(b) Listed alphabetically, there are two Cl^- *(dichloro)* and two en [*bis(ethylenediamine)*] as ligands. The complex ion is a cation, so the metal name is unchanged, but we specify its oxidation state because cobalt can have several. One NO_3^- balances the 1+ cation charge: with 2− for two Cl^- and 0 for two en, the metal must be *cobalt(III)*. The word *nitrate* follows a space: dichlorobis(ethylenediamine)cobalt(III) nitrate.

(c) The central metal ion is written first, followed by the neutral ligands and then (in alphabetical order) by the negative ligands. *Tetraammine* is four NH_3, *bromo* is one Br^-, *chloro* is one Cl^-, and *platinate(IV)* is Pt^{4+}, so the complex ion is $[Pt(NH_3)_4BrCl]^{2+}$. Its 2+ charge is the sum of 4+ for Pt^{4+}, 0 for four NH_3, 1− for one Br^-, and 1− for one Cl^-. To balance the 2+ charge, we need two Cl^- counter ions: $[Pt(NH_3)_4BrCl]Cl_2$.

(d) This compound consists of two different complex ions. In the cation, *hexaammine* is six NH_3 and *cobalt(III)* is Co^{3+}, so the cation is $[Co(NH_3)_6]^{3+}$. The 3+ charge is the sum of 3+ for Co^{3+} and 0 for six NH_3. In the anion, *tetrachloro* is four Cl^-, and *ferrate(III)* is Fe^{3+}, so the anion is $[FeCl_4]^-$. The 1− charge is the sum of 3+ for Fe^{3+} and 4− for four Cl^-. In the neutral compound, one 3+ cation is balanced by three 1− anions: $[Co(NH_3)_6][FeCl_4]_3$.

Check Reverse the process to be sure you obtain the name or formula asked for in the problem.

FOLLOW-UP PROBLEM 23.3 **(a)** What is the name of $[Cr(H_2O)_5Br]Cl_2$?
(b) What is the formula of barium hexacyanocobaltate(III)?

A Historical Perspective: Alfred Werner and Coordination Theory

The substances we now call coordination compounds had been known for almost 200 years when the young Swiss chemist Alfred Werner began studying them in the 1890s. He investigated a series of compounds such as the cobalt series shown in Table 23.10, each of which contains one cobalt(III) ion, three chloride ions, and a given number of ammonia molecules. At the time, which was 30 years before the idea of atomic orbitals was proposed, no structural theory could explain how compounds with similar, even identical, formulas could have widely different properties.

Werner measured the conductivity of each compound in aqueous solution to determine the total number of ions that became dissociated. He treated the solutions with excess $AgNO_3$ to precipitate released Cl^- ions as AgCl and thus determine the number of free Cl^- ions per formula unit. Previous studies had established that the NH_3 molecules were not free in solution. Werner's data, summarized in Table 23.10, could not be explained by the accepted, traditional formulas of the compounds. Other chemists had proposed "chain" structures, like those of organic compounds, to explain such data. For example, a proposed structure for $[Co(NH_3)_6]Cl_3$ was

```
NH3—Cl
|
Co—NH3—Cl
|
NH3—NH3—NH3—NH3—Cl
```

However, these models proved inadequate.

Werner's novel idea was the coordination complex, a central metal ion surrounded by a *constant total number* of covalently bonded molecules and/or anions. The coordination complex could be neutral or charged; if charged, it combined with oppositely charged counter ions, in this case Cl^-, to form the neutral compound.

Table 23.10 Some Coordination Compounds of Cobalt Studied by Werner

Traditional Formula	Werner's Data* Total Ions	Werner's Data* Free Cl^-	Modern Formula	Charge of Complex Ion
$CoCl_3{\cdot}6NH_3$	4	3	$[Co(NH_3)_6]Cl_3$	3+
$CoCl_3{\cdot}5NH_3$	3	2	$[Co(NH_3)_5Cl]Cl_2$	2+
$CoCl_3{\cdot}4NH_3$	2	1	$[Co(NH_3)_4Cl_2]Cl$	1+
$CoCl_3{\cdot}3NH_3$	0	0	$[Co(NH_3)_3Cl_3]$	—

*Moles per mole of compound.

Werner proposed two types of valence, or *combining ability,* for metal ions. *Primary valence,* now called *oxidation state,* is the positive charge on the metal ion that must be satisfied by an equivalent negative charge. In Werner's cobalt series, the primary valence is +3, and it is always balanced by three Cl^- ions. These anions can be bonded covalently to Co as part of the complex ion and/or associated with it as counter ions. *Secondary valence,* now called *coordination number,* is the constant total number of connections (anionic or neutral ligands) *within* the complex ion. The secondary valence in this series of Co compounds is 6.

As you can see, Werner's data are satisfied if the total number of ligands remains the same for each compound, even though the numbers of Cl^- ions and NH_3 molecules in the different complex ions vary. For example, the first compound, $[Co(NH_3)_6]Cl_3$, has a total of four ions: one $[Co(NH_3)_6]^{3+}$ and three Cl^-. All three Cl^- ions are free to form AgCl. The last compound, $[Co(NH_3)_3Cl_3]$, contains no separate ions.

Surprisingly, Werner was an organic chemist, and his work on coordination compounds, which virtually revolutionized his contemporaries' understanding of chemical bonding, was his attempt to demonstrate the unity of chemistry. For these pioneering studies, especially his prediction of optical isomerism (discussed next), Werner received the Nobel Prize in chemistry in 1913.

Isomerism in Coordination Compounds

Isomers are compounds with the same chemical formula but different properties. We discussed many aspects of isomerism in the context of organic compounds in Section 15.2; it may be helpful to review that section now. Figure 23.10 presents an overview of the most common types of isomerism in coordination compounds.

Figure 23.10 Important types of isomerism in coordination compounds.

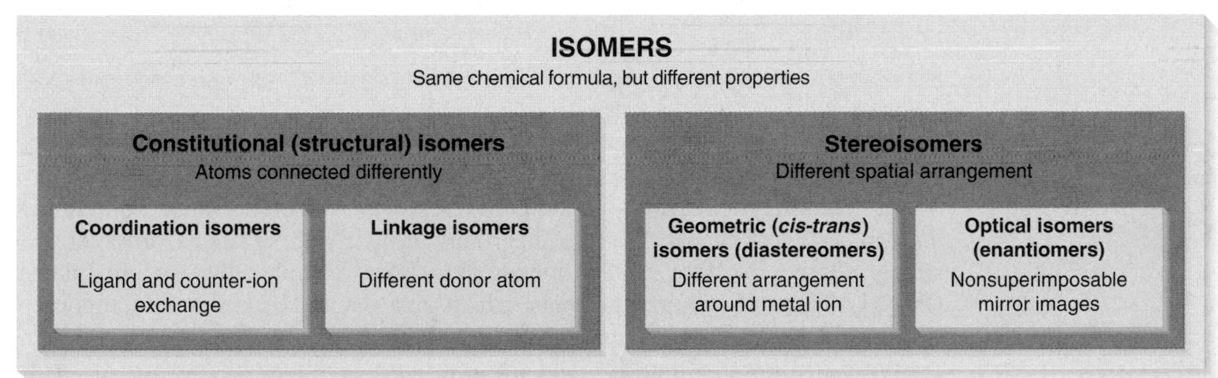

Constitutional Isomers: Same Atoms Connected Differently Two compounds with the same formula, but with the atoms connected differently, are called **constitutional (structural) isomers.** Coordination compounds exhibit the following two types of constitutional isomers: one involves a difference in the composition of the complex ion, the other in the donor atom of the ligand.

1. **Coordination isomers** occur when the composition of the complex ion changes but not that of the compound. One way this type of isomerism occurs is when ligand and counter ion exchange positions, as in $[Pt(NH_3)_4Cl_2](NO_2)_2$ and $[Pt(NH_3)_4(NO_2)_2]Cl_2$. In the first compound, the Cl^- ions are the ligands, and the NO_2^- ions are counter ions; in the second, the roles are reversed. Another way that this type of isomerism occurs is in compounds of two complex ions in which the two sets of ligands in one compound are reversed in the other, as in $[Cr(NH_3)_6][Co(CN)_6]$ and $[Co(NH_3)_6][Cr(CN)_6]$; note that NH_3 is a ligand of Cr^{3+} in one compound and of Co^{3+} in the other.

2. **Linkage isomers** occur when the composition of the complex ion remains the same but the attachment of the ligand donor atom changes. Some ligands can bind to the metal ion through *either of two donor atoms*. For example, the nitrite ion can bind through a lone pair on either the N atom (*nitro*, O_2N:) or one of the O atoms (*nitrito*, ONO:) to give linkage isomers, as in the orange compound pentaammine*nitro*cobalt(III) chloride $[Co(NH_3)_5(NO_2)]Cl_2$ *(below left)* and its red linkage isomer pentaammine*nitrito*cobalt(III) chloride $[Co(NH_3)_5(ONO)]Cl_2$ *(below right)*:

Nitro **isomer** ***Nitrito*** **isomer**

Another example is the cyanate ion, which can attach via a lone pair on the O atom (*cyanato*, NCO:) or the N atom (*isocyanato*, OCN:); the thiocyanate ion behaves similarly, attaching via the S atom or the N atom:

$[O{=}N{-}O]^-$ nitrite $[:\ddot{O}{=}C{=}\ddot{N}:]^-$ cyanate $[:\ddot{S}{=}C{=}\ddot{N}:]^-$ thiocyanate

Stereoisomers: Different Spatial Arrangements of Atoms **Stereoisomers** are compounds that have the same atomic connections but different spatial arrangements of the atoms. The two types we discussed for organic compounds, called *geometric* and *optical* isomers, are seen with coordination compounds as well:

1. **Geometric isomers** (also called ***cis-trans*** **isomers** and, sometimes, *diastereomers*) occur when atoms or groups of atoms are arranged differently in space relative to the central metal ion. For example, the square planar $[Pt(NH_3)_2Cl_2]$ has two arrangements, which give rise to two different compounds (Figure 23.11A). The isomer with identical ligands *next* to each other is *cis*-diamminedichloroplatinum(II), and the one with identical ligands *across* from each other is *trans*-diamminedichloroplatinum(II); their biological behaviors are remarkably different. Octahedral complexes also exhibit *cis-trans* isomerism (Figure 23.11B). The *cis* isomer of the $[Co(NH_3)_4Cl_2]^+$ ion has the two Cl^- ligands next to each other and is violet, whereas the *trans* isomer has these two ligands across from each other and is green.

2. **Optical isomers** (also called *enantiomers*) occur when a molecule and its mirror image cannot be superimposed (see Figures 15.8 to 15.10, pp. 627–628). Unlike other types of isomers, which have distinct physical properties, optical isomers are physically identical in all ways but one: *the direction in which they rotate the plane of polarized light*. Octahedral complex ions show many examples of optical isomerism, which we can observe by rotating one isomer and seeing if it is superimposable on the other isomer (its mirror image). For example, as you can see in Figure 23.12A, the two structures (I and II) of $[Co(en)_2Cl_2]^+$, the *cis*-dichlorobis(ethylenediamine)cobalt(III) ion, are mirror images of each other. Rotate structure I 180° around a vertical axis, and you obtain III. The Cl^- li-

Anticancer Geometric Isomers In the mid-1960s, Barnett Rosenberg and his colleagues found that *cis*-$[Pt(NH_3)_2Cl_2]$ (cisplatin) was a highly effective antitumor agent. This compound and several closely related platinum(II) complexes are still among the most effective treatments for certain types of cancer. The geometric isomer, *trans*-$[Pt(NH_3)_2Cl_2]$, however, has no antitumor effect. Cisplatin may work by lying within the cancer cell's DNA double helix, such that a donor atom on each strand replaces a Cl^- ligand and binds the platinum(II) strongly, preventing DNA replication (Section 15.6).

Table 23.11 Relation Between Absorbed and Observed Colors

Absorbed Color	λ (nm)	Observed Color	λ (nm)
Violet	400	Green-yellow	560
Blue	450	Yellow	600
Blue-green	490	Red	620
Yellow-green	570	Violet	410
Yellow	580	Dark blue	430
Orange	600	Blue	450
Red	650	Green	520

Foliage changing color in autumn.

by the chlorophyll. Xanthophylls absorb green and blue strongly, reflecting the bright yellows and reds of autumn (see photo).

Splitting of *d* Orbitals in an Octahedral Field of Ligands The crystal field model explains that the properties of complexes result from the splitting of *d*-orbital energies, which arises from electrostatic interactions between metal ion and ligands. The model assumes that a complex ion forms as a result of *electrostatic attractions between the metal cation and the negative charge of the ligands.* This negative charge is either partial, as in a polar neutral ligand like NH_3, or full, as in an anionic ligand like Cl^-. The ligands approach the metal ion along the mutually perpendicular *x*, *y*, and *z* axes, which minimizes the overall energy of the system.

Picture what happens as the ligands approach. Figure 23.17A shows six ligands moving toward a metal ion to form an octahedral complex. Let's see how the various *d* orbitals of the metal ion are affected as the complex forms. As the ligands approach, their electron pairs repel electrons in the five *d* orbitals. In the isolated metal ion, the *d* orbitals have equal energies despite their different orientations. In the electrostatic field of ligands, however, the *d* electrons are *repelled unequally because their orbitals have different orientations.* Because the ligands move along the *x*, *y*, and *z* axes, they approach *directly toward* the lobes of the $d_{x^2-y^2}$ and d_{z^2} orbitals (Figure 23.17B and C) but *between* the lobes of the d_{xy}, d_{xz}, and d_{yz} orbitals (Figure 23.17D to F). Thus, electrons in the $d_{x^2-y^2}$ and d_{z^2} orbitals experience *stronger* repulsions than those in the d_{xy}, d_{xz}, and d_{yz} orbitals.

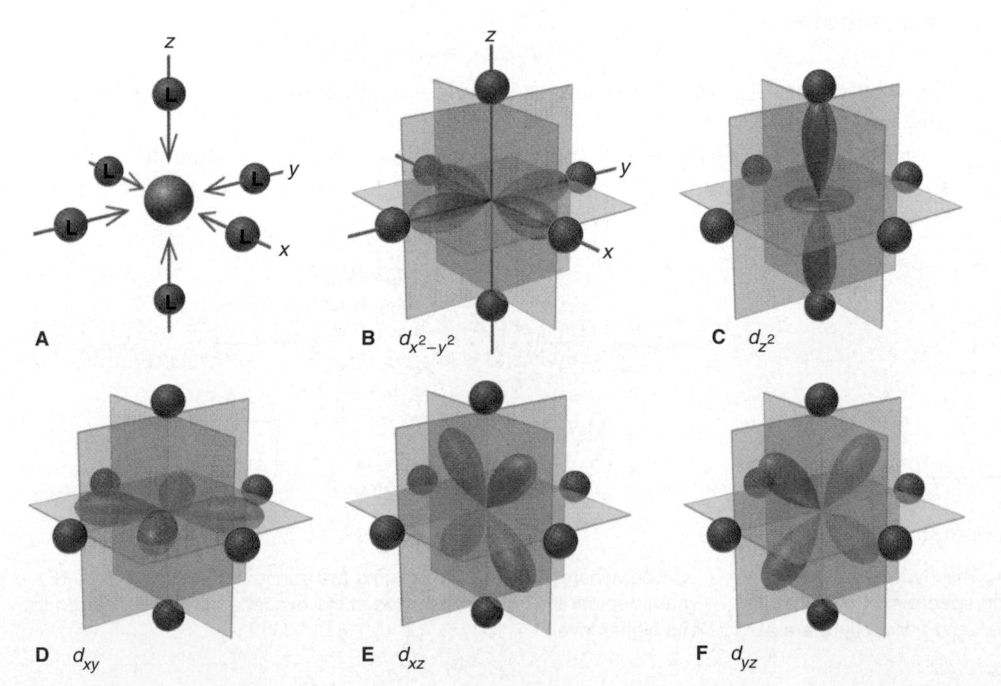

Figure 23.17 The five *d* orbitals in an octahedral field of ligands. The direction of ligand approach influences the strength of repulsions of electrons in the five metal *d* orbitals. **A,** We assume that ligands approach a metal ion along the three linear axes in an octahedral orientation. **B** and **C,** Lobes of the $d_{x^2-y^2}$ and d_{z^2} orbitals lie *directly in line* with the approaching ligands, so repulsions are stronger. **D** to **F,** Lobes of the d_{xy}, d_{xz}, and d_{yz} orbitals lie *between* the approaching ligands, so repulsions are weaker.

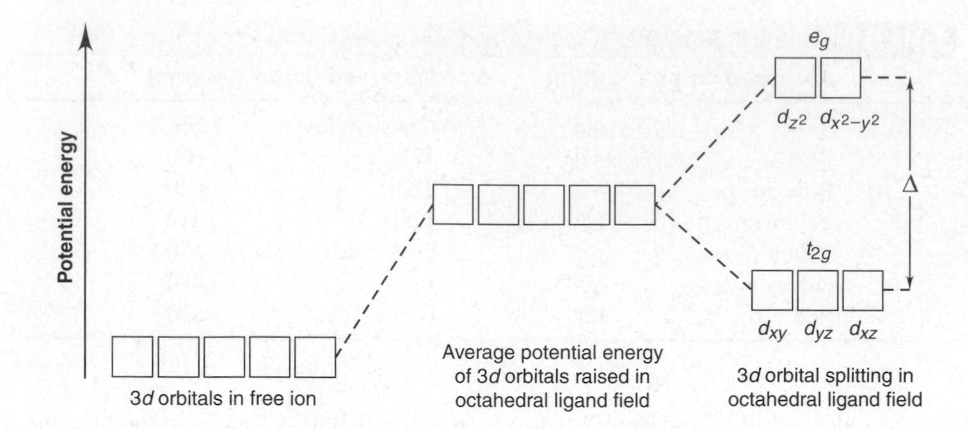

Figure 23.18 Splitting of *d*-orbital energies by an octahedral field of ligands. Electrons in the *d* orbitals of the free metal ion experience an *average* net repulsion in the negative ligand field that increases all *d*-orbital energies. Electrons in the d_{xy}, d_{yz}, and d_{xz} orbitals, which form the t_{2g} set, are repelled less than those in the $d_{x^2-y^2}$ and d_{z^2} orbitals, which form the e_g set. The energy difference between these two sets is the crystal field splitting energy, Δ.

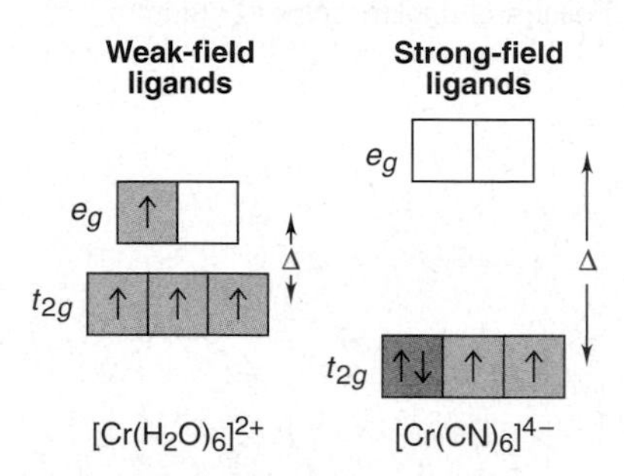

Figure 23.19 The effect of the ligand on splitting energy. Ligands interacting strongly with metal-ion *d* orbitals, such as CN^-, produce a larger Δ than those interacting weakly, such as H_2O.

An energy diagram of the orbitals shows that all five *d* orbitals are higher in energy in the forming complex than in the free metal ion because of repulsions from the approaching ligands, but *the orbital energies split, with two d orbitals higher in energy than the other three* (Figure 23.18). The two higher energy orbitals are called e_g **orbitals,** and the three lower energy ones are t_{2g} **orbitals.** (These designations refer to features of the orbitals that need not concern us here.)

The splitting of orbital energies is called the *crystal field effect,* and the difference in energy between the e_g and t_{2g} sets of orbitals is the **crystal field splitting energy (Δ).** Different ligands create crystal fields of different strength and, thus, cause the *d*-orbital energies to split to different extents. **Strong-field ligands** lead to a *larger* splitting energy (larger Δ); **weak-field ligands** lead to a *smaller* splitting energy (smaller Δ). For instance, H_2O is a weak-field ligand, and CN^- is a strong-field ligand (Figure 23.19). The magnitude of Δ relates directly to the color and magnetic properties of a complex.

Explaining the Colors of Transition Metals The remarkably diverse colors of coordination compounds are determined by the energy difference (Δ) between the t_{2g} and e_g orbital sets in their complex ions. When the ion absorbs light in the visible range, electrons are excited ("jump") from the lower energy t_{2g} level to the higher e_g level. Recall that the *difference* between two electronic energy levels in the ion is equal to the energy (and inversely related to the wavelength) of the absorbed photon:

$$\Delta E_{\text{electron}} = E_{\text{photon}} = h\nu = hc/\lambda$$

The substance has a color because only certain wavelengths of the incoming white light are absorbed.

Consider the $[Ti(H_2O)_6]^{3+}$ ion, which appears purple in aqueous solution (Figure 23.20). Hydrated Ti^{3+} is a d^1 ion, with the *d* electron in one of the three

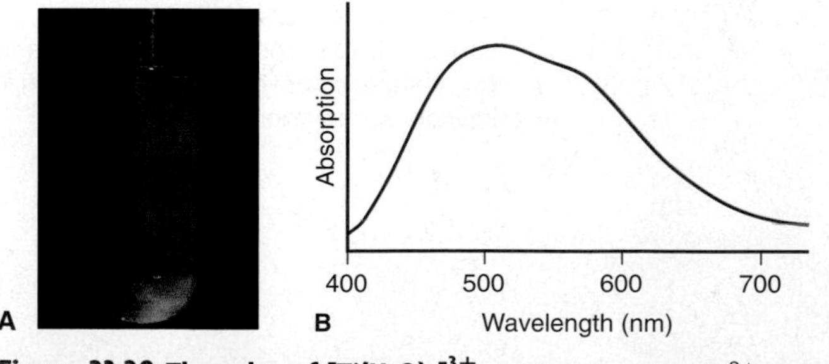

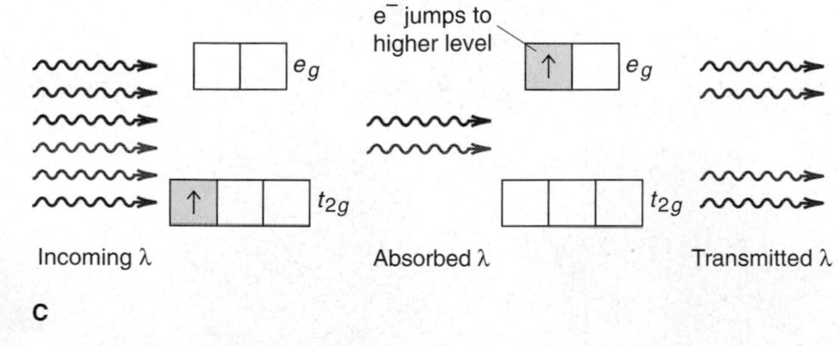

Figure 23.20 The color of $[Ti(H_2O)_6]^{3+}$. A, The hydrated Ti^{3+} ion is purple in aqueous solution. **B,** An absorption spectrum shows that incoming wavelengths corresponding to green and yellow light are absorbed, whereas other wavelengths are transmitted. **C,** An orbital diagram depicts the colors absorbed in the excitation of the *d* electron to the higher level.

lower energy t_{2g} orbitals. The energy difference (Δ) between the t_{2g} and e_g orbitals in this ion corresponds to the energy of photons spanning the green and yellow range. When white light shines on the solution, these colors of light are absorbed, and the electron jumps to one of the e_g orbitals. Red, blue, and violet light are transmitted, so the solution appears purple.

Absorption spectra show the wavelengths absorbed by a given metal ion with different ligands and by different metal ions with the same ligand. From such data, we relate the energy of the absorbed light to the Δ values, and two important observations emerge:

1. *For a given ligand, the color depends on the oxidation state of the metal ion.* A solution of $[V(H_2O)_6]^{2+}$ ion is violet, and a solution of $[V(H_2O)_6]^{3+}$ ion is yellow (Figure 23.21A).
2. *For a given metal ion, the color depends on the ligand.* Even a single ligand substitution can have a major effect on the wavelengths absorbed and, thus, the color, as you can see for two Cr^{3+} complex ions in Figure 23.21B.

Figure 23.21 Effects of the metal oxidation state and of ligand identity on color. **A,** Solutions of $[V(H_2O)_6]^{2+}$ (*left*) and $[V(H_2O)_6]^{3+}$ (*right*) ions have different colors. **B,** A change in even a single ligand can influence the color. The $[Cr(NH_3)_6]^{3+}$ ion is yellow-orange (*left*); the $[Cr(NH_3)_5Cl]^{2+}$ ion is purple (*right*).

The second observation allows us to rank ligands into a **spectrochemical series** with regard to their ability to split *d*-orbital energies. An abbreviated series, moving from weak-field ligands (small splitting, small Δ) to strong-field ligands (large splitting, large Δ), is shown in Figure 23.22. Using this series, we can predict the *relative* size of Δ for a series of octahedral complexes of the same metal ion. Although it is difficult to predict the actual color of a given complex, we can determine whether a complex will absorb longer or shorter wavelengths than other complexes in the series.

SAMPLE PROBLEM 23.5 Ranking Crystal Field Splitting Energies for Complex Ions of a Given Metal

Problem Rank the ions $[Ti(H_2O)_6]^{3+}$, $[Ti(NH_3)_6]^{3+}$, and $[Ti(CN)_6]^{3-}$ in terms of the relative value of Δ and of the energy of visible light absorbed.

Plan The formulas show that titanium's oxidation state is +3 in the three ions. From Figure 23.22, we rank the ligands in terms of crystal field strength: the stronger the ligand, the greater the splitting, and the higher the energy of light absorbed.

Solution The ligand field strength is in the order $CN^- > NH_3 > H_2O$, so the relative size of Δ and energy of light absorbed is

$$Ti(CN)_6^{3-} > Ti(NH_3)_6^{3+} > Ti(H_2O)_6^{3+}$$

FOLLOW-UP PROBLEM 23.5 Which complex ion absorbs visible light of higher energy, $[V(H_2O)_6]^{3+}$ or $[V(NH_3)_6]^{3+}$?

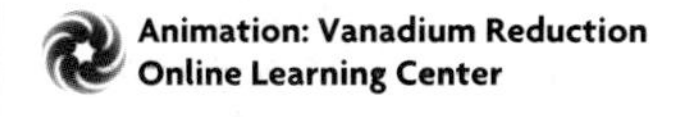

Animation: Vanadium Reduction
Online Learning Center

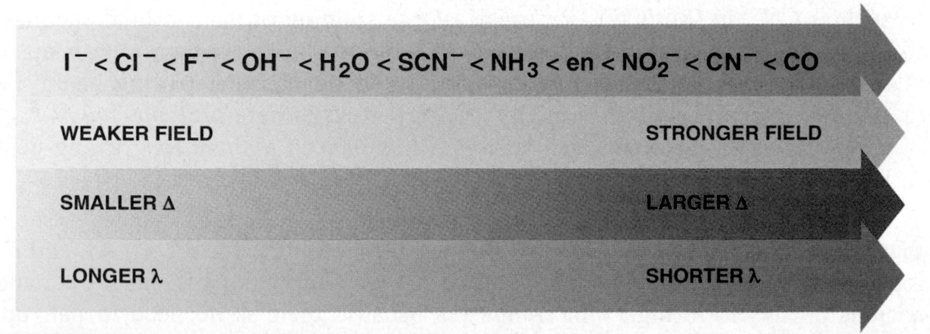

Figure 23.22 The spectrochemical series. As the crystal field strength of the ligand increases, the splitting energy (Δ) increases, so shorter wavelengths (λ) of light must be absorbed to excite electrons. Water is usually a weak-field ligand.

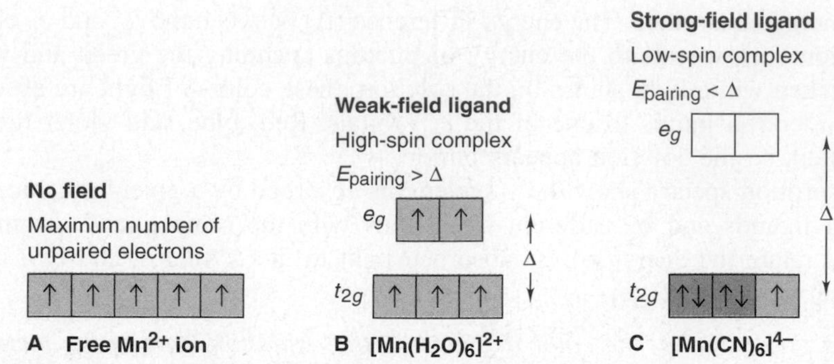

Figure 23.23 High-spin and low-spin complex ions of Mn^{2+}. **A,** The free Mn^{2+} ion has five unpaired electrons. **B,** Bonded to weak-field ligands (smaller Δ), Mn^{2+} still has five unpaired electrons (high-spin complex). **C,** Bonded to strong-field ligands (larger Δ), Mn^{2+} has only one unpaired electron (low-spin complex).

Explaining the Magnetic Properties of Transition Metal Complexes The splitting of energy levels influences magnetic properties by affecting the number of *unpaired* electrons in the metal ion's *d* orbitals. Based on Hund's rule, electrons occupy orbitals one at a time as long as orbitals of equal energy are available. When all lower energy orbitals are half-filled, the next electron can

- enter a half-filled orbital and pair up by overcoming a repulsive *pairing energy* ($E_{pairing}$), or
- enter an empty, higher energy orbital by overcoming the crystal field splitting energy (Δ).

Thus, *the relative sizes of $E_{pairing}$ and Δ determine the occupancy of the d orbitals.* The orbital occupancy pattern, in turn, determines the number of unpaired electrons and, thus, the paramagnetic behavior of the ion.

As an example, the isolated Mn^{2+} ion ([Ar] $3d^5$) has five unpaired electrons in 3*d* orbitals of equal energy (Figure 23.23A). In an octahedral field of ligands, the orbital energies split. The orbital occupancy is affected by the ligand in one of two ways:

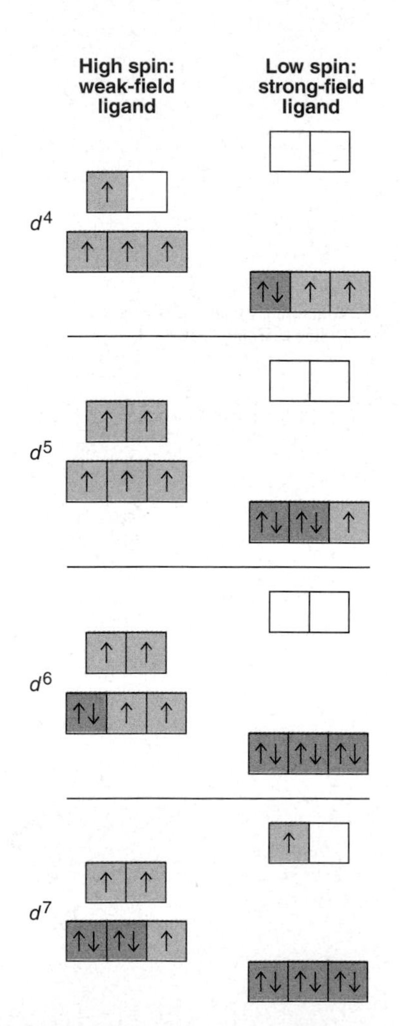

Figure 23.24 Orbital occupancy for high-spin and low-spin complexes of d^4 through d^7 metal ions.

- *Weak-field ligands and high-spin complexes.* Weak-field ligands, such as H_2O in $[Mn(H_2O)_6]^{2+}$, cause a *small* splitting energy, so it takes *less* energy for *d* electrons to jump to the e_g set than to pair up in the t_{2g} set. Therefore, the *d* electrons remain unpaired (Figure 23.23B). Thus, with weak-field ligands, the pairing energy is *greater* than the splitting energy ($E_{pairing} > \Delta$); therefore, *the number of unpaired electrons in the complex ion is the* ***same*** *as in the free ion.* Weak-field ligands create **high-spin complexes,** those with the *maximum* number of unpaired electrons.
- *Strong-field ligands and low-spin complexes.* In contrast, strong-field ligands, such as CN^- in $[Mn(CN)_6]^{4-}$, cause a *large* splitting of the *d*-orbital energies, so it takes *more* energy for electrons to jump to the e_g set than to pair up in the t_{2g} set (Figure 23.23C). With strong-field ligands, the pairing energy is *smaller* than the splitting energy ($E_{pairing} < \Delta$); therefore, *the number of unpaired electrons in the complex ion is* ***less*** *than in the free ion.* Strong-field ligands create **low-spin complexes,** those with *fewer* unpaired electrons.

Orbital diagrams for the d^1 through d^9 ions in octahedral complexes show that both high-spin and low-spin options are possible only for d^4, d^5, d^6, and d^7 ions (Figure 23.24). With three lower energy t_{2g} orbitals available, the d^1, d^2, and d^3 ions always form high-spin complexes because there is no need to pair up. Similarly, d^8 and d^9 ions always form high-spin complexes: because the t_{2g} set is filled with six electrons, the two e_g orbitals *must* have either two (d^8) or one (d^9) unpaired electron.

SAMPLE PROBLEM 23.6 Identifying Complex Ions as High Spin or Low Spin

Problem Iron(II) forms an essential complex in hemoglobin. For each of the two octahedral complex ions $[Fe(H_2O)_6]^{2+}$ and $[Fe(CN)_6]^{4-}$, draw an orbital splitting diagram, predict the number of unpaired electrons, and identify the ion as low spin or high spin.

Plan The Fe^{2+} electron configuration gives us the number of *d* electrons, and the spectrochemical series in Figure 23.22 shows the relative strengths of the two ligands. We draw the diagrams, separating the t_{2g} and e_g sets by a greater distance for the strong-field ligand. Then we add electrons, noting that a weak-field ligand gives the *maximum* number of unpaired electrons and a high-spin complex, whereas a strong-field ligand leads to electron pairing and a low-spin complex.

Solution Fe^{2+} has the [Ar] $3d^6$ configuration. According to Figure 23.22, H_2O produces smaller splitting than CN^-. The diagrams are shown in the margin. The $[Fe(H_2O)_6]^{2+}$ ion has four unpaired electrons (high spin), and the $[Fe(CN)_6]^{4-}$ ion has no unpaired electrons (low spin).

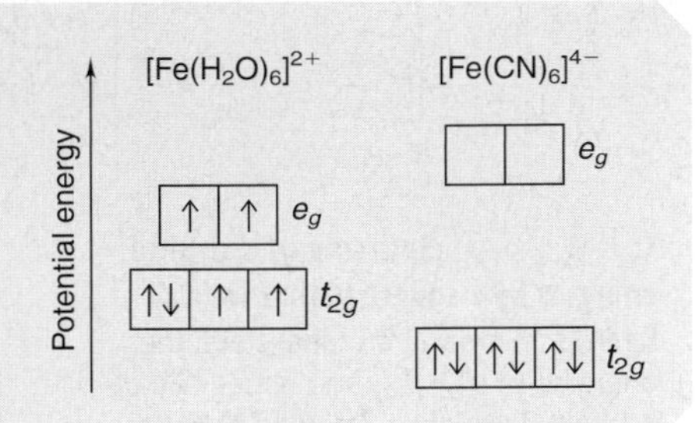

Comment 1. H_2O is a weak-field ligand, so it almost always forms high-spin complexes. **2.** These results are correct, but we cannot confidently predict the spin of a complex without having actual values for Δ and $E_{pairing}$. **3.** Cyanide ions and carbon monoxide are highly toxic because they interact with the iron cations in proteins.

FOLLOW-UP PROBLEM 23.6 How many unpaired electrons do you expect for $[Mn(CN)_6]^{3-}$? Is this a high-spin or low-spin complex ion?

Crystal Field Splitting in Tetrahedral and Square Planar Complexes Four ligands around a metal ion also cause *d*-orbital splitting, but the magnitude and pattern of the splitting depend on whether the ligands are in a tetrahedral or a square planar arrangement.

- *Tetrahedral complexes.* With the ligands approaching from the corners of a tetrahedron, none of the five *d* orbitals is directly in their paths (Figure 23.25). Thus, splitting of *d*-orbital energies is *less* in a tetrahedral complex than in an octahedral complex having the same ligands:

$$\Delta_{tetrahedral} < \Delta_{octahedral}$$

Minimal repulsions arise if the ligands approach the d_{xy}, d_{yz}, and d_{xz} orbitals closer than they approach the d_{z^2} and $d_{x^2-y^2}$ orbitals. This situation is the *opposite of the octahedral case,* and the relative *d*-orbital energies are reversed: the d_{xy}, d_{yz}, and d_{xz} orbitals become *higher* in energy than the d_{z^2} and $d_{x^2-y^2}$ orbitals. *Only high-spin tetrahedral complexes are known* because the magnitude of Δ is so small.

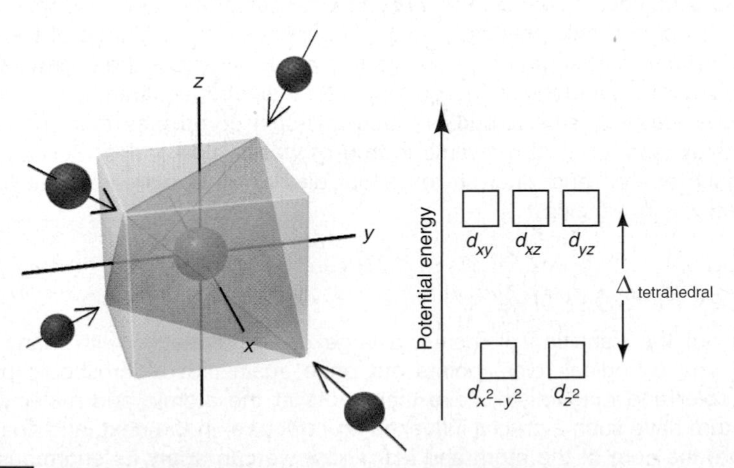

Figure 23.25 Splitting of *d*-orbital energies by a tetrahedral field of ligands. Electrons in d_{xy}, d_{yz}, and d_{xz} orbitals experience greater repulsions than those in $d_{x^2-y^2}$ and d_{z^2}, so the tetrahedral splitting pattern is the opposite of the octahedral pattern.

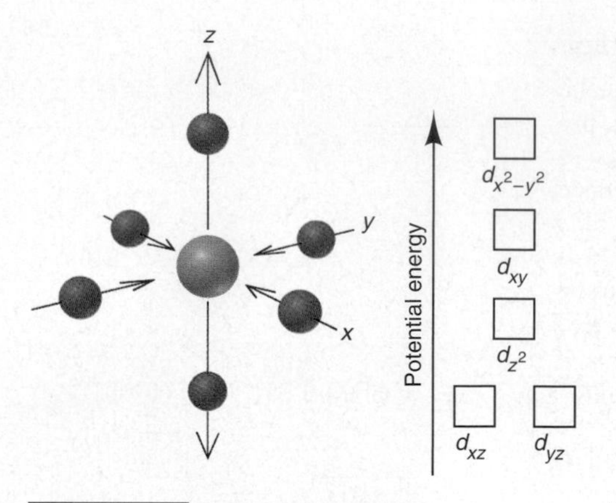

Figure 23.26 Splitting of *d*-orbital energies by a square planar field of ligands. In a square planar field, the energies of d_{xz}, d_{yz}, and especially d_{z^2} orbitals decrease relative to the octahedral pattern.

- *Square planar complexes*. The effects of the ligand field in the square planar case are easier to picture if we imagine starting with an octahedral geometry and then remove the two ligands along the *z*-axis, as depicted in Figure 23.26. With no *z*-axis interactions present, the d_{z^2} orbital energy decreases greatly, and the energies of the other orbitals with a *z*-axis component, the d_{xz} and d_{yz}, also decrease. As a result, the two *d* orbitals in the *xy* plane interact most strongly with the ligands, and because the $d_{x^2-y^2}$ orbital has its lobes *on* the axes, its energy is highest. As a consequence of this splitting pattern, square planar complexes with d^8 metal ions, such as $[PdCl_4]^{2-}$, are diamagnetic, with four pairs of *d* electrons filling the four lowest energy orbitals. Thus, as a general rule, *square planar complexes are low spin.*

At this point, a final word about bonding theories may be helpful. As you have seen in our discussions of several other topics, no one model is satisfactory in every respect. The VB approach offers a simple picture of bond formation but does not even attempt to explain color. The crystal field model predicts color and magnetic behavior beautifully but treats the metal ion and ligands as points of opposite charge and thus offers no insight about the covalent nature of metal-ligand bonding.

Despite its complexity, chemists now rely on a more refined model, called *ligand field–molecular orbital theory*. This theory combines aspects of the previous two models with MO theory (Section 11.3). We won't explore this model here, but it is a powerful predictive tool, yielding information on bond properties that result from the overlap of metal ion and ligand orbitals as well as information on the spectral and magnetic properties that result from the splitting of a metal ion's *d* orbitals.

In addition to their important chemical applications, complexes of the transition elements play vital roles in living systems, as the following Chemical Connections essay describes.

SECTION SUMMARY

Valence bond theory pictures bonding in complex ions as arising from coordinate covalent bonding between Lewis bases (ligands) and Lewis acids (metal ions). Ligand lone pairs occupy hybridized metal-ion orbitals to form complex ions with characteristic shapes.

Crystal field theory explains the color and magnetism of complexes. As the result of a surrounding field of ligands, the *d*-orbital energies of the metal ion split. The magnitude of this crystal field splitting energy (Δ) depends on the charge of the metal ion and the crystal field strength of the ligand. In turn, Δ influences the energy of the photon absorbed (color) and the number of unpaired *d* electrons (paramagnetism). Strong-field ligands create a large Δ and produce low-spin complexes that absorb light of higher energy (shorter λ); the reverse is true of weak-field ligands. Several transition metals, such as iron and zinc, are essential dietary trace elements that function in complexes within proteins.

Chapter Perspective

Our study of the transition elements, a large group of metals with many essential industrial and biological roles, points out once again that macroscopic properties, such as color and magnetism, have their roots at the atomic and molecular levels, which in turn have such a crucial influence on structure. In the next, and final, chapter we explore the core of the atom and learn how we can apply its enormous power.

CHEMICAL CONNECTIONS to Nutritional Science

Transition Metals as Essential Dietary Trace Elements

Living things consist primarily of water and complex organic compounds of four building-block elements: carbon, oxygen, hydrogen, and nitrogen (see Figure 2.18, p. 61). All organisms also contain seven other elements, known as *macro*nutrients because they occur in fairly high concentrations. In order of increasing atomic number, they are sodium, magnesium, phosphorus, sulfur, chlorine, potassium, and calcium. In addition, organisms contain a surprisingly large number of other elements in much lower concentrations, and most of these *micro*nutrients, or *trace elements,* are transition metals.

With the exception of scandium and titanium, all Period 4 transition elements are essential for organisms, and plants require molybdenum (from Period 5) as well. The transition metal ion usually occurs at a bend of a protein chain covalently bonded to surrounding amino acid groups whose N and O atoms act as ligands. Despite the structural complexity of biomolecules, the principles of bonding and *d*-orbital splitting are the same as in simple inorganic systems. Table B23.1 (on the next page) is a list of the Period 4 transition metals known, or thought, to be essential in human nutrition. In this discussion we focus on iron and zinc.

Iron plays a crucial role in oxygen transport in all vertebrates. The oxygen-transporting protein hemoglobin (Figure B23.1A) consists of four folded protein chains called *globins,* each cradling the iron-containing complex *heme.* Heme is a porphyrin, a complex derived from a metal ion and the tetradentate ring ligand known as *porphin.* Iron(II) is centered in the plane of the porphin ring, forming coordinate covalent bonds with four N lone pairs, resulting in a square planar complex. When heme is bound in hemoglobin (Figure B23.1B), the complex is *octahedral,* with the fifth ligand of iron(II) being an N atom from a nearby amino acid (histidine), and the sixth an O atom from either an O_2 (shown) or an H_2O molecule.

Hemoglobin exists in two forms, depending on the nature of the sixth ligand. In the blood vessels of the lungs, where O_2 concentration is high, heme binds O_2 to form *oxyhemoglobin,* which is transported in the arteries to O_2-depleted tissues. At these tissues, the O_2 is released and replaced by an H_2O molecule to form *deoxyhemoglobin,* which is transported in the veins back to the lungs. The H_2O is a weak-field ligand, so the d^6 ion Fe^{2+} in deoxyhemoglobin is part of a high-spin complex. Because of the relatively small *d*-orbital splitting, deoxyhemoglobin absorbs light at the red (low-energy) end of the spectrum and looks purplish blue, which accounts for the dark color of venous blood. On the other hand, O_2 is a strong-field ligand, so it increases the splitting energy, which gives rise to a low-spin complex. For this reason, oxyhemoglobin absorbs at the blue (high-energy) end of the spectrum, which accounts for the bright red color of arterial blood.

The position of the Fe^{2+} ion relative to the plane of the porphin ring also depends on this sixth ligand. Bound to O_2, Fe^{2+} is *in* the porphin plane; bound to H_2O, it moves *out of* the plane slightly. This tiny ($\sim$60 pm $= 6\times10^{-11}$ m) change in the position of Fe^{2+} on release or attachment of O_2 influences the shape of its globin chain, which in turn alters the shape of a neighboring globin chain, triggering the release or attachment of *its* O_2, and so on to the other two globin chains. The very survival of vertebrate life is the result of this cooperative "teamwork" by the four globin

(continued)

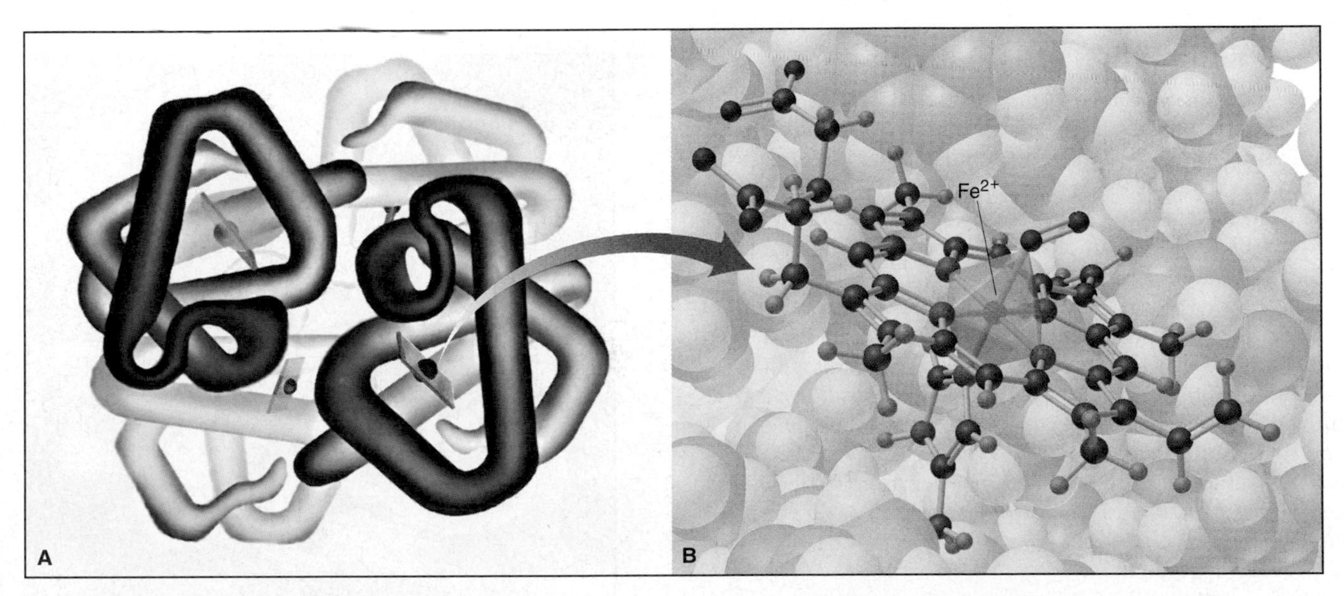

Figure B23.1 Hemoglobin and the octahedral complex in heme. **A,** Hemoglobin consists of four protein chains, each with a bound heme. (Illustration by Irving Geis. Rights owned by Howard Hughes Medical Institute. Not to be used without permission.) **B,** In oxyhemoglobin, the octahedral complex in heme has iron(II) at the center surrounded by the four N atoms of the porphin ring, a fifth N from histidine *(below),* and an O_2 molecule *(above).*

Table B23.1 Some Transition Metal Trace Elements in Humans

Element	Biomolecule Containing Element	Function of Biomolecule
Vanadium	Protein (?)	Redox couple in fat metabolism (?)
Chromium	Glucose tolerance factor	Glucose utilization
Manganese	Isocitrate dehydrogenase	Cell respiration
Iron	Hemoglobin and myoglobin	Oxygen transport
	Cytochrome *c*	Cell respiration; ATP formation
	Catalase	Decomposition of H_2O_2
Cobalt	Cobalamin (vitamin B_{12})	Development of red blood cells
Copper	Ceruloplasmin	Hemoglobin synthesis
	Cytochrome oxidase	Cell respiration; ATP formation
Zinc	Carbonic anhydrase	Elimination of CO_2
	Carboxypeptidase A	Protein digestion
	Alcohol dehydrogenase	Metabolism of ethanol

chains because it allows hemoglobin to pick up O_2 rapidly from the lungs and unload it rapidly in the tissues.

Carbon monoxide is highly toxic because it binds to the Fe^{2+} ion in heme about 200 times more tightly than O_2, thereby eliminating the heme group from functioning in the circulation. Like O_2, CO is a strong-field ligand and produces a bright red, "healthy" look in the individual. Because heme binding is an equilibrium process, CO poisoning can be reversed by breathing extremely high concentrations of O_2, which effectively displaces CO from the heme:

$$\text{heme—CO} + O_2 \rightleftharpoons \text{heme—}O_2 + \text{CO}$$

Porphin rings are among the most common biological ligands. Chlorophyll, the photosynthetic pigment of green plants, is a porphyrin with Mg^{2+} at the center of the porphin ring, and vitamin B_{12} has Co^{3+} at the center of a very similar ring system. Heme itself is found not only in hemoglobin, but also in proteins called *cytochromes* that are involved in energy metabolism (see Chemical Connections, pp. 947–948).

The zinc ion occurs in many enzymes, the protein catalysts of cells (see Chemical Connections, pp. 709–710). With its d^{10} configuration, Zn^{2+} has a tetrahedral geometry, typically with the N atoms of three amino acid groups at three of the positions, and the fourth position free to interact with the molecule whose reaction is being catalyzed (Figure B23.2). In every case studied, the Zn^{2+} ion acts as a Lewis acid, accepting a lone pair from the reactant as a key step in the catalytic process. Consider the enzyme carbonic anhydrase, which catalyzes the essential reaction between H_2O and CO_2 during respiration:

$$CO_2(g) + H_2O(l) \rightleftharpoons H^+(aq) + HCO_3^-(aq)$$

The Zn^{2+} ion at the enzyme's active site binds three histidine N atoms and the H_2O reactant as the fourth ligand. By withdrawing electron density from the O—H bonds, the Zn^{2+} makes the H_2O acidic enough to lose a proton. In the rate-determining step, the resulting bound OH^- ion attacks the partially positive C atom of CO_2 much more vigorously than could the lone pair of a free water molecule; thus, the reaction rate is higher. One reason the Cd^{2+} ion is toxic is that it competes with Zn^{2+} for fitting into the carbonic anhydrase active site.

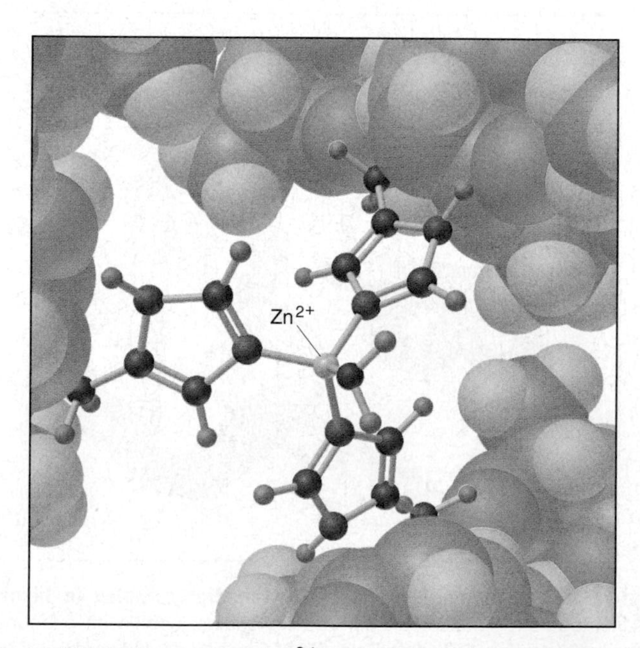

Figure B23.2 The tetrahedral Zn^{2+} complex in carbonic anhydrase.

For Review and Reference

(Numbers in parentheses refer to pages, unless noted otherwise.)

Learning Objectives

Relevant section and/or sample problem (SP) numbers appear in parentheses.

Understand These Concepts

1. The positions of the *d*- and *f*-block elements and the general forms of their atomic and ionic electron configurations (Section 23.1)
2. How atomic size, ionization energy, and electronegativity vary across a period and down a group of transition elements and how these trends differ from those of the main-group elements; why the densities of Period 6 transition elements are so high (Section 23.1)
3. Why the transition elements often have multiple oxidation states and why the +2 state is common (Section 23.1)
4. Why metallic behavior (prevalence of ionic bonding and basic oxides) decreases as oxidation state increases (Section 23.1)
5. Why many transition metal compounds are colored and paramagnetic (Section 23.1)
6. The common +3 oxidation state of lanthanides and the similarity in their M^{3+} radii; the radioactivity of actinides (Section 23.2)
7. How valence-state electronegativity explains why oxides become more covalent and acidic as the O.N. of the metal increases (Section 23.3)
8. Why Cr and Mn oxoanions are stronger oxidizing agents in acidic than in basic solutions (Section 23.3)
9. The role silver halides play in black-and-white photography (Section 23.3)
10. How the high density and low melting point of mercury account for its common uses; the toxicity of organomercury compounds (Section 23.3)
11. The coordination numbers, geometries, and ligand structures of complex ions (Section 23.4)
12. How coordination compounds are named and their formulas written (Section 23.4)
13. How Werner correlated the properties and structures of coordination compounds (Section 23.4)
14. The types of constitutional isomerism (coordination and linkage) and stereoisomerism (geometric and optical) of coordination compounds (Section 23.4)
15. How valence bond theory uses hybridization to account for the shapes of octahedral, square planar, and tetrahedral complexes (Section 23.5)
16. How crystal field theory explains that approaching ligands cause *d*-orbital energies to split (Section 23.5)
17. How the relative crystal-field strength of ligands (spectrochemical series) affects the *d*-orbital splitting energy (Δ) (Section 23.5)
18. How the magnitude of Δ accounts for the energy of light absorbed and, thus, the color of a complex (Section 23.5)
19. How the relative sizes of pairing energy and Δ determine the occupancy of *d* orbitals and, thus, the magnetic properties of complexes (Section 23.5)
20. How *d*-orbital splitting in tetrahedral and square planar complexes differs from that in octahedral complexes (Section 23.5)

Master These Skills

1. Writing electron configurations of transition metal atoms and ions (SP 23.1)
2. Using a partial orbital diagram to determine the number of unpaired electrons in a transition-metal atom or ion (SP 23.2)
3. Recognizing the structural components of complex ions (Section 23.4)
4. Naming and writing formulas of coordination compounds (SP 23.3)
5. Determining the type of stereoisomerism in complexes (SP 23.4)
6. Correlating a complex ion's shape with the number and type of hybrid orbitals of the central metal ion (Section 23.5)
7. Using the spectrochemical series to rank complex ions in terms of Δ and the energy of light absorbed (SP 23.5)
8. Using the spectrochemical series to determine if a complex is high or low spin (SP 23.6)

Key Terms

transition elements (1003)

Section 23.1

lanthanide contraction (1006)

Section 23.2

lanthanides (1010)
actinides (1010)
inner transition elements (1010)

Section 23.4

coordination compound (1017)
complex ion (1017)
ligand (1017)
counter ion (1017)
coordination number (1018)
donor atom (1019)
chelate (1020)
isomer (1023)
constitutional (structural) isomers (1023)
coordination isomers (1023)
linkage isomers (1024)
stereoisomers (1024)
geometric (*cis-trans*) isomers (1024)
optical isomers (1024)

Section 23.5

coordinate covalent bond (1026)
crystal field theory (1028)
e_g orbital (1030)
t_{2g} orbital (1030)
crystal field splitting energy (Δ) (1030)
strong-field ligand (1030)
weak-field ligand (1030)
spectrochemical series (1031)
high-spin complex (1032)
low-spin complex (1032)

Highlighted Figures and Tables

These figures (F) and tables (T) provide a review of key ideas.

Brief Solutions to Follow-up Problems

23.1 (a) Ag^+: $4d^{10}$; (b) Cd^{2+}: $4d^{10}$; (c) Ir^{3+}: $5d^6$
23.2 Three; Er^{3+} is [Xe] $4f^{11}$:

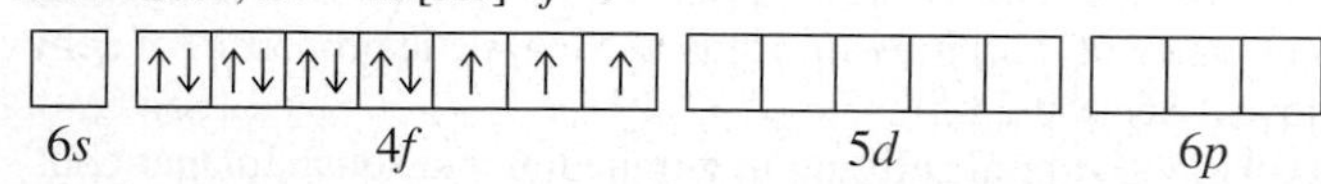

23.3 (a) Pentaaquabromochromium(III) chloride;
(b) $Ba_3[Co(CN)_6]_2$

23.4 Two sets of *cis-trans* isomers, and the two *cis* isomers are optical isomers.

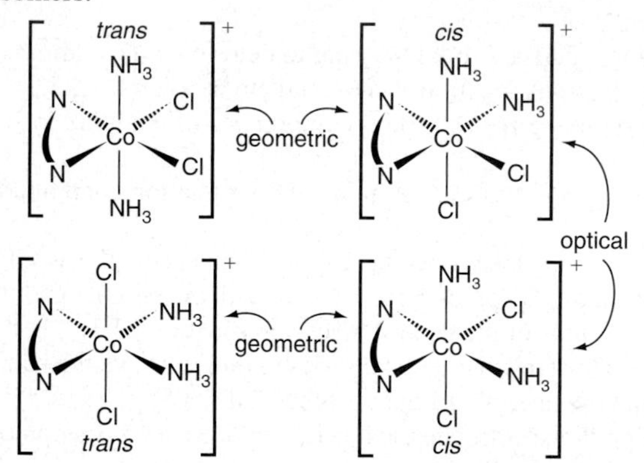

23.5 Both metal ions are V^{3+}; in terms of ligand field energy, $NH_3 > H_2O$, so $[V(NH_3)_6]^{3+}$ absorbs light of higher energy.

23.6 The metal ion is Mn^{3+}: [Ar] $3d^4$.

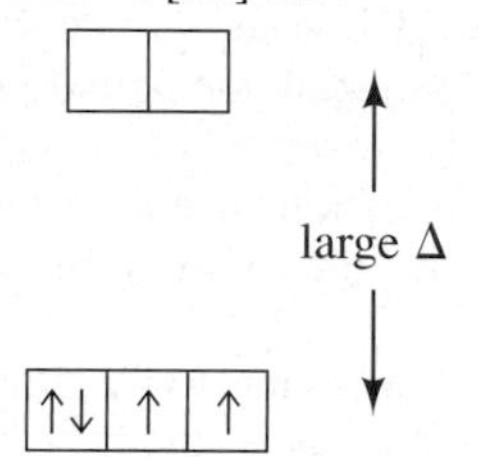

Two unpaired *d* electrons; low-spin complex

Problems

Problems with **colored** numbers are answered in Appendix E. Sections match the text and provide the numbers of relevant sample problems. Most offer Concept Review Questions, Skill-Building Exercises (grouped in pairs covering the same concept), and Problems in Context. Comprehensive Problems are based on material from any section or previous chapter.

Note: In these problems, the term *electron configuration* refers to the condensed, ground-state electron configuration.

Properties of the Transition Elements
(Sample Problem 23.1)

Concept Review Questions

23.1 How is the *n* value of the *d* sublevel of a transition element related to the period number of the element?

23.2 (a) Write the general electron configuration of a transition element in Period 5.
(b) Write the general electron configuration of a transition element in Period 6.

23.3 What is the general rule concerning the order in which electrons are removed from a transition metal atom to form an ion? Give an example from Group 5B(5). Name two types of measurements used to study electron configurations of ions.

23.4 What is the maximum number of unpaired *d* electrons that an atom or ion can possess? Give an example of an atom and an ion that have this number.

23.5 How does the variation in atomic size across a transition series contrast with the change across the main-group elements of the same period? Why?

23.6 (a) What is the lanthanide contraction?
(b) How does it affect atomic size down a group of transition elements?
(c) How does it influence the densities of the Period 6 transition elements?

23.7 (a) What is the range in electronegativity values across the first (3*d*) transition series?
(b) What is the range across Period 4 of main-group elements?
(c) Explain the difference between the two ranges.

23.8 (a) Explain the major difference between the number of oxidation states of most transition elements and that of most main-group elements.
(b) Why is the +2 oxidation state so common among transition elements?

23.9 (a) What difference in behavior distinguishes a paramagnetic substance from a diamagnetic one?
(b) Why are paramagnetic ions common among the transition elements but not the main-group elements?
(c) Why are colored solutions of metal ions common among the transition elements but not the main-group elements?

Skill-Building Exercises *(grouped in similar pairs)*

23.10 Using the periodic table to locate each element, write the electron configuration of (a) V; (b) Y; (c) Hg.

23.11 Using the periodic table to locate each element, write the electron configuration of (a) Ru; (b) Cu; (c) Ni.

23.12 Using the periodic table to locate each element, write the electron configuration of (a) Os; (b) Co; (c) Ag.

23.13 Using the periodic table to locate each element, write the electron configuration of (a) Zn; (b) Mn; (c) Re.

23.14 Give the electron configuration and the number of unpaired electrons for each of the following ions: (a) Sc^{3+}; (b) Cu^{2+}; (c) Fe^{3+}; (d) Nb^{3+}.

23.15 Give the electron configuration and the number of unpaired electrons for each of the following ions: (a) Cr^{3+}; (b) Ti^{4+}; (c) Co^{3+}; (d) Ta^{2+}.

23.16 What is the highest possible oxidation state for each of the following: (a) Ta; (b) Zr; (c) Mn?

23.17 What is the highest possible oxidation state for each of the following: (a) Nb; (b) Y; (c) Tc?

23.18 Which transition metals have a maximum oxidation state of +6?

23.19 Which transition metals have a maximum oxidation state of +4?

23.20 In which compound does Cr exhibit greater metallic behavior, CrF_2 or CrF_6? Explain.

23.21 VF_5 is a liquid that boils at 48°C, whereas VF_3 is a solid that melts above 800°C. Explain this difference in properties.

23.22 Is it more difficult to oxidize Cr or Mo? Explain.

23.23 Is MnO_4^- or ReO_4^- a stronger oxidizing agent? Explain.

23.24 Which oxide, CrO_3 or CrO, forms a more acidic aqueous solution? Explain.

23.25 Which oxide, Mn_2O_3 or Mn_2O_7, displays more basic behavior? Explain.

Problems in Context

23.26 The green patina of copper-alloy roofs of old buildings is the result of the corrosion (oxidation) of copper in the presence of O_2, H_2O, CO_2, and sulfur compounds. Silver and gold—the other stable members of Group 1B(11)—do not form this patina. Corrosion of copper and silver in the presence of sulfur and its compounds leads to the familiar black tarnish, but gold does not react with sulfur. This pattern is markedly different from that in Group 1A(1), where ease of oxidation *increases* down the group. What causes the different patterns in the two groups?

The Inner Transition Elements

(Sample Problem 23.2)

Concept Review Questions

23.27 What atomic property of the lanthanides leads to their remarkably similar chemical properties?

23.28 (a) What is the maximum number of unpaired electrons exhibited by an ion of a lanthanide?
(b) How does this number relate to occupancy of the 4*f* subshell?

23.29 Which of the actinides are radioactive?

Skill-Building Exercises *(grouped in similar pairs)*

23.30 Write the electron configurations of the following atoms and ions: (a) La; (b) Ce^{3+}; (c) Es; (d) U^{4+}.

23.31 Write the electron configurations of the following atoms and ions: (a) Pm; (b) Lu^{3+}; (c) Th; (d) Fm^{3+}.

23.32 Only a few of the lanthanides show any oxidation state other than +3. Two of these, europium (Eu) and terbium (Tb), are found near the middle of the series, and their unusual oxidation states can be associated with a half-filled *f* subshell.
(a) Write the electron configurations of Eu^{2+}, Eu^{3+}, and Eu^{4+}. Why is Eu^{2+} a common ion, whereas Eu^{4+} is unknown?
(b) Write the electron configurations of Tb^{2+}, Tb^{3+}, and Tb^{4+}. Would you expect Tb to show a +2 or a +4 oxidation state? Explain.

23.33 Cerium (Ce) and ytterbium (Yb) exhibit other oxidation states in addition to +3.
(a) Write the electron configurations of Ce^{2+}, Ce^{3+}, and Ce^{4+}.
(b) Write the electron configurations of Yb^{2+}, Yb^{3+}, and Yb^{4+}.
(c) In addition to the 3+ ions, the ions Ce^{4+} and Yb^{2+} are stable. Suggest a reason for this stability.

Problems in Context

23.34 One of the lanthanides displays the maximum possible number of unpaired electrons both for an atom and for a 3+ ion. Name the element and give the number of unpaired electrons in the atom and the ion.

Highlights of Selected Transition Metals

Concept Review Questions

23.35 What is the chemical reason that chromium is so useful for decorative plating on metals?

23.36 What is valence-state electronegativity? Use the concept to explain the change in acidity of the oxides of Mn with changing O.N. of the metal.

23.37 What property does manganese confer to steel?

23.38 What chemical property of silver leads to its use in jewelry and other decorative objects?

23.39 How is a photographic latent image different from the image you see on a piece of developed film?

23.40 Mercury has an unusual physical property and an unusual 1+ ion. Explain.

Problems in Context

23.41 When a basic solution of $Cr(OH)_4^-$ ion is slowly acidified, solid $Cr(OH)_3$ first precipitates out and then redissolves as excess acid is added. Assuming that $Cr(OH)_4^-$ actually exists as $Cr(H_2O)_2(OH)_4^-$, write equations that represent these two reactions.

23.42 Use the following data to determine if $Cr^{2+}(aq)$ can be prepared by the reaction of $Cr(s)$ with $Cr^{3+}(aq)$:

$$Cr^{3+}(aq) + e^- \longrightarrow Cr^{2+}(aq) \qquad E^0 = -0.41\ V$$
$$Cr^{3+}(aq) + 3e^- \longrightarrow Cr(s) \qquad E^0 = -0.74\ V$$
$$Cr^{2+}(aq) + 2e^- \longrightarrow Cr(s) \qquad E^0 = -0.91\ V$$

23.43 When solid CrO_3 is dissolved in water, the solution is orange rather than the yellow of H_2CrO_4. How does this observation indicate that CrO_3 is an acidic oxide?

23.44 Solutions of $KMnO_4$ are used commonly in redox titrations. The dark purple MnO_4^- ion serves as its own indicator, changing to the almost colorless Mn^{2+} as it is reduced. The end point occurs when a pale purple color remains as the $KMnO_4$ solution is added. If a sample that has reached this end point is allowed to stand for a long period of time, the color fades and a suspension of a small amount of brown, muddy MnO_2 appears. Use standard electrode potentials to explain this result.

Coordination Compounds

(Sample Problems 23.3 and 23.4)

Concept Review Questions

23.45 Describe the makeup of a complex ion, including the nature of the ligands and their interaction with the central metal ion. Explain how a complex ion can be positive or negative and how it occurs as part of a neutral coordination compound.

23.46 What electronic feature must a donor atom of any ligand possess?

23.47 What is the coordination number of a metal ion in a complex ion? How does it differ from oxidation number?

23.48 What structural feature is characteristic of a complex described as a chelate?

23.49 What geometries are associated with the coordination numbers 2, 4, and 6?

23.50 What are the coordination numbers of cobalt(III), platinum(II), and platinum(IV) in complexes?

23.51 In what sense is a complex ion the adduct of a Lewis acid-base reaction?

23.52 What does the ending *-ate* signify in a complex ion name?

23.53 In what order are the metal ion and ligands given in the name of a complex ion?

23.54 Is a linkage isomer a type of constitutional isomer or stereoisomer? Explain.

Skill-Building Exercises *(grouped in similar pairs)*

23.55 Give systematic names for the following formulas:
(a) $[Ni(H_2O)_6]Cl_2$ (b) $[Cr(en)_3](ClO_4)_3$ (c) $K_4[Mn(CN)_6]$

23.56 Give systematic names for the following formulas:
(a) $[Co(NH_3)_4(NO_2)_2]Cl$ (b) $[Cr(NH_3)_6][Cr(CN)_6]$
(c) $K_2[CuCl_4]$

23.57 What are the charge and coordination number of the central metal ion(s) in each compound of Problem 23.55?

23.58 What are the charge and coordination number of the central metal ion(s) in each compound of Problem 23.56?

23.59 Give systematic names for the following formulas:
(a) $K[Ag(CN)_2]$ (b) $Na_2[CdCl_4]$ (c) $[Co(NH_3)_4(H_2O)Br]Br_2$

23.60 Give systematic names for the following formulas:
(a) $K[Pt(NH_3)Cl_5]$ (b) $[Cu(en)(NH_3)_2][Co(en)Cl_4]$
(c) $[Pt(en)_2Br_2](ClO_4)_2$

23.61 What are the charge and coordination number of the central metal ion(s) in each compound of Problem 23.59?

23.62 What are the charge and coordination number of the central metal ion(s) in each compound of Problem 23.60?

23.63 Give formulas corresponding to the following names:
(a) Tetraamminezinc sulfate
(b) Pentaamminechlorochromium(III) chloride
(c) Sodium bis(thiosulfato)argentate(I)

23.64 Give formulas corresponding to the following names:
(a) Dibromobis(ethylenediamine)cobalt(III) sulfate
(b) Hexaamminechromium(III) tetrachlorocuprate(II)
(c) Potassium hexacyanoferrate(II)

23.65 What is the coordination number of the metal ion and the number of individual ions per formula unit in each of the compounds in Problem 23.63?

23.66 What is the coordination number of the metal ion and the number of individual ions per formula unit in each of the compounds in Problem 23.64?

23.67 Give formulas corresponding to the following names:
(a) Hexaaquachromium(III) sulfate
(b) Barium tetrabromoferrate(III)
(c) Bis(ethylenediamine)platinum(II) carbonate

23.68 Give formulas corresponding to the following names:
(a) Potassium tris(oxalato)chromate(III)
(b) Tris(ethylenediamine)cobalt(III) pentacyanoiodomanganate(II)
(c) Diamminediaquabromochloroaluminum nitrate

23.69 What is the coordination number of the metal ion and the number of individual ions per formula unit in each of the compounds in Problem 23.67?

23.70 What is the coordination number of the metal ion and the number of individual ions per formula unit in each of the compounds in Problem 23.68?

23.71 Which of these ligands can participate in linkage isomerism: (a) NO_2^-; (b) SO_2; (c) NO_3^-? Explain with Lewis structures.

23.72 Which of these ligands can participate in linkage isomerism: (a) SCN^-; (b) $S_2O_3^{2-}$ (thiosulfate); (c) HS^-? Explain with Lewis structures.

23.73 For any of the following that can exist as isomers, state the type of isomerism and draw the structures:
(a) $[Pt(CH_3NH_2)_2Br_2]$ (b) $[Pt(NH_3)_2FCl]$
(c) $[Pt(H_2O)(NH_3)FCl]$

23.74 For any of the following that can exist as isomers, state the type of isomerism and draw the structures:
(a) $[Zn(en)F_2]$ (b) $[Zn(H_2O)(NH_3)FCl]$
(c) $[Pd(CN)_2(OH)_2]^{2-}$

23.75 For any of the following that can exist as isomers, state the type of isomerism and draw the structures:
(a) $[PtCl_2Br_2]^{2-}$ (b) $[Cr(NH_3)_5(NO_2)]^{2+}$
(c) $[Pt(NH_3)_4I_2]^{2+}$

23.76 For any of the following that can exist as isomers, state the type of isomerism and draw the structures:
(a) $[Co(NH_3)_5Cl]Br_2$ (b) $[Pt(CH_3NH_2)_3Cl]Br$
(c) $[Fe(H_2O)_4(NH_3)_2]^{2+}$

Problems in Context

23.77 Chromium(III), like cobalt(III), has a coordination number of 6 in many of its complex ions. Compounds are known that have the traditional formula $CrCl_3 \cdot nNH_3$, where $n = 3$ to 6. Which of the compounds has an electrical conductivity in aqueous solution similar to that of an equimolar NaCl solution?

23.78 When $MCl_4(NH_3)_2$ is dissolved in water and treated with $AgNO_3$, 2 mol of AgCl precipitates immediately for each mole of $MCl_4(NH_3)_2$. Give the coordination number of M in the complex.

23.79 Palladium, like its group neighbor platinum, forms four-coordinate Pd(II) and six-coordinate Pd(IV) complexes. Write modern formulas for the complexes with these compositions:
(a) $PdK(NH_3)Cl_3$ (b) $PdCl_2(NH_3)_2$
(c) PdK_2Cl_6 (d) $Pd(NH_3)_4Cl_4$

Theoretical Basis for the Bonding and Properties of Complexes

(Sample Problems 23.5 and 23.6)

Concept Review Questions

23.80 (a) What is a coordinate covalent bond?
(b) Is it involved when $FeCl_3$ dissolves in water? Explain.
(c) Is it involved when HCl gas dissolves in water? Explain.

23.81 According to valence bond theory, what set of orbitals is used by a Period 4 metal ion in forming (a) a square planar complex; (b) a tetrahedral complex?

23.82 A metal ion is described as using a d^2sp^3 set of orbitals when forming a complex. What is the coordination number of the metal ion and the shape of the complex?

23.83 A complex in solution absorbs green light. What is the color of the solution?

23.84 What *two* possibilities of color absorption by a solution could give rise to an observed blue color?

23.85 (a) What is the crystal field splitting energy (Δ)?
(b) How does it arise for an octahedral field of ligands?
(c) How is it different for a tetrahedral field of ligands?

23.86 What is the distinction between a weak-field ligand and a strong-field ligand? Give an example of each.

23.87 Is a complex with the same number of unpaired electrons as the free gaseous metal ion termed high spin or low spin?

23.88 How do the relative magnitudes of $E_{pairing}$ and Δ affect the paramagnetism of a complex?

23.89 Why are there both high-spin and low-spin octahedral complexes but only high-spin tetrahedral complexes?

Skill-Building Exercises *(grouped in similar pairs)*

23.90 Give the number of *d* electrons (*n* of d^n) for the central metal ion in each of these species: (a) $[TiCl_6]^{2-}$; (b) $K[AuCl_4]$; (c) $[RhCl_6]^{3-}$.

23.91 Give the number of *d* electrons (*n* of d^n) for the central metal ion in each of these species: (a) $[Cr(H_2O)_6](ClO_3)_2$; (b) $[Mn(CN)_6]^{2-}$; (c) $[Ru(NO)(en)_2Cl]Br$.

23.92 Give the number of *d* electrons (*n* of d^n) for the central metal ion in each of these species: (a) $Ca[IrF_6]$; (b) $[HgI_4]^{2-}$; (c) $[Co(EDTA)]^{2-}$.

23.93 Give the number of *d* electrons (*n* of d^n) for the central metal ion in each of these species: (a) $[Ru(NH_3)_5Cl]SO_4$; (b) $Na_2[Os(CN)_6]$; (c) $[Co(NH_3)_4CO_3I]$.

23.94 Sketch the orientation of the orbitals relative to the ligands in an octahedral complex to explain the splitting and the relative energies of the d_{xy} and the $d_{x^2-y^2}$ orbitals.

23.95 The two e_g orbitals are identical in energy in an octahedral complex but have different energies in a square planar complex, with the d_{z^2} orbital being much lower in energy than the $d_{x^2-y^2}$. Explain with orbital sketches.

23.96 Which of these ions *cannot* form both high- and low-spin octahedral complexes: (a) Ti^{3+}; (b) Co^{2+}; (c) Fe^{2+}; (d) Cu^{2+}?

23.97 Which of these ions *cannot* form both high- and low-spin octahedral complexes: (a) Mn^{3+}; (b) Nb^{3+}; (c) Ru^{3+}; (d) Ni^{2+}?

23.98 Draw orbital-energy splitting diagrams and use the spectrochemical series to show the orbital occupancy for each of the following (assuming that H_2O is a weak-field ligand):
(a) $[Cr(H_2O)_6]^{3+}$ (b) $[Cu(H_2O)_4]^{2+}$ (c) $[FeF_6]^{3-}$

23.99 Draw orbital-energy splitting diagrams and use the spectrochemical series to show the orbital occupancy for each of the following (assuming that H_2O is a weak-field ligand):
(a) $[Cr(CN)_6]^{3-}$ (b) $[Rh(CO)_6]^{3+}$ (c) $[Co(OH)_6]^{4-}$

23.100 Draw orbital-energy splitting diagrams and use the spectrochemical series to show the orbital occupancy for each of the following (assuming that H_2O is a weak-field ligand):
(a) $[MoCl_6]^{3-}$ (b) $[Ni(H_2O)_6]^{2+}$ (c) $[Ni(CN)_4]^{2-}$

23.101 Draw orbital-energy splitting diagrams and use the spectrochemical series to show the orbital occupancy for each of the following (assuming that H_2O is a weak-field ligand):
(a) $[Fe(C_2O_4)_3]^{3-}$ ($C_2O_4^{2-}$ creates a weaker field than H_2O does.)
(b) $[Co(CN)_6]^{4-}$ (c) $[MnCl_6]^{4-}$

23.102 Rank the following complex ions in order of *increasing* Δ and energy of visible light absorbed: $[Cr(NH_3)_6]^{3+}$, $[Cr(H_2O)_6]^{3+}$, $[Cr(NO_2)_6]^{3-}$.

23.103 Rank the following complex ions in order of *decreasing* Δ and energy of visible light absorbed: $[Cr(en)_3]^{3+}$, $[Cr(CN)_6]^{3-}$, $[CrCl_6]^{3-}$.

23.104 A complex, ML_6^{2+}, is violet. The same metal forms a complex with another ligand, Q, that creates a weaker field. What color might MQ_6^{2+} be expected to show? Explain.

23.105 $[Cr(H_2O)_6]^{2+}$ is violet. Another CrL_6 complex is green. Can ligand L be CN^-? Can it be Cl^-? Explain.

Problems in Context

23.106 Octahedral $[Ni(NH_3)_6]^{2+}$ is paramagnetic, whereas planar $[Pt(NH_3)_4]^{2+}$ is diamagnetic, even though both metal ions are d^8 species. Explain.

23.107 The hexaaqua complex $[Ni(H_2O)_6]^{2+}$ is green, whereas the hexaammonia complex $[Ni(NH_3)_6]^{2+}$ is violet. Explain.

23.108 Three of the complex ions that are formed by Co^{3+} are $[Co(H_2O)_6]^{3+}$, $[Co(NH_3)_6]^{3+}$, and $[CoF_6]^{3-}$. These ions have the observed colors (listed in arbitrary order) yellow-orange, green, and blue. Match each complex with its color. Explain.

Comprehensive Problems

23.109 When neptunium (Np) and plutonium (Pu) were discovered, the periodic table did not include the actinides, so these elements were placed in Group 7B(7) and 8B(8). When americium (Am) and curium (Cm) were synthesized, they were placed in Group 8B(9) and 8B(10). However, during chemical isolation procedures, Glenn Seaborg and his group, who had synthesized these elements, could not find their compounds among other compounds of these groups, which led Seaborg to suggest they were part of a new inner transition series. (a) How do the electron configurations of these elements support Seaborg's suggestion? (b) The highest fluorides of Np and Pu are hexafluorides, as is the highest fluoride of uranium. How does this chemical evidence support the placement of Np and Pu as *inner* transition elements rather than transition elements?

23.110 How many different formulas are there for octahedral complexes with a metal M and four ligands A, B, C and D? Give the number of isomers for each formula and describe the isomers.

23.111 At one time, it was common to write the formula for copper(I) chloride as Cu_2Cl_2, instead of CuCl, analogously to Hg_2Cl_2 for mercury(I) chloride. Use electron configurations to explain why Hg_2Cl_2 is correct but so is CuCl.

23.112 Correct each name that has an error:
(a) $Na[FeBr_4]$, sodium tetrabromoferrate(II)
(b) $[Ni(NH_3)_6]^{2+}$, nickel hexaammine ion
(c) $[Co(NH_3)_3I_3]$, triamminetriiodocobalt(III)
(d) $[V(CN)_6]^{3-}$, hexacyanovanadium(III) ion
(e) $K[FeCl_4]$, potassium tetrachloroiron(III)

23.113 For the compound $[Co(en)_2Cl_2]Cl$, give:
(a) The coordination number of the metal ion
(b) The oxidation number of the central metal ion
(c) The number of individual ions per formula unit
(d) The moles of AgCl that precipitate immediately when 1 mol of compound is dissolved in water and treated with $AgNO_3$

23.114 Hexafluorocobaltate(III) ion is a high-spin complex. Draw the orbital-energy splitting diagram for its *d* orbitals.

23.115 A salt of each of the ions in Table 23.3 (p. 1009) is dissolved in water. A Pt electrode is immersed in each solution and connected to a 0.38-V battery. All of the electrolytic cells are run for the same amount of time with the same current.
(a) In which cell(s) will a metal plate out? Explain.
(b) Which cell will plate out the least mass of metal? Explain.

23.116 Criticize and correct the following statement: strong-field ligands always give rise to low-spin complexes.

23.117 Two major bidentate ligands used in analytical chemistry are bipyridyl (bipy) and *ortho*-phenanthroline (*o*-phen):

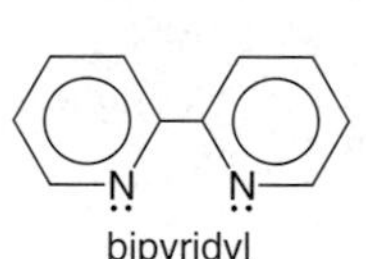

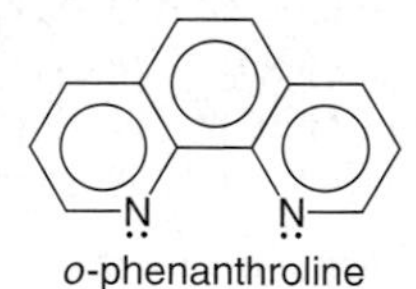

Draw structures and discuss the possibility of isomers for
(a) $[Pt(bipy)Cl_2]$ (b) $[Fe(o\text{-phen})_3]^{3+}$
(c) $[Co(bipy)_2F_2]^+$ (d) $[Co(o\text{-phen})(NH_3)_3Cl]^{2+}$

23.118 The following reaction is a key step in black-and-white photography (see p. 1015):

$$AgBr(s) + 2S_2O_3^{2-}(aq) \longrightarrow Ag(S_2O_3)_2^{3-}(aq) + Br^-(aq)$$

During fixing, 258 mL of hypo (sodium thiosulfate) was used. The hypo concentration was 0.1052 *M* before the AgBr reacted, and 0.0378 *M* afterward. How many grams of AgBr reacted?

23.119 The metal ion in platinum(IV) complexes, like that in cobalt(III) complexes, has a coordination number of 6. Many of these complexes occur with Cl^- ions and NH_3 molecules as ligands. Consider the following traditional (before the work of Werner) formulas for two coordination compounds:
(a) $PtCl_4{\cdot}6NH_3$ (b) $PtCl_4{\cdot}4NH_3$
For each of these compounds:
(1) Give the modern formula and charge of the complex ion.
(2) Predict moles of ions formed per mole of compound dissolved and moles of AgCl formed immediately with excess $AgNO_3$.

23.120 In 1940, when the elements Np and Pu were prepared, a controversy arose about whether the elements from Ac on were analogs of the transition elements (and related to Y through Mo) or analogs of the lanthanides (and related to La through Nd). The arguments hinged primarily on comparing observed oxidation states to those of the earlier elements. (a) If the actinides were analogs of the transition elements, what would you predict about the maximum oxidation state for U? For Np? (b) If the actinides were analogs of the lanthanides, what would you predict about the maximum oxidation state for U? For Pu?

23.121 In many species, a transition metal has an unusually high or low oxidation state. Write balanced equations for the following and find the oxidation state of the transition metal in the product:
(a) Iron(III) ion reacts with hypochlorite ion in basic solution to form ferrate ion (FeO_4^{2-}), Cl^-, and water.
(b) Potassium hexacyanomanganate(II) reacts with K metal to form $K_6[Mn(CN)_6]$.
(c) Heating sodium superoxide (NaO_2) with Co_3O_4 produces Na_4CoO_4 and O_2 gas.
(d) Vanadium(III) chloride reacts with Na metal under a CO atmosphere to produce $Na[V(CO)_6]$ and NaCl.
(e) Barium peroxide reacts with nickel(II) ions in basic solution to produce $BaNiO_3$.
(f) Bubbling CO through a basic solution of cobalt(II) ion produces $[Co(CO)_4]^-$, CO_3^{2-}, and water.
(g) Heating cesium tetrafluorocuprate(II) with F_2 gas under pressure gives Cs_2CuF_6.
(h) Heating tantalum(V) chloride with Na metal produces NaCl and Ta_6Cl_{15}, in which half of the Ta is in the +2 state.
(i) Heating cesium tetrafluoroargentate(II) with F_2 gas gives Cs_2AgF_6, in which half the silver is in the +3 state.
(j) Potassium tetracyanonickelate(II) reacts with hydrazine (N_2H_4) in basic solution to form $K_4[Ni_2(CN)_6]$ and N_2 gas.

23.122 For the permanganate ion, draw a Lewis structure that has the lowest formal charges.

23.123 The coordination compound $[Pt(NH_3)_2(SCN)_2]$ displays two types of isomerism. Name the types and give names and structures for the six possible isomers.

23.124 In the sepia "toning" of a black-and-white photograph, the image is converted to a rich brownish violet by placing the finished photograph in a solution of gold(III) ions, in which metallic gold replaces the metallic silver. Use Appendix D to explain the chemistry of this process.

23.125 An octahedral complex with three different ligands (A, B, and C) can have formulas with three different ratios of the ligands:
$[MA_4BC]^{n+}$, such as $[Co(NH_3)_4(H_2O)Cl]^{2+}$
$[MA_3B_2C]^{n+}$, such as $[Cr(H_2O)_3Br_2Cl]$
$[MA_2B_2C_2]^{n+}$, such as $[Cr(NH_3)_2(H_2O)_2Br_2]^+$
For each example, give the name, state the type(s) of isomerism present, and draw all isomers.

23.126 In black-and-white photography, what are the major chemical changes involved in exposing, developing, and fixing?

23.127 In $[Cr(NH_3)_6]Cl_3$, the $[Cr(NH_3)_6]^{3+}$ ion absorbs visible light in the blue-violet range, and the compound is yellow-orange. In $[Cr(H_2O)_6]Br_3$, the $[Cr(H_2O)_6]^{3+}$ ion absorbs visible light in the red range, and the compound is blue-gray. Explain these differences in light absorbed and colors of the compounds.

23.128 Dark green manganate salts contain the MnO_4^{2-} ion. The ion is stable in basic solution but disproportionates in acid to $MnO_2(s)$ and MnO_4^-. (a) What is the oxidation state of Mn in MnO_4^{2-}, MnO_4^-, and MnO_2? (b) Write a balanced equation for the reaction of MnO_4^{2-} in acidic solution.

23.129 Aqueous electrolysis of potassium manganate (K_2MnO_4) in basic solution is used for the production of tens of thousands of tons of potassium permanganate ($KMnO_4$) annually. (a) Write

thus providing a means for measuring its intensity. Two years later, a young doctoral student named Marie Sklodowska Curie began a search for other minerals that behaved like uranium in this way. She found that thorium minerals also emit radiation and discovered that *the intensity of the radiation is directly proportional to the concentration of the element in the mineral, not to the nature of the mineral or compound* in which the element occurs. Curie named the emissions *radioactivity* and showed that they are *unaffected by temperature, pressure, or other physical and chemical conditions.*

To her surprise, Curie found that certain uranium minerals were even more radioactive than pure uranium, which implied that they contained traces of one or more as yet unknown, highly radioactive elements. She and her husband, the physicist Pierre Curie, set out to isolate all the radioactive components in pitchblende, the principal ore of uranium. After months of painstaking chemical work, they isolated two extremely small, highly radioactive fractions, one that precipitated with bismuth compounds and another that precipitated with alkaline earth compounds. Through chemical and spectroscopic analysis, Marie Curie was able to show that these fractions contained two new elements, which she named polonium (after her native Poland) and radium. Polonium (Po; $Z = 84$), the most metallic member of Group 6A(16), lies to the right of bismuth in Period 6. Radium (Ra; $Z = 88$), which is the heaviest alkaline earth metal, lies under barium in Group 2A(2).

Purifying radium proved to be another arduous task. Starting with several tons of pitchblende residues from which the uranium had been extracted, Curie prepared compounds of the larger Group 2A(2) elements, continually separating minuscule amounts of radium compounds from enormously larger amounts of chemically similar barium compounds. It took her four years to isolate 0.1 g of radium chloride, which she melted and electrolyzed to obtain pure metallic radium.

Her Brilliant Career Marie Curie (1867–1934) is the only person to be awarded Nobel Prizes in two *different* sciences, one in physics in 1903 for her research into radioactivity and the other in chemistry in 1911 for the discovery of polonium and the discovery, isolation, and study of radium and its compounds.

During the next few years, Henri Becquerel, the Curies, and P. Villard in France and Ernest Rutherford and his coworkers in England studied the nature of radioactive emissions. Rutherford and his colleague Frederick Soddy observed that elements other than radium were formed when radium decayed. In 1902, they proposed that radioactive emission results in the change of one element into another. To their contemporaries, this idea sounded like a resurrection of alchemy and was met with disbelief and ridicule. We now know it to be true: under most circumstances, *when a nuclide of one element decays, it changes into a nuclide of a different element.*

These studies led to an understanding of the three most common types of radioactive emission:

- **Alpha particles** (symbolized α or ${}^{4}_{2}He$) are dense, positively charged particles identical to helium nuclei.
- **Beta particles** (symbolized β, β^-, or more usually ${}^{0}_{-1}\beta$) are negatively charged particles identified as high-speed electrons. (The emission of electrons from the nucleus may seem strange, but as you'll see shortly, β particles arise as a result of a nuclear reaction.)
- **Gamma rays** (symbolized as γ, or sometimes ${}^{0}_{0}\gamma$) are very high-energy photons, about 10^5 times as energetic as visible light.

The behavior of these three emissions in an electric field is shown in Figure 24.1. Note that α particles curve to a small extent toward the negative plate, β particles curve to a great extent toward the positive plate, and γ rays are not affected by the electric field. We'll discuss the effects of these emissions on matter later.

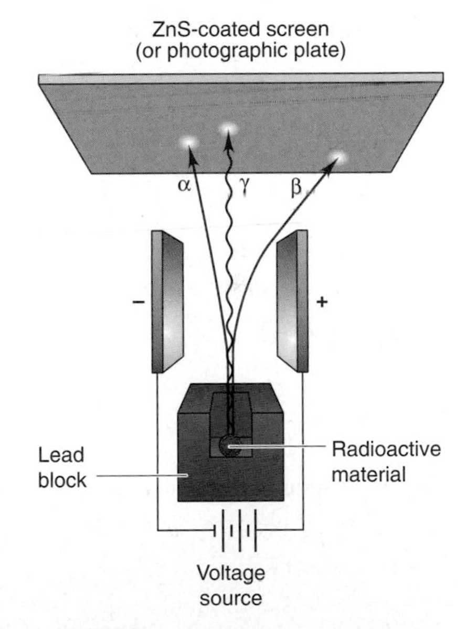

Figure 24.1 Three types of radioactive emissions in an electric field. Positively charged α particles curve toward the negative plate; negatively charged β particles curve toward the positive plate. The curvature is greater for β particles because they have much lower mass. The γ rays, uncharged high-energy photons, are unaffected by the field.

Types of Radioactive Decay; Balancing Nuclear Equations

When a nuclide decays, it forms a nuclide of lower energy, and the excess energy is carried off by the emitted radiation. The decaying, or reactant, nuclide is called the *parent;* the product nuclide is called the *daughter.* Nuclides can decay in several ways. As we discuss the major types of decay, which are summarized in Table 24.2, note the principle used to balance nuclear reactions: *the total Z (charge, number of protons) and the total A (sum of protons and neutrons) of the reactants equal those of the products:*

$$\frac{\text{Total } A}{\text{Total } Z} \text{ Reactants} = \frac{\text{Total } A}{\text{Total } Z} \text{ Products} \qquad \textbf{(24.1)}$$

1. **Alpha decay** involves the loss of an α particle ($^{4}_{2}He$) from a nucleus. For each α particle emitted by the parent nucleus, *A decreases by 4 and Z decreases by 2.* Every element that is heavier than lead (Pb; $Z = 82$), as well as a few lighter ones, exhibits α decay. In Rutherford's classic experiment that established the existence of the atomic nucleus (Section 2.4, pp. 47–48), radium was the source of the α particles that were used as projectiles. Radium undergoes α decay to yield radon (Rn; $Z = 86$):

$$^{226}_{88}Ra \longrightarrow \, ^{222}_{86}Rn + \, ^{4}_{2}He$$

Table 24.2 Modes of Radioactive Decay*

Mode	Emission	Decay Process	Change in A	Change in Z	Change in N
α Decay	α ($^{4}_{2}He$)	Reactant (parent) ⟶ Product (daughter) + α expelled	−4	−2	−2
β Decay†	$^{0}_{-1}\beta$	$^{1}_{0}n$ in nucleus ⟶ $^{1}_{1}p$ in nucleus + $^{0}_{-1}\beta$ β expelled	0	+1	−1
Positron emission†	$^{0}_{1}\beta$	$h\nu$ high-energy photon + nucleus with xp^{+} and yn^{0} ⟶ nucleus with $(x-1)p^{+}$ and $(y+1)n^{0}$ + $^{0}_{1}\beta$ positron expelled	0	−1	+1
Electron capture†	x-ray photon	$^{0}_{-1}e$ absorbed from low-energy orbital + $^{1}_{1}p$ in nucleus ⟶ $^{1}_{0}n$ in nucleus	0	−1	+1
γ Emission	$^{0}_{0}\gamma$	excited nucleus ⟶ stable nucleus + $^{0}_{0}\gamma$ γ photon radiated	0	0	0

*Neutrinos (ν) are involved in several of these processes but are not shown.

†Nuclear chemists consider β decay to be a more general process that includes three decay modes: negatron emission (which the text calls "β decay"), positron emission, and electron capture.

Note that the *A* value for Ra equals the sum of the *A* values for Rn and He (226 = 222 + 4), and that the *Z* value for Ra equals the sum of the *Z* values for Rn and He (88 = 86 + 2).

2. **Beta decay** involves the ejection of a β particle ($_{-1}^{0}\beta$) from the nucleus.* This change does not involve the expulsion of a β particle that was actually in the nucleus, but rather the *conversion of a neutron into a proton, which remains in the nucleus, and a β particle, which is expelled immediately:*

$$_{0}^{1}n \longrightarrow {}_{1}^{1}p + {}_{-1}^{0}\beta$$

As always, the totals of the *A* and the *Z* values for reactant and products are equal. Radioactive nickel-63 becomes stable copper-63 through β decay:

$$_{28}^{63}Ni \longrightarrow {}_{29}^{63}Cu + {}_{-1}^{0}\beta$$

Another example is the β decay of carbon-14, applied in radiocarbon dating:

$$_{6}^{14}C \longrightarrow {}_{7}^{14}N + {}_{-1}^{0}\beta$$

Note that *β decay results in a product nuclide with the same A but with Z one higher (one more proton) than in the reactant nuclide.* In other words, an atom of the element with the next *higher* atomic number is formed.

3. **Positron decay** involves the emission of a positron from the nucleus. A key idea of modern physics is that every fundamental particle has a corresponding *antiparticle,* another particle with the same mass but opposite charge. The **positron** (symbolized $_{1}^{0}\beta$; note the positive *Z*) is the antiparticle of the electron. Positron decay occurs through a process in which *a proton in the nucleus is converted into a neutron, and a positron is expelled.*† *Positron decay has the opposite effect of β decay, resulting in a daughter nuclide with the same A but with Z one lower (one fewer proton) than the parent;* thus, an atom of the element with the next *lower* atomic number forms. Carbon-11, a synthetic radioisotope, decays to a stable boron isotope through emission of a positron:

$$_{6}^{11}C \longrightarrow {}_{5}^{11}B + {}_{1}^{0}\beta$$

4. **Electron capture** occurs when the nucleus of an atom draws in an electron from an orbital of the lowest energy level. The net effect is that *a nuclear proton is transformed into a neutron:*

$$_{1}^{1}p + {}_{-1}^{0}e \longrightarrow {}_{0}^{1}n$$

(We use the symbol $_{-1}^{0}e$ to distinguish an orbital electron from a beta particle, symbol $_{-1}^{0}\beta$.) The orbital vacancy is quickly filled by an electron that moves down from a higher energy level, and that energy difference appears as an *x-ray photon.* Radioactive iron forms stable manganese through electron capture:

$$_{26}^{55}Fe + {}_{-1}^{0}e \longrightarrow {}_{25}^{55}Mn + h\nu \text{ (x-ray)}$$

Electron capture has the same net effect as positron decay (*Z* lower by 1, *A* unchanged), even though the processes are entirely different.

5. **Gamma emission** involves the radiation of high-energy γ photons from an excited nucleus. Recall that an atom in an excited *electronic* state reduces its energy by emitting photons, usually in the UV and visible ranges. Similarly, a nucleus in an excited state lowers its energy by emitting γ photons, which are of much higher energy (much shorter wavelength) than UV photons. Many nuclear processes leave the nucleus in an excited state, so *γ emission accompanies most other types of decay.* Several γ photons (γ rays) of different frequencies can be

The Little Neutral One A neutral particle called a *neutrino* (ν) is also emitted in many nuclear reactions, including the change of a neutron to a proton:

$$_{0}^{1}n \longrightarrow {}_{1}^{1}p + {}_{-1}^{0}\beta + \nu$$

Theory suggests that neutrinos have a mass much less than 10^{-4} times that of an electron, and that at least 10^{9} neutrinos exist in the universe for every proton. Neutrinos interact with matter so slightly that it would take a piece of lead 1 light-year thick to absorb them. We will not discuss them further, except to mention that experiments in Japan in the 1990s detected neutrinos and obtained evidence that they have mass. Using a cathedral-sized pool containing 50,000 tons of ultrapure water buried 1 mile underground in a zinc mine, an international team of scientists obtained results that suggest that neutrinos may account for a significant portion of the "missing" matter in the universe and may provide enough mass (and, thus, gravitational attraction) to prevent the universe from expanding forever.

*In formal nuclear chemistry terminology, *β decay* indicates a more general phenomenon that also includes positron emission and electron capture (see footnote to Table 24.2).

†The process, called *pair production,* involves a transformation of energy into matter. A high-energy (>1.63×10^{-13} J) photon becomes an electron and a positron simultaneously. The electron and a proton in the nucleus form a neutron, while the positron is expelled.

emitted from an excited nucleus as it returns to the ground state. Many of Marie Curie's experiments involved the release of γ rays, such as

$$^{238}_{92}\text{U} \longrightarrow {}^{234}_{90}\text{Th} + {}^{4}_{2}\text{He} + 2^{0}_{0}\gamma$$

Because γ rays have no mass or charge, *γ emission does not change A or Z.* Gamma rays also result when a particle and an antiparticle annihilate each other, as when an emitted positron meets an orbital electron:

$$^{0}_{1}\beta \text{ (from nucleus)} + {}^{0}_{-1}\text{e (outside nucleus)} \longrightarrow 2^{0}_{0}\gamma$$

SAMPLE PROBLEM 24.1 Writing Equations for Nuclear Reactions

Problem Write balanced equations for the following nuclear reactions:
(a) Naturally occurring thorium-232 undergoes α decay.
(b) Chlorine-36 undergoes electron capture.
Plan We first write a skeleton equation that includes the mass numbers, atomic numbers, and symbols of all the particles, showing the unknown particles as $^{A}_{Z}\text{X}$. Then, because the total of mass numbers and the total of charges on the left side and the right side must be equal, we solve for *A* and *Z*, and use *Z* to determine X from the periodic table.
Solution (a) Writing the skeleton equation:

$$^{232}_{90}\text{Th} \longrightarrow {}^{A}_{Z}\text{X} + {}^{4}_{2}\text{He}$$

Solving for *A* and *Z* and balancing the equation: For *A*, $232 = A + 4$, so $A = 228$. For *Z*, $90 = Z + 2$, so $Z = 88$. From the periodic table, we see that the element with $Z = 88$ is radium (Ra). Thus, the balanced equation is

$$^{232}_{90}\text{Th} \longrightarrow {}^{228}_{88}\text{Ra} + {}^{4}_{2}\text{He}$$

(b) Writing the skeleton equation:

$$^{36}_{17}\text{Cl} + {}^{0}_{-1}\text{e} \longrightarrow {}^{A}_{Z}\text{X}$$

Solving for *A* and *Z* and balancing the equation: For *A*, $36 + 0 = A$, so $A = 36$. For *Z*, $17 + (-1) = Z$, so $Z = 16$. The element with $Z = 16$ is sulfur (S), so we have

$$^{36}_{17}\text{Cl} + {}^{0}_{-1}\text{e} \longrightarrow {}^{36}_{16}\text{S}$$

Check Always read across superscripts and then across subscripts, with the yield arrow as an equal sign, to check your arithmetic. In part (a), for example, $232 = 228 + 4$, and $90 = 88 + 2$.

FOLLOW-UP PROBLEM 24.1 Write a balanced equation for the reaction in which a nuclide undergoes β decay and produces cesium-133.

Nuclear Stability and the Mode of Decay

There are several ways that an unstable nuclide *might* decay, but can we predict how it *will* decay? Indeed, can we predict *if* a given nuclide will decay at all? Our knowledge of the nucleus is much less than that of the atom as a whole, but some patterns emerge from observation of the naturally occurring nuclides.

The Band of Stability and the Neutron-to-Proton (*N/Z*) Ratio A key factor that determines the stability of a nuclide is the ratio of the number of neutrons to the number of protons, the ***N/Z* ratio,** which we calculate from $(A - Z)/Z$. For lighter nuclides, one neutron for each proton ($N/Z \approx 1$) is enough to provide stability. However, for heavier nuclides to be stable, the number of neutrons must exceed the number of protons, and often by quite a lot. But, if the *N/Z* ratio is either too high or not high enough, the nuclide is unstable and decays.

Figure 24.2A is a plot of number of neutrons vs. number of protons for the *stable* nuclides. The nuclides form a narrow **band of stability** that gradually increases from an *N/Z* ratio of 1, near $Z = 10$, to an *N/Z* ratio slightly greater than 1.5, near $Z = 83$ for ^{209}Bi. Several key points are as follows:

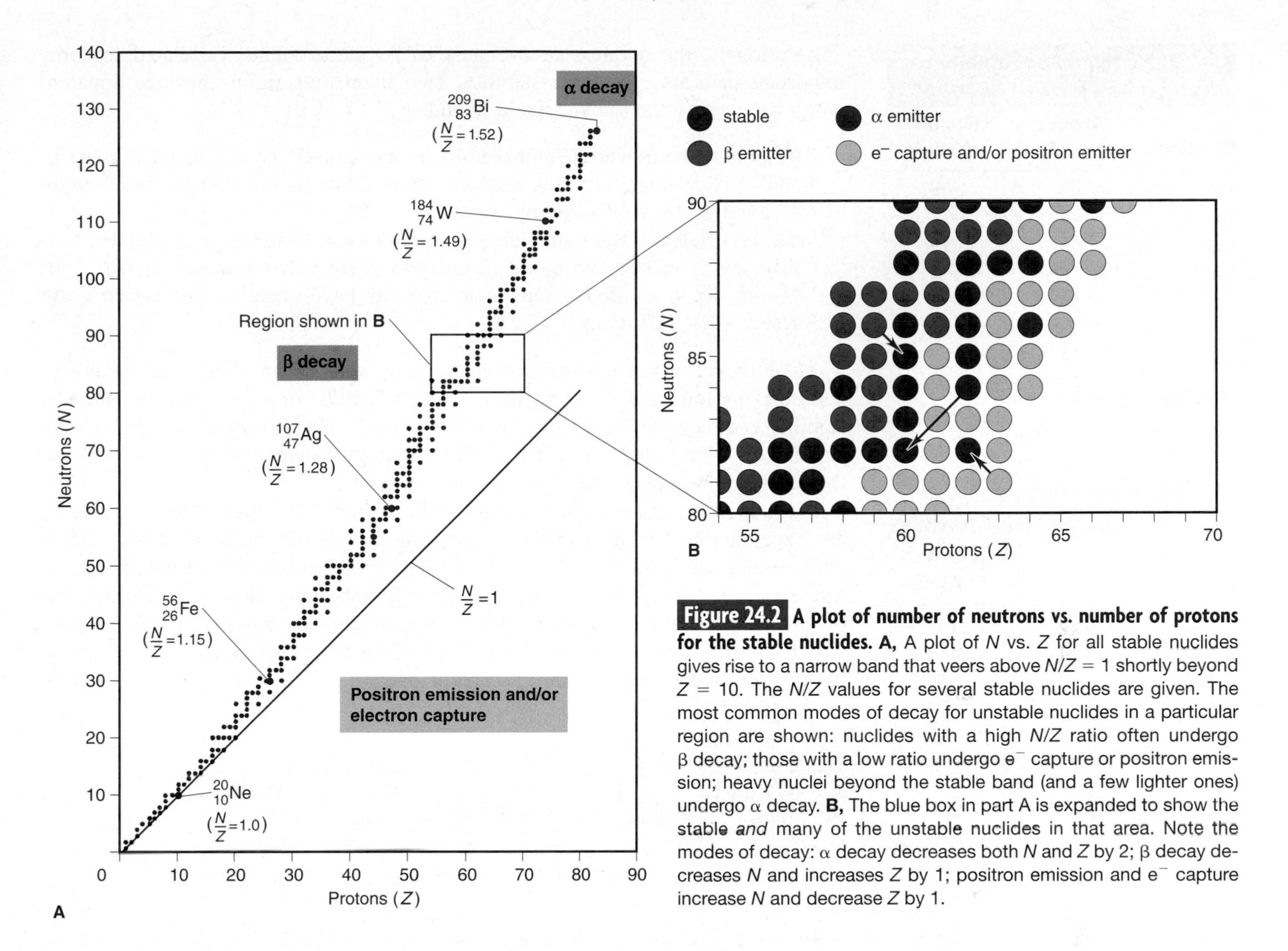

Figure 24.2 A plot of number of neutrons vs. number of protons for the stable nuclides. A, A plot of N vs. Z for all stable nuclides gives rise to a narrow band that veers above $N/Z = 1$ shortly beyond $Z = 10$. The N/Z values for several stable nuclides are given. The most common modes of decay for unstable nuclides in a particular region are shown: nuclides with a high N/Z ratio often undergo β decay; those with a low ratio undergo e⁻ capture or positron emission; heavy nuclei beyond the stable band (and a few lighter ones) undergo α decay. **B,** The blue box in part A is expanded to show the stable *and* many of the unstable nuclides in that area. Note the modes of decay: α decay decreases both N and Z by 2; β decay decreases N and increases Z by 1; positron emission and e⁻ capture increase N and decrease Z by 1.

- Very few stable nuclides exist with $N/Z < 1$; the only two are $^{1}_{1}H$ and $^{3}_{2}He$. For lighter nuclides, $N/Z \approx 1$: for example, $^{4}_{2}He$, $^{12}_{6}C$, $^{16}_{8}O$, and $^{20}_{10}Ne$ are particularly stable.
- The N/Z ratio of stable nuclides gradually increases as Z increases. No stable nuclide exists with $N/Z = 1$ for $Z > 20$. Thus, for $^{56}_{26}Fe$, $N/Z = 1.15$; for $^{107}_{47}Ag$, $N/Z = 1.28$; and for $^{184}_{74}W$, $N/Z = 1.49$.
- All nuclides with $Z > 83$ are unstable. Bismuth-209 is the heaviest stable nuclide. Therefore, the largest members of Groups 1A(1), 2A(2), 4A(14), 6A(16), 7A(17), and 8A(18) are radioactive, as are all the actinides and the elements of the fourth transition series (Period 7).

Stability and Nuclear Structure Given that protons are positively charged and neutrons uncharged, what holds the nucleus together? Nuclear scientists answer this question and explain the importance of the N/Z ratio in terms of two opposing forces. Electrostatic repulsive forces between protons would break the nucleus apart if not for the presence of an attractive force that exists between all nucleons (protons and neutrons) called the **strong force.** This force is about 1000 times stronger than the repulsive force but *operates only over the short distances within the nucleus.* Competition between the *attractive* strong force and the *repulsive* electrostatic force determines nuclear stability.

Table 24.3 Number of Stable Nuclides for Elements 48 to 54*

Element	Atomic No. (Z)	No. of Nuclides
Cd	**48**	**8**
In	49	2
Sn	**50**	**10**
Sb	51	2
Te	**52**	**8**
I	53	1
Xe	**54**	**9**

*Even *Z* shown in boldface.

Table 24.4 An Even-Odd Breakdown of the Stable Nuclides

Z	*N*	No. of Nuclides
Even	Even	157
Even	Odd	53
Odd	Even	50
Odd	Odd	7
	TOTAL	267

Curiously, the oddness or evenness of *N* and *Z* values is related to some important patterns of nuclear stability. Two interesting points become apparent when we classify the known stable nuclides:

- Elements with an even *Z* (number of protons) usually have a larger number of stable nuclides than elements with an odd *Z*. Table 24.3 demonstrates this point for cadmium ($Z = 48$) through xenon ($Z = 54$).
- Well over half the stable nuclides have *both* even *N* and even *Z* (Table 24.4). (Only seven nuclides with odd *N* and odd *Z* are either stable—$^{2}_{1}H$, $^{6}_{3}Li$, $^{10}_{5}B$, $^{14}_{7}N$—or decay so slowly that their amounts have changed little since Earth formed—$^{50}_{23}V$, $^{138}_{57}La$, and $^{176}_{71}Lu$.)

One model of nuclear structure that attempts to explain these findings postulates that protons and neutrons lie in *nucleon shells,* or energy levels, and that stability results from the *pairing* of like nucleons. This arrangement leads to the stability of even values of *N* and *Z*. (The analogy to electron energy levels and the stability that arises from electron pairing is striking.)

Just as noble gases—the elements with 2, 10, 18, 36, 54, and 86 electrons—are exceptionally stable because of their filled *electron* shells, nuclides with *N* or *Z* values of 2, 8, 20, 28, 50, 82 (and $N = 126$) are exceptionally stable. These so-called *magic numbers* are thought to correspond to the numbers of protons or neutrons in filled *nucleon* shells. A few examples are $^{50}_{22}Ti$ ($N = 28$), $^{88}_{38}Sr$ ($N = 50$), and the ten stable nuclides of tin ($Z = 50$). Some extremely stable nuclides have double magic numbers: $^{4}_{2}He$, $^{16}_{8}O$, $^{40}_{20}Ca$, and $^{208}_{82}Pb$ ($N = 126$).

SAMPLE PROBLEM 24.2 Predicting Nuclear Stability

Problem Which of the following nuclides would you predict to be stable and which radioactive: **(a)** $^{18}_{10}Ne$; **(b)** $^{32}_{16}S$; **(c)** $^{236}_{90}Th$; **(d)** $^{123}_{56}Ba$? Explain.

Plan In order to evaluate the stability of each nuclide, we determine the *N/Z* ratio from $(A - Z)/Z$, the value of *Z*, stable *N/Z* ratios (from Figure 24.2), and whether *Z* and *N* are even or odd.

Solution (a) Radioactive. The ratio $N/Z = \frac{18 - 10}{10} = 0.8$. The minimum ratio for stability is 1.0; so, despite even *N* and *Z*, this nuclide has too few neutrons to be stable.

(b) Stable. This nuclide has $N/Z = 1.0$ and $Z < 20$, with even *N* and *Z*. Thus, it is most likely stable.

(c) Radioactive. Every nuclide with $Z > 83$ is radioactive.

(d) Radioactive. The ratio $N/Z = 1.20$. For *Z* from 55 to 60, Figure 24.2A shows $N/Z \geq 1.3$, so this nuclide probably has too few neutrons to be stable.

Check By consulting a table of isotopes, such as the one in the *CRC Handbook of Chemistry and Physics,* we find that our predictions are correct.

FOLLOW-UP PROBLEM 24.2 Why is $^{31}_{15}P$ stable but $^{30}_{15}P$ unstable?

Predicting the Mode of Decay An unstable nuclide generally decays in a mode that shifts its *N/Z* ratio toward the band of stability. This fact is illustrated in Figure 24.2B on the preceding page, which expands a small region of Figure 24.2A to show all of the stable *and* many of the radioactive nuclides in that region, as well as their modes of decay. Note the following points, and then we'll apply them in a sample problem:

1. *Neutron-rich nuclides.* Nuclides with too many neutrons for stability (a high *N/Z*) lie above the band of stability. They undergo *β decay,* which converts a neutron into a proton, thus reducing the value of *N/Z*.

2. *Neutron-poor nuclides.* Nuclides with too few neutrons for stability (a low *N/Z*) lie below the band. They undergo *positron decay* or *electron capture,* both of which convert a proton into a neutron, thus increasing the value of *N/Z*.
3. *Heavy nuclides.* Nuclides with $Z > 83$ are too heavy to lie within the band and undergo *α decay,* which reduces their *Z* and *N* values by two units per emission. (Several lighter nuclides also exhibit α decay.)

SAMPLE PROBLEM 24.3 Predicting the Mode of Nuclear Decay

Problem Predict the nature of the nuclear change(s) each of the following radioactive nuclides is likely to undergo: **(a)** $^{12}_{5}B$; **(b)** $^{234}_{92}U$; **(c)** $^{74}_{33}As$; **(d)** $^{127}_{57}La$.

Plan We use the *N/Z* ratio to decide where the nuclide lies relative to the band of stability and how its ratio compares with others in the nearby region of the band. Then, we predict which of the decay modes just discussed will yield a product nuclide that is closer to the band.

Solution (a) This nuclide has an *N/Z* ratio of 1.4, which is too high for this region of the band. It will probably undergo β decay, increasing *Z* to 6 and lowering the *N/Z* ratio to 1.

(b) This nuclide is heavier than those close to it in the band of stability. It will probably undergo α decay and decrease its total mass.

(c) This nuclide, with an *N/Z* ratio of 1.24, lies in the band of stability, so it will probably undergo either β decay or positron emission.

(d) This nuclide has an *N/Z* ratio of 1.23, which is too low for this region of the band, so it will decrease *Z* by either positron emission or electron capture.

Comment Both possible modes of decay are observed for the nuclides in parts (c) and (d).

FOLLOW-UP PROBLEM 24.3 What mode of decay would you expect for **(a)** $^{61}_{26}Fe$; **(b)** $^{241}_{95}Am$?

Decay Series A parent nuclide may undergo a series of decay steps before a stable daughter nuclide forms. The succession of steps is called a **decay series,** or **disintegration series,** and is typically depicted on a gridlike display. Figure 24.3 shows the decay series from uranium-238 to lead-206. Numbers of neutrons (*N*) are plotted against numbers of protons (*Z*) to form the grid, which displays a series of α and β decays. The zigzag pattern is typical and occurs because α decay decreases both *N* and *Z*, whereas β decay decreases *N* but increases *Z*. Note that it is quite common for a given nuclide to undergo both types of decay. (Gamma decay accompanies many of these steps, but it does not affect the mass or type of the nuclide.) This decay series is one of three that occur in nature. All end with isotopes of lead whose nuclides all have one ($Z = 82$) or two ($N = 126$, $Z = 82$) magic numbers. A second series begins with uranium-235 and ends with lead-207, and a third begins with thorium-232 and ends with lead-208. (Neptunium-237 began a fourth series, but its half-life is so much less than the age of Earth that only traces of it remain today.)

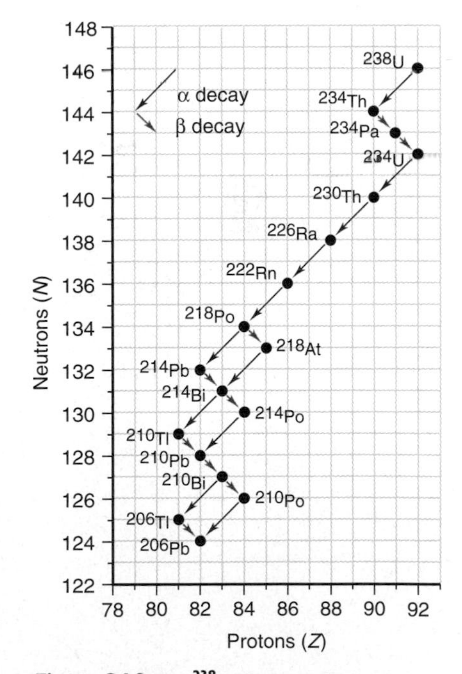

Figure 24.3 The ^{238}U decay series. Uranium-238 (*top right*) decays through a series of emissions of α or β particles to lead-206 (*bottom left*) in 14 steps.

SECTION SUMMARY

Nuclear reactions are not affected by reaction conditions or chemical composition and release much more energy than chemical reactions. A radioactive nuclide is unstable and may emit α particles ($^{4}_{2}He$ nuclei), β particles ($^{0}_{-1}\beta$; high-speed electrons), positrons ($^{0}_{1}\beta$), or γ rays ($^{0}_{0}\gamma$; high-energy photons) or may capture an orbital electron ($^{0}_{-1}e$). A narrow band of neutron-to-proton ratios (*N/Z*) includes those of all the stable nuclides. Radioactive decay allows an unstable nuclide to achieve a more stable *N/Z* ratio. Certain "magic numbers" of neutrons and protons are associated with very stable nuclides. By comparing a nuclide's *N/Z* ratio with those in the band of stability, we can predict that, in general, heavy nuclides undergo α decay, neutron-rich nuclides undergo β decay, and proton-rich nuclides undergo positron emission or electron capture. Three naturally occurring decay series all end in isotopes of lead.

24.2 THE KINETICS OF RADIOACTIVE DECAY

Chemical and nuclear systems both tend toward maximum stability. Just as the concentrations in a chemical system change in a predictable direction to give a stable equilibrium ratio, the type and number of nucleons in an unstable nucleus change in a predictable direction to give a stable N/Z ratio. As you know, however, the tendency of a chemical system to become more stable tells nothing about how long that process will take, and the same holds true for nuclear systems. In this section, we examine the kinetics of nuclear change; later, we'll examine the energetics of such change. To begin, a Tools of the Laboratory essay on the opposite page describes how radioactivity is detected and measured.

The Rate of Radioactive Decay

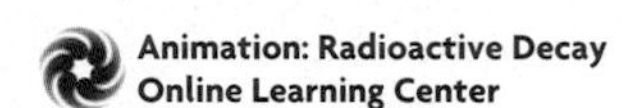
Animation: Radioactive Decay
Online Learning Center

Radioactive nuclei decay at a characteristic rate, regardless of the chemical substance in which they occur. The *decay rate,* or **activity ($\mathscr{A}$),** of a radioactive sample is the change in number of nuclei ($\mathscr{N}$) divided by the change in time (t). As we saw with chemical reaction rates, because the number of nuclei is *decreasing,* a minus sign precedes the expression for the decay rate:

$$\text{Decay rate } (\mathscr{A}) = -\frac{\Delta \mathscr{N}}{\Delta t}$$

The SI unit of radioactivity is the **becquerel (Bq);** it is defined as one disintegration per second (d/s): 1 Bq = 1 d/s. A much larger and more common unit of radioactivity is the **curie (Ci):** 1 curie equals the number of nuclei disintegrating each second in 1 g of radium-226:

$$1 \text{ Ci} = 3.70\times10^{10} \text{ d/s} \quad \textbf{(24.2)}$$

Because the curie is so large, the millicurie (mCi) and microcurie (μCi) are commonly used. We often express the radioactivity of a sample in terms of *specific activity,* the decay rate per gram.

An activity is meaningful only when we consider the large number of nuclei in a macroscopic sample. Suppose there are 1×10^{15} radioactive nuclei of a particular type in a sample and they decay at a rate of 10% per hour. Although any particular nucleus in the sample might decay in a microsecond or in a million hours, the *average* of all decays results in 10% of the entire collection of nuclei disintegrating each hour. During the first hour, 10% of the *original* number, or 1×10^{14} nuclei, will decay. During the next hour, 10% of the remaining 9×10^{14} nuclei, or 9×10^{13} nuclei, will decay. During the next hour, 10% of those remaining will decay, and so forth. Thus, for a large collection of radioactive nuclei, *the number decaying per unit time is proportional to the number present:*

$$\text{Decay rate } (\mathscr{A}) \propto \mathscr{N} \qquad \text{or} \qquad \mathscr{A} = k\mathscr{N}$$

where k is called the **decay constant** and is characteristic of each type of nuclide. The larger the value of k, the higher is the decay rate.

Combining the two rate expressions just given, we obtain

$$\mathscr{A} = -\frac{\Delta \mathscr{N}}{\Delta t} = k\mathscr{N} \quad \textbf{(24.3)}$$

Note that the activity depends only on $\mathscr{N}$ raised to the first power (and on the constant value of k). Therefore, *radioactive decay is a first-order process* (see Section 16.4). The only difference in the case of nuclear decay is that we consider the *number* of nuclei rather than their concentration.

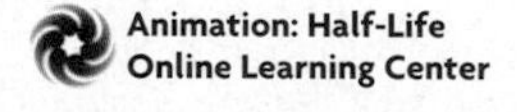
Animation: Half-Life
Online Learning Center

Half-Life of Radioactive Decay Decay rates are also commonly expressed in terms of the fraction of nuclei that decay over a given time interval. The **half-life ($t_{1/2}$)** of a nuclide is the time it takes for half the nuclei present in a sample to decay. *The number of nuclei remaining is halved after each half-life.* Thus, half-life has the same meaning for a nuclear change as for a chemical change (Section 16.4).

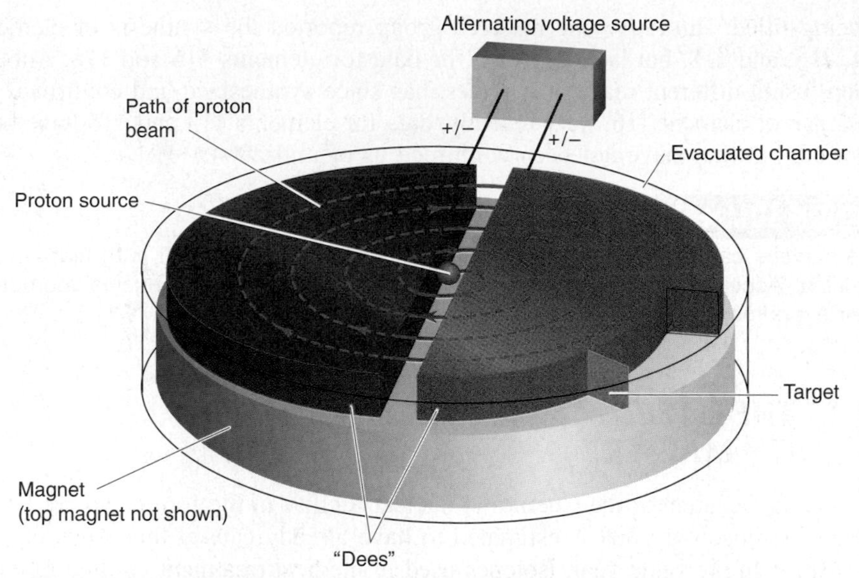

Figure 24.7 The cyclotron accelerator. When the positively charged particle reaches the gap between the two D-shaped electrodes ("dees"), it is repelled by one dee and attracted by the other. The particles move in a spiral path, so the cyclotron can be much smaller than a linear accelerator.

heavier particles, such as B, C, O, and Ne nuclei, several hundred million times faster, with correspondingly greater kinetic energies.

The *cyclotron* (Figure 24.7), invented by E. O. Lawrence in 1930, applies the principle of the linear accelerator but uses electromagnets to give the particle a spiral path, thus saving space. The magnets lie within an evacuated chamber above and below two "dees," open, D-shaped electrodes that function like the tubes in the linear design. The particle is accelerated as it passes from one dee, which is momentarily positive, to the other, which is momentarily negative. Its speed and radius increase until it is deflected toward the target nucleus. The *synchrotron* uses a synchronously increasing magnetic field to make the particle's path circular rather than spiral.

Accelerators have many applications, from producing radioisotopes used in medical applications to studying the fundamental nature of matter. Perhaps their most specific application for chemists is the synthesis of **transuranium elements,** those with atomic numbers higher than uranium, which is the heaviest naturally occurring element. Some reactions that were used to form several of these elements appear in Table 24.6. The transuranium elements include the remaining actinides (Z = 93 to 103), in which the 5*f* sublevel is being filled, and the elements in the fourth transition series (Z = 104 to 112), in which the 6*d* sublevel

The Powerful Bevatron The *bevatron,* used to study the physics of high-energy particle collisions, includes a linear section and a synchrotron section. The instrument at the Lawrence Berkeley Laboratory in California increases the kinetic energy of the particles by a factor of more than 6 billion. A beam of 10^{10} protons makes more than 4 million revolutions, a distance of 300,000 miles, in 1.8 s, attaining a final speed about 90% of the speed of light! Even more powerful bevatrons are in use at the Brookhaven National Laboratory in New York and at CERN, outside Geneva, Switzerland.

Table 24.6 Formation of Some Transuranium Nuclides

Reaction	Half-life of Product
$^{239}_{94}Pu + ^{4}_{2}He \longrightarrow ^{240}_{95}Am + ^{1}_{1}H + 2^{1}_{0}n$	50.9 h
$^{239}_{94}Pu + ^{4}_{2}He \longrightarrow ^{242}_{96}Cm + ^{1}_{0}n$	163 days
$^{244}_{96}Cm + ^{4}_{2}He \longrightarrow ^{245}_{97}Bk + ^{1}_{1}H + 2^{1}_{0}n$	4.94 days
$^{238}_{92}U + ^{12}_{6}C \longrightarrow ^{246}_{98}Cf + 4^{1}_{0}n$	36 h
$^{253}_{99}Es + ^{4}_{2}He \longrightarrow ^{256}_{101}Md + ^{1}_{0}n$	76 min
$^{252}_{98}Cf + ^{10}_{5}B \longrightarrow ^{256}_{103}Lr + 6^{1}_{0}n$	28 s

Naming the Transuranium Elements
The last naturally occurring element was named after Uranus, thought at the time to be the outermost planet; then, the first two artificial elements were named after the more recently discovered Neptune and Pluto. The next few elements were named after famous scientists, as in curium, and places, as in americium. But conflicting claims of discovery by scientists in different countries led to controversies about names for elements 104 and higher. To provide interim names until the disputes could be settled, the International Union of Pure and Applied Chemistry (IUPAC) adopted a system that uses the atomic number as the basis for a Latin name. Thus, for example, element 104 was named unnilquadium (un = 1, nil = 0, quad = 4, ium = element suffix), with the symbol Unq. After much compromise, the IUPAC has finalized these names: 104, rutherfordium (Rf); 105, dubnium (Db); 106, seaborgium (Sg); 107, bohrium (Bh); 108, hassium (Hs); 109, meitnerium (Mt); and 110, darmstadtium (Ds). Elements with atomic numbers 111 and higher have not yet been named.

is being filled. (In 1999, one research group reported the synthesis of elements 114, 116, and 118, but later retracted the data for elements 116 and 118. Another group, using different reactant nuclides, has since synthesized and confirmed the existence of element 116. Very recently, data for elements 113 and 115 have been reported, but they have not been confirmed as of mid-2004.)

SECTION SUMMARY

One nucleus can be transmuted to another through bombardment with high-energy particles. Accelerators increase the kinetic energy of particles in nuclear bombardment experiments and are used to produce transuranium elements.

24.4 THE EFFECTS OF NUCLEAR RADIATION ON MATTER

In 1986, an accident at the Chernobyl nuclear facility in the former Soviet Union released radioactivity that is estimated to have already caused thousands of cancer deaths. In the same year, isotopes used in medical treatment emitted radioactivity that prevented thousands of cancer deaths. In this section and the next, we examine the harmful and beneficial effects of radioactivity.

The key to both of these outcomes is that *nuclear changes cause chemical changes in surrounding matter.* In other words, even though the nucleus of an atom undergoes a reaction with little or no involvement of the atom's electrons, the emissions *do* affect the electrons of nearby atoms.

The Effects of Radioactive Emissions: Excitation and Ionization

Radioactive emissions interact with matter in two ways, depending on their energies:

- *Excitation.* In the process of **excitation,** radiation of relatively low energy interacts with an atom of a substance, which absorbs some of the energy and then re-emits it. Because electrons are not lost from the atom, the radiation that causes excitation is called **nonionizing radiation.** If the absorbed energy causes the atoms to move, vibrate, or rotate more rapidly, the material becomes hotter. Concentrated aqueous solutions of plutonium salts boil because the emissions excite the surrounding water molecules. (Polonium has therefore been suggested as a lightweight heat source, with no moving parts, for use on space stations.) Particles of somewhat higher energy excite electrons in other atoms to higher energy levels. As the atoms return to their ground state, they emit photons, often in the blue or UV region (see the description of scintillation counters in the Tools of the Laboratory essay, p. 1055).
- *Ionization.* In the process of **ionization,** radiation collides with an atom energetically enough to dislodge an electron:

$$\text{Atom} \xrightarrow{\text{ionizing radiation}} \text{ion}^+ + e^-$$

  A cation and a free electron result, and the number of such *cation-electron pairs* that are produced is directly related to the energy of the incoming radiation. The high-energy radiation that gives rise to this effect is called **ionizing radiation.** The free electron of the pair often collides with another atom and ejects a second electron (see the description of Geiger-Müller counters in the Tools of the Laboratory essay, p. 1055).

Effects of Ionizing Radiation on Living Matter

Whereas nonionizing radiation is relatively harmless, ionizing radiation has a destructive effect on living tissue. When the atom that was ionized is part of a biological macromolecule or membrane component, the results can be devastating.

Units of Radiation Dose and Its Effects To measure the effects of ionizing radiation, we need a unit for radiation dose. Units of radioactive decay, such as the becquerel and curie, measure the number of decay events in a given time but not their energy or absorption by matter. The number of cation-electron pairs produced in a given amount of living tissue is a measure of the energy absorbed by the tissue. The SI unit for such energy absorption is the **gray (Gy);** it is equal to 1 joule of energy absorbed per kilogram of body tissue: 1 Gy = 1 J/kg. A more widely used unit is the **rad (*r*adiation-*a*bsorbed *d*ose),** which is equal to 0.01 Gy:

$$1 \text{ rad} = 0.01 \text{ J/kg} = 0.01 \text{ Gy}$$

To measure actual tissue damage, we must account for differences in the strength of the radiation, the exposure time, and the type of tissue. To do this, we multiply the number of rads by a *r*elative *b*iological *e*ffectiveness (RBE) factor, which depends on the effect of a given type of radiation on a given tissue or body part. The product is the **rem (*r*oentgen *e*quivalent for *m*an),** the unit of radiation dosage equivalent to a given amount of tissue damage in a human:

$$\text{no. of rems} = \text{no. of rads} \times \text{RBE}$$

Doses are often expressed in millirems (10^{-3} rem). The SI unit for dosage equivalent is the **sievert (Sv).** It is defined in the same way as the rem but with absorbed dose in grays; thus, 1 rem = 0.01 Sv.

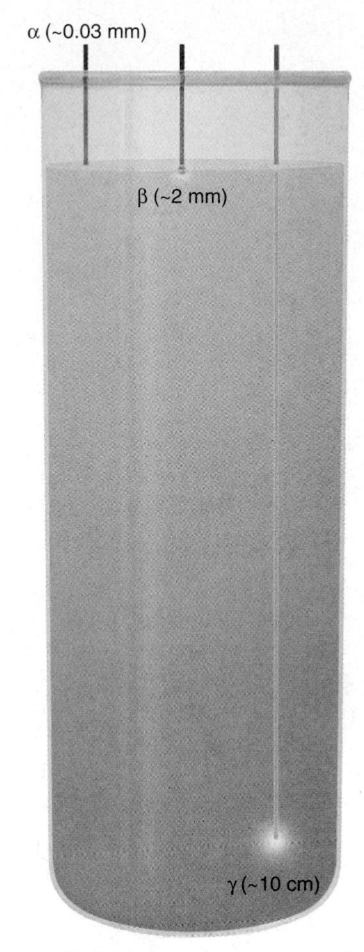

Figure 24.8 Penetrating power of radioactive emissions. Penetrating power is often measured in terms of the depth of water that stops 50% of the incoming radiation. (Water is the main component of living tissue.) Alpha particles, with the highest mass and charge, have the lowest penetrating power, and γ rays have the highest. (Average values of actual penetrating distances are shown.)

Penetrating Power of Emissions The effect on living tissue of a radiation dose depends on the penetrating power *and* ionizing ability of the radiation. Figure 24.8 depicts the differences in penetrating power of the three common emissions. Note, in general, that *penetrating power is inversely related to the mass and charge of the emission.* In other words, if a particle interacts strongly with matter, it penetrates only slightly, and vice versa:

- *α Particles.* Alpha particles are massive and highly charged, which means that they interact with matter most strongly of the three common types of emissions. As a result, they penetrate so little that a piece of paper, light clothing, or the outer layer of skin can stop α radiation from an external source. However, if ingested, an α emitter such as plutonium-239 causes grave localized damage through extensive ionization.
- *β Particles and positrons.* Beta particles and positrons have less charge and much less mass than α particles, so they interact less strongly with matter. Even though a given particle has less chance of causing ionization, a β (or positron) emitter is a more destructive external source because the particles penetrate deeper. Specialized heavy clothing or a thick (0.5 cm) piece of metal is required to stop these particles.
- *γ Rays.* Neutral, massless γ rays interact least with matter and, thus, penetrate most. A block of lead several inches thick is needed to stop them. Therefore, an external γ ray source is the most dangerous because the energy can ionize many layers of living tissue.

A Tragic Way to Tell Time in the Dark
In the early 20th century, wristwatch and clock dials were painted by hand with paint containing radium so they would glow in the dark. To write numbers clearly, the young women hired to apply the paint "tipped" fine brushes repeatedly between their lips. Small amounts of ingested $^{226}Ra^{2+}$ were incorporated into the bones of the women, along with normal Ca^{2+}, which led to numerous cases of bone fracture and jaw cancer.

Molecular Interactions How does the damage take place on the molecular level? When ionizing radiation interacts with a molecule, it causes the *loss of an electron from a bond or a lone pair.* The resulting charged species go on to form **free radicals,** molecular or atomic species with one or more unpaired electrons. As we've seen several times already, species with lone electrons are very reactive and tend to form electron pairs by bonding to other species. To do this, they attack bonds in other molecules, sometimes forming more free radicals.

When γ radiation strikes biological tissue, for instance, the most likely molecule to absorb it is water, which forms an electron and a water ion-radical:

$$H_2O + \gamma \longrightarrow H_2O\cdot^+ + e^-$$

The $H_2O\cdot^+$ and e^- collide with other water molecules to form free radicals:

$$H_2O\cdot^+ + H_2O \longrightarrow H_3O^+ + \cdot OH \quad \text{and} \quad e^- + H_2O \longrightarrow H\cdot + OH^-$$

These free radicals go on to attack more water molecules and surrounding biomolecules, whose bonding and structure, as you know (Section 15.6), are intimately connected with their function.

The double bonds in membrane lipids are particularly susceptible to free-radical attack:

$$H\cdot + RCH{=}CHR' \longrightarrow RCH_2{-}\dot{C}HR'$$

In this reaction, one electron of the π bond forms a C—H bond between one of the double-bonded carbons and the H·, and the other electron resides on the other carbon to form a free radical. Changes to lipid structure cause changes in membrane fluidity and other damage that, in turn, cause leakage of the cell and destruction of the protective fatty tissue around organs. Changes to critical bonds in enzymes lead to their malfunction as catalysts of metabolic reactions. Changes in the nucleic acids and proteins that govern the rate of cell division cause cancer. Genetic damage and mutations may occur when bonds in the DNA of sperm and egg cells are altered by free radicals.

Sources of Ionizing Radiation It is essential to keep the molecular effects of ionizing radiation in perspective. After all, we are continuously exposed to ionizing radiation from natural and artificial sources (Table 24.7). Indeed, life evolved in the presence of natural ionizing radiation, called **background radiation.** The

Table 24.7 Typical Radiation Doses from Natural and Artificial Sources

Source of Radiation	Average Adult Exposure
Natural	
Cosmic radiation	30–50 mrem/yr
Radiation from the ground	
From clay soil and rocks	~25–170 mrem/yr
In wooden houses	10–20 mrem/yr
In brick houses	60–70 mrem/yr
In concrete (cinder block) houses	60–160 mrem/yr
Radiation from the air (mainly radon)	
Outdoors, average value	20 mrem/yr
In wooden houses	70 mrem/yr
In brick houses	130 mrem/yr
In concrete (cinder block) houses	260 mrem/yr
Internal radiation from minerals in tap water and daily intake of food (^{40}K, ^{14}C, Ra)	~40 mrem/yr
Artificial	
Diagnostic x-ray methods	
Lung (local)	0.04–0.2 rad/film
Kidney (local)	1.5–3 rad/film
Dental (dose to the skin)	≤1 rad/film
Therapeutic radiation treatment	Locally ≤ 10,000 rad
Other sources	
Jet flight (4 h)	~1 mrem
Nuclear testing	<4 mrem/yr
Nuclear power industry	<1 mrem/yr
TOTAL AVERAGE VALUE	100–200 mrem/yr

same radiation that causes harmful mutations also causes beneficial mutations that, over time, allow organisms to adapt and species to change.

Background radiation has several sources. One source is *cosmic radiation,* which increases with altitude because of decreased absorption by the atmosphere. Thus, people in Denver absorb twice as much cosmic radiation as people in Los Angeles; even a jet flight involves measurable absorption. The sources of most background radiation are thorium and uranium minerals present in rocks and soil. Radon, the heaviest noble gas [Group 8A(18)], is a radioactive product of uranium and thorium decay, and its concentration in the air we breathe varies with type of local soil and rocks. About 150 g of K^+ ions is dissolved in the water in the tissues of an average adult, and 0.0118% of that amount is radioactive ^{40}K. The presence of these substances and of atmospheric $^{14}CO_2$ means that all food, water, clothing, and building materials are slightly radioactive.

The largest artificial source of radiation, and the easiest to control, is associated with medical diagnostic techniques, especially x-rays. The radiation dosage from nuclear testing and radioactive waste disposal is miniscule for most people, but exposures for those living near test sites, nuclear energy facilities, or disposal areas may be many times higher.

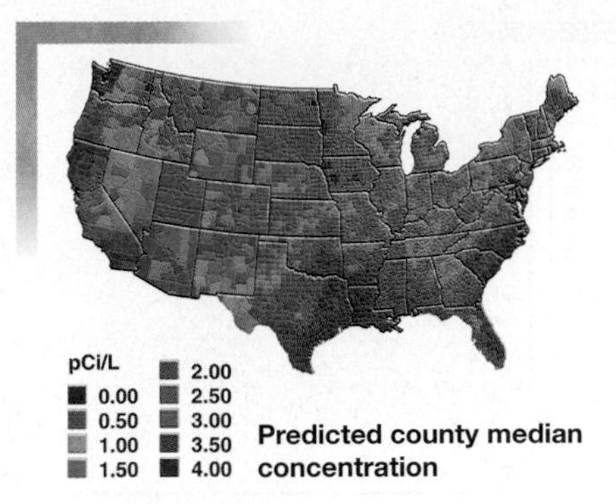

Risk of Radon Radon (Rn; $Z = 86$), the largest noble gas, is a natural decay product of uranium. Therefore, the uranium content of the local soil and rocks is a critical factor in the extent of the threat, but radon occurs everywhere in varying concentrations. Radon itself decays to radioactive nuclides of Po, Pb, and Bi, through α, β, and γ emission. These processes occur inside the body when radon is inhaled and pose a serious potential hazard. The emissions damage lung tissue, and the heavy-metal atoms formed aggravate the problem. The latest EPA estimates indicate that radon contributes to 15% of annual lung cancer deaths.

Assessing the Risk from Ionizing Radiation How much radiation is too much? To approach this question, we must ask several others: How strong is the exposure? How long is the exposure? Which tissue is exposed? Are offspring affected? One reason we lack clear data to answer these questions is that scientific ethical standards forbid the intentional exposure of humans in an experimental setting. However, accidentally exposed radiation workers and Japanese atomic bomb survivors have been studied extensively. Table 24.8 summarizes the immediate tissue effects on humans of an acute single dose of ionizing radiation to the whole body. The severity of the effects increases with dose; a dose of 500 rem will kill about 50% of the exposed population within a month.

Most data come from laboratory animals, whose biological systems may differ greatly from ours. Nevertheless, studies with mice and dogs show that lesions

Table 24.8 Acute Effects of a Single Dose of Whole-Body Irradiation

Dose (rem)	Effect	Lethal Dose: Population (%)	Lethal Dose: No. of Days
5–20	Possible late effect; possible chromosomal aberrations	—	—
20–100	Temporary reduction in white blood cells	—	—
50+	Temporary sterility in men (100+ rem = 1 yr duration)	—	—
100–200	"Mild radiation sickness": vomiting, diarrhea, tiredness in a few hours Reduction in infection resistance Possible bone growth retardation in children	—	—
300+	Permanent sterility in women	—	—
500	"Serious radiation sickness": marrow/intestine destruction	50–70	30
400–1000	Acute illness, early deaths	60–95	30
3000+	Acute illness, death in hours to days	100	2

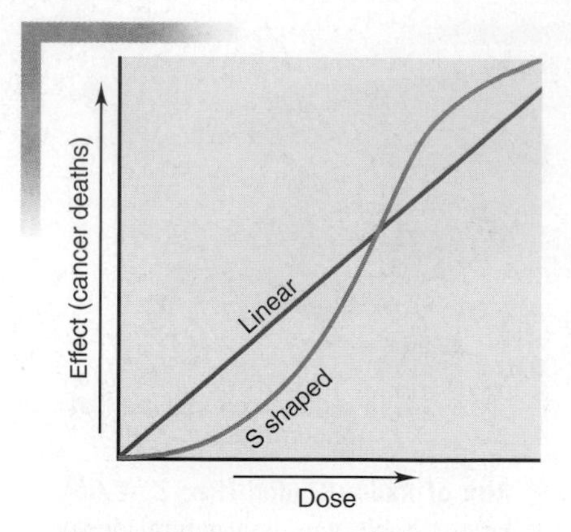

Modeling Radiation Risk There are two current models of effect vs. dose. The *linear response* model proposes that radiation effects, such as cancer risks, accumulate over time regardless of dose and that populations should not be exposed to any radiation above background levels. The *S-shaped response* model assumes an extremely low risk at low doses and advocates concern only at higher doses. If the linear model is more accurate, we should limit all excess exposure, but this would severely restrict medical diagnosis and research, military testing, and nuclear energy production.

and cancers appear after massive whole-body exposure, with rapidly dividing cells affected first. In an adult animal, these are cells of the bone marrow, organ linings, and reproductive organs, but many other tissues are affected in an immature animal or fetus. Studies in both animals and humans show an increase in the incidence of cancer from either a high, single exposure or a low, chronic exposure.

Reliable data on genetic effects are few. Pioneering studies on fruit flies show a linear increase in genetic defects with both dose and exposure time. However, in the mouse, whose genetic system is obviously much more similar to ours than is the fruit fly's, a total dose given over a long period created one-third as many genetic defects as the same dose given over a short period. Therefore, rate of exposure is a key factor. The children of atomic bomb survivors show higher-than-normal childhood cancer rates, implying that their parents' reproductive systems were affected.

SECTION SUMMARY

Relatively low-energy emissions cause excitation of atoms in surrounding matter, whereas high-energy emissions cause ionization. The effect of ionizing radiation on living matter depends on the quantity of energy absorbed and the extent of ionization in a given type of tissue. Radiation dose for the human body is measured in rem. Ionization forms free radicals, some of which proliferate and destroy biomolecular function. All organisms are exposed to varying quantities of natural ionizing radiation. Studies show that a large acute dose and a chronic small dose are both harmful.

24.5 APPLICATIONS OF RADIOISOTOPES

Our ability to detect minute amounts of radioisotopes makes them powerful tools for studying processes in biochemistry, medicine, materials science, environmental studies, and many other scientific and industrial fields. Such uses depend on the fact that *isotopes of an element exhibit* ***very*** *similar chemical and physical behavior*. In other words, except for having a less stable nucleus, a radioisotope has nearly the same chemical properties as a nonradioactive isotope of that element.* For example, the fact that $^{14}CO_2$ is utilized by a plant in the same way as $^{12}CO_2$ forms the basis of radiocarbon dating.

Radioactive Tracers: Applications of Nonionizing Radiation

Just think how useful it could be to follow a substance through a complex process or from one region of a system to another. A tiny amount of a radioisotope mixed with a large amount of the stable isotope can act as a **tracer,** a chemical "beacon" emitting nonionizing radiation that signals the presence of the substance.

Reaction Pathways Tracers help us choose from among possible reaction pathways. One well-studied example is the formation of an organic ester and water from a carboxylic acid and alcohol. Which portions of the reactants end up in the ester and which in the water? Figure 24.9 shows how ^{18}O-tracers answer the question: an ^{18}O-alcohol gives an ^{18}O-ester, but an ^{18}O-acid gives ^{18}O-water.

As another example, consider the reaction between periodate and iodide ions:

$$IO_4^-(aq) + 2I^-(aq) + H_2O(l) \longrightarrow I_2(s) + IO_3^-(aq) + 2OH^-(aq)$$

Is IO_3^- the result of IO_4^- reduction or I^- oxidation? When we add "cold" (nonradioactive) IO_4^- to a solution of I^- that contains some "hot" (radioactive, indi-

*Although this statement is generally correct, differences in isotopic mass *can* influence bond strengths and therefore reaction rates. Such behavior is called a *kinetic isotope effect* and is particularly important for isotopes of hydrogen—^1H, ^2H, and ^3H—because their masses differ by such large proportions. Section 22.4 discussed how the kinetic isotope effect is employed in the industrial production of heavy water, D_2O.

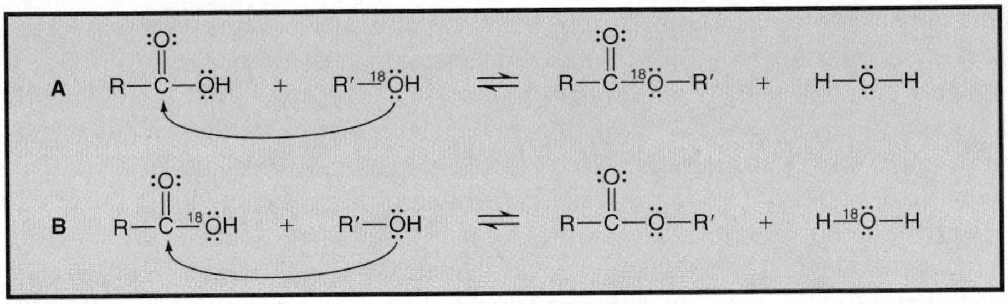

$$\textbf{A}\quad R{-}\overset{\overset{\displaystyle :O:}{\|}}{C}{-}\ddot{O}H \;+\; R'{-}^{18}\ddot{O}H \;\rightleftharpoons\; R{-}\overset{\overset{\displaystyle :O:}{\|}}{C}{-}^{18}\ddot{O}{-}R' \;+\; H{-}\ddot{O}{-}H$$

$$\textbf{B}\quad R{-}\overset{\overset{\displaystyle :O:}{\|}}{C}{-}^{18}\ddot{O}H \;+\; R'{-}\ddot{O}H \;\rightleftharpoons\; R{-}\overset{\overset{\displaystyle :O:}{\|}}{C}{-}\ddot{O}{-}R' \;+\; H{-}^{18}\ddot{O}{-}H$$

Figure 24.9 Which reactant contributes which group to the ester? An ester forms when a carboxylic acid reacts with an alcohol. To determine which reactant supplies the O atom in the —OR′ part of the ester group, acid and alcohol were labeled with the ^{18}O and used as tracers. **A,** When $R^{18}OH$ reacts with the unlabeled acid, the ester contains ^{18}O but the water doesn't. **B,** When $RCO^{18}OH$ reacts with the unlabeled alcohol, the water contains ^{18}O. Thus, the alcohol supplies the —OR′ part of the ester, and the acid supplies the RC=O part.

cated in red) $^{131}I^-$, we find that the I_2 is radioactive, not the IO_3^-:

$$IO_4^-(aq) + 2^{131}I^-(aq) + H_2O(l) \longrightarrow {}^{131}I_2(s) + IO_3^-(aq) + 2OH^-(aq)$$

These results show that IO_3^- forms through the reduction of IO_4^-, and that I_2 forms through the oxidation of I^-. To confirm this pathway, we add IO_4^- containing some hot $^{131}IO_4^-$ to a solution of cold I^-. As we expected, the IO_3^- is radioactive, not the I_2:

$$^{131}IO_4^-(aq) + 2I^-(aq) + H_2O(l) \longrightarrow I_2(s) + {}^{131}IO_3^-(aq) + 2OH^-(aq)$$

Thus, tracers act like "handles" we can "hold" to follow the changing reactants.

Far more complex pathways can be followed with tracers as well. The photosynthetic pathway, the most essential and widespread metabolic process on Earth, in which energy from sunlight is used to form the chemical bonds of glucose, has an overall reaction that looks quite simple:

$$6CO_2(g) + 6H_2O(l) \xrightarrow[\text{chlorophyll}]{\text{light}} C_6H_{12}O_6(s) + 6O_2(g)$$

However, the actual process is extremely complex, requiring 13 enzyme-catalyzed steps for each C atom from CO_2 incorporated; thus these steps must occur six times for each molecule of $C_6H_{12}O_6$ that forms. Melvin Calvin and his coworkers took seven years to determine the pathway, using ^{14}C in CO_2 as the tracer and paper chromatography as the means of separating the products formed after different times of light exposure. Calvin won the Nobel Prize in chemistry in 1961 for this remarkable achievement.

Tracers are used in many studies of biological function. Most recently, life in space has required answers to new questions. In an animal study of red blood cell loss during extended space flight, blood plasma volume was measured with ^{125}I-labeled albumin (a blood protein), and ^{51}Cr-labeled red blood cells were used to assess survival of blood cells. In another study, blood flow in skin under long periods of microgravity was monitored using injected ^{133}Xe.

Material Flow Tracers are used in studies of solid surfaces and the flow of materials. Metal atoms hundreds of layers deep within a solid have been shown to exchange with metal ions from the surrounding solution within a matter of minutes. Chemists and engineers use tracers to study material movement in semiconductor chips, paint, and metal plating, in detergent action, and in the process of corrosion, to mention just a few of many applications.

Hydrologic engineers use tracers to study the volume and flow of large bodies of water. By following radionuclides formed during atmospheric nuclear bomb tests (^3H in H_2O, $^{90}Sr^{2+}$, and $^{137}Cs^+$), scientists have mapped the flow of water from land to lakes and streams to oceans. Surface and deep ocean currents that circulate around the globe are also studied, as are the mechanisms of hurricane formation and the mixing of the troposphere and stratosphere. Industries employ tracers to study material flow during the manufacturing process, such as the flow of ore pellets in smelting kilns, the paths of wood chips and bleach in paper mills, the diffusion of fungicide into lumber, and in a particularly important application, the porosity and leakage of oil and gas wells in geological formations.

Spirit on the surface of Mars.

Activation Analysis A somewhat different use of tracers occurs in *neutron activation analysis* (NAA). In this method, neutrons bombard a nonradioactive sample, converting a small fraction of its atoms to radioisotopes, which exhibit characteristic decay patterns, such as γ-ray spectra, that reveal the elements present. Unlike chemical analysis, NAA leaves the sample virtually intact, so the method can be used to determine the composition of a valuable object or a very small sample. For example, a painting thought to be a 16th-century Dutch masterpiece was shown through NAA to be a 20th-century forgery, because a microgram-sized sample of its pigment contained much less silver and antimony than the pigments used by the Dutch masters. Forensic chemists use NAA to detect traces of ammunition on a suspect's hand or traces of arsenic in the hair of a victim of poisoning. In 2004, space scientists used NAA instrumentation in the *Spirit* and *Opportunity* robot vehicles to analyze the composition of Martian soils and rocks (see photo).

Automotive engineers employ NAA and γ-ray detectors to measure friction and wear of moving parts without having to take the engine apart. For example, when a steel surface that has been neutron-activated to form some radioactive ^{59}Fe moves against a second steel surface, the amount of radioactivity on the second surface indicates the amount of material rubbing off. The radioactivity appearing in a lubricant placed between the surfaces can demonstrate the lubricant's ability to reduce wear.

Table 24.9 Some Radioisotopes Used as Medical Tracers

Isotope	Body Part or Process
^{11}C, ^{18}F, ^{13}N, ^{15}O	PET studies of brain, heart
^{60}Co, ^{192}Ir	Cancer therapy
^{64}Cu	Metabolism of copper
^{59}Fe	Blood flow, spleen
^{67}Ga	Tumor imaging
^{123}I, ^{131}I	Thyroid
^{111}In	Brain, colon
^{42}K	Blood flow
^{81m}Kr	Lung
^{99}Tc	Heart, thyroid, liver, lung, bone
^{201}Tl	Heart muscle
^{90}Y	Cancer, arthritis

Medical Diagnosis The largest use of radioisotopes is in medical science. In fact, over 25% of U.S. hospital admissions are for diagnoses based on data from radioisotopes. Tracers with half-lives of a few minutes to a few days are employed to observe specific organs and body parts. For example, a healthy thyroid gland incorporates dietary I^- into iodine-containing hormones at a known rate. To assess thyroid function, the patient drinks a solution containing a trace amount of $Na^{131}I$, and a scanning monitor follows the uptake of $^{131}I^-$ into the thyroid (Figure 24.10A). Technetium-99 ($Z = 43$) is also used for imaging the thyroid (Figure 24.10B), as well as the heart, lungs, and liver. Technetium does not occur naturally, so the radioisotope (actually a metastable form, ^{99m}Tc) is prepared just before use from radioactive molybdenum:

$$^{99}_{42}\text{Mo} \longrightarrow \, ^{99m}_{43}\text{Tc} + \, ^{0}_{-1}\beta$$

Tracers are also used to measure physiological processes, such as blood flow. The rate at which the heart pumps blood, for example, can be observed by injecting ^{59}Fe, which concentrates in the hemoglobin of blood cells. Several radioisotopes used in medical diagnosis are listed in Table 24.9.

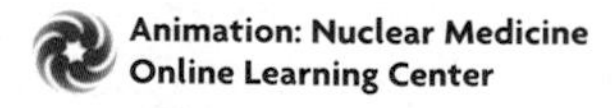

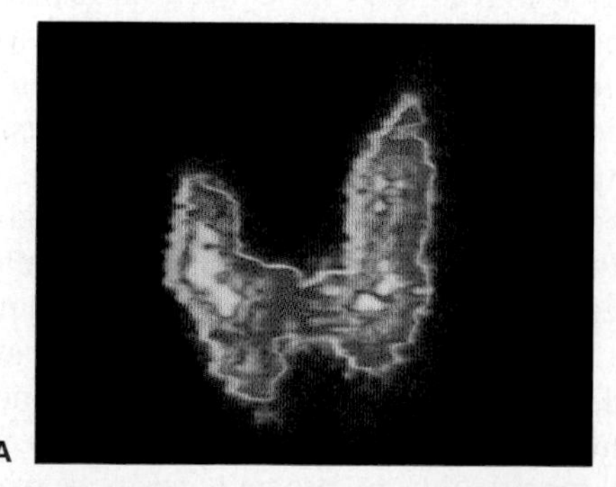
A

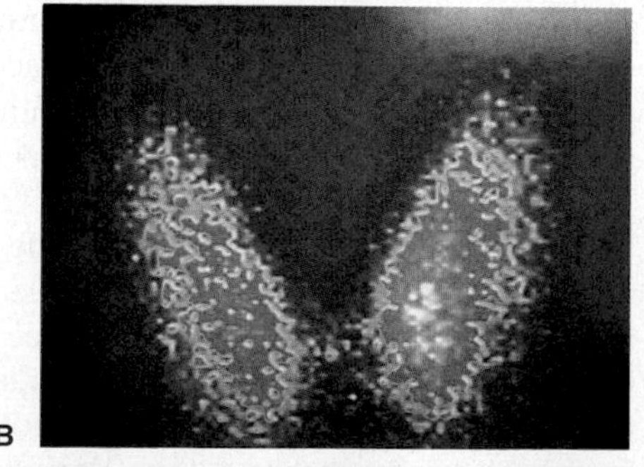
B

Figure 24.10 The use of radioisotopes to image the thyroid gland. Thyroid scanning is used to assess nutritional deficiencies, inflammation, tumor growth, and other thyroid-related ailments. **A,** In ^{131}I scanning, the thyroid gland absorbs $^{131}I^-$ ions whose β emissions expose a photographic film. The asymmetric image indicates disease. **B,** A ^{99}Tc scan of a healthy thyroid.

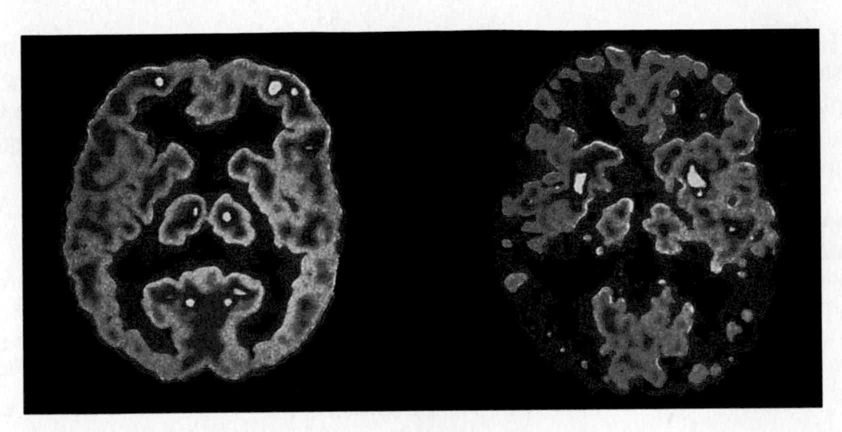

Figure 24.11 PET and brain activity. These PET scans show brain activity in a normal person (*left*) and in a patient with Alzheimer's disease (*right*). Red and yellow indicate relatively high activity within a region.

Positron-emission tomography (PET) is a powerful imaging method for observing brain structure and function. A biological substance is synthesized with one of its atoms replaced by an isotope that emits positrons. The substance is injected into a patient's bloodstream, from which it is taken up into the brain. The isotope emits positrons, each of which annihilates a nearby electron. In the annihilation process, two γ photons are emitted simultaneously 180° from each other:

$$^{0}_{1}\beta + {}^{0}_{-1}e \longrightarrow 2\,^{0}_{0}\gamma$$

An array of detectors around the patient's head pinpoints the sites of γ emission, and the image is analyzed by computer. Two of the isotopes used are ^{15}O, injected as $H_2{}^{15}O$ to measure blood flow, and ^{18}F bonded to a glucose analog to measure glucose uptake, which is a marker for energy metabolism. Among many fascinating PET findings are those that show how changes in blood flow and glucose uptake accompany normal or abnormal brain activity (Figure 24.11). In a recent nonmedical development, substances incorporating ^{11}C and ^{15}O are being investigated by PET to learn how molecules interact with and move along the surface of a catalyst.

Applications of Ionizing Radiation

To be used as a tracer, a radioisotope need emit only low-energy detectable radiation. Many other uses of radioisotopes, however, depend on the effects of high-energy, ionizing radiation.

The interaction between radiation and matter that causes cancer can also be used to eliminate it. Cancer cells divide more rapidly than normal cells, so radioisotopes that interfere with the cell-division process kill more cancer cells than normal ones. Implants of ^{198}Au or of a mixture of ^{90}Sr and ^{90}Y have been used to destroy pituitary and breast tumor cells, and γ rays from ^{60}Co have been used to destroy brain tumors.

Irradiation of food increases shelf life by killing microorganisms that cause food to rot (Figure 24.12), but the practice is quite controversial. Advocates point to the benefits of preserving fresh foods, grains, and seeds for long periods, whereas opponents suggest that irradiation might lower the food's nutritional content or produce harmful by-products. The increased use of antibiotics in animal feed has brought about an increased incidence of illness from newer, more resistant bacterial strains, providing a stronger argument for the use of irradiation. The United Nations has approved irradiation for potatoes, wheat, chicken, and strawberries, and the United States allows irradiation of chicken.

Figure 24.12 The increased shelf life of irradiated food.

Ionizing radiation has been used to control harmful insects. Captured males are sterilized by radiation and released to mate, thereby reducing the number of offspring. This method has been used to control the Mediterranean fruit fly in California and disease-causing insects, such as the tsetse fly and malarial mosquito, in other parts of the world.

SECTION SUMMARY

Radioisotopic tracers emit nonionizing radiation and have been used to study reaction mechanisms, material flow, elemental composition, and medical conditions. Ionizing radiation has been used to destroy cancerous tissue, kill organisms that spoil food, and control insect populations.

24.6 THE INTERCONVERSION OF MASS AND ENERGY

Most of the nuclear processes we've considered so far have involved radioactive decay, in which a nucleus emits one or a few small particles or photons to become a slightly lighter nucleus. Two other nuclear processes cause much greater changes. In nuclear **fission,** a heavy nucleus splits into two much lighter nuclei, emitting several small particles at the same time. In nuclear **fusion,** the opposite process occurs as two lighter nuclei combine to form a heavier one. Both fission and fusion release enormous quantities of energy. Let's take a look at the origins of this energy by first examining the change in mass that accompanies the breakup of a nucleus into its nucleons and then considering the energy that is equivalent to this mass change.

The Mass Defect

We have known for most of the 20th century that mass and energy are interconvertible. The traditional mass and energy conservation laws have been combined to state that *the total quantity of mass-energy in the universe is constant.* Therefore, when *any* reacting system releases or absorbs energy, there must be an accompanying loss or gain in mass.

This relation between mass and energy did not concern us earlier because the energy changes involved in breaking or forming chemical bonds are so small that the mass changes are negligible. When 1 mol of water breaks up into its atoms, for example, heat is absorbed:

$$H_2O(g) \longrightarrow 2H(g) + O(g) \qquad \Delta H^0_{rxn} = 2 \times \text{BE of O—H} = 934 \text{ kJ}$$

We find the mass that is equivalent to this energy from *Einstein's equation:*

$$E = mc^2 \quad \text{or} \quad \Delta E = \Delta mc^2 \quad \text{so} \quad \Delta m = \frac{\Delta E}{c^2} \qquad \textbf{(24.7)}$$

where Δm is the change in mass between the reactants and the products. Substituting the heat of reaction (in J/mol) for ΔE and the numerical value for c (2.9979×10^8 m/s), we obtain

$$\Delta m = \frac{9.34 \times 10^5 \text{ J/mol}}{(2.9979 \times 10^8 \text{ m/s})^2} = 1.04 \times 10^{-11} \text{ kg/mol} = 1.04 \times 10^{-8} \text{ g/mol}$$

(Units of kg/mol are obtained because the joule includes the kilogram: 1 J = 1 kg·m^2/s^2.) The mass of 1 mol of H_2O (reactant) is about 10 ng *less* than the combined masses of 2 mol of H and 1 mol of O (products), a change too small to measure with even the most sophisticated balance. Such minute mass changes when bonds break or form allow us to assume that mass is conserved in *chemical* reactions.

The much larger mass change that accompanies a *nuclear* process is related to the enormous energy required to bind the nucleus together or break it apart. Consider, for example, the change in mass that occurs when one ^{12}C nucleus breaks up into its nucleons: six protons and six neutrons. We calculate this change in mass by combining the mass of six H *atoms* and six neutrons and then subtracting the mass of one ^{12}C *atom*. This procedure cancels the masses of the electrons [six e^- (in six ^1H atoms) cancel six e^- (in one ^{12}C atom)]. The mass of one ^1H atom is 1.007825 amu, and the mass of one neutron is 1.008665 amu, so

$$\begin{aligned} \text{Mass of six } ^1\text{H atoms} &= 6.046950 \text{ amu} \\ \text{Mass of six neutrons} &= 6.051990 \text{ amu} \\ \hline \text{Total mass} &= 12.098940 \text{ amu} \end{aligned}$$

The mass of one ^{12}C atom is 12 amu (exactly). The difference in mass (Δm) is the total mass of the nucleons minus the mass of the nucleus:

$$\begin{aligned} \Delta m &= 12.098940 \text{ amu} - 12.000000 \text{ amu} \\ &= 0.098940 \text{ amu}/^{12}\text{C} = 0.098940 \text{ g/mol } ^{12}\text{C} \end{aligned}$$

Note that *the mass of the nucleus is* ***less*** *than the combined masses of its nucleons.* The mass decrease that occurs when nucleons are united into a nucleus is called the **mass defect.** The size of this mass change (9.89×10^{-2} g/mol) is nearly 10 million times that of the previous bond breakage (10.4×10^{-9} g/mol) and is easily observed on any laboratory balance.

Nuclear Binding Energy

Einstein's equation for the relation between mass and energy also allows us to find the energy equivalent of a mass defect. For ^{12}C, after converting grams to kilograms, we have

$$\begin{aligned} \Delta E = \Delta mc^2 &= (9.8940\times10^{-5} \text{ kg/mol})(2.9979\times10^8 \text{ m/s})^2 \\ &= 8.8921\times10^{12} \text{ J/mol} = 8.8921\times10^9 \text{ kJ/mol} \end{aligned}$$

This quantity of energy is called the **nuclear binding energy** for carbon-12. In general, the nuclear binding energy is the quantity of energy required to *break up 1 mol of nuclei into their individual nucleons:*

$$\text{Nucleus} + \text{nuclear binding energy} \longrightarrow \text{nucleons}$$

Thus, qualitatively, the nuclear binding energy is analogous to the sum of bond energies of a covalent compound or the lattice energy of an ionic compound. But, quantitatively, nuclear binding energies are typically several million times greater.

The Force That Binds Us According to current theory, the nuclear binding energy is related to the strong force, which holds nucleons together in a nucleus. There are three other fundamental forces: (1) the weak nuclear force, which is important in β decay, (2) the electrostatic force that we observe between charged particles, and (3) the gravitational force. Toward the end of his life, Albert Einstein tried unsuccessfully to develop a theory to explain how the four forces were really different aspects of one unified force that governs all nature. The 2004 Nobel Prize in physics was awarded to David J. Gross, H. David Politzer, and Frank Wilczek for their explanation of the strong force who, with others, may one day realize Einstein's dream.

We use joules to express the binding energy per mole of nuclei, but the joule is an impractically large unit to express the binding energy of a single nucleus. Instead, nuclear scientists use the **electron volt (eV),** the energy an electron acquires when it moves through a potential difference of 1 volt:

$$1 \text{ eV} = 1.602\times10^{-19} \text{ J}$$

Binding energies are commonly expressed in millions of electron volts, that is, in *mega–electron volts* (MeV):

$$1 \text{ MeV} = 10^6 \text{ eV} = 1.602\times10^{-13} \text{ J}$$

A particularly useful factor converts a given mass defect in atomic mass units to its energy equivalent in electron volts:

$$1 \text{ amu} = 931.5\times10^6 \text{ eV} = 931.5 \text{ MeV} \qquad \textbf{(24.8)}$$

Earlier we found the mass defect of the ^{12}C nucleus to be 0.098940 amu. Therefore, the binding energy per ^{12}C nucleus, expressed in MeV, is

$$\frac{\text{Binding energy}}{^{12}\text{C nucleus}} = 0.098940 \text{ amu} \times \frac{931.5 \text{ MeV}}{1 \text{ amu}} = 92.16 \text{ MeV}$$

We can compare the stability of nuclides of different elements by determining the *binding energy per nucleon.* For ^{12}C, we have

$$\text{Binding energy per nucleon} = \frac{\text{binding energy}}{\text{no. of nucleons}} = \frac{92.16 \text{ MeV}}{12 \text{ nucleons}} = 7.680 \text{ MeV/nucleon}$$

SAMPLE PROBLEM 24.6 Calculating the Binding Energy per Nucleon

Problem Iron-56 is an extremely stable nuclide. Compute the binding energy per nucleon for ^{56}Fe and compare it with that for ^{12}C (mass of ^{56}Fe atom = 55.934939 amu; mass of ^{1}H atom = 1.007825 amu; mass of neutron = 1.008665 amu).

Plan Iron-56 has 26 protons and 30 neutrons in its nucleus. We calculate the mass defect by finding the sum of the masses of 26 ^{1}H atoms and 30 neutrons and subtracting the given mass of 1 ^{56}Fe atom. Then we multiply Δm by the equivalent in MeV (931.5 MeV/amu) and divide by 56 (no. of nucleons) to obtain the binding energy per nucleon.

Solution Calculating the mass defect:

$$\begin{aligned}\text{Mass defect} &= [(26 \times \text{mass } ^{1}\text{H atom}) + (30 \times \text{mass neutron})] - \text{mass } ^{56}\text{Fe atom} \\ &= [(26)(1.007825 \text{ amu}) + (30)(1.008665 \text{ amu})] - 55.934939 \text{ amu} \\ &= 0.52846 \text{ amu}\end{aligned}$$

Calculating the binding energy per nucleon:

$$\text{Binding energy per nucleon} = \frac{0.52846 \text{ amu} \times 931.5 \text{ MeV/amu}}{56 \text{ nucleons}} = 8.790 \text{ MeV/nucleon}$$

An ^{56}Fe nucleus would require more energy to break up into its nucleons than would ^{12}C (7.680 MeV/nucleon), so ^{56}Fe is more stable than ^{12}C.

Check The answer is consistent with the great stability of ^{56}Fe. Given the number of decimal places in the values, rounding to check the math is useful only to find a *major* error. The number of nucleons (56) is an exact number, so we retain four significant figures.

FOLLOW-UP PROBLEM 24.6 Uranium-235 is the essential component of the fuel in nuclear power plants. Calculate the binding energy per nucleon for ^{235}U. Is this nuclide more or less stable than ^{12}C (mass of ^{235}U atom = 235.043924 amu)?

Fission or Fusion: Means of Increasing the Binding Energy Per Nucleon Calculations similar to Sample Problem 24.6 for other nuclides show that the binding energy per nucleon varies considerably. The essential point is that *the greater the binding energy per nucleon, the more stable the nuclide.*

Figure 24.13 shows a plot of the binding energy per nucleon vs. mass number. It provides information about nuclide stability and the two possible processes nuclides can undergo to form more stable nuclides. Nuclides with fewer than 10 nucleons have a relatively small binding energy per nucleon. The ^{4}He nucleus has an exceptionally large value, however, which is why it is emitted intact as an α particle. Above $A = 12$, the binding energy per nucleon varies from about 7.6 to 8.8 MeV.

The most important observation is that *the binding energy per nucleon peaks for elements with $A \approx 60$.* In other words, nuclides become more stable with increasing mass number up to around 60 nucleons and then become less stable with higher numbers of nucleons. The existence of a peak of stability suggests that there are two ways nuclides can increase their binding energy per nucleon:

- *Fission.* A heavier nucleus can *split into lighter ones (closer to $A \approx 60$)* by undergoing fission. The product nuclei have greater binding energy per nucleon (are more stable) than the reactant nucleus, and the difference in *energy is released.* Nuclear power plants generate energy through fission, as do atomic bombs (Section 24.7).

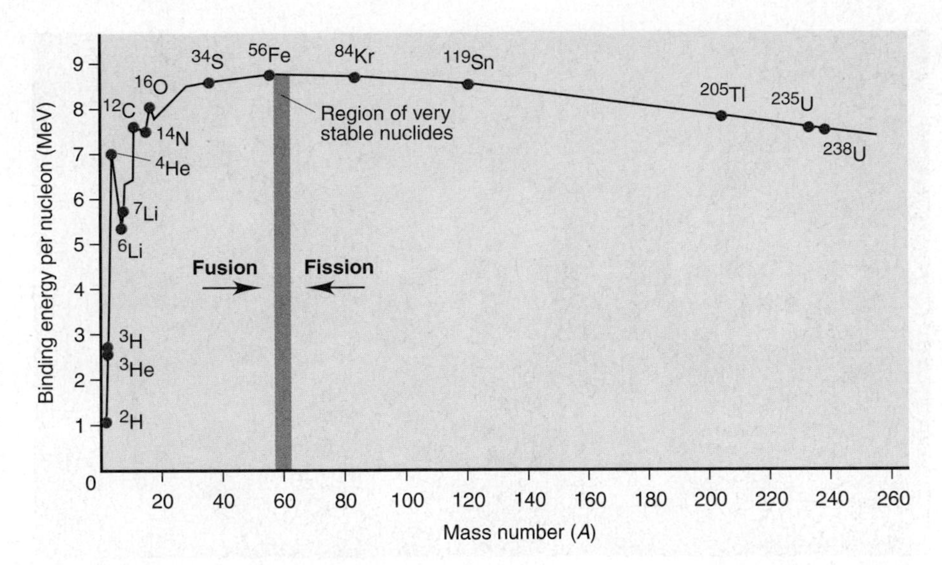

Figure 24.13 The variation in binding energy per nucleon. A plot of the binding energy per nucleon vs. mass number shows that nuclear stability is greatest in the region near ^{56}Fe. Lighter nuclei may undergo fusion to become more stable; heavier ones may undergo fission. Note the exceptional stability of ^{4}He among extremely light nuclei.

- *Fusion.* Lighter nuclei, on the other hand, can *combine to form a heavier one (closer to* $A \approx 60$*)* by undergoing fusion. Once again, the product is more stable than the reactants, and *energy is released.* The Sun and other stars generate energy through fusion, as do hydrogen bombs. In these examples and in all current research efforts for developing fusion as a useful energy source, hydrogen nuclei fuse to form the very stable helium-4 nucleus.

In the next section, we examine fission and fusion and the industrial energy facilities designed to utilize them.

SECTION SUMMARY

The mass of a nucleus is less than the sum of the masses of its nucleons by an amount called the mass defect. The energy equivalent to the mass defect is the nuclear binding energy, usually expressed in units of MeV. The binding energy per nucleon is a measure of nuclide stability and varies with the number of nucleons in a nuclide. Nuclides with $A \approx 60$ are most stable. Lighter nuclides can join (fusion) or heavier nuclides can split (fission) to become more stable.

24.7 APPLICATIONS OF FISSION AND FUSION

Of the many beneficial applications of nuclear reactions, the greatest is the potential for almost limitless amounts of energy, which is based on the multimillionfold increase in energy yield of nuclear reactions over chemical reactions. Our experience with nuclear energy from power plants in the late 20th century, however, has forced a realization that we must strive to improve ways to tap this energy source safely and economically. In this section, we discuss how fission and fusion occur and how we are applying them.

The Process of Nuclear Fission

During the mid-1930s, Enrico Fermi and coworkers bombarded uranium ($Z = 92$) with neutrons in an attempt to synthesize transuranium elements. Many of the

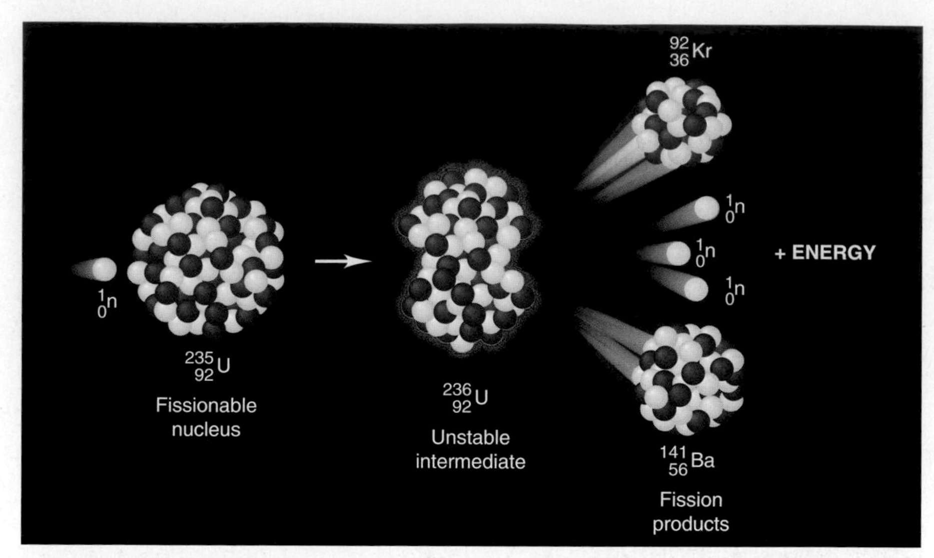

Figure 24.14 Induced fission of ^{235}U. A neutron bombarding a ^{235}U nucleus results in an extremely unstable ^{236}U nucleus, which becomes distorted in the act of splitting. In this case, which shows one of many possible splitting patterns, the products are ^{92}Kr and ^{141}Ba. Three neutrons and a great deal of energy are released also.

unstable nuclides produced were tentatively identified as having $Z > 92$, but other scientists were skeptical. Four years later, the German chemist Otto Hahn and his associate F. Strassmann showed that one of these unstable nuclides was an isotope of barium ($Z = 56$). The Austrian physicist Lise Meitner, a coworker of Hahn, and her nephew Otto Frisch proposed that barium resulted from the *splitting* of the uranium nucleus into *smaller* nuclei, a process that they called *fission* because of its similarity to the fission a biological cell undergoes during reproduction.

Lise Meitner (1878–1968) Until very recently, this extraordinary physicist received little of the acclaim she deserved. Meitner worked in the laboratory of the chemist Otto Hahn, and she was responsible for the discovery of protactinium (Pa; $Z = 91$) and numerous radioisotopes. After leaving Germany in advance of the Nazi domination, Meitner proposed the correct explanation of nuclear fission. In 1944 Hahn received the Nobel Prize in chemistry, but he did not even acknowledge Meitner in his acceptance speech. Today, most physicists believe Meitner should have received the prize. Despite controversy over names for elements 104 to 109, it was widely agreed that element 109 should be named meitnerium.

The ^{235}U nucleus can split in many different ways, giving rise to various daughter nuclei, but all routes have the same general features. Figure 24.14 depicts one of these fission patterns. Neutron bombardment results in a highly excited ^{236}U nucleus, which splits apart in 10^{-14} s. The products are two nuclei of unequal mass, two or three neutrons (average of 2.4), and a large quantity of energy. A single ^{235}U nucleus releases 3.5×10^{-11} J when it splits; 1 mol of ^{235}U (about $\frac{1}{2}$ lb) releases 2.1×10^{13} J—a billion times as much energy as burning $\frac{1}{2}$ lb of coal (about 2×10^{4} J)!

We harness the energy of nuclear fission, much of which appears as heat, by means of a **chain reaction,** illustrated in Figure 24.15: the two to three neutrons that are released by the fission of one nucleus collide with other fissionable nuclei and cause them to split, releasing more neutrons, which then collide with other nuclei, and so on, in a self-sustaining process. In this manner, the energy released increases rapidly because each fission event in a chain reaction releases two to three times as much energy as the preceding one.

Whether a chain reaction occurs depends on the mass (and thus the volume) of the fissionable sample. If the piece of uranium is large enough, the product neutrons strike another fissionable nucleus *before* flying out of the sample, and a chain reaction takes place. The mass required to achieve a chain reaction is called the **critical mass.** If the sample has less than the critical mass (called a *subcritical mass*), most of the product neutrons leave the sample before they have the opportunity to collide with and cause the fission of another ^{235}U nucleus, and thus a chain reaction does not occur.

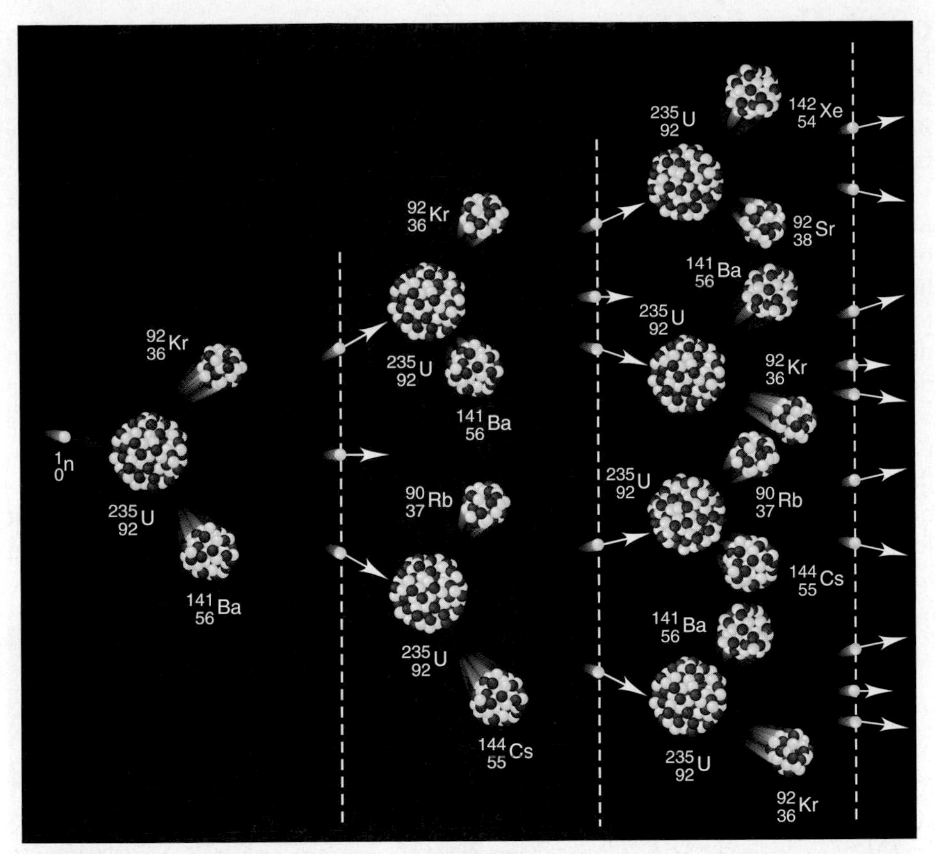

Figure 24.15 A chain reaction of ^{235}U. If a sample exceeds the critical mass, neutrons produced by the first fission event collide with other nuclei, causing their fission and the production of more neutrons to continue the process. Note that various product nuclei form. The vertical dashed lines identify succeeding "generations" of neutrons.

Uncontrolled Fission: The Atomic Bomb An uncontrolled chain reaction can be adapted to make an extremely powerful explosive, as several of the world's leading atomic physicists suspected just prior to the beginning of World War II. In August 1939, Albert Einstein wrote the president of the United States, Franklin Delano Roosevelt, to this effect, warning of the danger of allowing the Nazi government to develop this power first. It was this concern that led to the Manhattan Project, an enormous scientific effort to develop a bomb based on nuclear fission, which was initiated in 1941.* In August 1945, the United States detonated two atomic bombs over Japan, and the horrible destructive power of these bombs was a major factor in the surrender of the Japanese a few days later.

In an atomic bomb, small explosions of trinitrotoluene (TNT) bring subcritical masses of fissionable material together to exceed the critical mass, and the ensuing chain reaction brings about the explosion (Figure 24.16). The proliferation of nuclear power plants, which use fissionable materials to generate energy for electricity, has increased concern that more countries (and unscrupulous individuals) may have access to such material for making bombs. Since the devastating terrorist attacks of September 11, 2001 in the United States, this concern has been heightened. After all, only 1 kg of fissionable uranium was used in the bomb dropped on Hiroshima, Japan.

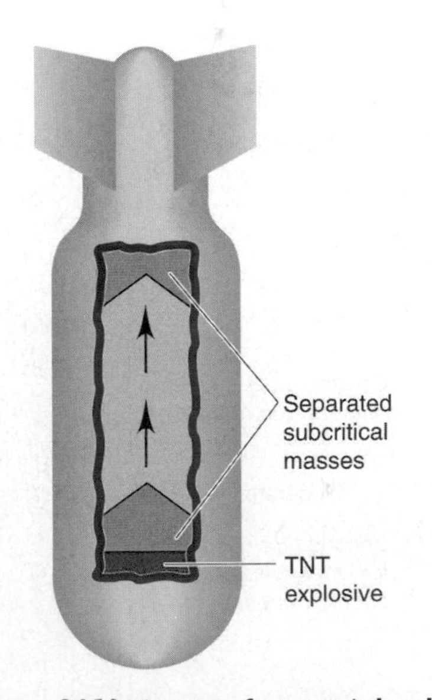

Figure 24.16 Diagram of an atomic bomb. Small TNT explosions bring subcritical masses together, and the chain reaction occurs.

*For an excellent scientific and historical account of the development of the atomic bomb, see R. Rhodes, *The Making of the Atomic Bomb,* New York, Simon and Schuster, 1986.

Controlled Fission: Nuclear Energy Reactors Controlled fission can produce electric power more cleanly than can the combustion of coal. Like a coal-fired power plant, *a nuclear power plant generates heat to produce steam, which turns a turbine attached to an electric generator.* In a coal plant, the heat is produced by burning coal; in a nuclear plant, it is produced by splitting uranium.

Heat generation takes place in the **reactor core** of a nuclear plant (Figure 24.17). The core contains the *fuel rods,* which consist of fuel enclosed in tubes of a corrosion-resistant zirconium alloy. The fuel is uranium(IV) oxide (UO_2) that has been *enriched* from 0.7% ^{235}U, the natural abundance of this fissionable isotope, to the 3% to 4% ^{235}U required to sustain a chain reaction. (Enrichment of nuclear fuel is the most important application of Graham's law; see the margin note, p. 205.) Sandwiched between the fuel rods are movable *control rods* made of cadmium or boron (or, in nuclear submarines, hafnium), substances that absorb neutrons very efficiently. When the control rods are moved between the fuel rods, the chain reaction slows because fewer neutrons are available to bombard uranium atoms; when they are removed, the chain reaction speeds up. Neutrons that leave the fuel-rod assembly collide with a *reflector,* usually made of a beryllium

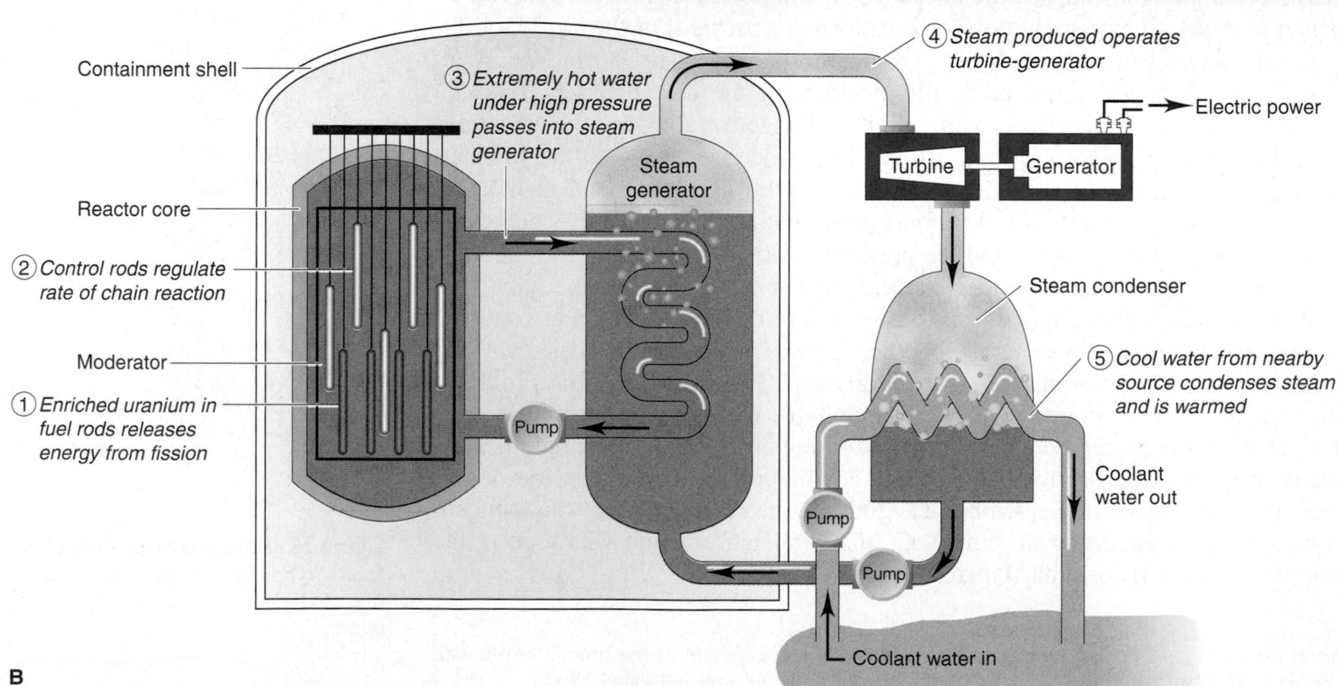

Figure 24.17 A light-water nuclear reactor. A, Photo of a facility showing the concrete containment shell and nearby water source. **B,** Schematic of a light-water reactor.

alloy, which absorbs very few neutrons. Reflecting the neutrons back to the fuel rods speeds the chain reaction.

Flowing around the fuel and control rods in the reactor core is the *moderator,* a substance that slows the neutrons, making them much better at causing fission than the fast ones emerging directly from the fission event. In most modern reactors, the moderator also acts as the *coolant,* the fluid that transfers the released heat to the steam-producing region. Because ^{1}H absorbs neutrons, *light-water reactors* use H_2O as the moderator; in heavy-water reactors, D_2O is used. The advantage of D_2O is that it absorbs very few neutrons, leaving more available for fission, so heavy-water reactors can use *unenriched* uranium. As the coolant flows around the encased fuel, pumps circulate it through coils that transfer its heat to the water reservoir. Steam formed in the reservoir turns the turbine that runs the generator. The steam is then condensed in large cooling towers (see Figure 13.22, p. 509) using water from a lake or river and returned to the water reservoir.

Some major accidents at nuclear plants have caused decidedly negative public reactions. In 1979, malfunctions of coolant pumps and valves at the Three-Mile Island facility in Pennsylvania led to melting of some of the fuel, serious damage to the reactor core, and the release of radioactive gases into the atmosphere. In 1986, a million times as much radioactivity was released when a cooling system failure at the Chernobyl plant in Ukraine caused a much greater melting of fuel and an uncontrolled reaction. High-pressure steam and ignited graphite moderator rods caused the reactor building to explode and expel radioactive debris. Carried by prevailing winds, the radioactive particles contaminated vegetables and milk in much of Europe. Health officials have evidence that thousands of people living near the accident have already or may eventually develop cancer from radiation exposure. The design of the Chernobyl plant was particularly unsafe because, unlike reactors in the United States and western Europe, the reactor was not enclosed in a massive, concrete containment building.

Despite potential safety problems, nuclear power remains an important source of electricity. In the late 1990s, nearly every European country employed nuclear power, and it is the major power source in some countries—Sweden creates 50% of its electricity this way and France almost 80%. Currently, the United States obtains about 20% of its electricity from nuclear power, and Canada slightly less. As our need for energy grows, safer reactors will be designed.

However, even a smoothly operating plant has certain inherent problems. The problem of *thermal pollution* is common to all power plants. Water used to condense the steam is several degrees warmer when returned to its source, which can harm aquatic organisms (Section 13.4). A more serious problem is *nuclear waste disposal.* Many of the fission products formed in nuclear reactors have long half-lives, and no satisfactory plan for their permanent disposal has yet been devised. Proposals to place the waste in containers and bury them in deep bedrock cannot possibly be field-tested for the thousands of years the material will remain harmful. Leakage of radioactive material into groundwater is a danger, and earthquakes can occur even in geologically stable regions. Despite studies indicating the proposed disposal site at Yucca Mountain, Nevada, may be too geologically active, the U.S. government recently approved the site. It remains to be seen whether we can operate fission reactors *and* dispose of the waste safely and economically.

"Breeding" Nuclear Fuel Uranium-235 is not an abundant isotope. One solution to a potential fuel shortage is a *breeder reactor,* designed to consume one type of nuclear fuel as it produces another. Fuel rods are surrounded by natural U_3O_8, which contains 99.3% *nonfissionable* ^{238}U atoms. As fast neutrons, formed during ^{235}U fission, escape the fuel rod, they collide with ^{238}U, transmuting it into ^{239}Pu, another fissionable nucleus:

$$^{238}_{92}U + ^{1}_{0}n \longrightarrow ^{239}_{92}U \quad (t_{1/2} \text{ of } ^{239}_{92}U = 23.5 \text{ min})$$

$$^{239}_{92}U \longrightarrow ^{239}_{93}Np + ^{0}_{-1}\beta \quad (t_{1/2} \text{ of } ^{239}_{93}Np = 2.35 \text{ days})$$

$$^{239}_{93}Np \longrightarrow ^{239}_{94}Pu + ^{0}_{-1}\beta \quad (t_{1/2} \text{ of } ^{239}_{94}Pu = 2.4\times10^{4} \text{ yr})$$

Although breeder reactors can make fuel as they operate, they are difficult and expensive to build, and ^{239}Pu is extremely toxic and long lived. Breeder reactors are not used in the United States, although several are operating in Europe and Japan.

The Promise of Nuclear Fusion

Nuclear fusion is the ultimate source of nearly all the energy on Earth because nearly all other sources depend, directly or indirectly, on the energy produced by nuclear fusion in the Sun. But the Sun and other stars generate more than energy; in fact, *all the elements larger than hydrogen were formed in fusion and decay processes within stars,* as the upcoming Chemical Connections essay describes.

CHEMICAL CONNECTIONS to Cosmology

Origin of the Elements in the Stars

How did the universe begin? Where did matter come from? How were the elements formed? Every culture has creation myths that address such questions, but only recently have astronomers, physicists, and chemists begun to offer a scientific explanation. The most accepted current model proposes that a sphere of unimaginable properties—diameter of 10^{-28} cm, density of 10^{96} g/mL (density of a nucleus $\approx 10^{14}$ g/mL), and temperature of 10^{32} K—exploded in a "Big Bang," for reasons not yet even guessed, and distributed its contents through the void of space. Cosmologists consider this moment the beginning of time.

One second later, the universe was an expanding mixture of neutrons, protons, and electrons, denser than rock and hotter than an exploding hydrogen bomb (about 10^{10} K). During the next few minutes, it became a gigantic fusion reactor creating the first atomic nuclei: ^2H, ^3He, and ^4He. After 10 minutes, more than 25% of the mass of the universe existed as ^4He, and only about 0.025% as ^2H. About 100 million years later, or about 15 billion years ago, gravitational forces pulled this cosmic mixture into primitive, contracting stars.

This account of the origin of the universe is based on the observation of spectra from the Sun, other stars, nearby galaxies, and cosmic (interstellar) dust. Spectral analysis of planets and chemical analysis of Earth and Moon rocks, meteorites, and cosmic-ray particles furnish data about isotope abundance. From these, a model has been developed for **stellar nucleogenesis,** the origin of the elements in the stars. The overall process occurs in several stages during a star's evolution, and the entire sequence of steps occurs only in very massive stars, having 10 to 100 times the mass of the Sun. Each step involves a contraction of the star that produces higher temperature and heavier nuclei. Such events are forming elements in stars today. The key stages in the process are shown in Figure B24.3 and described below:

1. *Hydrogen burning produces He.* The initial contraction of a star heats its core to about 10^7 K, at which point a fusion process called *hydrogen burning* begins, which produces helium from the abundant protons:

$$4\,^1_1H \longrightarrow\, ^4_2He + 2\,^0_1\beta + 2\gamma + \text{energy}$$

2. *Helium burning produces C, O, Ne, and Mg.* After several billion years of hydrogen burning, about 10% of the ^1H is consumed, and the star contracts further. The ^4He forms a dense core, hot enough (2×10^8 K) to fuse ^4He. The energy released during *helium burning* expands the remaining ^1H into a vast envelope: the star becomes a *red giant,* more than 100 times its original diameter. Within its core, pairs of ^4He nuclei (α particles) fuse into unstable ^8Be nuclei ($t_{1/2} = 7\times10^{-17}$ s). These collide with another ^4He to form stable ^{12}C. Then, further fusion with ^4He creates nuclei up to ^{24}Mg:

$$^{12}C \xrightarrow{\alpha}\, ^{16}O \xrightarrow{\alpha}\, ^{20}Ne \xrightarrow{\alpha}\, ^{24}Mg$$

3. *Elements through Fe and Ni form.* For another 10 million years, ^4He is consumed, and the heavier nuclei created form a core. This core contracts and heats, expanding the star into a *supergiant.* Within the hot core (7×10^8 K), *carbon and oxygen burning* occur:

$$^{12}C + \,^{12}C \longrightarrow\, ^{23}Na + \,^1H$$
$$^{12}C + \,^{16}O \longrightarrow\, ^{28}Si + \gamma$$

Absorption of α particles forms nuclei up to ^{40}Ca:

$$^{12}C \xrightarrow{\alpha}\, ^{16}O \xrightarrow{\alpha}\, ^{20}Ne \xrightarrow{\alpha}\, ^{24}Mg \xrightarrow{\alpha}$$
$$^{28}Si \xrightarrow{\alpha}\, ^{32}S \xrightarrow{\alpha}\, ^{36}Ar \xrightarrow{\alpha}\, ^{40}Ca$$

Further contraction and heating to a temperature of 3×10^9 K allow reactions in which nuclei release neutrons, protons, and α particles and then recapture them. As a result, nuclei with lower binding energies supply nucleons to create those with higher binding energies. This process, which takes only a few minutes, stops at iron ($A = 56$) and nickel ($A = 58$), the nuclei with the highest binding energies.

4. *Heavier elements form.* In very massive stars, the next stage is the most spectacular. With all the fuel consumed, the core collapses within a second. Many Fe and Ni nuclei break down into neutrons and protons. Protons capture electrons to form neutrons, and the entire core forms an incredibly dense *neutron star.* (An Earth-sized star that became a neutron star would fit in the Houston Astrodome!) As the core implodes, the outer layers explode in a *supernova,* which expels material throughout space. A supernova occurs an average of every few hundred years in each galaxy; the one shown in Figure B24.4 was observed from the southern hemisphere in 1987, about 160,000 years after the event occurred. The heavier elements are formed during supernova events and are found in *second-generation stars,* those that coalesce from interstellar ^1H and ^4He and the debris of exploded first-generation stars.

Heavier elements form through *neutron-capture* processes. In the *s-process,* a nucleus captures a neutron and emits a γ ray. Days, months, or even thousands of years after this event, the nucleus emits a β particle to form the next element, as in this conversion of ^{68}Zn to ^{70}Ge:

$$^{68}Zn \xrightarrow{n}\, ^{69}Zn \xrightarrow{\beta}\, ^{69}Ga \xrightarrow{n}\, ^{70}Ga \xrightarrow{\beta}\, ^{70}Ge$$

The stable isotopes of most heavy elements form by the s-process.

Less stable isotopes and those with A greater than 230 cannot form by the s-process because their half-lives are too short. These form by the *r-process* during the fury of the supernova. Multiple neutron captures, followed by multiple β decays, occur in a second, as when ^{56}Fe is converted to ^{79}Br:

$$^{56}_{26}Fe + 23\,^1_0n \longrightarrow\, ^{79}_{26}Fe \longrightarrow\, ^{79}_{35}Br + 9\,^{0}_{-1}\beta$$

We know from the heavy elements present in the Sun that it is at least a second-generation star presently undergoing hydrogen burning. Together with its planets, it was formed from the dust of exploded stars about 4.6×10^9 years ago. This means that many of the atoms on Earth, including some within you, came from exploded stars and are older than the Solar System itself!

Any theory of element formation must be consistent with the element abundances we observe (Section 22.1). Although local compositions, such as those of Earth and Sun, differ, large regions of the universe have, on average, similar compositions. Therefore, scientists believe that element forming reaches a dynamic equilibrium, which leads to *relatively constant amounts of the isotopes.*

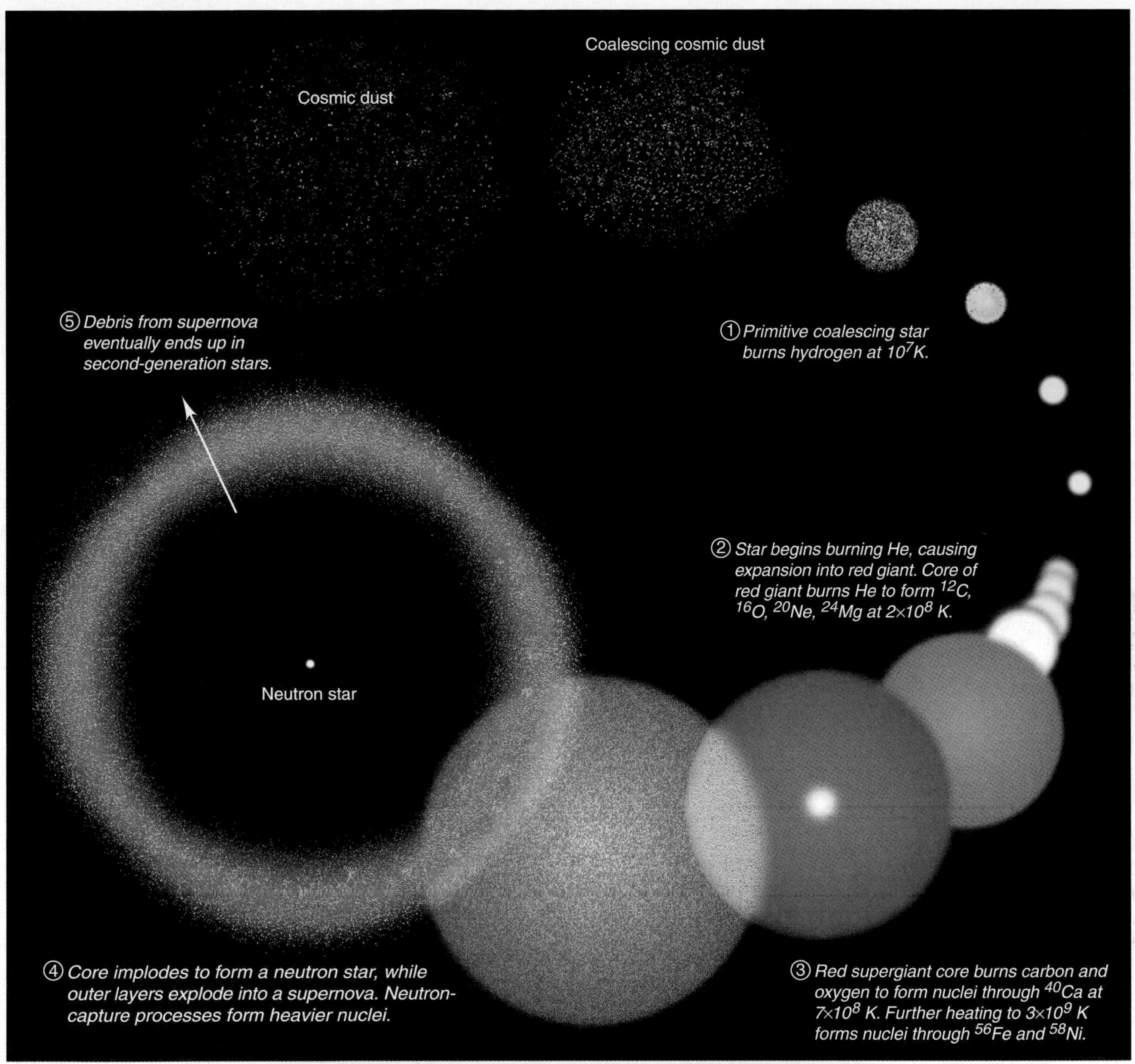

Coalescing cosmic dust
Cosmic dust
⑤ Debris from supernova eventually ends up in second-generation stars.
① Primitive coalescing star burns hydrogen at 10^7K.
② Star begins burning He, causing expansion into red giant. Core of red giant burns He to form ^{12}C, ^{16}O, ^{20}Ne, ^{24}Mg at 2×10^8 K.
Neutron star
④ Core implodes to form a neutron star, while outer layers explode into a supernova. Neutron-capture processes form heavier nuclei.
③ Red supergiant core burns carbon and oxygen to form nuclei through ^{40}Ca at 7×10^8 K. Further heating to 3×10^9 K forms nuclei through ^{56}Fe and ^{58}Ni.

Figure B24.3 Element synthesis in the life cycle of a star.

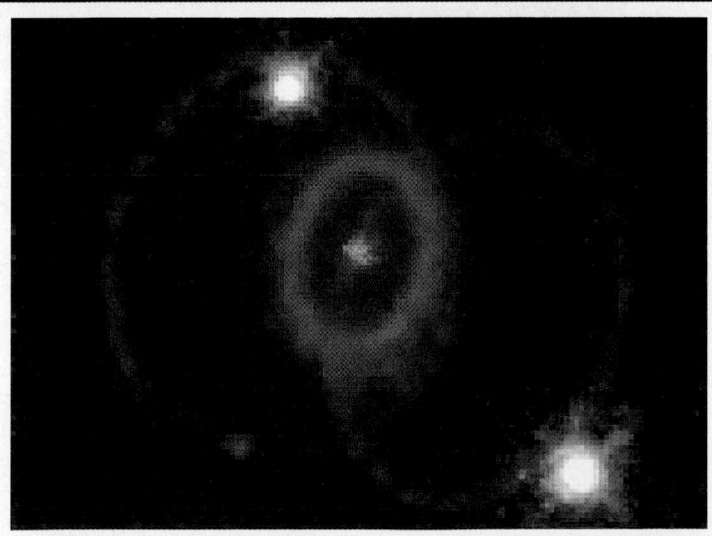

Figure B24.4 A view of Supernova 1987A.

Much research is being devoted to making nuclear fusion a practical, direct source of energy on Earth. To understand the advantages of fusion, let's consider one of the most discussed fusion reactions, in which deuterium and tritium react:

$$^{2}_{1}H + ^{3}_{1}H \longrightarrow ^{4}_{2}He + ^{1}_{0}n$$

This reaction produces 1.7×10^9 kJ/mol, an enormous quantity of energy with no radioactive by-products. Moreover, the reactant nuclei are relatively easy to come by. We obtain deuterium from the electrolysis of water (Section 22.4). In nature, tritium forms through the cosmic (neutron) irradiation of ^{14}N:

$$^{14}_{7}N + ^{1}_{0}n \longrightarrow ^{3}_{1}H + ^{12}_{6}C$$

However, this process results in a natural abundance of only $10^{-7}\%$ ^{3}H. More practically, tritium can be produced in nuclear accelerators by bombarding lithium-6 or by surrounding the fusion reactor itself with material containing lithium-6:

$$^{6}_{3}Li + ^{1}_{0}n \longrightarrow ^{3}_{1}H + ^{4}_{2}He$$

Thus, fusion seems very promising, at least in principle. However, some extremely difficult problems remain. Fusion requires enormous energy in the form of heat to give the positively charged nuclei enough kinetic energy to force themselves together. The fusion of deuterium and tritium, for example, occurs at practical rates at about 10^8 K, hotter than the Sun's core! How can such temperatures be achieved? The reaction that forms the basis of a *hydrogen, or thermonuclear, bomb* fuses lithium-6 and deuterium, with an atomic bomb inside the device providing the heat. Obviously, a power plant cannot begin operation by detonating atomic bombs.

Two research approaches are being used to achieve the necessary heat. In one, atoms are stripped of their electrons at high temperatures, which results in a gaseous *plasma*, a neutral mixture of positive nuclei and electrons. Because of the extreme temperatures needed for fusion, no *material* can contain the plasma. The most successful approach to date has been to enclose the plasma within a magnetic field. The *tokamak* design has a donut-shaped container in which a helical magnetic field confines the plasma and prevents it from contacting the walls (Figure 24.18). Scientists at the Princeton University Plasma Physics facility have achieved some success in generating energy from fusion this way. In another approach, the high temperature is reached by using many focused lasers to compress and heat the fusion reactants. In any event, as a practical, everyday source of energy, fusion still seems to be a long way off.

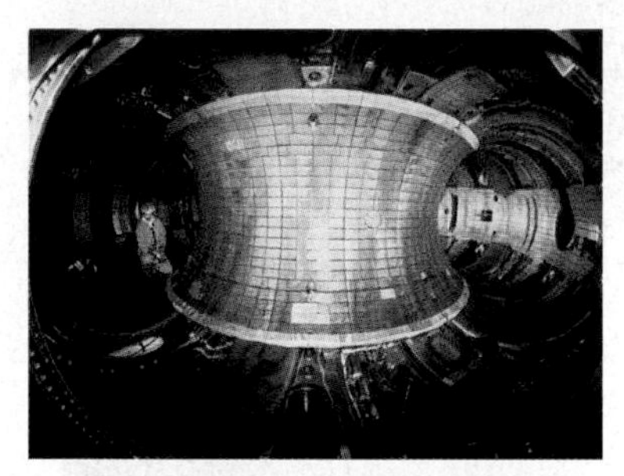

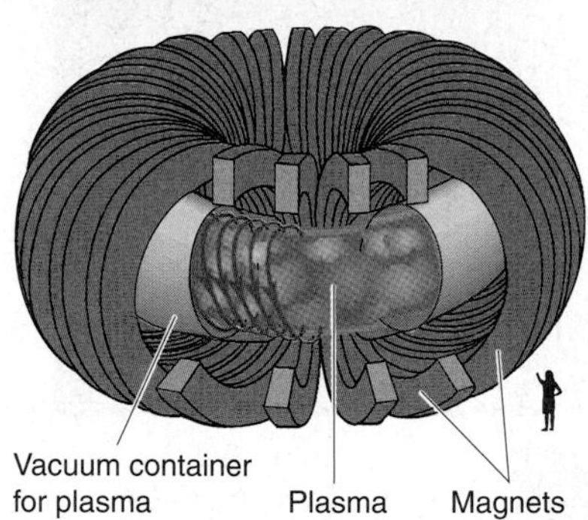

Figure 24.18 The tokamak design for magnetic containment of a fusion plasma. The donut-shaped chamber of the tokamak (photo, *top;* schematic, *bottom*) contains the plasma within a helical magnetic field.

SECTION SUMMARY

In nuclear fission, neutron bombardment causes a nucleus to split, releasing neutrons that split other nuclei to produce a chain reaction. A nuclear power plant controls the rate of the chain reaction to produce heat that creates steam, which is used to generate electricity. Potential hazards, such as radiation leaks, thermal pollution, and disposal of nuclear waste, remain current concerns. Nuclear fusion holds great promise as a source of clean abundant energy, but it requires extremely high temperatures and is not yet practical. The elements were formed through a complex series of nuclear reactions in evolving stars.

Chapter Perspective

With this chapter, our earlier picture of the nucleus as a static point of positive mass at the atom's core has changed radically. Now we picture a dynamic body, capable of a host of changes that involve incredible quantities of energy. Our attempts to apply the behavior of this minute system to benefit society have created some of the most fascinating and challenging fields in science today.

We began our investigation of chemistry 24 chapters ago, by seeing how the chemical elements and the products we make from them influence nearly every aspect of our material existence. Now we have come full circle to learn that these elements, whose patterns of behavior we have become familiar with yet still marvel at, are continually being born in the countless infernos twinkling in the night sky.

For you, the end of this course is a beginning—a chance to apply your new abilities to visualize molecular events and solve problems in whatever field you choose. For the science of chemistry, future challenges are great: What greener energy sources can satisfy our needs while sustaining our environment? What new products can feed, clothe, and house the world's people and maintain precious resources? How can we apply our new genetic insight to defend against cancer, AIDS, and other dreaded diseases? What new materials and technologies can make life more productive and meaningful? The questions are many, but the science of chemistry will always be one of our most powerful means of answering them.

For Review and Reference

(Numbers in parentheses refer to pages, unless noted otherwise.)

Learning Objectives

Relevant section and/or sample problem (SP) numbers appear in parentheses.

Understand These Concepts

1. How nuclear changes differ, in general, from chemical changes (Introduction)
2. The meanings of *radioactivity, nucleon, nuclide,* and *isotope* (Section 24.1)
3. Characteristics of three types of radioactive emissions: α, β, and γ (Section 24.1)
4. The various forms of radioactive decay and how each changes the values of A and Z (Section 24.1)
5. How the N/Z ratio and the even-odd nature of N and Z correlate with nuclear stability (Section 24.1)
6. How the N/Z ratio correlates with the mode of decay of an unstable nuclide (Section 24.1)
7. How a decay series combines numerous decay steps and ends with a stable nuclide (Section 24.1)
8. Why radioactive decay is a first-order process; the meanings of *decay rate* and *specific activity* (Section 24.2)
9. The meaning of *half-life* in the context of radioactive decay (Section 24.2)
10. How the specific activity of an isotope in an object is used to determine the object's age (Section 24.2)
11. How particle accelerators are used to synthesize new nuclides (Section 24.3)
12. The distinction between excitation and ionization and the extent of their effects on matter (Section 24.4)
13. The units of radiation dose; the effects on living tissue of various dosage levels; the inverse relationship between the mass and charge of an emission and its penetrating power (Section 24.4)
14. How ionizing radiation creates free radicals that damage tissue; sources and risks of ionizing radiation (Section 24.4)
15. How radioisotopes are used in research, analysis, and diagnosis (Section 24.5)
16. Why the mass of a nuclide is less than the sum of its nucleons' masses (mass defect) and how this mass difference is related to the nuclear binding energy (Section 24.6)
17. How nuclear stability is related to binding energy per nucleon (Section 24.6)
18. How unstable nuclides undergo either fission or fusion to increase their binding energy per nucleon (Section 24.6)
19. The current application of fission and potential application of fusion to produce energy (Section 24.7)

Master These Skills

1. Expressing the mass and charge of a particle with the ${}^{A}_{Z}X$ notation (Section 24.1; see also Section 2.5)
2. Using changes in the values of A and Z to write and balance nuclear equations (SP 24.1)
3. Using the N/Z ratio and the even-odd nature of N and Z to predict nuclear stability (SP 24.2)
4. Using the N/Z ratio to predict the mode of nuclear decay (SP 24.3)
5. Converting units of radioactivity (Section 24.2)
6. Calculating specific activity, decay constant, half-life, and number of nuclei (Section 24.2 and SP 24.4)
7. Estimating the age of an object from the specific activity and half-life of carbon-14 (SP 24.5)
8. Writing and balancing equations for nuclear transmutation (Section 24.3)
9. Calculating radiation dose and converting units (Section 24.4)
10. Calculating the mass defect and its energy equivalent in J and eV (Section 24.6)
11. Calculating the binding energy per nucleon and using it to compare stabilities of nuclides (SP 24.6)

Key Terms

Section 24.1
radioactivity (1046)
nucleon (1046)
nuclide (1046)
isotope (1046)
alpha (α) particle (1047)
beta (β) particle (1047)
gamma (γ) ray (1047)
alpha decay (1048)
beta decay (1049)
positron decay (1049)
positron (1049)
electron capture (1049)
gamma emission (1049)
N/Z ratio (1050)
band of stability (1050)
strong force (1051)
decay (disintegration) series (1053)

Section 24.2
activity ($\mathcal{A}$) (1054)
becquerel (Bq) (1054)
curie (Ci) (1054)
decay constant (1054)
half-life ($t_{1/2}$) (1054)
Geiger-Müller counter (1055)
scintillation counter (1055)
radioisotopic dating (1057)
radioisotope (1057)

Section 24.3
nuclear transmutation (1059)
deuteron (1060)
particle accelerator (1060)
transuranium element (1061)

Section 24.4
excitation (1062)
nonionizing radiation (1062)
ionization (1062)
ionizing radiation (1062)
gray (Gy) (1063)
rad (*r*adiation-*a*bsorbed *d*ose) (1063)
rem (*r*oentgen *e*quivalent for *m*an) (1063)
sievert (Sv) (1063)
free radical (1063)
background radiation (1064)

Section 24.5
tracer (1066)

Section 24.6
fission (1070)
fusion (1070)
mass defect (1071)
nuclear binding energy (1071)
electron volt (eV) (1071)

Section 24.7
chain reaction (1074)
critical mass (1074)
reactor core (1076)
stellar nucleogenesis (1078)

Key Equations and Relationships

24.1 Balancing a nuclear equation (1048):

$$^{\text{Total } A}_{\text{Total } Z}\text{Reactants} = {}^{\text{Total } A}_{\text{Total } Z}\text{Products}$$

24.2 Defining the unit of radioactivity (curie, Ci) (1054):

$$1 \text{ Ci} = 3.70{\times}10^{10} \text{ disintegrations per second (d/s)}$$

24.3 Expressing the decay rate (activity) for radioactive nuclei (1054):

$$\text{Decay rate } (\mathcal{A}) = -\frac{\Delta\mathcal{N}}{\Delta t} = k\mathcal{N}$$

24.4 Finding the number of nuclei remaining after a given time, $\mathcal{N}_t$ (1056):

$$\ln\frac{\mathcal{N}_0}{\mathcal{N}_t} = kt$$

24.5 Finding the half-life of a radioactive nuclide (1056):

$$t_{1/2} = \frac{\ln 2}{k}$$

24.6 Calculating the time to reach a given specific activity (age of an object in radioisotopic dating) (1058):

$$t = \frac{1}{k}\ln\frac{\mathcal{A}_0}{\mathcal{A}_t}$$

24.7 Using Einstein's equation and the mass defect to calculate the nuclear binding energy (1070):

$$\Delta E = \Delta mc^2$$

24.8 Relating the atomic mass unit to its energy equivalent in MeV (1071):

$$1 \text{ amu} = 931.5{\times}10^6 \text{ eV} = 931.5 \text{ MeV}$$

Highlighted Figures and Tables

These figures (F) and tables (T) provide a review of key ideas.

T24.1 Chemical vs. nuclear reactions (1045)
F24.1 Radioactive emissions in an electric field (1047)
T24.2 Modes of radioactive decay (1048)
F24.2 *N* vs. *Z* for the stable nuclides (1051)
F24.4 Decrease in number of ^{14}C nuclei over time (1056)
F24.13 The variation in binding energy per nucleon (1073)

Brief Solutions to Follow-up Problems

24.1 $^{133}_{54}\text{Xe} \longrightarrow {}^{133}_{55}\text{Cs} + {}^{0}_{-1}\beta$

24.2 Phosphorus-31 has a slightly higher *N/Z* ratio and an even *N* (16).

24.3 (a) $N/Z = 1.35$; too high for this region of band: β decay
(b) Mass too high for stability: α decay

24.4 $\ln \mathcal{A}_t = -kt + \ln \mathcal{A}_0$

$$= -\left(\frac{\ln 2}{15\text{ h}} \times 4.0 \text{ days} \times \frac{24\text{ h}}{1 \text{ day}}\right) + \ln(2.5{\times}10^9)$$

$$= 17.20$$

$$\mathcal{A}_t = 3.0{\times}10^7 \text{ d/s}$$

24.5 $t = \dfrac{1}{k}\ln\dfrac{\mathcal{A}_0}{\mathcal{A}_t} = \dfrac{5730 \text{ yr}}{\ln 2}\ln\left(\dfrac{15.3 \text{ d/min·g}}{9.41 \text{ d/min·g}}\right) = 4.02{\times}10^3 \text{ yr}$

The mummy case is about 4000 years old.

24.6 ^{235}U has 92 $^1_1$p and 143 $^1_0$n.

$$\Delta m = [(92 \times 1.007825 \text{ amu}) + (143 \times 1.008665 \text{ amu})] - 235.043924 \text{ amu} = 1.9151 \text{ amu}$$

$$\frac{\text{Binding energy}}{\text{nucleon}} = \frac{1.9151 \text{ amu} \times \dfrac{931.5 \text{ MeV}}{1 \text{ amu}}}{235 \text{ nucleons}} = 7.591 \text{ MeV/nucleon}$$

Therefore, ^{235}U is less stable than ^{12}C.

COMMON MATHEMATICAL OPERATIONS IN CHEMISTRY

In addition to basic arithmetic and algebra, four mathematical operations are used frequently in general chemistry: manipulating logarithms, using exponential notation, solving quadratic equations, and graphing data. Each is discussed briefly below.

MANIPULATING LOGARITHMS

Meaning and Properties of Logarithms

A *logarithm* is an exponent. Specifically, if $x^n = A$, we can say that the logarithm to the base x of the number A is n, and we can denote it as

$$\log_x A = n$$

Because logarithms are exponents, they have the following properties:

$$\log_x 1 = 0$$
$$\log_x (A \times B) = \log_x A + \log_x B$$
$$\log_x \frac{A}{B} = \log_x A - \log_x B$$
$$\log_x A^y = y \log_x A$$

Types of Logarithms

Common and natural logarithms are used frequently in chemistry and the other sciences. For *common* logarithms, the base (x in the examples above) is 10, but they are written without specifying the base; that is, $\log_{10} A$ is written more simply as $\log A$. The common logarithm of 1000 is 3; in other words, you must raise 10 to the 3rd power to obtain 1000:

$$\log 1000 = 3 \qquad \text{or} \qquad 10^3 = 1000$$

Similarly, we have

$$\log 10 = 1 \qquad \text{or} \qquad 10^1 = 10$$
$$\log 1{,}000{,}000 = 6 \qquad \text{or} \qquad 10^6 = 1{,}000{,}000$$
$$\log 0.001 = -3 \qquad \text{or} \qquad 10^{-3} = 0.001$$
$$\log 853 = 2.931 \qquad \text{or} \qquad 10^{2.931} = 853$$

The last example illustrates an important point about significant figures with all logarithms: the number of significant figures in the number equals the number of digits to the right of the decimal point in the logarithm. That is, the number 853 has three significant figures, and the logarithm 2.931 has three digits to the right of the decimal point.

To find a common logarithm with an electronic calculator, you simply enter the number and press the LOG button.

For *natural* logarithms, the base is the number e, which is 2.71828 . . . , and $\log_e A$ is written $\ln A$. The relationship between the common and natural logarithms is easily obtained: because

$$\log 10 = 1 \qquad \text{and} \qquad \ln 10 = 2.303$$

Therefore, we have

$$\ln A = 2.303 \log A$$

To find a natural logarithm with an electronic calculator, you simply enter the number and press the LN button. If your calculator does not have an LN button, enter the number, press the LOG button, and multiply by 2.303.

Antilogarithms

The *antilogarithm* is the number you obtain when you raise the base to the logarithm:

$$\text{antilogarithm (antilog) of } n \text{ is } 10^n$$

Using two of the earlier examples, the antilog of 3 is 1000, and the antilog of 2.931 is 853. To obtain the antilog with a calculator, you enter the number and press the 10^x button. Similarly, to obtain the natural antilogarithm, you enter the number and press the e^x button. [On some calculators, you enter the number and first press INV and then the LOG (or LN) button.]

USING EXPONENTIAL (SCIENTIFIC) NOTATION

Many quantities in chemistry are very large or very small. For example, in the conventional way of writing numbers, the number of gold atoms in 1 gram of gold is

59,060,000,000,000,000,000,000 atoms (to four significant figures)

As another example, the mass in grams of one gold atom is

0.0000000000000000000003272 g (to four significant figures)

Exponential (scientific) notation provides a much more practical way of writing such numbers. In exponential notation, we express numbers in the form

$$A \times 10^n$$

where A (the coefficient) is greater than or equal to 1 and less than 10 (that is, $1 \leq A < 10$), and n (the exponent) is an integer.

If the number we want to express in exponential notation is larger than 1, the exponent is positive ($n > 0$); if the number is smaller than 1, the exponent is negative ($n < 0$). The size of n tells the number of places the decimal point (in conventional notation) must be moved to obtain a coefficient A greater than or equal to 1 and less than 10 (in exponential notation). In exponential notation, 1 gram of gold contains 5.906×10^{22} atoms, and each gold atom has a mass of 3.272×10^{-22} g.

Changing Between Conventional and Exponential Notation

In order to use exponential notation, you must be able to convert to it from conventional notation, and vice versa.

1. To change a number from conventional to exponential notation, move the decimal point to the left for numbers equal to or greater than 10 and to the right for numbers between 0 and 1:

   75,000,000 changes to 7.5×10^7 (decimal point 7 places to the left)

   0.006042 changes to 6.042×10^{-3} (decimal point 3 places to the right)

2. To change a number from exponential to conventional notation, move the decimal point the number of places indicated by the exponent to the right for numbers with positive exponents and to the left for numbers with negative exponents:

   1.38×10^5 changes to 138,000 (decimal point 5 places to the right)

   8.41×10^{-6} changes to 0.00000841 (decimal point 6 places to the left)

3. An exponential number with a coefficient greater than 10 or less than 1 can be changed to the standard exponential form by converting the coefficient to the standard form and adding the exponents:

$$582.3\times10^6 \text{ changes to } 5.823 \times 10^2 \times 10^6 = 5.823\times10^{(2+6)} = 5.823\times10^8$$
$$0.0043\times10^{-4} \text{ changes to } 4.3 \times 10^{-3} \times 10^{-4} = 4.3\times10^{[(-3)+(-4)]} = 4.3\times10^{-7}$$

Using Exponential Notation in Calculations

In calculations, you can treat the coefficient and exponents separately and apply the properties of exponents (see earlier section on logarithms).

1. To multiply exponential numbers, multiply the coefficients, add the exponents, and reconstruct the number in standard exponential notation:

$$(5.5\times10^3)(3.1\times10^5) = (5.5 \times 3.1)\times10^{(3+5)} = 17\times10^8 = 1.7\times10^9$$
$$(9.7\times10^{14})(4.3\times10^{-20}) = (9.7 \times 4.3)\times10^{[14+(-20)]} = 42\times10^{-6} = 4.2\times10^{-5}$$

2. To divide exponential numbers, divide the coefficients, subtract the exponents, and reconstruct the number in standard exponential notation:

$$\frac{2.6\times10^6}{5.8\times10^2} = \frac{2.6}{5.8} \times 10^{(6-2)} = 0.45\times10^4 = 4.5\times10^3$$
$$\frac{1.7\times10^{-5}}{8.2\times10^{-8}} = \frac{1.7}{8.2} \times 10^{[(-5)-(-8)]} = 0.21\times10^3 = 2.1\times10^2$$

3. To add or subtract exponential numbers, change all numbers so that they have the same exponent, then add or subtract the coefficients:

$$(1.45\times10^4) + (3.2\times10^3) = (1.45\times10^4) + (0.32\times10^4) = 1.77\times10^4$$
$$(3.22\times10^5) - (9.02\times10^4) = (3.22\times10^5) - (0.902\times10^5) = 2.32\times10^5$$

SOLVING QUADRATIC EQUATIONS

A *quadratic equation* is one in which the highest power of x is 2. The general form of a quadratic equation is

$$ax^2 + bx + c = 0$$

where a, b, and c are numbers. For given values of a, b, and c, the values of x that satisfy the equation are called *solutions* of the equation. We calculate x with the quadratic formula:

$$x = \frac{-b \pm \sqrt{b^2 - 4ac}}{2a}$$

We commonly require the quadratic formula when solving for some concentration in an equilibrium problem. For example, we might have an expression that is rearranged into the quadratic equation

$$4.3x^2 + 0.65x - 8.7 = 0$$

Applying the quadratic formula, with $a = 4.3$, $b = 0.65$, and $c = -8.7$, gives

$$x = \frac{-0.65 \pm \sqrt{(0.65)^2 - 4(4.3)(-8.7)}}{2(4.3)}$$

The "plus or minus" sign ($\pm$) indicates that there are always two possible values for x. In this case, they are

$$x = 1.3 \quad \text{and} \quad x = -1.5$$

In any real physical system, however, only one of the values will have any meaning. For example, if x were $[H_3O^+]$, the negative value would mean a negative concentration, which has no meaning.

GRAPHING DATA IN THE FORM OF A STRAIGHT LINE

Visualizing changes in variables by means of a graph is a very useful technique in science. In many cases, it is most useful if the data can be graphed in the form of a straight line. Any equation will appear as a straight line if it has, or can be rearranged to have, the following general form:

$$y = mx + b$$

where y is the dependent variable (typically plotted along the vertical axis), x is the independent variable (typically plotted along the horizontal axis), m is the slope of the line, and b is the intercept of the line on the y axis. The intercept is the value of y when $x = 0$:

$$y = m(0) + b = b$$

The slope of the line is the change in y for a given change in x:

$$\text{Slope } (m) = \frac{y_2 - y_1}{x_2 - x_1} = \frac{\Delta y}{\Delta x}$$

The *sign* of the slope tells the *direction* of the line. If y increases as x increases, m is positive, and the line slopes upward with higher values of x; if y decreases as x increases, m is negative, and the line slopes downward with higher values of x. The *magnitude* of the slope indicates the *steepness* of the line. A line with $m = 3$ is three times as steep (y changes three times as much for a given change in x) as a line with $m = 1$.

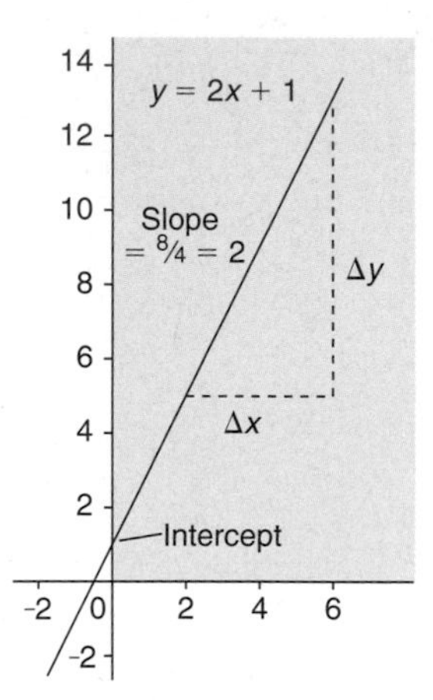

Figure A.1

Consider the linear equation $y = 2x + 1$. A graph of this equation is shown in Figure A.1. In practice, you can find the slope by drawing a right triangle to the line, using the line as the hypotenuse. Then, one leg gives Δy, and the other gives Δx. In the figure, $\Delta y = 8$ and $\Delta x = 4$.

At several places in the text, an equation is rearranged into the form of a straight line in order to determine information from the slope and/or the intercept. For example, in Chapter 16, we obtained the following expression:

$$\ln \frac{[A]_0}{[A]_t} = kt$$

Based on the properties of logarithms, we have

$$\ln [A]_0 - \ln [A]_t = kt$$

Rearranging into the form of an equation for a straight line gives

$$\begin{aligned} \ln [A]_t &= -kt + \ln [A]_0 \\ y &= mx + b \end{aligned}$$

Thus, a plot of $\ln [A]_t$ vs. t is a straight line, from which you can see that the slope is $-k$ (the negative of the rate constant) and the intercept is $\ln [A]_0$ (the natural logarithm of the initial concentration of A).

At many other places in the text, linear relationships occur that were not shown in graphical terms. For example, the conversion of temperature scales in Chapter 1 can also be expressed in the form of a straight line:

$$\begin{aligned} {}^\circ\text{F} &= \tfrac{9}{5}\,{}^\circ\text{C} + 32 \\ y &= mx + b \end{aligned}$$

STANDARD THERMODYNAMIC VALUES FOR SELECTED SUBSTANCES AT 298 K

Substance or Ion	ΔH_f° (kJ/mol)	ΔG_f° (kJ/mol)	S° (J/mol·K)
$e^-(g)$	0	0	20.87
Aluminum			
$Al(s)$	0	0	28.3
$Al^{3+}(aq)$	−524.7	−481.2	−313
$AlCl_3(s)$	−704.2	−628.9	110.7
$Al_2O_3(s)$	−1676	−1582	50.94
Barium			
$Ba(s)$	0	0	62.5
$Ba(g)$	175.6	144.8	170.28
$Ba^{2+}(g)$	1649.9	—	—
$Ba^{2+}(aq)$	−538.36	−560.7	13
$BaCl_2(s)$	−806.06	−810.9	126
$BaCO_3(s)$	−1219	−1139	112
$BaO(s)$	−548.1	−520.4	72.07
$BaSO_4(s)$	−1465	−1353	132
Boron			
B(β-rhombohedral)	0	0	5.87
$BF_3(g)$	−1137.0	−1120.3	254.0
$BCl_3(g)$	−403.8	−388.7	290.0
$B_2H_6(g)$	35	86.6	232.0
$B_2O_3(s)$	−1272	−1193	53.8
$H_3BO_3(s)$	−1094.3	−969.01	88.83
Bromine			
$Br_2(l)$	0	0	152.23
$Br_2(g)$	30.91	3.13	245.38
$Br(g)$	111.9	82.40	174.90
$Br^-(g)$	−218.9	—	—
$Br^-(aq)$	−120.9	−102.82	80.71
$HBr(g)$	−36.3	−53.5	198.59
Cadmium			
$Cd(s)$	0	0	51.5
$Cd(g)$	112.8	78.20	167.64
$Cd^{2+}(aq)$	−72.38	−77.74	−61.1
$CdS(s)$	−144	−141	71
Calcium			
$Ca(s)$	0	0	41.6
$Ca(g)$	192.6	158.9	154.78
$Ca^{2+}(g)$	1934.1	—	—
$Ca^{2+}(aq)$	−542.96	−553.04	−55.2
$CaF_2(s)$	−1215	−1162	68.87
$CaCl_2(s)$	−795.0	−750.2	114
$CaCO_3(s)$	−1206.9	−1128.8	92.9
$CaO(s)$	−635.1	−603.5	38.2
$Ca(OH)_2(s)$	−986.09	−898.56	83.39
$Ca_3(PO_4)_2(s)$	−4138	−3899	263
$CaSO_4(s)$	−1432.7	−1320.3	107
Carbon			
C(graphite)	0	0	5.686
C(diamond)	1.896	2.866	2.439
$C(g)$	715.0	669.6	158.0
$CO(g)$	−110.5	−137.2	197.5
$CO_2(g)$	−393.5	−394.4	213.7
$CO_2(aq)$	−412.9	−386.2	121
$CO_3^{2-}(aq)$	−676.26	−528.10	−53.1
$HCO_3^-(aq)$	−691.11	587.06	95.0
$H_2CO_3(aq)$	−698.7	−623.42	191
$CH_4(g)$	−74.87	−50.81	186.1
$C_2H_2(g)$	227	209	200.85
$C_2H_4(g)$	52.47	68.36	219.22
$C_2H_6(g)$	−84.667	−32.89	229.5
$C_3H_8(g)$	−105	−24.5	269.9
$C_4H_{10}(g)$	−126	−16.7	310
$C_6H_6(l)$	49.0	124.5	172.8
$CH_3OH(g)$	−201.2	−161.9	238
$CH_3OH(l)$	−238.6	−166.2	127
$HCHO(g)$	−116	−110	219
$HCOO^-(aq)$	−410	−335	91.6
$HCOOH(l)$	−409	−346	129.0
$HCOOH(aq)$	−410	−356	164
$C_2H_5OH(g)$	−235.1	−168.6	282.6
$C_2H_5OH(l)$	−277.63	−174.8	161
$CH_3CHO(g)$	−166	−133.7	266
$CH_3COOH(l)$	−487.0	−392	160
$C_6H_{12}O_6(s)$	−1273.3	−910.56	212.1
$C_{12}H_{22}O_{11}(s)$	−2221.7	−1544.3	360.24
$CN^-(aq)$	151	166	118
$HCN(g)$	135	125	201.7
$HCN(l)$	105	121	112.8
$HCN(aq)$	105	112	129
$CS_2(g)$	117	66.9	237.79
$CS_2(l)$	87.9	63.6	151.0
$CH_3Cl(g)$	−83.7	−60.2	234
$CH_2Cl_2(l)$	−117	−63.2	179

(Continued)

Substance or Ion	ΔH°_f (kJ/mol)	ΔG°_f (kJ/mol)	S° (J/mol·K)
$CHCl_3(l)$	−132	−71.5	203
$CCl_4(g)$	−96.0	−53.7	309.7
$CCl_4(l)$	−139	−68.6	214.4
$COCl_2(g)$	−220	−206	283.74
Cesium			
$Cs(s)$	0	0	85.15
$Cs(g)$	76.7	49.7	175.5
$Cs^+(g)$	458.5	427.1	169.72
$Cs^+(aq)$	−248	−282.0	133
$CsF(s)$	−554.7	−525.4	88
$CsCl(s)$	−442.8	−414	101.18
$CsBr(s)$	−395	−383	121
$CsI(s)$	−337	−333	130
Chlorine			
$Cl_2(g)$	0	0	223.0
$Cl(g)$	121.0	105.0	165.1
$Cl^-(g)$	−234	−240	153.25
$Cl^-(aq)$	−167.46	−131.17	55.10
$HCl(g)$	−92.31	−95.30	186.79
$HCl(aq)$	−167.46	−131.17	55.06
$ClO_2(g)$	102	120	256.7
$Cl_2O(g)$	80.3	97.9	266.1
Chromium			
$Cr(s)$	0	0	23.8
$Cr^{3+}(aq)$	−1971	—	—
$CrO_4^{2-}(aq)$	−863.2	−706.3	38
$Cr_2O_7^{2-}(aq)$	−1461	−1257	214
Copper			
$Cu(s)$	0	0	33.1
$Cu(g)$	341.1	301.4	166.29
$Cu^+(aq)$	51.9	50.2	−26
$Cu^{2+}(aq)$	64.39	64.98	−98.7
$Cu_2O(s)$	−168.6	−146.0	93.1
$CuO(s)$	−157.3	−130	42.63
$Cu_2S(s)$	−79.5	−86.2	120.9
$CuS(s)$	−53.1	−53.6	66.5
Fluorine			
$F_2(g)$	0	0	202.7
$F(g)$	78.9	61.8	158.64
$F^-(g)$	−255.6	−262.5	145.47
$F^-(aq)$	−329.1	−276.5	−9.6
$HF(g)$	−273	−275	173.67
Hydrogen			
$H_2(g)$	0	0	130.6
$H(g)$	218.0	203.30	114.60
$H^+(aq)$	0	0	0
$H^+(g)$	1536.3	1517.1	108.83
Iodine			
$I_2(s)$	0	0	116.14
$I_2(g)$	62.442	19.38	260.58
$I(g)$	106.8	70.21	180.67
$I^-(g)$	−194.7	—	—
$I^-(aq)$	−55.94	−51.67	109.4
$HI(g)$	25.9	1.3	206.33
Iron			
$Fe(s)$	0.	0.	27.3
$Fe^{3+}(aq)$	−47.7	−10.5	−293
$Fe^{2+}(aq)$	−87.9	−84.94	113
$FeCl_2(s)$	−341.8	−302.3	117.9
$FeCl_3(s)$	−399.5	−334.1	142
$FeO(s)$	−272.0	−251.4	60.75
$Fe_2O_3(s)$	−825.5	−743.6	87.400
$Fe_3O_4(s)$	−1121	−1018	145.3
Lead			
$Pb(s)$	0	0	64.785
$Pb^{2+}(aq)$	1.6	−24.3	21
$PbCl_2(s)$	−359	−314	136
$PbO(s)$	−218	−198	68.70
$PbO_2(s)$	−276.6	−219.0	76.6
$PbS(s)$	−98.3	−96.7	91.3
$PbSO_4(s)$	−918.39	−811.24	147
Lithium			
$Li(s)$	0	0	29.10
$Li(g)$	161	128	138.67
$Li^+(g)$	687.163	649.989	132.91
$Li^+(aq)$	−278.46	−293.8	14
$LiF(s)$	−616.9	−588.7	35.66
$LiCl(s)$	−408	−384	59.30
$LiBr(s)$	−351	−342	74.1
$LiI(s)$	−270	−270	85.8
Magnesium			
$Mg(s)$	0	0	32.69
$Mg(g)$	150	115	148.55
$Mg^{2+}(g)$	2351	—	—
$Mg^{2+}(aq)$	−461.96	−456.01	118
$MgCl_2(s)$	−641.6	−592.1	89.630
$MgCO_3(s)$	−1112	−1028	65.86
$MgO(s)$	−601.2	−569.0	26.9
$Mg_3N_2(s)$	−461	−401	88
Manganese			
$Mn(s, \alpha)$	0	0	31.8
$Mn^{2+}(aq)$	−219	−223	−84
$MnO_2(s)$	−520.9	−466.1	53.1
$MnO_4^-(aq)$	−518.4	−425.1	190
Mercury			
$Hg(l)$	0	0	76.027
$Hg(g)$	61.30	31.8	174.87
$Hg^{2+}(aq)$	171	164.4	−32
$Hg_2^{2+}(aq)$	172	153.6	84.5
$HgCl_2(s)$	−230	−184	144
$Hg_2Cl_2(s)$	−264.9	−210.66	196
$HgO(s)$	−90.79	−58.50	70.27
Nitrogen			
$N_2(g)$	0	0	191.5
$N(g)$	473	456	153.2
$N_2O(g)$	82.05	104.2	219.7
$NO(g)$	90.29	86.60	210.65
$NO_2(g)$	33.2	51	239.9
$N_2O_4(g)$	9.16	97.7	304.3
$N_2O_5(g)$	11	118	346
$N_2O_5(s)$	−43.1	114	178
$NH_3(g)$	−45.9	−16	193
$NH_3(aq)$	−80.83	26.7	110
$N_2H_4(l)$	50.63	149.2	121.2

(Continued)

Substance or Ion	ΔH_f° (kJ/mol)	ΔG_f° (kJ/mol)	S° (J/mol·K)
$NO_3^-(aq)$	−206.57	−110.5	146
$HNO_3(l)$	−173.23	−79.914	155.6
$HNO_3(aq)$	−206.57	−110.5	146
$NF_3(g)$	−125	−83.3	260.6
$NOCl(g)$	51.71	66.07	261.6
$NH_4Cl(s)$	−314.4	−203.0	94.6
Oxygen			
$O_2(g)$	0	0	205.0
$O(g)$	249.2	231.7	160.95
$O_3(g)$	143	163	238.82
$OH^-(aq)$	−229.94	−157.30	−10.54
$H_2O(g)$	−241.826	−228.60	188.72
$H_2O(l)$	−285.840	−237.192	69.940
$H_2O_2(l)$	−187.8	−120.4	110
$H_2O_2(aq)$	−191.2	−134.1	144
Phosphorus			
$P_4(s$, white)	0	0	41.1
$P(g)$	314.6	278.3	163.1
$P(s$, red)	−17.6	−12.1	22.8
$P_2(g)$	144	104	218
$P_4(g)$	58.9	24.5	280
$PCl_3(g)$	−287	−268	312
$PCl_3(l)$	−320	−272	217
$PCl_5(g)$	−402	−323	353
$PCl_5(s)$	−443.5	—	—
$P_4O_{10}(s)$	−2984	−2698	229
$PO_4^{3-}(aq)$	−1266	−1013	−218
$HPO_4^{2-}(aq)$	−1281	−1082	−36
$H_2PO_4^-(aq)$	−1285	−1135	89.1
$H_3PO_4(aq)$	−1277	−1019	228
Potassium			
$K(s)$	0	0	64.672
$K(g)$	89.2	60.7	160.23
$K^+(g)$	514.197	481.202	154.47
$K^+(aq)$	−251.2	−282.28	103
$KF(s)$	−568.6	−538.9	66.55
$KCl(s)$	−436.7	−409.2	82.59
$KBr(s)$	−394	−380	95.94
$KI(s)$	−328	−323	106.39
$KOH(s)$	−424.8	−379.1	78.87
$KClO_3(s)$	−397.7	−296.3	143.1
$KClO_4(s)$	−432.75	−303.2	151.0
Rubidium			
$Rb(s)$	0	0	69.5
$Rb(g)$	85.81	55.86	169.99
$Rb^+(g)$	495.04	—	—
$Rb^+(aq)$	−246	−282.2	124
$RbF(s)$	−549.28	—	—
$RbCl(s)$	−435.35	−407.8	95.90
$RbBr(s)$	−389.2	−378.1	108.3
$RbI(s)$	−328	−326	118.0
Silicon			
$Si(s)$	0	0	18.0
$SiF_4(g)$	−1614.9	−1572.7	282.4
$SiO_2(s)$	−910.9	−856.5	41.5
Silver			
$Ag(s)$	0	0	42.702
$Ag(g)$	289.2	250.4	172.892
$Ag^+(aq)$	105.9	77.111	73.93
$AgF(s)$	−203	−185	84
$AgCl(s)$	−127.03	−109.72	96.11
$AgBr(s)$	−99.51	−95.939	107.1
$AgI(s)$	−62.38	−66.32	114
$AgNO_3(s)$	−45.06	19.1	128.2
$Ag_2S(s)$	−31.8	−40.3	146
Sodium			
$Na(s)$	0	0	51.446
$Na(g)$	107.76	77.299	153.61
$Na^+(g)$	609.839	574.877	147.85
$Na^+(aq)$	−239.66	−261.87	60.2
$NaF(s)$	−575.4	−545.1	51.21
$NaCl(s)$	−411.1	−384.0	72.12
$NaBr(s)$	−361	−349	86.82
$NaOH(s)$	−425.609	−379.53	64.454
$Na_2CO_3(s)$	−1130.8	−1048.1	139
$NaHCO_3(s)$	−947.7	−851.9	102
$NaI(s)$	−288	−285	98.5
Strontium			
$Sr(s)$	0	0	54.4
$Sr(g)$	164	110	164.54
$Sr^{2+}(g)$	1784	—	—
$Sr^{2+}(aq)$	−545.51	−557.3	−39
$SrCl_2(s)$	−828.4	−781.2	117
$SrCO_3(s)$	−1218	−1138	97.1
$SrO(s)$	−592.0	−562.4	55.5
$SrSO_4(s)$	−1445	−1334	122
Sulfur			
S(rhombic)	0	0	31.9
S(monoclinic)	0.3	0.096	32.6
$S(g)$	279	239	168
$S_2(g)$	129	80.1	228.1
$S_8(g)$	101	49.1	430.211
$S^{2-}(aq)$	41.8	83.7	22
$HS^-(aq)$	−17.7	12.6	61.1
$H_2S(g)$	−20.2	−33	205.6
$H_2S(aq)$	−39	−27.4	122
$SO_2(g)$	−296.8	−300.2	248.1
$SO_3(g)$	−396	−371	256.66
$SO_4^{2-}(aq)$	−907.51	−741.99	17
$HSO_4^-(aq)$	−885.75	−752.87	126.9
$H_2SO_4(l)$	−813.989	−690.059	156.90
$H_2SO_4(aq)$	−907.51	−741.99	17
Tin			
Sn(white)	0	0	51.5
Sn(gray)	3	4.6	44.8
$SnCl_4(l)$	−545.2	−474.0	259
$SnO_2(s)$	−580.7	−519.7	52.3
Zinc			
$Zn(s)$	0	0	41.6
$Zn(g)$	130.5	94.93	160.9
$Zn^{2+}(aq)$	−152.4	−147.21	−106.5
$ZnO(s)$	−348.0	−318.2	43.9
$ZnS(s$, zinc blende)	−203	−198	57.7

Appendix C

EQUILIBRIUM CONSTANTS AT 298 K

Dissociation (Ionization) Constants (K_a) of Selected Acids

Name	Formula*	K_{a1}	K_{a2}	K_{a3}
Acetic acid	CH_3COOH	1.8×10^{-5}		
Acetylsalicylic acid†	$CH_3COOC_6H_4COOH$	3.6×10^{-4}		
Adipic acid	$HOOC(CH_2)_4COOH$	3.8×10^{-5}	3.8×10^{-6}	
Arsenic acid	H_3AsO_4	6×10^{-3}	1.1×10^{-7}	3×10^{-12}
Ascorbic acid	$H_2C_6H_6O_6$	1.0×10^{-5}	5×10^{-12}	
Benzoic acid	C_6H_5COOH	6.3×10^{-5}		
Carbonic acid	H_2CO_3	4.5×10^{-7}	4.7×10^{-11}	
Chloroacetic acid	$ClCH_2COOH$	1.4×10^{-3}		
Chlorous acid	$HClO_2$	1.1×10^{-2}		
Citric acid	$HOC(CH_2)_2(COOH)_3$	7.4×10^{-4}	1.7×10^{-5}	4.0×10^{-7}
Formic acid	$HCOOH$	1.8×10^{-4}		
Glyceric acid	$HOCH_2CH(OH)COOH$	2.9×10^{-4}		
Glycolic acid	$HOCH_2COOH$	1.5×10^{-4}		
Glyoxylic acid	$O{=}CHCOOH$	3.5×10^{-4}		
Hydrocyanic acid	HCN	6.2×10^{-10}		
Hydrofluoric acid	HF	6.8×10^{-4}		
Hydrosulfuric acid	H_2S	9×10^{-8}	1×10^{-17}	
Hypobromous acid	$HBrO$	2.3×10^{-9}		
Hypochlorous acid	$HClO$	2.9×10^{-8}		
Hypoiodous acid	HIO	2.3×10^{-11}		
Iodic acid	HIO_3	1.6×10^{-1}		
Lactic acid	$CH_3CH(OH)COOH$	1.4×10^{-4}		
Maleic acid	$HOOCCH{=}CHCOOH$	1.2×10^{-2}	4.7×10^{-7}	
Malonic acid	$HOOCCH_2COOH$	1.4×10^{-3}	2.0×10^{-6}	
Nitrous acid	HNO_2	7.1×10^{-4}		
Oxalic acid	$HOOCCOOH$	5.6×10^{-2}	5.4×10^{-5}	
Phenol	C_6H_5OH	1.0×10^{-10}		
Phenylacetic acid	$C_6H_5CH_2COOH$	4.9×10^{-5}		
Phosphoric acid	H_3PO_4	7.2×10^{-3}	6.3×10^{-8}	4.2×10^{-13}
Phosphorous acid	$HPO(OH)_2$	3×10^{-2}	1.7×10^{-7}	
Propanoic acid	CH_3CH_2COOH	1.3×10^{-5}		
Pyruvic acid	$CH_3C(O)COOH$	2.8×10^{-3}		
Succinic acid	$HOOCCH_2CH_2COOH$	6.2×10^{-5}	2.3×10^{-6}	
Sulfuric acid	H_2SO_4	Very large	1.0×10^{-2}	
Sulfurous acid	H_2SO_3	1.4×10^{-2}	6.5×10^{-8}	

*Acidic (ionizable) proton(s) shown in red.
†At 37°C in 0.15 *M* NaCl.

Dissociation (Ionization) Constants (K_b) of Selected Amine Bases

Name	Formula*	K_{b1}	K_{b2}
Ammonia	NH_3	1.76×10^{-5}	
Aniline	$C_6H_5NH_2$	4.0×10^{-10}	
Diethylamine	$(CH_3CH_2)_2NH$	8.6×10^{-4}	
Dimethylamine	$(CH_3)_2NH$	5.9×10^{-4}	
Ethanolamine	$HOCH_2CH_2NH_2$	3.2×10^{-5}	
Ethylamine	$CH_3CH_2NH_2$	4.3×10^{-4}	
Ethylenediamine	$H_2NCH_2CH_2NH_2$	8.5×10^{-5}	7.1×10^{-8}
Methylamine	CH_3NH_2	4.4×10^{-4}	
2-Methyl-2-propylamine	$(CH_3)_3CNH_2$	4.8×10^{-4}	
Piperidine	$C_5H_{10}NH$	1.3×10^{-3}	
1-Propylamine	$CH_3CH_2CH_2NH_2$	3.5×10^{-4}	
2-Propylamine	$(CH_3)_2CHNH_2$	4.7×10^{-4}	
1,3-Propylenediamine	$H_2NCH_2CH_2CH_2NH_2$	3.1×10^{-4}	3.0×10^{-6}
Pyridine	C_5H_5N	1.7×10^{-9}	
Triethylamine	$(CH_3CH_2)_3N$	5.2×10^{-4}	
Trimethylamine	$(CH_3)_3N$	6.3×10^{-5}	

*Basic nitrogen(s) shown in blue.

Dissociation (Ionization) Constants (K_a) of Some Hydrated Metal Ions

Free Ion	Hydrated Ion	K_a
Fe^{3+}	$Fe(H_2O)_6^{3+}(aq)$	6×10^{-3}
Sn^{2+}	$Sn(H_2O)_6^{2+}(aq)$	4×10^{-4}
Cr^{3+}	$Cr(H_2O)_6^{3+}(aq)$	1×10^{-4}
Al^{3+}	$Al(H_2O)_6^{3+}(aq)$	1×10^{-5}
Cu^{2+}	$Cu(H_2O)_6^{2+}(aq)$	3×10^{-8}
Pb^{2+}	$Pb(H_2O)_6^{2+}(aq)$	3×10^{-8}
Zn^{2+}	$Zn(H_2O)_6^{2+}(aq)$	1×10^{-9}
Co^{2+}	$Co(H_2O)_6^{2+}(aq)$	2×10^{-10}
Ni^{2+}	$Ni(H_2O)_6^{2+}(aq)$	1×10^{-10}

Formation Constants (K_f) of Some Complex Ions

Complex Ion	K_f
$Ag(CN)_2^-$	3.0×10^{20}
$Ag(NH_3)_2^+$	1.7×10^{7}
$Ag(S_2O_3)_2^{3-}$	4.7×10^{13}
AlF_6^{3-}	4×10^{19}
$Al(OH)_4^-$	3×10^{33}
$Be(OH)_4^{2-}$	4×10^{18}
CdI_4^{2-}	1×10^{6}
$Co(OH)_4^{2-}$	5×10^{9}
$Cr(OH)_4^-$	8.0×10^{29}
$Cu(NH_3)_4^{2+}$	5.6×10^{11}
$Fe(CN)_6^{4-}$	3×10^{35}
$Fe(CN)_6^{3-}$	4.0×10^{43}
$Hg(CN)_4^{2-}$	9.3×10^{38}
$Ni(NH_3)_6^{2+}$	2.0×10^{8}
$Pb(OH)_3^-$	8×10^{13}
$Sn(OH)_3^-$	3×10^{25}
$Zn(CN)_4^{2-}$	4.2×10^{19}
$Zn(NH_3)_4^{2+}$	7.8×10^{8}
$Zn(OH)_4^{2-}$	3×10^{15}

Solubility-Product Constants (K_{sp}) of Slightly Soluble Ionic Compounds

Name, Formula	K_{sp}
Carbonates	
Barium carbonate, $BaCO_3$	2.0×10^{-9}
Cadmium carbonate, $CdCO_3$	1.8×10^{-14}
Calcium carbonate, $CaCO_3$	3.3×10^{-9}
Cobalt(II) carbonate, $CoCO_3$	1.0×10^{-10}
Copper(II) carbonate, $CuCO_3$	$3\ \times10^{-12}$
Lead(II) carbonate, $PbCO_3$	7.4×10^{-14}
Magnesium carbonate, $MgCO_3$	3.5×10^{-8}
Mercury(I) carbonate, Hg_2CO_3	8.9×10^{-17}
Nickel(II) carbonate, $NiCO_3$	1.3×10^{-7}
Strontium carbonate, $SrCO_3$	5.4×10^{-10}
Zinc carbonate, $ZnCO_3$	1.0×10^{-10}
Chromates	
Barium chromate, $BaCrO_4$	2.1×10^{-10}
Calcium chromate, $CaCrO_4$	$1\ \times10^{-8}$
Lead(II) chromate, $PbCrO_4$	2.3×10^{-13}
Silver chromate, Ag_2CrO_4	2.6×10^{-12}
Cyanides	
Mercury(I) cyanide, $Hg_2(CN)_2$	$5\ \times10^{-40}$
Silver cyanide, $AgCN$	2.2×10^{-16}
Halides	
Fluorides	
Barium fluoride, BaF_2	1.5×10^{-6}
Calcium fluoride, CaF_2	3.2×10^{-11}
Lead(II) fluoride, PbF_2	3.6×10^{-8}
Magnesium fluoride, MgF_2	7.4×10^{-9}
Strontium fluoride, SrF_2	2.6×10^{-9}
Chlorides	
Copper(I) chloride, $CuCl$	1.9×10^{-7}
Lead(II) chloride, $PbCl_2$	1.7×10^{-5}
Silver chloride, $AgCl$	1.8×10^{-10}
Bromides	
Copper(I) bromide, $CuBr$	$5\ \times10^{-9}$
Silver bromide, $AgBr$	5.0×10^{-13}
Iodides	
Copper(I) iodide, CuI	$1\ \times10^{-12}$
Lead(II) iodide, PbI_2	7.9×10^{-9}
Mercury(I) iodide, Hg_2I_2	4.7×10^{-29}
Silver iodide, AgI	8.3×10^{-17}
Hydroxides	
Aluminum hydroxide, $Al(OH)_3$	$3\ \times10^{-34}$
Cadmium hydroxide, $Cd(OH)_2$	7.2×10^{-15}
Calcium hydroxide, $Ca(OH)_2$	6.5×10^{-6}
Cobalt(II) hydroxide, $Co(OH)_2$	1.3×10^{-15}
Copper(II) hydroxide, $Cu(OH)_2$	2.2×10^{-20}
Iron(II) hydroxide, $Fe(OH)_2$	4.1×10^{-15}
Iron(III) hydroxide, $Fe(OH)_3$	1.6×10^{-39}
Magnesium hydroxide, $Mg(OH)_2$	6.3×10^{-10}
Manganese(II) hydroxide, $Mn(OH)_2$	1.6×10^{-13}
Nickel(II) hydroxide, $Ni(OH)_2$	$6\ \times10^{-16}$
Zinc hydroxide, $Zn(OH)_2$	$3\ \times10^{-16}$
Iodates	
Barium iodate, $Ba(IO_3)_2$	1.5×10^{-9}
Calcium iodate, $Ca(IO_3)_2$	7.1×10^{-7}
Lead(II) iodate, $Pb(IO_3)_2$	2.5×10^{-13}
Silver iodate, $AgIO_3$	3.1×10^{-8}
Strontium iodate, $Sr(IO_3)_2$	3.3×10^{-7}
Zinc iodate, $Zn(IO_3)_2$	3.9×10^{-6}
Oxalates	
Barium oxalate dihydrate, $BaC_2O_4 \cdot 2H_2O$	1.1×10^{-7}
Calcium oxalate monohydrate, $CaC_2O_4 \cdot H_2O$	2.3×10^{-9}
Strontium oxalate monohydrate, $SrC_2O_4 \cdot H_2O$	5.6×10^{-8}
Phosphates	
Calcium phosphate, $Ca_3(PO_4)_2$	1.2×10^{-29}
Magnesium phosphate, $Mg_3(PO_4)_2$	5.2×10^{-24}
Silver phosphate, Ag_3PO_4	2.6×10^{-18}
Sulfates	
Barium sulfate, $BaSO_4$	1.1×10^{-10}
Calcium sulfate, $CaSO_4$	2.4×10^{-5}
Lead(II) sulfate, $PbSO_4$	1.6×10^{-8}
Radium sulfate, $RaSO_4$	$2\ \times10^{-11}$
Silver sulfate, Ag_2SO_4	1.5×10^{-5}
Strontium sulfate, $SrSO_4$	3.2×10^{-7}
Sulfides	
Cadmium sulfide, CdS	1.0×10^{-24}
Copper(II) sulfide, CuS	$8\ \times10^{-34}$
Iron(II) sulfide, FeS	$8\ \times10^{-16}$
Lead(II) sulfide, PbS	$3\ \times10^{-25}$
Manganese(II) sulfide, MnS	$3\ \times10^{-11}$
Mercury(II) sulfide, HgS	$2\ \times10^{-50}$
Nickel(II) sulfide, NiS	$3\ \times10^{-16}$
Silver sulfide, Ag_2S	$8\ \times10^{-48}$
Tin(II) sulfide, SnS	1.3×10^{-23}
Zinc sulfide, ZnS	2.0×10^{-22}

(b) $2N_2(g) + 5O_2(g) + 2H_2O(l) \longrightarrow 4HNO_3(g)$
(c) 5.62×10^3 t HNO_3
3.144(a) 0.027 g heme (b) 4.4×10^{-5} mol heme
(c) 2.4×10^{-3} g Fe (d) 2.9×10^{-2} g hemin **3.147**(a) 46.66 mass % N in urea; 31.98 mass % N in arginine; 21.04 mass % N in ornithine (b) 30.13 g N **3.149** 29.54% **3.151** (a) 84.3%
(b) 2.39 g ethylene **3.153**(a) 125 g salt (b) 65.6 L H_2O
3.155 44.6%

Chapter 4

4.2 ionic or polar covalent compounds **4.3** Ions must be present and they come from ionic compounds or from electrolytes such as acids and bases. **4.6** (2) **4.10** (a) Benzene is likely to be insoluble in water because it is nonpolar and water is polar.
(b) Sodium hydroxide, an ionic compound, is likely to be very soluble in water. (c) Ethanol (CH_3CH_2OH) is likely to be soluble in water because the alcohol group (—OH) will hydrogen bond with a water molecule. (d) Potassium acetate, an ionic compound, is likely to be very soluble in water. **4.12**(a) Yes, CsI is a soluble salt. (b) Yes, HBr is a strong acid. **4.14**(a) 0.74 mol
(b) 0.337 mol (c) 1.18×10^{-5} mol **4.16**(a) 3.3 mol
(b) 8.93×10^{-5} mol (c) 8.17×10^{-3} mol **4.18**(a) 0.245 mol Al^{3+}; 0.735 mol Cl^-; 1.48×10^{23} Al^{3+} ions; 4.43×10^{23} Cl^- ions
(b) 0.0848 mol Li^+; 0.0424 mol SO_4^{2-}; 5.11×10^{22} Li^+ ions; 2.55×10^{22} SO_4^{2-} ions (c) 3.78×10^{21} K^+ ions; 3.78×10^{21} Br^- ions; 6.28×10^{-3} mol K^+; 6.28×10^{-3} mol Br^-
4.20(a) 0.35 mol H^+ (b) 1.3×10^{-3} mol H^+ (c) 0.43 mol H^+
4.24 Spectator ions do not appear because they are not involved in the reaction and are present only to balance charge.
4.28 Assuming that the left beaker contains $AgNO_3$ (because it has gray Ag^+ ion), the right must contain NaCl. Then, NO_3^- is blue, Na^+ is brown, and Cl^- is green.
Molecular equation: $AgNO_3(aq) + NaCl(aq) \longrightarrow AgCl(s) + NaNO_3(aq)$
Total ionic equation: $Ag^+(aq) + NO_3^-(aq) + Na^+(aq) + Cl^-(aq) \longrightarrow AgCl(s) + Na^+(aq) + NO_3^-(aq)$
Net ionic equation: $Ag^+(aq) + Cl^-(aq) \longrightarrow AgCl(s)$
4.29(a) Molecular: $Hg_2(NO_3)_2(aq) + 2KI(aq) \longrightarrow Hg_2I_2(s) + 2KNO_3(aq)$
Total ionic: $Hg_2^{2+}(aq) + 2NO_3^-(aq) + 2K^+(aq) + 2I^-(aq) \longrightarrow Hg_2I_2(s) + 2K^+(aq) + 2NO_3^-(aq)$
Net ionic: $Hg_2^{2+}(aq) + 2I^-(aq) \longrightarrow Hg_2I_2(s)$
Spectator ions are K^+ and NO_3^-.
(b) Molecular: $FeSO_4(aq) + Ba(OH)_2(aq) \longrightarrow Fe(OH)_2(s) + BaSO_4(s)$
Total ionic: $Fe^{2+}(aq) + SO_4^{2-}(aq) + Ba^{2+}(aq) + 2OH^-(aq) \longrightarrow Fe(OH)_2(s) + BaSO_4(s)$
Net ionic: This is the same as the total ionic equation, because there are no spectator ions.
4.31(a) No precipitate will form. (b) A precipitate will form because silver ions, Ag^+, and iodide ions, I^-, will combine to form a solid salt, silver iodide, AgI. The ammonium and nitrate ions do not form a precipitate.
Molecular: $NH_4I(aq) + AgNO_3(aq) \longrightarrow AgI(s) + NH_4NO_3(aq)$
Total ionic: $NH_4^+(aq) + I^-(aq) + Ag^+(aq) + NO_3^-(aq) \longrightarrow AgI(s) + NH_4^+(aq) + NO_3^-(aq)$
Net ionic: $Ag^+(aq) + I^-(aq) \longrightarrow AgI(s)$
4.33(a) No precipitate will form.
(b) $BaSO_4$ will precipitate.
Molecular: $(NH_4)_2SO_4(aq) + BaCl_2(aq) \longrightarrow BaSO_4(s) + 2NH_4Cl(aq)$
Total ionic: $2NH_4^+(aq) + SO_4^{2-}(aq) + Ba^{2+}(aq) + 2Cl^-(aq) \longrightarrow BaSO_4(s) + 2NH_4^+(aq) + Cl^-(aq)$
Net ionic: $SO_4^{2-}(aq) + Ba^{2+}(aq) \longrightarrow BaSO_4(s)$
4.35 0.0389 *M* Pb^{2+} **4.37** 1.80 mass % Cl **4.43**(a) Formation of a gas, $SO_2(g)$, and of a nonelectrolyte (water) will cause the reaction to go to completion. (b) Formation of a precipitate $Ba_3(PO_4)_2(s)$, and of a nonelectrolyte (water) will cause the reaction to go to completion.
4.45(a) Molecular equation: $KOH(aq) + HI(aq) \longrightarrow KI(aq) + H_2O(l)$
Total ionic equation: $K^+(aq) + OH^-(aq) + H^+(aq) + I^-(aq) \longrightarrow K^+(aq) + I^-(aq) + H_2O(l)$
Net ionic equation: $OH^-(aq) + H^+(aq) \longrightarrow H_2O(l)$
The spectator ions are $K^+(aq)$ and $I^-(aq)$.
(b) Molecular equation: $NH_3(aq) + HCl(aq) \longrightarrow NH_4Cl(aq)$
Total ionic equation: $NH_3(aq) + H^+(aq) + Cl^-(aq) \longrightarrow NH_4^+(aq) + Cl^-(aq)$
NH_3, a weak base, is written in the molecular (undissociated) form. HCl, a strong acid, is written as dissociated ions. NH_4Cl is a soluble compound, because all ammonium compounds are soluble.
Net ionic equation: $NH_3(aq) + H^+(aq) \longrightarrow NH_4^+(aq)$
Cl^- is the only spectator ion.
4.47 Total ionic equation: $CaCO_3(s) + 2H^+(aq) + 2Cl^-(aq) \longrightarrow Ca^{2+}(aq) + 2Cl^-(aq) + H_2O(l) + CO_2(g)$
Net ionic equation: $CaCO_3(s) + 2H^+(aq) \longrightarrow Ca^{2+}(aq) + H_2O(l) + CO_2(g)$
4.49 0.05839 *M* CH_3COOH **4.58**(a) S has O.N. = +6 in SO_4^{2-} (i.e., H_2SO_4), and O.N. = +4 in SO_2, so S has been reduced (and I^- oxidized); H_2SO_4 acts as an oxidizing agent.
(b) The oxidation numbers remain constant throughout; H_2SO_4 transfers a proton to F^- to produce HF, so it acts as an acid.
4.60(a) +4 (b) +3 (c) +4 (d) −3 **4.62**(a) −1 (b) −2
(c) −3 (d) +3 **4.64**(a) −3 (b) +5 (c) +3 **4.66**(a) +6
(b) +3 (c) +7 **4.68**(a) MnO_4^- is the oxidizing agent; $H_2C_2O_4$ is the reducing agent. (b) Cu is the reducing agent; NO_3^- is the oxidizing agent. **4.70**(a) Oxidizing agent is NO_3^-; reducing agent is Sn. (b) Oxidizing agent is MnO_4^-; reducing agent is Cl^-. **4.72** S is in Group 6A(16), so its highest possible O.N. is +6 and its lowest possible O.N. is 6 − 8 = −2. (a) S = −2. The S can only increase its O.N. (oxidize), so S^{2-} can function only as a reducing agent. (b) S = +6. The S can only decrease its O.N. (reduce), so SO_4^{2-} can function only as an oxidizing agent.
(c) S = +4. The S can increase or decrease its O.N. Therefore, SO_2 can function as either an oxidizing or reducing agent.
4.74(a) $8HNO_3(aq) + K_2CrO_4(aq) + 3Fe(NO_3)_2(aq) \longrightarrow 2KNO_3(aq) + 3Fe(NO_3)_3(aq) + Cr(NO_3)_3(aq) + 4H_2O(l)$
Oxidizing agent is K_2CrO_4; reducing agent is $Fe(NO_3)_2$.
(b) $8HNO_3(aq) + 3C_2H_6O(l) + K_2Cr_2O_7(aq) \longrightarrow 2KNO_3(aq) + 3C_2H_4O(l) + 7H_2O(l) + 2Cr(NO_3)_3(aq)$
Oxidizing agent is $K_2Cr_2O_7$; reducing agent is C_2H_6O.
(c) $6HCl(aq) + 2NH_4Cl(aq) + K_2Cr_2O_7(aq) \longrightarrow 2KCl(aq) + 2CrCl_3(aq) + N_2(g) + 7H_2O(l)$
Oxidizing agent is $K_2Cr_2O_7$; reducing agent is NH_4Cl.
(d) $KClO_3(aq) + 6HBr(aq) \longrightarrow 3Br_2(l) + 3H_2O(l) + KCl(aq)$
Oxidizing agent is $KClO_3$; reducing agent is HBr.

4.76(a) 4.54×10^{-3} mol MnO_4^- (b) 0.0113 mol H_2O_2 (c) 0.386 g H_2O_2 (d) 2.80 mass % H_2O_2 (e) H_2O_2 **4.81** A combination reaction that is also a redox reaction is $2Mg(s) + O_2(g) \longrightarrow 2MgO(s)$. A combination reaction that is not a redox reaction is $CaO(s) + H_2O(l) \longrightarrow Ca(OH)_2(aq)$.
4.83 (a) $Ca(s) + 2H_2O(l) \longrightarrow Ca(OH)_2(aq) + H_2(g)$; displacement
(b) $2NaNO_3(s) \longrightarrow 2NaNO_2(s) + O_2(g)$; decomposition
(c) $C_2H_2(g) + 2H_2(g) \longrightarrow C_2H_6(g)$; combination
4.85(a) $2Sb(s) + 3Cl_2(g) \longrightarrow 2SbCl_3(s)$; combination
(b) $2AsH_3(g) \longrightarrow 2As(s) + 3H_2(g)$; decomposition
(c) $3Mn(s) + 2Fe(NO_3)_3(aq) \longrightarrow Mn(NO_3)_2(aq) + 2Fe(s)$; displacement
4.87(a) $Ca(s) + Br_2(l) \longrightarrow CaBr_2(s)$
(b) $2Ag_2O(s) \xrightarrow{\Delta} 4Ag(s) + O_2(g)$
(c) $Mn(s) + Cu(NO_3)_2(aq) \longrightarrow Mn(NO_3)_2(aq) + Cu(s)$
4.89(a) $N_2(g) + 3H_2(g) \longrightarrow 2NH_3(g)$
(b) $2NaClO_3(s) \xrightarrow{\Delta} 2NaCl(s) + 3O_2(g)$
(c) $Ba(s) + 2H_2O(l) \longrightarrow Ba(OH)_2(aq) + H_2(g)$
4.91 (a) $2Cs(s) + I_2(s) \longrightarrow 2CsI(s)$
(b) $2Al(s) + 3MnSO_4(aq) \longrightarrow Al_2(SO_4)_3(aq) + 3Mn(s)$
(c) $2SO_2(g) + O_2(g) \longrightarrow 2SO_3(g)$
(d) $C_3H_8(g) + 5O_2(g) \longrightarrow 3CO_2(g) + 4H_2O(g)$
(e) $2Al(s) + 3Mn^{2+}(aq) \longrightarrow 2Al^{3+}(aq) + 3Mn(s)$
4.93 315 g O_2; 3.95 kg Hg **4.95**(a) O_2 is in excess. (b) 0.117 mol Li_2O (c) 0 g Li, 3.49 g Li_2O, and 4.13 g O_2 **4.97** 56.7 mass % $KClO_3$ **4.99** 0.223 kg Fe **4.100** 99.9 g compound B, which is $FeCl_2$ **4.104** The reaction is $2NO + Br_2 \rightleftharpoons 2NOBr$ which can proceed in either direction. If NO and Br_2 are placed in a container, they will react to form NOBr, and NOBr will decompose to form NO and Br_2. Eventually, the concentrations of NO, Br_2, and NOBr adjust so that the rates of the forward and reverse reactions become equal, and equilibrium is reached.
4.106(a) $Fe(s) + 2H^+(aq) \longrightarrow Fe^{2+}(aq) + H_2(g)$
O.N.: 0 +1 +2 0
(b) 3.1×10^{21} Fe^{2+} ions
4.108 5.11 g C_2H_5OH; 2.49 L CO_2
4.110(a) $Ca^{2+}(aq) + C_2O_4^{2-}(aq) \longrightarrow CaC_2O_4(s)$
(b) $5H_2C_2O_4(aq) + 2MnO_4^-(aq) + 6H^+(aq) \longrightarrow 10CO_2(g) + 2Mn^{2+}(aq) + 8H_2O(l)$
(c) $KMnO_4$ (d) $H_2C_2O_4$ (e) 55.06 mass % $CaCl_2$
4.113 3.027 mass % $CaMg(CO_3)_2$ **4.115**(a) Step 1: oxidizing agent is O_2; reducing agent is NH_3. Step 2: oxidizing agent is O_2; reducing agent is NO. Step 3: oxidizing agent is NO_2; reducing agent is NO_2. (b) 1.2×10^4 kg NH_3 **4.119** 627 L air **4.121** 0.75 kg SiO_2; 0.26 kg Na_2CO_3; 0.18 kg $CaCO_3$ **4.124**(a) 4 mol IO_3^- (b) 12 mol I_2; IO_3^- is oxidizing agent and I^- is reducing agent. (c) 12.87 mass % thyroxine **4.126** 3.0×10^{-3} mol CO_2 (b) 0.11 L CO_2 **4.128**(a) $C_7H_5O_4Bi$ (b) $C_{21}H_{15}O_{12}Bi_3$ (c) $Bi(OH)_3(s) + 3HC_7H_5O_3(aq) \longrightarrow Bi(C_7H_5O_3)_3(s) + 3H_2O(l)$ (d) 0.490 mg $Bi(OH)_3$
4.130(a) Ethanol: $C_2H_5OH(l) + 3O_2(g) \longrightarrow 2CO_2(g) + 3H_2O(l)$
Gasoline: $2C_8H_{18}(l) + 25O_2(g) \longrightarrow 16CO_2(g) + 18H_2O(g)$
(b) 2.50×10^3 g O_2 (c) 1.75×10^3 L O_2 (d) 8.38×10^3 L air
4.132 yes **4.134**(a) Reaction (2) is a redox process. (b) 2.00×10^5 g Fe_2O_3; 4.06×10^5 g $FeCl_3$ (c) 2.09×10^5 g Fe; 4.75×10^5 g $FeCl_2$ (d) 0.313

Chapter 5

5.1(a) The volume of the liquid remains constant, but the volume of the gas increases to the volume of the larger container. (b) The volume of the container holding the gas sample increases when heated, but the volume of the container holding the liquid sample remains essentially constant when heated. (c) The volume of the liquid remains essentially constant, but the volume of the gas is reduced. **5.6** 979 cmH_2O **5.8**(a) 566 mmHg (b) 1.32 bar (c) 3.60 atm (d) 107 kPa **5.10** 0.9408 atm **5.12** 0.966 atm
5.18 At constant temperature and volume, the pressure of a gas is directly proportional to number of moles of the gas.
5.20(a) Volume decreases to one-third of the original volume. (b) Volume increases by a factor of 2.5. (c) Volume increases by a factor of 3. **5.22**(a) Volume decreases by a factor of 2. (b) Volume increases by a factor of 1.56. (c) Volume decreases by a factor of 4. **5.24** −42°C **5.26** 35.3 L **5.28** 0.061 mol Cl_2 **5.30** 0.674 g ClF_3 **5.33** no **5.35** Beaker is inverted for H_2 and upright for CO_2. The molar mass of CO_2 is greater than the molar mass of air, which, in turn, has a greater molar mass than H_2. **5.38** 5.86 g/L **5.40** 1.78×10^{-3} mol AsH_3; 3.48 g/L **5.42** 51.1 g/mol **5.44** 1.33 atm **5.48** C_5H_{12} **5.50**(a) 0.90 mol (b) 6.76 torr **5.51** 39.3 g P_4 **5.53** 41.2 g PH_3 **5.55** 0.0249 g Al **5.57** 286 mL SO_2 **5.59** 0.0997 atm SiF_4 **5.63** At STP, the volume occupied by a mole of any gas is the same. At the same temperature, all gases have the same average kinetic energy, resulting in the same pressure. **5.66**(a) $P_A > P_B > P_C$ (b) $E_A = E_B = E_C$ (c) $rate_A > rate_B > rate_C$ (d) total $E_A >$ total $E_B >$ total E_C (e) $d_A = d_B = d_C$. (f) collision frequency in A > collision frequency in B > collision frequency in C **5.67** 13.21 **5.69**(a) curve 1 (b) curve 1 (c) curve 1; fluorine and argon have about the same molar mass **5.71** 14.0 min **5.73** 4 atoms per molecule **5.75** negative deviations; $N_2 < Kr < CO_2$ **5.77** at 1 atm; because the pressure is lower. **5.80** 6.81×10^4 g/mol **5.83**(a) 22.5 atm (b) 21.2 atm **5.86**(a) 597 torr N_2; 159 torr O_2; 0.3 torr CO_2; 3.5 torr H_2O (b) 74.9 mol % N_2; 13.7 mol % O_2; 5.3 mol % CO_2; 6.2 mol % H_2O (c) 1.6×10^{21} molecules O_2 **5.88**(a) 4×10^2 mL (b) 0.013 mol N_2 **5.90** 35.7 L NO_2 **5.95** Al_2Cl_6 **5.97** 1.62×10^{-2} mol **5.101**(a) 1.95×10^3 g Ni (b) 3.5×10^4 g Ni (c) 63 m^3 CO **5.103**(a) 9 volumes of $O_2(g)$ (b) CH_5N **5.107** 4.86; 52.4 ft vapor to a depth of 73 ft **5.109** 6.07 g H_2O_2 **5.113** 4.94×10^{-3} g N_2 **5.117**(a) xenon (b) water vapor (c) mercury (d) water vapor **5.121** 17.2 g CO_2; 17.8 g Kr **5.126** 676 m/s Ne; 481 m/s Ar; 1.52×10^3 m/s He **5.128**(a) 0.055 g (b) 1.1 mL

5.130 (a)
$$\tfrac{1}{2}m\overline{u^2} = \frac{3}{2}\left(\frac{R}{N_A}\right)T$$
$$m\overline{u^2} = 3RT\left(\frac{R}{N_A}\right)T$$
$$\overline{u^2} = \frac{3RT}{mN_A}$$
$$u_{rms} = \sqrt{\frac{3RT}{\mathcal{M}}} \quad \text{where } \mathcal{M} = mN_A$$
(b)
$$\overline{E}_k = \tfrac{1}{2}m_1\overline{u_1^2} = \tfrac{1}{2}m_2\overline{u_2^2}$$
$$m_1\overline{u_1^2} = m_2\overline{u_2^2}$$
$$\frac{m_1}{m_2} = \frac{\overline{u_2^2}}{\overline{u_1^2}}; \text{ so } \frac{\sqrt{m_1}}{\sqrt{m_2}} = \frac{\overline{u_2}}{\overline{u_1}}$$

Substitute molar mass, $\mathcal{M}$, for m:

$$\frac{\sqrt{\mathcal{M}_1}}{\sqrt{\mathcal{M}_2}} = \frac{\text{rate}_2}{\text{rate}_1}$$

5.135 (a) 14.8 L CO_2 (b) $P_{H_2O} = 42.2$ torr; 3.7×10^2 torr $P_{O_2} = P_{CO_2}$ **5.140** 332 steps **5.142** 1.6 **5.146** $P_{total} =$ 0.323 atm; $P_{I_2} = 0.0332$ atm

Chapter 6

6.4 Increase: eating food, lying in the sun, taking a hot bath. Decrease: exercising, taking a cold bath, going outside on a cold day. **6.6** The amount of the change in internal energy is the same for heater and air conditioner. By the law of energy conservation, the change in energy of the universe is zero. **6.8** 0 J **6.10** 1.52×10^3 J **6.12**(a) 3.3×10^7 kJ (b) 7.9×10^6 kcal (c) 3.1×10^7 Btu **6.15** 9.3 h **6.17** Measuring the heat transfer at constant pressure is more convenient than measuring at constant volume. **6.19**(a) exothermic (b) endothermic (c) exothermic (d) exothermic (e) endothermic (f) endothermic (g) exothermic

6.22

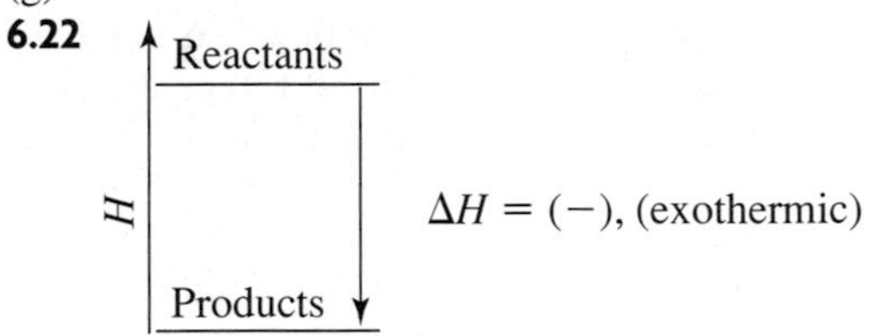

6.24(a) Combustion of methane: $CH_4(g) + 2O_2(g) \longrightarrow CO_2(g) + 2H_2O(g) +$ heat

$CH_4 + 2O_2$ (initial)

H $\Delta H = (-)$, (exothermic)

$CO_2 + 2H_2O$ (final)

(b) Freezing of water: $H_2O(l) \longrightarrow H_2O(s) +$ heat

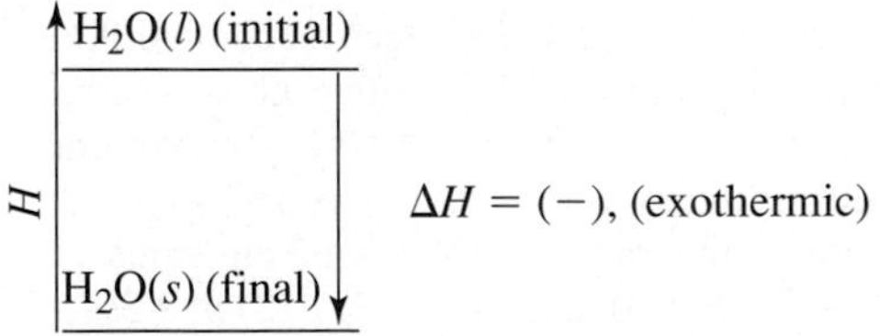

6.26(a) $C_2H_5OH(l) + 3O_2(g) \longrightarrow 2CO_2(g) + 3H_2O(g) +$ heat

$C_2H_6O + 3O_2$ (initial)

H $\Delta H = (-)$, (exothermic)

$2CO_2 + 3H_2O$ (final)

(b) $N_2(g) + 2O_2(g) +$ heat $\longrightarrow 2NO_2(g)$

$2NO_2$ (final)

H $\Delta H = (+)$, (endothermic)

$N_2 + 2O_2$ (initial)

6.29 To determine the specific heat capacity of a substance, you need its mass, the heat added (or lost), and the change in temperature. **6.31** Specific heat capacity is the quantity of heat required to raise one gram of a substance by one kelvin. Heat capacity is also the quantity of heat required for a one-kelvin temperature change, but it applies to an object instead of a specified amount of a substance. (a) Heat capacity, because the fixture is a combination of substances (b) Specific heat capacity, because the copper wire is a pure substance (c) Specific heat capacity, because the water is a pure substance **6.33** 4.0×10^3 J **6.35** 323°C **6.37** 77.5°C **6.39** 42°C **6.41** 57.0°C **6.48** The reaction has a positive ΔH_{rxn}, because this reaction requires the input of energy to break the oxygen-oxygen bond. **6.49** ΔH is negative; it is opposite in sign and half of the value for the vaporization of 2 mol of H_2O. **6.50**(a) exothermic (b) 20.2 kJ (c) -5.2×10^2 kJ (d) -12.6 kJ
6.52 (a) $\frac{1}{2}N_2(g) + \frac{1}{2}O_2(g) \longrightarrow NO(g)$; $\Delta H = 90.29$ kJ
(b) -4.51 kJ
6.54 -2.11×10^6 kJ
6.58(a) $C_2H_4(g) + 3O_2(g) \longrightarrow 2CO_2(g) + 2H_2O(l)$;
$\Delta H_{rxn} = -1411$ kJ
(b) 1.39 g C_2H_4
6.62 -110.5 kJ **6.63** -813.4 kJ **6.65** 44.0 kJ
6.67 $N_2(g) + 2O_2(g) \longrightarrow 2NO_2(g)$; $\Delta H_{rxn} = +66.4$ kJ; A = 1, B = 2, C = 3 **6.70** The standard heat of reaction, ΔH^0_{rxn}, is the enthalpy change for any reaction where all substances are in their standard states. The standard heat of formation, ΔH^0_f, is the enthalpy change that accompanies the formation of one mole of a compound in its standard state from elements in their standard states.
6.72(a) $\frac{1}{2}Cl_2(g) + Na(s) \longrightarrow NaCl(s)$
(b) $H_2(g) + \frac{1}{2}O_2(g) \longrightarrow H_2O(l)$
(c) no changes
6.73(a) $Ca(s) + Cl_2(g) \longrightarrow CaCl_2(s)$
(b) $Na(s) + \frac{1}{2}H_2(g) + C(\text{graphite}) + \frac{3}{2}O_2(g) \longrightarrow NaHCO_3(s)$
(c) $C(\text{graphite}) + 2Cl_2(g) \longrightarrow CCl_4(l)$
(d) $\frac{1}{2}H_2(g) + \frac{1}{2}N_2(g) + \frac{3}{2}O_2(g) \longrightarrow HNO_3(l)$
6.75(a) -1036.8 kJ (b) -433 kJ **6.77** -157.3 kJ/mol
6.80(a) 503.9 kJ (b) $-\Delta H_1 + 2\Delta H_2 = 504$ kJ
6.81(a) $C_{18}H_{36}O_2(s) + 26O_2(g) \longrightarrow 18CO_2(g) + 18H_2O(g)$
(b) $-10{,}488$ kJ (c) -36.9 kJ; -8.81 kcal (d) 8.81 kcal/g $\times$ 11.0 g = 96.9 kcal **6.83**(a) 23.6 L/mol initial; 24.9 L/mol final (b) 187 J (c) -1.2×10^2 J (d) 3.1×10^2 J (e) 310 J (f) $\Delta H = \Delta E + P\Delta V = \Delta E - w = (q + w) - w = q_P$
6.86(a) -65.2 kJ/mol (b) 1.558×10^4 kJ/kg SiC
6.90 -94 kJ/mol **6.93**(a) 1.1×10^3 J (b) 1.1×10^8 J (c) 1.3×10^2 mol CH_4 (d) \$0.53 **6.95** 721 kJ **6.103**(a) III > II > I (b) I > III > II **6.105**(a) $\Delta H^0_{rxn1} = -657.0$ kJ; $\Delta H^0_{rxn2} = 32.9$ kJ (b) -106.6 kJ **6.108**(a) 6.81×10^3 J (b) 243°C **6.109** -22.2 kJ **6.110**(a) 34 kJ/mol (b) -757 kJ **6.112**(a) -1.25×10^3 kJ (b) 2.24×10^3°C

Chapter 7

7.2 (a) x-ray < ultraviolet < visible < infrared < microwave < radio waves (b) radio < microwave < infrared < visible < ultraviolet < x-ray (c) radio < microwave < infrared < visible < ultraviolet < x-ray **7.5** The energy of an atom is not continuous, but quantized. It exists only in certain fixed amounts called *quanta.* **7.7** 312 m; 3.12×10^{11} nm; 3.12×10^{12} Å **7.9** 2.4×10^{-23} J **7.11** b < c < a **7.13** 1.3483×10^{9} nm; 1.3483×10^{8} Å **7.16**(a) 1.24×10^{15} s^{-1}; 8.21×10^{-19} J (b) 1.4×10^{15} s^{-1}; 9.0×10^{-19} J **7.18** Bohr's key assumption was that the electron in an atom does not radiate energy while in a stationary state, and it can move to a different orbit only by absorbing or emitting a photon whose energy is equal to the difference in energy between two states. These differences in energy correspond to the wavelengths in the known line spectra for the hydrogen atom. **7.20**(a) absorption (b) emission (c) emission (d) absorption **7.22** Yes, the predicted line spectra are accurate. The energies could be predicted from $E_n = \dfrac{-(Z^2)(2.18\times10^{-18}\text{ J})}{n^2}$, where Z is the atomic number for the atom or ion. The energy levels for Be^{3+} will be greater by a factor of 16 than those for the hydrogen atom. This means that the pattern of lines will be similar, but at different wavelengths. **7.23** 434.17 nm **7.25** 1875.6 nm **7.27** -2.76×10^{5} J/mol **7.29** d < a < c < b **7.31** $n = 4$ **7.34** 3.37×10^{-19} J; 203 kJ **7.37** Macroscopic objects do exhibit a wavelike motion, but the wavelength is too small for humans to perceive. **7.39**(a) 7.6×10^{-37} m (b) 1×10^{-35} m **7.41** 2.2×10^{-26} m/s **7.43** 3.75×10^{-36} kg **7.47** The total probability of finding the 1*s* electron in any distance *r* from the nucleus is greatest when the value of *r* is 0.529 Å. **7.48**(a) principal determinant of the electron's energy or distance from the nucleus (b) determines the shape of the orbital (c) determines the orientation of the orbital in three-dimensional space **7.49**(a) one (b) five (c) three (d) nine **7.51**(a) m_l: −2, −1, 0, +1, +2 (b) m_l: 0 (if $n = 1$, then $l = 0$) (c) m_l: −3, −2, −1, 0, +1, +2, +3

7.53

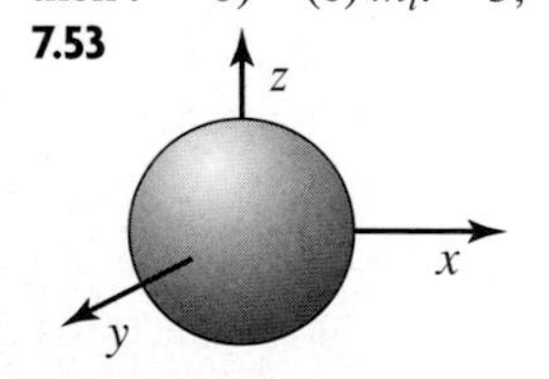

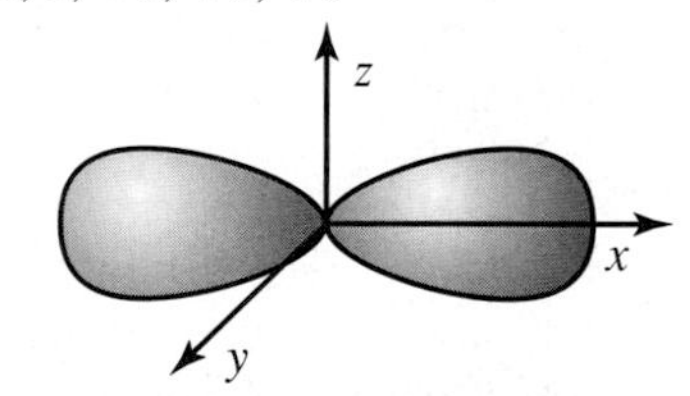

7.55

Sublevel	Allowable m_l	No. of orbitals
(a) d ($l = 2$)	−2, −1, 0, +1, +2	5
(b) p ($l = 1$)	−1, 0, +1	3
(c) f ($l = 3$)	−3, −2, −1, 0, +1, +2, +3	7

7.57(a) $n = 5$ and $l = 0$; one orbital (b) $n = 3$ and $l = 1$; three orbitals (c) $n = 4$ and $l = 3$; seven orbitals **7.59**(a) no; $n = 2$, $l = 1$, $m_l = -1$; $n = 2$, $l = 0$, $m_l = 0$ (b) allowed (c) allowed (d) no; $n = 5$, $l = 3$, $m_l = +3$; $n = 5$, $l = 2$, $m_l = 0$ **7.62**(a) $E = -(2.180\times10^{-18}\text{ J})(1/n^2)$. This is identical with the expression from Bohr's theory. (b) 3.028×10^{-19} J (c) 656.2 nm **7.63**(a) The attraction of the nucleus for the electrons must be overcome. (b) The electrons in silver are more tightly held by the nucleus. (c) silver (d) Once the electron is freed from the atom, its energy increases in proportion to the frequency of the light. **7.66** Li^{2+} **7.69**(a) 2 ⟶ 1 (b) 5 ⟶ 2 (c) 4 ⟶ 2 (d) 3 ⟶ 2 (e) 6 ⟶ 3 **7.71**(a) Ba; 462 nm (b) 278 to 292 nm **7.74**(a) 2.7×10^{2} s (b) 3.6×10^{8} m **7.75**(a) $l = 1$ (b) $l = 2$ (c) $l = 3$ (d) $l = 2$

7.77(a) $\Delta E = (-2.18\times10^{-18}\text{ J})\left(\dfrac{1}{\infty^2} - \dfrac{1}{n^2_{\text{initial}}}\right)Z^2\left(\dfrac{6.022\times10^{23}}{1\text{ mol}}\right)$ (b) 3.28×10^{7} J/mol (c) 205 nm (d) 22.8 nm **7.79**(a) 5.293×10^{-11} m (b) 5.293×10^{-9} m **7.81** 6.4×10^{27} photons **7.83**(a) no overlap (b) overlap (c) two (d) At longer wavelengths, the hydrogen spectrum begins to become a continuous band. **7.86**(a) 7.56×10^{-18} J; 2.63×10^{-8} m (b) 5.110×10^{-17} J; 3.890×10^{-9} m (c) 1.22×10^{-18} J; 1.63×10^{-7} m **7.88**(a) 1.87×10^{-19} J (b) 3.58×10^{-19} J **7.90**(a) red; green (b) 1.18 kJ (Sr); 1.17 kJ (Ba) **7.92**(a) This is the wavelength of maximum absorbance, so it gives the highest sensitivity. (b) ultraviolet region (c) 1.93×10^{-2} g vitamin A/g oil **7.97** 1.0×10^{18} photons/s **7.100** 3*s* ⟶ 2*p*; 3*d* ⟶ 2*p*; 4*s* ⟶ 2*p*; 3*p* ⟶ 2*s*

Chapter 8

8.1 Elements are listed in the periodic table in an ordered, systematic way that correlates with a periodicity of their chemical and physical properties. The theoretical basis for the table in terms of atomic number and electron configuration does not allow for a "new element" between Sn and Sb. **8.3**(a) predicted atomic mass = 54.23 amu (b) predicted melting point = 6.3°C **8.6** The quantum number m_s relates to just the electron; all the others describe the orbital. **8.9** Shielding occurs when inner electrons protect, or shield, outer electrons from the full nuclear attraction. The effective nuclear charge is the nuclear charge an electron actually experiences. As the number of inner electrons increases, the effective nuclear charge decreases. **8.11**(a) 6 (b) 10 (c) 2 **8.13**(a) 6 (b) 2 (c) 14 **8.16** Degenerate orbitals (those of identical energy) will be filled in such a way that the electron spins will have the same value. N: $1s^22s^22p^3$.

↑↓	↑↓	↑ ↑ ↑
1*s*	2*s*	2*p*

8.18 Main-group elements from the same group have similar outer electron configurations, and the (old) group number equals the number of outer electrons. Outer electron configurations vary in a periodic manner within a period, with each succeeding element having an additional electron. **8.20** The maximum number of electrons in any energy level n is $2n^2$, so the $n = 4$ energy level holds a maximum of $2(4^2) = 32$ electrons. **8.21**(a) $n = 5$, $l = 0$, $m_l = 0$, and $m_s = +\frac{1}{2}$ (b) $n = 3$, $l = 1$, $m_l = +1$, and $m_s = -\frac{1}{2}$ (c) $n = 5$, $l = 0$, $m_l = 0$, and $m_s = +\frac{1}{2}$ (d) $n = 2$, $l = 1$, $m_l = +1$, and $m_s = -\frac{1}{2}$

8.23(a) Rb: $1s^22s^22p^63s^23p^64s^23d^{10}4p^65s^1$
(b) Ge: $1s^22s^22p^63s^23p^64s^23d^{10}4p^2$
(c) Ar: $1s^22s^22p^63s^23p^6$

8.25(a) Cl: $1s^22s^22p^63s^23p^5$ (b) Si: $1s^22s^22p^63s^23p^2$ (c) Sr: $1s^22s^22p^63s^23p^64s^23d^{10}4p^65s^2$

8.27(a) Ti: [Ar] $4s^23d^2$

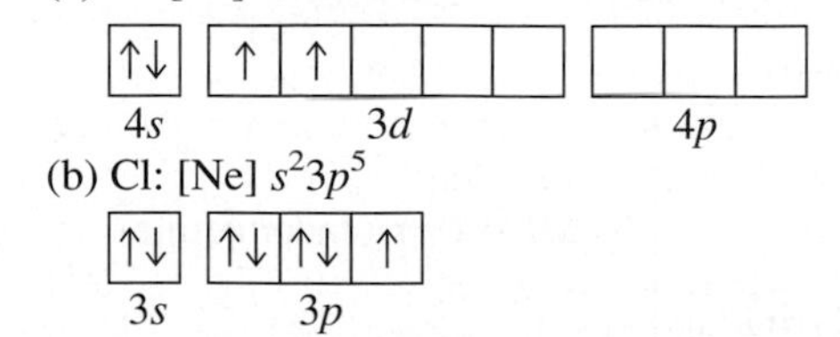

(b) Cl: [Ne] s^23p^5

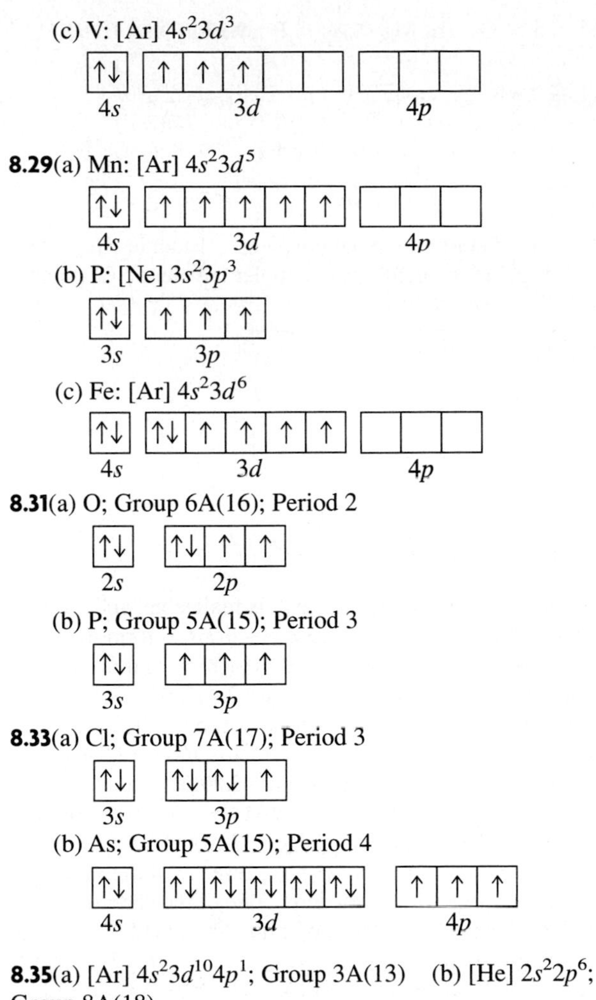

8.35(a) [Ar] $4s^2 3d^{10} 4p^1$; Group 3A(13) (b) [He] $2s^2 2p^6$; Group 8A(18)

8.37

	Inner Electrons	Outer Electrons	Valence Electrons
(a) O	2	6	6
(b) Sn	46	4	4
(c) Ca	18	2	2
(d) Fe	18	2	8
(e) Se	28	6	6

8.39(a) B; Al, Ga, In, and Tl (b) S; O, Se, Te, and Po (c) La; Sc, Y, and Ac **8.41**(a) C; Si, Ge, Sn, and Pb (b) V; Nb, Ta, and Db (c) P; N, As, Sb, and Bi
8.43 Na (first excited state): $1s^2 2s^2 2p^6 3p^1$

1s 2s 2p 3s 3p

8.46 Atomic size increases down a group. Ionization energy decreases down a group. These trends result because the outer electrons are more easily removed as the atom gets larger.
8.48 For a given element, successive ionization energies always increase. As each successive electron is removed, the positive charge on the ion increases, which results in a stronger attraction between the leaving electron and the ion. When a large jump between successive ionization energies is observed, the subsequent electron must come from a lower energy level. **8.50** A high IE_1 and a very negative EA_1 suggest that the elements are halogens, in Group 7A(17), which form 1− ions. **8.53**(a) K < Rb < Cs (b) O < C < Be (c) Cl < S < K (d) Mg < Ca < K

8.55(a) Ba < Sr < Ca (b) B < N < Ne (c) Rb < Se < Br (d) Sn < Sb < As **8.57** $1s^2 2s^2 2p^1$ (boron, B) **8.59**(a) Na (b) Na (c) Be **8.61** (1) Metals conduct electricity, nonmetals do not. (2) When they form stable ions, metal ions tend to have a positive charge, nonmetal ions tend to have a negative charge. (3) Metal oxides are ionic and act as bases, nonmetal oxides are covalent and act as acids. **8.62** Metallic character increases down a group and decreases toward the right across a period. These trends are the same as those for atomic size and opposite those for ionization energy. **8.64** Possible ions are 2+ and 4+. The 2+ ions form by loss of the outermost two *p* electrons, while the 4+ ions form by loss of these and the outermost two *s* electrons. **8.68**(a) Rb (b) Ra (c) I **8.70**(a) As (b) P (c) Be **8.72** acidic solution; $SO_2(g) + H_2O(l) \longrightarrow H_2SO_3(aq)$ **8.74**(a) Cl^-: $1s^2 2s^2 2p^6 3s^2 3p^6$ (b) Na^+: $1s^2 2s^2 2p^6$ (c) Ca^{2+}: $1s^2 2s^2 2p^6 3s^2 3p^6$ **8.76**(a) Al^{3+}: $1s^2 2s^2 2p^6$ (b) S^{2-}: $1s^2 2s^2 2p^6 3s^2 3p^6$ (c) Sr^{2+}: $1s^2 2s^2 2p^6 3s^2 3p^6 4s^2 3d^{10} 4p^6$ **8.78**(a) 0 (b) 3 (c) 0 (d) 1 **8.80** a, b, and d **8.82**(a) V^{3+}, [Ar] $3d^2$, paramagnetic (b) Cd^{2+}, [Kr] $4d^{10}$, diamagnetic (c) Co^{3+}, [Ar] $3d^7$, paramagnetic (d) Ag^+, [Kr] $4d^{10}$, diamagnetic **8.84** For palladium to be diamagnetic, all of its electrons must be paired. (a) You might first write the condensed electron configuration for Pd as [Kr] $5s^2 4d^8$. However, the partial orbital diagram is not consistent with diamagnetism.

5s 4d 5p

(b) This is the only configuration that supports diamagnetism, [Kr] $4d^{10}$.

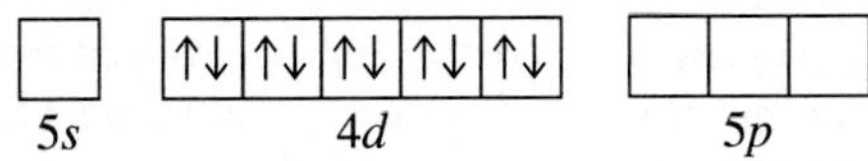

(c) Promoting an *s* electron into the *d* sublevel still leaves two electrons unpaired.

5s 4d 5p

8.86(a) $Li^+ < Na^+ < K^+$ (b) $Rb^+ < Br^- < Se^{2-}$ (c) $F^- < O^{2-} < N^{3-}$ **8.90** Ce: [Xe] $6s^2 4f^1 5d^1$; Ce^{4+}: [Xe]; Eu: [Xe] $6s^2 4f^7$; Eu^{2+}: [Xe] $4f^7$. Ce^{4+} has a noble-gas configuration; Eu^{2+} has a half-filled *f* subshell. **8.92**(a) Cl_2O, dichlorine monoxide (b) Cl_2O_3, dichlorine trioxide (c) Cl_2O_5, dichlorine pentaoxide (d) Cl_2O_7, dichlorine heptaoxide (e) SO_3, sulfur trioxide (f) SO_2, sulfur dioxide (g) N_2O_5, dinitrogen pentaoxide (h) N_2O_3, dinitrogen trioxide (i) CO_2, carbon dioxide (j) P_2O_5, diphosphorus pentaoxide **8.95**(a) $SrBr_2$, strontium bromide (b) CaS, calcium sulfide (c) ZnF_2, zinc fluoride (d) LiF, lithium fluoride **8.97**(a) 2009 kJ/mol (b) −549 kJ/mol **8.99** All ions except Fe^{8+} and Fe^{14+} are paramagnetic; Fe^+ and Fe^{3+} would be most attracted. **8.104** indium: 4.404×10^{-19} J; thallium: 3.713×10^{-19} J **8.107**(a) The first student chooses the lower bunk in the bedroom on the first floor. (b) There are seven students on the top bunk when the seventeenth student chooses. (c) The twenty-first student chooses a bottom bunk on the third floor in the largest room. (d) There are fourteen students in bottom bunks when the twenty-fifth student chooses.

Chapter 9

9.1(a) Greater ionization energy decreases metallic character. (b) Larger atomic radius increases metallic character. (c) Higher number of outer electrons decreases metallic character. (d) Larger effective nuclear charge decreases metallic character. **9.4**(a) Cs (b) Rb (c) As **9.6**(a) ionic (b) covalent (c) metallic **9.8**(a) covalent (b) ionic (c) covalent **9.10**(a) Rb· (b) ·Si· (c) :I· **9.12**(a) ·Sr· (b) :P· (c) :S: **9.14**(a) 6A(16); [noble gas] ns^2np^4 (b) 3(A)13; [noble gas] ns^2np^1 **9.17** Because the lattice energy is the result of electrostatic attractions between oppositely charged ions, its magnitude depends on several factors, including ionic size and ionic charge. For a particular arrangement of ions, the lattice energy increases as the charge on the ions increases and as their radii decrease. **9.20**(a) Ba^{2+}, [Xe]; Cl^-, [Ne] $3s^23p^6$, $:\ddot{Cl}:^-$; $BaCl_2$ (b) Sr^{2+}, [Kr]; O^{2-}, [He] $2s^22p^6$, $:\ddot{O}:^{2-}$; SrO (c) Al^{3+}, [Ne]; F^-, [He] $2s^22p^6$, $:\ddot{F}:^-$; AlF_3 (d) Rb^+, [Kr]; O^{2-}, [He] $2s^22p^6$, $:\ddot{O}:^{2-}$; Rb_2O. **9.22**(a) 2A(2) (b) 6A(16) (c) 1A(1) **9.24**(a) 3A(13) (b) 2A(2) (c) 6A(16) **9.26**(a) BaS; the charge on each ion is twice the charge on the ions in CsCl. (b) LiCl; Li^+ is smaller than Cs^+. **9.28**(a) BaS; Ba^{2+} is larger than Ca^{2+}. (b) NaF; the charge on each ion is less than the charge on Mg and O. **9.30** −788 kJ, the lattice energy for NaCl is less than that for LiF, because the Na^+ and Cl^- ions are larger than Li^+ and F^- ions. **9.33** −336 kJ **9.34** When two chlorine atoms are far apart, there is no interaction between them. As the atoms move closer together, the nucleus of each atom attracts the electrons of the other atom. The closer the atoms, the greater this attraction; however, the repulsions of the two nuclei and two electrons also increase at the same time. The final internuclear distance is the distance at which maximum attraction is achieved in spite of the repulsion. **9.35** The bond energy is the energy required to break the bond between H atoms and Cl atoms in one mole of HCl molecules in the gaseous state. Energy is needed to break bonds, so bond energy is always endothermic and $\Delta H^0_{\text{bond breaking}}$ is positive. The amount of energy needed to break the bond is released upon its formation, so $\Delta H^0_{\text{bond forming}}$ has the same magnitude as $\Delta H^0_{\text{bond breaking}}$ but is opposite in sign (always exothermic and negative). **9.39**(a) I—I < Br—Br < Cl—Cl (b) S—Br < S—Cl < S—H (c) C—N < C═N < C≡N **9.41**(a) C—O < C═O; the C═O bond (bond order = 2) is stronger than the C—O bond (bond order = 1). (b) C—H < O—H; O is smaller than C so the O—H bond is shorter and stronger than the C—H bond. **9.44** Less energy is required to break weak bonds. **9.46** Both are one-carbon molecules. Since methane contains fewer C—O bonds, it will have the greater heat of combustion per mole. **9.48** −168 kJ **9.50** −22 kJ **9.51** −59 kJ **9.52** Electronegativity increases from left to right and increases from bottom to top within a group. Fluorine and oxygen are the two most electronegative elements. Cesium and francium are the two least electronegative elements. **9.54** The H—O bond in water is polar covalent. A nonpolar covalent bond occurs between two atoms with identical electronegativities. A polar covalent bond occurs when the atoms have differing electronegativities. Ionic bonds result from electron transfer between atoms.

9.57(a) Si < S < O (b) Mg < As < P **9.59**(a) N > P > Si (b) As > Ga > Ca
9.61(a) N—B (arrow toward N) (b) N—O (arrow toward O) (c) C—S (none) (d) S—O (arrow toward O) (e) N—H (arrow toward N) (f) Cl—O (arrow toward O)
9.63 a, d, and e **9.65**(a) nonpolar covalent (b) ionic (c) polar covalent (d) polar covalent (e) nonpolar covalent (f) polar covalent; $SCl_2 < SF_2 < PF_3$
9.67(a) H—I < H—Br < H—Cl (b) H—C < H—O < H—F (c) S—Cl < P—Cl < Si—Cl
9.70(a) A shiny solid that conducts heat, is malleable, and melts at high temperatures. (b) Metals lose electrons to form positive ions, and metals form basic oxides. **9.74**(a) 800. kJ/mol, which is lower than the value in Table 9.2 (b) -2.417×10^4 kJ (c) 1690. g CO_2 (d) 65.2 L O_2 **9.76**(a) −125 kJ (b) yes, since ΔH^0_f is negative (c) −392 kJ (d) No, ΔH^0_f for $MgCl_2$ is much more negative than that for MgCl. **9.79**(a) 406 nm (b) 2.93×10^{-19} J (c) 1.87×10^4 m/s **9.83** C—Cl: 3.53×10^{-7} m; bond in O_2: 2.40×10^{-7} m **9.84** XeF_2: 132 kJ/mol; XeF_4: 150. kJ/mol; XeF_6: 146 kJ/mol **9.86**(a) The presence of the very electronegative fluorine atoms bonded to one of the carbons makes the C—C bond polar. This polar bond will tend to undergo heterolytic rather than homolytic cleavage. More energy is required to achieve heterolytic cleavage. (b) 1420 kJ **9.89** −13,286 kJ **9.91** 8.70×10^{14} s^{-1}; 3.45×10^{-7} m, which is in the ultraviolet region of the electromagnetic spectrum. **9.94**(a) $CH_3OCH_3(g)$: −326 kJ; $CH_3CH_2OH(g)$: −369 kJ (b) The formation of gaseous ethanol is more exothermic. (c) 43 kJ

Chapter 10

10.1 He and H cannot serve as central atoms in a Lewis structure. Both can have no more than two valence electrons. Fluorine needs only one electron to complete its valence level, and it does not have *d* orbitals available to expand its valence level. Thus, it can bond to only one other atom. **10.3** All the structures obey the octet rule except c and g.
10.5(a) SiF_4: Si bonded to four :F: atoms (b) :Cl—Se—Cl: (c) :F—C—F: with C═O
10.7(a) :F—P—F: with P—F below (b) H—O—C— with C═O (c) :S═C═S:
10.9(a) [O═N═O]$^+$
(b) :F: bonded to N, N bonded to O and ═O ⟷ :F: bonded to N, N bonded to ═O and O

10.11(a) $[:\ddot{N}{=}N{=}\ddot{N}:]^- \longleftrightarrow [:\ddot{\underset{..}{N}}{-}N{\equiv}N:]^- \longleftrightarrow [:N{\equiv}N{-}\ddot{\underset{..}{N}}:]^-$
(b) $[:\ddot{O}{=}\underset{..}{N}{-}\ddot{\underset{..}{O}}:]^- \longleftrightarrow [:\ddot{\underset{..}{O}}{-}\underset{..}{N}{=}\ddot{O}:]^-$

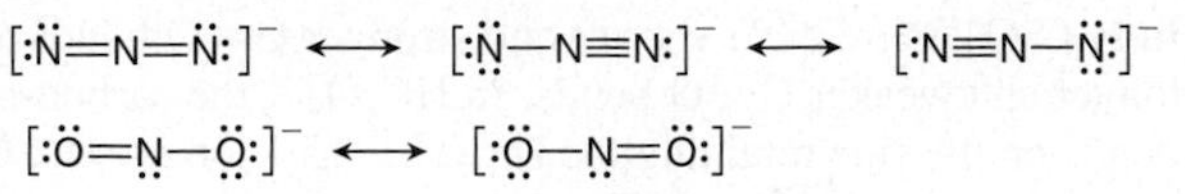

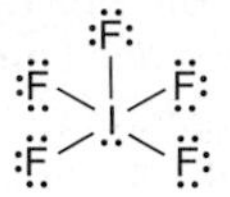

10.13(a) IF_5 (I bonded to five F atoms, lone pair on I) formal charges: I = 0, F = 0

(b) 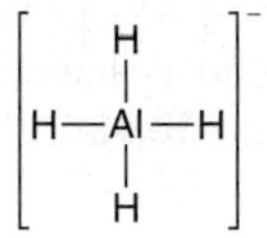 $[AlH_4]^-$ (Al bonded to four H atoms) formal charges: H = 0, Al = −1

10.15(a) $[:C{\equiv}N:]^-$ formal charges: C = −1, N = 0
(b) 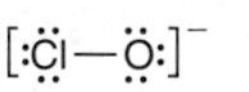 $[:\ddot{\underset{..}{Cl}}{-}\ddot{\underset{..}{O}}:]^-$ formal charges: Cl = 0, O = −1

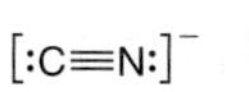

10.17(a) $[\ddot{O}{=}\ddot{Br}{=}\ddot{O}]^-$ with $:\ddot{\underset{..}{O}}:$ singly bonded to Br formal charges: Br = 0, doubly bonded O = 0, singly bonded O = −1

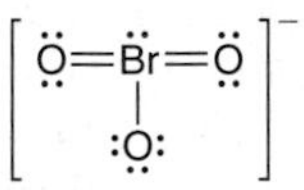

(b) 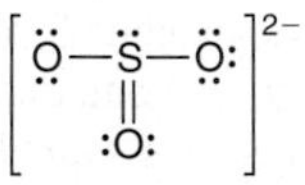 $[\ddot{\underset{..}{O}}{-}\ddot{S}{-}\ddot{\underset{..}{O}}:]^{2-}$ with $:\ddot{O}:$ doubly bonded to S formal charges: S = 0, singly bonded O = −1, doubly bonded O = 0

10.19(a) BH_3 has 6 valence electrons. (b) As has an expanded valence level with 10 electrons. (c) Se has an expanded valence level with 10 electrons.

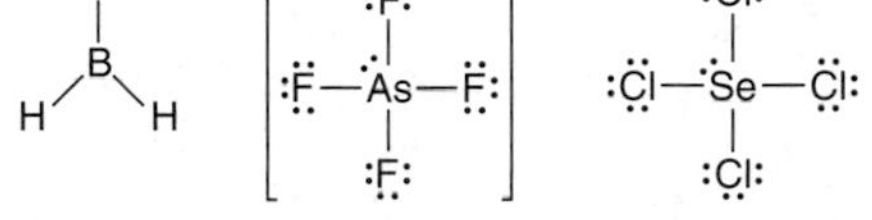

(a) BH_3 (B bonded to three H) (b) $[AsF_4]^-$ (c) $SeCl_4$

10.21(a) Br expands its valence level to 10 electrons. (b) I has an expanded valence level of 10 electrons. (c) Be has only 4 valence shell electrons.
(a) $:\ddot{\underset{..}{F}}{-}\ddot{Br}{-}\ddot{\underset{..}{F}}:$ with $:\ddot{\underset{..}{F}}:$ below Br (b) $[:\ddot{\underset{..}{Cl}}{-}\ddot{\underset{..}{I}}{-}\ddot{\underset{..}{Cl}}:]^-$ (c) $:\ddot{\underset{..}{F}}{-}Be{-}\ddot{\underset{..}{F}}:$

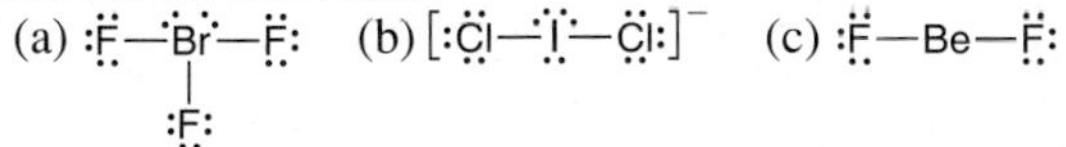

10.23

$:\ddot{\underset{..}{Cl}}{-}Be{-}\ddot{\underset{..}{Cl}}: + 2\,[:\ddot{\underset{..}{Cl}}:]^- \longrightarrow [BeCl_4]^{2-}$

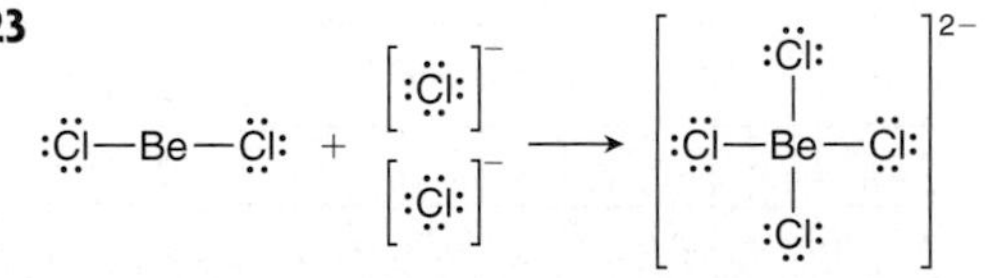

10.26 structure A **10.28** The molecular shape and the electron-group arrangement are the same when no lone pairs are present. **10.30** tetrahedral, AX_4; trigonal pyramidal, AX_3E; bent or V shaped, AX_2E_2

10.32(a) (b) (c) (d) (e) (f) [structures of X atoms around central A]

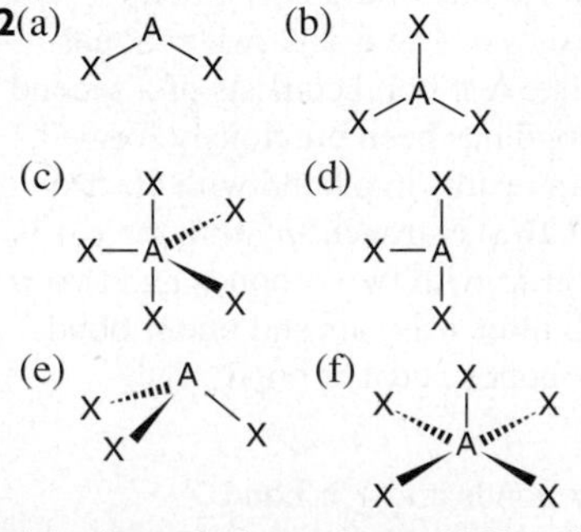

10.34(a) trigonal planar, bent, 120° (b) tetrahedral, trigonal pyramidal, 109.5° (c) tetrahedral, trigonal pyramidal, 109.5°

10.36(a) trigonal planar, trigonal planar, 120° (b) trigonal planar, bent, 120° (c) tetrahedral, tetrahedral, 109.5°
10.38(a) trigonal planar, AX_3, 120° (b) trigonal pyramidal, AX_3E, 109.5° (c) trigonal bipyramidal, AX_5, 90° and 120°
10.40(a) bent, 109.5°, less than 109.5° (b) trigonal bipyramidal, 90° and 120°, angles are ideal (c) see-saw, 90° and 120°, less than ideal (d) linear, 180°, angle is ideal **10.42**(a) C: tetrahedral, 109.5°; O: bent, $<$ 109.5° (b) N: trigonal planar, 120°
10.44(a) C in CH_3: tetrahedral, 109.5°; C with C=O: trigonal planar, 120°; O with H: bent, $<$ 109.5° (b) O: bent, $<$ 109.5°
10.46 $OF_2 < NF_3 < CF_4 < BF_3 < BeF_2$ **10.48**(a) The C and N each have three groups so the ideal angles are 120°, and the O has four groups so the ideal angle is 109.5°. The N and O have lone pairs so the angles are less than ideal. (b) All central atoms have four pairs, so the ideal angles are 109.5°. The lone pairs on the O reduce this value. (c) The B has three groups and an ideal bond angle of 120°. All the O's have four groups (ideal bond angles of 109.5°), two of which are lone pairs that reduce the angle.
10.51 PCl_5 $[PCl_4]^+$ $[PCl_6]^-$

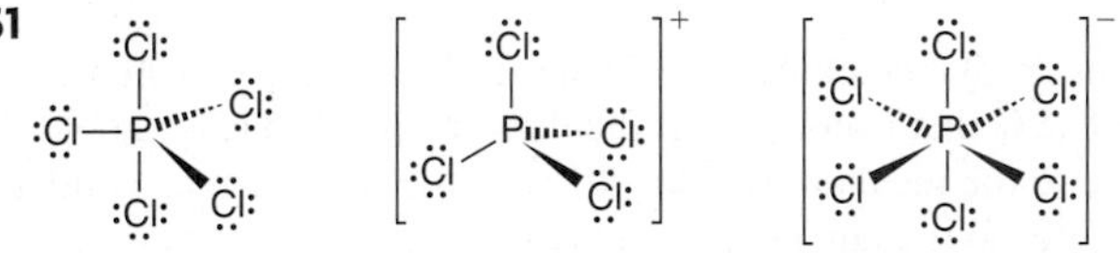

In the gas phase, PCl_5 is AX_5, so the shape is trigonal bipyramidal, and the bond angles are 120° and 90°. The $PCl_4{}^+$ ion is AX_4, so the shape is tetrahedral, and the bond angles are 109.5°. The $PCl_6{}^-$ ion is AX_6, so the shape is octahedral, and the bond angles are 90°. **10.52** Molecules are polar if they have polar bonds that are not arranged to cancel each other. A polar bond is present any time there is a bond between elements with differing electronegativities. **10.55**(a) CF_4 (b) BrCl and SCl_2
10.57(a) SO_2, because it is polar and SO_3 is not. (b) IF has a greater electronegativity difference between its atoms. (c) SF_4, because it is polar and SiF_4 is not. (d) H_2O has a greater electronegativity difference between its atoms.
10.59 X: $(Cl)(H)C{=}C(Cl)(H)$ Y: $(H)(H)C{=}C(Cl)(Cl)$ Z: $(Cl)(H)C{=}C(Cl)(H)$

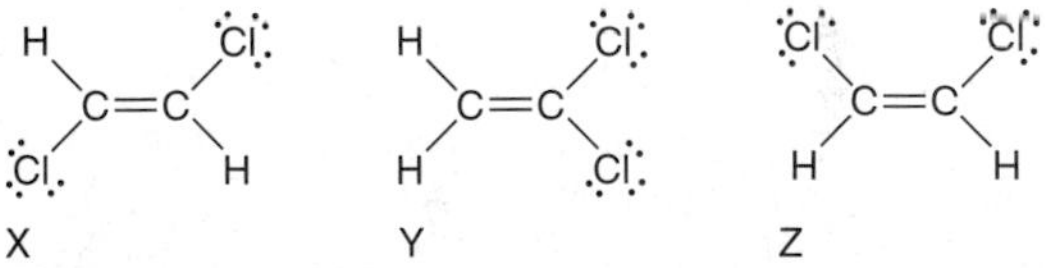

X Y Z

10.61 $H{-}\ddot{N}{-}\ddot{N}{-}H$ (each N bonded to H) $H{-}\ddot{N}{=}\ddot{N}{-}H$ $:N{\equiv}N:$

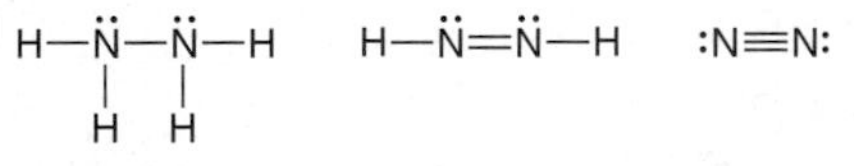

Hydrazine Diazene Nitrogen

(a) The single N—N bond (bond order = 1) is weaker and longer than the others. The triple bond (bond order = 3) is stronger and shorter than the others. The double bond (bond order = 2) has an intermediate strength and length.
(b) $\Delta H^0_{rxn} = -367$ kJ

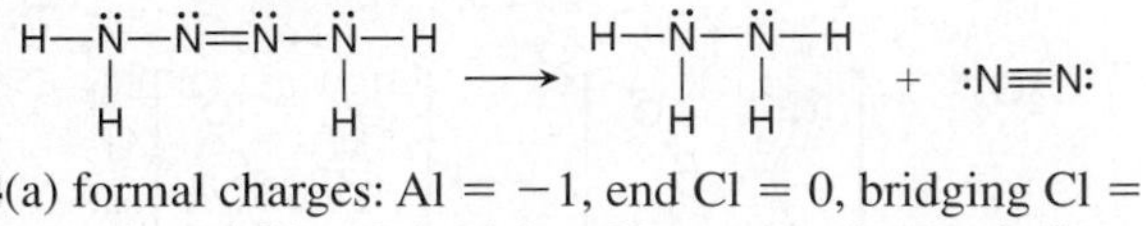

$H{-}\ddot{N}{-}\ddot{N}{=}\ddot{N}{-}\ddot{N}{-}H$ (terminal N's each bonded to H) $\longrightarrow$ $H{-}\ddot{N}{-}\ddot{N}{-}H$ (each N bonded to H) $+ :N{\equiv}N:$

10.64(a) formal charges: Al = −1, end Cl = 0, bridging Cl = +1; I = −1, end Cl = 0, bridging Cl = +1 (b) The iodine atoms are each AX_4E_2 and the shape around each is square planar. Placing these square planar portions adjacent gives a planar molecule.

10.67 :F̈—B(—F̈:)(—F̈:) + CH_3CH_2—Ö—CH_2CH_3 → :F̈—B⁻(—F̈:)(—F̈:)—Ö⁺(CH_2CH_3)—CH_2CH_3

trigonal planar; bent; B: tetrahedral, O: trigonal pyramidal

10.70

$H_3C(H)C{=}CH_2 + H_2O_2 \longrightarrow H_3C{-}C(H)(O){-}CH_2 + H_2O$

(a) In propylene oxide, the C's have ideal angles of 109.5°.
(b) The C that is not part of the three-membered ring should have close to the ideal angle. The atoms in the ring form an equilateral triangle, so the angles around the two C's in the ring are reduced from the ideal 109.5° to 60°.

10.72 Ö=Cl(=O)(=O)—Ö—Cl(=O)(=O)=Ö

The Cl—O—Cl bond angle is less than 109.5° because of the lone pairs. **10.76**(a) −1267 kJ/mol (b) −1226 kJ/mol (c) −1234.8 kJ/mol. The two answers differ by less than 10 kJ/mol. This is very good agreement since average bond energies were used to calculate answers a and b. (d) −37 kJ

10.79 H—Ö—C(=O)—C(=O)—Ö—H

10.82 CH_4: −409 kJ/mol O_2; H_2S: −398 kJ/mol O_2

10.84 (a) The OH species only has 7 valence electrons, which is less than an octet, and 1 electron is unpaired. (b) 426 kJ (c) 508 kJ **10.86**(a) The F atoms will substitute at the axial positions first. (b) PF_5 and PCl_3F_2

10.88 $H_2C_2O_4$: H—Ö—C(=O)—C(=O)—Ö—H

$HC_2O_4^-$: [resonance structures] ⟷ [resonance structures]$^-$

$C_2O_4^{2-}$: [resonance structures]$^{2-}$ ⟷ [ ]$^{2-}$ ⟷ [ ]$^{2-}$ ⟷ [ ]$^{2-}$

In $H_2C_2O_4$, there are 2 shorter and stronger C=O bonds and 2 longer and weaker C—O bonds. In $HC_2O_4^-$, the carbon-oxygen bonds on the side retaining the H remain as 1 long, weak C—O and 1 short, strong C=O. The carbon-oxygen bonds on the other side of the molecule have resonance forms with a bond order of 1.5, so they are intermediate in length and strength. In $C_2O_4^{2-}$, all the carbon-oxygen bonds have a bond order of 1.5. **10.91** 22 kJ **10.93** Trigonal planar molecules are nonpolar, so AY_3 cannot be that shape. Trigonal pyramidal molecules and T-shaped molecules are polar, so either could represent AY_3. **10.98**(a) 339 pm (b) 316 pm and 223 pm (c) 274 pm

Chapter 11

11.1(a) sp^2 (b) sp^3d^2 (c) sp (d) sp^3 (e) sp^3d **11.3** C has only 2*s* and 2*p* atomic orbitals, allowing for a maximum of four hybrid orbitals. Si has 3*s*, 3*p*, and 3*d* atomic orbitals, allowing it to form more than four hybrid orbitals. **11.5**(a) six, sp^3d^2 (b) four, sp^3 **11.7**(a) sp^2 (b) sp^2 (c) sp^2 **11.9** (a) sp^3 (b) sp^3 (c) sp^3 **11.11**(a) Si: one *s* and three *p* atomic orbitals form sp^3 hybrid orbitals. (b) C: one *s* and one *p* atomic orbitals form *sp* hybrid orbitals. **11.13**(a) S: one *s*, three *p*, and one *d* atomic orbital mix to form sp^3d hybrid orbitals. (b) N: one *s* and three *p* atomic orbitals mix to form sp^3 hybrid orbitals.

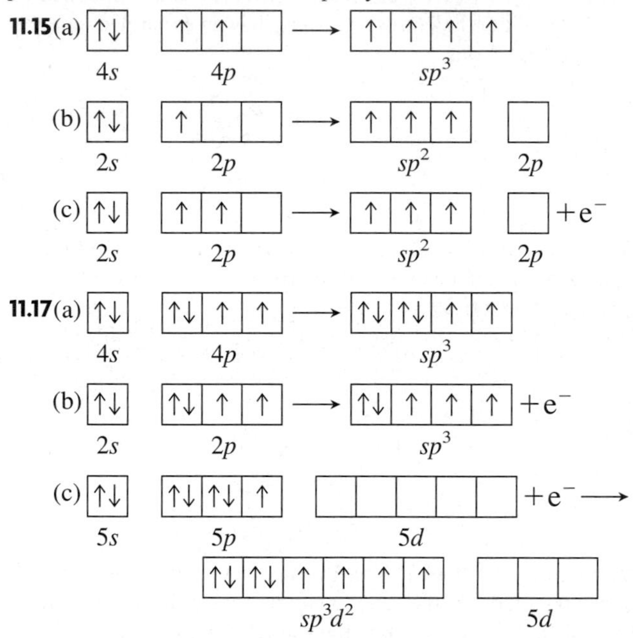

11.20(a) False. A double bond is one σ and one π bond. (b) False. A triple bond consists of one σ and two π bonds. (c) True (d) True (e) False. A π bond consists of a second pair of electrons after a σ bond has been previously formed. (f) False. End-to-end overlap results in a bond with electron density along the bond axis. **11.21**(a) Nitrogen sp^2 with three σ bonds and one π bond (b) Carbon *sp* with two σ bonds and two π bonds (c) Carbon sp^2 with three σ bonds and one π bond **11.23**(a) N: sp^2, forming 2 σ bonds and 1 π bond

:F̈—N̈=Ö:

(b) C: sp^2, forming 3 σ bonds and 1 π bond

(:F̈)(:F̈)C=C(F̈:)(F̈:)

(c) C: *sp*, forming 2 σ bonds and 2 π bonds

:N≡C—C≡N:

11.25

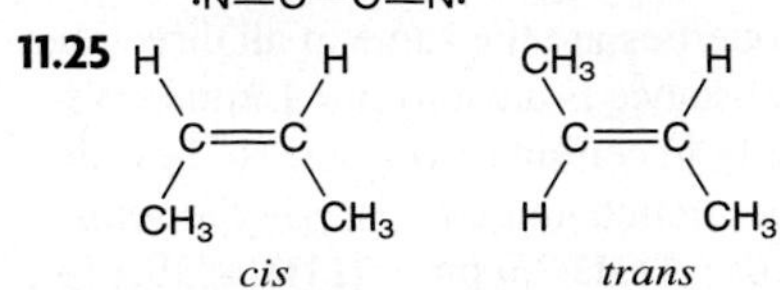

The single bonds are all σ bonds. The double bond is one σ bond and one π bond. **11.26** Four molecular orbitals form from the four *p* atomic orbitals. The total number of molecular orbitals must equal the number of atomic orbitals. **11.28**(a) Bonding MOs have lower energy than antibonding MOs. Lower energy = more stable (b) Bonding MOs do not have a nodal plane perpendicular to the bond. (c) Bonding MOs have higher electron density between nuclei than antibonding MOs. **11.30**(a) two (b) two (c) four

11.32

a) bonding $s + p$

antibonding $s - p$

b) bonding $p + p$

antibonding $p - p$

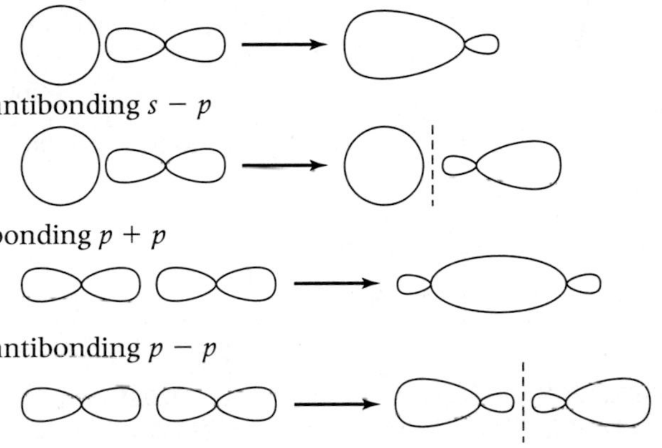

11.34(a) stable (b) paramagnetic (c) $(\sigma_{2s})^2(\sigma_{2s}^*)^1$ **11.36**(a) $C_2^+ < C_2 < C_2^-$ (b) $C_2^- < C_2 < C_2^+$ **11.40**(a) C (ring): sp^2; C (all others): sp^3; O (all): sp^3; N: sp^3 (b) 26 (c) 6 **11.42**(a) 17 (b) all carbons sp^2, the ring N is sp^2, the other N's are sp^3 **11.44**(a) B changes from sp^2 to sp^3. (b) P changes from sp^3 to sp^3d. (c) C changes from sp to sp^2. Two electron groups surround C in C_2H_2 and three electron groups surround C in C_2H_4. (d) Si changes from sp^3 to sp^3d^2. (e) no change for S **11.46** P: tetrahedral, sp^3; N: trigonal pyramid, sp^3; C_1 and C_2: tetrahedral, sp^3; C_3: trigonal planar, sp^2 **11.50** The central C is *sp* hybridized, and the other two C atoms are sp^2 hybridized.

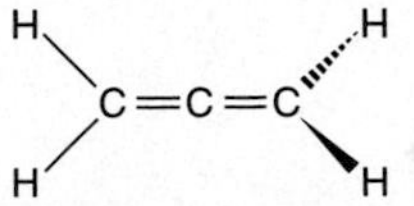

11.52(a) four (b) eight **11.56** Through resonance, the C—N bond gains some double-bond character, which hinders rotation about that bond.

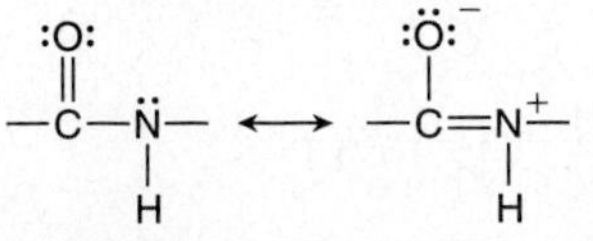

11.58(a) C in $—CH_3$: sp^3; all other C atoms: sp^2; O in two C—O bonds: sp^3; O in two C=O bonds: sp^2 (b) two (c) eight; one

Chapter 12

12.1 In a solid, the energy of attraction of the particles is greater than their energy of motion; in a gas, it is less. Gases have high compressibility and the ability to flow, while solids have neither. **12.4**(a) Because the intermolecular forces are only partially overcome when fusion occurs but need to be totally overcome in vaporization. (b) Because solids have greater intermolecular forces than liquids do. (c) $\Delta H_{vap} = -\Delta H_{cond}$ **12.5**(a) intermolecular (b) intermolecular (c) intramolecular (d) intramolecular **12.7**(a) condensation (b) fusion (c) vaporization **12.9** The gas molecules slow down as the gas is compressed. Therefore, much of the kinetic energy lost by the propane molecules is released to the surroundings. **12.13** At first, the vaporization of liquid molecules from the surface predominates, which increases the number of gas molecules and hence the vapor pressure. As more molecules enter the gas phase, gas molecules hit the surface of the liquid and "stick" more frequently, so the condensation rate increases. When the vaporization and condensation rates become equal, the vapor pressure becomes constant. **12.14** As intermolecular forces increase, (a) critical temperature increases (b) boiling point increases (c) vapor pressure decreases (d) heat of vaporization increases **12.18** because the condensation of the vapor supplies an additional 40.7 kJ/mol. **12.19** 4.16×10^3 J **12.21** 0.692 atm **12.23** 2×10^4 J/mol

12.25

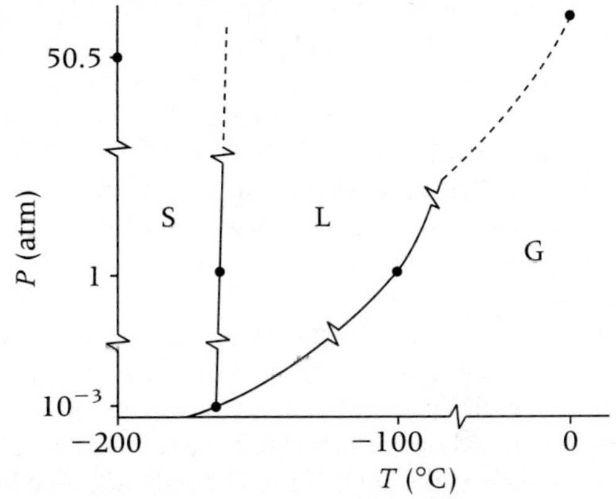

Solid ethylene is more dense than liquid ethylene.

12.28 42 atm **12.32** O is smaller and more electronegative than Se; so the electron density on O is greater, which attracts H more strongly. **12.34** All particles (atoms and molecules) exhibit dispersion forces, but the total force is weak in small molecules. Dipole-dipole forces in small polar molecules dominate the dispersion forces. **12.37**(a) hydrogen bonding (b) dispersion forces (c) dispersion forces **12.39**(a) dipole-dipole forces (b) dispersion forces (c) hydrogen bonding

12.41(a)

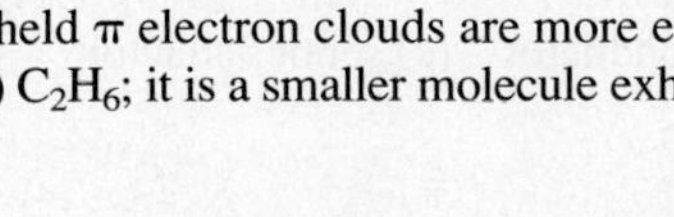

(b)

12.43(a) dispersion forces (b) hydrogen bonding (c) dispersion forces **12.45**(a) I^- (b) $CH_2{=}CH_2$ (c) H_2Se. In (a) and (c) the larger particle has the higher polarizability. In (b), the less tightly held π electron clouds are more easily distorted. **12.47**(a) C_2H_6; it is a smaller molecule exhibiting weaker dispersion forces than C_4H_{10}. (b) CH_3CH_2F; it has no H—F bonds, so it only exhibits dipole-dipole forces, which are weaker than the hydrogen bonds of CH_3CH_2OH. (c) PH_3; it has weaker intermolecular forces (dipole-dipole) than NH_3 (hydrogen bonding). **12.49**(a) Lithium chloride; it has ionic bonds versus dipole-dipole forces in HCl. (b) NH_3; it exhibits hydrogen bonding versus dipole-dipole forces in PH_3. (c) I_2; its molecules are more polarizable than Xe atoms because of their larger size. **12.51**(a) C_4H_8 (cyclobutane), because it is more compact than C_4H_{10}. (b) PBr_3; the dipole-dipole forces in PBr_3 are weaker than the ionic bonds in NaBr. (c) HBr; the dipole-dipole forces in HBr are weaker than the hydrogen bonds in water. **12.53** As atomic size decreases and electronegativity increases, the electron density of an atom increases. High electron density strengthens the attraction to an H atom on another molecule and makes its attached H atom more positive. Fluorine is the smallest of the three and the most electronegative, so its H bonds would be the strongest. Oxygen is smaller and more electronegative than nitrogen, so H bonds in water would be stronger than H bonds in ammonia. **12.57** The cohesive forces in water and mercury are stronger than the adhesive forces to the nonpolar wax on the floor. Weak adhesive forces result in spherical drops. The adhesive forces overcome the even weaker cohesive forces in the oil and so the oil drop spreads out. **12.59** Surface tension is defined as the energy needed to increase the surface area by a given amount, so units of energy per area are appropriate. **12.61** $CH_3CH_2CH_2OH < HOCH_2CH_2OH < HOCH_2CH(OH)CH_2OH$. More hydrogen bonding means more attraction between molecules so more energy is needed to increase surface area. **12.63** $CH_3CH_2CH_2OH < HOCH_2CH_2OH < HOCH_2CH(OH)CH_2OH$. More hydrogen bonding means more attraction between molecules, so they flow less easily. **12.68** Water is a good solvent for polar and ionic substances and a poor solvent for nonpolar substances. Water is a polar molecule and dissolves polar substances because their intermolecular forces are of similar strength. **12.69** A single water molecule can form four H bonds. The two hydrogen atoms each form one H bond to oxygen atoms on neighboring water molecules. The two lone pairs on the oxygen atom form H bonds with two hydrogen atoms on neighboring molecules. **12.72** Water exhibits strong capillary action, which allows it to be easily absorbed by the narrow spaces in the plant's roots and transported upward to the leaves. **12.78** simple cubic cell **12.81** The energy gap is the energy difference between the highest filled energy level (valence band) and the lowest unfilled energy level (conduction band). In conductors and superconductors, the energy gap is zero because the valence band overlaps the conduction band. In semiconductors, the energy gap is small. In insulators, the gap is large. **12.83** atomic mass and atomic radius **12.84**(a) face-centered cubic (b) body-centered cubic (c) face-centered cubic **12.86**(a) Tin, a metal, forms a metallic solid. (b) Silicon is in the same group as carbon, so it forms network covalent bonding. (c) Xenon is monatomic and forms an atomic solid. **12.88**(a) Metallic solid because nickel consists of metal atoms. (b) Molecular solid because F_2 exists as molecules. (c) Molecular solid because CH_3OH molecules are held together by hydrogen bonds. **12.90** four **12.92**(a) four Se^{2-} ions, four Zn^{2+} ions (b) 577.40 amu (c) 1.77×10^{-22} cm^3 (d) 5.61×10^{-8} cm **12.94**(a) insulator (b) conductor (c) semiconductor

12.96(a) Conductivity increases. (b) Conductivity increases. (c) Conductivity decreases. **12.98** 1.68×10^{-8} cm **12.105** A substance whose properties are the same in all directions is isotropic; otherwise, the substance is anisotropic. Liquid crystals have a degree of order only in certain directions, so they are anisotropic. **12. 111**(a) n-type semiconductor (b) p-type semiconductor **12.113** $n = 3.4\times10^3$ **12.115** 7.9 pm **12.118**(a) 19.8 torr (b) 0.0388 g
12.120(a)

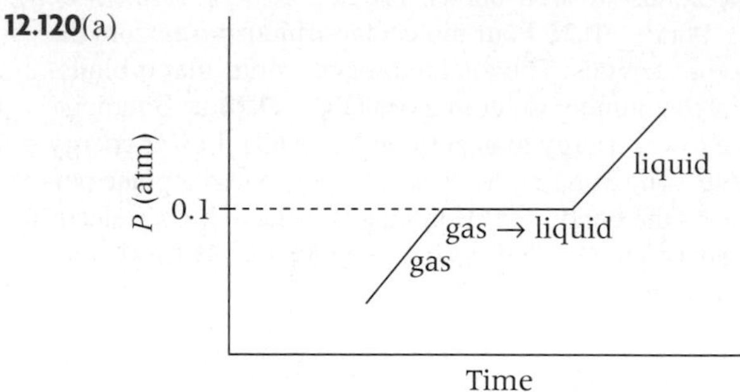

(b)

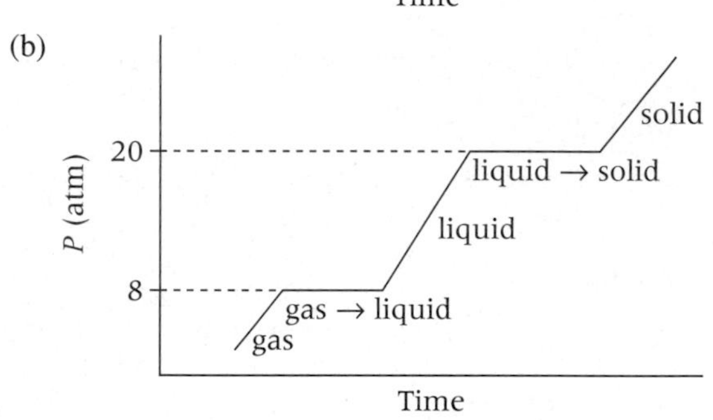

12.122 2.77×10^{-7} m; strikes on an edge **12.125** 259°C **12.127**(a) 2.52×10^{-2} atm (b) 5.26 L **12.130**(a) simple (b) 3.99 g/cm^3
12.134(a) furfuryl alcohol

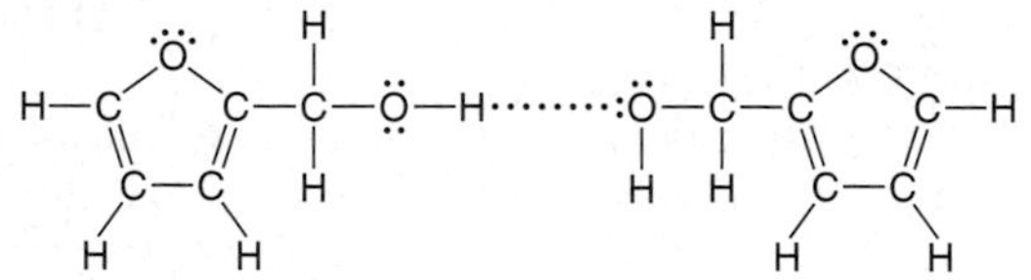

2-furoic acid

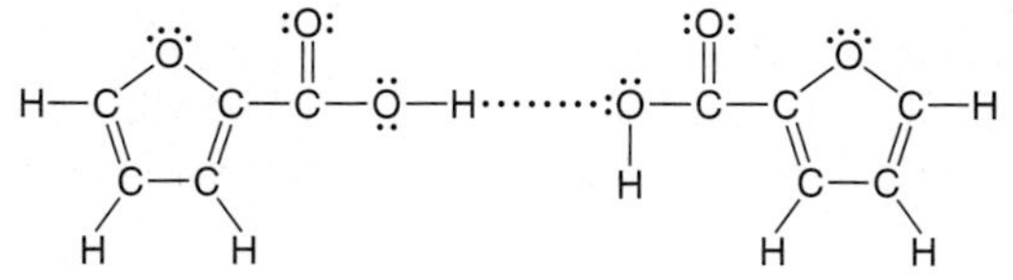

(b) furfuryl alcohol 2-furoic acid

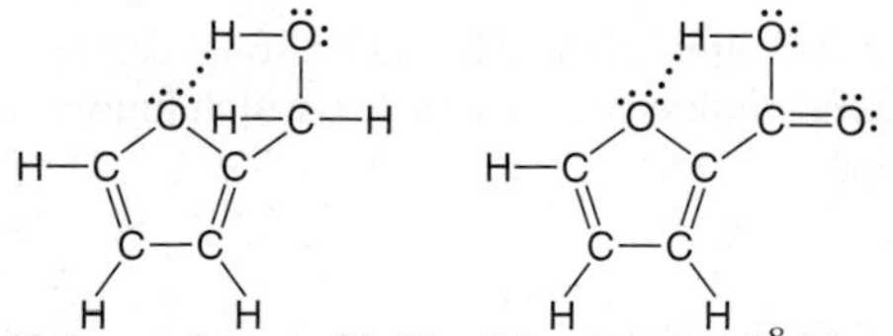

12.137(a) 49.3 metric tons H_2O (b) -1.11×10^8 kJ

12.139 2.9 g/m^3 **12.142**(a) 1.1 min (b) 10. min
(c)

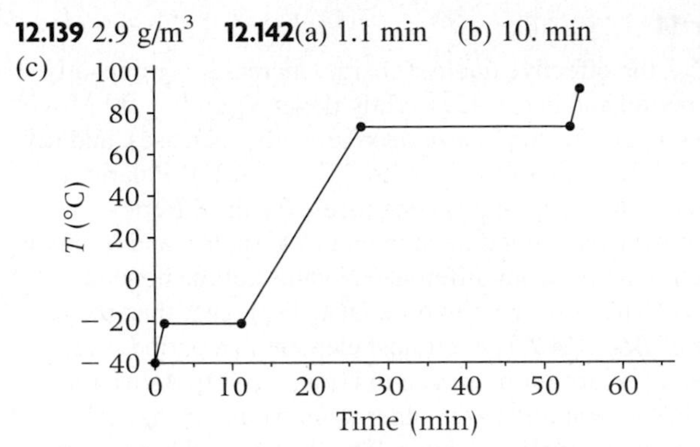

12.143 2.97×10^5 g BN **12.145**(a) $4r$ (b) $\sqrt{2}a$ (c) $a = 4r/\sqrt{3}$ (d) two (e) 0.68017 **12.147** 45.98 amu

Chapter 13

13.2 When a salt such as NaCl dissolves, ion-dipole forces cause the ions to separate, and many water molecules cluster around each ion in hydration shells. **13.4** Sodium stearate would be a more effective soap because the hydrocarbon chain in the stearate ion is longer than that in the acetate ion. The longer chain allows for more dispersion forces with grease molecules. **13.7**(a) KNO_3 is an ionic compound and is therefore more soluble in water. **13.9**(a) ion-dipole forces (b) hydrogen bonding (c) dipole–induced dipole forces **13.11**(a) hydrogen bonding (b) dipole–induced dipole forces (c) dispersion forces **13.13**(a) HCl(g), because the molecular interactions (dipole-dipole forces) in ether are like those in HCl but not like the ionic bonding in NaCl. (b) CH_3CHO(l), because the molecular interactions with ether (dipole-dipole) can replace those between CH_3CHO, but not the H bonds in water. (c) CH_3CH_2MgBr(s), because the molecular interactions (dispersion forces) are greater than between ether and the ions in $MgBr_2$. **13.16** Gluconic acid is soluble in water due to extensive hydrogen bonding from its —OH groups attached to five of its carbons. The dispersion forces in the nonpolar tail of caproic acid are more similar to the dispersion forces in hexane; thus, caproic acid is soluble in hexane. **13.18** The nitrogen-containing bases bond to their complementary bases. The flat, N-containing bases stack above each other, which allows extensive dispersion forces. The exterior negatively charged sugar-phosphate chains form ion-dipole and H bonds with water molecules in the aqueous surroundings, which also stabilizes the structure. **13.20** Dispersion forces are present between the nonpolar tails of the lipid molecules within the bilayer. Polar heads form H bonds and ion-dipole forces with the aqueous surroundings. **13.25** The energy change needed to separate the solvent ($\Delta H_{solvent}$), and that needed to mix the solvent and solute (ΔH_{mix}) combine to obtain $\Delta H_{solution}$.
13.29 Very soluble because a decrease in enthalpy and an increase in entropy both favor the formation of a solution.

13.30

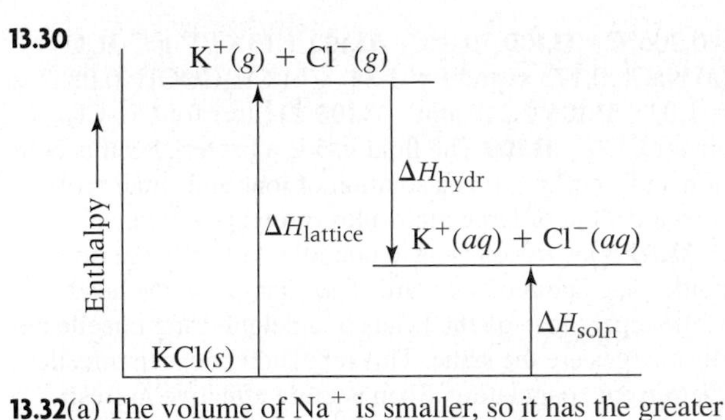

13.32(a) The volume of Na^+ is smaller, so it has the greater charge density. (b) Sr^{2+} has a larger ionic charge and a smaller volume, so it has the greater charge density. (c) Na^+ is smaller than Cl^-, so it has the greater charge density. (d) O^{2-} has a larger ionic charge with a similar ion volume, so it has the greater charge density. (e) OH^- has a smaller volume than SH^-, so it has the greater charge density. **13.34**(a) Na^+ (b) Sr^{2+} (c) Na^+ (d) O^{2-} (e) OH^- **13.36**(a) −704 kJ/mol (b) The K^+ ion contributes more because it is smaller and, therefore, has a greater charge density. **13.38**(a) increases (b) decreases (c) increases **13.41** Add a pinch of the solid solute to each solution. Addition of a "seed" crystal of solute to a supersaturated solution causes the excess solute to crystallize immediately, leaving behind a saturated solution. The solution in which the added solid solute dissolves is the unsaturated solution. The solution in which the added solid solute remains undissolved is the saturated solution. **13.44**(a) increase (b) decrease **13.46** (a) 0.0819 g O_2 (b) 0.0171 g O_2 **13.49** 0.20 mol/L **13.52**(a) molarity and % w/v or % v/v (b) parts-by-mass (% w/w) (c) molality **13.54** With just this information, you can convert between molality and molarity, but you need to know the molar mass of the solvent to convert to mole fraction. **13.56**(a) 1.24 M $C_{12}H_{22}O_{11}$ (b) 0.158 M $LiNO_3$ **13.58**(a) 0.0750 M NaOH (b) 0.31 M HNO_3 **13.60**(a) Add 4.22 g KH_2PO_4 to enough water to make 355 mL of aqueous solution. (b) Add 107 mL of 1.25 M NaOH to enough water to make 425 mL of solution. **13.62**(a) Weigh out 45.9 g KBr, dissolve it in about 1 L distilled water, and then dilute to 1.50 L with distilled water. (b) Measure 139 mL of the 0.244 M solution and add distilled water to make a total of 355 mL. **13.64**(a) 0.942 m glycine (b) 1.29 m glycerol **13.66** 3.09 m C_6H_6 **13.68**(a) Add 2.13 g $C_2H_6O_2$ to 298 g H_2O. (b) Add 0.0323 kg of 62.0% (w/w) HNO_3 to 0.968 kg H_2O to make 1.000 kg of 2.00% (w/w) HNO_3. **13.70**(a) 0.27 (b) 56 mass % (c) 21 m C_3H_7OH **13.72** 33.7 g CsCl; mole fraction = 7.67×10^{-3}; 6.74 mass % CsC **13.74** 5.11 m NH_3; 4.53 M NH_3; mole fraction = 0.0843 **13.76** 2.2 ppm Ca^{2+}, 0.66 ppm Mg^{2+} **13.80** It creates a "strong" current. A strong electrolyte dissociates completely in solution. **13.82** The boiling point is higher and the freezing point is lower for the solution compared to the solvent. **13.85** A dilute solution of an electrolyte behaves more ideally than a concentrated one. With increasing concentration, the effective concentration deviates from the molar concentration because of ionic attractions. Thus, 0.050 m NaF has a boiling point closer to its predicted value. **13.88**(a) strong electrolyte (b) strong electrolyte (c) nonelectrolyte (d) weak electrolyte **13.90**(a) 0.4 mol of solute particles (b) 0.14 mol (c) 3×10^{-4} mol (d) 0.07 mol **13.92**(a) CH_3OH in H_2O (b) H_2O in CH_3OH solution **13.94**(a) $\Pi_I = \Pi_{II} < \Pi_{III}$ (b) $bp_I = bp_{II} < bp_{III}$ (c) $fp_{III} < fp_I = fp_{II}$ (d) $vp_{III} < vp_I = vp_{II}$ **13.96** 23.36 torr

13.98 $-0.206°C$ **13.100** 79.0°C **13.102** 1.13×10^4 g $C_2H_6O_2$ **13.104**(a) NaCl: 0.173 *m* and $i = 1.84$ (b) CH_3COOH: 0.0837 *m* and $i = 1.02$ **13.106** 0.240 atm **13.108** 211 torr for CH_2Cl_2; 47.2 torr for CCl_4 **13.109** The fluid inside a bacterial cell is both a solution and a colloid. It is a solution of ions and small molecules and a colloid of large molecules (proteins and nucleic acids). **13.113** Soap micelles have nonpolar tails pointed inward and anionic heads pointed outward. The charges on the heads of one micelle repel those on the heads of a neighboring micelle because the charges are the same. This repulsion between micelles keeps them from coagulating. Soap is more effective in freshwater than in seawater because the divalent cations in seawater combine with the anionic heads to form a precipitate. **13.114** Foam comes from gas bubbles trapped in the cream. The best gas for making cream foam would be one that is not very soluble in the cream. Since nitrous oxide is better at making foam than carbon dioxide, nitrous oxide must be less soluble in the cream than carbon dioxide. Henry's law constant is larger for a gas that is more soluble, so the constant for carbon dioxide must be larger than the constant for nitrous oxide. **13.119** 3.5×10^9 L
13.126 (a) 89.9 g/mol (b) C_2H_5O; $C_4H_{10}O_2$.
(c) Forms H bonds

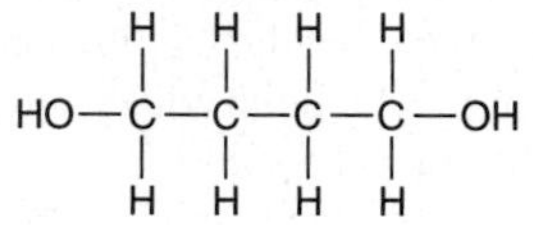

```
     H   H   H   H
     |   |   |   |
HO—C—C—C—C—OH
     |   |   |   |
     H   H   H   H
```

Does not form H bonds

```
   H       H       H   H
   |       |       |   |
H—C—O—C—O—C—C—H
   |       |       |   |
   H       H       H   H
```

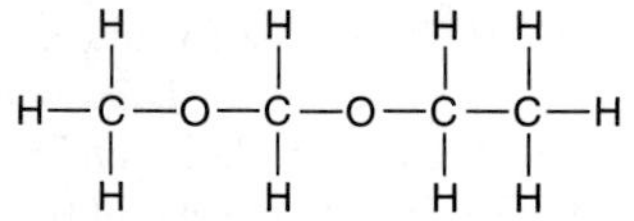

13.129(a) Mg^{2+} (b) Mg^{2+} (c) CO_3^{2-} (d) SO_4^{2-} (e) Fe^{3+} (f) Ca^{2+} **13.131**(a) 68 g/mol (b) 2.1×10^2 g/mol (c) The molar mass of CaN_2O_6 is 164.10 g/mol. This value is less than the 2.1×10^2 g/mol calculated when the compound is assumed to be a strong electrolyte and is greater than the 68 g/mol calculated when the compound is assumed to be a nonelectrolyte. Thus, the compound forms a non-ideal solution because the ions interact but do not dissociate completely in solution. (d) 2.4
13.135(a) 1.82×10^4 g/mol (b) 3.41×10^{-5}°C
13.136 9.4×10^5 ng/L **13.142** Weigh 3.11 g $NaHCO_3$ and dissolve in 247 g water. **13.146** $M = m$ (kg solvent/L solution) = $m \times d_{solution}$ Thus, for very dilute solutions molality × density = molarity. In an aqueous solution, the liters of solution have approximately the same value as the kg of solvent because the density of water is close to 1 kg/L, so $m = M$.
13.149(a) CH_4N_2O. The empirical formula mass is 60.06 g/mol. (b) 60. g/mol; CH_4N_2O. **13.152**(a) 0.088 cm^3/g H_2O (b) 1.5 cm^3/g H_2O·MPa (c) 0.74 cm^3/g H_2O, 11%
13.159(a) Yes, the phases of water can still coexist at some temperature and can therefore establish equilibrium. (b) The triple point would occur at a lower pressure and lower temperature because the dissolved air lowers the vapor pressure of the solvent. (c) Yes, this is possible because the gas-solid phase boundary exists below the new triple point. (d) No; at both temperatures, the presence of the solute lowers the vapor pressure of the liquid.

Chapter 14

14.2(a) Z_{eff}, the effective nuclear charge, increases significantly across a period and decreases slightly down a group. (b) Moving across a period, atomic size decreases, IE_1 increases, and EN increases, all because of the increased Z_{eff}. **14.3**(a) Polarity is the molecular property that is responsible for the difference in boiling points between iodine monochloride (polar) and bromine (nonpolar). It arises from different EN values of the bonded atoms. (b) The boiling point of polar ICl is higher than the boiling point of Br_2. **14.7** The leftmost element in a period is an alkali metal [except for hydrogen in Group 1A(1)]. As we move to the right, the bonding to the other atom would change from metallic to polar covalent to ionic. The IE, EA, and EN of the two atoms become increasingly different, causing changes in the character of the bond. **14.9** The elements on the left side in Period 3 form cations. The loss of electrons leads to an ion smaller than the atom. The elements on the right side form anions. The gain of electrons leads to an ion larger than the atom. **14.11**(a) The greater the electronegativity of the element, the more covalent the bonding is in its oxide. (b) The more electronegative the element, the more acidic the oxide is. **14.14** In general, network solids have very high melting and boiling points and are very hard, while molecular solids have low melting and boiling points and are soft. The properties of network solids reflect the necessity of breaking chemical bonds throughout the substances, whereas the properties of molecular solids reflect the weaker forces between individual molecules. **14.15**(a) $Mg < Sr < Ba$ (b) $Na < Al < P$ (c) $Se < Br < Cl$ (d) $Ga < Sn < Bi$
14.17(b) and (d) **14.19** $\ddot{O}=C=\ddot{O}$ **14.21** $CF_4 < SiCl_4 < GeBr_4$
14.23(a) $Na^+ < F^- < O^{2-}$ (b) $Cl^- < S^{2-} < P^{3-}$
14.25 $O < O^- < O^{2-}$
14.27(a) $CaO(s) + H_2O(l) \longrightarrow$
$Ca(OH)_2(s) \rightleftharpoons Ca^{2+}(aq) + 2OH^-(aq)$ (more basic)
$SO_3(g) + H_2O(l) \longrightarrow H_2SO_4(aq) \longrightarrow H^+(aq) + HSO_4^-(aq)$
(b) BaO (c) CO_2 (d) K_2O **14.29**(a) LiCl (b) PCl_3 (c) NCl_3
14.31(a) $Ar < Na < Si$ (b) $Cs < Rb < Li$ **14.33** The outermost electron is attracted by a smaller effective nuclear charge in Li because of shielding by the inner electrons, and it is farther from the nucleus in Li. Both of these factors lead to a lower ionization energy.
14.35(a) NH_3 will hydrogen bond.

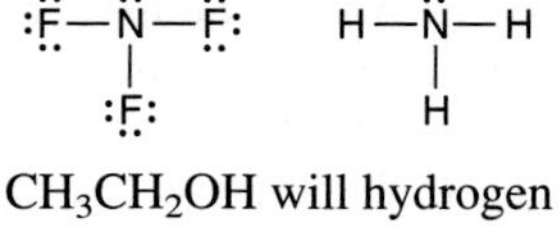

```
:F̤—N̈—F̤:     H—N̈—H
    |            |
   :F̤:           H
```

(b) CH_3CH_2OH will hydrogen bond.

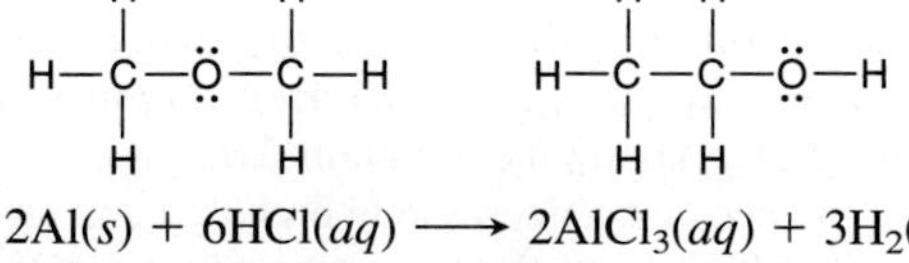

```
   H       H             H   H
   |       |             |   |
H—C—Ö—C—H        H—C—C—Ö—H
   |   ··  |             |   |
   H       H             H   H
```

14.37(a) $2Al(s) + 6HCl(aq) \longrightarrow 2AlCl_3(aq) + 3H_2(g)$
(b) $LiH(s) + H_2O(l) \longrightarrow LiOH(aq) + H_2(g)$
14.39(a) $NaBH_4$: +1 for Na, +3 for B, −1 for H
$Al(BH_4)_3$: +3 for Al, +3 for B, −1 for H
$LiAlH_4$: +1 for Li, +3 for Al, −1 for H
(b) tetrahedral

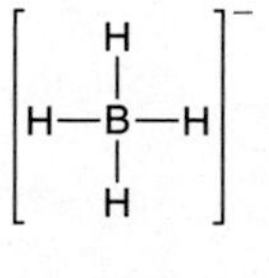

```
[     H    ]-
      |
   H—B—H
      |
[     H    ]
```

electron-rich O of water to the partially positive C. There is no such polarity in the alkene, so either C atom can be attacked.

$$CH_3-CH_2-C(=O)-CH_2-CH_3 + H_2O \xrightarrow{H^+} CH_3-CH_2-C(OH)(OH)-CH_2-CH_3$$

$$CH_3-CH_2-CH=CH-CH_3 + H_2O \xrightarrow{H^+} CH_3-CH_2-CH(OH)-CH_2-CH_3 + CH_3-CH_2-CH_2-CH(OH)-CH_3$$

15.56 Esters and acid anhydrides form through dehydration-condensation reactions, and water is the other product.

15.58(a) alkyl halide (b) nitrile (c) carboxylic acid (d) aldehyde

15.60(a) $CH_3-(CH=CH)-(CH_2-OH)$ alkene, alcohol

(b) $(Cl-CH_2)-C_6H_4-(C(=O)-OH)$ haloalkane, carboxylic acid

(c) cyclohexene ring with $C(=O)-NH-CH_3$: alkene, amide

(d) $(N\equiv C)-(CH_2-C(=O)-CH_3)$ nitrile, ketone

(e) cyclobutyl $-(C(=O)-O-CH_2)-CH_3$ ester

15.62 $H_3C-CH_2-CH_2-CH_2-CH_2-OH$

$H_3C-CH_2-CH_2-CH(OH)-CH_3$

$H_3C-CH(CH_3)-CH(OH)-CH_3$ $\quad H_3C-C(OH)(CH_3)-CH_2-CH_3$

$H_3C-CH_2-CH(OH)-CH_2-CH_3$ $\quad H_3C-CH(CH_3)-CH_2-CH_2-OH$

$H_3C-CH_2-CH(CH_3)-CH_2-OH$ $\quad H_3C-C(CH_3)_2-CH_2-OH$

15.64 $H_3C-CH_2-CH_2-CH_2-NH_2$ $\quad H_3C-CH_2-CH(NH_2)-CH_3$

$H_3C-CH(CH_3)-CH_2-NH_2$ $\quad H_3C-C(CH_3)_2-NH_2$

$H_3C-CH_2-NH-CH_2-CH_3$ $\quad H_3C-CH_2-CH_2-NH-CH_3$

$H_3C-CH_2-N(CH_3)-CH_3$ $\quad H_3C-CH(CH_3)-NH-CH_3$

15.66(a) $CH_3-C(=O)-CH_2-CH_3$ (b) $CH_3-CH(CH_3)-C(=O)-OH$

(c) cyclopentanone (five-membered ring with $=O$)

15.68(a) $CH_3-C(=O)-NH-CH_3$

(b) $CH_3-CH_2-CH_2-C(=O)-O-CH(CH_3)-CH_3$

(c) $H-C(=O)-O-CH_2-CH(CH_3)-CH_3$

15.70(a) $CH_3-(CH_2)_4-C(=O)-OH$ and $HO-CH_2-CH_3$

(b) $C_6H_5-C(=O)-OH$ and $HO-CH_2-CH_2-CH_3$

(c) CH_3-CH_2-OH and $HO-C(=O)-CH_2-CH_2-C_6H_5$

15.72(a) CH_3-CH_2-OH

$CH_3-CH_2-C(=O)-O-CH_2-CH_3$

(b) $CH_3-CH_2-CH(C\equiv N)-CH_3$

$CH_3-CH_2-CH(C(=O)OH)-CH_3$

15.74(a) hydrogen ion and propanoic acid (b) ethylamine **15.78** addition reactions and condensation reactions **15.81** Dispersion forces strongly attract the long, unbranched chains of high-density polyethylene (HDPE). Low-density polyethylene (LDPE) has branching in the chains that prevents packing and weakens the attractions. **15.83** An amine and a carboxylic acid react to form nylon; a carboxylic acid and an alcohol form a polyester.

15.84(a) $\left[\!-\mathrm{CH_2}-\mathrm{CHCl}-\!\right]_n$ (b) $\left[\!-\mathrm{CH_2}-\mathrm{CH(CH_3)}-\!\right]_n$

15.86 $n\mathrm{HO{-}C({=}O){-}C_6H_4{-}C({=}O){-}OH} + n\mathrm{HO{-}CH_2{-}CH_2{-}OH} \longrightarrow$

$\mathrm{HO}\left[\mathrm{C(=O)-C_6H_4-C(=O)-O-CH_2-CH_2-O}\right]_n\mathrm{H} + 2n - 1\ H_2O$

15.88(a) condensation (b) addition (c) condensation (d) condensation **15.90** The amino acid sequence in a protein determines its shape and structure, which determine its function. **15.93** The DNA base sequence determines the RNA base sequence, which determines the protein amino acid sequence.

15.94 (a) $-CH_3$ (b) $-CH_2$–(imidazolium ring, HN, $\overset{+}{N}H$) (c) $-CH_2-CH_2-S-CH_3$

15.96(a) $\mathrm{H_3\overset{+}{N}-CH(CH_2-C(=O)-O^-)-C(=O)-NH-CH(CH_2\text{-imidazolium})-C(=O)-NH-CH(CH_2\text{-indole})-C(=O)-O^-}$

(b) $\mathrm{H_3\overset{+}{N}-CH_2-C(=O)-NH-CH(CH_2-SH)-C(=O)-NH-CH(CH_2-C_6H_4-OH)-C(=O)-O^-}$

15.98(a) AATCGG (b) TCTGTA **15.100** ACAATGCCT; this sequence codes for three amino acids. **15.102**(a) Both R groups are from cysteine, which can form a disulfide bond (covalent bond). (b) Lysine and aspartic acid give a salt link. (c) Asparagine and serine will hydrogen bond. (d) Valine and phenylalanine interact through dispersion forces. **15.104**(a) Perform an acid-catalyzed dehydration of the alcohol followed by addition of Br_2. (b) Oxidize 1 mol of ethanol to acetic acid, then react 1 mol of acetic acid with 1 mol of ethanol to form the ester. **15.108** $CH_3-CH=CH-CH_3$; decolorization of bromine

15.110

$H_2N-CH_2-CH_2-CH_2-CH_2-CH_2-NH_2$

cadaverine

$H_2N-CH_2-CH_2-CH_2-CH_2-NH_2$

putrescine

$Br-CH_2CH_2-Br + 2CN^- \longrightarrow N{\equiv}C-CH_2CH_2-C{\equiv}N + 2Br^-$

$N{\equiv}C-CH_2CH_2-C{\equiv}N \xrightarrow{\text{reduction}} H_2N-CH_2CH_2CH_2CH_2-NH_2$

15.111(a)

(structure with labeled groups: alkene; ester; alkene; ester; ketone; carbons numbered 1–7)

(b) Carbon 1 is sp^2 hybridized. Carbon 2 is sp^3 hybridized. Carbon 3 is sp^3 hybridized. Carbon 4 is sp^2 hybridized. Carbon 5 is sp^3 hybridized. Carbons 6 and 7 are sp^2 hybridized.

(c) Carbons 2, 3, and 5 are chiral centers, as they are each bonded to four different groups.

15.113

(structure with labeled groups: aldehyde $H-C(=O)-$; ether $O-CH_3$; alcohol $-OH$)

The shortest carbon-oxygen bond is the double bond in the aldehyde group. **15.116** The resonance structures show that the bond between carbon and nitrogen will have some double bond character that restricts rotation around the bond.

$$\mathrm{-C(=\ddot{O}:)-\ddot{N}H-} \longleftrightarrow \mathrm{-C(-\ddot{O}:^-)=\overset{+}{N}H-}$$

15.120(a) Hydrolysis requires the addition of water to break the peptide bonds. (b) glycine, 4; alanine, 1; valine, 3; proline, 6; serine, 7; arginine, 49 (c) 10,700 g/mol **15.122** When 2-butanone is reduced, equal amounts of both isomers are produced because the reaction does not favor the production of one over the other.

Chapter 16

16.2 Reaction rate is proportional to concentration. An increase in pressure will increase the concentration, resulting in an increased reaction rate. **16.3** The addition of water will dilute the concentrations of all dissolved solutes, and the rate of the reaction will decrease. **16.5** An increase in temperature affects the rate of a reaction by increasing the number of collisions between particles, but more importantly, the energy of collisions increases. Both these factors increase the rate of reaction. **16.8**(a) The slope of the line joining any two points on a graph of concentration versus time gives the average rate between the two points. The closer the points, the closer the average rate will be to the instantaneous rate. (b) The initial rate is the instantaneous rate at the point on the graph where time = 0, that is, when reactants are mixed.

16.10

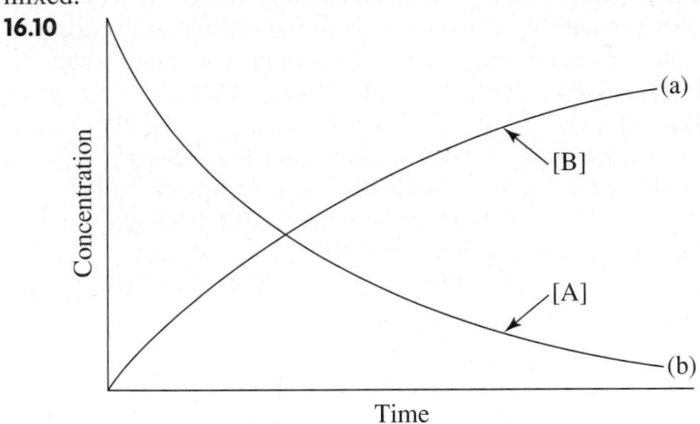

16.12(a) rate $= -\left(\frac{1}{2}\right)\frac{\Delta[AX_2]}{\Delta t}$

$= -\left(\frac{1}{2}\right)\frac{(0.0088\ M - 0.0500\ M)}{(20.0\text{ s} - 0\text{ s})}$

$= 0.0010\ M/\text{s}$

(b)

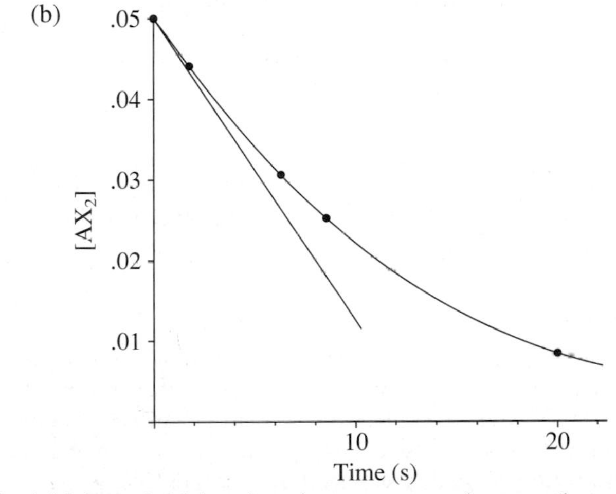

The initial rate is higher than the average rate because the rate will decrease as reactant concentration decreases.

16.14 rate $= -\frac{1}{2}\frac{\Delta[A]}{\Delta t} = \frac{\Delta[B]}{\Delta t} = \frac{\Delta[C]}{\Delta t}$; 4 mol/L·s

16.16 rate $= -\frac{\Delta[A]}{\Delta t} = -\frac{1}{2}\frac{\Delta[B]}{\Delta t} = \frac{\Delta[C]}{\Delta t}$; 0.2 mol/L·s

16.18 $2N_2O_5(g) \longrightarrow 4NO_2(g) + O_2(g)$

16.21 rate $= -\frac{\Delta[N_2]}{\Delta t} = -\frac{1}{3}\frac{\Delta[H_2]}{\Delta t} = \frac{1}{2}\frac{\Delta[NH_3]}{\Delta t}$

16.22(a) rate $= -\frac{1}{3}\frac{\Delta[O_2]}{\Delta t} = \frac{1}{2}\frac{\Delta[O_3]}{\Delta t}$ (b) 1.45×10^{-5} mol/L·s

16.23(a) k is the rate constant, the proportionality constant in the rate law; it is reaction and temperature specific. (b) m represents the order of the reaction with respect to [A], and n represents the order of the reaction with respect to [B]. The order of a reactant does not necessarily equal its stoichiometric coefficient in the balanced equation. (c) $L^2/mol^2\cdot min$ **16.25**(a) Rate doubles. (b) Rate decreases by a factor of four. (c) Rate increases by a factor of nine. **16.26** first order in BrO_3^-; first order in Br^-; second order in H^+; fourth order overall **16.28**(a) Rate doubles. (b) Rate is halved. (c) The rate increases by a factor of 16. **16.30** second order in NO_2; first order in Cl; third order overall **16.32**(a) Rate increases by a factor of 9. (b) Rate increases by a factor of 8. (c) Rate is halved. **16.34**(a) second order in A; first order in B (b) rate $= k[A]^2[B]$ (c) $5.00\times10^3\ L^2/mol^2\cdot min$ **16.36**(a) $time^{-1}$ (b) L/mol·time (c) $L^2/mol^2\cdot time$ (d) $L^{3/2}/mol^{3/2}\cdot time$ **16.39**(a) first order (b) second order (c) zero order **16.41** 7 s **16.43**(a) $k =$ 0.0660 min^{-1} (b) 21.0 min

16.45(a)

x-axis (time, s)	$[NH_3]$	y-axis (ln $[NH_3]$)
0	4.000 M	1.38629
1.000	3.986 M	1.38279
2.000	3.974 M	1.37977

(b) $t_{1/2} = 2\times10^2$ s

16.47 Measure the rate constant at a series of temperatures and plot ln k versus $1/T$. The slope of the line equals $-E_a/R$. **16.49** 0.033 s^{-1} **16.53** No, other factors that affect the fraction of collisions that lead to reaction are the energy and orientation of the collisions. **16.56** No, reaction is reversible and will eventually reach a state where the forward and reverse rates are equal. When this occurs, there are no concentrations equal to zero. Since some reactants are reformed from EF, the amount of EF will be less than 4×10^{-5} mol. **16.57** At the same temperature, both reaction mixtures have the same average kinetic energy, but not the same velocity. The trimethylamine molecule has greater mass than the ammonia molecule, so trimethylamine molecules will collide less often with HCl. Moreover, the bulky groups bonded to nitrogen in trimethylamine mean that collisions with HCl having the correct orientation occur less frequently. Therefore, the rate of the first reaction is greater. **16.58** 12 unique collisions **16.60** 2.96×10^{-18}

16.62(a)

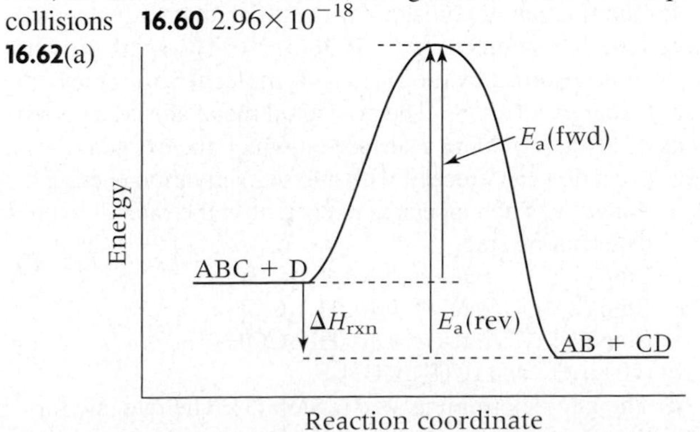

(b) 2.70×10^2 kJ/mol

(c)

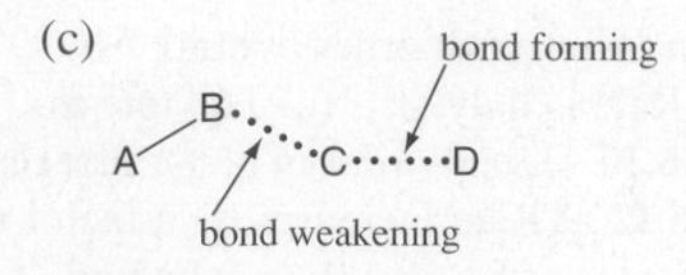

16.64(a) Because the enthalpy change is positive, the reaction is endothermic.

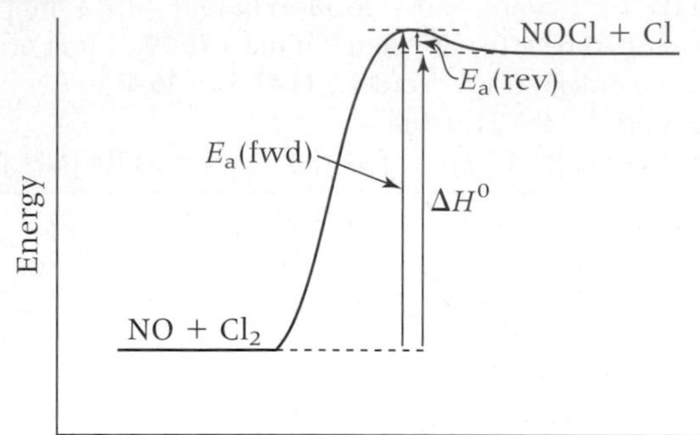

(b) 3 kJ (c) :C̈l·····C̈l·····N̈=Ö:

16.65 The rate of an overall reaction depends on the rate of the slowest step. The rate of the overall reaction will be slower than the average of the individual rates because the average includes faster rates as well. **16.69** The probability of three particles colliding with one another with the proper energy and orientation is much less than the probability for two particles. **16.70** No, the overall rate law must contain only reactants (no intermediates), and the overall rate is determined by the slow step.
16.72(a) $A(g) + B(g) + C(g) \longrightarrow D(g)$
(b) X and Y are intermediates.

(c)

Step	Molecularity	Rate Law
$A(g) + B(g) \rightleftharpoons X(g)$	bimolecular	$\text{rate}_1 = k_1[A][B]$
$X(g) + C(g) \longrightarrow Y(g)$	bimolecular	$\text{rate}_2 = k_2[X][C]$
$Y(g) \longrightarrow D(g)$	unimolecular	$\text{rate}_3 = k_3[Y]$

(d) yes (e) yes **16.74** The proposed mechanism is valid because the individual steps are chemically reasonable and add to give the overall equation. In addition, the rate law for the mechanism matches the observed rate law. **16.77** No. A catalyst changes the mechanism of a reaction to one with lower activation energy. Lower activation energy means a faster reaction. An increase in temperature does not influence the activation energy, but increases the fraction of collisions with sufficient energy to equal or exceed the activation energy. **16.78**(a) No. The spark provides energy that is absorbed by the H_2 and O_2 molecules to achieve the activation energy. (b) Yes. The powdered metal acts as a heterogeneous catalyst, providing a surface on which the oxygen-hydrogen reaction can proceed with a lower activation energy.
16.82(a) Water does not appear as a reactant in the rate-determining step.
(b) Step (1): $\text{rate}_1 = k_1[(CH_3)_3CBr]$
Step (2): $\text{rate}_2 = k_2[(CH_3)_3C^+]$
Step (3): $\text{rate}_3 = k_3[(CH_3)_3COH_2{}^+]$
(c) $(CH_3)_3C^+$ and $(CH_3)_3COH_2{}^+$
(d) The rate-determining step is Step (1). The rate law for this step agrees with the rate law observed with $k = k_1$.
16.85 4.61×10^4 J/mol **16.89**(a) Rate increases 2.5 times. (b) Rate is halved. (c) Rate decreases by a factor of 0.01. (d) Rate does not change. **16.92** second order **16.94** 7 times faster **16.96** 57 yr **16.99** (a) 0.21 h^{-1}; 3.3 h (b) 6.6 h (c) If the concentration of sucrose is relatively low, the concentration of water remains nearly constant even with small changes in the amount of water. This gives an apparent zero-order reaction with respect to water. Thus, the reaction is first order overall because the rate does not change with changes in the amount of water.
16.101(a) 0.68 M (b) 0.57 **16.104** 71 kPa **16.107** 7.3×10^3 J/mol **16.110**(a) 2.4×10^{-15} M (b) 2.4×10^{-11} mol/L·s **16.113**(a) 2.8 days (b) 7.4 days (c) 4.5 mol/m^3 **16.116**(a) $\text{rate}_1 = 1.7 \times 10^{-5}$ M s^{-1}; $\text{rate}_2 = 3.4 \times 10^{-5}$ M s^{-1}; $\text{rate}_3 = 3.4 \times 10^{-5}$ M s^{-1} (b) zero order with respect to $S_2O_8{}^{2-}$; first order with respect to I^- (c) 4.3×10^{-4} s^{-1} (d) rate = $(4.3 \times 10^{-4}\ s^{-1})$[KI] **16.119**(a) 7×10^4 cells/L (b) 2.0×10^1 min
16.122(a) Use the Monod equation.

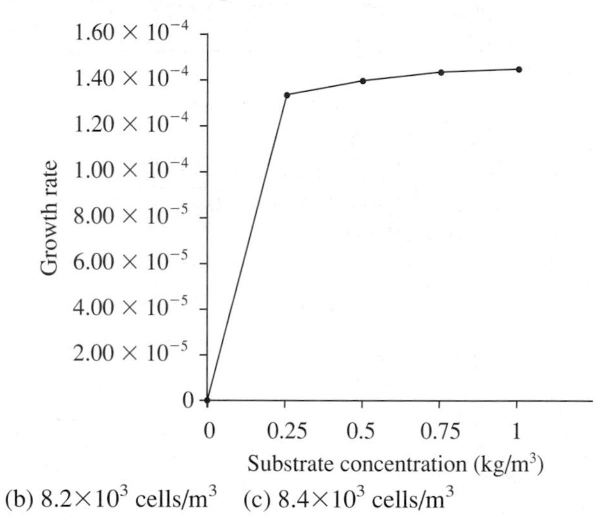

(b) 8.2×10^3 cells/m^3 (c) 8.4×10^3 cells/m^3

Chapter 17

17.1 If the change is one of concentrations, it results temporarily in more products and less reactants. After equilibrium is reestablished, the K_c remains unchanged because the ratio of products and reactants remains the same. If the change is one of temperature, [product] and K_c increase and [reactant] decreases.
17.5 This reaction is very exothermic, so it will probably have a large K. **17.7** The equilibrium constant expression is $K = [O_2]$. If the temperature remains constant, K remains constant. If the initial amount of Li_2O_2 present is sufficient to reach equilibrium, the amount of O_2 obtained will be constant.

17.8(a) $Q = \dfrac{[HI]^2}{[H_2][I_2]}$

Concentration
[HI]
[H_2]
Time

The value of Q increases as a function of time until it reaches the value of K. (b) no **17.11** Yes. If Q_1 is for the formation of

1 mol NH_3 from H_2 and N_2, and Q_2 is for the formation of NH_3 from H_2 and 1 mol of N_2, then $Q_2 = Q_1^2$.
17.12(a) $4NO(g) + O_2(g) \rightleftharpoons 2N_2O_3(g)$;
$$Q_c = \frac{[N_2O_3]^2}{[NO]^4[O_2]}$$
(b) $SF_6(g) + 2SO_3(g) \rightleftharpoons 3SO_2F_2(g)$;
$$Q_c = \frac{[SO_2F_2]^3}{[SF_6][SO_3]^2}$$
(c) $2SClF_5(g) + H_2(g) \rightleftharpoons S_2F_{10}(g) + 2HCl(g)$;
$$Q_c = \frac{[S_2F_{10}][HCl]^2}{[SClF_5]^2[H_2]}$$
17.14(a) $2NO_2Cl(g) \rightleftharpoons 2NO_2(g) + Cl_2(g)$;
$$Q_c = \frac{[NO_2]^2[Cl_2]}{[NO_2Cl]^2}$$
(b) $2POCl_3(g) \rightleftharpoons 2PCl_3(g) + O_2(g)$;
$$Q_c = \frac{[PCl_3]^2[O_2]}{[POCl_3]^2}$$
(c) $4NH_3(g) + 3O_2(g) \rightleftharpoons 2N_2(g) + 6H_2O(g)$;
$$Q_c = \frac{[N_2]^2[H_2O]^6}{[NH_3]^4[O_2]^3}$$
17.16(a) 7.9 (b) 3.2×10^{-5}
17.18(a) $2Na_2O_2(s) + 2CO_2(g) \rightleftharpoons 2Na_2CO_3(s) + O_2(g)$;
$$Q_c = \frac{[O_2]}{[CO_2]^2}$$
(b) $H_2O(l) \rightleftharpoons H_2O(g)$; $Q_c = [H_2O(g)]$
(c) $NH_4Cl(s) \rightleftharpoons NH_3(g) + HCl(g)$; $Q_c = [NH_3][HCl]$
17.20(a) $2NaHCO_3(s) \rightleftharpoons Na_2CO_3(s) + CO_2(g) + H_2O(g)$;
$Q_c = [CO_2][H_2O]$
(b) $SnO_2(s) + 2H_2(g) \rightleftharpoons Sn(s) + 2H_2O(g)$;
$$Q_c = \frac{[H_2O]^2}{[H_2]^2}$$
(c) $H_2SO_4(l) + SO_3(g) \rightleftharpoons H_2S_2O_7(l)$;
$$Q_c = \frac{1}{[SO_3]}$$
17.23(a) (1) $Cl_2(g) + F_2(g) \rightleftharpoons 2ClF(g)$
(2) $2ClF(g) + 2F_2(g) \rightleftharpoons 2ClF_3(g)$
overall: $Cl_2(g) + 3F_2(g) \rightleftharpoons 2ClF_3(g)$
$$\text{(b) } Q_{overall} = Q_1Q_2 = \frac{[ClF]^2}{[Cl_2][F_2]} \times \frac{[ClF_3]^2}{[ClF]^2[F_2]^2} = \frac{[ClF_3]^2}{[Cl_2][F_2]^3}$$
17.25 K_c and K_p are equal when $\Delta n_{gas} = 0$. **17.26**(a) smaller (b) Assuming that $RT > 1$ ($T > 12.2$ K), $K_p > K_c$ if there are more moles of products than reactants at equilibrium, and $K_p < K_c$ if there are more moles of reactants than products. **17.27**(a) 3 (b) −1 (c) 3 **17.29**(a) 3.2 (b) 28.5 **17.31**(a) 0.15 (b) 3.6×10^{-7} **17.33** The reaction quotient (Q) and equilibrium constant (K) are determined by the ratio [products]/[reactants]. When $Q < K$, the reaction proceeds to the right to form more products. **17.35** No, to the left **17.38** At equilibrium, equal concentrations of $CFCl_3$ and HCl exist, regardless of starting reactant concentrations, because the product coefficients are equal. **17.40**(a) The approximation applies when the change in concentration from initial concentration to equilibrium concentration is so small that it is insignificant; this occurs when K is small and initial concentration is large. (b) This approximation should not be used when the change in concentration is greater than 5%. This can occur when $[\text{reactant}]_{initial}$ is very small or when change in [reactant] is relatively large due to a large K. **17.41** 50.8
17.43

Concentration (M)	$PCl_5(g)$	$\rightleftharpoons$ $PCl_3(g)$	+ $Cl_2(g)$
Initial	0.075	0	0
Change	$-x$	$+x$	$+x$
Equilibrium	$0.075 - x$	x	x

17.45 28 atm **17.47** 0.33 atm **17.49** 3.5×10^{-3} M **17.51** $[I_2] =$ 0.00328 M; [HI] = 0.0152 M **17.53** $[I_2]_{eq} = [Cl_2]_{eq} = 0.0200$ M; $[ICl]_{eq} = 0.060$ M **17.55** 6.01×10^{-6} **17.58** Equilibrium position refers to the specific concentrations or pressures of reactants and products that exist at equilibrium, whereas equilibrium constant is the overall ratio of equilibrium concentrations or pressures. **17.59**(a) B, because the amount of product increases with temperature (b) A, because the lowest temperature will give the least product **17.62** A rise in temperature favors the forward direction of an endothermic reaction. The addition of heat makes K_2 larger than K_1. **17.63**(a) shifts toward products (b) shifts toward products (c) does not shift (d) shifts toward reactants **17.65**(a) more F and less F_2 (b) more C_2H_2 and H_2 and less CH_4 **17.67**(a) no effect (b) less H_2 and O_2 and more H_2O **17.69**(a) no change (b) increase volume
17.71(a) amount decreases (b) amount increases (c) amount increases (d) amount decreases **17.73** 2.0 **17.76**(a) lower temperature; higher pressure (b) Q decreases; no change in K (c) Reaction rates are lower at lower temperatures so a catalyst is used to speed up the reaction. **17.79**(a) 4.82×10^{-5} atm (b) 19.6 μg/L **17.81** 0.204 atm **17.84**(a) 3×10^{-3} atm (b) high pressure; low temperature (c) 2×10^5 (d) No, because water condenses at a higher temperature. **17.89**(a) 0.016 atm (b) $K_c = 5.6\times10^2$; $P_{SO_2} = 0.16$ atm **17.91** 12.5 g $CaCO_3$
17.95 Both concentrations increased by a factor of 2.2.
17.97(a) 3.0×10^{-14} atm (b) 0.013 pg CO/L **17.100**(a) 98.0% (b) 99.0% (c) 2.60×10^5 J/mol **17.102**(a) $2CH_4(g) + O_2(g) + 2H_2O(g) \rightleftharpoons 2CO_2(g) + 6H_2(g)$ (b) 1.76×10^{29} (c) 3.19×10^{23} (d) 48 atm **17.104**(a) 4.0×10^{-21} atm (b) 5.5×10^{-8} atm (c) 29 N atoms/L; 4.0×10^{14} H atoms/L (d) $N_2(g) + H(g) \longrightarrow NH(g) + N(g)$ **17.106**(a) $P_{N_2} = 31$ atm; $P_{H_2} = 93$ atm; $P_{total} = 174$ atm; (b) $P_{N_2} = 18$ atm; $P_{H_2} = 111$; $P_{total} = 179$ atm ; not a valid argument **17.108**(a) $P_{N_2} =$ 0.780 atm; $P_{O_2} = 0.210$ atm; $P_{NO} = 2.67\times10^{-16}$ atm (b) 0.990 atm (c) $K_c = K_p = 4.35\times10^{-31}$ **17.110**(a) 1.26×10^{-3} (b) 794 (c) −51.8 kJ (d) 1.2×10^4 J/mol **17.112**(a) 5 (b) 8 (c) Once sufficient F was produced, neither branch would proceed. (d) The lower branch would not proceed once sufficient F was made. **17.115**(a) 1.52 (b) 0.9626 atm (c) 0.2000 mol CO (d) 0.01128 M

Chapter 18

18.2 All Arrhenius acids contain hydrogen in their formula and produce hydronium ion (H_3O^+) in aqueous solution. All Arrhenius bases produce hydroxide ion (OH^-) in aqueous solution. Neutralization occurs when each H_3O^+ ion combines with an OH^- ion to form two molecules of H_2O. Chemists found the reaction of any strong base with any strong acid always produced 56 kJ/mol ($\Delta H = -56$ kJ/mol), which was consistent with Arrhenius' hypothesis describing neutralization. **18.4** Strong acids and bases dissociate completely into ions when dissolved in water. Weak acids and bases dissociate only partially. The characteristic property of all weak acids is that a significant number of the acid molecules are undissociated. **18.5**(a), (c), and (d) **18.7** (b) and (d)

18.9(a) $K_a = \frac{[CN^-][H_3O^+]}{[HCN]}$

(b) $K_a = \frac{[CO_3^{2-}][H_3O^+]}{[HCO_3^-]}$

(c) $K_a = \frac{[HCOO^-][H_3O^+]}{[HCOOH]}$

18.11(a) $K_a = \frac{[NO_2^-][H_3O^+]}{[HNO_2]}$

(b) $K_a = \frac{[CH_3COO^-][H_3O^+]}{[CH_3COOH]}$

(c) $K_a = \frac{[BrO_2^-][H_3O^+]}{[HBrO_2]}$

18.13 $CH_3COOH < HF < HIO_3 < HI$ **18.15**(a) weak acid (b) strong base (c) weak acid (d) strong acid **18.17**(a) strong base b) strong acid (c) weak acid (d) weak acid **18.22**(a) The acid with the smaller K_a (4×10^{-5}) has the higher pH, because less dissociation yields fewer hydronium ions. (b) The acid with the larger pK_a (3.5) has the high pH, because a larger pK_a means a smaller K_a. (c) Lower concentration (0.01 M) gives fewer hydronium ions. (d) A 0.1 M weak acid solution gives fewer hydronium ions. (e) The 0.1 M base solution has a lower concentration of hydronium ions. (f) The pOH = 6.0 because pH = 14.0 − 6.0 = 8.0 **18.23**(a) 12.05; basic (b) 11.09; acidic **18.25**(a) 2.298; acidic (b) −0.708; basic **18.27**(a) $[H_3O^+] = 1.7\times10^{-10}$ M, pOH = 4.22, $[OH^-] = 6.0\times10^{-5}$ M (b) pH = 3.57, $[H_3O^+] = 2.7\times10^{-4}$ M, $[OH^-] = 3.7\times10^{-11}$ M **18.29**(a) $[H_3O^+] = 1.7\times10^{-3}$ M, pOH = 11.23, $[OH^-] = 5.9\times10^{-12}$ M (b) pH = 8.82, $[H_3O^+] = 1.5\times10^{-9}$ M, $[OH^-] = 6.6\times10^{-6}$ M **18.31** 3.4×10^{-4} mol OH^-/L **18.33** 6×10^{-5} mol OH^- **18.36**(a) Rising temperature increases the value of K_w. (b) $K_w = 2.5\times10^{-14}$; pOH = 6.80; $[OH^-] = 1.6\times10^{-7}$ M **18.37** The Brønsted-Lowry theory defines acids as proton donors and bases as proton acceptors, while the Arrhenius definition looks at acids as containing ionizable hydrogen atoms and at bases as containing hydroxide ions. In both definitions, an acid produces hydronium ions and a base produces hydroxide ions when added to water. Ammonia and carbonate ion are two Brønsted-Lowry bases that are not Arrhenius bases because they do not contain hydroxide ions. Brønsted-Lowry acids must contain an ionizable hydrogen atom in order to be proton donors, so a Brønsted-Lowry acid is also an Arrhenius acid. **18.40** An amphoteric species can act as either an acid or a base. The dihydrogen phosphate ion, $H_2PO_4^-$, is an example.

18.41(a) $H_3PO_4(aq) + H_2O(l) \rightleftharpoons H_2PO_4^-(aq) + H_3O^+(aq)$;

$K_a = \frac{[H_3O^+][H_2PO_4^-]}{[H_3PO_4]}$

(b) $C_6H_5COOH(aq) + H_2O(l) \rightleftharpoons C_6H_5COO^-(aq) + H_3O^+(aq)$;

$K_a = \frac{[H_3O^+][C_6H_5COO^-]}{[C_6H_5COOH]}$

(c) $HSO_4^-(aq) + H_2O(l) \rightleftharpoons SO_4^{2-}(aq) + H_3O^+(aq)$;

$K_a = \frac{[H_3O^+][SO_4^{2-}]}{[HSO_4^-]}$

18.43(a) Cl^- (b) HCO_3^- (c) OH^- **18.45** (a) NH_4^+ (b) NH_3 (c) $C_{10}H_{14}N_2H^+$

18.47(a) $HCl + H_2O \rightleftharpoons Cl^- + H_3O^+$
acid base base acid
Conjugate acid-base pairs: HCl/Cl^- and H_3O^+/H_2O

(b) $HClO_4 + H_2SO_4 \rightleftharpoons ClO_4^- + H_3SO_4^+$
acid base base acid
Conjugate acid-base pairs: $HClO_4/ClO_4^-$ and $H_3SO_4^+/H_2SO_4$

(c) $HPO_4^{2-} + H_2SO_4 \rightleftharpoons H_2PO_4^- + HSO_4^-$
base acid acid base
Conjugate acid-base pairs: H_2SO_4/HSO_4^- and $H_2PO_4^-/HPO_4^{2-}$

18.49(a) $NH_3 + H_3PO_4 \rightleftharpoons NH_4^+ + H_2PO_4^-$
base acid acid base
Conjugate acid-base pairs: $H_3PO_4/H_2PO_4^-$ and NH_4^+/NH_3

(b) $CH_3O^- + NH_3 \rightleftharpoons CH_3OH + NH_2^-$
base acid acid base
Conjugate acid-base pairs: NH_3/NH_2^- and CH_3OH/CH_3O^-

(c) $HPO_4^{2-} + HSO_4^- \rightleftharpoons H_2PO_4^- + SO_4^{2-}$
base acid acid base
Conjugate acid-base pairs: HSO_4^-/SO_4^{2-} and $H_2PO_4^-/HPO_4^{2-}$

18.51(a) $OH^-(aq) + H_2PO_4^-(aq) \rightleftharpoons H_2O(l) + HPO_4^{2-}(aq)$
Conjugate acid-base pairs: $H_2PO_4^-/HPO_4^{2-}$ and H_2O/OH^-

(b) $HSO_4^-(aq) + CO_3^{2-}(aq) \rightleftharpoons SO_4^{2-}(aq) + HCO_3^-(aq)$
Conjugate acid-base pairs: HSO_4^-/SO_4^{2-} and HCO_3^-/CO_3^{2-}

18.53 $K_c > 1$: $HS^- + HCl \rightleftharpoons H_2S + Cl^-$
$K_c < 1$: $H_2S + Cl^- \rightleftharpoons HS^- + HCl$

18.55 $K_c > 1$ for both (a) and (b) **18.57** $K_c < 1$ for both (a) and (b) **18.59**(a) A strong acid is 100% dissociated, so the acid concentration will be very different after dissociation. (b) A weak acid dissociates to a very small extent, so the acid concentration before and after dissociation is nearly the same. (c) same as (b), but with the extent of dissociation greater. (d) same as (a) **18.60** No. HCl is a strong acid and dissociates to a greater extent than the weak acid CH_3COOH. The K_a of the acid, not the concentration of H_3O^+, determines the strength of the acid. **18.63** 1.5×10^{-5} **18.65** $[H_3O^+] = [NO_2^-] = 1.9\times10^{-2}$ M; $[OH^-] = 5.3\times10^{-13}$ M **18.67** $[H_3O^+] = [ClCH_2COO^-] = 0.038$ M; $[ClCH_2COOH] = 1.01$ M; pH = 1.42 **18.69**(a) $[H_3O^+] = 7.5\times10^{-3}$ M; pH = 2.12; $[OH^-] = 1.3\times10^{-12}$ M; pOH = 11.88 (b) 2.3×10^{-4} **18.71** 3.5×10^{-7} **18.73** (a) 2.47 (b) 11.41 **18.75** (a) 2.378 (b) 12.570 **18.77** 1.6% **18.79** $[H_3O^+] = [HS^-] = 9\times10^{-5}$ M; pH = 4.0; $[OH^-] = 1\times10^{-10}$ M; pOH = 10.0; $[H_2S] = 0.10$ M; $[S^{2-}] = 1\times10^{-17}$ M **18.82** 1.9% **18.83** All Brønsted-Lowry bases contain at least one lone pair of electrons, which binds an H^+ and allows the base to act as a proton acceptor.

18.86(a) $C_5H_5N(aq) + H_2O(l) \rightleftharpoons OH^-(aq) + C_5H_5NH^+(aq)$;

$K_b = \frac{[C_5H_5NH^+][OH^-]}{[C_5H_5N]}$

(b) $CO_3^{2-}(aq) + H_2O(l) \rightleftharpoons OH^-(aq) + HCO_3^-(aq)$;

$K_b = \frac{[HCO_3^-][OH^-]}{[CO_3^{2-}]}$

18.88(a) $HONH_2(aq) + H_2O(l) \rightleftharpoons OH^-(aq) + HONH_3^+(aq)$;
$K_b = \frac{[HONH_3^+][OH^-]}{[HONH_2]}$
(b) $HPO_4^{2-}(aq) + H_2O(l) \rightleftharpoons H_2PO_4^-(aq) + OH^-(aq)$;
$K_b = \frac{[H_2PO_4^-][OH^-]}{[HPO_4^{2+}]}$
18.90 11.71 **18.92** 11.34 **18.94**(a) 5.6×10^{-10} (b) 2.5×10^{-5}
18.96 (a) 12.04 (b) 10.77 **18.98** (a) 10.95 (b) 5.62
18.100 (a) 8.73 (b) 4.58 **18.102** $[OH^-] = 4.8\times10^{-4}$ *M*; pH = 10.68 **18.104** As a nonmetal becomes more electronegative, the acidity of its binary hydride increases. The electronegative nonmetal attracts the electrons more strongly in the polar bond, shifting the electron density away from H, thus making the H^+ more easily transferred to a water molecule to form H_3O^+.
18.107 Chlorine is more electronegative than iodine, and $HClO_4$ has more oxygen atoms than HIO. **18.108**(a) H_2SeO_4 (b) H_3PO_4 (c) H_2Te **18.110** (a) H_2Se (b) $B(OH)_3$ (c) $HBrO_2$ **18.112**(a) 0.05 *M* $Al_2(SO_4)_3$ (b) 0.1 *M* $PbCl_2$
18.114(a) 0.1 *M* $Ni(NO_3)_2$ (b) 0.1 *M* $Al(NO_3)_3$ **18.117** NaF contains the anion of the weak acid HF, so F^- acts as a base. NaCl contains the anion of the strong acid HCl.
18.119(a) $KBr(s) + H_2O(l) \longrightarrow K^+(aq) + Br^-(aq)$; neutral
(b) $NH_4I(s) + H_2O(l) \longrightarrow NH_4^+(aq) + I^-(aq)$
$NH_4^+(aq) + H_2O(l) \rightleftharpoons NH_3(aq) + H_3O^+(aq)$; acidic
(c) $KCN(s) + H_2O(l) \longrightarrow K^+(aq) + CN^-(aq)$
$CN^-(aq) + H_2O(l) \rightleftharpoons HCN(aq) + OH^-(aq)$; basic
18.121(a) $Na_2CO_3(s) + H_2O(l) \longrightarrow 2Na^+(aq) + CO_3^{2-}(aq)$
$CO_3^{2-}(aq) + H_2O(l) \rightleftharpoons HCO_3^-(aq) + OH^-(aq)$; basic
(b) $CaCl_2(s) + H_2O(l) \longrightarrow Ca^{2+}(aq) + 2Cl^-(aq)$; neutral
(c) $Cu(NO_3)_2(s) + H_2O(l) \longrightarrow Cu^{2+}(aq) + 2NO_3^-(aq)$
$Cu(H_2O)_6^{2+}(aq) + H_2O(l) \rightleftharpoons Cu(H_2O)_5OH^+(aq) + H_3O^+(aq)$; acidic
18.123(a) $SrBr_2(s) + H_2O(l) \longrightarrow Sr^{2+}(aq) + 2Br^-(aq)$; neutral
(b) $Ba(CH_3COO)_2(s) + H_2O(l) \longrightarrow Ba^{2+}(aq) + 2CH_3COO^-(aq)$
$CH_3COO^-(aq) + H_2O(l) \rightleftharpoons CH_3COOH(aq) + OH^-(aq)$; basic
(c) $(CH_3)_2NH_2Br(s) + H_2O(l) \longrightarrow (CH_3)_2NH_2^+(aq) + Br^-(aq)$
$(CH_3)_2NH_2^+(aq) + H_2O(l) \rightleftharpoons (CH_3)_2NH(aq) + H_3O^+(aq)$; acidic
18.125(a) $NH_4^+(aq) + H_2O(l) \rightleftharpoons NH_3(aq) + H_3O^+(aq)$
$PO_4^{3-}(aq) + H_2O(l) \rightleftharpoons HPO_4^{2-}(aq) + OH^-(aq)$;
$K_b > K_a$; basic
(b) $SO_4^{2-}(aq) + H_2O(l) \rightleftharpoons HSO_4^-(aq) + OH^-(aq)$;
Na^+ gives no reaction; basic
(c) $ClO^-(aq) + H_2O(l) \rightleftharpoons HClO(aq) + OH^-(aq)$;
Li^+ gives no reaction; basic
18.127(a) $Fe(NO_3)_2 < KNO_3 < K_2SO_3 < K_2S$
(b) $NaHSO_4 < NH_4NO_3 < NaHCO_3 < Na_2CO_3$
18.129 Since both bases produce OH^- ions in water, both bases appear equally strong. $CH_3O^-(aq) + H_2O(l) \longrightarrow OH^-(aq) + CH_3OH(aq)$ and $NH_2^-(aq) + H_2O(l) \longrightarrow OH^-(aq) + NH_3(aq)$. **18.131** Ammonia, NH_3, is a more basic solvent than H_2O. In a more basic solvent, weak acids such as HF act like strong acids and are 100% dissociated. **18.133** A Lewis acid is an electron pair acceptor while a Brønsted-Lowry acid is a proton donor. The proton of a Brønsted-Lowry acid fits the definition of a Lewis acid because it accepts an electron pair when it bonds with a base. All Lewis acids are not Brønsted-Lowry acids. A Lewis base is an electron pair donor and a Brønsted-Lowry base is a proton acceptor. All Brønsted-Lowry bases can be Lewis bases, and vice versa. **18.134**(a) No, $Zn(H_2O)_6^{2+}(aq) + 6NH_3(aq) \rightleftharpoons Zn(NH_3)_6^{2+} + 6H_2O(l)$; NH_3 is a weak Brønsted-Lowry base, but a strong Lewis base.
(b) cyanide ion and water (c) cyanide ion **18.137**(a) Lewis acid (b) Lewis base (c) Lewis acid (d) Lewis base **18.139**(a) Lewis acid (b) Lewis base (c) Lewis base (d) Lewis acid
18.141(a) Lewis acid: Na^+; Lewis base: H_2O
(b) Lewis acid: CO_2; Lewis base: H_2O
(c) Lewis acid: BF_3; Lewis base: F^-
18.143(a) Lewis (b) Brønsted-Lowry and Lewis (c) none (d) Lewis **18.147** 3.5×10^{-8} to 4.5×10^{-8} *M* H_3O^+; 5.2×10^{-7} to 6.6×10^{-7} *M* OH^- **18.148**(a) Acids vary in the extent of dissociation depending on the acid-base character of the solvent. (b) Methanol is a weaker base than water since phenol dissociates less in methanol than in water. (c) $C_6H_5OH(\text{solvated}) + CH_3OH(l) \rightleftharpoons CH_3OH_2^+(\text{solvated}) + C_6H_5O^-(\text{solvated})$
(d) $CH_3OH(l) + CH_3OH(l) \rightleftharpoons CH_3O^-(\text{solvated}) + CH_3OH_2^+(\text{solvated})$; $K = [CH_3O^-][CH_3OH_2^+]$ **18.151**(a) $SnCl_4$ is the Lewis acid; $(CH_3)_3N$ is the Lewis base (b) 5*d*
18.152 pH = 5.00, 6.00, 6.79, 6.98, 7.00 **18.155** H_3PO_4
18.159 3×10^{18} **18.161** (a) $[CH_3O^-] = 4\times10^{-9}$ *M* (b) 4×10^{-16}
18.163 10.43 **18.165** 2.41 **18.168** amylase, 2×10^{-7} *M*; pepsin, 1×10^{-2} *M*; trypsin, 3×10^{-10} *M* **18.172** 0.0227 **18.174**(a) Ca^{2+} does not react with water; $CH_3CH_2COO^-(aq) + H_2O(l) \rightleftharpoons CH_3CH_2COOH(aq) + OH^-(aq)$; basic (b) 9.03
18.179 4.5×10^{-5} **18.182**(a) The concentration of oxygen is higher in the lungs so the equilibrium shifts to the right. (b) In an oxygen-deficient environment the equilibrium shifts to the left to release oxygen. (c) A decrease in $[H_3O^+]$ shifts the equilibrium to the right. More oxygen is absorbed, but it will be more difficult to remove the O_2. (d) An increase in $[H_3O^+]$ shifts the equilibrium to the left. Less oxygen is bound to Hb, but it will be easier to remove it. **18.185**(a) 1.012 *M* (b) 0.004 *M* (c) 0.4%
18.186(a) 10.0 (b) The pK_b for the 3° amine group is much smaller than that for the aromatic ring, thus the K_b is significantly larger (yielding a much greater amount of OH^-). (c) 4.6 (d) 5.1

Chapter 19

19.2 The acid component neutralizes added base and the base component neutralizes added acid so the pH of the buffer solution remains relatively constant. The components of a buffer do not neutralize one another because they are a conjugate acid-base pair. **19.4** The pH of a buffer decreases only slightly with added H_3O^+. **19.7** The buffer range, the pH over which the buffer acts effectively, is greatest when the buffer-component ratio is 1; the range decreases as the component ratio deviates from 1.
19.9(a) Ratio and pH increase; added OH^- reacts with HA.
(b) Ratio and pH decrease; added H^+ reacts with A^-. (c) Ratio and pH increase; added A^- increases $[A^-]$. (d) Ratio and pH decrease; added HA increases [HA]. **19.11** $[H_3O^+] = 7.8\times10^{-6}$ *M*; pH = 5.11 **19.13** $[H_3O^+] = 5.5\times10^{-4}$ *M*; pH = 3.26
19.15 3.80 **19.17** 9.92 **19.19** 9.55 **19.21**(a) K_{a2} (b) 10.47
19.23 1.7 **19.25** 0.17 **19.27** 3.37 **19.29** 8.79 **19.31**(a) 4.91 (b) 0.66 g KOH **19.33**(a) $HCOOH/HCOO^-$ or $C_6H_5NH_2/C_6H_5NH_3^+$ (b) $H_2PO_4^-/HPO_4^{2-}$ or

$H_2AsO_4^-/HAsO_4^{2-}$ **19.35**(a) $H_3AsO_4/H_2AsO_4^-$ or $H_3PO_4/H_2PO_4^-$ (b) $C_5H_5N/C_5H_5NH^+$ **19.38** 1.6 **19.40** To see a distinct color in a mixture of two colors, you need one to have about 10 times the intensity of the other. For this to be the case, the concentration ratio [HIn]/[In$^-$] has to be greater than 10:1 or less than 1:10. This occurs when pH = pK_a − 1 or pH = pK_a + 1, respectively, giving a pH range of about two units. **19.42** The equivalence point in a titration is the point at which the number of moles of base equals the number of moles of acid. The endpoint is the point at which the added indicator changes color. If an appropriate indicator is selected, the endpoint is close to the equivalence point, but they are not usually the same. The endpoint, or color change, may precede or follow the equivalence point, depending on the indicator chosen. **19.44**(a) initial pH: *strong acid*–strong base < *weak acid*–strong base < strong acid–*weak base* (b) equivalence point: strong acid–*weak base* < *strong acid*–strong base < *weak acid*–strong base **19.46** At the center of the buffer region, the concentrations of weak acid and conjugate base are equal, so the pH = pK_a of the acid. **19.48** pH range from 7.3 to 9.3 **19.50**(a) bromthymol blue (b) thymol blue or phenolphthalein **19.52**(a) methyl red (b) bromthymol blue **19.54**(a) 1.00 (b) 1.48 (c) 3.00 (d) 4.00 (e) 7.00 (f) 10.00 (g) 11.96 **19.56**(a) 2.91 (b) 4.81 (c) 5.29 (d) 6.09 (e) 7.40 (f) 8.76 (g) 10.10 (h) 12.05 **19.58**(a) 59.0 mL and 8.54 (b) 45.2 mL and 7.14, total 90.4 mL and 9.69 **19.60**(a) 96.2 mL and 5.16 (b) 146 mL and 5.78 **19.63** Fluoride ion is the conjugate base of a weak acid and reacts with H_2O: $F^-(aq) + H_2O(l) \rightleftharpoons HF(aq) + OH^-(aq)$. As the pH increases, the equilibrium shifts to the left and [F$^-$] increases. As the pH decreases, the equilibrium shifts to the right and [F$^-$] decreases. The changes in [F$^-$] influence the solubility of CaF_2. Chloride ion is the conjugate base of a strong acid so it does not react with water and its concentration is not influenced by pH. **19.65** The compound precipitates. **19.66**(a) $K_{sp} = [Ag^+]^2[CO_3^{2-}]$ (b) $K_{sp} = [Ba^{2+}][F^-]^2$ (c) $K_{sp} = [Cu^{2+}][HS^-][OH^-]$ **19.68**(a) $K_{sp} = [Ca^{2+}][CrO_4^{2-}]$ (b) $K_{sp} = [Ag^+][CN^-]$ (c) $K_{sp} = [Ni^{2+}][HS^-][OH^-]$ **19.70** 1.3×10^{-4} **19.72** 2.8×10^{-11} **19.74**(a) 2.3×10^{-5} *M* (b) 4.2×10^{-9} *M* **19.76**(a) 1.7×10^{-3} *M* (b) 2.0×10^{-4} *M* **19.78**(a) $Mg(OH)_2$ (b) PbS (c) Ag_2SO_4 **19.80**(a) $CaSO_4$ (b) $Mg_3(PO_4)_2$ (c) $PbSO_4$ **19.82**(a) $AgCl(s) \rightleftharpoons Ag^+(aq) + Cl^-(aq)$. The chloride ion is the anion of a strong acid, so it does not react with H_3O^+. No change with pH. (b) $SrCO_3(s) \rightleftharpoons Sr^{2+}(aq) + CO_3^{2-}(aq)$. The strontium ion is the cation of a strong base so pH will not affect its solubility. The carbonate ion acts as a base: $CO_3^{2-}(aq) + H_2O(l) \rightleftharpoons HCO_3^-(aq) + OH^-(aq)$; also $CO_2(g)$ forms and escapes: $CO_3^{2-}(aq) + 2H_3O^+(aq) \longrightarrow CO_2(g) + 2H_2O(l)$. Therefore, the solubility of $SrCO_3$ will increase with addition of H_3O^+ (decreasing pH). **19.84**(a) $Fe(OH)_2(s) \rightleftharpoons Fe^{2+}(aq) + 2OH^-(aq)$. The OH^- ion reacts with added H_3O^+: $OH^-(aq) + H_3O^+(aq) \longrightarrow 2H_2O(l)$. The added H_3O^+ consumes the OH^-, driving the equilibrium toward the right to dissolve more $Fe(OH)_2$. Solubility increases with addition of H_3O^+ (decreasing pH). (b) $CuS(s) + H_2O(l) \rightleftharpoons Cu^{2+}(aq) + HS^-(aq) + OH^-(aq)$. Both HS^- and OH^- are anions of weak acids, so both ions react with added H_3O^+. Solubility increases with addition of H_3O^+ (decreasing pH). **19.86** yes **19.88** yes **19.93** No, because it indicates that a complex ion forms between the lead ion and hydroxide ions: $Pb^{2+}(aq) + nOH^-(aq) \rightleftharpoons Pb(OH)_n^{2-n}(aq)$ **19.94** $Hg(H_2O)_4^{2+}(aq) + 4CN^-(aq) \rightleftharpoons Hg(CN)_4^{2-}(aq) + 4H_2O(l)$ **19.96** $Ag(H_2O)_2^+(aq) + 2S_2O_3^{2-}(aq) \rightleftharpoons Ag(S_2O_3)_2^{3-}(aq) + 2H_2O(l)$ **19.98** 1×10^{-5} *M* **19.100** 7.0×10^{-17} *M* Zn^{2+}; 0.024 *M* $Zn(CN)_4^{2-}$; 0.053 *M* CN^- **19.102** 9.4×10^{-5} *M* **19.104**(a) $Fe(OH)_3$ (b) The two metal ions are separated by adding just enough NaOH to precipitate iron(III) hydroxide. (c) 2.0×10^{-7} *M* **19.106** 0.121 L of 2.00 *M* NaOH and 0.379 L of 0.200 *M* HCOOH **19.108**(a) 0.99 (b) Assuming the volumes are additive: 0.401 L of 1.0 *M* HCOOH and 0.199 L of 1.0 *M* NaOH (c) 0.34 *M* **19.110** 1.3×10^{-4} *M* **19.114**(a) 14 (b) 1 g **19.117**(a) 0.88 (b) 0.14 **19.119** TRIS = 0.260 *M*; 8.53 **19.121** Lower the pH below 6.6. **19.124** 8×10^{-5}
19.127(a)

V (mL)	pH	ΔpH/ΔV	$V_{average}$ (mL)
0.00	1.00		
		0.022	5.00
10.00	1.22		
		0.026	15.00
20.00	1.48		
		0.037	25.00
30.00	1.85		
		0.066	32.50
35.00	2.18		
		0.18	37.00
39.00	2.89		
		0.62	39.25
39.50	3.20		
		1.2	39.63
39.75	3.50		
		2.7	39.83
39.90	3.90		
		6	39.93
39.95	4.20		
		18	39.97
39.99	4.90		
		200	40.00
40.00	7.00		
		200	40.01
40.01	9.40		
		10	40.03
40.05	9.80		
		10	40.08
40.10	10.40		
		0.67	40.18
40.25	10.50		
		1.2	40.38
40.50	10.79		
		0.60	40.75
41.00	11.09		
		0.17	43.00
45.00	11.76		
		0.058	47.50
50.00	12.05		
		0.025	55.00
60.00	12.30		
		0.013	65.00
70.00	12.43		
		0.009	75.00
80.00	12.52		

(b)

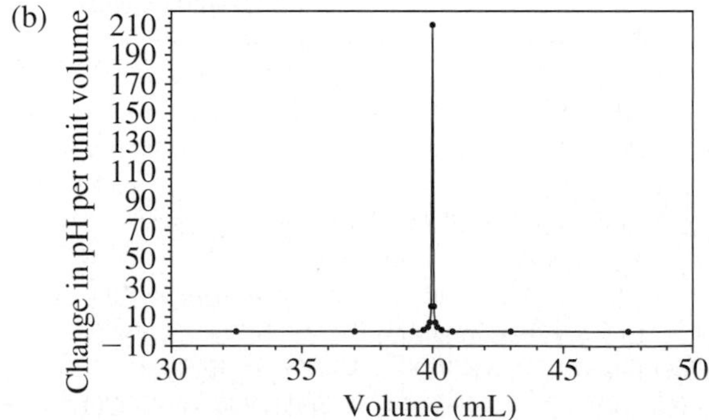

19.129 4.05

19.132 $K_b = \frac{[BH^+][OH^-]}{[B]}$

Rearranging to isolate $[OH^-]$:

$[OH^-] = K_b \frac{[B]}{[BH^+]}$

Taking the negative log:

$-\log [OH^-] = -\log K_b - \log \frac{[B]}{[BH^+]}$

Therefore, $pOH = pK_b + \log \frac{[BH^+]}{[B]}$

19.136 H_2CO_3/HCO_3^- and $H_2PO_4^-/HPO_4^{2-}$; 5.8 **19.139** 3.8 **19.141**(a) 58.2 mL (b) 7.4 mL (c) 6.30 **19.143** 170 mL **19.146**(a) 65 mol (b) 6.28 (c) 4.0×10^3 g **19.148** 5.68 **19.150** 3.9×10^{-9} μg Pb^{2+}/100 mL blood **19.152** No. NaCl will precipitate. **19.158**(a) A and D (b) $pH_A = 4.35$; $pH_B = 8.67$; $pH_C = 2.67$; $pH_D = 4.57$ (c) C, A, D, B (d) B

Chapter 20

20.2 A spontaneous process occurs by itself, whereas a nonspontaneous process requires a continuous input of energy to make it happen. It is possible to cause a nonspontaneous process to occur, but the process stops once the energy source is removed. A reaction that is nonspontaneous under one set of conditions may be spontaneous under a different set of conditions. **20.5** The transition from liquid to gas involves a greater increase in dispersal of energy and freedom of motion than does the transition from solid to liquid. **20.6** In an exothermic reaction, $\Delta S_{surr} > 0$. In an endothermic reaction, $\Delta S_{surr} < 0$. A chemical cold pack for injuries is an example of an application using a spontaneous endothermic process. **20.8**(a), (b), and (c) **20.10**(a) and (b) **20.12**(a) positive (b) negative (c) negative **20.14**(a) positive (b) positive (c) positive **20.16**(a) negative (b) negative (c) positive **20.18**(a) positive (b) negative (c) positive **20.20**(a) positive (b) negative (c) positive **20.22**(a) Butane. The double bond in 2-butene restricts freedom of rotation. (b) $Xe(g)$ because it has the greater molar mass (c) $CH_4(g)$. Gases have greater entropy than liquids. **20.24**(a) $C_2H_5OH(l)$ is a more complex molecule. (b) $KClO_3(aq)$. Ions in solution have their energy more dispersed than those in a solid. (c) $K(s)$, because it has a greater molar mass. **20.26**(a) diamond < graphite < charcoal. Freedom of motion is least in the network solid; more freedom between graphite sheets; most freedom in amorphous solid. (b) ice < liquid water < water vapor. Entropy increases as a substance changes from solid to liquid to gas. (c) O atoms < O_2 < O_3. Entropy increases with molecular complexity. **20.28**(a) $ClO_4^-(aq) > ClO_3^-(aq) > ClO_2^-(aq)$; decreasing molecular complexity (b) $NO_2(g) > NO(g) > N_2(g)$. N_2 has lower standard molar entropy because it consists of two of the same atoms; the other species have two different types of atoms. NO_2 is more complex than NO. (c) $Fe_3O_4(s) > Fe_2O_3(s) > Al_2O_3(s)$. Fe_3O_4 is more complex and more massive. Fe_2O_3 is more massive than Al_2O_3. **20.31** For a system at equilibrium, $\Delta S_{univ} = \Delta S_{sys} + \Delta S_{surr} = 0$. For a system moving to equilibrium, $\Delta S_{univ} > 0$. **20.32** $S^0_{Cl_2O(g)} = 2S^0_{HClO(g)} - S^0_{H_2O(g)} - \Delta S^0_{rxn}$ **20.33**(a) negative; $\Delta S^0 = -172.4$ J/K (b) positive; $\Delta S^0 = 141.6$ J/K (c) negative; $\Delta S^0 = -837$ J/K **20.35** $\Delta S^0 = 93.1$ J/K; yes, the positive sign of ΔS is expected because there is a net increase in the number of gas molecules. **20.37** $\Delta S^0 = -311$ J/K. Yes, the negative entropy change matches the decrease in moles of gas. **20.39** -75.6 J/K **20.41** -242 J/K **20.44** -97.2 J/K **20.46** A spontaneous process has $\Delta S_{univ} > 0$. Since the absolute temperature is always positive, ΔG_{sys} must be negative ($\Delta G_{sys} < 0$) for a spontaneous process. **20.48** ΔH^0_{rxn} is positive and ΔS^0_{sys} is positive. Melting is an example. **20.49** The entropy changes little within a phase. As long as the substance does not change phase, the value of ΔS^0 is relatively unaffected by temperature. **20.50**(a) -1138.0 kJ (b) -1379.4 kJ (c) -224 kJ **20.52**(a) -1138 kJ (b) -1379 kJ (c) -226 kJ **20.54**(a) Entropy decreases (ΔS^0 is negative) because the number of moles of gas decreases. The combustion of CO releases energy (ΔH^0 is negative). (b) -257.2 kJ or 257.3 kJ, depending on the method **20.56**(a) -0.409 kJ/mol·K (b) -197 kJ/mol **20.58**(a) $\Delta H^0_{rxn} = 90.7$ kJ; $\Delta S^0_{rxn} = 221$ J/K (b) at 38°C, $\Delta G^0 = 22.1$ kJ; at 138°C, $\Delta G^0 = 0.0$ kJ; at 238°C, $\Delta G^0 = -22.1$ kJ (c) For the substances in their standard states, the reaction is nonspontaneous at 38°C, near equilibrium at 138°C, and spontaneous at 238°C. **20.60** $\Delta H^0 = 30910$ J, $\Delta S^0 = 93.15$ J/K, $T = 331.8$ K **20.62**(a) $\Delta H^0_{rxn} = -241.826$ kJ, $\Delta S^0_{rxn} = -44.4$ J/K, $\Delta G^0_{rxn} = -228.60$ kJ (b) Yes. The reaction will become nonspontaneous at higher temperatures. (c) The reaction is spontaneous below 5.45×10^3 K. **20.64**(a) ΔG^0 is a relatively large positive value. (b) $K >> 1$. Q depends on initial conditions, not equilibrium conditions. **20.67** The standard free energy change, ΔG^0, applies when all components of the system are in their standard states; $\Delta G^0 = \Delta G$. **20.68**(a) 1.7×10^6 (b) 3.89×10^{-34} (c) 1.26×10^{48} **20.70**(a) 6.57×10^{173} (b) 4.46×10^{-15} (c) 3.46×10^4 **20.72** 4.89×10^{-51} **20.74** 3.36×10^5 **20.76** 2.7×10^4 J/mol; no **20.78**(a) 2.9×10^4 J/mol (b) The reverse direction, formation of reactants, is spontaneous so the reaction proceeds to the left. (c) 7.0×10^3 J/mol; the reaction proceeds to the left to reach equilibrium. **20.80**(a) no T (b) 163 kJ (c) 1×10^2 kJ/mol **20.83**(a) spontaneous (b) + (c) + (d) − (e) −, not spontaneous (f) − **20.87**(a) 2.3×10^2 (b) Administer oxygen-rich air to counteract the CO poisoning. **20.90** $-370.$ kJ **20.92**(a) $2N_2O_5(g) + 6F_2(g) \longrightarrow 4NF_3(g) + 5O_2(g)$ (b) $\Delta G^0_{rxn} = -569$ kJ (c) $\Delta G_{rxn} = -5.60\times10^2$ kJ/mol **20.95** $\Delta H^0_{rxn} = -137.14$ kJ; $\Delta S^0_{rxn} = -120.3$ J/K; $\Delta G^0_{rxn} = -101.25$ kJ **20.96**(a) $\Delta H^0_{rxn} = 470.5$ kJ; $\Delta S^0_{rxn} = 558.4$ J/K (b) The reaction will be spontaneous at high T, because the $-T\Delta S$ term will be larger in magnitude than ΔH. (c) no (d) 842.5 K **20.98**(a) yes, negative Gibbs free energy (b) Yes. It becomes spontaneous at 362.8 K. (c) 414 K. The temperature is different because the ΔH and ΔS values for N_2O_5 vary with physical state. **20.100**(a) 1.203×10^{23} molecules ATP/g glucose (b) 3.073×10^{23} molecules ATP/g tristearin **20.106**(a) 1.67×10^3 J/mol (b) 7.37×10^3 J/mol (c) -4.04×10^3 J/mol (d) 0.19 **20.110**(a) 465 K (b) 6.59×10^{-4} (c) The reaction rate is higher at the higher temperature. The shorter time required (kinetics) overshadows the lower yield (thermodynamics).

Chapter 21

21.1 Oxidation is the loss of electrons and results in a higher oxidation number; reduction is the gain of electrons and results in a lower oxidation number. **21.3** No, one half-reaction cannot take place independently because there is a transfer of electrons from

one substance to another. If one substance loses electrons, another substance must gain them. **21.6** To remove H^+ ions from an equation, add an equal number of OH^- ions to both sides to neutralize the H^+ ions and produce water. **21.8** Spontaneous reactions, $\Delta G_{sys} < 0$, take place in voltaic cells (also called galvanic cells). Nonspontaneous reactions, $\Delta G_{sys} > 0$, take place in electrolytic cells. **21.10**(a) Cl^- (b) MnO_4^- (c) MnO_4^- (d) Cl^- (e) from Cl^- to MnO_4^-
(f) $8H_2SO_4(aq) + 2KMnO_4(aq) + 10KCl(aq) \longrightarrow 2MnSO_4(aq) + 5Cl_2(g) + 8H_2O(l) + 6K_2SO_4(aq)$
21.12(a) $ClO_3^-(aq) + 6H^+(aq) + 6I^-(aq) \longrightarrow Cl^-(aq) + 3H_2O(l) + 3I_2(s)$
Oxidizing agent is ClO_3^- and reducing agent is I^-.
(b) $2MnO_4^-(aq) + H_2O(l) + 3SO_3^{2-}(aq) \longrightarrow 2MnO_2(s) + 3SO_4^{2-}(aq) + 2OH^-(aq)$
Oxidizing agent is MnO_4^- and reducing agent is SO_3^{2-}.
(c) $2MnO_4^-(aq) + 6H^+(aq) + 5H_2O_2(aq) \longrightarrow 2Mn^{2+}(aq) + 8H_2O(l) + 5O_2(g)$
Oxidizing agent is MnO_4^- and reducing agent is H_2O_2.
21.14(a) $Cr_2O_7^{2-}(aq) + 14H^+(aq) + 3Zn(s) \longrightarrow 2Cr^{3+}(aq) + 7H_2O(l) + 3Zn^{2+}(aq)$
Oxidizing agent is $Cr_2O_7^{2-}$ and reducing agent is Zn.
(b) $MnO_4^-(aq) + 3Fe(OH)_2(s) + 2H_2O(l) \longrightarrow MnO_2(s) + 3Fe(OH)_3(s) + OH^-(aq)$
Oxidizing agent is MnO_4^- and reducing agent is $Fe(OH)_2$.
(c) $2NO_3^-(aq) + 12H^+(aq) + 5Zn(s) \longrightarrow N_2(g) + 6H_2O(l) + 5Zn^{2+}(aq)$
Oxidizing agent is NO_3^- and reducing agent is Zn.
21.16(a) $4NO_3^-(aq) + 4H^+(aq) + 4Sb(s) \longrightarrow 4NO(g) + 2H_2O(l) + Sb_4O_6(s)$
Oxidizing agent is NO_3^- and reducing agent is Sb.
(b) $5BiO_3^-(aq) + 14H^+(aq) + 2Mn^{2+}(aq) \longrightarrow 5Bi^{3+}(aq) + 7H_2O(l) + 2MnO_4^-(aq)$
Oxidizing agent is BiO_3^- and is reducing agent is Mn^{2+}.
(c) $Pb(OH)_3^-(aq) + 2Fe(OH)_2(s) \longrightarrow Pb(s) + 2Fe(OH)_3(s) + OH^-(aq)$
Oxidizing agent is $Pb(OH)_3^-$ and reducing agent is $Fe(OH)_2$.
21.18(a) $5As_4O_6(s) + 8MnO_4^-(aq) + 18H_2O(l) \longrightarrow 20AsO_4^{3-}(aq) + 8Mn^{2+}(aq) + 36H^+(aq)$
Oxidizing agent is MnO_4^- and reducing agent is As_4O_6.
(b) $P_4(s) + 6H_2O(l) \longrightarrow 2HPO_3^{2-}(aq) + 2PH_3(g) + 4H^+(aq)$
P_4 is both the oxidizing agent and reducing agent.
(c) $2MnO_4^-(aq) + 3CN^-(aq) + H_2O(l) \longrightarrow 2MnO_2(s) + 3CNO^-(aq) + 2OH^-(aq)$
Oxidizing agent is MnO_4^- and reducing agent is CN^-.
21.21(a) $Au(s) + 3NO_3^-(aq) + 4Cl^-(aq) + 6H^+(aq) \longrightarrow AuCl_4^-(aq) + 3NO_2(g) + 3H_2O(l)$ (b) Oxidizing agent is NO_3^- and reducing agent is Au. (c) HCl provides chloride ions that combine with the gold(III) ion to form the stable $AuCl_4^-$ ion. **21.22**(a) A (b) E (c) C (d) A (e) E (f) E **21.25** An active electrode is a reactant or product in the cell reaction. An inactive electrode does not take part in the reaction and is present only to conduct a current. Platinum and graphite are commonly used as inactive electrodes. **21.26**(a) A (b) B (c) A (d) Hydrogen bubbles will form when metal A is placed in acid. Metal A is a better reducing agent than metal B, so if metal B reduces H^+ in acid, then metal A will also.

21.27(a) Oxidation: $Zn(s) \longrightarrow Zn^{2+}(aq) + 2e^-$
Reduction: $Sn^{2+}(aq) + 2e^- \longrightarrow Sn(s)$
Overall: $Zn(s) + Sn^{2+}(aq) \longrightarrow Zn^{2+}(aq) + Sn(s)$
(b)

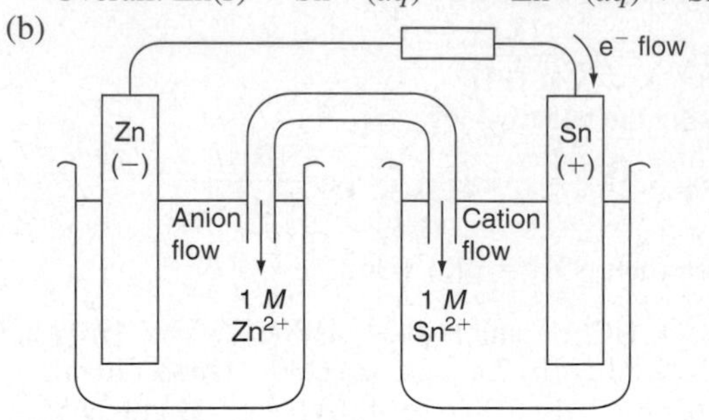

21.29(a) left to right (b) left (c) right (d) Ni (e) Fe (f) Fe (g) 1 *M* $NiSO_4$ (h) K^+ and NO_3^- (i) Fe (j) from right to left
(k) Oxidation: $Fe(s) \longrightarrow Fe^{2+}(aq) + 2e^-$
Reduction: $Ni^{2+}(aq) + 2e^- \longrightarrow Ni(s)$
Overall: $Fe(s) + Ni^{2+}(aq) \longrightarrow Fe^{2+}(aq) + Ni(s)$
21.31(a) Reduction: $Fe^{2+}(aq) + 2e^- \longrightarrow Fe(s)$
Oxidation: $Mn(s) \longrightarrow Mn^{2+}(aq) + 2e^-$
Overall: $Fe^{2+}(aq) + Mn(s) \longrightarrow Fe(s) + Mn^{2+}(aq)$
(b)

21.33(a) $Al(s) \mid Al^{3+}(aq) \parallel Cr^{3+}(aq) \mid Cr(s)$
(b) $Pt(s) \mid SO_2(g) \mid SO_4^{2-}(aq), H^+(aq) \parallel Cu^{2+}(aq) \mid Cu(s)$
21.36 A negative E^0_{cell} indicates that the redox reaction is not spontaneous, that is, $\Delta G^0 > 0$. The reverse reaction is spontaneous with $E^0_{cell} > 0$. **21.37** Similar to other state functions, E^0 changes sign when a reaction is reversed. Unlike ΔG^0, ΔH^0, and S^0, E^0 (the ratio of energy to charge) is an intensive property. When the coefficients in a reaction are multiplied by a factor, the values of ΔG^0, ΔH^0, and S^0 are multiplied by that factor. However, E^0 does not change because both the energy and charge are multiplied by the factor and thus their ratio remains unchanged.
21.38(a) Oxidation: $Se^{2-}(aq) \longrightarrow Se(s) + 2e^-$
Reduction: $2SO_3^{2-}(aq) + 3H_2O(l) + 4e^- \longrightarrow S_2O_3^{2-}(aq) + 6OH^-(aq)$
(b) $E^0_{anode} = E^0_{cathode} - E^0_{cell} = -0.57\ V - 0.35\ V = -0.92\ V$
21.40(a) $Br_2 > Fe^{3+} > Cu^{2+}$ (b) $Ca^{2+} < Ag^+ < Cr_2O_7^{2-}$
21.42(a) $Co(s) + 2H^+(aq) \longrightarrow Co^{2+}(aq) + H_2(g)$
$E^0_{cell} = 0.28$ V; spontaneous
(b) $2Mn^{2+}(aq) + 5Br_2(l) + 8H_2O(l) \longrightarrow 2MnO_4^-(aq) + 10Br^-(aq) + 16H^+(aq)$
$E^0_{cell} = -0.44$ V; not spontaneous
(c) $Hg_2^{2+}(aq) \longrightarrow Hg^{2+}(aq) + Hg(l)$
$E^0_{cell} = -0.07$ V; not spontaneous
21.44(a) $2Ag(s) + Cu^{2+}(aq) \longrightarrow 2Ag^+(aq) + Cu(s)$
$E^0_{cell} = -0.46$ V; not spontaneous
(b) $Cr_2O_7^{2-}(aq) + 3Cd(s) + 14H^+(aq) \longrightarrow 2Cr^{3+}(aq) + 3Cd^{2+}(aq) + 7H_2O(l)$
$E^0_{cell} = 1.73$ V; spontaneous

(c) $Pb(s) + Ni^{2+}(aq) \longrightarrow Pb^{2+}(aq) + Ni(s)$
$E^0_{cell} = -0.12$ V; not spontaneous
21.46 $3N_2O_4(g) + 2Al(s) \longrightarrow 6NO_2^-(aq) + 2Al^{3+}(aq)$
$E^0_{cell} = 0.867\text{ V} - (-1.66\text{ V}) = 2.53$ V
$2Al(s) + 3SO_4^{2-}(aq) + 3H_2O(l) \longrightarrow 2Al^{3+}(aq) + 3SO_3^{2-}(aq) + 6OH^-(aq)$
$E^0_{cell} = 2.59$ V
$SO_4^{2-}(aq) + 2NO_2^-(aq) + H_2O(l) \longrightarrow SO_3^{2-}(aq) + N_2O_4(g) + 2OH^-(aq)$
$E^0_{cell} = 0.06$ V
Oxidizing agents: $Al^{3+} < N_2O_4 < SO_4^{2-}$
Reducing agents: $SO_3^{2-} < NO_2^- < Al$
21.48 $2HClO(aq) + Pt(s) + 2H^+(aq) \longrightarrow Cl_2(g) + Pt^{2+}(aq) + 2H_2O(l)$
$E^0_{cell} = 0.43$ V
$2HClO(aq) + Pb(s) + SO_4^{2-}(aq) + 2H^+(aq) \longrightarrow Cl_2(g) + PbSO_4(s) + 2H_2O(l)$
$E^0_{cell} = 1.94$ V
$Pt^{2+}(aq) + Pb(s) + SO_4^{2-}(aq) \longrightarrow Pt(s) + PbSO_4(s)$
$E^0_{cell} = 1.51$ V
Oxidizing agent: $PbSO_4 < Pt^{2+} < HClO$
Reducing agent: $Cl_2 < Pt < Pb$
21.50 Yes; $C > A > B$ **21.53** $A(s) + B^+(aq) \longrightarrow A^+(aq) + B(s)$ with $Q = [A^+]/[B^+]$. (a) $[A^+]$ increases and $[B^+]$ decreases. (b) E_{cell} decreases. (c) $E_{cell} = E^0_{cell} - (RT/nF)\ln([A^+]/[B^+])$; $E_{cell} = E^0_{cell}$ when $(RT/nF)\ln([A^+]/[B^+]) = 0$. This occurs when $\ln([A^+]/[B^+]) = 0$, that is, $[A^+]$ equals $[B^+]$. (d) Yes, when $[A^+] > [B^+]$. **21.55** In a concentration cell, the overall reaction decreases the concentration of the more concentrated electrolyte because it is reduced in the cathode compartment. **21.56**(a) 3×10^{35} (b) 4×10^{-31} **21.58**(a) 1×10^{-67} (b) 6×10^{9} **21.60** (a) -2.03×10^5 J (b) 1.73×10^5 J **21.62**(a) -3.82×10^5 J (b) -5.6×10^4 J **21.64** $\Delta G^0 = -2.1\times10^4$ J; $E^0 = 0.22$ V **21.66** $E^0_{cell} = 0.056$ V; $\Delta G^0 = -1.1\times10^4$ J **21.68** 9×10^{-4} M **21.70** (a) 0.05 V (b) 0.60 M (c) $[Co^{2+}] = 0.91$ M; $[Ni^{2+}] = 0.09$ M **21.72** A; 0.085 V **21.74** Electrons flow from the anode, where oxidation occurs, to the cathode, where reduction occurs. The electrons always flow from the anode to the cathode no matter what type of battery. **21.76** A D-sized alkaline battery is much larger than an AAA-sized one, so the D-sized battery contains greater amounts of the cell components. The cell potential is an intensive property and does not depend on the amounts of the cell components. The total charge, however, depends on the amount of cell components so the D-sized battery produces more charge than the AAA-sized battery. **21.78** The Teflon spacers keep the two metals separated so the copper cannot conduct electrons that would promote the corrosion (rusting) of the iron skeleton. **21.81** Sacrificial anodes are made of metals with E^0 less than that of iron, -0.44 V, so they are more easily oxidized than iron. Only (b), (f), and (g) will work for iron. (a) will form an oxide coating that prevents further oxidation. (c) would react with groundwater quickly. **21.83** To reverse the reaction requires 0.34 V with the cell in its standard state. A 1.5 V cell supplies more than enough potential, so the cadmium metal is oxidized to Cd^{2+} and chromium plates out. **21.85** The oxidation number of N in NO_3^- is +5, the maximum O.N. for N. In the nitrite ion, NO_2^-, the O.N. of N is +3, so nitrogen can be further oxidized. **21.87**(a) Br_2 (b) Na **21.89** Anode is I_2 gas; cathode is magnesium (liquid). **21.91** Bromine gas forms at the anode; calcium metal forms at the cathode. **21.93** copper and bromine

21.95 iodine, zinc, and silver
21.97(a) Anode: $2H_2O(l) \longrightarrow O_2(g) + 4H^+(aq) + 4e^-$
Cathode: $2H_2O(l) + 2e^- \longrightarrow H_2(g) + 2OH^-(aq)$
(b) Anode: $2H_2O(l) \longrightarrow O_2(g) + 4H^+(aq) + 4e^-$
Cathode: $Sn^{2+}(aq) + 2e^- \longrightarrow Sn(s)$
21.99(a) Anode: $2H_2O(l) \longrightarrow O_2(g) + 4H^+(aq) + 4e^-$
Cathode: $NO_3^-(aq) + 4H^+(aq) + 3e^- \longrightarrow NO(g) + 2H_2O(l)$
(b) Anode: $2Cl^-(aq) \longrightarrow Cl_2(g) + 2e^-$
Cathode: $2H_2O(l) + 2e^- \longrightarrow H_2(g) + 2OH^-(aq)$
21.101(a) 2.93 mol e$^-$ (b) 2.83×10^5 C (c) 31.4 A **21.103** 0.282 g Ra **21.105** 1.10×10^4 s **21.107**(a) The sodium and sulfate ions make the water conductive so the current will flow through the water, facilitating electrolysis. Pure water, which contains very low (10^{-7} M) concentrations of H^+ and OH^-, conducts electricity very poorly. (b) The reduction of H_2O has a more positive half-potential than does the reduction of Na^+; the oxidation of H_2O is the only reaction possible because SO_4^{2-} cannot be oxidized. Thus, it is easier to reduce H_2O than Na^+ and easier to oxidize H_2O than SO_4^{2-}. **21.109** 44.2 g Zn **21.111**(a) 2.0×10^{11} C (b) 2.4×10^{11} J (c) 6.1×10^3 kg **21.114** 69.4 mass % Cu **21.116** (a) 13 days (b) 5.2×10^1 days (c) $310 **21.119**(a) 91 days (b) 50. g Ag (c) $8.90 **21.122**(a) Pb/Pb^{2+}: $E^0_{cell} = 0.13$ V; Cu/Cu^{2+}: $E^0_{cell} = 0.34$ V (b) The anode (negative electrode) is Pb. The anode in the other cell is platinum in the standard hydrogen electrode. (c) The precipitation of PbS decreases $[Pb^{2+}]$, which increases the potential. (d) -0.13 V **21.125** The three steps equivalent to the overall reaction $M^+(aq) + e^- \longrightarrow M(s)$ are
(1) $M^+(aq) \longrightarrow M^+(g)$ $\quad \Delta H$ is $-\Delta H_{hydration}$
(2) $M^+(g) + e^- \longrightarrow M(g)$ $\quad \Delta H$ is $-$IE
(3) $M(g) \longrightarrow M(s)$ $\quad \Delta H$ is $-\Delta H_{atomization}$
The energy for step 3 is similar for all three elements, so the difference in energy for the overall reactions depends on the values for $\Delta H_{hydration}$ and IE. The Li^+ ion has a much greater hydration energy than Na^+ and K^+ because it is smaller, with large charge density that holds the water molecules more tightly. The energy required to remove the waters surrounding the Li^+ offsets the lower ionization energy, making the overall energy for the reduction of lithium larger than expected. **21.127** The very high and very low standard electrode potentials involve extremely reactive substances, such as F_2 (a powerful oxidizer) and Li (a powerful reducer). These substances react directly with water because any aqueous cell with a voltage of more than 1.23 V has the ability to electrolyze water into hydrogen and oxygen.
21.129(a) 1.073×10^5 s (b) 1.5×10^4 kW·h (c) 6.8¢
21.131 F < D < E. If metal E and a salt of metal F are mixed, the salt is reduced, producing metal F because E has the greatest reducing strength of the three metals.
21.134(a) Cell I: 4 mol electrons; $\Delta G = -4.75\times10^5$ J
Cell II: 2 mol electrons; $\Delta G = -3.94\times10^5$ J
Cell III: 2 mol electrons; $\Delta G = -4.53\times10^5$ J
(b) Cell I: -13.2 kJ/g
Cell II: -0.613 kJ/g
Cell III: -2.62 kJ/g
Cell I has the highest ratio (most energy released per gram) because the reactants have very low mass, while Cell II has the lowest ratio because the reactants have large masses.
21.138(a) 9.6 g Cu (b) 0.38 M Cu^{2+}

21.140 $Sn^{2+}(aq) + 2e^- \longrightarrow Sn(s)$
$Cr^{3+}(aq) + e^- \longrightarrow Cr^{2+}(aq)$
$Fe^{2+}(aq) + 2e^- \longrightarrow Fe(s)$
$U^{4+}(aq) + e^- \longrightarrow U^{3+}(aq)$
21.144(a) 3.6×10^{-9} M (b) 1 M
21.146(a) Nonstandard cell:
$E_{waste} = E^0_{cell} - (0.0592\ V/1)\log[Ag^+]_{waste}$
Standard cell:
$E_{standard} = E^0_{cell} - (0.0592\ V/1)\log[Ag^+]_{standard}$
(b) $[Ag^+]_{waste} = \text{antilog}\left(\dfrac{E_{standard} - E_{waste}}{0.0592}\right)[Ag^+]_{standard}$
(c) $C_{Ag^+,\ waste} = \text{antilog}\left(\dfrac{E_{standard} - E_{waste}}{0.0592}\right)C_{Ag^+,\ standard}$, where
C is concentration in ng/L
(d) 900 ng/L
(e) $[Ag^+]_{waste} =$

$$\text{antilog}\left[\frac{(E_{standard} - E_{waste})\dfrac{nF}{2.303R} + T_{standard}\log[Ag^+]_{standard}}{T_{waste}}\right]$$

21.148(a) 1.08×10^3 C (b) 0.629 g Cd, 1.03 g NiO(OH), 0.202 g H_2O; total mass of reactants = 1.86 g (c) 14.0% **21.150** Li > Ba > Na > Al > Mn > Zn > Cr > Fe > Ni > Sn > Pb > Cu > Ag > Hg > Au. Metals with potentials lower than that of water (−0.83 V) can displace H_2 from water: Li, Ba, Na, Al, and Mn. Metals with potentials lower than that of hydrogen (0.00 V) can displace H_2 from acid: Li, Ba, Na, Al, Mn, Zn, Cr, Fe, Ni, Sn, and Pb. Metals with potentials greater than that of hydrogen (0.00 V) cannot displace H_2: Cu, Ag, Hg, and Au. **21.153**(a) 5.3×10^{-11} (b) 0.20 V (c) 0.43 V (d) 8.2×10^{-4} M NaOH **21.156**(a) -1.18×10^5 kJ (b) 1.20×10^4 L (c) 2.31×10^6 (d) 3.20×10^3 kW·h (e) 3.38×10^3 cents **21.158** 2.94

Chapter 22

22.2 Fe from Fe_2O_3; Ca from $CaCO_3$; Na from NaCl; Zn from ZnS **22.3**(a) Differentiation refers to the processes involved in the formation of Earth into regions (core, mantle, and crust) of differing composition. Substances separated according to their densities, with the more dense material in the core and the less dense in the crust. (b) O, Si, Al, and Fe (c) O **22.7** Plants produced O_2, slowly increasing the oxygen concentration in the atmosphere and creating an oxidative environment for metals. The oxygen-free decay of plant and animal material created large fossil fuel deposits. **22.9** Fixation refers to the process of converting a substance in the atmosphere into a form more readily usable by organisms. Carbon and nitrogen; fixation of carbon dioxide gas by plants and fixation of nitrogen gas by nitrogen-fixing bacteria. **22.12** Atmospheric nitrogen is fixed by three pathways: atmospheric, industrial, and biological. Atmospheric fixation requires high-temperature reactions (e.g., lightning) to convert N_2 into NO and other oxidized species. Industrial fixation involves mainly the formation of ammonia, NH_3, from N_2 and H_2. Biological fixation occurs in nitrogen-fixing bacteria that live on the roots of legumes. Human activity is an example of industrial fixation. It contributes about 17% of the nitrogen fixed. **22.14**(a) the atmosphere (b) Plants excrete acid from their roots to convert PO_4^{3-} ions into more soluble $H_2PO_4^-$ ions, which the plant can absorb. Through excretion and decay, organisms return soluble phosphate compounds to the cycle. **22.17**(a) 1.1×10^3 L (b) 4.2×10^2 m^3 **22.18**(a) The iron ions form an insoluble salt, $Fe_3(PO_4)_2$, that decreases the yield of phosphorus. (b) 8.8 t **22.20**(a) Roasting involves heating the mineral in air at high temperatures to convert the mineral to the oxide. (b) Smelting is the reduction of the metal oxide to the free metal using heat and a reducing agent such as coke. (c) Flotation is a separation process in which the ore is removed from the gangue by exploiting the difference in density in the presence of detergent. The gangue sinks to the bottom and the lighter ore-detergent mix is skimmed off the top. (d) Refining is the final step in the purification process to yield the pure metal. **22.25**(a) Slag is a by-product of steel-making and contains the impurity SiO_2. (b) Pig iron is the impure product of iron metallurgy (containing 3–4% C and other impurities). (c) Steel refers to iron alloyed with other elements to attain desirable properties. (d) The basic-oxygen process is used to purify pig iron and obtain carbon steel. **22.27** Iron and nickel are more easily oxidized and less easily reduced than copper. They are separated from copper in the roasting step and converted to slag. In the electrorefining process, all three metals are in solution, but only Cu^{2+} ions are reduced at the cathode to form Cu(s). **22.30** Le Châtelier's principle says that the system shifts toward formation of K as the gaseous metal leaves the cell. **22.31**(a) $E^0_{half\text{-}cell} = -3.05$ V, −2.93 V, and −2.71 V for Li^+, K^+, and Na^+, respectively. In all of these cases, it is energetically more favorable to reduce H_2O to H_2 than to reduce M^+ to M. (b) $2RbX + Ca \longrightarrow CaX_2 + 2Rb$, where $\Delta H = IE_1(Ca) + IE_2(Ca) - 2IE_1(Rb) = 929$ kJ/mol. Based on the IEs and positive ΔH for the forward reaction, it seems more reasonable that Rb metal will reduce Ca^{2+} than the reverse. (c) If the reaction is carried out at a temperature greater than the boiling point of Rb, the product mixture will contain gaseous Rb, which can be removed from the reaction vessel; this would cause a shift in equilibrium to form more Rb as product. (d) $2CsX + Ca \longrightarrow CaX_2 + 2Cs$, where $\Delta H = IE_1(Ca) + IE_2(Ca) - 2IE_1(Cs) = 983$ kJ/mol. This reaction is more unfavorable than for Rb, but Cs has a lower boiling point. **22.32**(a) 4.6×10^4 L (b) 1.26×10^8 C (c) 1.68×10^6 s **22.35**(a) Mg^{2+} is more difficult to reduce than H_2O, so $H_2(g)$ would be produced instead of Mg metal. $Cl_2(g)$ forms at the anode due to overvoltage. (b) The ΔH^0_f of $MgCl_2(s)$ is −641.6 kJ/mol. High temperature favors the reverse (endothermic) reaction, the formation of magnesium metal and chlorine gas. **22.37**(a) Sulfur dioxide is the reducing agent and is oxidized to the +6 state (SO_4^{2-}). (b) $HSO_4^-(aq)$ (c) $H_2SeO_3(aq) + 2SO_2(g) + H_2O(l) \longrightarrow Se(s) + 2HSO_4^-(aq) + 2H^+(aq)$ **22.42**(a) O.N. for Cu: in Cu_2S, +1; in Cu_2O, +1; in Cu, 0 (b) Cu_2S is the reducing agent, and Cu_2O is the oxidizing agent. **22.44**(a) 1.0×10^6 C (b) 1.2×10^3 A **22.47** $2ZnS(s) + C(graphite) \longrightarrow 2Zn(s) + CS_2(g)$; $\Delta G^0_{rxn} = 463$ kJ. Since ΔG^0_{rxn} is positive, this reaction is not spontaneous at standard-state conditions. $2ZnO(s) + C(s) \longrightarrow 2Zn(s) + CO_2(g)$; $\Delta G^0_{rxn} = 242.0$ kJ. This reaction is also not spontaneous, but is less unfavorable. **22.48** The formation of sulfur trioxide is very slow at ordinary temperatures. Increasing the temperature can speed up the reaction, but because the reaction is exothermic, increasing the temperature decreases the yield. Adding a catalyst increases the rate of the reaction, so a lower temperature can be used to enhance the yield. **22.51**(a) Cl_2, H_2, and NaOH

(b) The mercury-cell process yields higher purity NaOH, but releases some Hg, which is discharged into the environment. **22.52**(a) $\Delta G^0 = -142$ kJ; yes (b) The rate of the reaction is very slow at 25°C. (c) $\Delta G^0_{500} = -53$ kJ, so the reaction is spontaneous. (d) $K_{25} = 7.8\times10^{24} > K_{500} = 3.8\times10^3$ (e) 1.05×10^3 K **22.53** 7×10^2 lb Cl_2 **22.57**(a) $P_4O_{10}(s) + 6H_2O(l) \longrightarrow 4H_3PO_4(l)$ (b) 1.55 **22.59**(a) 9.006×10^9 g CO_2 (b) The 4.3×10^{10} g CO_2 produced by automobiles is much greater than that from the blast furnace. **22.61**(a) If $[OH^-] > 1.1\times10^{-4}$ *M* (i.e., if pH > 10.04), $Mg(OH)_2$ will precipitate. (b) 1 (To the correct number of significant figures, all the magnesium has precipitated.) **22.63**(a) K_{25}(step 1) $= 1\times10^{168}$; K_{25}(side rxn) $= 7\times10^{228}$; (b) K_{900}(step 1) $= 4.6\times10^{49}$; K_{900}(side rxn) $= 1.4\times10^{58}$ (c) $\$3.5\times10^7$
22.67 (1) $2H_2O(l) + 2FeS_2(s) + 7O_2(g) \longrightarrow 2Fe^{2+}(aq) + 4SO_4^{2-}(aq) + 4H^+(aq)$ increases acidity

(2) $4H^+(aq) + 4Fe^{2+}(aq) + O_2(g) \longrightarrow 4Fe^{3+}(aq) + 2H_2O(l)$

(3) $Fe^{3+}(aq) + 3H_2O(l) \longrightarrow Fe(OH)_3(s) + 3H^+(aq)$ increases acidity

(4) $8H_2O(l) + FeS_2(s) + 14Fe^{3+}(aq) \longrightarrow 15Fe^{2+}(aq) + 2SO_4^{2-}(aq) + 16H^+(aq)$ increases acidity

22.69 density of ferrite: 7.86 g/cm^3; density of austenite: 7.55 g/cm^3
22.71(a) Cathode: $Na^+(l) + e^- \longrightarrow Na(l)$
Anode: $4OH^-(l) \longrightarrow O_2(g) + 2H_2O(g) + 4e^-$
(b) 50%
22.74(a) $nCO_2(g) + nH_2O(l) \longrightarrow (CH_2O)_n(s) + nO_2(g)$ (b) 25 L (c) 7.6×10^4 L **22.75** 66 mg/L **22.77** 794 kg Na_3AlF_6 **22.78**(a) 20.4 min (b) 14 effusion steps **22.80**(a) 1.890 t Al_2O_3 (b) 0.3339 t C (c) 100% (d) 74% (e) 2.813×10^3 m^3 **22.85** Acid rain increases the leaching of phosphate into the groundwater, due to the protonation of PO_4^{3-} to form HPO_4^{2-} and $H_2PO_4^-$. As shown in calculations (a) and (b), solubility of $Ca_3(PO_4)_2$ increases from 6.4×10^{-7} *M* (in pure water) to 1.1×10^{-2} *M* (in acidic rainwater). (a) 6.4×10^{-7} *M* (b) 1.1×10^{-2} *M* **22.86**(a) 5.60 mol % (b) 237.4 g/mol **22.88** density of silver = 10.51 g/cm^3; density of sterling silver = 10.2 g/cm^3

Chapter 23

23.2(a) $1s^22s^22p^63s^23p^64s^23d^{10}4p^65s^24d^x$ (b) $1s^22s^22p^63s^23p^64s^23d^{10}4p^65s^24d^{10}5p^66s^24f^{14}5d^x$ **23.4** Five; examples are Mn, [Ar] $4s^23d^5$, and Fe^{3+}, [Ar] $3d^5$. **23.6**(a) The elements should increase in size as they increase in mass from Period 5 to Period 6. Because 14 additional elements lie between Periods 5 and 6, the effective nuclear charge increases significantly; so the atomic size decreases, or "contracts." This effect is significant enough that Zr^{4+} and Hf^{4+} are almost the same size but differ greatly in atomic mass. (b) The atomic size increases from Period 4 to Period 5, but stays fairly constant from Period 5 to Period 6. (c) Atomic mass increases significantly from Period 5 to Period 6, but atomic radius (and thus volume) increases slightly, so Period 6 elements are very dense. **23.9**(a) A paramagnetic substance is attracted to a magnetic field, while a diamagnetic substance is slightly repelled by one. (b) Ions of transition elements often have half-filled *d* orbitals whose unpaired electrons make the ions paramagnetic. Ions of main-group elements usually have a noble gas configuration with no partially filled levels. (c) Some *d* orbitals in the transition element ions are empty, which allows an electron from one *d* orbital to move to a slightly higher energy one. The energy required for this transition is small and falls in the visible wavelength range. All orbitals are filled in ions of main-group elements, so enough energy would have to be added to move an electron to the next principal energy level, not just another orbital within the same energy level. This amount of energy is very large and much greater than the visible range of wavelengths.
23.10(a) $1s^22s^22p^63s^23p^64s^23d^3$ (b) $1s^22s^22p^63s^23p^64s^23d^{10}4p^65s^24d^1$ (c) [Xe] $6s^24f^{14}5d^{10}$ **23.12**(a) [Xe] $6s^24f^{14}5d^6$ (b) [Ar] $4s^23d^7$ (c) [Kr] $5s^14d^{10}$ **23.14**(a) [Ar], no unpaired electrons (b) [Ar] $3d^9$, one unpaired electron (c) [Ar] $3d^5$, five unpaired electrons (d) [Kr] $4d^2$, two unpaired electrons **23.16**(a) +5 (b) +4 (c) +7 **23.18** Cr, Mo, and W **23.20** In CrF_2, because the chromium is in a lower oxidation state. **23.22** Atomic size increases slightly down a group of transition elements, but nuclear charge increases much more, so the first ionization energy generally increases. The reduction potential for Mo is lower, so it is more difficult to oxidize Mo than Cr. In addition, the ionization energy of Mo is higher than that of Cr, so it is more difficult to remove electrons from Mo. **23.24** CrO_3, with Cr in a higher oxidation state, yields a more acidic aqueous solution. **23.28**(a) seven (b) This corresponds to a half-filled *f* subshell. **23.30** (a) [Xe] $6s^25d^1$ (b) [Xe] $4f^1$ (c) [Rn] $7s^25f^{11}$ (d) [Rn] $5f^2$ **23.32**(a) Eu^{2+}: [Xe] $4f^7$; Eu^{3+}: [Xe] $4f^6$; Eu^{4+}: [Xe] $4f^5$ The stability of the half-filled *f* subshell makes Eu^{2+} most stable. (b) Tb^{2+}: [Xe] $4f^9$; Tb^{3+}: [Xe] $4f^8$; Tb^{4+}: [Xe] $4f^7$; Tb should show a +4 oxidation state because that gives the half-filled subshell. **23.34** Gd has the electron configuration [Xe] $6s^24f^75d^1$ with eight unpaired electrons. Gd^{3+} has seven unpaired electrons: [Xe] $4f^7$. **23.36** Valence-state electronegativity is the apparent electronegativity of an element in a given oxidation state. As Mn is bonded to more O atoms in its different oxides, its O.N. becomes more positive. In solution, an H_2O molecule is attracted to the increasingly positive Mn, and one of its protons becomes easier to lose. **23.40** Mercury exists in the liquid state at room temperature. Mercury occurs in the +1 oxidation state, unusual for its group, because two mercury 1+ ions form a dimer, Hg_2^{2+}, by sharing their unpaired 6*s* electrons in a covalent bond. **23.42** $2Cr^{3+}(aq) + Cr(s) \longrightarrow 3Cr^{2+}$; $E^0 = +0.50$ V; yes, the reaction is spontaneous. **23.44** $3Mn^{2+}(aq) + 2MnO_4^-(aq) + 2H_2O(l) \longrightarrow 5MnO_2(s) + 4H^+(aq)$; $E^0 = +0.28$ V; the reaction is spontaneous, so the Mn^{2+} will react with remaining MnO_4^- to form MnO_2. **23.47** The coordination number indicates the number of ligand atoms bonded to the metal ion. The oxidation number represents the number of electrons lost to form the ion. The coordination number is unrelated to the oxidation number. **23.49** 2, linear; 4, tetrahedral or square planar; 6, octahedral **23.52** The complex ion has a negative charge. **23.55**(a) hexaaquanickel(II) chloride (b) tris(ethylenediamine)chromium(III) perchlorate (c) potassium hexacyanomanganate(II) **23.57**(a) 2+, 6 (b) 3+, 6 (c) 2+, 6 **23.59**(a) potassium dicyanoargentate(I) (b) sodium tetrachlorocadmate(II) (c) tetraammineaquabromocobalt(III) bromide **23.61**(a) 1+, 2 (b) 2+, 4 (c) 3+, 6 **23.63**(a) $[Zn(NH_3)_4]SO_4$ (b) $[Cr(NH_3)_5Cl]Cl_2$ (c) $Na_3[Ag(S_2O_3)_2]$ **23.65**(a) 4, two ions (b) 6, three ions

(c) 2, four ions **23.67**(a) $[Cr(H_2O)_6]_2(SO_4)_3$ (b) $Ba[FeBr_4]_2$ (c) $[Pt(en)_2]CO_3$ **23.69**(a) 6, five ions (b) 4, three ions (c) 4, two ions **23.71**(a) The nitrite ion forms linkage isomers because it can bind to the metal ion through either the nitrogen or one of the oxygen atoms—both have a lone pair of electrons.

$[:\ddot{O}-\ddot{N}=\ddot{O}]^-$ ligand lone pair

(b) Sulfur dioxide molecules form linkage isomers because both the sulfur and the oxygen atoms can bind the central metal ion.

$\ddot{O}=\ddot{S}=\ddot{O}$ ligand lone pair

(c) Nitrate ions have three oxygen atoms, all with a lone pair that can bond to the metal ion, but all of the oxygen atoms are equivalent, so there are no linkage isomers.

$[:\ddot{O}-N=\ddot{O}\ (:\ddot{O}:)]^-$

23.73(a) geometric isomerism

Br, Br / Pt / CH_3NH_2, NH_2CH_3 (*cis*) and Br, NH_2CH_3 / Pt / CH_3NH_2, Br (*trans*)

(b) geometric isomerism

H_3N, NH_3 / Pt / Cl, F (*cis*) and H_3N, F / Pt / Cl, NH_3 (*trans*)

(c) geometric isomerism

H_2O, NH_3 / Pt / Cl, F and H_2O, NH_3 / Pt / F, Cl and H_2O, F / Pt / Cl, NH_3

23.75(a) geometric isomerism

$[Cl, Cl / Pt / Br, Br]^{2-}$ (*cis*) and $[Cl, Br / Pt / Br, Cl]^{2-}$ (*trans*)

(b) linkage isomerism

$[NH_3, H_3N, NH_3 / Cr / H_3N, NO_2, NH_3]^{2+}$ and $[NH_3, H_3N, NH_3 / Cr / H_3N, ONO, NH_3]^{2+}$

(c) geometric isomerism

$[NH_3, H_3N, I / Pt / H_3N, I, NH_3]^{2+}$ (*cis*) and $[NH_3, H_3N, I / Pt / I, NH_3, NH_3]^{2+}$ (*trans*)

23.77 The compound with the traditional formula is $CrCl_3{\cdot}4NH_3$; the actual formula is $[Cr(NH_3)_4Cl_2]Cl$. **23.79**(a) $K[Pd(NH_3)Cl_3]$ (b) $[PdCl_2(NH_3)_2]$ (c) $K_2[PdCl_6]$ (d) $[Pd(NH_3)_4Cl_2]Cl_2$ **23.81**(a) dsp^2 (b) sp^3 **23.84** absorption of orange or yellow **23.85**(a) The crystal field splitting energy (Δ) is the energy difference between the two sets of *d* orbitals that result from the bonding of ligands to a central transition metal atom. (b) In an octahedral field of ligands, the ligands approach along the *x*-, *y*-, and *z*- axes. The $d_{x^2-y^2}$ and d_{z^2} orbitals are located *along* the *x*-, *y*-, and *z*-axes, so ligand interaction is higher in energy. The other orbital-ligand interactions are lower in energy because the d_{xy}, d_{yz}, and d_{xz} orbitals are located *between* the *x*-, *y*-, and *z*-axes. (c) In a tetrahedral field of ligands, the ligands do not approach along the *x*-, *y*-, and *z*-axes. The ligand interaction is greater for the d_{xy}, d_{yz}, and d_{xz} orbitals and lesser for the $d_{x^2-y^2}$ and d_{z^2} orbitals. Therefore, the crystal field splitting is reversed, and the d_{xy}, d_{yz}, and d_{xz} orbitals are higher in energy than the $d_{x^2-y^2}$ and d_{z^2} orbitals. **23.88** If Δ is greater than $E_{pairing}$, electrons will pair their spins in the lower energy *d* orbitals before adding as unpaired electrons to the higher energy *d* orbitals. If Δ is less than $E_{pairing}$, electrons will add as unpaired electrons to the higher energy *d* orbitals before pairing in the lower energy *d* orbitals. **23.90**(a) no *d* electrons (b) eight *d* electrons (c) six *d* electrons **23.92**(a) five (b) ten (c) seven

23.94

y, x, $d_{x^2-y^2}$ y, x, d_{xy}

In an octahedral field of ligands, the ligands approach along the *x*, *y*, and *z* axes. The $d_{x^2-y^2}$ orbital is located along the *x* and *y* axes, so ligand interaction is greater. The d_{xy} orbital is offset from the *x* and *y* axes by 90°, so ligand interaction is less. The greater interaction of the $d_{x^2-y^2}$ orbital results in its higher energy. **23.96**(a) and (d)

23.98

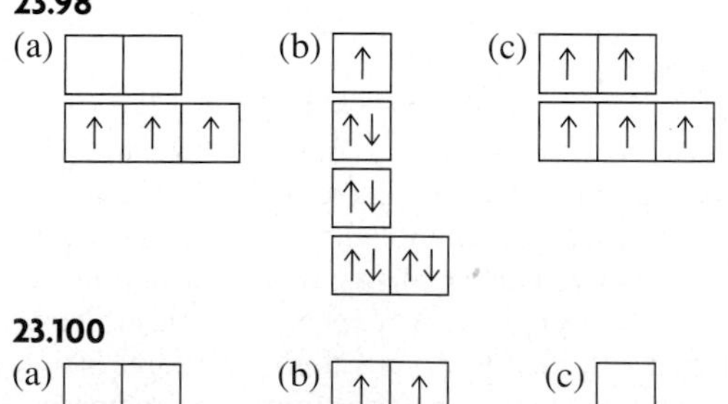

23.100

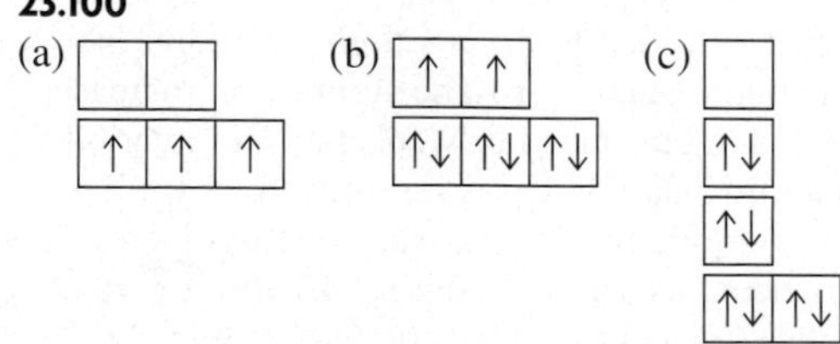

23.102 $[Cr(H_2O)_6]^{3+} < [Cr(NH_3)_6]^{3+} < [Cr(NO_2)_6]^{3-}$
23.104 A violet complex absorbs yellow-green light. The light absorbed by a complex with a weaker field ligand would be at a lower energy and higher wavelength. Light of lower energy than yellow-green light is yellow, orange, or red. The color observed would be blue or green. **23.107** The H_2O ligand is weaker than the NH_3 ligand. The weaker field ligand results in a lower splitting energy and absorbs a lower energy of visible light. The hexaaqua complex appears green because it absorbs red light. The hexaammine complex appears violet because it absorbs yellow light, which has higher energy (shorter λ) than red light.
23.111 Hg^+ is [Xe] $6s^1 4f^{14} 5d^{10}$ and Cu^+ is [Ar] $3d^{10}$. The mercury(I) ion has one electron in the 6*s* orbital that can form a covalent bond with the electron in the 6*s* orbital of another mercury(I) ion. In the copper(I) ion there are no electrons in the *s* orbital, so these ions cannot bond with one another.
23.113(a) 6 (b) +3 (c) two (d) 1 mol
23.123 Geometric (*cis-trans*) and linkage isomerism

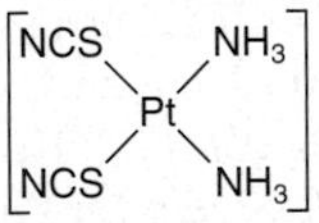

cis-diamminedithiocyanatoplatinum(II)

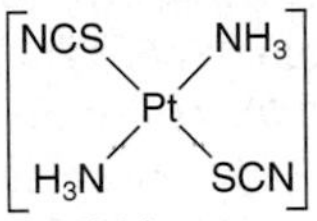

trans-diamminedithiocyanatoplatinum(II)

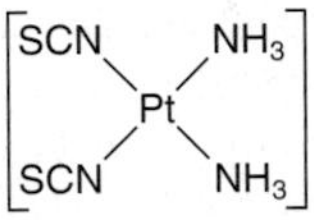

cis-diamminediisothiocyanatoplatinum(II)

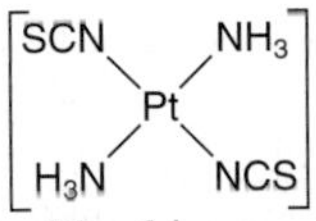

trans-diamminediisothiocyanatoplatinum(II)

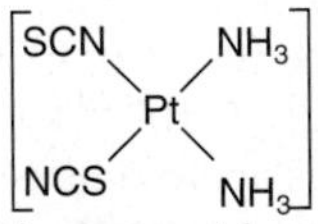

cis-diammineisothiocyanatothiocyanatoplatinum(II)

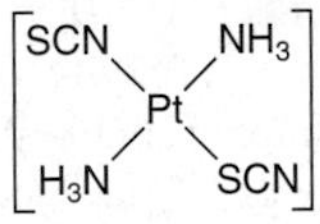

23.125 $[Co(NH_3)_4(H_2O)Cl]^{2+}$
tetraammineaquachlorocobalt(III) ion
2 geometric isomers

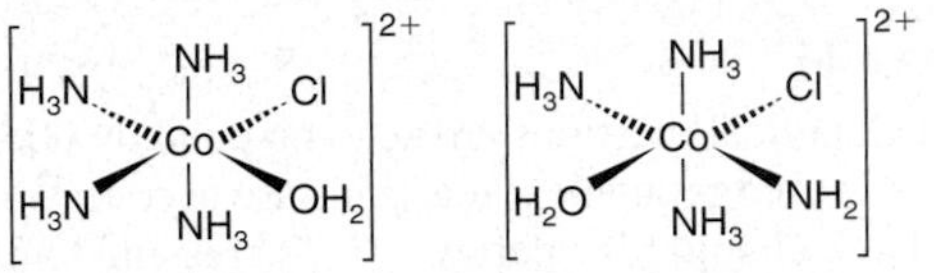

cis *trans*

$[Cr(H_2O)_3Br_2Cl]$ triaquadibromochlorochromium(III)
3 geometric isomers

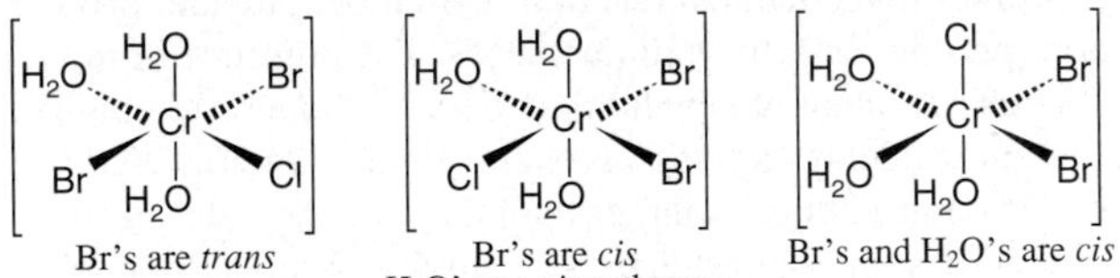

Br's are *trans*

Br's are *cis*
H_2O's are *cis* and *trans*

Br's and H_2O's are *cis*

$[Cr(NH_3)_2(H_2O)_2Br_2]^+$
diamminediaquadibromochromium(III) ion
6 isomers (5 geometric)

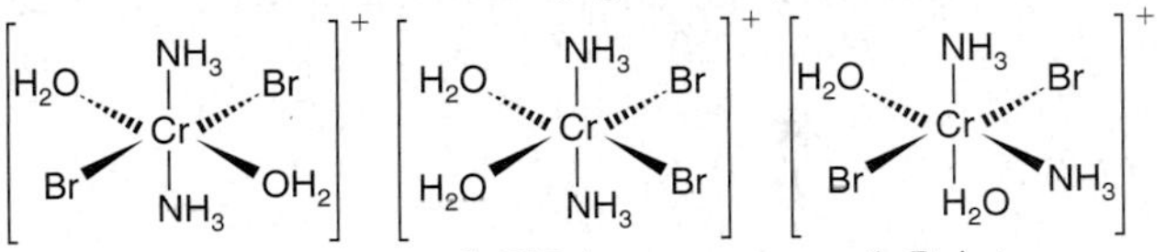

all ligands are *trans*

only NH_3 is *trans*

only Br is *trans*

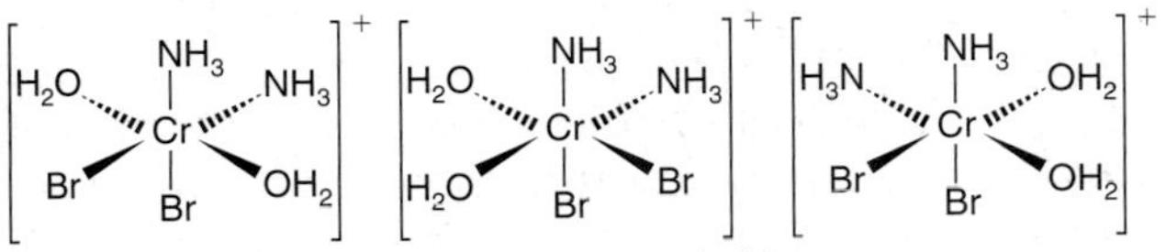

only H_2O is *trans*

optical isomers
all three ligands are *cis*

23.128(a) in MnO_4^{2-}, +6; in MnO_4^-, +7; in MnO_2, +4
(b) $3MnO_4^{2-}(aq) + 4H^+(aq) \longrightarrow 2MnO_4^-(aq) + MnO_2(s) + 2H_2O(l)$
23.135(a) no optical isomers (b) no optical isomers (c) no optical isomers (d) no optical isomers (e) exist as optical isomers
23.136 $Pt[P(C_2H_5)_3]_2Cl_2$

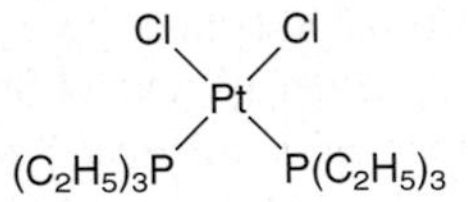

cis-dichlorobis(triethylphosphine)platinum(II)

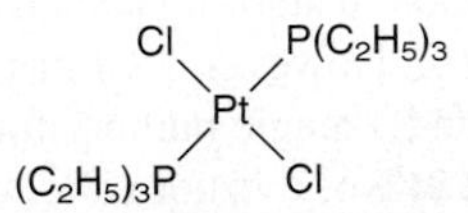

trans-dichlorobis(triethylphosphine)platinum(II)

23.140(a) The first reaction shows no change in the number of particles. In the second reaction, the number of reactant particles is greater than the number of product particles. A decrease in the number of particles means a decrease in entropy. Based on entropy change only, the first reaction is favored.

(b) The ethylenediamine complex will be more stable with respect to ligand exchange in water because the entropy change for that exchange is unfavorable.

Chapter 24

24.1(a) Chemical reactions are accompanied by relatively small changes in energy; nuclear reactions are accompanied by relatively large changes in energy. (b) Increasing temperature increases the rate of a chemical reaction but has no effect on a nuclear reaction. (c) Both chemical and nuclear reaction rates increase with higher reactant concentrations. (d) If the reactant is limiting in a chemical reaction, then more reactant produces more product and the yield increases. The presence of more radioactive reactant results in more decay product, so a higher reactant concentration increases the yield. **24.2**(a) 95.02% (b) The atomic mass is larger than the isotopic mass of ^{32}S. Sulfur-32 is the lightest isotope. **24.4**(a) Z down by 2, N down by 2 (b) Z up by 1, N down by 1 (c) no change in Z or N (d) Z down by 1, N up by 1; a different element is produced in all cases except (c). **24.6** A neutron-rich nuclide decays by beta decay. A neutron-poor nuclide undergoes positron decay or electron capture. **24.8**(a) $^{234}_{92}U \longrightarrow ^{4}_{2}He + ^{230}_{90}Th$ (b) $^{232}_{93}Np + ^{0}_{-1}e \longrightarrow ^{232}_{92}U$ (c) $^{12}_{7}N \longrightarrow ^{0}_{1}\beta + ^{12}_{6}C$ **24.10**(a) $^{27}_{12}Mg \longrightarrow ^{0}_{-1}\beta + ^{27}_{13}Al$ (b) $^{9}_{3}Li \longrightarrow ^{1}_{0}n + ^{8}_{3}Li$ (c) $^{103}_{46}Pd + ^{0}_{-1}e \longrightarrow ^{103}_{45}Rh$ **24.12**(a) $^{48}_{23}V \longrightarrow ^{48}_{22}Ti + ^{0}_{1}\beta$ (b) $^{107}_{48}Cd + ^{0}_{-1}e \longrightarrow ^{107}_{47}Ag$ (c) $^{210}_{86}Rn \longrightarrow ^{206}_{84}Po + ^{4}_{2}He$ **24.14**(a) $^{186}_{78}Pt + ^{0}_{-1}e \longrightarrow ^{186}_{77}Ir$ (b) $^{225}_{89}Ac \longrightarrow ^{221}_{87}Fr + ^{4}_{2}He$ (c) $^{129}_{52}Te \longrightarrow ^{129}_{53}I + ^{0}_{-1}\beta$ **24.16**(a) Appears stable because its N and Z values are both magic numbers, but its N/Z ratio (1.50) is too high; it is unstable. (b) Appears unstable because its Z value is an odd number, but its N/Z ratio (1.19) is in the band of stability, so it is stable. (c) Unstable because its N/Z ratio is too high. **24.18**(a) The N/Z ratio for ^{127}I is 1.4; it is stable. (b) The N/Z ratio for ^{106}Sn is 1.1; it is unstable because this ratio is too low. (c) The N/Z is 1.1 for ^{68}As. The ratio is within the range of stability, but the nuclide is most likely unstable because there is an odd number of both protons and neutrons. **24.20**(a) alpha decay (b) positron decay or electron capture (c) positron decay or electron capture **24.22**(a) beta decay (b) positron decay or electron capture (c) alpha decay **24.24** Stability results from a favorable N/Z ratio, even numbered N and/or Z, and the occurrence of magic numbers. The N/Z ratio of ^{52}Cr is 1.17, which is within the band of stability. The fact that Z is even does not account for the variation in stability because all isotopes of chromium have the same Z. However, ^{52}Cr has 28 neutrons, so N is both an even number and a magic number for this isotope only. **24.28** $4\,^{0}_{-1}\beta + 7\,^{4}_{2}He$ **24.31** No, it is not valid to conclude that $t_{1/2}$ equals 1 min because the number of nuclei is so small. Decay rate is an average rate and is only meaningful when the sample is macroscopic and contains a large number of nuclei. For the sample containing 6×10^{12} nuclei, the conclusion is valid. **24.33** 2.89×10^{-2} Ci/g **24.35** 1.4×10^{8} Bq/g **24.37** 1×10^{-12} h^{-1} **24.39** 2.31×10^{-7} yr^{-1} **24.41** 1.49 mg **24.43** 2.2×10^{9} yr **24.45** 1×10^{2} dpm **24.47** 1.5×10^{6} yr **24.50** Both γ radiation and neutron beams have no charge, but a neutron has a mass approximately equal to that of a proton. **24.52** Protons are repelled from the target nuclei due to interaction with like (positive) charges. Higher energy is required to overcome the repulsion. **24.53**(a) $^{10}_{5}B + ^{4}_{2}He \longrightarrow ^{1}_{0}n + ^{13}_{7}N$ (b) $^{28}_{14}Si + ^{2}_{1}H \longrightarrow ^{1}_{0}n + ^{29}_{15}P$ (c) $^{242}_{96}Cm + ^{4}_{2}He \longrightarrow 2^{1}_{0}n + ^{244}_{98}Cf$ **24.58** Ionizing radiation is more dangerous to children because their rapidly dividing cells are more susceptible to radiation than an adult's slowly dividing cells. **24.60**(a) 5.4×10^{-7} rad (b) 5.4×10^{-9} Gy **24.62**(a) 7.5×10^{-10} Gy (b) 7.5×10^{-5} mrem (c) 7.5×10^{-10} Sv **24.65** 2.45×10^{-3} rad **24.67** NAA does not destroy the sample, while chemical analyses do. Neutrons bombard a nonradioactive sample, inducing some atoms within the sample to be radioactive. The radioisotopes decay by emitting radiation characteristic of each isotope. **24.70** The oxygen isotope in the methanol reactant appears in the formaldehyde product. The oxygen isotope in the chromic acid reactant appears in the water product. The isotope traces the oxygen in methanol to the oxygen in formaldehyde. **24.73** Energy is released when a nuclide forms from nucleons. The nuclear binding energy is the amount of energy holding the nucleus together. This energy must be absorbed to break up the nucleus into nucleons and is released when nucleons come together. **24.75**(a) 1.861×10^{4} eV (b) 2.981×10^{15} J **24.77** 5.1×10^{11} J **24.79**(a) 7.976 MeV/nucleon (b) 127.6 MeV/atom (c) 1.231×10^{10} kJ/mol **24.81**(a) 8.768 MeV/nucleon (b) 517.3 MeV/atom (c) 4.99×10^{10} kJ/mol **24.85** Radioactive decay is a spontaneous process in which unstable nuclei emit radioactive particles and energy. Fission occurs as the result of high-energy bombardment of nuclei with small particles that cause the nuclei to break into smaller nuclides, radioactive particles, and energy. **24.88** The water serves to slow the neutrons so that they are better able to cause a fission reaction. Heavy water is a better moderator because it does not absorb neutrons as well as light water does, so more neutrons are available to initiate the fission process. However, D_2O does not occur naturally in great abundance, so its production adds to the cost of a heavy-water reactor. **24.91** In the stellar nucleogenesis model, the lighter nuclides (up to ^{24}Mg) are produced in a helium-burning stage in which heavier nuclei are formed by the fusion of helium nuclei, each with a mass of 4 amu. **24.95**(a) 1.1×10^{-29} kg (b) 9.8×10^{-13} J (c) 5.9×10^{8} kJ/mol. This is approximately 1 million times larger than a typical heat of reaction. **24.97** 7.6×10^{3} yr **24.101** 1.35×10^{-5} M **24.103** 6.3×10^{-2} **24.106**(a) 5.99 h (b) 21% **24.108**(a) 0.999 (b) 0.298 (c) 5.58×10^{-6} (d) Radiocarbon dating is more reliable for the fraction in (b) because a significant quantity of ^{14}C has decayed and a significant quantity remains. Therefore, a change in the amount of ^{14}C will be noticeable. For the fraction in (a), very little ^{14}C has decayed, and for (c) very little ^{14}C remains. In either case, it will be more difficult to measure the change, so the error will be relatively large. **24.110** 6.58 h **24.114** 4.904×10^{-9} L/h **24.117** 87 yr **24.119**(a) 0.15 Bq/L (b) 0.34 Bq/L (c) 4.6 days; a total of 13.1 days **24.122** 6.6 s **24.125** 1926 **24.128** 6.27×10^{5} eV, 6.05×10^{7} kJ/mol **24.131**(a) 2.07×10^{-17} J (b) 1.45×10^{7} H atoms (c) 1.4960×10^{-5} J (d) 1.4959×10^{-5} J (e) No, the Captain should continue using the current technology. **24.134**(a) 0.043 MeV, 2.9×10^{-11} m (b) 4.713 MeV **24.138**(a) 3.26×10^{3} days (b) 3.2×10^{-3} s (c) 2.78×10^{11} yr **24.140**(a) 1.80×10^{17} J (b) 6.15×10^{10} kJ (c) The procedure in part (b) produces more energy per kilogram of antihydrogen. **24.141** 9.316×10^{2} MeV **24.145** 0.357 days

Glossary

Numbers in parentheses refer to the page(s) on which a term is introduced and/or discussed.

A

absolute scale (also *Kelvin scale*) The preferred temperature scale in scientific work, which has absolute zero (0 K, or −273.15°C) as the lowest temperature. (23) [See also *kelvin (K)*.]

absorption spectrum The spectrum produced when atoms absorb specific wavelengths of incoming light and become excited from lower to higher energy levels. (269)

abundance The amount of an element in a particular region of the natural world. (961)

accuracy The closeness of a measurement to the actual value. (29)

acid In common laboratory terms, any species that produces H^+ ions when dissolved in water. (144) [See also *Brønsted-Lowry, classical (Arrhenius),* and *Lewis acid-base definitions*.]

acid anhydride A compound, sometimes formed by a dehydration-condensation reaction of an oxoacid, that yields two molecules of the acid when it reacts with water. (646)

acid-base buffer (also *buffer*) A solution that resists changes in pH when a small amount of either strong acid or strong base is added. (815)

acid-base indicator A species whose color is different in acid and in base, which is used to monitor the equivalence point of a titration or the pH of a solution. (146, 777, 824)

acid-base reaction Any reaction between an acid and a base. (144) (See also *neutralization reaction*.)

acid-base titration curve A plot of the pH of a solution of acid (or base) versus the volume of base (or acid) added to the solution. (824)

acid-dissociation (acid-ionization) constant (K_a) An equilibrium constant for the dissociation of an acid (HA) in H_2O to yield the conjugate base (A^-) and H_3O^+:

$$K_a = \frac{[H_3O^+][A^-]}{[HA]} \qquad (771)$$

actinides The Period 7 elements that constitute the second inner transition series (5*f* block), which includes thorium (Th; $Z = 90$) through lawrencium (Lr; $Z = 103$). (304, 1010)

activated complex (See *transition state*.)

activation energy (E_a) The minimum energy with which molecules must collide to react. (692)

active site The region of an enzyme formed by specific amino acid side chains at which catalysis occurs. (709)

activity ($\mathcal{A}$) (also *decay rate*) The change in number of nuclei ($\mathcal{N}$) of a radioactive sample divided by the change in time (t). (1054)

activity series of the metals A listing of metals arranged in order of decreasing strength of the metal as a reducing agent in aqueous reactions. (161)

actual yield The amount of product actually obtained in a chemical reaction. (114)

addition polymer (also *chain-reaction,* or *chain-growth, polymer*) A polymer formed when monomers (usually containing C=C) combine through an addition reaction. (651)

addition reaction A type of organic reaction in which atoms linked by a multiple bond become bonded to more atoms. (635)

adduct The product of a Lewis acid-base reaction characterized by the formation of a new covalent bond. (801)

adenosine triphosphate (ATP) A high-energy molecule that serves most commonly as a store and source of energy in organisms. (889)

alchemy An occult study of nature that flourished for 1500 years in northern Africa and Europe and resulted in the development of several key laboratory methods. (8)

alcohol An organic compound (ending, *-ol*) that contains a $-\overset{|}{\underset{|}{C}}-\ddot{\underset{..}{O}}-H$ functional group. (638)

aldehyde An organic compound (ending, *-al*) that contains the carbonyl functional group ($C=\ddot{\underset{..}{O}}$) in which the carbonyl C is also bonded to H. (643)

alkane A hydrocarbon that contains only single bonds (general formula, C_nH_{2n+2}). (623)

alkene A hydrocarbon that contains at least one C=C bond (general formula, C_nH_{2n}). (628)

alkyl group A saturated hydrocarbon chain with one bond available. (634)

alkyl halide (See *haloalkane*.)

alkyne A hydrocarbon that contains at least one C≡C bond (general formula, C_nH_{2n-2}). (631)

allotrope One of two or more crystalline or molecular forms of an element. In general, one allotrope is more stable than another at a particular pressure and temperature. (573)

alloy A mixture with metallic properties that consists of solid phases of two or more pure elements, a solid-solid solution, or distinct intermediate phases. (358, 494, 977)

alpha (α) decay A radioactive process in which an alpha particle is emitted from a nucleus. (1048)

alpha particle (α or ^4_2He) A positively charged particle, identical to a helium nucleus, that is one of the common types of radioactive emissions. (1047)

amide An organic compound that contains the $-\overset{:O:}{\overset{\|}{C}}-\underset{..}{\overset{|}{N}}-$ functional group. (646)

amine An organic compound (general formula, $-\overset{|}{\underset{|}{C}}-\overset{..}{\underset{|}{N}}-$) derived structurally by replacing one or more H atoms of ammonia with alkyl groups; a weak organic base. (640, 789)

amino acid An organic compound [general formula, $H_2N-CH(R)-COOH$] with at least one carboxyl and one amine group on the same molecule; the monomer unit of a protein. (495, 655)

amorphous solid A solid that occurs in different shapes because it lacks extensive molecular-level ordering of its particles. (449)

ampere (A) The SI unit of electric current; 1 ampere of current results when 1 coulomb flows through a conductor in 1 second. (945)
amphoteric Able to act as either an acid or a base. (315)
amplitude The height of the crest (or depth of the trough) of a wave; related to the intensity of the energy. (258)
angular momentum quantum number (*l*) (also *orbital-shape quantum number*) An integer from 0 to $n - 1$ that is related to the shape of an atomic orbital. (277)
anion A negatively charged ion. (57)
anode The electrode at which oxidation occurs in an electrochemical cell. Electrons are given up by the reducing agent and leave the cell at the anode. (909)
antibonding MO A molecular orbital formed when wave functions are subtracted from each other, which decreases electron density between the nuclei and leaves a node. Electrons occupying such an orbital destabilize the molecule. (410)
apatite A compound of general formula $Ca_5(PO_4)_3X$, where X is generally F, Cl, or OH. (970)
aqueous solution A solution in which water is the solvent. (73)
aromatic hydrocarbon A compound of C and H with one or more rings of C atoms (often drawn with alternating C—C and C=C bonds), in which there is extensive delocalization of π electrons. (632)
Arrhenius acid-base definition [See *classical (Arrhenius) acid-base definition*.]
Arrhenius equation An equation that expresses the exponential relationship between temperature and the rate constant: $k = Ae^{-E_a/RT}$. (692)
atmosphere The mixture of gases that extends from a planet's surface and eventually merges with outer space; the gaseous portion of Earth's crust. (207, 963) (For the unit, see *standard atmosphere*.)
atom The smallest particle of an element that retains the chemical nature of the element. A neutral, spherical entity composed of a positively charged central nucleus surrounded by one or more negatively charged electrons. (44)
atomic mass (also *atomic weight*) The average of the masses of the naturally occurring isotopes of an element weighted according to their abundances. (51)
atomic mass unit (amu) (also *dalton, Da*) A mass exactly equal to $\frac{1}{12}$ the mass of a carbon-12 atom. (51)
atomic number (*Z*) The unique number of protons in the nucleus of each atom of an element (equal to the number of electrons in the neutral atom). An integer that expresses the positive charge of a nucleus in multiples of the electronic charge. (50)
atomic orbital (also *wave function*) A mathematical expression that describes the motion of the electron's matter-wave in terms of time and position in the region of the nucleus. The term is used qualitatively to mean the region of space in which there is a high probability of finding the electron. (275)
atomic size A term referring to the atomic radius, one-half the distance between nuclei of identical bonded elements. (305, 543) (See also *covalent radius* and *metallic radius*.)
atomic solid A solid consisting of individual atoms held together by dispersion forces; the frozen noble gases are the only examples. (456)
atomic symbol (also *element symbol*) A one- or two-letter abbreviation for the English, Latin, or Greek name of an element. (50)
aufbau principle (also *building-up principle*) The conceptual basis of a process of building up atoms by adding one proton (and one or more neutrons) at a time to the nucleus and one electron around it to obtain the ground-state electron configurations of the elements. (296)
autoionization (also *self-ionization*) A reaction in which two molecules of a substance react to give ions. The most important example is for water:

$$2H_2O(l) \rightleftharpoons H_3O^+(aq) + OH^-(aq) \qquad (773)$$

average rate The change in concentration of reactants (or products) divided by a finite time period. (676)
Avogadro's law The gas law stating that, at fixed temperature and pressure, equal volumes of any ideal gas contain equal numbers of particles, and, therefore, the volume of a gas is directly proportional to its amount (mol): $V \propto n$. (187)
Avogadro's number A number (6.022×10^{23} to four significant figures) equal to the number of atoms in exactly 12 g of carbon-12; the number of atoms, molecules, or formula units in one mole of an element or compound. (87)
axial group An atom (or group) that lies above or below the trigonal plane of a trigonal bipyramidal molecule, or a similar structural feature in a molecule. (380)

B

background radiation Natural ionizing radiation, the most important form of which is cosmic radiation. (1064)
balancing coefficient (also *stoichiometric coefficient*) A numerical multiplier of all the atoms in the formula immediately following it in a chemical equation. (102)
band of stability The narrow band of stable nuclides that appears on a plot of number of neutrons versus number of protons for all nuclides. (1050)
band theory An extension of molecular orbital (MO) theory that explains many properties of metals, in particular, the differences in electrical conductivity of conductors, semiconductors, and insulators. (460)
barometer A device used to measure atmospheric pressure. Most commonly, a tube open at one end, which is filled with mercury and inverted into a dish of mercury. (179)
base In common laboratory terms, any species that produces OH^- ions when dissolved in water. (144) [See also *Brønsted-Lowry, classical (Arrhenius)*, and *Lewis acid-base definitions*.]
base-dissociation (base-ionization) constant (K_b) An equilibrium constant for the reaction of a base (B) with H_2O to yield the conjugate acid (BH^+) and OH^-:

$$K_b = \frac{[BH^+][OH^-]}{[B]} \qquad (788)$$

base pair Two complementary bases in mononucleotides that are H bonded to each other; guanine (G) always pairs with cytosine (C), and adenine (A) always pairs with thymine (T) (or uracil, U). (659)
base unit (also *fundamental unit*) A unit that defines the standard for one of the seven physical quantities in the International System of Units (SI). (16)
basic-oxygen process The method used to convert pig iron to steel, in which O_2 is blown over and through molten iron to oxidize impurities and decrease the content of carbon. (982)
battery A self-contained group of voltaic cells arranged in series. (932)

becquerel (Bq) The SI unit of radioactivity; 1 Bq = 1 d/s (disintegration per second). (1054)
bent shape (also *V shape*) A molecular shape that arises when a central atom is bonded to two other atoms and has one or two lone pairs; occurs as the AX_2E shape class (bond angle $< 120°$) in the trigonal planar arrangement and as the AX_2E_2 shape class (bond angle $< 109.5°$) in the tetrahedral arrangement. (378)
beta (β) decay A radioactive process in which a beta particle is emitted from a nucleus. (1049)
beta particle (β, β^-, or ${}^{0}_{-1}\beta$) A negatively charged particle identified as a high-speed electron that is one of the common types of radioactive emissions. (1047)
bimolecular reaction An elementary reaction involving the collision of two reactant species. (701)
binary covalent compound A compound that consists of atoms of two elements in which bonding occurs primarily through electron sharing. (68)
binary ionic compound A compound that consists of the oppositely charged ions of two elements. (57)
biomass conversion The process of applying chemical and biological methods to convert plant and/or animal matter into fuels. (245)
biosphere The living systems that have inhabited Earth. (963)
blast furnace A tower-shaped furnace made of brick material in which intense heat and blasts of air are used to convert iron ore and coke to iron metal and carbon dioxide. (980)
body-centered cubic unit cell A unit cell in which a particle lies at each corner and in the center of a cube. (450)
boiling point (bp or T_b) The temperature at which the vapor pressure of a gas equals the external (atmospheric) pressure. (433)
boiling point elevation (ΔT_b) The increase in the boiling point of a solvent caused by the presence of dissolved solute. (517)
bond angle The angle formed by the nuclei of two surrounding atoms with the nucleus of the central atom at the vertex. (376)
bond energy (BE) (also *bond strength*) The enthalpy change accompanying the breakage of a given bond in a mole of gaseous molecules. (341, 545)
bond length The distance between the nuclei of two bonded atoms. (342, 545)
bond order The number of electron pairs shared by two bonded atoms. (340, 545)
bonding MO A molecular orbital formed when wave functions are added to each other, which increases electron density between the nuclei. Electrons occupying such an orbital stabilize the molecule. (410)
bonding pair (also *shared pair*) An electron pair shared by two nuclei; the mutual attraction between the nuclei and the electron pair forms a covalent bond. (340)
Born-Haber cycle A series of hypothetical steps and their enthalpy changes needed to convert elements to an ionic compound and devised to calculate the lattice energy. (334)
Boyle's law The gas law stating that, at constant temperature and amount of gas, the volume occupied by a gas is inversely proportional to the applied (external) pressure: $V \propto 1/P$. (184)
branch A side chain appended to a polymer backbone. (476)
bridge bond (also *three-center, two-electron bond*) A covalent bond in which three atoms are held together by two electrons. (571)
Brønsted-Lowry acid-base definition A model of acid-base behavior based on proton transfer, in which an acid and a base are defined, respectively, as species that donate and accept a proton. (777)
buffer (See *acid-base buffer.*)
buffer capacity A measure of the ability of a buffer to resist a change in pH; related to the total concentrations and relative proportions of buffer components. (821)
buffer range The pH range over which a buffer acts effectively. (821)

C

calibration The process of correcting for systematic error of a measuring device by comparing it to a known standard. (30)
calorie (cal) A unit of energy defined as exactly 4.184 joules; originally defined as the heat needed to raise the temperature of 1 g of water 1°C (from 14.5°C to 15.5°C). (229)
calorimeter A device used to measure the heat released or absorbed by a physical or chemical process taking place within it. (236)
capillarity (or *capillary action*) A property that results in a liquid rising through a narrow space against the pull of gravity. (444)
carbon steel The steel that is produced by the basic-oxygen process, contains about 1% to 1.5% C and other impurities, and is alloyed with metals that prevent corrosion and increase strength. (982)
carbonyl group The C=O grouping of atoms. (643)
carboxylic acid An organic compound (ending, *-oic acid*) that contains the $-\overset{\overset{\Large :\ddot{O}:}{\|}}{C}-\ddot{\underset{\cdot\cdot}{O}}H$ group. (645)
catalyst A substance that increases the rate of a reaction without being used up in the process. (707)
cathode The electrode at which reduction occurs in an electrochemical cell. Electrons enter the cell and are acquired by the oxidizing agent at the cathode. (909)
cathode ray The ray of light emitted by the cathode (negative electrode) in a gas discharge tube; travels in straight lines, unless deflected by magnetic or electric fields. (46)
cation A positively charged ion. (57)
cell potential (E_{cell}) (also *electromotive force,* or *emf; cell voltage*) The potential difference between the electrodes of an electrochemical cell when no current flows. (914)
Celsius scale (formerly *centigrade scale*) A temperature scale in which the freezing and boiling points of water are defined as 0°C and 100°C, respectively. (23)
ceramic A nonmetallic material that is hardened by heating it to high temperatures and, in most cases, consists of silicate microcrystals suspended in a glassy cementing medium. (469)
chain reaction In nuclear fission, a self-sustaining process in which neutrons released by splitting of one nucleus cause other nuclei to split, which releases more neutrons, and so on. (1074)
change in enthalpy (ΔH) The change in internal energy plus the product of the constant pressure and the change in volume: $\Delta H = \Delta E + P\Delta V$; the heat lost or gained at constant pressure: $\Delta H = q_P$. (232)
charge density The ratio of the charge of an ion to its volume. (504)
Charles's law The gas law stating that at constant pressure, the volume occupied by a fixed amount of gas is directly proportional to its absolute temperature: $V \propto T$. (185)
chelate A complex ion in which the metal ion is bonded to a bidentate or polydentate ligand. (1020)

chemical bond The force that holds two atoms together in a molecule (or formula unit). (57)
chemical change (also *chemical reaction*) A change in which a substance is converted into a substance with different composition and properties. (3)
chemical equation A statement that uses chemical formulas to express the identities and quantities of the substances involved in a chemical or physical change. (101)
chemical formula A notation of atomic symbols and numerical subscripts that shows the type and number of each atom in a molecule or formula unit of a substance. (62)
chemical kinetics The study of the rates and mechanisms of reactions. (673)
chemical property A characteristic of a substance that appears as it interacts with, or transforms into, other substances. (3)
chemical reaction (See *chemical change*.)
chemistry The scientific study of matter and the changes it undergoes. (3)
chiral molecule One that is not superimposable on its mirror image; an optically active molecule. In organic compounds, a chiral molecule typically contains a C atom bonded to four different groups (asymmetric C). (627)
chlor-alkali process An industrial method that electrolyzes concentrated aqueous NaCl and produces Cl_2, H_2, and NaOH. (994)
chromatography A separation technique in which a mixture is dissolved in a fluid (gas or liquid), and the components are separated through differences in adsorption to (or solubility in) a solid surface (or viscous liquid). (75)
***cis-trans* isomers** (See *geometric isomers*.)
classical (Arrhenius) acid-base definition A model of acid-base behavior in which an acid is a substance that has H in its formula and produces H^+ in water, and a base is a substance that has OH in its formula and produces OH^- in water. (768)
Clausius-Clapeyron equation An equation that expresses the relationship between vapor pressure P of a liquid and temperature T:

$$\ln P = \frac{-\Delta H_{\text{vap}}}{R}\left(\frac{1}{T}\right) + C, \text{ where } C \text{ is a constant} \qquad (432)$$

coal gasification An industrial process for altering the large molecules in coal to sulfur-free gaseous fuels. (245)
colligative property A property of a solution that depends on the number, not the identity, of solute particles. (515) (See also *boiling point elevation, freezing point depression, osmotic pressure,* and *vapor pressure lowering*.)
collision frequency The average number of collisions per second that a particle undergoes. (206)
collision theory A model that explains reaction rate as the result of particles colliding with a certain minimum energy. (694)
colloid A suspension in which a solute-like phase is dispersed throughout a solvent-like phase. (527)
combustion The process of burning in air, often with release of heat and light. (9)
combustion analysis A method for determining the formula of a compound from the amounts of its combustion products; used commonly for organic compounds. (97)
common-ion effect The shift in the position of an ionic equilibrium away from an ion involved in the process that is caused by the addition or presence of that ion. (816)
complex (See *coordination compound*.)
complex ion An ion consisting of a central metal ion bonded covalently to molecules and/or anions called ligands. (844, 1017)
composition The types and amounts of simpler substances that make up a sample of matter. (2)
compound A substance composed of two or more elements that are chemically combined in fixed proportions. (40)
concentration A measure of the quantity of solute dissolved in a given quantity of solution. (117)
concentration cell A voltaic cell in which both compartments contain the same components but at different concentrations. (928)
condensation The process of a gas changing into a liquid. (426)
condensation polymer A polymer formed by monomers with two functional groups that are linked together in a dehydration-condensation reaction. (652)
conduction band In band theory, the empty, higher energy portion of the band of molecular orbitals into which electrons move when conducting heat and electricity. (462)
conductor A substance (usually a metal) that conducts an electric current well. (462)
conjugate acid-base pair Two species related to each other through the gain or loss of a proton; the acid has one more proton than its conjugate base. (779)
constitutional isomers (also *structural isomers*) Compounds with the same molecular formula but different arrangements of atoms. (625, 1023)
contact process An industrial process for the manufacture of sulfuric acid based on the catalyzed oxidation of SO_2. (992)
controlled experiment An experiment that measures the effect of one variable at a time by keeping other variables constant. (11)
conversion factor A ratio of equivalent quantities that is equal to 1 and used to convert the units of a quantity. (12)
coordinate covalent bond A covalent bond formed when one atom donates both electrons to give the shared pair and, once formed, is identical to any covalent single bond. (1026)
coordination compound (also *complex*) A substance containing at least one complex ion. (1017)
coordination isomers Two or more coordination compounds with the same composition in which the complex ions have different ligand arrangements. (1023)
coordination number In a crystal, the number of nearest neighbors surrounding a particle. (450) In a complex, the number of ligand atoms bonded to the central metal ion. (1018)
copolymer A polymer that consists of two or more types of monomer. (477)
core The dense, innermost region of Earth. (961)
core electrons (See *inner electrons*.)
corrosion The natural redox process that results in unwanted oxidation of a metal. (936)
coulomb (C) The SI unit of electric charge. One coulomb is the charge of 6.242×10^{18} electrons; one electron possesses a charge of 1.602×10^{-19} C. (914)
Coulomb's law A law stating that the electrostatic force associated with two charges A and B is directly proportional to the product of their magnitudes and inversely proportional to the square of the distance between them:

$$\text{electrostatic force} \propto \frac{\text{charge A} \times \text{charge B}}{(\text{distance})^2} \qquad (335)$$

counter ion A simple ion associated with a complex ion in a coordination compound. (1017)
coupling of reactions The pairing of reactions of which one releases enough free energy for the other to occur. (888)
covalent bond A type of bond in which atoms are bonded through the sharing of two electrons; the mutual attraction of the nuclei and an electron pair that holds atoms together in a molecule. (60, 339)
covalent bonding The idealized bonding type that is based on localized electron-pair sharing between two atoms with little difference in their tendencies to lose or gain electrons (most commonly nonmetals). (330, 544)
covalent compound A compound that consists of atoms bonded together by shared electron pairs. (57)
covalent radius One-half the distance between nuclei of identical covalently bonded atoms. (305)
critical mass The minimum mass needed to achieve a chain reaction. (1074)
critical point The point on a phase diagram above which the vapor cannot be condensed to a liquid; the end of the liquid-gas curve. (434)
crosslink A branch that covalently joins one polymer chain to another. (476)
crust The thin, light, heterogeneous outer layer of Earth. (961)
crystal defect Any of a variety of disruptions in the regularity of a crystal structure. (464)
crystal field splitting energy (Δ) The difference in energy between two sets of metal-ion *d* orbitals that results from electrostatic interactions with the surrounding ligands. (1030)
crystal field theory A model that explains the color and magnetism of coordination compounds based on the effects of ligands on metal-ion *d*-orbital energies. (1028)
crystalline solid Solid with a well-defined shape because of the orderly arrangement of the atoms, molecules, or ions. (449)
crystallization A technique used to separate and purify the components of a mixture through differences in solubility in which a component comes out of solution as crystals. (74)
cubic closest packing A crystal structure based on the face-centered cubic unit cell in which the layers have an *abcabc* . . . pattern. (452)
cubic meter (m^3) The SI-derived unit of volume. (17)
curie (Ci) The most common unit of radioactivity, defined as the number of nuclei disintegrating each second in 1 g of radium-226; 1 Ci = 3.70×10^{10} d/s (disintegrations per second). (1054)
cyclic hydrocarbon A hydrocarbon with one or more rings in its structure. (625)

D

***d* orbital** An atomic orbital with $l = 2$. (282)
dalton (Da) A unit of mass identical to *atomic mass unit*. (51)
Dalton's law of partial pressures A gas law stating that, in a mixture of unreacting gases, the total pressure is the sum of the partial pressures of the individual gases: $P_{total} = P_1 + P_2 + \cdots$. (196)
data Pieces of quantitative information obtained by observation. (11)
de Broglie wavelength The wavelength of a moving particle obtained from the de Broglie equation: $\lambda = h/mu$. (272)
decay constant The rate constant *k* for radioactive decay. (1054)
decay series (also *disintegration series*) The succession of steps a parent nucleus undergoes as it decays into a stable daughter nucleus. (1053)
degree of polymerization (*n*) The number of monomer repeat units in a polymer chain. (472)
dehydration-condensation reaction A reaction in which H and OH groups on two molecules react to form water as one of the products. (590)
delocalization (See *electron-pair delocalization.*)
density (*d*) An intensive physical property of a substance at a given temperature and pressure, defined as the ratio of the mass to the volume: $d = m/V$. (20)
deposition The process of changing directly from gas to solid. (427)
derived unit Any of various combinations of the seven SI base units. (16)
desalination A process used to remove large amounts of ions from seawater, usually by reverse osmosis. (530)
deuterons Nuclei of the stable hydrogen isotope deuterium, ^2H. (1060)
diagonal relationship Physical and chemical similarities between a Period 2 element and one located diagonally down and to the right in Period 3. (563)
diamagnetism The tendency of a species not to be attracted (or to be slightly repelled) by a magnetic field as a result of its electrons being paired. (318, 1009)
differentiation The geochemical process of forming regions in Earth based on differences in composition and density. (961)
diffraction The phenomenon in which a wave striking the edge of an object bends around it. A wave passing through a slit as wide as its wavelength forms a circular wave. (261)
diffusion The movement of one fluid through another. (205)
dimensional analysis (also *factor-label method*) A calculation method in which arithmetic steps are accompanied by the appropriate canceling of units. (14)
dipole-dipole force The intermolecular attraction between oppositely charged poles of nearby polar molecules. (438)
dipole–induced dipole force The intermolecular attraction between a polar molecule and the oppositely charged pole it induces in a nearby molecule. (491)
dipole moment (μ) A measure of molecular polarity; the magnitude of the partial charges on the ends of a molecule (in coulombs) times the distance between them (in meters). (387)
disaccharide An organic compound formed by a dehydration-condensation reaction between two simple sugars (monosaccharides). (654)
disintegration series (See *decay series.*)
dispersion force (also *London force*) The intermolecular attraction between all particles as a result of instantaneous polarizations of their electron clouds; the intermolecular force primarily responsible for the condensed states of nonpolar substances. (441)
disproportionation reaction A reaction in which a given substance is both oxidized and reduced. (588)
distillation A separation technique in which a more volatile component of a mixture vaporizes and condenses separately from the less volatile components. (74)
donor atom An atom that donates a lone pair of electrons to form a covalent bond, usually from ligand to metal ion in a complex. (1019)

doping Adding small amounts of other elements into the crystal structure of a semiconductor to enhance a specific property, usually conductivity. (464)
double bond A covalent bond that consists of two bonding pairs; two atoms sharing four electrons in the form of one σ and one π bond. (340, 407)
double helix The two intertwined polynucleotide strands held together by H bonds that form the structure of DNA (deoxyribonucleic acid). (501, 659)
Downs cell An industrial apparatus that electrolyzes molten NaCl to produce sodium and chlorine. (979)
dynamic equilibrium The condition at which the forward and reverse reactions are taking place at the same rate, so there is no net change in the amounts of reactants or products. (165)

E

e_g orbitals The set of orbitals (composed of $d_{x^2-y^2}$ and d_{z^2}) that results when the energies of the metal-ion *d* orbitals are split by a ligand field. This set is higher in energy than the other (t_{2g}) set in an octahedral field of ligands and lower in energy in a tetrahedral field. (1030)
effective collision A collision in which the particles meet with sufficient energy and an orientation that allows them to react. (696)
effective nuclear charge (Z_{eff}) The nuclear charge an electron actually experiences as a result of shielding effects due to the presence of other electrons. (294)
effusion The process by which a gas escapes from its container through a tiny hole into an evacuated space. (205)
elastomer A polymeric material that can be stretched and springs back to its original shape when released. (476)
electrochemical cell A system that incorporates a redox reaction to produce or use electrical energy. (903)
electrochemistry The study of the relationship between chemical change and electrical work. (903)
electrode The part of an electrochemical cell that conducts the electricity between the cell and the surroundings. (908)
electrolysis The nonspontaneous lysing (splitting) of a substance, often to its component elements, by supplying electrical energy. (941)
electrolyte A substance that conducts a current when it dissolves in water. (136, 515) A mixture of ions, in which the electrodes of an electrochemical cell are immersed, that conducts a current. (908)
electrolytic cell An electrochemical system that uses electrical energy to drive a nonspontaneous chemical reaction ($\Delta G > 0$). (908)
electromagnetic (EM) radiation (also *electromagnetic energy* or *radiant energy*) Oscillating, perpendicular electric and magnetic fields moving simultaneously through space as waves and manifested as visible light, x-rays, microwaves, radio waves, and so on. (257)
electromagnetic spectrum The continuum of wavelengths of radiant energy. (259)
electromotive force (emf) (See *cell potential*.)
electron (e^-) A subatomic particle that possesses a unit negative charge (1.60218×10^{-19} C) and occupies the space around the atomic nucleus. (49)
electron affinity (EA) The energy change (in kJ) accompanying the addition of one mole of electrons to one mole of gaseous atoms or ions. (312)
electron capture A type of radioactive decay in which a nucleus draws in an orbital electron, usually one from the lowest energy level, and releases energy. (1049)
electron cloud An imaginary representation of an electron's rapidly changing position around the nucleus over time. (276)
electron configuration The distribution of electrons within the orbitals of the atoms of an element; also the notation for such a distribution. (291, 542)
electron deficient Referring to a bonded atom, such as Be or B, that has fewer than eight valence electrons. (372)
electron density diagram (also *electron probability density diagram*) The pictorial representation for a given energy sublevel of the quantity ψ^2 (the probability density of the electron lying within a particular tiny volume) as a function of *r* (distance from the nucleus). (276)
electron-pair delocalization (also *delocalization*) The process by which electron density is spread over several atoms rather than remaining between two. (370)
electron-sea model A qualitative description of metallic bonding proposing that metal atoms pool their valence electrons into a delocalized "sea" of electrons in which the metal cores (metal ions) are submerged in an orderly array. (357)
electron volt (eV) The energy (in joules, J) that an electron acquires when it moves through a potential difference of 1 volt; $1 \text{ eV} = 1.602 \times 10^{-19}$ J. (1071)
electronegativity (EN) The relative ability of a bonded atom to attract shared electrons. (351, 543)
electronegativity difference (ΔEN) The difference in electronegativities between the atoms in a bond. (354)
electrorefining An industrial electrolytic process in which a sample of impure metal acts as the anode and a sample of the pure metal acts as the cathode. (977)
element The simplest type of substance with unique physical and chemical properties. An element consists of only one kind of atom, so it cannot be broken down into simpler substances. (39)
elementary reaction (also *elementary step*) A simple reaction that describes a single molecular event in a proposed reaction mechanism. (701)
elimination reaction A type of organic reaction in which C atoms are bonded to fewer atoms in the product than in the reactant, which leads to multiple bonding. (636)
emission spectrum The line spectrum produced when excited atoms return to lower energy levels and emit photons characteristic of the element. (269)
empirical formula A chemical formula that shows the lowest relative number of atoms of each element in a compound. (62)
enantiomers (See *optical isomers*.)
end point The point in a titration at which the indicator changes color. (147, 826)
end-to-end overlap Overlap of *s*, *p*, or hybrid atomic orbitals that yields a sigma (σ) bond. (545)
endothermic Occurring with an absorption of heat from the surroundings and therefore an increase in the enthalpy of the system ($\Delta H > 0$). (233)
energy The capacity to do work, that is, to move matter. (6) [See also *kinetic energy* (E_k) and *potential energy* (E_p).]
enthalpy (*H*) A thermodynamic quantity that is the sum of the internal energy plus the product of the pressure and volume. (238)
enthalpy diagram A graphic depiction of the enthalpy change of a system. (233)

enthalpy of hydration (ΔH_{hydr}) (See *heat of hydration*.)
enthalpy of solution (ΔH_{soln}) (See *heat of solution*.)
entropy (S) A thermodynamic quantity related to the number of ways the energy of a system can be dispersed through the motions of its particles. (505, 867)
environmental cycle The physical, chemical, and biological paths through which the atoms of an element move within Earth's crust. (966)
enzyme A biological macromolecule (usually a protein) that acts as a catalyst. (709)
enzyme-substrate complex (ES) The intermediate in an enzyme-catalyzed reaction, which consists of enzyme and substrate(s) and whose concentration determines the rate of product formation. (710)
equatorial group An atom (or group) that lies in the trigonal plane of a trigonal bipyramidal molecule, or a similar structural feature in a molecule. (380)
equilibrium constant (K) The value obtained when equilibrium concentrations are substituted into the reaction quotient. (725)
equivalence point The point in a titration when the number of moles of the added species is stoichiometrically equivalent to the original number of moles of the other species. (147, 825)
ester An organic compound that contains the $-\overset{\overset{\displaystyle :O\!:}{\|}}{C}-\ddot{\underset{\cdot\cdot}{O}}-\underset{|}{\overset{|}{C}}-$ group. (646)
exact number A quantity, usually obtained by counting or based on a unit definition, that has no uncertainty associated with it and, therefore, contains as many significant figures as a calculation requires. (29)
excitation The process by which a substance absorbs energy from low-energy radioactive particles, causing its electrons to move to higher energy levels. (1062)
excited state Any electron configuration of an atom or molecule other than the lowest energy (ground) state. (266)
exclusion principle A principle developed by Wolfgang Pauli stating that no two electrons in an atom can have the same set of four quantum numbers. The principle arises from the fact that an orbital has a maximum occupancy of two electrons and their spins are paired. (293)
exothermic Occurring with a release of heat to the surroundings and therefore a decrease in the enthalpy of the system ($\Delta H < 0$). (233)
expanded valence shell A valence level that can accommodate more than 8 electrons by using available *d* orbitals; occurs only for elements in Period 3 or higher. (373)
experiment A clear set of procedural steps that tests a hypothesis. (11)
extensive property A property, such as mass, that depends on the quantity of substance present. (20)
extraction A method of separating the components of a mixture based on differences in their solubility in a nonmiscible solvent. (74)

F

face-centered cubic unit cell A unit cell in which a particle occurs at each corner and in the center of each face of a cube. (450)
Faraday constant (F) The physical constant representing the charge of 1 mol of electrons: $F = 96{,}485$ C/mol e^-. (923)
fatty acid A carboxylic acid that has a long hydrocarbon chain and is derived from a natural source. (646)
filtration A method of separating the components of a mixture on the basis of differences in particle size. (74)
first law of thermodynamics (See *law of conservation of energy*.)
fission The process by which a heavier nucleus splits into lighter nuclei with the release of energy. (1070)
fixation A chemical/biochemical process that converts a gaseous substance in the environment into a condensed form that can be used by organisms. (967)
flame test A procedure for identifying the presence of metal ions in which a granule of a compound or a drop of its solution is placed in a flame to observe a characteristic color. (269)
flotation A metallurgical process in which oil and detergent are mixed with pulverized ore in water to create a slurry that separates the mineral from the gangue. (974)
formal charge The hypothetical charge on an atom in a molecule or ion, equal to the number of valence electrons minus the sum of all the unshared and half the shared valence electrons. (371)
formation constant (K_f) An equilibrium constant for the formation of a complex ion from the hydrated metal ion and ligands. (844)
formation equation An equation in which 1 mole of a compound forms from its elements. (242)
formula unit The chemical unit of a compound that contains the number and type of atoms (or ions) expressed in the chemical formula. (64)
fossil fuel Any fuel, including coal, petroleum, and natural gas, derived from the products of the decay of dead organisms. (245)
fraction by mass (also *mass fraction*) The portion of a compound's mass contributed by an element; the mass of an element in a compound divided by the mass of the compound. (42)
fractional distillation A process involving numerous vaporization-condensation steps used to separate two or more volatile components. (524)
free energy (G) A thermodynamic quantity that is the difference between the enthalpy and the product of the absolute temperature and the entropy: $G = H - TS$. (881)
free radical A molecular or atomic species with one or more unpaired electrons, which typically make it very reactive. (373, 1063)
freezing The process of cooling a liquid until it solidifies. (426)
freezing point depression (ΔT_f) A lowering of the freezing point of a solvent caused by the presence of dissolved solute particles. (519)
frequency (ν) The number of cycles a wave undergoes per second, expressed in units of 1/second, or s^{-1} [also called hertz (Hz)]. (258)
frequency factor (A) The product of the collision frequency Z and an orientation probability factor p that is specific for a reaction. (696)
fuel cell (also *flow battery*) A battery that is not self-contained and in which electricity is generated by the controlled oxidation of a fuel. (935)
functional group A specific combination of atoms, typically containing a carbon-carbon multiple bond and/or carbon-heteroatom bond, that reacts in a characteristic way no matter what molecule it occurs in. (619)
fundamental unit (See *base unit*.)
fusion (See *melting*.)

fusion (nuclear) The process by which light nuclei combine to form a heavier nucleus with the release of energy. (1070)

G

galvanic cell (See *voltaic cell.*)
gamma emission The type of radioactive decay in which gamma rays are emitted from an excited nucleus. (1049)
gamma (γ) ray A very high-energy photon. (1047)
gangue In an ore, the debris, such as sand, rock, and clay, attached to the mineral. (973)
gas One of the three states of matter. A gas fills its container regardless of the shape. (4)
Geiger-Müller counter An ionization counter that detects radioactive emissions through their ionization of gas atoms within the instrument. (1055)
genetic code The set of three-base sequences that is translated into specific amino acids during the process of protein synthesis. (659)
geometric isomers (also *cis-trans isomers* or *diastereomers*) Stereoisomers in which the molecules have the same connections between atoms but differ in the spatial arrangements of the atoms. The *cis* isomer has similar groups on the same side of a structural feature; the *trans* isomer has them on opposite sides. (629, 1024)
Graham's law of effusion A gas law stating that the rate of effusion of a gas is inversely proportional to the square root of its density (or molar mass):

$$\text{rate} \propto \frac{1}{\sqrt{\mathcal{M}}} \qquad (205)$$

gray (Gy) The SI unit of absorbed radiation dose; 1 Gy = 1 J/kg tissue. (1063)
green chemistry Field that is focused on developing methods to synthesize compounds efficiently and reduce or prevent the release of harmful products into the environment. (115)
ground state The electron configuration of an atom or ion that is lowest in energy. (266)
group A vertical column in the periodic table. (54)

H

Haber process An industrial process used to form ammonia from its elements. (755)
half-cell A portion of an electrochemical cell in which a half-reaction takes place. (910)
half-life ($t_{1/2}$) In chemical processes, the time required for half the initial reactant concentration to be consumed. (689) In nuclear processes, the time required for half the initial number of nuclei in a sample to decay. (1054)
half-reaction method A method of balancing redox reactions by treating the oxidation and reduction half-reactions separately. (904)
haloalkane (also *alkyl halide*) A hydrocarbon with one or more halogen atoms (X) in place of H; contains a $-\overset{|}{\underset{|}{C}}-\ddot{\underset{..}{X}}:$ group. (640)
hard water Water that contains large amounts of divalent cations, especially Ca^{2+} and Mg^{2+}. (529)
heat (q) The energy transferred between objects because of differences in their temperatures only; thermal energy. (22, 227)
heat capacity The quantity of heat required to change the temperature of an object by 1 K. (235)
heat of combustion (ΔH_{comb}) The enthalpy change occurring when 1 mol of a substance combines with oxygen in a combustion reaction. (234)
heat of formation (ΔH_f) The enthalpy change occurring when 1 mol of a compound is produced from its elements. (234)
heat of fusion (ΔH_{fus}) The enthalpy change occurring when 1 mol of a solid substance melts. (234, 427)
heat of hydration (ΔH_{hydr}) (also *enthalpy of hydration*) The enthalpy change occurring when 1 mol of a gaseous species is hydrated. The sum of the enthalpies from separating water molecules and mixing the gaseous species with them. (504)
heat of reaction (ΔH_{rxn}) The enthalpy change of a reaction. (233)
heat of solution (ΔH_{soln}) (also *enthalpy of solution*) The enthalpy change occurring when a solution forms from solute and solvent. The sum of the enthalpies from separating solute and solvent molecules and mixing them. (503)
heat of sublimation (ΔH_{subl}) The enthalpy change occurring when 1 mol of a solid substance changes directly to a gas. The sum of the heats of fusion and vaporization. (427)
heat of vaporization (ΔH_{vap}) The enthalpy change occurring when 1 mol of a liquid substance vaporizes. (234, 427)
heating-cooling curve A plot of temperature vs. time for a substance when heat is absorbed or released by the system at a constant rate. (428)
Henderson-Hasselbalch equation An equation for calculating the pH of a buffer system:

$$\text{pH} = \text{p}K_a + \log\left(\frac{[\text{base}]}{[\text{acid}]}\right) \qquad (820)$$

Henry's law A law stating that the solubility of a gas in a liquid is directly proportional to the partial pressure of the gas above the liquid: $S_{gas} = k_H \times P_{gas}$. (510)
Hess's law of heat summation A law stating that the enthalpy change of an overall process is the sum of the enthalpy changes of the individual steps of the process. (240)
heteroatom Any atom in an organic compound other than C or H. (619)
heterogeneous catalyst A catalyst that occurs in a different phase from the reactants, usually a solid interacting with gaseous or liquid reactants. (708)
heterogeneous mixture A mixture that has one or more visible boundaries among its components. (73)
hexagonal closest packing A crystal structure based on the hexagonal unit cell in which the layers have an *abab* . . . pattern. (452)
high-spin complex Complex ion that has the same number of unpaired electrons as in the isolated metal ion; contains weak-field ligands. (1032)
homogeneous catalyst A catalyst (gas, liquid, or soluble solid) that exists in the same phase as the reactants. (707)
homogeneous mixture (also *solution*) A mixture that has no visible boundaries among its components. (73)
homologous series A series of organic compounds in which each member differs from the next by a $-CH_2-$ (methylene) group. (623)
homonuclear diatomic molecule A molecule composed of two identical atoms. (412)
Hund's rule A principle stating that when orbitals of equal energy are available, the electron configuration of lowest energy

has the maximum number of unpaired electrons with parallel spins. (297)
hybrid orbital An atomic orbital postulated to form during bonding by the mathematical mixing of specific combinations of nonequivalent orbitals in a given atom. (400)
hybridization A postulated process of orbital mixing to form hybrid orbitals. (400)
hydrate A compound in which a specific number of water molecules are associated with each formula unit. (66)
hydration Solvation in water. (503)
hydration shell The oriented cluster of water molecules that surrounds an ion in aqueous solution. (490)
hydrocarbon An organic compound that contains only H and C atoms. (620)
hydrogen bond (H bond) A type of dipole-dipole force that arises between molecules that have an H atom bonded to a small, highly electronegative atom with lone pairs, usually N, O, or F. (439, 551)
hydrogenation The addition of hydrogen to a carbon-carbon multiple bond to form a carbon-carbon single bond. (708)
hydrolysis Cleaving a molecule by reaction with water, in which one part of the molecule bonds to the water —OH and the other to the water H. (647)
hydronium ion (H_3O^+) A proton covalently bonded to a water molecule. (768)
hydrosphere The liquid portion of Earth's crust. (963)
hypothesis A testable proposal made to explain an observation. If inconsistent with experimental results, a hypothesis is revised or discarded. (11)

I

ideal gas A hypothetical gas that exhibits linear relationships among volume, pressure, temperature, and amount (mol) at all conditions; approximated by simple gases at ordinary conditions. (183)
ideal gas law (also *ideal gas equation*) An equation that expresses the relationships among volume, pressure, temperature, and amount (mol) of an ideal gas: $PV = nRT$. (188)
ideal solution A solution whose vapor pressure equals the mole fraction of the solvent times the vapor pressure of the pure solvent; approximated only by very dilute solutions. (516) (See also *Raoult's law.*)
indicator (See *acid-base indicator.*)
induced-fit model A model of enzyme action that pictures the binding of the substrate as inducing the active site to change its shape and become catalytically active. (710)
infrared (IR) The region of the electromagnetic spectrum between the microwave and visible regions. (259)
infrared (IR) spectroscopy An instrumental technique for determining the types of bonds in a covalent molecule by measuring the absorption of IR radiation. (345)
initial rate The instantaneous rate occurring as soon as the reactants are mixed, that is, at $t = 0$. (677)
inner electrons (also *core electrons*) Electrons that fill all the energy levels of an atom except the valence level; electrons also present in atoms of the previous noble gas and any completed transition series. (303)
inner transition elements The elements of the periodic table in which *f* orbitals are being filled; the lanthanides and actinides. (303, 1010)
instantaneous rate The reaction rate at a particular time, given by the slope of a tangent to a plot of reactant concentration vs. time. (677)
insulator A substance (usually a nonmetal) that does not conduct an electric current. (463)
integrated rate law A mathematical expression for reactant concentration as a function of time. (686)
intensive property A property, such as density, that does not depend on the quantity of substance present. (20)
interhalogen compound A compound consisting entirely of halogens. (602)
intermolecular forces (also *interparticle forces*) The attractive and repulsive forces among the particles—molecules, atoms, or ions—in a sample of matter. (425)
internal energy (E) The sum of the kinetic and potential energies of all the particles in a system. (226)
interstitial hydride The substance formed when metals absorb H_2 into the spaces between their atoms. (555)
ion A charged particle that forms from an atom (or covalently bonded group of atoms) when it gains or loses one or more electrons. (57)
ion-dipole force The intermolecular attractive force between an ion and a polar molecule (dipole). (437)
ion exchange A process of softening water by exchanging one type of ion (usually Ca^{2+}) for another (usually Na^+) by binding the ions on a specially designed resin. (529)
ion–induced dipole force The intermolecular attractive force between an ion and the dipole it induces in the electron cloud of a nearby particle. (490)
ion pair A pair of ions that form a gaseous ionic molecule; sometimes formed when a salt boils. (338)
ion-product constant for water (K_w) The equilibrium constant for the autoionization of water:

$$K_w = [H_3O^+][OH^-] \qquad (773)$$

ionic atmosphere A cluster of ions of net opposite charge surrounding a given ion in solution. (525)
ionic bonding The idealized type of bonding based on the attraction of oppositely charged ions that arise through electron transfer between atoms with large differences in their tendencies to lose or gain electrons (typically metals and nonmetals). (330, 544)
ionic compound A compound that consists of oppositely charged ions. (57, 551)
ionic radius The size of an ion as measured by the distance between the centers of adjacent ions in a crystalline ionic compound. (320)
ionic solid A solid whose unit cell contains cations and anions. (457)
ionization The process by which a substance absorbs energy from high-energy radioactive particles and loses an electron to become ionized. (1062)
ionization energy (IE) The energy (in kJ) required to remove completely one mole of electrons from one mole of gaseous atoms or ions. (309, 543)
ionizing radiation The high-energy radiation that forms ions in a substance by causing electron loss. (1062)
isoelectronic Having the same number and configuration of electrons as another species. (316)

isomer One of two or more compounds with the same molecular formula but different properties, often as a result of different arrangements of atoms. (100, 625, 1023)
isotopes Atoms of a given atomic number (that is, of a specific element) that have different numbers of neutrons and therefore different mass numbers. (50, 1046)
isotopic mass The mass (in amu) of an isotope relative to the mass of carbon-12. (51)

J

joule (J) The SI unit of energy; $1\ J = 1\ kg{\cdot}m^2/s^2$. (229)

K

kelvin (K) The SI base unit of temperature. The kelvin is the same size as the Celsius degree. (23)
Kelvin scale (See *absolute scale*.)
ketone An organic compound (ending, *-one*) that contains a carbonyl group bonded to two other C atoms, $-\overset{|}{\underset{|}{C}}-\overset{:O:}{\overset{\|}{C}}-\overset{|}{\underset{|}{C}}-$. (644)
kilogram (kg) The SI base unit of mass. (19)
kinetic energy (E_k) The energy an object has because of its motion. (6)
kinetic-molecular theory The model that explains gas behavior in terms of particles in random motion whose volumes and interactions are negligible. (200)

L

lanthanide contraction The additional decrease in atomic and ionic size, beyond the expected trend, caused by the poor shielding of the increasing nuclear charge by *f* electrons in the elements following the lanthanides. (1006)
lanthanides (also *rare earths*) The Period 6 (4*f*) series of inner transition elements, which includes cerium (Ce; $Z = 58$) through lutetium (Lu; $Z = 71$). (304, 1010)
lattice The three-dimensional arrangement of points created by choosing each point to be at the same location within each particle of a crystal; thus, the lattice consists of all points with identical surroundings. (450)
lattice energy ($\Delta H^\circ_{lattice}$) The enthalpy change (always positive) that accompanies a solid ionic compound forming separate gaseous ions. (334)
law (See *natural law*.)
law of chemical equilibrium (also *law of mass action*) The law stating that when a system reaches equilibrium at a given temperature, the ratio of quantities that make up the reaction quotient has a constant numerical value. (726)
law of conservation of energy (also *first law of thermodynamics*) A basic observation that the total energy of the universe is constant: $\Delta E_{universe} = \Delta E_{system} + \Delta E_{surroundings} = 0$. (229)
law of definite (or constant) composition A mass law stating that, no matter what its source, a particular compound is composed of the same elements in the same parts (fractions) by mass. (42)
law of mass action (See *law of chemical equilibrium*.)
law of mass conservation A mass law stating that the total mass of substances does not change during a chemical reaction. (41)
law of multiple proportions A mass law stating that if elements A and B react to form two compounds, the different masses of B that combine with a fixed mass of A can be expressed as a ratio of small whole numbers. (43)
Le Châtelier's principle A principle stating that if a system in a state of equilibrium is disturbed, it will undergo a change that shifts its equilibrium position in a direction that reduces the effect of the disturbance. (745)
leaching A hydrometallurgical process that extracts a metal selectively, usually through formation of a complex ion. (974)
level (also *shell*) A specific energy state of an atom given by the principal quantum number *n*. (278)
leveling effect The inability of a solvent to distinguish the strength of an acid (or base) that is stronger than the conjugate acid (or conjugate base) of the solvent. (800)
Lewis acid-base definition A model of acid-base behavior in which acids and bases are defined, respectively, as species that accept and donate an electron pair. (800)
Lewis electron-dot symbol A notation in which the element symbol represents the nucleus and inner electrons, and surrounding dots represent the valence electrons. (331)
Lewis structure (also *Lewis formula*) A structural formula consisting of electron-dot symbols, with lines as bonding pairs and dots as lone pairs. (366)
ligand A molecule or anion bonded to a central metal ion in a complex ion. (844, 1017)
like-dissolves-like rule An empirical observation stating that substances having similar kinds of intermolecular forces dissolve in each other. (490)
limiting reactant (also *limiting reagent*) The reactant that is consumed when a reaction occurs and therefore the one that determines the maximum amount of product that can form. (110)
line spectrum A series of separated lines of different colors representing photons whose wavelengths are characteristic of an element. (265) (See also *emission spectrum*.)
linear arrangement The geometric arrangement obtained when two electron groups maximize their separation around a central atom. (377)
linear shape A molecular shape formed by three atoms lying in a straight line, with a bond angle of 180° (shape class AX_2 or AX_2E_3). (377, 381)
linkage isomers Coordination compounds with the same composition but with different ligand donor atoms linked to the central metal ion. (1024)
lipid Any of a class of biomolecules, including fats and oils, that are soluble in nonpolar solvents. (647)
lipid bilayer An extended sheetlike double layer of phospholipid molecules that forms in water and has the charged heads of the molecules on the surfaces of the bilayer and the polar tails within the interior. (499)
liquid One of the three states of matter. A liquid fills a container to the extent of its own volume and thus forms a surface. (4)
liquid crystal A substance that flows like a liquid but packs like a crystalline solid at the molecular level. (466)
liter (L) A non-SI unit of volume equivalent to 1 cubic decimeter ($0.001\ m^3$). (17)
lithosphere The solid portion of Earth's crust. (963)
lock-and-key model A model of enzyme function that pictures the enzyme active site and the substrate as rigid shapes that fit together as a lock and key, respectively. (710)
lone pair (also *unshared pair*) An electron pair that is part of an atom's valence shell but not involved in covalent bonding. (340)

low-spin complex Complex ion that has fewer unpaired electrons than in the free metal ion because of the presence of strong-field ligands. (1032)

M

macromolecule (See *polymer.*)
magnetic quantum number (m_l) (also *orbital-orientation quantum number*) An integer from $-l$ through 0 to $+l$ that specifies the orientation of an atomic orbital in the three-dimensional space about the nucleus. (277)
manometer A device used to measure the pressure of a gas in a laboratory experiment. (181)
mantle A thick homogeneous layer of Earth's internal structure that lies between the core and the crust. (961)
mass The quantity of matter an object contains. Balances are designed to measure mass. (19)
mass-action expression (See *reaction quotient.*)
mass defect The mass decrease that occurs when nucleons combine to form a nucleus. (1071)
mass fraction (See *fraction by mass.*)
mass number (A) The total number of protons and neutrons in the nucleus of an atom. (50)
mass percent (also *mass %* or *percent by mass*) The fraction by mass expressed as a percentage. (42) A concentration term [% (w/w)] expressed as the mass in grams of solute dissolved per 100. g of solution. (512)
mass spectrometry An instrumental method for measuring the relative masses of particles in a sample by creating charged particles and separating them according to their mass-charge ratio. (51)
matter Anything that possesses mass and occupies volume. (2)
mean free path The average distance a molecule travels between collisions at a given temperature and pressure. (206)
melting (also *fusion*) The change of a substance from a solid to a liquid. (426, 1070)
melting point (mp or T_f) The temperature at which the solid and liquid forms of a substance are at equilibrium. (434)
metabolic pathway A biochemical reaction sequence that flows in one direction and in which each reaction is enzyme catalyzed. (754)
metal A substance or mixture that is relatively shiny and malleable and is a good conductor of heat and electricity. In reactions, metals tend to transfer electrons to nonmetals and form ionic compounds. (56)
metallic bonding An idealized type of bonding based on the attraction between metal ions and their delocalized valence electrons. (330, 544) (See also *electron-sea model.*)
metallic radius One-half the distance between the nuclei of adjacent individual atoms in a crystal of an element. (305)
metallic solid A solid whose individual atoms are held together by metallic bonding. (459)
metalloid (also *semimetal*) An element with properties between those of metals and nonmetals. (56)
metallurgy The branch of materials science concerned with the extraction and utilization of metals. (973)
metathesis reaction (also *double-displacement reaction*) A reaction in which atoms or ions of two compounds exchange bonding partners. (143)
meter (m) The SI base unit of length. The distance light travels in a vacuum in 1/299,792,458 second. (17)
milliliter (mL) A volume (0.001 L) equivalent to 1 cm^3. (17)
millimeter of mercury (mmHg) A unit of pressure based on the difference in the heights of mercury in a barometer or manometer. Renamed the *torr* in honor of Torricelli. (181)
mineral The portion of an ore that contains the element of interest, a mineral is a naturally occurring, homogeneous, crystalline inorganic solid, with a well-defined composition. (973)
miscible Soluble in any proportion. (490)
mixture A group of two or more elements and/or compounds that are physically intermingled. (40, 489)
MO bond order One-half the difference between the number of electrons in bonding and antibonding MOs. (411)
model (also *theory*) A simplified conceptual picture based on experiment that explains how an aspect of nature occurs. (11)
molality (m) A concentration term expressed as number of moles of solute dissolved in 1000 g (1 kg) of solvent. (511)
molar heat capacity (C) The quantity of heat required to change the temperature of 1 mol of a substance by 1 K. (235)
molar mass ($\mathcal{M}$) (also *gram-molecular weight*) The mass of 1 mol of entities (atoms, molecules, or formula units) of a substance, in units of g/mol. (89)
molar solubility The solubility expressed in terms of amount (mol) of dissolved solute per liter of solution. (834)
molarity (M) A concentration term expressed as the moles of solute dissolved in 1 L of solution. (117)
mole (mol) The SI base unit for amount of a substance. The amount that contains a number of objects equal to the number of atoms in exactly 12 g of carbon-12. (87)
mole fraction (X) A concentration term expressed as the ratio of moles of one component of a mixture to the total moles present. (196, 513)
molecular compounds Compounds that consist of individual molecules and whose physical state depends on intermolecular forces. (551)
molecular equation A chemical equation showing a reaction in solution in which reactants and products appear as intact, undissociated compounds. (141)
molecular formula A formula that shows the actual number of atoms of each element in a molecule. (62, 96)
molecular mass (also *molecular weight*) The sum (in amu) of the atomic masses of a formula unit of a compound. (70)
molecular orbital (MO) An orbital of given energy and shape that extends over a molecule and can be occupied by no more than two electrons. (410)
molecular orbital (MO) diagram A depiction of the relative energy and number of electrons in each MO, as well as the atomic orbitals from which the MOs form. (411)
molecular orbital (MO) theory A model that describes a molecule as a collection of nuclei and electrons in which the electrons occupy orbitals that extend over the entire molecule. (409)
molecular polarity The overall distribution of electronic charge in a molecule, determined by its shape and bond polarities. (387)
molecular shape The three-dimensional structure defined by the relative positions of the atomic nuclei in a molecule. (376)
molecular solid A solid held together by intermolecular forces between individual molecules. (456)
molecularity The number of reactant particles involved in an elementary step. (701)
molecule A structure consisting of two or more atoms that are chemically bound together and behave as an independent unit. (40)

monatomic ion An ion derived from a single atom. (57)
monomer A small molecule, linked covalently to others of the same or similar type to form a polymer, on which the repeat unit of the polymer is based. (472)
mononucleotide A monomer unit of a nucleic acid, consisting of an N-containing base, a sugar, and a phosphate group. (500, 658)
monosaccharide A simple sugar; a polyhydroxy ketone or aldehyde with three to nine C atoms. (501, 654)

N

nanotechnology The science and engineering of nanoscale (1–50 nm) systems. (477)
natural law (also *law*) A summary, often in mathematical form, of a universal observation. (11)
Nernst equation An equation stating that the voltage of an electrochemical cell under any conditions depends on the standard cell voltage and the concentrations of the cell components:

$$E_{cell} = E^0_{cell} - \frac{RT}{nF} \ln Q \qquad (926)$$

net ionic equation A chemical equation of a reaction in solution in which spectator ions have been eliminated to show the actual chemical change. (141)
network covalent solid A solid in which all the atoms are bonded covalently. (459, 551)
neutralization reaction An acid-base reaction that yields water and a solution of a salt; when a strong acid reacts with a stoichiometrically equivalent amount of a strong base, the solution is neutral. (144, 768)
neutron (n^0) An uncharged subatomic particle found in the nucleus, with a mass slightly greater than that of a proton. (49)
nitrile An organic compound containing the $-C\equiv N:$ group. (649)
node A region of an orbital where the probability of finding the electron is zero. (281)
nonbonding MO A molecular orbital that is not involved in bonding. (417)
nonelectrolyte A substance whose aqueous solution does not conduct an electric current. (138, 515)
nonionizing radiation Radiation that does not cause loss of electrons. (1062)
nonmetal An element that lacks metallic properties. In reactions, nonmetals tend to bond with each other to form covalent compounds or accept electrons from metals to form ionic compounds. (56)
nonpolar covalent bond A covalent bond between identical atoms such that the bonding pair is shared equally. (353)
nuclear binding energy The energy required to break 1 mole of nuclei of an element into individual nucleons. (1071)
nuclear magnetic resonance (NMR) spectroscopy An instrumental technique used to determine the molecular environment of a given type of nucleus, most often ^1H, from its absorption of radio waves in a magnetic field. (634)
nuclear transmutation The induced conversion of one nucleus into another by bombardment with a particle. (1059)
nucleic acid An unbranched polymer consisting of mononucleotides that occurs as two types, DNA and RNA (deoxyribonucleic and ribonucleic acids), which differ chemically in the nature of the sugar portion of the mononucleotides. (500, 658)
nucleon A subatomic particle that makes up a nucleus; a proton or neutron. (1046)
nucleus The tiny central region of the atom that contains all the positive charge and essentially all the mass. (48)
nuclide A nuclear species with specified numbers of protons and neutrons. (1046)
***N/Z* ratio** The ratio of the number of neutrons to the number of protons, a key factor that determines the stability of a nuclide. (1050)

O

observation A fact obtained with the senses, often with the aid of instruments. Quantitative observations provide data that can be compared objectively. (11)
occurrence (also *source*) The form(s) in which an element exists in nature. (965)
octahedral arrangement The geometric arrangement obtained when six electron groups maximize their space around a central atom; when all six groups are bonding groups, the molecular shape is octahedral (AX_6; ideal bond angle = 90°). (381)
octet rule The observation that when atoms bond, they often lose, gain, or share electrons to attain a filled outer shell of eight electrons. (332)
optical isomers (also *enantiomers*) A pair of stereoisomers consisting of a molecule and its mirror image that cannot be superimposed on each other. (627, 1024)
optically active Able to rotate the plane of polarized light. (628)
orbital diagram A depiction of electron number and spin in an atom's orbitals by means of arrows in a series of small boxes, lines, or circles. (296)
ore A naturally occurring compound or mixture of compounds from which an element can be profitably extracted. (965)
organic compound A compound in which carbon is nearly always bonded to itself and to hydrogen, and often to other elements. (617)
organometallic compound An organic compound in which carbon is bonded covalently to a metal atom. (644)
osmosis The process by which solvent flows through a semipermeable membrane from a dilute to a concentrated solution. (520)
osmotic pressure (Π) The pressure that results from the inability of solute particles to cross a semipermeable membrane. The pressure required to prevent the net movement of solvent across the membrane. (520)
outer electrons Electrons that occupy the highest energy level (highest *n* value) and are, on average, farthest from the nucleus. (303)
overall (net) equation A chemical equation that is the sum of two or more balanced sequential equations in which a product of one becomes a reactant for the next. (109)
overvoltage The additional voltage, usually associated with gaseous products, that is required above the standard cell voltage to accomplish electrolysis. (943)
oxidation The loss of electrons by a species, accompanied by an increase in oxidation number. (151)
oxidation number (O.N.) (also *oxidation state*) A number equal to the magnitude of the charge an atom would have if its shared electrons were held completely by the atom that attracts them more strongly. (151)

oxidation number method A method for balancing redox reactions in which the change in oxidation numbers is used to determine balancing coefficients. (154)
oxidation-reduction reaction (also *redox reaction*) A process in which there is a net movement of electrons from one reactant (reducing agent) to another (oxidizing agent). (150)
oxidation state (See *oxidation number.*)
oxidizing agent The substance that accepts electrons in a reaction and undergoes a decrease in oxidation number. (151)
oxoanion An anion in which an element is bonded to one or more oxygen atoms. (66)

P

***p* block** The main-group elements that lie to the right of the transition elements in the periodic table and whose *p* orbitals are being filled. (542)
***p* orbital** An atomic orbital with $l = 1$. (281)
packing efficiency The percentage of the available volume occupied by atoms, ions, or molecules in a unit cell. (452)
paramagnetism The tendency of a species with unpaired electrons to be attracted by an external magnetic field. (318, 1009)
partial ionic character An estimate of the actual charge separation in a bond (caused by the electronegativity difference of the bonded atoms) relative to complete separation. (354)
partial pressure The portion of the total pressure contributed by a gas in a mixture of gases. (195)
particle accelerator A device used to impart high kinetic energies to nuclear particles. (1060)
pascal (Pa) The SI unit of pressure; $1\ Pa = 1\ N/m^2$. (181)
penetration The process by which an outer electron moves through the region occupied by the core electrons to spend part of its time closer to the nucleus; penetration increases the average effective nuclear charge for that electron. (295)
percent by mass (mass %) (See *mass percent.*)
percent yield (% yield) The actual yield of a reaction expressed as a percent of the theoretical yield. (114)
period A horizontal row of the periodic table. (54)
periodic law A law stating that when the elements are arranged by atomic number, they exhibit a periodic recurrence of properties. (291)
periodic table of the elements A table in which the elements are arranged by atomic number into columns (groups) and rows (periods). (54)
pH The negative common logarithm of $[H_3O^+]$. (775)
phase A physically distinct portion of a system. (425)
phase change A physical change from one phase to another, usually referring to a change in physical state. (425)
phase diagram A diagram used to describe the stable phases and phase changes of a substance as a function of temperature and pressure. (434)
phlogiston theory An incorrect theory of combustion proposing that a burning substance releases the undetectable material phlogiston. (9)
photoelectric effect The observation that when monochromatic light of sufficient frequency shines on a metal, an electric current is produced. (263)
photon A quantum of electromagnetic radiation. (263)
photovoltaic cell A device capable of converting light directly into electricity. (247)
physical change A change in which the physical form (or state) of a substance, but not its composition, is altered. (2)
physical property A characteristic shown by a substance itself, without interacting with or changing into other substances. (2)
pi (π) bond A covalent bond formed by sideways overlap of two atomic orbitals that has two regions of electron density, one above and one below the internuclear axis. (407)
pi (π) MO A molecular orbital formed by combination of two atomic (usually *p*) orbitals whose orientations are perpendicular to the internuclear axis. (413)
Planck's constant (*h*) A proportionality constant relating the energy and the frequency of a photon, equal to 6.626×10^{-34} J·s. (262)
plastic A material that, when deformed, retains its new shape. (476)
polar covalent bond A covalent bond in which the electron pair is shared unequally, so the bond has partially negative and partially positive poles. (353)
polar molecule A molecule with an unequal distribution of charge as a result of its polar bonds and shape. (135)
polarimeter A device used to measure the rotation of plane-polarized light by an optically active compound. (628)
polarizability The ease with which a particle's electron cloud can be distorted. (440)
polyatomic ion An ion in which two or more atoms are bonded covalently. (66)
polymer (also *macromolecule*) An extremely large molecule that results from the covalent linking of many simpler molecular units (monomers). (472)
polyprotic acid An acid with more than one ionizable proton. (786)
polysaccharide A macromolecule composed of many simple sugars linked covalently. (501, 654)
positron ($^{0}_{1}\beta$) The antiparticle of an electron. (1049)
positron decay A type of radioactive decay in which a positron is emitted from a nucleus. (1049)
potential energy (E_p) The energy an object has as a result of its position relative to other objects or because of its composition. (6)
precipitate The insoluble product of a precipitation reaction. (141)
precipitation reaction A reaction in which two soluble ionic compounds form an insoluble product, a precipitate. (141)
precision (also *reproducibility*) The closeness of a measurement to other measurements of the same phenomenon in a series of experiments. (29)
pressure (*P*) The force exerted per unit of surface area. (179)
pressure-volume work (*PV* work) A type of work in which a volume change occurs against an external pressure. (228)
principal quantum number (*n*) A positive integer that specifies the energy and relative size of an atomic orbital. (277)
probability contour A shape that defines the volume around an atomic nucleus within which an electron spends a given percentage of its time. (277)
product A substance formed in a chemical reaction. (102)
property A characteristic that gives a substance its unique identity. (2)
protein A natural, linear polymer composed of any of about 20 types of amino acid monomers linked together by peptide bonds. (495, 655)
proton (p^+) A subatomic particle found in the nucleus that has a unit positive charge (1.60218×10^{-19} C). (49)

proton acceptor A substance that accepts an H^+ ion; a Brønsted-Lowry base. (778)
proton donor A substance that donates an H^+ ion; a Brønsted-Lowry acid. (777)
pseudo–noble gas configuration The $(n-1)d^{10}$ configuration of a *p*-block metal atom that has emptied its outer energy level. (316)
pure substance (See *substance*.)

Q

qualitative analysis The separation and identification of the ions in an aqueous mixture. (851)
quantum A packet of energy equal to $h\nu$. The smallest quantity of energy that can be emitted or absorbed. (262)
quantum mechanics The branch of physics that examines the wave motion of objects on the atomic scale. (275)
quantum number A number that specifies a property of an orbital or an electron. (262)

R

rad (radiation-absorbed dose) The quantity of radiation that results in 0.01 J of energy being absorbed per kilogram of tissue; 1 rad = 0.01 J/kg tissue = 10^{-2} Gy. (1063)
radial probability distribution plot The graphic depiction of the total probability distribution (sum of ψ^2) of an electron in the region near the nucleus. (277)
radioactivity The emissions resulting from the spontaneous disintegration of an unstable nucleus. (1046)
radioisotope An isotope with an unstable nucleus that decays through radioactive emissions. (1057)
radioisotopic dating A method for determining the age of an object based on the rate of decay of a particular radioactive nuclide. (1057)
radius of gyration (R_g) A measure of the size of a coiled polymer chain as the average distance from the center of mass of the chain to its outside edge. (474)
random coil The shape adopted by most polymer chains and caused by random rotation about the bonds joining the repeat units. (473)
random error Human error that occurs in all measurements and results in values *both* higher and lower than the actual value. (29)
Raoult's law A law stating that the vapor pressure of a solution is directly proportional to the mole fraction of solvent: $P_{\text{solvent}} = X_{\text{solvent}} \times P^0_{\text{solvent}}$. (516)
rare earths (See *lanthanides*.)
rate constant (k) The proportionality constant that relates the reaction rate to reactant (and product) concentrations. (680)
rate-determining step (also *rate-limiting step*) The slowest step in a reaction mechanism and therefore the step that limits the overall rate. (702)
rate law (also *rate equation*) An equation that expresses the rate of a reaction as a function of reactant (and product) concentrations. (679)
reactant A starting substance in a chemical reaction. (102)
reaction energy diagram A graph that shows the potential energy of a reacting system as it progresses from reactants to products. (698)
reaction intermediate A substance that is formed and used up during the overall reaction and therefore does not appear in the overall equation. (701)
reaction mechanism A series of elementary steps that sum to the overall reaction and is consistent with the rate law. (700)
reaction order The exponent of a reactant concentration in a rate law that shows how the rate is affected by changes in that concentration. (680)
reaction quotient (Q) (also *mass-action expression*) A ratio of terms for a given reaction consisting of product concentrations multiplied together and divided by reactant concentrations multiplied together, each raised to the power of their balancing coefficient. The value of Q changes until the system reaches equilibrium, at which point it equals K. (726)
reaction rate The change in the concentrations of reactants (or products) with time. (675)
reactor core The part of a nuclear reactor that contains the fuel rods and generates heat from fission. (1076)
redox reaction (See *oxidation-reduction reaction*.)
reducing agent The substance that donates electrons in a redox reaction and undergoes an increase in oxidation number. (151)
reduction The gain of electrons by a species, accompanied by a decrease in oxidation number. (151)
refraction A phenomenon in which a wave changes its speed and therefore its direction as it passes through a phase boundary. (260)
rem (roentgen equivalent for man) The unit of radiation dosage for a human based on the product of the number of rads and a factor related to the biological tissue; 1 rem = 10^{-2} Sv. (1063)
resonance hybrid The weighted average of the resonance structures of a molecule. (370)
resonance structure (also *resonance form*) One of two or more Lewis structures for a molecule that cannot be adequately depicted by a single structure. Resonance structures differ only in the position of bonding and lone electron pairs. (369)
reverse osmosis A process for preparing drinkable water that uses an applied pressure greater than the osmotic pressure to remove ions from an aqueous solution, typically seawater. (530)
rms (root-mean-square) speed (u_{rms}) The speed of a molecule having the average kinetic energy; very close to the most probable speed. (204)
roasting A pyrometallurgical process in which metal sulfides are converted to oxides. (974)
round off The process of removing digits based on a series of rules to obtain an answer with the proper number of significant figures (or decimal places). (27)

S

***s* block** The main-group elements that lie to the left of the transition elements in the periodic table and whose *s* orbitals are being filled. (542)
***s* orbital** An atomic orbital with $l = 0$. (280)
salt An ionic compound that results from a classical acid-base reaction. (145)
salt bridge An inverted U tube containing a solution of nonreacting electrolyte that connects the compartments of a voltaic cell and maintains neutrality by allowing ions to flow between compartments. (910)
saturated hydrocarbon A hydrocarbon in which each C is bonded to four other atoms. (623)

saturated solution A solution that contains the maximum amount of dissolved solute at a given temperature in the presence of undissolved solute. (507)
scanning tunneling microscopy An instrumental technique that uses electrons moving across a minute gap to observe the topography of a surface on the atomic scale. (455)
Schrödinger equation An equation that describes how the electron matter-wave changes in space around the nucleus. Solutions of the equation provide allowable energy levels of the H atom. (275)
scientific method A process of creative thinking and testing aimed at objective, verifiable discoveries of the causes of natural events. (10)
scintillation counter A device used to measure radioactivity through its excitation of atoms and their subsequent emission of light. (1055)
second (s) The SI base unit of time. (24)
second law of thermodynamics A law stating that a process occurs spontaneously in the direction that increases the entropy of the universe. (870)
seesaw shape A molecular shape caused by the presence of one equatorial lone pair in a trigonal bipyramidal arrangement (AX_4E). (380)
selective precipitation The process of separating ions through differences in the solubility of their compounds with a given precipitating ion. (850)
self-ionization (See *autoionization.*)
semiconductor A substance whose electrical conductivity is poor at room temperature but increases significantly with rising temperature. (463)
semimetal (See *metalloid.*)
semipermeable membrane A membrane that allows solvent, but not solute, to pass through. (520)
shared pair (See *bonding pair.*)
shell (See *level.*)
shielding The ability of other electrons, especially inner ones, to lessen the nuclear attraction for an outer electron. (294)
SI unit A unit composed of one or more of the base units of the Système International d'Unités, a revised metric system. (16)
side reaction An undesired chemical reaction that consumes some of the reactant and reduces the overall yield of the desired product. (114)
side-to-side overlap Overlap of *p* with *p* (or sometimes with *d*) atomic orbitals that leads to a pi (π) bond. (545)
sievert (Sv) The SI unit of human radiation dosage; 1 Sv = 100 rem. (1063)
sigma (σ) bond A type of covalent bond that arises through end-to-end orbital overlap and has most of its electron density along the bond axis. (407)
sigma (σ) MO A molecular orbital that is cylindrically symmetrical about an imaginary line that runs through the nuclei of the component atoms. (411)
significant figures The digits obtained in a measurement. The greater the number of significant figures, the greater the certainty of the measurement. (26)
silicate A type of compound found throughout rocks and soil and consisting of repeating —Si—O groupings and, in most cases, metal cations. (578)
silicone A type of synthetic polymer containing —Si—O repeat units, with organic groups and crosslinks. (578)
simple cubic unit cell A unit cell in which a particle occurs at each corner of a cube. (450)
single bond A bond that consists of one electron pair. (340)
slag A molten waste product formed in a blast furnace by the reaction of acidic silica with a basic metal oxide. (981)
smelting Heating a mineral with a reducing agent, such as coke, to obtain a metal. (975)
soap The salt of a fatty acid and usually a Group 1A(1) or 2A(2) hydroxide. (498)
solid One of the three states of matter. A solid has a fixed shape that does not conform to the container shape. (4)
solubility (*S*) The maximum amount of solute that dissolves in a fixed quantity of a particular solvent at a specified temperature when excess solute is present. (490)
solubility-product constant (K_{sp}) An equilibrium constant for the dissolving of a slightly soluble ionic compound in water. (833)
solute The substance that dissolves in the solvent. (117, 490)
solution (See *homogeneous mixture.*)
solvated Surrounded closely by solvent molecules. (136)
solvation The process of surrounding a solute particle with solvent particles. (136, 503)
solvent The substance in which the solute(s) dissolve. (117, 490)
source (See *occurrence.*)
***sp* hybrid orbital** An orbital formed by the mixing of one *s* and one *p* orbital of a central atom. (400)
sp^2 hybrid orbital An orbital formed by the mixing of one *s* and two *p* orbitals of a central atom. (402)
sp^3 hybrid orbital An orbital formed by the mixing of one *s* and three *p* orbitals of a central atom. (402)
sp^3d hybrid orbital An orbital formed by the mixing of one *s*, three *p*, and one *d* orbital of a central atom. (403)
sp^3d^2 hybrid orbital An orbital formed by the mixing of one *s*, three *p*, and two *d* orbitals of a central atom. (404)
specific heat capacity (*c*) The quantity of heat required to change the temperature of 1 gram of a substance by 1 K. (235)
spectator ion An ion that is present as part of a reactant but is not involved in the chemical change. (141)
spectrochemical series A ranking of ligands in terms of their ability to split *d*-orbital energies. (1031)
spectrophotometry A group of instrumental techniques that create an electromagnetic spectrum to measure the atomic and molecular energy levels of a substance. (269)
speed of light (*c*) A fundamental constant giving the speed at which electromagnetic radiation travels in a vacuum: $c = 2.99792458\times10^8$ m/s. (258)
spin quantum number (m_s) A number, either $+\frac{1}{2}$ or $-\frac{1}{2}$, that indicates the direction of electron spin. (293)
spontaneous change A change that occurs by itself, that is, without an ongoing input of energy. (864)
square planar shape A molecular shape (AX_4E_2) caused by the presence of two axial lone pairs in an octahedral arrangement. (381)
square pyramidal shape A molecular shape (AX_5E) caused by the presence of one lone pair in an octahedral arrangement. (381)
standard atmosphere (atm) The average atmospheric pressure measured at sea level, defined as 1.01325×10^5 Pa. (181)
standard cell potential (E°_{cell}) The potential of a cell measured with all components in their standard states and no current flowing. (915)

standard electrode potential ($E^0_{\text{half-cell}}$) (also *standard half-cell potential*) The standard potential of a half-cell, with the half-reaction written as a reduction. (915)
standard entropy of reaction (ΔS^0_{rxn}) The entropy change that occurs when all components are in their standard states. (875)
standard free energy change (ΔG^0) The free energy change that occurs when all components are in their standard states. (882)
standard free energy of formation (ΔG^0_f) The standard free energy change that occurs when 1 mol of a compound is made from its elements. (883)
standard half-cell potential (See *standard electrode potential.*)
standard heat of formation (ΔH^0_f) The enthalpy change that occurs when 1 mol of a compound forms from its elements, with all substances in their standard states. (242)
standard heat of reaction (ΔH^0_{rxn}) The enthalpy change that occurs during a reaction, with all substances in their standard states. (242)
standard hydrogen electrode (See *standard reference half-cell.*)
standard molar entropy (S^0) The entropy of 1 mol of a substance in its standard state. (871)
standard molar volume The volume of 1 mol of an ideal gas at standard temperature and pressure: 22.4141 L. (187)
standard reference half-cell (also *standard hydrogen electrode*) A specially prepared platinum electrode immersed in 1 M $H^+(aq)$ through which H_2 gas at 1 atm is bubbled. $E^0_{\text{half-cell}}$ is defined as 0 V. (916)
standard states A set of specifications used to compare thermodynamic data: 1 atm for gases behaving ideally, 1 M for dissolved species, or the pure substance for liquids and solids. (242)
standard temperature and pressure (STP) The reference conditions for a gas:

$$0°\text{C } (273.15 \text{ K}) \text{ and } 1 \text{ atm } (760 \text{ torr}) \quad (187)$$

state function A property of the system determined by its current state, regardless of how it arrived at that state. (230)
state of matter One of the three physical forms of matter: solid, liquid, or gas. (4)
stationary state In the Bohr model, one of the allowable energy levels of the atom in which it does not release or absorb energy. (265)
steel An alloy of iron with small amounts of carbon and usually other metals. (980)
stellar nucleogenesis The process by which elements are formed in the stars through nuclear fusion. (1078)
stereoisomers Molecules with the same connections of atoms but different orientations of groups in space. (627, 1024) (See also *geometric isomers* and *optical isomers.*)
stoichiometric coefficient (See *balancing coefficient.*)
stoichiometry The study of the mass-mole-number relationships of chemical formulas and reactions. (87)
strong-field ligand A ligand that causes larger crystal field splitting energy and therefore is part of a low-spin complex. (1030)
strong force An attractive force that exists between all nucleons and is about 100 times stronger than the electrostatic repulsive force. (1051)
structural formula A formula that shows the actual number of atoms, their relative placement, and the bonds between them. (62)
structural isomers (See *constitutional isomers.*)
sublevel (also *subshell*) An energy substate of an atom within a level. Given by the n and l values, the sublevel designates the size and shape of the atomic orbitals. (278)
sublimation The process by which a solid changes directly into a gas. (427)
substance (also *pure substance*) A type of matter, either an element or a compound, that has a fixed composition. (39)
substitution reaction An organic reaction that occurs when an atom (or group) from one reactant substitutes for one in another reactant. (636)
substrate A reactant that binds to the active site in an enzyme-catalyzed reaction. (709)
superconductivity The ability to conduct a current with no loss of energy to resistive heating. (463)
supersaturated solution An unstable solution in which more solute is dissolved than in a saturated solution. (507)
surface tension The energy required to increase the surface area of a liquid by a given amount. (443)
surroundings All parts of the universe other than the system being considered. (225)
suspension A heterogeneous mixture containing particles that are distinct from the surrounding medium. (527)
syngas Synthesis gas; a combustible mixture of CO and H_2. (245)
synthetic natural gas (SNG) A gaseous fuel mixture, mostly methane, formed from coal. (245)
system The defined part of the universe under study. (225)
systematic error A type of error producing values that are all either higher or lower than the actual value, often caused by faulty equipment or a consistent fault in technique. (29)

T

t_{2g} orbitals The set of orbitals (composed of d_{xy}, d_{yz}, and d_{xz}) that results when the energies of the metal-ion d orbitals are split by a ligand field. This set is lower in energy than the other (e_g) set in an octahedral field and higher in energy in a tetrahedral field. (1030)
T shape A molecular shape caused by the presence of two equatorial lone pairs in a trigonal bipyramidal arrangement (AX_3E_2). (380)
temperature (T) A measure of how hot or cold a substance is relative to another substance. (22)
tetrahedral arrangement The geometric arrangement formed when four electron groups maximize their separation around a central atom; when all four groups are bonding groups, the molecular shape is tetrahedral (AX_4; ideal bond angle 109.5°). (379)
theoretical yield The amount of product predicted by the stoichiometrically equivalent molar ratio in the balanced equation. (114)
theory (See *model.*)
thermochemical equation A chemical equation that shows the heat of reaction for the amounts of substances specified. (238)
thermochemistry The branch of thermodynamics that focuses on the heat involved in chemical reactions. (225)
thermodynamics The study of heat (thermal energy) and its interconversions. (225)
thermometer A device for measuring temperature that contains a fluid that expands or contracts within a graduated tube. (22)
third law of thermodynamics A law stating that the entropy of a perfect crystal is zero at 0 K. (871)

titration A method of determining the concentration of a solution by monitoring its reaction with a solution of known concentration. (146)
torr A unit of pressure identical to 1 mmHg. (181)
total ionic equation A chemical equation for an aqueous reaction that shows all the soluble ionic substances dissociated into ions. (141)
tracer A radioisotope that signals the presence of the species of interest by emitting nonionizing radiation. (1066)
transition element (also *transition metal*) An element that occupies the *d* block of the periodic table; one whose *d* orbitals are being filled. (301, 1003)
transition state (also *activated complex*) An unstable species formed in an effective collision of reactants that exists momentarily when the system is highest in energy and that can either form products or re-form reactants. (697)
transition state theory A model that explains how the energy of reactant collision is used to form a high-energy transitional species that can change to reactant or product. (694)
transuranium element An element with atomic number higher than that of uranium ($Z = 92$). (1061)
trigonal bipyramidal arrangement The geometric arrangement formed when five electron groups maximize their separation around a central atom. When all five groups are bonding groups, the molecular shape is trigonal bipyramidal (AX_5; ideal bond angles, axial-center-equatorial 90° and equatorial-center-equatorial 120°). (380)
trigonal planar arrangement The geometric arrangement formed when three electron groups maximize their separation around a central atom. (377)
trigonal planar shape A molecular shape (AX_3) formed when three atoms around a central atom lie at the corners of an equilateral triangle; ideal bond angle, 120°. (377)
trigonal pyramidal shape A molecular shape (AX_3E) caused by the presence of one lone pair in a tetrahedral arrangement. (379)
triple bond A covalent bond that consists of three bonding pairs, two atoms sharing six electrons; one σ and two π bonds. (340)
triple point The pressure and temperature at which three phases of a substance are in equilibrium. In a phase diagram, the point at which three phase-transition curves meet. (435)
Tyndall effect The scattering of light by a colloidal suspension. (528)

U

ultraviolet (UV) region The region of the electromagnetic spectrum between the visible and the x-ray regions. (259)
uncertainty A characteristic of every measurement that results from the inexactness of the measuring device and the necessity of estimating when taking a reading. (25)
uncertainty principle The principle stated by Werner Heisenberg that it is impossible to know simultaneously the exact position and velocity of a particle; the principle becomes important only for particles of very small mass. (274)
unimolecular reaction An elementary reaction that involves the decomposition or rearrangement of a single particle. (701)
unit cell The smallest portion of a crystal that, if repeated in all three directions, gives the crystal. (450)
universal gas constant (*R*) A proportionality constant that relates the energy, amount of substance, and temperature of a system; $R = 0.0820578$ atm·L/mol·K = 8.31447 J/mol·K. (188)
unsaturated hydrocarbon A hydrocarbon with at least one carbon-carbon multiple bond; one in which at least two C atoms are bonded to fewer than four atoms. (628)
unsaturated solution A solution in which more solute can be dissolved at a given temperature. (507)
unshared pair (See *lone pair.*)

V

V shape (See *bent shape.*)
valence band In band theory, the lower energy portion of the band of molecular orbitals, which is filled with valence electrons. (462)
valence bond (VB) theory A model that attempts to reconcile the shapes of molecules with those of atomic orbitals through the concepts of orbital overlap and hybridization. (399)
valence electrons The electrons involved in compound formation; in main-group elements, the electrons in the valence (outer) level. (303)
valence-shell electron-pair repulsion (VSEPR) theory A model explaining that the shapes of molecules and ions result from minimizing electron-pair repulsions around a central atom. (376)
van der Waals constants Experimentally determined positive numbers used in the van der Waals equation to account for the intermolecular attractions and molecular volume of real gases. (212)
van der Waals equation An equation that accounts for the behavior of real gases. (212)
van der Waals radius One-half of the closest distance between the nuclei of identical nonbonded atoms. (437)
vapor pressure (also *equilibrium vapor pressure*) The pressure exerted by a vapor at equilibrium with its liquid in a closed system. (431)
vapor pressure lowering (ΔP) The lowering of the vapor pressure of a solvent caused by the presence of dissolved solute particles. (516)
vaporization The process of changing from a liquid to a gas. (426)
variable A quantity that can have more than a single value. (11) (See also *controlled experiment.*)
viscosity A measure of the resistance of a liquid to flow. (445)
volatility The tendency of a substance to become a gas. (74)
volt (V) The SI unit of electric potential: 1 V = 1 J/C. (914)
voltage (See *cell potential.*)
voltaic cell (also *galvanic cell*) An electrochemical cell that uses a spontaneous reaction to generate electric energy. (908)
volume (*V*) The space occupied by a sample of matter. (17)
volume percent [% (v/v)] A concentration term defined as the volume of solute in 100. volumes of solution. (512)

W

wastewater (also *sewage*) Used water, usually containing industrial and/or residential waste, that is treated before being returned to the environment. (530)
water softening The process of removing hard-water ions, such as Ca^{2+} and Mg^{2+}, from water. (529)
wave function (See *atomic orbital.*)

wave-particle duality The principle stating that both matter and energy have wavelike and particle-like properties. (274)

wavelength (λ) The distance between any point on a wave and the corresponding point on the next wave, that is, the distance a wave travels during one cycle. (258)

weak-field ligand A ligand that causes smaller crystal field splitting energy and therefore is part of a high-spin complex. (1030)

weight The force exerted by a gravitational field on an object. (19)

work (w) The energy transferred when an object is moved by a force. (227)

X

x-ray diffraction analysis An instrumental technique used to determine spatial dimensions of a crystal structure by measuring the diffraction patterns caused by x-rays impinging on the crystal. (455)

Z

zone refining A process used to purify metals and metalloids in which impurities are removed from a bar of the element by concentrating them in a thin molten zone. (521, 977)

Credits

Chapter 1

Figure 1.1A: © Paul Morrell/Stone/Getty Images; *1.1B, p. 3(top left):* © McGraw-Hill Higher Education/Stephen Frisch Photographer; *p. 3(bottom left):* © Ruth Melnick; *p. 3(top right, middle right, bottom right):* © McGraw-Hill Higher Education/Stephen Frisch Photographer; *p. 4:* NASA Media Resource Center; *1.4:* © Giraudon/Art Resource, NY; *p. 9:* © Culver Pictures; *1.5A:* © PhotoDisc; *1.5B:* © George Haling/Photo Researchers; *1.5C:* © Phanie Agency/Photo Researchers; *1.5D:* © Lehtikuva/Heikki Saukkomaa/Wide World Photos; *1.7:* © Loren Santow/Stone/Getty Images; *1.9A:* © McGraw-Hill Higher Education/Stephen Frisch Photographer; *1.9B:* © Brandtech Scientific; *1.13:* © Steve Jefferts (Time and Frequency)/National Institute of Standards and Technology; *p. 25:* © North Wind Picture Archives; *1.14B:* © Hart Scientific; *1.15(both):* © McGraw-Hill Higher Education/Stephen Frisch Photographer; *B1.1A:* © Bruce Forester/Stone/Getty Images; *B1.1B:* © Bob Daemmrich/Daemmrich Photography; *B1.1C:* © Cindy Yamanaka/National Geographic Image Collection

Chapter 2

Opener: © Museum of London Archaeology Service; *2.2, 2.2B, p. 41(all):* © McGraw-Hill Higher Education/Stephen Frisch Photographer; *2.3(top):* © Sleeping Children, after 1859–1874: William Henry Rinehart, American (1825–1874), Marble, overall 15 × 18 × 37 in. (38.1 × 45.7 × 94 cm), Gift of Daniel and Jessie Lie Farver, Roy Taylor, and Mary E. Moore Gift, 1984.268/Museum of Fine Arts, Boston; *2.3(bottom):* © Wards Earth Science Catalog/Wards Natural Science Establishment; *p. 44:* © Oesper Collection, University of Cincinnati; *p. 46:* © Michael Dalton/Fundamental Photographs; *B2.2D:* © James Holmes/Oxford Centre for Molecular Sciences/SPL/Photo Researchers; *B2.2E:* © Craig Hendrickson/National High Magnetic Field Laboratory; *p. 53:* © Hulton/Getty Images; *2.11(all), 2.12A(both):* © McGraw-Hill Higher Education/Stephen Frisch Photographer; *2.17:* © Dane S. Johnson/Visuals Unlimited; *2.21A&B, B2.3, B2.4, B2.9:* © McGraw-Hill Higher Education/Stephen Frisch Photographer

Chapter 3

Opener: © McGraw-Hill Higher Education/Stephen Frisch Photographer; *3.1A&B, 3.2, p. 95:* © McGraw-Hill Higher Education/Stephen Frisch Photographer; *p. 98:* © Ruth Melnick; *3.8(all), 3.13(all), 3.14(both):* © McGraw-Hill Higher Education/Stephen Frisch Photographer

Chapter 4

Opener: © David Taylor/SPL/Photo Researchers; *4.3(all), 4.5(both):* © McGraw-Hill Higher Education/Stephen Frisch Photographer; *4.6:* © Richard Megna/Fundamental Photographs; *4.7:* © McGraw-Hill Higher Education/Stephen Frisch Photographer; *p. 144(both), 4.8A–C, 4.10, p. 155, 4.14(both), 4.15(all), 4.16(all), 4.17(all), 4.18, 4.19(both):* © McGraw-Hill Higher Education/Stephen Frisch Photographer

Chapter 5

Figures 5.1A–C, 5.2A, 5.2A–B, 5.9: © McGraw-Hill Higher Education/Stephen Frisch Photographer; *p. 193:* © PhotoDisc/Vol. 25; *p. 199:* © McGraw-Hill Higher Education/Stephen Frisch Photographer; *p. 209:* © Julian Baum/SPL/Photo Researchers; *p. 220:* © Romilly Lockyer/Getty Images

Chapter 6

Opener: © Dave Spier/Visuals Unlimited; *p. 225:* © David Malin/David Malin Images; *6.1:* © McGraw-Hill Higher Education/Stephen Frisch Photographer; *p. 229:* © AIP Emilio Segre Visual Archives, E. Scott Barr Collection/American Institute of Physics

Chapter 7

Opener: © PhotoDisc Website; *p. 259:* © Ralph Wetmore/Getty Images; *p. 261:* © Michael Marten/SPL/Photo Researchers; *7.6A:* © Dr. E. R. Degginger/Color-Pic; *7.6B:* © Phil Degginger/Color-Pic; *7.6C:* © Paul Dance/Getty Images; *B7.1A(both):* © McGraw-Hill Higher Education/Stephen Frisch Photographer; *B7.1B:* Washington DC Convention & Visitors Association; *p. 271:* © SPL/Photo Researchers; *p. 272:* © Dennis Kunkel/Phototake; *7.14A–B:* PSSC Physics © 1965, Education Development Center, Inc.; D.C. Heath & Company/Education Development Center, Inc.; *p. 275:* © Mark Oliphant, AIP Emilio Segre Visual Archives, Margrethe Bohr Collection/American Institute of Physics

Chapter 8

Opener: © Adam Jones/Visuals Unlimited; *p. 291:* © Corbis; *p. 292:* © Oesper Collection, University of Cincinnati; *8.10A–B, p. 314, 8.23(all):* © McGraw-Hill Higher Education/Stephen Frisch Photographer

Chapter 9

Opener: © McGraw-Hill Higher Education/C. P. Hammond Photographer; *p. 332:* © Bancroft Library, University of California; *9.5A–B, 9.8A, 9.9A–C, 9.14, 9.24:* McGraw-Hill Higher Education/Stephen Frisch Photographer; *p. 358:* © Diane Hirsch/Fundamental Photographs; *9.27A:* © McGraw-Hill Higher Education/Stephen Frisch Photographer

Chapter 10

Opener: M. C. Escher's "Sun and Moon" © 2004 The M. C. Escher Company–Baarn–Holland. All Rights Reserved; *p. 373:* © Richard Megna/Fundamental Photographs; *10.2A(all), 10.6(both):* © McGraw-Hill Higher Education/Stephen Frisch Photographer

Chapter 11

Figure 11.22: © Richard Megna/Fundamental Photographs

Chapter 12

Page 426: © Michael Lustbader/Photo Researchers; *p. 427:* © Jill Birschbach Photo Services; *12.8:* © Richard Megna/McGraw-Hill Higher Education; *12.20A–B:* © McGraw-Hill Higher Education/Stephen Frisch Photographer; *12.22B:* © Scott Camazine/Photo Researchers; *p. 444:* © Yoav Levy/Phototake; *p. 446(water strider):* © Nuridsany & Perennou/Photo Researchers; *p. 446(leaf):* © Rod Planck/Photo Researchers; *p. 446(child):* © Ruth Melnick; *p. 446(oil):* © Chris Sorensen Photography; *p. 446(bubbles):* © Phil Jude/SPL/Photo Researchers; *12.23:* © Tom Pantages; *12.25A:*

© Paul Silverman/Fundamental Photographs; *12.25B:* © Scovil Photography; *12.25C:* © Mark Schneider/Visuals Unlimited; *p. 452:* © Michael Heron/Woodfin Camp & Associates; *B12.3(left):* © Science VU/LLL/Visuals Unlimited; *B12.3(right):* © Digital Instruments/VEECO/Photo Researchers; *12.38:* © AT&T Bell Labs/SPL/Photo Researchers; *p. 464:* NASA Johnson Space Center; *12.44A:* © J. R. Factor/Photo Researchers; *12.44B:* © James Dennis/Phototake; *p. 476:* © Sue Klemens/Stock Boston; *12.50A:* © Felice Frankel; *12.50B:* © Delft University of Technology/Photo Researchers

Chapter 13

Opener: © Jeff Smith/Getty Images; *p. 489:* © A. B. Dowsett/SPL/Photo Researchers; *13.4A:* © Ruth Melnick; *13.4B:* © Dr. E. R. Degginger/Color-Pic; *p. 495:* © MacDonald Photography/Root Resources; *13.20A–C:* © McGraw-Hill Higher Education/Stephen Frisch Photographer; *13.22:* © Sheila Terry/SPL/Photo Researchers; *p. 510:* © Peter Scoones/Photo Researchers; *13.24:* © Darwin Dale/Photo Researchers; *13.25A–C:* © McGraw-Hill Higher Education/Stephen Frisch Photographer; *p. 521(top):* © Traverse City Record-Eagle, John L. Russell/AP/Wide World Photos; *p. 521(middle left):* © OSF/Doug Allan/Animals, Animals; *p. 521(middle right):* © Chris Sorensen Photography; *p. 521(bottom):* © Westinghouse/Visuals Unlimited; *p. 522(top row):* © David M. Phillips/Photo Researchers; *p. 522(middle right):* © Kenneth W. Fink/Root Resources; *p. 522(middle left):* © John Lei/Stock Boston; *p. 522(bottom):* © Osentoski & Zoda/Envision; *13.32A–B:* © E. R. Degginger/Color-Pic; *13.32B:* © E. R. Degginger/Color-Pic; *B13.2A:* © McGraw-Hill Higher Education/Stephen Frisch Photographer; *B13.3A:* © Robert Essel/Corbis; *p. 531:* © Earth Satellite Corporation/SPL/Photo Researchers

Interchapter

Opener: © Richard Megna/Fundamental Photographs; *p. 548(all):* © McGraw-Hill Higher Education/Stephen Frisch Photographer

Chapter 14

Page 556(all), 557(all), p. 558, p. 560(all), p. 561(top): © McGraw-Hill Higher Education/Stephen Frisch Photographer; *p. 561(bottom):* © Visuals Unlimited; *p. 564(all), p. 565(top):* © McGraw-Hill Higher Education/Stephen Frisch Photographer; *p. 565(middle left):* © Scovil Photography; *p. 565(middle right):* © Scott Camazine/Photo Researchers; *p. 565(bottom):* © Annebicque Bernard/Corbis Sygma; *p. 568(all):* © McGraw-Hill Higher Education/Stephen Frisch Photographer; *p. 569(top):* © Scott Camazine/Photo Researchers; *p. 569(middle):* © E. R. Degginger/Color-Pic; *p. 569(bottom left):* © Phil Degginger/Color-Pic; *p. 569(bottom right):* © AT&T Bell Labs/SPL/Photo Researchers; *p. 570:* © McGraw-Hill Higher Education/Stephen Frisch Photographer; *14.13A:* Courtesy MER Corporation; *14.13B:* © S. C. Tsang/SPL/Photo Researchers; *p. 574(all):* © McGraw-Hill Higher Education/Stephen Frisch Photographer; *p. 575(top):* © Charles Winters/Photo Researchers; *p. 575(middle):* © Charles Falco/Photo Researchers; *p. 575(bottom left):* © Phil Jude/SPL/Photo Researchers; *p. 575(bottom right):* © McGraw-Hill Higher Education/Pat Watson Photographer; *14.14A–B:* © McGraw-Hill Higher Education/Stephen Frisch Photographer; *p. 580(bottom), p. 581(top and middle):* © Scott Camazine/Photo Researchers; *p. 581(bottom):* © Tom Pantages; *p. 582(top):* © Photri-Microstock; *p. 582(middle):* © Mark Richards/PhotoEdit; *p. 582(bottom):* Boston Public Library/Rare Books Department. Courtesy of the Trustees; *p. 584(all):* © McGraw-Hill Higher Education/Stephen Frisch Photographer; *p. 585(left):* © Stone/Getty Images; *p. 585(right, bottom):* © McGraw-Hill Higher Education/Pat Watson Photographer; *p. 592(all):* © McGraw-Hill Higher Education/Stephen Frisch Photographer; *p. 593(top):* © Chris Sorensen Photography; *p. 593(bottom):* © Gregory Dimijian/Photo Researchers; *p. 596:* © Bruce Forester/Stone/Getty Images; *14.26(all):* © McGraw-Hill Higher Education/Stephen Frisch Photographer; *14.27:* © Wards Earth Science Catalog/Wards Natural Science Establishment; *14.30B(both), p. 600(all):* © McGraw-Hill Higher Education/Stephen Frisch Photographer; *p. 601(left):* © Scovil Photography; *p. 601(right):* © NASA/Photo Researchers; *14.33:* © E.R. Degginger/Color-Pic; *p. 606(all):* © McGraw-Hill Higher Education/Stephen Frisch Photographer

Chapter 15

Page 617: © SPL/Photo Researchers; *B15.5:* © Scott Camazine/Photo Researchers; *15.13A–B:* © Richard Megna/McGraw-Hill Higher Education; *p. 647:* © McGraw-Hill Higher Education/Pat Watson Photographer; *15.26:* © E. R. Degginger/Color-Pic; *p. 654:* © McGraw-Hill Higher Education/Pat Watson Photographer; *15.30A:* © J. Gross/SPL/Photo Researchers; *15.30B:* © E. R. Degginger/Color-Pic

Chapter 16

Opener: © STUDIO CARLO DANI/Animals Animals/Earth Scenes; *16.2B:* © Ruth Melnick; *16.2A:* © Crown Copyright/Health & Safety Lab/Photo Researchers; *16.2C:* © Paul Silverman/Fundamental Photographs; *16.2D:* © Ruth Melnick; *16.3A–B:* © McGraw-Hill Higher Education/Stephen Frisch Photographer; *p. 675:* © Mike Powell/Getty Images; *B16.1:* © Varian Inc.; *B16.2, B16.3:* © McGraw-Hill Higher Education/Stephen Frisch Photographer; *p. 702:* © Angie Norwood Browne/Getty Images; *B16.6(all):* National Oceanic and Atmospheric Administration–NOAA

Chapter 17

Opener: © Beth A. Keiser/AP/Wide World Photos; *17.1A–D:* © McGraw-Hill Higher Education/Stephen Frisch Photographer; *p. 729(top):* © PhotoDisc/Vol. 25; *p. 729(bottom):* Noranda Inc. Horne Smelter; *p. 746:* © Kennan Ward 2004; *B17.3:* © Grant Heilman Photography

Chapter 18

Opener: © Richard Megna/Fundamental Photographs; *p. 767(both), 18.3, 18.7A–B, p. 797(all):* © McGraw-Hill Higher Education/Stephen Frisch Photographer

Chapter 19

Opener: © M. B. Angelo/Photo Researchers; *19.1A–B, 19.2A–B, 19.6:* © McGraw-Hill Higher Education/Stephen Frisch Photographer; *19.11:* © Jackie Lewin, Royal Free Hospital/Photo Researchers; *p. 832:* © McGraw-Hill Higher Education/Stephen Frisch Photographer; *p. 835:* © Scovil Photography; *19.12A–B:* © McGraw-Hill Higher Education/Stephen Frisch Photographer; *p. 838:* © CNRI/Science Source/Photo Researchers; *19.13:* © McGraw-Hill Higher Education/Stephen Frisch Photographer; *B19.1:* © Maurizio Lanini/Corbis Images; *B19.2:* © K. R. Downey Photography; *p. 841:* © McGraw-Hill Higher Education/Stephen Frisch Photographer; *B19.4:* © Will & Deni McIntyre/Corbis Images; *B19.5:* © NYC Parks Archive/Fundamental Photographs; *B19.5:* © Kristen Brochmann/Fundamental Photographs; *p. 847:* © Phil Degginger/Color-Pic; *19.16, 19.18, 19.19A–C:* © McGraw-Hill Higher Education/Stephen Frisch Photographer

Chapter 20

Opener: © Richard A. Cooke/Corbis Images; *20.1A–B:* © McGraw-Hill Higher Education/Stephen Frisch Photographer; *20.4:* Alder and Wainwright, "Molecular Motion," October 1959/E. O. Lawrence

Berkeley National Laboratory; *B20.1:* © Bettmann/Corbis; *B20.2:* © Vanderbilt University-Clinical Nutrition Research; *p. 881:* © SPL/ Photo Researchers; *p. 883:* © Karl Weatherly/PhotoDisc

Chapter 21

Opener: © Sam Fried/Photo Researchers; *p. 903:* © AP/Wide World Photos; *21.1, 21.2, 21.4(both), 21.5B:* © McGraw-Hill Higher Education/Stephen Frisch Photographer; *p. 915:* © Richard Megna/Fundamental Photographs; *21.8:* © McGraw-Hill Higher Education/ Stephen Frisch Photographer; *p. 926:* © AIP Emilio Segre Visual Archives/American Institute of Physics; *p. 930:* © Science VU/ Look Up/Visuals Unlimited; *p. 931:* Courtesy of Dr. Gerhard Dahl/University of Miami School of Medicine; *21.13:* © Chris Sorensen Photography; *21.14:* © McGraw-Hill Higher Education/Pat Watson Photographer; *21.15:* Courtesy of the Guidant Corporation; *21.17:* © McGraw-Hill Higher Education/Stephen Frisch Photographer; *21.18A:* © AP/Wide World Photos; *21.20A:* © F. J. Dias/Photo Researchers; *21.21:* © David Weintraub/Photo Researchers; *21.25:* © Chris Sorensen Photography; *21.26, p. 944:* © McGraw-Hill Higher Education/Stephen Frisch Photographer; *p. 945:* © The Royal Institution, London, UK/Bridgeman Art Library

Chapter 22

Opener: © Space Imaging LLC; *22.3A:* © Doug Sherman/Geofile; *22.3B:* © University of Cambridge, Department of Earth Sciences; *p. 964:* © Ruth Melnick; *p. 965:* © Robert Holmes/Corbis; *p. 974:* © Patrick W. Grace/Photo Researchers; *22.12:* © Richard Megna/ Fundamental Photographs; *p. 979:* © Christine L. Case/Visuals Unlimited; *p. 980:* © Remi Benali/Getty Images; *22.17B:* Courtesy Bethlehem Steel Corporation; *p. 982:* © Lester Lefkowitz/Tech Photo, Inc.; *22.18B:* © Tom Hollyman/Photo Researchers; *p. 984:* © Lain Remi/Getty Images; *22.20(airplane):* © Rob Crandall/Rainbow; *22.20(house):* © ALCOA/Alcoa Technical Center; *22.20(aluminum):* © E. R. Degginger/Color-Pic; *p. 986(bottom):* © Hank Morgan/Rainbow; *p. 988:* © Ken Whitmore/Getty Images; *p. 990:* © McGraw-Hill Higher Education/Stephen Frisch Photographer; *22.22C:* © Nathan Benn/Woodfin Camp & Associates

Chapter 23

Opener: © E. R. Degginger/Color-Pic; *23.2(all), 23.5A–B, 23.6:* © McGraw-Hill Higher Education/Stephen Frisch Photographer; *p. 1012:* © E. R. Degginger/Color-Pic; *23.7B:* © Richard Megna/Fundamental Photographs; *p. 1013(left):* © McGraw-Hill Higher Education/Stephen Frisch Photographer; *p. 1013(right):* © E. R. Degginger/ Color-Pic; *p. 1014:* © Institute of Oceanographic Sciences/ NERC/SPL/Photo Researchers; *23.8(crystals):* © F. S. Judd Research Laboratories; *23.8(negative):* © Kathy Shorr; *p. 1017:* © The Granger Collection; *p. 1024(both):* © Richard Megna/Fundamental Photographs; *p. 1029:* © Ruth Melnick; *23.20A:* © McGraw-Hill Higher Education/Pat Watson Photographer; *23.21A–B:* © McGraw-Hill Higher Education/Stephen Frisch Photographer

Chapter 24

Opener: Courtesy Sandia National Laboratories; *p. 1047:* © W. F. Meggers Collection/American Institute of Physics; *B24.1:* © Hank Morgan/Rainbow; *B24.2:* © Hewlett-Packard; *p. 1058:* © Gamma/Getty Images; *p. 1059:* © Getty Images; *24.6B:* © Stanford Linear Accelerator Center; *p. 1060:* © Hulton Archive/Getty Images; *24.10A:* © Scott Camazine/Photo Researchers; *24.10B:* © T. Youssef/Custom Medical Stock; *p. 1068:* Jet Propulsion Laboratory/NASA; *24.11:* © Dr. Robert Friedland/SPL/Photo Researchers; *24.12:* © Dr. Dennis Olson/Meat Lab, Iowa State University, Ames, IA; *p. 1074:* © Emilio Segre Visual Archives/American Institute of Physics; *24.17A:* © Albert Copley/Visuals Unlimited; *B24.4:* Dr. Christopher Burrows, ESA/STScI and NASA; *24.18:* © Dietmar Krause/Princeton Plasma Physics Lab

Index

Page numbers followed by *f* indicate figures; *mn*, marginal notes; *n*, footnotes; and *t*, tables. Page numbers preceded by "A-" indicate pages in the *Appendix*. Page numbers preceded by "G-" indicate pages in the *Glossary*.

A

B

D

E

F

G

H

I

J

K

L

N

O

P

Q

R

S

X

Y

Z